AF631469

Microwave Radar and Radiometric Remote Sensing

Microwave Radar and Radiometric Remote Sensing

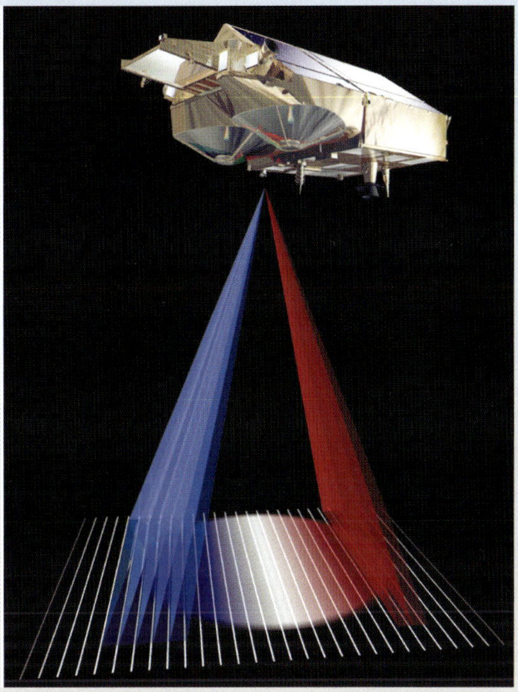

Fawwaz T. Ulaby **David G. Long**

WITH CONTRIBUTIONS BY

William Blackwell **Charles Elachi** **Adrian Fung** **Chris Ruf**
Kamal Sarabandi **Howard Zebker** **Jakob van Zyl**

THE UNIVERSITY OF MICHIGAN PRESS
2014
Ann Arbor

Copyright © 2014 by Fawwaz Ulaby
All rights reserved

This book may not be reproduced, in whole or in part, including illustrations, in any form (beyond that copying permitted by Sections 107 and 108 of the U.S. Copyright Law and except by reviewers for the public press), without written permission from the publisher.

Published in the United States of America by
The University of Michigan Press
Manufactured in the United States of America
Printed on acid-free paper

2017 2016 2015 2014 4 3 2 1

A CIP catalog record for this book is available from the British Library.

ISBN 978-0-472-11935-6

Dedicated to the memory of
four colleagues:

Richard K. Moore University of Kansas

Jin Au Kong Massachusetts Institute of Technology

Hans J. Liebe Institute of Telecommunication Sciences

Erwin Schanda University of Bern

Each has left an indelible mark on the
field of microwave remote sensing.

Preface

Designed to cover the theory, systems, and applications of microwave remote sensing, this textbook is written with the postgraduate student in mind. Even though nine coauthors contributed to its creation, it was developed sequentially—not in parallel—to insure consistency in style, symbols, and notation. The many intentional interconnections between the book's sections and chapters are intended to create a sense of clarity and cohesiveness. One of our goals was to limit the length of the book to no more than 1000 pages, and we nearly succeeded.

Special Features

(1) Book Homepage: www.MRS.eecs.umich.edu

The website provides access to:

- Computer Codes — see item (2) below.
- Interactive Modules — see item (3) below.
- URLs for spaceborne microwave sensors.
- High-resolution images.

(2) Computer Codes: To avail the user the opportunity to simulate and evaluate electromagnetic interactions with natural materials, we developed or acquired (with permission) MATLAB codes for computing reflection, absorption, scattering, and emission by a variety of different media, including atmospheric constituents, ocean surfaces, sea ice, snow cover, and bare and vegetation-covered surfaces. These are downloadable files with clear instructions and examples. Multiple codes can be combined together to suit the intended application.

(3) Interactive Modules: For quick visual analyses, the Interactive Modules allow the user to enter a set of parameter values and then observe the results in the form of graphical presentations. The modules execute the MATLAB codes and present their results on a computer screen.

Acknowledgments

This book could not have been possible had it not been for the energy and talent of several special individuals, each deserving of singular recognition:

Richard Carnes: Applying his legendary facility with LaTeX, Richard—who is a highly talented piano performer—typed and retyped the manuscript several times, always blending the precision of a compositor with the eye of an artist.

James Green: A recent Michigan graduate with a degree in architecture, James drafted the nearly 1000 colorful figures appearing in the book. His careful attention to detail is responsible for the consistency in style, symbols, and annotation. He compiled and meticulously error-checked the more than 1000 references in the bibliography.

Dr. Adib Nashashibi: The book's 50 MATLAB-based computer codes were all developed by Adib. He also used the codes to generate many of the model plots contained in the book.

Dr. Leland Pierce and Ms. Janice Richards: Transforming the MATLAB codes into user-friendly interactive modules required Leland's and Janice's talents for coding and execution in JavaScript.

Jane Whitcomb: A highly talented graduate student, Jane provided solutions to about half of the end-of-chapter problems.

Generating a 1000-page error-free manuscript is next to impossible, but minimizing the number of errors is possible. We are grateful to the following graduate students for their thorough inspection of every word, figure, and equation: Michael Benson, David Chen, Mark Crocket, Rachel Kroodsma, Victor Lee, Nathan Madsen, Kyra Moon, Steven Reeves, Craig Stringham, and Joe Winkler. Because of their reviews, many errors were identified and corrected.

Concluding Remark

Writing this book has been an endeavor of pure love. We enjoyed writing it; we hope you enjoy using it.

On behalf of the authors,

Fawwaz Ulaby
January 2014

Photo Credits

Page	Figure	Source
—	Ch. 1	NASA
11	1-8	David Long
12	1-9	David Sandwell and Walter Smith
13	—	NASA
15	1-10	David Long
19	1-14	William Johnson and Li Li, Naval Research Laboratory
27	—	NASA
28	1-19	Sandia National Laboratories
29	1-20	JPL / NASA
30	1-21	Environmental Research Institute of Michigan
31	1-22	U.S. National Sea Ice Data Center
32	1-23	NOAA
33	1-24	JPL / NASA
34	Ch. 2	Adapted from GNU Free Documentation License image
74	Ch. 3	NASA
177	5-13	Sandia National Laboratories
189	5-23	Fawwaz Ulaby
191	5-25	Fawwaz Ulaby
193	5-26	Chris Oliver and Shaun Quegan
210	5-38	JPL / NASA
211	5-39	JPL / NASA
219	5-45	JPL / NASA
220	5-46	JPL / NASA
221	5-48	JPL / NASA
221	5-47	JPL / NASA
224	5-50	JPL / NASA
262	Ch. 7	ESA / AOES Medialab
303	7-35	ESA / AOES Medialab
307	7-38	Chris Ruf
366	Ch. 9	Northrop-Grumman Electronic Systems
401	9-30	Northrop-Grumman Electronic Systems
402	9-31	Northrop-Grumman Electronic Systems
411	9-43	NASA
510	11-48	Roger Lang and Peggy O'Neill
542	11-87	Prasad Gogenini
552	Ch. 12	JPL / NASA
584	12-34	ESA / AOES Medialab
585	12-37	JPL / NASA
598	12-54	U.S. National Snow and Ice Data Center
604	Ch. 13	JPL / NASA
670	Ch. 14	Sandia National Laboratories
727	14-42	JPL / NASA
728	14-43	David Long
728	14-44	Sandia National Laboratories
742	Ch. 15	Alberto Moreira
768	15-23	JPL / NASA
769	15-25	NASA
770	15-26	GNU Free Documentation License (USGS and NASA data)
773	15-30	JPL / NASA
776	15-33	German Aerospace Center (DLR)
782	Ch. 16	David Long
799	16-13	David Long
801	16-16	David Long
832	16-46	Canadian Space Agency
832	16-47	Canadian Space Agency
832	16-48	Canadian Space Agency
833	16-49	David Long
836	16-50	U.S. National Snow and Ice Data Center: Ted Scambos, Terry Haran, Alice O'Connor, Rob Bauer
850	16-70	David Long
850	16-71	David Long
851	16-72	David Long
853	16-75	David Long
857	16-80	David Long
857	16-81	David Long
858	16-82	U.S. National Snow and Ice Center
858	16-84	Canadian Space Agency
859	16-85	European Space Agency
859	16-86	European Space Agency
864	Ch. 17	European Space Agency
880	17-13	NASA
886	17-18	European Space Agency
888	17-19	European Space Agency
890	Ch. 18	University of Illinois
898	18-10	National Oceanic and Atmospheric Administration (NOAA)
904	18-18	NASA
918	18-36	University of Illinois

Computer Codes

The following is a list of the MATLAB-based computer codes available at the book website, **mrs.eecs.umich.edu**:

Code 2.1 Polarization State

Code 2.2 Propagation Constant and Intrinsic Impedance

Code 2.3 Reflection and Transmission at Planar Boundary

Code 2.4 Reflection by Two-Layer Composite

Code 3.1 Array Factor for Uniformly Spaced Array

Code 3.2 Array Factor for Uniformly Spaced Array with Linear Phase Distribution

Code 4.1 Relative Dielectric Constant of Pure Water

Code 4.2 Relative Dielectric Constant of Saline Water

Code 4.3 Relative Dielectric Constant of Pure Ice

Code 4.4 TVB Dielectric Model for Heterogeneous Mixtures

Code 4.5 Relative Dielectric Constant of Dry Snow

Code 4.6 Relative Dielectric Constant of Wet Snow

Code 4.7 Relative Dielectric Constant of Soils

Code 4.8 Relative Dielectric Constant of Vegetation Material

Code 5.1 σ^0 Polarization Synthesis of PolSAR Images

Code 6.1 Emissivity of Homogeneous Medium with Planar Boundary

Code 6.2 Emissivity of Two-Layer Composite

Code 8.1 Attenuation Coefficient of Water Vapor

Code 8.2 Attenuation Coefficient of Oxygen

Code 8.3 Gaseous Attenuation Coefficient

Code 8.4 Water-Cloud Extinction Coefficient

Code 8.5 Ice-Cloud Extinction Coefficient

Code 8.6 Rain Extinction Coefficient

Code 8.7 Optical Thickness and Transmissivity—Clear Atmosphere

Code 8.8 Optical Thickness and Transmissivity—Cloudy Atmosphere

Code 8.9 Zenith Atmospheric Opacity and Transmissivity—Clear Atmosphere

Code 8.10 Zenith Atmospheric Opacity and Transmissivity—Cloudy Atmosphere

Code 8.11 TSKY—Clear Atmosphere

Code 8.12 Mie and Rayleigh Scattering

Code 10.1 I^2EM Backscattering from Single-Scale Random Surface

Code 10.2 I^2EM Backscattering from Multiscale Random Surface

Code 10.3 I^2EM Bistatic Scattering from Single-Scale Random Surface

Code 10.4 I^2EM Backscattering from Periodic Sinusoidal Surface

Contents

CHAPTER 1

Introduction

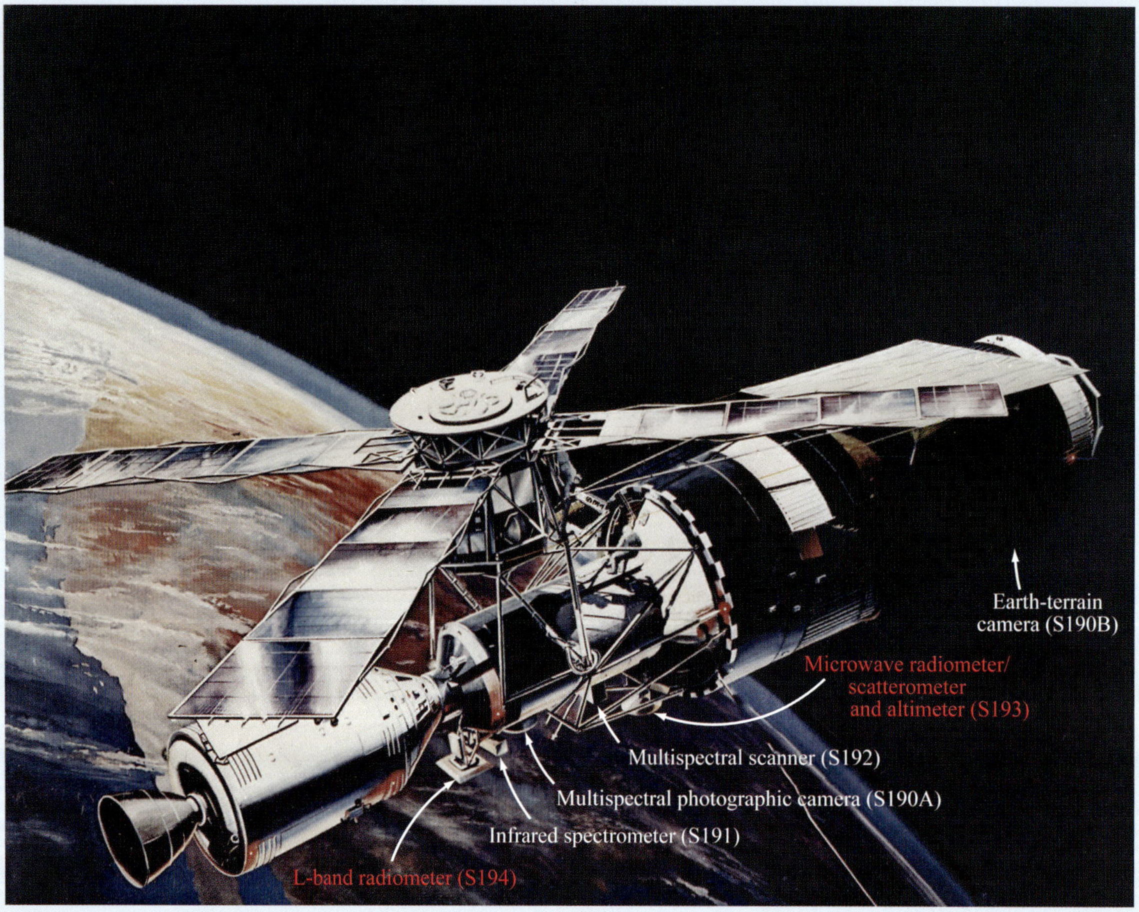

Skylab

CONTENTS

1-1 Why Microwaves for Remote Sensing?

Compared with aerial photography, which has been in use since the late 1800s, microwave remote sensing is relatively new, having been used in civilian applications only since the early 1960s. Given the success of photography, why use microwaves for remote sensing?

> ▶ Microwaves provide a different and unique view that offers new information about Earth's environment that often can be obtained in no other way. ◀

Microwaves can penetrate clouds—and to some extent rain—and do not rely on the sun as the source for illumination. These attributes allow sensing of the Earth independently of the time of day and under almost all weather conditions. Figure 1-1 illustrates the attenuating effect of clouds on radio transmission between space and the Earth surface as a function of electromagnetic wavelength. Clouds that are dense enough to completely obscure the ground, and thus preclude aerial or satellite optical imaging, have little effect on microwaves. Ice clouds have almost no effect at any wavelength longer than 1 cm, while water clouds have a significant effect only for wavelengths less than 2 cm. Rain has a greater effect than clouds, but as shown in Fig. 1-2, it is important only for wavelengths less than 4 cm. Active (radar) microwave sensors are thus able to observe the Earth's surface under almost all conditions.

> ▶ For passive microwave remote sensing (***radiometry***) the attenuation effects of clouds and rain can obscure the surface at some frequencies. However, this same attenuation provides an observable signal that can be exploited by microwave radiometers to gather information about key geophysical parameters of interest such as total liquid water fraction, cloud droplet size distribution, atmospheric temperature, and rain intensity. ◀

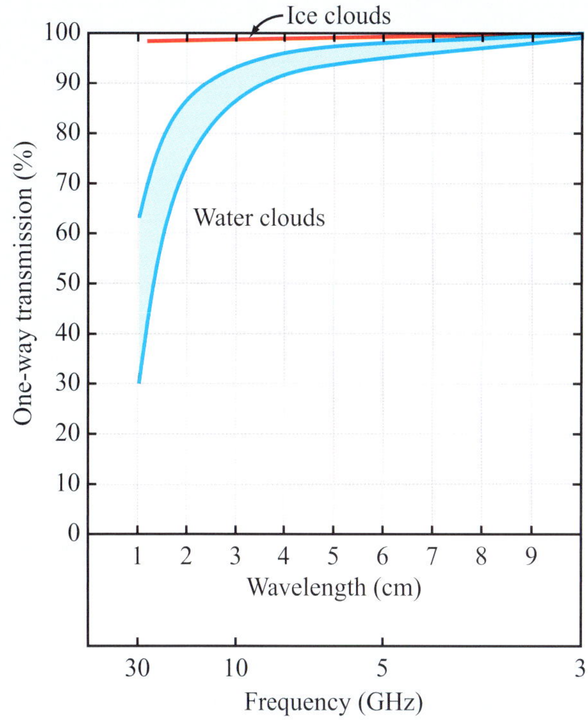

Figure 1-1: Attenuating effects of clouds on radio transmission from space to ground.

At longer wavelengths, radiometers are able to observe the surface—particularly in the polar regions—through clouds, whereas mapping such areas in the visible spectrum is much more difficult.

> ▶ Another reason for the use of microwaves is that they are able to penetrate more deeply into vegetation than optical waves can. ◀

A pictorial illustration of the relation between penetration depth and wavelength is shown in Fig. 1-3. The extent of penetration into vegetation depends upon the moisture content and density of the vegetation canopy, as well as upon the wavelength of the microwaves. Longer wavelengths penetrate much better than shorter wavelengths. Hence, shorter wavelengths yield information about the upper layers

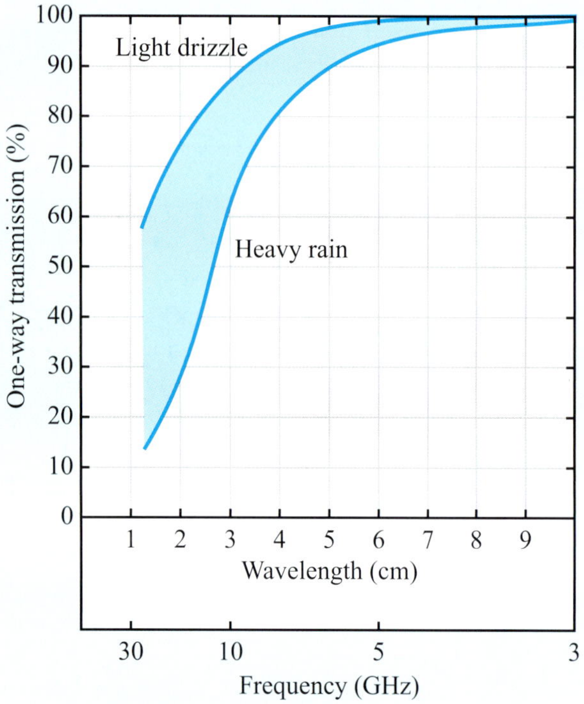

Figure 1-2: Attenuating effects of rain on radio transmission from space to ground.

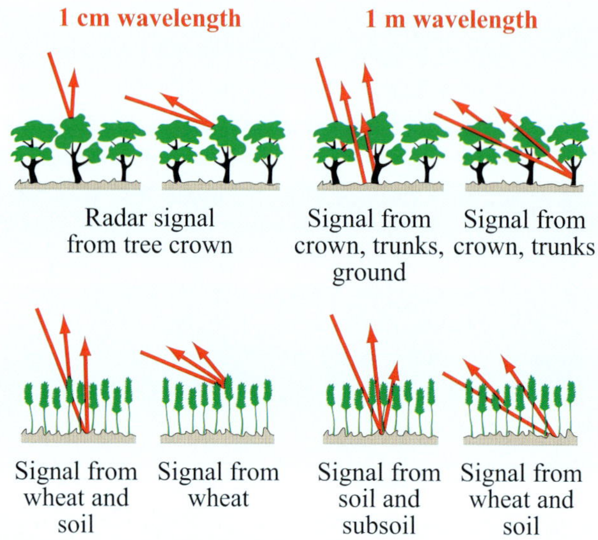

Figure 1-3: Illustration of radar signal penetration through and scattering by vegetation. There is more penetration if vegetation is dry than if wet.

of the vegetation, while longer wavelengths yield information about the lower layers and the ground beneath. Moreover, microwaves are able to penetrate significantly into the ground itself. Figure 1-4 illustrates the depth at which the microwave signal decreases by 63 percent for three different frequencies and three different types of soil. For dry soil the penetration depth at the lower microwave frequencies can be substantial, whereas penetration is limited for wet soil.

> ▶ A third, and perhaps the most important, reason for the use of microwaves is that the information available from microwaves is different from that available in the visible and infrared regions of the electromagnetic spectrum. ◀

In particular, microwave scattering from a natural surface is related to both the electrical (bulk-dieletric)

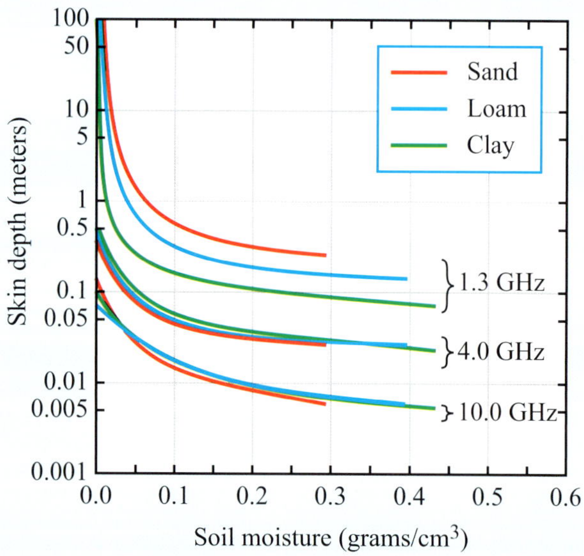

Figure 1-4: Skin depth as a function of soil volumetric water content, frequency, and soil type [from Ulaby et al., 1974].

and geometric (roughness) properties of the surface. With suitably designed sensors, it is thus possible to infer information about surface geometry as well as the electrical properties of the surface. Further, microwave-derived properties complement the information obtainable from visible and infrared radiation, facilitating studies of geometric, bulk-dielectric, and molecular-resonance properties of a surface feature.

1-2 A Brief Overview of Microwave Sensors

Microwave remote sensing instruments can be divided into two broad classes: ***passive***, known as ***radiometers***, and ***active***, known as ***radars*** (Fig. 1-5). Both include antennas and receivers, but radars differ from radiometers in that they include a transmitter as well. Both classes of sensors have been used on aircraft and spacecraft to study the Earth as well as other planets. These broad classes can be subdivided into subclasses based on their general operating characteristics and functions.

Active microwave sensors can be grouped into five general classes: ***synthetic-aperture radar*** (SAR) systems, ***side-looking airborne radar*** (SLAR), ***scatterometers***, ***altimeters***, and ***meteorological radars***. As suggested by the name, SAR employs synthetic-aperture antenna-processing techniques, whereas the other sensor systems typically use real-aperture antennas. ***Inverse synthetic-aperture radar*** (ISAR) is used for ground-based sensing of extra-terrestrial bodies.

▶ SAR and SLAR systems are designed to create images from moving platforms. ◀

Such systems operate by transmitting modulated pulses and using Doppler/range processing to construct backscatter images. SARs provide the highest resolution, but are significantly more complex than other sensors. Scatterometers measure radar backscatter very precisely, but they typically have lower resolutions than SARs. Altimeters are specialized radars designed to measure platform height, although other information can be extracted from the echo. Weather radars are specially designed scatterometers with ranging capability that measure rainfall and other meteorological phenomena. ***Interferometric SAR*** (InSAR or IFSAR) is a special SAR configuration used to measure topography and the motion of surfaces. Surveillance radars, which are designed for aircraft, ship, and missile detection and/or tracking, are treated in more detail in dedicated radar texts such as Richards et al. (2010) and Skolnik (1980).

▶ ***Radiometers*** are passive microwave sensors that observe thermal emission of microwave signals. ◀

The emission is related to the physical temperature and electrical properties of the sensed surface or volume, with modulation by the intervening atmosphere. Since they contain no transmitters, radiometers typically consume much less power. They frequently operate over a broader frequency range (bandwidth) than radar sensors and primarily rely on real-aperture antennas, although ***synthetic-aperture radiometers*** (interferometers) are also used for both Earth remote sensing and astronomical observation (Burke and Graham-Smith, 2002; Goodberlet, 2000; Le Vine, 1999; Ruf et al., 1988; Thompson et al., 1994; Wohlleben et al., 1991). ***Sounders*** are specialized radiometers designed for extracting vertical profiles of atmospheric parameters.

1-3 A Short History of Microwave Remote Sensing

1-3.1 Radar

The very first successful attempt at generating radio waves was conducted at a frequency of 200 MHz, which, by today's frequency allocation nomenclature, is in the UHF band and very close to the lower edge of the microwave band (300 MHz to 300 GHz). In 1886, Heinrich Hertz experimentally tested Maxwell's electromagnetic theory with resonators and demonstrated that reflections from various metallic and nonmetallic objects can indeed be detected.

Radar was first used to detect ships in 1903 by Hülsmeyer, who obtained a patent for his idea in 1904

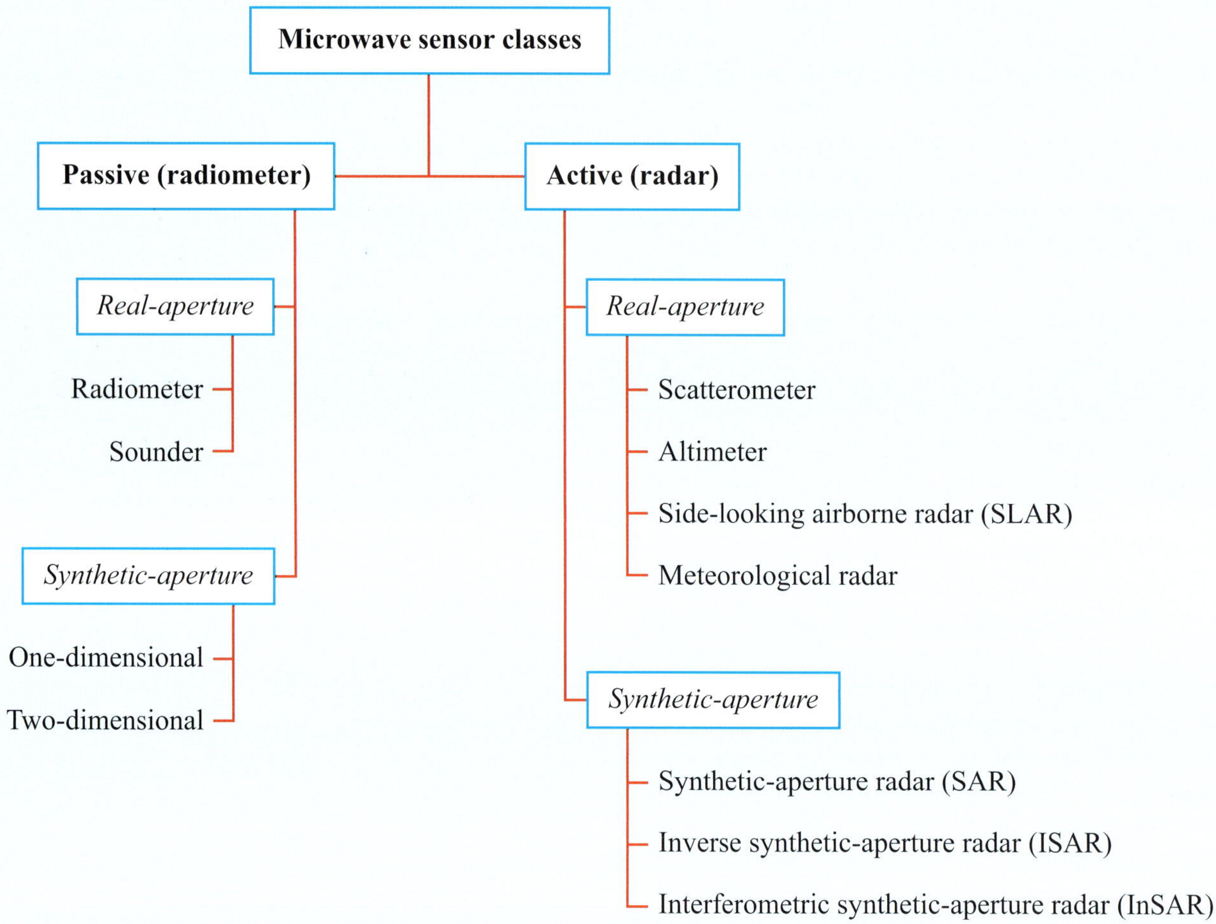

Figure 1-5: Major classes of microwave remote sensors.

(Hülsmeyer, 1904). Marconi (1922) predicted the use of radio for detection, but did not develop it. A. H. Taylor and others at the U.S. Naval Research Laboratory were pioneers in the development of radar as a tool for detecting ships and aircraft. The first experiment was conducted with a continuous-wave system in 1922 (Skolnik, 1980).

Interestingly, the first ***pulsed radar*** was used for remote sensing experiments in ionospheric sounding in 1925 (Breit and Tuve, 1926). Taylor, Breit, and Tuve transmitted pulses from a radio station and then observed the pulses that were reflected by the ionosphere at a separate location a few miles distant so as to measure the height of the ionosphere. The development of pulse radars for detecting objects closer to the Earth did not begin until the early 1930s (Taylor et al., 1934) and the first successful pulsed radar at the U.S. Naval Research Laboratory (NRL) operated in 1936. About this same time, pulse radars were developed in Great Britain under Sir Robert Watson-Watt (1957). The first ***airborne radars*** to be used extensively prior to World War II were frequency-modulated ***altimeters*** operating at about 400 MHz (Matsuo, 1938).

By the time of World War II, all the major participants had pulse-radar systems developed for the purpose of locating aircraft and/or ships. The airborne radars deployed during the early part of World War II operated at long wavelengths (Rowe, 1948). Shortly thereafter, the MIT Radiation Laboratory (MIT Rad. Lab., 1948–52) in the United States and comparable laboratories in other countries demonstrated airborne radars operating at microwave frequencies that were capable of producing images of the ground. By 1946, radars producing images of the ground were in extensive use in the 3, 10, and 24 GHz bands. Other types of radar systems also were developed and widely used during World War II, including altimeters and radar ***proximity fuses***. Proximity fuses permitted the detonation of artillery shells in the immediate vicinity of an aircraft in flight, thus damaging it without the necessity of a direct hit by a shell. It also permitted the bursting of shells well above the ground, thereby dispersing their destructive effect so that, again, a direct hit was not required.

Early airborne imaging radars required rotating antennas to scan the surface. The antennas were limited in size and thus restricted the image resolution. In the 1950s, a new type of radar, the side-looking airborne radar (SLAR), was developed. With the SLAR, finer-resolution images could be achieved because the antenna could be quite long since it was fixed parallel to the fuselage of the aircraft.

▶ SLAR scanning was achieved by a fixed beam pointed to the side, with the aircraft's motion moving the beam across the land, hence producing a scanned image. Images were recorded on long strips of film. ◀

Through the 1950s many different SLAR systems were developed, including systems that operated at 10 GHz (3 cm wavelength), 16 GHz (1.9 cm wavelength) and 35 GHz (8.6 mm wavelength). With these high frequencies and with antennas that were as long as 15 m, very fine-resolution images could be produced, extending to quite long ranges. By 1960, extensive imagery had been produced, but were unavailable on an unclassified basis; however, in 1964, images produced by the 8.6 mm AN / APQ-97 system were declassified. This data set clearly demonstrated the power of radar imaging as a tool for Earth observation and led to further development of imaging radars.

▶ In 1952, Wiley developed an airborne radar based on what he called a "Doppler beam-sharpening" system. This was the first of what is now known as synthetic-aperture radar (SAR). ◀

Wiley's first system operated at 75 MHz and looked to the side of the aircraft with a beam pointed about 45° ahead (this is now called "squint mode") (Cutrona, 1970). An independent development of a Doppler-processing radar was started in 1954 at the University of Illinois (Sherwin et al., 1962), where it was termed "synthetic-aperture." This work was transferred to the University of Michigan in 1956 (Cutrona et al., 1961). During the late 1950s and early 1960s, classified development of synthetic-aperture radar systems took place at the University of Michigan and at various companies. The first unclassified papers describing this new type of imaging radar appeared in 1961 (Cutrona et al., 1961, 1962). At the same time, similar developments were proceeding in other countries, including the USSR (Reutov and Mikhaylov, 1970), France, and the United Kingdom. However, papers in the open literature reporting these developments did not appear until after the 1961 papers in the U.S.

▶ A synthetic-aperture radar permitted the production of side-looking radar images whose pixel dimension in the along-track direction was independent of distance from the radar and could be much smaller than was possible for a SLAR with a feasibly short antenna. ◀

This represented a major step forward in improved resolution for airborne radars and made feasible the concept of fine-resolution spaceborne imaging radar. SAR and SLAR are discussed in Chapters 13 and 14.

The first scientific applications of imaging radar were to geological studies, but it was recognized from the beginning that many other land-use, water-resource, and vegetation studies were possible with imaging radars.

The first major radar mapping project was conducted by the U.S. Army and the Government of Panama in Oriente Province, Panama, in 1967 (Viksne et al., 1969). During the 1970s multiple-polarization research synthetic-aperture systems at 1.25 GHz and about 9 GHz were flown by the Environmental Research Institute of Michigan (ERIM) (Rawson and Smith, 1974) and the Jet Propulsion Laboratory (JPL) (Schaber et al., 1980). The production of multiple-polarization, multiple-frequency images has contributed significantly to the knowledge of the capabilities of imaging radar in remote sensing. This then led to fully-polarimetric imaging radars (Ulaby and Elachi, 1990; Cloude and Pottier, 1996; Touzi et al., 2004). The development of interferometric SAR systems has enabled the generation of global topography maps from space (Farr et al., 2007; Hanssen, 2001; Zebker and Goldstein, 1986). Technology for building airborne SARs has continued to evolve, resulting in the reduction of sensor size and improved imaging resolution. Spatial resolutions finer than 10 cm have been achieved (Doerry et al., 2005).

The use of radar to observe weather conditions began with early ground-based radars whose operators noticed that the height of the "grass" (the reflection from ocean waves) was related to wind conditions. Experiments showed that rain, snow, and hail could be mapped with radar. In the U.S., the first weather radars were developed in the 1950s. The first civilian Doppler radar specifically deployed to study the morphology of storms was set up at the National Severe Storms Laboratory in Oklahoma in 1971 (Cobb, 2004). Other meteorological (weather) radars followed, which led to the deployment of the ***NEXRAD*** network of meteorological radars, which eventually achieved (essentially) complete coverage of the U.S. Other nations have deployed similar networks. These radars measure rain and wind at multiple heights, and provide storm warnings (Meischner, 2004). They are familiar to the general public from weather forecast broadcasts.

Spaceborne radars

Radars were first used in space for spacecraft docking, but have been extensively used for imaging, altimetry, and wind measurement. While spaceborne radars have been developed and flown by the military, we focus on civilian applications. Table 1-1 summarizes the history of imaging radars flown on civilian spacecraft. The first spaceborne imaging radars were used to study cloud-covered Venus and have been used to study the moon Titan (West et al., 2009). Beginning in the late 1970s the Soviet Union developed and flew a series of SLARs to observe Earth as part of its ***Okean missions***, which also included scanning radiometers. These early radars had very low resolution.

Proposals for spaceborne synthetic-aperture Earth-observation radars were made in the early 1960s, but the first such radar to fly in space was on the oceanographic satellite ***Seasat*** that was launched in June 1978; see Fig. 1-6 (Jordan, 1980). Operating for only three months, this system produced many millions of square kilometers of high-resolution radar imagery. Several shuttle SAR missions were flown through the 1980s, 1990s, and early 2000s, including the ***Shuttle Radar Topography Mission*** (SRTM), which used ***radar interferometry*** to measure global surface topography (Farr et al., 2007). Many other spaceborne SAR systems have been flown or are in various planning stages.

Two other classes of radars have been flown extensively on spacecraft: scatterometers and altimeters.

► A ***scatterometer*** (the term was coined in 1965 by R. K. Moore) is a device used for measuring the radar scattering coefficient quantitatively. ◄

Scattering-coefficient measurements were made during World War II (Kerr and Goldstein, 1951; Davies and MacFarlane, 1946) and continued thereafter (e.g., MacDonald, 1956; Rouse, 1969). A summary of many of the scattering measurements made prior to 1975 is provided by Long (1975).

The flight of the S-193 scatterometer onboard ***Skylab*** in 1974 was conducted to measure scattering from space. It demonstrated the ability of radar to infer near-surface wind speed from space (Moore and Young, 1977). The success of this experiment led to the flight of the scatterometer (SASS) on the Seasat oceanographic satellite launched in 1978 (Grantham et al., 1977). This multiantenna fan-beam Doppler scatterometer was the

Table 1-1: Overview of spaceborne SAR sensors and their main characteristics [Moreira et al., 2013].

Sensor	Operation	Frequency Band (Polarization)	Comments	Institution, Country
Seasat	1978	L (hh)	First civilian SAR satellite, operation for only ca. three months	NASA/JPL, USA
SIRA-A, SIR-B	1981, 1984	L (hh)	Shuttle radar missions	NASA/JPL, USA
ERS-1/2	1991–2000/ 1995–2011	C (vv)	European Remote Sensing Satellites (first European SAR satellites)	ESA, Europe
J-ERS-1	1992–1998	L (hh)	Japanese Earth Resource Satellite (first Japanese SAR satellite)	JAXA, Japan
SIR-C/X-SAR	April and October 1994	L & C (quad) X (vv)	Shuttle imaging radar mission, first demonstration of spaceborne multifrequency SAR	NASA/JPL, USA DLR, Germany ASI, Italy
Radarsat-1	1995–today	C (hh)	First Canadian SAR satellite, swath width of up to 500 km with Scan SAR imaging mode	CSA, Canada
SRTM	Feb. 2000	C (hh+vv) and X (vv)	Shuttle Radar Topography Mission, first spaceborne interferometric SAR	NASA/JPL, USA DLR, Germany ASI, Italy
ENVISAT/ASAR	2002–2012	C (dual)	First SAR satellite with Transmit/ Receive module technology, swath width up to 400 km	ESA, Europe
ALOS/PALSAR	2006–2011	L (quad)	Advanced Land Observing Satellite (Daichi), swath width up to 360 km	JAXA, Japan
TerraSAR-X/ TanDEM-X	2007–today 2010–today	X (quad)	First bi-static radar in space, resolution up to 1 m, global topography available by end of 2014	DLR/Astrium, Germany
Radarsat-2	2007–today	C (quad)	Resolution up to: 1 m ×3 m (azimuth × range), swath width up to 500 km	CSA, Canada
COSMO-SkyMed-1/4	2007 . . . 2010–today	X (dual)	Constellation of four satellites, up to 1 m resolution	ASI/MID, Italy
RISAT-1/2	2008–today	C (quad)	Follow-on satellite (RISAT-1a) to be launched in 2016, RISAT-3 (L-band) in development	ISRO, India
HJ-1C	2012–today	S (vv)	Constellation of four satellites, first satellite launched in 2012	CRESDA/CAST/ NRSCC, China
Kompsat-5	Launch scheduled in 2013	X (dual)	Korea Multi-Purpose Satellite 5, resolution up to 1 m	KARI, Korea
PAZ	Launch scheduled in 2013	X (quad)	Constellation with TerraSAR-X and TanDEM-X planned	CDTI, Spain
ALOS-2	Launch scheduled in 2013	L (quad)	Resolution up to: 1 m ×3 m (azimuth × range), swath width up to 490 km	JAXA, Japan
Sentinel-1a/1b	Launch scheduled in 2013/2015	C (dual)	Constellation of two satellites, swath width up to 400 km	ESA, Europe
Radarsat Constellation-1/2/3	Launch scheduled in 2017	C (dual)	Constellation of three satellites, swath width up to 500 km	CSA, Canada
SAOCOM-1/2	Launch scheduled in 2014/2015	L (quad)	Constellation of two satellites, fully polarimetric	CONAE, Argentina

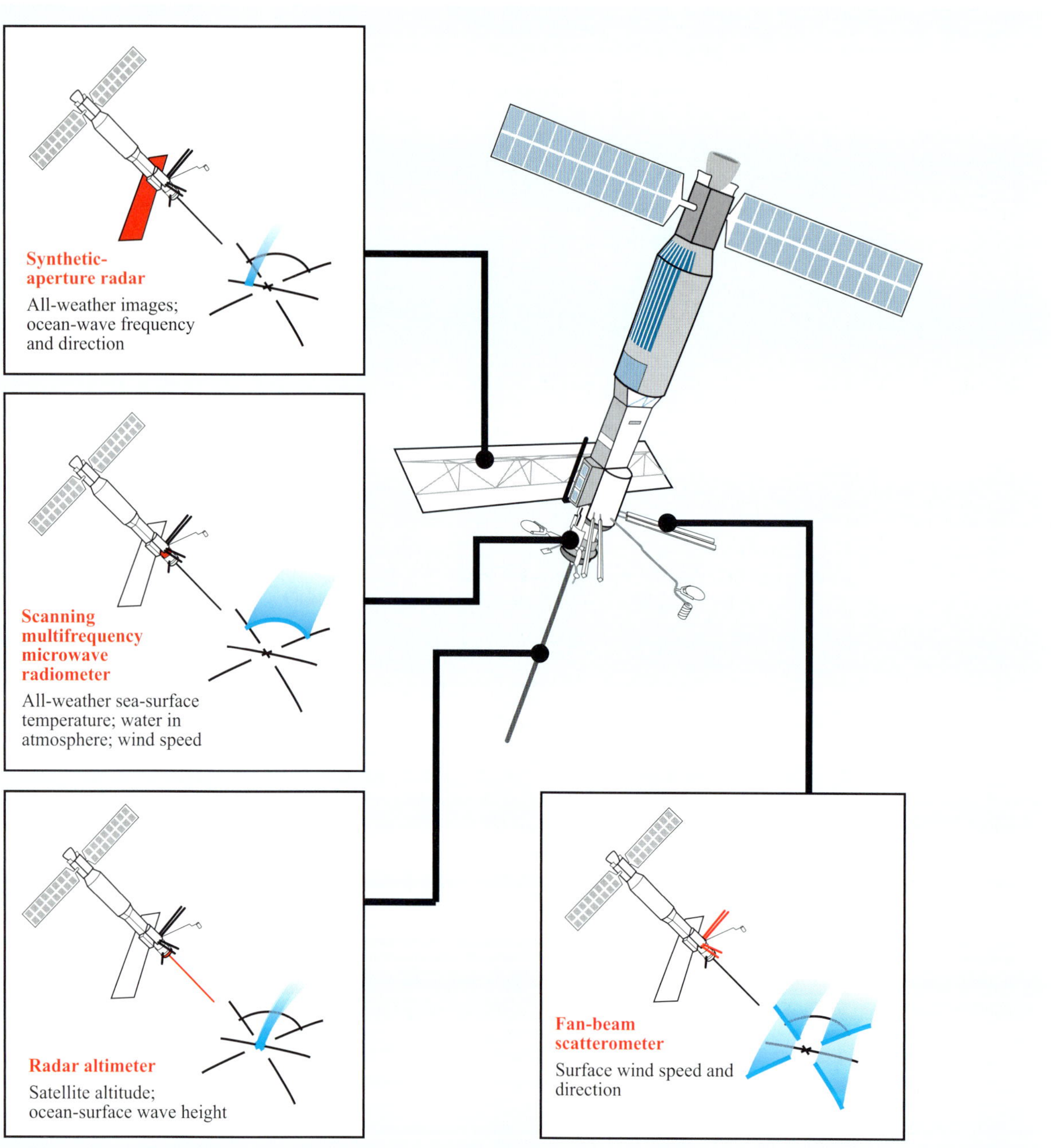

Figure 1-6: The Seasat-1 satellite, launched in 1978, with an illustration of the antenna illumination patterns for each of its microwave sensors.

Table 1-2: Operational scatterometers on spacecraft.

	SASS	ESCAT	NSCAT	SeaWinds	ASCAT	OSCAT
Frequency	14.6 GHz	5.3 GHz	13.995 GHz	13.6 GHz	5.3 GHz	13.6 GHz
Antenna azimuths	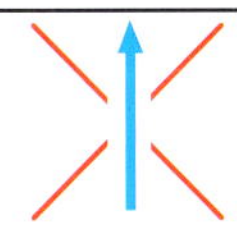	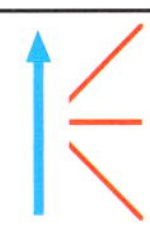	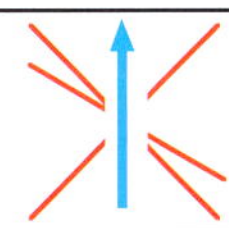	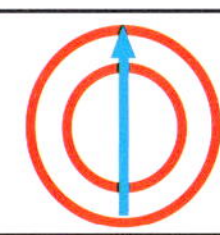	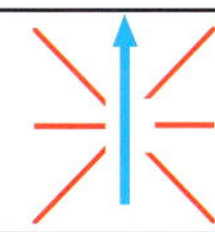	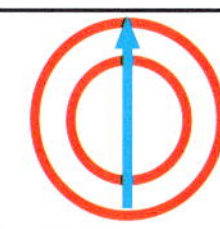
Polarizations	vv and hh	vv only	vv and hh	h-outer/h-inner	v only	v-outer/h-inner
Beam resolution	Fixed Doppler	Range gate	Variable Doppler	Pencil-beam	Range gate	Pencil-beam
Science modes	Many	SAR wind	Wind only	Wind/hi-res	Wind only	Wind/hi-res
Resolution (σ^0)	Normally 50 km	50 km	25 km	Egg: 25×35 km Slice: 6×25 km	25/50 km	Egg: 30×68 km Slice: 5×30 km
Swath, km	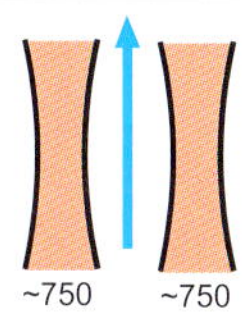	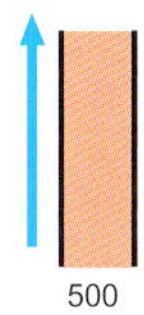	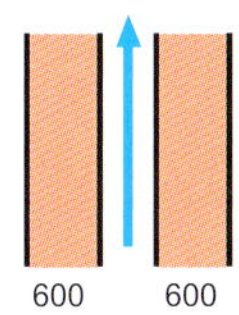	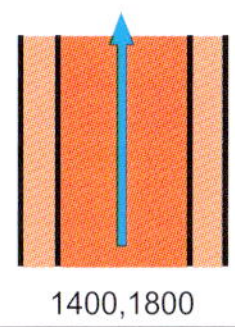	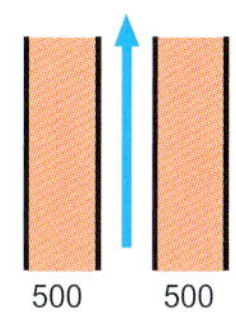	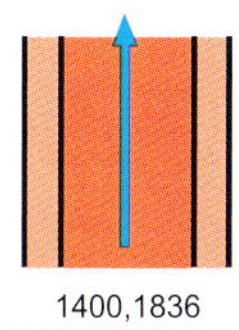
Incidence angles	0°–70°	18°–59°	17°–60°	46° & 54.4°	25°–65°	49° & 57°
Daily coverage	Variable	< 41%	78%	92%	65%	> 90%
Mission & dates	Seasat: 6/78–10/78	ERS-1: 92–96 ERS-2: 95–01	ADEOS-I: 8/96–6/97	QuikSCAT: 6/99-11/09 ADEOS II: 1/02–10/02	METOPD: 6/2007– METOPB: 9/2012–	OceanSat-2: 10/09–

first scatterometer designed for more than just research, and was operationally used for wind observation. It led a long series of operational scatterometer missions designed for wind observation (see Table 1-2). The Seasat scatterometer used four orthogonal fan beams [see Fig. 1-7(a)], to collect measurements at various incidence and azimuth angles of the same location on the surface, which enabled estimation of near-surface wind speed and direction. A number of follow-on instruments, such as the ***ERS-1/2*** scatterometer (1991-2011) (Attema, 1991), the ***NASA Scatterometer*** (NSCAT) (1996–1997) (Naderi et al., 1991), and the ***advanced scatterometer*** (ASCAT) (2008-present) (Figa-Saldaña et al., 2002), have used fan-beam antennas and employed either Doppler filters and/or range gating. The ***SeaWinds scatterometer*** (1999–2010 and 2006) (Spencer et al., 2000, 2003) was the first spaceborne wind scatterometer to use a dual-rotating beam. It was followed by the ***Indian Oceansat-2*** scatterometer (Parmar et al., 2006).

Wind scatterometers have become a standard instrument for satellite wind measurement. The scattering measurements from these sensors have been extensively used for other applications such as the mapping of sea-ice type and extent, detection of icebergs, and measuring soil moisture and vegetation density (Liu, 2002). A global view of the radar backscatter of the Earth's surface as observed by the SeaWinds scatterometer is shown in Fig. 1-8. Advanced capability scatterometers with higher resolution and dual-frequency have been proposed for the future (e.g., Long et al., 2009). The ***Soil Moisture Active Passive*** (SMAP) mission planned for launch in 2014–15 will employ a low-resolution L-band scatterometer/imager in conjunction with an L-band radiometer to improve the accuracy of soil moisture measurements (Entekhabi et al., 2010).

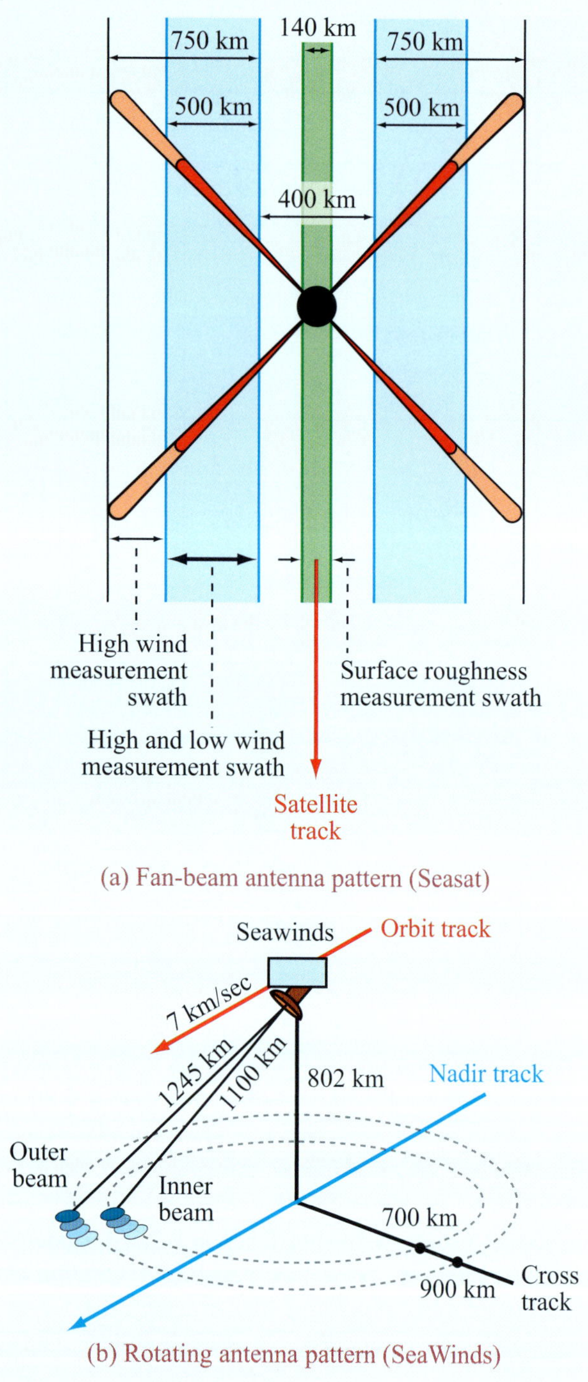

Figure 1-7: (a) Fan-beam antenna pattern for the Seasat scatterometer. (b) Rotating antenna pattern for the SeaWinds scatterometer. These are representative examples of the two major types of spaceborne scatterometers, fan-beam and scanning pencil-beam.

Satellite altimeters have been widely used for ocean studies and have become essential tools for the study of ocean circulation (Fu and Cazenave, 2001). Altimeters have also been used for measuring the topography of other planets such as perpetually cloud-covered Venus. Table 1-3 summarizes the history of satellite altimeters. Satellite altimeters provide very precise measurements of the distance from the satellite to the surface, and often include radiometers to enable correction for atmospheric delay. Coupled with precise orbit determination, very accurate surface topography can be derived. This information is useful for determining the geoid (the gravitational field of the Earth). Ocean topography measurements enable mapping of undersea topography and derivation of ocean currents. Wind speed can also be estimated. Figure 1-9 illustrates the derived shape of the geoid measured by the ***ERS-1 altimeter***. Planned developments in altimetry include a wide-swath system (Rodriguez and Pollard, 1999; Pollard et al., 2002). By examining the shape of the pulse return to a spaceborne radar altimeter over the sea, one also may establish the height of the waves beneath the spacecraft (Sandwell, 1991).

The ***Tropical Rain Mapping Mission*** (TRMM) ***Precipitation Radar*** is the first meteorological radar to be launched in space for mapping rain distributions. Launched in 1997 in an equatorial orbit and operating at Ku band, the TRMM PR used an electronically-scanned antenna to measure rainfall over a 50 km swath at a 4 to 5 km horizontal and 250 m vertical resolution (Kozu et al., 2001). The ***CloudSat*** satellite launched in 2006 carried a 94 GHz radar designed for vertically profiling clouds and precipitation along its nadir track (Tanelli et al., 2008).

1-3.2 Radiometers

> ▶ All matter radiates electromagnetic energy as a consequence of the movement of atoms and molecules in its material. ◀

A material also may absorb and/or reflect energy incident upon it. The spectral, polarization, and angular variations of the radiation emitted, absorbed, and

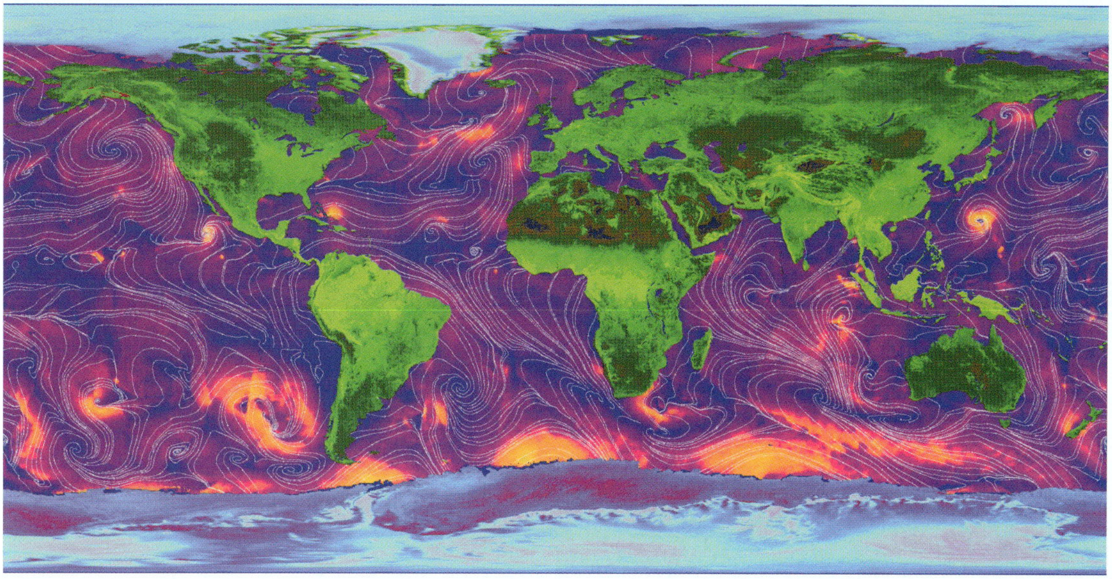

Figure 1-8: Global backscatter over land and ice derived from the SeaWinds spaceborne scatterometer. Wind speed and direction derived from ocean backscatter is shown over the ocean.

Table 1-3: History of civilian remote sensing altimeters on spacecraft.

Years	Spacecraft/Instrument	Band	Height accuracy[a] (cm)
1973	Skylab/S193	Ku	–
1975	Geos3	C	–
1978	Seasat/ALT[b]	Ku	10
1983	Venera 15/16[b]	S (8 cm)	
1985–1990	Geosat[d]	Ku	
1978–1992	Pioneer Venus-1/ORAD[b]	S (1.757 GHz)	
1990–1992	Magellan/ALT[b]	S (2.385 GHz)	3000
1991–1996	ERS-1/RA[d]	Ku	
1995–2011	ERS-2/RA[d]	Ku	
1981–2006	TOPEX-Poseidon/SSALT, NRA[d]	C, Ku	5
1997–	Cassini/ALT[b]	Ku	100
1998–2008	GFO[d]	C, Ku	
2001–	Jason-1/Poseidon-2[d]	C, Ku	
2002–2008	Envisat/RA 2[d]	Ku	
2006–	CloudSat/CPR[e]	W	5000
2008–	Jason-2/Poseidon-3[d]	C, Ku	2.5
2010–	Cryosat-2/Siral[c]	Ku	
2011–	HaiYang-2[d]	C, Ku	
2013	Saral/AltiK[d]a	Ka	

[a]Reported/estimated based on post processing; [b]Designed for planetary topography measurement; [c]Designed primarily for ice-sheet elevation and sea-ice freeboard; [d]Primarily designed for ocean topography; [e]Designed for cloud profiling.

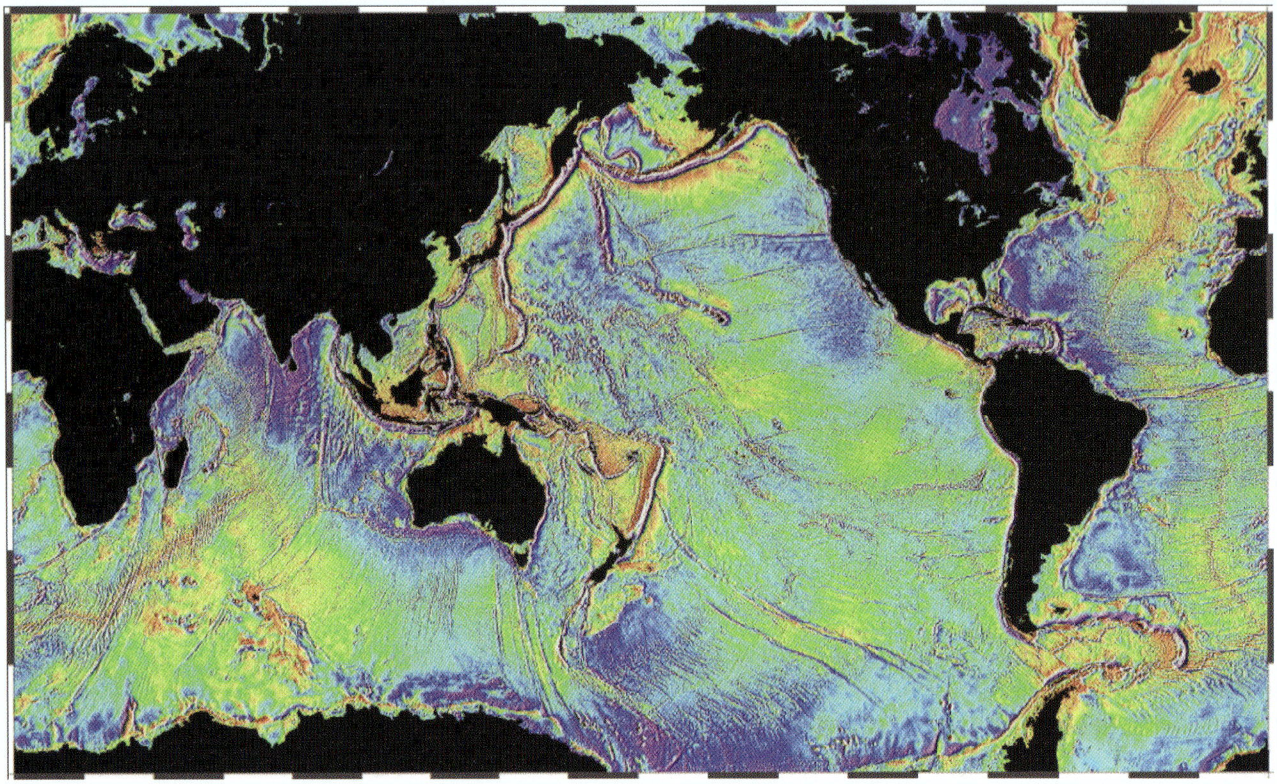

Figure 1-9: Shape of the geoid derived from spaceborne altimetry data by the ERS-1 altimeter. [Courtesy of D. Sandwell and W. Smith.]

scattered by a medium are governed by the geometrical configuration of the surface and interior of the medium, and by the spatial distributions of its temperature and dielectric properties. By measuring the electromagnetic energy radiated by a material, it is possible to infer some of these properties.

▶ Radiometry is the measurement of electromagnetic radiation. A ***microwave radiometer*** is a highly sensitive receiver capable of measuring low levels of microwave radiation. ◀

When a scene, such as terrain, is observed by a microwave radiometer (via its antenna beam), the radiation received by the antenna is partly due to self-emission by the scene and partly due to reflected radiation originating from the surroundings, such as the atmosphere. Through proper choice of the radiometer instrument parameters (wavelength, polarization, and viewing angle), it is sometimes possible to establish useful relations between the magnitude of the energy received by the radiometer and specific terrestrial or atmospheric parameters of interest. For example, observations of bare soil surfaces have shown that the radiometric response in the 20 to 30 cm wavelength range is influenced strongly by the soil moisture content (Schmugge, 1978). Such a dependence can be used over large areas to remotely map soil moisture content, an important physical parameter in many hydrological, agricultural, and meteorological applications. In many cases, multiwavelength observations can be processed to estimate more than one geophysical variable or to predict the profile of a particular variable. Satellite-borne radiometer systems employ multiwavelength radiometric observations to generate global maps of atmospheric temperature profiles and of total water vapor and liquid water over the ocean.

Microwave radiometric techniques were first developed in the 1930s and 1940s to measure electromagnetic energy of extraterrestrial origin—antennas were pointed away from the Earth. Terrestrial microwave radiometric remote sensing had its beginnings in the late 1950s. Using a 4.3 mm wavelength radiometer, which had been developed to measure the solar temperature and atmospheric attenuation, Straiton's group at the University of Texas made radiometric observations of several terrestrial materials, such as water, wood, grass, and asphalt (Straiton et al., 1958). These observations represent the first deliberate effort in which the radiometer antenna was pointed "downward" rather than "upward."

Since then, the science of microwave radiometry has established itself as an integral part of the general field of environmental remote sensing. Numerous investigations have been conducted to evaluate the use of passive microwave sensors for meteorological, hydrological, oceanographic, and military applications. Experience with microwave radiometric data acquisition from spacecraft dates back to December 14, 1962, when the sensors aboard the ***Mariner 2*** spacecraft provided the first close observation of the planet Venus (Barrett and Lilley, 1963). Among the sensors flown aboard Mariner 2 was a two-channel microwave radiometer operating at wavelengths of 1.35 and 1.9 cm. Passive observations of the planet Earth were initiated in 1968 by the Russian satellite ***Cosmos 243***, which was equipped with four microwave radiometers (Basharinov et al., 1971). The 1978 Seasat-1 spacecraft included a microwave radiometer, as well as several different microwave radars, as shown in Fig. 1-6 (Barrick and Swift, 1980).

▶ Since 1968, microwave radiometers have been carried aboard many spacecraft (Table 1-4) and have become a standard tool for Earth observation, particularly over the ocean. ◀

Early radiometers had a limited number of channels, but recent radiometers have many channels, which may include fully polarimetric capability. Figure 1-10 illustrates global brightness temperature measurements of the Earth collected by the ***Advanced Microwave Scanning Radiometer*** (AMSR). The linear spatial resolution of these radiometer images is comparatively coarse, on the order of 25 to 50 km, which is typical of many spaceborne radiometers flown to date (Table 1-4). This is a consequence of the fact that the spatial resolution of a radiometer is governed by the antenna beamwidth and the platform altitude. Since the beamwidth is approximately equal to the reciprocal of the number of wavelengths across the antenna, very narrow beamwidths require very large antenna sizes. This problem is particularly acute for low frequency radiometer systems, which are desirable for measuring sea surface salinity and soil moisture. To overcome this limitation, ***synthetic-aperture radiometer systems***, also known as ***interferometers***, have been developed to improve radiometric imaging resolution. Interferometers were originally developed for use in radio astronomy (Ryle, 1952; Ryle and Hewish, 1960).

Airborne synthetic-aperture radiometers were operated as early as 1990 (Le Vine, 1999). The first synthetic-aperture radiometer to fly in space was the ***Soil Moisture and Ocean Salinity*** (SMOS) mission launched in 2009. Using 69 antennas arrayed along three

AMSR

Table 1-4: History of microwave radiometry on spacecraft.

Year(s)	Spacecraft Instrument Acronym	Frequencies (GHz)	Antenna type	Swath width (km)	Resolution[b] (km)
1962	Mariner 2 Venus Flyby	15.8, 22.4	mechanically scanned parabola	planetary	1300[a]
1968	Cosmos 243	3.5, 8.8, 22.3, 37	horns		13
1972	Nimbus 5 / ESMR	19.3	electrically scanned array	3000	25
	Nimbus 5 / NEMS	22.2, 31.4, 53.6, 54.9, 58.8	lens-loaded horns	nadir	200
1973	Skylab / S193	13.9	mechanically scanned parabola	180	16
	Skylab / S194	1.4	phased-array	nadir	115
1974	Meteor	37			
1975	Nimbus 6 / ESMR	19.3	electrically scanned array	3000	20×43
	Nimbus 6 / SCAMS	22.2, 31.6, 52.8, 53.8, 55.4	rotating hyperbolic mirrors	2700	150
1978	DMSP / SSM/T	50.5, 54.2, 54.3, 54.9, 58.4, 58.8, 59.4	rotating mirror	1600	175
1978, '78	TIRO N/MSU	50.3, 53.7, 55.09, 57.6	rotating mirror	2300	110
1978–87	Seasat 1 / SMMR	6.6. 10.7, 18, 21, 37	oscillating offset parabolic reflector	600	14×21
1978–	Nimbus 7 /SMMR	6.6. 10.7, 18, 21, 37	oscillating offset parabolic reflector	800	18×27
1987, '90, '91 '97, '99	DMSP /SSM/I	19.35, 22.2, 37, 85.5	spinning offset parabolic reflector	1400	15×13
1989	COBE	31.56, 53, 90	differential horns	full-sky	
1997–	TRMM TMI	10.7, 19.3, 21.3, 37, 85.5	spinning offset parabolic reflector	878	6
1998, '00, '02, '05	DMSP / AMSU-A/B	23.8, 89, 31.4	spinning offset reflector	1650	45
2001–11	Aqua / AMSR-E	6.9, 10.7, 18.7, 23.8, 36.5, 89	spinning offset parabolic reflector	1445	4×6
2001	WMAP	22, 30, 40, 60, 90	differential horns	full-sky	
2002	ADEOS-II / AMSR	6.9, 10.7, 18.7, 23.8, 36.5, 89	spinning offset parabolic reflector	1400	4×6
2003–	Coriolis / WindSat	6.68, 10.7, 18.7, 23.8, 37.0	spinning offset parabolic reflector	1000/400 (fore/aft)	8×13
2009	SMOS	1.41	synthetic aperture	1500	40
2011	Aquarius	1.413	pushbroom offset parabolic reflector	340	62×68
2014–15	SMAP	1.41	conically scanning mesh reflector	1000	39×47

[a]Courtesy of L. King, NASA/GSFC; [b]resolution of highest frequency

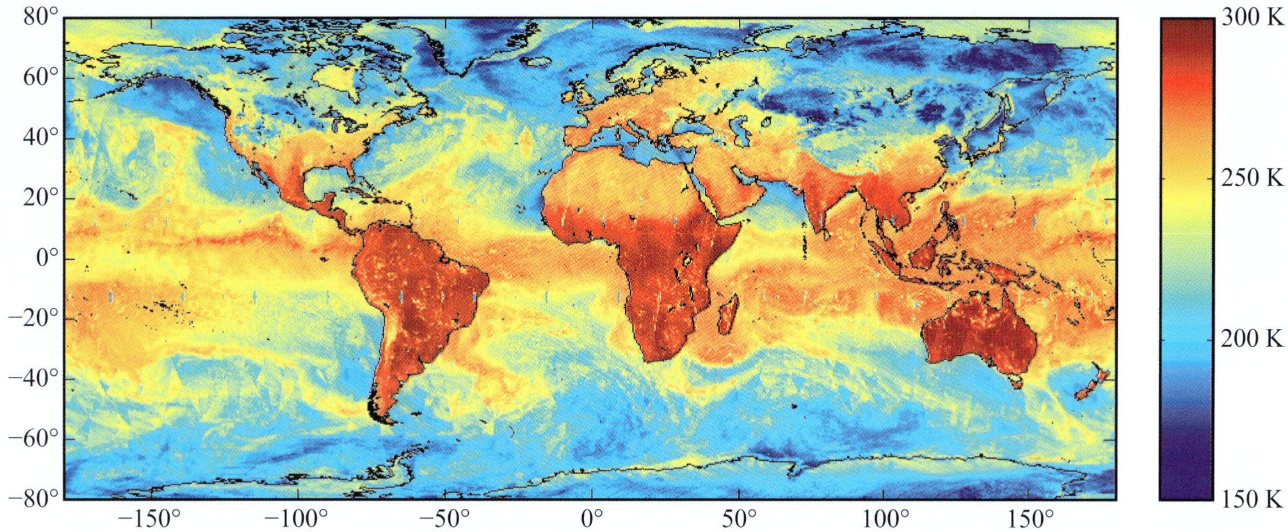

(a) AMSR-E, 89.0 GHz h-polarized

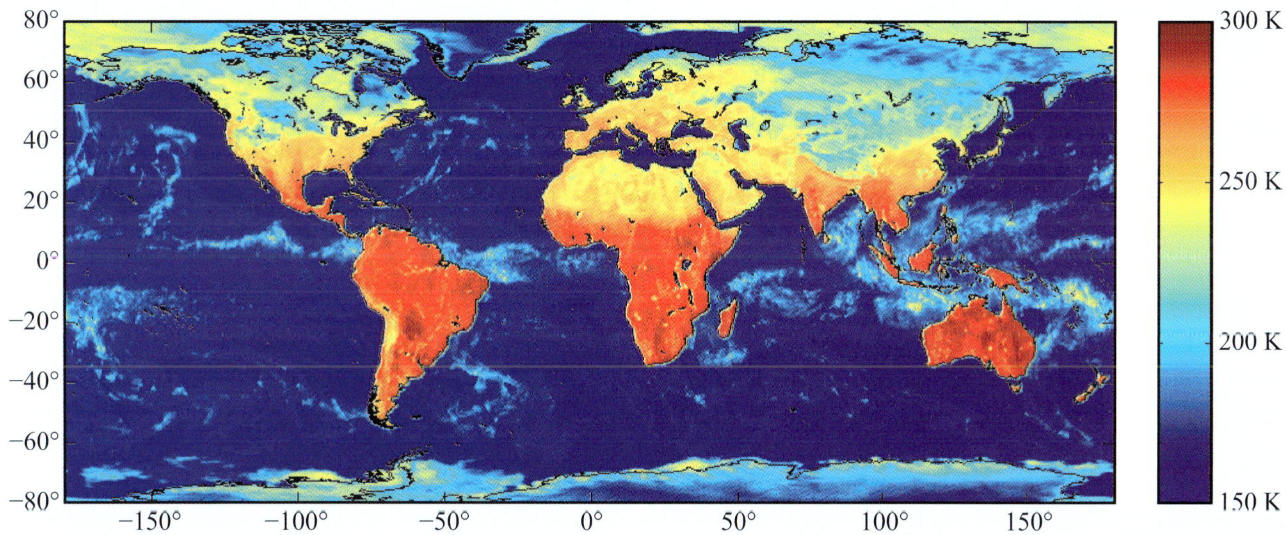

(b) AMSR-E, 36.5 GHz h-polarized

Figure 1-10: Passive microwave images of the measured brightness temperature of the globe collected by the Advanced Microwave Sensor Radiometer on the Earth Observation System (AMSR-E): (a) 89 GHz h-pol, (b) 36.5 GHz h-pol.

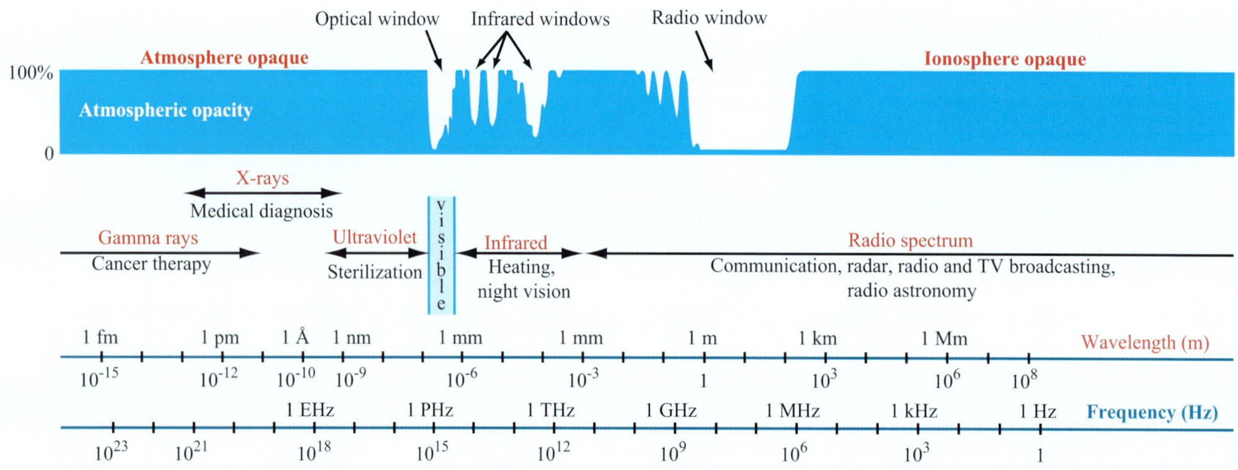

Figure 1-11: The electromagnetic spectrum. Atmospheric opacity for the Earth is shown along the top.

arms spanning 8 m, SMOS collected L-band radiometry measurements that could be processed to an effective spatial resolution of 35 km over a 1000 km wide swath (Kerr et al., 2010; Font et al., 2000).

1-4 The Electromagnetic Spectrum

Figure 1-11 provides a spectral plot of ***atmospheric opacity*** under clear-sky conditions. The electromagnetic spectrum spans over many decades in frequency (or wavelength). The lowest frequencies (longest wavelengths) constitute the radio spectrum, which includes the microwave portion of the spectrum. At higher frequencies (shorter wavelengths) above the radio-frequency spectrum lies the infrared spectrum, followed by the visible range. Above the visible spectrum lies the ultraviolet spectrum and, overlapping it, the X-ray spectrum. Finally, at the highest frequencies are the gamma rays. Remote sensing is practiced across most of the electromagnetic spectrum. However, our primary interest in this book is in the radio portion of the spectrum, and specifically in the microwave region. The atmospheric opacity, the ability of the signal to penetrate the atmosphere, is plotted along the top of Fig. 1-11. ***Atmospheric transmissivity*** (the inverse of

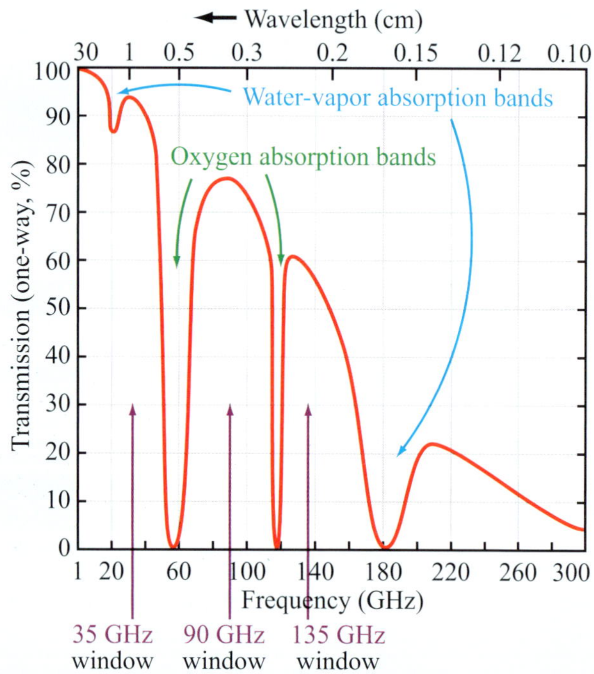

Figure 1-12: Percentage transmission through the Earth's atmosphere, along the vertical direction, under clear sky conditions.

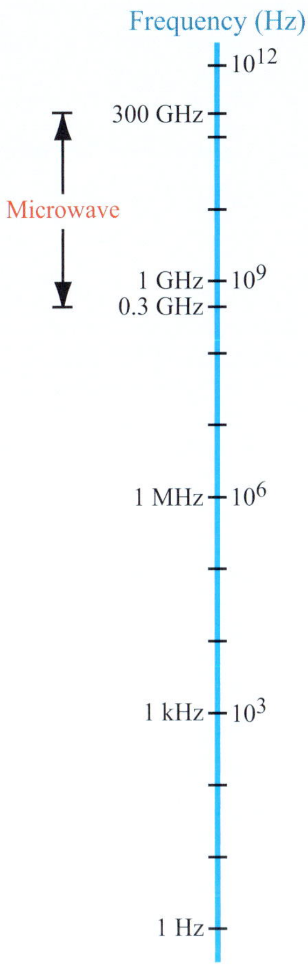

Band	Applications
Extremely High Frequency (30–300 GHz)	Radar, advanced communication systems, remote sensing, radio astronomy
Super High Frequency (3–30 GHz)	Radar, satellite communication systems, aircraft navigation, radio astronomy, remote sensing
Ultra High Frequency (300 MHz – 3 GHz)	TV broadcasting, radar, radio astronomy, microwave ovens, cellular telephone
Very High Frequency (30–300 MHz)	TV and FM broadcasting, mobile radio communication, air traffic control
High Frequency (3–30 MHz)	Shortwave broadcasting
Medium Frequency (300 kHz – 3 MHz)	AM broadcasting
Low Frequency (30–300 kHz)	Radio beacons, weather broadcast stations for air navigation
Very Low Frequency (3–30 kHz)	Navigation and position location
Ultra Low Frequency (300 Hz – 3 kHz)	Audio signals on telephone
Super Low Frequency (30–300 Hz)	Ionospheric sensing, electric power distribution, submarine communication
Extremely Low Frequency (3–0 Hz)	Detection of buried metal objects
	Magnetotelluric sensing of the Earth's structure

Figure 1-13: The RF spectrum and some of its applications.

opacity) over the microwave band is shown in Fig. 1-12. Atmospheric opacity and transmissivity play a key role in frequency selection for remote sensing within or through the atmosphere.

For example, because of water vapor absorption near 22 and 183 GHz and oxygen absorption near 58 and 119 GHz, these frequencies are used almost exclusively for radiometric observations of the atmosphere. In contrast, because at frequencies below 20 GHz the atmosphere is essentially transparent, and in atmospheric "windows" signal attenuation is tolerable, sensors observing the Earth's surface through the atmosphere are operated at frequencies below 20 GHz and at those within the transmission windows shown in Fig. 1-12. Short-range radar applications such as vehicle anticollision radars operate at 77 GHz where atmospheric attenuation helps minimize interference from distant radars.

Figure 1-13 summarizes some of the applications of the radio spectrum, including communication, navigation, broadcasting, radar, and passive remote sensing. Since the spectrum is a finite resource, these services must share the spectrum, so national governments coordinate with international organizations such as the International Telecommunication Union (ITU) for this purpose. Letter designations adopted

Table 1-5: Common band designations (GHz).

Band	Frequency Range (GHz)
P	0.225–0.390
L	0.390–1.550
S	1.550–4.20
C	4.20–5.75
X	5.75–10.9
K	10.9–36.0
Q	36.0–46.0
V	46.0–56.0
W	56.0–100

Table 1-6: Radar frequency allocations for remote sensing (GHz) summarized from the 1992 WARC. (All are shared with other services.)

1.215–1.300, 3.1–3.3, 5.25–5.35, 8.55–8.65
9.50–9.80, 13.4–14.0, 17.2–17.3, 24.05–24.25
35.5–35.6, 78.0–79.0

Examples of Other Allocations	
Radar altimeter	4.2–4.4
Doppler navigator	8.8, 13.25–13.40
Meteorological radar	5.6–5.65, 9.3–9.5
Coastal radar	5.35–5.65, 9.0–9.2, 10.0–10.55
Ship radar	5.46–5.47, 9.3–9.5, 14–14.3, 24.25–25.25, 31.8–33.4

internationally by the ITU are shown for decade regions of the spectrum.

► By convention, the ***microwave region*** encompasses the UHF, SHF, and EHF bands, extending from 0.3 to 300 GHz (1 m to 1 mm in wavelength). ◄

Numerous schemes of letter designation for bands within the microwave region exist. A commonly used set of these schemes is indicated in Table 1-5. The various bands also have official ITU subdesignations not identified here. Common usage, however, describes the lower part of ***K-band*** (by convention 10.9 to 22 GHz) as Ku-band, whereas frequencies above 22 GHz in the K-band are normally designated Ka-band. These designations are used commonly in spite of the existence of more precise definitions.

Radars may be found in all of the bands, with the possible exception of the Q- and V-bands, with most remote sensing radars operating at L-band or higher frequencies. Formal administrative approval (from the FCC in the U.S. or from agencies with a similar function in other countries; international frequency allocation is done at infrequent World Administrative Radio Conferences) in the form of a license is required for radar operation within specifically allocated frequency bands. Table 1-6 lists some of these allocations along with some selected non-remote-sensing allocations. Sharing between radar remote-sensing systems and other radars is usually not permitted. Thus, the designer of a remote-sensing radar system cannot arbitrarily choose a frequency and use it.

Radiometers are also found throughout the microwave spectrum. The measurement precision of a microwave radiometer is proportional to $(B\tau)^{-1/2}$, where B is the receiver bandwidth and τ is the integration time. Since the radiometer is on a moving platform and may be scanning to achieve a wide swath, τ is typically constrained. Thus, to achieve precision radiometry it is desirable to use as wide a bandwidth as possible. However, since radiometers are very sensitive to interference, they usually use protected frequencies exclusively allocated for passive sensing. Allocations for these purposes are summarized in Table 1-7. Passive microwave systems share frequency allocations with radio astronomy, and these frequencies are protected against transmission of any kind. However, since some of the radio-astronomy bands are too narrow for airborne and spaceborne passive sensors, additional frequencies have been allocated for them on a shared basis (see Table 1-7), but users of such systems must be aware of the likelihood that some points on the globe will be inaccessible for their sensing because of interference from ground transmitters. No licenses are required for passive microwave remote sensing, though the designer must be prepared to deal with possible interference, even when operating within allocated passive-only bands. As an illustration, Fig. 1-14 shows a map of the

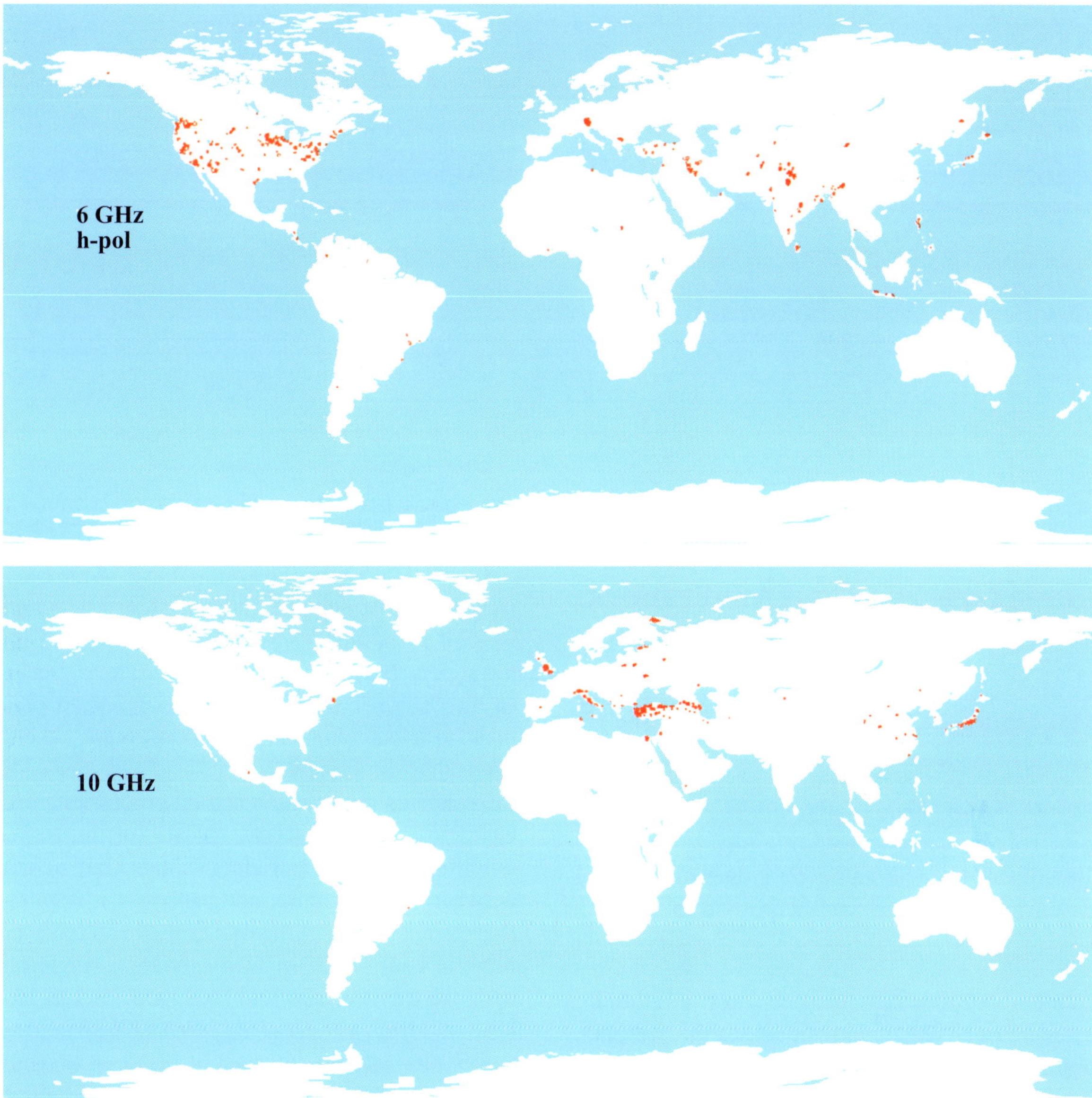

Figure 1-14: Intensity of radio frequency interference (RFI) observed by the WindSat polarimetric radiometer at 6 GHz h-pol and 10 GHz polarimetric. [Images courtesy of William Johnson and Li Li, Naval Research Laboratory.]

location of interference observed in 2008 by the WindSat polarimetric radiometer.

Table 1-7: Passive-sensor frequency allocations (GHz).

0.404–0.406a	15.35–15.40a	89–92a
1.370–1.400s	21.2–21.4p	100–102p
1.400–1.427s	22.21–22.5s	105–116a
1.6605–1.6684p	23.36–24.0a	116–126p
2.600–2.640s	31.3–31.5a	150–151p
4.2–4.4s	31.5–31.8p	164–168a
4.80–4.99s	36–37p	174.5–176.5p
6.425–7.250s	50.2–50.4p	200–202p
10.60–10.68p	51.4–54.25a	217–231a
10.68–10.70a	54.25–58.2p	235–238p
15.20–15.35s	64–65a	250–242a

a: protected for radio astronomy—no transmitters allowed;
p: shared, with primary use for services having transmitters;
s: shared, with secondary use for services having transmitters.

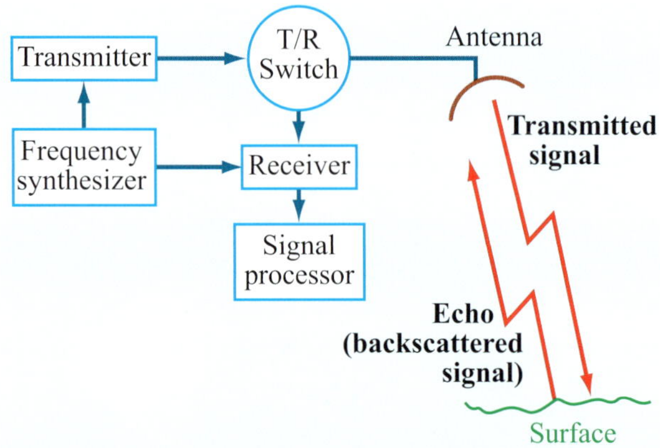

Figure 1-15: Basic elements of a remote-sensing radar system.

1-5 Basic Operation and Applications of Radar

1-5.1 Operation of Remote-Sensing Radars

▶ The term ***radar*** stands for radio detection and ranging. This term was coined at a time when the detection of ships and aircraft and the determination of their ranges was the primary purpose of radar. ◀

Today some radio devices measure only amplitude; others measure only speed. In many cases the presence of a target, such as the ground, is known; thus, "detection" may be unnecessary. The term "radar" is now applied to any radio device in which a transmitter illuminates a reflecting or scattering surface or object and a receiver measures some property of the reflected signal. Fundamentally, a radar can determine (1) direction (by the pointing of the antenna), (2) distance (by measuring the time of flight of the signal to and from the target), (3) speed (by measuring the Doppler shift of the echo), and (4) radar scattering cross section (by comparison of the energy of the echo return to that of the transmitted signal).

Figure 1-15 illustrates the basic components of a typical remote sensing radar system. The ***transmitter*** is used to produce a signal that is coupled into space through a transmitting antenna. The signal goes out to the "target," from which it is reflected or scattered back to a receiving antenna that couples the scattered wave into the receiver circuits. The ***receiver*** amplifies the echoed signal, which is then processed to extract the characteristics of interest. In all but the very simplest amplitude-measuring radars, some kind of correlation between the transmitter and processor is required so that the characteristics of the transmitted and received signals may be compared. In many radars a single antenna is used for transmission and reception. In this case, some kind of a ***transmit-receive switch*** (TR switch) is used to switch the antenna alternately between transmitter and receiver, although with some radars, isolation may be achieved by the use of a ***microwave circulator***. Radar system design and operation is described in more detail in Chapters 13–16.

As previously noted, radar systems used for remote sensing fall into broad categories: ***imaging radars*** (SAR and SLAR), and nonimaging radars such scatterometers, altimeters, and meteorological radars. Many remote-sensing applications call for an imaging radar, but specialized applications, such as measuring winds at sea, use scatterometers. Though classified as a

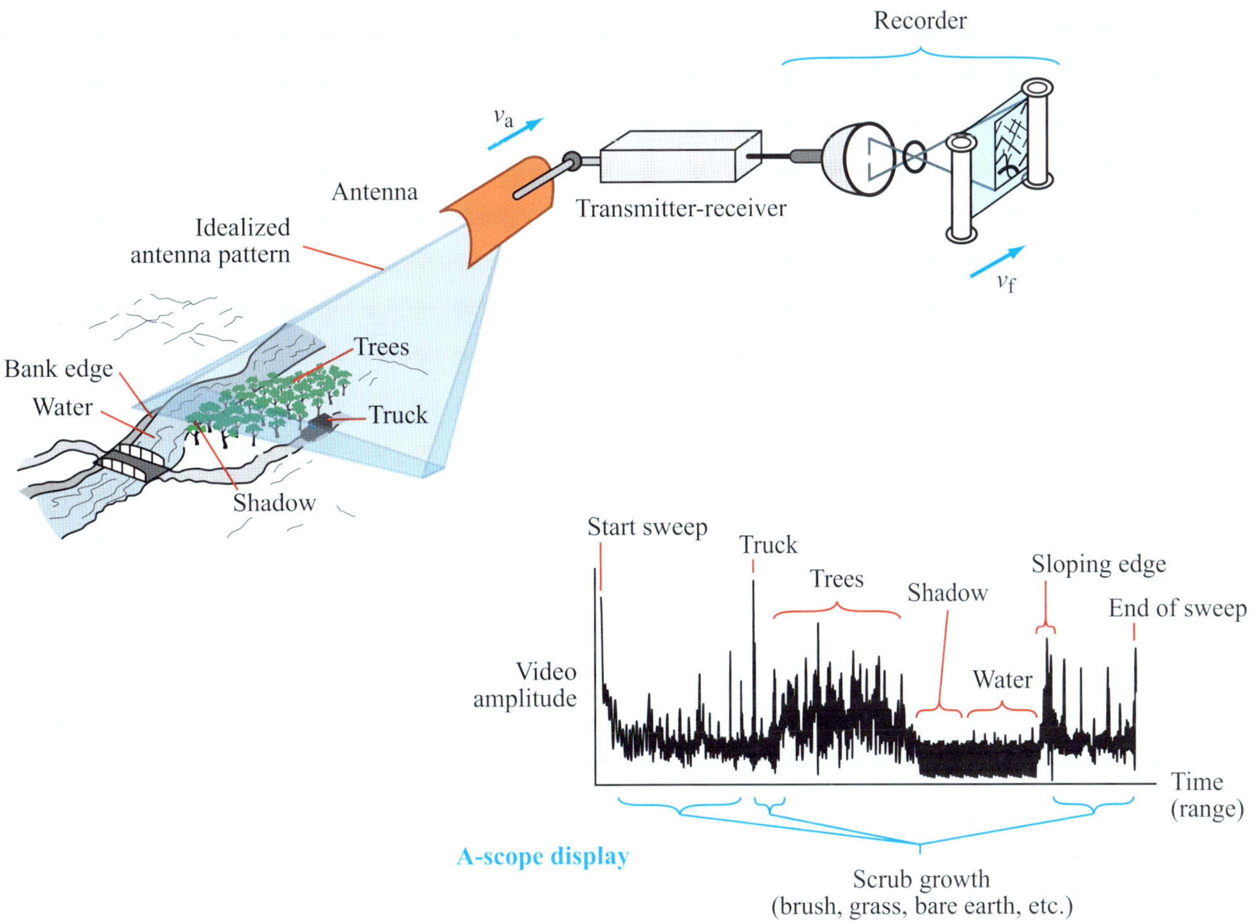

Figure 1-16: Real-aperture SLAR imaging technique.

nonimaging sensor, spaceborne scatterometers can be used to generate low-resolution imagery.

In side-looking airborne radar (SLAR) imaging, the antenna points to the side with a beam that is wide vertically and narrow horizontally. The image is produced by motion of the aircraft past the area being covered, as illustrated in Fig. 1-16. In this simple pulsed system, a short pulse is transmitted from the airborne radar, primarily within the horizontally narrow beam shown. When the pulse strikes a target, a signal returns to the aircraft.

▶ The time delay associated with this received signal, as with other pulse radars, gives the distance between target and radar. ◀

The picture illustrates a typical return at a particular instant in the flight, with strong signals coming from clumps of trees and bridges and no signal coming from the radar "shadow" of a stream bed. When a single pulse is transmitted, the return signal can be displayed on an oscilloscope as shown in Fig. 1-16. In early SLAR systems, the intensity of the signal was recorded on a moving strip of film that moved synchronously with the motion of the aircraft, so that as the aircraft moved forward the film also moved. As the aircraft moved, the series of intensity lines formed an image, a two-dimensional picture of the radar-return from the surface. In practice, film-based recording has significant limitations and modern systems store the data digitally.

▶ Side-looking airborne radars can be divided into two groups: real-aperture systems that depend on the beamwidth determined by the actual antenna, and synthetic-aperture systems that depend upon platform motion and signal processing to achieve a much narrower beamwidth in the along-track direction than that attainable with the real antenna alone. ◀

The customary nomenclature used is "SLAR" for the real-aperture system and "SAR" for the synthetic-aperture radar (SAR) system, although the latter is also a type of a side-looking airborne radar.

The along-track resolution of a SLAR is proportional to the product of the antenna beamwidth and the distance to the target. Thus, the resolution of a SLAR varies with cross-track distance and large antennas are required for fine resolution. A major advantage of SAR over SLAR is that the resolution of the SAR in the along-track direction is proportional to the length of the synthetic antenna and is independent of the distance. (This key result is derived in Chapter 14.) Thus, a SAR may be used in either an aircraft or a spacecraft without compromising the along-track resolution. Unlike SLAR, fine resolution for a SAR requires a *short* antenna rather than a long one! Figure 1-17 illustrates how SAR combines data collected from multiple antenna positions to synthesize a longer effective antenna, thereby achieving high along-track resolution. Many modern SAR systems include multiple polarization or polarimetric capabilities that yield information on scattering mechanisms and enhance the utility of the SAR image data (Ulaby and Elachi, 1990; Cloude and Pottier, 1996; Touzi et al., 2004).

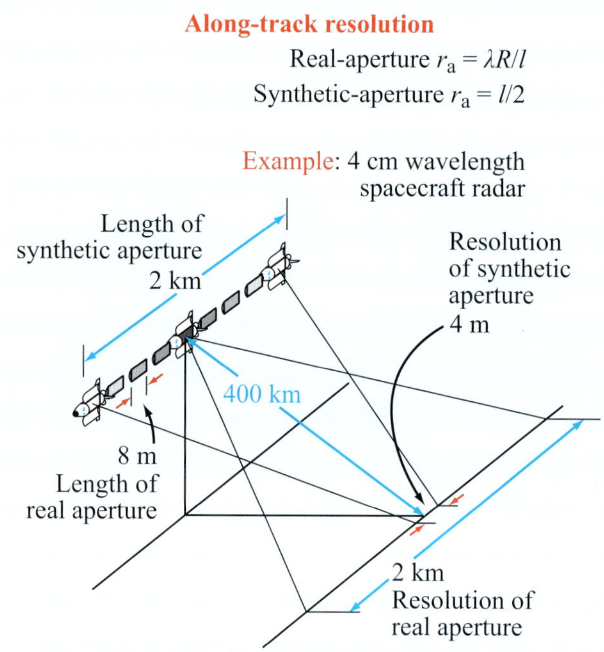

Figure 1-17: An illustration of how synthetic aperture works.

The most commonly used nonimaging remote-sensing radars include meteorological radars, scatterometers, and altimeters. ***Wind scatterometers*** collect backscatter measurements of the same location at multiple aspect angles and/or frequencies, generally over a wide swath. Accurate backscatter measurements are the essential function of a scatterometer. Careful attention to thermal design and internal calibration schemes are required to achieve the desired accuracy over a time period of up to a decade for spaceborne systems. For ocean measurement they must make multiple measurements over a diversity of incidence and/or azimuth angles. This is accomplished with either multiple fan-beams (Naderi et al., 1991) or with rotating pencil-beam antennas (Spencer et al., 1997), as illustrated in Fig. 1-7. For fan-beam antennas, Doppler processing and/or range gating is used to obtain along-beam resolution.

▶ Satellite altimeters measure the distance between the platform and the surface with a measurement accuracy on the order of 1 cm. ◀

While radar range measurement accuracy is critical, high precision orbit determination is also required to achieve this level of accuracy over the ocean (Zieger et al., 1991). The need to avoid aliasing from tides leads to the use of unusual orbits for altimeters, such as the 1600 km orbit employed by the ***TOPEX/Poseidon*** mission (Fu et al., 1994; Fu and Cazenave, 2001). The shape of the pulse return can be used to extract information about surface waves, and the mean backscatter provides information about wind-induced surface roughness.

Meteorological radars measure radar backscatter and ***Doppler shift*** as a function of distance to observe and map precipitation and wind. They often combine multiple beams in rotating antennas along with range gating to collect data in three dimensions (Cobb, 2004; Meischner, 2004).

1-5.2 Applications of Remote-Sensing Radars

Remote-sensing radar has become a standard tool for Earth observation and new applications are being continually developed. Table 1-8 summarizes some common applications of airborne and spaceborne imaging radars (SLAR and SAR) in remote sensing.

> ▶ The primary application of nonimaging radars to remote sensing is in oceanography and polar ice, where spaceborne radar altimeters and scatterometers are used. ◀

The altimeter has been used to measure the detailed characteristics of the geoid, to establish characteristics of tides and even of current flows by very precise measurement of the deflection of the ocean associated with the currents (Fu and Cazenave, 2001; Leitao et al., 1978), and to determine the height of waves directly beneath the spacecraft (Gower, 1979). Such information aids in predicting El Niño and La Niña events. Radar scatterometers enable accurate determination of the wind speed and direction over the ocean, as well as rain rate (Naderi et al., 1991). Scatterometers have also been used for sea-ice mapping and iceberg tracking as well as soil moisture and vegetation. Reconstruction-based processing techniques (Early and Long, 2001) have been developed to enable scatterometers to be used as imaging sensors, which has helped expand their applications.

Radar imaging has been used extensively in support of many geologic applications. It is used in identifying

Table 1-8: Selected applications of airborne/spaceborne imaging radar in remote sensing.

Geology:
- Structure
- Lithology

Hydrology:
- Soil moisture
- Watershed mapping
- Flood mapping
- Mapping of surface water (ponds, lakes, rivers)
- Snow mapping

Agriculture:
- Crop mapping
- Agricultural-practice monitoring
- Identifying field boundaries
- Monitoring growth and harvest progress
- Identifying stress areas
- Rangeland monitoring
- Water problems–same as hydrology

Forests:
- Monitoring cutting
- Mapping fire damage
- Identifying stress areas
- Vegetation density

Cartography:
- Topographic mapping
- Land-use mapping
- Monitoring land-use changes, urban development, etc.

Polar regions (cryosphere):
- Monitoring and mapping sea ice
- Detecting and tracking icebergs
- Mapping glacial ice sheets
- Monitoring glacial changes, including measuring velocity

Ocean:
- Measuring wave spectra
- Monitoring oil spills
- Monitoring ship traffic and fishing fleets
- Wind speed and direction measurement
- Rain
- Clouds
- Measuring currents
- Undersea mapping

geologic structures with the goal of mineral exploration and general geologic mapping; of lesser importance is the direct determination of lithologies. This is effective in desert regions where the rocks are exposed, but less so in heavily vegetated regions where the radar signal comes largely from the vegetation, since differences in the underlying rock cannot be detected except by surrogates in the vegetation signature.

While scatterometers have been used for soil moisture mapping, imaging radar has not yet been used operationally for soil moisture determination, though experiments have demonstrated that SAR images do provide useful soil moisture information (Ulaby et al., 1974, 1979, 1982a; Oh et al., 1992; Dubois et al., 1995a). Radar images have been used for mapping floods, a task made easy by the strong contrast between a body of water and a land surface. Missions such as SMAP are planned to provide global soil moisture mapping from space (Entekhabi et al., 2010). This mission will also enable mapping the extent, water content, and state of snow cover, which are important in determining future runoff and whether to forecast floods or the amount of water that will be available for irrigation or hydroelectric power generation (Stiles and Ulaby, 1980a).

Radar is particularly useful in polar regions because of the long period of darkness and the extensive presence of ice fogs, both of which limit the use of sensors at visible wavelengths. Radar's ability to map the characteristics of sea ice has been well demonstrated (Page and Ramseier, 1975).

Spaceborne SARs have been used for lake and ice monitoring and have been used extensively to map glacial ice and determine the speed of glacier flow in Greenland, Antarctica, and other regions (Jezek, 2003). SAR images are used operationally to map sea ice and to detect and track icebergs in the polar regions. Computer-based techniques for spatially combining and processing scatterometer measurements have made it possible to produce scatterometer (and radiometer) imagery suitable for many applications. While scatterometer images are of much lower resolution than the images generated by SLARs and SARs, they are finding important applications in climate research, particularly in the polar regions.

The Seasat SAR first demonstrated that certain characteristics of wave patterns can be monitored on a global scale (Beal, 1980). Since then, most spaceborne SARs now include a "wave mode" to measure the ocean surface wave spectrum and detect and track oil spills (Kraus et al., 1977; Moncrief, 1980; Lehner et al., 2000). Because the radar signal from the ocean's surface is largely determined by capillary waves, which are dampened by the presence of oil, a dark area in an ocean image can be an indication of an oil layer. Though of lower resolution, scatterometer images can also be used for mapping large oil spills (Lindsley and Long, 2012). The ability of radar to monitor ship traffic, both from the shore and from other ships, is well known and is being exploited by spaceborne SARs such as ***TerraSAR-X*** (Moreira and Bamler, 2010).

1-6 Basic Operation and Applications of Radiometers

1-6.1 Radiometer Operation

Unlike the signal received by a radar, which consists of energy scattered back by the target after it has been illuminated by the radar's transmitting antenna, the "transmission source" for a radiometer is the target itself, and the radiometer is merely a passive receiver. The energy received by a radiometer is due to radiation emitted and/or reflected by the scene and collected by the antenna. The total-power radiometer shown in Fig. 1-18 consists of an antenna, an amplifying predetection section, a detector, an integrator (low-pass filter) and a recording device.

As discussed in detail in Chapter 6, the power P emitted by an object in thermodynamic equilibrium is a function of its physical temperature T, and in the microwave region P is directly proportional to the ***brightness temperature*** T_{B},

$$P = kT_{\mathrm{B}}B, \tag{1.1}$$

where k is Boltzmann's constant and B is the radiometer bandwidth. If the material has a constant physical

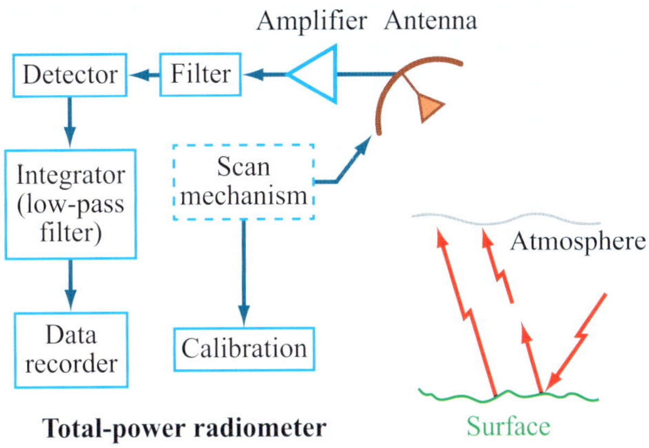

Figure 1-18: Key elements of a microwave radiometer.

temperature T, the material is said to have an emissivity $e = T_B/T$ where e varies between 0 for a perfectly nonemitting material and 1 for a perfect emitter known as a ***blackbody***.

Ideally, if the scene observed by the antenna beam is characterized by a uniform brightness temperature T_B (representing radiation in the direction of the antenna), then the observed antenna power is the same as P. In reality, however, the observed power represents all radiation incident upon the antenna, integrated over all possible directions and weighted according to the antenna directional pattern. Other factors play an important role, including the effects of the atmosphere and self emission by the antenna structure, as well as noise generated internal to the receiver. These factors are considered in Chapters 6 and 7.

While it was stated earlier that a radiometer is merely a passive receiver, a radiometer receiver differs from a traditional radar or communications receiver in two key respects. First, whereas the input signal that is processed by a traditional receiver may be phase-coherent and nearly monochromatic, the natural radiation emitted by material media is phase-incoherent and extends over the entire electromagnetic spectrum. That is, it is "noiselike" in character and similar to the noise power generated by the receiver components.

The second difference relates to the ***signal-to-noise*** ratio, S_n, at the receiver output. In traditional receivers, faithful extraction of the information contained in the received signal necessitates that $S_n \gg 1$ to differentiate the signal from the fluctuating component of the noise. This condition usually is achieved through a combination of amplitude and waveform design of the transmitted signal and the application of signal-processing techniques in the receiver section, although scatterometers often use methods similar to those used in radiometers (Chapter 7). However for radiometers, the radiometric signal to be measured usually is much smaller than the receiver noise power. Highly sensitive receivers with processing that averages in time and frequency are required to extract precise measurements of T_B. High stability and precise, frequent calibration are also required. To aid in this, several different receiver configurations have been developed, some of which are discussed in Chapter 7.

Typical radiometers measure T_B of both vertically polarized and horizontally polarized signals, each of which conveys information about the scene. As explained later in Chapter 7, in ***polarimetric radiometry*** all four Stokes polarization parameters are measured (Gasiewski and Kunkee, 1993; Tinbergen, 1996) from which more information about the observed scene can be inferred.

1-6.2 Applications of Microwave Radiometry

▶ Microwave radiometry is used for (1) astronomical studies, (2) military applications and (3) environmental monitoring. ◀

Although both radars and radiometers are employed in radio astronomy, Earth-based radars are limited to observations of the sun and nearby targets such as the planets and moons within the solar system. Radiometers, on the other hand, have been used to measure the radio emission from numerous objects in our galaxy, as well as from objects in other galaxies.

The primary military use of radiometers has been for detecting or locating metal objects. As was mentioned earlier, the term used for characterizing the emission by the scene observed by the radiometer (through its antenna beam) is the brightness temperature T_B, which may vary from zero kelvins (for a nonemitting medium) to a maximum value equal to the physical temperature T_0 of the scene (for a perfect emitter known as a blackbody). Equivalently, the emissivity e ($= T_B/T_0$) varies between zero and unity. Theoretically, a perfectly conducting material, such as a metal object, has zero emissivity, thereby making it easy to differentiate from the Earth's background (the emissivity for land surfaces is rarely < 0.3 and is often > 0.7). Although metal objects do not self-emit, their radiometric temperatures are not identically zero because they reflect downward-emitted sky radiation. The radiometric contrast between a ***field of view*** (FOV) with and without a metal target contained in it is a function of the ***beam-fill factor*** (ratio of the cross-sectional area of the metal target to the area of the antenna footprint on the ground). The size of the footprint (i.e., the spatial resolution) is governed by the distance between the radiometer and the ground and by the antenna beamwidths, the latter being in turn determined by the size of the antenna and the microwave frequency. Because of its inherently limited spatial resolution even at high microwave frequencies, a missile-borne microwave radiometer used for locating military targets can detect successfully the presence of the target only if the target is within a few hundred meters. A study conducted by Deitz and Constantine (1979), for example, shows that a 35 GHz radiometer at an altitude of up to 100 m above the ground can easily detect a tank against the signal fluctuation due to the ground surrounding it. This altitude range may be increased by a factor of 3 to 4 by operating at higher microwave frequencies, such as the atmospheric-window frequencies located at 94 and 140 GHz, because, for a given antenna size, the FOV varies as the square of the height-to-frequency ratio, $(h/f)^2$. Operationally, microwave radiometers are used in conjunction with other superior spatial-resolution sensors such as radars and infrared scanners, which serve as the prime sensors for detecting the presence of military targets. Microwave radiometers are used as the prime sensors in the final stage as a missile approaches a target area, since no warning signal is emitted by a radiometer. Because the microwave radiometer is a wideband incoherent receiver, its output fluctuates less as the antenna beam scans over a relatively homogeneous terrain than does the output from a "coherent" radar system. This smaller fluctuation translates into a lower probability of false alarm for the radiometer if the target's brightness temperature T_B is different from that of the background.

Table 1-9: Selected applications of airborne/spaceborne microwave radiometers in remote sensing.

Hydrology:
- Soil moisture
- Watershed surface drainage
- Flood mapping
- Mapping of surface water (ponds, lakes, rivers)
- Snowcover extent, snow water equivalent, snow wetness

Agriculture:
- Soil moisture distribution for crop yield estimation and irrigation scheduling
- Delineation of freeze-thaw boundaries

Polar regions (cryosphere):
- Monitoring and mapping sea ice and sea-ice type
- Mapping glacial ice sheets
- Monitoring ice-sheet melt conditions

Ocean:
- Measuring surface wind speed
- Measuring surface temperature
- Measuring surface salinity
- Monitoring oil spills
- Measuring and mapping rain

Meteorology and climatology (primarily over the ocean):
- Temperature profile
- Integrated water vapor
- Water vapor profile
- Liquid water (rain)

Stratosphere, mesosphere, and lower thermosphere:
- Atmospheric temperature profile
- Magnetic field profile
- Abundance of atmospheric gases

In addition to their use in radio astronomy and for military applications, microwave radiometers have found extensive use in several geoscientific fields, including meteorology, oceanography, and hydrology (see Table 1-9). The coarse spatial resolution of satellite microwave radiometers that have flown on satellites to date (see Table 1-4) limits their use to applications compatible with resolutions of the order of 15 km or larger. This means that, with the exception of a few land applications, most radiometric applications are associated with observations of oceanographic parameters or atmospheric parameters over the ocean. These parameters include sea surface temperature and wind speed (Wilheit, 1979), sea-ice type and concentration (Zwally and Gloersen, 1977), atmospheric water-vapor content, liquid-water content, and temperature profiles (Staelin et al., 1973; Wilheit, 1979), rainfall rates (Wilheit et al., 1977), and salinity (Pampaloni and Paloscia, 2000), among others (Narvekar et al., 2007).

Over land, radiometers have proven useful in soil-moisture determination (Newton and Rouse, 1980) and snow accumulation (Chang et al., 1982). These applications are sensitive to soil-surface roughness and (to a lesser extent) vegetation cover. For this reason soil-moisture systems use lower frequencies, which are less sensitive to these confounding parameters. The SMAP mission will include an active sensor operating at essentially the same frequency as the radiometer in order to estimate and thus remove the vegetation contribution (Entekhabi et al., 2010).

Ocean wind speed and direction measurement are currently the primary application of polarimetric radiometry. The wave structure leads to azimuthal variations in the Stokes brightness temperature from which the vector wind creating the wave structure can be estimated using measurements from only a single azimuth direction (Yueh et al., 1994, 1995; Gaiser et al., 2004).

1-7 Image Examples

To introduce the reader to the types of products generated by active and passive microwave sensors, we present Figs. 1-19 through 1-24. These are typical image products that display either the signal backscattered or emitted by a scene, or geophysical quantities derived from them. These six illustrative examples, along with Figs. 1-8, 1-9, and 1-10, are merely the tip of the iceberg, with many more included in later chapters. High-resolution versions of these images are available on the book's website (www.mrs.eecs.umich.edu).

In 1978, the Seasat satellite carried the first civilian SAR in space.

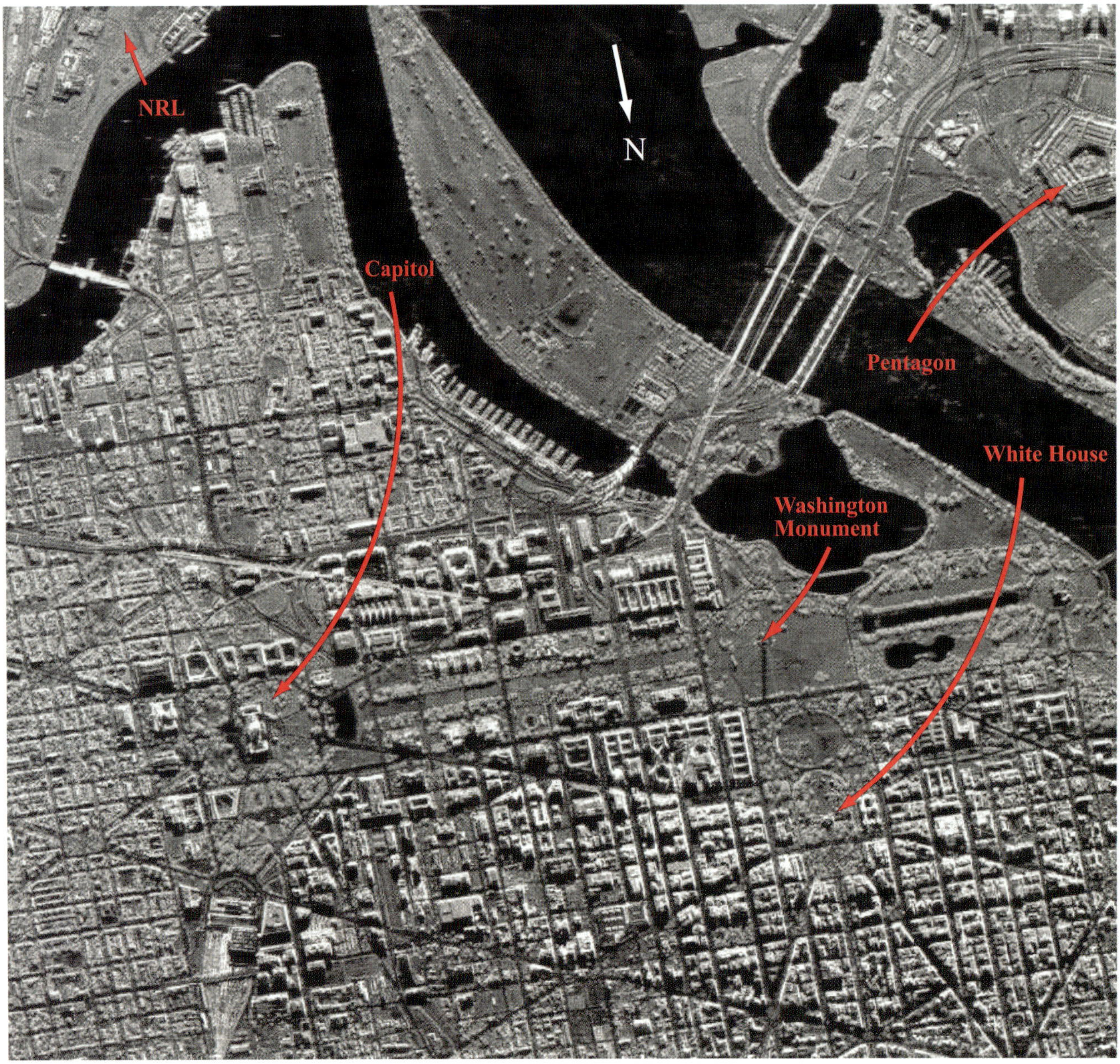

Figure 1-19: This figure presents a Ku-band SAR image collected over Washington, D.C. The SAR illumination is from the top. Man-made features, such as buildings and bridges, stand out clearly in this radar backscatter image. The backscatter intensity of water is much smaller than that of buildings and vegetation, so water shows up black in the image. Individual trees can be distinguished in some areas, and the pattern of roads and urban development is clearly apparent. The Capitol building is left of center with the Mall extending to the right. Right of center is the Washington Monument, though only the shadow of the obelisk is readily apparent in the image. Just below the monument are the Ellipse and the White House. The Naval Research Laboratory is at the upper left. [Courtesy of Sandia National Laboratory.]

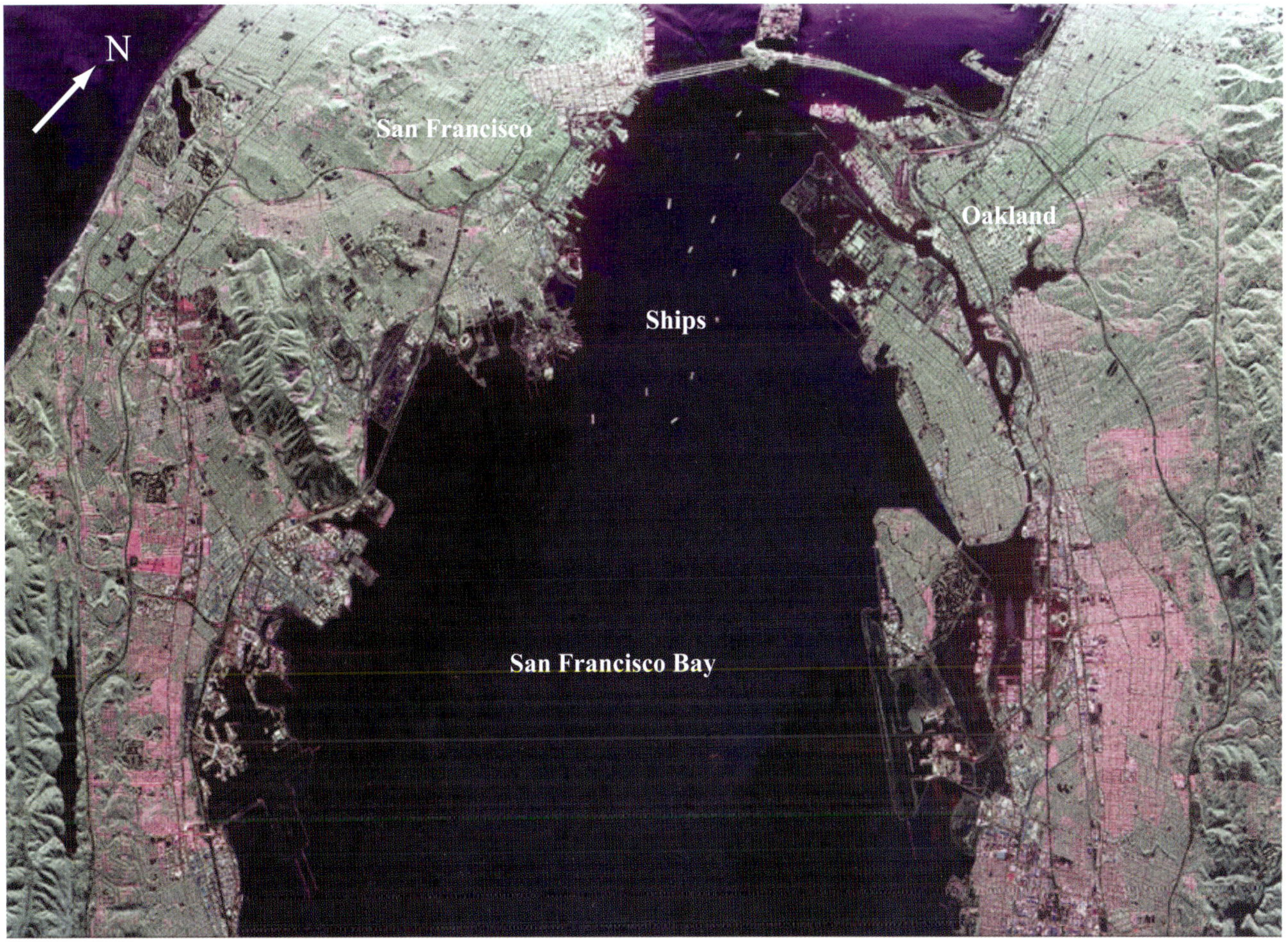

Figure 1-20: SAR image of a portion the San Francisco Bay area. Color describes the polarimetric backscatter for each pixel. The orientation of streets and roads in urban areas alters the polarimetric response in these areas, so the color is related to road orientation relative to the radar view direction, which is from the top. Water has low backscatter and appears dark. The Pacific Ocean is the dark area at the left, while San Francisco Bay is in the center. Bright objects in the bay are ships. The city of San Francisco, at the upper left, is connected to Oakland via the San Francisco-Oakland Bay Bridge, which passes through Yerba Buena Island at the top center. San Bruno Mountain is visible left of center and south of San Francisco. San Andreas Lake lies along the San Andreas fault at the western edge of the mountains at the lower left. [Courtesy of the Jet Propulsion Laboratory.]

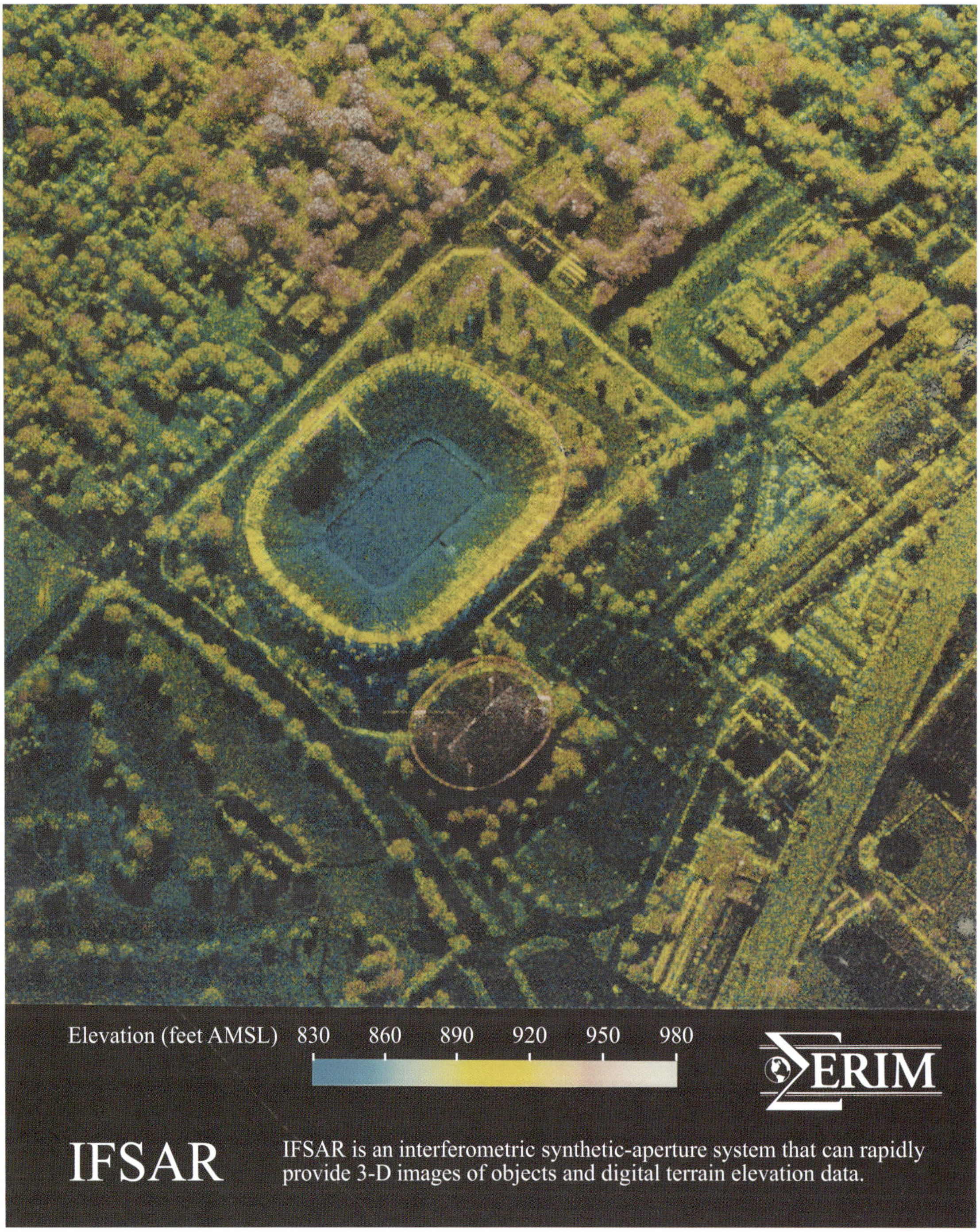

Figure 1-21: This SAR image is overlaid on a height map (color) derived using an interferometric SAR (IFSAR) technique. The image is one of the earliest demonstrations of SAR interferometry. Radar illumination is from the top. The University of Michigan football stadium is visible in the image center. Note that the playing field is depressed relative to the stands. The heights of individual trees growing on the rise in the image center can be seen. [Courtesy of the Environmental Research Institute of Michigan.]

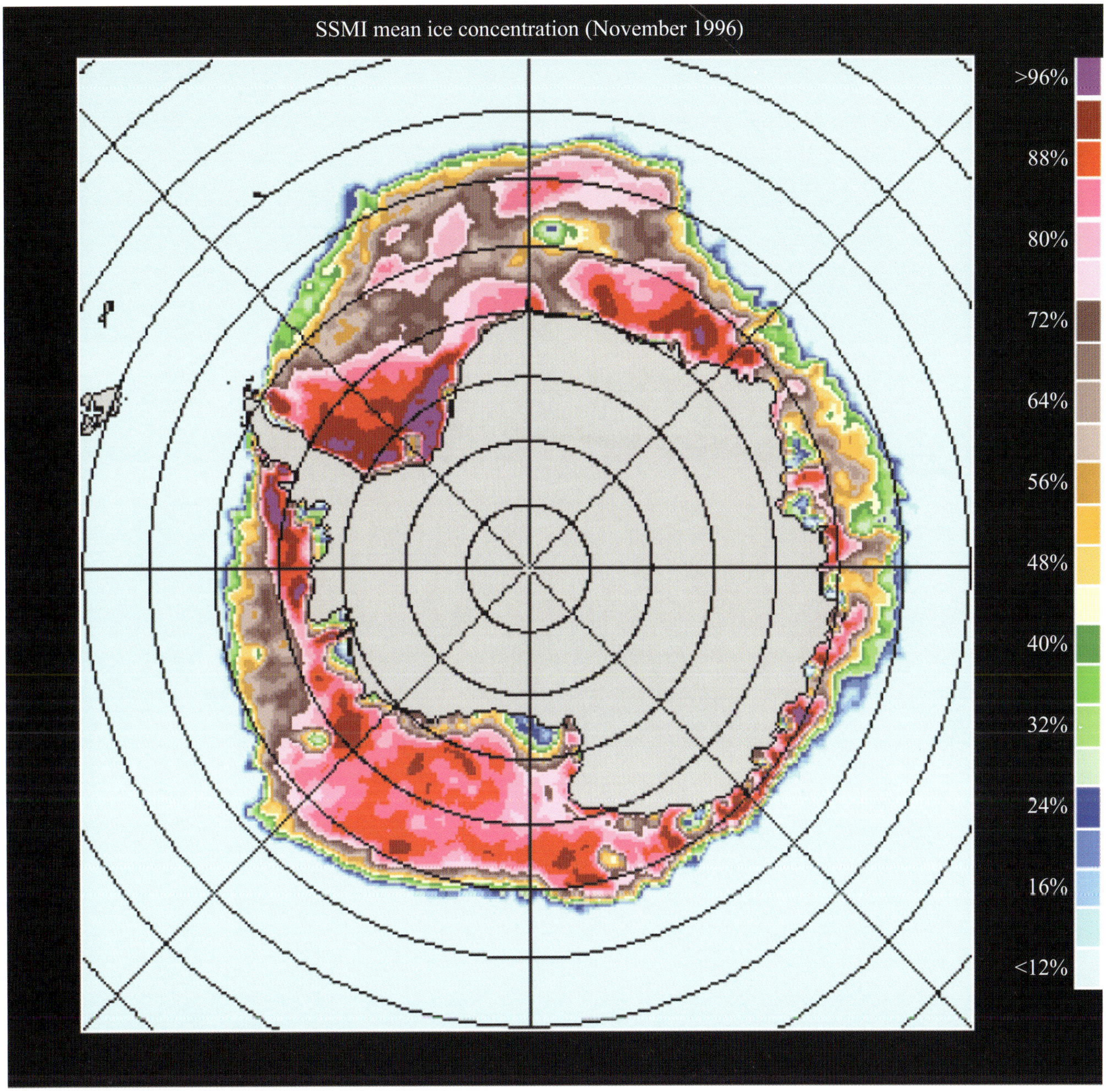

Figure 1-22: This map of sea-ice concentration in Antarctica is based on passive multichannel microwave measurements. For each pixel, the pixel color denotes the fractional amount of the pixel area covered by sea ice, with the remaining area of the pixel being open ocean. Land and near-land areas are masked off. Sea-ice coverage is an important factor in Earth's climate heat balance. [Courtesy U.S. National Sea Ice Data Center.]

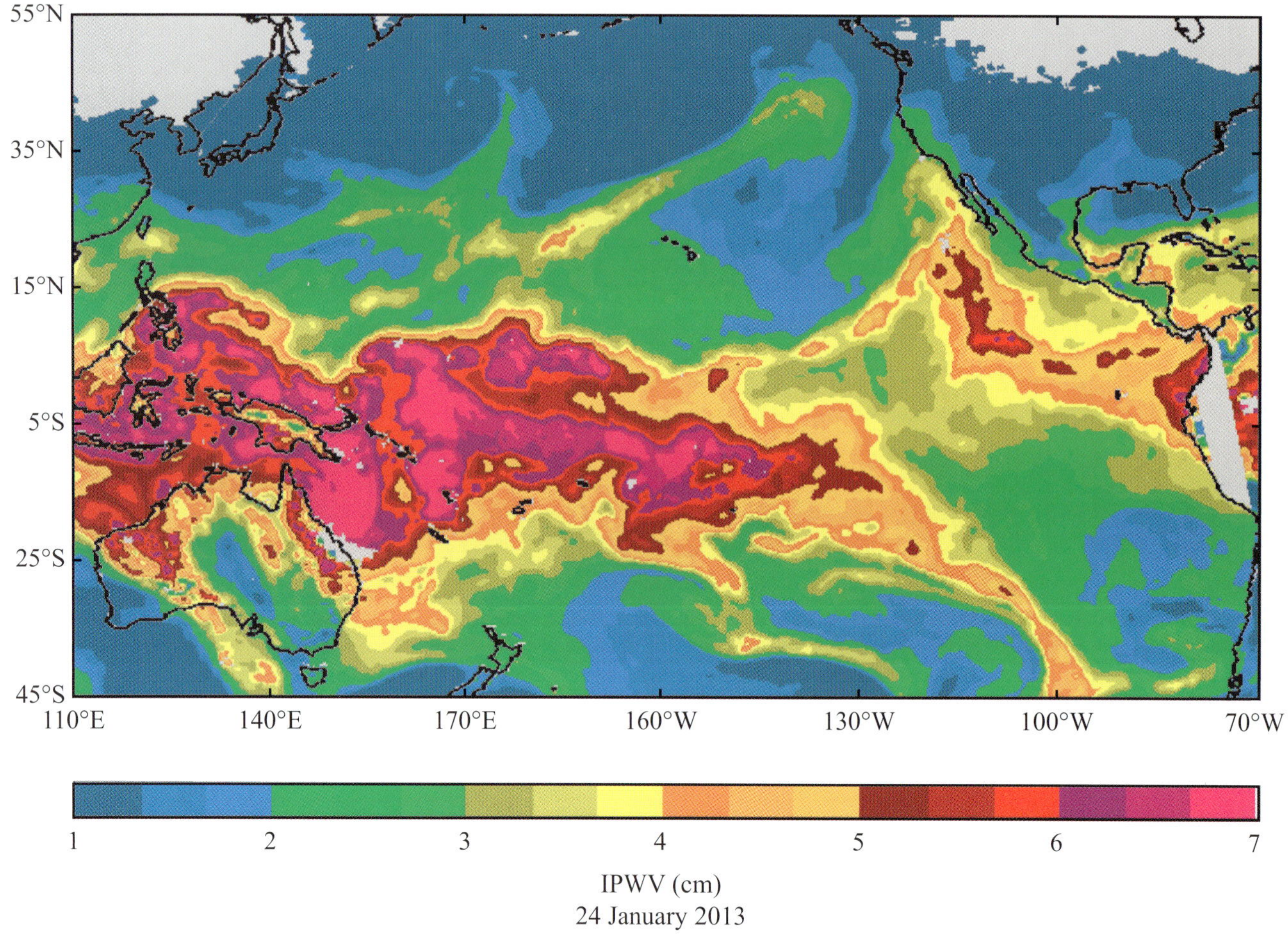

Figure 1-23: A map of integrated precipitable water vapor (IPWV). This is an important parameter used for predicting heavy precipitation. It also affects the heat balance of the atmosphere and is thus critical in climate modeling. This 6-hour-average map is derived from passive multichannel microwave observations from multiple satellite platforms and sensors whose measurements have been histogram matched (Kidder and Jones, 2007). The high values of IPWV along the southern tropical convergence zone south of the equator are indicative of high evaporation and convective rain. IPWV is measured in centimeters of equivalent precipitable water. [Courtesy of National Oceanic and Atmospheric Administration.]

Figure 1-24: This spaceborne radar image shows part of the Mojave Desert in the vicinity of Barstow, California. [Courtesy of NASA/JPL.]

CHAPTER

2

Electromagnetic Wave Propagation and Reflection

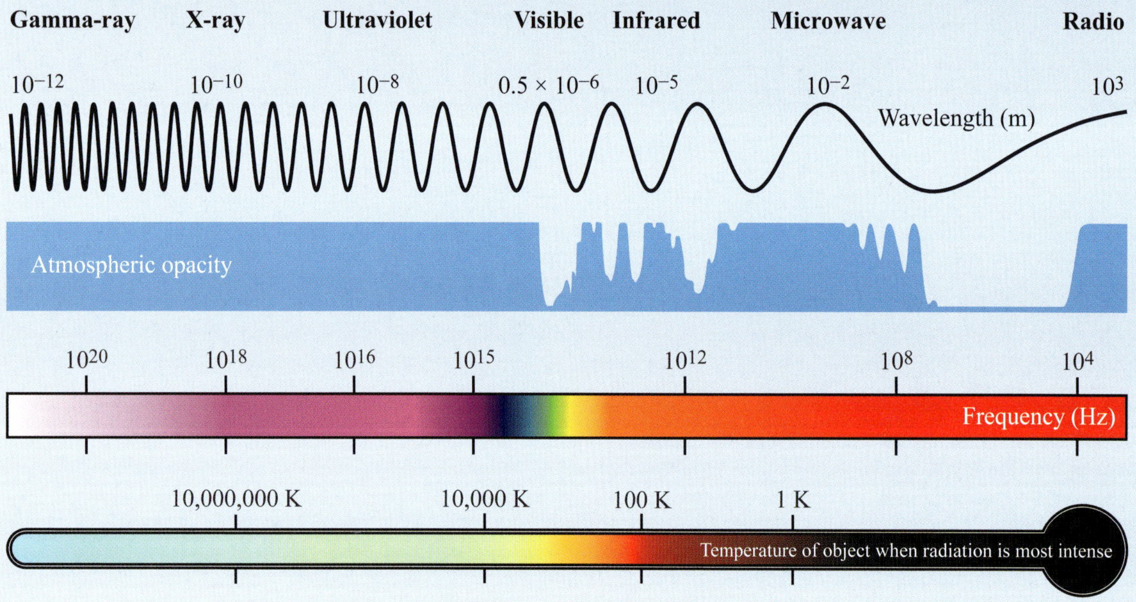

Electromagnetic spectrum

CONTENTS

Overview

To understand how active and passive microwave sensors operate and how the ***electromagnetic*** (EM) quantities they measure are transformed into biological, hydrological, and geophysical information about Earth's surface and atmosphere, it is imperative that we have a solid understanding of EM theory and *how EM waves interact with material media.* The interaction may involve ***scattering***, ***absorption***, ***transmission***, and ***emission***, or combinations thereof. Accordingly, this chapter provides a review of EM waves and their properties, EM propagation in lossy and lossless media, and wave reflection and transmission at planar boundaries. This review will allow us in future chapters to develop models for scattering by boundaries with rough surfaces and by volumes containing particles (such as water droplets in clouds, rain and snow, and leaves and needles in vegetation-covered canopies).

In a parallel treatment, Chapter 6 establishes the connection between the EM energy naturally emitted by a surface or medium (such as the atmosphere) and the dielectric and geometrical properties of that surface or medium. The scattered EM energy measured by a radar and the emitted energy measured by a radiometer are such that they depend not only on the properties of the scene, but also on the properties of the microwave sensor, namely its EM wavelength, observation direction relative to the scene, and the polarization configuration of its antenna(s). We will keep these ***sensor parameters*** in mind throughout our forthcoming discussions.

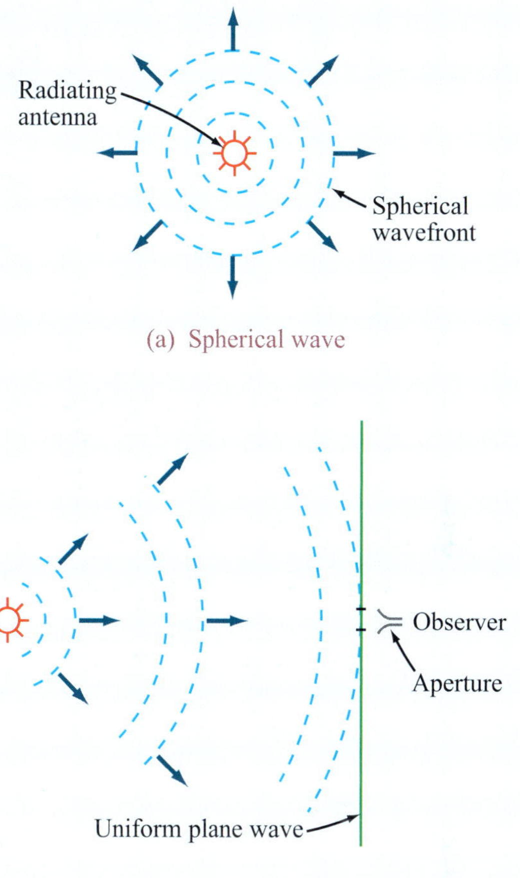

Figure 2-1: Waves radiated by an EM source, such as a lightbulb or an antenna, have spherical wavefronts, as in (a); to a distant observer, however, the wavefront across the observer's aperture appears approximately planar, as in (b).

2-1 EM Plane Waves

A time-varying electric field induces a magnetic field and, conversely, a time-varying magnetic field induces an electric field. This cyclic pattern often results in electromagnetic waves propagating through free space and in material media. When a wave propagates through a homogeneous medium without interacting with obstacles or material interfaces, it is said to be ***unbounded***. Light waves emitted by the sun and radio transmissions by antennas are good examples. Unbounded waves may propagate in both lossless and lossy media. Waves propagating in a ***lossless medium*** (e.g., air and perfect dielectrics) do not attenuate. When propagating in a ***lossy medium*** (material with nonzero conductivity, such as water), part of the power carried by an EM wave gets converted into heat.

A wave produced by a localized source, such as an antenna, expands outwardly in the form of a ***spherical wave***, as depicted in Fig. 2-1(a). Even though an antenna may radiate more energy along some directions than along others, the spherical wave travels at the same speed in all directions. To an observer very far away from the source, however, the ***wavefront*** of the spherical wave appears approximately *planar*, as if it were part of a ***uniform plane wave*** with identical properties at all points in the plane tangent to the wavefront [Fig. 2-1(b)]. Plane waves are easily described using a Cartesian coordinate system, which is mathematically easier to work with than the spherical coordinate system needed to describe spherical waves.

2-1.1 Constitutive Parameters

Regardless of its material composition, a small differential volume is characterized by four electromagnetic ***constitutive parameters***:

- $\varepsilon'\varepsilon_0$ = ***electrical permittivity*** (F/m),[†]
- μ = ***magnetic permeability*** (H/m),
- ρ_v = ***volume charge density*** (C/m^3),
- σ = ***conductivity*** (S/m)

where $\varepsilon_0 = 8.85 \times 10^{-12}$ F/m is the permittivity of free space and ε' is the ***permittivity of the material* relative relative *to that of free space***.

> ▶ We denote the ***relative permittivity*** by the symbol ε', instead of ε, so we may use the latter in future sections to represent the complex dielectric constant of the material under sinusoidal time-varying conditions. ◀

Examples of constitutive parameters:

(a) Free space:
$\varepsilon' = 1$
$\mu = \mu_0 = 4\pi \times 10^{-7}$ H/m
$\rho_\mathrm{v} = 0$
$\sigma = 0$

[†]Abbreviations for units of physical quantities are given in Appendix A.

(b) Pure dielectric:
ε': medium-specific
$\mu = \mu_0$ (all materials except ferromagnetic)
$\rho_\mathrm{v} = 0$
$\sigma = 0$

(c) Conducting medium:
ε': medium-specific
$\mu = \mu_0$ (except ferromagnetic)
ρ_v may or may not be zero
$\sigma \neq 0$

As we shall see shortly, an EM wave can propagate through a dielectric medium (including free space) with no loss of energy (zero attenuation). In contrast, a conducting medium absorbs part of the energy carried by an EM wave traveling through it, thereby attenuating it.

2-1.2 Maxwell's Equations

In a homogeneous, isotropic medium, the differential form of Maxwell's equations is given by

$$\nabla \cdot \mathbf{E} = \frac{\rho_\mathrm{v}}{\varepsilon'\varepsilon_0} \qquad \textbf{(Gauss's law)}, \tag{2.1a}$$

$$\nabla \times \mathbf{E} = -\mu\,\frac{\partial \mathbf{H}}{\partial t} \qquad \textbf{(Faraday's law)}, \tag{2.1b}$$

$$\nabla \cdot \mathbf{H} = 0 \qquad \textbf{(Gauss's law for magnetism)}, \tag{2.1c}$$

$$\nabla \times \mathbf{H} = \mathbf{J} + \varepsilon'\varepsilon_0\,\frac{\partial \mathbf{E}}{\partial t} \qquad \textbf{(Ampère's law)}, \tag{2.1d}$$

where $\mathbf{E}$ is the ***electric field intensity*** (in V/m), $\mathbf{H}$ is the ***magnetic field intensity*** (in A/m), and $\mathbf{J}$ is the ***current density*** (in A/m^2) flowing through the medium.

Time-varying electric and magnetic fields ($\mathbf{E}$ and $\mathbf{H}$) and their sources (the charge density ρ_v and current density $\mathbf{J}$) generally depend on the spatial coordinates (x,y,z) and the time variable t. However, if their time variation is sinusoidal with angular frequency ω, then these quantities can be represented by a phasor that depends on (x,y,z) only. The vector phasor $\mathbf{E}(x,y,z)$ and the instantaneous field $\mathbf{E}(x,y,z;t)$ it describes are related as

$$\mathbf{E}(x,y,z;t) = \mathfrak{Re}\left[\mathbf{E}(x,y,z)\,e^{j\omega t}\right], \tag{2.2}$$

where $j = \sqrt{-1}$. Similar definitions apply to ρ_v and $\mathbf{J}$. For time-harmonic quantities, differentiation in the time domain corresponds to multiplication by $j\omega$ in the phasor domain. Maxwell's equations (Eq. (2.1)) assume the following form in the phasor domain:

$$\nabla \cdot \mathbf{E} = \rho_v/\varepsilon'\varepsilon_0, \tag{2.3a}$$

$$\nabla \times \mathbf{E} = -j\omega\mu\mathbf{H}, \tag{2.3b}$$

$$\nabla \cdot \mathbf{H} = 0, \tag{2.3c}$$

$$\nabla \times \mathbf{H} = \mathbf{J} + j\omega\varepsilon'\varepsilon_0\mathbf{E}. \tag{2.3d}$$

2-1.3 Complex Permittivity

In a medium with conductivity σ, the conduction current density $\mathbf{J}$ is related to $\mathbf{E}$ by $\mathbf{J} = \sigma\mathbf{E}$. Assuming no other current flows in the medium, Eq. (2.3d) may be written as

$$\begin{aligned}\nabla \times \mathbf{H} &= \mathbf{J} + j\omega\varepsilon'\varepsilon_0\mathbf{E} \\ &= (\sigma + j\omega\varepsilon'\varepsilon_0)\mathbf{E} = j\omega\varepsilon_0\left(\varepsilon' - j\,\frac{\sigma}{\omega\varepsilon_0}\right)\mathbf{E}.\end{aligned} \tag{2.4}$$

By defining the ***complex dielectric constant*** ε as

$$\varepsilon = \varepsilon' - j\frac{\sigma}{\omega\varepsilon_0}, \tag{2.5}$$

Eq. (2.4) can be rewritten as

$$\nabla \times \mathbf{H} = j\omega\varepsilon\varepsilon_0\mathbf{E}. \tag{2.6}$$

Taking the divergence of both sides of Eq. (2.6), and recalling that the divergence of the curl of any vector field vanishes (i.e., $\nabla \cdot \nabla \times \mathbf{H} = 0$), it follows that $\nabla \cdot (j\omega\varepsilon\varepsilon_0\mathbf{E}) = 0$, or $\nabla \cdot \mathbf{E} = 0$. Comparing this with Eq. (2.3a) implies that $\rho_v = 0$. This result is a consequence of the assumption that the only current that may exist in the medium is due to conduction. Upon replacing Eq. (2.3d) with Eq. (2.6) and setting $\rho_v = 0$ in Eq. (2.3a), Maxwell's equations become

$$\nabla \cdot \mathbf{E} = 0, \tag{2.7a}$$

$$\nabla \times \mathbf{E} = -j\omega\mu\mathbf{H}, \tag{2.7b}$$

$$\nabla \cdot \mathbf{H} = 0, \tag{2.7c}$$

$$\nabla \times \mathbf{H} = j\omega\varepsilon\varepsilon_0\mathbf{E}. \tag{2.7d}$$

The complex dielectric constant ε given by Eq. (2.5) is often written in terms of a real part ε' and an imaginary part ε''. Thus,

$$\varepsilon = \varepsilon' - j\frac{\sigma}{\omega\varepsilon_0} = \varepsilon' - j\varepsilon'', \tag{2.8}$$

with

$$\varepsilon'' = \frac{\sigma}{\omega\varepsilon_0}. \tag{2.9}$$

Since ε'' is associated with heat loss in the material, it is called the ***dielectric loss factor*** of the material. For a lossless medium with $\sigma = 0$, it follows that $\varepsilon'' = 0$ and $\varepsilon = \varepsilon'$.

2-1.4 Wave Equations

To derive wave equations for $\mathbf{E}$ and $\mathbf{H}$ and solve them to obtain explicit expressions for $\mathbf{E}$ and $\mathbf{H}$ as a function of the spatial variables (x, y, z), we take the curl of both sides of Eq. (2.7b) and then use appropriate substitutions to obtain the ***homogeneous wave equation*** for $\mathbf{E}$, namely

$$\nabla^2\mathbf{E} + \omega^2\mu\varepsilon\varepsilon_0\mathbf{E} = 0. \tag{2.10}$$

In Cartesian coordinates the Laplacian of $\mathbf{E}$ is given by

$$\nabla^2\mathbf{E} = \left(\frac{\partial^2}{\partial x^2} + \frac{\partial^2}{\partial y^2} + \frac{\partial^2}{\partial z^2}\right)\mathbf{E}. \tag{2.11}$$

By defining the ***propagation constant*** γ as

$$\gamma^2 = -\omega^2\mu\varepsilon\varepsilon_0, \tag{2.12}$$

Eq. (2.10) can be written as

$$\nabla^2\mathbf{E} - \gamma^2\mathbf{E} = 0. \tag{2.13}$$

In deriving Eq. (2.13), we started by taking the curl of both sides of Eq. (2.7b). If we reverse the process, that is, if we start by taking the curl of both sides of Eq. (2.7d) and then use Eq. (2.7b) to eliminate **E**, we obtain a wave equation for **H**:

$$\nabla^2 \mathbf{H} - \gamma^2 \mathbf{H} = 0. \tag{2.14}$$

Since the wave equations for **E** and **H** are of the same form, so are the forms of their solutions.

2-2 Plane-Wave Propagation in Lossless Media

The properties of an electromagnetic wave, such as its phase velocity u_p and wavelength λ, depend on the angular frequency ω and the medium's three constitutive parameters: ε', μ, and σ. If the medium is ***nonconducting*** ($\sigma = 0$), the wave does not suffer any attenuation as it travels and hence the medium is said to be ***lossless***. Because in a lossless medium $\varepsilon = \varepsilon'$, Eq. (2.12) becomes

$$\gamma^2 = -\omega^2 \mu \varepsilon' \varepsilon_0. \tag{2.15}$$

For lossless media, it is customary to define the ***wavenumber*** k as

$$k = \omega \sqrt{\mu \varepsilon' \varepsilon_0}\,. \tag{2.16}$$

In view of Eq. (2.15), $\gamma^2 = -k^2$ and Eq. (2.13) becomes

$$\nabla^2 \mathbf{E} + k^2 \mathbf{E} = 0. \tag{2.17}$$

2-2.1 Uniform Plane Waves

For an electric field phasor expressed in a Cartesian coordinate system as

$$\mathbf{E} = \hat{\mathbf{x}} E_x + \hat{\mathbf{y}} E_y + \hat{\mathbf{z}} E_z, \tag{2.18}$$

substitution of Eq. (2.11) into Eq. (2.17) gives

$$\left(\frac{\partial^2}{\partial x^2} + \frac{\partial^2}{\partial y^2} + \frac{\partial^2}{\partial z^2}\right)(\hat{\mathbf{x}} E_x + \hat{\mathbf{y}} E_y + \hat{\mathbf{z}} E_z) + k^2(\hat{\mathbf{x}} E_x + \hat{\mathbf{y}} E_y + \hat{\mathbf{z}} E_z) = 0. \tag{2.19}$$

To satisfy Eq. (2.19), each vector component on the left-hand side of the equation must vanish. Hence,

$$\left(\frac{\partial^2}{\partial x^2} + \frac{\partial^2}{\partial y^2} + \frac{\partial^2}{\partial z^2} + k^2\right) E_x = 0, \tag{2.20}$$

and similar expressions apply to E_y and E_z.

▶ A ***uniform plane wave*** is characterized by electric and magnetic fields that have uniform properties at all points across an infinite plane. ◀

If this happens to be the x–y plane, then **E** and **H** do not vary with x and y. Hence, $\partial E_x/\partial x = 0$ and $\partial E_x/\partial y = 0$, and Eq. (2.20) reduces to

$$\frac{d^2 E_x}{dz^2} + k^2 E_x = 0. \tag{2.21}$$

Similar expressions apply to E_y, H_x, and H_y. The remaining components of **E** and **H** are zero; that is, $E_z = H_z = 0$.

▶ A plane wave has no electric- or magnetic-field components along its direction of propagation. ◀

For the phasor quantity E_x, the general solution of the ordinary differential equation given by Eq. (2.21) is

$$E_x(z) = E_x^+(z) + E_x^-(z) = E_{x0}^+ e^{-jkz} + E_{x0}^- e^{jkz}, \tag{2.22}$$

where E_{x0}^+ and E_{x0}^- are constants to be determined from boundary conditions. The first term in Eq. (2.22), containing the negative exponential e^{-jkz}, represents a wave with amplitude E_{x0}^+ traveling in the $+z$ direction. Likewise, the second term (with e^{jkz}) represents a wave with amplitude E_{x0}^- traveling in the $-z$ direction. Assume for the time being that **E** only has a component along x (i.e., $E_y = 0$) and that E_x is associated with a wave traveling in the $+z$ direction only (i.e., $E^-_{x0} = 0$). Under these conditions,

$$\mathbf{E}(z) = \hat{\mathbf{x}} E^+_x(z) = \hat{\mathbf{x}} E^+_{x0} e^{-jkz}. \tag{2.23}$$

To find the magnetic field **H** associated with this wave, we apply Eq. (2.7b) with $\mathbf{E}_y = \mathbf{E}_z = 0$. The process leads to

$$H_y(z) = H^+_{y0} e^{-jkz}, \tag{2.24a}$$

with

$$H^+_{y0} = \frac{k}{\omega\mu} E^+_{x0}, \tag{2.24b}$$

and the other components of **H** are zero (i.e., $H_x = H_z = 0$). The ***intrinsic impedance*** of a lossless medium is defined as

$$\eta = \frac{\omega\mu}{k} = \frac{\omega\mu}{\omega\sqrt{\mu\varepsilon'\varepsilon_0}} = \sqrt{\frac{\mu}{\varepsilon'\varepsilon_0}} \quad (\Omega), \tag{2.25}$$

where we used the expression for k given by Eq. (2.16). In view of Eq. (2.25), the electric and magnetic fields of a $+z$-propagating plane wave with **E**-field along $\hat{\mathbf{x}}$ are:

$$\mathbf{E}(z) = \hat{\mathbf{x}} E^+_x(z) = \hat{\mathbf{x}} E^+_{x0} e^{-jkz}, \tag{2.26a}$$

$$\mathbf{H}(z) = \hat{\mathbf{y}} \frac{E^+_x(z)}{\eta} = \hat{\mathbf{y}} \frac{E^+_{x0}}{\eta} e^{-jkz}. \tag{2.26b}$$

▶ Because the electric and magnetic fields are perpendicular to each other, and both are perpendicular to the direction of wave travel, this wave is said to be ***transverse electromagnetic*** (TEM) (Fig. 2-2). ◀

In the general case, E^+_{x0} is a complex quantity with magnitude $|E^+_{x0}|$ and phase angle ϕ^+. That is,

$$E^+_{x0} = |E^+_{x0}| e^{j\phi^+}. \tag{2.27}$$

The instantaneous electric and magnetic fields therefore are

$$\begin{aligned}\mathbf{E}(z,t) &= \mathfrak{Re}\left[\mathbf{E}(z)\, e^{j\omega t}\right] \\ &= \hat{\mathbf{x}} |E^+_{x0}| \cos(\omega t - kz + \phi^+) \quad \text{(V/m)},\end{aligned} \tag{2.28a}$$

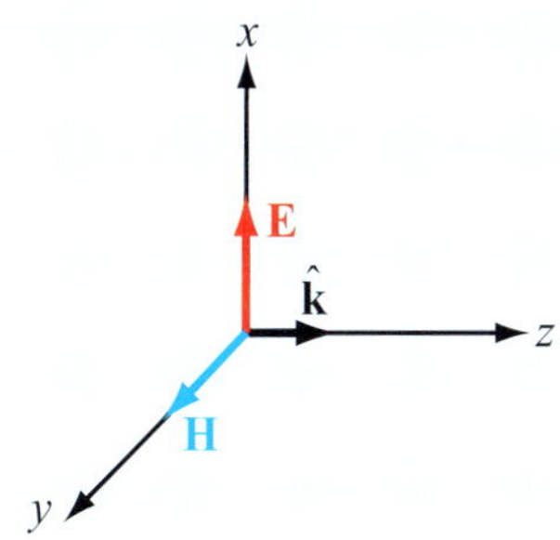

Figure 2-2: A transverse electromagnetic (TEM) wave propagating in the direction $\hat{\mathbf{k}} = \hat{\mathbf{z}}$. For all TEM waves, $\hat{\mathbf{k}}$ is parallel to $\mathbf{E} \times \mathbf{H}$.

and

$$\begin{aligned}\mathbf{H}(z,t) &= \mathfrak{Re}\left[\mathbf{H}(z)\, e^{j\omega t}\right] \\ &= \hat{\mathbf{y}} \frac{|E^+_{x0}|}{\eta} \cos(\omega t - kz + \phi^+) \quad \text{(A/m)}.\end{aligned} \tag{2.28b}$$

Because $\mathbf{E}(z,t)$ and $\mathbf{H}(z,t)$ exhibit the same functional dependence on z and t, they are said to be ***in phase***; when the amplitude of one of them reaches a maximum, the amplitude of the other does so too. The fact that **E** and **H** are in phase is characteristic of waves propagating in lossless media.

The ***phase velocity*** of the wave is

$$u_\text{p} = \frac{\omega}{k} = \frac{\omega}{\omega\sqrt{\mu\varepsilon'\varepsilon_0}} = \frac{1}{\sqrt{\mu\varepsilon'\varepsilon_0}} \quad \text{(m/s)}, \tag{2.29}$$

and its wavelength is

$$\lambda = \frac{2\pi}{k} = \frac{u_\text{p}}{f} \quad \text{(m)}. \tag{2.30}$$

In vacuum, $\varepsilon' = 1$ and $\mu = \mu_0$, and the phase velocity u_p and the intrinsic impedance η given by Eq. (2.25) are

$$u_\text{p} = c = \frac{1}{\sqrt{\mu_0\varepsilon_0}} = 299{,}792{,}458 \approx 3 \times 10^8 \quad \text{(m/s)}, \tag{2.31}$$

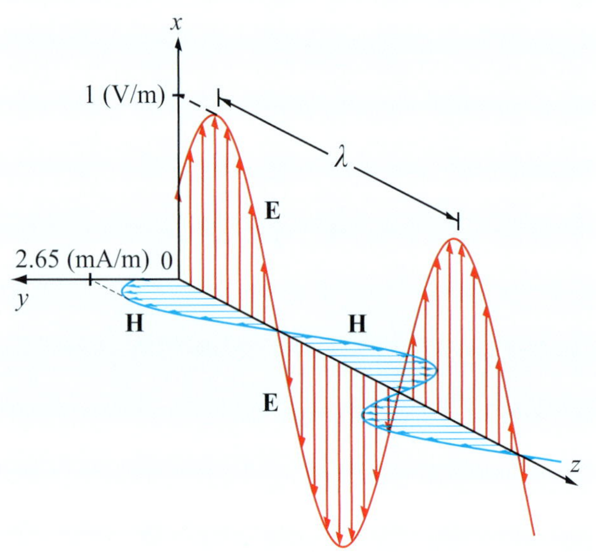

Figure 2-3: Spatial variations of **E** and **H** at $t = 0$ for the plane wave defined by Eq. (2.33).

$$\eta = \eta_0 = \sqrt{\frac{\mu_0}{\varepsilon_0}} = 377\ (\Omega) \approx 120\pi \quad (\Omega), \tag{2.32}$$

where c is the velocity of light and η_0 is called the ***intrinsic impedance of free space***.

Figure 2-3 displays the profiles of **E** and **H** for a 1 MHz EM wave traveling in air, with

$$\mathbf{E}(z,t) = \hat{\mathbf{x}} \cos\left(2\pi \times 10^6 t - \frac{2\pi z}{300} + \frac{\pi}{3}\right) \quad (\text{V/m}), \tag{2.33a}$$

$$\begin{aligned}\mathbf{H}(z,t) &= \hat{\mathbf{y}}\, \frac{E(z,t)}{\eta_0} \\ &= \hat{\mathbf{y}}\, 2.65 \cos\left(2\pi \times 10^6 t - \frac{2\pi z}{300} + \frac{\pi}{3}\right) \quad (\text{mA/m}).\end{aligned} \tag{2.33b}$$

2-2.2 General Relation between E and H

It can be shown that, for any uniform plane wave traveling in an arbitrary direction denoted by the unit vector $\hat{\mathbf{k}}$, the electric and magnetic field phasors **E** and **H** are related as

$$\mathbf{H} = \frac{1}{\eta}\, \hat{\mathbf{k}} \times \mathbf{E}, \tag{2.34a}$$

$$\mathbf{E} = -\eta\, \hat{\mathbf{k}} \times \mathbf{H}. \tag{2.34b}$$

▶ The following right-hand rule applies: when we rotate the four fingers of the right hand from the direction of **E** toward that of **H**, the thumb points in the direction of wave travel, $\hat{\mathbf{k}}$. ◀

The relations given by Eqs. (2.34a and b) are valid not only for lossless media, but for lossy media as well. As shown later in Section 2-4, the expression for η of a lossy medium is different from that given by Eq. (2.25). As long as the expression used for η is appropriate for the medium in which the wave is traveling, the relations given by Eqs. (2.34a and b) always hold.

Let us apply Eq. (2.34a) to the wave given by Eq. (2.26a). The direction of propagation $\hat{\mathbf{k}} = \hat{\mathbf{z}}$ and $\mathbf{E} = \hat{\mathbf{x}} E_x^+(z)$. Hence,

$$\mathbf{H} = \frac{1}{\eta} \hat{\mathbf{k}} \times \mathbf{E} = \frac{1}{\eta} (\hat{\mathbf{z}} \times \hat{\mathbf{x}})\, E_x^+(z) = \hat{\mathbf{y}}\, \frac{E_x^+(z)}{\eta}, \tag{2.35}$$

which is the same as the result given by Eq. (2.26b).

In general, a uniform plane wave traveling in the $+z$ direction may have both x and y components, in which case **E** is given by

$$\mathbf{E} = \hat{\mathbf{x}} E_x^+(z) + \hat{\mathbf{y}} E_y^+(z), \tag{2.36a}$$

and the associated magnetic field is

$$\mathbf{H} = \hat{\mathbf{x}} H_x^+(z) + \hat{\mathbf{y}} H_y^+(z). \tag{2.36b}$$

Application of Eq. (2.34a) gives

$$\mathbf{H} = \frac{1}{\eta}\, \hat{\mathbf{z}} \times \mathbf{E} = -\hat{\mathbf{x}}\, \frac{E_y^+(z)}{\eta} + \hat{\mathbf{y}}\, \frac{E_x^+(z)}{\eta}. \tag{2.37}$$

By equating Eq. (2.36b) to Eq. (2.37), we have

$$H_x^+(z) = -\frac{E_y^+(z)}{\eta}, \qquad H_y^+(z) = \frac{E_x^+(z)}{\eta}. \tag{2.38}$$

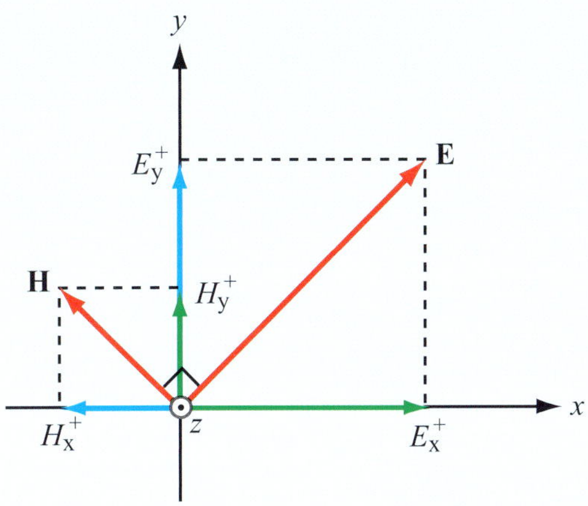

Figure 2-4: The wave $(\mathbf{E},\mathbf{H})$ is equivalent to the sum of two waves, one with fields (E_x^+, H_y^+) and another with (E_y^+, H_x^+), with both traveling in the $+z$ direction.

These results are illustrated in Fig. 2-4. The wave may be considered to be the sum of two waves, one with electric and magnetic components (E_x^+, H_y^+), and another with components (E_y^+, H_x^+). In general, a TEM wave may have an electric field in any direction in the plane orthogonal to the direction of wave travel, and the associated magnetic field is also in the same plane and its direction is dictated by Eq. (2.34a).

2-3 Wave Polarization in a Lossless Medium

▶ The ***polarization*** of a uniform plane wave describes the locus traced by the tip of the **E** vector (in the plane orthogonal to the direction of propagation) at a given point in space as a function of time. ◀

In the most general case, the locus of the tip of **E** is an ellipse, and the wave is said to be ***elliptically polarized***. Under certain conditions, the ellipse may degenerate into a circle or a straight line, in which case the ***polarization state*** is called ***circular*** or ***linear***, respectively.

It was shown in Section 2-2 that the z components of the electric and magnetic fields of a z-propagating plane wave are both zero. Hence, in the most general case, the electric field phasor $\mathbf{E}(z)$ of a $+z$-propagating plane wave may consist of an x component, $\hat{\mathbf{x}}\, E_x(z)$, and a y component, $\hat{\mathbf{y}}\, E_y(z)$, or

$$\mathbf{E}(z) = \hat{\mathbf{x}}E_x(z) + \hat{\mathbf{y}}E_y(z), \tag{2.39}$$

with

$$E_x(z) = E_{x0}e^{-jkz}, \tag{2.40a}$$

$$E_y(z) = E_{y0}e^{-jkz}, \tag{2.40b}$$

where E_{x0} and E_{y0} are the amplitudes of $E_x(z)$ and $E_y(z)$, respectively. For the sake of simplicity, the plus sign superscript has been suppressed; the negative sign in e^{-jkz} is sufficient to remind us that the wave is traveling in the positive z direction.

The two amplitudes E_{x0} and E_{y0} are, in general, complex quantities, each characterized by a magnitude and a phase angle. The phase of a wave is defined relative to a reference state, such as $z = 0$ and $t = 0$ or any other combination of z and t. As will become clear from the discussion that follows, the polarization of the wave described by Eqs. (2.39) and (2.40) depends on the phase of E_{y0} relative to that of E_{x0}, but not on the absolute phases of E_{x0} and E_{y0}. Hence, for convenience, we assign E_{x0} a phase of zero and denote the phase of E_{y0}, relative to that of E_{x0}, as δ. Thus, δ is the *phase difference* between the y- and x components of **E**. Accordingly, we define E_{x0} and E_{y0} as

$$E_{x0} = a_x, \tag{2.41a}$$

$$E_{y0} = a_y e^{j\delta}, \tag{2.41b}$$

where $a_x = |E_{x0}| \geq 0$ and $a_y = |E_{y0}| \geq 0$ are the magnitudes of E_{x0} and E_{y0}, respectively. Thus, by definition, a_x and a_y may not assume negative values. Using Eqs. (2.41a) and (2.41b) in Eqs. (2.40a) and (2.40b), the total electric field phasor is

$$\mathbf{E}(z) = (\hat{\mathbf{x}}a_x + \hat{\mathbf{y}}a_y e^{j\delta})e^{-jkz}, \tag{2.42}$$

and the corresponding instantaneous field is

$$\begin{aligned}\mathbf{E}(z,t) &= \mathfrak{Re}\left[\mathbf{E}(z)\,e^{j\omega t}\right] \\ &= \hat{\mathbf{x}}a_x\cos(\omega t - kz) \\ &\quad + \hat{\mathbf{y}}a_y\cos(\omega t - kz + \delta).\end{aligned} \tag{2.43}$$

When characterizing an electric field at a given point in space, two attributes of particular interest are its magnitude and direction. The magnitude of $\mathbf{E}(z,t)$ is

$$\begin{aligned}|\mathbf{E}(z,t)| &= [E_x^2(z,t) + E_y^2(z,t)]^{1/2} \\ &= [a_x^2\cos^2(\omega t - kz) \\ &\quad + a_y^2\cos^2(\omega t - kz + \delta)]^{1/2}.\end{aligned} \tag{2.44}$$

The electric field $\mathbf{E}(z,t)$ has components along the x- and y directions. At a specific position z, the direction of $\mathbf{E}(z,t)$ is characterized by its ***inclination angle*** $\tau(z,t)$, defined with respect to the x axis and given by

$$\tau(z,t) = \tan^{-1}\left[\frac{E_y(z,t)}{E_x(z,t)}\right]. \tag{2.45}$$

In the general case, both the intensity of $\mathbf{E}(z,t)$ and its direction are functions of z and t. Next, we examine some special cases.

2-3.1 Linear Polarization

▶ A wave is said to be linearly polarized if for a fixed z, the tip of $\mathbf{E}(z,t)$ traces a straight line segment as a function of time. This happens when $E_x(z,t)$ and $E_y(z,t)$ are ***in phase*** (i.e., $\delta = 0$) or ***out of phase*** ($\delta = \pi$). ◀

Under these conditions Eq. (2.43) simplifies to

$$\mathbf{E}(z,t) = (\hat{\mathbf{x}}a_x + \hat{\mathbf{y}}a_y)\cos(\omega t - kz) \quad \textbf{(in phase)}, \tag{2.46a}$$

$$\mathbf{E}(z,t) = (\hat{\mathbf{x}}a_x - \hat{\mathbf{y}}a_y)\cos(\omega t - kz) \quad \textbf{(out of phase)}. \tag{2.46b}$$

Let us examine the out-of-phase case. The field's magnitude is

$$|\mathbf{E}(z,t)| = [a_x^2 + a_y^2]^{1/2}\left|\cos(\omega t - kz)\right|, \tag{2.47a}$$

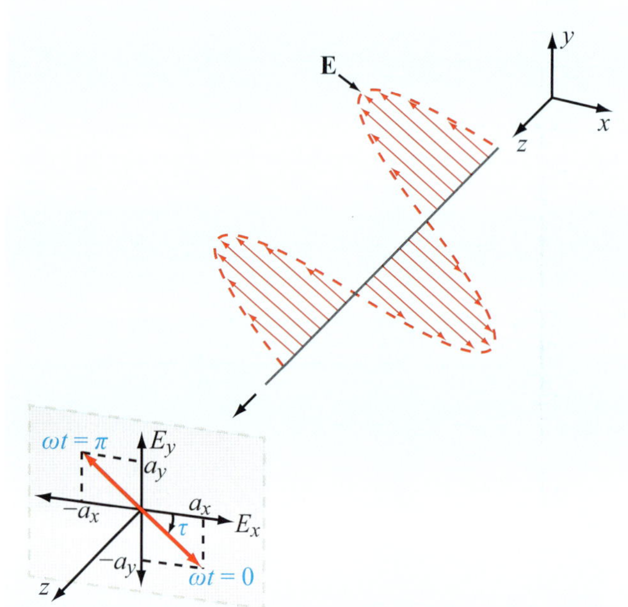

Figure 2-5: Linearly polarized wave traveling in the $+z$ direction.

and the inclination angle is

$$\begin{aligned}\tau(z,t) &= \tan^{-1}\left[\frac{-a_y\cos(\omega t - kz)}{a_x\cos(\omega t - kz)}\right] \\ &= \tan^{-1}\left(\frac{-a_y}{a_x}\right) \qquad \textbf{(out of phase)}.\end{aligned} \tag{2.47b}$$

We note that τ is independent of both z and t. Figure 2-5 displays the line segment traced by the tip of $\mathbf{E}$ at $z = 0$ over a half of a cycle. The trace would be the same at any other value of z as well. At $z = 0$ and $t = 0$, $|E(0,0)| = [a_x^2 + a_y^2]^{1/2}$. The length of the vector representing $\mathbf{E}(0,t)$ decreases to zero at $\omega t = \pi/2$. The vector then reverses direction and increases in magnitude to $[a_x^2 + a_y^2]^{1/2}$ in the second quadrant of the x–y plane at $\omega t = \pi$. Since τ is independent of both z and t, $\mathbf{E}(z,t)$ maintains a direction along the line making an angle τ with the x axis, while oscillating back and forth across the origin.

If $a_y = 0$, then $\tau = 0°$ or $180°$, and the wave is x polarized; conversely, if $a_x = 0$, then $\tau = 90°$ or $-90°$, and the wave is y polarized.

2-3.2 Circular Polarization

We now consider the special case when the magnitudes of the x- and y components of $\mathbf{E}(z)$ are equal, and the phase difference $\delta = \pm\pi/2$. For reasons that will become evident shortly, the wave polarization is called ***left-hand circular*** when $\delta = \pi/2$, and ***right-hand circular*** when $\delta = -\pi/2$.

Left-hand circular (LHC) polarization

For $a_x = a_y = a$ and $\delta = \pi/2$, Eqs. (2.42) and (2.43) become

$$\begin{aligned}\mathbf{E}(z) &= (\hat{\mathbf{x}}a + \hat{\mathbf{y}}ae^{j\pi/2})e^{-jkz} \\ &= a(\hat{\mathbf{x}} + j\hat{\mathbf{y}})e^{-jkz}, \qquad (2.48a)\\ \mathbf{E}(z,t) &= \mathfrak{Re}\left[\mathbf{E}(z)\,e^{j\omega t}\right] \\ &= \hat{\mathbf{x}}a\cos(\omega t - kz) + \hat{\mathbf{y}}a\cos(\omega t - kz + \pi/2) \\ &= \hat{\mathbf{x}}a\cos(\omega t - kz) - \hat{\mathbf{y}}a\sin(\omega t - kz). \qquad (2.48b)\end{aligned}$$

The corresponding field magnitude and inclination angle are

$$\begin{aligned}|\mathbf{E}(z,t)| &= \left[E_x^2(z,t) + E_y^2(z,t)\right]^{1/2} \\ &= [a^2\cos^2(\omega t - kz) + a^2\sin^2(\omega t - kz)]^{1/2} \\ &= a \qquad (2.49a)\end{aligned}$$

and

$$\begin{aligned}\tau(z,t) &= \tan^{-1}\left[\frac{E_y(z,t)}{E_x(z,t)}\right] \\ &= \tan^{-1}\left[\frac{-a\sin(\omega t - kz)}{a\cos(\omega t - kz)}\right] \\ &= -(\omega t - kz). \qquad (2.49b)\end{aligned}$$

We observe that the magnitude of $\mathbf{E}$ is independent of both z and t, whereas $\tau(z,t)$ depends on both variables. These functional dependencies are the converse of those for the linear polarization case.

At $z = 0$, Eq. (2.49b) gives $\tau = -\omega t$; the negative sign implies that the inclination angle decreases as time increases. As illustrated in Fig. 2-6(a), the tip of $\mathbf{E}(t)$ traces a circle in the x–y plane and rotates in a clockwise direction as a function of time (when viewing the wave approaching). Such a wave is called ***left-hand circularly polarized***, because when the thumb of the left hand points along the direction of propagation (the z direction in this case), the other four fingers curl in the direction of rotation of $\mathbf{E}$.

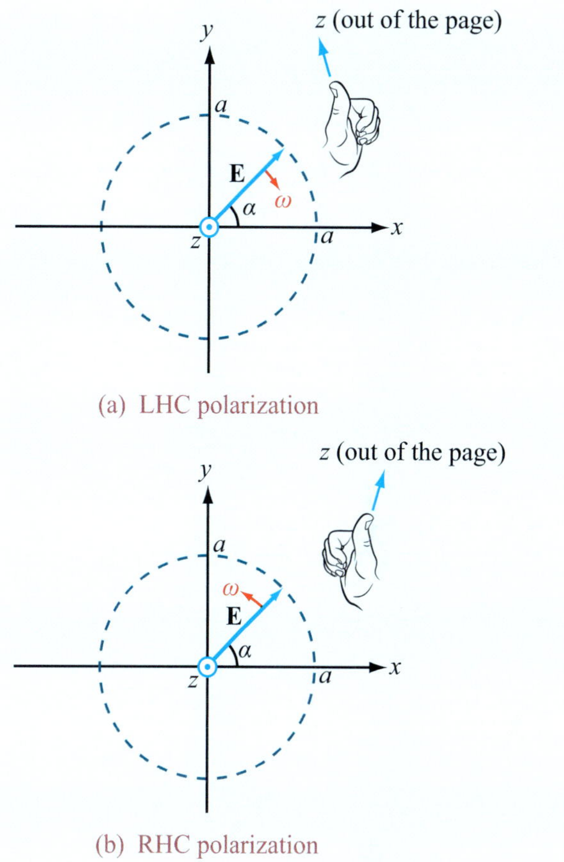

Figure 2-6: Circularly polarized plane waves propagating in the $+z$ direction (out of the page).

Right-hand circular (RHC) polarization

For $a_x = a_y = a$ and $\delta = -\pi/2$, we have

$$|\mathbf{E}(z,t)| = a, \qquad \tau = (\omega t - kz). \qquad (2.50)$$

The trace of $\mathbf{E}(0,t)$ as a function of t is shown in Fig. 2-6(b). For RHC polarization, the fingers of the

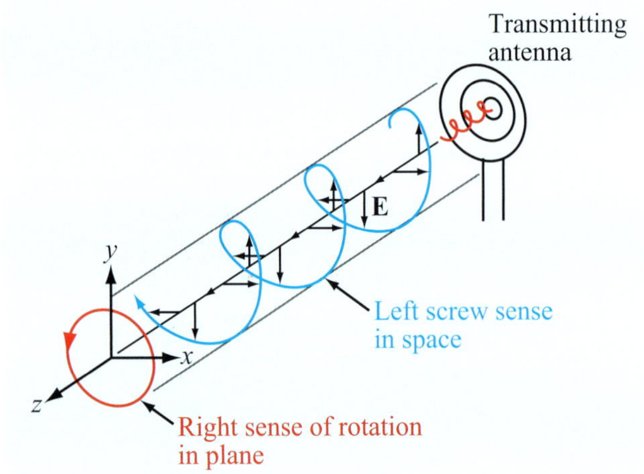

Figure 2-7: Right-hand circularly polarized wave radiated by a helical antenna.

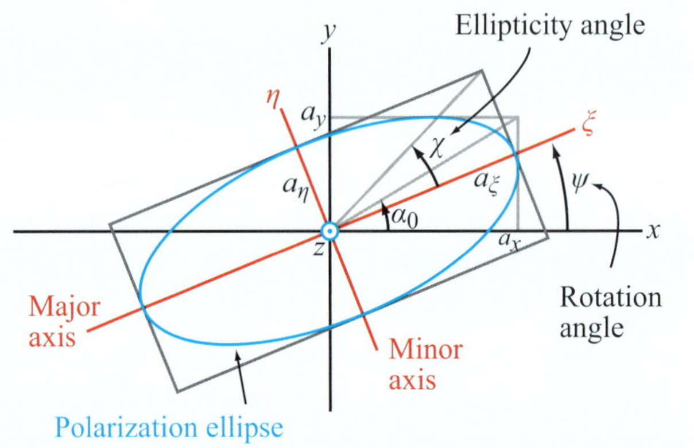

Figure 2-8: Polarization ellipse in the x–y plane, with the wave traveling in the z direction (out of the page).

right hand curl in the direction of rotation of $\mathbf{E}$ when the thumb is along the propagation direction. Figure 2-7 depicts a right-hand circularly polarized wave radiated by a helical antenna.

▶ Polarization handedness is defined by the rotation of $\mathbf{E}$ as a function of time in a fixed plane orthogonal to the direction of propagation, which is opposite of the direction of rotation of $\mathbf{E}$ as a function of distance at a fixed point in time. ◀

2-3.3 Elliptical Polarization

Plane waves that are not linearly or circularly polarized are elliptically polarized. That is, the tip of $\mathbf{E}(z,t)$ traces an ellipse in the plane perpendicular to the direction of propagation. The shape of the ellipse and the field's handedness (left-hand or right-hand) are determined by the values of the ratio (a_y/a_x) and the phase difference δ.

The polarization ellipse shown in Fig. 2-8 has its major axis with length a_ξ along the ξ direction and its minor axis with length a_η along the η direction. The ***rotation angle*** ψ is defined as the angle between the major axis of the ellipse and a reference direction, chosen here to be the x axis, with ψ being bounded within the range $-\pi/2 \le \psi \le \pi/2$. The shape of the ellipse and its handedness are characterized by the ***ellipticity angle*** χ, defined as

$$\tan\chi = \pm\frac{a_\eta}{a_\xi} = \pm\frac{1}{R}, \tag{2.51}$$

with the plus sign corresponding to left-handed rotation and the minus sign corresponding to right-handed rotation. The limits for χ are $-\pi/4 \le \chi \le \pi/4$. The quantity $R = a_\xi/a_\eta$ is called the ***axial ratio*** of the polarization ellipse, and it varies between 1 for circular polarization and ∞ for linear polarization. The polarization angles ψ and χ are related to the wave parameters a_x, a_y, and δ by (Born and Wolf, 1965)

$$\tan 2\psi = (\tan 2\alpha_0)\cos\delta \quad (-\pi/2 \le \psi \le \pi/2), \tag{2.52a}$$

$$\sin 2\chi = (\sin 2\alpha_0)\sin\delta \quad (-\pi/4 \le \chi \le \pi/4), \tag{2.52b}$$

where α_0 is an ***auxiliary angle*** defined by

$$\tan\alpha_0 = \frac{a_y}{a_x} \qquad \left(0 \le \alpha_0 \le \frac{\pi}{2}\right). \tag{2.53}$$

Sketches of the polarization ellipse are shown in Fig. 2-9 for various combinations of the angles (ψ,χ). The ellipse reduces to a circle for $\chi = \pm 45^\circ$ and to a line for

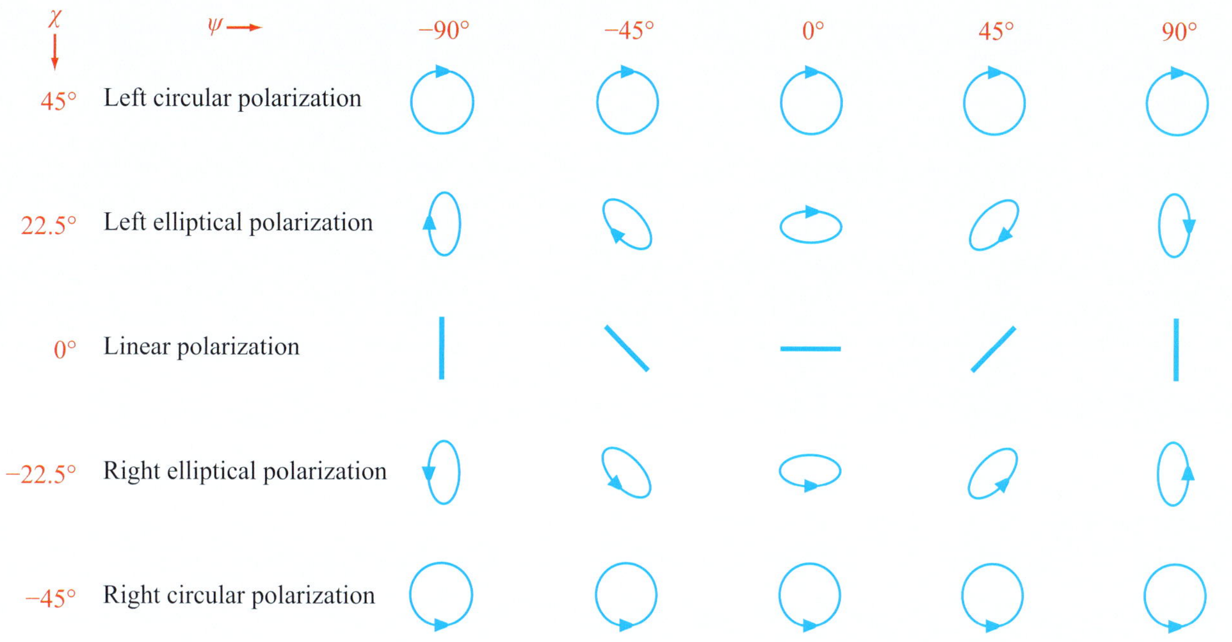

Figure 2-9: Polarization states for various combinations of the polarization angles (ψ, χ) for a wave traveling out of the page.[†]

$\chi = 0$. *Positive values of χ, corresponding to $\sin\delta > 0$, are associated with left-handed rotation, and negative values of χ, corresponding to $\sin\delta < 0$, are associated with right-handed rotation.*

Since the magnitudes a_x and a_y are, by definition, nonnegative numbers, the ratio a_y/a_x may vary between zero for an x-polarized linear polarization and ∞ for a y-polarized linear polarization. Consequently, the angle α_0 is limited to the range $0 \le \alpha_0 \le 90°$. Application of Eq. (2.52a) leads to two possible solutions for the value of ψ, both of which fall within the defined range from $-\pi/2$ to $\pi/2$. The correct choice is governed by the following rule:

$$\psi > 0 \text{ if } \cos\delta > 0,$$
$$\psi < 0 \text{ if } \cos\delta < 0.$$

▶ In summary, the sign of the rotation angle ψ is the same as the sign of $\cos\delta$ and the sign of the ellipticity angle χ is the same as the sign of $\sin\delta$. ◀

2-4 Plane-Wave Propagation in Lossy Media

To examine wave propagation in a lossy (conducting) medium, we return to the wave equation given by Eq. (2.13),

$$\nabla^2\mathbf{E} - \gamma^2\mathbf{E} = 0,$$

with

$$\gamma^2 = -\omega^2\mu\varepsilon\varepsilon_0 = -\omega^2\mu\varepsilon_0(\varepsilon' - j\varepsilon''), \qquad (2.54)$$

where $\varepsilon'' = \sigma/\omega\varepsilon_0$. Since γ is complex, we express it as

$$\gamma = \alpha + j\beta, \qquad (2.55)$$

[†] Computer Code 2.1 (see page vii).

where α is the medium's ***attenuation constant*** and β its ***phase constant***. The combination of Eqs. (2.54) and (2.55) leads to

$$\alpha = -\omega\sqrt{\mu\varepsilon_0}\ \mathfrak{Im}\left\{\sqrt{\varepsilon}\right\}, \tag{2.56a}$$

$$\beta = \omega\sqrt{\mu\varepsilon_0}\ \mathfrak{Re}\left\{\sqrt{\varepsilon}\right\}. \tag{2.56b}$$

Alternatively, by replacing γ with $(\alpha + j\beta)$ in Eq. (2.54), we obtain

$$\begin{aligned}(\alpha + j\beta)^2 &= (\alpha^2 - \beta^2) + j2\alpha\beta \\ &= -\omega^2\mu\varepsilon'\varepsilon_0 + j\omega^2\mu\varepsilon''\varepsilon_0.\end{aligned} \tag{2.57}$$

The rules of complex algebra require the real and imaginary parts on one side of an equation to equal the real and imaginary parts on the other side, which leads to

$$\begin{aligned}\alpha &= \omega\left\{\frac{\mu_0\varepsilon'\varepsilon_0}{2}\left[\sqrt{1+\left(\frac{\varepsilon''}{\varepsilon'}\right)^2} - 1\right]\right\}^{1/2} \\ &= \frac{2\pi}{\lambda_0}\left\{\frac{\varepsilon'}{2}\left[\sqrt{1+\left(\frac{\varepsilon''}{\varepsilon'}\right)^2} - 1\right]\right\}^{1/2} \quad \text{(Np/m)},\end{aligned} \tag{2.58a}$$

$$\begin{aligned}\beta &= \omega\left\{\frac{\mu_0\varepsilon'\varepsilon_0}{2}\left[\sqrt{1+\left(\frac{\varepsilon''}{\varepsilon'}\right)^2} + 1\right]\right\}^{1/2} \\ &= \frac{2\pi}{\lambda_0}\left\{\frac{\varepsilon'}{2}\left[\sqrt{1+\left(\frac{\varepsilon''}{\varepsilon'}\right)^2} + 1\right]\right\}^{1/2} \quad \text{(Np/m)},^\dagger\end{aligned} \tag{2.58b}$$

where $\lambda_0 = c/f$ is the wavelength in free space.

> ▶ We set $\mu = \mu_0$ because the natural materials encountered in remote sensing are nonmagnetic. This includes water, ice, soil, and vegetation, among many others. ◀

† **Computer Code 2.2.**

For a uniform plane wave with electric field $\mathbf{E} = \hat{\mathbf{x}}\, E_x(z)$ traveling along the $+z$ direction, the wave equation given by Eq. (2.13) reduces to

$$\frac{d^2 E_x(z)}{dz^2} - \gamma^2 E_x(z) = 0, \tag{2.59}$$

and its solution gives

$$\mathbf{E}(z) = \hat{\mathbf{x}} E_{x0} e^{-\gamma z} = \hat{\mathbf{x}} E_{x0} e^{-\alpha z} e^{-j\beta z}. \tag{2.60}$$

The associated magnetic field $\mathbf{H}$ can be determined by applying Eq. (2.3b): $\nabla \times \mathbf{E} = -j\omega\mu\mathbf{H}$, or using Eq. (2.34a): $\mathbf{H} = (\hat{\mathbf{k}} \times \mathbf{E})/\eta_c$, where η_c is the ***intrinsic impedance of the lossy medium***. Both approaches give

$$\mathbf{H}(z) = \hat{\mathbf{y}}\, H_y(z) = \hat{\mathbf{y}}\, \frac{E_x(z)}{\eta_c} = \hat{\mathbf{y}}\, \frac{E_{x0}}{\eta_c} e^{-\alpha z} e^{-j\beta z}, \tag{2.61}$$

where

$$\begin{aligned}\eta_c &= \sqrt{\frac{\mu_0}{\varepsilon\varepsilon_0}} = \sqrt{\frac{\mu_0}{\varepsilon'\varepsilon_0}}\left(1 - j\frac{\varepsilon''}{\varepsilon'}\right)^{-1/2} \\ &= \frac{\eta_0}{\sqrt{\varepsilon'}}\left(1 - j\,\frac{\varepsilon''}{\varepsilon'}\right)^{-1/2} \quad (\Omega),^\dagger\end{aligned} \tag{2.62}$$

where η_0 is the intrinsic impedance of free space. We noted earlier that in a lossless medium, $\mathbf{E}(z,t)$ is in phase with $\mathbf{H}(z,t)$; however, this property no longer holds true in a lossy medium because η_c is complex.

From Eq. (2.60), the magnitude of $E_x(z)$ is given by

$$|E_x(z)| = |E_{x0} e^{-\alpha z} e^{-j\beta z}| = |E_{x0}| e^{-\alpha z}, \tag{2.63}$$

which decreases exponentially with z at a rate dictated by the attenuation constant α. Since $H_y = E_x/\eta_c$, the magnitude of H_y also decreases as $e^{-\alpha z}$. As the field attenuates, part of the energy carried by the electromagnetic wave is converted into heat due to conduction in the medium. As the wave travels through a distance $z = \delta_s$ with

$$\delta_s = \frac{1}{\alpha} \quad \text{(m)}, \tag{2.64}$$

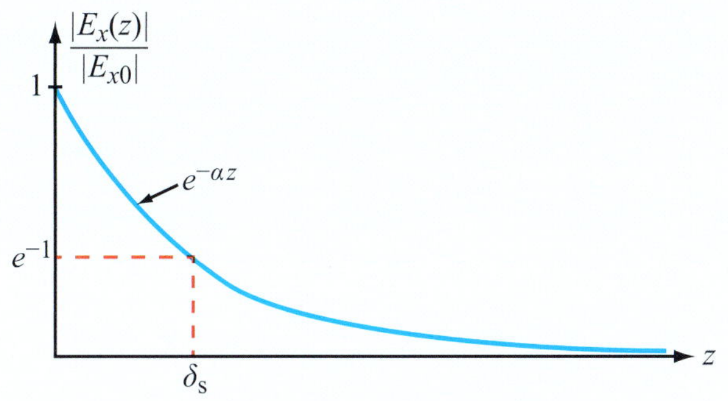

Figure 2-10: Attenuation of the magnitude of $E_x(z)$ with distance z. The skin depth δ_s is the value of z at which $|E_x(z)|/|E_{x0}| = e^{-1}$, or $z = \delta_s = 1/\alpha$.

the wave magnitude decreases by a factor of $e^{-1} \approx 0.37$ (Fig. 2-10). At depth $z = 3\delta_s$, the field magnitude is less than 5% of its initial value, and at $z = 5\delta_s$, it is less than 1%.

▶ This distance δ_s, called the ***skin depth*** of the medium, characterizes how deep an electromagnetic wave can penetrate into a conducting medium. ◀

In a perfect dielectric, $\sigma = 0$ and $\varepsilon'' = 0$; use of Eq. (2.58a) yields $\alpha = 0$ and therefore $\delta_s = \infty$. Thus, in free space, a plane wave can propagate indefinitely with no loss in magnitude. On the other extreme, in a perfect conductor, $\sigma = \infty$ and use of Eq. (2.58a) leads to $\alpha = \infty$ and hence $\delta_s = 0$. Thus, the electric field is confined to the surface of a perfect conductor.

The expressions given by Eqs. (2.58a), (2.58b), and (2.62) for α, β, and η_c are valid for any linear, isotropic, and homogeneous medium. For a perfect dielectric ($\sigma = 0$), these expressions reduce to those for the lossless case (Section 2-2), wherein $\alpha = 0$, $\beta = k = \omega\sqrt{\mu_0\varepsilon'\varepsilon_0}$, and $\eta_c = \eta$. For a lossy medium, the ratio $\varepsilon''/\varepsilon' = \sigma/\omega\varepsilon'\varepsilon_0$, which appears in all these expressions, plays an important role in classifying how lossy the medium is. When $\varepsilon''/\varepsilon' \ll 1$, the medium is considered a ***low-loss dielectric***, and when $\varepsilon''/\varepsilon' \gg 1$, it is considered a ***good conductor***. In practice, the medium may be regarded as a low-loss dielectric if $\varepsilon''/\varepsilon' < 10^{-2}$, as a good conductor if $\varepsilon''/\varepsilon' > 10^2$, and as a ***quasi conductor*** if $10^{-2} \le \varepsilon''/\varepsilon' \le 10^2$. For low-loss dielectrics and good conductors, the expressions given by Eq. (2.58) can be significantly simplified, as shown next.

2-4.1 Low-Loss Dielectric

From Eq. (2.54), the general expression for γ in a nonmagnetic medium is

$$\gamma = j\omega\sqrt{\mu_0\varepsilon'\varepsilon_0}\left(1 - j\frac{\varepsilon''}{\varepsilon'}\right)^{1/2}. \tag{2.65}$$

For $|x| \ll 1$, the function $(1-x)^{1/2}$ can be approximated by the first two terms of its binomial series; that is, $(1-x)^{1/2} \approx 1 - x/2$. By applying this approximation to Eq. (2.65) for a low-loss dielectric with $x = j\varepsilon''/\varepsilon'$ and $\varepsilon''/\varepsilon' \ll 1$, we obtain

$$\begin{aligned}\gamma &\approx j\omega\sqrt{\mu_0\varepsilon'\varepsilon_0}\left(1 - j\,\frac{\varepsilon''}{2\varepsilon'}\right)\\ &\approx j\,\frac{2\pi}{\lambda_0}\,\sqrt{\varepsilon'}\left(1 - j\,\frac{\varepsilon''}{2\varepsilon'}\right).\end{aligned} \tag{2.66}$$

The real and imaginary parts of Eq. (2.66) are

$$\alpha \approx \frac{\pi\varepsilon''}{\lambda_0\sqrt{\varepsilon'}} \qquad \text{(Np/m)}, \tag{2.67a}$$

$$\beta \approx \frac{2\pi}{\lambda_0}\sqrt{\varepsilon'} \qquad \text{(rad/m)}. \tag{2.67b}$$

We note that the expression for β is the same as that for the wavenumber k of a lossless medium. Applying the binomial approximation $(1-x)^{-1/2} \approx (1 + x/2)$ to Eq. (2.62) leads to

$$\eta_c \approx \frac{\eta_0}{\sqrt{\varepsilon'}}\left(1 + j\,\frac{\varepsilon''}{2\varepsilon'}\right). \tag{2.68a}$$

In practice, because $\varepsilon''/\varepsilon' < 10^{-2}$ for low-loss materials, the second term in Eq. (2.68a) often is ignored. Thus,

$$\eta_c \approx \frac{\eta_0}{\sqrt{\varepsilon'}}, \tag{2.68b}$$

which is the same as Eq. (2.25) for a lossless nonmagnetic material.

2-4.2 Good Conductor

When $\varepsilon''/\varepsilon' > 100$, Eqs. (2.58a), (2.58b), and (2.62) can be approximated as

$$\alpha \approx \frac{\pi\sqrt{2\varepsilon''}}{\lambda_0} \qquad \text{(Np/m)}, \tag{2.69a}$$

$$\beta = \alpha \approx \frac{\pi\sqrt{2\varepsilon''}}{\lambda_0} \qquad \text{(rad/m)}, \tag{2.69b}$$

$$\eta_c \approx \sqrt{\frac{j\mu_0}{\varepsilon''\varepsilon_0}} = \frac{(1+j)\eta_0}{\sqrt{2\varepsilon''}} \qquad (\Omega). \tag{2.69c}$$

In Eq. (2.69c), we used the relation $\sqrt{j} = (1+j)/\sqrt{2}$. For a perfect conductor with $\sigma = \infty$, these expressions yield $\alpha = \beta = \infty$, and $\eta_c = 0$.

Expressions for the propagation parameters in various types of media are summarized in Table 2-1.

2-5 Electromagnetic Power Density

This section deals with the flow of power carried by an electromagnetic wave. For any wave with time-domain electric field $\mathbf{E}(t)$ and associated magnetic field $\mathbf{H}(t)$, the ***Poynting vector*** $\mathbf{S}(t)$ is defined as

$$\mathbf{S}(t) = \mathbf{E}(t) \times \mathbf{H}(t) \qquad (\text{W/m}^2). \tag{2.70}$$

The unit of $\mathbf{S}(t)$ is (V/m) $\times$ (A/m) = (W/m^2), and the direction of $\mathbf{S}(t)$ is along the wave's direction of propagation. Thus, $\mathbf{S}(t)$ represents the power per unit area (or power density) carried by the wave.

In practice, the quantity of greater interest is the ***average power density*** of the wave, $\mathbf{S}$, which is the time-average value of $\mathbf{S}(t)$:

$$\mathbf{S} = \tfrac{1}{2}\mathfrak{Re}\,[\mathbf{E} \times \mathbf{H}^*] \qquad (\text{W/m}^2). \tag{2.71}$$

If the wave is incident upon an aperture of area A with outward surface unit vector $\hat{\mathbf{n}}$ as shown in Fig. 2-11, then

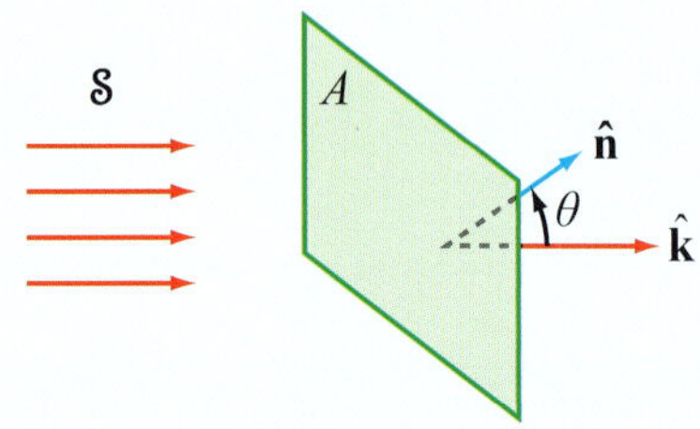

Figure 2-11: EM power flow through an aperture.

the total power that flows through or is intercepted by the aperture is

$$P = \int_A \mathbf{S} \cdot \hat{\mathbf{n}}\, dA \qquad \text{(W)}. \tag{2.72}$$

For a uniform plane wave propagating in a direction $\hat{\mathbf{k}}$ that makes an angle θ with $\hat{\mathbf{n}}$, $P = \mathcal{S}A\cos\theta$, where $\mathcal{S} = |\mathbf{S}|$.

2-5.1 Plane Wave in a Lossless Medium

Recall that the general expression for the electric field of a uniform plane wave with arbitrary polarization traveling in the $+z$ direction is

$$\begin{aligned} \mathbf{E}(z) &= \hat{\mathbf{x}}E_x(z) + \hat{\mathbf{y}}E_y(z) \\ &= (\hat{\mathbf{x}}E_{x0} + \hat{\mathbf{y}}E_{y0})e^{-jkz}, \end{aligned} \tag{2.73}$$

where, in the general case, E_{x0} and E_{y0} are complex quantities. The magnitude of $\mathbf{E}$ is

$$|\mathbf{E}| = (\mathbf{E}\cdot\mathbf{E}^*)^{1/2} = [|E_{x0}|^2 + |E_{y0}|^2]^{1/2}. \tag{2.74}$$

The phasor magnetic field associated with $\mathbf{E}$ is obtained by applying Eq. (2.34a):

$$\begin{aligned} \mathbf{H}(z) &= (\hat{\mathbf{x}}H_x + \hat{\mathbf{y}}H_y)e^{-jkz} \\ &= \frac{1}{\eta}\,\hat{\mathbf{z}} \times \mathbf{E} = \frac{1}{\eta}\,(-\hat{\mathbf{x}}E_{y0} + \hat{\mathbf{y}}E_{x0})e^{-jkz}. \end{aligned} \tag{2.75}$$

The wave can be considered as the sum of two waves, one comprising fields (E_x, H_y) and another comprising fields (E_y, H_x). Use of Eqs. (2.73) and (2.75) in

Table 2-1: Expressions for α, β, η_c, u_p, and λ for various types of nonmagnetic media.[†]

	Any Medium	Lossless Medium ($\sigma = 0$)	Low-Loss Medium ($\varepsilon''/\varepsilon' \ll 1$)	Good Conductor ($\varepsilon''/\varepsilon' \gg 1$)	Units
$\alpha =$	$\omega \left[\frac{\mu_0 \varepsilon' \varepsilon_0}{2} \left[\sqrt{1 + \left(\frac{\varepsilon''}{\varepsilon'} \right)^2} - 1 \right] \right]^{1/2}$	0	$\frac{\pi \varepsilon''}{\lambda_0 \sqrt{\varepsilon'}}$	$\frac{\pi \sqrt{2\varepsilon''}}{\lambda_0}$	(Np/m)
$\beta =$	$\omega \left[\frac{\mu_0 \varepsilon' \varepsilon_0}{2} \left[\sqrt{1 + \left(\frac{\varepsilon''}{\varepsilon'} \right)^2} + 1 \right] \right]^{1/2}$	$\frac{2\pi \sqrt{\varepsilon'}}{\lambda_0}$	$\frac{2\pi \sqrt{\varepsilon'}}{\lambda_0}$	$\frac{\pi \sqrt{2\varepsilon''}}{\lambda_0}$	(rad/m)
$\eta_c =$	$\sqrt{\frac{\mu_0}{\varepsilon' \varepsilon_0}} \left(1 - j \frac{\varepsilon''}{\varepsilon'} \right)^{-1/2}$	$\frac{\eta_0}{\sqrt{\varepsilon'}}$	$\frac{\eta_0}{\sqrt{\varepsilon'}}$	$\frac{(1+j)\eta_0}{\sqrt{2\varepsilon''}}$	(Ω)
$u_p =$	ω/β	$c/\sqrt{\varepsilon'}$	$c/\sqrt{\varepsilon'}$	$c\sqrt{2/\varepsilon''}$	(m/s)
$\lambda =$	$2\pi/\beta = u_p/f$	u_p/f	u_p/f	u_p/f	(m)

Notes: In practice, a material is considered a low-loss medium if $\varepsilon''/\varepsilon' < 0.01$ and a good conducting medium if $\varepsilon''/\varepsilon' > 100$; $c = 3 \times 10^8$ m/s; $\eta_0 = 377\ \Omega$.

Eq. (2.71) leads to

$$\mathcal{S} = \hat{\mathbf{z}} \frac{1}{2\eta} (|E_{x0}|^2 + |E_{y0}|^2) = \hat{\mathbf{z}} \frac{|\mathbf{E}|^2}{2\eta} \qquad (\text{W/m}^2), \qquad (2.76)$$

which states that power flows in the z direction with average power density equal to the sum of the average power densities of the (E_x, H_y) and (E_y, H_x) waves. Note that, because $\mathcal{S}$ depends only on η and $|\mathbf{E}|$, waves characterized by different polarizations carry the same amount of average power as long as their electric fields have the same magnitudes.

2-5.2 Plane Wave in a Lossy Medium

The expressions given by Eqs. (2.60) and (2.61) characterize the electric and magnetic fields of an x-polarized plane wave propagating along the z direction in a lossy medium with propagation constant $\gamma = \alpha + j\beta$. By extending these expressions to the more general case of a wave with components along both x and y, we have

$$\mathbf{E}(z) = \hat{\mathbf{x}} E_x(z) + \hat{\mathbf{y}} E_y(z) = (\hat{\mathbf{x}} E_{x0} + \hat{\mathbf{y}} E_{y0}) e^{-\alpha z} e^{-j\beta z}, \qquad (2.77a)$$

$$\mathbf{H}(z) = \frac{1}{\eta_c} (-\hat{\mathbf{x}} E_{y0} + \hat{\mathbf{y}} E_{x0}) e^{-\alpha z} e^{-j\beta z}, \qquad (2.77b)$$

where η_c is the intrinsic impedance of the lossy medium. Application of Eq. (2.71) gives

$$\mathcal{S}(z) = \frac{1}{2} \mathfrak{Re} [\mathbf{E} \times \mathbf{H}^*] = \frac{\hat{\mathbf{z}}(|E_{x0}|^2 + |E_{y0}|^2)}{2} e^{-2\alpha z} \mathfrak{Re} \left(\frac{1}{\eta_c^*} \right). \qquad (2.78)$$

By expressing η_c in polar form as

$$\eta_c = |\eta_c| e^{j\theta_\eta}, \qquad (2.79)$$

[†] Computer Code 2.1.

Eq. (2.78) can be rewritten as

$$\begin{aligned} \mathbf{S}(z) &= \hat{\mathbf{z}}\,\frac{|E(0)|^2}{2|\eta_c|}\,e^{-2\alpha z}\cos\theta_\eta \\ &= \hat{\mathbf{z}}S_0 e^{-2\alpha z} \qquad (\text{W/m}^2), \end{aligned} \tag{2.80}$$

where $|E(0)|^2 = [|E_{x0}|^2 + |E_{y0}|^2]^{1/2}$ is the magnitude of $\mathbf{E}(z)$ at $z = 0$.

► Whereas the fields $\mathbf{E}(z)$ and $\mathbf{H}(z)$ decay with z as $e^{-\alpha z}$, the power density $\mathbf{S}$ decreases as $e^{-2\alpha z}$. ◄

When a wave propagates through a distance $z = \delta_s = 1/\alpha$, the magnitudes of its electric and magnetic fields decrease to $e^{-1} \approx 37\%$ of their initial values, and its average power density decreases to $e^{-2} \approx 14\%$ of its initial value.

2-5.3 Decibel Scale for Power Ratios

The unit for power P is watts (W). In many engineering problems, the quantity of interest is the ratio of two power levels, P_1 and P_2, such as the received and transmitted powers for a radar system, and often the ratio P_1/P_2 may vary over several orders of magnitude. The decibel (dB) scale is logarithmic, thereby providing a convenient representation of the power ratio, particularly when numerical values of P_1/P_2 are plotted against some variable of interest. If

$$G = \frac{P_1}{P_2}, \tag{2.81}$$

then

$$G\,[\text{dB}] = 10\log G = 10\log\left(\frac{P_1}{P_2}\right) \qquad (\text{dB}). \tag{2.82}$$

The ***attenuation rate***, representing the rate of decrease of the magnitude of $S(z)$ as a function of propagation distance, is defined as

$$\begin{aligned} A &= 10\log\left[\frac{S(z)}{S(0)}\right] \\ &= 10\log(e^{-2\alpha z}) \\ &= -20\alpha z \log e \\ &= -8.68\alpha z = -\alpha\,[\text{dB/m}]\,z \qquad (\text{dB}), \end{aligned} \tag{2.83}$$

where

$$\alpha\,[\text{dB/m}] = 8.68\alpha\,[\text{Np/m}]. \tag{2.84}$$

We also note that, since $S(z)$ is directly proportional to $|\mathbf{E}(z)|^2$,

$$A = 10\log\left[\frac{|E(z)|^2}{|E(0)|^2}\right] = 20\log\left[\frac{|E(z)|}{|E(0)|}\right] \qquad (\text{dB}). \tag{2.85}$$

Table 2-2 compares selected values of G with the corresponding values of G [dB].

Table 2-2: Power ratios in natural numbers and in decibels.

G	G [dB]
10^x	$10x$ dB
4	6 dB
2	3 dB
1	0 dB
0.5	−3 dB
0.25	−6 dB
0.1	−10 dB
10^{-3}	−30 dB

2-6 Wave Reflection and Transmission at Normal Incidence

For convenience, we divide our treatment of wave reflection by, and transmission through, planar boundaries into three parts: in this section we confine our discussion to the normal-incidence case depicted in Fig. 2-12(a), in

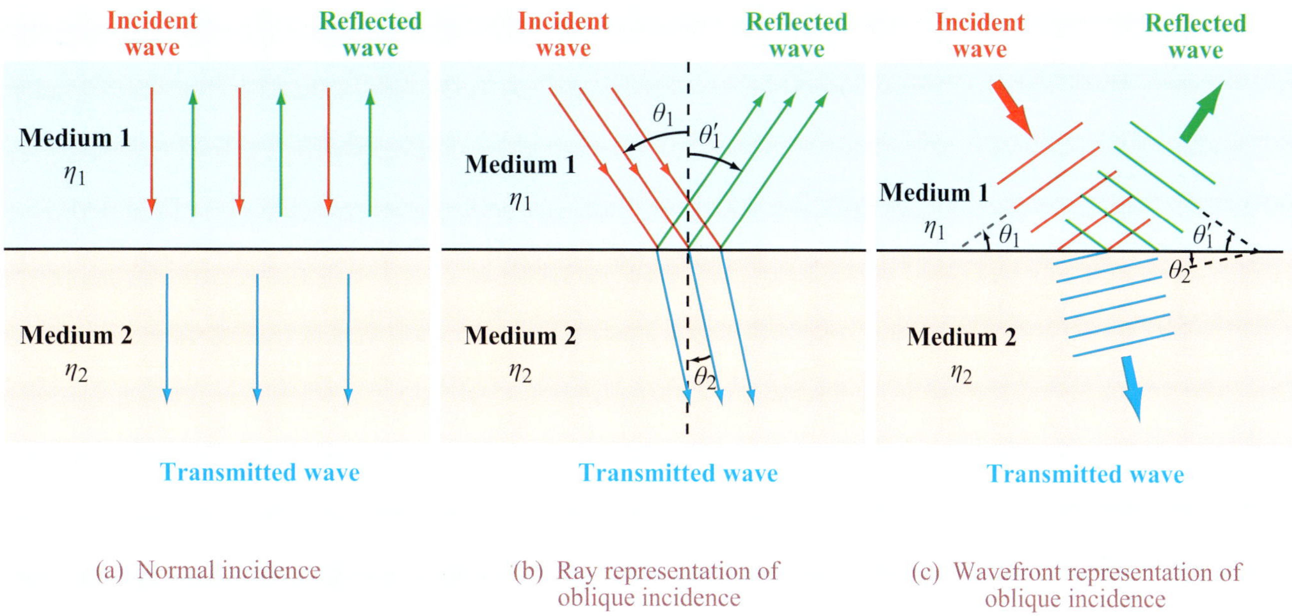

Figure 2-12: Ray representation of wave reflection and transmission at (a) normal incidence and (b) oblique incidence, and (c) wavefront representation of oblique incidence.

Section 2-7 to 2-8 we examine the more general oblique-incidence case depicted in Fig. 2-12(b), and we do so for two different wave polarizations, and in Sections 2-9 and 2-10 we treat special cases of particular interest in remote sensing.

Before proceeding, however, we should explain the notion of rays and wavefronts, and the relationship between them, as both are used throughout this chapter to represent electromagnetic waves. A ***ray*** is a line representing the direction of flow of electromagnetic energy carried by a wave, and therefore it is parallel to the propagation unit vector $\hat{\mathbf{k}}$. A ***wavefront*** is a surface across which the phase of a wave is constant; it is perpendicular to vector $\hat{\mathbf{k}}$. Hence, rays are perpendicular to wavefronts. The ray representation of wave incidence, reflection, and transmission shown in Fig. 2-12(b) is equivalent to the wavefront representation depicted in Fig. 2-12(c). The two representations are complimentary; the ray representation is easier to use in graphical illustrations, whereas the wavefront representation provides greater physical insight into what happens to a wave when it encounters a discontinuous boundary. Both representations are used in forthcoming discussions.

2-6.1 Boundary between Lossless Media

A planar boundary located at $z = 0$ (Fig. 2-13) separates two lossless, homogeneous, dielectric media. Medium 1 has relative permittivity ε_1' and permeability μ_0 and fills the half-space $z \geq 0$. Medium 2 has relative permittivity ε_2' and permeability μ_0 and fills the half-space $z \leq 0$. A y-polarized plane wave, with electric and magnetic fields $(\mathbf{E}^{\mathrm{i}}, \mathbf{H}^{\mathrm{i}})$ propagates in medium 1 along direction $\hat{\mathbf{k}}_{\mathrm{i}} = -\hat{\mathbf{z}}$ toward medium 2. Reflection and transmission at the boundary at $z = 0$ result in a reflected wave, with electric and magnetic fields $(\mathbf{E}^{\mathrm{r}}, \mathbf{H}^{\mathrm{r}})$, traveling along direction $\hat{\mathbf{k}}_{\mathrm{r}} = \hat{\mathbf{z}}$ in medium 1, and a transmitted wave, with electric and magnetic fields $(\mathbf{E}^{\mathrm{t}}, \mathbf{H}^{\mathrm{t}})$, traveling along direction $\hat{\mathbf{k}}_{\mathrm{t}} = -\hat{\mathbf{z}}$ in medium 2. The expressions for the electric and magnetic fields of the three linearly polarized waves (Fig. 2-13) are given

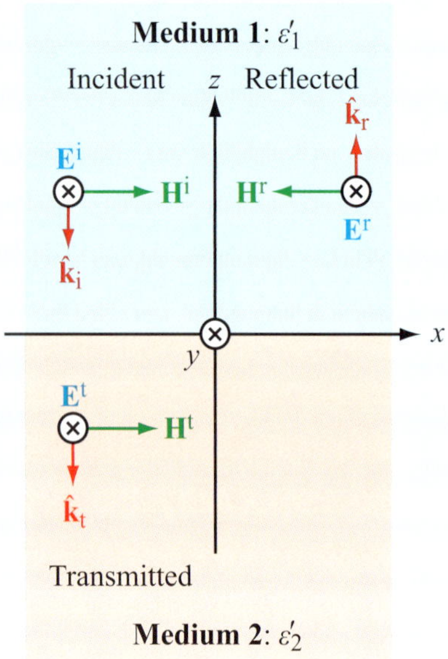

Figure 2-13: Two dielectric media separated by the x–y plane.

in the left-hand section of Table 2-3. The ***total fields*** in media 1 and 2 are

$$\mathbf{E}_1 = \mathbf{E}^i + \mathbf{E}^r, \tag{2.86a}$$

$$\mathbf{H}_1 = \mathbf{H}^i + \mathbf{H}^r, \tag{2.86b}$$

$$\mathbf{E}_2 = \mathbf{E}^t, \tag{2.86c}$$

$$\mathbf{H}_2 = \mathbf{H}^t. \tag{2.86d}$$

Note that superscripts i, r, and t are used to denote incident, reflected, and transmitted, and subscripts 1 and 2 denote the total fields in media 1 and 2. The amplitudes of the incident, reflected, and transmitted electric fields at $z = 0$ (the boundary between the two media) are E_0^i, E_0^r, and E_0^t, respectively. The wavenumber and intrinsic impedance of medium 1 are $k_1 = 2\pi\sqrt{\varepsilon_1'}/\lambda_0$ and $\eta_1 = \eta_0/\sqrt{\varepsilon_1'}$, and those for medium 2 are $k_2 = 2\pi\sqrt{\varepsilon_2'}/\lambda_0$ and $\eta_2 = \eta_0/\sqrt{\varepsilon_2'}$.

The amplitude E_0^i is imposed by the source responsible for generating the incident wave, and therefore is assumed known. Our goal is to relate E_0^r and E_0^t to E_0^i. We do so by applying EM boundary conditions for the total electric and magnetic fields at $z = 0$. These boundary conditions state that the tangential component of the total electric field is always continuous across a boundary between two contiguous media, and in the absence of current sources at the boundary, the same is true for the total magnetic field. In the present case, the electric and magnetic fields of the incident, reflected, and transmitted waves are all tangential to the boundary. Application of boundary conditions at $z = 0$, which also is called ***phase matching***, leads to

$$E_0^r = \left(\frac{\eta_2 - \eta_1}{\eta_2 + \eta_1}\right) E_0^i = \rho E_0^i, \tag{2.87a}$$

$$E_0^t = \left(\frac{2\eta_2}{\eta_2 + \eta_1}\right) E_0^i = \tau E_0^i, \tag{2.87b}$$

where

$$\rho = \frac{E_0^r}{E_0^i} = \frac{\eta_2 - \eta_1}{\eta_2 + \eta_1} \quad \textbf{(normal incidence)}, \tag{2.88a}$$

$$\tau = \frac{E_0^t}{E_0^i} = \frac{2\eta_2}{\eta_2 + \eta_1} \quad \textbf{(normal incidence)}. \tag{2.88b}$$

The quantities ρ and τ are called the ***Fresnel reflection*** and ***transmission coefficients***. For lossless dielectric media, η_1 and η_2 are real; consequently, both ρ and τ are also real. From Eqs. (2.88a) and (2.88b), it is easy to show that ρ and τ are interrelated by

$$\tau = 1 + \rho \qquad \textbf{(normal incidence)}. \tag{2.89}$$

Using

$$\eta_1 = \frac{\eta_0}{\sqrt{\varepsilon_1'}},$$

$$\eta_2 = \frac{\eta_0}{\sqrt{\varepsilon_2'}},$$

in Eq. (2.88a) leads to

$$\rho = \frac{\sqrt{\varepsilon_1'} - \sqrt{\varepsilon_2'}}{\sqrt{\varepsilon_1'} + \sqrt{\varepsilon_2'}}. \tag{2.90}$$

Table 2-3: Expressions for EM fields in lossless and lossy media under normal incidence.†

Lossless Media	Lossy Media
$\mathbf{E}^{\rm i} = \hat{\mathbf{y}} E_0^{\rm i} e^{jk_1 z}, \quad \mathbf{H}^{\rm i} = \hat{\mathbf{x}} \dfrac{E_0^{\rm i}}{\eta_1} e^{jk_1 z}$	$\mathbf{E}^{\rm i} = \hat{\mathbf{y}} E_0^{\rm i} e^{\gamma_1 z}, \quad \mathbf{H}^{\rm i} = \hat{\mathbf{x}} \dfrac{E_0^{\rm i}}{\eta_{\rm c_1}} e^{\gamma_1 z}$
$\mathbf{E}^{\rm r} = \hat{\mathbf{y}} \rho E_0^{\rm i} e^{-jk_1 z}, \quad \mathbf{H}^{\rm r} = -\hat{\mathbf{x}} \rho \dfrac{E_0^{\rm i}}{\eta_1} e^{-jk_1 z}$	$\mathbf{E}^{\rm r} = \hat{\mathbf{y}} \rho E_0^{\rm i} e^{-\gamma_1 z}, \quad \mathbf{H}^{\rm r} = -\hat{\mathbf{x}} \rho \dfrac{E_0^{\rm i}}{\eta_{\rm c_1}} e^{-\gamma_1 z}$
$\mathbf{E}^{\rm t} = \hat{\mathbf{y}} \tau E_0^{\rm i} e^{jk_2 z}, \quad \mathbf{H}^{\rm t} = \hat{\mathbf{x}} \tau \dfrac{E_0^{\rm i}}{\eta_2} e^{jk_2 z}$	$\mathbf{E}^{\rm t} = \hat{\mathbf{y}} \tau E_0^{\rm i} e^{\gamma_2 z}, \quad \mathbf{H}^{\rm t} = \hat{\mathbf{x}} \tau \dfrac{E_0^{\rm i}}{\eta_{\rm c_2}} e^{\gamma_2 z}$
Phase matching: $(\mathbf{E}^{\rm i} + \mathbf{E}^{\rm r})\big\vert_{z=0} = \mathbf{E}^{\rm t}\big\vert_{z=0}$	Phase matching: $(\mathbf{E}^{\rm i} + \mathbf{E}^{\rm r})\big\vert_{z=0} = \mathbf{E}^{\rm t}\big\vert_{z=0}$
$(\mathbf{H}^{\rm i} + \mathbf{H}^{\rm r})\big\vert_{z=0} = \mathbf{H}^{\rm t}\big\vert_{z=0}$	$(\mathbf{H}^{\rm i} + \mathbf{H}^{\rm r})\big\vert_{z=0} = \mathbf{H}^{\rm t}\big\vert_{z=0}$
$\rho = \dfrac{\eta_2 - \eta_1}{\eta_2 + \eta_1}, \quad \tau = 1 + \rho$	$\rho = \dfrac{\eta_{\rm c_2} - \eta_{\rm c_1}}{\eta_{\rm c_2} + \eta_{\rm c_1}}, \quad \tau = 1 + \rho$
$k_1 = \dfrac{2\pi}{\lambda_0} \sqrt{\varepsilon_1'}, \quad k_2 = \dfrac{2\pi}{\lambda_0} \sqrt{\varepsilon_2'}$	$\gamma_1 = \alpha_1 + j\beta_1, \quad \gamma_2 = \alpha_2 + j\beta_2$
$\eta_1 = \dfrac{\eta_0}{\sqrt{\varepsilon_1'}}, \quad \eta_2 = \dfrac{\eta_0}{\sqrt{\varepsilon_2'}}$	$\eta_{\rm c_1} = \dfrac{\eta_0}{\sqrt{\varepsilon_1}}, \quad \eta_{\rm c_2} = \dfrac{\eta_0}{\sqrt{\varepsilon_2}}$
Notes: $\varepsilon_1 = \varepsilon_1' - j\varepsilon_1''$; $\varepsilon_2 = \varepsilon_2' - j\varepsilon_2''$.	

Complete expressions for the fields and power densities in media 1 and 2 are available in the left-hand side of Table 2-3.

Using Eq. (2.71), the net average power density flowing in medium 1 is

$$\begin{aligned}\mathcal{S}_1(z) &= \tfrac{1}{2}\mathfrak{Re}[\mathbf{E}_1(z) \times \mathbf{H}_1^*(z)] \\ &= \tfrac{1}{2}\mathfrak{Re}\left[\hat{\mathbf{y}} E_0^{\rm i}(e^{jk_1 z} + \rho e^{-jk_1 z}) \right. \\ &\qquad \left. \times\, \hat{\mathbf{x}} \frac{E_0^{\rm i*}}{\eta_1} (e^{-jk_1 z} - \rho^* e^{jk_1 z})\right],\end{aligned}$$

which leads to

$$\mathcal{S}_1(z) = -\hat{\mathbf{z}} \frac{|E_0^{\rm i}|^2}{2\eta_1} (1 - |\rho|^2). \tag{2.91}$$

The first and second terms inside the bracket in Eq. (2.91) represent the average power density of the incident and reflected waves, respectively. Thus,

$$\mathcal{S}_1 = \mathcal{S}^{\rm i} + \mathcal{S}^{\rm r}, \tag{2.92a}$$

with

$$\mathcal{S}^{\rm i} = -\hat{\mathbf{z}} \frac{|E_0^{\rm i}|^2}{2\eta_1}, \tag{2.92b}$$

$$\mathcal{S}^{\rm r} = \hat{\mathbf{z}} |\rho|^2 \frac{|E_0^{\rm i}|^2}{2\eta_1} = \hat{\mathbf{z}} |\rho|^2 \mathcal{S}^{\rm i}. \tag{2.92c}$$

Even though ρ is purely real when both media are lossless dielectrics, we chose to treat it as complex, thereby providing in Eq. (2.92c) an expression that is also valid when medium 2 is conducting.

†Computer Code 2.3.

The average power density of the transmitted wave in medium 2 is

$$\begin{aligned}\mathbf{\mathcal{S}}^{\mathrm{t}}(z) = \mathbf{\mathcal{S}}_2(z) &= \tfrac{1}{2}\Re\mathfrak{e}[\mathbf{E}_2(z)\times\mathbf{H}_2^*(z)] \\ &= \tfrac{1}{2}\Re\mathfrak{e}\left[\hat{\mathbf{y}}\tau E_0^{\mathrm{i}}e^{jk_2z}\times\hat{\mathbf{x}}\tau^*\,\frac{E_0^{\mathrm{i}*}}{\eta_2}\,e^{-jk_2z}\right] \\ &= -\hat{\mathbf{z}}|\tau|^2\,\frac{|E_0^{\mathrm{i}}|^2}{2\eta_2}\,. \end{aligned} \tag{2.93}$$

Through the use of Eqs. (2.88a) and (2.88b), it can be easily shown that for lossless media

$$\frac{\tau^2}{\eta_2} = \frac{1-\rho^2}{\eta_1} \qquad \textbf{(lossless media)}, \tag{2.94}$$

which leads to

$$\mathcal{S}_1 = \mathcal{S}_2.$$

This result is expected from considerations of power conservation.

2-6.2 Boundary between Lossy Media

In Section 2-6.1 we considered a plane wave in a lossless medium incident normally on a planar boundary of another lossless medium. We now generalize our expressions to lossy media. In a medium with constitutive parameters $(\varepsilon', \mu, \sigma)$, the propagation constant $\gamma = \alpha + j\beta$ and the intrinsic impedance η_c are both complex. General expressions for α, β, and η_c are given in Table 2-1. If media 1 and 2 have constitutive parameters $(\varepsilon_1', \mu_0, \sigma_1)$ and $(\varepsilon_2', \mu_0, \sigma_2)$, the expressions for the electric and magnetic fields in media 1 and 2 can be obtained from those for the lossless case listed in the left-hand section of Table 2-3 by replacing jk with γ and η with η_c. The results are listed in the right-hand section of Table 2-3.

2-7 Wave Reflection and Transmission at Oblique Incidence

In the preceding section we examined reflection and transmission of plane waves that are normally incident upon a planar interface between two different media. We now consider the oblique-incidence case depicted in Fig. 2-14, and for simplicity we start by assuming all media to be lossless, and later we extend our results to lossy media. The $z = 0$ plane forms the boundary between media 1 and 2 with constitutive parameters (ε_1', μ_0) and (ε_2', μ_0), respectively. The two lines in Fig. 2-14 with direction $\hat{\mathbf{k}}_\mathrm{i}$ represent rays drawn normal to the wavefront of the incident wave, and those along directions $\hat{\mathbf{k}}_\mathrm{r}$ and $\hat{\mathbf{k}}_\mathrm{t}$ are similarly associated with the reflected and transmitted waves. The ***angles of incidence***, ***reflection***, and ***transmission*** (or ***refraction***), defined with respect to the normal to the boundary (the z axis), are θ_1, θ_1', and θ_2, respectively. These three

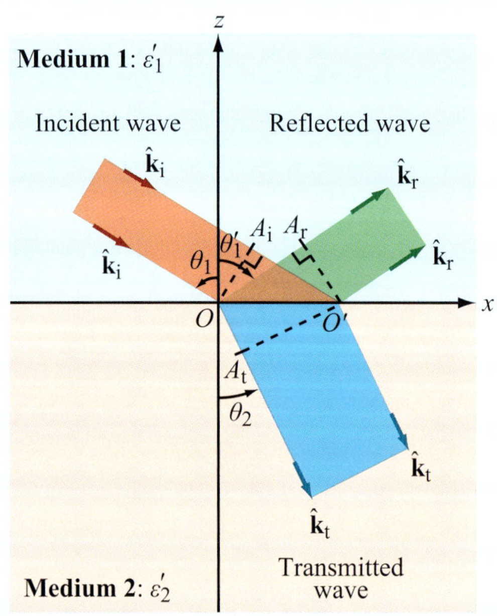

Figure 2-14: Wave reflection and refraction at a planar boundary between different media.

angles are interrelated by ***Snell's laws***:

$$\theta_1 = \theta_1' \quad \textbf{(Snell's law of reflection)},$$

$$\frac{\sin\theta_2}{\sin\theta_1} = \frac{u_{p_2}}{u_{p_1}} = \sqrt{\frac{\varepsilon_1'}{\varepsilon_2'}} \qquad (2.95)$$

(Snell's law of refraction),

where u_{p_1} and u_{p_2} are the phase velocities in media 1 and 2, respectively.

> ▶ ***Snell's law of reflection*** states that the angle of reflection equals the angle of incidence, and ***Snell's law of refraction*** provides a relation between $\sin\theta_2$ and $\sin\theta_1$ in terms of the ratio of the phase velocities. ◀

The ***index of refraction*** of a medium, n, is defined as the ratio of the phase velocity in free space (i.e., the speed of light c) to the phase velocity in the medium. Thus,

$$n = \frac{c}{u_p} = \sqrt{\frac{\mu_0\varepsilon_0\varepsilon'}{\mu_0\varepsilon_0}} = \sqrt{\varepsilon'} \,. \qquad (2.96)$$

In view of Eq. (2.96), Eq. (2.95) may be rewritten as

$$\frac{\sin\theta_2}{\sin\theta_1} = \frac{n_1}{n_2} \,. \qquad (2.97)$$

Usually, materials with higher densities have higher permittivities. Air, with $\varepsilon' = 1$, has an index of refraction $n_0 = 1$. Since for nonmagnetic materials $n = \sqrt{\varepsilon'}$, *a material is often referred to as more dense than another material if it has a greater index of refraction.*

At normal incidence ($\theta_1 = 0$), Eq. (2.97) gives $\theta_2 = 0$, as expected. At oblique incidence $\theta_2 < \theta_1$ when $n_2 > n_1$ and $\theta_2 > \theta_1$ when $n_2 < n_1$.

> ▶ If a wave is incident on a more dense medium [Fig. 2-15(a)], the transmitted wave refracts inwardly (toward the z axis), and the opposite is true if a wave is incident on a less dense medium [Fig. 2-15(b)]. ◀

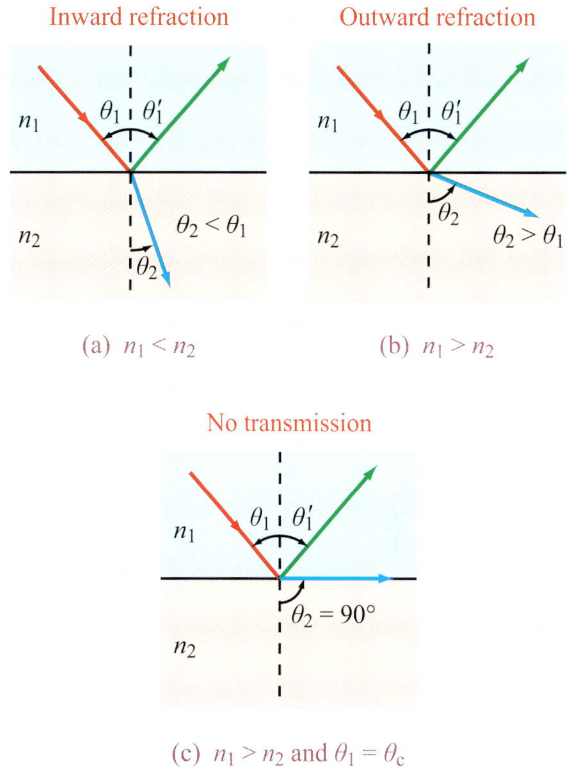

Figure 2-15: Snell's laws state that $\theta_1' = \theta_1$ and $\sin\theta_2 = (n_1/n_2)\sin\theta_1$. Refraction is (a) inward if $n_1 < n_2$ and (b) outward if $n_1 > n_2$; and (c) the refraction angle is 90° if $n_1 > n_2$ and θ_1 is equal to or greater than the critical angle $\theta_c = \sin^{-1}(n_2/n_1)$.

A case of particular interest is when $\theta_2 = \pi/2$, as shown in Fig. 2-15(c); in this case, the refracted wave flows along the surface and no energy is transmitted into medium 2. The value of the angle of incidence θ_1 corresponding to $\theta_2 = \pi/2$ is called the ***critical angle*** θ_c and is obtained from Eq. (2.97) as

$$\sin\theta_c = \frac{n_2}{n_1}\sin\theta_2\bigg|_{\theta_2=\pi/2} = \frac{n_2}{n_1} \qquad (2.98a)$$

$$= \sqrt{\frac{\varepsilon_2'}{\varepsilon_1'}} \,. \qquad (2.98b)$$

(critical angle)

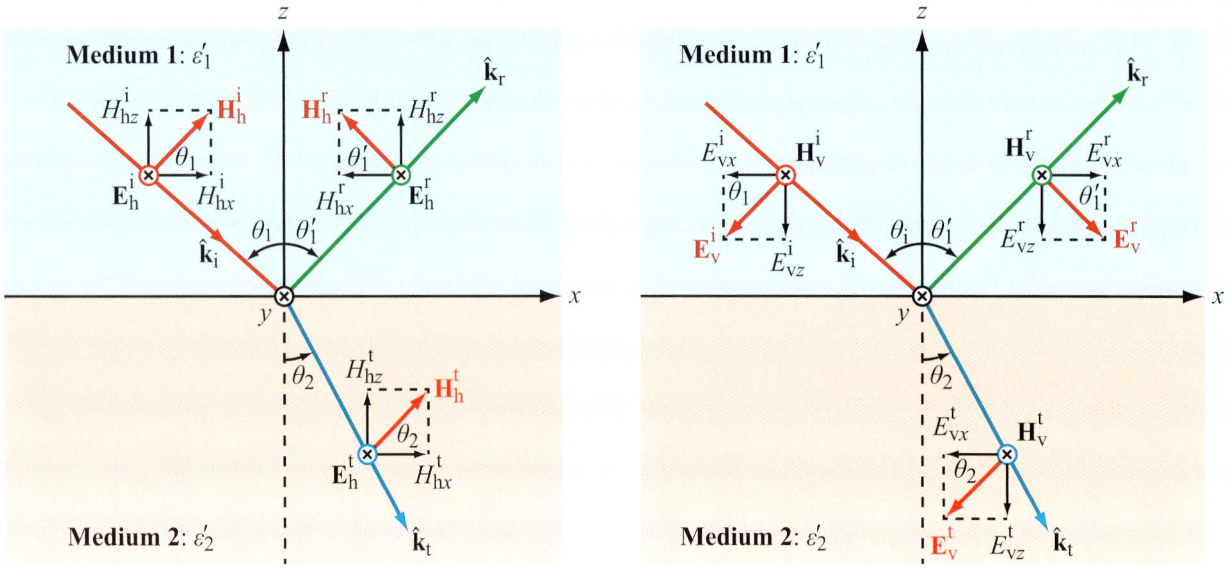

(a) Horizontal polarization

(b) Vertical polarization

Figure 2-16: The plane of incidence is the plane containing the direction of wave travel, $\hat{\mathbf{k}}_i$, and the surface normal to the boundary. In the present case the plane of incidence containing $\hat{\mathbf{k}}_i$ and $\hat{\mathbf{z}}$ coincides with the plane of the paper. A wave is (a) perpendicularly polarized (also called ***horizontally polarized***) when its electric field vector is perpendicular to the plane of incidence and (b) parallel polarized (also called ***vertically polarized***) when its electric field vector lies in the plane of incidence.

If θ_1 exceeds θ_c, the incident wave is totally reflected, and the refracted wave becomes a nonuniform ***surface wave*** that travels along the boundary between the two media. This wave behavior is called ***total internal reflection***.

For normal incidence, the reflection and transmission coefficients ρ and τ at a boundary between two media are independent of the polarization of the incident wave, as both the electric and magnetic fields of a normally incident plane wave are tangential to the boundary regardless of the wave polarization. This is not the case for obliquely incident waves traveling at an angle $\theta_1 \neq 0$ with respect to the normal to the interface. In what follows, the ***plane of incidence*** *is defined as the plane containing the normal to the boundary and the direction of propagation of the incident wave.* A wave of arbitrary polarization may be described as the superposition of two orthogonally polarized waves, one with its electric field parallel to the plane of incidence (***parallel polarization***) and the other with its electric field perpendicular to the plane of incidence (***perpendicular polarization***). These two polarization configurations are shown in Fig. 2-16, in which the plane of incidence is coincident with the x–z plane.

▶ In remote sensing, the polarization with **E** perpendicular to the plane of incidence is also called ***horizontal polarization*** because **E** is parallel to Earth's surface, and that with **E** parallel to the plane of incidence is called ***vertical polarization*** because in this case it is the magnetic field that is parallel to Earth's surface. ◀

For the general case of a wave with an arbitrary polarization, it is common practice to decompose the incident wave $(\mathbf{E}^{\rm i}, \mathbf{H}^{\rm i})$ into a horizontally polarized component $(\mathbf{E}^{\rm i}_{\rm h}, \mathbf{H}^{\rm i}_{\rm h})$ and a vertically polarized component $(\mathbf{E}^{\rm i}_{\rm v}, \mathbf{H}^{\rm i}_{\rm v})$. Then, after determining the reflected waves $(\mathbf{E}^{\rm r}_{\rm h}, \mathbf{H}^{\rm r}_{\rm h})$ and $(\mathbf{E}^{\rm r}_{\rm v}, \mathbf{H}^{\rm r}_{\rm v})$ due to the two incident components, the reflected waves are added together to give the total reflected wave $(\mathbf{E}^{\rm r}, \mathbf{H}^{\rm r})$ corresponding to the original incident wave. A similar process can be used to determine the total transmitted wave $(\mathbf{E}^{\rm t}, \mathbf{H}^{\rm t})$.

2-7.1 Horizontal Polarization—Lossless Media

In the normal-incidence configuration depicted in Fig. 2-13, $\mathbf{E}^{\rm i}$, $\mathbf{H}^{\rm i}$, and $\hat{\mathbf{k}}_{\rm i}$ of the incident wave pointed along $\hat{\mathbf{y}}$, $\hat{\mathbf{x}}$, and $-\hat{\mathbf{z}}$, respectively. For the horizontally polarized incident wave depicted in Fig. 2-16(a), $\mathbf{E}^{\rm i}_{\rm h}$ continues to be along $\hat{\mathbf{y}}$, but $\mathbf{H}^{\rm i}_{\rm h}$ and $\hat{\mathbf{k}}_{\rm i}$ point along new directions. Accordingly, the electric and magnetic fields of the incident plane wave are given by

$$\mathbf{E}^{\rm i}_{\rm h} = \hat{\mathbf{y}} E^{\rm i}_{\rm h0} e^{-jk_1(x\sin\theta_1 - z\cos\theta_1)},$$

$$\mathbf{H}^{\rm i}_{\rm h} = (\hat{\mathbf{x}}\cos\theta_1 + \hat{\mathbf{z}}\sin\theta_1)\,\frac{E^{\rm i}_{h0}}{\eta_1}\, e^{-jk_1(x\sin\theta_1 - z\cos\theta_1)}.$$

Similar expressions can be written for the reflected and transmitted wave (and are given in Table 2-4). Boundary conditions require that the tangential component of the total electric field in medium 1 (i.e., the sum of $\mathbf{E}^{\rm i}_{\rm h}$ and $\mathbf{E}^{\rm r}_{\rm h}$) be equal to the tangential component of $\mathbf{E}^{\rm t}$ at the boundary ($z = 0$). In this case all three electric fields are along $\hat{\mathbf{y}}$ and, therefore, tangential to the x–y boundary. A similar boundary condition applies to the $\hat{\mathbf{x}}$ components of the magnetic fields. Application of these boundary conditions leads to the following expressions for the reflection and transmission coefficients in the horizontal polarization case:

$$\rho_{\rm h} = \frac{E^{\rm r}_{\rm h0}}{E^{\rm i}_{\rm h0}} = \frac{\eta_2\cos\theta_1 - \eta_1\cos\theta_2}{\eta_2\cos\theta_1 + \eta_1\cos\theta_2}, \tag{2.99a}$$

$$\tau_{\rm h} = \frac{E^{\rm t}_{\rm h0}}{E^{\rm i}_{\rm h0}} = \frac{2\eta_2\cos\theta_1}{\eta_2\cos\theta_1 + \eta_1\cos\theta_2}. \tag{2.99b}$$

These two coefficients, which formally are known as the ***Fresnel reflection and transmission coefficients for horizontal polarization***, are related by

$$\tau_{\rm h} = 1 + \rho_{\rm h}. \tag{2.100}$$

If medium 2 is a perfect conductor ($\eta_2 = 0$), Eqs. (2.99a) and (2.99b) reduce to $\rho_{\rm h} = -1$ and $\tau_{\rm h} = 0$, respectively, which means that the incident wave is totally reflected by the conducting medium.

In view of Eq. (2.95), the expression for $\rho_{\rm h}$ can be written as

$$\rho_{\rm h} = \frac{\cos\theta_1 - \sqrt{(\varepsilon'_2/\varepsilon'_1) - \sin^2\theta_1}}{\cos\theta_1 + \sqrt{(\varepsilon'_2/\varepsilon'_1) - \sin^2\theta_1}}. \tag{2.101}$$

Since $(\varepsilon_2/\varepsilon_1) = (n_2/n_1)^2$, this expression can also be written in terms of the indices of refraction n_1 and n_2.

2-7.2 Vertical Polarization

Due to the dual nature of the electromagnetic fields, the expressions for the vertical polarization case can be obtained from the horizontal polarization case by replacing $\mathbf{E}$ with $\mathbf{H}$ and $\mathbf{H}$ with $-\mathbf{E}$. The process leads to the expressions given in Table 2-4. Of particular note are the reflection and transmission coefficients,

$$\rho_{\rm v} = \frac{E^{\rm r}_{\rm v0}}{E^{\rm i}_{\rm v0}} = \frac{\eta_2\cos\theta_2 - \eta_1\cos\theta_1}{\eta_2\cos\theta_2 + \eta_1\cos\theta_1},$$

$$= \frac{\left(\dfrac{\varepsilon'_2}{\varepsilon'_1} - \sin^2\theta_1\right)^{1/2} - \left(\dfrac{\varepsilon'_2}{\varepsilon'_1}\right)\cos\theta_1}{\left(\dfrac{\varepsilon'_2}{\varepsilon'_1} - \sin^2\theta_1\right)^{1/2} + \left(\dfrac{\varepsilon'_2}{\varepsilon'_1}\right)\cos\theta_1}, \tag{2.102a}$$

and

$$\tau_{\rm v} = \frac{E^{\rm t}_{\rm v0}}{E^{\rm i}_{\rm v0}} = (1 + \rho_{\rm v})\,\frac{\cos\theta_1}{\cos\theta_2}. \tag{2.102b}$$

To illustrate the angular variations of the magnitudes of $\rho_{\rm h}$ and $\rho_{\rm v}$, Fig. 2-17 shows plots for waves incident in air onto three different types of dielectric surfaces: dry

Table 2-4: Expressions for EM fields in lossless and lossy* media under normal incidence.†

Horizontal Polarization	Vertical Polarization
$\mathbf{E}_\mathrm{h}^\mathrm{i} = \hat{\mathbf{y}} E_\mathrm{h0}^\mathrm{i} e^{-jk_1(x\sin\theta_1 - z\cos\theta_1)}$	$\mathbf{E}_\mathrm{v}^\mathrm{i} = (-\hat{\mathbf{x}}\cos\theta_1 - \hat{\mathbf{z}}\sin\theta_1) E_\mathrm{v0}^\mathrm{i} e^{-jk_1(x\sin\theta_1 - z\cos\theta_1)}$
$\mathbf{H}_\mathrm{h}^\mathrm{i} = (\hat{\mathbf{x}}\cos\theta_1 + \hat{\mathbf{z}}\sin\theta_1)\,\dfrac{E_\mathrm{h0}^\mathrm{i}}{\eta_1}\, e^{-jk_1(x\sin\theta_1 - z\cos\theta_1)}$	$\mathbf{H}_\mathrm{v}^\mathrm{i} = \hat{\mathbf{y}}\,\dfrac{E_\mathrm{v0}^\mathrm{i}}{\eta_1}\, e^{-jk_1(x\sin\theta_1 - z\cos\theta_1)}$
$\mathbf{E}_\mathrm{h}^\mathrm{r} = \hat{\mathbf{y}} \rho_\mathrm{h} E_\mathrm{h0}^\mathrm{i} e^{-jk_1(x\sin\theta_1 + z\cos\theta_1)}$	$\mathbf{E}_\mathrm{v}^\mathrm{r} = (\hat{\mathbf{x}}\cos\theta_1 - \hat{\mathbf{z}}\sin\theta_1)\rho_\mathrm{v} E_\mathrm{v0}^\mathrm{i} e^{-jk_1(x\sin\theta_1 + z\cos\theta_1)}$
$\mathbf{H}_\mathrm{h}^\mathrm{r} = (-\hat{\mathbf{x}}\cos\theta_1 + \hat{\mathbf{z}}\sin\theta_1)\rho_\mathrm{h}\,\dfrac{E_\mathrm{h0}^\mathrm{i}}{\eta_1}\, e^{-jk_1(x\sin\theta_1 + z\cos\theta_1)}$	$\mathbf{H}_\mathrm{v}^\mathrm{r} = \hat{\mathbf{y}}\rho_\mathrm{v}\,\dfrac{E_\mathrm{v0}^\mathrm{i}}{\eta_1}\, e^{-jk_1(x\sin\theta_1 + z\cos\theta_1)}$
$\mathbf{E}_\mathrm{h}^\mathrm{t} = \hat{\mathbf{y}} \tau_\mathrm{h} E_\mathrm{h0}^\mathrm{i} e^{-jk_2(x\sin\theta_2 - z\cos\theta_2)}$	$\mathbf{E}_\mathrm{v}^\mathrm{t} = (-\hat{\mathbf{x}}\cos\theta_2 - \hat{\mathbf{z}}\sin\theta_2)\tau_\mathrm{v} E_\mathrm{v0}^\mathrm{i} e^{-jk_2(x\sin\theta_2 - z\cos\theta_2)}$
$\mathbf{H}_\mathrm{h}^\mathrm{t} = (\hat{\mathbf{x}}\cos\theta_2 + \hat{\mathbf{z}}\sin\theta_2)\tau_\mathrm{h}\,\dfrac{E_\mathrm{h0}^\mathrm{i}}{\eta_2}\, e^{-jk_2(x\sin\theta_2 - z\cos\theta_2)}$	$\mathbf{H}_\mathrm{v}^\mathrm{t} = \hat{\mathbf{y}}\tau_\mathrm{v}\,\dfrac{E_\mathrm{v0}^\mathrm{i}}{\eta_1}\, e^{-jk_2(x\sin\theta_2 - z\cos\theta_2)}$
$\rho_\mathrm{h} = \dfrac{\eta_2\cos\theta_1 - \eta_1\cos\theta_2}{\eta_2\cos\theta_1 + \eta_1\cos\theta_2} = \dfrac{\cos\theta_1 - \sqrt{(\varepsilon_2'/\varepsilon_1') - \sin^2\theta_1}}{\cos\theta_1 + \sqrt{(\varepsilon_2'/\varepsilon_1') - \sin^2\theta_1}}$	$\rho_\mathrm{v} = \dfrac{\eta_2\cos\theta_2 - \eta_1\cos\theta_1}{\eta_2\cos\theta_2 + \eta_1\cos\theta_1} = \dfrac{\left[\dfrac{\varepsilon_2'}{\varepsilon_1'} - \sin^2\theta_1\right]^{1/2} - \left(\dfrac{\varepsilon_2'}{\varepsilon_1'}\right)\cos\theta_1}{\left[\dfrac{\varepsilon_2'}{\varepsilon_1'} - \sin^2\theta_1\right]^{1/2} + \left(\dfrac{\varepsilon_2'}{\varepsilon_1'}\right)\cos\theta_1}$
$\tau_\mathrm{h} = 1 + \rho_\mathrm{h}$	$\tau_\mathrm{v} = (1 + \rho_\mathrm{v})\,\dfrac{\cos\theta_1}{\cos\theta_2}$
$k_1 = \dfrac{2\pi}{\lambda_0}\sqrt{\varepsilon_1}$ $\quad \eta_1 = \dfrac{\eta_0}{\sqrt{\varepsilon_1}}$	$\eta_2 = \dfrac{\eta_0}{\sqrt{\varepsilon_2}}$
$k_2 = \dfrac{2\pi}{\lambda}\sqrt{\varepsilon_2}$ $\quad k_2\sin\theta_2 = k_1\sin\theta_1$	
$\cos\theta_2 = \left[1 - \left(\dfrac{k_1}{k_2}\sin\theta_1\right)^2\right]^{1/2}$	
*Notes: (1) Lossless medium: $\varepsilon = \varepsilon'$; (2) lossy medium: $\varepsilon = \varepsilon' - j\varepsilon''$ and jk should be replaced with $\gamma = j(2\pi\sqrt{\varepsilon' - j\varepsilon''})/\lambda_0$.	

soil ($\varepsilon_2' = 3$), wet soil ($\varepsilon_2' = 25$), and water ($\varepsilon_2' = 81$). For each of the surfaces, (a) $\rho_\mathrm{h} = \rho_\mathrm{v}$ at normal incidence ($\theta_1 = 0$), as expected, (b) $|\rho_\mathrm{h}| = |\rho_\mathrm{v}| = 1$ at ***grazing incidence*** ($\theta_1 = 90^\circ$), and (c) ρ_v goes to zero at an angle called the ***Brewster angle***. For nonmagnetic materials, the Brewster angle exists only for vertical polarization, and its value depends on the ratio $(\varepsilon_2'/\varepsilon_1')$.

► At the Brewster angle, the vertically polarized component of the incident wave is totally transmitted into medium 2. ◄

From Eq. (2.102a), $\rho_\mathrm{v} = 0$ when

$$\eta_1\cos\theta_1 = \eta_2\cos\theta_2. \tag{2.103}$$

Combining Eq. (2.103) with Snell's law for nonmagnetic media, namely $\sin\theta_2 = (\eta_2/\eta_1)\sin\theta_1$, and then denot-

†Computer Code 2.3.

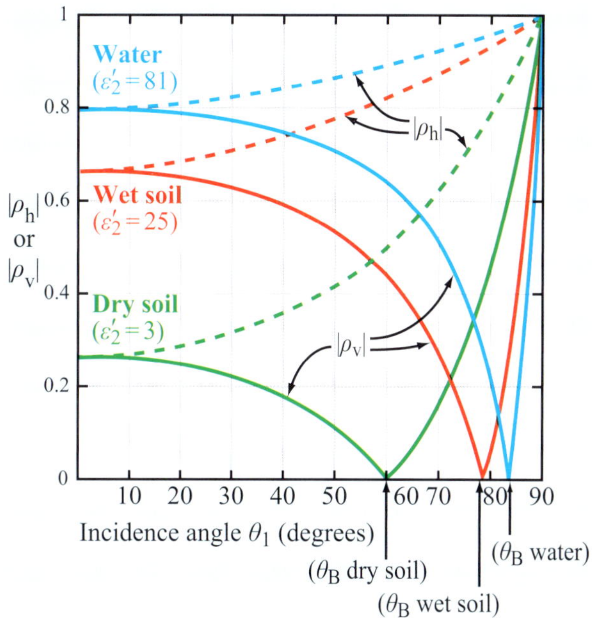

Figure 2-17: Plots for $|\rho_h|$ and $|\rho_v|$ as a function of θ_1 for a dry soil surface, a wet-soil surface, and a water surface. For each surface, $|\rho_v| = 0$ at the Brewster angle.

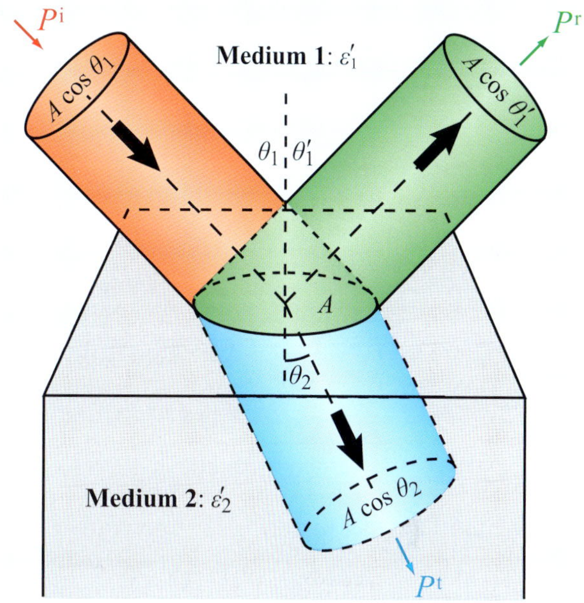

Figure 2-18: Reflection and transmission of an incident circular beam illuminating a spot of size A on the interface.

ing θ_1 as θ_B, leads to

$$\theta_B = \sin^{-1}\sqrt{\frac{1}{1+(\varepsilon_1'/\varepsilon_2')}} = \tan^{-1}\sqrt{\frac{\varepsilon_2'}{\varepsilon_1'}}. \qquad (2.104)$$

The Brewster angle is also called the ***polarizing angle***. This is because, if a wave composed of both horizontal and vertical polarization components is incident upon a nonmagnetic surface at the Brewster angle θ_B, the vertically polarized component is totally transmitted into the second medium, and only the horizontally polarized component is reflected by the surface.

2-8 Reflectivity and Transmissivity

The reflection and transmission coefficients are ratios of the reflected and transmitted electric field amplitudes to the amplitude of the incident electric field. We now examine power ratios, starting with the horizontal polarization case. Figure 2-18 shows a circular beam of electromagnetic energy incident upon the boundary between two contiguous, lossless media. The area of the spot illuminated by the beam is A, and the incident, reflected, and transmitted beams have electric-field amplitudes E^i_{h0}, E^r_{h0}, and E^t_{h0}, respectively. The average magnitudes of the power densities carried by the incident, reflected, and transmitted beams are

$$\mathcal{S}^i_h = \frac{|E^i_{h0}|^2}{2\eta_1}, \qquad \mathcal{S}^r_h = \frac{|E^r_{h0}|^2}{2\eta_1}, \qquad \mathcal{S}^t_h = \frac{|E^t_{h0}|^2}{2\eta_2},$$

where η_1 and η_2 are the intrinsic impedances of media 1 and 2, respectively. The cross-sectional areas of the incident, reflected, and transmitted beams are

$$A_i = A\cos\theta_1, \qquad A_r = A\cos\theta_1', \qquad A_t = A\cos\theta_2,$$

and the corresponding average powers carried by the beams are

$$P^i_h = \mathcal{S}^i_h A_i = \frac{|E^i_{h0}|^2}{2\eta_1} A\cos\theta_1, \qquad (2.105a)$$

$$P_{\mathrm{h}}^{\mathrm{r}} = \mathbb{S}_{\mathrm{h}}^{\mathrm{r}} A_{\mathrm{r}} = \frac{|E_{\mathrm{h0}}^{\mathrm{r}}|^2}{2\eta_1} A \cos\theta_1', \tag{2.105b}$$

$$P_{\mathrm{h}}^{\mathrm{t}} = \mathbb{S}_{\mathrm{h}}^{\mathrm{t}} A_{\mathrm{t}} = \frac{|E_{\mathrm{h0}}^{\mathrm{t}}|^2}{2\eta_2} A \cos\theta_2. \tag{2.105c}$$

► The ***reflectivity*** Γ (also called ***reflectance*** in optics) is defined as the ratio of the reflected power to the incident power. ◄

The reflectivity for horizontal polarization is then

$$\Gamma^{\mathrm{h}} = \frac{P_{\mathrm{h}}^{\mathrm{r}}}{P_{\mathrm{h}}^{\mathrm{i}}} = \frac{|E_{\mathrm{h0}}^{\mathrm{r}}|^2 \cos\theta_1'}{|E_{\mathrm{h0}}^{\mathrm{i}}|^2 \cos\theta_1} = \left|\frac{E_{\mathrm{h0}}^{\mathrm{r}}}{E_{\mathrm{h0}}^{\mathrm{i}}}\right|^2, \tag{2.106}$$

where we used the fact that $\theta_1' = \theta_1$, in accordance with Snell's law of reflection. The ratio of the reflected to incident electric field amplitudes, $|E_{\mathrm{h0}}^{\mathrm{r}}/E_{\mathrm{h0}}^{\mathrm{i}}|$, is equal to the magnitude of the reflection coefficient ρ_{h}. Hence,

$$\Gamma^{\mathrm{h}} = |\rho_{\mathrm{h}}|^2, \tag{2.107}$$

and, similarly, for vertical polarization

$$\Gamma^{\mathrm{v}} = \frac{P_{\mathrm{v}}^{\mathrm{r}}}{P_{\mathrm{v}}^{\mathrm{i}}} = |\rho_{\mathrm{v}}|^2. \tag{2.108}$$

► The ***transmissivity*** $\mathbb{T}$ (or ***transmittance*** in optics) is defined as the ratio of the transmitted power to incident power. ◄

That is,

$$\mathbb{T}^{\mathrm{h}} = \frac{P_{\mathrm{h}}^{\mathrm{t}}}{P_{\mathrm{h}}^{\mathrm{i}}} = \frac{|E_{\mathrm{h0}}^{\mathrm{t}}|^2}{|E_{\mathrm{h0}}^{\mathrm{i}}|^2} \frac{\eta_1}{\eta_2} \frac{A\cos\theta_2}{A\cos\theta_1} = |\tau_{\mathrm{h}}|^2 \left(\frac{\eta_1 \cos\theta_2}{\eta_2 \cos\theta_1}\right), \tag{2.109a}$$

$$\mathbb{T}^{\mathrm{v}} = \frac{P_{\mathrm{v}}^{\mathrm{t}}}{P_{\mathrm{v}}^{\mathrm{i}}} = |\tau_{\mathrm{v}}|^2 \left(\frac{\eta_1 \cos\theta_2}{\eta_2 \cos\theta_1}\right). \tag{2.109b}$$

► The incident, reflected, and transmitted waves have to obey the law of conservation of power. ◄

In fact, in many cases the transmitted electric field is larger than the incident electric field. However, conservation of power requires that the incident power equals the sum of the reflected and transmitted powers. That is, for horizontal polarization,

$$P_{\mathrm{h}}^{\mathrm{i}} = P_{\mathrm{h}}^{\mathrm{r}} + P_{\mathrm{h}}^{\mathrm{t}}, \tag{2.110}$$

which leads to

$$\Gamma^{\mathrm{h}} + \mathbb{T}^{\mathrm{h}} = 1, \tag{2.111a}$$

$$\Gamma^{\mathrm{v}} + \mathbb{T}^{\mathrm{v}} = 1, \tag{2.111b}$$

Table 2-5 provides a summary of the general expressions for ρ, τ, Γ, and $\mathbb{T}$ for both normal and oblique incidence.

2-9 Oblique Incidence onto a Lossy Medium

Consider a plane EM wave incident in medium 1 upon a planar boundary between medium 1 and medium 2, with medium 1 being a lossless dielectric and medium 2 a lossy medium. Water is highly lossy at microwave frequencies (Section 4-2), so lakes and oceans are perfect examples of lossy media. Medium 1 is characterized by constitutive parameters (ε_1' and μ_0) and medium 2 is characterized by ($\varepsilon_2 = \varepsilon_2' - j\varepsilon_2''$ and μ_0). In turn,

$$k_1 = \frac{2\pi}{\lambda_0}\sqrt{\varepsilon_1'}\,, \qquad \eta_1 = \frac{\eta_0}{\sqrt{\varepsilon_1'}}\,, \qquad \textbf{(medium 1)} \tag{2.112a}$$

$$\gamma_2 = \alpha_2 + j\beta_2, \qquad \eta_{c_2} = \frac{\eta_0}{\sqrt{\varepsilon_2}}\,, \qquad \textbf{(medium 2)} \tag{2.112b}$$

with α_2 and β_2 as given by the expressions in Table 2-1.

Formally, the only modifications we need to make to the expressions given earlier in Table 2-4 (for the

Table 2-5: Expressions for ρ, τ, Γ, and $\mathbb{T}$ for wave incidence from a lossless medium with intrinsic impedance η_1 onto a second lossless medium with intrinsic impedance η_2. Angles θ_1 and θ_2 are the angles of incidence and transmission, respectively.†

Property	Normal Incidence $\theta_1 = \theta_2 = 0$	Horizontal Polarization	Vertical Polarization
Reflection coefficient	$\rho = \dfrac{\eta_2 - \eta_1}{\eta_2 + \eta_1}$	$\rho_h = \dfrac{\eta_2 \cos\theta_1 - \eta_1 \cos\theta_2}{\eta_2 \cos\theta_1 + \eta_1 \cos\theta_2}$	$\rho_v = \dfrac{\eta_2 \cos\theta_2 - \eta_1 \cos\theta_1}{\eta_2 \cos\theta_2 + \eta_1 \cos\theta_1}$
Transmission coefficient	$\tau = \dfrac{2\eta_2}{\eta_2 + \eta_1}$	$\tau_h = \dfrac{2\eta_2 \cos\theta_1}{\eta_2 \cos\theta_1 + \eta_1 \cos\theta_2}$	$\tau_v = \dfrac{2\eta_2 \cos\theta_1}{\eta_2 \cos\theta_2 + \eta_1 \cos\theta_1}$
Relation of ρ to τ	$\tau = 1 + \rho$	$\tau_h = 1 + \rho_h$	$\tau_v = (1 + \rho_v)\dfrac{\cos\theta_1}{\cos\theta_2}$
Reflectivity	$\Gamma = \lvert\rho\rvert^2$	$\Gamma^h = \lvert\rho_h\rvert^2$	$\Gamma^v = \lvert\rho_v\rvert^2$
Transmissivity	$\mathbb{T} = \lvert\tau\rvert^2 \left(\dfrac{\eta_1}{\eta_2}\right)$	$\mathbb{T}^h = \lvert\tau_h\rvert^2 \dfrac{\eta_1 \cos\theta_2}{\eta_2 \cos\theta_1}$	$\mathbb{T}^v = \lvert\tau_v\rvert^2 \dfrac{\eta_1 \cos\theta_2}{\eta_2 \cos\theta_1}$
Relation of Γ to $\mathbb{T}$	$\mathbb{T} = 1 - \Gamma$	$\mathbb{T}^h = 1 - \Gamma^h$	$\mathbb{T}^v = 1 - \Gamma^v$
Notes: $\sin\theta_2 = \sqrt{\varepsilon'_1/\varepsilon'_2}\sin\theta_1$; $\eta_1 = \eta_0/\sqrt{\varepsilon'_1}$; $\eta_2 = \eta_0/\sqrt{\varepsilon'_2}$.			

case where both media are lossless) are to replace η_2 with η_{c_2} and jk_2 with γ_2. The new expressions are listed in Table 2-6. For the incident and reflected fields in medium 1, the expressions remain unchanged, but those for the transmitted fields in the lossy medium 2 now involve $\gamma_2 \sin\theta_2$ and $\gamma_2 \cos\theta_2$. Snell's law (for both lossless and lossy media) requires the tangential components of the phase factors (which in the present case are the x components) to match at the boundary. That is,

$$\gamma_2 \sin\theta_2 = \gamma_1 \sin\theta_1 = jk_1 \sin\theta_1, \tag{2.113}$$

where we have replaced γ_1 with jk_1 because medium 1 is lossless. We still need an expression for $\cos\theta_2$ in terms of θ_1 and the constitutive parameters of the two media. To that end, we use Eq. (2.113) to obtain

$$\cos\theta_2 = [1 - \sin^2\theta_2]^{1/2}$$

$$= \left[1 - \left(\frac{jk_1}{\gamma_2}\sin\theta_1\right)^2\right]^{1/2}$$

$$= \left[1 - \left(\frac{jk_1}{\alpha_2 + j\beta_2}\sin\theta_1\right)^2\right]^{1/2}. \tag{2.114}$$

► Clearly $\cos\theta_2$ is a complex quantity, which means that θ_2 is no longer a real angle in the traditional sense. ◄

Nevertheless, for the purpose of computing ρ, τ, and the electric and magnetic fields in the transmitted medium, the expression given by Eq. (2.114) is all that is needed. Figure 2-19 displays angular plots of $|\rho_v|$ and $|\rho_h|$ for incidence in air upon a material with $\varepsilon_2 = \varepsilon'_2 - j\varepsilon''_2$. The plots contrast three surfaces, all with $\varepsilon'_2 = 50$, but with very different values for the loss factor ε''_2.

As noted in connection with Eq. (2.114), θ_2 is a complex quantity whose sine and cosine functions

† Computer Code 2.3.

Table 2-6: Oblique incidence in a lossless medium onto a lossy medium.[†]

Horizontal Polarization	Vertical Polarization
$\mathbf{E}_\text{h}^\text{i} = \hat{\mathbf{y}} E_\text{h0}^\text{i} e^{-jk_1(x\sin\theta_1 - z\cos\theta_1)}$	$\mathbf{E}_\text{v}^\text{i} = (-\hat{\mathbf{x}}\cos\theta_1 - \hat{\mathbf{z}}\sin\theta_1) E_\text{v0}^\text{i} e^{-jk_1(x\sin\theta_1 - z\cos\theta_1)}$
$\mathbf{H}_\text{h}^\text{i} = (\hat{\mathbf{x}}\cos\theta_1 + \hat{\mathbf{z}}\sin\theta_1) \dfrac{E_\text{h0}^\text{i}}{\eta_1} e^{-jk_1(x\sin\theta_1 - z\cos\theta_1)}$	$\mathbf{H}_\text{v}^\text{i} = \hat{\mathbf{y}} \dfrac{E_\text{v0}^\text{i}}{\eta_1} e^{-jk_1(x\sin\theta_1 - z\cos\theta_1)}$
$\mathbf{E}_\text{h}^\text{r} = \hat{\mathbf{y}} \rho_\text{h} E_\text{h0}^\text{i} e^{-jk_1(x\sin\theta_1 + z\cos\theta_1)}$	$\mathbf{E}_\text{v}^\text{r} = (\hat{\mathbf{x}}\cos\theta_1 - \hat{\mathbf{z}}\sin\theta_1) \rho_\text{v} E_\text{v0}^\text{i} e^{-jk_1(x\sin\theta_1 + z\cos\theta_1)}$
$\mathbf{H}_\text{h}^\text{r} = (-\hat{\mathbf{x}}\cos\theta_1 + \hat{\mathbf{z}}\sin\theta_1) \rho_\text{h} \dfrac{E_\text{h0}^\text{i}}{\eta_1} e^{-jk_1(x\sin\theta_1 + z\cos\theta_1)}$	$\mathbf{H}_\text{v}^\text{r} = \hat{\mathbf{y}} \rho_\text{v} \dfrac{E_\text{v0}^\text{i}}{\eta_1} e^{-jk_1(x\sin\theta_1 + z\cos\theta_1)}$
$\mathbf{E}_\text{h}^\text{t} = \hat{\mathbf{y}} \tau_\text{h} E_\text{h0}^\text{i} e^{-\gamma_2(x\sin\theta_2 - z\cos\theta_2)}$	$\mathbf{E}_\text{v}^\text{t} = (-\hat{\mathbf{x}}\cos\theta_2 - \hat{\mathbf{z}}\sin\theta_2) \tau_\text{v} E_\text{v0}^\text{i} e^{-\gamma_2(x\sin\theta_2 - z\cos\theta_2)}$
$\mathbf{H}_\text{h}^\text{t} = (\hat{\mathbf{x}}\cos\theta_2 + \hat{\mathbf{z}}\sin\theta_2) \tau_\text{h} \dfrac{E_\text{h0}^\text{i}}{\eta_2} e^{-\gamma_2(x\sin\theta_2 - z\cos\theta_2)}$	$\mathbf{H}_\text{v}^\text{t} = \hat{\mathbf{y}} \tau_\text{v} \dfrac{E_\text{v0}^\text{i}}{\eta_1} e^{-\gamma_2(x\sin\theta_2 - z\cos\theta_2)}$
$\rho_\text{h} = \dfrac{\eta_2\cos\theta_1 - \eta_1\cos\theta_2}{\eta_2\cos\theta_1 + \eta_1\cos\theta_2}$	$\rho_\text{v} = \dfrac{\eta_2\cos\theta_2 - \eta_1\cos\theta_1}{\eta_2\cos\theta_2 + \eta_1\cos\theta_1}$
$\tau_\text{h} = 1 + \rho_\text{h}$	$\tau_\text{v} = (1 + \rho_\text{v}) \dfrac{\cos\theta_1}{\cos\theta_2}$

Notes:

$k_1 = \dfrac{2\pi}{\lambda_0}\sqrt{\varepsilon_1'}$ $\qquad \eta_1 = \dfrac{\eta_0}{\sqrt{\varepsilon_1'}}$ $\qquad \eta_2 = \dfrac{\eta_0}{\sqrt{\varepsilon_2}}$

$\gamma_2 = \alpha_2 + j\beta_2 = j\dfrac{2\pi}{\lambda_0}\sqrt{\varepsilon_2}$ $\qquad \gamma_2\sin\theta_2 = k_1\sin\theta_1$ $\qquad \varepsilon_2 = \varepsilon_2' - j\varepsilon_2''$

$$\cos\theta_2 = \left[1 - \left(\frac{jk_1}{\gamma_2}\sin\theta_1\right)^2\right]^{1/2} = \left[1 - \left(\frac{\varepsilon_1'}{\varepsilon_2' - j\varepsilon_2''}\right)\sin^2\theta_1\right]^{1/2}$$

provide the necessary phase matching at the interface between the two media. The ***real angle*** of transmission χ_2 is given by (Stratton, 1941):

$$\chi_2 = \tan^{-1}\left[\frac{\sqrt{2}\,k_1\sin\theta_1}{[(p^2+q^2)^{1/2}+q]^{1/2}}\right],$$

where

$$p = 2\alpha_2\beta_2$$

and

$$q = \beta_2^2 - \alpha_2^2 - k_1^2\sin^2\theta_1.$$

If medium 2 is lossless ($\alpha_2 = 0$), the expression for χ_2 reduces to $\chi_2 = \theta_2$.

[†]**Computer Code 2.3.**

2-10 Oblique Incidence onto a Two-Layer Composite

The formulations presented in the preceding section allow us to compute the reflection by and transmission through an ocean surface when observed by a downward-looking airborne or spaceborne microwave sensor. For a nadir-looking ($\theta_1 = 0$) altimeter, the reflected pulse measured by its receiver is proportional to the reflection coefficient ρ of the surface. Side-looking radars measure a quantity called the ***backscattering coefficient*** of the surface, σ^0, which is related to ρ_v and ρ_h of the surface. A radiometer measures the emissivity of the surface, which is related to $(1 - |\rho_\text{v}|^2)$

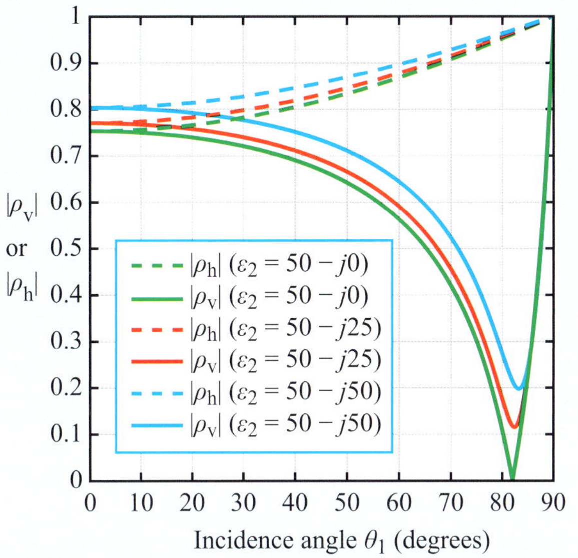

Figure 2-19: Angular plots of $|\rho_v|$ and $|\rho_h|$ for incidence in air upon a material with $\varepsilon_2 = \varepsilon_2' - j\varepsilon_2''$.

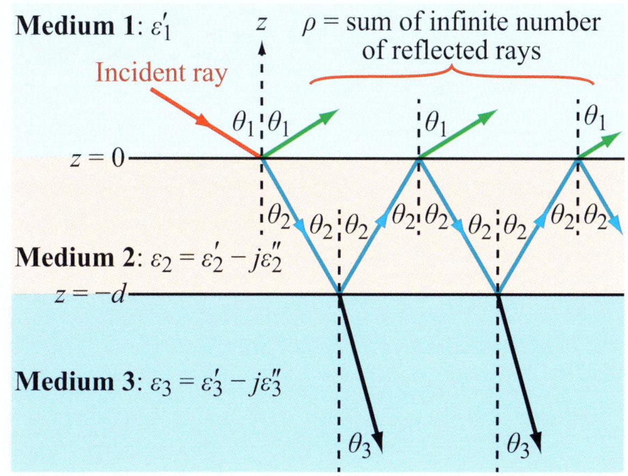

Figure 2-20: Multiple reflections in a two-layer composite.

for vertical polarization and to $(1 - |\rho_h|^2)$ for horizontal polarization. So for all of these sensor configurations, it is important to be able to model the reflection coefficient of lossless, low-loss, and lossy media.

A specific scenario of interest is when a lake or ocean water surface is covered by a uniform layer of ice or oil (due to spillage by oil tankers). A similar scenario occurs when a soil surface is covered by a layer of snow. At microwave frequencies, oil is a low-loss material and so is lake ice. Sea ice may or may not behave like a low-loss medium depending on its type and the specific microwave frequency used by the sensor. Snow is a low-loss medium when dry and lossy when wet. To compute the reflectivity of such two-layer composites, we model the problem as a three-layer configuration:

Medium 1: Air
Medium 2: Intervening uniform layer of thickness d (ice, oil, or snow)
Medium 3: Water (lossy)

Before we proceed with the development of an appropriate mathematical model for ρ_v and ρ_h, however, we should examine the physics of the multilayer reflection process. Figure 2-20 depicts a uniform plane wave represented by a ray incident at angle θ_1 in medium 1. In addition to the reflection at the boundary between media 1 and 2, part of the incident ray crosses the boundary and propagates down to the second boundary that separates medium 2 from medium 3, thereupon part of it is reflected back up towards the upper boundary. The upward-propagating ray in medium 2 gets both partly reflected by the upper boundary and partly transmitted across into the top medium. The multiple reflection process continues indefinitely, with each successive reflection carrying less energy than its predecessor.

Under steady state conditions, the reflection coefficient in medium 1 is equal to the sum of all of the electric fields of the upward directed rays in medium 1, divided by the amplitude of the electric field of the incident wave.

2-10.1 Input Parameters

(a) Constitutive Parameters (Fig. 2-20)

- **Medium 1:** ε_1', μ_0, and $\sigma_1 = 0$

$$\alpha_1 = 0, \qquad \gamma_1 = jk_1 = j\,\frac{2\pi}{\lambda_0}\sqrt{\varepsilon_1'}\,, \tag{2.115a}$$

$$\eta_1 = \frac{\eta_0}{\sqrt{\varepsilon_1'}} \tag{2.115b}$$

- **Medium 2:** $\varepsilon_2 = \varepsilon_2' - j\varepsilon_2''$ and μ_0

$$\gamma_2 = \alpha_2 + j\beta_2 \quad \text{(see Table 2-1)}, \tag{2.116a}$$

$$\eta_2 = \frac{\eta_0}{\sqrt{\varepsilon_2}}\,. \tag{2.116b}$$

If medium 2 is low-loss (i.e., $\varepsilon_2''/\varepsilon_2' \ll 1$), then

$$\alpha_2 \approx \frac{\pi}{\lambda_0}\,\frac{\varepsilon_2''}{\sqrt{\varepsilon_2'}}\,, \qquad \beta_2 \approx \frac{2\pi}{\lambda_0}\sqrt{\varepsilon_2'}\,. \tag{2.116c}$$

- **Medium 3:** $\varepsilon_3 = \varepsilon_3' - j\varepsilon_3''$ and μ_0

$$\gamma_3 = \alpha_3 + j\beta_3 \quad \text{(see Table 2-1)}, \tag{2.117}$$

$$\eta_3 = \frac{\eta_0}{\sqrt{\varepsilon_3}}\,. \tag{2.118}$$

(b) Snell's Law Phase-Matching Condition

- $\gamma_1 \sin\theta_1 = \gamma_2 \sin\theta_2 = \gamma_3 \sin\theta_3$ (2.119a)

$$\cos\theta_2 = \left[1 - \left(\frac{\gamma_1}{\gamma_2}\sin\theta_1\right)^2\right]^{1/2} \tag{2.119b}$$

$$\approx \left[1 - \left(\frac{\beta_1}{\beta_2}\sin\theta_1\right)^2\right]^{1/2} \tag{2.119c}$$

if medium 2 is low loss

$$\cos\theta_3 = \left[1 - \left(\frac{\gamma_1}{\gamma_3}\sin\theta_1\right)^2\right]^{1/2}. \tag{2.120}$$

2-10.2 Propagation Matrix Method

Reflection and transmission by a layered medium consisting of N layers with known constitutive parameters can be computed by defining electric and magnetic fields in each layer and then applying boundary conditions at all $(N-1)$ boundaries. Each such application lends itself into a matrix that relates the fields above a boundary to those below it. Successive multiplication of the $(N-1)$ matrices can lead to computing both the reflection coefficient in medium 1 of the N-layer configuration, as well as the transmission coefficient in medium N (Tsang et al., 1985, pp. 25–31; Balanis, 1989, pp. 229–236). We now apply the matrix method to the three-layer configuration shown in Fig. 2-21, and we initially do so for horizontal polarization.

From Table 2-6, the electric field of a horizontally polarized wave incident in medium 1 (Fig. 2-21) is given by the general form

$$\mathbf{E}_1^- = \hat{\mathbf{y}} A_1 e^{-jk_1(x\sin\theta_1 - z\cos\theta_1)}, \tag{2.121}$$

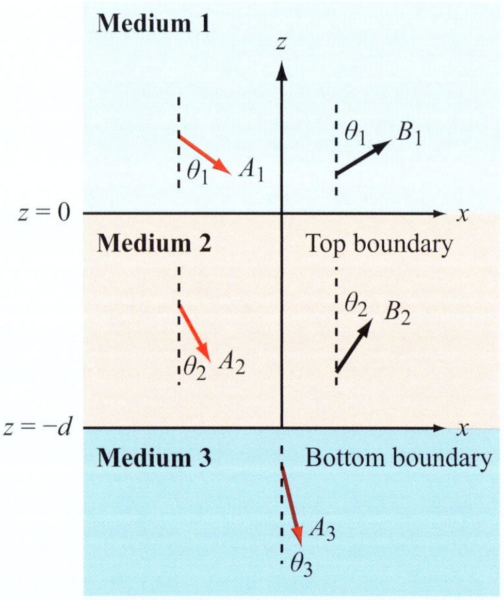

Figure 2-21: Electric field amplitudes of downward and upward propagating waves in media 1, 2 and 3.

where A_1 is the amplitude of $\mathbf{E}_1^-$ at $x = 0$ and $z = 0$, the subscript 1 denotes that $\mathbf{E}_1^-$ is in medium 1, and the superscript $(-)$ denotes that it belongs to a downward-propagating wave. Similarly, the electric field of the upward-propagating wave in medium 1 is

$$\mathbf{E}_1^+ = \hat{\mathbf{y}} B_1 e^{-jk_1(x\sin\theta_1 + z\cos\theta_1)}, \tag{2.122}$$

where B_1 represents the amplitude of the sum total of the electric fields of all multiple-reflection components as they exist in medium 1, and the superscript $(+)$ denotes that $\mathbf{E}_1^+$ represents an upward propagating wave. The total electric field in medium 1 is

$$\begin{aligned}\mathbf{E}_1 &= \mathbf{E}_1^- + \mathbf{E}_1^+ \\ &= \hat{\mathbf{y}}(A_1 e^{jk_1 z\cos\theta_1} + B_1 e^{-jk_1 z\cos\theta_1})e^{-jk_1 x\sin\theta_1}.\end{aligned} \tag{2.123}$$

Application of Eq. (2.34a) leads to the associated magnetic fields $\mathbf{H}_1^-$ and $\mathbf{H}_1^+$ and to their sum:

$$\mathbf{H}_1 = \hat{\mathbf{x}} H_{1x} + \hat{\mathbf{z}} H_{1z}, \tag{2.124}$$

with

$$H_{1x} = \frac{\cos\theta_1}{\eta_1}(A_1 e^{jk_1 z\cos\theta_1} - B_1 e^{-jk_1 z\cos\theta_1})e^{-jk_1 x\sin\theta_1}, \tag{2.125a}$$

$$H_{1z} = \frac{\sin\theta_1}{\eta_1}(A_1 e^{jk_1 z\cos\theta_1} + B_1 e^{-jk_1 z\cos\theta_1})e^{-jk_1 x\sin\theta_1}. \tag{2.125b}$$

Extending the formulation to medium 2 entails replacing A_1 and B_1 with A_2 and B_2, respectively, jk_1 with γ_2, and η_1 with η_{c_2}. The process leads to

$$\begin{aligned}\mathbf{E}_2 &= \mathbf{E}_2^- + \mathbf{E}_2^+ \\ &= \hat{\mathbf{y}}(A_2 e^{\gamma_2 z\cos\theta_2} + B_2 e^{-\gamma_2 z\cos\theta_2})e^{-\gamma_2 x\sin\theta_2},\end{aligned} \tag{2.126}$$

$$\mathbf{H}_2 = \hat{\mathbf{x}} H_{2x} + \hat{\mathbf{z}} H_{2z}, \tag{2.127}$$

with

$$H_{2x} = \frac{\cos\theta_2}{\eta_{c_2}}(A_2 e^{\gamma_2 z\cos\theta_2} - B_2 e^{-\gamma_2 z\cos\theta_2})e^{-\gamma_2 x\sin\theta_2}, \tag{2.128a}$$

$$H_{2z} = \frac{\sin\theta_2}{\eta_{c_2}}(A_2 e^{\gamma_2 z\cos\theta_2} + B_2 e^{-\gamma_2 z\cos\theta_2})e^{-\gamma_2 x\sin\theta_2}. \tag{2.128b}$$

Medium 3 contains only a downward-propagating wave. Hence,

$$\mathbf{E}_3 = \hat{\mathbf{y}}(A_3 e^{\gamma_3 z\cos\theta_3})e^{-\gamma_3 x\sin\theta_3}, \tag{2.129}$$

$$\mathbf{H}_3 = \hat{\mathbf{x}} H_{3x} + \hat{\mathbf{z}} H_{3z}, \tag{2.130}$$

with

$$H_{3x} = \left(\frac{\cos\theta_3}{\eta_{c_3}} A_3 e^{\gamma_3 z\cos\theta_3}\right) e^{-\gamma_3 x\sin\theta_3}, \tag{2.131a}$$

$$H_{3z} = \left(\frac{\sin\theta_3}{\eta_{c_3}} A_3 e^{\gamma_3 z\cos\theta_3}\right) e^{-\gamma_3 x\sin\theta_3}. \tag{2.131b}$$

In view of Eq. (2.119a) and keeping in mind that in the present case $\gamma_1 = jk_1$ because medium 1 is lossless, it follows that

$$\underbrace{e^{-jk_1 x\sin\theta_1}}_{\text{medium 1}} = \underbrace{e^{-\gamma_2 x\sin\theta_2}}_{\text{medium 2}} = \underbrace{e^{-\gamma_3 x\sin\theta_3}}_{\text{medium 3}}. \tag{2.132}$$

Hence, the x component of the phase in all of the expressions given by Eqs. (2.123) through (2.131) are identical (which is another way to state Snell's law).

At the top boundary (at $z = 0$), the tangential components of $\mathbf{E}$ and $\mathbf{H}$ must be continuous. Consequently,

$$\mathbf{E}_1\big|_{z=0} = \mathbf{E}_2\big|_{z=0},$$

which, upon equating Eq. (2.123) to Eq. (2.126) and setting $z = 0$, leads to

$$A_1 + B_1 = A_2 + B_2. \tag{2.133}$$

For the magnetic field, continuity at $z = 0$ applies to its x component. That is,

$$H_{1x}\big|_{z=0} = H_{2x}\big|_{z=0},$$

which leads to

$$\frac{\cos\theta_1}{\eta_1}(A_1 - B_1) = \frac{\cos\theta_2}{\eta_2}(A_2 - B_2). \tag{2.134}$$

Similarly, at the bottom boundary (at $z = -d$),

$$\mathbf{E}_2\big|_{z=-d} = \mathbf{E}_3\big|_{z=-d} \quad \text{and} \quad H_{2x}\big|_{z=-d} = H_{3x}\big|_{z=-d},$$

which leads to

$$(A_2 e^{-\gamma_2 d\cos\theta_2} + B_2 e^{\gamma_2 d\cos\theta_2}) = A_3 e^{-\gamma_3 d\cos\theta_3}, \quad (2.135)$$

$$\frac{\cos\theta_2}{\eta_2}(A_2 e^{-\gamma_2 d\cos\theta_2} - B_2 e^{\gamma_2 d\cos\theta_2}) = \frac{A_3\cos\theta_3}{\eta_3}\, e^{-\gamma_3 d\cos\theta_3}. \quad (2.136)$$

The effective reflection coefficient of the combined two-layer structure (media 2 and 3) is given by

$$\rho = \frac{B_1}{A_1}. \quad (2.137)$$

Simultaneous solution of Eqs. (2.133) through (2.136) leads to

$$\rho = \frac{\rho_{12} + \rho_{23} e^{-2\gamma_2 d\cos\theta_2}}{1 + \rho_{12}\rho_{23} e^{-2\gamma_2 d\cos\theta_2}} \quad \textbf{(h or v polarization)},^{\dagger} \quad (2.138)$$

where ρ_{12} is the reflection coefficient at the top boundary for incidence in semi-infinite medium 1 upon semi-infinite medium 2 (i.e., as if it were the only boundary). Similarly, ρ_{23} is the reflection coefficient at the bottom boundary for incidence in medium 2 upon medium 3. For horizontal polarization, the expressions for ρ_{12} and ρ_{23} are given by

$$\left.\begin{aligned} \rho_{12} &= \frac{\eta_2\cos\theta_1 - \eta_1\cos\theta_2}{\eta_2\cos\theta_1 + \eta_1\cos\theta_2} \\ \rho_{23} &= \frac{\eta_3\cos\theta_2 - \eta_2\cos\theta_3}{\eta_3\cos\theta_2 + \eta_2\cos\theta_3} \end{aligned}\right\} \quad \textbf{(h polarization)}. \quad (2.139)$$

For the vertically polarized case:

$$\left.\begin{aligned} \rho_{12} &= \frac{\eta_2\cos\theta_2 - \eta_1\cos\theta_1}{\eta_2\cos\theta_2 + \eta_1\cos\theta_1} \\ \rho_{23} &= \frac{\eta_3\cos\theta_3 - \eta_2\cos\theta_2}{\eta_3\cos\theta_3 + \eta_2\cos\theta_2} \end{aligned}\right\} \quad \textbf{(v polarization)}. \quad (2.140)$$

Figure 2-22 displays two examples:

[†] **Computer Code 2.4.**

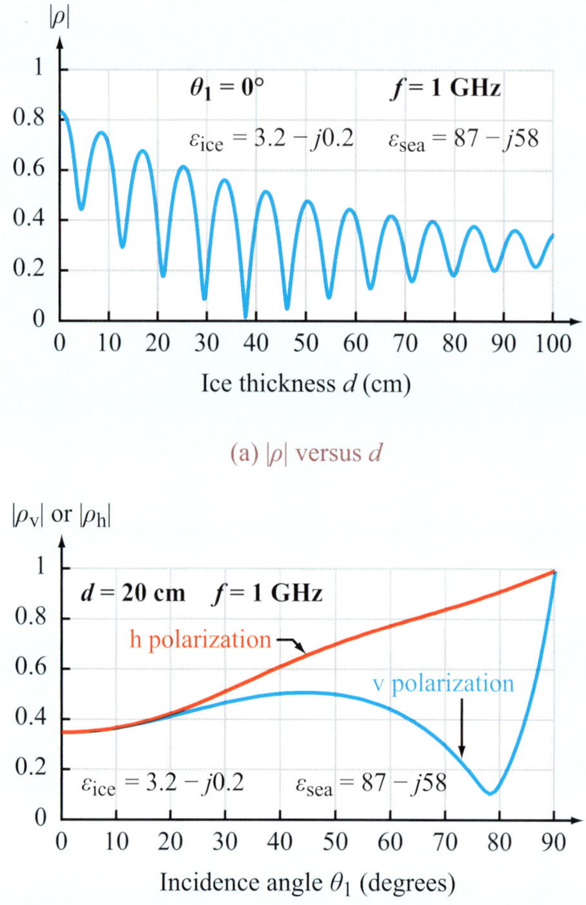

Figure 2-22: Plots of (a) $|\rho|$ as a function of ice-layer thickness d for normal incidence and (b) $|\rho_v|$ and $|\rho_h|$ as a function of incidence angle θ_1 for a 20 cm ice layer over the ocean surface.

(a) a plot of the magnitude of the normal-incidence reflection coefficient, $|\rho|$, as a function of d, where d is the thickness of an ice layer over the ocean surface, and

(b) plots of $|\rho_v|$ and $|\rho_h|$ as a function of incidence angle θ_1 for an ice layer of thickness $d = 20$ cm.

The oscillatory behavior exhibited by $|\rho|$ as a function of d in Fig. 2-22(a) has maxima and minima corresponding to, respectively, constructive and destructive interference of the multiple reflections that occur in the ice layer.

2-10.3 Multiple Reflection Method

The expression for ρ given by Eq. (2.138) was obtained by applying boundary conditions at the upper and lower boundaries of the middle layer in Fig. 2-21. We can gain additional insight by deriving Eq. (2.138) a second time by tracking all the multiple reflections and transmissions that occur at the two boundaries. We demonstrate the process for h polarization, and we suppress the x component of the phase factor since, as noted earlier, it is the same for all electric and magnetic fields in all three media.

Figure 2-23 depicts the various reflection, transmission, and propagation mechanisms encountered by a wave incident in medium 1 upon medium 2 at incidence angle θ_1. They include:

- $\rho_{12} =$ reflection coefficient at the boundary between media 1 and 2, for incidence in medium 1 at angle θ_1.
- $\rho_{21} =$ reflection coefficient at the boundary between media 1 and 2, for incidence in medium 2 at angle θ_2. Note that $\rho_{21} = -\rho_{12}$.
- $\tau_{12} =$ transmission coefficient from medium 1 to medium 2 when incidence is at angle θ_2. Note that $\tau_{12} = 1 + \rho_{12}$ for h polarization and $\tau_{12} = (1 + \rho_{12})\cos\theta_1/\cos\theta_2$ for v polarization.
- $\tau_{21} =$ transmission coefficient from medium 2 to medium 1 when incidence is at angle θ_2. Note that $\tau_{21} = 1 + \rho_{21}$ for h polarization and $\tau_{21} = (1 + \rho_{21})\cos\theta_2/\cos\theta_1$ for v polarization.
- $\mathcal{L} = e^{-\gamma_2 d\cos\theta_2} =$ propagation factor in medium 2 between its top boundary and bottom boundary (or between its bottom boundary and top boundary) along angle θ_2.
- $\rho_{23} =$ reflection coefficient at the boundary between media 2 and 3 for incidence in medium 2 at angle θ_2.

For an incident field with $E_0^{\mathrm{i}} = 1$ V/m, the first reflection in Fig. 2-24 contributes ρ_{12}, the second one is $\tau_{21}\rho_{23}\mathcal{L}^2\tau_{12}$, and so on. Summing up all of the upward-moving waves in medium 1 gives

$$\begin{aligned}\rho &= \rho_{12} + \tau_{21}\rho_{23}\mathcal{L}^2\tau_{12} + \tau_{21}\rho_{23}^2\rho_{21}\mathcal{L}^4\tau_{12} + \cdots \\ &= \rho_{12} + \tau_{21}\tau_{12}\rho_{23}\mathcal{L}^2(1 + x + x^2 + \cdots),\end{aligned} \tag{2.141}$$

where $x = \rho_{21}\rho_{23}\mathcal{L}^2$. We note that the magnitudes of ρ_{21}, ρ_{23}, and $\mathcal{L}^2$ are all smaller than 1. Hence, $x < 1$. If we make the substitutions

$$\tau_{21} = 1 + \rho_{21} = 1 - \rho_{12},$$
$$\tau_{12} = 1 + \rho_{12},$$

and

$$\frac{1}{1-x} = 1 + x + x^2 + \cdots,$$

Eq. (2.141) becomes

$$\rho = \rho_{12} + \frac{(1-\rho_{12})(1+\rho_{12})\rho_{23}\mathcal{L}^2}{1-\rho_{21}\rho_{23}\mathcal{L}^2}. \tag{2.142}$$

Upon making the substitutions $\rho_{21} = -\rho_{12}$ and $\mathcal{L}^2 = e^{-2\gamma_2 d\cos\theta_2}$ and simplifying the expression, we obtain

$$\rho = \frac{\rho_{12} + \rho_{23}e^{-2\gamma_2 d\cos\theta_2}}{1 + \rho_{12}\rho_{23}e^{-2\gamma_2 d\cos\theta_2}},^{\dagger} \tag{2.143}$$

which is identical with the result obtained in the previous subsection in the form of Eq. (2.138). The reflectivity of the two-layer composite is $\Gamma = |\rho|^2$.

PROBLEMS

2.1 Determine (a) the direction of propagation, (b) phase velocity, (c) wavelength, (d) relative permittivity of the material, and (e) the electric field phasor of an EM wave propagating in a nonmagnetic material, given that the wave's magnetic field is

$$\mathbf{H} = \hat{\mathbf{x}}6\cos(10^9 t - 20y) \qquad \text{(mA/m)}.$$

2.2 Determine (a) the direction of propagation, (b) phase velocity, (c) wavelength, (d) relative

† Computer Code 2.4.

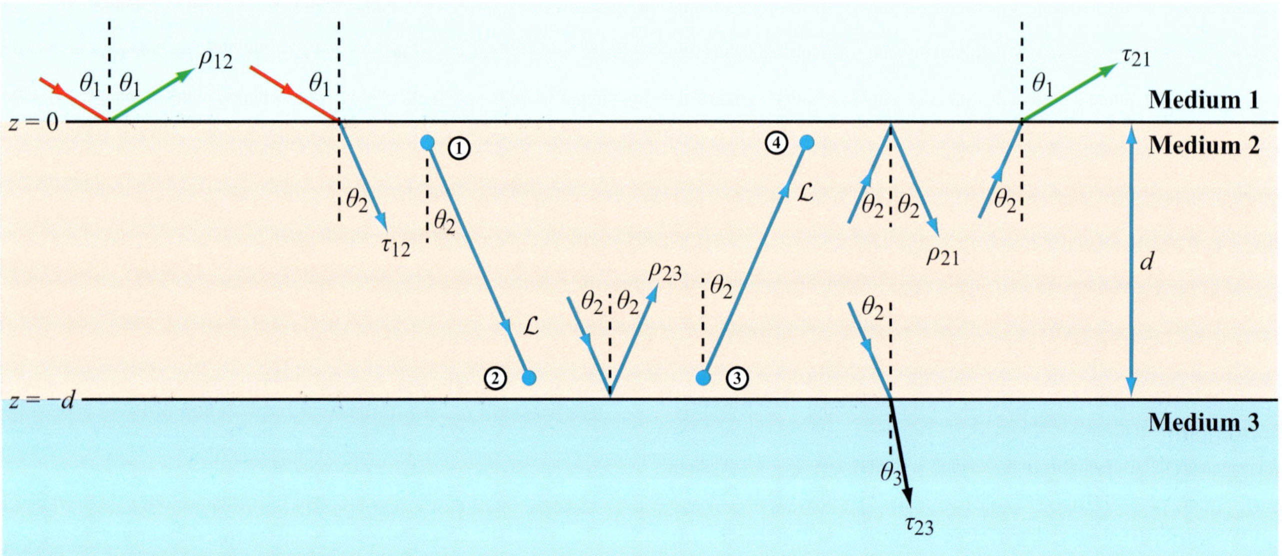

Figure 2-23: Reflection, transmission, and propagation mechanisms.

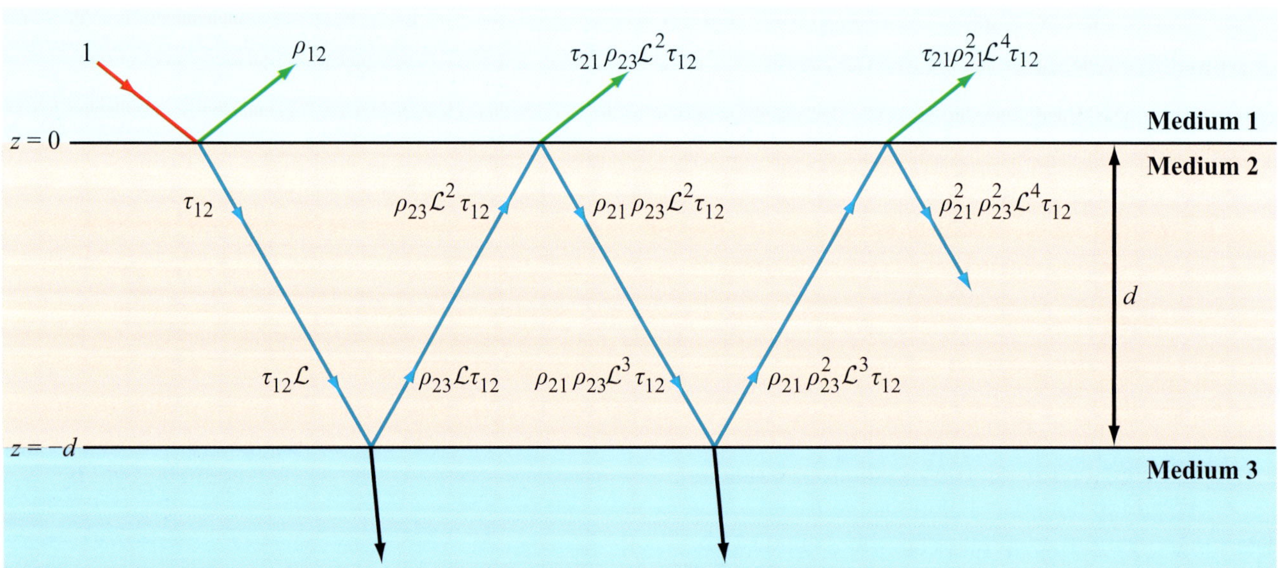

Figure 2-24: Multiple reflection process.

permittivity of the material, and (e) the electric field phasor of an EM wave propagating in a nonmagnetic material, given that the wave's magnetic field is

$$\mathbf{H} = \hat{\mathbf{z}}60\cos(10^7t + 0.1x - 60^\circ) \qquad \text{(mA/m)}.$$

2.3 Provide general expressions for the electric and magnetic fields of a 10 GHz plane wave traveling in the $+x$ direction in a lossless nonmagnetic medium with relative permittivity $\varepsilon' = 4$. The electric field is polarized along the z direction, its peak value is 12 V/m, and its intensity is 8 V/m at $t = 0$ and $x = 0.5$ cm.

2.4 Provide general expressions for the electric and magnetic fields of a 30 GHz plane wave traveling in the $+z$ direction in a lossless nonmagnetic medium with relative permittivity $\varepsilon' = 16$. The electric field is polarized along the z direction, its peak value is 10 V/m, and its intensity is 6 V/m at $t = 0$ and $z = 2$ cm.

2.5 The electric field of a plane EM wave propagating in a nonmagnetic medium is given by

$$\mathbf{E} = \hat{\mathbf{x}}6\cos(2\times 10^7t - 0.4y) + \hat{\mathbf{z}}3\sin(2\times 10^7t - 0.4y) \qquad \text{(V/m)}.$$

Determine: (a) λ, (b) ε', and (c) $\mathbf{H}$.

2.6 The electric field of a plane EM wave propagating in a nonmagnetic medium is given by

$$\mathbf{E} = \hat{\mathbf{y}}12\sin(2\times 10^7t - 0.1x) + \hat{\mathbf{z}}4\cos(2\times 10^7t - 0.1x) \qquad \text{(V/m)}.$$

Determine: (a) λ, (b) ε', and (c) $\mathbf{H}$.

2.7 The electric field of a plane wave propagating in a nonmagnetic material with $\varepsilon' = 9$ is given by

$$\mathbf{E} = \hat{\mathbf{x}}100\cos(1.2\pi\times 10^{10}t + kz) \qquad \text{(V/m)}.$$

Determine: f, u_p, λ, k, η, and $\mathbf{H}$.

2.8 An LHC-polarized wave with a modulus of 2 (V/m) is traveling in free space in the negative z direction. Write the expression for the wave's electric field vector, given that the wavelength is 6 cm.

2.9 For a wave characterized by the electric field

$$\mathbf{E}(z,t) = \hat{\mathbf{x}}a_x\cos(\omega t - kz) + \hat{\mathbf{y}}a_y\cos(\omega t - kz + \delta)$$

identify the polarization state, determine the polarization angles (ψ, χ), and sketch the locus of $\mathbf{E}(0,t)$ for each of the following cases:

(a) $a_x = 3$ V/m, $a_y = 4$ V/m, and $\delta = 0$
(b) $a_x = 3$ V/m, $a_y = 4$ V/m, and $\delta = 180^\circ$
(c) $a_x = 3$ V/m, $a_y = 3$ V/m, and $\delta = 45^\circ$
(d) $a_x = 3$ V/m, $a_y = 4$ V/m, and $\delta = -135^\circ$

2.10 The electric field of a uniform plane wave propagating in free space is given by

$$\mathbf{E} = (\hat{\mathbf{x}} + j\hat{\mathbf{y}})60e^{-j\pi z/6} \qquad \text{(V/m)}.$$

Specify the modulus and direction of the electric field intensity at the $z = 0$ plane at $t = 0$, 5, and 10 ns.

2.11 A linearly polarized plane wave of the form $\mathbf{E} = \hat{\mathbf{x}}a_xe^{-jkz}$ can be expressed as the sum of an RHC polarized wave with magnitude a_R, and an LHC polarized wave with magnitude a_L. Prove this statement by finding expressions for a_R and a_L in terms of a_x.

2.12 The electric field of an elliptically polarized plane wave is given by

$$\mathbf{E}(z,t) = [-\hat{\mathbf{x}}20\sin(\omega t - kz - 60^\circ) + \hat{\mathbf{y}}30\cos(\omega t - kz)] \qquad \text{(V/m)}.$$

Determine: (a) the polarization angles (ψ, χ), and (b) the direction of rotation.

2.13 Compare the polarization states of each of the following pairs of plane waves:

(a) Wave 1: $\mathbf{E}_1 = \hat{\mathbf{x}}4\cos(\omega t - kz) + \hat{\mathbf{y}}4\sin(\omega t - kz)$.
Wave 2: $\mathbf{E}_2 = \hat{\mathbf{x}}4\cos(\omega t + kz) + \hat{\mathbf{y}}4\sin(\omega t + kz)$.
(b) Wave 1: $\mathbf{E}_1 = \hat{\mathbf{x}}4\cos(\omega t - kz) - \hat{\mathbf{y}}4\sin(\omega t - kz)$.
Wave 2: $\mathbf{E}_2 = \hat{\mathbf{x}}4\cos(\omega t + kz) - \hat{\mathbf{y}}4\sin(\omega t + kz)$.

2.14 Plot the locus of $\mathbf{E}(0,t)$ for a plane wave with

$$\mathbf{E}(z,t) = \hat{\mathbf{x}}\sin(\omega t + kz) + \hat{\mathbf{y}}2\cos(\omega t + kz).$$

Determine the polarization state from your plot.

2.15 For each of the following combinations of parameters, determine if the material is a low-loss dielectric, a quasi conductor, or a good conductor, and then calculate α, β, λ, u_p, and η_c:

(a) Glass with $\mu = \mu_0$, $\varepsilon' = 5$, and $\sigma = 10^{-12}$ S/m at 10 GHz.

(b) Animal tissue with $\mu = \mu_0$, $\varepsilon' = 12$, and $\sigma = 0.3$ S/m at 100 MHz.

(c) Wood with $\mu = \mu_0$, $\varepsilon' = 3$, and $\sigma = 10^{-4}$ S/m at 1 kHz.

2.16 Dry soil is characterized by $\varepsilon' = 2.5$, $\mu = \mu_0$, and $\sigma = 10^{-4}$ (S/m). At each of the following frequencies, determine if dry soil may be considered a good conductor, a quasi conductor, or a low-loss dielectric, and then calculate α, β, λ, μ_p, and η: (a) 60 Hz, (b) 1 kHz, (c) 1 MHz, (d) 1 GHz.

2.17 In a medium characterized by $\varepsilon' = 9$, $\mu = \mu_0$, and $\sigma = 0.1$ S/m, determine the phase angle by which the magnetic field leads the electric field at (a) 1 MHz, (b) 1 GHz, and (c) 10 GHz.

2.18 Generate a plot for the skin depth δ_s versus frequency for seawater for the range from 1 kHz to 10 GHz (use log-log scales). The constitutive parameters of seawater are $\mu = \mu_0$, $\varepsilon' = 80$, and $\sigma = 4$ S/m.

2.19 Wet soil is characterized by $\mu = \mu_0$, $\varepsilon' = 9$, and $\sigma = 5 \times 10^{-4}$ S/m. Ignoring reflection at the air–soil boundary, if the amplitude of an incident wave is 10 V/m at the surface of a wet soil medium, at what depth will it be down to 1 mV/m if (a) $f = 1$ MHz, (b) 1 GHz, and (c) 10 GHz?

2.20 Based on wave attenuation and reflection measurements conducted at 1 MHz, it was determined that the intrinsic impedance of a certain medium is $28.1\angle 45^\circ$ (Ω) and the skin depth is 2 m. Determine: (a) the conductivity of the material, (b) the wavelength in the medium, and (c) the phase velocity.

2.21 The electric field of a plane wave propagating in a nonmagnetic medium is given by

$$\mathbf{E} = \hat{\mathbf{z}}\, 25e^{-30x} \cos(2\pi \times 10^9 t - 40x) \qquad \text{(V/m)}.$$

Obtain the corresponding expressions for $\mathbf{H}(t)$ and $\mathbf{S}(t)$.

2.22 The electric field of a plane wave propagating in a nonmagnetic medium is given by

$$\mathbf{E} = \hat{\mathbf{x}} 10e^{-15z} \cos(4\pi \times 10^9 t - 80z) + \hat{\mathbf{y}} 20e^{-15z} \cos(4\pi \times 10^9 t - 80z + 30^\circ) \quad \text{(V/m)}.$$

Determine the average power density carried by the wave.

2.23 A wave traveling in a nonmagnetic medium with $\varepsilon' = 9$ is characterized by an electric field given by

$$\mathbf{E} = [\hat{\mathbf{y}}\, 3\cos(\pi \times 10^7 t + kx) - \hat{\mathbf{z}}\, 2\cos(\pi \times 10^7 t + kx)]e^{-0.2x} \qquad \text{(V/m)}.$$

Determine the direction of wave travel and average power density carried by the wave.

2.24 The electric-field phasor of a uniform plane wave traveling downward in water is given by

$$\mathbf{E} = \hat{\mathbf{x}}\, 5e^{-0.2z} e^{-j0.2z} \qquad \text{(V/m)},$$

where $\hat{\mathbf{z}}$ is the downward direction and $z = 0$ is the water surface. If $\sigma = 4$ S/m, (a) obtain an expression for the average power density, and (b) determine the depth at which the power density has been reduced by 40 dB.

2.25 The amplitudes of an elliptically polarized plane wave traveling in a lossless, nonmagnetic medium with $\varepsilon' = 4$ are $H_{y0} = 3$ (mA/m) and $H_{z0} = 4$ (mA/m). Determine the average power flowing through an aperture in the y–z plane if its area is 20 m^2.

2.26 At microwave frequencies, the power density considered safe for human exposure is 1 (mW/cm^2). A radar radiates a wave with an electric field amplitude E that decays with distance as $E(R) = (3{,}000/R)$ (V/m), where R is the distance in meters. What is the radius of the unsafe region?

2.27 Consider the imaginary rectangular box shown in Fig. P2.27. A wave traveling in the medium has electric and magnetic fields

$$\mathbf{E} = \hat{\mathbf{x}}\, 100e^{-20y} \cos(2\pi \times 10^9 t - 40y) \qquad \text{(V/m)},$$

$$\mathbf{H} = -\hat{\mathbf{z}}0.64e^{-20y}\cos(2\pi \times 10^9 t - 40y - 36.85^\circ) \quad \text{(A/m)}.$$

The box has dimensions $a = 1$ cm, $b = 2$ cm, and $c = 0.5$ cm. Determine: (a) the net time-average power entering the box, (b) the time-average power exiting the box, and (c) the time-average power absorbed by the box.

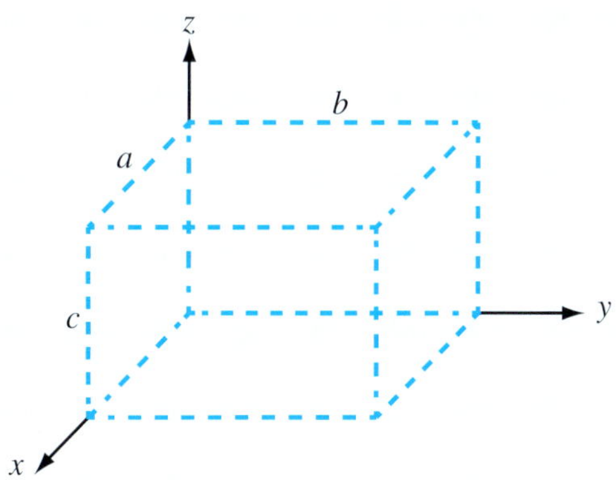

Figure P2.27: Imaginary rectangular box of Problem 2.27.

2.28 A team of scientists is designing a radar as a probe for measuring the depth of the ice layer over the Antarctic land mass. In order to measure a detectable echo due to the reflection by the ice–rock boundary, the thickness of the ice sheet should not exceed three skin depths. If $\varepsilon' = 3$ and $\varepsilon'' = 10^{-2}$ for ice and if the maximum anticipated ice thickness in the area under exploration is 1.2 km, what frequency range is usable with the radar?

2.29 A plane wave traveling in medium 1 with $\varepsilon'_1 = 2.25$ is normally incident upon medium 2 with $\varepsilon'_2 = 4$. Both media are made of nonmagnetic, nonconducting materials. If the electric field of the incident wave is given by

$$\mathbf{E}^{\mathrm{i}} = \hat{\mathbf{y}}8\cos(6\pi \times 10^9 t - 30\pi x) \quad \text{(V/m)}.$$

(a) Obtain time-domain expressions for the electric and magnetic fields in each of the two media.

(b) Determine the average power densities of the incident, reflected and transmitted waves.

2.30 A plane wave traveling in a medium with $\varepsilon'_1 = 9$ is normally incident upon a second medium with $\varepsilon'_2 = 4$. Both media are made of nonmagnetic, nonconducting materials. If the magnetic field of the incident plane wave is given by

$$\mathbf{H}^{\mathrm{i}} = \hat{\mathbf{z}}2\cos(2\pi \times 10^9 t - ky) \quad \text{(A/m)}.$$

(a) Obtain time-domain expressions for the electric and magnetic fields in each of the two media.

(b) Determine the average power densities of the incident, reflected, and transmitted waves.

2.31 A 200 MHz LHC-polarized plane wave with an electric field modulus of 5 V/m is normally incident in air upon a dielectric medium with $\varepsilon' = 4$, and occupies the region defined by $z \geq 0$.

(a) Write an expression for the electric field phasor of the incident wave, given that the field is a positive maximum at $z = 0$ and $t = 0$.

(b) Calculate the reflection and transmission coefficients.

(c) Write expressions for the electric field phasors of the reflected wave, the transmitted wave, and the total field in the region $z \leq 0$.

(d) Determine the percentages of the incident average power reflected by the boundary and transmitted into the second medium.

2.32 The three regions shown in Fig. P2.32 contain perfect dielectrics. For a wave in medium 1, incident normally upon the boundary at $z = -d$, what combination of ε'_2 and d produces no reflection? Express your answers in terms of ε'_1, ε'_3 and the oscillation frequency of the wave, f.

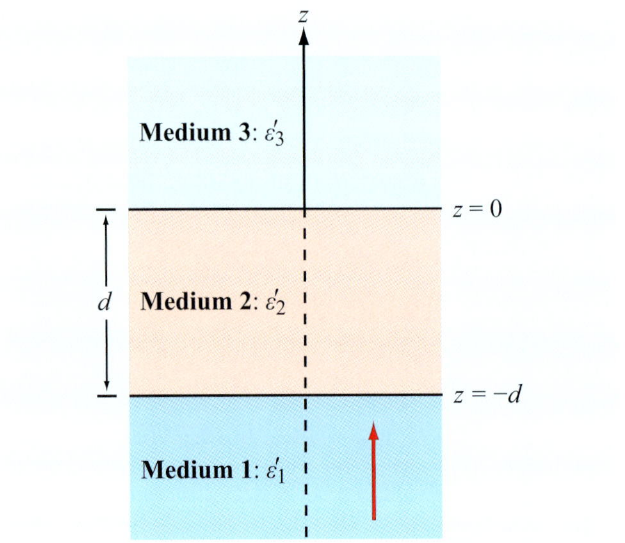

Figure P2.32: Dielectric layers for Problem 2.32.

2.33 A plane wave in air with

$$\mathbf{E}^{\mathrm{i}} = \hat{\mathbf{y}}\, 20e^{-j(3x+4z)} \qquad \text{(V/m)}$$

is incident upon the planar surface of a dielectric material, with $\varepsilon' = 4$, occupying the half-space $z \geq 0$. Determine:

(a) The polarization of the incident wave.

(b) The angle of incidence.

(c) The time-domain expressions for the reflected electric and magnetic fields.

(d) The time-domain expressions for the transmitted electric and magnetic fields.

(e) The average power density carried by the wave in the dielectric medium.

2.34 Repeat Problem 2.33 for a wave in air with

$$\mathbf{H}^{\mathrm{i}} = \hat{\mathbf{y}}\, 2 \times 10^{-2} e^{-j(8x+6z)} \qquad \text{(A/m)}$$

incident upon the planar boundary of a dielectric medium ($z \geq 0$) with $\varepsilon' = 9$.

2.35 A plane wave in air with

$$\mathbf{E}^{\mathrm{i}} = (\hat{\mathbf{x}}9 - \hat{\mathbf{y}}4 - \hat{\mathbf{z}}6)e^{-j(2x+3z)} \qquad \text{(V/m)}$$

is incident upon the planar surface of a dielectric material, with $\varepsilon' = 2.25$, occupying the half-space $z \geq 0$. Determine:

(a) The incidence angle θ_1.

(b) The frequency of the wave.

(c) The field $\mathbf{E}^{\mathrm{r}}$ of the reflected wave.

(d) The field $\mathbf{E}^{\mathrm{t}}$ of the wave transmitted into the dielectric medium.

(e) The average power density carried by the wave into the dielectric medium.

2.36 A parallel-polarized plane wave is incident from air onto a dielectric medium with $\varepsilon' = 9$ at the Brewster angle. What is the refraction angle?

2.37 A perpendicularly polarized wave in air is obliquely incident upon a planar glass–air interface at an incidence angle of 30°. The wave frequency is 600 THz (1 THz $= 10^{12}$ Hz), which corresponds to green light, and the index of refraction of the glass is 1.6. If the electric field amplitude of the incident wave is 50 V/m, determine the following:

(a) The reflection and transmission coefficients.

(b) The instantaneous expressions for **E** and **H** in the glass medium.

2.38 A parallel-polarized beam of light in air with an electric field amplitude of 10 (V/m) is incident on polystyrene with $\mu = \mu_0$ and $\varepsilon' = 2.6$. If the incidence angle at the air–polystyrene planar boundary is 50°, determine the following:

(a) The reflectivity and transmissivity.

(b) The power carried by the incident, reflected, and transmitted beams if the spot on the boundary illuminated by the incident beam is 1 m^2 in area.

2.39 A 50 MHz RHC-polarized plane wave with an electric field modulus of 30 V/m is normally incident in air upon a dielectric medium with $\varepsilon' = 9$ and occupying the region defined by $z \geq 0$.

(a) Write an expression for the electric field phasor of the incident wave, given that the inclination angle is zero at $z = 0$ and $t = 0$.

(b) Calculate the reflection and transmission coefficients.

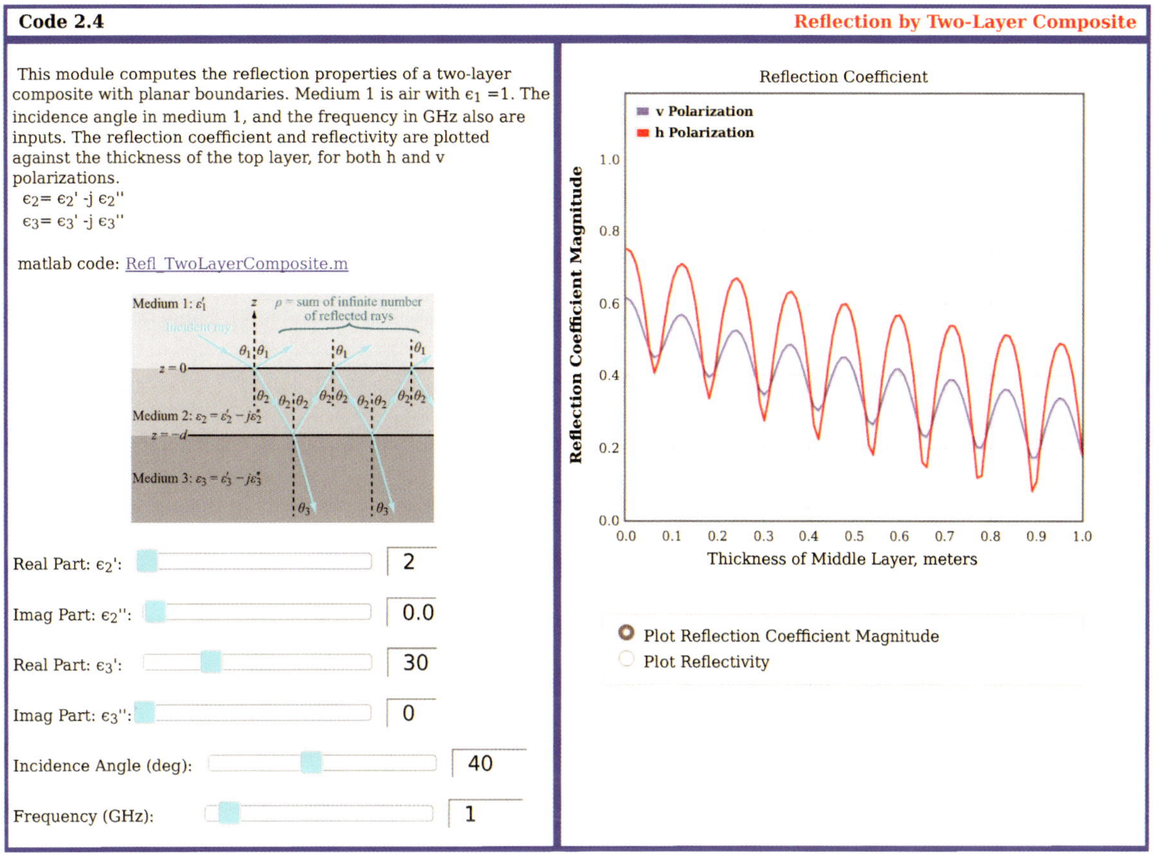

An example of one of the interactive modules available at the book website: mrs.eecs.umich.edu.

(c) Write expressions for the electric field phasors of the reflected wave, the transmitted wave, and the total field in the region $z \leq 0$.

(d) Determine the percentages of the incident average power reflected by the boundary and transmitted into the second medium.

2.40 A horizontal sand layer of thickness d overlays a semi-infinite medium of wet soil. If $\varepsilon_{\text{sand}} = (3 - j0.1)$ and $\varepsilon_{\text{soil}} = (20 - j5)$, compute and plot the magnitude of the reflection coefficient, as a function of d from 0 to 1 m, at normal incidence at 2 GHz.

2.41 For the same sand-over-soil configuration of Problem 2.40, compute and plot the magnitude of the h-polarized reflection coefficient as a function of incidence angle from 0 to 90° for $d = 20$ cm. Keep f at 2 GHz.

2.42 At 20 GHz, the permittivity of ocean water at 20 °C is $\varepsilon_{\text{water}} = (36 - j30)$ and the permittivity of crude oil is $\varepsilon_{\text{oil}} = (2.1 - j0.1)$. The oil layer is of thickness d and overlays the water surface. Compute and plot the magnitude of the normal-incidence reflection coefficient as a function of d from 0 to 30 mm in 0.1 mm increments.

2.43 Repeat Problem 2.42 for horizontal polarization at an incidence angle of 50°.

2.44 Derive an expression for the transmission coefficient $\tau = A_3/A_1$ of the three-medium configuration treated in Section 2-10.2.

CHAPTER

3

Remote-Sensing Antennas

12 m long SIR-C radar antenna carried by the Space Shuttle

CONTENTS

Overview

The ***antenna*** of a microwave remote-sensing system is one of the most important elements of the overall system. It defines the ground area illuminated by its beam, and through its properties—radiation pattern, gain, radiation efficiency, polarization purity, etc.—it influences the quality of the data measured by the sensor. This book assumes that the reader has general familiarity with how antennas transmit and receive EM waves. Accordingly, the first part of this chapter provides a general review of antennas, and the latter part focuses on the type of antennas commonly used by microwave sensing systems, both active and passive.

► An antenna is a transducer that converts a guided wave propagating on a transmission line into an electromagnetic wave propagating in an unbounded medium (usually free space), or vice versa. ◄

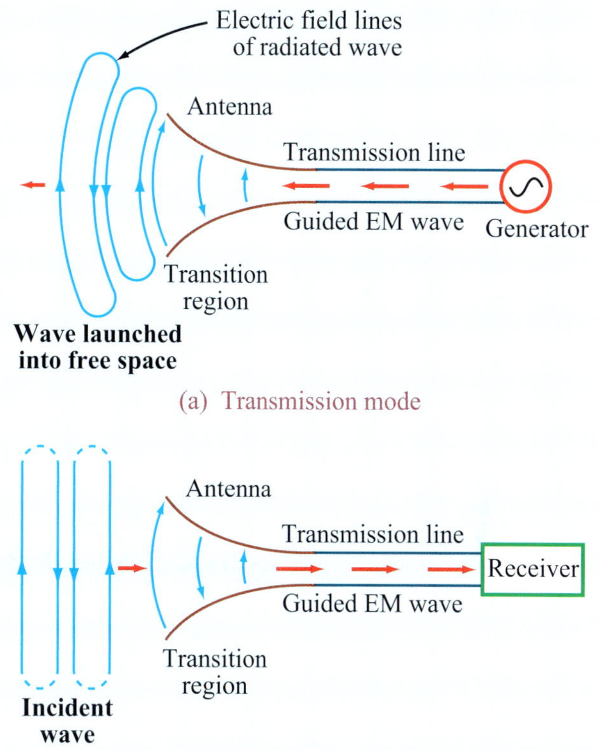

Figure 3-1: Antenna as a transducer between a guided electromagnetic wave and a free-space wave, for both transmission and reception.

Figure 3-1 shows how a wave is launched by a hornlike antenna, with the horn acting as the transition segment between the waveguide and free space. Antennas are made in various shapes and sizes (Fig. 3-2). The radiation and impedance properties of an antenna are governed by its shape, size, and material properties, and its dimensions are usually measured in units of λ of the wave it is launching or receiving; a 1 m long dipole antenna operating at a wavelength $\lambda = 2$ m exhibits the same properties as a 1 cm long dipole operating at $\lambda = 2$ cm. Hence, in most of our discussions in this chapter, we refer to antenna dimensions in wavelength units.

► Most antennas are ***reciprocal*** devices, exhibiting the same radiation pattern for transmission as for reception. ◄

Reciprocity

The directional function characterizing the *relative* distribution of power radiated by an antenna is known as the ***antenna radiation pattern***, or simply the ***antenna pattern***. An ***isotropic*** antenna is a hypothetical antenna that radiates equally in all directions, and it is often used as a reference radiator when describing the radiation properties of real antennas.

Reciprocity means that if in the transmission mode a given antenna transmits in direction A 100 times the power it transmits in direction B, then when used in the reception mode it is 100 times more sensitive to electromagnetic radiation incident from direction A than from B. All the antennas shown in Fig. 3-2 obey the reciprocity law, but not all antennas are reciprocal devices. Reciprocity may not hold for some solid-state antennas composed of nonlinear semiconductors or ferrite materials, as discussed in Section 3-12. The

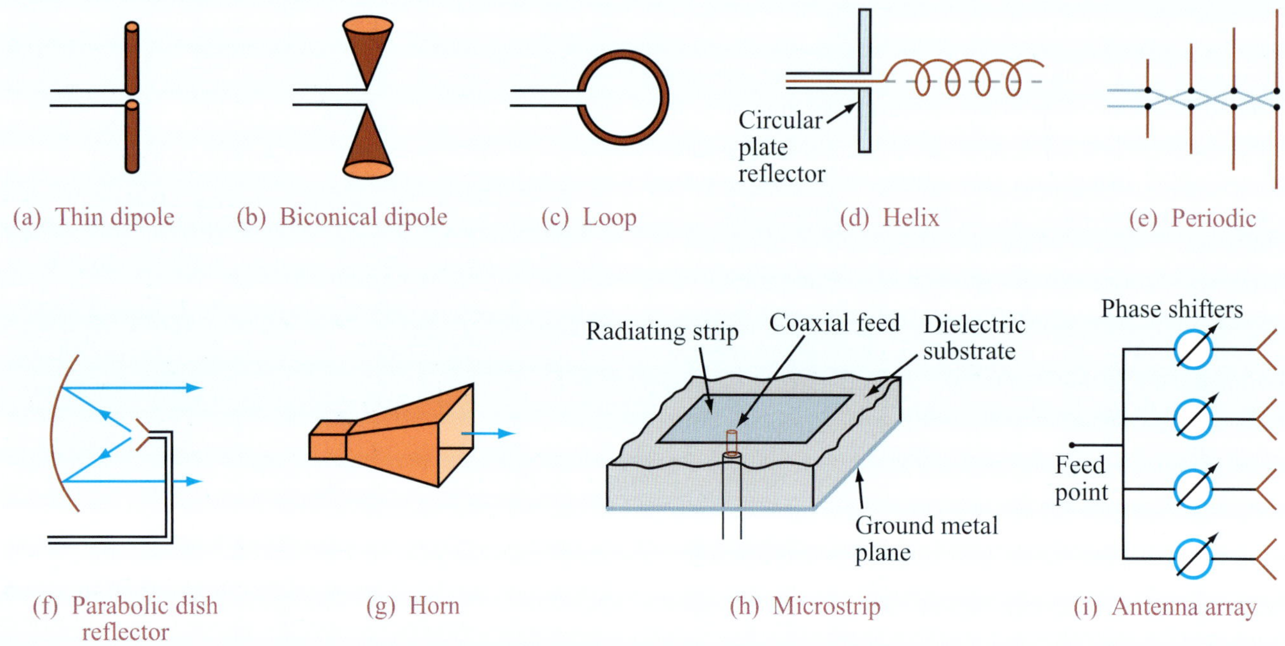

Figure 3-2: Various types of antennas.

reciprocity property is very convenient because it allows us to compute the radiation pattern of an antenna in the transmission mode, even when the antenna is intended to operate as a receiver. Reciprocity is assumed throughout Sections 3-1 to 3-11.

> ▶ As a reciprocal device, an antenna operating in the receiving mode extracts from an incident wave only that component of the wave whose electric field matches the antenna polarization state. ◀

Another important reciprocity property is ***antenna impedance***, which pertains to the transfer of power from a generator to the antenna when the antenna is used as a transmitter and, conversely, the transfer of power from the antenna to a load when the antenna is used as a receiver. Throughout this chapter, it is assumed that the antenna is properly matched to the transmission line connected to its terminals, thereby avoiding reflections and their associated problems.

Radiation sources

Radiation sources fall into two categories: conduction currents and aperture fields. The dipole and loop antennas [Fig. 3-2(a) and (c)] are examples of current sources; the time-varying currents flowing in the conducting wires give rise to the radiated electromagnetic fields. A horn antenna [Fig. 3-2(g)] is an example of the second group because the electric and magnetic fields across the horn's aperture serve as the sources of the radiated fields. The aperture fields are themselves induced by time-varying currents on the surfaces of the horn's walls, and therefore ultimately all radiation is due to time-varying currents. The choice of currents or apertures as the sources is merely a computational convenience arising from the structure of the antenna. We briefly examine the radiation processes associated with both types of sources.

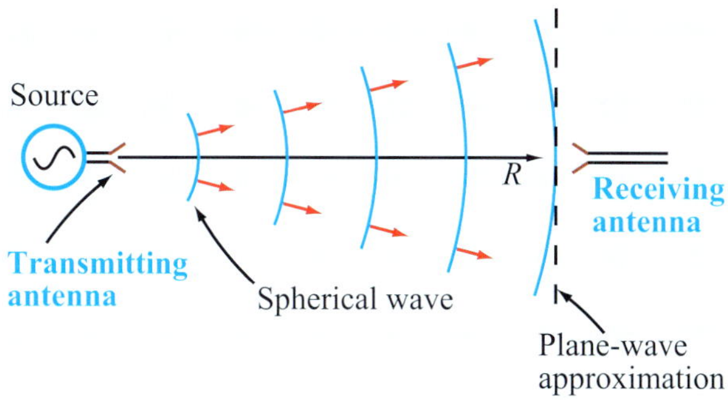

Figure 3-3: Far-field plane-wave approximation.

Far-field region

The wave radiated by a point source is spherical in nature, with the wavefront expanding outward at a rate equal to the phase velocity u_{p} (or the velocity of light c if the medium is free space). If R, the distance between the transmitting antenna and the receiving antenna, is sufficiently large such that the wavefront across the receiving aperture may be considered planar (Fig. 3-3), then the receiving aperture is said to be in the ***far-field*** (or ***far-zone***) region of the transmitting point source. This region is of particular significance because for most applications, the location of the observation point is indeed in the far-field region of the antenna. The far-field plane-wave approximation allows the use of certain mathematical approximations that simplify the computation of the radiated field and, conversely, provide convenient techniques for synthesizing the appropriate antenna structure that gives rise to the desired far-field antenna pattern.

By convention, the ***far-field distance*** is defined as the distance R from the antenna at which the maximum error between the phase of the spherical wave radiated by the antenna and the phase of its plane-wave approximation is $\pi/8$. For an antenna whose longest linear dimension is d, the far-field distance is

$$R \geq 2d^2/\lambda \qquad \textbf{(far-field distance)}, \qquad (3.1)$$

where λ is the wavelength of the wave radiated by the antenna.

Antenna arrays

When multiple antennas operate together, the combination is called an ***antenna array*** [Fig. 3-2(i)], and the array as a whole behaves as if it were a single antenna. By controlling the magnitude and phase of the signal feeding each antenna, it is possible to *shape the radiation pattern* of the array and to *electronically steer the direction of the beam.* These topics are treated in Sections 3-8 to 3-10.

3-1 The Hertzian Dipole

By regarding a linear antenna as consisting of a large number of infinitesimally short conducting elements, each of which is so short that current may be considered uniform over its length, the field of the entire antenna may be obtained by integrating the fields from all these differential antennas, with the proper magnitudes and phases taken into account. A short linear antenna of length $l \ll \lambda$ is known as a ***Hertzian dipole***.

The wire oriented along the z direction in Fig. 3-4 carries a sinusoidally varying current given by

$$i(t) = I_0 \cos \omega t = \mathfrak{Re}[I_0 e^{j\omega t}] \qquad \text{(A)}, \qquad (3.2)$$

where I_0 is the current amplitude and ω is the angular frequency. From Eq. (3.2), the phasor current $I = I_0$. Even though the current has to go to zero at the two ends of the dipole, we treat it as constant across its entire length.

To characterize the directional character of the radiated power at a fixed distance R from the antenna, antenna pattern plots usually are presented in a spherical coordinate system (Fig. 3-5). Its variables, R, θ, and ϕ, are called the ***range***, ***zenith angle***, and ***azimuth angle***, respectively. In the far-field region, the expressions for the electric and magnetic fields of the wave radiated by the Hertzian dipole in a medium with wavenumber k are given by

$$\mathbf{E} = \hat{\boldsymbol{\theta}}\, \frac{jI_0 lk\eta_0}{4\pi} \left(\frac{e^{-jkR}}{R}\right) \sin\theta \quad \text{(V/m)}, \qquad (3.3)$$

$$\mathbf{H} = \hat{\boldsymbol{\phi}}\, \frac{E_\theta}{\eta_0} \qquad \text{(A/m)}. \qquad (3.4)$$

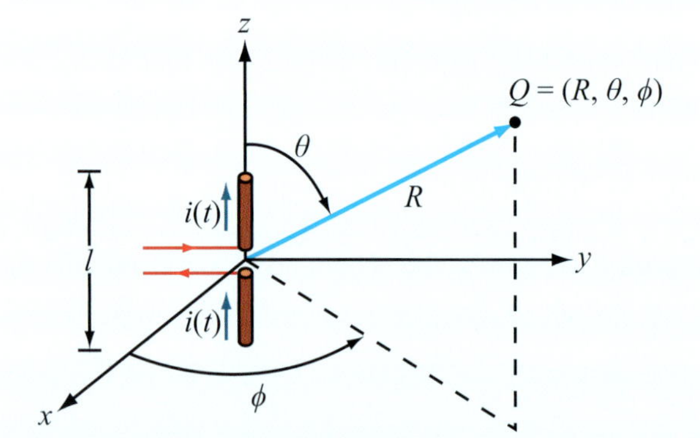

Figure 3-4: Short dipole placed at the origin of a spherical coordinate system.

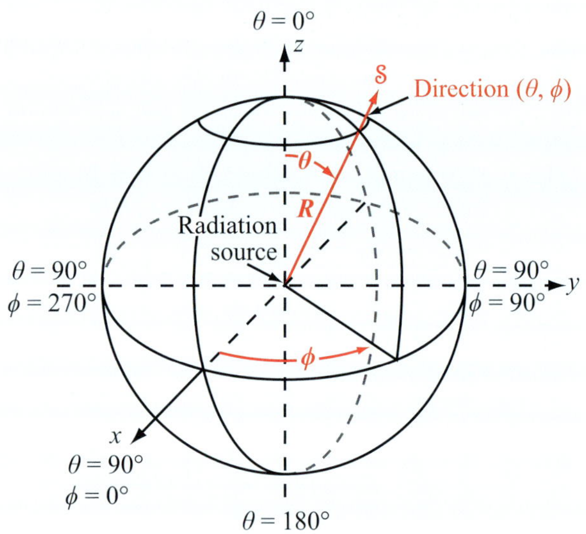

Figure 3-5: Spherical coordinate system.

At the observation point Q (Fig. 3-4), the wave is similar to a uniform plane wave with its electric and magnetic fields in phase, related by the intrinsic impedance of the medium η_0, and their directions orthogonal to each other and to the direction of propagation $\hat{\mathbf{R}}$. Both fields are proportional to $\sin\theta$ and independent of ϕ (which is expected from symmetry considerations).

Given **E** and **H**, the ***time-averaged power density*** of the radiated wave is obtained by applying Eq. (2.71); that is,

$$\mathcal{S} = \tfrac{1}{2}\mathfrak{Re}\,(\mathbf{E}\times H^*) \qquad (\text{W/m}^2). \tag{3.5}$$

For the short dipole, use of Eqs. (3.3) and (3.4) yields

$$\mathcal{S} = \hat{\mathbf{R}}\,\mathcal{S}(R,\theta), \tag{3.6}$$

with

$$\begin{aligned}\mathcal{S}(R,\theta) &= \left(\frac{\eta_0 k^2 I_0^2 l^2}{32\pi^2 R^2}\right)\sin^2\theta \\ &= \mathcal{S}_0 \sin^2\theta \qquad (\text{W/m}^2).\end{aligned} \tag{3.7}$$

The directional pattern of an antenna is described by its ***normalized radiation intensity*** $F(\theta,\phi)$, defined as the ratio of the power density $\mathcal{S}(R,\theta,\phi)$ at a specified range R to $\mathcal{S}_{\max}$, the maximum value of $\mathcal{S}(R,\theta,\phi)$ at the same range,

$$F(\theta,\phi) = \frac{\mathcal{S}(R,\theta,\phi)}{\mathcal{S}_{\max}} \qquad \textbf{(dimensionless)}. \tag{3.8}$$

For the Hertzian dipole, the $\sin^2\theta$ dependence in Eq. (3.7) indicates that the radiation is maximum in the broadside direction ($\theta = \pi/2$), corresponding to the azimuth plane, and is given by

$$\begin{aligned}\mathcal{S}_{\max} = \mathcal{S}_0 &= \frac{\eta_0 k^2 I_0^2 l^2}{32\pi^2 R^2} \\ &= \frac{15\pi I_0^2}{R^2}\left(\frac{l}{\lambda}\right)^2 \qquad (\text{W/m}^2),\end{aligned} \tag{3.9}$$

where use was made of the relations $k = 2\pi/\lambda$ and $\eta_0 \approx 120\pi$. We observe that $\mathcal{S}_{\max}$ is directly proportional to I_0^2 and l^2 (with l measured in wavelengths), and that it decreases with distance as $1/R^2$.

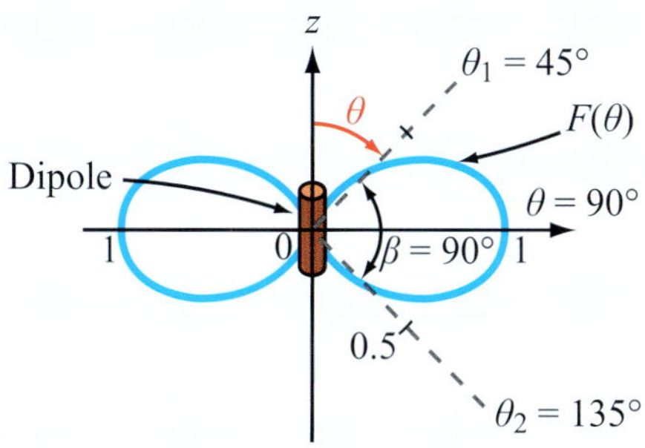

(a) Elevation pattern

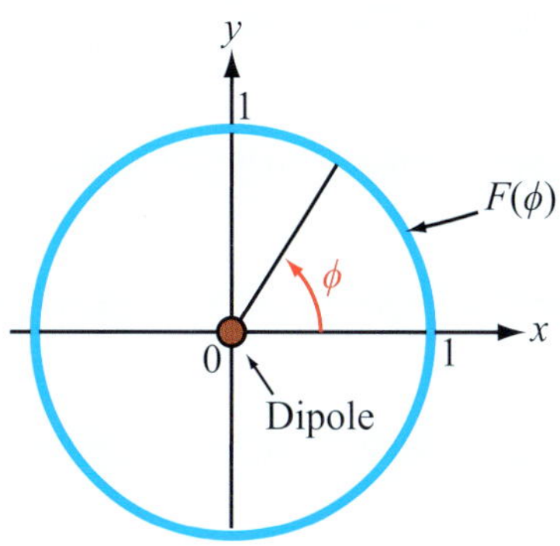

(b) Azimuth pattern

Figure 3-6: Radiation patterns of a short dipole.

From the definition of the normalized radiation intensity given by Eq. (3.8), it follows that

$$F(\theta,\phi) = F(\theta) = \sin^2\theta. \qquad (3.10)$$

Plots of $F(\theta)$ are shown in Fig. 3-6 in both the elevation plane (the θ plane) and the azimuth plane (ϕ plane).

▶ No energy is radiated by the short dipole along the direction of the dipole axis and maximum radiation ($F = 1$) occurs in the ***broadside direction*** ($\theta = 90°$). Since $F(\theta)$ is independent of ϕ, the pattern is doughnut-shaped in θ–ϕ space. ◀

3-2 Antenna Radiation Characteristics

▶ An ***antenna pattern*** describes the far-field directional properties of an antenna when measured at a fixed distance from the antenna. By virtue of reciprocity, a receiving antenna has the same directional antenna pattern as the pattern that it exhibits when operated in the transmission mode. ◀

In general, the antenna pattern is a three-dimensional plot that displays the strength of the radiated field or power density as a function of direction, with direction being specified by the zenith angle θ and the azimuth angle ϕ.

Consider a transmitting antenna placed at the origin of the observation sphere shown in Fig. 3-7. The differential power radiated by the antenna through an elemental area dA is

$$dP_{\text{rad}} = \mathbf{S}\cdot d\mathbf{A} = \mathbf{S}\cdot\hat{\mathbf{R}}\,dA = S\,dA \qquad \text{(W)}, \qquad (3.11)$$

where S is the radial component of the time-average power density $\mathbf{S}$. In the far-field region of any antenna, $\mathbf{S}$ is always in the radial direction. In a spherical

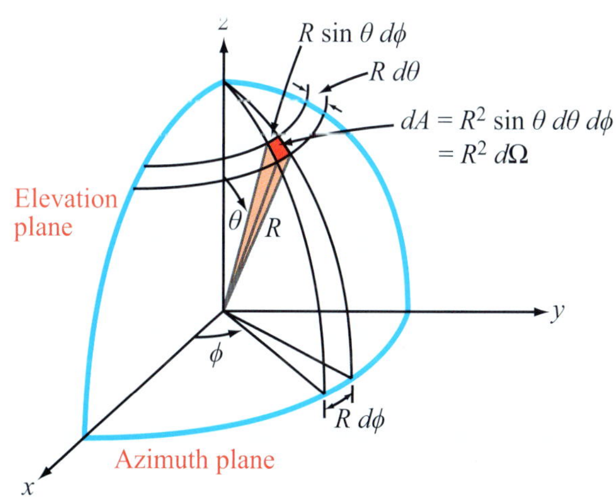

Figure 3-7: Definition of solid angle $d\Omega = \sin\theta\, d\theta\, d\phi$.

coordinate system,

$$dA = R^2 \sin\theta \, d\theta \, d\phi, \qquad (3.12)$$

and the ***solid angle*** $d\Omega$ associated with dA, defined as the subtended area divided by R^2, is given by

$$d\Omega = \frac{dA}{R^2} = \sin\theta \, d\theta \, d\phi \quad \text{(sr)}. \qquad (3.13)$$

Note that whereas a planar angle is measured in radians and the angular measure of a complete circle is 2π (rad), a solid angle is measured in ***steradians*** (sr), and the angular measure for a spherical surface is $\Omega = (4\pi R^2)/R^2 = 4\pi$ (sr). The solid angle of a hemisphere is 2π (sr).

Using the relation $dA = R^2 \, d\Omega$, dP_{rad} can be rewritten as

$$dP_{\text{rad}} = R^2 \, \mathcal{S}(R,\theta,\phi) \, d\Omega. \qquad (3.14)$$

The total power radiated by an antenna through a spherical surface at a fixed distance R is obtained by integrating Eq. (3.14) over that surface:

$$\begin{aligned} P_{\text{rad}} &= R^2 \int_{\phi=0}^{2\pi} \int_{\theta=0}^{\pi} \mathcal{S}(R,\theta,\phi) \sin\theta \, d\theta \, d\phi \\ &= R^2 \mathcal{S}_{\max} \int_{\phi=0}^{2\pi} \int_{\theta=0}^{\pi} F(\theta,\phi) \sin\theta \, d\theta \, d\phi \\ &= R^2 \mathcal{S}_{\max} \iint_{4\pi} F(\theta,\phi) \, d\Omega \quad \text{(W)}, \qquad (3.15) \end{aligned}$$

where $F(\theta,\phi)$ is the normalized radiation intensity defined by Eq. (3.8). The 4π symbol under the integral sign is used as an abbreviation for the indicated limits on θ and ϕ. Formally, P_{rad} is called the ***total radiated power***.

3-2.1 Antenna Pattern

Each specific combination of the zenith angle θ and the azimuth angle ϕ denotes a specific direction in the spherical coordinate system of Fig. 3-7.

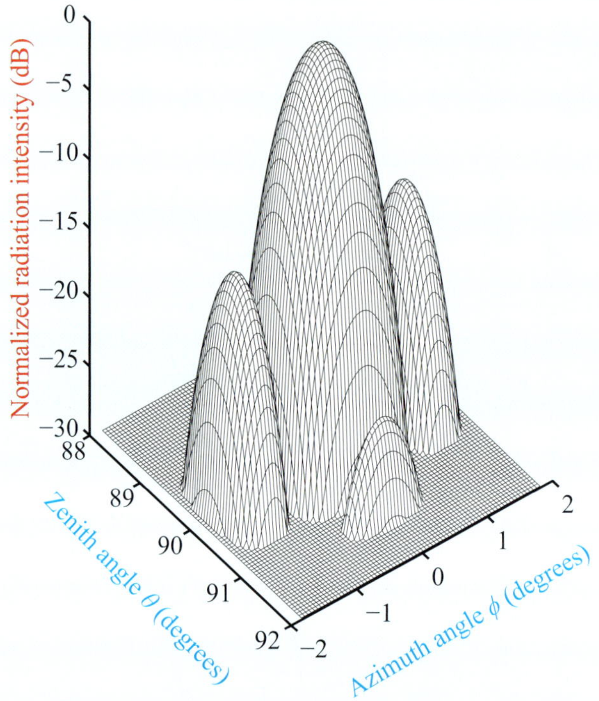

Figure 3-8: Three-dimensional pattern of a narrow-beam antenna.

▶ The normalized radiation intensity $F(\theta,\phi)$ characterizes the directional pattern of the energy radiated by the antenna. ◀

A plot of $F(\theta,\phi)$ as a function of both θ and ϕ constitutes a three-dimensional pattern, an example of which is shown in Fig. 3-8.

Often, it is of interest to characterize the variation of $F(\theta,\phi)$ in the form of two-dimensional plots in specific planes in the spherical coordinate system. The two planes most commonly specified for this purpose are the elevation and azimuth planes. The ***elevation plane***, also called the θ plane, is a plane corresponding to a constant value of ϕ. For example, $\phi = 0$ defines the x–z plane and $\phi = 90^\circ$ defines the y–z plane, both of which are elevation planes (Fig. 3-7). A plot of $F(\theta,\phi)$ versus θ in either of these planes constitutes a two-dimensional pattern in the elevation plane. This is not to imply,

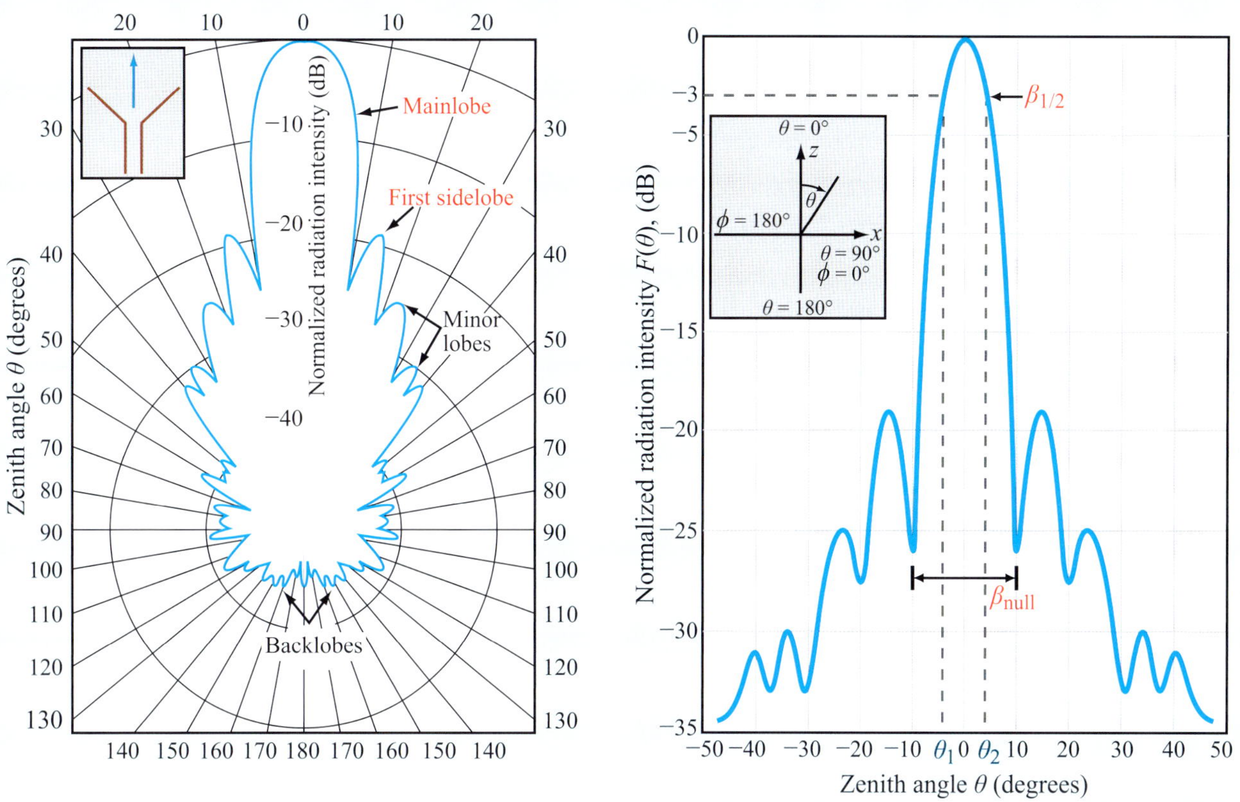

(a) Polar diagram (b) Rectangular plot

Figure 3-9: Representative plots of the normalized radiation pattern of a microwave antenna in (a) polar form and (b) rectangular form.

however, that the elevation-plane pattern is necessarily the same in all elevation planes.

The ***azimuth plane***, also called the ϕ plane, is specified by $\theta = 90^\circ$ and corresponds to the x–y plane. The elevation and azimuth planes are often called the two ***principal planes*** of the spherical coordinate system.

Some antennas exhibit highly directive patterns with narrow beams, in which case it is often convenient to plot the antenna pattern on a decibel scale by expressing F in decibels:

$$F\text{ [dB]} = 10 \log F.$$

As an example, the antenna pattern shown in Fig. 3-9(a) is plotted on a decibel scale in polar coordinates, with intensity as the radial variable. This format permits a convenient visual interpretation of the directional distribution of the ***radiation lobes***.

Another format commonly used for inspecting the pattern of a narrow-beam antenna is the rectangular display shown in Fig. 3-9(b), which permits the pattern to be easily expanded by changing the scale of the horizontal axis. These plots represent the variation in only one plane in the observation sphere, the $\phi = 0$ plane. Unless the pattern is symmetrical in ϕ, additional plots are required to define the overall variation of $F(\theta, \phi)$ with θ and ϕ.

Strictly speaking, the polar angle θ is always positive, being defined over the range from 0° (z direction) to 180° (z direction), and yet the θ axis in Fig. 3-9(b) is shown to have both positive and negative values. This is not a contradiction, but rather a different form of plotting antenna patterns. The right-hand half of the plot represents the variation of F (dB) with θ as θ is increased in a clockwise direction in the x–z plane [see inset in Fig. 3-9(b)], corresponding to $\phi = 0$, whereas the left-hand half of the plot represents the variation of F (dB) with θ as θ is increased in a counterclockwise direction at $\phi = 180°$. Thus, a negative θ value simply denotes that the direction (θ, ϕ) is in the left-hand half of the x–z plane.

The pattern shown in Fig. 3-9(a) indicates that the antenna is fairly directive, since most of the energy is radiated through a narrow sector called the ***mainlobe***. In addition to the mainlobe, the pattern exhibits several ***sidelobes*** and ***backlobes*** as well. For most applications, these extra lobes are considered undesirable because they represent wasted energy for transmitting antennas and potential interference directions for receiving antennas.

3-2.2 Beam Dimensions

For an antenna with a single mainlobe, the ***pattern solid angle*** Ω_p describes the equivalent width of the mainlobe of the antenna pattern (Fig. 3-10). It is defined as the integral of the normalized radiation intensity $F(\theta, \phi)$ over a sphere:

$$\Omega_p = \iint_{4\pi} F(\theta, \phi) \, d\Omega \qquad \text{(sr)}. \tag{3.16}$$

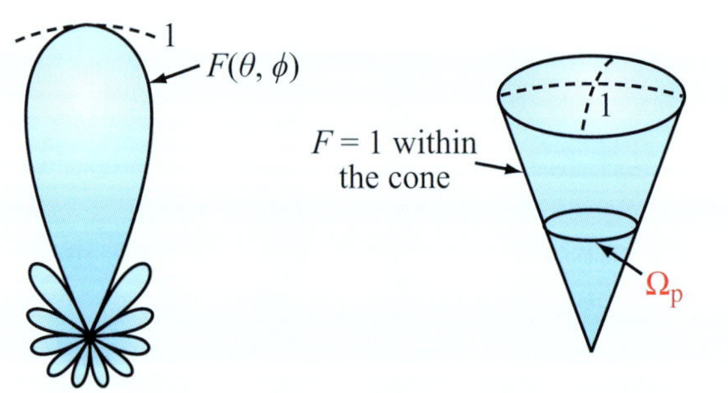

(a) Actual pattern (b) Equivalent solid angle

Figure 3-10: The pattern solid angle Ω_p defines an equivalent cone over which all the radiation of the actual antenna is concentrated with uniform intensity equal to the maximum of the actual pattern.

▶ For an isotropic antenna with $F(\theta, \phi) = 1$ in all directions, $\Omega_p = 4\pi$ (sr). ◀

The pattern solid angle characterizes the directional properties of the three-dimensional radiation pattern. To characterize the width of the mainlobe in a given plane, the term used is ***beamwidth***. The ***half-power beamwidth***, or simply the beamwidth β, is defined as the angular width of the mainlobe between the two angles at which the magnitude of $F(\theta, \phi)$ is equal to half of its peak value (or -3 dB on a decibel scale). For example, for the pattern displayed in Fig. 3-9(b), β is given by

$$\beta = \theta_2 - \theta_1, \tag{3.17}$$

where θ_1 and θ_2 are the ***half-power angles*** at which $F(\theta, 0) = 0.5$ (with θ_2 denoting the larger value and θ_1 denoting the smaller one, as shown in the figure). If the pattern is symmetrical and the peak value of $F(\theta, \phi)$ is at $\theta = 0$, then $\beta = 2\theta_2$. For the short-dipole pattern shown earlier in Fig. 3-6(a), $F(\theta)$ is maximum at $\theta = 90°$; θ_2 is at $135°$, and θ_1 is at $45°$. Hence, $\beta = 135° - 45° = 90°$. The beamwidth β is also known as the ***3 dB beamwidth***. In addition to the half-power beamwidth, other beam dimensions may be of interest for certain applications, such as the ***null beamwidth*** β_{null}, which is the angular width between the first nulls on the two sides of the peak [Fig. 3-9(b)].

3-2.3 Antenna Directivity

The ***peak directivity*** D of an antenna is defined as the ratio of its maximum normalized radiation intensity, F_{max} (which by definition is equal to 1), to the average value of $F(\theta,\phi)$ over all directions (4π space):

$$D = \frac{F_{\rm max}}{F_{\rm av}} = \frac{1}{\dfrac{1}{4\pi}\displaystyle\iint_{4\pi} F(\theta,\phi)\, d\Omega} = \frac{4\pi}{\Omega_{\rm p}} \quad \textbf{(dimensionless)}. \tag{3.18}$$

Here $\Omega_{\rm p}$ is the pattern solid angle defined by Eq. (3.16). Thus, the narrower $\Omega_{\rm p}$ of an antenna pattern is, the greater is the directivity. For an isotropic antenna, $\Omega_{\rm p} = 4\pi$; hence, its directivity $D_{\rm iso} = 1$.

By using Eq. (3.15) in Eq. (3.18), D can be expressed as

$$D = \frac{4\pi R^2 \mathcal{S}_{\rm max}}{P_{\rm rad}} = \frac{\mathcal{S}_{\rm max}}{\mathcal{S}_{\rm av}}, \tag{3.19}$$

where $\mathcal{S}_{\rm av} = P_{\rm rad}/(4\pi R^2)$ is the average value of the radiated power density and is equal to the total power radiated by the antenna, $P_{\rm rad}$, divided by the surface area of a sphere of radius R.

▶ Since $\mathcal{S}_{\rm av} = \mathcal{S}_{\rm iso}$, where $\mathcal{S}_{\rm iso}$ is the power density radiated by an isotropic antenna, D represents the ratio of the maximum power density radiated by the antenna to the power density radiated by an isotropic antenna, both measured at the same range R and excited by the same amount of input power. ◀

Usually, D is expressed in decibels:* D [dB] $= 10 \log D$.

For an antenna with a single mainlobe pointing in the z direction as shown in Fig. 3-11, $\Omega_{\rm p}$ may be approximated as the product of the half-power beamwidths β_{xz} and β_{yz} (in radians):

$$\Omega_{\rm p} \approx \beta_{xz}\beta_{yz}, \tag{3.20}$$

and therefore

$$D = \frac{4\pi}{\Omega_{\rm p}} \approx \frac{4\pi}{\beta_{xz}\beta_{yz}}. \tag{3.21}$$

*A note of caution: Even though we often express certain dimensionless quantities in decibels, we should always convert their decibel values to natural values before using them in the relations given in this chapter.

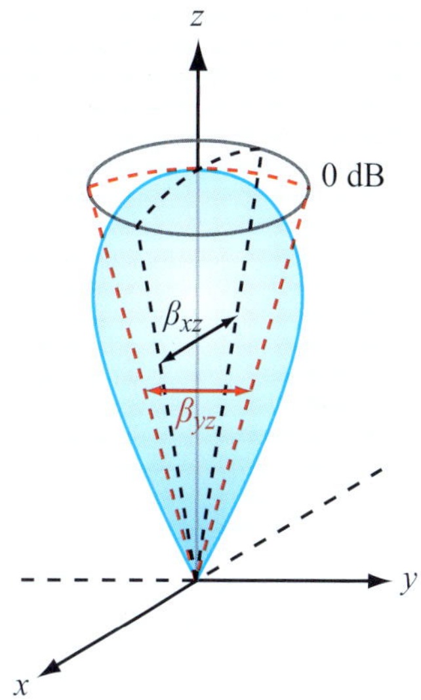

Figure 3-11: The solid angle of a unidirectional radiation pattern is approximately equal to the product of the half-power beamwidths in the two principal planes; that is, $\Omega_{\rm p} \approx \beta_{xz}\beta_{yz}$.

Although approximate, this relation provides a useful method for estimating the antenna directivity from measurements of the beamwidths in the two orthogonal planes whose intersection is the axis of the mainlobe.

For a Hertzian dipole, application of Eq. (3.18) with $F(\theta) = \sin^2\theta$ [from Eq. (3.10)] gives

$$D = \frac{4\pi}{\displaystyle\iint_{4\pi} F(\theta,\phi)\sin\theta\, d\theta\, d\phi} = \frac{4\pi}{\displaystyle\int_{\phi=0}^{2\pi}\int_{\theta=0}^{\pi} \sin^3\theta\, d\theta\, d\phi} = \frac{4\pi}{8\pi/3} = 1.5$$

or, equivalently, 1.76 dB.

3-2.4 Antenna Gain

Of the total power $P_{\rm t}$ (transmitter power) supplied to the antenna, a part, $P_{\rm rad}$, is radiated out into space, and the remainder, $P_{\rm loss}$, is dissipated as heat in the antenna structure. The ***radiation efficiency*** ξ is defined as the ratio of P_{rad} to P_t:

$$\xi = \frac{P_{rad}}{P_t} \qquad \textbf{(dimensionless)}. \tag{3.22}$$

The ***peak gain*** of an antenna is defined as

$$G = \frac{4\pi R^2 \mathcal{S}_{max}}{P_t}, \tag{3.23}$$

which is similar in form to the expression given by Eq. (3.19) for the directivity D except that it is referenced to the input power supplied to the antenna, P_t, rather than to the radiated power P_{rad}. In view of Eq. (3.22),

$$G = \xi D \qquad \textbf{(dimensionless)}. \tag{3.24}$$

▶ The gain accounts for ohmic losses in the antenna material, whereas the directivity does not. For a lossless antenna, $\xi = 1$ and $G = D$. ◀

3-2.5 Radiation Efficiency

To a transmission line connected between a generator supplying power P_t on one end and an antenna on the other end, the antenna is merely a load with ***input impedance*** $\mathbf{Z}$. If the line is lossless and properly matched to the antenna, all of P_t is transferred to the antenna. In general, $\mathbf{Z}$ consists of a resistive component R and a reactive component X:

$$\mathbf{Z} = R + jX. \tag{3.25}$$

The resistive component is defined as equivalent to a resistor R that would consume an average power P_t when the amplitude of the ac current flowing through it is I_0,

$$P_t = \tfrac{1}{2} I_0^2 R. \tag{3.26}$$

Since $P_t = P_{rad} + P_{loss}$, it follows that R can be defined as the sum of a ***radiation resistance*** R_{rad} and a ***loss resistance*** R_{loss},

$$R = R_{rad} + R_{loss}, \tag{3.27}$$

with

$$P_{rad} = \tfrac{1}{2} I_0^2 R_{rad}, \tag{3.28a}$$

$$P_{loss} = \tfrac{1}{2} I_0^2 R_{loss}, \tag{3.28b}$$

where I_0 is the amplitude of the sinusoidal current exciting the antenna. As defined earlier, the radiation efficiency is the ratio of P_{rad} to P_t, or

$$\xi = \frac{P_{rad}}{P_t} = \frac{P_{rad}}{P_{rad} + P_{loss}} = \frac{R_{rad}}{R_{rad} + R_{loss}}. \tag{3.29}$$

The radiation resistance R_{rad} can be calculated by integrating the far-field power density over a sphere to obtain P_{rad} and then equating the result to Eq. (3.28a).

3-2.6 Effective Area of a Receiving Antenna

So far, antennas have been treated as directional radiators of energy. Now, we shall examine the reverse process, namely how a receiving antenna extracts energy from an incident wave and delivers it to a load. The ability of an antenna to capture energy from an incident wave of power density $\mathcal{S}_i$ (W/m^2) and convert it into an ***intercepted power*** P_{int} (W) for delivery to a matched load (Fig. 3-12) is characterized by the ***effective area*** A_e:

$$A_e = \frac{P_{int}}{\mathcal{S}_i} \qquad (\text{m}^2). \tag{3.30}$$

Other commonly used names for A_e include ***effective aperture*** and ***receiving cross section***. It can be

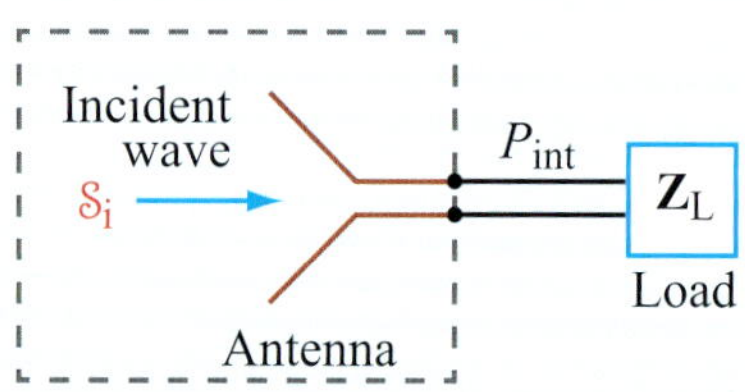

Figure 3-12: Receiving antenna represented by an equivalent circuit.

shown (Ulaby et al., 2010) that for any antenna, the directivity D, pattern solid angle Ω_p, and effective aperture A_e are related by

$$\Omega_p = \frac{\lambda^2}{A_e}, \tag{3.31a}$$

$$D = \frac{4\pi}{\Omega_p} = \frac{4\pi}{\lambda^2} A_e, \tag{3.31b}$$

which implies that a highly directional antenna (with narrow beamwidths) has a large effective area (measured in λ^2 units) and, conversely, an antenna with a broad beam has a small effective area. The concept of effective area may or may not bear a meaningful connection to the ***physical aperture*** A_p of the antenna, depending on whether or not the antenna has a definable physical aperture. For dipoles and many other wire antennas, the antenna directivity D does not depend on the wire diameter, so it is theoretically possible to make the wire diameter infinitesimally small, thereby reducing any definition of the antenna's cross-sectional area to zero, without affecting either its D or A_e.

If the antenna has a readily definable physical aperture A_p, as in the case of horn and dish antennas, it is useful to relate A_e to A_p through the ***aperture efficiency*** η_a as follows:

$$A_e = \eta_a A_p. \tag{3.32}$$

3-3 Friis Transmission Formula

The two antennas shown in Fig. 3-13 are part of a free-space communication link, with the separation between the antennas, R, being large enough for each to be in the far-field region of the other. The transmitting and receiving antennas have ***effective areas*** A_t and A_r and ***radiation efficiencies*** ξ_t and ξ_r, respectively. Our objective is to find a relationship between P_t, the power supplied to the transmitting antenna, and P_{rec}, the power delivered to the receiver. As always, we assume that both antennas are impedance-matched to their respective transmission lines. Initially, we consider the case where

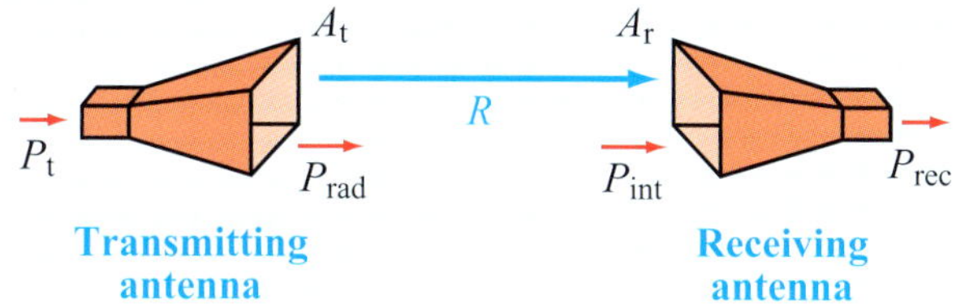

Figure 3-13: Transmitter–receiver configuration.

the two antennas are oriented such that the peak of the radiation pattern of each antenna points in the direction of the other.

We start by treating the transmitting antenna as a lossless isotropic radiator. The power density incident upon the receiving antenna at a distance R from an isotropic transmitting antenna is simply equal to the transmitter power P_t divided by the surface area of a sphere of radius R:

$$\mathcal{S}_{iso} = \frac{P_t}{4\pi R^2}. \tag{3.33}$$

The real transmitting antenna is neither lossless nor isotropic. Hence, the received power density $\mathcal{S}_r$ due to the real antenna is

$$\mathcal{S}_r = G_t \mathcal{S}_{iso} = \xi_t D_t \mathcal{S}_{iso} = \frac{\xi_t D_t P_t}{4\pi R^2}. \tag{3.34}$$

Through the gain $G_t = \xi_t D_t$, ξ_t accounts for the fact that only part of the power P_t supplied to the antenna is radiated out into space, and D_t accounts for the directivity of the transmitting antenna (in the direction of the receiving antenna). Moreover, by Eq. (3.31b), the peak directivity D_t is related to A_t by $D_t = 4\pi A_t/\lambda^2$. Hence, Eq. (3.34) becomes

$$\mathcal{S}_r = \frac{\xi_t A_t P_t}{\lambda^2 R^2}. \tag{3.35}$$

On the receiving-antenna side, the power intercepted by the receiving antenna is equal to the product of the incident power density $\mathcal{S}_r$ and the effective area A_r:

$$P_{int} = \mathcal{S}_r A_r = \frac{\xi_t A_t A_r P_t}{\lambda^2 R^2}. \tag{3.36}$$

The power delivered to the receiver, P_{rec}, is equal to the intercepted power P_{int} multiplied by the radiation efficiency of the receiving antenna, ξ_r. Hence, $P_{rec} = \xi_r P_{int}$, which leads to the result

$$\frac{P_{rec}}{P_t} = \frac{\xi_t \xi_r A_t A_r}{\lambda^2 R^2} = G_t G_r \left(\frac{\lambda}{4\pi R}\right)^2. \tag{3.37}$$

▶ This relation is known as the ***Friis transmission formula***, and P_{rec}/P_t is called the ***power transfer ratio***. ◀

If the two antennas are not oriented in the direction of maximum power transfer, Eq. (3.37) assumes the general form

$$\frac{P_{rec}}{P_t} = \left(\frac{\lambda}{4\pi R}\right)^2 G_t(\theta_t, \phi_t)\, G_r(\theta_r, \phi_r), \tag{3.38}$$

where $G_t(\theta_t, \phi_t)$ is the gain of the transmitting antenna at angles (θ_t, ϕ_t) corresponding to the direction of the receiving antenna (as seen by the antenna pattern of the transmitting antenna), and a similar definition applies to $G_r(\theta_r, \phi_r)$ for the receiving antenna.

For a pair of lossless isotropic antennas with $G_t = G_r = 1$,

$$\frac{P_{rec}}{P_t} = \left(\frac{\lambda}{4\pi R}\right)^2 = \frac{1}{L_{FS}} \qquad \textbf{(lossless isotropic antennas)},$$

where $L_{FS} = (4\pi R/\lambda)^2$ is called the ***free-space propagation loss***. In terms of L_{FS}, Eq. (3.38) can be rewritten as

$$\frac{P_{rec}}{P_t} = \frac{1}{L_{FS}}\, G_t(\theta_t, \phi_t)\, G_r(\theta_r, \phi_r). \tag{3.39}$$

The Friis transmission formula can also be applied to calculate the power reflected by a specular surface. In Fig. 3-14(a), the transmitter and receiver are oriented along the specular direction relative to the reflecting surface ($\theta_i = \theta_r$) and their antennas have the same polarization state. Using image theory, the configuration in Fig. 3-14(b) is equivalent to that in Fig. 3-14(b). In the latter, the range between the transmitter and the receiver is $(R_1 + R_2)$. Also, reflection by the surface modifies the power density by the factor $\Gamma^p(\theta)$, the p-polarized Fresnel reflectivity (with $p =$ v or h). Incorporating these

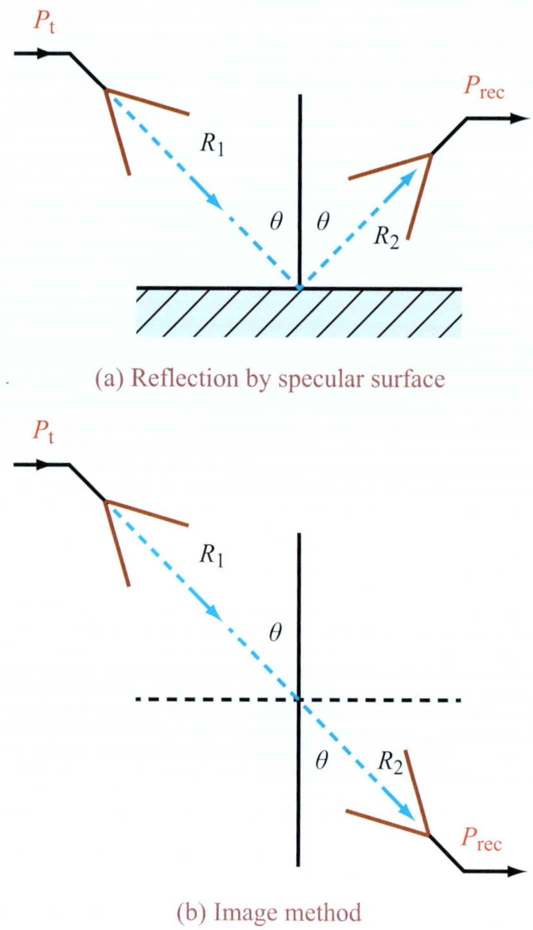

Figure 3-14: Reflection by a specular surface is equivalent to direct transmission along a path of length $(R_1 + R_2)$, but modified by the reflectivity of the surface $\Gamma^p(\theta)$.

modifications into Eq. (3.37) leads to

$$\frac{P_{\text{rec}}}{P_{\text{t}}} = G_{\text{t}}G_{\text{r}} \left[\frac{\lambda}{4\pi(R_1+R_2)} \right]^2 \Gamma^p(\theta), \quad (p = \text{h or v}). \tag{3.40}$$

3-4 Radiation by Large-Aperture Antennas

For wire antennas, the sources of radiation are the infinitesimal current elements comprising the current distribution along the wire, and the total radiated field at a given point in space is equal to the sum, or integral, of the fields radiated by all the elements. A parallel scenario applies to aperture antennas, except that now the source of radiation is the electric-field distribution across the aperture. Consider the horn antenna shown in Fig. 3-15(a). It is connected to a source through a coaxial transmission line, with the outer conductor of the line connected to the metal body of the horn and the inner conductor made to protrude, through a small hole, partially into the throat end of the horn, thereby providing a transition from coax to waveguide. The protruding conductor acts as a monopole antenna, generating waves that radiate along the short waveguide segment and then outwardly toward the horn's aperture. The electric field of the wave arriving at the aperture, which may vary as a function of x_a and y_a over the horn's aperture, is called the ***electric-field aperture distribution*** or ***illumination***, $E_a(x_a, y_a)$. Inside the horn, wave propagation is guided by the horn's geometry, but as the wave transitions from a guided wave into an unbounded wave, every point of its wavefront serves as a source of spherical secondary wavelets. The aperture may then be represented as a distribution of isotropic radiators. At a distant point Q, the combination of all the waves arriving from all of these radiators constitutes the total wave observed by a receiver placed at that point.

The radiation process described for the horn antenna is equally applicable to any aperture upon which an electromagnetic wave is incident. For example, in the case of the parabolic reflector shown in Fig. 3-15(b), it can be described in terms of an imaginary aperture representing the electric-field distribution across a plane in front of the reflector.

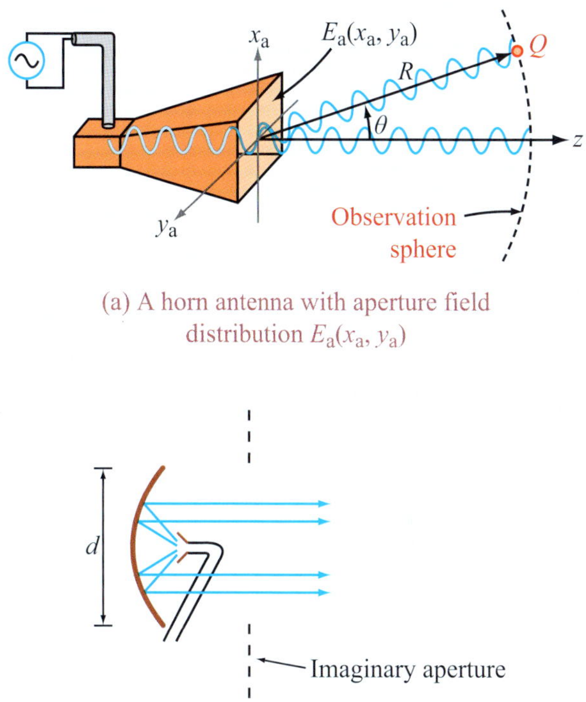

Figure 3-15: (a) A horn antenna with aperture field distribution $E_a(x_a, y_a)$, and (b) radiation by a parabolic dish reflector illuminated by a small horn antenna.

Two types of mathematical formulations are available for computing the electromagnetic fields of waves radiated by apertures. The first is a ***scalar formulation*** based on Kirchhoff's work and the second is a ***vector formulation*** based on Maxwell's equations. In this section, we limit our presentation to the scalar diffraction technique, in part because of its inherent simplicity and also because it is applicable to a wide range of practical applications.

▶ The key requirement for the validity of the scalar formulation is that the antenna aperture be at least several wavelengths long along each of its principal dimensions. ◀

A distinctive feature of such an antenna is its high directivity and correspondingly narrow beam, which makes it attractive for microwave remote-sensing systems. The frequency range commonly used for such systems is the 1 to 300 GHz band. Because the corresponding wavelength range is 30 cm to 1 mm, respectively, it is quite practical to construct and use antennas (in this frequency range) with aperture dimensions that are many wavelengths in size.

The x_a–y_a plane in Fig. 3-16, denoted ***aperture plane*** A, contains an aperture with an electric field distribution $E_a(x_a, y_a)$. For the sake of convenience, the opening has been chosen to be rectangular in shape, with dimensions l_x along x_a and l_y along y_a, even though the formulation we are about to discuss is general enough to accommodate any two-dimensional aperture distribution, including those associated with circular and elliptical apertures. At a distance z from plane A in Fig. 3-16, we have an observation plane O with axes (x, y). The two planes have parallel axes and are separated by a distance z. Moreover, z is sufficiently large that any point Q in the observation plane is in the far-field region of the aperture. To satisfy the far-field condition, it is necessary that

$$R \geq 2d^2/\lambda \qquad \textbf{(far-field distance)}, \tag{3.41}$$

where, as noted earlier in connection with Eq. (3.1), d is the longest linear dimension of the radiating aperture.

The position of observation point Q is specified by the range R between the center of the aperture and point Q and by the angles θ and ϕ (Fig. 3-16), which jointly define the direction of the observation point relative to the coordinate system of the aperture. The z axis is orthogonal to the plane containing the antenna aperture. Also, θ usually is called the ***elevation angle***.

The electric field phasor of the wave incident upon point Q is denoted $E(R, \theta, \phi)$. ***Kirchhoff's scalar diffraction theory*** provides the following relationship between the radiated field $E(R, \theta, \phi)$ and the aperture illumination $E_a(x_a, y_a)$:

$$E(R, \theta, \phi) = \frac{j}{\lambda}\left(\frac{e^{-jkR}}{R}\right) h(\theta, \phi), \tag{3.42}$$

where

$$h(\theta, \phi) = \iint_{-\infty}^{\infty} E_a(x_a, y_a) \cdot \exp\left[jk \sin\theta (x_a \cos\phi + y_a \sin\phi)\right] dx_a \, dy_a, \tag{3.43}$$

where $k = 2\pi/\lambda$ is the wave number. We shall refer to $h(\theta, \phi)$ as the ***form factor*** of $E(R, \theta, \phi)$. Its integral is written with infinite limits, with the understanding that $E_a(x_a, y_a)$ is identically zero outside the aperture. The spherical propagation factor (e^{-jkR}/R) accounts for wave propagation between the center of the aperture and the observation point, and $h(\theta, \phi)$ represents an integration of the exciting field $E_a(x_a, y_a)$ over the extent of the aperture, taking into account [through the exponential function in Eq. (3.43)] the approximate deviation in distance between R and s, where s is the distance to any point (x_a, y_a) in the aperture plane (see Fig. 3-16).

▶ In Kirchhoff's scalar formulation, the polarization direction of the radiated field $E(R, \theta, \phi)$ is the same as that of the aperture field $E_a(x_a, y_a)$. ◀

The power density of the radiated wave is given by

$$\mathcal{S}(R, \theta, \phi) = \frac{|E(R, \theta, \phi)|^2}{2\eta_0} = \frac{|h(\theta, \phi)|^2}{2\eta_0 \lambda^2 R^2}. \tag{3.44}$$

3-5 Rectangular Aperture with Uniform Field Distribution

In the general case, $E_a(x_a, y_a)$ is a complex quantity consisting of an amplitude distribution and a phase distribution over the aperture. As we will see in future

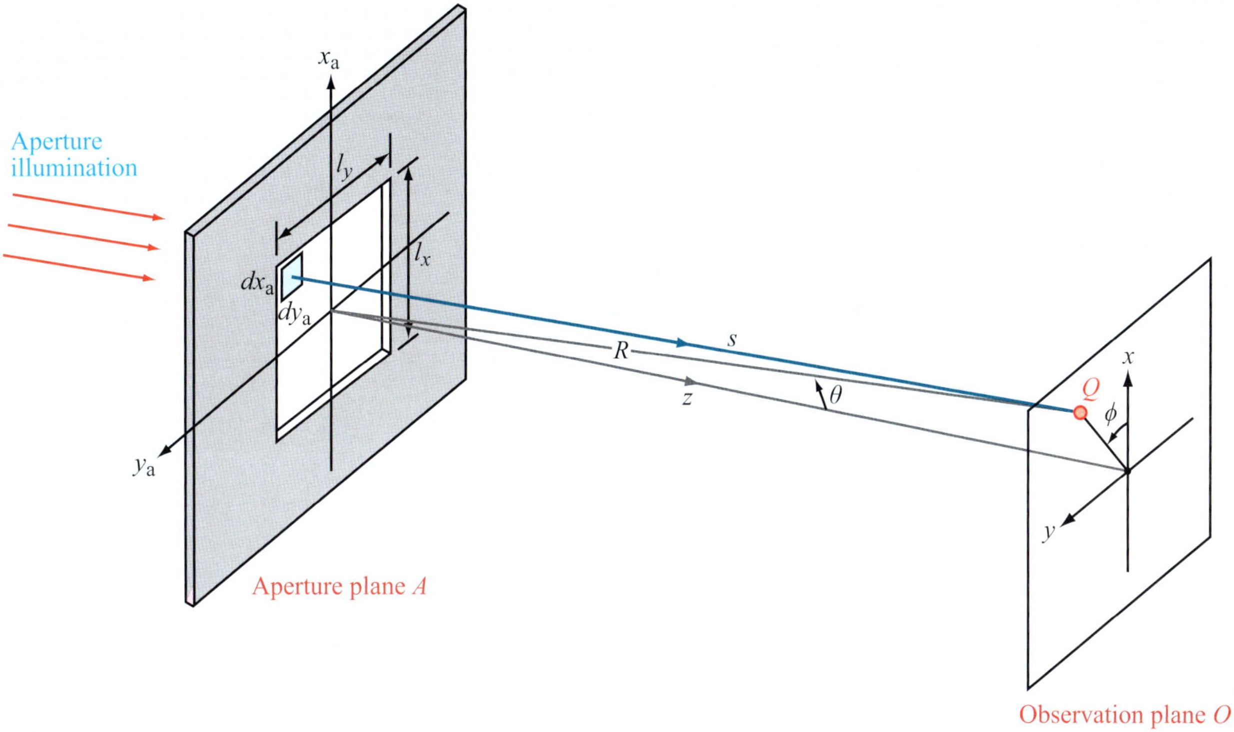

Figure 3-16: Radiation by an aperture in the x_a–y_a plane at $z = 0$.

sections, the two distributions govern both the shape and direction of the antenna pattern.

To illustrate the scalar diffraction technique, consider a rectangular aperture of height l_x and width l_y, both at least a few wavelengths long. The aperture is excited by a ***uniform field distribution*** (i.e., constant value) given by

$$E_a(x_a, y_a) = \begin{cases} E_0 & \text{for } -l_x/2 \leq x_a \leq l_x/2 \\ & \text{and } -l_y/2 \leq y_a \leq l_y/2, \\ 0 & \text{otherwise.} \end{cases} \quad (3.45)$$

Inserting Eq. (3.45) into Eq. (3.43) gives

$$h(\theta, \phi) = \int_{y_a=-l_y/2}^{l_y/2} \int_{x_a=-l_x/2}^{l_x/2} E_0 \exp[jk \sin\theta \cos\phi \; x_a] \cdot \exp[jk \sin\theta \sin\phi \; x_a] \, dx_a \, dy_a. \quad (3.46)$$

In preparation for performing the integration, we introduce the intermediate variables u and v defined as

$$u = k \sin\theta \cos\phi = \frac{2\pi}{\lambda} \sin\theta \cos\phi, \quad (3.47a)$$

$$v = k \sin\theta \sin\phi = \frac{2\pi}{\lambda} \sin\theta \sin\phi. \quad (3.47b)$$

Use of Eq. (3.47) in Eq. (3.46) leads to

$$h(\theta, \phi) = E_0 \int_{-l_x/2}^{l_x/2} e^{jux_a} \, dx_a \int_{-l_y/2}^{l_y/2} e^{jvy_a} \, dy_a. \quad (3.48)$$

After performing the integration in Eq. (3.48), replacing u and v with the expressions given in Eq. (3.47), and finally inserting the expression for $h(\theta, \phi)$ into

Eq. (3.42), we obtain the result

$$E(R,\theta,\phi) = \frac{jE_0A_p}{\lambda}\left(\frac{e^{-jkR}}{R}\right)\,\mathrm{sinc}\left(\frac{\pi l_x}{\lambda}\sin\theta\cos\phi\right) \cdot \mathrm{sinc}\left(\frac{\pi l_y}{\lambda}\sin\theta\sin\phi\right), \qquad (3.49)$$

where $A_p = l_x l_y$ and the sinc function for any argument t is defined as[†]

$$\mathrm{sinc}\,t = \frac{\sin t}{t}. \qquad (3.50)$$

3-5.1 Antenna Pattern in x–y Plane

To illustrate the result represented by Eq. (3.49), let us examine the x–z elevation pattern of the antenna at a fixed range **R**. Setting $\phi = 0$ in Eq. (3.49) and then using Eq. (3.44) leads to

$$\mathcal{S}(R,\theta) = \mathcal{S}_0\,\mathrm{sinc}^2(\pi l_x \sin\theta/\lambda) \quad (x\text{–}z \text{ plane}), \qquad (3.51)$$

where $\mathcal{S}_0 = E_0^2 A_p^2/(2\eta_0\lambda^2 R^2)$.

▶ The sinc function is maximum when its argument is zero; i.e., $\mathrm{sinc}(0) = 1$. ◀

This occurs when $\theta = 0$. Hence, at a fixed range R, $\mathcal{S}_{max} = \mathcal{S}(\theta{=}0) = \mathcal{S}_0$. The normalized radiation intensity is then given by

$$F(\theta) = \frac{\mathcal{S}(R,\theta)}{\mathcal{S}_{max}} = \mathrm{sinc}^2\left(\frac{\pi l_x}{\lambda}\sin\theta\right) = \mathrm{sinc}^2(\pi\gamma) \quad (x\text{–}z \text{ plane}). \qquad (3.52)$$

Figure 3-17 shows $F(\theta)$ plotted (on a decibel scale) as a function of the intermediate variable $\gamma = (l_x/\lambda)\sin\theta$. The pattern exhibits nulls at nonzero integer values of γ.

[†]This is one of two common definitions for the sinc function; the other is $\mathrm{sinc}\,t = \sin(\pi t)/(\pi t)$. In this chapter we use the definition given by Eq. (3.50) exclusively.

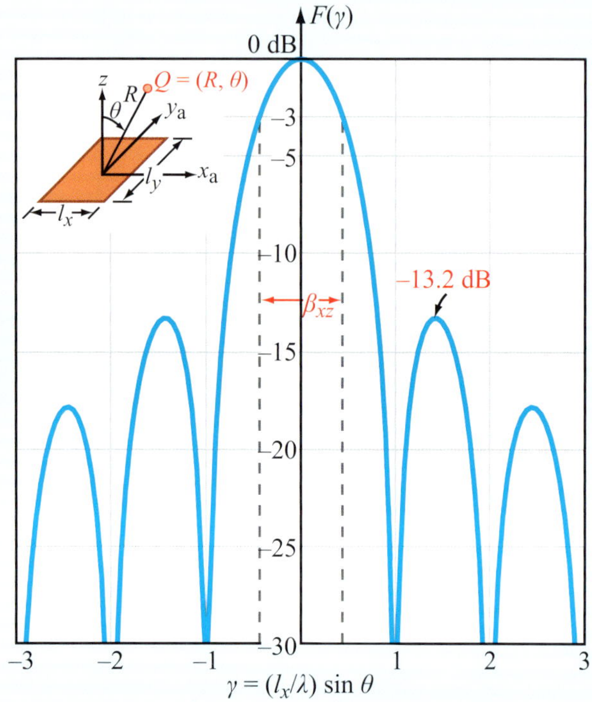

Figure 3-17: Normalized radiation pattern of a uniformly illuminated rectangular aperture in the x–z plane ($\phi = 0$).

3-5.2 Beamwidth

The normalized radiation intensity $F(\theta)$ is symmetrical in the x–z plane, and its maximum is along the boresight direction ($\theta = 0$ in this case). Its half-power beamwidth $\beta_{xz} = \theta_2 - \theta_1$, where θ_1 and θ_2 are the values of θ at which $F(\theta,0) = 0.5$ (or -3 dB on a decibel scale), as shown in Fig. 3-17. Since the pattern is symmetrical with respect to $\theta = 0$, $\theta_1 = -\theta_2$ and $\beta_{xz} = 2\theta_2$. The angle θ_2 can be obtained from a solution of

$$F(\theta_2) = \mathrm{sinc}^2\left(\frac{\pi l_x}{\lambda}\sin\theta_2\right) = 0.5. \qquad (3.53)$$

Solution of Eq. (3.53) yields the result

$$\frac{\pi l_x}{\lambda}\sin\theta_2 = 1.39, \qquad (3.54)$$

or

$$\sin\theta_2 = 0.44\,\frac{\lambda}{l_x}\,. \tag{3.55}$$

A fundamental condition for the validity of scalar diffraction theory is that the aperture dimensions are much larger than the wavelength λ; that is, $\lambda/l_x \ll 1$. Hence, θ_2 is a small angle, the approximation $\sin\theta_2 \approx \theta_2$ is appropriate, and the half-power beamwidth of a rectangular aperture is

$$\beta_{xz} = 2\theta_2 \approx 2\sin\theta_2 = 0.88\frac{\lambda}{l_x} \qquad \text{(rad)}. \tag{3.56a}$$

A similar solution for the y–z plane ($\phi = \pi/2$) gives

$$\beta_{yz} = 0.88\,\frac{\lambda}{l_y} \qquad \text{(rad)}. \tag{3.56b}$$

> ▶ The uniform aperture distribution ($E_a = E_0$ across the aperture) gives a far-field pattern with the narrowest possible beamwidth. ◀

The first sidelobe level is 13.2 dB below the peak value (see Fig. 3-17), which is equivalent to 4.8% of the peak value. If the intended application calls for a pattern with a lower sidelobe level (to avoid interference with signals from sources along directions outside the main beam of the antenna pattern), this can be accomplished by using a ***tapered aperture distribution*** (Sections 3-7 and 3-8), one in which the electric field intensity is weighted over the aperture with a maximum at the center and decreases toward the edges.

> ▶ A tapered distribution provides a pattern with lower sidelobes, but the mainlobe becomes wider. ◀

As discussed later in Section 3-7, the steeper the taper, the lower are the sidelobes and the wider is the mainlobe. In general, the beamwidth in a given plane, say the x–z plane, is given by

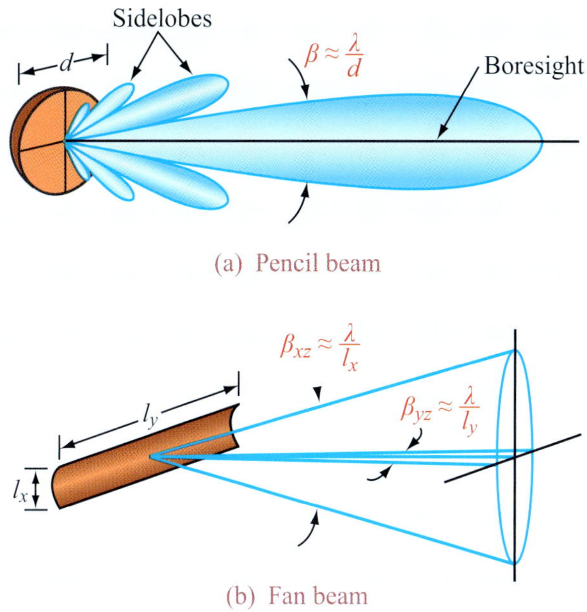

Figure 3-18: Radiation patterns of (a) a circular reflector and (b) a cylindrical reflector (sidelobes not shown).

$$\beta_{xz} = k_x\frac{\lambda}{l_x}\,, \tag{3.57}$$

where k_x is a constant related to the steepness of the taper. For a uniform distribution with no taper, $k_x - 0.88$, whereas for a highly tapered distribution, $k_x \approx 2$. In the typical case, $k_x \approx 1$.

To illustrate the relationship between the antenna dimensions and the corresponding beam shape, we show in Fig. 3-18 the radiation patterns of a circular reflector and a cylindrical reflector. The circular reflector has a circularly symmetric pattern, whereas the pattern of the cylindrical reflector has a narrow beam in the azimuth plane corresponding to its long dimension and a wide beam in the elevation plane corresponding to its narrow dimension. For a circularly symmetric antenna pattern, the beamwidth β is related to the diameter d by the approximate relation $\beta \approx \lambda/d$.

3-5.3 Directivity and Effective Area

In Section 3-2.3, we derived an approximate expression [Eq. (3.21)] for the antenna directivity D in terms of the half-power beamwidths β_{xz} and β_{yz} for antennas characterized by a single major lobe whose boresight is along the z direction:

$$D \approx \frac{4\pi}{\beta_{xz}\beta_{yz}}. \tag{3.58}$$

If we use the approximate relations $\beta_{xz} \approx \lambda/l_x$ and $\beta_{yz} \approx \lambda/l_y$, we obtain

$$D \approx \frac{4\pi l_x l_y}{\lambda^2} = \frac{4\pi A_p}{\lambda^2}. \tag{3.59}$$

More precisely, for any antenna, its directivity is related to its effective area A_e by Eq. (3.31b):

$$D = \frac{4\pi A_e}{\lambda^2}. \tag{3.60}$$

► As a first-order approximation, the effective area of an aperture antenna is equal to its physical aperture; that is, $A_e \approx A_p$. ◄

To compute D accurately, it is necessary to execute the following steps:

1. For a specified aperture distribution $E_a(x_a, y_a)$, use Eqs. (3.42) and (3.43) to determine the radiated field $E(R, \theta, \phi)$.

2. Compute power density $\mathcal{S}(R, \theta, \phi) = |E|^2/2\eta_0$, and then determine the direction at which $\mathcal{S}$ is at its maximum value (at a fixed R). Call it $\mathcal{S}_{max}$.

3. Obtain the normalized radiation intensity

$$F(\theta, \phi) = \left. \frac{\mathcal{S}(R, \theta, \phi)}{\mathcal{S}_{max}} \right|_{@ \text{ same } R}.$$

4. Use Eq. (3.18) to compute D.

5. Use Eq. (3.31b) to compute the effective area A_e.

6. The aperture efficiency is then given by $\xi_a = A_e/A_p$, where A_p is the physical aperture of the antenna.

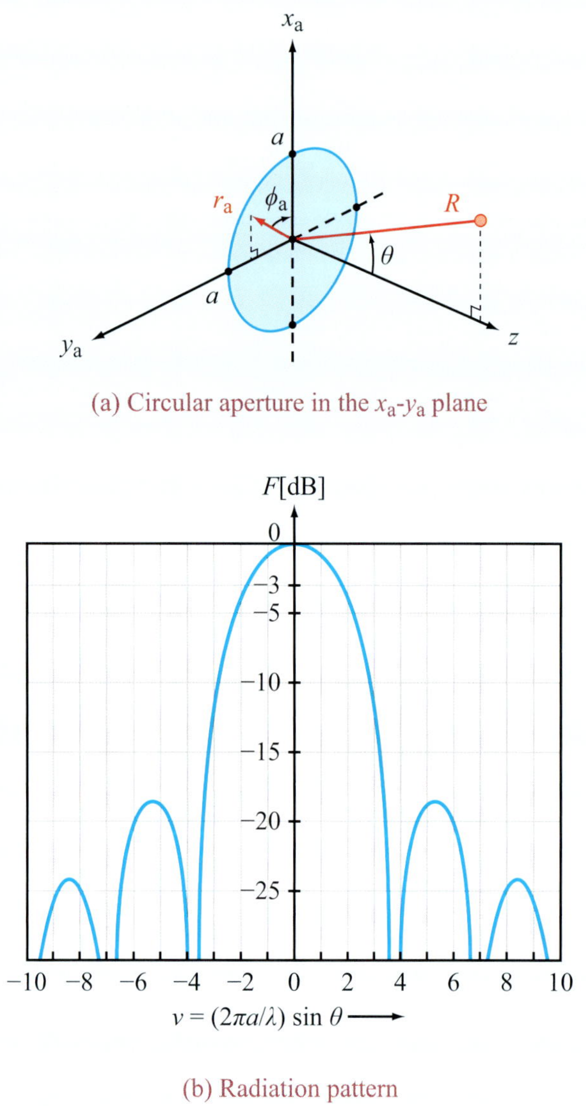

Figure 3-19: Circular aperture and its radiation pattern.

3-6 Circular Aperture with Uniform Field Illumination

For circularly symmetric aperture illuminations, the far-field patterns can be found most conveniently if we first express x_a and y_a in terms of the polar coordinates r_a and ϕ_a (Fig. 3-19a) as follows:

$$x_a = r_a \cos\phi_a, \quad (3.61a)$$

$$y_a = r_a \sin\phi_a. \quad (3.61b)$$

Substituting the above equations in Eq. (3.43) and replacing the elemental area $dx_a\, dy_a$ with its polar-coordinate equivalent $r_a\, dr_a\, d\phi_a$, the form factor $h(\theta,\phi)$ becomes

$$h(\theta,\phi) = \int_0^{2\pi}\int_0^{\infty} E_a(r_a) \times \exp[jkr_a \sin\theta(\cos\phi\cos\phi_a + \sin\phi\sin\phi_a)]r_a\, dr_a\, d\phi_a, \quad (3.62)$$

or equivalently,

$$h(\theta,\phi) = \int_0^{2\pi}\int_0^{\infty} E_a(r_a) \times \exp[jkr_a \sin\theta\cos(\phi-\phi_a)]r_a\, dr_a\, d\phi_a. \quad (3.63)$$

Using the Bessel-function identity

$$J_0(x) = \frac{1}{2\pi}\int_0^{2\pi} \exp[jx\cos(\phi-\phi_a)]\, d\phi_a,$$

where J_0 is a zero-order Bessel function of the first kind, the expression given by Eq. (3.63) can be simplified to:

$$h(\theta,\phi) = h(\theta) = 2\pi \int_0^{\infty} E_a(r_a)\, J_0(kr_a\sin\theta)\, r_a\, dr_a. \quad (3.64)$$

Note that because of the circular symmetry, the dependence on the angles ϕ and ϕ_a has disappeared. Evaluation of Eq. (3.64) as a function of θ leads to the radiation pattern shown in Fig. 3-19(b), plotted as a function of the variable $v = (2\pi a/\lambda)\sin\theta$. The width of the mainlobe between first nulls can be shown to be

$$\beta_{null} = 2[\sin^{-1}(0.61\lambda/a)] \approx 1.22\lambda/a = 2.44\lambda/d, \quad (3.65)$$

where $d = 2a$ is the aperture diameter.

In a similar fashion, the half-power beamwidth can be obtained by setting $F(v) = 0.5$, which leads to $v = 1.57$ and

$$\beta_{1/2} = 2[\sin^{-1}(0.25\lambda/a)] \approx 0.5\lambda/a = \lambda/d. \quad (3.66)$$

Following the same procedure used earlier in connection with the rectangular aperture, the maximum directivity of the circular aperture can be shown to be

$$D = \frac{4\pi A_p}{\lambda^2}, \quad (3.67)$$

and in terms of $\beta_{1/2}$,

$$D = 0.78\,\frac{4\pi}{\beta_{1/2}^2}. \quad (3.68)$$

3-7 Nonuniform-Amplitude Illumination

By controlling the shape of the amplitude illumination, the antenna designer can optimize certain features of interest such as the antenna directivity, the sidelobe level, or the location of the nulls in the radiation pattern. Since these features are interrelated, the optimization process is determined by the intended application.

To illustrate the influence of the amplitude illumination on the far-field pattern of a rectangular aperture, we consider some commonly used distributions. For simplicity, we limit our discussion to the x–z plane ($\phi = 0$), uniform-phase illumination, and unity-amplitude illumination in the y_a direction. That is, we define $E_a(x_a, y_a)$ as a product of ***separable distributions***,

$$E_a(x_a, y_a) = E_1(x_a)\, E_2(y_a), \quad (3.69)$$

and then we set $E_2(y_a) = 1$.

Introducing the new variables

$$x_1 = \frac{2}{l_x}\, x_a, \quad (3.70a)$$

$$u_1 = \frac{\pi l_x}{\lambda}\sin\theta, \quad (3.70b)$$

the form factor $h(\theta,\phi)$ given by Eq. (3.43) becomes

$$h(\theta) = \frac{l_x l_y}{2}\int_{x_1=-1}^{1} E_1(x_1)\, e^{ju_1x_1}\, dx_1 \qquad (at\ \phi = 0). \quad (3.71)$$

Table 3-1: **Radiation patterns produced by various types of amplitude distributions along x_1 over a rectangular aperture $l_x \times l_y$.**

	Amplitude distribution[a]	Relative directivity[b]	Sidelobe level (dB)[c]	Half-power (−3 dB) beamwidth (radians)[d]
Cosine: $E_1(x_1) = \cos^n(\pi x_1/2)$				
	$n = 0^{\text{e}}$	1.00	13.2	$0.88\lambda/l_x$
	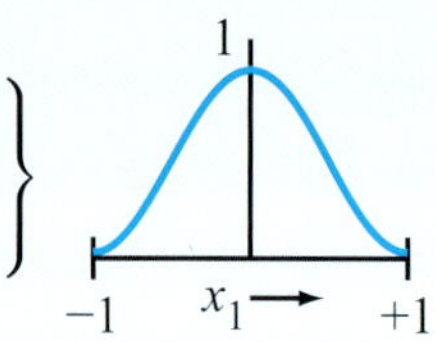			
	$n = 1$	0.81	23	$1.20\lambda/l_x$
	2	0.67	32	$1.45\lambda/l_x$
	3	0.58	40	$1.66\lambda/l_x$
	4	0.52	48	$1.94\lambda/l_x$
Parabolic: $E_1(x_1) = 1 - (1-\Delta)x_1^2$				
	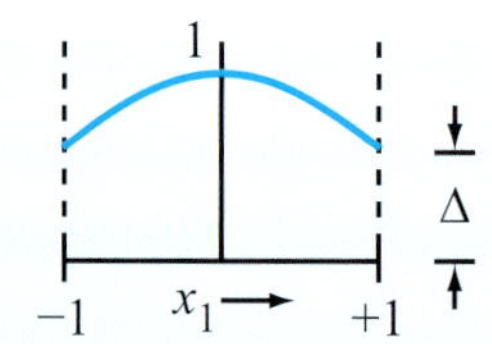			
	$\Delta = 1.0^{\text{e}}$	1.00	13.2	$0.88\lambda/l_x$
	0.8	0.99	15.8	$0.92\lambda/l_x$
	0.5	0.97	17.1	$0.97\lambda/l_x$
	0	0.83	20.6	$1.15\lambda/l_x$
Triangular: $E_1(x_1) = 1 - \lvert x_1 \rvert$				
	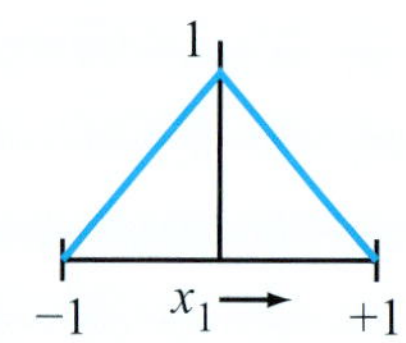	0.75	26.4	$1.28\lambda/l_x$

[a]The variable $x_1 = (2/l_x)x_a$, and $|x_1| \le 1$. [b]Relative to a uniform aperture distribution. [c]Below maximum intensity. [d]In x–z plane. [e]Same as uniform-distribution case.

Antenna radiation performance usually is described in terms of its normalized radiation intensity $F(\theta)$, which is proportional to $|h(\theta)|^2$.

Table 3-1 provides a summary of three important characteristics of $|h(u_1)|^2$ for several classes of amplitude illumination $E_1(x_1)$. These characteristics are the directivity D relative to the uniform illumination case, the sidelobe level in dB below peak intensity, and the half-power beamwidth $\beta_{1/2}$. Complete expressions for $h(u_1)$ for these and other illuminations are available in the literature (Silver, 1949; Hansen, 1964).

For the first class of illumination in Table 3-1, $E_1(x_1) = \cos^n(\pi x_1/2)$, the illumination decreases from 1.0 at the center of the aperture ($x_1 = 0$) to zero at the edges ($x_1 = \pm 1$) for $n > 1$. As n increases, faster roll-off illumination tapers are produced, resulting in lower sidelobe levels, at the expense, however, of lower directivities and wider beamwidths. This is true also for the second class of illumination included in Table 3-1

Table 3-2: Radiation patterns produced by the amplitude distribution function $E(r_1) = (1 - r_1^2)^n$ over a circular aperture of diameter d.

n	Relative directivity[a]	Sidelobe level (dB)[b]	Half-power (−3 dB) beamwidth (radians)
0 (uniform)	1.00	17.6	$1.02\lambda/d$
1	0.75	24.6	$1.27\lambda/d$
2	0.55	30.6	$1.47\lambda/d$
3	0.45	36.1	$1.65\lambda/d$

The variable $r_1 = (2/d)r_a$ and $0 \le r_1 \le 1$. [a]Relative to a uniform aperture distribution. [b]Below maximum intensity.

as well as for the illumination given in Table 3-2 for circular apertures.

Note that for the amplitude distributions listed in Table 3-1, the taper occurs along x_a only. If an identical taper occurs along y_a as well, the radiation pattern exhibits similar first sidelobe levels and half power beamwidths in the y–z plane (with $\beta_{1/2}$ being inversely proportional to l_y), and the relative directivity becomes the squared value of the value listed in Table 3-1. For example, if $E_a(x_a, y_a)$ has a cosine taper with $n = 1$ along both x_a and y_a, the relative directivity becomes $(0.81)^2 = 0.66$. Also, the effective aperture becomes approximately

$$A_e \approx \frac{\lambda^2}{\beta_{xz}\beta_{yz}} = \frac{\lambda^2}{(1.2\lambda/\lambda_x)(1.2\lambda/\lambda_y)} \approx 0.69 l_x l_y = 0.69 A_p.$$

3-8 Beam Efficiency

Consider a square aperture with $l_x = l_y = l$ situated in the x_a–y_a plane. The aperture dimension l is at least several wavelengths long; consequently, the radiation pattern is highly directional and has a narrow mainlobe. Figure 3-20 displays plots of the radiation pattern in the x–z plane for two types of aperture illuminations, a uniform illumination and a cosine distribution with $n = 1$ (see Table 3-1). In each case, the same illumination is applied along aperture axis x_a as along y_a. Hence, the radiation patterns are the same in the x–z and y–z planes.

The uniform-illumination radiation pattern, labeled $F_u(\gamma)$ in Fig. 3-20, is an exact replica of the pattern shown earlier in Fig. 3-17. Its half-power beamwidth is $\beta_u = \beta_{xz} = \beta_{yz} = 0.88\lambda/l$. If $l = 10\lambda$, $\beta_u = 0.088$ rad $\approx 5°$. The first sidelobe is 13.2 dB below the maximum value of $F_u(\gamma)$, corresponding to a level of only 4.8%, which means that the power radiated by such an antenna along the direction corresponding to the peak of the first sidelobe is less than 5% of that radiated along the main-beam axis. This may be quite adequate for many applications, but not for microwave radiometry.

The radiation pattern labeled $F_c(\gamma)$ in Fig. 3-20 corresponds to a cosine-tapered illumination with $n = 1$ along both x_a and y_a. Its half-power beamwidth is $\beta_c = 1.2\lambda/l$, which is 36% wider than that of the

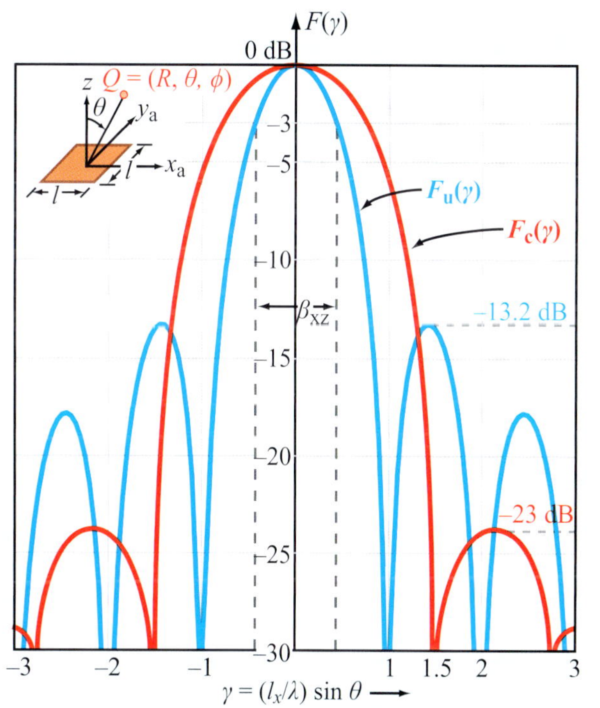

Figure 3-20: Comparison of the antenna pattern $F_u(\gamma)$ of a uniformly illuminated aperture with $F_c(\gamma)$ for a cosine-tapered illuminated aperture.

uniformly illuminated aperture, but its first sidelobe is at 23 dB below the peak (or 0.5%), which is 10 times better than that of the uniformly illuminated aperture. Lowering the sidelobes is equivalent to "squeezing" their pattern content into the mainlobe, causing it to become wider.

For an antenna with radiation pattern $F(\theta,\phi)$, the ***beam efficiency*** η_b is defined as

$$\eta_b = \frac{\text{power radiated in cone with half-angle } \theta_l}{\text{total power radiated in } 4\pi \text{ sr}} = \frac{\displaystyle\int_0^{2\pi}\int_0^{\theta_l} F(\theta,\phi)\sin\theta\, d\theta\, d\phi}{\displaystyle\int_0^{2\pi}\int_0^{\pi} F(\theta,\phi)\sin\theta\, d\theta\, d\phi} = \frac{\Omega_m}{\Omega_p}, \tag{3.72}$$

where Ω_m and Ω_p are the ***main-beam solid angle*** and ***pattern solid angle***, respectively, and angle θ_l is selected such that the cone encompassed within it is a good approximation of the mainlobe of the antenna pattern. Typically, θ_l is defined as the angle between the beam axis and the first minimum.

For $F_u(\gamma)$ in Fig. 3-20, θ_l corresponds to $\gamma_l = 1$, and since $\sin\theta \approx \theta$, it follows that

$$\theta_{l_u} = \frac{\lambda}{l} \qquad \textbf{(uniform illumination)}. \tag{3.73a}$$

The first null of the tapered-illumination radiation pattern occurs at $\gamma_l = 1.5$. Hence,

$$\theta_{l_c} = 1.5\,\frac{\lambda}{l}\,. \tag{3.73b}$$

(cosine illumination with $n = 1$)

To illustrate with an example, let us consider a uniformly illuminated square antenna with $l = 10\lambda$. Computing Eq. (3.72) with $\theta_l = 0.1$ rad and

$$\begin{aligned} F(\theta,\phi) &= \text{sinc}^2\left(\frac{\pi l_x}{\lambda}\sin\theta\cos\phi\right)\text{sinc}^2\left(\frac{\pi l_y}{\lambda}\sin\theta\sin\phi\right) \\ &= \text{sinc}^2(10\pi\sin\theta\cos\phi)\,\text{sinc}^2(10\pi\sin\theta\sin\phi), \end{aligned} \tag{3.74}$$

leads to $\eta_b \approx 0.8$, or 80%. This means that 80% of the total power radiated by the antenna is radiated along directions within the mainlobe, and the remaining 20% is lost to other directions. Conversely, if the antenna is operating in a receiving mode and power is incident upon it uniformly from all directions, 20% of the total power it receives comes in from directions outside the mainlobe.

A similar evaluation carried out for the cosine-tapered illumination with $n = 1$ leads to a beam efficiency $\eta_b \approx 93\%$, and for $n = 2$, $\eta_b \approx 98\%$.

► The application of steep illumination tapers can raise the beam efficiency to ≈ 100%, at the expense of a wider mainlobe and a smaller directivity. ◄

Nevertheless, as we discuss later in Chapter 6, high beam efficiency is critically important for making accurate measurements with a microwave radiometer.

3-9 Antenna Arrays

► When two or more antennas are used together, the combination is called an ***antenna array***. ◄

Even though an array need not consist of similar radiating elements, most arrays actually use identical elements, such as dipoles, slots, horn antennas, or parabolic dishes. The antenna elements of an array may be arranged in various configurations, but the most common are the linear one-dimensional configuration—wherein the elements are arranged along a straight line—and the two-dimensional lattice configuration in which the elements sit on a planar grid. The desired shape of the far-field radiation pattern of the array can be synthesized by controlling the relative amplitudes of the array elements' excitations.

▶ Through the use of electronically controlled solid-state phase shifters, the beam direction of the antenna array can be steered electronically by controlling the relative phases of the array elements. ◀

This flexibility of the array antenna has led to numerous applications, including ***electronic steering*** and ***multiple-beam generation***.

The purpose of this and the next two sections is to introduce the reader to the basic principles of array theory and design techniques used in shaping the antenna pattern and steering the mainlobe. The presentation is confined to the one-dimensional linear array with equal spacing between adjacent elements.

A linear array of N identical radiators is arranged along the z axis as shown in Fig. 3-21. The radiators are fed by a common oscillator through a branching network. In each branch, an attenuator (or amplifier) and phase shifter are inserted in series to control the amplitude and phase of the signal feeding the antenna element in that branch.

In the far-field region of any radiating element, the ***element electric-field intensity*** $E_e(R,\theta,\phi)$ may be expressed as a product of two functions: the spherical propagation factor e^{-jkR}/R, which accounts for the dependence on the range R, and $f_e(\theta,\phi)$, which accounts for the directional dependence of the element's electric field. Thus, for an isolated element, the radiated field is

$$E_e(R,\theta,\phi) = \frac{e^{-jkR}}{R} f_e(\theta,\phi), \tag{3.75}$$

and the corresponding power density $\mathcal{S}_e$ is

$$\mathcal{S}_e(R,\theta,\phi) = \frac{1}{2\eta_0}|E_e(R,\theta,\phi)|^2 = \frac{1}{2\eta_0 R^2}|f_e(\theta,\phi)|^2. \tag{3.76}$$

For the array shown in Fig. 3-21(b), the far-zone field due to element i at observation point Q at range R_i is

$$E_i(R_i,\theta,\phi) = A_i \frac{e^{-jkR_i}}{R_i} f_e(\theta,\phi), \tag{3.77}$$

where $A_i = a_i e^{j\psi_i}$ is a ***complex feeding coefficient*** representing the amplitude a_i and phase ψ_i of the

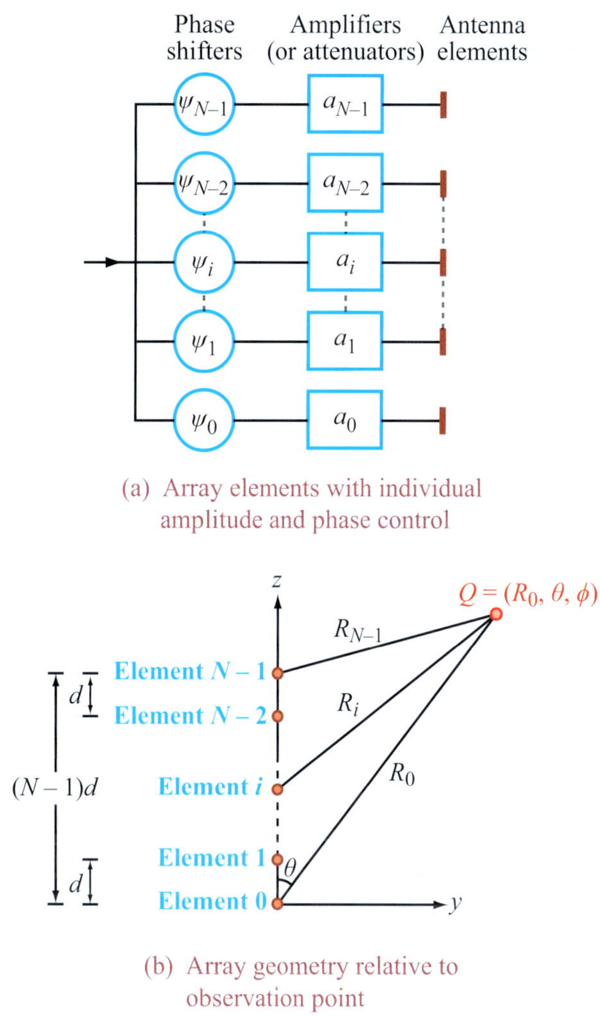

Figure 3-21: Linear-array configuration and geometry.

excitation giving rise to E_i, relative to the phase of a reference excitation. In practice, the excitation of one of the elements is used as reference. Note that R_i and A_i may be different for different elements in the array, but $f_e(\theta,\phi)$ is the same for all of the antenna elements because they are identical, and hence exhibit identical directional patterns.

The total field at the observation point $Q(R_0, \theta, \phi)$ is the sum of the fields due to the N elements:

$$E(R_0, \theta, \phi) = \sum_{i=0}^{N-1} E_i(R_i, \theta, \phi)$$
$$= \left[\sum_{i=0}^{N-1} A_i \frac{e^{-jkR_i}}{R_i} \right] f_e(\theta, \phi), \quad (3.78)$$

where R_0 denotes the range of Q from the center of the coordinate system, chosen to be at the location of the zeroth element. To satisfy the far-field condition given by Eq. (3.41) for an array of length $l = (N-1)d$, where d is the interelement spacing, the range R_0 should be sufficiently large to satisfy

$$R_0 \geq \frac{2l^2}{\lambda} = \frac{2(N-1)^2 d^2}{\lambda} . \quad (3.79)$$

This condition allows us to ignore differences in the distances from Q to the individual elements as far as the magnitudes of the radiated fields are concerned. Thus, we can set $R_i = R_0$ in the denominator in Eq. (3.78) for all i. With regard to the phase part of the propagation factor, we can use the parallel-ray approximation given by

$$R_i \approx R_0 - z_i \cos\theta = R_0 - id \cos\theta, \quad (3.80)$$

where $z_i = id$ is the distance between the ith element and the zeroth element (Fig. 3-22). Employing these two approximations in Eq. (3.78) leads to

$$E(R_0, \theta, \phi) = f_e(\theta, \phi) \left(\frac{e^{-jkR_0}}{R_0} \right) \left[\sum_{i=0}^{N-1} A_i e^{jikd\cos\theta} \right], \quad (3.81)$$

and the corresponding array-antenna power density is given by

$$\mathcal{S}(R_0, \theta, \phi) = \frac{1}{2\eta_0} |E(R_0, \theta, \phi)|^2$$
$$= \frac{1}{2\eta_0 R_0^2} |f_e(\theta, \phi)|^2 \left| \sum_{i=0}^{N-1} A_i e^{jikd\cos\theta} \right|^2 .$$

Use of Eq. (3.76) leads to

$$\mathcal{S}(R_0, \theta, \phi) = \mathcal{S}_e(R_0, \theta, \phi) \left| \sum_{i=0}^{N-1} A_i e^{jikd\cos\theta} \right|^2 . \quad (3.82)$$

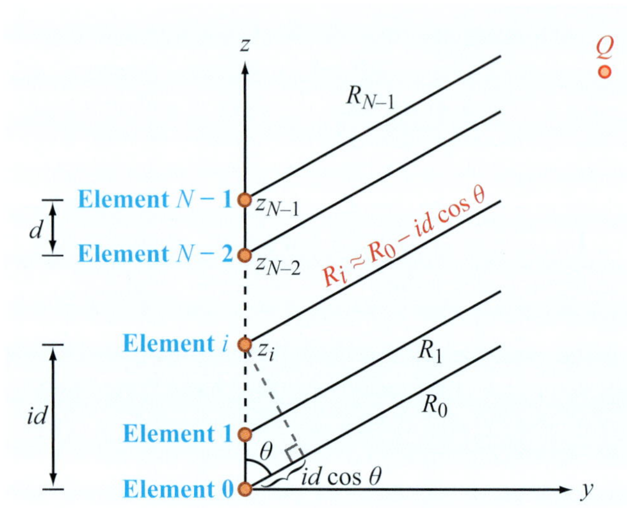

Figure 3-22: The rays between the elements and a far-away observation point are approximately parallel lines. Hence, the distance $R_i \approx R_0 - id\cos\theta$.

This expression is a product of two factors. The first factor, $\mathcal{S}_e(R_0, \theta, \phi)$, is the power density of the energy radiated by an individual element, and the second, called the ***array factor***, is a function of the positions of the individual elements and their feeding coefficients, but not a function of the specific type of radiator used.

> ▶ The array factor represents the far-field radiation intensity of the N elements, had the elements been isotropic radiators. ◀

Denoting the ***array factor*** by

$$F_a(\theta) = \left| \sum_{i=0}^{N-1} A_i e^{jikd\cos\theta} \right|^2 , \quad (3.83)$$

the power density of the antenna array is then written as

$$\mathcal{S}(R_0, \theta, \phi) = \mathcal{S}_e(R_0, \theta, \phi)\, F_a(\theta). \quad (3.84)$$

This equation demonstrates the ***pattern multiplication principle***. It allows us to find the far-field power density of the antenna array by first computing the far-field power pattern with the array elements replaced with isotropic radiators, which yields the array factor $F_a(\theta)$, and then multiplying the result by $\mathcal{S}_e(R_0, \theta, \phi)$, the power density for a single element (which is the same for all elements).

The feeding coefficient A_i is, in general, a complex amplitude consisting of an amplitude factor a_i and a phase factor ψ_i:

$$A_i = a_i e^{j\psi_i}. \tag{3.85}$$

Insertion of Eq. (3.85) into Eq. (3.83) leads to

$$F_a(\theta) = \left| \sum_{i=0}^{N-1} a_i e^{j\psi_i} e^{jikd\cos\theta} \right|^2 .^{\dagger} \tag{3.86}$$

The array factor is governed by two input functions: the ***array amplitude distribution*** given by the a_i's and the ***array phase distribution*** given by the ψ_i's. The amplitude distribution serves to control the shape of the array radiation pattern, while the phase distribution can be used to steer its direction.

3-10 N-Element Array with Uniform Phase Distribution

We now consider an array of N elements with equal spacing d and equal-phase excitations; that is, $\psi_i = \psi_0$ for $i = 1, 2, \ldots, (N-1)$. Such an array of ***in-phase elements*** is sometimes referred to as a ***broadside array*** because the main beam of the radiation pattern of its array factor is always in the direction broadside to the array axis. From Eq. (3.86), the array factor is given by

$$\begin{aligned} F_a(\theta) &= \left| e^{j\psi_0} \sum_{i=0}^{N-1} a_i e^{jikd\cos\theta} \right|^2 \\ &= |e^{j\psi_0}|^2 \left| \sum_{i=0}^{N-1} a_i e^{jikd\cos\theta} \right|^2 = \left| \sum_{i=0}^{N-1} a_i e^{jikd\cos\theta} \right|^2 . \end{aligned} \tag{3.87}$$

† **Computer Code 3.1.**

The phase difference between the fields radiated by adjacent elements is

$$\gamma = kd\cos\theta = \frac{2\pi d}{\lambda}\cos\theta. \tag{3.88}$$

In terms of γ, Eq. (3.87) takes the compact form

$$F_a(\gamma) = \left| \sum_{i=0}^{N-1} a_i e^{ji\gamma} \right|^2 \quad \textbf{(uniform phase)}. \tag{3.89}$$

3-10.1 Uniform Amplitude Distribution

For a uniform amplitude distribution with $a_i = 1$ for $i = 0, 1, \ldots, (N-1)$, Eq. (3.89) becomes

$$F_a(\gamma) = |1 + e^{j\gamma} + e^{j2\gamma} + \cdots + e^{j(N-1)\gamma}|^2. \tag{3.90}$$

This geometric series can be rewritten (Ulaby et al., 2010) in the compact form

$$F_a(\gamma) = |f_a(\gamma)|^2, \tag{3.91}$$

with

$$f_a(\gamma) = e^{j(N-1)\gamma/2} \, \frac{\sin(N\gamma/2)}{\sin(\gamma/2)} . \tag{3.92}$$

After multiplying $f_a(\gamma)$ by its complex conjugate, we obtain the following result for the array factor:

$$F_a(\gamma) = \frac{\sin^2(N\gamma/2)}{\sin^2(\gamma/2)} . \tag{3.93}$$

(uniform amplitude and phase)

From Eq. (3.90), $F_a(\gamma)$ is maximum when all terms are 1, which occurs when $\gamma = 0$ (or equivalently, $\theta - \pi/2$). Moreover, $F_a(0) = N^2$. Hence, the normalized array factor is given by

$$\begin{aligned} F_{an}(\gamma) &= \frac{F_a(\gamma)}{F_{a,\max}} = \frac{\sin^2(N\gamma/2)}{N^2 \sin^2(\gamma/2)} \\ &= \frac{\sin^2\left(\dfrac{N\pi d}{\lambda}\cos\theta\right)}{N^2 \sin^2\left(\dfrac{\pi d}{\lambda}\cos\theta\right)} . \end{aligned} \tag{3.94}$$

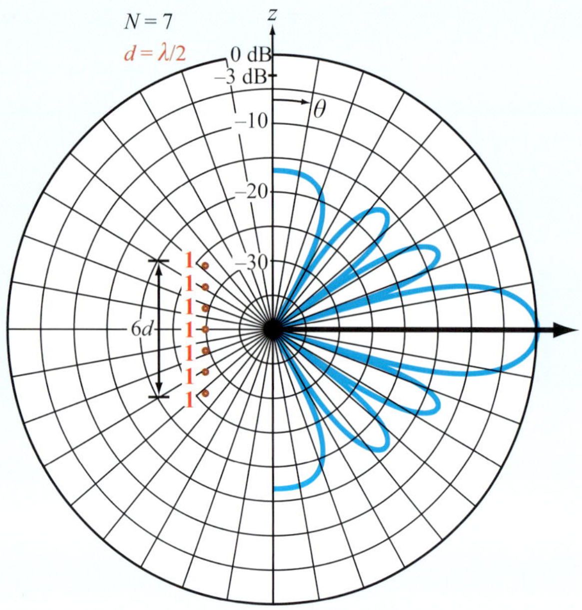

(a) Uniform distribution, $d = \lambda/2$

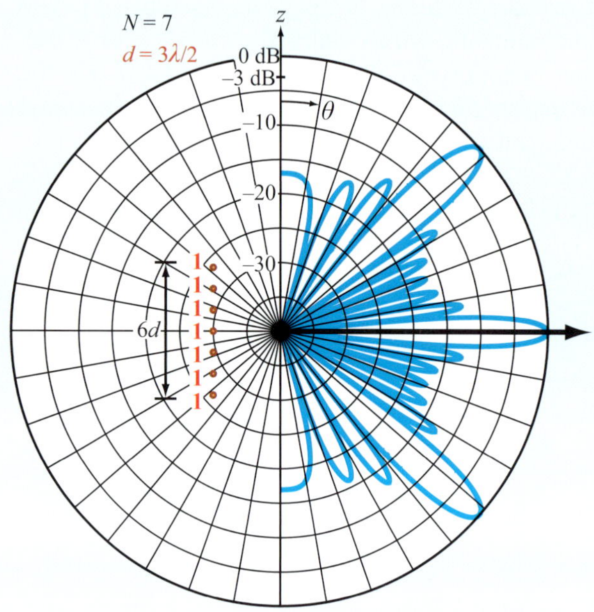

(b) Uniform distribution, $d = 3\lambda/2$

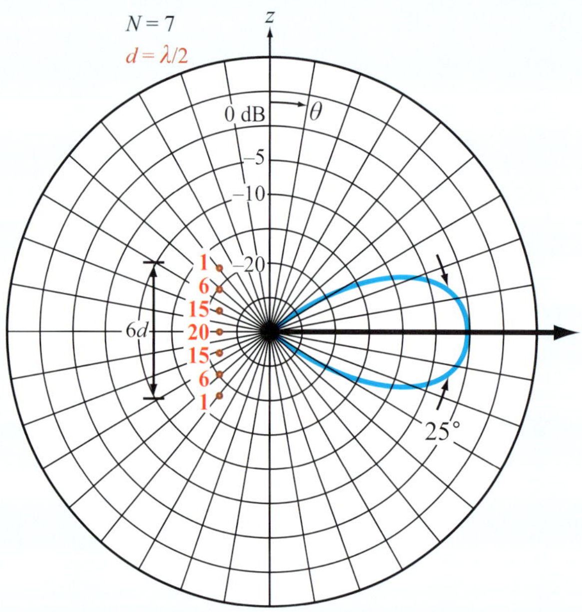

(c) Binomial distribution

Figure 3-23: Normalized array patterns of 7-element arrays: (a) uniform distribution with $d = \lambda/2$, (b) uniform distribution with $d = 3\lambda/2$, and (c) binomial distribution with $d = \lambda/2$.

A polar plot of $F_{an}(\theta)$ is shown in Fig. 3-23(a) for $N = 7$ and $d = \lambda/2$. The reader is reminded that this is a plot of the radiation pattern of the array factor alone; the pattern for the antenna array is equal to the product of this pattern and that of a single element, as discussed earlier in connection with the pattern multiplication principle.

3-10.2 Grating Lobes

Even though Eq. (3.94) is defined for all values of γ, only the interval $0 \leq \theta \leq \pi$ pertains to the real antenna situation. This interval is called the ***visible*** region of the antenna array. The array pattern shown in Fig. 3-23(a) for the uniform array with $d = \lambda/2$ consists of a single pattern, whereas if the interelement spacing is increased to $d = 3\lambda/2$, the pattern gets repeated three times, as shown in Fig. 3-23(b). The multiple patterns, called ***grating lobes***, are introduced by the repetitive behavior of the sine function in the denominator of Eq. (3.94). The grating lobes are spaced 2π apart along the γ axis. To avoid grating lobes, it is necessary that

$$|\gamma| = \left| \left(\frac{2\pi d}{\lambda} \right) \cos\theta \right| \leq 2\pi.$$

Over the visible region, $0 \leq \theta \leq \pi$, the above inequality is satisfied if $d \leq \lambda$. For $d = \lambda$, two grating lobes exist, one on each side of the mainlobe, with their peaks (of the same amplitude as that of the mainlobe) pointing along the two directions of the array axis. Since the radiation pattern of the array antenna is the product of the array factor and the radiation pattern of an individual element, grating lobes may be unimportant if the element pattern has small values (relative to its peak value) in those directions. *When phase delay is introduced between adjacent array elements at their inputs in order to steer the antenna beam into a direction other than broadside (Section 3-11), avoiding grating lobes requires that $d \leq \lambda/2$.* This condition is analogous to the Nyquist sampling theorem in communication theory (Ulaby and Yagle, 2013).

3-10.3 Binomial Distribution

In some applications, it may be preferable to have a less directive array factor than that of the uniform array, but with lower sidelobe levels. To reduce sidelobe levels, several nonuniform amplitude distributions have been proposed. Among these is the ***binomial distribution***, so designated because the a_i's, the amplitudes of array elements, are proportional to the binomial coefficients:

$$a_i = \frac{(N-1)!}{i!(N-i-1)!}$$

with $i = 0, 1, \ldots, N-1$. To illustrate with an example, the relative amplitudes of a seven-element binomial array are $a_0 = a_6 = 1$, $a_1 = a_5 = 6$, $a_2 = a_4 = 15$, and $a_3 = 20$. With these coefficients, the array factor defined by Eq. (3.89) becomes

$$F_a(\gamma) = \left|1 + 6e^{j\gamma} + 15e^{j2\gamma} + 20e^{j3\gamma} + 15e^{j4\gamma} + 6e^{j5\gamma} + e^{j6\gamma}\right|^2. \tag{3.95}$$

Transferring the array phase reference from the edge to the center by factoring out $e^{j3\gamma}$, the array factor can be expressed in terms of cosine functions of γ:

$$\begin{aligned} F_a(\gamma) &= \left|e^{j3\gamma}(e^{-j3\gamma} + 6e^{-j2\gamma} + 15e^{-j\gamma} + 20 + 15e^{j\gamma} + 6e^{j2\gamma} + e^{j3\gamma})\right|^2 \\ &= (20 + 30\cos\gamma + 12\cos 2\gamma + 2\cos 3\gamma)^2. \end{aligned}$$

The pattern maximum occurs at $\gamma = 0$, and the corresponding peak value is $F_a(0) = (64)^2 = 4096$. The normalized array factor, $F_{an}(\gamma) = F_a(\gamma)/F_a(0)$, is shown in Fig. 3-23(c) for $d = \lambda/2$. The binomial array exhibits no minor lobes, but its half-power beamwidth is substantially larger than that of the uniform array.

3-11 Electronic Scanning of Arrays

The discussion in the preceding section was concerned with uniform-phase arrays, in which the phases of the feeding coefficients, ψ_0 to ψ_{N-1}, are all equal. In this section, we examine the use of phase delay between adjacent elements as a tool to ***electronically steer*** the direction of the array-antenna beam from broadside at $\theta = 90^\circ$ to any desired angle θ_0. By eliminating the need to mechanically steer an antenna to change its beam's

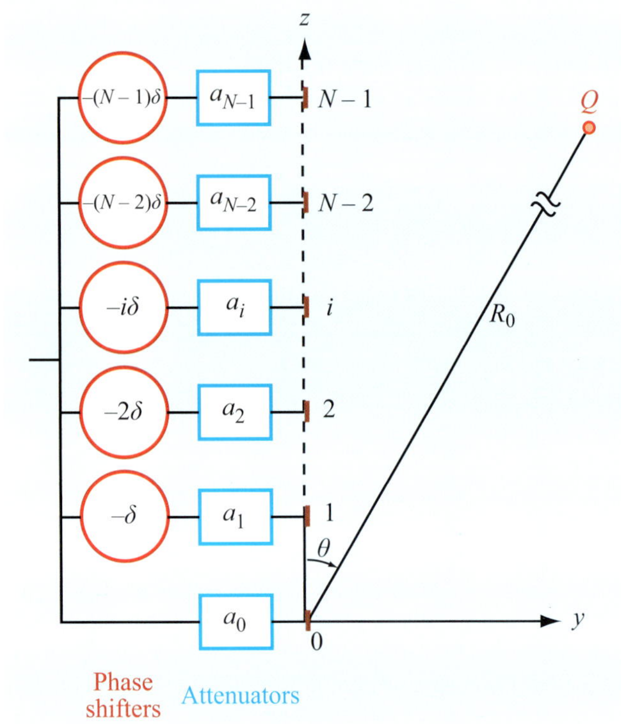

Figure 3-24: The application of linear phase.

direction, electronic steering allows beam scanning at very fast rates.

▶ Electronic steering is achieved by applying a ***linear phase distribution*** across the array: $\psi_0 = 0$, $\psi_1 = -\delta$, $\psi_2 = -2\delta$, etc. ◀

As shown in Fig. 3-24, the phase of the ith element, relative to that of the zeroth element, is

$$\psi_i = -i\delta, \tag{3.96}$$

where δ is the ***incremental phase delay*** between adjacent elements. Use of Eq. (3.96) in Eq. (3.86) leads to

$$\begin{aligned} F_{\rm a}(\theta) &= \left| \sum_{i=0}^{N-1} a_i e^{-ji\delta} e^{jikd\cos\theta} \right|^2 \\ &= \left| \sum_{i=0}^{N-1} a_i e^{ji(kd\cos\theta-\delta)} \right|^2 \\ &= \left| \sum_{i=0}^{N-1} a_i e^{ji\gamma'} \right|^2 = F_{\rm a}(\gamma'),^{\dagger} \end{aligned} \tag{3.97}$$

where we introduced a new variable given by

$$\gamma' = kd\cos\theta - \delta. \tag{3.98}$$

For reasons that become clear later, we define the phase shift δ in terms of an angle θ_0, which we call the ***scan angle***, as follows:

$$\delta = kd\cos\theta_0. \tag{3.99}$$

Hence, γ' becomes

$$\gamma' = kd(\cos\theta - \cos\theta_0). \tag{3.100}$$

The array factor given by Eq. (3.97) has the same functional form as the array factor developed earlier for the uniform-phase array [see Eq. (3.89)], except that γ is replaced with γ'. Hence:

▶ For any amplitude distribution across an array, the array factor $F_{\rm a}(\gamma')$ when excited by a linear-phase distribution can be obtained from $F_{\rm a}(\gamma)$, the expression developed for the array assuming a uniform-phase distribution, by replacing γ with γ'. ◀

If the amplitude distribution is symmetrical with respect to the array center, the array factor $F_{\rm a}(\gamma')$ is maximum when its argument $\gamma' = 0$. When the phase is uniform ($\delta = 0$), this condition corresponds to the direction

† **Computer Code 3.2.**

$\theta = 90^\circ$, which is why the uniform-phase arrangement is called a broadside array. According to Eq. (3.100), in a linearly phased array, $\gamma' = 0$ when $\theta = \theta_0$. Thus, by applying linear phase across the array, the array pattern is shifted along the $\cos\theta$ axis by an amount $\cos\theta_0$, and the direction of maximum radiation is ***steered*** from the broadside direction ($\theta = 90^\circ$) to the direction $\theta = \theta_0$. To steer the beam all the way to the ***end-fire direction*** ($\theta = 0$), the incremental phase shift δ should be equal to kd radians.

To illustrate the process with an example, consider the case of the N-element array excited by a uniform-amplitude distribution. Its normalized array factor is given by Eq. (3.94). Upon replacing γ with γ', we have

$$F_{\text{an}}(\gamma') = \frac{\sin^2(N\gamma'/2)}{N^2\sin^2(\gamma'/2)}, \tag{3.101}$$

with γ' as defined by Eq. (3.100). For an array with $N = 10$ and $d = \lambda/2$, plots of the mainlobe of $F_{\text{an}}(\theta)$ are shown in Fig. 3-25 for $\theta_0 = 0^\circ$, 45°, and 90°. We note that the half-power beamwidth increases as the array beam is steered from broadside to end fire.

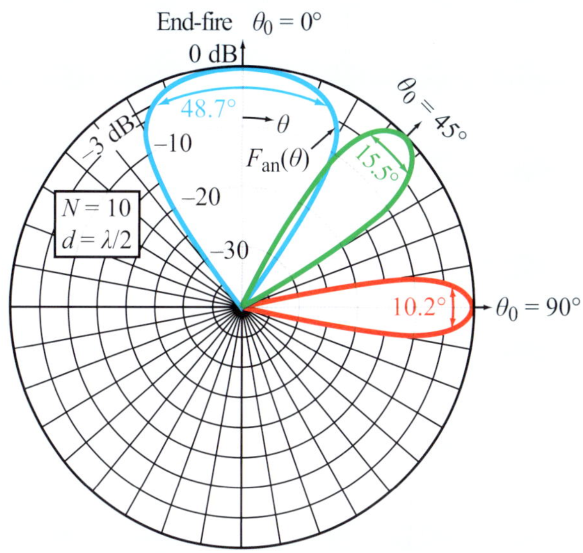

Figure 3-25: Normalized array pattern of a 10-element array with $\lambda/2$ spacing between adjacent elements. All elements are excited with equal amplitude. Through the application of linear phase across the array, the main beam can be steered from the broadside direction ($\theta_0 = 90^\circ$) to any scan angle θ_0. Equiphase excitation corresponds to $\theta_0 = 90^\circ$.

3-12 Antenna Types

Antennas are constructed in a wide variety of geometrical configurations and feeding arrangements. A specific antenna type usually is chosen on the basis of its radiation pattern, gain, polarization, impedance, and bandwidth. In some applications, geometry also may be important. For example, the antenna used for a side-looking airborne radar (SLAR) needs to be long in the direction of flight (so as to produce a narrow beam in the along-track direction) and narrow in height (so as to illuminate a wide swathwidth on the ground). Although both the waveguide slot array antenna and the cylindrical reflector antenna fed by a dipole array can provide the desired illumination pattern, the waveguide slot array or microstrip patch array (Section 3-12.3) are more suitable aerodynamically.

A thorough treatment of the various types of antennas used in microwave remote sensing would constitute a sizable volume in itself. Therefore, we have chosen to consider below two basic types of antennas—the electromagnetic horn and the slot antenna—as examples with which to illustrate how the techniques discussed in previous sections may be applied to determine their radiation properties. More detailed treatments of these and other types of antennas, including reflector, log-periodic, and lens antennas, are contained in the books and handbooks listed at the end of the book.

3-12.1 Horn Antennas

Electromagnetic horns are widely used in microwave communication and remote-sensing systems. Small horns with modest gain are used commonly as primary antennas to illuminate ("feed") large apertures such as reflector-type antennas, while large horns with high gain are employed on their own in some applications.

The horn antenna provides a gradual transition between a waveguide and free space. If the waveguide

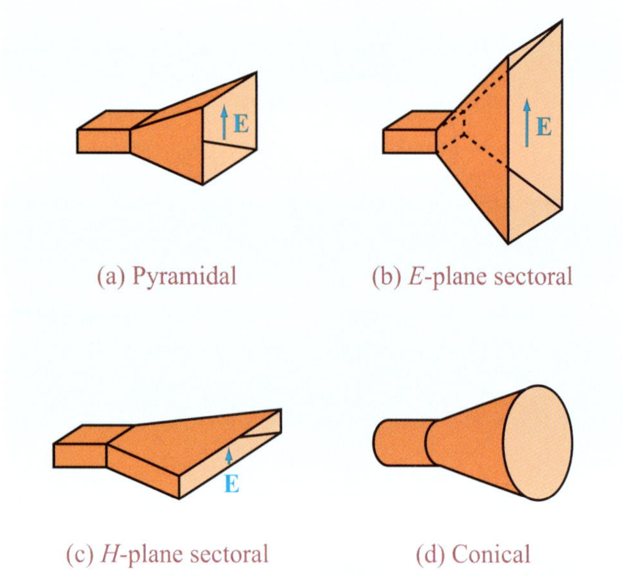

Figure 3-26: Commonly used types of horn antennas.

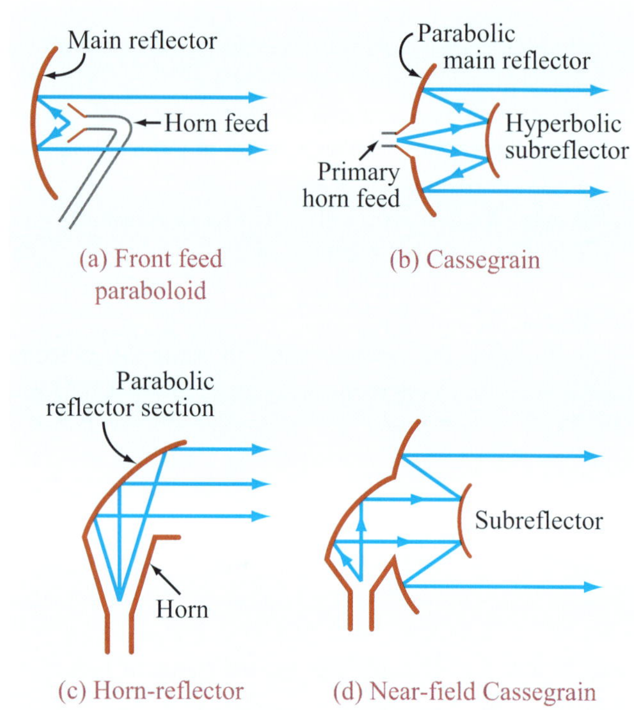

Figure 3-27: Horn-reflector antennas.

dimensions are such that it can support only the dominant propagation mode, then by gradually flaring the terminal section of the waveguide, the excitation of higher-order modes may be avoided. Thus, a horn can provide simultaneously single-mode propagation (which is difficult to achieve with an oversize open-ended waveguide) as well as a large radiating aperture (compared to the single-mode waveguide). Furthermore, horns are broader-band antennas than dipole and slot antennas.

Horns are constructed in a variety of shapes; some of the more frequently used types are illustrated in Fig. 3-26. The combination of size and shape of a horn dictates its directivity, impedance, shape of radiation pattern, and polarization properties. The horn shown in Fig. 3-26(a) is known as a ***pyramidal horn***. It is often called a "standard-gain horn" because it is commonly used in antenna measurements as a reference antenna of known gain. A pyramidal horn flared in only one plane is known as a ***sectoral horn***. If the waveguide is flared to increase the aperture dimension in the direction of the electric field [Fig. 3-26(b)], it is referred to as an ***E-plane sectoral horn***, and if the flaring is in the direction of the magnetic field, it is an ***H-plane sectoral horn*** [Fig. 3-26(c)]. In the nonflared direction, the sectoral horn usually has the same dimension as the waveguide connected to its throat. Another horn configuration is the ***conical horn*** [Fig. 3-26(d)], which, by virtue of its axial symmetry, produces a symmetrical radiation pattern. Horn antennas also have been combined with reflector antennas in a variety of different configurations, some of which are illustrated in Fig. 3-27.

Rectangular horns

Expressions for the radiation from open-ended waveguides and horns of rectangular aperture were derived by Barrow and Chu (1939) and Chu (1940), and are given by Risser (in Silver, 1949). They used Maxwell's equations in spherical coordinates to solve for the field at the aperture and then applied the theory of vector diffraction to compute the far-zone field. An equivalent approach using cylindrical coordinates is available in

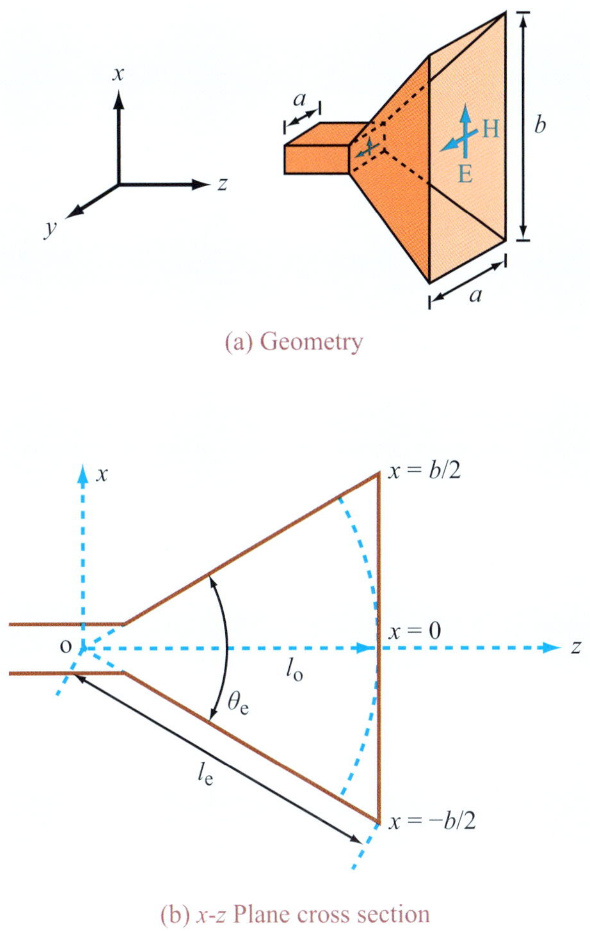

Figure 3-28: E-plane sectoral horn geometry and coordinates.

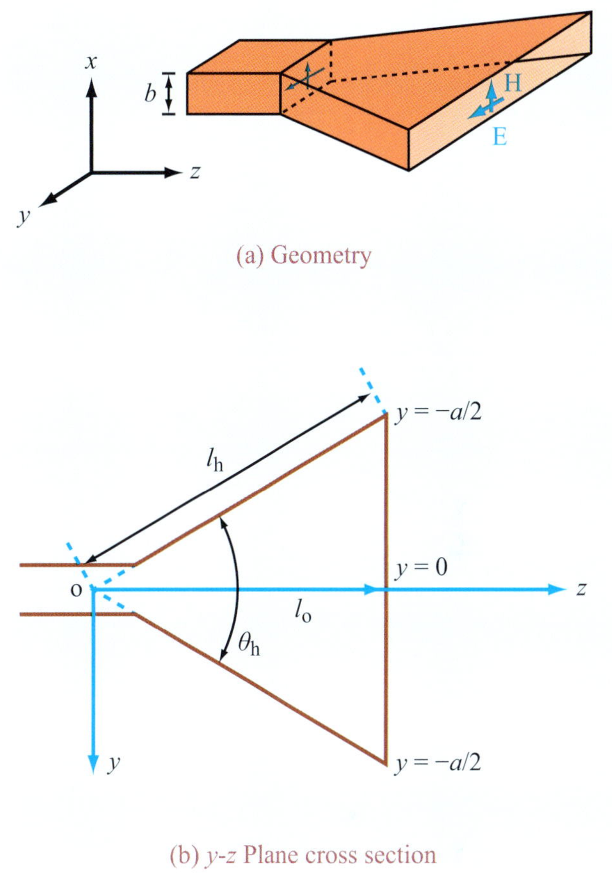

Figure 3-29: H-plane sectoral horn geometry and coordinates.

Compton and Collin (1969a). The resultant expressions are fairly involved; hence, they are not reproduced in this presentation. Instead, we devote the remainder of this section to discussions of directivity as a function of horn dimensions.

The geometry of the E-plane sectoral horn is shown in Fig. 3-28. The angle θ_e is called the ***flare angle***, and l_e is the ***slant length***. A similar geometry applies to the H-plane sectoral horn, as shown in Fig. 3-29.

The directivity D of rectangular horns, including the pyramidal horn and the E-plane and H-plane sectoral horns, was computed by Schelkunoff (1943) for the TE_{10} mode on the basis of simplified theoretical calculations using a scalar formulation approach. Despite its theoretical simplicity, Schelkunoff's derivations provide results that have been found to be quite accurate in practice (Jakes, 1951).

E-plane sectoral horns

Schelkunoff's expression for the directivity of an E-plane horn is given by

$$D_e = \frac{64al_e}{\pi\lambda b}\left[C^2(\tau) + S^2(\tau)\right], \tag{3.102}$$

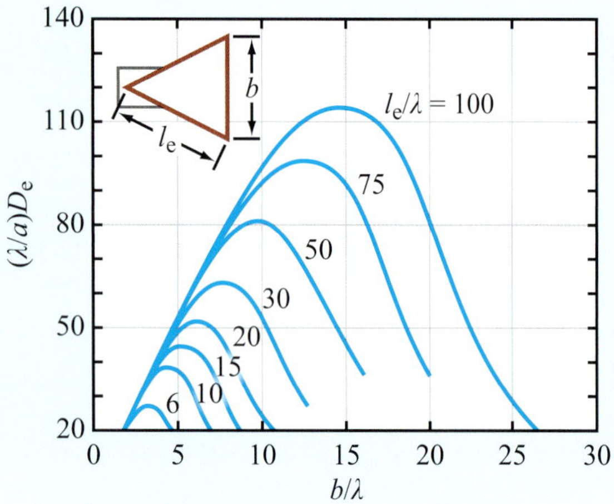

Figure 3-30: Directivity of E-plane sectoral horns with aperture height b and width a (based on Fig. 16.4 of Schelkunoff and Friis, 1952).

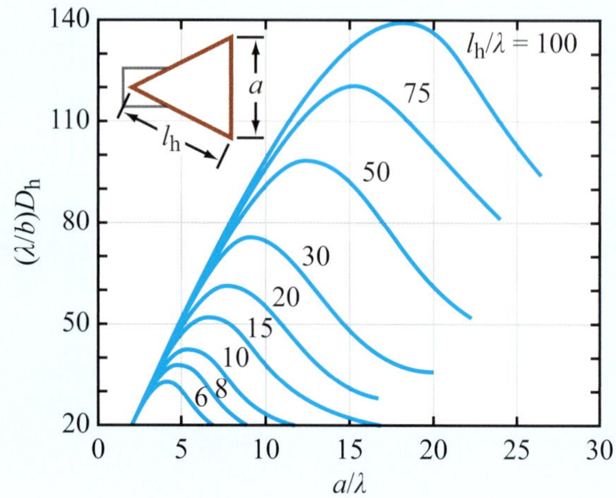

Figure 3-31: Directivity of H-plane horns of width a and height b (based on Fig. 16.3 of Schelkunoff and Friis, 1952).

where a, b, and l_e are as defined in Fig. 3-28, and C and S are the ***Fresnel integrals*** given by

$$C(\tau) = \int_0^\tau \cos\left(\frac{\pi t^2}{2}\right) dt, \tag{3.103a}$$

$$S(\tau) = \int_0^\tau \sin\left(\frac{\pi t^2}{2}\right) dt, \tag{3.103b}$$

with

$$\tau = \frac{b}{\sqrt{2\lambda l_e}}. \tag{3.103c}$$

Plots of D_e as a function of b/λ are shown in Fig. 3-30 for several values of the slant length ℓ_e. Note that for a given slant length, the directivity increases with the aperture height b until it reaches a maximum value, and then beyond this optimum aperture height the directivity decreases. This behavior is due to the nonuniform phase distribution across the aperture.

For a given value of l_e, the horn is said to be optimum if b is chosen so that the directivity D_e is maximum. This occurs when $l_e - l_0 = \lambda/4$ (Jakes, 1961). Under this condition, b and l_e are related by

$$b \approx \sqrt{2 l_e \lambda}. \tag{3.104}$$

In the E plane (x–z plane in Fig. 3-28), $\beta_{1/2} \approx \lambda/b$ radians and $\beta_{\text{null}} \approx 2\lambda/b$ for the optimum E-plane sectoral horn.

H-plane sectoral horns

The directivity D_h of the H-plane horn as given by Schelkunoff (1943) is

$$D_h = \frac{4\pi b s_h}{\lambda a}\{[C(u) - C(v)]^2 + [S(u) - S(v)]^2\}, \tag{3.105}$$

where

$$u = \frac{1}{\sqrt{2}}\left(\frac{\sqrt{\lambda l_h}}{a} + \frac{a}{\sqrt{\lambda l_h}}\right), \tag{3.106a}$$

$$v = \frac{1}{\sqrt{2}}\left(\frac{a}{\sqrt{\lambda l_h}} - \frac{\sqrt{\lambda l_h}}{a}\right). \tag{3.106b}$$

Plots of D_h as a function of a/λ are shown in Fig. 3-31 for several values of the slant length l_h. The curves are similar to those of Fig. 3-30 for the E-plane horn. The H-plane sectoral horn of Fig. 3-29 is optimum when $l_h - l_0 = 3\lambda/8$, which leads to the approximate relation (Jakes, 1961)

$$a \approx \sqrt{3 l_h \lambda}. \tag{3.107}$$

Pyramidal horns

The E-plane and H-plane sectoral horns have narrow beams in the flared direction and wide beams in the unflared direction. Such a beam is sometimes called a fan beam. If the application calls for narrow beams in both directions, the horn can be flared along both axes, in which case we have a pyramidal horn. Thus, a pyramidal horn may be regarded as a superposition of an E-plane and an H-plane sectoral horn.

The directivity of the pyramidal horn, D_p (Schelkunoff, 1943), may be expressed in terms of the directivities D_e and D_h as follows:

$$D_p = \frac{\pi\lambda^2}{32ab} D_e D_h, \tag{3.108}$$

where D_e and D_h are given by Eq. (3.102) and Eq. (3.105), respectively. That is, D_p may be obtained by multiplying the product of the normalized directivities $(\lambda/a)D_e$ and $(\lambda/b)D_h$ obtained from Figs. 3-30 and 3-31 by the factor $(\pi/32)$.

Conical horns

A conical horn [Fig. 3-26(d)] usually is connected to a circular waveguide excited with a TE_{11} mode. Its performance is similar to that of the pyramidal horn in that, for a fixed length, its directivity increases as a function of the aperture diameter d up to a certain optimum value. This is demonstrated in Fig. 3-32, which contains theoretical curves that were derived by Gray and Schelkunoff and reported by King (1950). Note that the family of curves is for different values of the ***axial length*** l_0, unlike the plots in Figs. 3-30 and 3-31, where the parameter was the slant length. Experimental measurements made over a wide range of values of l_0 and d were found to be in excellent agreement with the calculated values. The dashed line of Fig. 3-32 describes the dimensions of the optimum horn. The conical horn is optimum (maximum directivity for a fixed slant length l_c) when its diameter is (Jakes, 1961)

$$d \approx \sqrt{3 l_c \lambda}\,, \tag{3.109}$$

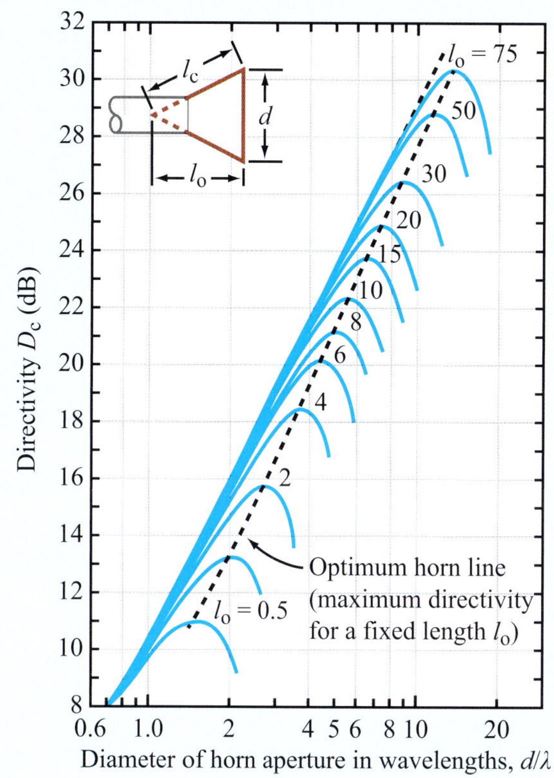

Figure 3-32: The directivity of a conical horn as a function of axial length and aperture diameter.

which is the same criterion as for the H-plane sectoral horn. Under this condition, the directivity in dB is given by

$$D_c\ [\text{dB}] = 20\log\left(\frac{\pi d}{\lambda}\right) - 2.82\ \text{dB}, \tag{3.110}$$

(optimum horn)

and its effective area is equal to 52 percent of its physical aperture area: $A_e = 0.52A_p$, where $A_p = \pi d^2/4$.

Lens-corrected horns

According to Eqs. (3.109) and (3.110), an optimum horn with a 26 dB directivity should have a slant length $l_c \approx 26\lambda$ and a diameter $d \approx 8.8\lambda$. If $\lambda = 3$ cm

($f = 10$ GHz), for example, these dimensions translate into a 78 cm long horn with an aperture diameter of 26.4 cm. For many applications, such long lengths are undesirable. If l_c were to be limited to a certain value, so would d and the directivity D_c. Because of the quadratic phase error in the aperture, increasing the aperture diameter (by increasing the flare angle) beyond its optimum value results in a loss in directivity, rather than gain. This limitation on the maximum flare angle for a given horn length has led to the development of the ***lens-corrected horn***.

By mounting a lens in the horn aperture (Fig. 3-33) to correct the phase distribution across the aperture, the limitation on the maximum flare angle can be removed, thereby allowing the use of large apertures to increase the directivity. This technique has been implemented successfully in both rectangular and conical horns. Without the lens, the phase shift between the vertex 0 and the point $x = x_1$ on the aperture (Fig. 3-33) is larger than it is to the point $x = 0$ (cone axis) because of the longer travel time. Through proper design of the lens shape and its index of refraction, the total phase shift between 0 and x_1 can be made equal to the phase-shift between 0 and the point $x = 0$ for any value of x_1 on the aperture. Effectively, the wave propagating along the cone axis has to traverse a longer thickness of the lens (in which the velocity of propagation is lower than that in air) than the wave traveling to x_1, thereby arriving in phase at the aperture.

Since the index of refraction of the lens is different from that of air, only part of the energy incident upon it is transmitted into the lens, and the remainder is reflected. Multiple reflections from the surface of the lens and the surface of the cone can result in amplitude and phase variations across the aperture. To reduce these reflections, several techniques have been developed (Compton and Collin, 1969a). The wave impedance of the lens can be matched to that of free space by covering the lens surface with a dielectric layer of appropriate thickness and index of refraction, or by cutting slots of appropriate thickness and depth into the surface of the lens (Collin, 1959).

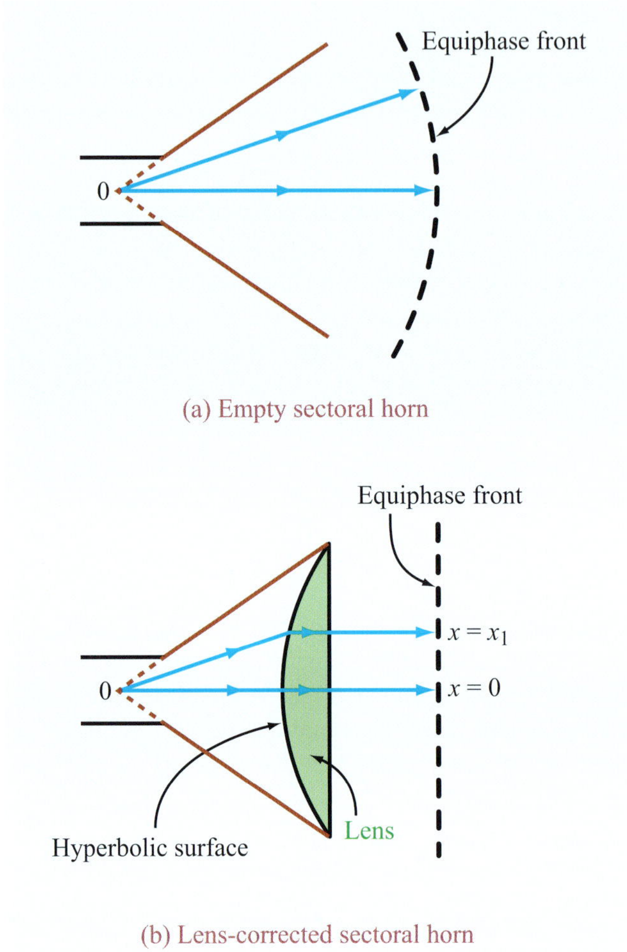

Figure 3-33: A hyperbolic lens of the appropriate index of refraction can transform the phase front to a uniform distribution across the aperture.

3-12.2 Slot Antennas

In this section we consider the far-field radiation properties of a rectangular slot antenna, followed by a brief discussion of waveguide slot arrays.

Examples of slots cut into the walls of a rectangular waveguide are shown in Fig. 3-34. When a slot interrupts the flow of transverse surface currents, it couples energy from the waveguide to free space. Figure 3-35 shows the distribution of currents flowing along the inner walls of a rectangular waveguide propagating a

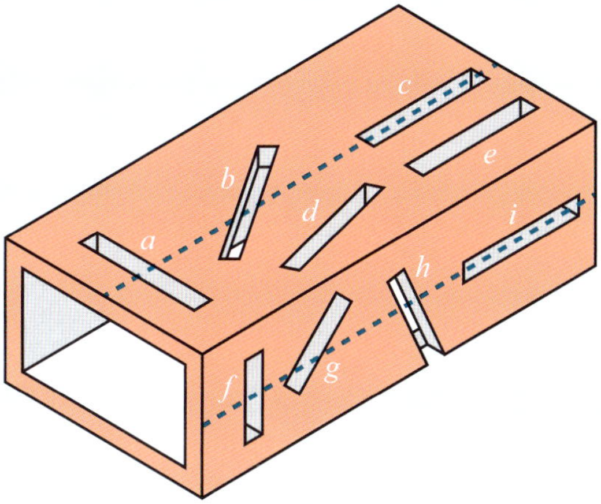

Figure 3-34: Various types of slots cut in the walls of a rectangular waveguide. Slots c and f do not radiate, because they do not interrupt the flow of surface current (see Fig. 3-35).

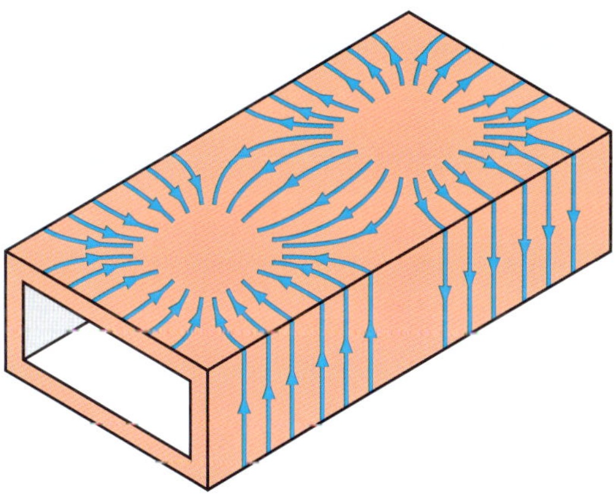

Figure 3-35: The flow of surface currents in the walls of a rectangular waveguide excited with a TE_{10} mode.

TE_{10} mode. Among the slots shown in Fig. 3-34, the only nonradiating slots are those labeled c and f (their axes are parallel to the direction of current-flaw). The magnitude of the energy coupled by a slot is a function of the waveguide dimensions; the slot dimensions, orientation, and position; and the density of the current interrupted by the slot.

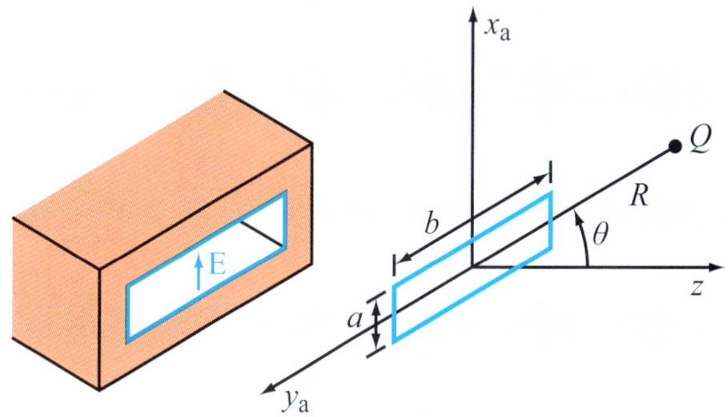

Figure 3-36: Rectangular slot with a tangential electric field in the $\hat{\mathbf{x}}_a$ direction.

To demonstrate the approach used for computing the far-zone fields of a slot antenna, we consider the simple case of a half-wave rectangular slot excited by an aperture electric field that is orthogonal to the slot axis, as shown in Fig. 3-36. More complicated configurations are beyond the scope of this presentation and are not discussed here; the interested reader is referred to Compton and Collin (1969b).

Half-wave slot

In reference to the coordinate system shown in Fig. 3-36, the tangential electric field in the aperture plane is zero everywhere except in the region of the slot. If $b = \lambda/2$ and $a \ll \lambda$, the slot behaves like a half-wave dipole with a singular beam centered along the z axis.

Waveguide slot array

We now extend the theory of linear arrays and the techniques used to provide shaped beams and electronic scanning to an array of slot radiators cut in one of the walls of a rectangular waveguide.

The choice of the geometrical configuration of the slotted waveguide array is dictated in part by the desire to minimize mutual coupling between the individual slots. Mutual coupling can influence the radiation properties of the slots as well as change their input impedance, thereby causing reflections toward the feed structure, which in turn modify the amplitude and

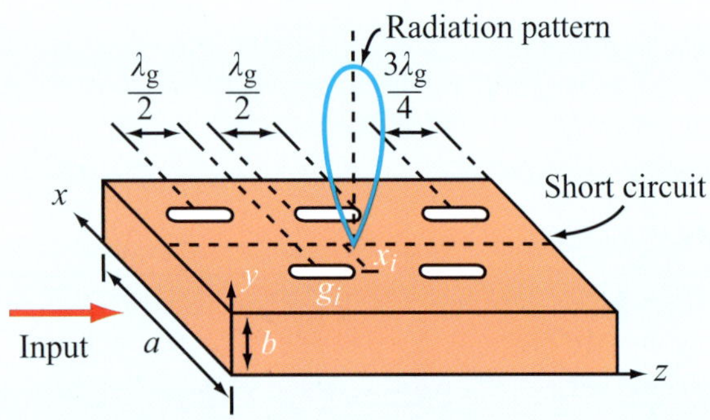

(a) Resonant slot array, with maximum radiation in broadside direction (y)

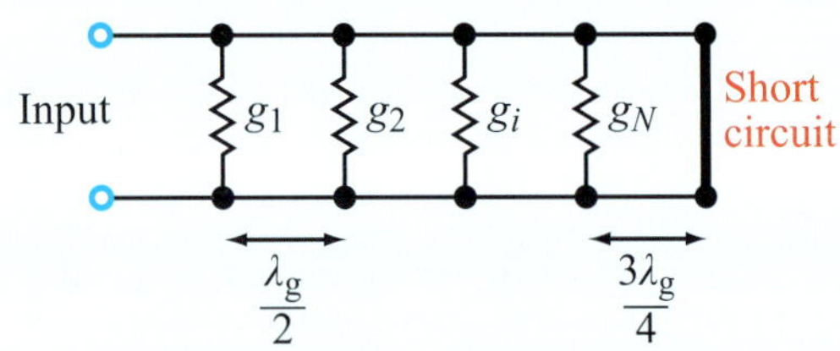

(b) Transmission-line equivalent circuit

Figure 3-37: The resonant slot array.

phase distributions among the array elements. An array consisting of longitudinal slots cut in the broad side of the waveguide (Fig. 3-37) has been found to exhibit negligible mutual coupling between adjacent slots (Ehrlich and Short, 1954). For the purpose of this introductory presentation, the discussion to follow is limited to this type of configuration, and the rectangular waveguide is assumed to be excited in the dominant TE_{10} mode.

Longitudinal slot arrays are commonly divided into two types: the ***resonant array*** and the ***nonresonant array***. A brief description of each follows.

Resonant array

The resonant array [Fig. 3-37(a)] is an in-phase array configured for broadside operation at a single frequency. For the slots to be excited in phase, they must be spaced one guide wavelength λ_g apart. In order to avoid grating lobes in the visible region of the array, the spacing between elements must be less than one free-space wavelength λ. Since in air-filled waveguides $\lambda_g > \lambda$, the λ_g spacing required for in-phase excitation results in grating lobes. The solution to this problem is to arrange the slots $\lambda_g/2$ apart with adjacent elements on opposite sides of the longitudinal axis of the waveguide, as shown in Fig. 3-37(a). The $\lambda_g/2$ spacing provides a phase difference of π between adjacent elements, and because the x component of the surface current flows in opposite directions on the two sides of the longitudinal axis, an additional phase difference of π is introduced in the excitation between adjacent slots. Thus, the slots are excited in phase, and if the guide dimensions are such that $\lambda_g/2 < \lambda$, grating lobes do not occur.

The transmission-line equivalent of the slot array is shown in Fig. 3-37(b). Associated with the ith slot is a ***shunt conductance*** g_i, whose magnitude is given by (Compton and Collin, 1969b)

$$g_i = K \sin^2\left(\frac{\pi x_i}{a}\right), \tag{3.111}$$

where K is a constant and x_i is the displacement of the slot from the longitudinal axis as shown in Fig. 3-37(a). The constant K is related to the waveguide dimensions a and b and to the wavelengths λ and λ_g. For a given voltage across the transmission line, the power radiated by a given slot is proportional to its conductance g_i, and hence to the displacement x_i. Thus, a desired amplitude distribution of the element feeding coefficients can be achieved by specifying the displacements x_i of the array slots. Since the amplitude distribution is usually defined in terms of the excitation field (or current) rather than power, the slot displacements should be chosen so that the desired relative amplitudes are proportional to $\sqrt{g_i}$, or $\sin(\pi x_i/a)$. Finally, by terminating the waveguide in a short circuit at a distance of $3\lambda_g/4$ from the last slot, such a termination translates to an open circuit at the last slot, thereby having no effect on its conductance. To satisfy this condition and to maintain the excitations of the slots in phase with one another, the array design is configured for operation at a single frequency, which has led to the name "resonant array." Consequently, a resonant slot array has a small bandwidth, typically on the order of $\pm 50\%/N$, where N is the number of slots.

Nonresonant array

Unlike the fixed-frequency broadside resonant array, the nonresonant array usually is designed for operation over a large bandwidth so that the direction of the mainlobe of the antenna array can be steered electronically by changing the frequency of the input signal. The array shown in Fig. 3-38 is terminated in a matched load to avoid reflections. Since part of the input power now is absorbed in the matched load, the nonresonant array is less efficient than the resonant array, but on the other hand, the use of the matched load allows operation over a much wider frequency range than is possible with the resonant array.

A constant spacing d between adjacent slots results in progressive (linear) phase shift along the array. If the wavelength of the input signal is such that d corresponds to $\lambda_g/2$, then the elements are excited in phase and the resultant beam is in the broadside direction ($\theta_0 = \pi/2$ in Fig. 3-38). In the general case, however, the beam angle θ_0 is governed by the relationship

$$\cos\theta_0 = \frac{\lambda}{\lambda_g} - \frac{\lambda}{2d}, \tag{3.112}$$

and for the TE_{10} mode in an unloaded waveguide

$$\frac{\lambda}{\lambda_g} = \sqrt{1 - \left(\frac{\lambda}{2a}\right)^2}. \tag{3.113}$$

Thus, the nonresonant slot array is amenable to frequency scanning wherein the mainlobe of the array is made to scan in the y–z plane by changing the frequency of the input signal. For more details regarding the design of slot arrays, the reader is referred to Compton and Collin (1969b), Balanis (2008), and Croswell (2007).

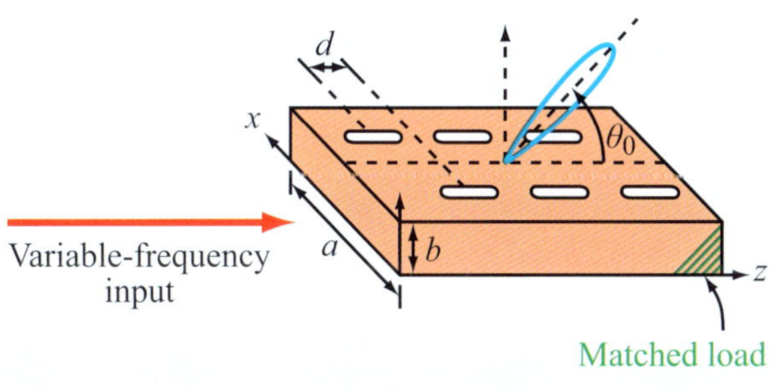

Figure 3-38: With the nonresonant slot array, the main beam may be steered in direction by varying the frequency of the input signal.

3-12.3 Microstrip Antennas

A microstrip antenna consists of a flat metallic patch separated from a ground conducting plane by a thin, uniform, nearly lossless dielectric substrate. Figure 3-39 shows a cross-sectional view of a microstrip antenna fed by a coaxial transmission line, with the center conductor of the coaxial line connected to the patch and the outer conductor connected to the ground plane. Alternatively, the microstrip antenna can be fed by other types of transmission lines, such as the microstrip line shown in Fig. 3-40. Because it can be inexpensively fabricated using printed circuit technology, both singly and in arrays, the microstrip antenna dominates the mobile phone market and is used extensively in many microwave applications. Disadvantages include low power-handling capability and a narrow bandwidth ($\approx$ 1–5%).

The patch of the microstrip antenna is typically rectangular or circular in shape, although other shapes are also used. For a rectangular patch, its length L is usually in the range $\lambda_0/3 < L < \lambda_0/2$, where λ_0 is the free-space wavelength. Its width W may vary over a wider range, typically $0.25\lambda_0 \le W \le 0.75\lambda_0$. The height h of the dielectric substrate, which influences both the bandwidth and radiation efficiency of the antenna, is usually in the range $0.003\lambda_0 \le h \le 0.05\lambda_0$. The rationale for these ranges will become evident in the next few paragraphs.

The radiation pattern of the microstrip patch antenna consists primarily of a broad single lobe, similar to that of a half-wave dipole. A typical example is shown in Fig. 3-41. If configured into a one-dimensional array, as depicted by Fig. 3-42, the overall pattern becomes much narrower in the plane containing the array axis and the normal to the antenna surface, and if arrayed into a two-dimensional lattice, the beam can be narrowed into a conical beam. To avoid grating lobes, the spacing between adjacent patches should be less than $\lambda_0/2$. Also, in the ***corporate feed network*** shown in

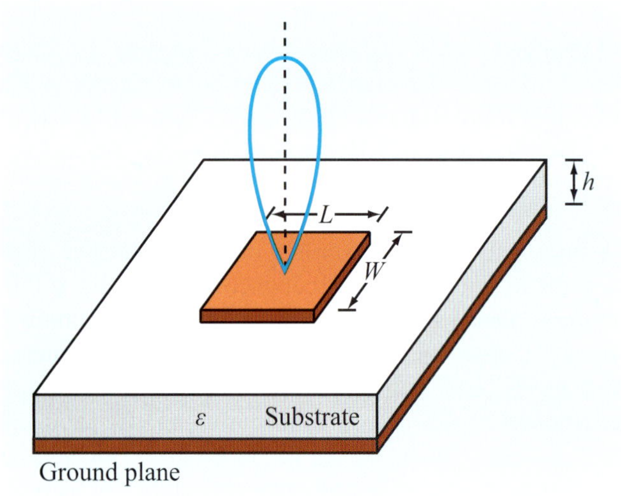

(a) Microstrip antenna

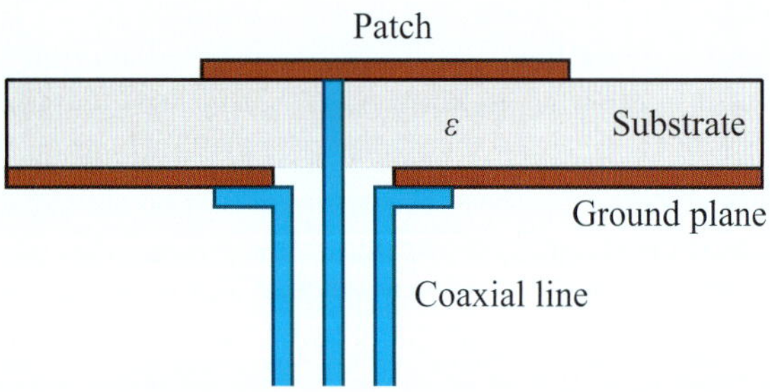

(b) Side view with coaxial feed line

Figure 3-39: (a) Top view of a rectangular patch antenna and (b) side view showing how it connects to a coaxial transmission line.

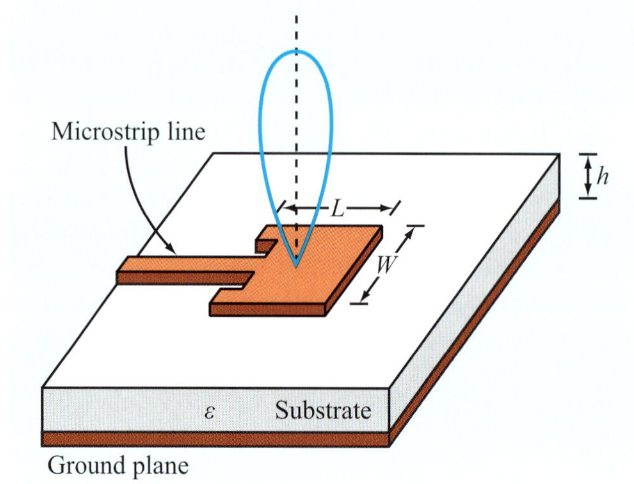

Figure 3-40: Microstrip patch antenna fed by a microstrip transmission line.

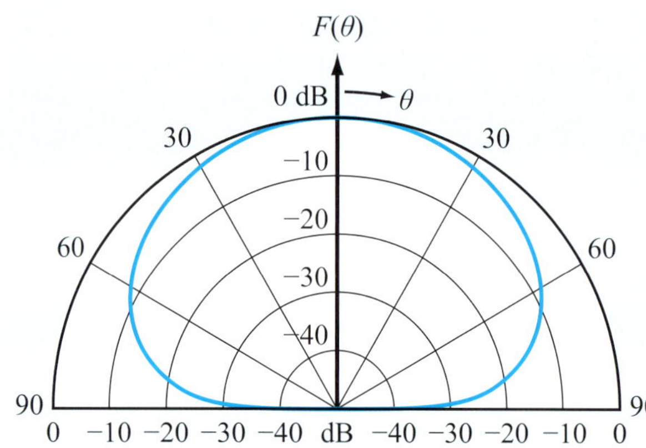

Figure 3-41: Far-field patterns of a rectangular patch on an infinite ground plane (Jackson, 2007).

Fig. 3-42, quarter-wave transformers are used to match the impedances of the individual antenna elements to a common 50-Ω microstrip transmission line.

Using electromagnetic simulation models (Jackson and Oliner, 2007; Balanis, 2008) it is possible to select the antenna and substrate dimensions so that the antenna input impedance $\mathbf{Z}_i$ provides a good match to the transmission line feed at a specific frequency of interest. In the example shown in Fig. 3-43, $\mathbf{Z}_i = (50 + j0)$ Ω at $f = 1.575$ GHz, which is an ideal match to a 50-Ω transmission line, but $\mathbf{Z}_i$ varies rapidly with frequency. A small deviation away from 1.575 GHz causes both the real and imaginary components of $\mathbf{Z}_i$ to change significantly. The operational fractional bandwidth is defined as

$$\text{BW} = \frac{f_2 - f_1}{f_r}, \tag{3.114}$$

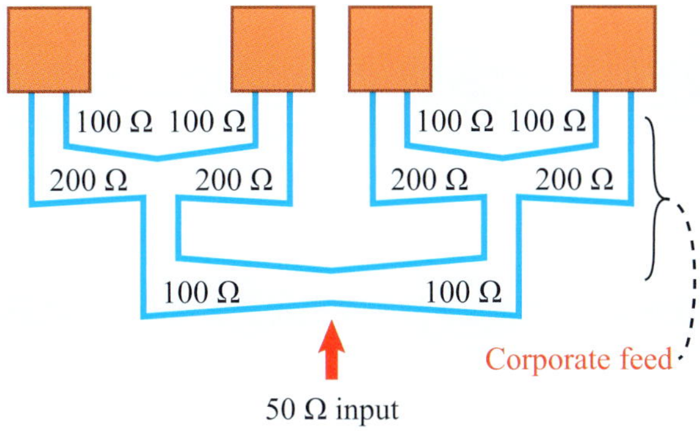

Figure 3-42: Tapered lines to match 100-Ω patches to a 50-Ω line (R. E. Munson, 1974).

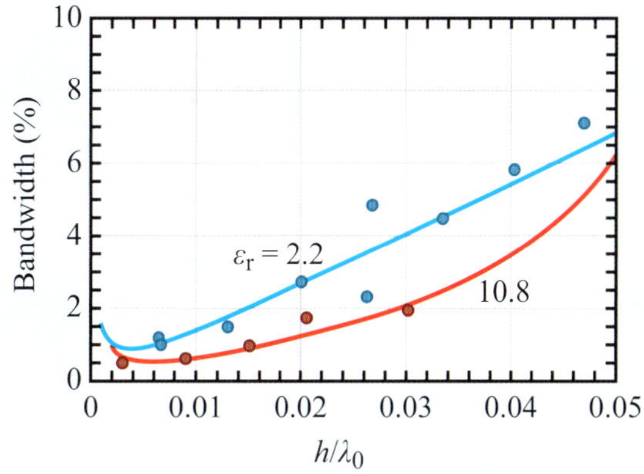

(a) Bandwidth

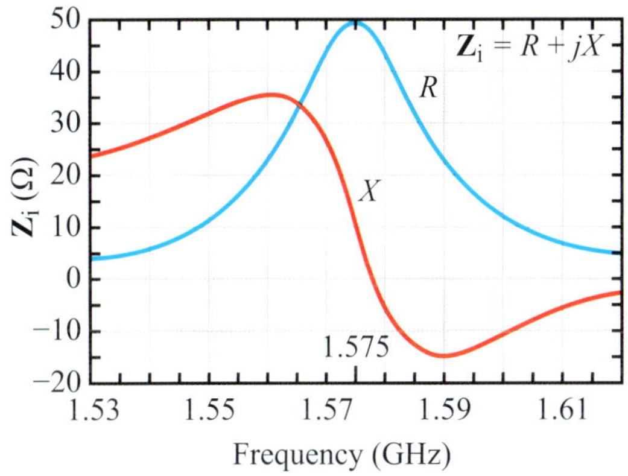

Figure 3-43: Input impedance $\mathbf{Z}_i$ of a microstrip antenna with $L = 6.255$ cm, $W = 9.383$ cm, $h = 0.1524$ cm, and $\varepsilon_r = 2.2$ (Jackson, 2007).

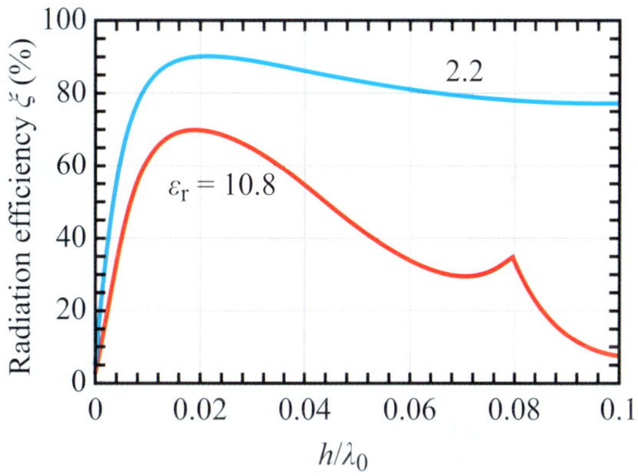

(b) Radiation efficiency

Figure 3-44: (a) Bandwidth and radiation efficiency versus the normalized substrate thickness for a moderate-permittivity substrate and a high-permittivity substrate (Jackson, 2007).

where f_r is the resonant frequency (1.575 GHz, in the present case), and f_1 and f_2 are the frequencies below and above f_r at which the impedance mismatch between the antenna and the transmission line is characterized by a standing wave ratio of a specified value, typically $S = 2$. Figure 3-44(a) displays the variation of percent fractional bandwidth as a function of h/λ_0, where h is the height of the dielectric substrate. The bandwidth increases with h/λ_0. However, beyond $h/\lambda_0 = 0.015$, the radiation efficiency ξ exhibits a decreasing trend with h/λ_0 [Fig. 3-44(b)]. Ultimately, the antenna dimensions and the height and permittivity of its substrate are selected by the antenna designer through a trade-off analysis to suit the intended application.

3-13 Active Antennas

As previously noted, while most antennas obey reciprocity, a particularly important class of antennas may not: those that contain active elements. Antennas containing active electronic components are known as

active antennas. In this context, an active antenna should not be confused with the "active" elements in a Yagi-Uda dipole array antenna (Yagi, 1928). In a Yagi-Uda array, some of the elements are not connected to the antenna feed but are parasitic elements designed to improve the antenna directivity. The passive elements are termed "directors" or "relectors" depending on their place in the array. In such antennas, the antenna elements connected to the antenna feed are commonly referred to as the active elements (Pozar, 1997). Old-style television antennas are good examples of Yagi-Uda antennas. In this book we define an active antenna as one that contains active electronic elements.

An active antenna is generally an array antenna in which transmit and receive amplifiers are distributed throughout the array. Figure 3-45 illustrates the difference between a conventional phased array antenna and an active array antenna. In a conventional phased array, the array elements are connected to phase shifters fed through a corporate (divided) feed and a ***transmit/receive*** (T/R) switch that selects between a ***high-power transmit amplifier*** (HPA) and a ***low-noise receive amplifier*** (LNA), which are connected to the upconverter and downconverter, respectively. (Up- and downconverters are discussed in Chapters 13 and 14.)

In the active antenna diagrammed in Fig. 3-45(b), each array element is fed from a dedicated transmit/receive module (TRM). These are connected by a corporate feed to a transmit/receive switch connected to the upconverter and the downconverter. The same upconverter and downconverter designs can be used in both the phased-array and the active antenna array. A diagram of a typical TRM is shown in Fig. 3-46, though some TRMs do not contain all components. For example, a receive-only active antenna would obviously not contain transmit amplifiers. Each TRM includes its own LNA and a power amplifier (PA), as well as T/R switches or circulators. In some applications a calibration feedback circuit is included to enable feeding some of the transmitted signal into the LNA (see Chapter 13). TRMs may contain phase shifters and variable-gain amplifiers to permit dynamic beam steering and beam shaping. The TRM includes power conditioning for the electronics and control logic for

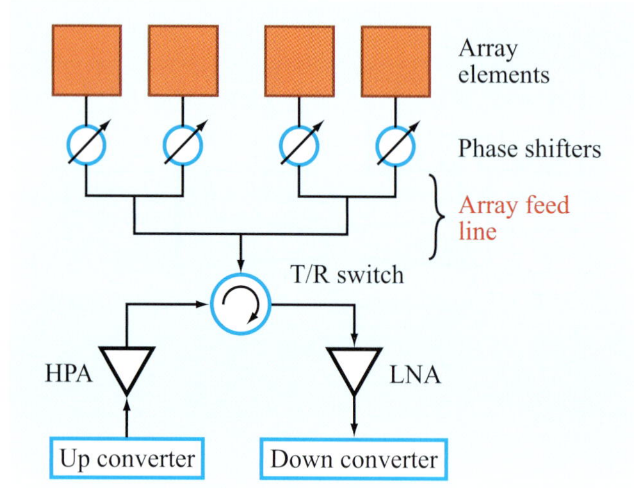

(a) Conventional array

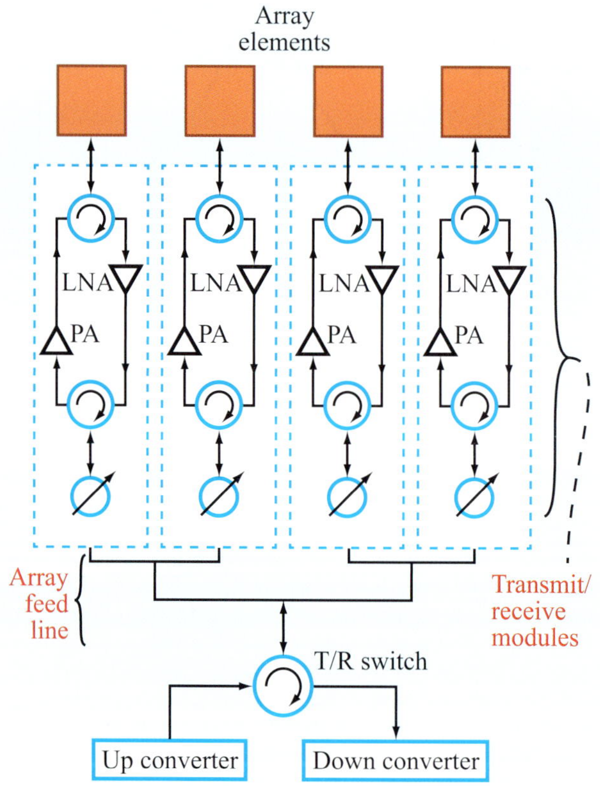

(b) Active array

Figure 3-45: Comparison of (a) conventional and (b) active array system.

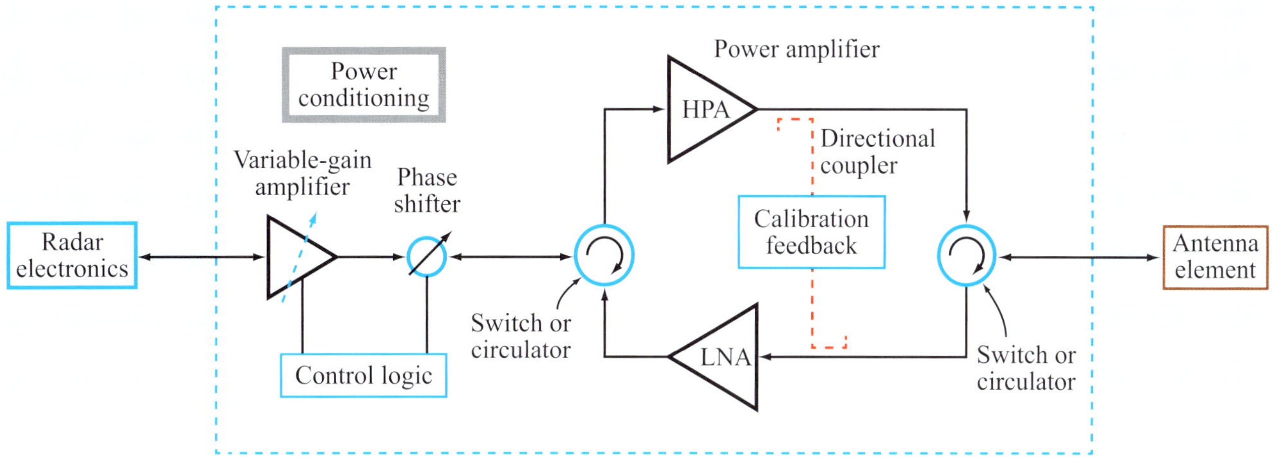

Figure 3-46: Design of an active transmit/receive module.

setting the gain of the variable-gain amplifier and phase of the phase shifter and for controlling the T/R switches and the calibration circuit.

3-13.1 Advantages of Active Antennas

Given the significant complexity that active antennas have in comparison with conventional passive antennas, why would one want to use an active antenna? In the following we explore some of the trade-offs between active and passive antennas.

Note that for a conventional antenna, the radiated power must be generated by one large HPA, and the feed structure and the phase shifters must be designed to handle high power levels. Also, the heat generated by the HPA is localized. In contrast, in an active antenna array, signal amplification is spread over multiple smaller amplifiers and the feeding and phase shifting occurs at much-smaller signal levels. These can reduce component cost and size, enable more rapid phase switching, and produce increased component reliability. The shorter signal paths between the antenna elements and the amplifiers result in less signal loss and thus in improved signal-to-noise ratio (SNR). Mass production of identical TRMs lower their cost. The savings can be used to reduce their size, weight, and power efficiency.

We also note that heat dissipation from the distributed amplifiers may be simplified and require less weight, a particular concern for spaceborne radar systems. The trade-off is the requirement that the TRMs must either have matched phase and gain or contain calibration and adjustment circuitry to ensure the achievement of the desired antenna performance. Due to thermal variations in amplifier gain and phase shifter performance, self-calibration and correction is essential in spaceborne systems.

Active antenna systems generally have higher fault tolerance: the failure of an amplifier has less effect in an active antenna. There is a trade-off, however, because there are many more individual electronic components in an active antenna system that can fail than in a system with a conventional passive antenna with fewer electronic elements. Due to the potentially higher mass of an active antenna compared with a passive one, the supporting structure requirements for the antenna system may be higher for an active antenna than for a passive antenna of the same performance. Finally, another disadvantage of an active antenna is that there is less room for high performance front-end filters (those between the antenna and LNA) in the TRM to minimize interference from undesired signals, so the front-end filter performance may be reduced, which can result in increased sensitivity to interference.

Though higher in cost and complexity, the improved performance of active antennas have led to their use in spaceborne SAR systems such as ASAR on Envisat (Desnos et al., 2000), RADARSAT-2 (Riendeau and Grenier, 2007), and TerraSAR-X (Buckreuss et al., 2008). As technology continues to improve the performance of TRMs, while shrinking their size and cost, it is expected that active antennas will become increasingly more common.

3-13.2 Digital Beamforming with Active Antennas

The active phased-array antenna considered thus far forms only a single antenna beam based on analog summation of the phase-shifted receive signals in the feed, as illustrated in Fig. 3-47(a). After summation, the signal is amplified, filtered, downconverted, and digitized with an analog-to-digital (A/D) converter for digital processing. This requires only a single LNA, downconverter, and A/D converter. An alternative architecture, known as ***digital beamforming***, is illustrated in Fig. 3-47(b).

In a digital beamformer each antenna array element has a dedicated receiver that includes an LNA, filtering, downconversion, and an A/D. The digital data stream from each A/D is then summed in a digital summer processor. Any required phase shift or time delay is included in the digital summer output to produce precisely the same signal as in the analog beamformer. However, using a digital processor, multiple summations, each with a different phase distribution. This produces multiple simultaneous beams which can be steered independently. These can be used to provide wider coverage at the expense of increased complexity, cost, and data rate. Digital processing provides finer and faster control of beam steering and beam shaping than conventional analog beamforming. Digital beamforming offers the same SNR advantage as other active antennas with short antenna element-to-LNA paths. Elimination of the analog phase shifters and feed structure reduces antenna mass and simplifies calibration of the array. Finally, a digital beamformer has improved dynamic range since the signal level at the LNA is smaller. The coherent signal increase due to coherent summation over the aperture now occurs in the digital processor. The resulting increase in dynamic range can be handled digitally.

Note that digital beamforming can also be used for the transmit function by replacing A/Ds with digital-to-analog converters (D/As), the downconverter with an upconverter, and LNAs with HPAs. A single transmit beam can be formed and it can be steered rapidly. Using multiple orthogonal signals makes it possible to generate multiple simultaneous transmit beams, but specialized

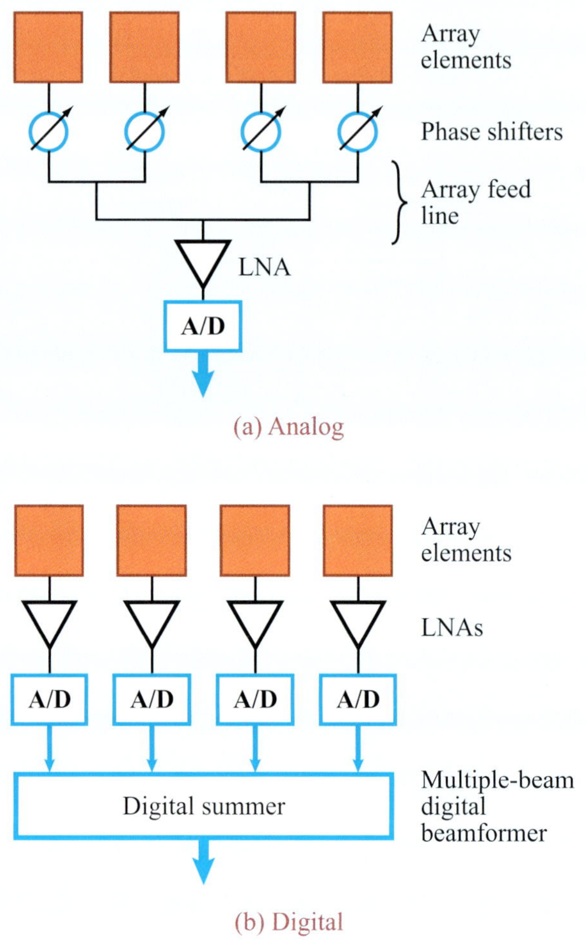

Figure 3-47: Comparison of (a) analog and (b) digital beamforming.

digital processing of the received signal is required (Li and Stoica, 2009).

PROBLEMS

3.1 Determine the following:

(a) The direction of maximum radiation.

(b) Directivity.

(c) Beam solid angle.

(d) Half-power beamwidth in the x–z plane.

for an antenna whose normalized radiation intensity is given by

$$F(\theta,\phi) = \begin{cases} 1 & \text{for } 0 \le \theta \le 60^\circ \text{and } 0 \le \phi \le 2\pi \\ 0 & \text{elsewhere.} \end{cases}$$

Suggestion: Sketch the pattern prior to calculating the desired quantities.

3.2 Repeat Problem 3.1 for an antenna with

$$F(\theta,\phi) = \begin{cases} \sin^2\theta\cos^2\phi & \text{for } 0 \le \theta \le \pi \\ & \text{and } -\pi/2 \le \phi \le \pi/2 \\ 0 & \text{elsewhere} \end{cases}$$

3.3 An antenna with a pattern solid angle of 1.5 (sr) radiates 60 W of power. At a range of 1 km, what is the maximum power density radiated by the antenna?

3.4 An antenna with a radiation efficiency of 90% has a directivity of 7.0 dB. What is its gain in decibels?

3.5 The radiation pattern of a circular parabolic-reflector antenna consists of a circular major lobe with a half-power beamwidth of 3° and a few minor lobes. Ignoring the minor lobes, obtain an estimate for the antenna directivity in dB.

3.6 The normalized radiation intensity of a certain antenna is given by

$$F(\theta) = \exp(-20\theta^2) \qquad \text{for } 0 \le \theta \le \pi,$$

where θ is in radians. Determine:

(a) The half-power beamwidth.

(b) The pattern solid angle.

(c) The antenna directivity.

3.7 A 3 GHz line-of-sight microwave communication link consists of two lossless parabolic dish antennas, each 1 m in diameter. If the receive antenna requires 10 nW of receive power for good reception and the distance between the antennas is 40 km, how much power should be transmitted?

3.8 Consider the communication system shown in Fig. P3.8, with all components properly matched. If $P_t = 10$ W and $f = 6$ GHz:

(a) What is the power density at the receiving antenna (assuming proper alignment of antennas)?

(b) What is the received power?

(c) If the receiver noise power is $P_n = KT_{sys}B$, where K is Boltzmann's constant (1.38×10^{-23} J/K), $T_{sys} = 1000$ K is the receiver system temperature, and the receiver bandwidth is $B = 20$ MHz, what is the signal-to-noise ratio in decibels?

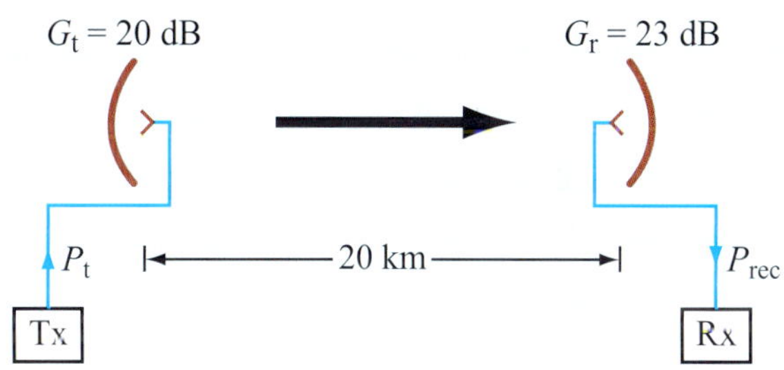

Figure P3.8: Communication system of Problem 3.8.

3.9 A uniformly illuminated aperture is of length $l_x = 20\lambda$. Determine the beamwidth between first nulls in the x–z plane.

3.10 The 10 dB beamwidth is the beam size between the angles at which $F(\theta)$ is 10 dB below its peak value. Determine the 10 dB beamwidth in the x–z plane for a uniformly illuminated aperture with length $l_x = 10\lambda$.

3.11 A uniformly illuminated rectangular aperture situated in the x–y plane is 2 m high (along x) and 1 m wide (along y). If $f = 10$ GHz, determine the following:

(a) The beamwidths of the radiation pattern in the elevation plane (x–z plane) and the azimuth plane (y–z plane).

(b) The antenna directivity D in decibels.

3.12 An antenna with a circular aperture has a circular beam with a beamwidth of 3° at 20 GHz.

(a) What is the antenna directivity in dB?

(b) If the antenna area is doubled, what will be the new directivity and new beamwidth?

(c) If the aperture is kept the same as in (a), but the frequency is doubled to 40 GHz, what will the directivity and beamwidth become then?

3.13 A 94 GHz automobile collision-avoidance radar uses a rectangular-aperture antenna placed above the car's bumper. If the antenna is 15 cm in length and 5 cm in height, determine the following:

(a) Its elevation and azimuth beamwidths.

(b) The horizontal extent of the beam at a distance of 300 m.

3.14 A microwave telescope consisting of a very sensitive receiver connected to a 100 m parabolic-dish antenna is used to measure the energy radiated by astronomical objects at 20 GHz. If the antenna beam is directed toward the moon and the moon extends over a planar angle of 0.5° from Earth, what fraction of the moon's cross section will be occupied by the beam?

3.15 Derive an expression for the far-zone normalized radiation pattern of a rectangular aperture with dimensions a and b, illuminated by

(a)

$$E_a(x_a, y_a) = \begin{cases} E_0 \cos\left(\dfrac{\pi x_a}{a}\right) & \text{for } |x_a| \le \dfrac{a}{2} \text{ and } |y_a| \le \dfrac{b}{2}, \\ 0 & \text{otherwise,} \end{cases}$$

(b)

$$E_a(x_a, y_a) = \begin{cases} \dfrac{E_0}{2}\left[1 + \cos\left(\dfrac{\pi x_a}{a}\right)\right] & \text{for } |x_a| \le \dfrac{a}{2} \text{ and } |y_a| \le \dfrac{b}{2}, \\ 0 & \text{otherwise.} \end{cases}$$

(c) Plot the x–z-plane radiation patterns of the uniformly illuminated aperture and the above two apertures. Compare half-power beamwidths and sidelobe levels.

3.16 Find the far-field radiation pattern of a Gaussianly illuminated rectangular aperture, with

$$E_a(x_a, y_a) = \begin{cases} \dfrac{1}{\sqrt{2\pi\sigma^2}} \exp\left(\dfrac{-\pi x_1^2}{2\sigma^2}\right) & \text{for } |x_1| \le 1 \text{ and } |y_1| \le 1, \\ 0 & \text{otherwise,} \end{cases}$$

where $x_1 = 2x_a/a$ and $y_1 = 2y_a/b$.

3.17 A rectangular aperture is illuminated by the function

$$E_a(x_a, y_a) = \begin{cases} E_0\, f(x_1) & \text{for } |x_1| \le 1 \text{ and } |y_1| \le 1, \\ 0 & \text{otherwise,} \end{cases}$$

where $f(x_1)$ is shown below, and x_1 and y_1 are as defined in Problem 3.16.

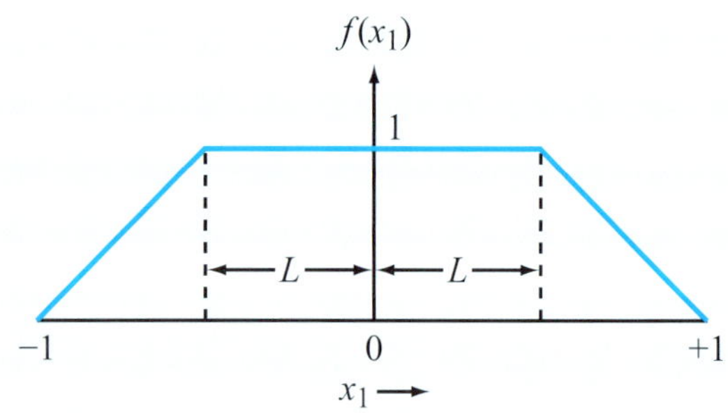

Figure P3.17: Figure for Problem 3.17.

(a) Express $f(x_1)$ as a convolution of two rectangular functions: $f(x_1) = f_1(x_1) * f_2(x_1)$.

(b) Use the convolution property of the Fourier transform to obtain the normalized radiation pattern of the aperture.

3.18 A two-element array consisting of two isotropic antennas separated by a distance d along the z axis is placed in a coordinate system whose z axis points eastward and whose x axis points toward the zenith. If a_0 and a_1 are the amplitudes of the excitations of the antennas at $z = 0$ and at $z = d$, respectively, and if δ is the phase of the excitation of the antenna at $z = d$ relative to that of the other antenna, find the array factor and plot the pattern in the x–z plane for the following:

(a) $a_0 = a_1 = 1,\ \delta = \pi/4$, and $d = \lambda/2$

(b) $a_0 = 1,\ a_1 = 2,\ \delta = 0$, and $d = \lambda$

(c) $a_0 = a_1 = 1,\ \delta = -\pi/2$, and $d = \lambda/2$

(d) $a_0 = 1,\ a_1 = 2,\ \delta = \pi/4$, and $d = \lambda/2$

(e) $a_0 = 1,\ a_1 = 2,\ \delta = \pi/2$, and $d = \lambda/4$

3.19 Find and plot the normalized array factor and determine the half-power beamwidth for a five-element linear array excited with equal phase and a uniform amplitude distribution. The interelement spacing is $3\lambda/4$.

3.20 A five-element, equally spaced linear array with $d = \lambda/2$ is excited with uniform phase and an amplitude distribution given by the binomial distribution

$$a_i = \frac{(N-1)!}{i!(N-i-1)!}, \qquad i = 0, 1, \ldots, (N-1),$$

where N is the number of elements. Develop an expression for the array factor.

3.21 A three-element linear array of isotropic sources aligned along the z axis has an interelement spacing of $\lambda/4$ (Fig. P3.21). The amplitude excitation of the center element is twice that of the bottom and top elements, and the phases are $-\pi/2$ for the bottom element and $\pi/2$ for the top element, relative to that of the center element. Determine the array factor and plot it in the elevation plane.

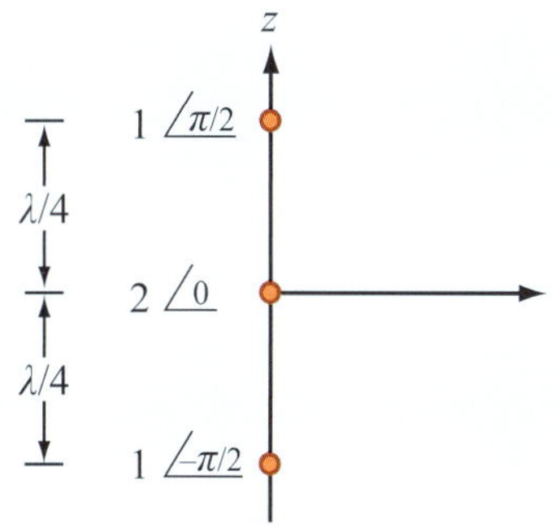

Figure P3.21: Three-element array of Problem 3.21.

3.22 An eight-element linear array with $\lambda/2$ spacing is excited with equal amplitudes. To steer the main beam to a direction 60° below the broadside direction, what should be the incremental phase delay between adjacent elements? Also, give the expression for the array factor and plot the pattern.

3.23 A linear array arranged along the z axis consists of 12 equally spaced elements with $d = \lambda/2$. Choose an appropriate incremental phase delay δ so as to steer the main beam to a direction 30° above the broadside direction. Provide an expression for the array factor of the steered antenna and plot the pattern. From the pattern, estimate the beamwidth.

3.24 Design a 3 cm wavelength optimum pyramidal horn, given the constraint that the axial length l_0 cannot exceed 30 cm. Find its maximum directivity D_p.

3.25 A square aperture with dimensions much larger than λ is illuminated by a cosine amplitude distribution with $n = 1$ along both dimensions, thereby generating an approximately azimuthally symmetric radiation pattern. Use the x–z pattern to calculate the beam efficiency of the antenna.

3.26 A square aperture with dimensions much larger than λ is illuminated by a cosine amplitude distribution with $n = 2$ along both dimensions, thereby generating an approximately azimuthally symmetric radiation pattern. Use the x–z pattern to calculate the beam efficiency of the antenna.

3.27 A square aperture with dimensions much larger than λ is illuminated by a triangular amplitude distribution along both dimensions, thereby generating an approximately azimuthally symmetric radiation pattern. Use the x–z pattern to calculate the beam efficiency of the antenna.

3.28 Designers of a 5.4 GHz SAR are choosing between an active antenna and a conventional array. To achieve the desired beam pattern, the aperture size is 15 m $\times$ 1.4 m. For a two-dimensional array of microstrip patches with $\lambda/2$ element spacing and assuming one TRM per patch with a TRM transmit power of 1 W/element, what is the minimum HPA output power required from a conventional array to achieve the same peak radiated power supplied by the active array? If the TRM transmit efficiency is 50% and the HPA efficiency is 75% but the HPA feed loss is 3 dB, what is the peak power required for each case?

3.29 How is the antenna pattern affected by the failure of a single TRM in a 16-element linear array with spacing $d = \lambda/2$? For simplicity, assume that the failure occurs in one of the end elements and the beam is steered to broadside. Sketch the difference in the antenna pattern before and after the failure.

3.30 Assume a plane-wave signal incident upon an 8-element linear phased array ($d = \lambda/2$) at an angle of 30°. Compute the signal level of the output of the antenna feed when (a) the antenna is steered to broadside and (b) when it is steered at 30° toward the signal source. What does your analysis indicate about the dynamic range of a phased-array antenna?

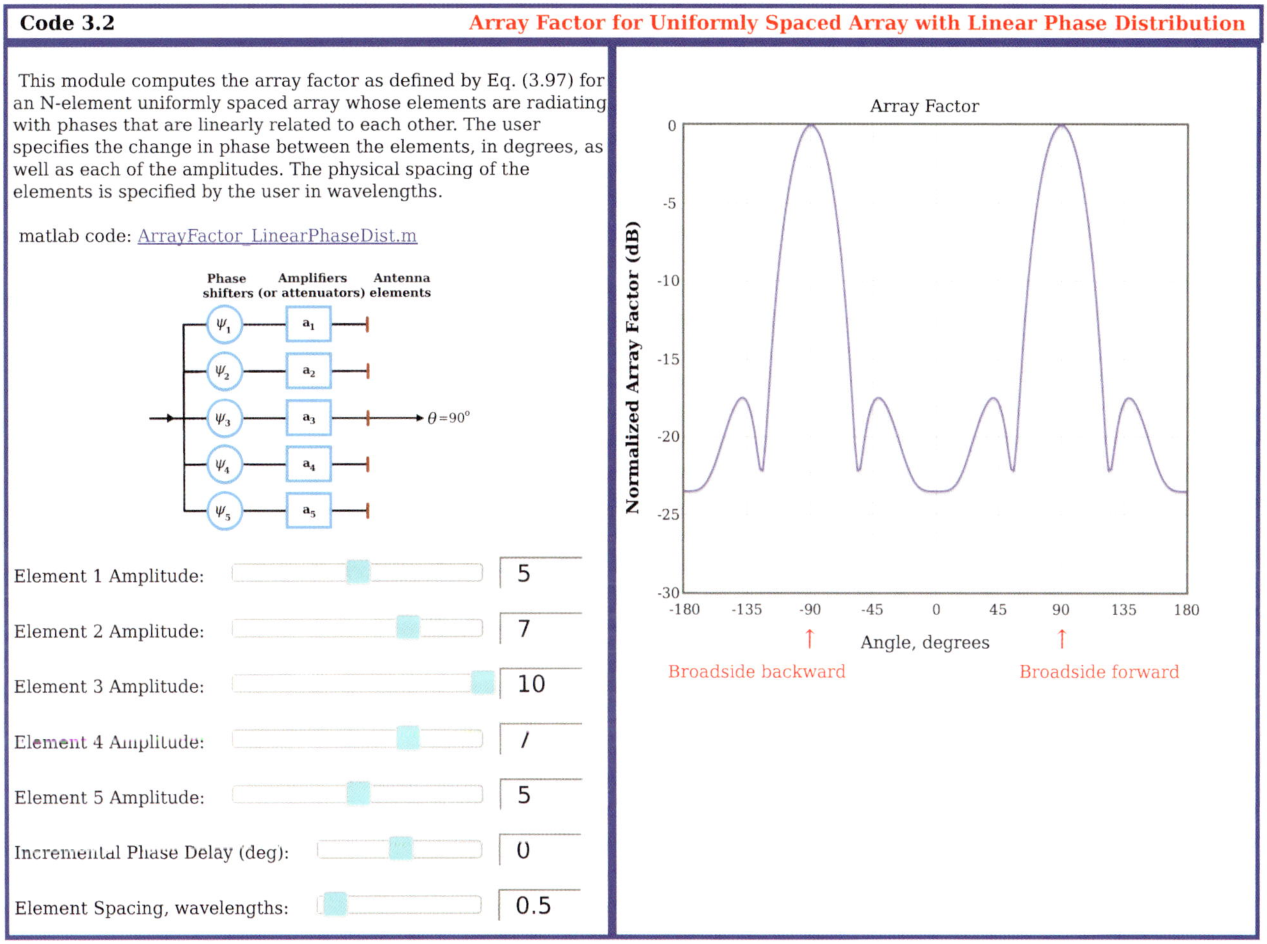

An example of one of the interactive modules available at the book website: mrs.eecs.umich.edu.

CHAPTER

4

Microwave Dielectric Properties of Natural Earth Materials

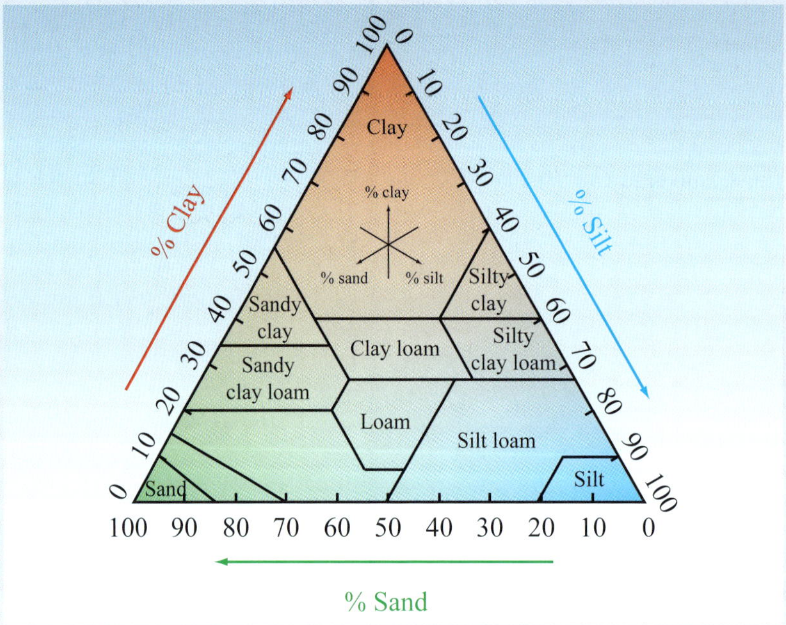

Soil textural triangle

CONTENTS

Overview

This chapter is a sourcebook of representative models and measured data of the microwave dielectric properties of natural earth materials. These materials can generally be classified into one of the following dielectric groups: (a) ***homogeneous substances***, (b) ***electrolytic solutions***, or (c) ***heterogeneous mixtures***. Only pure water and ice belong to group (a). Ionic salts or sugars dissolved in water produce an electrolytic solution whose microwave dielectric properties may be markedly different from those of pure water. Section 4-2 therefore examines the dielectric properties of both pure and saline water, and Section 4-3 treats the case for pure ice.

Heterogeneous mixtures include: sea ice (a mixture of ice crystals, air pockets, and liquid brine inclusions); snow (a mixture of air, ice particles, and sometimes liquid water inclusions); dry soil (a mixture of soil particles and air); wet soil; vegetation; and other materials. For a medium consisting of a ***host material*** with an intrinsic relative dielectric constant ε_h, within which is contained a homogeneous random concentration of ellipsoidal particles (***inclusions***) of material with an intrinsic relative dielectric constant ε_i, the relative dielectric constant of the ***mixture***, ε_m, is said to consist of two components,

$$\varepsilon_m(x,y,z;\hat{\mathbf{p}}) = \varepsilon_m(\hat{\mathbf{p}}) + \varepsilon_f(x,y,z;\hat{\mathbf{p}}), \tag{4.1}$$

where $\varepsilon_m(\hat{\mathbf{p}}) = \langle \varepsilon_m(x,y,z;\hat{\mathbf{p}}) \rangle$ is the ***effective or average*** value of the dielectric constant of the medium and ε_f is the ***fluctuating component***. Thus, $\varepsilon_m(\hat{\mathbf{p}})$ is independent of position within the medium but may be a function of the polarization unit vector $\hat{\mathbf{p}}$, which denotes the direction of the electric field of the incident wave relative to the geometry of the ellipsoids if the ellipsoids have a preferred orientation. If the orientation is random, ε_m becomes independent of the direction of the incident field. The fluctuating component ε_f accounts for the deviation from the average value $\varepsilon_m(\hat{\mathbf{p}})$ at any point (x,y,z) within the medium. Its average value $\langle \varepsilon_f \rangle = 0$.

The propagation constant of the medium, γ, is governed by the average dielectric constant $\varepsilon_m(\hat{\mathbf{p}})$, as discussed below. The significance of the fluctuating component ε_f, or rather of its statistical spatial distribution, has to do with volume scattering in the medium. Volume scattering is treated in Chapter 11 and therefore is not addressed in this chapter. Instead, we focus our attention on methods of relating the average dielectric constant ε_m of the mixture to the dielectric properties and volume fractions of the constituents. A brief review of dielectric mixing models is given in Section 4-4, followed in succeeding sections by a survey of measured dielectric constant data and applicable models for several heterogeneous mixtures of interest, including sea ice, snow, rocks and powders, soils, and vegetation. Microwave dielectric measurement techniques are not covered in this chapter; the interested reader is referred to Bussey (1967), Nicolson and Ross (1970), Weir (1974), Jones (1976), Campbell (1978), Stuchly and Stuchly (1980), Hallikainen et al. (1986), and El-Rayes and Ulaby (1987).

Throughout this book, the symbol ε refers to the ***average relative dielectric constant of the material*** under consideration, and its first subscript denotes the name of the material (except for ε_0, the permittivity of free space). In general, ε is complex, consisting of a real part, ε', and an imaginary part, ε'',

$$\varepsilon = \varepsilon' - j\varepsilon'', \tag{4.2}$$

where ε' is the ***relative permittivity*** of the material and ε'' is its ***dielectric loss factor***.

The dielectric constant ε is related to the complex ***index of refraction*** n through

$$\varepsilon = n^2. \tag{4.3}$$

With n defined as

$$n = n' - jn'', \tag{4.4}$$

it follows that

$$\varepsilon' = (n')^2 - (n'')^2, \qquad \varepsilon'' = 2n'n'', \tag{4.5}$$

and, conversely,

$$n' = \mathfrak{Re}\{\sqrt{\varepsilon}\}, \qquad n'' = -\mathfrak{Im}\{\sqrt{\varepsilon}\}. \tag{4.6}$$

For a plane wave propagating in a lossy medium in the z direction, the electric-field intensity at a distance z is given by

$$E(z) = E_0 \exp(-\gamma z), \tag{4.7}$$

where E_0 is the field intensity at $z = 0$ and

$$\gamma = \alpha + j\beta, \tag{4.8}$$

where γ, α, and β are the ***propagation***, ***absorption***, and ***phase*** constants of the medium. They are related to n and ε by

$$\alpha = k_0 n'' = -k_0 \,\mathfrak{Im}\{\sqrt{\varepsilon}\} \quad \text{Np/m}, \tag{4.9a}$$

$$\beta = k_0 n' = k_0 \,\mathfrak{Re}\{\sqrt{\varepsilon}\} \quad \text{rad/m}, \tag{4.9b}$$

where $k_0 = 2\pi/\lambda_0$ is the wave number in free space, and λ_0 is the free-space wavelength in meters.

Ignoring scattering losses in the medium, the power density $\mathcal{S}(z)$ at a point z is given by

$$\mathcal{S}(z) = \mathcal{S}_0 \exp(-\kappa_a z), \tag{4.10}$$

where κ_a, the ***power absorption coefficient***, is related to α by

$$\kappa_a = 2\alpha. \tag{4.11}$$

A related quantity of interest in remote sensing is the ***penetration depth*** δ_p. For a scatter-free medium,

$$\delta_p = 1/\kappa_a \qquad \text{(m)}, \tag{4.12a}$$

and if $\varepsilon''/\varepsilon' \ll 1$,

$$\delta_p \approx \frac{\sqrt{\varepsilon'}}{k_0 \varepsilon''}\,. \tag{4.12b}$$

Note that δ_p is related to the ***skin depth*** δ_s by $\delta_p = \delta_s/2$.

4-1 Pure-Water Single-Debye Dielectric Model ($f \leq 50$ GHz)

For pure (distilled) water with no dissolved salts, we offer two dielectric models, a relatively simple model applicable to frequencies $f \leq 50$ GHz, and a more elaborate model applicable to frequencies as high as 1000 GHz. The latter is a special case of the saline water model of Section 4-2.

Until the mid-1990s, the general consensus within the microwave community was that the dielectric constant of water obeys the single-relaxation Debye model for polar molecules (Hasted, 1973), namely

$$\varepsilon_w = \varepsilon_{w\infty} + \frac{\varepsilon_{w0} - \varepsilon_{w\infty}}{1 + j2\pi f \tau_w}\,, \tag{4.13}$$

where

$\varepsilon_{w0} =$ static dielectric constant (at $f = 0$), dimensionless,

$\varepsilon_{w\infty} =$ high-frequency dielectric constant (as $f \to \infty$), dimensionless,

$\tau_w =$ relaxation time constant (s)

$f =$ frequency (Hz).

Rationalizing Eq. (4.13) and equating it to $\varepsilon_w = \varepsilon'_w - j\varepsilon''_w$ leads to

$$\varepsilon'_w = \varepsilon_{w\infty} + \frac{\varepsilon_{w0} - \varepsilon_{w\infty}}{1 + (2\pi f \tau_w)^2}\,, \tag{4.14a}$$

$$\varepsilon''_w = \frac{2\pi f \tau_w(\varepsilon_{w0} - \varepsilon_{w\infty})}{1 + (2\pi f \tau_w)^2}\,. \tag{4.14b}$$

In addition to their dependence on frequency, ε'_w and ε''_w also are temperature-dependent because ε_{w0}, τ_w, and (possibly) $\varepsilon_{w\infty}$ are all functions of the water temperature T.

The magnitude of the high-frequency dielectric constant $\varepsilon_{w\infty}$ was determined by Lane and Saxton (1952) to be

$$\varepsilon_{w\infty} = 4.9. \tag{4.15}$$

Contradictory views have been expressed on the temperature dependence of $\varepsilon_{w\infty}$; but the dependence is so weak that for computational purposes $\varepsilon_{w\infty}$ may be considered a constant equal to 4.9.

The relaxation time of pure water is given by

$$2\pi\tau_w(T) = 1.1109 \times 10^{-10} - 3.824 \times 10^{-12} T + 6.938 \times 10^{-14} T^2 - 5.096 \times 10^{-16} T^3, \tag{4.16}$$

where T is in °C. This expression was obtained by Stogryn (1971) by fitting a polynomial to experimental data. A related term used in the literature is the ***relaxation frequency*** f_0, where

$$f_0 = (2\pi\tau_w)^{-1}. \tag{4.17}$$

The relaxation frequency of pure water lies in the microwave region; f_0 (0 °C) $\approx$ 8.9 GHz and f_0 (20 °C) $\approx$ 16.7 GHz. From Eq. (4.14), it can be shown that ε''_w has its maximum value at $f = f_0$.

Using dielectric measurements made at 1.43 GHz and 2.65 GHz, Klein and Swift (1977) generated the following regression fit for $\varepsilon_{w0}(T)$:

$$\varepsilon_{w0}(T) = 88.045 - 0.4147T + 6.295 \times 10^{-4}T^2 + 1.075 \times 10^{-5}T^3. \tag{4.18}$$

We shall refer to the group of expressions given by Eqs. (4.14) through (4.18) as the ***single-Debye dielectric model*** (or SD^{2}M) for pure water.

The plots in Fig. 4-1 depict the frequency responses of ε'_w and ε''_w of pure water at $T = 0$ °C and 20 °C. They cover the frequency range from 0 to 50 GHz. We observe that the two sets of spectral plots have the same general shape, but their absolute levels are different at the two different temperatures, and the relaxation frequency (frequency at which ε''_w is a maximum) shifts from 8.9 GHz at 0 °C to 16.7 GHz at 20 °C.

Values computed using SD^{2}M are found to be in excellent agreement with experimental measurements conducted at frequencies below 50 GHz across the temperature range $0 \leq T \leq 30$ °C. The error is less than 5%, and at frequencies below 10 GHz it is within 1%.

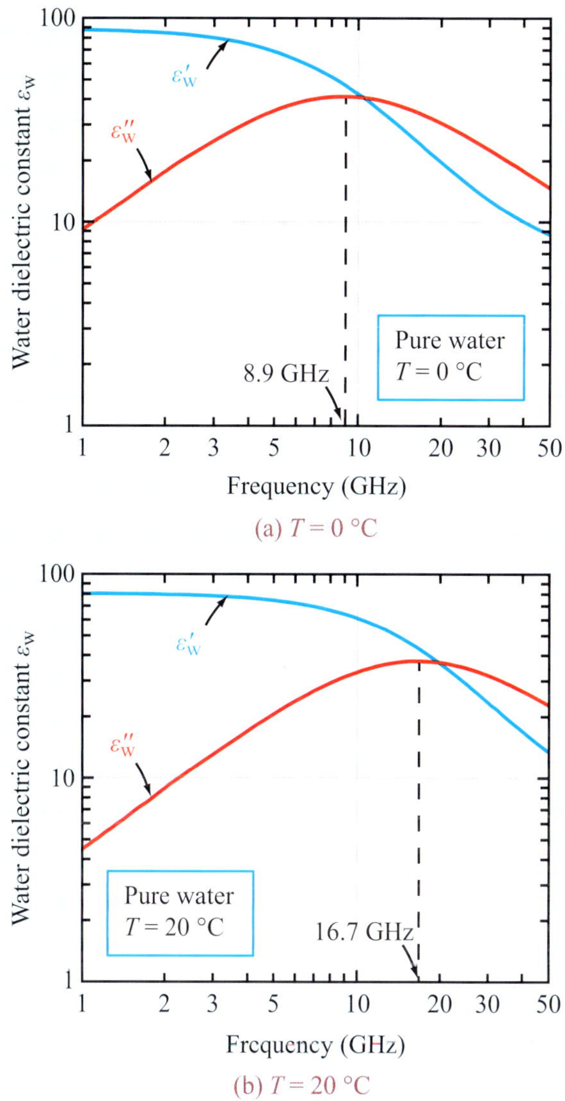

Figure 4-1: Microwave spectra of the permittivity and dielectric loss factor of pure water at (a) 0 °C and (b) 20 °C.

4-2 Saline-Water Double-Debye Dielectric Model ($f \leq 1000$ GHz)

†A review of the literature suggests that the most accurate model available at the present time for computing the dielectric constant of water is based on a ***double-Debye dielectric model*** (D3M) developed by William Ellison and reported in the book edited by Mätzler (2006, pp. 431–455). The D3M, which was developed for seawater (and reduces to a model for pure water when the water salinity is set to zero) is given by

$$\varepsilon'_w = \varepsilon_{w\infty} + \frac{\varepsilon_{w0} - \varepsilon_{w1}}{1 + (2\pi f \tau_{w1})^2} + \frac{\varepsilon_{w0} - \varepsilon_{w\infty}}{1 + (2\pi f \tau_{w2})^2}, \tag{4.19a}$$

†Computer Code 4.1 (pure water); Computer Code 4.2 (saline water).

$$\varepsilon''_{\rm w} = \frac{2\pi f \tau_{\rm w1}(\varepsilon_{\rm w0} - \varepsilon_{\rm w1})}{1 + (2\pi f \tau_{\rm w1})^2} + \frac{2\pi f \tau_{\rm w2}(\varepsilon_{\rm w0} - \varepsilon_{\rm w\infty})}{1 + (2\pi f \tau_{\rm w2})^2} + \frac{\sigma_{\rm i}}{2\pi\varepsilon_0 f}, \quad (4.19b)$$

where ε_0 is the permittivity of free space. The expression given by Eq. (4.19a) contains two relaxation terms, one with a relaxation time constant $\tau_{\rm w1}$ and another with $\tau_{\rm w2}$. The same is true for $\varepsilon''_{\rm w}$, which implies that the spectral plot of $\varepsilon''_{\rm w}$ should exhibit maxima at two relaxation frequencies,

$$f_{01} = \frac{1}{2\pi\tau_{\rm w1}}, \qquad f_{02} = \frac{1}{2\pi\tau_{\rm w2}}. \quad (4.20)$$

At $T = 0$ °C, $f_{01} = 8.9$ GHz and $f_{02} = 201.8$ GHz. The expression for $\varepsilon''_{\rm w}$ also includes a term proportional to the ionic conductivity $\sigma_{\rm i}$ of the water solution when the solution contains dissolved salts. For pure water, $\sigma_{\rm i} = 0$.

As noted earlier in Section 4-1, the dielectric spectrum of water exhibits a relaxation frequency in the microwave region ($f_0 = 8.9$ GHz at 0 °C). The suggestion to add a second relaxation frequency was first made in the 1950s, but it was not seriously considered until several new experimental measurements were conducted in the mid-1990s (Barthel et al., 1990; Kindt et al., 1996; Stogryn et al., 1996; Rønne et al., 1997). Using the double-Debye model and all available, credible dielectric data for pure and saline water, Ellison (in Mätzler, 2006, pp. 431–455) generated empirical functions for all of the parameters in Eq. (4.19). The applicable range of conditions is:

$$0 \le T \le 30^\circ\text{C},$$
$$0 \le S \le 40‰ \text{ or } 40 \text{ psu},$$
$$0 \le f \le 1000 \text{ GHz},$$

where S is the water ***salinity***, defined as the total mass of solid salt in grams dissolved in one kilogram of solution. Thus, S is in parts per thousand (‰) on a weight basis. Instead of using the symbol (‰), salinity has been given the equivalent unit of psu, which stands for ***practical salinity unit***, with 1 psu = 1 (‰). For pure water, $S = 0$.

The parameter functions are:

$$\varepsilon_{\rm w0} = 87.85306 \cdot \exp\{-0.00456992T - a_1 S - a_2 S^2 - a_3 ST\}, \quad (4.21a)$$

$$\varepsilon_{\rm w1} = a_4 \exp\{-a_5 T - a_6 S - a_7 ST\}, \quad (4.21b)$$

$$\tau_{\rm w1} = (a_8 + a_9 S) \exp\left(\frac{a_{10}}{T + a_{11}}\right) \text{ ns}, \quad (4.21c)$$

$$\tau_{\rm w2} = (a_{12} + a_{13} S) \exp\left(\frac{a_{14}}{T + a_{15}}\right) \text{ ns}, \quad (4.21d)$$

$$\varepsilon_{\rm w\infty} = a_{16} + a_{17} T + a_{18} S, \quad (4.21e)$$

$$\sigma_{\rm i} = \sigma(T, 35) \cdot P(S) \cdot Q(T, S), \quad (4.21f)$$

where

$$\sigma(T, 35) = 2.903602 + 8.607 \cdot 10^{-2} T + 4.738817 \cdot 10^{-4} T^2 - 2.991 \cdot 10^{-6} T^3 + 4.3041 \cdot 10^{-9} T^4, \quad (4.21g)$$

$$P(S) = S\, \frac{37.5109 + 5.45216S + 0.014409S^2}{1004.75 + 182.283S + S^2}, \quad (4.21h)$$

$$Q(T, S) = 1 + \frac{\alpha_0 (T - 15)}{T + \alpha_1}, \quad (4.21i)$$

$$\alpha_0 = \frac{6.9431 + 3.2841S - 0.099486S^2}{84.85 + 69.024S + S^2}, \quad (4.21j)$$

$$\alpha_1 = 49.843 - 0.2276S + 0.00198S^2. \quad (4.21k)$$

Coefficients a_1 to a_{18} are listed in Table 4-1. According to Ellison (in Mätzler, 2006, p. 454), this semiempirical model represents the dielectric constant "of pure water to within 1% over the frequency range 0–20 GHz, to within 3% over 30–100 GHz, and to within 5% over 100–1000 GHz." He also adds that for seawater, the model is accurate to within 3% (of experimental data) over the frequency range 3–105 GHz. As there are no reported experimental measurements of the dielectric constant of seawater at frequencies greater than 105 GHz, it is not possible to ascertain the accuracy of the model at $f > 105$ GHz.

Figures 4-2(a) and 4-2(b) display the frequency responses of $\varepsilon'_{\rm w}$ and $\varepsilon''_{\rm w}$ for pure water and seawater

Table 4-1: Values of coefficients in Eq. (4.21) of D3M (Double-Debye dielectric model) [Mätzler, 2006].

$a_1 = 0.46606917\text{E}-02$
$a_2 = -0.26087876\text{E}-04$
$a_3 = -0.63926782\text{E}-05$
$a_4 = 0.63000075\text{E}+01$
$a_5 = 0.26242021\text{E}-02$
$a_6 = -0.42984155\text{E}-02$
$a_7 = 0.34414691\text{E}-04$
$a_8 = 0.17667420\text{E}-03$
$a_9 = -0.20491560\text{E}-06$
$a_{10} = 0.58366888\text{E}+03$
$a_{11} = 0.12684992\text{E}+03$
$a_{12} = 0.69227972\text{E}-04$
$a_{13} = 0.38957681\text{E}-06$
$a_{14} = 0.30742330\text{E}+03$
$a_{15} = 0.12634992\text{E}+03$
$a_{16} = 0.37245044\text{E}+01$
$a_{17} = 0.92609781\text{E}-02$
$a_{18} = -0.26093754\text{E}-01$

at 20 °C, calculated using the D3M expressions in Eqs. (4.19) and (4.21). The average salinity of seawater is $S = 32.54$ psu. Whereas the relaxation phenomenon at $f_{01} = 16.7$ GHz is clearly visible in the plots of Fig. 4-2, the effect is much more subtle at $f_{02} = 281.4$ GHz. Moreover, ε'_{w} of pure water and ε'_{sw} of seawater are approximately the same over the entire spectrum from 1 to 1000 GHz, and so are the loss factors ε''_{w} and ε''_{w} at frequencies above the relaxation frequency f_{01}, but $\varepsilon''_{\text{sw}}$ of seawater is much larger than ε''_{w} of pure water at frequencies below f_{01}.

For the sake of completeness, the reader is referred to another equally credible model developed by Meissner and Wentz (2004). The model uses the same double-Debye expressions given by Eq. (4.19) and relies on basically the same experimental data used by Ellison (Mätzler, 2006), but the forms of the parameter functions are different from those developed by Ellison.

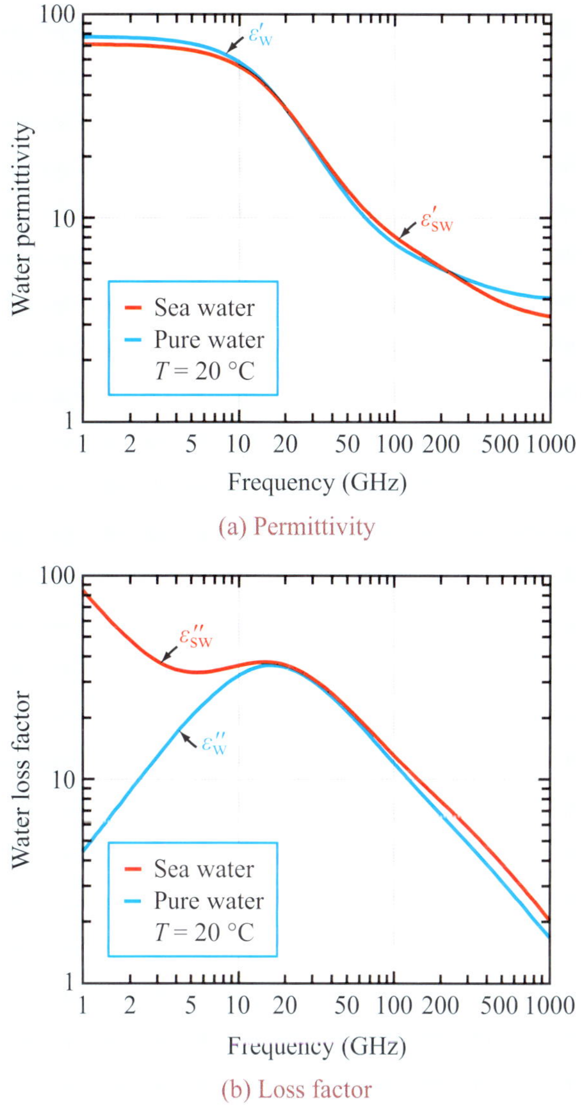

Figure 4-2: Microwave spectra of (a) the permittivities and (b) loss factors of pure water (ε_{w}) and seawater (ε_{sw}) at $T = 20$ °C. Sea salinity is 32.54 psu.

4-3 Dielectric Constant of Pure Ice

Unlike liquid water, whose relaxation frequencies lie in the microwave region, the relaxation frequency of pure ice f_{i0} occurs in the kilohertz region. Hence, in the microwave region (where f is on the order of 10^9 Hz), the quantity

$$2\pi f \tau_i = \frac{f}{f_{i0}} \gg 1.$$

Consequently, for ice the Debye model expressions of the form given by Eq. (4.14) simplify to

$$\varepsilon_i' \approx \varepsilon_{i\infty}, \tag{4.22a}$$

$$\varepsilon_i'' \approx \frac{\varepsilon_{i0} - \varepsilon_{i\infty}}{2\pi f \tau_i} = \frac{\alpha_0}{f}, \tag{4.22b}$$

where $\alpha_0 = (\varepsilon_{i0} - \varepsilon_{i\infty})/2\pi\tau_i$. According to Mätzler and Wegmüller (1987), ε_i' is essentially independent of frequency from 10 MHz to 300 GHz, and it exhibits a weak temperature dependence of the form:

$$\varepsilon_i' = 3.1884 + 9.1 \times 10^{-4} T \quad (-40^\circ\text{C} \leq T \leq 0), \tag{4.23}$$

where T is the temperature in °C. In practice, we can ignore the temperature dependence and use $\varepsilon_i' \approx 3.2$ across the entire microwave spectrum.

In Eq. (4.22b), ε_i'' varies as $1/f$ and the coefficient α_0 is a function of only one physical variable, namely the temperature T. However, ice exhibits an infrared absorption spectrum that includes a nonresonant term that varies as f. This led Hufford (1991) to propose the model†

$$\varepsilon_i'' = \frac{\alpha_0}{f} + \beta_0 f, \tag{4.24}$$

where α_0 and f are in GHz and β_0 is in $(\text{GHz})^{-1}$. As explained in detail by Mätzler (2006, pp. 456–460), the coefficients α_0 and β_0 are given by the semiempirical expressions

$$\alpha_0 = (0.00504 + 0.0062\theta) \cdot \exp(-22.1\theta) \qquad (\text{GHz}), \tag{4.25a}$$

$$\beta_0 = \frac{B_1}{T_K} \frac{\exp(b/T_K)}{[\exp(b/T_K) - 1]^2} + B_2 f^2 + \exp[-9.963 + 0.0372(T_K - 273.16)] \quad (\text{GHz}^{-1}), \tag{4.25b}$$

† Computer Code 4.3.

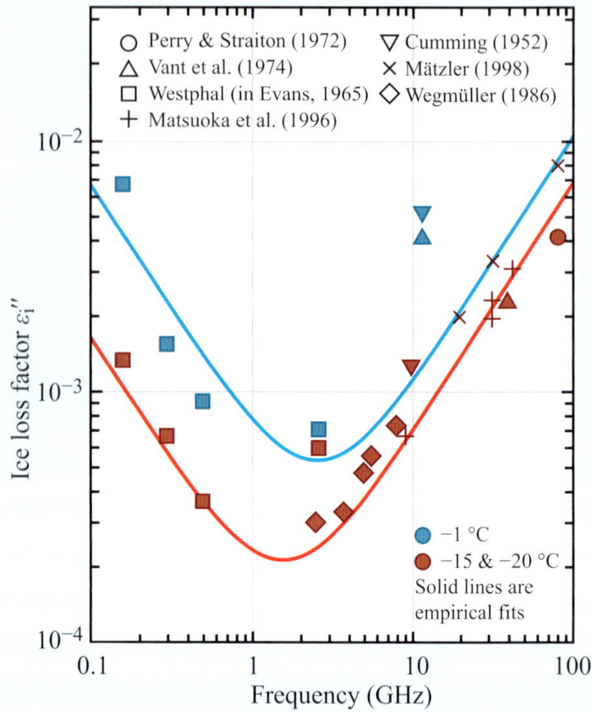

Figure 4-3: Loss factor of pure ice.

$$\theta = \frac{300}{T_K} - 1, \qquad T_K \text{ in K}, \tag{4.25c}$$

$$B_1 = 0.0207 \text{ K/GHz}, \qquad b = 335 \text{ K},$$

and

$$B_2 = 1.16 \times 10^{-11} \text{ GHz}^{-3}.$$

Figure 4-3 displays plots of the model for ε_i'' at 272 K (−1 °C) and 253 K (−20 °C). Also shown are data reported in the literature by various research groups. Both the data and the model exhibit a minimum in the neighborhood of 1 GHz. Some of the variability in the data is attributed to the difficulty associated with making accurate measurements of ε_i'' due to its small magnitude.

Fresh water ice often contains ionic impurities, such as dissolved salts, which causes its ε_i'' to increase significantly in comparison with that for pure ice. According to the data reported by Wegmüller (1986) and reproduced in Fig. 4-4, ε_i'' of an impure ice sample with $S = 13$ psu is 2–8 times larger than that of a

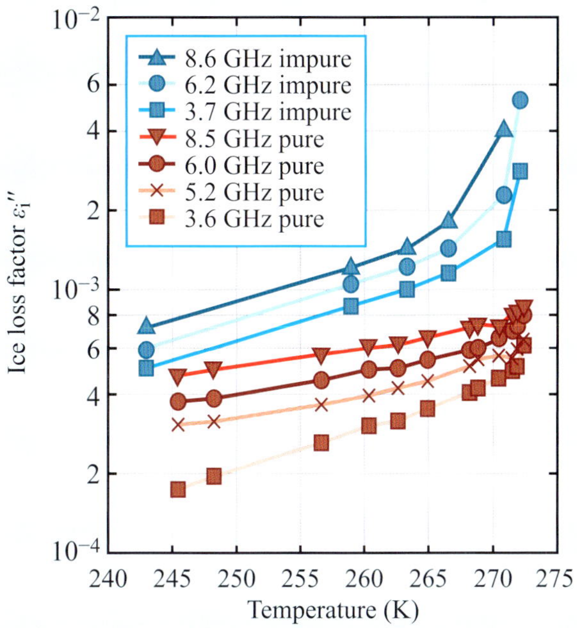

Figure 4-4: Loss factor ε_i'' of pure and slightly saline ($S = 13$ psu) ice as a function of temperature [Wegmüller, 1986].

pure ice sample; the relative increase depends on the temperature T, the frequency f, and the salinity S.

4-4 Dielectric Mixing Models for Heterogeneous Materials

The equivalent dielectric constant of a heterogeneous material (mixture) consisting of two or more substances is related to the dielectric constants of the individual substances, their volume fractions, their spatial distributions, and their orientations relative to the direction of the incident electric-field vector. Usually the substance with the highest volume fraction is regarded as the ***host material***, or continuous medium, and the other substances are regarded as ***inclusions***.

To relate the average dielectric constant of a mixture to the factors mentioned above, it is necessary to relate the average electric field (within the mixture as a whole) to the fields within the inclusions. If the inclusions are randomly dispersed within the host material, it is not possible to derive an exact solution for the fields within the inclusions because the mutual interactions of the inclusions (through their polarization fields) are dependent upon their positions relative to each other. Tinga and his associates (1973) provide a concise historical review of the various approximations proposed in the literature for solving the interaction problem, and Sihvola (1999) provides a comprehensive treatment of electromagnetic mixing formulas for computing the equivalent dielectric constant of a heterogeneous mixture. These approximations include (a) ignoring short-range interactions (between inclusions) altogether by restricting the validity of the dielectric mixing model to only those mixtures characterized by a low concentration of inclusions, (b) assigning an ***effective dielectric constant*** ε^* to the immediate surroundings of an included particle in an attempt to account for short-range interactions (de Loor, 1956; 1968), and (c) accounting for first-order inclusion interactions by solving Maxwell's equations with appropriate boundary conditions (Tinga et al., 1973). Also, *it is assumed in all cases that the dimensions of the inclusions are much smaller than the wavelength of the radiation propagating in the mixture medium*, thereby allowing the treatment to ignore volume scattering in the material.

4-4.1 Randomly Oriented Ellipsoidal Inclusions

A host material containing only one type of inclusion is called a ***two-phase mixture***. Consider the case where the inclusions are ellipsoidal particles of identical shape and size, randomly dispersed within the host material and randomly oriented with respect to the direction of the electric field of the incident wave. Each ellipsoid has dimensions $2a$, $2b$, and $2c$ along its major axes (Fig. 4-5), and the host and inclusion materials have isotropic dielectric constants ε_h and ε_i, respectively. The inclusion concentration is defined by the ***inclusion volume fraction*** v_i,

$$v_i = \frac{4}{3}\pi abcN, \qquad 0 \le v_i \le 1, \tag{4.26}$$

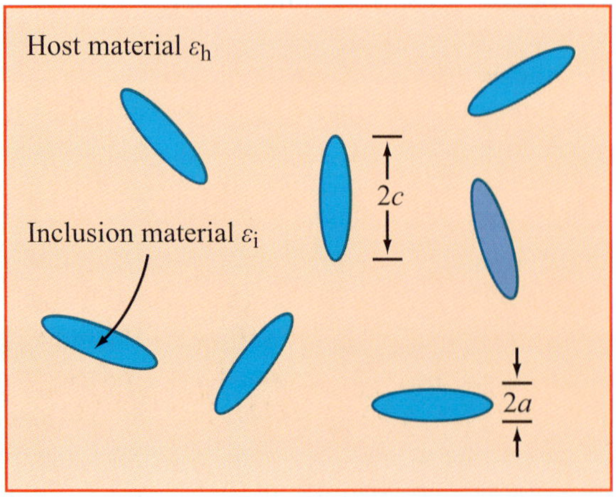

Figure 4-5: Randomly oriented ellipsoidal inclusions of identical size, shape, and dielectric constant ε_i, dispersed in a host material with dielectric constant ε_h. The ellipsoidal particles have semiaxes a, b, and c.

where N is the number of ellipsoids per m^3. The volume fraction of the host material is $v_h = 1 - v_i$.

4-4.2 Polder–van Santen/de Loor Formulas

Extending the early work of Polder and van Santen (1946) on the dielectrics of mixtures of solids, de Loor (1968) developed the following formula for the equivalent dielectric constant of a two-phase mixture containing randomly oriented ellipsoidal particles:

$$\varepsilon_m = \varepsilon_h + \frac{v_i}{3}(\varepsilon_i - \varepsilon_h) \sum_{u=a,b,c} \left[\frac{1}{1 + A_u \left(\dfrac{\varepsilon_i}{\varepsilon^*} - 1 \right)} \right], \tag{4.27}$$

where ε^* is an ***effective dielectric constant*** for the region immediately surrounding an included particle. For small inclusion volume fractions ($v_i \leq 0.1$), de Loor argues that short-range particle interactions may be ignored, in which case ε^* may be set equal to ε_h, the dielectric constant of the host material. For higher values of v_i, particle interactions may be accounted for, at least in part, by regarding the particle as being surrounded by the mixture rather than the host material. In that case, de Loor assigns $\varepsilon^* = \varepsilon_m$.

The quantity A_u inside the summation in Eq. (4.27) is the ***depolarization factor*** of the ellipsoid along its u axis ($u = a$, b, or c) and is given by (Landau and Lifshitz, 1975):

$$A_u = \frac{abc}{2} \int_0^\infty \frac{ds}{(s+u^2)[(s+a^2)(s+b^2)(s+c^2)]^{1/2}}, \quad u = a,\ b \text{ or } c. \tag{4.28}$$

The sum of the three depolarization factors is

$$A_a + A_b + A_c = 1. \tag{4.29}$$

For some specific ellipsoidal configurations of interest, the integral form of Eq. (4.28) can be simplified into closed-form solutions.

(a) A_u for Prolate Spheroids: For a prolate spheroid (needle-like) with $a = b$ (symmetric about the c axis) and $c > a$, Eq. (4.28) can be shown to lead to (Kerker, 1969)

$$A_c = \frac{1-e^2}{2e^3} \left[\ln\left(\frac{1+e}{1-e} \right) - 2e \right], \tag{4.30a}$$

$$A_a = A_b = \frac{1-A_c}{2}, \tag{4.30b}$$

where e is the ellipsoid eccentricity,

$$e = \left[1 - \left(\frac{a}{c} \right)^2 \right], \qquad c > a. \tag{4.31}$$

(b) A_u for Spheres: For $a = b = c$,

$$A_a = A_b = A_c = \frac{1}{3}. \tag{4.32}$$

(c) A_u for Oblate Spheroids: For an oblate spheroid (disc-like) with $a = b$ and $c < a$, Eq. (4.28) has the closed-form solution

$$A_c = \frac{1}{e^2} \left[1 - \frac{(1-e^2)^{1/2}}{e} \sin^{-1} e \right], \tag{4.33a}$$

$$A_a = A_b = \frac{1-A_c}{2}, \tag{4.33b}$$

where the eccentricity e of an oblate spheroid is defined as

$$e = \left[1 - \left(\frac{c}{a}\right)^2\right]^{1/2}, \qquad c < a. \tag{4.34}$$

Next, we consider Eq. (4.27) for specific shapes of inclusions.

(a) ε_m of Circular Disc Inclusions: For thin circular discs [Fig. 4-6(a)] with $a = b$ and $c \ll a$:

$$A_a = A_b = 0, \qquad A_c = 1, \tag{4.35a}$$

and Eq. (4.27) simplifies to

$$\varepsilon_m = \varepsilon_h + \frac{v_i}{3}\,(\varepsilon_i - \varepsilon_h)\left(2 + \frac{\varepsilon^*}{\varepsilon_i}\right). \tag{4.35b}$$

(b) ε_m of Spherical Inclusions: For spheres with $a = b = c$ [Fig. 4-6(b)]:

$$A_a = A_b = A_c \tag{4.36a}$$

and

$$\varepsilon_m = \varepsilon_h + 3v_i\varepsilon^*\,\frac{\varepsilon_i - \varepsilon_h}{\varepsilon_i + 2\varepsilon^*}. \tag{4.36b}$$

(c) ε_m of Needle Inclusions: For long, narrow needles [Fig. 4-6(c)] with $a = b$ and $c \gg a$:

$$A_a - A_b - 0.5, \qquad A_c - 0, \tag{4.37a}$$

and

$$\varepsilon_m = \varepsilon_h + v_i\,\frac{(\varepsilon_i - \varepsilon_h)(5\varepsilon^* + \varepsilon_i)}{3(\varepsilon_i + \varepsilon^*)}. \tag{4.37b}$$

► In all cases,

$$\varepsilon_m \to \varepsilon_h \text{ as } v_i \to 0,$$
$$\varepsilon_m \to \varepsilon_i \text{ as } v_i \to 1, \text{ if } \varepsilon^* = \varepsilon_m.$$

In the intermediate range, setting $\varepsilon^* = \varepsilon_h$ tends to underestimate the value of ε_m (when compared with experimental data) and setting $\varepsilon^* = \varepsilon_m$ tends to overestimate ε_m. ◄

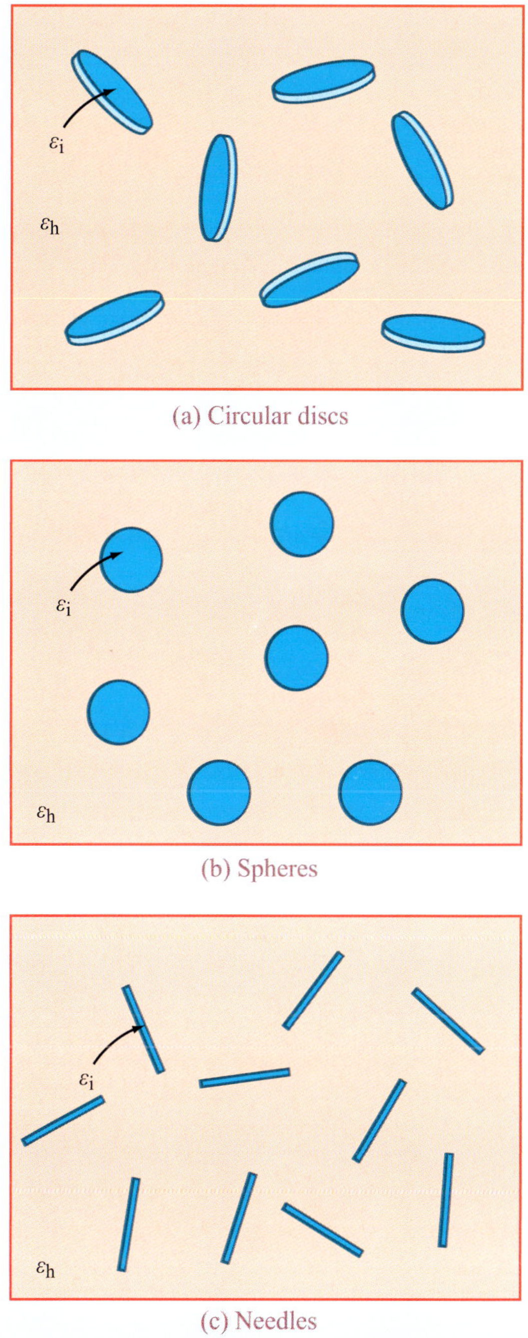

Figure 4-6: Examples of randomly distributed ellipsoidal configurations.

4-4.3 Tinga-Voss-Blossey (TVB) Formulas

The general dielectric formulation given by Tinga et al. (1973) pertains to randomly dispersed confocal ellipsoids consisting of an inner ellipsoid with dielectric constant ε_i surrounded by a shell of another material having a dielectric constant ε_h. In the derivation given by Tinga et al. (1973), the two-phase mixture is compelled to take on the geometry of two confocal ellipsoids with an inner ellipsoid representing the inclusion dielectric material and a "fictitious" outer ellipsoidal shell representing the host material (Fig. 4-7). The dimensions of the fictitious outer ellipsoid are such that the volume of the shell surrounding the inner ellipsoid is equal to the total volume of the host material in the mixture divided by the total number of ellipsoids. The inner ellipsoid has semiaxes a_2, b_2, and c_2 and a volume $V_2 = (4/3)\pi a_2 b_2 c_2$; similarly, the outer ellipsoid has semiaxes a_1, b_1, and c_1 and volume $V_1 = (4/3)\pi a_1 b_1 c_1$. Because of the chosen confocal geometry,

$$a_1^2 - a_2^2 = b_1^2 - b_2^2 = c_1^2 - c_2^2 = k^2, \tag{4.38}$$

where k^2 is a constant whose value is determined from the condition that the inclusion volume fraction v_i is

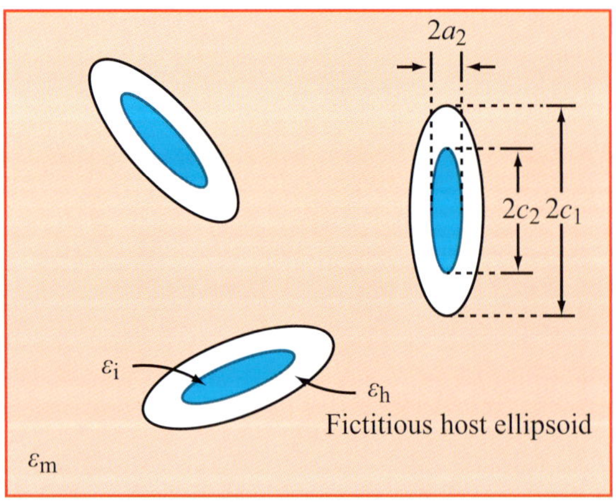

Figure 4-7: Two-phase confocal ellipsoidal TVB model (Tinga et al., 1973) in which each ellipsoidal inclusion is surrounded by a shell of host material whose thickness is governed by the volume fraction of the host material.

given by

$$v_i = \frac{V_2}{V_1}. \tag{4.39}$$

For randomly oriented inclusions, the equivalent dielectric constant of the two-phase mixture is given by[†]

$$\varepsilon_m = \varepsilon_h + \frac{v_i}{3}(\varepsilon_i - \varepsilon_h) \cdot \sum_{u=a,b,c} \frac{1}{1 + (A_{u2} - A_{u1} v_i)\left(\dfrac{\varepsilon_i}{\varepsilon_h} - 1\right)}, \tag{4.40}$$

where A_{u1} and A_{u2} are the depolarization factors of the outer and inner ellipsoids, respectively,

$$A_{u_j} = \frac{a_j b_j c_j}{2} \int_0^\infty \frac{ds}{(s+u_j^2)[(s+a_j^2)(s+b_j^2)(s+c_j^2)]^{1/2}}$$

$$u_j = a_j,\ b_j,\ \text{or}\ c_j; \quad j = 1 \text{ or } 2. \tag{4.41}$$

(a) ε_m of Circular Disc Inclusions: For thin circular discs with $a_1 = b_1$, $a_2 = b_2$, $c_1 \ll a_1$, and $c_2 \ll a_2$,

$$A_{a_1} = A_{a_2} = A_{b_1} = A_{b_2} = 0,$$
$$A_{c_1} = A_{c_2} = 1,$$

and

$$\varepsilon_m = \varepsilon_h + \frac{v_i}{3}(\varepsilon_i - \varepsilon_h)\left[\frac{2\varepsilon_i(1-v_i) + \varepsilon_h(1+2v_i)}{v_i\varepsilon_h + (1-v_i)\varepsilon_i}\right]. \tag{4.42}$$

(b) ε_m of Spherical Inclusions: For spheres,

$$A_{a_1} = A_{a_2} = A_{b_1} = A_{b_2} = A_{c_1} = A_{c_2} = \frac{1}{3}$$

and

$$\varepsilon_m = \varepsilon_h + \frac{3v_i\varepsilon_h(\varepsilon_i - \varepsilon_h)}{(2\varepsilon_h + \varepsilon_i) - v_i(\varepsilon_i - \varepsilon_h)}. \tag{4.43}$$

(c) ε_m of Needle Inclusions: For long, narrow needles,

$$A_{a_1} = A_{a_2} = A_{b_1} = A_{b_2} = 0.5,$$
$$A_{c_1} = A_{c_2} = 0,$$

and

[†] **Computer Code 4.4.**

$$\varepsilon_m = \varepsilon_h + \frac{v_i}{3}(\varepsilon_i - \varepsilon_h)\left[\frac{\varepsilon_h(5+v_i)+(1-v_i)\varepsilon_i}{\varepsilon_h(1+v_i)+\varepsilon_i(1-v_i)}\right]. \tag{4.44}$$

▶ In all cases,

$$\varepsilon_m \to \varepsilon_h \text{ as } v_i \to 0,$$

$$\varepsilon_m \to \varepsilon_i \text{ as } v_i \to 1,$$

and the attractive feature of the ***Tinga-Voss-Blossey*** (TVB) model is that the unknown quantity, ε_m, is only on one side of the expressions given by Eqs. (4.42) to (4.44). ◀

Figure 4-8 displays plots for ε'_m and ε''_m as a function of the inclusion volume fraction v_i for spherical inclusions. The plots include the two de Loor formulas with $\varepsilon^* = \varepsilon_h$ and $\varepsilon^* = \varepsilon_m$ and the formula given by Eq. (4.43) based on the TVB model. The plots are for a host material with $\varepsilon_h = 1 - j0$ (air) and inclusions with $\varepsilon_i = 10 - j1$. The TVB model plot for ε'_m is intermediate between the two de Loor plots, and generally provides a better fit to experimental data. The same is true for ε''_m.

4-4.4 Other Dielectric Mixing Formulas

The literature contains numerous models and formulas for describing the dielectric behavior of mixtures. Comprehensive reviews are available in the paper by Tinga et al. (1973) and the books by Van Beek (1967) and Sihvola (1999). Most of these formulas are special cases of formulas given in earlier subsections.

We occasionally find it useful in forthcoming sections to use the formula

$$\varepsilon_m^\alpha = \varepsilon_h^\alpha + v_i(\varepsilon_i^\alpha - \varepsilon_h^\alpha), \tag{4.45}$$

where α is a constant. For $\alpha = 1$, the preceding formula is known as the ***linear model***, for $\alpha = 1/2$ as the ***refractive model*** (since $\varepsilon^{1/2} = n$ is the refractive index), and for $\alpha = 1/3$ as the ***cubic model***.

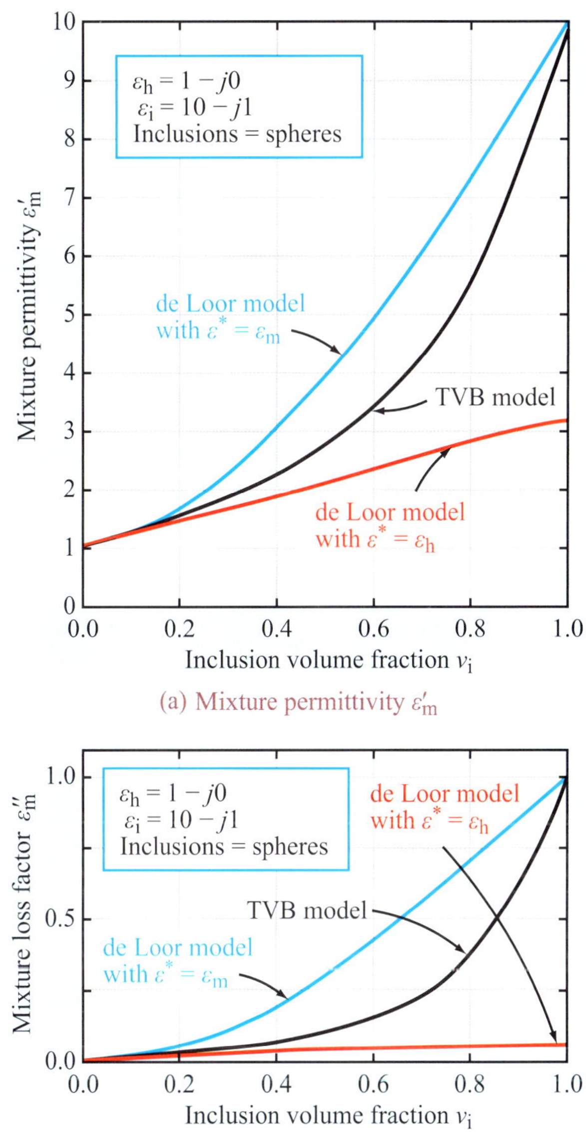

(a) Mixture permittivity ε'_m

(b) Mixture loss factor ε''_m

Figure 4-8: Comparison of three dielectric mixing formulas for spherical inclusions.

4-5 Sea Ice

Sea ice is a much more complicated medium, both structurally and electromagnetically, than fresh-water ice. Sea ice is a heterogeneous mixture of liquid brine inclusions and air pockets interspersed within the ice medium. The ***brine inclusions***, containing salt and water, exercise a strong influence on the complex dielectric constant of the mixture, because of their high complex-dielectric constant when compared with that of ice.

The growth rate of the ice layer governs both the shape and concentration of the brine inclusions. Sea ice generally is divided into three categories on the basis of age: (a) ***young ice***, which typically is less than 30 cm thick, (b) ***first-year ice***, whose thickness usually is between 30 cm and 2 m, and (c) ***multiyear ice***, whose thickness usually exceeds 2 m (Vant et al., 1978). The first two ice categories are similar in structure and often are treated as a single category. One of the major differences between multiyear ice and the other ice categories is their brine concentrations. A salinity profile of first-year ice typically decreases from about 5–16 psu near the surface to about 4–5 psu in the bulk of the ice and then increases rapidly to about 30 psu near the ice-water interface. In contrast, the salinity of multiyear ice usually is less than 1 psu in the surface layer and about 2–3 psu in the bulk portion. These values are representative of sea ice in the Arctic Ocean, where the salinity of liquid water is of the order of 32 psu. Close to land, the water's salinity can be much lower; in the Gulf of Finland, for example, the salinity of water is 6 psu and in the Gulf of Bothnia only 2 psu (Hallikainen, 1980). Sea-ice salinity in such areas is correspondingly lower than that of sea ice in arctic areas.

In general, the complex ***dielectric constant of sea ice***, ε_{si}, is a function of the following parameters: (a) the complex dielectric constant of pure ice, ε_{i}, (b) the complex dielectric constant of the brine pockets or inclusions, ε_{b}, (c) the ***fraction of brine by volume***, v_{b}, and (d) the shape and orientation of the brine pockets or inclusions relative to the direction of the electric field of the wave propagating in the ice medium. Indirectly, ε_{si} also is a function of the ice temperature T [through parameters (a)–(c)] and the ice salinity S_{i} (through v_{b}). Expressions and graphs for ε_{i} are available in Section 4-3. The other parameters are discussed next.

4-5.1 Dielectric Constant of Brine

Liquid brine is the same as saline water; therefore, the same basic expressions defining ε_{w} in Section 4-2 may be used to compute the complex dielectric constant of brine, ε_{b}, except for two modifications:

(1) The salinity of liquid brine is governed by its temperature. In sea ice, liquid brine exists in the form of pockets and inclusions, so a drop in temperature causes additional water to freeze, thereby increasing the brine-salt concentration. The ***brine salinity*** S_{b} (in psu) is related to the temperature T (°C) by the following empirical expressions (Assur, 1960; Poe et al., 1972):

$$\begin{aligned} S_{\text{b}} &= 1.725 - 18.756T - 0.3964T^2, \\ &\quad -8.2 \leq T \leq -2°\text{C}; \\ S_{\text{b}} &= 57.041 - 9.929T - 0.16204T^2 - 0.002396T^3, \\ &\quad -22.9 \leq T \leq -8.2°\text{C}; \\ S_{\text{b}} &= 242.94 + 1.5299T + 0.0429T^2, \\ &\quad -36.8 \leq T \leq -22.9°\text{C}; \\ S_{\text{b}} &= 508.18 + 14.535T + 0.2018T^2, \\ &\quad -43.2 \leq T \leq -36.8°\text{C}. \end{aligned} \tag{4.46}$$

A plot of S_{b} as a function of T is shown in Fig. 4-9.

The link between S_{b} and T means that the dielectric constant of brine ε_{b} is determined by the electromagnetic frequency f and the temperature T only.

(2) Some of the numerical expressions given in Section 4-2 were formulated with seawater in mind; therefore, they are not always applicable when salinities extend beyond 35 psu. Hence, the expressions developed by Stogryn (1971) for NaCl solutions are used instead because they cover the range $0 \leq S \leq 157$ psu. Even though sea-ice brine contains other salts in addition to NaCl, NaCl is the principal component; therefore, the expressions developed for pure NaCl solutions are assumed applicable. In terms of Stogryn's (1971) formulation, ε'_{b} and ε''_{b} are given by

$$\varepsilon'_{\text{b}} = \varepsilon_{\text{w}\infty} + \frac{\varepsilon_{\text{b}0} - \varepsilon_{\text{w}\infty}}{1 + (2\pi f \tau_{\text{b}})^2}, \tag{4.47a}$$

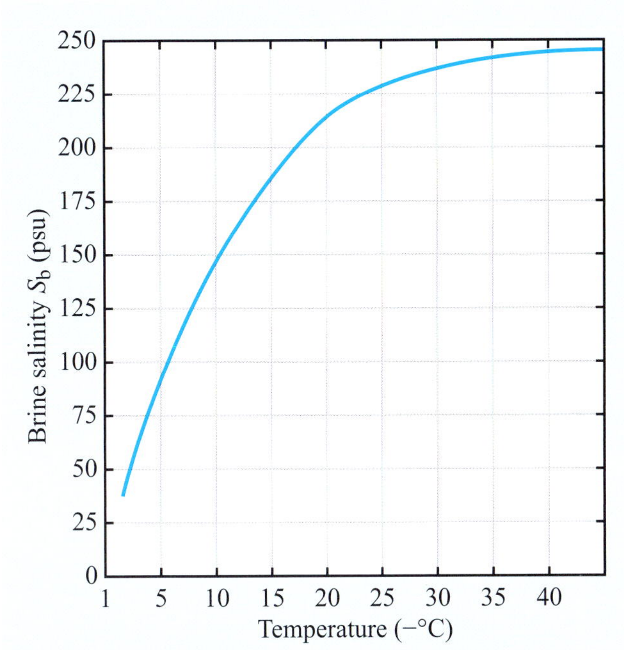

Figure 4-9: Brine salinity in sea ice as a function of negative temperature computed using Eq. (4.46).

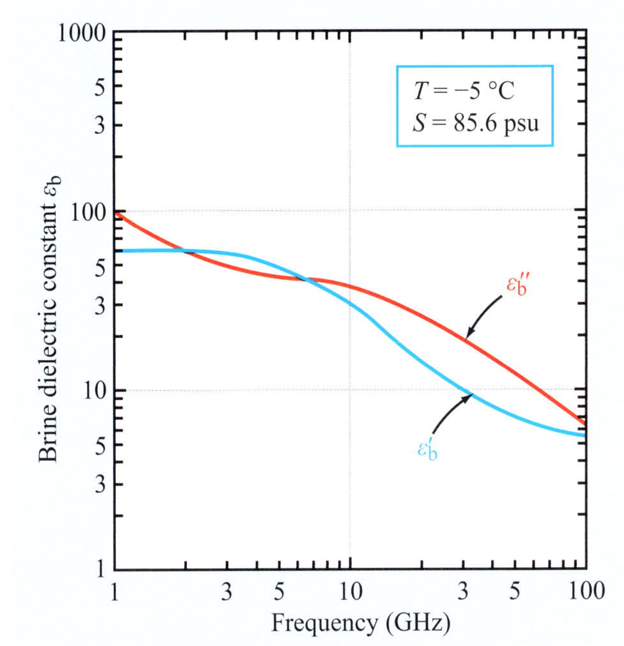

Figure 4-10: Dielectric constant of liquid brine as a function of frequency, calculated using the expressions in Section 4-5.1.

$$\varepsilon_b'' = (2\pi f \tau_b) \frac{(\varepsilon_{b0} - \varepsilon_{w\infty})}{1 + (2\pi f \tau_b)^2} + \frac{\sigma_b}{2\pi f \varepsilon_0}, \qquad (4.47b)$$

where $\varepsilon_{w\infty} = 4.9$,

$$\varepsilon_{b0}(T, N_b) = \varepsilon_{b0}(T, 0)\, a_1(N_b), \qquad (4.48a)$$

$$\tau_b(T, N_b) = \tau_b(T, 0)\, b_1(T, N_b), \qquad (4.48b)$$

$$\sigma_b(T, N_b) = \sigma_b(25, N_b)\, c_1(\Delta, N_b). \qquad (4.48c)$$

The preceding functions are given by the following expressions:

$$\varepsilon_{b0}(T, 0) = \varepsilon_{w0}(T), \text{ as defined by Eq. (4.18)},$$

$$a_1(N_b) = 1.0 - 0.255N_b + 5.15 \times 10^{-2} N_b^2 - 6.89 \times 10^{-3} N_b^3, \qquad (4.49a)$$

$$2\pi\tau_b(T, 0) = 2\pi\tau_w(T), \text{ as defined by Eq. (4.16)},$$

$$b_1(T, N_b) = 1.0 + 0.146 \times 10^{-2} T N_b - 4.89 \times 10^{-2} N_b - 2.97 \times 10^{-2} N_b^2 + 5.64 \times 10^{-3} N_b^3, \qquad (4.49b)$$

$$\sigma_b(25, N_b) = N_b(10.39 - 2.378N_b + 0.683N_b^2 - 0.135N_b^3 + 1.01 \times 10^{-2} N_b^4), \qquad (4.49c)$$

$$c_1(\Delta, N_b) = 1.0 - 1.96 \times 10^{-2}\Delta + 8.08 \times 10^{-5}\Delta^2 - N_b\Delta[3.02 \times 10^{-5} + 3.92 \times 10^{-5}\Delta + N_b(1.72 \times 10^{-5} - 6.58 \times 10^{-6}\Delta)], \qquad (4.49d)$$

$$N_b = S_b[1.707 \times 10^{-2} + 1.205 \times 10^{-5} S_b + 4.058 \times 10^{-9} S_b^2]. \qquad (4.49e)$$

In Eq. (4.49), N_b is the ***normality*** of the brine solution and $\Delta = (25 - T)$ in °C.

An example depicting the frequency dependence of ε_b' and ε_b'' is displayed in Fig. 4-10 for $T = -5$ °C. In comparison with the values discussed in the previous section for fresh-water ic, ε_b' is larger than ε_i' by more than an order of magnitude and ε_b'' is larger than ε_i'' by more than three orders of magnitude. Hence, the presence of brine pockets and inclusions in ice can exercise a very strong influence on the dielectric properties of the mixture, even for very small values of the ***brine volume fraction*** v_b.

4-5.2 Brine Volume Fraction

The volume fraction of brine in sea ice is given by

$$v_b = \frac{S_i}{S_b}\frac{\rho_i}{\rho_b}, \tag{4.50}$$

where S_i and S_b are the salinities of the sea-ice mixture and of the brine component, respectively; ρ_i is the density of pure ice; and ρ_b is the density of the brine. As was indicated in the previous section, S_b is a function of the temperature T only. The brine density ρ_b also is temperature-dependent, although its relative sensitivity to T is much weaker than that of S_b. Using the experimental data reported by Assur in 1960, Frankenstein and Garner (1967) generated the following empirical expression:

$$v_b = 10^{-3} S_i \left(-\frac{49.185}{T} + 0.532 \right), \tag{4.51}$$
$$-0.5°\text{C} \geq T \geq -22.9°\text{C},$$

where S_i is the salinity of the sea ice. Figure 4-11 is a plot of v_b/S_i for the above temperature range.

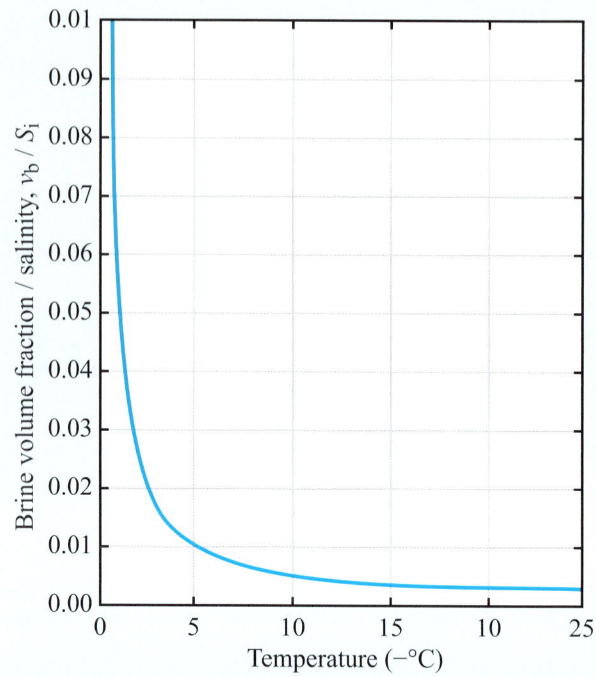

Figure 4-11: Plot of the ratio v_b/S_i (where S_i is in psu) as a function of negative temperature, based on Eq. (4.51).

4-5.3 Dielectric Properties

Due to the structural complexity and nonuniformity of sea ice, it has been difficult to measure its dielectric constant over a wide range of the governing parameters, namely the temperature T, brine volume fraction v_b and electromagnetic frequency f. These are in addition to the structural parameters, most notably the shapes and orientations of the brine inclusions. According to the review by Hallikainen (1992), measurements had been conducted at several frequencies between 0.1 GHz and 40 GHz, but most were at frequencies below 10 GHz. Some of the ice samples were simulated under laboratory conditions while others were extracted from natural sea ice in the Bering Sea and the Beaufort Sea. In view of all of these factors, no credible model exists for the dielectric constant of sea ice, so our approach in the present section is to present selected samples of reported experimental data so as to give the reader an overview, albeit incomplete, of the dielectric behavior of sea ice.

Variation with T

▶ According to measured data, the real part of the dielectric constant of sea ice, ε'_{si}, varies with several sea-ice parameters; however, over the 1 to 40 GHz range it appears to be bounded within the range $2.5 \leq \varepsilon'_{si} \leq 8$. ◀

The temperature dependence of ε_{si} is illustrated in Figs. 4-12 and 4-13 (from Vant et al., 1974) for three types of ice: (a) frazil ice, which usually comprises the top layer of first-year ice and is characterized by spherical brine pockets, (b) columnar ice, which usually

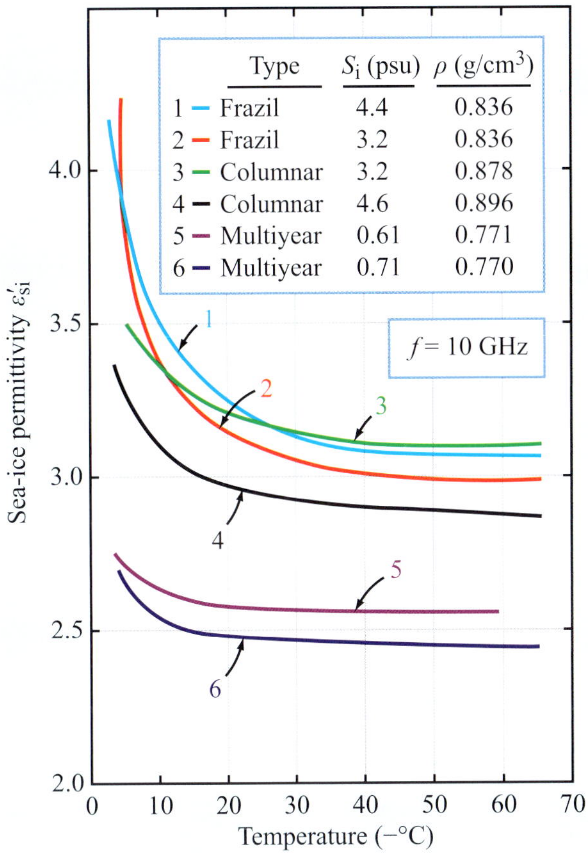

Figure 4-12: Temperature variation of the permittivity of three types of sea ice at 10 GHz [from Vant et al., 1974].

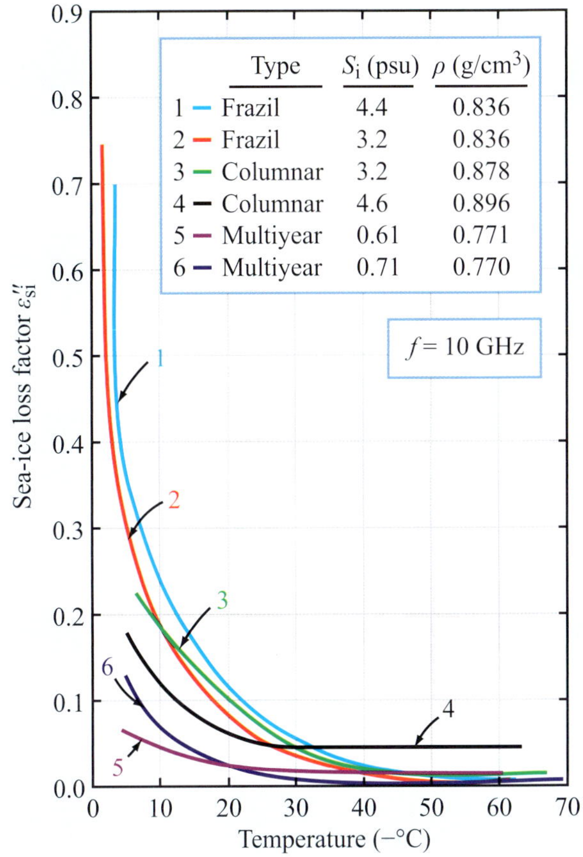

Figure 4-13: Temperature variation of the loss factor for three types of sea ice at 10 GHz [Vant et al., 1974].

occurs beneath a layer of frazil ice in first-year ice and is characterized by elongated brine pockets, and (c) multiyear ice. In Fig. 4-14, dielectric data reported by various investigators through 1992 are grouped into frequency bands: (a) 0.9–1 GHz, (b) 4–4.75 GHz, and (c) 9.5–16 GHz.

Variation with v_b

In their investigation of the dielectric properties of laboratory-simulated sea ice, Arcone et al. (1986) examined the dependence of ε'_{si} and ε''_{si} on the brine volume fraction v_b over a wide range extending between 5 psu and 100 psu. The ice was grown from water with a salinity in the 23–25 psu range, which is characteristic of first-year sea ice. Their results, measured at 4.75 GHz, are shown in Fig. 4-15. We observe that ε'_{si} increases by a factor of about 30% as v_b is increased from about zero to 100 psu, but this rate of increase pales by comparison with that for ε''_{si}, which increases by more than two orders of magnitude (ε''_i of pure ice at 4.75 GHz is on the order of 3×10^{-4}, compared with $\varepsilon''_{si} \approx 0.5$ at $v_b = 100$ psu).

Differences in the propagation properties of pure ice and sea ice may be illustrated through a comparison of their attenuation coefficients. At 10 GHz, the attenuation coefficient of pure (fresh-water) ice, α_i, is on the order of 1 dB m^{-1}. In contrast, the attenuation

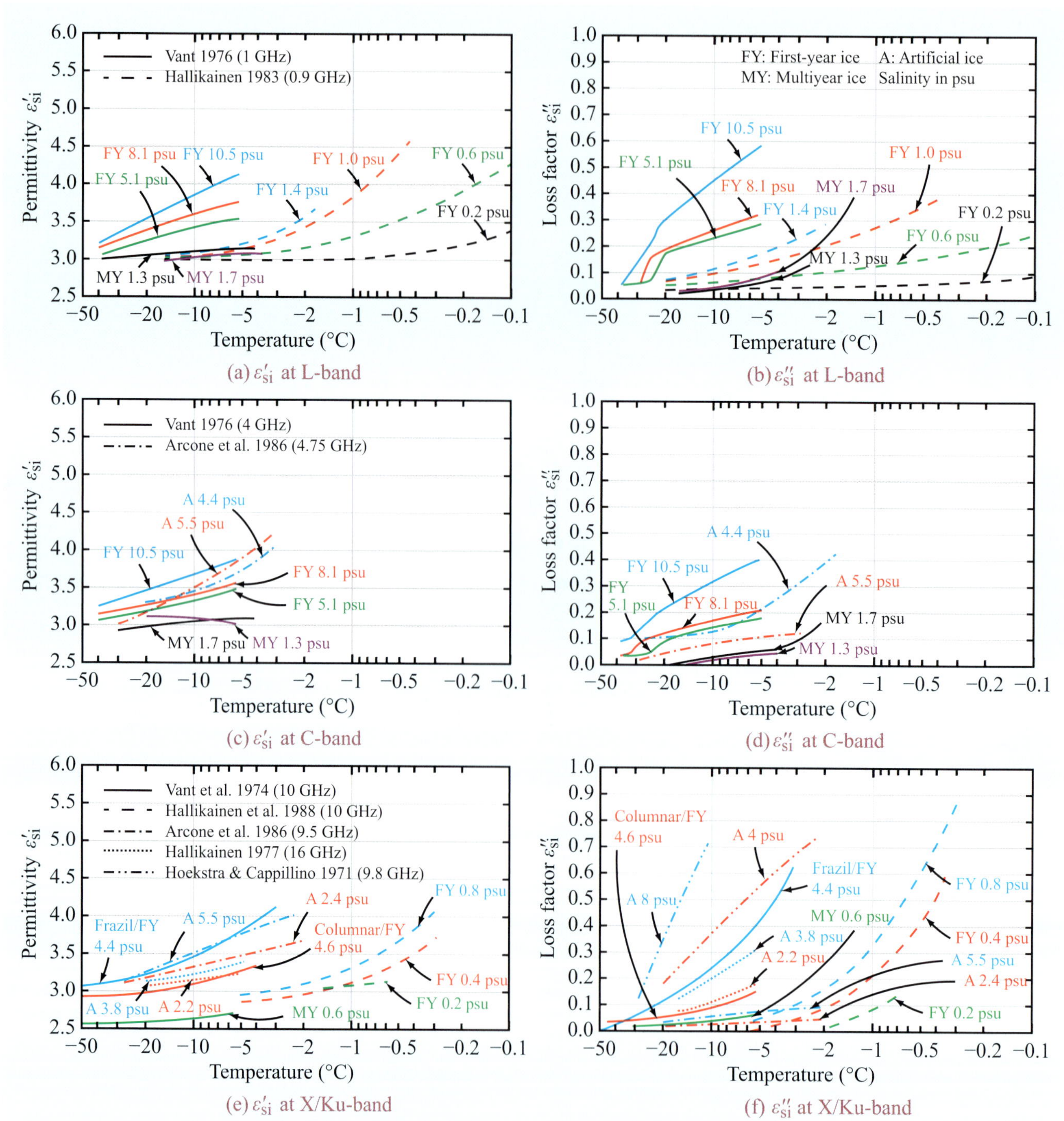

(a) ε'_{si} at L-band

(b) ε''_{si} at L-band

(c) ε'_{si} at C-band

(d) ε''_{si} at C-band

(e) ε'_{si} at X/Ku-band

(f) ε''_{si} at X/Ku-band

Figure 4-14: Measured permittivity and loss factor of sea ice at L-, C-, and X/Ku-bands [Hallikainen, 1992].

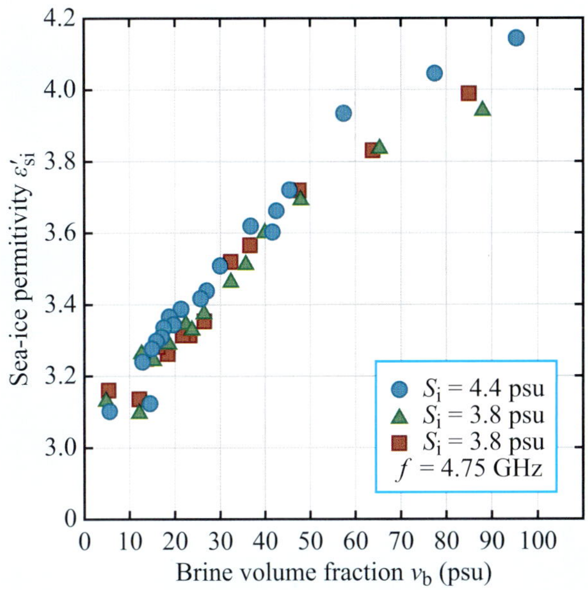

(a) Sea-ice permitivity ε'_{si}

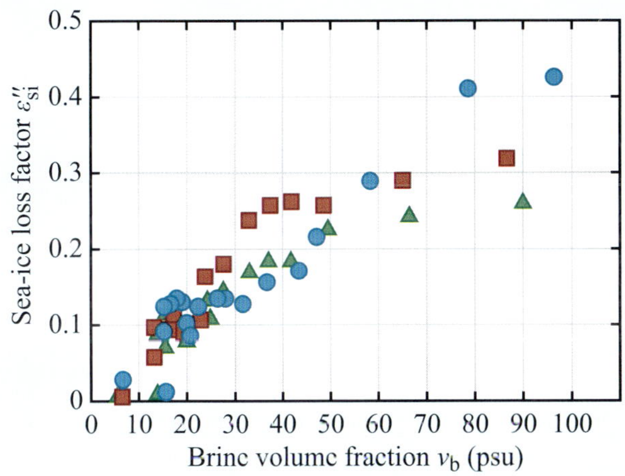

(b) Sea-ice loss factor ε''_{si}

Figure 4-15: Measured permittivity and loss factor of simulated sea ice at 4.7 GHz, plotted as a function of brine volume fraction [Arcone et al., 1986].

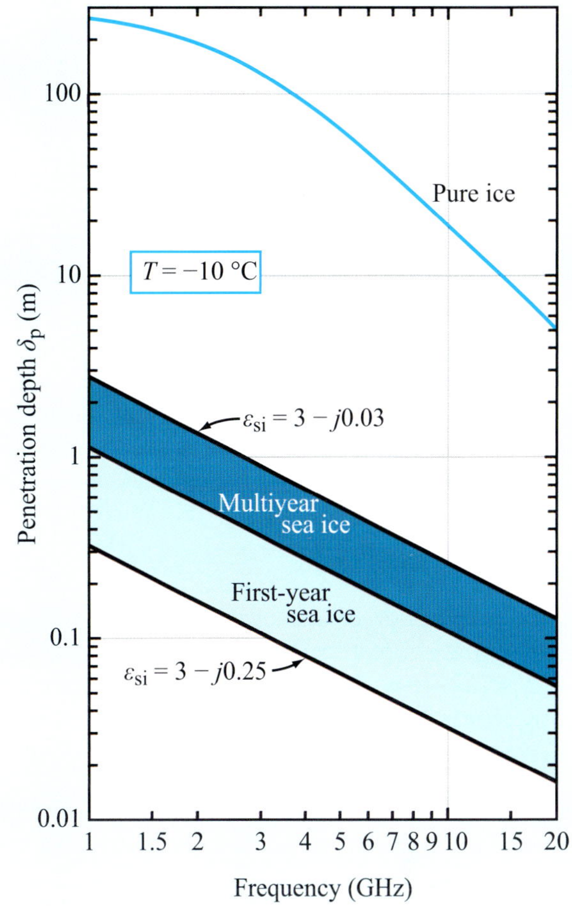

Figure 4-16: Calculated penetration depth for pure ice, first-year sea ice, and multiyear sea ice.

coefficient of sea ice may reach values in excess of 200 dB m^{-1}, depending upon the temperature and salinity of the ice (Finkelstein et al., 1970; Glushnev et al., 1976). This point may be illustrated further by using the curves shown in Fig. 4-16, which depict the spectral variation of the penetration depth δ_p at $T = -10$ °C for representative cases of pure ice, first-year ice, and multiyear ice. The curves were calculated using Eq. (4.12) as follows:

(a) For pure ice, $\varepsilon'_i = 3.18$ at $T = -10$ °C and it is independent of frequency; ε''_i is given graphically in Fig. 4-3 for $T = -1$ °C and -20 °C. Assuming an approximately linear temperature dependence, the values of ε''_i at $T = -10$ °C were obtained by interpolation between the two curves.

(b) For first-year ice at $T = -10$ °C, the dielectric constant was taken to be $\varepsilon_{\rm si} = 3.3 - j0.25$. The real part was chosen on the basis of Fig. 4-12, which shows that $\varepsilon'_{\rm si}$ of first-year ice varies between 3.1 and 3.5 at -10 °C. According to Fig. 4-13, the imaginary part of the relative dielectric constant of first-year ice varies between 0.1 and 0.25 (at 10 GHz and $T = -10$ °C for the cases shown). The bottom curve in Fig. 4-16 is intended to represent a high-loss example of first-year ice. Hence, $\varepsilon''_{\rm si} = 0.25$ was chosen as a representative value for this case. Additionally, in view of the fact that no consistent, discernible frequency dependence has been observed over the 1–20 GHz region for the dielectric constant of "natural" sea ice, the 10 GHz values of $\varepsilon_{\rm si}$ noted above were assumed to be representative of the 1–20 GHz frequency range. The curve defining the upper boundary of the shaded area denoted as "first-year sea ice" in Fig. 4-16 corresponds to a dielectric constant $\varepsilon_{\rm si} = 3 - j0.07$, which is taken to be representative of the boundary between first-year and multiyear sea-ice conditions.

(c) The shaded area in Fig. 4-16 corresponding to "multiyear sea ice" is bounded by curves calculated for $\varepsilon_{\rm si} = 3 - j0.07$ and $\varepsilon_{\rm si} = 3 - j0.03$. The imaginary parts were based on Fig. 4-13.

The curves for $\varepsilon_{\rm si}$ given in this section clearly show that both $\varepsilon'_{\rm si}$ and $\varepsilon''_{\rm si}$ decrease rapidly with increasing negative temperature for the temperature range between 0 °C and -10 °C, and decrease at a lower rate at higher negative temperatures. Moreover, the normalized temperature sensitivity of the real part, $|\partial\varepsilon'_{\rm si}/\partial T|/\varepsilon'_{\rm si}$, is several times smaller in magnitude than the normalized temperature sensitivity of the imaginary part, $|\partial\varepsilon''_{\rm si}/\partial T|/\varepsilon''_{\rm si}$. This means that the penetration depth, $\delta_{\rm p} \approx \sqrt{\varepsilon'_{\rm si}}\,(k_0\varepsilon''_{\rm si})^{-1}$, increases with increasing negative temperature. Thus, if the curves in Fig. 4-16 for first-year and multiyear sea ice at $T = -10$ °C are used as a reference, the corresponding values of $\delta_{\rm p}$ at temperatures closer to 0 °C are 2 to 3 times smaller and values at -30 °C are 3 to 10 times larger.

4-6 Dielectric Constant of Snow

We divide our treatment of the dielectric properties of snow into two subsections, one for ***dry snow***, which is a mixture of ice and air and contains no free (liquid) water, and ***wet snow***, which does contain free water.

4-6.1 Dry Snow

The dielectric constant of dry snow, $\varepsilon_{\rm ds} = \varepsilon'_{\rm ds} - j\varepsilon''_{\rm ds}$, depends on three quantities: (a) the dielectric constant of air, $\varepsilon_{\rm air} = 1 - j0$, (b) the dielectric constant of ice, $\varepsilon_{\rm i} = \varepsilon'_{\rm i} - j\varepsilon''_{\rm i}$, and the ice volume factor $v_{\rm i}$. The volume fraction of ice in snow is related to the ***snow density*** $\rho_{\rm s}$ (in g/cm^3) through

$$v_{\rm i} = \frac{\rho_{\rm s}}{0.9167}\,, \tag{4.52}$$

where 0.9167 g/cm^3 is the density of pure ice.

Permittivity of dry snow

As was noted in Section 4-3 in connection with Eq. (4.23), $\varepsilon'_{\rm i}$ of ice is independent of frequency across the entire microwave band and its temperature sensitivity is so small as to be ignored. Hence, we assign it the value

$$\varepsilon'_{\rm i} = 3.17,$$

which corresponds to Eq. (4.23) at $T = -20$ °C. If we use $\varepsilon_{\rm h} = 1$ (for air) in the Tinga-Voss-Blossey (TVB) model for spherical inclusions given by Eq. (4.43), we obtain the following expression for the dielectric constant of the mixture, which now is dry snow,

$$\varepsilon_{\rm ds} = 1 + \frac{3v_{\rm i}(\varepsilon_{\rm i} - 1)}{(2+\varepsilon_{\rm i}) - v_{\rm i}(\varepsilon_{\rm i} - 1)}\,, \tag{4.53}$$

where $\varepsilon_{\rm i} = \varepsilon'_{\rm i} - j\varepsilon''_{\rm i}$ is the dielectric constant of ice. Our goal is to derive a single, but fairly accurate, expression for $\varepsilon'_{\rm ds}$. Since $\varepsilon''_{\rm i} \ll \varepsilon'_{\rm i}$, we ignore $\varepsilon''_{\rm i}$ altogether and we use $\varepsilon_{\rm i} = 3.17$ in Eq. (4.53). The result is

$$\varepsilon'_{\rm ds} \approx \frac{1 + 0.84v_{\rm i}}{1 - 0.42v_{\rm i}}\,. \tag{4.54}$$

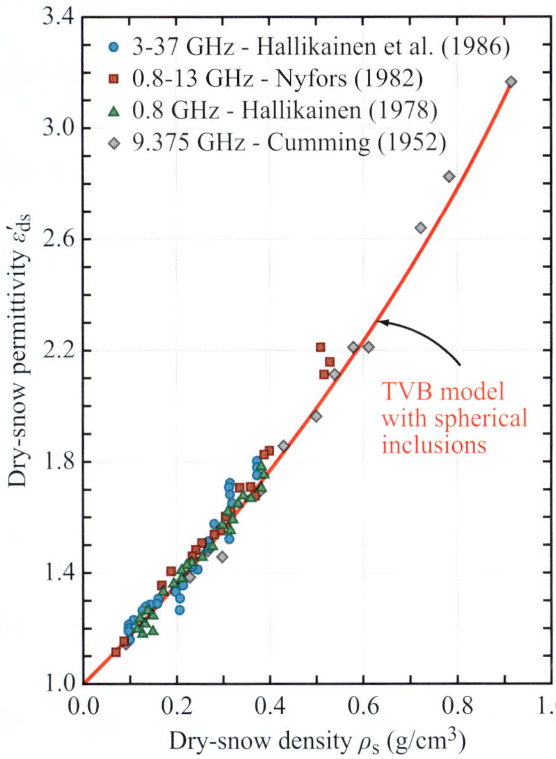

Figure 4-17: Measured permittivity of dry snow as a function of snow density.

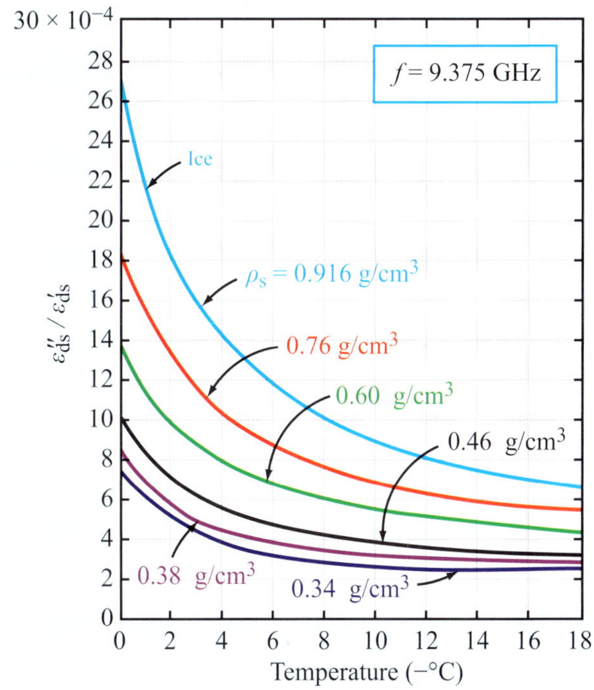

Figure 4-18: The loss tangent of dry snow as a function of temperature at a frequency of 9.375 GHz [from Cumming, 1952].

For $v_i = 0$ (pure air), $\varepsilon'_{ds} = 1$; and for $v_i = 1$ (pure ice), $\varepsilon'_{ds} = 3.17$, thereby confirming that the formula predicts the correct values at the two extreme values of v_i. Figure 4-17 shows a plot of Eq. (4.54), together with experimental measurements of ε'_{ds} made over a wide frequency range extending from 0.8 GHz to 37 GHz. The TVB model represented by Eq. (4.54) provides a good fit to the experimental data.

An equally good fit to the data is provided by the empirical expression (Mätzler, 2006)[†]

$$\varepsilon'_{ds} = \begin{cases} 1 + 1.4667 v_i + 1.435 v_i^3 & \text{for } 0 \le v_i \le 0.45, \\ (1 + 0.4759 v_i)^3 & \text{for } v_i \ge 0.45. \end{cases} \tag{4.55}$$

[†] **Computer Code 4.5.**

Loss factor of dry snow

As was stated earlier, ε'_{ds} is essentially independent of both temperature and frequency in the microwave region. In contrast, the loss factor of dry snow ε''_{ds} is strongly dependent on both parameters. The family of plots in Fig. 4-18 depict the variation of $\varepsilon''_{ds}/\varepsilon'_{ds}$ as a function of negative temperature at 9.375 GHz, with ρ_s as a parameter. The variation of $\varepsilon''_{ds}/\varepsilon'_{ds}$ with frequency is shown in Fig. 4-19, based on measurements from 0.8 GHz to 12.6 GHz. The family of plots exhibits a minimum in the 1–2 GHz range, similar to the minimum exhibited by ε''_i of ice in Fig. 4-3. Since dry snow is a mixture of ice and air, ε''_{ds} should mimic the temperature and frequency behavior of ε''_i. To obtain a model expression for $\varepsilon''_{ds}/\varepsilon''_i$, we proceed as follows:

(1) In the TVB model expression given by Eq. (4.53), we set $\varepsilon_{ds} = \varepsilon'_{ds} - j\varepsilon''_{ds}$ and $\varepsilon_i = \varepsilon'_i - j\varepsilon''_i$.

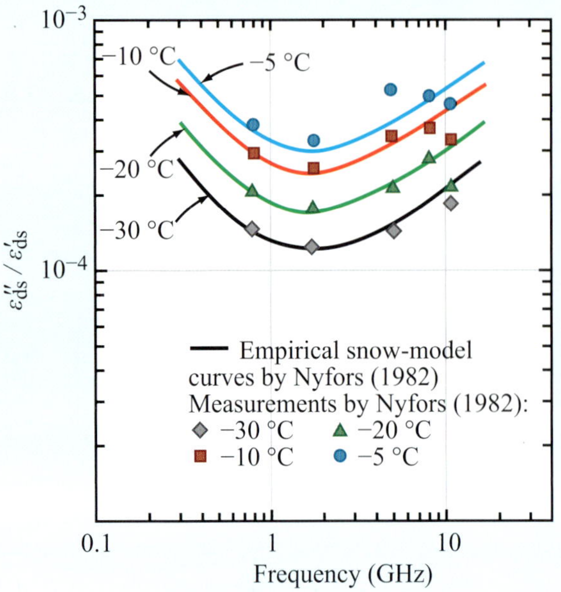

Figure 4-19: The loss tangent of dry snow as a function of frequency with temperature as parameter. Snow density is 0.45 g/cm^3 [from Nyfors, 1982].

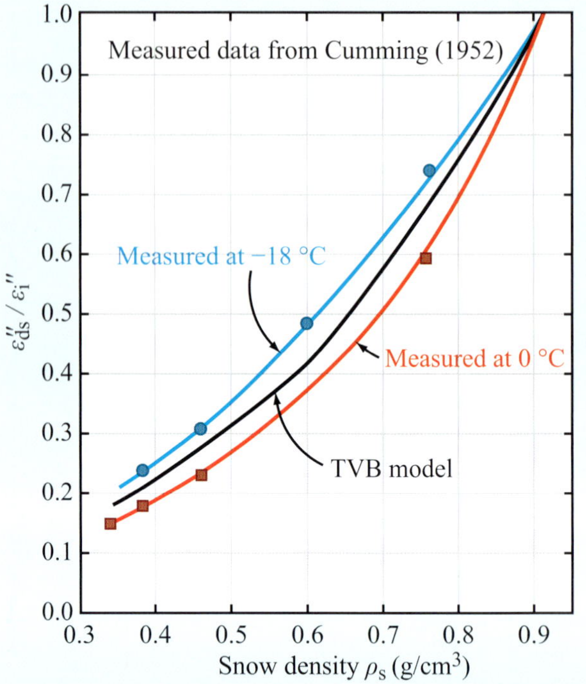

Figure 4-20: Comparison of Eq. (4.57) with measured data.

(2) We multiply the numerator and denominator of the second term on the right-hand side of the equation by the complex conjugate of its denominator so as to obtain an expression of the form $\varepsilon'_{ds} - j\varepsilon''_{ds} = a - jb$.

(3) We then equate ε''_{ds} to the term b, and we simplify it by ignoring terms of small magnitude when appropriate.

The process leads to (Hallikainen et al., 1986):

$$\varepsilon''_{ds} = \frac{9v_i\varepsilon''_i}{[(2+v_i)+\varepsilon'_i(1-v_i)]^2}\,. \tag{4.56}$$

Upon setting $\varepsilon'_i = 3.17$, the expression simplifies to

$$\varepsilon''_{ds} = \frac{0.34v_i\varepsilon''_i}{(1-0.42v_i)^2}\,. \tag{4.57}$$

A plot of $\varepsilon''_{ds}/\varepsilon''_i$ as a function of ρ_s is shown in Fig. 4-20. Also shown are plots extracted from Fig. 4-18 for $T = 0$ °C and -18 °C. The theoretical curve due to the TVB model is intermediate between the 0 °C and −18 °C data curves. According to Eq. (4.57), $\varepsilon''_{ds}/\varepsilon''_i$ should be temperature-independent, which means that the 0 °C and −18 °C plots should be identical. The fact that they are not is attributed to measurement inaccuracy.

To determine ε''_{ds} from Eq. (4.57), we need to know the value of ε''_i of the ice from which the snow is made with a reasonable degree of accuracy. As noted earlier in connection with Fig. 4-4, ε''_i can vary significantly with the amount of impurity present in the ice; a salinity as low as 1 psu is sufficient to cause ε''_i to double in value. To reconcile the data in Figs. 4-18 and 4-19 with the model given by Eq. (4.57), it is necessary to use values for ε''_i that are multiple times those reported for pure ice in Fig. 4-3.

4-6.2 Wet Snow

In the proximity of 0 °C, snow can support water in liquid form. Because of the high dielectric constant of liquid water compared with that of air and ice, the presence of liquid water strongly affects the dielectric constant of the snow mixture. The liquid-water content of snow (or ***snow wetness***) m_v is the volume fraction of liquid water in the snow mixture, and usually it is expressed as a percentage. Occasionally snow wetness is expressed as a percentage of liquid water by weight, w_w. These two quantities are related by the approximate expression $m_v = w_w \rho_{ws}$, where ρ_{ws} is the density of wet snow, which is approximately equal to the density of dry snow ρ_s because the densities of water (1 g/cm^3) and pure ice (0.9167 g/cm^3) are fairly close to one another.

An extensive investigation of the dielectric behavior of natural snow was conducted by Hallikainen et al. (1983, 1984, 1986) using a free-space transmission technique. The permittivity and the dielectric loss factors of wet snow, ε'_{ws} and ε''_{ws}, were measured for 110 snow samples at nine frequencies between 4 and 18 GHz. Of these, 62 samples were also measured at 3 GHz and 37 GHz. Thus, although their results cover the range between 3 and 37 GHz, they are more heavily influenced by the 4–18 GHz measurements. The range of m_v covered by the samples extended between 1 and 12 percent. The measurement apparatus employed a pair of horn antennas used for transmission and reception and was arranged for wave propagation along the vertical direction. By operating the apparatus from a location adjacent to a snow-covered field, it was possible to remove snow slabs from the field while causing minimal disturbance to the snow's geometry. Each slab—approximately 50 cm in diameter, a few centimeters thick, and supported by a thin styrofoam sheet—was placed between the two antennas in order to measure the amplitude and phase of its transmission coefficient (from which ε'_{ws} and ε''_{ws} were determined).

In addition to measuring the complex dielectric constant of wet snow ($m_v > 1\%$), Hallikainen et al. (1986) measured the permittivity of dry snow for the range of snow density between 0.09 and 0.38 g/cm^3. The following linear fit to the data was obtained:

$$\varepsilon'_{ds} = 1.0 + 1.832\rho_s, \tag{4.58}$$

with a linear correlation coefficient of 0.96 and an rms error of 0.04.

When they analyzed the dielectric measurements for wet snow, Hallikainen et al. found that the analysis became simpler when the dependence of ε'_{ws} on snow density was (approximately) removed through conversion to the incremental permittivity $\Delta\varepsilon'_{ws}$,

$$\Delta\varepsilon'_{ws} = \varepsilon'_{ws} - 1.0 - 1.832\rho_s. \tag{4.59}$$

This is not the case for the dielectric loss factor ε''_{ws} because it is almost entirely an incremental quantity introduced by the presence of liquid water.

Typical examples of the variations of $\Delta\varepsilon'_{ws}$ and ε''_{ws} with m_v are shown in Fig. 4-21. The data points are based on measurements of ε'_{ws} and ε''_{ws}, whereas the solid curve was calculated using an empirical, modified ***Debye-like model*** given by[†]

$$\varepsilon'_{ws} = A + \frac{Bm_v^x}{1+(f/f_0)^2}, \tag{4.60a}$$

$$\varepsilon''_{ws} = \frac{C(f/f_0)m_v^x}{1+(f/f_0)^2}, \tag{4.60b}$$

where f_0 is an effective relaxation frequency of wet snow. The constants A, B, C, f_0, and x were determined by fitting 955 measurements for ε'_{ws}, and an identical number for ε''_{ws}, to the above expressions. The results, which pertain to the ranges $3 \le f \le 37$ GHz, $0.09 \le \rho_s \le 0.38$ g/cm^3, and $1 \le m_v \le 12\%$, are

$$A = A_1(1.0 + 1.83\rho_s + 0.02m_v^{1.015}) + B_1, \tag{4.61a}$$

$$B = 0.073A_1, \tag{4.61b}$$

$$C = 0.073A_2, \tag{4.61c}$$

$$x = 1.31, \tag{4.61d}$$

$$f_0 = 9.07 \text{ GHz}, \tag{4.61e}$$

$$A_1 = 0.78 + 0.03f - 0.58 \times 10^{-3} f^2, \tag{4.61f}$$

$$A_2 = 0.97 - 0.39f \times 10^{-2} + 0.39 \times 10^{-3} f^2, \tag{4.61g}$$

$$B_1 = 0.31 - 0.05f + 0.87 \times 10^{-3} f^2, \tag{4.61h}$$

where f is in GHz. The importance of the frequency-dependent constants is primarily at frequencies above

[†] **Computer Code 4.6.**

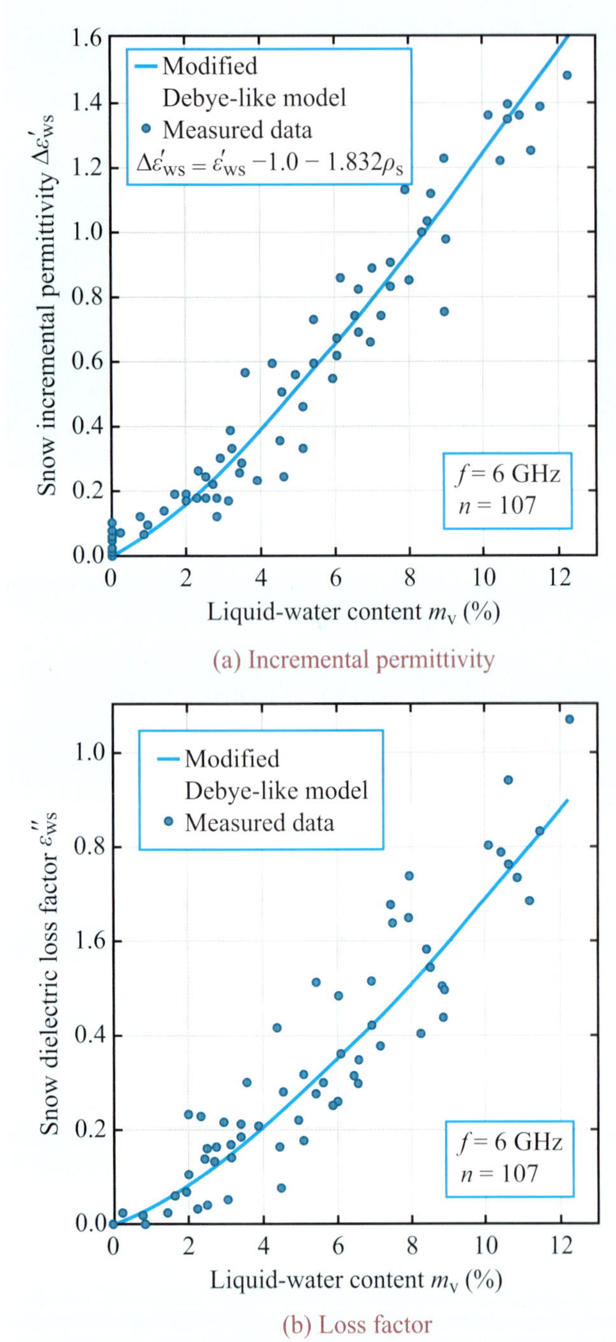

Figure 4-21: Real and imaginary parts of the relative dielectric constant of snow as a function of liquid-water content at 6 GHz [from Hallikainen et al., 1986].

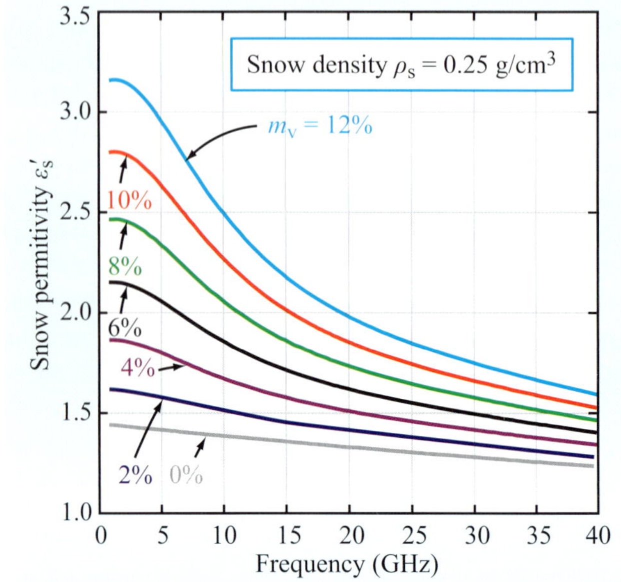

Figure 4-22: Spectral variation of the permittivity of wet snow with snow wetness as a parameter. The plots are based on the modified Debye-like model given by Eq. (4.60).

15 GHz; below 15 GHz we may set $A_1 = A_2 = 1.0$ and $B_1 = 0$.

The value of f_0 is slightly higher than the relaxation frequency of water at 0 °C, $f_0 = 8.9$ GHz. This is in agreement with the results of previous studies concerning heterogeneous mixtures containing water. Both theoretically and experimentally, de Loor (1968) found that the relaxation frequency of such mixtures is always equal to or higher than that of the relaxing component.

The curves shown in Fig. 4-21, which were calculated using the expressions given by Eq. (4.60) for $f = 6$ GHz, appear to provide a reasonable fit to the data for both $\Delta\varepsilon'_{ws}$ and ε''_{ws}. For the data-set as a whole, Hallikainen et al. (1986) evaluated the linear correlation coefficients between the measured and calculated values of the dielectric constant of snow, and found them to be on the order of 0.99 for $\Delta\varepsilon'_{ws}$ and 0.98 for $\Delta\varepsilon''_{ws}$.

Figures 4-22 to 4-24 show the effect of liquid-water content on the dielectric behavior of wet snow between

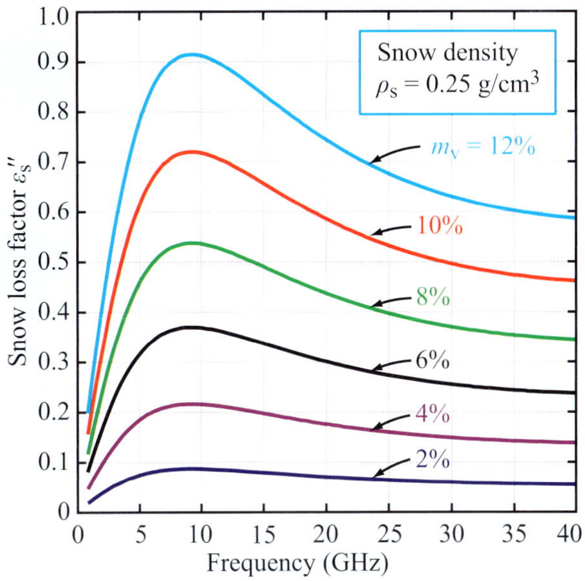

Figure 4-23: Spectral variation of the loss factor of wet snow with wetness as a parameter. The plots are based on the modified Debye-like model given by Eq. (4.60).

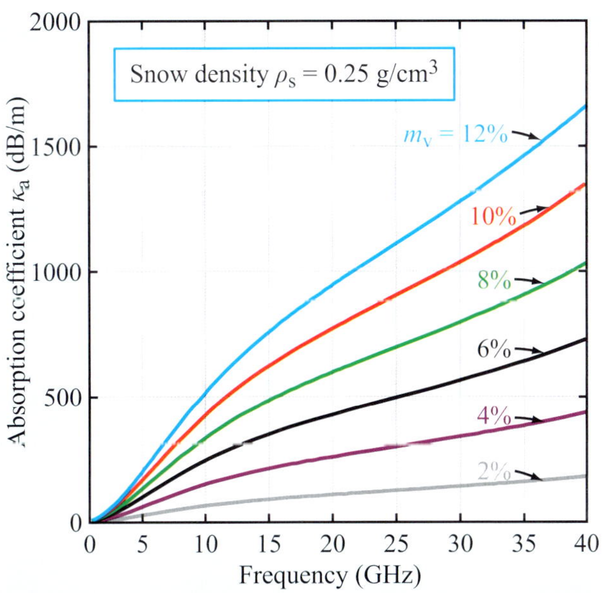

Figure 4-24: Spectral variation of the attenuation coefficient of wet snow with wetness as a parameter. The curves are based on the modified Debye-like model.

3 GHz and 37 GHz. The results are shown for a dry snow density $\rho_{ds} = 0.25$ g/cm^3, which was the average observed value during this study. The values of ε'_{ws} and ε''_{ws} were calculated using the modified Debye-like equations (4.60) and (4.61). As shown in Fig. 4-23, ε''_{ws} reaches its maximum value at 9.0 GHz. This agrees with the relaxation frequency of water at 0 °C. The sharp increase in ε''_{ws} between 3 GHz and 6 GHz is due to the large increase in the magnitude of ε''_w for water over this frequency range. The attenuation constant of wet snow increases practically linearly with frequency up to 15 GHz. Between 18 GHz and 37 GHz, the average slope is slightly smaller.

4-7 Dielectric Constant of Dry Rocks

An extensive study of the dielectric properties of rocks and powders was conducted by Campbell and Ulrichs (1969), in which they made measurements of 36 types of rocks, in both solid and powder forms, at 450 MHz and 35 GHz. Two decades later, several measurement techniques were used to investigate the dielectric properties of 80 rock samples across the 0.5–18 GHz range (Ulaby et al., 1990a). A summary of the results of these two studies follows.

4-7.1 Powdered Rocks

The bulk density of rocks, ρ_b in g/cm^3, varies between 1 g/cm^3 for rock material crushed into powder form to about 3.4 for some high density rocks like peridotites. The data set reported by Campbell and Ulrichs (1969) included dielectric measurements for 25 different types of powdered rocks, all at a density of 1 g/cm^3.

▶ Their results show that the relative permittivity ε'_p of the powdered rocks varied over a narrow range between 1.9 and 2.1, with a mean value around 2.0. Thus, when normalized to the same density, all rocks have approximately the same permittivity $\varepsilon'_p \approx 2.0$, regardless of rock type. ◀

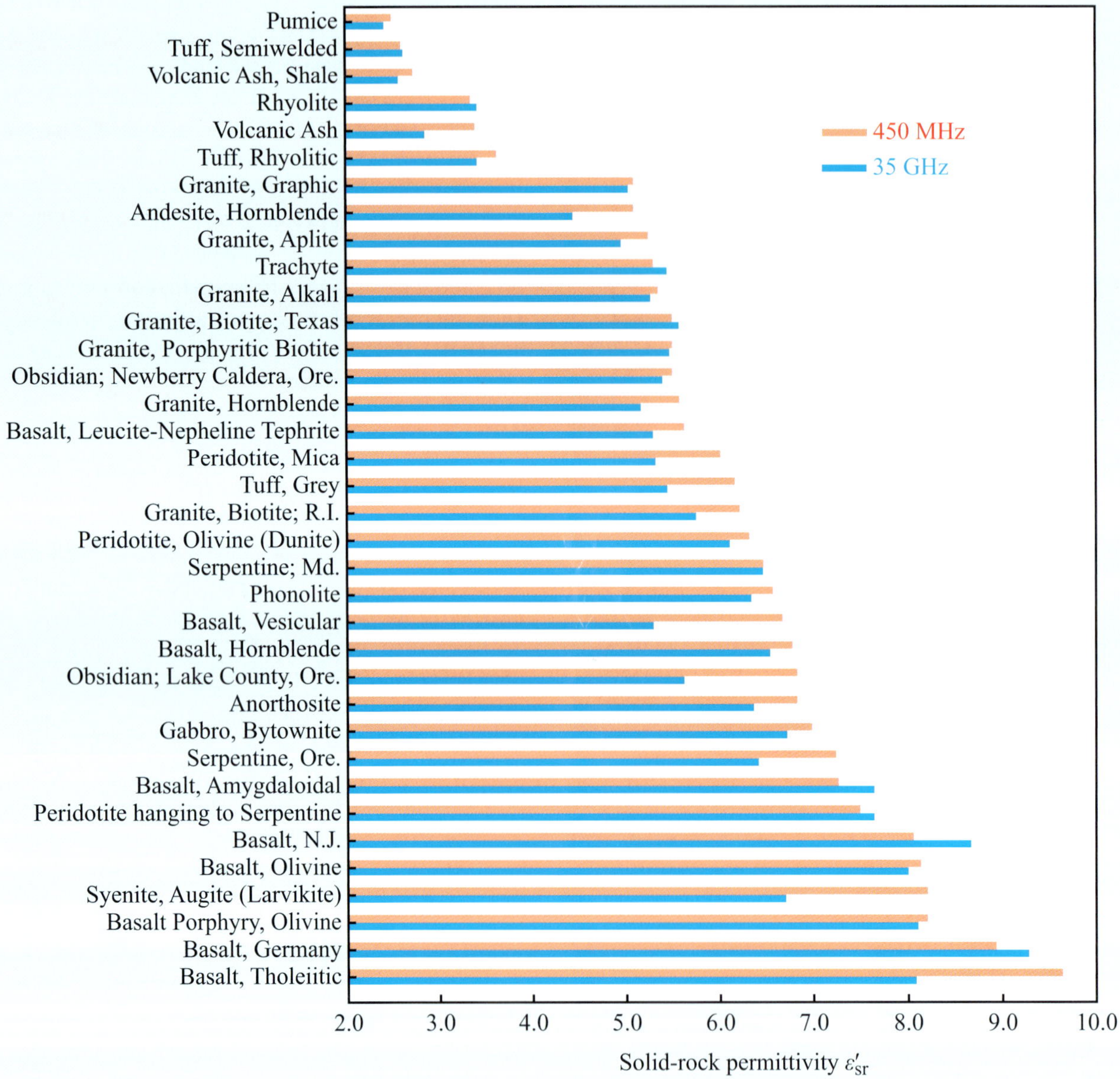

Figure 4-25: A comparison of the permittivities of solid rocks at 450 MHz and 35 GHz [from Campbell and Ulrichs, 1969].

4-7.2 Solid Rocks

According to the data shown in Fig. 4-25, among 36 types of solid rocks, ε'_{sr} varies between 2.5 and 9.6, and the values measured at 450 MHz are not statistically different from those measured at 35 GHz, suggesting that ε'_{sr} is frequency-independent in the microwave band. This conclusion is supported by the spectra shown in Fig. 4-26 for four typical examples of the 80 rock types examined by Ulaby et al. (1990a). Dielectric measurements made as a function of temperature revealed that ε'_{sr} is temperature-independent also.

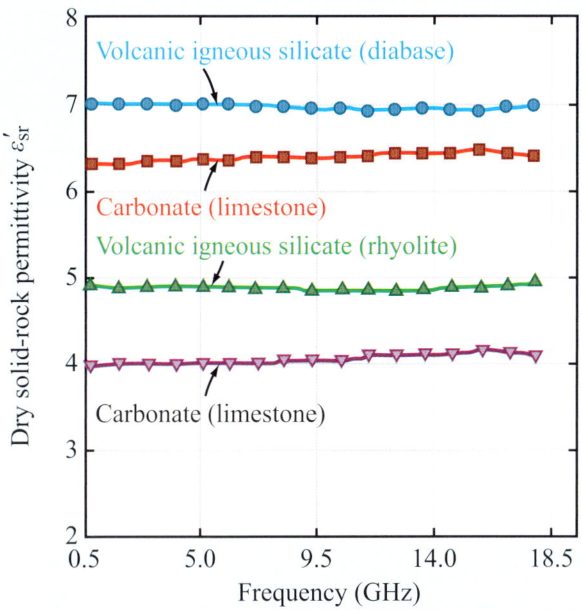

Figure 4-26: Typical examples of measured permittivity spectra [Ulaby et al., 1990a].

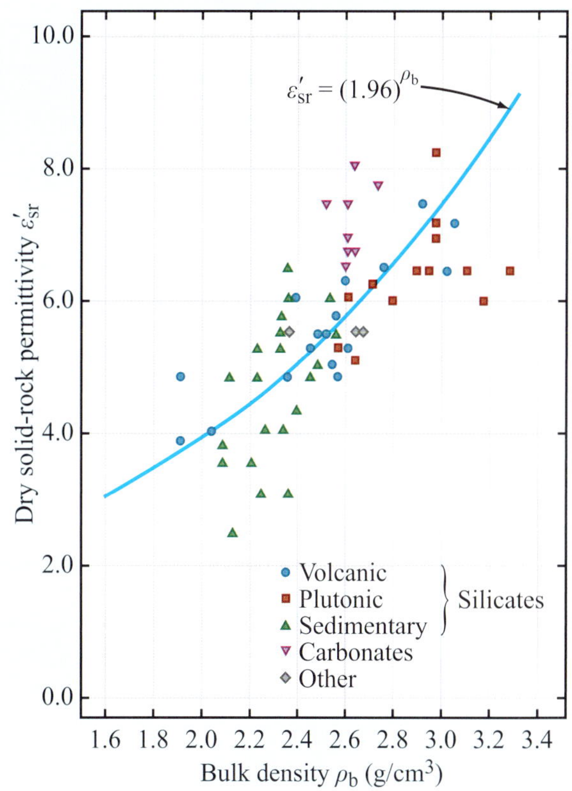

Figure 4-27: Variation of $\varepsilon'_{\rm sr}$ with rock bulk density [Ulaby et al., 1990a].

The variation of $\varepsilon'_{\rm sr}$ with bulk density $\rho_{\rm b}$ is displayed in Fig. 4-27. The figure includes measured data as well as an empirical fit given by

$$\varepsilon'_{\rm sr} = (\varepsilon'_{\rm p})^{\rho_{\rm b}} = 2^{\rho_{\rm b}}, \tag{4.62}$$

where $\varepsilon'_{\rm p} = 2$ is the average permittivity of powdered rocks. The deviation from a smooth variation with $\rho_{\rm b}$ is attributed to differences in mineralogy among the various types of rocks.

Unlike the frequency-independent behavior of $\varepsilon'_{\rm sr}$, the loss factor $\varepsilon''_{\rm sr}$ exhibits a definite variation with frequency for all rocks tested by Ulaby et al. (1990a). Typical examples are shown in Fig. 4-28. The frequency variation of $\varepsilon''_{\rm sr}$ is attributed, at least in part, to the dispersion spectrum of bound water within the rock samples.

Analysis of the measured data of $\varepsilon''_{\rm sr}$ led to the following conclusions:

(a) No apparent correlation between $\varepsilon''_{\rm sr}$ and bulk density $\rho_{\rm b}$.

(b) At 1.6 GHz, the range of $\varepsilon''_{\rm sr}$ among the 80 different types of rocks extended between 0.002 and 0.24, and the corresponding range at 16 GHz extended between 0.002 and 0.18. Similar ranges of values were reported by Vaccaneo et al. (2004) at L-band.

(c) Most rock types exhibited a frequency variation of the form

$$\varepsilon''_{\rm sr} = a + \frac{b}{f}, \tag{4.63}$$

where a and b are constants specific to the rock type under consideration and f is frequency. No apparent correlation could be established between coefficients a and b and bulk density $\rho_{\rm b}$.

In summary, for individual rock types, $\varepsilon''_{\rm sr}$ exhibits a predictable variation similar to that of the plots shown in

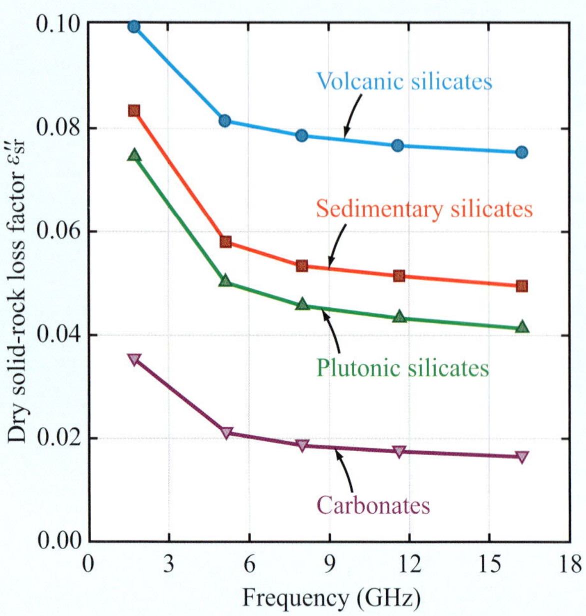

Figure 4-28: Variation of $\varepsilon''_{\text{sr}}$ with frequency for four types of rocks [Ulaby et al., 1990a].

Fig. 4-28, but the magnitude of $\varepsilon''_{\text{sr}}$ does not appear to be related to bulk density.

4-8 Dielectric Constant of Soils

4-8.1 Dry Soil

In the absence of liquid water, the microwave dielectric constant of soil, $\varepsilon_{\text{soil}}$, is essentially independent of both temperature and frequency. Moreover, $2 \lesssim \varepsilon'_{\text{soil}} \lesssim 4$ and $\varepsilon''_{\text{soil}} < 0.05$. Based on experimental measurements of several soil types, Dobson et al. (1985) determined that $\varepsilon'_{\text{soil}}$ can be modeled as

$$\varepsilon'_{\text{soil}} = (1 + 0.44\rho_{\text{b}})^2 \qquad \textbf{(dry soil)}, \tag{4.64}$$

where ρ_{b} is the soil bulk density.

4-8.2 Wet Soil

A wet-soil medium is a mixture of soil particles, air pockets, and liquid water. The water contained in the soil usually is divided into two fractions: (a) ***bound water***, and (b) ***free water***. Bound water refers to water molecules that are contained in the first few molecular layers surrounding the soil particles and therefore are tightly held by the soil particles due to the influence of matric and osmotic forces. Because the matric forces acting on a water molecule decrease rapidly with increasing distance from the soil particle, water molecules that are located several molecular layers away from soil particles are able to move within the soil medium with relative ease, and hence are referred to as "free."

Dividing the water into bound and free fractions is only an approximate description of the actual distribution of water molecules in the soil medium and is based on a somewhat arbitrary criterion for establishing the transition point between bound and free water layers. The amount of water contained in the molecular layer adjoining the soil particles is directly proportional to the total surface area of the soil particles contained in a unit volume. The total surface area of the particles is, in turn, a function of the size distribution and mineralogy of the soil particles. For the most part, a soil is assigned to a ***textural class*** on the basis of its ***particle-size distribution***.

Soil particles are classified as sand, silt, or clay, according to size (Fig. 4-29). Because all soils contain a distribution of particle sizes, it is convenient to classify a soil by the weight-percent of the soil within each specific size category (sand, silt, or clay). The relative percentages of the three size categories then determine the soil's textural class (Fig. 4-30 shows the U.S. Department of Agriculture triaxial classification system). Soil type refers to a subdivision of one of the soil textural classes shown in Fig. 4-30.

The two terms commonly used to characterize the moisture content of a soil sample are ***volumetric***

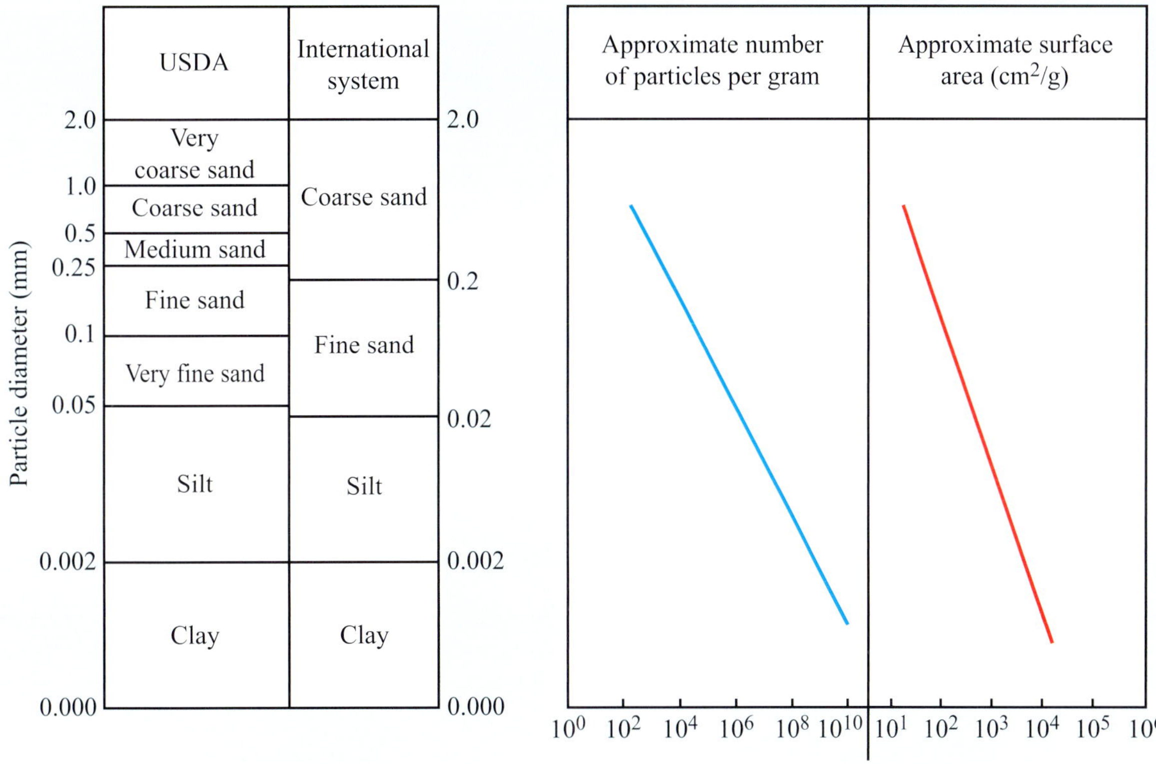

Figure 4-29: Particle-size classes, number of particles, and external surface area based on spherical particles of density 2.65 g/cm^3 [modified from Birkeland, 1974].

moisture $m_{\rm v}$ and ***gravimetric moisture*** $m_{\rm g}$,

$$m_{\rm v} = \frac{V_{\rm w}}{V_{\rm t}} = \frac{V_{\rm w}}{V_{\rm dry}} = \frac{W_{\rm w}\rho_{\rm b}}{W_{\rm dry}\rho_{\rm w}} = \frac{W_{\rm w}\rho_{\rm b}}{W_{\rm dry}} \ \text{cm}^3/\text{cm}^3 \text{ or g/cm}^3, \tag{4.65a}$$

and

$$m_{\rm g} = \frac{W_{\rm w}}{W_{\rm dry}} \times 100 = 100\,\frac{m_{\rm v}}{\rho_{\rm b}} \ (\%), \tag{4.65b}$$

where $W_{\rm w}$ and $W_{\rm dry}$ are the weights of the water in the sample and of the dry sample, respectively; $V_{\rm w}$ is the water volume; $V_{\rm t}$ is the total volume of the sample, which includes the volumes of air, soil, and water and is equal to the volume of the dry sample (assuming that when water is added to the sample, it fills air pockets but does not increase the total volume); $\rho_{\rm b}$ is the bulk density of the dry soil sample; and $\rho_{\rm w} = 1$ g/cm^3 is the density of water. Although $m_{\rm v}$, being a fraction, is unitless, it usually is expressed in cm^3/cm^3 or g/cm^3.

In general, a soil medium is electromagnetically a four-component dielectric mixture consisting of air, bulk soil, bound water, and free water. Due to the intensity of the forces acting upon it, a bound water molecule interacts with an incident electromagnetic wave in much the same way that it does when it is in the form of ice, thereby exhibiting a dielectric dispersion spectrum that is very different from that of free water. The complex dielectric constants of bound and free water are functions of the electromagnetic frequency f, the physical temperature T, and the salinity S. Hence, the dielectric constant of the soil mixture is, in general, a function of (a) f, T, and S, (b) the total volumetric

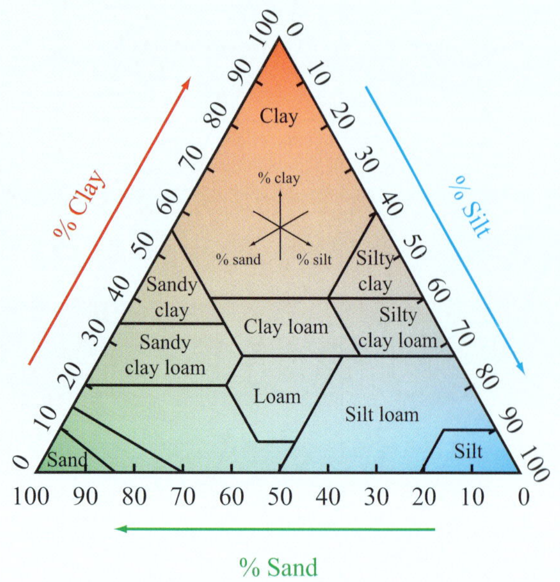

Figure 4-30: Soil textured classification triangle.

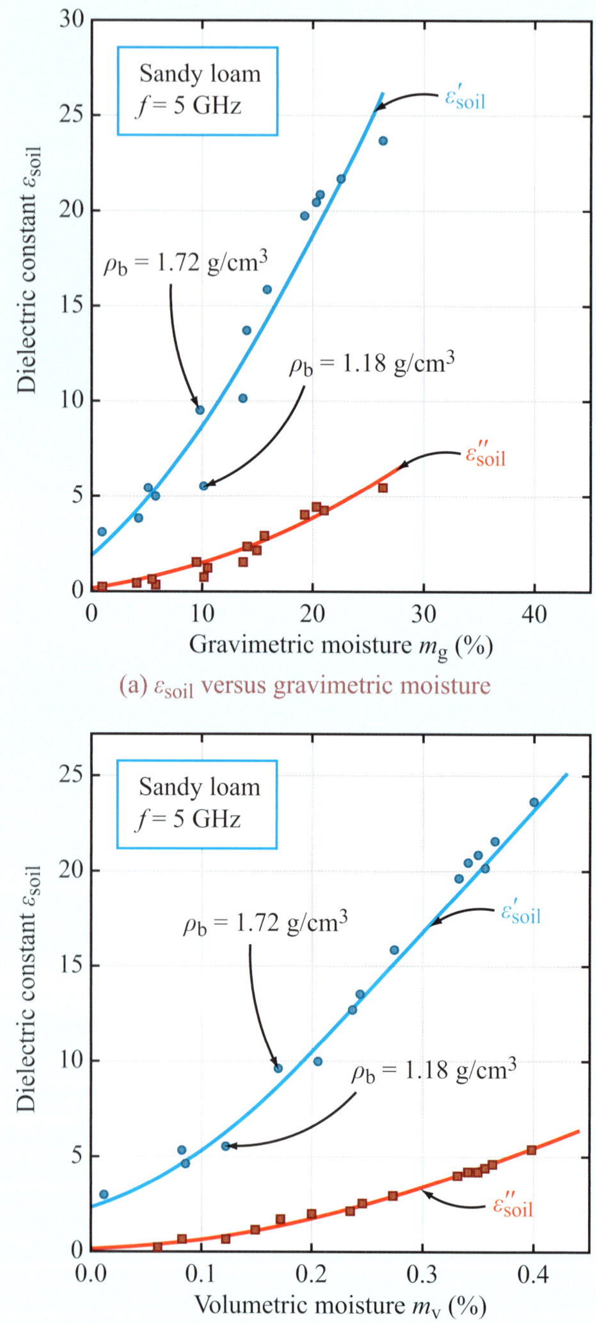

(a) ε_{soil} versus gravimetric moisture

(b) ε_{soil} versus volumetric moisture

Figure 4-31: Comparison of measured soil dielectric constant data plotted as a function of (a) gravimetric moisture content and (b) volumetric moisture content [from Hallikainen et al., 1985].

water content m_v, (c) the relative fractions of bound and free water, which are related to the particle-size distribution (or soil texture), (d) the bulk soil density ρ_b, (e) the shape of the soil particles, and (f) the shape of the water inclusions. We will endeavor to describe ε_{soil} by a relatively straightforward model that accounts for most of the dependence on this long list of variables.

Numerous studies have been conducted to determine the dielectric behavior of soil-water mixtures, most notably those reported by Wang et al. (1980), Dobson et al. (1985), Hallikainen et al. (1985), Roth et al. (1990), and Peplinski et al. (1995). The measured dielectric data usually are plotted as a function of either m_v or m_g, and occasionally, of both.

From the standpoint of dielectric mixing models, the volumetric measure is preferred because the dielectric constant of the soil-water mixture is a function of the water volume-fraction in the mixture. This preference is demonstrated by Fig. 4-31, in which a greater degree of scattering about the regression curve is apparent for the plots of ε'_{soil} and ε''_{soil} versus m_g [Fig. 4-31(a)] than for those plotted versus m_v [Fig. 4-31(b)]. Additionally, measurements made for two soil samples

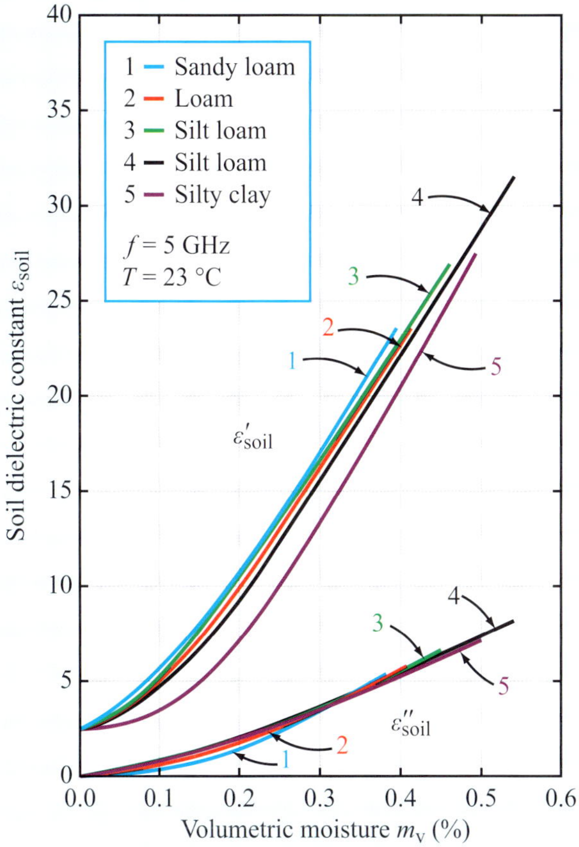

Figure 4-32: Measured dielectric constant for five soils at 5 GHz [Hallikainen et al., 1985].

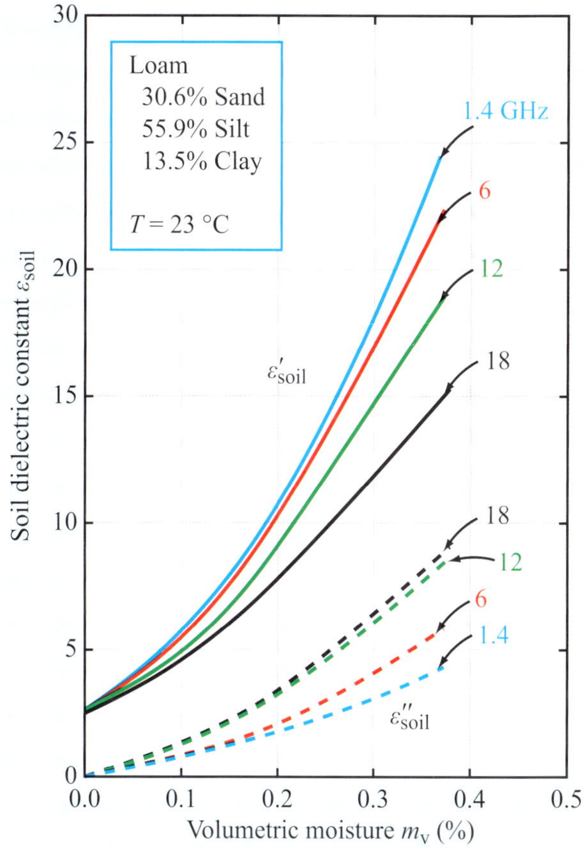

Figure 4-33: Measured dielectric constant as a function of volumetric moisture content for a loamy soil at four microwave frequencies [Hallikainen et al., 1985].

with approximately the same m_g, but significantly different bulk densities, resulted in significantly different values for ε_{soil}. However, samples with the same m_v but different bulk densities resulted in dielectric values that were approximately the same for both samples.

The curves shown in Fig. 4-32 depict the measured variation of ε_{soil} with m_v for five soil types at 5 GHz. Similar responses were measured at other frequencies between 1.4 and 18 GHz. The indicated moisture range for each soil type extends between $m_v = 0$ and the highest moisture content that can be supported by that soil type without producing drainage.

Figure 4-33 illustrates the role of frequency for one of the soil types measured by Hallikainen et al. (1985), which indicates that ε'_{soil} decreases with increasing frequency between 1.4 and 18 GHz, whereas ε''_{soil} increases with increasing frequency. The character of the frequency response of ε_{soil} is similar to that of water (Section 4-1). The dependence of ε_{soil} on frequency is more clearly displayed in Fig. 4-34, which shows the spectra for ε'_{soil} and ε''_{soil} for several values of m_v.

The dependence of ε_{soil} on temperature was examined by Hoekstra and Delaney (1974) at 10 GHz and by Hallikainen et al. (1984b) over a wide frequency range extending from 3 GHz to 37 GHz. Above 0 °C, ε'_{soil} and ε''_{soil} are weakly dependent on temperature, but as T crosses the freezing temperature of water, both ε'_{soil}

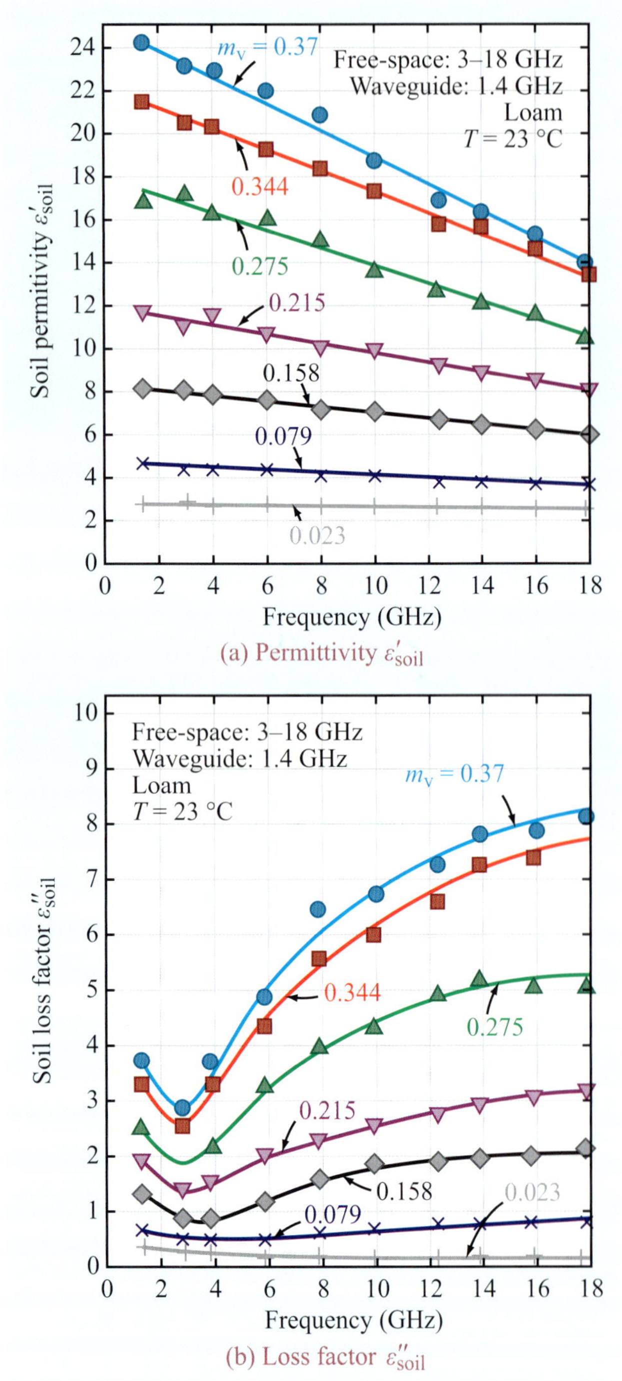

Figure 4-34: Measured (a) permittivity and (b) loss factor of loamy soil as a function of frequency with volumetric moisture content as a parameter [Hallikainen et al., 1985].

and ε''_{soil} change markedly, as illustrated by the data in Fig. 4-35.

To model the dielectric behavior of soils at microwave frequencies, Dobson et al. (1985) used a four-component model consisting of solid soil material, air, bound water, and free water. Using a refractive-like model along the lines of Eq. (4.45), together with representative values for the density of solid soil material, they developed the following semiempirical model for ε_{soil}:[†]

$$\varepsilon'_{soil} = [1 + 0.66\rho_b + m_v^{\beta_1} (\varepsilon'_w)^\alpha - m_v]^{1/\alpha}, \tag{4.66a}$$

$$\varepsilon''_{soil} = m_v^{\beta_2} \varepsilon''_w, \tag{4.66b}$$

where ρ_b is the bulk density of the soil in g/cm^3 (if the value of ρ_b is unknown, it is recommended that the typical value of 1.7 g/cm^3 be used), m_v is the volumetric moisture content in g/cm^3, and ε_w is the dielectric constant of water. In the model, ε_w is given by the single Debye model of Section 4-1, with an added conductivity term for ε''_w. That is,

$$\varepsilon'_w = \varepsilon_{w\infty} + \frac{\varepsilon_{w0} - \varepsilon_{w\infty}}{1 + (2\pi f \tau_w)^2}, \tag{4.67a}$$

$$\varepsilon''_w = \frac{2\pi f \tau_w (\varepsilon_{w0} - \varepsilon_{w\infty})}{1 + (2\pi f \tau_w)^2} + \left(\frac{2.65 - \rho_b}{2.65 m_v}\right) \frac{\sigma}{2\pi\varepsilon_0 f}. \tag{4.67b}$$

The expressions for ε_{w0}, $\varepsilon_{w\infty}$, and τ_w are given in Section 4-1, and $\varepsilon_0 = 8.854 \times 10^{-12}$ F/m. The exponents α, β_1, and β_2, and the effective conductivity σ are related to the soil properties. Their empirically determined expressions are:

$$\alpha = 0.65 \tag{4.68a}$$

$$\beta_1 = 1.27 - 0.519S - 0.152C, \tag{4.68b}$$

$$\beta_2 = 2.06 - 0.928S - 0.255C, \tag{4.68c}$$

$$\sigma = -1.645 + 1.939\rho_b - 2.256S + 1.594C, \tag{4.68d}$$

where S and C represent the mass fractions of sand and clay, respectively (i.e., $0 \le S \le 1$ and the same range applies to C).

[†]**Computer Code 4.7.**

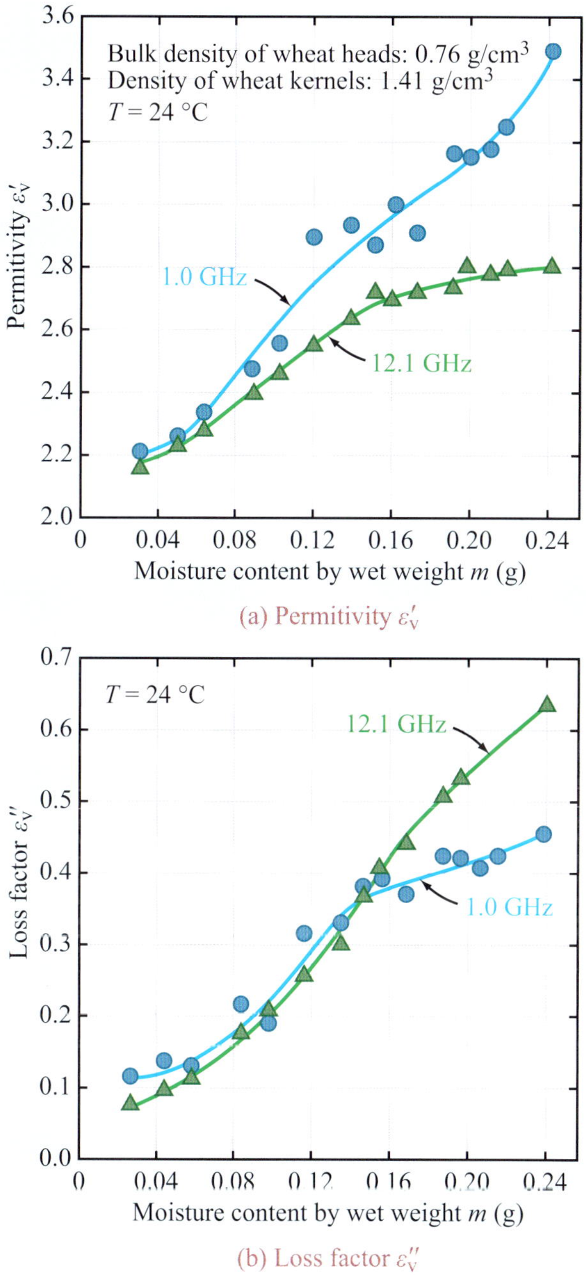

Figure 4-37: Measured dielectric constant of red winter wheat heads as a function of moisture content [from Nelson and Stetson, 1976].

available dielectric data for grains and seeds may not be directly applicable to remote-sensing problems, they can nonetheless supplement higher moisture data.

The moisture content of a vegetation material, such as leaves or stalks, is measured on a ***gravimetric*** wet-weight basis (m_g) or on a ***volume basis*** (m_v). The two quantities are related by

$$m_g = \frac{m_v}{m_v + (1 - m_v)\rho_s}, \tag{4.71a}$$

$$m_v = \frac{\rho_s m_g}{1 - m_g(1 - \rho_s)}, \tag{4.71b}$$

where ρ_s is the dry density of the solid material. A typical value of ρ_s for leaves is 0.3 g/cm^3. The range of m_g extends from 0 to about 0.9, and the corresponding range of m_v is 0 to 0.7 (for $\rho_s \approx 0.3$).

Using a waveguide transmission technique, Ulaby and Jedlicka (1984) measured the dielectric properties of leaves and stalks of corn and wheat over the 1.1 to 8.4 GHz range. Examples are shown in Fig. 4-38 for corn leaves at three microwave frequencies. As discussed shortly, among the important parameters governing the behavior of ε'_v and ε''_v is the salinity S of the fluid samples extracted from the vegetation material. Figure 4-38 displays dielectric data measured for corn leaves with $S = 11$ psu.

In a follow-up study, dielectric measurements were conducted over the 0.2 to 20 GHz range for various types of vegetation material (El-Rayes and Ulaby, 1987; Ulaby and El-Rayes, 1987). An example illustrating the spectra of ε'_v and ε''_v at various moisture levels is shown in Fig. 4-39. The data were used in the development of the dielectric model introduced in Section 4-9.2. Data measured for corn stalks and other types of vegetation were found to exhibit a moisture dependence very similar to that for corn leaves.

Pressure techniques were used to extract samples of the fluids contained in some of the vegetation material, so as to measure the dielectric spectrum of the fluid. Figure 4-40 displays the measured spectrum for the fluid contained in a freshly cut corn stalk. Also shown in the figure is the dielectric spectrum computed for saline water with a salinity of 7 psu, which was the measured salinity of the extracted fluid. The measured and calculated spectra are in very close agreement.

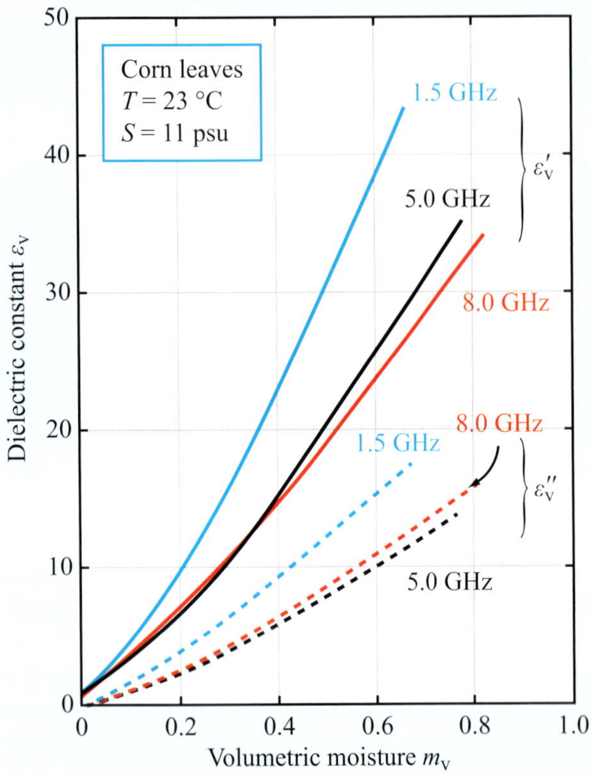

Figure 4-38: Measured moisture dependence of the dielectric constant of corn leaves at 1.5, 5.0, 8.0 GHz [Ulaby and Jedlicka, 1984].

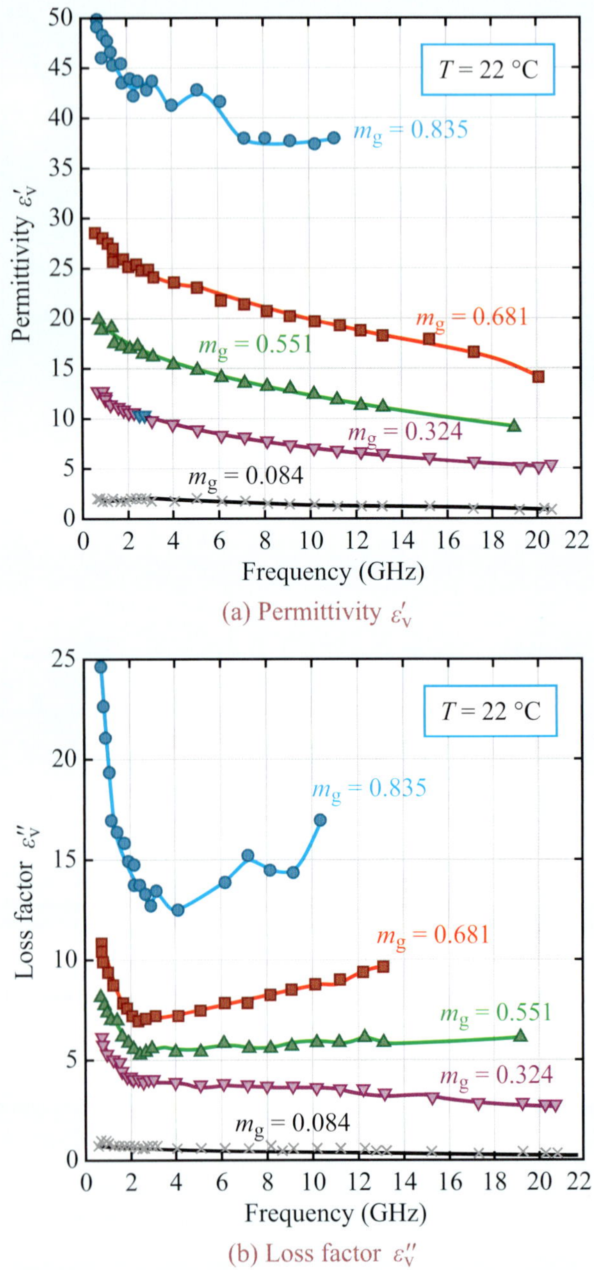

Figure 4-39: Family of dielectric spectra for corn leaves at $T = 22$ °C and various moisture contents [El-Rayes and Ulaby, 1987].

The role of temperature was examined by measuring the dielectric constant as the temperature was slowly lowered from room temperature at 22 °C down to -32 °C. At a salinity of 5 psu, the freezing temperature of the vegetation fluid should be around -0.3 °C. Hence, the expectation was that the water in the corn leaves would freeze when the temperature reached between 0 and -1 °C. The results shown in Fig. 4-41 suggest otherwise; the fluid in the corn leaf goes through a supercooling state down to about -6 °C, and then it freezes instantaneously.

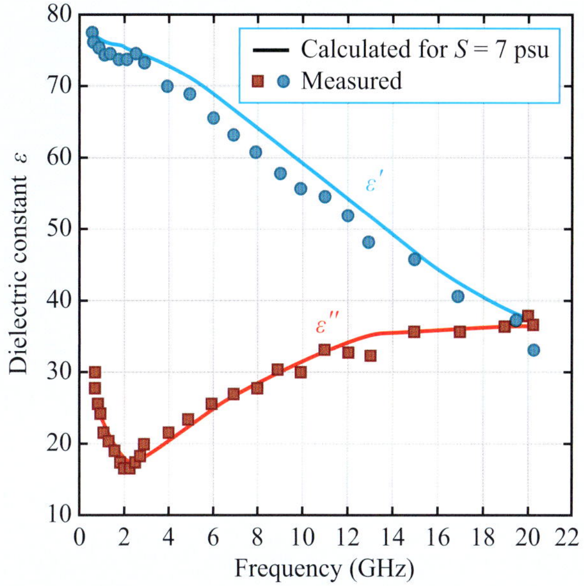

Figure 4-40: Measured and calculated dielectric spectra of the fluid extracted from corn stalks [El-Rayes and Ulaby, 1987].

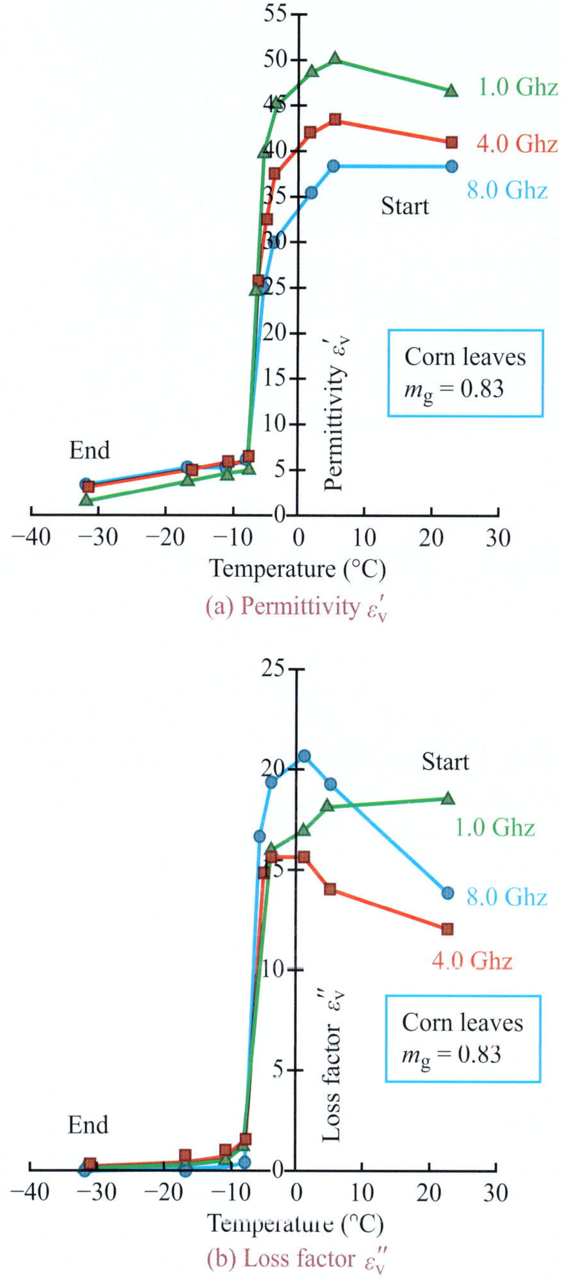

Figure 4-41: Variation of ε_r of corn leaves with decreasing temperature from 22 °C down to −32 °C [El-Rayes and Ulaby, 1987].

4-9.2 Dielectric Model

In most dielectric models of a solid material containing water (Section 4-4), the mixture is treated as an inhomogeneous medium consisting of discrete water particles dispersed in the bulk material (as host). The water particles are assigned a shape (such as spherical or cylindrical) in order to match the model to experimental data. Such models are unrealistic for vegetation because the water inside a leaf is closer to being spatially continuous than discrete. Accordingly, Ulaby and El-Rayes (1987) introduced a linear model for the dielectric constant of vegetation in the form of an additive mixture of three components:[†]

$$\varepsilon_v = \varepsilon_r + v_{fw}\varepsilon_w + v_{bw}\varepsilon_b, \tag{4.72}$$

where ε_r is a nondispersive ***residual component*** to be determined empirically, ε_w and ε_b are the complex dielectric constants of ***free water*** and ***bound water***,

[†] **Computer Code 4.8.**

respectively, and v_{fw} and v_{bw} are their associated volume fractions. The real and imaginary parts of ε_v are

$$\varepsilon'_v = \varepsilon_r + v_{fw}\varepsilon'_w + v_{bw}\varepsilon'_b, \tag{4.73a}$$

$$\varepsilon''_v = v_{fw}\varepsilon''_w + v_{bw}\varepsilon''_b. \tag{4.73b}$$

Free water

For the free-water component, the expressions for ε'_w and ε''_w given by Eqs. (4.14a) and (4.14b) are quite adequate, but because the free water in vegetation includes low concentrations of salts and sugars (salinity S seldom exceeds 15 psu), the expression for ε''_w requires the addition of a conductivity term. That is,

$$\varepsilon'_w = \varepsilon_{w\infty} + \frac{\varepsilon_{w0} - \varepsilon_{w\infty}}{1+(f/f_0)^2} \tag{4.74a}$$

$$\varepsilon''_w = \frac{(f/f_0)(\varepsilon_{w0} - \varepsilon_{w\infty})}{1+(f/f_0)^2} + \frac{\sigma_i}{2\pi\varepsilon_0 f}, \tag{4.74b}$$

where f_0 is the relaxation frequency of free water. The ionic conductivity σ_i is a function of the salinity S and the temperature T, and its functional form is given by Eq. (4.21f). At $T = 22$ °C, the relaxation frequency of water is 18 GHz, and the expressions for ε'_w and ε''_w become

$$\varepsilon'_w = 4.9 + \frac{74.4}{1+(f/18)^2}, \tag{4.75a}$$

$$\varepsilon''_w = \frac{74.4(f/18)}{1+(f/18)^2} + \frac{18\sigma_i}{f}, \tag{4.75b}$$

where f is in GHz and

$$\sigma_i \approx 0.17S - 0.0013S^2 \quad \text{(S/m)}. \tag{4.75c}$$

Bound water

A water molecule is considered to be in the "bound" state if its response to a suddenly applied electric field is inhibited because of some force acting on it. When in the bound state, the relaxation time τ of a water molecule is longer than when in the free state, but it is difficult to quantitatively relate the type of force and its magnitude to the increase in τ. To develop a model for the dielectric constant of the bound-water component in the vegetation mixture, Ulaby and El-Rayes (1987) conducted dielectric measurements for sucrose-water mixtures. Sucrose was chosen because it is a good example of the organic substances present in vegetation, and the binding arrangement of sucrose-water molecules is well known, thereby allowing them to compute the bound water concentration. Dielectric measurements conducted from 0.2 to 20 GHz for a sucrose-water solution at room temperature (22 °C) were fitted to a Cole-Cole dispersion equation of the form

$$\varepsilon'_b - j\varepsilon''_b = 2.9 + \frac{55}{1+(jf/0.18)^{0.5}}. \tag{4.76}$$

The factor 0.18 represents the relaxation frequency of bound water in GHz, which is exactly two orders of magnitude smaller than that for free water. In a Cole-Cole equation, the exponent of (f/f_0) may be smaller than 1. In the present case, the value 0.5 was found to provide an optimum fit to the measured data, as demonstrated by Fig. 4-42. Rationalizing Eq. (4.76) leads to

$$\varepsilon'_b = 2.9 + \frac{55(1+\sqrt{f/0.36})}{(1+\sqrt{f/0.36})^2 + (f/0.36)}, \tag{4.77a}$$

$$\varepsilon''_b = + \frac{55\sqrt{f/0.36}}{(1+\sqrt{f/0.36})^2 + (f/0.36)}, \tag{4.77b}$$

with f in GHz.

Empirical fits

By fitting the measured dielectric data to the dielectric model, the remaining quantities in Eq. (4.73) were found to assume the forms

$$\varepsilon_r = 1.7 - 0.74m_g + 6.16m_g^2, \tag{4.78a}$$

$$v_{fw} = m_g(0.55m_g - 0.076), \tag{4.78b}$$

$$v_{bw} = \frac{4.64m_g^2}{1+7.36m_g^2}. \tag{4.78c}$$

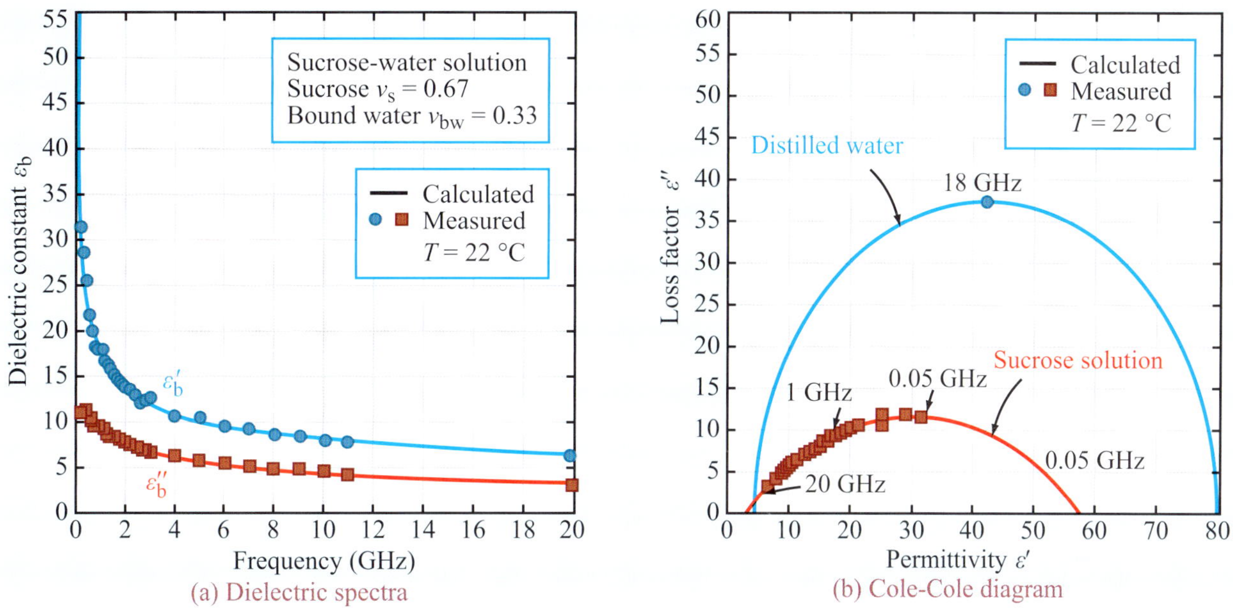

Figure 4-42: (a) Measured dielectric spectrum of sucrose-water solution and theoretical spectrum calculated according to Eq. (4.76). (b) Cole-Cole diagrams for distilled water and sucrose solution.

The residual dielectric constant ε_r varies from 1.7 for dry vegetation to 4.5 for $m_g = 0.7$. Since it represents the bulk vegetation material, it should not depend on m_g, but including a dependence on m_g leads to a better fit. Examples illustrating the dielectric spectra of corn leaves are shown in Fig. 4-43, and the variation of ε_v with m_g is displayed in Fig. 4-44. Even though the model was developed on the basis of dielectric measurements for corn leaves, it appears to provide a good fit to other types of vegetation as well. An example is shown in Fig. 4-45 for rubber leaves (Chuah et al., 1995).

PROBLEMS

4.1 Use Computer Code 4.1 to determine the microwave frequency at which the dielectric loss factor of pure water at 10 °C is a maximum in the range 1 to 30 GHz.

4.2 Use Computer Code 4.2 to generate plots for the relative permittivity and dielectric loss factor of saline water, at $T = 20$ °C and $f = 3$ GHz, as a function of salinity over the range from 0 to 40 psu.

4.3 Use Computer Code 4.3 to generate a plot of the dielectric loss factor of pure ice at 1 GHz as a function of temperature over the range from −1 °C to −30 °C.

4.4 For pure ice at $T = -20$ °C, below what frequency is the penetration depth greater than:

(a) 1 m?

(b) 10 m?

(c) 100 m?

4.5 Randomly oriented ice inclusions with

$$\varepsilon_i = 3.2 - j0.02$$

are present in an air background. Compute and plot the equivalent TVB permittivity and dielectric loss factor of the mixture as a function of the inclusion volume fraction, assuming the inclusions to be:

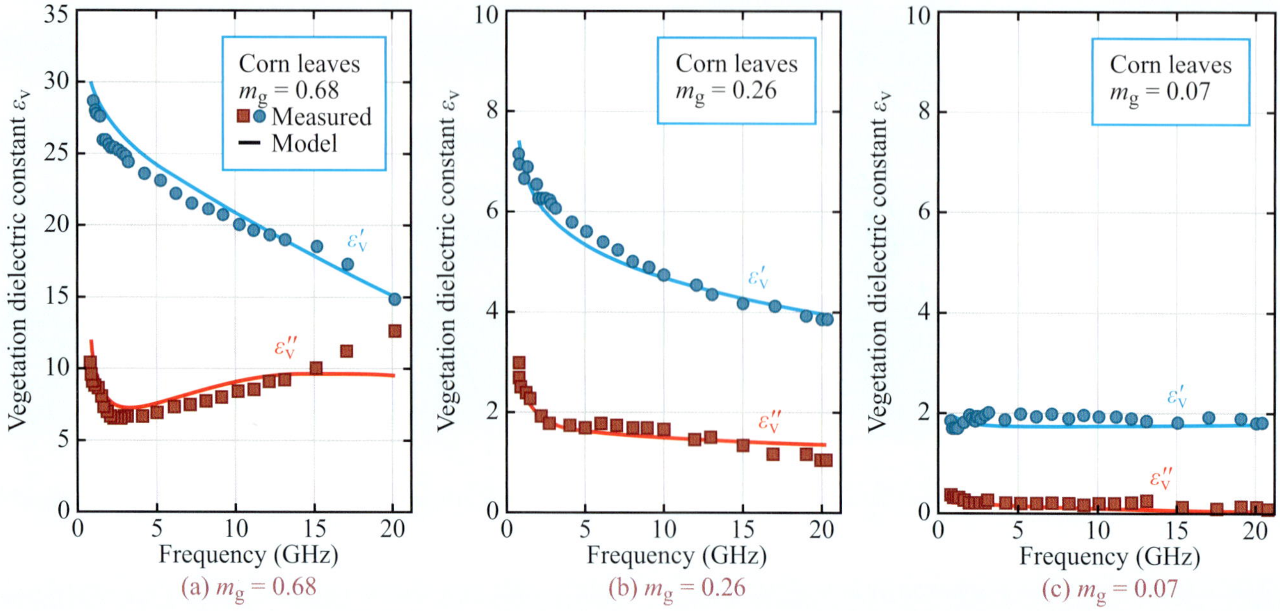

Figure 4-43: Comparison of the dielectric spectrum predicted by the vegetation model with measured data at (a) $m_g = 0.68$, (b) $m_g = 0.26$, and (c) $m_g = 0.07$.

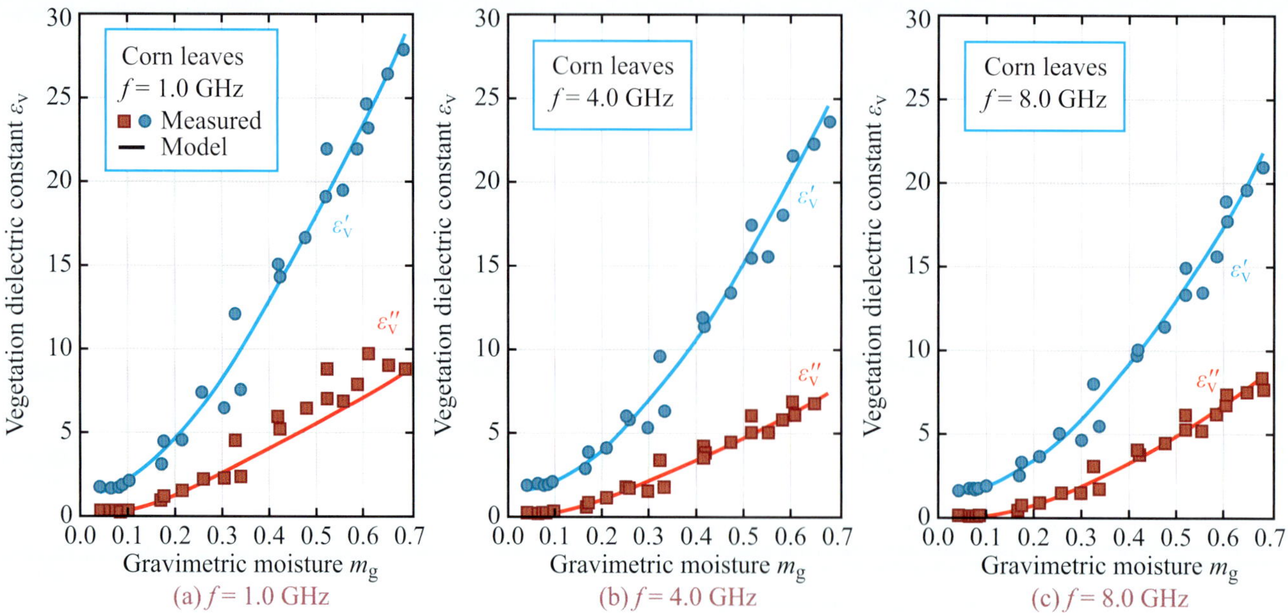

Figure 4-44: Comparison of the dielectric variation with moisture as predicted by the vegetation model with data at (a) 1 GHz, (b) 4 GHz, and (c) 8 GHz.

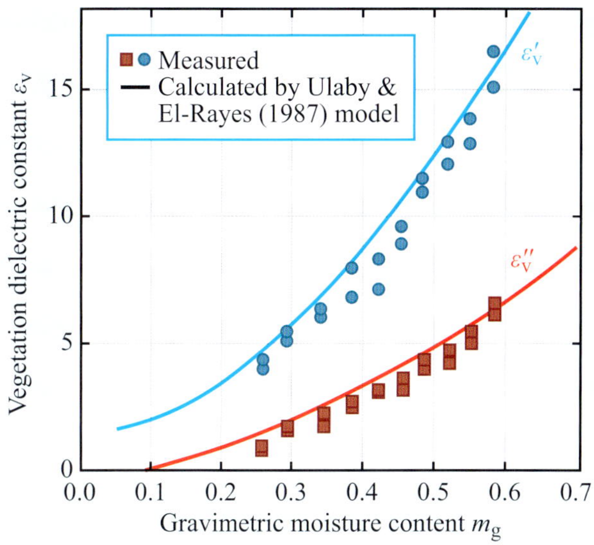

Figure 4-45: Dielectric constant of rubber leaves at 10.0 GHz [Chuah et al., 1995].

(a) circular discs,

(b) spheres,

(c) needles.

Suggestion: Use Computer Code 4.4.

4.6 A certain medium is modeled as a mixture of randomly oriented ellipsoids in a background of air. The ellipsoid dimensions are $2a_1 = 2b_1 = 0.12$ mm and $2c_1 = 1.76$ mm, and the dielectric constant of the ellipsoid material is $\varepsilon_i = 3.2 - j0.1$. Use the TVB model to develop an expression for the dielectric constant of the mixture, and then plot it as a function of the inclusion volume fraction.

4.7 Use the refractive model given by Eq. (4.45) with $\alpha = 1/2$ to compute the real and imaginary parts of the dielectric constant of a mixture composed of ice with $\varepsilon_i = 3.2 - j0.02$ in an air background. Plot your results as a function of the inclusion volume fraction.

4.8 Based on the data in Fig. 4-4, we estimate ε''_i of slightly impure ice at $T = -1$ °C and $f = 8.6$ GHz to be $\approx 4 \times 10^{-3}$. Use this value to compute the loss factor of dry snow and then to compute:

(a) ε''_{ds} as a function of ice volume fraction v_i.

(b) the penetration depth as a function of v_i.

4.9 Use Computer Code 4.6 to compute and plot the penetration depth in wet snow as a function of snow wetness (from 0 to 12%). The snow density is 0.4 g/cm^3 and the frequency is 10 GHz.

4.10 Use Computer Code 4.7 to compute and plot the penetration depth in wet soil as a function of volumetric soil moisture (from 3% to 30%) at 1.4 GHz. Assume $T = 20$ °C, the sand fraction $S = 30\%$, and the clay fraction $C = 50\%$.

4.11 Use Computer Code 4.8 to generate a Cole-Cole plot [similar to that in Fig. 4-42(b)] for vegetation with a gravimetric moisture content of 0.5. Assume the salinity of the vegetation fluid to be 7 psu.

CHAPTER

5

Radar Scattering

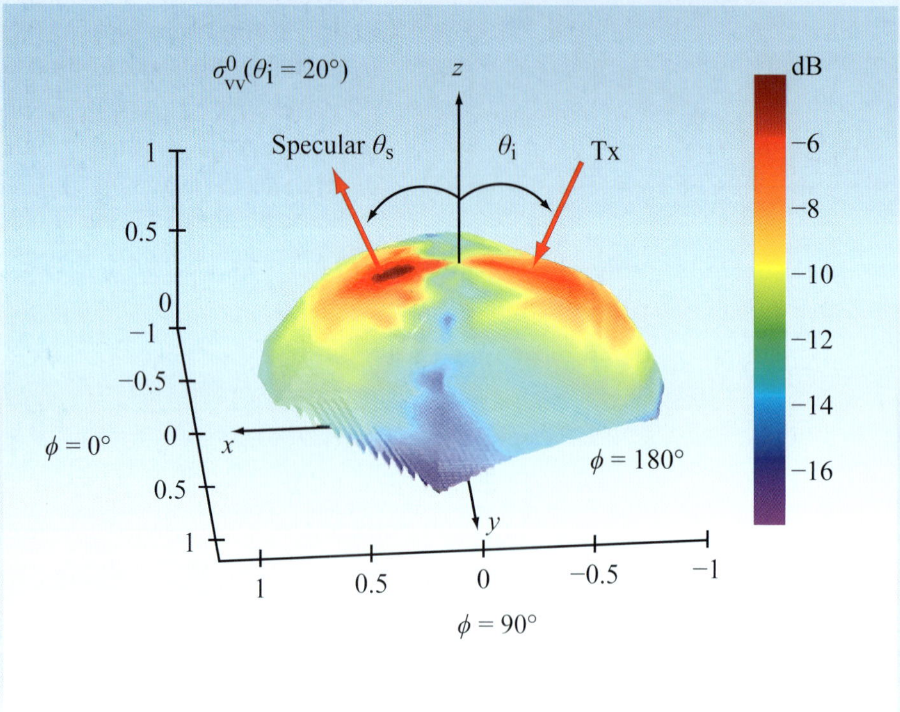

Bistatic scattering

CONTENTS

Overview

This chapter establishes the framework and associated terminology for how electromagnetic waves are scattered by point and distributed targets, the role of wave polarization, and how polarimetric radar can capture the full polarimetric response of the target. In the language of radar, any object "seen" by its antenna beam is called a "target," it is called a ***point target*** if it is self-contained—such as a car or aircraft, and it is called a ***distributed target*** if it is terrain-like.

5-1 Wave Polarization in a Spherical Coordinate System

The polarization of a plane wave describes the shape traced by the tip of its electric field in a plane orthogonal to the direction of propagation, as the wave passes through that plane. In Section 2-3, we examined the properties of linear, circular, and elliptical polarizations for a wave traveling in the z direction in a Cartesian (x,y,z) system. Orthogonality required $\mathbf{E}$ to reside in the x–y plane. When modeling radar scattering from the earth's surface, it is more convenient to represent the earth's surface by the x–y plane and to represent the polarization vector $\mathbf{E}$ of waves incident upon or scattered from the surface in terms of the spherical angles θ and ϕ. For a plane EM wave traveling in the direction $\hat{\mathbf{k}}$, as shown in Fig. 5-1, its electric field phasor $\mathbf{E}$ consists of a ***horizontal polarization*** component, $\hat{\mathbf{h}}E_\text{h}$, and a ***vertical polarization*** component, $\hat{\mathbf{v}}E_\text{v}$, defined such that the coordinate system $(\hat{\mathbf{k}},\hat{\mathbf{v}},\hat{\mathbf{h}})$ coincides with $(\hat{\mathbf{R}},\hat{\boldsymbol{\theta}},\hat{\boldsymbol{\phi}})$ of a standard spherical coordinate system. Thus, in a medium with wave number $k=2\pi/\lambda$,

$$\mathbf{E} = (\hat{\mathbf{v}}E_\text{v} + \hat{\mathbf{h}}E_\text{h})e^{-jk\hat{\mathbf{k}}\cdot\hat{\mathbf{R}}}, \tag{5.1}$$

with

$$\hat{\mathbf{k}} = \hat{\mathbf{x}}\sin\theta\cos\phi + \hat{\mathbf{y}}\sin\theta\sin\phi + \hat{\mathbf{z}}\cos\theta, \tag{5.2a}$$

$$\hat{\mathbf{h}} = \frac{\hat{\mathbf{z}}\times\hat{\mathbf{k}}}{|\hat{\mathbf{z}}\times\hat{\mathbf{k}}|} = -\hat{\mathbf{x}}\sin\phi + \hat{\mathbf{y}}\cos\phi, \tag{5.2b}$$

and

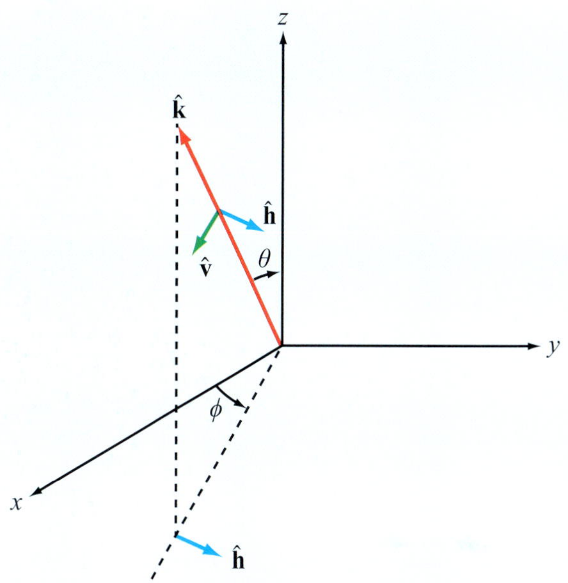

Figure 5-1: Spherical coordinate system for a wave traveling in the $\hat{\mathbf{k}}$ direction.

$$\hat{\mathbf{v}} = \hat{\mathbf{h}}\times\hat{\mathbf{k}} = \hat{\mathbf{x}}\cos\theta\cos\phi + \hat{\mathbf{y}}\cos\theta\sin\phi - \hat{\mathbf{z}}\sin\theta. \tag{5.2c}$$

In future sections, we use the matrix notation for $\mathbf{E}$ as

$$\mathbf{E} = \begin{bmatrix} E_\text{v} \\ E_\text{h} \end{bmatrix}, \tag{5.3}$$

where it is understood that E_v and E_h are the ***complex amplitudes*** of $\mathbf{E}$ along $\hat{\mathbf{v}}$ and $\hat{\mathbf{h}}$, respectively, and associated with both components is a ***propagation phase factor*** $e^{-jk\hat{\mathbf{k}}\cdot\hat{\mathbf{R}}}$.

In Section 2-3, for a wave traveling in the z direction, its wave polarization was defined by the magnitudes and phases of its electric field complex amplitudes E_x and E_y. Conversion to the spherical coordinate system entails the simple change of subscripts from (x,y) to (v,h) in all of the expressions given in Section 2-3. Specifically,

$$E_\text{v} = a_\text{v}, \tag{5.4a}$$

$$E_\text{h} = a_\text{h}e^{j\delta}, \tag{5.4b}$$

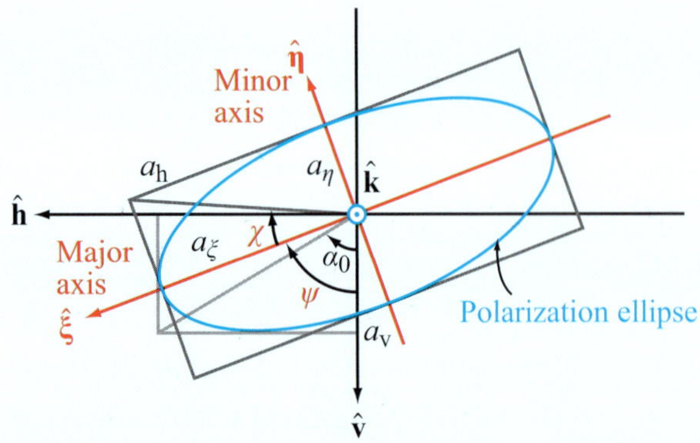

Figure 5-2: Polarization ellipse in the v–h plane for a wave traveling in the $\hat{\mathbf{k}}$ direction.

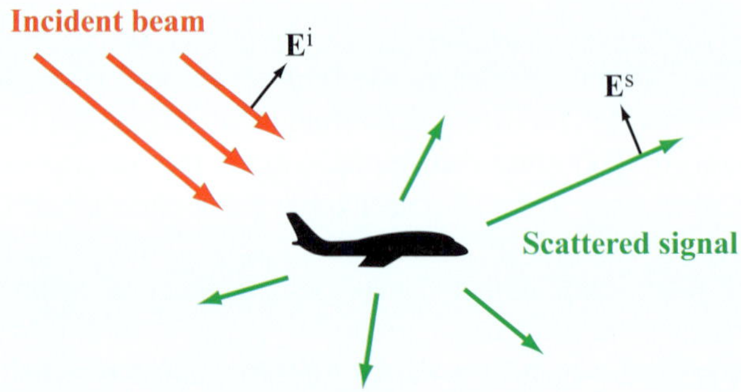

Figure 5-3: EM beam incident upon an object is scattered along many directions.

where $a_v = |E_v|$, $a_h = |E_h|$, and for convenience we assigned E_v a phase of zero. Thus, δ is the phase difference between the h and v components of $\mathbf{E}$.

If a_v, a_h, and δ all have nonzero values, the trace of the tip of $\mathbf{E}$ as a function of time (at a fixed R) forms an ellipse with rotation angle ψ and ellipticity angle χ (Fig. 5-2) given by

$$\tan 2\psi = (\tan 2\alpha_0)\cos\delta = \frac{2a_v a_h}{a_v^2 - a_h^2}\cos\delta \qquad (5.5a)$$

$$(-\pi/2 \le \psi \le \pi/2),$$

$$\sin 2\chi = (\sin 2\alpha_0)\sin\delta = \frac{2a_h a_v}{a_h^2 + a_v^2}\sin\delta \qquad (5.5b)$$

$$(-\pi/4 \le \chi \le \pi/4),$$

where α_0 is an auxiliary angle defined by

$$\tan\alpha_0 = \frac{a_h}{a_v}\,. \qquad (5.6)$$

For $\delta = 0$ or π, the ellipse collapses into a line and the wave polarization becomes linear, and if $\delta = \pm\pi/2$ and $a_v = a_h$, the polarization becomes circular (see Section 2-3).

5-2 Scattering Coordinate Systems

When illuminated by a plane wave with electric field $\mathbf{E}^i$, an object (Fig. 5-3) absorbs part of the energy it intercepts and scatters the remainder along many directions. If the object is a perfect conductor, it can scatter energy but not absorb any of it. Associated with each scattered wave is an electric field $\mathbf{E}^s$ whose magnitude and direction (polarization) are related to the incident field $\mathbf{E}^i$, the shape and orientation of the object relative to the incident and scattered directions, and the electrical properties of the object (permittivity ε' and conductivity σ).

In its simplest form, a radar is an instrument that uses a transmit antenna to illuminate an object, as depicted by Fig. 5-4(a), and then intercepts and measures part of the scattered energy. The scenario shown in Fig. 5-4(a), in which the transmitter and receiver are at different locations, is called a ***bistatic radar*** configuration. Most radars, however, operate in the ***monostatic radar*** configuration shown in Fig. 5-4(b) wherein the transmitter and receiver are collocated, and in most cases the same antenna is used for both transmission and reception. The monostatic configuration also is known as the ***backscatter configuration*** because the receiver measures the energy *scattered back* towards the radar.

Scattering calculations are performed in two major types of coordinate systems, the ***forward scatter alignment*** (FSA) convention and the ***backscatter alignment*** (BSA) convention. In both cases, the electric fields

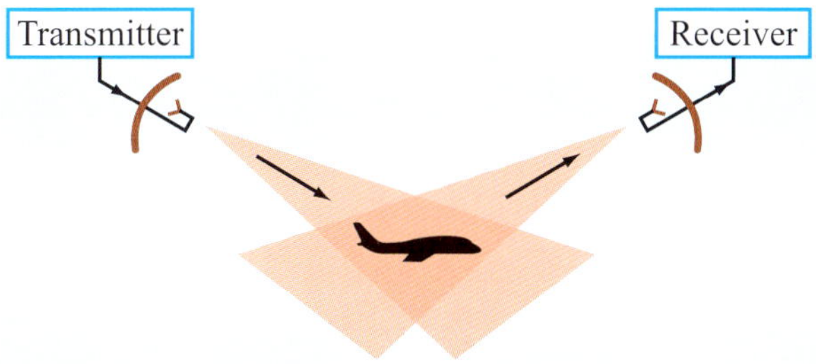

(a) Bistatic radar

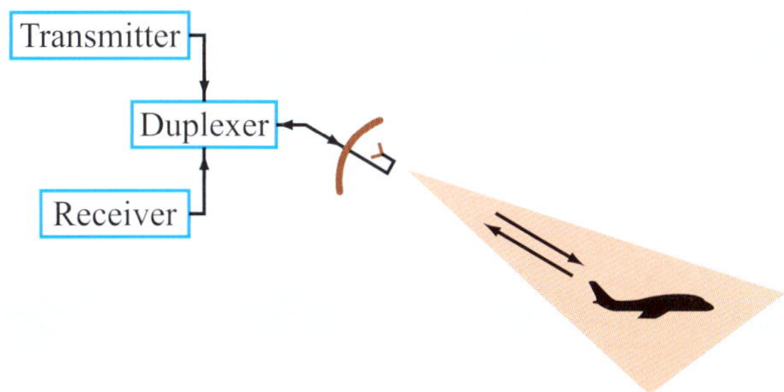

(b) Monostatic radar

Figure 5-4: The duplexer used by a monostatic radar allows the antenna to be shared between the transmitter and the receiver.

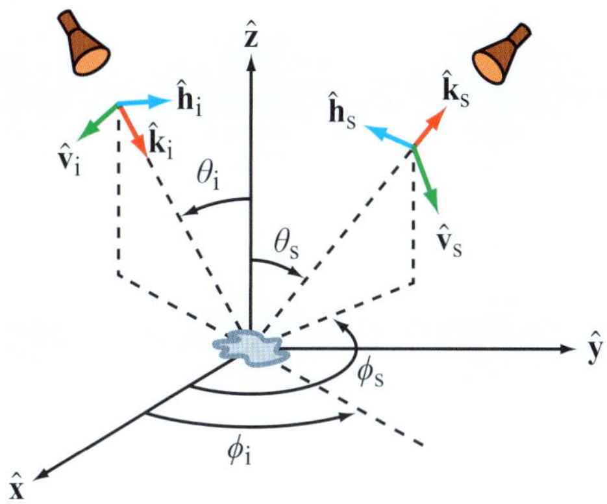

(a) FSA convention

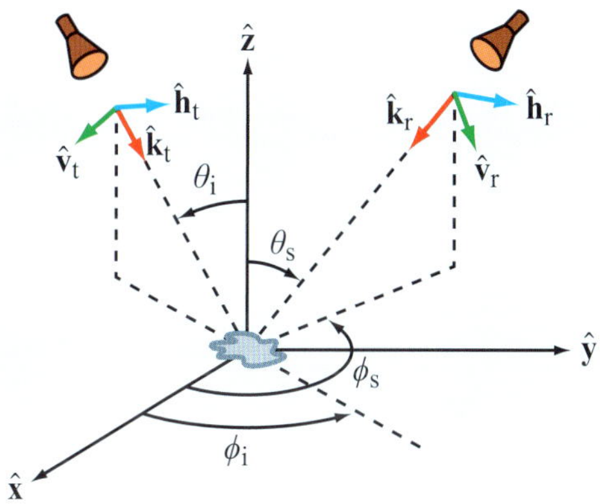

(b) BSA convention

Figure 5-5: Coordinate systems and scattering geometry for (a) the forward scattering alignment convention and (b) the backscatter alignment (BSA) convention.

of the incident and scattered waves are expressed in local coordinate systems centered on the transmitting and receiving antennas, respectively. All coordinate systems are defined in terms of a global coordinate system centered inside the scattering object, as shown in Fig. 5-5.

Both conventions are used in the literature, with the FSA being the preferred system for problems involving bistatic scattering from particles and inhomogeneous media and the BSA being the preferred system for calculating the radar backscatter from a given target or medium. Because both conventions are used in this book, we define and interrelate their individual coordinate systems in the following subsections.

5-2.1 Forward Scattering Alignment (FSA) Convention

FSA is a "wave-oriented" convention in that the directions of the vertical and horizontal unit vectors, $\hat{\mathbf{v}}$ and $\hat{\mathbf{h}}$, are always defined with respect to the direction of propagation of the wave, $\hat{\mathbf{k}}$. The coordinate system $(\hat{\mathbf{k}}, \hat{\mathbf{v}}, \hat{\mathbf{h}})$ is chosen to coincide with $(\hat{\mathbf{R}}, \boldsymbol{\theta}, \boldsymbol{\phi})$ in a standard spherical coordinate system, as shown earlier in Section

5-1. Thus, for a wave incident in the direction $\hat{\mathbf{k}}_i$ and scattered in the direction $\hat{\mathbf{k}}_s$, as shown in Fig. 5-5(a), the unit vectors are

$$\hat{\mathbf{k}}_i = \hat{\mathbf{x}}\cos\phi_i \sin\theta_i + \hat{\mathbf{y}}\sin\phi_i \sin\theta_i - \hat{\mathbf{z}}\cos\theta_i, \tag{5.7a}$$

$$\hat{\mathbf{h}}_i = \frac{\hat{\mathbf{z}}\times\hat{\mathbf{k}}_i}{|\hat{\mathbf{z}}\times\hat{\mathbf{k}}_i|} = -\hat{\mathbf{x}}\sin\phi_i + \hat{\mathbf{y}}\cos\phi_i, \tag{5.7b}$$

$$\hat{\mathbf{v}}_i = \hat{\mathbf{h}}_i \times \hat{\mathbf{k}}_i = -\hat{\mathbf{x}}\cos\phi_i \cos\theta_i - \hat{\mathbf{y}}\sin\phi_i \cos\theta_i - \hat{\mathbf{z}}\sin\theta_i, \tag{5.7c}$$

$$\hat{\mathbf{k}}_s = \hat{\mathbf{x}}\cos\phi_s \sin\theta_s + \hat{\mathbf{y}}\sin\phi_s \sin\theta_s + \hat{\mathbf{z}}\cos\theta_s, \tag{5.7d}$$

$$\hat{\mathbf{h}}_s = \frac{\hat{\mathbf{z}}\times\hat{\mathbf{k}}_s}{|\hat{\mathbf{z}}\times\hat{\mathbf{k}}_s|} = -\hat{\mathbf{x}}\sin\phi_s + \hat{\mathbf{y}}\cos\phi_s, \tag{5.7e}$$

$$\hat{\mathbf{v}}_s = \hat{\mathbf{h}}_s \times \hat{\mathbf{k}}_s = \hat{\mathbf{x}}\cos\phi_s \cos\theta_s + \hat{\mathbf{y}}\sin\phi_s \cos\theta_s - \hat{\mathbf{z}}\sin\theta_s. \tag{5.7f}$$

► In the FSA convention, the forward scattering direction corresponds to $\theta_s = \pi - \theta_i$ and $\phi_s = \phi_i$, and the backscattering direction corresponds to $\theta_s = \theta_i$ and $\phi_s = \phi_i + \pi$. ◄

For forward scattering, we have $\hat{\mathbf{k}}_s = \hat{\mathbf{k}}_i$, $\hat{\mathbf{v}}_s = \hat{\mathbf{v}}_i$, and $\hat{\mathbf{h}}_s = \hat{\mathbf{h}}_i$. In contrast, for backscattering, the relations are $\hat{\mathbf{k}}_s = -\hat{\mathbf{k}}_i$, $\hat{\mathbf{v}}_s = \hat{\mathbf{v}}_i$, $\hat{\mathbf{h}}_s = -\hat{\mathbf{h}}_i$. The direction of propagation $\hat{\mathbf{k}}_s$ is opposite to $\hat{\mathbf{k}}_i$, as expected, but only one of the two polarization vectors reverses direction upon backscattering.

5-2.2 Backscatter Alignment (BSA) Convention

In contrast to the FSA convention, in the BSA convention the polarization unit vectors are defined with respect to the radar antennas. This is in accordance with the standard convention that defines the polarization state of an antenna to be the polarization of the wave radiated by the antenna, even when it is used as a receiving antenna. Consequently, for the backscatter case under the BSA convention, the vertical and horizontal unit polarization vectors of the scattered wave are identical with their counterparts of the incident wave. The local coordinate system unit vectors are [Fig. 5-5(b)]:

$$\hat{\mathbf{k}}_t = \hat{\mathbf{k}}_i = \hat{\mathbf{x}}\cos\phi_i \sin\theta_i + \hat{\mathbf{y}}\sin\phi_i \sin\theta_i - \hat{\mathbf{z}}\cos\theta_i, \tag{5.8a}$$

$$\hat{\mathbf{h}}_t = \hat{\mathbf{h}}_i = -\hat{\mathbf{x}}\sin\phi_i + \hat{\mathbf{y}}\cos\phi_i, \tag{5.8b}$$

$$\hat{\mathbf{v}}_t = \hat{\mathbf{v}}_i = -\hat{\mathbf{x}}\cos\phi_i \cos\theta_i - \hat{\mathbf{y}}\sin\phi_i \cos\theta_i - \hat{\mathbf{z}}\sin\theta_i, \tag{5.8c}$$

$$\hat{\mathbf{k}}_r = -\hat{\mathbf{k}}_s = -[\hat{\mathbf{x}}\cos(\phi_s)\sin(\theta_s) + \hat{\mathbf{y}}\sin\phi_s \sin\theta_s + \hat{\mathbf{z}}\cos\theta_s], \tag{5.8d}$$

$$\hat{\mathbf{h}}_r = -\hat{\mathbf{h}}_s = \hat{\mathbf{x}}\sin\phi_s - \hat{\mathbf{y}}\cos\phi_s, \tag{5.8e}$$

$$\hat{\mathbf{v}}_r = \hat{\mathbf{v}}_s = \hat{\mathbf{x}}\cos\phi_s \cos\theta_s + \hat{\mathbf{y}}\sin\phi_s \cos\theta_s - \hat{\mathbf{z}}\sin\theta_s, \tag{5.8f}$$

where the subscripts t and r refer to the transmitting (or incident) and receiving antenna orientations. Because the two coordinates systems, $(\hat{\mathbf{k}}_t, \hat{\mathbf{v}}_t, \hat{\mathbf{h}}_t)$ and $(\hat{\mathbf{k}}_r, \hat{\mathbf{v}}_r, \hat{\mathbf{h}}_r)$, are coincident when the two antennas are collocated, the BSA convention has been the preferred system in the area of radar polarimetry (van Zyl et al., 1987a, b; Zebker et al., 1987).

To distinguish easily between the coordinate systems of the two conventions, we choose the subscripts i and s (which denote incident and scattered) for the FSA and the subscripts t and r (which denote transmitted and received) for the BSA convention. We must keep in mind, however, that the i coordinate system and the t coordinate systems are identical. To denote whether a given symbol represents a quantity that has been defined according to the FSA or BSA convention, we adopt the rule that a quantity is defined according to the FSA (or BSA) convention if its symbol has (or does not have) a tilde ($\sim$) over it.

► Thus, the scattering matrix $\widetilde{\mathbf{S}}$ introduced in the next section, for example, is defined in the FSA convention, whereas $\mathbf{S}$ is defined in the BSA convention. ◄

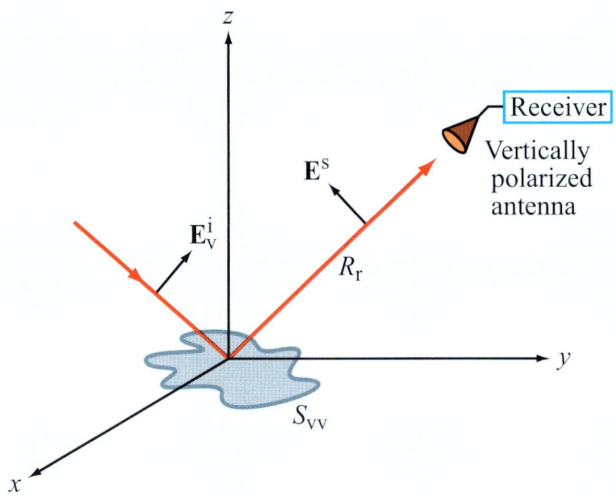

Figure 5-6: A vertically polarized incident wave E_{v}^{i}, defined at the location of the scattering object, generates a scattered wave with $\mathbf{E}^{s}$ defined at range R_r from the object. A vertically polarized receiving antenna measures $E_{v}^{s} = (e^{-jkR_r}/R_r)\widetilde{S}_{vv}E_{v}^{i}$.

5-3 Scattering Matrix

5-3.1 FSA Convention

Consider a small scattering object illuminated by a plane wave with incident electric field $\mathbf{E}^{i} = \hat{\mathbf{v}}_i E_{v}^{i}$ (Fig. 5-6). That is, $\mathbf{E}^{i}$ is vertically polarized. Moreover, *E_{v}^{i} is defined at the location of the object*. In the far zone of the scattering object, the scattered wave at a range R_r from the object is an outgoing spherical wave that can be approximated by a plane wave over the relatively small aperture of the receiving antenna. In general, the electric field $\mathbf{E}^{s}$ of the scattered wave consists of components along $\hat{\mathbf{v}}_s$ and $\hat{\mathbf{h}}_s$. If the antenna is vertically polarized, it will intercept only the vertically polarized component E_{v}^{s}, which is related to E_{v}^{i} by

$$E_{v}^{s} = \left(\frac{e^{-jkR_r}}{R_r}\right)\widetilde{S}_{vv}\,E_{v}^{i} \qquad \textbf{(vv polarization)}, \tag{5.9}$$

where (e^{-jkR_r}/R_r) is the spherical propagation factor and $\widetilde{S}_{vv}$ is the vv-polarized ***scattering amplitude*** of the scattering object, defined in the FSA convention.

> ▶ Scattering amplitude is denoted as:
>
> $\widetilde{S}$ in FSA convention,
> S in BSA convention. ◀

Similarly, had the receiving antenna been horizontally polarized instead, it would have measured the horizontal polarization component of $\mathbf{E}^{s}$, namely

$$E_{h}^{s} = \left(\frac{e^{-jkR_r}}{R_r}\right)\widetilde{S}_{hv}\,E_{v}^{i} \qquad \textbf{(hv polarization)}. \tag{5.10}$$

> ▶ Note that the first subscript of the scattering amplitude $\widetilde{S}_{pq}$ (where p and q may each be either v or h) refers to the polarization component of the scattered wave intercepted by the receive antenna and the second subscript refers to the polarization of the incident wave. ◀

In the general case, the incident wave may have both $\hat{\mathbf{h}}$ and $\hat{\mathbf{v}}$ polarization components, and the same is true for the scattered wave. That is,

$$\mathbf{E}^{i} = \hat{\mathbf{v}}_i E_{v}^{i} + \hat{\mathbf{h}}_i E_{h}^{i}, \tag{5.11a}$$

$$\mathbf{E}^{s} = \hat{\mathbf{v}}_s E_{v}^{s} + \hat{\mathbf{h}}_s E_{h}^{s}, \tag{5.11b}$$

and the two fields are related by

$$\begin{bmatrix} E_{v}^{s} \\ E_{h}^{s} \end{bmatrix} = \left(\frac{e^{-jkR_r}}{R_r}\right)\begin{bmatrix} \widetilde{S}_{vv} & \widetilde{S}_{vh} \\ \widetilde{S}_{hv} & \widetilde{S}_{hh} \end{bmatrix}\begin{bmatrix} E_{v}^{i} \\ E_{h}^{i} \end{bmatrix}. \tag{5.12}$$

The four scattering amplitudes characterize the scattering behavior of the object for the four possible combinations of the v and h polarization orientations of the incident and scattered fields. Each may be a real or complex quantity and may be a function of not only the target shape, size, orientation, permittivity, and conductivity, but also of the illumination and scattering angles, (θ_i, ϕ_i) and (θ_s, ϕ_s), respectively.

Equation (5.12) can be cast in the abbreviated form

$$\mathbf{E}^{s} = \left(\frac{e^{-jkR_r}}{R_r}\right)\widetilde{\mathbf{S}}\,\mathbf{E}^{i}, \tag{5.13}$$

where $\widetilde{\mathbf{S}}$ is the ***scattering matrix*** of the target in the FSA convention,

$$\widetilde{\mathbf{S}} = \begin{bmatrix} \widetilde{S}_{\mathrm{vv}} & \widetilde{S}_{\mathrm{vh}} \\ \widetilde{S}_{\mathrm{hv}} & \widetilde{S}_{\mathrm{hh}} \end{bmatrix} \qquad \textbf{(FSA)}. \tag{5.14}$$

As stated earlier, each component of $\widetilde{\mathbf{S}}$ is a function of the illumination and scattering angles. That is, for any polarization combination pq,

$$\begin{aligned} \widetilde{S}_{pq} &= \widetilde{S}_{pq}(\theta_i, \phi_i;\ \theta_{\mathrm{s}}, \phi_{\mathrm{s}}; \theta_j, \phi_j), \qquad p,q = \mathrm{v \ or \ h}, \\ &= \lim_{R_{\mathrm{r}}\to\infty} \left[R_{\mathrm{r}} e^{-jkR_{\mathrm{r}}} \left(\frac{E^{\mathrm{s}}_p}{E^{\mathrm{i}}_q} \right) \right], \end{aligned} \tag{5.15}$$

where (θ_j, ϕ_j) are the orientation angles of the object and the limit is used to denote that R_{r} is in the far-zone region of the scattering object.

5-3.2 BSA Convention

In the BSA convention, the transmitted and received fields are expressed in the form

$$\mathbf{E}^{\mathrm{t}} = \hat{\mathbf{v}}_{\mathrm{t}} E^{\mathrm{t}}_{\mathrm{v}} + \hat{\mathbf{h}}_{\mathrm{t}} E^{\mathrm{t}}_{\mathrm{h}}, \tag{5.16a}$$

$$\mathbf{E}^{\mathrm{r}} = \hat{\mathbf{v}}_{\mathrm{r}} E^{\mathrm{r}}_{\mathrm{v}} + \hat{\mathbf{h}}_{\mathrm{r}} E^{\mathrm{r}}_{\mathrm{h}}, \tag{5.16b}$$

with the unit vectors defined by Eq. (5.8). Upon comparing the expressions for the FSA and BSA local coordinate systems, the following becomes evident:

$$\mathbf{E}^{\mathrm{i}} = \mathbf{E}^{\mathrm{t}}, \tag{5.17}$$

$$\mathbf{E}^{\mathrm{s}} = \begin{pmatrix} 1 & 0 \\ 0 & -1 \end{pmatrix} \mathbf{E}^{\mathrm{r}}. \tag{5.18}$$

Therefore, the fields $\mathbf{E}^{\mathrm{t}}$ and $\mathbf{E}^{\mathrm{r}}$ are related by

$$\mathbf{E}^{\mathrm{r}} = \frac{e^{-jkR_{\mathrm{r}}}}{R_{\mathrm{r}}} \mathbf{S}\, \mathbf{E}^{\mathrm{t}}, \tag{5.19}$$

where

$$\mathbf{S} = \begin{pmatrix} S_{\mathrm{vv}} & S_{\mathrm{vh}} \\ S_{\mathrm{hv}} & S_{\mathrm{hh}} \end{pmatrix} \qquad \textbf{(BSA)} \tag{5.20}$$

is the scattering matrix of the object, defined according to the BSA convention. If we use Eqs. (5.17) and (5.18), we obtain the relationship

$$\widetilde{\mathbf{S}} = \begin{pmatrix} 1 & 0 \\ 0 & -1 \end{pmatrix} \mathbf{S}, \tag{5.21}$$

which implies that

$$\begin{aligned} \widetilde{S}_{\mathrm{vv}} &= S_{\mathrm{vv}}, & \widetilde{S}_{\mathrm{hh}} &= -S_{\mathrm{hh}}, \\ \widetilde{S}_{\mathrm{vh}} &= S_{\mathrm{vh}}, & \widetilde{S}_{\mathrm{hv}} &= -S_{\mathrm{hv}}. \end{aligned}$$

In general, $\widetilde{S}_{\mathrm{hv}}$ and $\widetilde{S}_{\mathrm{vh}}$ may not bear a simple relationship to one another, but in the backscatter direction ($\hat{\mathbf{k}}_{\mathrm{s}} = -\hat{\mathbf{k}}_{\mathrm{i}}$), the ***reciprocity theorem*** of electromagnetic scattering (Tsang et al., 1985) asserts that

$$\widetilde{S}_{\mathrm{vh}} = -\widetilde{S}_{\mathrm{hv}} \qquad \textbf{(backscatter)} \tag{5.22a}$$

and

$$S_{\mathrm{vh}} = S_{\mathrm{hv}} \qquad \textbf{(backscatter)}. \tag{5.22b}$$

5-3.3 Stokes Parameters and Mueller Matrix

The relationship given by Eq. (5.12) relates the h- and v-polarized components of the electric field $\mathbf{E}^{\mathrm{s}}$ of the wave scattered by an object with scattering matrix $\widetilde{\mathbf{S}}$ to the polarized components of $\mathbf{E}^{\mathrm{i}}$ of the incident wave illuminating the object. Often, the quantities of greater interest are the specific intensities of the incident and scattered waves. To that end, we introduce the ***vector specific intensity*** of an incident wave as

$$\mathbf{I}^{\mathrm{i}} = \begin{bmatrix} I^{\mathrm{i}}_{\mathrm{v}} \\ I^{\mathrm{i}}_{\mathrm{h}} \\ U^{\mathrm{i}} \\ V^{\mathrm{i}} \end{bmatrix} = \begin{bmatrix} |E^{\mathrm{i}}_{\mathrm{v}}|^2 \\ |E^{\mathrm{i}}_{\mathrm{h}}|^2 \\ 2\mathfrak{Re}(E^{\mathrm{i}}_{\mathrm{v}} E^{\mathrm{i}\,*}_{\mathrm{h}}) \\ 2\mathfrak{Im}(E^{\mathrm{i}}_{\mathrm{v}} E^{\mathrm{i}\,*}_{\mathrm{h}}) \end{bmatrix} \Big/ \eta, \tag{5.23a}$$

where η is the intrinsic impedance of the medium. The elements of $\mathbf{I}_{\mathrm{i}}$ are called the ***Stokes parameters***, of which $I^{\rm i}_{\rm v}$ and $I^{\rm i}_{\rm h}$ represent the specific intensities of the v- and h-polarized components of $\mathbf{I}^{\rm i}$, and the last two terms, namely $U^{\rm i}$ and $V^{\rm i}$, jointly account for the phase difference between $E^{\rm i}_{\rm v}$ and $E^{\rm i}_{\rm h}$.

A similar definition with superscript i replaced with superscript s applies to the vector specific intensity of the scattered wave:

$$\mathbf{I}^{\rm s} = \begin{bmatrix} I^{\rm s}_{\rm v} \\ I^{\rm s}_{\rm h} \\ U^{\rm s} \\ V^{\rm s} \end{bmatrix} = \begin{bmatrix} |E^{\rm s}_{\rm v}|^2 \\ |E^{\rm s}_{\rm h}|^2 \\ 2\mathfrak{Re}(E^{\rm s}_{\rm v} E^{\rm s*}_{\rm h}) \\ 2\mathfrak{Im}(E^{\rm s}_{\rm v} E^{\rm s*}_{\rm h}) \end{bmatrix} \Big/ \eta. \tag{5.23b}$$

By using Eq. (5.12) to evaluate the four quantities in Eq. (5.23b) and then relating their expressions to the four quantities defined in Eq. (5.23a) for the incident wave, we obtain the following relationship:

$$\mathbf{I}^{\rm s} = \frac{1}{R_{\rm r}^2}\, \widetilde{\mathbf{M}}\, \mathbf{I}^{\rm i} \qquad \textbf{(FSA)}, \tag{5.24a}$$

where $R_{\rm r}$ is the range between the scattering object and the location at which $\mathbf{I}^{\rm s}$ is measured. The 4×4 matrix $\widetilde{\mathbf{M}}$ is the ***modified Mueller matrix*** of the scattering object, defined in the forward scattering alignment (FSA) convention:

$$\widetilde{\mathbf{M}} = \begin{bmatrix} |\widetilde{S}_{\rm vv}|^2 & |\widetilde{S}_{\rm vh}|^2 & \mathfrak{Re}(\widetilde{S}_{\rm vv}\widetilde{S}^*_{\rm vh}) & -\mathfrak{Im}(\widetilde{S}_{\rm vv}\widetilde{S}^*_{\rm vh}) \\ |\widetilde{S}_{\rm hv}|^2 & |\widetilde{S}_{\rm hh}|^2 & \mathfrak{Re}(\widetilde{S}_{\rm hv}\widetilde{S}^*_{\rm hh}) & -\mathfrak{Im}(\widetilde{S}_{\rm hv}\widetilde{S}^*_{\rm hh}) \\ 2\mathfrak{Re}(\widetilde{S}_{\rm vv}\widetilde{S}^*_{\rm hv}) & 2\mathfrak{Re}(\widetilde{S}_{\rm vh}\widetilde{S}^*_{\rm hh}) & \mathfrak{Re}(\widetilde{S}_{\rm vv}\widetilde{S}^*_{\rm hh} + \widetilde{S}_{\rm vh}\widetilde{S}^*_{\rm hv}) & -\mathfrak{Im}(\widetilde{S}_{\rm vv}\widetilde{S}^*_{\rm hh} - \widetilde{S}_{\rm vh}\widetilde{S}^*_{\rm hv}) \\ 2\mathfrak{Im}(\widetilde{S}_{\rm vv}\widetilde{S}^*_{\rm hv}) & 2\mathfrak{Im}(\widetilde{S}_{\rm vh}\widetilde{S}^*_{\rm hh}) & \mathfrak{Im}(\widetilde{S}_{\rm vv}\widetilde{S}^*_{\rm hh} + \widetilde{S}_{\rm vh}\widetilde{S}^*_{\rm hv}) & \mathfrak{Re}(\widetilde{S}_{\rm vv}\widetilde{S}^*_{\rm hh} - \widetilde{S}_{\rm vh}\widetilde{S}^*_{\rm hv}) \end{bmatrix}. \tag{5.24b}$$

The scattering amplitudes of the scattering object are defined for the specific combination of the six angles listed explicitly in Eq. (5.15).

The significance of the relationships given by Eq. (5.24) is that it is possible to construct a receiver that can measure all of the elements of the matrix $\widetilde{\mathbf{M}}$, without having to measure the phase angles of any of the scattering amplitudes directly. The Stokes vector–Mueller matrix formulation is at the heart of the radiative transfer models used in Chapters 6, 8, 9, and 12 to compute polarimetric scattering and emission by surfaces and volumes.

Before closing this section, we should note that in the backscatter alignment convention (BSA), the counterpart of Eq. (5.24) takes the form

$$\mathbf{I}^{\rm r} = \frac{1}{R_{\rm r}^2}\, \mathbf{M}\, \mathbf{I}^{\rm t} \qquad \textbf{(BSA)} \tag{5.25a}$$

with $\mathbf{I}^{\rm t} = \mathbf{I}^{\rm i}$,

$$\mathbf{I}^{\rm r} = \begin{bmatrix} I^{\rm r}_{\rm v} \\ I^{\rm r}_{\rm h} \\ U^{\rm r} \\ V^{\rm r} \end{bmatrix} = \begin{bmatrix} |E^{\rm r}_{\rm v}|^2 \\ |E^{\rm r}_{\rm h}|^2 \\ 2\mathfrak{Re}(E^{\rm r}_{\rm v} E^{\rm r*}_{\rm h}) \\ 2\mathfrak{Im}(E^{\rm r}_{\rm v} E^{\rm r*}_{\rm h}) \end{bmatrix} \Big/ \eta. \tag{5.25b}$$

and

$$\mathbf{M} = \begin{bmatrix} |S_{\rm vv}|^2 & |S_{\rm vh}|^2 & \mathfrak{Re}(S_{\rm vv}S^*_{\rm vh}) & -\mathfrak{Im}(S_{\rm vv}S^*_{\rm vh}) \\ |S_{\rm hv}|^2 & |S_{\rm hh}|^2 & \mathfrak{Re}(S_{\rm hv}S^*_{\rm hh}) & -\mathfrak{Im}(S_{\rm hv}S^*_{\rm hh}) \\ 2\mathfrak{Re}(S_{\rm vv}S^*_{\rm hv}) & 2\mathfrak{Re}(S_{\rm vh}S^*_{\rm hh}) & \mathfrak{Re}(S_{\rm vv}S^*_{\rm hh} + S_{\rm vh}S^*_{\rm hv}) & -\mathfrak{Im}(S_{\rm vv}S^*_{\rm hh} - S_{\rm vh}S^*_{\rm hv}) \\ 2\mathfrak{Im}(S_{\rm vv}S^*_{\rm hv}) & 2\mathfrak{Im}(S_{\rm vh}S^*_{\rm hh}) & \mathfrak{Im}(S_{\rm vv}S^*_{\rm hh} + S_{\rm vh}S^*_{\rm hv}) & \mathfrak{Re}(S_{\rm vv}S^*_{\rm hh} - S_{\rm vh}S^*_{\rm hv}) \end{bmatrix}. \tag{5.25c}$$

Even though the forms of $\mathbf{I}^{\rm r}$ and $\mathbf{M}$ are the same as those of $\mathbf{I}^{\rm s}$ and $\widetilde{\mathbf{M}}$, respectively, $\mathbf{I}^{\rm r} \neq \mathbf{I}^{\rm s}$ and $\widetilde{\mathbf{M}} \neq \mathbf{M}$. In $\mathbf{I}^{\rm r}$, for example, $E^{\rm r}_{\rm h}$ is defined along $\hat{\mathbf{h}}_{\rm r}$ and in $\mathbf{I}^{\rm s}$, $E^{\rm s}_{\rm h}$ is defined along $\hat{\mathbf{h}}_{\rm s}$, but $\hat{\mathbf{h}}_{\rm s} = -\hat{\mathbf{h}}_{\rm r}$.

5-4 Radar Equation

Figure 5-7 depicts a bistatic radar system observing a target at a range $R_{\rm t}$ from the transmitter and a range $R_{\rm r}$ from the receiver. The radar transmitter injects power $P^{\rm t}_q$ into a q-polarized transmit antenna. At the receiver,

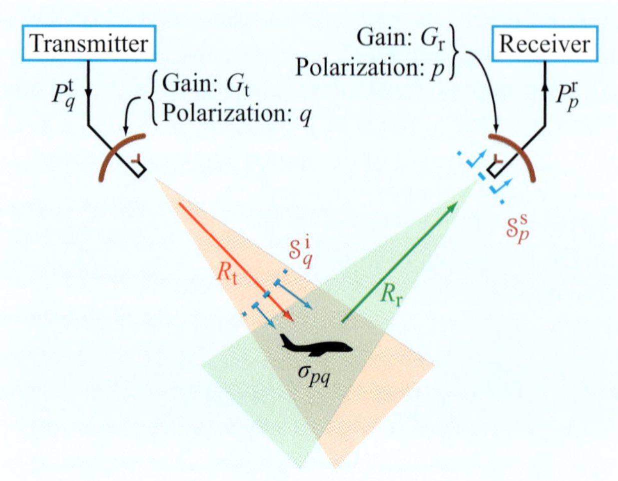

Figure 5-7: Power density $\mathcal{S}_q^i$ of the incident wave is defined at the location of the target, whereas $\mathcal{S}_p^s$ of the scattered energy is defined at the location of the receive antenna.

a p-polarized receive antenna delivers power P_p^r to the receiver.

▶ A pq-polarized radar (with p denoting the polarization of the receive antenna and q denoting the polarization of the transmit antenna, in that order) measures hh scattering by the target if $p = q = \text{h}$; it measures hv scattering if $p = \text{h}$ and $q = \text{v}$; and so on. ◀

At the target location, the power density illuminating the target is

$$\mathcal{S}_q^i = \frac{P_q^t}{4\pi R_t^2} G_t \qquad (\text{W/m}^2), \qquad (5.26)$$

where $(P_q^t/4\pi R_t^2)$ represents the power density that would have been radiated had the antenna been isotropic, and G_t is the gain of the transmit antenna in the direction of the target. The power intercepted by the target is scattered (reradiated) along many directions, but our interest is in only the power radiated toward the receive antenna, and in only the component of that power with polarization state p, because that is the only component that will be captured by the p-polarized receive antenna. The p-polarized power reradiated by the target, denoted P_p^{rer}, is related to the power density $\mathcal{S}_q^i$ incident on the target by

$$P_p^{\text{rer}} = \sigma_{pq} \mathcal{S}_q^i, \qquad (5.27a)$$

where σ_{pq} is the pq-polarized ***radar cross section*** (RCS) of the target for the specific combination of incident and scattered directions. Combining Eqs. (5.26) and (5.27a) gives

$$P_p^{\text{rer}} = \frac{P_q^t G_t}{4\pi R_t^2} \sigma_{pq} \qquad (\text{W}). \qquad (5.27b)$$

The reradiated power spreads over a spherical surface, resulting in scattered power density $\mathcal{S}_p^s$ incident upon the receive antenna. At range R_r from the target

$$\mathcal{S}_p^s = \frac{P^{\text{rer}}}{4\pi R_r^2} = \frac{P_q^t G_t}{(4\pi R_t R_r)^2} \sigma_{pq} \qquad (\text{W/m}^2). \qquad (5.28a)$$

With an effective area A_r and radiation efficiency ξ_r, the receive antenna intercepts and delivers (to the receiver) power P_p^r given by

$$P_p^r = \xi_r A_r \mathcal{S}_q^s = \frac{P_q^t G_t \xi_r A_r}{(4\pi R_t R_r)^2} \sigma_{pq}. \qquad (5.28b)$$

By incorporating the relations $A_r = (\lambda^2/4\pi) D_r$ and $G_r = \xi_r D_r$ [Eqs. (3.24) and (3.31b)], we obtain the ***bistatic radar equation***

$$\frac{P_p^r}{P_q^t} = \frac{G_t G_r \lambda^2}{(4\pi)^3 R_t^2 R_r^2} \sigma_{pq} \qquad \textbf{(point-target bistatic radar equation).} \qquad (5.29a)$$

Under monostatic conditions, $R_t = R_r = R$, and if the radar uses the same antenna for both the transmit and receive functions (but with polarization q when transmitting and polarization p when receiving), Eq. (5.29a) simplifies to

$$\frac{P_p^r}{P_q^t} = \frac{G^2 \lambda^2}{(4\pi)^3 R^4} \sigma_{pq} \qquad \textbf{(point-target monostatic radar equation),} \qquad (5.29b)$$

where $G = G_t = G_r$.

Note that the range dependence in the radar equation varies as $1/R^4$, the product of two one-way propagation processes.

The expressions given by Eq. (5.29) pertain to a target whose physical dimensions are such that the solid angle it subtends (from where the radar is located) is much smaller than the solid angle of the radar beams [Fig. 5-4(b)]. We call such a target a ***point target***, even though it may have a complex geometry and a nonuniform scattering pattern. Its backscattering *strength* is defined by its RCS σ_{pq}, which has units of area. If we take the ratio of Eq. (5.28a) to Eq. (5.26), we obtain the result

$$\sigma_{pq} = \lim_{R_\mathrm{r}\to\infty} \left(4\pi R_\mathrm{r}^2 \, \frac{\mathcal{S}_p^\mathrm{s}}{\mathcal{S}_q^\mathrm{i}} \right), \tag{5.30}$$

where the limit has been added to emphasize the condition that the range R_r at which the scattered power density $\mathcal{S}_p^\mathrm{s}$ is measured is in the far-zone region of the target. Power densities $\mathcal{S}_p^\mathrm{s}$ and $\mathcal{S}_q^\mathrm{i}$ are related to the electric fields of their respective waves by $\mathcal{S}_p^\mathrm{s} = |E_p^\mathrm{s}|^2/2\eta_0$ and $\mathcal{S}_q^\mathrm{i} = |E_q^\mathrm{s}|^2/2\eta_0$, where η_0 is the intrinsic impedance of air. In view of Eq. (5.15), σ_{pq} is related to the scattering amplitude S_{pq} by

$$\sigma_{pq} = 4\pi|\widetilde{S}_{pq}|^2. \tag{5.31}$$

Models for computing $\widetilde{S}_{pq}$, and by extension σ_{pq}, of simple point targets, such as spheres and cylinders, are available in Section 5-11. For complex targets, like ships and aircraft, highly sophisticated computer simulation techniques beyond the scope of this book are needed to compute σ_{pq}.

5-5 Scattering from Distributed Targets

The monostatic radar equation given by Eq. (5.29b) for a point target may be extended to the distributed target shown in Fig. 5-8 by integrating the backscattered power over the illuminated area A:

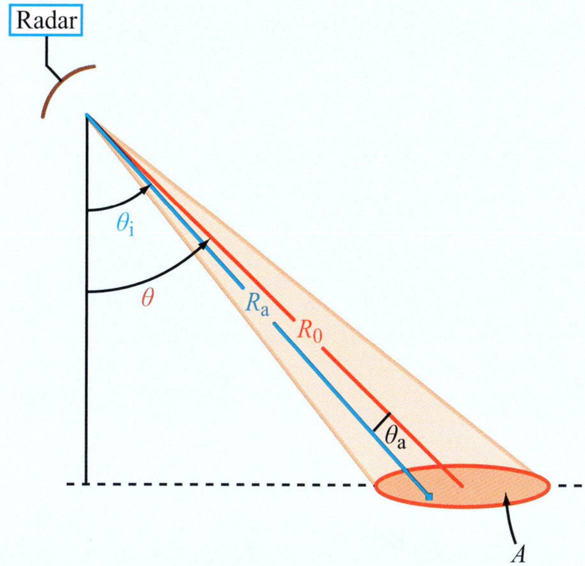

Figure 5-8: Illumination geometry for a distributed target.

$$P_p^\mathrm{r}(\theta) = \iint_A \frac{P_q^\mathrm{t}\, G^2(\theta_\mathrm{a},\phi_\mathrm{a})\,\lambda^2}{(4\pi)^3 R_\mathrm{a}^4} \cdot \sigma_{pq}^0 \, dA \tag{5.32}$$

where θ is the ***incidence angle*** of the boresight direction of the antenna, relative to normal incidence, $(\theta_\mathrm{a},\phi_\mathrm{a})$ define the direction relative to boresight of a point within A, and R_a is the range to it. The polarization indices p and q are included in the above expression to show the connection between the q-polarized transmitted power, the p-polarized received power, and the pq-polarized ***backscattering cross section per unit area*** σ_{pq}^0, which is defined as the backscattering cross section σ_{pq} of a distributed target of horizontal area A, normalized with respect to A,

$$\sigma_{pq}^0 = \sigma_{pq}/A. \tag{5.33a}$$

Often, σ^0 is referred to as the (differential) ***backscattering coefficient*** or ***radar reflectivity***.

As discussed later in Section 5-6, the power backscattered by a distributed target with *seemingly uniform* properties, such as a bare soil surface, exhibits large variations from one illuminated area to the next.

The reasons for these variations and the steps usually taken to reduce them are discussed in Sections 5-6 to 5-9. For the present, however, we should amend the definition of σ^0_{pq} given by Eq. (5.33a) to the more correct and formal definition

$$\sigma^0_{pq} = \frac{1}{A}\left\langle \sigma_{pq} \right\rangle, \tag{5.33b}$$

where $\langle\ \rangle$ denotes the ensemble-average operation over many measurements of σ_{pq}, corresponding to independent illuminated areas of the distributed target.

▶ That is, σ^0_{pq} is, by definition, the *average value* of the backscattering cross section of the distributed target, normalized to the illumination area of the antenna beam. ◀

Accordingly, the ensemble-average operation should apply to the left-hand side of Eq. (5.32). That is,

$$\left\langle P^{\rm r}_p(\theta) \right\rangle = \iint\limits_A \frac{P^{\rm t}_q\, G^2(\theta_{\rm a},\phi_{\rm a})\, \lambda^2}{(4\pi)^3 R^4_{\rm a}} \cdot \sigma^0_{pq}\, dA. \tag{5.34}$$

5-5.1 Narrow-Beam Scatterometer

If the distributed target has uniform properties across the area illuminated by the antenna beam, and if the beam is sufficiently narrow that $\sigma^0(\theta_{\rm i})$ at the local angle of incidence $\theta_{\rm i}$ may be regarded as constant over the angular extent of the beam, then Eq. (5.34) may be simplified to

$$\begin{aligned}\left\langle P^{\rm r}_p(\theta) \right\rangle &= \frac{P^{\rm t}_q \lambda^2}{(4\pi)^3}\, \sigma^0_{pq}(\theta) \iint\limits_{A_{\rm i}} \frac{G^2(\theta_{\rm a},\phi_{\rm a})\, dA}{R^4_{\rm a}} \\ &= \frac{P^{\rm t}_q \lambda^2}{(4\pi)^3}\, \sigma^0_{pq}(\theta) \cdot I, \end{aligned} \tag{5.35}$$

where

$$I = \iint\limits_A \frac{G^2(\theta_{\rm a},\phi_{\rm a})\, dA}{R^4_{\rm a}} \tag{5.36}$$

is called the ***illumination integral***. Often, an additional approximation is made, namely $R_{\rm a} \approx R_0$ over the solid angle of the antenna beam, in which case $R_{\rm a}$ can be taken outside the integral, and the antenna pattern is replaced with an equivalent pattern of gain G_0 and effective width β. These approximations lead to

$$I \approx \frac{G^2_0}{R^4_0}\, A$$

and

$$\left\langle P^{\rm r}_p(\theta) \right\rangle \approx \left[\frac{P^{\rm t}_q \lambda^2 G^2_0 A}{(4\pi)^3 R^4_0}\right] \cdot \sigma^0_{pq}(\theta), \tag{5.37}$$

(narrow-beam scatterometer)

where A is the area illuminated by the equivalent beam of the antenna. We should note that because the integration in Eq. (5.36) involves G^2, rather than just G, β is the effective beamwidth of the product pattern G^2. For a symmetrical Gaussian pattern $G(\theta_{\rm a})$, the effective beamwidth is approximately equal to its half-power (3 dB) beamwidth $\beta_{1/2}$. The half-power beamwidth of $G^2(\theta_{\rm a})$ is $\beta = \beta_{1/2}/\sqrt{2}$.

5-5.2 Imaging Radar

The assumptions made in deriving Eq. (5.37) are well suited to the imaging radar case because the dimensions of an illuminated ground cell are very small relative to the distance R between the radar and the cell. Figures 5-9 and 5-10 show the imaging geometries associated with real-aperture and synthetic-aperture radars when used in the side-looking mode. A real-aperture radar (Chapter 14) uses an antenna that is wide horizontally and narrow vertically. The horizontal beamwidth $\beta_{\rm h}$ is typically 1° or less, and the azimuth resolution is

$$r_{\rm a} = \beta_{\rm h} R \approx \frac{\lambda R}{l}, \tag{5.38}$$

where R is the slant range and l is the antenna length. Resolution in the range or across-track dimension is achieved by transmitting a narrow pulse of length τ.

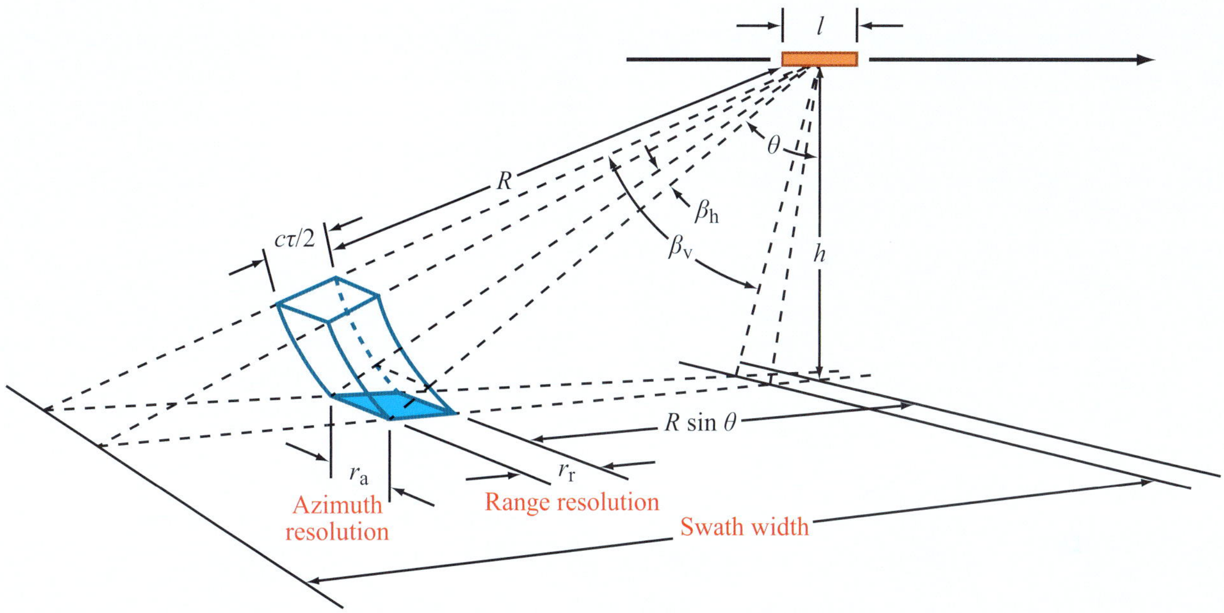

Figure 5-9: Real-aperture radar (RAR) geometry and resolved surface area.

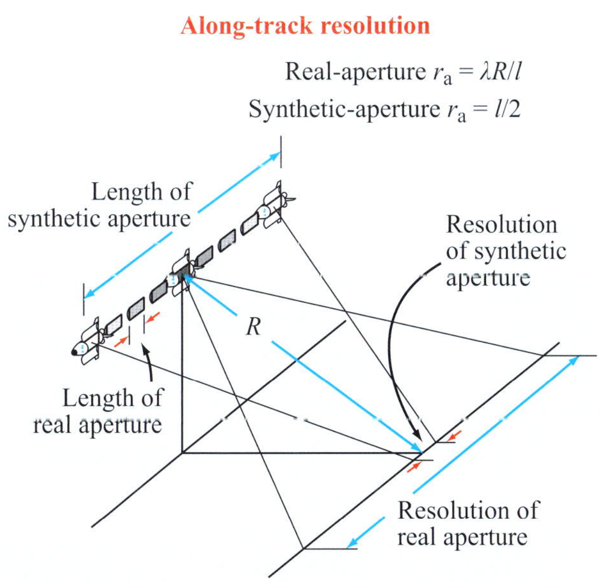

Figure 5-10: An illustration of how synthetic aperture works.

The image is produced by recording the returns from successive pulses as the aircraft moves past the area being covered. The ground-range resolution is independent of R and is given by

$$r_r = \frac{c\tau}{2\sin\theta} . \tag{5.39}$$

Inserting $A = r_a r_r$ in Eq. (5.37) leads to the following form of the radar equation for the real-aperture radar:

$$\langle P_p^r(\theta)\rangle = \left[\frac{P_q^t \lambda^3\, G^2(\theta)\, c\tau}{2l(4\pi)^3 R^3 \sin\theta}\right] \cdot \sigma_{pq}^0(\theta), \tag{5.40}$$

(real-aperture radar)

where $G(\theta)$ is the antenna gain in the direction θ.

The antenna used in a synthetic-aperture radar (SAR) system is usually shorter in the along-track (azimuth) direction, but successive pulses are processed together to "focus" the azimuth beam of the antenna down to a very narrow beam corresponding to a very long synthetic array (Chapter 14). Theoretically, the cell dimension in the along-track direction can be made independent of R and as small as

$$r_{\text{a}} = \frac{l}{2}, \tag{5.41}$$

where l is the along-track length of the antenna. When this is achieved, the SAR is said to be ***fully focused***. In the across-track direction, resolution is achieved in the same way as with the real-aperture radar, so Eq. (5.39) applies. For a fully focused SAR, the radar equation becomes

$$\langle P_p^{\text{r}}(\theta)\rangle = \left[\frac{P_q^{\text{t}}\lambda^2\, G^2(\theta)\, lc\tau}{4(4\pi)^3R^4\sin\theta}\right]\cdot\sigma_{pq}^0(\theta). \tag{5.42}$$

(synthetic-aperture radar)

In summary, the power received due to backscatter from an illuminated cell that subtends a small solid angle as viewed from the radar is directly proportional to the backscattering coefficient σ^0, regardless of the specific type of radar used. Thus, the radar equations for the narrow-beam scatterometer, the real-aperture radar, and the synthetic-aperture radar can all be written in the compact form

$$\langle P_p^{\text{r}}(\theta)\rangle = K_{pq}\,\sigma_{pq}^0(\theta), \tag{5.43}$$

(transmit q / receive p polarization)

where K_{pq} may be regarded as a system constant representing the quantity in the square bracket in Eqs. (5.37), (5.40), or (5.42).

► In radar remote sensing, the differential scattering coefficient σ_{pq}^0 is the critical link between the radar as a sensing instrument and the intended application. ◄

By calibrating the radar to determine K_{pq}, the measured power can be used to determine the reflectivity σ_{pq}^0. Then, from models relating σ_{pq}^0 to the biophysical or geophysical properties of the terrain, we can generate estimates of those biophysical or geophysical properties. Often the process involves multiple measurements of σ_{pq}^0 corresponding to different combinations of polarizations and microwave frequencies.

5-5.3 Specific Intensities for Distributed Target

In Section 5-3.3, the vector specific intensity of the wave incident upon a point target, $\mathbf{I}^{\text{i}}$, and the vector specific intensity of the wave scattered by it, $\mathbf{I}^{\text{s}}$, are represented by Eqs. (5.23a) and (5.23b). The representations for $\mathbf{I}^{\text{i}}$ and $\mathbf{I}^{\text{s}}$ are identical in form, despite the fact that $\mathbf{I}^{\text{i}}$ is associated with a plane wave and $\mathbf{I}^{\text{s}}$ is associated with a spherical wave. The forms of $\mathbf{I}^{\text{i}}$ and $\mathbf{I}^{\text{s}}$ are identical because the target is treated as a point.

For a distributed target, the scattering coefficient is defined as the average scattering cross section per unit area for plane-wave incidence and spherical-wave scattering by the distributed target. Hence, Eq. (5.23a) continues to provide a valid representation of the vector specific intensity of the incident wave, but the representation for the vector specific intensity of the scattered wave needs to be defined in terms of the net electric field $\mathbf{E}^{\text{s}}$, representing the vector sum of all fields scattered from an illuminated area A subtending a solid angle $d\Omega_{\text{s}}$ given by

$$d\Omega_{\text{s}} = \frac{A\cos\theta_{\text{s}}}{R_{\text{r}}^2}, \tag{5.44}$$

where R_{r} is the range from the surface to the receiver and θ_{s} is the angle between the surface normal and the scattering direction [Fig. 5-5(a)]. Accordingly, $\mathbf{I}^{\text{i}}$ and $\mathbf{I}^{\text{s}}$ are defined as

$$\mathbf{I}^{\text{i}} = \begin{bmatrix} I_{\text{v}}^{\text{i}} \\ I_{\text{h}}^{\text{i}} \\ U^{\text{i}} \\ V^{\text{i}} \end{bmatrix} = \begin{bmatrix} |E_{\text{v}}^{\text{i}}|^2 \\ |E_{\text{h}}^{\text{i}}|^2 \\ 2\mathfrak{Re}(E_{\text{v}}^{\text{i}}E_{\text{h}}^{\text{i}\,*}) \\ 2\mathfrak{Im}(E_{\text{v}}^{\text{i}}E_{\text{h}}^{\text{i}\,*}) \end{bmatrix} \Big/ \eta, \tag{5.45a}$$

$$\mathbf{I}^{\text{s}}\, d\Omega_{\text{s}} = \begin{bmatrix} I_{\text{v}}^{\text{s}} \\ I_{\text{h}}^{\text{s}} \\ U^{\text{s}} \\ V^{\text{s}} \end{bmatrix} d\Omega_{\text{s}} = \begin{bmatrix} |E_{\text{v}}^{\text{s}}|^2 \\ |E_{\text{h}}^{\text{s}}|^2 \\ 2\mathfrak{Re}(E_{\text{v}}^{\text{s}}E_{\text{h}}^{\text{s}*}) \\ 2\mathfrak{Im}(E_{\text{v}}^{\text{s}}E_{\text{h}}^{\text{s}*}) \end{bmatrix} \Big/ \eta. \tag{5.45b}$$

Note the distinction between the Stokes parameters of $\mathbf{I}^{\mathrm{i}}$ and $\mathbf{I}^{\mathrm{s}}$. For the vertical-polarization Stokes parameters,

$$I_{\mathrm{v}}^{\mathrm{i}} = \frac{|E_{\mathrm{v}}^{\mathrm{i}}|^2}{\eta}, \tag{5.46a}$$

whereas

$$I_{\mathrm{v}}^{\mathrm{s}}\, d\Omega_{\mathrm{s}} = \frac{|E_{\mathrm{v}}^{\mathrm{s}}|^2}{\eta}. \tag{5.46b}$$

For the simple case where both the transmitting and receiving antennas are vertically polarized, the scattering coefficient is defined as

$$\begin{aligned}\sigma_{\mathrm{vv}}^0 &= \frac{4\pi R_{\mathrm{r}}^2}{A}\, \frac{\langle |E_{\mathrm{v}}^{\mathrm{s}}|^2\rangle}{|E_{\mathrm{v}}^{\mathrm{i}}|^2} \\ &= \frac{4\pi R_{\mathrm{r}}^2}{A}\, \frac{\langle I_{\mathrm{v}}^{\mathrm{s}}\rangle\, d\Omega_{\mathrm{s}}}{I_{\mathrm{v}}^{\mathrm{i}}} \\ &= \frac{4\pi R_{\mathrm{r}}^2}{A}\, \frac{\langle I_{\mathrm{v}}^{\mathrm{s}}\rangle}{I_{\mathrm{v}}^{\mathrm{i}}} \cdot \frac{A\cos\theta_{\mathrm{s}}}{R_{\mathrm{r}}^2} = 4\pi\cos\theta_{\mathrm{s}}\, \frac{\langle I_{\mathrm{v}}^{\mathrm{s}}\rangle}{I_{\mathrm{v}}^{\mathrm{i}}}. \end{aligned} \tag{5.47}$$

The modified definition of $\mathbf{I}^{\mathrm{s}}$ does not change the relationship between σ_{vv}^0 and the scattering amplitude S_{vv}. From Eq. (5.12),

$$E_{\mathrm{v}}^{\mathrm{s}} = \frac{e^{-ikR_{\mathrm{r}}}}{R_{\mathrm{r}}}\, S_{\mathrm{vv}} E_{\mathrm{v}}^{\mathrm{i}}, \tag{5.48}$$

which leads to

$$\sigma_{\mathrm{vv}}^0 = \frac{4\pi R_{\mathrm{r}}^2}{A}\, \frac{\langle |E_{\mathrm{v}}^{\mathrm{s}}|^2\rangle}{|E_{\mathrm{v}}^{\mathrm{i}}|^2} = \frac{4\pi}{A}\, \langle |S_{\mathrm{vv}}|^2\rangle. \tag{5.49}$$

If the radar is polarimetric, i.e., capable of measuring the scattering matrix $\widetilde{\mathbf{S}}$, not only can it determine σ_{pq}^0 for hh, hv, vv, and vh polarizations, but through the application of polarization synthesis (Section 5-10), it is also possible to synthesize images corresponding to the transmission and reception of any combination of polarizations, including circular and elliptical.

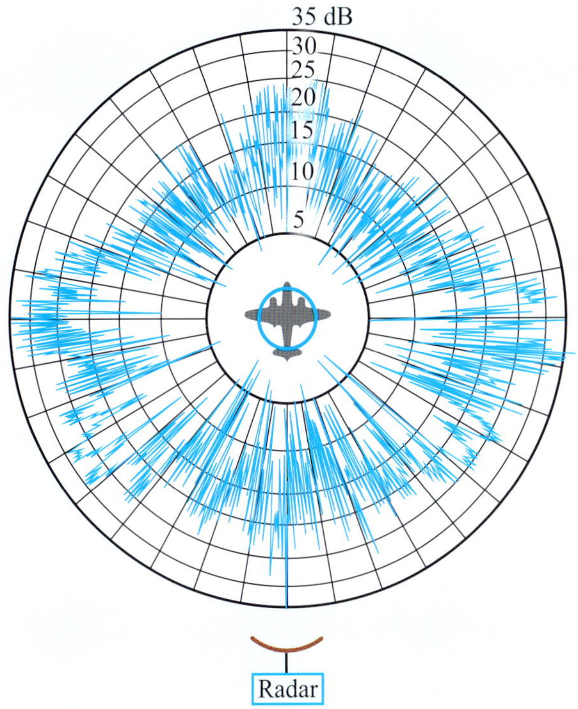

Figure 5-11: Measured azimuthal pattern of the radar backscatter from a B-26 airplane at 10 cm wavelength [from Ridenour, 1947, courtesy McGraw-Hill Book Company].

5-6 RCS Statistics

As noted earlier, the scattering amplitude S_{pq} of a target is dependent not only on the incident and scattered directions, but also on its shape and orientation. The question is: how sensitive is S_{pq}, or σ_{pq}, to the orientation of the target. Obviously, it depends on the shape of the target. For a sphere, its RCS is independent of its orientation, but that is not true for most targets. To illustrate with a real example, let us consider the pattern shown in Fig. 5-11. The pattern represents the backscatter cross section σ for a B-26 aircraft as a function of aspect angle. The data were obtained by mounting the aircraft on a turntable and then observing the backscattered signal measured by a 3 GHz radar as the turntable was rotated about its axis. The most striking feature of the scattering pattern is the strong sensitivity of σ to aspect angle; σ changes by an order of magnitude (10 dB) in response to a change of a fraction of a degree in aspect angle, and by two orders of magnitude (20 dB) in response to a change of a full degree. The full dynamic range exceeds 30 dB (1000 : 1)!

The aircraft consists of a number of scattering elements, each characterized by a scattering pattern that reflects its particular shape and orientation. The composite backscatter in a given direction is the phasor sum of the waves backscattered from the individual elements, which includes the propagation phase delay to and from each element. The peaks in the pattern shown in Fig. 5-11 are associated with constructive interference of the backscattered phasor and the nulls are associated with destructive interference.

Although an aircraft has a distinctly different geometry from that of a distributed terrain surface, the nature of the scattering process is fundamentally the same in both cases. The aircraft has a ***deterministic scattering pattern*** because it has a well-defined geometry, whereas a distributed target has a ***random scattering pattern*** (Fig. 5-12) because its scattering elements are randomly distributed. Both patterns exhibit large fluctuations. The pattern shown in Fig. 5-12(b) was generated from radar backscatter measurements made by a 35 GHz truck-mounted scatterometer as the truck was driven across an asphalt surface with the antenna beam pointing downward along the aft direction at an incidence angle of 40° relative to normal incidence [Fig. 5-12(a)]. The antennas were mounted atop a telescopic boom at a height of 10.3 m above the asphalt surface. The sampling rate was such that the footprints (on the asphalt surface) corresponding to adjacent samples were totally independent (no overlap). The vertical axis in Fig. 5-12(b) represents the ratio of the received power to the average value computed for all 1000 measurements, expressed in dB. The measured values extend over a range of 50.2 dB, or $10^5 : 1$.

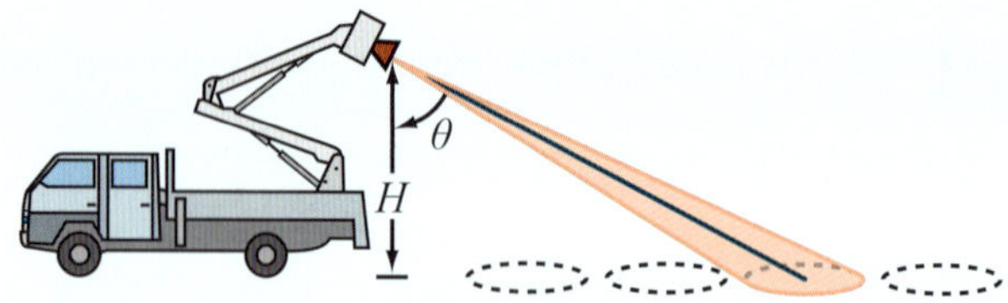

(a) Beam movement

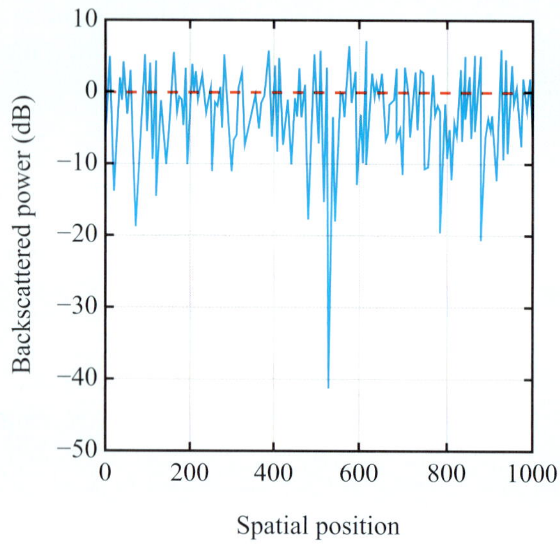

(b) Backscattering variation

Figure 5-12: The sketch in (a) shows how the measurements [shown in (b)] of the backscatter from an asphalt surface were acquired. The incidence angle was 40°, the frequency 35 GHz, the platform height 10.4 m, and the polarization vv [from Ulaby et al., 1988a].

▶ The implication of the results displayed in Fig. 5-12 is that the backscattered signal measured by a terrain-looking radar exhibits fluctuations from one spot to the next, "even when the terrain is the same." An image of a seemingly uniform bare soil field, for example, exhibits large pixel-to-pixel intensity variations, often referred to as ***image speckle*** (Fig. 5-13). To extract information from a radar image, it is necessary to understand the statistics of image speckle and how to reduce it. ◀

Figure 5-13: Image speckle refers to the pixel-to-pixel variation of image tone. This is a 1 m × 1 m resolution Ku-band SAR image of an agricultural area [Sandia National Lab].

5-7 Rayleigh Fading Model

The term ***radar clutter*** was introduced during World War II to denote unwanted radar echoes from extended targets such as rain, land, and water. When a downward-looking radar is used for the purpose of detecting or tracking a ***hard target***, such as a bridge or a vehicle, the terrain background represents a source of interference because the energy backscattered from the terrain *clutters* the fidelity of the desired signal. In the 1950s and 1960s, ***radar terrain clutter*** encompassed all aspects of scattering from terrain surfaces [Barton, 1975]. This included theoretical models for electromagnetic interaction with surfaces and volumes, experimental measurements of the backscattering cross section per unit area for various types of terrain, and the statistics associated with the scattering process. Research in this field was conducted primarily in support of military applications. The advent of high resolution imaging radar in the 1950s led to the development of ***radar remote sensing***, which evolved into a major scientific discipline with wide-ranging civilian, military, and environmental application. In many of these application, the terrain-scattered radar signal, which previously was regarded as a source of interference, is itself the information carrier because the information sought is derived from knowledge of how waves scatter from rough surfaces and inhomogeneous media. Because of this wider scope of the intended applications of radar scattering, in today's literature the generic term ***radar scattering from distributed targets*** has become the standard term used (instead of terrain clutter) to refer to the topics previously classified as radar clutter.

From the standpoint of detection and tracking applications, terrain scattering is viewed in statistical terms because the objective is to determine the degradation in the false alarm probability (associated

Random Variables

For any random variable x, defined over the range x_1 to x_2 and characterized by a probability density function (pdf) $p(x)$:

$$\overline{x} = \langle x \rangle = \int_{x_1}^{x_2} x\, p(x)\, dx = \text{mean value of } x$$

$$\overline{x^2} = \langle x^2 \rangle = \int_{x_1}^{x_2} x^2\, p(x)\, dx = \text{second moment of } x$$

$$s_x^2 = \langle x^2 \rangle - \langle x \rangle^2 = \text{variance of } x$$

$$s_x = \text{standard deviation of } x$$

$$\beta_x = \left(\frac{s_x}{\overline{x}}\right)^2 = \text{normalized variance of } x$$

$$P(x \leq x') = \int_{x_1}^{x'} p(x)\, dx$$

$$= \text{cumulative distribution of } x \leq x'$$

with detecting a target) caused by interference of the radar return from the target with that contributed by the terrain background. Consequently, the term radar clutter has evolved into the more specific term ***radar clutter statistics*** and is used in today's literature to denote the ***probability density function*** (*pdf*) associated with a statistical scattering process whose mean value is σ^0, the backscattering cross section per unit area. As we shall see later, σ^0 of terrain varies as a function of two sets of parameters: (1) the sensor parameters, namely the wavelength (or frequency), the incidence angle (defined here with respect to normal incidence), and the polarization configurations of the transmitting and receiving antennas, and (2) the terrain parameters, which include its dielectric properties and geometrical characteristics. Hence, in general the pdf is both sensor- and terrain type-specific. Actually, the pdf is also terrain condition-specific because a forest, for example, undergoes changes in both its dielectric properties and geometry as a function of time.

A distributed target usually consists of a large number of randomly distributed scatterers. When a distributed target is illuminated by a coherent electromagnetic wave, the magnitude of the backscattered signal is equal to the phasor sum of the returns from all of the scatterers illuminated by the incident beam. The backscattered signal is a random variable because the dielectric and geometrical properties of a terrain surface are random variables.

▶ Two ground patches with the same statistical properties may produce backscattered signals with different magnitudes because the individual scatterers in the two patches have different locational arrangements. This variability in the magnitude of the backscattered signal is referred to as ***signal fading*** or ***signal scintillation***. ◀

To characterize the fluctuation statistics associated with a terrain surface of uniform electromagnetic properties, the usual approach is to model the surface as an ensemble of independent, randomly located scatterers, all of comparable scattering strengths. Such a model leads to the result that the amplitude of the backscattered signal is ***Rayleigh-distributed***. If the return is dominated by backscatter from one or a few strong scatterers, the fading process is characterized by the ***Rice distribution*** (Raemer, 1997). Some experimental observations support the Rayleigh behavior (Bush and Ulaby, 1975; Ulaby et al., 1986b; Weinstock, 1965) whereas others, particularly those measured for complex terrain categories, are in closer agreement with the ***log-normal distribution***, the ***Weibull distribution*** (Kashihara et al., 1984; Schleher, 1976; Valenzuela and Laing, 1972), or other more complicated distributions (Jao, 1984). In this chapter we limit our treatment to the Rayleigh clutter model.

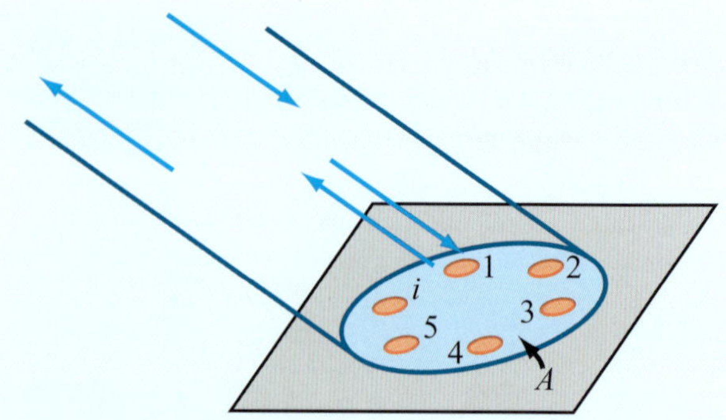

Figure 5-14: The illuminated area A contains N_s randomly distributed scatterers.

5-7.1 Underlying Assumptions

The Rayleigh clutter model used for radar scattering from an area-extended (distributed) target is essentially the same as the model used for random noise and is based on the same mathematical assumptions. A review of these assumptions will prove useful in later sections.

The sketch shown in Fig. 5-14 depicts a radar beam illuminating an area A of an area-extended target. The illuminated area contains N_s point scatterers, designated by the index $i = 1, 2, \ldots, N_s$. For simplicity, we shall confine our present discussion to the backscatter case. The electric field at the input of the receiving antenna due to backscatter by the ith scatterer may be expressed as

$$E_i(t) = K_i E_{i0} \cos(\omega t - 2kR_i + \theta_i), \tag{5.50}$$

where E_{i0} is the scattering magnitude and θ_i is the scattering phase of the ith scatterer, R_i is the range from the antenna to the scatterer, $k = 2\pi/\lambda$ is the wavenumber, and K_i is a system constant that accounts for several radar system factors including propagation losses to and from the scatterer and antenna gain in the direction of the scatterer. The phasor-domain counterpart of Eq. (5.50) is

$$\mathbf{E}_i = K_i E_{i0} e^{j\phi_i}, \tag{5.51}$$

where ϕ_i, the phase angle of $\mathbf{E}_i$, is

$$\phi_i = \theta_i - 2kR_i. \tag{5.52}$$

Assumption 1: The scatterers are statistically independent. This assumption allows us to express the total instantaneous field due to the N_s scatterers contained in the area A as a simple sum,

$$\mathbf{E} = \sum_{i=1}^{N_s} K_i E_{i0} e^{j\phi_i}, \tag{5.53}$$

and it implies that interaction effects between adjacent scatterers may be ignored.

Assumption 2: The maximum range extent of the target $\Delta R = |R_i - R_j|_{\max}$ is much smaller than the mean range to the target area A, and the antenna gain is uniform across A. This allows us to set $K_i = K$ for all i. For convenience, we shall set $K = 1$. Hence,

$$\mathbf{E} = \sum_{i=1}^{N_s} E_{i0} e^{j\phi_i}. \tag{5.54}$$

The total field $\mathbf{E}$ is a vector sum of N_s phasors. If we express these phasors graphically (Fig. 5-15), with the first one starting at the origin and the successive ones each starting at the tip of the preceding one, the result is a vector from the origin to the tip of the last phasor. The length of this vector (which is called the ***voltage envelope***) and its phase angle are denoted E_e and ϕ, respectively. Thus,

$$\mathbf{E} = E_e e^{j\phi} = \sum_{i=1}^{N_s} E_{i0} e^{j\phi_i}. \tag{5.55}$$

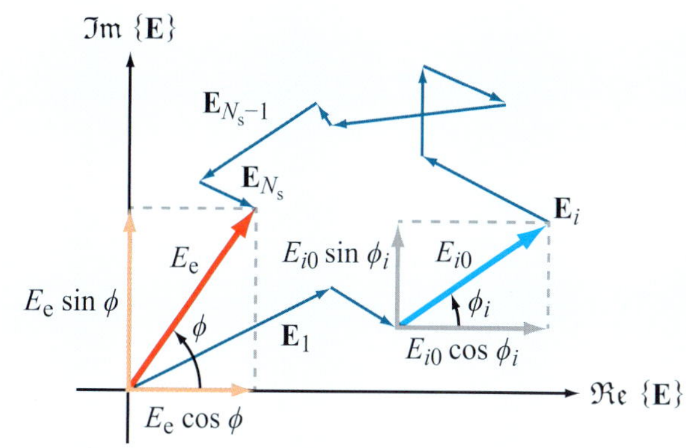

Figure 5-15: The vector $\mathbf{E} = E_e e^{j\phi}$ is the phasor sum of N_s fields.

Phasor $E_e e^{j\phi}$ has Cartesian components E_x and E_y (Fig. 5-15) given by

$$E_x = E_e \cos\phi = \sum_{i=1}^{N_s} E_{i0}\cos\phi_i, \tag{5.56a}$$

$$E_y = E_e \sin\phi = \sum_{i=1}^{N_s} E_{i0}\sin\phi_i. \tag{5.56b}$$

Assumption 3: N_s is a large number, which allows us to use the central-limit theorem, thereby treating E_x and E_y as normally distributed random variables with means

$$\overline{E}_x = \langle E_x \rangle = \sum_{i=1}^{N_s} \langle E_{i0}\cos\phi_i \rangle, \tag{5.57a}$$

$$\overline{E}_y = \langle E_y \rangle = \sum_{i=1}^{N_s} \langle E_{i0}\sin\phi_i \rangle. \tag{5.57b}$$

Through computer simulation, it can be demonstrated that N_s need not be exceedingly large to satisfy the central-limit theorem; in practice, N_s as small as 10 is sufficient (approximately).

Assumption 4: The scattering amplitude E_{i0} and phase ϕ_i are independent random variables. This condition is easily satisfied if E_{i0} is independent of the range R_i (and hence of the phase ϕ_i given by Eq. (5.52)), which would be the case if the scatterers are randomly distributed in range because ϕ_i is dependent on the range R_i (through Eq. (5.52)), whereas E_{i0} is not.

Assumption 5: The phase ϕ_i is uniformly distributed over the range $[0, 2\pi]$. To satisfy this condition, it is not only necessary that the scatterers be randomly distributed in range, but the maximum range extent of the target ΔR must also be several wavelengths across.

Assumption 6: No one individual scatterer produces an electric field of magnitude commensurate with the resultant field from all other scatterers. In other words, the field **E** is not dominated by one (or a few) very strong scatterer(s). If this condition is not satisfied, the Rayleigh noise-like statistics do not apply and the statistics developed by Rice (Raemer, 1997) for one or more large signals contained in a background of noise should be used instead.

Based on Assumptions 3–6, Eq. (5.57a) becomes

$$\overline{E}_x = \sum_{i=1}^{N_s} \langle E_{i0} \rangle \langle \cos \phi_i \rangle = \sum_{i=1}^{N_s} \langle E_{i0} \rangle \left[\frac{1}{2\pi} \int_0^{2\pi} \cos \phi_i \, d\phi_i \right] = 0, \qquad (5.58)$$

because the average value of $\cos \phi_i$ is zero. Similarly, $\overline{E}_y = 0$. Additionally, the mean value of the product $E_x E_y$ is

$$\langle E_x E_y \rangle = \sum_{i=1}^{N_s} \sum_{i=1}^{N_s} \langle E_{i0}^2 \rangle \left[\frac{1}{2\pi} \int_0^{2\pi} \cos \phi_i \sin \phi_i \, d\phi_i \right] = 0, \qquad (5.59)$$

which implies that E_x and E_y are uncorrelated. For normally distributed random variables, this also implies statistical independence. Hence, their joint ***probability density function*** (pdf) is equal to the product of their individual pdfs, and also recalling that E_x and E_y have zero means, the joint pdf becomes

$$p(E_x, E_y) = \frac{1}{\sqrt{2\pi}\, s} \, e^{-E_x^2/2s^2} \cdot \frac{1}{\sqrt{2\pi}\, s} \, e^{-E_y^2/2s^2} = \frac{1}{2\pi s^2} \, e^{-(E_x^2 + E_y^2)/2s^2}, \qquad (5.60)$$

(Gaussian pdf)

where s is the ***standard deviation*** of E_x and E_y. Our true interest is in E_e, the envelope of the total electric field, so we convert the pdf given by Eq. (5.60) into a pdf in the polar variables E_e and ϕ through the transformation

$$p(E_x, E_y) \, dE_x \, dE_y = p(E_e, \phi) \, dE_e \, d\phi. \qquad (5.61)$$

Noting that

$$E_e^2 = E_x^2 + E_y^2 \qquad (5.62)$$

and equating a differential area in (E_e, ϕ) space with its equivalent in (E_x, E_y) space, namely

$$E_e \, dE_e \, d\phi = dE_x \, dE_y, \qquad (5.63)$$

we obtain the pdf

$$p(E_e, \phi) = \frac{E_e}{2\pi s^2} \, e^{-E_e^2/2s^2}, \qquad 0 \leq E_e \leq \infty \quad \text{and} \quad 0 \leq \phi \leq 2\pi. \qquad (5.64)$$

The phase angle ϕ is uniformly distributed over $[0, 2\pi]$. Hence, the pdf for E_e alone is

$$p(E_e) = \int_0^{2\pi} p(E_e, \phi) \, d\phi = \frac{E_e}{s^2} \, e^{-E_e^2/2s^2}, \qquad 0 \leq E_e \leq \infty. \qquad (5.65)$$

(Rayleigh pdf)

The pdf given by Eq. (5.65) is known as the ***Rayleigh distribution***.

The mean value of E_e is

$$\overline{E}_e = \int_0^{\infty} E_e \, p(E_e) \, dE_e = \int_0^{\infty} \frac{E_e^2}{s^2} \, e^{-E_e^2/2s^2} \, dE_e = \sqrt{\frac{\pi}{2}} \, s. \qquad (5.66)$$

The variable E_e represents the magnitude of the cumulative electric field at the receiving antenna prior to detection by the receiver. Most receivers use either linear detection or square-law detection to convert the input signal to an output voltage. As we see in the following subsections, the output voltage of the linear receiver also is characterized by a Rayleigh pdf, but that of the square-law receiver is characterized by an exponential pdf. Both cases, however, belong to the Rayleigh clutter model because the underlying pdf, $p(E_e)$, is Rayleigh.

▶ The terms ***Rayleigh fading*** and ***Rayleigh clutter*** refer to the model and the assumptions leading to Eq. (5.65) and should not be confused with the form of the pdf of the detected voltage. ◀

Between the input to the radar receiver, represented by the signal E_e (due to backscatter from a given pixel on the ground), and the final product in the form of a radar image, the signal undergoes two basic transformations, namely detection and image formation. Most radar receivers use either:

(a) Linear detection: the tone (digital value) of the processed image is directly proportional to E_e, in which case the image is called a voltage or ***amplitude image***. In the present treatment the image tone is represented by the symbol V.

(b) Square-law detection: the image tone is proportional to $|E_e|^2$, which is equivalent to power. The image is called an ***intensity image*** and its tone is represented by the symbol I.

The two detection options are discussed in the next two subsections.

5-7.2 Linear Detection

If the radar receiver uses a linear detector, its output voltage V is directly proportional to E_e; that is,

$$V = K_1 E_e = K_1 \overline{E}_e \frac{E_e}{\overline{E}_e} = K_1 \overline{E}_e f, \qquad (5.67)$$

where K_1 is a system constant and f is a ***normalized fading random variable for voltage*** given by

$$f = \frac{E_e}{\overline{E}_e}\,. \qquad (5.68)$$

The mean value of f is

$$\overline{f} = \frac{\overline{E}_e}{\overline{E}_e} = 1. \qquad (5.69)$$

The mean field $\overline{E}_e$ is related to σ^0, the mean value of the backscattering cross section per unit area of the distributed target, by

$$\overline{E}_e = K_2(\sigma^0)^{1/2}, \qquad (5.70)$$

where K_2 is another system constant. Use of Eq. (5.70) in Eq. (5.67) gives

$$V = K_1\, K_2(\sigma^0)^{1/2}\, f. \qquad (5.71)$$

In view of Eq. (5.69), the mean value of the linearly detected voltage measured by the radar is

$$\overline{V} = K_1\, K_2(\sigma^0)^{1/2}\, \overline{f} = K_1\, K_2(\sigma^0)^{1/2}, \qquad (5.72)$$

where the averaging process is performed by adding many independent observations of the distributed target. For an imaging radar, this can be realized by adding together the voltages corresponding to many pixels. The value of constant K_1K_2 can be established through calibration, thereby providing a direct mechanism for relating $\overline{V}$ to σ^0.

▶ Because the output voltage V is a product of its mean value $\overline{V} = K_1\, K_2(\sigma^0)^{1/2}$ and the random variable f, the process is sometimes referred to as a ***multiplicative noise model***. ◀

From Eqs. (5.71) and (5.72), it follows that

$$f = \frac{V}{\overline{V}}\,. \qquad (5.73)$$

The two relevant quantities are σ^0 and f; σ^0 represents the "strength" of the average backscatter from the distributed target and f accounts for variations among different observations (or image pixels).

The fading random variable f is linearly related to E_e through Eq. (5.68). Using the expression for $p(E_e)$ given by Eq. (5.65), in combination with the area relation

$$p(f)\, df = p(E_e)\, dE_e, \qquad (5.74)$$

leads to the pdf

$$p(f) = \frac{\pi f}{2} \exp\left(-\frac{\pi}{4} f^2\right), \qquad f \geq 0. \qquad (5.75)$$

(Rayleigh pdf)

The mean values of f and f^2 are

$$\overline{f} = 1, \quad (5.76a)$$

$$\overline{f^2} = \frac{4}{\pi}, \quad (5.76b)$$

and the corresponding standard deviation s_f is

$$s_f = [\overline{f^2} - \overline{f}^2]^{1/2} = \left(\frac{4}{\pi} - 1\right)^{1/2} = 0.523. \quad (5.76c)$$

The associated ***cumulative distribution*** $P(f \leq f')$, where f' is some threshold level of interest, is

$$P(f \leq f') = \int_0^{f'} p(f)\, df$$
$$= 1 - \exp\left(-\frac{\pi}{4} f'^2\right), \qquad f' \geq 0. \quad (5.76d)$$

From Eq. (5.73) and the area relation $p(V)\, dV = p(f)\, df$, it is easy to show that the pdf $p(V)$ of the linearly detected voltage is

$$p(V) = \frac{\pi V}{2\overline{V}^2} \exp\left[-\frac{\pi}{4\overline{V}^2} V^2\right], \qquad V \geq 0. \quad (5.77a)$$

The mean value $\overline{V}$ is related to the reflectivity σ^0 by Eq. (5.72), and the ***standard-deviation-to-mean ratio*** is

$$\frac{s_V}{\overline{V}} = 0.523. \quad (5.77b)$$

5-7.3 Square-Law Detection

The voltage output of a square-law detector is directly proportional to the power of the input signal, rather than to its electric intensity E_e. Hence, within a proportionality constant, the output, represented by the symbol I (for intensity), is

$$I = K_3 E_e^2 = K_3 \overline{E_e^2}\, \frac{E_e^2}{\overline{E_e^2}} = K_3 K_4 \sigma^0 F, \quad (5.78)$$

where

$$F = \frac{E_e^2}{\overline{E_e^2}} = \frac{I}{\overline{I}} \quad (5.79)$$

is the ***normalized fading random variable for power***. Ignoring the constant $K_3 K_4$ and using the area relation $p(E_e)\, dE_e = p(I)\, dI = p(F)\, dF$ leads to the exponential pdf:

$$p(I) = \frac{1}{\overline{I}} e^{-I/\overline{I}}, \qquad p(F) = e^{-F}, \quad (5.80a)$$

(exponential pdf)

with

$$\overline{I} = \sigma^0, \qquad \overline{F} = 1, \quad (5.80b)$$

$$\frac{s_I}{\overline{I}} = 1, \qquad s_F = 1, \quad (5.80c)$$

and

$$P(I \leq I') = 1 - e^{-I'/\overline{I}}, \qquad P(F \leq F') = 1 - e^{-F'}. \quad (5.80d)$$

Note that for the exponential distribution, the ***standard-deviation-to-mean ratio*** is $s_I/\overline{I} = s_F/\overline{F} = 1$.

5-7.4 Interpretation

What do these statistics tell us? To answer this question, we start by examining Fig. 5-16(a), which shows plots of $p(f)$ and $p(F)$ for the Rayleigh and exponential distributions, respectively, and Fig. 5-16(b), which shows the corresponding cumulative distributions. We observe that the range of fading associated with the distributions is very large. That is, if we take a single sample of the signal from a Rayleigh-distributed or exponentially distributed ensemble, we have very little chance of selecting a value close to the mean. Let us illustrate this with a specific example. According to the Rayleigh distribution in Fig. 5-16(b), the value of f exceeded 5 percent of the time is 1.95 (relative to the mean) and that exceeded 95 percent of the time is 0.25. In decibels, these levels correspond to +5.8 dB and −11.9 dB, respectively. If we select a sample at random, the probability is 90 percent (95 to 5 range) that its value will be within the range extending from 11.9 dB below the true mean to 5.8 dB above the true mean. We may think of this as the ***90 percent confidence interval***

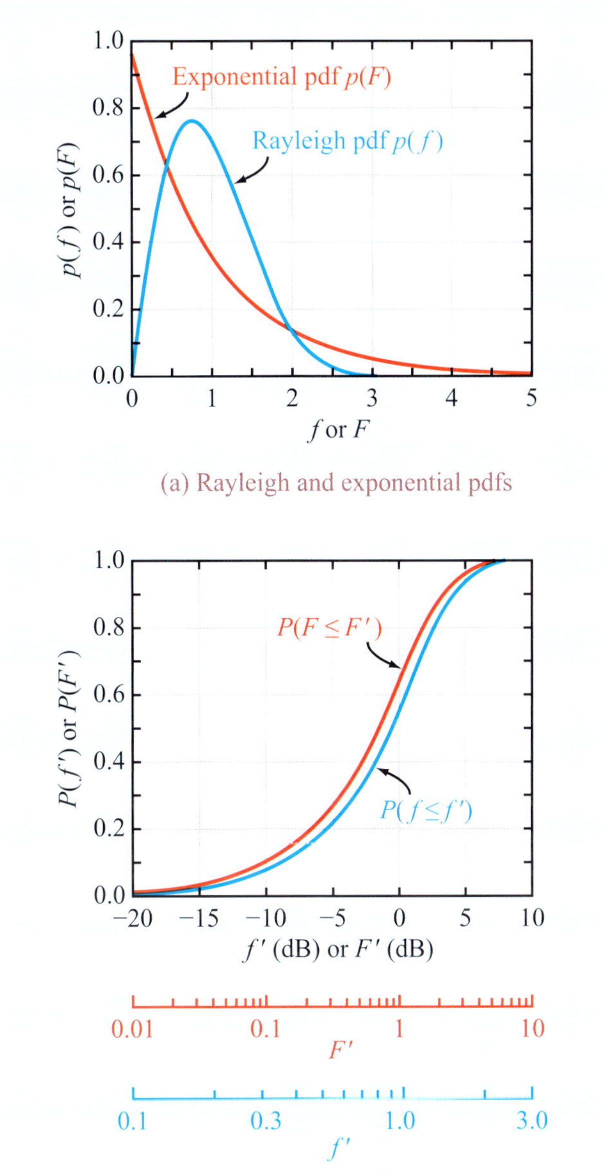

Figure 5-16: Plot of (a) probability density functions and (b) cumulative distributions for f and F.

associated with our measurement. The important point to note here is the fact that this interval (17.7 dB) is very large!

The situation is not much different when square-law detection is used: the 5 percent and 95 percent levels of the cumulative distribution for the exponential pdf are +4.8 and −12.9 dB, also totaling 17.7 dB. For the data presented earlier in Fig. 5-12, the data points that fall within the −12.9 dB to +4.8 dB range constitute 90.8 percent of the total, which is in close agreement with the 90 percent figure predicted by the exponential distribution. Furthermore, the measured pdf of F, shown in Fig. 5-17, exhibits a good fit to the exponential distribution.

5-8 Multiple Independent Samples

To reduce the uncertainty of a radar measurement of the backscatter from a terrain surface, it is necessary to average many independent samples together. Independence requires that the distribution of scatterers be statistically different between one observation and another. If a radar transmits multiple pulses, and then records the returns from the same target with identical scatterers at identical locations, the returns will not be independent. In fact, the backscattered signal is the same for all pulses. The noise generated by the receiver is likely to be random in time, so averaging multiple pulse returns improves the signal-to-noise ratio, but the returns are not statistically independent.

An easy way to increase the number of independent samples N contained in an estimate of the radar backscatter is through spatial averaging, which amounts to *trading spatial resolution for improved radiometric resolution*. The image generated by replacing N pixels with a new pixel bearing their average value is called an ***N*-look image**.

5-8.1 *N*-Look Amplitude Image

If N randomly selected samples of a Rayleigh-distributed voltage V are averaged together, the average value V_N has the following properties:

$$V_N = \frac{1}{N}\sum_{i=1}^{N} V_i = K_1\, K_2(\sigma^0)^{1/2} \left[\frac{1}{N}\sum_{i=1}^{N} f_i\right] = K_1\, K_2(\sigma^0)^{1/2}\, f_N, \qquad (5.81)$$

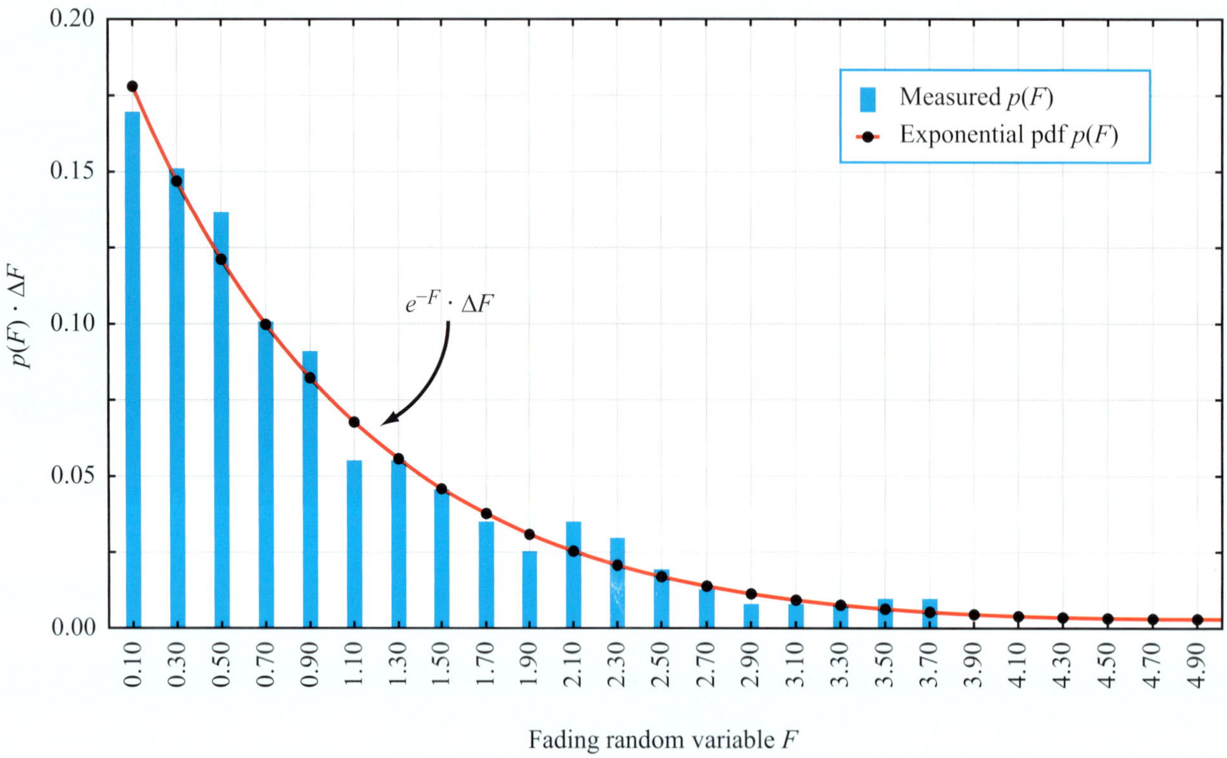

Figure 5-17: The measured pdf of the backscatter from asphalt [corresponding to the data in Fig. 5-12(b)] is found to be in good agreement with the exponential pdf based on the Rayleigh fading model.

where we define

$$f_N = \frac{1}{N} \sum_{i=1}^{N} f_i \tag{5.82}$$

as the fading random variable corresponding to the average of N independent samples. Its properties are

$$\overline{f}_N = 1, \tag{5.83a}$$

$$s_{f_N} = \frac{0.523}{\sqrt{N}}, \tag{5.83b}$$

and its pdf may be obtained by N-successive convolutions of the Rayleigh pdf given by Eq. (5.75). Plots of $p(f_N)$ are shown in Fig. 5-18(a) for several values of N. As expected, as N increases, the distribution becomes more peaked and narrow (the standard deviation decreases as $N^{-1/2}$) and eventually approaches the Gaussian pdf.

5-8.2 *N*-Look Intensity Image

If the receiver uses square-law detection, the average received power due to N independent samples is

$$I_N = K_3 K_4 \sigma^0 F_N \tag{5.84}$$

with

$$F_N = \frac{I_N}{\overline{I}_N} = \frac{1}{N} \sum_{i=1}^{N} F_i.$$

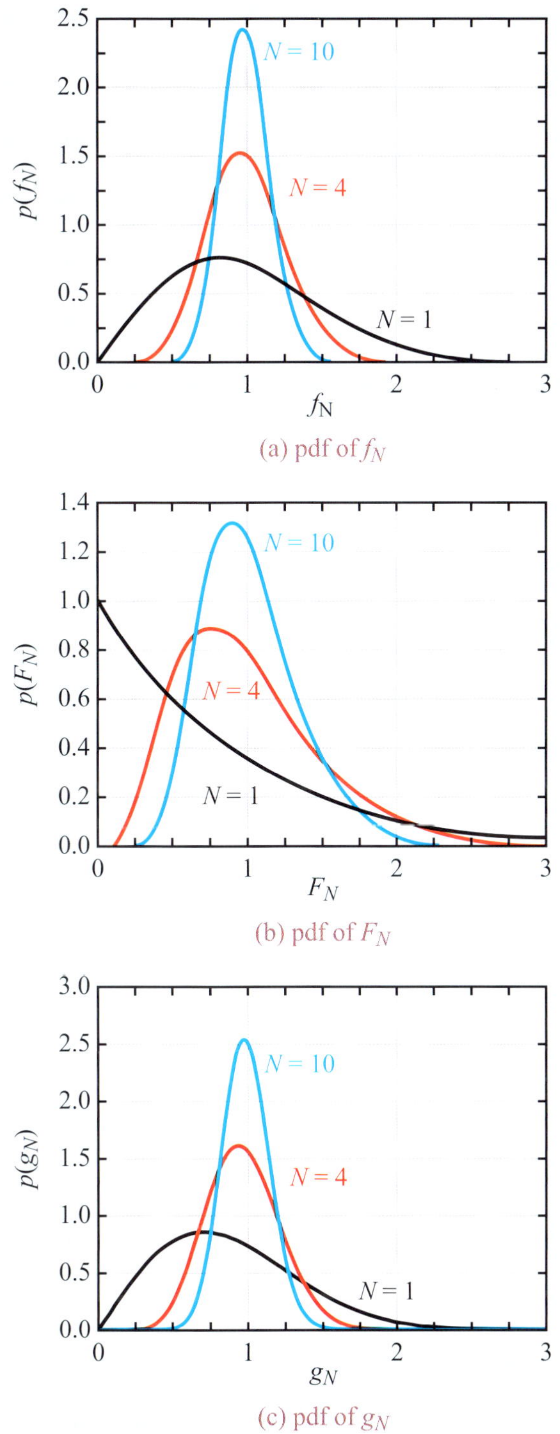

Figure 5-18: Probability density function for $N = 1, 4,$ and 10 for (a) f_N (linear deduction), (b) F_N (square-law detection), and (c) g_N (square-root intensity).

Because the averaging process is linear, $\bar{I} = \bar{I}_N$. The statistical attributes of F_N are:

$$\overline{F}_N = 1, \qquad \overline{F_N^2} = 1 + \frac{1}{N}, \tag{5.85a}$$

$$s_{F_N} = \left[\overline{F_N^2} - \overline{F}_N^2\right]^{1/2} = \frac{1}{\sqrt{N}}, \tag{5.85b}$$

and its pdf is a ***χ^2 distribution with $2N$ degrees of freedom***:

$$p(F_N) = \frac{F_N^{N-1} N^N e^{-NF_N}}{(N-1)!}, \qquad F_N \geq 0. \tag{5.86}$$

The corresponding pdf for I_N is

$$p(I_N) = \frac{p(F_N)}{\bar{I}_N}. \tag{5.87}$$

Plots of $p(F_N)$ are shown in Fig. 5-18(b) for various values of N, and Fig. 5-19 shows plots of the 5% and 95% levels versus N. The 5% level is a plot of F'_N and represents the value of F'_N exceeded only 5% of the time; i.e., $P(F_N \geq F'_N) = 0.05$. A similar definition applies to the 95% level.

5-8.3 N-Look Square-Root Intensity Image

In some radar systems, square-law detection is followed by (a) averaging N image pixels together and then (b) performing a square-root operation to compress the dynamic range. The tone (digital value) T of the resultant image is related to I by

$$T_N = \left[\frac{1}{N}\sum_{i=1}^{N} I_i\right]^{1/2}$$

$$= [I_N]^{1/2} = \left[\bar{I}_N \frac{I_N}{\bar{I}_N}\right]^{1/2} = \bar{I}_N^{1/2} F_N^{1/2} = \bar{T}_N\, g_N, \tag{5.88}$$

where g_N is the fading random variable for the N-look square-root intensity image,

$$g_N = F_N^{1/2}. \tag{5.89}$$

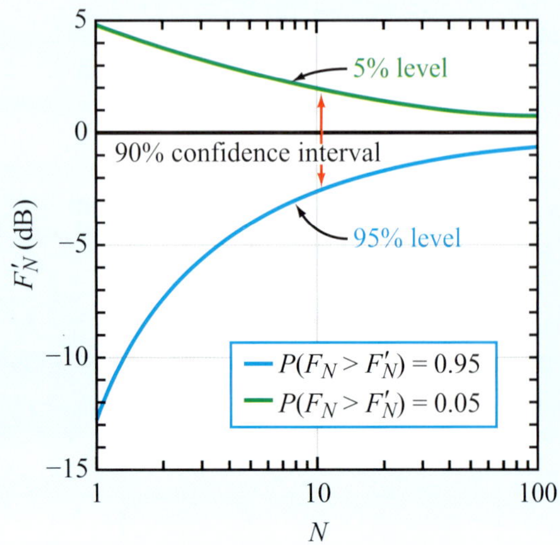

Figure 5-19: Plots of the 5% and 95% cumulative distribution levels versus N. The vertical spacing between these two curves is a measure of the "confidence interval" associated with a measurement of the radar backscatter.

The pdf of F_N is given by Eq. (5.86), so to obtain the pdf of g_N we use the density relation

$$p(g_N)\, dg_N = p(F_N)\, dF_N,$$

which leads to

$$p(g_N) = \frac{2g_N^{2N-1} N^N e^{-Ng_N^2}}{(N-1)!}, \qquad g_N \geq 0. \tag{5.90}$$

The mean values of g_N and its variance are given by

$$\overline{g}_N = \frac{\Gamma(N+1/2)}{N^{1/2}\,\Gamma(N)}, \tag{5.91a}$$

$$s_{g_N}^2 = 1 - \overline{g}_N^2, \tag{5.91b}$$

where $\Gamma(\;)$ is the Gamma function:

$$\Gamma(z) = \int_0^\infty t^{z-1} e^{-t}\, dt. \tag{5.92}$$

As $N \to \infty$, $\overline{g}_N \to 1$.

If a random variable x has a mean value $\overline{x}$ and a standard deviation s_x, and if a second random variable y is related to x by a smooth function $y = h(x)$, then the mean $\overline{y}$ and the standard deviation s_y of y may be related to $\overline{x}$ and s_x by the following approximate relations (Papoulis, 1965; pp. 151–152):

$$\overline{y} \approx y(x) + y''(x)\,\frac{s_x^2}{2}, \tag{5.93a}$$

$$s_y^2 \approx [y'(x)]^2 s_x^2, \tag{5.93b}$$

where $y' = dy/dx$ and all quantities are evaluated at $x = \overline{x}$. The application of these relations to $g_N = F_N^{1/2}$, evaluated at $F_N = \overline{F}_N = 1$, leads to

$$\overline{g}_N \approx 1 - \frac{1}{8N}, \tag{5.94a}$$

$$s_{g_N} \approx \frac{1}{2\sqrt{N}}. \tag{5.94b}$$

The relative error between the exact and approximate values of $\overline{g}_N$ and s_{g_N} is less than 1% if $N \geq 4$. Figure 5-18(c) displays plots of $p(g_N)$ as a function of g_N for $N = 1$, 4, and 10. The plots for the square-root intensity image are very similar to those for the amplitude image [Fig. 5-18(a)]. The similarity is also evident in the plots shown in Fig. 5-20, which depict the variation of the standard-deviation-to-mean ratio as a function of N for the three types of images. In all cases, significant improvement is realized by increasing N from 1 to 4 looks, followed by a more moderate rate of improvement as N is increased further.

5-8.4 Spatial Resolution vs. Radiometric Resolution

Part (a) of Fig. 5-21 depicts a synthesized single-look ($N = 1$) amplitude radar image with each pixel corresponding to a ground resolution of 1 m × 1 m. The image terrain consists of 3 square parcels of land, one covered with trees, another with grass, and the third with bare soil. Each parcel is imaged into $16 \times 16 = 256$ pixels.

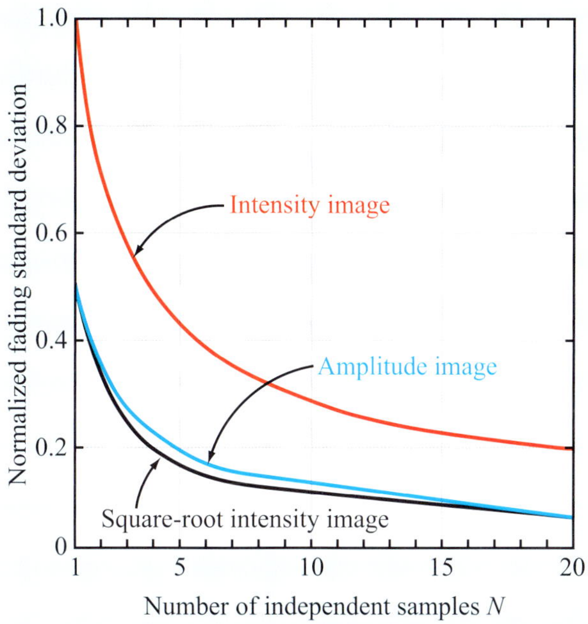

Figure 5-20: Variation of $s_{f_N}/\overline{f}_N$ (amplitude image), $s_{F_N}/\overline{F}_N$ (intensity image), and $s_{g_N}/\overline{g}_N$ (square-root intensity image) as a function of the number of looks (independent samples) N.

By averaging neighboring pixels together in groups of 4, we obtain the image in Fig. 5-21(b), which contains only $8 \times 8 = 64$ pixels with each pixel representing the backscatter from a 2 m $\times$ 2 m cell. The associated number of looks is $N = 4$. Upon repeating the process once again, we obtain the 16-look image in Fig. 5-21(c).

▶ Fundamentally, increasing N is equivalent to trading off spatial resolution for improved radiometric resolution. ◀

This statement is true when discrete measurements (corresponding to discrete resolution cells) are averaged together after detection, as well as when the averaging process is an integral part of the detection process. When generating a SAR image, steps may be taken during the image formation process to reduce image speckle. The fading or speckle pattern of the resultant image can be reduced by processing N separate segments of the total SAR bandwidth and then averaging these separate ***looks*** together. The process, which is referred to as ***mixed-integration*** or ***multiple-looking***, produces images with $N \geq 1$. This gain in the magnitude of N, however, is achieved at the expense of spatial resolution because the range resolution of the SAR image is inversely proportional to the bandwidth.

For illustration purposes, let us assume that the hypothetical image represented by Fig. 5-21 is an amplitude image with $K_1 K_2 = 1$ [in Eq. (5.72)] and that the average backscatter values for the trees, grass, and soil parcels are 36, 16, and 4 m^2/m^2, respectively. For an amplitude image, the standard-deviation-to-mean ratio is given by

$$\frac{s}{\sqrt{\sigma^0}} = \frac{0.523}{\sqrt{N}} ,$$

where for an amplitude image, $\overline{V} = \sqrt{\sigma^0}$. Hence, for $N = 1$, the range for trees corresponding to ± 1 standard deviation (relative to the mean) extends from

$$\overline{V}_\text{t} - s_\text{t} = (1 - 0.523)\sqrt{\sigma_\text{t}^0} = 0.477 \times \sqrt{36} \approx 2.9 \text{ m}^2/\text{m}^2$$

to

$$\overline{V}_\text{t} + s_\text{t} = (1 + 0.523)\sqrt{\sigma_\text{t}^0} = 1.523 \times \sqrt{36} \approx 9.1 \text{ m}^2/\text{m}^2.$$

This range, along with corresponding ranges for grass and soil, is shown in Fig. 5-22(a). Because of the large overlap among the three ranges, if we do not know the identities of the pixels and we were to pick one at random and try to determine its identity (trees, grass, or soil) on the basis of its amplitude, the probability that we would classify it correctly is very low.

The potential for correct classification improves as we consider Fig. 5-22(b), representing an image with $N = 4$, because less overlap exists among the three vertical ranges. And for the 16-look image, there is no overlap between the three ranges, so classification accuracy should be much higher. We should note, however, that even with $N = 16$ looks, some of the pixels may be misclassified because the criterion we used is based on the ± 1 standard deviation range, which captures a significant part of the pdf, but not all of it.

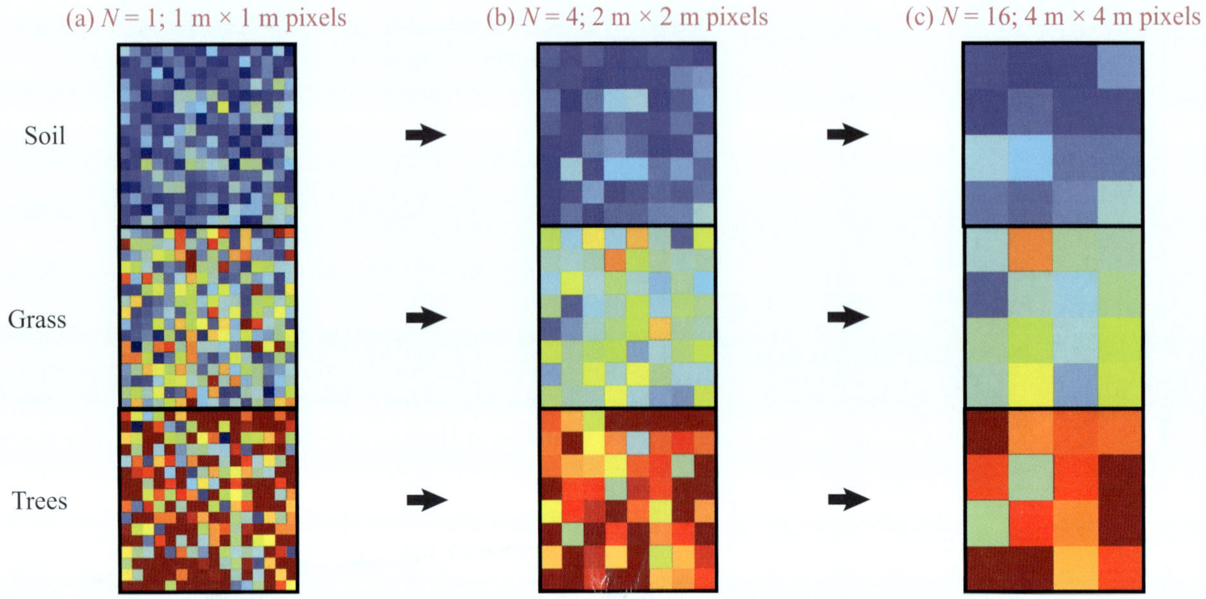

Figure 5-21: Simulated amplitude images of three distributed targets: trees with $\sigma_t^0 = 36\ \mathrm{m}^2/\mathrm{m}^2$, grass with $\sigma_g^0 = 16\ \mathrm{m}^2/\mathrm{m}^2$, and soil with $\sigma_s^0 = 4\ \mathrm{m}^2/\mathrm{m}^2$. Averaging multiple pixels trades off spatial resolution for improved radiometric resolution (less pixel-to-pixel variation).

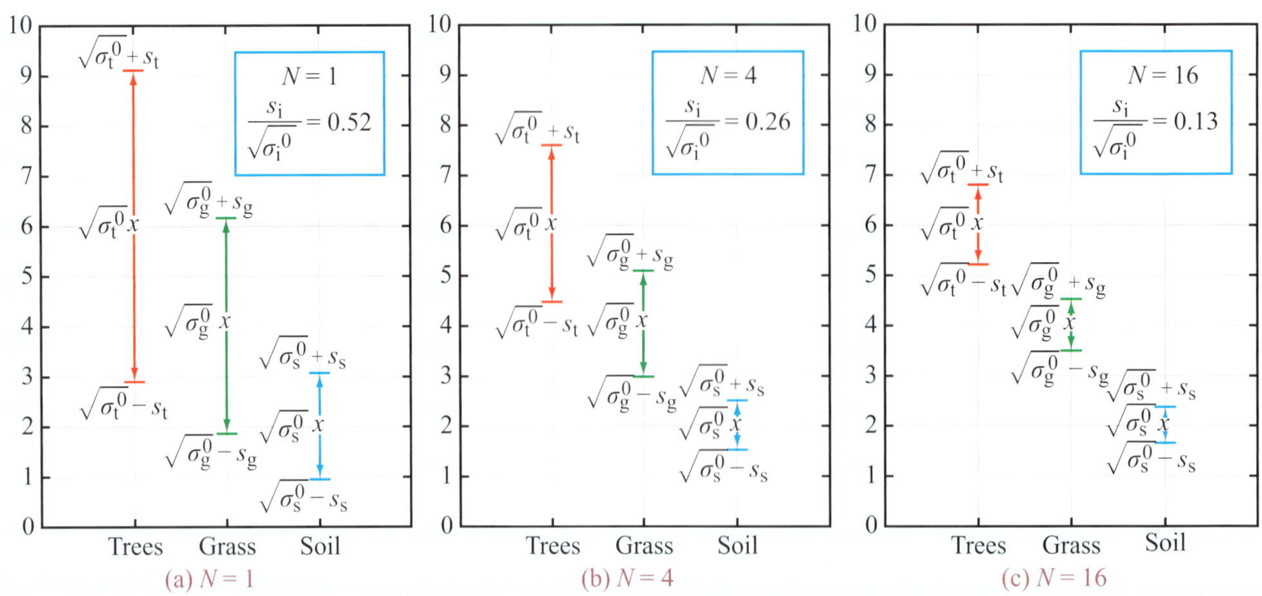

Figure 5-22: As the number of independent samples N is increased from 1 to 16, the confidence intervals around the means decrease by $1/\sqrt{N}$.

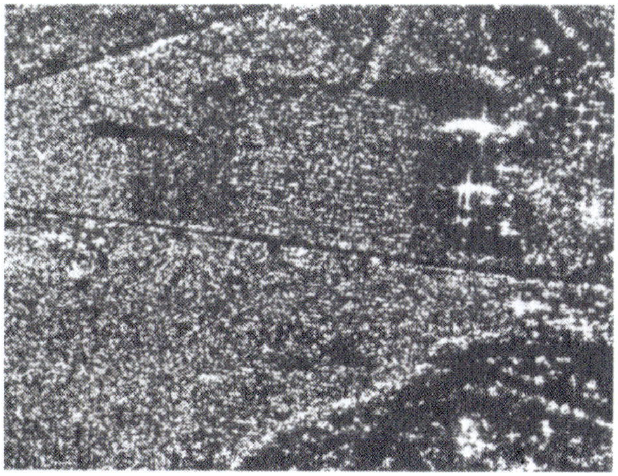

(a) 6 m × 6 m, $N = 1$

(b) 6 m × 6 m, $N = 4$

(c) 6 m × 6 m, $N = 12$

(d) 1.5 m × 2.1 m, $N = 1$

Figure 5-23: The four-image set provides a visual illustration of the relationships between image speckle, number of looks N, spatial resolution, and interpretability.

The image shown in Fig. 5-23(d), which is a fully focused SAR image with $N = 1$, has an image resolution capability corresponding to a ground resolution area approximately 1.5 m×2.1 m. From the raw data of this image, three other images were generated, all having the same ground resolution of 6 m×6 m. The image in Fig. 5-23(a) was generated by subsampling the raw data in both the azimuth and range dimensions to degrade the resolution to 6 m×6 m. Hence, no subsequent averaging was employed and N remained equal to 1. Partial and maximum possible averaging (or multilooking) was used in the generation of the images in parts (b) and (c) of Fig. 5-23 such that the spatial resolution was maintained at 6 m × 6 m.

The relationship between image speckle and the magnitude of N is quite *visible* in Fig. 5-23(a)–(c).

Despite the fact that the spatial resolution is the same for all three images, some features that are clearly distinguishable in Fig. 5-23(c) are totally indistinguishable in Fig. 5-23(a). An equally significant observation can be surmised from a comparison of the 6 m×6 m image with $N = 12$ [Fig. 5-23(c)] with the 1.5 m×2.1 m image with $N = 1$ [Fig. 5-23(d)].

▶ From an *interpretability* point of view, the two images offer comparable *resolving power* despite the fact that the area of the resolution cell of the high-resolution image is 12 times smaller than that of the other image. Thus, the resolution quality of an image is governed by both the spatial resolution of the ground cell represented by an image pixel and the number of independent samples (looks) N "contained" in the measurement of the power backscattered from the ground cell. ◀

5-8.5 Applicability of the Rayleigh Fading Model

Does the Rayleigh fading model provide an appropriate approach to characterize the statistics of radar backscatter from terrain? The answer is a qualified yes. If the assumptions underlying the Rayleigh fading model are reasonably satisfied, the available experimental evidence suggests that the Rayleigh model is quite applicable (Bush and Ulaby, 1975; Ulaby et al., 1986b; Ulaby et al., 1988a). Terrain targets satisfying the Rayleigh assumption include bare ground surfaces, agricultural fields, dense forest canopies, and snow-covered ground. In all cases the target must have stationary statistics, which requires that its "local-average" electromagnetic properties be uniform across the extent of the target.

Rayleigh fading is inapplicable for a sparse forest observed by a high-resolution radar because the high spatial variations in tree density at the scale of the radar resolution violate the stationarity assumption. Thus, a very important parameter governing applicability of Rayleigh statistics to backscatter from terrain is the size of the radar resolution cell relative to the spatial frequency spectrum characterizing the scattering from the terrain target under consideration.

▶ An urban scene is another target class or condition for which Rayleigh statistics may not apply. ◀

If the resolution cell size is such that the backscatter is likely to be dominated by the return from one or a few strong scatterers, such as a building or a corner reflector formed by two intersecting flat surfaces, the Rayleigh clutter model is no longer applicable.

These observations are borne out by the histograms shown in Fig. 5-24, which were generated from 4-look Seasat SAR data for image segments containing water, forest, and urban scenes. The image was generated by a square-root intensity process. The continuous curves are plots of the pdf $p(T_4)$ given by

$$p(T_4) = \frac{1}{\bar{I}^{1/2}}\, p(g_4),$$

where $p(g_4)$ is the pdf of the speckle random variable given by Eq. (5.90). We observe that the histogram for water is in reasonably close agreement with the theoretical pdf based on Rayleigh fading, the histogram for the forest category exhibits some departure from the theoretical pdf, and the histogram of the urban scene is markedly different from the theoretical pdf. These differences are attributable to the *textural variations* of the image, which are not accounted for in the theoretical model.

5-9 Image Texture and Despeckle Filtering

5-9.1 Image Texture

A radar image of a distributed target that obeys Rayleigh fading statistics is said to be ***textureless***; the image exhibits the usual pixel-to-pixel tonal variations characteristic of speckle, but it has no other spatial patterns.

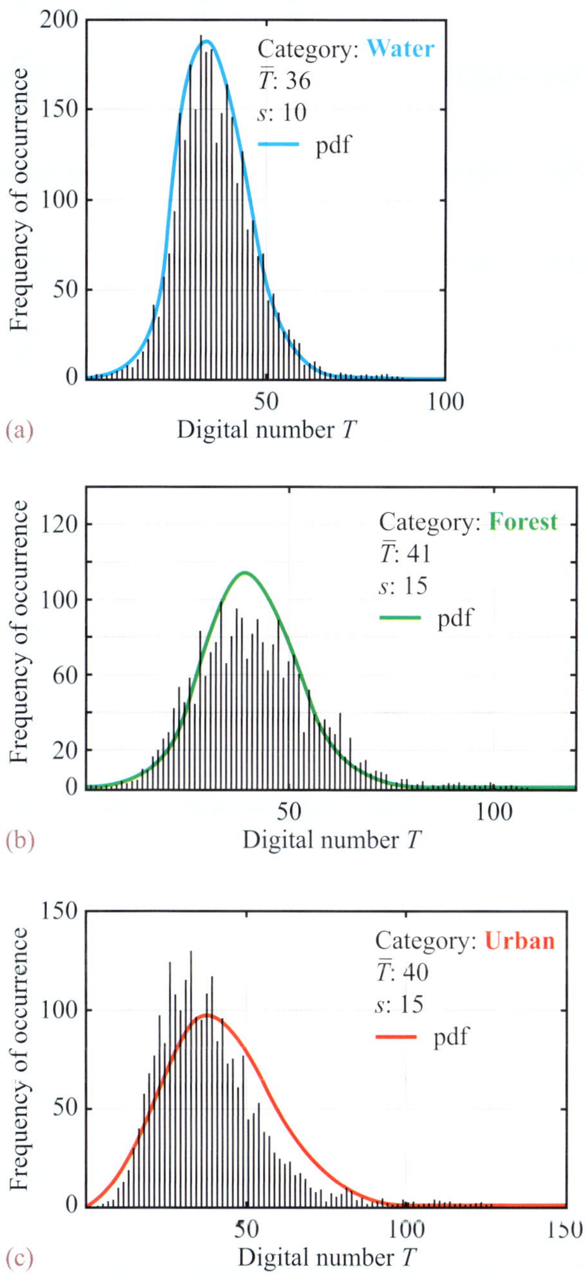

Figure 5-24: Histograms of image tone T (digital number) of square-root-intensity Seasat SAR images of three categories of distributed targets. The water category provides a good fit to the theoretical pdf given by Eq. (5.90); the forest category exhibits a significant departure due to textural variations and the urban category exhibits the greatest departure from theory. Urban scenes often have bright, dominant scatterers, which violates Assumption 6 of the Rayleigh model.

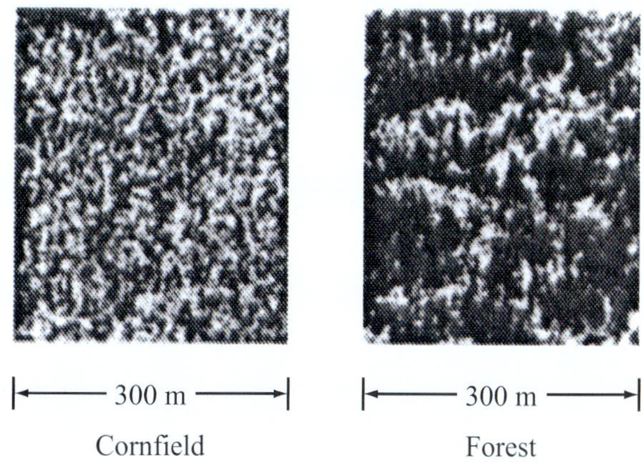

Figure 5-25: The pixel-to-pixel variation in the cornfield image is almost entirety due to Rayleigh fading (speckle), whereas the forest-parcel image includes spatial patterns (in addition to speckle) associated with the density variation of trees.

> ▶ ***Image texture*** refers to the intrinsic spatial variability of image tone, beyond that caused by speckle, due to corresponding spatial variabilities in the physical and/or electromagnetic properties of the distributed target. ◀

Consider the two image segments shown in Fig. 5-25, one of which belongs to an agricultural field (corn) and the other belongs to a forest parcel. A radar image of a cornfield is essentially textureless if the image pixel contains many rows of corn, whereas that of the forest parcel is highly textured, typical of a sparse forest that may have a dense canopy in some areas and mostly open spaces in others.

Image texture is a useful property that can be used in computer classification of radar images, particularly when attempting to discriminate between different classes of distributed targets with comparable average tones (reflectivities). Image textural attributes in common use include:

(a) the ***texture autocorrelation function***, which is a measure of the degree of similarity between a given pixel and its neighboring pixels.

(b) The ***texture standard deviation***, representing pixel-to-pixel variability over and above that due to speckle.

(c) Orientation or periodic textural patterns, if either exists.

In addition to the numerous useful papers available in the literature on radar image texture (Collins and Huang, 1998; Glaister et al., 2013; Collins and Allan, 2009; De Grandi et al., 2007), a highly recommended source is the authoritative book by Oliver and Quegan (2004), which not only discusses image texture in great detail, but also provides a comprehensive review of the various types of filters available for despeckling radar images.

5-9.2 Despeckling Filters

> ▶ An ideal ***despeckling filter*** is able to remove speckle from a radar image while preserving texture, and to do so without sacrificing spatial resolution. No such filter exists, but several of the available filters provide products that are far superior to the original speckled images. ◀

In their book on *Understanding SAR Images* (2004), Oliver and Quegan describe despeckle filters as "RCS reconstruction filters," meaning that a despeckle filter should generate an image of σ^0, the radar cross section per unit area of the imaged scene, without the speckle pattern.

Figure 5-26(a) displays a 2-look airborne SAR image with a spatial resolution of 3 m $\times$ 3 m. Given that N is only 2, the image exhibits the standard speckle appearance. The remaining six images are images filtered using a variety of different image processing algorithms. They include:

(a) Multilook Filter [Fig. 5-26(b)]: The image intensity I of a given pixel in an N-look image is assigned a value equal to μ_I, where μ_I is the average intensity of M^2 pixels contained in an $M \times M$ window centered at the given pixel. The image in Fig. 5-26(b) was generated by applying multilook averaging over an 11×11 window. With this type of primitive filter, the averaging process removes the speckle, but the image has a blurry appearance. Even though the number of pixels in the image has not changed, the averaging process, in effect, degrades the spatial resolution of the image. In practice this type of filter is of limited utility.

(b) Minimum Mean-Square Error (MMSE) Filter [Fig. 5-26(c)]: Instead of replacing the intensity I of a given pixel with μ_I (the average intensity over the $M \times M$ window), it is replaced with an intensity I' that incorporates information about the local statistics within the $M \times M$ window. Intensity I' is related to I and μ_I by

$$I' = \mu_I + \alpha(I - \mu_I), \tag{5.95}$$

where α is a parameter that gauges the degree of uniformity within the $M \times M$ window (Oliver and Quegan, 2004). For a window containing uniform terrain with constant σ, α is set to 0, in which case $I' = \mu_I$. On the other extreme, if the window contains multiple classes of targets with very different σ values, α is assigned a value of 1; consequently, $I' = I$, which means that the pixel intensity remains unchanged. To estimate the appropriate value of α for each pixel, a minimum-mean-square-error criterion is applied to the ***multiplicative intensity model*** given by Eq. (5.84), namely

$$I = \sigma F, \tag{5.96}$$

where I is the N-look intensity of the pixel, σ is its true reflectivity, and F is the fading random variable for an N-look intensity image (for the sake of simplicity, the system constant has been set to 1 and the subscript N has been suppressed).

If the window contains a uniform target with a constant σ, it follows that the ***normalized variance*** β_I of I is equal to the normalized variance β_F of F:

$$\beta_I = \left(\frac{s_I}{\mu_I}\right)^2 = \left(\frac{s_F}{\mu_F}\right)^2 = \frac{1}{N} \qquad \textbf{(uniform } \sigma\textbf{)}, \tag{5.97}$$

where in the last step we used Eq. (5.85). The values of the mean intensity μ_I and standard deviation s_I are estimated using the M^2 intensity values of the pixels in the window centered at the pixel under consideration.

However, if the window contains more than one class of targets, σ is no longer uniform across it, which means that it should be treated as a random variable with

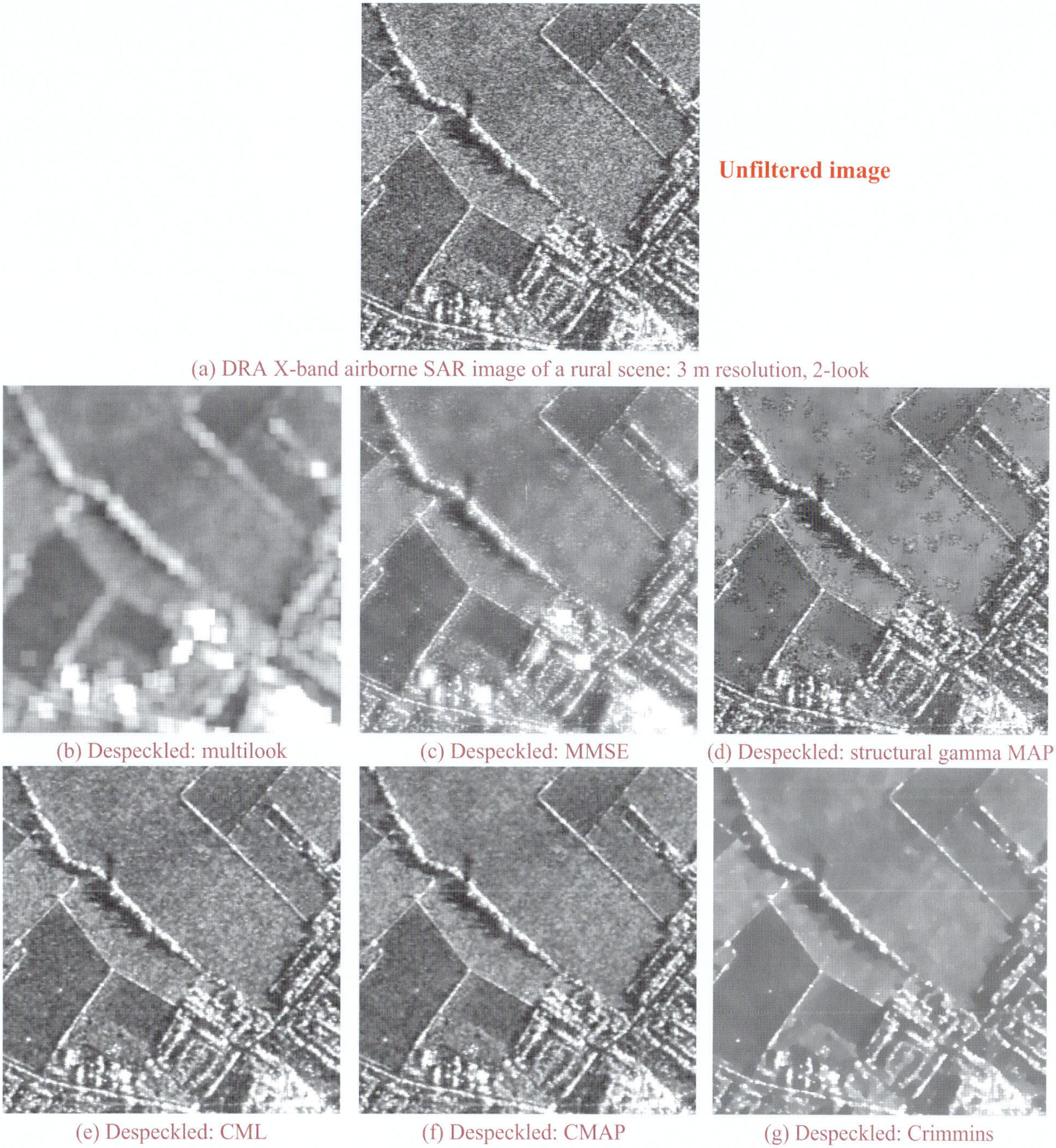

Figure 5-26: The 3 m × 3 m resolution 2-look image in (a) was subjected to six different despeckling filters [courtesy of Oliver and Quegan].

mean μ_σ and standard deviation s_σ. Statistically, σ and F_N are independent random variables. Hence,

$$\mu_I = \mu_\sigma \mu_F, \tag{5.98a}$$

$$\mu_{I^2} = \mu_{\sigma^2} \mu_{F^2}, \tag{5.98b}$$

and the normalized variance of I is

$$\beta_I = \left(\frac{s_I}{\mu_I}\right)^2 = \frac{\mu_{I^2} - \mu_I^2}{\mu_I^2} = \frac{\mu_{\sigma^2}\mu_{F^2} - \mu_\sigma^2 \mu_F^2}{\mu_\sigma^2 \mu_F^2}. \tag{5.99}$$

In view of the moments $\mu_F = 1$ and $\mu_{F^2} = 1 + 1/N$, Eq. (5.99) becomes

$$\beta_I = \left(\frac{\mu_{\sigma^2}}{\mu_\sigma^2}\right)\left(1 + \frac{1}{N}\right) - 1. \tag{5.100}$$

The normalized variance of σ is

$$\beta_\sigma = \left(\frac{s_\sigma}{\mu_\sigma}\right)^2 = \frac{\mu_{\sigma^2} - \mu_\sigma^2}{\mu_\sigma^2} = \left(\frac{\mu_{\sigma^2}}{\mu_\sigma^2}\right) - 1. \tag{5.101}$$

The parameter α is defined as the ratio of the reflectivity variance β_σ to the intensity variance β_I,

$$\alpha = \frac{\beta_\sigma}{\beta_I} = \frac{\beta_I - 1/N}{\beta_I\left(1 + \frac{1}{N}\right)} \qquad \textbf{(any } \boldsymbol{\sigma} \textbf{ variation)}. \tag{5.102}$$

The expression given by Eq. (5.102) is obtained by combining the expressions in Eqs. (5.100) and (5.101) to eliminate the ratio $(\mu_{\sigma^2}/\mu_\sigma^2)$. The intensity variance β_I is calculated using the M^2 intensity values. For an image window of a textureless distributed target (σ = constant), β_I should be close to $1/N$, in which case $\alpha \approx 0$. In contrast, for a textured window, $\beta_I > 1/N$.

Application of the MMSE filter to the original image shown in Fig. 5-26(a) leads to the image in Fig. 5-26(c). The filter is successful in suppressing speckle while preserving image features such as roads and boundaries.

(c)–(e) Bayesian Filters [Fig. 5-26(d) to (f)]: The filters used in generating the despeckled images shown in Fig. 5-26(d) to (f) all are based on Bayesian statistics, which uses conditional probabilities in combination with assumed pdf models for the reflectivity σ. For more details, the reader is referred to the book by Oliver and Quegan (2004) and its comprehensive list of references.

5-10 Coherent and Noncoherent Scattering

5-10.1 Surface Roughness

Consider the ground surface shown in Fig. 5-27(a). Let us assume that the surface height $z(x,y)$ varies randomly and isotropically with x and y, which means that $z(x,y)$ has the same statistical properties across any straight line segment of the ground surface. An example of a one-dimensional height profile for such a random surface is shown in Fig. 5-27(b). We label it $z(x)$. Upon digitizing the profile and arranging its values into "height bins" relative to the mean surface $\langle z \rangle = 0$, we obtain the ***height probability density function*** $p(z)$ displayed in Fig. 5-27(c). The measured pdf looks approximately Gaussian in shape, which is true for most random surfaces. That is,

$$p(z) = \frac{1}{\sqrt{2\pi s^2}}\, e^{-z^2/2s^2}, \tag{5.103}$$

where s^2 is the variance of surface heights (and since $\langle z \rangle = 0$, s is the standard deviation).

Given $p(z)$, we can calculate several statistical attributes of the random surface, including:

(a) The height standard deviation s given by

$$s = \langle z^2 \rangle^{1/2} = \left[\int_{-\infty}^{\infty} z^2\, p(z)\, dz\right]^{1/2}, \tag{5.104}$$

which also is called the ***rms height***, and

(b) the ***surface correlation function*** defined by

$$\rho(\xi) = \frac{\langle z(x,y)\, z(x',y') \rangle}{s^2}, \tag{5.105}$$

where $u = x - x'$, $v = y - y'$, and $\xi = \sqrt{u^2 + v^2}$ is the lateral separation between the two locations, (x,y) and (x',y') on the surface. The correlation function $\rho(\xi)$ is a measure of the degree of correlation between the surface at location (x,y) and the surface at location (x',y').

The correlation function for the random surface under consideration is shown in Fig. 5-27(d). As expected, the correlation function decreases with ξ, so that if the spacing between two locations is greater than a

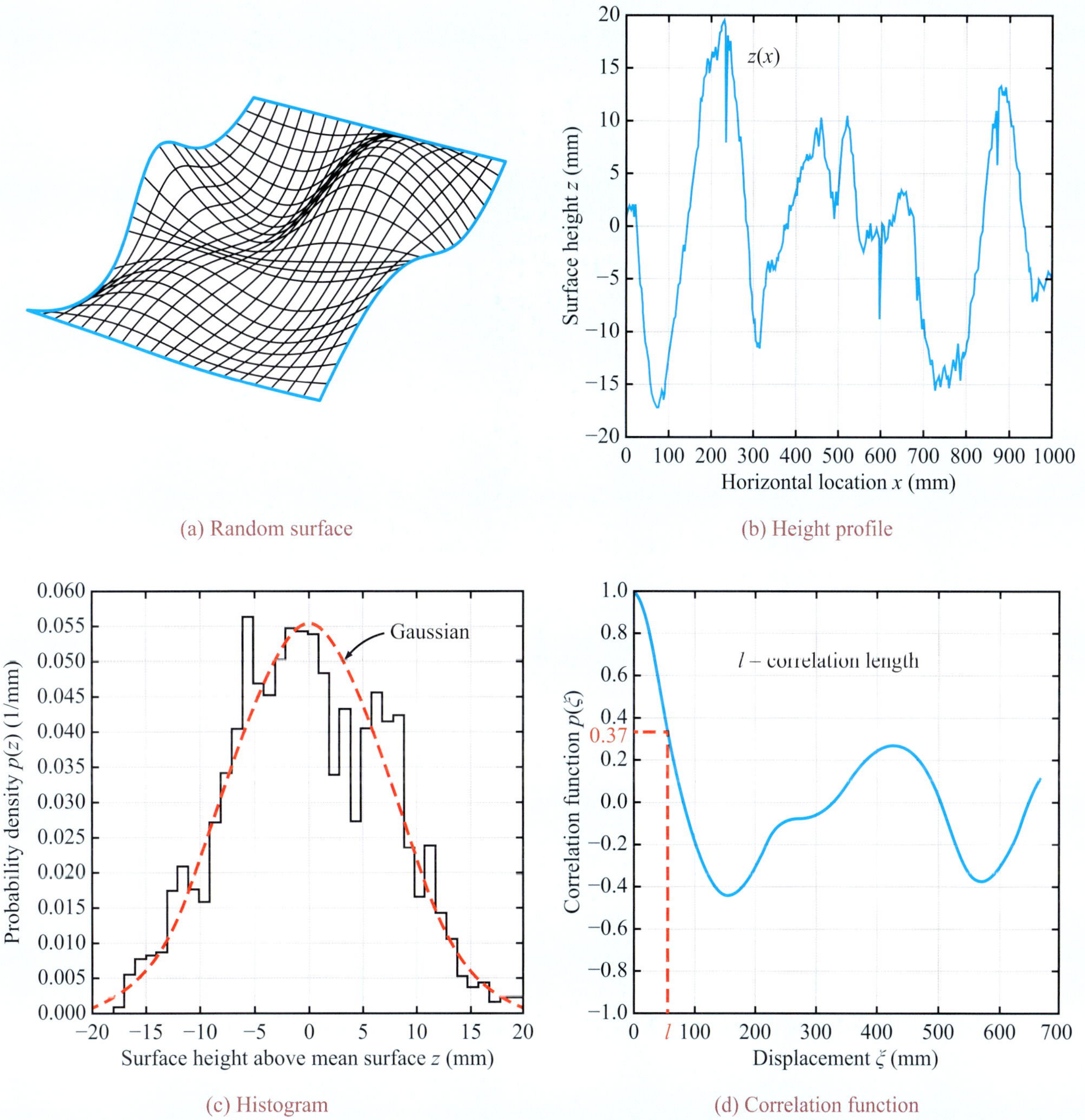

Figure 5-27: Random, isotropic surface $z(x,y)$: (a) pictorial view, (b) measured height profile $z(x)$, (c) pdf of digitized height profile, and (d) autocorrelation function $\rho(\xi)$, where ξ is the displacement between two points on the surface.

certain distance called the ***correlation length*** l, then the heights at the two locations are considered statistically uncorrelated. The correlation length l is defined as the separation $\xi = l$ for which

$$\rho(\xi) = e^{-1} \qquad (\text{at } \xi = l). \tag{5.106}$$

For the correlation function displayed in Fig. 5-27(d), $l = 52.5$ mm. Because the measurement precision of $z(x,y)$ is limited by the tool (such as a laser profiler) used to profile the surface height, the correlation function $\rho(\xi)$ exhibits noise-like oscillations beyond about $\xi = 100$ mm. Negative values of $\rho(\xi)$ are an artifact of the measurement process.

As discussed later in Chapter 10, models for electromagnetic scattering by random rough surfaces involve the use of assumed functional forms for the correlation function $\rho(\xi)$. The two most common forms are the ***exponential correlation function*** $\rho_e(\xi)$ and the ***Gaussian correlation function*** $\rho_G(\xi)$, defined as:

$$\rho_e(\xi) = e^{-|\xi|/l} \qquad \textbf{(exponential)}, \tag{5.107a}$$

$$\rho_G(\xi) = e^{-\xi^2/l^2} \qquad \textbf{(Gaussian)}, \tag{5.107b}$$

where l is the correlation length of the surface. Plots of both functions are shown in Fig. 5-28 as a function of the normalized displacement ξ/l. Experimental measurements of $\rho(\xi)$ for random soil surfaces tend to favor the applicability of the exponential correlation function (Nashashibi and Ulaby, 2007).

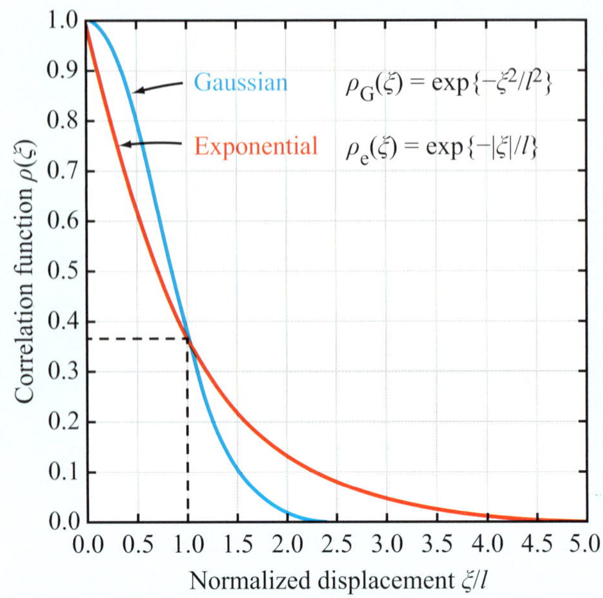

Figure 5-28: Exponential and Gaussian correlation functions.

5-10.2 Bistatic Scattering

Electromagnetically, the roughness of a surface is measured relative to the EM wavelength λ. For a surface with rms height s, its ***electromagnetic roughness*** ks is

$$ks = \frac{2\pi}{\lambda} s \qquad \textbf{(EM roughness)}. \tag{5.108}$$

In the general case, bistatic scattering involves four angles: incidence and scattering angles θ_i and θ_s and azimuth angles ϕ_i and ϕ_s (Fig. 5-29), but for a random isotropic surface, the x and y axes are arbitrary, so only one azimuth angle is sufficient, namely $\phi = \phi_s - \phi_i$. In terms of θ_i, θ_s, and ϕ, we define the following terms:

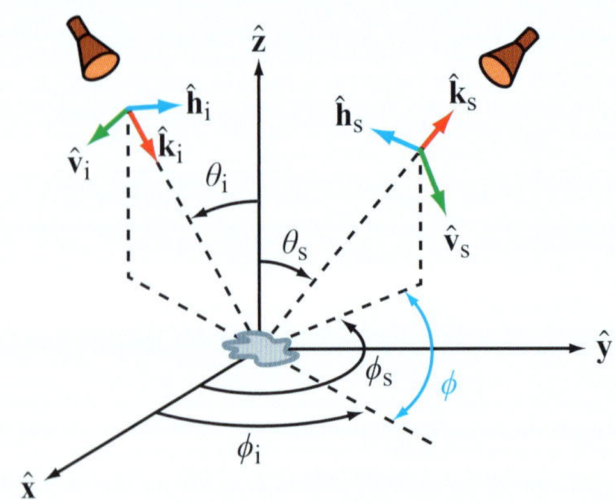

Figure 5-29: Bistatic-scattering coordinate system.

(a) Specular direction: $\theta_i = \theta_s$ and $\phi = 0$,

(b) Plane of incidence: $\phi = 0$,

(c) Out of the plane of incidence: $\phi \neq 0$,

(d) Backscattering direction: $\theta_i = \theta_s$ and $\phi = 180°$.

For a perfectly smooth surface with rms height $s = 0$, an incident EM wave with polarization p (where $p = \mathrm{h}$ or v) is reflected along the specular direction and the reflected power P_p^{r} is related to the incident power P_p^{i} by Eq. (2.108), namely

$$P_p^{\mathrm{r}} = \Gamma^p P_p^{\mathrm{i}} \qquad (p = \mathrm{v \ or \ h \ polarization}), \qquad (5.109\mathrm{a})$$

(specular surface)

where Γ^p is the p-polarized Fresnel reflectivity of a perfectly flat surface. Figure 5-30 depicts the scattering patterns for three surfaces with varying degrees of electromagnetic roughness. The pattern for the perfectly smooth surface is essentially a delta function, as shown in Fig. 5-30(a). The component of the scattering pattern along the specular direction is called the ***coherent component*** because the reflected wave has a uniform phase front. The scattering pattern of the perfectly smooth surface consists of only a coherent component.

If the surface is slightly rough with ks on the order of 0.1, the scattering pattern continues to be dominated by the coherent component along the specular direction, as shown in Fig. 5-30(b), but the pattern also includes a ***noncoherent component*** along all other directions. In that case, the power reflected along the specular direction is related to the incident power by

$$P_p^{\mathrm{r}} = \Gamma_{\mathrm{coh}}^p P_p^{\mathrm{i}}, \qquad (5.109\mathrm{b})$$

(nonspecular surface)

where Γ_{coh}^p is the ***coherent reflectivity***. The relationship between Γ_{coh}^p and Γ^p is introduced shortly.

Furthermore, the incoherent component contains not only waves with the same polarization as that of the incident wave, but also waves with the orthogonal polarization. That is, if the incident wave is h polarized, the scattering noncoherent component consists of both h-polarized (***co-pol***) and v-polarized (***cross-pol***) components. If the surface roughness is increased to $ks > 2$, the coherent component becomes negligibly small and the noncoherent component becomes dominant along all directions, including the specular direction [Fig. 5-30(c)].

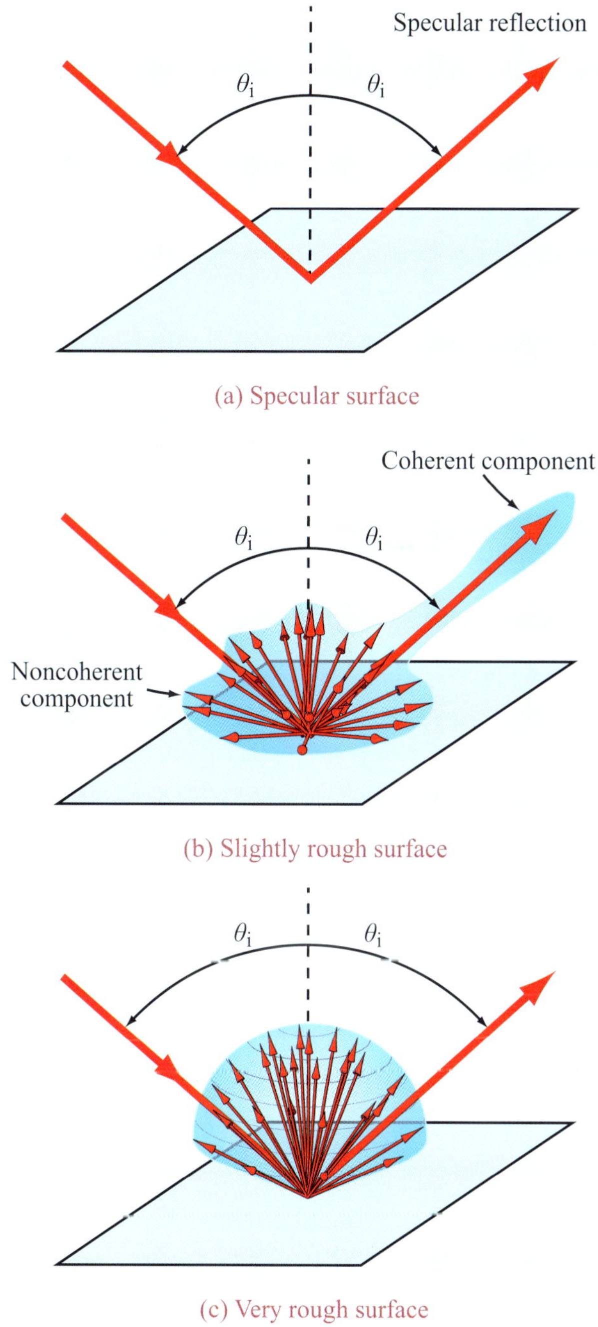

Figure 5-30: The bistatic-scattering pattern consists of a coherent component along the specular direction and a noncoherent component along all directions. For a perfectly smooth surface, only the coherent component exists, and at the opposite extreme, for a very rough surface the coherent component becomes negligibly small in comparison with the noncoherent component.

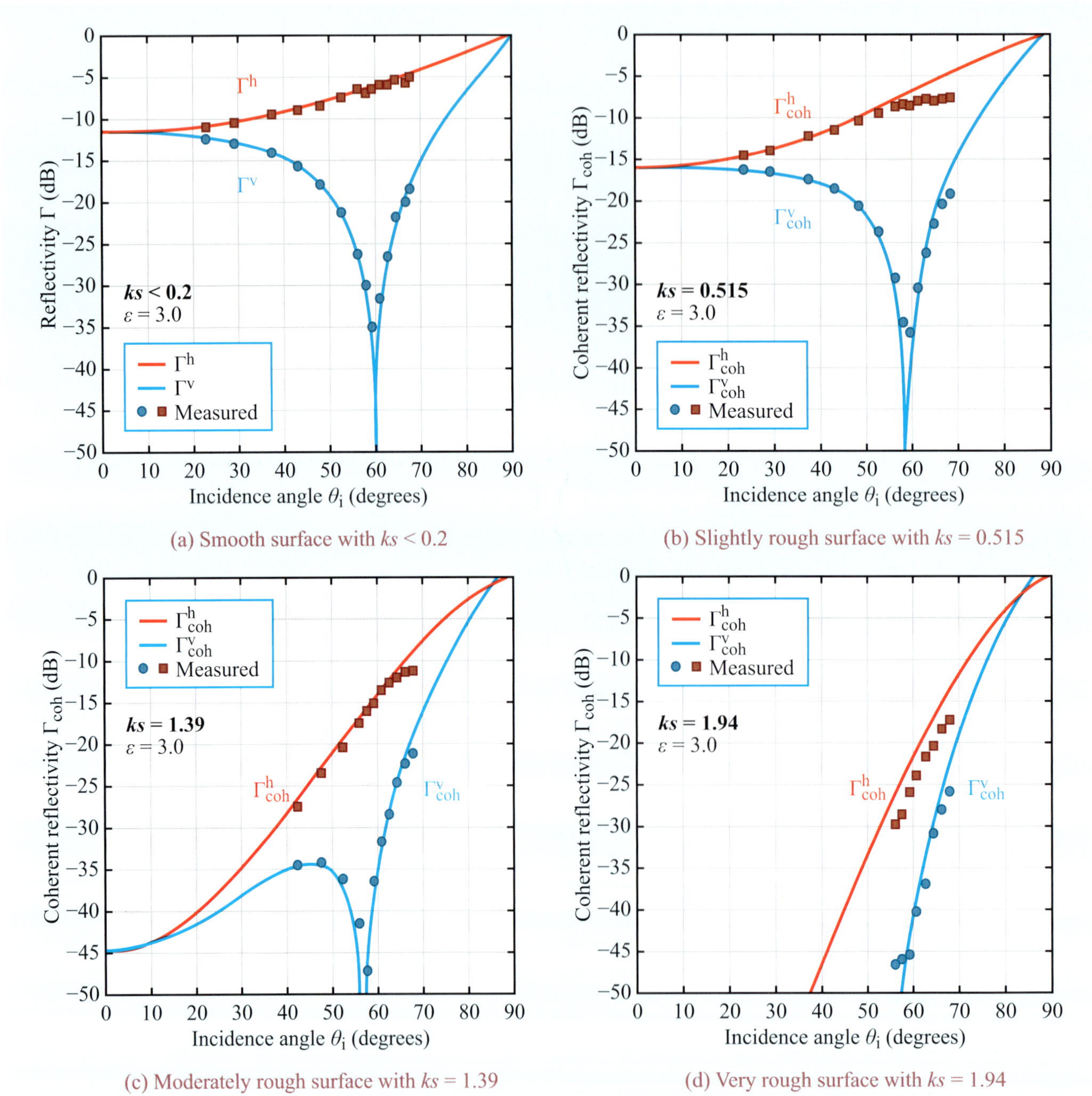

(a) Smooth surface with $ks < 0.2$

(b) Slightly rough surface with $ks = 0.515$

(c) Moderately rough surface with $ks = 1.39$

(d) Very rough surface with $ks = 1.94$

Figure 5-31: Measured and calculated reflectivities for h and v polarizations for four surfaces with different EM roughnesses [De Roo and Ulaby, 1994].

5-10.3 Specular Reflectivity

Using an X-band bistatic radar facility, De Roo and Ulaby (1994) measured the bistatic-scattering pattern for several sand surfaces with varying roughnesses. The smoothest surface they measured had an electromagnetic roughness $ks < 0.2$, and the electromagnetic roughness of the roughest random surface was $ks = 1.94$, which is about an order of magnitude greater than that of the smoothest surface. Figure 5-31(a) displays the measured specular reflectivities Γ^{v} and Γ^{h} of the smoothest surface as a function of the incidence angle θ_{i}. Also shown are plots calculated on the basis of the Fresnel reflection coefficient equation given by Eqs. (2.101) and (2.102a) for a specular surface and the relations $\Gamma^{\text{v}} = |\rho_{\text{v}}|^2$ and $\Gamma^{\text{h}} = |\rho_{\text{h}}|^2$. The sand dielectric permittivity used in the calculation is $\varepsilon = 3$. We observe that the measured data are in very close agreement with the calculated plots, implying that a surface may be considered perfectly flat if its electromagnetic roughness $ks < 0.2$.

The three remaining figures in Fig. 5-31 pertain to much rougher surfaces. The calculated plots, which are in close agreement with the measured reflectivities, are based on the relationship

$$\Gamma^p_{\text{coh}} = \Gamma^p \, e^{-4\psi^2} \qquad (p = \text{v or h}), \tag{5.110}$$

where

$$\psi = ks\cos\theta_{\text{i}} = \frac{2\pi}{\lambda}\, s\cos\theta_{\text{i}}, \tag{5.111}$$

and Γ^p is the Fresnel reflectivity for polarization p given by the expressions in Table 2-5. To appreciate the sensitivity of Γ^p_{coh} to ks, let us consider two roughness cases, both at $\theta_{\text{i}} = 45^\circ$:

(a) Relatively smooth surface with $ks = 0.2$:

$$\Gamma^p_{\text{coh}} = 0.92\,\Gamma^p.$$

(b) Very rough surface with $ks = 2$:

$$\Gamma^p_{\text{coh}} = 3.35 \times 10^{-4}\Gamma^p.$$

> ▶ Increasing the EM roughness from 0.2 to 2 causes the reflectivity to decrease by more than three orders of magnitude. ◀

The variation of $\Gamma^p_{\text{coh}}/\Gamma^p$ with $\psi = ks\cos\theta_{\text{i}}$ is shown in Fig. 5-32(a), and angular plots of $\Gamma^{\text{v}}_{\text{coh}}$ are shown in Fig. 5-32(b) for multiple values of the electromagnetic roughness parameter ks. Note that the vertical axis is in dB, so the observed reduction in level as a function of increasing roughness is very significant, and much more so at angles below the Brewster angle than above it.

The expression given by Eq. (5.110) is called the zeroth-order physical-optics solution for reflection by a rough surface (De Roo and Ulaby, 1994). Whereas the model provides very good agreement to experimental observations, it does not predict a rather subtle shift in the location of the Brewster angle (Saillard and Maystre, 1990; Greffet, 1992). De Roo and Ulaby (1994) demonstrated that the location of the Brewster angle can be predicted with excellent accuracy by using a second-order physical-optics model, which also is predicted by the I^{2}EM model presented in Section 10-3.

5-10.4 Bistatic-Scattering Coefficient

In the preceding subsection, we examined the dependence of the coherent reflectivity Γ_{coh} of a continuous two-dimensional surface on surface roughness. Now, we extend the discussion to the bistatic-scattering coefficient of the surface, σ^0. The two quantities are related, but distinct.

The q-polarized coherent reflectivity $\Gamma^q_{\text{coh}}(\theta_{\text{i}})$ is an idealized quantity defined for a continuous, infinite, two-dimensional surface; its definition relies on plane-wave incidence and plane-wave reflection. For an area A of the surface, $\Gamma^q_{\text{coh}}(\theta_{\text{i}})$ is given by Eq. (5.109a) as

$$\Gamma^q_{\text{coh}}(\theta_{\text{i}}) = \frac{P^{\text{r}}_q}{P^{\text{i}}_q} = \frac{S^{\text{r}}_q A\cos\theta_{\text{i}}}{S^{\text{i}}_q A\cos\theta_{\text{i}}} = \frac{S^{\text{r}}_q}{S^{\text{i}}_q} \qquad (q = \text{v or h}), \tag{5.112}$$

where S^{i}_q and S^{r}_q are the q-polarized power densities of the incident and reflected plane waves. In contrast, the bistatic-scattering cross section σ_{pq} is defined by Eq. (5.27a) as

$$\sigma_{pq} = \frac{P^{\text{rer}}_p}{S^{\text{i}}_q}, \tag{5.113}$$

where P^{rer}_p is the p-polarized power reradiated by the target. For a distributed surface of area A, its scattering

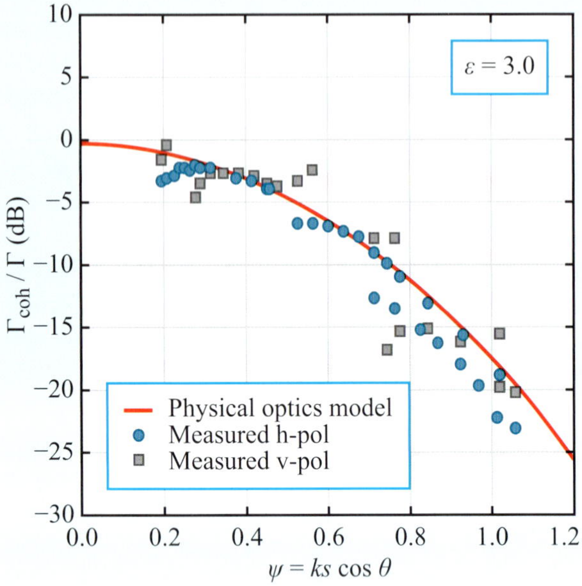

(a) $\Gamma_{\text{coh}} / \Gamma$ vs. roughness parameter

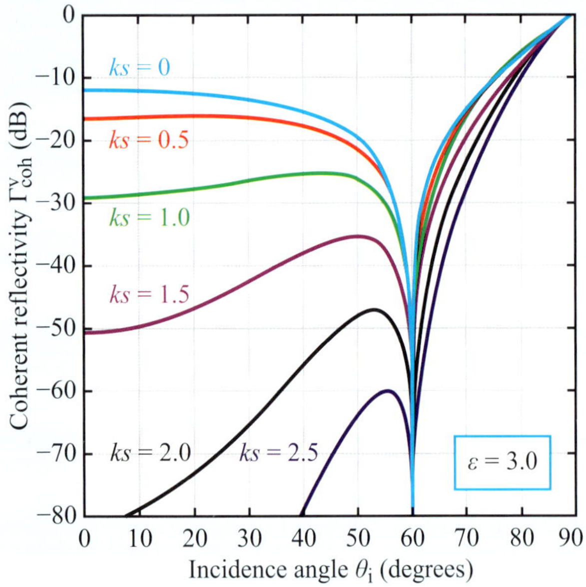

(b) $\Gamma^{\text{v}}_{\text{coh}}(\theta)$ at various surface roughnesses

Figure 5-32: Variation of (a) $\Gamma_{\text{coh}}/\Gamma$ with $\psi = ks\cos\theta$ and (b) $\Gamma^{\text{v}}_{\text{coh}}$ with θ_{i} for various values of ks [De Roo and Ulaby, 1994].

coefficient is $\sigma^0 = \sigma/A$. Also, the reradiated power spreads spherically, so at a range R_{r} from the surface, the power density of the scattered wave is $\mathbb{S}^{\text{s}}_p = P^{\text{rer}}/(4\pi R_{\text{r}}^2)$. Hence,

$$\sigma^0_{pq} = \frac{1}{A}\,\frac{P^{\text{rer}}_p}{\mathbb{S}^{\text{i}}_q} = \left(\frac{4\pi R_{\text{r}}^2}{A}\right)\frac{\mathbb{S}^{\text{s}}_p}{\mathbb{S}^{\text{i}}_q}\,. \tag{5.114}$$

Coherent component

In general, the *total* co-polarized bistatic-scattering coefficient σ^0_{pp} consists of a ***coherent component*** $\sigma^0_{pp_{\text{coh}}}$ and an ***incoherent component*** $\sigma^0_{pp_{\text{inc}}}$:

$$\sigma^0_{pp}(\theta_{\text{i}},\phi_{\text{i}};\theta_{\text{s}},\phi_{\text{s}}) = \sigma^0_{pp_{\text{coh}}}(\theta_{\text{i}}) + \sigma^0_{pp_{\text{inc}}}(\theta_{\text{i}},\phi_{\text{i}};\theta_{\text{s}},\phi_{\text{s}}),$$
$$p = \text{h or v}, \tag{5.115}$$

where $(\theta_{\text{i}},\phi_{\text{i}})$ and $(\theta_{\text{s}},\phi_{\text{s}})$ define the incident and scattering directions as shown in Fig. 5-29.

> ▶ The cross-polarized component does not have a coherent component. ◀

The coherent component, which exists only along a narrow angular range around the specular direction, is given by

$$\sigma^0_{pp_{\text{coh}}}(\theta_{\text{i}}) = 4\pi\cos\theta_{\text{i}}\,\Gamma^p(\theta_{\text{i}})\,e^{-4\psi^2} \cdot \delta(\cos\theta_{\text{i}} - \cos\theta_{\text{s}})\,\delta(\phi_{\text{i}} - \phi_{\text{s}}), \tag{5.116}$$

where ψ is the roughness parameter defined by Eq. (5.111). The presence of the delta functions in Eq. (5.116) asserts that $\sigma^0_{pp_{\text{coh}}}$ is nonzero only when $\theta_{\text{i}} = \theta_{\text{s}}$ and $\phi_{\text{i}} = \phi_{\text{s}}$. The factor $4\pi\cos\theta_{\text{i}}$ is associated with the definition of σ^0. To demonstrate the validity of Eq. (5.116), consider the case of a flat surface with no roughness. For $s = 0$, $\psi = 0$ and so the incoherent component is zero, in which case σ^0_{pp} reduces to $\sigma^0_{pp_{\text{coh}}}$,

$$\begin{aligned}\sigma^0_{pp}(\theta_{\text{i}}) &= \sigma^0_{pp_{\text{coh}}}(\theta_{\text{i}}) \\ &= 4\pi\cos\theta_{\text{i}}\,\Gamma^p(\theta_{\text{i}}) \\ &\quad\cdot \delta(\cos\theta_{\text{i}} - \cos\theta_{\text{s}})\,\delta(\phi_{\text{i}} - \phi_{\text{s}}).\end{aligned} \tag{5.117}$$

At a distance R_r from the flat surface, the total p-polarized scattered (reflected) power is related to the scattered power density $\mathcal{S}_p^\mathrm{s}$ by

$$P_p^\mathrm{s} = \iint\limits_{\text{upper hemisphere}} \mathcal{S}_p^\mathrm{s} R_\mathrm{r}^2 \, d\Omega_\mathrm{s}. \tag{5.118}$$

Using Eq. (5.114), with $p = q$, and Eq. (5.117) in Eq. (5.118) leads to

$$\begin{aligned} P_p^\mathrm{s} &= \iint \frac{\mathcal{S}_p^\mathrm{i} A}{4\pi R_\mathrm{r}^2} \, \sigma_{pp}^0 R_\mathrm{r}^2 \, d\Omega_\mathrm{s} \\ &= \iint \frac{\mathcal{S}_p^\mathrm{i} A}{4\pi} \, (4\pi \cos\theta_\mathrm{i}) \, \Gamma^p(\theta_\mathrm{i}) \\ &\qquad \cdot \delta(\cos\theta_\mathrm{i} - \cos\theta_\mathrm{s}) \, \delta(\phi_\mathrm{i} - \phi_\mathrm{s}) \, d\Omega_\mathrm{s} \\ &= P_p^\mathrm{i} \, \Gamma^p(\theta_\mathrm{i}), \end{aligned} \tag{5.119}$$

where we used the relation $P_p^\mathrm{i} = \mathcal{S}_p^\mathrm{i} A \cos\theta_\mathrm{i}$ for the p-polarized power incident upon the surface. The result given by Eq. (5.119) is equivalent to Eq. (5.109a), the basic equation that defines the Fresnel reflectivity Γ^p as the ratio of the power reflected from a perfectly flat surface to the incident power.

Incoherent component

▶ Increasing the roughness of a surface causes a transfer of energy from the coherent component to the incoherent component, as well as to increased depolarization. ◀

The redistribution of energy across the upper hemisphere depends on the geometrical, dielectric, and statistical properties of the surface, as discussed in more detail in Chapter 10.

Figure 5-33 displays a 3-D bistatic-scattering pattern of the scattering coefficient σ_hh^0 for a sand surface whose rms height s is estimated to be smaller than 0.1 cm. The measurements were made by a 35 GHz bistatic radar, so the corresponding electromagnetic roughness is $ks = (2\pi s/\lambda) = 0.73$, which makes the surface slightly rough electromagnetically. The incidence angle is 60° and both the transmit and receive antennas are horizontally polarized. We observe that the pattern has a peak along the direction $\theta_\mathrm{s} = \theta_\mathrm{i} = 60°$ and $\phi = 0$, corresponding to the specular direction. This peak is the coherent bistatic-scattering component $\sigma_{hh_\mathrm{coh}}^0$. Another example, shown in Fig. 5-34 for a rough surface with $ks = 3.28$, includes 3-D patterns for the vv-polarized scattering coefficient σ_vv^0 and the ***depolarized ratio*** defined as

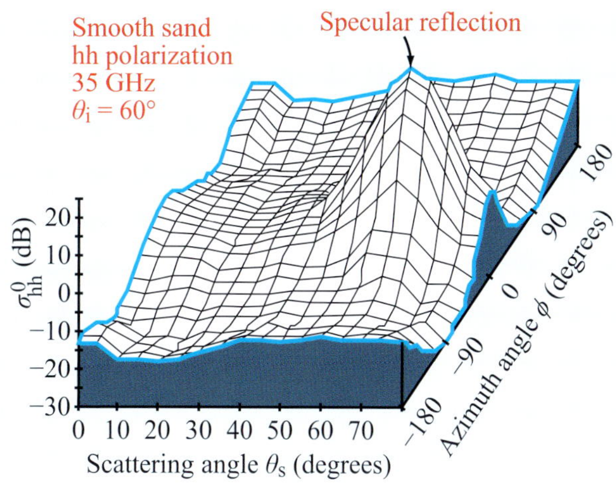

Figure 5-33: Measured bistatic-scattering pattern for a sand surface with $ks = 0.73$ at 35 GHz [Ulaby et al., 1988d].

$$\chi = \frac{\sigma_\mathrm{hv}^0 + \sigma_\mathrm{vh}^0}{\sigma_\mathrm{hh}^0 + \sigma_\mathrm{vv}^0}. \tag{5.120}$$

We note that maximum depolarization occurs in the plane $\phi = 90°$, which is orthogonal to the plane of incidence.

5-10.5 Backscattering Response of a Smooth Surface

In the backscattering direction, the total co-polarized backscattering coefficient at incidence angle θ_i consists of a coherent component $\sigma_{pp_\mathrm{coh}}^0(\theta_\mathrm{i})$, which is important only at or near normal incidence, and an incoherent

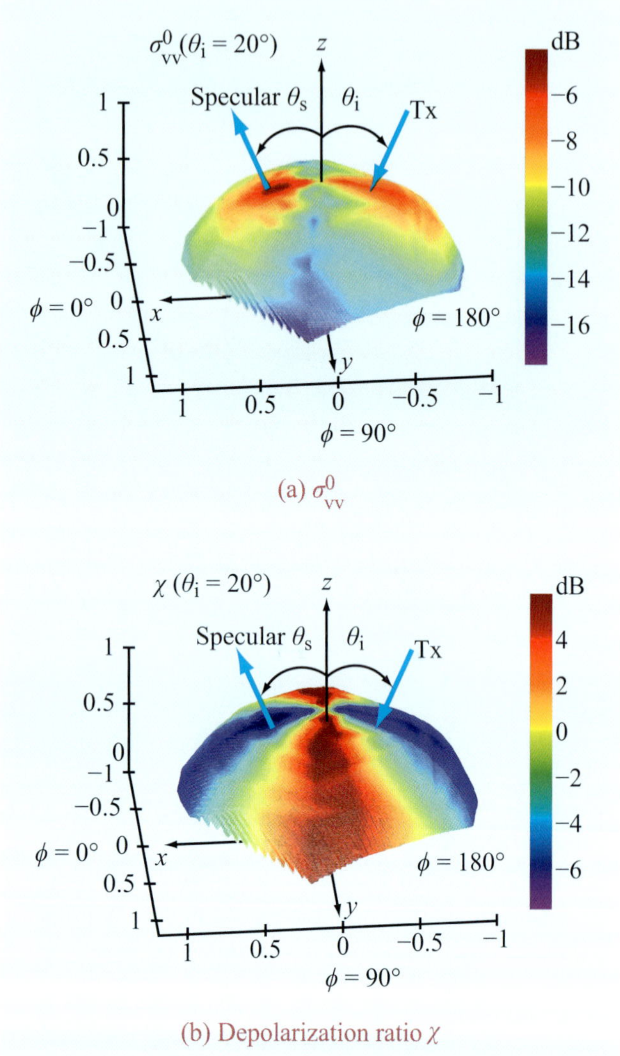

Figure 5-34: Measured bistatic-scattering patterns for a rough surface: (a) σ^0_{vv} and (b) χ [Nashashibi and Ulaby, 2007].

component $\sigma^0_{pp_{inc}}(\theta_i)$, which is important at all angles:

$$\sigma^0_{pp}(\theta_i) = \sigma^0_{pp_{coh}}(\theta_i) + \sigma^0_{pp_{inc}}(\theta_i). \tag{5.121}$$

For a p-polarized antenna pointed along a direction θ_i relative to normal incidence, the coherent backscattering coefficient is given by Eq. (5.116), which treats the coherent component as a pure delta function. In reality, $\sigma^0_{pp_{coh}}$ has a finite beamwidth surrounding the specular direction, so we use the approximate expression (Ulaby et al., 1983)

$$\sigma^0_{pp_{coh}} = \frac{\Gamma^p(\theta_i)}{\beta_c^2} \exp(-4k^2s^2) \exp\left(\frac{-\theta_i^2}{\beta_c^2}\right), \tag{5.122}$$

where $\Gamma^p(\theta_i)$ is the p-polarized Fresnel reflectivity, $k = 2\pi/\lambda$, s is the surface rms height, and β_c is an effective width of the angular extent of $\sigma^0_{pp_{coh}}$. Both θ_i and β_c are in radians, and in the derivation leading to Eq. (5.122), it was assumed that the range is sufficiently large to satisfy the far-field criterion for the antenna and θ_i is sufficiently small to permit making the approximations $\sin\theta_i \approx \theta_i$ and $\cos\theta_i \approx 1$. The factor $\exp(-4k^2s^2)$ accounts for the reduction of the Fresnel reflectivity due to surface roughness, consistent with Eqs. (5.110) and (5.111).

For a specular surface with $s = 0$, the expression given by Eq. (5.122) leads to the same result for the power received from a specular surface at normal incidence as does the application of the image method to the Friis transmission formula given by Eq. (3.39). In the image method, the transmit antenna "sees" its image in the reflection and hence may be modeled as two communication antennas with the same gain G_0, separated by a distance $2h$, where h is the height of the antenna above the surface. The received power in this image case, when modified by the surface nadir reflectivity Γ_0, is given by

$$P^r = \frac{P^t G_0^2 \lambda^2\, \Gamma_0}{(4\pi)^2 (2h)^2} \qquad \textbf{(Friis transmission formula)}. \tag{5.123}$$

For a distributed target with only a coherent scattering coefficient $\sigma^0_{pp_{coh}}$, the radar equation given by Eq. (5.32) becomes

$$P^r = \frac{P^t\lambda^2}{(4\pi)^3} \iint \frac{G^2(\theta_i,\phi_i)}{R^4}\, \sigma^0_{pp_{coh}}\, dA. \tag{5.124}$$

For a specular surface with $s = 0$, $\sigma^0_{pp_{coh}}(\theta_i)$ is given by

$$\sigma^0_{pp_{coh}}(\theta_i) = \frac{\Gamma^p(\theta_i)}{\beta_c^2} \exp\left(\frac{-\theta_i^2}{\beta_c^2}\right). \tag{5.125}$$

The magnitude of β_c is on the order of $1°$ or smaller. Hence, the contribution to the integral of Eq. (5.124) is insignificant for θ_i beyond a few degrees, in which case the quantity dA/R^4 may be approximated as follows:

$$\frac{dA}{R^4} = \frac{(R\,d\theta_i)(R\sin\theta_i\,d\phi_i)}{R^4} \approx \frac{\theta_i\,d\theta_i\,d\phi_i}{h^2}, \quad (5.126)$$

where we set $R \approx h$ and $\sin\theta_i \approx \theta_i$. With $G(\theta_i,\phi_i) \approx G_0$ over the effective region of the integration in Eq. (5.124), inserting Eqs. (5.125) and (5.126) in Eq. (5.124) gives

$$P^r = \frac{P^t G_0^2 \lambda^2}{(4\pi)^3 h^2} \int_{\phi_i=0}^{2\pi} \int_{\theta_i=0}^{\pi/2} \Gamma^p(\theta_i)\, \frac{e^{-\theta_i^2/\beta_c^2}}{\beta_c^2}\, \theta_i\,d\theta_i\,d\phi_i. \quad (5.127)$$

Since β_c is on the order of $1°$, $\Gamma^p(\theta_i) \approx \Gamma^p(0) = \Gamma_0$, and the integration of the Gaussian-like function leads to

$$P^r = \frac{P^t G_0^2 \lambda^2}{(4\pi)^2 (2h)^2}\,\Gamma_0, \quad (5.128)$$

which is identical to Eq. (5.123).

Having established that the model for $\sigma^0_{pp_{coh}}(\theta_i)$ given by Eq. (5.122) provides a reasonable mathematical representation for the coherent component of the backscattering coefficient, we now show in Fig. 5-35 plots of $\sigma_{pp_{coh}}(\theta_i)$, $\sigma_{pp_{inc}}(\theta_i)$, and their sum. The angular plot of $\sigma^0_{pp_{coh}}(\theta_i)$, which has an effective beamwidth $\beta_c = 0.51°$, was obtained from the measured values of $\sigma^0_{pp}(\theta_i)$ and the radar antenna pattern $G(\theta_i,\phi_i)$ by performing a deconvolution process. The roughness of the observed surface is $ks = 0.3$, which defines it as a relatively smooth surface. We note that at nadir, $\sigma^0_{pp_{coh}}$ is about 6 dB higher than that of the incoherent component $\sigma^0_{pp_{inc}}$, but beyond a few degrees from nadir, $\sigma^0_{pp_{coh}}$ becomes negligibly small. Since most imaging radars operate at incidence angles far away from nadir, the coherent component usually is inconsequential. The only situation of concern is when the nadir return through an antenna sidelobe interferes with the intended signal from the scene illuminated by the mainbeam of the antenna.

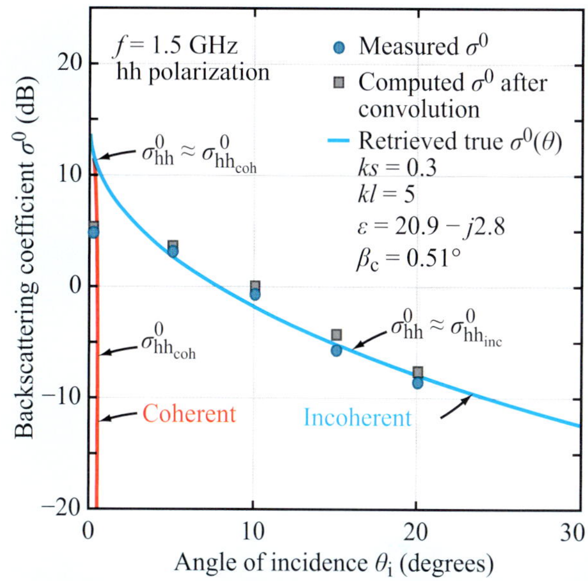

Figure 5-35: Measured and retrieved backscattering angular pattern for a surface with $ks = 0.3$ at 1.5 GHz [Ulaby et al., 1983].

5-11 Polarization Synthesis

In Section 5-3 we defined the scattering matrix of a target in the backscattering alignment (BSA) convention as

$$\mathbf{S} = \begin{bmatrix} S_{vv} & S_{vh} \\ S_{hv} & S_{hh} \end{bmatrix}, \quad (5.129)$$

and we explained how the four elements of $\mathbf{S}$ can be measured by a radar with polarization agility. Element S_{hv}, for example, corresponds to transmitting with a v-polarized antenna and receiving with a h-polarized antenna. The typical implementation of a radar polarimeter involves transmitting a wave of one polarization and receiving echoes in two orthogonal polarizations simultaneously. This is followed by transmitting a wave with a second polarization, and again receiving echoes with both polarizations simultaneously as shown in Fig. 5-36.

Before these measurements of the scattering matrix elements are useful for further quantitative analysis and comparison to models, the measured values must be calibrated so that the final results can be converted to normalized radar cross sections. Polarimetric radar measurements require additional corrections; not only the amplitudes, but also the relative phases between

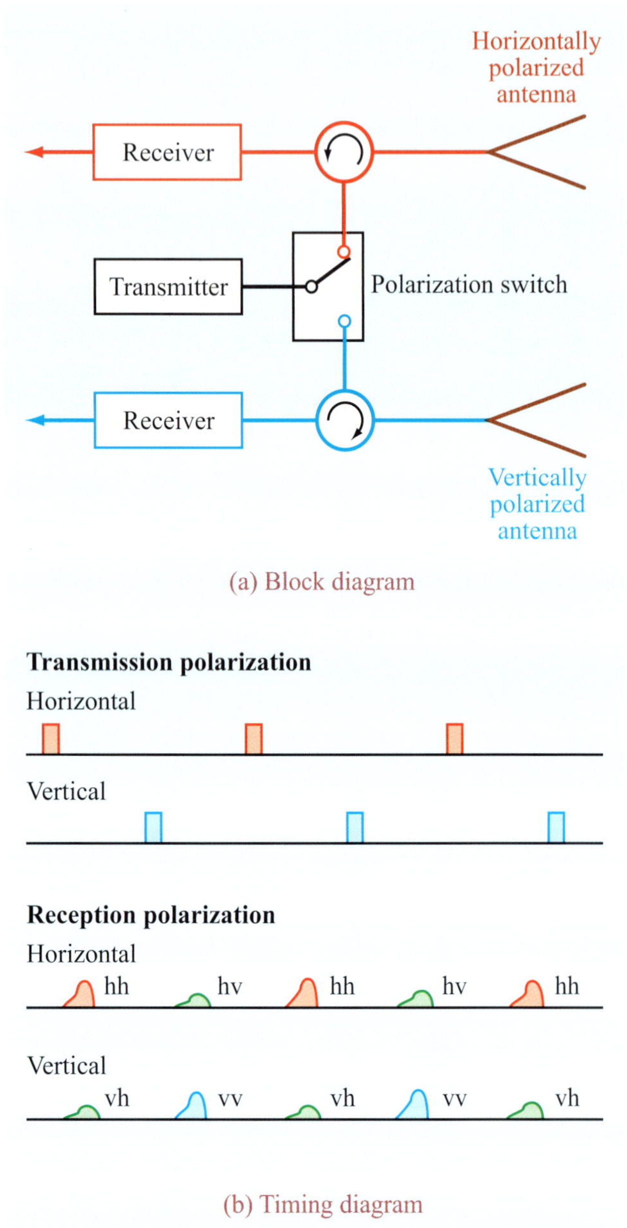

Figure 5-36: Calibration of polarimetric radar. A polarimetric radar is implemented by alternately transmitting signals out of horizontally and vertically polarized antennas and receiving at both polarizations simultaneously. Two pulses are needed to measure all the elements in the scattering matrix [van Zyl and Kim, 2011].

channels must be calibrated. Polarimetric radar calibration is the subject of Section 13-14.

For a target with scattering matrix **S**, its *pq*-polarized radar cross section (RCS) is given by

$$\sigma_{pq} = 4\pi|S_{pq}|^2 \qquad (p,q = \text{h or v}). \tag{5.130}$$

(point target)

> ▶ The scattering matrix representation is applicable not only to point targets, but to distributed targets as well. A ***polarimetric*** imaging radar with polarization agility can generate four images, one for each of the four elements of the scattering matrix. ◀

Since S_{pq} is a complex quantity, each pixel in each of the four images is represented by a magnitude and associated phase angle. The RCS of a given ground resolution cell is related to S_{pq} by Eq. (5.130), and if the distributed target consists of a large number of resolution cells N_c, its reflectivity σ^0_{pq} is obtained from

$$\sigma^0_{pq} = \frac{4\pi}{AN_c}\sum_{i=1}^{N_c}|S^i_{pq}|^2$$
$$= \frac{4\pi}{A}\left\langle|\mathbf{S}_{pq}|^2\right\rangle \qquad (p,q = \text{h or v}), \tag{5.131}$$

(distributed target)

where S^i_{pq} is the *pq*-polarized scattering amplitude of the *i*th pixel and *A* is the area of the corresponding resolution cell. Since, according to Eq. (5.22b), $S_{\text{vh}} = S_{\text{hv}}$, it follows that $\sigma^0_{\text{vh}} = \sigma^0_{\text{hv}}$, which means that a polarimetric imaging radar generates three distinct complex images, not four.

At this juncture, we stop to pose the following questions:

Q$_1$: Do hh, vv, and hv images of terrain contain the same or different information about the terrain?

A$_1$: As we will learn in a later section, the backscatter from a terrain surface or volume (such as vegetation) depends on two sets of parameters, the physical (and hence dielectric) properties of the terrain surface or volume and the sensor parameters, namely the wavelength, incidence angle, and transmit/receive polarization mode. Consequently, two terrain classes that may appear almost indistinguishable in an hh-polarized image may exhibit markedly different tones in an hv-polarized image. So the simple answer is: yes, different polarization modes carry different information and the combination of images at multiple polarizations offers distinct advantages in many applications.

Q$_2$: A polarimetric radar measures images of S_{pq}, which is a complex quantity. If the images are converted into σ^0_{pq} images, what happens to the phase of S_{pq}? Does the phase carry useful information?

A$_2$: In converting an $S_{\rm vv}$ image into a $\sigma^0_{\rm vv}$ image, only the magnitude of $S_{\rm vv}$ is used, so the phase is unnecessary. The absolute phase of $S_{\rm vv}$ is not important, but the phase of $S_{\rm vv}$ relative to the phases of $S_{\rm hh}$ and $S_{\rm hv}$ is very important. The relative phases, particularly the co-pol phase $\phi_{\rm c}$, carry information about the target.

Q$_3$: What is polarization synthesis?

A$_3$: Given the matrix **S** for a point or distributed target, it is possible to generate the response that the radar would measure for any specified combination of linear, circular, or elliptical polarizations. If a certain application calls for measuring the RCS of a target (or the reflectivity of terrain) using a v-polarized antenna for transmission and a right-hand circularly polarized antenna for reception, polarization synthesis provides the analytical tools for computing that RCS from the scattering matrix **S** without having to actually build a radar with that specific transmit/receive antenna polarization configuration.

5-11.1 RCS Polarization Response

Using the matrix notation given by Eq. (5.3), the electric field of a propagating EM wave is given by

$$\mathbf{E} = \begin{bmatrix} E_{\rm v} \\ E_{\rm h} \end{bmatrix} = \begin{bmatrix} a_{\rm v} \\ a_{\rm h} e^{j\delta} \end{bmatrix}. \tag{5.132}$$

Because the polarization state of the wave depends on the phase of $E_{\rm h}$ relative to that of $E_{\rm v}$—rather than on the absolute values of the two phases—we set $\delta_{\rm v} = 0$ and define δ as the phase difference between $E_{\rm h}$ and $E_{\rm v}$. Furthermore, the magnitudes are defined such that $a_{\rm h} \geq 0$ and $a_{\rm v} \geq 0$, and their ratio is related to the auxiliary angle α by

$$\tan\alpha = \frac{a_{\rm h}}{a_{\rm v}}. \tag{5.133}$$

Per Eq. (5.5), each pair of values of angles (α, δ) defines a polarization ellipse with polarization angles (ψ, χ) given by

$$\tan 2\psi = (\tan 2\alpha)\cos\delta \qquad (-\pi/2 \leq \psi \leq \pi/2), \tag{5.134a}$$

$$\sin 2\chi = (\sin 2\alpha)\sin\delta \qquad (-\pi/4 \leq \chi \leq \pi/4), \tag{5.134b}$$

with the additional constraint that $\psi > 0$ if $\cos\delta > 0$ and $\psi < 0$ if $\cos\delta < 0$.

Conversely, angles (α, δ) are derived from (ψ, χ) through

$$\cos 2\alpha = (\cos 2\chi)(\cos 2\psi), \tag{5.135a}$$

$$\tan\delta = \frac{\tan 2\chi}{\sin 2\psi}. \tag{5.135b}$$

A transmit antenna radiating a wave with a polarization state defined by angles $(\alpha_{\rm t}, \delta_{\rm t})$, or equivalently, polarization angles $(\psi_{\rm t}, \chi_{\rm t})$, is said to have an ***antenna polarization vector*** $\mathbf{p}^{\rm t}$ given by

$$\mathbf{p}^{\rm t} = \frac{\mathbf{E}_{\rm t}}{|\mathbf{E}_{\rm t}|} = \frac{1}{\sqrt{a^2_{\rm v_t} + a^2_{\rm h_t}}} \begin{bmatrix} a_{\rm v_t} \\ a_{\rm h_t} e^{j\delta_{\rm t}} \end{bmatrix} = \begin{bmatrix} \cos\alpha_{\rm t} \\ \sin\alpha_{\rm t}\, e^{j\delta_{\rm t}} \end{bmatrix}. \tag{5.136}$$

Similarly, in the BSA convention, for a receive antenna configured to intercept waves with polarization states defined by $(\alpha_{\rm r}, \delta_{\rm r})$, or equivalently by polarization angles $(\psi_{\rm r}, \chi_{\rm r})$, its antenna polarization vector is

$$\mathbf{p}^{\rm r} = \begin{bmatrix} \cos\alpha_{\rm r} \\ \sin\alpha_{\rm r}\, e^{j\delta_{\rm r}} \end{bmatrix}. \tag{5.137}$$

If a radar with transmit and receive antenna polarization vectors $\mathbf{p}^{\rm t}$ and $\mathbf{p}^{\rm r}$—both defined in the BSA convention—is used to observe a target with scattering matrix $\mathbf{S}$, the radar would measure a radar cross section σ_{rt} given by (Kennaugh, 1951)

$$\sigma_{\text{rt}}(\psi_{\text{r}}, \chi_{\text{r}}; \psi_{\text{t}}, \chi_{\text{t}}) = 4\pi |\mathbf{p}^{\text{r}} \cdot \mathbf{S}\, \mathbf{p}^{\text{t}}|^2. \quad (5.138)$$

(point target)

This is known as the ***polarization synthesis equation***. It allows us to compute the RCS σ of a point target, or σ/A of a distributed target with resolution cell area A, for any specified combination of transmit/receive antenna polarizations. As an example, let us consider a large metal sphere with radius $a \gg \lambda$. Its scattering matrix in the BSA convention is given by

$$\mathbf{S} = \frac{a}{2}\begin{pmatrix} 1 & 0 \\ 0 & 1 \end{pmatrix} \qquad \textbf{(metal sphere)}. \quad (5.139)$$

In view of its symmetry, the sphere does not generate cross-polarized components ($S_{\text{hv}} = S_{\text{vh}} = 0$) and it has the same scattering amplitudes for hh and vv polarizations. From among the infinite number of combinations of transmit/receive polarizations that can be synthesized using Eq. (5.138), two subsets of specific interest are:

(a) The ***co-polarized response***: which encompasses the full range of $(\psi_{\text{t}}, \chi_{\text{t}})$ such that the antenna polarizations are always the same: $\mathbf{p}^{\text{t}} = \mathbf{p}^{\text{r}}$, $\psi_{\text{t}} = \psi_{\text{r}}$, and $\chi_{\text{t}} = \chi_{\text{r}}$. The normalized plot of σ as a function of ψ_{t} and χ_{t} is called the co-pol response. The co-pol response for the metal sphere is shown in Fig. 5-37(a).

(b) The ***cross-polarized response***: receive antenna polarization is always orthogonal to the transmit antenna polarization. If the transmit antenna is v-polarized ($\mathbf{p}^{\text{t}} = \hat{\mathbf{v}}$), the receive antenna is h-polarized ($\mathbf{p}^{\text{r}} = \hat{\mathbf{h}}$); if $\mathbf{p}^{\text{t}}$ represents a RHC-polarization state, then $\mathbf{p}^{\text{r}}$ represents a LHC-polarization state, and so on. The cross-pol response of the metal sphere is displayed in Fig. 5-37(a) also.

Parts (b) through (f) of Fig. 5-37 display the co-polarized and cross-polarized responses for the following point targets:

(1) Dihedral metal reflector aligned with respect to the radar for maximum return. Its scattering matrix is given by Jasik (1961) as

$$\mathbf{S} = \frac{k_0 ab}{\pi}\begin{pmatrix} -1 & 0 \\ 0 & 1 \end{pmatrix}. \quad (5.140)$$

where $k_0 = 2\pi/\lambda$ is the wavenumber in free space.

(2) Trihedral metal reflector aligned with respect to the radar for maximum return (radar beam points towards the trihedral's vortex). Its scattering matrix is given (Ruck et al., 1970) as

$$\mathbf{S} = \frac{k_0 l^2}{\sqrt{2}\,\pi}\begin{pmatrix} 1 & 0 \\ 0 & 1 \end{pmatrix}. \quad (5.141)$$

Since the form of $\mathbf{S}$ of the trihedral reflector is the same as that of the sphere, the polarization responses of the two point targets look identical.

(3) Thin conducting cylinder of length l (with $l \ll \lambda$) and radius a (with $a \ll \lambda$), oriented at an angle μ (measured in the clockwise direction) relative to the unit vector $\hat{\mathbf{h}}_{\text{t}}$ of the transmit antenna. Its scattering matrix is given by Ruck et al. (1970) as

$$\mathbf{S} = \frac{k_0^2 l^3}{3[\ln(4l/a) - 1]}\begin{pmatrix} \sin^2\mu & -\sin\mu\cos\mu \\ -\sin\mu\cos\mu & \cos^2\mu \end{pmatrix}. \quad (5.142)$$

The responses shown in Fig. 5-37(d), (e), and (f) correspond to $\mu = 0$ (horizontal wire), $\mu = 45^\circ$, and $\mu = 90^\circ$ (vertical wire), respectively.

5-11.2 Distributed Targets

Polarization synthesis also can be applied to radar images. A terrain-imaging polarimetric radar generates *complex* images of S_{vv}, S_{hh}, and S_{hv} (or S_{vh} since $S_{\text{hv}} = S_{\text{vh}}$). For a distributed target occupying N_{c} cells each of area A, its polarization synthesis equation is given by

$$\begin{aligned} \sigma^0_{\text{rt}}(\psi_{\text{r}}, \chi_{\text{r}}; \psi_{\text{t}}, \chi_{\text{t}}) &= \frac{4\pi}{AN_{\text{c}}}\sum_{i=1}^{N_{\text{c}}} |\mathbf{p}^{\text{r}} \cdot \mathbf{S}_i \mathbf{p}^{\text{t}}|^2 \\ &= \frac{4\pi}{A}\left\langle |\mathbf{P}^{\text{r}} \cdot \mathbf{S}\mathbf{p}^{\text{t}}|^2 \right\rangle, ^{\dagger} \end{aligned} \quad (5.143)$$

(distributed target)

† Computer Code 5.1.

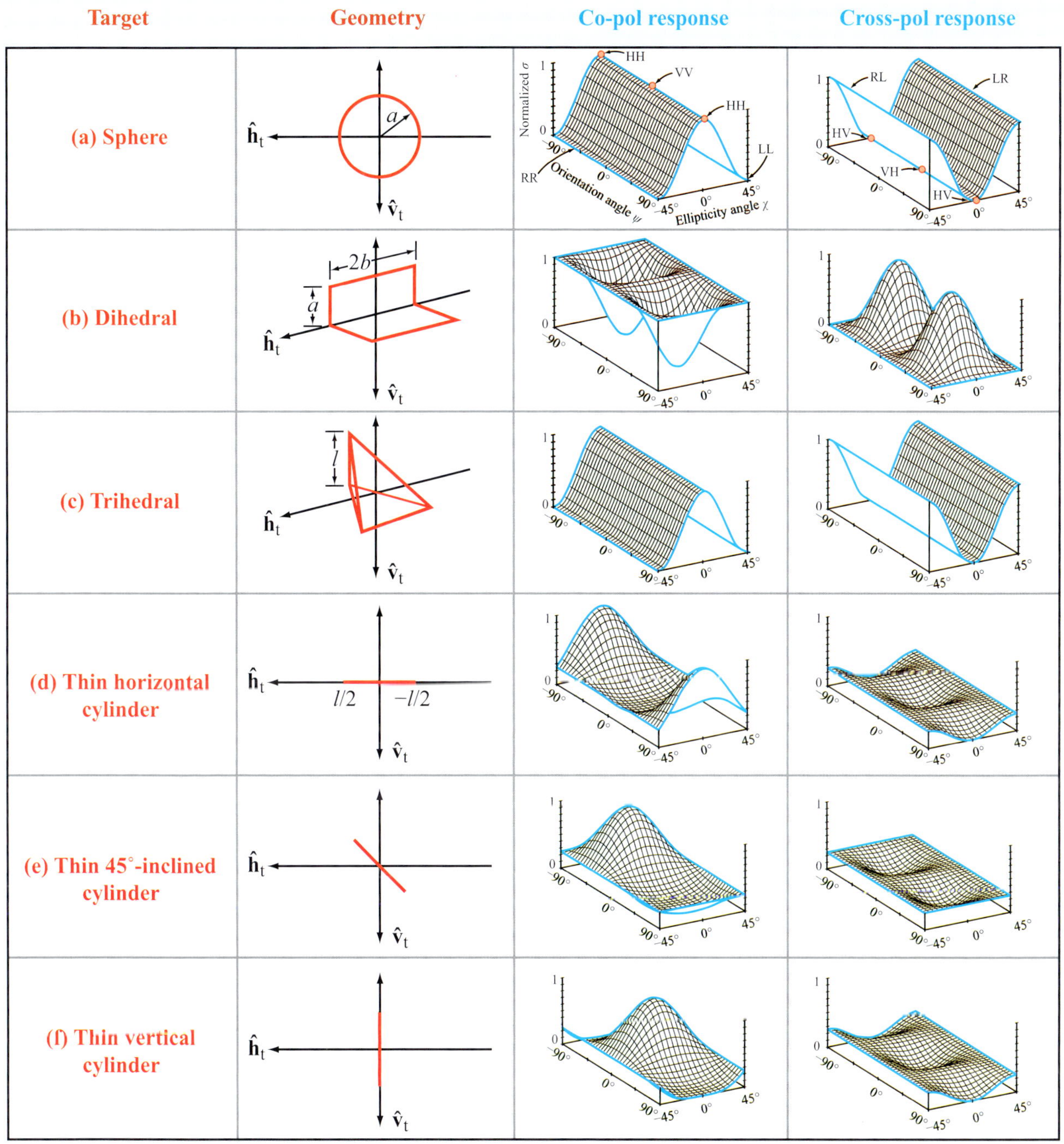

Figure 5-37: Co-pol and cross-pol responses of 6 point targets. LR stands for a left-hand (receive)/right-hand (transmit) polarization configuration.

where $\mathbf{S}_i$ is the scattering matrix of the ith cell. Of course, N_c should be a large number in order for the computation to yield an accurate estimate of the true value of σ^0_{rt}.

5-11.3 Mueller Matrix Approach

Instead of synthesizing the measured backscatter N_c times and then averaging the sum to obtain σ^0, as outlined by Eq. (5.143), it is possible to use a computationally more efficient approach based on the modified Mueller matrix representation of the polarized scattering by a target, instead of the scattering matrix $\mathbf{S}$. To show the procedure, we start by reintroducing the ***modified Stokes vector*** given by Eq. (5.23a) and then defining its elements in terms of the polarization angles given by Eqs. (5.5) and (5.6). For the transmit Stokes vector, the process leads to (Ulaby and van Zyl, Ch. 1, in Ulaby and Elachi, 1990):

$$\mathbf{I}^t(\psi,\chi) = \begin{bmatrix} |E^t_v|^2 \\ |E^t_h|^2 \\ 2\Re\mathfrak{e}(E^t_v E^{t*}_h) \\ 2\Im\mathfrak{m}(E^t_v E^{t*}_h) \end{bmatrix} \Bigg/ \eta = \begin{bmatrix} a^{t\,2}_v \\ a^{t\,2}_h \\ 2a^t_v a^t_h \cos\delta_t \\ 2a^t_v a^t_h \sin\delta_t \end{bmatrix} \Bigg/ \eta = \begin{bmatrix} \frac{1}{2}(1+\cos 2\psi_t \cos 2\chi_t) \\ \frac{1}{2}(1-\cos 2\psi_t \cos 2\chi_t) \\ \sin 2\psi_t \cos 2\chi_t \\ \sin 2\chi_t \end{bmatrix} I^t_0, \tag{5.144}$$

where the ***total intensity*** $I^t_0 = (a^{t\,2}_v + a^{t\,2}_h)/\eta$. The four elements of the vector $\mathbf{I}^t$ are defined in terms of the same quantities that define $\mathbf{E}$ in Eq. (5.132), namely a_v, a_h, and δ.

▶ Whereas the elements of $\mathbf{E}$ are electric field components and one of them may be either real or complex, the elements of $\mathbf{I}^t$ represent power and all are real quantities. ◀

Similarly, the modified Stokes vector of the received wave is related to polarization angles (ψ_r, χ_r) by

$$\mathbf{I}^r = \begin{bmatrix} \frac{1}{2}(1+\cos 2\psi_r \cos 2\chi_r) \\ \frac{1}{2}(1-\cos 2\psi_r \cos 2\chi_r) \\ \sin 2\psi_r \cos 2\chi_r \\ \sin 2\chi_r \end{bmatrix} I^r_0. \tag{5.145}$$

The modified Stokes vectors of the transmitted and received waves are related through Eq. (5.25a), namely

$$\mathbf{I}^r = \frac{1}{R^2_r} \mathbf{M}\,\mathbf{I}^t, \tag{5.146}$$

where R_r is the range between the target and the radar antenna, and $\mathbf{M}$ is the modified Mueller matrix of the target, defined in the BSA coordinate system. From Eq. (5.25c),

$$\mathbf{M} = \begin{bmatrix} |S_{vv}|^2 & |S_{vh}|^2 & \Re\mathfrak{e}(S_{vv}S^*_{vh}) & -\Im\mathfrak{m}(S_{vv}S^*_{vh}) \\ |S_{hv}|^2 & |S_{hh}|^2 & \Re\mathfrak{e}(S_{hv}S^*_{hh}) & -\Im\mathfrak{m}(S_{hv}S^*_{hh}) \\ 2\Re\mathfrak{e}(S_{vv}S^*_{hv}) & 2\Re\mathfrak{e}(S_{vh}S^*_{hh}) & \Re\mathfrak{e}(S_{vv}S^*_{hh} + S_{vh}S^*_{hv}) & -\Im\mathfrak{m}(S_{vv}S^*_{hh} - S_{vh}S^*_{hv}) \\ 2\Im\mathfrak{m}(S_{vv}S^*_{hv}) & 2\Im\mathfrak{m}(S_{vh}S^*_{hh}) & \Im\mathfrak{m}(S_{vv}S^*_{hh} + S_{vh}S^*_{hv}) & \Re\mathfrak{e}(S_{vv}S^*_{hh} - S_{vh}S^*_{hv}) \end{bmatrix}. \tag{5.147}$$

The 16 elements of $\mathbf{M}$ are real quantities related to various combinations of the elements of $\mathbf{S}$.

For a point target, the following polarization synthesis equation is applicable (Kennaugh, 1951; van Zyl et al., 1987a):

$$\sigma_{rt}(\psi_r,\chi_r;\,\psi_t,\chi_t) = 4\pi \mathbf{I}^r_n \cdot \mathbf{Q}\,\mathbf{M}\,\mathbf{I}^t_n, \tag{5.148}$$

(point target)

where $\mathbf{I}^r_n = \mathbf{I}^r/I^r_0$ and $\mathbf{I}^t_n = \mathbf{I}^t/I^t_0$, and

$$\mathbf{Q} = \begin{bmatrix} 1 & 0 & 0 & 0 \\ 0 & 1 & 0 & 0 \\ 0 & 0 & 1/2 & 0 \\ 0 & 0 & 0 & -1/2 \end{bmatrix}. \tag{5.149}$$

In the case of a distributed target for which N_c measurements of $\mathbf{M}$ are available, the reflectivity σ^0 can be synthesized for any transmit and receive antenna polarizations from[†]

$$\sigma^0_{\mathrm{rt}}(\psi_{\mathrm{r}},\chi_{\mathrm{r}};\psi_{\mathrm{t}},\chi_{\mathrm{t}}) = \frac{4\pi}{A}\,\mathbf{I}^{\mathrm{r}}_{\mathrm{n}}\cdot\mathbf{Q}\,\langle\mathbf{M}\rangle\,\mathbf{I}^{\mathrm{t}}_{\mathrm{n}}, \tag{5.150}$$

(distributed target)

where

$$\langle\mathbf{M}\rangle = \frac{1}{N_{\mathrm{c}}}\sum_{i=1}^{N}\mathbf{M}_i. \tag{5.151}$$

Because the averaging over the index i is performed prior to applying Eq. (5.150), this approach has been found to be computationally superior to the direct synthesis approach presented in Section 5-11.2.

Figure 5-38 shows a number of synthesized images for the San Francisco Bay area in California. The data were acquired with the NASA/JPL AIRSAR system at L-band. The radar look direction for these images is from left to right. The Golden Gate Bridge is the linear feature at the top middle of the image. The Golden Gate Park is the large rectangle about 1/3 from the bottom of the image. Note the strong pixel-to-pixel variation in the relative return from the urban areas in the image (top left). These are related to the orientation of the buildings and the streets relative to the radar look direction (Zebker et al., 1987). When the streets are oriented nearly orthogonal to the radar look direction, the dominant scattering mechanism is that of a dihedral corner reflector formed by a street as the horizontal surface, and the front face of a building as the vertical surface. Double reflections tend to result in strong signals as evidenced by the bright areas in the city. Also note that the contrast between the urban areas and the vegetated areas, such as Golden Gate Park, is maximized using the 45-degree linear polarization for transmit, and 135-degree linear polarization for receive, as shown by the image on the right in the middle row. This is the polarization that generally maximizes the contrast between areas that exhibit double-reflection scattering and those that exhibit either odd numbers of reflections, or diffuse scattering (see Chapter 7 of Ulaby and Elachi (1990) for a more detailed discussion).

In a second example, we show in Fig. 5-39 a similar set of synthesized images for a portion of the Florida coastline, acquired at L-band with the NASA/JPL UAVSAR system. The large body of water is part of Choctawhatchee Bay, and the urban areas to the left of the bay in the image are the cities of Valparaiso and Niceville. The bright line near the top right of the image is where the Mid Bay Bridge Road crosses Choctawhatchee Bay. This set of images shows that for natural vegetation, there is generally less variation in radar backscatter with changing polarization than with urban areas. The exception is the case where there is standing water under vegetation, where, at long wavelengths, double reflections from the water surface to the tree trunks and back to the radar tend to dominate the scattered signals. In this image, the bright area near the top left of the hh image (top left image in the series) is such an area. Note how there is more contrast between the vegetation with standing water and the surrounding vegetation in the hh image than in the vv image (center image top row). Double reflections from dielectric surfaces tend to be stronger at hh than at vv, because the Fresnel reflection coefficient is stronger for horizontal polarization than for vertical polarization. Natural vegetation with nearly uniform orientation shows little difference between hh and vv scattering. In fact, if the vegetation is uniformly randomly oriented, hh and vv scattering would be equal. The HV image (rightmost image in top row) does not show much contrast between the areas with standing water under the vegetation and the surrounding vegetation. This indicates that the cross-polarized return is mostly due to the randomly oriented parts of the vegetation canopy. Note, however, how the contrast between the vegetated and nonvegetated areas is the best in the hv image when comparing across the top row. This is expected because the cross-polarized return is mostly due to the randomly oriented parts of the vegetation canopy. Also note that the hv image shows subtle differences in the vegetated areas on the left edge of the image, where areas that were previously cleared and have since experienced regrowth are visible

[†]Computer Code 5.1.

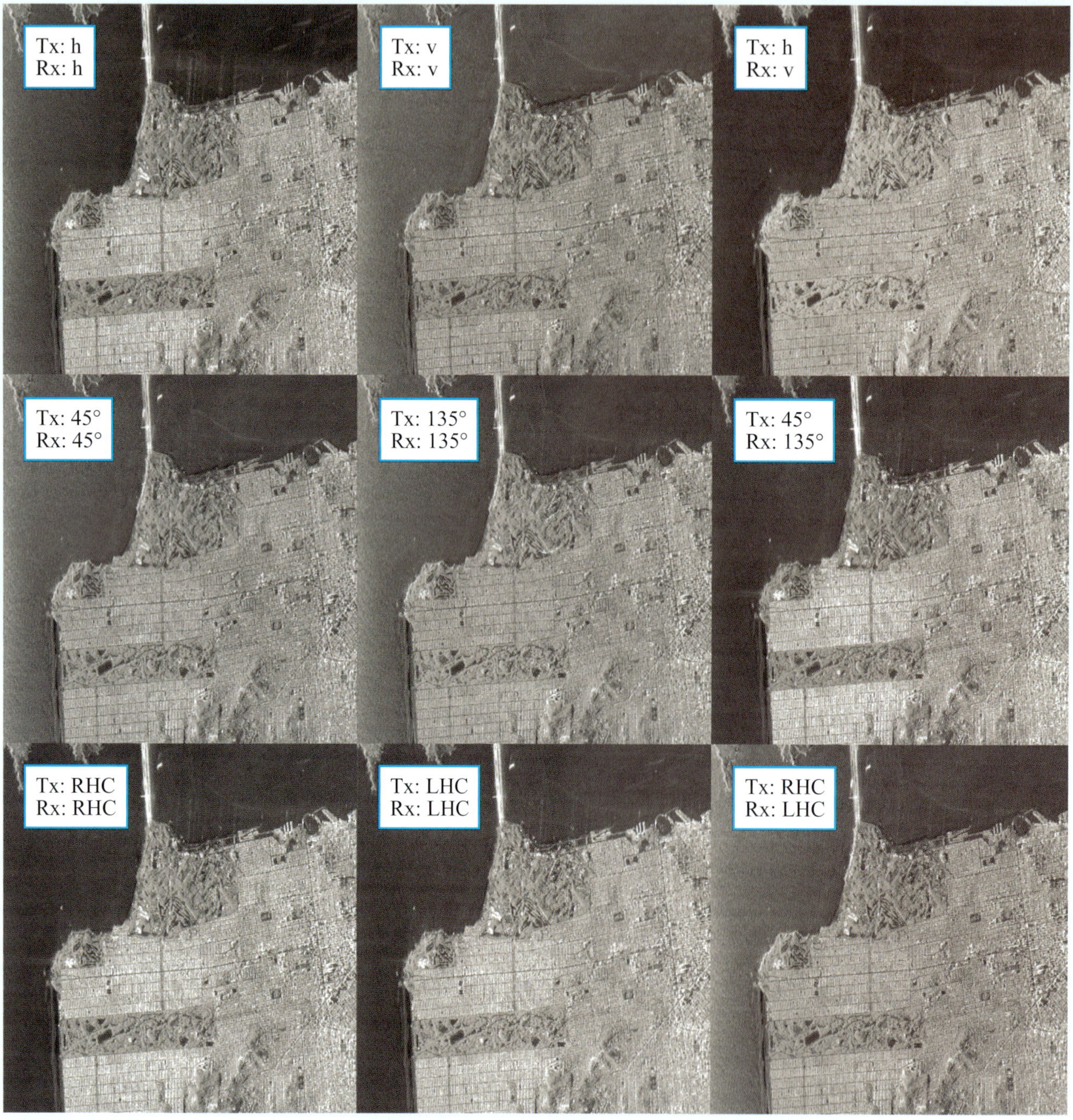

Figure 5-38: This series of L-band images of San Francisco were synthesized from a single polarimetric image acquired by the NASA/JPL AIRSAR system. The nine images include various combinations of co-polarized (transmit and receive polarizations are the same) and cross-polarized (transmit and receive polarizations are orthogonal) images. Note the relative change in brightness between the city of San Francisco, the ocean, and the Golden Gate Park, which is the large rectangle about 1/3 from the bottom of the images. The radar illumination is from the left.

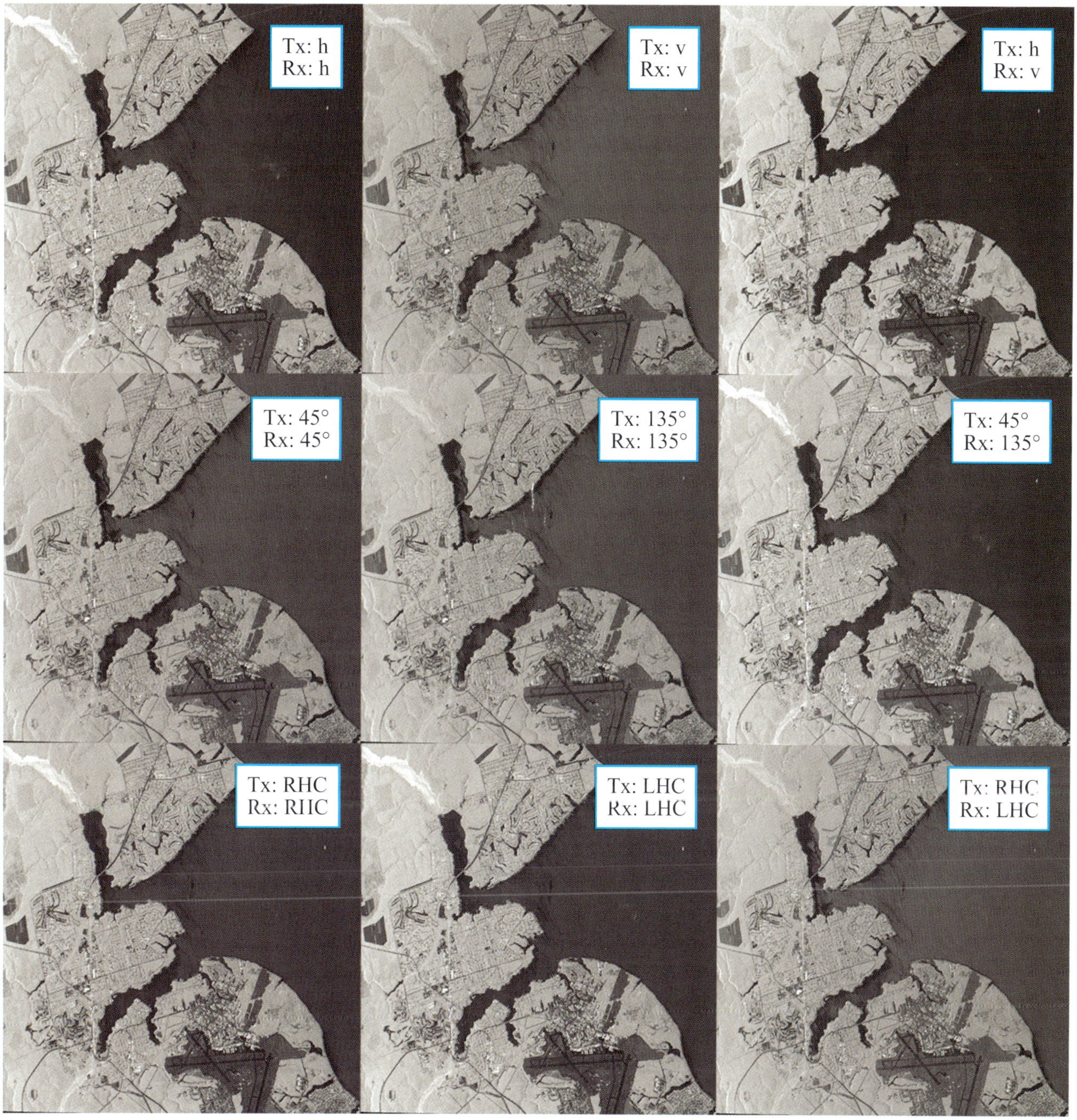

Figure 5-39: This series of L-band images of the Florida Coast were synthesized from a single polarimetric image acquired by the NASA/JPL UAVSAR system. Radar illumination is from the left. The large body of water is part of Choctawhatchee Bay. See the text for additional description of the area.

near the top on the left of the hv image. As in the case of the San Francisco image, the best contrast between the areas with standing water under the canopy and the surrounding vegetation is found when using the 45/135-degree cross-polarized linear polarizations, as can be seen in the center image of the right-hand column. Finally, the co-polarized circularly polarized images combine the characteristics of the hv image in that they show reasonable contrast between vegetated and nonvegetated areas, and like the 45/135-degree linear cross-polarized image, show good contrast between the areas with standing water under the vegetation and the surrounding areas.

5-12 Polarimetric Scattering Statistics

Consider a textureless distributed target (such as a random soil surface) imaged by a polarimetric radar. The target extent encompasses a large number of independent resolution cells N_c. Each cell is represented by a 1-look scattering matrix **S** from which we can generate hh-, vv-, and hv-polarized images by applying Eq. (5.131). As noted previously, $S_{hv} = S_{vh}$ in the backscatter mode.

If the radar uses linear detection, the tone of the hh-polarized amplitude image is proportional to $|S_{hh}|$ and the pdf of $|S_{hh}|$ (for the N_c image pixels) is given by the Rayleigh distribution of Eq. (5.75) with $f = |S_{hh}|/\langle|S_{hh}|\rangle$. The same statistics apply to the vv-polarized and hv-polarized images. Similarly, square-law detection generates an hh image whose tone is proportional to $|S_{hh}|^2$ and whose associated pdf is the exponential function given by Eq. (5.80a) with $F = |S_{hh}|^2/\langle|S_{hh}|^2\rangle$. *Our goal in this section is to examine the phase statistics of* **S**.

In the backscatter mode, the scattering matrix of an image pixel,

$$\mathbf{S} = \begin{pmatrix} S_{vv} & S_{vh} \\ S_{hv} & S_{hh} \end{pmatrix} \tag{5.152}$$

consists of

$$S_{vv} = |S_{vv}|e^{j\phi_{vv}} = X_{vv} + jY_{vv}, \tag{5.153a}$$

$$S_{hh} = |S_{hh}|e^{j\phi_{hh}} = X_{hh} + jY_{hh}, \tag{5.153b}$$

$$S_{hv} = |S_{hv}|e^{j\phi_{hv}} = X_{hv} + jY_{hv}. \tag{5.153c}$$

In total, **S** consists of six real quantities, namely the three magnitudes ($|S_{vv}|$, $|S_{hh}|$, and $|S_{hh}|$) and three phase angles (ϕ_{vv}, ϕ_{hh}, ϕ_{hv}), or equivalently, of the three real parts (X_{vv}, X_{hh}, and X_{hv}) and three imaginary parts (Y_{vv}, Y_{hh}, and Y_{hv}) of the scattering amplitudes. In Section 5-7.1, we stated that for a textureless distributed target, the real and imaginary parts of the backscattered electric field—E_x and E_y defined by Eq. (5.56)—are both Gaussian-distributed. Within a calibration constant, $E_x = X_{vv}$ and $E_y = Y_{vv}$ (for vv polarization), so X_{vv} and Y_{vv} should be Gaussian-distributed also. The plots in Fig. 5-40 are pdfs of X_{vv} and Y_{vv} generated from 1000 radar measurements acquired by a track-mounted polarimetric scatterometer at 4.75 GHz. The target is a random soil surface. The two pdfs are indeed zero-mean Gaussian, and they have the same standard deviation (within measurement error). Similar results were observed for the hh- and hv-polarized data.

Individually, the phase angles ϕ_{vv}, ϕ_{hh}, and ϕ_{hv} are uniformly distributed over $[-\pi, \pi]$, and therefore they contain no information about the geometrical and dielectric properties of the distributed targets. However, their differences

$$\phi_c = \phi_{hh} - \phi_{vv} \qquad \textbf{(co-pol phase)}, \tag{5.154a}$$

and

$$\phi_x = \phi_{hv} - \phi_{vv} \qquad \textbf{(cross-pol phase)}, \tag{5.154b}$$

may or may not be uniformly distributed and may in fact be related to the target's properties. For most distributed targets, S_{vv} and S_{hh} are dominated by ***first-order scattering*** (direct backscatter with no multiple reflections), whereas S_{hv} and S_{vh} are due to second- and higher-order scattering (two or more reflections involving two or more scatterers). Consequently, S_{hv} and S_{vv} are uncorrelated random variables, and the phase difference ϕ_x is indeed uniformly distributed, as shown in Fig. 5-41(a). That is,

$$\langle S_{hv}S^*_{hh}\rangle = \langle S_{hv}S^*_{vv}\rangle = 0 \tag{5.154c}$$

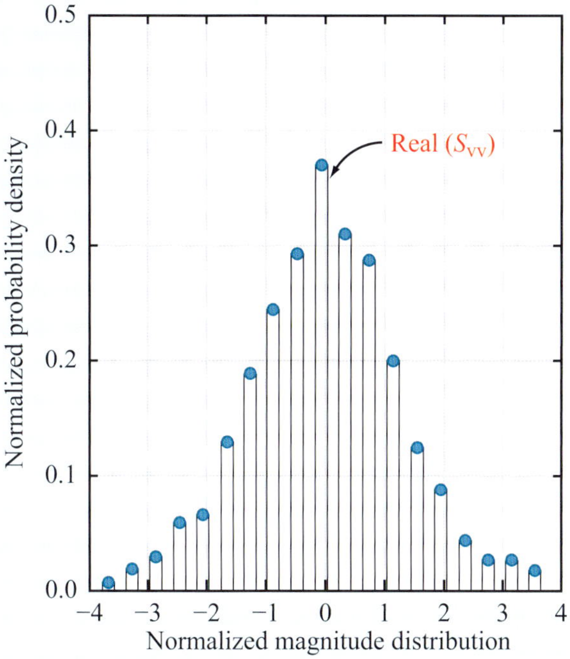

(a) $\text{Re}(S_{vv}) = X_{vv}$

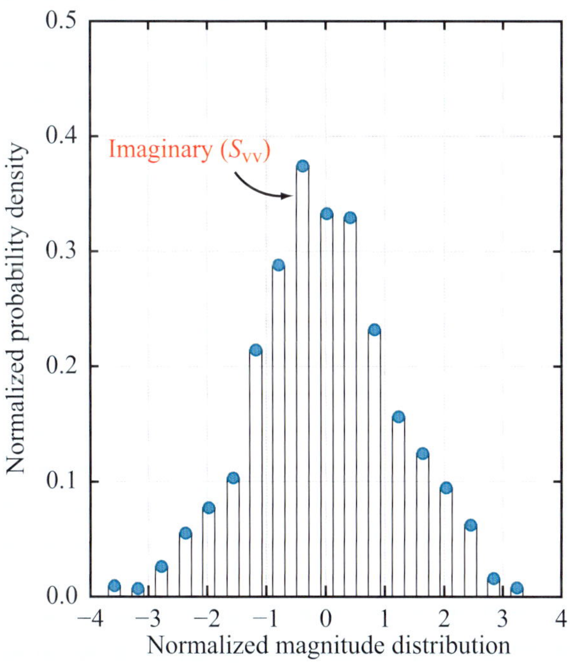

(b) $\text{Im}(S_{vv}) = Y_{vv}$

Figure 5-40: Measured histograms of the real and imaginary parts of S_{vv} for a soil surface observed by a 4.75 GHz radar scatterometer [Sarabandi, 1992].

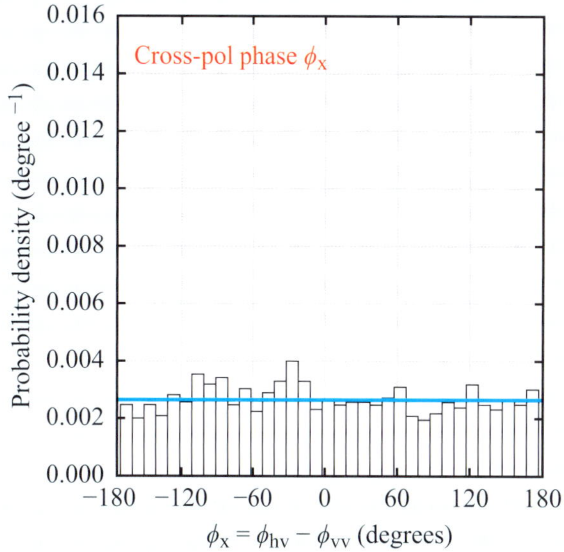

(a) Cross-pol phase

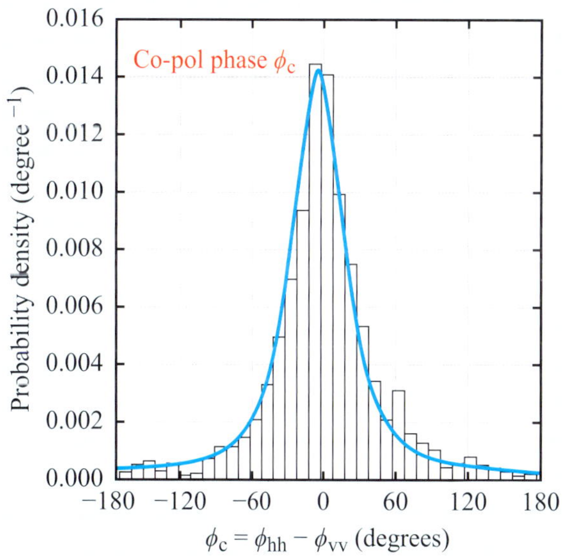

(b) Co-pol phase

Figure 5-41: Measured histograms of the cross-pol phase difference ϕ_x and the co-pol phase difference ϕ_c of a slightly rough soil surface [Sarabandi, 1992].

and

$$p(\phi_x) = \frac{1}{2\pi}, \qquad -\pi \le \phi \le \pi. \tag{5.154d}$$

In contrast, $p(\phi_c)$, the pdf of the co-pol phase difference ϕ_c, is approximately Gaussian, as shown in Fig. 5-41(b). Furthermore, as we show shortly, the mean value and variance of ϕ_c are directly related to the elements of the average Mueller matrix of the distributed target, $\langle \mathbf{M} \rangle$.

Based on the assumption that the co-polarized and cross-polarized scattering amplitudes are statistically uncorrelated, it follows that 8 of the 16 elements of the Mueller matrix given by Eq. (5.147) have zero average values. Hence,

$$\langle \mathbf{M} \rangle = \begin{bmatrix} M_{11} & M_{12} & 0 & 0 \\ M_{21} & M_{22} & 0 & 0 \\ 0 & 0 & M_{33} & M_{34} \\ 0 & 0 & M_{43} & M_{44} \end{bmatrix}, \tag{5.155}$$

with

$$M_{11} = \langle |S_{vv}|^2 \rangle, \qquad M_{22} = \langle |S_{hh}|^2 \rangle, \tag{5.156a}$$

$$M_{12} = M_{21} = \langle |S_{hv}|^2 \rangle, \tag{5.156b}$$

$$M_{33} = \langle |S_{vv}||S_{hh}| \cos\phi_c \rangle + M_{12}, \tag{5.156c}$$

$$M_{43} = -M_{34} = \langle |S_{vv}||S_{hh}| \sin\phi_c \rangle, \tag{5.156d}$$

$$M_{44} = \langle |S_{vv}||S_{hh}| \cos\phi_c \rangle - M_{12}. \tag{5.156e}$$

Using Rayleigh fading statistics, it can be shown that the pdf of the co-pol phase difference ϕ_c is given by (Sarabandi, 1992; Ulaby et al., 1992):

$$p(\phi_c) = \frac{a}{2\pi b} \left\{ 1 + c \left[\frac{\pi}{2} + \tan^{-1} c \right] \right\}, \tag{5.157}$$

where

$$a = 1 - \beta^2 \tag{5.158a}$$

$$b = 1 - \beta^2 \cos^2(\phi_c - \phi_0), \tag{5.158b}$$

$$c = \frac{\beta}{\sqrt{b}} \cos(\phi_c - \phi_0), \tag{5.158c}$$

$$\beta = \frac{1}{2} \left[\frac{(M_{33} + M_{44})^2 + (M_{34} - M_{43})^2}{M_{11} M_{22}} \right]^{1/2}, \quad 0 \le \beta \le 1, \tag{5.158d}$$

$$\phi_0 = \tan^{-1} \left(\frac{M_{34} - M_{43}}{M_{33} + M_{44}} \right), \qquad -\pi \le \phi_0 \le \pi. \tag{5.158e}$$

The parameter ϕ_0 is the value of ϕ_c at which $p(\phi_c)$ is a maximum, so it is called the ***highest likelihood co-pol phase difference***, and the parameter β is called the ***degree of co-pol correlation*** and it controls the width of the pdf. Figure 5-42 displays plots of $p(\phi_c)$ for various combinations of ϕ_0 and β, and Fig. 5-43 compares a histogram of measured values with the pdf based on Eq. (5.157) for a grassy area observed by a 1.2 GHz polarimetric SAR in April 1990.

Whereas a bare soil surface or a grassy field usually exhibits a pdf $p(\phi_c)$ with a mean value $\langle \phi_0 \rangle \approx 0$, $\langle \phi_0 \rangle$ may be much larger for vegetation canopies composed of plants with vertical stalks or trees with vertical trunks, particularly when imaged at large incidence angles. In Fig. 5-44, plots of $p(\phi_c)$ are shown for two fields of mature corn, one observed by a 1.2 GHz polarimetric SAR at an incidence angle of 19° and the other at 50°. Their pdfs have significantly different peak values, widths, and mean values.

In summary, a lot of evidence exists to suggest that the co-pol phase difference ϕ_c contains information about the physical and dielectric properties of a distributed target, but insufficient progress has been made to date to utilize such information in support of specific applications.

5-13 Polarimetric Analysis Tools

Even though the foundational principles of radar polarimetry were established well before 1980 (Kennaugh, 1951; Huynen, 1970; Boerner, 1981; Kostinski and Boerner, 1986), the implementation of the polarimetric radar concept in an imaging mode was not realized until 1985. In that year, the Jet Propulsion Laboratory successfully designed, constructed, and demonstrated the capabilities of the first polarimetric SAR to fly on an aircraft (van Zyl, 1985). Since then, the field of

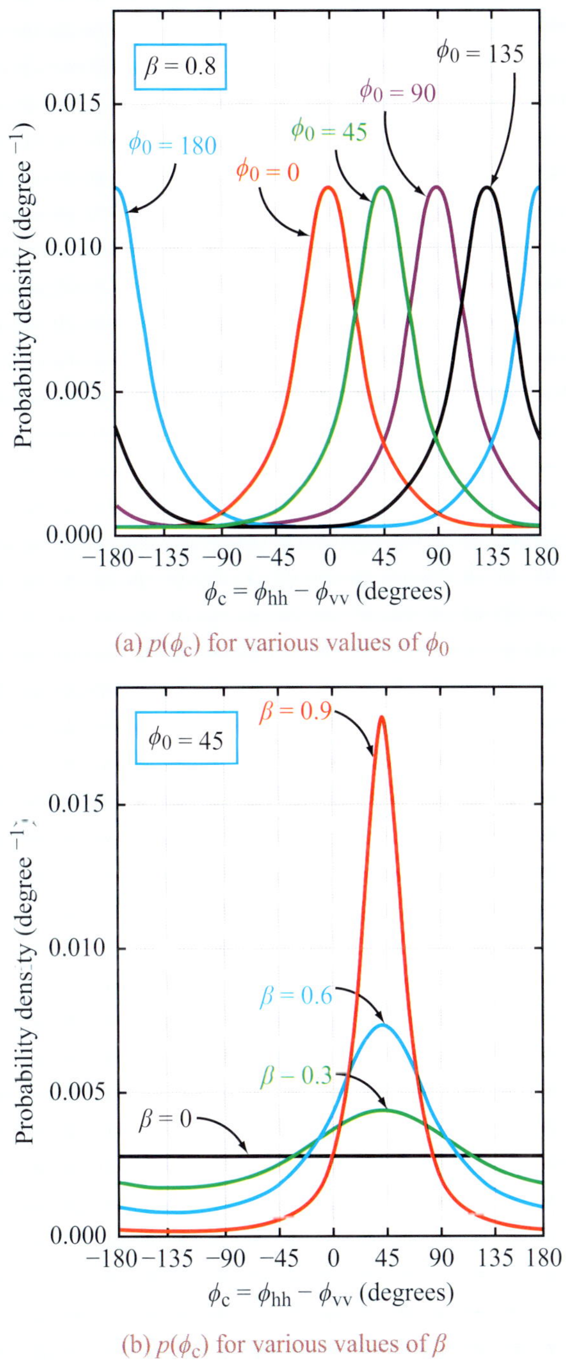

Figure 5-42: Plots of the pdf $p(\phi_c)$ of the co-pol phase difference ϕ_c for various combinations of ϕ_0 and β.

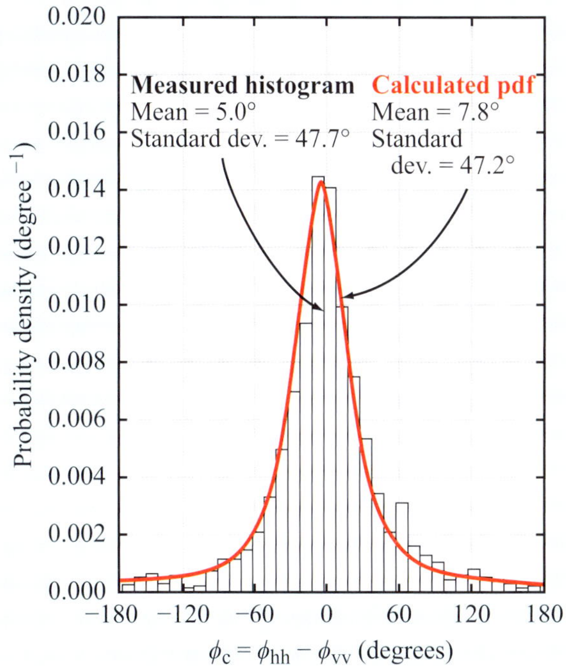

Figure 5-43: Comparison of the measured histogram of ϕ_c extracted from a 1.2 GHz SAR image with $p(\phi_c)$ given by Eq. (5.157).

radar polarimetry has expanded exponentially; several polarimetric SAR systems have been flown in space, numerous articles and books have been published on the subject (Ulaby and Elachi, 1990; Mott, 2007; Lee and Pottier, 2009; van Zyl and Kim, 2011), and numerous techniques and analysis tools have been developed to enhance target detection (Ioannidis and Hammers, 1979; Swartz et al., 1988) or extract information about an imaged scene from its polarimetric radar observations. In this section, we highlight some of these analysis tools and illustrate them with examples.

5-13.1 Scattering Covariance Matrix

Each pixel in a polarimetric radar image of a given scene is represented by a scattering matrix of the form given by

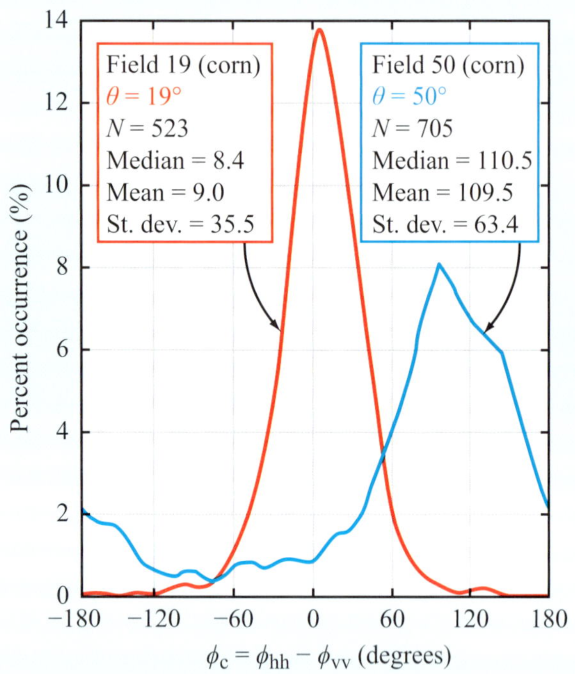

Figure 5-44: Measured histograms of ϕ_c for two cornfields, extracted from a 1.2 GHz SAR image.

Eq. (5.152), namely

$$\mathbf{S} = \begin{pmatrix} S_{vv} & S_{vh} \\ S_{hv} & S_{hh} \end{pmatrix}. \tag{5.159}$$

In Section 5-11, we demonstrated that it is possible to synthesize an image for any specified combination of transmit and receive polarizations. The polarization synthesis capability allows us to generate images that better discriminate between certain terrain types or better accentuate the appearance of certain terrain features such as boundaries. However, given that the number of possible polarization combinations is limitless, the process of identifying the *best* combination for a given application and a given imaged scene becomes a laborious search process. Fortunately, it is possible to mathematically solve for the optimum polarization combination that maximizes the backscatter from a certain type of target, or for the combination that maximizes the contrast between two types of targets. The details are available in the book by van Zyl and Kim (2011).

In the present section, we examine certain polarimetric attributes that have been found to provide useful information about polarimetrically imaged terrain. These attributes include:

(a) The ***target entropy*** (Cloude, 1992a; Cloude and Pottier, 1995), which describes the degree of randomness exhibited by the polarized return among the different pixels of a target type (such as a bare surface, a forest, or an urban area).

(b) The ***polarimetric pedestal*** (Durden et al., 1990), which is a measure of the degree of depolarization generated by a target.

(c) The ***radar vegetation index*** (RVI) introduced by Kim and van Zyl (2001) as a parameter with which to characterize the sizes of scattering cylinders (such as branches and needles in a vegetation canopy) relative to the wavelength λ.

Computing these or other polarimetric attributes is a multistep process that starts by computing the ***average covariance matrix*** $\langle \mathbf{C} \rangle$. For an image segment composed of N_c pixels of a distributed scene, with scattering matrices $\mathbf{S}_i$ where $i = 1, 2, \ldots, N_c$, the scattering coefficient is given by Eq. (5.143) as

$$\begin{aligned} \sigma^0_{rt}(\psi_r, \chi_r; \psi_t, \chi_t) &= \frac{4\pi}{AN_c} \sum_{i=1}^{N_c} |\mathbf{p}^r \cdot \mathbf{S}_i \mathbf{p}^t|^2 \\ &= \frac{4\pi}{A} \left\langle |\mathbf{P}^r \cdot \mathbf{S}\mathbf{p}^t|^2 \right\rangle, \end{aligned} \tag{5.160}$$

where $\mathbf{p}^t$ and $\mathbf{p}^r$ are the antenna polarization vectors defined by Eqs. (5.136) and (5.137):

$$\mathbf{p}^t = \begin{bmatrix} p^t_v \\ p^t_h \end{bmatrix} = \begin{bmatrix} \cos\alpha_t \\ \sin\alpha_t \, e^{j\delta_t} \end{bmatrix}, \tag{5.161a}$$

$$\mathbf{p}^r = \begin{bmatrix} p^r_v \\ p^r_h \end{bmatrix} = \begin{bmatrix} \cos\alpha_r \\ \sin\alpha_r \, e^{j\delta_r} \end{bmatrix}. \tag{5.161b}$$

The polarization states of the transmit and receive antennas are specified by the values of α_t, α_r, δ_t, δ_r, as described in Section 5-11.1. These four parameters specify, in turn, the polarization angles ψ_r, χ_r, ψ_t, and χ_t.

Covariance matrix in bistatic mode

By expanding the expression inside the magnitude sign in Eq. (5.160), it can be shown that

$$\mathbf{p}^{\mathrm{r}} \cdot \mathbf{S}_i \mathbf{p}^{\mathrm{t}} = \begin{bmatrix} p_{\mathrm{v}}^{\mathrm{r}} p_{\mathrm{v}}^{\mathrm{t}} \\ p_{\mathrm{v}}^{\mathrm{r}} p_{\mathrm{h}}^{\mathrm{t}} \\ p_{\mathrm{h}}^{\mathrm{r}} p_{\mathrm{v}}^{\mathrm{t}} \\ p_{\mathrm{h}}^{\mathrm{r}} p_{\mathrm{h}}^{\mathrm{t}} \end{bmatrix} \cdot \begin{bmatrix} S_{\mathrm{vv}} \\ S_{\mathrm{vh}} \\ S_{\mathrm{hv}} \\ S_{\mathrm{hh}} \end{bmatrix} = \mathbf{A} \cdot \mathbf{T}, \tag{5.162}$$

where

$$\mathbf{A} = \begin{bmatrix} p_{\mathrm{v}}^{\mathrm{r}} p_{\mathrm{v}}^{\mathrm{t}} \\ p_{\mathrm{v}}^{\mathrm{r}} p_{\mathrm{h}}^{\mathrm{t}} \\ p_{\mathrm{h}}^{\mathrm{r}} p_{\mathrm{v}}^{\mathrm{t}} \\ p_{\mathrm{h}}^{\mathrm{r}} p_{\mathrm{h}}^{\mathrm{t}} \end{bmatrix} \tag{5.163a}$$

and

$$\mathbf{T} = \begin{bmatrix} S_{\mathrm{vv}} \\ S_{\mathrm{vh}} \\ S_{\mathrm{hv}} \\ S_{\mathrm{hh}} \end{bmatrix}. \tag{5.163b}$$

It then follows that

$$\begin{aligned} |\mathbf{p}^{\mathrm{r}} \cdot \mathbf{S}_i \mathbf{p}^{\mathrm{t}}|^2 &= (\mathbf{A} \cdot \mathbf{T})(\mathbf{T} \cdot \mathbf{A}) = \mathbf{A} \cdot \mathbf{T}\mathbf{T}^{\dagger}\mathbf{A}^* \\ &= \mathbf{A} \cdot \mathbf{C}\mathbf{A}^*, \end{aligned} \tag{5.164}$$

where the symbol $\dagger$ denotes transpose complex conjugate, and the matrix

$$\mathbf{C} = \mathbf{T}\mathbf{T}^{\dagger} = \begin{bmatrix} S_{\mathrm{vv}}S_{\mathrm{vv}}^* & S_{\mathrm{vv}}S_{\mathrm{vh}}^* & S_{\mathrm{vv}}S_{\mathrm{hv}}^* & S_{\mathrm{vv}}S_{\mathrm{hh}}^* \\ S_{\mathrm{vh}}S_{\mathrm{vv}}^* & S_{\mathrm{vh}}S_{\mathrm{vh}}^* & S_{\mathrm{vh}}S_{\mathrm{hv}}^* & S_{\mathrm{vh}}S_{\mathrm{hh}}^* \\ S_{\mathrm{hv}}S_{\mathrm{vv}}^* & S_{\mathrm{hv}}S_{\mathrm{vh}}^* & S_{\mathrm{hv}}S_{\mathrm{hv}}^* & S_{\mathrm{hv}}S_{\mathrm{hh}}^* \\ S_{\mathrm{hh}}S_{\mathrm{vv}}^* & S_{\mathrm{hh}}S_{\mathrm{vh}}^* & S_{\mathrm{hh}}S_{\mathrm{hv}}^* & S_{\mathrm{hh}}S_{\mathrm{hh}}^* \end{bmatrix} \tag{5.165}$$

is called the ***covariance matrix*** of the scattering object or surface.

Covariance matrix in backscatter mode

In the general bistatic-scattering case, S_{vh} may be different from S_{hv}, but in the monostatic backscatter case, the reciprocity theorem dictates that $S_{\mathrm{vh}} = S_{\mathrm{hv}}$ [see Eq. (5.22b)]. Accordingly, vectors **A** and **T** can be reduced to 3×1 and the matrix **C** can be reduced to 3×3. These new versions are given by

$$\mathbf{A} = \begin{bmatrix} p_{\mathrm{v}}^{\mathrm{r}} p_{\mathrm{v}}^{\mathrm{t}} \\ \frac{1}{\sqrt{2}} (p_{\mathrm{v}}^{\mathrm{r}} p_{\mathrm{h}}^{\mathrm{t}} + p_{\mathrm{h}}^{\mathrm{r}} p_{\mathrm{v}}^{\mathrm{t}}) \\ p_{\mathrm{h}}^{\mathrm{r}} p_{\mathrm{h}}^{\mathrm{t}} \end{bmatrix}, \tag{5.166a}$$

$$\mathbf{T} = \begin{bmatrix} S_{\mathrm{vv}} \\ \sqrt{2}\, S_{\mathrm{vh}} \\ S_{\mathrm{hh}} \end{bmatrix}, \tag{5.166b}$$

and

$$\mathbf{C} = \begin{bmatrix} S_{\mathrm{vv}}S_{\mathrm{vv}}^* & \sqrt{2}\, S_{\mathrm{vv}}S_{\mathrm{vh}}^* & S_{\mathrm{vv}}S_{\mathrm{hh}}^* \\ \sqrt{2}\, S_{\mathrm{vh}}S_{\mathrm{vv}}^* & 2S_{\mathrm{vh}}S_{\mathrm{vh}}^* & \sqrt{2}\, S_{\mathrm{vh}}S_{\mathrm{hh}}^* \\ S_{\mathrm{hh}}S_{\mathrm{vv}}^* & \sqrt{2}\, S_{\mathrm{hh}}S_{\mathrm{vh}}^* & S_{\mathrm{hh}}S_{\mathrm{hh}}^* \end{bmatrix}. \tag{5.166c}$$

In view of Eqs. (5.160) and (5.164), the backscattering coefficient can be expressed in terms of the ***average scatterer covariance matrix*** $\langle \mathbf{C} \rangle$ as follows:

$$\begin{aligned} \sigma_{\mathrm{rt}}^0(\psi_{\mathrm{r}}, \chi_{\mathrm{r}};\ \psi_{\mathrm{t}}, \chi_{\mathrm{t}}) &= \frac{4\pi}{AN_{\mathrm{c}}} \sum_{i=1}^{N_{\mathrm{c}}} \mathbf{A} \cdot \mathbf{C}_i \mathbf{A}^* \\ &= \frac{4\pi}{AN_{\mathrm{c}}} \mathbf{A} \cdot \left(\sum_{i=1}^{N_{\mathrm{c}}} \mathbf{C}_i \right) \mathbf{A}^* \\ &= \frac{4\pi}{A} \mathbf{A} \cdot \langle \mathbf{C} \rangle \mathbf{A}. \end{aligned} \tag{5.167}$$

Given a segment of a polarimetric radar image with N_{c} pixels (representing a quasi-uniform distributed target such as a forest canopy), the measured scattering matrix of pixel i, $\mathbf{S}_i$, is used in Eq. (5.166c) to compute the corresponding covariance matrix $\mathbf{C}_i$. Upon averaging all of the $\mathbf{C}_i$ matrices, we obtain $\langle \mathbf{C} \rangle$. If N_{c} is sufficiently large, the off-diagonal terms of $\langle \mathbf{C} \rangle$ should be negligibly small and can be set equal to zero. This is because for a distributed target the co-pol and cross-pol components of **S** are uncorrelated, as was shown earlier in Section 5-12. Thus,

$$\langle \mathbf{C} \rangle = \begin{bmatrix} \langle S_{\mathrm{vv}}S_{\mathrm{vv}}^* \rangle & 0 & \langle S_{\mathrm{vv}}S_{\mathrm{hh}}^* \rangle \\ 0 & 2 \langle S_{\mathrm{vh}}S_{\mathrm{vh}}^* \rangle & 0 \\ \langle S_{\mathrm{hh}}S_{\mathrm{vv}}^* \rangle & 0 & \langle S_{\mathrm{hh}}S_{\mathrm{hh}}^* \rangle \end{bmatrix}. \tag{5.168}$$

This is called the covariance matrix of a distributed target with ***reflection symmetry***.

The covariance matrix given by Eq. (5.168) is ***Hermitian*** and positive semidefinite, which means that none of its ***eigenvalues*** are negative. This follows from the way in which we constructed the covariance matrix.

▶ Hermitian matrices have the property that all their ***eigenvalues*** are real numbers and their ***eigenvectors*** form an orthonormal set. ◀

Cloude (1992b) was the first to use the orthonormality of the eigenvectors of the covariance matrix (in the context of radar polarimetry) to propose the ***decomposition*** of the covariance matrix in terms of its eigenvalues and eigenvectors, namely that $\langle \mathbf{C} \rangle$ can be expressed as

$$\langle \mathbf{C} \rangle = \sum_{j=1}^{3} \lambda_j \hat{\mathbf{e}}_j \hat{\mathbf{e}}_j^{\dagger}, \tag{5.169}$$

where λ_1 to λ_3 are the eigenvalues and $\hat{\mathbf{e}}_1$ to $\hat{\mathbf{e}}_3$ are the eigenvectors.

5-13.2 Eigenvector Decomposition

The decomposition expressed by Eq. (5.169) is unique. That is, since the eigenvectors of the covariance matrix are orthogonal, they form a natural basis in which to express the scattering. Note that in this decomposition, the eigenvector contains the information about the scattering mechanism, while the eigenvalue contains the information about the relative strength of that scattering mechanism compared to the others. Solution of the determinant of the matrix equation $\langle \mathbf{C} \rangle - \lambda \mathbf{I} = 0$, where $\mathbf{I}$ is the identity matrix, leads to

$$\lambda_1 = \frac{1}{2}\left\{ \langle S_{\text{hh}} S_{\text{hh}}^* \rangle + \langle S_{\text{vv}} S_{\text{vv}}^* \rangle + \sqrt{\left(\langle S_{\text{hh}} S_{\text{hh}}^* \rangle - \langle S_{\text{vv}} S_{\text{vv}}^* \rangle \right)^2 + 4 |\langle S_{\text{hh}} S_{\text{vv}}^* \rangle|^2} \right\}, \tag{5.170a}$$

$$\lambda_2 = \frac{1}{2}\left\{ \langle S_{\text{hh}} S_{\text{hh}}^* \rangle + \langle S_{\text{vv}} S_{\text{vv}}^* \rangle - \sqrt{\left(\langle S_{\text{hh}} S_{\text{hh}}^* \rangle - \langle S_{\text{vv}} S_{\text{vv}}^* \rangle \right)^2 + 4 |\langle S_{\text{hh}} S_{\text{vv}}^* \rangle|^2} \right\}, \tag{5.170b}$$

$$\lambda_3 = 2 \langle S_{\text{hv}} S_{\text{hv}}^* \rangle . \tag{5.170c}$$

All of these are real numbers, as expected for a Hermitian matrix.

The corresponding three eigenvectors are

$$\hat{\mathbf{e}}_1 = \frac{1}{\sqrt{\left[\langle S_{\text{vv}} S_{\text{vv}}^* \rangle - \langle S_{\text{hh}} S_{\text{hh}}^* \rangle + \sqrt{\Delta} \right]^2 + 4 |\langle S_{\text{hh}} S_{\text{vv}}^* \rangle|^2}} \cdot \begin{pmatrix} 2 \langle S_{\text{hh}} S_{\text{vv}}^* \rangle \\ 0 \\ \langle S_{\text{vv}} S_{\text{vv}}^* \rangle - \langle S_{\text{hh}} S_{\text{hh}}^* \rangle + \sqrt{\Delta} \end{pmatrix}, \tag{5.171a}$$

$$\hat{\mathbf{e}}_2 = \frac{1}{\sqrt{\left[\langle S_{\text{vv}} S_{\text{vv}}^* \rangle - \langle S_{\text{hh}} S_{\text{hh}}^* \rangle - \sqrt{\Delta} \right]^2 + 4 |\langle S_{\text{hh}} S_{\text{vv}}^* \rangle|^2}} \cdot \begin{pmatrix} 2 \langle S_{\text{hh}} S_{\text{vv}}^* \rangle \\ 0 \\ \langle S_{\text{vv}} S_{\text{vv}}^* \rangle - \langle S_{\text{hh}} S_{\text{hh}}^* \rangle - \sqrt{\Delta} \end{pmatrix}, \tag{5.171b}$$

$$\hat{\mathbf{e}}_3 = \begin{pmatrix} 0 \\ 1 \\ 0 \end{pmatrix}. \tag{5.171c}$$

In these expressions, we used the shorthand notation

$$\Delta = \left(\langle S_{\text{vv}} S_{\text{vv}}^* \rangle - \langle S_{\text{hh}} S_{\text{hh}}^* \rangle \right)^2 + 4 |\langle S_{\text{hh}} S_{\text{vv}}^* \rangle|^2 . \tag{5.172}$$

There are a couple of interesting properties to point out about the eigenvalues and eigenvectors in this case. First, eigenvalue λ_3 and its associated eigenvector $\hat{\mathbf{e}}_3$ depend on the cross-polarized return only, and the other two are purely made up of the co-polarized returns. This has led to the interpretation that the third eigenvector and its associated eigenvalue indicate the amount of diffuse scattering. The second point is that the first two eigenvectors show that the co-polarized phase of these two eigenvectors are 180 degrees different. Furthermore, their co-polarized amplitude ratios are the inverse of each other. It is easily shown from the expressions given by Eq. (5.171) that (van Zyl and Kim, 2011)

$$\frac{2 \langle S_{\text{hh}} S_{\text{vv}}^* \rangle}{\langle S_{\text{vv}} S_{\text{vv}}^* \rangle - \langle S_{\text{hh}} S_{\text{hh}}^* \rangle + \sqrt{\Delta}} = \frac{\langle S_{\text{vv}} S_{\text{vv}}^* \rangle - \langle S_{\text{hh}} S_{\text{hh}}^* \rangle - \sqrt{\Delta}}{2 \langle S_{\text{hh}} S_{\text{vv}}^* \rangle^*} . \tag{5.173}$$

This means that the first two eigenvectors represent scattering matrices that are related to odd and even numbers of reflections. Without looking explicitly at the eigenvectors, however, it is not obvious as to which eigenvalue to associate with which scattering mechanism. Also, we should point out that these eigenvectors do not necessarily represent pure single or double reflections. The phase relationship between the two eigenvectors means that if one eigenvector is more similar to single (odd) reflections, i.e., its phase is closer to zero than to 180 degrees, then the other eigenvalue is associated with double (even) reflections.

5-13.3 Useful Polarimetric Parameters

Target entropy

There are several ways by which we can estimate the amount of randomness exhibited in the state of the polarization of the wave scattered by the various resolution cells of the distributed target. Cloude (1992b), for example, defined ***target entropy*** as

$$H_T = -\sum_{i=1}^{3} P_i \log_3 P_i, \tag{5.174}$$

where

$$P_i = \frac{\lambda_i}{\lambda_1 + \lambda_2 + \lambda_3}. \tag{5.175}$$

As pointed out by Cloude, the target entropy H_T is a measure of target disorder, with $H_T = 1$ for highly random targets with equal eigenvalues, and $H_T = 0$ for simple, single, nonrandom targets.

Polarimetric pedestal

Figure 5-45 displays co-pol responses for a point target, namely a sphere, and three distributed targets representing an ocean surface, an urban area, and a city park. For the point target, the co-pol response is computed by applying the polarization synthesis equation given by Eq. (5.138) for all possible polarizations (such that the transmit and receive polarizations are always the same). A similar process

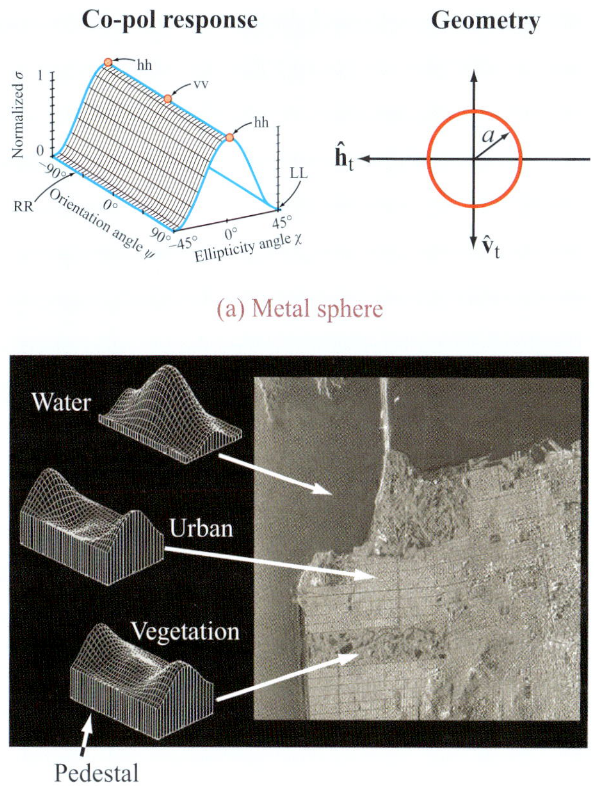

(a) Metal sphere

(b) San Francisco

Figure 5-45: Co-pol responses for (a) a metal sphere and (b) three areas in the San Francisco image.

applies to each of the three different distributed targets, except that for each polarization the process is repeated for all pixels of the distributed target and then averaged according to Eq. (5.143). Note that the minimum response for the point target is zero. It occurs when the receive/transmit polarization combination is RR or LL, where R and L stand for right-hand circular and left-hand circular. The co-pol response in Fig. 5-45(a) has no "pedestal," whereas the co-pol responses of the distributed targets do. The pedestal is small for the ocean surface and large for the urban and the city park. Distributed targets that generate significant multiple scattering have significant pedestal heights.

Durden et al. (1990) showed that measuring the pedestal height is equivalent to measuring the ratio of

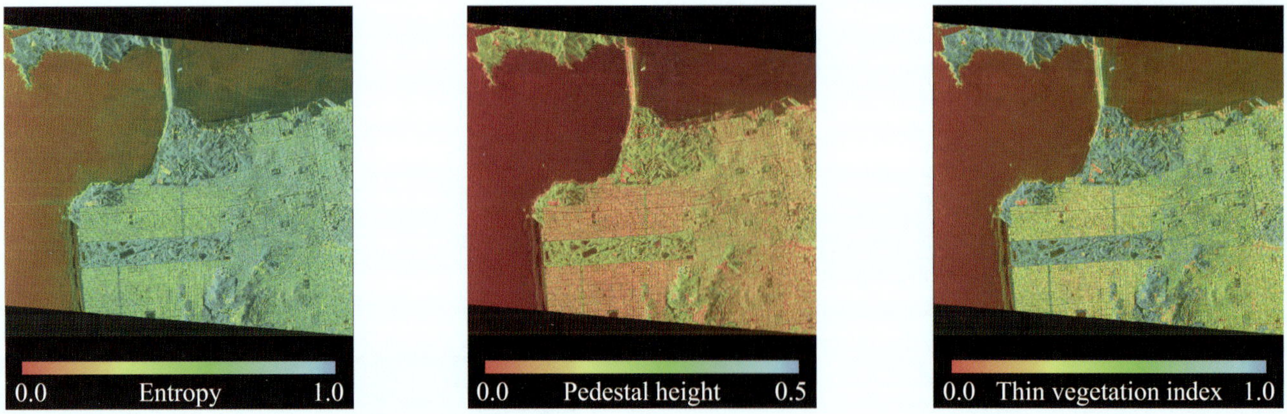

(a) Entropy (b) Pedestal (c) RVI

Figure 5-46: Radar images of San Francisco showing the three measures of scattering randomness: (a) entropy scaled from 0 (black) to 1 (white); (b) pedestal height scaled from 0 (black) to 0.5 (white); and (c) the radar vegetation index scaled from 0 (black) to 1 (white).

the minimum eigenvalue to the maximum eigenvalue:

$$P_{\rm H} = \frac{\min(\lambda_1, \lambda_2, \lambda_3)}{\max(\lambda_1, \lambda_2, \lambda_3)} . \quad (5.176)$$

Radar vegetation index

Kim and van Zyl (2001) introduced a parameter for volume scattering media, such as vegetation canopies, and they named it the ***radar vegetation index*** (RVI). The RVI is defined as

$$\text{RVI} = \frac{4\min(\lambda_1, \lambda_2, \lambda_3)}{\lambda_1 + \lambda_2 + \lambda_3} . \quad (5.177)$$

By modeling a vegetation canopy as a collection of randomly oriented cylinders of length l and diameter d, Kim and van Zyl (2001) evaluated the RVI as a function of d/λ, where λ is the EM wavelength. Their results indicate that RVI approaches 1 when $d/\lambda \ll 1$ and approaches 0 when d approaches l (that is, when the cylinder becomes like a disc). Hence, the RVI provides information about the canopy architecture (at the needle and small branch scale) that may prove useful in distinguishing between different types of canopies.

5-13.4 Image Examples

The three polarimetric parameters introduced in the preceding subsection are different measures of ***polarization randomness*** of a distributed target. The similarities between the three polarization randomness parameters are illustrated in Fig. 5-46 using the image of San Francisco introduced earlier. From the polarization responses displayed in Fig. 5-45, we expect a small amount of randomness for the ocean but a more significant amount for the vegetation-dominated area in the Golden Gate Park. As noted earlier, the height of the pedestal of the polarization response is a measure of polarization randomness. Except for scaling differences, the three images in Fig. 5-46 convey similar information about the imaged scene. The entropy image shows less color variation than the other two, and the pedestal height image shows the greatest dynamic range. The choice among the three parameters as terrain type discriminators is somewhat arbitrary.

Next, we investigate the RVI at three different frequencies over a portion of the Black Forest in

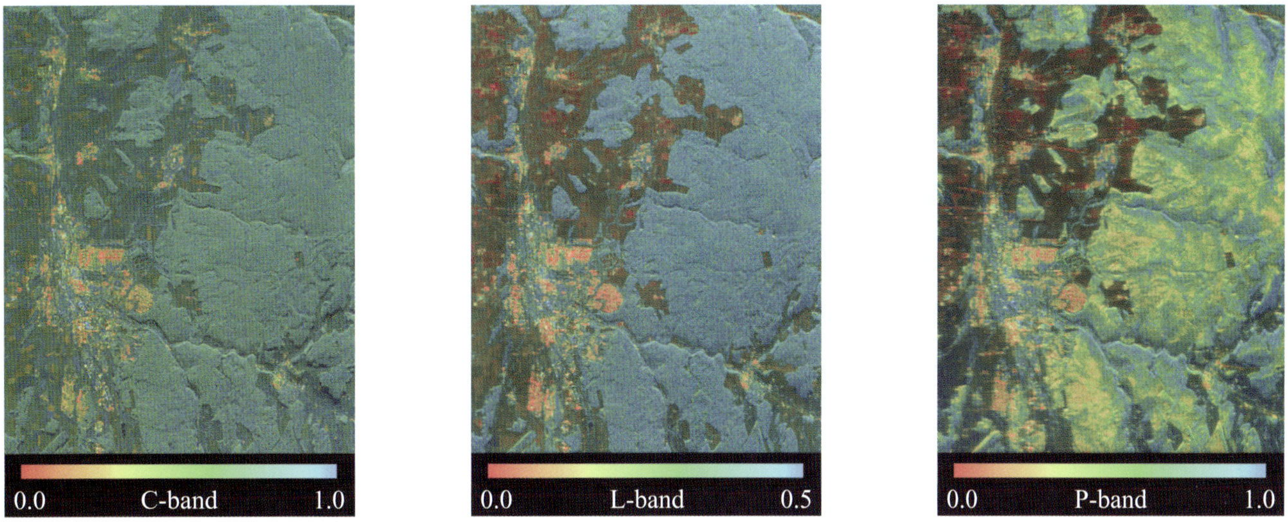

(a) C-band (b) L-band (c) P-band

Figure 5-48: RVI images of the area shown in Fig. 5-47 at three different frequencies. The RVI is scaled from 0 (black) to 1 (white). Note that the L-band RVI in the forested area is higher than the C-band RVI, while the C-band RVI is higher than the others in the agricultural areas..

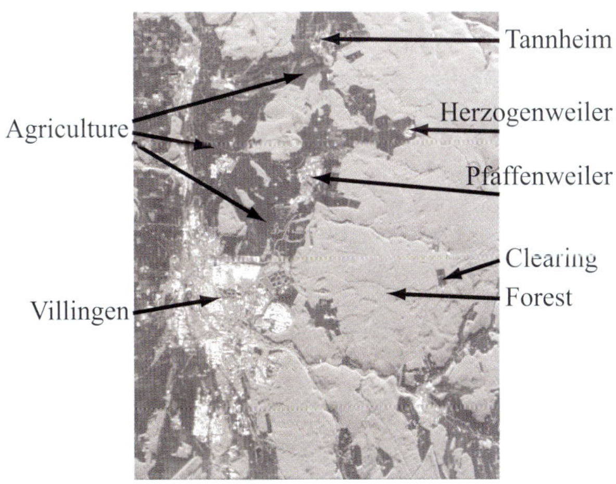

Figure 5-47: L-band total-power image of a portion of the Black Forest in Germany acquired with the NASA/JPL AIRSAR system in the summer of 1991. Radar illumination is from the top.

Germany acquired with the NASA/JPL AIRSAR system during the summer of 1991. The L-band image is shown for reference in Fig. 5-47. The bright feature in the left portion of the image is the town of Vilingen. The brighter right-hand portion of the image is a mixed forest consisting of spruce (*Picea abies*), pine (*Pinus sylvestris*) and fir (*Abies alba*) trees. The dry weight biomass ranges up to 50 kg/m^2. The darker areas in the upper portion and to the left in the image are mostly agricultural fields with varying amounts of biomass depending on the crop type and maturity of the plants.

Figure 5-48 compares the radar vegetation index at the three frequencies (C-band, L-band and P-band) of the AIRSAR system. The C-band image shows much higher values of the RVI in the agricultural areas, because the shorter wavelength is more sensitive to the smaller biomass in these fields. Fields with increased values of RVI generally had more vegetation on them at the time of acquisition. The forested area shows a high level of RVI at C-band, but somewhat less than some of the agricultural fields. The L-band image shows

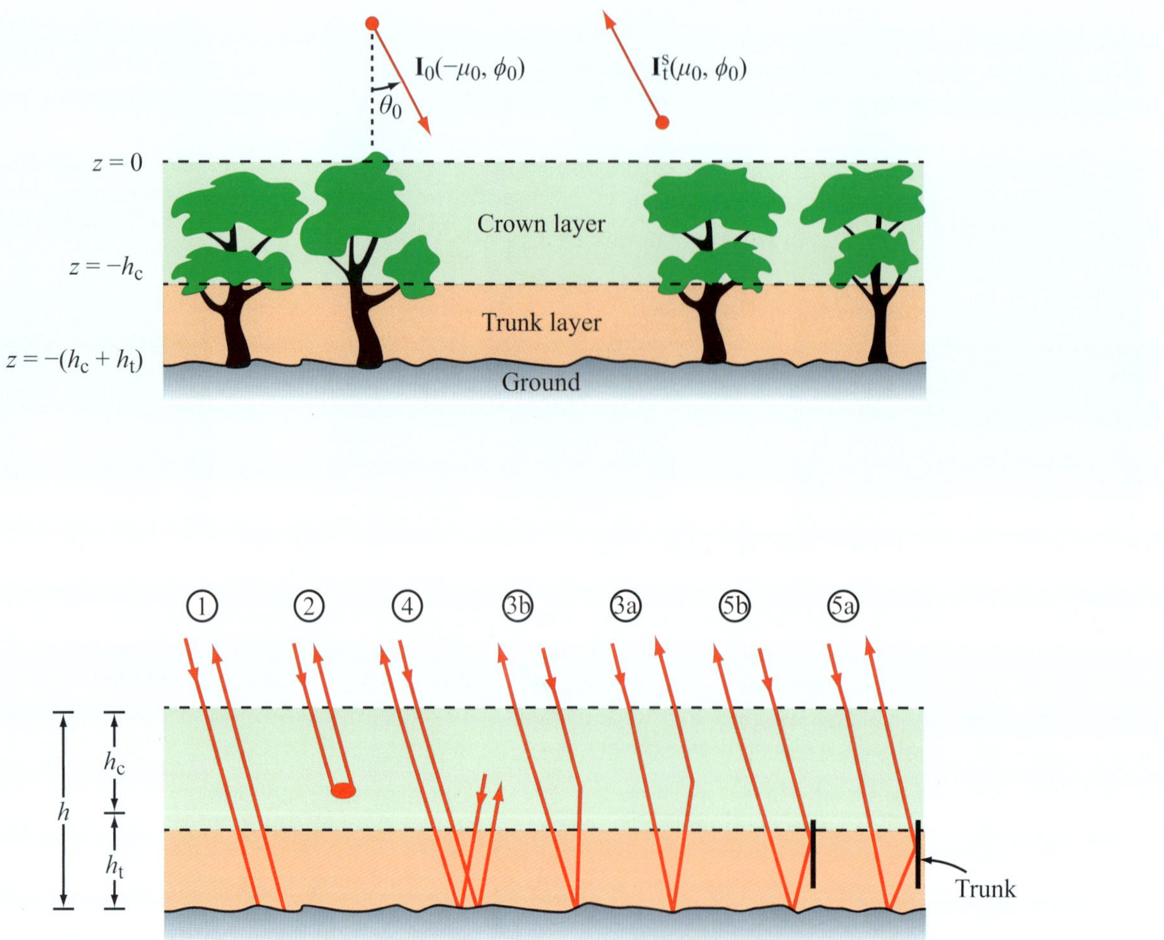

Figure 5-49: Scattering mechanisms for a forest canopy.

lower values of the RVI in the agricultural areas than the C-band image. This is consistent with the notion that L-band signals interact less with the relatively short vegetation than the shorter wavelength C-band signals. At L-band, the returns from the agricultural areas are probably still dominated by scattering from the underlying ground surface. The forest, on the other hand, shows a larger value of the RVI in the L-band than in the C-band image.

5-13.5 Freeman-Durden Decomposition

As discussed later in Section 11-13.3, the radar energy backscattered by a forest canopy contains contributions due to (a) direct scattering by the underlying ground surface, as depicted by ray 1 in Fig. 5-49, (b) direct and indirect scattering by the crown layer, depicted by rays 2, 3a, 3b, and 4 in Fig. 5-49, and (c) scattering by the ground-tree trunk combination, depicted by rays 5a and 5b. The ground-trunk scattering mechanism is an example of ***dihedral corner reflection***. Another example of this type of reflection is the reflection at the corner between a ground surface and the vertical wall of a building. Accordingly, Freeman and Durden (1998) proposed a three-component decomposition for the covariance matrix of Eq. (5.168):

$$\langle \mathbf{C} \rangle = f_{\rm s} \langle \mathbf{C}_{\rm s} \rangle + f_{\rm d} \langle \mathbf{C}_{\rm d} \rangle + f_{\rm v} \langle \mathbf{C}_{\rm v} \rangle, \qquad (5.178)$$

where $f_{\rm s}$, $f_{\rm d}$, and $f_{\rm v}$ are the relative surface, dihedral, and volume contributions, and $\langle \mathbf{C}_{\rm s} \rangle$, $\langle \mathbf{C}_{\rm d} \rangle$, and $\langle \mathbf{C}_{\rm v} \rangle$ are the associated covariance matrices. By applying certain assumptions about the sizes and orientation distributions of the volume scatterers in the crown layer, they were able to decompose the measured covariance matrix into its three components, as well as compute $f_{\rm s}$, $f_{\rm d}$, and $f_{\rm v}$ for every group of image pixels over which $\langle \mathbf{C} \rangle$ was measured. An example of their results is shown in Fig. 5-50 for the Black Forest area. The relative surface, dihedral, and volume contributions ($f_{\rm s}$, $f_{\rm d}$, and $f_{\rm v}$) are displayed in blue, red, and green, respectively. The images in Fig. 5-50 indicate that volume scattering (represented by $f_{\rm v}$ in green) dominated in the vegetated areas, particularly at C-band where the penetration through the canopy is the smallest. The dihedral-corner mechanism (represented by $f_{\rm d}$ in red) is the dominant mechanism in the urban areas at all three frequencies. At P-band, where penetration through vegetation is the greatest, we see significant areas in yellow, which is a mixture of red and blue, indicating significant contributions due to both surface scattering (blue) and ground-trunk (red) reflections.

PROBLEMS

5.1 The scattering matrix of a dihedral corner reflector in the BSA convention is given by Eq. (5.140). Compute the corresponding modified Mueller matrix in the FSA convention.

5.2 Relate the modified Mueller matrix in the FSA convention to the modified Mueller matrix in the BSA convention.

5.3 The scattering matrix of a thin conducting cylinder is given by Eq. (5.142). Compute σ_{pq} for $pq =$ vv, vh, hv, and hh for a cylinder oriented at 45°. The cylinder is 3 cm long, its radius is 0.1 cm, and the frequency is 1 GHz.

5.4 A linear-detection radar is used to measure the backscatter from a large bare-soil field. Based on a large number of measurements, it was determined that the average value of the electric field amplitude is 0.2 V/m. Determine the most-probable single-observation electric field amplitude.

5.5 For a single radar observation, compute the lower and upper bounds in dB of the 75% range of fading for:

(a) Linear detection.

(b) Square-law detection.

5.6 Repeat Problem 5.5 for the average of three independent samples.

5.7 How many independent samples are required to reduce the 90% confidence interval for square-law detection to within ±0.3 dB?

5.8 Compute the normalized fading standard deviation for an $N = 2$ image generated by:

(a) Amplitude detection.

(b) Intensity detection.

(c) Square-root intensity detection.

5.9 An airborne sidelooking radar traveling at 200 m/s is used to measure the backscatter from a distributed target. The radar uses a 2 m long antenna, operates at 1 cm wavelength, and its detection mode is square-law. The target range is 10 km.

(a) Determine the 90% range of fading after averaging 4, 10, 100, and 1000 samples.

(b) As the radar beam sweeps across each resolution cell of width $r_{\rm a}$, the number of independent samples inherent in the measured backscatter is $N_{\rm a} = 2\beta_{\rm h}^2 R/\lambda$ (from Eq. (14.123) in Chapter 14). Determine the distances that the radar has to travel so as to realize a total of $N = 4$, 10, 100, and 1000 samples.

5.10 The vertical ranges shown in Fig. 5-22(c) for $N = 16$ show no overlap between them. Does it then follow that it should always be possible to separate pixels of trees, grass, and soil from each other? Explain.

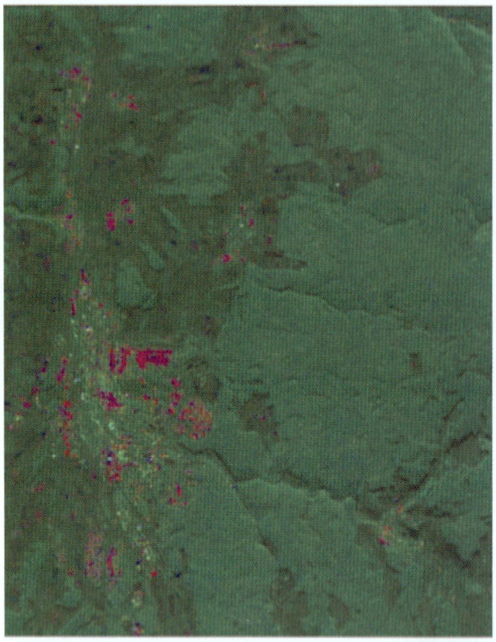

(a) C-band

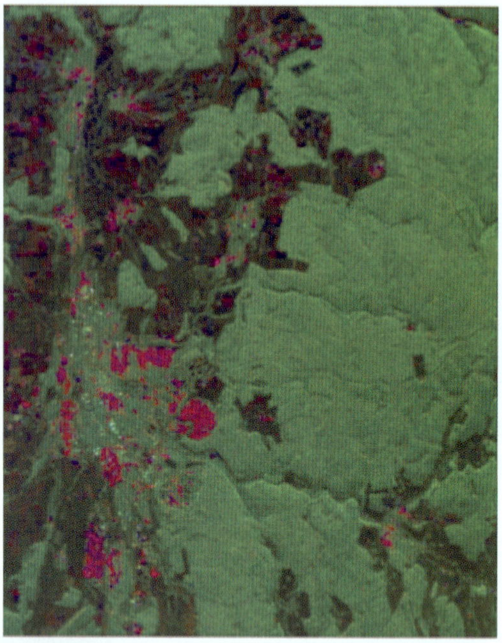

(b) L-band

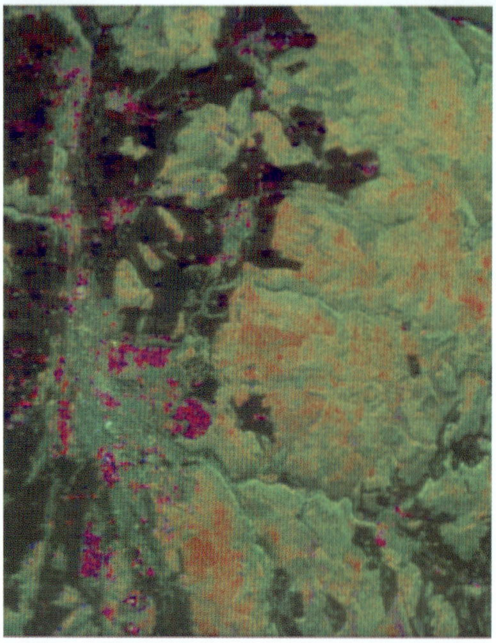

(c) P-band

Figure 5-50: Results of the Freeman-Durden decomposition for the Black Forest image at (a) C-band, (b) L-band, and (c) P-band. Surface, dihedral-corner, and volume scattering components are displayed in blue, red, and green colors, respectively.

5.11 What information about a random surface do the rms height and the correlation length convey? Explain with examples of smooth and rough surfaces.

5.12 A plane wave is incident at 30° upon a random surface with an rms height of 1 cm. Compute the ratio of the coherent reflectivity to the specular reflectivity at:

(a) $f = 0.5$ GHz.

(b) $f = 2$ GHz.

(c) $f = 10$ GHz.

5.13 Compute the hv-polarized radar cross section of a thin metal cylinder oriented at 30°, given that its length is 1 cm and its radius is 0.1 cm. The frequency is 3 GHz.

5.14 Select a polarimetric SAR image and then apply Computer Code 5.1 to generate images for the following combinations of receive/transmit polarizations:

(a) RHC/RHC.

(b) RHC/LHC.

(c) 45° Linear/LHC.

5.15 The average Mueller matrix of a soil surface measured by a polarimetric scatterometer is given by

$$\langle \mathbf{M} \rangle = \begin{bmatrix} 1.0 & 0.03 & 0 & 0 \\ 0.03 & 0.77 & 0 & 0 \\ 0 & 0 & 0.77 & -0.11 \\ 0 & 0 & 0.11 & 0.77 \end{bmatrix}.$$

Compute and plot the co-pol phase pdf.

5.16 The average covariance matrix of a grass-covered surface measured by a polarimetric radar in the BSA convention is given by

$$\langle \mathbf{C} \rangle = \begin{bmatrix} 1.0 & 0 & 0.7 \\ 0 & 0.06 & 0 \\ 0.7 & 0 & 0.9 \end{bmatrix}.$$

Compute:

(a) The target entropy.

(b) The pedestal height.

(c) The radar vegetation index.

CHAPTER 6

Microwave Radiometry and Radiative Transfer

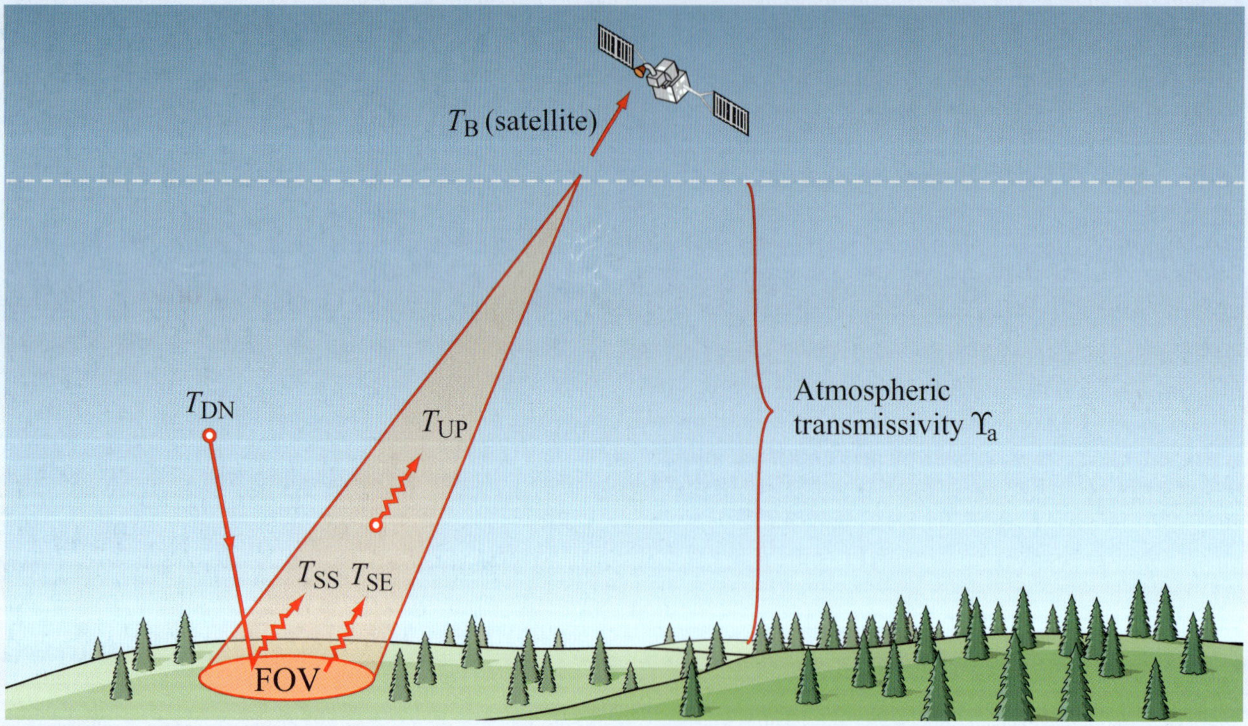

Microwave emission

CONTENTS

Overview

The bulk of the energy received by the planet Earth is in the form of solar electromagnetic radiation. Part of the incident solar energy is scattered and absorbed by the Earth's atmosphere, and the remainder is transmitted to the Earth's surface. Of the latter, a part is scattered outward and the remainder is absorbed. According to thermodynamic principles, absorption of electromagnetic energy by a material medium is a transformation into thermal energy, which is accompanied by a rise in the thermometric temperature of the material. The reverse process, that of ***thermal emission***, serves to create a balance between absorbed solar radiation and radiation emitted by the Earth's surface and its atmosphere. These transformation processes are at the heart of the ***theory of radiative transfer***.

Radiometry is a field of science and engineering concerned with the measurement of electromagnetic radiation. This chapter begins by introducing radiometric quantities of interest, and then proceeds to treat the radiometric measurement process, with special emphasis on the microwave region of the electromagnetic spectrum. Along the way, several topics are discussed, including blackbody radiation, radiative transfer, and terrestrial emission and scattering.

As was mentioned earlier in Chapter 5, the introduction of ***radar polarimetry*** in the mid-1980s led to the development of new remote sensing applications and better algorithms for retrieving certain geophysical parameters of interest. Radar polarimetry was enabled by advances in antenna and measurement technologies. These same advances led to the development of ***radiometric polarimetry***; a dual-polarized radiometer antenna measures not only the magnitudes of the polarized electric fields of the radiation emitted by the scene, but it also measures the phase difference between them. As we will see in this and later chapters, the phase difference contains information related to the geophysical properties of the scene. This is particularly true for the estimation of wind vectors over the ocean surface. The second part of this chapter is devoted to radiometric polarimetry, and so is the second part of Chapter 7 on radiometric receivers.

6-1 Radiometric Quantities

The term ***radiometry*** means the measurement of incoherent radiant electromagnetic energy. As will be discussed in future sections, all material media (gases, liquids, solids, and plasma) radiate (emit) electromagnetic energy. In this section, we define the radiometric quantities of interest and develop the relation between the directional distribution of power incident upon an antenna and the power measured at its output terminals. The nomenclature, symbols, and units used to express radiometric quantities, which often are different from those used in optical and infrared radiometry, are chosen from the standpoint of microwave engineering. Table 6-1 provides a list of these quantities along with their optical counterparts.

6-2 Thermal Radiation

6-2.1 Quantum Theory of Radiation

All substances at a finite absolute temperature radiate electromagnetic energy. Atomic gases radiate electromagnetic waves at discrete frequencies, or wavelengths; that is, they have ***line spectra***. According to quantum theory, each spectral line in the radiation spectrum of an atomic gas corresponds to a specific transition by an electron from one atomic energy level to another (lower) energy level. When the transition is between quantum level $\mathcal{E}_1$ and $\mathcal{E}_2$, the frequency f of the emitted radiation (photon) is given by Bohr's equation

$$f = \frac{\mathcal{E}_1 - \mathcal{E}_2}{h}, \tag{6.1}$$

where h is Planck's constant. Quantum theory specifies the quantum energy levels of an atom and the allowed transitions between them, thereby defining the line spectrum of an atomic gas.

Energy incident upon an atom can be absorbed by the atom to provide the necessary energy to move an electron to a higher energy level, provided the frequency of the incident wave satisfies the Bohr condition. Thus, the absorption spectrum of a gas is identical to its emission spectrum. As discussed later, this principle

Table 6-1: **Standard units, symbols, and defining equations for fundamental radiometric quantities.**

Microwave terminology	Optical terminology	Symbol	Defining equation	Unit	Abbreviation
Energy	Radiant energy	$\mathcal{E}$		joule	J
Power	Radiant flux	P	$P = \partial\mathcal{E}/\partial t$	watt	W
Power (or flux) density	Radiant flux	$\mathcal{S}$	$\mathcal{S} = \partial P/\partial A$	watt per square meter	Wm^{-2}
Brightness intensity	Radiance	I	$I = \partial^2 P/\partial\Omega\,\partial A$	watt per steradian per square meter	$\text{W sr}^{-1}\text{m}^{-2}$
Emissivity	Emissivity	e	$e = I/I_{\text{blackbody}}$	(unitless)	
Reflectivity	Reflectance	Γ	$\Gamma = P^{\text{r}}/P^{\text{i}}$	(unitless)	
Absorptivity	Absorptance	a	$a = P^{\text{a}}/P^{\text{i}}$	(unitless)	
Transmissivity	Transmittance	$\mathbb{T}$	$\mathbb{T} = P^{\text{t}}/P^{\text{i}}$	(unitless)	

Superscripts: i = incident, r = reflected, a = absorbed, and t = transmitted.

applies also to more complicated structures including molecular gases, liquids, and solids.

Spontaneous emission of radiation by an atom is caused by collision with another atom or particle. The probability that a collision (and hence emission) takes place is governed by the density of atoms and the kinetic energies of their random motions. And since the kinetic (or heat) energy of a substance is defined in terms of its absolute temperature, it follows that the intensity of energy radiated by a substance increases with its temperature.

Associated with a molecule consisting of two or more atoms is a set of vibrational and rotational modes describing the motion of the atoms relative to one another. These modes define a corresponding set of allowable energy levels. With contributions from vibrational, rotational, and electronic transitions, the spectra of molecules again consist of discrete lines, but the number of lines is much larger than for atoms. In the visible region of the spectrum, the spectral lines of some molecular gases are so close together that it is difficult to resolve individual lines, especially for molecules consisting of a large number of atoms. Translational molecule motion introduces Doppler shifts that further blur the observed spectral lines.

As we go from gases to liquids and solids, the interaction between particles increases and the radiation (or absorption) spectrum becomes more complicated.

► A liquid or solid may be regarded as an enormous molecule with a correspondingly increased number of degrees of freedom, which leads to such a large number of closely spaced spectral lines that the radiation spectrum becomes effectively continuous, with all frequencies being radiated. ◄

The radiation mechanisms described above for atoms and molecules are based on Planck's quantum theory, which he introduced in 1901 in his derivation of the ***blackbody radiation law***. Prior to Planck's work, attempts by nineteenth-century physicists to derive an expression for the radiation spectrum of solids were unsuccessful because of their reliance on classical mechanics. The key to Planck's solution was his assumption that the emitted radiation occurs only in discrete energy quanta. It is this assumption that marks the origin of the quantum theory.

The remainder of this section is devoted to further discussion of radiation from solids (and liquids). Microwave spectra of atmospheric gases are discussed in Chapter 8.

6-2.2 Planck's Blackbody Radiation Law

The concept of a blackbody radiator is of fundamental importance to gaining a good understanding of how real materials emit thermal energy because the emission spectrum of a blackbody represents a reference, relative to which the radiant emittance of any material can be expressed.

▶ In general, of the radiation incident upon the surface of a solid (or liquid) substance, a certain fraction is absorbed and the remainder is reflected. A ***blackbody*** is defined as an idealized, perfectly opaque material that absorbs all the incident radiation at all frequencies, reflecting none. ◀

The quantum-mechanical model of a blackbody may be described as consisting of such a large number of quantized energy levels, with a correspondingly large number of allowable transitions, that any photon, whatever its energy or frequency, is absorbed when incident upon the blackbody.

▶ A body in thermodynamic equilibrium emits to its environment the same amount of energy it absorbs from its environment. Hence, in addition to being a perfect absorber, *a blackbody also is a perfect emitter*. ◀

In practice, a good approximation to a blackbody is a hollow body with a small opening. Any radiation entering the container through the opening is absorbed or undergoes many reflections on the inside surfaces, so that all the original energy effectively is absorbed before it can be reflected back out through the opening. If the container is maintained at some fixed temperature T, its inner surfaces emit and absorb photons at the same rate, and therefore the energy leaking out through the opening resembles radiation by a blackbody in thermodynamic equilibrium. At microwave frequencies, good approximations to ideal blackbodies are the highly absorbing materials used in the construction of anechoic chambers.

According to Planck's radiation law, a blackbody radiates uniformly in all directions with a ***spectral brightness intensity*** (also called spectral specific intensity) I_f given by

$$I_f = \frac{2hf^3}{c^2}\left(\frac{1}{e^{hf/kT}-1}\right) \quad \textbf{(Planck's law)}, \tag{6.2}$$

where I_f is in $\text{Wm}^{-2}\text{sr}^{-1}\text{Hz}^{-1}$, h is Planck's constant ($= 6.63 \times 10^{-34}$ joules·s), f is frequency (Hz), k is Boltzmann's constant ($= 1.38 \times 10^{-23}$ joule K^{-1}), T is the blackbody's absolute temperature (K), and c is the velocity of light in vacuum ($= 3 \times 10^8$ m/s). The only two variables appearing in Eq. (6.2), which includes intensities of two orthogonal polarizations, are f and T. A family of curves displaying I_f as a function of frequency is shown in Fig. 6-1 (with T as a parameter). Logarithmic scales are used on both axes so that a wide range of values can be accommodated. The curves exhibit two interesting features:

(a) raising the temperature T causes the overall level of the corresponding curve to increase, and

(b) the frequency at which I_f is maximum increases with T.

Over a narrow frequency interval df centered at f, the ***brightness intensity*** is given by

$$dI = I_f \, df. \tag{6.3}$$

It is sometimes of interest to express the spectral brightness intensity in terms of I_λ rather than I_f, where I_λ is the power per unit area per unit solid angle per unit wavelength. Over a wavelength interval $d\lambda$ corresponding to the frequency interval df above, the area under the curve contains the same brightness intensity dI. Thus,

$$dI = I_\lambda \, d\lambda. \tag{6.4}$$

Equating Eq. (6.3) and Eq. (6.4) and noting that

$$df = -\frac{c}{\lambda^2}\, d\lambda, \tag{6.5}$$

the following expression for I_λ is obtained:

$$I_\lambda = \frac{2hc^2}{\lambda^5}\left(\frac{1}{e^{hc/\lambda kT}-1}\right). \tag{6.6}$$

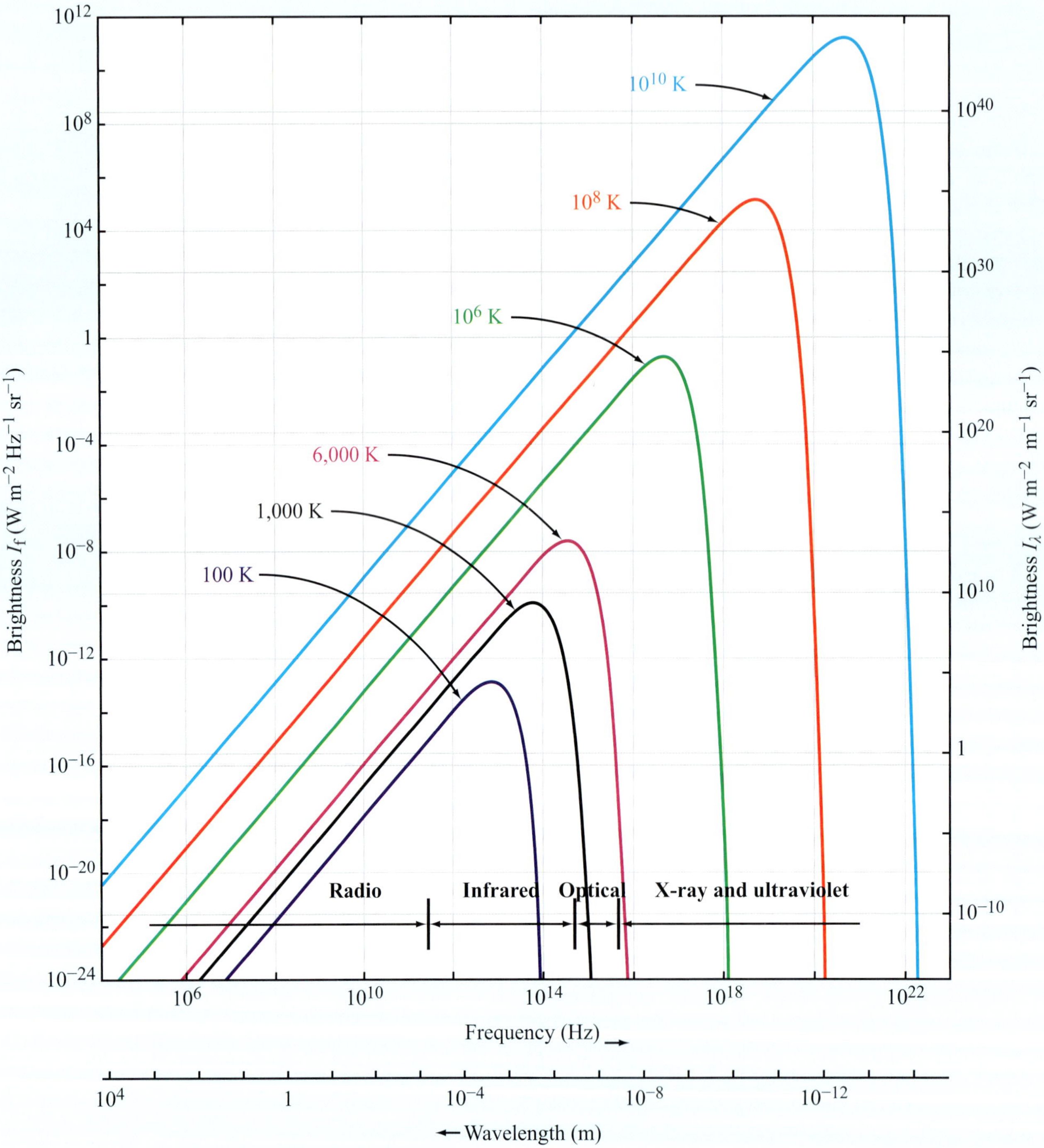

Figure 6-1: Planck's radiation law [adapted from Kraus, 1966].

The negative sign in Eq. (6.5) has been ignored because it merely reflects the fact that f and λ increase in opposite directions and it has no bearing on the magnitudes of $d\lambda$ and df. The curves shown in Fig. 6-1 are also plots of I_λ measured along the right-hand ordinate scale.

To fully appreciate what the blackbody spectral brightness intensity I_f represents, let us consider the situation shown in Fig. 6-2. By definition, I_f is the power (in watts) emitted by a blackbody source of surface area 1 m^2, through a solid angle of 1 sr, and over a bandwidth of 1 Hz. In Fig. 6-2, the blackbody source has an area A_s, and our objective is to determine the amount of ***spectral power*** P_f intercepted by a receiving aperture of area A_r at a range R from the source. Spectral power is power contained in a bandwidth of 1 Hz. Accordingly, its units are W Hz^{-1}.

The receiving aperture A_r subtends a solid angle Ω_r. Hence,

$$P_f = I_f A_s \Omega_r \qquad (\text{W Hz}^{-1}). \tag{6.7}$$

From the perspective of the receiving aperture, the source area A_s subtends a solid angle Ω_s given by

$$\Omega_s = \frac{A_s}{R^2}, \tag{6.8}$$

where R is the range between the receiving aperture and the source. Similarly,

$$\Omega_r = \frac{A_r}{R^2}. \tag{6.9}$$

The combination of Eq. (6.8) and Eq. (6.9) allows us to rewrite Eq. (6.7) as

$$P_f = I_f A_r \Omega_s. \tag{6.10}$$

In this new form, P_f is proportional to the product of the area of the receiving aperture and the angular extent of the source. When passed through a receiver (filter) with a bandwidth extending from frequency f_1 to frequency f_2, the total captured power (Fig. 6-2) is

$$P = \int_{f_1}^{f_2} P_f \, df = A_r \Omega_s \int_{f_1}^{f_2} I_f \, df. \tag{6.11}$$

6-2.3 The Rayleigh–Jeans Law

Figure 6-3 displays a plot of Planck's law for a blackbody at $T = 300$ K, together with a dashed line that provides an excellent approximation to Planck's law, as long as the frequency f is well below f_{max}, where f_{max} is the frequency at which I_f is a maximum. This low-frequency approximation is known as the ***Rayleigh–Jeans law***.

If $x \ll 1$, the expression

$$\begin{aligned} e^x - 1 &= \left(1 + x + \frac{x^2}{2!} + \cdots\right) - 1 \\ &\approx x \qquad (\text{for } x \ll 1). \end{aligned} \tag{6.12}$$

Hence, for $hf/kT \ll 1$ in Eq. (6.2), the expression for I_f simplifies to

$$I_f \approx \frac{2kT}{\lambda^2}. \tag{6.13}$$

(Rayleigh–Jeans's law)

▶ The Rayleigh–Jeans approximation is very useful in the microwave region: it is mathematically simpler than Planck's law and yet its fractional deviation from Planck's exact expression is less than 1% if

$$\lambda T > 0.77 \text{ m K}, \tag{6.14a}$$

or equivalently,

$$f/T < 3.9 \times 10^8 \text{ Hz K}^{-1}. \tag{6.14b}$$

◀

For a blackbody at a room temperature of 300 K, the above inequalities will hold if $\lambda > 2.57$ mm, or equivalently $f < 117$ GHz, which covers the entire radio region and most of the usable part of the microwave spectrum. At 300 GHz, the fractional deviation is about 3 percent. On the other hand, the energy radiated by the cosmic background surrounding our galaxy has an equivalent temperature of 2.7 K, which does not satisfy the condition for the Rayleigh–Jeans approximation in the microwave region of the spectrum. When a

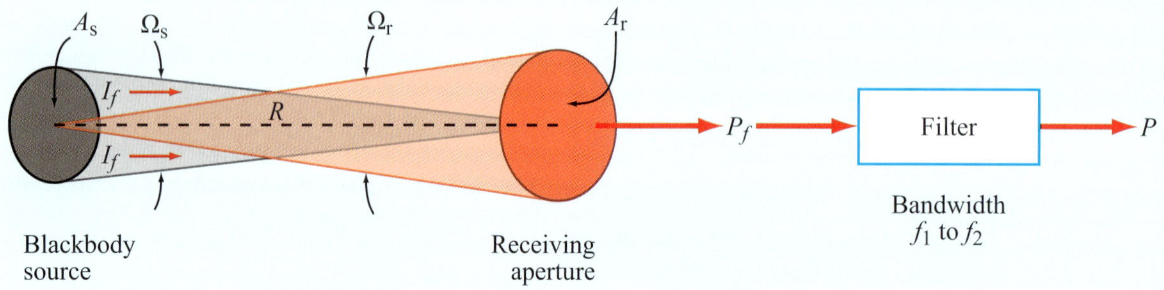

Figure 6-2: Geometry for power received from a blackbody source.

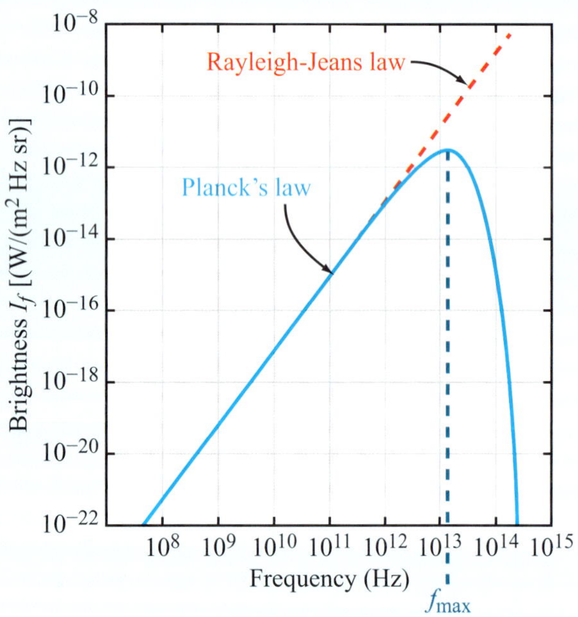

Figure 6-3: Comparison of Planck's law with its low-frequency approximation (Rayleigh–Jeans law) at 300 K.

radiometer or radio telescope is pointed towards the cosmos, it is said to be looking at ***cold space***.

6-3 Power-Temperature Correspondence

Consider the arrangement shown in Fig. 6-4 for a lossless receive antenna with effective aperture A_r, surrounded by a blackbody of spectral brightness intensity I_f. The antenna is characterized by a radiation pattern $F(\theta, \phi)$. To account for the directionality of the antenna, we revise Eq. (6.10) into a differential form that expresses the differential spectral power dP_f received

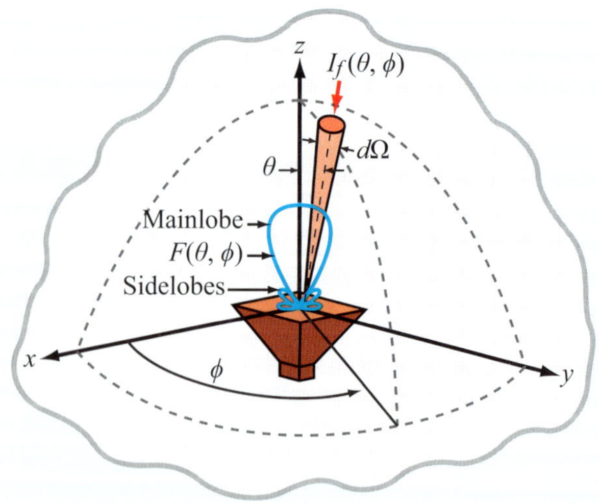

Figure 6-4: Blackbody spectral brightness I_f incident on an antenna with effective aperture A_r and radiation pattern $F(\theta, \phi)$.

through a differential solid angle $d\Omega$ along the direction (θ,ϕ), relative to the boresight direction of the antenna, as

$$dP_f = I_f A_r\, F(\theta,\phi)\, d\Omega. \tag{6.15}$$

The total amount of power received by the antenna from all directions over the frequency range f_1 to f_2 is

$$P = A_r \int_{f_1}^{f_2} \iint_{4\pi} I_f\, F(\theta,\phi)\, d\Omega\, df, \tag{6.16}$$

where the integration is over a complete solid angle of 4π (i.e., for $\theta = 0$ to π and $\phi = 0$ to 2π). The energy emitted by a blackbody is unpolarized, and the same is true in general for the Earth's atmosphere and many other media. This means that on average, half of the energy is polarized along any specified polarization direction and the other half is polarized along the direction orthogonal to it. If we choose a vertical/horizontal pair of orthogonal polarization directions, and if we observe the polarization direction of the energy radiated by a blackbody as a function of time and then decompose it into vertical and horizontal components, we discover that, on average, half of the total energy is vertically polarized and the other half is horizontally polarized. We obtain the same result if we choose the orthogonal pair of polarization states to be right-hand circular and left-hand circular, or any two orthogonal elliptically polarized states.

At any instant in time, a typical antenna can intercept EM energy along a single polarization direction. Consequently, regardless of its polarization state (as a function of time), the antenna can detect, on average, only $1/2$ of the energy incident upon its aperture. Hence, we introduce a factor of $1/2$ in Eq. (6.16), which leads to

$$P = \frac{1}{2} A_r \int_{f_1}^{f_2} \iint_{4\pi} I_f\, F(\theta,\phi)\, d\Omega\, df. \tag{6.17}$$

(polarized antenna)

In the microwave region, under the Rayleigh–Jeans approximation given by Eq. (6.13), Eq. (6.17) becomes

$$P_{bb} = \frac{1}{2} A_r \int_{f_1}^{f_2} \iint_{4\pi} \frac{2kT}{\lambda^2}\, F(\theta,\phi)\, d\Omega\, df, \tag{6.18}$$

where the subscript bb has been added to denote a blackbody enclosure. If the detected power is limited to a narrow bandwidth $B = (f_2 - f_1)$ such that I_f is approximately constant over B (i.e., $B \ll f$), then Eq. (6.18) simplifies to

$$P_{bb} = kTB\, \frac{A_r}{\lambda^2} \iint_{4\pi} F(\theta,\phi)\, d\Omega. \tag{6.19}$$

The integral in Eq. (6.19) is recognized as the pattern solid angle Ω_p, defined by Eq. (3.16) as

$$\iint_{4\pi} F(\theta,\phi)\, d\Omega = \Omega_p, \tag{6.20}$$

and Ω_p is related to the effective aperture A_r through Eq. (3.31a),

$$\Omega_p = \frac{\lambda^2}{A_r}\,. \tag{6.21}$$

In view of these relationships, Eq. (6.19) becomes

$$P_{bb} = kTB. \tag{6.22}$$

This result is of fundamental significance in microwave remote sensing; the direct linear relationship between power and temperature has led to the interchangeable use of the two terms.

A result analogous to Eq. (6.22) was derived by Nyquist in 1928 for a resistor at a temperature T [Fig. 6-5(b)]. He showed that the noise power P_n available at the resistor terminals is

$$P_n = kTB. \tag{6.23}$$

From the standpoint of an ideal receiver of bandwidth B, the antenna connected to its input terminals is equivalent to a resistance R_{rad}, called the ***antenna radiation resistance*** (see Section 3-2.5). In both cases shown in Fig. 6-5 the receiver is connected to a "resistor," but in the case of the real resistor of Fig. 6-5(b) the noise power available at its output terminals is determined by the physical temperature of the resistor, whereas for the antenna the power at its output terminals is determined by the temperature of the blackbody

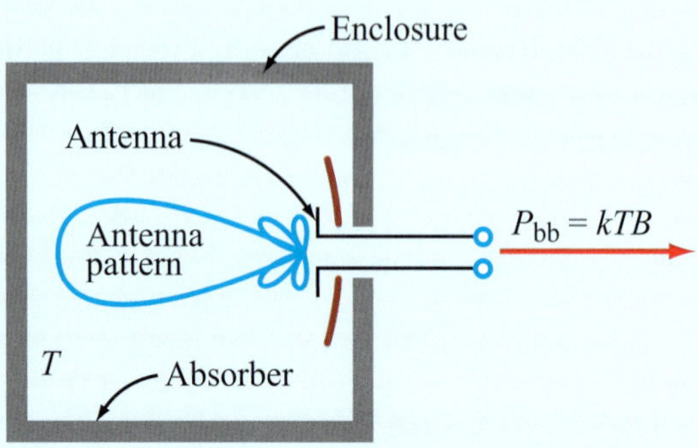

(a) Antenna inside a blackbody enclosure

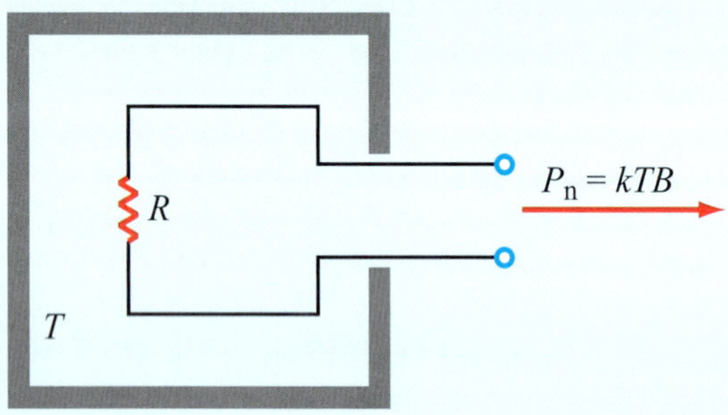

(b) Resistor at temperature T

Figure 6-5: The power delivered by (a) an antenna placed inside a blackbody enclosure of temperature T is equal to the power delivered by (b) a resistor maintained at the same temperature.

enclosure [Fig. 6-5(a)], regardless of how close or far the enclosure walls are from the antenna. Moreover, the physical temperature of the antenna structure has no bearing on its output power (as long as the antenna is lossless).

6-4 Radiation by Natural Materials

6-4.1 Brightness Temperature

A blackbody is an idealized body that, when in thermodynamic equilibrium at a temperature T, radiates at least as much energy as any other body at the same temperature. A blackbody is also a perfect absorber.

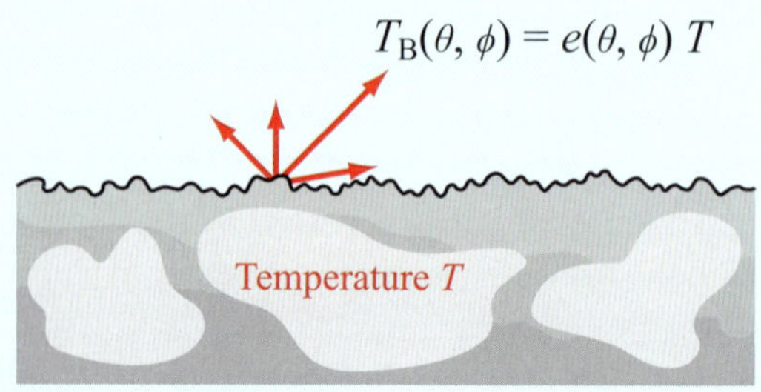

Figure 6-6: Brightness temperature of a semi-infinite isothermal medium.

Real materials, sometimes referred to as ***gray bodies***, emit less than a blackbody does and do not necessarily absorb all the energy incident upon them. In the microwave region, the singularly polarized brightness intensity I_{bb} of a blackbody at a temperature T is

$$I_{bb} = I_f B = \frac{kT}{\lambda^2} B, \tag{6.24}$$

where we have used Eq. (6.13) and multiplied it by a factor of $1/2$. Consider the semi-infinite material shown in Fig. 6-6. If its brightness, which may be direction-dependent, is $I(\theta, \phi)$ and its physical temperature is T, a ***blackbody equivalent radiometric**** ***temperature*** may be defined so that $I(\theta, \phi)$ can assume a form similar to Eq. (6.24). Such a temperature is called the ***brightness temperature*** $T_B(\theta, \phi)$ of the material, and it is related to the brightness intensity $I(\theta, \phi)$ by

$$I(\theta, \phi) = \frac{k}{\lambda^2} T_B(\theta, \phi)\, B. \tag{6.25}$$

The ratio of the brightness intensity $I(\theta, \phi)$ of the material to that of a blackbody at the same temperature is defined as the ***emissivity*** $e(\theta, \phi)$:

$$e(\theta, \phi) = \frac{I(\theta, \phi)}{I_{bb}} = \frac{T_B(\theta, \phi)}{T}. \tag{6.26}$$

*To avoid confusion between physical and radiometric temperatures, the latter always carries uppercase subscripts, while the former carries either no subscript or lowercase subscripts.

▶ Since $I(\theta,\phi) \leq I_{\text{bb}}$, it follows that the emissivity is constrained to the range $0 \leq e(\theta,\phi) \leq 1$. Thus, the brightness temperature $T_B(\theta,\phi)$ of a material always is smaller than or equal to its physical temperature T. When $e < 1$, the material is said to have a "cooler" brightness temperature than its physical temperature. ◀

Implied in this definition of emissivity is the assumption that the material is homogeneous and of uniform temperature. Departures from these conditions are considered in later sections.

▶ Because of the one-to-one correspondence between the brightness temperature $T_B(\theta,\phi)$ of a material and the thermal energy it radiates (represented by $I(\theta,\phi)$), we will henceforth use $T_B(\theta,\phi)$ as a surrogate for $I(\theta,\phi)$. That is, when we use a phrase of the form "as $T_B(\theta,\phi)$ propagates between the material and an observing antenna" we are actually referring to the propagation of the emitted brightness intensity $I(\theta,\phi)$ between the material and the antenna. ◀

6-4.2 Brightness Temperature Distribution

In the preceding subsection we defined the brightness temperature T_B of a material as the temperature of an equivalent blackbody that emits the same brightness as the material under consideration. In practice, when a radiometer is used to observe the emission from a certain material, such as the Earth's surface, neither the radiometer nor the surface are isolated from radiation by other sources, such as the atmosphere.

Let us consider the diagram shown in Fig. 6-7(a). The radiation incident upon the antenna from any specific direction may contain components originating from several different sources. Specifically, we have:

(a) self-emitted radiation from the terrain, represented by the ***surface emission*** brightness temperature $T_{SE}(\theta,\phi)$,

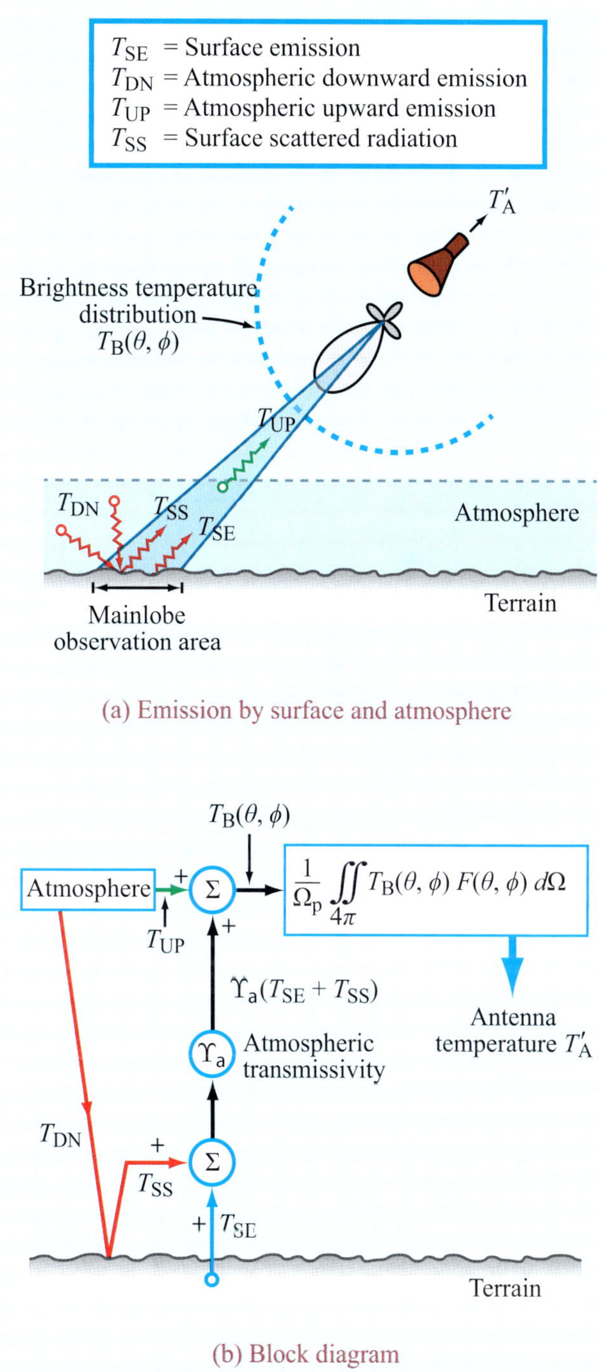

Figure 6-7: Relationships between (lossless) antenna temperature T'_A, brightness temperature $T_B(\theta,\phi)$ incident on the antenna and surface emitted brightness temperature T_{SE}: (a) schematic representation; (b) block-diagram representation.

(b) self-emitted ***upward radiation*** from the atmosphere, represented by brightness temperature $T_{\text{UP}}(\theta,\phi)$, and

(c) ***downward-emitted*** atmospheric radiation $[T_{\text{DN}}(\theta,\phi)]$ that gets scattered (reflected) by the terrain in the direction of the antenna. In Fig. 6-7, the part of T_{DN} scattered by the surface is labeled $T_{\text{SS}}(\theta,\phi)$, where the subscript SS stands for ***surface-scattered***.

In addition to acting as a radiation source, the atmosphere also attenuates EM energy passing through it. Consequently, both $T_{\text{SE}}(\theta,\phi)$ and $T_{\text{SS}}(\theta,\phi)$ are attenuated as they propagate through the intervening atmosphere between the terrain and the antenna. We account for the attenuation by multiplying $(T_{\text{SE}}+T_{\text{SS}})$ by the atmospheric transmissivity $\Upsilon_{\text{a}}(\theta,\phi)$. Hence, the net brightness temperature T_{B}, representing the energy incident upon the antenna, is given by

$$T_{\text{B}} = T_{\text{UP}} + \Upsilon_{\text{a}}(T_{\text{SE}} + T_{\text{SS}}). \tag{6.27}$$

Models for computing T_{SE}, T_{DN}, T_{UP}, T_{SS}, and Υ_{a} are presented in forthcoming sections of this chapter or future chapters. Our purpose at the present time is to establish the framework [Fig. 6-7(b)] and associated terminology so as to relate the output power delivered by the antenna to the various sources of emission present in the scene observed by the antenna through its radiation pattern $F(\theta,\phi)$. To that end, we introduce two new terms, the ***brightness temperature distribution*** $T_{\text{B}}(\theta,\phi)$ and the (lossless) ***antenna temperature*** T'_{A}. The latter, which is defined and discussed in the next subsection, represents the power delivered by a perfectly lossless antenna (radiation efficiency $=1$) to a radiometer receiver. The effects of antenna losses are dealt with in Section 6-5.

The brightness temperature distribution $T_{\text{B}}(\theta,\phi)$ represents the brightness distribution $I_{\text{i}}(\theta,\phi)$ of the energy incident upon the antenna [Fig. 6-7(a)]. It incorporates all of the sources of energy in the scene including reflections and attenuation effects. If the atmosphere in Fig. 6-7 is lossless, it follows that:

$$\Upsilon_{\text{a}} = 1, \qquad T_{\text{UP}} = T_{\text{DN}} = T_{\text{SS}} = 0,$$

and

$$T_{\text{B}} = T_{\text{SE}} \qquad \textbf{(lossless atmosphere)}.$$

6-4.3 Antenna Temperature

From Eq. (6.18), the power at the output terminals of a lossless antenna placed inside a blackbody enclosure is given by

$$P_{\text{bb}} = \frac{1}{2}\, A_{\text{r}} B \iint_{4\pi} \frac{2kT}{\lambda^2}\, F(\theta,\phi)\, d\Omega. \tag{6.28}$$

The expression applies to a narrow-band receiver with $B \ll f$. For a nonblackbody enclosure emitting a brightness distribution represented by $T_{\text{B}}(\theta,\phi)$, the analogous expression is

$$P = \frac{1}{2}\, A_{\text{r}} B \iint_{4\pi} \frac{2k}{\lambda^2}\, T_{\text{B}}(\theta,\phi)\, F(\theta,\phi)\, d\Omega. \tag{6.29a}$$

Use of Eq. (6.21) in Eq. (6.29a) leads to

$$P = \frac{kB}{\Omega_{\text{p}}} \iint_{4\pi} T_{\text{B}}(\theta,\phi)\, F(\theta,\phi)\, d\Omega = kB\, \frac{\displaystyle\iint_{4\pi} T_{\text{B}}(\theta,\phi)\, F(\theta,\phi)\, d\Omega}{\displaystyle\iint_{4\pi} F(\theta,\phi)\, d\Omega}, \tag{6.29b}$$

where in the last step we replaced the pattern solid angle Ω_{p} with its defining equation, namely Eq. (6.20).

The ***receiver transfer function*** of a microwave radiometer is established by measuring the output voltage V_{out} as a function of the physical temperature T of a matched resistor placed at the receiver input (in place of the antenna), as shown in Fig. 6-8. The process relies on the relationship expressed by Eq. (6.23), which states that the noise power P_{n} delivered by a resistor is proportional to its physical temperature. By extension, corresponding to the power P provided by the antenna to the receiver (as a consequence of observing a scene with brightness temperature distribution $T_{\text{B}}(\theta,\phi)$), a resistor-equivalent ***lossless-antenna temperature*** T'_{A} can

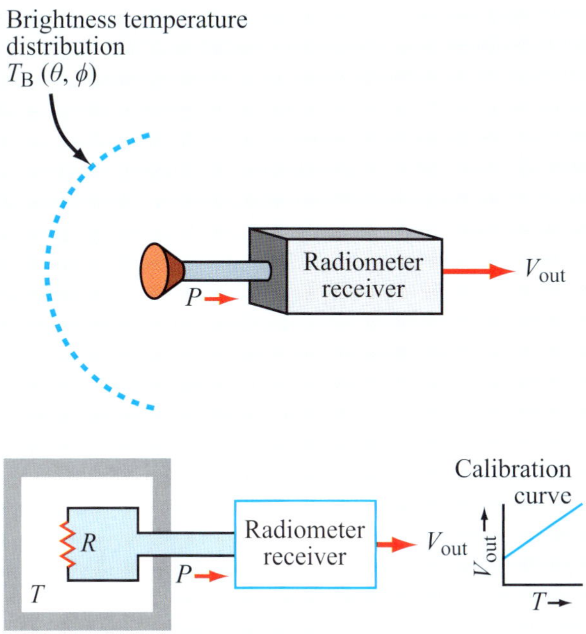

Figure 6-8: The power received by an antenna is equivalent to the noise power delivered by a matched resistor.

be defined such that the noise power delivered by the resistor at that temperature is equal to P. Thus,

$$P = kT'_A B. \tag{6.30}$$

Equating Eqs. (6.29b) and (6.30) leads to

$$T'_A = \frac{\iint\limits_{4\pi} T_B(\theta,\phi)\, F(\theta,\phi)\, d\Omega}{\iint\limits_{4\pi} F(\theta,\phi)\, d\Omega}. \tag{6.31}$$

(lossless antenna)

▶ The (lossless) antenna temperature T'_A is equal to the integrated value of the weighted brightness temperature distribution $T_B(\theta,\phi)$, with the antenna radiation pattern acting as the weighting function, normalized to the integral of the weighting function $F(\theta,\phi)$. The integrations encompass all directions (4π solid angle). ◀

For a blackbody enclosure at temperature T, the brightness temperature of the enclosure is simply $T_B(\theta,\phi) = T$ for all directions (θ,ϕ), in which case Eq. (6.31) reduces to $T'_A = T$.

6-5 Antenna Efficiency Considerations

6-5.1 Beam Efficiency

A microwave radiometer may use a single antenna or an array of antenna elements to observe or image a scene of interest. In either case, the ideal antenna pattern $F(\theta,\phi)$ is a narrow pencil beam with no sidelobes. In reality, most antennas have sidelobes and backlobes, so in addition to the thermal radiation received through the main beam of the antenna pattern, the antenna also receives contributions through the remainder of the pattern (Fig. 6-9). To evaluate the significance of these undesirable contributions, let us divide the integral in the numerator of Eq. (6.31) into two parts, one representing the mainlobe contribution and the other representing contributions received from directions outside the mainlobe:

$$T'_A = \frac{\iint\limits_{\text{mainlobe}} T_B(\theta,\phi)\, F(\theta,\phi)\, d\Omega}{\iint\limits_{4\pi} F(\theta,\phi)\, d\Omega} + \frac{\iint\limits_{4\pi-\text{mainlobe}} T_B(\theta,\phi)\, F(\theta,\phi)\, d\Omega}{\iint\limits_{4\pi} F(\theta,\phi)\, d\Omega}. \tag{6.32}$$

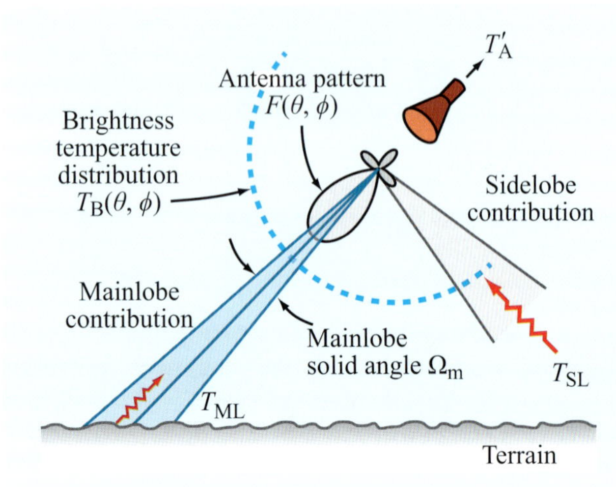

Figure 6-9: T_{ML} and T_{SL} are the mainlobe and sidelobe contributions to antenna temperature T'_A.

The integration in the numerator of the first term is limited to the mainlobe (down to the first minimum), in contrast with the integration in the second term, which includes all 4π solid angle, except for the mainlobe.

Next, we introduce T_{ML}, the average ***mainlobe brightness temperature***, and T_{SL}, the average ***sidelobe brightness temperature***, defined as

$$T_{ML} = \frac{\displaystyle\iint_{\text{mainlobe}} T_B(\theta,\phi)\, F(\theta,\phi)\, d\Omega}{\displaystyle\iint_{\text{mainlobe}} F(\theta,\phi)\, d\Omega}, \tag{6.33}$$

and

$$T_{SL} = \frac{\displaystyle\iint_{4\pi-\text{mainlobe}} T_B(\theta,\phi)\, F(\theta,\phi)\, d\Omega}{\displaystyle\iint_{4\pi-\text{mainlobe}} F(\theta,\phi)\, d\Omega}. \tag{6.34}$$

In Eq. (6.33), the numerator and denominator have the same integration limits, namely over the mainlobe. A similar definition applies to T_{SL} in Eq. (6.34), but over the angular range (4π minus mainlobe).

An antenna's ***pattern solid angle*** Ω_p and mainlobe solid angle Ω_m are defined as

$$\Omega_p = \iint_{4\pi} F(\theta,\phi)\, d\Omega \tag{6.35a}$$

and

$$\Omega_m = \iint_{\text{mainlobe}} F(\theta,\phi)\, d\Omega, \tag{6.35b}$$

and their ratio is the ***beam efficiency*** η_b:

$$\eta_b = \frac{\Omega_m}{\Omega_p}. \tag{6.36}$$

In view of Eqs. (6.33) and (6.36), Eq. (6.32) can be cast in the form

$$T'_A = \eta_b T_{ML} + (1-\eta_b) T_{SL}. \tag{6.37}$$

For an ideal antenna with only a mainlobe and no sidelobes, $\eta_b = 1$, in which case Eq. (6.37) reduces to $T'_A = T_{ML}$.

6-5.2 Radiation Efficiency

The temperature T'_A represents the power delivered by a lossless receiving antenna at its output terminals. In practice, however, antennas are not lossless. As was noted earlier in Section 3-2.5, part of the energy received (or transmitted) by an antenna is absorbed by the antenna material and converted to heat. When an antenna is operated in a transmission mode, its radiation efficiency ξ is the ratio of the total power radiated by the antenna to the power supplied to it by a transmission source. In essence, ξ is a power transmission coefficient. Similarly, when the antenna is operated as a receiver, ξ is the ratio of the power it delivers to the receiver to the total power intercepted by it.

To account for the effects of antenna ohmic losses on the received power, we define the ***antenna temperature*** T_A as the antenna temperature in the general case, whether or not the antenna is lossless. As we show later in Section 7-2.4, a lossy device that absorbs EM energy also is a radiator of EM energy. Hence, an antenna with $\xi < 1$ not only delivers less power to the receiver, but also contributes noise power of its own. An antenna with radiation efficiency ξ has an absorption coefficient $(1-\xi)$. Under thermodynamic equilibrium the emissivity of the antenna is equal to its absorption coefficient and the power it radiates towards the receiver is equivalent to a noise temperature T_N given by

$$T_N = (1-\xi)T_0, \tag{6.38}$$

where T_0 is the physical temperature of the antenna.

The relationship between T_A (for any antenna) and T'_A (for a lossless antenna) is

$$T_A = \xi T'_A + T_N = \xi T'_A + (1-\xi)T_0. \tag{6.39}$$

Inserting Eq. (6.37) into Eq. (6.39) gives

$$T_A = \xi \eta_b T_{ML} + \xi(1-\eta_b)T_{SL} + (1-\xi)T_0. \tag{6.40}$$

For an ideal antenna with radiation efficiency $\xi = 1$ and main-beam efficiency $\eta_b = 1$, Eq. (6.40) reduces to $T_A = T_{ML}$.

> ▶ The goal of the radiometer designer is to select an antenna with radiation and main-beam efficiencies that are as close to 1 as possible. ◀

6-5.3 Radiometer Measurement Ambiguity

The power delivered by an antenna to a radiometer receiver is

$$P = kT_AB, \tag{6.41}$$

where T_A is the antenna temperature given by Eq. (6.40). As discussed later in Chapter 7, the receiver output voltage V_{out} can be calibrated to "read" temperature. Hence, T_A may be regarded as the measured quantity, and T_{ML}, the average brightness temperature of the resolution cell delineated by the main beam, as the quantity to be estimated from the measured value of T_A. Solving Eq. (6.40) for T_{ML} gives

$$T_{ML} = \left(\frac{1}{\xi\eta_b}\right)T_A - \left(\frac{1-\eta_b}{\eta_b}\right)T_{SL} - \left(\frac{1-\xi}{\xi\eta_b}\right)T_0. \tag{6.42}$$

This is a linear equation of the form

$$T_{ML} = aT_A + b, \tag{6.43}$$

where a is a ***scaling factor***,

$$a = \frac{1}{\xi\eta_b}, \tag{6.44a}$$

and b is a ***bias term*** given by

$$b = -\left(\frac{1-\eta_b}{\eta_b}\right)T_{SL} - \left(\frac{1-\xi}{\xi\eta_b}\right)T_0. \tag{6.44b}$$

If ξ, η_b, T_{SL}, and T_0 are known quantities, T_{ML} can be easily recovered from the measured value of T_A. The antenna parameters ξ and η_b can be determined using the measurement techniques covered in Section 7-11, and T_0 can be monitored by attaching a thermistor to the skin of the antenna. The only remaining quantity is the sidelobe temperature T_{SL}. Unfortunately, determining the value of T_{SL} is not so straightforward. Recall from Eq. (6.34) that T_{SL} represents the emission by the scene, integrated over all directions outside the mainlobe of the antenna pattern. The apparent temperature of the scene may vary over a wide range of values, depending on whether the sidelobes are pointing in the direction of the atmosphere, water, or land. For an Earth-pointing airborne or spaceborne radiometer, the scene changes constantly (except perhaps over the ocean) as the radiometer flies along its trajectory, which means that the ***sidelobe factor*** $(1-\eta_b)T_{SL}/\eta_b$ (second term in Eq. (6.42)) is an unknown time-varying bias. To appreciate the significance of this bias term, we display in Fig. 6-10 plots of the sidelobe factor as a function of the magnitude of T_{SL} for each of several values of the beam efficiency η_b. For typical land and ocean scenes, T_{SL} may vary between 100 K and 300 K. If the radiometer uses a standard communication antenna with $\eta_b = 0.8$, the sidelobe factor may vary between 25 K

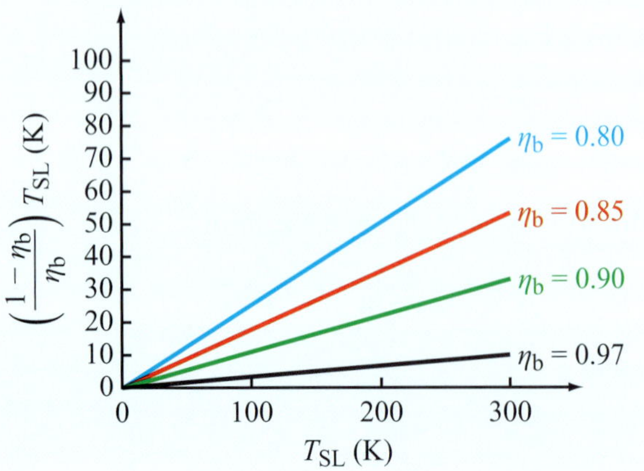

Figure 6-10: Sidelobe factor as a function of the incident sidelobe brightness temperature T_{SL} for each of several values of the beam efficiency η_b.

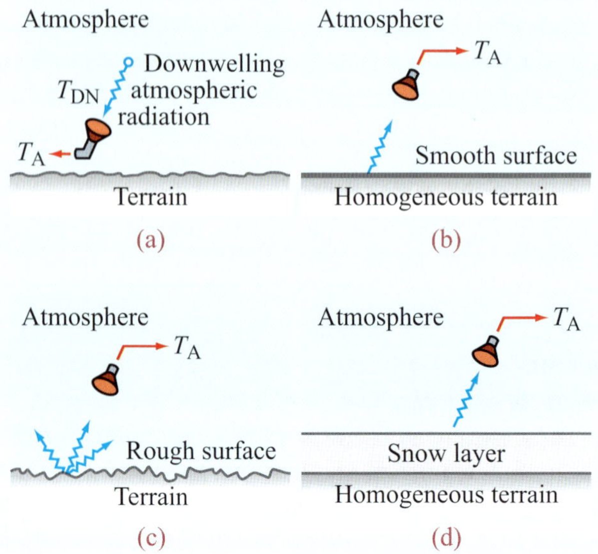

Figure 6-11: Examples of configurations of interest in radiometric remote sensing: (a) upward-looking radiometer, and downward-looking radiometer with (b) smooth-surface boundary, (c) rough-surface boundary, and (d) two-layer terrain.

and 75 K (corresponding to T_{SL} varying between 100 K and 300 K). Most applications require a radiometric accuracy on the order of 1 K, so the 50-K uncertainty is unacceptable. To reduce the uncertainty to an acceptable level, it is critically important that the radiometer antenna be designed to have exceedingly low sidelobe levels, by increasing η_b to as close to 1 as possible. This is realized by controlling the amplitude distribution of the electric field across the antenna aperture. Of course, increasing η_b causes the mainlobe to enlarge, which reduces the angular resolution of the antenna beam and the corresponding spatial resolution (at a given range R) of the radiometer. This is a fundamental trade-off in radiometer system design.

Antennas can be designed to have beam efficiencies as high as 0.99. By so doing, the variation in the sidelobe factor is substantially reduced. From our previous example, it is reduced down to a narrow range extending from 1 K to 3 K, compared with 25 K to 75 K for an antenna with $\eta_b = 0.8$. If the sidelobe factor is assigned a value of 2 K, the maximum absolute error becomes ± 1 K.

▶ The accuracy of a radiometric measurement of T_{ML}, the brightness temperature of the scene observed by the mainlobe of the antenna, is strongly influenced by the accuracies with which the radiation and beam efficiencies of the antenna, ξ and η_b, are known, and by how close η_b is to 1. ◀

6-6 Theory of Radiative Transfer

Having established the connection between the antenna temperature measured by the radiometer and the brightness energy incident upon it, we now examine the nature of the energy emitted by the medium observed by the antenna.

The four scenarios shown in Fig. 6-11 depict:

(a) an *upward-looking* microwave radiometer measuring the downwelling radiation emitted by the atmosphere,

(b) the reverse of scenario (a), namely a *downward-looking* radiometer observing a smooth surface through the atmosphere,

(c) the same scenario as (b) except now the surface is *electromagnetically rough*, which alters the nature of the emission by the ground surface, and

(d) the same as scenario (b), except now the soil surface is covered by a *layer of snow*.

These are a few of the many scenarios that an upward-looking or downward-looking radiometer may encounter in microwave remote sensing. The purpose of the present and future sections in this chapter is to develop formulations that can relate the brightness temperature T_B (representing the brightness intensity I) incident upon the radiometer's antenna aperture to the physical and electromagnetic properties of the scene, including the intervening atmospheric layer between the terrain and the radiometer antenna. These formulations, which are sufficiently general and flexible to apply to all the scenarios shown in Fig. 6-11 and to many variations thereof, are obtained by solving the ***equation of radiative transfer*** under a variety of different conditions.

The interaction between EM radiation and matter involves two processes: ***extinction*** and ***emission***. When radiation traversing a medium is reduced in intensity due to absorption and/or scattering by the medium, the process is called extinction, and when the medium adds energy of its own to the original beam of energy, the process is called emission. Usually the interaction consists of both processes occurring simultaneously.

6-6.1 Equation of Radiative Transfer

Consider the propagation of EM energy in the direction $\hat{\mathbf{R}}$ in Fig. 6-12. The imaginary infinitesimal cylindrical volume has cross section dA and length (thickness) dR. The cylinder is located in a spherical coordinate system with its center at range vector $\mathbf{R}$. The material contained in the cylinder and its surroundings may be a gas, a liquid, or a solid. Incident normally upon the lower face of the cylinder is brightness intensity $I(\mathbf{R}, \hat{\mathbf{R}})$, where $\mathbf{R}$ denotes the location of the cylinder and $\hat{\mathbf{R}}$ denotes the direction of the brightness. Exiting from

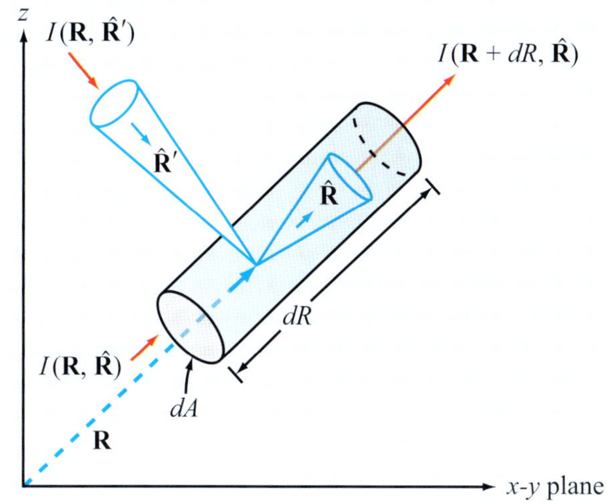

Figure 6-12: Radiation transfer through an infinitesimal cylinder.

the cylinder's upper face along the same direction $\hat{\mathbf{R}}$ is brightness intensity $I(\mathbf{R}+d\mathbf{R}, \hat{\mathbf{R}})$. Recall that intensity is energy per unit area dA, radiated within a solid angle $d\Omega$, during an interval of 1 s, and containing EM waves over a frequency interval extending over a bandwidth B centered at f (with $B \ll f$). The loss in intensity (the amount removed from $I(\mathbf{R}, \hat{\mathbf{R}})$) due to propagation over thickness dR is

$$dI_{\text{extinction}}(\mathbf{R}, \hat{\mathbf{R}}) = \kappa_e \, I(\mathbf{R}, \hat{\mathbf{R}}) \, dR, \tag{6.45}$$

where κ_e, measured in nepers/m (Np/m), is the ***extinction coefficient*** of the medium containing the imaginary cylinder. The extinction coefficient is also known as the ***power attenuation*** coefficient. The extincted energy may be due to absorption by the material, scattering by particles contained in it, or both. Loss due to scattering refers to changing the direction of the incident energy into new directions away from the original direction $\hat{\mathbf{R}}$ of $I(\mathbf{R}, \hat{\mathbf{R}})$. Absorption and scattering are linear processes. Hence, the extinction coefficient κ_e may be expressed as the sum of an ***absorption coefficient*** κ_a and a ***scattering coefficient*** κ_s:

$$\kappa_e = \kappa_a + \kappa_s \qquad \text{(Np/m)}. \tag{6.46}$$

In addition to attenuating the incident brightness, the medium in the cylinder also generates brightness of its own and adds it to the brightness propagating along $\hat{\mathbf{R}}$. The emitted brightness intensity is given by

$$dI_{\text{emission}}(\mathbf{R},\hat{\mathbf{R}}) = [\kappa_a J_a(\mathbf{R},\hat{\mathbf{R}}) + \kappa_s J_s(\mathbf{R},\hat{\mathbf{R}})]\, dR, \tag{6.47}$$

where J_a and J_s are ***source functions*** that account for thermal emission and scattering in the direction $\hat{\mathbf{R}}$. Source function J_a is called an ***absorption source function*** because under conditions of local thermodynamic equilibrium, thermal emission has to be equal to absorption (Kirchhoff's law). Otherwise, the temperature of the medium would change. Fortunately, when a radiometer is used to measure the emission from a scene of interest, it integrates the emission over a very short interval, typically on the order of milliseconds to seconds. Over such short durations, the physical temperature of the scene remains essentially constant, thereby satisfying the condition of thermodynamic equilibrium.

The scattering source function $J_s(\mathbf{R},\hat{\mathbf{R}})$ accounts for radiation scattered in direction $\hat{\mathbf{R}}$ (by the material in the cylinder) due to radiation incident upon the cylinder from all other directions (Fig. 6-12). It is given by

$$J_s(\mathbf{R},\hat{\mathbf{R}}) = \frac{1}{\kappa_s} \iint\limits_{4\pi} \psi(\hat{\mathbf{R}},\hat{\mathbf{R}}')\, I(\mathbf{R},\hat{\mathbf{R}}')\, d\Omega', \tag{6.48}$$

where $I(\mathbf{R},\hat{\mathbf{R}}')$ is the brightness incident from direction $\hat{\mathbf{R}}'$, a portion of which gets redirected along the direction $\hat{\mathbf{R}}$, thereby adding to the original brightness incident upon the cylinder in that direction. For every combination of directions $(\hat{\mathbf{R}},\hat{\mathbf{R}}')$, the redirection fraction is governed by the ***scattering phase function*** $\psi(\hat{\mathbf{R}},\hat{\mathbf{R}}')$, which is related to the size (relative to λ), shape, orientation and spatial distribution, and constitutive properties of the particles contained within the cylindrical volume. As we discuss later in Section 11-5.5, $\psi = \kappa_s/4\pi$ for volumes containing particles that scatter the energy incident upon them isotropically.

Oftentimes, κ_a and κ_s are expressed as

$$\kappa_a = (1-a)\kappa_e, \qquad \kappa_s = a\kappa_e, \tag{6.49}$$

where a is the single-scattering albedo defined as

$$a = \frac{\kappa_s}{\kappa_e}\,. \tag{6.50}$$

Using these definitions, Eq. (6.47) becomes

$$\begin{aligned} dI_{\text{emission}}(\mathbf{R},\hat{\mathbf{R}}) &= \kappa_e[(1-a)\, J_a(\mathbf{R},\hat{\mathbf{R}}) \\ &\qquad + a\, J_s(\mathbf{R},\hat{\mathbf{R}})]\, dR \\ &= \kappa_e\, J(\mathbf{R},\hat{\mathbf{R}})]\, dR, \end{aligned} \tag{6.51}$$

where we combined the source terms into a ***total source function*** $J(\mathbf{R},\hat{\mathbf{R}})$ given by

$$J(\mathbf{R},\hat{\mathbf{R}}) = (1-a)\, J_a(\mathbf{R},\hat{\mathbf{R}}) + a\, J_s(\mathbf{R},\hat{\mathbf{R}}). \tag{6.52}$$

Propagation over thickness dR results in a change in intensity dI given by

$$\begin{aligned} dI(\hat{\mathbf{R}},\hat{\mathbf{R}}) &= I_{\text{emission}}(\mathbf{R},\hat{\mathbf{R}}) - I_{\text{extinction}}(\mathbf{R},\hat{\mathbf{R}}) \\ &= \kappa_e\, J(\mathbf{R},\hat{\mathbf{R}})\, dR - \kappa_e\, I(\mathbf{R},\hat{\mathbf{R}})\, dR \\ &= \kappa_e\, dR[J(\mathbf{R},\hat{\mathbf{R}}) - I(\mathbf{R},\hat{\mathbf{R}})]. \end{aligned} \tag{6.53}$$

The dimensionless product $\kappa_e\, dR$ is often abbreviated as

$$d\tau = \kappa_e\, dR, \tag{6.54}$$

and $d\tau$ is called an increment of ***optical depth***. A more appropriate name is ***electromagnetic depth***, because it is applicable across the entire EM spectrum. Nevertheless, for historical reasons, we continue to use the term "optical" even when treating radiative transfer situations in other parts of the spectrum.

Using Eq. (6.54) in Eq. (6.53) leads to the ***radiative transfer equation***:

$$\frac{dI}{d\tau} + I = J \quad \textbf{(radiative transfer equation)}, \tag{6.55}$$

where it is understood that I and J are defined at location $\mathbf{R}$ and for propagation in direction $\hat{\mathbf{R}}$.

6-6.2 Brightness-Temperature Equation

Per Eq. (6.25), we convert intensity I to brightness temperature T_B,

$$I(\mathbf{R},\hat{\mathbf{R}}) = \frac{k}{\lambda^2}\, T_B(\mathbf{R},\hat{\mathbf{R}})\, B. \tag{6.56}$$

As noted earlier, Kirchhoff's law states that under conditions of local thermodynamic equilibrium, thermal emission has to be equal to absorption, which leads to the conclusion that the absorption source function J_a is isotropic and given by the Rayleigh–Jeans approximation to Planck's law. That is,

$$J_a(\mathbf{R}) = \frac{k}{\lambda^2}\, T(\mathbf{R})\, B, \tag{6.57}$$

where $T(\mathbf{R})$ is the kinetic (physical) temperature of the medium at location $\mathbf{R}$. Because $J_a(\mathbf{R})$ is isotropic, it is not dependent on $\hat{\mathbf{R}}$; thermal emission by an infinitesimal volume is the same along all directions.

In an analogous fashion, the scattering source function J_s can be defined in terms of a ***volume scattered radiometric temperature*** $T_{VS}(\mathbf{R},\hat{\mathbf{R}})$ as follows:

$$J_s(\mathbf{R},\hat{\mathbf{R}}) = \frac{k}{\lambda^2}\, T_{VS}(\mathbf{R},\hat{\mathbf{R}})\, B. \tag{6.58}$$

Replacing $I(\mathbf{R},\hat{\mathbf{R}})$ and $J_s(\mathbf{R},\hat{\mathbf{R}})$ in Eq. (6.48) with their surrogates, $T_B(\mathbf{R},\hat{\mathbf{R}})$ and $T_{VS}(\mathbf{R},\hat{\mathbf{R}})$, leads to

$$T_{VS}(\mathbf{R},\hat{\mathbf{R}}) = \frac{1}{\kappa_s} \iint_{4\pi} \psi(\hat{\mathbf{R}},\hat{\mathbf{R}}')\, T_B(\mathbf{R},\hat{\mathbf{R}}')\, d\Omega. \tag{6.59}$$

Use of Eqs. (6.56) to (6.58) in Eqs. (6.52) and (6.55) leads to

$$\frac{dT_B}{d\tau} + T_B = (1-a)T + aT_{VS}. \tag{6.60}$$

(brightness temperature radiative transfer equation)

The brightness-related quantities (T_B, T and T_{VS}) are defined at location $\mathbf{R}$ for propagation in direction $\hat{\mathbf{R}}$.

6-6.3 Brightness Temperature of a Stratified Medium

Consider the stratified medium shown in Fig. 6-13, in which the radiative transfer properties of the medium are functions of z only. That is, a, T, T_{VS} and τ do not vary with x and y (or equivalently, with θ and ϕ). The figure depicts a microwave radiometer at height H above the surface of a lower medium. The upper hemisphere may be the Earth's atmosphere and the lower hemisphere may be land or water. The radiometer is pointed at an incidence angle θ relative to the vertical axis.

Our plan is to solve Eq. (6.60) to obtain an expression for the brightness temperature $T_B(\theta,H)$ incident upon the radiometer antenna (the azimuth angle ϕ is assumed constant throughout our forthcoming discussion.)

For the stratified medium in Fig. 6-13, Eq. (6.54) becomes

$$d\tau = \kappa_e\, dR = \kappa_e \sec\theta\, dz. \tag{6.61}$$

For propagation along the θ direction between two points located at (θ,z_1) and (θ,z_2), as illustrated in Fig. 6-13, the optical thickness $\tau(z_1,z_2)$ is obtained by

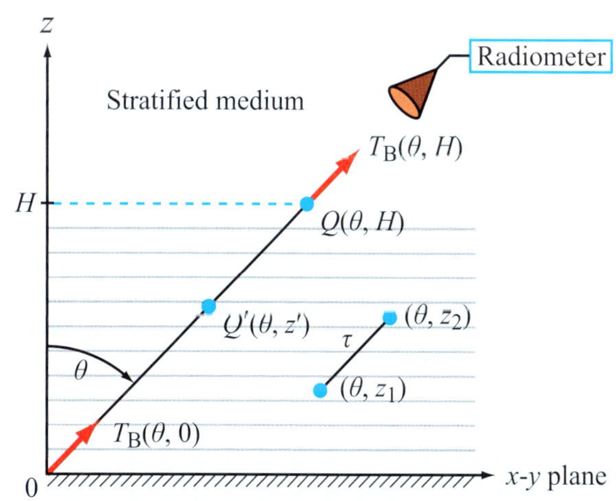

Figure 6-13: Geometry for brightness temperature radiative transfer equation.

integrating $d\tau$ between z_1 and z_2:

$$\tau(z_1, z_2) = \sec\theta \int_{z_1}^{z_2} \kappa_e \, dz.$$

For the specific path from the origin at $z = 0$ to an arbitrary point at $Q'(\theta, z')$, the optical thickness is

$$\tau(0, z') = \sec\theta \int_0^{z'} \kappa_e \, dz.$$

The differential equation given by Eq. (6.60) is applicable at any point in space. If at location (θ, z'), we multiply all of its terms by $e^{\tau(0,z')}$, we have

$$\begin{aligned} &\frac{dT_B(\theta, z')}{d\tau} \, e^{\tau(0,z')} + T_B(\theta, z') \, e^{\tau(0,z')} \\ &\quad = [(1-a)\, T(z') + aT_{VS}(\theta, z')] e^{\tau(0,z')}, \end{aligned} \tag{6.62}$$

where $d\tau = \kappa_e \sec\theta \, dz'$. The sum of the two terms on the left-hand side is recognized as

$$\begin{aligned} &\frac{dT_B(\theta, z')}{d\tau} \, e^{\tau(0,z')} + T_B(\theta, z') \, e^{\tau(0,z')} \\ &\quad = \frac{d}{d\tau} [T_B(\theta, z') \, e^{\tau(0,z')}], \end{aligned} \tag{6.63}$$

which leads to

$$\begin{aligned} &\frac{d}{d\tau} [T_B(\theta, z') \, e^{\tau(0,z')}] \\ &\quad = [(1-a)\, T(z') + aT_{VS}(\theta, z')] e^{\tau(0,z')}. \end{aligned} \tag{6.64}$$

Integrating both sides from $\tau(0,0)$ to $\tau(0,z)$ gives

$$\begin{aligned} &\int_0^{\tau(0,z)} \frac{d}{d\tau} [T_B(\theta, z') \, e^{\tau(0,z')}] \, d\tau \\ &\quad = \int_0^{\tau(0,z)} [(1-a)\, T(z') + aT_{VS}(\theta, z')] e^{\tau(0,z')} \, d\tau, \end{aligned} \tag{6.65}$$

which leads to the solution

$$\begin{aligned} &T_B(\theta, z') \, e^{\tau(0,z')} \Big|_{z'=0}^{z'=z} \\ &= \int_0^{z'=z} [(1-a)\, T(z') + aT_{VS}(\theta, z')] e^{\tau(0,z')} \kappa_e \sec\theta \, dz' \end{aligned}$$

or

$$\begin{aligned} T_B(\theta, z) &= T_B(\theta, 0) \, e^{-\tau(0,z)} \\ &\quad + \int_0^z [(1-a)\, T(z') + aT_{VS}(\theta, z')] \\ &\qquad \cdot e^{-\tau(z',z)} \kappa_e \sec\theta \, dz'. \end{aligned} \tag{6.66}$$

In the last step, we divided all terms by $e^{\tau(0,z)}$ and used the equality $\tau(0,z') - \tau(0,z) = -\tau(z',z)$, which follows from

$$\begin{aligned} &e^{\tau(0,z')} \cdot e^{-\tau(0,z)} \\ &\quad = \exp\left[\int_0^{z'} \kappa_e \sec\theta \, dz\right] \cdot \exp\left[-\int_0^{z} \kappa_e \sec\theta \, dz\right] \\ &\quad = \exp\left[-\int_{z'}^{0} \kappa_e \sec\theta \, dz - \int_0^{z} \kappa_e \sec\theta \, dz\right] \\ &\quad = \exp\left[-\int_{z'}^{z} \kappa_e \sec\theta \, dz\right] = e^{-\tau(z',z)}. \end{aligned} \tag{6.67}$$

At height $z = H$, Eq. (6.66) becomes

$$T_B(\theta, H) = T_B(\theta, 0) \, e^{-\tau(0,H)} + T_{UP}(\theta, H), \tag{6.68}$$

where $T_{UP}(\theta, H)$ is the ***upward emission contribution*** of the upper hemisphere,

$$\begin{aligned} T_{UP}(\theta, H) &= \sec\theta \int_0^H [(1-a)\, T(z') + aT_{VS}(\theta, z')] \\ &\qquad \cdot e^{-\tau(z',H)} \kappa_e \, dz'. \end{aligned} \tag{6.69}$$

Let us consider the physical meaning of each of the contributions to $T_B(\theta, H)$.

(a) Brightness temperature $T_B(\theta, 0)$ represents the upward-propagating energy emerging from the surface (x–y plane) along the θ direction. It may consist of thermal energy emitted by the lower hemisphere as well as downward propagating energy that gets reflected (scattered) by the surface into upward propagation along the θ direction. We will examine models for $T_B(\theta, 0)$ in future sections.

As $T_B(\theta,0)$ travels from the surface at $Q(\theta,0)$ to the radiometer at $Q(\theta,H)$, it gets attenuated due to absorption and scattering along the intervening path. The ***one-way atmospheric transmissivity*** of the layer between $z=0$ and $z=H$ along direction θ is

$$\Upsilon = e^{-\tau_0(0,H)\sec\theta} = \exp\left[-\int_0^H \kappa_e \sec\theta \, dz\right], \quad (6.70)$$

where τ_0 is τ at $\theta = 0$.

(b) The path between the surface and location $Q(\theta,H)$ contains two emission sources, one due to thermal emission and the other due to scattering. Differential thickness $\sec\theta \, dz'$, at location z', has a ***differential brightness temperature***

$$(1-a)\kappa_e \, T(z')\sec\theta \, dz' = \kappa_a \, T(z')\sec\theta \, dz', \quad (6.71)$$

where we used the relationship given by Eq. (6.49), namely $\kappa_a = (1-a)\kappa_e$. This emitted radiation is reduced in magnitude by the propagation factor $e^{-\tau(z',H)}$, which accounts for the attenuation between location $Q(\theta,z')$ and the location of the radiometer at $Q(\theta,H)$.

The second term inside the square bracket in Eq. (6.69) represents the energy scattered by the differential thickness, along the θ direction toward the radiometer. It is given by

$$a\kappa_e \, T_{VS}(z')\sec\theta \, dz' = \kappa_s \, T_{VS}(z')\sec\theta \, dz'. \quad (6.72)$$

This contribution also is attenuated by the same propagation factor as it propagates between locations $Q(\theta,z')$ and $Q(\theta,H)$.

▶ In general, the extinction coefficient κ_e consists of an absorption component κ_a and a scattering component κ_s. In the absence of absorption, the process is referred to as one of ***perfect scattering***, and if absorption is the only extinction process, we have ***perfect absorption***. ◀

Chandrasekhar (1960), in his book *Radiative Transfer*, provides an extensive treatment of the equation of transfer for the atmosphere under a variety of conditions, with special emphasis on problems involving a perfectly scattering medium. Microwave interactions with terrain and atmospheric media, however, rarely encounter situations characterized by perfect scattering. When both scattering and absorption are present, the general solution for $T_B(\theta,H)$ as given by Eqs. (6.68) and (6.69) requires the evaluation of an integral involving the scattering radiometric temperature $T_{VS}(\theta,z')$, which itself is determined by an integral involving T_B for all directions over 4π solid angle, as in Eq. (6.59). Thus, the solution at every point in the medium depends on the interactions at every other point, which leads to highly complicated formulations. If the single scattering albedo $a \ll 1$, the complexity of the problem is reduced immensely, as discussed next.

6-6.4 Brightness Temperature of a Scatter-Free Medium

In a scatter-free medium, $\kappa_s = 0$ and $a = 0$, in which case Eq. (6.68) simplifies to

$$T_B(\theta,H) = T_B(\theta,0)\, e^{-\tau(0,H)} + \int_0^H \kappa_a(z') \, T(z')\sec\theta \, e^{-\tau(z',H)} \, dz', \quad (6.73)$$

where now

$$\tau(z_1,z_2) = \int_{z_1}^{z_2} \kappa_a \sec\theta \, dz'. \quad (6.74)$$

Under clear sky conditions, the Earth's atmosphere is free from scattering in the microwave region. When a cloud and/or rain is present, scattering by the water droplets may or may not be negligible depending upon the density and drop-size distribution of the water droplets relative to the wavelength. Generally, scattering effects may be ignored for most atmospheric weather conditions if the frequency is below 10 GHz. For further discussion of atmospheric absorption and scattering, the reader is referred to Chapter 8.

Absorption within a material medium is governed by the average conductivity of the medium, as discussed earlier in Chapter 2, while scattering is governed by the degree of spatial inhomogeneity and/or anisotropy of the dielectric properties of the medium, measured in wavelength units.

Scattering by particles contained inside a medium often is referred to as ***volume scattering***, which should not be confused with ***surface scattering*** by an interface between two dissimilar media, such as the boundary between a soil medium and the atmosphere. To explain the mechanics of volume scattering, let us consider the example of dry snow. The diameters of the ice particles in snow typically are on the order of 0.1–5 mm in size. If the wavelength of the propagating wave is much larger than the dimensions of the ice particles (and of the average distance between particles), the medium appears electromagnetically homogeneous and no appreciable scattering takes place. But if the wavelength is on the same order of magnitude as the dimensions of the ice particles, the spatial inhomogeneity (due to the 3.2-to-1.0 ratio of the relative permittivity of the ice particles to that of the air background) leads to volume scattering. Thus, at 1 GHz ($\lambda = 30$ cm in air and $30/\sqrt{3.2} \approx 16.8$ cm in snow), snow may be considered purely absorptive, while at 30 GHz ($\lambda = 5.59$ mm in snow) both absorption and scattering have to be considered. Another terrain cover characterized by dielectric inhomogeneities is vegetation, and in the case of row crops, the dielectric also is anisotropic.

In the next few sections, we examine the solution of the radiative transfer equation for scatter-free media with plane boundaries. Surface scattering from irregular boundaries and volume scattering in inhomogeneous media are considered in later chapters.

6-6.5 Upwelling and Downwelling Atmospheric Brightness Temperatures

From Eq. (6.73), the second term represents the ***upwelling atmospheric brightness temperature*** [Fig. 6-14(a)] in the absence of scattering. That is,

$$T_{UP}(\theta, H) = \sec\theta \int_0^H \kappa_a(z')\, T(z')\, e^{-\tau(z',H)}\, dz'. \qquad (6.75)$$

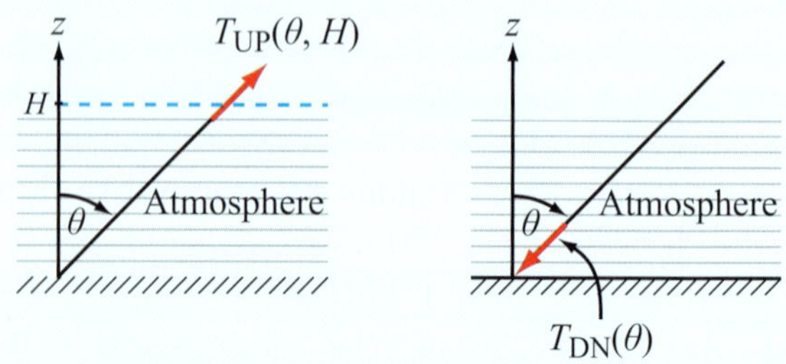

(a) T_{UP} (b) T_{DN}

Figure 6-14: Upward and downward emission contributions of a plane-stratified atmosphere.

If the height H of the downward-looking radiometer is sufficiently higher than the physical extent of the atmosphere, it may be replaced with ∞, since $\kappa_a(z') = 0$ for free space. Models for $\kappa_a(z')$ and $T(z')$ are available in Chapter 8.

A ground-based upward-looking radiometer [Fig. 6-14(b)] observes a ***downwelling atmospheric brightness temperature*** $T_{DN}(\theta)$ given by

$$T_{DN}(\theta) = \sec\theta \int_0^\infty \kappa_a(z')\, T(z')\, e^{-\tau(0,z')}\, dz', \qquad (6.76)$$

where the atmosphere is considered semi-infinite in height.

The physical meaning of the above equation is quite simple; the energy emitted by a stratum at height z' and of vertical thickness dz' (slant thickness $\sec\theta\, dz'$) is proportional to $\sec\theta\ \kappa_a(z')\ T(z')\ dz'$, which, after propagation down to the surface, is reduced by the factor $\exp[-\tau(0,z')]$ due to absorption by the intervening layers. Thus, of two adjacent layers with comparable temperature and absorption coefficients, the layer closer to the ground provides the larger contribution to $T_{DN}(\theta)$.

According to Eqs. (6.75) and (6.76), $T_{UP}(\theta,H)$ and $T_{DN}(\theta)$ of a stratified atmosphere are governed by the vertical profiles of $T(z)$ and $\kappa_a(z)$. For $\kappa_a(z) = 0$ (lossless atmosphere), $T_{UP} = T_{DN} = 0$, as would be expected from Kirchhoff's law. On the other hand, if

κ_a is large, do T_{UP} and T_{DN} have upper limits? As the medium becomes more absorptive, it approaches a perfect blackbody, in which case its brightness temperature becomes equal to its physical temperature. For a medium with a nonuniform temperature profile, the exact value of the brightness temperature depends on the shape of the vertical profiles of its temperature and absorption coefficient, but never exceeds the maximum physical temperature in the profile.

For the special case of a planar, homogeneous atmosphere (or cloud) with $T(z) = T_0$ and $\kappa_a(z) = \kappa_{a0}$ over the range $z = 0$ to $z = H$, where T_0 and κ_{a0} are constants, the expression for $T_{DN}(\theta, H)$ becomes

$$T_{DN}(\theta,H) = \sec\theta \int_0^H \kappa_{a0} T_0 e^{-\kappa_{a0} z' \sec\theta}\, dz'$$
$$= T_0[1 - e^{-\tau_0(0,H)\sec\theta}], \tag{6.77}$$

where $\tau_0(0,H)$ is the zenith optical thickness of the atmospheric layer. It can be shown that $T_{UP}(\theta,H)$ is given by the same expression. If the optical thickness of the cloud, $\tau(0,H)$, is very large, so that the second term inside the bracket of Eq. (6.77) is negligible compared with unity, the expression reduces to

$$T_{DN} = T_0.$$

The variation of κ_a with microwave frequency is discussed in Chapter 8 for atmospheric gases, clouds, and rain. Also, calculated and measured values of κ_a, T_{DN}, and other related parameters are presented there.

6-7 Terrain Brightness Temperature

Consider a downward-looking radiometer in air pointed at an angle θ_1 relative to the z axis, as shown in the three scenarios of Fig. 6-15. In the first scenario, the boundary is a specular surface and the lower hemisphere is occupied by a lossy homogeneous material (such as land or water). The second scenario also has a specular surface, but the lower medium has a nonuniform temperature profile $T(z)$. Finally, in the scenario shown in Fig. 6-15(c), the lower medium is homogeneous, but the boundary is a rough surface. Our goal in this section

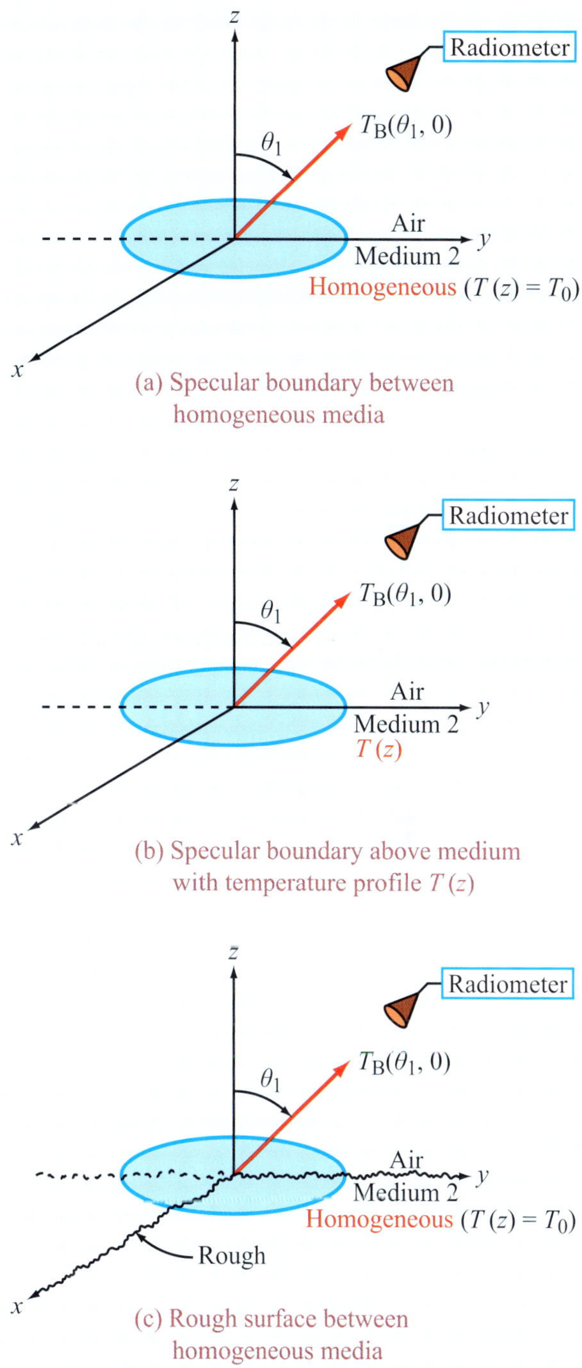

Figure 6-15: Three scenarios for medium 2 and its surface boundary.

is to derive expressions for the brightness temperature $T_B(\theta_1, 0)$ observed by the radiometer for each of the three scenarios.

6-7.1 Brightness Transmission Across a Specular Boundary

Recall that brightness I represents power per unit area per solid angle, so in Fig. 6-16, h-polarized power P_2^h incident upon area A along angle θ_2 from below is

$$P_2^h(\theta_2) = I_2^h(\theta_2)\, A\cos\theta_2\, d\Omega_2. \tag{6.78}$$

At this stage, we confine the treatment to horizontal polarization, and later we generalize the results to vertical polarization. From Snell's law, a wave incident in medium 2 at angle θ_2 is partially transmitted into medium 1 at angle θ_1 such that

$$\sqrt{\varepsilon_1}\sin\theta_1 = \sqrt{\varepsilon_2}\sin\theta_2 \qquad \textbf{(Snell's law)}. \tag{6.79}$$

where ε_1 and ε_2 are the relative dielectric constants (which may be complex) of media 1 and 2, respectively,

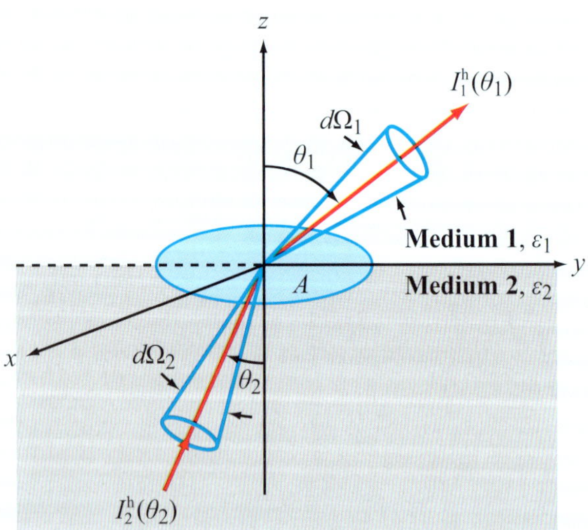

Figure 6-16: Brightness transmission across a planar boundary.

and both media are assumed nonmagnetic. The power transmitted into medium 1 with brightness $I_1^h(\theta_1)$ is

$$P_1^h(\theta_1) = I_1^h(\theta_1)\, A\cos\theta_1\, d\Omega_1. \tag{6.80}$$

Differentiating Eq. (6.79) gives

$$\sqrt{\varepsilon_1}\cos\theta_1\, d\theta_1 = \sqrt{\varepsilon_2}\cos\theta_2\, d\theta_2. \tag{6.81}$$

Transmission across the boundary refracts angle θ_2 into θ_1, but has no effect on the azimuth angle ϕ. That is, $\phi_1 = \phi_2$ and $d\phi_1 = d\phi_2$. Multiplying the left-hand side of Eq. (6.81) by $d\phi_1$ and the right-hand side by $d\phi_2$, and then multiplying the result by Eq. (6.79) leads to

$$\varepsilon_1\cos\theta_1\, d\Omega_1 = \varepsilon_2\cos\theta_2\, d\Omega_2. \tag{6.82}$$

After taking the ratio of Eq. (6.80) to Eq. (6.78) and incorporating Eq. (6.82), we have

$$\begin{aligned}\frac{P_1^h(\theta_1)}{P_2^h(\theta_2)} &= \frac{I_1^h(\theta_1)\, A\cos\theta_1\, d\Omega_1}{I_2^h(\theta_2)\, A\cos\theta_2\, d\Omega_2} \\ &= \frac{I_1^h(\theta_1)\cos\theta_2\, d\Omega_2}{I_2^h(\theta_2)\cos\theta_2\, d\Omega_2}\cdot\frac{\varepsilon_2}{\varepsilon_1} = \frac{I_1^h(\theta_1)}{I_2^h(\theta_2)}\cdot\frac{\varepsilon_2}{\varepsilon_1}.\end{aligned} \tag{6.83}$$

According to Section 2-8, the power ratio on the left-hand side of Eq. (6.83) is equal to the transmissivity $\mathbb{T}_{21}^h$ for incidence in medium 2 at angle θ_2 and transmission into medium 1 at angle θ_1:

$$\frac{P_1^h(\theta_1)}{P_2^h(\theta_2)} = \mathbb{T}_{21}^h. \tag{6.84}$$

Furthermore,

$$I_1^h(\theta_1) = \frac{kB}{\lambda_1^2}\, T_{B_1}^h(\theta_1) = \frac{kB}{\lambda_0^2}\, \varepsilon_1\, T_{B_1}^h(\theta_1) \tag{6.85a}$$

and

$$I_2^h(\theta_2) = \frac{kB}{\lambda_2^2}\, T_{B_2}^h(\theta_1) = \frac{kB}{\lambda_0^2}\, \varepsilon_2\, T_{B_2}^h(\theta_2), \tag{6.85b}$$

where B is the receiver bandwidth, $T_{B_1}^h(\theta_1)$ and $T_{B_2}^h(\theta_2)$ are respectively the brightness temperatures corresponding to brightnesses $I_1^h(\theta_1)$ and $I_2^h(\theta_2)$, $\lambda_1 = \lambda_0/\sqrt{\varepsilon_1}$ and $\lambda_2 = \lambda_0/\sqrt{\varepsilon_2}$ are respectively the wavelengths in media 1 and 2, and λ_0 is the wavelength in free space. Incorporating Eqs. (6.84) and (6.85) in Eq. (6.83) leads to

$$\begin{aligned} T^{\rm h}_{B_1}(\theta_1) &= \mathbb{T}^{\rm h}_{21}\, T^{\rm h}_{B_2}(\theta_2) \\ &= [1-\Gamma^{\rm h}_{21}(\theta_2)]\, T^{\rm h}_{B_2}(\theta_2) \\ &= [1-\Gamma^{\rm h}_{12}(\theta_1)]\, T^{\rm h}_{B_2}(\theta_2), \end{aligned} \tag{6.86}$$

where we used the relationships $\mathbb{T}^{\rm h}_{21}(\theta_2) = 1-\Gamma^{\rm h}_{21}(\theta_2)$ and $\Gamma^{\rm h}_{12}(\theta_1) = \Gamma^{\rm h}_{21}(\theta_2)$. Note that $\Gamma^{\rm h}_{21}(\theta_2)$ is the Fresnel reflectivity for incidence in medium 2 at angle θ_2, whereas $\Gamma^{\rm h}_{12}(\theta_1)$ is the Fresnel reflectivity for incidence in medium 1 at angle θ_1.

For vertical polarization, a similar derivation leads to

$$T^{\rm v}_{B_1}(\theta_1) = [1-\Gamma^{\rm v}_{12}(\theta_1)]\, T^{\rm v}_{B_2}(\theta_2). \tag{6.87}$$

From Section 2-8, the reflectivities for incidence in medium 1 at angle θ_1 are given by

$$\Gamma^{\rm h}_{12}(\theta_1) = \left| \frac{\sqrt{\varepsilon_1}\cos\theta_1 - \sqrt{\varepsilon_2}\cos\theta_2}{\sqrt{\varepsilon_1}\cos\theta_1 + \sqrt{\varepsilon_2}\cos\theta_2} \right|^2, \tag{6.88a}$$

$$\Gamma^{\rm v}_{12}(\theta_1) = \left| \frac{\sqrt{\varepsilon_1}\cos\theta_2 - \sqrt{\varepsilon_2}\cos\theta_1}{\sqrt{\varepsilon_1}\cos\theta_2 + \sqrt{\varepsilon_2}\cos\theta_1} \right|^2, \tag{6.88b}$$

where, as defined by Eq. (2.8), ε_1 and ε_2 are the complex relative dielectric constants of media 1 and 2, respectively. In the present case, medium 1 is air, so $\varepsilon_1 = 1$. Also, θ_2 is related to θ_1 by Snell's law, given by Eq. (6.79).

6-7.2 Emission by a Specular Surface

We established in the previous subsection that the upwelling brightness temperature observed in the upper medium is related to the upwelling brightness temperature incident upon the boundary in the lower medium by the transmissivity of the boundary, in accordance with Snell's laws. Next, for the scenario shown in Fig. 6-17, we determine $T^{\rm h}_{B_2}(\theta_2)$ by applying Eq. (6.75), but we change the integration limits for z' from $(0,H)$ to $(-\infty,0)$. Hence,

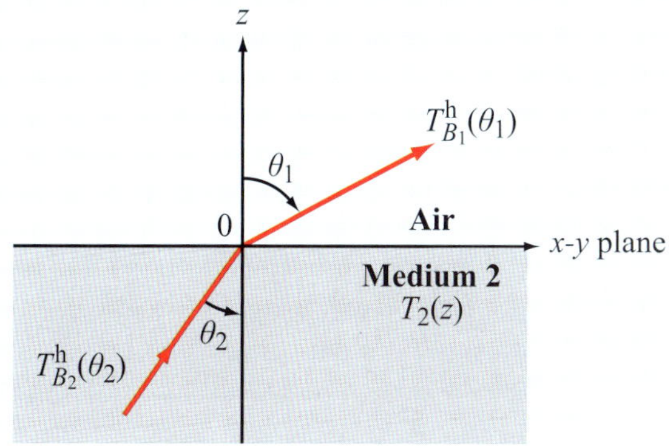

Figure 6-17: Medium 2 has a temperature-depth profile $T_2(z)$.

$$T^{\rm h}_{B_2}(\theta_2) = \sec\theta_2 \int_{-\infty}^{0} \kappa_{\rm a}(z')\, T_2(z')\, e^{-\tau(z',0)}\, dz', \tag{6.89}$$

where $T_2(z')$ is the (physical) temperature profile of medium 2 and $\kappa_{\rm a}(z')$ is its absorption coefficient. The exponential inside the integral is

$$e^{-\tau(z',0)} = \exp\left[-\int_{z'}^{0} \kappa_{\rm a} \sec\theta_2\, dz' \right]. \tag{6.90}$$

For a short interval z' to 0 over which $\kappa_{\rm a}$ is a constant,

$$e^{-\tau(z',0)} = e^{\kappa_{\rm a} z' \sec\theta_2} = e^{-\kappa_{\rm a}|z'|\sec\theta_2} \tag{6.91}$$

where we used the fact that z' is a negative number (lower hemisphere). The exponent is the product of $-\kappa_{\rm a}$ and the slant thickness $z'\sec\theta_2$. From Chapter 2, we know that the magnitude of the electric field of an EM wave traveling in a lossy medium over a slant thickness $|z'|\sec\theta_2$ gets attenuated as $e^{-\alpha|z'|\sec\theta_2}$, and the power carried by the wave gets attenuated as $e^{-2\alpha|z'|\sec\theta_2}$, where α is the attenuation constant of the medium. Complete expressions for α are available in Section 2-4 for both lowloss and lossy media. The conclusion from this discussion is that

$$\kappa_{\rm a} = 2\alpha. \tag{6.92}$$

If $T_2(z')$ and the constitutive parameters of medium 2 are known, evaluation of Eq. (6.89) can be performed numerically. Since none of the quantities involved is polarization specific, we can write for medium 2

$$T^{\rm h}_{B_2}(\theta_2) = T^{\rm v}_{B_2}(\theta_2). \tag{6.93}$$

▶ Whereas below the boundary, the emission within medium 2 is randomly polarized, it is the process of transmission across the boundary that determines what fraction is transmitted per each polarization. ◀

For a homogeneous medium with uniform temperature profile $T_2(z') = T_0$, Eq. (6.89) simplifies to

$$T^{\rm h}_{B_2}(\theta_2) = T^{\rm v}_{B_2}(\theta_2) = T_0, \tag{6.94}$$

in which case the brightness temperature observed by a downlooking radiometer situated in the upper medium becomes

$$T^p_{B_1}(\theta_1) = [1 - \Gamma^p_{12}(\theta_1)]T_0, \qquad p = \text{h or v polarization}, \tag{6.95}$$

where T_0 is the physical temperature of the surface. In reality, neither land nor water media have uniform temperature profiles, so in effect T_0 represents the physical temperature of the uppermost surface layer responsible for the bulk of the contributions to $T_{B_2}(\theta_2)$.

The ***specular emissivity*** of medium 2 is given by

$$e^p(\theta_1) = \frac{T^p_{B_1}(\theta_1)}{T_0} = 1 - \Gamma^p_{12}(\theta_1),^\dagger \qquad p = \text{h or v polarization}. \tag{6.96}$$

Figure 6-18 displays angular plots of $e^{\rm h}(\theta_1)$ and $e^{\rm v}(\theta_1)$ for three types of media.

6-7.3 Emissivity of a Rough Surface

A p-polarized plane wave incident upon a perfectly flat surface of a homogeneous medium is specularly reflected as shown in Fig. 6-19(a), and the reflected power is related to the incident power by the specular reflectivity given by Eq. (6.88). Now consider the surface shown in Fig. 6-19(b); superimposed on the flat surface are small height irregularities. If these are on the order of the wavelength λ or larger, the incident power is scattered by the surface in many directions. Part of the scattered power is reflected in the ***specular direction*** and retains the ***phase-coherence*** characteristic of reflection off a plane surface, and the remainder, known as the ***diffuse component***, is ***phase-incoherent***. Moreover, part of the diffuse scattering bears the same polarization state as the incident power, and the remainder is polarized in the orthogonal state. In like manner, the self-emitted radiation incident upon the surface from the underlying medium [Fig. 6-19(c)] is transmitted across the boundary into many different directions. Consequently, the brightness temperature $T^p_{\rm B}(\theta_1)$ of the medium (as seen from above) is composed of contributions that are incident upon the surface from below along many different directions [Fig. 6-19(d)], and in both polarization states.

Kirchhoff's radiation law stipulates that under conditions of thermodynamic equilibrium, the net exchange of energy between two adjoining media has to be zero, which requires the p-polarized emissivity $e^p(\theta,\phi)$ of medium 2 (into medium 1) along direction (θ,ϕ) to be equal to the p-polarized ***absorptivity*** $a^p(\theta,\phi)$ of medium 2 (from medium 1) for incidence from the same direction:

$$e^p(\theta,\phi) = a^p(\theta,\phi). \tag{6.97}$$

A corollary statement applies for emission by medium 1 into medium 2.

For wave incidence in medium 1 upon medium 2, the power absorbed by medium 2 is equal to the incident power minus the power scattered by the surface boundary. Consider an h-polarized wave with electric field $E^{\rm i}_{\rm h}$ incident upon a surface of area A along direction $(\theta_{\rm i},\phi_{\rm i})$ as shown in Fig. 6-20. In general, the target area scatters part of the incident power along all directions $(\theta_{\rm s},\phi_{\rm s})$ in the upper hemisphere, with some of the power being h-polarized and the remainder v-polarized. The

† Computer Code 6.1.

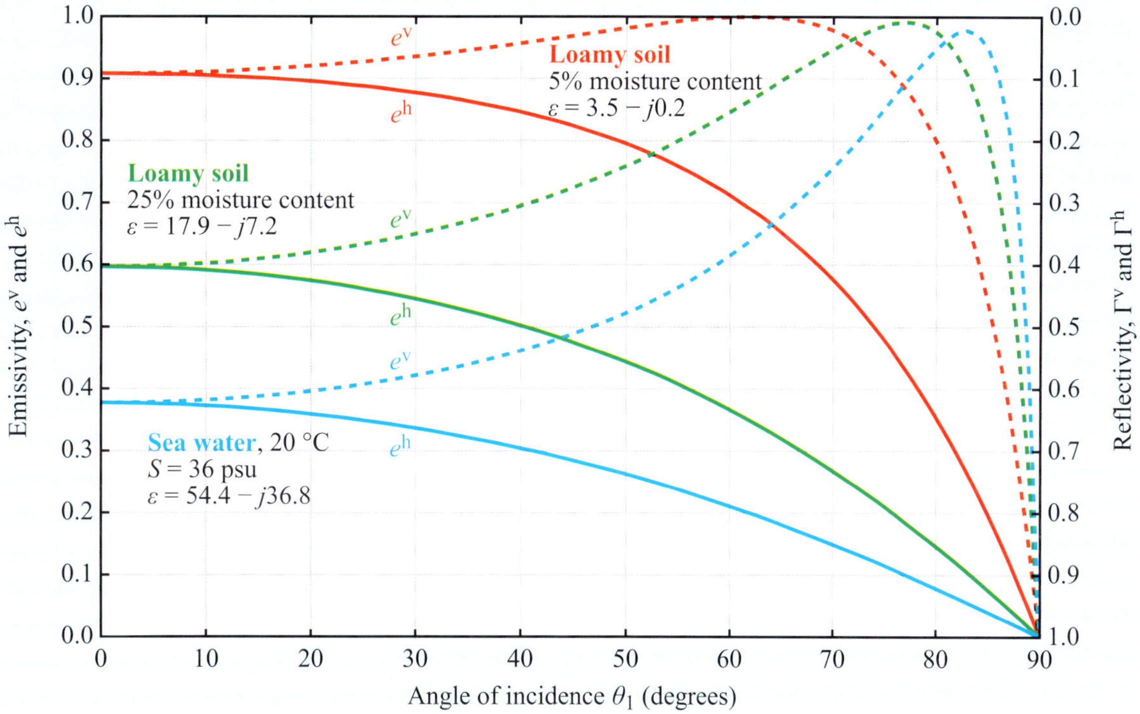

Figure 6-18: Calculated reflectivities and emissivities of a specular surface for h and v polarizations at 10 GHz.

power density of the incident wave is

$$\mathcal{S}^{\text{i}}_{\text{h}} = \frac{1}{2\eta_1} \, |E^{\text{i}}_{\text{h}}|^2, \tag{6.98}$$

where η_1 is the intrinsic impedance of medium 1. The corresponding incident power intercepted by the area A is

$$P^{\text{i}}_{\text{h}} = \mathcal{S}^{\text{i}}_{\text{h}} A \cos\theta = \frac{1}{2\eta_1} \, |E^{\text{i}}_{\text{h}}|^2 A \cos\theta_{\text{i}}. \tag{6.99}$$

The electric field of the wave scattered along $(\theta_{\text{s}}, \phi_{\text{s}})$ consists of an h-polarized component E^{s}_{h} and a v-polarized component E^{s}_{v}. The corresponding power density of the scattered wave is

$$\mathcal{S}^{\text{s}} = \mathcal{S}^{\text{s}}_{\text{h}} + \mathcal{S}^{\text{s}}_{\text{v}} = \frac{1}{2\eta_1} \left[|E^{\text{s}}_{\text{h}}|^2 + |E^{\text{s}}_{\text{v}}|^2 \right]. \tag{6.100}$$

From Section 5-3.1, E^{s}_{h} and E^{s}_{v} at a distance R_{r} from the scattering surface are related to the incident electric field E^{i}_{h} by

$$E^{\text{s}}_{\text{h}} = \left(\frac{e^{-jkR_{\text{r}}}}{R_{\text{r}}} \right) \widetilde{S}_{\text{hh}} E^{\text{i}}_{\text{h}}, \tag{6.101a}$$

$$E^{\text{s}}_{\text{v}} = \left(\frac{e^{-jkR_{\text{r}}}}{R_{\text{r}}} \right) \widetilde{S}_{\text{vh}} E^{\text{i}}_{\text{h}}, \tag{6.101b}$$

where $\widetilde{S}_{\text{hh}}$ and $\widetilde{S}_{\text{vh}}$ are the hh- and vh-scattering amplitudes of the surface in the FSA convention. Using Eq. (6.101) in Eq. (6.100) leads to

$$\begin{aligned} \mathcal{S}^{\text{s}} &= \frac{1}{2\eta_1 R_{\text{r}}^2} \left[|\widetilde{S}_{\text{hh}}|^2 + |\widetilde{S}_{\text{vh}}|^2 \right] |E^{\text{i}}_{\text{h}}|^2 \\ &= \frac{A}{8\pi\eta_1 R_{\text{r}}^2} \, [\sigma^0_{\text{hh}} + \sigma^0_{\text{vh}}] |E^{\text{i}}_{\text{h}}|^2, \end{aligned} \tag{6.102}$$

where we used the relations $\sigma_{pq} = 4\pi|\widetilde{S}_{pq}|^2$ and $\sigma^0_{pq} = \sigma_{pq}/A$. At range R_{r}, the total power scattered by

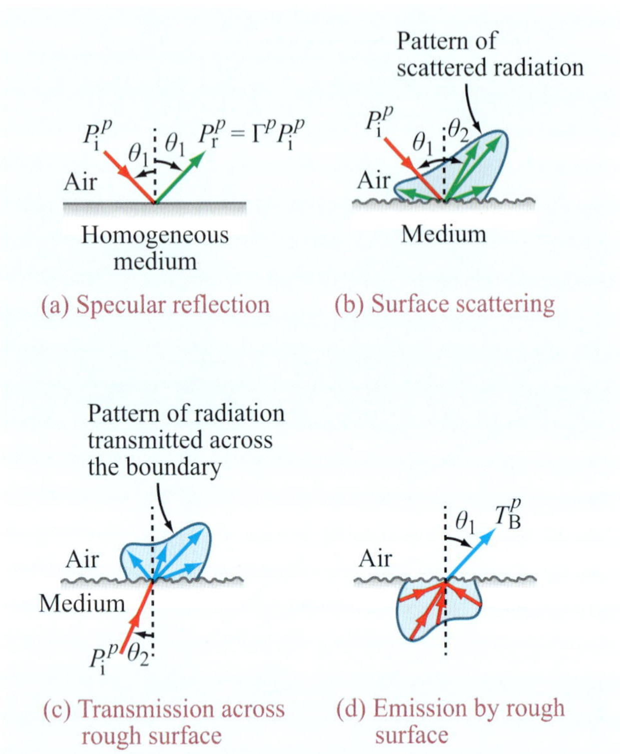

Figure 6-19: Specular- and rough-surface scattering and emission.

the surface into the upper hemisphere is

$$\begin{aligned}
P^{\mathrm{s}} &= \iint_{\text{upper hemisphere}} \mathcal{S}^{\mathrm{s}} R_{\mathrm{r}}^2 \, d\Omega_{\mathrm{s}} \\
&= \frac{A}{8\pi\eta_1 R_{\mathrm{r}}^2} \int_{\phi_{\mathrm{s}}=0}^{2\pi} \int_{\theta_{\mathrm{s}}=0}^{\pi/2} (\sigma_{\mathrm{hh}}^0 + \sigma_{\mathrm{vh}}^0) |E_{\mathrm{h}}^{\mathrm{i}}|^2 R_{\mathrm{r}}^2 \, d\Omega_{\mathrm{s}} \\
&= \frac{P_{\mathrm{i}}}{4\pi \cos\theta_{\mathrm{i}}} \int_{\phi_{\mathrm{s}}=0}^{2\pi} \int_{\theta_{\mathrm{s}}=0}^{\pi/2} [\sigma_{\mathrm{hh}}^0(\theta_{\mathrm{i}}, \phi_{\mathrm{i}}; \theta_{\mathrm{s}}, \phi_{\mathrm{s}}) \\
&\qquad + \sigma_{\mathrm{vh}}^0(\theta_{\mathrm{i}}, \phi_{\mathrm{i}}; \theta_{\mathrm{s}}, \phi_{\mathrm{s}})] \sin\theta_{\mathrm{s}} \, d\theta_{\mathrm{s}} \, d\phi_{\mathrm{s}},
\end{aligned} \tag{6.103}$$

where we used $d\Omega_{\mathrm{s}} = \sin\theta_{\mathrm{s}} \, d\theta_{\mathrm{s}} \, d\phi_{\mathrm{s}}$ and the expression for P_{i} given by Eq. (6.99). The h-polarized emissivity of the surface is equal to the h-polarized absorptivity, which, in turn, is equal to 1 minus the fractional power scattered by the surface. Thus,

$$\begin{aligned}
e^{\mathrm{h}}(\theta_{\mathrm{i}}, \phi_{\mathrm{i}}) &= 1 - \frac{1}{4\pi \cos\theta_{\mathrm{i}}} \\
&\times \int_{\phi_{\mathrm{s}}=0}^{2\pi} \int_{\theta_{\mathrm{s}}=0}^{\pi/2} [\sigma_{\mathrm{hh}}^0(\theta_{\mathrm{i}}, \phi_{\mathrm{i}}; \theta_{\mathrm{s}}, \phi_{\mathrm{s}}) + \sigma_{\mathrm{vh}}^0(\theta_{\mathrm{i}}, \phi_{\mathrm{i}}; \theta_{\mathrm{s}}, \phi_{\mathrm{s}})] \\
&\cdot \sin\theta_{\mathrm{s}} \, d\theta_{\mathrm{s}} \, d\phi_{\mathrm{s}}.
\end{aligned} \tag{6.104}$$

As discussed in Section 5-10.4, the co-polarized bistatic-scattering coefficient consists of a coherent component that exists along only the specular direction and an incoherent component that exists along all directions. Replacing σ_{hh}^0 in Eq. (6.104) with Eqs. (5.115) and (5.116) leads to

$$\begin{aligned}
e^{\mathrm{h}}(\theta_{\mathrm{i}}, \phi_{\mathrm{i}}) &= 1 - \Gamma^{\mathrm{h}}(\theta_{\mathrm{i}}) \, e^{-4\psi^2} - \frac{1}{4\pi \cos\theta_{\mathrm{i}}} \\
&\times \int_{\phi_{\mathrm{s}}=0}^{2\pi} \int_{\theta_{\mathrm{s}}=0}^{\pi/2} [\sigma_{\mathrm{hh_{inc}}}^0(\theta_{\mathrm{i}}, \phi_{\mathrm{i}}; \theta_{\mathrm{s}}, \phi_{\mathrm{s}}) \\
&\qquad + \sigma_{\mathrm{vh}}^0(\theta_{\mathrm{i}}, \phi_{\mathrm{i}}; \theta_{\mathrm{s}}, \phi_{\mathrm{s}})] \, d\Omega_{\mathrm{s}},
\end{aligned} \tag{6.105}$$

where $\psi = ks\cos\theta_{\mathrm{i}}$, s is the surface rms height, $\Gamma^{\mathrm{h}}(\theta_{\mathrm{i}})$ is the h-polarized specular reflectivity of the surface, and *it is implicitly understood that the bistatic-scattering coefficients inside the integral of Eq. (6.105) account for only the incoherent part of the scattering by the surface.* The v-polarized expressions for the emissivity $e^{\mathrm{v}}(\theta_{\mathrm{i}}, \phi_{\mathrm{i}})$ can be obtained from Eqs. (6.104) and (6.105) by interchanging v and h.

The expression given by Eq. (6.105) provides an electromagnetic link between the emissivity measured by a radiometer and the bistatic-scattering coefficients measured by radar. To evaluate Eq. (6.105) for a specific surface of interest, we need to have expressions that relate the bistatic-scattering coefficients of that surface to its dielectric and roughness properties, as discussed in Chapter 10.

The formulation leading to Eq. (6.105) can also be used to relate the p-polarized surface-scattered brightness temperature $T_{\mathrm{SS}}^{p}(\theta, \phi)$ to the unpolarized downward emitted atmospheric temperature $T_{\mathrm{DN}}(\theta_{\mathrm{s}}, \phi_{\mathrm{s}})$ incident from all directions $(\theta_{\mathrm{s}}, \phi_{\mathrm{s}})$ in the upper hemisphere:

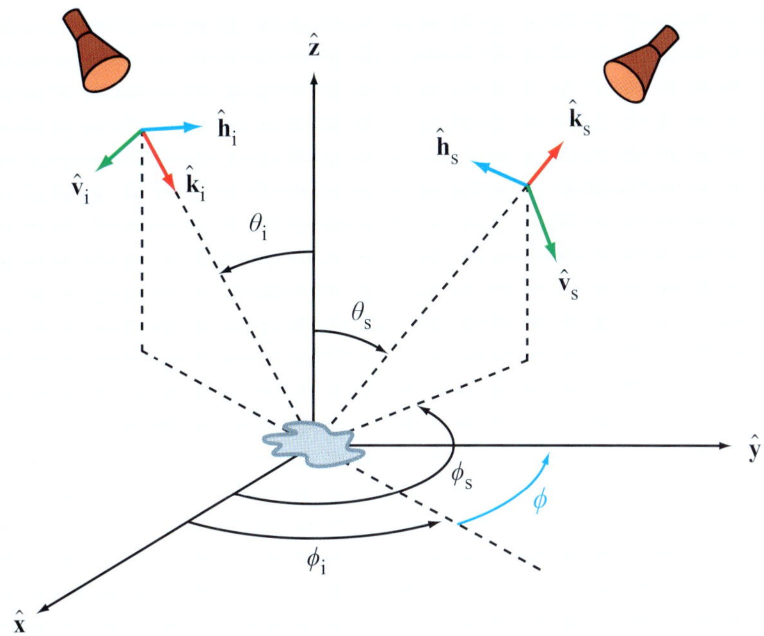

Figure 6-20: Bistatic-scattering coordinate system.

$$T^p_{\text{SS}}(\theta_\text{i},\phi_\text{i}) = \Gamma^p(\theta_\text{i})\,T_{\text{DN}}(\theta_\text{i})\,e^{-4\psi^2} + \frac{1}{4\pi\cos\theta_\text{i}} \times \int_{\theta_\text{s}=0}^{\pi/2}\int_{\phi_\text{s}=0}^{2\pi}[\sigma^0_{pp}(\theta_\text{i},\phi_\text{i};\,\theta_\text{s},\phi_\text{s}) + \sigma^0_{pq}(\theta_\text{i},\phi_\text{i};\,\theta_\text{s},\phi_\text{s})] \cdot T_{\text{DN}}(\theta_\text{s},\phi_\text{s})\,d\Omega_\text{s}, \quad p,q = \text{h or v}. \tag{6.106}$$

6-7.4 Extreme Surface Conditions

Specular surface

The energy scattered by a perfectly flat surface consists of only the specularly reflected, coherent component. Thus, the incoherent component of $\sigma^0_{pp_\text{inc}}$ is zero and so is the rms height s. Consequently, Eqs. (6.105) and (6.106) reduce to

$$e^\text{h}(\theta_\text{i},\phi_\text{i}) = 1 - \Gamma^\text{h}(\theta_\text{i}), \tag{6.107a}$$

(specular surface)

and

$$T^\text{h}_{\text{SS}}(\theta_\text{i},\phi_\text{i}) = \Gamma^\text{h}(\theta_\text{i})\,T_{\text{DN}}(\theta_\text{i}). \tag{6.107b}$$

Similar expressions apply to the v polarization. Not surprisingly, Eq. (6.107a) is a reconfirmation of the result obtained earlier in the form of Eq. (6.96).

Lambertian surface

A perfectly rough surface, also known as a ***Lambertian surface***, scatters energy uniformly (per unit projected area) in all directions. The coherent component is zero and the angular variation of the sum $(\sigma^0_{\text{hh}_\text{inc}} + \sigma^0_{\text{vh}})$ is dependent solely on the product $\cos\theta_\text{i}\cos\theta_\text{s}$:

$$\sigma^0_{\text{hh}}(\theta_\text{i},\phi_\text{i};\,\theta_\text{s},\phi_\text{s}) + \sigma^0_{\text{vh}}(\theta_\text{i},\phi_\text{i};\,\theta_\text{s},\phi_\text{s}) = \sigma^0_0\cos\theta_\text{i}\cos\theta_\text{s}, \tag{6.108}$$

where σ^0_0 is a constant related to the dielectric properties of the scattering surface. The corresponding emissivity

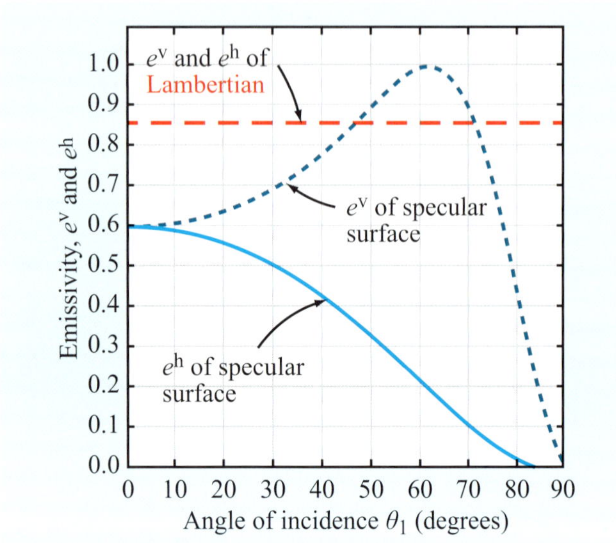

Figure 6-21: Comparison of the polarized emissivity of a specular surface with that of a very rough (Lambertian) surface.

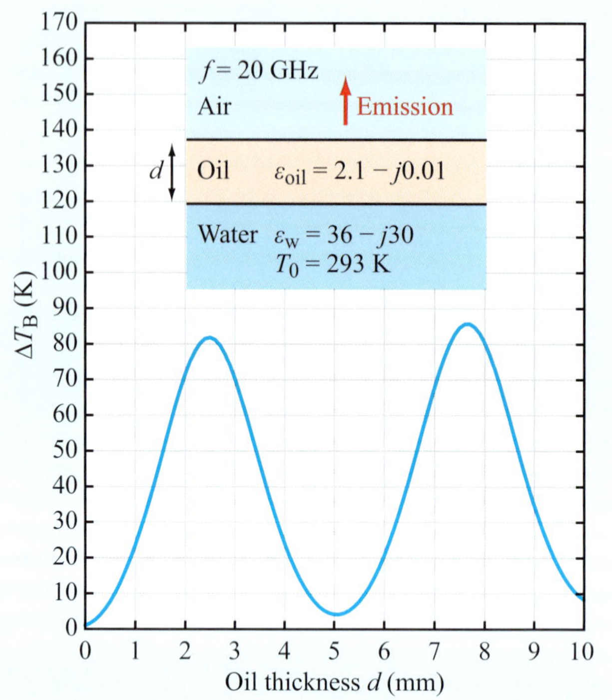

Figure 6-22: Increase in brightness temperature as a function of oil thickness at 20 GHz at nadir ($\theta_1 = 0$).

is

$$e^{\rm h}(\theta_{\rm i}, \phi_{\rm i}) = 1 - \frac{1}{4\pi \cos\theta_{\rm i}} \times \int_0^{2\pi} \int_0^{\pi/2} \sigma_0^0 \cos\theta_{\rm i} \cos\theta_{\rm s} \sin\theta_{\rm s} \, d\theta_{\rm s} \, d\phi_{\rm s},$$

which leads to

$$e^{\rm h}(\theta_{\rm i}, \phi_{\rm i}) = 1 - \frac{\sigma_0^0}{4} \qquad \textbf{(Lambertian)}, \tag{6.109}$$

which is polarization- and angle-independent. Figure 6-21 compares the angular emissivity responses of the specular and Lambertian surfaces. Most real surfaces exhibit angular responses in-between these two extreme roughness conditions.

6-7.5 Emissivity of a Two-Layer Composite

The two-layer composite diagrammed in Fig. 6-22 is viewed by a downward-looking radiometer pointed at angle θ_1 relative to the vertical. Medium 1 is air with permittivity ε_0, and media 2 and 3 are lossy dielectrics with ε_{c_2} and ε_{c_3}, respectively. Both boundaries at $z = 0$ and $z = -d$ are planar, and all media are at the same physical temperature T_0. Since all media have the same physical temperature, the concept of emissivity is applicable and the brightness temperature of the two-layer composite can be expressed in the form

$$T_{\rm B}^p(\theta_1) = e^p(\theta_1)\, T_0 \qquad (p = \text{v or h polarization}). \tag{6.110}$$

Given that the three media are in thermodynamic equilibrium (same physical temperature), Kirchhoff's radiation law mandates that the rates at which they exchange energies are such that the net change is zero for each of them. In medium 1, the downward incident p-polarized energy (at angle θ_1) is partially reflected specularly (back into medium 1) and the remainder is transmitted across the boundary at $z = 0$ into the two-layer composite. The reflected component is proportional to the p-polarized reflectivity $\Gamma^p(\theta_1)$ and the transmitted component is proportional to $\mathbb{T}^p(\theta_1) = 1 - \Gamma^p(\theta_1)$. Kirchhoff's law requires that the two-layer composite emit into medium 1 an amount of energy equal to that it received from medium 1. Hence, the emissivity of the two-layer composite is†

$$\begin{aligned} e^p(\theta_1) &= 1 - \Gamma^p(\theta_1) \\ &= 1 - |\rho^p(\theta_1)|^2 \\ &= 1 - \left| \frac{\rho_{12} + \rho_{23} e^{-2\gamma_2 d \cos\theta_2}}{1 + \rho_{12}\rho_{23} e^{-2\gamma_2 d \cos\theta_2}} \right|^2, \end{aligned} \tag{6.111}$$

where $\rho^p(\theta_1)$, the reflection coefficient of the two-layer composite, is given by Eq. (2.138), and ρ_{12} and ρ_{23} are given by Eq. (2.139) and Eq. (2.140) for both h and v polarizations. Additional related expressions are given in Section 2-10.1.

To illustrate the utility of the model given by Eq. (6.111), let us consider an oil layer of thickness d overlaying the ocean surface. A 20 GHz ***nadir-looking*** (downward at $\theta_1 = 0$) radiometer is used to map the extent and measure the thickness of the oil (dumped by an oil tanker). At 20 GHz and $T_0 = 293$ K (20 °C), the permittivity of the ocean water is $\varepsilon_w = (36 - j30)\varepsilon_0$ and the permittivity of oil is $\varepsilon_{oil} = (2.1 - j0.01)\varepsilon_0$. Figure 6-22 displays a plot of the increase in brightness temperature as a function of oil thickness,

$$\Delta T_B = (e_{oil} - e_w)T_0, \tag{6.112}$$

where e_{oil} is the emissivity of the two-layer composite, computed using Eq. (6.111) with $\theta_1 = 0$, and e_w is the emissivity of the water surface with no oil covering it. As d increases from zero to 2.5 mm, ΔT_B increases by about 80 K, and then it decreases by about the same amount as d increases further to 5 mm. These are steep slopes, indicative of strong sensitivity to d, but the overall response curve poses an ambiguity problem (same ΔT_B for multiple different thicknesses). Fortunately, the ambiguity can be resolved by using a dual frequency radiometer as discussed later in Section 18-5.

† **Computer Code 6.2.**

Thermal emission is fundamentally phase incoherent, meaning that the phases of two EM waves radiated at a given instant in time by two neighboring differential volumes within a material bear no specific relationship to one another. The difference between the two phase angles is a random variable with a uniform distribution over the range 0 to 2π (all values are equally probable). Yet, the emission by a two-layer (or multilayer) composite may exhibit attributes of a phase coherent signal. An example is the oscillatory behavior of ΔT_B as a function of d in Fig. 6-22, which is due to constructive and destructive interference of the multiple reflections within the oil layer. Moreover, phase coherence is integral to the derivation leading to Eq. (6.111), which was used in calculating the curve shown in Fig. 6-22. The explanation for what may appear as a contradiction is that the emission in the water layer is indeed phase incoherent, but once it is transmitted into the oil layer, propagation within and multiple reflections by the upper and lower boundaries of the oil layer generate multiple contributions that bear specific phase relationships to one another. In essence, the middle layer imposes coherence onto the incoherent emission generated by the bottom layer. Similarly, the emission by every infinitesimal volume within the middle layer assumes coherence properties as it undergoes multiple reflections within that layer.

6-8 Downward-Looking Satellite Radiometer

The scenario shown in Fig. 6-23 depicts an orbiting radiometer above the Earth's atmosphere, with its antenna beam pointed at a ground cell of area A. Through the application of appropriate calibration and correction algorithms, the antenna temperature T_A measured by the radiometer is converted into a high precision estimate of T_B^p(satellite), the p-polarized brightness temperature representing the energy incident upon the radiometer antenna. The brightness temperature T_B^p(satellite) is, in turn, related to the surface-emitted brightness temperature of the ground cell, T_{SE}^p, by

$$T_B^p(\text{satellite}) = \Upsilon_a(T_{SE}^p + T_{SS}^p) + T_{UP}, \tag{6.113}$$

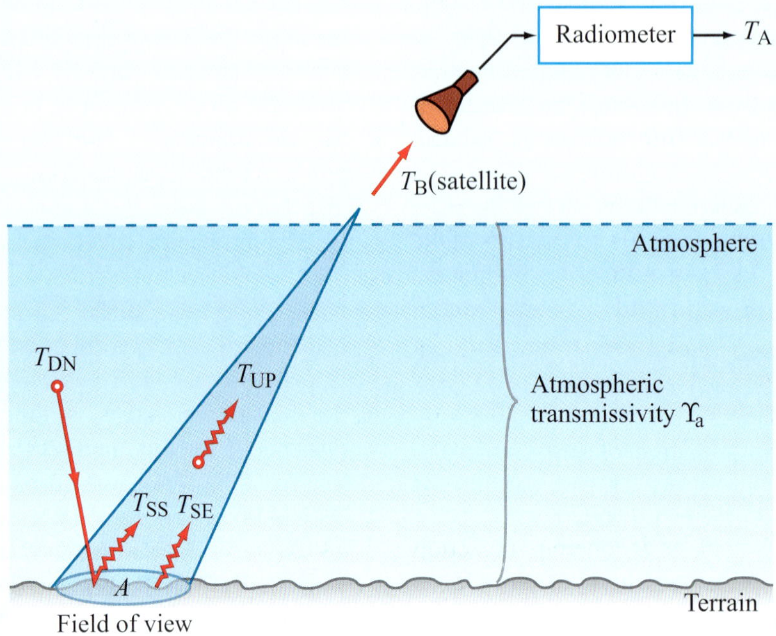

Figure 6-23: Satellite-borne downward-looking radiometer.

where Υ_a is the one-way transmissivity of the atmosphere as defined by Eq. (6.70), T^p_{SS} is the p-polarized upward-reflected surface-scattered fraction of the downward-emitted atmospheric temperature T_{DN}, as defined by Eq. (6.106), and T_{UP} is the upward-emitted atmospheric temperature. To obtain a good estimate of T^p_{SE} (so as to relate it to the biophysical properties of the ground cell), it is necessary to factor out Υ_a, T^p_{SS}, and T_{UP}. This is usually done from knowledge of the state of the atmosphere and the type of surface cover present in the ground cell. As discussed later in Chapters 8 and 9, radiometric observations made at carefully selected microwave frequencies can be used to characterize the profiles of atmospheric temperature, pressure, and water vapor density as a function of height, as well as to measure important attributes of clouds when present. Knowledge of these atmospheric parameters is then used in radiative transfer models to estimate Υ_a, T^p_{SS}, and T_{UP}. At frequencies in the 1 to 10 GHz band, $\Upsilon_a \approx 1$ and both T^p_{SS} and T_{UP} are only a few degrees in magnitude, so the process of converting T^p_B(satellite) to T^p_{SE} is fairly straightforward.

The expression given by Eq. (6.113) is equally applicable to a downward-looking airborne radiometer except that in that case Υ_a and T_{UP} pertain to only the atmospheric layer between the ground and the airborne radiometer.

6-9 Polarimetric Radiometry

Individual polarization components of the brightness temperature often contain independent and complementary information about their thermal emission source(s). For example, simultaneous measurements of the h- and v-polarized brightness temperatures of the ocean surface at the proper frequency can be used to estimate both the physical temperature and wind-induced roughness of the surface. The p-polarized brightness temperature is given by

$$T^p_B = K \left\langle |E_p|^2 \right\rangle, \qquad (6.114)$$

where E_p is the associated p-polarized electric field incident on the radiometer antenna, $K = \lambda_0^2/(k\eta_0 B)$ is a scale factor to convert units of $(\mathrm{V/m})^2$ to kelvins, λ_0 is the wavelength and B is the radiometer bandwidth, k is Boltzmann's constant, and η_0 is the intrinsic impedance of free space.

Traditional dual-linear polarized radiometers measure vertical ($p = \mathrm{v}$) and horizontal ($p = \mathrm{h}$) components of the brightness temperature. Fully polarimetric radiometers measure T_B^v and T_B^h as well as the third and fourth Stokes parameters of brightness temperature, respectively T_B^3 and T_B^4. The third and fourth Stokes parameters are here designated as T_B^3 and T_B^4, rather than the conventional U and V notation used in Section 5-3.3, to avoid confusion with the vertical polarization component. Measurement of the third and fourth Stokes parameters can be performed by either coherent or incoherent detection (Ishimaru, 1991). Coherent detection implies that a direct cross correlation is performed between the vertical and horizontal components of the incident electric field, as described by $\langle E_\mathrm{v} E_\mathrm{h}^* \rangle$, where the asterisk denotes complex conjugation.

Whereas $\langle E_\mathrm{v} E_\mathrm{v}^* \rangle$ and $\langle E_\mathrm{h} E_\mathrm{h}^* \rangle$ are real quantities, $\langle E_\mathrm{v} E_\mathrm{h}^* \rangle$ is in general a complex quantity. Hence, we introduce brightness temperature components T_B^3 and T_B^4, defined as

$$T_\mathrm{B}^3 = 2K\, \mathfrak{Re} \langle E_\mathrm{v} E_\mathrm{h}^* \rangle, \tag{6.115a}$$

$$T_\mathrm{B}^4 = 2K\, \mathfrak{Im} \langle E_\mathrm{v} E_\mathrm{h}^* \rangle. \tag{6.115b}$$

When these two components are added to the two traditional components, T_B^v and T_B^h, we end up with the ***polarimetric brightness temperature vector***

$$\mathbf{T}_\mathrm{B} = \begin{bmatrix} T_\mathrm{B}^\mathrm{v} \\ T_\mathrm{B}^\mathrm{h} \\ T_\mathrm{B}^3 \\ T_\mathrm{B}^4 \end{bmatrix} = K \begin{bmatrix} \langle E_\mathrm{v} E_\mathrm{v}^* \rangle \\ \langle E_\mathrm{h} E_\mathrm{h}^* \rangle \\ 2\mathfrak{Re} \langle E_\mathrm{v} E_\mathrm{h}^* \rangle \\ 2\mathfrak{Im} \langle E_\mathrm{v} E_\mathrm{h}^* \rangle \end{bmatrix}. \tag{6.116}$$

Incoherent detection of T_B^3 and T_B^4 relies on the following relationships between the various polarization components of brightness temperatures:

$$T_\mathrm{B}^3 = T_\mathrm{B}^\mathrm{P} - T_\mathrm{B}^\mathrm{M}, \tag{6.117a}$$

$$T_\mathrm{B}^4 = T_\mathrm{B}^\mathrm{L} - T_\mathrm{B}^\mathrm{R}, \tag{6.117b}$$

where T_B^P, T_B^M, T_B^L, and T_B^R, correspond to components of the brightness temperature at polarizations $p = \mathrm{P}$ ($+45^\circ$ slant linear), M (-45° slant linear), L (left-hand circular), and R (right-hand circular). Hardware implementations of coherent and incoherent detection are discussed in Section 7-14.

6-10 Stokes Parameters and Periodic Structures

Estimation of ocean-surface wind speed by traditional v- and h-pol microwave radiometers, using linearly polarized brightness temperatures only, is a well-established technique (e.g., Wilheit and Chang, 1980; Goodberlet et al., 1989). Early investigations into the azimuthal anisotropy of microwave emission by the ocean surface identified wind direction as the cause (Bespalova et al., 1982; Etkin et al., 1991; Wentz, 1992). The results suggest that the microwave brightness temperature is sensitive to both wind speed and direction. However, traditional v- and h-pol observations, while responsive to wind direction, cannot independently estimate it due to azimuthal ambiguities in the sensitivity. Therefore, their wind-speed retrievals tend to have a direction-dependent component of error. In subsequent investigations, it was determined that additional information about wind direction is contained in the third and fourth Stokes parameters of the emission (e.g., Yueh et al., 1995; Gasiewski and Piepmeier, 1996; Skou and Laursen, 1998). Periodic, anisotropic patterns in the surface shape and emissivity introduce partial correlation between the otherwise uncorrelated v- and h-pol electric fields associated with thermal emission from the surface. Analytical surface-scattering models are able to predict some of the observed polarimetric emission characteristics of an ocean surface modified by directional wind roughening (Yueh, 1997; Zhang and Johnson, 2001; Johnson and Cai, 2002). However, these models remain only an approximation, as they contain modeling uncertainties due to their treatment of the surface roughness spectrum and the nonisotropic distribution of foam and white caps. As a result, empirical geophysical model functions, derived from polarimetric radiometer observations made over a wide range of wind speeds and directions, combined with *in-situ* near-surface measurements of the wind vector, are often used as the forward model in operational wind vector retrieval algorithms. As discussed in more detail in Chapter 18, this is the approach taken by WindSat, the first fully polarimetric microwave radiometer (Gaiser et al., 2004) to fly in space.

Partially polarized thermal emission from the Arctic and Antarctic ice sheets has also been observed. Li et al. (2008) show strong time- and azimuthal dependences for both T_B^3 and T_B^4, which appear to be related to the seasonal development and decay of sastrugi (parallel wavelike ridges) in dry snow regions and to snowmelt in the wet-snow regions. These cyclic patterns alter the subsurface snow grain size and introduce subsurface ice lenses and layering that have distinct, periodic spatial distributions. It is these subsurface features that produce the partial correlation between linearly polarized components of thermal emission necessary to generate significant nonzero T_B^3 and T_B^4.

PROBLEMS

6.1 Consider a blackbody at 300 K observed by a 10 GHz radiometer. What is the fractional deviation between the spectral brightness intensity computed using the Rayleigh-Jeans approximation versus Planck's law?

6.2 From Earth, the (planar) angle subtended by the sun is 0.5°. A 1 cm wavelength radio telescope antenna pointed at the sun measured an antenna temperature of 1174 K. The antenna effective area is 0.4 m^2, its physical temperature is 300 K, its beam efficiency is 1.0, and its radiation efficiency is 0.8.

(a) What is the antenna temperature that the radio telescope would have measured had the antenna been lossless?

(b) Taking into account the relative solid angle subtended by the sun relative to the pattern solid angle of the antenna, determine the brightness temperature of the sun.

6.3 A radiometer is used to observe terrain surfaces with brightness temperatures ranging from 100 K to 300 K. The antenna radiation efficiency is 0.9 and its physical temperature T_0 is always measured. The sidelobe brightness temperature is unknown, but it has been calculated to be somewhere between 100 K and 200 K. Corresponding to this uncertainty in T_{SL}, what should the minimum value of the beam efficiency be in order that the percentage error of the estimated value of the terrain brightness temperature be less than:

(a) 3%?

(b) 1%?

6.4 The absorption coefficient of a water cloud is given by the approximate expression

$$\kappa_a \approx 2.4 \times 10^{-4} f^2 m_v \qquad \text{(Np/km)},$$

where f is in GHz and m_v is the water content in g/m^3. An upward-looking zenith-pointing radiometer is observing a 2 km thick cloud with a uniform physical temperature of 275 K and a uniform water content of 15 g/m^3. Compute and plot the brightness temperature of the cloud as a function of frequency between 1 and 30 GHz. The cloud may be treated as a nonscattering medium.

6.5 Repeat Problem 6.4, but now place the radiometer above the cloud and point it downward along the nadir direction. Also, an ocean surface exists below the cloud and its upward-radiated brightness temperature is 150 K.

6.6 Repeat Problem 6.5 for $\theta = 60°$, while keeping the ocean brightness temperature the same.

6.7 A water cloud with a 2 km vertical thickness is characterized by a temperature profile of the form

$$T(z) = 300(1 - 4 \times 10^{-2} z) \qquad \text{(K)},$$

where z is the height in km above the base of the cloud. Compute T_{UP} and T_{DN} at $\theta = 60°$ for

(a) $\kappa_a = 10^{-4}$ Np/km.

(b) $\kappa_a = 10^{-2}$ Np/km.

(c) $\kappa_a = 1$ Np/km.

Are T_{UP} and T_{DN} the same? If not, why not?

6.8 One of the sensors flown aboard Skylab was a nadir-looking, 21 cm wavelength radiometer. The radiometer antenna had an approximately Gaussian radiation pattern given by

$$F(\theta) = \exp\left[-2.77\left(\frac{\theta}{\beta}\right)^2\right],$$

where β, the half-power beamwidth, is 15°. The ground track of one of Skylab's passes was a line perpendicular to the South Carolina coastline, as shown in Fig. P6.8. Assuming T_B to be 250 K for land and 90 K for the Atlantic Ocean, plot the observed antenna temperature as a function of time between 27.5 and 29.5 minutes. You may assume that (1) the antenna is lossless, (2) the atmosphere is lossless, (3) sidelobe contributions are negligible, and (4) T_B is angle-of-incidence independent over the extent of the beam for both land and ocean. Skylab's orbit was at an altitude of 435 km above sea level and its ground velocity was 7.65 km/s. *Hint:* Assume the radiation pattern to consist of one mainlobe 30° wide and no sidelobes. Also, for small angles, use the approximations $\sin\theta \approx \theta$ and $\cos\theta \approx 1$.

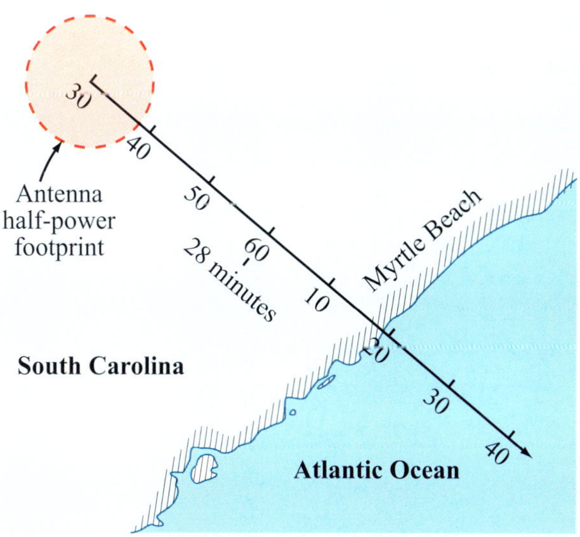

Figure P6.8: Skylab ground track over South Carolina coastline.

6.9 Compute and plot the h-polarized and v-polarized specular emissivities as a function of incidence angle for a sea surface at 1.4 GHz. Assume $T_0 = 300$ K and $S = 32$ psu. *Suggestion:* Use Computer Codes 4.2 and 6.1.

6.10 Compute and plot the specular nadir emissivity of the sea surface as a function of frequency between 1 and 30 GHz. Assume $T_0 = 300$ K and $S = 32$ psu. *Suggestion:* Use Computer Codes 4.2 and 6.1.

6.11 Compute the h-polarized 10 GHz emissivity as a function of incidence angle for a two-layer composite consisting of an ice layer of thickness $d = 1$ cm and $\varepsilon_{ice} = 3.2 - j0.1$, overlaying a water surface with $\varepsilon_w = 55 - j30$. Assume both layers have the same temperature.

6.12 Compute the 10 GHz emissivity of a two-layer composite consisting of an ice layer of thickness d and $\varepsilon_{ice} = 3.2 - j0.1$, overlaying a water surface with $\varepsilon_w = 55 - j30$, and plot it as a function of d between 0 and 2 cm.

6.13 A commonly used model for the azimuthal dependence of polarized thermal emission by the ocean surface is given by

$$T_B^v = a_0 + a_1\cos\theta + a_2\cos 2\theta, \qquad (6.118a)$$

$$T_B^h = b_0 + b_1\cos\theta + b_2\cos 2\theta, \qquad (6.118b)$$

$$T_B^3 = c_1\sin\theta + c_2\sin 2\theta, \qquad (6.118c)$$

$$T_B^4 = d_2\sin 2\theta, \qquad (6.118d)$$

where θ is the relative angle between the azimuthal pointing direction of the radiometer's antenna and the wind direction. At a frequency of 19.35 GHz and an incidence angle of 50° with a wind speed of 12 m/s, the model coefficients are $a_0 = 172$ K, $a_1 = 1.5$ K, $a_2 = 0.95$ K, $b_0 = 113$ K, $b_1 = 0.5$ K, $b_2 = -1.0$ K, $c_1 = -1.25$ K, $c_2 = -1.7$ K, and $d_2 = 0.5$ K.

(a) An attempt is made to retrieve the relative wind direction by measurement of only T_B^v and T_B^h. Can it be uniquely determined from these two measurements for all possible values of θ?

(b) Now assume that measurements are made by a fully polarimetric radiometer of all four Stokes parameters: T_B^v, T_B^h, T_B^3, and T_B^4. If the measurements are perfect (i.e., if they are predicted exactly by the forward model given in Eq. (6.118), so that the instrument calibration is perfect, and if there is no additive noise present), then there will be a unique set of measurements for every value of the relative wind direction θ. However, perfect measurements are not possible. Find the pair of relative wind directions that are at least 30° apart that have the smallest difference between measurements. The solution can be found numerically. Define the difference as the rms difference between the four Stokes parameters at the two angles. What is the minimum rms difference? In practice, one would want the ΔT of the measurements to be significantly smaller than this value.

(c) In practice, it is common to make measurements at a single azimuthal pointing angle only, and to accept a certain amount of uncertainty due to measurement noise. Suppose a suitable wind-direction retrieval algorithm exists that incorporates the forward model sensitivity of T_B to θ represented by Eq. (6.118). We are interested in finding the sensitivity of the wind-direction solution to small perturbations in the brightness temperature. Derive an approximate expression for the standard deviation of the retrieved value of θ if all four polarization components of T_B are perturbed by uncorrelated additive Gaussian noise with standard deviation ΔT. Plot the standard deviation of the retrieved value of θ for the case $\Delta T = 0.3$ K. At which values of θ are the retrievals most and least sensitive to measurement noise?

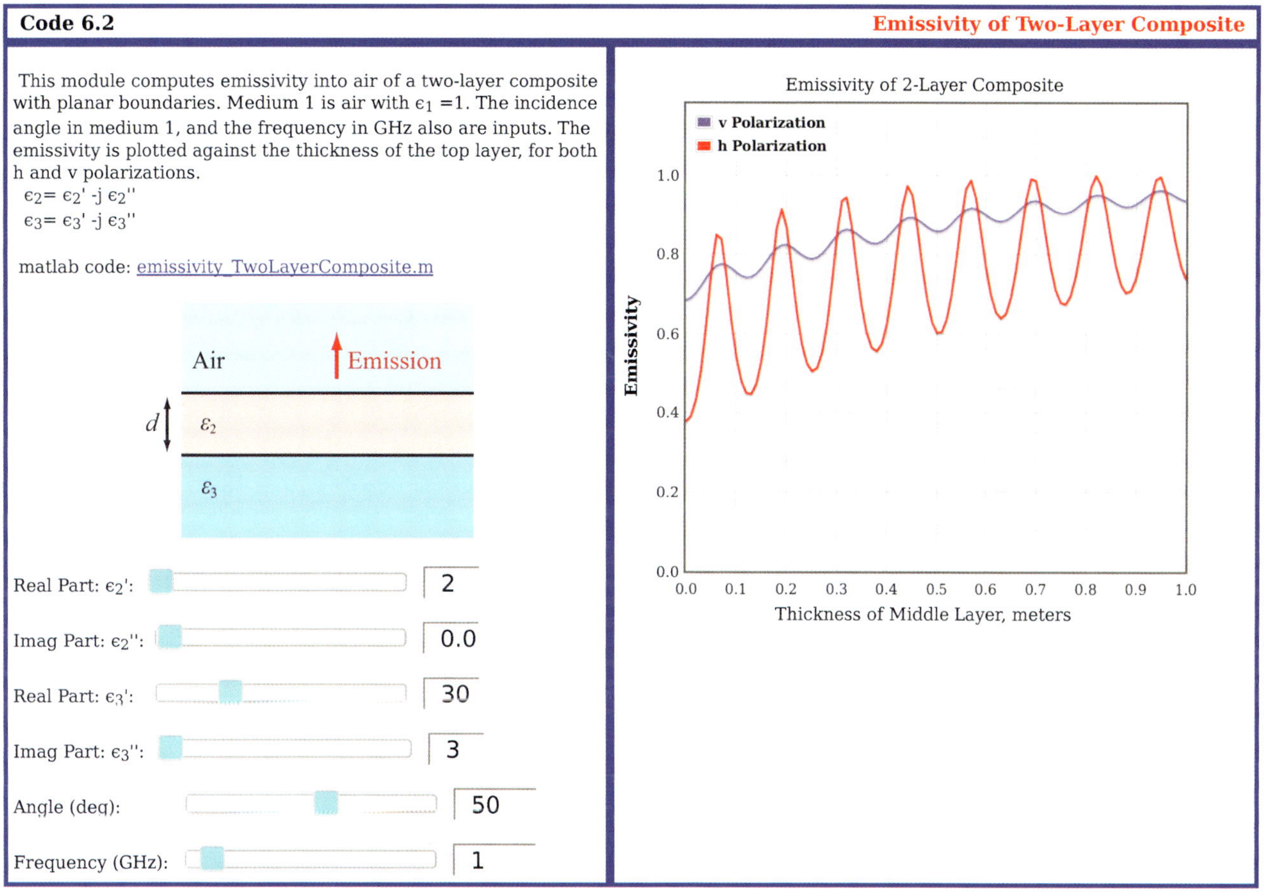

An example of one of the interactive modules available at the book website: mrs.eecs.umich.edu.

CHAPTER 7

Microwave Radiometric Systems

SMOS antenna array in orbit

CONTENTS

Overview

Radiometers are highly sensitive receivers designed to measure thermal electromagnetic emission by material media. In Chapter 6, the theory of radiative transfer was used to relate the electromagnetic properties of the scene observed by an antenna to the power P_A delivered by the antenna to the receiver: $P_A = kT_AB$, where k is Boltzmann's constant and $B = \Delta f$ is the bandwidth of the receiver. The antenna temperature T_A incorporates the intensity of the radiation incident upon the antenna (weighted by the antenna directional pattern) as well as self-emitted energy contributed by the antenna structure itself. The function of a radiometer is to measure T_A; however, T_A represents the average value of a fluctuating noise-like signal. Hence, strictly speaking, a radiometer can only provide an *estimate* of T_A; therefore, not only is the radiometer transfer function (relating T_A to the output voltage V_{out}) of interest, but so is the precision with which T_A can be estimated. The measurement precision of T_A is referred to as the ***radiometric sensitivity*** or ***radiometric resolution*** ΔT. This chapter examines the operation and performance characteristics of several different types of radiometers, as well as several related topics including calibration techniques, imaging considerations, polarimetric systems, and digital radiometry.

7-1 Equivalent Noise Temperature

In a conductor, electrons are always in a state of random motion and their average kinetic energy is proportional to T, the temperature of the conductor. These random motions of the electrons produce fluctuations in electric current, which in turn produce voltage fluctuations. If we were to measure the voltage $\upsilon(t)$ across a conductor of resistance R through an ideal rectangular filter of bandwidth B, we would observe an output similar to that shown in Fig. 7-1. Sources that emit incoherent energy are called ***noise sources***. Hence, voltage $\upsilon(t)$ is called the ***noise voltage*** generated by the resistor, and since it is related to the temperature of the conductor, it is called ***thermal noise***.

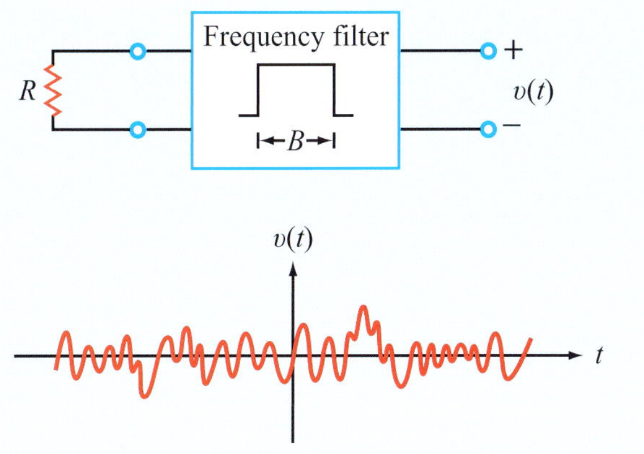

Figure 7-1: Random variation of noise voltage across a resistor.

▶ Whereas the average (or mean) value of $\upsilon(t)$ is zero, its rms value, V_{rms}, is not. ◀

In 1928, Nyquist showed that at the output of a unity gain rectangular filter,

$$V_{rms}^2 = \langle \upsilon^2(t) \rangle = 4RkTB, \tag{7.1}$$

where k is Boltzmann's constant, T is the physical temperature of the resistor, and B is the bandwidth of the filter.

The average noise power delivered by a *noisy* resistor at a temperature T is determined by replacing the noisy resistor with an equivalent circuit consisting of a voltage generator V_{rms}, in series with an impedance $\mathbf{Z} = R + jX$ consisting of a noise-free resistance R and a reactance X, as shown in Fig. 7-2. The reactance X accounts for the combination of the self-inductance of the resistor and the possible capacitance that may exist between the resistor's terminals. The power transferred from the equivalent voltage generator to a load $\mathbf{Z}_L = R_L + jX_L$ is maximum when $\mathbf{Z}_L$ is matched to the generator impedance; that is, when $\mathbf{Z}_L = \mathbf{Z}^* = R - jX$, or equivalently, $R = R_L$ and $X = -X_L$. Under such a matched condition, the ***average noise power*** dissipated

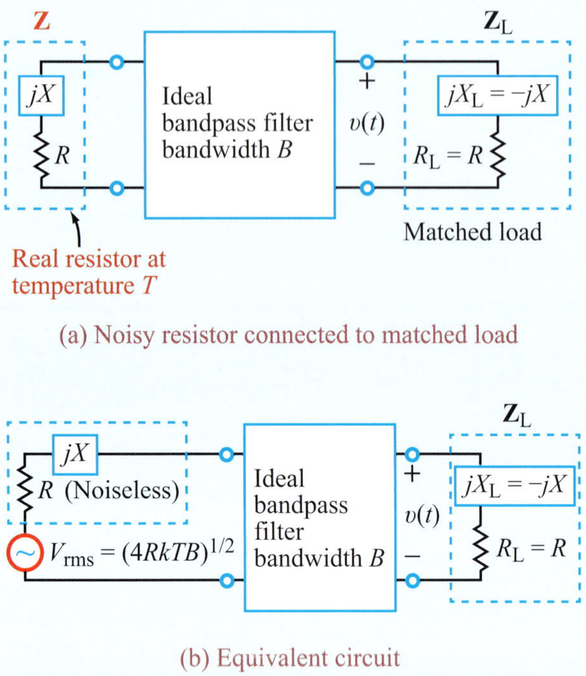

Figure 7-2: (a) Noisy resistor connected to a matched load, and (b) its equivalent circuit. The average power delivered to R_L is $P_n = kTB$.

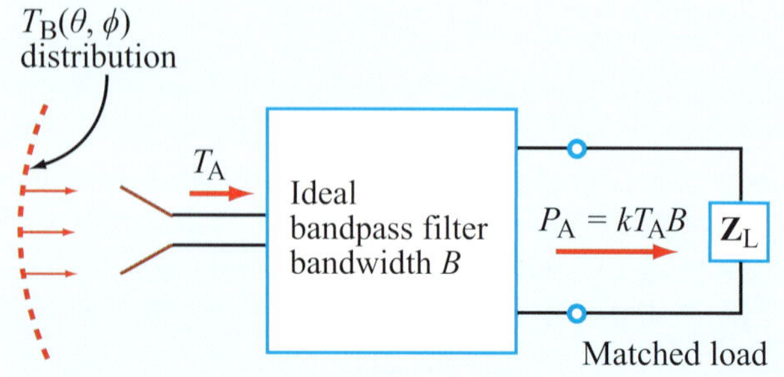

Figure 7-3: The average power delivered by an antenna with radiometric antenna temperature T_A to a matched load is $P_A = kT_AB$.

in R_L is

$$P_n = I_{rms}^2 R = \left(\frac{V_{rms}}{2R}\right)^2 R = \frac{V_{rms}^2}{4R} = kTB. \tag{7.2}$$

The power-temperature correspondence expressed by Eq. (7.2) is identical in form to the power delivered by a lossless antenna enclosed in a chamber of constant temperature T. The similarity can be extended further to any antenna, lossless or not, and the antenna need not be confined to a constant-temperature chamber (Fig. 7-3). In view of Eq. (6.41), which states that the average power delivered by an antenna with antenna temperature T_A (representing the energy incident upon its aperture) is $P_A = kT_AB$, it follows that a resistor of physical temperature T_A is equivalent to an antenna with antenna temperature T_A; both deliver the same amount of average power to a matched load. This equivalence between an antenna and a resistor provides a convenient tool for calibrating microwave radiometers (Section 7-11).

Whereas thermal-noise generation is a universal property of all matter, it is not the only type of random noise encountered in electronic circuits and devices. Other types include quantum noise, shot noise, and flicker noise. Quantum noise, which is associated with discrete energy levels of atoms and molecules, is insignificant in comparison with thermal noise unless the EM frequency is very high or the temperature is very low; quantum noise may be neglected as long as the Rayleigh-Jeans approximation holds (Section 6-2.3). Shot noise arises from the discrete nature of current flow in electronic devices such as diodes and transistors, while flicker noise arises from surface irregularities in cathodes and semiconductors.

According to Eq. (7.2), the (time) average thermal noise power per unit bandwidth is dependent on only the physical temperature of the resistor and is independent of both the operating frequency and the resistance R. This property of thermal-noise can be extended to define an ***equivalent output noise temperature*** T_E^o for any noise source, regardless of the mechanism responsible for the generated noise. If P_n^o is the output noise power of a nonthermal noise source over a bandwidth B, its equivalent output noise temperature is defined such that

$$P_n^o = kT_E^o B. \tag{7.3}$$

7-2 Characterization of Noise

7-2.1 Noise Figure

The ***noise figure*** F of a linear two-port device (or system) is a measure of the degradation in signal-to-noise ratio between its input and output ports, due to noise addition by the device. For the device shown in Fig. 7-4(a),

$$F = \frac{P_\text{s}^\text{i}/P_\text{n}^\text{i}}{P_\text{s}^\text{o}/P_\text{n}^\text{o}}, \tag{7.4}$$

where P_s^i is the available input signal power, P_n^i is the available input noise, and similarly, P_s^o and P_n^o are the output signal and noise powers, respectively. Noise figure F is defined for a specific input noise power, namely that power that would be provided by a resistor matched to the input port of the device and whose temperature is $T_0 = 290$ K ($\approx$ room temperature). That is,

$$P_\text{n}^\text{i} = kT_0B. \tag{7.5}$$

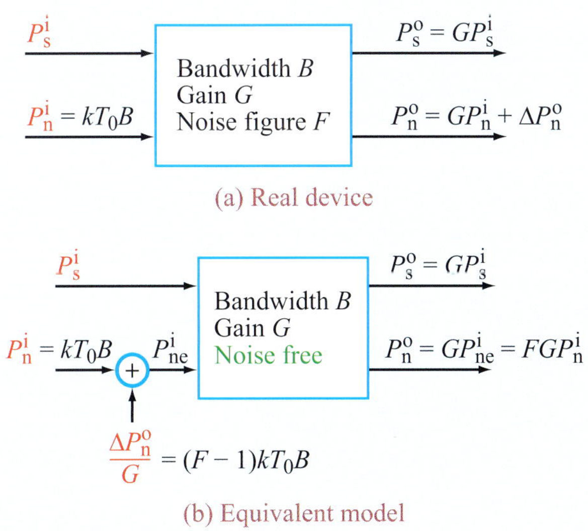

Figure 7-4: A noisy device can be replaced by a noise-free device if the input noise is increased by the noise figure of the device: (a) noisy device; (b) equivalent representation of (a) in terms of a noise-free device.

If over the bandwidth B the average power gain of the device is G, then

$$P_\text{s}^\text{o} = GP_\text{s}^\text{i}, \tag{7.6a}$$

$$P_\text{n}^\text{o} = GP_\text{n}^\text{i} + \Delta P_\text{n}^\text{o}, \tag{7.6b}$$

where $\Delta P_\text{n}^\text{o}$ is the noise power generated by the device itself. Hence,

$$F = \frac{P_\text{s}^\text{i}}{P_\text{s}^\text{o}} \cdot \frac{P_\text{n}^\text{o}}{P_\text{n}^\text{i}} = \frac{1}{G} \cdot \frac{GkT_0B + \Delta P_\text{n}^\text{o}}{kT_0B} = 1 + \frac{\Delta P_\text{n}^\text{o}}{GkT_0B}. \tag{7.7}$$

> ▶ The noise figure of any system is always larger than or equal to 1. For an ideal noise-free system, $\Delta P_\text{n}^\text{o} = 0$ and $F = 1$. ◀

Occasionally, F may be expressed in decibels:

$$F\ [\text{dB}] = 10 \log F. \tag{7.8}$$

From Eq. (7.7), $\Delta P_\text{n}^\text{o}$ is given by

$$\Delta P_\text{n}^\text{o} = (F-1)GkT_0B, \tag{7.9}$$

and the total output noise power is

$$P_\text{n}^\text{o} = GP_\text{n}^\text{i} + \Delta P_\text{n}^\text{o} = GkT_0B + (F-1)GkT_0B = FGkT_0B. \tag{7.10}$$

Since the device self-generates $\Delta P_\text{n}^\text{o}$ at its output terminals, we can model it as a noise-free device with an equivalent input noise source with power output $\Delta P_\text{n}^\text{o}/G$ at its input terminals, so that after amplification by gain G, it generates noise power $\Delta P_\text{n}^\text{o}$. This is in addition to amplifying input noise power $P_\text{n}^\text{i} = kT_0B$ by G. In Fig. 7-4(b), we show an equivalent circuit in which the device is represented by a noise-free equivalent of the same gain and bandwidth, and its self-generated noise is represented by a fictitious source with power

$$\Delta P_\text{n}^\text{i} = \Delta P_\text{n}^\text{o}/G = (F-1)kT_0B. \tag{7.11}$$

▶ The process of replacing a device or a system with a noise-free equivalent and adding a fictitious noise source at its input terminals so as to account for the noise self-generated by the device or system is called ***referring the noise to the input terminals.*** ◀

This ***referral*** process allows us to characterize the overall performance of a cascaded system in terms of the noise figures of its individual subsystems (Section 7-3).

7-2.2 Equivalent Input Noise Temperature

For a linear two-port device (or system), the internally generated noise power $\Delta P_\mathrm{n}^\mathrm{o}$ should be independent of both the signal and the noise at the input terminals of the device. But according to Eq. (7.9), $\Delta P_\mathrm{n}^\mathrm{o}$ is a function of T_0, the input noise temperature. This artificial dependence on T_0 follows from the form of the definition of the noise figure F, which specifies the input noise power as $P_\mathrm{n}^\mathrm{i} = kT_0B$. To avoid confusion, the definition of noise figure has been standardized by choosing $T_0 = 290$ K since most radio-frequency communication systems usually operate at a room temperature in the neighborhood of 290 K.

The ***equivalent input noise temperature*** T_E^i offers an alternative approach for characterizing the noise performance of devices and systems. In fact, it is the preferred approach in microwave radiometry because radiometer receivers are often cooled to temperatures much below room temperature in order to reduce their self-generated noise. It is an attractive approach because T_E^i depends upon only the parameters of the device, not P_n^i.

In Fig. 7-5(a), the noisy device under consideration is connected to a fictitious resistor R maintained at 0 K at its input terminals and a matched load impedance $\mathbf{Z}_\mathrm{L}$ at its output terminals. The fictitious resistor at absolute zero temperature does not radiate, so the output noise delivered to the matched load is entirely due to the noise self-generated by the device; i.e., $\Delta P_\mathrm{n}^\mathrm{o}$. In part (b) of the same figure, the device is replaced with a noise-free equivalent and the resistor's temperature is assigned a temperature T_E^i, defined such that the noise-free equivalent device with gain G generates the same amount of noise output $\Delta P_\mathrm{n}^\mathrm{o}$. That is,

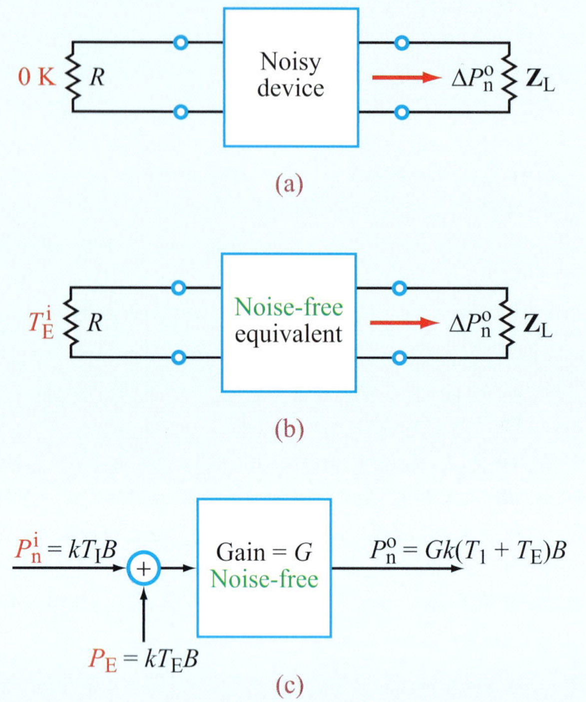

Figure 7-5: Definition and representation of equivalent input noise temperature T_E: (a) noisy device with input connected to a fictitious resistor at 0 K temperature; (b) equivalent noise-free device with input connected to a fictitious resistor at temperature T_E; (c) device-generated noise referred to its input terminals.

$$\Delta P_\mathrm{n}^\mathrm{o} = GkT_\mathrm{E}^\mathrm{i}B. \tag{7.12}$$

▶ From here on forward, we simplify our notation by deleting the superscript on T_E^i; we refer to $T_\mathrm{E} = T_\mathrm{E}^\mathrm{i}$ as simply the ***noise temperature*** of the device, while keeping in mind that it is referred to its input terminals. ◀

If the actual noise power available at the input to the device, P_n^i, is characterized by an equivalent noise temperature $T_\mathrm{I} = P_\mathrm{n}^\mathrm{i}/kB$, then the total input noise temperature of the equivalent noise-free device

[Fig. 7-5(c)] is $(T_{\mathrm{I}}+T_{\mathrm{E}})$, and the corresponding noise power at the output is

$$P_{\mathrm{n}}^{\mathrm{o}} = Gk(T_{\mathrm{I}}+T_{\mathrm{E}})B. \qquad (7.13)$$

The noise temperature T_{E} can be related to the noise figure F by equating Eqs. (7.9) and (7.12) , which leads to

$$T_{\mathrm{E}} = (F-1)T_0. \qquad (7.14)$$

While it again seems like T_{E} depends on T_0, from Eq. (7.9), $(F-1)T_0 = \Delta P_{\mathrm{n}}^{\mathrm{o}}/(GkB)$, which means that T_{E} is actually independent of T_0. Even though both quantities, F and T_{E}, describe the same property of a device or system, F is used more commonly for describing the noise performance of conventional receivers, whereas T_{E} is preferred for low-noise devices and systems, particularly if they operate at cooled-down temperatures.

7-2.3 Noise Temperature of a Cascaded System

We now extend the concept of noise temperature defined for an individual device to a complete system consisting of N cascaded subsystems. We first consider two subsystems connected in series, both with the same bandwidth, but with different noise temperatures and power gains, as shown in Fig. 7-6(a). From the definition of equivalent input noise temperature, each subsystem can be replaced with a noise-free equivalent subsystem with a noise generator added at its input terminals, as shown in Fig. 7-6(b). For subsystem 1 with noise temperature T_{E_1} (referred to its input terminals), the noise generator power is

$$P_{\mathrm{E}_1} = kT_{\mathrm{E}_1}B, \qquad (7.15a)$$

and similarly,

$$P_{\mathrm{E}_2} = kT_{\mathrm{E}_2}B \qquad (7.15b)$$

for subsystem 2. The total noise power at the output of the two-subsystem cascade in Fig. 7-6(b) is

$$\begin{aligned} P_{\mathrm{n}}^{\mathrm{o}} &= G_1G_2P_{\mathrm{n}}^{\mathrm{i}} + G_1G_2P_{\mathrm{E}_1} + G_2P_{\mathrm{E}_2} \\ &= kG_1G_2\left(T_{\mathrm{I}} + T_{\mathrm{E}_1} + \frac{T_{\mathrm{E}_2}}{G_1}\right)B, \qquad (7.16) \end{aligned}$$

where T_{I} is the (real) noise temperature at the input to the system.

Next, we replace the two cascaded noise-free subsystems with an equivalent, single, noise-free system of gain G_1G_2 and input noise temperature T_{E}, defined such that $P_{\mathrm{n}}^{\mathrm{o}}$ is the same for the cascaded subsystems and for their single-system equivalent. Comparison of the expressions for $P_{\mathrm{n}}^{\mathrm{o}}$ in Figs. 7-6(b) and (c) leads to

$$T_{\mathrm{E}} = T_{\mathrm{E}_1} + \frac{T_{\mathrm{E}_2}}{G_1}. \qquad (7.17)$$

By extension, the noise temperature of a system consisting of N cascaded subsystems is

$$T_{\mathrm{E}} = T_{\mathrm{E}_1} + \frac{T_{\mathrm{E}_2}}{G_1} + \frac{T_{\mathrm{E}_3}}{G_1G_2} + \cdots + \frac{T_{\mathrm{E}_N}}{G_1G_2\ldots G_{N-1}}, \qquad (7.18)$$

where T_{E_1} is the input noise temperature of the first subsystem (located at the input terminals of the overall system), T_{E_2} belongs to the next subsystem, and so on to T_{E_N} of the final subsystem.

7-2.4 Noise Temperature of a Lossy Two-Port Device

An antenna with a radiation efficiency $\xi < 1$ qualifies as a lossy two-port device; it has an input and an output, and by virtue of the fact that its radiation efficiency is smaller than 1, it is lossy because it absorbs part of the energy passing through it. A transmission line connecting an antenna to a ***radio frequency*** (RF) amplifier also is a lossy two-port device, and from the standpoint of radiative transfer, so is the Earth's atmosphere. If a lossy device absorbs EM energy (radiated by other sources), the law of thermodynamic equilibrium mandates that it also radiate an equal amount of energy.

Consider the two-port device shown in Fig. 7-7. Its ambient (physical) temperature is T_{p} and it is connected to matched loads on both sides, and both loads are maintained at the same temperature T_{p}. Thus, the entire system is at temperature T_{p}. The device has a

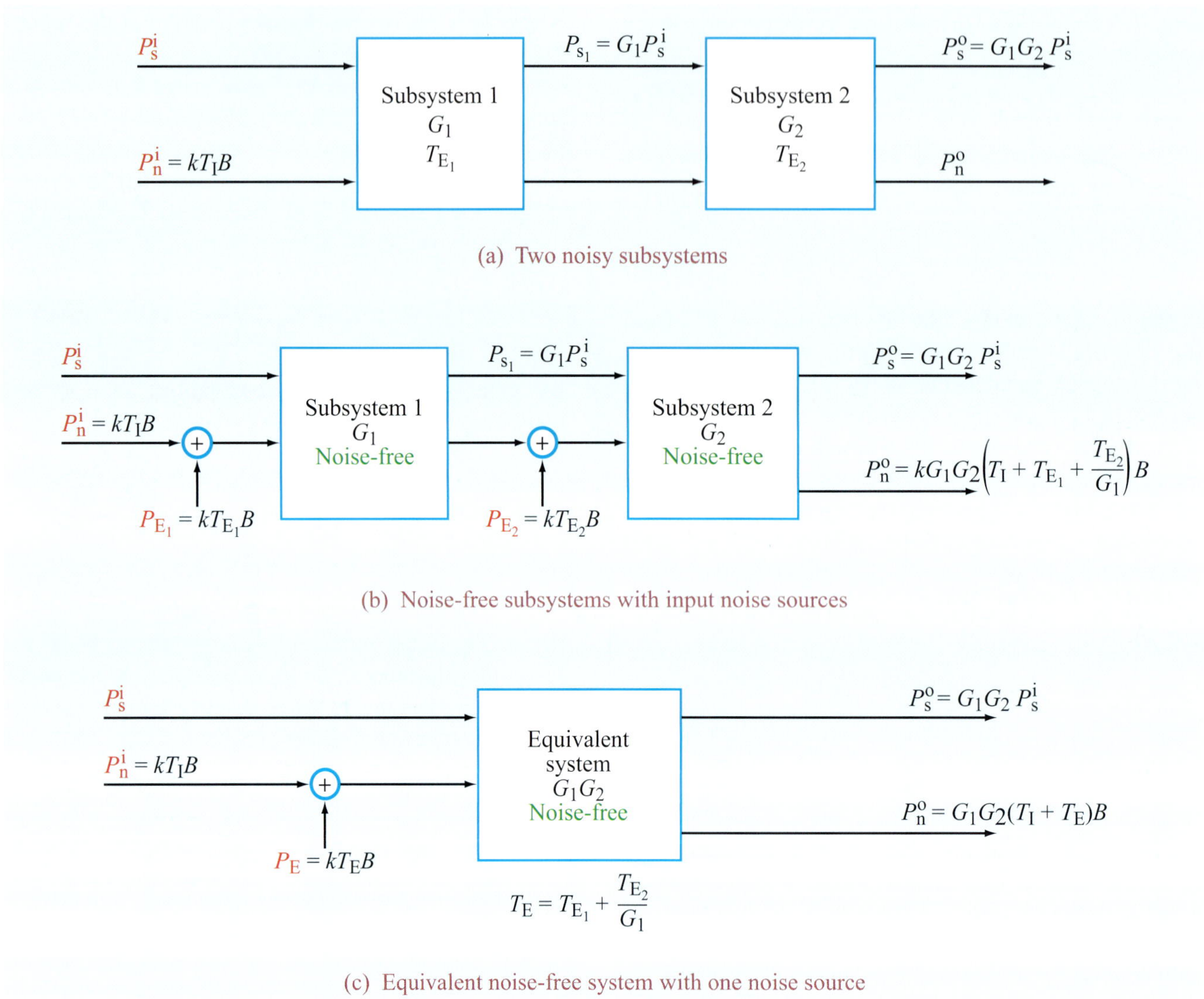

Figure 7-6: Effective input noise temperature of two subsystems in cascade: (a) two noisy subsystems in cascade; (b) equivalent representation of (a) in terms of noise-free subsystems; (c) equivalent single-system representation of (b).

loss factor L, which is by definition the inverse of ***power gain***:

$$L = \frac{1}{G} = \frac{P^i}{P^o}, \tag{7.19}$$

where P^i and P^o are, respectively, the input and output power. For a reciprocal two-port device, propagation can occur in either direction, so the definition given by Eq. (7.19) applies to either port serving as the input and the other as the output.

Since the total system is in thermodynamic equilibrium (all elements at the same physical temperature T_p), the flow of noise power into load 2 (crossing the reference plane from the left in Fig. 7-7) should be equal to the flow of power from load 2 (crossing the reference plane from the right). The former is designated P_n^o and

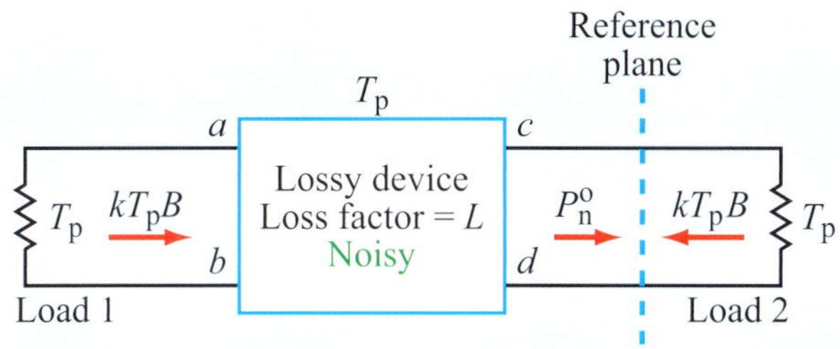

Figure 7-7: Noise generated in a lossy device of loss factor L and physical temperature T_p.

the latter is equal to kT_pB. Hence,

$$P_n^o = kT_pB. \tag{7.20}$$

Noise power P_n^o at terminals (c, d) consists of the noise generated by load 1, namely kT_pB, attenuated by the factor L as it passes through the device, plus noise ΔP_n^o generated internally by the device:

$$P_n^o = \frac{1}{L} kT_pB + \Delta P_n^o. \tag{7.21}$$

Equating Eqs. (7.20) and (7.21) gives

$$\Delta P_n^o = \left(1 - \frac{1}{L}\right) kT_pB. \tag{7.22}$$

This is the noise power generated by the device at its output terminals, and it is proportional to a noise temperature T_E^o (referred to the same output terminals) given by

$$T_E^o = \frac{\Delta P_n^o}{kB} = \left(1 - \frac{1}{L}\right) T_p. \tag{7.23}$$

Equivalently, we can define a noise temperature $T_E = T_E^i$ referred to its input terminals as

$$T_E = T_E^i = LT_E^o = (L-1)T_p. \tag{7.24}$$

For an antenna, its radiation efficiency ξ is the ratio of the power it radiates into space to that provided by a transmitter, or the power it delivers to a receiver to that it captures by its aperture. In both cases, ξ is a ratio of

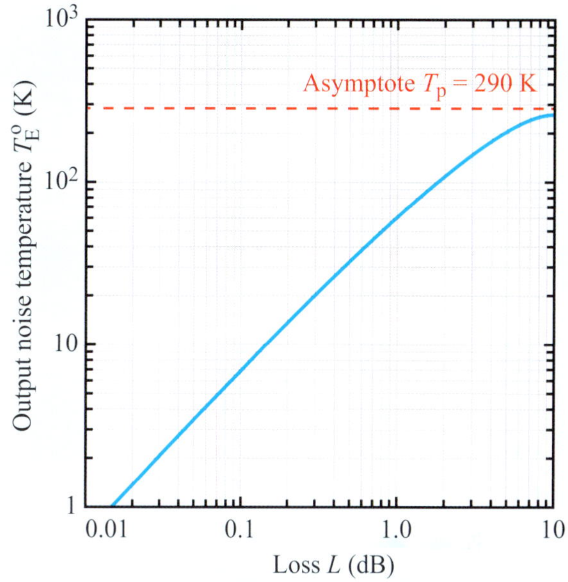

Figure 7-8: Output noise generated by a lossy device with a physical temperature $T_p = 290$ K.

output power to input power, which is the inverse of the definition of the loss factor L given by Eq. (7.19). Hence,

$$\xi = \frac{1}{L} \qquad \textbf{(antenna)}, \tag{7.25}$$

and the amount of noise power it generates by self-emission is

$$\Delta P_n^o = \left(1 - \frac{1}{L}\right) kT_pB = (1-\xi)kT_pB. \tag{7.26}$$

The antenna radiates noise power ΔP_n^o into free space, as well as towards the receiver.

To appreciate the dependence of T_E^o on L, we show in Fig. 7-8 a plot of T_E^o as a function of L, with the latter expressed in dB. We also consider the example of a transmission line of loss L connecting an antenna to a low-noise receiver with equivalent input noise temperature T_{REC}, as shown in Fig. 7-9. The overall noise performance of the transmission line and receiver combination can be represented by an equivalent input

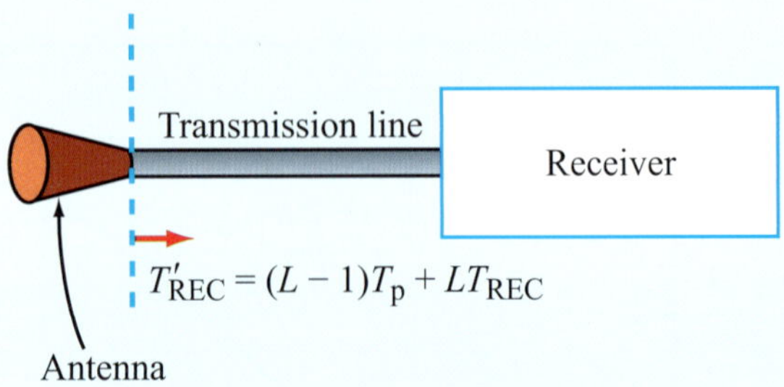

Figure 7-9: T'_{REC} is the input noise temperature of an equivalent noise-free transmission line and receiver combination. T_{REC} is the receiver input noise temperature alone, T_p is the physical temperature of the transmission line and L is its loss factor.

noise temperature T'_{REC} referred to the antenna terminals, thereby treating the transmission line and receiver as noise-free. For the two-stage system,

$$T'_{REC} = T_{E_1} + \frac{T_{E_2}}{G_1}. \tag{7.27}$$

In the present case, T_{E_1} is the input noise temperature of the transmission line and is given by Eq. (7.24), $T_{E_2} = T_{REC}$, and $G_1 = 1/L$. Hence,

$$T'_{REC} = (L-1)T_p + LT_{REC}. \tag{7.28}$$

If the low-noise receiver is characterized by $T_{REC} = 50$ K and if $T_p = 290$ K and $L = 1.12$ (= 0.5 dB), $T'_{REC} = 90.8$ K. Hence, even a loss factor as low as 0.5 dB can degrade the noise performance of the overall receiver system by approximately a factor of 2 (compared with a receiver connected to the antenna directly). Had the receiver been of the conventional type, with a typical noise temperature $T_{REC} = 1000$ K, the noise added by the transmission line would have been of minor significance, since in that case $T'_{REC} = 1155$ K, an increase of only 11.6%.

7-3 Receiver and System Noise Temperatures

7-3.1 Receiver Alone

The front-end of a superheterodyne receiver usually consists of an RF amplifier connected to a mixer that translates the carrier frequency f_{RF} of the signal at the output of the RF amplifier into an intermediate frequency $f_{IF} = |f_{RF} - f_{LO}|$, where f_{LO} is the frequency of the local oscillator. The signal at the output of the mixer is then fed into an ***intermediate frequency*** (IF) amplifier, as shown in Fig. 7-10. From the standpoint of noise, the overall performance of the receiver is dominated by the performance of the RF amplifier, mixer, and IF amplifier, usually in that order.

The local oscillator serves to shift the carrier frequency and usually has no role so far as noise is concerned. Hence, the superheterodyne receiver qualifies as a cascaded system, in which case we can apply Eq. (7.18) to compute its (input) receiver noise temperature T_{REC}:

$$T_{REC} = T_{RF} + \frac{T_M}{G_{RF}} + \frac{T_{IF}}{G_{RF}G_M} + \cdots, \tag{7.29}$$

where T_{RF} and G_{RF} are the noise temperature and gain of the RF amplifier, and similar definitions apply to the mixer and IF amplifier. In Eq. (7.29), the noise contributions of the stages following the IF amplifier are not included explicitly because, as we see shortly, they are insignificant in comparison with the contributions of the front-end stages. To demonstrate through an example, let us consider the following list of specifications for a typical receiver operating at a center frequency of 1.5 GHz:

$$T_{RF} = 200 \text{ K}, \qquad G_{RF} = 1000 \text{ (30 dB)},$$
$$T_M = 1200 \text{ K}, \qquad G_M = 200 \text{ (23 dB)},$$
$$T_{IF} = 100 \text{ K}, \qquad G_{IF} = 1000 \text{ (30 dB)}.$$

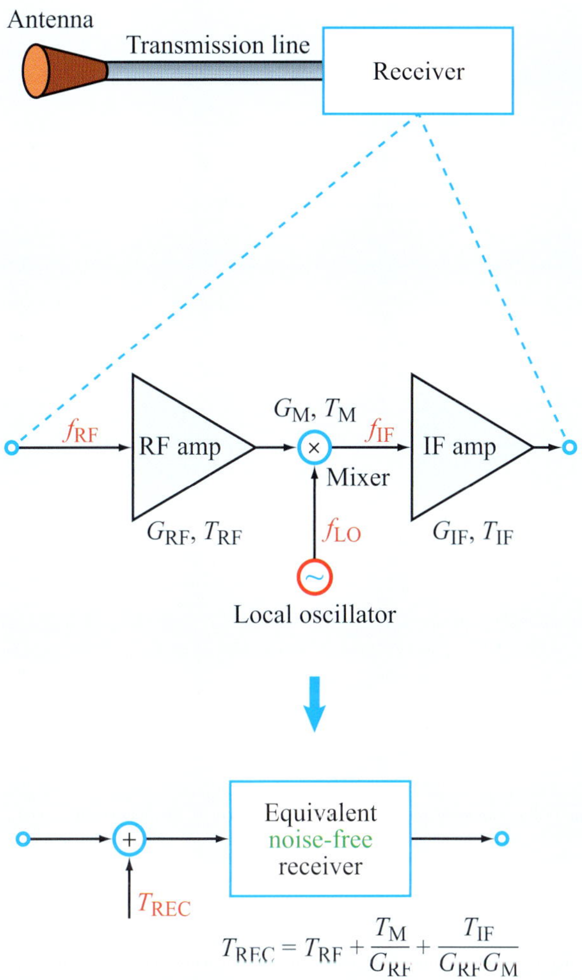

Figure 7-10: Front-end segment of superheterodyne receiver.

Inserting these values in Eq. (7.29) gives

$$\begin{aligned} T_{\text{REC}} &= 200 + \frac{1200}{1000} + \frac{100}{1000 \times 200} + \cdots \\ &= 200 + 1.2 + 5 \times 10^{-4} \\ &\approx 201.2 \text{ K} \qquad \textbf{(with RF amplifier)}. \end{aligned}$$

▶ For all practical purposes, the receiver noise temperature is equal to the noise temperature of the first stage, the RF amplifier. With regard to noise performance, the first stage of a receiver is the most critical, provided its power gain is much larger than 1. ◀

Without an RF amplifier, the receiver noise temperature would be

$$\begin{aligned} T_{\text{REC}} &= T_{\text{M}} + \frac{T_{\text{IF}}}{G_{\text{M}}} \\ &= 1200 + \frac{100}{200} \\ &\approx 1200 \text{ K} \qquad \textbf{(without RF amplifier)}. \end{aligned}$$

7-3.2 Total System Including Antenna

Given a receiver with a noise temperature T_{REC} (referred to its input terminals), we now consider the overall system including the antenna, the transmission line, and the receiver (Fig. 7-11). We define P_{SYS} as the total noise power flowing toward the receiver from the junction between the antenna and the transmission line, with the noise generated by all stages to the right of the junction referred to that junction. That is, the transmission line and receiver are treated as noise-free and their combined noise temperature T'_{REC} is ***referred***

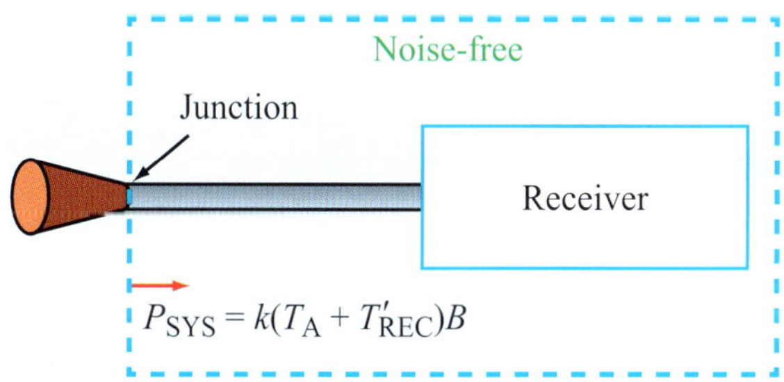

Figure 7-11: Equivalent input system noise power incorporates noise generated by the receiver, transmission line, antenna self-emission, and emission by the scene observed by the antenna.

to the antenna junction. The second component of P_{SYS} is the power delivered by the antenna, P_A. Hence,

$$\begin{aligned} P_{SYS} &= P_A + P'_{REC} \\ &= k(T_A + T'_{REC})B \\ &= kT_{SYS}B, \end{aligned} \tag{7.30}$$

with T_{SYS} defined as the ***system noise temperature***,

$$\begin{aligned} T_{SYS} &= T_A + T'_{REC} \\ &= \xi T'_A + (1-\xi)T_0 + (L-1)T_{tl} + LT_{REC}. \end{aligned} \tag{7.31}$$

In Eq. (7.31), the first pair of terms represents T_A as given by Eq. (6.39), the second pair represents T'_{REC} as given by Eq. (7.28), T_0 and T_{tl} are the physical temperatures of the antenna and transmission line,* and L is the loss factor of the transmission line.

To a radiometer receiver, P_{SYS} represents the "signal" at the input terminals of an equivalent noise-free receiving system (transmission line and receiver). That is, the receiver output voltage is proportional to P_{SYS}. By comparing the output voltage due to P_{SYS} with the output voltage due to a matched load (in place of the antenna) of known physical temperature, a radiometer provides an estimate of T'_A, the antenna temperature of the scene under observation (provided T_0 and ξ are known). Details of the technique and the precision associated with the estimate of T'_A are given in future sections.

In the case of a radar, P_{SYS} represents the noise power at the input to the receiver, while the input signal is the received power P_r given by the radar equation (Chapter 5). The input signal-to-noise ratio is $S_n = P_r/P_{SYS}$. This ratio is used to establish the transmitter power and antenna parameters required for a given performance specification. If through the application of signal-processing techniques the signal-to-noise ratio at the receiver output is better than at the input, then the effective input signal-to-noise ratio is S_n multiplied by the improvement factor. Considerations of radar-system design and performance are given in Chapter 13.

*We use upper-case subscripts for T to denote radiometric temperatures and lower-case subscripts to denote physical temperatures.

7-4 Radiometer Operation

The function of a radiometer is to measure the antenna radiometric temperature T_A, representing the radiation power delivered by the antenna to the receiver. The measurement process is characterized by two important attributes: (1) accuracy and (2) precision. Accuracy is associated with the mean value of a process, whereas precision is associated with the uncertainty relative to the mean.

7-4.1 Measurement Accuracy

The transfer function of a radiometer receiver can be established by measuring the output voltage as a function of the noise temperature of a noise source connected to the receiver input terminals, instead of the antenna. Alternatively, calibration is achieved through the use of a switch placed at a point as close to the antenna as possible. The scheme is shown in Fig. 7-12(a). Also, if the radiometer employs square-law detection, which most radiometers do, its average output voltage is linearly related to the noise temperature of the input source. Linearity allows us to establish the calibration line of the radiometer by measuring its output voltage, V_{out}, for each of two calibration temperatures, as illustrated in Fig. 7-12(b). After performing the calibration (with the switch in position 2), the switch is moved to position 1 to connect the antenna to the receiver. The calibration line allows the user to convert a measured value of V_{out} into a value for T_A. Ideally, the two calibration temperatures, T_1 and T_2, should be selected to be close to the two extreme ends of the expected range of T_A, so as to minimize slope errors. Assuming that the calibration measurements of V_1 and V_2 [Fig. 7-12(b)] are made with a high degree of measurement precision, the absolute accuracy of the measurement of T_A is then dictated by the accuracy with which the absolute values of the calibration temperatures T_1 and T_2 are known. If a passive device, such as a matched load (resistor), is used as the calibration

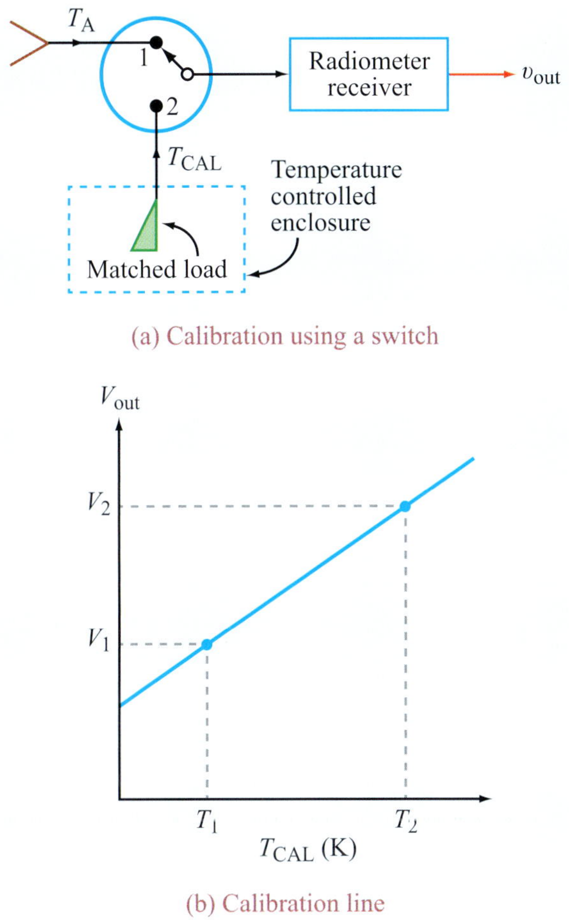

(a) Calibration using a switch

(b) Calibration line

Figure 7-12: Calibration of a microwave radiometer.

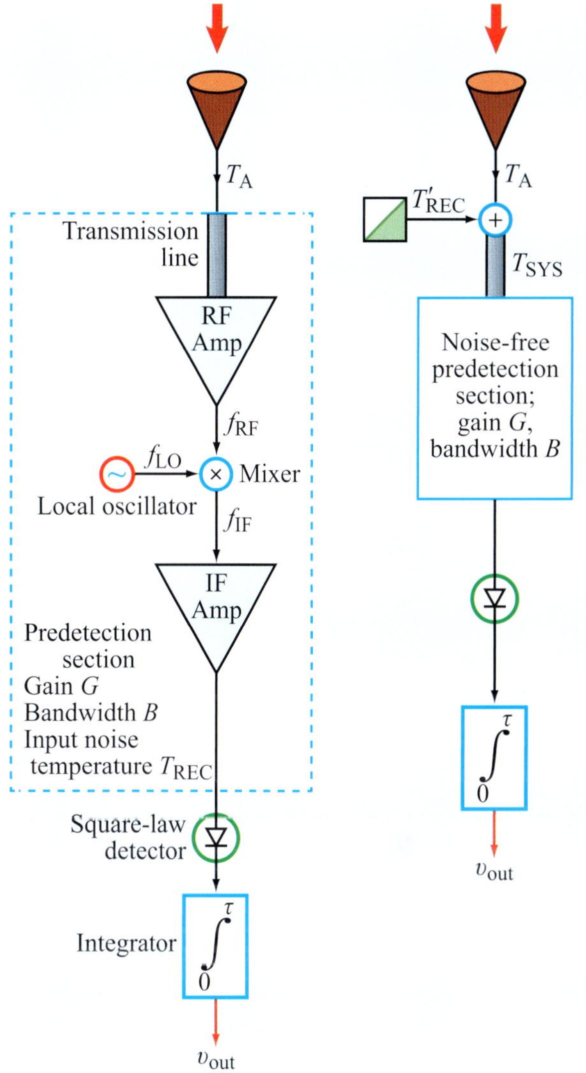

(a) Total-power receiver (b) Equivalent receiver

Figure 7-13: The representation in (b) replaces the predetection section with a noise-free equivalent and refers the receiver noise to the antenna terminals.

source, its temperature accuracy can (in practice) be maintained to within a fraction of 1 K by controlling the temperature of its environment. This is true so long as there are no impedance mismatches in either the antenna-switch-receiver path or the load-switch-receiver path. Impedance mismatches contribute to reflections, which can compromise the measurement accuracy of T_A, as discussed in Section 7-11.

7-4.2 Total-Power Radiometer

Among the various designs of radiometer receivers (several of which are discussed in future sections), the ***total power radiometer*** is considered the reference standard, primarily because of its simple architecture. In Fig. 7-13(a), the antenna is connected to a ***superheterodyne receiver*** of ***predetection bandwidth*** B and total predetection power gain G. The predetection section includes all subsystems between the antenna and the input terminals of the square-law detector.

The power delivered by the antenna is broadband noise extending over a wider bandwidth than the receiver bandwidth B. The function of the radio-frequency (RF) amplifier is to filter the input signal by amplifying the frequency components contained in a bandwidth $B_{RF} \geq B$, centered at the RF frequency f_{RF}. The mixer and intermediate frequency (IF) amplifier translate the RF band of signals from f_{RF} to f_{IF} and provide further amplification.

▶ In practice, the IF stage has a narrower bandwidth than that of the RF stage, so the predetection bandwidth B is effectively determined by the bandpass characteristics of the IF stage. For simplicity, only RF and IF amplifiers are shown in Fig. 7-13(a) to define the passbands. In practice, bandpass filters are usually included in series with the amplifiers to more accurately and stably set the passbands. ◀

A system that translates signals over a bandwidth B from RF to IF is called a ***single-sideband receiver***.

Consider a receiver with a local oscillator frequency $f_{LO} = 10$ GHz and $f_{IF} = 100$ MHz. Also, the IF amplifier has an approximately rectangular frequency response with a bandwidth $B = 10$ MHz (Fig. 7-14). The mixer can generate an IF signal centered at 100 MHz by mixing the local oscillator signal with either or both of two RF signals, one at f_{RF_1}, and another at f_{RF_2}, where

$$f_{RF_1} = f_{LO} - f_{IF} = (10 - 0.1) \text{ GHz} = 9.9 \text{ GHz},$$
$$f_{RF_2} = f_{LO} + f_{IF} = (10 + 0.1) \text{ GHz} = 10.1 \text{ GHz}.$$

With a single-sideband receiver, only one of the two RF signals is allowed to pass through the RF amplifier, whereas without an RF amplifier (or with a broadband RF amplifier whose passband extends over both RF frequencies), the IF signal consists of the sum of two signals, one originally at f_{RF_1} and another originally at f_{RF_2}. Since each signal is of bandwidth B, the total input power for such a ***double-sideband*** receiver [Fig. 7-14(b)] is twice the input power received by the single-sideband configuration. Even though the signal power at the input to the double-sideband receiver is larger by a factor 2 in comparison with that for the single-sideband receiver, the absence of a low-noise RF amplifier may raise the receiver noise temperature by a much larger factor (see Section 7-3.1). Furthermore, if the total available frequency band is fixed, the single-sideband receiver can be designed to use the entire bandwidth, whereas in the double-sideband configuration, part of the available bandwidth is not used.

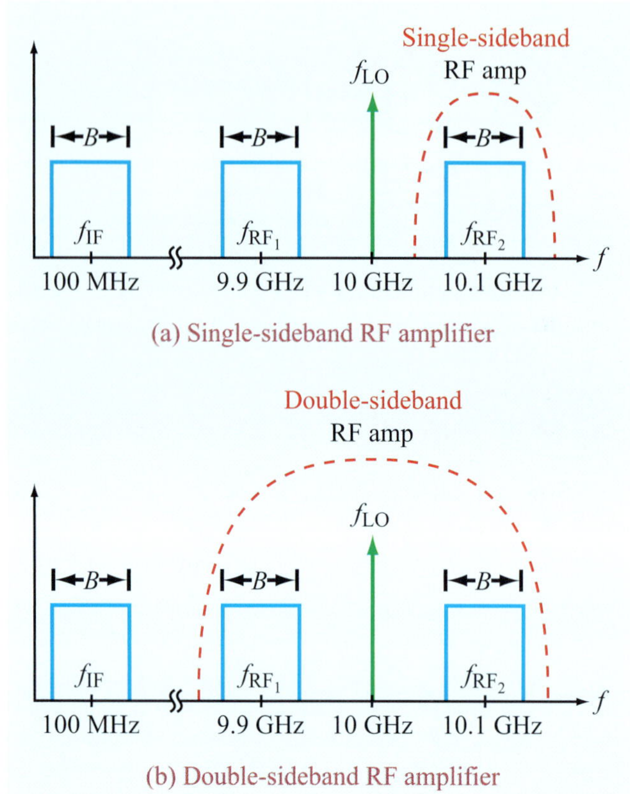

Figure 7-14: A local oscillator can mix with RF spectra centered at either f_{RF_1}, f_{RF_2}, or both, to generate the same bandwidth(s) at f_{IF}. The determination is made by the spectrum of the RF amplifier.

▶ Generally speaking, low-noise RF amplifiers are readily available at frequencies below 100 GHz, moderately available at frequencies in the 100–300 GHz range, and seldom available at higher frequencies. Availability is dictated by technology development and cost. ◀

To relate the input signal, represented by antenna temperature T_A, to the output voltage υ_{out}, we show in Fig. 7-13(b) an equivalent representation of the single-sideband total power radiometer in which the predetection section of the receiver (including the transmission line) is replaced with the combination of a noise-free receiver and an input noise source with output power $P'_{REC} = kT'_{REC}B$, where T'_{REC} is the receiver noise temperature referred to the antenna terminals. The sum of the two sources is the total system noise power P_{SYS}, with a corresponding system noise temperature T_{SYS},

$$P_{SYS} = P_A + P'_{REC} = kT_{SYS}B \tag{7.32}$$

with

$$T_{SYS} = T_A + T'_{REC}. \tag{7.33}$$

Noise is generated by a random process of collisions between the particles in a material medium, with the consequent spontaneous emission of EM radiation by its atoms and molecules. Consequently, a time plot of the voltage or power at any stage in the predetection section of the receiver would exhibit random fluctuations over a wide range of values (Fig. 7-15). Whereas it is not possible to predict the instantaneous values of these voltages and powers, their average values can be ascertained with no ambiguity. At the output of the IF amplifier, the ***average noise power*** is

$$\overline{P}_{IF} = GkT_{SYS}B, \tag{7.34}$$

where G is the power gain of the predetection section. In Eq. (7.34), we added an overbar to P_{IF} to emphasize that $\overline{P}_{IF}$ is the time average value of the instantaneous power $p_{IF}(t)$.

Often, the detection bandwidth B of a microwave radiometer is much smaller than the IF frequency f_{IF}. Hence, the instantaneous IF voltage is approximately a sinusoidal signal at frequency f_{IF} with a time-varying envelope $\upsilon_e(t)$, as shown in Fig. 7-15. That is,

$$\upsilon_{IF}(t) = \upsilon_e(t)\cos[2\pi f_{IF}t + \phi(t)], \tag{7.35}$$

where $\phi(t)$ is a random phase angle. For noise, the envelope $\upsilon_e(t)$ and phase angle $\phi(t)$ are statistically independent random variables. Hence, the mean value of $\upsilon_{IF}(t)$ is

$$\overline{V}_{IF} = \langle \upsilon_{IF}(t)\rangle = \langle \upsilon_e(t)\rangle \langle \cos[2\pi f_{IF}t + \phi(t)]\rangle = 0. \tag{7.36}$$

This is because $\phi(t)$ is uniformly distributed over $[-\pi,\pi]$, so the average value of the cosine function is zero.

Within a multiplicative constant, the instantaneous IF power $p_{IF}(t)$ is related to the instantaneous IF voltage $\upsilon_{IF}(t)$ by

$$\begin{aligned} p_{IF}(t) &= \upsilon_{IF}^2(t) \\ &= \upsilon_e^2(t)\cos^2[2\pi f_{IF}t + \phi(t)] \\ &= \upsilon_e^2(t)\left\{\frac{1}{2} + \frac{1}{2}\cos[4\pi f_{IF}t + 2\phi(t)]\right\}. \end{aligned} \tag{7.37}$$

The time-average value of $p_{IF}(t)$ is

$$\begin{aligned} \overline{P}_{IF} &= \langle p_{IF}(t)\rangle \\ &= \frac{1}{2}\langle \upsilon_e^2(t)\rangle + \frac{1}{2}\langle \upsilon_e^2(t)\rangle\langle\cos[4\pi f_{IF}t + 2\phi(t)]\rangle \\ &= \frac{1}{2}\overline{V_e^2}. \end{aligned} \tag{7.38}$$

The square-law detector in the block diagram of Fig. 7-15 is a slow-response circuit, which means that its output voltage $\upsilon_d(t)$ can follow only the slow variations of the IF voltage, namely those associated with the envelope $\upsilon_e(t)$, but not those associated with the cosine term. Moreover, $\upsilon_d(t)$ is proportional to $\upsilon_e^2(t)$. Hence,

$$\upsilon_d(t) = \upsilon_e^2(t), \tag{7.39}$$

where, for the sake of simplicity, we assigned the proportionality constant a value of 1. In view of

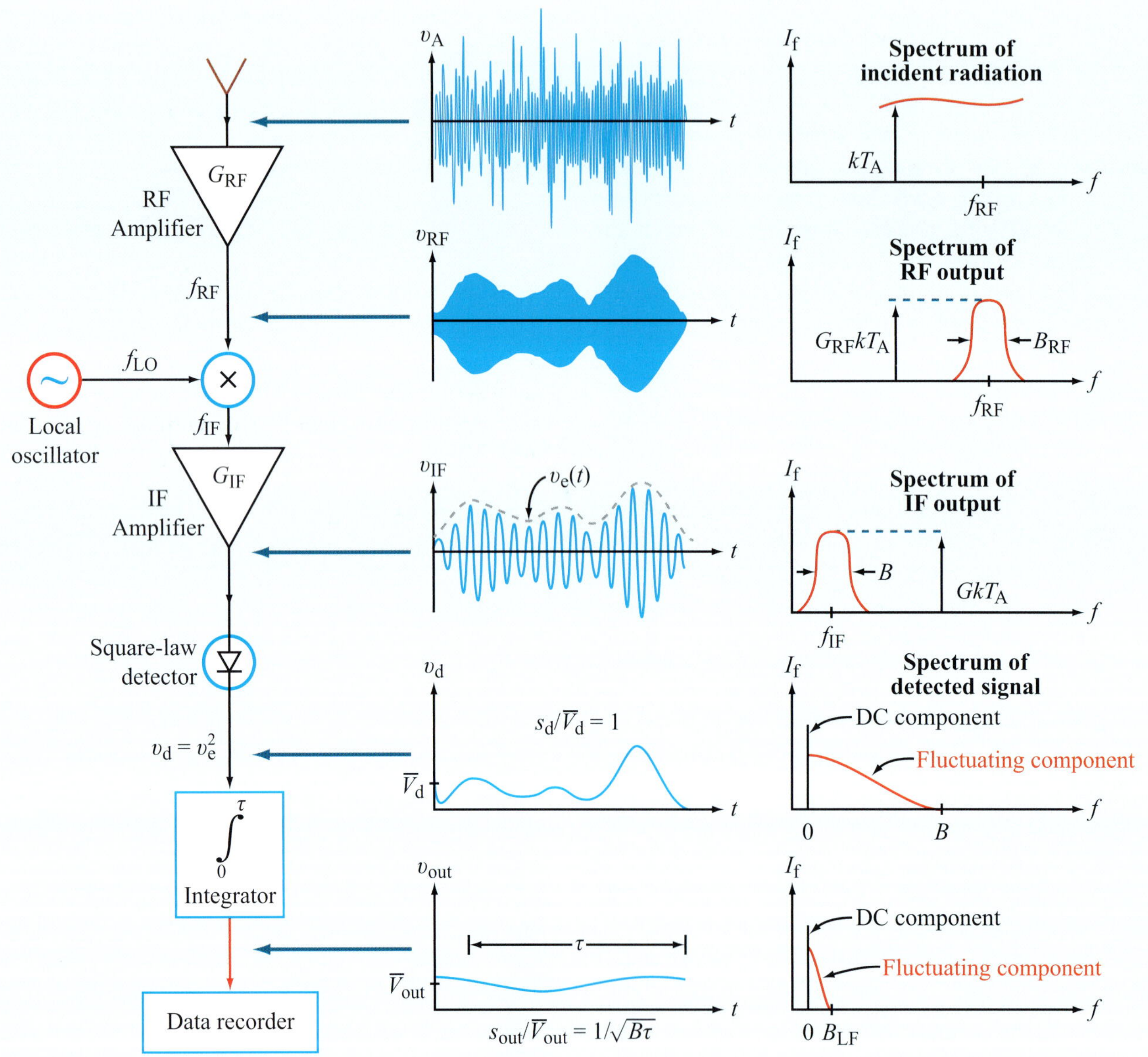

Figure 7-15: Total-power radiometer with a superheterodyne receiver. The signal voltage and corresponding spectrum are shown at various stages. Quantity I_f is the spectral density and B_{LF} is the bandwidth of the low-pass filter (integrator).

Eqs. (7.34), (7.38), and (7.39), it follows that the average value of the detected voltage is

$$\overline{V}_{\rm d} = \overline{V_{\rm e}^2} = 2GkT_{\rm SYS}B. \tag{7.40}$$

To realize the averaging process, the radiometer uses an integrator with a time constant τ and voltage gain $g_{\rm I}$. In analog circuits, an integrator in the time domain is equivalent to a low-pass RC filter in the frequency domain. Its output voltage is

$$\upsilon_{\rm out}(t) = \frac{g_{\rm I}}{\tau}\int_{t-\tau}^{t} \upsilon_{\rm d}(t')\,dt'. \tag{7.41}$$

As we show in the next subsection, if $B\tau \gg 1$,

$$\upsilon_{\rm out}(t) \approx \overline{V}_{\rm out} = g_{\rm I}\overline{V}_{\rm d} = G_{\rm s}T_{\rm SYS} \qquad (B\tau \gg 1), \tag{7.42}$$

where $G_{\rm s}$ is an overall system gain factor given by

$$G_{\rm s} = 2g_{\rm I}GkB. \tag{7.43}$$

Combining Eqs. (7.33) and (7.42) leads to

$$T_{\rm A} = \frac{\overline{V}_{\rm out}}{G_{\rm s}} - T'_{\rm REC}, \tag{7.44}$$

which provides a linear relationship between the unknown antenna temperature $T_{\rm A}$ and the measured output voltage $\overline{V}_{\rm out}$. The system constants can be determined through calibration, as discussed later in Section 7-11.

Most modern radiometers use an analog-to-digital (A/D) converter to digitize the detected signal $\upsilon_{\rm d}(t)$, followed by a digital integrator to perform the operation equivalent to that described by Eq. (7.41). As shown in Fig. 7-15, the spectrum of $\upsilon_{\rm d}(t)$ extends from dc to B, so to satisfy Shannon's sampling theorem, $\upsilon_{\rm d}(t)$ should be sampled at a rate of at least $2B$ (Hz) prior to the A/D operation.

7-4.3 Radiometric Resolution

Because for a real radiometer the integration time τ in Eq. (7.41) is finite, the value of $T_{\rm A}$ it measures is not the true value of $T_{\rm A}$, but rather an estimate of that value. Associated with that estimate is an uncertainty ΔT. If the measured value is 236 K and the measurement uncertainty (*precision*) is $\Delta T = 2$ K, then we report the measurement as $T_{\rm A} = 236\text{ K} \pm 2$ K. In practical terms, ΔT is the smallest change detectable by the measurement system. Hence, it is called the ***radiometric sensitivity*** or ***radiometric resolution*** of the system.

How is ΔT related to the parameters of the radiometer receiver and how do we quantify it? Obviously, the integration time τ is very relevant since we already know that $\Delta T \to 0$ as $\tau \to \infty$, but other system parameters must play a role as well.

To derive an expression for ΔT, we start by examining the statistical nature of the signal detected by the receiver. The signal is represented by $T_{\rm SYS}$, which includes multitudes of photons emitted by the atmosphere and terrain viewed by the antenna beam, an undeterminable number of currents spontaneously generated within the electronic components comprising the receiver, and similar currents generated by the conductors in the antenna and transmission line. All of these sources are noise-like in character, meaning that they occur spontaneously, for infinitesimally short durations, and independently of each other. They constitute a large number of uncorrelated sources, so at any instant in time, the voltage at the output of the IF amplifier can be represented as the sum of $N_{\rm s}$ sinusoids, all oscillating at the same IF frequency $f_{\rm IF}$, but each having its own amplitude and phase angle. If the voltage due to the ith source is

$$\upsilon_i(t) = V_i\cos(2\pi f_{\rm IF}t + \phi_i), \tag{7.45}$$

where V_i is its amplitude and ϕ_i is its phase angle, then the IF voltage is the sum due to all $N_{\rm s}$ sources,

$$\upsilon_{\rm IF}(t) = \sum_{i=1}^{N_{\rm s}} V_i\cos(2\pi f_{\rm IF}t + \phi_i). \tag{7.46}$$

The phasor-domain counterpart of Eq. (7.46) is

$$\mathbf{V}_{\rm IF} = V_{\rm e}e^{j\phi} = \sum_{i=1}^{N_{\rm s}} V_i e^{j\phi_i}, \tag{7.47}$$

where $V_{\rm e}$ is the envelope voltage and ϕ is its associated phase angle. The statistics associated with the sum of

N_s noise sources are identical with the Rayleigh fading statistics we examined in Chapter 5 for radar scattering from a large number of randomly distributed scatterers. Hence, V_e is characterized by a Rayleigh pdf of the form given by Eq. (5.65).

In the radiometer receiver (Fig. 7-15), the IF voltage output is fed into a square-law detector with output voltage

$$V_d = V_e^2. \tag{7.48}$$

Following the square-law detection steps outlined in Section 5-7 for radar scattering leads to the following pdf for V_d:

$$p(V_d) = \frac{1}{\overline{V}_d}\, e^{-V_d/\overline{V}_d}, \tag{7.49}$$

with an associated standard-deviation-to-mean ratio

$$\frac{s_d}{\overline{V}_d} = 1 \qquad \textbf{(before integration)}. \tag{7.50}$$

The detected voltage $\upsilon_d(t)$ consists of a dc (average) component $\overline{V}_d$ and a fluctuating component, as illustrated by the waveform in Fig. 7-15. By Eq. (7.40), the dc component is directly proportional to the system temperature T_{SYS}. The standard deviation s_d, which is the rms value of the fluctuating component, represents the statistical uncertainty associated with the measurement of T_{SYS} (through $\overline{V}_d$). According to Eq. (7.50), $s_d = \overline{V}_d$, which means that the uncertainty is so large that it makes a radiometer (without an integrator) useless and impractical. The obvious solution to the problem is to filter out the high-frequency fluctuations of the detected voltage, which is equivalent to averaging $\upsilon_d(t)$ over some interval of time τ. This is precisely the function of the integrator shown in Fig. 7-15.

The voltage $\upsilon_{out}(t)$ at the output of the integrator (Fig. 7-15) also consists of a dc component $\overline{V}_{out}$ and it is also accompanied by a fluctuating component, but the fluctuating component exhibits a much gentler profile than that of the detected voltage $\upsilon_d(t)$. Integrating a random signal of bandwidth B over a time τ leads to a reduction of its variance (normalized to the square of its mean value) by the factor $N = B\tau$. That is, the ratio $s_{out}^2/\overline{V}_{out}^2$ at the integrator output is related to $s_d^2/\overline{V}_d^2$ at the integrator input by

$$\frac{s_{out}^2}{\overline{V}_{out}^2} = \frac{s_d^2}{\overline{V}_d^2} \cdot \frac{1}{B\tau}. \tag{7.51}$$

In view of Eq. (7.50), it follows that

$$\frac{s_{out}}{\overline{V}_{out}} = \frac{1}{\sqrt{B\tau}} \qquad \textbf{(after integration)}. \tag{7.52}$$

Since $\overline{V}_{out} = G_s T_{SYS}$, Eq. (7.52) is equivalent to

$$\frac{\Delta T_{SYS}}{T_{SYS}} = \frac{1}{\sqrt{B\tau}}, \tag{7.53}$$

where ΔT_{SYS} is the standard deviation associated with the measured (estimated) value of T_{SYS}.

Operationally, ΔT_{SYS} may be viewed as the minimum deviation that T_{SYS} has to undergo so as to produce a detectable change at the radiometer output, where detectable change is defined as a change in the dc level of the output voltage equal to the standard deviation s_{out}.

Recalling that $T_{SYS} = T_A + T'_{REC}$ and that T'_{REC} (the receiver input noise temperature) is independent of the radiation incident upon the antenna, Eq. (7.53) may be rewritten as

$$\Delta T = \Delta T_{SYS} = \frac{T_{SYS}}{\sqrt{B\tau}} = \frac{T_A + T'_{REC}}{\sqrt{B\tau}}, \tag{7.54}$$

where ΔT is regarded as the minimum (statistically) detectable change in the radiometric antenna temperature T_A of the observed scene.

► The expression given by Eq. (7.54) defines the ***radiometric sensitivity*** (or ***resolution***) of an ideal total power radiometer with no gain fluctuations. ◄

As discussed in the next section, gain fluctuations are an important limitation, so to emphasize the importance of the no-gain-fluctuation condition, ΔT is denoted ΔT_{IDEAL} and rewritten as

$$\Delta T_{\text{IDEAL}} = \frac{T_{\text{SYS}}}{\sqrt{B\tau}} \qquad \left(\textbf{ideal total-power radiometer}\right) \tag{7.55}$$

Before concluding this section, we should remark on how to determine the bandwidth B for a real predetection section with a nonuniform transfer function. In his classic paper on radio-telescope receivers, Tiuri (1964) showed that a predetection section with power transfer function $H(f)$ has an equivalent predetection bandwidth B given by

$$B = \frac{\left[\int_0^\infty H(f)\,df\right]^2}{\int_0^\infty H^2(f)\,df}. \tag{7.56}$$

7-5 Effects of Receiver Gain Variations

As stated earlier, the expression for ΔT given by Eq. (7.55) accounts for only the measurement uncertainty due to noise fluctuations and does not incorporate receiver gain fluctuations. For convenience, Eq. (7.55) is repeated in the form

$$\Delta T_{\text{N}} = \frac{T_{\text{SYS}}}{\sqrt{B\tau}}, \tag{7.57a}$$

where the subscript N denotes *noise*-caused uncertainty.

The output voltage of the total-power radiometer is directly proportional to G_{s}, the system gain factor defined by Eq. (7.43), which was assumed to be constant in the derivation leading to Eq. (7.55). In practice, this is a fair assumption for the postdetection stages, but may not be valid for the predetection power gain G. Gain variations in the predetection section arise primarily from the RF amplifier and secondarily from the mixer and IF amplifier.

Since V_{out} is linearly related to the product $G_{\text{S}}T_{\text{SYS}}$, an increase in G_{S} by ΔG_{S} will be misinterpreted by the output as an increase in T_{SYS} by $\Delta T_{\text{SYS}} = T_{\text{SYS}}(\Delta G_{\text{S}}/G_{\text{S}})$. Long-term (slow) variations of G_{S}, with periods on the order of minutes, may be factored out (approximately) by calibrating the radiometer output voltage against known input noise sources as frequently as is practicable. However, calibration does not eliminate short-term (fast) gain variations that occur over intervals smaller than the period between successive calibrations. Statistically, the rms uncertainty in T_{SYS} due to system gain variations is

$$\Delta T_{\text{G}} = T_{\text{SYS}}\left(\frac{\Delta G_{\text{S}}}{\overline{G}_{\text{S}}}\right), \tag{7.57b}$$

where $\overline{G}_{\text{S}}$ is the average system power gain and ΔG_{S} is the rms value of its fluctuating component.

Since the noise uncertainty ΔT_{N} and the gain uncertainty ΔT_{G} are caused by unrelated mechanisms, they are statistically independent, in which case the total rms uncertainty is given by

$$\begin{aligned}\Delta T &= [(\Delta T_{\text{N}})^2 + (\Delta T_{\text{G}})^2]^{1/2}\\ &= T_{\text{SYS}}\left[\frac{1}{B\tau} + \left(\frac{\Delta G_{\text{S}}}{\overline{G}_{\text{S}}}\right)^2\right]^{1/2}.\end{aligned} \tag{7.58}$$

(total-power radiometer)

▶ This expression defines the radiometer sensitivity of the total-power radiometer, incorporating the effects of both noise and gain variations. ◀

To gain an appreciation for the relative significance of the two sources of measurement uncertainty, let us consider the following example. A total-power radiometer operating at a center frequency of 30 GHz is characterized by the following parameters: $T'_{\text{REC}} = 600$ K, $B = 100$ MHz, $\tau = 0.01$ s, and a ***normalized gain variation*** $\Delta G_{\text{S}}/\overline{G}_{\text{S}} = 10^{-2}$. For an antenna temperature T_{A} in the neighborhood of 300 K, the above values lead to the following results:

$$\Delta T_{\text{N}} = 0.9 \text{ K}, \qquad \Delta T_{\text{G}} = 9 \text{ K}, \qquad \Delta T = 9.05 \text{ K}.$$

That is, the radiometer sensitivity is, in essence, governed by gain variations. The desired sensitivity in remote-sensing observations usually is on the order of 1 K or less. To reduce ΔT of the above radiometer to 1 K, the product $T_{\text{SYS}}(\Delta G_{\text{S}}/\overline{G}_{\text{S}})$ has to be reduced by a factor of 20. Typically, the gain variation factor for standard low-noise microwave amplifiers is between 10^{-2} and 10^{-4}; therefore they are not well suited for use in total-power receivers. Gain variations may be reduced by an order of magnitude or better by controlling the sources that cause these variations, namely, power-supply voltages and environmental temperature variations. This may provide an acceptable solution for a total-power radiometer operating in the centimeter wavelength range, but at millimeter wavelengths it is difficult to construct highly stable receivers with $\Delta G_{\text{S}}/\overline{G}_{\text{S}} \leq 10^{-4}$.

One possible solution to the gain variation problem is to use a receiver with no RF amplifier. In this case, the receiver noise temperature T'_{REC} may be as large as 3000 K or more, but without the RF amplifier, values of $\Delta G_{\text{S}}/\overline{G}_{\text{S}}$ as low as 10^{-4} to 10^{-5} are achievable and therefore the product $T_{\text{SYS}}(\Delta G_{\text{S}}/\overline{G}_{\text{S}})$ may be reduced to about 0.1 K. Other approaches are discussed in the following sections.

7-6 Dicke Radiometer

Studies conducted to evaluate the gain fluctuations of RF amplifiers reveal the following observations: (1) the power spectral density of G_{S} (fluctuation spectrum) decreases with increasing f as $1/f$ or faster, (2) the bulk of the fluctuation spectrum lies at frequencies below 1 Hz, and (3) practically no fluctuations with frequencies above 1 kHz exist. Through the example cited in the previous section, it was concluded that gain variations are often the limiting factor to achieving high radiometric resolutions (small values of ΔT).

▶ To reduce the effects of gain fluctuations in the radiometer receiver, Dicke (1946) introduced a modulation technique that has since become known as the ***Dicke radiometer***. ◀

A block diagram of the Dicke radiometer is shown in Fig. 7-16. It is basically a total-power radiometer with two additional features: (1) a switch, which has become known as the ***Dicke switch***, connected at the receiver input (at a point as close to the antenna as possible) and used to *modulate* the receiver input signal, and (2) a ***synchronous demodulator*** (also called synchronous detector) placed in between the square-law detector and the integrator. The predetection section consists of the RF amplifier, mixer, and IF amplifier and is characterized by a predetection power gain G and bandwidth B.

The modulation consists of periodically switching the receiver input between the antenna and a constant (***reference***) noise source (Fig. 7-16) at a switching rate higher than the highest significant spectral component in the gain variation spectrum. That is, the switching rate f_{s} is chosen so that over a period of one switching cycle (typically between 1 and 20 ms) the system gain G_{s} essentially remains constant, and therefore *identical* for the half-cycle during which the receiver is connected to the antenna and the half-cycle during which the receiver is connected to the reference source. For square-wave modulation, the dc components of the detected outputs corresponding to the antenna and reference-source powers are, respectively,

$$\overline{V}_{\text{d}}^{\text{Ant}} = 2GkB(T_{\text{A}} + T'_{\text{REC}}) \qquad \text{for } 0 \leq t \leq \tau_{\text{s}}/2, \quad (7.59)$$

$$\overline{V}_{\text{d}}^{\text{Ref}} = 2GkB(T_{\text{REF}} + T'_{\text{REC}}) \qquad \text{for } \tau_{\text{s}}/2 \leq t \leq \tau_{\text{s}}, \quad (7.60)$$

where T_{REF} is the reference-source noise temperature and τ_{s} $(= 1/f_{\text{s}})$ is the period of one switching cycle. The receiver noise temperature T'_{REC} includes noise contributions from the input switch.

Superimposed on the dc voltages are ac components due to noise and gain fluctuations. The synchronous demodulator (Fig. 7-16) consists of a switch that operates in synchronism with the input Dicke switch, followed by two unity-gain amplifiers (in parallel), with opposite polarity, one to receive $\upsilon_{\text{d}}^{\text{Ant}}$ and the other to receive $\upsilon_{\text{d}}^{\text{Ref}}$. The unity-gain amplifier outputs are then summed and fed into the low-pass filter (integrator). The switching period τ_{s} is much shorter than the integration time τ, so over the duration τ, the integrator spends

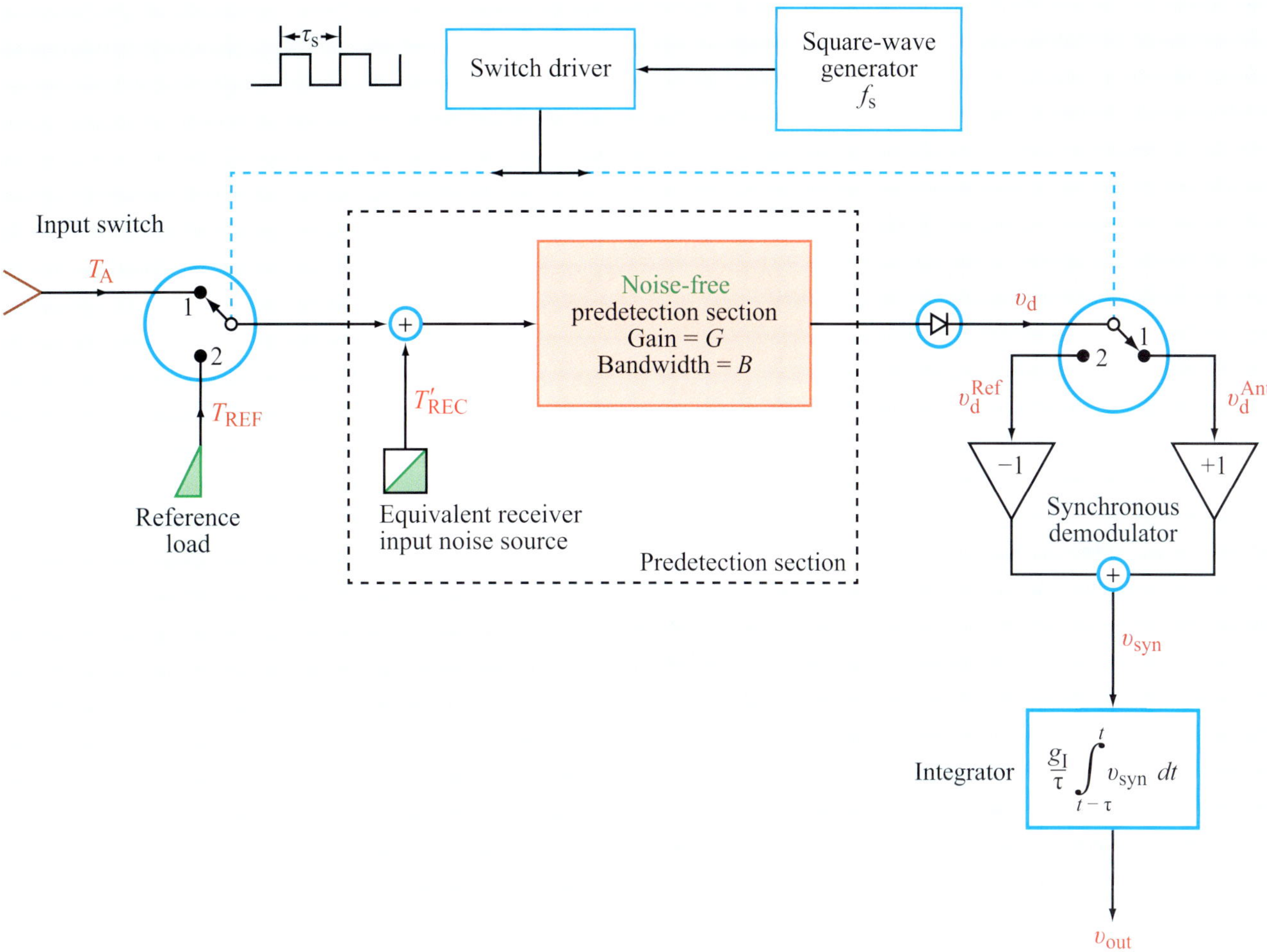

Figure 7-16: Functional block diagram of a Dicke radiometer.

half of its time integrating $\upsilon_d^{Ant}(t)$ and the other half integrating $\upsilon_d^{Ref}(t)$. Hence, the output of the integrator is equivalent to

$$\upsilon_{out}(t) = \frac{g_I}{\tau}\left[\int_{t-\tau}^{t-\tau/2} \upsilon_d^{Ant}(t)\, dt - \int_{t-\tau/2}^{t} \upsilon_d^{Ref}(t)\, dt\right]. \tag{7.61}$$

With most of the ac fluctuations removed by the integration, the output voltage consists of a dc component, $\overline{V}_{out}$, and a relatively very small fluctuating component with rms value s_{out}. Thus,

$$\begin{aligned}\overline{V}_{out} &= \tfrac{1}{2}\, g_I(\overline{V}_d^{Ant} - \overline{V}_d^{Ref}) \\ &= g_I GkB(T_A - T_{REF}) = \tfrac{1}{2}\, G_S(T_A - T_{REF}).\end{aligned} \tag{7.62}$$

In addition to filtering out most of the fluctuating component of $\upsilon_{out}(t)$, the low-pass filter also filters out the ac components at f_s, and its higher harmonics (due to the square-wave modulation). That is, the switching rate f_s, must be much higher than the low-pass filter bandwidth B_{LF}. The relationship between f_s and B_{LF} may also be considered from the viewpoint of the sampling theorem. Bandwidth B_{LF} represents the range of frequencies of the fluctuating component of the input signal retained in the output voltage, and f_s represents the frequency at which the input signal is sampled. To satisfy the sampling theorem, $f_s \geq 2B_{LF}$.

▶ According to Eq. (7.62), the dc output voltage is proportional to the difference between T_A and T_{REF}, and is independent of the receiver noise temperature T'_{REC}. This is the key attribute of the Dicke radiometer. ◀

To derive an expression for the radiometric resolution ΔT of the Dicke radiometer, we start by rewriting Eq. (7.62) in the following form:

$$\overline{V}_{out} = \frac{1}{2} G_S[(T_A + T'_{REC}) - (T_{REF} + T'_{REC})]. \quad (7.63)$$

The fluctuating component of $v_{out}(t)$ consists of three contributions:

(a) Gain variations: from Eq. (7.57b), the gain uncertainty is

$$\Delta T_G = (T_A - T_{REF}) \left(\frac{\Delta G_S}{\overline{G}_S} \right). \quad (7.64)$$

(b) Noise variations of $(T_A + T'_{REC})$: integration for a period $\tau/2$ (antenna is observed during only half of the integration time τ) leads to a noise uncertainty

$$\Delta T_N^{Ant} = \frac{T_A + T'_{REC}}{\sqrt{B\tau/2}} = \frac{\sqrt{2}\,(T_A + T'_{REC})}{\sqrt{B\tau}}. \quad (7.65)$$

(c) Noise variations of $(T_{REF} + T'_{REC})$:

$$\Delta T_N^{Ref} = \frac{\sqrt{2}\,(T_{REF} + T'_{REC})}{\sqrt{B\tau}}. \quad (7.66)$$

Assuming that the above three uncertainties are statistically independent, the total radiometric resolution is given by

$$\Delta T = [(\Delta T_G)^2 + (\Delta T_N^{Ant})^2 + (\Delta T_N^{Ref})^2]^{1/2}, \quad (7.67)$$

Table 7-1: Nimbus 5 electrically scanning microwave radiometer (ESMR) flight model parameters (*Nimbus 5 User's Guide*) [NASA, 1972].

Parameter	Value
Antenna	
Antenna type	Phased array
Aperture size	83.3 cm ×85.5 cm
Half-power beamwidth	1.4° × 1.4° (at nadir)
Beam efficiency	90–92.7%
Beam scan angle	±50°
Antenna loss	1.7 dB
Polarization	Horizontal
Radiometer	
Center frequency	19.35 GHz
Predetection bandwidth	200 MHz
Mixer noise figure	6.5 dB
ΔT_{min} (for $\tau = 47$ ms)	1.5 K
Absolute accuracy	2 K
Dynamic range	50–330 K
Calibration	
(a) Reference load	338 K
(b) Ambient load	Local ambient
(c) Sky horn antenna	3 K

which reduces to

$$\Delta T = \left[\frac{2(T_A + T'_{REC})^2 + 2(T_{REF} + T'_{REC})^2}{B\tau} + \left(\frac{\Delta G_S}{\overline{G}_S} \right)^2 (T_A - T_{REF})^2 \right]^{1/2}. \quad (7.68)$$

(unbalanced Dicke radiometer)

For reasons that become obvious later, the expression given by Eq. (7.68) is referred to as the radiometric resolution of the ***unbalanced Dicke radiometer***. Examples of satellite Dicke radiometers that operated in the unbalanced mode include the 19.35 GHz Nimbus 5 and the 37 GHz Nimbus 6 electrically scanning microwave radiometer (ESMR) systems; parameters of the Nimbus 5 ESMR are summarized in Table 7-1.

Before we proceed further, let us compare ΔT of the unbalanced Dicke radiometer with that of the total-power radiometer. For $B = 100$ MHz, $\tau = 1$ s,

$T'_{\text{REC}} = 700$ K, and $\Delta G_{\text{S}}/G_{\text{S}} = 10^{-2}$, Eq. (7.58) gives

$$\Delta T(\text{Total Power}) \approx \begin{cases} 7 \text{ K} & \text{for } T_{\text{A}} = 0 \text{ K}, \\ 10 \text{ K} & \text{for } T_{\text{A}} = 300 \text{ K}. \end{cases}$$

If we choose a reference noise source with temperature $T_{\text{REF}} = 300$ K, Eq. (7.68) gives

$$\Delta T(\text{unbalanced Dicke}) \approx \begin{cases} 3 \text{ K} & \text{for } T_{\text{A}} = 0 \text{ K}, \\ 0.2 \text{ K} & \text{for } T_{\text{A}} = 300 \text{ K}. \end{cases}$$

Overall, the radiometric resolution of the unbalanced Dicke radiometer is superior to that of the total-power radiometer. Of particular note is the condition $T_{\text{A}} = T_{\text{REF}}$, because when this condition is met, the second term inside the square brackets of Eq. (7.68) goes to zero, thereby eliminating the effects of gain variations altogether. *When $T_{\text{A}} = T_{\text{REF}}$ the radiometer is said to be* ***balanced***, in which case Eq. (7.68) reduces to

$$\Delta T = \frac{2(T_{\text{A}} + T'_{\text{REC}})}{\sqrt{B\tau}} = 2\,\Delta T_{\text{IDEAL}}, \tag{7.69}$$

(balanced Dicke radiometer)

or twice the theoretical sensitivity of an ideal total-power radiometer (free of gain variations). The factor of 2 comes about from the fact that T_{A} is observed for only half the time. Techniques used for continuously maintaining the Dicke radiometer in a balanced state are discussed in the next section.

Sometimes, practical considerations dictate that the square wave at the output of the square-law detector be amplified prior to feeding it into the synchronous demodulator. A video amplifier is used for this purpose. To preserve the square-wave shape of the detected signal, the video amplifier must be capable of providing equal amplification to all of the major harmonics of the square-wave signal, which means that the bandpass of the video amplifier should extend from below f_{s} to at least $5f_{\text{s}}$, and preferably to as high as $10f_{\text{s}}$. A square wave consists of only odd harmonics, with the amplitude of the first harmonic equal to $4/\pi$ times the amplitude of the square wave, and the amplitude of the nth harmonic equal to $1/n$ times the amplitude of the first harmonic. This places an additional requirement on the necessary dynamic range of the video amplifier and makes it vulnerable to noise saturation. To avoid this problem, some Dicke radiometers employ a narrow-bandpass amplifier tuned to f_{s}, but whose bandwidth exceeds the bandwidth of the low-pass filter, B_{LF}. With only the first harmonic of the square wave reaching the synchronous demodulator, the dc output of the synchronous demodulator is smaller than when the full square wave is used, which leads to a reduction in radiometric sensitivity (larger ΔT) by a factor of $(\pi\sqrt{2}/4) = 1.11$, or approximately 11% (Tiuri, 1964). The simplification in system design and specification brought about by the sinusoidal demodulation, with a relatively small cost in terms of loss in radiometric sensitivity, has made the square-wave modulated, sine-wave-demodulated Dicke radiometer a popular choice. The merits of other modulation and demodulation waveforms have been considered also (Colvin, 1961; McGillem and Seling, 1963) but in terms of the radiometric sensitivity ΔT, using square-wave modulation and demodulation gives the best results.

7-7 Balancing Techniques

A Dicke radiometer is said to be ***balanced*** if

$$\overline{V}_{\text{d}}^{\text{Ant}} - \overline{V}_{\text{d}}^{\text{Ref}} = 0, \tag{7.70}$$

where $\overline{V}_{\text{d}}^{\text{Ant}}$ is the dc component of the detected voltage during the half cycle that the receiver input switch is connected to the antenna, and a similar definition applies to $\overline{V}_{\text{d}}^{\text{Ref}}$. In the balanced condition, $\overline{V}_{\text{syn}} = 0$ and consequently $\overline{V}_{\text{out}} = 0$. With reference to Eqs. (7.59) and (7.60), the balanced condition may be realized by either: (1) adjusting T_{REF} to equal T_{A} (or vice versa) prior to the input switch, or (2) controlling the predetection gain of $\overline{V}_{\text{d}}^{\text{Ant}}$ and $\overline{V}_{\text{d}}^{\text{Ref}}$ separately, in such a manner that the two voltages are made equal. This section presents examples of feedback configurations proposed to realize the balanced condition in a Dicke radiometer automatically.

7-7.1 Reference-Channel Control Method

Figure 7-17 is a block diagram of a ***null-balancing Dicke radiometer*** that employs a feedback loop to control the magnitude of T_{REF} so that it continuously balances the antenna temperature T_A (Machin et al., 1952). This is realized by feeding the integrator output into a control circuit that applies the necessary voltage (or current) to an electronically controlled variable attenuator to maintain a null at the integrator output. In this case, the control voltage V_c becomes the output voltage to be recorded, since it is related to T_{REF}, which, in the balanced condition, is equal to T_A. The radiometer is calibrated by measuring V_c as a function of the noise temperature T_{CAL} of a calibration noise source introduced in lieu of the antenna (or through the use of a calibration switch as shown in Fig. 7-12). In general, the calibration curve may be nonlinear, thereby necessitating that V_c be measured over a wide range of values of T_{CAL}.

Returning to Fig. 7-17, the reference noise temperature T_{REF} is in part due to the noise generated by the noise source and in part due to self-emission by the attenuator,

$$T_{REF} = \frac{T_N}{L} + \left(1 - \frac{1}{L}\right) T_0, \tag{7.71}$$

where T_N is the noise temperature of the source, L is the loss factor of the attenuator, and T_0 is its physical temperature. As a function of L, the limiting values of T_{REF} are $T_{REF} = T_N$ for $L = 1$ (no attenuation) and $T_{REF} = T_0$ for L very large. Conceptually, T_N and T_0 can be specified so that the range between them covers the range of values over which T_A is expected to vary. In practice, however, this specification cannot be realized with passive noise sources unless cryogenic cooling is used (the use of active "cold" sources is discussed below). To illustrate with an example, consider a radiometer system specified to operate over the range $50\text{ K} \leq T_A \leq 300\text{ K}$. The upper limit of T_A can be accommodated easily by maintaining the environmental temperature of the front end of the radiometer, including the attenuator, at $T_0 = 300$ K, which for L very large, gives $T_{REF} = T_A = 300$ K. To accommodate the lower limit of T_A, T_N has to be assigned a value of 50 K (or lower). If a matched load is used as the noise source, it has to be physically cooled to the desired noise temperature. Moreover, for real, variable attenuators, the loss factor L cannot be reduced all the way to unity, and therefore the temperature of the matched load has to be lower than 50 K in order to compensate for the noise power generated by the attenuator. Cryogenic techniques sometimes are used in ground-based radiometer systems to reduce the receiver noise temperature, and therefore they can be used to cool the matched load of the radiometer system. However, in airborne and spaceborne operations, cryogenic cooling usually is avoided because of the large amounts of power required to maintain the cryogenic fluids at the desired temperature.

Traditionally, active noise sources are used to provide noise temperatures in excess of their ambient temperatures, but they can also be designed to behave like "cold" sources with output noise temperatures that are much smaller than their ambient temperatures (Frater and Williams, 1981).

7-7.2 Antenna-Channel Noise-Injection Method

The configuration shown in Fig. 7-18 was proposed by Goggins (1967) to achieve null balancing through the use of noise injection into the line connecting the antenna to the input switch. The front end of the radiometer—including the reference load, the variable attenuator, and the directional coupler (as well as other RF devices)—is enclosed in a constant-temperature chamber at a temperature T_0 slightly higher than the highest specified value for T_A (for reasons that become obvious later). In other words, when the input switch is connected to terminal 2 (Fig. 7-18), the input noise temperature is $T_{REF} = T_0 =$ constant. To balance the reference load, a sufficient amount of noise power is injected into the antenna port through a directional coupler so that the ***supplemented*** input noise temperature T_A^s appearing at the input switch (terminal 1 in Fig. 7.19) is equal to T_{REF},

$$T_A^s = T_{REF} = T_0. \tag{7.72}$$

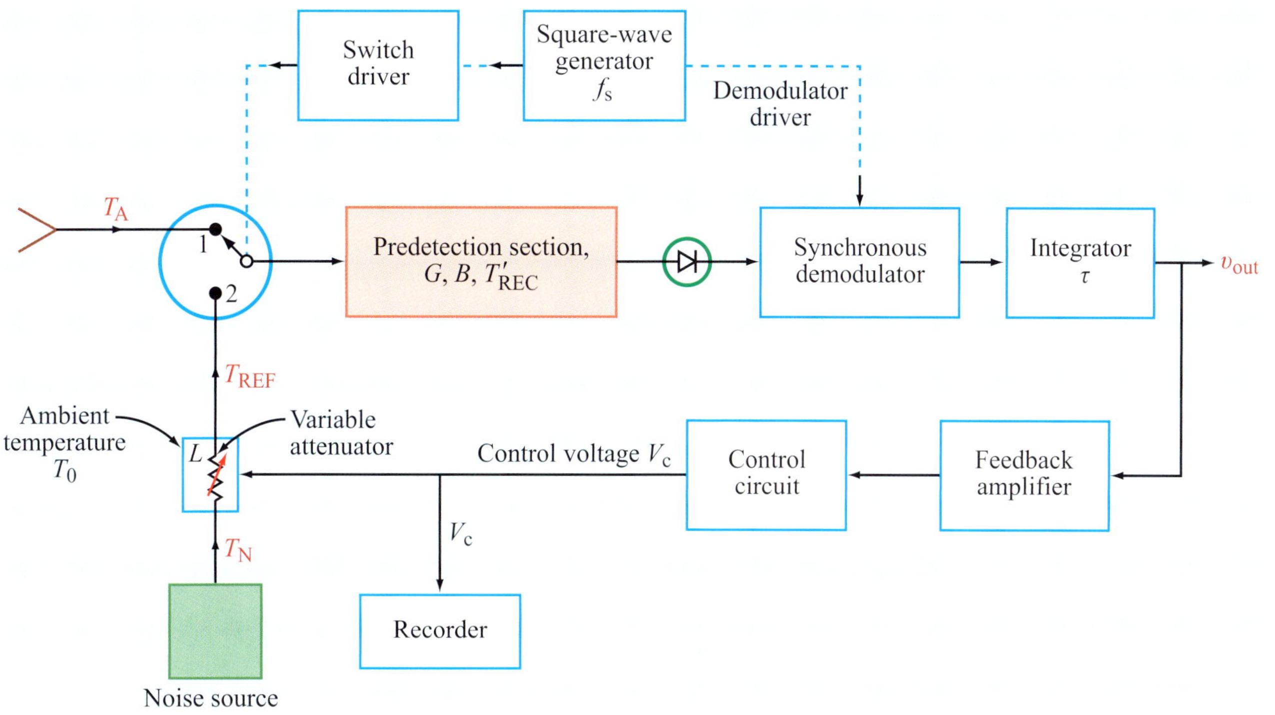

Figure 7-17: Dicke radiometer using feedback to control the reference noise temperature T_{REF}.

The amount of injected power is controlled by the variable attenuator, which itself is controlled by a feedback network. The noise temperature T_A^s is related to the antenna noise temperature T_A through

$$T_A^s = \left(1 - \frac{1}{F_c}\right) T_A + \frac{T'_N}{F_c}, \tag{7.73}$$

where F_c (≥ 1) is the coupling factor of the directional coupler, and T'_N is the noise temperature of the injected noise at the input to the directional coupler and is given by

$$T'_N = \frac{T_N}{L} + \left(1 - \frac{1}{L}\right) T_0. \tag{7.74}$$

Combining the above three expressions leads to

$$L = \frac{T_N - T_0}{(F_c - 1)(T_0 - T_A)}. \tag{7.75}$$

The radiometer output indicator is the attenuator control voltage V_c. If V_c is scaled and linearized so that $V_c = 1/L$, then Eq. (7.75) becomes a linear relationship between V_c and T_A:

$$V_c = \frac{F_c - 1}{T_N - T_0}(T_0 - T_A). \tag{7.76}$$

Suppose the radiometer is specified to operate over the range 50 K $\leq T_A \leq$ 300 K. Let us choose $T_0 = 310$ K, and let us use a 20 dB directional coupler ($F_c = 100$) and an avalanche noise diode having an output power noise temperature T_N of 50,000 K. From Eq. (7.75), L has to vary between 1.9 ($\approx$ 2.9 dB) and 50 ($\approx$ 17 dB) to cover the specified temperature range. This dynamic range of 14.1 dB can easily be accommodated by a *PIN* diode attenuator, which is a current-driven device capable of more than 40 dB in linear dynamic range. The minimum loss factor (usually called insertion loss) of a *PIN* diode is typically about 2 dB. Thus, the values chosen for T_0, F_c, and T_N result in a range of values for L that is compatible with the capabilities of a *PIN* diode. Had we chosen T_0 to be exactly equal to $T_A(\max) = 300$ K rather

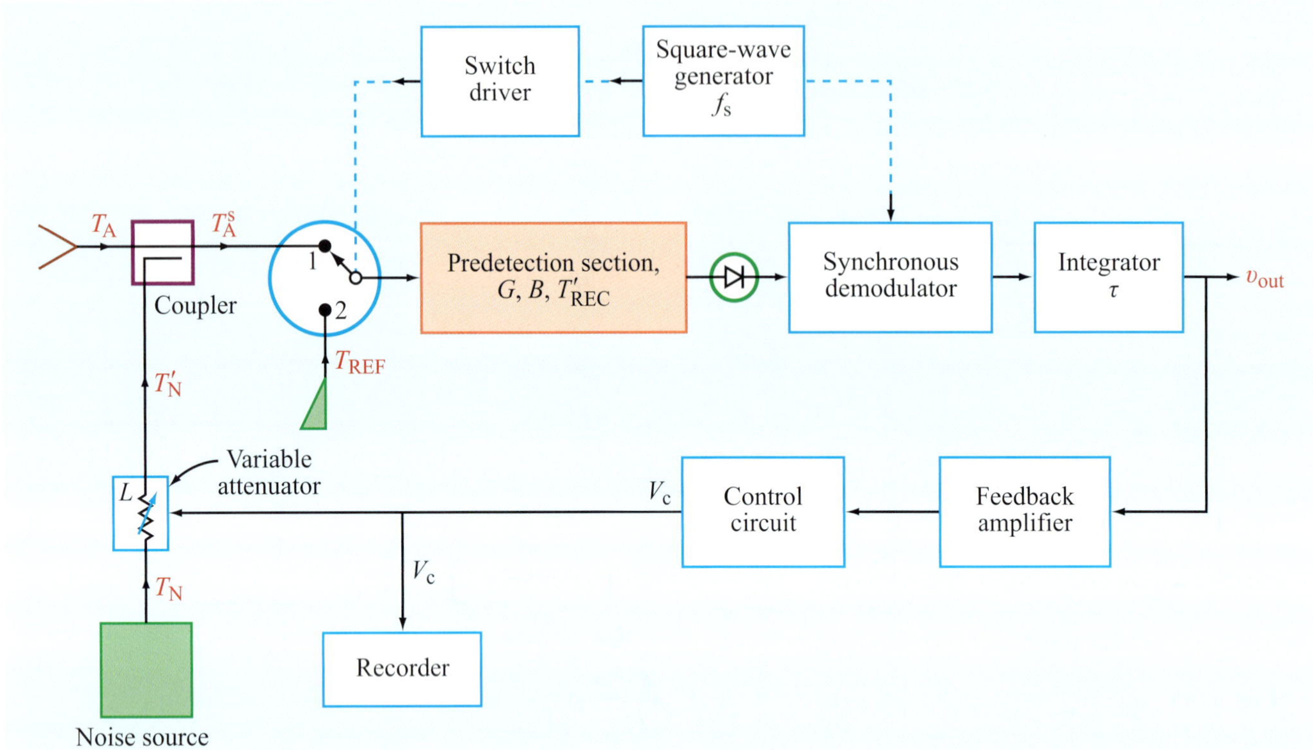

Figure 7-18: Balanced Dicke radiometer using feedback to control the level of the injected noise temperature T'_N to maintain the condition $T_A^s = T_{REF}$.

than 310 K, L would have had to have been infinite to produce null balancing.

Since the noise temperature seen by the input switch is always equal to T_0, the sensitivity is given by Eq. (7.69) with $T_A = T_0$,

$$\Delta T = \frac{2(T_0 + T'_{REC})}{\sqrt{B\tau}} . \tag{7.77}$$

7-7.3 Pulsed Noise-Injection Method

Diode attenuators are most stable in their extreme ON and OFF states. When operated in only one or the other of these two states (by controlling the magnitude and polarity of the bias voltage), a *PIN* diode attenuator becomes an ON/OFF switch; in the ON position it causes minimum attenuation (≈ 2.0 dB), and in the OFF position it causes maximum attenuation (typically 60 dB). Use is made of this diode property in the configuration shown in Fig. 7-19 (Hardy et al., 1974), which basically is the same as that shown in Fig. 7-18 except that now the injected noise is generated in the form of narrow rectangular pulses rather than continuously. The feedback control circuit drives a voltage-controlled oscillator (VCO), which in turn drives a pulse generator. The output of the pulse generator consists of narrow rectangular pulses, each of length τ_p, with a repetition period $\tau_R = 1/f_R$, where f_R is the pulse repetition frequency (Fig. 7-20).

When the diode switch is in the OFF position (no pulse), we have maximum attenuation, which, in effect, masks the noise generated by the noise diode. The resultant value of T'_N is designated T'_{OFF}. When the *PIN* diode is switched by a pulse to its ON position, we have

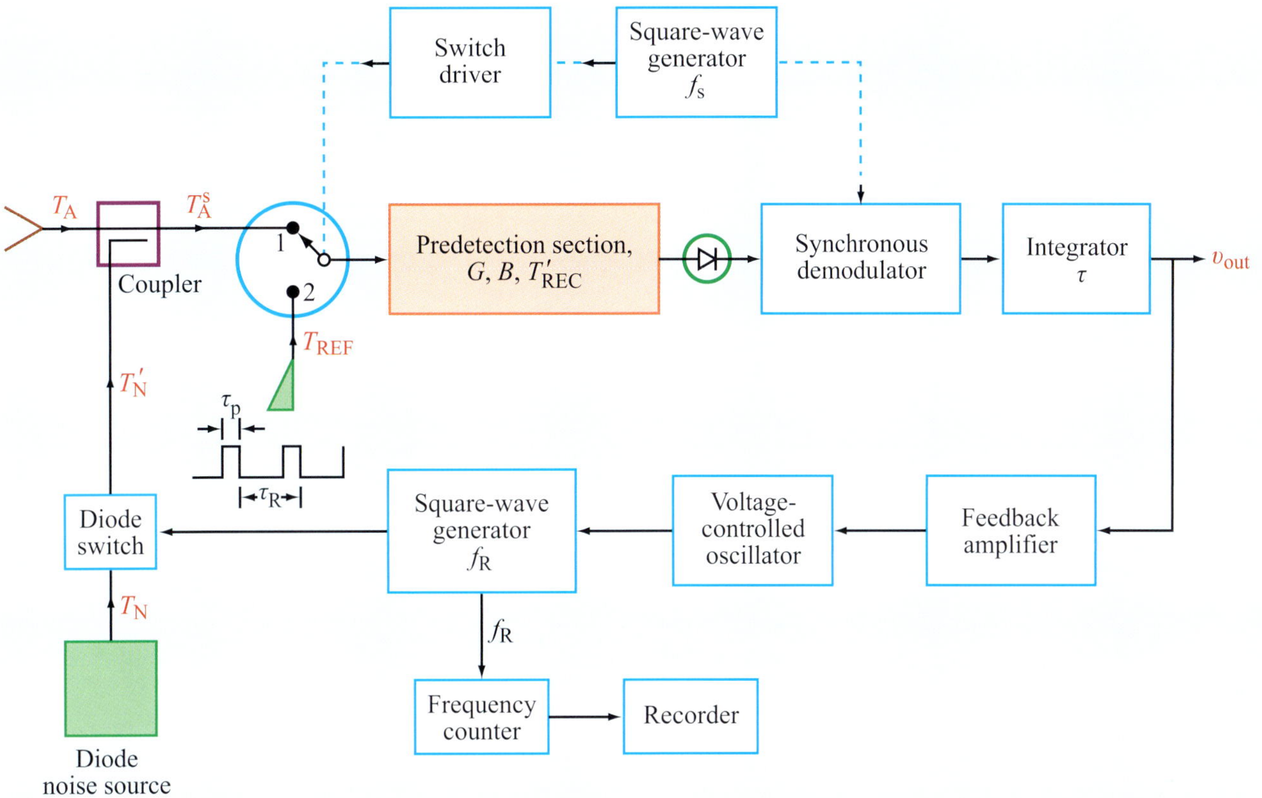

Figure 7-19: Balanced Dicke radiometer, using pulsed noise-injection to maintain $T_A^s = T_{REF}$. The output indicator of T_A is the pulse repetition frequency f_R.

minimum attenuation, or a large value of T'_N, which we designate T'_{ON}. Thus, over a repetition period τ_R

$$T'_N = \begin{cases} T'_{ON} & \text{for } 0 \le t \le \tau_p, \\ T'_{OFF} & \text{for } \tau_p \le t \le \tau_R. \end{cases} \tag{7.78}$$

The pulse repetition frequency f_R should be much higher than f_s, the square wave modulation frequency of the input switch, so that many pulses are generated during the period $\tau_s/2$. The average value of T'_N is

$$\overline{T}'_N = \tau_p f_R T'_{ON} + (1 - \tau_p f_R) T'_{OFF}, \tag{7.79}$$

where $\tau_p f_R$ is the total time, during one second, that the diode switch is ON.

If τ_p is kept constant, f_R can be controlled by the feedback loop to provide the necessary value of $\overline{T}'_N$ (to inject into the antenna port via the directional coupler) so that a null condition always is maintained at the integrator output. The radiometer output indicator is a frequency counter that measures f_R, which is linearly related to T_A by

$$f_R = \frac{(F_c T_0 - T'_{OFF}) - (F_c - 1) T_A}{\tau_p (T'_{ON} - T'_{OFF})} . \tag{7.80}$$

If during its OFF position, the attenuator loss factor is high enough to reduce the noise-diode contribution to a negligible value, then T'_{OFF} becomes approximately equal to T_0, and Eq. (7.80) reduces to

$$f_R = \frac{(F_c - 1)(T_0 - T_A)}{\tau_p (T'_{ON} - T_0)} . \tag{7.81}$$

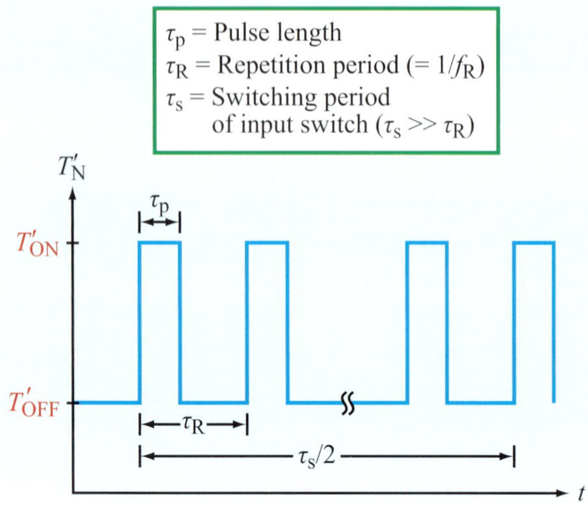

Figure 7-20: Timing diagram showing the ON and OFF states of the noise temperature T'_N that gets injected through the directional coupler of Fig. 7-19.

Table 7-2: 2.65 GHz satellite prototype ocean-temperature sensing radiometer using a pulsed noise-injection scheme [Hardy et al., 1974].

Antenna	
Antenna type	Multimode pyramidal horn
Aperture size	35.6 cm ×35.6 cm
Beam efficiency	98%
Beam scan angle	±50°
Polarization	Circular
Radiometer	
Center frequency	2.65 GHz
Predetection bandwidth	100 MHz
Receiver noise temperature	60 K
Switching frequency	50 Hz
Diode excess noise	36 dB
Pulse width	40 μs
Pulse-generator frequency range	0–12 kHz
ΔT_{min} for 1 s integration time	0.15 K

Since T_0 is known, it is sufficient to measure f_R at one value of a known noise temperature of a calibration source in order to determine the value of the quantity multiplying $(T_0 - T_A)$ in Eq. (7.81). The radiometric sensitivity of this type of radiometer is given by Eq. (7.77).

The parameters of a 2.65 GHz pulsed noise-injection Dicke radiometer are listed in Table 7-2. The system was developed as a satellite prototype sensor for measuring the surface temperature of the ocean (Hardy et al., 1974).

7-7.4 Gain-Modulation Method

In the methods considered thus far, the balanced condition was realized by adjusting the noise temperature of the reference channel to equal the noise temperature of the antenna channel, or vice versa, at the input switch. Another way to affect balance is by modulating the gain of the IF output voltage prior to the square-law detector. In the configuration shown in Fig. 7-21 (Orhaug and Waltman, 1962), the modulation consists of switching the IF output between a constant attenuator and a variable attenuator for the alternate half cycles of the input switch. The attenuator switch is driven by the switch-driver in synchronism with the input switch. The function of the feedback loop is to control the variable attenuator so that its output voltage (corresponding to the reference channel) is equal to the output voltage of the fixed attenuator (corresponding to the antenna channel). That is, if the loss factor of the variable attenuator, L_v, is adjusted to maintain the condition

$$\frac{1}{L_0}\,(T_A + T'_{REC}) = \frac{1}{L_v}\,(T_{REF} + T'_{REC}), \qquad (7.82)$$

then the dc value of the output voltage is zero. The system is calibrated by relating the control voltage V_c to the noise temperature of a calibration noise source replacing the antenna. If the voltage V_c is scaled and linearized so that $V_c = 1/L_v$, then V_c becomes linearly related to T_A:

$$V_c = \frac{1}{L_0(T_{REF} + T'_{REC})}\,(T_A + T'_{REC}). \qquad (7.83)$$

Conceptually, gain modulation may be applied before or after the mixer, but in practice it is easier to control

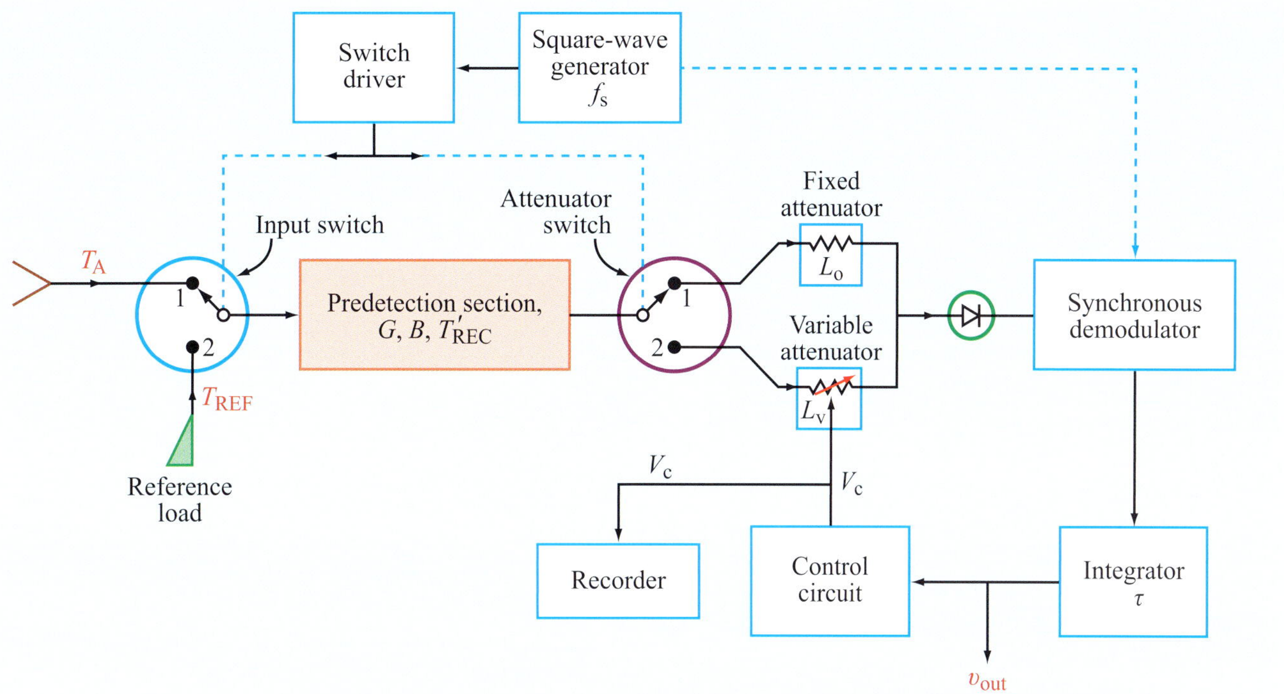

Figure 7-21: Dicke radiometer using gain modulation of the IF output to achieve null balance ($\upsilon_{out} = 0$).

the loss factor (or gain) of a device operating in the IF frequency range than to control a microwave device.

One drawback of the gain-modulation technique is that slow variations in the receiver noise temperature T'_{REC} (between calibrations) result in a measurement error of the absolute value of T_A. This is due to the dependence of V_c on T'_{REC}. In contrast, the techniques employing temperature control to balance the radiometer are insensitive to drifts in T'_{REC}. Another limitation of the gain-modulation technique pertains to the amplitude of the gain modulation, L_v/L_0. If this factor is large, the gain modulation is liable to produce excessive output fluctuations. Thus, for low-noise receivers (relatively small T'_{REC}), the gain-modulation method is useful over a narrow range of the difference $(T'_{REC} - T_A)$.

7-8 Automatic-Gain-Control (AGC) Techniques

Automatic gain control (AGC) is a feedback technique used for stabilizing the gain of receiver systems. In continuous AGC, the output voltage of the receiver is compared with a reference voltage on a continuous basis, and the difference between the two voltages is used to adjust the gain of the receiver so that the output voltage is maintained at a constant level. Continuous AGC is inapplicable to radiometer receivers because the AGC action removes all variations, including those due to the signal (T_A) that the radiometer is intended to measure. To eliminate the dependence of the AGC voltage on the antenna temperature, Seling (1964) introduced the ***sampled-AGC technique*** in which the detector output voltage is monitored by the AGC feedback circuit during only the half cycles (of the square-wave Dicke switching period) for which the Dicke switch is connected to the (constant-temperature) reference load. Successful operation of the sampled-AGC technique depends upon meeting certain constraints that dictate the choice of the AGC bandwidth, as discussed by Seling (1964).

The sampled-AGC technique shares a drawback with the gain-modulated radiometer (Section 7-7.4) in that slow variations in the receiver noise temperature (between calibrations) are compensated for as if they were gain variations, thereby introducing a bias error in the measurement accuracy of T_A.

The sampled-AGC approach was extended by Hach (1966, 1968), who developed a ***two-reference-temperature AGC radiometer***. This radiometer has several attractive features: (1) insensitivity to system gain variations, (2) insensitivity to receiver noise-temperature variations, and (3) the ability to provide continuous calibration. Hach's concept was adopted by General Electric (1973) in the construction of the radiometer section of the RADSCAT system that was flown aboard the Skylab satellite in 1973.

7-9 Noise-Adding Radiometer

The noise-adding radiometer (Ohm and Snell, 1963; Batelaan et al., 1973) removes the effects of gain variations, but without the use of a Dicke switch. Square-wave noise is coupled into the receiver input from a noise diode driven by a constant-rate square-wave generator, as shown in Fig. 7-22. The output voltage of the square-law detector is synchronously detected (demodulated) at the same rate, and a voltage ratio Y is formed, whose average value is given by

$$\overline{Y} = \frac{\overline{V}_1}{\overline{V}_2 - \overline{V}_1} = \frac{T_A + T'_{REC}}{T''_N}, \tag{7.84}$$

where $\overline{V}_1$ and $\overline{V}_2$ are the average square-law detector output voltage corresponding to the half cycle (of the switching period) during which the noise diode is OFF and ON, respectively, and T''_N is the added noise to receiver input during the diode ON half cycle. The ratio meter is followed by a low-pass filter to reduce noise fluctuations. For a unity-gain low-pass filter, we have

$$\overline{V}_{out} = \overline{Y} = \frac{T_A + T'_{REC}}{T''_N}. \tag{7.85}$$

The measurement accuracy of T_A is independent of system gain variations, but it is linked directly to the stability of the receiver noise temperature T'_{REC} and to the excess noise temperature T''_N.

The theoretical sensitivity of the noise-adding radiometer is given by (Batelaan et al., 1973)

$$\Delta T = \frac{2(T_A + T'_{REC})}{\sqrt{B\tau}} \left[1 + \frac{2(T_A + T'_{REC})}{T''_N}\right]. \tag{7.86}$$

The quantity multiplying the square bracket is recognized as twice the radiometric sensitivity of the ideal radiometer. Hence,

$$\Delta T(\text{noise-adding}) = 2\Delta T_{IDEAL} \left[1 + \frac{2T_{SYS}}{T''_N}\right]. \tag{7.87}$$

The absence of an input switch is an attractive feature of the noise-adding radiometer, particularly in low-noise receivers. A Dicke switch usually adds about 7–75 K to the receiver noise temperature T'_{REC}. In star-tracking and astronomical research, the brightness temperature of some targets may be only a few kelvins, which necessitates the use of low-noise receivers with noise temperatures on the order of a few tens of kelvins in order to achieve radiometric sensitivities on the order of 0.01–0.1 K. For such cases, the absence of an input switch is highly advantageous.

7-10 Summary of Radiometer Properties

For easy reference, Table 7-3 provides a summary of the input-output relationship and of the radiometric sensitivity for most of the receiver configurations discussed in the previous sections. The output indication I_{out} is related linearly to the antenna radiometric temperature T_A through

$$I_{out} = a(T_A + b), \tag{7.88}$$

where a and b are constants and I_{out} is either V_{out} (integrator output voltage), V_c (control voltage), or f_R (pulse-repetition frequency).

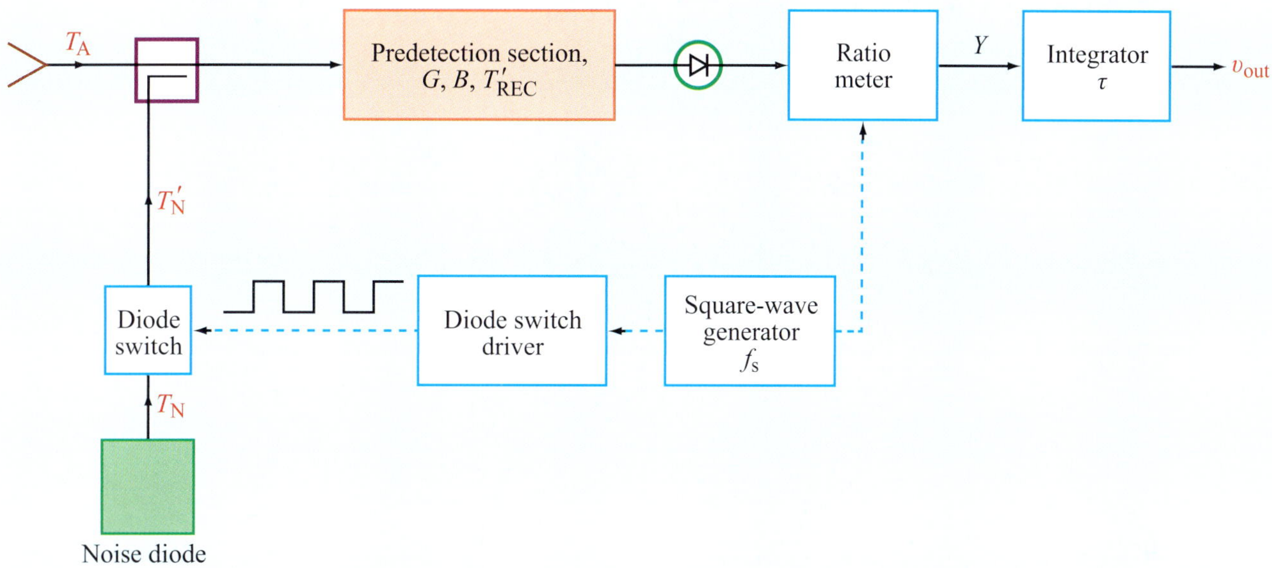

Figure 7-22: Noise-adding radiometer.

Since, in general, the radiometric sensitivity ΔT is a function of T_A, Table 7-3 considers ΔT for each of two values of T_A, namely, $T_A = 0$ K and $T_A = T_{REF} = T_0$, where T_0 (≈ 310 K) is the environmental temperature of the radiometer front end. In both cases, ΔT is normalized to ΔT_{IDEAL}, the sensitivity of the ideal total-power radiometer given by Eq. (7.55),

$$\Delta T_{IDEAL} = \frac{T_A + T'_{REC}}{\sqrt{B\tau}} . \tag{7.89}$$

7-11 Radiometer Calibration Techniques

Radiometer calibration consists of two steps. The first step involves relating the receiver output indication I_{out} (voltage, count, deflection, etc.) to the antenna temperature at the radiometer input. This usually is accomplished by measuring the output indication as a function of the noise temperature T_{CAL} of a calibration source connected to the radiometer input (in place of the antenna). The resultant relationship between the output indicator I_{out} and T_{CAL} provides the scale factors necessary for relating the output to the antenna temperature T_A (when the radiometer receiver is connected to the antenna). We shall refer to this first step as ***receiver calibration***.

The second step involves relating T_A to the radiative properties of the scene under observation. According to Eq. (6.40), T_A consists of three components: (1) energy received through the antenna mainbeam at the polarization of interest, which is the quantity of interest, (2) energy received from directions outside the antenna mainbeam (sidelobe contributions) and from the mainbeam at a polarization orthogonal to that of interest, and (3) thermal energy emitted by the antenna structure itself. To evaluate the significance of the latter two components and to factor out (partially) their influence on T_A, it is necessary to know the radiative properties of the antenna with a high degree of accuracy. Standard test procedures have been defined for the measurement of antenna properties (IEEE, 1979), but for one of the properties of interest, namely the radiation efficiency ξ, the standard procedures do not provide (in practice) the level of accuracy desired for the application of radiometric corrections. Alternate

Table 7-3: Summary of system transfer functions and sensitivities of different types of radiometers. The radiometer output indication I_{out} is related to the input antenna temperature T_A by $I_{out} = a(T_A + b)$, and the sensitivity ΔT is defined relative to ΔT_{IDEAL} of the ideal radiometer given by Eq. (7.55).

Radiometer type	*Output indicator* I_{out}	a	b	$\Delta T/\Delta T_{IDEAL}$ *for* $T_A = 0$ K	$\Delta T/\Delta T_{IDEAL}$ *for* $T_A = T_0$
Ideal (Fig. 7-15)	V_{out}	G_S	T'_{REC}	1	1
Total power (Fig. 7-15)	V_{out}	G_S	T'_{REC}	$\left[1+B\tau\left(\frac{\Delta G_S}{G_S}\right)^2\right]^{1/2}$	$\left[1+B\tau\left(\frac{\Delta G_S}{G_S}\right)^2\right]^{1/2}$
Dicke (unbalanced) (Fig. 7-16)	V_{out}	$\frac{G_S}{2}$	$-T_0$	$\sqrt{2}\left[\left(\frac{T_0+T'_{REC}}{T'_{REC}}\right)^2+1+\frac{B\tau}{2}\left(\frac{\Delta G_S}{G_S}\right)^2\left(\frac{T_0}{T'_{REC}}\right)^2\right]^{1/2}$	2
Balanced Dicke* with noise injection (Fig. 7-18)	V_c	$-\left(\frac{F_c-1}{T_N-T_0}\right)$	$-T_0$	$2\left(\frac{T_0}{T'_{REC}}+1\right)$	2
Balanced Dicke* with pulsed noise injection (Fig. 7-19)	f_R	$-\left[\frac{F_c-1}{\tau_p(T'_{ON}-T_0)}\right]$	$-T_0$	$2\left(\frac{T_0}{T'_{REC}}+1\right)$	2
Noise adding* (Fig. 7-22)	V_{out}	$\frac{1}{T''_N}$	T'_{REC}	$2\left(\frac{2T'_{REC}}{T''_N}+1\right)$	$2\left(\frac{T_0+2T'_{REC}}{T''_N}+1\right)$

*$T_{REF} = T_0$

measurement techniques developed specifically for microwave radiometric applications are discussed in Section 7-11.4 under the heading **Antenna Calibration**.

7-11.1 Receiver Calibration

Most radiometer receivers are linear systems in the sense that the output indication I_{out} is directly proportional to the antenna temperature T_A:

$$I_{out} = a(T_A + b). \quad (7.90)$$

Hence, it is sufficient to measure I_{out} for each of two known values of T_A to determine the constants a and b. For some radiometer configurations, the constant b is given in terms of known constant temperature(s); $b = -T_{REF} = -T_0$ for the three Dicke radiometer types given in Table 7-3, and $b = T'_{REC}$ for the noise-adding radiometer. In such cases, it is sufficient to perform a single calibration measurement so as to determine the value of the constant a. In practice, however, it is advisable to calibrate the radiometer at more than one value of the input noise temperature, and at least one point should be for a calibration noise temperature lower than 100 K.

Figure 7-23 shows a calibration noise source placed at the radiometer input. Corresponding to measurements with calibration noise temperatures T^h_{cal} and T^c_{cal} (where the superscripts h and c stand for hot and cold, respectively), the radiometer output indicator records values of

$$I^h_{out} = a(T^h_{cal} + b), \quad (7.91a)$$

$$I^c_{out} = a(T^c_{cal} + b). \quad (7.91b)$$

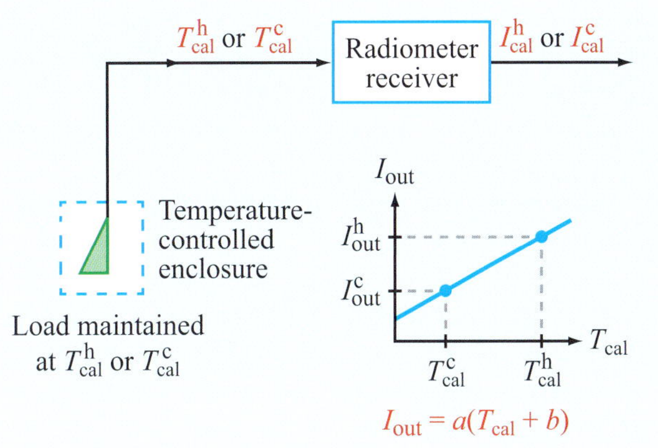

Figure 7-23: Calibration of a radiometer receiver.

The above two equations lead to the solutions

$$a = \frac{I^{\rm h}_{\rm out} - I^{\rm c}_{\rm out}}{T^{\rm h}_{\rm cal} - T^{\rm c}_{\rm cal}}, \qquad (7.92a)$$

$$b = \frac{I^{\rm c}_{\rm out} T^{\rm h}_{\rm cal} - I^{\rm h}_{\rm out} T^{\rm c}_{\rm cal}}{I^{\rm h}_{\rm out} - I^{\rm c}_{\rm out}}. \qquad (7.92b)$$

7-11.2 Calibration Sources

Calibration noise sources consist of the following four types:

(a) A matched load (resistor) whose temperature is held at a constant, known value. A ***hot load*** usually is enclosed in a temperature-controlled oven, and a ***cold load*** usually is immersed in a Dewar flask containing a boiling cryogen. Liquid nitrogen (N_2), whose boiling temperature is 77.36 K at sea-level barometric pressure, is commonly used for this purpose.

(b) Materials with known emissivities, such as the highly absorbing, nonreflective materials used in anechoic chambers. These materials can be constructed to have emissivities close to unity, and therefore their brightness temperatures are approximately equal to their physical temperatures. Low brightness-temperature values are obtained by saturating the absorbing material in a boiling cryogen. The calibration process involves measuring the emission by the material with the antenna beam directed at it. As the process requires knowledge of the antenna properties (radiation and beam efficiencies), we defer further discussion until Section 7-11.5.

(c) The two preceding calibration methods usually are used to calibrate a radiometer only before and/or after a measurement program. Using the matched-load method during an aircraft measurement mission requires the availability of a cryogen for the duration of the flight. Radiometers employing balanced Dicke receiver configurations can be made sufficiently stable that calibration against external sources is needed only occasionally, thereby avoiding the cryogenic refrigeration problem. Most modern satellite-borne radiometer systems use a third type of cold-calibration source, namely, outer space. When the antenna (or an auxiliary antenna) is pointed at ***cosmic space***, it observes a brightness temperature of 2.7 K. Typically, a rotating feed with a fixed reflector is used, so that each rotation includes a deep-space calibration measurement.

(d) A fourth type of calibration noise source may be added to the above list, namely, solid-state circuits designed to exhibit noise levels corresponding to very low equivalent noise temperatures. An example is the ***COLDFET*** (cold FET) discussed by Frater and Williams (1981).

7-11.3 Effects of Impedance Mismatches

Throughout the foregoing discussions it was assumed that every RF component and transmission line contained in the radiometer front end is perfectly matched to whatever it was connected to. In other words, no reflections due to impedance mismatches exist. In practice, it may be possible to reduce reflections to small levels through the use of impedance-matching techniques, but they cannot always be eliminated altogether.

The effects of impedance mismatches on the radiometer measurement accuracy may be divided into two groups: (1) mismatches within the receiver and (2) mismatches between the antenna (or calibration source) and the receiver input (Fig. 7-24). For a reflection-free radiometer, the output indication $I_{\rm out}$ is linearly related to the input noise temperature $T_{\rm IN}$, which represents the net power delivered to the receiver due to noise power

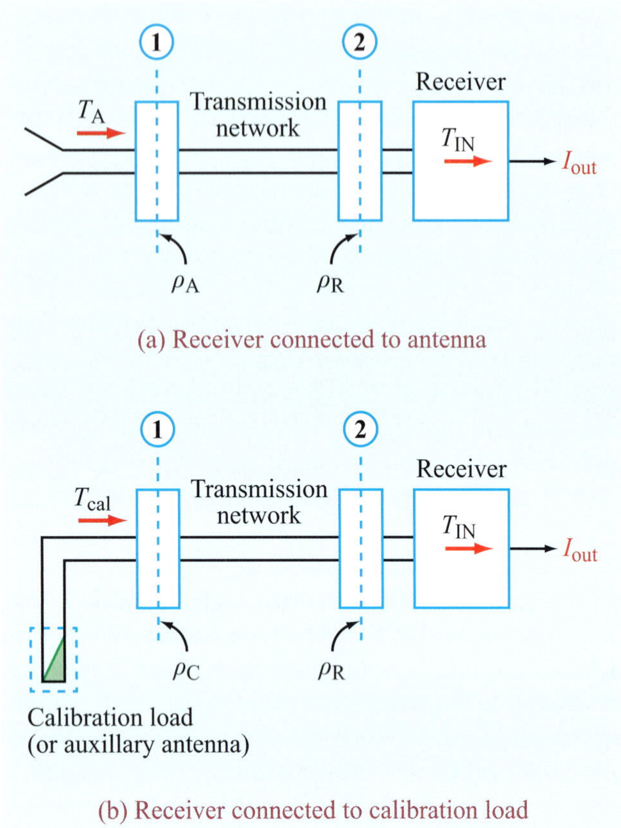

Figure 7-24: Radiometer receiver connected to (a) antenna and (b) calibration source. The ρ's are voltage reflection coefficients looking into the indicated ports.

supplied by the antenna (or calibration source) and the line in between. That is,

$$I_{out} = a_1 T_{IN} + b_1, \tag{7.93}$$

where a_1 and b_1 are constants. Mismatched components within the receiver give rise to reflection coefficients that end up modifying the values of a_1 and b_1. As long as the magnitudes and phases of these reflection coefficients remain constant, and as long as the noise temperatures of the components remain constant, the effects of mismatches may be factored out in the calibration process. However, the reflection coefficients of components often are susceptible to temperature variations. Therefore, not only is it important to maintain the environmental temperature of the radiometer front end constant, but it is equally important that the absolute value of this temperature be the same (as closely as possible) during the radiometer operating mode and during the calibration mode. That is, if a_1 and b_1 are maintained constant, I_{out} will be related accurately to T_{IN}.

This brings us to the second step, namely, relating T_{IN} to the antenna temperature T_A or to the noise temperature of the calibration source, T_{cal}. If the voltage reflection coefficient as seen looking into the antenna, ρ_A, is equal to the voltage reflection coefficient (in magnitude and phase) as seen looking into the calibration load, ρ_C, the problem is an easy one. In this case, the calibration constants obtained by calibrating I_{out} against T_{cal} will be the same as the constants in the relation between I_{out} and T_A, assuming that the same line used to connect the antenna to the radiometer also is used during calibration. In the general case, however, $\rho_A \neq \rho_C$, which means that the relationship between I_{out} and T_{cal} will not be exactly the same as the relationship between I_{out} and T_A; therefore, if the calibration equation is used (without corrections) to predict T_A-values, the predicted values may be in error. The magnitude of the error varies from values on the order of 1 K if $|\rho_C| < 0.05$ and $|\rho_A| < 0.05$ and to much larger values if the magnitudes of ρ_C and ρ_A are larger and very different.

Under matched conditions, T_{IN} is related to T_A by

$$T_{IN} = \frac{1}{L}\,T_A + \left(1 - \frac{1}{L}\right) T_0, \tag{7.94}$$

where L is the loss factor of the transmission network connecting the antenna to the receiver and T_0 is its physical temperature. The effects of mismatches on receiver calibration have been examined by Wells et al. (1964), Miller et al. (1967), and Otoshi (1968), among others. If ρ_R (looking into the receiver) and ρ_A (looking into the antenna) are nonzero, the expression for T_{IN} includes new terms that account for the multiple reflections (between ports 1 and 2 of the transmission network) of both the noise power delivered by the antenna (T_A) and the noise power generated by the receiver (T_{REC}). Figure 7-25 displays plots of the

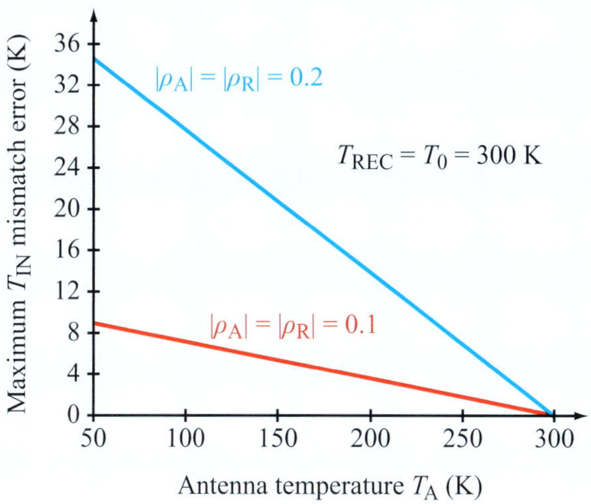

Figure 7-25: Variation of maximum mismatch errors with antenna temperature.

maximum mismatch error that can occur between the true value of T_{IN} and the reflection-free value. Even for the plot corresponding to $T_{REC} = 300$ K and small reflection coefficients, $|\rho_A| = |\rho_R| = 0.1$, the mismatch error can be as large as 9 K for $T_A = 50$ K. Increasing the magnitudes of the reflection coefficients to 0.2 raises the mismatch error to 34 K.

7-11.4 Antenna Calibration

In the preceding material, we discussed calibration methods for transforming the radiometer output indicator to an antenna temperature T_A, where T_A represents the noise powers delivered by the antenna. Now, we discuss a second transformation, namely that between T_A and T_{ML}, the mainlobe brightness temperature of the scene observed by the antenna. To that end, use is made of Eq. (6.40):

$$T_A = \xi \eta_b T_{ML} + \xi(1 - \eta_b) T_{SL} + (1 - \xi) T_0, \quad (7.95)$$

where ξ is the antenna radiation efficiency, η_b is the antenna beam efficiency, and T_0 is the physical temperature of the antenna. Note that Eq. (7.95) assumes that the antenna is sensitive only to the polarization of interest. In practice, the mainbeam sensitivity to orthogonally polarized signals must also be included in the expression for T_A, and corrected for during antenna calibration in a manner similar to that of the unwanted sidelobe contribution. Brightness temperatures T_{ML} and T_{SL} are the mainlobe and sidelobe contributions, and are given by Eqs. (6.33) and (6.34), respectively.

With T_A measured by the radiometer, the objective is to estimate T_{ML}. In the ideal case of a lossless antenna ($\xi = 1$) with a radiation pattern consisting of only one main lobe ($\eta_b = 1$), the above expression reduces to

$$T_A = T_{ML}.$$

In the real case, however, exact determination of T_{ML} necessitates that ξ, η_b, and T_{SL} be known. Techniques for measuring the antenna parameters ξ and η_b are discussed below, but T_{SL} is not a measurable quantity, nor is it a constant, since it depends on the distribution of radiation incident upon the antenna from directions outside the main lobe.

As was discussed in Section 6-5.3, the magnitude of the error associated with the estimated value of T_{ML} due to the lack of knowledge of T_{SL} is a function of η_b only. Therefore, in order to minimize this source of error, it is essential that the radiometer antenna be characterized by a beam efficiency that is as close to unity as possible. High beam efficiency is achieved by suppressing the side lobes of the radiation pattern, which is accomplished by properly tapering the aperture distribution, as discussed in Chapter 3. A higher beam efficiency, however, means a wider main beam, or equivalently, a lower aperture efficiency. The trade-off between beam efficiency η_b and aperture efficiency η_a is shown in Fig. 7-26. For radiometric remote sensing, this is equivalent to a trade-off between radiometric resolution and angular (or spatial) resolution.

From Eqs. (3.72), (3.18), and (3.24) in Chapter 3, the antenna beam efficiency η_b and radiation efficiency ξ

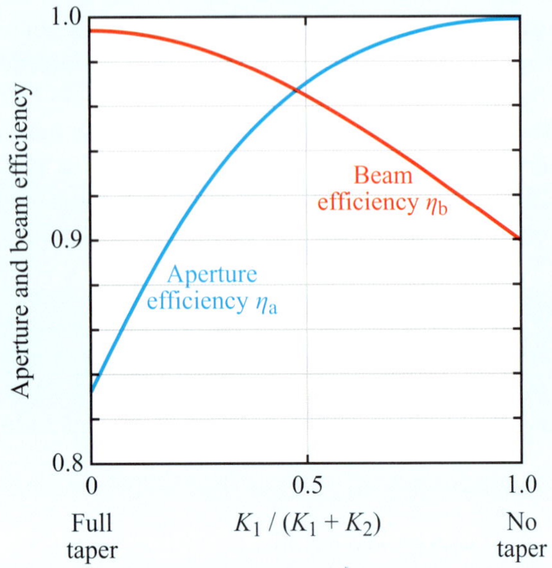

Figure 7-26: Beam and aperture efficiencies for a one-dimensional aperture as a function of taper [after Nash (1964)]. The aperture distribution is $E_a(x_a) = K_1 + K_2(1 - x_l^2)$, where $x_l = 2x_a/l$ and l is the aperture length.

are given by

$$\eta_b = \frac{\iint\limits_{\text{mainlobe}} F(\theta,\phi)\, d\Omega}{\iint\limits_{4\pi} F(\theta,\phi)\, d\Omega}, \tag{7.96}$$

$$\xi = \frac{G}{D} = \frac{4\pi G}{\iint\limits_{4\pi} F(\theta,\phi)\, d\Omega}, \tag{7.97}$$

where $F(\theta,\phi)$ is the antenna radiation pattern, and D and G are its peak directivity and peak gain, respectively. If the radiation pattern $F(\theta,\phi)$ is accurately known for all directions θ,ϕ over 4π solid angle, the determination of η_b becomes a straightforward task. Similarly, ξ also can be determined provided G is known. In practice, however, whereas G is one of the easiest antenna parameters to measure, the problem with the proposed approach has to do with accuracy and cost; if the denominator in Eqs. (7.96) and (7.97) is in error by only 1%, it is liable to result in an error of several kelvins in the estimated value of T_{SL}, and in order to have good accuracy it is necessary to measure the complete two-dimensional variation of $F(\theta,\phi)$ with a measurement precision of at least -60 dB relative to its peak value, which is likely to be a costly operation. Alternative, and more accurate, methods for measuring ξ are described next. Once ξ has been determined, η_b can be easily computed from

$$\eta_b = \frac{\xi}{4\pi G} \iint\limits_{\text{mainlobe}} F(\theta,\phi)\, d\Omega. \tag{7.98}$$

As was mentioned earlier, G is an easily measurable quantity, and so is $F(\theta,\phi)$ over the narrow angular range covered by the mainlobe. In some cases, it is difficult to define exactly the extent of the main lobe, and so η_b is quoted as the beam efficiency for a certain angular range (such as within two half-power beamwidths of the main-beam center) or for a range of $F(\theta,\phi)$ relative to the maximum (such as down to the -20 dB level).

7-11.5 Cryoload Technique

The aperture of the antenna shown in Fig. 7-27 is placed directly over a box containing microwave-absorbing material. The porous absorbing material is saturated with liquid nitrogen as described by Hardy et al. (1973). The absorbing material is characterized by a very small reflection coefficient (Fig. 7-28) and therefore acts like a perfect absorber (emitter) with a brightness temperature equal to its physical temperature. A voltage reflection coefficient of -40 dB is equivalent to a magnitude of 0.01.

With the antenna aperture viewing an essentially constant brightness temperature distribution T_B (from all directions), the antenna temperature T_A measured by the microwave radiometer is given by

$$T_A = \xi T_B + (1 - \xi) T_0, \tag{7.99}$$

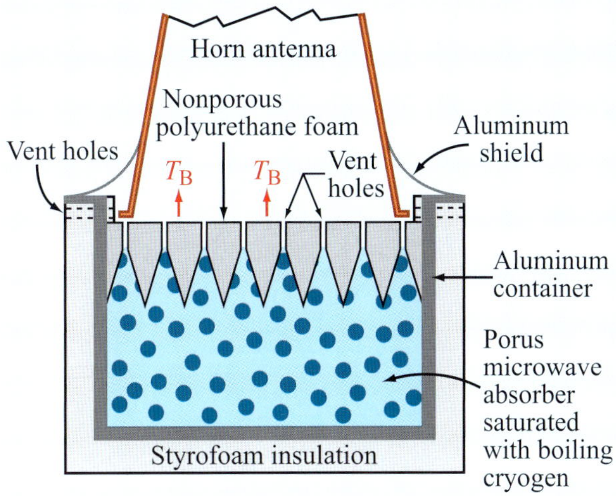

Figure 7-27: Construction of cryoload for calibration of radiometer antenna [after Hardy et al., 1974].

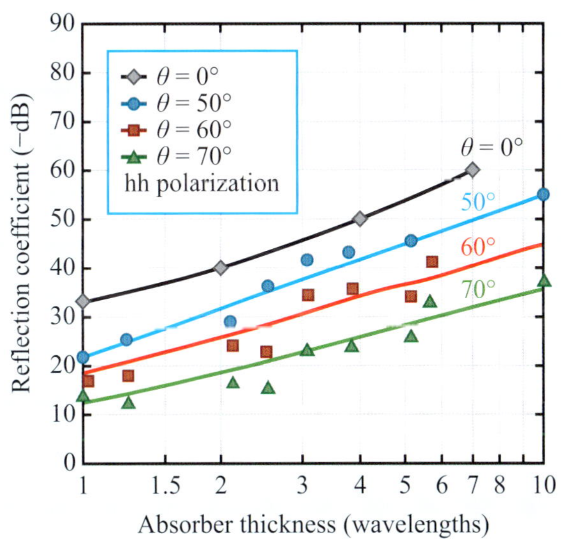

Figure 7-28: Characteristics of RF absorbers [after Emerson, 1973].

where ξ is the radiation efficiency of the antenna and T_0 is its physical temperature. Solving for ξ, we have

$$\xi = \frac{T_0 - T_A}{T_0 - T_B}, \tag{7.100}$$

where in this case $T_B = 77.36$ K (the boiling temperature of liquid nitrogen at sea level).

The cryoload calibration technique has several attractive features: it is accurate, repeatable, and relatively inexpensive. According to Hardy et al. (1974), who used the cryoload technique to calibrate a 2.65 GHz radiometer, ξ can be measured with an absolute accuracy of about $\pm 0.1\%$, and according to Blume (1977) the calibration repeatability had an rms value of 0.7 K and an average deviation of 0.03 K. A slightly modified version of the above technique was used to calibrate a small-aperture (10 cm × 10 cm) horn antenna at 86.1 GHz (Ulich, 1977).

The use of the cryoload calibration technique has been limited to relatively small antenna apertures (of the order of 1 m per side). For large antennas, the ***bucket technique*** has proved useful.

7-11.6 Bucket Technique

The antenna shown in Fig. 7-29 is placed inside a large metal bucket whose dimensions are sufficiently large that it can be safely assumed that there is no mutual coupling between the antenna and the bucket. The indicated dimensions are those of the New Mexico State University antenna calibration bucket, which was constructed on a mountain to avoid emissions from the surrounding terrain (Carver, 1975). With the antenna

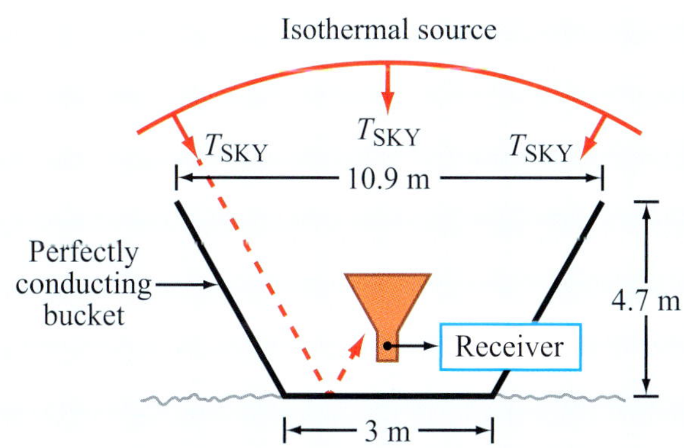

Figure 7-29: The bucket method for measuring the radiation efficiency of an antenna [after Carver, 1975].

main beam pointed in the zenith direction, the antenna temperature T_A measured by the radiometer is given by

$$T_A = \xi T'_A + (1-\xi)T_0, \qquad (7.101)$$

where T'_A is the antenna temperature for a lossless antenna and is equal to the integrated brightness temperature of the sky:

$$T'_A = \frac{\iint_{4\pi} T_{SKY}(\theta,\phi)\, F(\theta,\phi)\, d\Omega}{\iint_{4\pi} F(\theta,\phi)\, d\Omega}. \qquad (7.102)$$

The surface of the perfectly reflecting metal bucket has an emissivity of zero. Hence, the radiation received by the antenna is due entirely to atmospheric emission, the major portion of which is received directly from above and the rest of which is received through the side lobes after being reflected by the bucket walls. Assuming that $T_{SKY}(\theta,\phi)$ is approximately constant over the angular range subtended by the main lobe and the first few significant side lobes, Eq. (7.102) reduces to

$$T'_A = T_{SKY}(\theta = 0°).$$

Using values of meteorological parameters provided by a weather station located next to the bucket, $T_{SKY}(\theta = 0°)$ can be calculated using the atmospheric-emission formulas given in Chapter 8. With T'_A computed, and T_0 and T_A measured, ξ is determined from Eq. (7.101).

7-12 Imaging Considerations

According to Slater (1980), "the limit of resolution of an optical system is reached when, according to a given criterion, the system can just separate the elements of a well-defined test object, such as a double star in astronomy, the line of a grating in microscopy, and the bars of a bar target in photography." Several standard methods are used for measuring the resolving power of optical systems (Slater, 1980), but no equivalent methods or standard targets have been adopted for microwave systems.

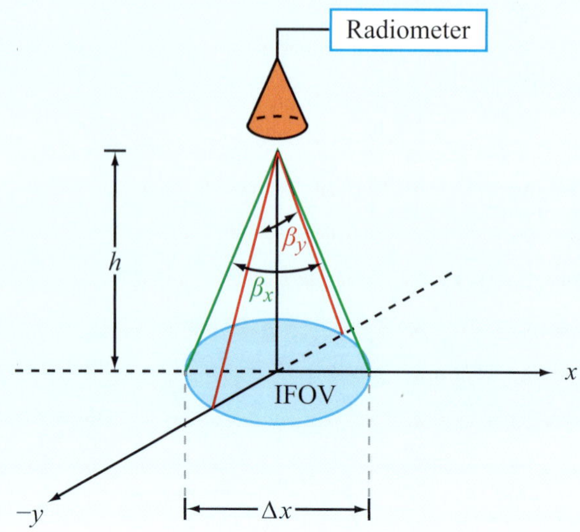

Figure 7-30: The Instantaneous Field of View (IFOV) for a nadir-pointing antenna with beamwidths β_x and β_y. The antenna platform is at a height h above the ground.

The most frequently used definition for the spatial resolution of a microwave radiometer is given in terms of the ***instantaneous field of view*** (IFOV) corresponding to the half-power beamwidths. The IFOV defines the area on the ground covered by the antenna main beam. For the configuration shown in Fig. 7-30, the spatial resolutions Δx and Δy are given by

$$\Delta x = \beta_x h, \qquad (7.103a)$$

$$\Delta y = \beta_y h, \qquad (7.103b)$$

where h is the height of the antenna platform, and β_x and β_y are the half-power beamwidths in the x and y directions, respectively. Most microwave radiometric systems use antennas with circularly symmetric patterns; i.e., $\beta_x = \beta_y = \beta$.

As was discussed in Chapter 3,

$$\beta = k\,\frac{\lambda}{l} \qquad \text{radians}, \qquad (7.104)$$

where l is the length (and width) of a square aperture or the diameter of a circular aperture, and k is a constant for a given antenna configuration. Usually k is between

0.88 (for a uniformly illuminated aperture) and about 2 (for steeply tapered illuminations) (see Tables 3-1 and 3-2). For calculation purposes, a value of $k = 1.5$ is appropriate for an antenna with a high beam efficiency.

7-12.1 Scanning Configurations

Radiometric imaging of a scene of interest is accomplished by scanning the main beam of the antenna; for a moving platform, scanning in the cross-track dimension is sufficient to produce an image. Both mechanical and electronic (beam-steering) scanning techniques are used in microwave radiometry. In mechanical scanning, the direction of the antenna beam is changed by mechanical rotation or angular movement of the radiating aperture of the antenna system. Examples are shown in Fig. 7-31; they include (a) the simple configuration in which the entire antenna structure is made to scan in angle, (b) a configuration where scanning is achieved by rotating a reflector (mirror) back and forth while maintaining the antenna in a fixed position, and (c) a parabolic torus configuration consisting of a fixed reflector and a spun feed.

Phased-array antennas are used to steer the direction of the antenna beam electronically, without the involvement of mechanical motion in the scanning process (see Section 3-11). If the antenna is to scan in only one direction, it is sufficient to control the relative phase in one dimension, as illustrated in Fig. 7-32. Another advantage over mechanically scanned antennas is the high scanning-speed capability of phased arrays. Electronic beam steering has some drawbacks, however: phased arrays are more complex, more expensive, heavier, and lossier than single-antenna structures of comparable size. The higher antenna losses of phased arrays are due to the phase shifters used to control the phase of each individual feed line. Usually ferrite or *PIN* diode phase shifters are used.

Among the scanning microwave radiometer systems that have been flown on the Nimbus satellite series, the 19.35 GHz (Nimbus 5) and 37 GHz (Nimbus 6) electrically scanning microwave radiometers (ESMRs) employed phased-array antennas, while the Nimbus 6 scanning microwave spectrometer (SCAMS) and the Nimbus 7 scanning multichannel microwave radiometer

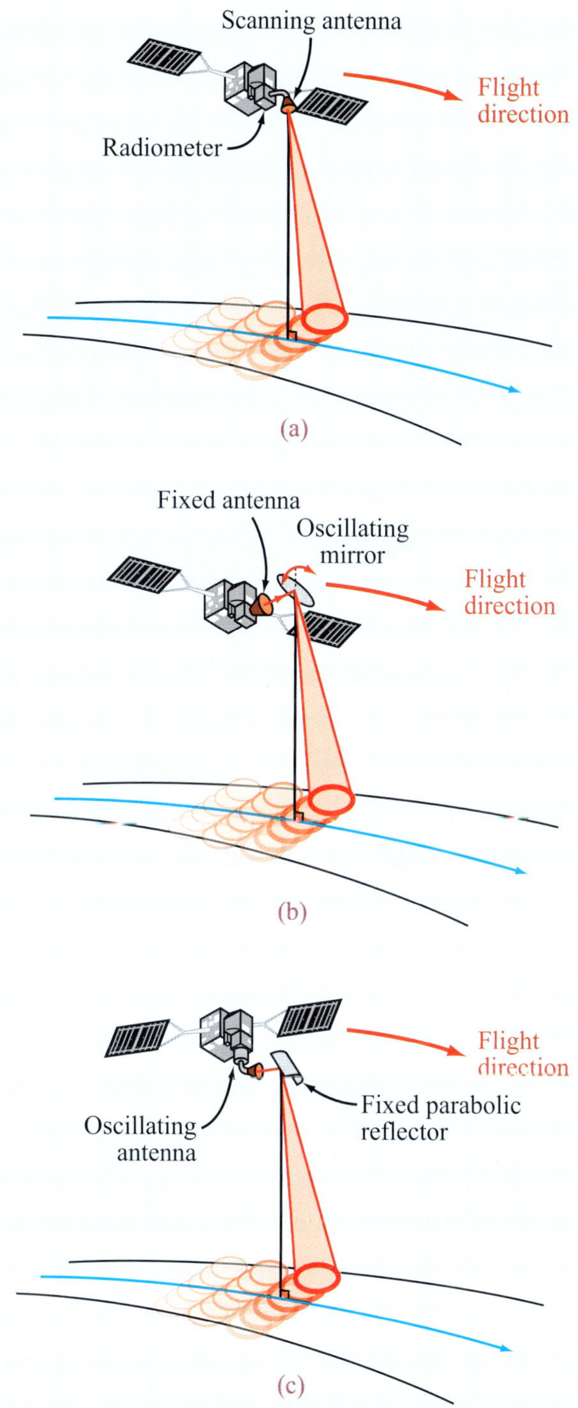

Figure 7-31: Mechanical scanning configurations: (a) scanning antenna; (b) fixed antenna and oscillating reflector; (c) fixed parabolic reflector and oscillating antenna feed.

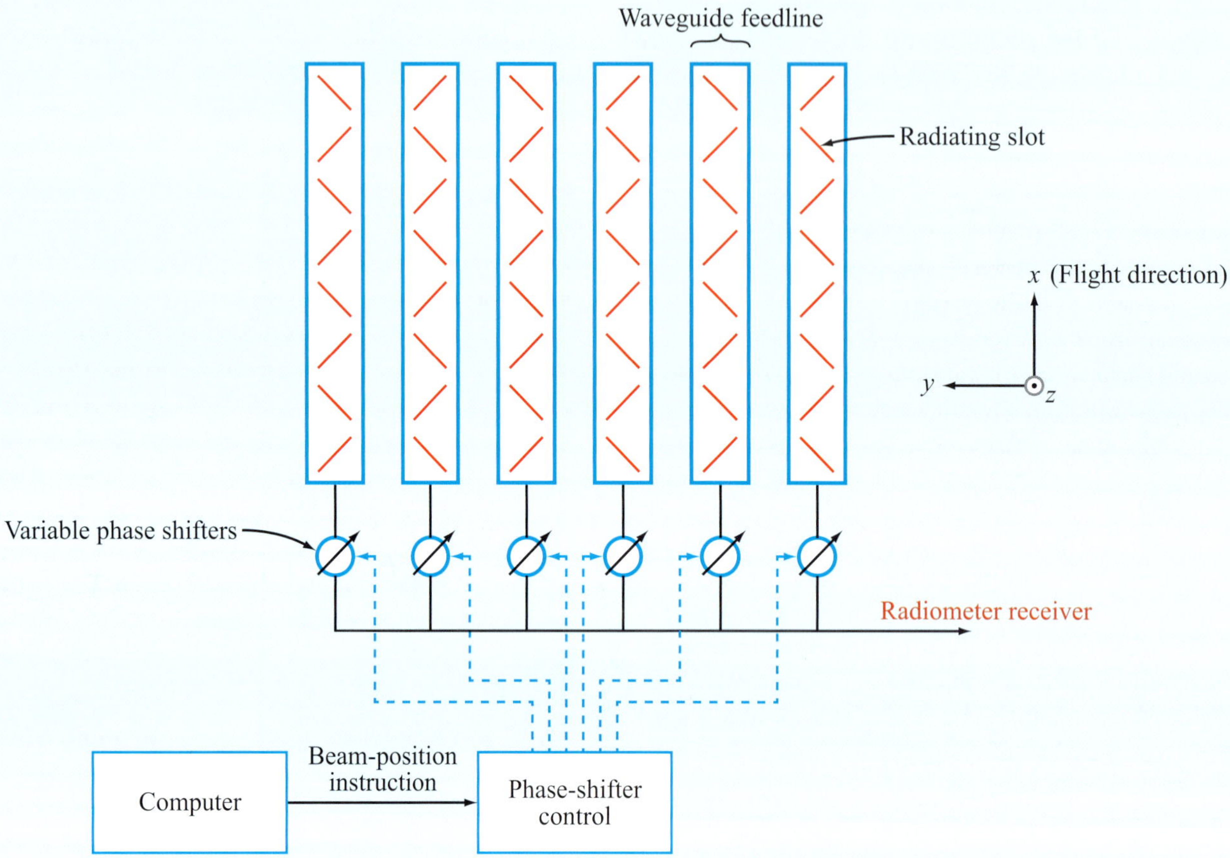

Figure 7-32: Phased-array antenna configuration for electronic scanning of the main beam in the y–z plane.

(SMMR) employed mechanically scanned antenna configurations.

Figure 7-33 illustrates the two most commonly used viewing configurations. In the first configuration [Fig. 7-33(a)] the antenna beam scans in a plane approximately perpendicular to the direction of motion. The angle of incidence varies between $\theta = 0^\circ$ for the nadir beam and $\theta = \theta_s$ at the edge of the scanline. Also, the shape of the IFOV changes from a circle (for a circular antenna pattern) at nadir to an ellipse with its long axis in the y direction. For some antenna scanning configurations, the direction of the polarization vector also is a function of the beam position within the scanline.

The major advantage of the viewing configuration shown in Fig. 7-33(b) is that the angle of incidence remains approximately constant as the beam scans in azimuth along a *conical surface ahead of the radiometer platform*. An example of a conical scan system is the 37 GHz ESMR, which scans in azimuth $\pm 35^\circ$ about the forward direction at a constant tilt angle of 45° with respect to the direction of motion. With the Earth's curvature taken into account, the angle of incidence at the Earth's surface varies between 49.6° for the beam position corresponding to an azimuth angle of zero and 50.8° at the edge of the scanline; in other words, the angle of incidence remains essentially constant across the image generated by such a scanning arrangement.

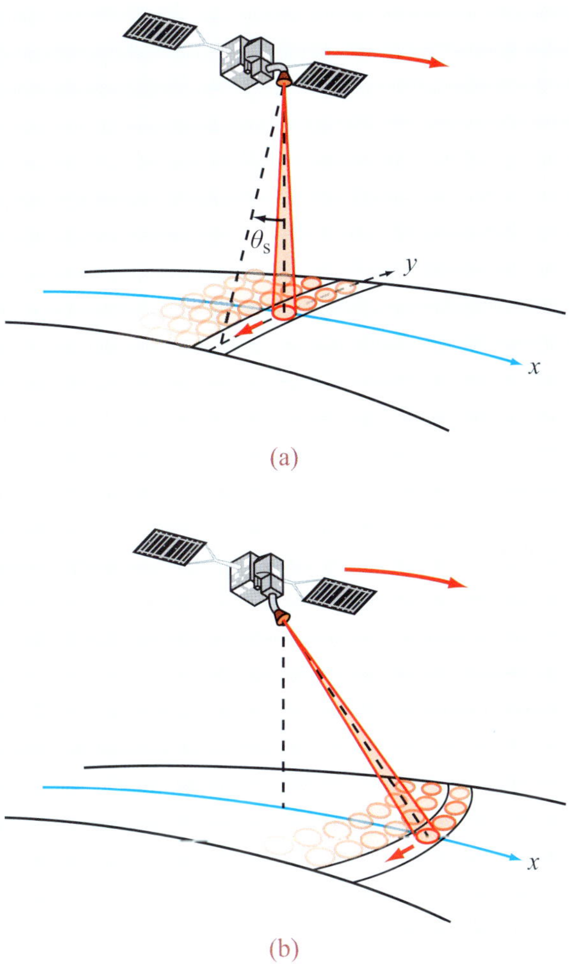

Figure 7-33: Radiometric imaging by (a) cross-track scanning in the plane normal to the direction of flight, and (b) conical scanning.

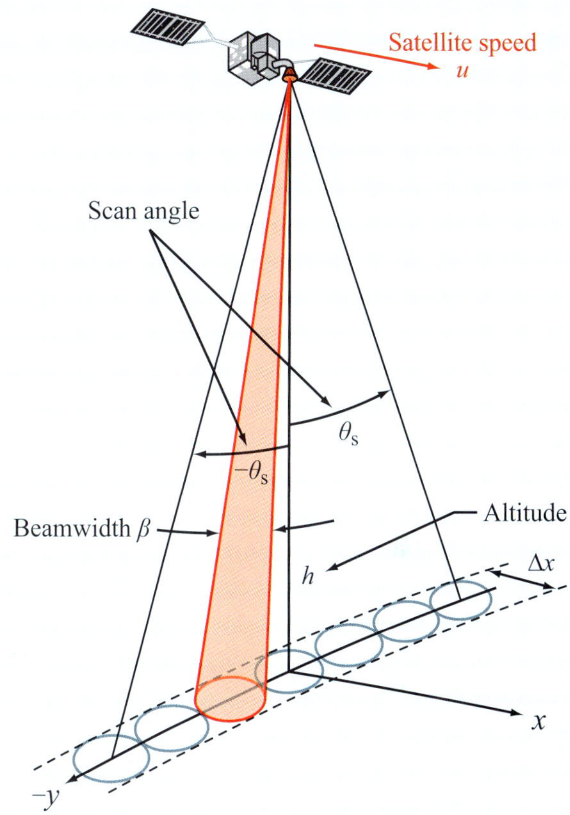

Figure 7-34: Geometry of airborne scanning microwave radiometer [after McGillem and Seling (1963)].

7-12.2 Radiometer Uncertainty Principle

We pointed out in preceding sections that, for a given integration time τ, there is a trade-off between spectral resolution (the predetection bandwidth B) and the radiometric resolution ΔT. For most radiometer systems, ΔT may be expressed in the general form

$$\Delta T = \frac{M}{\sqrt{B\tau}}, \tag{7.105}$$

where the radiometer figure of merit M is a constant for a given system temperature and receiver configuration. For a stationary radiometer (with respect to the scene), there are no fundamental constraints imposed on how long τ may be. The situation is different for a moving platform.

To relate the parameters of a scanning system to τ, consider the simple case shown in Fig. 7-34. The platform is at a height h above the ground and is moving with a speed u in the x direction. The radiometer antenna scans between $+\theta_s$ ($+y$ direction) and $-\theta_s$ ($-y$ direction) in a direction transverse to the flight direction. The forward motion of the platform provides a line-by-line scanning format. The time it takes to travel through one nadir beamwidth in the longitudinal (forward) direction is given by

$$t_l = \frac{\Delta x}{u} = \frac{\beta h}{u} \,. \tag{7.106}$$

Ignoring the reset time involved in steering the beam from the end of one scanline ($\theta = \theta_s$) to start a new scanline, and assuming that one transverse scan is completed per beamwidth of forward motion (i.e., in time t_l), the angular scanning rate is

$$\omega = \frac{2\theta_s}{t_l} \qquad \text{rad/s.} \tag{7.107}$$

If we assume further that the antenna beamwidth remains constant as the beam is made to scan between $-\theta_s$ and $+\theta_s$ (which is valid only if the scanning is mechanical or if the effective aperture remains constant), then the time it takes to scan through one beamwidth in the transverse direction is

$$\tau_d = \frac{\beta}{\omega} = \frac{t_l \beta}{2\theta_s} \,. \tag{7.108}$$

This time τ_d is called the ***dwell time*** because it is equal to the time that a point on the ground is observed by the antenna beam. Using Eq. (7.106), τ_d can be expressed in terms of the spatial resolution Δx:

$$\tau_d = \frac{(\Delta x)^2}{2u\theta_s h} \,, \tag{7.109}$$

where θ_s is in radians.

Now let us suppose that the radiometer beam crosses over a sharp boundary (in either direction) between two areas of very different emission characteristics. If the radiometer integration time τ is much smaller than τ_d, it takes the radiometer output approximately τ_d seconds to make the transition between the two levels corresponding to the two dissimilar areas. If, on the other hand, $\tau \gg \tau_d$, it takes the radiometer much longer than τ_d seconds to register the change due to the sharp boundary, which is equivalent to having an effective spatial resolution that is much larger than Δx. Hence, from the standpoint of radiometric resolution it is desirable to have τ as long as possible, and from the standpoint of spatial resolution it is desirable to have $\tau \ll \tau_d$. The optimum choice is dependent upon the nature of the application for which the radiometer is used and on other system parameters. In the general case, however, a compromise solution is to set

$$\tau = \tau_d. \tag{7.110}$$

For $\tau = \tau_d$ inserting Eq. (7.109) into Eq. (7.105) leads to

$$\Delta T \cdot \Delta x \cdot B^{1/2} = M(2u\theta_s h)^{1/2}. \tag{7.111}$$

This ***radiometer uncertainty equation*** states that for a given radiometer configuration (i.e., M), flight parameters (h and u), and angular scan range ($2\theta_s$), the product of the radiometric uncertainty ΔT, the spatial uncertainty Δx, and the square root of the spectral uncertainty ($B^{1/2}$) is a constant. Similar expressions may be developed for more complex scanning configurations, but the basic idea remains the same, namely that the three types of resolutions (uncertainties) are interrelated, and therefore improving one of them is likely to degrade one or both of the other two (unless the flight parameters and/or the scanning configuration are changed).

7-13 Interferometric Aperture Synthesis

A particular type of phased array antenna, which relies on software beamforming, is based on a technique known as ***interferometric aperture synthesis***.

> ▶ Its objective is to synthesize a large effective antenna aperture—with a correspondingly narrow beamwidth and high angular resolution—using a distributed network of much smaller physical antennas. ◀

The method was originally developed in the 1950s for use in radio astronomy (Ryle, 1952; Ryle and Hewish, 1960). The 1974 Nobel Prize in Physics was awarded in recognition of the importance of this new type of imager, and of the role it played in the original discovery of the pulsar. Interferometric aperture synthesis has

Figure 7-35: SMOS antenna array deployed in orbit.

since matured into an established tool of astronomical research (Napier et al., 1983).

The radio astronomy technique has been adopted for use in Earth remote sensing as ***Microwave Interferometric Radiometry*** (MIR). MIR was originally proposed as a means for improving the spatial resolution of passive microwave imagers that operate at low microwave frequencies, in order to avoid the large, mechanically steered physical aperture that would otherwise be required (Ruf et al., 1988). Early demonstrations of the method were made by airborne prototypes (Tanner and Swift, 1993; Ruf and Principe, 2003). The first spaceborne MIR example, the ***Microwave Imaging Radiometer with Aperture Synthesis*** (MIRAS) on the ***Soil Moisture and Ocean Salinity*** (SMOS) satellite, was launched by the European Space Agency in November 2009 (Kerr et al., 2010). An artist's rendition of the fully deployed satellite in orbit is shown in Fig. 7-35. In the figure, the two solar panel arrays extend out diametrically from the central spacecraft bus, and the three antenna array arms below them are rotationally arranged in 120° increments.

The fundamental building block of a MIR sensor is the correlation radiometer, which performs a complex correlation between the electric fields received by two antennas that are separated apart in space. The spatial cross-correlation of the electric field generated by a distant source is given by (van Cittert, 1934; Zernike, 1938)

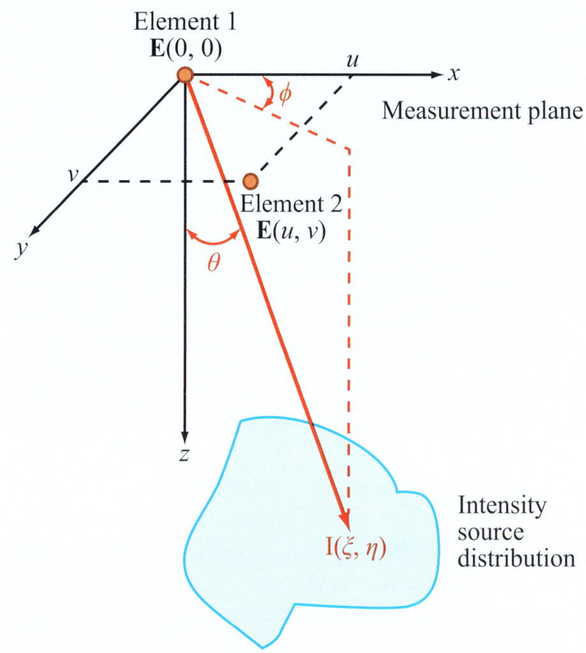

Figure 7-36: Relationship between measurement and source coordinates used in the van Cittert–Zernike theorem.

$$\langle \mathbf{E}(0,0)\,\mathbf{E}^*(u,v)\rangle \propto \iint_{\xi^2+\eta^2\leq 1} I(\xi,\eta)\, e^{-j2\pi(u\xi+v\eta)}\, d\xi\, d\eta, \qquad (7.112)$$

where (u,v) denote the coordinates of spatial separation between the two **E**-fields of the waves arriving at elements 1 and 2 in Fig. 7-36, measured in units of wavelengths, the coordinates $\xi = \sin\theta\cos\phi$ and $\eta = \sin\theta\sin\phi$ denote the direction cosines pointing from the **E**-field measurement plane to the source, I is the intensity of the source, and the integration is over the half space in which the source lies. The relationship between the measurement coordinate system (u,v) and the source coordinate system (ξ,η) is illustrated in Fig. 7-36. Equation (7.112) is often referred to as the ***van Cittert–Zernike theorem***. It highlights the Fourier transform relationship that exists between the spatial cross-correlation of an **E**-field and the source distribution that produced it.

With MIR sensors, an image of the intensity distribution $I(\xi,\eta)$ is generated in software by performing an inverse Fourier transformation of the cross-correlation measurements. Alias-free image reconstruction requires that cross-correlation measurements be made at specific intervals in the (u,v) measurement plane. The proper (u,v) sampling depends on the field of view over which the source distribution is to be imaged, as specified by the Nyquist sampling theorem (Ulaby and Yagle, 2013). If image reconstruction extends over the full half-space ($z > 0$ in Fig. 7-36), then measurements are required in half integer multiples of u and v. This is equivalent to the case of a conventional phased array arrangement in which the antenna elements are spaced in half-wavelength intervals in order for the array factor to be free of grating lobes over a complete half-space (Section 3-10.2). If image reconstruction is restricted to a narrower field of view, the samples in the (u,v) plane can be spaced further apart.

▶ In general, the size of a synthesized MIR aperture equals the greatest distance between pairs of antenna elements. ◀

For Earth remote sensing of an extended microwave brightness temperature distribution, the van Cittert–Zernike theorem can be rewritten in terms of the cross-correlation between two bandpass signals received by antennas separated apart by (u,v) as given by Corbella et al. (2004):

$$V(u,v) \propto \frac{1}{kBG} \iint_{\xi^2+\eta^2\leq 1} \left\{ \frac{T_{\mathrm{B}}(\xi,\eta) - T_{\mathrm{REC}}}{\sqrt{1-\xi^2-\eta^2}}\, F(\xi,\eta) \right\} \cdot R\left(\frac{u\xi+v\eta}{f_0}\right) e^{-j2\pi(u\xi+v\eta)}\, d\xi\, d\eta, \qquad (7.113)$$

where V, the so-called ***"visibility" function***, is the cross-correlation between the signals received by the two antennas, k is Boltzmann's constant, B is the (receiver) noise bandwidth, G is the power gain, T_{REC} is the noise temperature of the two receivers (assumed common), F is the power radiation pattern of the two antennas (assumed common), R is called the ***fringe washing function***, and it accounts for decorrelation of the thermal noise signal due to differences in time-of-flight from the source to the two antennas, and f_0 is the center frequency of the receiver passband. The source to be imaged has the brightness temperature distribution $T_{\mathrm{B}}(\xi,\eta)$. If the propagation time between antennas is much less than the inverse of the noise bandwidth, as is often the case, then the fringe washing effect is negligible, in which case $R \approx 1$.

In practice, MIR sensors are implemented using an array of small antenna elements configured so that the cross-correlation of all possible pairs among them provides the necessary sampling of the (u,v) measurement plane. Instruments are generally divided into two classes: ***one-dimensional imagers***, with fan beam antenna element patterns, and ***two-dimensional imagers***, with flood beam patterns. In the one dimensional case, the antenna elements are configured along a line and the (u,v) measurement plane collapses to one of its axes. The image reconstruction process correspondingly collapses to a one-dimensional inverse Fourier transform of the visibility function (Tanner and Swift, 1993).

An example of the design of a one-dimensional instrument is shown in Fig. 7-37. Thermal emission signals incident on the instrument enter the antenna through a low-loss kapton radome in the upper left corner of the diagram. The antenna subsystem consists of a linear array of 14 elements, each of which is a 1 m long slotted waveguide with a resonance operating frequency of 10.7 GHz. The elements are configured in a minimum-redundancy linear array for which the set of distances between all possible antenna pairs includes every integer multiple of one-half wavelength from 1 through 68. This arrangement is necessary to ensure alias-free Nyquist sampling of the visibility function (Ruf, 1993). The signals exiting each of the 14 antennas pass through directional couplers that add a broadband calibration signal, which is generated by a noise diode. A single noise source is used and its output is fanned out to the 14 signal paths, so that a controlled and

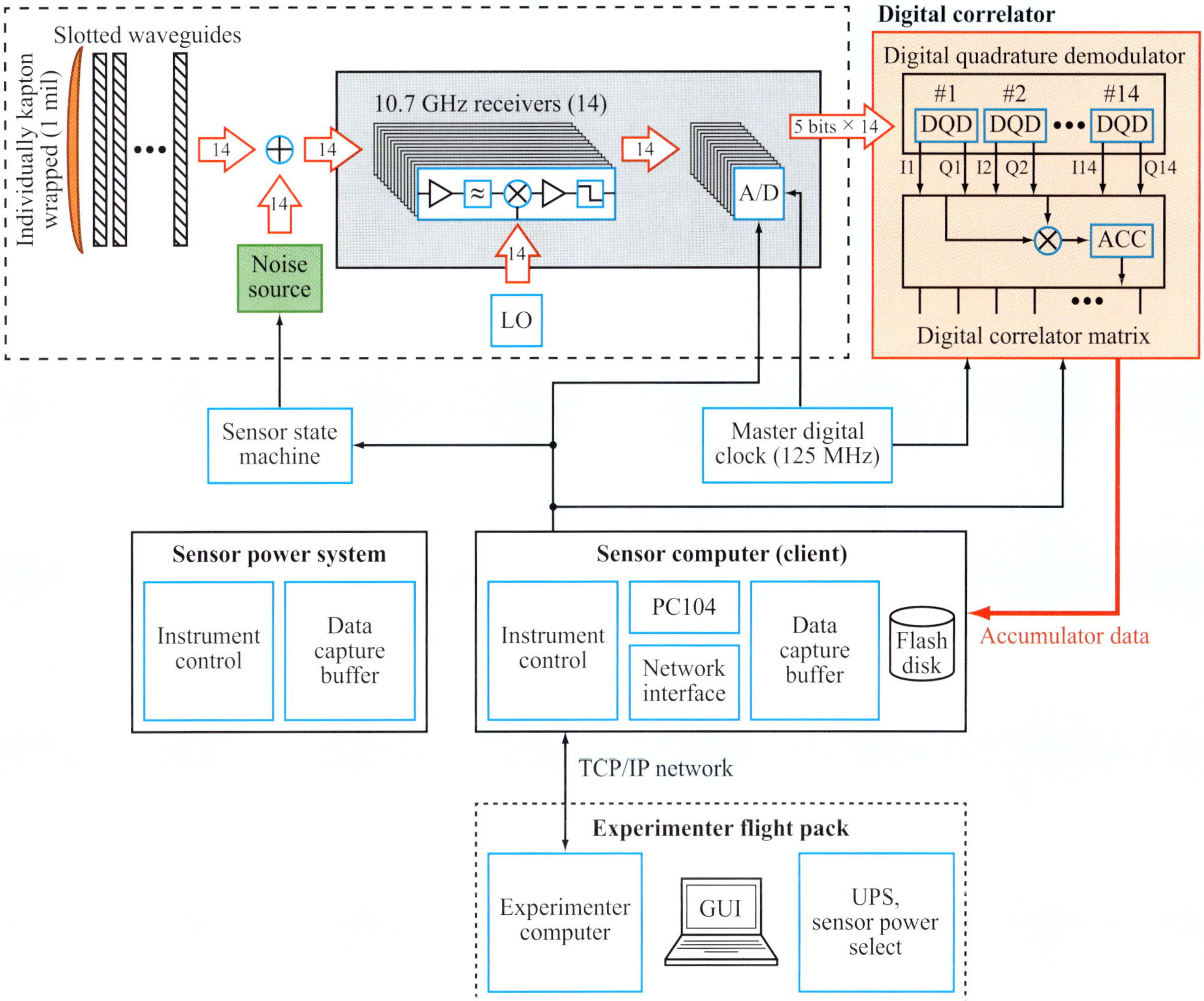

Figure 7-37: Functional block diagram of a one-dimensional-synthesis MIR sensor with 14 fan-beam antennas that are combined to synthesize a 68-element phased array using software beam steering [Ruf and Principe, 2003].

repeatable degree of correlation is added between each of the 14 signals.

Next, the signals are passed to a bank of 14 radiometer receivers. Each receiver contains a front-end Dicke switch with an ambient blackbody reference position, gain and frequency downconversion stages, and RF and IF band definition and sideband suppression filters. The final stage of the receiver is an analog-to-digital converter. Digitization is performed on the undetected IF signals in order to preserve the relative phase information that is present between the 14 signals.

The 14 digitized signals are then passed to a digital correlator subsystem. The first stage of digital signal processing is a bank of 14 quadrature demodulators.

They produce In-phase and Quadrature (I and Q) baseband components of the incoming IF signals using sine and cosine versions of the IF carrier as local oscillator inputs to downconversion mixers. The second stage of the correlator consists of a matrix of multipliers for all possible pairs of signals coming from different antenna elements, followed by accumulators that integrate and store the multiplication products. The multiplication matrix includes all combinations of I and Q components of the signals. The sample of the complex visibility function corresponding to the nth and mth antenna elements in the array is derived from the correlation data products according to

$$V_{nm} = (\langle I_n I_m \rangle + \langle Q_n Q_m \rangle) + j(\langle I_n Q_m \rangle - \langle Q_n I_m \rangle), \tag{7.114}$$

where I_n and Q_n are the I and Q components of the baseband signal from the nth antenna element, and $\langle\cdot\rangle$ denotes integration by the accumulator.

7-13.1 Image Reconstruction

Reconstruction of the brightness temperature distribution from measurements of the visibility function can, in principle, be carried out by an inverse Fourier transform. To see this, Eq. (7.113) may be rewritten as

$$V(u,v) = \int_0^{2\pi} \int_0^{\pi} T'_{\rm B}(\theta,\phi)\, e^{-j2\pi(u\sin\theta\cos\phi + v\sin\theta\sin\phi)} \cdot \sin\theta\, d\theta\, d\phi, \tag{7.115}$$

where the visibility function is assumed to be properly scaled to account for the leading terms outside the integral in Eq. (7.113), and where $T'_{\rm B}$ is the modified brightness temperature defined by collecting together the terms other than the exponential kernel in the integrand of Eq. (7.113). Fourier inversion of Eq. (7.115) follows as

$$T'_{\rm B}(\theta,\phi) = \int_{-\infty}^{\infty} \int_{-\infty}^{\infty} V(u,v) \cdot e^{j2\pi(u\sin\theta\cos\phi + v\sin\theta\sin\phi)}\, du\, dv, \tag{7.116}$$

where it is implicitly assumed that visibility measurements are available at all possible spatial frequency components, u and v. In practice, visibility measurements correspond to discrete baseline separations between pairs of elements in the antenna array. If the antenna element radiation pattern, $F(\xi,\eta)$ in Eq. (7.113), is broad enough to allow imaging over the entire half-space in front of the MIR instrument, then antenna pairs are spaced at integer multiples of one-half wavelength up to a maximum that is determined by the greatest distance between pairs of elements. The resulting truncated and discretized version of the image reconstruction algorithm follows as

$$\widehat{T'_{\rm B}}(\theta,\phi) = \sum_{n=-N}^{N} \sum_{m=-M}^{M} V(u_n, v_m)\, e^{j2\pi(u_n\sin\theta\cos\phi + v_m\sin\theta\sin\phi)}, \tag{7.117}$$

where $\widehat{T'_{\rm B}}$ is the reconstructed estimate of the true modified brightness temperature, $u_n = n/2$ for $n = 0$, ± 1, ± 2, ..., $\pm N$ and $v_m = m/2$ for $m = 0$, ± 1, ± 2, ..., $\pm M$. Inserting Eq. (7.115) into Eq. (7.117), passing the summation operations through the integrals, and collapsing the summations gives

$$\widehat{T'_{\rm B}}(\theta,\phi) = \int_0^{2\pi} \int_0^{\pi} T'_{\rm B}(\theta',\phi')\, F^u_{\rm a}(\theta,\phi;\,\theta',\phi') \cdot F^v_{\rm a}(\theta,\phi;\,\theta',\phi')\, \sin\theta'\, d\theta'\, d\phi', \tag{7.118}$$

where $F^u_{\rm a}$ and $F^v_{\rm a}$ are antenna array factors (see Section 3-9) given by

$$F^u_{\rm a}(\theta,\phi;\,\theta',\phi') = \frac{\sin\left[\pi\,\frac{2N+1}{2}\,(\sin\theta\cos\phi - \sin\theta'\cos\phi')\right]}{\sin\left[\frac{\pi}{2}\,(\sin\theta\cos\phi - \sin\theta'\cos\phi')\right]}, \tag{7.119a}$$

$$F^v_{\rm a}(\theta,\phi;\,\theta',\phi') = \frac{\sin\left[\pi\,\frac{2M+1}{2}\,(\sin\theta\sin\phi - \sin\theta'\sin\phi')\right]}{\sin\left[\frac{\pi}{2}\,(\sin\theta\sin\phi - \sin\theta'\sin\phi')\right]}. \tag{7.119b}$$

Equation (7.118) is essentially a spatial low-pass filter of the true brightness temperature distribution. The resolution of the reconstructed image can be defined as the angular distance between first nulls of the array factors in the two principal planes, or

$$\theta_{\text{null-to-null}} = \sin^{-1}\left(\frac{4}{2N+1}\right) \quad \text{in the } \phi' = 0 \text{ plane,} \tag{7.120a}$$

$$\theta_{\text{null-to-null}} = \sin^{-1}\left(\frac{4}{2M+1}\right) \quad \text{in the } \phi' = \pi/2 \text{ plane.} \tag{7.120b}$$

For large arrays ($N, M \gg 1$), these expressions can be simplified using the approximation $\sin\theta \approx \theta$. Recognizing that N and M are the maximum number of half wavelengths between pairs of antenna elements, and that the null-to-null beamwidth is approximately twice the half-power beamwidth of an antenna pattern, the expressions for angular resolution become

$$\theta_{\text{half-power}} \approx \left(\frac{\lambda}{D_u}\right) \quad \text{in the } \phi' = 0 \text{ plane,} \tag{7.121a}$$

$$\theta_{\text{half-power}} \approx \left(\frac{\lambda}{D_v}\right) \quad \text{in the } \phi' = \pi/2 \text{ plane,} \tag{7.121b}$$

where D_u and D_v are the maximum spacings between antenna elements along the u and v axes of the array (see Fig. 7-36).

Image reconstruction by Fourier inversion is an idealized approach that is useful to illustrate the fundamental relationships between brightness temperature and visibility, and between the MIR instrument architecture and its angular resolution. However, in practice an empirical pseudo-inverse approach is more often used instead. This allows the image reconstruction process to better account for such nonideal characteristics of the measurements as differences in the radiation patterns of the individual antenna array elements and variations in the relative phase and amplitude transfer functions of the different pairs of receivers. A pseudo-inverse is constructed directly from measurements of the MIR antenna interference patterns. The interference patterns are visibility measurements made as a point source is moved across the field of view of the instrument. Examples of interference patterns for the instrument portrayed in Figs. 7-37 and 7-38 are shown in Fig. 7-39 for three antenna spacings: (1) the "zeroth spacing," which is essentially the power radiation pattern of a single antenna array element; (2) the half-wavelength

(a) MIR instrument

(b) Instrument mounted on NASA aircraft

Figure 7-38: Photographs of the instrument described schematically in Fig. 7-37: (a) in its handling fixture and (b) mounted on the NASA DC-8 aircraft.

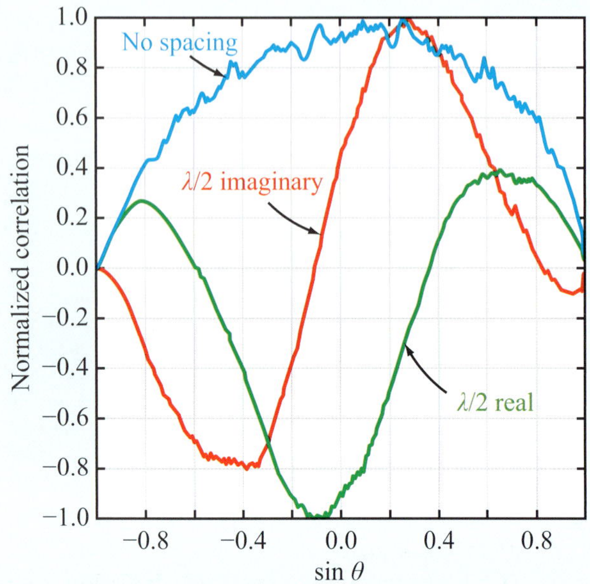

(a) No spacing and λ/2 spacing

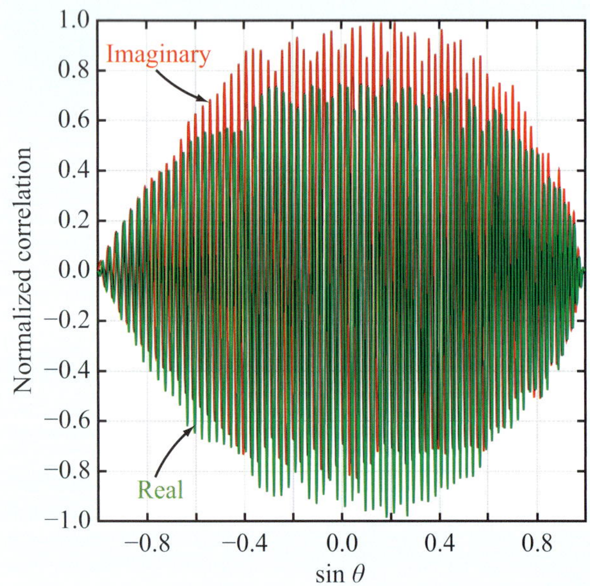

(b) 34-λ spacing

Figure 7-39: Complex interference patterns measured for the "zeroth" and half-wavelength spacings of the one-dimensional synthesis MIR sensor shown in Fig. 7-38: (a) The zeroth (real part only) and half-wavelength spacing, and (b) the 34-wavelength spacing.

spacing, with both real and imaginary components included to illustrate the sine and cosine components of the exponential term in Eq. (7.115); and (3) the 34-wavelength spacing, which is the largest spacing between antenna elements—hence the highest spatial frequency interference pattern, which determines the angular resolution of the overall imager.

The image reconstruction algorithm is constructed by assembling all of the interference patterns into an impulse response matrix **G** defined by

$$G_{ij} = V_{i=nm}(\text{point source at location } \theta_j), \qquad (7.122)$$

where i indexes all possible pairs of antenna elements, θ_j is the angular location of the point source in the instrument field of view, and V_{nm} is the measured visibility at that point. Each row of **G** corresponds to one of the antenna pairs and each column corresponds to one position in the field of view. Measurement of the "G-matrix" is typically done under controlled conditions in an anechoic chamber, and the effects of thermal emission by the walls of the chamber are removed by taking the difference between visibility measurements made with the point noise source on and off. Once the G-matrix is known, the vector of visibility measurements that would be made when viewing a general brightness temperature distribution, $T_B(\theta)$, can be described by

$$\mathbf{V} = \mathbf{G}\mathbf{T}_B. \qquad (7.123)$$

Here, $\mathbf{T}_B$ is the vector representation of the angular brightness temperature distribution. Its elements are aligned to the coordinates θ_j used to index the columns of the G-matrix. The image reconstruction algorithm follows as the pseudo-inverse of Eq. (7.123), or

$$\widehat{\mathbf{T}_B} = \mathbf{G}^\dagger(\mathbf{G}\mathbf{G}^\dagger)^{-1}\mathbf{V}, \qquad (7.124)$$

where $\widehat{\mathbf{T}_B}$ is the estimate of the true brightness temperature distribution. The synthesized antenna pattern and effective resolution are found by combining Eqs. (7.123) and (7.124), which leads to

$$\widehat{\mathbf{T}_B} = \mathbf{G}^\dagger(\mathbf{G}\mathbf{G}^\dagger)^{-1}\mathbf{G}\mathbf{T}_B = \mathbf{A}\mathbf{T}_B, \qquad (7.125)$$

where the matrix $\mathbf{A} = \mathbf{G}^\dagger(\mathbf{G}\mathbf{G}^\dagger)^{-1}\mathbf{G}$ represents the synthesized antenna pattern in matrix form. Its jth row corresponds to the antenna pattern steered to the angle θ_j in the image.

The above example describes a one-dimensional MIR instrument and its image reconstruction. An example of a two-dimensional instrument is MIRAS, shown earlier in Fig. 7-35. MIRAS operates at 1.4 GHz with 69 half-wave dipole antenna elements deployed along three 4.5 m arms. Its synthesized aperture is 9 m in diameter and it is steerable in software across the field-of-view defined by the individual dipole patterns. The MIRAS image reconstruction algorithm is a direct extension of the one-dimensional approach. Rows of the G-matrix correspond to all possible pairs of the 69 antenna elements, and columns correspond to all possible angular locations in the two-dimensional field of view of the image.

7-13.2 MIR Radiometric Sensitivity

The radiometric sensitivity ΔT of a total-power radiometer was described in Section 7-4 as the standard deviation of measurements made by an ideal, gain-stable, radiometer. The ΔT is a function of system noise temperature, bandwidth, and integration time. A similar relationship exists for the standard deviation of the individual visibility measurements from which a MIR image of brightness temperature is constructed. The ΔV expressions are given by (Ruf et al., 1988)

$$\Delta V_{\mathrm{r}} = \sqrt{\frac{T_{\mathrm{SYS}}^2 + V_{\mathrm{r}}^2 - V_{\mathrm{i}}^2}{2B\tau}}, \qquad (7.126\mathrm{a})$$

$$\Delta V_{\mathrm{i}} = \sqrt{\frac{T_{\mathrm{SYS}}^2 + V_{\mathrm{i}}^2 - V_{\mathrm{r}}^2}{2B\tau}}, \qquad (7.126\mathrm{b})$$

where $V = V_{\mathrm{r}} + jV_{\mathrm{i}}$ is the complex visibility function and $T_{\mathrm{SYS}} = T_{\mathrm{A}} + T'_{\mathrm{REC}}$, where T_{A} is the average brightness temperature over the entire field of view of a single MIR antenna array element. The reduction in measurement noise by an additional factor of $\sqrt{2}$ in Eq. (7.126), relative to the ΔT of a total power radiometer, results from the fact that visibility measurements are made with a correlating radiometer, for which the system noise temperature is largely uncorrelated between channels.

The standard deviation of individual pixels in the MIR brightness temperature image is found by a propagation-of-errors analysis of the image reconstruction algorithm, given by either Eq. (7.117) or, if image reconstruction is performed using the G-matrix method, Eq. (7.124). As a simplifying case, consider the central pixel in the image. The central pixel is located at $\theta = \phi = 0$, for which the reconstruction algorithm given by Eq. (7.117) collapses to

$$\widehat{T'_{\mathrm{B}}}(\theta = 0,\ \phi = 0) = \sum_{n=-N}^{N} \sum_{m=-M}^{M} V(u_n, v_m). \qquad (7.127)$$

The standard deviation of each visibility measurement, $V(u_n, v_m)$, is given by Eq. (7.126). In most cases, the higher spatial-frequency visibility components are much smaller than the system noise temperature and can be ignored. In that case, the standard deviation of the reconstructed image becomes

$$\Delta T_{\mathrm{MIR}} = \sqrt{\frac{T_{\mathrm{SYS}}^2 (2N+1)(2M+1)}{2B\tau}}. \qquad (7.128)$$

In large imaging systems, for which $N, M \gg 1$, the sensitivity of the reconstructed image is significantly poorer (i.e., there is a much larger standard deviation) than that of the individual visibility measurements. This decrease in sensivitity, relative to that of a conventional real aperture radiometer, can be largely recovered in practice, provided all visibility measurements are made simultaneously. If a real aperture radiometer were to generate an image of the brightness temperature distribution in the same time and with the same angular resolution, by scanning a single narrow antenna beam across the field of view of the MIR instrument, its available dwell time on each pixel would be τ divided by the total number of pixels. Since the number of independent pixels is roughly comparable to the total number of visibility measurements, $(2N+1)(2M+1)$, the ΔT associated with each pixel in the real aperture image is similar to that of the MIR instrument.

7-14 Polarimetric Radiometer

The p-polarized brightness temperature was defined previously in Section 6-9 in terms of the associated electric field E_p as

$$T_{\mathrm{B}}^{p} = K\left\langle |E_p|^2 \right\rangle, \tag{7.129}$$

where K is the scale factor $\lambda_0^2/k\eta_0 B$. In practice, the "norm squared" operator, $|\ |^2$ in Eq. (7.129), is performed by a square-law detector diode followed by a low-pass filter, and the expectation operator $\langle\ \rangle$ is approximated by a video integrator circuit. In more modern digital radiometers, these operations are performed by digital signal processing modules that square and integrate the signal following its digitization.

7-14.1 Coherent Detection

As noted in Section 6-9, the third and fourth Stokes parameters of the brightness temperature can be measured using either a coherent- or incoherent-detection radiometer. Coherent detection is performed by complex cross correlation of the v- and h-polarized components of the associated electric field:

$$T_{\mathrm{B}}^{3} + jT_{\mathrm{B}}^{4} = 2K\left\langle E_{\mathrm{v}}E_{\mathrm{h}}^{*} \right\rangle. \tag{7.130}$$

The coherent approach has been used successfully by a number of polarimetric radiometers (e.g., Gasiewski and Piepmeier, 1996; Ruf et al., 2006a). An example is shown in Fig. 7-40. Coherent detection has several attractive features. If the complex correlation is performed digitally, then that portion of the instrument will have nonlinearity, gain ripple, and amplitude and phase imbalances that are highly repeatable over time and are, therefore, easier to calibrate and remove. In addition, only one complex-correlation channel is required, as opposed to the four-channel system required in the incoherent detection case (discussed below). This can significantly simplify the antenna and receiver architectures, resulting in lower size, mass, power, complexity and cost.

7-14.2 Incoherent Detection

Incoherent detection relies on the following relationships between the various polarimetric components of the brightness temperature and the associated electric field components:

$$T_{\mathrm{B}}^{\mathrm{P}} = \frac{T_{\mathrm{B}}^{\mathrm{v}} + T_{\mathrm{B}}^{\mathrm{h}}}{2} + K\,\Re\mathfrak{e}\left\{\left\langle E_{\mathrm{v}}E_{\mathrm{h}}^{*}\right\rangle\right\}, \tag{7.131a}$$

$$T_{\mathrm{B}}^{\mathrm{M}} = \frac{T_{\mathrm{B}}^{\mathrm{v}} + T_{\mathrm{B}}^{\mathrm{h}}}{2} - K\,\Re\mathfrak{e}\left\{\left\langle E_{\mathrm{v}}E_{\mathrm{h}}^{*}\right\rangle\right\}, \tag{7.131b}$$

$$T_{\mathrm{B}}^{\mathrm{L}} = \frac{T_{\mathrm{B}}^{\mathrm{v}} + T_{\mathrm{B}}^{\mathrm{h}}}{2} + K\,\Im\mathfrak{m}\left\{\left\langle E_{\mathrm{v}}E_{\mathrm{h}}^{*}\right\rangle\right\}, \tag{7.131c}$$

$$T_{\mathrm{B}}^{\mathrm{R}} = \frac{T_{\mathrm{B}}^{\mathrm{v}} + T_{\mathrm{B}}^{\mathrm{h}}}{2} - K\,\Im\mathfrak{m}\left\{\left\langle E_{\mathrm{v}}E_{\mathrm{h}}^{*}\right\rangle\right\}, \tag{7.131d}$$

where $T_{\mathrm{B}}^{\mathrm{P}}$ ($T_{\mathrm{B}}^{\mathrm{M}}$) is the brightness temperature at a $+45^\circ$ (-45°) slant linear polarization and $T_{\mathrm{B}}^{\mathrm{L}}$ ($T_{\mathrm{B}}^{\mathrm{R}}$) is the brightness temperature at a left-hand (right-hand) circular polarization. Combining Eqs. (7.130) and (7.131), it can be shown that

$$T_{\mathrm{B}}^{3} = T_{\mathrm{B}}^{\mathrm{P}} - T_{\mathrm{B}}^{\mathrm{M}} \tag{7.132a}$$

and

$$T_{\mathrm{B}}^{4} = T_{\mathrm{B}}^{\mathrm{L}} - T_{\mathrm{B}}^{\mathrm{R}}. \tag{7.132b}$$

Brightness temperatures $T_{\mathrm{B}}^{\mathrm{P}}$, $T_{\mathrm{B}}^{\mathrm{M}}$, $T_{\mathrm{B}}^{\mathrm{L}}$, and $T_{\mathrm{B}}^{\mathrm{R}}$ can all be measured either directly or indirectly. In the direct approach, a conventional total-power radiometer is used in combination with antennas of the appropriate polarizations. Such an instrument design architecture typically requires a relatively large antenna feed array plus one radiometer receiver for each of the six individual polarization states to be measured.

The total-power approach was successfully used by WindSat on board the Coriolis satellite platform (Gaiser et al., 2004). WindSat measures the full polarimetric Stokes brightness temperature vector at 10.7, 18.7 and 37.0 GHz and vertical and horizontal polarizations only at 6.8 and 23.8 GHz. For the fully polarimetric channels, one antenna feed orthomode transducer (OMT) is used to form vertical and horizontal linear polarizations, a second OMT forms $+45^\circ$ and -45° slant linear polarizations, and a third OMT

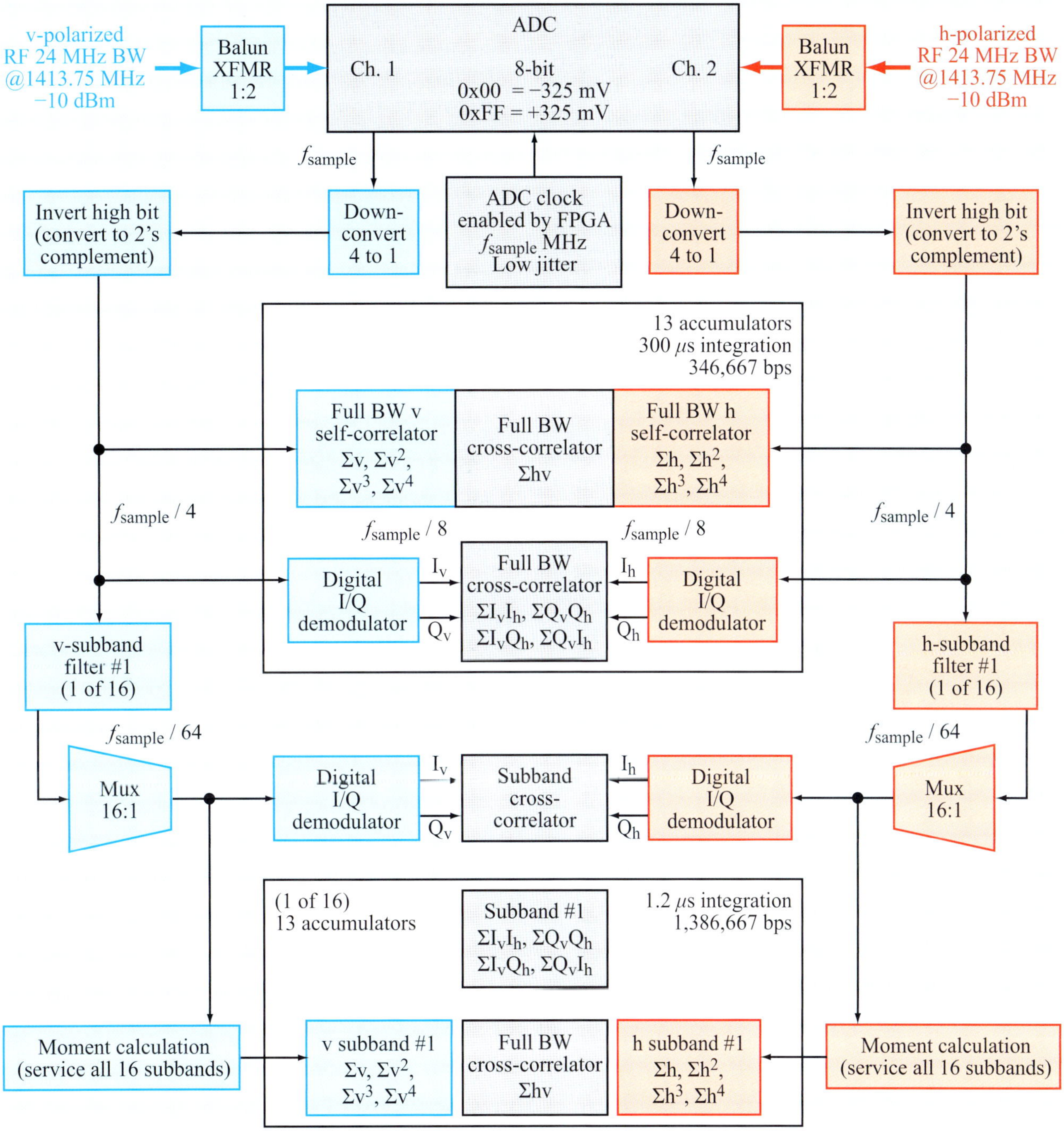

Figure 7-40: Fully polarimetric radiometer architecture using coherent detection and digital correlation.

forms left-hand and right-hand circular polarizations. Generation of the third and fourth Stokes brightness temperatures in this manner has several advantages. It permits the use of standard, square-law, detector-based radiometer receivers that are well understood in terms of component and subsystem design and system calibration. The six individual polarization brightness temperatures are calibrated before and after Earth observations by measurements pointed at an ambient blackbody load and a cold (~ 2.7 K) space reflector. This calibration method is useful because it reduces biases in the calibration of T_B^3 and T_B^4. Calibration biases that are common to T_B^P and T_B^M or to T_B^L and T_B^R cancel out in Eq. (7.132). The single largest contributor to T_B biases for many radiometers is inaccurate accounting for antenna sidelobe contributions. Mechanical and electrical symmetries in the WindSat antenna design between both pairs of channels that are differenced in Eq. (7.132) give rise to a common mode bias that is removed by the differencing operation. The residual biases that remain in T_B^3 and T_B^4 should generally be smaller than those in the original six polarization channels. However, even errors as small as a few tenths of kelvins can be problematic for wind-direction retrieval algorithms (Chapter 18).

In the indirect measurement approach, an incoherent detection radiometer architecture composed of standard v- and h-pol radiometer RF front ends is used in conjunction with a polarization-combining hybrid network to measure the required polarization components of brightness temperature, T_B^P, T_B^M, T_B^L, and T_B^R. The estimates of T_B^3 and T_B^4 then follow as in the total power case. This indirect (hybrid) approach has been successfully used by airborne and ground-based polarimetric radiometers (Yueh et al., 1995; Sollner and Suss, 1996). An example of the architecture is shown in Fig. 7-41. The principal advantage of the hybrid approach is the simpler antenna design, which requires only a single dual-linear polarization pair of channels.

7-15 Calibration of Polarimetric Radiometers

Traditional noncorrelating radiometers measure the total microwave noise power present in the receiver by integrating the square of the signal originating from a single antenna at a single polarization. Often referred to as total-power radiometers, they are traditionally tested and calibrated using targets that provide two or more known brightness temperatures. The most common total-power calibration targets are blackbody radiators whose brightness temperatures are equal to their physical temperatures (Section 7-11.5). The hardware design requirements for such a calibration standard generally reduce to that of providing a known and stable thermal environment for the blackbody material. The availability of brightness temperature calibration standards has greatly facilitated the development, testing and calibration of total power radiometers

Fully polarimetric radiometers would similarly benefit from a suitable calibration standard. In fact, the need for accurate calibration is in this case arguably more pressing because of the added complexity of the hardware and the much-smaller partially correlated signal levels. The correlated brightness between E_v and E_h in a fully polarimetric radiometer is typically about two orders of magnitude smaller than either T_B^v or T_B^h. A suitable fully polarimetric standard needs to generate E_v and E_h signals that are themselves stable (to calibrate T_B^v and T_B^h) as well as provide a known and stable level of complex partial correlation between them (to calibrate T_B^3 and T_B^4). No device as simple as a blackbody load can provide such signals. Two approaches have most often been used to generate fully polarimetric calibration signals. The first is a modification to the traditional blackbody load, made capable of producing important subsets of the range of partial correlations possible between two microwave noise signals (Gasiewski and Kunkee, 1993; Lahtinen et al., 2003). This polarized load consists of two traditional blackbody absorbers maintained at different physical temperatures. The microwave noise power emitted by them is combined quasi-optically with a polarizing grid. The relative alignment between the

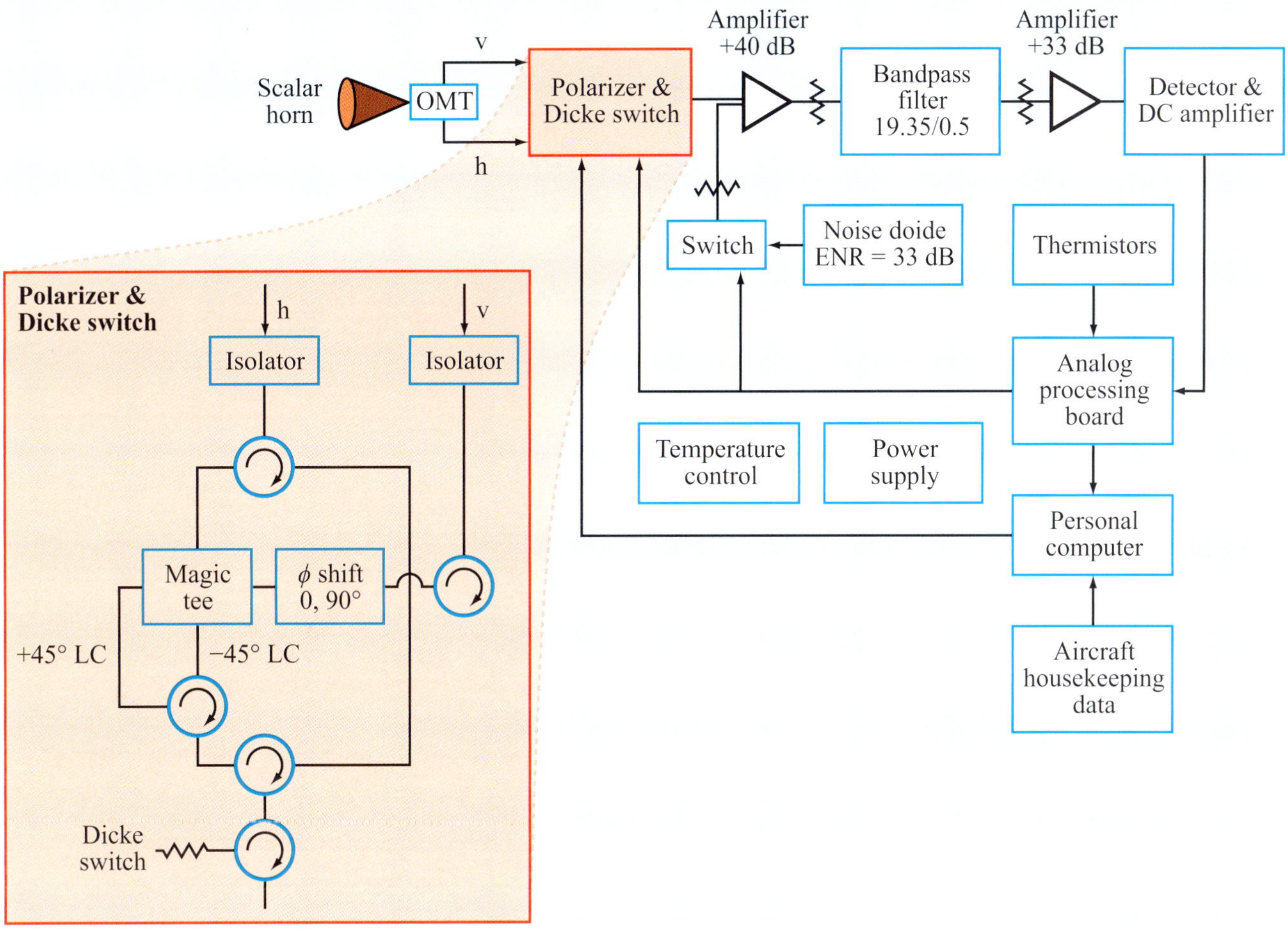

Figure 7-41: Fully polarimetric radiometer architecture using incoherent detection and polarization forming hybrids [after Yueh et al., 1995].

polarizing grid and the polarization reference axes of a radiometer's antenna determine the apparent polarization of the emission by the load. Variations in the relative alignment provide a set of calibration references against which to verify sensor performance in a manner roughly analogous to total-power radiometer calibration. A schematic example of one such polarized load is shown in Fig. 7-42.

The quasi-optical polarized load has several limitations. It was developed specifically for use with polarimetric radiometers and cannot be easily adapted to more general use by other types of correlating radiometers. Two practical issues with the design of a polarized load are also noteworthy. Very precise mechanical alignment is required to guarantee known and stable polarimetric brightness temperatures. Thermal design and control can also be challenging. Two relatively large blackbody loads are required to be in close proximity to one another while maintained at significantly different physical temperatures that are well known and stable.

An alternative polarimetric calibration approach relies on the generation of a pair of partially correlated noise sources using a dual channel ***arbitrary waveform***

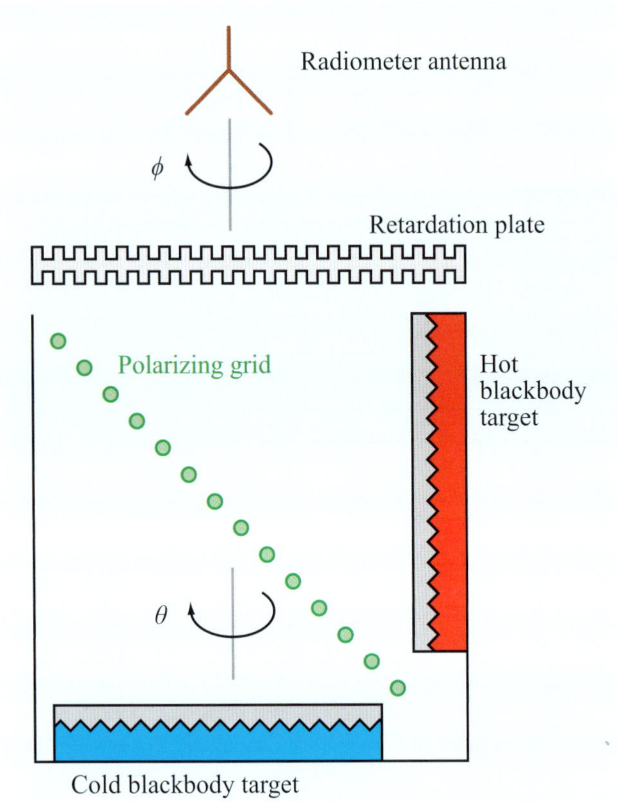

Figure 7-42: Quasi-optically combined polarimetric blackbody calibration target [after Lahtinen et al., 2003].

generator (AWG), combined with active cold load upconverters. This approach has been used for the calibration of both polarimetric and interferometric radiometers (Ruf and Li, 2003; Peng and Ruf, 2008). The AWG contains a pair of high-resolution ***digital-to-analog converters*** (DACs) that are synchronously triggered by a common clock. The digital inputs to the DACs are stored as a pair of ***lookup tables*** (LUTs) in memory. For purposes of radiometric calibration, it is common to pre-load the LUTs with realizations of noise-like Gaussian distributed signals with the desired noise power spectrum and partial correlation statistics. It is common to cycle the pointer to the LUT back to the beginning after it reaches the end. In this manner, the output signal resembles a cyclostationary random process, with periodicity determined by the length of the LUT (i.e., the available memory in the AWG) and clock rate. The clock rate also determines the maximum possible bandwidth of the output signal. Use of an AWG provides a significant amount of freedom in choosing the output analog calibration waveforms. For example, it is possible to control the noise bandwidth and variance of the noise, its spectral shape (i.e., its "color"), and the quadrature components of partial correlation (i.e., the complex correlation) between the pair of noise signals. With this flexibility, pairs of signals can be generated that individually resemble pre-detected v- and h-pol voltages corresponding to arbitrary values of T_B^v and T_B^h and with arbitrary complex correlations (i.e., arbitrary values for T_B^3 and T_B^4). The functional block diagram for such a calibration system is shown in Fig. 7-43. Whereas this type of polarimetric calibration source is well suited to calibrating and characterizing the electronics stages of a radiometer, it is important to note that the calibration process does not incorporate the antenna portion of the radiometer. This type of source is typically coupled into the radiometer in place of the antenna.

7-15.1 Forward Model for a Fully Polarimetric Radiometer

The instrument forward model for a fully polarimetric radiometer describes the relationship between its raw data and the input Stokes parameters of the brightness temperature. The model accounts for the net cascaded gain of individual polarization channels, the cross-coupling between the channels, and the measurement offsets that are not dependent on the input Stokes parameters (e.g., due to internally generated receiver noise and detector biases). Calibration of the radiometer consists of inverting its forward model. Details of the forward model will vary depending on the particular radiometer design. Examples appropriate for several coherent and incoherent detection radiometer architectures are considered here.

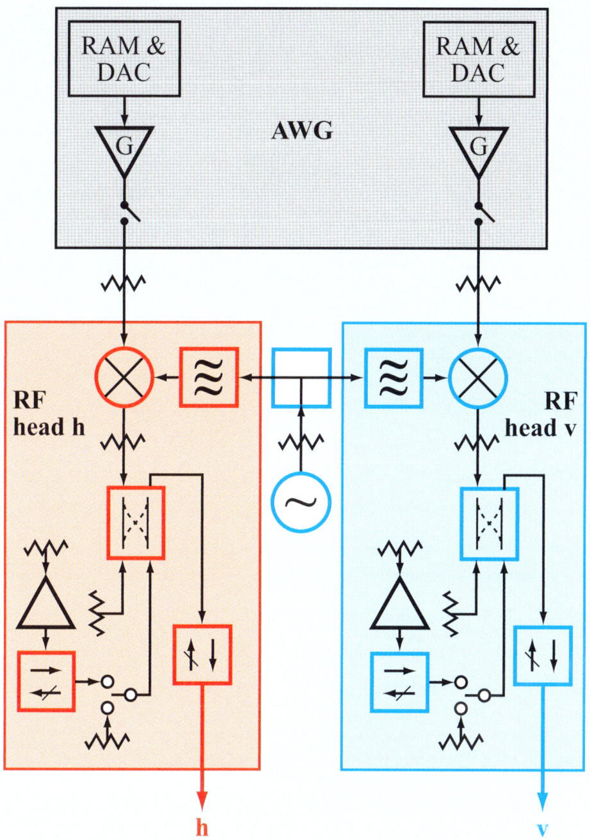

Figure 7-43: Digitally programmable fully polarimetric calibration source [after Peng and Ruf, 2008].

Incoherent detection

For a fully polarimetric radiometer using incoherent detection, the forward model is given by

$$\begin{pmatrix} C_\mathrm{v} \\ C_\mathrm{h} \\ C_\mathrm{P} \\ C_\mathrm{M} \\ C_\mathrm{L} \\ C_\mathrm{R} \end{pmatrix} = \begin{pmatrix} G_\mathrm{vv} & G_\mathrm{vh} & G_\mathrm{v3} & G_\mathrm{v4} \\ G_\mathrm{hv} & G_\mathrm{hh} & G_\mathrm{h3} & G_\mathrm{h4} \\ G_\mathrm{Pv} & G_\mathrm{Ph} & G_\mathrm{P3} & G_\mathrm{P4} \\ G_\mathrm{Mv} & G_\mathrm{Mh} & G_\mathrm{M3} & G_\mathrm{M4} \\ G_\mathrm{Lv} & G_\mathrm{Lh} & G_\mathrm{L3} & G_\mathrm{L4} \\ G_\mathrm{Rv} & G_\mathrm{Rh} & G_\mathrm{R3} & G_\mathrm{R4} \end{pmatrix} \cdot \begin{pmatrix} T_\mathrm{B}^\mathrm{v} \\ T_\mathrm{B}^\mathrm{h} \\ T_\mathrm{B}^3 \\ T_\mathrm{B}^4 \end{pmatrix} + \begin{pmatrix} O_\mathrm{v} \\ O_\mathrm{h} \\ O_\mathrm{P} \\ O_\mathrm{M} \\ O_\mathrm{L} \\ O_\mathrm{R} \end{pmatrix}, \tag{7.133}$$

where C_x ($x = \mathrm{v, h, P, M, L}$ and R) are the raw radiometer measurement "counts" for each of the six polarimetric channels, T_B^y ($y = \mathrm{v, h}$, 3 and 4) are the input Stokes parameters, G_{xy} are the gain matrix elements that relate input signals to output measurements, and O_x are offsets in the measurements that are independent of the input signals.

The 30 unknown parameters that need to be determined in this forward model are G_{xy} and O_x. Some partially polarimetric radiometers are designed to measure only T_B^v, T_B^h and T_B^3. They omit left- and right-hand circular polarization channels. Consequently, their appropriate forward model is a simplified version of Eq. (7.133) that omits C_x, O_x, and G_{xy} for $x = \mathrm{L}$ and R. In this case, the number of unknown parameters to be determined is reduced to 20. Note that the input signal T_B^4 should not be omitted from the reduced forward model because it might still influence the measurements if there is sufficient polarization leakage present.

Coherent detection

The forward model for a fully polarimetric radiometer using coherent detection is given by

$$\begin{pmatrix} C_\mathrm{v} \\ C_\mathrm{h} \\ C_3 \\ C_4 \end{pmatrix} = \begin{pmatrix} G_\mathrm{vv} & G_\mathrm{vh} & G_\mathrm{v3} & G_\mathrm{v4} \\ G_\mathrm{hv} & G_\mathrm{hh} & G_\mathrm{h3} & G_\mathrm{h4} \\ G_\mathrm{3v} & G_\mathrm{3h} & G_{33} & G_{34} \\ G_\mathrm{4v} & G_\mathrm{4h} & G_{43} & G_{44} \end{pmatrix} \cdot \begin{pmatrix} T_\mathrm{B}^\mathrm{v} \\ T_\mathrm{B}^\mathrm{h} \\ T_\mathrm{B}^3 \\ T_\mathrm{B}^4 \end{pmatrix} + \begin{pmatrix} O_\mathrm{v} \\ O_\mathrm{h} \\ O_3 \\ O_4 \end{pmatrix}. \tag{7.134}$$

In this case, there are 20 unknown parameters (G_{xy} and O_x) to be determined. If the radiometer uses coherent detection without a fourth Stokes channel (i.e., measures only the in-phase component of the correlation between v- and h-pol signals), then the forward model becomes a simplified version of Eq. (7.134) in which C_4, G_{4y} and O_4 are omitted. In this case, there are 15 unknown parameters to be determined.

7-15.2 Forward Model for the Polarimetric Calibration Source

A forward model for the fully polarimetric calibration source shown in Fig. 7-43 should account for relevant properties of its active components and passive interconnects. The most critical of these include the DACs and amplifiers inside the AWG, the upconversion mixer installed in the RF head, and the interconnecting transmission lines. The strength of the signal generated by the DAC and the gain of the subsequent stages affect the output signal level of the calibration source. In addition, the correlation between v- and h-pol calibration signals is affected by the phase imbalance between the two channels of the calibration source. These nonideal properties can lead to calibration errors if not properly accounted for. In order to do so, a parameterized forward model for the calibration source is constructed and parameters of the model are solved for simultaneously with those of the radiometer under test. Doing so requires that the proper variety of calibration signals be used.

The strength of the signal generated by the AWG is controlled in two ways. The relative signal level is determined by the values loaded into its LUT (the LUT defines only a normalized version of the signal). The absolute signal strength is determined by an AWG voltage gain setting, G. Signals generated by the calibration source contain one component generated by the AWG and a second component originating from either an ambient matched load or an active cold load. The brightness temperature of the AWG signal, referenced to the output of the RF head, can be expressed as

$$T_{\mathrm{AWG},\,p} = k_p(G_p^2 T_{\mathrm{N}} + O_{\mathrm{AWG},\,p}), \tag{7.135}$$

where k_p is an overall scale factor, G_p is the voltage gain of the AWG channel, T_{N} is the nominal brightness temperature generated by the AWG, and $O_{\mathrm{AWG},\,p}$ is a possible offset brightness temperature for the p-polarized channel of the calibration source.

The total brightness temperature is the sum of the AWG component plus the component from either the ambient load or the active cold load. The complete forward model for the calibration source is given by

$$\begin{aligned}
T_{\mathrm{B}}^{\mathrm{v}} &= k_{\mathrm{v}}(G_{\mathrm{v}}^2 T_{\mathrm{N}} + O_{\mathrm{AWG},\,\mathrm{v}}) + T_{Y,\mathrm{v}},\\
T_{\mathrm{B}}^{\mathrm{h}} &= k_{\mathrm{h}}(G_{\mathrm{h}}^2 T_{\mathrm{N}} + O_{\mathrm{AWG},\,\mathrm{h}}) + T_{Y,\mathrm{h}},\\
T_{\mathrm{B}}^{3} &= 2\cdot\sqrt{k_{\mathrm{v}}(G_{\mathrm{v}}^2 T_{\mathrm{N}} + O_{\mathrm{AWG},\,\mathrm{v}})\cdot k_{\mathrm{h}}(G_{\mathrm{h}}^2 T_{\mathrm{N}} + O_{\mathrm{AWG},\,\mathrm{h}})}\\
&\quad\cdot|\rho|\cos(\theta+\Delta),\\
T_{\mathrm{B}}^{4} &= 2\cdot\sqrt{k_{\mathrm{v}}(G_{\mathrm{v}}^2 T_{\mathrm{N}} + O_{\mathrm{AWG},\,\mathrm{v}})\cdot k_{\mathrm{h}}(G_{\mathrm{h}}^2 T_{\mathrm{N}} + O_{\mathrm{AWG},\,\mathrm{h}})}\\
&\quad\cdot|\rho|\sin(\theta+\Delta),
\end{aligned} \tag{7.136}$$

where $T_{Y,p}$ is the brightness temperature of the $Y =$ ***ambient*** or ***cold*** background reference load for the p-polarized channel, $|\rho|$ and θ are the magnitude and phase of the correlation coefficient between the two channels, and Δ is the electrical path length imbalance between the two channels (resulting in a phase offset in the correlation, where a positive value for Δ means that the electrical path length of the h-pol channel is larger than that of the v-pol channel). Parameters of the forward model that can be controlled and varied by the instrument designer include the AWG gain settings $(G_{\mathrm{v}}, G_{\mathrm{h}})$ and the complex correlation coefficient $(|\rho|, \theta)$. Parameters of the forward model that are not controllable and need to be determined as part of the calibration procedure include the gain imbalances $(k_{\mathrm{v}}, k_{\mathrm{h}})$, the AWG channel offsets $(O_{\mathrm{AWG},\,\mathrm{v}}, O_{\mathrm{AWG},\,\mathrm{h}})$, and the interchannel phase imbalance Δ.

7-15.3 Calibration by Inversion of the Forward Models

The forward models for both the calibration source and radiometer under test can be combined together into a composite forward model that relates the brightness temperatures that are programmed into the calibration source to the output digital counts that are measured by the radiometer under test. In order to simultaneously retrieve the gain matrix and offsets of the radiometer, together with the forward model parameters of the calibration source, a suitable set of test signals is required. The test signals are generated by varying calibration source settings ($|\rho|$, θ, G_{v} and G_{h} in Eq. (7.136)) and hardware switch positions (AWG signal either on or off, background T_{Y} source either ambient or cold load).

An example of a combination of settings that is capable of uniquely solving for all unknown model parameters except Δ is given in Table 7-4. The test vectors t_1, t_4, t_7, t_{10}, and t_{13} are signals from the AWG and are added onto the ColdFET. Those test vectors are linearly independent of one another and

Table 7-4: Fully polarimetric calibration source test set.

Test vector	ρ	θ	AWG G_v	AWG G_h	Background signal	T_Y
t_1	0	0	0.17	0.17	on	cold
t_2	0	0	0.17	0.17	off	cold
t_3	0	0	0.17	0.17	off	ambient
t_4	0	0	0.25	0.17	on	cold
t_5	0	0	0.25	0.17	off	cold
t_6	0	0	0.25	0.17	off	ambient
t_7	0	0	0.17	0.25	on	cold
t_8	0	0	0.17	0.25	off	cold
t_9	0	0	0.17	0.25	off	ambient
t_{10}	1	0	0.25	0.25	on	cold
t_{11}	1	0	0.25	0.25	off	cold
t_{12}	1	0	0.25	0.25	off	ambient
t_{13}	1	45°	0.25	0.25	on	cold
t_{14}	1	45°	0.25	0.25	off	cold
t_{15}	1	45°	0.25	0.25	off	ambient

are varied by selecting different values for the relative AWG gains and the correlation coefficient magnitude and phase. The brightness temperature entering the radiometer under test is a function of the strength of each channel and the correlation coefficient magnitude and phase. Because the calibration source is not ideal and the absolute strength from the AWG is unknown, two additional test vectors with known but different brightness temperatures are required to calibrate the calibration source signal strength. The two additional test vectors are t_2 and t_3 in Table 7-4. There is some redundancy of information provided by this set of test vectors. The precision and robustness of the retrieved forward model parameters are both improved by overconstraining the system of equations from which they are derived.

All elements of the radiometer gain matrix and offset vector, and all parameters of the calibration source forward model except for its interchannel phase imbalance, Δ, can be estimated from the measurements using a standard over-constrained, nonlinear, iterative minimization technique. The phase imbalance can be estimated by manually cross-swapping the interconnecting cables between the calibration source and the radiometer under test. This is the one step in the calibration procedure that is not fully automated. Fortunately, in practice the channel phase imbalance tends to be extremely stable over time and so is not in need of frequent recalibration.

The complete set of unknown parameters in the composite forward model for both the calibration source and the radiometer under test can be combined together into a single state vector $\mathbf{X}$ given by

$$\mathbf{X} = \{k_v,\ k_h,\ O_{\text{AWG, v}},\ O_{\text{AWG, h}},\ \Delta,\ G_{mn},\ \text{and } O_m\}, \tag{7.137}$$

where $m =$ v, h, P, M, L, R for a fully polarimetric incoherent detection radiometer, while $m =$ v, h, 3, and 4 for a fully polarimetric coherent detection radiometer, and $n =$ v, h, 3, and 4 in both cases. The individual elements of $\mathbf{X}$ are defined in Eq. (7.133), Eq. (7.134), and Eq. (7.136). All of the radiometer measurements made during the calibration test can be similarly combined into a measurement vector $\mathbf{C}$ given by

$$\mathbf{C} = f(\mathbf{X}), \tag{7.138}$$

where each element of $\mathbf{C}$ is the measurement count for one of the six polarimetric radiometer channels at one of the fifteen calibration source settings listed in Table 7-4, and where f represents the complete forward model.

The estimation of $\mathbf{X}$ given $\mathbf{C}$ is performed by inverting equation Eq. (7.138). Specifically, a first guess at $\mathbf{X}$ is made, denoted by $\hat{\mathbf{X}}$. The Jacobian matrix $\mathbf{J}$ of partial derivatives of f with respect to elements of $\mathbf{X}$ evaluated at $\hat{\mathbf{X}}$ is next computed numerically. An updated estimate of $\hat{\mathbf{X}}$ is then made according to

$$\hat{\mathbf{X}}1 = \hat{\mathbf{X}} + (\mathbf{J}^{\mathrm{T}} \cdot \mathbf{J})^{-1} \cdot \mathbf{J}^{\mathrm{T}} \cdot [\mathbf{C} - f(\hat{\mathbf{X}})], \tag{7.139}$$

where T denotes matrix transpose. This process is iterated until changes in $\hat{\mathbf{X}}$ become acceptably small and $\hat{\mathbf{X}}$ has converged to the value that minimizes the norm-squared difference between $\mathbf{C}$ and $f(\hat{\mathbf{X}})$.

The calibration source interchannel phase imbalance Δ is determined by manually cross-swapping the interconnecting cables between the source and the radiometer under test. When cross-swapped, the v-pol output port of the calibration source is connected to the h-pol input port of the radiometer, and the h-pol output port of the calibration source is connected to the v-pol input port of the radiometer. Cross-swapping changes the sign, but not the magnitude, of Δ in the forward model. In both the standard and cross-swapped positions, the radiometer parameters that are retrieved should be the same since the radiometer itself has not changed. Some of the radiometer parameters (e.g., G_{vv} and G_{hh}) are not sensitive to either the radiometer or the source phase imbalance. Some of the other parameters (e.g., G_{P3} and G_{P4}, or G_{M3} and G_{M4} in Eq. (7.133), or G_{33} and G_{34} in Eq. (7.134)) are sensitive to the source imbalance and so will vary with Δ. Using both the standard and cross-swapped measurements, the assumed value of Δ is varied incrementally over all possible values and, for each value, a new estimate of G_{xy} ($x =$ P, M, L, R, 3, or 4; $y = 3$ or 4) is made using this iterative method. In this way, a value of Δ is found for which the retrieved estimate of G_{xy} ($x =$ P, M, L, R, 3, or 4; $y = 3$ or 4) is the same in both the normal and cross-swapped cable positions. Two possible solutions are obtained in this manner and the proper one is selected based on a first guess at the solution using component-level characteristics of the calibration source.

7-16 Digital Radiometers

Digital radiometers replace as much as possible of the conventional analog circuitry with an Analog-to-Digital Converter (ADC) followed by a high-speed ***digital signal processing*** (DSP) stage.

> ▶ The conversion to digital technology improves the long term stability of the receiver and offers independence from temperature variations, thereby improving and simplifying the radiometer calibration. ◀

Digital technology can also add capabilities to a radiometer that would be much more difficult (and often cost prohibitive) to include otherwise. For example, the signal bandwidth can be digitally subdivided into many well-isolated subbands. This enables fine-resolution spectral analysis of the signal. It can also simplify the removal of radio frequency interference (RFI), which tends to have a highly localized spectrum. The DSP stage can also be programmed to detect other moments of the signal amplitude, in addition to the second moment that is detected by a conventional analog square-law detector diode. In particular, the second and fourth central moments have been shown to enable extremely sensitive detection of RFI (Ruf et al., 2006b). A third significant advantage of the digital approach is that it simplifies the design of fully polarimetric radiometers. If the analog front-end stage of a radiometer has two orthogonal linearly-polarized channels, a DSP back end-stage can be used to perform the complex correlation between them, in addition to performing individual square-law detection of each channel. This results in coherent-detection measurements of the third and fourth Stokes parameters of the brightness temperature with negligible additional complexity to the analog front ends. The incoherent-detection method using analog circuitry, on the other hand, would require a very significant modification to the analog front-end in order to realize the correlation process. Examples of the two approaches are given in Section 7-14. Massively parallel digital correlators are the standard back-end processing stage used by microwave interferometric radiometers (see Section 7-13).

The digitization of an analog signal x can be represented by the mapping

$$d_m(x) = a_n, \qquad \text{if } x \in V_n \qquad \text{for } n = 1, \ldots, m, \quad (7.140)$$

where m is the number of quantization levels, a_n is the digitized value assigned to the nth level, and V_n defines the range of values of x that map onto a_n. For example, a convenient mapping with nonnegative analog signals digitized by an N-bit digitizer, for which $m = 2^N$, is

$$a_n = \frac{n - \frac{1}{2}}{2^N}$$

and

$$V_N = (a_N \pm 2^{-N-1})/x_{\max},$$

where $x_{\max}$ is the maximum value of x. In radiometry, signals are typically digitized at the pre-detection IF stage, at which point they are zero-mean gaussian distributed voltages with variance proportional to the system noise temperature, or $s_x^2 = KT_{\text{SYS}}\text{B}$. It is common practice to use very coarse digitization, with quantization levels $m = 2$ or $m = 4$ (i.e., one- or two-bit digitizers). In this case, the most common mappings are given by

$$d_2(x) = \begin{cases} -1 & \text{if } x < 0, \\ 1 & \text{if } x \geq 0, \end{cases} \tag{7.141a}$$

$$d_4(x) = \begin{cases} -a_2 & \text{if } x < -V_0, \\ -a_1 & \text{if } -V_0 \leq x < 0, \\ a_1 & \text{if } 0 \leq x < V_0, \\ a_2 & \text{if } V_0 \leq x, \end{cases} \tag{7.141b}$$

where, in Eq. (7.141b), the so-called "Cooper two-bit" mapping of $a_1 = 1$ and $a_2 = 3$ is often used to produce a uniform increment of the digitized values between successive quantization levels (Cooper, 1970).

The DSP stage of a digital radiometer estimates certain statistics of the digitized signal, from which the corresponding statistics of the analog signal are derived. In the case of a pair of signals, such as the v- and h-pol channels of a polarimetric radiometer, the DSP stage estimates joint statistics of the two signals. Consider the case where x_{v} and x_{h} represent the v- and h-pol analog signals prior to digitization. The joint probability distribution of their amplitudes is given by

$$p(x_{\text{v}}, x_{\text{h}}) = \frac{1}{2\pi s_{\text{v}} s_{\text{h}} \sqrt{1 - r^2}} \cdot \exp\left[-\frac{\left(\frac{x_{\text{v}}}{s_{\text{v}}}\right)^2 + \left(\frac{x_{\text{h}}}{s_{\text{h}}}\right)^2 - 2r\left(\frac{x_{\text{v}} x_{\text{h}}}{s_{\text{v}} s_{\text{h}}}\right)}{2(1 - r^2)} \right], \tag{7.142}$$

where r is the partial correlation between the two signals and s_p^2 is the variance of the p-polarized channel ($p = \text{v}$ or h), which is proportional to the system noise temperature:

$$s_p^2 = KT_{\text{SYS}}^p \text{B} = K(T_{\text{B}}^p + T_{\text{REC}}^p)\text{B}. \tag{7.143}$$

The partial correlation can be estimated directly from the statistics of x_{v} and x_{h} by

$$r = \frac{\langle x_{\text{v}} x_{\text{h}}^* \rangle}{\sqrt{s_{\text{v}}^2 s_{\text{h}}^2}}. \tag{7.144}$$

The third and fourth Stokes parameters of brightness temperature can then be derived from the partial correlation according to

$$T_{\text{B}}^3 + jT_{\text{B}}^4 = r\sqrt{T_{\text{SYS}}^{\text{v}} T_{\text{SYS}}^{\text{h}}}, \tag{7.145}$$

where the system noise temperature is determined directly from the measured variance.

With a digital radiometer, the necessary statistics of the digitized signals are computed by its DSP stage. The partial correlation of the digitized signals then follows as

$$r_{\text{digital}} = \frac{\langle d(x_{\text{v}})\, d(x_{\text{h}}^*) \rangle}{\sqrt{\langle d^2(x_{\text{v}}) \rangle \langle d^2(x_{\text{h}}) \rangle}}. \tag{7.146}$$

A mapping from r_{digital} to r is required prior to scaling the unitless correlation value r to units of kelvins according to Eq. (7.145). That mapping is, in general, nonlinear and dependent on details of the digitization. For the simplest case of the one-bit digitizer specified by Eq. (7.141a), the mapping is given by (Weinreb, 1961)

$$r = \sin\left(\frac{\pi}{2}\, r_{\text{digital}}\right). \tag{7.147}$$

For more complicated digitizers, the mapping of the correlation is often determined numerically and implemented as a lookup table.

With analog radiometers, a square-law detector is used to measure the variance of the pre-detected signal, or the covariance between two analog signals in the case of analog correlators. With a digital radiometer, the output products are obtained by computing the variance of, or covariance between, a population of digital samples. The standard error associated with estimating the variance or covariance from a finite population plays an identical role as the radiometric uncertainty ΔT defined in Section 7-4. With a one-bit digitizer, the standard error is approximately 1.57 times that of the analog detector. With one-bit digitization of a zero-mean signal, the single quantization threshold is always set at the zero-crossing of the analog signal. With more than two levels of quantization, the choice of digitization thresholds used to define the boundaries between quantization levels is variable and can be optimized to minimize the standard error associated with estimates of the relevant statistics (Hagen and Farley, 1973; Ruf, 1995).

PROBLEMS

7.1 An antenna is connected to a receiver system consisting of a transmission line with a loss factor of 1.5 dB; an RF amplifier with a noise figure of 7 dB and gain of 20 dB, followed by a mixer-preamplifier with a noise figure of 8 dB and conversion gain of 6 dB; and finally, an IF amplifier with a noise figure of 6 dB and gain of 40 dB. The entire receiver is maintained at an environmental temperature of 290 K.

(a) Find the overall noise figure and effective noise temperature of the receiver system.

(b) Find the overall noise figure and effective noise temperature of the receiver system with RF amplifier and transmission line interchanged.

7.2 The receiver of Problem 7.1(a) is connected to an antenna with a radiation efficiency of 0.9. If the calculated antenna temperature of the scene under observation is $T_A = 100$ K, and if the antenna physical temperature is 290 K, what is the system noise temperature at the antenna terminals?

7.3 In a superheterodyne receiver with local oscillator frequency f_{LO} and IF bandwidth B centered at f_{IF}, the IF output spectrum corresponds to two RF spectral bands as illustrated in Fig. 7-14. If an RF amplifier is used (ahead of the mixer) whose passband allows only one of the two RF bands to reach the mixer input, the system is referred to as a single-sideband receiver. On the other hand, if the mixer input contains both RF bands (centered at RF_1 and RF_2), the IF output will include input signals and noise from a total RF bandwidth of $2B$. This latter situation, which is known as double-sideband reception, exists if no RF amplifier (or RF filter) is used or if the RF amplifier passband extends from below $f_{RF_1} - B/2$ to above $f_{RF_2} + B/2$, thereby amplifying both RF bands.

If F_{SSB} is the single-sideband noise figure of a mixer-preamplifier assembly, show that for the double-sideband receiver, the double-sideband noise figure F_{DSB} of the mixer-preamplifier is given by

$$F_{DSB} = \frac{F_{SSB} + 1}{2}$$

and that the corresponding equivalent noise temperatures are related by

$$T_{DSB} = \frac{T_{SSB}}{2} .$$

7.4 In the absence of gain variations, the sensitivity of a total-power radiometer with a system temperature of 1000 K is 1.5 K. Including gain variations, the sensitivity is 2 K. What is the normalized gain variation?

7.5 The pulsed noise-injection radiometer of Fig. 7-19 drives the PIN diode switch with 40 μs pulses at a pulse repetition frequency f_R. The diode noise source has an ***excess noise ratio*** G (dB) = 23 dB. That is, $T_N = T_0 + GT_0$, with G in natural units. The loss factor of the PIN diode switch is 2 dB in the ON state and 60 dB in the OFF state. The directional coupler has a 20 dB coupling factor and the entire radiometer is enclosed in a chamber maintained at 320 K. What should the range of f_R be in order to maintain a balanced condition over the range 50 K $\leq T_A \leq$ 300 K?

7.6 A 1 GHz balanced Dicke radiometer with a 100 MHz bandwidth is to be flown on a satellite at an altitude of 600 km with an average speed of 7.5 km/s. The radiometer uses a 10 m diameter antenna and the receiver is characterized by $T'_{REC} = 1000$ K and $T_{REF} = T_0 = 300$ K. The radiometer integration time is chosen to be equal to 0.1 of the dwell time of the antenna beam for a point on the ground. If the antenna is fixed so that its main beam is always pointed in the nadir direction, what will ΔT be?

7.7 Suppose the antenna of Problem 7.6 is made to scan between $-20°$ and $+20°$ relative to nadir in the plane orthogonal to the flight direction. What is ΔT?

7.8 An airborne scanning radiometer operating in the imaging mode shown in Fig. 7-34 has a radiometric resolution of 1 K and a spatial resolution of 80 m. If a longer antenna is used so as to improve the spatial resolution to 40 m, by what factor should the speed of the aircraft be changed so as to maintain the same radiometric resolution?

7.9 The one-dimensional MIR sensor described in Fig. 7-37 uses 14 antennas to synthesize a 68-element phased array. It does so by spacing the antennas along a line at specific integer multiples of one-half wavelength at its 10.7 GHz frequency of operation. Let $S = \{ a_n \mid n = 1, \ldots, 14 \}$ be the ordered set of integers describing the locations of the antennas in units of one-half wavelength. Define

$$D(S) = \{ |a_n - a_m| \mid m = 1, \ldots, 13;\ n = m+1, \ldots, 14 \}$$

as the set of differences generated by S. The set S is said to be a complete difference basis if $D(S)$ contains every integer from 1 to $(a_{14} - a_1)$ at least once. Find the integers in S for which $D(S) = \{ 1, \ldots, 68 \}$ is complete.

7.10 For the MIR sensor of Problem 7.9, can you find some other complete difference basis with 14 elements that generates a difference set, $D(S)$, with a larger maximum element than 68?

7.11 The fringe washing function in Eq. (7.113) accounts for the decorrelation of a noise signal when correlated with a time delayed version of itself. Assume a noise signal has a flat power spectral density over a bandwidth $f_c \pm B/2$, where f_c is the center frequency, B is the bandwidth, and $B \ll f_c$. Derive an expression for the correlation of this signal with a time delayed version of itself, and evaluate it at time delays of $0.1/B$, $0.5/B$ and $5/B$.

7.12 Prove Eqs. (7.131a)-(7.131d).

7.13 Prove the following relationships between the polarimetric components of the brightness temperature:

(a) $T_B^3 = 2T_B^P - T_B^v - T_B^h$.

(b) $T_B^3 = T_B^v + T_B^h - 2T_B^M$.

(c) $T_B^4 = 2T_B^L - T_B^v - T_B^h$.

(d) $T_B^4 = T_B^v + T_B^h - 2T_B^R$.

7.14 Fully polarimetric radiometers using incoherent detection typically measure T_B^P, T_B^M, T_B^L, and T_B^R and use Eqs. (7.132a and b) to compute T_B^3 and T_B^4. However, since T_B^v and T_B^h also are measured in practice, it should be possible to compute T_B^3 and T_B^4 using just those two and T_B^P and T_B^L (see Problem 7.13), without measuring T_B^M and T_B^R. This reduces the number of channels required, which would make the radiometer smaller, lower power and cheaper. Why do you think this is not done in practice?

7.15 An AWG-based polarimetric calibration source is used. Its LUT has 16 MB of sample memory that is clocked at 1.25 GHz. It is operated in a cyclical mode, in which the pointer continuously cycles back to the beginning of the 16 MB table of numbers when it reaches the end.

(a) If the LUT is populated with independent realizations of a zero-mean Gaussian-distributed random number generator, what will be the power spectrum of the analog output signal? Assume the output is Nyquist filtered by an ideal low-pass filter with a cut-off frequency of 600 MHz.

(b) In order to introduce partial correlation into the two channels of the AWG, their LUTs can be populated with linear combinations of two independent Gaussian-distributed random number sequences. Starting with two such sequences, both with unity standard deviations, what linear combinations produce one sequence with a variance of 100 K, another with a variance of 200 K, and a covariance between them of 5 K? This could be used, for example, to produce ocean-like calibration test signals with $T_B^h = 100$ K, $T_B^v = 200$ K, and $T_B^3 = 5$ K.

7.16 Derive an analytical expression for the two-bit digital correlation between signals with an analog correlation of 10% using the Cooper two-bit mapping as a function of the ratio V_0/s_x, where V_0 is the quantization threshold in Eq. (7.141b) and s_x is the standard deviation of the two signals being digitized. Plot the digital correlation versus V_0/s_x. The correlator has a minimum sensitivity to signal strength or gain fluctuations when the slope of this plot is zero. Indicate that desired operating point.

7.17 Suppose a zero-mean Gaussian noise signal is digitized with a one-bit digitizer, but the zero value of the analog signal is slightly offset from the zero threshold of the digitizer. In this case, the digitization mapping given by Eq. (7.141a) can be rewritten as

$$d_2(x) = \begin{cases} -1 & \text{if } x < a, \\ 1 & \text{if } x \geq a, \end{cases}$$

where $a \neq 0$. This is referred to as a ***null offset*** in the digitizer. Suppose a second, uncorrelated, zero-mean Gaussian noise signal is digitized by another one-bit digitizer with a slightly different null offset. Since the two analog signals are statistically uncorrelated, the numerical correlation of the two digitized signals should be zero. Derive an expression for the digital correlation as a function of the two null-offset values. Your expression should go to zero as the two null offsets both go to zero.

CHAPTER

8

Microwave Interaction with Atmospheric Constituents

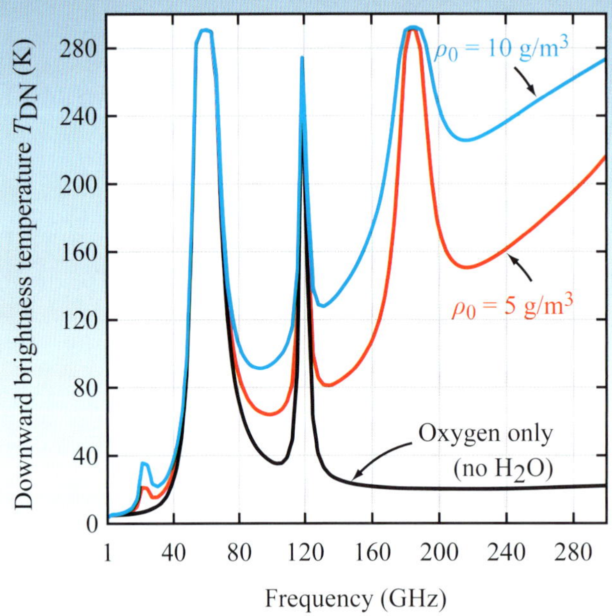

Zenith brightness temperature

CONTENTS

Overview

The microwave transmission spectrum of the Earth's atmosphere offers a wide range of conditions, from near-total transparency to near-total opacity. In the 1–15 GHz range, the atmosphere is practically transparent, even in the presence of clouds and moderate rainfall rates. Consequently, this frequency range is highly suitable for radar observation of land and ocean surfaces from satellite platforms. If, on the other hand, the objective is to measure and/or map atmospheric constituents, the radar and radiometer microwave frequencies should be selected such that the atmospheric constituents of interest exhibit significant scattering or absorption (and hence emission) at those frequencies. The atmospheric absorption spectrum includes resonances due to water vapor (at 22.235 GHz and 183.31 GHz) and oxygen (in the 50–70 GHz range and at 118.75 GHz). From multifrequency radiometric observations made by orbiting satellites at and near the absorption maxima, it is possible to generate height profiles for water vapor and atmospheric temperature. Between the absorption maxima, the spectrum includes a number of low-attenuation ***atmospheric windows***—most notably the ***35 GHz window***—that also are suitable for terrain observations.

The practice of remote sensing may be divided into two interrelated activities: the solution of the ***direct problem*** and the implementation of the ***inverse problem***. As portrayed in Fig. 8-1, associated with each of the two problems are certain inputs and outputs. In the case of the direct problem (also known as the ***forward model***), the inputs are intended to provide complete characterization of all relevant atmospheric parameters and boundary conditions, and the output is the brightness temperature $T_B(f, \theta)$ computed for a ground-based upward-looking radiometer or a satellite borne downward-looking radiometer operating at microwave frequency f and oriented along angle θ relative to the vertical axis. The inputs include the vertical profiles of temperature, pressure, water vapor density, and the attributes of clouds and rain if present; the boundary conditions for an upward-looking radiometer may include extraterrestrial downward emission from our galaxy; and the boundary conditions for a downward-looking radiometer include upward-emitted terrestrial radiation. Connecting the inputs to the output is made possible by a radiative transfer model that accounts for all absorption, emission, and scattering mechanisms by all constituents present along the path viewed by the radiometer antenna. The accuracy with which the model can predict the true value of $T_B(f, \theta)$ is dependent on the accuracies of the input data, boundary conditions, and

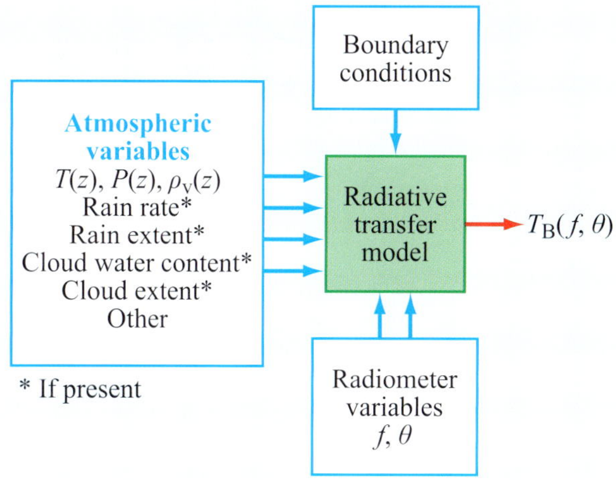

(a) Direct problem

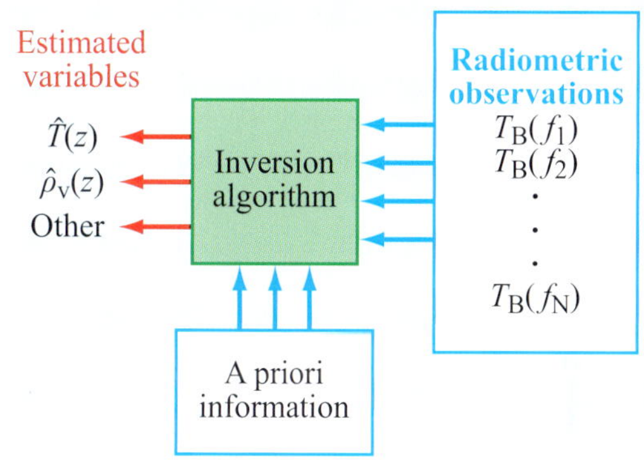

(b) Inverse problem

Figure 8-1: Elements of the (a) direct and (b) inverse problems in upward-looking atmospheric remote sensing.

the model itself. The direct problem (forward model) is the subject of the present chapter.

In Chapter 9, we treat the inverse problem, which entails the use of measured values of $T_B(f,\theta)$, usually made at several frequencies f or angles θ, as inputs to an inversion algorithm designed to produce estimates of one or more of the atmospheric variables of interest (as outputs). Another topic deferred to Chapter 9 is ***limb sounding***, which refers to observing the Earth's atmosphere along a direction tangential to the atmosphere, rather than upward or downward.

8-1 Standard Atmosphere

Atmospheric pressure $P(z)$ and ***atmospheric density*** $\rho_{\text{air}}(z)$ decrease approximately exponentially with increasing geometric height z (Fig. 8-2) as does the water-vapor density $\rho_v(z)$, although its variation with height is somewhat irregular and is strongly dependent on time of day, season, geographic location, and atmospheric activity. The variation of ***atmospheric temperature*** with height, $T(z)$, exhibits a cyclic pattern that subdivides the Earth's atmosphere into a number of atmospheric layers. As a guide, we show in Fig. 8-3 the height profile of $T(z)$ as defined by a reference model known as the ***International Standard Atmosphere (ISA)***, which provides a common reference for how atmospheric parameters—including temperature, pressure, and density—vary with altitude. For altitudes up to 32 km, the ISA models are identical to those defined according to the widely used 1976 ***U.S. Standard Atmosphere***. At 30 km, atmospheric density is only 1.5% of its sea-level value, so for computational purposes associated with downward-looking or upward-looking microwave sensors, the international and U.S. standard atmospheric models are practically identical.

The height profile shown in Fig. 8-3 consists of four layers: the ***troposphere***, ***stratosphere***, ***mesosphere***, and ***thermosphere***. Adjacent layers are separated by a boundary defined as the height at which the temperature gradient, dT/dz, changes from a nonzero magnitude to zero. If dT/dz is negative, its magnitude is called the ***temperature lapse rate***. In the Standard Atmosphere model, the lapse rate in the troposphere is 6.5 K/km.

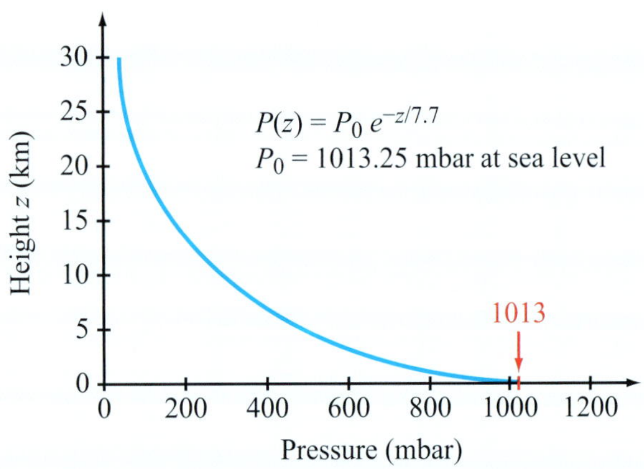

Figure 8-2: Atmospheric pressure decreases exponentially with height.

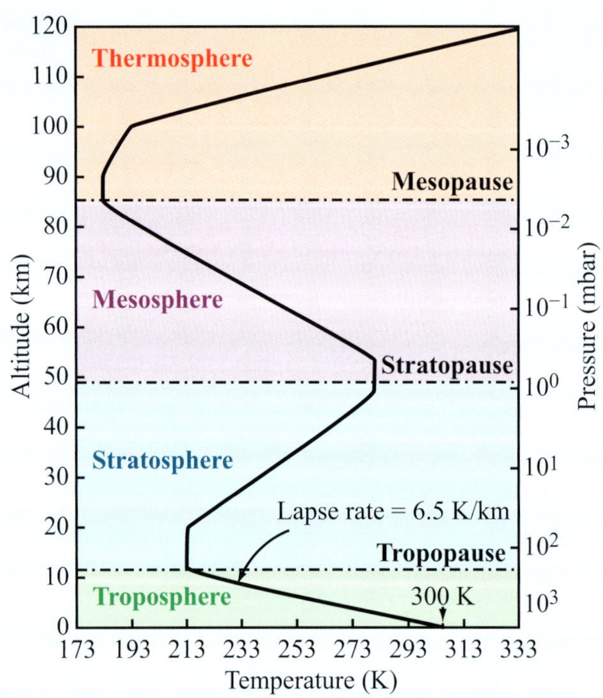

Figure 8-3: Atmospheric temperature height profile according to the Standard Atmosphere model.

At its upper boundary, known as the ***tropopause***, dT/dz changes abruptly to zero. According to the standard-model atmospheres, the tropopause is at a height of 11 km, but in reality it varies according to latitude, season, and weather activity. In general, it is between 8 and 10 km in arctic regions in winter and between 16 and 18 km in the tropics.

The next layer above the troposphere is the ***stratosphere***, which extends to the ***stratopause*** at a height of 47 km. The third layer, extending between the stratopause and approximately 80 km, is the ***mesosphere***. In this layer, the temperature decreases to the lowest value of the 120 km profile, and it occurs at its upper boundary, known as the ***mesopause***.

The composition of the atmosphere and the molecular weight of its air remain approximately constant up to the mesopause. Above the mesopause, the density becomes very small and the composition changes due to dissociation, which makes direct measurement of air temperature impossible. The temperature profile shown in Fig. 8-3 for the ***thermosphere*** is modeled on the basis of temperature values inferred from measurements of other atmospheric parameters. There is no upper limit to the thermosphere.

8-1.1 Atmospheric Composition

▶ Except for water-vapor variations, the relative composition of the atmosphere essentially is constant up to 90 km above sea level. ◀

The sea-level composition is given in Table 8-1 for contaminant-free air containing no water vapor. The principal constituents are molecular nitrogen and molecular oxygen, the molecular weight of the standard atmosphere is 28.9644, and the sea-level density of dry air is $\rho_{\text{air}}(0) = 1.225$ kg/m^3.

Table 8-1: Composition of clean, dry atmospheric air near sea level [from Hering, 1965].

Constituent gas	Gas symbol	Content (% by volume)	Molecular weight
Nitrogen	N_2	78.084	28.0134
Oxygen	O_2	20.9476	31.9988
Argon	Ar	0.932	39.948
Carbon dioxide	CO_2	0.0314	44.00995
Neon	Ne	0.001818	20.183
Helium	He	0.000524	4.0026
Krypton	Kr	0.000114	83.80
Xenon	Xe	0.0000087	131.30
Hydrogen	H_2	0.00005	2.01594
Methane	CH_4	0.0002	16.04303
Nitrous oxide	N_2O	0.00005	44.0128
Ozone	O_3	Summer: 0–7 $\times 10^{-6}$	47.9982
		Winter: 0–2 $\times 10^{-6}$	47.9982
Sulfur dioxide	SO_2	0 to 0.0001	64.0628
Nitrogen dioxide	NO_2	0 to 0.000002	46.0055
Ammonia	NH_3	0 to trace	17.03061
Carbon monoxide	CO	0 to trace	28.01055
Iodine	I_2	0 to 0.000001	253.8088

8-1.2 Temperature Profile

For the U.S. or ***International Standard Atmosphere*** (ISA), the atmospheric temperature $T(z)$ is given by

$$T(z) = \begin{cases} T_0 - az, & 0 \leq z \leq 11 \text{ km}, \\ T(11), & 11 \text{ km} \leq z \leq 20 \text{ km}, \\ T(11) + (z - 20), & 20 \text{ km} \leq z \leq 32 \text{ km}, \end{cases} \tag{8.1}$$

where z is height above sea level in km, T_0 is the sea-level atmospheric temperature and $T(11)$ is the atmospheric temperature at $z = 11$ km, both in kelvins, and a is the ***temperature lapse rate*** in the lower 11 km of the atmosphere. For the U.S. and ISA Standard Atmosphere models, $T_0 = 288.15$ K and $a = 6.5$ K/km.

8-1.3 Density Profile

The density of dry air decreases exponentially with altitude:

$$\rho_{\text{air}}(z) = 1.225e^{-z/H_1} \qquad (\text{kg/m}^3) \qquad (0 \le z \le 10 \text{ km}), \tag{8.2}$$

where z is the altitude (km) and $H_1 = 9.5$ km is the ***density scale height***. The above expression is in excellent agreement (less than 3 percent error) with that defined by the Standard Atmosphere model for the lowermost 10 km of the atmosphere, but deviates from the tabulated values at higher altitudes. A better fit to the tabulated values for altitudes up to 30 km is provided by the expression

$$\rho_{\text{air}}(z) = 1.225e^{-z/H_2}[1 + 0.3\sin(z/H_2)] \text{ kg/m}^3, \tag{8.3}$$

where $H_2 = 7.3$ km.

8-1.4 Pressure Profile

Assuming air to be an ideal gas, its equation of state is

$$P = \rho_{\text{air}}\text{R}T/M = \rho_{\text{air}}R_{\text{a}}T, \tag{8.4}$$

where P is the atmospheric pressure in millibars, R is the universal gas constant, T is the kinetic (actual) temperature in kelvins, and M is the molecular weight of air. If the density ρ_{air} is in kg/m^3, the gas constant for air is $R_{\text{a}} = 2.87$. Hence, using

$$P(z) = 2.87\rho_{\text{air}}(z)\, T(z), \tag{8.5}$$

$P(z)$ can be found for any altitude z between sea level and 30 km via Eqs. (8.1) and (8.3).

Alternatively, the pressure profile can be fitted to an exponential of the form

$$P(z) = P_0 e^{-z/H_3} \qquad (\text{mbar}), \tag{8.6}$$

where P_0 is the sea-level pressure and H_3 is the ***pressure scale height***. For the Standard Atmosphere, $P_0 = 1013.25$ mbar and $H_3 = 7.7$ km. The above expression provides values that are within 3 percent of the tabulated Standard Atmosphere values for $z \le 10$ km.

8-1.5 Water-Vapor Density Profile

The amount of water vapor contained in the atmosphere is a function of several weather parameters, but it is especially dependent on the atmospheric temperature. At sea level, the water-vapor density ρ_{v} can vary from 10^{-2} g/m^3 in very cold, dry climates to as much as 30 g/m^3 in hot, humid climates. The average surface value used by the U.S. Standard Atmosphere model is $\rho_0 = 7.72$ g/m^3 for middle latitudes. The altitude profile of $\rho_{\text{v}}(z)$ is usually described by a decreasing exponential,

$$\rho_{\text{v}}(z) = \rho_0 e^{-z/H_4} \qquad (\text{g/m}^3), \tag{8.7}$$

where the scale height H_4 is typically chosen to be between 2 and 2.5 km. The ***total mass of water vapor*** contained in a vertical column of unit cross section is

$$M_{\text{v}} = \int_0^\infty \rho_{\text{v}}(z)\, dz = \rho_0 H_4. \tag{8.8}$$

For $\rho_0 = 7.72$ g/m^3, and $H_4 = 2$ km, $M_{\text{v}} = 15.44$ kg/m^2 or 1.54 g/cm^2.

8-2 Absorption and Emission by Gases

8-2.1 Electromagnetic Interaction with Individual Molecules

The total internal energy $\mathcal{E}$ of an isolated molecule consists of three types of energy states,

$$\mathcal{E} = \mathcal{E}_{\text{e}} + \mathcal{E}_{\text{v}} + \mathcal{E}_{\text{r}}, \tag{8.9}$$

where $\mathcal{E}_{\text{e}}$ = ***electronic*** energy, $\mathcal{E}_{\text{v}}$ = ***vibrational*** energy, and $\mathcal{E}_{\text{r}}$ = ***rotational*** energy. These energy states are quantized; they can assume discrete values that are specified by one or more quantum numbers. Corresponding to each possible electronic state, there are a number of possible vibrational states, and corresponding to each possible vibrational state there are a number of possible rotational states. Rotational energy is associated with rotational motions of the atoms about their equilibrium positions.

Radiation is absorbed (or emitted) when a transition takes place from a lower (or higher) energy state to a higher (or lower) energy state. The frequency f_0 of the absorbed (or emitted) quantum is given by the Bohr formula,

$$f_0 = \frac{\mathcal{E}_m - \mathcal{E}_l}{h}, \tag{8.10}$$

where h is Planck's constant and $\mathcal{E}_m$ and $\mathcal{E}_l$ are the internal energies of the higher and lower molecular states, respectively. The transition may involve changes of electronic, vibrational, or rotational energy, or any combination of the three types. The absorption spectrum due to a single transition is called an ***absorption line***. The energy differences are largest between different electronic states, being typically from 2 to 10 eV,* followed by differences between vibrational states of the same electronic state, which typically vary between 0.1 and 2 eV; and the smallest energy differences are associated with differences between rotational states belonging to the same electronic and vibrational state. Changes in rotational energy alone are usually on the order of 10^{-4} to 5×10^{-2} eV. Transitions between pure rotational states (same $\mathcal{E}_e$ and $\mathcal{E}_v$ but different $\mathcal{E}_r$'s) give rise to rotation lines that occur in the microwave and far-infrared portions of the spectrum. Since energy changes between pure vibrational states (same $\mathcal{E}_e$) are several orders of magnitude larger than energy changes between pure rotational states, vibrational transitions never occur alone, but are always accompanied by many rotational transitions. This condition givens rise to a group of lines, usually referred to as a vibration-rotation band. Such bands, corresponding to values of $\Delta\mathcal{E}$ in the 0.1–2-eV range, occur between the red part of the visible spectrum and the thermal infrared region.

The large energy differences associated with electronic transitions (compared with pure vibrational or rotational transitions) usually result in complex band systems (in the visible and ultraviolet parts of the spectrum) involving simultaneous changes of all three types of energy.

*1 eV= 1.6×10^{-19} J.

Absorption of electromagnetic energy by molecules in the gaseous state usually involves the interaction of the electric or magnetic field of the incident wave with an electric or magnetic dipole or quadrupole moment of the molecules.

► Of the various gases in the Earth's atmosphere, oxygen and water vapor are the only constituents that exhibit significant absorption bands in the microwave spectrum. ◄

The oxygen molecule has a permanent magnetic moment. Magnetic interaction with the incident field produces a family of rotation lines in the vicinity of 60 GHz and an isolated line at 118.75 GHz. Water vapor, on the other hand, is a polar molecule with an electric dipole. Electric interaction with the incident field produces rotation lines at 22.235 GHz, 183.31 GHz, and several frequencies in the submillimeter-wave region (above 300 GHz).

8-2.2 The Shape of a Spectral Line

According to the preceding discussion, the absorption (or emission) spectrum of a molecule consists of sharply defined frequency lines [Fig. 8-4(a)] corresponding to transitions between sharply defined (quantized) energy levels of the molecule. Such a spectrum would be characteristic of an isolated, undisturbed, and stationary molecular system. In reality, however, the molecules are in constant motion, interacting and colliding with one another, and colliding with other material objects (such as dust particles). These disturbances cause the energy levels to vary in width, which results in spectral lines with finite widths, as illustrated in Fig. 8-4(b). The increase in ***linewidth*** is called ***line broadening***. Among the various sources of spectral line broadening, ***pressure broadening***—which arises from collisions between molecules—is the most important for atmospheric absorption in the microwave region of the spectrum.

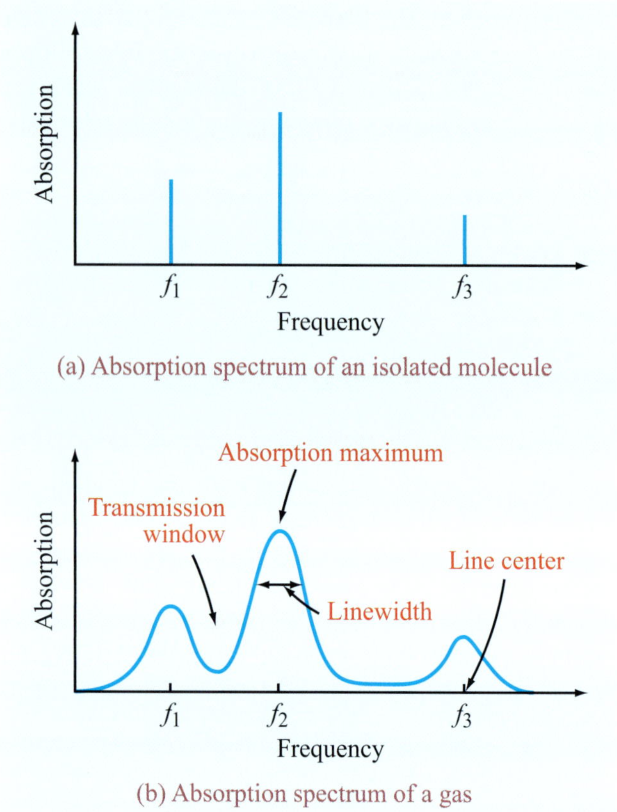

Figure 8-4: Absorption spectrum of (a) a single isolated molecule and (b) a gas containing many molecules.

8-2.3 Absorption Spectrum

In Chapter 2, we related the propagation constant γ of a lossy medium to the medium's complex dielectric constant ε through Eq. (2.54),

$$\gamma = j\omega\sqrt{\mu_0\varepsilon_0}\,\sqrt{\varepsilon} = j\,\frac{\omega}{c}\,\sqrt{\varepsilon}\,, \tag{8.11}$$

where we set $\mu = \mu_0$ (because the medium of interest is the atmosphere and none of its constituents are ferromagnetic) and we used the relation $\sqrt{\mu_0\varepsilon_0} = 1/c$. The dielectric constant is related to the index of refraction of the medium by $\varepsilon = n^2$, and since ε is complex, so is n. Hence,

$$\sqrt{\varepsilon} = n = n' - jn''. \tag{8.12}$$

Because n' of air is close to 1 (at sea level $n \approx 1.0003$), it is customary to describe the atmosphere in terms of a related quantity called the ***refractivity*** N, where

$$n = 1 + 10^{-6}N. \tag{8.13}$$

As n' decreases from about 1.0003 at sea level to 1 at very high altitudes, the corresponding decrease in N is from 300 to zero. Thus, working with N is simply more convenient than working with n.

Since $n = n' - jn''$, we define $N = N' - jN''$, and the corresponding real and imaginary parts are related by

$$n' = 1 + 10^{-6}N', \tag{8.14a}$$

$$n'' = 10^{-6}N''. \tag{8.14b}$$

The deviation of n' from 1.0 is due to the presence of atmospheric gases, causing the density of air to be slightly larger than zero. The imaginary part, n'', accounts for absorption by atmospheric constituents, namely gases, clouds and rain. In the present section, our focus is on absorption by atmospheric gases.

Inserting Eq. (8.12) in Eq. (8.11) leads to

$$\gamma = \frac{j2\pi f}{3\times 10^8}\,(n' - jn''), \tag{8.15}$$

and by equating Eq. (8.15) to $\gamma = \alpha + j\beta$, we obtain the following expressions for the attenuation coefficient α and phase constant β:

$$\alpha = \Re\mathfrak{e}[\gamma] = \frac{\omega}{c}\,n'' = 10^{-6}\,\frac{\omega}{c}\,N'', \tag{8.16a}$$

$$\beta = \Im\mathfrak{m}[\gamma] = \frac{\omega}{c}\,n' = \frac{\omega}{c}(1 + 10^{-6}N'). \tag{8.16b}$$

The power density carried by a wave propagating in the z direction (for example) varies as

$$\mathcal{S}(z) = \mathcal{S}_0 e^{-2\alpha z} = \mathcal{S}_0 e^{-\kappa_a z}, \tag{8.17}$$

where $\kappa_a = 2\alpha$ is the ***power absorption coefficient***. The units for κ_a are Np/m, but it is common practice to express it in dB/km by multiplying it by 4.34×10^3. The conversion process leads to

$$\begin{aligned}\kappa_a(\text{dB/km}) &= 2\alpha(\text{dB/km}) \\ &= 2\times 4.34\times 10^3\times 10^{-6}\times \frac{2\pi f}{3\times 10^8}\, N'' \\ &= 0.182 f N'' \qquad (\text{dB/km}),\end{aligned} \tag{8.18}$$

with f expressed in GHz. Our next step is to relate N'' to the microwave line spectra of atmospheric gases.

As stated earlier, for upward-looking and downward-looking radiometry, the only atmospheric gases that have significant absorption/emission lines in the microwave spectrum are oxygen and water vapor. Since each spectral line has tails that extend (conceptually) over the entire electromagnetic spectrum, the absorption coefficient at any microwave frequency f is the sum of the contributions of all lines, including those outside the microwave spectrum. In practice, we exclude all lines whose cumulative contribution to κ_a is less than 1% or a similar threshold.

In the 1980s, Hans Liebe and his collaborators at the Institute of Telecommunication Sciences of the U.S. National Telecommuncation and Information Administration assembled a model and a computer code that computes the real and imaginary parts of the refractivity N of the Earth's atmosphere for the frequency range from 1 to 1000 GHz (Liebe et al., 1985; 1993). Their model, which incorporates measurements and models reported by numerous investigators, accounts for absorption by 44 oxygen spectral lines spanning the range from 50.47 GHz to 834.14 GHz, and 34 water vapor spectral lines spanning the range from 22.235 GHz to 1780 GHz, in addition to residual contributions of water vapor lines above 1 THz. The formal name of the model and associated computer code is the ***millimeter-wave propagation model*** (MPM).[†]

For the clear-sky atmosphere, MPM computes the ***gaseous absorption coefficient*** κ_g as the sum of the the ***oxygen absorption coefficient*** κ_{O_2} and the ***water vapor absorption coefficient*** κ_{H_2O}:

$$\kappa_g = \kappa_{O_2} + \kappa_{H_2O}, \qquad (\text{dB/km}) \tag{8.19a}$$

[†]MPM is contained in Computer Codes 8.1 to 8.12 (Appendix D).

Table 8-2: Input/output parameters of the MPM code.

Input Parameter	Units	Range
Ambient temperature t	°C	−100 to +50 °C
Barometric pressure P	mb	10^{-5} to 1013 mb
H_2O partial pressure P_{H_2O}	mb	0 to $P_{H_2O}(\max)^*$
Microwave frequency f	GHz	1 to 1000 GHz
Output Products		
κ_{O_2}	dB/km	
κ_{H_2O}	dB/km	
$\kappa_g = \kappa_{O_2} + \kappa_{H_2O}$	dB/km	

$^*P_{H_2O}(\max)$ is the maximum possible partial pressure of water vapor at specified values of T and P.

with

$$\kappa_{O_2} = 0.182 f N''_{O_2} = 0.182 f \sum_{j=1}^{44} S_i F_i \qquad (\text{dB/km}) \tag{8.19b}$$

and

$$\begin{aligned}\kappa_{H_2O} &= 0.182 f N''_{H_2O} \\ &= 0.182 f\left[\sum_{j=1}^{34} S_j F_j + S_c F_c\right] \qquad (\text{dB/km}).\end{aligned} \tag{8.19c}$$

In Eq. (8.19b), the summation is over 44 oxygen spectral lines, each characterized by a ***center frequency*** f_i, a ***line strength*** S_i, and a line ***shape function*** F_i. A similar arrangement applies to the 34 water vapor lines and the equivalent line representing the continuum above 1 THz. The line strengths and linewidths are functions of the ambient temperature T, barometric pressure P, and partial pressure of water vapor P_{H_2O}. The input/output parameters of the MPM code are listed in Table 8-2.

8-2.4 Oxygen Spectrum

Below 350 GHz, the absorption spectrum of oxygen consists of 37 significant absorption lines spread out over the 50–69 GHz frequency range, known

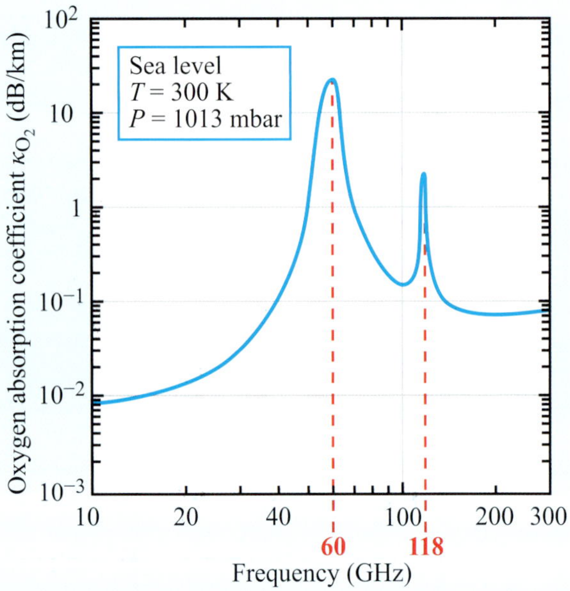

Figure 8-5: Absorption spectrum of oxygen at sea level.

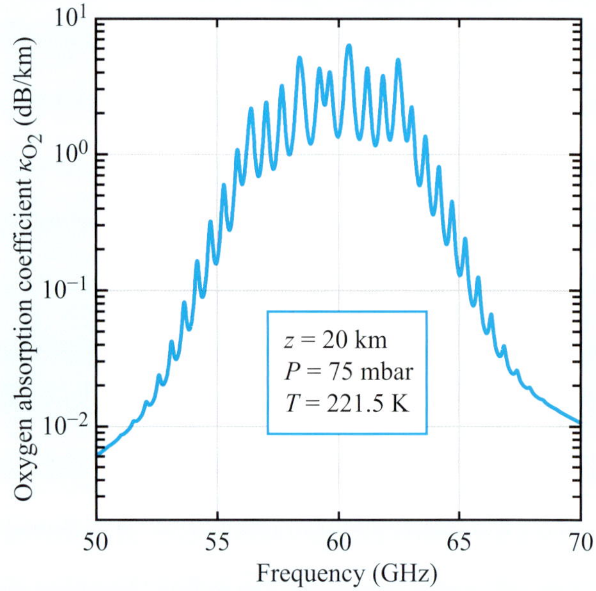

Figure 8-6: Absorption spectrum of the 60 GHz oxygen complex at an altitude of 20 km.

collectively as the ***60 GHz*** oxygen complex, and an additional line at 118.75 GHz. At barometric pressures characteristic of the lower part of the Earth's atmosphere, pressure broadening causes the complex of lines to blend together, forming a continuous band centered at around 60 GHz. This is evident in the spectrum shown in Fig. 8-5 for sea level conditions.† The absorption coefficient κ_{O_2} is the rate of absorption that an EM wave experiences if propagating horizontally through a perfectly dry atmosphere (zero relative humidity) at sea level. We note that whereas $\kappa_{O_2} \approx 0.01$ dB/km at 20 GHz, it is greater by more than three orders of magnitude at the center of the 60 GHz oxygen complex. Thus, the choice of frequency is critically important in remote sensing. If the goal is to measure an atmospheric parameter that is closely connected to the absorption spectrum of oxygen—such as atmospheric temperature, the choice of radiometer frequencies is likely to include the center of the oxygen complex and its "wings." On the other hand, a microwave sensor intended for ocean or land applications is designed to operate at a frequency at which the atmosphere is as transparent as possible.

The width of an oxygen spectral line is directly proportional to the partial pressure of oxygen, which is 0.21 of the dry air pressure (because oxygen constitutes 21% of air content by volume). Air pressure decreases exponentially with height, so the widths of the individual oxygen lines within the 60 GHz complex become narrower at higher altitudes. At an altitude of 20 km, the pressure is on the order of 6% of its value at sea level. Consequently, the lines are much narrower at 20 km and the oxygen spectrum becomes more resolvable, as shown in Fig. 8-6.

8-2.5 Water-Vapor Spectrum

In the microwave region (1–300 GHz), water vapor has two significant rotational absorption lines at 22.235 GHz and 183.31 GHz. There are, however, numerous other lines at frequencies just above this region that also contribute to the microwave absorption spectrum. The closest is at 325.15 GHz. Figure 8-7 displays the

†**Computer Code 8.2.**

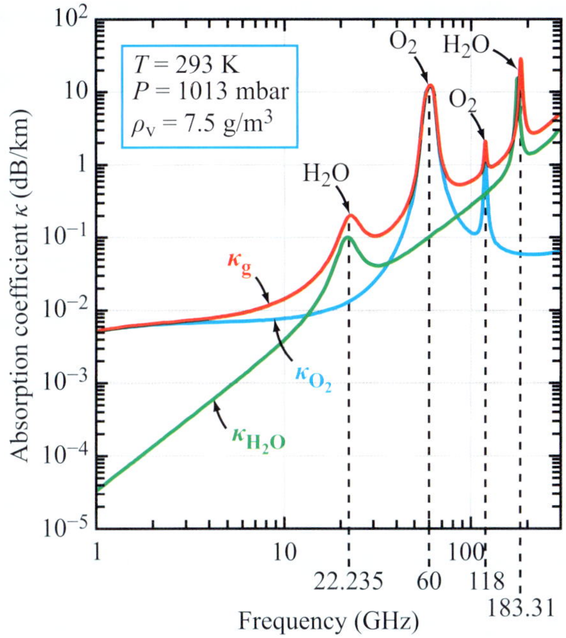

Figure 8-7: Absorption spectra of oxygen, water vapor, and their sum at sea level.

spectrum of κ_{H_2O} at sea level over the frequency range from 1 to 300 GHz.†

8-2.6 Total Gaseous Spectrum

In addition to oxygen and water vapor, other atmospheric gases and pollutants have absorption lines in the microwave spectrum. These include O_3, SO_2, NO_2, and N_2O, among others, but their relative concentrations at sea level are so small that their cumulative contribution to the microwave gaseous absorption spectrum is negligible in comparison to the contributions of oxygen and water vapor. At high altitudes, however, their contributions become discernible and measurable by microwave radiometers whose antenna beams are pointed tangentially to the Earth's atmosphere. Such an observation mode is called ***limb sounding*** and is discussed in Section 9-11.

† Computer Code 8.1.

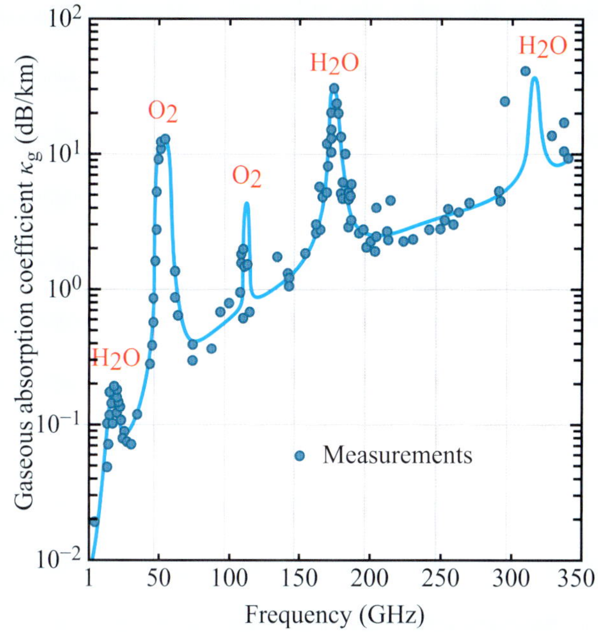

Figure 8-8: Microwave absorption spectrum due to atmospheric gases, $\kappa_g(f)$, at sea level for surface conditions $P_0 = 1013$ mb, $T_0 = 293$ K, and $\rho_0 = 7.5$ g/m^3. Solid curves are calculated according to theory, and dots are measured values [Crane, 1981].

For the purposes of upward-looking and downward-looking microwave remote sensing, the gaseous absorption coefficient κ_g is simply the sum of κ_{O_2} and κ_{H_2O}. In addition to plots of individual spectra for κ_{O_2} and κ_{H_2O}, Fig. 8-7 also displays the spectrum of their sum. Comparison of theory with measured data is shown in Fig. 8-8.†

8-3 Opacity of the Clear Atmosphere

For zenith angles up to 70°, and in some cases up to 85°, the atmosphere may be considered to be ***horizontally stratified***, but for spaceborne systems, spherical geometries are needed at such large angles. Nevertheless, we shall adopt the horizontal stratification model because of its simplicity, in which case the

† Computer Code 8.3.

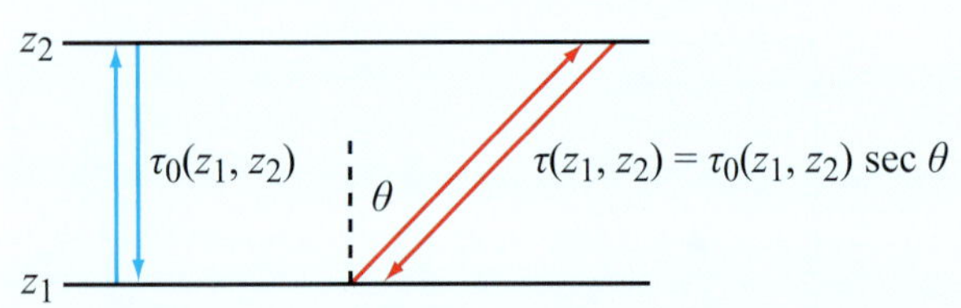

Figure 8-9: For a horizontally stratified atmosphere, the optical thickness $\tau(z_1,z_2)$ for propagation along angle θ is related to the zenith optical thickness $\tau_0(z_1,z_2)$ by $\sec\theta$.

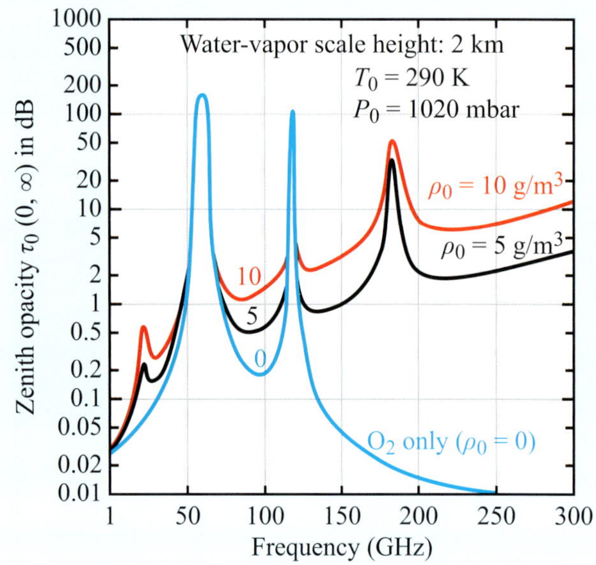

Figure 8-10: Zenith opacity of the atmosphere for three water-vapor conditions.

oblique optical thickness of a layer with uniform properties extending between heights z_1 and z_2 is[†]

$$\tau(z_1,z_2) = \tau_0(z_1,z_2)\sec\theta, \tag{8.20}$$

where θ defines the direction of propagation relative to the zenith (Fig. 8-9) and $\tau_0(z_1,z_2)$ is the ***zenith optical thickness***, defined in terms of the absorption coefficient $\kappa_a(z)$ as

$$\tau_0(z_1,z_2) = \int_{z_1}^{z_2} \kappa_a(z)\,dz. \tag{8.21}$$

The units of τ_0 are nepers (Np) or dB, depending on whether the absorption coefficient κ_a is expressed in nepers per unit length or dB per unit length, respectively. Under clear-sky conditions, $\kappa_a = \kappa_g$, where κ_g is the gaseous absorption coefficient given by Eq. (8.19a).

If $(z_1,z_2) = (0,\infty)$, which encompasses the entire atmosphere, $\tau_0(0,\infty)$ is then called the ***zenith opacity*** τ_0.

Ignoring the minor contributions of ozone and other minor gases, the zenith opacity τ_0 is determined by three atmospheric-height profiles, namely those of temperature, the partial pressure of oxygen, and the partial pressure of water vapor. Figure 8-10 shows the microwave spectra of $\tau_0(0,\infty)$ for surface water-vapor densities of 0, 5, and 10 g/m^3, all calculated for the Standard Atmosphere model assuming a 2 km scale height for water vapor. As expected, $\tau_0(0,\infty)$ has a strong sensitivity to the surface water-vapor density ρ_0, except within the bands surrounding the oxygen absorption resonances centered at 60 GHz and 118.75 GHz. Comparison with experimental observations is provided in Fig. 8-11.

The concentrations of oxygen and water vapor decrease exponentially as a function of height above the Earth's surface, with scale heights of 7.7 km for oxygen and 2 to 2.5 km for water vapor (see (Section 7-1). Consequently, the bulk of the attenuation represented by the opacity τ_0 occurs within the first few kilometers of the lower atmosphere. This fact is demonstrated by the five curves in Fig. 8-12, which depict the microwave spectra of the zenith optical thickness $\tau_0(z,\infty)$ corresponding to five discrete values of z, namely 0, 4, 8, 12, and 16 km. At the peak of the 22.235 GHz water-vapor line, for example, the attenuation by the layer between the surface and $z = 4$ km (difference in dB between the two curves) comprises more than 85 percent of the zenith opacity. Figure 8-13 examines $\tau(z,\infty)$ in the 50–70 GHz band. While it is relatively difficult to resolve the individual oxygen lines in the spectrum of the $\tau_0(0,\infty)$, the lines are

[†]Computer Code 8.7.

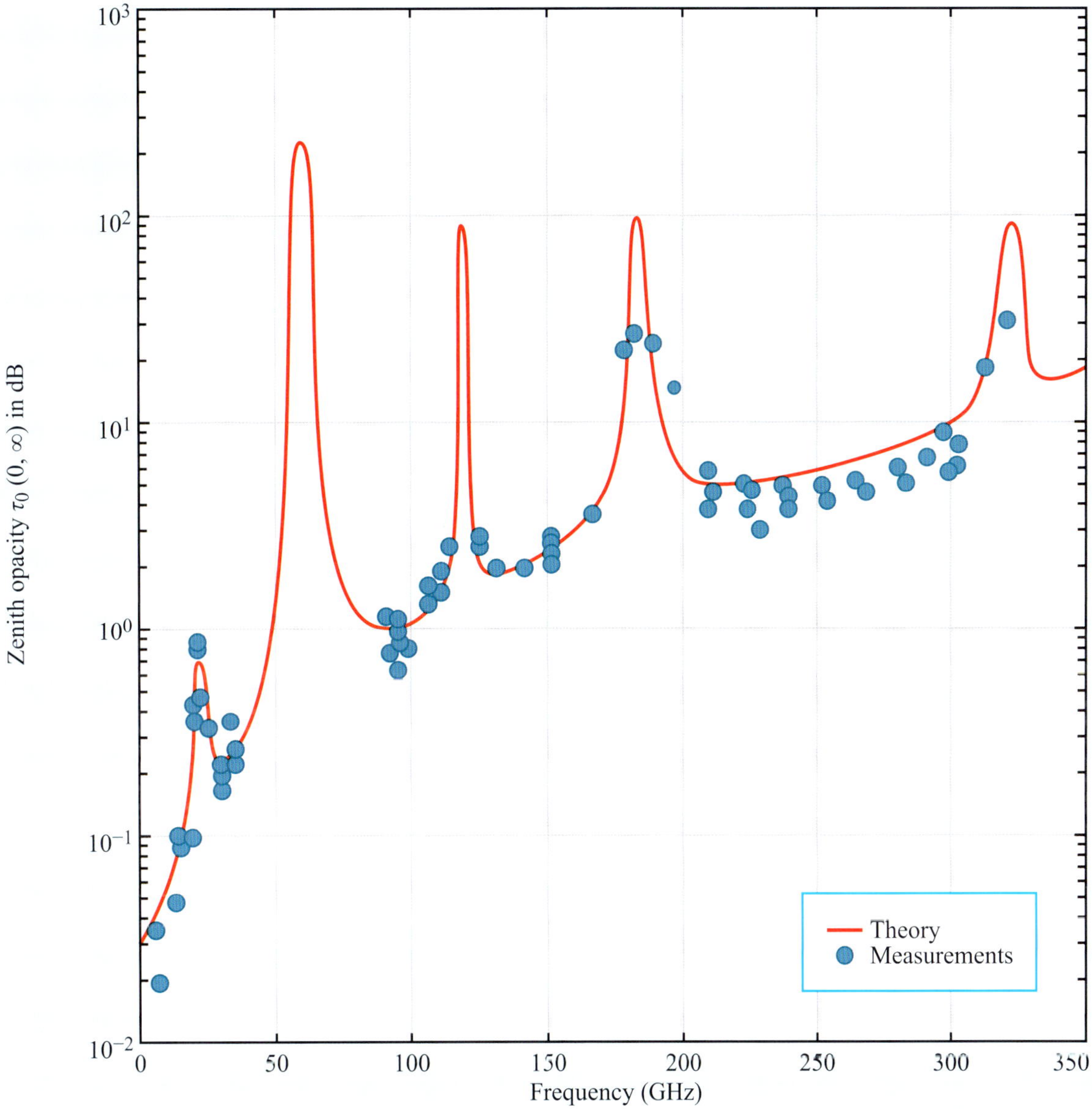

Figure 8-11: Comparison of calculated spectrum of the zenith opacity $\tau_0(0,\infty)$ for surface conditions $T_0 = 293$ K, $P_0 = 1013$ mbar, and $\rho_0 = 7.5$ g/m^3 with experimental observations [Crane, 1981].

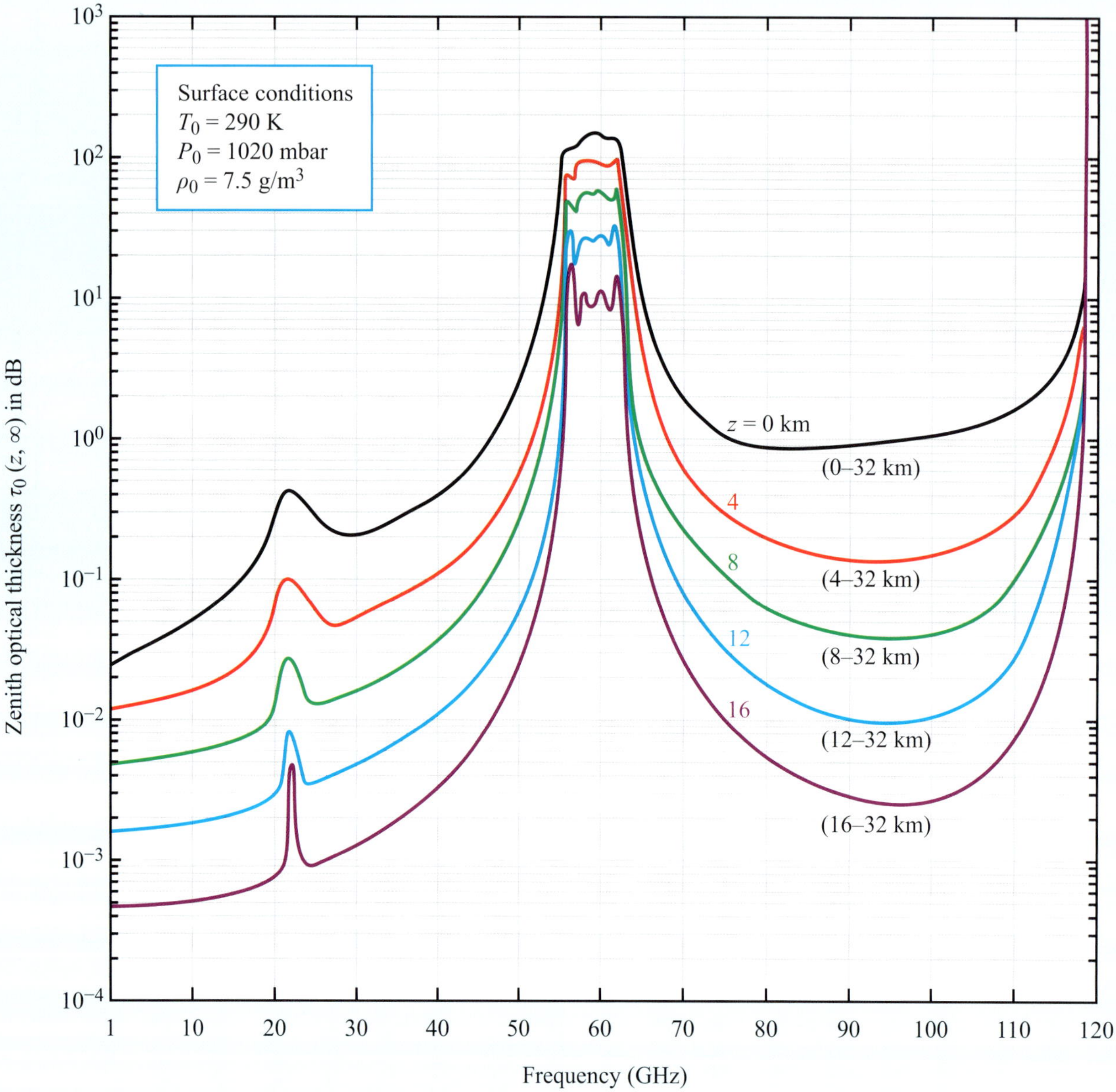

Figure 8-12: Zenith optical thickness $\tau_0(z,\infty)$ for five values of z [from Smith, 1982].

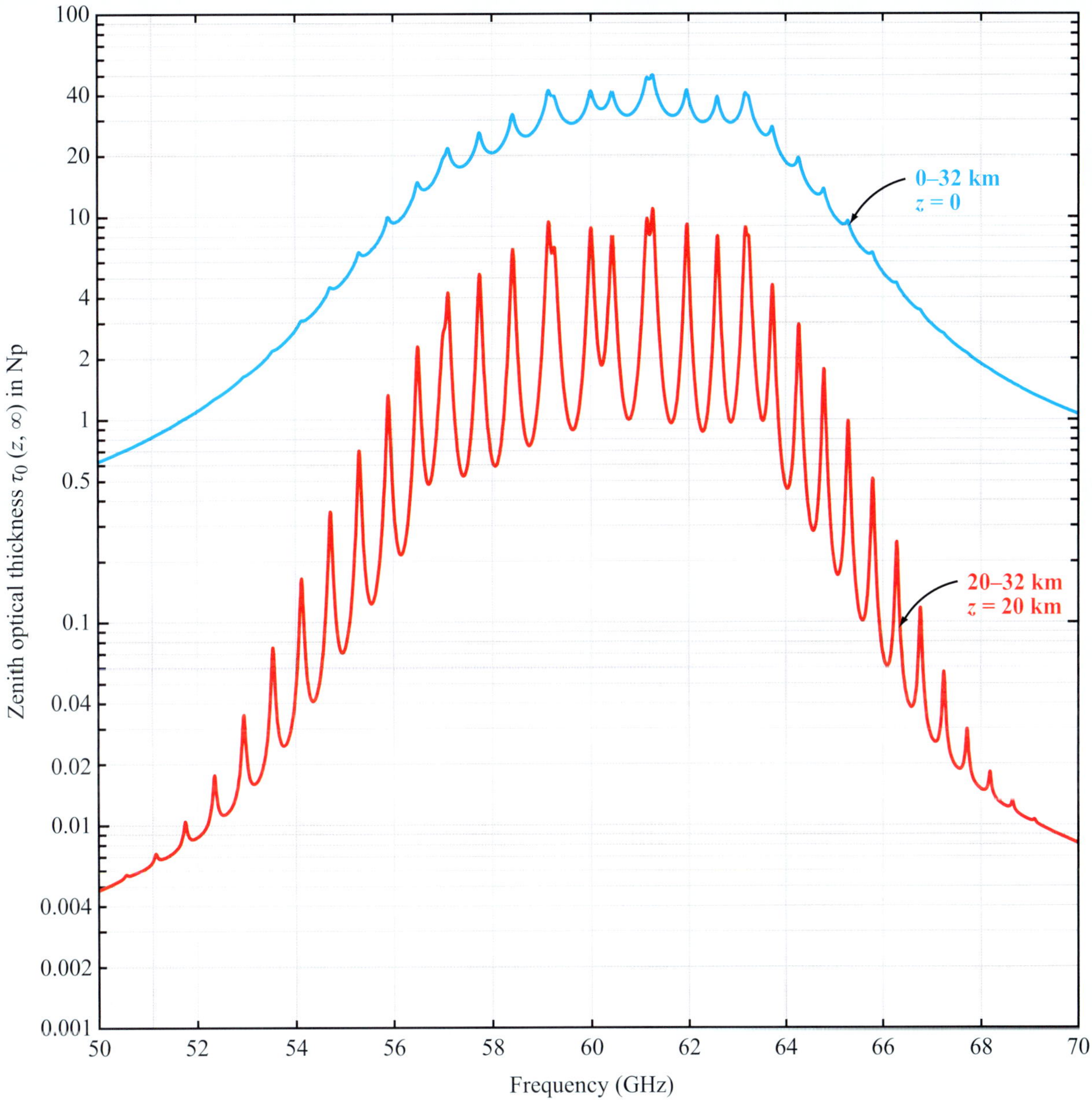

Figure 8-13: Computed zenith optical thickness $\tau_0(z, \infty)$ in the 50 to 70 GHz oxygen band, for $z = 0$ (sea level) and 20 km. Note that τ_0 is in Np [τ_0 (dB) $= 4.34\tau_0$ (Np)].

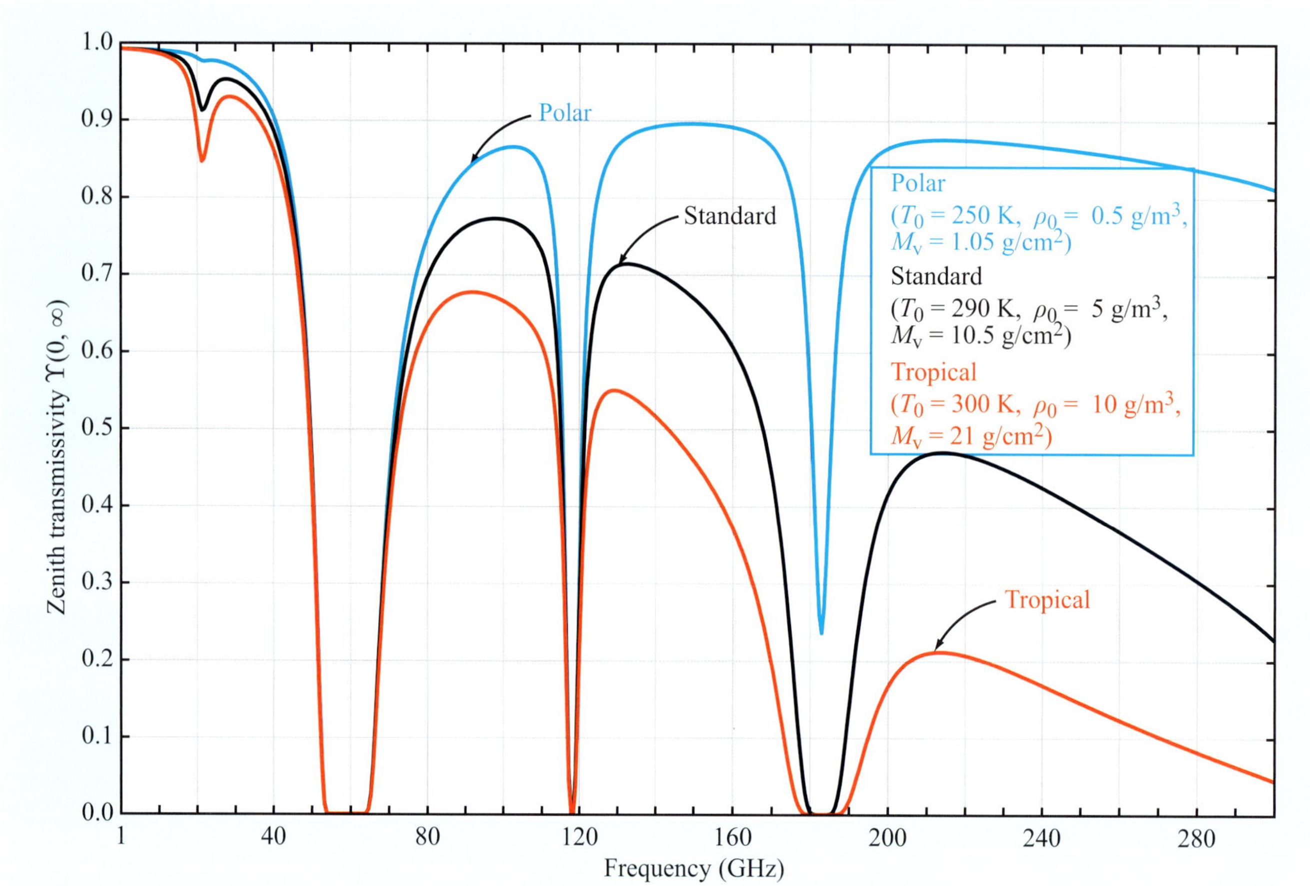

Figure 8-14: Atmosphere transmissivity as characterized by different surface temperatures T_0 and integrated water-vapor content M_v.

quite resolvable in the spectrum of τ_0 (20 km, ∞) because the lower atmospheric density above 20 km results in weaker pressure-broadening of the lines.

The ***oblique transmissivity*** of the stratified atmosphere is related to the zenith opacity by

$$\Upsilon(0,\infty) = e^{-\tau_0(0,\infty)\sec\theta} = [\Upsilon_0(0,\infty)]^{\sec\theta}, \qquad (8.22a)$$

where

$$\Upsilon_0(0,\infty) = e^{-\tau_0(0,\infty)} \qquad (8.22b)$$

is the ***zenith transmissivity***.

► When calculating the transmissivity or brightness temperature of a medium, the optical thickness τ should be expressed in Np; to convert dB to nepers, use $\tau(\mathrm{Np}) = (1/4.34)\tau(\mathrm{dB})$. ◄

Figure 8-14 displays the microwave spectrum of $\Upsilon_0(0,\infty)$ for three different atmospheres,† characterized by different surface temperatures T_0 and total integrated water-vapor contents M_v (see definition given by Eq. (8.8)). In the window bands, the dry polar atmosphere is far more transparent than the humid atmosphere of the tropics.

†**Computer Code 8.9.**

8-4 Emission by the Clear Atmosphere

For an upward-looking radiometer, the downward emission by the atmosphere (excluding cosmic background radiation) is denoted $T_{DN}(f,\theta)$ and is given by Eq. (6.76) as

$$T_{DN}(f,\theta) = \sec\theta \int_0^\infty \kappa_a(f,z)\, T(z)\, e^{-\tau_0(0,z)\sec\theta}\, dz, \tag{8.23}$$

where $\kappa_a(f,z)$ and $T(z)$ are the absorption coefficient and thermometric temperature, respectively, of the atmosphere at height z. In the absence of clouds and rain, $\kappa_a = \kappa_g$, the gaseous absorption coefficient. Figure 8-15 presents plots of T_{DN} for $\theta = 0$ (zenith-looking) for the same model atmosphere and conditions that pertain to Fig. 8-10. As expected, the high attenuation levels at and around the resonant frequencies of the oxygen absorption lines at 60 and 118 GHz cause T_{DN} to "saturate" at a value only slightly lower than the surface temperature of 290 K. The same is true for the

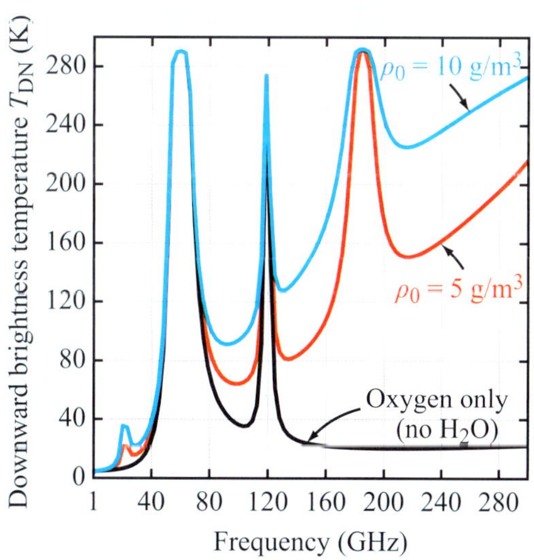

Figure 8-15: Zenith brightness temperature of the atmosphere T_{DN} for 0, 5, and 10 g/m^3 surface water vapor (2.1 km scale height) computed using the Standard Atmosphere model with surface temperature 290 K.

183 and 323 GHz water-vapor absorption lines (except for $\rho = 0$).

Since emission is linked to absorption, the spectrum of the radiometric temperature T_{DN} is similar in shape to the atmospheric opacity spectrum of Fig. 8-10, particularly in those regions of the spectrum where τ_0 does not exceed 2 to 3 dB. However, in the high attenuation regions, around the 60 GHz oxygen complex and the 118.75 and 183.31 GHz lines, the atmosphere behaves like a blackbody with a radiometric temperature approximately equal to the weighted mean of the atmospheric temperature profile, where the weighting function accounts for the relative contributions of the various atmospheric layers to T_{DN}.

In addition to the downwelling radiation represented by T_{DN}, an upward-looking radiometer receives a relatively small amount of energy due to cosmic and galactic radiation incident upon the atmosphere from the top (Fig. 8-16). Thus, the apparent temperature of the sky, usually referred to as the ***sky radiometric temperature***, T_{SKY}, consists of

$$T_{SKY}(\theta) = T_{EXTRA} e^{-\tau_0 \sec\theta} + T_{DN}(\theta) \tag{8.24a}$$

and

$$T_{EXTRA} = T_{COS} + T_{GAL}, \tag{8.24b}$$

where τ_0 is the total zenith atmospheric opacity, and T_{COS} and T_{GAL} are the ***cosmic*** and ***galactic*** brightness temperatures, respectively, whose sum we call the ***extraterrestrial brightness temperature***, T_{EXTRA}. The cosmic radiation contribution is independent of frequency and the zenith angle θ, and has the constant value $T_{COS} = 2.7$ K. The galactic contribution is due to radiation from our own galaxy. Its magnitude is direction-dependent due to the nonuniformity of radiation emitted by different parts of the galaxy; T_{GAL} is a maximum in the direction of the center of the galaxy and a minimum in the direction of the galactic pole. Plots of $T_{SKY}(\theta)$ are shown in Fig. 8-17 (solid curves) for several values of θ, calculated for minimum galactic emission.[†] Also shown is a dashed line corresponding to T_{GAL}(max), which arises from a narrow belt extending about 3° in the galactic plane and concentrated toward

[†] **Computer Code 8.11.**

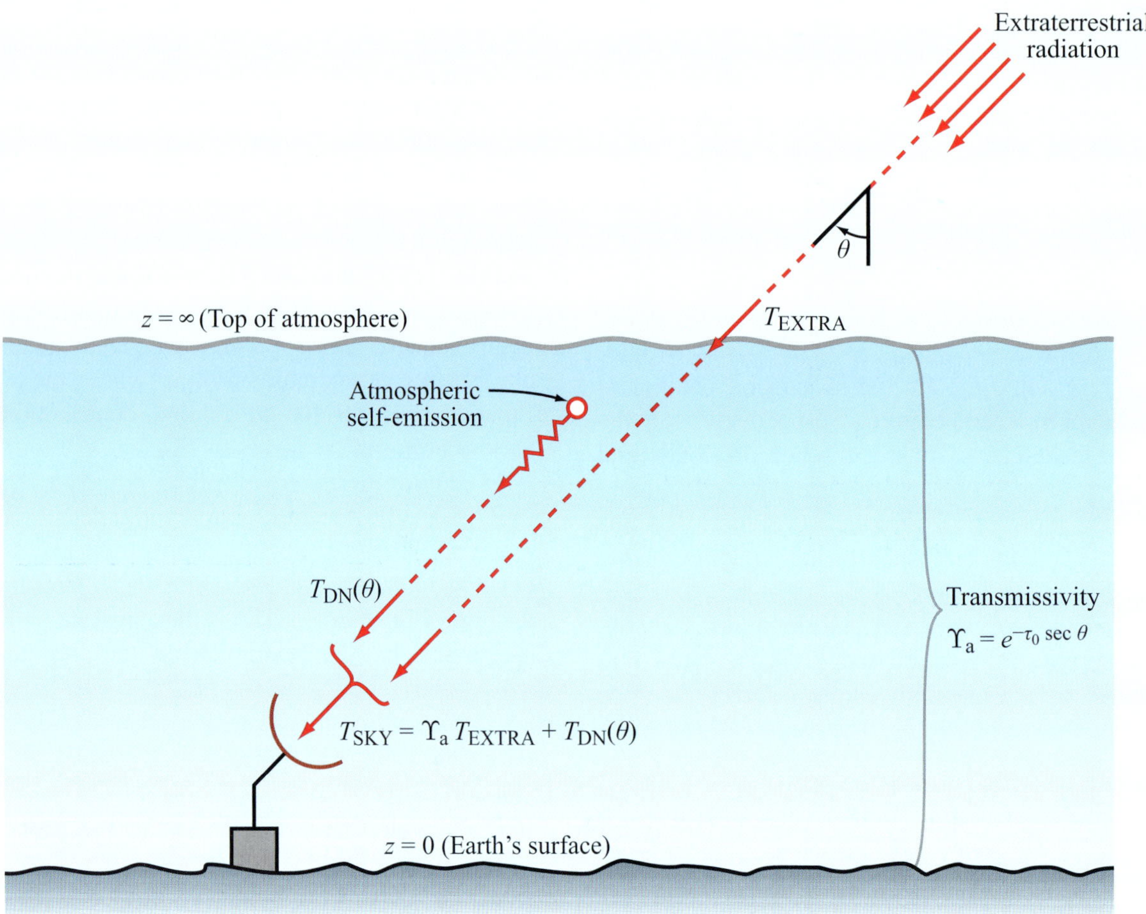

Figure 8-16: The radiometric temperature measured by an upward-looking radiometer includes contributions from our galaxy.

the galactic center. The frequency dependence of T_{GAL} varies between $f^{-2.5}$ and f^{-3}, depending on the specific region of the galaxy. Above about 5 GHz, T_{GAL} may be neglected in comparison to T_{DN}. Below 1 GHz, however, the contribution of T_{GAL} may not be ignored.

The wide range of values that T_{GAL} may assume (between $T_{\text{GAL}}(\text{min})$ and $T_{\text{GAL}}(\text{max})$) poses a serious problem for the use of microwave radiometry for Earth observations at frequencies below 1 GHz. This factor, together with the problem of interference from man-made radio sources (primarily radio and television transmitters) and the poor spatial resolution attainable with reasonable-size antennas, is responsible for the very limited use of frequencies below 1 GHz for radiometric Earth observations.

In the above discussion, contributions of point sources, such as the sun, were not included. The brightness temperature of the "quiet" sun decreases rapidly with increasing frequency from about 10^6 K at 100 MHz down to about 10^4 K at 10 GHz. Above 10 GHz, the solar brightness temperature decreases at a slower rate towards the value of 6000 K, which remains approximately constant above about 30 GHz. When sunspots and flares occur, the solar

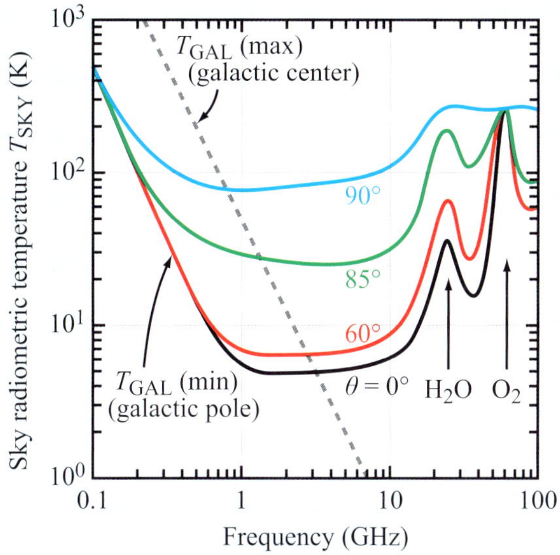

Figure 8-17: The sky radiometric temperature T_{SKY} includes contributions from our galaxy, which may be ignored if $f > 5$ GHz.

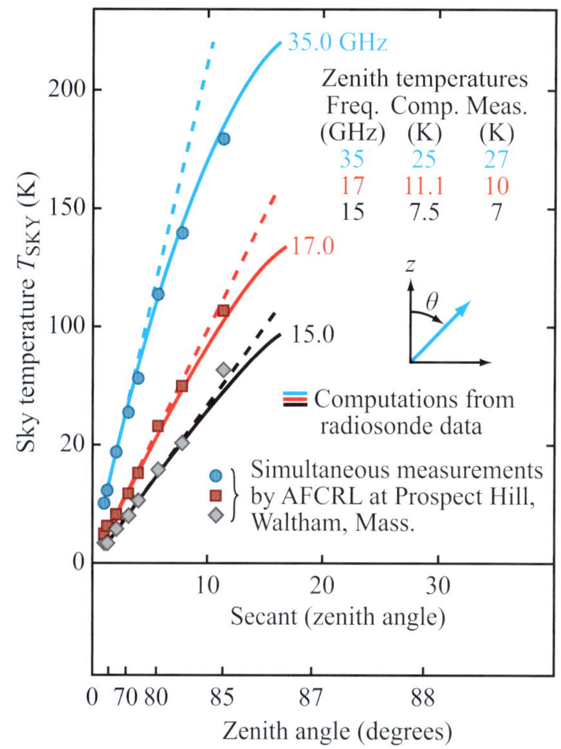

Figure 8-18: Comparison between measured and computed sky temperature as a function of $\sec\theta$, where θ is the zenith angle [Crane, 1971].

brightness temperature may increase by several orders of magnitude. The contribution received by an upward-looking radiometer depends on the direction of the sun relative to the antenna pattern and on the brightness temperature of the sun.

The atmospheric emission plots shown in Figs. 8-15 and 8-17 are for downwelling radiation such as would be observed by an upward-looking radiometer located at sea level. Ignoring the cosmic and galactic contributions, which are quite small in comparison with atmospheric emission above 5 GHz, the plots of $T_{DN}(\theta)$ are within a few degrees of what would be calculated for $T_{UP}(\theta)$, the upwelling atmospheric radiation that would be observed by a downward-looking radiometer located above the Earth's atmosphere (excluding emission from the Earth's surface). Hence, the curves may be used as approximate representatives of $T_{UP}(\theta)$.

It was stated earlier that for computational purposes, the approximation of horizontal atmospheric stratification (which led to the $\sec\theta$ dependence in Eq. (8.23)) may be used up to about 70° from the zenith. This statement is verified by the excellent agreement shown in Fig. 8-18 between measured values of $T_{SKY}(\theta)$ and the results of computations based on the above approximation. The agreement is good even for angles as large as 82°.

8-5 Extinction by Hydrometeors

The treatment thus far has been limited to clear-sky conditions, where absorption and emission are due solely to atmospheric gases. The interaction of electromagnetic radiation with hydrometeors (such as those in clouds, fog, snow, and rain) is different in nature and may involve both absorption and scattering. In this and the sections that follow, we first consider the interaction with a single particle and then proceed to treat the case of a large number of particles within the volume of interest.

▶ The volume extinction coefficient is governed by the density, shape, size distribution, and dielectric properties of the particles contained in the volume. ◀

In extending the results applicable to a single particle to the ensemble containing many particles, the particles usually are assumed to be randomly distributed within the volume, and therefore the contributions of the individual particles can be summed assuming an incoherent process (by considering the phases of the fields scattered by the particles as random). In addition to the above assumptions, one further assumption will be made in the discussions below, namely, that the particles are spherical, which is a reasonable assumption for most atmospheric water and ice droplets. Although models have been developed (Oguchi, 1973) to account for the nonspherical nature of raindrops, the differences between their results and those obtained for spherical droplets usually are smaller than the statistical uncertainties due to other parameters such as the drop-size distribution.

8-5.1 Electromagnetic Interaction with Individual Spherical Particles

If $\mathcal{S}_i$ is the power density (W/m^2) of an electromagnetic wave incident upon a suspended material particle of geometrical cross-sectional area A, then a fraction of the incident power is absorbed by the particle, and an additional fraction is scattered by the particle in all directions. The ratio of absorbed power P_a to incident power density $\mathcal{S}_i$ is known as the ***absorption cross section*** Q_a:

$$Q_a = \frac{P_a}{\mathcal{S}_i} \quad \text{m}^2, \tag{8.25a}$$

and the ratio of Q_a to the physical cross section A is known as the ***absorption efficiency factor*** ξ_a. For a spherical particle of radius r, $A = \pi r^2$ and therefore

$$\xi_a = \frac{Q_a}{\pi r^2}. \tag{8.25b}$$

If the incident plane wave is traveling in the z direction and if $\mathcal{S}_s(\theta,\phi)$ is the power density of the radiation scattered in the direction (θ,ϕ) at a distance R from the particle, the total power scattered by the particle is the integral of $\mathcal{S}_s(\theta,\phi)$ over the spherical surface of radius R centered on the particle:

$$P_s = \iint_{4\pi} \mathcal{S}_s(\theta,\phi)\, R^2\, d\Omega. \tag{8.26}$$

The ***scattering cross section*** Q_s and ***scattering efficiency factor*** ξ_s are defined as

$$Q_s = \frac{P_s}{\mathcal{S}_i} \quad \text{m}^2, \tag{8.27a}$$

$$\xi_s = \frac{Q_s}{\pi r^2}. \tag{8.27b}$$

The total power removed from the incident wave is $P_a + P_s$, and the corresponding ***extinction (or attenuation) cross section*** Q_e and efficiency ξ_e are:

$$Q_e = Q_a + Q_s, \tag{8.28a}$$

$$\xi_e = \xi_a + \xi_s. \tag{8.28b}$$

The scattered power given by Eq. (8.26) is the total power scattered by the particle in all possible directions. Of particular interest in radar meteorology is the power density scattered in the backward direction towards the radiation source. This is $\mathcal{S}_b = \mathcal{S}_s$ at $\theta = \pi$. In this connection, the ***radar backscattering cross section*** σ_b is defined such that σ_b multiplied by the incident power density $\mathcal{S}_i$ is equal to the total power radiated by an equivalent isotropic radiator, so that at a distance R from the scatterer, $\mathcal{S}_b$ is given by

$$\mathcal{S}_b = \frac{\mathcal{S}_i \sigma_b}{4\pi R^2}. \tag{8.29}$$

From Eq. (8.29), it follows that σ_b is given by

$$\sigma_b = 4\pi R^2 \frac{\mathcal{S}_b}{\mathcal{S}_i} \quad (\text{m}^2), \tag{8.30a}$$

(backscattering cross section)

and the corresponding ***backscattering efficiency*** ξ_b is

$$\xi_b = \frac{\sigma_b}{\pi r^2}. \tag{8.30b}$$

8-5.2 Mie Scattering

The solution for the scattering and absorption of electromagnetic waves by a dielectric sphere of arbitrary radius r is due to Mie (1908). His derivation, which has been restated by several authors, including Ishimaru (1991), is formulated in terms of two parameters, the ***normalized circumference*** χ and the relative index of refraction n:

$$\chi = k_\text{b} r = \frac{2\pi r}{\lambda_\text{b}} = \frac{2\pi r}{\lambda_0} \sqrt{\varepsilon'_\text{b}} \tag{8.31a}$$

and

$$n = \frac{n_\text{p}}{n_\text{b}} = \left(\frac{\varepsilon_\text{p}}{\varepsilon_\text{b}}\right)^{1/2} = \varepsilon^{1/2}, \tag{8.31b}$$

where k_b is the wave number in the background medium, ε'_b is the real part of the ***relative dielectric constant*** of the background medium, λ_b is the wavelength in the background medium, λ_0 is the free-space wavelength, and n_p and n_b are the complex indices of refraction of the particle (sphere) material and of the background medium, respectively, and ε_p and ε_b are the corresponding ***complex dielectric constants***. When the background medium is air, as is true in the atmosphere, we have $\varepsilon'_\text{b} = 1$, $n_\text{b} = 1$, and $\lambda_\text{b} = \lambda_0$. However, for other remote-sensing applications of Mie scattering, the background need not be air.

The results of Mie's solution lead to expressions for the scattering and extinction efficiencies of the sphere in the form of converging series, given by[†]

$$\xi_\text{s}(n,\chi) = \frac{2}{\chi^2} \sum_{l=1}^{\infty} (2l+1)(|a_l|^2 + |b_l|^2), \tag{8.32a}$$

$$\xi_\text{e}(n,\chi) = \frac{2}{\chi^2} \sum_{l=1}^{\infty} (2l+1)\, \mathfrak{Re}\{a_l + b_l\}, \tag{8.32b}$$

where a_l and b_l, known as the ***Mie coefficients***, are functions of n and χ. The formal expressions for the Mie coefficients involve Bessel functions of complex argument. For computational purposes, Deirmendjian (1969) developed an iterative procedure using the recursion formulas for Bessel functions. His development leads to the expressions

$$a_l = \frac{\left(\dfrac{A_l}{n} + \dfrac{l}{\chi}\right) \mathfrak{Re}\{W_l\} - \mathfrak{Re}\{W_{l-1}\}}{\left(\dfrac{A_l}{n} + \dfrac{l}{\chi}\right) W_l - W_{l-1}} \tag{8.33a}$$

and

$$b_l = \frac{\left(nA_l + \dfrac{l}{\chi}\right) \mathfrak{Re}\{W_l\} - \mathfrak{Re}\{W_{l-1}\}}{\left(nA_l + \dfrac{l}{\chi}\right) W_l - W_{l-1}}, \tag{8.33b}$$

where

$$W_l = \left(\frac{2l-1}{\chi}\right) W_{l-1} - W_{l-2} \tag{8.34}$$

with

$$W_0 = \sin\chi + j\cos\chi, \tag{8.35a}$$

$$W_{-1} = \cos\chi - j\sin\chi, \tag{8.35b}$$

and

$$A_l = -\frac{l}{n\chi} + \left[\frac{l}{n\chi} - A_{l-1}\right]^{-1} \tag{8.36}$$

with

$$A_0 = \cot n\chi. \tag{8.37}$$

The index of refraction n is in general complex,

$$n = n' - jn'', \tag{8.38}$$

with

$$n' = \mathfrak{Re}\{\varepsilon^{1/2}\} = \left[\frac{\varepsilon' + \sqrt{(\varepsilon')^2 + (\varepsilon'')^2}}{2}\right]^{1/2}, \tag{8.39a}$$

$$n'' = -\mathfrak{Im}\{\varepsilon^{1/2}\} = \frac{\varepsilon''}{2n'}. \tag{8.39b}$$

The above computational form of the Mie coefficients, which involves no approximations, is substantially simpler to use for machine processing than the traditional expressions involving Bessel functions of the complex argument.

[†] **Computer Code 8.12.**

Mie's solution for the field scattered by a spherical particle leads to the following expression for the ***backscattering efficiency*** ξ_b

$$\xi_b = \frac{\sigma_b}{\pi r^2} = \frac{1}{\chi^2} \left| \sum_{l=1}^{\infty} (-1)^l (2l+1)(a_l - b_l) \right|^2 . \quad (8.40)$$

The Mie coefficients a_l and b_l are computed using the recursion procedure outlined earlier.

8-5.3 Rayleigh Approximation

If the particle size is much smaller than the wavelength of the incident wave, so that $|n\chi| \ll 1$, the Mie expressions for ξ_s and ξ_e may be shown to reduce to simple well-known expressions, known as the ***Rayleigh approximations*** (van de Hulst, 1957). Specifically, if only the most significant terms of the series expansion of ξ_s and ξ_e are retained, the Mie expressions take the form

$$\xi_s = \tfrac{8}{3}\, \chi^4 |K|^2 + \cdots \quad (8.41)$$

and

$$\xi_e = 4\chi\, \Im\mathfrak{m}\{-K\} + \tfrac{8}{3}\, \chi^4 |K|^2 + \cdots , \quad (8.42)$$

where K is a complex quantity defined in terms of the complex index of refraction n (of the sphere to the background medium),

$$K = \frac{n^2-1}{n^2+2} = \frac{\varepsilon - 1}{\varepsilon + 2} . \quad (8.43)$$

The absorption efficiency ξ_a is obtained from Eqs. (8.41) and (8.42):

$$\xi_a = \xi_e - \xi_s = 4\chi\, \Im\mathfrak{m}\{-K\}. \quad (8.44)$$

The corresponding scattering and absorption cross sections are

$$Q_s = \frac{2\lambda^2}{3\pi}\, \chi^6 |K|^2 , \quad (8.45a)$$

$$Q_a = \frac{\lambda^2}{\pi}\, \chi^3\, \Im\mathfrak{m}\{-K\}. \quad (8.45b)$$

Because Q_s varies as χ^6 and Q_a varies as χ^3, Q_a is usually much larger than Q_s in the Rayleigh region ($\chi \ll 1$) unless the particle is made of a very weakly absorbing material (i.e., $n'' \ll n'$ such that $\Im\mathfrak{m}\{-K\} \ll |K|^2$).

> ▶ For most computational purposes, the Rayleigh approximations yield acceptable accuracies if $|n\chi| < 0.5$. ◀

Comparison of the Rayleigh scattering values with the exact values obtained from the Mie expressions is discussed in a later section for water and ice droplets.

In the Rayleigh region, the backscattering efficiency given by Eq. (8.40) simplifies to (Kerr, 1951)

$$\xi_b = 4\chi^4 |K|^2 , \qquad |n\chi| < 0.5, \quad (8.46)$$

which is known as ***Rayleigh's backscattering law***. If $|n| \gg 1$ but finite, $|K| \approx 1$. The corresponding expression for ξ_b is

$$\xi_b \approx 4\chi^4 \qquad \text{for finite } |n| \gg 1 \text{ and } |n\chi| < 0.5. \quad (8.47)$$

For a sphere of infinite n, such as a perfectly conducting metal sphere, Eq. (8.46) is not applicable, because it is the result of an approximation of the Mie expression for $|n\chi| < 0.5$. A solution obtained directly from the Mie coefficients a_l and b_l for $|n| = \infty$ and $\chi \ll 1$ leads to (Kerr, 1951)

$$\xi_b = 9\chi^4 \qquad \text{for } |n| = \infty \text{ and } |\chi| \ll 1. \quad (8.48)$$

(conducting sphere)

8-5.4 Frequency Response

The Mie parameter χ is given by

$$\chi = \frac{2\pi r}{\lambda_0} = \frac{2\pi r}{u_p}\, f,$$

where u_p is the phase velocity of the EM wave and f is its frequency. If the particle is in air, $u_p = c$. For a fixed particle radius r, increasing χ is equivalent to

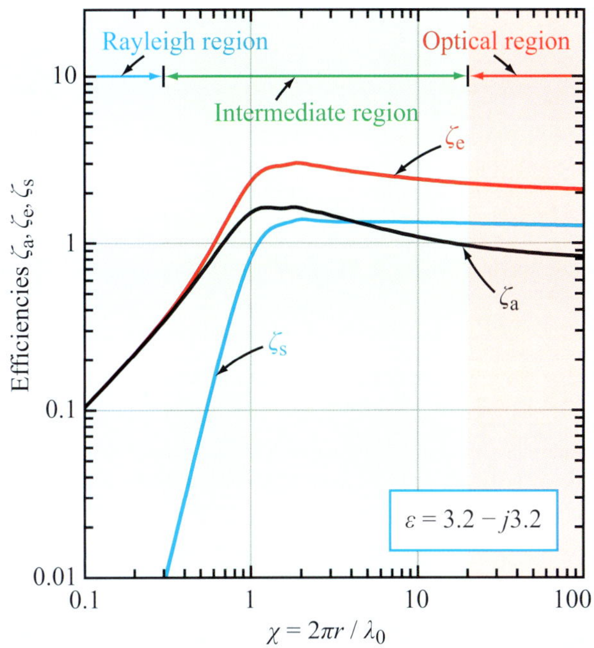

Figure 8-19: Calculated extinction, absorption, and scattering efficiencies as a function of χ for a sphere with with $\varepsilon = 3.2(1 - j1)$.

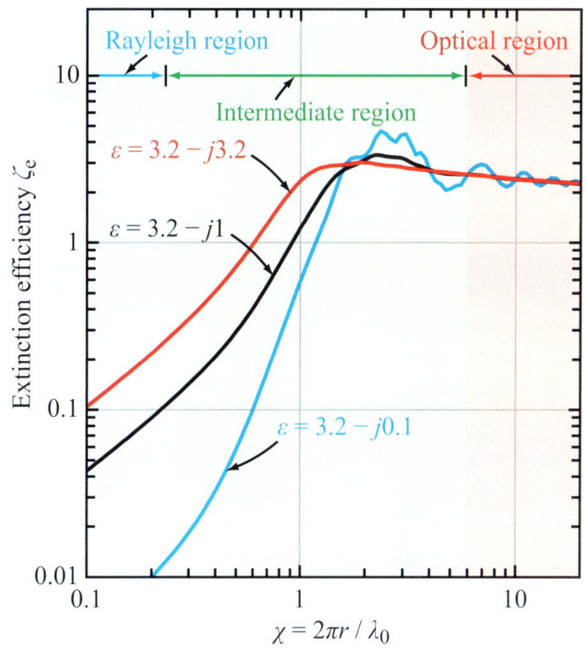

Figure 8-20: The role of $\varepsilon''/\varepsilon'$.

increasing f. Hence, we sometimes refer to plots of the extinction, absorption, scattering, and backscattering efficiencies as a function of χ as their ***frequency responses***.

Figure 8-19 displays the frequency responses of ξ_e, ξ_a, and ξ_s for a particle with $\varepsilon = 3.2(1 - j1)$. The χ-range is divided into three regions: (a) a ***low-frequency (Rayleigh) region*** defined (in the preceding subsection) as $|n|\chi < 0.5$, (b) a high-frequency (optical) region, and (c) an intermediate region. We discern the following behavior:

(1) In the low-frequency region, $\xi_s \ll \xi_a$, so $\xi_e = \xi_s + \xi_a \approx \xi_a$.

(2) In the low-frequency region, ξ_e varies as χ^4.

(3) In the intermediate region, ξ_s becomes comparable in magnitude to ξ_a, and eventually exceeds it in the high-frequency region.

The role of $\varepsilon''/\varepsilon'$ is explored in Fig. 8-20, which displays plots of the extinction efficiency ξ_e for three particles, all with the same $\varepsilon' = 3.2$, but with different values of ε''. We observe that:

(1) The ratio $\varepsilon''/\varepsilon'$ is very important in the low-frequency region (because ξ_e is dominated by absorption, and absorption is strongly dependent on $\varepsilon''/\varepsilon'$), but less so in the intermediate and high-frequency regions.

(2) The extinction efficiency exhibits an oscillatory behavior in the intermediate-frequency region if $\varepsilon''/\varepsilon'$ is small.

For radar-related applications, the quantities of interest are the extinction cross section Q_e and the backscattering cross section σ_b. Figure 8-21 displays a plot of the backscattering efficiency $\xi_b = \sigma_b/(\pi r^2)$ as a function of χ for a perfectly conducting sphere with $|n| = \infty$. The condition we used earlier for defining the acceptable upper limit of the Rayleigh region—namely, $|n|\chi < 0.5$—no longer applies to the perfectly conducting sphere. Comparison of ξ_b computed using the Mie expression with that computed using the Rayleigh approximation reveals that the Rayleigh

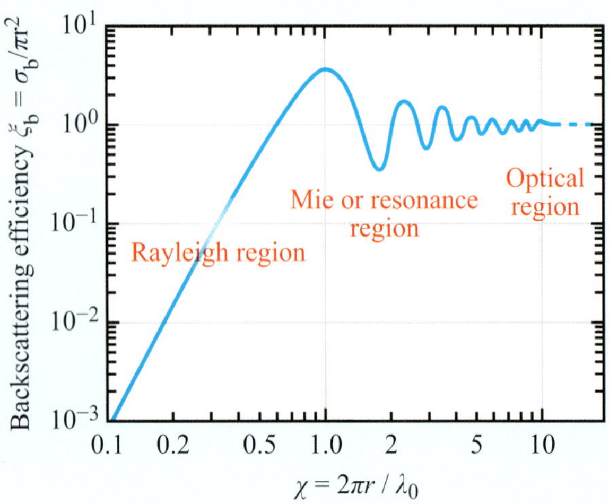

Figure 8-21: Backscattering efficiency of a perfectly conducting sphere.

approximation incurs an error of less than a few percent if

$\chi < 0.7$ **(Rayleigh region for conducting sphere)**.

In the Rayleigh region, ξ_b varies as χ^4, and in the intermediate region it exhibits an oscillatory behavior. For any sphere of index of refraction n, the optical limit of ξ_b is

$$\lim_{\chi \to \infty} \xi_b(\chi) = \left| \frac{n-1}{n+1} \right|^2 . \tag{8.49}$$

For a perfectly conducting sphere with $|n| = \infty$, the optical limit is $\xi_b = 1$.

8-6 Dielectric Properties of Hydrometeors

According to the previous section, the scattering and absorption characteristics of a spherical particle present in air are governed by three parameters: (a) the EM wavelength λ_0, (b) the complex index of refraction of the particle, n (or equivalently, the particle's complex dielectric constant $\varepsilon = n^2$), and (c) the particle radius r. This section provides an overview of the dielectric properties of water, ice, and snow, extracted from Chapter 4.

8-6.1 Water Particles

The complex dielectric constant of pure water (containing no salts or other solvents) is a function of only two parameters, the EM frequency f and the water temperature T. Figure 8-22 displays the spectra of ε'_w and ε''_w at 0 °C and 20 °C for the 1–50 GHz range. From dc to about 1 GHz, the spectrum of the dielectric permittivity $\varepsilon'_w \approx 88$ at 0 °C (and ≈ 92 at 20 °C). Beyond 1 GHz, ε'_w transitions into a much-lower high-frequency plateau at a level $\varepsilon'_w \approx 4$ (as $f \to \infty$). The transition passes through an intermediate point at which the dielectric loss factor ε''_w is a maximum. The frequency corresponding to the peak value of ε''_w is called the ***relaxation frequency*** f_0 of the water molecule. At $T = 0$ °C, $f_0 = 8.9$ GHz, and it increases with increasing temperature; at 20 °C, $f_0 = 16.7$ GHz.

> ▶ Because both ε'_w and ε''_w have large magnitudes in the 1–50 GHz range, a water particle exhibits strong scattering and strong absorption in this part of the microwave spectrum. ◀

8-6.2 Ice Particles

The microwave relative permittivity of pure ice, ε'_i, is much smaller in magnitude than that of pure water. At 1 GHz, for example, $\varepsilon'_i \approx 3.2$, compared with $\varepsilon'_w \approx 90$ for water. Furthermore, according to both theory and experimental measurements (see Section 4-3), ε'_i is essentially independent of frequency over a wide range extending between 10 MHz and 300 GHz, and it is very weakly dependent on temperature. So, for all practical purposes, ε'_i is a constant:

$$\varepsilon'_i \approx 3.2. \tag{8.50}$$

The dielectric loss factor of pure ice, ε''_i, is at least two orders of magnitude smaller than ε'_i (for $f < 100$ GHz), but it exhibits both a frequency dependence and a temperature dependence. According to the model-calculated plots and the accompanying experimental

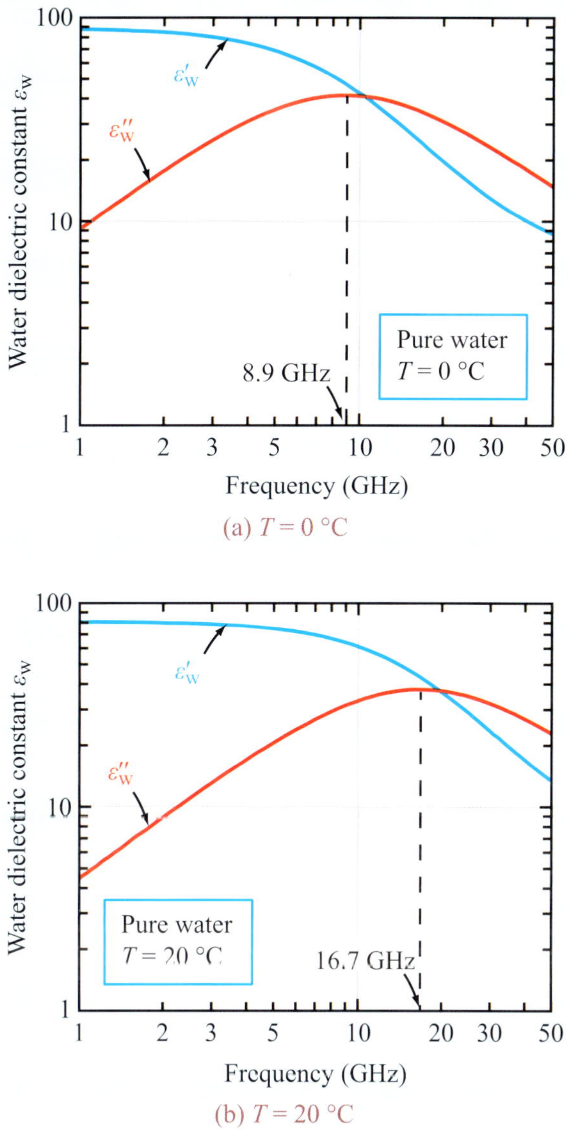

Figure 8-22: Dielectric spectrum of pure water at (a) 0 °C and (b) 20 °C.

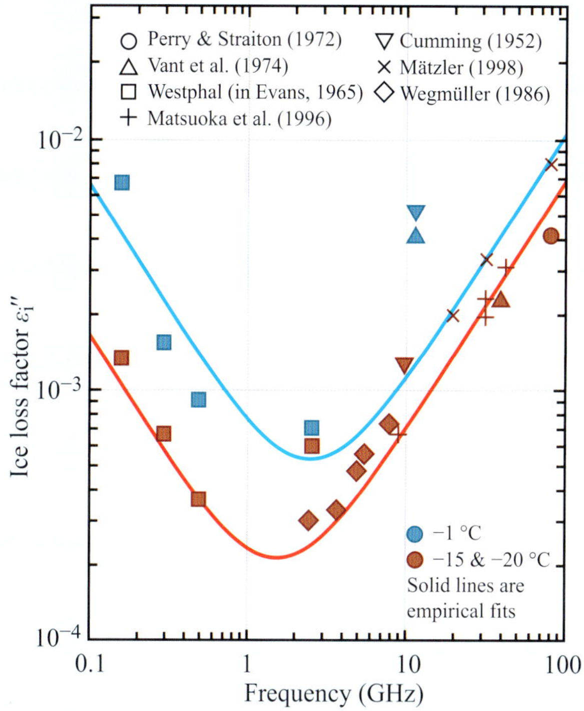

Figure 8-23: Loss factor of pure ice.

data shown in Fig. 8-23, ε''_i exhibits a minimum in the 1–5 GHz range (and the location of the minimum depends on the temperature T).

For a dielectric material with $\varepsilon'' < \varepsilon'$, the absorption coefficient is given by the approximate expression

$$\kappa_a \approx k_0 \frac{\varepsilon''}{\sqrt{\varepsilon'}} . \tag{8.51}$$

Because ε''_i of ice is < 0.01 in the microwave region, ice is a weak absorber of microwaves, in stark contrast with water.

8-6.3 Snowflakes

According to the review by Atlas (1964) on scattering by spheroids, the scattering and absorption efficiencies of an ice particle are only weakly dependent on the shape of the particle. Hence, a snowflake, nonspherical in shape, may be treated using the Rayleigh expression for a spherical particle of the same mass provided the Rayleigh criterion applies.

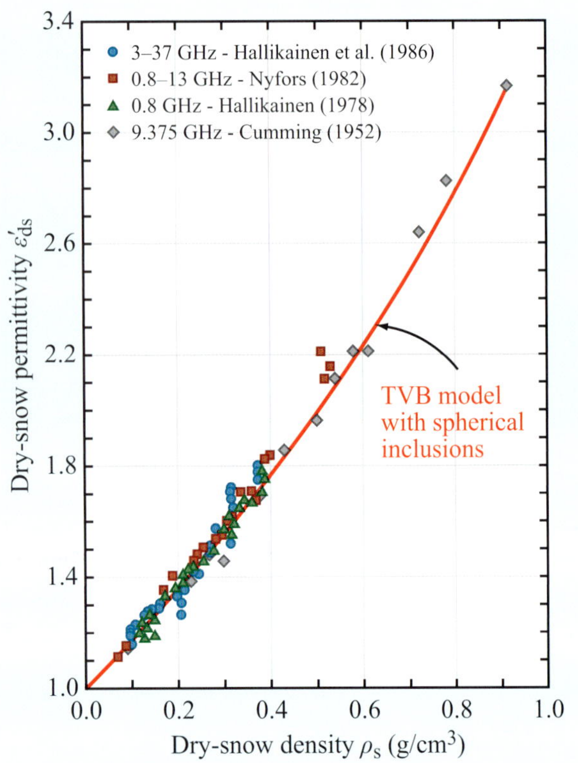

Figure 8-24: Measured permittivity of dry snow as a function of snow density [see Section 4-6].

Snowflakes are a mixture of air and ice crystals. The density of pure ice is $\rho_i = 0.9167$ g/cm^3 and the density of a snowflake, ρ_s, usually is between 0.05 and 0.3 g/cm^3. Since $\varepsilon'_{air} = 1$ and $\varepsilon'_i \approx 3.2$, if we model snowflakes as equivalent spheres of ice surrounded by air, we obtain the relationship given by Eq. (4.54) for permittivity of dry snow:

$$\varepsilon'_{ds} = \frac{1 + 0.92\rho_s}{1 - 0.46\rho_s}. \tag{8.52}$$

A plot of Eq. (8.52) is shown in Fig. 8-24, together with experimental measurements made across the 0.8 to 37 GHz range (the reader is referred to Section 4-6 for more details).

The same snow dielectric model leads to the following approximate relationship for the ratio of the loss factor ε''_{ds} to the permittivity ε'_{ds}:

$$\frac{\varepsilon''_{ds}}{\varepsilon'_{ds}} = \frac{0.37\rho_s}{(1 - 0.46\rho_s)^2}, \tag{8.53}$$

where ρ_s is the snow density. For falling snow, ρ_s is the average density of snowflakes.

8-7 Extinction and Backscattering by Clouds, Fog, or Haze

In a cloud or rain volume, the scatterers (particles) usually are assumed to be randomly distributed within the volume, so that there are no coherent phase relationships between the fields scattered or emitted by the individual particles, thereby allowing the use of incoherent scattering theory to compute the absorption and scattering by a volume containing many particles. Additionally, the concentration of particles usually is small enough to support the assumption that shadowing of one particle by another may be ignored. These two assumptions lead to the conclusion that the total scattering cross section of a given volume is equal to the algebraic sum of the scattering cross sections of all the individual particles contained in that volume. Similar statements apply to the absorption and backscattering cross sections.

The ***volume scattering coefficient*** κ_s is the total scattering cross section per unit volume, and its units are $(\text{Np/m}^3) \times \text{m}^2 = \text{Np/m}$. The range of sizes of the particles contained in a cloud or rain mass usually is described by a continuous function known as the ***drop-size distribution*** $p(r)$, which defines the partial concentration of particles per unit volume and per unit increment of the radius r. Hence κ_s is given by

$$\kappa_s = \int_{r_1}^{r_2} p(r)\, Q_s(r)\, dr, \tag{8.54}$$

where $Q_s(r)$ is the scattering cross section of a sphere of radius r, and r_1 and r_2 are the lower and upper limits, respectively, of drop radii contained in the cloud.

For numerical computation, it sometimes is convenient to express κ_s in terms of the ***scattering***

Table 8-3: Properties of standard cloud models [from Fraser et al., 1975].

Descriptive cloud name	Cloud base (m)	Height top (m)	Mass density liquid H_2O (g/m^3)	Mode radius of distribution r_c (μm)	Shape parameters α	γ	Principal composition
Cirrostratus, midlatitude	5000.	7000.	0.10	40.0	6.0	0.5	Ice
Low-lying stratus	500.	1000.	0.25	10.0	6.0	1.0	Water
Fog layer	0.	50.	0.15	20.0	7.0	2.0	Water
Haze, heavy	0.	1500.	10^{-3}	0.05	1.0	0.5	Water
Fair-weather cumulus	500.	1000.	0.50	10.0	6.0	0.5	Water
Cumulus congestus	1600.	2000.	0.80	20.0	5.0	0.3	Water

efficiency $\xi_s = Q_s/\pi r^2$ and the dimensionless parameter $\chi = 2\pi r/\lambda_0$. Hence, Eq. (8.54) assumes the form

$$\kappa_s = \frac{\lambda_0^3}{8\pi^2}\int_0^\infty \chi^2\, p(\chi)\,\xi_s(\chi)\,d\chi, \qquad (8.55)$$

where the limits on the integral cover the entire range of possible values of χ, it being understood that $p(\chi) = 0$ for $r < r_1$ or $r > r_2$. Similar expressions may be written for the volume ***absorption***, ***extinction***, and ***backscattering*** coefficients κ_a, κ_e, and σ_v by replacing ξ_s in Eq. (8.55) with ξ_a, ξ_e, and ξ_b, respectively. The volume backscattering coefficient σ_v often is called the ***radar reflectivity***.

8-7.1 Drop-Size Distribution

Cloud nomenclature and models are defined according to several cloud parameters, the most significant of which are: (1) the ***liquid-water content per unit volume***, m_v (g/m^3), (2) the drop-size distribution $p(r)$, (3) the principal ***composition*** (water, ice, or rain), and (4) the height of the cloud base above the ground. Table 8-3 provides a list of properties for each of six cloud models, which include ice and water clouds, fog, and haze. Three of the listed parameters can be related to the drop-size distribution formulated by Deirmendjian (1969):

$$p(r) = ar^\alpha \exp(-br^\gamma), \qquad 0 \le r \le \infty, \qquad (8.56)$$

which vanishes at $r = 0$ and $r = \infty$. Deirmendjian calls it a ***modified gamma distribution***, since it reduces to the gamma distribution when $\gamma = 1$. For a given drop-size distribution, a, α, b, and γ are positive, real constants and are related to the physical properties of the cloud. For convenience in mathematical manipulations, the constant α is specified to be an integer.

The total number of particles per unit volume, N_v, is obtained by integrating $p(r)$ over all values of r:

$$N_v = \int_0^\infty p(r)\,dr = a\int_0^\infty r^\alpha \exp(-br^\gamma)\,dr$$
$$= \frac{a\Gamma(x)}{\gamma b^x} \qquad (m^{-3}), \qquad (8.57)$$

where

$$x = \frac{\alpha+1}{\gamma} \qquad (8.58)$$

and $\Gamma(\ )$ is the standard gamma function given by

$$\Gamma(n) = (n-1)! \qquad \text{for } n = \text{positive integer}, \qquad (8.59a)$$

and

$$\Gamma(x) = \int_0^\infty e^{-t} t^{x-1}\,dt, \qquad (8.59b)$$

for $x =$ positive, but not an integer.

Another parameter of interest is the mode radius of the distribution, r_c, defined as the radius at which $p(r)$ is maximum. Upon differentiating $p(r)$ with respect to r and setting the result equal to zero, the solution for r gives (Deirmendjian, 1969)

$$r_c^\gamma = \frac{\alpha}{b\gamma}, \qquad (8.60)$$

and the corresponding maximum density of the distribution is

$$p(r_c) = ar_c^{\alpha} \exp\left(-\frac{\alpha}{\gamma}\right). \qquad (8.61)$$

The ***water content*** (or mass density) of the cloud, m_v (g/m^3), is equal to the fractional volume occupied by the particles, V_p, multiplied by the density of water (= 10^6 g/m^3). The fractional volume V_p is obtained by multiplying $p(r)$ by $4\pi r^3/3$ and integrating. Thus,

$$m_v = \tfrac{4}{3} \times 10^6 a\pi \int_0^\infty r^{\alpha+3} \exp(-br^\gamma)\, dr$$
$$= \frac{4 \times 10^6 a\pi}{3\gamma b^y}\, \Gamma(y) \qquad (\text{g/m}^3), \qquad (8.62)$$

where

$$y = \frac{\alpha + 4}{\gamma}. \qquad (8.63)$$

By virtue of Eqs. (8.60) and (8.62), the distribution function $p(r)$ is completely specified by m_v, r_c, α, and γ. The ***normalized drop-size distribution*** $p_n(r)$ is the ratio of $p(r)$ to its peak value $p(r_c)$:

$$p_n(r) = \frac{p(r)}{p(r_c)}$$
$$= \left(\frac{r}{r_c}\right)^{\alpha} \exp\left\{-\frac{\alpha}{\gamma}\left[\left(\frac{r}{r_c}\right)^{\gamma} - 1\right]\right\}. \qquad (8.64)$$

Plots of $p_n(r)$ as a function of r are shown in Fig. 8-25 for two types of water clouds and for fog and haze. As formulated, the distributions are defined for all values of r between 0 and ∞, although clouds do not contain particles with radii much larger than r_c. The justification is based on the fact that $p_n(r)$ is very small for $r \gg r_c$. If desired, the distribution may be truncated at the largest value of r.

For a given drop-size distribution $p(r)$, the volume absorption, scattering and extinction coefficients, κ_a, κ_s, and κ_e (= $\kappa_a + \kappa_s$), and the volume backscattering coefficient σ_v can be obtained through expressions of the form given by Eq. (8.55), namely:

$$\kappa = \frac{\lambda_0^3}{8\pi^2} \int_0^\infty \chi^2\, p(\chi)\, \xi(\chi)\, d\chi, \qquad (8.65)$$

where $\chi = 2\pi r/\lambda_0$, and κ is κ_s, κ_a, κ_e, or σ_v corresponding to ξ_s, ξ_a, ξ_e, or ξ_b, respectively, in the integrand. In the general case, the ξ's are the Mie efficiencies defined in Section 8-5. However, if the wavelength λ_0 is such that the Rayleigh approximations are valid over the range of r of the droplets contained in the cloud, simpler expressions are obtained, as shown next.

8-7.2 Mie versus Rayleigh

The Rayleigh approximation yields results accurate to within approximately 1% of the Mie calculations for the absorption, scattering, extinction, and backscattering cross sections of a dielectric sphere of radius r and complex index of refraction n if

$$|n|\chi \lesssim 0.5 \qquad \textbf{(Rayleigh approximation)}, \qquad (8.66)$$

where $\chi = 2\pi r/\lambda_0$. What are the implications of this statement for clouds, fog, and rain?

The Mie extinction and scattering efficiencies, ξ_e and ξ_s, are shown in Figs. ref7-26 to 8-28 (from Fraser et al., 1975) as a function of drop radius r between 10 μm and 10,000 μm (= 10 mm). The three figures, corresponding to 3, 30, and 300 GHz, also include dashed lines representing the Rayleigh extinction efficiency. The heavy horizontal lines indicate the ranges of drop radii characteristic of two types of water clouds and a rain cloud with a rain rate of 25.4 mm/hr.

▶ At 3 GHz, the Rayleigh approximation is certainly applicable for the water clouds and is approximately valid for the rain cloud, while at 30 GHz the Rayleigh approximation is valid for the two water clouds only, and at 300 GHz, it is valid for the fair-weather cloud only. ◀

8-7.3 The Rayleigh Volume Extinction Coefficient

It was shown in Section 8-5 that, for individual particles, the Rayleigh approximation is applicable if $|n|\chi \leq 0.5$.

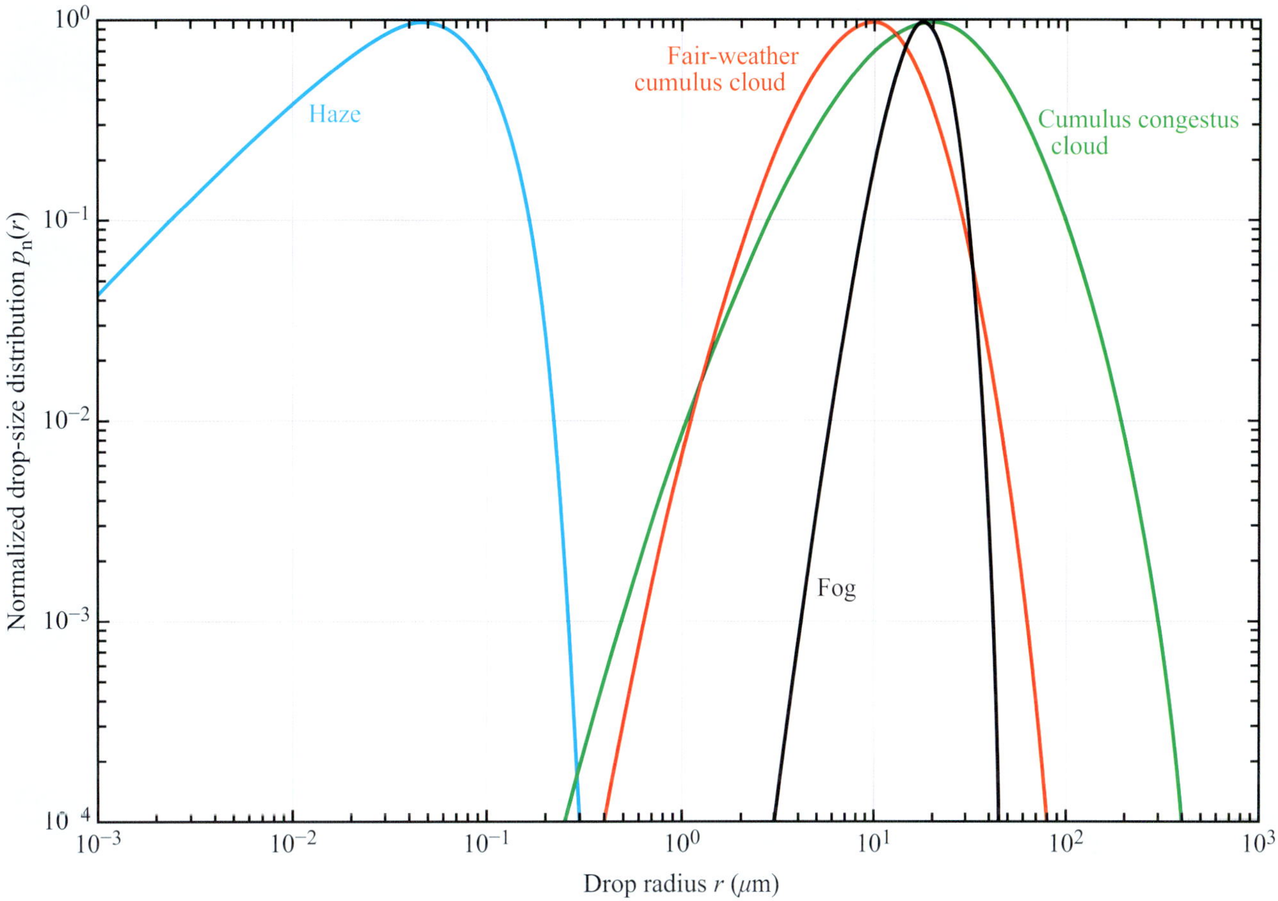

Figure 8-25: Normalized drop-size distributions for four cloud types. The cloud parameters are given in Table 8-3.

Most water clouds (except rain-bearing clouds) do not contain particles with radii larger than 0.1 mm. For the values of $|n_\text{w}| = |\varepsilon_\text{w}^{1/2}|$ corresponding to the plots shown in Fig. 8-22, the Rayleigh approximation applies to frequencies up to about 50 GHz. Ice clouds may contain particles with radii up to about 0.2 mm, but the refractive index of ice is smaller than that of water. The combination of these two factors leads to the conclusion that the Rayleigh criterion is applicable up to about 70 GHz for ice clouds.

For water and ice spheres, the refractive index is such that the absorption cross section Q_a is much larger than the scattering cross section Q_s in the Rayleigh region (see Section 8-5.4). The ***cloud volume extinction coefficient*** κ_e is therefore approximately equal to the volume absorption cross sections of all particles contained in a unit volume of the cloud:

$$\kappa_\text{e} = \sum_{i=1}^{N_\text{v}} Q_\text{a}(r_i), \tag{8.67}$$

where N_v is the number of particles per unit volume and r_i is the radius of the ith particle. Using Eq. (8.45b) for $Q_\text{a}(r_i)$, namely

$$Q_\text{a}(r_i) = \frac{8\pi^2}{\lambda_0}\, r_i^3\, \Im\text{m}\{-K\}, \tag{8.68}$$

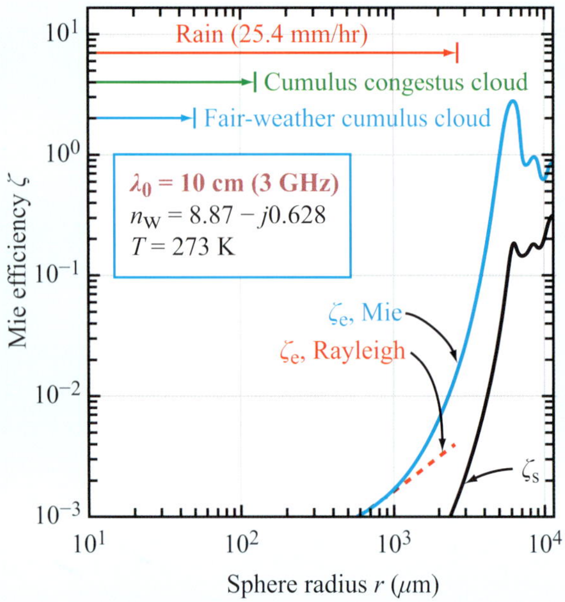

Figure 8-26: Mie efficiency factors for scattering and extinction by a water sphere as a function of a drop radius, at 3 GHz [Fraser et al., 1975]. Horizontal arrows indicate range of drop radii.

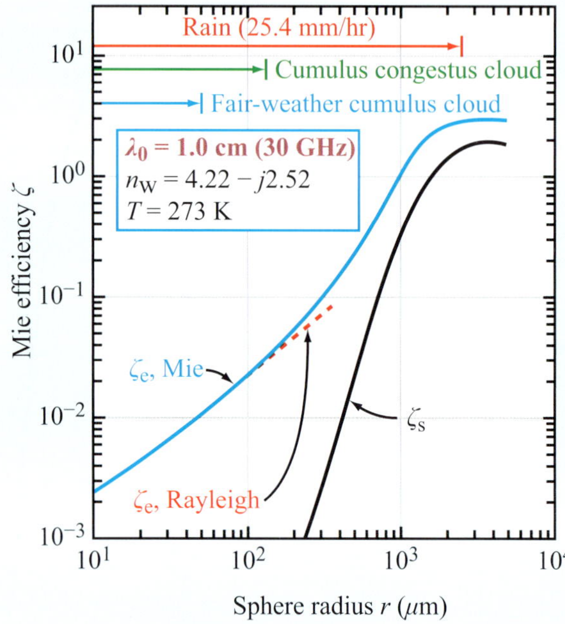

Figure 8-27: Mie efficiency factors for scattering and extinction by a water sphere as a function of drop radius, at 30 GHz [Fraser et al., 1975]. Horizontal arrows indicate drop radii.

Eq. (8.67) becomes

$$\kappa_e = \frac{8\pi^2}{\lambda_0}\mathfrak{Im}\{-K\}\sum_{i=1}^{N_v} r_i^3. \qquad (8.69)$$

The factor K is related to the relative dielectric constant of the droplet through Eq. (8.43).

The ***water content*** of the cloud is

$$m_v = 10^6 \sum_{i=1}^{N_v} \frac{4\pi}{3}\, r_i^3, \qquad (8.70)$$

where the multiplication factor 10^6 g/m^3 accounts for the density of water. Comparison of Eq. (8.69) with Eq. (8.70) leads to[†]

$$\kappa_e = \kappa_L m_v \qquad \text{(Np/m)}, \qquad (8.71a)$$

[†] **Computer Codes 8.4 and 8.5.**

where κ_L is a ***liquid extinction coefficient*** (corresponding to $m_v = 1$ g/m^3) given by

$$\kappa_L = \frac{6\pi}{\lambda_0}\mathfrak{Im}\{-K\}\times 10^{-6}. \qquad (8.71b)$$

In Eq. (8.71b) λ_0 is in meters. If λ_0 is expressed in cm and κ_e is expressed in dB/km by multiplying κ_e (Np/m) by 4.34×10^3, Eq. (8.71a) and Eq. (8.71b) become

$$\kappa_e = \kappa_L m_v \qquad \text{(dB/km)} \qquad (8.72)$$

and

$$\kappa_L = 0.434\,\frac{6\pi}{\lambda_0}\,\mathfrak{Im}\{-K\} \qquad [\text{dB/km/(g/m}^3)], \quad (8.73)$$

with λ_0 in cm. Plots of κ_L as a function of frequency are shown in Fig. 8-29 for water and ice clouds at several temperatures. We note that κ_L increases rapidly

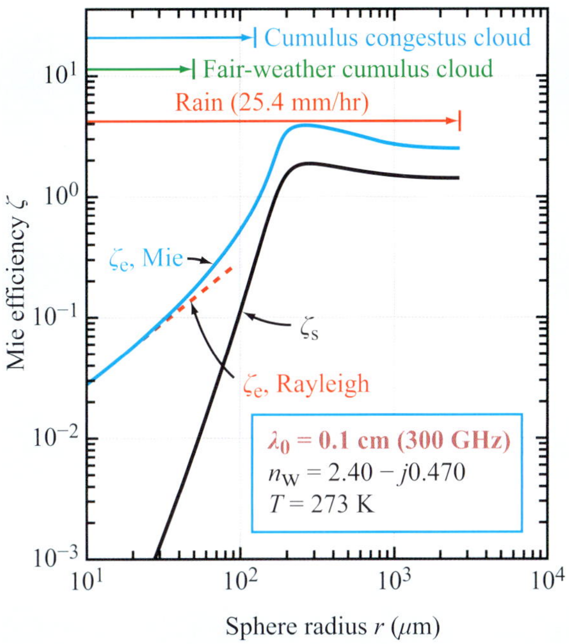

Figure 8-28: Mie efficiency factors for scattering and extinction by a water sphere as a function of drop radius, at 300 GHz [Fraser et al., 1975]. Horizontal arrows indicate range of drop radii.

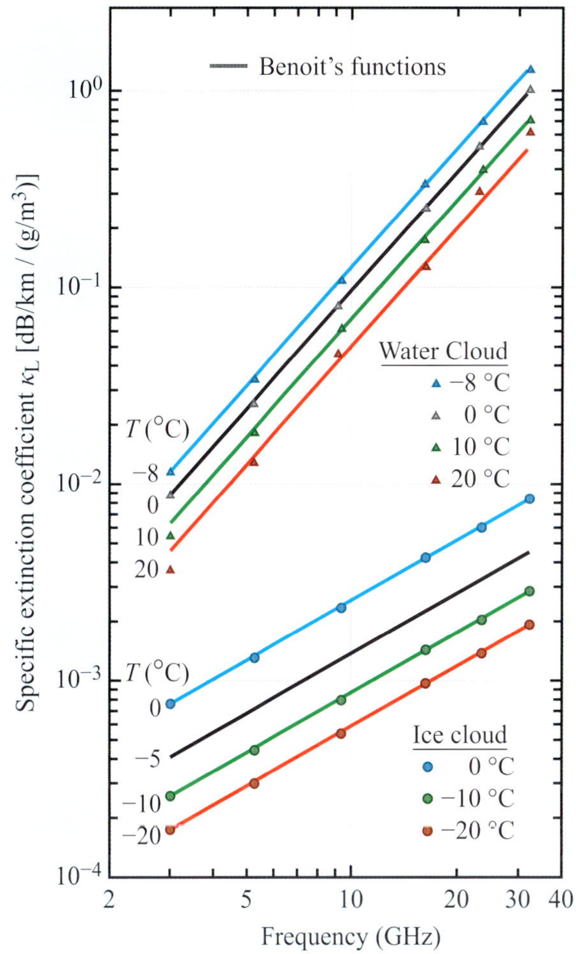

Figure 8-29: Attenuation coefficient for clouds and fog computed using Eq. (8.73) and the dielectric models in Chapter 4.

with increasing frequency f and increases slowly with decreasing temperature.

▶ The graphs also show that κ_L of ice clouds is one to two orders of magnitude smaller than that of water clouds and fog. This difference is attributed to the fact that ice spheres are considerably less lossy than water spheres at the same frequency. ◀

8-7.4 Cloud Attenuation Above 50 GHz

The form of Eq. (8.73) leads to a linear plot for κ_L versus f on a log-log scale, as shown in Fig. 8-29. However, this expression is not valid for frequencies much higher than 50 GHz.

Tsang et al. (1977) used the Mie expressions for spherical particles to compute the scattering and extinction coefficients for a water cloud at 273 K as a function of frequency. The cloud is characterized by a Deirmendjian drop-size distribution of the form given by Eq. (8.56), with a mode radius $r_c = 20$ μm, a water content $m_v = 0.8$ g/m^3, and distribution parameters $\alpha = 5$ and $\gamma = 0.3$. In Table 8-3 these specifications define a ***cumulus congestus*** cloud. The results of Tsang's computations, which are displayed in Fig. 8-30, show that the ***single-scattering albedo*** $a = \kappa_s/\kappa_e$ is indeed negligible for frequencies below 50 GHz. At higher frequencies or for clouds with larger particles,

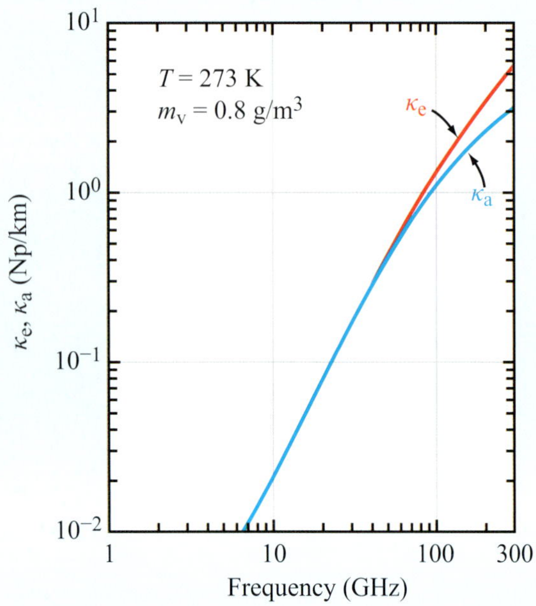

Figure 8-30: Calculated Mie extinction and absorption coefficients of a cloud with water content of 0.8 g/m^3 [from Tsang et al., 1977].

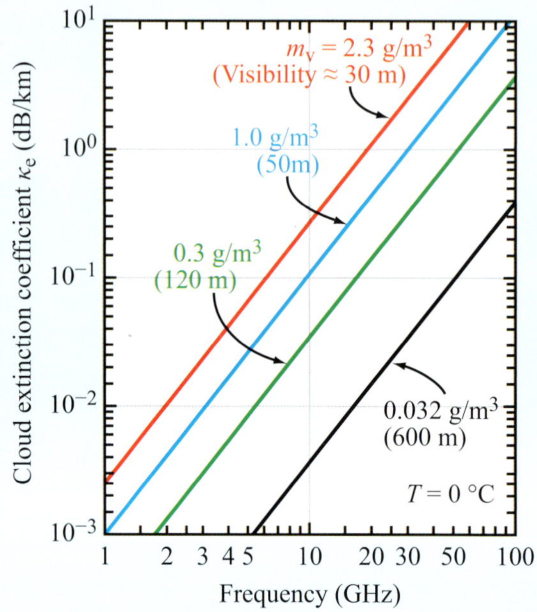

Figure 8-31: Extinction coefficient of clouds or fog for several water contents (and corresponding visibilities) at 0 °C [adapted from Barton, 1974].

however, scattering effects become quiet significant. Note that the linear dependence of κ_e on f (on a log-log scale) continues with frequency up to about 100 GHz, then at higher frequencies κ_e exhibits a shallower slope due to the weaker frequency dependence of the dielectric constant of water.

Before concluding this section, it is relevant to compare the attenuation by clouds and fog in the microwave region with the visible region of the electromagnetic spectrum. Figure 8-31 shows the attenuation coefficient κ_e (dB/km) as a function of frequency for several values of the cloud water content m_v. Corresponding to each value of m_v, a "visibility" range R_v is indicated. ***Visibility*** is defined as the greatest distance at which it is possible to identify (with the unaided eye) a dark object against the sky at the horizon (in the daytime). Mathematically, R_v may be defined as the range over which the total attenuation in the visible region ($\lambda = 0.4$–$0.7\ \mu$m) is approximately 18 dB, which corresponds to a transmission coefficient slightly larger than 1 percent. Thus,

$$\kappa_e R_v \approx 18 \text{ dB}. \tag{8.74}$$

For $R_v = 600$ m, $\kappa_e \approx 30$ dB/km in the visible region, compared to about 0.3 dB/km at 30 GHz.

8-7.5 Volume Backscattering Coefficient

In the Rayleigh region, the backscattering cross section of an individual spherical particle of radius r is given by

$$\sigma_b = \pi r^2 \xi_b = \frac{64\pi^5}{\lambda_0^4} r^6 |K|^2 \qquad (\text{m}^2), \tag{8.75}$$

where Eq. (8.46) was used for ξ_b. Summing over N_v particles per unit volume, the cloud ***volume backscattering coefficient*** σ_v is

$$\sigma_v = \sum_{i=1}^{N_v} \sigma_b(r_i) = \frac{64\pi^5}{\lambda_0^4} |K|^2 \sum_{i=1}^{N_v} r_i^6 \qquad (\text{m}^{-1}). \tag{8.76}$$

The ***reflectivity factor*** Z is defined as

$$Z = \sum_{i=1}^{N_v} d_i^6, \tag{8.77}$$

where $d_i = 2r_i$ is the diameter of the ith particle. Use of Eq. (8.77) leads to

$$\sigma_v = \frac{\pi^5}{\lambda_0^4} |K|^2 Z. \tag{8.78}$$

With d_i expressed in meters, Z has the dimensions of meters to the sixth power per cubic meter. Upon converting Z from m^6 per m^3 to mm^6 per m^3 and expressing λ_0 in centimeters (rather than meters), Eq. (8.78) becomes

$$\sigma_v = 10^{-10} \frac{\pi^5}{\lambda_0^4} |K|^2 Z \qquad (m^{-1}). \tag{8.79}$$

For water clouds, the reflectivity factor Z_w was related by Atlas (1964) to the cloud water content m_v (g/m^3) through the expression

$$Z_w = 4.8 \times 10^{-2} m_v^2 \qquad (mm^6/m^3). \tag{8.80}$$

Ice crystals in ice clouds can attain sizes an order of magnitude larger than water droplets found in water clouds. Hence, the reflectivity factor Z_i of an ice cloud may be several orders of magnitude larger than that of a water cloud with the same liquid water content m_v. Atlas (1964) derived the following relation for the reflectivity Z_i of an ice cloud:

$$Z_i = 9.21 \times 10^3 m_v^4. \tag{8.81}$$

Upon inserting Eq. (8.80) in Eq. (8.79), and similarly inserting Eq. (8.81) in Eq. (8.79), we obtain the following expressions for the volume backscattering coefficients of water clouds and ice clouds in the Rayleigh region:

$$\sigma_{vw} = \frac{1.47}{\lambda_0^4} \times 10^{-9} |K_w|^2 m_v^2 \quad (m^{-1}), \tag{8.82a}$$

(water cloud)

$$\sigma_{vi} = \frac{2.82}{\lambda_0^4} \times 10^{-9} |K_i|^2 m_v^2 \quad (m^{-1}). \tag{8.82b}$$

(ice cloud)

It should be noted that σ_v has the dimension of reciprocal length (m^{-1}), but it represents physically the backscattering cross section (m^2) per unit volume (m^3). As can be seen from Eq. (8.82), the magnitude of σ_v is quite small in the microwave region. The water content of a cloud typically is less than 1 g/m^3 and rarely exceeds 4 g/m^3. The factor $|K|^2$ for water varies between 0.89 and 0.93 over the 0–20 °C temperature range and 1–10 cm wavelength range. Hence, for $\lambda_0 = 1$ cm and $m_v = 1$ g/m^3, we have $\sigma_{vw} \approx 1.3 \times 10^{-9}$ m^{-1}.

► For ice, $|K_i|^2 \approx 0.2$, which is about 4.5 times smaller than $|K_w|^2$, but because of the much larger value of Z_i (in comparison with Z_w), ice clouds are much more readily detectable by radar. ◄

8-8 Extinction and Backscattering by Rain

Raindrops are typically two orders of magnitude larger in diameter than cloud droplets. Hence, whereas the Rayleigh approximation is valid up to about 50 GHz for most cloud types or even up to 300 GHz for fair-weather clouds, for rain the validity of the Rayleigh approximation is limited to rainfall rates of less than 10 millimeters per hour in the centimeter wavelength range, and to very low rainfall rates at frequencies above 30 GHz. Hence, in the general case, Mie scattering should be employed for computing absorption and scattering in a rain volume even at the lower microwave frequencies.

8-8.1 Drop-Size Distribution

The drop-size distribution for rain has been determined by several investigators, including Laws and Parsons (1943), Wexler (1948), Marshall and Palmer (1948), and Best (1950). Among these, the most widely used in the literature are the Laws-Parsons and Marshall-Palmer distributions, which compare reasonably well with one another. Graphical presentations are given in Fig. 8-32 for three different rainfall rates.

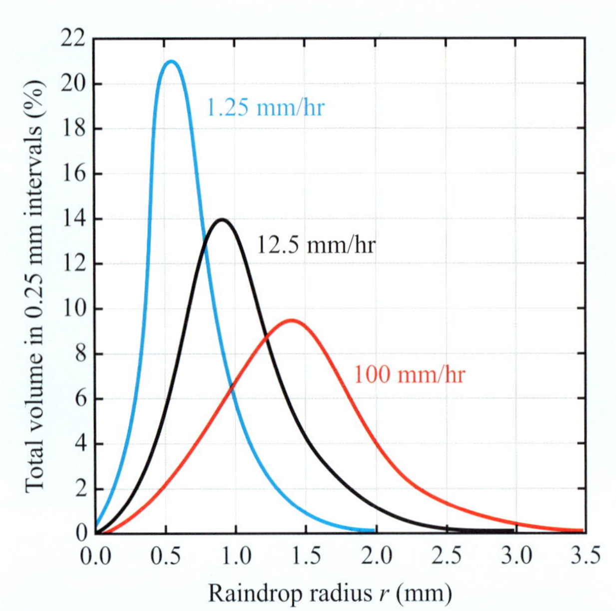

Figure 8-32: Laws-Parsons raindrop-size distributions [from Chu and Hogg, 1968].

Using drop-size distributions measured at the ground surface for rainfall intensities between 1 and 23 mm/hr, Marshall and Palmer (1948) developed the following expression for the distribution of drop diameter d:

$$p(d) = N_0 e^{-bd}, \tag{8.83}$$

where $p(d)$ is the number of drops of diameter d (in m) per unit volume per unit drop-diameter interval, $N_0 = 8.0 \times 10^6/\text{m}^4$, and b is related to the ***rainfall rate*** R_r (mm/hr) by

$$b = 4100 R_\text{r}^{-0.21}. \tag{8.84}$$

► Measurements of raindrop-size distributions show large variations for the same location, rain type, and rain rate. Consequently, drop-size distribution models should be regarded as representative of average, rather than individual, rainfall conditions. ◄

Drop-size distributions measured for falling snow have been reported be Sekhon and Srivastava (1970).

8-8.2 Volume Extinction Coefficient

For a given drop-size distribution $p(r)$, where r is the drop radius, the volume extinction coefficient κ_e of rain can be computed using Eq. (8.65):

$$\kappa_\text{e} = \frac{\lambda_0^3}{8\pi^2} \int_0^\infty \chi^2 \, p(\chi) \, \xi_\text{e}(\chi) \, d\chi, \tag{8.85}$$

where, as previously defined, $\chi = 2\pi r/\lambda_0$, and $\xi_\text{e}(\chi)$ is the Mie extinction coefficient given by Eq. (8.32b).

Figure 8-33 displays plots of κ_e as a function of frequency for several values of the rain rate R_r. The plots were calculated using the Laws and Parsons drop-size

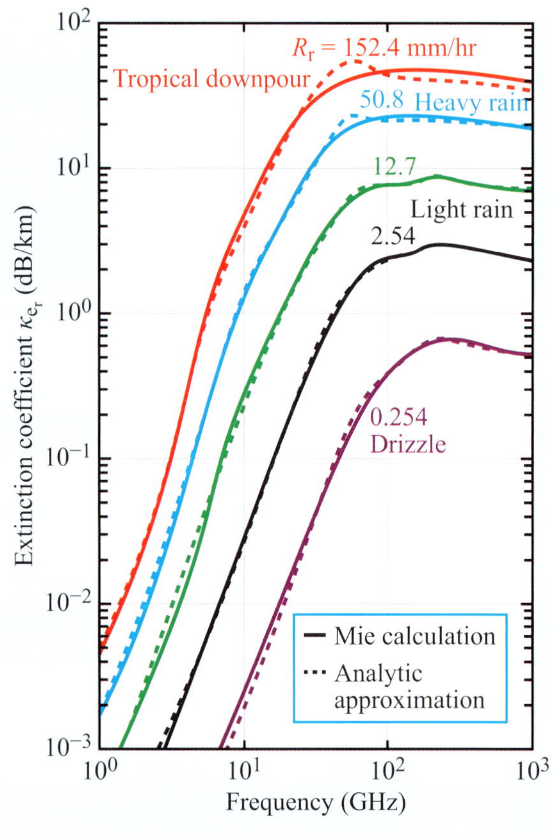

Figure 8-33: Comparison of Mie calculations with the analytic approximation given by Eqs. (8.86) and (8.87) for the extinction coefficient of rain at several rain rates [from Olsen et al., 1978].

distribution. Also shown in Fig. 8-33 are plots based on simple analytic expressions generated by regression analysis (Olsen et al., 1978; CCIR report, 1981). The analytic expression is[†]

$$\kappa_e = \kappa_1 R_r^b \qquad \text{(dB/km)}, \tag{8.86}$$

with R_r in mm/hr. The ***specific extinction coefficient*** κ_1 (for $R_r = 1$ mm/hr) and the exponent b are given by the regression equations:

$$\kappa_1(f) = \begin{cases} 6.39 \times 10^{-5} f^{2.03} & \text{for } f < 2.9 \text{ GHz}, \\ 4.21 \times 10^{-5} f^{2.42} & \text{for } 2.9 \leq f < 54 \text{ GHz}, \\ 4.09 \times 10^{-2} f^{0.699} & \text{for } 54 \leq f < 180 \text{ GHz}, \\ 3.38 f^{-0.151} & \text{for } 180 \text{ GHz} \leq f, \end{cases} \tag{8.87a}$$

and

$$b(f) = \begin{cases} 0.851 f^{0.158} & \text{for } f < 8.5 \text{ GHz}, \\ 1.41 f^{-0.0779} & \text{for } 8.5 \leq f < 25 \text{ GHz}, \\ 2.63 f^{-0.272} & \text{for } 25 \leq f < 164 \text{ GHz}, \\ 0.616 f^{0.0126} & \text{for } 164 \text{ GHz} \leq f. \end{cases} \tag{8.87b}$$

The plots based on the approximate analytic expressions agree reasonably well with those based on the Mie calculations.

For many investigators, including Olsen et al. (1978), the impetus to generate practical expressions for κ_e originates in terrestrial and Earth-space communication concerns. The quantity of greatest interest is the total attenuation resulting from combined absorption and scattering, while the relative contributions of each of these two processes is secondary. In radiometric remote sensing, however, emission from a scattering medium is different from that of a nonscattering medium, even if the total attenuation is the same (the effects of scattering on the brightness temperature of rain are discussed in the next section). At frequencies below 10 GHz and for rain rates of only a few millimeters per hour, the scattering coefficient κ_s is much smaller than the absorption coefficient κ_a. Hence, $\kappa_e = \kappa_s + \kappa_a \approx \kappa_a$, and the single-scattering albedo $a = \kappa_s/\kappa_e$ is close to zero. Figure 8-34, from Tsang et al. (1977), shows plots of κ_e and κ_a as a function of frequency for a rain layer with a modest precipitation rate of 12 mm/hr. It is clear from those plots that κ_s (which is equal to the difference between the two curves) cannot be ignored at frequencies above 10 GHz. This point is illustrated further by the plots in Fig. 8-35, which depict the spectral variation of the single-scattering albedo for three rain rates.

For the sake of comparison, we show in Fig. 8-36 plots of the extinction coefficient as a function of frequency for both rain and dry snow at two precipitation rates.

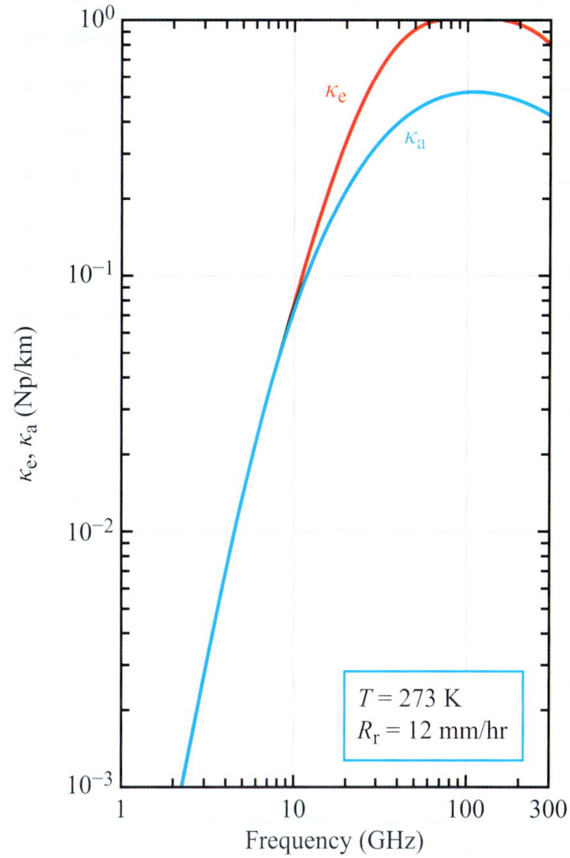

Figure 8-34: Calculated Mie extinction and absorption coefficients of rain characterized by a precipitation rate of 12 mm/hr [from Tsang et al., 1977].

[†] Computer Code 8.6.

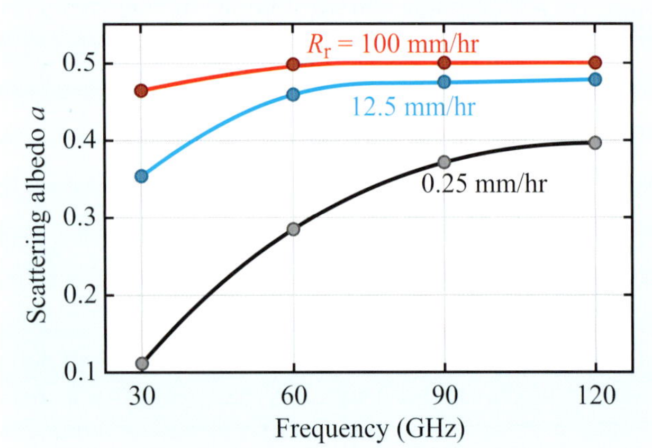

Figure 8-35: Spectral variation of Mie scattering albedo $a = \kappa_s/\kappa_e$ for rain at three rain rates [data from Ishimaru and Cheung, 1980].

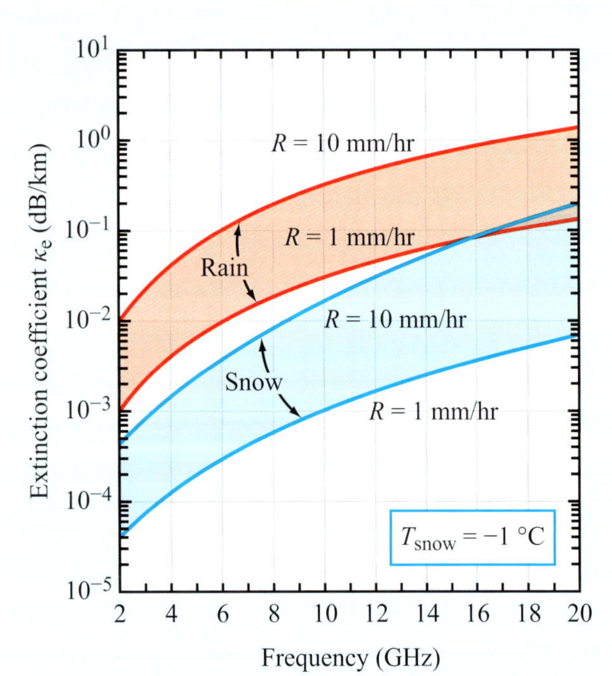

Figure 8-36: Extinction coefficient of rain and snow as a function of frequency [adapted from Barton, 1974].

► For the same precipitation rate, the extinction coefficient of dry snow is about 20 to 50 times smaller than that of rain. This is because the dielectric loss factor of dry snow is one to two orders of magnitude smaller than that of water. In contrast, the extinction coefficient of melting snow can be substantially larger than that of rain. Since the regions in the atmosphere where snow melting occurs produce high reflectivities on radar images, these regions are called ***bright bands***. They are usually found at altitudes just below the 0 °C isotherm. ◄

8-8.3 Volume Backscattering Coefficient

Formally, the volume backscattering coefficient of rain, σ_v, may be computed using Eq. (8.85) with κ_e replaced by σ_v and ξ_e replaced by the Mie backscattering efficiency ξ_b, as defined by Eq. (8.40). The practice adopted in the literature, however, has been to model σ_v using the form obtained when the Rayleigh approximation is valid. The Rayleigh volume backscattering coefficient is given by the same expression developed earlier for clouds, namely Eq. (8.79):

$$\sigma_v = 10^{-10}\,\frac{\pi^5}{\lambda_0^4}\,|K_w|^2 Z \qquad (\mathrm{m}^{-1}), \tag{8.88}$$

where λ_0 (cm) is the wavelength, $|K_w|$ is a function of the refractive index of water and is given by Eq. (8.43), and Z (mm^6/m^3) is the reflectivity factor defined by Eq. (8.77). For rain, Eq. (8.88) is adequate at frequencies below about 10 GHz. The relationship between Z and the rain rate R_r depends on the drop-size distribution. A widely used expression developed by Marshall and Palmer (1948) is given by

$$Z = 200 R_r^{1.6}, \qquad f < 10 \text{ GHz}, \tag{8.89}$$

where Z is in mm^6 per m^3 and R_r is in mm/hr.

At frequencies above 10 GHz, the practice has been to continue to use the form of Eq. (8.88) by empirically determining an effective (or equivalent) reflectivity factor Z_e that fits the form of Eq. (8.88).

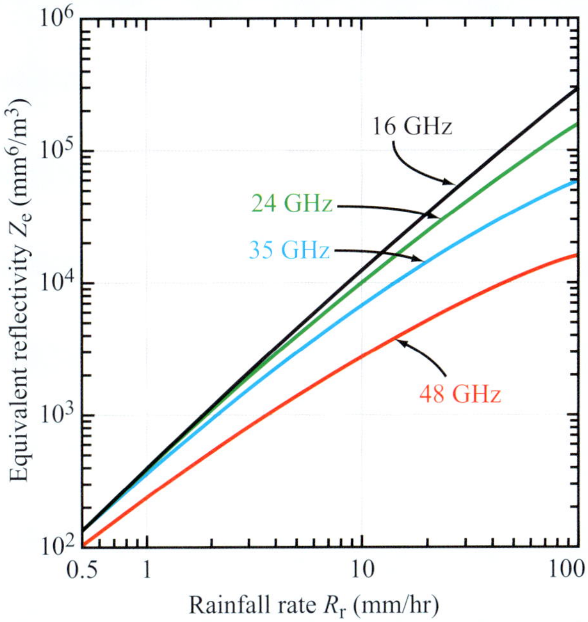

Figure 8-37: Relationships between equivalent radar reflectivity factor Z_e and rainfall rate for 0 °C [Wexler and Atlas, 1963].

Figure 8-37 shows Z_e as a function of rain-rate R_r at various frequencies. The relationship between Z_e and R_r was determined by first computing σ_v at 0 °C using Mie scattering and the Marshall-Palmer (1948) drop-size distribution, and then calculating Z_e from the expression

$$Z_e = \frac{\lambda_0^4 \sigma_v \times 10^{10}}{\pi^4 |K_w|^2} . \qquad (8.90)$$

8-9 Radar Equation for Meteorology

For a short-pulse radar, the received power due to backscattering from a group of volume-distributed incoherent scatterers is given by

$$P_r = \frac{P_t G_0^2 \lambda_0^2}{(4\pi)^3 R^4} \sigma e^{-2\tau}, \qquad (8.91)$$

where P_t is the peak transmitted power, G_0 is the antenna gain along the beam axis, R is the range to the scattering volume, τ is the total one-way path attenuation (Np), and σ is the radar cross section of the scattering volume (in m^2). The attenuation τ is the sum of the extinction coefficients due to atmospheric gases, clouds, and precipitation (rain or snow), integrated along the path between the radar and the scattering volume at range R:

$$\tau = \int_0^R (\kappa_g + \kappa_{e_c} + \kappa_{e_p}) \, dr \qquad \text{(Np)}, \qquad (8.92)$$

where κ_{e_p} is equal to κ_{e_r} for rain or equal to κ_{e_s} for snow, whichever is present. The radar cross section σ is equal to the backscattering cross section per unit volume, σ_v, integrated over the volume contributing to the received power P_r. For a narrow-beam antenna with half-power beamwidths β_θ and β_ϕ, the contributing volume at a range R is

$$V = \pi \left(\frac{R\beta_\theta}{2} \right) \left(\frac{R\beta_\phi}{2} \right) \left(\frac{c\tau_p}{2} \right), \qquad (8.93)$$

where c is the velocity of light and τ_p is the pulse length, assumed sufficiently short for R to be considered constant over the depth range $c\tau_p/2$ of the scattering volume. If σ_v is approximately uniform over V, then $\sigma = \sigma_v V$ and P_r becomes

$$P_r = \left[\frac{P_t G_0^2 \lambda_0^2 \beta_\theta \beta_\phi c \tau_p e^{-2\tau}}{32(4\pi R)^2} \right] \sigma_v . \qquad (8.94)$$

8-10 Emission by Clouds and Rain

Downward emission by atmospheric gases was discussed in Section 8-4. For moderate cloud cover and rain-rate conditions, the atmosphere may be considered a nonscattering medium in the centimeter and part of the millimeter wavelength regions. In that case, the brightness temperature expression given by Eq. (8.23) may be used:

$$T_{DN}(\theta) = \sec\theta \int_0^\infty \kappa_a(z) \, T(z) \, e^{-\tau_0(0,z)\sec\theta} \, dz, \qquad (8.95)$$

where T_{DN} represents the downwelling atmospheric radiation at angle θ from the vertical, $T(z)$ is the thermometric temperature of the atmosphere at height z, and $\tau_0(0,z)$ is the zenith optical thickness of the layer between the surface and height z,

$$\tau_0(0,z) = \int_0^z \kappa_a(z)\,dz \qquad \text{(Np)}.$$

The quantity $\kappa_a(z)$ is the total absorption coefficient at height z (km) consisting of contributions due to atmospheric gases, clouds (if present), and precipitation (if present). It is given by

$$\kappa_a(z) = \kappa_g(z) + \kappa_{a_c}(z) + \kappa_{a_p}(z) \qquad \text{(Np/km)}, \quad (8.96)$$

where the subscripts g, c, and p refer to gases, clouds, and precipitation, respectively. The expression for κ_g is given by Eq. (8.19a) in dB/km; therefore, it should be divided by the factor 4.34 to convert it to Np/km before using it in Eq. (8.96). For clouds, the Rayleigh approximation for computing the volume absorption coefficient κ_{a_c} [as given by Eq. (8.71)] is valid up to about 50 GHz, and for fair-weather clouds it is applicable throughout the microwave region.

If the antenna main beam is not pointing at the sun, the contributions of extraterrestrial radiation sources can be neglected above 10 GHz, in which case the sky radiometric temperature $T_{SKY}(\theta)$, as defined by Eq. (8.24a), reduces to $T_{DN}(\theta)$:

$$T_{SKY}(\theta) = T_{DN}(\theta).$$

Weger (1960) computed $T_{SKY}(\theta)$ at several wavelengths for each of three atmospheric conditions, which he describes as follows:

Case 1: ***Clear sky*** with sea-level water-vapor content $\rho_0 = 7.5$ g/m^3 and temperature 290 K.

Case 2: Uniform moderate ***cloud cover*** between 900 and 1800 m with condensed-water content 0.3 g/m^3, superimposed on Case 1.

Case 3: Uniform ***moderate rain*** between 0 and 900 m with a low precipitation rate of 4 mm/hr, superimposed on Case 2.

At 10 GHz [Fig. 8-38(a)], T_{SKY} of the clear sky is on the order of a few kelvins, and the addition of cloud

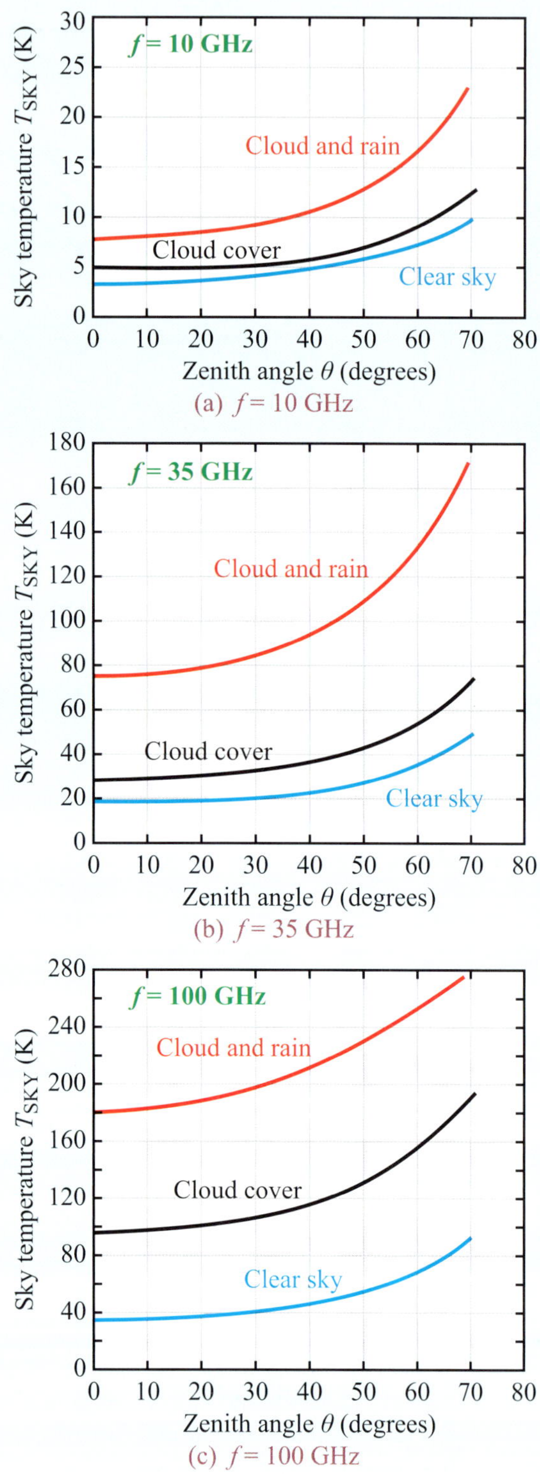

Figure 8-38: Sky radiometric temperature as a function of zenith angle θ for three atmospheric conditions at (a) 10 GHz, (b) 35 GHz, and (c) 100 GHz [from Weger, 1960].

cover, with or without rain, increases T_{SKY} by only a few degrees. At 35 GHz, $T_{SKY} = 20$ K at nadir under clear-sky conditions; the addition of cloud cover raises T_{SKY} to 30 K; and adding precipitation raises it some more to 77 K. Finally, at 100 GHz, $T_{SKY} = 40$ K at nadir under clear-sky conditions; the presence of cloud cover causes T_{SKY} to increase to about 100 K and the addition of rain on top of that increases T_{SKY} to 180 K. The three examples illustrate the important role of frequency with regard to the sensitivity of T_{SKY} to atmospheric conditions.

8-11 Error Sources and Estimation Statistics

The process culminating in the deployment of operational satellite sensors capable of generating useful data products about Earth's environment consists of three major steps:

(a) Solution of the direct problem: ***forward model***.

(b) Solution of the inverse problem: ***inversion algorithm***.

(c) Extending the observing platform from ground-based to satellite-based.

Active and passive microwave sensors measure quantities that are proportional to respectively, the scattering amplitude S (or $|S|^2$ in the case of non-polarimetric radars) and the brightness temperature T_B of the scene observed by their antenna beams. Using a combination of theoretical model simulations and experimental observations (usually carried out by sensors on ground-based or airborne platforms), relationships are established between T_B and the physical parameters of the atmospheric volumes or land and ocean surfaces contributing to the emission measured by the radiometer. These relationships constitute the forward model and represent a solution of the direct problem. A similar process applies in the radar case.

The understanding gained from the solution of the direct problem invariably leads to the development of inversion algorithms designed to estimate or infer the state of the physical parameters of interest on the basis of the measured value(s) of T_B (which may consist of one or more channel(s) due to the use of multifrequency, multipolarization, multiangle, or multitemporal observations). The inverse algorithm constitutes the solution of the inverse problem. The third step in this evolutionary process is to extend the applicability of the algorithms to satellite altitudes so they may be used by similarly configured spaceborne sensors.

Integral to the preceding three-step process are the following critical questions:

(a) With what precision can the ground-based or airborne sensor estimate the physical quantity of interest, and how does that precision compare with the precision required by the intended application?

(b) How is the measurement precision impacted by the atmosphere as the platform is raised to orbital altitudes? If the estimation accuracy and precision are distorted by atmospheric effects, are there correction algorithms available to remove the distortions?

(c) Is the field of view (spatial resolution) of the satellite-borne sensor significantly different from those of the ground-based and airborne sensors used in the development of the estimation algorithm? If yes, what impact does the difference in FOV have on the estimation process?

Questions (b) and (c) are very important, but the answers are application-specific, so it is best to address those questions in future chapters in conjunction with our discussions of atmospheric, oceanic, and land applications. In contrast, question (a) is generic to any estimation process in which the value of a quantity estimated by a sensor is compared with the value of that same quantity estimated ***in situ*** (on site). Hence, we devote the remainder of this section to an examination of how the various sources of measurement error contribute to the overall estimation precision.

8-11.1 Error Sources

Let us assume we are interested in mapping the spatial variation of a geophysical parameter g, which may represent an atmospheric variable (such as the average

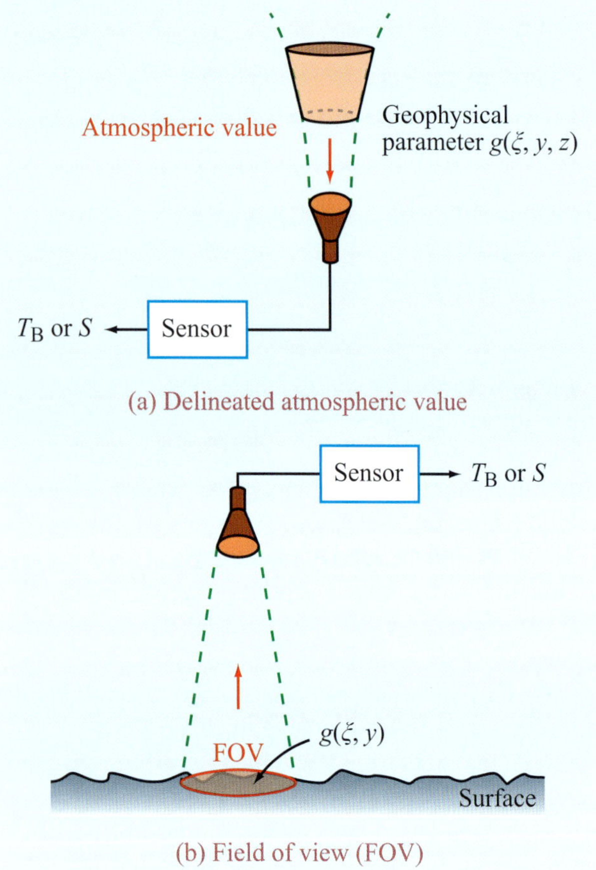

Figure 8-39: Sensor output T_B or S is proportional to (a) the delineated atmospheric volume or (b) field of view (FOV).

temperature, pressure, or cloud water content within a specified atmospheric volume), an ocean parameter (e.g., average salinity within the sensor's FOV), or a land parameter (e.g., soil moisture content or vegetation biomass within the FOV). In each case, the estimation process involves three manifestations of the geophysical variable g, namely:

g_a = the ***actual*** (true) value of g. For the atmospheric volume shown in Fig. 8-39(a), g_a is the value of g averaged over the said volume, and for the ocean sensor depicted in Fig. 8-39(b), g_a is the average value of g across the FOV. For many geophysical parameters, it is not feasible to determine g_a with absolute precision. The FOV of a typical satellite-borne radiometer, for example, is on the order of 300 km^2. If g is water salinity, determining its actual value g_a would require removing the top few-centimeter thick layer of the ocean surface from the entire 300 km^2 area!

x = the ***in-situ*** measured value of g. Within a designated atmospheric volume or across an FOV, x is determined by measuring g at N_s sampling locations. Thus, $x = (\sum_{j=1}^{N_s} x_j)/N_s$, where x_j is the *in-situ* value measured at location j.

y = the ***sensor-estimated*** value of g. This is the value inferred by the radar or radiometer on the basis of an inversion algorithm.

The *in-situ* values x_j are measured by a probe or similar device, such as a radiosonde (for measuring atmospheric parameters) or a soil-moisture dielectric circuit. The *in-situ* measurement process is subject to at least the following two sources of error:

(a) an instrument or process error associated with the *in-situ* probe, and

(b) sampling error associated with the fact that the variable g is measured at a discrete number of locations. If the expected spatial variability of g is known (or can be ascertained with a high degree of confidence), the sampling error can be eliminated by sampling the volume or FOV at a spatial rate greater than twice the highest frequency component of the spatial spectrum of g (i.e., by satisfying the Nyquist sampling criterion).

The **in-situ *error*** due to all error sources associated with the *in-situ* measurement process is

$$\varepsilon_x = x - g_a. \tag{8.97}$$

Assuming the *in-situ* probe is well calibrated and the associated error sources are random in nature, we can treat ε_x as a zero-mean Gaussian random variable with mean $\langle \varepsilon_x \rangle = 0$ and **in-situ *variance***

$$s_x^2 = \langle \varepsilon_x^2 \rangle = \langle (x - g_a)^2 \rangle. \tag{8.98}$$

A similar story applies to y. The error between the value estimated by the remote sensor (radar or radiometer) and the actual value is

$$\varepsilon_y = y - g_a, \tag{8.99}$$

and the error sources include instrument and algorithm errors. For a well-calibrated sensor and a properly configured inversion algorithm, there should be no significant bias between y and g_a. Hence, we again assume that ε_y is a zero-mean random variable with mean $\langle \varepsilon_y \rangle = 0$ and ***algorithm variance***

$$s_y^2 = \langle (y - g_a)^2 \rangle. \tag{8.100}$$

Since the *in-situ* measurement process and the remote-sensor inference process are statistically unrelated, x and y are uncorrelated. Hence, the variance between them is the sum of their individual variances:

$$s_{yx}^2 = s_x^2 + s_y^2. \tag{8.101}$$

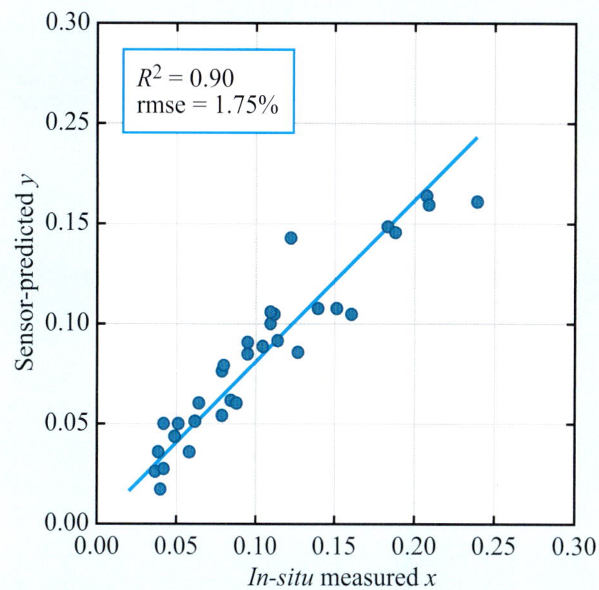

Figure 8-40: Linear regression between y, the remote sensor algorithm-predicted value of a geophysical parameter, and x, its *in-situ* sampled value.

8-11.2 Model Validation

To validate an inversion algorithm and quantify its statistical performance, the standard approach is to compare x with y under as many different relevant conditions and across as wide a range of the expected span of g as possible. The results of such a hypothetical test are shown in Fig. 8-40 for a variable with a span that extends from 0 to 0.25. Linear regression between x and y leads to a model of the form

$$y = ax + b. \tag{8.102}$$

Ideally, the bias term b should be zero and the slope term a should be 1. This is because y and x are different estimates of the same variable g_a. In most cases, the assumption is that the *in-situ* value x is more accurate (absolute value) and more precise (smaller uncertainty) than y, so if $b \neq 0$ and $a \neq 1$, the algorithm gets modified so as to generate a 1 : 1 line between x and y in Fig. 8-40.

The scatter of the data pairs relative to the 1 : 1 line is caused by the combination of errors in x and in y. Linear regression provides two statistical measures of the goodness of fit between x and y:

R^2 = coefficient of determination (square of the correlation coefficient between x and y), and

rmse = root-mean-square-error, also called ***standard error of estimate*** (SEE), which is an estimate of the standard deviation s_{yx}.

The value of the rmse is commonly regarded as the ***precision*** with which the remote-sensing algorithm is capable of measuring the geophysical variable g.

8-11.3 The Curse of Remote Sensing

From Eq. (8.101), we note that s_{yx}^2 has two components: s_y^2, which is related to the remote sensor and its associated inversion algorithm, and s_x^2, which is related to the *in-situ* sampled estimate of the geophysical variable. If the sampling is sufficiently intensive so as to render $s_x^2 \ll s_y^2$, then the rmse computed by linear regression is due (almost) entirely to the precision s_y of the inversion algorithm. Unfortunately, that is not always the case. Occasionally, the process of validating a certain remote sensing algorithm runs into a state of ambiguity known as ***ground truth***!

Consider a satellite radiometer designed to measure ocean salinity. How do we validate its estimation algorithm? The footprint of the radiometer's antenna beam on the ocean surface is a 20 km diameter circle. How do we "ground-truth" the true average salinity of such a 314 km^2 area? How many samples should be taken? How far apart? And how do we combine their statistical variability with the statistical precision of the ocean sensors used on buoys and ships to measure the salinity? Obviously, the *in-situ* measured "ground-truth" value of the salinity is only an estimate of the true average salinity of the area viewed by the radiometer beam. And consequently, the estimated precision with which the radiometer can measure salinity is constrained by the precision of the "ground-truth" data used in validating its algorithm. Occasionally, we encounter situations in which the estimate provided by the remote sensor is more accurate than the estimate based on the *in-situ* measurements. That is, $s_y^2 < s_x^2$.

As we move from ocean to land, the issue of spatial diversity assumes even greater significance; vegetation cover and topography vary at scales of meters or tens of meters, compared with kilometers for ocean surface parameters. High-resolution sensors with image pixel dimensions on the order of meters or tens of meters are better suited for land applications than sensors with kilometers or tens of kilometer resolutions. But even with high-resolution sensors, there is no getting away from the "ground-truth" ***curse of remote sensing***.

PROBLEMS

8.1 For Earth's atmosphere, the total air mass contained in a semiinfinite vertical column of 1 m^2 cross-sectional area is 1.034×10^4 kg. How does this value compare with the value obtained on the basis of the density profile defined by:

(a) Eq. (8.2)?

(b) Eq. (8.3)?

8.2 For the water-vapor density profile given by Eq. (8.7), what fraction of the total mass of water M_v is contained in the lowermost 5 km layer of the atmosphere?

8.3 Use Computer Codes 8.2 to compute and plot the absorption spectrum of oxygen (30–300 GHz) at an altitude of 3 km, using the Standard Atmosphere model.

8.4 Use Computer Code 8.2 to compute and plot the absorption spectrum of oxygen in the 50–70 GHz band at an altitude of 20 km, using the Standard Atmosphere model.

8.5 Use Computer Code 8.1 to compute and plot the absorption spectrum of water vapor over the range from 1 to 200 GHz at sea level for a humid climate with $T_0 = 310$ K, $\rho_0 = 12$ g/m^3 and $P_0 = 1013$ mbar.

8.6 Compute and plot the zenith transmissivity for a clear atmosphere as a function of sea-level water-vapor density (0 to 12 g/m^3) at:

(a) 1 GHz.

(b) 10 GHz.

(c) 22.235 GHz.

(d) 35 GHz.

Use a Standard Atmosphere with $T_0 = 288.15$ K and $P_0 = 1013$ mbar.

8.7 Given a Standard Atmosphere with $\rho_0 = 3$ g/m^3, compute and plot the zenith transmissivity over the frequency range from 1 to 150 GHz.

8.8 Compute and plot the zenith sky brightness temperature as a function of sea-level water-vapor density (over the range 0 to 12 g/m^3) at

(a) 1 GHz.

(b) 10 GHz.

(c) 22.235 GHz.

(d) 35 GHz.

Use a Standard Atmosphere with $T_0 = 288.15$ K and $P_0 = 1013$ mbar.

8.9 Develop an empirical relationship between the zenith transmissivity of the atmosphere and its zenith sky brightness temperature at 10 GHz by computing each of the two quantities at multiple values of the

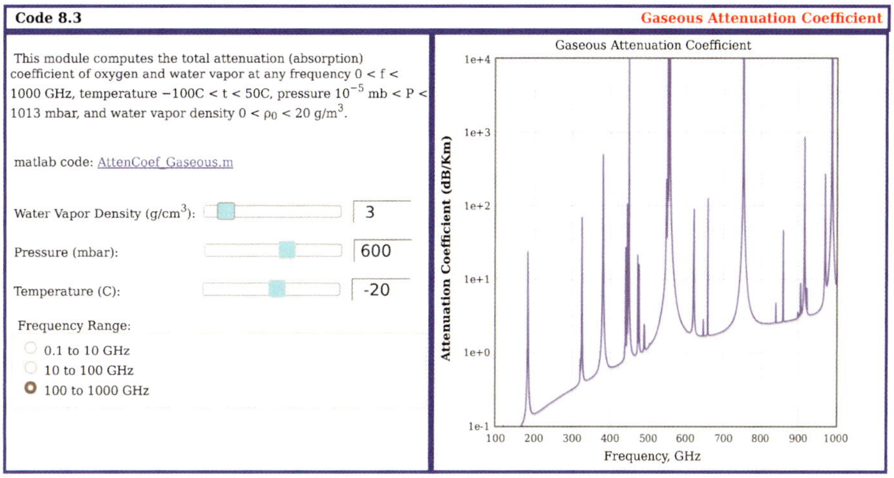

An example of one of the interactive modules available at the book website: mrs.eecs.umich.edu.

sea-level water-vapor density ρ_0 (from 0 to 12 g/m^3). Assume $T_0 = 288.15$ K and $P_0 = 1013$ mbar.

8.10 Repeat Problem 8.9 at $f = 35$ GHz.

8.11 For a sphere with dielectric constant $\varepsilon = 50 - j10$ and of radius $r = 0.1$ mm, compute and compare the following Mie and Rayleigh efficiencies over the range 1 to 100 GHz:

(a) Extinction.

(b) Scattering.

(c) Absorption.

(d) Backscattering.

In each case specify the frequency f (if it exists) below which the deviation between the Mie and Rayleigh values is less than 1%.

8.12 Repeat Problem 8.11 for an ice sphere of the same size but with $\varepsilon_i = 3.2 - j10^{-3}$.

8.13 According to Computer Code 4.1, the dielectric constant of pure water is a function of only two variables, the temperature T and the frequency f. Suppose T is a constant at 20 °C. Compute the Mie and Rayleigh extinction efficiencies for a water sphere of radius $r = 0.05$ mm, and plot them as a function of f over the range from 1 to 100 GHz.

8.14 Use Computer Code 8.4 to compute and plot the extinction coefficient of a water cloud as a function of frequency over the range from 1 to 100 GHz, assuming the cloud is at 10 °C, has a water content of 0.7 g/m^3, and its droplets have a radius $r = 0.01$ mm.

8.15 Repeat Problem 8.14 for an ice cloud with the same water content but larger particles ($r = 0.1$ mm), assuming that $\varepsilon_{ice} = 3.2 - j10^{-3}$ across the 1–100 GHz range.

8.16 Use Computer Code 8.6 to compute and plot the extinction coefficient of rain as a function of rainfall rate from 0.1 to 150 mm/hr at:

(a) 1 GHz.

(b) 10 GHz.

(c) 35 GHz.

(d) 100 GHz.

8.17 For a Standard Atmospheric model with sea-level parameters $T_0 = 300$ K, $P_0 = 1013$ mbar, and $\rho_0 = 5$ g/m^3, compute and plot the zenith sky temperature as a function of frequency (1 to 100 GHz), given that the atmosphere contains a water cloud extending between 3 km and 5 km. The cloud water content is 1.2 g/m^3.

CHAPTER 9

Radiometric Sounding of the Atmosphere

SSMIS instrument

CONTENTS

Overview

Important attributes of the atmospheric state can be inferred from remote microwave measurements through the fundamental mechanisms of absorption and scattering discussed in the preceding chapter. Atmospheric absorption near water-vapor resonances at 22.235 and 183.31 GHz allows the retrieval of humidity profiles, and oxygen resonances near 50–70 GHz and at 118.75 GHz allow the retrieval of temperature profiles. Spectral lines in the millimeter-wave region have also been suggested for the retrieval of temperature (424.76 GHz) and water-vapor (325.15 and 380.20 GHz) profiles and also for the sensing of hydrometeor characteristics and precipitation intensity and type (Gasiewski, 1992). These observations can be obtained from ground-based sensors scanning upward across the sky or from airborne or space-based sensors in low-Earth or geostationary orbit scanning downward across the Earth (Fig. 9-1). Characteristics of the sensor viewing geometry also impact the quantities that can be measured usefully and their fidelity.

There is a vast and growing assortment of applications that benefit from atmospheric measurements obtained by microwave and millimeter-wave sounding instruments spanning research areas that transcend meteorology, oceanography, geology, and ecology. Measurements of the atmosphere using microwave radiometry have provided a benchmark climate record of temperature trends dating back to the ***Microwave Sounding Unit*** (MSU), which began operation in 1979. MSU was followed by the ***Advanced Microwave Sounding Unit*** (AMSU), which began operation in 1998. The ***Advanced Technology Microwave Sounder*** (ATMS) is the first in a series of new cross-track scanning sounders developed for the ***Joint Polar Satellite System*** (JPSS). ATMS was launched on October 28, 2011 on the Suomi National Polar Partnership satellite. The ***Special Sensor Microwave Imager/Sounder*** (SSMIS) was launched in 2003 and provides conically scanned capabilities to observe atmospheric temperature up into the mesosphere as well as surface phenomena such as near-surface wind speed and sea-surface temperature. The first ***Microwave Limb Sounder*** (MLS) aboard the Upper Atmosphere Research Satellite (UARS) was launched in 1991. The MLS provided measurements of stratospheric ClO, O_3, and H_2O, upper tropospheric H_2O vapor and cloud ice, stratospheric HNO and temperature variances associated with atmospheric gravity waves. The NASA ***Precipitation and All-Weather Temperature and Humidity*** (PATH) mission is in formulation by NASA and aims to place a microwave sounder in geostationary orbit for high-revisit observations of severe-weather and hydrologic parameters (Lambrigtsen, 2010).

9-1 Atmospheric Weighting Function

As stated in the opening section of Chapter 8, in remote sensing we deal with two interrelated problems: the ***forward problem*** and the ***inverse problem***. Chapter 8 addressed the forward problem, represented diagrammatically by part (a) of Fig. 9-2. The present chapter addresses the inverse problem: given radiometric observations of the atmosphere at multiple microwave frequencies, the goal of the solution of the inverse problem is to use those observations—together with other ancillary information—to estimate the state of the atmosphere [Fig. 9-2(b)]. Estimating the state of the atmosphere includes generating height profiles for atmospheric temperature, pressure, and water vapor, measuring cloud liquid-water content and rain rate, and profiling the concentrations of atmospheric gases of interest. The estimation process is realized through the application of inversion algorithms that are formulated in terms of ***atmospheric weighting functions***. A weighting function is specific to the physical quantity being estimated (such as the temperature or pressure profile) and the observation geometry (upward-looking, downward-looking, or limb-sounding).

9-1.1 Upward-Looking Temperature Weighting Function

By way of introducing the concept of a weighting function, we start by considering the case of an upward-looking radiometer observing the emission by the atmosphere under conditions that satisfy the Rayleigh-Jeans approximation (Section 6-2.3). Afterwards, we

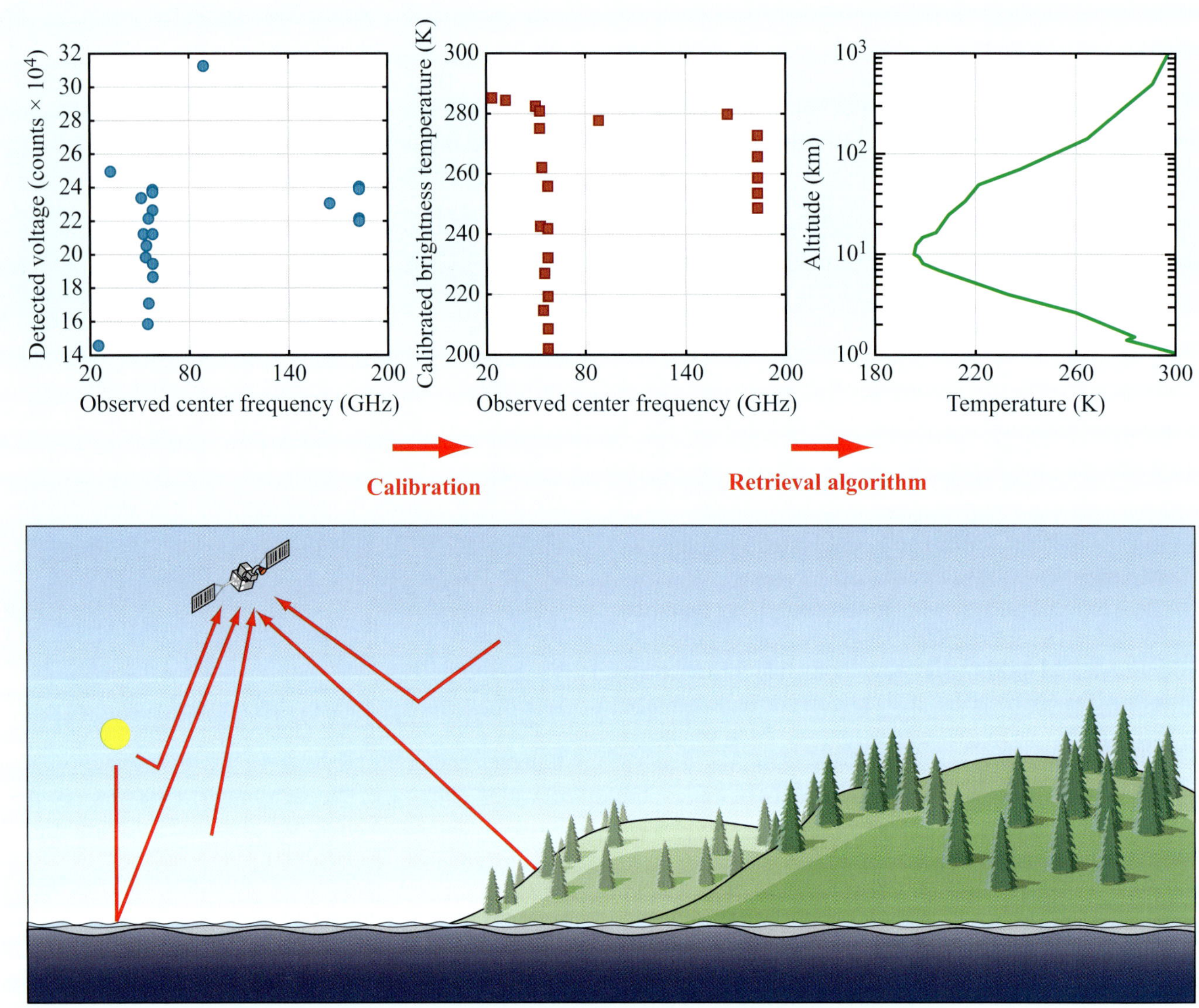

Figure 9-1: Typical measurement scenario for spaceborne atmospheric remote sensing. Electromagnetic radiation that reaches the sensor is emitted by the sun, atmosphere, surface, clouds, and cosmic background. This radiation can also be reflected or scattered by the surface, atmosphere, or clouds; some of these interactions are indicated with arrows in the figure. The spectral radiances measured by the sensor are related to geophysical quantities such as the vertical temperature profile of the atmosphere. An appropriate retrieval algorithm is necessary to convert these radiances into a geophysical quantity of interest.

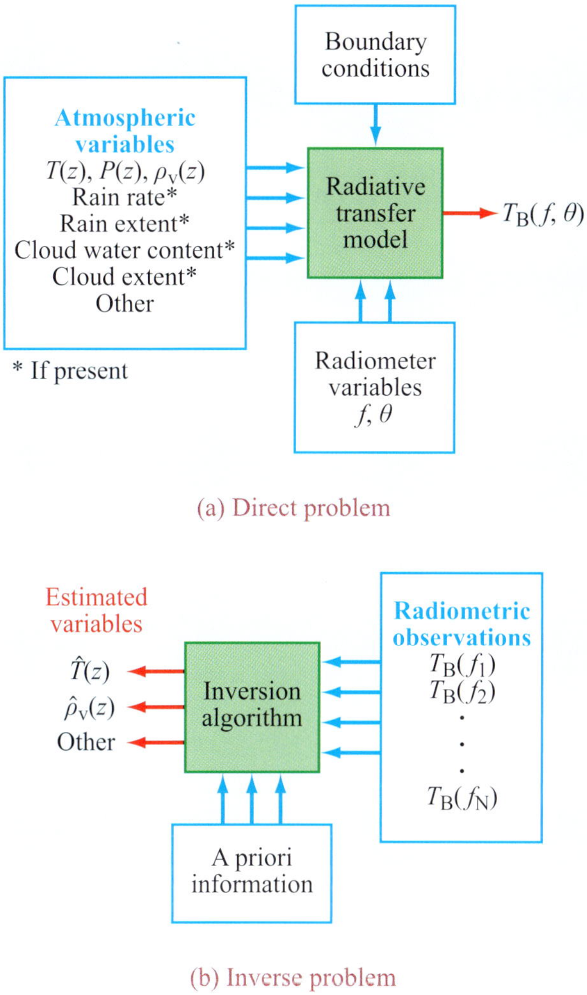

Figure 9-2: Elements of the (a) direct and (b) inverse problems in atmospheric remote sensing.

consider the satellite downward-looking configuration.

For an upward-directed antenna beam (Fig. 9-3), the sky brightness temperature is given by Eq. (8.24a) as

$$T_{\mathrm{SKY}}(f,\theta) = T_{\mathrm{EXTRA}} e^{-\tau_0 \sec\theta} + T_{\mathrm{DN}}(f,\theta), \qquad (9.1)$$

where T_{EXTRA} accounts for the galactic and cosmic radiation originating from outside of the Earth's atmosphere, τ_0 is the zenith opacity, and $T_{\mathrm{DN}}(f,\theta)$ is the ***downward emitted brightness temperature*** of the atmosphere at frequency f and zenith angle θ. For $f \geq 10$ GHz, T_{EXTRA} may be set equal to 2.7 K, or (for first-order calculations) the first term in Eq. (9.1) may be neglected altogether. Hence, the observed brightness temperature is essentially $T_{\mathrm{DN}}(f,\theta)$, which is given by Eq. (8.23) as

$$T_{\mathrm{DN}}(f,\theta) = \sec\theta \int_0^\infty \kappa_{\mathrm{a}}(f,z)\, T(z)\, e^{-\tau_0(0,z)\sec\theta}\, dz, \qquad (9.2)$$

where $\kappa_{\mathrm{a}}(f,z)$ and $T(z)$ are the absorption coefficient and thermometric temperature, respectively, of the atmosphere at height z. The formulation applies to a nonscattering, horizontally stratified atmosphere. Also, in general $\kappa_{\mathrm{a}}(f,z)$ consists of multiple components:

$$\kappa_{\mathrm{a}}(f,z) = \kappa_{\mathrm{g}}(f,z) + \kappa_{\mathrm{c}}(f,z) + \kappa_{\mathrm{p}}(f,z), \qquad (9.3)$$

where the subscripts refer to the gaseous, cloud, and precipitation absorption coefficients (if clouds and rain are present at height z). The ***zenith optical thickness*** $\tau_0(0,z)$ of the atmospheric layer between the surface and height z is related to $\kappa_{\mathrm{a}}(f,z)$ by

$$\tau_0(0,z) = \int_0^z \kappa_{\mathrm{a}}(f,z')\, dz' \qquad \text{(Np)}. \qquad (9.4)$$

For the purpose of formulating the inverse problem, let us assume that we are interested in remotely sensing (i.e., retrieving) the atmospheric temperature profile $T(z)$ from radiometric observations of the sky in the absence of clouds and rain. If we rewrite Eq. (9.2) in the form

$$T_{\mathrm{DN}}(f,\theta) = \int_0^\infty W_{\mathrm{T}}(f,\theta,z)\, T(z)\, dz, \qquad (9.5)$$

where $W_{\mathrm{T}}(f,\theta,z)$ is the ***temperature weighting function***,

$$W_{\mathrm{T}}(f,\theta,z) = \kappa_{\mathrm{a}}(f,z)\, e^{-\tau_0(0,z)\sec\theta} \sec\theta, \qquad (9.6)$$

we obtain a convolution integral involving a ***temperature weighting function*** $W_{\mathrm{T}}(f,z)$ and the desired

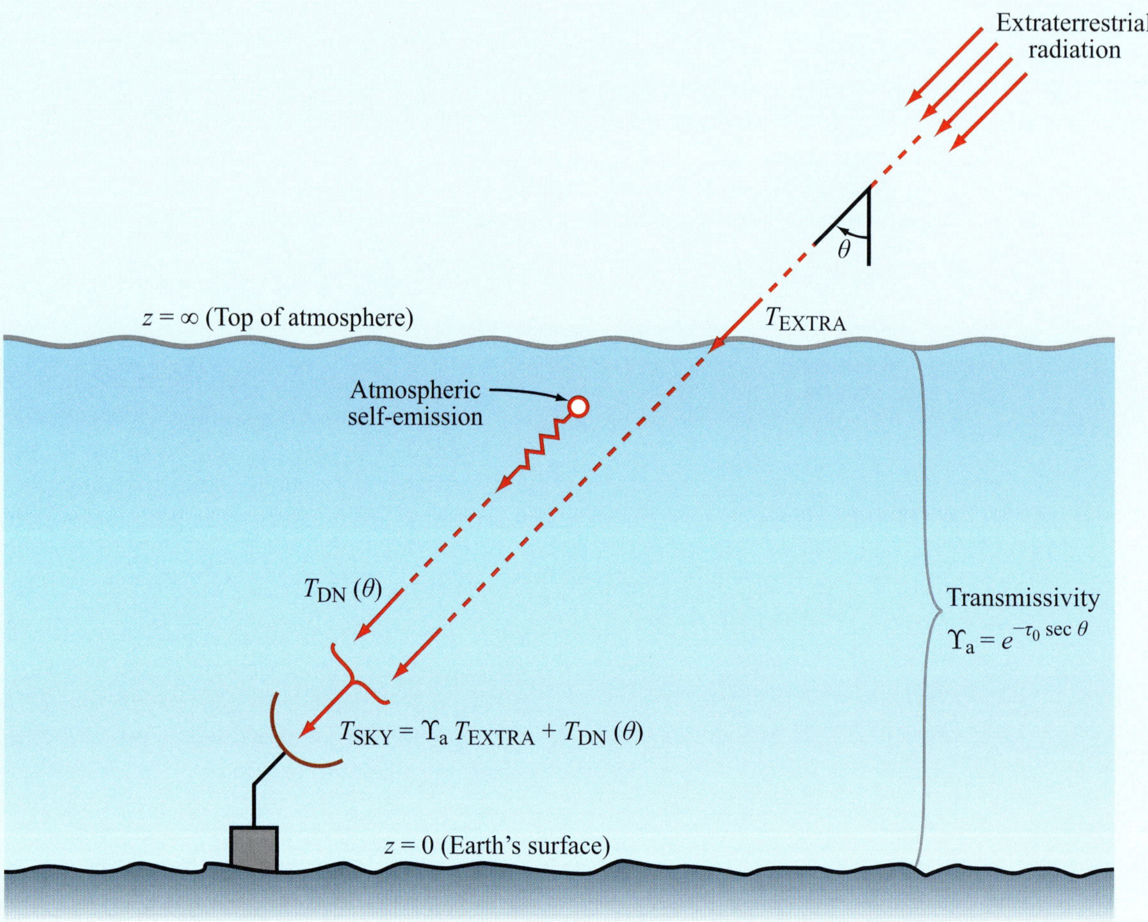

Figure 9-3: The radiometric temperature measured by an upward-looking radiometer includes contributions from our galaxy.

temperature profile $T(z)$. In this case the integral equation given by Eq. (9.5) is linear in $T(z)$, assuming $W(f,z)$ is independent of $T(z)$. In practice, $T_{DN}(f)$ is measured at a discrete number of frequencies f_i, and the objective of the inversion technique is to find a function $T(z)$ that, when substituted in Eq. (9.5), yields values of $T_{DN}(f_i)$ that are approximately equal to the measured values.

In Chapter 6, we related the ***oblique transmissivity*** $\Upsilon(0,z)$ of the layer extending between its lower boundary at $z = 0$ and its top boundary at height z (along the direction defined by the zenith angle θ) to the ***oblique optical thickness*** of the layer, $\tau(0,z)$ through Eq. (6.70):

$$\Upsilon(0,z) = e^{-\tau(0,z)} = e^{-\tau_0(0,z)\sec\theta}, \tag{9.7}$$

where

$$\tau(0,z) = \tau_0(0,z)\sec\theta = \int_0^z \kappa_a(z')\sec\theta\, dz'. \tag{9.8}$$

For an infinitesimal segment dz,

$$d\tau = \kappa_a(z)\sec\theta\, dz. \tag{9.9}$$

The partial derivative of $\Upsilon(0,z)$ with respect to z is

$$\begin{aligned}\frac{\partial\Upsilon(0,z)}{\partial z} &= \frac{\partial\Upsilon(0,z)}{\partial\tau}\cdot\frac{\partial\tau}{\partial z}\\ &= -e^{-\tau(0,z)}\cdot\kappa_a(z)\sec\theta\\ &= -\kappa_a(z)\, e^{-\tau_0(0,z)\sec\theta}\sec\theta, \qquad (9.10)\end{aligned}$$

where we used the relation $\tau(0,z) = \tau_0(0,z)\sec\theta$. Comparison of Eq. (9.10) with Eq. (9.6) leads to the conclusion that

$$W_T(f,\theta,z) = -\frac{\partial\Upsilon(0,z)}{\partial z}. \qquad (9.11)$$

Thus, the temperature weighting function W_T is the derivative of the atmospheric transmittance with respect to altitude z. Figure 9-4(a) displays vertical profiles of the weighting function $W_T(f,\theta,z)$ at three discrete frequencies, namely 52.85, 53.85, and 55.45 GHz, all for $\theta = 0$ (zenith direction). The profiles, which were calculated for the standard model atmosphere described in Section 8-1, are normalized to unity at the surface. The 55.45 GHz profile decreases more rapidly with increasing height than the lower-frequency profiles, which means that the surface layer between $z = 0$ and $z = 1$ km, for example, has the greatest influence (in terms of relative contribution or weight) on T_{DN} (55.45 GHz, 0) and the least influence on T_{DN} (52.85 GHz, 0). Also, at 55.45 GHz, T_{DN} is determined almost totally by the emission from the first three to four kilometers of the lower atmosphere. In contrast, the magnitude of the weighting function at 52.85 GHz remains significant up to $z = 10$ km and even higher.

Another way to obtain multiple weighting functions with different height profiles is to vary the observation angle θ while keeping f constant. The family of plots shown in Fig. 9-4(b) consists of temperature weighting functions calculated for a stratified standard atmosphere at 54.5 GHz. As before, each profile has been normalized to unity at the surface by dividing $W_T(f,\theta,z)$ by its value at $z = 0$, which is $\kappa_a(0)\sec\theta$. Westwater et al. (1975) have shown that it is possible to combine, linearly, weighting profiles of the type and shape shown in Figs. 9-4(a) and (b) in order to generate

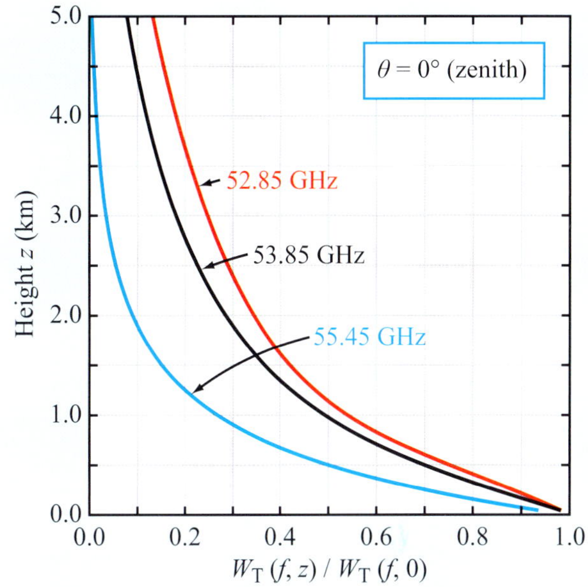

(a) Single-angle multifrequency

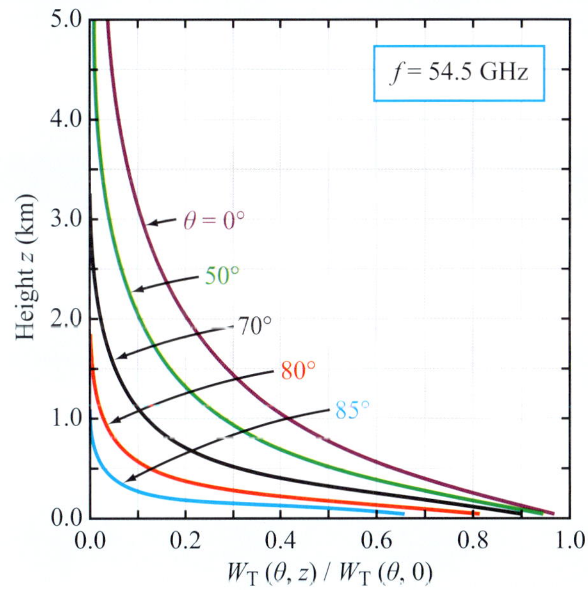

(b) Single-frequency multiangle

Figure 9-4: Normalized temperature weighting function for ground-based sensing at (a) several Nimbus 6/SCAMS frequencies along the zenith ($\theta = 0$), and (b) several angles θ at 54.5 GHz (from Westwater and Decker, 1977).

synthetic weighting functions having peaks at various altitudes.

There is a very good reason for selecting frequencies in the oxygen-absorption band for the temperature-retrieval problem. If f is close to either 60 GHz or 118.75 GHz, the absorption coefficient $\kappa_a(z)$, and in turn $\tau(0,z)$, is dominated by the oxygen-absorption coefficient $\kappa_{O_2}(z)$. Because $\kappa_{O_2}(z)$ is strongly dependent on the partial pressure of oxygen in the atmosphere, $P_{O_2}(z)$, and because oxygen has a uniform and time-invariant mixing ratio in the atmosphere ($P_{O_2}(z) = 0.21P(z)$), it follows that $W_T = W_T(f,\theta,P(z))$. That is, W_T depends on the height z through the pressure $P(z)$. However, the height variation of atmospheric pressure $P(z)$ is generally well behaved. This line of reasoning, in addition to the fact that the oxygen-absorption coefficient is only weakly dependent on temperature, underlies the choice of the oxygen bands as a means of retrieving T(z). The choice of frequencies f for a radiometer system usually is made on the basis of the shapes of the corresponding temperature weighting functions $W_T(f,\theta,P(z))$.

Even though the variation of $P(z)$ with z is fairly predictable, it is not unchangeable. In fact, at some altitudes the atmospheric pressure may vary by as much as 10 percent or more as a function of time or location. For this reason, Eqs. (9.5) and (9.6) are sometimes expressed as a function of the pressure P rather than of the height z. Actually, because P decreases exponentially with increasing height z, it is preferable to express the two equations in terms of $\ln P$. This is easily accomplished by taking a ratio of the ***hydrostatic equation***

$$dP = -\rho g\, dz \tag{9.12}$$

to the ***equation of state*** for air

$$P = \frac{\rho RT}{M} \tag{9.13}$$

to obtain

$$\frac{dP}{P} = d(\ln P) = -\frac{gM}{RT}\, dz, \tag{9.14}$$

where P, ρ, and T are the atmospheric pressure, density, and temperature, respectively; M is the molecular weight of air; g is the acceleration of gravity; and R is the universal gas constant. Employing Eq. (9.14) in conjunction with

$$\frac{\partial\Upsilon(P_s,P)}{\partial(\ln P)} = \frac{\partial\Upsilon(0,z)}{\partial z}\cdot\frac{\partial z}{\partial(\ln P)} \tag{9.15}$$

leads to

$$T_{DN}(f,\theta) = \int_{-\infty}^{\ln P_s} W_T(f,\theta,P)\, T(P)\, d(\ln P) \tag{9.16}$$

and

$$W_T(f,\theta,P) = \frac{\partial\Upsilon(P_s,P)}{\partial(\ln P)}\,, \tag{9.17}$$

where $P_s = P(0)$ is the atmospheric pressure at the surface. As we see later, most temperature-retrieval algorithms are configured to provide the profile of T as a function of P; i.e., as $T(P)$. From $T(P)$, it is possible to relate P to z, and hence to obtain $T(z)$. Alternatively, the height variation of P may be characterized by a model atmosphere suitable to the specific location and season of interest.

9-1.2 Downward-Looking Temperature Weighting Function

For a satellite-borne downward-looking radiometer, the upward-emitted brightness temperature includes multiple contributions (as diagrammed in Fig. 6-7):

$$T_B(\text{satellite}) = T_{UP}(F,\theta) + T_{SKY}(F,\theta)\,\Gamma\,\Upsilon(0,\infty) + T_{SE}\,\Upsilon(0,\infty), \tag{9.18}$$

where T_{SKY} is the downward-emitted radiation given by Eq. (9.1), Γ is the reflectivity of the ground (or ocean) surface, $\Upsilon(0,\infty)$ is the one-way transmissivity of the atmosphere along direction θ, and T_{SE} is the brightness temperature of the surface when observed in the absence of the atmosphere. At frequencies such that the atmosphere is fairly lossy, the single most important contribution is $T_{UP}(f,\theta)$, which is given by

$$\begin{aligned} T_{UP}(f,\theta) &= \sec\theta \int_0^\infty \kappa_a(f,z)\, T(z)\, e^{-\tau_0(z,\infty)\sec\theta}\, dz \\ &= \int_0^\infty W_T(f,\theta,z)\, T(z)\, dz, \end{aligned} \tag{9.19}$$

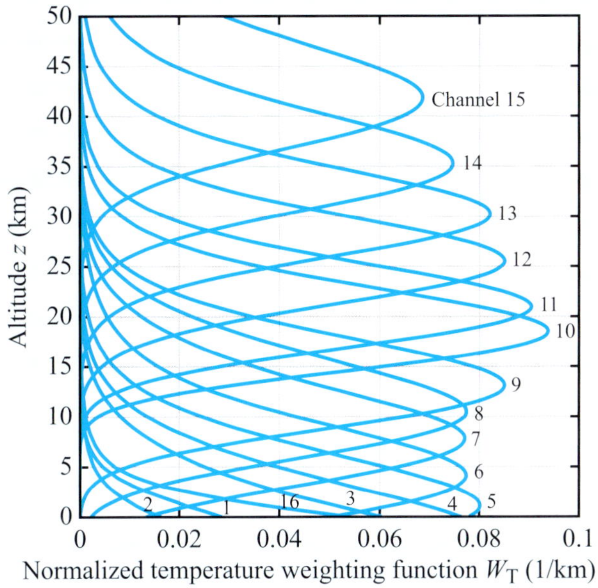

Figure 9-5: The temperature weighting functions for ATMS channels 1–16 are shown for nadir incidence over a nonreflective surface assuming the 1976 US Standard Atmosphere.

where now the temperature weighting function is

$$W_T(f,\theta,z) = \kappa_a(f,z)\, e^{-\tau_0(z,\infty)\sec\theta}\sec\theta. \tag{9.20}$$

Note that the temperature weighting function for the downward-looking radiometer involves $\tau_0(z,\infty)$, which is the zenith optical depth for the atmospheric layer between height z and the satellite at $z = \infty$. This is in contrast with W_T of the upward-looking radiometer given by Eq. (9.6), which involves $\tau_0(0,z)$ instead, where $\tau_0(0,z)$ pertains to the atmospheric layer between height z and the radiometer at the ground surface.

The temperature weighting functions for ATMS channels 1–16 are shown in Fig. 9-5. The altitude resolution of the temperature weighting functions is related to the rate at which absorption increases with depth in the atmosphere. For this reason, weighting functions of channels located close to line centers are less sharply peaked in altitude than those farther out on the wings. Likewise, the weighting function of a given channel tends to become slightly sharper with increasing incidence angle because it peaks at lower pressure, where the lines are narrower. This latter effect is sometimes offset by other effects such as changing magnetic field direction or temperature lapse rate. The weighting functions of narrowband channels very close to line centers depend on the terrestrial magnetic field strength and orientation because of the magnetic dipole moment of diatomic oxygen, as discussed by Rosenkranz and Staelin (1988) and Stogryn (1989a,b). At pressures higher than a few millibar, pressure broadening overcomes the magnetic (Zeeman) effect. In the case of conical scanning (discussed in Section 9-9.2), the observed line frequency will shift during a scan due to the Doppler effect associated with the component of satellite motion along the line of sight (Rosenkranz et al., 1997).

9-2 Data Representation

Before proceeding with the application of mathematical inversion techniques to geophysical retrieval problems, we first examine the statistical nature of the parameters involved, as it may be possible to substantially simplify the inversion methodology by exploiting statistical properties of the data to be represented. This in turn can have several beneficial consequences. Reducing the number of free parameters in the inversion usually results in a more stable, better-conditioned retrieval operator that is less sensitive to data artifacts, such as sensor noise or interfering atmospheric or surface features. Computational efficiency is also improved by reducing retrieval complexity, and for the case of statistical estimation methods, a smaller training ensemble is generally required for simple retrieval operators with fewer free parameters. Furthermore, surface emissivity variability can degrade retrievals of atmospheric parameters, and data representations that facilitate the isolation and removal of this interfering signature can be very useful.

For the following analyses, we assume that a noisy observation of a ***random radiance vector*** $\widetilde{\mathbf{R}}$ is related to some ***atmospheric state vector*** $\mathbf{S}$ through a

forward model $\mathbf{f}(\cdot)$ as follows:

$$\widetilde{\mathbf{R}} = \mathbf{f}(\mathbf{S}) + \mathbf{\Psi} = \mathbf{R} + \mathbf{\Psi}, \tag{9.21}$$

where $\mathbf{\Psi}$ is a ***random noise vector*** (that may depend on $\mathbf{S}$), and $\mathbf{R}$ is the ***"noise-free" radiance*** observation. The retrieval seeks to estimate the state vector $\mathbf{S}$, given an observation $\widetilde{\mathbf{R}}$, where we use $\widehat{\mathbf{S}}(\widetilde{\mathbf{R}})$ to denote the estimate of $\mathbf{S}$ given an observation $\widetilde{\mathbf{R}}$. In a typical case, the elements of $\widetilde{\mathbf{R}}$ are multifrequency brightness temperature observations and $\mathbf{f}(\mathbf{S})$ is the forward model relating T_B to the vertical profiles of relevant atmospheric parameters [as in Eq. (9.2)].

9-2.1 Analysis of the Information Content of Atmospheric Sounding Data

The information content of a measurement can be defined in a number of ways. Two common scalar metrics that are used to measure the information contained in a measurement are the Shannon information content (Shannon, 1948) and the number of degrees of freedom present in the signal (Rodgers, 2000), both of which are related to the eigenvalues of data covariance matrices.

Shannon information content

The Shannon definition of information content depends on the entropy of the underlying probability density functions (pdfs) that characterize the measurement. The ***entropy*** of a continuous pdf $p(\widetilde{\mathbf{R}})$ can be defined as

$$H(p) = -\int p(\widetilde{\mathbf{R}}) \log \left[p(\widetilde{\mathbf{R}}) \right] \, d\widetilde{\mathbf{R}}. \tag{9.22}$$

The base of the logarithm is usually taken to be 2, in which case the units of entropy are ***bits***, or e, in which case the units of entropy are ***nats***. The information content of a measurement in the Shannon sense can be defined as the ***reduction of entropy*** upon making a measurement of $\widetilde{\mathbf{R}}$:

$$I(\mathbf{R}, \widetilde{\mathbf{R}}) = H[p(\mathbf{R})] - H[p(\mathbf{R}|\widetilde{\mathbf{R}})] \tag{9.23}$$

or equivalently

$$I(\mathbf{R}, \widetilde{\mathbf{R}}) = H[p(\widetilde{\mathbf{R}})] - H[p(\widetilde{\mathbf{R}}|\mathbf{R})]. \tag{9.24}$$

For example, the entropy of $\widetilde{\mathbf{R}}$ can be calculated assuming a multivariate Gaussian distribution

$$p(\widetilde{\mathbf{R}}) = \frac{1}{(2\pi)^{N/2} |\mathbf{C}_{\widetilde{\mathbf{R}}\widetilde{\mathbf{R}}}|^{1/2}} \exp\left\{ -\frac{1}{2} \widetilde{\mathbf{R}}^T \mathbf{C}^{-1}_{\widetilde{\mathbf{R}}\widetilde{\mathbf{R}}} \widetilde{\mathbf{R}} \right\}, \tag{9.25}$$

where $\mathbf{C}_{\widetilde{\mathbf{R}}\widetilde{\mathbf{R}}}$ is the covariance of $\widetilde{\mathbf{R}}$. For such a distribution the entropy is given by

$$\begin{aligned} H[p(\widetilde{\mathbf{R}})] &= \sum_{i=1}^{N} \log(2\pi e \lambda_i)^{1/2} \\ &= N \log(2\pi e)^{1/2} + \frac{1}{2} \log \left(\prod_{i=1}^{N} \lambda_i \right) \\ &= N \log(2\pi e)^{1/2} + \frac{1}{2} \log |\mathbf{C}_{\widetilde{\mathbf{R}}\widetilde{\mathbf{R}}}| \\ &= c_1 + \frac{1}{2} \log |\mathbf{C}_{\mathbf{RR}} + \mathbf{C}_{\mathbf{\Psi\Psi}}|. \end{aligned} \tag{9.26}$$

Here, λ_i are the eigenvalues of $C_{\widetilde{\mathbf{R}}\widetilde{\mathbf{R}}}$ and N is the number of elements of $\mathbf{R}$. It can be shown that the volume of an ellipsoid describing a surface of constant probability is proportional to the square root of the product of the eigenvalues $\{\lambda_i\}$ of $\mathbf{C}_{\widetilde{\mathbf{R}}\widetilde{\mathbf{R}}}$ (Strang, 1980). Therefore, the entropy of the pdf is related to the volume inside a surface of constant probability. When a measurement is made, this "volume of uncertainty" decreases. The information content is a measure of the factor by which it decreases, a generalization of the scalar concept of signal-to-noise ratio (Rodgers, 2000).

The information content of a measurement of $\widetilde{\mathbf{R}}$ can be calculated under the assumption of Gaussianity. The covariance of $\widetilde{\mathbf{R}}$ before the measurement is $\mathbf{C}_{\mathbf{RR}} + \mathbf{C}_{\mathbf{\Psi\Psi}}$ while the covariance of $\widetilde{\mathbf{R}}$ after the measurement is $\mathbf{C}_{\mathbf{\Psi\Psi}}$. These values are used in Eq. (9.24), which leads to:

$$\begin{aligned} I(\mathbf{R}, \widetilde{\mathbf{R}}) &= H[p(\widetilde{\mathbf{R}})] - H[p(\widetilde{\mathbf{R}}|\mathbf{R})] \\ &= \frac{1}{2} \log |\mathbf{C}_{\mathbf{RR}} + \mathbf{C}_{\mathbf{\Psi\Psi}}| - \frac{1}{2} \log |\mathbf{C}_{\mathbf{\Psi\Psi}}| \\ &= \frac{1}{2} \log |\mathbf{C}^{-1}_{\mathbf{\Psi\Psi}} (\mathbf{C}_{\mathbf{RR}} + \mathbf{C}_{\mathbf{\Psi\Psi}})| \\ &= \frac{1}{2} \log |\mathbf{C}^{-1/2}_{\mathbf{\Psi\Psi}} \mathbf{C}_{\mathbf{RR}} \mathbf{C}^{-1/2}_{\mathbf{\Psi\Psi}} + \mathbf{I}| \\ &= \frac{1}{2} \log |\widetilde{\mathbf{C}}_{\mathbf{RR}} + \mathbf{I}|, \end{aligned} \tag{9.27}$$

where

$$\widetilde{\mathbf{C}}_{\mathbf{RR}} \triangleq \mathbf{C}_{\Psi\Psi}^{-1/2}\mathbf{C}_{\mathbf{RR}}\mathbf{C}_{\Psi\Psi}^{-1/2} \tag{9.28}$$

is the ***whitened covariance matrix*** describing **R**. Note that Eq. (9.27) can be easily calculated from the eigenvalues of $\widetilde{\mathbf{C}}_{\mathbf{RR}}$:

$$I(\mathbf{R},\widetilde{\mathbf{R}}) = \frac{1}{2}\sum_i \log(1+\lambda_i). \tag{9.29}$$

Degrees of freedom

Another measure of the information contained within a measurement is the number of degrees of freedom (DOF), where a degree of freedom can be loosely defined as an independent component of $\widetilde{\mathbf{R}}$ that contains some information about **R**, and the uncertainty of this information is smaller than the measurement error of the component. For example, if $\widetilde{\mathbf{R}}$ is prewhitened and projected onto the eigenvectors of $\widetilde{\mathbf{C}}_{\mathbf{RR}}+\mathbf{I}$, where **I** is the identity matrix, the eigenvalues of $\widetilde{\mathbf{C}}_{\mathbf{RR}}$ give the signal-to-noise ratio (SNR) of each uncorrelated component. It is intuitive that the number of degrees of freedom should be the same as the number of components with SNR greater than or about equal to one. Furthermore, it is convenient to make a distinction between a degree of freedom due to signal (DOF_s) and a degree of freedom due to noise (DOF_n). The previous description assumed implicitly that the degree of freedom was due to signal. If there are N elements of **R**, we require

$$\text{DOF}_\text{s} + \text{DOF}_\text{n} = N. \tag{9.30}$$

Rodgers (2000) offers the following definitions for DOF_s and DOF_n:

$$\text{DOF}_\text{s} = \text{tr}(\widetilde{\mathbf{C}}_{\mathbf{RR}}[\widetilde{\mathbf{C}}_{\mathbf{RR}}+\mathbf{I}]^{-1}) = \sum_i \frac{\lambda_i^2}{(1+\lambda_i^2)} \tag{9.31a}$$

and

$$\text{DOF}_\text{n} = \text{tr}([\widetilde{\mathbf{C}}_{\mathbf{RR}}+\mathbf{I}]^{-1}) = \sum_i \frac{1}{(1+\lambda_i^2)}\,, \tag{9.31b}$$

where "tr" denotes the trace of the matrix. These definitions do not necessarily yield integer values; for example, a component with SNR = 1 would contribute $\frac{1}{2}$ to DOF_s and $\frac{1}{2}$ to DOF_n. Equation (9.30) is satisfied, nevertheless. As a final comment, it is interesting to note the similarity between Eq. (9.29) and Eq. (9.31a), both of which depend only on the eigenvalues of $\widetilde{\mathbf{C}}_{\mathbf{RR}}$.

An example

A radiative transfer model (Rosenkranz, 1995) was used to simulate the 22-channel Advanced Technology Microwave Sounder measuring upwelling thermal emission at the top of the atmosphere. Random noise was added to the simulated measurements. Figure 9-6 shows the number of degrees of freedom for the Shannon and Rodgers conventions; both show that cumulative information reaches an asymptote after approximately 15 eigenvalues. The implication is that the radiance spectra can be compressed with negligible information loss by simply setting to zero eigenvalues with order higher than 15. Note also that the high-order eigenvalues

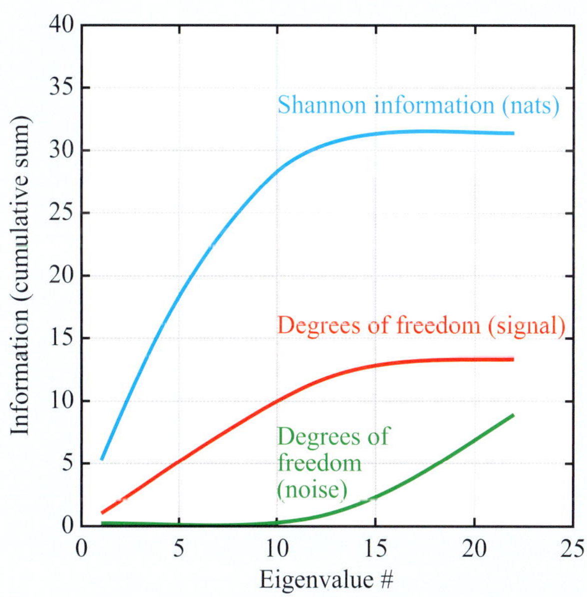

Figure 9-6: The information content of the Advanced Technology Microwave Sounder using the Shannon and Degrees of Freedom (DOF) metrics as a function of eigenvalue number.

are dominated by random noise, as the $\mathrm{DOF_n}$ curve increases linearly after the 15th eigenvalue. Therefore, in addition to the more compact representation of the radiance information afforded by the first 15 eigenvalues, the noise component has been reduced substantially (Rodgers, 1996).

9-2.2 Principal Components Analysis (PCA)

We now demonstrate how the information content in the radiance spectrum can be recovered through the application of a linear transform of minimal rank. A random vector (of atmospheric radiance intensity observations at N frequencies, for example) given by

$$\mathbf{R} = \begin{bmatrix} R_{\nu_1} \\ R_{\nu_2} \\ \vdots \\ R_{\nu_N} \end{bmatrix} \tag{9.32}$$

can be decomposed into a vector $\mathbf{I}_r$ of r statistically independent components (where $1 \leq r \leq N$):

$$\mathbf{I}_r = \begin{bmatrix} I_1 \\ I_2 \\ \vdots \\ I_r \end{bmatrix} = \begin{bmatrix} f_1(\mathbf{R}) \\ f_2(\mathbf{R}) \\ \vdots \\ f_r(\mathbf{R}) \end{bmatrix} = \mathbf{f}_r(\mathbf{R}), \tag{9.33}$$

where $\mathbf{f}_r(\mathbf{R})$ is a continuous (usually nonlinear) function. The radiance vector $\mathbf{R}$ may be reconstructed from the independent components of $\mathbf{I}_r$ (possibly with some distortion) as follows:

$$\widehat{\mathbf{R}}_r = \begin{bmatrix} g_1(\mathbf{I}_r) \\ g_2(\mathbf{I}_r) \\ \vdots \\ g_N(\mathbf{I}_r) \end{bmatrix} = \mathbf{g}_r(\mathbf{I}_r), \tag{9.34}$$

where $\mathbf{g}_r$ is a continuous (usually nonlinear) function. The vector-valued functions $\mathbf{f}_r(\cdot)$ and $\mathbf{g}_r(\cdot)$ are usually chosen to minimize some scalar-valued cost function

$$C(\mathbf{R} - \widehat{\mathbf{R}}_r) \tag{9.35}$$

over $1 \leq r \leq N$. Note that $\widehat{\mathbf{R}}_r = \mathbf{R}$ for $r = N$, and possibly for $r < N$ if the elements of $\mathbf{R}$ are statistically dependent.

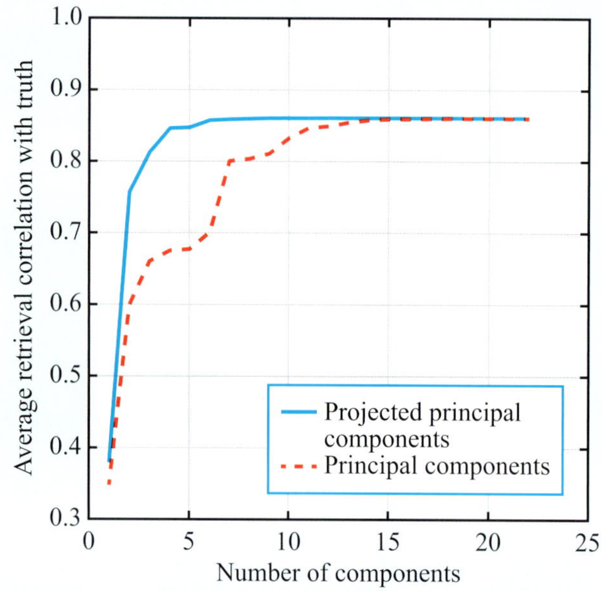

Figure 9-7: Temperature-retrieval accuracy (expressed as the average correlation of the retrieval with truth over the troposphere) derived from principal components (dashed curve) and projected principal component (solid curve).

The functions $\mathbf{f}_r(\cdot)$ and $\mathbf{g}_r(\cdot)$ and the statistical moments of $\mathbf{I}_r$ provide a measure of the statistical structure of $\mathbf{R}$.

If the cost function to be minimized is the expected value of the sum of the squares of the error of $\mathbf{R} - \widehat{\mathbf{R}}_r$, that is,

$$C(\cdot) = E[(\mathbf{R} - \widehat{\mathbf{R}}_r)^T (\mathbf{R} - \widehat{\mathbf{R}}_r)], \tag{9.36}$$

then the elements of $\mathbf{I}_r$ are called the ***principal components*** of $\mathbf{R}$.

Several methods are available for constructing vector $\mathbf{I}_r$, depending on whether $\mathbf{f}_r(\cdot)$ and $\mathbf{g}_r(\cdot)$ are linear or nonlinear functions (Kramer, 1991; Tan and Mavrovouniotis, 1995; Slone, 1995). The details are beyond the scope of this book, but it is instructive to illustrate the relationship between retrieval precision and the number of principal components. Figure 9-7 shows temperature-profile retrieval skill, expressed as the averaged correlation coefficient between the estimated and true tropospheric profile over an ensemble, for two principal component transforms. The retrieval error is shown as a function of the number of principal components that were retained during the retrieval. The Projected Principal Components Transform performance exceeds that of a Principal Components Transform for any number of coefficients.

9-3 Inversion Techniques

If the integral of Eq. (9.5) is approximated as a summation over m layers each of height Δz, the observed radiometric temperature $T_{\rm DN}(f_i)$ at frequencies f_i can be written as

$$T_{\rm DN}(f_i) = \sum_{j=1}^{m} W(f_i, z_j)\, T(z_j), \qquad i = 1, 2, \ldots, n, \tag{9.37}$$

where n is the total number of frequencies at which observations are made, $T(z_j)$ is the average temperature over the height interval Δz centered at z_j, and

$$W(f_i, z_j) = \int_{z_j - \Delta/2}^{z_j + \Delta/2} W(f_i, z)\, dz. \tag{9.38}$$

To simplify the notation, let us replace the variables f_i and z_j with their respective subscripts,

$$T_{\rm DN}(i) = \sum_{j=1}^{m} W(i, j)\, T(j), \qquad i = 1, 2, \ldots, n, \tag{9.39}$$

and further compact the expression into the form

$$\mathbf{T}_{\rm DN} = \mathbf{WT}, \tag{9.40}$$

where $\mathbf{T}_{\rm DN}$ and $\mathbf{T}$ are vectors of dimension n and m respectively, and $\mathbf{W}$ is an $(n \times m)$ matrix. The vector $\mathbf{T}_{\rm DN}$ represents the observations, assumed to be noise-free for the time being; $\mathbf{W}$ is a weighting matrix, also presumed known; and $\mathbf{T}$, the atmospheric temperature profile, is the unknown vector. Typically, the number of observation frequencies n is as small as three or four, whereas the desired number of vertical layers m is much larger, because the larger the m, the greater the vertical resolution (i.e., the layers are narrower). With $m > n$,

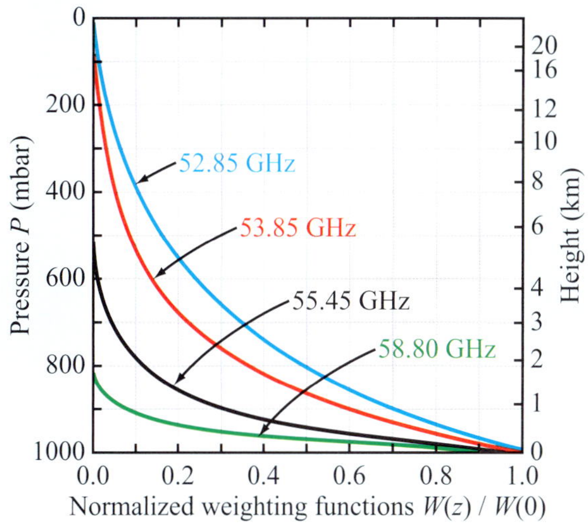

Figure 9-8: Normalized weighting functions for upward-looking zenith observations at four frequencies in the 50 to 60 GHz oxygen band [from Westwater and Grody, 1980].

Eq. (9.40) has an infinite number of solutions for $\mathbf{T}$, and therein lies the problem. In addition, the real situation is characterized by the following departures from the ideal conditions assumed or implied above:

(a) The measurements $\mathbf{T}_{\rm DN}$ are not noise-free because of instrument noise introduced in the measurement process.

(b) The mathematical models defining $\mathbf{T}_{\rm DN}$ and $\mathbf{W}$ are not exact because of the physical and mathematical assumptions and approximations inherent in the derivations leading to the final expressions.

(c) The form of $W(f, z)$ implies that the weighting function depends on f and z only. In reality, $W(f, z)$ depends on z through the absorption coefficient $\kappa_{\rm a}(z)$, which in turn depends on z through the atmospheric variables $T(z)$, $\rho_{\rm v}(z)$ and $P(z)$. Thus, $W(f, z)$ is a function of all three atmospheric variables, and its magnitude changes as their magnitudes change. The curves shown in Fig. 9-8 are for a specific set of atmospheric profiles. If clouds (or rain) are present in the atmosphere, the situation becomes even more complicated. However, solutions incorporating a correction for the presence of clouds do exist and are used in inversion algorithms.

On the one hand, the above factors seem to pose a real challenge, namely that of retrieving the profile $T(z)$ from measurements of $T_{DN}(f)$ made at only a few frequencies. On the other hand, the counterbalance that renders the problem solvable is the ***a priori information*** available about the "character" of the atmosphere for a given geographic location and a given time of year. Rodgers (1976a, 1977) calls such *a priori* information "***virtual measurements***," because they provide specifications about the profile just as actual measurements do. Rodgers (1977) states that "to solve an inverse problem, the real measurements and the virtual measurements must contain enough pieces of information to determine all the parameters of the profile with an accuracy adequate for the applications." The virtual measurements may include historical statistics about the temperature profile, constraints imposed by atmospheric physics, and any other information that, if integrated into the inversion algorithm, would narrow the range of values that $T(z)$ can have. The degree to which virtual measurements are incorporated in an inversion algorithm depends to some extent on the underlying structure of the inversion method.

9-3.1 General Formulation

Instead of showing how inversion techniques can be applied to solve Eq. (9.40) and retrieve $T(z)$, it is more useful to consider these techniques in connection with more general forms that may include bias or error terms. Generally, two basic forms of integral equations are encountered in radiative transfer problems, linear and nonlinear.

Linear form

If the brightness temperature measured by a radiometer, $T_B(f_i)$, is linearly related to the unknown or ***sought function*** $g(z)$,

$$T_B(f_i) = T_{BGD}(f_i) + \int_0^\infty W(f_i, z)\, g(z)\, dz, \qquad (9.41)$$
$$i = 1, 2, \dots, n,$$

then the physical model is considered to be linear. For numerical calculations, it may be expressed in the discrete form

$$\mathbf{T}_B = \mathbf{T}_{BGD} + \mathbf{W}\mathbf{g}, \qquad (9.42)$$

where $\mathbf{T}_B$ and $\mathbf{T}_{BGD}$ are vectors of order n, $\mathbf{g}$ is of order m, and $\mathbf{W}$ is an $(n \times m)$ matrix. The radiometric temperature $T_{BGD}(f_i)$ accounts for emission by the background (cosmic space for an upward-looking ground-based radiometer, and the Earth's surface for a downward-looking satellite-borne radiometer). The function $W(f_i, z)$ is commonly called the ***kernel*** or ***weighting function*** of the sought function $g(z)$. If the observations are made at an angle other than $\theta = 0°$(zenith or nadir depending on the platform), we need only replace z with $z \sec\theta$ in Eq. (9.41), assuming that horizontal stratification is applicable. If $\mathbf{T}_{BGD}$ is constant or can be accurately determined from calculations or other measurements, Eq. (9.42) can be rewritten as

$$\mathbf{T}_M = \mathbf{W}\mathbf{g}, \qquad (9.43)$$

where

$$\mathbf{T}_M = \mathbf{T}_B - \mathbf{T}_{BGD} \qquad (9.44)$$

is a vector representing the ***input measurements***. Equation Eq. (9.43) has the same form as Eq. (9.40).

In most atmospheric-profile-retrieval problems, the unknown function $g(z)$ is not completely unknown. Statistical records exist from which a time average or ***mean profile*** $\langle g(z) \rangle$ can be generated. The objective of the inversion algorithm becomes to determinate (estimate) the amount of ***departure from the mean profile***, $g'(z)$, where

$$g'(z) = g(z) - \langle g(z) \rangle\,. \qquad (9.45)$$

The brightness temperature corresponding to the mean profile $\langle g(z) \rangle$ is, from Eq. (9.41)

$$\langle T_B(f_i) \rangle = T_{BGD}(f_i) + \int_0^\infty W(f_i, z)\, \langle g(z) \rangle\, dz, \qquad (9.46)$$
$$i = 1, 2, \dots, n.$$

Upon subtracting Eq. (9.46) from Eq. (9.41), and defining

$$T'_B(f_i) = T_B(f_i) - \langle T_B(f_i) \rangle\,, \qquad (9.47)$$

we obtain

$$T'_{\mathrm{B}}(f_i) = \int_0^\infty W(f_i,z)\, g'(z)\, dz, \qquad i = 1,2,\ldots,n. \tag{9.48}$$

whose discrete equivalent is given by

$$\mathbf{T}'_{\mathrm{B}} = \mathbf{W}\mathbf{g}'. \tag{9.49}$$

We implicitly assumed that $T_{\mathrm{BGD}}(f_i)$ is time-independent; consequently, it was canceled out in the subtraction. The assumption may be valid for upward-looking observations as long as no point sources, such as the sun, pass through the antenna beam. Even then, the assumption is only partially valid because $T_{\mathrm{BGD}}(f_i)$ is continuously modulated by clouds or rain. For downward-looking observations, $T_{\mathrm{BGD}}(f_i)$ usually is not a constant because the emission by the background (the Earth's surface) can vary spatially as well as temporally. These variations, if not taken into account, can lead to errors in the retrieved values of $g(z)$. The relative magnitudes of the errors, however, are strongly dependent on the frequency f_i.

Nonlinear form

In some cases it may not be possible to develop a formulation that is linear in the unknown function $g(z)$, but it may be possible to develop a nonlinear model of the form

$$T_{\mathrm{B}}(f_i) = T_{\mathrm{BGD}}(f_i) + \int_0^\infty W_{\mathrm{I}}(f_i,z)\, I[f_i, g(z)]\, dz, \tag{9.50}$$
$$i = 1,2,\ldots,n,$$

where $I[f_i, g(z)]$ is now the sought function from which $g(z)$ can be determined, and $W_{\mathrm{I}}(f_i,z)$ is a weighting function for $I[f_i, g(z)]$. The nonlinear form is not as commonly encountered in the microwave region as it is in the infrared region. The infrared equivalent of the linear integral equation given by Eq. (9.5), for example, involves the Planck brightness function (Section 6-2), which is strongly nonlinear in T in the infrared region, but is approximately linear for most of the microwave region.

The discrete form of Eq. (9.50) is

$$\mathbf{T}_{\mathrm{M}} = \mathbf{W}_{\mathrm{I}}\mathbf{I}. \tag{9.51}$$

Basically, the same solution methods are used to retrieve $g(z)$ in both the linear and nonlinear cases, except that in the latter case the function $I(f_i, g(z))$ must be retrieved first. This is not to imply, however, that the accuracy of the retrieved profile $g(z)$ is necessarily the same in both cases. Measurement errors and errors in the propagation equations are likely to result in different accuracies for $g(z)$ in the linear and nonlinear models. As discussed later, one of the most successful inversion methods is the statistical regression technique, which does not require knowledge of the model given by Eq. (9.43), but does require that the relationship between $\mathbf{T}_{\mathrm{M}}$ and $\mathbf{g}$ be linear. If the model is nonlinear and unknown, one may resort to using nonlinear regression routines based on neural networks (Blackwell and Chen, 2009).

If the nonlinear model is only moderately nonlinear, it can be linearized by Taylor expansion. Because, for any z, the departure from the mean profile, $g'(z)$, usually is much smaller than $g(z)$, the nonlinear function $I[f_i, g(z)]$ can be expanded in a Taylor series about the mean profile $\langle g(z)\rangle$,

$$I[f_i, g(z)] = I[f_i, \langle g(z)\rangle] + a(f_i,z)[g(z) - \langle g(z)\rangle] + \cdots, \tag{9.52}$$

where

$$a(f_i,z) = \left.\frac{\partial I[f_i, g(z)]}{\partial g(z)}\right|_{g(z)=\langle g(z)\rangle}. \tag{9.53}$$

The constant $a(f_i,z)$ is the derivative of $I[f_i, g(z)]$ with respect to $g(z)$, evaluated at $g(z)$ equal to its mean value. If terms higher than first order are ignored, and Eq. (9.52) is inserted in Eq. (9.50), the latter becomes

$$T_{\mathrm{B}}(f_i) = T_{\mathrm{BDG}}(f_i) + \int_0^\infty W_{\mathrm{I}}(f_i,z) \cdot \left\{ I[f_i, \langle g(z)\rangle] + a(f_i,z)\, g'(z) \right\} dz. \tag{9.54}$$

Recognizing that the brightness temperature corresponding to the mean profile $\langle g(z)\rangle$ is

$$\langle T_{\mathrm{B}}(f_i)\rangle = T_{\mathrm{BGD}}(f_i) + \int_0^\infty W_{\mathrm{I}}(f_i,z)\, I[f_i, \langle g(z)\rangle]\, dz, \tag{9.55}$$

the combination of the above two equations leads to

$$\begin{aligned}T'_{\rm B}(f_i) &= T_{\rm B}(f_i) - \langle T_{\rm B}(f_i)\rangle \\ &= \int_0^\infty W_{\rm I}(f_i,z)\, a(f_i,z)\, g'(z)\, dz \\ &= \int_0^\infty W(f_i,z)\, g'(z)\, dz, \qquad i = 1,2,\ldots,n,\end{aligned} \tag{9.56}$$

where

$$W(f_i,z) = W_{\rm I}(f_i,z)\, a(f_i,z)$$

is now the weighting function for $g'(z)$. The discrete form of Eq. (9.56), which is now linear in $g'(z)$, is

$$\mathbf{T}'_{\rm B} = \mathbf{W}\mathbf{g}'. \tag{9.57}$$

The remainder of this section provides a brief review of the inversion methods commonly used to retrieve profiles of atmospheric variables from remote sensing observations. For a more detailed examination of inversion methods, the reader is referred to the excellent tutorial reviews by Westwater et al. (1968), and Rodgers (1976b), and to the papers published in the proceedings of two technical workshops on the subject of inversion methods in remote sensing, both edited by Deepak (1977, 1980). Additional reviews are available in the books by Janssen (1993) and Blackwell and Chen (2009).

9-3.2 Least-Squares Solution of the Ill-Posed Problem

Let us consider the physical model given by Eq. (9.43),

$$\mathbf{T}_{\rm M} = \mathbf{W}\mathbf{g}, \tag{9.58}$$

which relates a measurable vector $\mathbf{T}_{\rm M}$ of order n to an unknown vector $\mathbf{g}$ of order m through a known weighting matrix $\mathbf{W}$ of order $(n \times m)$. In practice, we cannot measure the true $\mathbf{T}_{\rm M}$ exactly because of experimental error, which may include both measurement error and modeling errors, the latter being a result of the approximations and assumptions underlying the model given by Eq. (9.43). Let us denote the observations as $\hat{\mathbf{T}}_{\rm M}$, where

$$\hat{\mathbf{T}}_{\rm M} = \mathbf{T}_{\rm M} + \boldsymbol{\varepsilon} = \mathbf{W}\mathbf{g} + \boldsymbol{\varepsilon} \tag{9.59}$$

and $\boldsymbol{\varepsilon}$ is the error vector.

The first requirement that usually needs to be satisfied in order to invert Eq. (9.59) is for the number of measurements, n, to be at least equal to and preferably greater than the number of unknowns, m. The standard solution (estimate) $\hat{\mathbf{g}}$ of Eq. (9.59) is given by Franklin (1968) as

$$\hat{\mathbf{g}} = (\mathbf{W}^\dagger\mathbf{W})^{-1}\mathbf{W}^\dagger\hat{\mathbf{T}}_{\rm M}, \tag{9.60}$$

where $\mathbf{W}^\dagger$ is the ***transpose*** of $\mathbf{W}$. It is called the ***least-squares solution*** because it minimizes the sum of the squares of the differences between the measurements $\hat{\mathbf{T}}_{\rm M}$ and the values of $\mathbf{T}_{\rm M}$ calculated from the model on the basis of the solution; i.e.,

$$\sum_{i=1}^{n}\left[\hat{\mathbf{T}}_{\rm M}(i) - \sum_{j=1}^{m}\mathbf{W}(i,j)\,\hat{\mathbf{g}}(j)^2\right]^2 = \text{minimum}. \tag{9.61}$$

For $m = n$, Eq. (9.60) reduces to

$$\hat{\mathbf{g}} = \mathbf{W}^{-1}\hat{\mathbf{T}}_{\rm M}. \tag{9.62}$$

The least-squares solution is not suitable for most atmospheric retrieval problems for two reasons. First, the number of observations n is usually small; in the 1970s satellite radiometers consisted of only 5 channels, and more modern versions use on the order of 10 channels, compared with the desire to estimate a temperature profile with m on the order of 20 or more layers. To satisfy the condition $m \leq n$, we must divide the vertical profile of the atmosphere into five or fewer layers. In most cases, a profile with so coarse a vertical resolution is simply not useful.

Even if the first limitation could be alleviated by increasing the number of observations (frequencies or angles), the least-squares solution is handicapped by the problem of instability. The solution is considered unstable if the error associated with the estimate $\hat{\mathbf{g}}$ is unacceptably large. The instability problem arises from the fact that the weighting function $\mathbf{W}$ is typically a smooth function of j (altitude), and hence its elements are not very different from each other, which in turn means that the matrix $\mathbf{W}^\dagger\mathbf{W}$ is ill-conditioned with respect to matrix inversion. That is, the elements of its inverse, $(\mathbf{W}^\dagger\mathbf{W})^{-1}$, may be so large that a small experimental error $\boldsymbol{\varepsilon}$ in $\mathbf{T}_\mathrm{M}$ is amplified many times by $(\mathbf{W}^\dagger\mathbf{W})^{-1}$. This was illustrated very well by Ishimaru (1978) through a simple example in which he demonstrated that if the magnitudes of adjacent elements of a (2×2) matrix $\mathbf{W}$ are different from one another by as little as 10^{-3} of either element, an experimental error of 1 percent in the value of the measured function leads to an error of the order of 2000 percent in the unknown function $\hat{\mathbf{g}}$! A problem having an unstable solution is termed an ***ill-posed problem*** (Franklin, 1970).

9-3.3 Constrained Linear Inversion Method

The least-squares method was extended by Phillips (1962) and Twomey (1965) by introducing a smoothing function to constrain the shape of the profile $g(z)$. The method also was developed independently by Tikhonov (1963), who called it the "regularization of incorrectly posed problems." Today the method is known by several different names, including the ***smoothing method***, the ***regularization method***, the ***Twomey-Phillips method***, and the ***Twomey-Tikhonov method***.

The solution of Eq. (9.58) has the form

$$\hat{\mathbf{g}} = (\mathbf{W}^\dagger\mathbf{W} + \gamma\mathbf{H})^{-1}\mathbf{W}^\dagger\hat{\mathbf{T}}_\mathrm{M}, \tag{9.63}$$

where $\mathbf{H}$ is an $(m \times m)$ ***smoothing matrix*** that must be chosen in such a way as to describe or incorporate the smoothness of the function g, and γ is a Lagrangian multiplier that usually is determined empirically. The choice of γ determines the stability of the solution. For details, the reader is referred to the articles by Turchin et al. (1971), Deschamps and Cabayan (1972), Rodgers (1976b), and Westwater (in Janssen, 1993).

9-3.4 Optimal Estimation Method

As was alluded to earlier, the key to successful retrieval of an unknown vector $\mathbf{g}$ of order m using observations $\hat{\mathbf{T}}_\mathrm{M}$ of order n (with $n \ll m$) is to supplement the observations with sufficient ***virtual measurements*** (a priori information) to regularize the ill-posed problem. One such form of a priori information is the mean profile $\langle\mathbf{g}\rangle$ and its covariance matrix $\mathbf{S}_\mathbf{g}$,

$$\mathbf{S}_\mathbf{g} = \left\langle \mathbf{g}\mathbf{g}^\dagger \right\rangle, \tag{9.64}$$

both of which are presumed to be known (or calculable) from the available historical records of $\mathbf{g}$. If $\mathbf{g}$ represents the atmospheric temperature profile $T(z)$, for example, a representative ensemble of radiosonde-measured temperature profiles can provide $\langle\mathbf{g}\rangle$ and its covariance $\mathbf{S}_\mathrm{g}$. In addition, let us assume that the error covariance matrix $\mathbf{S}_{\boldsymbol{\varepsilon}} = \left\langle \boldsymbol{\varepsilon}\boldsymbol{\varepsilon}^\dagger \right\rangle$ of the experimental error vector $\boldsymbol{\varepsilon}$ also is known (or may be determined through calibration, as discussed in connection with the statistical inversion method of the next subsection).

Using Eq. (9.59) the covariance matrix of $\hat{\mathbf{T}}_\mathrm{M}$ is

$$\begin{aligned}
\mathbf{S}_{\hat{\mathbf{T}}_\mathrm{M}} &= \left\langle \mathbf{T}_\mathrm{M}\mathbf{T}_\mathrm{M}^\dagger \right\rangle \\
&= \left\langle (\mathbf{W}\mathbf{g}+\boldsymbol{\varepsilon})(\mathbf{W}\mathbf{g}+\boldsymbol{\varepsilon})^\dagger \right\rangle \\
&= \left\langle (\mathbf{W}\mathbf{g}+\boldsymbol{\varepsilon})(\mathbf{g}^\dagger\mathbf{W}^\dagger+\boldsymbol{\varepsilon}^\dagger) \right\rangle \\
&= \left\langle \mathbf{W}\mathbf{g}\mathbf{g}^\dagger\mathbf{W}^\dagger \right\rangle + \left\langle \boldsymbol{\varepsilon}\mathbf{g}^\dagger\mathbf{W}^\dagger \right\rangle + \left\langle \mathbf{W}\mathbf{g}\boldsymbol{\varepsilon}^\dagger \right\rangle + \left\langle \boldsymbol{\varepsilon}\boldsymbol{\varepsilon}^\dagger \right\rangle \\
&= \mathbf{W}\mathbf{S}_\mathbf{g}\mathbf{W}^\dagger + \mathbf{S}_{\boldsymbol{\varepsilon}},
\end{aligned} \tag{9.65}$$

where it was assumed that the error $\boldsymbol{\varepsilon}$ has a zero mean ($\langle\boldsymbol{\varepsilon}\rangle = 0$) and is statistically independent of $\mathbf{g}$. That is, $\left\langle \boldsymbol{\varepsilon}\mathbf{g}^\dagger\mathbf{W}^\dagger \right\rangle = \langle\boldsymbol{\varepsilon}\rangle \left\langle \mathbf{g}^\dagger\mathbf{W}^\dagger \right\rangle = 0$, and similarly, $\left\langle \mathbf{W}\mathbf{g}\boldsymbol{\varepsilon}^\dagger \right\rangle = \langle\mathbf{W}\mathbf{g}\rangle \left\langle \boldsymbol{\varepsilon}^\dagger \right\rangle = 0$.

The optimal estimation method finds the "most likely" value of $\mathbf{g}$, based on the combination of a virtual measurement $\langle\mathbf{g}\rangle$ and a real measurement through $\hat{\mathbf{T}}_\mathrm{M}$, and on their associated covariance matrices, $\mathbf{S}_\mathbf{g}$ and $\mathbf{S}_{\hat{\mathbf{T}}_\mathrm{M}}$. Rodgers (1976b) explains the concept by means of a simple example: suppose we have two independent measurements x_1 and x_2 of a scalar quantity x, and suppose the associated variances are s_1 and s_2, respectively. The combined estimate of x is then

$$\hat{x} = \left(\frac{x_1}{s_1} + \frac{x_2}{s_2} \right) s_{\hat{x}},$$

where $s_{\hat{x}}$ is the variance of $\hat{x}$,

$$s_{\hat{x}} = \left(\frac{1}{s_1} + \frac{1}{s_2} \right)^{-1}.$$

The optimal estimate of $\mathbf{g}$ may be obtained by generalizing the method to vectors and considering the vector equivalents of x_1 and x_2 to be $\langle \mathbf{g} \rangle$ and $(\mathbf{D}\hat{\mathbf{T}}_\mathrm{M})$, where $\mathbf{D}$ is any exact solution such that $\mathbf{WD} = \mathbf{I}$, the identity matrix (Strand and Westwater, 1968; Westwater and Strand, 1968; Rodgers, 1976b).

The solution of the ill-posed problem given by Eq. (9.59) is

$$\hat{\mathbf{g}} = \langle \mathbf{g} \rangle + \mathbf{S_g W}^\dagger (\mathbf{W S_g W}^\dagger + \mathbf{S}_{\boldsymbol{\varepsilon}})^{-1} (\hat{\mathbf{T}}_\mathrm{M} - \mathbf{W} \langle \mathbf{g} \rangle). \tag{9.66}$$

The covariance matrix $\mathbf{S}_{(\mathbf{g}-\hat{\mathbf{g}})}$ of $(\mathbf{g}-\hat{\mathbf{g}})$, whose diagonal elements provide an estimate of the accuracy of $\mathbf{g}$, is given by (Rodgers, 1976b) as

$$\mathbf{S}_{(\mathbf{g}-\hat{\mathbf{g}})} = \mathbf{S_g} - \mathbf{S_g W}^\dagger (\mathbf{W S_g W}^\dagger + \mathbf{S}_{\boldsymbol{\varepsilon}})^{-1} \mathbf{W S_g}. \tag{9.67}$$

Rodgers (1976b) provides a sequential estimation procedure for arriving at $\mathbf{g}$ and $\mathbf{S}_{(\mathbf{g}-\hat{\mathbf{g}})}$ without having to perform any matrix inversions.

9-3.5 Statistical Inversion Method

The ***statistical inversion method*** is perhaps the most widely applied inversion technique in remote sensing. According to Westwater, one of the pioneers in the development of techniques for retrieving the profiles of atmospheric variables, "statistical inversion appears capable of extracting maximum information from the measurements" (Westwater and Decker, 1977). The underlying basis of the method and its formal solution are identical to the optimal estimation method discussed in the previous section, if the variables $\mathbf{T}_\mathrm{M}$ and $\mathbf{g}$ have a Gaussian distribution. The method used to implement the solution is different, however.

Assuming that the observations $\hat{\mathbf{T}}_\mathrm{M}$ are linearly related to the unknown function $\mathbf{g}$ through

$$\hat{\mathbf{T}}_\mathrm{M} = \mathbf{W g} + \boldsymbol{\varepsilon} \tag{9.68}$$

and that $\langle \boldsymbol{\varepsilon} \rangle = 0$, we can write

$$\hat{\mathbf{T}}'_\mathrm{M} = \mathbf{W g}' + \boldsymbol{\varepsilon}, \tag{9.69}$$

where the prime denotes departure from the mean value,

$$\hat{\mathbf{g}}' = \hat{\mathbf{g}} - \langle \mathbf{g} \rangle \tag{9.70}$$

and

$$\hat{\mathbf{T}}'_\mathrm{M} = \hat{\mathbf{T}}_\mathrm{M} - \langle \hat{\mathbf{T}}_\mathrm{M} \rangle = \hat{\mathbf{T}}_\mathrm{M} - \mathbf{W} \langle \mathbf{g} \rangle. \tag{9.71}$$

The statistical inversion method, which is closely related to multiple regression, provides a solution, $\mathbf{g}'$, that is linearly related to the observations $\hat{\mathbf{T}}'_\mathrm{M}$ through a ***predictor matrix*** $\mathbf{D}$,

$$\hat{\mathbf{g}}' = \mathbf{D}\hat{\mathbf{T}}'_\mathrm{M}. \tag{9.72}$$

The matrix $\mathbf{D}$, whose elements are the regression coefficients, is given by (Franklin, 1970; Ishimaru, 1978):

$$\mathbf{D} = E\left\{ \mathbf{g}' (\hat{\mathbf{T}}'_\mathrm{M})^\dagger \right\} E\left\{ (\hat{\mathbf{T}}'_\mathrm{M})(\hat{\mathbf{T}}'_\mathrm{M})^\dagger) \right\}^{-1}, \tag{9.73}$$

which is found by minimizing the expected value of the variance of the error in the estimate:

$$E\left\{ (\mathbf{g}-\hat{\mathbf{g}})^\dagger (\mathbf{g}-\hat{\mathbf{g}}) \right\} = \text{minimum}. \tag{9.74}$$

For this reason, the method also is known as the ***minimum-variance method***.

It can be shown (Franklin, 1970; Ishimaru, 1978) that given Eq. (9.69), evaluation of the quantities in Eq. (9.73) leads to

$$\hat{\mathbf{g}}' = \mathbf{S}_{\mathbf{g}'} \mathbf{W}^\dagger (\mathbf{W S}_{\mathbf{g}'} \mathbf{W}^\dagger + \mathbf{S}_{\boldsymbol{\varepsilon}})^{-1} \hat{\mathbf{T}}'_\mathrm{M}, \tag{9.75}$$

which basically is identical to Eq. (9.66), if $\langle \mathbf{g} \rangle$ and $\langle \hat{\mathbf{T}}_\mathrm{M} \rangle$ are set equal to zero in the latter.

In practice, given a joint distribution (data set) of radiometric measurements $\hat{\mathbf{T}}_\mathrm{M}$, and corresponding values of $\mathbf{g}$ measured independently (by radiosonde, for example), the statistical quantities given in Eq. (9.73) can be found empirically by using standard regression programs. Thus, the elements of $\mathbf{D}$ are found without the need to know anything about the weighting function $\mathbf{W}$. The availability of the joint distribution essentially serves to ***calibrate*** the linear relation between $\hat{\mathbf{g}}'$ and $\hat{\mathbf{T}}'_\mathrm{M}$. For purposes of error analysis, the joint distribution can be used to find $\mathbf{S_g}$ and $\mathbf{S}_{\hat{\mathbf{T}}_\mathrm{M}}$, and if $\mathbf{W}$ is known, then Eq. (9.65) can be used to find $\mathbf{S}_{\boldsymbol{\varepsilon}}$. Here $\mathbf{S}_{\boldsymbol{\varepsilon}}$ is the total error covariance, part of which is due to the instrument. Given this latter part, which can be obtained from instrument calibration, an estimate of model error

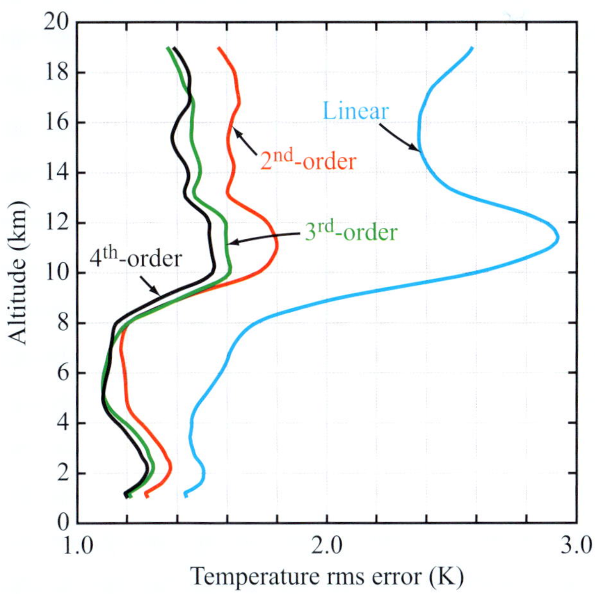

Figure 9-9: RMS temperature-retrieval error for linear regression and nonlinear (polynomial) regression for ATMS over ocean.

can be projected. If a joint distribution of measurements is not available, a simulated data set can be generated using the model defined by Eq. (9.69), together with any information available about the statistics of both the independent variable **g** and the measurement process.

Both linear and nonlinear regression techniques are used to estimate the profile of the geophysical variable **g**. In some cases, physical nonlinearities can be better addressed by nonlinear estimators. As an example, Fig. 9-9 provides a performance comparison (for estimating the temperature profile $T(z)$) between a linear regression estimator and three nonlinear estimators of varying orders. Clearly, the nonlinear estimators outperform the linear estimator, in some cases by a factor of 2.

9-3.6 Backus-Gilbert Synthetic-Averaging Inversion Method

The statistical inversion method is an effective retrieval technique provided that (a) sufficient statistical knowledge of the measurement errors and the unknown function **g** is available to render the problem a well-posed one, and (b) the error covariance $\mathbf{S}_{(\mathbf{g}-\hat{\mathbf{g}})}$ of the solution is acceptable. Let us assume that the method was applied to a given profile-retrieval problem and that condition (b) was not satisfied; i.e., that the trace of the matrix $\mathbf{S}_{(\mathbf{g}-\hat{\mathbf{g}})}$, which is a measure of overall solution quality, was unacceptably large. The error variance can be reduced by spatial averaging in the vertical direction; i.e., by degrading the resolution of the retrieved profile. There is no provision in the statistical method, however, to trade off resolution against noise to obtain the optimum compromise for a particular application. Such a trade-off feature is available in the ***Backus-Gilbert inversion method***, which was originally developed for application to inverse problems involving seismic sensing of the solid Earth (Backus and Gilbert, 1970) and later applied to atmospheric remote-sensing problems (Conrath, 1972; Westwater and Cohen, 1973).

Given the model

$$T_{\mathrm{M}}(f_i) = \int_0^\infty W_i(z)\, g(z)\, dz, \qquad i = 1,2,\dots,n, \quad (9.76)$$

and observations

$$\hat{T}_{\mathrm{M}}(f_i) = T_{\mathrm{M}}(f_i) + \varepsilon_i, \qquad i = 1,2,\dots,n, \quad (9.77)$$

at frequency f_i, with experimental error ε_i characterized by $\langle \varepsilon_i \rangle = 0$, the Backus-Gilbert method provides the solution

$$\hat{g}(z) = \sum_{i=1}^{n} a_i(z)\, \hat{T}_{\mathrm{M}}(f_i). \quad (9.78)$$

The z-dependent coefficients $a_i(z)$ form an $n \times 1$ column vector given by

$$\mathbf{a}(z) = \frac{\mathbf{R}^{-1}(z)\, \mathbf{U}}{\mathbf{U}^\dagger\, \mathbf{R}^{-1}(z)\, \mathbf{U}}, \quad (9.79)$$

where $\mathbf{R}(z)$ is a function that is linearly related to a weighted sum of the error covariance matrix $\mathbf{S}_\varepsilon$ and an $n \times n$ matrix $\mathbf{S}(z)$ and is given by

$$\mathbf{R}(z) = \alpha\, \mathbf{S}(z) + (1-\alpha) r \mathbf{S}_\varepsilon. \quad (9.80)$$

The element U_i of the $n \times 1$ column vector $\mathbf{U}$ is equal to the total integral of the corresponding weighting function,

$$U_i = \int_0^\infty W_i(z)\, dz, \tag{9.81}$$

and the matrix $\mathbf{S}$ has elements

$$S_{kl}(z) = 12 \int_0^\infty (z - z')^2\, W_k(z')\, W_l(z')\, dz'. \tag{9.82}$$

The factor r in Eq. (9.80) is a constant inserted so that the two terms have the same physical dimensions, and α is a ***trade-off parameter***. By varying α, whose range is between 0 and 1, emphasis can be shifted from minimizing the estimate variance (by choosing $\alpha = 0$) at the expense of resolution, to maximizing resolution ($\alpha = 1$) at the expense of measurement accuracy. The degree of emphasis between the two considerations is chosen to suit the particular application.

Although it is clear from Eq. (9.80) that the magnitude of the parameter α controls the relative weights of the two terms whose sum defines $\mathbf{R}$, it is not clear how the choice of α is related to resolution and variance. This is because we furnished the result of the Backus-Gilbert solution without deriving it. The connection between α and resolution and variance can be shown easily through the following steps:

(a) For a given set of weighting functions $W_i(z)$, $i = 1, 2, \ldots, n$, the matrix $\mathbf{S}(z)$ is completely specified by Eq. (9.82).

(b) For a given error covariance matrix $\mathbf{S}_{\boldsymbol{\varepsilon}}$, the trade-off parameter α specifies the matrix $\mathbf{R}(z)$, which in turn specifies the column vector $\mathbf{a}(z)$ through Eq. (9.79), because the vector $\mathbf{U}$ has already been specified by the weighting functions in Eq. (9.81).

(c) With $\mathbf{a}(z)$ specified, the estimate $\hat{g}(z)$ of the unknown function is made through a linear sum of the measurements $\hat{T}_{\rm M}(f_i)$, each weighted by the corresponding coefficient $a_i(z)$.

(d) If we ignore the random error ε and substitute Eq. (9.76) into Eq. (9.78), we obtain the result

$$\begin{aligned}
\hat{g}(z) &= \sum_{i=1}^{n} a_i(z)\, T_{\rm M}(f_i) \\
&= \sum_{i=1}^{n} a_i(z) \int_0^\infty W_i(z')\, g(z')\, dz' \\
&= \int_0^\infty \sum_{i=1}^{n} a_i(z)\, W_i(z')\, g(z')\, dz' \\
&= \int_0^\infty A(z,z')\, g(z')\, dz',
\end{aligned} \tag{9.83}$$

where $A(z,z')$, called the ***averaging kernel***, is defined as

$$A(z,z') = \sum_{i=1}^{n} a_i(z)\, W_i(z'). \tag{9.84}$$

Thus, according to Eq. (9.83), the estimate $\hat{g}(z)$ at a height z is a weighted average of the profile $g(z)$ over the entire height of the atmosphere, and the weighting function $A(z,z')$ is governed by the coefficients $a_i(z)$, which in turn are governed by the trade-off parameter α.

By substituting Eq. (9.77) into Eq. (9.78) and computing the variance $\sigma^2_{\hat{g}}(z)$ of the estimate $\hat{g}(z)$, it can be easily shown that

$$\sigma^2_{\hat{g}}(z) = \mathbf{a}^\dagger(z)\, \mathbf{S}_{\boldsymbol{\varepsilon}}\, \mathbf{a}(z), \tag{9.85}$$

where $\mathbf{S}_{\boldsymbol{\varepsilon}} = \langle \boldsymbol{\varepsilon}\boldsymbol{\varepsilon}^\dagger \rangle$ is the error covariance matrix.

Thus, the coefficients $a_i(z)$ control the variance $\sigma^2_{\hat{g}}$, as well as the averaging kernel $A(z,z')$, whose shape controls the effective resolution associated with the estimate $\hat{g}(z)$. The ***effective resolution width*** is called the ***spread*** and is given by

$$s(z) = 12 \int_0^\infty (z - z')^2\, A^2(z,z')\, dz' \tag{9.86}$$

The spread $s(z)$ has units of z and is a measure of the spread of $A(z,z')$ about $z' = z$.

To illustrate how the various parameters relate to each other, let us consider the following example. Figure 9-10 shows a set of eight weighting functions $W_i(z)$

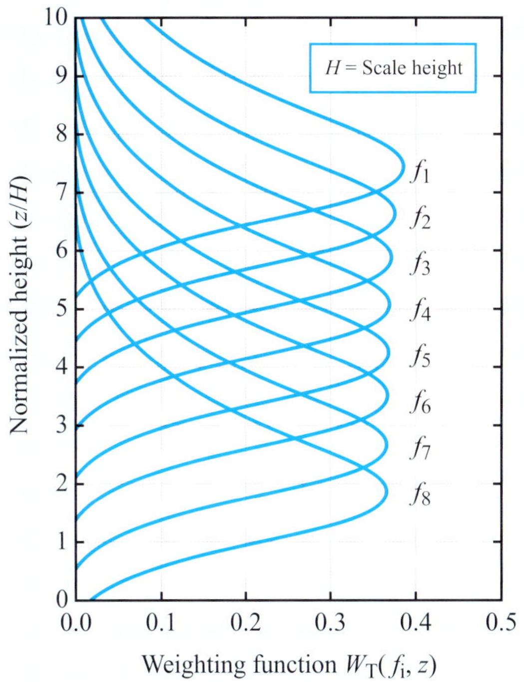

Figure 9-10: A set of idealized weighting functions normalized to unit area (adapted from Rodgers, 1976b).

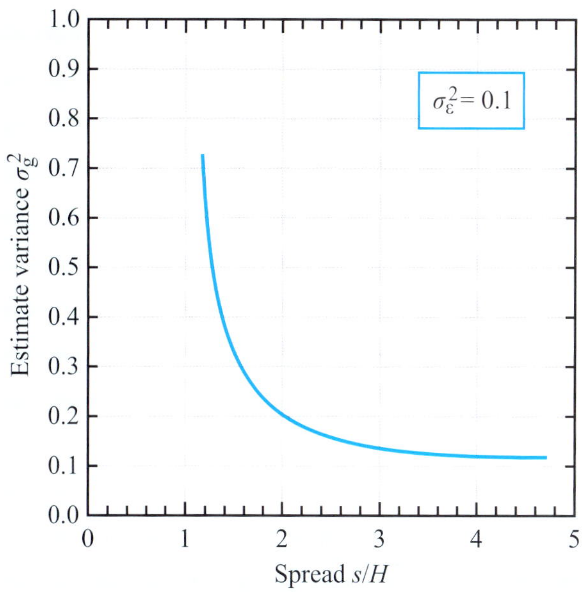

Figure 9-11: Trade-off between estimation accuracy and vertical resolution for the idealized set of weighting functions (Fig. 9-10) at a height of $z = 5H$ (from Rodgers, 1976b).

plotted as a function of z/H, where H is a (constant) scale height. Eight measurements are made, $\hat{\mathbf{T}}_{\mathrm{M}}(f_i)$ with $i = 1, 2, \ldots, 8$, and associated with each is a zero-mean random error ε_i whose variance $\langle \varepsilon_i^2 \rangle = \sigma_\varepsilon^2$ for all i. Assume $\sigma_\varepsilon^2 = 0.1$ and choose the dimensionality factor r in Eq. (9.80) to be unity for convenience. The trade-off between measurement accuracy and resolution is illustrated in Figs. 9-11 and 9-12. The curve shown in Fig. 9-11 is a plot of the estimate variance $\sigma_{\hat{g}}^2$ as a function of the spread s at $z = 5H$, with α as parameter.

From Fig. 9-12, the best resolution (minimum spread) is obtained for $\alpha = 1$ and is equal to $1.15H$, but the corresponding variance is $\sigma_{\hat{g}}^2 = 0.72$. On the other hand, if $\alpha = 0$, $\sigma_{\hat{g}}^2 \approx 0.1$, but $s = 4.8H$. The averaging kernels $A(z, z')$ associated with $\alpha = 0$ and $\alpha = 1$ are shown in Fig. 9-12, along with others for intermediate values of α.

9-3.7 Retrievals Based on Neural Networks

Artificial neural networks, or neural nets, are computational structures inspired by biological networks of densely connected neurons, each of which is capable only of simple computations. Just as biological neural networks are capable of learning from their environment, neural nets are able to learn from the presentation of training data, as the free parameters (weights and biases) are adaptively tuned to fit the training data. Neural nets can be used to learn and compute functions for which the analytical relationships between inputs and outputs are unknown and/or computationally complex and are therefore useful for pattern recognition, classification, and function approximation. Neural nets are particularly appealing for the inversion of atmospheric remote sensing data, where relationships are commonly nonlinear and non-Gaussian, and the physical processes may not be well understood.

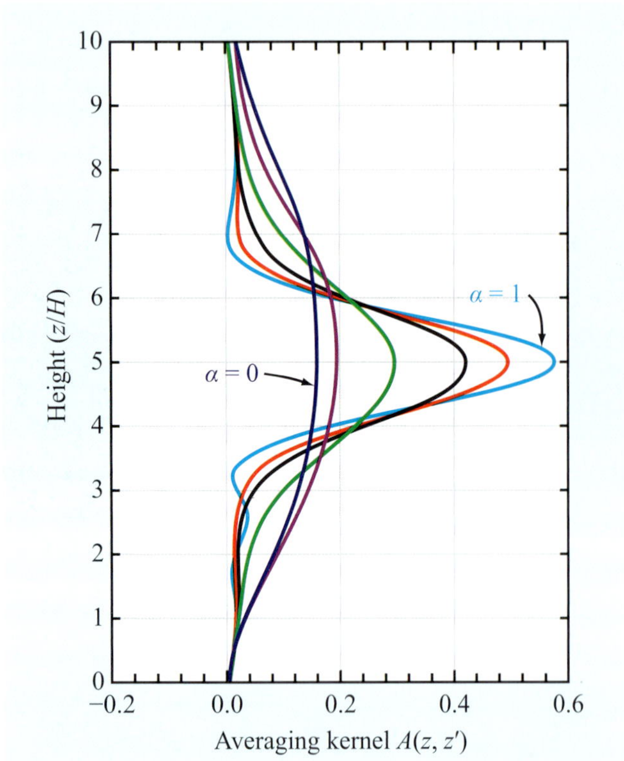

Figure 9-12: Averaging kernels at $z = 5H$ for $\alpha = 0$, $\alpha = 1$, and intermediate values (adapted from Rodgers, 1976b).

Feedforward multilayer perceptron (FFMLP) neural networks are often used due to their simplicity, flexibility, and ease of use. Multilayer neural networks most often consist of an input layer, one or more nonlinear hidden layers, and a linear output layer. The mathematical functions implemented by FFMLP nets are continuous and differentiable, which greatly simplifies training and error analysis. But perhaps the most useful attribute of neural nets is their scalability; a network with a sufficient number of weights and biases is capable of approximating a bounded, continuous function to an arbitrary level of precision over a finite domain (Hornik et al., 1989). Therefore, neural networks can be used as universal function approximators.

The example of nonlinear regression shown in Fig. 9-9 illustrated that a linear combination of simple, nonlinear functions (polynomials) could be used to construct powerful nonlinear estimators. However, no guidance was provided for the selection of these functions or the determination of how many should be used. Neural networks are constructed in a similar fashion: simple computational elements (nodes) are connected in a way that allows complex functions to be synthesized by adding nodes to the network. The distinctive features of neural networks in this context are the "activation functions" applied by each of the nodes, the number of nodes, a detailed description of the node connections in the network (the topology), and the algorithm used to derive the weights and biases (training). The interested reader is referred to Haykin (2008), Bishop (2007), and Blackwell and Chen (2009) for a more detailed treatment.

9-4 Temperature-Profile Retrieval from Ground-Based Observations

In their 1963 paper on the microwave spectrum of atmospheric oxygen, Meeks and Lilley (1963) suggested that it might be possible to remotely acquire information about the structure of the atmosphere from measurements of the emission spectrum of oxygen. Two years later, Westwater (1965) published a paper in which he examined this thesis for the upward-looking case, and developed a mathematical procedure for inverting the emission measurements to infer the temperature profile of the atmosphere. Moreover, he evaluated the sources and magnitudes of the errors expected to be associated with the inversion technique. Several studies have ensued since then, including those by Westwater and Strand (1968), Waters (1971), Snider (1972), Westwater et al. (1975), Decker et al. (1978, 1982), Westwater (in Janssen, 1993) and Blackwell and Chen (2009). Thus, extensive experience has been gained in the use of measurement techniques, retrieval methods, and other elements associated with the remote-sensing problem. In this section, we present the basic steps involved in the temperature-profile retrieval process and highlight some of the experience gained to date.

9-4.1 Single-Frequency Multiangle Observations

Figure 9-13 is an example of a temperature profile $\hat{T}(z)$ inferred from angular scanning measurements recorded by a single-frequency radiometer operating at 54.5 GHz. Also shown in the figure are the mean temperature profile $\langle T(z) \rangle$ (calculated from historical records) and the radiosonde profile $T_{\rm rs}(z)$ recorded at the same time as the radiometric observations leading to $\hat{T}(z)$. Overall, the temperature profile $\hat{T}(z)$, which was inferred from six angular measurements (at $\theta = 0°$, $30°$, $60°$, $75°$, $80°$, and $85°$) using the statistical inversion method of Section 9-3.5, provides a good estimate of the "true" profile $T_{\rm rs}(z)$, but fails to predict the temperature inversion at $z \approx 2.7$ km.

It should be noted that the temperature profile retrieved by a microwave radiometer is qualitatively different from that measured by a radiosonde. A balloon-carried radiosonde makes point-specific measurements, while a radiometer makes volume-specific measurements. Furthermore, the radiometer averages out spatial variations in T if they occur over distances smaller than its spatial resolution capabilities.

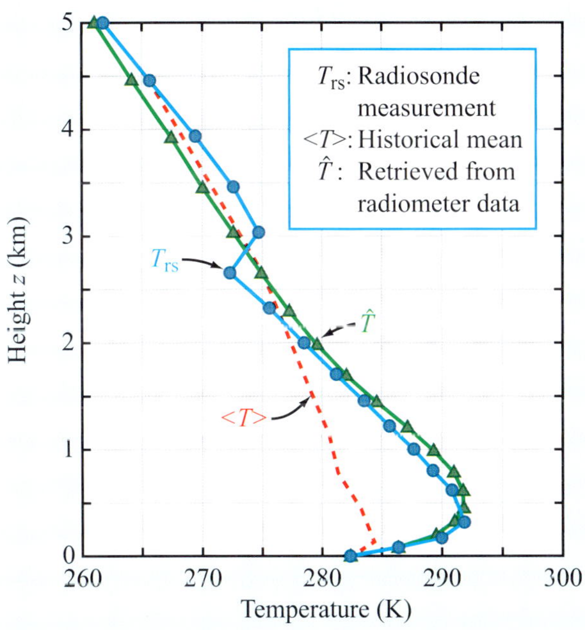

Figure 9-13: Typical example of a temperature profile derived from single-frequency (54.5 GHz) angular scanned radiometer data at Cincinnati, Ohio. $T_{\rm rs}$ = radiosonde profile, $\langle T \rangle$ = mean profile for constrained surface conditions, $\hat{T}$ = inferred profile (from Westwater and Decker, 1977).

Apart from the a priori information provided as input to the inversion algorithm, the radiometric measurements $\hat{T}_{\rm DN}$ are the only inputs used to estimate the profile $T(z)$. Thus, the inversion process does not take into consideration the variations in $\hat{T}_{\rm DN}$ resulting from atmospheric variables other than the temperature profile $T(z)$. In addition, the statistical inversion technique used to obtain $\hat{T}(z)$ is based on the assumption that the direct problem is linear in form, which means that the weighting function $W_{\rm T}(f,\theta,z)$ is independent of $T(z)$. In practice, it is not possible to select a frequency f such that $W_{\rm T}(f,\theta,z)$ is totally independent of deviations from the mean profiles for temperature, pressure, and water-vapor density. Variations in $W_{\rm T}(f,\theta,z)$ will in turn cause variations in the observations $\hat{T}_{\rm DN}(f,\theta)$.

9-4.2 Multifrequency Single-Angle Observations

One of the obvious limitations of the multiangle retrieval scheme is its requirement that the atmosphere be horizontally stratified. A multifrequency, single angle system, on the other hand, has the distinct advantage that all measurements are made for the same atmospheric volume, provided the antennas are chosen so that their beams are approximately equal in size at all of the radiometer frequencies.

The ***scanning microwave spectrometer*** (SCAMS) is part of the payload carried into orbit by the Nimbus-6 satellite in 1975. SCAMS is a five-channel radiometer with three channels in the 50–60 GHz oxygen band, one at the center of the 22.235 GHz water-vapor line, and one at 31.65 GHz, which is in an atmospheric-window band. A ground-based version of the SCAMS was used to retrieve temperature and water-vapor profiles from zenith radiometric measurements at the five SCAMS frequencies (Decker et al., 1978). Since water-vapor retrievals are discussed in the next section, we

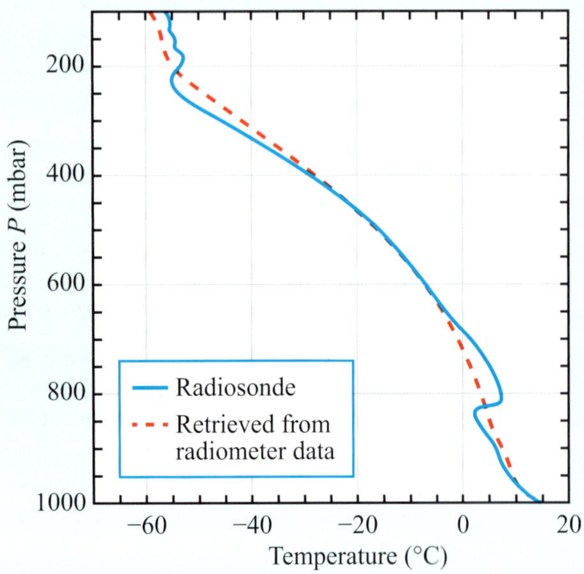

Figure 9-14: An example of a temperature profile retrieved from SCAMS radiometric data (dashed line) compared with concurrent radiosondes (solid line) (from Decker et al., 1978).

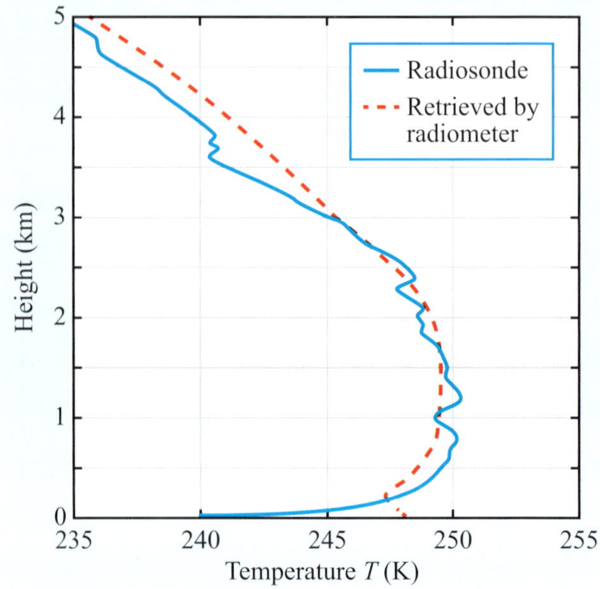

Figure 9-15: Typical example of radiometric temperature retrievals compared with the radiosonde measured in Barrow, Alaska on March 11, 2007 [Cimini et al., 2010].

shall concern ourselves here with temperature retrievals only. The temperature weighting functions at the three SCAMS oxygen frequencies were shown earlier in Fig. 9-4(b). A typical example of the retrieved temperature profiles $T(z)$ is shown in Fig. 9-14, obtained by applying a statistical inversion to the radiometric measurements.

Overall, the retrieved temperature profiles have good accuracies, on the order of 1–2 K rms (relative to the radiosonde measurements) from the surface to about 600 mbar ($\approx$ 4 km) (Decker et al., 1978). One obvious deficiency in the retrieval algorithm is that it is incapable of retrieving rapid variations in temperature, such as temperature inversions, with height. This is not surprising when we consider that the profile is essentially being retrieved from four numbers: the three microwave measurements at the oxygen frequencies, and the measured surface temperature. The nonoxygen channels do not contribute directly to the retrieval of the temperature profile, but can be used to correct for water-vapor and cloud effects when necessary.

By increasing the number of channels in the 50 to 70 GHz range and by exploiting more robust inversion techniques, it is possible to retrieve the temperature profile with better precision, but it continues to be difficult to retrieve temperature inversions at high altitudes. Figure 9-15 shows a profile of atmospheric temperature in the Arctic retrieved by a six-channel radiometer, in combination with observations at six elevation angles, adding up to a total of 36 observations (Cimini et al., 2010). The temperature profile measured by a radiosonde is included for comparison. Closer agreement between the two profiles is observed over the atmospheric layer between the surface and 3 km than at higher elevations.

9-4.3 Pressure Height

The temperature profile $T(P)$ can be used to determine the ***pressure height*** z_P, which is defined as the height

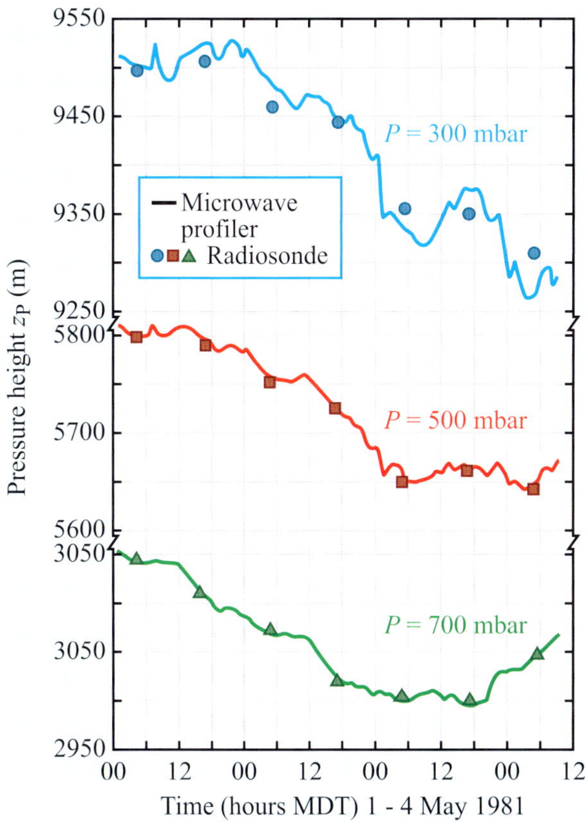

Figure 9-16: Time series of 700, 500, and 300 mbar pressure heights, recorded continuously by the microwave radiometer system, and twice daily radiosonde measurements (Decker et al., 1982).

corresponding to a given pressure level P. Integrating both sides of Eq. (9.14) from the surface at $z = 0$ (and corresponding pressure P_s) to the height z_P (and corresponding pressure P) gives

$$z_P = -\frac{R}{Mg}\int_{\ln P_s}^{\ln P} T(P)\, d(\ln P). \tag{9.87}$$

Figure 9-16 shows a time series of the 700, 500, and 300 mbar pressure heights, all derived from radiometric measurements made by a six-channel radiometer consisting of four oxygen channels in the 50–60 GHz band, a water-vapor channel, and a 31.65 GHz window channel (Decker et al., 1982). Based on a comparison of the pressure height as derived from the microwave radiometer data with that measured by the radiosonde, the rms difference was found to be 3 m, 8.6 m, and 24 m for the 700, 500, and 300 mbar levels, respectively. The analysis consisted of 70 to 80 cases (Decker et al., 1982).

9-5 Water-Vapor Profile Retrieval from Ground-Based Observations

In principle, radiometric data can be used to infer the height profile of the water-vapor density $\rho_v(z)$ using the same statistical inversion techniques referred to earlier in connection with the atmospheric temperature profile $T(z)$. In selecting the radiometer frequencies, the following criteria are important: (a) the radiometric temperature $T_{DN}(f)$ at selected frequencies should be strongly sensitive to $\rho_v(z)$ and only very weakly sensitive (or totally insensitive) to other atmospheric variables such as the temperature profile $T(z)$, and (b) the water-vapor weighting functions $W_\rho(f,z)$ at the selected frequencies should have height profiles sufficiently different to minimize redundancy. These criteria can be met conceptually by choosing frequencies that lie at the peak and shoulders of either the 22.235 GHz or the 183.31 GHz water-vapor absorption lines. Each of these cases is discussed in some detail below. First, however, an important difference between the problems posed by the retrieval of the temperature profile and those posed by the water-vapor density profile should be pointed out. Whereas the temperature profile $T(z)$ has a basically stable structure and a relative variability about the mean profile $\langle T(z)\rangle$ of the order of 0.2, the water-vapor density may exhibit large variations as a function of both space and time. For a fixed location, the ***total integrated water content*** M_v can vary by a factor of 30 during a given year (Westwater, 1979). Estimating the profile of a parameter that may vary over a wide range of values is clearly a more demanding task than estimating the profile of a parameter whose profile is more narrowly bounded.

For simplicity, we limit the discussion to zenith observations, i.e., $\theta = 0$. The ***weighting function for***

water vapor $W_\rho(f,z)$ is defined through the standard form

$$T_{\rm DN}(f) = \int_0^\infty W_\rho(f,z)\,\rho_{\rm v}(z)\,dz, \tag{9.88}$$

where $T_{\rm DN}(f)$ is the downward-emitted brightness temperature along the zenith direction and $\rho_{\rm v}(z)$ is the water-vapor profile. Upon equating Eq. (9.88) to Eq. (9.2) for $\theta = 0$, we obtain

$$W_\rho(f,z) = \kappa_{\rm a}(f,z)\,\frac{T(z)}{\rho_{\rm v}(z)}\,e^{-\tau_0(0,z)}, \tag{9.89}$$

and its units are K·(g/m^3)/km. The absorption coefficient $\kappa_{\rm a}(f,z)$ accounts for absorption by oxygen and water vapor. At or near the peak of a water-vapor line, $\kappa_{\rm a}$ is dominated by the water-vapor component, $\kappa_{\rm H_2O}$. This is certainly true for the strong 183.31 GHz line, and also true for the relatively weak 22.235 GHz line, except in dry climates. Consider Fig. 8-7 of the previous chapter: at $f = 22.235$ GHz, $\kappa_{\rm H_2O}$ is larger than $\kappa_{\rm O_2}$ by about a factor of 10 (for $\rho_0 = 7.5$ g/m^3). In a dry climate with, let us say, $\rho_0 = 1$ g/m^3, the ratio would be only about 2, and hence the total absorption coefficient $\kappa_{\rm a}$ is not as strongly dominated by water vapor. When $\kappa_{\rm a}$ can indeed be approximated as $\kappa_{\rm a} \approx \kappa_{\rm H_2O}$, the explicit dependence of $W_\rho(f,z)$ on $\rho_{\rm v}(z)$, as indicated in Eq. (9.89), disappears, because $\kappa_{\rm H_2O}$ is directly proportional to $\rho_{\rm v}(z)$.

Height profiles of $W_\rho(f,z)$ are shown in Fig. 9-17 at several microwave frequencies, including the resonance frequencies at 22.235 and 183.31 GHz and several line-shoulder frequencies, as well as at two window frequencies (31.4 and 140 GHz). The window frequencies usually are included in order to estimate cloud attenuation (when clouds are present) as discussed in the next section. Because of the weakness of the 22.235 GHz line relative to the other microwave water-vapor and oxygen lines, the weighting functions at 21.0 and 22.235 GHz do not vary rapidly with height. Consequently, radiometric measurements at or near the 22.235 GHz line cannot produce water-vapor profiles with good vertical resolution. As an illustration, consider the water-vapor profiles shown

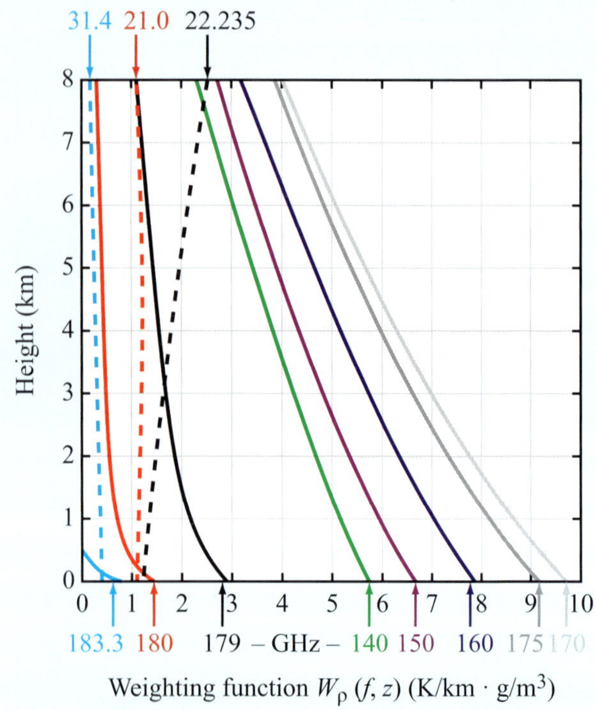

Figure 9-17: Weighting functions for frequencies around the 22.235 GHz line (dashed) and the 183.31 GHz line (solid) [from Askne and Skoog, 1983].

in Fig. 9-18(a). These profiles were retrieved from radiometric observations at five microwave frequencies, all made along the zenith direction. The associated weighting functions are shown in Fig. 9-18(b). The radiometer system used to retrieve the profiles was a ground-based version of the SCAMS discussed in connection with temperature-profile retrievals in Section 9-4.2.

The retrieval accuracy (rms of the difference between retrieved and radiosonde profiles) was found by Decker et al. (1978) to be slightly better than 1 g/m^3 at a height of about 500 m and to improve with increasing altitude. No error exists at the surface because the retrieved profiles were constrained by available surface measurements. A similar water-vapor retrieval study was reported by Skoog et al. (1982), who used radiometer frequencies of 31.65 GHz and 21.0 GHz,

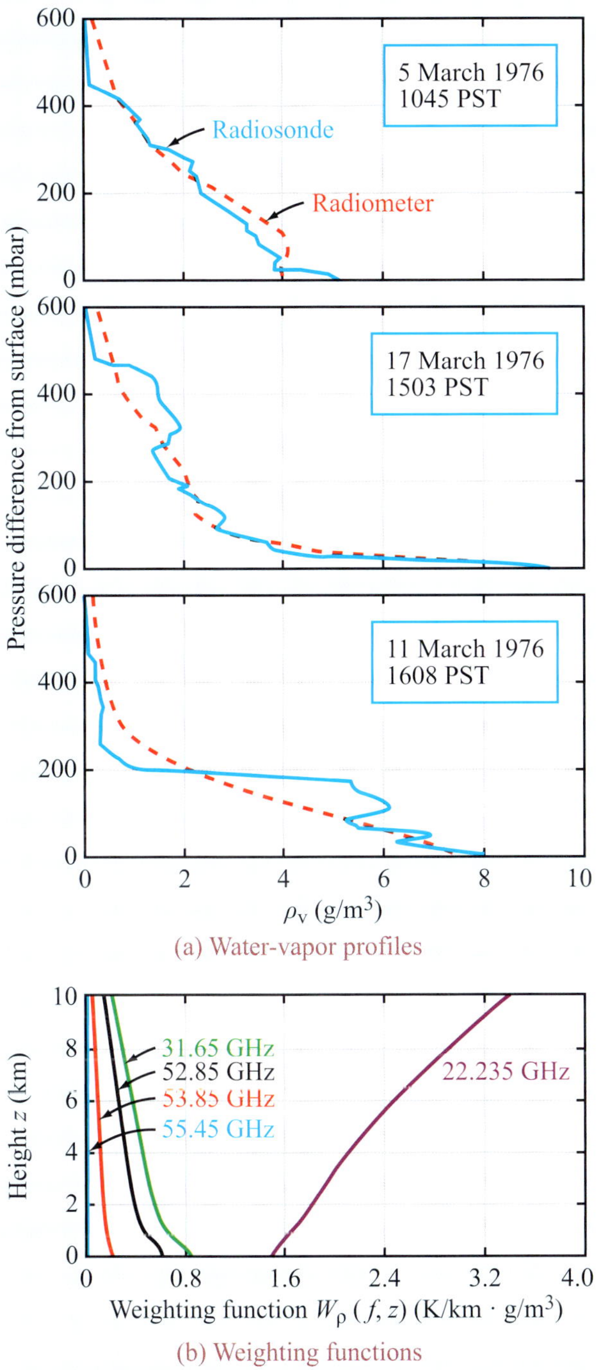

Figure 9-18: (a) Examples of retrieved profiles (dashed lines) of $\rho_v(z)$ compared with concurrent radiosonde profiles (solid lines) and (b) water-vapor weighting functions at the five SCAMS frequencies [from Westwater and Decker, 1977].

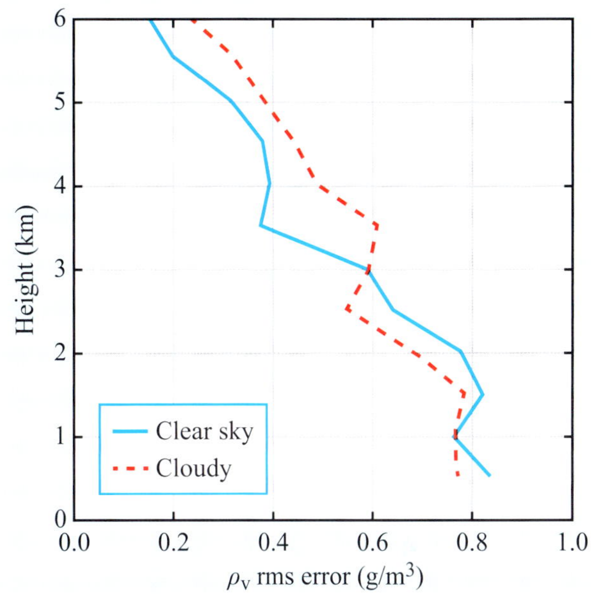

Figure 9-19: Water-vapor profiling accuracy: rms differences between retrieved and radiosonde profiles for 22 clear-sky measurements (solid line) and for 16 measurements during cloudy conditions (dashed line) [from Askne and Skoog, 1983].

rather than the line-center frequency of 22.235 GHz used by Decker et al. (1978). Skoog et al. (1982) reported profiling accuracies (Fig. 9-19) that were similar both in shape and in terms of their variations with height but were somewhat better in magnitude than those reported by Decker et al. (1978).

These results indicate that the water-vapor profiles retrieved from radiometric measurements using the 22.235 GHz water-vapor line are not sufficiently accurate for the technique to be considered for operational use. It has been suggested that a multifrequency radiometer system operating at frequencies near the strong 183.31 GHz water-vapor line is needed for accurate water-vapor profiling (Hogg, 1980; Askne and Skoog, 1983; Wilheit, 1990). Figure 9-20 shows plots of water-vapor density weighting functions, normalized to unity at the surface, for several frequencies in the 140 to 183.3 GHz region. According to a simulation study using these weighting functions (Askne and

Skoog, 1983), the profiling accuracy can be improved by 40 percent (over that provided by a two-frequency radiometer operating at 21 and 31.4 GHz) through the addition of at least one frequency around 175 GHz.

Since these early explorations, numerous studies were conducted to simulate, measure, and retrieve the atmospheric profile of water vapor using multiple channels at and near the 183.31 GHz water-vapor absorption line (Wilheit, 1990; Lutz et al., 1991; Wang et al., 1993; Wang et al., 1997; Kuo et al., 1994; Racette et al., 1996; Pazmany, 2007; Cimini et al., 2007a; Racette et al., 2005; Mattioli et al., 2010). One example of such a system is the HAMSTRAD (***H_2O Antarctica Microwave Stratospheric and Tropospheric Radiometer***), which is a ground-based radiometer developed by a European team to measure the water-vapor profile over Concordia Station in Antarctica (Ricaud et al., 2010a, b). It includes 7 channels in the 50–60 GHz band, which are used to profile $T(z)$, and an additional 13 channels in the 169–197 GHz band, selected to retrieve water-vapor density $\rho_v(z)$. The system (Fig. 9-21) was initially deployed and validated in the Pyrenees in France before transporting it to Antarctica. Both locations are very cold and extremely dry environments; whereas in temperate climates ρ_v typically is in the range 3–10 g/m^3, in Antarctica ρ_v seldom exceeds 1 g/m^3. Profiling ρ_v in such a dry environment requires an accuracy on the order of 0.05 g/m^3. Figure 9-22 displays profiles retrieved by HAMSTRAD on three occasions, representing relatively dry, medium, and slightly wet Antarctic atmospheres. Also shown are profiles recorded concurrently by radiosondes. In all three cases, good overall agreement is apparent between the pairs of profiles. A one-week-long temporal profile of ρ_v is displayed in Fig. 9-23.

9-6 Retrieval of Integrated Precipitable Water Vapor (IPWV) from Ground-Based Observations

In this section we examine how radiometric measurements can be used to estimate the ***integrated precipitable water vapor*** (IPWV) of the atmosphere and

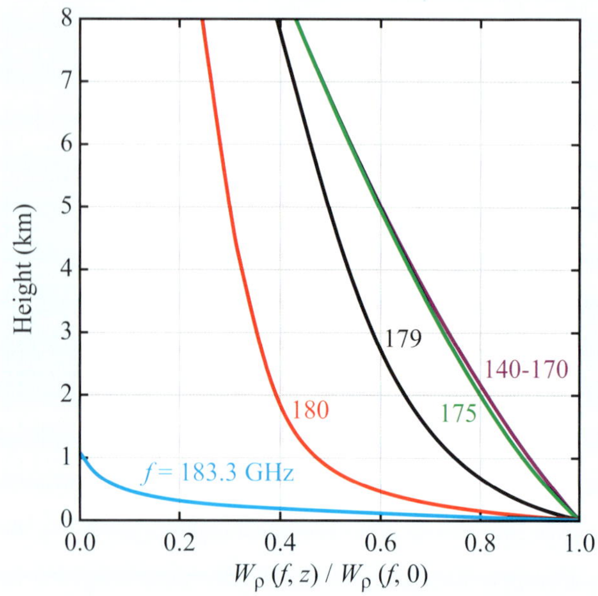

Figure 9-20: Normalized water-vapor weighting functions for frequencies in the 140 to 183.3 GHz band [from Askne and Skoog, 1983].

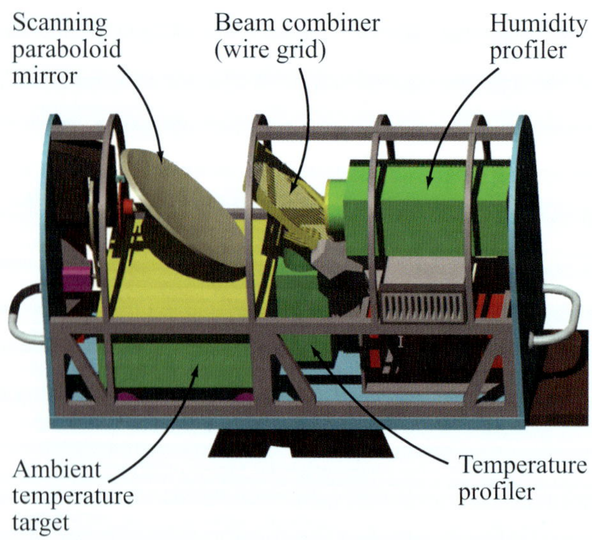

Figure 9-21: Internal structure of the HAMSTRAD radiometer [Ricaud et al., 2010a].

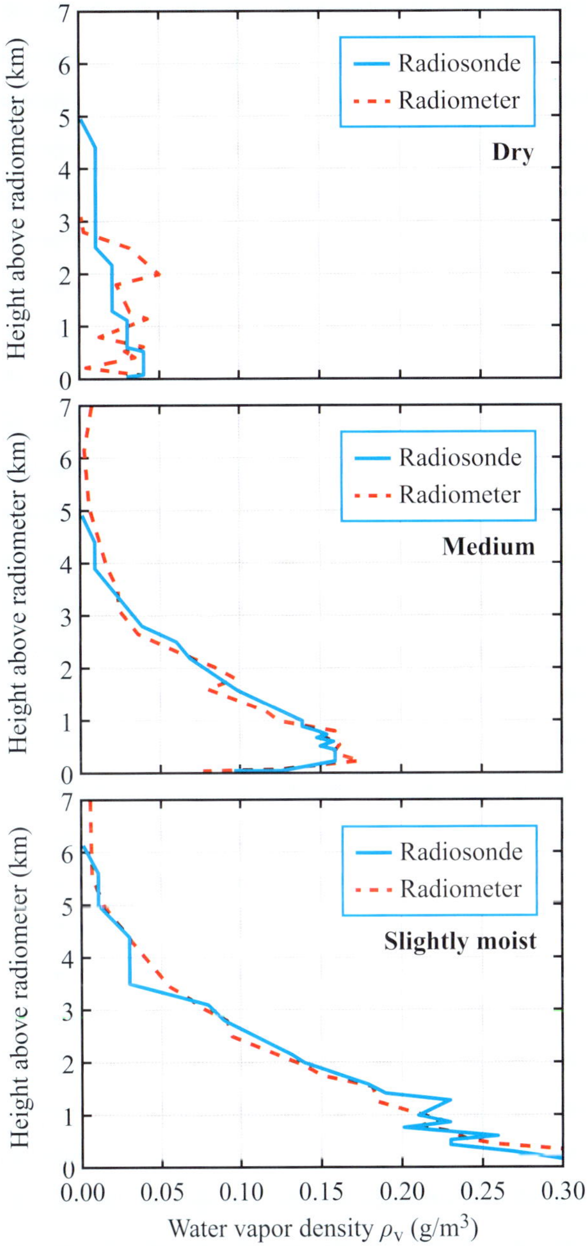

Figure 9-22: Height profiles for water-vapor density ρ_v at Concordia Station, Antarctica [Ricaud et al., 2010a].

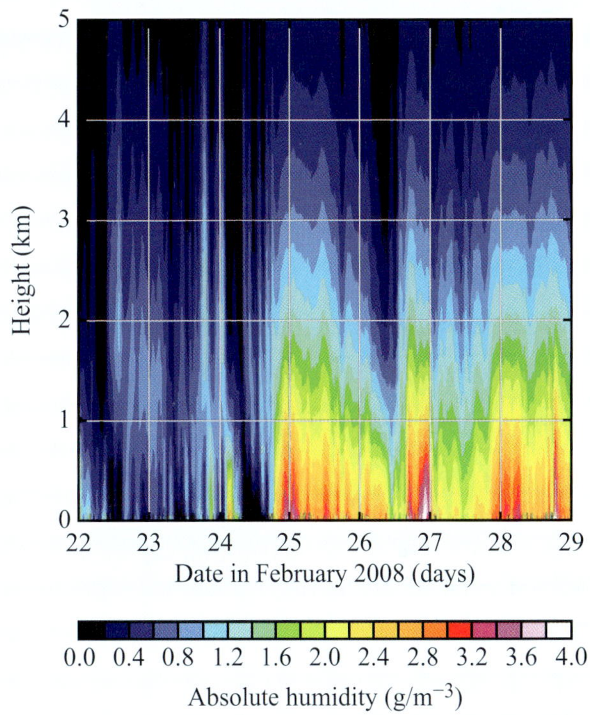

Figure 9-23: Temporal evolution of the water-vapor density as measured by HAMSTRAD in February 2008 [Ricaud et al., 2010a].

the ***liquid-water path*** (LWP) of nonprecipitating clouds. The measurement of these atmospheric parameters is useful in meteorological research and in astronomical and military applications (Decker et al., 1973; Gorelik et al., 1973; Westwater, 1978; Goldfinger, 1980).

The total mass of atmospheric water vapor contained in a vertical column of unit cross section (1 cm^2) was defined by Eq. (8.8) as

$$M_v = \int_0^\infty \rho_v(z)\, dz \qquad (\text{g/cm}^2), \tag{9.90}$$

where $\rho_v(z)$ is the water-vapor density at height z. If water vapor having this particular mass were to precipitate inside a tube having a cross section of 1 cm^2, the height of the liquid water within the tube would be

$$h_\mathrm{v} = \frac{M_\mathrm{v}}{\rho_\mathrm{L}} = \int_0^\infty \rho_\mathrm{v}(z)\,dz \quad \textbf{cm of precipitable water}, \tag{9.91}$$

where $\rho_\mathrm{L} = 1$ g/cm^3 is the density of liquid water. This height h_v is the ***integrated precipitable water vapor*** (IPWV) of the atmosphere.

Similarly, for a cloud of vertical thickness H (km) and uniform water content m_v (g/m^3), the ***liquid-water path*** (LWP) of the cloud would be measured in terms of the height h_L, which is defined as

$$h_\mathrm{L} = \frac{m_\mathrm{v}}{\rho_\mathrm{L}} H \qquad \frac{\mathrm{g/m^3 \cdot km}}{\mathrm{g/cm^3}} = \text{mm of liquid water}. \tag{9.92}$$

These two quantities, h_v and h_L, can be estimated from microwave radiometric observations at two frequencies selected such that the emitted radiation is particularly sensitive to water vapor at one of the frequencies and to cloud liquid-water content at the other. For measuring h_v, the optimum frequency is the one for which the water-vapor density weighting function has a constant profile with height. If at a frequency f_1, $W_\rho(f_1, z) \approx W_\rho(f_1)$, then Eq. (9.88) becomes

$$T_\mathrm{DN}(f_1) = W_\rho(f_1) \int_0^\infty \rho_\mathrm{v}(z)\,dz = W_\rho(f_1)\,h_\mathrm{v}. \tag{9.93}$$

This provides a radiometric measurement that is directly proportional to h_v. An example of a weighting function whose profile is approximately height-independent is the 21.0 GHz function plotted in Fig. 9-17. Such a profile, however, represents average (mean) conditions over a certain period of time. For a given location and at a given point in time, the actual profile of the water-vapor density weighting function may differ significantly from this mean profile in response to variations in temperature and water-vapor density. Hence, relying on a single radiometric channel to estimate h_v may not provide acceptably accurate results. Also, the presence of clouds within the volume observed by the antenna beam contributes to the uncertainty in h_v, unless the emission by the clouds is monitored separately. Hence, an additional channel should be chosen (with a center frequency in the 30–32 GHz window) to perform that function.

Figure 9-24 displays values of the brightness temperature T_B measured by upward-looking radiometers, plotted as a function of h_v. The measurements, reported by Payne et al. (2008), were performed at four frequencies in the neighborhood of the 22 GHz water-vapor line and at another four frequencies in the neighborhood of the 183 GHz line. The response of T_B to h_v is linear for the first set and nonlinear, but with a much larger slope, for the second set. Note the relatively small amount of scatter of the data points relative to the smooth curves.

The ***Ground-Based Scanning Radiometer*** (GSR) is a system capable of making radiometric observations at 25 channels in the 50–380 GHz range, including 7 channels in the 183 GHz band (Cimini et al., 2007b). It can scan in elevation from 15° relative to the horizon) to 165°. A photograph of the system is shown in Fig. 9-25. According to the analysis reported by Cimini et al. (2007a), operating in the millimeter- and submillimeter-wavelength bands yields sensitivity "improvement factors from 1.5 to 69 for precipitable water vapor (IPWV) and 3 to 4 for cloud liquid water when compared to 20–30 GHz radiometers." Not only is the sensitivity greater, but so is the retrieval accuracy. This is illustrated by the time series shown in Fig. 9-26, which includes two retrievals, one based on the GSR system and another based on observations made by an older-generation microwave radiometer (MWR) that operates at only 23.8 and 31.4 GHz. The time series are compared with direct measurements based on radiosonde recordings. The results indicate that the GSR-retrieval values of h_v are consistently in agreement with the radiosonde measurements, but not so for the MWR. Among 126 observations, the rms error is 0.03 cm for h_v when retrieved by the GSR (relative to the sonde), compared with 0.04 cm when retrieved by the MWR (Fig. 9-27). It is noteworthy that these are excellent precisions, particularly in a very dry environment (Barrow, Alaska) in which the maximum value of h_v seldom exceeds 1.5 cm.

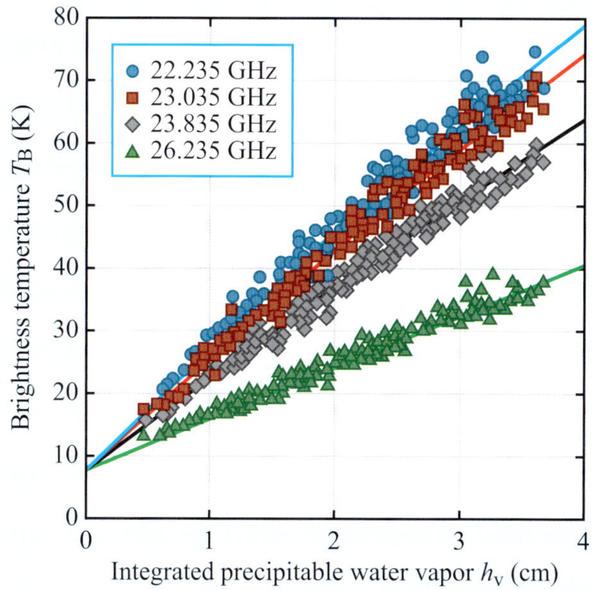

(a) 22 GHz band

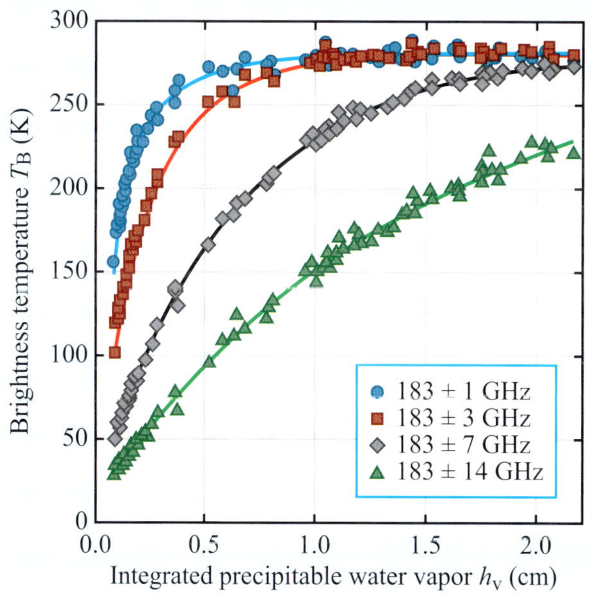

(b) 183 GHz band

Figure 9-24: Measured brightness temperatures plotted against radiosonde IPWV values for cloud-free cases [Payne et al., 2008].

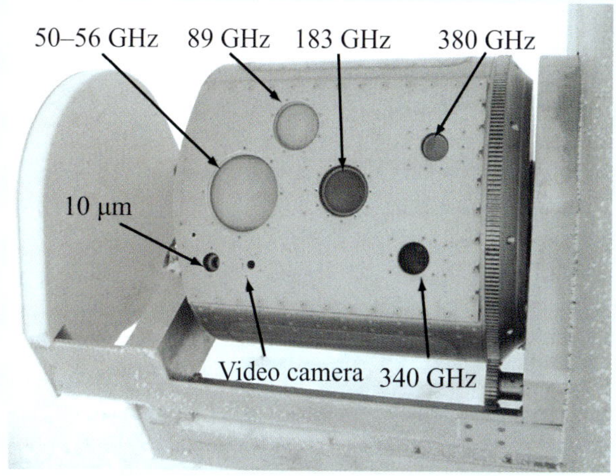

Figure 9-25: Picture of the GSR scanhead with the lens antennas indicated for each radiometer. The lens antennas were designed to have approximately 3.5° beamwidth for 55 and 89 GHz radiometers and 1.8° for 183, 340, and 380 GHz radiometers [Cimini et al., 2007b].

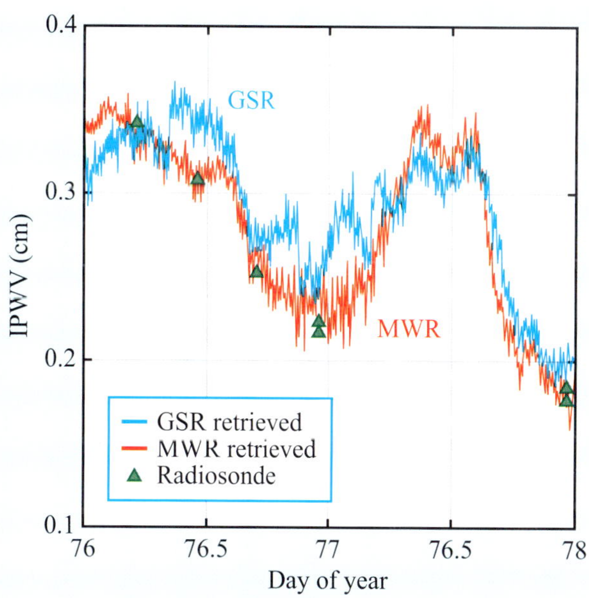

Figure 9-26: Time series of IPWV as retrieved by GSR and MWR [Cimini et al., 2007a].

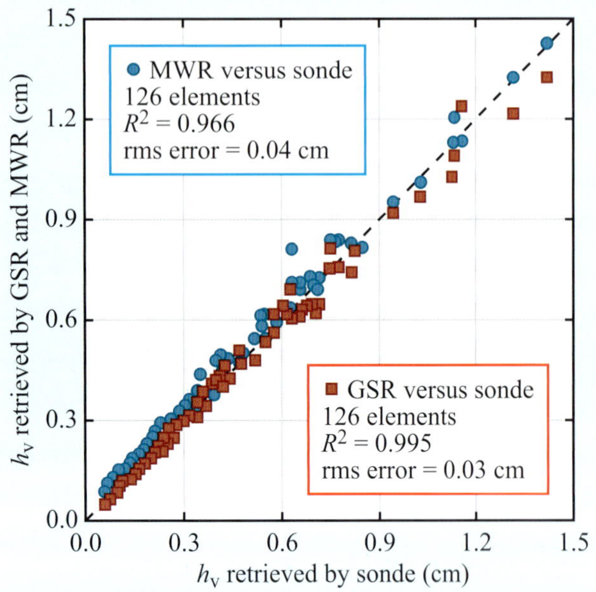

Figure 9-27: Scatter plots of remotely sensed versus *in-situ* IPWV measurements [Cimini et al., 2007a].

9-7 Retrieval of Cloud Liquid-Water Path (LWP) from Ground-Based Observations

9-7.1 Physical Basis

From Eqs. (9.1) and (9.2), the upward-looking brightness temperature of the atmosphere along the zenith direction ($\theta = 0$) is given by

$$T_B = T_{\cos} e^{-\tau_0} + \int_0^\infty \kappa_a(z)\, T(z)\, e^{-\tau_0(0,z)}\, dz, \qquad (9.94)$$

where $T_{\cos} = 2.7$ K, $\tau_0 = \tau_0(0,\infty)$, and

$$\tau_0(0,z) = \int_0^z \kappa_a(z')\, dz'. \qquad (9.95)$$

These expressions pertain to a nonscattering atmosphere, which includes both clear-sky conditions and nonprecipitating clouds when present. In general, $\kappa_a(z)$ accounts for absorption by oxygen, water vapor, and cloud liquid-water content:

$$\kappa_a(z) = \kappa_{O_2}(z) + \kappa_{H_2O}(z) + \kappa_c(z). \qquad (9.96)$$

Furthermore, $\kappa_{H_2O}(z)$ is approximately proportional to the water-vapor density $\rho_v(z)$ and $\kappa_c(z)$ is proportional to the cloud ***liquid-water content*** $m_v(z)$ [as demonstrated by Eq. (8.71a)], which allows us to express Eq. (9.96) as

$$\kappa_a(z) = \kappa_{O_2} + \kappa_v\, \rho_v(z) + \kappa_L\, m_v(z), \qquad (9.97)$$

where κ_v and κ_L are frequency-dependent constants. With $\kappa_a(z)$ and $\kappa_{O_2}(z)$ expressed in Np/km and both $\rho_v(z)$ and $m_v(z)$ expressed in g/m^3, the units for κ_v and κ_L are (Np/km)/(g/m^3).

Using Eq. (9.97), the expression for the zenith opacity τ_0 is

$$\begin{aligned}\tau_0 &= \int_0^\infty \kappa_a(z)\, dz \\ &= \int_0^\infty \kappa_{O_2}(z)\, dz + \int_0^\infty \kappa_v\, \rho_v(z)\, dz + \int_0^\infty \kappa_L\, m_v(z)\, dz.\end{aligned} \qquad (9.98)$$

In view of the definitions for IPWV and LWP given by Eqs. (9.91) and (9.92),

$$\tau_0 = \tau_d + \kappa_v h_v + \kappa_L h_L, \qquad (9.99)$$

where τ_d is the dry component of the zenith opacity,

$$\tau_d = \int_0^\infty \kappa_{O_2}(z)\, dz. \qquad (9.100)$$

According to Eq. (9.99), the zenith opacity consists of a ***dry component***, a ***vapor component***, and a ***liquid component***.

The second term in Eq. (9.94) is the downward-emitted brightness temperature T_{DN}, exclusive of the cosmic radiation component. The ***effective (or mean) radiating temperature*** T_m is defined as (Wu, 1979):

$$T_m = \frac{\int_0^\infty \kappa_a(z)\, T(z)\, e^{-\tau_0(0,z)}\, dz}{\int_0^\infty \kappa_a(z)\, e^{-\tau_0(0,z)}\, dz}. \qquad (9.101)$$

Note that the numerator is the same as T_{DN}. Furthermore, the denominator can be abbreviated as follows:

$$\int_0^\infty \kappa_a(z)\, e^{-\tau_0(0,z)}\, dz = \int_0^{\tau_0(0,\infty)} e^{-\tau_0(0,z)}\, d\tau_0(0,z) = [1-e^{-\tau_0(0,\infty)}] = [1-e^{-\tau_0}], \tag{9.102}$$

where, as noted earlier, $\tau_0 = \tau_0(0,\infty)$. Upon combining Eqs. (9.101) and (9.102) with Eq. (9.94), we obtain

$$T_B = T_{\cos} e^{-\tau_0} + T_m[1-e^{-\tau_0}]. \tag{9.103}$$

In Eq. (9.103), both τ_0 and T_m are frequency dependent. Solving for τ_0, we have

$$\tau_0 = \ln\left(\frac{T_m - T_{\cos}}{T_m - T_B}\right). \tag{9.104a}$$

From Fig. 8-12, we note that $\tau_0 \leq 0.4$ dB for $f < 35$ GHz, which is equivalent to $0.4/4.34 \approx 0.09$ Np. Since $\tau_0 \ll 1$, we can use the approximation $e^{-\tau_0} \approx 1-\tau_0$, in which case Eq. (9.103) leads to

$$\tau_0 = \frac{T_B - T_{\cos}}{T_m - T_{\cos}}, \qquad \text{for } \tau_0 \ll 1 \text{ Np}. \tag{9.104b}$$

Based on 24 radiosonde profiles of the atmosphere above Point Mugu, California, Wu (1979) developed a simple relationship of the form

$$T_m = aT_s, \tag{9.105}$$

where T_s is the surface physical temperature in kelvins and a is a frequency-dependent constant. He found that $a \approx 0.95$ at several frequencies between 20 and 24.5 GHz, and $a \approx 0.94$ at the window frequency of 31.4 GHz. The standard deviation associated with such an estimate of T_m was less than 3.5 K.

► This means that the opacity τ_0 can be determined from T_B directly. ◄

If measurements of T_B were to be performed at two microwave frequencies, such as 21 GHz to measure h_v and 31 GHz to maximize sensitivity to h_L, leading to estimates $\tau_0(f_1)$ and $\tau_0(f_2)$, then

$$\tau_0(f_1) = \tau_d(f_1) + \kappa_v(f_1)\, h_v + \kappa_L(f_1)\, h_L, \tag{9.106a}$$

$$\tau_0(f_2) = \tau_d(f_2) + \kappa_v(f_2)\, h_v + \kappa_L(f_2)\, h_L. \tag{9.106b}$$

At each of the two frequencies, τ_d can be computed using the MPM code described in Section 8-2.3 (in conjunction with a Standard Atmosphere model and directly measured surface values of T, ρ_v, and P), and κ_v and κ_L are constants (see Eq. (8.71b) for κ_L). Hence, the combination of Eqs. (9.106a) and (9.106b) can be solved to determine h_v and h_L:

$$h_v = \frac{\tau_1\, \kappa_L(f_2) - \tau_2\, \kappa_L(f_1)}{\kappa_v(f_1)\, \kappa_L(f_2) - \kappa_v(f_2)\, \kappa_L(f_1)}, \tag{9.107a}$$

$$h_L = \frac{-\tau_1\, \kappa_v(f_2) + \tau_2\, \kappa_v(f_1)}{\kappa_v(f_1)\, \kappa_L(f_2) - \kappa_v(f_2)\, \kappa_L(f_1)}, \tag{9.107b}$$

where

$$\tau_1 = \tau_0(f_1) - \tau_d(f_1), \tag{9.108a}$$

$$\tau_2 = \tau_0(f_2) - \tau_d(f_2). \tag{9.108b}$$

9-7.2 Statistical Inversion

For $\tau_0 \ll 1$ Np, the linear form of τ_0 given by Eq. (9.104b) can be combined with Eq. (9.99) to generate linear retrieval algorithms of the form

$$h_v = a_0 + a_1\, T_B(f_1) + a_2\, T_B(f_2), \tag{9.109a}$$

$$h_L = b_0 + b_1\, T_B(f_1) + b_2\, T_B(f_2), \tag{9.109b}$$

where $T_B(f_1)$ and $T_B(f_2)$ are the measured brightness temperatures, and the a values and b values are constant coefficients (for a specific (f_1, f_2) combination) determined in terms of the various specific absorption coefficients defined previously. Alternatively, these constant coefficients can be found empirically through the application of statistical linear-regression analysis. For example, using data from Denver, Colorado, Guiraud et al. (1979) established the following relationships for a dual-frequency radiometer system operating at $f_1 = 20.6$ GHz and $f_2 = 31.6$ GHz:

$$h_v = -0.19 + 0.118 T_B(f_1) - 0.0560 T_B(f_2) \quad \text{(cm)}, \tag{9.110a}$$

$$h_L = -0.018 + 0.00114 T_B(f_1) + 0.00284 T_B(f_2) \quad \text{(cm)}. \tag{9.110b}$$

The coefficient of $T_B(f_1)$ in the bottom equation has been corrected for a misprint that appeared in the original paper by Guiraud et al. (1979).

The performance of an h_v retrieval technique can be evaluated by comparing the retrieved values to the values obtained from radiosonde data. This, however, cannot be done for h_L, because a radiosonde cannot measure the liquid properties of the atmosphere. In order to verify the capabilities of the radiometric technique as an estimator of the integrated liquid water of nonprecipitating clouds, Snider et al. (1980) obtained direct measurements of the cloud opacity τ_L at $f = 28$ GHz. This was accomplished by using a ground receiver to measure the signal level of the transmission from the Comstar 3 satellite. Part (a) of Fig. 9-28 shows a 3.5-hour record of $\tau_L = \kappa_L h_L$, part (b) shows the corresponding increase in T_B as clouds traversed the antenna beam, and part (c) shows the inferred values of h_L. The triangles in Fig. 9-28(c) represent values inferred from the radiometric technique, which consisted of (1) calculating τ_1 and τ_2 from measurements of T_B at $f_1 = 20.6$ GHz and $f_2 = 30.6$ GHz, and (2) then applying the empirically determined algorithm

$$h_v = -0.01 + 26.97\tau_1 - 11.77\tau_2 \quad \text{(cm)}, \tag{9.111a}$$

$$h_L = -0.01 - 0.229\tau_1 + 0.563\tau_2 \quad \text{(cm)}. \tag{9.111b}$$

The solid line in Fig. 9-28(c) represents values of h_L obtained from the attenuation measurements of the satellite signal and the triangles are based on Eq. (9.111b). The agreement between the two sets is quite good.

9-8 Estimation of Propagation Delay

A typical value for the ***atmospheric index of refraction*** at sea level is $n' = 1.0003$, which means that the velocity of propagation is very close to c, the velocity of light in vacuum. Nonetheless, over long distances the deviation from unity for n' can lead to significant differences in propagation time delay. It is standard practice to

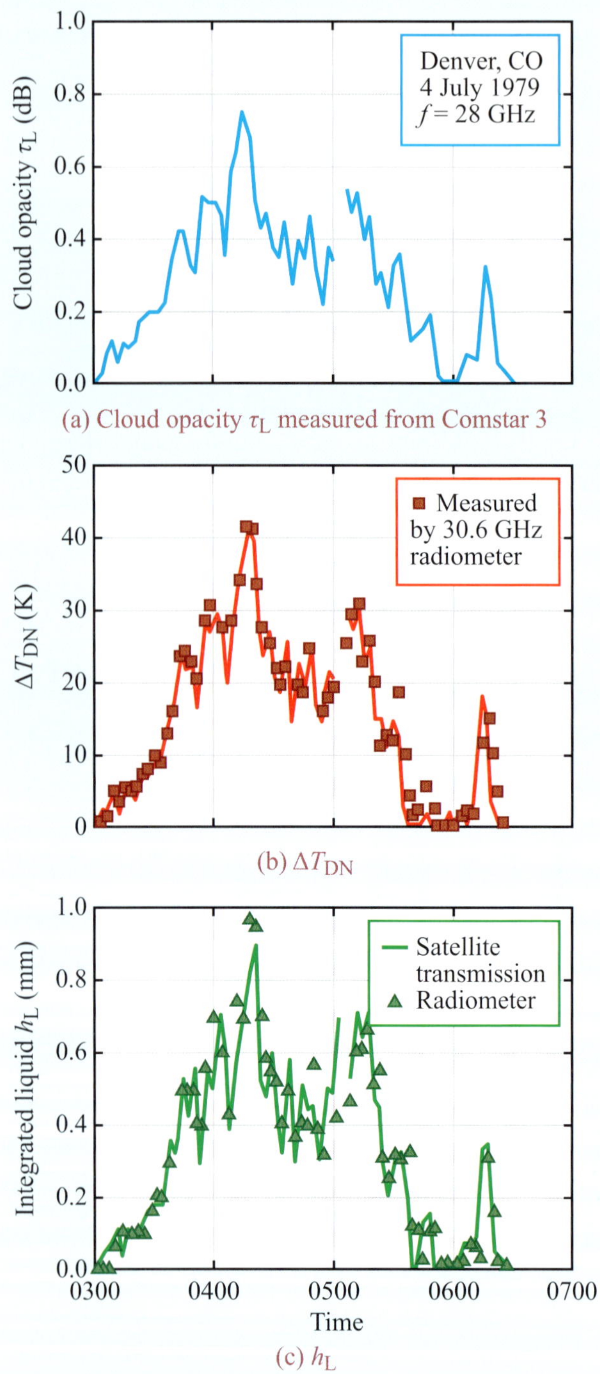

Figure 9-28: (a) Increase in attenuation of the signal received from the Comstar 3 satellite while a cloud traversed the antenna beam, (b) corresponding increases in radiometric temperatures and (c) inferred integrated cloud liquid water [from Snider et al., 1980].

define the ***refractivity*** N as the excess index of refraction relative to that of free space. As a practical measure, the difference is magnified by a factor of 10^6. Thus, from Eq. (8.13),

$$N' = (n' - 1) \times 10^6. \tag{9.112}$$

Corresponding to the typical value for n' given earlier, $N' = 300$. However, the magnitude of N is governed by the temperature, pressure, and water-vapor density of the medium; therefore, it can vary as a function of space (both horizontally and in elevation) and time. For a radio signal propagating through the Earth's atmosphere between a point on the surface and a point outside the atmosphere, the electrical path length is longer than the physical path length between the two points by a distance termed the ***excess electrical path length***, ΔR. This excess path length, which is introduced by the atmosphere and is proportional to the integral of the profile of N' along the designated path, is approximately 220 to 270 cm for propagation paths along the vertical direction (Pandey and Kakar, 1983), and longer for propagation along slanting paths. The ability to measure ΔR accurately is important in many applications, including radioastronomical observations using very long baseline interferometry to observe small radio sources (Schaper et al., 1970; Moran and Rosen, 1981; Mathur et al., 1970), missile and spacecraft tracking and navigation (Black, 1978), and radar altimetry (Chapter 17). As we see shortly, N' consists of two terms: a relatively constant ***"dry" term*** that does not involve water vapor, and a highly variable ***"wet" term*** that is directly proportional to the water-vapor density ρ_v. In this section we discuss how radiometric observations can be used to measure the component of the excess electrical path length attributable to ρ_v.

The refractivity N' is given by the model of Smith and Weintraub (1953),

$$N' = N'_d + N'_v, \tag{9.113}$$

which provides the following expressions for the dry term N'_d and wet term N'_v:

$$N'_d = 77.6 \frac{P}{T} \tag{9.114a}$$

and

$$N'_v = 3.73 \times 10^5 \frac{e}{T^2} = 1720 \frac{\rho_v}{T}, \tag{9.114b}$$

where P(mbar) is the total pressure, T(K) is the temperature, e(mbar) is the ***partial pressure of water vapor***, ρ_v(g/m^3) is the water-vapor density, and the following relation was used:

$$\rho_v = \frac{217e}{T}. \tag{9.115}$$

Because P, T, and ρ_v vary with the height z above the surface, so do N'_d and N'_v. At the surface, N'_d typically varies between 250 and 300, and N'_v varies between about 10 and 200 (Moran and Rosen, 1981). Actually N'_v consists of two components: a nondispersive component given by Eq. (9.114b), and a dispersive component associated with the microwave water-vapor lines. Below 100 GHz, however, the dispersive component is less than 1 percent of the nondispersive component (Liebe, 1981).

The total ***tropospheric excess electrical path length*** consists of a ***dry path length*** ΔR_d and a ***wet path length*** ΔR_v:

$$\Delta R_{\text{Tropo}} = \Delta R_d + \Delta R_v.$$

The dry path length is associated with the profile of $N'_d(r)$ and is given by

$$\Delta R_d = 10^{-6} \int_0^\infty N'_d(r)\, dr, \tag{9.116}$$

where r defines the propagation path, and for a stratified atmosphere, $dr = \sec\theta\, dz$. The magnitude of ΔR_d is on the order of 230 cm. According to Goldfinger (1980), Hopfield's model (1971) can accurately predict the profile $N'_d(r)$ from surface weather data alone, and Moran and Rosen (1981) state that ΔR_d at zenith can "be estimated with an uncertainty of less than 1 cm provided the (surface) pressure is measured to an accuracy of a few millibars." Thus, the problem of estimating ΔR_{Tropo} reduces to that of accurately estimating ΔR_v. The ***wet path length*** is given by

$$\Delta R_v = 10^{-6} \int_0^\infty N'_v(r)\, dr = 1.72 \times 10^{-3} \int_0^\infty \frac{\rho_v(r)}{T(r)}\, dr, \tag{9.117}$$

where Eq. (9.114b) was used. With ρ_v in g/m^3 and T in kelvins, ΔR_v has the same units as r. Conventional techniques used to determine ΔR_v rely on the evaluation of the above integral either by (a) assuming specific models for the profiles $\rho_v(z)$ and $T(z)$ and augmenting them with measured surface values, or (b) measuring $\rho_v(z)$ and $T(z)$ with radiosondes. The former approach generally is not very successful, and the latter usually is adequate but only for the zenith case. If the application of interest calls for the estimation of ΔR_v along a slant path whose direction forms an angle θ with the zenith, the radiosonde technique may still be used, provided the assumption that atmospheric variables are horizontally stratified remains valid. Clearly, this assumption is not always valid, especially for large zenith angles ($\theta \geq 80^\circ$). In contrast, when a radiometer antenna is pointed along the slant path of interest, it observes the same atmospheric mass for which excess path length is to be measured; therefore the assumption of horizontal stratification is not necessary. As we show later, the brightness temperature of the radiometer can be directly related to ΔR_v.

Let us examine Eq. (9.117) while, without loss of generality, we limit our discussion to the zenith case. The water-vapor density $\rho_v(z)$ usually decreases exponentially with height and is characterized by a scale height of 2.2 km (Section 8-1.5). Thus, $\rho_v(z)$ is essentially zero at heights above 10 km. Below this height, $T(z)$ can be modeled by Eq. (8.1), namely as

$$T(z) = T_0 - az, \qquad 0 \leq z \leq 11 \text{ km},$$

where T_0 is the surface temperature and a is the temperature lapse rate ($a = 6.5$ K/km for the U.S. Standard Atmosphere). Upon expanding $T^{-1}(z)$ in Eq. (9.117) in a Taylor series, as proposed by Goldfinger (1980), we obtain

$$\frac{1}{T(z)} = \left[T_0\left(1 - \frac{az}{T_0}\right)\right]^{-1} \approx \frac{1}{T_0}\left(1 + \frac{az}{T_0} + \cdots\right).$$

Upon inserting the first two terms of the above expansion into Eq. (9.117), the latter becomes

$$\Delta R_v \approx \frac{1.72 \times 10^{-3}}{T_0}\left[\int_0^\infty \rho_v(z)\, dz + \frac{a}{T_0}\int_0^\infty \rho_v(z)\, dz\right]. \tag{9.118}$$

We recognize the first integral as h_v, the integrated precipitable water vapor (IPWV) given by Eq. (9.91), which, as noted in Section 9-6, can be estimated by a dual-frequency microwave radiometer with a small rms uncertainty. Furthermore, Goldfinger (1980) has shown that the magnitude of the second term inside the square bracket is on the order of 4 percent of that of the first term (for $T_0 = 300$ K). Formally, the first integral in the above expression is in units of g/m^2. To replace the integral with h_v, and to express h_v and ΔR_v in cm units, we need to multiply the above expression by 10^6. Incorporating these modifications, including the replacement of the second term inside the square bracket in Eq. (9.118) with $0.04\, h_v$, we obtain

$$\Delta R_v \approx \frac{1.79 \times 10^3}{T_s}\, h_v, \qquad \text{(cm)}. \tag{9.119}$$

Microwave radiometric measurements may be used to estimate h_v from which ΔR_v can be estimated; alternatively, ΔR_v can be estimated directly from a linear algorithm of the form

$$\Delta R_v = a_0 + a_1\, T_B(f_1) + a_2\, T_B(f_2). \tag{9.120}$$

According to a simulation analysis reported by Wu (1979), it should be possible to estimate ΔR_v with an rms error of 0.3 cm at zenith and an error of less than 2 cm at a zenith angle of 80°. The frequency combination of 20.3 and 31.4 GHz is one of three recommended by Wu (1979) to achieve such accuracies. A similar simulation study was conducted by Pandey and Kakar (1983) for a downward-looking radiometer. They concluded that the optimum frequency pair was 16 and 21 GHz, and that the corresponding rms error was 0.54 cm at nadir. Based on zenith measurements at 19 and 22 GHz, Moran and Rosen (1981) reported that the rms error in ΔR_v was 1.5 cm, which is about 3 to 5 times larger than the results obtained from the simulation studies. Nonetheless, this 1.5 cm error is significantly smaller than the 3.2 cm rms error associated with the estimates based on surface meteorological parameters. A scatter plot is shown in Fig. 9-29, comparing radiometer-derived values of ΔR_v with values calculated from radiosonde observations. The radiometric algorithm used by Moran and Rosen (1981) is of the form given by Eq. (9.120). For

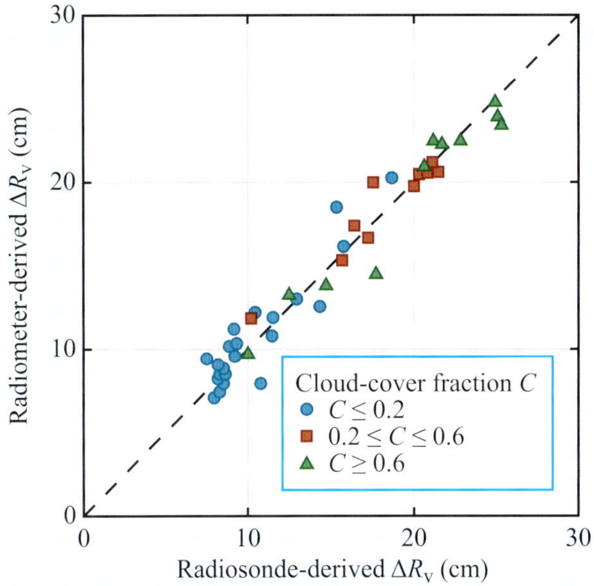

Figure 9-29: Radiometer-derived wet path lengths versus wet path lengths derived from radiosonde data. The symbols denote different cloud-cover conditions [from Moran and Rosen, 1981].

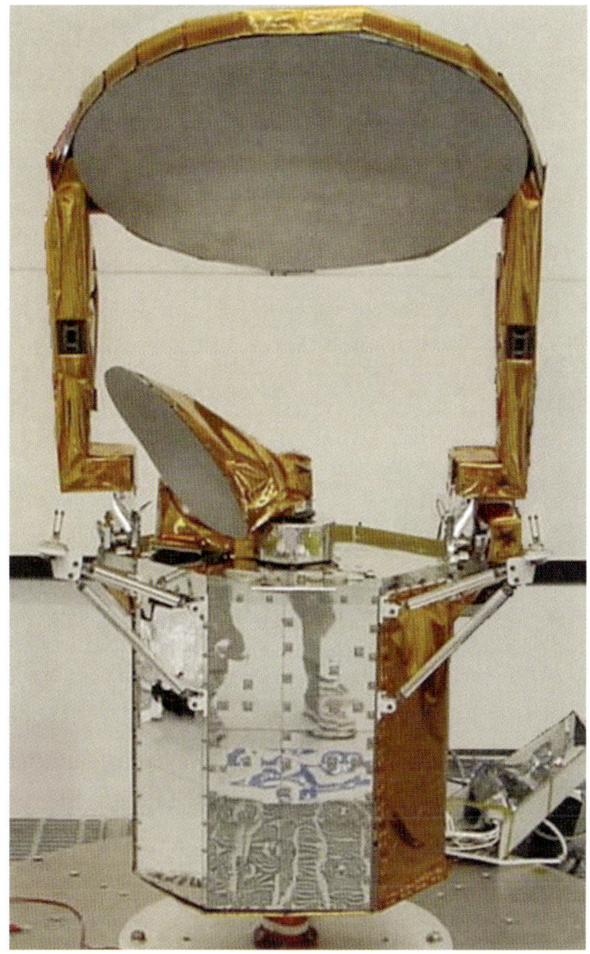

Figure 9-30: The deployed SSMIS instrument showing the main canister, reflector antenna, and calibration load assembly [Kunkee et al., 2008]. The sensor is approximately 1.2 m tall when deployed with a mass of approximately 96 kg.

$f_1 = 19$ GHz and $f_2 = 22.2$ GHz, statistical linear regression using 45 data points results in the following values for the regression coefficients: $a_0 = -4.1$ cm, $a_1 = -0.33$ cm/K, and $a_2 = 0.67$ cm/K.

9-9 Space-Based Atmospheric Sounding Radiometers

Mechanically-scanned microwave sensors viewing the Earth from space have historically utilized one of three possible scan modes: cross-track scanning, in which the angle of Earth incidence varies and passes through nadir; conical scanning, which has a fixed angle of incidence and a vertical axis of rotation; and limb scanning, which has a narrow range of near-tangent incidence angles.

The ***Special Sensor Microwave Imager Sounder*** (SSMIS) conical-scanning instrument shown in Fig. 9-30 was first launched in October 2003 aboard the Air Force Defense Meteorological Satellite Program (DMSP) F-16 Spacecraft. The SSMIS combines the imaging capabilities of the heritage DMSP conically scanning ***Special Sensor Microwave/Imager*** (SSM/I) sensor with the cross-track microwave sounder's ***Special Sensor Microwave Temperature and Special Sensor Microwave Humidity Sounder***, SSM/T-2 into a single conically scanning 24-channel instrument with extended

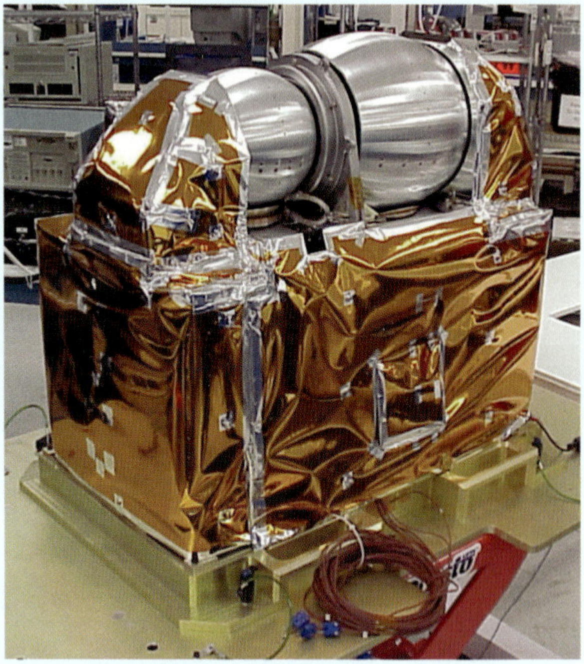

Figure 9-31: The Advanced Technology Microwave Sounder (ATMS) ready for integration on the Suomi NPP satellite. The sensor dimensions are $70 \times 60 \times 40$ cm and mass is 74.1 kg. [Courtesy of Northrop Grumman Electronic Systems.]

Table 9-1: AMSU instrument characteristics.

Channel number	Center frequency (GHz)	Band-width (MHz)	Radiometric sensitivity ΔT
1	23.80	251	0.30
2	31.40	161	0.30
3	50.30	161	0.40
4	52.80	380	0.25
5	53.59 ± 0.115	168	0.25
6	54.40	380	0.25
7	54.94	380	0.25
8	55.50	310	0.25
9	$57.29 = f_0$	310	0.25
10	$f_0 \pm 0.217$	76	0.40
11	$f_0 \pm 0.322 \pm 0.048$	34	0.40
12	$f_0 \pm 0.322 \pm 0.022$	15	0.60
13	$f_0 \pm 0.322 \pm 0.010$	8	0.80
14	$f_0 \pm 0.322 \pm 0.004$	3	1.20
15	89.00	2000	0.50
16	89.00	5000	2.00
17	150	4000	2.00
18	183 ± 1	1000	2.00
19	183 ± 3	2000	2.00
20	183 ± 7	4000	2.00

sounding capability to profile the mesosphere (Kunkee et al., 2008).

The ***Advanced Technology Microwave Sounder*** (ATMS, shown in Fig. 9-31) is a cross-track scanning sounder with 22 channels. ATMS provides sounding observations needed to retrieve profiles of atmospheric temperature and moisture for weather forecasting as well as continuity of these measurements for climate monitoring purposes. A global image of ATMS brightness temperatures observed near 23.8 GHz is shown in Fig. 9-32. Like the long heritage of its predecessors, ATMS combines all the channels of the preceding AMSU–A1, AMSU–A2, and AMSU–B (Table 9-1) sensors into a single package with considerable savings in mass, power, and volume. ATMS, together with the ***Crosstrack Infrared Sounder*** (CrIS), forms the ***Crosstrack Infrared and Microwave Sounding Suite*** (CrIMSS). Clouds are mostly opaque in the infrared part of the spectrum while they are largely transparent at microwave frequencies. Therefore, microwave and infrared sounders operate synergistically to jointly exploit the cloud penetration of the microwave sounder and the relatively high vertical resolution of the infrared sounder. The combination of microwave and hyperspectral infrared sounding was first demonstrated from space with the ***Atmospheric Infrared Sounder*** (AIRS) and AMSU launched by NASA on the Aqua research satellite in 2002 (Aumann et al., 2003a, b).

Given adequate antenna size to take advantage of their respective geometries, cross-track and conical scanning modes typically offer better horizontal spatial resolution, whereas limb scanning offers better vertical resolution. The relative merits of cross-track versus conical scanning involve the following considerations, covered in great detail by Rosenkranz et al. (1997) and summarized below.

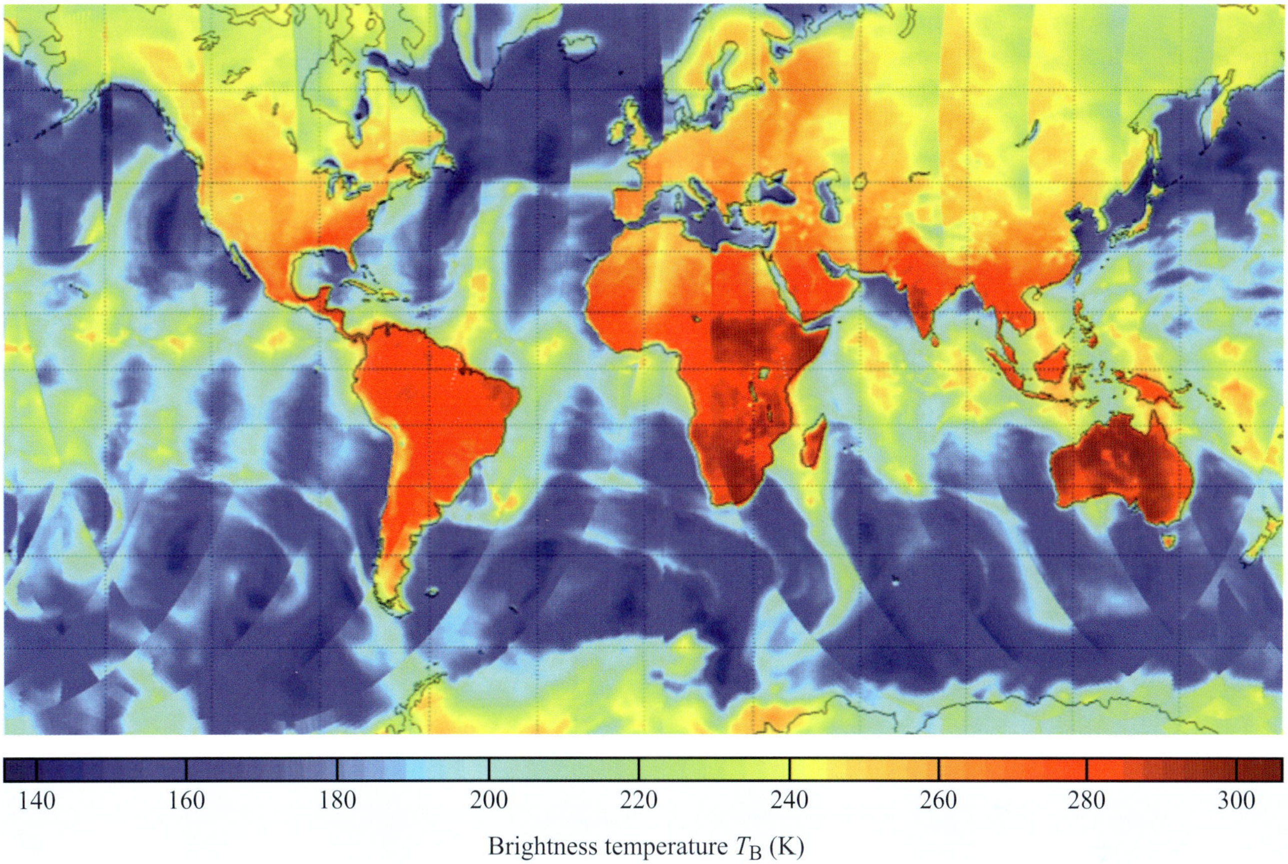

Figure 9-32: A global mosaic of ATMS 23.8 GHz brightness temperatures observed on December 16, 2011.

9-9.1 Vertical Shifting of the Weighting Functions

The pressure corresponding to the peak of the temperature weighting function of a given channel decreases as $\sqrt{\cos\theta}$, where θ is measured from the vertical at the surface (Grody, 1993). Because a cross-track instrument should provide good retrievals over its entire range of incidence angles, its frequencies are chosen to obtain weighting functions that penetrate to the bottom of the atmosphere at the maximum incidence angle. Furthermore, as a cross-track instrument scans toward nadir, its lowest sounding channels become more sensitive to variations in surface emissivity because of the decreasing atmospheric path length. However, the impact of this greater sensitivity on retrieval accuracy depends on how these channels are used in the retrieval algorithm. In an optimum algorithm, the shift of all weighting functions to lower altitudes would cause a compensating lesser weight to be given, at any fixed altitude, to the channel that peaks closest to the surface. Lastly, imaging of surface parameters with microwaves is best done with dual-polarization at oblique incidence (among other reasons, to separate the effects of near-ocean-surface wind speed from other parameters), as discussed by Gaiser et al. (2004).

9-9.2 Swath Width

Conical-scanning systems typically view Earth at an angle of approximately 50° from the vertical at the surface in part to maximize the sensitivity of the radiance observations to surface parameters of interest, such as the near-surface wind speed. Furthermore, the swath is typically constrained by blockages due to the mechanical assembly holding the antenna reflector (see Fig. 9-30), satellite appendages, and so forth to an active scan angle of approximately ±70° in front of the spacecraft. These considerations limit the swath width of conical scanning systems to approximately 1700 km. The ***Special Sensor Microwave Imager Sounder*** (SSMIS) scan geometry is shown in Fig. 9-33. Cross-track scanning systems such as AMSU and ATMS observe out to angles approaching 50° or more. The maximum ATMS scan angle is approximately 52.8° (see Fig. 9-34), resulting in a swath that is contiguous at the equator.

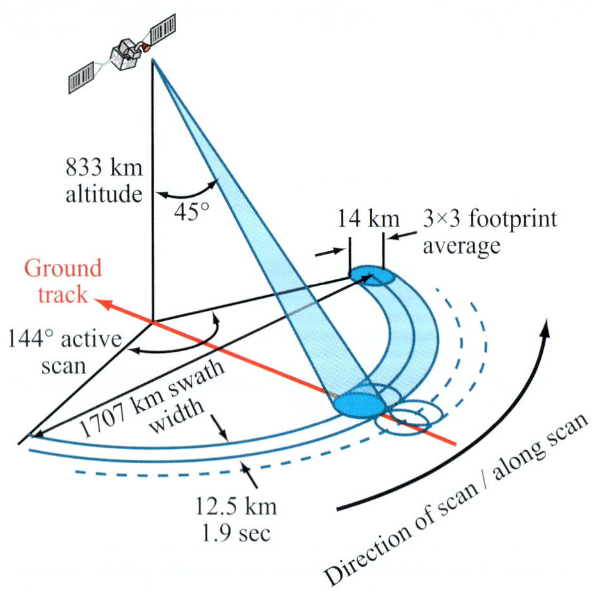

Figure 9-33: SSMIS scan geometry showing direction of active scan, swath width, ground track, and footprint averages (Kunkee et al., 2008).

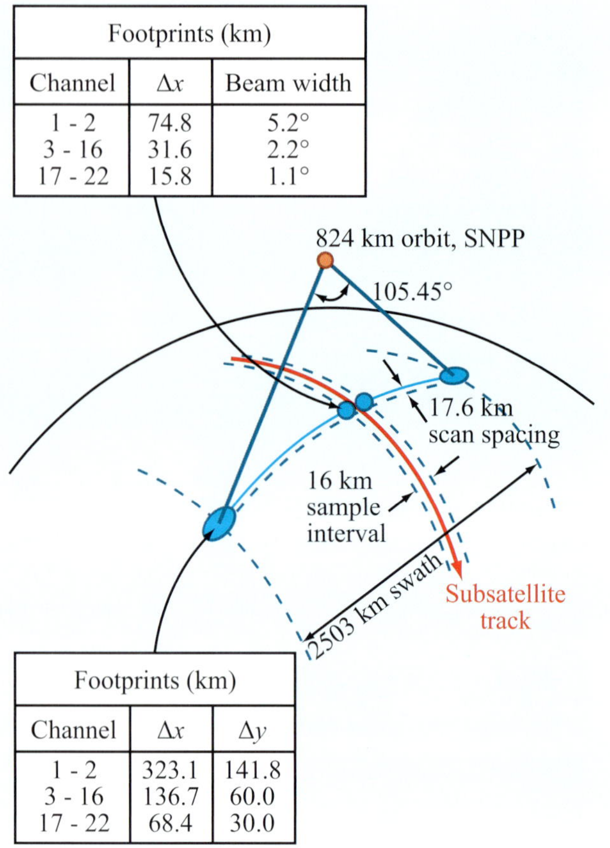

Figure 9-34: Scan pattern of the Advanced Technology Microwave Sounder. [Courtesy of Northrop Grumman Electronic Systems.]

9-9.3 Cloud Sensitivity

Lower-atmosphere channels become more sensitive to clouds and precipitation as the sensor viewing angle decreases because the shielding effect due to the opacity of the atmosphere overlying the clouds decreases. However, the profiling accuracy near nadir is less impacted by clouds because the opacity of a planar cloud (in a simple case) would be greater by the factor $\sec\theta$.

9-9.4 Calibration

Cross-track scanning instruments can be calibrated by pointing the beam at cold space and at a target within the instrument. With a conical scan, the main reflector never points in those directions, and a subreflector is typically moved into position to direct energy from cold space into the instrument antenna feed horns, and a calibration target is also moved in front of the feed horns. This precludes a complete "through- the-antenna" calibration as the primary reflector transfer function is not calibrated during this procedure. Therefore, pre-launch corrections are needed to properly characterize the primary reflector for system calibration.

9-9.5 Modeling and Retrieval Algorithm Complexity

Because a cross-track scan measures a range of angles, the retrieval algorithm and associated forward model are necessarily more complicated than for a fixed angle. This consideration does not bear directly on the accuracy of the retrievals, but if the algorithm yields biased results, the biases may be angle-dependent. Note also that the footprint size for a conical scanning sounder is constant, thereby simplifying data modeling and processing.

9-9.6 Compatibility with Other Sensors

Sounding with infrared wavelengths is possible only in clear air or through broken clouds, and cloud-clearing algorithms decrease in accuracy as the percentage of cloud cover increases (Blackwell, 2011). Given some vertical cloud extent, this percentage will tend to increase at oblique angles. Hence, a cross-track scan is preferred for infrared sounders. If microwave channels are to be combined with infrared channels in a retrieval algorithm, greater complexity is involved in registering a conical microwave sounder, because its slant path through the atmosphere is different from that of the cross-track infrared sounder, and thus the mapping would depend on altitude.

9-10 Atmospheric Sounding by Downward-Looking Radiometers

9-10.1 Brightness Temperature

The brightness temperature observed by a satellite-borne radiometer viewing the Earth at an angle θ relative to the nadir direction (Fig. 9-35) is given by Eq. (9.18) as

$$T_B = T_{UP} + \Upsilon[T_{SE} + T_{SS}], \tag{9.121}$$

where, in addition to the upward-emitted atmospheric radiation represented by T_{UP}, the signal includes contributions related to the ground surface. Upon setting

$$T_{SE} = e_s T_s, \tag{9.122a}$$

and

$$T_{SS} = \Gamma_s T_{DN} \approx (1 - e_s) T_{DN}, \tag{9.122b}$$

where e_s, Γ_s, and T_s are the surface emissivity, reflectivity, and physical temperature, respectively, we have

$$\begin{aligned} T_B &= T_{UP} + \Upsilon[e_s T_s + (1 - e_s) T_{DN}] \\ &= \sec\theta \int_0^\infty \kappa_a(z)\, T(z)\, e^{-\tau_0(z,\infty)\sec\theta}\, dz \\ &\quad + e_s T_s e^{-\tau_0(0,\infty)\sec\theta} \\ &\quad + (1 - e_s) \sec\theta\, e^{-\tau_0(0,\infty)\sec\theta} \\ &\quad \cdot \int_0^\infty \kappa_a(z)\, T(z)\, e^{-\tau_0(0,z)\sec\theta}\, dz. \end{aligned} \tag{9.123}$$

Since atmospheric emission is part of the first and third terms in Eq. (9.123), we can combine them into the form

$$T_B = T_{BGD} + \int_0^\infty W_T(z)\, T(z)\, dz, \tag{9.124}$$

where

$$T_{BGD} = e_s T_s e^{-\tau_0(0,\infty)\sec\theta}, \tag{9.125a}$$

and

$$\begin{aligned} W_T(z) &= \sec\theta\ \kappa_a(z) \\ &\cdot [e^{-\tau_0(z,\infty)\sec\theta} + (1 - e_s) e^{-\tau_0(0,\infty)\sec\theta} e^{-\tau_0(0,z)\sec\theta}]. \end{aligned} \tag{9.125b}$$

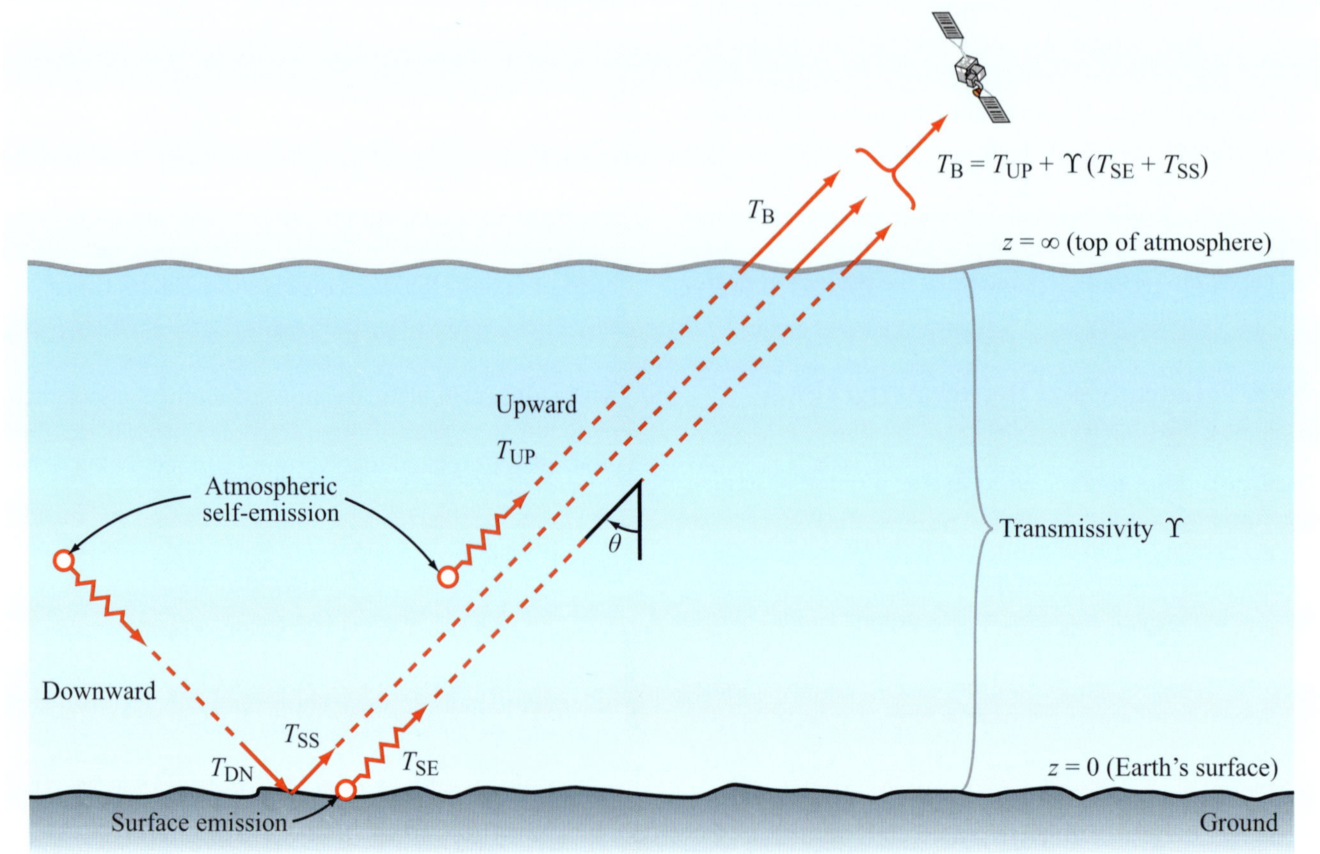

Figure 9-35: Satellite-borne radiometer observing the Earth at a nadir angle θ.

Figure 9-36 shows plots of the temperature weighting function at three oxygen frequencies. Two sets of plots are shown, one for a land surface with an assumed emissivity of 1.0 and the other for a water surface with an emissivity of 0.5. Each set includes curves for nadir viewing ($\theta = 0$) and for $\theta = 53.3^\circ$, which corresponds to the extreme scan position for SCAMS. Note that the 55.45 GHz weighting function is identical for both land and water, whereas the weighting functions of the lower frequency channels are influenced by surface effects at altitudes corresponding to pressures greater than 700 mbar.

9-10.2 Examples of Retrieved Parameters

A typical example of temperature-profile retrieval is shown in Fig. 9-37 (Liu and Weng, 2005). Very good agreement is observed between the profile retrieved by the 20-channel AMSU instrument (Table 9-1) and the profile recorded by a radiosonde, at all heights between the surface and 12.5 km (P at 12.5 km is ≈ 220 mbar). Also indicated in the figure are values for the relative humidity retrieved by AMSU's three channels with center frequencies surrounding the 183 GHz water-vapor line. Other atmospheric parameters retrieved by the AMSU instrument include integrated precipitable water vapor (IPWV), cloud liquid-water path (LWP), cloud ice-water path, and surface rain rate. The image

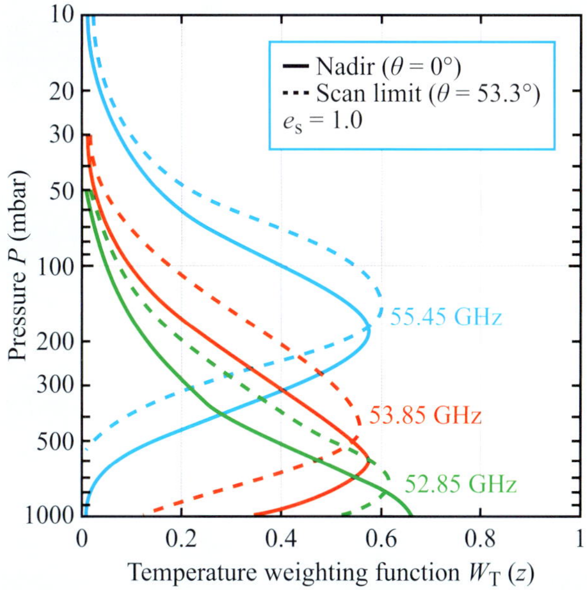

(a) Land surface

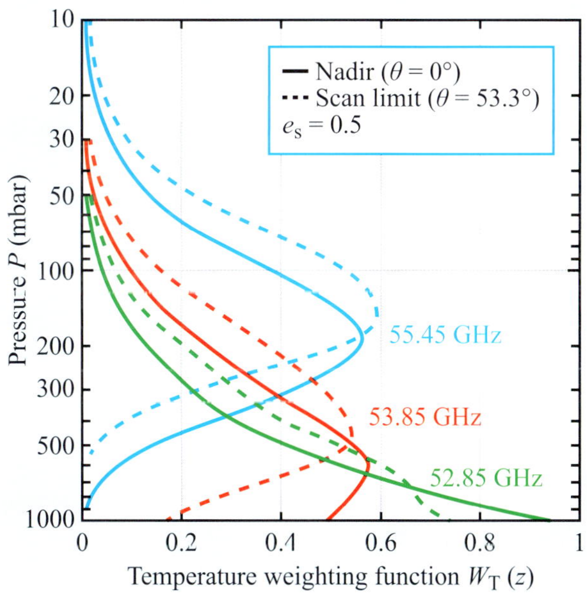

(b) Water surface

Figure 9-36: SCAMS temperature weighting functions calculated for nadir- and scan-limit positions, for land and sea surfaces [from Grody, 1978].

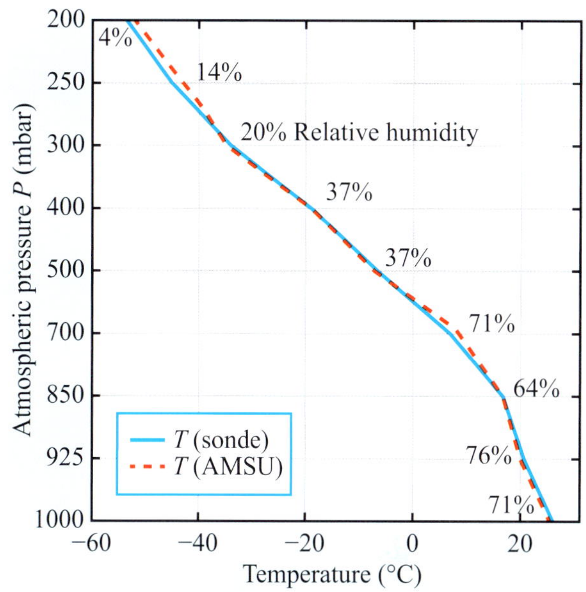

Figure 9-37: Vertical distributions of the atmospheric temperatures from (solid line) dropsondes and (dashed line) the retrievals from AMSU measurements. The values in percentage represent measured relative humidity [Liu and Weng, 2005].

in Fig. 9-38 displays IPWV on a global scale and the four images in Fig. 9-39 display retrieved parameters based on observations made on 12 September, 2003 for Hurricane Isabel.

9-11 Atmospheric Limb Sounding

Microwave limb sounders measure atmospheric parameters by observing thermal emission as the instrument field of view is scanned through the atmospheric limb (Waters, 1993).

► Microwave limb-sounding observations are particularly useful in the upper atmosphere for applications spanning stratospheric ozone chemistry to climate implications of upper tropospheric humidity. ◄

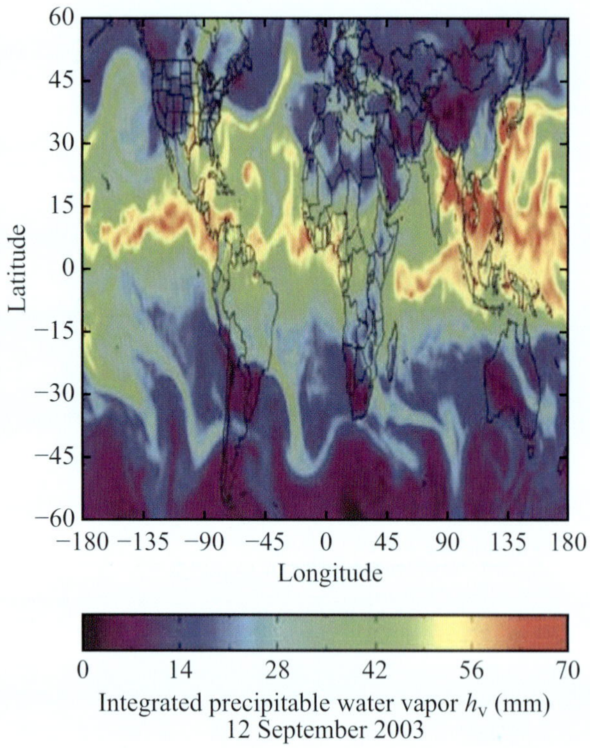

Figure 9-38: Global distribution of integrated precipitable water vapor retrieved from AMSU [Liu and Weng, 2005].

The scanning geometry is shown in Fig. 9-40 for a radiometer on a spaceborne platform at orbital altitude H above the Earth's surface. The angle θ is the ***elevation angle*** of the observation path relative to the satellite horizon, and the ***tangent height*** h_T is defined as the altitude above the surface where the observation path makes a perpendicular intersection with the radius vector of the Earth. For a given orbital altitude H, the tangent height is controlled by θ. The scanning is typically performed in the upward direction to yield a more vertical tangent point locus of the observation path (Waters et al., 2006). Tangent heights spanning zero to approximately 100 km are typically observed, thereby providing vertical distributions of a host of atmospheric chemical species, cloud ice, temperature, and geopotential height. The ***Microwave Limb Sounder*** (MLS) began operating on the NASA Aura satellite in 2004, and the ***Superconducting Submillimeter-Wave Limb-Emission Sounder*** (SMILES) on the Japanese Experimental Module of the International Space Station (ISS) began operating in 2009 (Kikuchi et al., 2010).

The limb-scanning technique offers many advantages relative to the cross-track and conical scanning geometries previously discussed. The atmospheric path length is longer, thereby providing greater sensitivity to trace gases with relatively low concentrations, including many molecules present in the upper atmosphere. Microwave limb sounders can exploit narrowband spectroscopy to spectrally resolve emission lines at all altitudes, which allows measurements of very weak lines in the presence of nearby strong lines (Waters et al., 2006). The viewing geometry permits a vertical resolution approaching 1 km with spatial resolution of a few hundred kilometers along the line of sight. Atmospheric emission is measured relative to a uniform cold sky background, thereby avoiding ambiguities due to surface emissivity uncertainties. Engineering challenges include the need for very accurate pointing knowledge, as the brightness temperature change with boresight elevation can exceed 100 K per degree. Product accuracy is far more sensitive to the relative pointing of the various frequency bands in the instrument than to the absolute value of the pointing relative to the Earth, so careful boresighting of the various radiometers is required (Cofield and Stek, 2006). Furthermore, narrow antenna beamwidths are required to optimize the vertical resolution of the measurements.

9-11.1 Fundamental Considerations

It is clear from Fig. 9-41 that no part of the signal originates below the tangent point (for an infinitesimally small beamwidth), and it can be shown that the path length along the line of sight at a given tangent height z_k is much greater for the tangent layer (k) than for any other layer. In addition to this geometrical factor, the density of the atmospheric species whose emission is being measured usually decreases with increasing altitude. Because of these two factors, most of the radiation measured by the radiometer originates in a narrow vertical layer within about 2 km of the ray's tangent point, if the atmosphere is not opaque (Staelin,

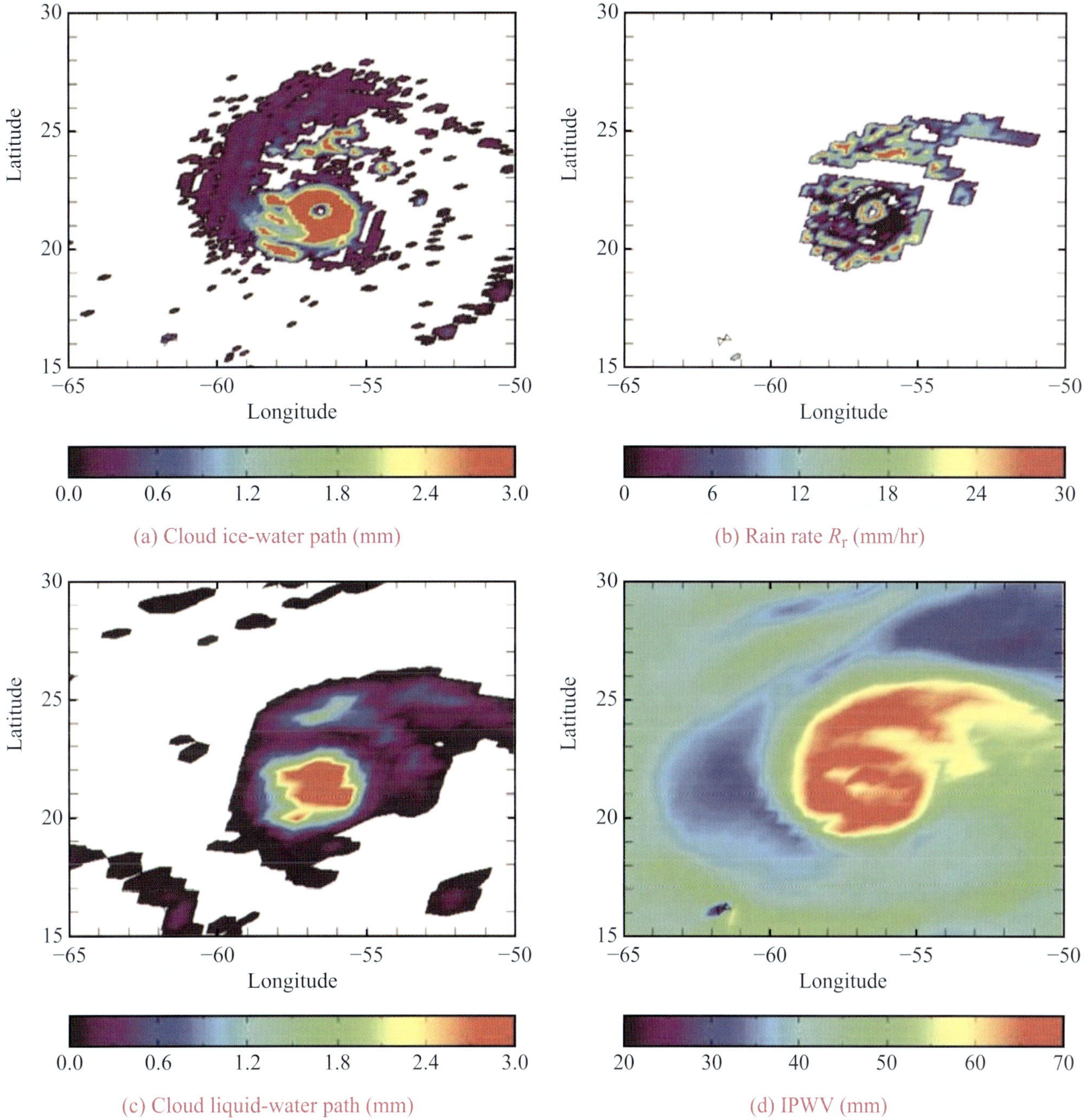

Figure 9-39: AMSU-retrieved (a) ice-water path, (b) surface rain rate, (c) cloud liquid water, and (d) integrated precipitable water vapor (IPWV) for Hurricane Isabel on September 12, 2003 [Liu and Weng, 2005].

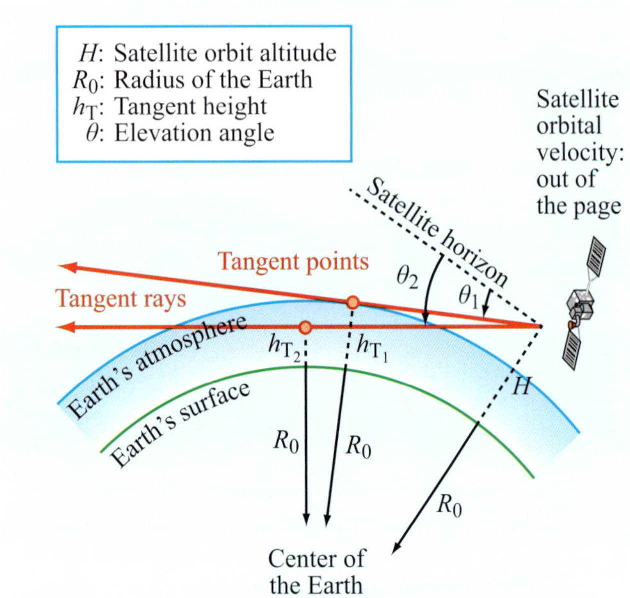

Figure 9-40: Limb-sounding geometry for a sensor at height H above Earth viewing at angle θ below the local horizontal resulting in a tangent height h_t (not drawn to scale).

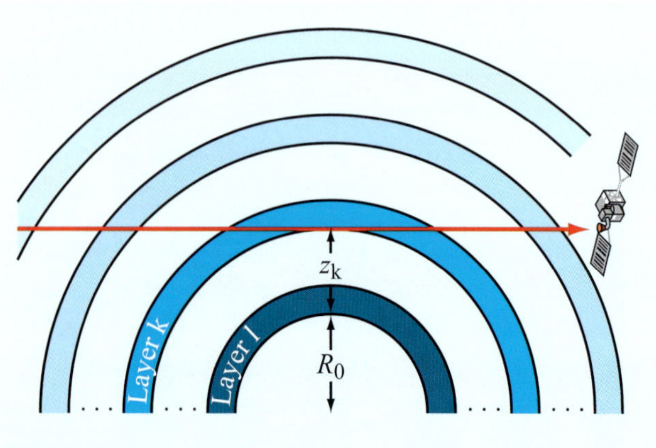

Figure 9-41: The path length in layer k is greater than in any other layer (not drawn to scale).

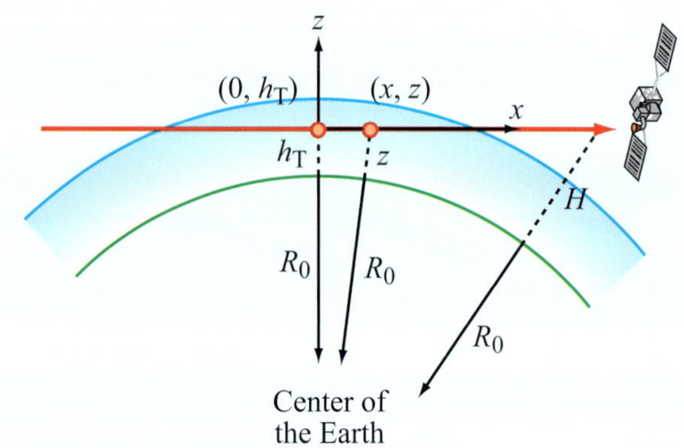

Figure 9-42: Geometry for relating x to z (not drawn to scale).

1977). This is equivalent to saying that limb-sounding observations are characterized by very narrow weighting functions.

Assuming that (a) the atmosphere is spherically stratified and spherically symmetric, (b) atmospheric refraction can be neglected, and (c) the emitting atmospheric species is in local thermodynamic equilibrium, the brightness temperature detected by a radiometer pointing toward the horizon at a tangent height h_T (Fig. 9-42) can be written as

$$T_B(h_T) = T_{cos} e^{-\tau(-\infty,\infty)} + \int_{-\infty}^{\infty} T(x)\, \kappa_a(x)\, e^{-\tau(x,\infty)}\, dx, \tag{9.126}$$

where $T_{cos} = 2.7$ K is the cosmic (background) brightness temperature, $T(x)$ is the atmospheric temperature at a point x along the tangent path, $\kappa_a(x)$ is the absorption coefficient, and $\tau(x,\infty)$ is the optical thickness between points x and ∞ (satellite position),

$$\tau(x,\infty) = \int_x^{\infty} \kappa_a(x)\, dx. \tag{9.127}$$

From the geometry of Fig. 9-42, we can relate the distance x to the altitude z above the Earth's surface,

$$x^2 = (R_0 + z)^2 - (R_0 + h_T)^2, \tag{9.128}$$

where R_0 is the Earth's radius. Through this relation, Eq. (9.126) can be transformed into an integral over z,

$$T_B(h_T) = T_{cos} e^{-2\tau(h_T,\infty)} + \int_{h_T}^{\infty} T(z)\, W_T(z)\, dz, \tag{9.129}$$

where the ***temperature weighting function*** is given by

$$W_T(z) = \kappa_a(z)\, e^{-\tau(z,\infty)}[1 + e^{-2\tau(h_T,z)}]\, f(z) \qquad (9.130)$$

and

$$\tau(z_1,z_2) = \int_{z_1}^{z_2} \kappa_a(z)\, f(z)\, dz. \qquad (9.131)$$

The function $f(z)$ defines the derivative of x with respect to z,

$$f(z) = \frac{dx}{dz} = (R_0+z)[(R_0+z)^2 - (R_0+h_T)^2]^{-1/2}, \qquad (9.132)$$

and the presence of two terms in Eq. (9.130) is the result of the fact that x is a double-valued function of z.

If the frequency f is close to the absorption line frequency f_0 of a given atmospheric species, and if the absorption coefficient $\kappa_a(z)$ is dominated by that species, it then follows that the brightness temperature $T_B(h_T)$ is a function of the profile of the ***volume mixing ratio*** $v(z)$ of that species. This ratio is defined by $v(z) = N(z)/N_{air}(z)$, where $N(z)$ and $N_{air}(z)$ are the number densities for the species and for air, respectively.

9-11.2 The NASA Aura Microwave Limb Sounder

The Microwave Limb Sounder is shown in Fig. 9-43. MLS measures thermal emission near 118, 190, 240, 640, and 2500 GHz using a three-reflector offset antenna system that vertically scans the limb. Atmospheric limb scan and radiometric calibration are performed routinely every 25 seconds for all bands. Vertical profiles are retrieved every 165 km along the suborbital track, covering 82°S to 82°N latitudes on each orbit. The MLS temperature weighting functions for the 118, 190, 240, and 640 GHz channels are shown in Fig. 9-44. These plots were calculated using Eq. (9.130), in conjunction with a nonzero antenna beam of a 1.6 m diameter antenna. The MLS product suite is summarized in Fig. 9-45 indicating a broad range of chemical families, as well as temperature, geopotential height, and volcanic sulfur dioxide. MLS measurements of temperature and six atmospheric gases are shown in Fig. 9-46 for a tangent height of 18 km.

Figure 9-43: The Earth Observing System Microwave Limb Sounder was launched July 15, 2004 on the NASA Aura satellite (Waters et al., 2006).

Potential improvements being considered for a follow-on microwave limb-sounding mission include the addition of azimuth scanning to provide a wide measurement swath, improved vertical resolution by a factor of approximately two, and measurement capabilities for NO, NO_2, HDO, and $H_2^{18}O$ (Livesey et al., 2008).

9-12 Global Precipitation Mapping Using Atmospheric Sounding Observations

Sounding observations made simultaneously in transparent and opaque water-vapor and oxygen microwave absorption bands can be used to retrieve precipitation

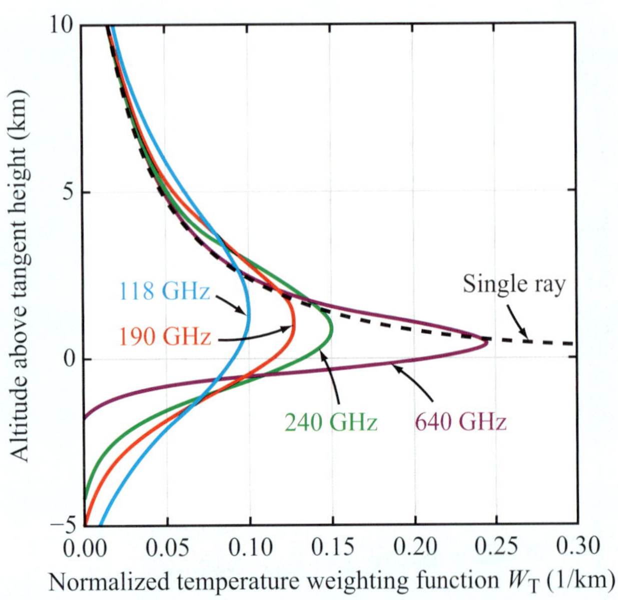

Figure 9-44: Normalized temperature weighting functions at 50 km tangent height for a limb sounder in 705 km orbit. All curves assume that the absorption coefficient is proportional to pressure. The spectral and spatial characteristics of the Microwave Limb Sounder [Waters et al., 2006; Cofield and Stek, 2006; Jarnot et al., 2006] are assumed.

intensity (Ferraro et al., 2005; Vila et al., 2007; Chen and Staelin, 2003; Surussavadee and Staelin, 2010a,b) using the brightness contrast between hydrometers and the radiatively cool ocean. Over land surfaces, scattering from millimeter-wave channels due to ice crystals in convective clouds provides a brightness temperature signature that is correlated with rainfall rate, as the concentration and size of ice particles increases with heavier rainfall. These signatures, however, are less sensitive for detecting nonconvective "warm rain" processes. Furthermore, interfering signatures from surfce emissivity variations must be carefully addressed. Recent associated research thrusts involve improved radiative transfer and electromagnetic modeling and advanced validation techniques (Surussavadee and Staelin, 2008; 2010c).

Spaceborne sounding systems, such as AMSU, ATMS, and SSMIS, provide uniform global coverage relative to ground-based precipitation monitoring systems. The use of opaque bands also reduces error due to surface emissivity uncertainties, and recent work has shown useful skill for both snow-covered land and sea ice (Surussavadee and Staelin, 2009). Advantages relative to current satellite-borne radars include wider swath width, which allows global coverage every 12 hours from a single satellite. Recent work combining the active and passive microwave sensors on board the Tropical Rainfall Measuring Mission (TRMM) has shown improvement by adjusting the radar solution to simultaneously match the radiometer observations (Kummerow et al., 2011). Fusion of infrared and microwave observations have also been shown to provide useful precipitation-retrieval skill (Ebert et al., 2007).

9-12.1 Physical Foundation: Attenuation and Scattering

The two principal physical mechanisms that permit the measurement of rain with microwave radiometers are absorption and scattering (Wilheit, 1986). Attenuation measurements at frequencies below the 22 GHz water-vapor line are routinely used over the ocean. These measurements represent observations of the liquid hydrometeors themselves and thereby can be considered direct measurements of the rainfall (Wilheit et al., 1991). Scattering measurements at frequencies above the 60 GHz oxygen band respond to hydrometeor characteristics (size, abundance, phase, etc.) and are indirectly related to rainfall intensity. We further point out that transformations of microwave brightness temperatures can be used to better isolate the signals associated with liquid water attenuation and ice scattering in a precipitating cloud system and for decoupling both signals from variations in the atmospheric and oceanic background signals (Petty, 1994). Recent work has exploited similar approaches to improve precipitation-retrieval accuracies over ocean and land (Surussavadee and Staelin, 2008).

Airborne observations made by the NPOESS ***Airborne Sounder Testbed-Microwave*** (NAST-M) instrument illustrating the attenuation and scattering phenom-

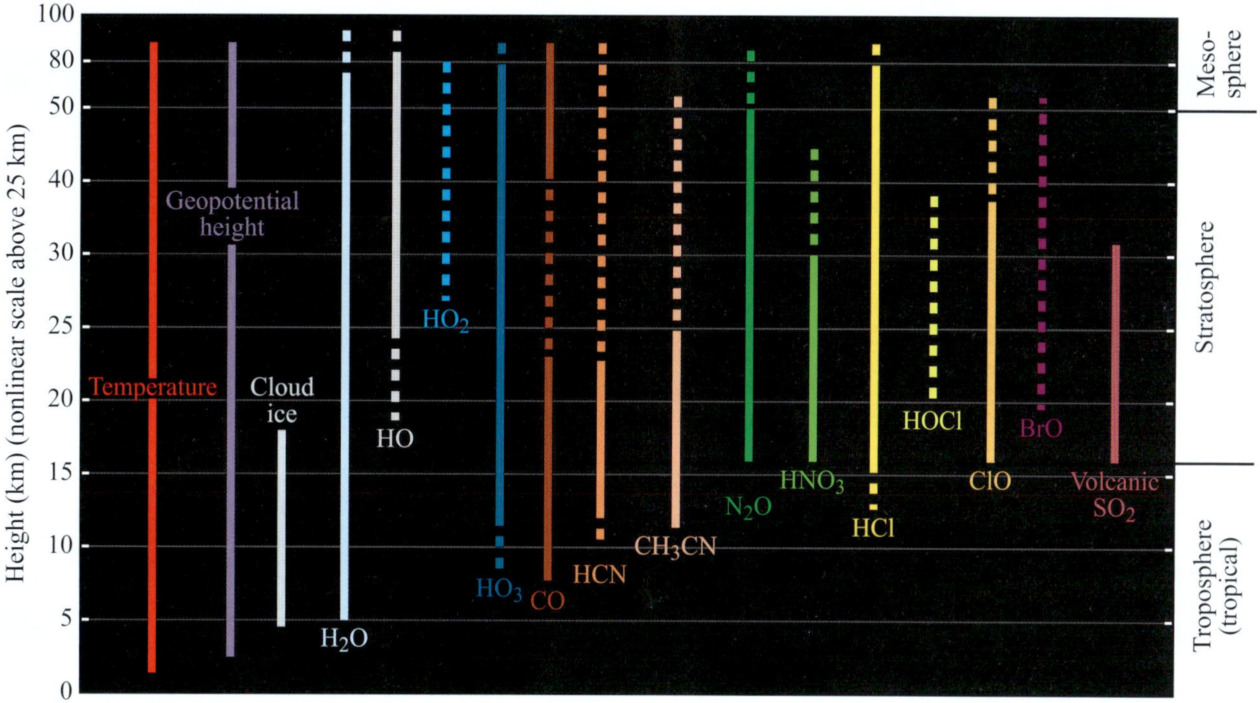

Figure 9-45: Microwave Limb Sounder products. Solid lines indicate generally useful precision for individual profiles. Dotted lines indicate that zonal (or other) averages are generally needed to obtain useful precision [Waters et al., 2006].

ena are shown in Fig. 9-47. Data were collected in four bands near 50–57, 118.75, 183.31, and 424.76 GHz during the Pacific THOR*pex* (THe Observing-system Research and *p*redictability *ex*periment) Observing System Test (PTOST 2003). The four columns of Fig. 9-47 correspond to four oceanic convective cells overflown on March 14, 2003 by the ER-2 at 20 km altitude over the North Pacific. The top four rows of the figure correspond to channels in the four bands that have weighting functions peaking at 14 km and have comparable cloud-free brightness temperatures. Only the innermost 11 scan angles are plotted; they span a swath approximately 32 km wide. The cloud-free baseline brightness temperature for each channel was subtracted to yield a ***cloud-perturbation image*** that simplifies interband comparisons.

The differences between the bands are striking. The 52 GHz images clearly respond strongly only to the narrow convective cores of these cells where typically icy hydrometeors are sufficiently large to have strong scattering signatures (more than approximately 2 mm in diameter). The apparent width of these cores is approximately 9 km. The apparent diameters of the cells near 118 GHz are markedly larger because Rayleigh scattering (ice or liquid) is inversely proportional to the fourth power of wavelength. This trend continues as the frequency progresses to 183 and 425 GHz, for which these cell diameters approach approximately 1617 km. A related trend continues for the minimum brightness temperatures at each of the cell tops. In order of increasing frequency, these minima are approximately 6, 18, 20–30, and 24–40 K below their baselines in all cases (Leslie and Staelin, 2004).

Of particular interest from a severe-weather perspective is convective precipitation, which is generally characterized by ascending saturated air masses and

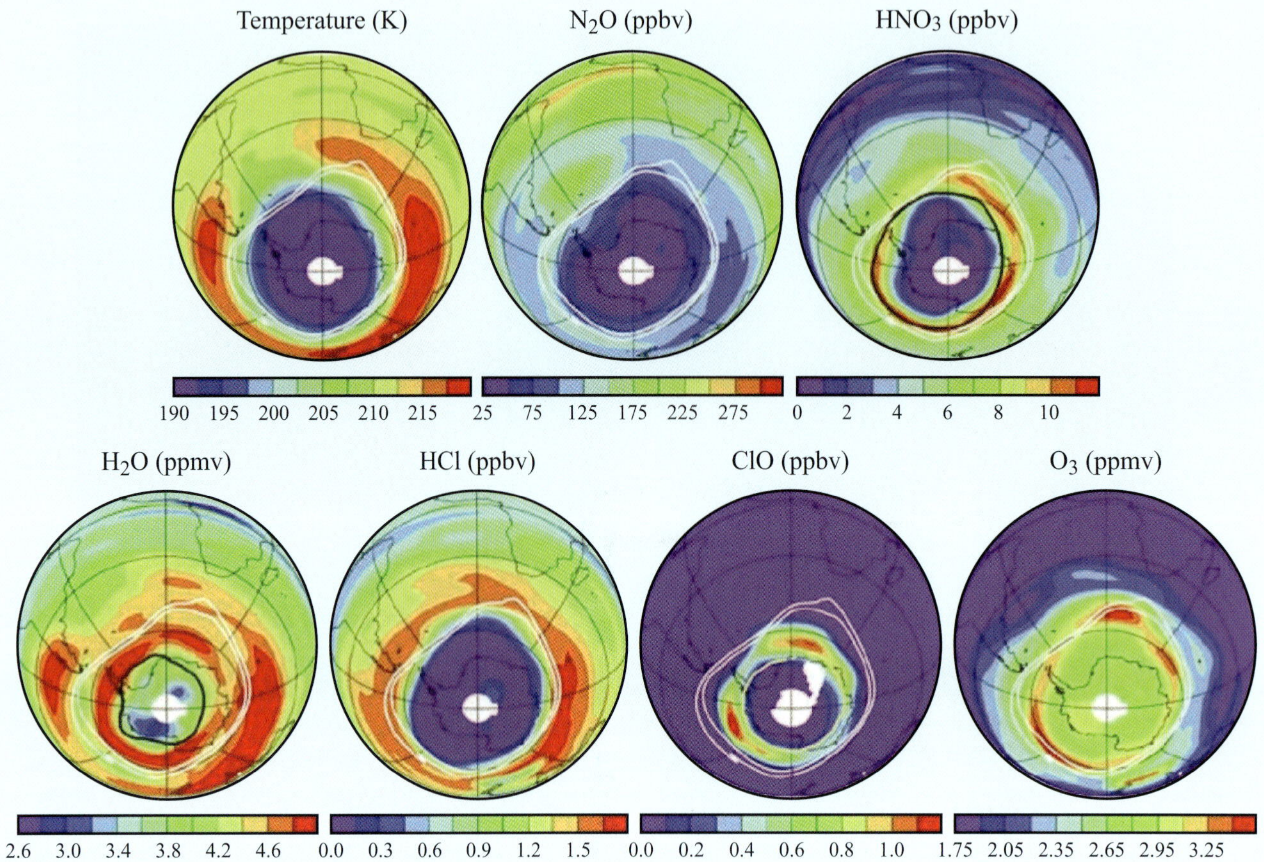

Figure 9-46: MLS measurements on August 3, 2004 for $h_T \approx 18$ km. Center of white area is the south pole [Waters et al., 2006].

wet adiabatic temperature profiles (Staelin and Chen, 2000). Retrievals of precipitation near 183 GHz rely on cold brightness-temperature signatures due to hydrometeor scattering and absorption. Transparent (window) channels near 23.8, 31.4, 89.0, and 150 GHz are useful because of their sensitivity to water-vapor absorption over the ocean and to hydrometeor scattering. The synergistic use of channels near 60 GHz and 183 GHz allows the retrieval of temperature and water-vapor profiles, both of which play important roles in the development of hydrometeors and are thus correlated to some degree with precipitation rate. Average precipitation rates are related to absolute humidity multiplied by vertical wind. In contrast to precipitation retrievals from centimeter-wave observations, which mostly employ dual-polarized window channels below 90 GHz viewed at large constant zenith angles, to estimate atmospheric absorption and water paths (Wilheit et al., 2003), precipitation retrievals from millimeter-wave spectrometers rely more on frequency-dependent and size-dependent scattering signatures from icy hydrometeors of the cold cosmic background radiation.

The sensitivity of each channel to varying altitudes reveals information about the three-dimensional structure of a precipitating cloud and vertical updraft velocity, and the sensitivity of each frequency band to different ranges of particle sizes reveals information about the

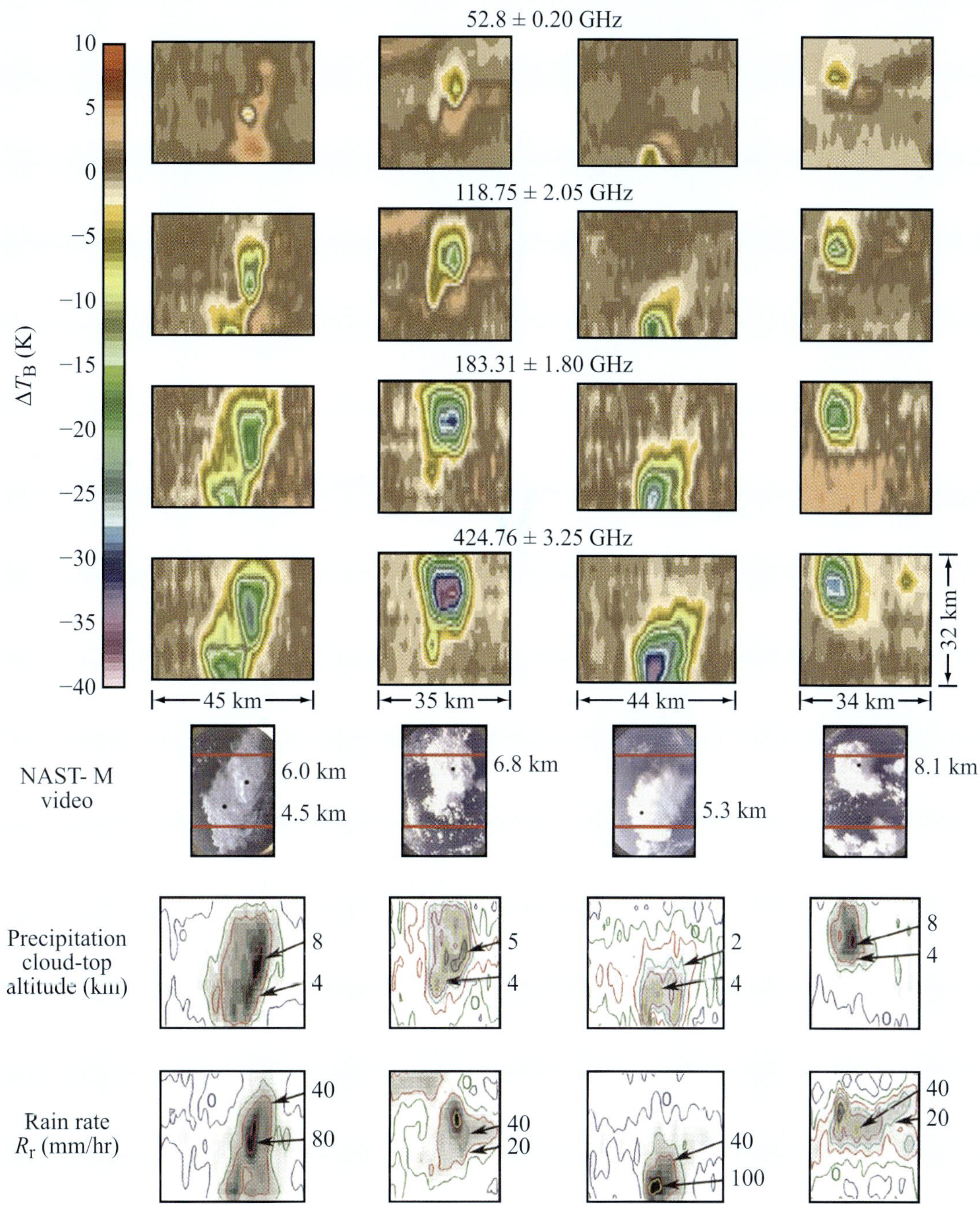

Figure 9-47: Convective cell comparison with the NAST-M spectrometers between channels sharing similar clear-air temperature weighting functions. NAST-M video images of the clouds, precipitation cell-top altitude retrievals, and rain-rate retrievals are also presented. Observations were taken during the PTOST campaign, March 14, 2003 [Leslie and Staelin, 2004].

particle-size distribution. Vertical wind is related to cell-top altitudes, altitude profiles of hydrometeor size distributions, and cell-top albedos. Cell-top altitudes can be sensed by "altitude slicing" in both the 60- and 183 GHz bands. Cell tops observed by millimeter waves are defined by hydrometeors with approximate diameters of 1–5 mm, which dominate millimeter-wave scattering. These cell tops can lie well below those observed at optical frequencies. Only strong convective cell tops rise high enough to be visible for the most opaque channels in the 183 GHz band, and cell tops deep down in the atmosphere can only be observed at the most transparent frequencies. Since scattering depends strongly on ice particle-size distribution and abundance, altitude profiles of hydrometeor size distributions can be sensed by differences in scattering cross sections in the 60- and 183 GHz bands. Only large icy hydrometeors associated with strong vertical wind and strong precipitation scatter strongly in the 60 GHz band. Scattering from smaller icy hydrometeors associated with lighter vertical wind and lighter precipitation dominate in the 183 GHz band. Icy hydrometeors affecting the 60 and 183 GHz bands are generally large enough to fall, and scattering by water droplets and cloud ice has much less effect.

One difficulty of the use of opaque channels for precipitation intensity estimation is that some precipitation may evaporate prior to reaching the surface, thereby complicating comparisons with radars and rain gauges. This challenge has been addressed by recent work to develop near-surface evaporation corrections to further improve the performance of passive millimeter-wave precipitation retrievals (Surussavadee and Staelin, 2011).

9-12.2 Example: Precipitation Retrieval Using ATMS

The ***ATMP-1 precipitation retrieval algorithm*** (Surussavadee et al., 2012) developed for ATMS employs neural networks trained and evaluated using a validated global reference physical model NCEP/MM5/TBSCAT/F(λ) (Surussavadee and Staelin, 2006). Scan angle and surface effects and other interfering signals are filtered out using principal component analysis. ATMP-1 performs well for all retrieved parameters including surface precipitation rate, water paths for rainwater, snow, graupel, the sum of rainwater, snow, and graupel, and peak vertical wind. ATMP-1 estimates are useful for surface precipitation rates higher than 1 mm/hr, hydrometeor water-paths higher than 0.1 mm, and peak vertical wind velocities higher than 0.1 m/s (Surussavadee and Staelin, 2006). A global rain-rate retrieval showing Hurricane Sandy encroaching the eastern coast of the United States on October 29, 2012 is shown in Fig. 9-48.

9-13 GPS Radio Occultation

In astronomy, the term ***occultation*** refers to the concealment of a celestial body, such as a planet, by a second body passing in front of it. An eclipse is an example of an occultation event. ***Radio occultation*** involves radio waves, and the first application of radio occultation occurred in 1964 when the spacecraft Mariner 4 flew along a path behind Mars with respect to Earth (Melbourne, 2004). As the spacecraft moved behind Mars, the radio-link signal between the spacecraft and Earth receiving stations passed through and was modulated by the atmospheric limb of Mars. The modulation caused both amplitude and phase delay variations of the radio signal, and similar variations were observed as the spacecraft emerged from the other site of Mars. These observations led to the development of a height profile for the atmospheric density of Mars. Similar studies followed for Venus, Saturn, and other planets and moons.

In the 1990s, the radio occultation technique was used in conjunction with the Global Positioning System (GPS) to evaluate its utility for profiling the temperature and water-vapor density of the Earth's atmosphere by measuring the signal transmitted through the atmosphere between a GPS satellite transmitter and an orbiting satellite receiver (Kursinski et al., 1997; Rocken et al., 1997). Within a decade, several satellite radio occultation missions were configured and launched (Hajj et al., 2004; Wickert et al., 2006; Heise et al., 2006; Healy et al., 2007), which today provide atmospheric temperature and humidity profiles with

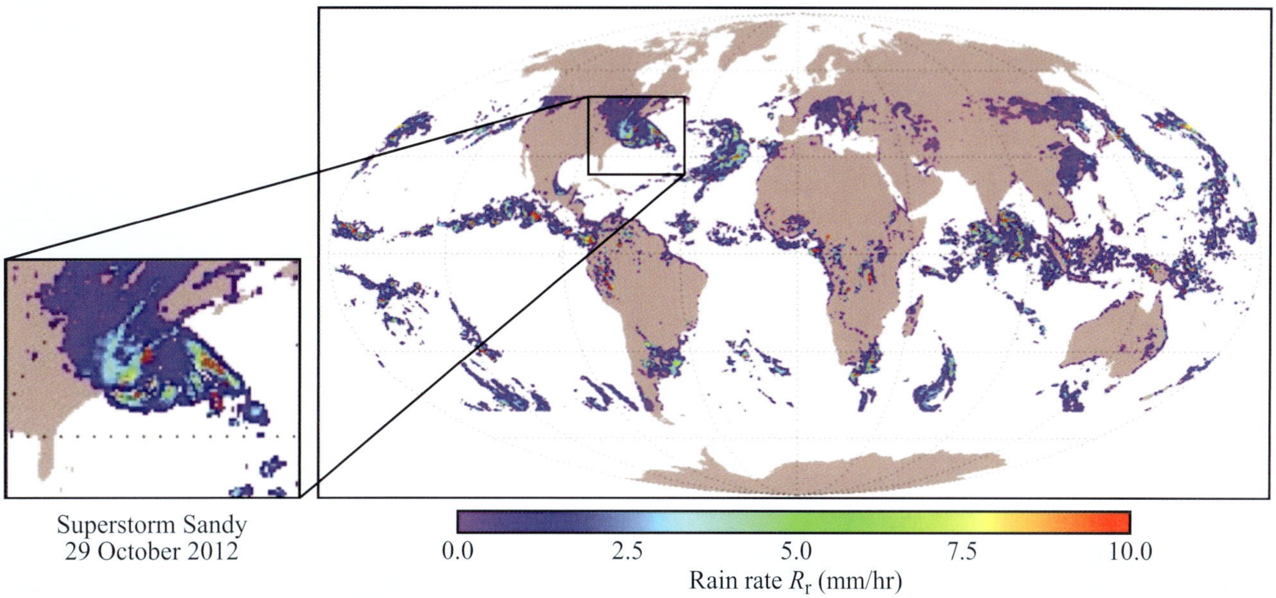

Figure 9-48: Global ATMS retrieval of rain rate (mm/hr) on October 29, 2012.

very high precision and excellent vertical resolution. A brief history of the radio occultation technique is available in the article by Yunck et al. (2000) and an extensive treatment of its underlying physics is available in the monograph by Melbourne (2004). To date, nine low-Earth orbiting satellite systems have been launched, including the ***Challenging Minisatellite Payload*** (CHAMP), the ***Gravity Recovery and Climate Experiment*** (GRACE), and the ***Constellation Observation System for Meteorology, Ionosphere, and Climate*** (COSMIC). Collectively, these systems have recorded over 6 million radio occultations, with over half of the total by COSMIC alone.

In the radio occultation technique, a low-Earth orbiting satellite, such as CHAMP, compares two received signals, one from an ***occulting GPS satellite*** and another from a reference GPS satellite (Fig. 9-49). As the occulted signal passes through Earth's atmospheric limb, it bends by an angle α, whereas the path of the reference signal is essentially unaffected by the atmosphere. The ***bending angle*** α is determined from the Doppler shift of the carrier frequency. The measurements

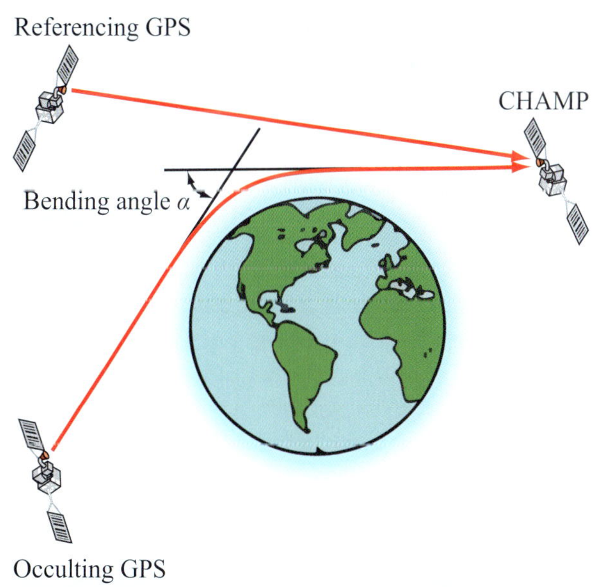

Figure 9-49: Occultation scenario for a low-Earth orbiter [Wickert et al., 2002].

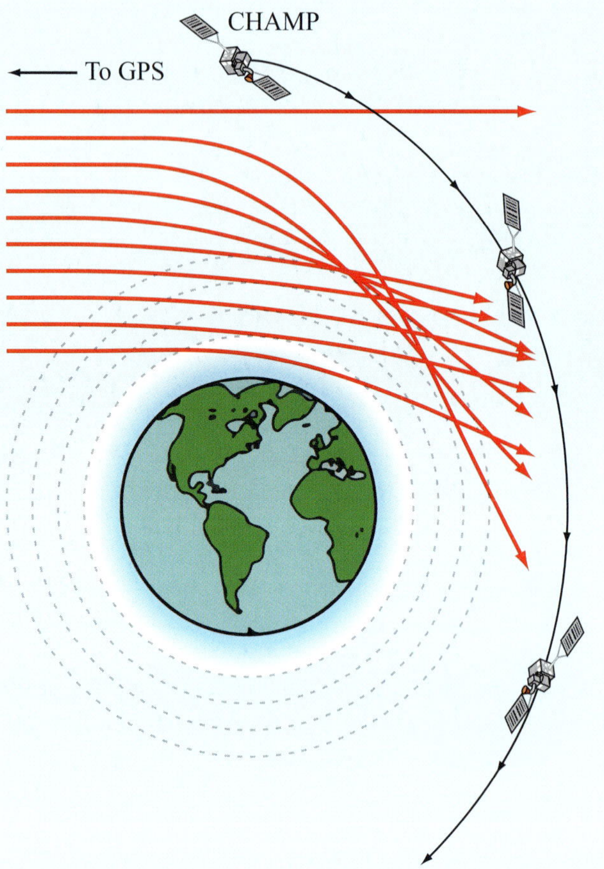

Figure 9-50: An ordered set of rays from an increasing refractive gradient with depth [Melbourne, 2004].

are made as a function of time as CHAMP moves below the horizon, which corresponds to traversing through different spherical layers of Earth's atmosphere (Fig. 9-50). From the recorded time record of Doppler shift, it is possible to generate a refractivity profile for the atmosphere and ionosphere, which is then used in an inversion algorithm to estimate the temperature and humidity profiles of the atmosphere. A typical example is shown in Fig. 9-51; the atmospheric pressure range extends from about 1000 mbars at the surface to 1 mbar at an altitude of approximately 50 km.

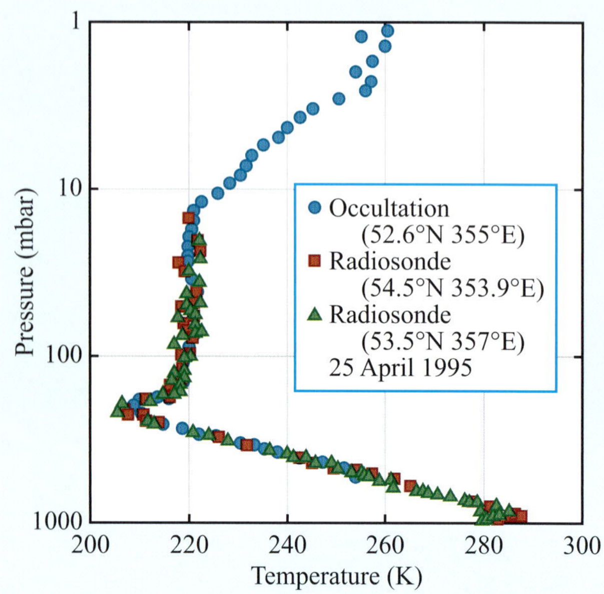

Figure 9-51: GPS temperature profile compared with two radiosondes (Yunck, 2002).

PROBLEMS

9.1 Use the Standard Atmosphere model and appropriate Computer Codes from Chapter 8 to compute and plot the normalized temperature weighting function W_T as a function of height z (from 0 to 30 km) for a ground-based zenith-pointed radiometer operating at:

(a) 50 GHz.

(b) 52 GHz.

(c) 53 GHz.

(d) 55 GHz.

(e) 60 GHz.

9.2 Repeat Problem 9.1 for a satellite-borne nadir-pointed radiometer. Ignore the presence and contributions of the Earth's surface.

9.3 Repeat Problem 9.2, but limit it to 53 GHz, and perform the computations at $\theta = 0°$, $20°$, $30°$, $50°$, and $60°$.

9.4 Use the Standard Atmospheric model with $\rho_0 = 7.5$ g/m^3 and appropriate Computer Codes from

Chapter 8 to compute and plot the normalized water-vapor weighting function as a function of height z (from 0 to 10 km) for a ground-based zenith-pointed radiometer operating at:

(a) 20 GHz.

(b) 35 GHz.

(c) 170 GHz.

(d) 175 GHz.

(e) 180 GHz.

9.5 Repeat Problem 9.4 for a satellite-borne nadir-pointed radiometer. Ignore the presence and contributions of the Earth's surface.

9.6 Given zenith-looking radiometers operating at frequencies of 21 GHz, 35 GHz, and 180 GHz, which one would provide the "best" estimate of the integrated precipitable water vapor h_v? Simulate the process by relating T_{DN} to h_v using a Standard Atmosphere with $T_0 = 300$ K and $P_0 = 1013$ mbar, but with a varying ρ_0 (from 1 to 12 g/m^3). Explain your rationale for "best performance."

9.7 Given a zenith-looking 35 GHz radiometer, simulate the results reported by Wu (1979) as described in Section 9-7.1. Specifically:

(a) Using a Standard Atmosphere, create 25 profiles consisting of the combination of 5 different surface temperatures: $T_s = 275$, 285, 295, 305, and 315 K and five different water-vapor densities ρ_0: 2, 5, 7, 10, and 12 g/m^3.

(b) For each profile, compute the mean radiating temperature T_m and brightness temperature T_B.

(c) Compute τ_0 and then relate it to T_s and ρ_0 using linear regression.

(d) Use linear regression to relate T_m to T_s.

9.8 Using a Standard Atmosphere with $T_0 = 300$ K, $P_0 = 1013$ mbar, and $\rho_0 = 10$ g/m^3:

(a) Compute the excess dry, wet, and total path lengths along the zenith direction.

(b) Independently, compute ΔR_v using Eq. (9.119), with h_v computed using Eq. (9.91).

(c) What is the percentage error between the values of ΔR_v computed in (a) and (b)?

9.9 In reference to the normalized weighting functions shown in Fig. 9-44:

(a) Reproduce the plots, taking into account that the 1.6 m diameter antenna has a nonzero beamwidth.

(b) Repeat the process for the 240 GHz channel by comparing the profiles of the weighting functions for different size antennas, namely 1.6 m, 3 m, and 5 m diameters.

9.10 Show that the rate of change of h_T with θ is $\sqrt{(R_0 + H)^2 - (R_0 + h_T)^2}$ km/radian.

9.11 Calculate the field-of-view width needed by a microwave limb sounder orbiting at 705 km altitude to provide 3 km vertical resolution at a 50 km tangent height. What is the elevation angle for this observation?

CHAPTER

10

Surface-Scattering Models and Land Observations

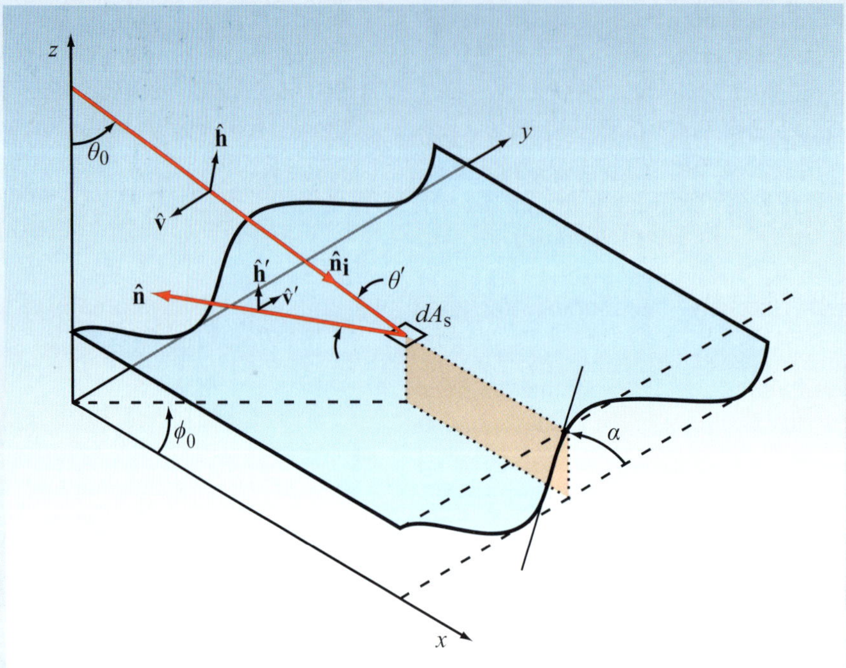

Surface scattering

CONTENTS

Overview

The backscattered signal intercepted by a radar observing a ground pixel is a result of either ***surface scattering***, ***volume scattering***, or some combination of both. Scattering by the air-soil interface in Fig. 10-1(a) generates surface scattering, which may be the result of ***single-scattering*** by facets oriented such that their surface normal points towards the radar, ***multiple scattering*** involving reflections by multiple facets, or by resonant effects with particular sinusoidal components of the surface shape. A vegetation canopy [Fig. 10-1(b)] is considered a volume-scattering medium because the layer between the soil surface and the top of the canopy contains many individual scatterers in the form of leaves, needles, and branches. The backscatter from vegetation often includes a component due to scattering by the soil surface as well as multiple scattering involving both the soil surface and the canopy constituents. The third example diagrammed in Fig. 10-1(c) is a snow layer over a soil surface. In this case, the backscatter may include scattering by the upper surface if it is sufficiently rough, volume scattering by the ice crystals within the air background, surface scattering by the lower soil surface, and multiple scattering involving the two interfaces and the snow volume between them.

The focus of this chapter is surface scattering, to be followed with a treatment of volume scattering in Chapter 11. We offer the reader an overview of scattering models for surfaces, supplemented with numerous examples of experimental measurements. The intent is to illustrate the backscatter behavior as a function of the radar wave parameters—namely, the wavelength, incidence angle and receive/transmit polarization configuration—and the surface's geometrical and dielectric properties. We should caution, however, that this is not a chapter about the mathematical techniques used in the development of scattering models. Such an endeavor would consume one or more book volumes all on its own. We refer the reader, instead, to the books by Tsang et al. (1985), Jin (1993), and Fung (1994).

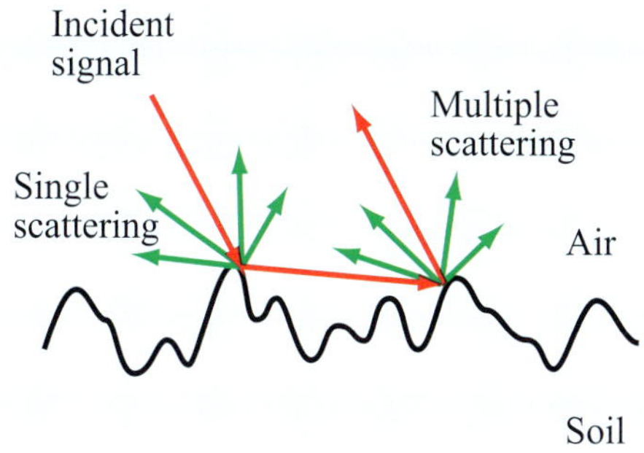

(a) Random surface

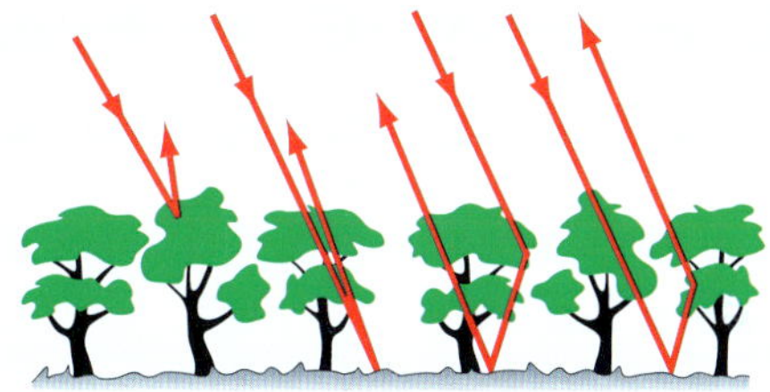

(b) Vegetation canopy

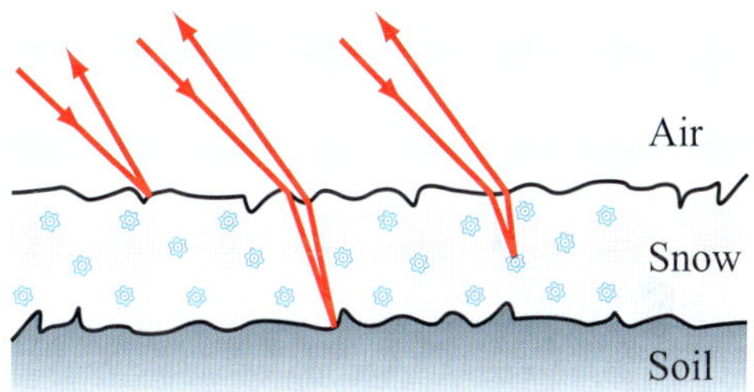

(c) Snow layer containing ice crystals

Figure 10-1: Scattering from a random surface, a vegetation canopy, and snow-covered soil.

10-1 The Role of Scattering Models

The geometry of most natural terrain is a superposition of two factors, a quasi-deterministic shape and a random shape variation. The surface of a soil surface has a mean boundary, added to which are random height deviations. A tree canopy contains scatterers with specific shapes, but their sizes and orientations are characterized by probability distributions. Similar statements apply to other types of terrain. The combination of the statistical nature of terrain targets and the complexity of their shapes makes the task of modeling bistatic scattering from terrain a daunting endeavor. Even a bare soil surface is a challenge to model; capturing the exact profile of the surface geometry is almost impossible, and even if we could do that so as to test the validity of a theoretical model, extending the results to other surfaces (about which we often have little or no information) would be very difficult if not impossible. *So, how do we approach the task of developing scattering models?* The answer is:

(1) We try to describe the terrain in terms of statistical distributions—such as height and slope distributions for a random surface, size and orientation distributions for volume scatterers, etc., and

(2) We perform a mathematical averaging process equivalent to (a) generating a large number of synthesized surfaces and volumes based on the assumed statistics, (b) computing for each synthesized target the scattering cross section for the radar wave parameters of interest, and then (c) performing an ensemble average. Step 2 is equivalent to implementing a Monte-Carlo simulation, but it is carried out mathematically by injecting the assumed statistical distributions into the scattering formulation and then calculating the mean radar cross section.

> ▶ Hence, scattering models of terrain are, at best, good approximations of the true scattering process experienced by a real radar observing a real terrain surface or volume. They serve as guides to explain experimental observations and as predictors of how the radar scattering coefficient σ^0 is likely to behave as a function of a particular terrain parameter of interest. ◀

We should note that scattering models are useful not only for radar remote sensing, but also for modeling and understanding the terrain emission observed by a microwave radiometer. As noted in Section 6-7.3, the emissivity of terrain is computed by integrating the bistatic scattering coefficient over all scattering directions [Eq. (6.105)].

10-2 Surface Parameters

As a prelude to the presentations contained in succeeding sections of this chapter, we now introduce (or reintroduce) the principal terminology associated with the physical and statistical properties of natural surfaces.

A given surface that may "appear" very rough to an optical wave may appear very smooth to a microwave. This is because the ***degree of roughness***, or simply the ***roughness***, of a random surface is characterized in terms of statistical parameters, measured in units of wavelength. The two fundamental parameters used to characterize surface roughness are the ***standard deviation of the surface height variation*** (or ***rms height***) s and the surface correlation length l. A third surface parameter that also is important to some surface models is the ***rms slope*** m. In all cases, these statistical parameters are associated with the ***random component*** of surface height, relative to a reference surface. The reference surface may be the unperturbed surface of a deterministic periodic pattern (as in the case of row-tilled soil surfaces or wind-driven ocean waves) as shown in Fig. 10-2(a), or it may be the ***mean surface*** if only random variations exist [Fig. 10-2(b)]. The periodic-surface case is discussed in some detail in Section 10-4.2, so we limit our present discussion to nonperiodic random surfaces.

10-2.1 rms Height

For a random surface whose mean is coincident with the x-y plane, its height $z(x,y)$ above the x-y plane usually is characterized by a Gaussian probability density function $p(z)$ given by

$$p(z) = \frac{1}{\sqrt{2\pi s^2}}\, e^{-z^2/2s^2}, \qquad (10.1)$$

where s is the ***rms height***:

$$s = \langle z^2 \rangle^{1/2} = \left[\int_{-\infty}^{\infty} z^2\, p(z)\, dz \right]^{1/2}. \qquad (10.2)$$

As noted earlier in Section 5-10.1, the Gaussian assumption is a very good approximation for most natural random surfaces.

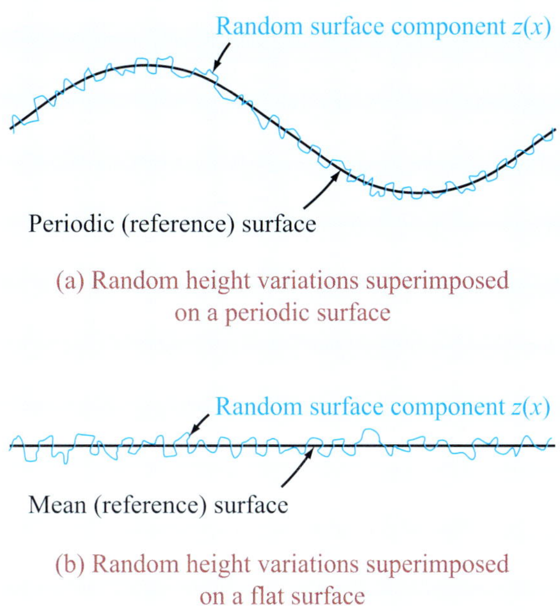

Figure 10-2: Two configurations of height variations: (a) random height variations superimposed on a periodic surface, and (b) random height variations superimposed on a flat surface.

Alternatively, if $z(x,y)$ of a statistically representative segment of the surface, of dimensions L_x and L_y, centered at the origin, is available, the mean height $\overline{z}$ and second moment $\overline{z^2}$ can be calculated from

$$\overline{z} = \frac{1}{L_x L_y} \int_{-L_x/2}^{L_x/2} \int_{-L_y/2}^{L_y/2} z(x,y)\, dx\, dy, \tag{10.3a}$$

and

$$\overline{z^2} = \frac{1}{L_x L_y} \int_{-L_x/2}^{L_x/2} \int_{-L_y/2}^{L_y/2} z^2(x,y)\, dx\, dy. \tag{10.3b}$$

The rms height is then given by

$$s = (\overline{z^2} - \overline{z}^2)^{1/2}. \tag{10.4}$$

If the surface is azimuthally symmetrical, a one-dimensional height profile $z(x)$ is sufficient, in which case the integration over y is not necessary. In practice, the profile is digitized into discrete values $z_i(x_i)$ at an appropriate spacing Δx. If the height variation Δz corresponding to a horizontal segment Δx is much smaller than the EM wavelength λ, Δz exercises little or no appreciable effect on the reflection by the surface of segment Δx. As a rule of thumb, the spacing Δx should be chosen such that $\Delta x \leq 0.1\lambda$.

The rms height of the discrete one-dimensional case is

$$s = \left[\frac{1}{N-1}\left(\sum_{i=1}^{N} z_i^2 - N\overline{z}^2\right)\right]^{1/2}, \tag{10.5}$$

where N is the number of samples and

$$\overline{z} = \frac{1}{N}\sum_{i=1}^{N} z_i. \tag{10.6}$$

10-2.2 Surface Correlation Length

The correlation function of the continuous random surface, $\rho(\xi)$, was defined by Eq. (5.105). For the discrete, one-dimensional case,

$$\rho(\xi) = \frac{\displaystyle\sum_{i=1}^{N+1-j} z_i\, z_{j+i-1}}{\displaystyle\sum_{i=1}^{N} z_i^2}, \tag{10.7}$$

where $\xi = (j-1)\,\Delta x$ and j is an integer ≥ 1. The ***correlation length*** l is the value of ξ at which $\rho(\xi) = e^{-1}$. Figure 10-3, which was introduced earlier in Chapter 5, provides a graphical description of $z(x)$, $p(z)$, and $\rho(\xi)$ for a real surface.

For a perfectly smooth surface, $l = \infty$.

10-2.3 rms Slope

The slope of $z(x)$ at location x is given by

$$Z_x(x) = \lim_{\Delta x \to 0} \frac{z(x+\Delta x) - z(x)}{\Delta x}. \tag{10.8a}$$

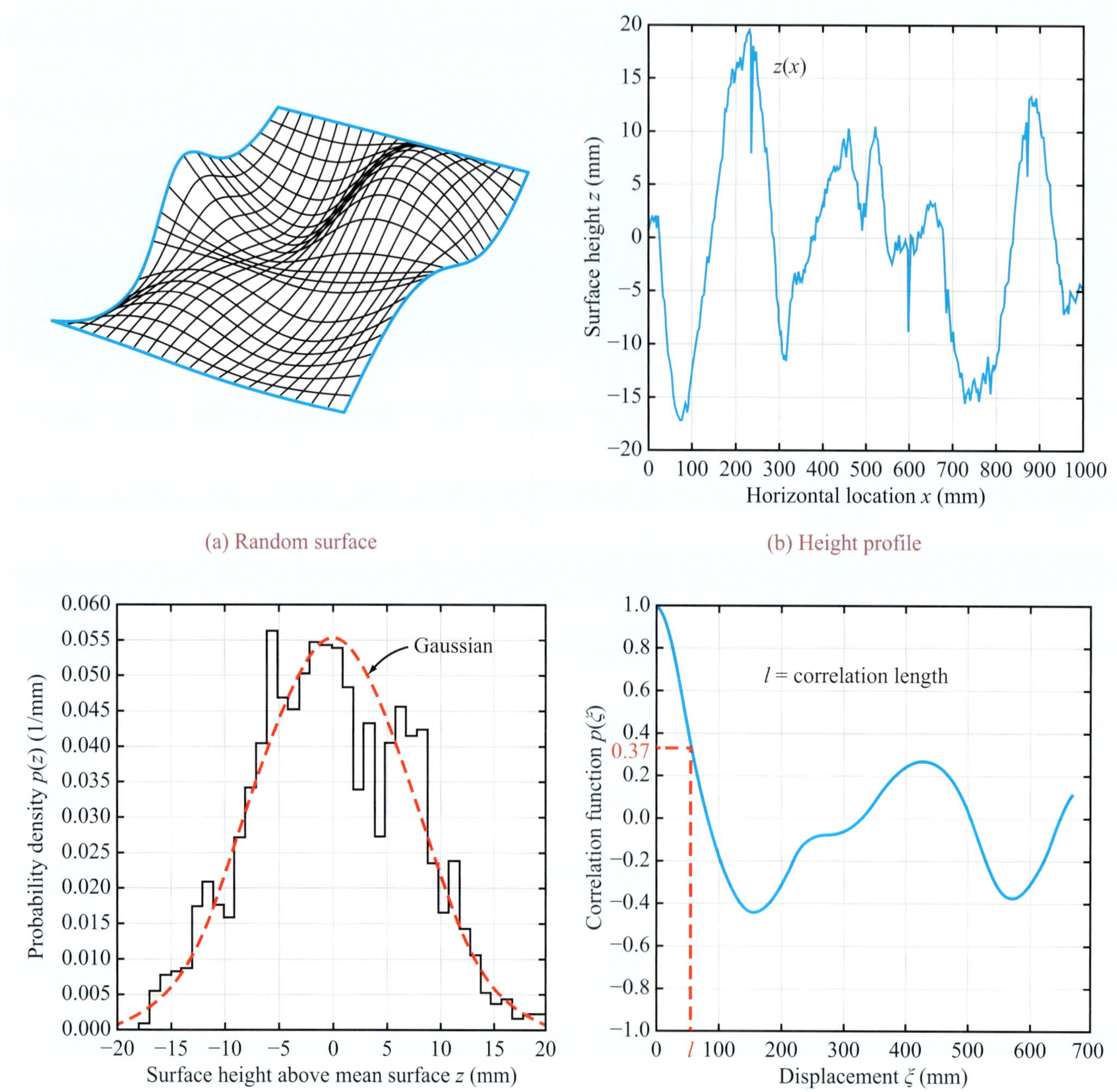

(a) Random surface

(b) Height profile

(c) Histogram

(d) Correlation function

Figure 10-3: Random, isotropic surface $z(x,y)$: (a) pictorial view, (b) measured height profile $z(x)$, (c) pdf of digitized height profile, and (d) autocorrelation function $\rho(\xi)$, where ξ is the displacement between two points on the surface.

The ensemble average of Z_x^2 is

$$\langle Z_x^2 \rangle = \lim_{\Delta x \to 0} \left\langle \frac{z^2(x+\Delta x) - 2z(x)\, z(x+\Delta x) + z^2(x)}{(\Delta x)^2} \right\rangle$$
$$= \lim_{\Delta x \to 0} \left[\frac{s^2 - 2s^2 \rho(\Delta x) + s^2}{(\Delta x)^2} \right]$$
$$= 2s^2 \lim_{\Delta x \to 0} \left[\frac{1 - \rho(\Delta x)}{(\Delta x)^2} \right], \tag{10.8b}$$

where s is the rms height and $\rho(\Delta x)$ is the correlation function for separation Δx. To find the limit, we expand $\rho(\Delta x)$ about $\Delta x = 0$ in a Taylor series. Because $\rho(\Delta x)$ is an even function, its first derivative $\rho'(0) = 0$, in which case

$$\langle Z_x^2 \rangle = 2s^2 \lim_{\Delta x \to 0} \left[\frac{1 - [1 + \rho''(0)\, (\Delta x)^2/2 + \cdots]}{(\Delta x)^2} \right]$$
$$\approx -s^2 \rho''(0), \tag{10.9}$$

where $\rho''(0)$ is the second derivative of the surface correlation function $\rho(\xi)$ evaluated at $\xi = 0$. The ***rms slope*** is

$$m = \langle Z_x^2 \rangle^{1/2} = [-s^2 \rho''(0)]^{1/2}. \tag{10.10}$$

Note that the correlation function is an even function. Hence, $\rho'(0) = 0$ and $\rho''(0)$ is a negative quantity.

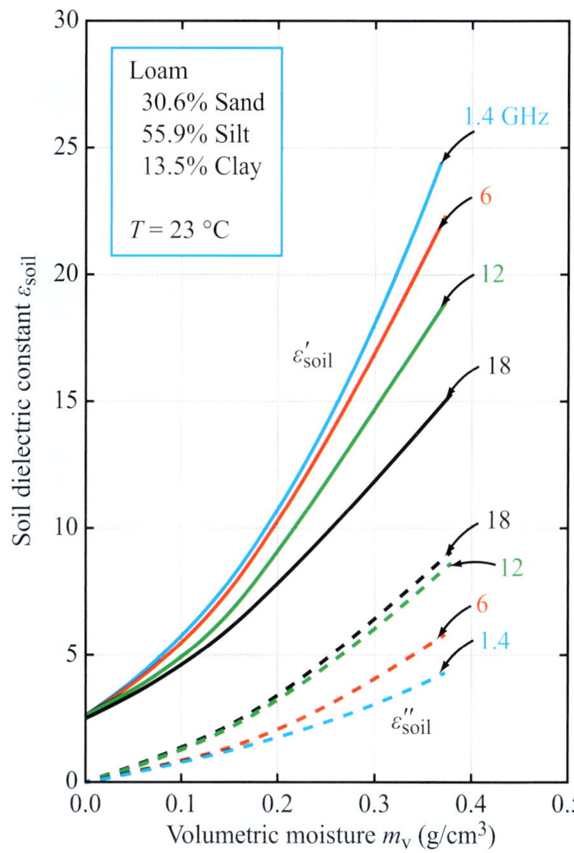

Figure 10-4: Measured dielectric constant as a function of volumetric moisture content for a loamy soil at four microwave frequencies [Hallikainen et al., 1985].

10-2.4 Fresnel Reflection Coefficient

The scattering coefficients $\sigma^0_{\rm vv}$, $\sigma^0_{\rm hh}$, and $\sigma^0_{\rm hv}$ of a surface are related to the dielectric constant of the surface material, ε, through the Fresnel reflectivities $\Gamma_{\rm h}$, and $\Gamma_{\rm v}$. To first order, $\sigma^0_{\rm vv}$ is proportional to $\Gamma_{\rm v}$, $\sigma^0_{\rm hh}$ is proportional to $\Gamma_{\rm h}$, and $\sigma^0_{\rm hv}$ is related to the product $(\Gamma_{\rm v}\Gamma_{\rm h})^{1/2}$. For soils, $\varepsilon_{\rm soil}$ is strongly dependent on the soil's volumetric moisture content $m_{\rm v}$. Figure 10-4, extracted from Chapter 4, illustrates the variation of $\varepsilon_{\rm soil}$ with $m_{\rm v}$ at four microwave frequencies. In part (a) of Fig. 10-5, we display the angular variations of $\Gamma_{\rm v}(\theta)$ and $\Gamma_{\rm h}(\theta)$ at two moisture contents, and in parts (b) and (c) of the same figure we display the responses of $\Gamma_{\rm v}$ and $\Gamma_{\rm h}$, respectively, to $m_{\rm v}$ at three incidence angles. We note that at $\theta = 30^\circ$, for example, $\Gamma_{\rm v}$ and $\Gamma_{\rm h}$ exhibit a dynamic range of about 4 : 1 (or 6 dB) as $m_{\rm v}$ is increased from 0.05 g/cm^3 to 0.35 g/cm^3.

10-2.5 Smooth-Surface Criteria

Under what conditions may we consider a surface to be electromagnetically *smooth*?

The geometry shown in Fig. 10-6(a) represents two rays incident normally upon a surface. Had the surface been perfectly flat, the electric fields of the two reflected rays would have been in-phase, but because the rough surface is higher at point B

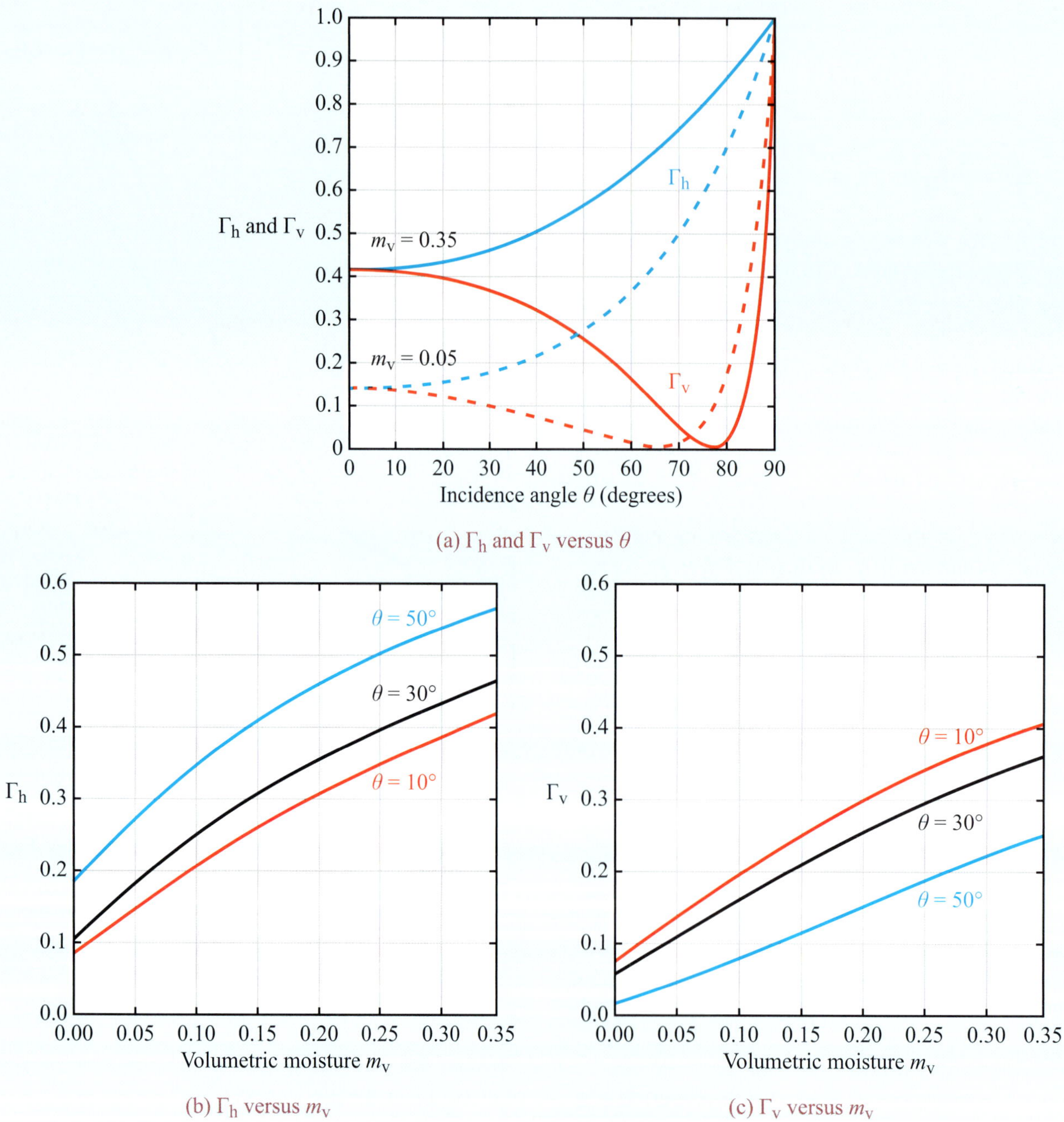

Figure 10-5: Variation of reflectivity Γ with θ and m_v for a loamy soil at 1.5 GHz.

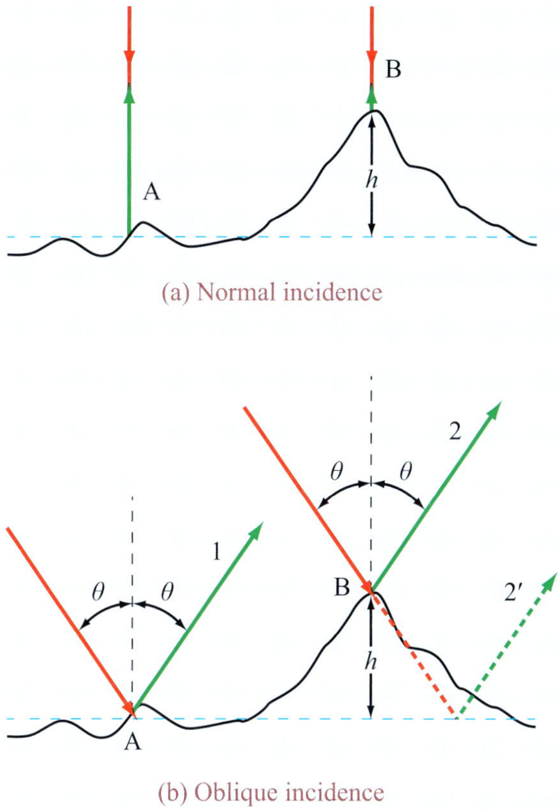

Figure 10-6: At (a) normal incidence, the phase difference between the rays reflected from A and B is $\Delta\phi = 2kh$ and at (b) oblique incidence $\Delta\phi = 2kh\cos\theta$.

than at point A by a height h, the field of the ray reflected by point B travels a shorter distance by $2h$. The associated phase difference is $\Delta\phi = 2kh = 4\pi h/\lambda$, where $k = 2\pi/\lambda$ is the wavenumber. For oblique incidence at angle θ [Fig. 10-6(b)],

$$\Delta\phi = 2kh\cos\theta = \frac{4\pi h}{\lambda}\cos\theta. \tag{10.11}$$

The ***Rayleigh roughness criterion*** states that the surface may be considered smooth if $\Delta\phi < \pi/2$, which corresponds to

$$h < \frac{\lambda}{8\cos\theta}.$$

For a random surface with rms surface height s, h may be replaced with s:

$$s < \frac{\lambda}{8\cos\theta} \quad \text{or} \quad ks < 0.8 \quad (\text{at } \theta = 0). \tag{10.12a}$$

(Rayleigh criterion)

The Rayleigh criterion is useful as a first-order classifier of surface roughness or smoothness, but for modeling the scattering and emission behavior of natural surfaces in the microwave region—where the wavelength λ usually is of the order of the rms height s, a more stringent criterion is needed. For this purpose, we adopt the criterion used to define the far-field distance of an antenna, which requires the maximum phase difference between the electric fields of rays coming from the center and the edge of the antenna to be less than $\pi/8$ radians. Such a requirement, which we call the ***Fraunhofer roughness criterion***, leads to the condition

$$s < \frac{\lambda}{32\cos\theta} \quad \text{or} \quad ks < 0.2 \quad (\text{at } \theta_i = 0) \tag{10.12b}$$

(Fraunhofer criterion)

for a surface to be considered smooth. The Fraunhofer criterion is consistent with the results reported earlier in Section 5-10.3 on the coherent reflectivity of a random surface. The dependence of the backscattering coefficient σ^0 on surface roughness is discussed in forthcoming sections of the present chapter, and the dependence of the emissivity on roughness is examined in Chapter 12.

10-3 Surface-Scattering Models

Serious efforts to develop mathematical models for randomly rough surfaces began in the 1950s and culminated in two approximate models. The first one, developed by Rice (1951) for slightly rough surfaces whose rms heights and correlation lengths are both smaller than the incident wavelength, became known as the ***small perturbation model***. Its validity conditions are:

$$\left.\begin{array}{l} ks < 0.3 \\ kl < 3 \\ \dfrac{s}{l} < 0.3 \end{array}\right\} \quad \textbf{(small perturbation model)},$$

where s is the rms height, l is the correlation length and $k = 2\pi/\lambda$.

The second approximate model, known as the ***Kirchhoff scattering model***, was developed by Beckman and Spizzichino (1963) to describe EM scattering by surfaces with gentle undulations whose average horizontal dimensions are large compared with λ. To evaluate the integral representing the total field scattered by the surface, two types of approximations have been used, one for $ks = (2\pi s/\lambda) \geq 3$, which is then known as the ***geometric optics model***, and another for $ks < 3$, which is known as the ***physical optics model***. The validity conditions include additional constraints related to the average radius of curvature and rms slope of the undulating surface. When $ks \geq 3$ and the radius of curvature of the roughness scale is much larger than a wavelength, the scattering process is dominated by reflection. The correlation function (introduced shortly in Section 10-3.1) can be approximated by the first two terms of its Taylor expansion in the scattered field integral, and this leads to the ***geometric optics model***. When $ks < 3$ and the radius of curvature of the roughness is still larger than a wavelength, then locally both reflection and diffraction are important. In this case, the correlation function cannot be approximated and the resulting model is known as the ***physical optics model***.

▶ The Kirchhoff model deals with roughness scales that are larger than a wavelength in horizontal extent, while the perturbation model deals with roughness scales that are small relative to the radar wavelength. Clearly, a better surface-scattering model is needed, one that has no restriction on the size of the surface roughness scales. ◀

Across the microwave band, which extends over the wavelength range from 1 mm ($f = 300$ GHz) to 30 cm ($f = 1$ GHz), natural surfaces encompass a wide range of roughness scales. Consequently, the aforementioned 1960s scattering models are applicable over limited segments of the roughness scale of interest. Over the next three decades, many attempts were made to improve the models and extend their regions of validity, culminating in the development of an ***integral equation model*** (IEM) (Fung et al., 1992) capable of bridging the gaps between its various predecessors. The applicability of the model, however, is limited to the backscatter direction. Over the next decade, Fung and his colleagues extended the range of applicability of the IEM to bistatic scattering (Fung et al., 2002). The new model is referred to in this book as the ***improved IEM scattering model***, or I^2EM for short.

As mentioned earlier, the emissivity of a rough surface is computed by integrating the total bistatic scattering over the upper hemisphere. The availability of the I^2EM makes such a computation viable, as discussed in more detail in Chapter 12.

10-3.1 I^2EM Parameters

The I^2EM model is a mathematically sophisticated computational algorithm with which one can compute the backscattering or bistatic scattering coefficient of a random surface with any specified degree of roughness for any combination of receive and transmit wave polarizations. The computation includes multiple scattering contributions and accounts for shadowing effects. Generally, multiple scattering from a surface is small (in comparison with single scattering) except for surfaces with large rms slopes, which is uncommon for natural surfaces. When multiple scattering contributions, which require a two-dimensional integration, are excluded, the model reduces to a relatively simple algebraic form. Figure 10-7 provides in block-diagram format the list of surface and radar wave parameters required to specify the model to compute σ^0. Theoretically, correlation functions used in the model should be differentiable at the origin. Hence, the commonly used exponential correlation function is not acceptable. However, for low to medium frequency calculations this property of the correlation function can be relaxed, because the computed result does not depend on this property until

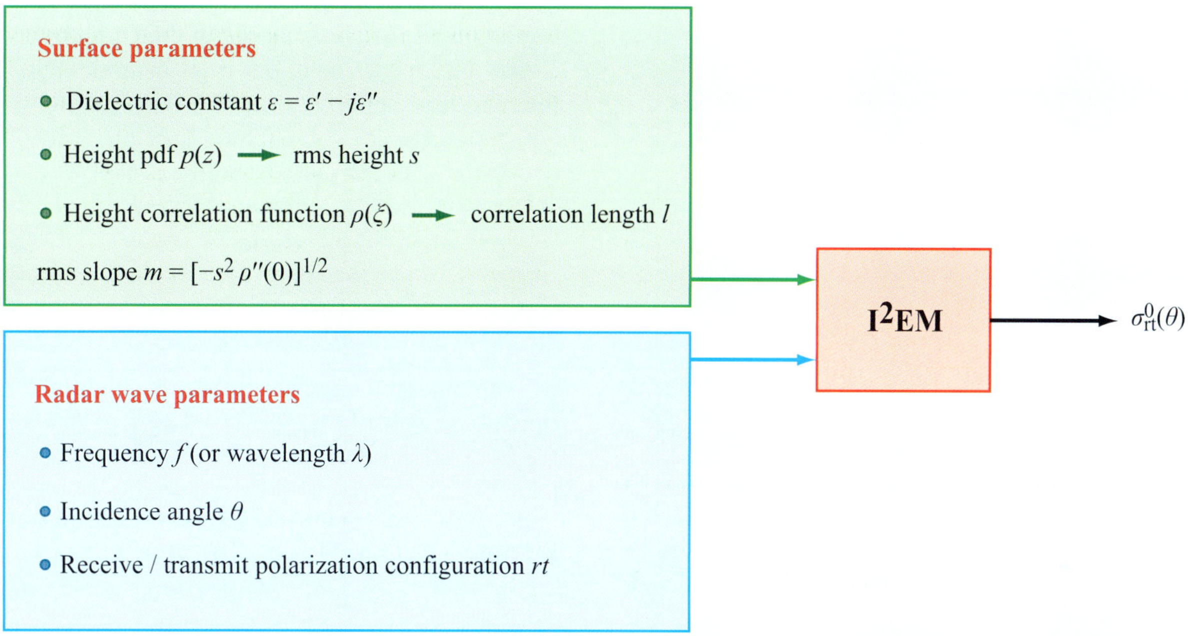

Figure 10-7: Block-diagram representation of the surface and radar input parameters for the I²EM model in the backscatter mode. For bistatic scattering, θ becomes θ_i and the scattered direction is specified by θ_s and ϕ, where ϕ is the azimuth angle between the incident and scattered directions.

we get close to the geometric-optics condition, where all roughness scales are large compared with the incident wavelength. In practice, we rarely reach this high-frequency condition, because natural surfaces normally have many roughness scales. As we shorten the incident wavelength, the wave simply responds to smaller scales. Some of the ***correlation functions*** reported in the literature include:†

Exponential: $$\rho(\xi) = e^{-|\xi|/l} \tag{10.13a}$$

Gaussian: $$\rho(\xi) = e^{-\xi^2/l^2} \tag{10.13b}$$

***x*-exponential**: $$\rho(\xi) = e^{-(|\xi|/l)^x}. \tag{10.13c}$$

***x*-power**: $$\rho(\xi) = \frac{1}{[1 + (\xi^2/l^2)]^x} \tag{10.13d}$$

The four correlation functions represent different approaches on how to characterize the statistical correlation between the height $z(x)$ at location x with the height $z(x+\xi)$ at a location separated from the first location by a distance ξ. Measuring the height profile $z(x)$ of a random surface with a high degree of vertical accuracy across a long horizontal segment and along many different horizontal directions is a difficult and laborious process. Based on the relatively few such measurements reported in the literature for natural surfaces, the exponential correlation function appears to provide the best fit, with the Gaussian a close second. For both the exponential and Gaussian correlation functions, l represents the correlation distance of the random surface, defined as the separation ξ at which $\rho(\xi) = 1/e$ (Section 5-10.1). The x-exponential correlation function is a two-parameter function (x and l) that reduces to the exponential function for $x = 1$ and to the Gaussian for $x = 2$ [Fig. 10-8(a)]. Having two fitting parameters, the x-exponential function provides greater flexibility in matching the model to correlation functions based on measured surface height profiles. Moreover, the value of l in the x-exponential correlation function

† **Computer Code 10.1.**

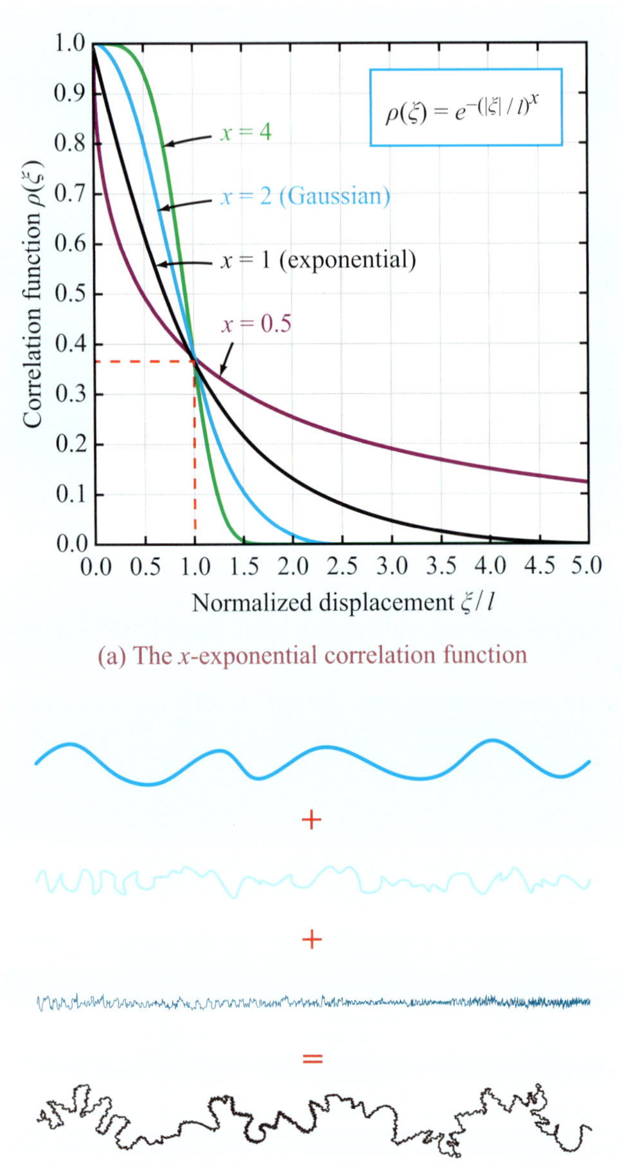

(a) The x-exponential correlation function

(b) Multiscale roughness

Figure 10-8: (a) x-exponential correlation function and (b) three-scale random surface.

continues to represent the usual definition associated with correlation distance. The last correlation function listed in Eq. (10.13) is the x-power function, which also has two fitting parameters, but in this case l no longer represents the usual definition of the correlation length.

For the x-power and Gaussian correlation functions, the rms slope defined by Eq. (10.10) is given by

$$m = \sqrt{3}\,\frac{s}{l} \quad (x\text{-power with } x = 1.5), \tag{10.14a}$$

$$m = \sqrt{2}\,\frac{s}{l} \quad (\text{Gaussian}). \tag{10.14b}$$

10-3.2 Multiscale Surfaces

A real random surface consists of multiple, or continuous, scales of roughness, with the smaller scales "riding" on top of the larger scales. Figure 10-8(b) depicts a ***three-scale surface*** consisting of a large-scale undulation with correlation length l_1 and associated rms height s_1, a medium-scale perturbation with l_2 and s_2, and a small-scale high-spatial frequency variation with l_3 and s_3. If we were to measure the height profile $z(x)$ of the composite surface with a sensor (such as a laser beam) capable of measuring $z(x)$ with a precision superior to the smallest scale (i.e., with a vertical resolution $\Delta z \ll s_3$ and a horizontal resolution $\Delta x \ll l_3$), and if the surface correlation is Gaussian, the correlation function derived from the measured record would have the functional form†

$$\rho(\xi) = \frac{1}{s^2}\left[s_1^2 e^{-\xi^2/l_1^2} + s_2^2 e^{-\xi^2/l_2^2} + s_3^2 e^{-\xi^2/l_3^2}\right], \tag{10.15a}$$

where s, the equivalent rms height of the composite surface, is related to the individual rms heights by

$$s^2 = s_1^2 + s_2^2 + s_3^2. \tag{10.15b}$$

The surface scattering process is subject to a ***wavelength filtering*** effect that, in effect, determines which scales of roughness are important and which are not. Let us examine, qualitatively, what happens as λ is varied from $\lambda \gg l_1$ to $\lambda \ll s_3$. When λ is much longer than s_1 and l_1 of the largest scale, the surface "appears" to the incident wave as if it were perfectly flat, in which case the wave reflects coherently along the specular direction relative to the mean surface, and no incoherent

†**Computer Code 10.2.**

scattering occurs along any direction. As the wavelength is made shorter so that it approaches the dimensions of s_1 and l_1 of the largest scale, incoherent scattering does occur, but the shorter scales do not play a significant role. This is what is meant by wavelength filtering. As the dimension of λ approaches the dimensions of s_2 and l_2 of the medium-scale roughness, the scattering process becomes responsive primarily to the medium-scale statistics. Further reduction of λ toward the small scale leads to diminution of the role of the medium- and large-scale roughnesses (except for defining the mean slope upon which the small-scale roughness resides) and to an increased role for the small-scale roughness. An illustration of the scattering behavior just described is shown in Fig. 10-9, where we show backscattering computations at 2, 4, and 9 GHz for a surface with three roughness scales. It is clear from the figure that as either the incident angle or frequency increases, the smaller roughness scales become increasingly important.

At microwave frequencies, λ is on the order of centimeters to a few tens of centimeters. Hence, the centimeter scale is the scale of roughness of primary importance from the standpoint of surface scattering. At $f > 10$ GHz, the millimeter scale starts becoming important, but unfortunately it is very difficult to measure millimeter-scale roughness of natural surfaces.

▶ Hence, in practice we tend to characterize surfaces by only their centimeter-scale roughness statistics, which occasionally leads to inconsistencies between model-calculated values and experimentally observed values of the backscattering coefficient. The scattering by the real surface involves scattering at multiple scales, whereas the model-calculated scattering is based on single-scale statistics. This is not because the model cannot compute multiscale statistics, but rather because the vertical precision and horizontal spacing of the measured roughness profile are inadequate for recording the millimeter-scale roughness component of the real profile. ◀

The mathematical details of I^{2}EM are outside the scope of this book, but its utility as a guide and predictor of scattering behavior is very much of interest. Hence, in the next few subsections, we use it to examine the roles of surface roughness and dielectric constant and to compare model predictions with experimental measurements.

10-3.3 Role of Correlation Function

In Fig. 10-10 we display angular plots for σ^0_{vv}, σ^0_{hh}, and σ^0_{hv} computed using I^{2}EM for two surfaces with identical parameters, except for their correlation functions. The two surfaces have the same correlation length of 10 cm, but one is characterized by an exponential correlation function, as defined by Eq. (10.13a), and the other is characterized by the Gaussian correlation function of Eq. (10.13b). We observe that for each of the three polarization combinations, σ^0 of the Gaussian surface exhibits a much steeper angular response than does that of the exponential surface. At $\theta = 70^\circ$, σ^0_{hh} (Gaussian) is about 38 dB lower in magnitude than σ^0_{hh} (exponential), and for hv polarization the difference is about 50 dB. These are very large differences; 50 dB corresponds to five orders of magnitude!

▶ Experimental measurements of σ^0 for soil surfaces with comparable surface conditions (rms height s, correlation length l, and dielectric constant ε) are much closer in level to the predictions of the I^{2}EM for a surface with an exponential correlation function than for a Gaussian. ◀

10-3.4 Role of rms Height s

To illustrate the effect that the rms height s has on the level of σ^0, we used I^{2}EM to generate angular plots of σ^0_{hh}, σ^0_{vv}, and σ^0_{hv} at 3 GHz for two surfaces (Fig. 10-11, one with $s = 0.5$ cm (with a corresponding $ks \approx 0.3$) and a second one with $s = 1.5$ cm (with a corresponding $ks \approx 0.9$). At angles greater than 20°, the co-polarized backscattering coefficients are on the order of 8–15 dB higher for the rougher surface than for the smoother surface, and the difference in level is closer to 17 dB for σ^0_{hv}. In general, at angles greater than 20°, all three

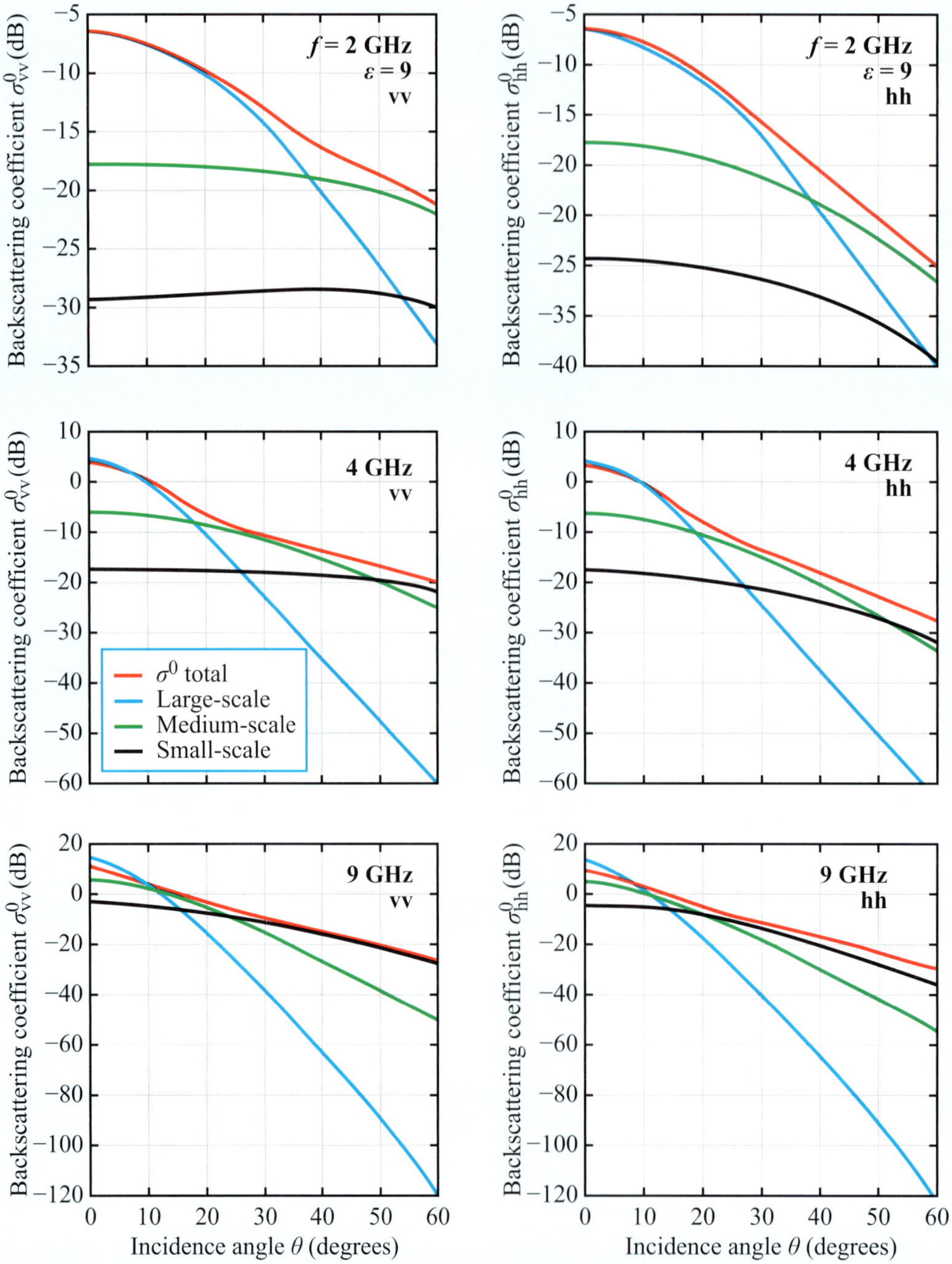

Figure 10-9: In backscattering, the effective wavelength is $\lambda_e = \lambda/(2\sin\theta)$. As λ_e is shortened due to an increase in either frequency or angle, the smaller roughness scales become more effective in scattering. The surface has three roughness scales: large-scale with $(s_1, l_1) = (0.4\text{ cm},\ 7\text{ cm})$, medium-scale with $(s_2, l_2) = (0.25\text{ cm},\ 3\text{ cm})$, and small-scale with $(s_3, l_3) = (0.13\text{ cm},\ 1.5\text{ cm})$. The plots show the individual contributions of the three roughness scales, as well as the total.

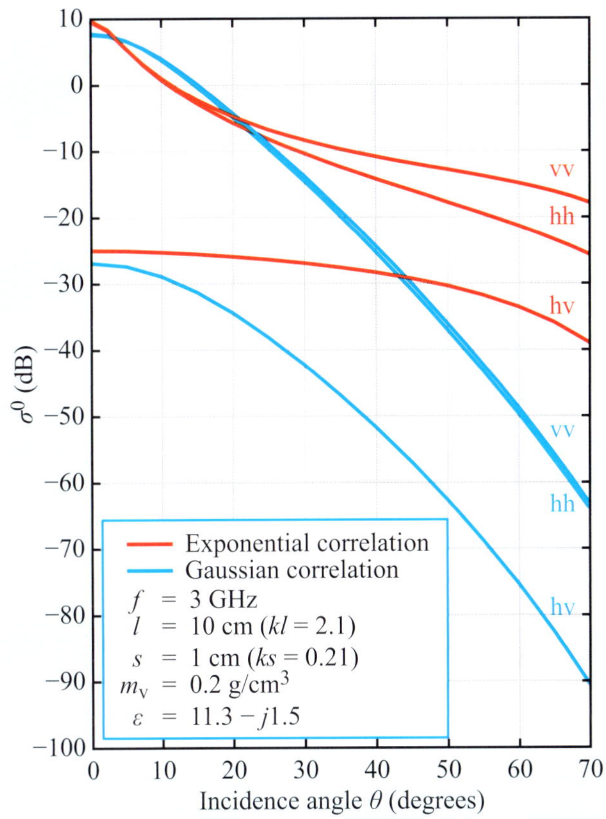

Figure 10-10: I²EM comparison of the backscattering coefficients for two surfaces with identical surface parameters, but different correlation functions.

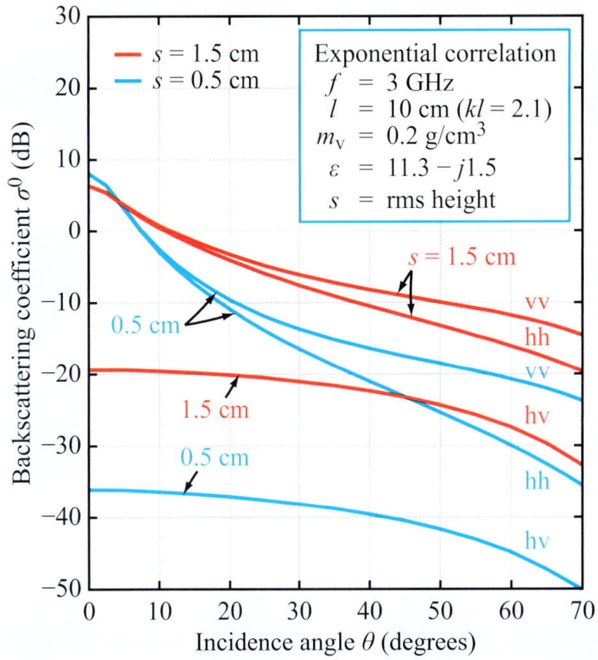

Figure 10-11: I²EM backscatter response to two surfaces with different rms heights.

backscattering coefficients increase with increasing ks (but at different rates) up to about $ks \approx 2$, beyond which their responses approach constant levels (as discussed later in Section 10-3.6).

10-3.5 Role of Correlation Length l

The correlation length l exercises a weaker influence on σ^0 compared with the rms height s; increasing l from 2 cm to 16 cm in Fig. 10-12 causes σ^0 to drop in level by about 5–6 dB for vv and hv polarizations, and by about the same amount for hh polarization (not included in Fig. 10-12 to avoid unnecessary crowding of curves).

10-3.6 Role of Dielectric Constant

At 3 GHz, the real part of the dielectric constant of loamy soil increases with volumetric moisture content from about 3 at $m_v \approx 0$ to 20.6 at $m_v = 0.35$ g/cm^3. Figure 10-13(a) displays multipolarization angular plots of σ^0 and Fig. 10-13(b) displays plots of σ^0 as a function of m_v at $\theta = 30°$. Over the full range of m_v, σ^0_{vv} increases by 8 dB, σ^0_{hh} increases 6 dB, and σ^0_{hv} increases 13 dB. Comparable rates are observed at other angles of incidence.

10-3.7 Role of Polarization Ratios

The ***co-polarized*** and ***cross-polarized*** ratios are defined as

$$p = \frac{\sigma^0_{hh}}{\sigma^0_{vv}}, \quad \text{or } p(\text{dB}) = \sigma^0_{hh}(\text{dB}) - \sigma^0_{vv}(\text{dB}), \tag{10.16a}$$

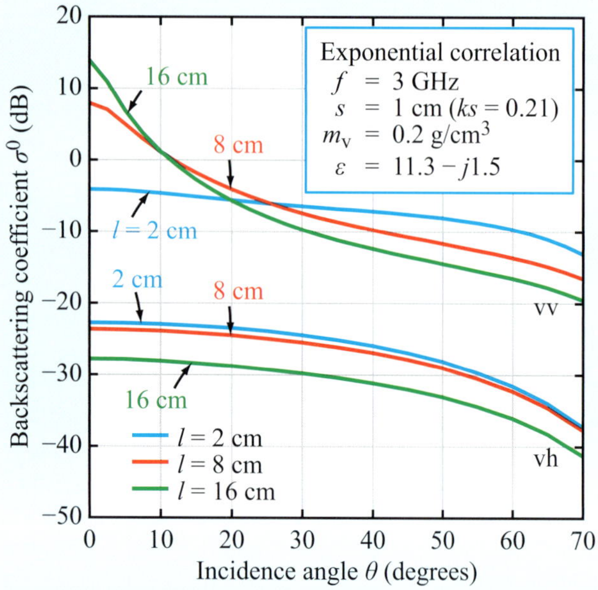

Figure 10-12: The role of the correlation length l is demonstrated by three surfaces with $l = 2$ cm, 8 cm, and 16 cm, all computed using I^{2}EM.

and

$$q = \frac{\sigma^0_{\mathrm{hv}}}{\sigma^0_{\mathrm{vv}}}, \quad \text{or} \quad q(\mathrm{dB}) = \sigma^0_{\mathrm{hv}}(\mathrm{dB}) - \sigma^0_{\mathrm{vv}}(\mathrm{dB}). \tag{10.16b}$$

The plots shown in Fig. 10-14 were computed using I^{2}EM at $\theta = 30°$ for a surface with an rms height $s = 1$ cm and correlation length $l = 10$ cm. The backscattering coefficients were calculated as a function of the electromagnetic roughness ks by incrementally increasing the frequency from 1 GHz to 15 GHz. We note the following general trends:

(a) The co-pol ratio p(dB) increases by about 3 dB as ks is increased from 0.25 to 2. As ks is increased beyond 2, p(dB) approaches 0 dB, or $\sigma^0_{\mathrm{hh}} = \sigma^0_{\mathrm{vv}}$.

(b) The co-pol ratio has a higher level for the lower moisture content of 0.05 g/cm^3 than for the higher moisture content of 0.35 g/cm^3.

(c) The cross-pol ratio q(dB) increases by about 18 dB as ks is increased from 0.25 to 3.

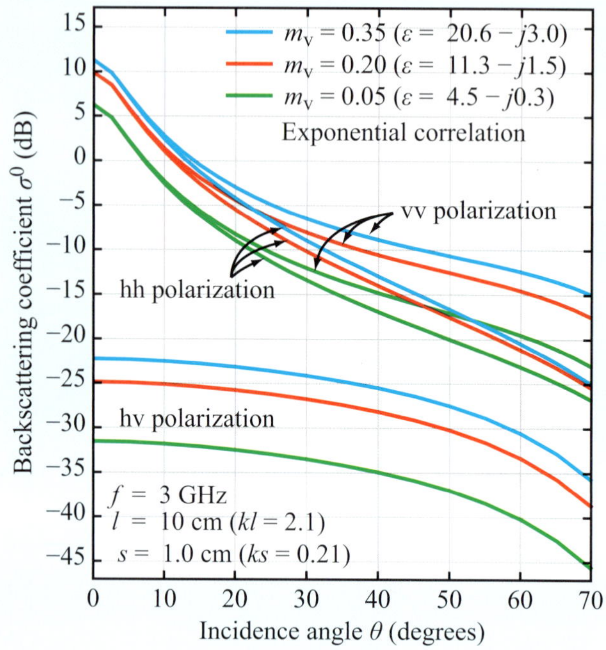

(a) $\sigma^0(\theta)$ at three values of m_v

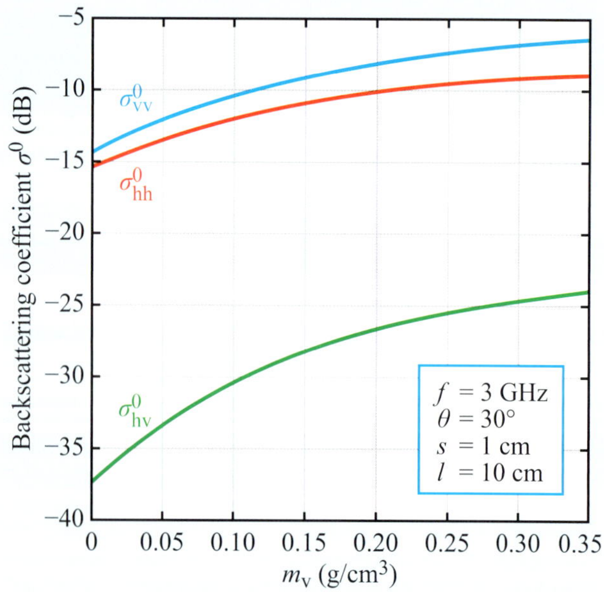

(b) σ^0 versus m_v

Figure 10-13: The role of soil moisture: (a) angular response of σ^0 for three surfaces with different soil moisture content, (b) σ^0 versus m_v at 30°.

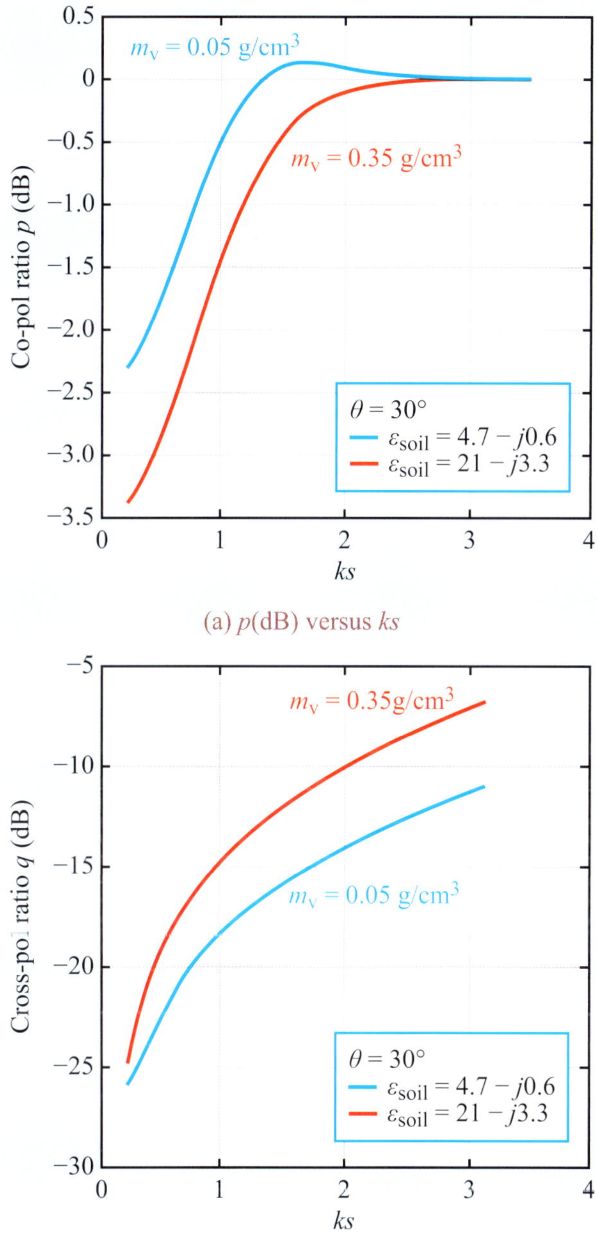

(a) p(dB) versus ks

(b) q(dB) versus ks

Figure 10-14: I^{2}EM-computed plots of (a) the co-pol ratio p(dB) and (b) the cross-pol ratio q(dB) as a function of ks.

(d) For q, the role of soil moisture is opposite that for p; q increases with increasing soil moisture, whereas p decreases.

The foregoing general trends of I^{2}EM are all confirmed by the experimental observations presented in Section 10-5.

10-3.8 Comparison with Experimental Backscattering Measurements

By way of illustration, we show in Fig. 10-15 angular plots for σ^0_{vv} and σ^0_{hh} calculated using I^{2}EM, together with experimental measurements reported by Oh et al. (1992) for a relatively smooth random surface with an rms height $s = 0.4$ cm. Good overall agreement is observed between the model-calculated curves and the experimental measurements, at both 1.5 GHz and 4.75 GHz.

A similar set of data is shown in Fig. 10-16 for a rougher surface with $s = 1.12$ cm. Again, the model and the experimental data are in close agreement.

10-3.9 Comparison with Experimental Bistatic Measurements

So far in this section, we examined the behavior of I^{2}EM for the backscattering case only. The model, however, is equally applicable to bistatic scattering, which is critical for computing the emissivity of random surfaces. Recall from Section 6-7.3 that the emissivity of a surface given by Eq. (6.105) is proportional to an integration of the bistatic scattering coefficient over the upper hemisphere. For an isotropic random surface, bistatic scattering involves three angles, the incidence angle θ_i, the scattering angle θ_s, and the azimuth angle $\phi = \phi_s - \phi_i$.

In Fig. 10-17 we display plots of σ^0_{vv} and σ^0_{hh} as a function of θ_i in the ***incidence plane***, defined as $\theta_s = \theta_i$ and $\phi = 0$. The plots correspond to three surfaces with different roughnesses, all measured at 10 GHz by De Roo and Ulaby (1994). Not only does I^{2}EM[†] provide computed values that are in close agreement with the experimental measurements, but it

[†] **Computer Code 10.3.**

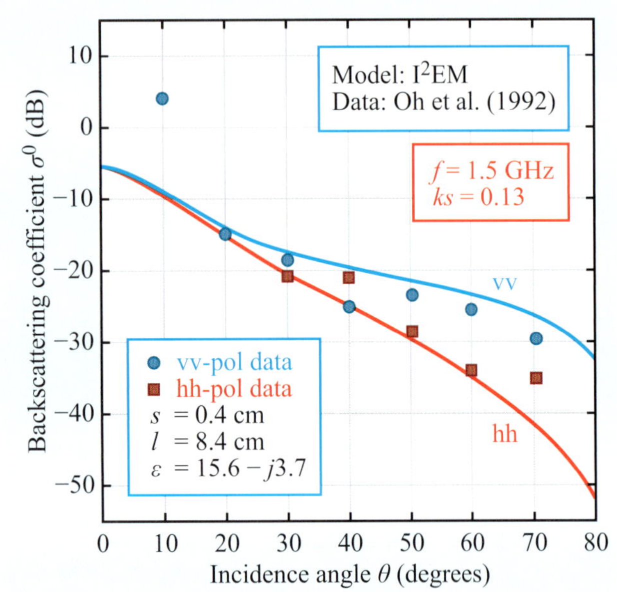

(a) Smooth surface at 1.5 GHz

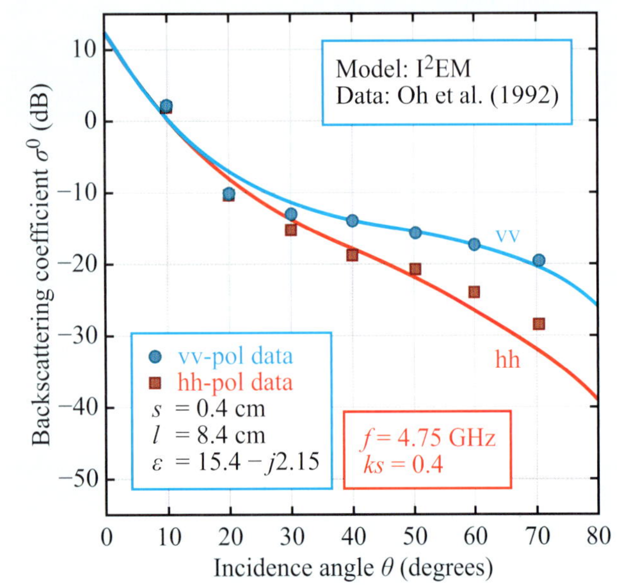

(b) Smooth surface at 4.75 Ghz

Figure 10-15: Comparison of I²EM-calculated values of σ^0 with experimental data measured for a smooth surface with $s = 0.4$ cm.

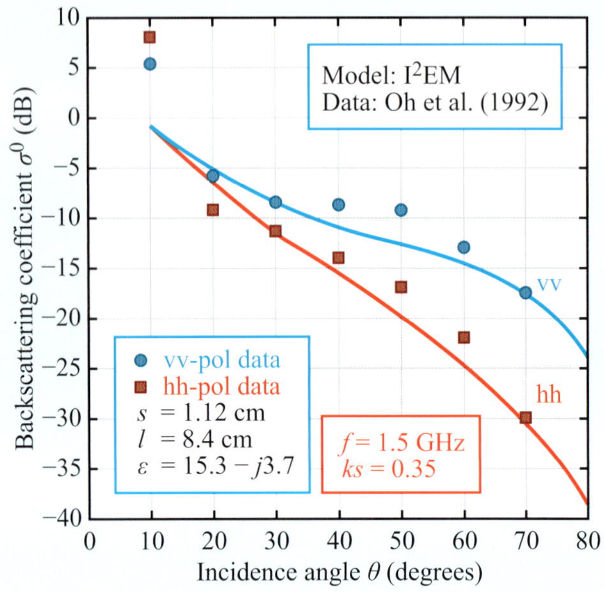

(a) Rough surface at 1.5 Ghz

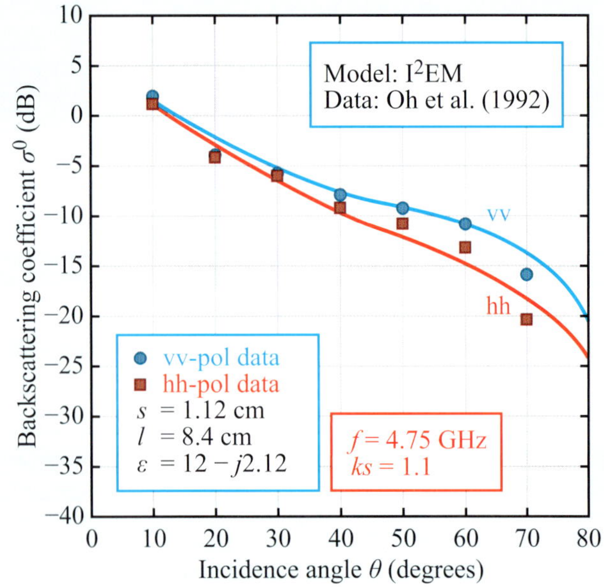

(b) Rough surface at 4.75 GHz

Figure 10-16: Comparison of I²EM-calculated values of σ^0 with experimental data measured for a rough surface with $s = 1.12$ cm.

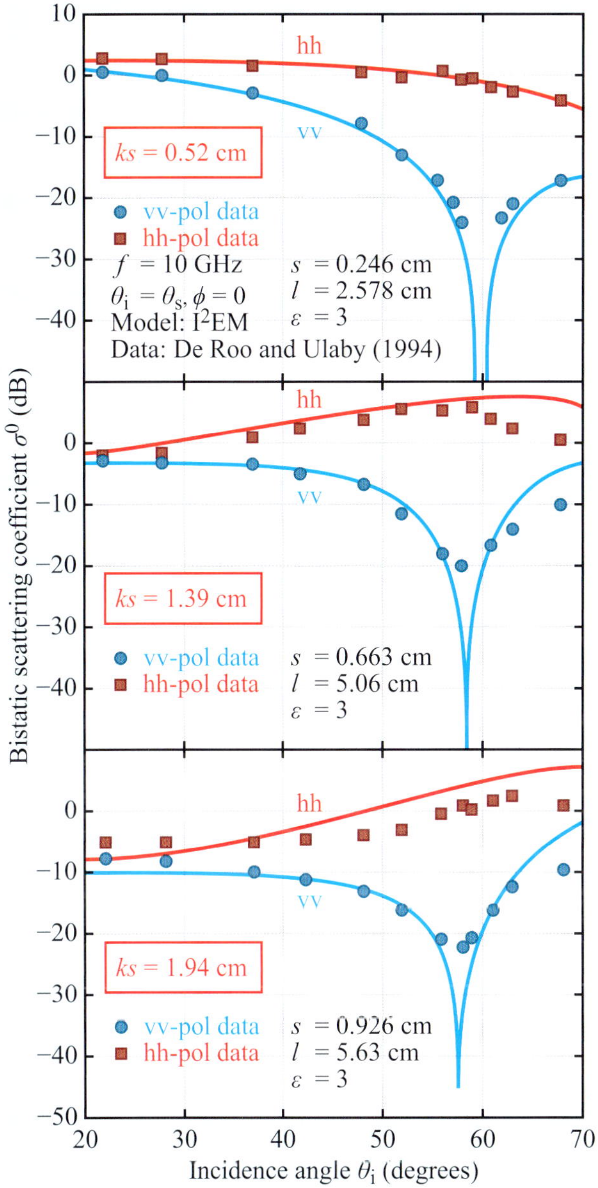

Figure 10-17: Comparison of I^{2}EM-computed bistatic scattering coefficient with measurements made in the incidence plane ($\theta_i = \theta_s$ and $\phi = 0$) for three surfaces with different roughnesses.

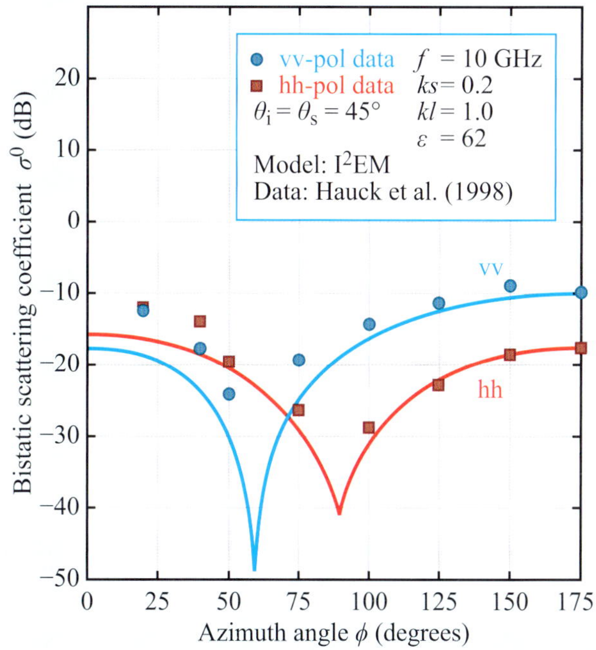

Figure 10-18: Comparison of I^{2}EM-computed bistatic scattering coefficient with measurements made as a function of the azimuth angle ϕ for $\theta_i = \theta_s = 45°$.

also correctly predicts the shift in the location of the Brewster angle from about 60° for the smoothest surface with $s = 0.246$ cm (or $ks = 0.515$) to about 57° for the roughest surface with $s = 0.926$ cm (or $ks = 1.94$).

The variations of σ^0_{hh} and σ^0_{vv} with the azimuth angle ϕ are illustrated in Fig. 10-18 for a relatively smooth random surface with $s = 0.2$ cm observed at 4.775 GHz (Hauck et al., 1998). The incidence and scattering angles are held constant at $\theta_i = \theta_s = 45°$. I^{2}EM correctly predicts the levels of σ^0_{vv} and σ^0_{hh} and their variations with ϕ.

Our last example, presented in Fig. 10-19, pertains to the variations of σ^0_{hh} and σ^0_{vv} as a function of the scattering angle θ_s, while holding the other two angles constant at $\theta_i = 20°$ and $\phi = 0$. The measured data, which were reported by the European Microwave Signature Laboratory of the Joint Research Center in Italy on its website (wmw-emsl.jrc.it), were made for an artificially generated random surface with a Gaussian

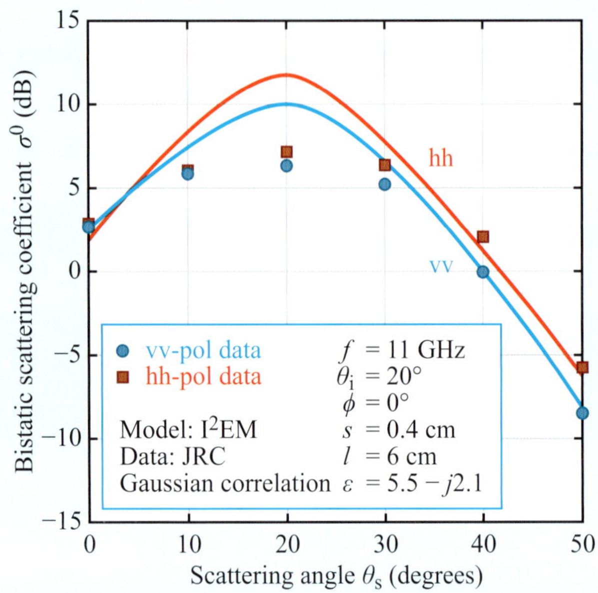

Figure 10-19: Comparison of I²EM-computed bistatic scattering coefficients with measurements as a function of the scattering angle θ_s.

correlation function, $s = 0.4$ cm and $l = 6$ cm. The surface was measured at 11 GHz and 13 GHz; only the 11 GHz data are shown in Fig. 10-19, as the 13 GHz data are almost identical, which is not surprising.

10-3.10 Applicability of Surface Scattering Models

Nature is far too complex for us to be able to describe it in terms of relatively simple models and mathematical functions. A natural random surface consists of multiple scales of roughness, so any description of its statistics in terms of a simple correlation function is certainly only an approximation, and even "direct" measurements of its rms height and correlation length are only estimates of its true statistical properties. The applicability of a model, be it theoretical or empirical, to a real surface is, at best, as good as the accuracies with which we know its statistics.

Models help us understand the observations made by our microwave sensors, and they guide our approach in the development of inversion techniques to estimate physical parameters—such as soil moisture content and surface roughness—from microwave observations.

10-4 Scattering by Random and Periodic Surfaces

As discussed in Section 5-12.5, for a monostatic radar, the coherent scattering component is important only over a narrow angular range of about 1° surrounding nadir incidence, and then only if the surface is relatively smooth ($ks \lesssim 0.5$). In this section, we examine the roles of surface roughness, and dielectric constant for periodic and nonperiodic surfaces.

10-4.1 Backscattering by Nonperiodic Random Surfaces

Surface roughness

We start by illustrating the angular variation of σ^0 for a statistically uniform surface of a dry asphalt parking lot. The measurements shown in Fig. 10-20 were conducted at 8.6 GHz, 17 GHz, and 35.6 GHz for hh, vv, and hv polarizations. The estimated value of the surface rms height is $s < 0.5$ cm. We note that:

(a) At all frequencies, the hh- and vv-polarized responses are essentially identical between normal incidence and $\theta \approx 10°$ and then they diverge as a function of θ, with σ^0_{vv} maintaining a higher level than σ^0_{hh}.

(b) The cross-polarized backscattering coefficient σ^0_{hv} exhibits a cosine-like angular dependence.

(c) Increasing the frequency from 8.6 GHz to 35.6 GHz is equivalent to increasing the electromagnetic roughness by a factor of 4. Increasing the roughness causes the angular plots of σ^0_{hh} and σ^0_{vv} to come closer together. Also, the level difference between σ^0_{hh} and σ^0_{hv} becomes narrower (more depolarization).

A similar set of measurements, conducted at the same frequencies and over the same angular range, are shown in Fig. 10-21 for a much rougher surface with an estimated rms height of 1.5 cm. The corresponding value of ks at 8.6 GHz is 2.7, which classifies the surface

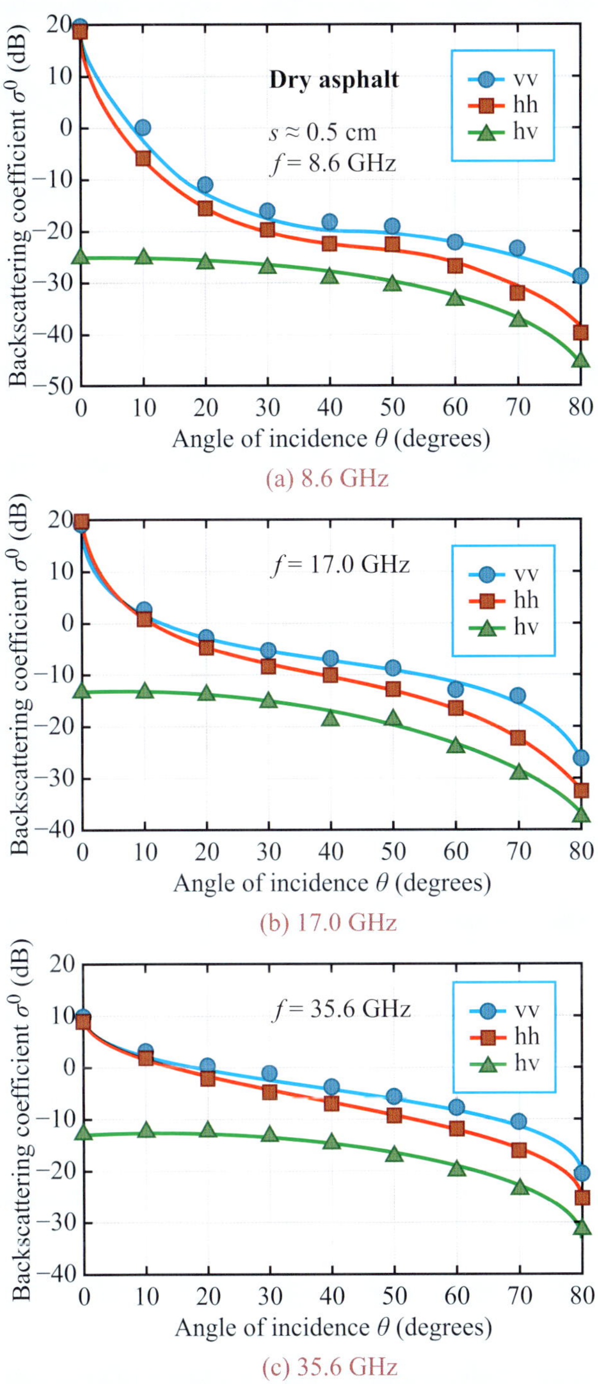

Figure 10-20: Measured backscattering coefficient of a dry asphalt surface at three microwave frequencies. The estimated value of the rms height is $s = 0.5$ cm.

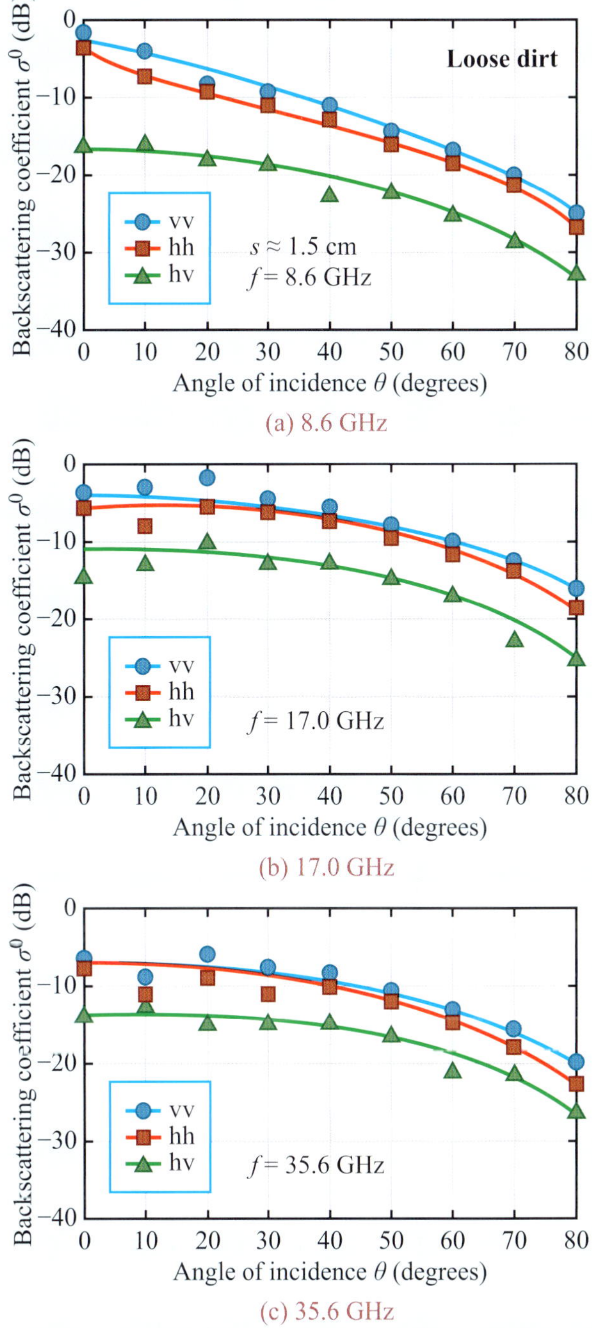

Figure 10-21: Measured backscattering coefficient of loose dirt with $s \approx 1.5$ cm at three microwave frequencies.

as electromagnetically very rough, and even more so at the higher frequencies. Consequently, (1) there is no evidence of the existence of a significant coherent component, (2) $\sigma^0_{\rm hh} \approx \sigma^0_{\rm vv}$ across the entire angular range, and (3) $\sigma^0_{\rm hv}$ is on the order of 5–10 dB below the level of the co-polarized backscattering coefficients.

Moisture content

Since (a) σ^0 is, to first order, proportional to the Fresnel reflectivity Γ, (b) Γ is dependent on the dielectric constant of a soil surface $\varepsilon_{\rm soil}$, and (c) $\varepsilon_{\rm soil}$ is dependent on the volumetric moisture content $m_{\rm v}$, it is not surprising that σ^0 also is soil moisture dependent. The data shown in Fig. 10-22 indicate that for a relatively smooth surface ($ks = 0.35$), σ^0 (in dB) increases approximately linearly with $m_{\rm v}$ at a rate of 0.24 dB/0.01 g/cm^3. In contrast, for the rough surface with $ks = 1.3$, the slope is 0.17 dB/0.01 g/cm^3. The data shown in Fig. 10-22 represent the early stage of a long investigation to develop a viable algorithm for estimating soil moisture content using an imaging radar system. The investigation culminated in the development of the inversion algorithm described in Section 10-5.

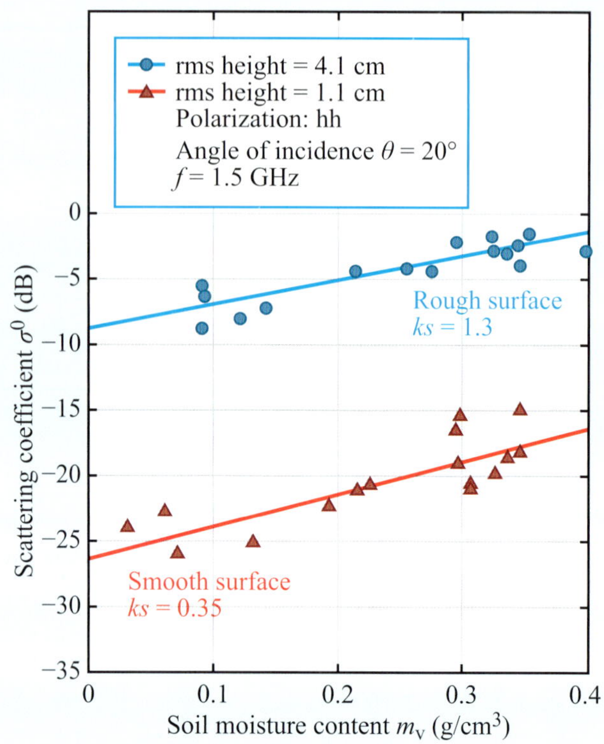

Figure 10-22: Measured 1.5 GHz response of $\sigma^0_{\rm hh}$ to $m_{\rm v}$ for a smooth surface with $ks = 0.35$ and a rougher surface with $ks = 1.3$ [from Ulaby et al., 1978].

10-4.2 Backscattering by Periodic Surfaces

Several types of agricultural crops are commonly planted in parallel rows, either in a rectangular format or in concentric rings (e.g., some types of center-pivot irrigated fields). In this subsection we examine the dependence of σ^0 on ***look direction***, where look direction is defined in terms of the azimuth angle ϕ_0 shown in Fig. 10-23, defined as the angle between the ground projection of the antenna beam axis and the x axis. The periodic variation is along the y direction, so $\phi_0 = 0$ is called the ***parallel look direction*** and $\phi_0 = 90°$ is called the ***perpendicular look direction***.

▶ One approach to calculating the backscattering coefficient of a composite surface is to assume that the scattering is caused exclusively by the random part of the surface and that the periodic component acts as a modulator of the local mean slope of the random surface component. ◀

A description of such a two-scale scattering model is given in Ulaby et al. (1982c). Another approach, which treats the random and periodic components together, was reported by Eom and Fung (1984).

Geometrical considerations

The sketch in Fig. 10-24(a) depicts a narrow-beam antenna pointed at a soil surface located in the x-y

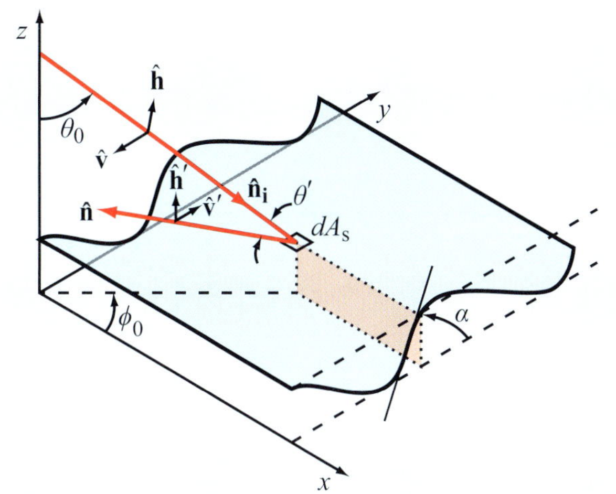

Figure 10-23: Parallel-look direction corresponds to $\phi_0 = 0$, and perpendicular-look direction corresponds to $\phi_0 = 90°$.

plane. The soil surface consists of ***random height variations*** denoted by $z(x,y)$ superimposed on a much larger, ***deterministic, one-dimensional periodic height variation*** denoted by

$$\mathcal{Z}(x,y) = \mathcal{Z}(y) = \mathcal{Z}(y+m\Lambda),$$

where m is any integer and Λ is the spatial period (row spacing) of the periodic pattern in the y direction. For a point (x,y) within the illuminated area in Fig. 10-24(a), the angle of incidence $\theta(x,y)$ is the angle between $-\hat{\mathbf{z}}$ and $\hat{\mathbf{n}}_i$, where $\hat{\mathbf{n}}_i$ is the unit vector in the propagation direction from the antenna to point (x,y).

Imaging radar

For an imaging radar, the resolution cell subtends an angular beamwidth so small that the angle of incidence θ is essentially constant for all points within the cell. This situation is depicted in Fig. 10-24(b). Furthermore, it is assumed that the dimensions of the cell are much larger than the spatial period Λ, so that many spatial cycles are included within the cell. (For simplicity of illustration, the resolution cell in Fig. 10-24(b) includes only part of one period). Recalling that the backscattering coefficient σ^0 of a resolution cell is equal to the radar cross section of the cell, σ, normalized by the physical cross-sectional area of the cell, A, the above assumption that the resolution cell contains many spatial periods leads to the conclusion that σ^0 may be determined by performing the calculation for a single rectangular strip within A whose dimensions are Λ in the y direction and some arbitrary length L_s in the x direction, with $L_s \gg l$,

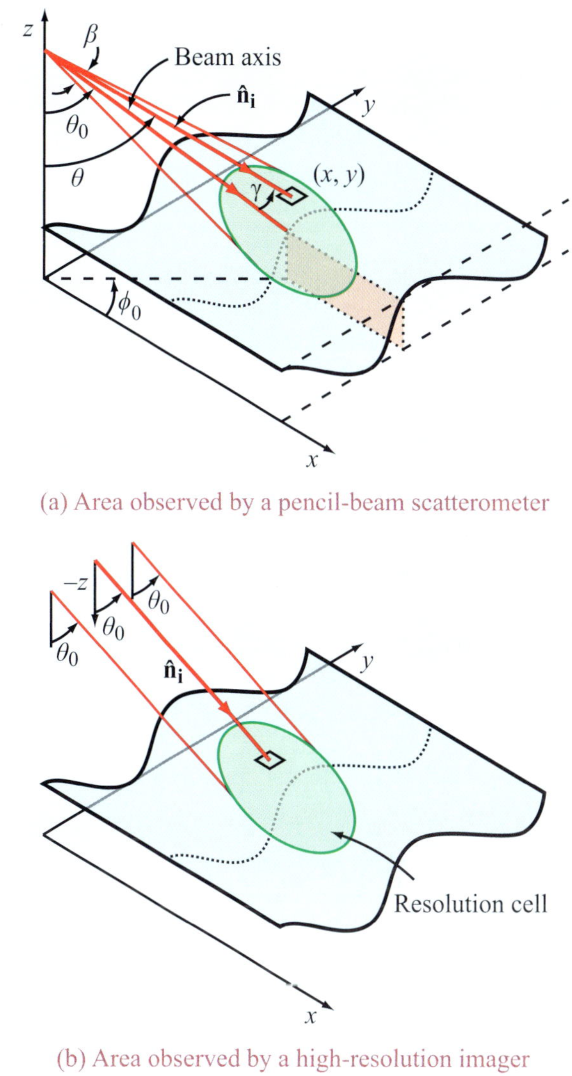

Figure 10-24: Area as observed by (a) a pencil-beam scatterometer and (b) a high-resolution imager.

where l is the correlation length of the random surface $z(x,y)$.

For a differential area $dx\,dy$ with center at (x,y) in the x-y plane, the corresponding area of the soil surface is given by

$$dA_s(x,y) = \left[\frac{dx}{\cos\alpha(y)}\right] dy, \tag{10.17}$$

where α (Fig. 10-23) is the angle whose tangent is equal to the slope of the periodic surface $\mathcal{Z}(y)$:

$$\tan\alpha(y) = \frac{d\mathcal{Z}(y)}{dy}. \tag{10.18}$$

For a linearly polarized antenna with a receive-transmit configuration denoted by pq, the ***observed backscattering coefficient*** $\sigma^0_{pq}(\theta_0,\phi_0)$ may be expressed as the incoherent integration over the rectangular strip

$$\sigma^0_{pq}(\theta_0,\phi_0) = \frac{\displaystyle\int_{y=0}^{\Lambda}\int_{x=0}^{L_s} \sigma^{ss}_{pq}(\theta')\,dA_s(x,y)}{\displaystyle\int_{y=0}^{\Lambda}\int_{x=0}^{L_s} dx\,dy}, \tag{10.19}$$

where $\sigma^{ss}_{pq}(\theta')$ is the backscattering coefficient of the differential area $dA_s(x,y)$, with the same polarization configuration as that of the radar antenna, and θ' is the local angle of incidence (Fig. 10-23). The superscript ss denotes the ***smaller-scale*** random roughness. The scattering by the differential area $dA_s(x,y)$ is governed by its surface characteristics and by the direction of the incident unit vector $\hat{\mathbf{n}}_i$ [Fig. 10-24(a)] relative to the local vector normal to the periodic surface $\hat{\mathbf{n}}$.

Upon inserting Eq. (10.17) into Eq. (10.19) and canceling out the dependence on x, Eq. (10.19) becomes

$$\sigma^0_{pq}(\theta_0,\phi_0) = \frac{1}{\Lambda}\int_0^{\Lambda} \sigma^{ss}_{pq}(\theta')\sec\alpha(y)\,dy. \tag{10.20}$$

Conical-beam scatterometer

For a circularly symmetric antenna beam with an antenna radiation pattern $F(\gamma)$, where γ is the angle away from the beam axis in Fig. 10-24(a), the variation in antenna gain across the illuminated cell may be accounted for by modifying Eq. (10.19) to

$$\sigma^0_{pq}(\theta_0,\phi_0) = \frac{\displaystyle\iint_{\text{main beam}} F^2(\gamma)\,\sigma^{ss}_{pq}(\theta')\sec\alpha(y)\,dx\,dy}{\displaystyle\iint_{\text{main beam}} F^2(\gamma)\,dx\,dy}. \tag{10.21}$$

Similar expressions may be derived for other antenna-beam configurations.

Sinusoidal periodic surface

To compute the integral given by Eq. (10.20), we need an expression for $\sigma^{ss}_{pq}(\theta')$ and the means to relate θ' and α to y. From the geometry in Fig. 10-23,

$$\theta' = \cos^{-1}(-\hat{\mathbf{n}}_i.\hat{\mathbf{n}}), \tag{10.22}$$

where unit vectors $\hat{\mathbf{n}}_i$ and $\hat{\mathbf{n}}$ are related to the incident direction (θ_0,ϕ_0) and the profile of the periodic surface as follows:

$$\hat{\mathbf{n}}_i = \hat{\mathbf{x}}\sin\theta_0\cos\phi_0 + \hat{\mathbf{y}}\sin\theta_0\sin\phi_0 - \hat{\mathbf{z}}\cos\theta_0, \tag{10.23}$$

$$\hat{\mathbf{n}} = (-\hat{\mathbf{x}}Z_x - \hat{\mathbf{y}}Z_y + \hat{\mathbf{z}})D_0, \tag{10.24}$$

where

$$D_0 = (1 + Z_x^2 + Z_y^2)^{-1/2},$$

$$Z_x = \frac{d\mathcal{Z}}{dx}, \qquad Z_y = \frac{d\mathcal{Z}}{dy}. \tag{10.25}$$

For a one-dimensional sinusoidal periodic surface that varies with y only, the periodic component is given by

$$\mathcal{Z}(y) = A\sin\left(\frac{2\pi y}{\Lambda}\right), \tag{10.26a}$$

and its slopes along x and y are

$$Z_x = \frac{d\mathcal{Z}}{dx} = 0, \tag{10.26b}$$

$$Z_y = \frac{d\mathcal{Z}}{dx} = \left(\frac{2\pi A}{\Lambda}\right)\cos\left(\frac{2\pi y}{\Lambda}\right), \tag{10.26c}$$

and

$$\alpha(y) = \tan^{-1}(Z_y). \tag{10.26d}$$

σ^0:	Averaged over one or more spatial periods, as would be observed by a radar
σ^{s}:	Small-scale roughness without periodic surface
σ^{ss}:	Small-scale roughness, modified by periodic surface

Small-scale random roughness $\sigma_{pq}^{\mathrm{ss}}(\theta')$

Because of the periodic surface, the local coordinate system (x', y', z') of a differential area dA_{s} is, in general, different from the reference coordinate system (x, y, z). Consequently, the directions of the polarization vectors $\hat{\mathbf{h}}$ and $\hat{\mathbf{v}}$ in the reference (radar) coordinate system are not identical to $\hat{\mathbf{h}}'$ and $\hat{\mathbf{v}}'$ in the local coordinate system. To relate the two systems, we use the following relationships (from Ulaby et al., 1982a):

$$\hat{\mathbf{z}}' = \hat{\mathbf{n}}, \tag{10.27a}$$

$$\begin{aligned}\hat{\mathbf{y}}' = \frac{\hat{\mathbf{n}} \times \hat{\mathbf{n}}_{\mathrm{i}}}{|\hat{\mathbf{n}} \times \hat{\mathbf{n}}_{\mathrm{i}}|} &= D_1[\hat{\mathbf{x}}(Z_y \cos\theta_0 - \sin\theta_0 \sin\phi_0) \\ &+ \hat{\mathbf{y}}(\sin\theta_0 \cos\phi_0 - Z_x \cos\theta_0) \\ &+ \hat{\mathbf{z}} \sin\theta_0 (Z_y \cos\phi_0 - Z_x \sin\phi_0)]\end{aligned} \tag{10.27b}$$

$$\begin{aligned}\hat{\mathbf{x}}' &= \hat{\mathbf{y}}' \times \hat{\mathbf{z}}' \\ &= \{\hat{\mathbf{x}}[\sin\theta_0 \cos\phi_0 - Z_x \cos\theta_0 \\ &\quad + \sin\theta_0 (Z_y \cos\phi_0 - Z_x \sin\phi_0) Z_y] \\ &\quad + \hat{\mathbf{y}}[\sin\theta_0 (Z_x \sin\phi_0 - Z_y \cos\phi_0) Z_x \\ &\quad + \sin\theta_0 \sin\phi_0 - Z_y \cos 0_0] \\ &\quad + \hat{\mathbf{z}}[(\sin\theta_0 \sin\phi_0 - Z_y \cos\theta_0) Z_y \\ &\quad + (\sin\theta_0 \cos\phi_0 - Z_x \cos\theta_0) Z_x]\} D_0 D_1,\end{aligned} \tag{10.27c}$$

where

$$\begin{aligned}D_1 = \{&\sin^2\theta_0 + Z_x^2(\cos^2\theta_0 + \sin^2\theta_0 \sin^2\phi_0) \\ &+ Z_y^2(\cos^2\theta_0 + \sin^2\theta_0 \cos^2\phi_0) \\ &- 2\sin\theta_0 \cos\theta_0 (Z_x \cos\phi_0 \\ &+ Z_y \sin\phi_0 + Z_x Z_y \tan\theta_0 \sin\phi_0 \cos\phi_0)\}^{-1/2}.\end{aligned} \tag{10.27d}$$

Note that the expressions given by Eq. (10.27) are for the general case where $Z_x \neq 0$. For a one-dimensional periodic surface with periodicity along the y direction, $Z_x = 0$.

The horizontal and vertical polarization vectors in the local and the reference frames are defined as follows:

$$\hat{\mathbf{h}}' = \hat{\mathbf{y}}', \tag{10.28a}$$

$$\hat{\mathbf{v}}' = -\hat{\mathbf{x}}' \cos\theta' - \hat{\mathbf{z}}' \cos\theta' \tag{10.28b}$$

$$\hat{\mathbf{h}} = -\hat{\mathbf{x}} \sin\phi_0 + \hat{\mathbf{y}} \cos\phi_0, \tag{10.28c}$$

$$\hat{\mathbf{v}} = -\hat{\mathbf{x}} \cos\theta_0 \cos\phi_0 - \hat{\mathbf{y}} \cos\theta_0 \sin\phi_0 - \hat{\mathbf{z}} \sin\theta_0. \tag{10.28d}$$

From Eq. (5.30), the backscattering coefficient $\sigma_{pq}^0 = \sigma_{pq}/A$ is given by

$$\sigma_{pq}^0 = \frac{4\pi R_{\mathrm{r}}^2}{A} \frac{\mathrm{S}_p^{\mathrm{s}}}{\mathrm{S}_q^{\mathrm{i}}} = \frac{4\pi R_{\mathrm{r}}^2}{A} \frac{|E_p^{\mathrm{s}}|^2}{|E_q^{\mathrm{i}}|^2} = \frac{4\pi}{A} |S_{pq}|^2, \tag{10.29}$$

where S_{pq} is the scattering amplitude of the target for incidence with polarization q and reception with polarization p. The derivation leading to the scattering coefficient of the random surface, $\sigma_{pq}^{\mathrm{ss}}(\theta')$, encounters products of scattering amplitudes, necessitating the introduction of the generalized scattering coefficient

$$\sigma_{mnpq}^0 = \frac{4\pi}{A} \,\mathfrak{Re}(S_{mn} S_{pq}^*). \tag{10.30}$$

In terms of this new definition and the polarization unit vectors defined by Eq. (10.28), the backscattering coefficients of the random surface are given by

$$\begin{aligned}\sigma_{\mathrm{vv}}^{\mathrm{ss}}(\theta') &= (\hat{\mathbf{v}} \cdot \hat{\mathbf{v}}')^4 \, \sigma_{\mathrm{vv}}^{\mathrm{s}}(\theta') + (\hat{\mathbf{v}} \cdot \hat{\mathbf{h}}')^4 \, \sigma_{\mathrm{hh}}^{\mathrm{s}}(\theta') \\ &\quad + 2(\hat{\mathbf{v}} \cdot \hat{\mathbf{h}}')^2 (\hat{\mathbf{v}} \cdot \hat{\mathbf{v}}')^2 \, \sigma_{\mathrm{vvhh}}^{\mathrm{s}}(\theta'),\end{aligned} \tag{10.31a}$$

$$\begin{aligned}\sigma_{\mathrm{hh}}^{\mathrm{ss}}(\theta') &= (\hat{\mathbf{h}} \cdot \hat{\mathbf{v}}')^4 \, \sigma_{\mathrm{vv}}^{\mathrm{s}}(\theta') + (\hat{\mathbf{h}} \cdot \hat{\mathbf{h}}')^4 \, \sigma_{\mathrm{hh}}^{\mathrm{s}}(\theta') \\ &\quad + 2(\hat{\mathbf{h}} \cdot \hat{\mathbf{h}}')^2 (\hat{\mathbf{h}} \cdot \hat{\mathbf{v}}')^2 \, \sigma_{\mathrm{vvhh}}^{\mathrm{s}}(\theta'),\end{aligned} \tag{10.31b}$$

and

$$\begin{aligned}\sigma_{\mathrm{vh}}^{\mathrm{ss}}(\theta') &= (\hat{\mathbf{v}} \cdot \hat{\mathbf{v}}')^2 (\hat{\mathbf{v}}' \cdot \hat{\mathbf{h}})^2 \, \sigma_{\mathrm{vv}}^{\mathrm{s}}(\theta') \\ &\quad + (\hat{\mathbf{v}} \cdot \hat{\mathbf{h}}')^2 (\hat{\mathbf{h}}' \cdot \hat{\mathbf{h}})^2 \, \sigma_{\mathrm{hh}}^2(\theta') \\ &\quad + 2(\hat{\mathbf{v}} \cdot \hat{\mathbf{v}}')(\hat{\mathbf{v}}' \cdot \hat{\mathbf{h}})(\hat{\mathbf{v}} \cdot \hat{\mathbf{h}}')(\hat{\mathbf{h}}' \cdot \hat{\mathbf{h}}) \, \sigma_{\mathrm{vvhh}}^{\mathrm{s}}(\theta') \\ &\quad + [(\hat{\mathbf{v}} \cdot \hat{\mathbf{v}}')(\hat{\mathbf{h}}' \cdot \hat{\mathbf{h}}) + (\hat{\mathbf{v}} \cdot \hat{\mathbf{h}}')(\hat{\mathbf{v}}' \cdot \hat{\mathbf{h}})]^2 \, \sigma_{\mathrm{hv}}^{\mathrm{s}}(\theta').\end{aligned} \tag{10.31c}$$

Here, σ^{ss}_{vv} is the vv-polarized coefficient of the smaller-scale roughness, defined in the antenna coordinate system, whereas σ^{s}_{vv} is the vv-polarized backscattering coefficient of the same random surface, but without the periodic surface. An examination of Eq. (10.31a) shows that the dominant term is $(\hat{\mathbf{v}} \cdot \hat{\mathbf{v}}')^4\, \sigma^{s}_{vv}(\theta')$. Similarly, the dominant term in Eq. (10.31b) is $(\hat{\mathbf{h}} \cdot \hat{\mathbf{h}}')^4\, \sigma^{s}_{hh}(\theta')$. In Eq. (10.31c), the first three terms tend to cancel and the last term is representative of $\sigma^{ss}_{vh}(\theta')$.

Model behavior

Before we examine the dependence of the backscattering coefficient of a periodic surface on the parameters of that surface, let us review the results obtained so far. Equation (10.20) expresses the backscattering coefficient $\sigma^0_{pq}(\theta_0, \phi_0)$, as observed by an imaging radar illuminating the periodic surface at an angle of incidence θ_0 (measured relative to the mean surface) and an azimuth angle ϕ_0, in terms of a line integral involving $\sigma^{ss}_{pq}(\theta')$, which is the backscattering coefficient of a differential area whose surface normal is at an angle θ' relative to the observation direction. The integral is an incoherent summation of backscattering contributions from all elemental areas observed by the radar over one spatial period Λ.

An incident wave with a linear (horizontal or vertical) polarization orientation (as defined in the antenna reference frame) generally has both $\hat{\mathbf{h}}$ and $\hat{\mathbf{v}}$ components in the local coordinate system of the differential area dA_s. Consequently, the backscattering coefficient $\sigma^{ss}_{hh}(\theta')$ of the area dA_s is not only related to $\sigma^{s}_{hh}(\theta')$, the backscattering coefficient of dA_s in its own coordinate system, but also includes contributions related to the backscattering coefficients $\sigma^{s}_{vv}(\theta')$ and $\sigma^{s}_{vvhh}(\theta')$ of the area.

The backscattering coefficients having the single s superscript characterize the backscattering properties of the random surface in the absence of the periodic surface. To illustrate the effect of the periodic surface and look angle on the hh-backscattering coefficient, we apply the model to two surfaces with identical sinusoidal surfaces, but with very different random components. The angular plot of the backscattering coefficient of the random component $\sigma^{s}_{hh}(\theta')$ is displayed in Fig. 10-25

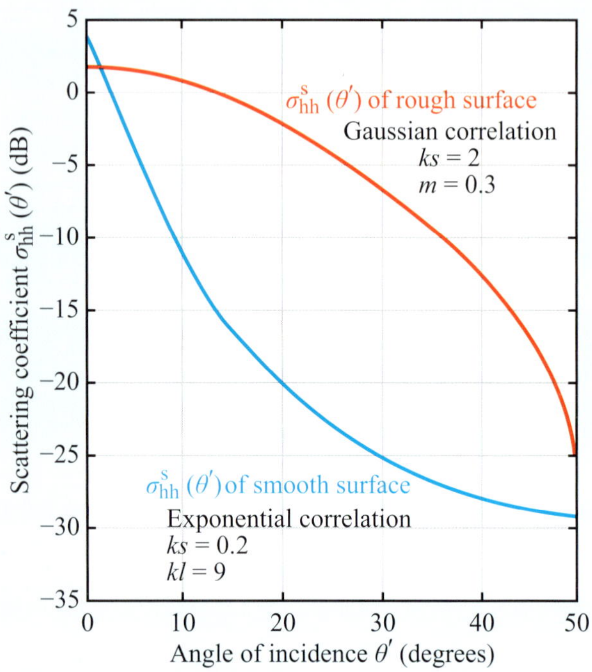

Figure 10-25: Angular plots of $\sigma^{s}_{hh}(\theta')$ for smooth and rough random surfaces in the absence of the periodic surface.

for both surfaces. The smooth random surface, which is characterized by a roughness $ks = 0.2$, exhibits an angular variation of 28 dB between $\theta' = 0$ and $\theta' = 30°$. In contrast, the rough surface with $ks \approx 2$ exhibits a much smaller angular variation of only 10 dB over the same angular range.

As stated earlier in connection with the expression for $\sigma^{ss}_{hh}(\theta')$ given by Eq. (10.31b), the dominant term is the second term. That is,

$$\sigma^{ss}_{hh}(\theta') \approx (\hat{\mathbf{h}} \cdot \hat{\mathbf{h}}')^4\, \sigma^{s}_{hh}(\theta').$$

Computing $\sigma^0_{hh}(\theta_0, \phi_0)$ at incidence angle θ_0 and azimuth look-angle ϕ_0 entails performing the integration in Eq. (10.20). For each value of the integration variable y, (1) Eq. (10.26) provides values for Z_x, Z_y, and α, which then (2) specify $\hat{\mathbf{n}}_i$ and $\hat{\mathbf{n}}$, which in turn specify θ' by Eq. (10.22). Also, use of Eqs. (10.27) and (10.28) specifies $\hat{\mathbf{h}}$ and $\hat{\mathbf{h}}'$.

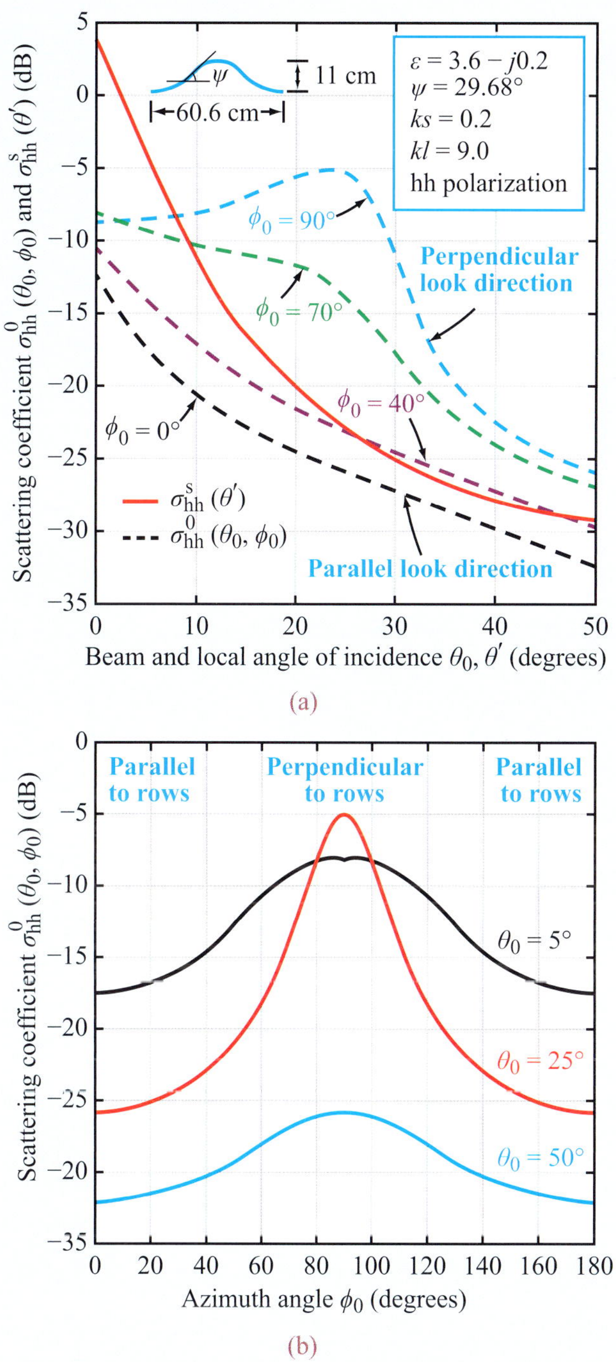

Figure 10-26: Periodic surface with relatively smooth random undulations. (a) Variation of the scattering coefficient $\sigma^0_{\text{hh}}(\theta_0, \phi_0)$ as a function of the beam angle of incidence θ_0 for several values of the azimuth angle ϕ_0, and (b) variation of $\sigma^0_{\text{hh}}(\theta_0, \phi_0)$ with the azimuth angle ϕ_0.

Figure 10-26(a) displays plots for $\sigma^0_{\text{hh}}(\theta_0, \phi_0)$ as a function of θ_0 for specific values of ϕ_0, and Fig. 10-26(b) displays plots as a function of ϕ_0 at specific values of θ_0, all for a smooth random surface with $ks = 0.2$. The periodic surface is sinusoidal with an amplitude $A = 5.5$ cm and a spatial period $\Lambda = 60.6$ cm. The greatest contrast is between $\phi_0 = 0$ (parallel look direction) and $\phi_0 = 90^\circ$ (perpendicular look direction). At $\theta_0 = 25^\circ$, the variation of $\sigma^0_{\text{hh}}(\theta_0, \phi_0)$ with ϕ_0 exceeds 20 dB!

Similar plots are shown in Fig. 10-27 for a surface with the same periodic structure, but with a much rougher random-surface component. Not surprisingly, $\sigma^0_{\text{hh}}(\theta_0, \phi_0)$ is much less sensitive to look direction.

Experimental observations

Variations in σ^0 due to look direction have been observed by scatterometers (Batlivala and Ulaby, 1976; Ulaby and Bare, 1979; Bradley and Ulaby, 1981; Fenner et al., 1981; Ulaby et al., 1982c) and registered on radar imagery (Batlivala and Ulaby, 1976; MacDonald and Waite, 1978; Blanchard and Chang, 1983). An example is shown in Fig. 10-28, which displays plots of σ^0 versus time as measured by airborne scatterometers passing over a rectangular wheat-stubble field with row configurations as illustrated. The presence of wheat stubble above the soil surface has almost no discernible effect on the backscattering behavior of the soil background. Hence, for all practical purposes, at microwave frequencies a wheat-stubble field may be regarded as a bare soil surface. Over the triangular sections at each end of the field, the radar look direction is orthogonal to row direction ($\phi_0 = 90^\circ$), whereas over the middle section of the field, the radar look direction is parallel to row direction ($\phi_0 = 0^\circ$). The plots shown are for observations at $\theta_0 = 20^\circ$ and include like-polarized (hh and vv) plots at 1.6, 4.75, and 13.3 GHz and hv-polarized plots at 1.6 and 4.75 GHz. For the like-polarized plots, the change in the level of σ^0 across the boundary between $\phi_0 = 0^\circ$ and $\phi_0 = 90^\circ$ is largest at 1.6 GHz and smallest at 13.3 GHz. This behavior is predictable, because the random roughness component becomes increasingly rough as the frequency increases. A rougher random surface exhibits a σ^{s} versus θ' curve

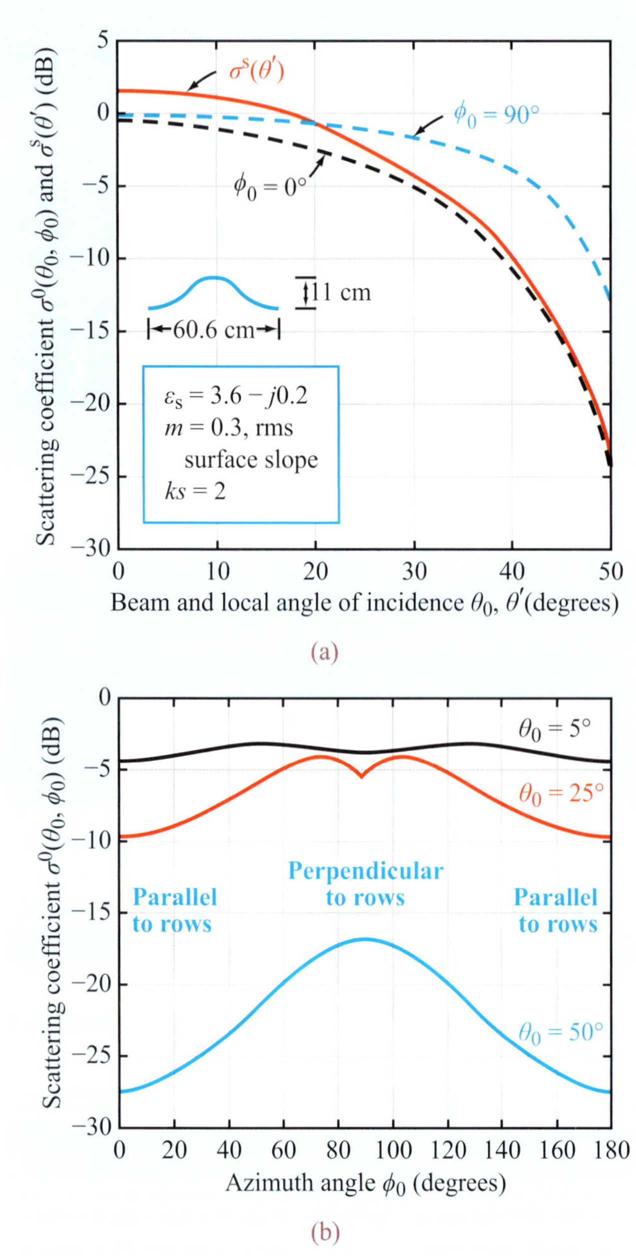

Figure 10-27: Periodic surface with rough random undulations. (a) Variation of the scattering coefficient $\sigma^0_{\text{hh}}(\theta_0, \phi_0)$ as a function of the beam angle of incidence θ_0 for several values of the azimuth angle ϕ_0, and (b) variation of $\sigma^0_{\text{hh}}(\theta_0, \phi_0)$ with the azimuth angle ϕ_0.

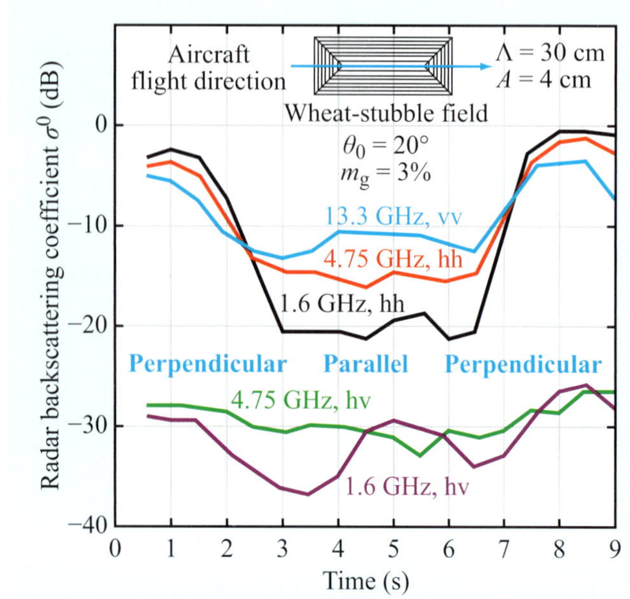

Figure 10-28: Scatterometer time response for a wheat-stubble field [Ulaby et al., 1982c].

with a slower decay rate, which results in reduced sensitivity to look direction.

► In contrast to the like-polarization scattering coefficient, the cross-polarization scattering coefficient shows a much weaker sensitivity to row direction. ◄

Information about the cross-polarization isolation performance of the antennas suggests that part of the observed increase in σ^0_{hv} across the boundaries between the $\phi = 0^\circ$ and $\phi = 90^\circ$ regions is actually due to the coupling of part of the like-polarized return to the cross-polarized receiving antenna. Thus, the actual sensitivity of σ^0_{hv} to row direction may be even weaker than indicated in Fig. 10-28. Based on theoretical considerations, σ^0_{hv} should indeed exhibit a very weak dependence on ϕ_0 because, as noted in connection with Figs. 10-20 and 10-21, σ^s_{hv} is weakly dependent on incidence angle, particularly in the region between $\theta' = 0^\circ$ and 40°.

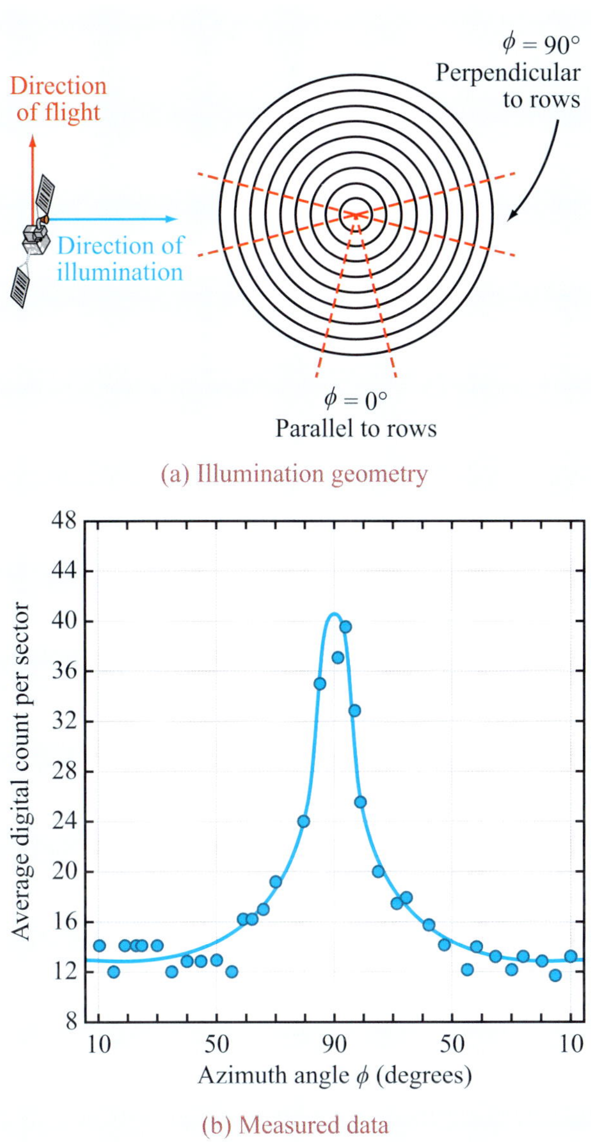

Figure 10-29: Graphic illustration of the difference between Seasat SAR returns as influenced by tillage direction [from Blanchard and Chang, 1983].

Figure 10-29(a) shows the geometry associated with the radar observation of a circularly tilled field. Assuming that the radar platform is moving northward and the radar beam is pointing eastward, the narrow biconical section of the field (whose axis is east-west) contains rows that are approximately orthogonal to the radar look direction ($\phi_0 = 90°$). The plot in Fig. 10-29(b) displays the variation in image intensity, averaged along radial vectors originating at the center of the field, as a function of the azimuth angle ϕ_0 for a field with a circularly tilled surface (Blanchard and Chang, 1983). The data, which were obtained from a digital Seasat-A SAR image of the field, show that the backscattered power (which is proportional to σ^0) is highest for pixels corresponding to $\phi_0 \approx 90°$ and lowest for those corresponding to $\phi_0 \approx 0°$. These results are in good agreement with the theoretical plots shown in Fig. 10-26.

10-5 PRISM (Polarimetric Radar Inversion for Soil Moisture)

Both theoretical models and experimental observations indicate that for a random surface observed by a radar at an incidence angle θ and microwave frequency f:

(a) The magnitudes of the surface backscattering coefficients σ^0_{vv}, σ^0_{hh}, and σ^0_{hv} are governed primarily by the rms height s and moisture content m_{v}, and secondarily by the correlation length l.

(b) At incidence away from nadir, the co-pol ratio $p = \sigma^0_{\text{hh}}/\sigma^0_{\text{vv}}$ of a relatively smooth surface (with $ks \approx 0.1$) is on the order of 0.1 to 0.4 (depending on θ and m_{v}), and increases with ks to 1 as ks increases to beyond 2.

(c) For a perfectly smooth surface, $\sigma^0_{\text{hv}} = 0$. Increasing surface roughness causes the cross-pol ratio $q = \sigma^0_{\text{hv}}/\sigma^0_{\text{vv}}$ to increase monotonically with ks, reaching a plateau as ks exceeds 2.

These considerations suggest that it should be possible to retrieve the values of s and m_{v} of a random surface from the measurement of the three scattering coefficients. To develop such a model, a University of Michigan team conducted field experiments using a mobile, truck-mounted, multifrequency, polarimetric scatterometer, supported by a laser-profiling system to measure the height profiles of the observed surfaces and a microwave dielectric probe to measure the dielectric constant and moisture content of the soil medium (Oh et al., 1992). The study included four surfaces with roughnesses extending from $s = 0.32$ cm to $s = 3.02$ cm, each observed under dry and wet soil conditions. The objective of the study was not only to quantify the radar backscatter response to surface roughness and soil moisture content, but also to develop a semiempirical inversion model to estimate surface roughness and soil moisture content from multipolarized radar observations. The inversion algorithm that emerged from the 1992 study—which we call the ***Polarimetric Radar Inversion for Soil Moisture-1***, or ***PRISM-1*** for short—is discussed in Section 10-5.2. A few years later, additional experiments were conducted using a variety of radar instruments including airborne polarimetric SARs, which led to a slightly modified version of PRISM-1, which we refer to as ***PRISM-2***. Its details are covered in Section 10-5.3.

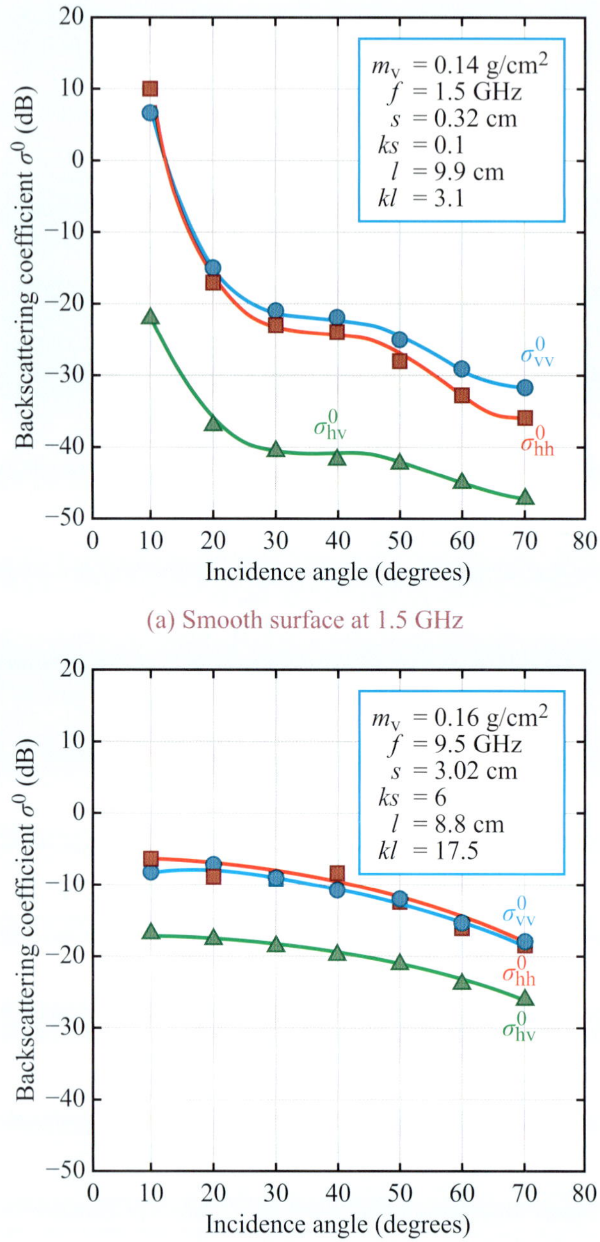

(a) Smooth surface at 1.5 GHz

(b) Very rough surface at 9.5 GHz

Figure 10-30: Measured backscattering coefficient of (a) a smooth surface with $s = 0.32$ cm at 1.5 GHz and (b) a rough surface with $s = 3.02$ cm at 9.5 GHz.

10-5.1 Co-Pol and Cross-Pol Ratios

As a prelude to developing an inversion model, it is important to analyze the behavior of σ^0 as a function of θ, ks, m_v, and f.

The role of roughness is evident by comparing the experimental data displayed in Fig. 10-30(a) for a smooth surface (with $s = 0.32$ cm) at 1.5 GHz ($ks = 0.1$) with those in Fig. 10-30(b) for a rough surface (with $s = 3.02$ cm) at 9.5 GHz ($ks = 6$). The volumetric moisture contents of the two surfaces are essentially the same (0.14 for the smooth surface and 0.16 for the rough surface). The co-polarized backscattering coefficients ($\sigma^0_{hh}(\theta)$ and $\sigma^0_{vv}(\theta)$) of the smooth surface exhibit steep angular responses ($\approx$ 45 dB drop between $\theta = 10^\circ$ and $\theta = 70^\circ$), compared with much gentler responses for the rough surface ($\approx$ 12 dB drop between $\theta = 10^\circ$ and $\theta = 70^\circ$). The over 30 dB difference in the angular slope translates to a factor of over 1000 in m^2/m^2. That is, the angular decay rate, on average, is about 1000 times greater for the smooth surface [Fig. 10-30(a)] than for the rough surface [Fig. 10-30(b)].

Co-polarized ratio

Based on the data in Fig. 10-30 and other data reported by Oh et al. (1992), the ***co-polarized ratio***, defined as

$$p = \frac{\sigma^0_{\mathrm{hh}}}{\sigma^0_{\mathrm{vv}}}, \tag{10.32}$$

is observed to always be smaller than or equal to 1, and it approaches 1 as ks becomes large. In dB,

$$p(\mathrm{dB}) = \sigma^0_{\mathrm{hh}}(\mathrm{dB}) - \sigma^0_{\mathrm{vv}}(\mathrm{dB}), \tag{10.33}$$

which in Fig. 10-30 is the difference between the levels of $\sigma^0_{\mathrm{hh}}(\mathrm{dB})$ and $\sigma^0_{\mathrm{vv}}(\mathrm{dB})$. The measured variation of $p(\mathrm{dB})$ with ks is shown in Fig. 10-31 for dry and wet soil surfaces observed at $\theta = 50^\circ$. We observe that p depends on both ks and ε, but it approaches 0 dB and becomes essentially independent of the dielectric constant ε for $ks \geq 2$.

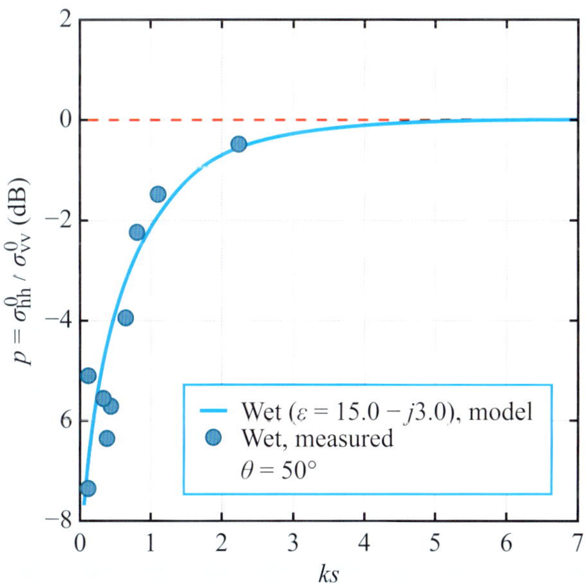

Figure 10-31: Measured and modeled variation of the co-pol ratio p with EM roughness ks.

Oh et al. (1992) developed the following empirical model† for the co-polarized ratio p:

$$p = \frac{\sigma^0_{\mathrm{hh}}}{\sigma^0_{\mathrm{vv}}} = \left[1 - \left(\frac{2\theta}{\pi}\right)^{\alpha} e^{-ks}\right]^2, \tag{10.34}$$

where θ is the incidence angle in radians, $k = 2\pi/\lambda$, s is the rms height, and

$$\alpha = \frac{1}{3\Gamma_0}, \tag{10.35}$$

with Γ_0 representing the surface (Fresnel) reflectivity at normal incidence,

$$\Gamma_0 = \left|\frac{1-\sqrt{\varepsilon}}{1+\sqrt{\varepsilon}}\right|^2. \tag{10.36}$$

As noted in Section 4-8, the dielectric constant ε of the soil surface is strongly dependent on the volumetric moisture content m_{v}.

Cross-polarized ratio

The ***cross-polarized ratio***, defined as

$$q = \frac{\sigma^0_{\mathrm{hv}}}{\sigma^0_{\mathrm{vv}}}, \tag{10.37}$$

exhibits a strong dependence on ks, as shown in Fig. 10-32, and can be modeled empirically as

$$q = 0.23\Gamma_0^{1/2}[1 - e^{-ks}]. \tag{10.38}$$

Using the empirical models developed for p and q, the following models were developed for σ^0_{vv}, σ^0_{hh}, and σ^0_{hv}:

$$\sigma^0_{\mathrm{vv}} = 0.7\left[1 - e^{-0.65(ks)^{1.8}}\right] \frac{\cos^3\theta}{\sqrt{p}} [\Gamma_{\mathrm{v}}(0) + \Gamma_{\mathrm{h}}(\theta)], \tag{10.39a}$$

$$\sigma^0_{\mathrm{hh}} = p\sigma^0_{\mathrm{vv}}, \tag{10.39b}$$

$$\sigma^0_{\mathrm{hv}} = q\sigma^0_{\mathrm{vv}}. \tag{10.39c}$$

† Computer Code 10.5.

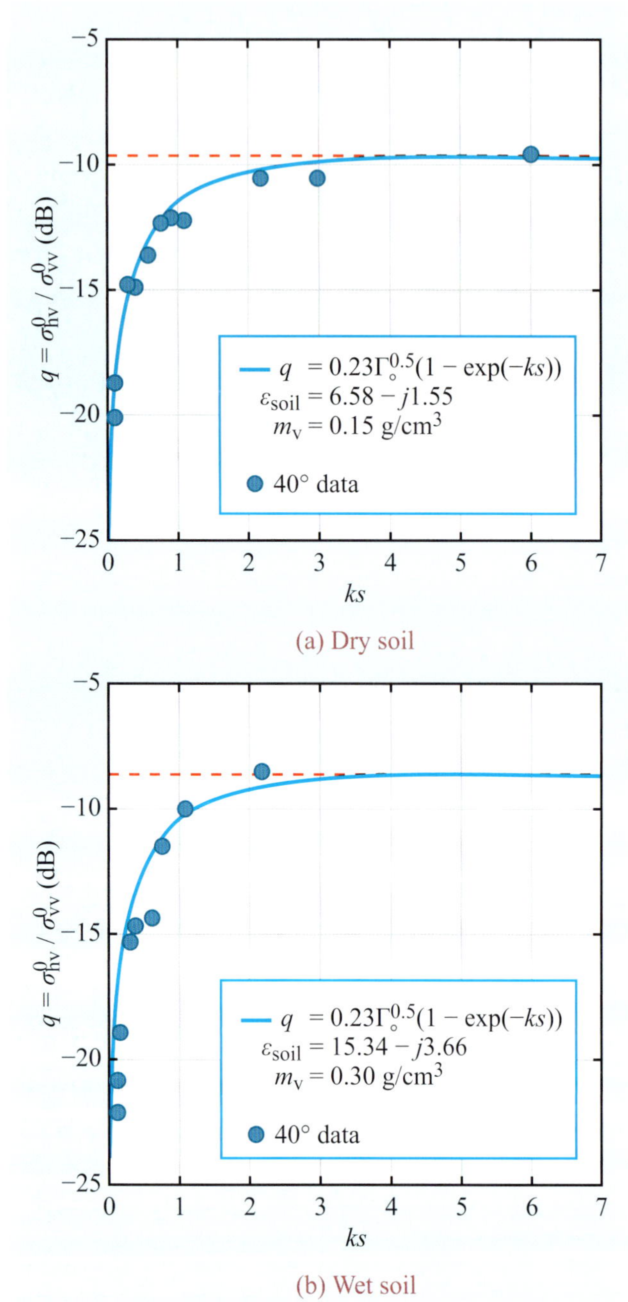

Figure 10-32: Measured and modeled variation of the cross-pol ratio q with EM roughness ks for (a) a dry and (b) a wet soil surface [Oh et al., 1992].

The expressions given by Eqs. (10.34) to (10.39) constitute the ***forward model*** of PRISM-1.

10-5.2 PRISM-1

Using the models for p and q, a nonlinear inversion model (PRISM-1) was developed to estimate the values of ks, ε', ε'', and m_v. An example is shown in Fig. 10-33 for $f = 1.25$ GHz and $\theta = 30°$. The combination of p and q provides values for s and m_v. Comparisons between estimated values and measured values are provided in Fig. 10-34. The rms error for m_v is 0.04 g/cm^3, which is as good as the rms error associated with measuring m_v directly under field conditions. To measure m_v for an area extensive target, several soil samples are acquired from different spatial locations and processed to determine their mean value and associated standard deviation. Even under very uniform soil conditions, the spatial variability alone (i.e.,

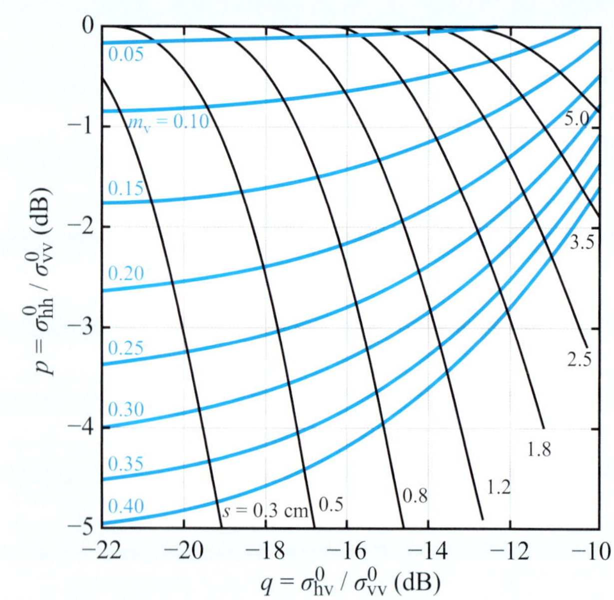

Figure 10-33: PRISM-1 uses the co-pol ratio p and cross-pol ratio q at any specified frequency/angle combination to estimate the values of the rms height s and volumetric moisture m_v. This algorithm is for $f = 1.25$ GHz and $\theta = 30°$.

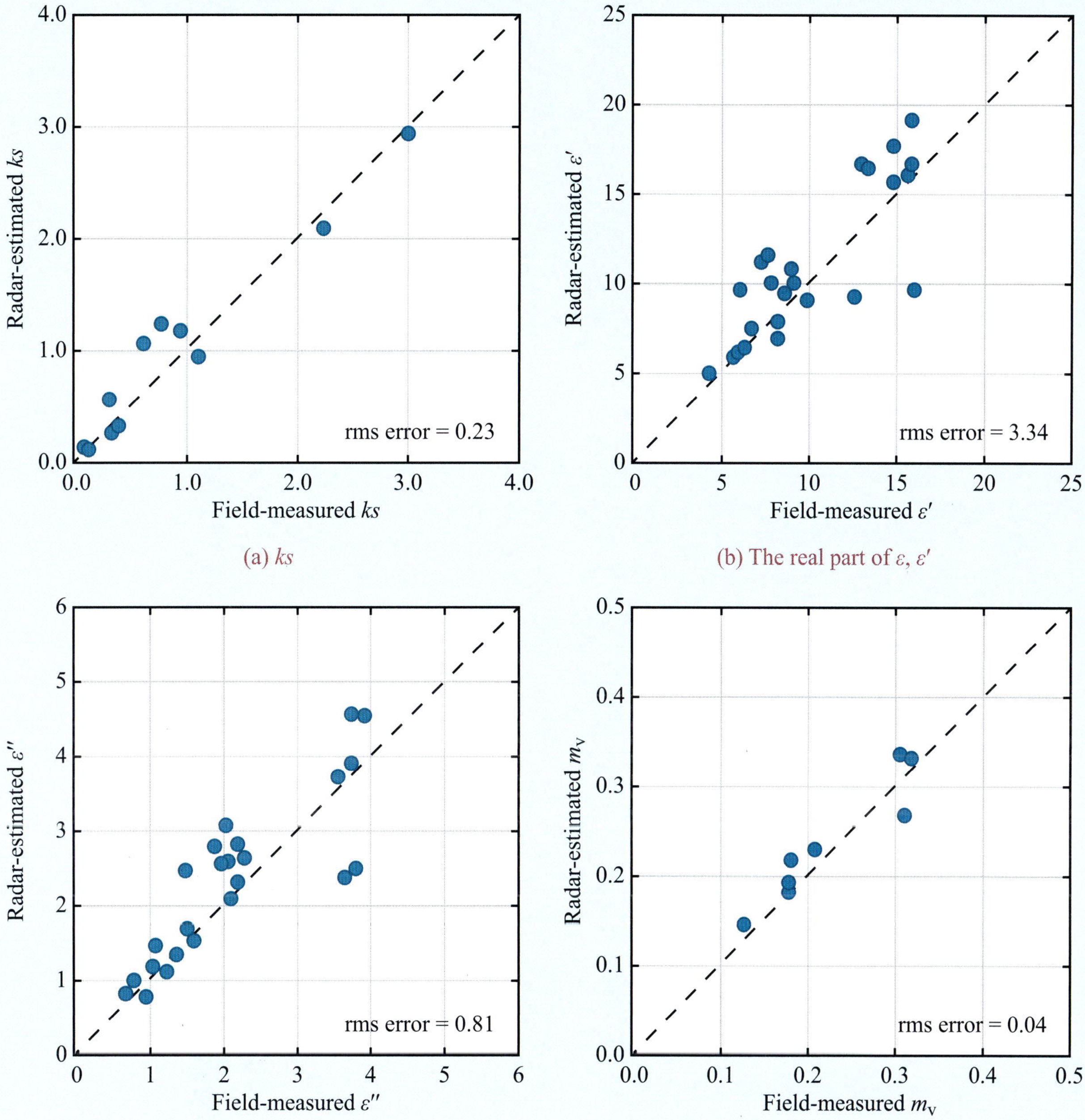

(a) ks

(b) The real part of ε, ε'

(c) The imaginary part of ε, ε''

(d) The volumetric moisture content m_v

Figure 10-34: Comparison of radar-estimated values (based on PRISM-1) of ks, ε', ε'', and m_v with corresponding values measured directly [Oh et al., 1992].

ignoring measurement error of the dielectric probe used to measure m_v) is on the order of 0.04 g/cm^3. This means that the precision associated with the values estimated by the radar is comparable to the natural spatial variability of m_v.

10-5.3 PRISM-2

Even though the 1992 empirical model for σ^0 provides very good fits to the measured data and the inverse model provides accurate soil moisture estimates, we should keep in mind that the project represents a single experiment involving only four different soil surfaces. Over the next decade, many more experiments were conducted, using not only truck-mounted and indoor scatterometers, but also several flights by the JPL AirSAR over test sites in Michigan, Oklahoma, and California. The combined data set of the seven experiments encompassed approximately 40 bare soil fields, with ks values extending over the range from 0.13 to 6.98 and volumetric moisture values covering the range from 0.04 to 0.29 g/cm^3 (Oh et al., 2002).

Using the expanded data set, a new semiempirical model was developed (PRISM-2), but this time m_v was used explicitly as an independent variable, rather than through ε and Γ as was done previously in the model given by Eqs. (10.34), (10.38), and (10.39). The 2004 model (Oh, 2004) takes the form

$$\sigma^0_{\text{vh}} = 0.11 m_v^{0.7} (\cos\theta)^{2.2} \left[1 - e^{-0.32(ks)^{1.8}}\right], \quad (10.40\text{a})$$

$$p = \frac{\sigma^0_{\text{hh}}}{\sigma^0_{\text{vv}}} = 1 - \left(\frac{2\theta}{\pi}\right)^{0.35 m_v^{-0.65}} \cdot e^{-0.4(ks)^{1.4}}, \quad (10.40\text{b})$$

$$q = \frac{\sigma^0_{\text{hv}}}{\sigma^0_{\text{vv}}} = 0.095(0.13 + \sin 1.5\theta)^{1.4} \left[1 - e^{-1.3(ks)^{0.9}}\right]. \quad (10.40\text{c})$$

In this model, the cross-polarized ratio q is a function of only θ and ks. Its variation with ks is shown in Fig. 10-35(a) for $\theta = 45°$. Parts (b) and (c) depict the variations of σ^0_{vh} and p with m_v and ks.

Using the model expressions of Eq. (10.40), Oh (2004) developed an inverse model to generate estimates of ks and m_v from the multipolarized radar measurements. Figure 10-36 compares field-measured values

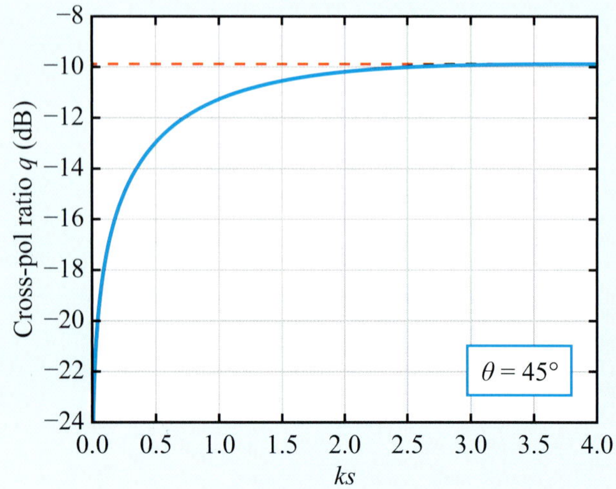

(a) q versus ks

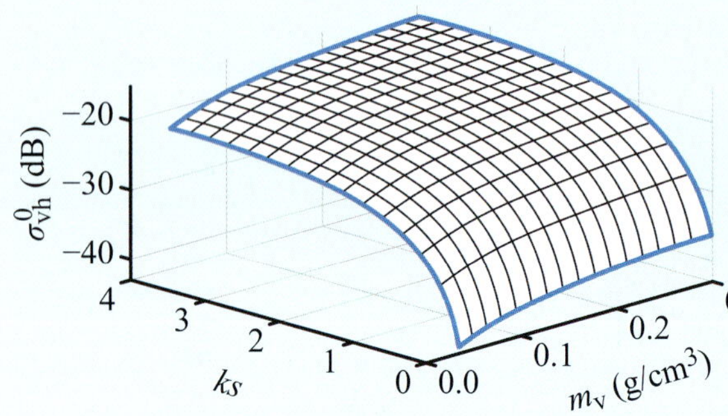

(b) Mesh plot of σ^0_{vh} against ks and m_v

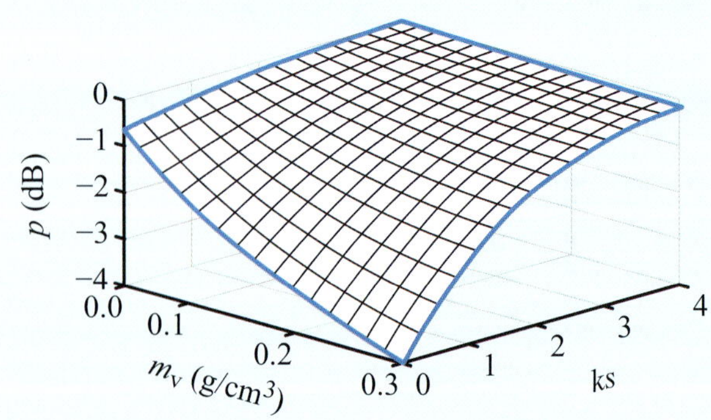

(c) Mesh plot of p against m_v and ks

Figure 10-35: PRISM-2 plots of (a) the cross-pol ratio q as a function of ks at $\theta = 45°$, (b) σ^0_{hv}, and (c) co-pol ratio p.

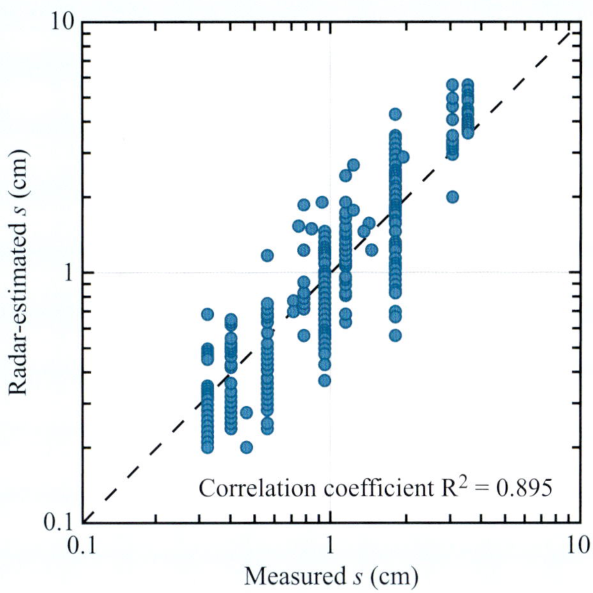

(a) rms height s

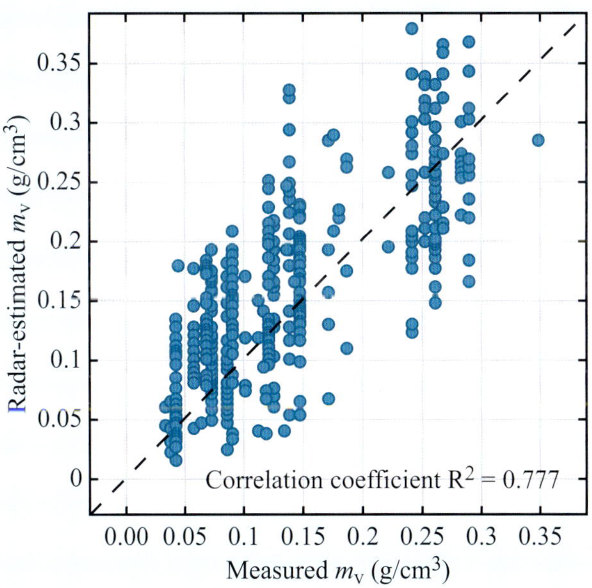

(b) Volumetric moisture content m_v

Figure 10-36: Comparison of radar-estimated values of s and m_v using PRISM-2 with corresponding values measured directly. Each data point represents a single frequency/angle combination [Oh, 2004].

of s and m_v with values estimated using the inverse model applied to 414 data vectors composed of σ^0_{hh}, σ^0_{hv}, and σ^0_{vv} at 1 angle × 1 frequency × 1 soil-surface condition. For each field-measured value of ks or m_v, multiple estimated values are generated by the inverse model—corresponding to the various frequency/angle combinations of available radar observations. Part of the variability along the vertical axes in Figs. 10-36(a) and (b) is due to the imprecision of the inversion model, but some of the variability is due to intercalibration among the various radars and field campaigns. When the multiple estimates of ks generated from the multiple frequency/angle combinations are averaged together, much of the variability disappears. The result is shown in Fig. 10-37.

In addition to the I^{2}EM theoretical model described in Section 10-3 and the empirical models presented in the present section, the literature contains other theoretical and empirical models for radar scattering by random surfaces, two of which are presented in forthcoming sections. Intercomparison of some of the available models was performed by Baghdadi et al. (2011). Also, concerted efforts have been devoted to advancing the development of soil moisture inversion algorithms that incorporate simultaneous radar and radiometer observations (Njoku et al., 2002; Narayan et al., 2006; Zhan et al., 2006). A major challenge faced by such an approach is the large difference in the spatial-resolution capabilities of the two sensors: tens of kilometers for a spaceborne radiometer compared with only meters to tens of meters for a SAR. Estimating soil moisture with microwave radiometry is discussed in Chapter 12.

10-6 SMART (Soil Moisture Assessment Radar Technique)

Dubois et al. (1995a, 1995b) developed a soil moisture inversion model that we have named the ***Soil Moisture Assessment Radar Technique***, or SMART for short. The model is based on a data set that combines the University of Michigan measurements made by the LCX POLARSCAT (Tassoudji et al., 1989) and the measurements made by the RASAM systems (Wegmüller, 1993) at the University of Berne in

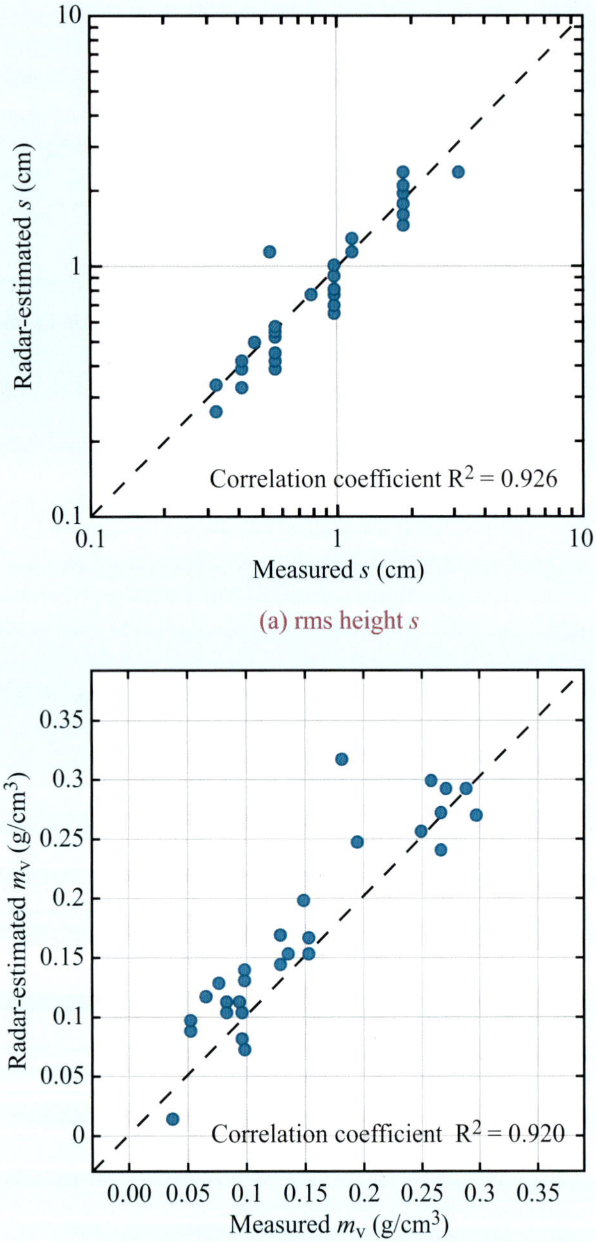

(a) rms height s

(b) Volumetric moisture content m_{v}

Figure 10-37: To reduce the scattering in Fig. 10-36 due to intercalibration among the various radars and measurement campaigns, multifrequency/angle radar estimates are averaged together [Oh, 2004].

Switzerland. The first of these two data sets led to the development of the PRISM model described in the preceding section. It included measurements at three microwave frequencies (1.25, 4.75, and 9.5 GHz) for multiple surfaces with rms height s varying from 0.32 cm to 3.02 cm. The second data set included measurements at six frequencies between 2.5 and 11 GHz for surfaces with rms height varying from 0.57 cm to 1.12 cm. Both data sets included soil moisture and soil density measurements.

Using these data sets for $\theta \geq 30^\circ$, the SMART model offers the following empirical relationships:¶†

$$\sigma^0_{\mathrm{hh}} = 10^{-2.75} \cdot \frac{\cos^{1.5}\theta}{\sin^5\theta} \cdot 10^{0.028\varepsilon' \tan\theta} (ks\sin\theta)^{1.4} \lambda^{0.7}, \tag{10.41a}$$

$$\sigma^0_{\mathrm{vv}} = 10^{-2.35} \cdot \frac{\cos^{3}\theta}{\sin^3\theta} \cdot 10^{0.046\varepsilon' \tan\theta} (ks\sin\theta)^{1.1} \lambda^{0.7}, \tag{10.41b}$$

where ε' is the real part of the soil dielectric constant, $k = 2\pi/\lambda$, s is the rms height in cm, and λ is the wavelength in cm. The units of σ^0_{hh} and σ^0_{vv} are m^2/m^2. Figure 10-38 displays a family of plots of σ^0_{vv} (dB) against the co-pol ratio $p = \sigma^0_{\mathrm{hh}}/\sigma^0_{\mathrm{vv}}$, with ks and ε' as parameters. The calculations are for $\theta = 45^\circ$. As noted by van Zyl and Kim (2011), the co-pol ratio, based on the model expressions, can exceed 1 for large values of ks, which contradicts theoretical expectations. Hence, the model applicability is better for surfaces with $ks \lesssim 1.2$ than for rougher surfaces.

Aside from the radar parameters θ and λ, both of which are known, the only unknown parameters in Eq. (10.41) are ε' and s. Measured radar values for σ^0_{hh} and σ^0_{vv} are used to determine ε' and s, and then a soil dielectric model (Section 4-8) is used to relate ε' to the soil moisture content m_{v}.

To demonstrate the viability of the SMART model, Dubois et al. (1995a, 1995b) isolated the term $(ks\sin\theta)$ on the right-hand side of Eq. (10.41) and then converted

¶The original paper by Dubois et al. (1995a) included printing errors that were noted in a later correction (1995b).

†Computer Code 10.7.

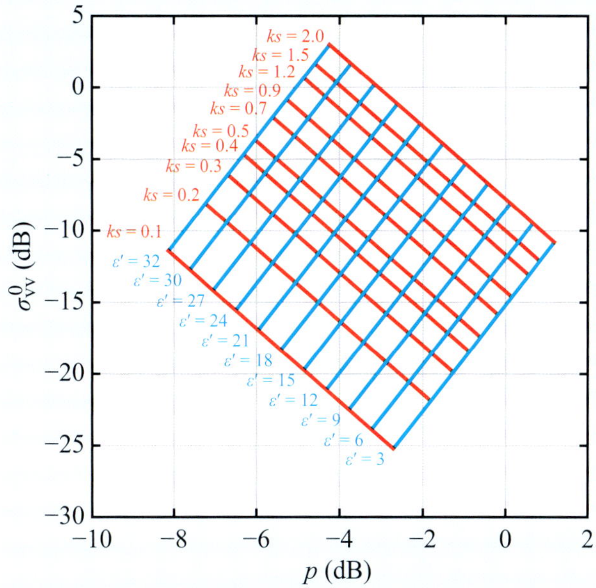

Figure 10-38: SMART model algorithm.

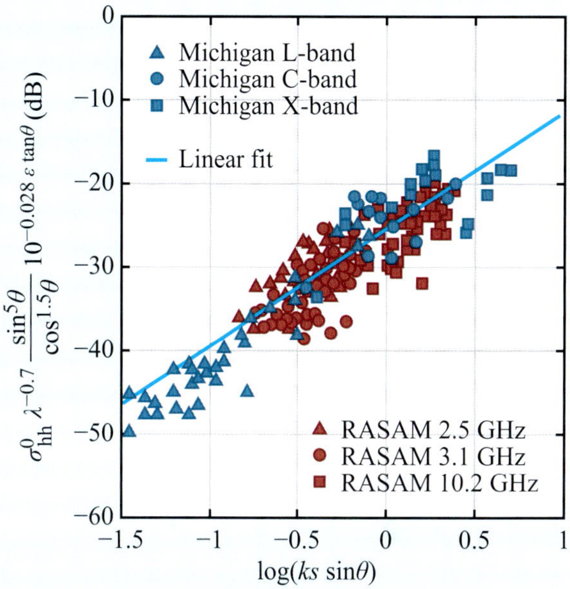

Figure 10-39: Plot of $10\log[\sigma^0_{\text{hh}}\lambda^{-0.7}(\sin^5\theta/\cos^{1.5}\theta) \times 10^{-0.028\varepsilon'\tan\theta}]$ as a function of $\log(ks\sin\theta)$ for the POLARSCAT and RASAM scatterometer data [Dubois et al., 1995a, b].

both sides to dB, which leads to

$$\begin{aligned} 10\log\left[\sigma^0_{\text{hh}}\lambda^{-0.7}\cdot\frac{\sin^5\theta}{\cos^{1.5}\theta}\cdot 10^{-0.028\varepsilon'\tan\theta}\right] &= 10\log[10^{-2.75}(ks\sin\theta)^{1.4}] \\ &= -27.5 + 14\log(ks\sin\theta). \end{aligned} \tag{10.42}$$

Figure 10-39 displays the calculated values of the left-hand side of Eq. (10.42), plotted against $\log(ks\sin\theta)$, for the combined data set. The linear form of Eq. (10.42) is represented by the straight line shown in Fig. 10-39. Application of the model in an inverse mode to estimate m_v and s from σ^0_{hh} and σ^0_{vv} leads to the results shown in Fig. 10-40. The reported rms error is on the order of 4% for m_v and 0.4 cm for s. Application of the inverse model to L-band images recorded by airborne and spaceborne imaging radars led to the results displayed in Fig. 10-41.

The inversion equations are given by (van Zyl and Kim, 2011):[†]

$$\begin{aligned} \varepsilon' = \frac{1}{3.36\tan\theta}\{&14\sigma^0_{\text{vv}}\,(\text{dB}) \quad 11\sigma^0_{\text{hh}}\,(\text{dB}) \\ &+ 26.5 - 255\log_{10}(\cos\theta) \\ &- 130\log_{10}(\sin\theta) - 21\log_{10}(\lambda)\}, \end{aligned} \tag{10.43a}$$

and

$$\begin{aligned} \log_{10}(ks) = &-0.083\sigma^0_{\text{vv}}\,(\text{dB}) + 0.137\sigma^0_{\text{hh}}\,(\text{dB}) + 1.807 \\ &+ 0.446\log_{10}(\cos\theta) + 3.345\log_{10}(\sin\theta) \\ &- 0.375\log_{10}(\lambda) \end{aligned} \tag{10.43b}$$

with λ in cm.

The SMART model has several attractive features: (a) its expressions are straightforward and easy to use, (b) it relies on only hh and vv polarizations (whereas

[†] **Computer Code 10.8.**

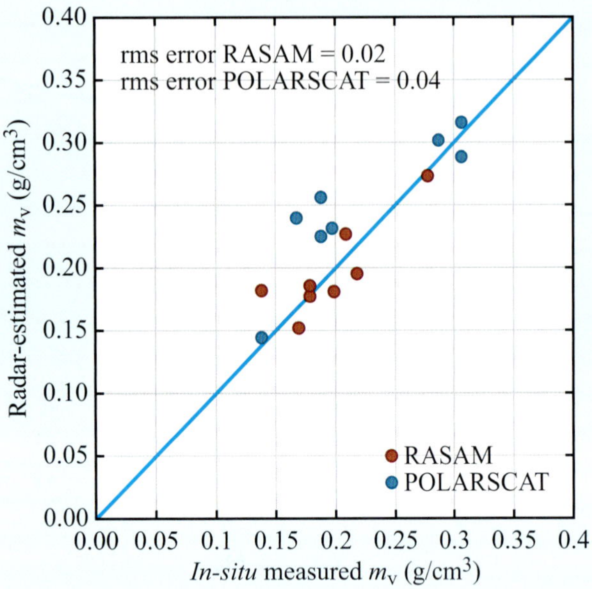

(a) Soil moisture versus *in-situ* measured soil moisture

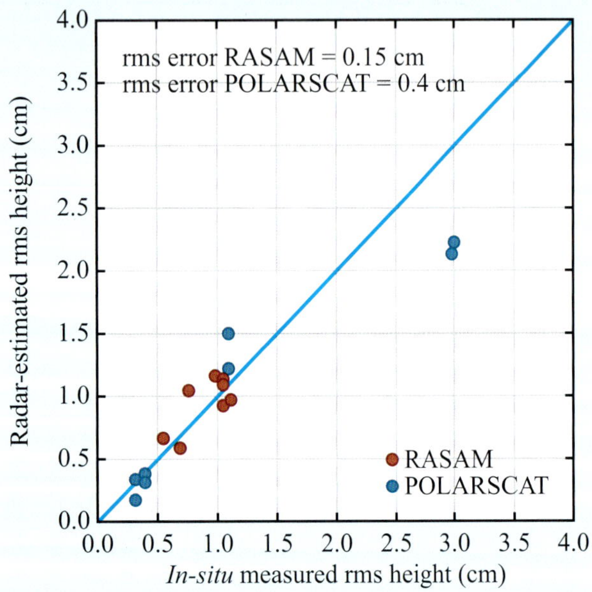

(b) rms height versus *in-situ* measured rms height

Figure 10-40: Estimated (a) soil moisture versus *in-situ* measured soil moisture and (b) rms height versus *in-situ* measured rms height [Dubois et al., 1995a].

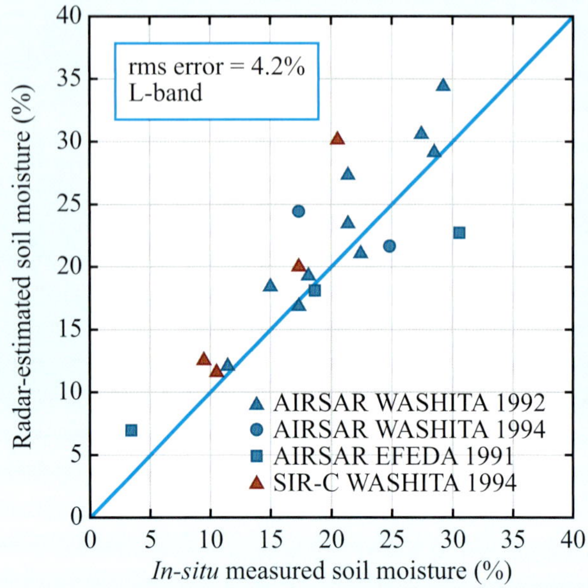

Figure 10-41: Radar-estimated soil moisture versus *in-situ* measured soil moisture for bare-soil fields imaged by the JPL AIRSAR and SIR-C [Dubois et al., 1995a].

the PRISM model uses those two plus σ^0_{hv}), and (c) it provides estimates of m_v and s with good accuracies. Consequently, it has been applied at numerous sites to map soil moisture content on the basis of multipolarized radar imagery (Kseneman and Gleich, 2009; Truong-Loï et al., 2009; Merzouki et al., 2011; Prakash et al., 2012).

10-7 Model Comparisons

EM scattering models are of three varieties. The simplest in form and expression are the semiempirical, and, in the present context, they include the PRISM model of Section 10-5 and the SMART model of Section 10-6. They tend to be of hybrid construction based on experimental data but guided by trends predicted by theoretical models.

The second variety consists of theoretical models that are based on a firmer theoretical foundation, but whose analytical solutions require the imposition of various assumptions to support the use of mathematical approximations. Often, their solutions require complex integrations, but their computational demands are relatively modest. The I^2EM model of Section 10-3 is a perfect example.

The third variety encompasses numerical solutions of Maxwell's equations in 3-D, which are regarded as more exact than the approximate theoretical models, but they are computationally very demanding. In a recent study, Huang et al. (2010) used a method-of-moments technique to compute about 6000 numerical solutions of σ° for hh and vv polarizations and of the emissivity e, also for h and v polarizations, as a function of several surface variables. The surface variables included s/λ, l/λ, and ε, where s and l are the rms height and correlation length of the random surface, and ε is its dielectric constant. All computations were performed at $\theta = 40°$. Their results for σ° are summarized in Figs. 10-42 and 10-43.

Part (a) of Fig. 10-42 displays three curves of σ^0_{vv} as a function of ks, together with six data points. The data points correspond to the numerical method-of-moment 3-D (NMM3D) computation reported by Huang et al. (2010), and the three curves represent the I^2EM, PRISM-1, and SMART models. A similar set for σ^0_{hh} is displayed in part (b) of the same figure, and similar comparisons are provided in Fig. 10-43 as a function of ε'.

In a study similar to that reported by Huang et al. (2010), Baghdadi et al. (2011) evaluated the I^2EM, PRISM, and SMART models relative to measurements made by the high-resolution TerraSAR-X imaging radar for several sites in France and Tunisia.

10-8 Concluding Observations

Since the publication of the first report examining the radar response to soil moisture (Ulaby, 1974), hundreds of studies have been conducted to quantify the nature of the response and to use that knowledge to develop inversion techniques so as to generate soil-moisture distribution maps on a large scale. We now have the basic tools—embodied in models like I^2EM, PRISM, and SMART, but we still need to develop an algorithm capable of generating the desired soil-moisture maps

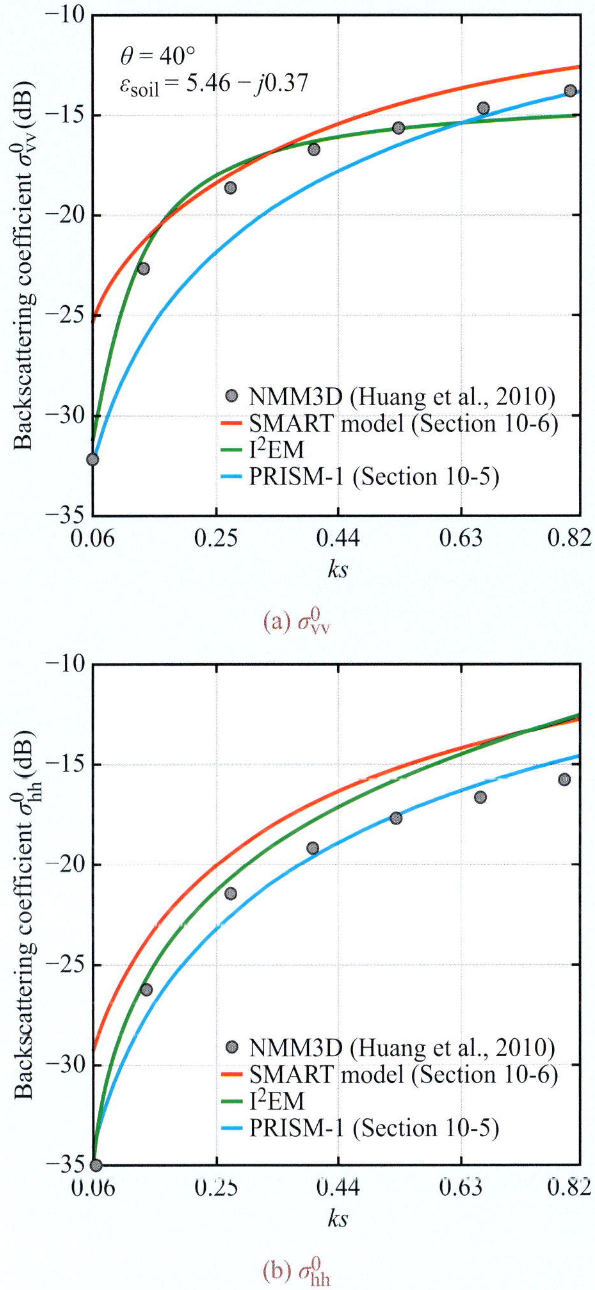

Figure 10-42: Comparison of numerical 3-D computed values with I^2EM, PRISM-1, and SMART models for (a) vv polarization and (b) hh polarization.

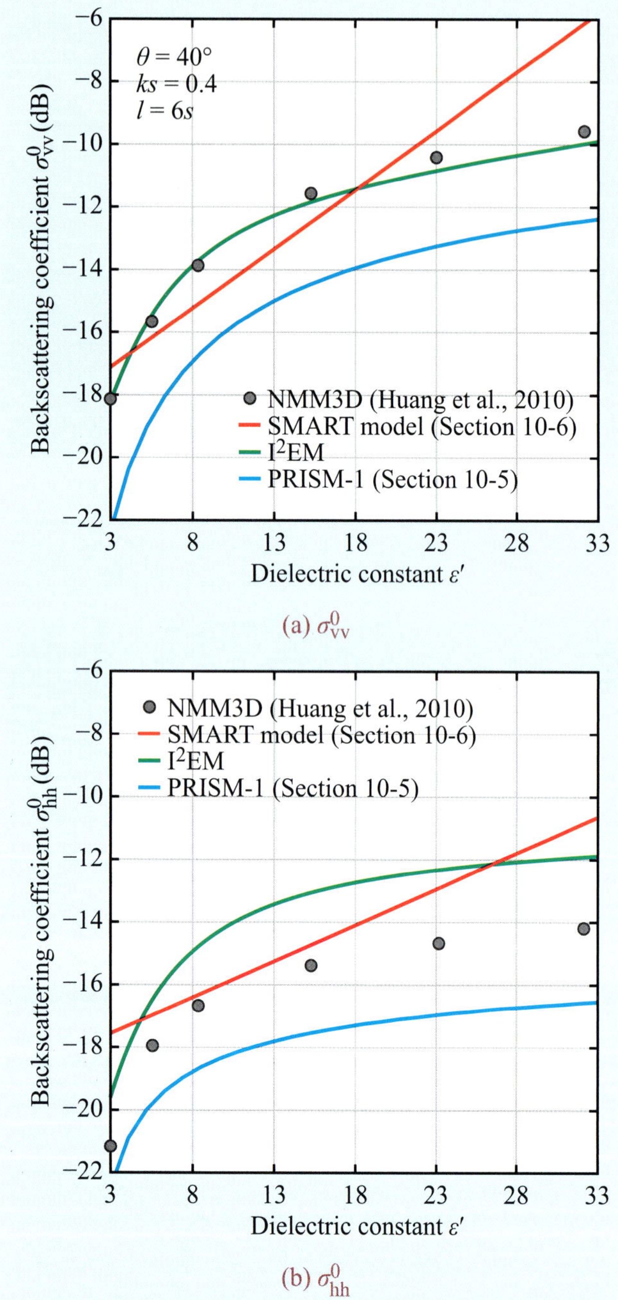

Figure 10-43: Comparison of numerical 3-D computed values with I^{2}EM, PRISM-1, and SMART models for (a) vv and (b) hh polarizations.

with an acceptable degree of accuracy, on the order of 3–4%. The challenge is two-dimensional. The first dimension has to do with our definition of the soil volumetric moisture content m_v. In some field studies, m_v is measured over the top 1 cm of the soil surface, in others it is over 2 cm or as much as 5 cm. The moisture actually has a profile that may decrease or increase with depth, further complicating the definition of what we mean by m_v. Perhaps the answer to this issue is to designate the m_v predicted by the radar inversion models as a ***radar-equivalent moisture content*** and then let the users of the soil-moisture information (hydrologist, climate modelers, farmers, etc.) figure out how best to use it for their intended applications.

The second challenge has to do with the oversimplification of our description of surface roughness. According to I^{2}EM, the surface-height statistics are governed by the form of the correlation function, the rms height, and the correlation length l, and if the surface-height profile includes multiple scales of roughness, these aforementioned parameters apply to each roughness scale. The PRISM and SMART models rely on ks as the sole representative of roughness. Since different roughness scales respond differently to different wavelengths, a two-frequency polarimetric radar (such as an L-band plus a C-band) is more likely to provide a higher-accuracy estimate of m_v than a single-frequency system can. This was certainly the conclusion reached by the study reported by Oh (2004); compare the results displayed in Fig. 10-36 with those in Fig. 10-37! To sense soil moisture under vegetation cover (as discussed in Chapter 12), it may be necessary to add a third (or even a fourth) lower frequency channel.

PROBLEMS

10.1 Suppose a random surface is characterized by three scales of roughness. Use the results displayed in Fig. 10-9 to explain which roughness scale is the most important and under what conditions. Consider the ratio s_i/λ_e where s_i is the rms height of the ith scale and λ_e is the effective wavelength.

10.2 Consider a three-scale random surface with a Gaussian correlation function. The surface roughness parameters are $s_1 = 1$ cm, $l_1 = 15$ cm, $s_2 = 0.4$ cm, $l_2 = 5$ cm, $s_3 = 0.2$ cm, and $l_3 = 2.5$ cm. The dielectric constant is $\varepsilon = 12$. Compute the backscattering coefficient for hh polarization at $\theta = 30°$ and plot it as a function of ks, where k is the wavenumber and s is the equivalent rms height of the surface. The range of ks should extend from 0.1 to 1.

10.3 Figure 10-13 contains a plot for σ^0_{hv} as a function of m_{v} at $f = 3$ GHz. Generate similar plots for the same surface at 1.5 GHz and 4.5 GHz.

10.4 Consider a one-dimensional sinusoidal surface with $A = 8$ cm and $\Lambda = 40$ cm. Superimposed on the periodic surface is a relatively smooth random surface with backscattering coefficient (in the absence of the periodic surface) given by

$$\sigma^{\mathrm{s}}_{\mathrm{vv}} = e^{-13.2\theta'} \qquad (\mathrm{m^2/m^2}),$$

where θ' is the incidence angle in radians. Compute and plot the backscattering coefficient that would be observed by an imaging radar (with a spatial resolution many times greater than Λ) at incidence angle $\theta_0 = 20°$. Your plot should show the variation with the azimuth angle ϕ_0 (0 to 180°).

10.5 Repeat Problem 10.4 for a relatively rough random surface with

$$\sigma^{\mathrm{s}}_{\mathrm{vv}} = e^{-2\theta'} \qquad (\mathrm{m^2/m^2}).$$

10.6 Use Computer Codes 4.7 and 10.5 to compute the PRISM-1 model variations with m_{v} (0 to 0.3 g/cm^3) for:

(a) σ^0_{vv}

(b) σ^0_{hh}

(c) σ^0_{hv}

(d) co-pol and cross-pol ratios.

Assume $f = 3$ GHz, $\theta = 45°$, and $s = 1.5$ cm.

10.7 Use Computer Codes 4.7 and 10.7 to generate SMART model variations with m_{v} (0 to 0.3 g/cm^3) for:

(a) σ^0_{vv}

(b) σ^0_{hh}

Assume $f = 3$ GHz, $\theta = 45°$, and $s = 1.5$ cm.

10.8 A 1.5 GHz polarimetric radar is used to image bare-soil fields at an incidence angle of 30°. The soil dielectric constant is given by the approximate model

$$\varepsilon' = 57m_{\mathrm{v}} + 3,$$
$$\varepsilon'' = 11m_{\mathrm{v}},$$

where m_{v} is in g/cm^3. Use the PRISM-1 forward model to generate a database for σ^0_{hh} and σ^0_{vv} for various combinations of m_{v} and the rms height s over the ranges $0.05 \leq m_{\mathrm{v}} \leq 0.35$ g/cm^3 and $0.2 \leq s \leq 2$ cm. Use the database to develop an inverse model. Evaluate your model performance by performing a linear-regression analysis to compare the values of m_{v} predicted by the model with the true values of m_{v}.

CHAPTER

11

Volume-Scattering Models and Land Observations

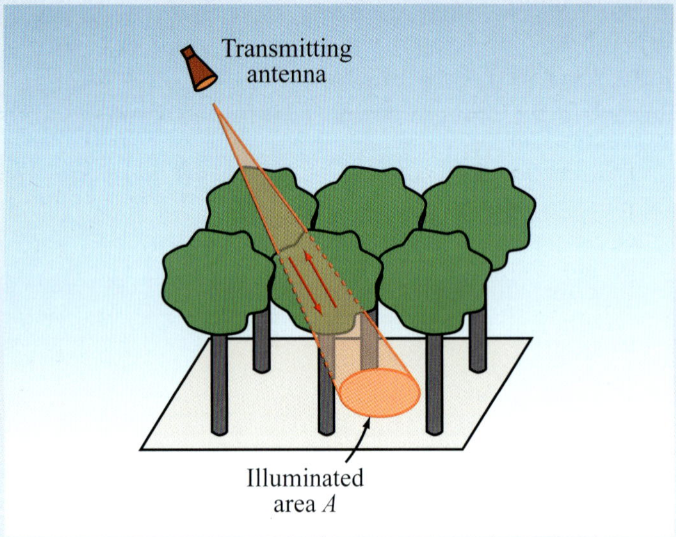

Volume scattering

CONTENTS

The Nature of Volume Scattering

In contrast with surface scattering, which occurs at a continuous surface interface between two media with different dielectric properties, volume scattering is caused by discrete particles present in an otherwise homogeneous background dielectric medium. The sketch shown in Fig. 11-1 depicts a layer of dry snow above a soil medium. The snow consists of ice crystals suspended in an air background. The total signal scattered by the snow-soil composite includes scattering components due to surface scattering by the air-snow interface and by the snow-soil interface, volume scattering by the ice crystals, and multiple scattering involving both surface and volume scattering. A similar scenario applies to the vegetation canopy shown in Fig. 11-2, except that the top of the canopy does not have a distinct boundary. To compute the scattering at the diffuse air-canopy boundary, we represent the canopy in terms of an equivalent, homogeneous dielectric medium. However, because the vegetation material contained in the canopy constitutes less than 1% of the canopy by volume, the real part of the effective dielectric constant of the canopy is only slightly larger than that of air. Consequently, we treat the air-canopy boundary as a nonscattering diffuse boundary.

▶ When modeling or analyzing radar scattering or radiometric emission by a volume-scattering medium, it is prudent to consider the following factors:

(a) The size distribution of the scatterers, relative to the wavelength λ.

(b) The 3-D orientation distribution of the scatterers.

(c) The shape distributions of the scatterers.

(d) The dielectric constant of the scatterers (for each type of scatterer, if more than one type exists in the scattering volume). ◀

These factors determine:

(a) Whether or not the shape of a scatterer is important so far as its scattering pattern is concerned. Large scatterers (relative to λ) have distinct scattering patterns, whereas the scattering patterns of very small scatterers can be approximated by the Rayleigh phase function (Section 11-6.5) regardless of their shapes.

(b) Whether or not volume scattering by the layer can be limited to single scattering, thereby simplifying the computation.

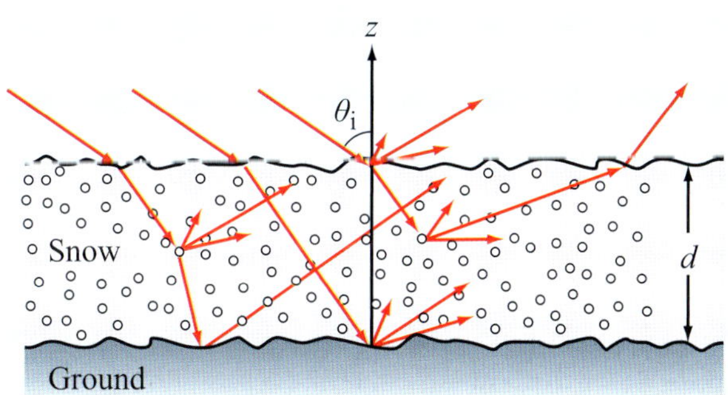

Figure 11-1: Scattering by a snow layer.

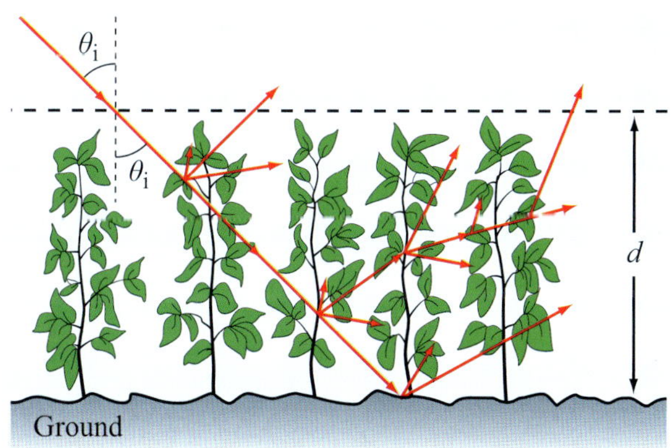

Figure 11-2: Volume scattering by a vegetation canopy.

(c) Whether or not the extinction coefficient of the medium is dominated by absorption (relative to scattering).

(d) Whether or not the volume exhibits azimuthal symmetry.

To help the reader develop an understanding of the interrelationships between these various factors and their consequences, we begin with heuristic descriptions of the scattering process for simple cases, followed by an overview presentation of the wave propagation properties of various media of interest. The combination of the two topics provides a useful background for our discussion of radiative transfer theory in Sections 11-6 and 11-7.

11-1 Heuristic Single-Scattering Model for Vegetation

Consider the vegetation canopy depicted in Fig. 11-3.

> ▶ A ***single-scattering*** backscatter model accounts for scattering contributions that involve single scattering by the canopy volume, but may include double, single, or no scattering by the underlying ground surface. ◀

These contributions include:

(a) single backscattering by the ground surface, taking into account the two-way transmission through the canopy (denoted "ray 1" in Fig. 11-3).

(b) single direct backscattering by the canopy constituents (denoted "ray 2" in Fig. 11-3).

(c) combination of single bistatic scattering (reflection) by the ground followed by single bistatic scattering by vegetation elements such that the final direction is towards the radar, or the reverse of the scattering sequence (denoted "ray 3a" and "ray 3b", respectively, in Fig. 11-3).

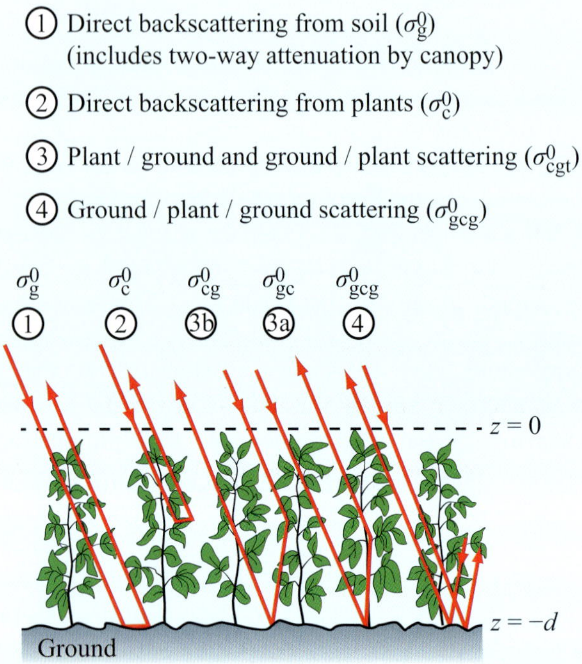

Figure 11-3: Single-scattering contributions in a vegetation canopy.

(d) transmission through the canopy, followed by specular reflection by the ground surface, followed by backscatter by the vegetation volume, followed by another specular reflection by the ground surface ("ray 4" in Fig. 11-3).

For a differential canopy volume, the ***p-polarized extinction coefficient*** κ_e^p accounts for absorption and scattering losses experienced by a wave propagating through it. That is,

$$\kappa_e^p = \kappa_a^p + \kappa_s^p. \tag{11.1}$$

Because the canopy may be nonuniform in height and nonisotropic azimuthally, and because its constituents (leaves, stalks, etc.) have specific shapes and orientations, κ_e^p is, in general, height-dependent (between the top of the canopy at $z = 0$ and the bottom of the canopy at depth $z = -d$), direction-dependent, and polarization-dependent. *For the sake of this simple heuristic model,*

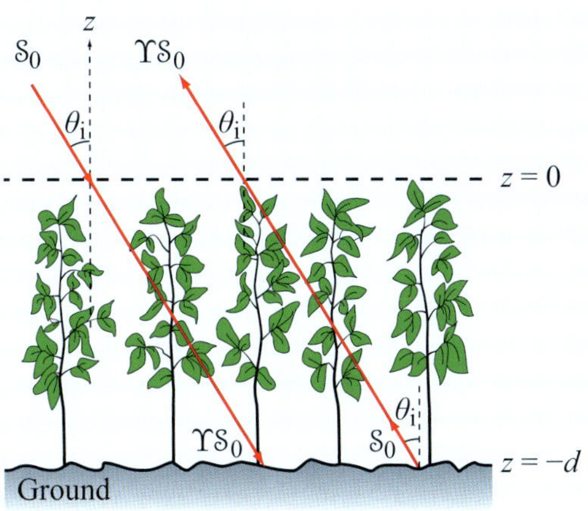

Figure 11-4: Transmissivity Υ accounts for transmission loss through the canopy along angle θ_i.

we assume that κ_s^p, κ_a^p, and κ_e^p are uniform as a function of z within the canopy.

For propagation downward along incidence angle θ_i relative to nadir, or upward along angle θ_i relative to $-z$ (Fig. 11-4), the p-polarized one-way oblique ***transmittivity*** of the canopy, Υ_p, is given by

$$\Upsilon_p = e^{-\tau_p}, \tag{11.2}$$

where τ_p is the p-polarized ***attenuation (optical depth)*** of the canopy,

$$\tau_p = \kappa_e^p d \sec\theta_i \qquad \text{(Np)}. \tag{11.3}$$

Scattering Terminology

In the forthcoming derivation, we use multiple scattering-related quantities. For the sake of clarity, we offer the following definitions:

$\sigma^0_{\text{name}_{pq}}$**:** The superscript 0 identifies this quantity as a ***scattering coefficient*** (scattering cross section per unit area), measured in m^2/m^2; its "name" subscript refers to the abbreviated name of the associated scattering mechanism (see Fig. 11-3); and pq refers to the receive/transmit polarizations (with $p, q =$ v or h).

$\sigma^{\text{back}}_{\text{v}_{pq}}$**:** The subscript v denotes this quantity as the ***volume backscattering coefficient*** (scattering cross section per unit volume), measured in $m^2/m^3 = m^{-1}$; and the superscript "back" denotes that the incident and scattering directions are exact opposites.

$\sigma^{\text{bist}}_{\text{v}_{pq}}$**:** This quantity also is a ***volume scattering coefficient***, but it is for bistatic scattering.

$\sigma^{\text{back}}_{pq}$**:** ***Backscattering cross section*** of a single particle (m^2).

$\sigma^{\text{bist}}_{pq}$**:** ***Bistatic scattering cross section*** of a single particle (m^2).

11-1.1 Direct Ground Contribution

If in the absence of the vegetation layer the backscattering coefficient of the ground surface is $\sigma^0_{\text{s}_{pq}}$, the presence of the vegetation simply modifies the direct ground contribution by a factor of $\Upsilon_p\Upsilon_q$,

$$\sigma^0_{\text{g}_{pq}}(\theta_i) = \Upsilon_p\Upsilon_q\, \sigma^0_{\text{s}_{pq}}(\theta_i) \qquad (m^2/m^2), \tag{11.4}$$
$$(p, q = \text{v or h polarization}).$$

Since $\sigma^0_{\text{s}_{pq}}$ is, in general, polarization and angle dependent, so is $\sigma^0_{\text{g}_{pq}}$.

11-1.2 Direct Volume Contribution—The Cloud Model

To model the scattering contribution of the canopy volume $\sigma^0_{\text{c}_{pq}}(\theta_i)$, Attema and Ulaby (1978) treated the canopy as an equivalent "water cloud" consisting of identical scatterers that are uniformly distributed throughout the volume, as illustrated in Fig. 11-5. Ignoring multiple scattering, the pq-polarized ***volume backscattering coefficient*** $\sigma^{\text{back}}_{\text{v}_{pq}}$ of the vegetation medium is

$$\sigma^{\text{back}}_{\text{v}_{pq}} = N_{\text{v}}\sigma^{\text{back}}_{pq} \qquad (m^{-1}), \tag{11.5}$$

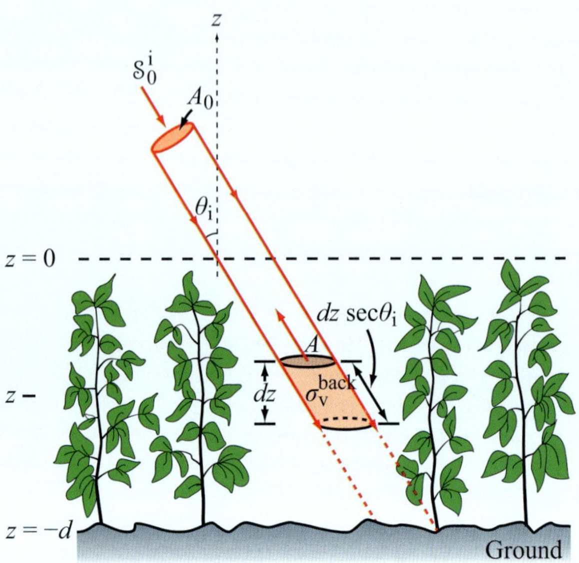

Figure 11-5: Direct backscatter contribution by the canopy volume.

where N_v is the ***number density*** (number of scattering particles per unit volume [in m^{-3})] and σ_{pq}^{back} is the pq-polarized backscattering cross section of a single particle.

With reference to Fig. 11-5, for a q-polarized incident plane wave with power density S_0^i before intercepting the canopy, the q-polarized power density across an area A of a differential volume within the canopy at a lateral distance $z \sec\theta_i$ from the surface is

$$S_q^i(z) = S_0^i \cos\theta_i \, e^{\kappa_e^q z \sec\theta_i} \qquad (W/m^2), \qquad (11.6)$$

where κ_e^q is the extinction coefficient for q polarization. Note that the top of the canopy is denoted $z = 0$ and the ground is denoted $z = -d$, so within the canopy layer, $z \leq 0$. The $\cos\theta$ factor accounts for the reduction in power density between that incident upon A_0 and that incident upon the larger area $A = A_0/\cos\theta_i$, and the exponential function represents the one-way transmissivity between the canopy surface and the differential volume at depth z.

Given that $\sigma_{v_{pq}}^{back}$ is the backscattering cross section per unit volume and the volume of the differential volume is $A \sec\theta_i \, dz$, the total backscattering cross section of the differential volume is

$$d\sigma_{pq}^{back} = \sigma_{v_{pq}}^{back} A \sec\theta_i \, dz \qquad (m^2). \qquad (11.7)$$

The differential p-polarized power reradiated by the differential volume in the backscatter direction, after propagation from z to the top of the canopy, is

$$\begin{aligned} dP_p^{rer} &= S_q^i(z)\, d\sigma_{pq}^{back}\, e^{\kappa_e^p z \sec\theta_i} \\ &= S_0^i A \cos\theta_i \, \sigma_{v_{pq}}^{back}\, e^{2(\kappa_e^q + \kappa_e^p) z \sec\theta_i} \sec\theta_i \, dz. \quad (11.8) \end{aligned}$$

Integrating dP_p^{rer} over the full depth of the canopy yields

$$\begin{aligned} P_p^{rer} &= \int_{-d}^{0} S_0^i A \cos\theta_i \, \sigma_{v_{pq}}^{back} e^{(\kappa_e^p + \kappa_e^q) z \sec\theta_i} \sec\theta_i \, dz \\ &= \frac{S_0^i A \, \sigma_{v_{pq}}^{back}}{(\kappa_e^p + \kappa_e^q)\sec\theta_i} \left[1 - e^{-(\kappa_e^p + \kappa_e^q) d \sec\theta_i}\right] \\ &= \frac{S_0^i A \cos\theta_i \, \sigma_{v_{pq}}^{back}}{\kappa_e^p + \kappa_e^q} \left[1 - \Upsilon_p \Upsilon_q\right], \qquad (11.9) \end{aligned}$$

where Υ is the one-way transmissivity defined by Eq. (11.2).

At a range R_r to the receive antenna, the p-polarized power density of the scattered wave is

$$S_p^s = \frac{P_p^{rer}}{4\pi R_r^2} .$$

Application of the standard definition for the radar cross section given by Eq. (5.30) and then dividing by A to obtain the pq-polarized ***canopy backscattering coefficient*** $\sigma_{c_{pq}}^0$, leads to

$$\begin{aligned} \sigma_{c_{pq}}^0(\theta_i) &= \frac{4\pi R_r^2}{A} \frac{S_p^s}{S_0^i} \\ &= \frac{\sigma_{v_{pq}}^{back} \cos\theta_i}{\kappa_e^p + \kappa_e^q} (1 - \Upsilon_p \Upsilon_q) \qquad (m^2/m^2). \end{aligned} \qquad (11.10)$$

If σ_v^{back} and κ_e are polarization-independent and do not generate cross polarization, then $\sigma_c^0(\theta_i)$ is the same for hh and vv polarization and zero for hv polarization. For real canopies, σ_c^0 is not zero for hv polarization, and if the canopy contains vertical stalks, σ_c^0 may not have the same magnitude for hh and vv polarizations.

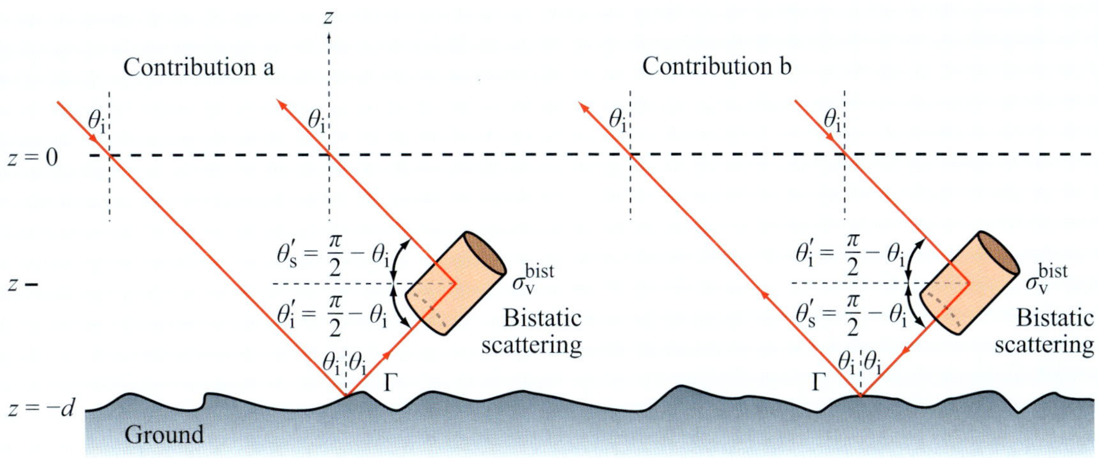

Figure 11-6: The two volume-ground single-scattering contributions.

11-1.3 Canopy-Ground Contributions

To compute the two components of the canopy ground contribution, namely the one involving ground reflection followed by canopy bistatic scattering in the direction of the radar, and the other involving the sequence in reverse order (Fig. 11-6), we treat the ground boundary as a ***quasi-specular*** surface. By quasi-specular, we mean that we limit reflection to the specular direction, but we account for the reduction in the magnitude of the Fresnel reflectivity caused by surface roughness. That is, we use Eq. (5.110) to describe the p-polarized coherent reflectivity $\Gamma^p_{\rm coh}(\theta_{\rm i})$ in terms of the p-polarized Fresnel reflectivity for a specular surface $\Gamma^p(\theta_{\rm i})$,

$$\Gamma^p_{\rm coh}(\theta_{\rm i}) = \Gamma^p(\theta_{\rm i})\, e^{-4k^2s^2\cos^2\theta_{\rm i}}, \qquad (p = \text{v or h}), \tag{11.11}$$

where $k = 2\pi/\lambda$ is the wavenumber and s is the rms height of the ground surface.

▶ In the material that follows, we use the symbol Γ to represent the surface reflectivity, with the implicit understanding that Γ should be replaced with $\Gamma_{\rm coh}$ if the electromagnetic roughness $ks > 0.2$. ◀

Under the specular-reflection condition, all propagation occurs along downward and upward directions that make an angle $\theta_{\rm i}$ relative to $\hat{\mathbf{z}}$ or $-\hat{\mathbf{z}}$. Consider scattering contribution "a" in Fig. 11-6. After propagation through the canopy down to the ground surface and reflection along the specular direction, the wave propagates all the way to the top of the canopy and beyond. Along the way, part of the upward propagating energy gets bistatically scattered in the direction of the radar. The fraction of energy so scattered is proportional to the ***volume bistatic scattering coefficient*** $\sigma^{\rm bist}_{{\rm v}_{pq}}(\theta'_{\rm s},\phi'_{\rm s};\theta'_{\rm i},\phi'_{\rm i})$, where $(\theta'_{\rm i},\phi'_{\rm i})$ and $(\theta'_{\rm s},\phi'_{\rm s})$ define the incident and scattered directions relative to the differential volume under consideration. Under the specular-direction condition, $\phi'_{\rm s} = \phi'_{\rm i} = 0$ and $\theta'_{\rm i} = \theta'_{\rm s} = \pi/2 - \theta_{\rm i}$. Repeat of the steps we used earlier to derive the expression for $\sigma^0_{\rm c}$ given by Eq. (11.10) leads to the following expression for the ***ground-canopy backscattering coefficient***:

$$\sigma^0_{{\rm gc}_{pq}}(\theta_{\rm i}) = \sigma^{\rm bist}_{{\rm v}_{pq}} \cdot d\Gamma^q \Upsilon_q \Upsilon_p, \tag{11.12}$$

where d is the height of the canopy and the volume bistatic scattering coefficient $\sigma^{\rm bist}_{{\rm v}_{pq}}$ is defined for $\theta'_{\rm i} = \theta'_{\rm s} = \pi/2 - \theta_{\rm i}$.

Similarly, for contribution "b" in Fig. 11-6,

$$\sigma^0_{{\rm cg}_{pq}}(\theta_{\rm i}) = \sigma^{\rm bist}_{{\rm v}_{pq}} \cdot d\Gamma^p \Upsilon_p \Upsilon_q. \tag{11.13}$$

The sum of the two components is denoted the ***total canopy ground contribution***:

$$\sigma^0_{\text{cgt}_{pq}}(\theta_\text{i}) = \sigma^0_{\text{gc}_{pq}}(\theta_\text{i}) + \sigma^0_{\text{cg}_{pq}}(\theta_\text{i}) = \sigma^{\text{bist}}_{\text{v}_{pq}} \cdot d[\Gamma^p + \Gamma^q]\Upsilon_p\Upsilon_q. \qquad (11.14)$$

For $p = q$ (i.e., for hh or vv polarization), Eq. (11.14) simplifies to

$$\sigma^0_{\text{cgt}_{pp}}(\theta_\text{i}) = 2\sigma^{\text{bist}}_{\text{v}_{pp}} \cdot d\Gamma^p\Upsilon_p^2 \qquad (p = \text{h or v}).$$

(incoherent addition) (11.15a)

Our treatment is based on radiative transfer theory, which deals with power quantities instead of electric fields. Hence, different contributions are added incoherently (with regard to phase). Since the two contributions undergo the same propagation and reflection sequence, but in reverse order, they should have the same amplitude and phase, in which case their electric fields should be added coherently.

Coherent addition increases the magnitude of σ^0_{cgt} for hh and vv polarizations by a factor of 2. Thus,

$$\sigma^0_{\text{cgt}_{pp}}(\theta_\text{i}) = 4\sigma^{\text{bist}}_{\text{v}_{pp}} \cdot d\Gamma^p\Upsilon_p^2 \qquad (p = \text{h or v}).$$

(coherent addition) (11.15b)

This increase in the level of the co-polarized total canopy-ground contribution is known as ***bistatic scattering enhancement***.

11-1.4 Ground-Canopy-Ground Contribution

Term 2 in Fig. 11-3 represents upward backscatter by the canopy volume. Term 4 represents backscatter by the canopy volume along the downward direction, in addition to double reflections by the ground surface and an extra two-way propagation through the canopy. Hence, the ground-canopy-ground contribution is

$$\sigma_{\text{gcg}_{pq}}(\theta_\text{i}) = \Upsilon_p\Upsilon_q\Gamma^p\Gamma^q\,\sigma^0_{\text{c}_{pq}}(\theta_\text{i}) = \frac{\sigma^{\text{back}}_{\text{v}_{pq}}\cos\theta_\text{i}}{\kappa_\text{e}^p + \kappa_\text{e}^q}\,\Gamma^p\Gamma^q(\Upsilon_p\Upsilon_q - \Upsilon_p^2\Upsilon_q^2). \qquad (11.16)$$

11-1.5 Single-Scattering Radiative Transfer Model

The total single-scattering backscattering coefficient is the sum of the four contributions derived in the preceding subsections:

$$\begin{aligned}\sigma^0_{pq} &= \sigma^0_{\text{g}_{pq}} + \sigma^0_{\text{c}_{pq}} + \sigma^0_{\text{cgt}_{pq}} + \sigma^0_{\text{gcg}_{pq}} \\ &= \Upsilon_p\Upsilon_q\,\sigma^0_{\text{s}_{pq}}(\theta_\text{i}) \\ &\quad + \frac{\sigma^{\text{back}}_{\text{v}_{pq}}\cos\theta_\text{i}}{\kappa_\text{e}^p + \kappa_\text{e}^q}\,(1 - \Upsilon_p\Upsilon_q)(1 + \Gamma^p\Gamma^q\Upsilon_p\Upsilon_q) \\ &\quad + n\,\sigma^{\text{bist}}_{\text{v}_{pq}}\,d(\Gamma^p + \Gamma^q)\Upsilon_p\Upsilon_q, \end{aligned} \qquad (11.17)$$

with

$$n = \begin{cases} 1 & \text{for hv polarization and also for hh and vv under the incoherent addition assumption,} \\ 2 & \text{for hh and vv polarizations with coherent addition.} \end{cases}$$

We refer to Eq. (11.17) as the ***single-scattering radiative-transfer*** (S^2RT) model because it is identical with the formal solution of the radiative transfer equation obtained later in Section 11-5.

The absence of a vegetation layer implies that $d = 0$ and $\Upsilon = 1$, in which case Eq. (11.17) reduces to $\sigma^0 = \sigma^0_{\text{s}_{pq}}$. At the other extreme, for a very dense canopy with high attenuation, $\Upsilon \approx 0$, in which case the canopy appears like a semi-infinite layer with

$$\sigma^0_{pq}(\theta_\text{i}) = \frac{\sigma^{\text{back}}_{\text{v}_{pq}}\cos\theta_\text{i}}{\kappa_\text{e}^p + \kappa_\text{e}^q}, \qquad (\Upsilon \to 0). \qquad (11.18)$$

(semi-infinite canopy)

11-2 Isotropic and Rayleigh Scatterers

11-2.1 Canopy Elements as Isotropic Scatterers

For isotropic scatterers, the scattering pattern is uniform along all directions for both hh and vv polarizations. Hence,

$$\sigma_{\rm v}^{\rm back} = \sigma_{\rm v}^{\rm bist} = \kappa_{\rm s}, \tag{11.19}$$

where κ_s is the scattering component of the extinction coefficient κ_e. Also, $\Upsilon_p = \Upsilon_q$. Using Eq. (11.19) in Eq. (11.17) leads to

$$\begin{aligned}\sigma_{pp}^0(\theta_{\rm i}) &= \Upsilon^2\,\sigma_{s_{pp}}^0(\theta_{\rm i}) \\ &\quad + \frac{a\cos\theta_{\rm i}}{2}\,(1-\Upsilon^2)(1+\Gamma^2\Upsilon^2)+4\kappa_{\rm s}d\Gamma\Upsilon^2, \\ &\quad (p = \text{h or v}),\end{aligned} \tag{11.20}$$

where $\Gamma = \Gamma^p$ and we introduced the ***single-scattering albedo*** a, defined as

$$a = \frac{\kappa_{\rm s}}{\kappa_{\rm e}}. \tag{11.21}$$

11-2.2 Canopy Elements as Rayleigh Scatterers

In Section 8-5.3, we discussed the scattering and absorption properties of small spheres under the Rayleigh approximation (which we discuss further in Section 11-5.5). The scattering pattern of a Rayleigh particle depends on the incident and scattered directions, but for hh and vv polarizations and for the incident and scattered directions shown in Fig. 11-6,

$$\sigma_{\rm v}^{\rm back} = \sigma_{\rm v}^{\rm bist} = \frac{3}{2}\,\kappa_{\rm s} \qquad \textbf{(hh and vv polarizations)}, \tag{11.22}$$

which leads to

$$\begin{aligned}\sigma_{pp}^0(\theta_{\rm i}) &= \Upsilon^2\,\sigma_{s_{pp}}^0(\theta_{\rm i}) \\ &\quad + \frac{3a}{4}\cos\theta_{\rm i}(1-\Upsilon^2)(1+\Gamma^2\Upsilon^2) \\ &\quad + 3n\kappa_{\rm s}d\Gamma\Upsilon^2, \qquad (p = \text{h or v}).\end{aligned} \tag{11.23}$$

We call this the ***S^2RT model with Rayleigh particles***, or ***S^2RT/R*** for short.[†]

The relative contributions of the four components of σ^0 are illustrated in Fig. 11-7 for three Rayleigh canopies: a lowloss canopy with $\Upsilon_0 = 0.8$, a medium-loss canopy with $\Upsilon_0 = 0.3$, and a high-loss canopy with $\Upsilon_0 = 0.1$. All three canopies have a single-scattering albedo $a = 0.1$ and an extinction coefficient $\kappa_e = 1$ Np/m, but their heights vary from 0.22 m for the low-loss canopy to 2.3 m for the most lossy. The underlying soil surface is slightly rough ($ks = 0.63$) and its volumetric soil moisture is 0.2 g/cm^3. As expected, the ground component $\sigma_{g_{pq}}^0$ is the dominant contributor, particularly near normal incidence, for the low-loss canopy, whereas the volume component σ_c^0 is the dominant contributor for the high-loss canopy.

11-3 Heuristic Single-Scattering Model for Snow-Covered Ground

The snow layer depicted in Fig. 11-8 is of depth d and contains uniformly distributed scatterers (ice crystals) in a background of air. The process of computing the backscatter from the snow-ground combination is identical with what we had done in the preceding section, except for modifications that incorporate the presence of the air-snow boundary. These modifications include:

(a) Backscatter by the air-snow boundary, represented by a backscattering coefficient $\sigma_{\rm as}^0$. This term is proportional to the air-snow reflectivity, which (except for very wet snow conditions) is very small in magnitude because the dielectric constant of the snow layer is only slightly larger than that of air. Hence, this term usually is negligibly small in comparison with the other backscatter contributions.

(b) Transmission from air to snow across the air-snow boundary. This happens twice, on the way down into the snow layer and later on the way out, and can be accounted for by $\mathbb{T}^p\mathbb{T}^q$, where $\mathbb{T}^p$ is the p-polarized air-snow boundary transmissivity defined in Table 2-5.

[†] **Computer Code 11.1.**

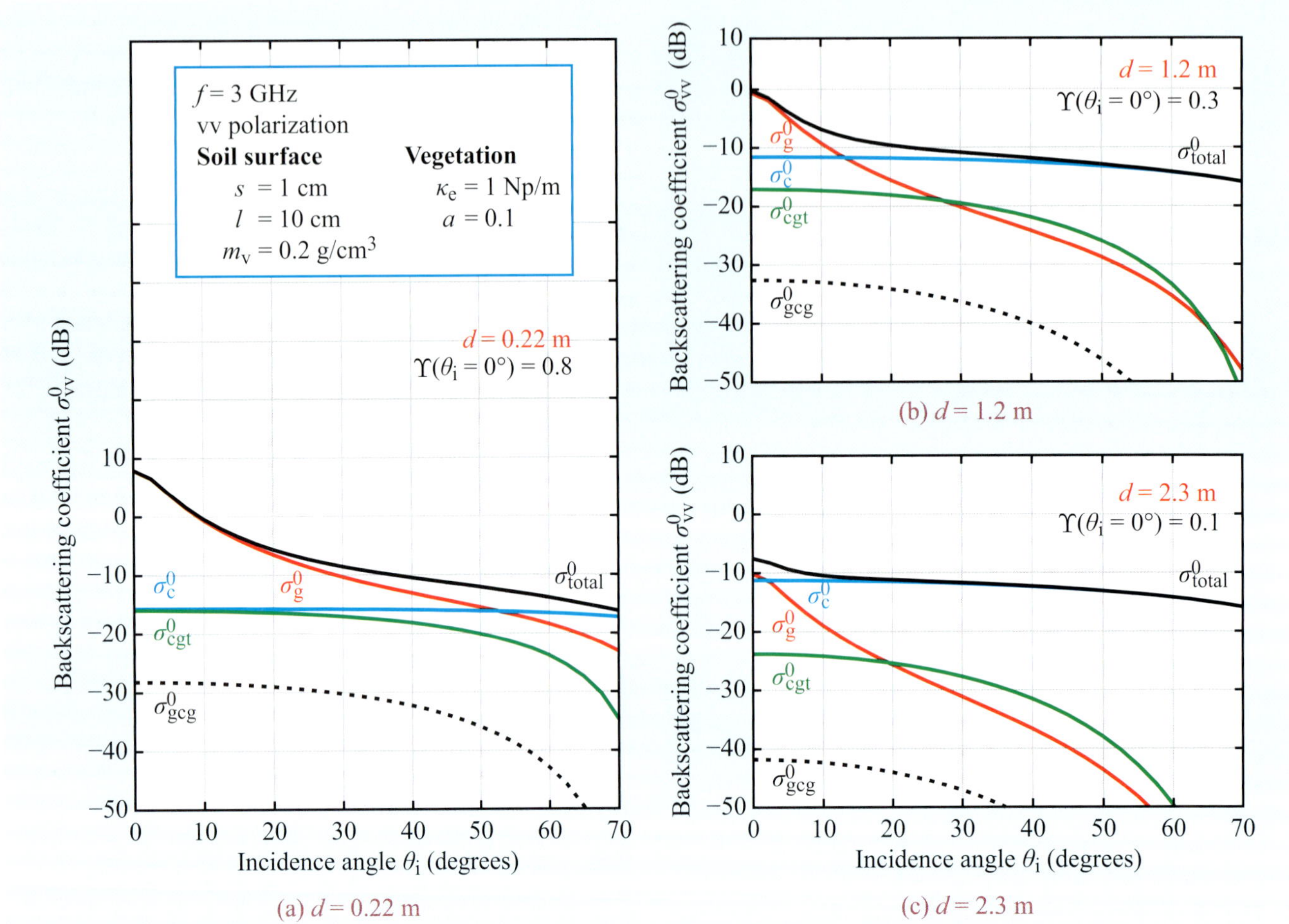

Figure 11-7: Computed backscattering by a canopy containing Rayleigh particles. The canopy heights are: (a) 0.22 m, (b) 1.2 m, and (c) 2.3 m.

Also, because of refraction from incidence angle θ_i to angle θ_i', θ_i in the expression for Υ given by Eqs. (11.2) and (11.3) should be replaced with θ_i'. Hence,

$$\sigma^0_{pq}(\theta_i) = \mathbb{T}^p(\theta_i)\,\mathbb{T}^q(\theta_i)\left[\Upsilon_p\Upsilon_q\,\sigma^0_{\text{sg}_{pq}}(\theta_i') + \frac{\sigma^{\text{back}}_{\text{v}_{pq}}(\theta_i')\cos\theta_i'}{\kappa^p_e+\kappa^q_e}(1-\Upsilon_p\Upsilon_q)(1+\Gamma^p\Gamma^q\Upsilon_p\Upsilon_q) + n\sigma^{\text{bist}}_{\text{v}_{pq}}\,d(\Gamma^p+\Gamma^q)\Upsilon_p\Upsilon_q\right] + \sigma^0_{\text{as}_{pq}}(\theta_i), \quad (11.24)$$

with

$$n = \begin{cases} 1 & \text{for hv polarization and also for hh and vv under the incoherent addition assumption,} \\ 2 & \text{for hh and vv polarizations with coherent addition,} \end{cases}$$

where σ^0_{sg} is the snow-ground backscattering coefficient and σ^0_{as} is the air-snow backscattering coefficient. The

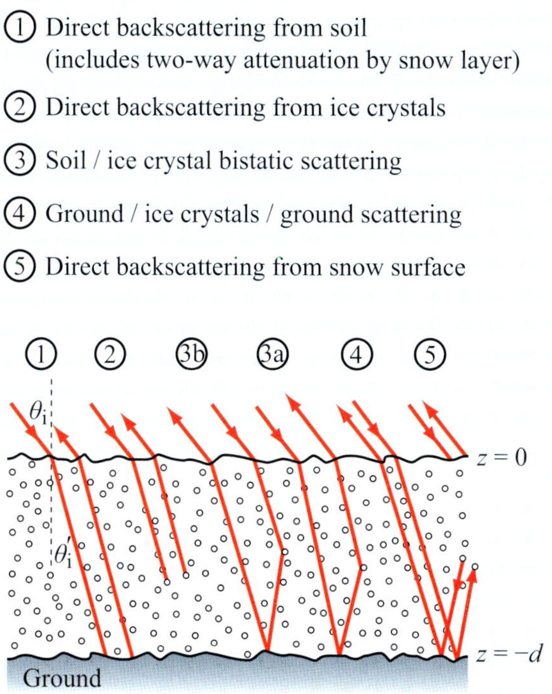

Figure 11-8: Single-scattering contributions for snow-over-ground.

formulation given by Eq. (11.24) is an ***S²RT/R model for a layer with a distinct upper boundary.***†

11-4 Penetration Depth

The power density $\mathcal{S}(z)$ of an EM wave propagating in the z direction in a medium characterized by an ***extinction coefficient*** κ_e is given by

$$\mathcal{S}(z) = \mathcal{S}(0)\, e^{-\kappa_e z},$$

where $\mathcal{S}(0)$ is the power density at a reference location denoted by $z = 0$, and

$$\kappa_e = \kappa_a + \kappa_s.$$

The ***volume absorption coefficient*** κ_a is related to the effective dielectric constant ε of the medium as

$$\kappa_a = 2k_0 n'' = -2k_0\, \Im\mathfrak{m}\left\{\sqrt{\varepsilon}\right\}, \qquad (11.25)$$

†Computer Code 11.2.

where $k_0 = 2\pi/\lambda_0$ with λ_0 being the free-space wavelength, and n is the complex index of refraction. If the medium is composed of multiple constituents—such as wet snow, which is a mixture of air, ice, and liquid water—then ε is the equivalent dielectric constant of the mixture. Strictly speaking, the concept of "effective" or "equivalent" dielectric constant for an inhomogeneous medium is applicable only if the constituents (ice crystals and small water pockets, for example) are much smaller in size compared with λ_0. When this condition is satisfied, it also follows that the ***volume scattering coefficient*** κ_s is much smaller than κ_a, in which case $\kappa_e \approx \kappa_a$.

Often, the equivalent-dielectric models of Chapter 4 are used to obtain a first-order, approximate estimate of ε even when the particle's dimensions are not much smaller than λ. A useful parameter in this connection is the ***single-scattering albedo*** of the medium, defined by Eq. (11.21). The contribution of scattering loss to the total extinction may be ignored if $a < 0.01$.

The ***penetration depth*** of a medium with extinction coefficient κ_e is defined as the depth $\delta_p = z$ at which $\mathcal{S}(z) = \mathcal{S}(0)\, e^{-1} = 0.37\mathcal{S}(0)$. Hence,

$$\delta_p = \frac{1}{\kappa_e} \qquad \text{(m)}.$$

For a scatter-free medium, $\delta_p = \delta_s/2$, where δ_s is the skin depth. The factor of 2 arises from the fact that δ_s is defined for the electric field amplitude $E(z)$, whereas δ_p is defined for the power density $\mathcal{S}(z)$, which is proportional to $|E(z)|^2$.

11-5 Radiative Transfer Theory

Two distinct theoretical approaches are available for computing multiple scattering by an inhomogeneous medium: ***analytical theory*** and ***radiative transfer theory*** (also called ***transport theory***). The analytical theory approach, which starts with Maxwell's equations and wave equations, can (in principle) account for all multiple scattering, diffraction, and interference effects, but it is mathematically complicated and computationally demanding.

▶ In contrast, the radiative-transfer approach deals with the transport of energy through a medium containing particles and assumes that there is no correlation between the fields scattered by different particles. This assumption allows the incoherent addition of the powers associated with multiple scattering contributions, rather than the addition of their electric fields. ◀

To justify this assumption, the scattering particles should be randomly distributed in location, and the average spacing between them should be sufficiently large (measured in wavelengths) as to ignore mutual coupling between them. Fortunately, most natural media that generate multiple scattering when EM waves pass through them do obey the random distribution requirement. The second requirement, which pertains to the spacing between adjacent scattering particles (such as between two ice crystals in a layer of snow or between two adjacent leaves in a vegetation canopy), is not always satisfied. Nevertheless, with few exceptions, the radiative transfer model has provided results in good agreement with experimental observations.

For an unbounded medium containing spherical particles (such as the atmosphere when viewed by an upward-looking sensor), the polarization state of the wave propagating through the medium becomes irrelevant. Consequently, radiative transfer can be formulated in terms of the scalar specific intensity I, as was done in Section 6-6. For radar, such a formulation leads to a scalar radiative transfer equation of the form

$$\frac{dI(\mathbf{R},\hat{\mathbf{s}})}{ds} = -\kappa_e\, I(\mathbf{R},\hat{\mathbf{s}}) - \kappa_{a,b}\, I(\mathbf{R},\hat{\mathbf{s}}) + \iint_{4\pi} \psi(\hat{\mathbf{s}},\hat{\mathbf{s}}')\, I(\mathbf{R},\hat{\mathbf{s}}')\, d\Omega', \qquad (11.26)$$

which is analogous to the combination of Eqs. (6.48), (6.52), and (6.53), without the self-emission contribution $\kappa_a J_a$. This ***scalar radiative transfer equation*** states that propagation through a differential volume of length ds in direction $\hat{\mathbf{s}}$ entails losses (negative signs) and gains (positive signs). The loss due to extinction (absorption and scattering) by the particles is accounted for by κ_e, and the loss due to absorption by the background material (if not lossless) is accounted for by the background absorption coefficient $\kappa_{a,b}$. The last term on the right-hand side of Eq. (11.26) accounts for energy incident from all directions $\hat{\mathbf{s}}'$ (due to bistatic scattering by other volumes) that gets redirected along the original direction $\hat{\mathbf{s}}$ of intensity I (Fig. 11-9). For passive remote sensing, the formulation includes an additional source term that accounts for self emission by the differential volume (Chapter 12).

Most terrestrial media are not composed of spherical particles, which means that scattering by the particles is, in general, wave-polarization dependent. Also, the scattering medium is bounded by a lower surface (ground) and sometimes by an upper surface. Reflection by and transmission across surface boundaries also are polarization dependent. Hence, we should generalize Eq. (11.26) to the ***vector radiative transfer equation***

$$\frac{d\mathbf{I}(\mathbf{R},\hat{\mathbf{s}})}{ds} = -\boldsymbol{\kappa}_e\, \mathbf{I}(\mathbf{R},\hat{\mathbf{s}}) - \kappa_{a,b}\, \mathbf{I}(\mathbf{R},\hat{\mathbf{s}}) + \iint_{4\pi} \boldsymbol{\psi}(\hat{\mathbf{s}},\hat{\mathbf{s}}')\, \mathbf{I}(\mathbf{R},\hat{\mathbf{s}}')\, d\Omega', \qquad (11.27)$$

where $\mathbf{I}$ is now a 4×1 vector representing the vector specific intensity of a plane wave with electric field

$$\mathbf{E} = (\hat{\mathbf{v}} E_v + \hat{\mathbf{h}} E_h) e^{-jk\hat{\mathbf{s}}.\mathbf{R}}, \qquad (11.28)$$

where $\hat{\mathbf{s}}$ is the propagation direction and $\mathbf{R}$ is a vector from the center of the coordinate system to the location of the differential volume under consideration. From Eq. (5.23a),

$$\mathbf{I} = \begin{bmatrix} I_v \\ I_h \\ U \\ V \end{bmatrix} = \begin{bmatrix} \langle |E_v|^2 \rangle \\ \langle |E_h|^2 \rangle \\ \langle 2\mathfrak{Re}(E_v E_h^*) \rangle \\ \langle 2\mathfrak{Im}(E_v E_h^*) \rangle \end{bmatrix} \Big/ \eta, \qquad (11.29)$$

where $\langle\ \rangle$ denotes an ensemble average over many independent observations of the distributed target.

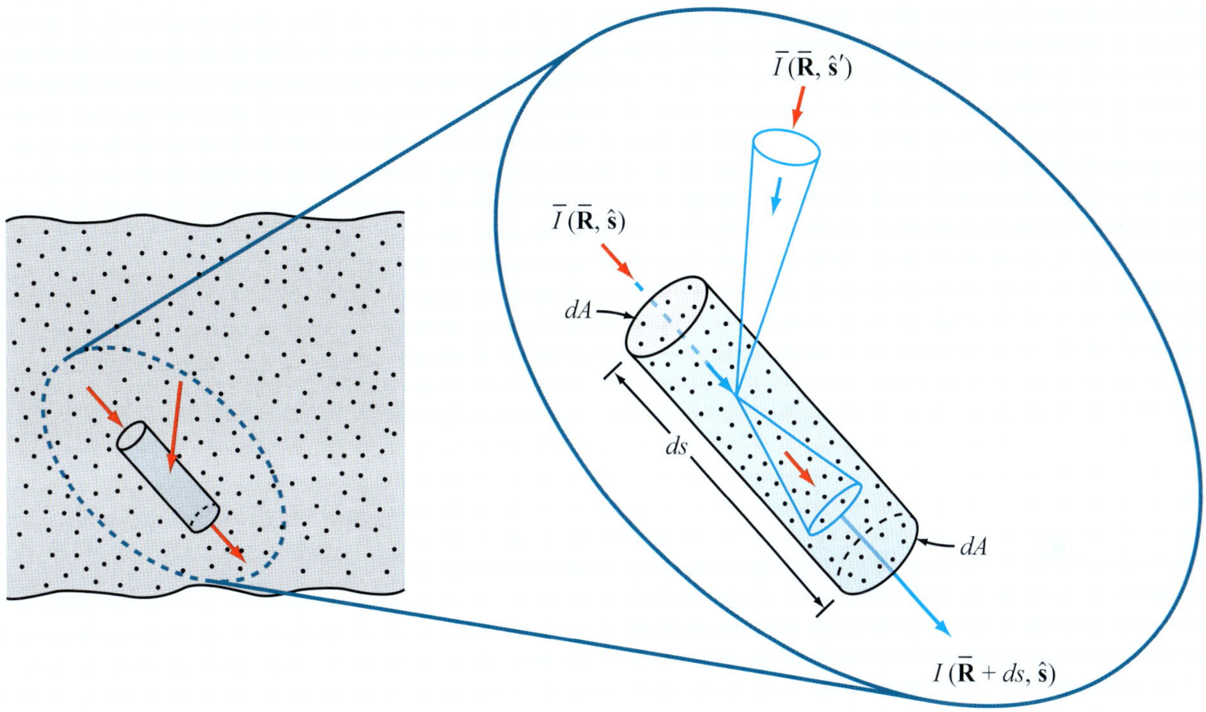

Figure 11-9: Propagation and scattering in a differential volume containing randomly positioned scatterers.

The other quantities printed in bold in Eq. (11.27) are the ***extinction matrix*** $\boldsymbol{\kappa}_\text{e}$ and the ***phase matrix*** $\boldsymbol{\psi}$, both of which are 4×4 matrices. Their definitions follow.

11-5.1 Extinction Matrix

For a single particle, the ***optical theorem*** provides a useful relationship for computing the extinction cross section Q_e in terms of elements of the scattering matrix of the particle for scattering in the forward direction, defined by $\theta_\text{s} = \theta_\text{i}$ and $\phi_\text{s} = \phi_\text{i}$. If the incident wave is p polarized, where $p = \text{v}$ or h, the extinction cross section of a particle with orientation direction (θ_j, ϕ_j) is

$$Q_\text{e}^p = \frac{4\pi}{k} \,\mathfrak{Im}[\widetilde{S}_{pp}(\theta_i, \phi_i; \theta_i, \phi_i; \theta_j, \phi_j)], \qquad (11.30)$$

where $k = 2\pi/\lambda$ and $\widetilde{S}_{pp}$ is the pp-scattering amplitude of the particle, defined in the forward scattering alignment (FSA) convention (Section 5-3.1).

The p-polarized extinction coefficient of a medium containing a random distribution of particles may be obtained by performing an ensemble average over the orientations of the particles. Thus,

$$\kappa_\text{e} = N_\text{v} \,\langle Q_\text{e}^p \rangle, \qquad (11.31)$$

where N_v is the number of particles per unit volume. For spherical particles, κ_e is the same for both polarizations and the extinction matrix is given by the diagonal form

$$\boldsymbol{\kappa}_\text{e} = \begin{bmatrix} \kappa_\text{e} & 0 & 0 & 0 \\ 0 & \kappa_\text{e} & 0 & 0 \\ 0 & 0 & \kappa_\text{e} & 0 \\ 0 & 0 & 0 & \kappa_\text{e} \end{bmatrix} \quad \textbf{(spheres)}. \qquad (11.32)$$

In the general case when the particles are nonspherical, the extinction matrix assumes the form (Ishimaru and

Cheung, 1980)

$$\boldsymbol{\kappa}_{\mathrm{e}} = \begin{bmatrix} 2\Re e(m_{\mathrm{vv}}) & 0 & \Re e(m_{\mathrm{vh}}) & \Im m(m_{\mathrm{vh}}) \\ 0 & 2\Re e(m_{\mathrm{hh}}) & \Re e(m_{\mathrm{hv}}) & -\Im m(m_{\mathrm{hv}}) \\ 2\Re e(m_{\mathrm{hv}}) & \Re e(m_{\mathrm{vh}}) & \Re e(m_{\mathrm{vh}}+m_{\mathrm{hh}}) & \Im m(m_{\mathrm{hh}}-m_{\mathrm{vv}}) \\ -2\Im m(m_{\mathrm{hv}}) & \Im m(m_{\mathrm{vh}}) & \Im m(m_{\mathrm{vh}}-m_{\mathrm{hh}}) & \Re e(m_{\mathrm{vv}}+m_{\mathrm{hh}}) \end{bmatrix}, \tag{11.33}$$

where

$$m_{pq} = j\left[\frac{2\pi}{k}\right] N_{\mathrm{v}} \left\langle \widetilde{S}_{pq} \right\rangle, \qquad (p,q = \mathrm{v\ or\ h}). \tag{11.34}$$

11-5.2 Phase Matrix

From Eq. (5.12) in Section 5-3, the electric field $\mathbf{E}^{\mathrm{s}}$ of the spherical wave scattered by a particle with scattering matrix $\widetilde{\mathbf{S}}$ (defined in the FSA convention) is related to $\mathbf{E}^{\mathrm{i}}$ of the incident plane wave by

$$\begin{bmatrix} E^{\mathrm{s}}_{\mathrm{v}} \\ E^{\mathrm{s}}_{\mathrm{h}} \end{bmatrix} = \left(\frac{e^{-jkR_{\mathrm{r}}}}{R_{\mathrm{r}}}\right) \begin{bmatrix} \widetilde{S}_{\mathrm{vv}} & \widetilde{S}_{\mathrm{vh}} \\ \widetilde{S}_{\mathrm{hv}} & \widetilde{S}_{\mathrm{hh}} \end{bmatrix} \begin{bmatrix} E^{\mathrm{i}}_{\mathrm{v}} \\ E^{\mathrm{i}}_{\mathrm{h}} \end{bmatrix}. \tag{11.35}$$

In the same section, we showed that the corresponding vector specific intensities, $\mathbf{I}^{\mathrm{s}}$ and $\mathbf{I}^{\mathrm{i}}$, are related by Eq. (5.24a), namely

$$\mathbf{I}^{\mathrm{s}} = \frac{1}{R_{\mathrm{r}}^2} \widetilde{\mathbf{M}}\, \mathbf{I}^{\mathrm{i}}, \tag{11.36}$$

where $\widetilde{\mathbf{M}}$ is the ***modified Mueller matrix***, repeated here for convenience:

$$\widetilde{\mathbf{M}} = \begin{bmatrix} |\widetilde{S}_{\mathrm{vv}}|^2 & |\widetilde{S}_{\mathrm{vh}}|^2 & \Re e(\widetilde{S}_{\mathrm{vv}}\widetilde{S}^*_{\mathrm{vh}}) & -\Im m(\widetilde{S}_{\mathrm{vv}}\widetilde{S}^*_{\mathrm{vh}}) \\ |\widetilde{S}_{\mathrm{hv}}|^2 & |\widetilde{S}_{\mathrm{hh}}|^2 & \Re e(\widetilde{S}_{\mathrm{hv}}\widetilde{S}^*_{\mathrm{hh}}) & -\Im m(\widetilde{S}_{\mathrm{hv}}\widetilde{S}^*_{\mathrm{hh}}) \\ 2\Re e(\widetilde{S}_{\mathrm{vv}}\widetilde{S}^*_{\mathrm{hv}}) & 2\Re e(\widetilde{S}_{\mathrm{vh}}\widetilde{S}^*_{\mathrm{hh}}) & \Re e(\widetilde{S}_{\mathrm{vv}}\widetilde{S}^*_{\mathrm{hh}}+\widetilde{S}_{\mathrm{vh}}\widetilde{S}^*_{\mathrm{hv}}) & -\Im m(\widetilde{S}_{\mathrm{vv}}\widetilde{S}^*_{\mathrm{hh}}-\widetilde{S}_{\mathrm{vh}}\widetilde{S}^*_{\mathrm{hv}}) \\ 2\Im m(\widetilde{S}_{\mathrm{vv}}\widetilde{S}^*_{\mathrm{hv}}) & 2\Im m(\widetilde{S}_{\mathrm{vh}}\widetilde{S}^*_{\mathrm{hh}}) & \Im m(\widetilde{S}_{\mathrm{vv}}\widetilde{S}^*_{\mathrm{hh}}+\widetilde{S}_{\mathrm{vh}}\widetilde{S}^*_{\mathrm{hv}}) & \Re e(\widetilde{S}_{\mathrm{vv}}\widetilde{S}^*_{\mathrm{hh}}-\widetilde{S}_{\mathrm{vh}}\widetilde{S}^*_{\mathrm{hv}}) \end{bmatrix}. \tag{11.37}$$

▶ One of the fundamental assumptions of radiative transfer theory is that in a medium containing a random distribution of particles, the waves scattered by the particles are random in phase, thereby allowing the addition of the multiple waves incoherently (without regard to their individual phases). ◀

That is, the Stokes parameters of the mixture are the sum of the respective Stokes parameters of the separate waves. For the general case where the medium may contain particles of different shapes, sizes, and orientations, Eq. (11.36) should be replaced with

$$\mathbf{I}^{\mathrm{s}}(\theta_{\mathrm{s}},\phi_{\mathrm{s}}) = \frac{1}{R_{\mathrm{r}}^2} \boldsymbol{\psi}(\theta_{\mathrm{s}},\phi_{\mathrm{s}};\theta_{\mathrm{i}},\phi_{\mathrm{i}})\, \mathbf{I}^{\mathrm{i}}(\theta_{\mathrm{i}},\phi_{\mathrm{i}}), \tag{11.38}$$

where $\boldsymbol{\psi}$ is the ***phase matrix*** given by

$$\begin{aligned} \boldsymbol{\psi}(\theta_{\mathrm{s}},\phi_{\mathrm{s}};\theta_{\mathrm{i}},\phi_{\mathrm{i}}) &= N_{\mathrm{v}} \left\langle \widetilde{\mathbf{M}}(\theta_{\mathrm{s}},\phi_{\mathrm{s}};\theta_{\mathrm{i}},\phi_{\mathrm{i}};x_1,x_2,\ldots,x_n) \right\rangle \\ &= N_{\mathrm{v}} \int \widetilde{\mathbf{M}}(\theta_{\mathrm{s}},\phi_{\mathrm{s}};\theta_{\mathrm{i}},\phi_{\mathrm{i}};x_1,x_2,\ldots,x_n) \\ &\qquad \times p(x_1,x_2,\ldots,x_n)\, dx_1\, dx_2 \ldots dx_n, \end{aligned} \tag{11.39}$$

and $p(x_1,x_2,\ldots,x_n)$ is the joint probability density function over the shape, size, and orientation parameters $(x_1,x_2,\ldots,x_n)$.

11-5.3 Scattering and Absorption Cross Sections

The backscattering cross section of a particle is defined by Eq. (5.30), but the form also applies to the general case of bistatic scattering. Thus, for plane-wave incidence along direction $\hat{\mathbf{i}}$, corresponding to angles $(\theta_{\mathrm{i}},\phi_{\mathrm{i}})$, and spherical-wave scattering along direction $\hat{\mathbf{s}}$, corresponding to angles $(\theta_{\mathrm{s}},\phi_{\mathrm{s}})$, the pq-polarized bistatic scattering cross section is given by

$$\sigma_{pq}(\hat{\mathbf{s}},\hat{\mathbf{i}}) = 4\pi R_{\mathrm{r}}^2 \frac{\mathcal{S}^{\mathrm{s}}_p(\hat{\mathbf{s}})}{\mathcal{S}^{\mathrm{i}}_q(\hat{\mathbf{i}})}, \qquad (p,q = \mathrm{h\ or\ v}), \tag{11.40}$$

where R_{r} is the distance to the receiver (assumed to be sufficiently large as to satisfy the far-field criterion),

$\mathcal{S}_p^{\mathrm{s}}(\hat{\mathbf{s}})$ is the p-polarized power density of the scattered wave and $\mathcal{S}_q^{\mathrm{i}}(\hat{\mathbf{i}})$ is the q-polarized power density of the incident wave. The p-polarized (total) ***scattering cross section*** of the particle, Q_{s}^p, is defined as the ratio of the total power scattered by the particle along all directions (which includes both h- and v-polarized waves), divided by the p-polarized incident power density:

$$\begin{aligned} Q_{\mathrm{s}}^p = \frac{P^{\mathrm{s}}}{\mathcal{S}_p^{\mathrm{i}}(\hat{\mathbf{i}})} &= \frac{\displaystyle\iint \left[\mathcal{S}_{\mathrm{h}}^{\mathrm{s}}(\hat{\mathbf{s}}') + \mathcal{S}_{\mathrm{v}}^{\mathrm{s}}(\hat{\mathbf{s}}')\right] dA'}{\mathcal{S}_p^{\mathrm{i}}(\hat{\mathbf{i}})} \\ &= \frac{\displaystyle\iint_{4\pi} \left[\mathcal{S}_{\mathrm{h}}^{\mathrm{s}}(\hat{\mathbf{s}}') + \mathcal{S}_{\mathrm{v}}^{\mathrm{s}}(\hat{\mathbf{s}}')\right] R_{\mathrm{r}}^2 \, d\Omega_{\mathrm{s}}'}{\mathcal{S}_p^{\mathrm{i}}(\hat{\mathbf{i}})} \\ &= \frac{1}{4\pi} \iint_{4\pi} \left[\sigma_{\mathrm{h}p}(\hat{\mathbf{s}}',\hat{\mathbf{i}}) + \sigma_{\mathrm{v}p}(\hat{\mathbf{s}}',\hat{\mathbf{i}})\right] d\Omega_{\mathrm{s}}'. \end{aligned} \tag{11.41}$$

In view of the relationship given by Eq. (5.31), namely

$$\sigma_{pq}(\hat{\mathbf{s}}',\hat{\mathbf{i}}) = 4\pi \left|\widetilde{S}_{pq}(\hat{\mathbf{s}}',\hat{\mathbf{i}})\right|^2 \qquad (p = \mathrm{h} \text{ or } \mathrm{v}), \tag{11.42}$$

the expression for Q_{s}^p becomes

$$Q_{\mathrm{s}}^p = \iint_{4\pi} \left[\left|\widetilde{S}_{\mathrm{h}p}(\hat{\mathbf{s}}',\hat{\mathbf{i}})\right|^2 + \left|\widetilde{S}_{\mathrm{v}p}(\hat{\mathbf{s}}',\hat{\mathbf{i}})\right|^2\right] d\Omega_{\mathrm{s}}'. \tag{11.43}$$

If the volume has N_{v} randomly distributed particles per unit volume, then the scattering coefficient of the medium is

$$\kappa_{\mathrm{s}}^p = N_{\mathrm{v}} \left\langle Q_{\mathrm{s}}^p \right\rangle. \tag{11.44}$$

The absorption coefficient of the medium is

$$\kappa_{\mathrm{a}}^p = \kappa_{\mathrm{e}}^p - \kappa_{\mathrm{s}}^p = N_{\mathrm{v}} \left\langle Q_{\mathrm{e}}^p - Q_{\mathrm{s}}^p \right\rangle, \qquad (p = \mathrm{h} \text{ or } \mathrm{v}), \tag{11.45}$$

where Q_{e}^p is the extinction cross section given by Eq. (11.30). The p-polarized ***single scattering albedo*** is

$$a^p = \frac{\kappa_{\mathrm{s}}^p}{\kappa_{\mathrm{e}}^p}. \tag{11.46}$$

If the background medium in which the particles are embedded is not air or a lossless dielectric, we can account for the ***background absorption coefficient*** $\kappa_{\mathrm{a,b}}$ as follows:

$$\kappa_{\mathrm{a,b}} = 2k_0 n_{\mathrm{b}}'' = -2k_0 \,\mathfrak{Im}\left\{\sqrt{\varepsilon_{\mathrm{b}}}\right\}, \tag{11.47}$$

where n_{b}'' is the imaginary part of the complex index of refraction of the background medium and $k_0 = 2\pi/\lambda_0$. The ***total extinction matrix*** of the medium becomes

$$\boldsymbol{\kappa}_{\mathrm{e,t}} = \boldsymbol{\kappa}_{\mathrm{e}} + \boldsymbol{\kappa}_{\mathrm{a,b}}, \tag{11.48}$$

where $\boldsymbol{\kappa}_{\mathrm{e}}$ is the extinction matrix of the particles given by Eq. (11.33), and $\boldsymbol{\kappa}_{\mathrm{a,b}} = \kappa_{\mathrm{a,b}}\mathbf{U}$ and $\mathbf{U}$ is a 4×4 unit matrix.

11-5.4 Applicability Conditions

The radiative transfer equation is formulated on the basis of energy balance. Thus, the phase changes of the scattered wave and its cross-correlation terms are ignored in the solution of the transfer equation. Furthermore, the phase function used in practice comes from the averaging of the magnitude squared of the far-zone scattering amplitudes caused by an individual scatterer or its equivalent. Hence, for the transfer method to apply, the spacing between scatterers in a discrete medium or adjacent inhomogeneities in a continuous medium must be large. An experimental study by Vasalos (1969) shows that for the transfer equation to be applicable, the spacing between scatterers must be larger than $\lambda/3$ and $0.4d$, where λ is the wavelength in the host medium and d is the diameter of the scatterer. Vasalos considered volume fractions as high as 0.295, d/λ ratios from 0.186 to 1.2, and optical thicknesses in the 0.01–3211 range. Additional studies by Hottel et al. (1971) show that a smaller spacing-to-wavelength ratio ($l/\lambda = 0.117$) can be acceptable if the d/λ ratio is larger than 0.23 at a volume fraction of 0.2195 or less. Theoretical studies on the relationship between the radiative transfer approach and the wave approach using Maxwell's equations have been carried out by Fante (1981). Related studies have also been reported by Ishimaru and Kuga (1982).

11-5.5 Phase Matrix of Simple Objects

With the definitions given in the preceding subsections for the phase and extinction matrices, we can, in principle, apply a suitable numerical technique to solve the vector radiative transfer equation given by Eq. (11.27) for a medium with any specified particle distribution and boundary conditions. Because of multiple scattering, each elemental volume within the inhomogeneous medium is coupled to every other elemental volume, and wave propagation in a given direction is coupled to wave propagation along all other directions. Obtaining a numerical solution is doable, but computationally intensive. To avoid the computational challenge, we have two basic options, namely:

(a) to limit the solution to first-order or second-order scattering, as illustrated later in Section 11-6, or

(b) to treat the particles as isotropic point scatterers or as equivalent spherical Rayleigh scatterers, thereby simplifying the forms of the extinction and phase matrices. To justify doing so requires that the wavelength λ be much larger than the particle size. Under such simplifying conditions, it is possible to obtain exact numerical solutions for the vector radiative transfer equation.

Isotropic scatterers

Isotropic point scatterers are characterized by a simple scalar phase function given by

$$\psi = \frac{\kappa_s}{4\pi} = a\,\frac{\kappa_e}{4\pi}, \qquad (11.49)$$

where $a = \kappa_s/\kappa_e$ is the single scattering albedo. Note that ψ is independent of both polarization and incidence angle.

Spherical Rayleigh scatterers

If an ice crystal is not spherical in shape but small in size relative to λ, it may be modeled as an equivalent sphere with the same size and mass as the actual crystal. In principle, this equivalency can also be applied to other types of scattering particle, such as small leaves and needles in vegetation canopies. Scattering by a sphere of radius r and relative index of refraction n can be computed using the Rayleigh approximation, which yields acceptable accuracies so long as the product of n and the parameter $\chi = kr = 2\pi r/\lambda_b$ satisfies the condition

$$|n\chi| = \left|\frac{2\pi n}{\lambda_b}\,r\right| < 0.5, \qquad (11.50)$$

where λ_b is the wavelength in the background medium and $n = n_p/n_b$ (with n_p and n_b being the indices of refraction of the particle and background, respectively). The phase matrix of a random medium containing spherical Rayleigh scatterers with incidence along (θ_i, ϕ_i) and scattering along (θ_s, ϕ_s), as shown in Fig. 11-10, is given by (Tsang et al., 1985, pp. 157–158):

$$\boldsymbol{\psi}(\phi_s,\theta_s;\phi_i,\theta_i) = \begin{bmatrix} \psi_{11} & \psi_{12} & \psi_{13} & 0 \\ \psi_{21} & \psi_{22} & \psi_{23} & 0 \\ \psi_{31} & \psi_{32} & \psi_{33} & 0 \\ 0 & 0 & 0 & \psi_{44} \end{bmatrix}, \qquad (11.51)$$

where

$$\begin{aligned}
\psi_{11} &= b[\sin^2\theta_s \sin^2\theta_i + \cos^2\theta_s\cos^2\theta_i\cos^2(\phi_s-\phi_i) \\
&\qquad + 2\sin\theta_s\sin\theta_i\cos\theta_s\cos\theta_i\cos(\phi_s-\phi_i)],\\
\psi_{12} &= b\cos^2\theta_s\sin^2(\phi_s-\phi_i),\\
\psi_{13} &= b[\cos\theta_s\sin\theta_s\sin\theta_i\sin(\phi_s-\phi_i)\\
&\qquad + \cos^2\theta_s\cos\theta_i\sin(\phi_s-\phi_i)\cos(\phi_s-\phi_i)],\\
\psi_{21} &= b\cos^2\theta_i\sin^2(\phi_s-\phi_i),\\
\psi_{22} &= b\cos^2(\phi_s-\phi_i),\\
\psi_{23} &= -b\cos\theta_s\sin(\phi_s-\phi_i)\cos(\phi_s-\phi_i),\\
\psi_{31} &= b[-2\sin\theta_s\sin\theta_i\cos\theta_i\sin(\phi_s-\phi_i)\\
&\qquad - 2\cos\theta_s\cos^2\theta_i\cos(\phi_s-\phi_i)\sin(\phi_s-\phi_i)],\\
\psi_{32} &= 2b\cos\theta_s\sin(\phi_s-\phi_i)\cos(\phi_s-\phi_i)],\\
\psi_{33} &= b\{\sin\theta_s\sin\theta_i\cos(\phi_s-\phi_i)\\
&\qquad + \cos\theta_s\cos\theta_i[\cos^2(\phi_s-\phi_i)-\sin^2(\phi_s-\phi_i)]\},\\
\psi_{44} &= b[\sin\theta_s\sin\theta_i\cos(\phi_s-\phi_i)+\cos\theta_s\cos\theta_i],\\
b &= \frac{3}{8\pi}\,\kappa_s,
\end{aligned}$$

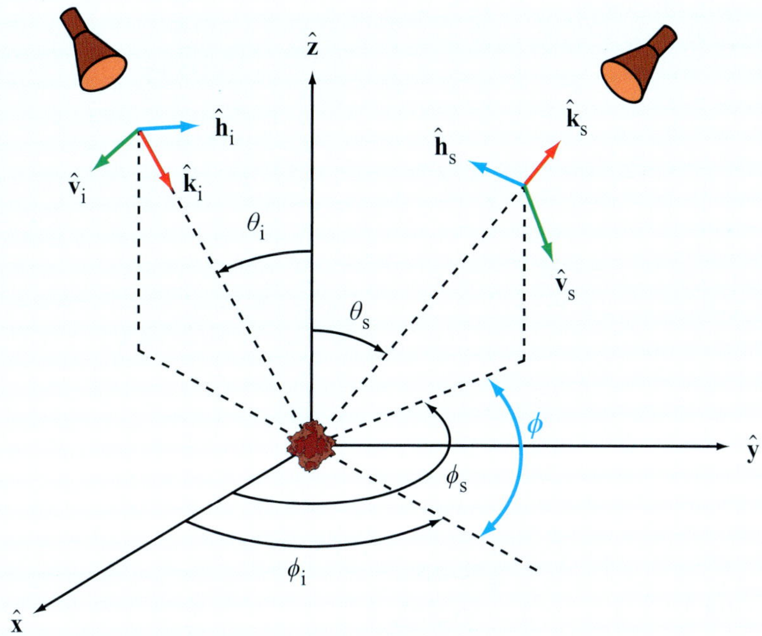

Figure 11-10: Incident and scattering directions for a point scatterer.

and κ_s is the scattering coefficient.

From Eq. (8.45a),

$$\kappa_s = N_v Q_s = \frac{2\lambda^2}{3\pi} N_v \chi^6 |K|^2 = \frac{8\pi}{3} N_v k^4 r^6 |K|^2 = 2\upsilon k^4 r^3 |K|^2, \qquad (11.52)$$

where $k = 2\pi/\lambda_b$, N_v is the number density of particles (in #/m^3), υ is the ***volume fraction*** of scatterers,

$$\upsilon = \left(\frac{4}{3}\pi r^3\right) N_v, \qquad (11.53)$$

and K is defined by Eq. (8.43) as

$$K = \frac{n^2-1}{n^2+2} = \frac{\varepsilon - 1}{\varepsilon + 2}, \qquad (11.54)$$

where $\varepsilon = n^2 = \varepsilon_p/\varepsilon_b$.

The phase matrix describes the bistatic scattering pattern of Rayleigh particles, while accounting for the polarization directions of the incident and scattered waves. In the ***plane of incidence*** ($\phi_s = \phi_i$), all nondiagonal terms are zero, and the diagonal terms simplify to

$$\psi_{11} = \frac{3\kappa_s}{8\pi}(\sin\theta_s \sin\theta_i + \cos\theta_s \cos\theta_i)^2, \qquad (11.55a)$$

$$\psi_{22} = \frac{3\kappa_s}{8\pi}, \qquad (11.55b)$$

$$\psi_{33} = \psi_{44} = \frac{3\kappa_s}{8\pi}(\sin\theta_s \sin\theta_i + \cos\theta_s \cos\theta_i). \qquad (11.55c)$$

Moreover, if $\theta_s = \theta_i$ (in addition to $\phi_s = \phi_i$),

$$\psi_{11} = \psi_{22} = \psi_{33} = \psi_{44} = \frac{3\kappa_s}{8\pi}. \qquad (11.56)$$

From Eq. (8.45a), the absorption coefficient is given by

$$\kappa_a = N_v Q_a = \frac{\lambda^2}{\pi} N_v \chi^3 \, \mathfrak{Im}[-K] = 3\upsilon k \, \mathfrak{Im}[-K] = 3\upsilon k \frac{\varepsilon''}{|\varepsilon+2|^2}. \qquad (11.57)$$

The extinction coefficient due to the particles is $\kappa_e = \kappa_s + \kappa_a$, and the extinction matrix is diagonal with each element equal to κ_e, as in Eq. (11.32).

11-5.6 Boundary Conditions for a Planar Interface

When the specific intensity impinges upon the boundary separating two media with different dielectric constants, part of the incident energy is reflected and the remaining part is transmitted across the boundary. In this section, we consider the boundary conditions relating the Stokes parameters of the specific intensities in the two media.

If the boundary is a planar dielectric interface, scattering at the boundary occurs in only the specular direction and does not include a diffuse component. If the boundary is rough, however, the incident wave is scattered in all directions at the boundary and the incident intensity is coupled into all of the reflected and transmitted directions. However, if the scattering albedo of the particles contained in the scattering layer is larger than about 0.3, the numerical approach to solving the radiative transfer equation is the only viable technique for obtaining reasonably accurate results; the iterative-solution approach is inadequate for $a \gtrsim 0.3$. Nevertheless, in this book, we limit our discussion to the smooth-boundary case. For examination of the boundary conditions associated with a rough surface, we refer the interested reader to the article by Fung and Eom (1981), who investigated radar scattering from a layer of Rayleigh particles over a rough surface by using the Kirchhoff approximation, and to the books by Tsang et al. (1985) and Fung and Chen (2010).

For a plane wave incident upon a planar boundary from medium 1 to medium 2, as shown in Fig. 11-11, the expressions for the Fresnel reflection and transmission coefficients and associated reflectivities and transmissivities given in Table 2-5 can be written in the form:

$$\rho_h = \frac{n_1 \cos\theta_1 - n_2 \cos\theta_2}{n_1 \cos\theta_1 + n_2 \cos\theta_2}, \tag{11.58a}$$

$$\rho_v = \frac{n_1 \cos\theta_2 - n_2 \cos\theta_1}{n_1 \cos\theta_2 + n_2 \cos\theta_1}, \tag{11.58b}$$

$$\tau_h = 1 + \rho_h, \qquad \tau_v = (1 + \rho_v)\,\frac{\cos\theta_1}{\cos\theta_2} \tag{11.58c}$$

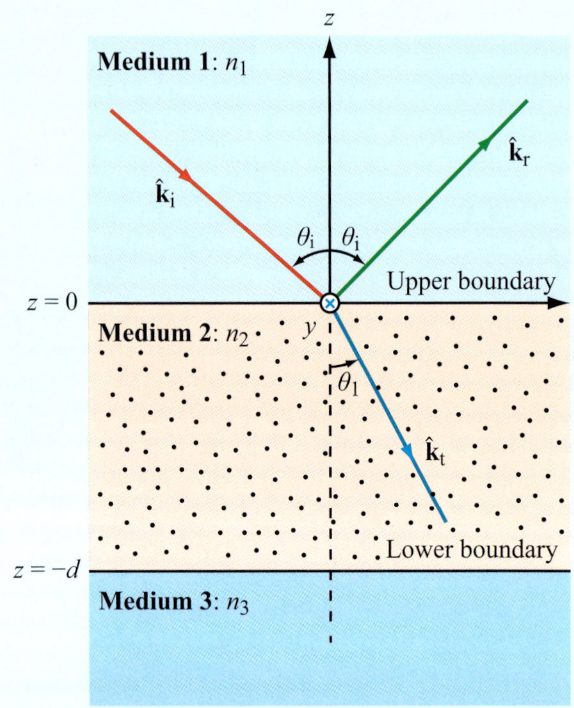

Figure 11-11: Incidence in medium 1 with index of refraction n_1 upon medium 2 with index of refraction n_2.

$$\Gamma^h = |\rho_h|^2, \qquad \Gamma^v = |\rho_v|^2, \tag{11.58d}$$

$$\mathbb{T}^h = 1 - \Gamma^h, \qquad \mathbb{T}^v = 1 - \Gamma^v, \tag{11.58e}$$

where n_1 and n_2 are the indices of refraction of media 1 and 2, respectively, and angles θ_1 and θ_2 are related by Snell's law:

$$n_2 \sin\theta_2 = n_1 \sin\theta_1. \tag{11.59}$$

The vector specific intensity $\mathbf{I}$ of a propagating wave is defined in Eq. (11.29) in terms of the h- and v-polarized components of the wave's electric field. Using superscripts $\mathbf{I}^i$, $\mathbf{I}^r$, and $\mathbf{I}^t$ to denote the incident, reflected, and transmitted specific intensities, the relationships between them are given by Ishimaru (1978) as

$$\mathbf{I}^r = \mathbb{R}\mathbf{I}^i, \tag{11.60a}$$

and

$$\mathbf{I}^t = \mathbb{T}\mathbf{I}^i, \tag{11.60b}$$

where $\mathbb{R}$ and $\mathbb{T}$ are ***reflectivity*** and ***transmissivity matrices*** given by

$$\mathbb{R} = \begin{bmatrix} \Gamma^{\mathrm{v}} & 0 & 0 & 0 \\ 0 & \Gamma^{\mathrm{h}} & 0 & 0 \\ 0 & 0 & \Re\mathrm{e}(\rho_{\mathrm{v}}\rho_{\mathrm{h}}^*) & -\Im\mathrm{m}(\rho_{\mathrm{v}}\rho_{\mathrm{h}}^*) \\ 0 & 0 & \Im\mathrm{m}(\rho_{\mathrm{v}}\rho_{\mathrm{h}}^*) & \Re\mathrm{e}(\rho_{\mathrm{v}}\rho_{\mathrm{h}}^*) \end{bmatrix} \tag{11.61a}$$

and

$$\mathbb{T} = \frac{n_2^3 \cos\theta_2}{n_1^3 \cos\theta_1} \begin{bmatrix} |\tau_{\mathrm{v}}|^2 & 0 & 0 & 0 \\ 0 & |\tau_{\mathrm{v}}|^2 & 0 & 0 \\ 0 & 0 & \Re\mathrm{e}(\tau_{\mathrm{v}}\tau_{\mathrm{h}}^*) & -\Im\mathrm{m}(\tau_{\mathrm{v}}\tau_{\mathrm{h}}^*) \\ 0 & 0 & \Im\mathrm{m}(\tau_{\mathrm{v}}\tau_{\mathrm{h}}^*) & \Re\mathrm{e}(\tau_{\mathrm{v}}\tau_{\mathrm{h}}^*) \end{bmatrix}. \tag{11.61b}$$

11-6 Iterative Solution of the Radiative Transfer Equation

Except for some special cases, the vector radiative transfer equation does not have an analytical solution. The approximations and numerical techniques used for solving the radiative transfer equation include the ***iterative-solution technique*** and the ***discrete-ordinate technique*** (Chandrasekhar, 1960; Ishimaru, 1978; Tsang et al., 1985; Ulaby et al., 1986a; Fung, 1994). In the iterative approach, the radiative transfer equation is cast into an integral form, and then solved iteratively to obtain the zero-, first-, and second-order solutions. In principle, we can obtain an accurate solution by iterating many times. In practice, however, iteration beyond the second-order solution is not necessary.

The discrete-ordinate eigenanalysis technique provides good numerical solutions for the radiative transfer equation (Ishimaru et al., 1982; Ulaby et al., 1986a) and has been used extensively in the literature. In this technique, the specific intensity and phase matrix are first discretized into a finite number of directions by means of quadrature, and the resulting matrix equation is solved by eigenanalysis.

In addition to these, several other techniques have been applied to solve the radiative transfer equation, including the invariant-embedding method and the finite-difference method (Tsang et al., 1985). The major limitation of these numerical techniques is the fact that they are computationally very demanding, particularly if the angular variation of the phase matrix is very complex. However, if the scattering albedo a of the particles contained in the scattering layer is larger than about 0.3, the numerical approach to solving the radiative transfer equation is the only viable technique for obtaining reasonably accurate results; the iterative-solution approach is inadequate for $a \gtrsim 0.3$. Nevertheless, in this book, we limit our presentation to the iterative method.

11-6.1 Iterative-Solution Method

The iterative method is applicable when the random medium is ***weakly scattering***; that is, when the albedo a is small. The procedure starts by computing the ***zero-order solution***, which ignores scattering, except for its contribution to extinction. The zero-order solution then serves as a ***source function*** for computing the first-order solution, and then the first-order solution becomes the source function for the second-order solution.

To illustrate the mechanics of the iterative solution method, we consider the simple three-layer case shown in Fig. 11-12, in which:

(a) the middle layer contains uniformly distributed spherical particles, which reduces the extinction matrix $\boldsymbol{\kappa}_{\mathrm{e}}$ into the product of a unitary diagonal matrix and the scalar extinction coefficient κ_{e},

(b) the upper boundary is diffuse, which means that no reflection occurs at the top interface at $z = 0$, and

(c) the bottom boundary is flat, which means that only specular reflection occurs at the lower interface at $z = -d$.

The incident specific intensity is $\mathbf{I}_0^{\mathrm{i}}(-\mu_{\mathrm{i}}, \phi_{\mathrm{i}})$, where $\mu_{\mathrm{i}} = \cos\theta_{\mathrm{i}}$ and $-\mu_{\mathrm{i}} = \cos(\pi - \theta_{\mathrm{i}})$. Thus, μ_{i} denotes an upward propagating wave and $-\mu_{\mathrm{i}}$ denotes a downward propagating wave. The total backscattered intensity is denoted $\mathbf{I}^{\mathrm{s}}(\mu_{\mathrm{i}}, \phi_{\mathrm{i}} + \pi)$, which is related to the incident intensity $\mathbf{I}_0^{\mathrm{i}}$ by

$$\mathbf{I}^{\mathrm{s}}(\mu_{\mathrm{i}}, \phi_{\mathrm{i}} + \pi) = \boldsymbol{\mathcal{T}}_{\mathrm{t}}(\mu_{\mathrm{i}}, \phi_{\mathrm{i}} + \pi; -\mu_{\mathrm{i}}, \phi_{\mathrm{i}})\, \mathbf{I}_0^{\mathrm{i}}(-\mu_{\mathrm{i}}, \phi_{\mathrm{i}}), \tag{11.62}$$

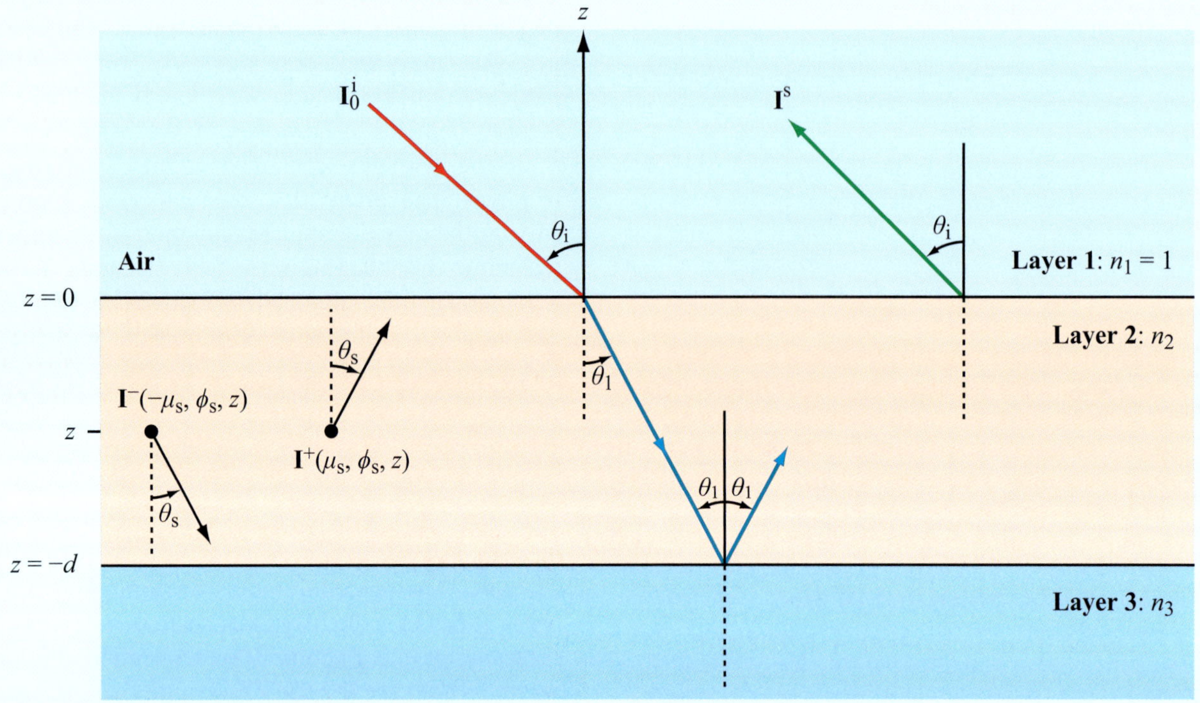

Figure 11-12: Incident vector intensity $\mathbf{I}_0^{\mathrm{i}}$, downward and upward propagating intensities $\mathbf{I}^-$ and $\mathbf{I}^+$, and backscattered intensity $\mathbf{I}^{\mathrm{s}}$.

where $\mathcal{T}_{\mathrm{t}}(\mu_{\mathrm{i}}, \phi_{\mathrm{i}} + \pi; -\mu_{\mathrm{i}}, \phi_{\mathrm{i}})$ is a ***total backscattering transformation matrix***. The goal of the radiative-transfer solution is to obtain an expression for $\mathcal{T}_{\mathrm{t}}$ for downward incidence along $(-\mu_{\mathrm{i}}, \phi_{\mathrm{i}})$ and upward backscattering along $(\mu_{\mathrm{i}}, \phi_{\mathrm{i}} + \pi)$. However, because the bottom boundary is assumed to be flat, the radiative transfer model does not account for the attenuated direct backscatter from the bottom surface.

Given $\mathcal{T}_{\mathrm{t}}$, the backscattering coefficients for vv, hh, and hv polarizations are given by

$$\sigma^0_{\mathrm{vv}}(\theta_{\mathrm{i}}) = \Upsilon^2\, \sigma^0_{\mathrm{s_{vv}}}(\theta_{\mathrm{i}}) + 4\pi\cos\theta_{\mathrm{i}}[\mathcal{T}_{\mathrm{t}}]_{11}, \tag{11.63a}$$

$$\sigma^0_{\mathrm{hh}}(\theta_{\mathrm{i}}) = \Upsilon^2\, \sigma^0_{\mathrm{s_{hh}}}(\theta_{\mathrm{i}}) + 4\pi\cos\theta_{\mathrm{i}}[\mathcal{T}_{\mathrm{t}}]_{22}, \tag{11.63b}$$

$$\sigma^0_{\mathrm{hv}}(\theta_{\mathrm{i}}) = \Upsilon^2\, \sigma^0_{\mathrm{s_{hv}}}(\theta_{\mathrm{i}}) + 4\pi\cos\theta_{\mathrm{i}}[\mathcal{T}_{\mathrm{t}}]_{21}, \tag{11.63c}$$

where Υ^2 is the two-way transmissivity of the middle layer and $\sigma^0_{\mathrm{s}_{pq}}$ is the backscattering coefficient of the bottom surface in the absence of the middle layer. To compute the backscattering coefficient for other receive/transmit polarization combinations of interest, we can use the polarization synthesis technique of Section 5-11.3. The parallel to the formulation given by Eq. (5.150) is

$$\sigma^0_{\mathrm{rt}}(\psi_{\mathrm{r}}, \chi_{\mathrm{r}}; \psi_{\mathrm{t}}, \chi_{\mathrm{t}}) = \Upsilon^2\, \sigma^0_{\mathrm{s_{rt}}}(\theta_{\mathrm{i}}) + 4\pi\cos\theta_{\mathrm{i}}\; \mathbf{I}^{\mathrm{r}}_{\mathrm{n}} \cdot \mathbf{Q}\, \mathcal{T}_{\mathrm{t}} \mathbf{I}^{\mathrm{t}}_{\mathrm{n}}, \tag{11.64}$$

where $\mathbf{I}^{\mathrm{r}}_{\mathrm{n}}$ and $\mathbf{I}^{\mathrm{t}}_{\mathrm{n}}$ describe the polarizations of the received and transmitted waves and are defined by Eq. (5.145), and $\mathbf{Q}$ is a 4×4 matrix given by Eq. (5.149).

11-6.2 Upward- and Downward-Propagating Intensities

We start by designating $\mathbf{I}^{+}(\mu_{\mathrm{s}}, \phi_{\mathrm{s}}, z)$ as the ***upward-propagating intensity*** at depth z, along direction $(\mu_{\mathrm{s}}, \phi_{\mathrm{s}})$. Similarly, $\mathbf{I}^{-}(-\mu_{\mathrm{s}}, \phi_{\mathrm{s}}, z)$ is the ***downward-propagating intensity*** along $(-\mu_{\mathrm{s}}, \phi_{\mathrm{s}})$. Inside the middle layer, the intensities $\mathbf{I}^+$ and $\mathbf{I}^-$ must satisfy the coupled radiative transfer equations:

$$\frac{d}{dz}\,\mathbf{I}^+(\mu_s,\phi_s,z) = -\frac{\kappa_e}{\mu_s}\,\mathbf{I}^+(\mu_s,\phi_s,z) + \mathfrak{F}^+(\mu_s,\phi_s,z), \tag{11.65a}$$

$$-\frac{d}{dz}\,\mathbf{I}^-(-\mu_s,\phi_s,z) = -\frac{\kappa_e}{\mu_s}\,\mathbf{I}^-(-\mu_s,\phi_s,z) + \mathfrak{F}^-(-\mu_s,\phi_s,z). \tag{11.65b}$$

Note that $d\mathbf{I}^-/dz$ is preceded by a minus sign in Eq. (11.65b). This is because the direction of propagation is along $-\mu_s$. This pair of equations is obtained from Eq. (11.27) by (a) replacing the lateral distance ds with $dz \sec\theta_s$, (b) using Eq. (11.32) for $\boldsymbol{\kappa}_e$, thereby, in essence, replacing $\boldsymbol{\kappa}_e$ with the scalar extinction coefficient κ_e, (c) either deleting the background absorption term in Eq. (11.27) because the background dielectric in the middle layer is lossless, in which case $\kappa_{a,b} = 0$, or including $\kappa_{a,b}$ in κ_e if the background dielectric is not lossless, and (d) denoting the term that includes the phase matrix as $\mathfrak{F}^+$ for upward propagation and denoting it as $\mathfrak{F}^-$ for downward propagation. Thus, $\mathfrak{F}^+$ and $\mathfrak{F}^-$ are ***source functions*** that account for directing the energy incident upon an elemental volume from all directions into directions (θ_s,ϕ_s) and $(\pi-\theta_s,\phi_s)$, respectively. They are given by

$$\mathfrak{F}^+(\mu_s,\phi_s,z) = \frac{1}{\mu_s}\left[\int_0^{2\pi}\!\!\int_0^1 \boldsymbol{\psi}(\mu_s,\phi_s;\mu',\phi')\,\mathbf{I}^+(\mu',\phi',z)\,d\Omega' + \int_0^{2\pi}\!\!\int_0^1 \boldsymbol{\psi}(\mu_s,\phi_s;-\mu',\phi')\,\mathbf{I}^-(-\mu',\phi',z)\,d\Omega'\right], \tag{11.66a}$$

$$\mathfrak{F}^-(-\mu_s,\phi_s,z) = \frac{1}{\mu_s}\left[\int_0^{2\pi}\!\!\int_0^1 \boldsymbol{\psi}(-\mu_s,\phi_s;\mu',\phi')\,\mathbf{I}^+(\mu',\phi',z)\,d\Omega' + \int_0^{2\pi}\!\!\int_0^1 \boldsymbol{\psi}(-\mu_s,\phi_s;-\mu',\phi')\,\mathbf{I}^-(-\mu',\phi',z)\,d\Omega'\right], \tag{11.66b}$$

where

$$d\Omega' = d\mu'\,d\phi' = \sin\theta'\,d\theta'\,d\phi'$$

and $\boldsymbol{\psi}(\mu_s,\phi_s;\mu',\phi')$ is the phase matrix of the middle layer.

The boundary conditions are

$$\mathbf{I}^-(-\mu_s,\ \phi_s,\ z=0) = \mathbf{I}_0^{\mathrm{i}}\,\delta(\mu_s-\mu_{\mathrm{i}})\,\delta(\phi_s-\phi_{\mathrm{i}}), \tag{11.67a}$$

$$\mathbf{I}^+(\mu_s,\ \phi_s,\ z=-d) = \mathbb{R}(\mu_s)\,\mathbf{I}^-(-\mu_s,\ \phi_s,\ z=-d), \tag{11.67b}$$

where $\mathbf{I}_0^{\mathrm{i}}$ is the specific intensity incident in medium 1 along the direction $(\mu_{\mathrm{i}},\phi_{\mathrm{i}})$ and $\mathbb{R}(\mu_s)$ is the reflectivity matrix at $z=-d$ at angle μ_s. The expression for $\mathbb{R}(\mu_s)$ is given by Eq. (11.61a). Repeat of the steps used in Section 6-6.3 leads to the following general solution for Eqs. (11.65a and b):

$$\mathbf{I}^+(\mu_s,\phi_s,z) = e^{-\kappa_e(z+d)/\mu_s}\,\mathbf{I}^+(\mu_s,\ \phi_s,\ z=-d) + \int_{-d}^{z} e^{-\kappa_e(z-z')/\mu_s}\,\mathfrak{F}^+(\mu_s,\ \phi_s,\ z')\,dz', \tag{11.68a}$$

$$\mathbf{I}^-(-\mu_s,\phi_s,z) = e^{\kappa_e z/\mu_s}\,\mathbf{I}^-(-\mu_s,\ \phi_s,\ z=0) + \int_{z}^{0} e^{\kappa_e(z-z')/\mu_s}\,\mathfrak{F}^-(-\mu_s,\ \phi_s,\ z')\,dz', \tag{11.68b}$$

By applying the boundary conditions given by Eqs. (11.67a and b), we have

$$\mathbf{I}^+(\mu_s,\phi_s,z) = e^{-\kappa_e(z+d)/\mu_s}\,\mathbb{R}(\mu_{\mathrm{i}})\,e^{-\kappa_e d/\mu_{\mathrm{i}}}\mathbf{I}_0^{\mathrm{i}}\,\delta(\mu_s-\mu_{\mathrm{i}})\,\delta(\phi_s-\phi_{\mathrm{i}}) \cdot e^{-\kappa_e(z+d)/\mu_s}\,\mathbb{R}(\mu_s)\int_{-d}^{z} e^{\kappa_e(-d-z')/\mu_s}\,\mathfrak{F}^-(\ \ \mu_s,\phi_s,z')\,dz' + \int_{-d}^{z} e^{-\kappa_e(z-z')/\mu_s}\,\mathfrak{F}^+(-\mu_s,\phi_s,z')\,dz', \tag{11.69a}$$

$$\mathbf{I}^-(-\mu_s,\phi_s,z) = e^{\kappa_e z/\mu_{\mathrm{i}}}\mathbf{I}_0^{\mathrm{i}}\,\delta(\mu_s-\mu_{\mathrm{i}})\,\delta(\phi_s-\phi_{\mathrm{i}}) + \int_{z}^{0} e^{\kappa_e(z-z')/\mu_s}\,\mathfrak{F}^-(-\mu_s,\phi_s,z')\,dz'. \tag{11.69b}$$

In the iterative approach, the solutions are expanded into a series of perturbation orders:

$$\mathbf{I}(\mu,\phi,z)=\mathbf{I}_0(\mu,\phi,z)+\mathbf{I}_1(\mu,\phi,z)+\mathbf{I}_2(\mu,\phi,z)+\cdots, \tag{11.70}$$

where $-1<\mu<1$. Because the scattering is treated as an iterative parameter, the perturbation orders have one-to-one correspondence with the multiple scattering process. The solution $\mathbf{I}_0$ represents reduced incident intensity, $\mathbf{I}_1$ represents the single scattering solution, and $\mathbf{I}_2$ represents the double scattering solution. The iterative solution method is applicable only when the medium's single scattering albedo $a=\kappa_s/\kappa_e$ is small (i.e., $a\lesssim 0.2$).

11-6.3 Zeroth-Order Solution

The zeroth-order solution is obtained by setting $\mathfrak{F}^+=0$ and $\mathfrak{F}^-=0$ in Eqs. (11.69a and b):

$$\mathbf{I}_0^-(-\mu_s,\phi_s,z)=e^{\kappa_e z/\mu_i}\mathbf{I}_0^i\,\delta(\mu_s-\mu_i)\,\delta(\phi_s-\phi_i), \tag{11.71a}$$

$$\mathbf{I}_0^+(\mu_s,\phi_s,z)=e^{-\kappa_e(z+d)/\mu_i}\,\mathbb{R}\,e^{-\kappa_e d/\mu_i}\mathbf{I}_0^i\cdot\delta(\mu_s-\mu_i)\,\delta(\phi_s-\phi_i). \tag{11.71b}$$

The impulse functions in Eq. (11.71a) specify the direction of the incident wave, and the impulse functions in Eq. (11.71b) specify that reflection by the lower boundary occurs only along the specular direction.

> ▶ The zeroth-order solution is simply the ***reduced incident intensity***, which attenuates exponentially inside the medium. Scattering is not included, except for its contribution to extinction. ◀

11-6.4 First-Order Solution

Use of the zeroth-order solutions given by Eq. (11.71) in Eq. (11.66) leads to the zeroth-order source functions:

$$\begin{aligned}
&\mathfrak{F}_0^+(\mu_s,\phi_s,z)\\
&=\frac{1}{\mu_s}\left[\int_0^{2\pi}\int_0^1\boldsymbol{\psi}(\mu_s,\phi_s;\mu',\phi')\,\mathbf{I}_0^+(\mu',\phi',z)\,d\Omega'\right.\\
&\qquad\left.\int_0^{2\pi}\int_0^1\boldsymbol{\psi}(\mu_s,\phi_s;-\mu',\phi')\,\mathbf{I}_0^-(-\mu',\phi',z)\,d\Omega'\right]\\
&=\frac{1}{\mu_s}\left[\int_0^{2\pi}\int_0^1\boldsymbol{\psi}(\mu_s,\phi_s;\mu',\phi')\,e^{-\kappa_e(z+d)/\mu_i}\,\mathbb{R}(\mu_i)\right.\\
&\qquad\cdot e^{-\kappa_e d/\mu_i}\mathbf{I}_0^i\,\delta(\mu'-\mu_i)\,\delta(\phi'-\phi_i)\,d\Omega'\\
&\qquad+\int_0^{2\pi}\int_0^1\boldsymbol{\psi}(\mu_s,\phi_s;-\mu',\phi')\,e^{\kappa_e z/\mu_i}\,\mathbf{I}_0^i\\
&\qquad\left.\cdot\,\delta(\mu'-\mu_i)\,\delta(\phi'-\phi_i)\,d\Omega'\right]\\
&=\left[e^{-\kappa_e(z+2d)/\mu_i}\,\boldsymbol{\psi}(\mu_s,\phi_s;\mu_i,\phi_i)\,\mathbb{R}(\mu_i)\right.\\
&\qquad\left.+e^{\kappa_e z/\mu_i}\,\boldsymbol{\psi}(\mu_s,\phi_s;-\mu_i,\phi_i)\,\mathbb{R}(\mu_i)\right]\frac{\mathbf{I}_0}{\mu_s},
\end{aligned} \tag{11.72a}$$

$$\begin{aligned}
&\mathfrak{F}_0^-(-\mu_s,\phi_s,z)\\
&=\frac{1}{\mu_s}\left[\int_0^{2\pi}\int_0^1\boldsymbol{\psi}(-\mu_s,\phi_s;\mu',\phi')\,\mathbf{I}_0^+(\mu',\phi',z)\,d\Omega'\right.\\
&\qquad\left.\int_0^{2\pi}\int_0^1\boldsymbol{\psi}(-\mu_s,\phi_s;-\mu',\phi')\,\mathbf{I}_0^-(-\mu',\phi',z)\,d\Omega'\right]\\
&=\frac{1}{\mu_s}\left[\int_0^{2\pi}\int_0^1\boldsymbol{\psi}(-\mu_s,\phi_s;\mu',\phi')\,e^{-\kappa_e(z+d)/\mu_i}\,\mathbb{R}(\mu_i)\right.\\
&\qquad\cdot e^{-\kappa_e d/\mu_i}\mathbf{I}_0^i\,\delta(\mu'-\mu_i)\,\delta(\phi'-\phi_i)\,d\Omega'\\
&\qquad+\int_0^{2\pi}\int_0^1\boldsymbol{\psi}(-\mu_s,\phi_s;-\mu',\phi')\,e^{\kappa_e z/\mu_i}\,\mathbf{I}_0^i\\
&\qquad\left.\cdot\,\delta(\mu'-\mu_i)\,\delta(\phi'-\phi_i)\,d\Omega'\right]\\
&=\left[e^{-\kappa_e(z+2d)/\mu_i}\,\boldsymbol{\psi}(-\mu_s,\phi_s;\mu_i,\phi_i)\,\mathbb{R}(\mu_i)\right.\\
&\qquad\left.+e^{\kappa_e z/\mu_i}\,\boldsymbol{\psi}(-\mu_s,\phi_s;-\mu_i,\phi_i)\,\mathbb{R}(\mu_i)\right]\frac{\mathbf{I}_0}{\mu_s}.
\end{aligned} \tag{11.72b}$$

To obtain the first-order solutions for the upward- and downward-propagating intensities, $\mathbf{I}_1^+$ and $\mathbf{I}_1^-$, we insert the zeroth-order source functions given by Eq. (11.72) into Eq. (11.69):

$$\begin{aligned}
\mathbf{I}_1^+(\mu_s,\phi_s,z) = {} & \\
& e^{-\kappa_e(z+2d)/\mu_i}\,\mathbb{R}(\mu_i)\,\mathbf{I}_0^i\,\delta(\mu_s-\mu_i)\,\delta(\phi_s-\phi_i) \\
& + e^{-\kappa_e(z+d)/\mu_s}\,\mathbb{R}(\mu_s)\int_{-d}^{z} e^{\kappa_e(-d-z')/\mu_s} \\
& \quad \cdot \Big[e^{-\kappa_e(z'+2d)/\mu_i}\,\boldsymbol{\psi}(-\mu_s,\phi_s;\mu_i,\phi_i)\,\mathbb{R}(\mu_i) \\
& \qquad + e^{\kappa_e z'/\mu_i}\,\boldsymbol{\psi}(-\mu_s,\phi_s;-\mu_i,\phi_i)\Big]\frac{\mathbf{I}_0^i}{\mu_s}\,dz' \\
& + \int_{-d}^{z} e^{-\kappa_e(z-z')/\mu_s} \\
& \quad \cdot \Big[e^{-\kappa_e(z'+2d)/\mu_i}\,\boldsymbol{\psi}(\mu_s,\phi_s;\mu_i,\phi_i)\,\mathbb{R}(\mu_i) \\
& \qquad + e^{\kappa_e z'/\mu_i}\,\boldsymbol{\psi}(\mu_s,\phi_s;-\mu_i,\phi_i)\Big]\frac{\mathbf{I}_0^i}{\mu_s}\,dz' \qquad (11.73a)
\end{aligned}$$

and

$$\begin{aligned}
\mathbf{I}_1^-(-\mu_s,\phi_s,z) = {} & e^{\kappa_e z/\mu_i}\mathbf{I}_0^i\,\delta(\mu_s-\mu_i)\,\delta(\phi_s-\phi_i) \\
& + \int_{z}^{0} e^{\kappa_e(z-z')/\mu_s} \\
& \quad \cdot \Big[e^{-\kappa_e(z'+2d)/\mu_i}\,\boldsymbol{\psi}(-\mu_s,\phi_s;\mu_i,\phi_i)\,\mathbb{R}(\mu_i) \\
& \qquad + e^{\kappa_e z'/\mu_i}\,\boldsymbol{\psi}(-\mu_s,\phi_s;-\mu_i,\phi_i)\Big]\frac{\mathbf{I}_0^i}{\mu_s}\,dz'. \qquad (11.73b)
\end{aligned}$$

The expressions given by Eq. (11.73) are applicable at any height z within the middle layer and for any direction (μ_s,ϕ_s) for $\mathbf{I}_1^+$ and $(-\mu_s,\phi_s)$ for $\mathbf{I}_1^-$. To compute the backscattering coefficient $\sigma^0_{pq}(\theta_i)$, we should set $z=0$, $\mu_s=\mu_i$ and $\phi_s=\pi+\phi_i$ in Eq. (11.73a), which leads to

$$\begin{aligned}
\mathbf{I}_1^+(\theta_i,\phi_i,0) = {} & e^{-2\kappa_e d/\mu_i}\,\mathbb{R}(\mu_i)\,\mathbf{I}_0^i\,\delta(\phi_s-\phi_i=\pi) \\
& + \Bigg[e^{-4\kappa_e d/\mu_i}\,\mathbb{R}(\mu_i)\,\boldsymbol{\psi}(-\mu_i,\phi_i+\pi;\,\mu_i,\phi_i)\,\mathbb{R}(\mu_i) \\
& \quad \cdot \int_{-d}^{0} e^{-2\kappa_e z'/\mu_i}\,dz' \\
& + e^{-2\kappa_e d/\mu_i}\;\boldsymbol{\psi}(-\mu_i,\phi_i+\pi;\,-\mu_i,\phi_i)\,\mathbb{R}(\mu_i)\int_{-d}^{0} dz' \\
& + e^{-2\kappa_e d/\mu_i}\,\mathbb{R}(\mu_i)\,\boldsymbol{\psi}(\mu_i,\phi_i+\pi;\,\mu_i,\phi_i)\int_{-d}^{0} dz' \\
& + \boldsymbol{\psi}(\mu_i,\phi_i+\pi;\,-\mu_i,\phi_i)\int_{-d}^{0} e^{2\kappa_e z'/\mu_i}\,dz'\Bigg]\frac{\mathbf{I}_0^i}{\mu_i} \\
= {} & \Upsilon^2\,\mathbb{R}(\theta_i)\,\mathbf{I}_0^i\,\delta(\phi_s-\phi_i=\pi) \\
& + \Bigg[\Upsilon^2\,\mathbb{R}(\theta_i)\,\boldsymbol{\psi}(-\mu_i,\phi_i+\pi;\,\mu_i,\phi_i)\,\mathbb{R}(\theta_i)\,\frac{\mu_i}{2\kappa_e}\,(1-\Upsilon^2) \\
& + d\,\Upsilon^2\,\boldsymbol{\psi}(-\mu_i,\phi_i+\pi;\,-\mu_i,\phi_i)\,\mathbb{R}(\theta_i) \\
& + d\,\Upsilon^2\,\mathbb{R}(\theta_i)\,\boldsymbol{\psi}(\mu_i,\phi_i+\pi;\,\mu_i,\phi_i) \\
& \quad + \boldsymbol{\psi}(\mu_i,\phi_i+\pi;\,-\mu_i,\phi_i)\,\frac{\mu_i}{2\kappa_e}\,(1-\Upsilon^2)\Bigg]\frac{\mathbf{I}_0^i}{\mu_i}, \qquad (11.74)
\end{aligned}$$

where Υ is the one-way transmissivity of the middle layer, defined by Eq. (11.2).

The five terms of Eq. (11.74) are diagrammed in Fig. 11-13:

① The first term represents the transfer of the incident power through the middle layer down to the interface between layers 2 and 3, followed by specular reflection and propagation through layer 2 and beyond. This term does not contribute to the backscatter, except at normal incidence.

② The second term includes single volume backscatter downward by the middle layer and double specular reflection by the boundary at $z=-d$.

③ The third and fourth terms each includes single bistatic scattering by the middle layer and single reflection by the lower boundary.

④ The last term represents single volume backscatter upwards by the middle layer and no involvement of the lower boundary.

After deletion of the first term in Eq. (11.74) and comparison with Eq. (11.62), we obtain the transformation matrix

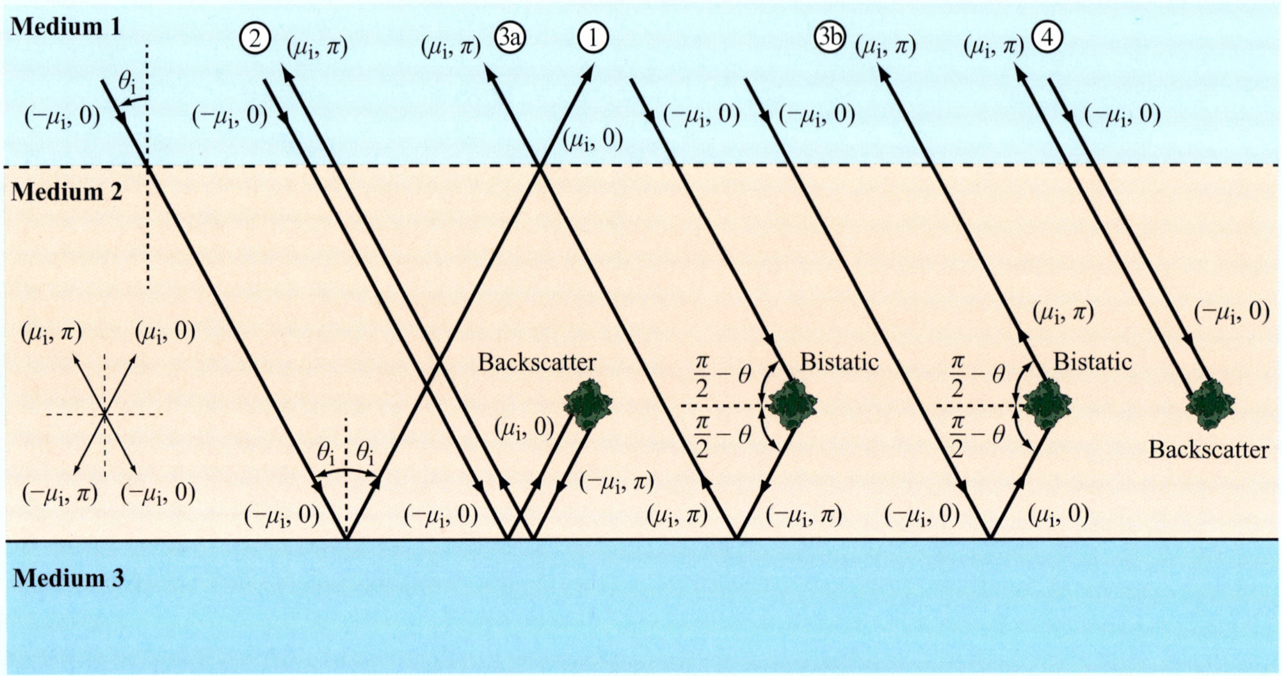

Figure 11-13: Volume scattering mechanisms.

$$\begin{aligned}
\mathcal{T}_{\mathrm{t}}(\mu_{\mathrm{i}},\phi_{\mathrm{i}}+\pi;\,-\mu_{\mathrm{i}},\phi_{\mathrm{i}}) = {} \\
&\frac{(1-\Upsilon^2)}{2\kappa_{\mathrm{e}}}\,\big[\boldsymbol{\psi}(\mu_{\mathrm{i}},\phi_{\mathrm{i}}+\pi;\,-\mu_{\mathrm{i}},\phi_{\mathrm{i}}) \\
&+\Upsilon^2\,\mathbb{R}(\theta_{\mathrm{i}})\,\boldsymbol{\psi}(-\mu_{\mathrm{i}},\phi_{\mathrm{i}}+\pi;\,\mu_{\mathrm{i}},\phi_{\mathrm{i}})\,\mathbb{R}(\theta_{\mathrm{i}})\big] \\
&+\frac{d\,\Upsilon^2}{\cos\theta_{\mathrm{i}}}\,\big[\boldsymbol{\psi}(-\mu_{\mathrm{i}},\phi_{\mathrm{i}}+\pi;\,-\mu_{\mathrm{i}},\phi_{\mathrm{i}})\,\mathbb{R}(\theta_{\mathrm{i}}) \\
&+\mathbb{R}(\theta_{\mathrm{i}})\boldsymbol{\psi}(\mu_{\mathrm{i}},\phi_{\mathrm{i}}+\pi;\,\mu_{\mathrm{i}},\phi_{\mathrm{i}})\big]. \qquad (11.75)
\end{aligned}$$

Because the lower boundary is assumed to be a perfectly flat surface, the radiative-transfer solution does not include the attenuated direct backscatter contribution of the lower surface, so it needs to be added. Hence, to synthesize the backscattering coefficient $\sigma^0_{\mathrm{rt}}(\theta_{\mathrm{i}})$ for a specified receive/transmit polarization configuration, the expression for $\mathcal{T}_{\mathrm{t}}$ should be used in Eq. (11.64).

11-6.5 Rayleigh Scatterers

If the middle layer is composed of Rayleigh particles and our interest is in vv polarization, we use Eq. (11.63a) with $[\mathcal{T}_{\mathrm{t}}]_{11}$, which corresponds to Eq. (11.75) after replacing $\mathbb{R}$ with Γ_{v} and $\boldsymbol{\psi}$ with ψ_{11}. From Eq. (11.56), $\psi_{11} = 3\kappa_{\mathrm{s}}/8\pi$ for all four combinations of incident and scattered directions contained in Eq. (11.75). Hence,

$$\begin{aligned}
\sigma^0_{\mathrm{vv}}(\theta_{\mathrm{i}}) &= \Upsilon^2\,\sigma^0_{\mathrm{s}_{vv}}(\theta_{\mathrm{i}}) + 4\pi\cos\theta_{\mathrm{i}}\,[\mathcal{T}_{\mathrm{t}}]_{11} \\
&= \Upsilon^2\,\sigma^0_{\mathrm{s}_{\mathrm{vv}}}(\theta_{\mathrm{i}}) + 4\pi\cos\theta_{\mathrm{i}}\;\frac{(1-\Upsilon^2)}{2\kappa_{\mathrm{e}}}\cdot\frac{3\kappa_{\mathrm{s}}}{8\pi} \\
&\quad + 4\pi\cos\theta_{\mathrm{i}}\;\Upsilon^2\Gamma^2\,\frac{(1-\Upsilon^2)}{2\kappa_{\mathrm{e}}}\cdot\frac{3\kappa_{\mathrm{s}}}{8\pi} \\
&\quad + 4\pi\cos\theta_{\mathrm{i}}\cdot\frac{d\,\Upsilon^2}{\cos\theta_{\mathrm{i}}}\left(2\Gamma^2\cdot\frac{3\kappa_{\mathrm{s}}}{8\pi}\right)
\end{aligned}$$

$$= \Upsilon^2\, \sigma^0_{\mathrm{s_{vv}}}(\theta_\mathrm{i}) + \frac{3}{4}\, a\cos\theta_\mathrm{i}\, (1-\Upsilon^2)(1+\Gamma^2\Upsilon^2) + 3\kappa_\mathrm{s} d\, \Gamma\Upsilon^2, \qquad (11.76)$$

(incoherent addition)

where $\Gamma = \Gamma_\mathrm{v}$ everywhere in Eq. (11.76). We refer to this expression as the ***single-scattering radiative transfer*** model with Rayleigh scatterers (S^2RT/R) under the incoherent addition assumption.[†]

As was mentioned earlier in Section 11-1, the radiative transfer model adds all energy sources incoherently, without regard to phase interference, if present. The last term in Eq. (11.76) represents the incoherent addition of terms "3a" and "3b" in Fig. 11-13, which (for hh and vv polarizations) undergo the same phase delays due to propagation, the same phase changes due to specular reflection by the underlying ground surface, and probably the same phase changes introduced by bistatic scattering by the same differential volume. Hence, the two contributions should be in phase, in which case the magnitude of the last term in Eq. (11.76) should be increased by a factor of 2. Consequently,

$$\sigma^0_{\mathrm{vv}}(\theta_\mathrm{i}) = \Upsilon^2\, \sigma^0_{\mathrm{s_{vv}}}(\theta_\mathrm{i}) + \frac{3}{4}\, a\cos\theta_\mathrm{i}(1-\Upsilon^2)(1+\Gamma^2\Upsilon^2) + 6\kappa_\mathrm{s} d\, \Gamma\,\Upsilon^2.$$

(coherent addition, vv-pol) (11.77)

We refer to this expression as the S^2RT/R under the coherent addition assumption.

The single-scattering expression for σ^0_{vv} is identical with the expression given by Eq. (11.23) based on the heuristic model developed earlier in Section 11-1. It is extendable to hh polarization by replacing $\sigma^0_{\mathrm{s_{vv}}}(\theta_\mathrm{i})$ with $\sigma^0_{\mathrm{s_{hh}}}(\theta_\mathrm{i})$ and Γ^v with Γ^h. For hv polarization, Rayleigh particles do not generate cross polarization; hence none of the volume scattering terms contribute, in which case Eq. (11.77) simplifies to

$$\sigma^0_{\mathrm{hv}}(\theta_\mathrm{i}) = \Upsilon^2\, \sigma^0_{\mathrm{s_{hv}}}(\theta_\mathrm{i}). \qquad (11.78)$$

(hv polarization)

[†] Computer Code 11.1.

11-6.6 Distinct Upper Boundary

For a snow layer over ground or sea ice over water, we incorporate the presence of the distinct upper boundary by modifying the expression for σ^0 using the same steps outlined earlier in Section 11-3. Hence, for vv polarization

$$\sigma^0_{\mathrm{vv}}(\theta_\mathrm{i}) = \mathbb{T}^2_{12}(\theta_\mathrm{i}) \Bigg[\Upsilon^2\, \sigma^0_{23}(\theta'_\mathrm{i}) + \frac{3}{4}\, a\cos\theta'_\mathrm{i}\, (1-\Upsilon^2)(1+\Gamma^2_{23}(\theta'_\mathrm{i})\,\Upsilon^2) + 6\kappa_\mathrm{s} d\, \Gamma_{23}(\theta'_\mathrm{i})\,\Upsilon^2 \Bigg] + \sigma^0_{12}(\theta_\mathrm{i}), \qquad (11.79)$$

where $\mathbb{T}_{12}(\theta_\mathrm{i})$ is the v-polarized boundary transmissivity between media 1 and 2 (with medium 2 being the middle layer containing snow or sea ice), Υ is the one-way transmissivity of the middle layer along direction θ', $\sigma^0_{23}(\theta'_\mathrm{i})$ is the vv-polarized backscattering coefficient of the lower boundary separating media 2 and 3, $\sigma^0_{12}(\theta_\mathrm{i})$ is the backscattering coefficient of the upper boundary separating media 1 and 2, and $\Gamma_{23}(\theta'_\mathrm{i})$ is the v-polarized reflectivity of the lower boundary. A similar expression can be developed for hh polarization.[†]

11-7 Approximate Form of S^2RT/R Model

In the absence of information about the roughness and moisture content of the lower boundary, it is difficult to assign lower and upper bounds on its backscattering coefficient σ^0_s. Imaging radar seldom operates at incidence angles below 20° or above 50°. Within this angular range, a careful review of experimental data and model predictions leads to the conclusion that for hh and vv polarizations σ^0_s of a random soil surface does not exceed 0 dB (or 1 $\mathrm{m^2/m^2}$). The same upper limit applies to the overwhelming surface-roughness conditions associated with water surfaces. The lower bound, however, may be as low as −30 dB (or 10^{-3}

[†] Computer Code 11.2.

m^2/m^2) or lower for both land and water surfaces. Hence, we should keep this range in mind as we examine the volume scattering terms in Eq. (11.77) under various scenarios.

The last term in Eq. (11.77) includes the product $\kappa_s d$, which can be written as

$$\kappa_s d = \frac{\kappa_s}{\kappa_e}\,(\kappa_e d) = a\,\frac{\tau}{\sec\theta}\,, \tag{11.80}$$

where $a = \kappa_s/\kappa_e$ and, from Eq. (11.3), $\tau = \kappa_e d\sec\theta$. Also, from Eq. (11.2),

$$\Upsilon = e^{-\tau}, \qquad \text{or } \tau = -\ln\Upsilon. \tag{11.81}$$

Hence,

$$\kappa_s d = -a\cos\theta_i\,(\ln\Upsilon). \tag{11.82}$$

Let us consider Eq. (11.77) for pp polarization (with $p=$ v or h) for a vegetation canopy over a ground surface, and let us rewrite it in the form

$$\sigma^0_{pp}(\theta_i) = \sigma^0_g(\theta_i) + \sigma^0_c(\theta_i) + \sigma^0_{gcg}(\theta_i) + \sigma^0_{cg}(\theta_i), \tag{11.83}$$

where

$$\sigma^0_g(\theta_i) = \Upsilon^2\,\sigma^0_s(\theta_i) \tag{11.84a}$$

(ground contribution),

$$\sigma^0_c(\theta_i) = \frac{3}{4}\,a\cos\theta_i\,(1-\Upsilon^2) \tag{11.84b}$$

(direct canopy contribution),

$$\sigma^0_{gcg}(\theta_i) = \Gamma^2(\theta_i)\,\Upsilon^2\,\sigma^0_c(\theta_i) \tag{11.84c}$$

(ground-canopy-ground contribution),

$$\sigma^0_{cgt}(\theta_i) = -6a\cos\theta_i\,\Gamma(\theta_i)\,\Upsilon^2\,(\ln\Upsilon), \tag{11.84d}$$

(total ground-canopy contribution).

By applying the relationship given by Eq. (11.82), the three volume-related contributions (σ^0_c, σ^0_{gcg}, and σ^0_{cgt}) are now defined in terms of four quantities, namely: θ_i, $\Gamma(\theta_i)$, a, and Υ. Also, all three contributions are directly proportional to the albedo a. To gain some insight into the relative magnitudes of the three contributions, let us fix θ_i at 30° and evaluate the three contributions under six scenarios corresponding to three values of Υ (namely, 0.8, 0.5, and 0.1) and two soil moisture conditions (dry and wet). From Fig. 10-7:

$$\left.\begin{aligned}\Gamma^v &= 0.06\\ \Gamma^h &= 0.08\end{aligned}\right\}\ \text{at } \theta = 30^\circ \text{ and } m_v = 0,$$

$$\left.\begin{aligned}\Gamma^v &= 0.36\\ \Gamma^h &= 0.46\end{aligned}\right\}\ \text{at } \theta = 30^\circ \text{ and } m_v = 0.35 \text{ g/cm}^3.$$

Table 11-1 displays the magnitudes of the ratios $\sigma^0_{gcg}/\sigma^0_c$ and $\sigma^0_{cgt}/\sigma^0_c$ for the various aforementioned scenarios. Under all scenarios but one, σ^0_{gcg} is at least one order of magnitude smaller than σ^0_c. The one exception is when the soil is very wet and the canopy is only slightly lossy ($\Upsilon = 0.8$); for vv polarization, $\sigma^0_{gcg} = 0.135\sigma^0_c$, which may seem significant, but it is only 4% of the sum of σ^0_c and σ^0_{cgt}. Hence, the contribution σ^0_{gcg} may be ignored altogether, in which case Eq. (11.77) simplifies to

$$\sigma^0_{pp}(\theta_i) = \Upsilon^2\,\sigma^0_{s_{pp}}(\theta_i) + \frac{3}{4}\,a\cos\theta_i\left[(1-\Upsilon^2) - 8\Gamma^p(\theta_i)\,\Upsilon^2(\ln\Upsilon)\right]. \tag{11.85}$$

Note that because $0 \le \Upsilon \le 1$, it follows that $\ln(\Upsilon) \le 0$.

11-7.1 Applicability of the Single-Scattering Model

At this juncture, the reader may ask the following question: "How applicable is the single-scattering Rayleigh model to real media such as vegetation canopies, snow cover, and sea ice?" The simple answer is: The applicability of the Rayleigh model depends on the sizes of the scattering elements relative to λ, and on their orientations and the average spacing between them. However, even when the model is not applicable, its form often is used as a starting point for generating applicable semiempirical models.

The model derived in the previous subsections is predicated on several limiting assumptions, namely that:

Table 11-1: Ratios $\sigma^0_{gcg}/\sigma^0_c$ and $\sigma^0_{cgt}/\sigma^0_c$ computed for a Rayleigh layer at $\theta_i = 30°$ for hh and vv polarizations under dry and wet soil conditions for three values of the canopy one-way transmissivity Υ.

	$m_v = 0$ (dry)			$m_v = 0.35$ g/cm³ (very wet)		
	$\Upsilon = 0.8$	0.5	0.1	$\Upsilon = 0.8$	0.5	0.1
$\sigma^0_{gcg}/\sigma^0_c$, hh-pol	4.1×10^{-3}	1.6×10^{-3}	6.4×10^{-5}	0.135	5.3×10^{-2}	2.1×10^{-3}
$\sigma^0_{gcg}/\sigma^0_c$, vv-pol	2.3×10^{-3}	9×10^{-4}	3.6×10^{-5}	8.3×10^{-2}	3.2×10^{-2}	1.3×10^{-3}
$\sigma^0_{cgt}/\sigma^0_c$, hh-pol	0.40	6.2×10^{-2}	1.5×10^{-2}	2.3	0.36	8.6×10^{-2}
$\sigma^0_{cgt}/\sigma^0_c$, vv-pol	0.30	4.7×10^{-2}	1.1×10^{-2}	1.7	0.28	6.6×10^{-2}

(a) the lower boundary is perfectly flat, thereby limiting the volume scattering contributions made by the middle layer to specific directions that simultaneously satisfy the specular reflection condition by the lower boundary and that the incidence and backscatter directions are exactly opposite of one another,

(b) the scattering constituents of the middle layer are sufficiently small in size relative to λ so as to satisfy the Rayleigh phase-function criterion,

(c) second- and higher-order scattering contributions are significantly smaller than first-order contributions and, therefore, may be ignored.

In the microwave region, the underlying surface is not electromagnetically smooth, and in the case of a vegetation canopy, the leaves and branches are comparable to or larger than λ in size. A flat leaf, for example, exhibits an elaborate scattering pattern similar to the radiation pattern of an antenna aperture of the same size and shape, as illustrated by the experimental measurements shown in Fig. 11-14. Furthermore, the scattering pattern changes if the leaf is not flat (Sarabandi et al., 1988). Wheat and corn stalks and tree branches are modeled as dielectric cylinders, which exhibit strong polarization-dependent scattering. In a tree canopy, the scattering elements have size, shape and orientation distributions, all of which may vary as a function of height between the ground surface and the top of the canopy.

To address these physical attributes of the scattering medium, several theoretically sophisticated and computationally demanding models were developed and used to compare their predictions with experimental observations (Ulaby et al., 1990b; Sun and Ranson, 1995; Lin and Sarabandi, 1999a,b; Marliani et al., 2002). Because in most cases the various size, shape, and orientation distributions are either unknown or only rudimentary estimates of them are available, numerous assumptions and parameter adjustments are made in

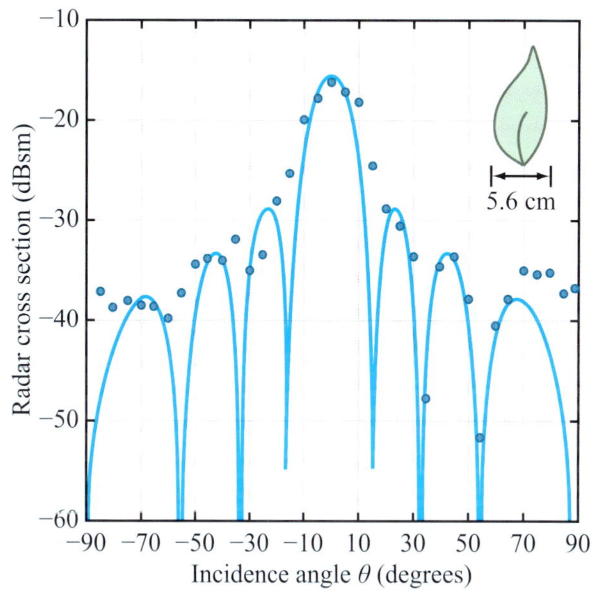

Figure 11-14: Radar scattering pattern of a leaf as a function of illumination angle relative to normal incidence [Senior et al., 1987].

computational models so as to arrive at good agreement with experimental observations. The other problem with many-parameter models is that they are rather difficult to invert.

Even though the scattering pattern of an individual leaf or branch in a canopy is strongly dependent on the direction and polarization of the illumination incident upon it, the scattering pattern of an ensemble of such leaves and branches of various sizes, shapes, and orientations is not. The randomness "destroys" most of the sensitivity to these physical parameters. Consequently, the simple single-scattering Rayleigh model often is a viable contender in many volume scattering situations of interest. One of its attracting features is that it consists of only three parameters, the extinction coefficient κ_e, the albedo a, and the depth of the layer d. These, in addition to the roughness and dielectric properties of the underlying surface, are sufficient to compute σ^0_{vv} and σ^0_{hh} of a vegetation canopy. The average dielectric constant of the canopy layer may be assumed to be ≈ 1. For a snow or ice layer, the average dielectric constant may depart significantly from 1, in which case it needs to be added to the total list of parameters. We should also note that even though one of its implicit assumptions is that the scattering layer is a weakly scattering medium (small albedo a), the model is often applied with no restriction on the value of a (between 0 and 1).

For the most part, the single-scattering Rayleigh model provides good agreement with experimental observations for many scene conditions, but only for hh and vv polarizations: the only contributions to σ^0_{hv} are due to scattering by the lower and upper surfaces. This implies that for a high-loss vegetation canopy, $\sigma^0_{hv} = 0$, which certainly is not the case.

11-7.2 Comparison with Experimental Observations

Using the combination of the I^{2}EM model to compute surface scattering and the S^{2}RT/R (single-scattering radiative transfer model with Rayleigh particles) to compute volume scattering, model parameters were selected to fit the model to experimental data for vegetation over soil, snow over soil, and sea ice over ocean water. Examples are displayed in Figs. 11-15 to 11-17.

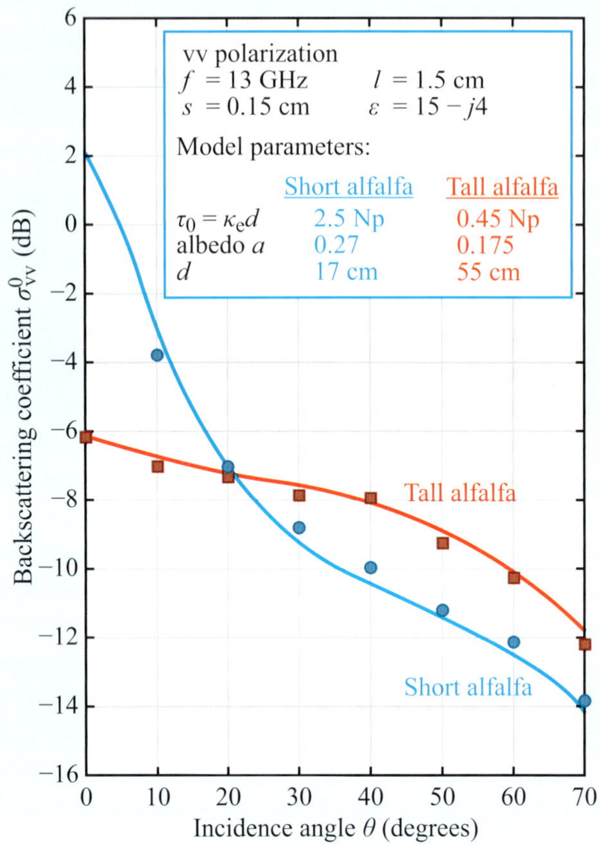

Figure 11-15: Fitting the S^{2}RT/R model to backscattering data measured for a short alfalfa canopy and a tall alfalfa canopy.

The alfalfa measurements include two canopies, a 17 cm tall canopy with a zenith optical thickness $\tau = 0.45$ Np, and a 55 cm tall canopy with $\tau = 2.5$ Np. The second canopy is not only three times taller, but the vegetation density also is greater. As expected, the backscatter response of the short canopy is dominated by the soil-surface contribution, particularly at incidence angles below 40°. In contrast, the gentle angular response of the tall canopy suggests that the canopy masks the soil contribution and the backscatter is dominated by the direct canopy term σ^0_c.

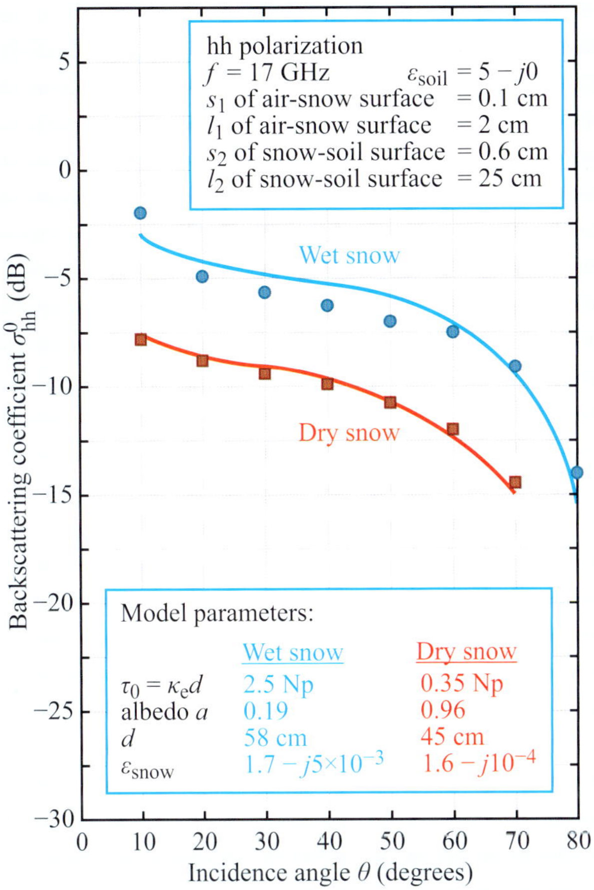

Figure 11-16: Fitting the S^2RT/R model to dry and wet snow data measured at 17 GHz.

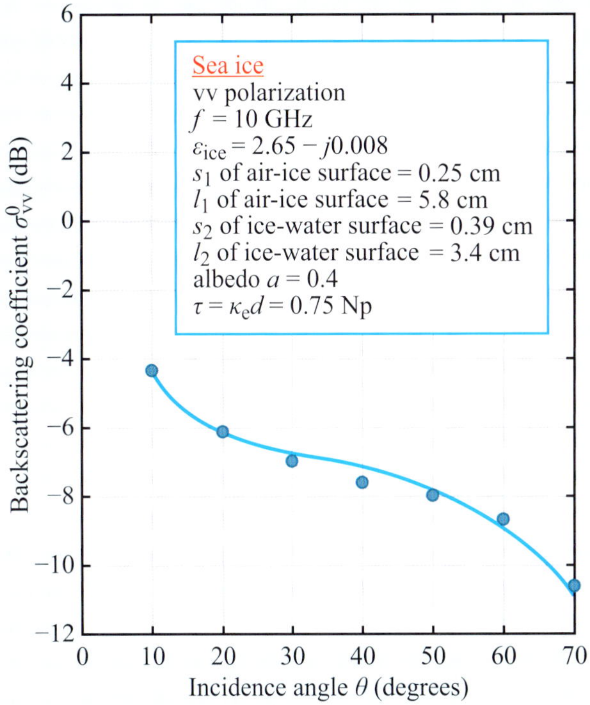

Figure 11-17: Fitting S^2RT/R model to sea-ice data at 10 GHz. Data reported by Onstott (1990).

Dry snow is modeled as low-loss ice crystals in a background of air. At 17 GHz, scattering by the ice crystals dominates over absorption. Consequently, the model parameters chosen for matching the measured data in Fig. 11-16(a) to the model include an albedo $a = 0.96$ and $\tau = 0.35$ Np. In contrast, the data measured for wet snow [Fig. 11-16(b)] required values of $a = 0.19$ and $\tau = 2.5$ Np because wet snow is a lossy medium.

The 10 GHz data shown in Fig. 11-17 for sea ice was reported by Onstott (1990). The model parameters include $a = 0.4$ and $\tau = 0.75$ Np. As a volume scattering medium, the ice layer is intermediate between those of dry and wet snow.

11-8 Radar Observations of Vegetation Canopies

The purpose of this section is to provide an overview, based on models and experimental observations, of how the backscattering coefficient responds to the various physical parameters of a vegetation canopy. Before we embark on such an overview we summarize our understanding of the penetration and propagation properties of vegetation canopies and the soil surface underneath them.

11-8.1 Penetration Depth in Soil

For perfectly dry soil, $2 \lesssim \varepsilon'_{soil} \lesssim 4$ and $\varepsilon''_{soil} < 0.05$. The exact values of ε'_{soil} and ε''_{soil} depend on the soil density and composition. For $\varepsilon'_{soil} = 3$ and $\varepsilon''_{soil} = 0.05$, the penetration depth at 1 GHz is $\delta_p = 1.65$ m. This

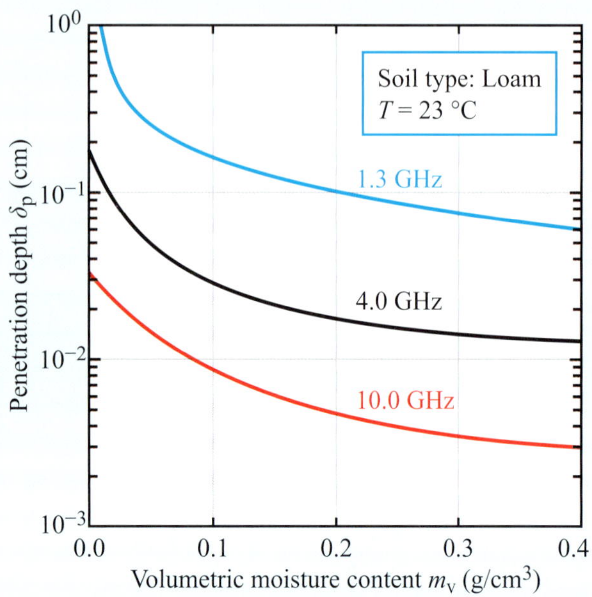

Figure 11-18: Penetration depth in loamy soil as a function of volumetric moisture content.

means that for dry desert sand, signal penetration of a few meters is quite feasible at 1 GHz, which explains why the L-band Shuttle Imaging Radar-A (SIR-A) was capable of detecting and mapping subsurface topography in the Egyptian desert (Schaber et al., 1986).

Increasing the soil volumetric moisture content m_v leads to an exponential-like decrease in the magnitude of δ_p, as shown in Fig. 11-18.

11-8.2 Propagation Properties of Cultural Vegetation

A wheat canopy may be divided into three constituents: (1) leaves, (2) stalks, and (3) heads. For modeling the radiation transfer both within and through the canopy, it may be divided into an upper layer comprising the heads and a lower layer comprising the leaves and stalks. For another crop, such as corn, the canopy may be approximated as an upper layer consisting primarily of leaves and a lower layer consisting of stalks. Similarly, other configurations may be appropriate for other crop types. In modeling or interpreting the emission and scattering from a crop canopy, it is useful to know the propagation characteristics of the canopy as a whole, as well as those of its constituent parts.

The propagation properties of a vegetation canopy may be computed using either a deterministic approach or a statistical one. The ***deterministic approach*** requires that the absorption and scattering cross sections be known for each leaf, stalk, branch, etc., in the canopy and that their locations and orientations and the degree of interaction between them also be known. Clearly, such a problem is rather difficult to solve when the canopy constituents are comparable to or larger than λ in size. Moreover, the size, orientation, shape, and location information usually is not available. Furthermore, a vegetation canopy exhibits some degree of randomness, which means that the canopy architecture contained in a given radar resolution cell is statistically different from those in other resolution cells.

In contrast, the ***statistical approach*** usually (1) assumes that the canopy constituents (leaves, etc.) are randomly distributed in space, either within the canopy as a whole or within a subregion of the canopy (such as in the upper part of the canopy if its plants do not grow leaves in the lower part), (2) ignores interactions between canopy constituents, and (3) uses statistical distribution functions to describe the shapes, sizes, and orientations of the canopy constituents. If these parameters vary with height between the ground surface and the top of the canopy, the canopy is modeled as a series of horizontal layers each with its own statistical distributions.

For an elemental canopy volume—defined such that it is small enough so that the density, type, etc., of canopy constituents is approximately uniform across it and yet large enough to contain a representative sample of constituents, its extinction coefficient κ_e may, in general, be a function of height z above the ground, incidence angle θ, wave polarization p, and azimuth angle ϕ if the canopy architecture is not azimuthally symmetrical. Thus,

$$\kappa_e = \kappa_e(\theta, \phi, z, p).$$

The corresponding ***attenuation*** (***optical thickness***) of a canopy of height h is

$$\tau = \int_0^h \kappa_e z \sec\theta. \tag{11.86}$$

Computing κ_e for an elemental canopy volume requires knowledge of all of the shape, size, and orientation distributions for each type of canopy constituent, as well as scattering and absorption models for each type of constituent. Stalks and branches usually are modeled as dielectric cylinders, leaves are modeled as dielectric plates, and so on. Because these scattering elements are comparable to λ in size, the scattering models tend to be mathematically complex and computationally demanding. Examples of such models are available in Ulaby and Elachi (1990) and Ulaby et al. (1990b).

The propagation properties of a vegetation canopy can be measured experimentally by transmitting a signal through the canopy. The sketch shown in Fig. 11-19(a) illustrates the setup used by Allen and Ulaby (1984) to measure the optical thickness of a wheat canopy. The transmitting antenna was situated above the canopy, and its boresight direction was at an angle of incidence of 60° and an azimuth angle of 90° relative to the row direction of the crop. The receiving antenna was mounted on a motor-drive platform that could be moved vertically from 23 cm to 132 cm above the soil surface. The total canopy height was 105 cm, of which the heads comprised about 8 cm. The receiving antenna was connected to a detector, which in turn was connected to a power meter. The height of the receiving antenna was converted to a voltage through a potentiometer and was used to drive the x axis of an x-y plotter with the y axis denoting the strength of the received power. After measuring the received power as a function of receiving-antenna height, the system was calibrated by repeating the measurements with no canopy present between the antennas. The net result was a curve of the type shown in Fig. 11-19(b), which displays the measured cumulative attenuation (optical thickness) of the canopy as a function of the path length s as defined in the inset. The average slope of the curve, which is about 4.6 Np/m, represents the slant extinction coefficient of the wheat stalks at $\theta = 60°$ for vertical polarization. Repetition of

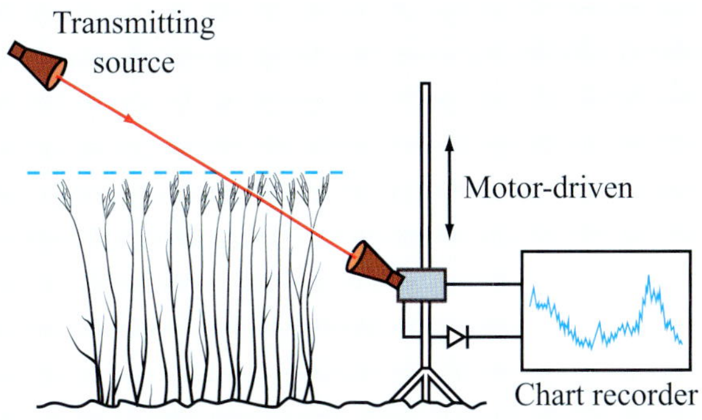

(a) Sketch showing measurement setup

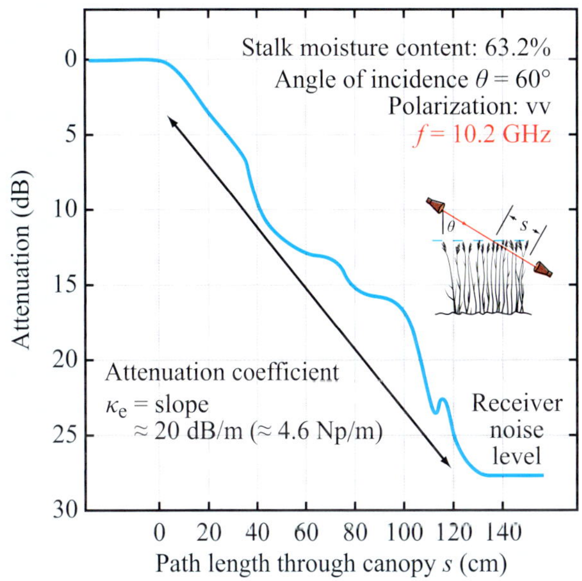

(b) Stalk attenuation

Figure 11-19: Sketch showing the setup used by Allen and Ulaby (1984) to measure canopy attenuation as a function of depth. The receiving antenna was a 5 cm diameter horn.

this type of experiment at five additional locations within the wheat field resulted in an average overall extinction coefficient of 6.91 Np/m (or 30 dB/m) and an associated standard deviation of 2.8 Np/m.

Most attempts to measure the extinction coefficient of vegetation canopies have relied on either (a) measuring the power transmitted through the canopy as detected by a receiver placed beneath the canopy, or (b) measuring the power reflected from a standard target (such as a trihedral reflector) placed beneath the canopy. In both cases, the attenuation is computed by comparing the power intensity with and without the canopy. The transmission approach measures the one-way total attenuation of the canopy, whereas the reflection approach measures the two-way attenuation.

When using the one-way transmission method, placement of the receiving antenna in close proximity of the vegetation may cause near-field interactions that may distort the gain and pattern of the antenna, thereby leading to inaccurate estimates of attenuation. Similar effects can occur in the two-way transmission method; the scattering pattern of the standard target can be distorted by the close presence of the vegetation. Another source of measurement uncertainty is the spatial inhomogeneity of the vegetation canopy; large fluctuations (due to location) are observed among measurements made through different transects.

Both sources of uncertainty can be reduced through spatial averaging. In the configuration depicted in Fig. 11-20, the receiving antenna and detector are placed on plastic rails close to the ground surface. The antenna assembly is tied to a rope and synchronized with the position of the radar on the truck by means of pulleys. As the track moved alongside the field, the assembly moved as well. Using this technique, a long record of the power transmitted through the canopy can be measured, from which an average value with a small standard deviation can be computed. Based on measurements performed for corn and wheat canopies at 10.2 GHz, Ulaby and Jedlicka (1984) determined that the standard deviation associated with the average value of the total attenuation is on the order of 1 dB (or equivalently, 0.23 Np). Figures 11-21(a) and (b) show the temporal variation of canopy attenuation τ (in dB) over the full growing season for canopies of soybeans and corn. The measurements were made at 10.2 GHz using v-polarized transmit and receive antennas. The incidence angle is 52° and the propagation direction is orthogonal to the row direction.

In both cases, the measured total attenuation τ is small during both the early and final stages of plant growth and is a maximum during the period that begins after the plants have attained their full height and ends when the plants begin to lose their moisture.

Also shown in Figs. 11-21(a) and (b) are plots based on relatively simple models. For soybeans, the model relates the total attenuation τ (in Np) to the canopy height and ***volume fractions*** occupied by leaves and stalks, υ_ℓ and υ_{st}, as follows:

$$\tau = \tau_\ell + \tau_{st} = (\kappa_\ell + \kappa_{st}) h \sec\theta, \tag{11.87}$$

where *κ_ℓ and κ_{st} account for absorption by leaves and stalks* by treating the canopy as a superposition of two canopies (Fig. 11-22), one composed of leaves and has a ***leaf canopy effective index of refraction*** $n_{c,\ell}$, and another composed of stalks and has a ***stalk canopy effective index of refraction*** $n_{c,st}$. The canopy leaf and stalk indices of refraction are related to the ***indices of refraction of the leaf and stalk material***, n_ℓ and n_{st}, via the refractive model given by Eq. (4.45) with $\alpha = 1/2$. That is,

$$n_{c,\ell} = (1 - \upsilon_\ell) n_{air} + \upsilon_\ell n_\ell, \tag{11.88a}$$

$$n_{c,st} = (1 - \upsilon_{st}) n_{air} + \upsilon_{st} n_{st}, \tag{11.88b}$$

where $n_{air} = 1$, and υ_ℓ and υ_{st} are the volume fractions of leaves and stalks contained in the canopy. Both υ_ℓ and υ_{st} vary as a function of time, but throughout the growth cycle their magnitudes are on the order of 0.01. Hence, Eqs. (11.88a) and (11.88b) yield values for $n'_{c,\ell}$ and $n'_{c,st}$ that are only slightly larger than that of the air background, namely $n_{air} = 1$. Hence, the equations simplify to

$$n''_{c,\ell} \approx \upsilon_\ell n''_\ell, \tag{11.89a}$$

$$n''_{c,st} \approx \upsilon_{st} n''_{st}. \tag{11.89b}$$

Finally, the extinction coefficients are related to the canopy leaf and stalk indices of refraction by

$$\kappa_e = \frac{2\pi}{\lambda_0} (n''_{c,\ell} + n''_{c,st}) = \frac{2\pi}{\lambda_0} (\upsilon_\ell n''_\ell + \upsilon_{st} n''_{st}). \tag{11.90}$$

For each date on which direct transmission measurements were made, Ulaby and Jedlicka (1984) measured

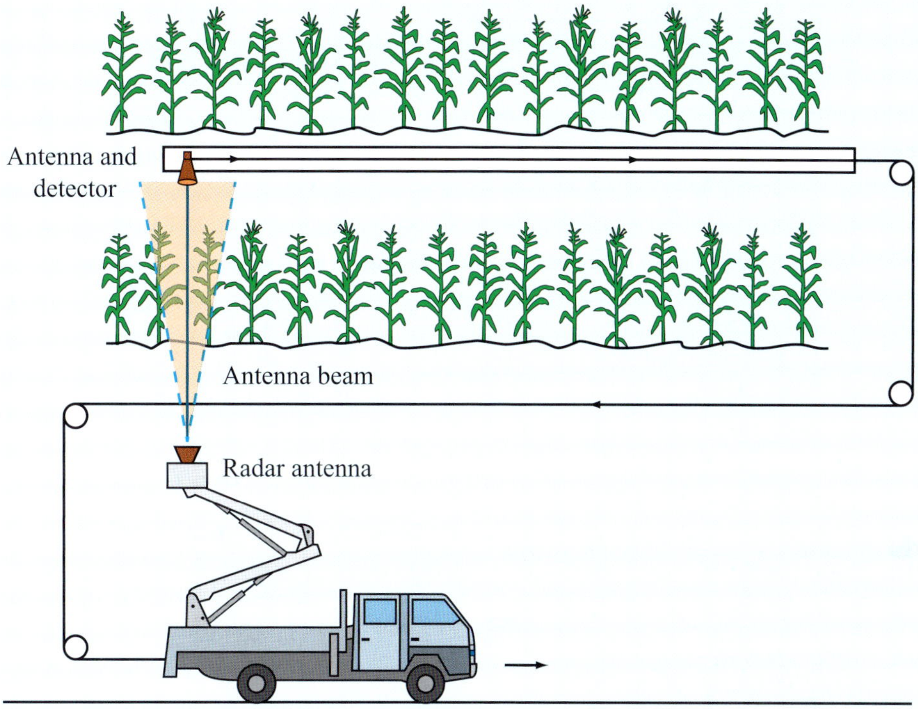

Figure 11-20: Attenuation measurement setup [Ulaby and Jedlicka, 1984].

υ_ℓ and υ_{st} of the canopy by removing plants from a representative section of the canopy and estimating their leaf and stalk volume fractions, and they also used a dielectric measurement technique to determine the value of ε_ℓ and ε_{st} (and hence $n_\ell = \sqrt{\varepsilon_\ell}$ and $n_{st} = \sqrt{\varepsilon_{st}}$). Good agreement is observed between the measured and model-calculated values shown in Fig. 11-21(a), despite the fact that (a) the model completely ignored scattering by the leaves and stalks and (b) treated the canopy constituents as if they are very small relative to λ, which they are not.

A similar attempt was made for a corn canopy, but the simple model did not provide good agreement with the measured values of τ. The curve labeled "calculated attenuation" in Fig. 11-21(b), which accounts for absorption by only the leaves of the corn plants, underestimates the value of τ. Including absorption by the stalks, as was done previously for the soybeans canopy, leads to overestimation of the value of τ. The apparent inapplicability of the model to the corn canopy is attributed to the fact that the corn stalks have an ordered spacing and orientation, whereas the refractive model assumes the inclusions to be not only small in size, but also randomly oriented within the mixture volume.

11-8.3 Extinction by a Canopy Containing Stalks

To model the propagation properties of a canopy with vertical stalks, Ulaby et al. (1987) developed a canopy model in which the stalks are represented by vertically oriented, lossy dielectric cylinders and the leaves are treated as randomly oriented discs. The model computes absorption and scattering by the cylinders, as well as absorption by the leaves. To evaluate the validity of the model, they also measured the transmission loss for a mature corn canopy at incidence angles of 20°,

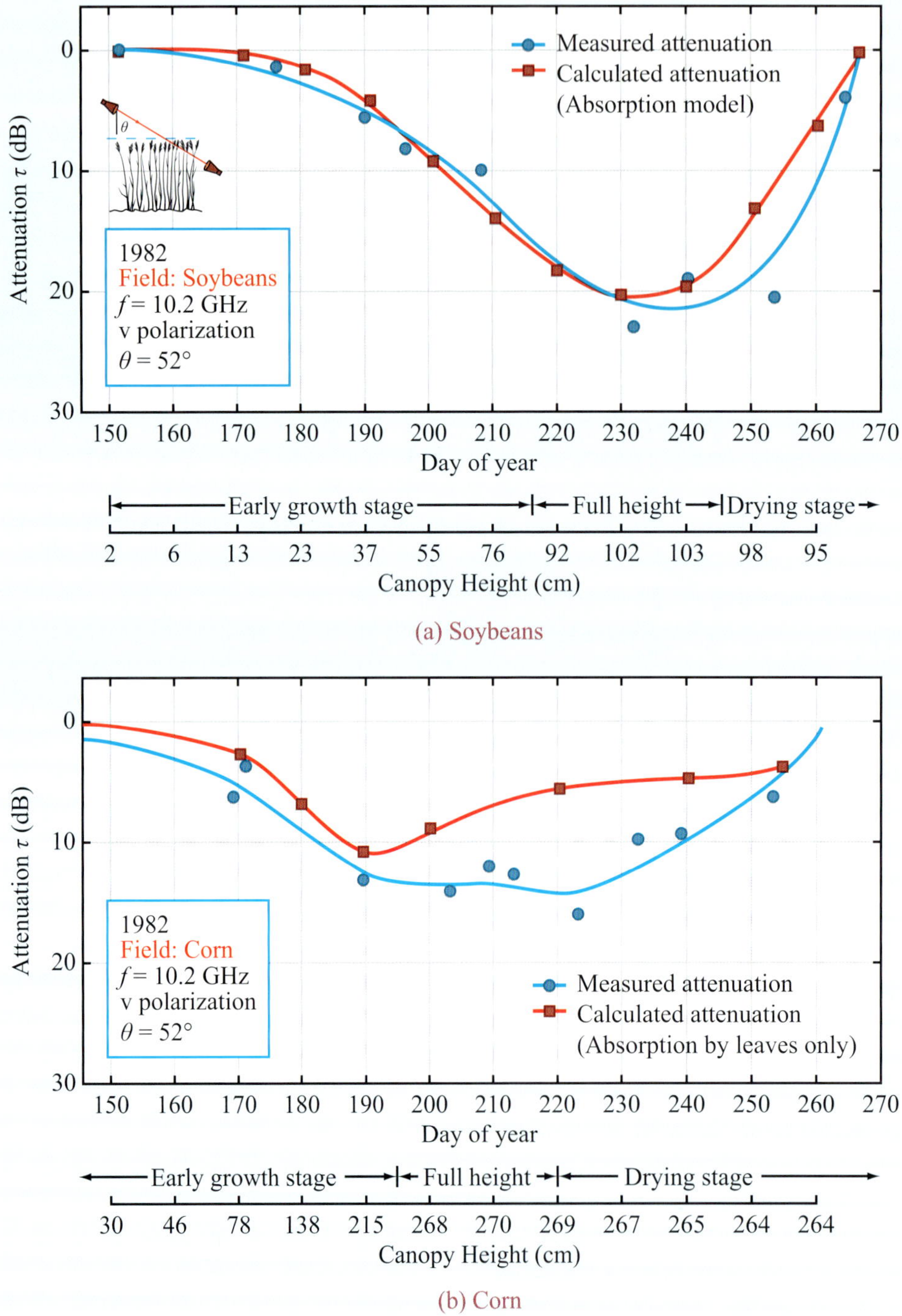

Figure 11-21: Measured and calculated one-way attenuation of soybean and corn canopies at 10.2 GHz [Ulaby and Jedlicka, 1984].

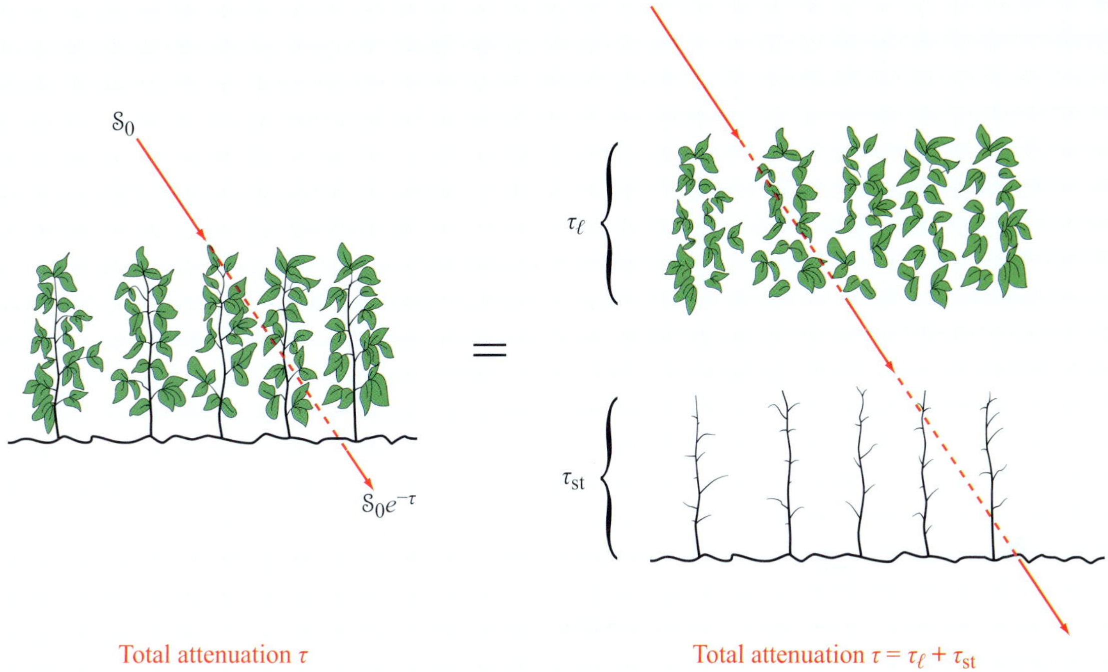

Figure 11-22: A vegetation canopy consisting of leaves and stalks is modeled as the sum of two equivalent canopies, one containing leaves only and another containing the stalks.

40°, 60°, and 90° using the configurations depicted in Fig. 11-23. The transmitters and receivers used dual-polarized antennas with center frequencies at 1.62, 4.75, and 10.2 GHz.

Using the vegetation dielectric model given in Section 4-9.2 and the volumetric moisture content measured for samples of leaves and stalks, it was possible to compute the dielectric spectra displayed in Fig. 11-24 and to use them to compute the extinction cross section for a single stalk, and by extension, the extinction coefficient of the canopy. The stalks, which were about 2.7 m tall, varied in diameter from 2.8 cm at the base to 0.6 cm at the top. In the model, the stalks were assumed to have a uniform diameter of 1.7 cm. Figure 11-25 displays the spectra for the absorption, scattering, and extinction "widths," normalized to the cylinder diameter d, when illuminated along a direction normal to the cylinder axis. Note the strong sensitivity to the polarization direction, particularly at frequencies below 3 GHz. Scattering and absorption by a dielectric cylinder are strongly dependent on the diameter d relative to λ in the cylinder. For the specified value of the stalk dielectric constant given in Fig. 11-25, the wavelength λ at 1 GHz is approximately $30\text{ cm}/4.2 \approx 7.1$ cm. Hence, $d/\lambda \approx 1.7/7.1 \approx 0.24$. As f is increased to 10 GHz, d/λ increases to 2.4. When the cylinder diameter becomes very large relative to λ, scattering and absorption become insensitive to the direction of the polarization vector.

By extending the model from a single dielectric cylinder to a medium of parallel cylinders illuminated at any incidence angle θ_i, it was possible to compute the extinction coefficient κ_e of the corn canopy and to compare the model predictions with experimental measurements. An example is shown in Fig. 11-26(a) for transmission at 1.62 GHz. For an h-polarized wave, κ_e is less than 0.2 dB/m at all incidence angles, rendering the canopy almost transparent. In contrast, the extinction

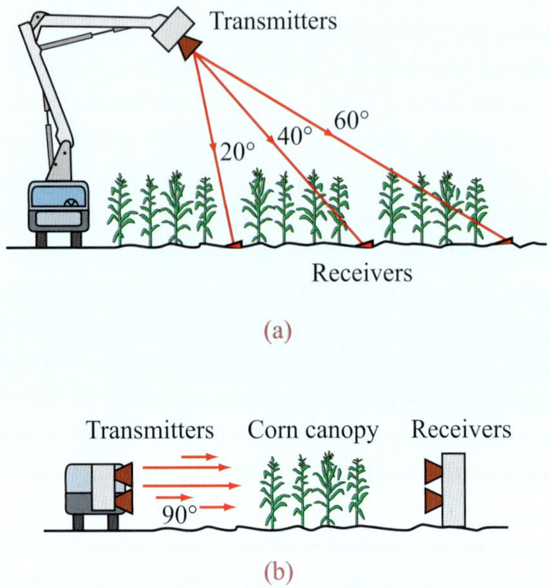

Figure 11-23: Configurations used for transmission measurements at (a) 20°, 40°, 60°, and (b) 90° [Ulaby et al., 1987].

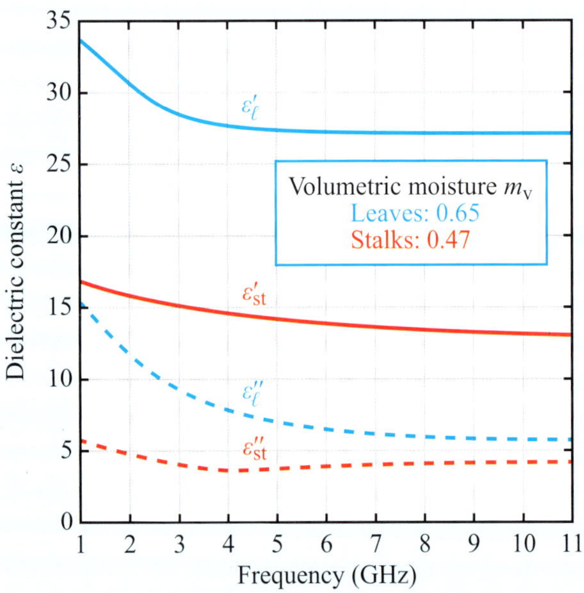

Figure 11-24: Dielectric constant of corn leaves and stalks as a function of frequency [Ulaby et al., 1987].

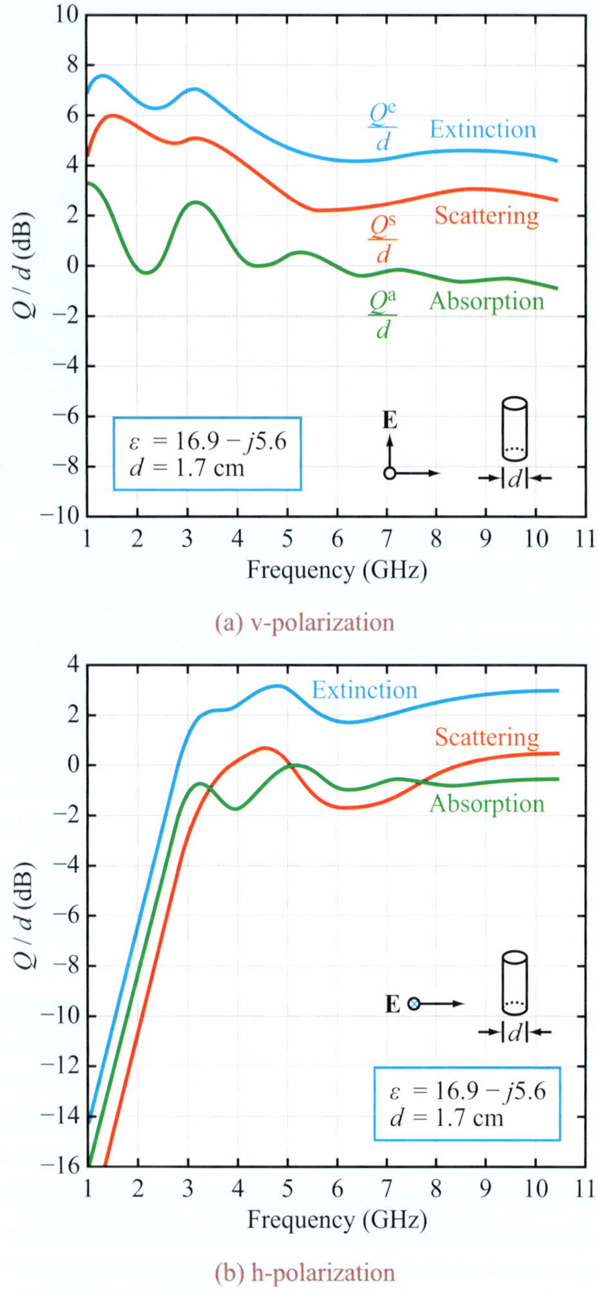

Figure 11-25: Calculated spectral variation of the normalized absorption, scattering, and extinction widths for an infinite cylinder illuminated with (a) a v-polarized wave and (b) an h-polarized wave [Ulaby et al., 1987].

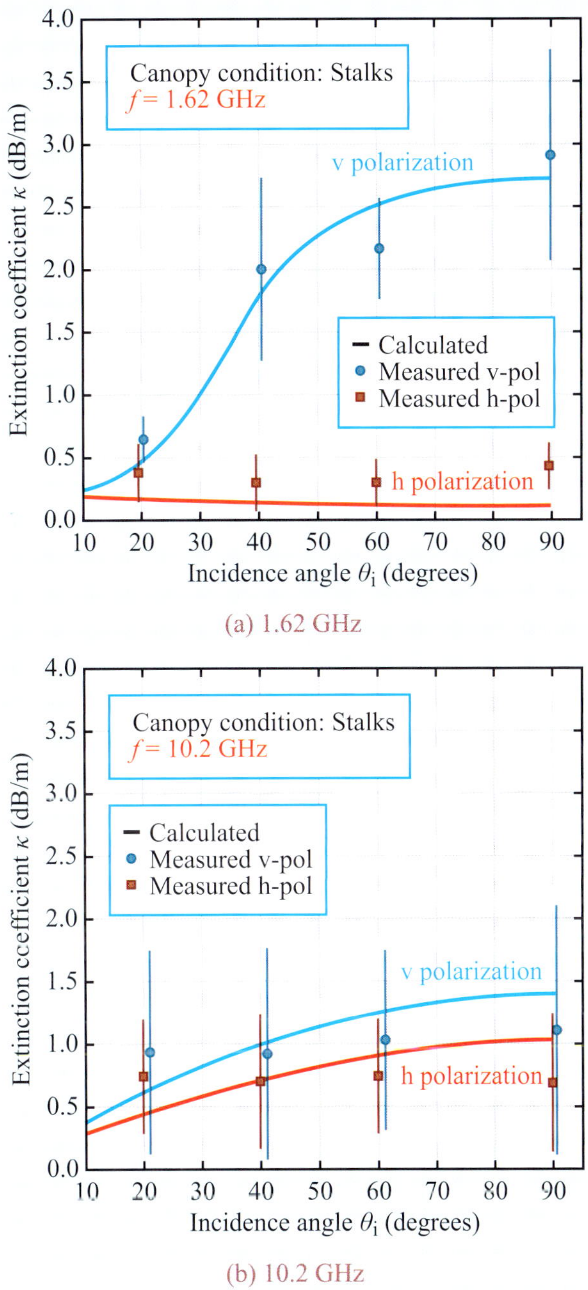

Figure 11-26: Comparison of calculated extinction coefficient for stalks with measured values at (a) 1.6 GHz and (b) 10.2 GHz.

coefficient of a v-polarized wave is strongly dependent on θ_i, increasing from about 0.2 dB at near normal incidence to as high as 2.5 dB/m at grazing incidence ($\theta_i = 90°$).

For a homogeneous medium with a complex index of refraction $n = n' - jn''$, the attenuation coefficient is given by

$$\alpha = \frac{2\pi}{c} f n'', \tag{11.91}$$

where c is the velocity of light. Hence, if n does not vary with the frequency f, α should increase linearly with f. Whereas this is true for a homogeneous medium, as well as approximately so for a medium containing dielectric particles that are sufficiently smaller than λ (thereby representing the medium in terms of an equivalent dielectric medium along the lines outlined in Section 4-4), the stated frequency dependence of α is not always valid. Comparison of the results shown in Fig. 11-26(b) for $f = 10.2$ GHz with those shown in Fig. 11-26(a) for $f = 1.62$ GHz demonstrates that the extinction coefficient for h polarization is indeed larger at the higher frequency, but the reverse is true for v polarization.

11-8.4 Role of Soil Surface Contribution

The volumetric moisture content of a green leaf is typically in the 70% range, in contrast to that of a yellow leaf at the end of the growing season, which may be as low as 10%. According to the leaf dielectric model in Section 4-9, the dielectric constant ε_v is strongly dependent on the leaf's gravimetric moisture content m_g. At 10 GHz, for example, $\varepsilon_v \approx 20 - j8$ for $m_g = 0.7$, compared with $\varepsilon_v \approx 1.9 - j0.2$ for $m_g = 0.1$. Correspondingly, the extinction coefficient is much smaller for the dry vegetation than for the green vegetation. Hence, when a radar was used to measure the angular response of a wheat field at 9 GHz immediately before and after harvesting the wheat plants, no statistically discernible change was observed. The measured plots of σ^0 are shown in Fig. 11-27. A relatively dry canopy of wheat is essentially "transparent" to microwaves at all angles of incidence up to at least 60°.

Figure 11-28 shows the results obtained from a

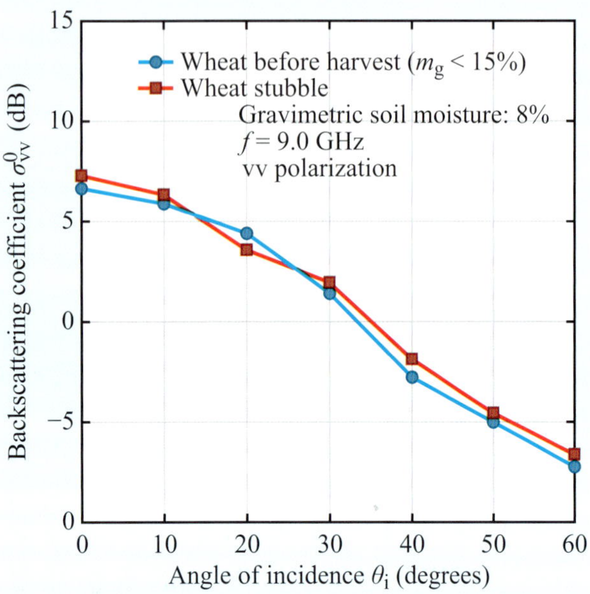

Figure 11-27: A canopy of dry wheat plants is "transparent" to microwaves; no discernible change in σ^0 is observed following the cutting and removal of the wheat plants [from Lopes, 1983].

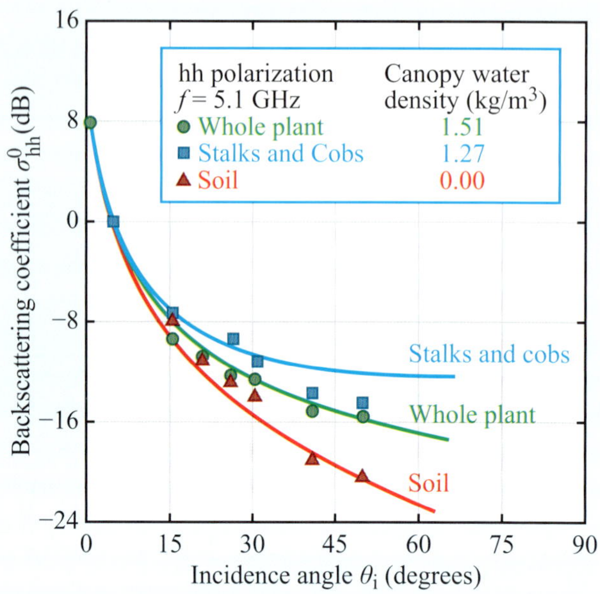

Figure 11-28: Measured σ^0 of a fully mature corn canopy in three consecutive stages of defoliation; all measurements were made on the same day.

defoliation experiment conducted on a corn canopy. A 5.1 GHz scatterometer was used to measure $\sigma^0_{hh}(\theta)$ for:

(a) a mature corn canopy under natural conditions,

(b) the same canopy after removing its leaves, but keeping the stalks and cobs in place, and

(c) the bare soil remaining after removal of stalks and cobs.

Based on the data displayed in Fig. 11-28, we deduce the following conclusions:

(1) At incidence angles $\theta_i \lesssim 20°$, the total canopy backscattering coefficient is dominated by the soil surface contribution.

(2) As θ_i increases beyond 30°, the plant becomes the dominant contributor to the total σ^0 of the canopy.

(3) For $\theta_i \gtrsim 30°$, the level of the curve labeled "stalks and cobs only" is higher than that corresponding to that labeled "whole plant." This implies that the backscattering contribution of the corn stalks and cobs is very significant, but that the attenuation by the leaves present in the upper part of the canopy reduces the stalk contribution to a level comparable to or smaller than the backscattering contribution by the leaves.

These observations suggest that a uniform canopy model in which all of its constituents (leaves, stalks, and cobs) are treated as if they have the same extinction and scattering properties, is not applicable in this case. The constituents should not be modeled as isotropic or Rayleigh particles because their dimensions are comparable to or longer than λ, and their orientations are certainly not randomly distributed.

To illustrate the relative importance of the soil contribution, we show in Fig. 11-29 pairs of angular plots for four vegetation canopies, all observed at 4.25 GHz. For each crop, the experimental observations were made under almost identical canopy conditions, but under very different soil-moisture conditions. Note that the observation dates between the dry-soil and wet-

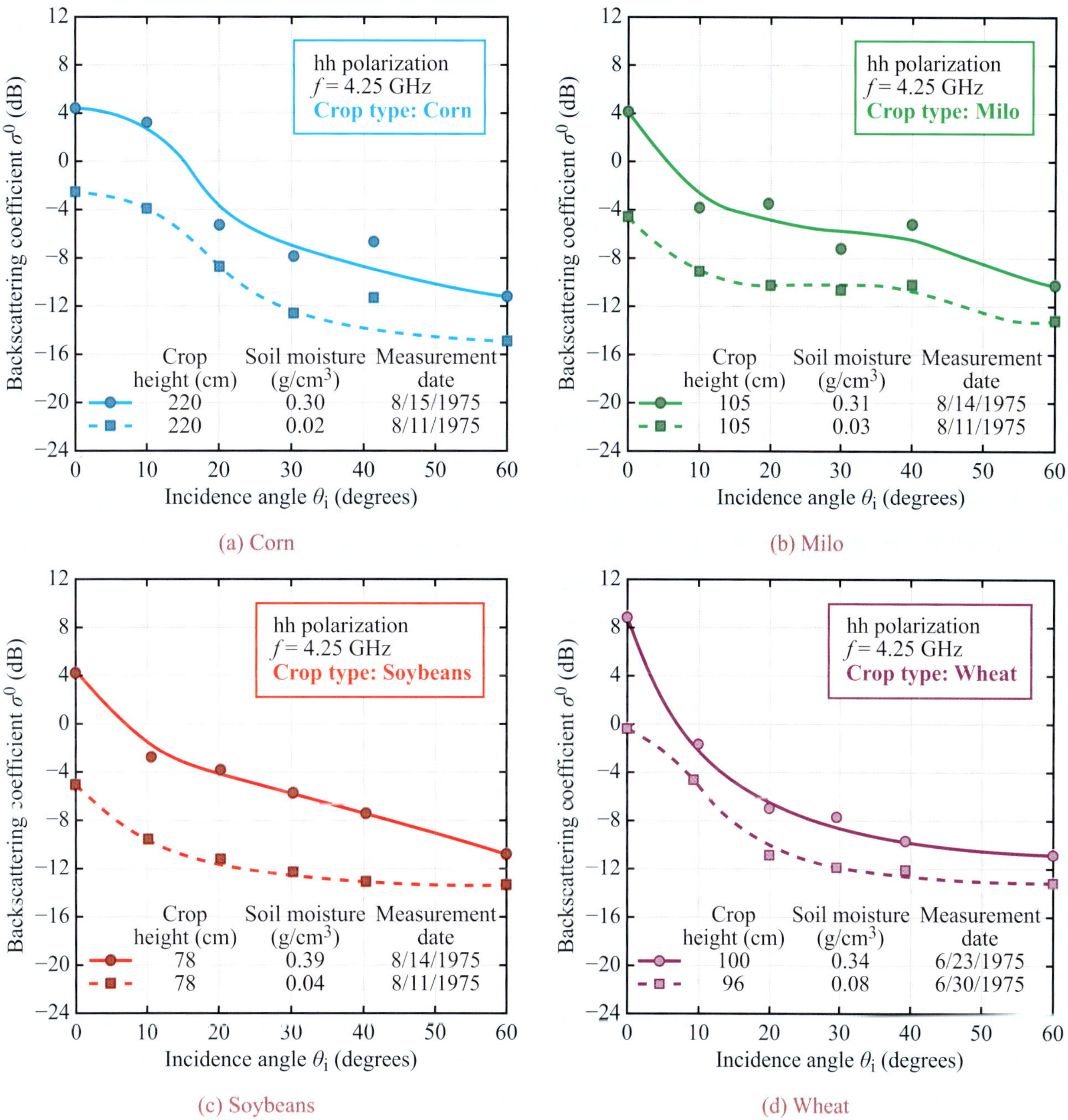

Figure 11-29: Angular response of scattering coefficient for wet and dry soil conditions at 4.25 GHz for (a) corn, (b) milo, (c) soybeans, and (d) wheat.

soil conditions are only a few days apart. The dates corresponding to the wet-soil conditions are only one day after a heavy rain that soaked the soil surface and increased its volumetric moisture content from $m_v \approx 0.04$ g/m^3 to $m_v \approx 0.35$ g/m^3.

Because the canopies were essentially unchanged between the two observation conditions, we attribute the difference in level between the two curves entirely to the underlying soil surface. In general, the wet-soil σ^0 is higher than the dry-soil σ^0 by about 7–8 dB at nadir, decreasing to about 3–4 dB at 60°.

11-8.5 σ^0 Relationship to Leaf-Area Index

Consider the plots displayed in Fig. 11-30. The left-hand vertical scale pertains to σ^0 of a corn canopy, plotted as a function of time over the growing season (approximately 100 days). The units of σ^0 are m^2/m^2 (rather than dB). The right-hand scale pertains to the temporal variation of the canopy's green ***leaf-area index*** (LAI), defined as the total single-sided surface area of all leaves contained in the canopy over a unit ground area. For the corn canopy of Fig. 11-30, the LAI $L(t)$ increases from zero at planting to a maximum of about 3.5 when the plants reach full maturity (known as Hanway stage 5), and then slowly decreases back to zero by the end of the growth cycle. Note the general similarity between the shapes of the temporal plots of $\sigma^0(t)$ and $L(t)$. Similar patterns also were observed for sorghum and wheat canopies.

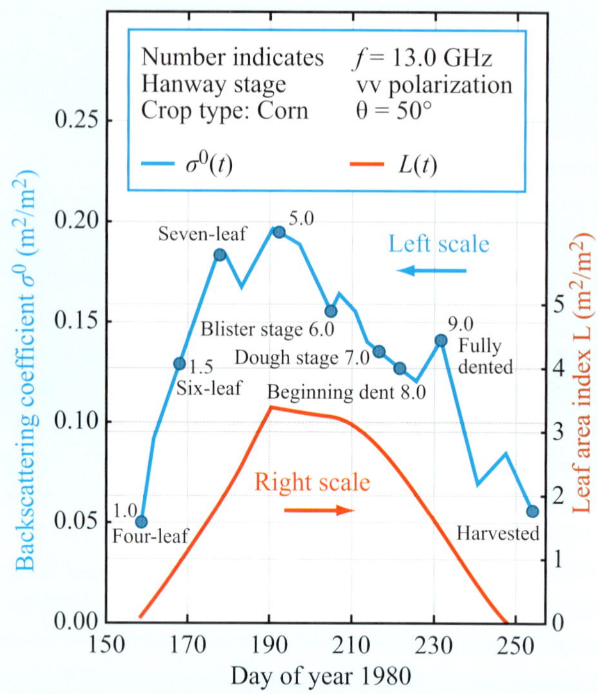

Figure 11-30: Measured temporal patterns of the backscattering coefficient σ^0 and leaf-area index (LAI) of a corn field. Stage of growth is indicated by the Hanway (1971) stage [from Ulaby et al., 1984]. Note that σ^0 is expressed in (m^2/m^2) instead of in dB.

To some extent, the leaf-area index is a surrogate for the amount of water contained in the canopy constituents, which is the single most important parameter influencing the attenuation by and scattering from vegetation. Also, even though the leaf-area index pertains to green leaves only, its temporal variation is correlated with the temporal variation of the amount of water contained in the stocks. With the understanding that the leaf area index $L(t)$ serves as an approximate representative of the canopy mass, Ulaby et al. (1984) used the form of the single-scattering model of Section 11-1 to develop a semiempirical model of the form

$$\sigma^0 = A_1 L^n (1 - e^{-bL}) + A_2 e^{-bL}, \tag{11.92}$$

where A_1, A_2, and b are empirically determined constants, and $n = 0$ for corn and sorghum and 1 for wheat. The model parameters were determined using only the radar observations at times t such that $L \geq 0.2$. The results, displayed in Fig. 11-31, confirm the general applicability of the relatively simple semiempirical model.

Another example demonstrating high correlation between σ^0 and L is shown in Fig. 11-32 for savannas observed by the SeaWinds 13.6 GHz scatterometer that flew on board of the QuikSCAT satellite (Table 1-2). The incidence angle is $\approx 50°$. The radar data shown in Fig. 11-32, which covers approximately six years (Yang et al., 2011), is highly correlated with the LAI observations made by optical sensors aboard the MODIS satellite.

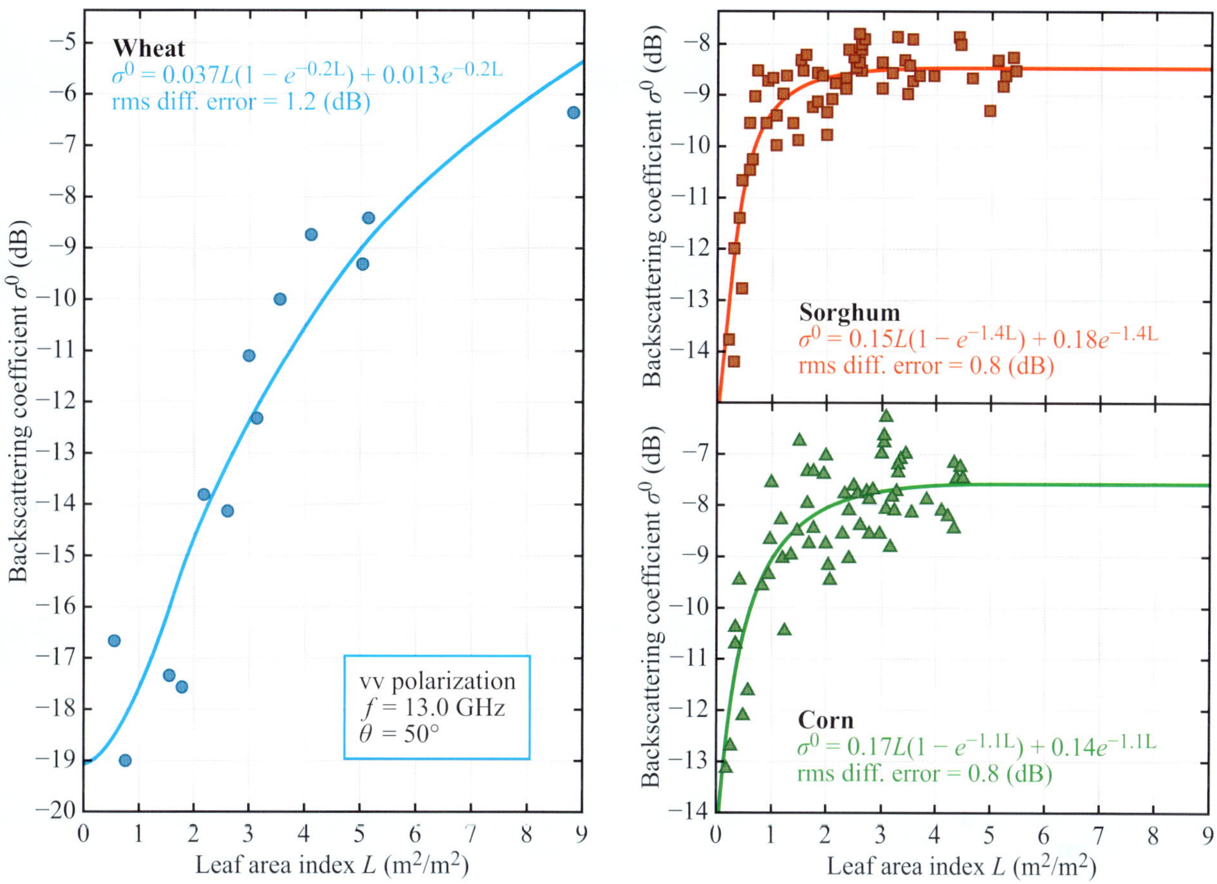

(a) Wheat

(b) Sorghum, (c) Corn

Figure 11-31: Comparison of measured backscattering coefficient of (a) corn, (b) wheat, and (c) sorghum with that computed on the basis of the single-parameter (LAI) model.

11-8.6 σ^0 Relationship to Canopy Water Content

As stated earlier, the (green) leaf-area index L should be highly correlated to the ***integrated canopy volume water content*** M_w measured in kg/m^2 and defined as the total mass of water contained in a vertical column of the canopy per unit ground surface area. That is,

$$M_w = M_f - M_d \qquad (\text{kg/m}^2), \tag{11.93}$$

where M_f and M_d are the weights of the biomass fresh and after drying, respectively, above 1 m^2 of the ground surface. In essence, M_w represents the columnar water content of the ***cloud-equivalent*** canopy.

Figure 11-33 displays the results of measurements reported by Macelloni et al. (2001) relating L to M_f, the fresh biomass/m^2, for five crops. Over most of the growing season, M_f is much larger than M_d, so the relationship between L and the canopy volume water content M_w would be only slightly shifted from that shown in Fig. 11-33.

Building on studies aimed at relating σ^0 to biophysical parameters of vegetation canopies (Jin and Liu, 1997; Bouvet et al., 2009; Kim et al., 2000; Maity

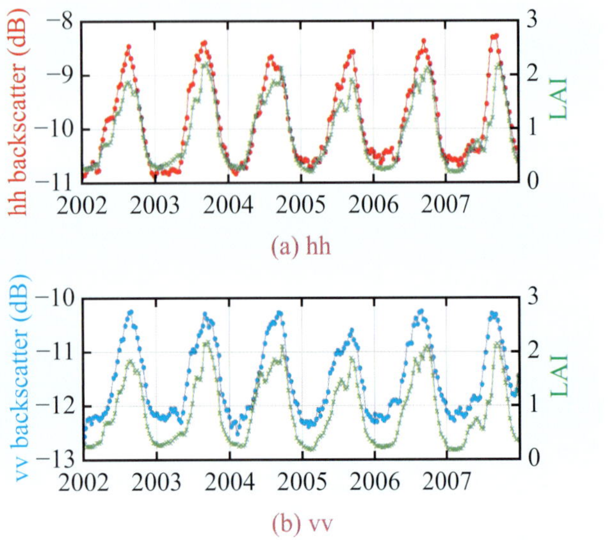

Figure 11-32: The temporal evolution of QuikSCAT backscatter (left axis) at the Savannas (10.528N, 3.976W) from 2002 to 2007, and the green time series is LAI (right axis) [from Yang et al., 2011].

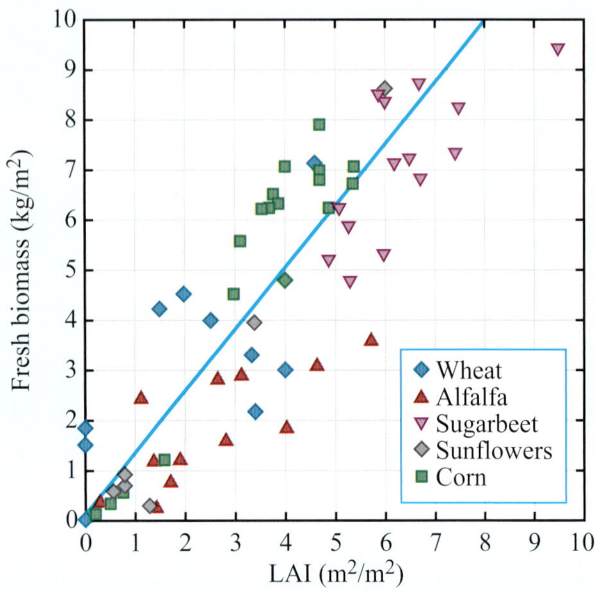

Figure 11-33: Relationship between canopy fresh biomass M_{f} (kg/m^2) and the leaf area index L (m^2/m^2) of various crop types [from Macelloni et al., 2001].

et al., 2004; Inoue et al., 2002; Oh et al., 2009), Kim et al. (2012) conducted extensive multitemporal experiments for fields planted in rice and soybeans using a ground-based multifrequency polarimetric scatterometer system. The system's center frequencies are 1.27 GHz (L-band), 5.3 GHz (C-band), and 9.65 GHz (X-band). The fields were observed over the entire growth period of the rice and soybeans plants, and all radar observations were made at an incidence angle of 40°.

Figure 11-34 displays plots of the temporal variations of LAI and volumetric moisture content M_{w} for fields planted in rice and soybeans. As expected, the two quantities are highly correlated to one another, for both types of crops. In part (a) of Fig. 11-35, we display plots of $\sigma^0(t)$ for rice over the growing season for vv, hh, and hv polarizations. In part (b) of the same figure we display a plot of $M_{\mathrm{w}}(t)$ and a temporal plot of a radar-related quantity called the ***radar vegetation index*** (RVI), which was proposed by Kim and Van Zyl (2009) as a parameter for separating vegetation-covered surfaces from bare-soil surfaces. For a canopy composed of randomly oriented thin cylinders, the RVI expression given by Eq. (5.177) reduces to (Kim and van Zyl, 2001):

$$R_{\mathrm{VI}} = \frac{8\sigma^0_{\mathrm{hv}}}{\sigma^0_{\mathrm{hh}} + \sigma^0_{\mathrm{vv}} + 2\sigma^0_{\mathrm{hv}}} \,. \tag{11.94}$$

We observe that R_{VI} and M_{v} exhibit similar temporal patterns, for both rice and soybeans. The soybeans data are shown in Fig. 11-36.

The study by Kim et al. (2012) concludes with the plots shown in Fig. 11-37, in which M_{w} is plotted against R_{VI} for both rice and soybeans combined. The sensitivity (slope) of R_{VI} to M_{w} is highest at L-band and lowest at X-band.

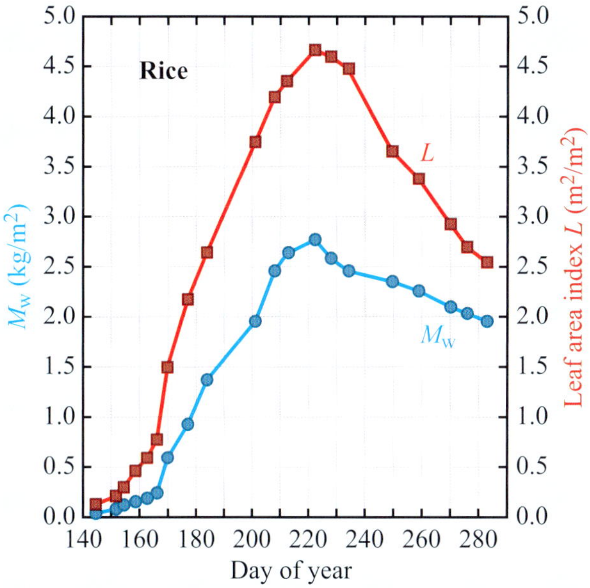

(a) Rice

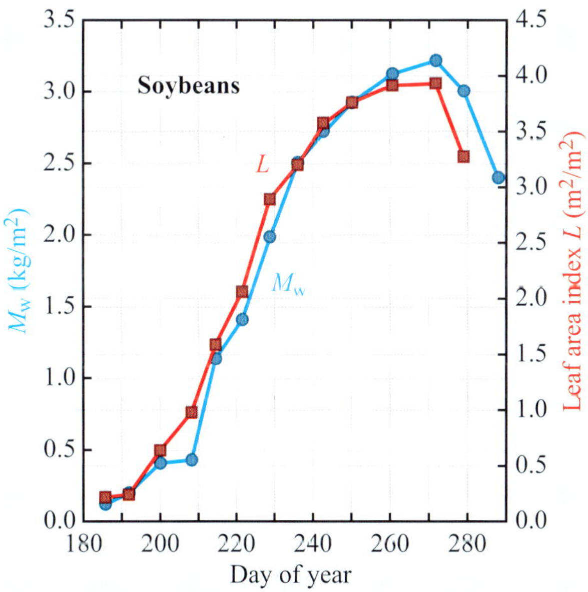

(b) Soybeans

Figure 11-34: Temporal variations of leaf area index L and canopy water content M_w for fields planted in (a) rice and (b) soybeans [Kim et al., 2012].

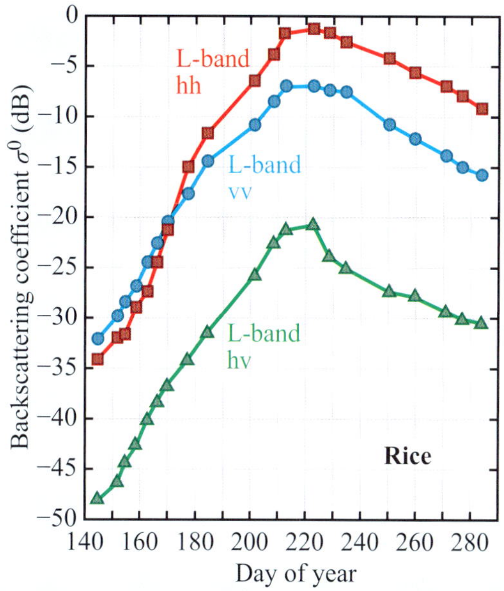

(a) $\sigma^0(t)$

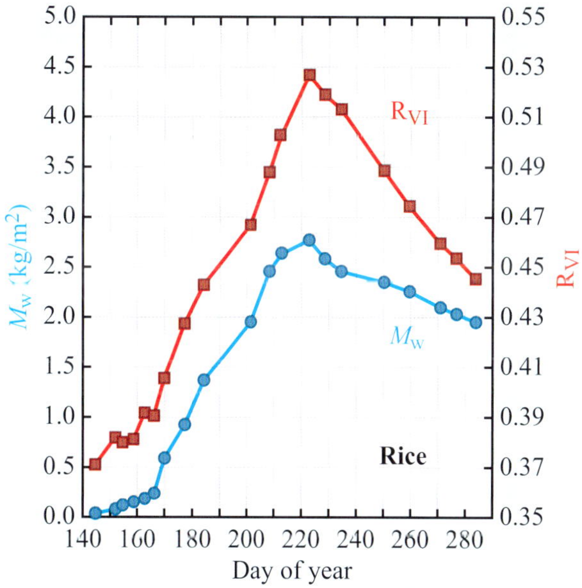

(b) $M_w(t)$ and $R_{VI}(t)$

Figure 11-35: Temporal variations of (a) σ^0 and (b) radar vegetation index R_{VI} and canopy water content M_w for a field planted in rice [Kim et al., 2012].

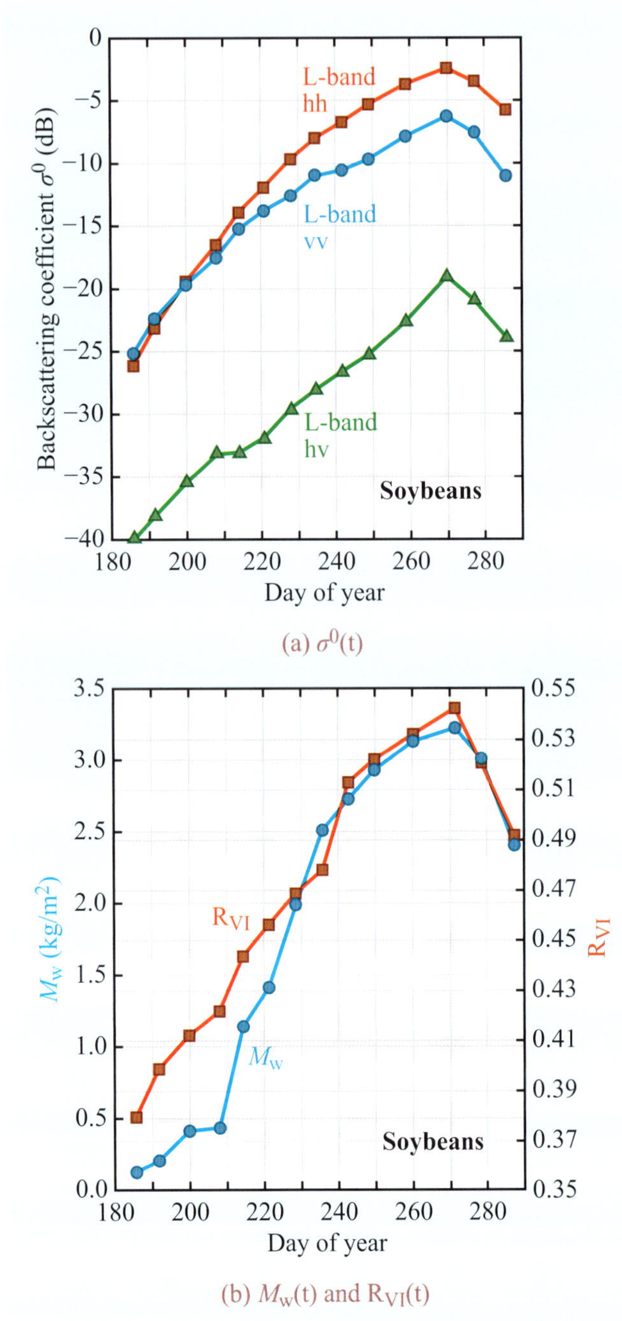

Figure 11-36: Temporal variations of (a) σ^0 and (b) radar vegetation index R_{VI} and canopy water content M_w for a field planted in soybeans [Kim et al., 2012].

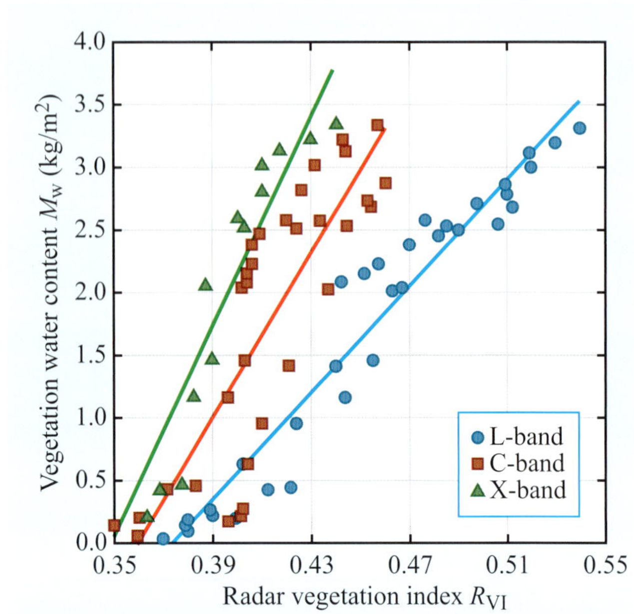

Figure 11-37: Relationship between vegetation water content M_w and L-, C-, and X-band R_{VI} for rice and soybean combined [Kim et al., 2012].

11-9 Soil-Moisture Inversion Example

To explore the possibility of developing an inversion model to estimate the soil moisture content underneath a vegetation canopy from radar observations, De Roo et al. (2001) used a truck-mounted radar to measure the backscatter from two adjacent soybean canopies over the entire growing season of 1996. Fully polarimetric L-band (at 1.25 GHz) and C-band (at 5.4 GHz) scatterometers observed the canopy at an incidence angle of 45° and an azimuth angle of 45° with respect to the row direction. Approximately 60 radar data sets were acquired over a temporal span of 130 days, covering a wide range of conditions, extending from 0.02 kg/m^2 to 0.97 kg/m^2 in vegetation water content M_w, 3% to 26% in volumetric soil moisture content m_v, and 12 cm to 63 cm in canopy depth d. The measured temporal history of m_v is shown in Fig. 11-38. The vertical bars around the mean values of m_v represent the ±1 standard deviation. The goal of the study is to estimate m_v from the radar observations, while keeping in mind that the

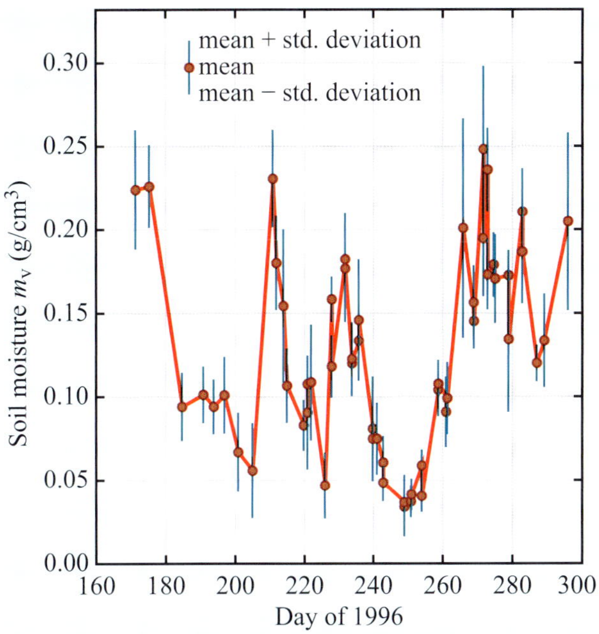

Figure 11-38: Temporal history of volumetric soil moisture content m_v under the soybean canopy. Each moisture value is based on measurements made by a soil-moisture probe at multiple locations [De Roo et al., 2001].

in-situ-measured value of m_v has an error on the order of ± 0.03.

11-9.1 The Direct Model

The investigation reported by De Roo et al. (2001) adopted the single-scattering coherent model given by Eq. (11.17) as

$$\begin{aligned}\sigma^0_{pq}(\theta_i) &= \Upsilon_p \Upsilon_q\, \sigma^0_{s_{pq}}(\theta_i)\\ &+ \frac{\sigma^{\text{back}}_{v_{pq}} \cos\theta_i}{\kappa^p_e + \kappa^q_e}(1-\Upsilon_p\Upsilon_q)(1+\Gamma^p\Gamma^q\Upsilon_p\Upsilon_q)\\ &+ n\sigma^{\text{bist}}_{v_{pq}}\, d(\Gamma^p+\Gamma^q)\Upsilon_p\Upsilon_q \qquad (p,q = \text{v or h}),\end{aligned} \tag{11.95}$$

where Γ^p is the p-polarized soil surface reflectivity at incidence angle θ_i (defined in Table 2-5), $\sigma^0_s(\theta_i)$ is the pq-polarized backscattering coefficient of the soil surface, σ^{back}_v and σ^{bist}_v are the pq-polarized volume backscatter and bistatic scattering coefficients of the soybeans canopy, and Υ_p is the canopy one-way transmissivity,

$$\Upsilon_p = e^{-\kappa^p_e d \sec\theta_i}. \tag{11.96}$$

The index n is 1 for hv polarization and 2 for hh and vv polarizations. The soil-surface backscattering coefficient $\sigma^0_{s_{pq}}$ is modeled by the semiempirical expressions given by Eqs. (10.37) to (10.39), which rely on two physical parameters, namely the rms surface height s and the soil moisture content m_v. The vegetation-related quantities were assumed to depend on the canopy ***water content per unit height***, $\rho_w = M_w/d$, as follows:

$$\sigma^{\text{back}}_{v_{pq}} = a^{pq}_2 \rho_w, \tag{11.97a}$$

$$\sigma^{\text{bist}}_{v_{pq}} = a^{pq}_3 \rho_w, \tag{11.97b}$$

$$\kappa^{pq}_e = \frac{\kappa^p_e + \kappa^q_e}{2} = a^{pq}_4 \sqrt{\rho_w}\,, \tag{11.97c}$$

$$\Upsilon_p \Upsilon_q = e^{-2\kappa^{pq}_e \sec\theta_i}, \tag{11.97d}$$

where the coefficients a^{pq}_i for $i = 2$, 3, and 4 are free parameters to be determined empirically. In addition, the expression given by Eq. (11.95) is multiplied by an empirically determined scaling constant a^{pq}_1.

At each frequency (1.25 GHz or 5.4 GHz) and polarization configuration (hh, vv, or hv), the known or measured quantities are θ_i, d, ρ_w, and $\sigma^0_{pq}(\theta_i)$, and the unknown parameters are a^{pq}_i, for $i = 1$ to 4, and the surface rms height s. To determine s, radar measurements were made early in the growing season and then used in PRISM-1 to invert for the value of s, which was found to be 2.8 cm. Using an error minimization routine, the values of the other four unknown parameters were selected such that the multiparameter error between the measured and computed values of σ^0 is a minimum. The result of the parameter-fitting process led to the values given in Table 11-2. Among the six frequency-polarization combinations, the rms error between measured and computed values of σ^0 is between 0.55 dB and 0.81 dB, which is comparable to the precision of the radar measurements. The parameter P is an indicator of the

Table 11-2: Best-fit free parameters for the semiempirical soybean model [De Roo et al., 2001].

f (GHz)	Pol	a_2	a_3	a_4	a_1	rms error	max error	P
		m^2/kg	m^2/kg	Np/(kg/m)$^{1/2}$	dB	dB	dB	%
5.4	hh	0.151	$= a_2$	0.341	2.50	0.71	1.79	23
5.4	vv	0.170	$= a_2$	0.484	3.47	0.73	1.82	16
5.4	hv	0.051	$= a_2$	0.948	5.16	0.55	1.26	84
1.25	hh	0.0	0.132	0.126	3.39	0.81	1.38	1.5
1.25	vv	0.0025	0.0605	0.0	4.78	0.62	1.81	49
1.25	hv	0.0	0.0351	0.125	5.06	0.70	1.95	15

goodness-of-fit of the model. A good fit requires P to be greater than 5%. In Table 11-2, $P > 5\%$ for five of the six frequency-polarization combinations. Comparisons of the semiempirical model calculations with the measured data are shown in Figs. 11-39 and 11-40 for L-band and C-band, respectively. Overall, the model tracks the temporal variations of the radar measurements quite well.

11-9.2 The Inverse Model

The goal of the inverse model is to use the radar data to obtain reliable estimates of the volumetric soil moisture content m_v and the canopy water content M_w. Because the direct model is highly nonlinear in M_w, it is very difficult to invert it analytically. The alternative is to explore empirical relations involving ratios of backscattering coefficients. The process led to the following inverse models:

Soil Moisture:

$$m_v = 0.234 + 0.024\sigma^0_{L-vv} - 0.014(\sigma^0_{C-hv} - \sigma^0_{C-vv}) \quad (g/cm^3),$$

$$\text{rmse} = 1.75\%, \qquad R^2 = 0.90, \tag{11.98}$$

with all three σ^0 values expressed in dB. Note that the ratio of two quantities in natural units becomes the difference between their values when their values are converted to dB.

Canopy Water Content:

$$M_w = 3.84\left(\frac{\sigma^0_{L-hv}}{\sigma^0_{L-vv}}\right)^{0.97} \quad (kg/m^2),$$

$$\text{rmse} = 0.068 \text{ kg/m}^2, \qquad R^2 = 0.87, \tag{11.99}$$

and in this case, σ^0 is in (m^2/m^2).

Comparison of radar-derived values with *in-situ* measurements is provided in Fig. 11-41.

11-10 Look-Direction Dependence

In Section 9-4.2 we examined the dependence of σ^0 on the radar look direction for periodic bare soil surfaces, wherein we defined the ***look direction*** as the azimuth angle ϕ_0 between the ground projection of the antenna beam axis and the row direction. Parallel-look direction is designated $\phi_0 = 0$ and perpendicular-look direction is designated $\phi_0 = 90°$. Now we consider a similar situation for row crops.

Using truck-mounted multifrequency scatterometers, Batlivala and Ulaby (1976) and Ulaby and Bare (1979) conducted several experiments to determine the variation of σ^0 with incidence angle θ at $\phi_0 = 0$ and $\phi_0 = 90°$ for fields planted in soybeans and wheat. Samples of their results are shown in Figs. 11-42 and 11-43 for a soybeans canopy observed at 1.1 GHz and 4.25 GHz, respectively. Strong sensitivity is observed for hh polarization at 1.1 GHz, whereas independence of look direction is observed at 4.25 GHz and higher

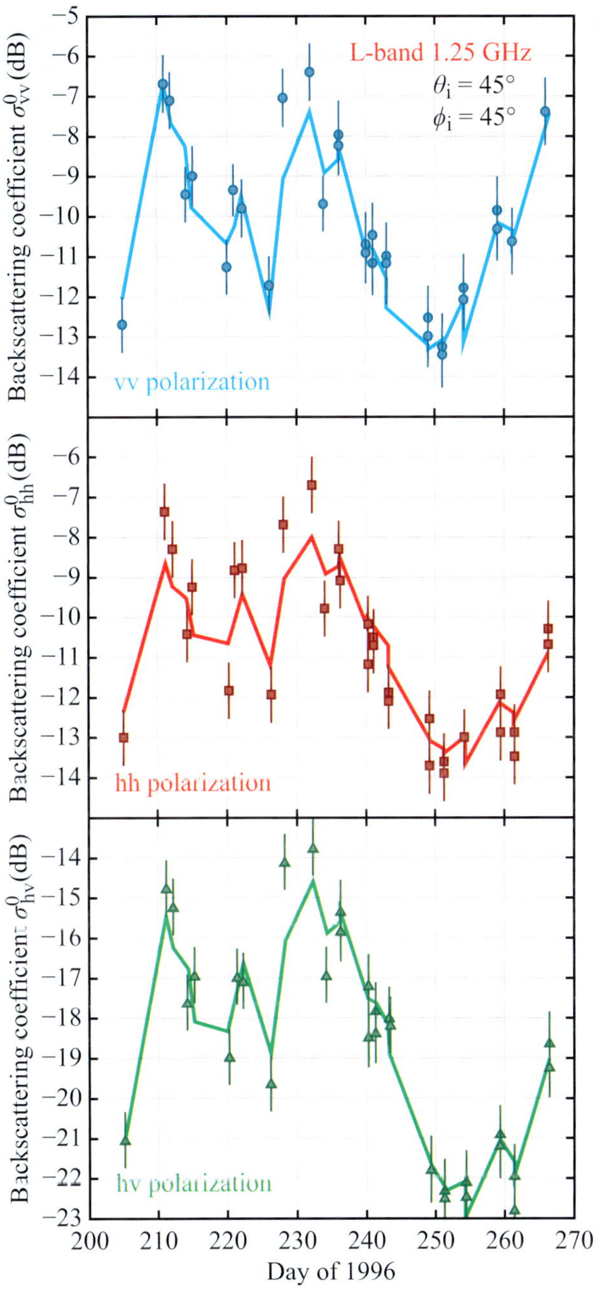

Figure 11-39: Comparison of the semiempirical model with the measured data at L-band [De Roo et al., 2001].

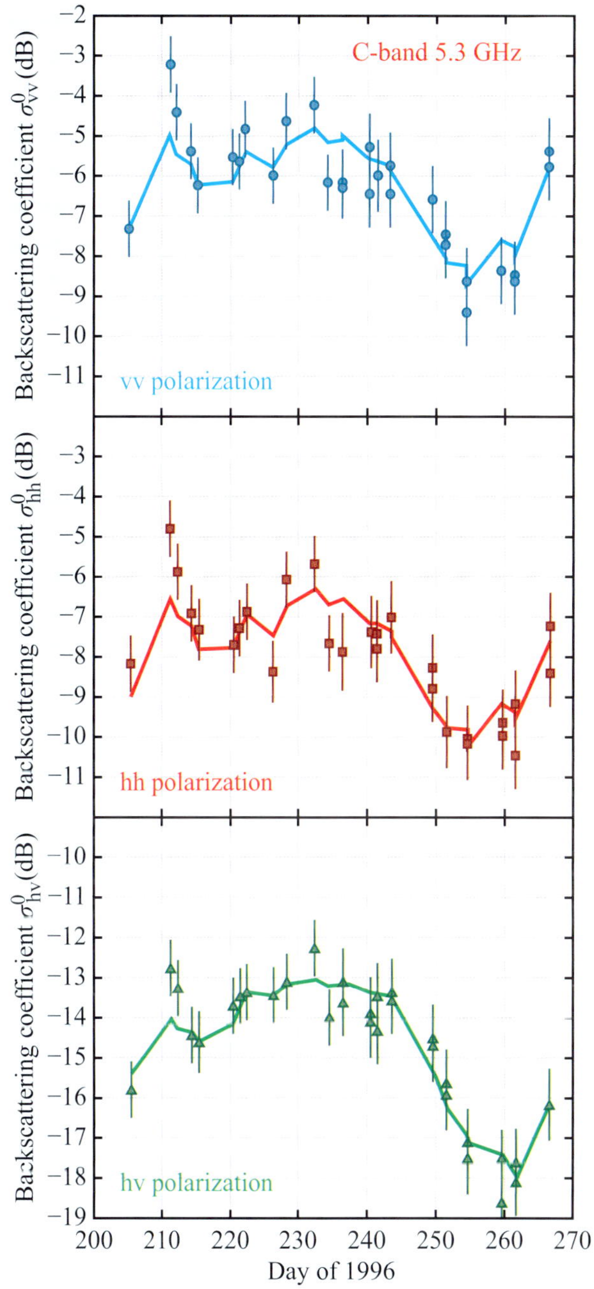

Figure 11-40: Comparison of the semiempirical model with the measured data at C-band [De Roo et al., 2001].

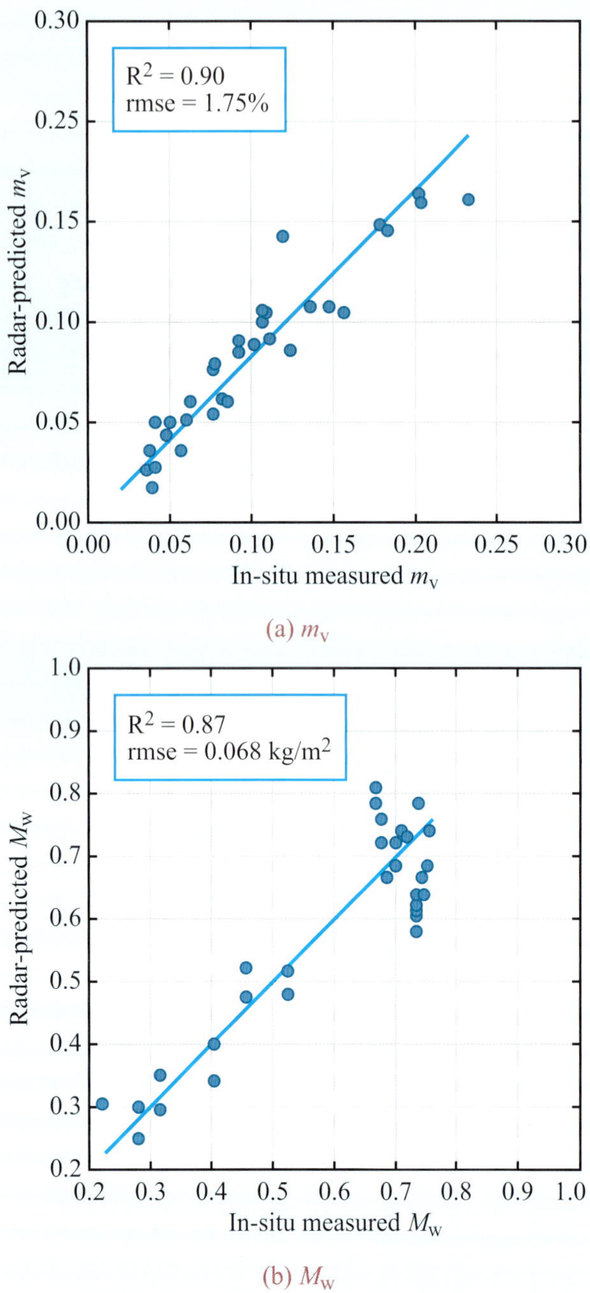

Figure 11-41: Comparison of (a) *in-situ* measured soil moisture with inverted soil moisture derived from radar measurements, and (b) similarly for canopy water content M_w [De Roo et al., 2001].

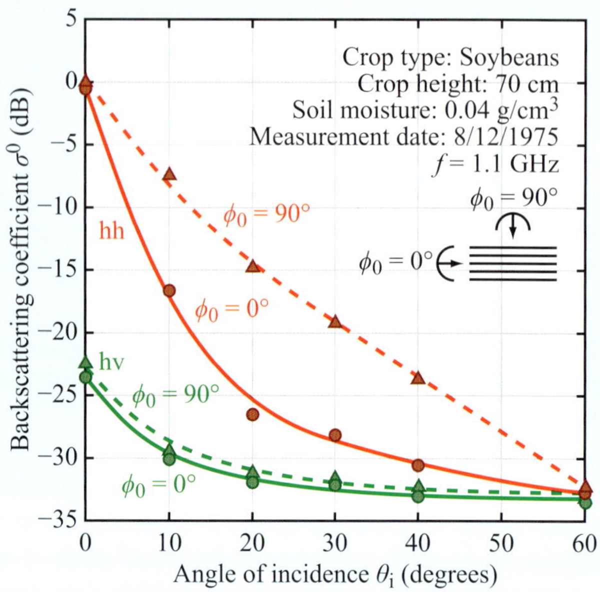

Figure 11-42: Comparison of the angular responses of $\sigma^0_\perp$ ($\phi_0 = 90°$) and $\sigma^0_\parallel$ ($\phi_0 = 0$) of a soybeans canopy at 1.1 GHz.

frequencies. Also, for hv polarization, σ^0 appears to be independent of look direction at all microwave frequencies. This observation for hv is consistent with the base-soil results discussed in Section 10-4.2, as well as with the attenuation data reported in Figs. 11-25 and 11-26. Similar conclusions were reached for wheat and other canopies observed at various stages of growth.

► For fields planted in wheat, soybeans, and corn, σ^0_{hh} and σ^0_{vv} exhibit no sensitivity to radar look direction (relative to row direction) so long as $f \gtrsim 4$ GHz. For cross polarization, σ^0_{hv} exhibits no sensitivity to look direction at all frequencies even as low as 1 GHz. ◄

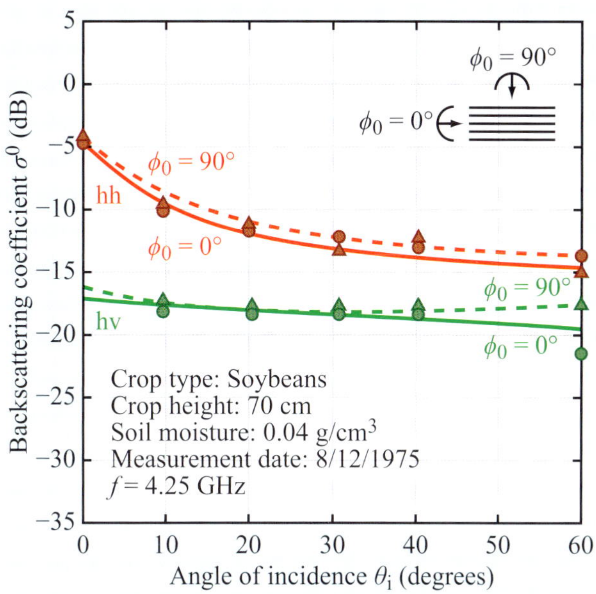

Figure 11-43: Comparison of the angular responses of $\sigma^0_\perp$ ($\phi_0 = 90°$) and $\sigma^0_\parallel$ ($\phi_0 = 0$) of a soybeans canopy at 4.25 GHz.

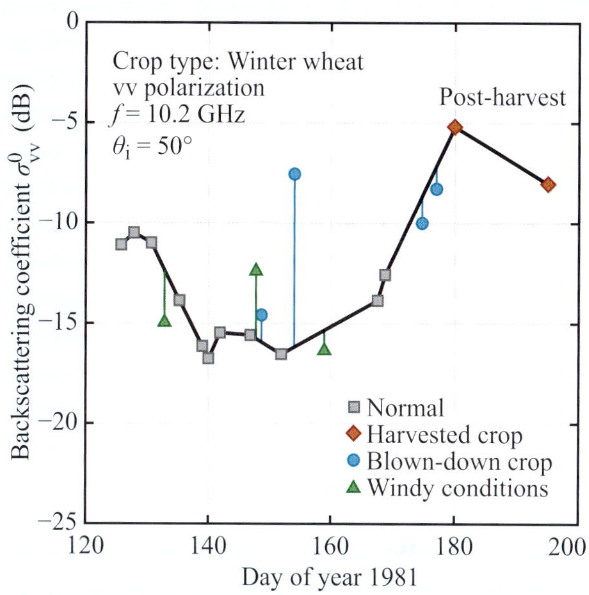

Figure 11-44: The solid curve connects backscattering measurement points made for a wheat canopy under "normal" environmental conditions. Deviations from the normal pattern caused by winds are denoted ▲, and those related to rain-caused "blowdown" by ●.

11-11 Effects of Dew, Wind, and Other Environmental Factors

The backscattering models and experimental data presented in the preceding sections pertain to "normal environmental and crop developmental conditions." Ordinarily, such models are based on certain assumptions about canopy geometry, which if violated are likely to render the model invalid. For example, logically, a wheat-canopy model is based on the assumption that most of the wheat stalks in the canopy are vertically oriented, which is a reasonable assumption under normal conditions. However, in the presence of strong winds, the stalks are likely to be bent away from the vertical by the force of the wind, thereby violating the model's initial assumption. Moreover, the effects of abnormal environmental factors usch as strong winds, heavy rain, and hail on σ^0 of a vegetation canopy are difficult to incorporate into the design of a backscattering model because of the associated complexity of the canopy's geometry. Hence, such effects may be regarded as deviations from normal behavior, as illustrated in Fig. 11-44. The plot consisting of line segments that connect points denoted by solid squares (■) represents the temporal variation of σ^0_{vv} for a field of winter wheat observed during the latter part of the wheat's growth cycle. These points are regarded as "normal" data according to the criteria that (a) relatively calm winds (i.e., with gusts of less than 25 mph) prevailed at the time of measurement and (b) the canopy's geometry appeared normal to visual inspection. The points denoted by triangles (▲) represent observations made for wind conditions under which the wheat plants were bent away from their normal, vertical orientation, and the points denoted by solid circles (●) represent observations of a canopy partially damaged by heavy rain, strong wind, or hail, which are collectively referred to as "blown-down" canopies.

The presence of water droplets on the leaf surfaces

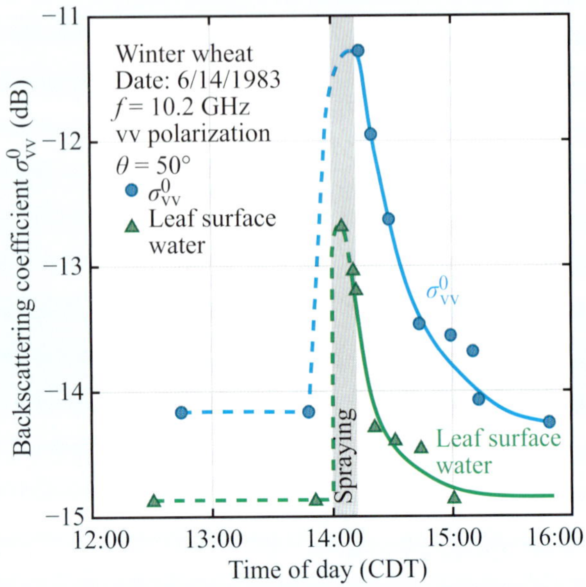

Figure 11-45: The presence of water droplets on the leaves of wheat plants can have a significant effect on σ^0 [Allen et al., 1984].

of plants—dew, for example—can modify the scattering behavior of the canopy. To evaluate this effect, Allen et al. (1984) conducted an experiment in which σ^0_{vv} of a winter wheat canopy was measured prior to and immediately after the canopy had been sprayed with water (Fig. 11-45). Conductivity plates were attached to the wheat stalks to monitor the amount of water (which the leaves retain in the form of surface droplets) present as a function of time following spraying. The conductivity plates have thermal properties similar to those of vegetation and thereby act as leaf analogs from the standpoint of the evaporation rate of surface water. The plots in Fig. 11-45 indicate that σ^0_{vv} increased by about 3 dB as a result of the spraying and then decreased to its original level over a period of about two hours as the surface water of the leaves evaporated. The experiment demonstrated that the effect of leaf-surface water on σ^0_{vv} may be significant, although its variation with frequency, angle, and polarization has not yet been established.

11-12 Radar Backscattering from Tree Canopies

11-12.1 Propagation Properties of Forest Canopies

Because a forest canopy is an inhomogeneous, random, and possibly anisotropic medium, proper characterization of its propagation properties entails measuring the attenuation loss and phase delay through many paths, so as to establish the mean values and associated statistics. The quantities of interest are the attenuation (optical depth) τ_{h} and τ_{v} for horizontal and vertical polarization, and the ***polarization phase difference*** $\Delta\phi = \phi_{\text{v}} - \phi_{\text{h}}$, where ϕ_{v} and ϕ_{h} are the one-way phase delays for v and h polarizations.

To demonstrate how to measure these quantities, Ulaby et al. (1990b) used a 1.6 GHz polarimetric scatterometer mounted atop a 19 m high platform, looking down at a dense, 13.7 m tall canopy of pine trees. The incidence angle was 40°. Figure 11-46(a) depicts the layout. The truck boom was rotated in a conical scan arrangement, thereby maintaining the range to the ground constant for all measurements. A total of 30 measurements were made, approximately 1 m apart. The first four locations were outside the canopy, and the remaining locations were inside the canopy. At each location, two scattering matrix measurements were made, one with a trihedral reflector placed in the illuminated area and another without the trihedral reflector. The dimensions of the antenna footprint on the ground were 2.3 m in the azimuth direction and 2.5 m in the range direction. The front surface of the trihedral reflector had 85 cm edges. The combination of 30 scattering matrix measurements, with and without the trihedral reflector, allowed for extracting values for the horizontal and vertical polarized attenuations τ_{h} and τ_{v} (both in dB), as well as $\Delta\phi = \phi_{\text{v}} - \phi_{\text{h}}$. The distributions and mean values are shown in Fig. 11-47.

The attenuation values extend over a wide range, from $\tau \approx 0$ for an unobstructed view of the radar beam through the canopy, to about 4.6 Np (or approximately 20 dB) for a beam intercepted by several trees. The average one-way attenuation $\tau_{\text{av}} \approx 2$ Np, which

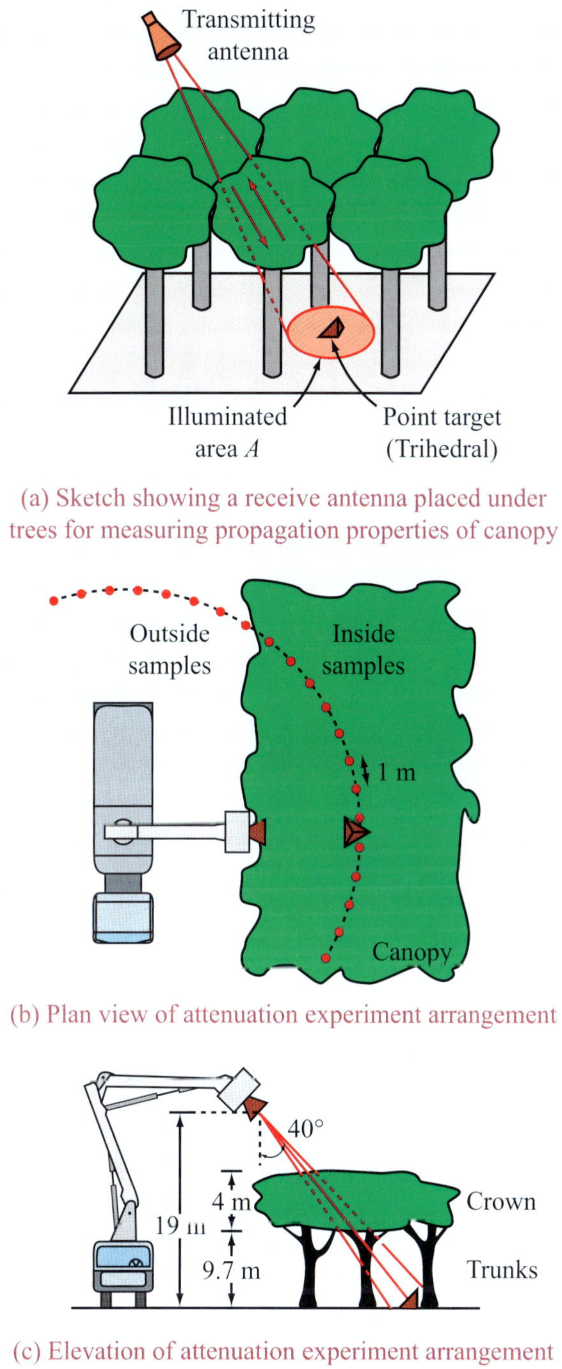

Figure 11-46: Tree attenuation experiment setup used by Ulaby et al. (1990b) at 1.6 GHz.

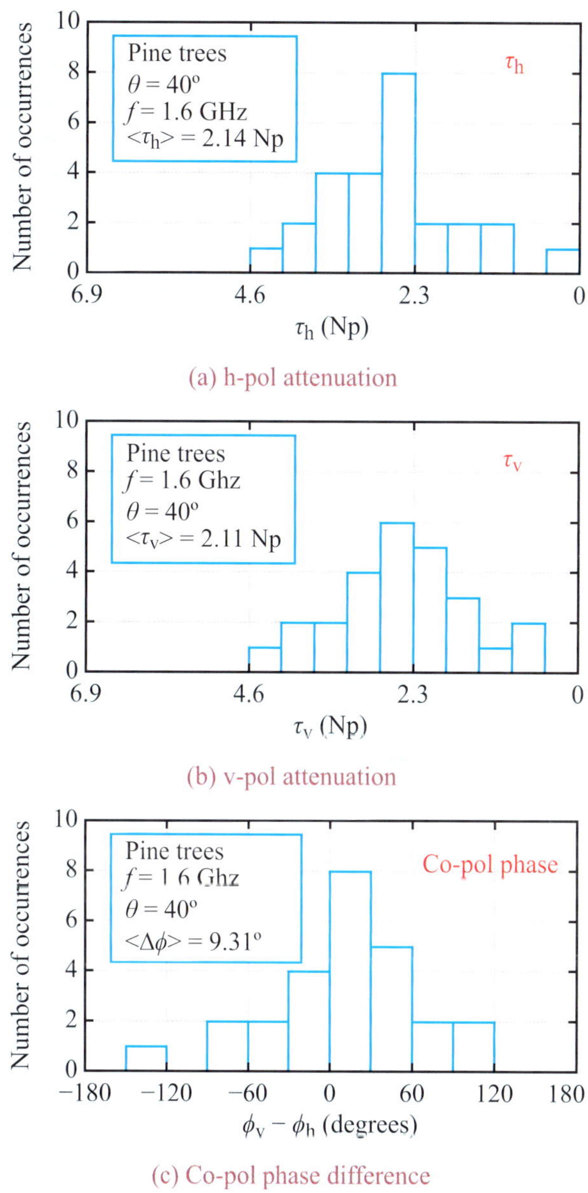

Figure 11-47: Measured distribution of one-way attenuations τ_h and τ_v, and phase delay difference $\Delta\phi$ [Ulaby et al., 1990b].

corresponds to a two-way transmissivity

$$\Upsilon^2 = e^{-2\tau_{\text{av}}} = e^{-4} \approx 0.02.$$

Hence, except possibly along incidence directions very close to normal incidence, even at the low microwave frequency of 1.6 GHz, the attenuation by the canopy is sufficiently high as to reduce the average ground backscattering contribution to a level much smaller than that of the canopy's direct backscattering contribution. In general, τ_{av} increases with frequency, so the aforementioned conclusion should be valid not only at L-band, but also across the microwave spectrum ($f \geq 1$ GHz). Obviously, this assertion does not apply to short or sparse canopies, but when it does, the canopy appears like a semi-infinite lossy medium, in which case Eq. (11.18) is applicable:

$$\sigma^0_{pq}(\theta) = \frac{\sigma^{\text{back}}_{\text{v}_{pq}} \cos\theta}{\kappa^p_{\text{e}} + \kappa^q_{\text{e}}}, \qquad (p, q = \text{h or v}).$$

For a particular tree canopy, the volume backscattering coefficient $\sigma^{\text{back}}_{\text{v}_{pq}}$ and the extinction coefficients κ^p_{e} and κ^q_{e} are likely to have constant values over a wide range of the incidence angle θ. Hence, σ^0_{pq} should exhibit a simple cosine dependence, or equivalently, $\gamma_{pq} = \sigma^0_{pq}/\cos\theta$ should exhibit a flat response. This is indeed the case, as shown in the next subsection.

In a similar study reported by Kurum et al. (2009), which also used a truck-mounted platform (Fig. 11-48), time-domain analysis of the recorded backscattered signal was used to measure the signal attenuation τ of a tree canopy composed of 11–14 m tall trees with approximately 13 kg/m^2 in above-ground biomass. The measurements were conducted at 1.25 GHz, along four incidence angles (15° to 45° in 10° increments), and for two canopy conditions: full-canopy and leaf-drop (no leaves). Their results, which were reported in dB, were converted to Np and are plotted in Fig. 11-49. These are average values based on observations made along 28 different azimuth orientations. As expected, τ_{av} increases with incidence angle for both wave polarizations and both canopy conditions. In general, τ_{av} is slightly larger for v polarization, and the leaf-drop attenuation is about 10% smaller than its full-canopy counterpart.

Figure 11-48: 1.25 GHz radar system deployed over a stand of Paulownia trees [Kurum et al., 2009].

11-12.2 Angular and Frequency Response of σ^0

Most tree canopies are spatially nonuniform, consisting of a random distribution of trees (or groups of trees) and open spaces. Hence, to obtain a representative value for the backscattering coefficient σ^0 of a tree canopy, it is important that sufficient spatial and temporal averaging be performed during the measurement process to ensure that the measured value of σ^0 is not adversely influenced by wind or local variations of the canopy cover fraction. This consideration is particularly important for high-resolution observations (such as those made using a low-altitude, narrowbeam scatterometer or a high-resolution imager) at angles close to normal incidence.

As noted earlier, if we model a tree canopy as a

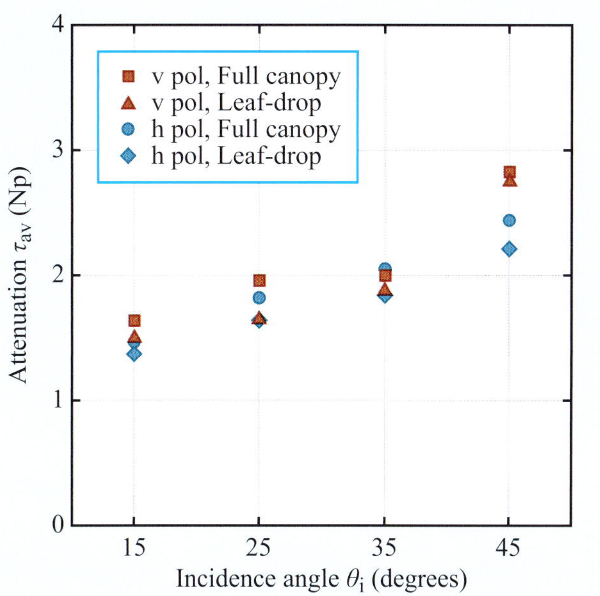

Figure 11-49: Average one-way attenuations through a canopy of Paulownia trees under full-canopy and leaf-drop conditions. Tree height ≈ 11–14 m and above-ground biomass ≈ 13 kg/m^2 [Kurum et al., 2009].

lossy scattering volume consisting of a collection of random scatterers of varying sizes and orientation, its backscattering coefficient σ^0 should exhibit an angular dependence approximately equal to $\cos\theta$, which is equivalent to $\gamma(\theta) = \sigma^0(\theta)/\cos\theta =$ constant. Such an angular variation is indeed typical of most observations reported in the literature; the examples given in Fig. 11-50 for deciduous trees at 3.3 GHz and 8.6 GHz show that $\gamma(\theta)$ is essentially constant (within 1 to 2 dB) for $\theta \geq 20^\circ$ for all linear polarization configurations. In general, deciduous trees in leaf exhibit a σ^0 level higher than that of defoliated trees (leaf-drop) by typically 2 to 8 dB, depending on angle, frequency, and polarization (Figs. 11-51 and 11-52). These observations are based on measurements made over the 1–18 GHz band. Similar observations were noted by Trebits et al. (1978) on the basis of their measurements at 9.4, 35, and 95 GHz.

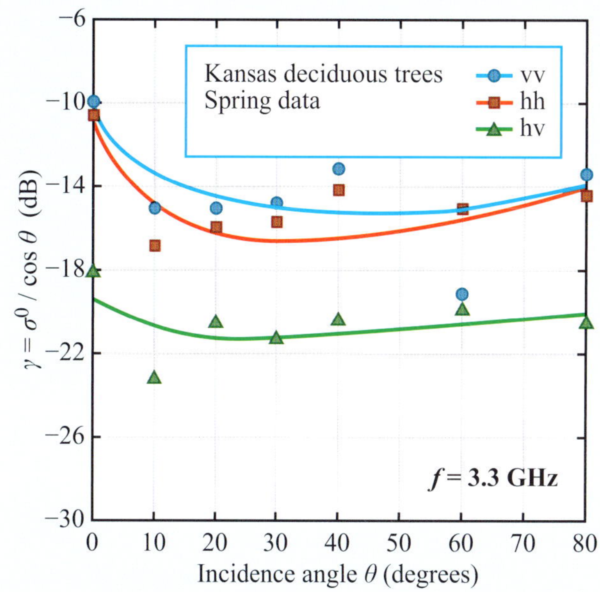

(a) 3.3 GHz

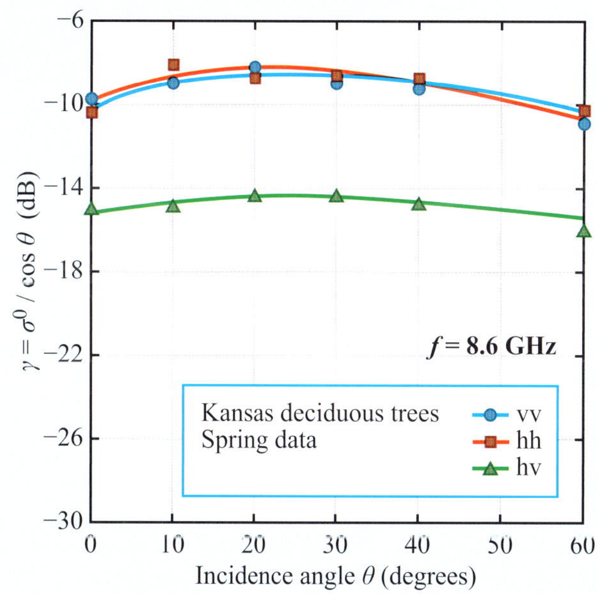

(b) 8.6 GHz

Figure 11-50: Angular variation of γ at 3.3 and 8.6 GHz as a function of θ for deciduous trees with leaves [from Bush et al., 1976]. Note that γ (dB) $= 10\log(\sigma^0/\cos\theta)$.

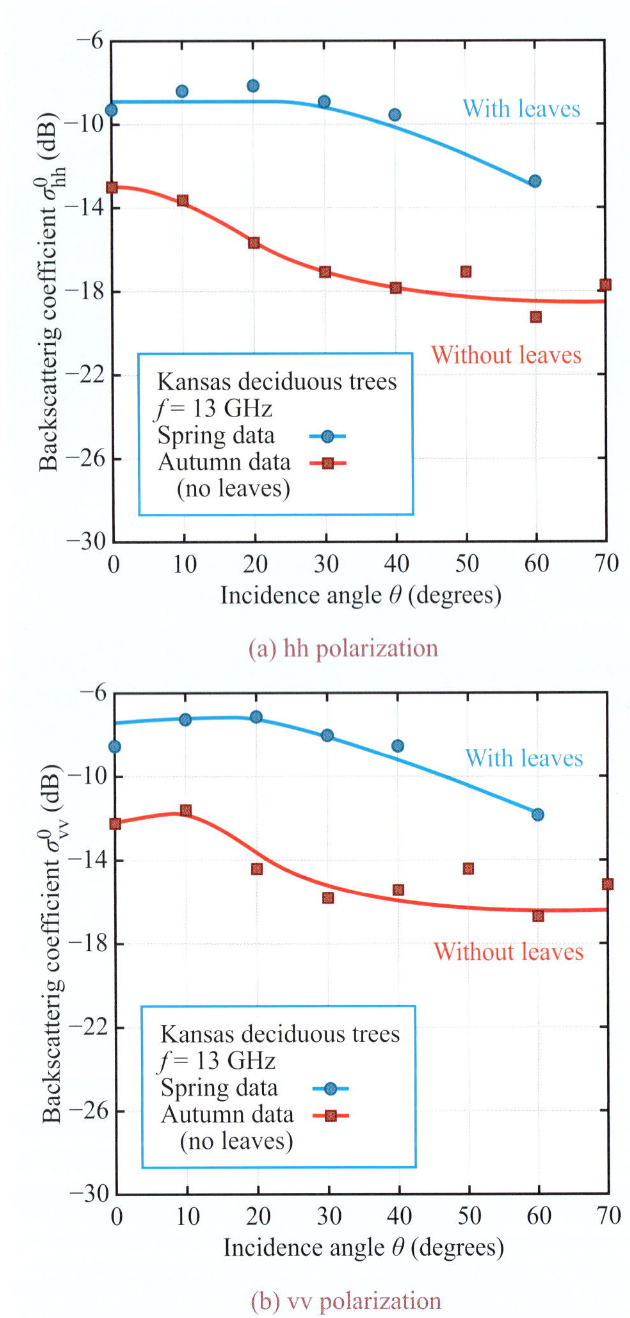

Figure 11-51: Angular variation of σ^0_{hh} of trees measured at 13.0 GHz in the spring and autumn [from Bush et al., 1976].

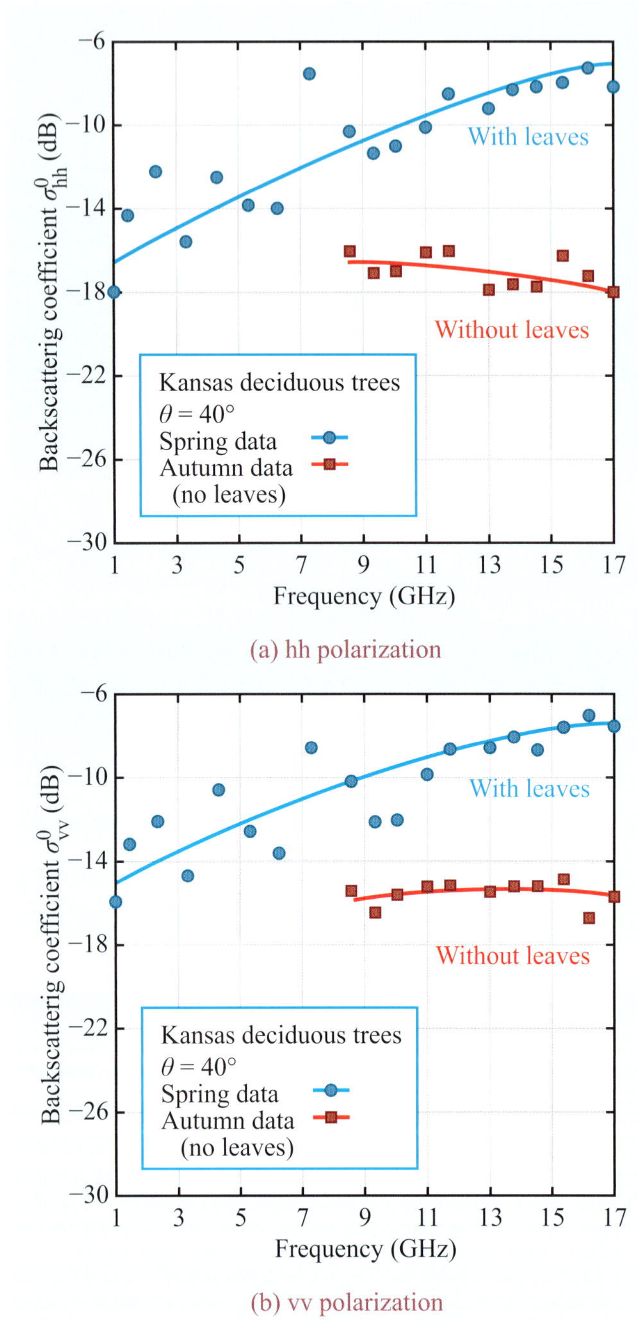

Figure 11-52: Spectral responses of (a) σ^0_{hh} and (b) σ^0_{vv} of trees measured at 40° [Bush et al., 1976].

11-12.3 MIMICS

The purpose of this subsection is to demonstrate that the single-scattering solution of the radiative transfer equation is perfectly suitable for modeling the backscatter from forest canopies, provided sufficient care is taken in characterizing the physical and dielectric properties of the canopy and underlying soil surface. We do so by offering a brief overview of MIMICS: the ***Mi***chigan ***Mi***crowave ***C***anopy ***S***cattering model. The first generation of MIMICS, known as MIMICS-1, was developed by Ulaby et al. (1988c) for canopies with continuous crown layers. Later generations addressed discontinuous canopies (McDonald and Ulaby, 1993) and periodic canopies (Whitt and Ulaby, 1994). We limit our present discussion to MIMICS-1 (continuous canopies).

MIMICS-1 is a generalized version of the single-scattering solution given in Section 11-6; it is fully polarimetric, which means that all scattering and extinction processes are defined in terms of 4×4 matrices. It models a forest canopy as two distinct horizontal vegetation layers over a dielectric ground surface (Fig. 11-53), the top layer contains the tree crowns, has vertical extent h_c, and consists of an appropriate combination of leaves and branches. The bottom layer contains the tree trunks and has height h_t. The leaves are modeled as flat circular disks of radius a and orientation (θ, ϕ), where θ is the angle between the zenith direction and the vector normal to the leaf surface. Similar definitions apply to the branches and the trunks, both of which are modeled as dielectric cylinders.

Walnut canopy experiment

In 1987, a field experiment was carried out at the Kearney Agricultural Center in Fresno, California, using the University of Michigan POLARSCAT system. The site contained separate orchards of water-stressed and unstressed plots of walnut trees. This study pertains to the orchard with unstressed trees. The trees were irrigated in the evening with the amount of water equivalent to 100% of the water that evaporated from the canopy during the day, as determined by meteorological measurements made at a nearby weather station. The radar experiments were designed to investigate the backscatter response to diurnal variations in canopy condition. Two experiments were conducted. The first is a set of multiangle measurements (40°–55°) recorded over a short span of 2 hours. The second set consisted of a three-day measurement series during which the walnut canopy was observed continuously at $\theta = 55°$.

Canopy architecture

Statistical sampling of the canopy geometry was performed for eight trees. It included measuring the length, diameter at midlength, and zenith and azimuth orientation angles for all branches with diameters greater than 2 cm. The number and size classes of all lateral branches were also recorded. All branch segments were numbered so that the tree skeletons could be reconstructed from these observations. In addition, the smaller branches with diameters less than 4 cm were statistically subsampled by class size. Four sample classes were considered (0–1, 1–2, 2–3, and 3–4 cm diameters). To adapt the branch geometry data for input to MIMICS, the orchard is divided into distinct crown and trunk layers with heights 2.5 m and 1.7 m, respectively. These heights correspond to the observed canopy architecture. The branches are then divided into four size classes. Figure 11-54 is a sketch of the geometry of an individual tree, showing the four branch classes and the leaves. The larger branches tend to be located in the lower portion of the canopy and are therefore considered to be part of the trunk layer. This layer consists of the trunk-branch size class and includes the tree trunks and all branches with diameters greater than 4.0 cm. The remaining three branch size classes are distributed throughout the crown layer.

In general, the larger branches tend to have mostly vertical orientations whereas the smaller branches exhibit no preferred orientation. Plots of the branch orientation pdfs are shown in Fig. 11-55. The branches in the trunk layer were assigned a $\cos^6\theta$ distribution so that the mean value of θ is 0. The primary branches were assigned a distribution of $\sin^4(2\theta)$ so that θ has a mean value of 45°. The secondary branches and higher-order stems were assigned spherical distribution functions $(\sin\theta)$ so that they show no preferred pointing direction

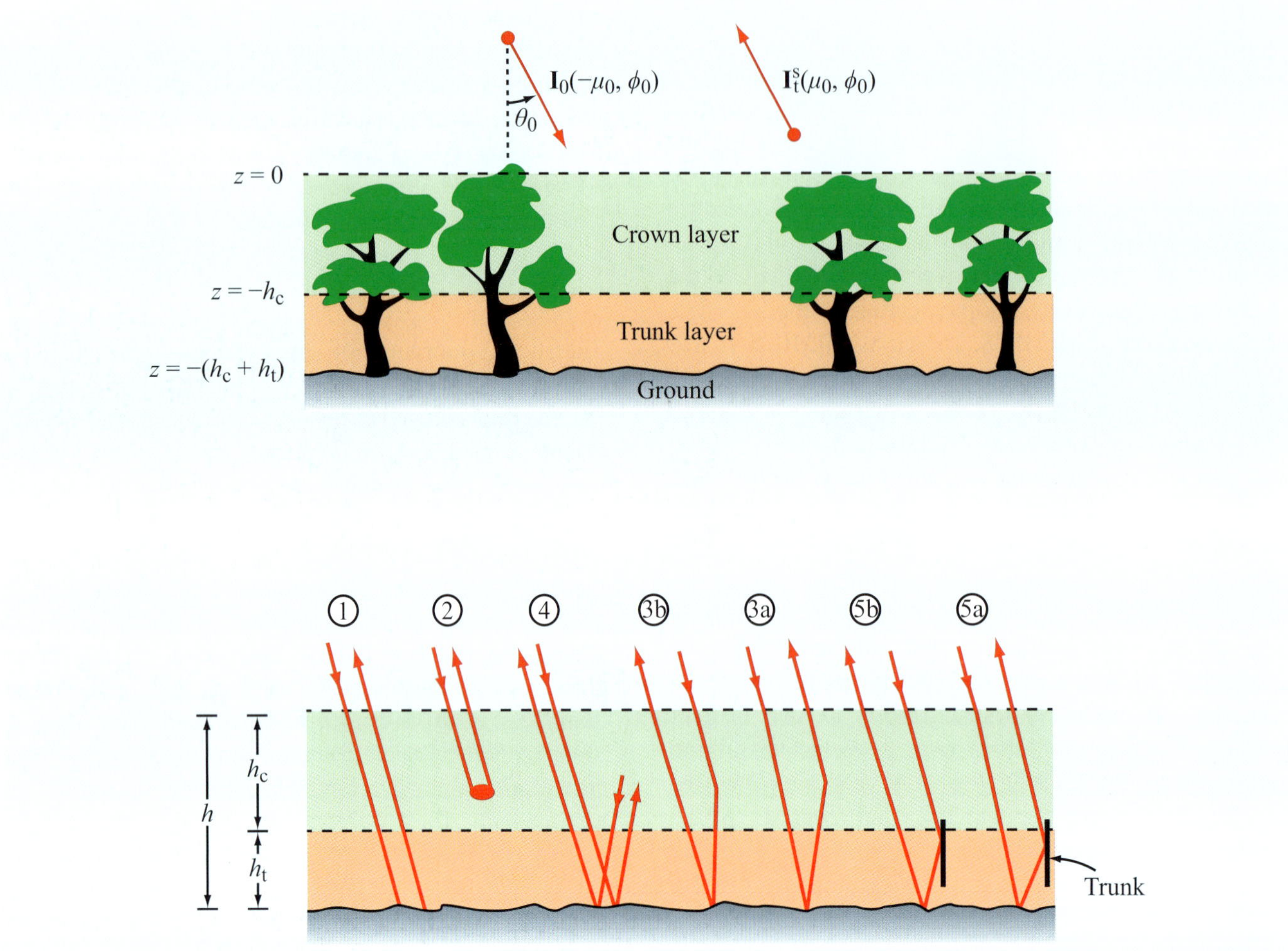

Figure 11-53: Illustration of geometry used to model tree canopy and scattering contributions obtained by MIMICS-1 [Ulaby et al., 1990b].

in the elemental solid angle $d\Omega = \sin\theta \; d\theta \; d\phi$. This implies that these branches are oriented such that their axis directions are uniformly distributed on a spherical surface.

Canopy dielectric constant

Observations of the relative dielectric constant of soil and vegetation were made *in situ* at 1.2 GHz using a field portable dielectric probe. Observations were made of the soil surface and tree trunks. Trunk measurements included both the exterior bark and the interior sapwood. The vegetation measurements were made using coaxial probe tips designed specifically for insertion into the tree boles. A statistically insignificant amount of dielectric data were obtained for the vegetation in the crown layer. However, the dielectric behavior of these constituents may be inferred from observations of other canopy physiological parameters—and the models applied here to predict the relative dielectric constant do in fact agree with the few recorded observations.

The dielectric properties of the tree boles were seen to vary dramatically with time and exhibit a repetitive

Leaf density	652 leaves / m^3
Average leaf diameter	5.0 cm
Average leaf thickness	0.02 cm
Leaf area index	3.2
Leaf orientation pdf	spherical

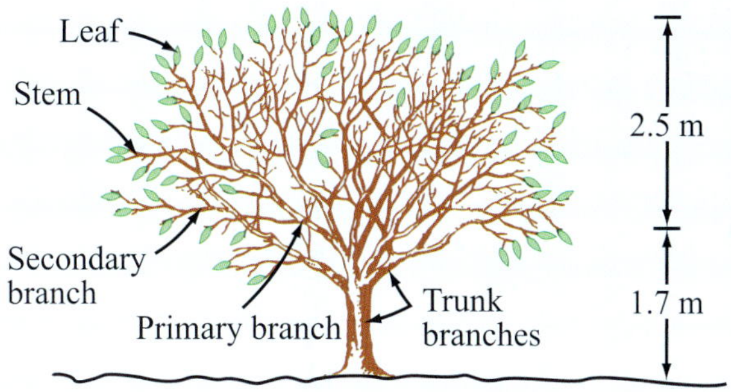

Characteristic	Branch size class			
	Trunk branches	Crown branches		
		Primary	Secondary	Stems
Max. diam. (cm)	-	4.0	0.9	0.4
Min. diam. (cm)	4.0	0.9	0.4	-
Ave. diam. (cm)	7.3	1.9	0.6	0.1
Ave. length (cm)	92.8	35.8	10.9	5.0
Density (# / m^3)	0.13	1.55	1.41	900
Orientation $f(\theta)$	$\cos^6\theta$	$\sin^4 2\theta$	$\sin\theta$	$\sin\theta$

Figure 11-54: Illustration of walnut tree showing four branch classes and leaves [McDonald et al., 1991].

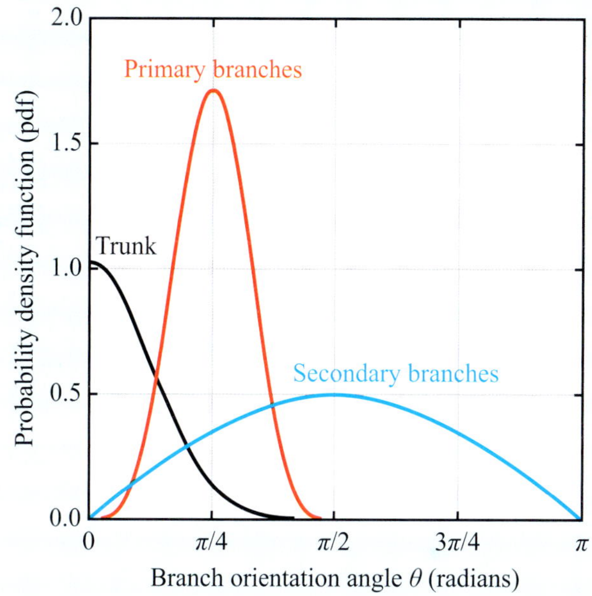

Figure 11-55: Branch orientation probability distribution functions [McDonald et al., 1991].

diurnal pattern. Figure 11-56 shows a piecewise fit to the measured values of the dielectric constant. The numbers on the time axis correspond to midnight on that day of August. The dielectric constant is seen to reach a peak near daybreak at about 6:00 am. Shortly thereafter, the values decrease rapidly until a minimum is reached at about 12 noon. Beginning at about 7:00 pm, the dielectric constant increases until the maximum is reached again. These trends are consistent with data observed throughout the entire course of the experiment. Note that ε_v' varies between a low of 10 and a high of 75, close to that of water.

Soil dielectric constant

The soil was monitored on an hourly basis. Each observation consisted of a minimum of 15 independent, randomly selected samples used to calculate the mean $\varepsilon_{\text{soil}}$ of the surface layer. Because of spatial variations associated with the locations of the sprinkler heads used to irrigate the soil, the measured dielectric data exhibited a high degree of scatter that made it difficult to estimate an effective soil dielectric. Therefore the soil surface area that received irrigation and remained mostly wet was analyzed separately from the area that received no irrigation and remained mostly dry (Fig. 11-57). The overall effective dielectric behavior was then estimated by combining these results.

Angular response of σ^0

To test the MIMICS-1 model, it was run as a function of θ for $f = 1.25$ GHz using the geometrical information given in Figs. 11-54 and 11-55 and the dielectric constant data listed in conjunction with

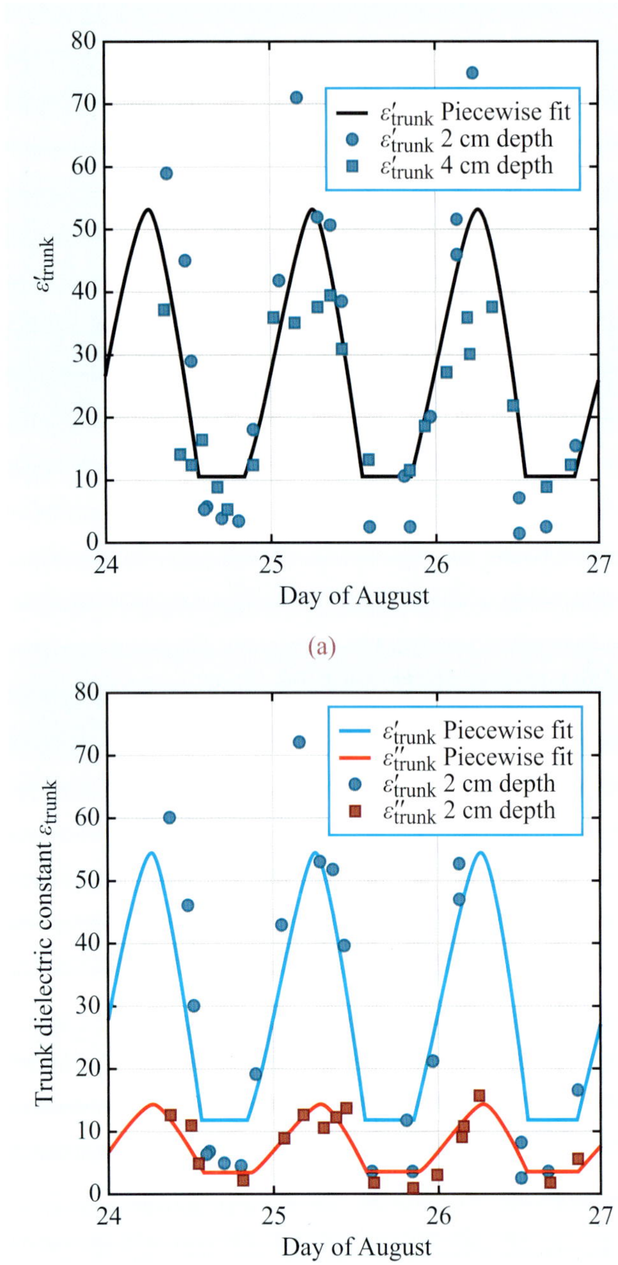

Figure 11-56: Comparison of periodic piecewise fit to measured trunk dielectric constant data: (a) for two insertion depths and (b) for real and imaginary parts at 2 cm insertion point [McDonald et al., 1991].

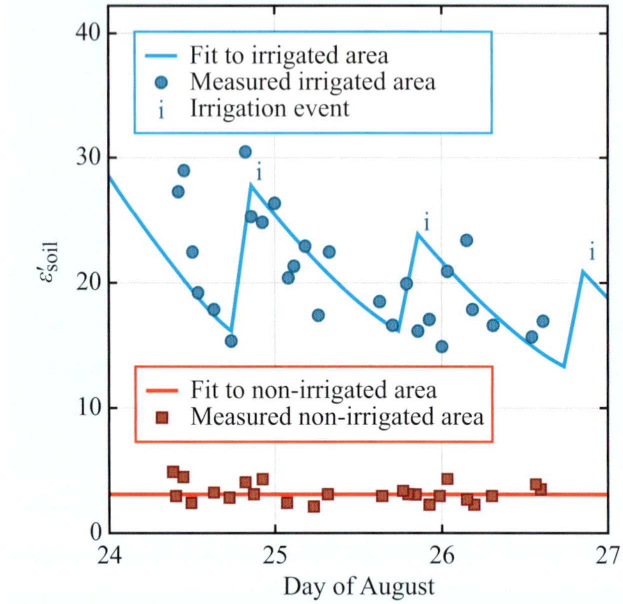

Figure 11-57: Behavior of soil dielectric constant. Fits to measured dielectric constants of irrigated and nonirrigated areas. The (i) indicates beginning of 2.5-hour irrigation period [McDonald et al., 1991].

Fig. 11-58. These dielectric values were measured at the approximate time of the multiangle experiment. The comparison of modeled and measured data shown in Fig. 11-58 demonstrates very good agreement between MIMICS-1 and the backscattering values measured by the scatterometer.

Temporal response of σ^0

Having established that MIMICS-1 successfully predicts canopy backscatter as a function of angle, the model was run at a constant incidence angle, $\theta = 55°$, while varying the canopy dielectric parameters so as to simulate the variations seen over the three-day diurnal experiment. Figure 11-59 presents the computed backscatter along with the measured values of canopy backscatter for the like- and cross-polarizations. In each case, MIMICS successfully predicts the appropriate level of the measured data together with the decreasing trend in backscatter observed over the three-day period.

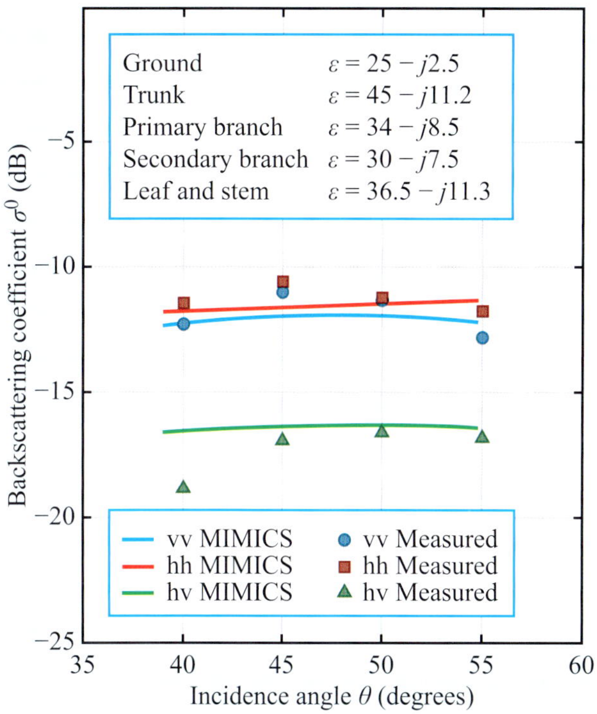

Figure 11-58: Comparison of MIMICS with recorded data for the multiangle data set [McDonald et al., 1991].

Furthermore, MIMICS-1 predicts the 1–2 dB dip seen in σ^0_{vv} and σ^0_{hv} in the early afternoon of each day.

11-12.4 Canopy Biophysical Parameters

Forest canopies are characterized by elaborate architectures that may vary not only by species, but also by age. The most important biophysical parameters of forest canopies include:

- ***Tree height*** $h = h_t + h_c$
- ***Trunk height*** h_t (m)
- ***Crown depth*** h_c (m)
- ***Tree (or stock) density*** N: # of trunks/hectare (ha) or trunks/m^2
- ***Trunk diameter*** d (averaged over the trunk height)

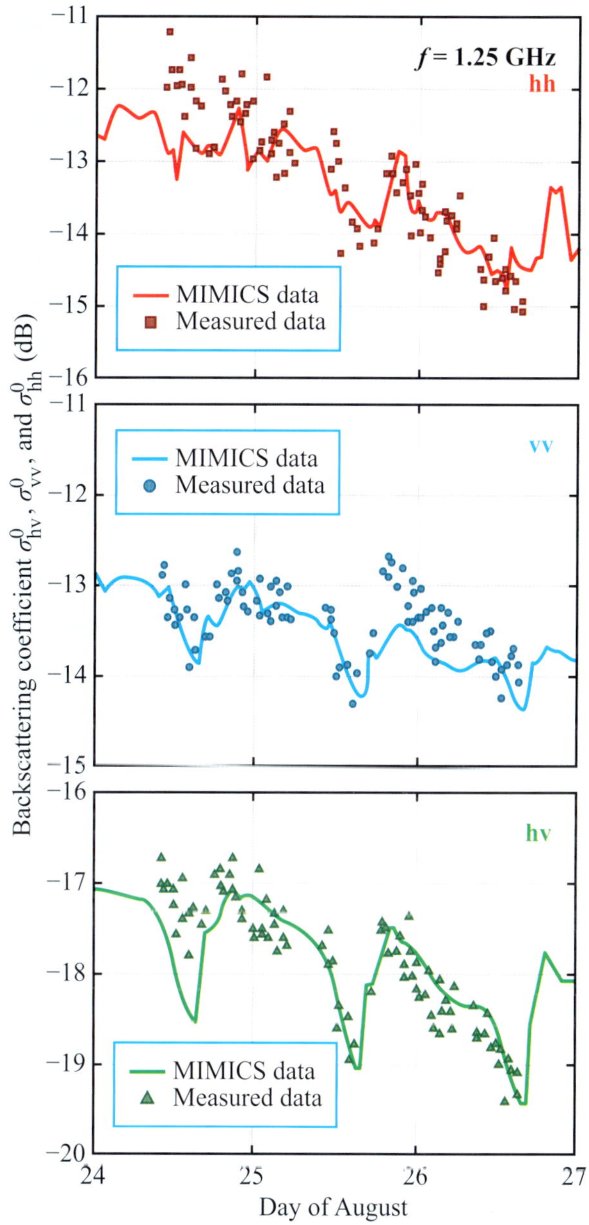

Figure 11-59: Measured and predicted backscatter during three-day diurnal experiment: (a) hh-polarized response, (b) vv-polarized response, (c) hv-polarized response [McDonald et al., 1991].

- ***Trunk (or stem) volume*** $V_t = Nh_t(\pi d^2/4)$, in m^3/m^2 (or m^3/ha where 1 m^3/ha $= 10^{-4}$ m^3/m^2)
- ***Stem (or trunk) biomass*** $B_t = \rho_t V_t =$ trunk *dry* mass per unit area (kg/m^2). Sometimes biomass is expressed in tons per hectare (tons/ha); the conversion factors are 1 ton $= 10^3$ kg and 1 ha $= 10^4$ m^2. Hence, 1 ton/ha $= 0.1$ kg/m^2.
- ***Basal area*** $A_b = V_t/h_t$, in m^2/m^2 or m^2/ha
- ***Crown biomass*** B_c: kg/m^2 of crown section
- ***Above ground standing dry biomass*** $B_d = B_t + B_c$

The term ***stem*** often is used as a generic term for trunk. The canopy architecture describes the size, shape, and orientation distributions of the canopy constituents. A popular model called AMAP is the subject of an excellent article by Castel et al. (2001) in which the authors connect the statistical information provided by AMAP (for many tree species) to the parameters commonly used in radiative transfer models for computing σ^0. Also, for many forest species, empirical ***allometric equations*** have been developed to relate height to diameter of trunks, and those two dimensions to biomass. If height and diameter were perfectly correlated for a given species, then the trunk biomass B_t can be uniquely determined from only the trunk height h_t or its diameter d. This, however, is not always the case; competition, natural disturbance processes, and silvicultural management modify tree density over time. Consequently, B_t does not necessarily vary in a monotonic fashion with either h_t or d.

Figure 11-60 presents a structural classification model for vegetation based on six classes. The distinction between classes is based on the presence or absence of woody stems, and leaf size and shape. Thus, the proposed model was developed from the perspective of how a polarized radar signal is expected to interact with a vegetation canopy (Dobson et al., 1996).

11-12.5 Backscatter Response to Forest Parameters

In this subsection, we explore answers to the following direct and *inverse* questions:

(a) How does σ^0 respond to the biophysical parameters introduced in the preceding subsection?

(b) What radar wave parameters are best suited for estimating tree height and stem and crown biomasses?

The answer to the first question depends not only on the radar wave parameters (incidence angle θ, frequency f, and polarization configuration), but also on the tree species and the degree of spatial uniformity of the canopy. The literature includes two types of reports, one in which σ^0 is plotted against a biophysical parameter of a single species, and another in which multiple tree species are combined together. As we see shortly, plots representing the combination of multiple species exhibit more scatter of the data about the trend lines than single-species data.

Ideally, we may want to develop a radar system that can measure trunk biomass with usable accuracy regardless of tree species. Realistically, however, this is not likely to succeed; the vast diversity of canopy architectures suggests that a more viable approach entails a two-step process:

(1) a land-cover classification step that identifies the species of individual forest parcels, followed by

(2) a biophysical parameter estimation step in which tree species-specific inversion algorithms are used.

Let us briefly examine the expected roles of the incidence angle θ, the frequency f, and receive/transmit polarization configuration pq (with p and $q =$ h or v).

Incidence angle

Because the range resolution of a SAR is proportional to $1/\sin\theta$, it is impractical to operate it at θ smaller than about 20°. By operating at $\theta \geq 20°$, we also avoid the issues associated with the near-nadir steep angular variation of σ^0 of the underlying surface. On the other hand, the transmissivity of the canopy, Υ, decreases with increasing θ because of the longer (slant) pathlength through the canopy. At very large values of θ (such as $\theta > 75°$), the canopy approaches an electromagnetically semi-infinite medium, which may provide a good estimate of the crown biomass, but not of the tree height or trunk biomass. Hence, most

	Herbaceous		Woody			
	Blade-like	Broadleaf	Shrubs	Trees		
Growth form	 (e.g. grass)	 (e.g. soybeans)	 (e.g. alder)	Excurrent 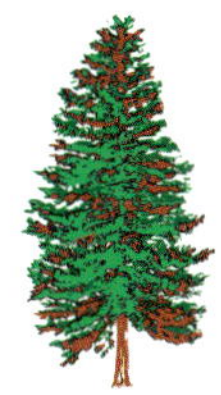 Gymnosperms (e.g. pine)	Decurrent  Angiosperms Dicotyledons (e.g. oak)	Columnar  Angiosperms Monocotyledons (e.g. palm)
Structural characteristics Trunks	None	None	Many small trunks with characteristic orientations	Conical trunk with layered dielectric	Cylindrical, forked trunk with layered dielectric	Cylindrical trunk of homogeneous dielectric
Branches	Non-woody stalks or stems	Non-woody stems	Many small branches and stems	Branch size and orientation vary with height	Branches forked with few horizontal elements	None
Foliage	Blade-like erectophile	Broad leaves	Blade-like or broad leaves	Needles	Broad leaves	Blade-like clump at top of trunk

Figure 11-60: Description of simple structural classes of vegetation [Dobson et al., 1996].

imaging radar systems operate within the angular range $20^\circ \leq \theta \leq 60^\circ$.

Frequency

The data reported earlier in Figs. 11-47 and 11-49 indicate that for a forest populated with trees on the order of 10–15 m in height, the two-way transmissivity at L-band is on the order of 10^{-2}. This means that by operating at C-band or a higher frequency, it may be possible to obtain a useful estimate of the biomass contained in the upper part (crown section) of the canopy, but lower-frequency channels are needed to measure trunk biomass.

Polarization

In general, the canopy volume is a much stronger depolarizing medium than is the underlying surface. Consequently, σ^0_{hv} usually exhibits a greater dynamic range as a function of canopy height or biomass than do σ^0_{hh} or σ^0_{vv}.

This behavior is evident in the plots shown in Fig. 11-61 for a forest parcel of jack pine observed by the L-band channel of the SIR-C SAR. Over the indicated range of the canopy biomass B_{d}, the dynamic range of σ^0_{hv} is about 10 dB, compared with only 5 dB for σ^0_{hh} and 3 dB for σ^0_{vv}.

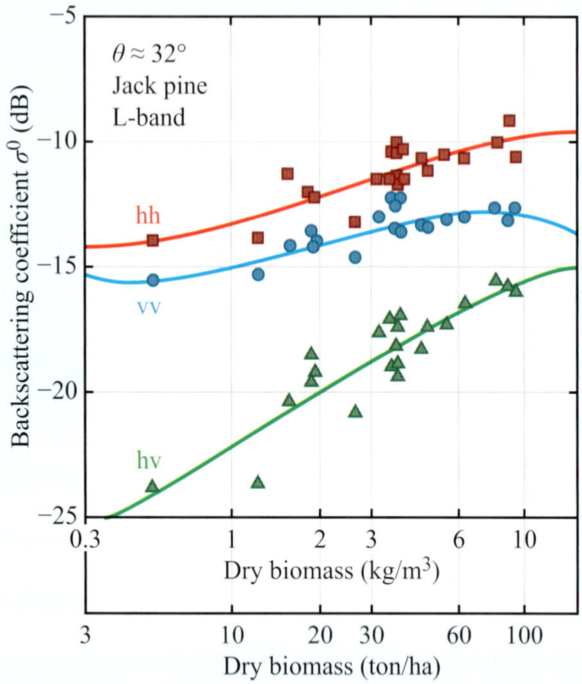

Figure 11-61: SIR-C measured radar response to dry biomass B_d for jack pine canopies. Biomass scale is logarithmic [Dobson et al., 1995a].

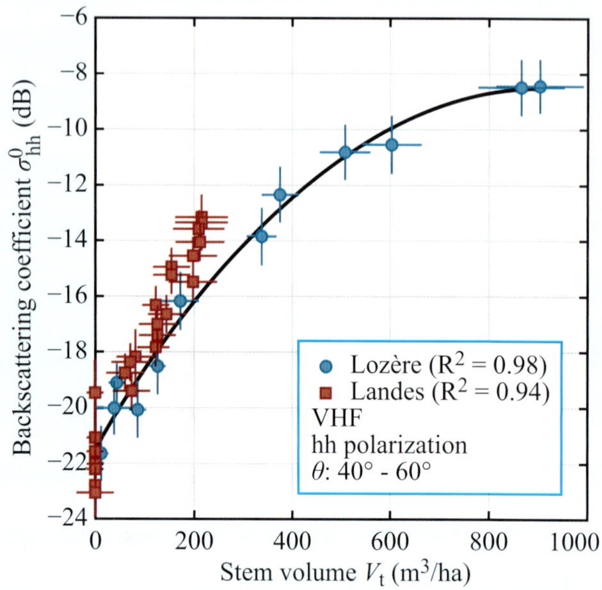

Figure 11-62: VHF backscatter coefficient versus stem volume for two French test sites [Melon et al., 2001].

11-12.6 σ^0 Response at VHF-Band

The CARABAS SAR system—which is an acronym for ***C***oherent ***A***ll-***Ra***dio ***Ba***nd ***S***ensing—is an airborne SAR that operates over the 20–90 MHz range, which is in the lower part of the VHF band. The SAR spatial resolution is on the order of 3 m (Gustavsson et al., 1996). Even though a system at such a low frequency cannot be used from a space platform (because of propagation loss through the ionosphere), it is nevertheless attractive as an airborne imaging system because its long wavelength affords greater penetration through forest canopies.

The CARABAS system was used to image forest test sites in Sweden and France (Manninen and Ulander, 2001; Melon et al., 2001) to determine the response of σ^0 to trunk volume V_t. The results obtained from two French test sites are shown in Fig. 11-62. In both cases, the data exhibit excellent fits to the individual trend lines, primarily because of the special uniformity of the forest parcels. According to the authors of the study (Melon et al., 2001):

> *The Landes forest in southwestern France is almost entirely formed of quasi-uniform large stands of maritime pine* [Pinus pinaster *(Ait.)*] *over sandy terrain and is managed in a consistent fashion which ensures that the canopy is homogeneous. Most of the stands are artificially sown, generally in row[s] of 4 m spacing. The rows follow an east-west direction on the test sites. This flat plantation includes clear-cuts and a range of age classes from seedling to the harvest age of 46 years, corresponding to a maximum stem volume of about 215 m^3/ha.*
>
> *The second test site is the Lozère forest in southern France, [. . .] Austrian pine stands* (Pinus nigra nigrae) *made of even-aged trees ranging from 0 to 130 years old, [. . .]*

presenting a very large range of stem volume, up to 900 m^3/ha.

We note that σ^0_{hh} exhibits a dynamic range of about 14 dB between $V_t = 0$ and $V_t = 900$ m^2ha. A similar response was observed for the test site in Sweden.

11-12.7 σ^0 Response at P- and L-Bands

PiSAR is a Japanese airborne, polarimetric, L-band SAR with a 4-look, 2 m resolution. The system was flown over a test site in northern Japan to evaluate its capability as an estimator of forest biomass. The site included stands of conifers and other types of trees. Figure 11-63 displays measured values of σ^0 for hh, vv, and hv polarizations, plotted against three biophysical parameters: the tree height h, the basal area A_b, and the total biomass B_d. The observed dynamic range is about 5 dB for vv polarization, 6 dB for hh polarization, and 7.5 dB for hv polarization. For all polarizations, the general trend of σ^0 as a function of canopy biomass consists of a monotonic increase up to about 100 ton/ha (or equivalently, 10 kg/m^2), followed by a leveling off to a saturation level characteristic of a *semi-infinite* canopy. The scatter around the general trend is attributed, in part, to the different tree species.

Similar results are evident in Fig. 11-64, which displays σ^0 as a function of canopy biomass B_d at L-band and P-band. As expected, the P-band response exhibits a wider dynamic range and a higher saturation level. However, because of the large amount of data scatter around the trend line, it is difficult to quantitatively ascertain the dynamic range of B_d that can be estimated by the radar. The data scatter is attributed to the wide variety of trees and shrubs imaged by the NASA AIRSAR system.

The E-SAR system is a dual-frequency (P- and L-band) airborne system operated by the German Aerospace Center (DLR). The center frequencies of its P- and L-band signals are 340 MHz and 1.3 GHz, respectively. Based on imagery acquired by E-SAR for a test site in the south of Sweden, Sandberg et al. (2011) generated the results shown in Fig. 11-65, which encompass a total of 68 forest stands. The primary tree species were Norway spruce, Scots pine, and birch. To remove most of the angular dependence on θ, the backscattering coefficient σ^0 was converted to $\gamma = \sigma^0/\cos\theta$. The dynamic range of γ is highest for hh polarization at P-band ($\approx$ 12 dB) and smallest for vv polarization at L-band ($\approx$ 2.5 dB). The latter is not surprising, but the former is; in all other studies, σ^0_{hv} (or γ_{hv}) exhibited a wider dynamic range than did σ^0_{hh} (or γ_{hh}).

Given the 6-channel data (3 polarization $\times$ 2 frequencies), the authors of the study (Sandberg et al., 2011) attempted to develop inversion algorithms to estimate the canopy biomass B_d from the measured values of γ. Their best results, shown in Fig. 11-66, use a combination of γ_{hv} and γ_{hh}, both at P-band. The correlation coefficient associated with their regression model is $R^2 \approx 0.8$.

11-13 SIR-C/X-SAR Case Study

The SIR-C/X-SAR, which flew onboard of the Space Shuttle in April of 1994, consisted of three radar instruments: fully polarimetric SARs at L- and C-band and a single-polarization vv-channel at X-band. As such, SIR/C/X-SAR provided a unique opportunity to image terrestrial surfaces at three different microwave frequencies. In preparation for the Space Shuttle mission, the Radar Remote Sensing team at the University of Michigan selected a test site at the eastern end of the Upper Peninsula of Michigan, located near Raco, Michigan. Over a three-year period, the Michigan team conducted a comprehensive survey of the approximately 40 km $\times$ 24 km Raco supersite. This section presents a summary of their observations and conclusions (Dobson et al., 1995).

11-13.1 Raco Supersite Description

The site contained many boreal forest species and several temperate species. Homogeneous forest stands usually exceed 10 ha in size. The SIR-C/X-SAR images were processed to 25 m $\times$ 25 pixels (each representing approximately 7.7 looks at L- and C-bands and 6.5 looks at X band). A 10-ha stand is equivalent to an area

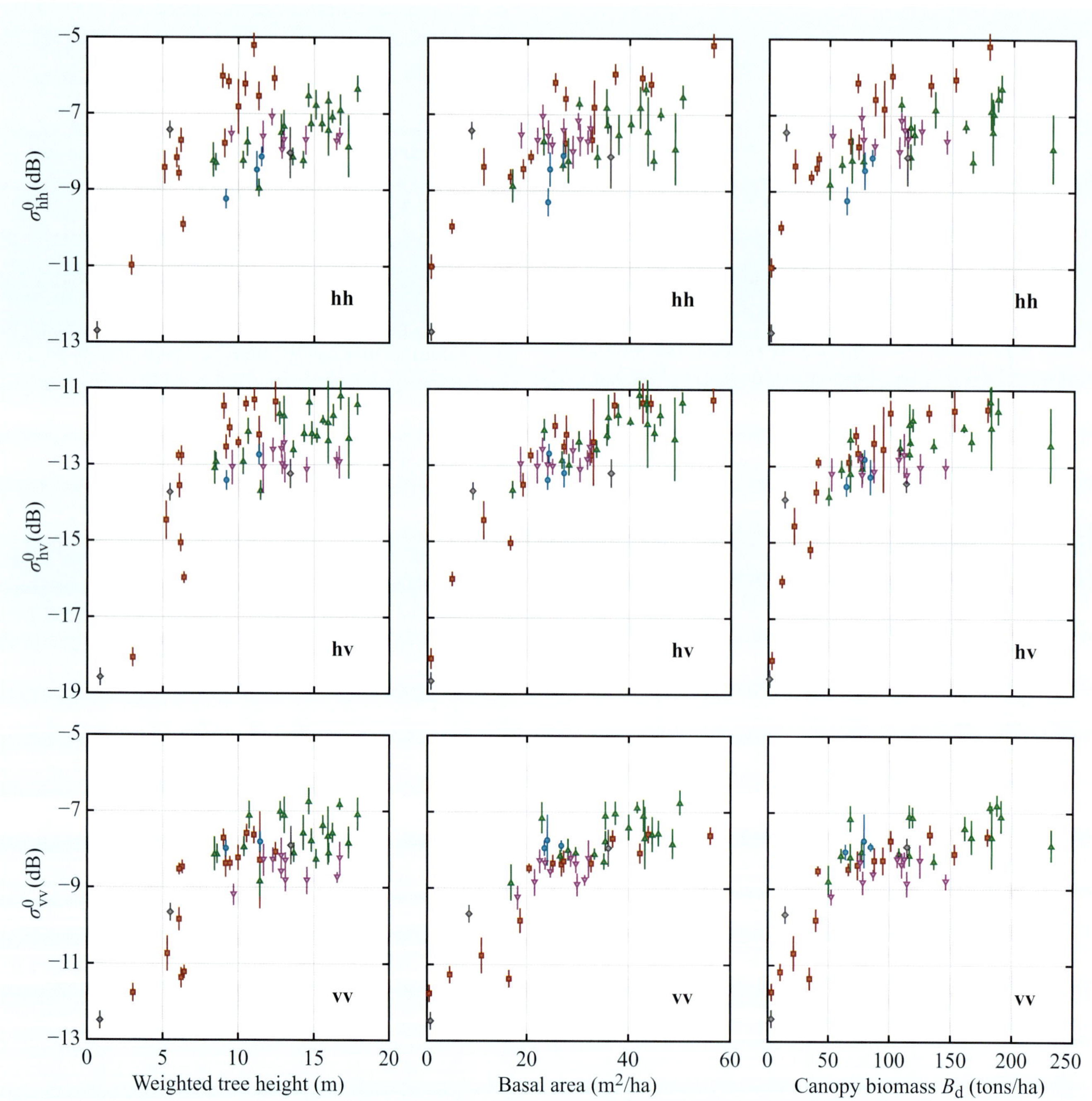

Figure 11-63: Measured L-band backscattering coefficient as a function of tree height, basal area, and canopy dry biomass. Colors represent different tree species, range of incidence angle is 43°–48° [Watanabe et al., 2006].

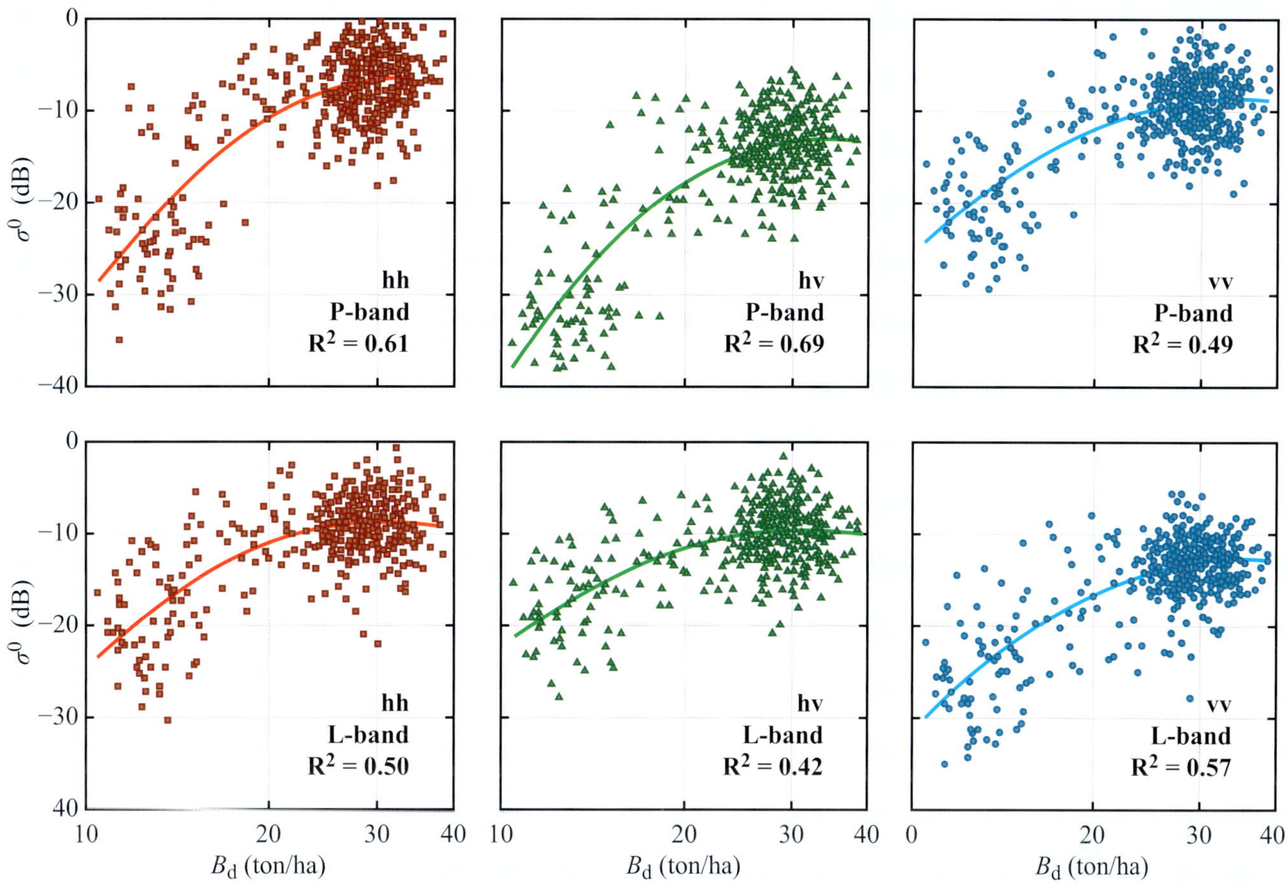

Figure 11-64: Backscattering coefficient at P- and L-band versus B_d with B_d on a log scale. Incidence angle range is 26°–63° [Saatchi et al., 2007].

1000 m × 100 m, which would contain 40 × 4 = 160 image pixels.

The forest communities included:

(a) Upland Conifers: they have an ***excurrent architecture*** (see Fig. 11-60) with large branches and long needles, and they include jack pine, red pine, and eastern pine.

(b) Lowland Conifer: they have excurrent architecture with small branches and short needles, and they include black spruce, white spruce, northern white cedar and tamarack.

(c) Northern Hardwoods: they have a ***decurrent architecture*** (see Fig. 11-60) with large branches, and they include sugar maple, red maple, American beech, eastern hemlock, and balsam fir.

(d) Lowland Hardwoods: they have a decurrent architecture with small branches, and they include red maple, white birch, yellow birch, and trembling aspen.

In addition to the preceding four forest categories, the test site included permanent clearings, hayfields, and a large agricultural region in the southeastern portion of

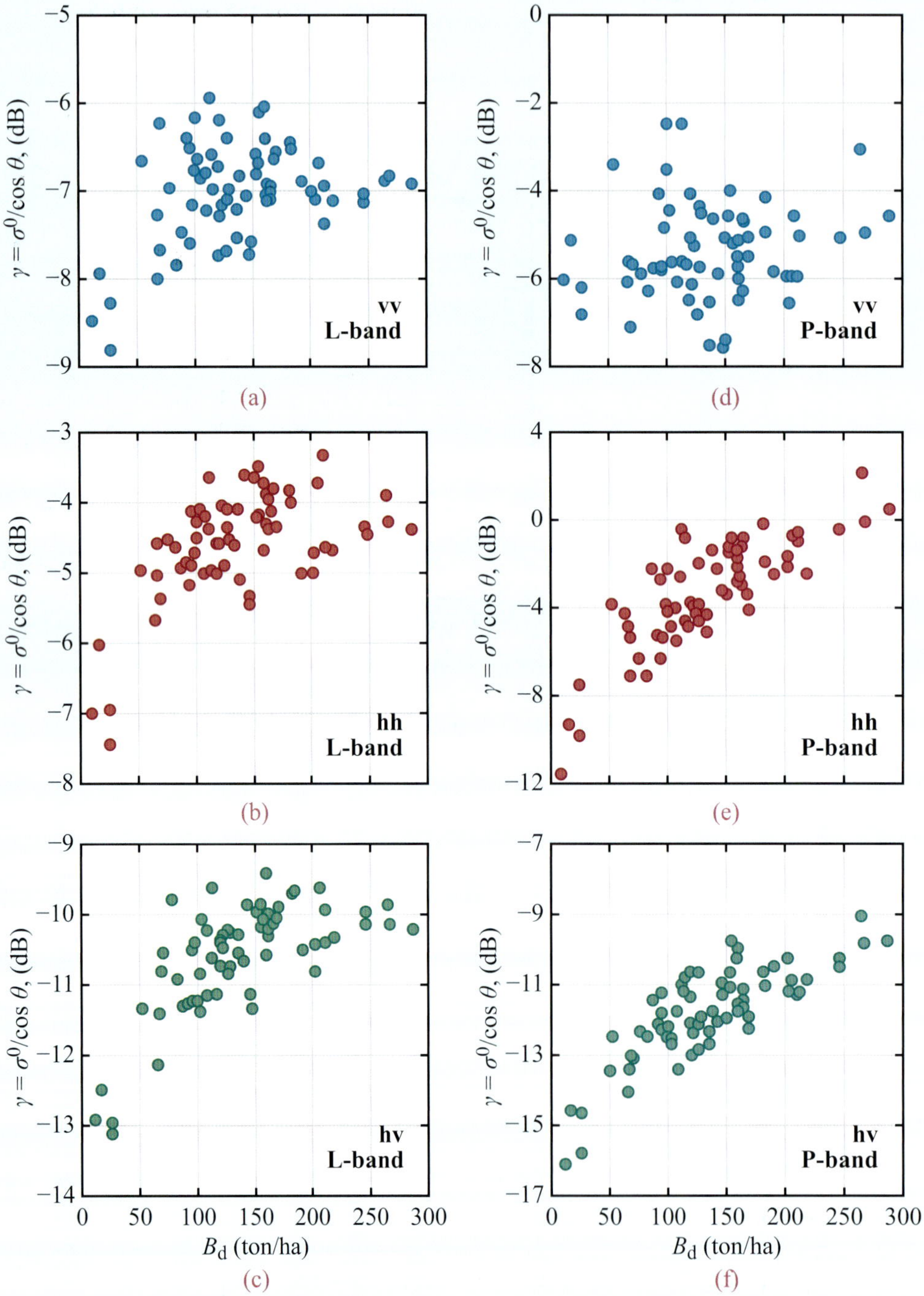

Figure 11-65: Measured backscattering coefficient of forest canopies at L- and P-band, plotted versus dry biomass B_d. Angular range is 28°–50° [Sandberg et al., 2011].

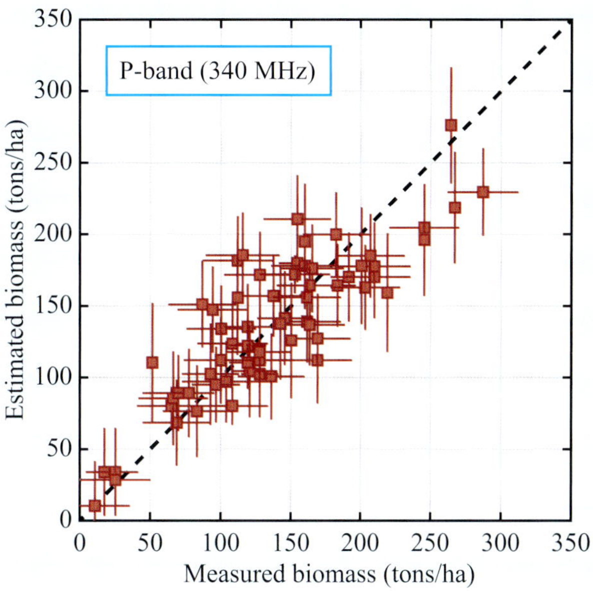

Figure 11-66: Result of biomass estimation using P-band data from 2 May 2007 based on the multipolarized regression model

$$B_d = [\beta_0 + \beta_1 \gamma_{hv}\ (\text{dB}) + \beta_2 \gamma_{hh}\ (\text{dB})]^2,$$

where β_0, β_1, and β_2 are empirically determined constants [Sandberg et al., 2011].

the site. These nonforested areas were designated as either:

(e) Short vegetation, or

(f) Surfaces.

A stratified sample of each of the primary forest communities was selected for biometric survey. A total of over 60 forest stands were selected, each covering an area of 200 m × 200 m. Using GPS with 3-D differential processing, multiple transects were selected within each sampled area to characterize the upper stratum (height > 5 m), the middle stratum (1 m $< h <$ 5 m) and the lower stratum ($h < 1$ m). The diameters of all live and dead stems were recorded by species to the nearest 0.1 cm. Numerous other measurements were performed including trunk diameter, trunk biomass, trunk density, crown biomass, etc. The total number of trees so sampled was on the order of 64,000!!

11-13.2 Land-Cover Classification

The block diagram in Fig. 11-67 depicts the processing steps applied to each channel of the SAR data. Following full calibration of the SAR images, a digital elevation model is used to mask areas where layover or shadow prohibit analysis. This is part of the ***image segmentation*** step in Fig. 11-67. The remainder of the scene is classified in a three-level hierarchical manner:

(a) ***Level 1***: The goal of the first SAR classifier, which was developed using training data of known classes, is used to differentiate the scene into four general structural classes: urban, surfaces, short vegetation, and tall vegetation (forest).

(b) ***Level 2***: The Level-1 tall vegetation class is further subdivided into subcategories on the basis of the structural characteristics of the vegetation.

(c) ***Level 3***: Following classification into the four forest classes, class-specific inversion algorithms are used to estimate the biophysical properties of the forest canopies.

To reduce the effects of image speckle, all images were post-processed using a 5 × 5 averaging window, thereby increasing the number of looks from ≈ 7.7 to ≈ 200 at L- and C-bands, and from ≈ 6.5 to ≈ 160 at X-band. The three-level classification algorithms were performed on the post-processed images.

The image shown in Fig. 11-68(a) is a false-color composite of the Raco supersite; the primary colors are assigned as follows: red = L-band hv-pol, green = C-band hv-pol, and blue = X-band vv-pol. The intent of the color composite is to demonstrate that multipolarized radar observations contain a great deal more information about the imaged terrain than the intensity of a single-polarized channel.

The Level-2 classification results are presented in Table 11-3 and displayed in the image shown in Fig. 11-68(b). The average classification accuracy is 94%.

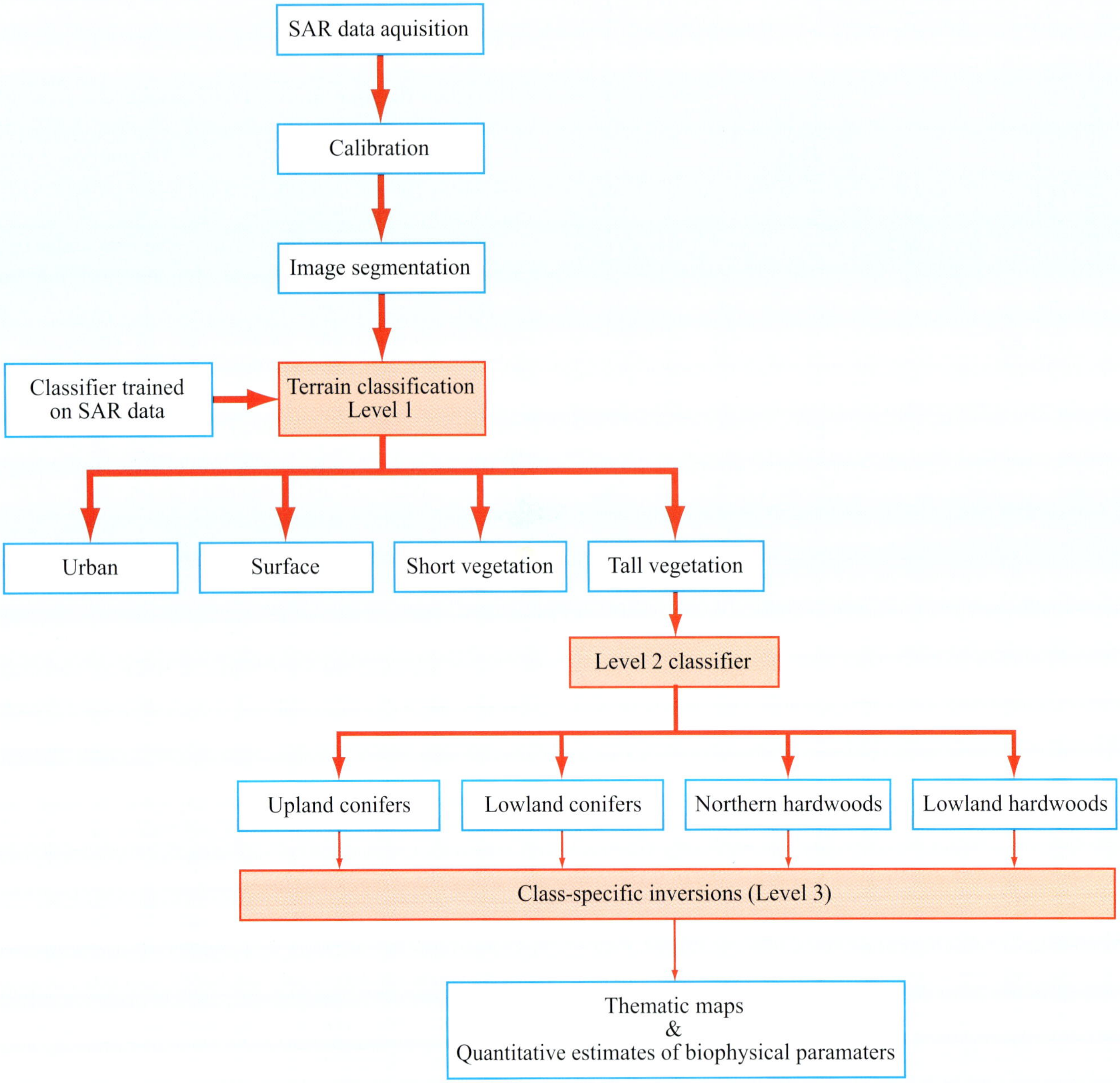

Figure 11-67: Three-level classifier for landcover classification and biophysical parameter estimation [Dobson et al., 1995b].

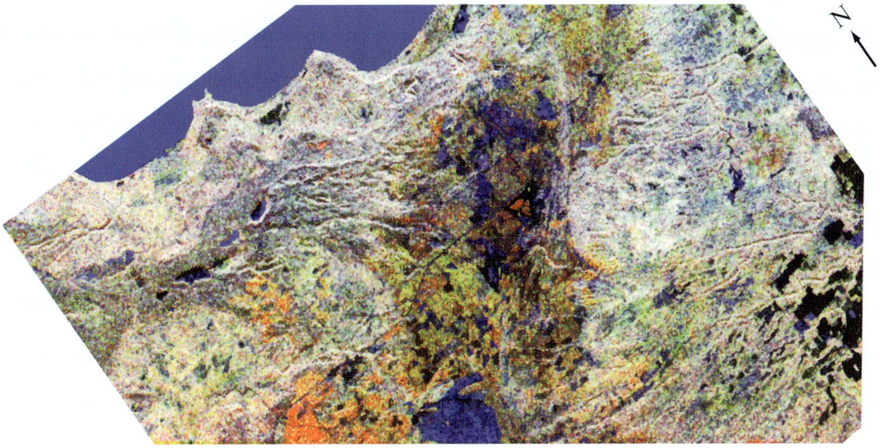

(a) Composite SIR-C / X-SAR image

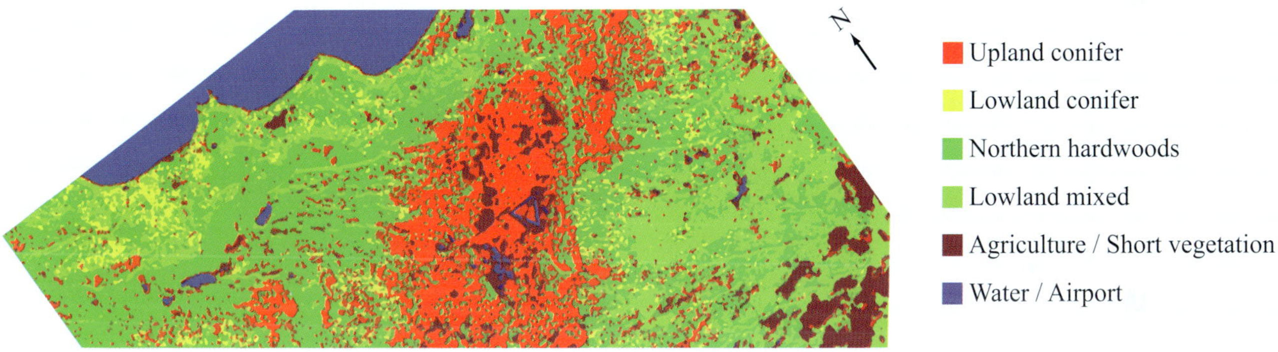

(b) Classified image

Figure 11-68: (a) Composite SIR-C/X-SAR image. Data take 22.2, SRL-2, 1 October 1994. Incidence angle = 22.2°; red, L-band hv; green, C-band hv; blue, X-band vv, (b) Classified image [Dobson et al., 1995b].

11-13.3 Estimation of Forest Biophysical Parameters

In order to decouple the errors associated with the biophysical retrievals of Level-3 from those associated with the Level-2 classifier, only those pixels that were correctly classified by the Level-2 classifier were subjected to the inversion algorithms. Each co-registered post-processed image pixel is represented by a vector of length 9, of which 7 represent the image intensities at X-band (vv polarization) and three each at L- and C-bands (vv, hh, and hv polarizations), and the remaining 2 represent the co-polarized phase differences at L- and C-bands. Using a variety of empirically generated algorithms that utilize various combinations of the 9 elements of the image pixels, Dobson et al. (1995b) were able to develop estimators for tree height h, basal area A_b, trunk biomass B_t, crown biomass B_c, and total biomass B_d, for each of five different forest communities, and to do so with a high degree of accuracy. The results are displayed in

Table 11-3: Level-2 classification results [Dobson et al., 1995b].

		PERCENT CLASSIFIED AS (independent test region)					
Structural class	Type	Tall vegetation				Short vegetation	Surface
	Tree form	Excurrent		Decurrent		NA	NA
	Branch size	Large	Small	Large	Small		
	Foliage type	Long-needle	Short-needle	None	None		
Local floristic community		Upland conifers	Lowland conifers	Northern hardwoods	Lowland hardwoods	Grasses, sedges, shrubs	Bare soil, paved, water
TRUE CLASS	Upland conifers	94.53	0.01	5.38	0	0.08	0
	Lowland conifers	4.53	92.88	2.41	0.11	0.07	0
	Northern hardwoods	0.87	7.81	90.30	1.02	0	0
	Lowland hardwoods	0	5.49	6.11	88.40	0	0
	Short vegetation	0.85	0.01	1.41	0.02	96.38	1.33
	Surface	0	0	0	0	1.05	98.95

Figs. 11-69 and 11-70. The indicated vertical error bars represent ± 1 standard deviation of the "ground-truth" in-situ measurements, and the horizontal error bars represent the ± 1 standard deviation of SAR-related uncertainties, which include co-registration errors of the SAR images and calibration errors. The excellent agreement between the *in-situ* measured values and the SAR-estimated values is a testimonial for why the world needs to have multifrequency polarimetric SAR systems orbiting the earth on a continuing basis.

Finally, by way of illustrating the type of information products that such as a system is capable of generating, we show in Fig. 11-71 images displaying:

(a) Total above-ground biomass B_d.

(b) Below-ground biomass contained in the root system, estimated using allometric relations (Bergen et al., 1998).

(c) Total above- and below-ground biomass.

11-14 Propagation Properties of Snow

A layer of dry snow is a dielectric medium consisting of ***ice crystals*** in an ***air background***. Wet snow is somewhat more complicated, as we see later in this section. Assuming that the ice particles in snow are spherical, the ***absorption***, ***scattering***, and ***extinction*** cross section of each ice particle (Q_a, Q_s, and $Q_\text{e} = Q_\text{a} + Q_\text{s}$), respectively) may be calculated in terms of the particle radius r and the relative dielectric constant ε_i using the ***Mie expressions*** given in Section 8-5.2.[†] Actually, the expressions are formulated in terms of the parameters

$$\chi = k_\text{b} r = \frac{2\pi r}{\lambda_\text{b}} = \frac{2\pi r}{\lambda_0}\sqrt{\varepsilon'_\text{b}} \tag{11.100a}$$

and

$$n = \frac{n_\text{i}}{n_\text{b}} = \left(\frac{\varepsilon_\text{i}}{\varepsilon_\text{b}}\right)^{1/2} = \varepsilon^{1/2}, \tag{11.100b}$$

where k_b is the wave number in the background medium; ε'_b is the relative permittivity of the background medium; n_i, n_b, and n are the complex indices of refraction of the particle (ice), the background medium, and the particle relative to the background medium, respectively; and ε_i, ε_b, and ε, are the corresponding complex dielectric constants. In the case of dry snow, the background is air; hence $\varepsilon'_\text{b} = 1$ and $\varepsilon = \varepsilon_\text{i}$.

For reasons that will become clear later, scattering in both dry and wet snow is attributed to the ice particles

[†] **Computer Codes 11.3 and 11.4.**

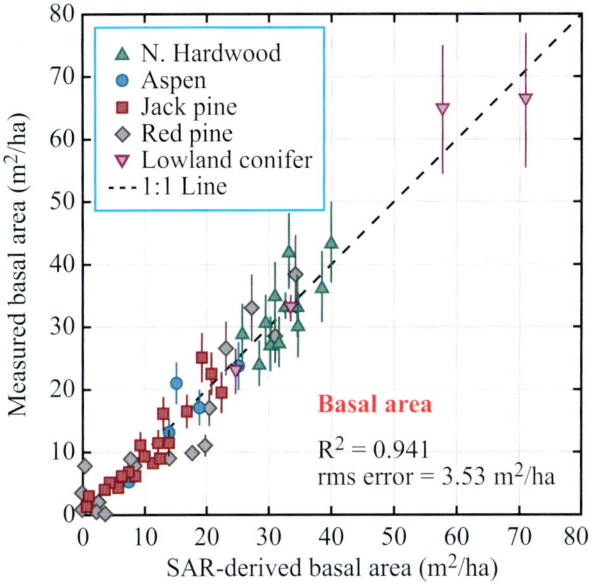

(a) Basal area

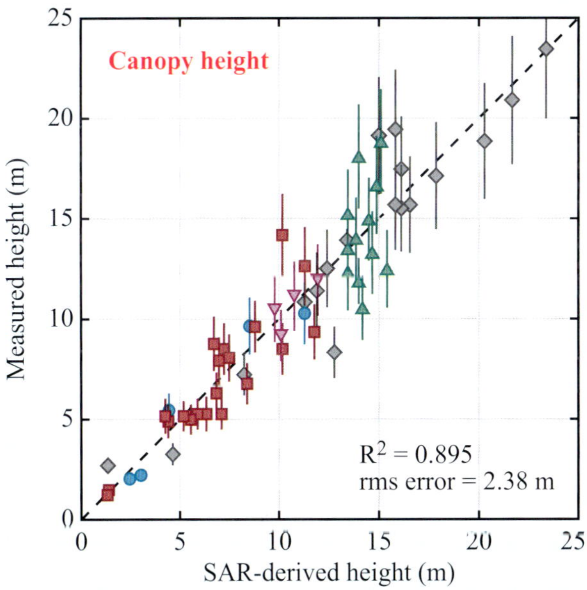

(b) Canopy height

Figure 11-69: *in-situ* measured basal area and tree height versus SAR-derived values based on SIR-C/X-SAR measurements [Dobson et al., 1995b].

in the snow. Ignoring the mutual interaction between the ice particles, κ_s is the sum of the scattering cross sections of all ice spheres contained in a unit volume:

$$\kappa_s = \sum_{j=1}^{N_v} Q_s(r_j, n), \tag{11.101}$$

where N_v is the number density of ice spheres in m^{-3}.

In the general case, the absorption coefficient κ_a consists of two components,

$$\kappa_a = \kappa_{ai} + \kappa_{ab}, \tag{11.102}$$

where κ_{ai} accounts for absorption by the ***ice spheres*** and κ_{ab} accounts for absorption by the ***background***. These quantities are given by

$$\kappa_{ai} = \sum_{j=1}^{N_v} Q_a(r_j, n), \tag{11.103}$$

and

$$\kappa_{ab} = 2k_0(1 - v_i)n''_b = 2k_0(1 - v_i)\left|\Im\mathfrak{m}\{\sqrt{\varepsilon_b}\}\right|, \tag{11.104}$$

where $k_0 = 2\pi/\lambda_0$ and v_i is the volume fraction of ice particles in the mixture. For dry snow, the background is air, and $n''_b = 0$. Hence, $\kappa_{ab} = 0$ and $\kappa_a = \kappa_{ai}$. For wet snow, the background is taken to be the air-water mixture (as discussed later). Hence, $n''_b \neq 0$.

If $\varepsilon''_b/\varepsilon'_b \ll 1$, which is the case for low values of the liquid-water content m_v, κ_{ab} can be calculated using an approximate form of Eq. (11.104), namely,

$$\kappa_{ab} \approx k_0(1 - v_i)\,\frac{\varepsilon''_b}{\sqrt{\varepsilon'_b}}\,. \tag{11.105}$$

For a given ***drop-size distribution*** $p(r)$, κ_{ai} and κ_s may be obtained by evaluating the integral in the expression

$$\kappa = \int_0^\infty p(r)\,Q(r,n)\,dr, \tag{11.106}$$

where κ is κ_{ai} or κ_{si} corresponding to Q_a or Q_s, respectively. The units of $p(r)$ are drops per m^3 per unit increment of drop radius r. If the drop-size distribution

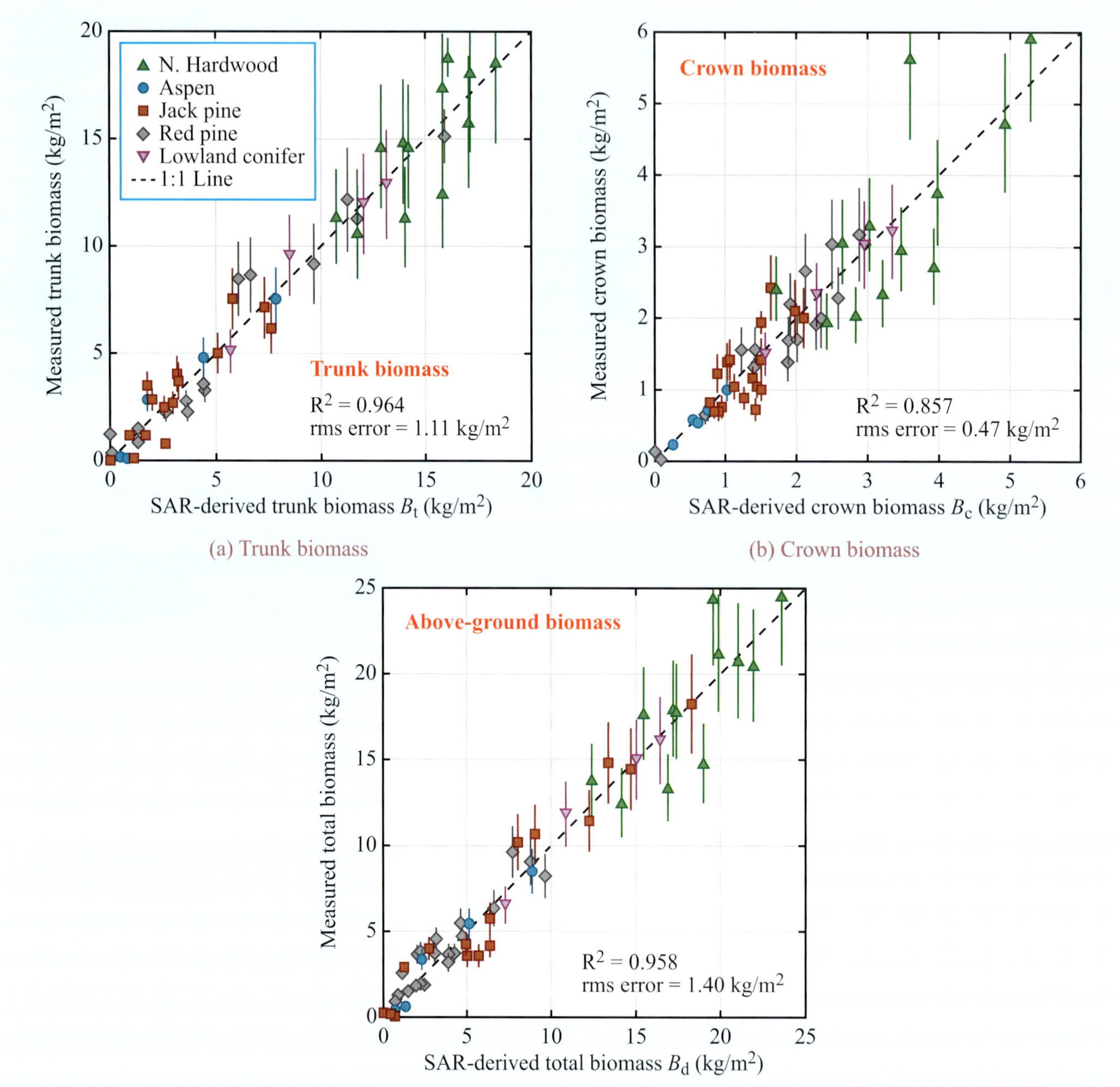

(a) Trunk biomass

(b) Crown biomass

(c) Total above-ground biomass

Figure 11-70: Measured *in-situ* versus SAR-derived values for (a) trunk biomass, (b) crown biomass, and (c) total above-ground biomass [Dobson et al., 1995b].

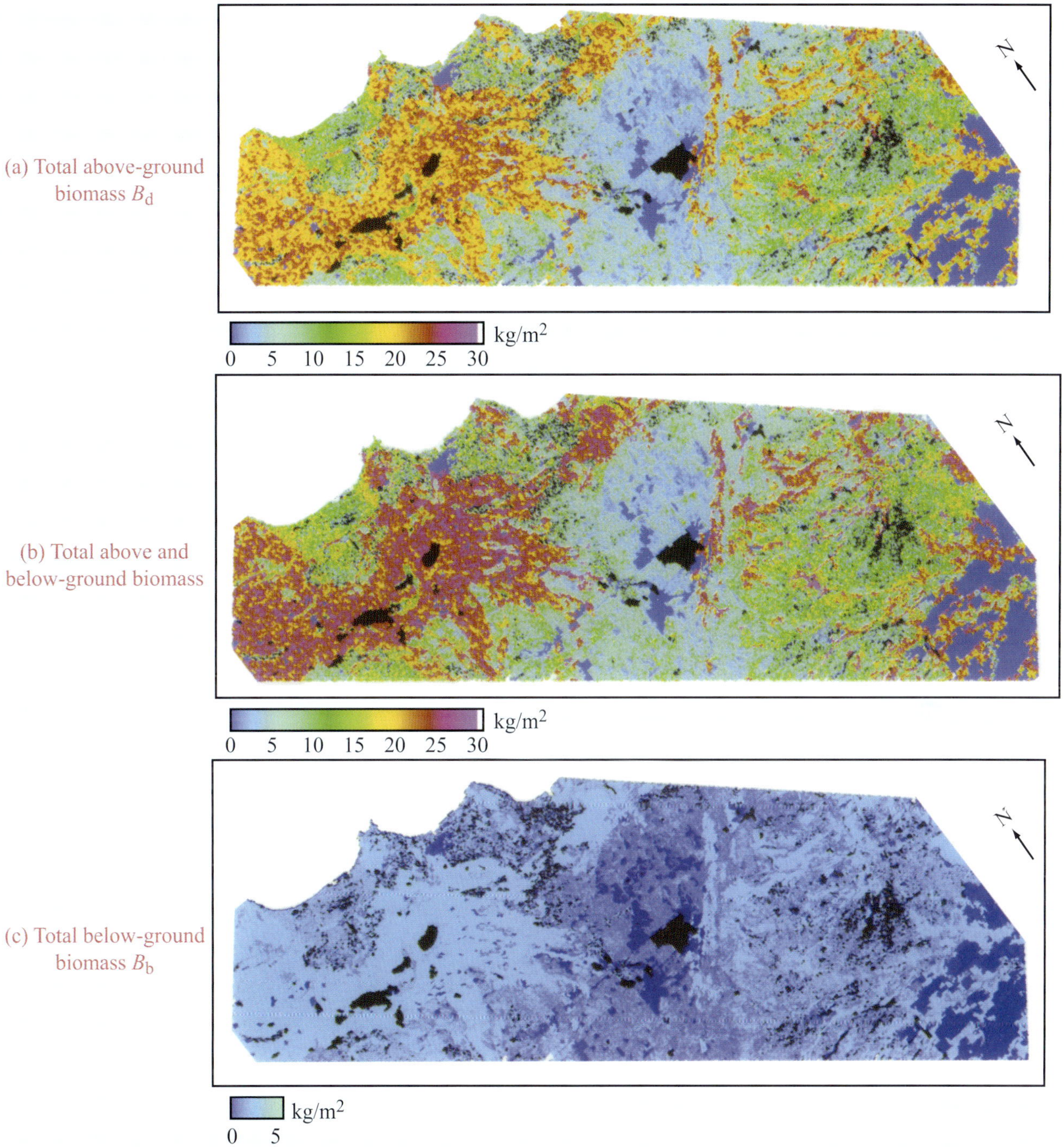

Figure 11-71: False-color images depicting the distribution of above-ground biomass, below-ground biomass, and total biomass for the Raco supersite [Dobson et al. (1995b) and Bergen et al. (1998)].

is equal to zero for r larger than some radius r_{max}, and if the wavelength λ_b ($= \lambda_0$ for dry snow) is such that $\chi_{max} = 2\pi r_{max}/\lambda_b \leq 0.3$, the Mie expressions for the cross sections Q_a and Q_s reduce to the Rayleigh expressions given in Section 8-5.3. This not only simplifies the computation of all quantities concerned, but also leads to an expression for κ_{ai} that is independent of $p(r)$. The ***Rayleigh*** expression for $Q_a(r_i, n)$ for an ice particle of radius r_i is

$$Q_a(r_i, n) = \frac{8\pi^2}{\lambda_b} r_i^3 \,\mathfrak{Im}\{-K\}, \tag{11.107}$$

where K is a complex quantity defined by Eq. (8.43) in terms of the particle-to-background index of refraction n,

$$K = \frac{n^2 - 1}{n^2 + 2} = \frac{\varepsilon - 1}{\varepsilon + 2}\,. \tag{11.108}$$

Inserting Eq. (11.107) into the discrete form for κ_{ai} given by Eq. (11.103) leads to

$$\kappa_{ai} = \frac{8\pi^2}{\lambda_b} \,\mathfrak{Im}\{-K\} \sum_{j=1}^{N} r_j^3. \tag{11.109}$$

The density of dry snow ρ_s (g/cm^3) is equal to the volume fraction occupied by the ice particles, v_i, multiplied by the ***density of ice*** ($\rho_i = 0.9167$ g/cm^3),

$$\rho_s = \rho_i \sum_{j=1}^{N_v} \frac{4\pi}{3} r_j^3. \tag{11.110}$$

Incorporating Eq. (11.110) into Eq. (11.109) leads to

$$\kappa_{ai} = \frac{6\pi}{\lambda_b} \frac{\rho_s}{\rho_i} \,\mathfrak{Im}\{-K\}. \tag{11.111}$$

(Rayleigh approximation)

The absorption and scattering coefficients are model parameters used in the radiative transfer equation to compute the scattering and emission of a layer of snow. When such computations are performed, often $p(r)$ is not known for the snow layer under consideration; however, it may be possible to estimate the mean or typical particle radius instead. If this typical radius is r, and all the particles are assumed to be of this size,

$$\kappa(r) = N_v(r)\, Q(r, n) \tag{11.112}$$

for κ_{ai} and κ_{si} (corresponding to Q_a and Q_s), and the number density $N(r)$ is determined from the density ρ_s by

$$N_v(r) = \frac{\rho_s}{\rho_i} \cdot \left(\frac{4}{3}\pi r^3\right)^{-1} = \frac{v_i}{V_i(r)}\,, \tag{11.113}$$

where $v_i = \rho_s/\rho_i$ is the ***volume fraction*** of ice in dry snow, and $V_i(r)$ is the volume of a single ice particle of radius r.

► If the Rayleigh approximation is appropriate, we use Eq. (11.107) for $Q(r, n)$, but if the particle radius r is too large to satisfy the condition $|n\chi| < 0.5$ [where n and χ are as defined by Eq. (8.31)], then the Mie solution is called for, in which case $Q = (\pi r^2)\xi$, and ξ is as given in Section 8-5.2. ◄

11-14.1 Dry Snow

The dielectric properties of ice are presented in Section 4-3. The real part ε_i' is independent of temperature and frequency (in the microwave region) and is equal to $\varepsilon_i' \approx 3.18$. The dielectric loss factor ε_i'' is a function of both temperature and frequency, and its variation is given in Fig. 4-3. With $n_b = 1.0$, the only additional parameters to be specified for calculating κ_{ai} and κ_s [using Eqs. (11.103) and (11.101)] are the snow density ρ_s, the frequency f, and the particle radius r.

Figure 11-72 shows spectral plots of κ_s, κ_a, and κ_e for a dry snow medium containing spherical ice crystals with $r = 0.5$ mm. All computations were performed using Mie expressions.

► Below about 5 GHz $\kappa_s \ll \kappa_a$, and the albedo $a = \kappa_s/(\kappa_s + \kappa_a)$ is very small. On the other hand, at frequencies above 20 GHz, scattering becomes the dominant component of the total extinction loss of the dry snow medium. ◄

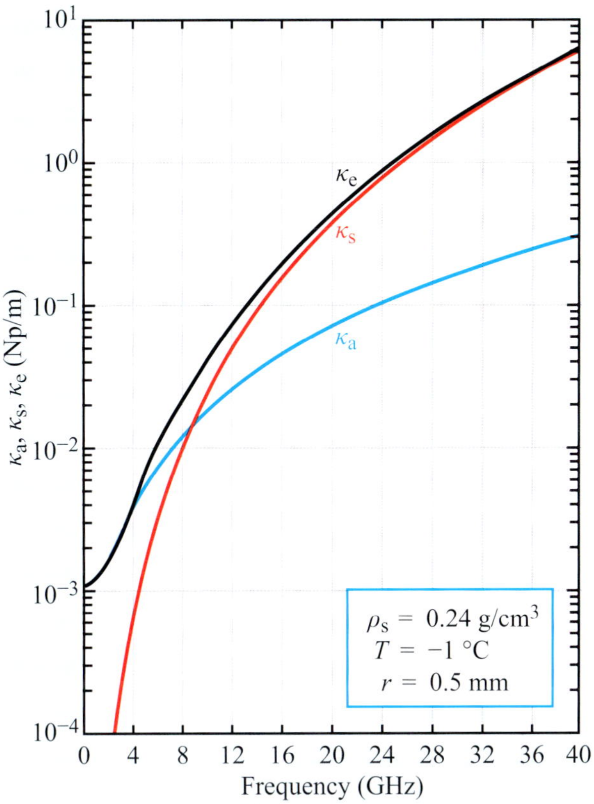

Figure 11-72: Mie-computed absorption, scattering and extinction coefficients for dry snow containing 1 mm diameter ice spheres, plotted as a function of frequency.†

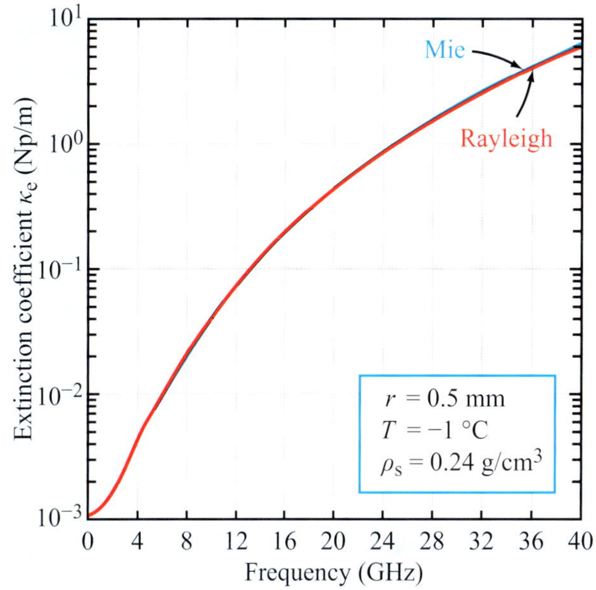

Figure 11-73: Comparison of the Rayleigh approximation with the Mie computation for κ_e of dry snow.

The variation of the extinction coefficient κ_e with the microwave frequency f is illustrated in Fig. 11-73 for snow containing ice particles with $r = 0.5$ mm. The Mie and Rayleigh plots are almost identical across the complete frequency range shown in the figure, which extends to 40 GHz.

▶ Based on this and similar comparisons performed for larger particles, the Rayleigh approximation for κ_e is valid up to 40 GHz for particles as large as 2 mm in radius. For much larger particles with $r = 5$ mm, the Rayleigh approximation provides reasonably accurate results up to 15 GHz. ◀

Plots of the penetration depth δ_p are displayed in Fig. 11-74 for ice particles with $r = 0.5$ mm, 2 mm, and 5 mm. All computations were performed using Mie expressions. We note that δ_p may vary from a few centimeters to tens of meters depending on the particle size and the microwave frequency.

Of course, the snow density ρ_s is important also, since it defines the numerical density of the particles in the snow medium. The plots in Fig. 11-75 indicate that at 37 GHz, κ_s, κ_a, and κ_e vary linearly with ρ_s for $r = 0.5$ mm.

The scattering albedo a of the medium is one of the key parameters determining snow scattering and emission. It is governed by the ice-particle size relative

†Computer Code 11.3.

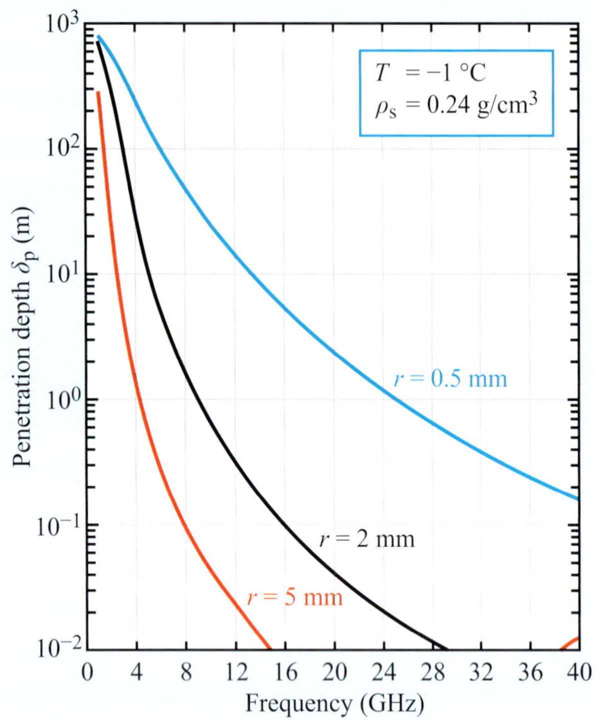

Figure 11-74: Mie calculated microwave spectra of the penetration depth of dry snow.

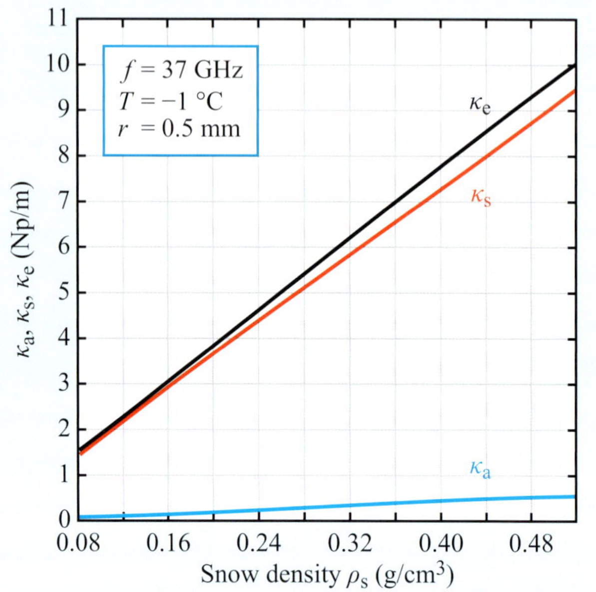

Figure 11-75: Mie-computed dependence of κ_a, κ_s, and κ_e on ρ_s for dry snow at 37 GHz.[†]

to λ. Plots of a as a function of the frequency f are shown in Fig. 11-76 for three sizes of particle radius. The albedo increases rapidly with increasing frequency f for frequencies corresponding to the size parameter $\chi < 0.5$, and at a slower rate for larger values of χ.

11-14.2 Wet Snow

When water in liquid form is present in snow, the snow medium becomes a mixture of ice particles, water droplets, and air. The volumetric water content of snow, m_v, which also is called ***snow wetness***, usually does not exceed a few percent, and the water droplets usually are much smaller than the ice particles.

▶ Hence, we assume that scattering in the medium is caused mainly by the ice particles, and that the background medium may be regarded as a mixture of water droplets and air. ◀

Assuming that the water droplets are randomly oriented ellipsoids of dielectric constant ε_w and that the host material is air, the dielectric constant of the background mixture (air-water) is, from Eq. (4.27), given by

$$\varepsilon_b = 1 + \frac{m_v}{3}(\varepsilon_w - 1) \sum_{u=a,b,c} \left[\frac{1}{1 + A_u \left(\dfrac{\varepsilon_w}{\varepsilon_b} - 1 \right)} \right], \tag{11.114}$$

where A_a, A_b, and A_c are the three ***depolarization factors*** of the ellipsoidal water droplet. Complete expressions for ε_w are given in Section 4-1, and according to the study reported by Ambach and Denoth

[†]**Computer Code 11.3.**

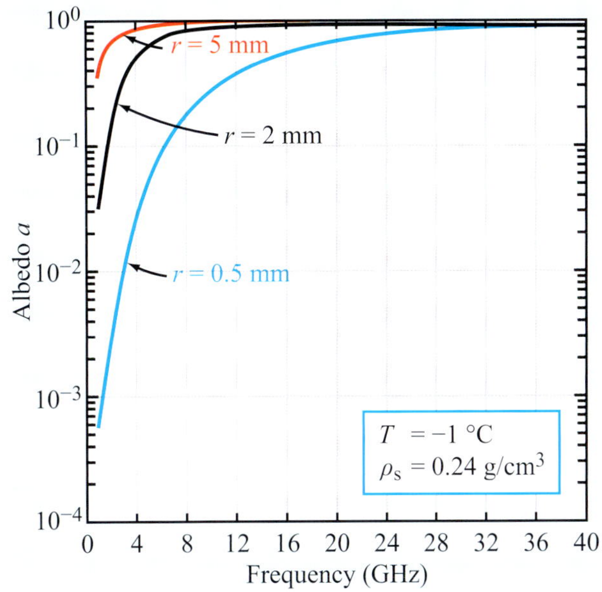

Figure 11-76: Calculated spectral variation of the albedo a of dry snow for three sizes of Mie spheres.

(1980), appropriate values one might assign to the depolarization factors are $A_c = 0.88$ and $A_a = A_b = 0.06$. With these parameters specified, it should be easy to solve Eq. (11.114) to determine the background dielectric constant ε_b for any value of the snow wetness m_v (note that ε_b appears on both sides of Eq. (11.114)).

Calculating κ_{ai} and κ_{si} for wet snow involves using the same expressions as for dry snow, except that ε_b in Eq. (11.100) is no longer equal to 1.0. In addition, we must use Eq. (11.104) to calculate the absorption coefficient of the background medium, which was equal to zero for dry snow.

The solid curves in Fig. 11-77 are plots of κ_e versus m_v at 4 GHz, 16 GHz, and 37 GHz. The data points are based on transmission measurements made for snow samples of varying wetness. Observations of the snow samples under a microscope indicate that a typical value for r is 0.5 mm; hence, this was the value used in calculating the curves in Fig. 11-77. Upon comparing the calculated curves with the measured data, we observe that the curves underpredict the measurements at 4 GHz, are comparable in level to the measurements at 16 GHz, and predict a level generally higher than the measured values at 37 GHz.

When snow melts, the absorption coefficient of the background medium, κ_{ab}, increases from zero for dry snow to a value larger than κ_s for $m_v \gtrsim 1\%$. Thus, the presence of a very small amount of liquid water in the snow mixture is sufficient to reduce the magnitude of the scattering albedo a to a very small value. This behavior is illustrated graphically in Fig. 11-78(a) for $r = 0.5$ mm. For $m_v \gtrsim 2\%$, the albedo is so small in magnitude that its dependence on r becomes inconsequential. Thus, when wet, snow becomes a nonscattering medium. The variation of the penetration depth δ_p as a function of m_v is shown in Fig. 11-78(b) for 4, 10, and 37 GHz.

11-15 Backscattering Behavior of Dry Snow

In the general case, the backscattering coefficient of snow-covered terrain may consist of (a) a direct contribution from the air-snow boundary, (b) a two-way attenuated contribution from the snow-ground boundary, (c) direct backscatter from the snow volume, and (d) indirect contributions resulting from the interaction between the snow volume and either or both boundaries of the snow layer. These various contributions are incorporated into the radiative-transfer formulations given in Sections 11-3 and 11-6.

Because the dielectric mismatch at the air-snow boundary is small for dry snow, the reflection coefficient is quite small, which, in turn, means that the snow-surface backscattering contribution may be neglected in comparison with the other contributions, except perhaps at normal incidence. This also means that multiple reflections involving the upper and lower boundaries of the snow layer may also be ignored. This contention is supported by the radar scatterometer observations reported in Fig. 11-79(a), which indicate that σ^0 is essentially insensitive to snow surface roughness when the snow layer is dry. The plots in Fig. 11-79(a)

†Computer Code 11.4.

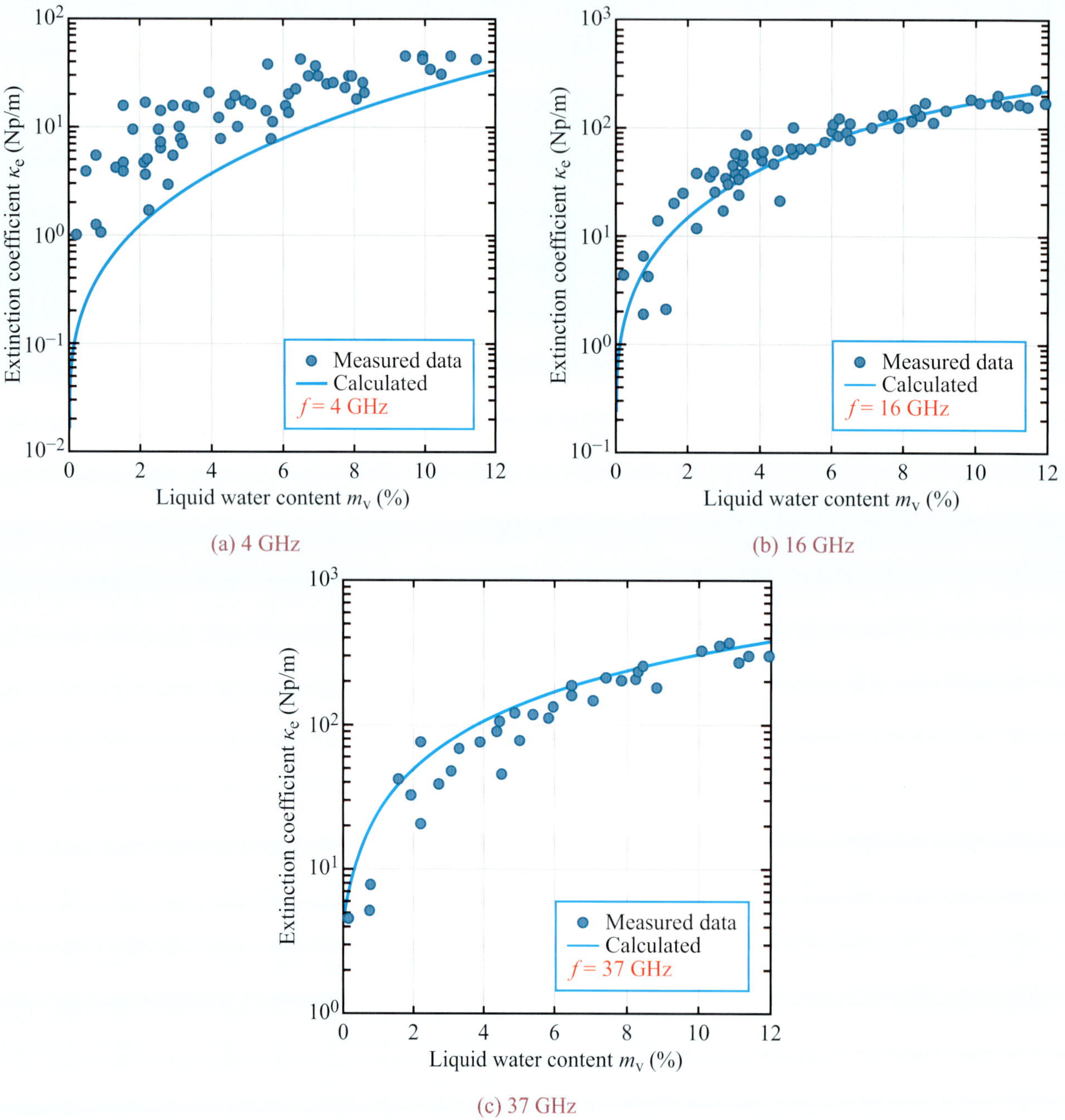

Figure 11-77: Comparison of the measured values of κ_e with the curves calculated for $\rho_s = 0.24$ g/cm^3, $r = 0.5$ mm, $A_a = A_b = 0.06$.

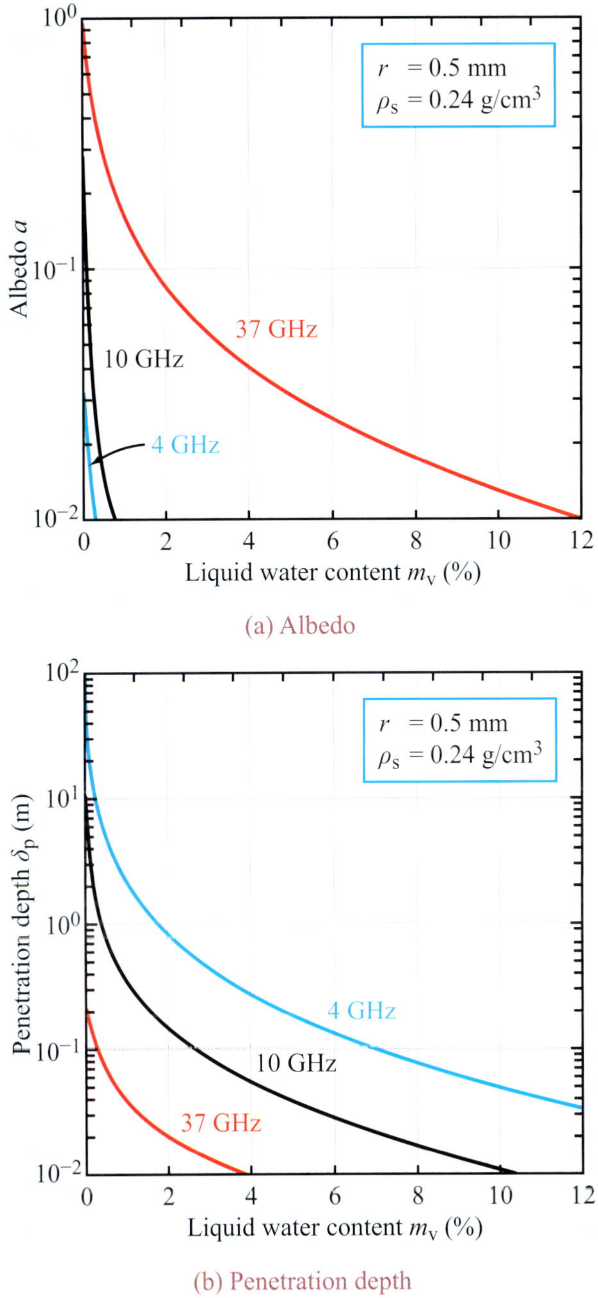

Figure 11-78: Mie-computed variation of snow (a) albedo and (b) penetration depth with liquid-water content.†

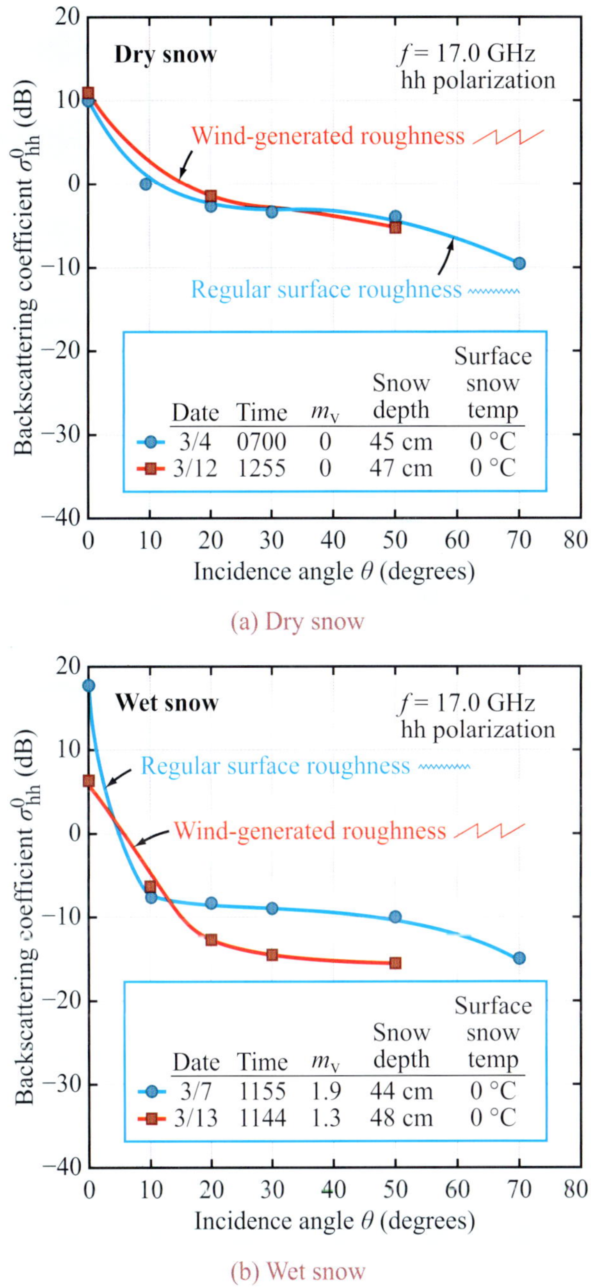

Figure 11-79: Change in snow surface roughness has no effect on σ^0 when the snow is dry, but it does when the snow is wet [Stiles and Ulaby, 1980b].

compare σ^0 for dry snow with a "regular surface" characterized by high-frequency spatial variations with small amplitudes (on the order of 0.5 cm) with a "wind-generated surface" characterized by large smooth facets (about 30 cm long) connected by 2 cm high ridges, forming a sawtooth-like pattern. This latter surface condition was caused by strong southerly winds. The lack of any significant difference between the two curves in Fig. 11-72(a) indicates that σ^0 is insensitive to snow surface roughness when the snow is dry. This conclusion is based not only on the 17 GHz data shown in Fig. 11-79, but also on data measured at several other frequencies over the 1 to 35.6 GHz range.

In contrast to its lack of sensitivity to surface roughness for dry snow, σ^0 exhibits a significant response to surface roughness when the snow is wet. Figure 11-79(b) shows angular plots of σ^0 for essentially the same snowpack as before, except that now the top 5 cm of the snow layer contains some water in liquid form. Wet snow has a higher dielectric constant and lower albedo [Fig. 11-78(a)] than dry snow. Consequently, not only is the general angular pattern of $\sigma^0(\theta)$ different for wet snow, but so is the sensitivity of σ^0 to snow surface roughness.

11-15.1 Radiative Transfer Model

We open this subsection with Fig. 11-80, which displays angular plots of $\sigma^0_{\rm hh}$ for a 58 cm deep snowpack, measured at multiple frequencies between 1.6 GHz and 35.6 GHz. Based on the penetration depth and albedo plots shown in Figs. 11-74 and 11-76, we can surmise that at 1.6 GHz, the backscatter contribution of the snow layer is very small, so the measured plot of σ^0 is due primarily, if not entirely, to the underlying ground surface, as if the snow layer is totally absent. On the other end of the spectral range of the plots shown in Fig. 11-80, namely at 35.6 GHz, the 58 cm snow depth is equal to $\approx 5\delta_{\rm p}$ (where $\delta_{\rm p}$ is the penetration depth) and the albedo $a \gtrsim 0.9$. Hence, electromagnetically the snow layer is semi-infinite in depth; the backscatter is due to volume scattering in primarily the upper part of the snow layer.

The radiative transfer theory introduced in Section 11-6 is perfectly suitable for modeling the backscatter

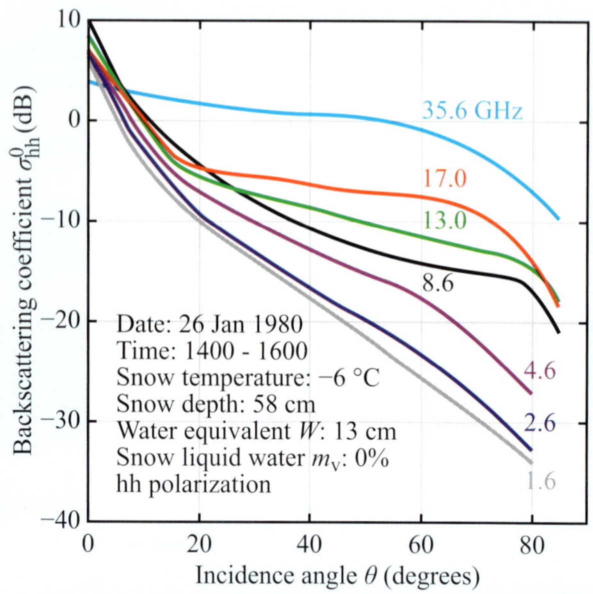

Figure 11-80: Angular patterns of measured backscattering coefficient of dry snow [Stiles et al., 1981].

from a snow layer above a rough-surface snow-ground boundary. However, the iterative-solution method is capable of providing solutions of the radiative transfer equation (with reasonably accuracy) only if the albedo $a \lesssim 0.3$. Otherwise, a numerical solution is warranted. According to Fig. 11-76, for $r = 0.5$ mm, $a > 0.3$ if $f > 12$ GHz. Hence, a numerical solution is called for at frequencies above 12 GHz, and for snowpacks with larger ice particles the upper frequency limit of the iterative-solution method is lower still.

The theoretical behavior of σ^0 as a function of snow depth is illustrated by the families of plots displayed in Figs. 11-81 and 11-82. All computations were performed using the numerical technique outlined in the books by Fung (1994), Tsang et al. (1985), and Ulaby et al. (1986a). The plots in Fig. 11-81 illustrate the important role of the microwave frequency f. At 1.2 GHz, σ^0 is totally insensitive to snow depth d; the snow layer is essentially transparent to a 1.2 GHz EM wave, even for a layer as deep as 1.8 m. The sensitivity of σ^0 to d increases with increasing f. At 35 GHz, σ^0 exhibits a strong response to d between $d = 0$ and

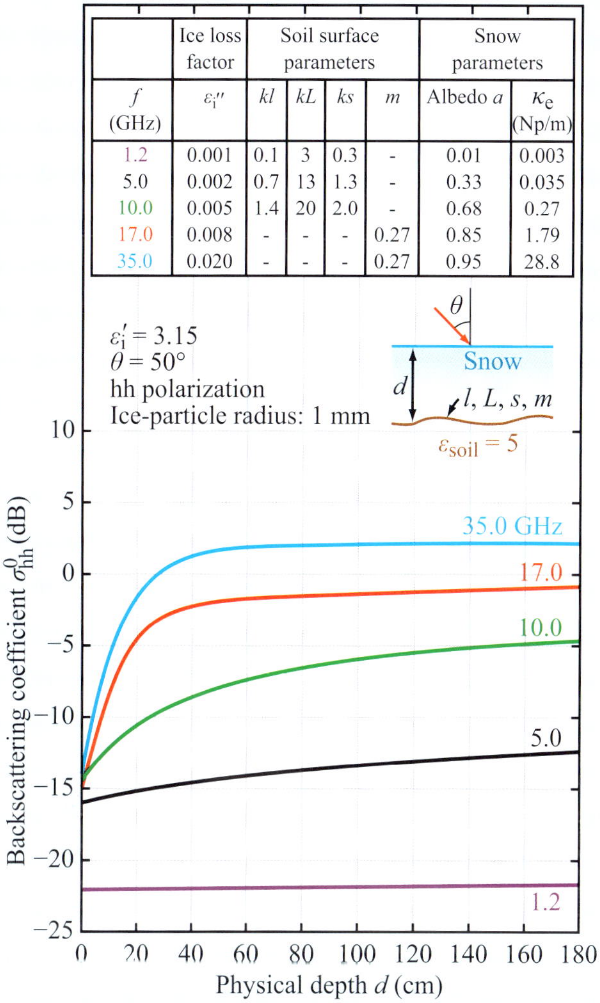

Figure 11-81: Calculated variation of σ^0_{hh} with d for a dry snow layer containing ice particles 1 mm in radius. Note the important role of frequency.

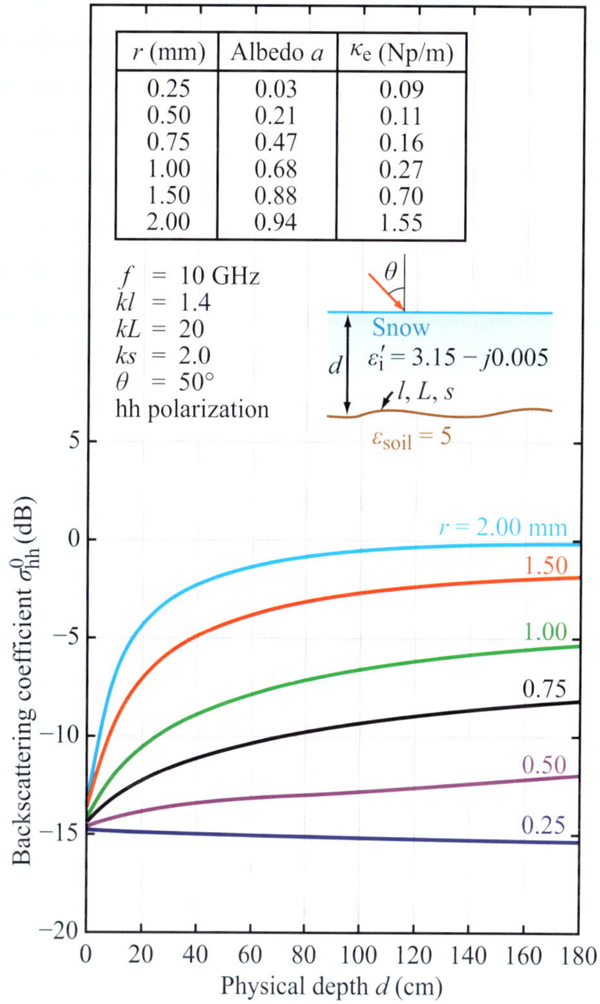

Figure 11-82: Calculated variation of σ^0_{hh} for a layer of dry snow over a dry soil surface as a function of the layer depth d, with the ice-crystal radius r as parameter. All calculations are for hh polarization at 10 GHz and $\theta = 50°$.

$d \approx 40$ cm, beyond which σ^0 approaches a saturation level.

The role of the ice-crystal radius is illustrated in Fig. 11-82. Here, $f = 10$ GHz and the variable parameter is r. Small crystals have small albedos and do not generate much scattering. Conversely, large crystals generate a great deal of scattering.

▶ The key parameter influencing volume scattering in snow is r/λ, where r is the crystal radius and λ is the wavelength. If $r/\lambda < 10^{-2}$, the snow layer appears electromagnetically invisible if its thickness $d \leq 2$ m. Conversely, the layer appears like a semi-infinite medium if $d \gtrsim 3000(r/\lambda)$. ◀

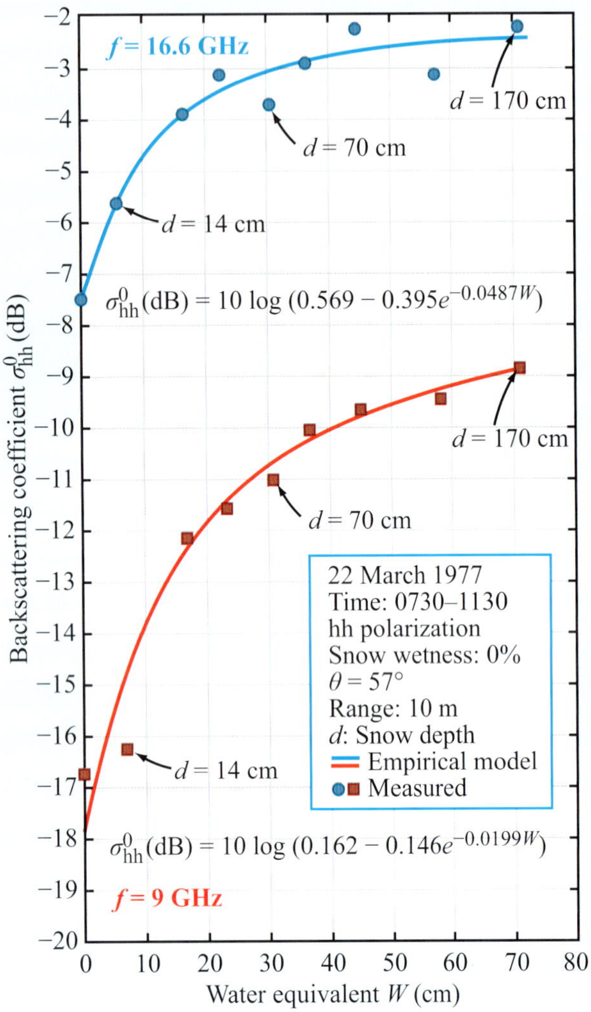

Figure 11-83: Response of backscattering coefficient σ^0 to snow water equivalent for dry snow at 9 and 16.6 GHz [Stiles and Ulaby, 1980a].

The general trends predicted by the theoretical plots of Fig. 11-81 are evident in the experimental measurements displayed in Fig. 11-83. The backscattering coefficient σ^0 was measured as a function of d by piling up dry snow in layers (Stiles and Ulaby, 1980a). The experiment was conducted at 9 and 16.6 GHz at an angle of incidence $\theta = 57^\circ$. The measured data was empirically fitted with the following expressions:

$$\sigma^0 = c_1 - c_2 e^{-c_3 W} \qquad (\text{m}^2/\text{m}^2), \tag{11.115}$$

where c_1 to c_3 are constants and W is the ***snow water equivalent*** of the snow layer,

$$W = \int_0^d \rho_s \, dz. \tag{11.116a}$$

Here, ρ_s is the snow density in g/cm^3. For a layer with uniform density,

$$W = \rho_s d. \tag{11.116b}$$

The water equivalent is the height in cm of liquid water contained in a vertical column of the snow layer (after melting) with a horizontal cross section of 1 cm^2. The average density of the snow used in the experiment leading to Fig. 11-83 is $\rho_s = 0.42$ g/cm^3.

11-15.2 Role of Snow-Ground Interface

The dynamic range of σ^0 as a function of snow depth d extends between $\sigma_1^0 = \sigma_s^0$, representing the backscattering coefficient of the ground surface under snow-free conditions ($d \approx 0$), and the "saturation" level σ_2^0 representing the semi-infinite case. The lower bound, σ_1^0, is determined by the dielectric constant and roughness of the ground surface, and the upper bound, σ_2^0, is determined by the albedo of the snow layer. Figure 11-84 displays measurements made by a 13.9 GHz airborne system called POLSCAT. The measurements were made at three different sites. Whereas in all three cases the data appears to confirm the general pattern of σ^0 versus d, the starting levels of σ^0 at $d = 0$ among the three sites are different by as much as 8 dB. These differences are attributed to differences in ground surface conditions.

Figure 11-85 shows another example depicting the role of the snow-ground interface. Radiative-transfer calculations are compared with measurements made as a function of f from 1 to 18 GHz. The measurements shown are for $\theta = 20^\circ$ and 50°. Excellent agreement is observed between theory and measurements when the ground surface is treated as a rough surface with $\sigma_s^0 \neq 0$,

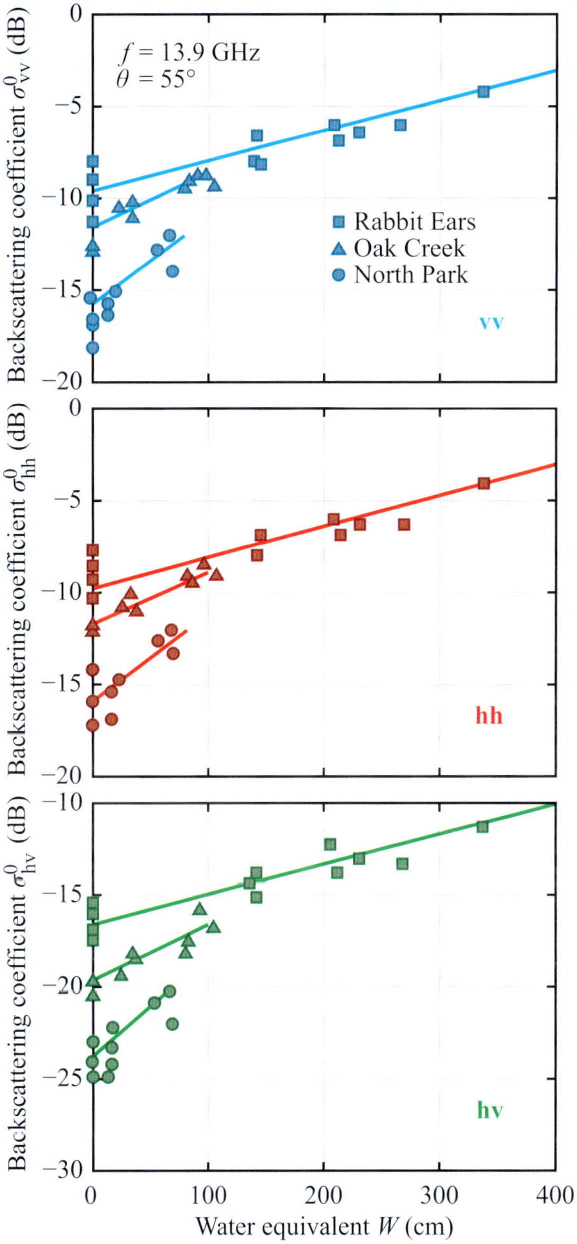

Figure 11-84: Comparison of 13.9 GHz POLSCAT radar data for three test sites [Yueh et al., 2009].

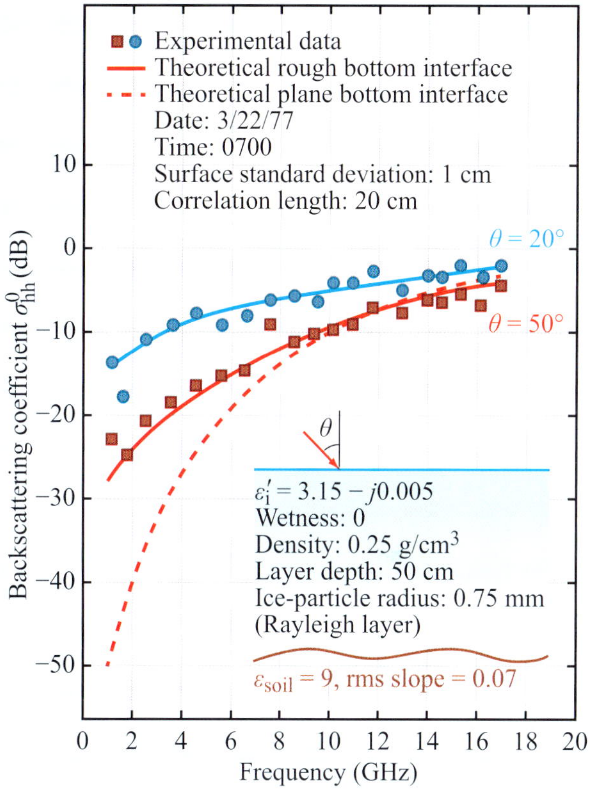

Figure 11-85: Comparison of theory and measurement for a layer of dry snow over the 1 to 18 GHz frequency band at 20° and 50° incidence angles.

but the agreement is quite poor at the low frequency end if we treat the ground surface as perfectly flat. For the purpose of computing the bistatic ground-snow and snow-ground contributions, the specular-surface assumption remains a reasonable approximation, but we should not altogether ignore the direct backscatter from the ground surface.

For deep snow with ice crystals of comparable size, the backscattering coefficient should be independent of the underlying ground surface. The plots in Fig. 11-86 were measured at very different sites, one in Colorado and the other in the Swiss Alps. The radar frequencies were approximately the same, the snow layer was sufficiently deep in both cases as to mask the underlying

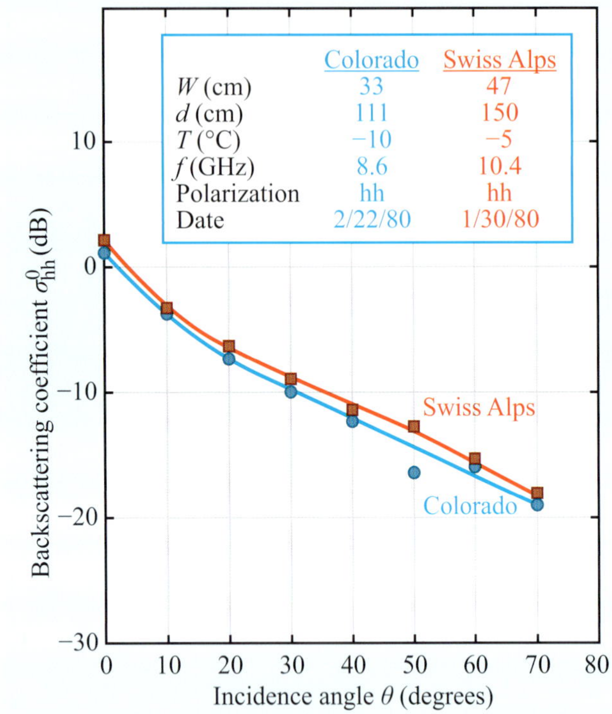

Figure 11-86: Comparison of deep-snow σ^0 measurements made at a site in the Swiss Alps (courtesy of C. Mätzler and E. Schanda, Univ. of Bern) with measurements made at a site in Colorado [from Ulaby et al., 1982d.].

surface, and so the measured angular responses of σ^0 of the two sites are almost identical.

11-15.3 Measuring Snow Thickness Over Sea Ice

Using an ultrawideband (2–8 GHz) FMCW radar mounted on a sled, a University of Kansas team evaluated the capability of the system for measuring the thickness of the snow layer over sea ice in Antarctica (Kanagaratnam et al., 2007). Figure 11-87(a) is a photograph of the FMCW radar system and Fig. 11-87(b) displays a typical example of the radar echo observed by the system. The first peak corresponds to the return from the air-snow boundary and the second peak corresponds to the snow-ice boundary. The

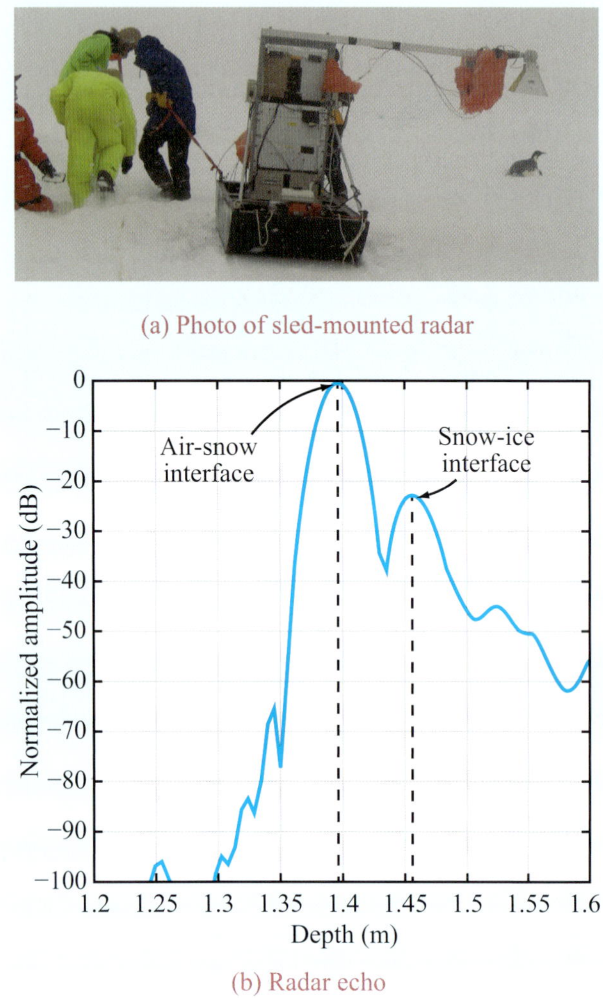

Figure 11-87: (a) University of Kansas sled-mounted FMCW radar in Antarctica and (b) typical example of a radar echo [Kangaratnam et al., 2007].

snow layer thickness is computed from the time delay and the phase velocity u_p. The latter is calculated from $u_p = c/\sqrt{\varepsilon'_{snow}}$, and ε'_{snow} is determined using standard equations relating ε'_{snow} to the snow density ρ_s. The results of measurements conducted over a snow-thickness range from 4 cm to 85 cm are displayed in Fig. 11-88. By sweeping the signal frequency over a 6 GHz bandwidth (from 2 to 8 GHz), the theoretical

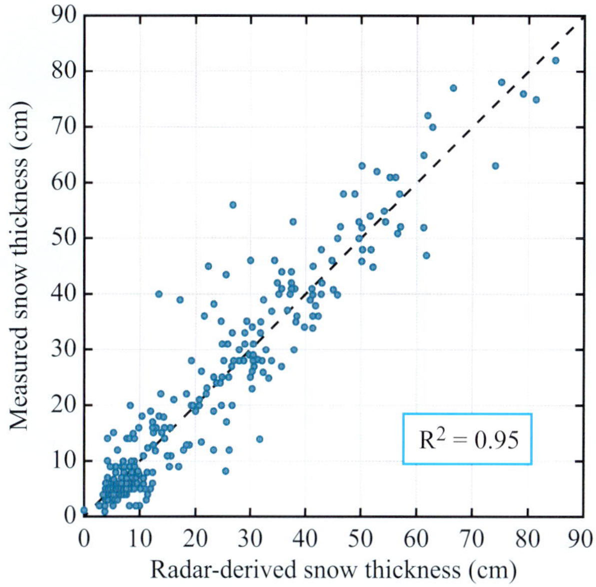

Figure 11-88: Comparison of radar-derived snow thickness with manually measured snow thickness [Kangaratnam et al., 2007].

range resolution of the FMCW system is

$$\Delta R = \frac{c}{2B} = \frac{3 \times 10^8}{2 \times 6 \times 10^9} = 0.025 \text{ m} = 2.5 \text{ cm} \qquad \text{(in air)},$$

and

$$\Delta R = \frac{2.5 \text{ cm}}{\sqrt{\varepsilon'_{\text{snow}}}} \qquad \text{(in snow)}.$$

The sled-mounted experiments were conducted in 2003 and reported in 2007. In 2008, a helicopter-mounted version of the FMCW radar [Fig. 11-89(a)] was used to demonstrate the snow-thickness measurement capability from an airborne platform. The experimental results, shown in Fig. 11-89(b), also exhibit good correlation between *in-situ* measurements of snow thickness and radar-derived values of snow thickness. The range of snow thickness that was encountered in Antarctica in 2008, however, was significantly narrower than that of 2003.

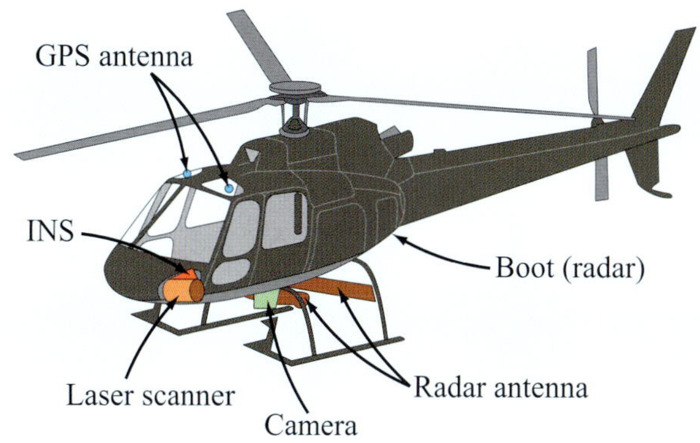

(a) Helicopter-mounted FMCW radar

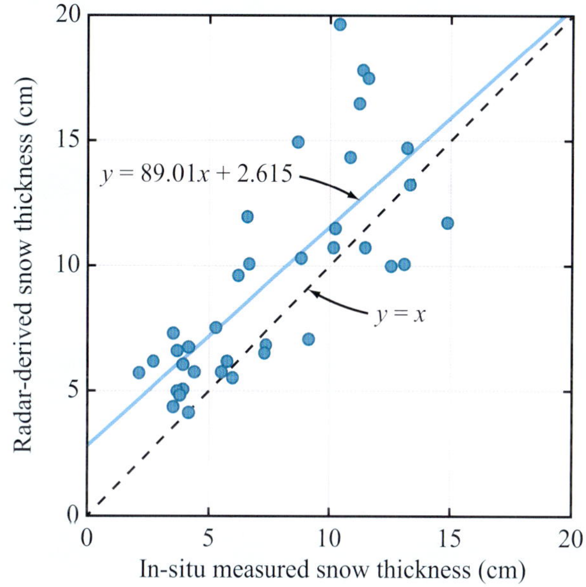

(b) Radar-derived vs in-situ measurements

Figure 11-89: (a) Helicopter-mounted FMCW radar and (b) results of measurement campaign [Galin et al., 2012].

11-16 Backscattering Behavior of Wet Snow

While the roughness of the snow surface exercises a negligible effect on the backscattering from snow when the snow layer is dry, it has a significant impact on σ^0 when the snow is wet. This behavior, which is demonstrated by the curves introduced earlier in Fig. 11-79, is related to the average dielectric permittivity of snow. The relative permittivity for dry snow with a typical density of 0.25 g/cm^3 is about 1.5, compared with 2.6 at 6 GHz for snow containing 10 percent liquid water. The corresponding Fresnel reflectivities at normal incidence are 0.01 for dry snow and 0.06 for wet snow; i.e., the reflectivity of wet snow is larger than that of dry snow by a factor of 6. Consequently, when constructing a backscattering model for wet snow, the upper snow boundary is treated as a rough surface, in contrast to the dry-snow case in which it was regarded as a planar interface.

In addition to causing the permittivity of the snow layer to increase, the presence of liquid water in the snow volume causes a dramatic increase in the dielectric loss factor of the layer, $\varepsilon''_{\rm ws}$, as detailed in Section 4-6.2. This, in turn, results in a large increase in the absorption coefficient, thereby reducing the relative importance of the ground contribution. Hence, to reduce the complexity of the single-scattering radiative transfer model of Section 11-6, the snow-ground interface usually is assumed to be a plane boundary.

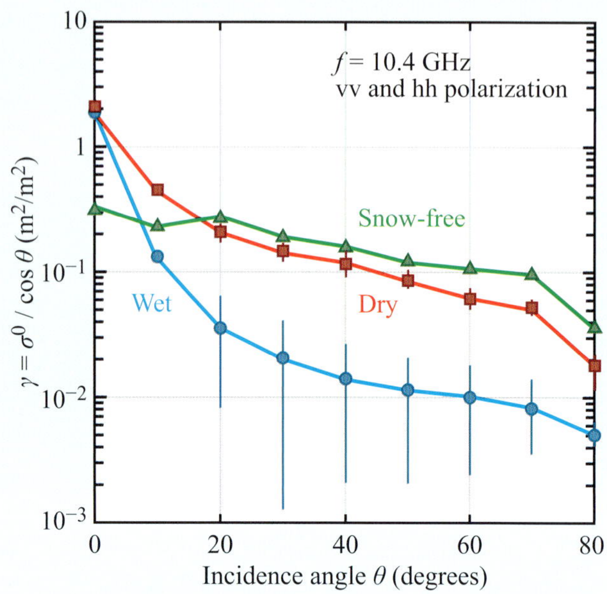

Figure 11-90: Mean values of vv- and hh-polarized backscattering coefficients averaged over all wet and dry snow, and over the snow-free situations, respectively. The standard deviations show the variability of the snow data [from Mätzler and Schanda, 1984].

11-16.1 Angular Dependence

To assess the potential of radar for mapping snow conditions, Mätzler and Schanda (1984) compared the measured angular response of $\gamma(\theta)$ for snow-free ground (Fig. 11-90) with responses for dry and wet snow, all measured at 10.4 GHz. The curves for dry and wet snow in Fig. 11-90 represent mean values measured over a wide range of snow conditions, and the vertical bars represent ± 1 standard deviation relative to the calculated means. The data in Fig. 11-90 indicate that for $\theta \geq 20°$ it should be easy to discriminate (using radar) between areas covered with dry snow and areas covered with wet snow, because of the large difference between σ^0 of dry snow and that of wet snow.

As an illustration of the role of the microwave frequency f with respect to both the shape of the angular response $\sigma^0(\theta)$ and sensitivity to snow wetness, we show in Fig. 11-91 data measured at 2.6 GHz and 35.6 GHz for a snow layer 27 cm deep. The 2.6 GHz curves show no sensitivity to the liquid-water content $m_{\rm v}$. Furthermore, they exhibit an angular dynamic range of about 35 dB (between normal incidence and $\theta = 70°$). In contrast, at 35.6 GHz, $\sigma^0(\theta)$ of dry snow is about 8 dB higher than that of wet snow at all angles $\theta \geq 10°$. Also $\sigma^0(\theta)$ exhibits a much narrower angular dynamic range at 35.6 GHz, particularly when the snow is dry.

11-16.2 Frequency Dependence

The spectral curves in Fig. 11-92 illustrate the importance of the frequency f with regard to the sensitivity of σ^0 to snow wetness $m_{\rm v}$. For a snowpack 48 cm deep, a change in wetness from zero to 1.26 percent results in a drop in level of only 1 dB at 1 GHz, but as f increases, the two curves diverge, and the difference between them increases to a maximum of 15 dB at 35.6 GHz.

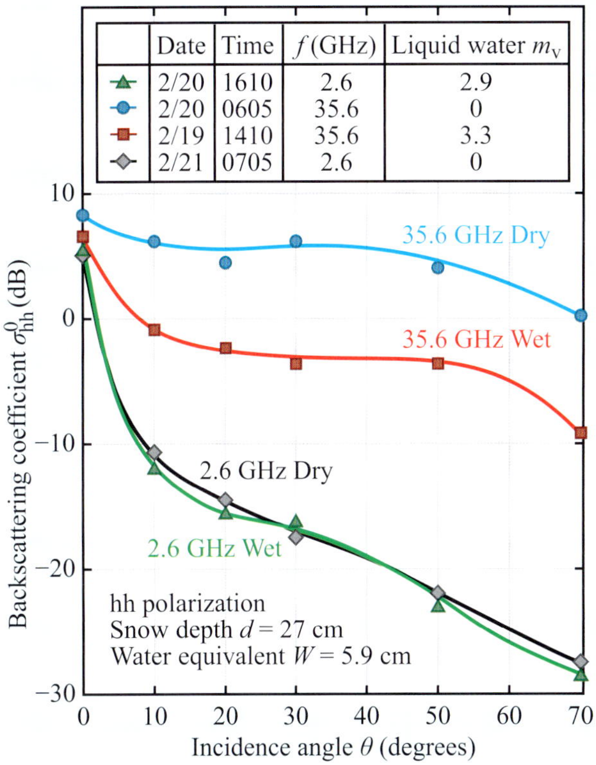

Figure 11-91: Angular dependence of σ^0 for wet and dry snow at 2.6 GHz and 35.6 GHz [from Stiles and Ulaby, 1980a].

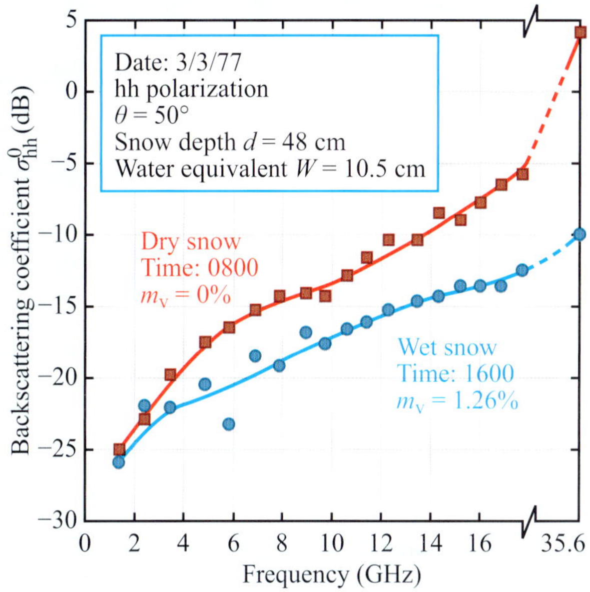

Figure 11-92: Spectral response of σ^0 for wet and dry snow [from Ulaby and Stiles, 1981].

11-16.3 Wetness Dependence

According to the data shown in Fig. 11-93, σ^0 exhibits an approximately linear response to m_v up to about 5 percent by volume. The measurements were made at 17 GHz for the depth and water equivalent values indicated alongside the curves. Similar measurements made at 8 GHz indicate that the response is linear over an even wider range of m_v (Ulaby and Stiles, 1981). In general, in the 1 to 35 GHz region, the magnitude of the sensitivity of σ^0 to m_v (i.e., $|d\sigma^0/dm_v|$ over the quasi-linear range of σ^0 versus m_v) increases with increasing frequency; however, concurrently, the detectable range of m_v (for which the σ^0 response is quasi-linear) decreases with increasing frequency. Thus, the choice of frequency should be dictated by whether it is more important to detect the presence of water in liquid form in the snowpack or to actually measure its percentage over as wide a range as possible.

11-16.4 Diurnal Variations

A convenient way to evaluate the response of σ^0 to snow wetness variations is to monitor σ^0 over a 24-hour diurnal cycle under clear-sky conditions. Figure 11-94 shows the results of such an experiment performed at $\theta = 50°$. During the daylight hours, the wetness m_v (of the top 5 cm layer) increased from zero at 0800 hours to a maximum of 1.26 percent by volume at 1500 hours and then decreased back to zero in the late afternoon. The observed σ^0 diurnal patterns are essentially mirror images of the m_v pattern, with some possible time-shifting between the σ^0 and m_v patterns. The magnitude of the dip in σ^0 increases from about 2 dB at 2.6 GHz to 15 dB at 35.6 GHz.

Under overcast conditions at below freezing temperatures, diurnal observations show no variation in σ^0. However, solar illumination under clear-sky conditions

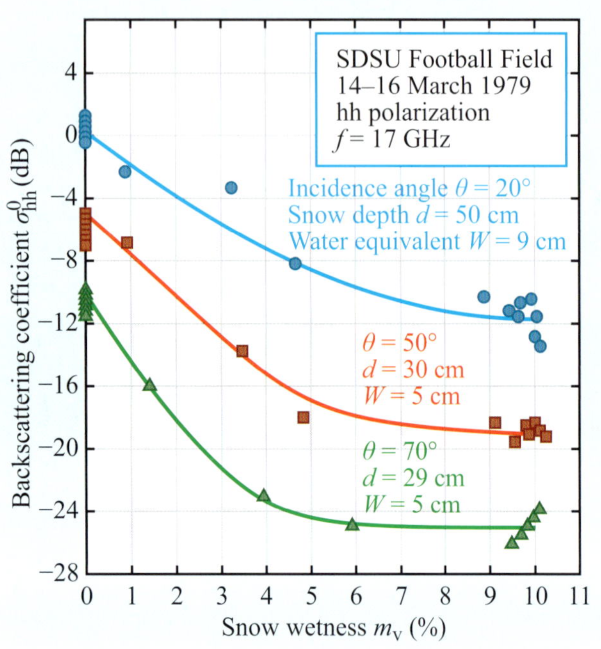

Figure 11-93: Measured response of σ^0 to snow liquid-water content at 17 GHz [from Ulaby and Stiles, 1981].

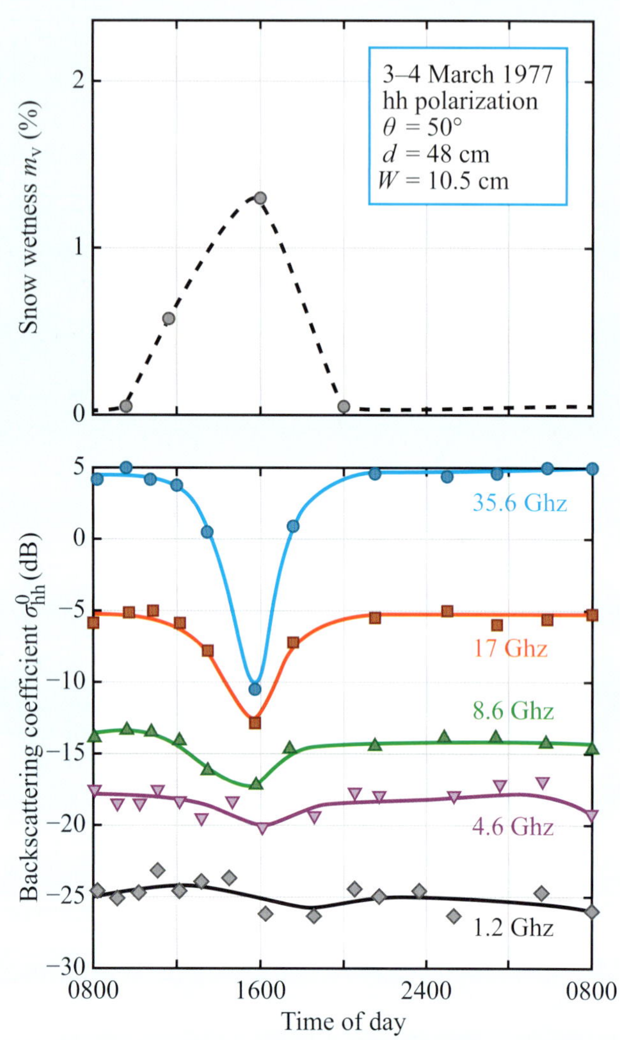

Figure 11-94: Diurnal patterns of liquid-water content m_v and backscattering coefficient σ^0 at several microwave frequencies [from Stiles and Ulaby, 1980a].

may cause partial melting of the snow's surface layer, even though the air temperature may be below freezing.

11-16.5 Seasonal Variations

Using a truck-mounted radar system, backscattering observations were made at a snow test site in Colorado for a period of six weeks (Ulaby and Stiles, 1981). The results for $f = 8.6$ GHz and $\theta = 50°$ are presented in Fig. 11-95 in the form of temporal plots of σ^0, the snow water equivalent W, and the liquid-water content m_v. In the figure, the boundaries of the temporal plot of σ^0 appear to form an envelope extending between two lines designated σ^0_{max} and σ^0_{min}. When the top layer of the snow is wet, it masks the contributions of the lower layers, and hence σ^0_{min} is essentially independent of the water equivalent W (so long as W is larger than some minimum value that equals 2 to 3 penetration depths in wet snow) and varies with m_v in a fashion similar to that indicated in Fig. 11-93. For dry snow, σ^0 increases with increasing W. This dependence is responsible for the observed positive slope of σ^0_{max}.

Although σ^0 varies with both m_v and W, it is possible to separate the causes of the changes in σ^0 by making observations both during the day and at night. At night, the snowpack is likely to be cold and dry; therefore, multidate nighttime observations may be used to update the status of W. The magnitude of the decrease in the

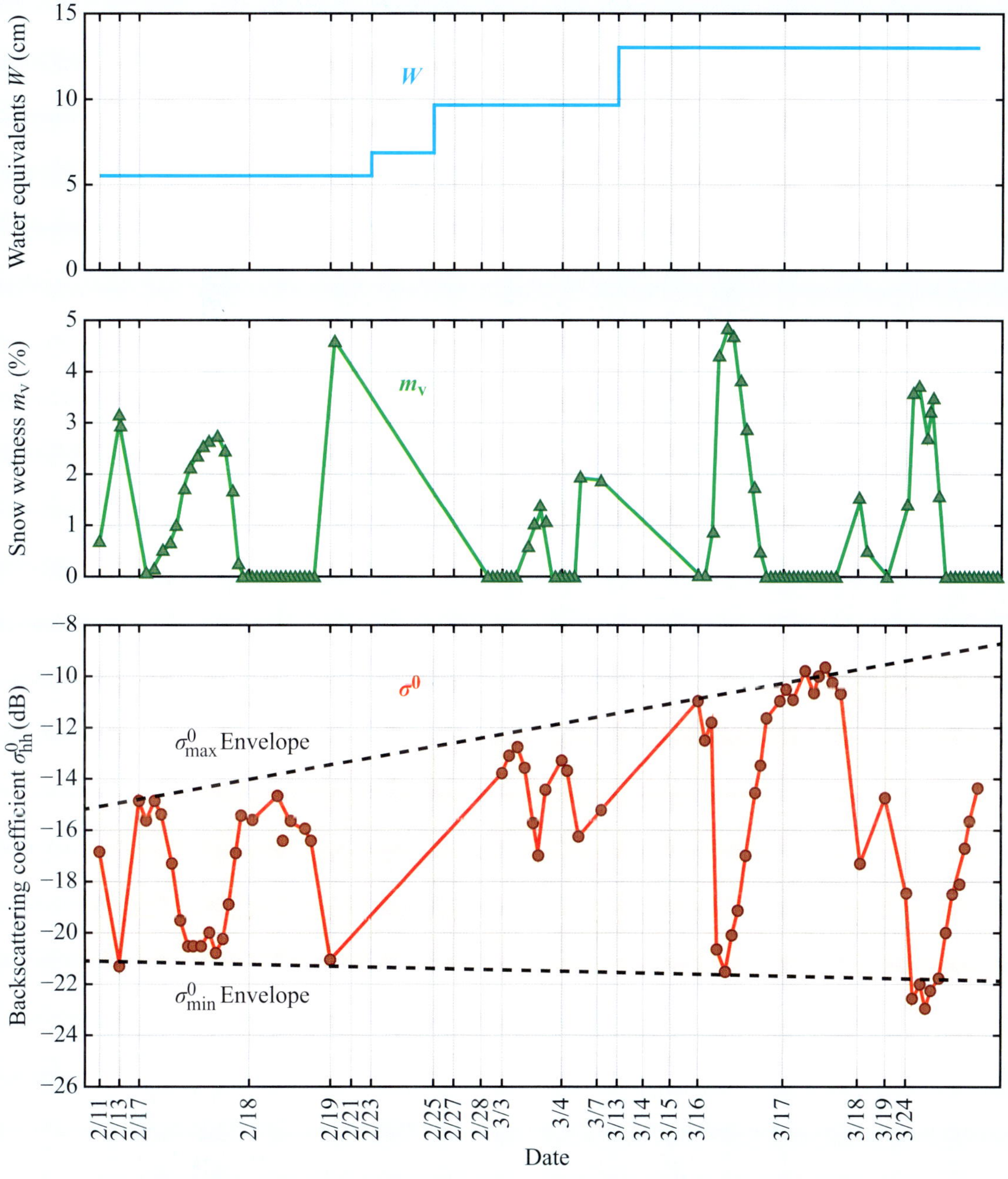

Figure 11-95: Temporal variation of σ^0 for Colorado snow test site [from Ulaby and Stiles, 1981].

level of σ^0 observed during daytime hours following a nighttime observation should then be indicative of the degree of wetness of the snowpack.

It is important to note that although the total increase in σ^0_{max} over the six-week period was only about 5 dB (in response to an increase in W of about 7 cm and a corresponding increase in depth of about 25 cm), such a change is quite significant when based on multidate observations of the same snowpack scene. During nighttime observations, the snow-layer and underlying ground parameters (other than W) influencing σ^0 are likely to exhibit much smaller temporal variations for a given ground cell than spatial variations on a given date. Thus, the use of change-detection techniques, in which a radar image of a given scene is "subtracted" from an image made earlier, may prove quite useful in estimating the water equivalent W of snowpacks over large areas.

11-16.6 Millimeter-Wave Observations

Our discussion thus far has dealt with the backscattering behavior of snow in the 1 to 35 GHz band. Far fewer studies have been reported involving measurements at higher frequencies. Examples illustrating the behavior of σ^0 for wet and dry snow at 35, 98, and 140 GHz are given in Fig. 11-96. As f increases from 35 GHz to 140 GHz, we observe the following patterns:

(a) The level of σ^0_{hv} approaches that of σ^0_{vv}, indicating the increased generation of cross-polarized energy as a result of the increase in the size of the ice crystals (in the snow volume) relative to λ.

(b) The sensitivity of σ^0 to wetness decreases. At 140 GHz, σ^0 is almost completely insensitive to m_v. This behavior is attributed to the albedo of the snow medium. At 35 GHz, the albedo of dry snow is on the order of 0.8, and when the snow is wet ($m_v \geq 1\%$) the albedo decreases to 0.3 or less (see Fig. 11-78(a)]. At 140 GHz, on the other hand, the albedo of dry snow is very close to 1.0 (because of the much larger sizes of the ice crystals relative to λ), and the presence of water in liquid form has little influence on the albedo because the permittivity of liquid water at 140 GHz is around 5,

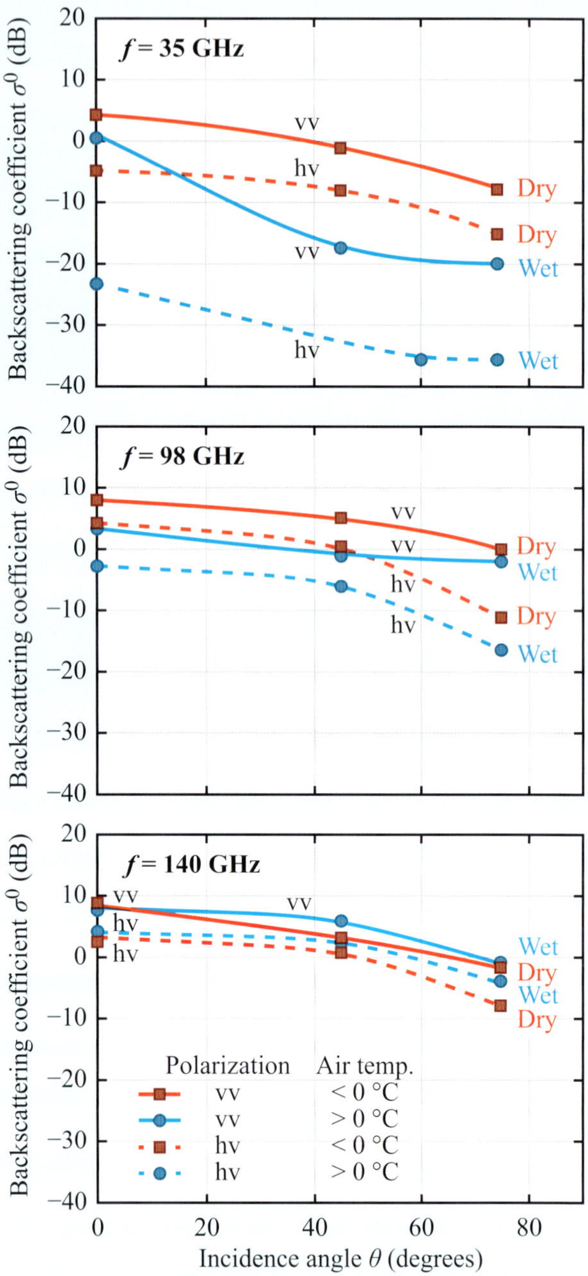

Figure 11-96: Measured angular variation of σ^0 for dry and wet snow at (a) 35 GHz, (b) 98 GHz, and 140 GHz [from Hayes et al., 1979].

which is not drastically different from that of the ice crystals, $\varepsilon_i' = 3.15$.

Applying a numerical solution to the radiative transfer equation, Kuga et al. (1991) evaluated the theoretical response of σ^0 to snow wetness at 35, 95, and 140 GHz. They modeled a 45 cm deep snowpack with an ice volume fraction of 0.4 (or an equivalent snow density $\rho_s = 0.37$ g/cm^3), containing ice particles characterized by a normal size distribution with a mean diameter of 1 mm and a standard deviation of 0.2 mm. The snow surface was assumed to have a rms slope $m = 0.5$. Their results for hh and hv polarizations are displayed in Fig. 11-97. We observe that:

(a) The cross-pol scattering coefficient σ^0_{hv} is more sensitive to snow wetness than the co-polarized component σ^0_{hh}.

(b) The sensitivity to snow wetness decreases with frequency; among the three millimeter-wave frequencies, the dynamic range of σ^0 due to m_v changing from 0 to 6% is highest at 35 GHz and lowest at 140 GHz.

The theoretical predictions of the radiative transfer model are confirmed by the experimental results and associated model calculations shown in Figs. 11-98 and 11-99 (Ulaby et al., 1991). The first figure displays plots of the recorded air temperature, snow surface temperature, and average snow wetness of the top 5 cm snow layer. Figure 11-99 contains plots of the measured diurnal variation of σ^0_{hh} at 35, 95, and 140 GHz, as well as model-calculated values. The calculated values are based on a hybrid first-order-numerical model applied to a snowpack with a snow-wetness depth profile that incorporates the thawing and refreezing processes that a snowpack undergoes over a diurnal cycle. Very good agreement is observed between the model-calculated and radar-measured diurnal patterns at all three millimeter-wave frequencies.

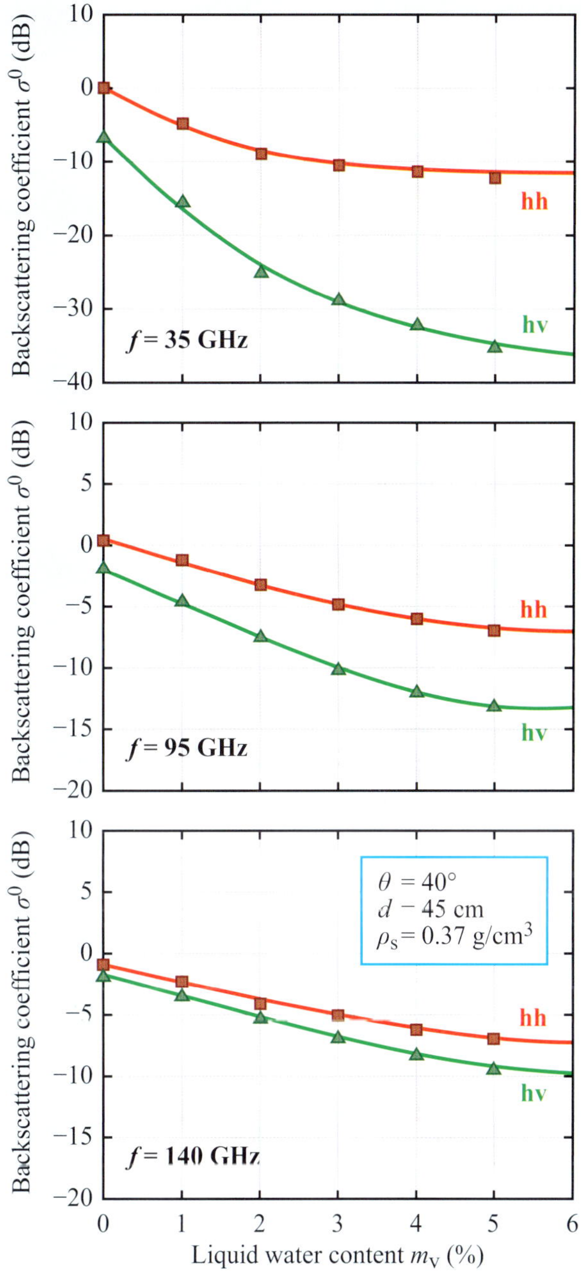

Figure 11-97: Backscattering coefficient σ^0 versus m_v at 35, 95, and 140 GHz [Kuga et al., 1991].

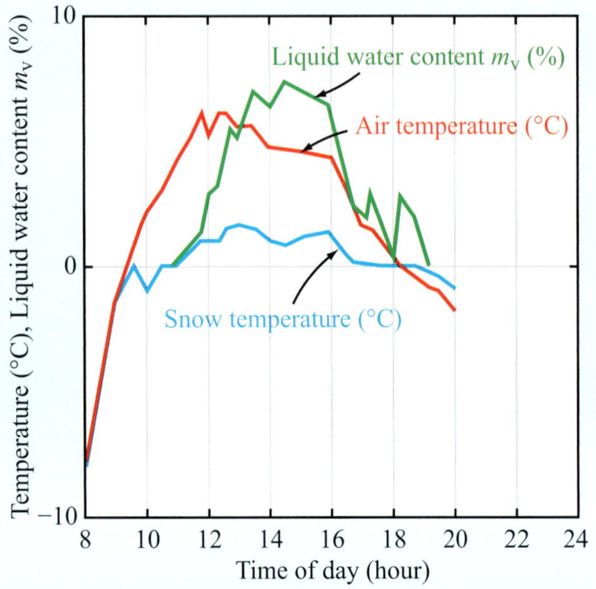

Figure 11-98: Temporal variations of air and snow temperatures and snow liquid-water content [Ulaby et al., 1991].

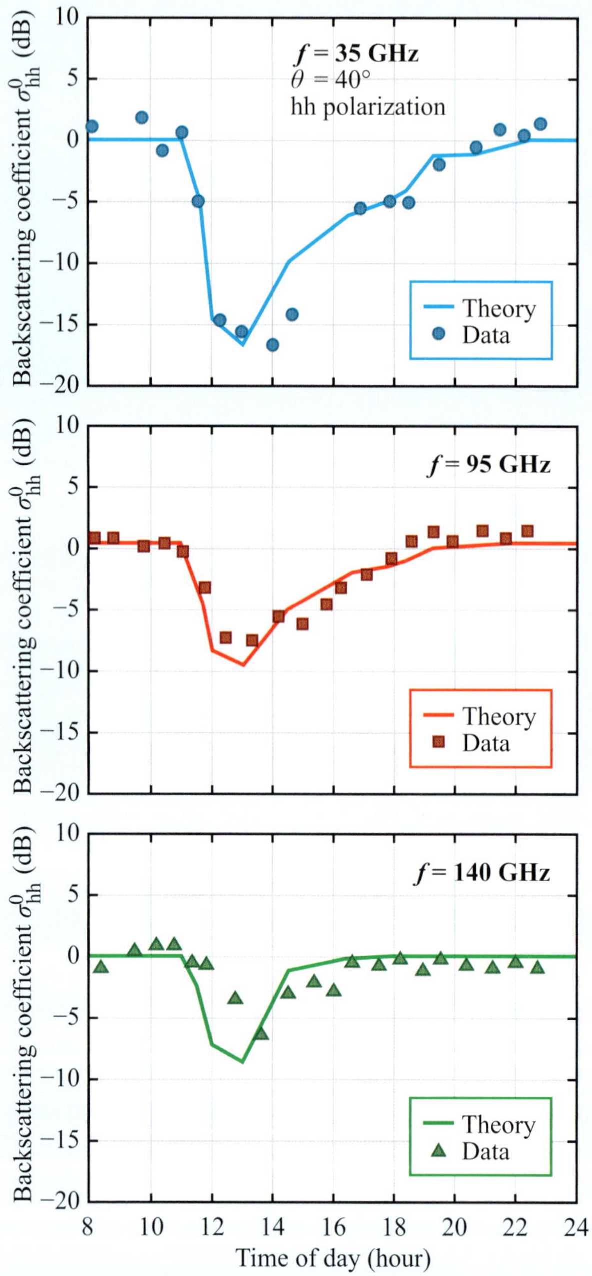

Figure 11-99: Comparison of the diurnal variation of the measured backscattering coefficient (relative to dry snow) with that using a hybrid first-order–numerical radiative transfer model [Ulaby et al., 1991].

PROBLEMS

Canopy for Problems 11.1 to 11.3

A vegetation canopy is characterized by the following parameters:

albedo $a = 0.03f$, with f in GHz ($1 \leq f \leq 20$ GHz),

extinction coefficient $\kappa_e = 0.3f$ (Np/m), with f in GHz ($1 \leq f \leq 20$ GHz).

The underlying soil is loam and has a random surface with an rms height of 1 cm. Assume PRISM-1 is applicable.

11.1 Compute the backscattering coefficients for hh and vv polarizations at $\theta_i = 30°$ and $f = 3$ GHz as a function of the canopy thickness d (0 to 2 m) for:

(a) a dry soil surface with $m_v = 0.05$ g/cm^3.

(b) a wet soil surface with $m_v = 0.3$ g/cm^3.

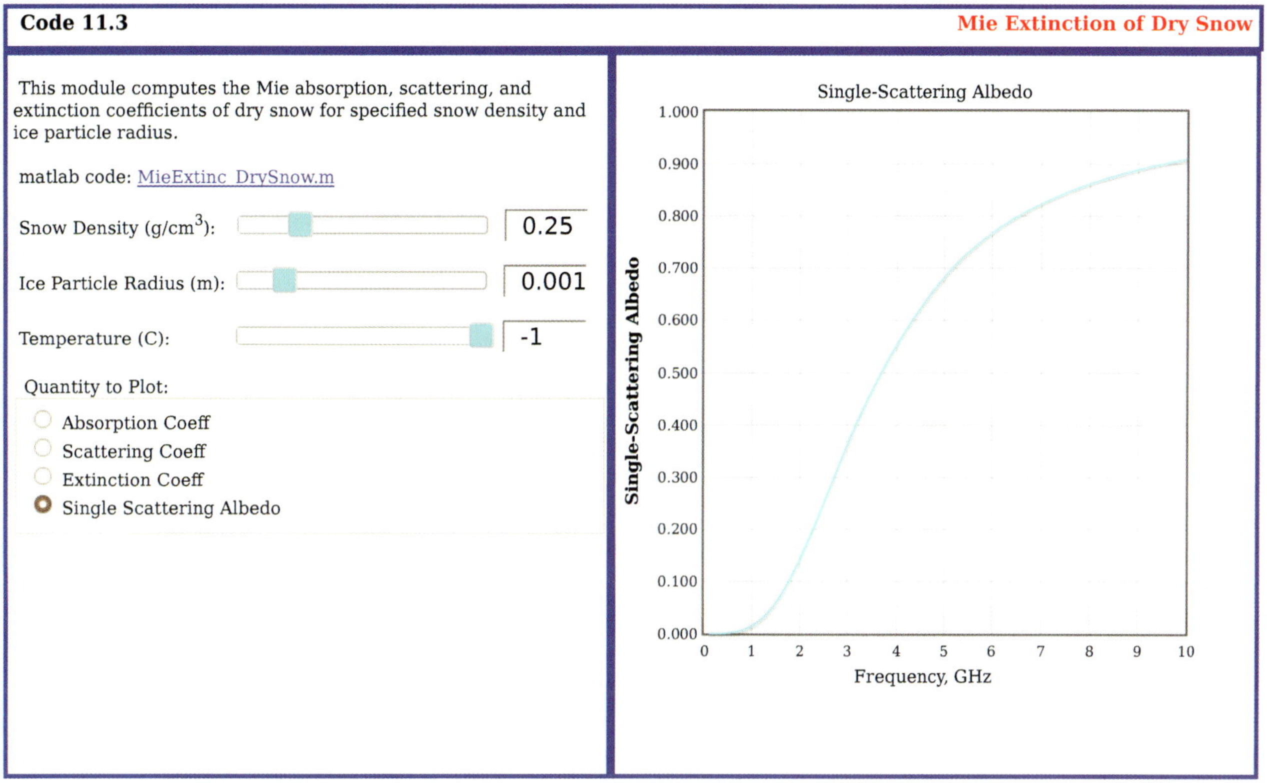

An example of one of the interactive modules available at the book website: mrs.eecs.umich.edu.

11.2 Compute the hh-polarized backscattering coefficient at $\theta_i = 30^\circ$ as a function of volumetric soil moisture content (0 to 0.3 g/cm^3) for a 1 m tall canopy at:

(a) 1 GHz.

(b) 3 GHz.

(c) 10 GHz.

In each case, determine the radar sensitivity to soil moisture by computing the slope at $m_v = 0.2$ g/cm^3.

11.3 Compute the hh-polarized backscattering coefficient at $\theta_i = 30^\circ$ as a function of frequency (1 to 10 GHz) for a 1 m tall canopy above a soil surface with:

(a) $m_v = 0.05$ g/cm^3.

(b) $m_v = 0.30$ g/cm^3.

11.4 A layer of dry snow of depth d overlays a soil surface. In the absence of snow, the hh-polarized backscattering coefficient of the soil surface at $\theta = 30^\circ$ is given by

$$\sigma^0_{\text{soil}} = 0.01 + 0.001 f,$$

where f is in GHz ($1 \leq f \leq 30$ GHz).

The snow density is 0.5 g/cm^3 and at 10 GHz, the snow albedo is 0.2 and the extinction coefficient is 0.15 Np/m. Assuming that the snow layer has a flat top surface and that it can be treated as a Rayleigh layer, compute and plot the hh-polarized backscattering coefficient at 30° as a function of d from 0 to 2 m.

11.5 Repeat Problem 11.4 for wet snow, keeping all parameters the same except for changing the albedo to 0.01 and the extinction coefficient to 1.2 Np/m.

CHAPTER

12

Emission Models and Land Observations

SMAP sensor

CONTENTS

Overview

Satellite-borne microwave radiometers have been providing information about atmospheric and oceanic parameters in either a fully operational or a quasi-operational mode for several decades. This, however, has not been the case for land parameters. The primary constraints for land applications include spatial resolution and relative complexity.

(a) Spatial resolution

The spatial resolution, or field of view (FOV), of a satellite radiometer is dictated by the orbital altitude, the incidence angle, the wavelength, and antenna size. Among the majority of radiometers flown to date, the FOV has ranged between about 5 km for radiometers operating in the 100 GHz range to about 50 km for L-band radiometers operating at 1.4 GHz. Except for localized, small-scale phenomena, most atmospheric and oceanic parameters (such as temperature, pressure, windspeed, etc.) vary horizontally at scales on the order of tens of kilometers. Furthermore, the frequency channels used in support of atmospheric and oceanic applications are at 19 GHz and up, so the radiometer FOVs at those frequencies are comparable in size with the spatial scale of the atmospheric and oceanic variables of interest. Consequently, microwave radiometers have been highly successful at providing useful information about the state of the atmosphere and ocean at those scales.

Land parameters vary at scales of tens of meters to tens of kilometers, depending on the specific variable. For snow-related applications—which include snow-cover mapping, snow water equivalent estimation, and the detection of the onset of snowmelt, their spatial scales are on the order of kilometers to tens of kilometers, and so are the FOVs of the AMSR (advanced microwave scanning radiometer) channels used operationally in support of these applications. The snow products generated by the AMSR algorithms have a grid size of 25 km.

To map the spatial distribution of soil moisture with passive microwave remote sensing, the physics of the emission process dictates the use of very low microwave frequencies. The FOV of the 1.4 GHz SMOS (***soil moisture and ocean salinity***) radiometer (launched in 2009) is around 40 km. Unfortunately, such a large FOV is one to two orders of magnitude larger than the scale of spatial variations characteristic of soil moisture and the vegetation covering it.

(b) Relative complexity

Among the three media—land, ocean, and atmosphere, the atmosphere is generally the best "behaved." Atmospheric variables, such as temperature and pressure, have predictable average profiles with height, so the task of remote sensing is to measure the deviations from these nominal profiles. Observations made by multifrequency microwave radiometers are used in inversion algorithms to estimate these deviations with very good accuracy (Chapter 9).

Ocean dynamics are generally more complex than those in the atmosphere. Nevertheless, inversion algorithms that rely on multifrequency radiometric observations have been used for many years to provide useful information about sea state, sea-ice concentration, and other ocean surface attributes (Chapter 18).

In contrast, land is by far the most complex; surface features—such as vegetation cover—vary widely in shape, orientation, and density, as do their dielectric properties. Sensing soil moisture content for bare and vegetation-covered soils has been a prime goal of the remote-sensing research community for several decades. Extensive experimental observations and model simulations have been conducted to develop good understanding of the emission process and the relationship between the observed brightness temperature and the various properties of the soil surface and vegetation cover above it. Setting the spatial resolution issue aside for the moment, it should be possible to design a microwave radiometer to generate accurate estimates of soil moisture, but such a radiometer should include multiple-frequency channels in the L-, S-, and C-bands (see Table 1-7 for passive microwave frequency allocations). Otherwise, it is next to impossible to segregate with a single channel the dependence of the brightness temperature on the various soil and vegetation cover parameters. Current plans call for the launch of a

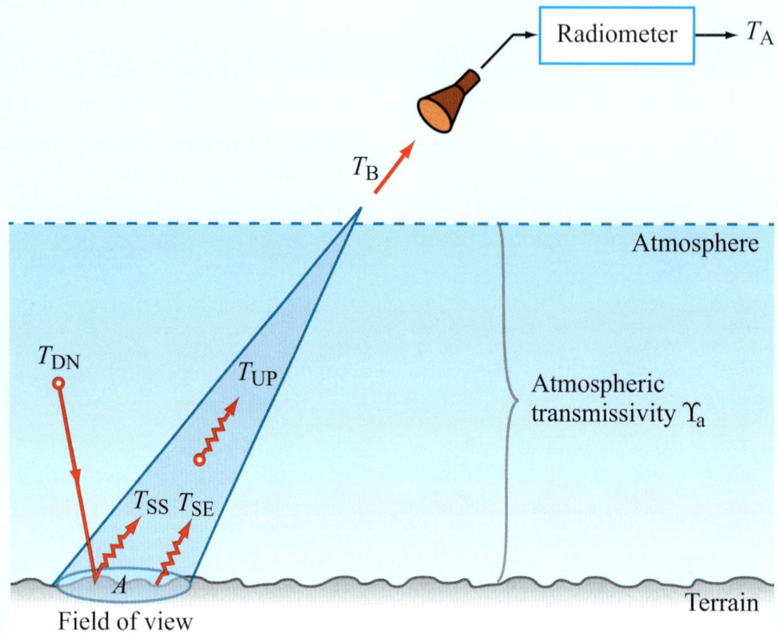

Figure 12-1: The brightness temperature T_B incident upon the radiometer antenna includes emission contributions from the atmosphere and the ground cell observed by the antenna field of view.

Soil Moisture Active and Passive (SMAP) instrument in 2014 (Entekhabi et al., 2010). The instrument includes a 1.41 GHz radiometer and a 1.26 GHz synthetic-aperture radar. The intent is to merge the 40 km resolution radiometer images with the 1 to 3 km resolution radar images to generate soil moisture maps at 10 km resolution. Only time will tell if the proposed single-band active/passive approach proves successful.

This chapter presents a review of microwave radiometric emission for bare soil surfaces, vegetation-covered surfaces, snow cover, and ice in freshwater lakes.

12-1 Emissivity and Reflectivity

Consider the downward-looking radiometer depicted in Fig. 12-1. As noted earlier in Section 6-8, a calibration algorithm is used to convert the antenna temperature T_A measured by the radiometer receiver into the p-polarized brightness temperature T_B incident upon the antenna. Our interest is in T^p_{SE}, *the p-polarized brightness temperature due to* ***surface emission*** *by the cell viewed by the beam of the p-polarized antenna.* The ground cell may be a bare soil surface, an ocean surface, a vegetation-covered soil surface, or any other type of terrestrial surface. Here, T^p_{SE} represents the emission by the ground cell when viewed from above, in the absence of the atmosphere.

In the general case, T^p_B and T^p_{SE} are related by Eq. (6.113),

$$T^p_B = \Upsilon_a(T^p_{SE} + T^p_{SS}) + T_{UP} \qquad (p = \text{v or h}), \qquad (12.1)$$

where Υ_a is the ***transmissivity*** of the atmospheric layer between the ground cell and the antenna, T_{UP} is the ***upwelling atmospheric emission*** by the same layer, and T^p_{SS} is the fraction of the downward-emitted atmospheric radiation that gets ***surface scattered*** by the ground cell into the direction of the radiometer. For an isothermal ground cell at a physical temperature T_0,

$$T_{\text{SE}}^p = e^p T_0, \tag{12.2}$$

where e^p is the p-polarized ***emissivity***. From Section 6-7.3, the expression for e^p is given by

$$e^p(\theta_i, \phi_i) = 1 - \Gamma_{\text{tot}}^p(\theta_i, \phi_i), \tag{12.3}$$

(rough-surface emissivity)

where $\Gamma_{\text{tot}}^p(\theta_i, \phi_i)$ is the ***total reflectivity*** of the surface, which includes a ***coherent component*** $\Gamma_{\text{coh}}^p(\theta_i, \phi_i)$ and an ***incoherent component*** $\Gamma_{\text{inc}}^p(\theta_i, \phi_i)$:

$$\Gamma_{\text{tot}}^p(\theta_i, \phi_i) = \Gamma_{\text{coh}}^p(\theta_i) + \Gamma_{\text{inc}}^p(\theta_i, \phi_i), \tag{12.4}$$

with

$$\Gamma_{\text{coh}}^p(\theta_i) = \Gamma^p(\theta_i)\, e^{-4\psi^2}, \tag{12.5}$$

$$\Gamma_{\text{inc}}^p(\theta_i, \phi_i) = \frac{1}{4\pi\cos\theta_i} \times \int_{\phi_s=0}^{2\pi}\int_{\theta_s=0}^{\pi/2} [\sigma_{pp}^\circ(\theta_i, \phi_i; \theta_s, \phi_s) + \sigma_{pq}^\circ(\theta_i, \phi_i; \theta_s, \phi_s)]\, d\Omega_s, \tag{12.6}$$

where $\psi = ks\cos\theta_i$, s is the surface rms height, $k = 2\pi/\lambda$, $\Gamma^p(\theta_i)$ is the Fresnel specular reflectivity of the surface at incidence angle θ_i, and σ_{pp}° and σ_{pq}° are the incoherent co-polarized and cross-polarized bistatic scattering coefficients of the ground cell.

Similarly, the surface scattered brightness temperature is

$$T_{\text{SS}}^p(\theta_i, \phi_i) = \Gamma_{\text{coh}}^p(\theta_i)\, T_{\text{DN}}(\theta_i) + \int_{\theta_s=0}^{\pi/2}\int_{\phi_s=0}^{2\pi} [\sigma_{pp}^0(\theta_i, \phi_i; \theta_s, \phi_s) + \sigma_{pq}^0(\theta_i, \phi_i; \theta_s, \phi_s)] \cdot T_{\text{DN}}(\theta_s, \phi_s)\, d\Omega_s. \tag{12.7}$$

Even though $T_{\text{DN}}(\theta_s, \phi_s)$ is different along different angles θ_s (because the extent of the atmospheric layer contributing to T_{DN} is greater along directions away from nadir than along the nadir direction), it is often approximated as a constant equal to its value along θ_i. Setting $T_{\text{DN}}(\theta_s, \phi_s) = T_{\text{DN}}(\theta_i)$ in Eq. (12.7) leads to

$$T_{\text{SS}}(\theta_i, \phi_i) = \Gamma_{\text{tot}}^p\, T_{\text{DN}}(\theta_i). \tag{12.8}$$

Using Eqs. (12.3) and (12.8) in Eq. (12.1) gives

$$T_{\text{B}}^p = \Upsilon_a[(1 - \Gamma_{\text{tot}}^p)T_0 + \Gamma_{\text{tot}}^p T_{\text{DN}}] + T_{\text{UP}}. \tag{12.9}$$

To determine Γ_{tot}^p from T_{B}^p, we need good estimates of T_0, Υ_a, T_{DN}, and T_{UP}. The physical temperature is estimated from either recorded weather data or from satellite observations made by the 37 GHz radiometric channel on the TRMM or WindSat satellites (see Table 1-4). Examples of studies documenting the relationship between the surface temperature T_0 and the brightness temperature observed by the 37 GHz radiometer onboard of Nimbus-7 are given by Owe and Van de Griend (1991) and by Van de Griend (2001). The other quantities can be estimated using atmospheric inversion models of the type described in Chapter 9. The bottom line is that we should be able to obtain a high-accuracy estimate of the integrated reflectivity $\Gamma_{\text{tot}}^p(\theta_i)$ from the value of T_{B}^p recorded by the satellite radiometer.

The two primary land-related applications of microwave radiometry are soil-moisture sensing and the estimation of snow water-equivalent. As we see later, the preferred frequency for soil moisture sensing (if only a single-frequency channel is available) is 1.4 GHz; it is a compromise between two considerations: spatial resolution and vegetation effects. The size of the antenna-beam footprint (field of view or resolution) decreases with increasing frequency, but so does the masking effect of vegetation. For snow, to measure its water equivalent with good sensitivity, the frequency should be in the 10 to 20 GHz range. Under "standard" weather conditions, as defined in Section 8-3, the three atmospheric quantities appearing in Eq. (12.9) have the following nominal values:

	1 GHz	**10 GHz**	**20 GHz**
Υ_a	~ 0.98	~ 0.95	~ 0.80
$T_{\text{UP}} \approx T_{\text{DN}}$	5 K	6 K	18 K

Using available weather information, it should be possible to estimate Υ_a to within an error of ± 0.01 and T_{DN} and T_{UP} to within an error of ± 1 K.

Reflectivity Terminology

To avoid confusion, we should elaborate on the use of the term "reflectivity":

$\Gamma^p(\theta_i)$ = p-polarized Fresnel reflectivity of a specular surface for incidence angle θ_i, and given by the expressions for Γ^v and Γ^h in Table 2-5.

$\Gamma^p_{coh}(\theta_i) = \Gamma^p e^{-(2ks\cos\theta_i)^2}$ = coherent reflectivity of a surface with rms height s and wavenumber $k = 2\pi/\lambda$.

$\Gamma^p_{inc}(\theta_i, \phi_i)$ = p-polarized incoherent reflectivity, which incorporates scattering along direction (θ_i, ϕ_i) due to incidence from all directions (θ_s, ϕ_s) in the upper hemisphere.

$\Gamma^p_{tot}(\theta_i, \phi_i) = \Gamma^p_{coh} + \Gamma^p_{inc}$ = p-polarized total reflectivity.

$e^p(\theta_i, \phi_i) = 1 - \Gamma^p_{tot}(\theta_i, \phi_i)$ = p-polarized emissivity.

12-2 Emission by a Specular Surface

The simplest possible soil-emission configuration to consider is that of a homogeneous, isothermal soil medium (constant-depth profiles for both its physical temperature T and dielectric constant ε_{soil}) with a planar air-soil boundary. For a specular surface, the incoherent reflectivity is zero, and upon setting $s = 0$, the coherent reflectivity reduces to the Fresnel reflectivity Γ. Hence, the p-polarized brightness temperature of the soil medium when observed from air at an angle θ_i is†

$$T^p_{Bsoil}(\theta_i) = e^p_{soil}(\theta_i)\, T_0 = [1 - \Gamma^p(\theta_i)]T_0, \qquad (12.10)$$

(specular surface)

†**Computer Code 12.1.**

where T_0 is the physical temperature of the soil medium.

For a real soil medium, the physical temperature varies with depth, as does the dielectric constant ε_{soil}. Except for extremely dry soil conditions (volumetric moisture $m_v < 0.02$ g/cm^3), the penetration depth at 1 GHz and higher frequencies does not exceed a few centimeters. Hence, T_0 in Eq. (12.10) represents an effective value for the top few centimeters of the soil medium. The same is true for the dielectric constant ε_{soil} and, in turn, for the soil moisture m_v. Several studies were conducted in the 1970s and 1980s to evaluate the sensitivity of T_{Bsoil} to soil temperature and dielectric profiles (Njoku and Kong, 1977; Wilheit, 1978; Burke et al., 1979; Schmugge and Choudhury, 1981). The results indicate that even for very steep profiles, the deviation of T_{Bsoil} from the value computed assuming a uniform profile is only a few degrees (Ulaby et al., 1986a).

In Chapter 11, we presented plots displaying the dependence of ε_{soil} on m_v and plots of the specular reflectivities as a function of moisture at multiple angles of incidence. For thc convenience of the reader, these figures are reintroduced here as Fig. 12-2.

12-3 Rough-Surface Emissivity

12-3.1 I^{2}EM

In Section 10-7, we reported on a study that compared values of the backscattering coefficient σ^0 computed according to the improved integral equation method (I^{2}EM) with 3-D method-of-moment numerical computations. A similar investigation was carried out for the polarized emissivity e^p_{soil}. The investigative team (Chen et al., 2003), which evaluated the response of e^p_{soil} to m_v for p = v and h polarizations at $\theta_i = 30°$, considered rough surfaces with the roughness parameter ks varying from a very smooth surface with $ks = 0.22$ to a very rough surface with $ks = 2.576$. They also varied the surface correlation length l over a correspondingly wide range of values. We extract from this important study two major findings:

(a) Among four different analytical models (the small-perturbation model (SPM), the geometric-optics

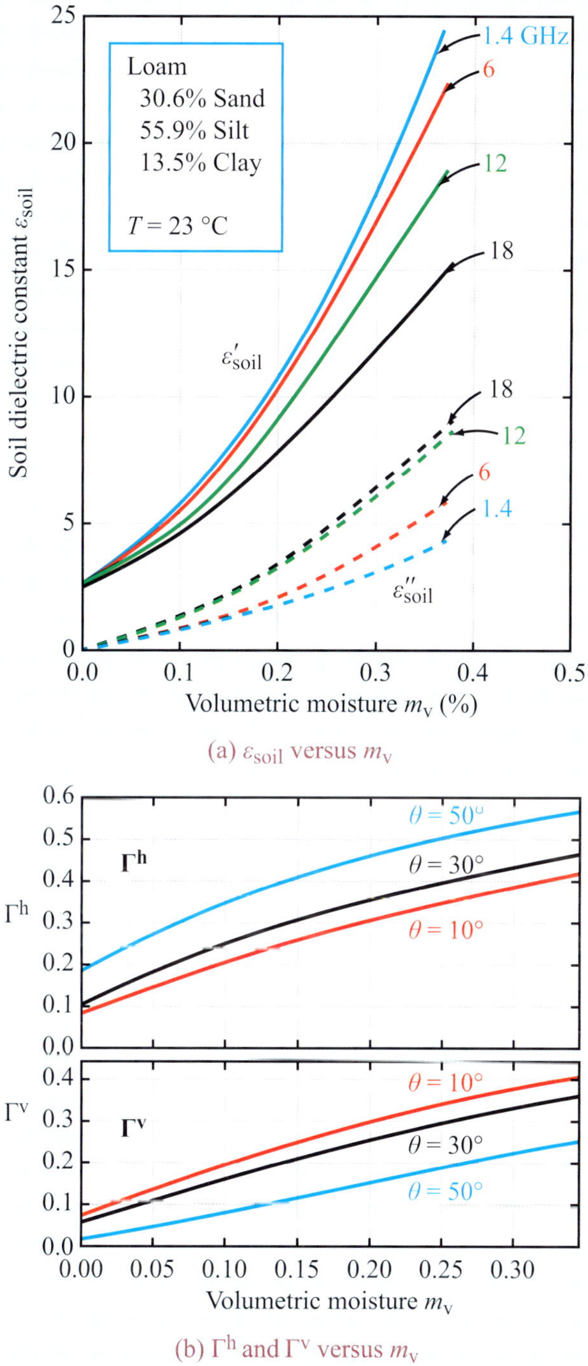

Figure 12-2: Variations of (a) ε_{soil} and (b) Γ^h and Γ^v with m_v for loamy soil.

model (GOM), the traditional integral-equation method (IEM), and the I^{2}EM), the only model that consistently provided excellent overall agreement with the numerical technique is I^{2}EM. This provides further confirmation for the applicability of I^{2}EM, not only for backscattering, but for bistatic scattering as well.[†]

(b) Figure 12-3 displays the variations of e^h_{soil} and e^v_{soil} with m_v for a smooth surface and a slightly rough surface, both computed using I^{2}EM at $\theta = 40°$.

The role of surface roughness is also evident in the experimental measurements shown in Fig. 12-4 for a wet soil surface observed at 1.4 GHz under three different soil surface-roughness conditions.

12-3.2 Semiempirical Models

Increasing the roughness of a surface generally increases its emissivity relative to that of a specular surface with the same dielectric constant. The change in emissivity (or total reflectivity) can be computed using Eqs. (12.3), (12.5), and (12.6), with the bistatic scattering coefficients in Eq. (12.6) computed using the I^{2}EM model of Section 10-3. The key surface parameters are the dielectric constant ε_{soil}, the rms height s, the correlation function $\rho(\xi)$, and the correlation length l. Alternatively, we can develop semiempirical models for the total reflectivity $\Gamma^p_{tot}(\theta_i)$, as discussed next.

Form of semiempirical model

The notion of describing the emissivity by a semiempirical model was first proposed by Choudhury et al. (1979) and later refined by other researchers (Wang and Choudhury, 1981; Shi et al., 2005; Escorihuela et al., 2007; Wigneron et al., 2011). The basic structure of the semiempirical model is

$$e^p_{soil}(\theta_i) = 1 - \Gamma^p_{rs}(\theta_i), \qquad (12.11a)$$

where Γ^p_{rs} is an ***effective rough-surface reflectivity*** given by

$$\Gamma^p_{rs}(\theta_i) = [(1-Q)\,\Gamma^p(\theta_i) + Q\,\Gamma^q(\theta_i)]e^{-h'\cos^n\theta_i},$$
$$(p = \text{h and } q = \text{v, or vice versa}), \qquad (12.11b)$$

[†]**Computer Code 12.2.**

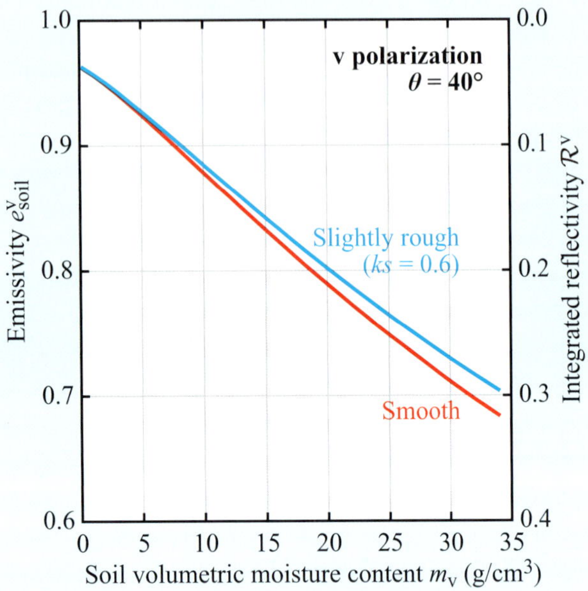

(a) v polarization

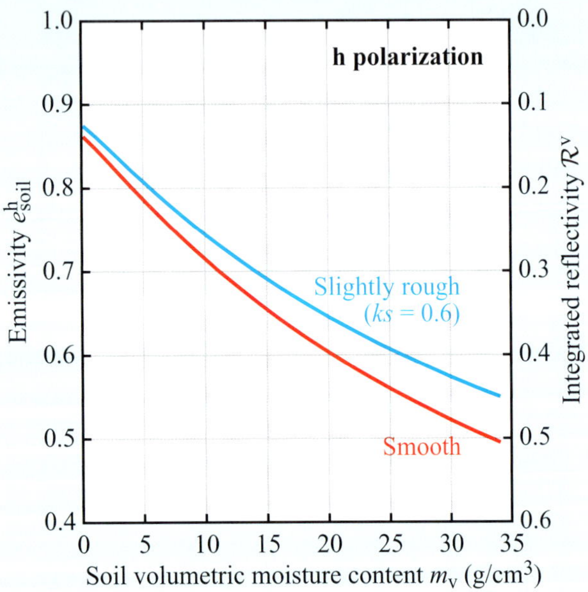

(b) h polarization

Figure 12-3: Variations of e^{v} and e^{h} at $\theta = 40°$ with m_v for two different surface roughnesses, both computed using I^{2}EM.

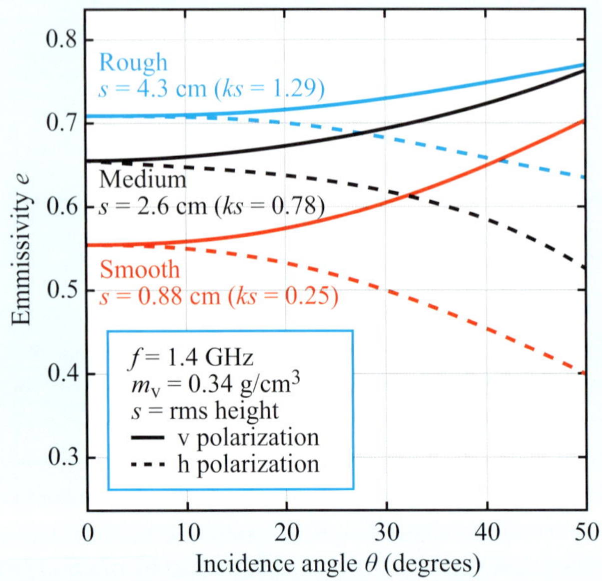

Figure 12-4: Angular patterns of the emissivity measured at 1.4 GHz for three bare-soil fields with different surface roughnesses [Newton and Rouse, 1980].

where Γ^{p} and Γ^{q} are Fresnel reflectivities, Q is a ***polarization mixing factor***, h' is an ***equivalent rms height***, and n is an ***angular exponent*** (which was treated as polarization-independent in early versions of the model, but then generalized into a polarization-dependent exponent in later versions). The rationale for Q is that when an h-polarized radiometer observes the surface, most of the h-polarized energy emitted by the surface originates as h-polarized energy incident upon the surface from below, but a small part of the h-polarized emitted energy originates as v-polarized energy incident upon the surface from below which gets converted to h-polarized energy through the process of transmission across the rough surface. Hence, the h-polarized rough-surface reflectivity Γ^{h}_{rs} includes a factor $(1 - Q)$ multiplying the h-polarized specular reflectivity Γ^{h} and a factor Q multiplying the v-polarized specular reflectivity Γ^{v}. *For a perfectly smooth surface,* $Q = 0$.

The factor $\exp[-h' \cos^{n} \theta_i]$ was selected by Choudhury et al. (1979) so as to resemble the function $e^{-4\psi^2}$ of the coherent reflectivity given by Eq. (12.5). Finally, the exponent n is a constant intended to describe the angular dependence of $\Gamma^p_{\rm rs}(\theta_{\rm i})$, beyond the angular variations inherent to $\Gamma^{\rm h}(\theta_{\rm i})$ and $\Gamma^{\rm v}(\theta_{\rm i})$.

To quantify the values of the three roughness-related parameters (Q, h', n) for various combinations of polarizations and microwave frequencies, Wang et al. (1983) used a data set composed of brightness-temperature measurements made at 1.4, 5, and 10.7 GHz, for both h and v polarizations. The measurements were acquired for three fields with very different surface roughnesses, but almost identical moisture contents. Figure 12-5 displays the angular patterns measured for the smoothest field with surface rms height $s = 0.21$ cm and for the roughest field with $s = 2.45$ cm. The process of minimizing the statistical error between the measured values of the emissivity and those calculated using the model given by Eq. (12.11) led to the values of Q and h' listed in Table 12-1. Moreover, the best fit they established for the angular coefficient n was $n = 0$, which corresponds to no dependence on $\theta_{\rm i}$ (beyond the dependencies included in $\Gamma^{\rm v}$ and $\Gamma^{\rm h}$). Using these tabulated parameters, they obtained reasonable agreement between the emissivity values calculated according to the model given by Eq. (12.11) and the emissivity values derived from the brightness temperatures recorded by their ground-based radiometers.

Over the next 30 years, numerous experiments, model simulations, and analyses were performed to refine the semiempirical model given by Eq. (12.11) by developing empirical expressions for the roughness parameters (Q, h', n) in terms of the frequency f (or wavelength λ) and surface roughness parameters. Because the sensitivity of the emissivity to surface roughness increases with increasing microwave frequency, and because the masking effect of vegetation cover also increases with increasing frequency, the L-band (1.4 GHz) has become the frequency of choice for mapping soil moisture with radiometry. Accordingly, numerous investigations have focused on modeling the emission by both bare-soil and vegetation-covered soil at 1.4 GHz. Moreover, two satellite missions have been configured with the express goal of mapping the distribution of soil moisture with

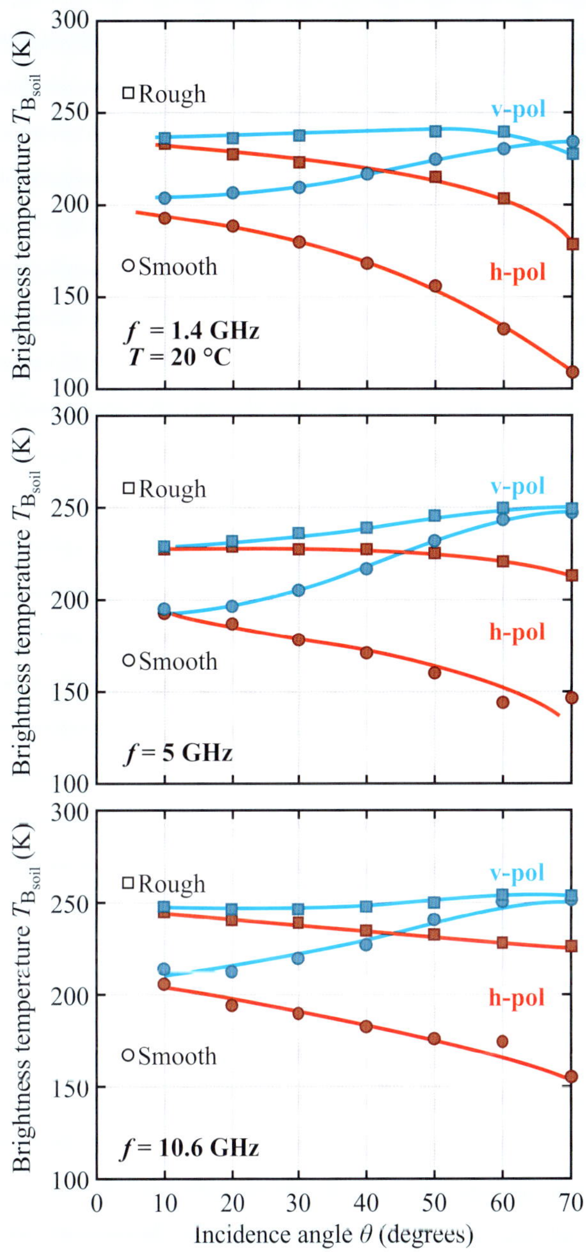

Figure 12-5: Measured brightness temperatures as a function of incidence angle at (a) 1.4 GHz, (b) 5 GHz, and (c) 10.7 GHz. The soil temperature for both smooth and rough fields was ~20 °C. The volumetric soil moisture content for the smooth field was ~0.25 g/cm^3 and for the rough field was ~0.26 g/cm^3 in the top 0 to 10 cm layer [Wang et al., 1983].

Table 12-1: Parameters derived from regression analyses (Wang et al., 1983).

f (GHz)	Field No.	Roughness Parameters				Statistics	
		s	ks	h'	Q	R^2	SEE (g/cm^3)
1.4	1	0.21 cm	0.06	0	0	0.95	0.026
	2	0.73 cm	0.21	0.10	0.01	0.97	0.019
	3	2.45 cm	0.72	0.53	0.12	0.83	0.024
5	1	0.21 cm	0.22	0	0	0.99	0.020
	2	0.73 cm	0.76	0.05	0.15	0.96	0.029
	3	2.45 cm	2.57	0.58	0.28	0.88	0.025
10.7	1	0.21 cm	0.47	0	0	0.97	0.025
	2	0.73 cm	1.64	0.11	0.20	0.94	0.031
	3	2.45 cm	5.50	0.60	0.30	0.81	0.029

microwave radiometers. A synopsis of the two missions is given later in Section 12-9. For the remainder of the present section, we examine the results of studies aimed at using and refining the semiempirical model given by Eq. (12.11) at L-band specifically, and at other frequencies more generally.

12-3.3 Model Parameters at L-Band

From Table 12-1, we note that at 1.4 GHz, $Q = 0$ for the smooth field, 0.01 for the field with a medium-rough surface, and 0.12 for the field with the very rough surface. Based on the investigation reported by Wang et al. (1983) and several follow-up investigations, the general consensus is that as a first-order approximation, Q can be set equal to zero for all surfaces, without incurring significant error. As we discuss shortly, the results of a more recent study suggest that Q can be empirically related to h' (instead of setting it equal to zero), yielding estimates of m_v with smaller errors.

PORTOS-93 is a large data set comprising brightness-temperature measurements made by ground-based radiometers at a site in France (Wigneron et al., 2001). The study area included 7 bare-soil fields covering a wide range of roughness conditions: s extended between 0.476 cm and 5.94 cm, and l extended between 3.15 cm and 20.6 cm. The fields were irrigated and then allowed to dry out, during which multitemporal radiometric observations were made at multiple microwave frequencies and at incidence angles of 10° to 40° in 10° increments. Using the L-band part of the data set, two promising formulations were proposed for (Q, h', n), both obtained by minimizing the error between measured or model-computed values of emissivity and the empirical model given by Eq. (12.11), for both $p =$ v and $p =$ h, at all four angles of incidence and for all moisture conditions represented in the data set.

(a) 2011 formulation

The study by Wingeron et al. (2011) led to the following parameterization:

- $Q = 0$
- $$h' = \left(\frac{9.4s}{8.865s + 2.29}\right)^6 \qquad (12.12a)$$

  or

  $$h' = 1.39m = 1.39\left(\frac{s}{l}\right) \qquad (12.12b)$$
- $$n = \begin{cases} 0.5 & \text{for h-pol,} \\ -1.5 & \text{for v-pol.} \end{cases} \qquad (12.12c)$$

The rms surface height s and correlation length l are in centimeters and m is the rms slope. With this formulation, surface roughness is characterized by a single parameter, either s or m, depending on whether h' is defined by Eq. (12.12a) or Eq. (12.12b), respectively.

The plot in Fig. 12-6(a) shows the variation of h' with s. The solid curve is an empirical fit given by Eq. (12.12a). The dashed curve represents the functional form of the roughness factor in the expression for the coherent reflectivity, namely $h' = 4\psi^2 = (2ks\cos\theta_i)^2$. The expression is suitable for $s < 1$ cm, but it overestimates the effective value of h' at values of s exceeding 1 cm.

In Fig. 12-6(b), we display the results reported by Wigneron et al. (2011) for the variation of h' with the rms slope of the surface, $m = s/l$. The linear relationship between h' and m is given by Eq. (12.12b). From the standpoint of scattering theory, it stands to reason that s and l are both important roughness parameters, and for a surface with a Gaussian height correlation function, the critical parameter is indeed the rms slope s/l. A similar dependence on m was proposed by Chen et al. (2010).

The final parameter evaluated in the study by Wigneron et al. (2011) is the power factor n. The results for h polarization are shown in Fig. 12-7(a). Except for the wide vertical spread at around $s = 0.5$ cm, most of the values of n are between 0 and 1. Consequently, most users assign $n_h = 0.5$ for h polarization. The vertical spread at around $s = 0.5$ cm is irrelevant because when s is that small, $h' \approx 0$ [see Fig. 12-6(a)], in which case the boundary becomes effectively a specular surface (the factor $e^{-h'\cos^{n}\theta_i} \approx 1$). A similar conclusion applies to v polarization [Fig. 12-7(b)], except that in that case a more appropriate value is $n_v = -1.5$.

(b) 2013 formulation

The results of a numerically generated data set based on the finite element method (FEM), together with reexamination of the PORTOS-93 data set, led Lawrence et al. (2013) to a model that does not set $Q = 0$ nor assign fixed values to n, and yet remains a single-parameter roughness model (in addition to soil moisture content). The equivalent rms height h' was found to vary with s, l, and $m = s/l$, but the relationship yielding the highest correlation was with an artificial surface

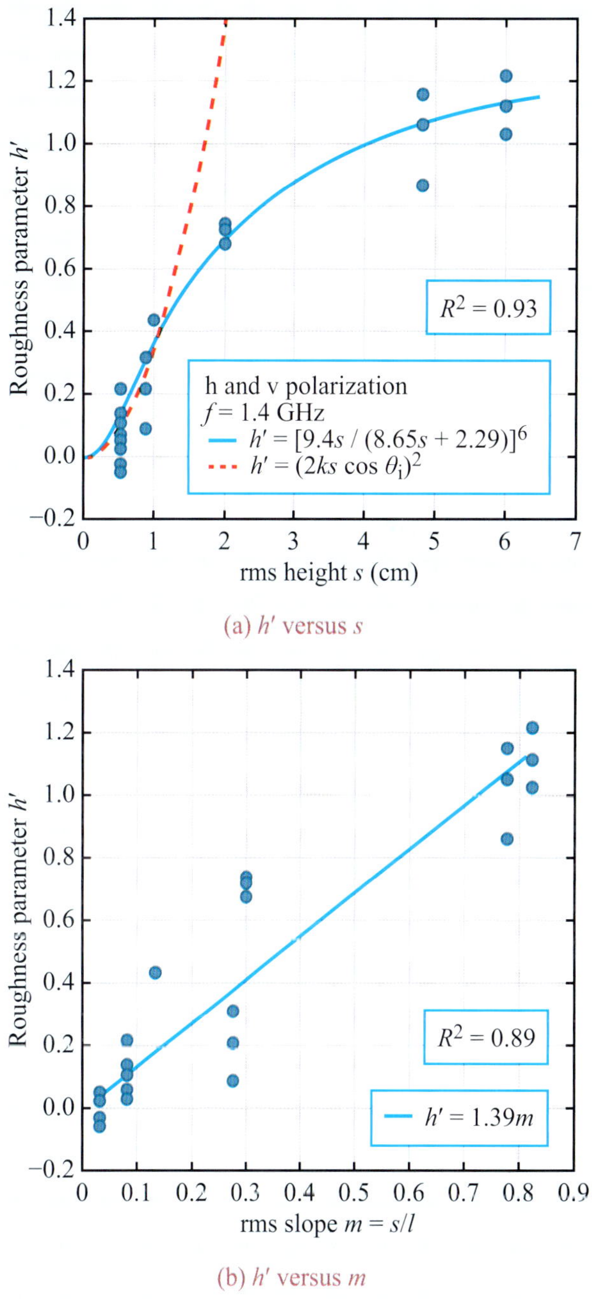

Figure 12-6: Retrieved values of h' plotted as a function of (a) the surface rms height s and (b) the rms slope $m = s/l$ [Wigneron et al., 2011].

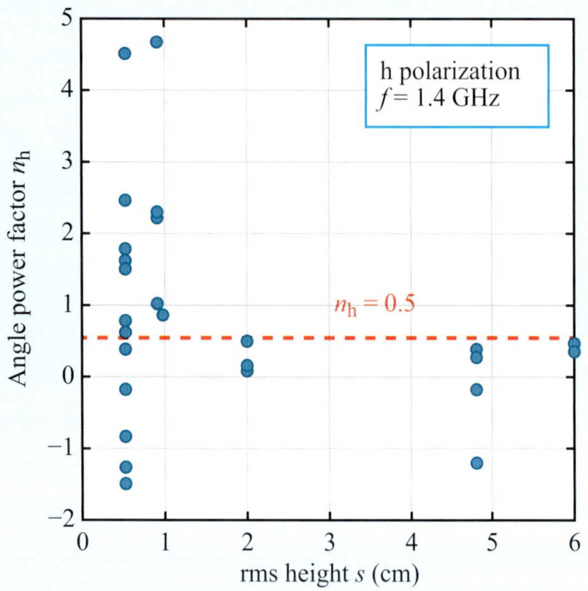

(a) h-pol

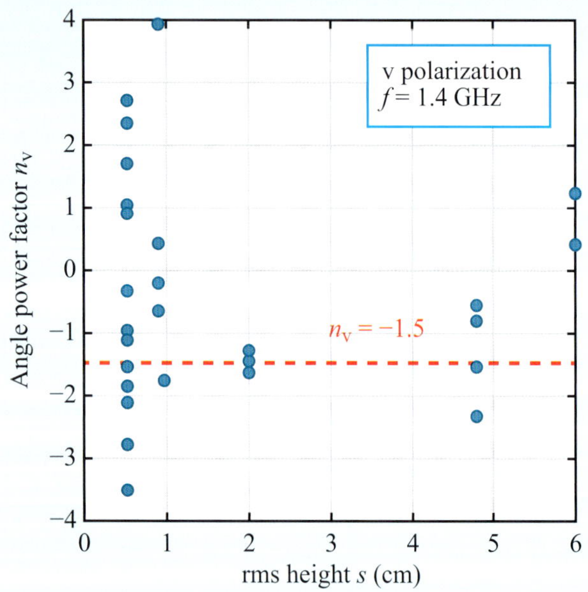

(b) v-pol

Figure 12-7: Retrieved values of (a) n_h and (b) n_v as a function of surface rms height s [Wigneron et al., 2011].

parameter defined as

$$Z_s = \frac{s^2}{l} \quad \text{(cm)}.$$

The solid line in Fig. 12-8(a) describes the relationship

$$\bullet\ h' = \begin{cases} 2.615\left(1 - \exp\left(\frac{-Z_s}{2.473}\right)\right) & \text{for } Z_s < 1.235 \text{ cm}, \\ 1.0279 & \text{for } Z_s > 1.235 \text{ cm}. \end{cases} \tag{12.13}$$

The remaining roughness parameters are related to h' by:

$$\bullet\ Q = 0.1771h', \tag{12.14a}$$

$$\bullet\ n = \begin{cases} 1.615\left(1 - \exp\left(\frac{-h'}{0.359}\right)\right) - 0.238 & \text{for v-pol}, \\ 0.767h' - 0.099 & \text{for h-pol}. \end{cases} \tag{12.14b}$$

These relationships are shown graphically in Fig. 12-8(b).

Application of the empirical relations† given by Eqs. (12.13) and (12.14) to the PORTOS-93 data set led to the results shown in Fig. 12-9, which compare the values of m_v retrieved from the measured brightness temperatures with the *in-situ* values measured in the field. These results presume that Z_s is known; in practice, however, it is not known.

12-3.4 Model Parameters at Other Frequencies

The Advanced Microwave Scanning Radiometer–Earth Observing System, or ***AMSRE*** for short, carries 6 radiometers that image the earth's surface at an incidence angle of 55°. Their center frequencies (see Table 1-4) extend from 6.92 GHz to 89 GHz. In an effort to characterize the expected emission from bare-soil fields at the first 5 radiometer frequencies (6.92 10.7, 18.7, 23.8, and 36.4 GHz), Shi et al. (2005) used a combination of emissivities measured in the PORTOS-93 field campaign (Wigneron et al., 2001) and emissivities computed using I^{2}EM. They then

†**Computer Code 12.3.**

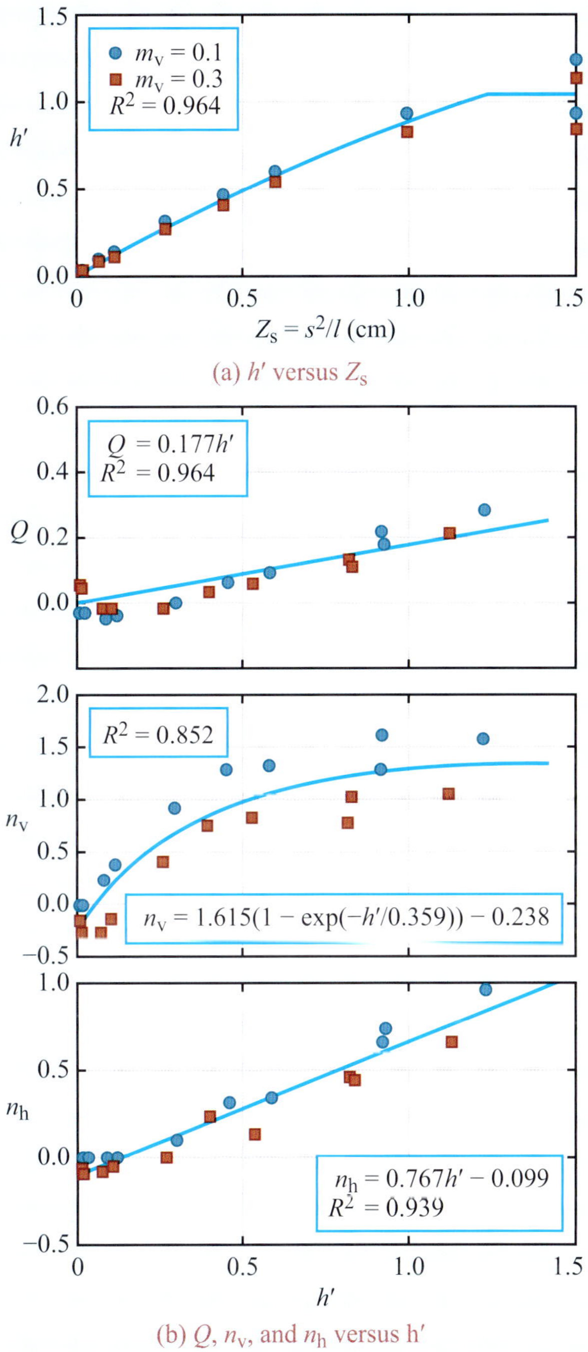

Figure 12-8: Semiempirical model fits given by Eqs. (12.13) and (12.14) [Lawrence et al., 2013] .

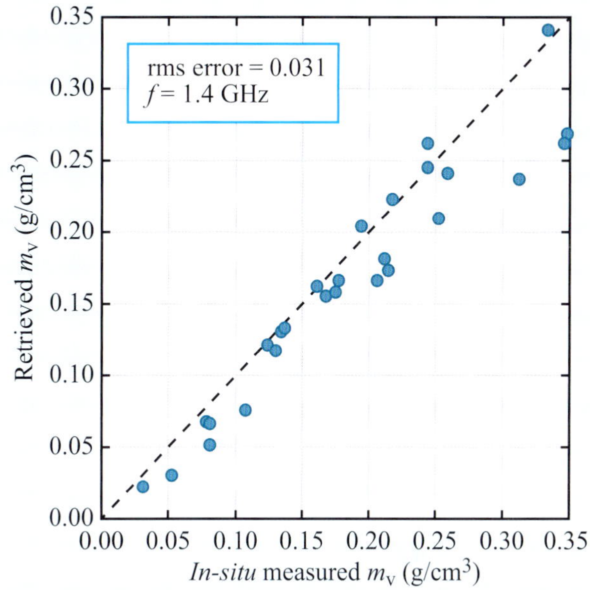

Figure 12-9: Comparison of *in-situ* values of m_v with radiometer-retrieved values using the PORTOS-93 measured values of emissivity and the model given by Eqs. (12.13) and (12.14) [Lawrence et al., 2013].

subjected the combined data set to the development of a semiempirical model given by

$$\Gamma^p_{rs}(\theta_i) = (1 - Q_p)\, \Gamma^p(\theta_i) + Q_p\, \Gamma^q(\theta_i), \qquad (12.15)$$

where $p = \text{v}$ and $q = \text{h}$, or vice versa. In this model, Q_p is a polarization-specific roughness parameter that depends on the microwave frequency f and the surface rms slope $m = s/l$. Initially, they developed the model specifically for the 10.7 GHz channel and arrived at the following expression for Q_p:

$$\left.\begin{aligned} Q_v(10.7) &= 3.22 + 2.45 \log m - 6.67m \\ Q_h(10.7) &= 5.60 + 3.10 \log m - 9.38m \end{aligned}\right\} \text{ at 10.7 GHz.} \qquad (12.16)$$

Plots of $Q_v(10.7)$ and $Q_h(10.7)$ as a function of m are shown in Fig. 12-10.

Next, expressions for $Q_p(f)$ at other frequencies were developed, in terms of the values of Q_p at 10.7 GHz:

$$Q_v(f) \approx Q_v(10.7), \qquad (12.17a)$$

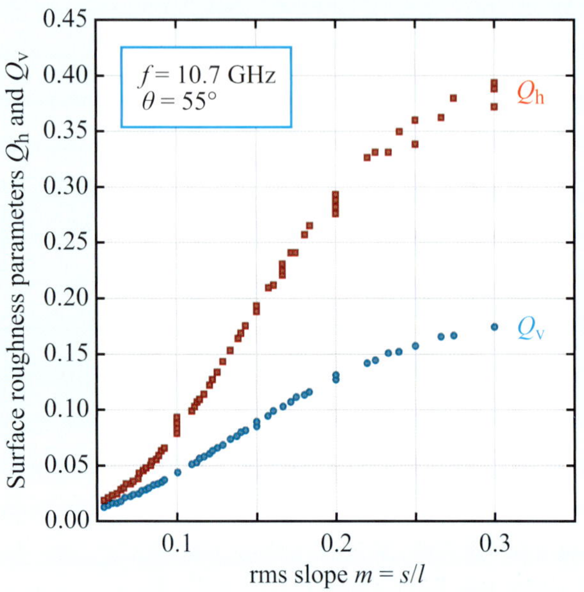

Figure 12-10: Relationships of Q_v and Q_h to rms slope $m = s/l$ at 10.7 GHz and 55° incidence [Shi et al., 2005].

and

$$Q_h(f) \approx \beta_h\, Q_h(10.7), \tag{12.17b}$$

$$\text{with} \quad \beta_h = \begin{cases} 0.958 \text{ at } 6.9 \text{ GHz}, \\ 1.013 \text{ at } 18.7 \text{ GHz}, \\ 1.015 \text{ at } 23.8 \text{ GHz}, \\ 1.026 \text{ at } 36.4 \text{ GHz}. \end{cases} \tag{12.17c}$$

The validity of the relationships given by Eqs. (12.16) and (12.17) was confirmed by an investigation carried out independently by a research team at the University of Tokyo in Japan (Kuria et al., 2006).

12-4 Emission by a Periodic Surface

This section on the emission by a periodic surface is analogous to Section 10-4.2 on the backscattering coefficient of a periodic surface. In fact, we have reproduced Fig. 10-23 as Fig. 12-11 to reintroduce the geometry of the periodic surface. To keep the formulation manageable, we restrict our consideration to

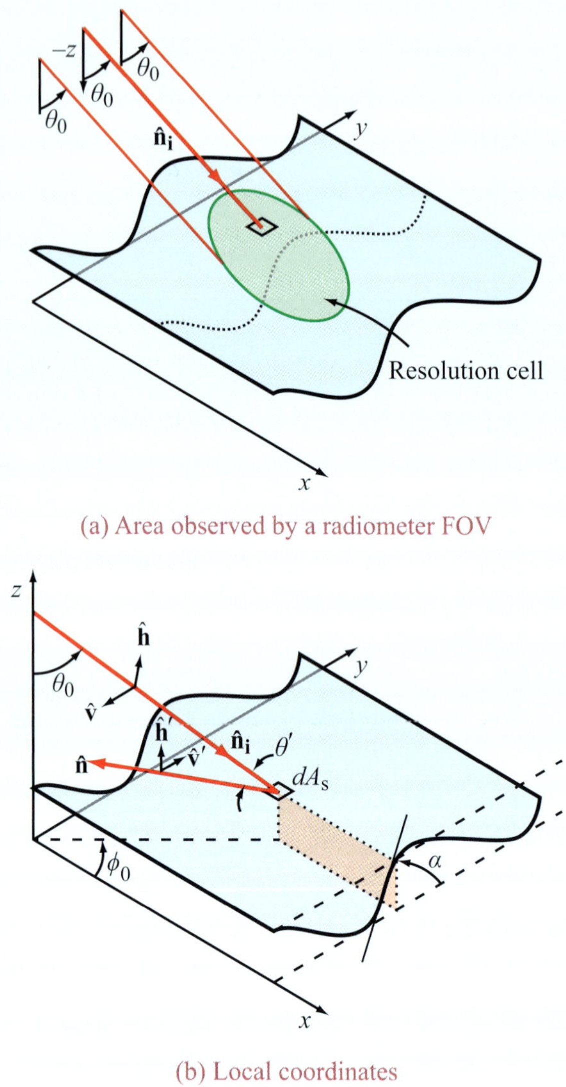

Figure 12-11: Periodic surface geometry.

a one-dimensional sinusoidal surface with height $\mathcal{Z}(y)$ given by

$$\mathcal{Z}(y) = A\left[1 + \cos\left(\frac{2\pi y}{\Lambda}\right)\right], \tag{12.18}$$

with amplitude A and spatial period Λ. With angles (θ_0, ϕ_0) as defined in Fig. 12-11, the p-polarized brightness temperature of the periodic surface is given by

$$T_{\rm B}(\theta_0,\phi_0;p)=\frac{\displaystyle\iint_{\Omega_{\rm M}} T_{\rm B}^{\rm SS}(\theta';p)\,F(\theta,\phi)\,d\Omega}{\displaystyle\iint_{\Omega_{\rm M}} F(\theta,\phi)\,d\Omega_{\rm f}}, \quad (12.19)$$

where $\Omega_{\rm M}$ is the main-beam solid angle of the antenna pattern, $F(\theta,\phi)$ is the normalized antenna gain along direction (θ,ϕ) relative to boresight, and $T_{\rm B}^{\rm SS}(\theta';p)$ is the p-polarized brightness temperature of the local ***small-scale surface*** with surface normal $\hat{\mathbf{n}}$ at angle θ' relative to the negative of the incident direction $\hat{\mathbf{n}}_{\rm i}$. In both the numerator and denominator, we treat $F(\theta,\phi)=1$ inside the main beam. To relate $T_{\rm B}(\theta_0,\phi_0;p)$ to the periodic surface given by Eq. (12.18), we convert the differential solid angles to differential areas. For a differential area dA of the undulating surface, the corresponding solid angle is

$$d\Omega=\frac{dA}{R^2}\cos\theta', \quad (12.20)$$

where R is the range between the radiometer antenna and the surface. In terms of $dx\,dy$,

$$dA=\frac{dx\,dy}{\cos\alpha}=dx\,dy\sec\alpha, \quad (12.21)$$

where α is the slope angle [Fig. 12-11(b)] associated with dA. Hence,

$$d\Omega=\frac{dx\,dy}{R^2}\sec\alpha\cos\theta'. \quad (12.22)$$

Both α and θ' are functions of y.

The denominator in Eq. (12.19) accounts for the solid angle of the main beam, so it represents the gain variation across a ***flat surface*** with differential solid angle $d\Omega_{\rm f}$ given by

$$d\Omega_{\rm f}=\frac{dA}{R^2}\cos\theta_0=\frac{dx\,dy}{R^2}\cos\theta_0. \quad (12.23)$$

Using Eqs. (12.22) and (12.23) in Eq. (12.19) (with $F(\theta,\phi)=1$) leads to

$$T_{\rm B}(\theta_0,\phi_0;p)=\frac{\displaystyle\int T_{\rm B}^{\rm SS}(\theta';p)\,\sec\alpha\cos\theta'\,dy}{\displaystyle\int\cos\theta_0\,dy}. \quad (12.24)$$

Assuming that the antenna footprint contains many spatial periods Λ, we can limit the integration to a single period, in which case Eq. (12.24) simplifies to

$$T_{\rm B}(\theta_0,\phi_0;p)=\frac{1}{\Lambda\cos\theta_0}\int_0^\Lambda T_{\rm B}^{\rm SS}(\theta';p)\,\sec\alpha\cos\theta'\,dy.^\dagger \quad (12.25)$$

For a flat surface with $A=0$, it follows that $\alpha=0$ and $\theta'=\theta_0$ for every point on the surface, in which case Eq. (12.25) reduces to $T_{\rm B}=T_{\rm B}^{\rm SS}$, as expected.

Suppose we have a nonundulating surface (either flat or with random roughness) with v-polarized and h-polarized brightness-temperature angular functions $T_{\rm B}(\theta';{\rm v})$ and $T_{\rm B}(\theta';{\rm h})$, respectively, where angle θ' defines the incidence direction relative to the surface normal. Next, suppose we superimpose this given surface onto the one-dimensional sinusoidal periodic surface given by Eq. (12.18). For a tilted differential surface area, the polarization directions $\hat{\mathbf{v}}'$ and $\hat{\mathbf{h}}'$ in its local coordinate system are, in general, different from $\hat{\mathbf{v}}$ and $\hat{\mathbf{h}}$ in the antenna coordinate system. The expressions for $\hat{\mathbf{v}}$, $\hat{\mathbf{h}}$, $\hat{\mathbf{v}}'$, and $\hat{\mathbf{h}}'$ are given by Eq. (10.28) in terms of θ_0, ϕ_0, and θ'. Additionally, θ' and α are given by Eqs. (10.22) and (10.26d), and $T_{\rm B}^{\rm SS}(\theta';{\rm h})$ and $T_{\rm B}^{\rm SS}(\theta';{\rm v})$ are given by (Ulaby et al., 1986a)

$$\begin{aligned}T_{\rm B}^{\rm SS}(\theta';{\rm h})&=(\hat{\mathbf{h}}\cdot\hat{\mathbf{h}}')^2\,T_{\rm B}(\theta';{\rm h})+(\hat{\mathbf{h}}\cdot\hat{\mathbf{v}}')^2\,T_{\rm B}(\theta';{\rm v}),\\ T_{\rm B}^{\rm SS}(\theta';{\rm v})&=(\hat{\mathbf{v}}\cdot\hat{\mathbf{h}}')^2\,T_{\rm B}(\theta';{\rm h})+(\hat{\mathbf{v}}\cdot\hat{\mathbf{v}}')^2\,T_{\rm B}(\theta';{\rm v}),\end{aligned} \quad (12.26)$$

where, as stated earlier, $T_{\rm B}(\theta';{\rm h})$ and $T_{\rm B}(\theta';{\rm v})$ pertain to the random surface in the absence of the periodic pattern. Identical expressions were obtained by Wang et al. (1980) through a different solution method.

† **Computer Code 12.4.**

To illustrate the effect of the periodic undulation of T_B, we present in Fig. 12-12 plots of $T_B(\theta_0)$ for a specular surface, and for two sinusoidal surfaces with the same spatial period $\Lambda = 100$ cm, but with different amplitudes. The plots in part (a) of the figure correspond to $\phi_0 = 0°$ (parallel look direction) and those in part (b) correspond to $\phi_0 = 90°$ (perpendicular look direction). An interesting feature of the periodic surface is that at nadir ($\theta_0 = 0$), the horizontally polarized brightness temperature at $\phi_0 = 0°$ is equal to the vertically polarized brightness temperature at $\phi_0 = 90°$; i.e., $T_B(0°, 0°; \text{h}) = T_B(0°, 90°; \text{v})$. In both cases, the electric-field vector points along the same direction (y direction in Fig. 12-11) relative to the geometry of the periodic pattern. The converse also is true: $T_B(0°, 90°; \text{h}) = T_B(0°, 0°; \text{v})$.

Experimental measurements demonstrating the effect of row direction were reported by Wang et al. (1980) for a sinusoidal surface with $A = 10$ cm and $\Lambda = 95$ cm. Their results are presented in Fig. 12-13. The measured data is in reasonably good agreement with theory.

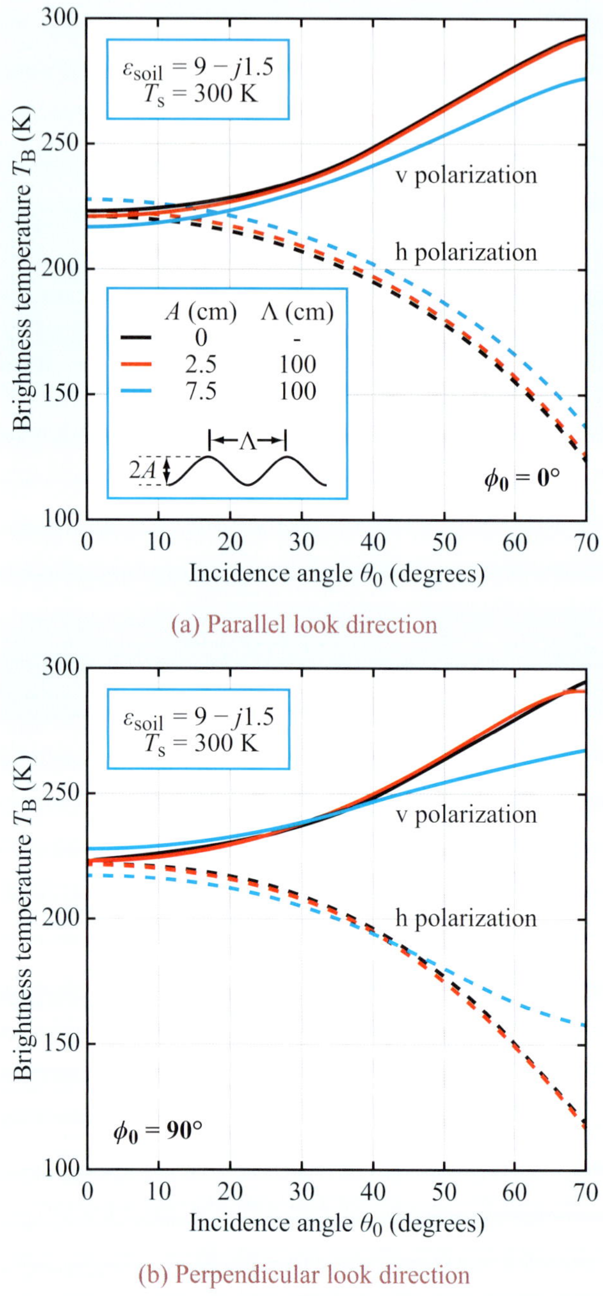

Figure 12-12: Calculated brightness temperature for sinusoidal surfaces with no random roughness. The look direction is (a) parallel to the rows and (b) perpendicular to the rows.

12-5 Radiative Transfer Equation for Vegetation-Covered Ground

The vegetation layer depicted in Fig. 12-14 is of depth d, with $z = 0$ denoting the top of the layer and $z = -d$ denoting the underlying ground surface. Our goal is to derive an expression for $T_{B_{veg}}(\theta_i)$, the brightness temperature of the ground cell observed by a radiometer pointed along angle θ_i. This brightness temperature accounts for only the emission by the vegetation-ground combination, with no regard for atmospheric effects. The transmission and emission contributions of the atmosphere are incorporated through Eq. (12.1).

Assuming the vegetation and underlying soil surface have the same physical temperature T_0, Eqs. (12.2) and (12.3) can be used to define the emissivity $e_{veg}(\theta_i)$ corresponding to $T_{B_{veg}}(\theta_i)$:

$$e_{veg}(\theta_i) = \frac{T_{B_{veg}}(\theta_i)}{T_0} . \qquad (12.27)$$

Of course, $T_{B_{veg}}$ and e_{veg} are associated with h polarization or v polarization.

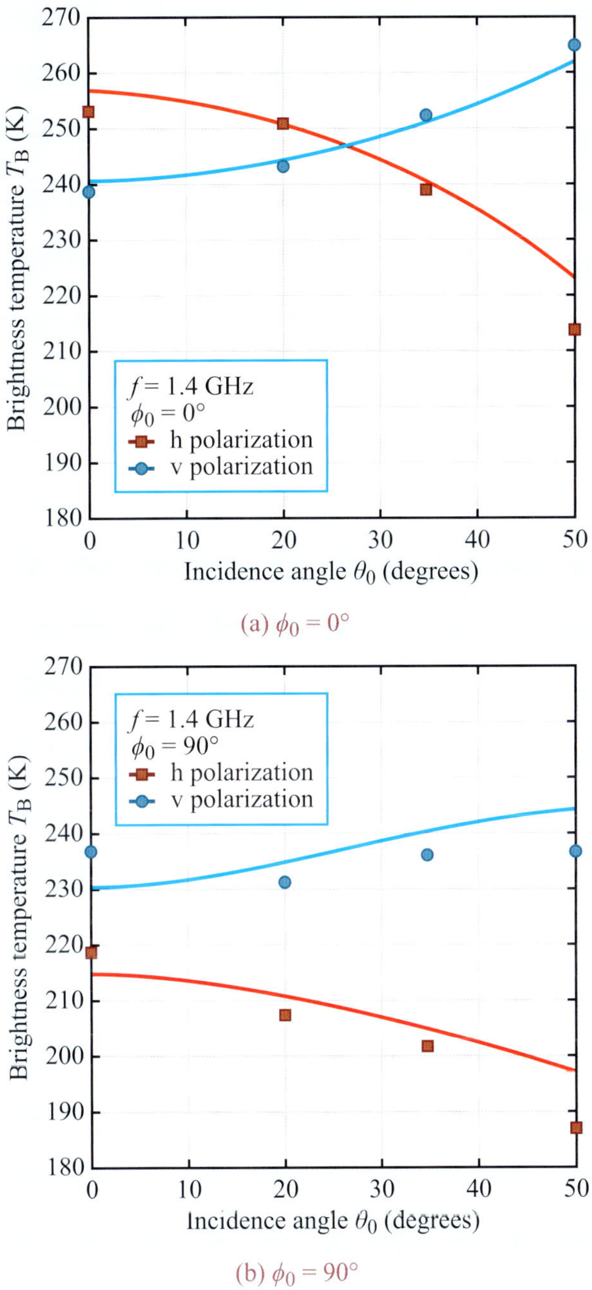

Figure 12-13: Comparison of measured and calculated brightness temperatures for soil with a gravimetric moisture of 30 percent, a sinusoidal periodic surface with a peak-to-peak amplitude of 20 cm, and a period of 95 cm [from Wang et al., 1980].

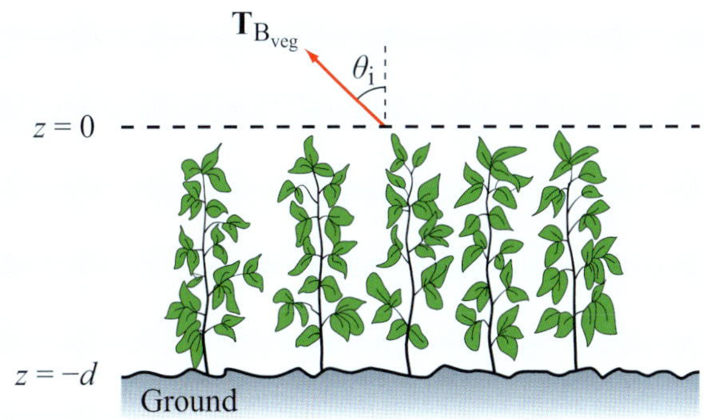

Figure 12-14: Microwave emission by a vegetation layer over a ground surface.

12-5.1 Scalar Radiative Transfer Equation

The radiative transfer formulation for the emission case is fundamentally similar to that for the radar case except that now we need to add a self-emission term to account for the emission by the vegetation layer. Also, the boundary conditions at $z = 0$ and $z = -d$ are different.

To keep the presentation relatively simple, we limit our derivation to the scalar case. Upon adding a self-emission term $\kappa_a J_a$ to the pair of equations given by Eq. (11.65), we have

$$\frac{d}{dz} I^+(\mu_s, \phi_s, z) = -\frac{\kappa_e}{\mu_s} I^+(\mu_s, \phi_s, z) + \frac{\kappa_a}{\mu_s} J_a(\mu_s, \phi_s, z) + \mathcal{F}^+(\mu_s, \phi_s, z), \tag{12.28a}$$

$$-\frac{d}{dz} I^-(-\mu_s, \phi_s, z) = -\frac{\kappa_e}{\mu_s} I^-(-\mu_s, \phi_s, z) + \frac{\kappa_a}{\mu_s} J_a(-\mu_s, \phi_s, z) + \mathcal{F}^-(-\mu_s, \phi_s, z), \tag{12.28b}$$

where I^+ and I^- are the ***upward-propagating*** and ***downward-propagating intensities*** in the middle layer (Fig. 12-15) and $\mu_s = \cos\theta_s$. *We have implicitly assumed that the propagation parameters of the canopy are polarization-independent.*

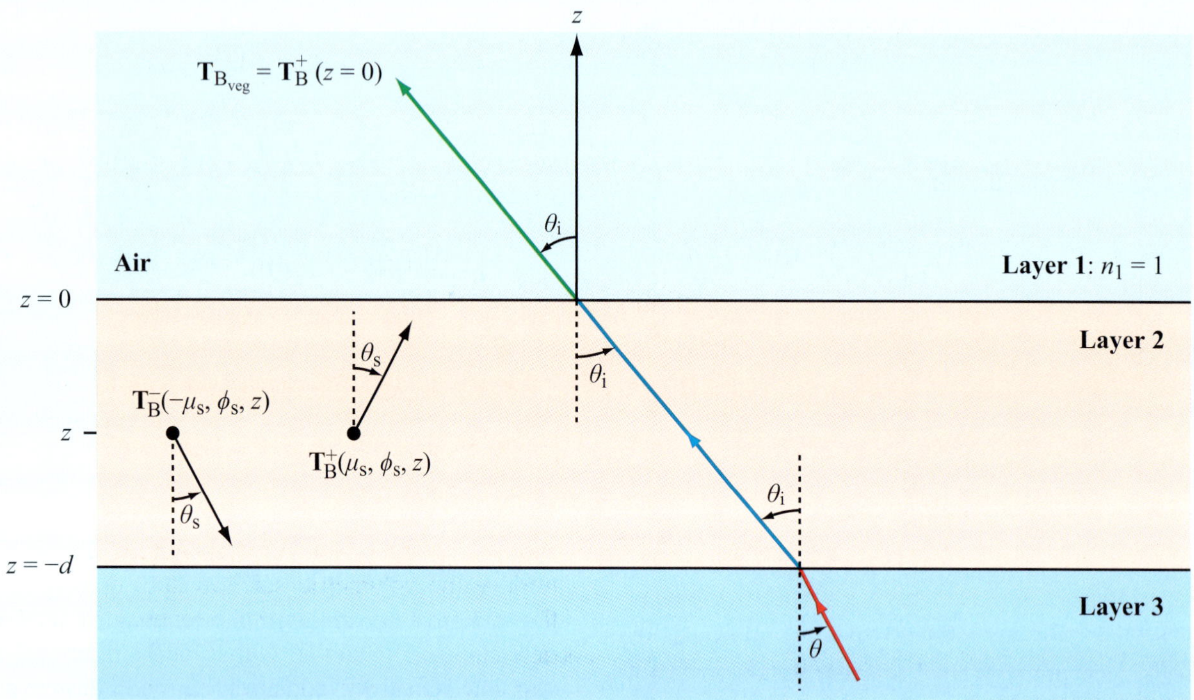

Figure 12-15: Downward- and upward-propagating brightness temperatures T_{B}^{+} and T_{B}^{-}.

Under the Rayleigh-Jeans approximation, Eqs. (6.56) and (6.57) are applicable, which allows us to express Eq. (12.28) in terms of the upward-propagating brightness temperature T_{B}^{+} and downward-propagating brightness temperature T_{B}^{-}, and the physical temperature T_0 (which is assumed to be uniform within the vegetation layer and the same as the soil's physical temperature):

$$\frac{d}{dz}\, T_{\mathrm{B}}^{+}(\mu_s,\phi_s,z) = -\frac{\kappa_e}{\mu_s}\, T_{\mathrm{B}}^{+}(\mu_s,\phi_s,z) + \frac{\kappa_a}{\mu_s}\, T_0 + \mathcal{F}^{+}(\mu_s,\phi_s,z), \tag{12.29a}$$

$$-\frac{d}{dz}\, T_{\mathrm{B}}^{-}(-\mu_s,\phi_s,z) = -\frac{\kappa_e}{\mu_s}\, T_{\mathrm{B}}^{-}(-\mu_s,\phi_s,z) + \frac{\kappa_a}{\mu_s}\, T_0 + \mathcal{F}^{-}(-\mu_s,\phi_s,z), \tag{12.29b}$$

The source functions $\mathcal{F}^{+}$ and $\mathcal{F}^{-}$ are given by

$$\mathcal{F}^{+}(\mu_s,\phi_s,z) = \frac{1}{\mu_s}\left[\int_0^{2\pi}\!\!\int_0^1 \psi(\mu_s,\phi_s;\mu',\phi')\, T_{\mathrm{B}}^{+}(\mu',\phi',z)\, d\Omega' + \int_0^{2\pi}\!\!\int_0^1 \psi(\mu_s,\phi_s;-\mu',\phi')\, T_{\mathrm{B}}^{-}(-\mu',\phi',z)\, d\Omega'\right], \tag{12.30a}$$

$$\mathcal{F}^{-}(-\mu_s,\phi_s,z) = \frac{1}{\mu_s}\left[\int_0^{2\pi}\!\!\int_0^1 \psi(-\mu_s,\phi_s;\mu',\phi')\, T_{\mathrm{B}}^{+}(\mu',\phi',z)\, d\Omega' + \int_0^{2\pi}\!\!\int_0^1 \psi(-\mu_s,\phi_s;-\mu',\phi')\, T_{\mathrm{B}}^{-}(-\mu',\phi',z)\, d\Omega'\right], \tag{12.30b}$$

Following the basic steps contained in Section 6-6.3, the differential equations can be converted into integral solutions:

$$T_B^+(\mu_s,z) = e^{-\kappa_e(z+d)/\mu_s}\, T_B^+(\mu_s,-d) + \int_{-d}^{z} e^{-\kappa_e(z-z')/\mu_s} \left[\frac{\kappa_a}{\mu_s}\, T_0 + \mathcal{F}^+(\mu_s,z')\right] dz', \tag{12.31a}$$

$$T_B^-(-\mu_s,z) = e^{\kappa_e z/\mu_s}\, T_B^-(-\mu_s,0) + \int_{z}^{0} e^{\kappa_e(z-z')/\mu_s} \left[\frac{\kappa_a}{\mu_s}\, T_0 + \mathcal{F}^-(-\mu_s,z')\right] dz', \tag{12.31b}$$

where, for simplicity, we suppressed the angle ϕ_s. Except for the self-emission term involving $\kappa_a T_0/\mu_s$, this pair of equations parallels the pair given by Eq. (11.68) for the radar case.

12-5.2 Boundary Conditions

Since the effective index of refraction of a vegetation layer is nearly 1, there is no abrupt discontinuity at the air-vegetation boundary, and hence no reflection-related boundary conditions to contend with at $z=0$. In Eq. (12.31b), $T_B^-(-\mu_s,0)$ represents downward-propagating radiation at the air-vegetation boundary. Since our treatment is confined to radiation sources emanating from the vegetation layer or the soil underneath it (i.e., no downwelling atmospheric contributions), and since no reflection occurs at the air-vegetation boundary, it follows that

$$T_B^-(-\mu_s,0) = 0. \tag{12.32}$$

At the quasi-specular lower boundary at $z=-d$, the p-polarized brightness temperatures T_B^+ and T_B^- are related by

$$T_B^+(\mu_s,-d) = \Gamma^p_{\text{coh}}(\mu_s)\, T_B^-(-\mu_s,-d) + e^p_{\text{soil}}(\mu_s)\, T_0, \tag{12.33}$$

where $\Gamma^p_{\text{coh}}(\mu_s)$ is the p-polarized coherent reflectivity of the soil surface at angle θ_s [see Eq. (12.5)], and $e^p_{\text{soil}}(\mu_s)$ is the emissivity of the soil surface, given by

$$e^p_{\text{soil}}(\mu_s) = 1 - \Gamma^p_{\text{tot}}(\mu_s), \tag{12.34}$$

with $\Gamma^p_{\text{tot}}(\mu_s)$ being the total reflectivity of the soil surface given by Eq. (12.4). For a specular surface with Fresnel reflectivity $\Gamma^p(\mu_s)$,

$$\Gamma^p_{\text{tot}}(\mu_s) = \Gamma^p_{\text{coh}}(\mu_s) = \Gamma^p(\mu_s). \tag{12.35}$$

(specular surface)

12-5.3 Weakly Scattering Medium

At frequencies above 10 GHz, the presence of a vegetation layer masks most of the emission contributed by the underlying surface. Exceptions include very short, very sparse, or very dry vegetation covers. To retrieve the volumetric soil moisture content m_v, it is then necessary to operate at a frequency, or combination of frequencies, in the 1 to 5 GHz band. Because vegetation elements contain water and because water is very lossy at the low end of the microwave spectrum, the albedo of the vegetation layer seldom exceeds 0.2, thereby qualifying the vegetation layer to be treated as a ***weakly scattering*** medium.

For a weakly scattering medium, we may ignore contributions related directly to the volume scattering coefficient κ_s, which is equivalent to setting the scattering phase function $\psi = 0$. This, in turn, leads to setting the source functions $\mathcal{F}^+ = \mathcal{F}^- = 0$. The consequent solution is known as the ***zeroth-order solution***.

Setting $T_B^-(-\mu_s,0) = 0$ and $\mathcal{F}^-(-\mu_s,z') = 0$ in Eq. (12.31b) and then evaluating the expression for $z=-d$ leads to:

$$\begin{aligned} T_B^-(-\mu_s,-d) &= \int_{-d}^{0} e^{-\kappa_e(d+z')/\mu_s}\, \frac{\kappa_a T_0}{\mu_s}\, dz' \\ &= \frac{\kappa_a T_0}{\mu_s}\, e^{-\kappa_e d/\mu_s} \int_{-d}^{0} e^{-\kappa_e z'/\mu_s}\, dz' \\ &= \frac{\kappa_a T_0}{\kappa_e}\,(1-e^{-\kappa_e d/\mu_s}) \\ &= (1-a)T_0(1-\Upsilon_{\text{veg}}), \end{aligned} \tag{12.36}$$

where we used the relations

$$a = \kappa_s/\kappa_e, \tag{12.37a}$$

$$\frac{\kappa_a}{\kappa_e} = 1 - a, \tag{12.37b}$$

and

$$\Upsilon_{veg} = e^{-\kappa_e d/\mu_s}. \tag{12.37c}$$

Next, we use Eq. (12.36) in Eq. (12.33) to obtain

$$T_B^+(\mu_s, -d) = \Gamma^p_{coh}(\mu_s)\,[(1-a)T_0(1-\Upsilon_{veg})] + e^p_{soil}(\mu_s)\,T_0. \tag{12.38}$$

Finally, we set $\mathcal{F}^+ = 0$ and insert Eq. (12.38) into Eq. (12.31a), and we evaluate it at $z = 0$ and $\theta_s = \theta_i$:

$$\begin{aligned} T_B^+(\theta_i, 0) &= e^{-\kappa_e d/\mu_i}\{\Gamma^p_{coh}(\mu_i)\,[(1-a)T_0(1-\Upsilon_{veg})] + e^p_{soil}T_0\} \\ &\quad + \int_{-d}^{0} e^{\kappa_e z'/\mu_i}\,\frac{\kappa_a}{\mu_i}\,T_0\,dz' \\ &= \Upsilon_{veg}\{\Gamma^p_{coh}(1-a)(1-\Upsilon_{veg})T_0 + e^p_{soil}T_0\} \\ &\quad + (1-a)T_0(1-\Upsilon_{veg}). \end{aligned} \tag{12.39}$$

Upon setting

$$T_B^+(\theta_i, 0) = T^p_{B_{veg}}(\theta_i) = e^p_{veg}(\theta_i)\,T_0, \tag{12.40}$$

we obtain the following expression for the p-polarized ***zeroth-order radiative transfer*** (ZRT) emissivity $e^p_{veg}(\theta_i)$:[†]

$$e^p_{veg}(\theta_i) = (1 + \Gamma^p_{coh}(\theta_i)\,\Upsilon_{veg})(1-a)(1-\Upsilon_{veg}) + \Upsilon_{veg}e^p_{soil}(\theta_i), \tag{12.41}$$

(ZRT model)

with Υ_{veg} given by Eq. (12.37c),

$$\Gamma^p_{coh} = \Gamma^p e^{-(2ks\cos\theta_i)^2}, \tag{12.42}$$

and

$$e^p_{soil} = 1 - \Gamma^p_{tot}, \tag{12.43}$$

where $k = 2\pi/\lambda$, s is the rms height, Γ^p is the Fresnel reflectivity of the soil surface, and Γ^p_{tot} is the integrated reflectivity given by

$$\Gamma^p_{tot} = \Gamma^p_{coh} + \Gamma^p_{inc}, \tag{12.44}$$

with the incoherent reflectivity Γ^p_{inc} defined by the integral expression given by Eq. (12.6). At low microwave frequencies, Γ^p_{tot} may be approximated as consisting of only a coherent-like component that can be modeled by the semiempirical rough surface expression given by Eq. (12.11b):

$$\begin{aligned} \Gamma^p_{tot}(\theta_i) &= \Gamma^p_{rs}(\theta_i) \\ &= [(1-Q)\,\Gamma^p(\theta_i) + Q\,\Gamma^q(\theta_i)]e^{-h'\cos^n\theta_i}. \end{aligned} \tag{12.45}$$

(p = h and q = v, or vice versa)

A similar approximation is applied to Eq. (12.42):

$$\Gamma^p_{coh}(\theta_i) = \Gamma^p(\theta_i)\,e^{-h'\cos^n\theta_i}. \tag{12.46}$$

For a perfectly specular soil surface with Fresnel reflectivity $\Gamma^p(\theta_i)$, the expressions for the coherent reflectivity and soil emissivity reduce to $\Gamma^p_{coh} = \Gamma^p$ and $e^p_{soil}(\theta_i) = 1 - \Gamma^p(\theta_i)$, in which case Eq. (12.41) simplifies to

$$e^p_{veg}(\theta_i) = (1 + \Gamma^p(\theta_i)\,\Upsilon_{veg})(1-a)(1-\Upsilon_{veg}) + [1 - \Gamma^p(\theta_i)]\Upsilon_{veg}. \tag{12.47}$$

(ZRT model with specular soil boundary)

The components of Eq. (12.41) include three contributions:

(a) $\Upsilon_{veg}e^p_{soil}$: emission by the soil surface (represented by e^p_{soil}), reduced by Υ_{veg}, the one-way transmissivity of the canopy.

(b) $(1-a)(1-\Upsilon_{veg})$: upward emission by the slanted column extending between the ground surface and the top of the canopy, along direction θ_i.

[†] **Computer Code 12.5.**

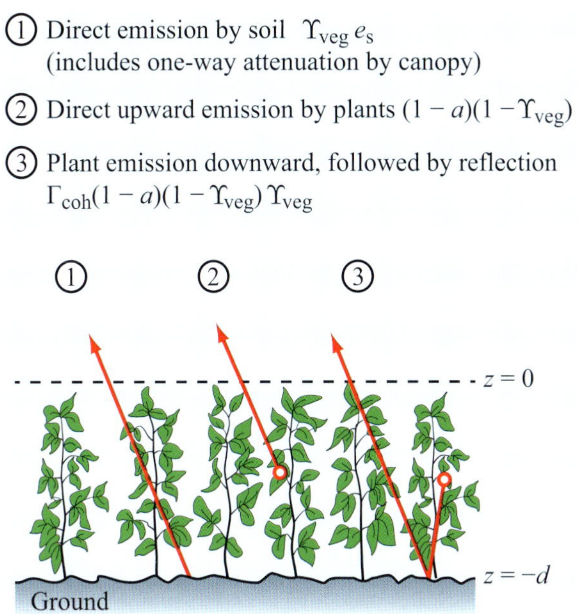

Figure 12-16: Emission contributions in a weakly scattering vegetation canopy.

(c) $\Gamma^p_{coh}(1-a)(1-\Upsilon_{veg})\Upsilon_{veg}$: downward emission by a slanted column, followed by reflection by the ground surface and then transmission through the vegetation layer to the upper boundary.

The three contributions are diagrammed in Fig. 12-16.

For an opaque vegetation layer with $\Upsilon_{veg} \approx 0$, the emissivity reduces to

$$e^p_{veg}(\theta_i) = 1 - a. \tag{12.48}$$

(opaque canopy with $\Upsilon \ll 1$)

Variations on the radiative transfer solution include:

(a) extending the model to first-order by using the zeroth-order solutions for T^+_B and T^-_B as the source functions $\mathcal{F}^+$ and $\mathcal{F}^-$ respectively, and then proceeding to solve for a new pair of T^+_B and T^-_B.

(b) solving the radiative transfer equations in vector form and incorporating 4×4 scattering phase functions $\boldsymbol{\psi}$, thereby accounting for polarization-dependent scattering and absorption in the vegetation layer.

(c) solving the radiative transfer equation numerically because the scattering albedo $a > 0.4$.

Relevant sources include the books by Tsang et al. (1985), Ulaby et al. (1986a), and Fung (1994).

If $T_{DN}(\theta_i)$ due to downwelling sky radiation is incident upon the canopy from above, propagation through the vegetation layer to the ground surface entails multiplication by Υ_{veg}. When followed by quasi-specular reflection and retransmission to the air-vegetation boundary, the net contribution is

$$T^p_{SS} = \Upsilon^2_{veg}\Gamma^p_{coh} T_{DN}. \tag{12.49}$$

Upon inserting the expression for T^p_{SS} in Eq. (12.1) and using Eq. (12.41) for e^p_{veg}, we obtain the following expression for the brightness temperature representing the energy incident upon the radiometer antenna

$$T^p_B = \Upsilon_a\{[(1+\Gamma^p_{coh}\Upsilon_{veg})(1-a)(1-\Upsilon_{veg}) + \Upsilon_{veg}e^p_{soil}]T_0 + \Upsilon^2_{veg}\Gamma^p_{coh}T_{DN}\} + T_{UP}, \tag{12.50}$$

where Υ_a and T_{UP} pertain to the atmospheric layer between the ground cell and the radiometer antenna.

12-6 ZRT Model for Layer with Distinct Upper Boundary

For a layer with a district upper boundary, such as snow over ground or ice over water, reflections (and refractions) occur at both boundaries of the middle layer. If we treat the boundaries as quasi-specular, the brightness temperature representing the upward emission along direction θ_i (Fig. 12-17) is given by (Ulaby et al., 1981a, p. 243):†

† Computer Code 12.6.

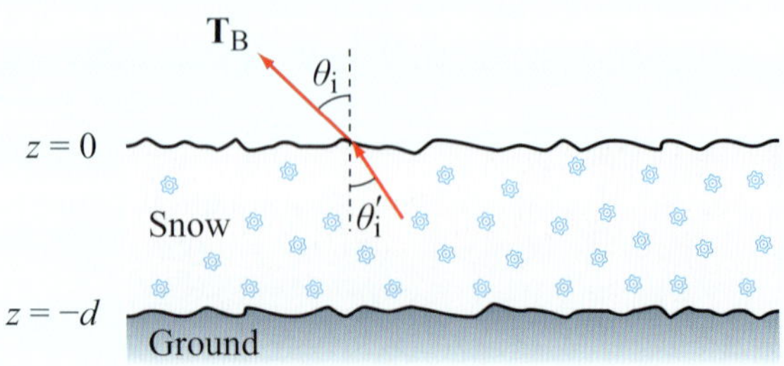

Figure 12-17: Middle layer, representing snow over soil or ice over water, with distinct boundaries.

$$T_B^p(\theta_i) = \left(\frac{1-\Gamma_{12}^p}{1-\Gamma_{12}^p\Gamma_{23}^p\Upsilon^2}\right) \cdot [(1+\Gamma_{23}^p\Upsilon)(1-a)(1-\Upsilon)T_2 + (1-\Gamma_{23}^p)\Upsilon T_3], \tag{12.51}$$

where $\Gamma_{12}^p(\theta_i)$ is the p-polarized reflectivity of the upper boundary, $\Gamma_{23}^p(\theta_i')$ is the reflectivity of the lower boundary at angle θ_i', where θ_i' is the refraction angle in the middle layer and is related to θ_i by Snell's law, and T_2 and T_3 are the physical temperatures of the middle and bottom layers, respectively. The front term multiplying the square-bracketed term accounts for the multiple reflections by the two boundaries.

For a vegetation layer, the upper boundary is diffuse, which means that $\Gamma_{12}^p \approx 0$. Moreover, if $T_2 = T_3 = T_0$, Eq. (12.51) leads to the following expression for the emissivity of the vegetation-ground combination

$$\begin{aligned} e_{\rm veg}^p(\theta_i) &= \frac{T_B^p(\theta_i)}{T_0} \\ &= (1+\Gamma^p(\theta_i)\,\Upsilon_{\rm veg})(1-a)(1-\Upsilon_{\rm veg}) \\ &\quad + [1-\Gamma^p(\theta_i)]\Upsilon_{\rm veg}, \end{aligned} \tag{12.52}$$

where we replaced Γ_2 with $\Gamma^p(\theta_i)$ and added the subscript "veg" to Υ. The expression given by Eq. (12.52) is identical with the ZRT model expression given by Eq. (12.47). Thus, the ZRT vegetation model is a simplified version of the more general model given by Eq. (12.51).

12-7 Applicability of the ZRT Vegetation Model

Upon inserting the semiempirical model given by Eqs. (12.45) and (12.46) into Eq. (12.52), the ZRT vegetation model becomes

$$\begin{aligned} e_{\rm veg}^p(\theta_i) &= (1+\Upsilon_{\rm veg}\Gamma^p e^{-\psi'})(1-a)(1-\Upsilon_{\rm veg}) \\ &\quad + \Upsilon_{\rm veg}\left\{1-[(1-Q)\Gamma^p + Q\Gamma^q]e^{-\psi'}\right\}, \end{aligned} \tag{12.53}$$

where $\psi' = h'\cos^n\theta_i$.

For a specified set of radiometer parameters (frequency f, incidence angle θ_i, and antenna polarization p), the emissivity is a function of the following scene parameters:

(a) Soil parameters:

- Soil dielectric constant: $\varepsilon_{\rm soil}$
- Soil surface roughness: Q, h', and n

(b) Vegetation parameters:

- Single-scattering albedo: a
- Nadir optical thickness $\tau_0 = \kappa_e d$.

If the operating frequency is 1.4 GHz, the number of surface roughness parameters can be reduced from 3 to 1 by adopting the empirical model proposed by Lawrence et al. (2013), as described earlier in Section 12-3.3. The model provides empirical expressions for Q and n in terms of h'. Consequently, the total number of scene parameters is reduced to 3: h', a, and τ_0.

To evaluate the sensitivity of the emissivity $e_{\rm veg}^p(\theta_i)$ to the various soil and vegetation parameters, while simultaneously keeping the analysis manageable, let us limit the analysis to the lower end of the microwave spectrum (1 to 5 GHz band). After examining the ZRT model for a vegetation layer over a specular soil surface, we extend our examination to a medium-rough soil-surface scenario.

12-7.1 Model Behavior for Specular Soil Surface

For $a = 0.05$ and a specular soil surface with $\Gamma^p_{\text{coh}} = \Gamma^p$, the number of scene parameters reduces to two: $\varepsilon_{\text{soil}}$ and τ_0. The angular variation of $e^{\text{h}}_{\text{veg}}(\theta_{\text{i}})$ is plotted in Fig. 12-18 with the attenuation τ_0 as a variable parameter. Parts (a) and (b) are for a relatively dry and a very wet soil surface, respectively. The nadir optical depth is varied between zero (no vegetation) and 1 Np. For reference, at L-band $\tau_0 \approx 0.7$ Np for a mature corn canopy, so $\tau_0 > 1$ Np is associated with denser canopies or higher frequencies.

We deduce from the plots in Fig. 12-18 that:

(a) The emissivity increases with increasing optical thickness. This is because the emission contributed by the addition of the vegetation canopy is greater than the reduction to the soil emissivity.

(b) For all vegetation-covered cases, as θ_{i} approaches $90°$, e^{h} approaches $(1 - a)$.

(c) Because the emissivity of wet soil is much lower than that for dry soil, the dynamic range between the angular plot for $\tau_0 = 0$ and that for $\tau_0 = 1$ Np in Fig. 12-18(b) is much greater than the corresponding pair of plots in Fig. 12-18(a), but in both cases the vegetation effectively masks the underlying soil surface when τ exceeds 1 Np.

12-7.2 Model Behavior for Moderately Rough Soil Surface

In parts (a) and (b) of Fig. 12-19, we display the response of the emissivity to soil moisture m_{v} for a soil medium with a smooth and a moderately rough surface, respectively. In each case, we use τ_0 as a parameter. The general behavior is the same for the smooth and rough soil surfaces, except that the slope of the emissivity plot for the smooth soil surface is steeper than that for the rough surface. If we approximate the curves in Fig. 12-19 as straight lines, we can compute the ***soil-moisture sensitivity*** as

$$S_{m_{\text{v}}} = \left| \frac{\partial e_{\text{veg}}}{\partial m_{\text{v}}} \right| \approx \left| \frac{e_{\text{veg}}(m_{\text{v}} = 0.35) - e_{\text{veg}}(m_{\text{v}} = 0.05)}{0.35 - 0.05} \right|. \tag{12.54}$$

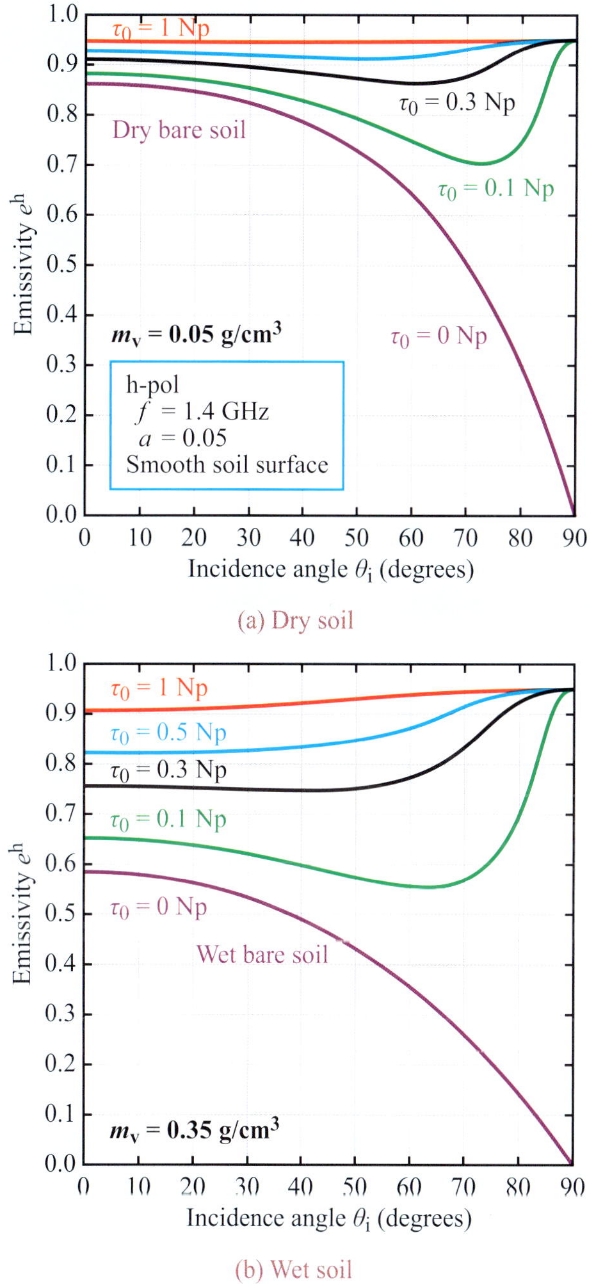

Figure 12-18: Angular plots of the h-polarized emissivity for (a) a dry soil surface and (b) a very wet soil surface, covered with vegetation of nadir optical thickness τ_0. The soil surface is perfectly smooth.

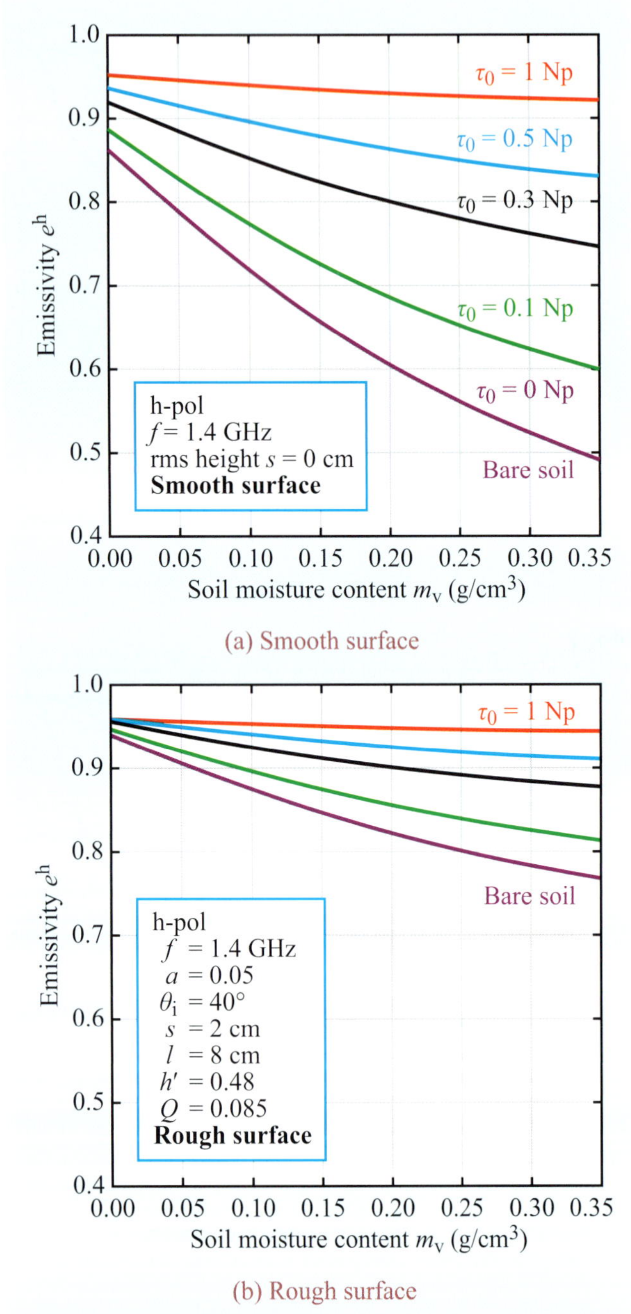

Figure 12-19: Variation of h-polarized emissivity with soil moisture content for (a) a smooth soil surface and (b) a moderately rough soil surface.

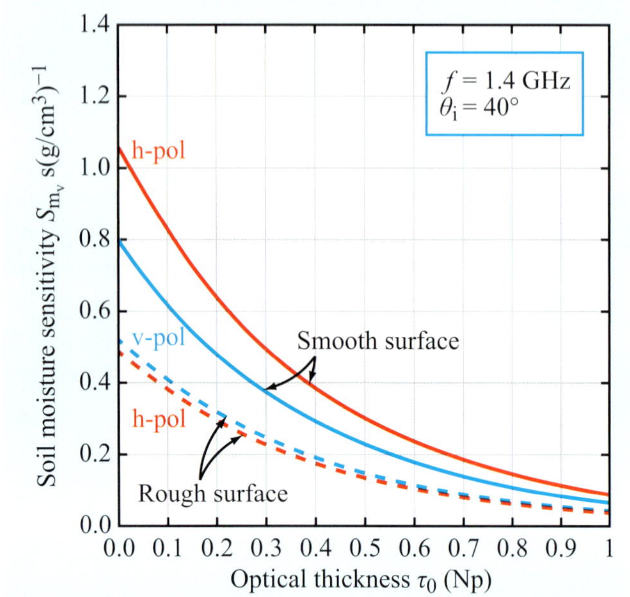

Figure 12-20: Reduction in soil-moisture sensitivity due to vegetation. The plots are based on approximating the family of curves in Fig. 12-19 with straight lines.

The exercise leads to the plots shown in Fig. 12-20. For the smooth-surface case, S_{m_v} decreases from 1.05 for h-pol and 0.8 for v-pol at $\tau_0 = 0$ to less than 0.1 at $\tau_0 = 1$ Np. The corresponding reduction in soil moisture sensitivity for the rough soil surface is from 0.5 at $\tau_0 = 0$ to 0.05 at $\tau_0 = 1$ Np.

12-7.3 Experimental Observations

The data shown in Fig. 12-21 were recorded by an L-band truck-mounted radiometer. Over the relatively short measurement period, the soil moisture content varied between 0.08 g/cm^3 and 0.39 g/cm^3, while the vegetation canopy parameters (physical temperature, height, and water content) remained approximately constant. The measurements were made at $\theta_i = 0°$ (nadir-looking), which means that $\Gamma^v = \Gamma^h$, thereby rendering Q in Eq. (12.53) irrelevant and setting $\psi' = h'$. The continuous curve in Fig. 12-21 is based on the ZRT model with $a = 0.04$ and $h' = 0.08$. These two parameters were treated as free parameters whose values

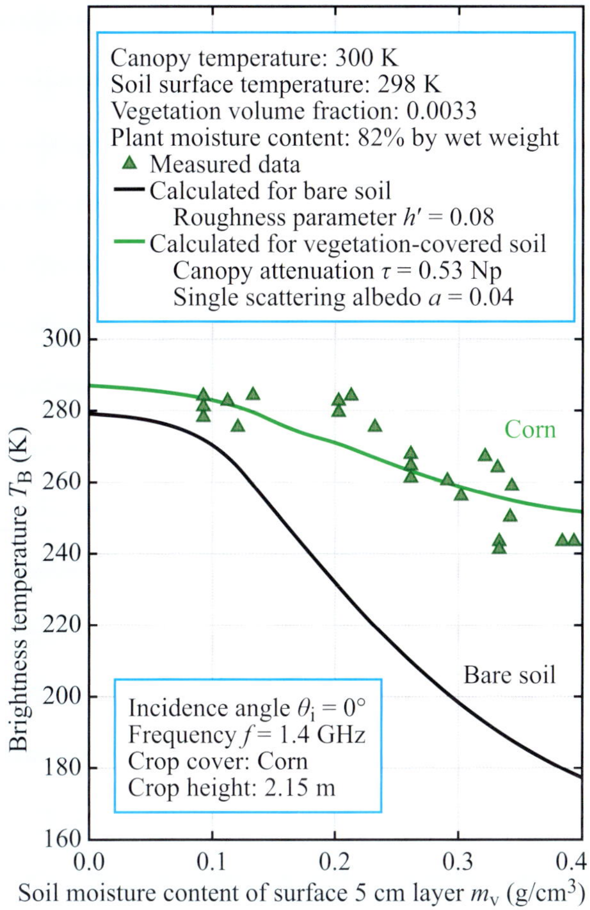

Figure 12-21: Variation of nadir looking brightness temperature with soil moisture content for fields of mature corn [Ulaby et al., 1981b].

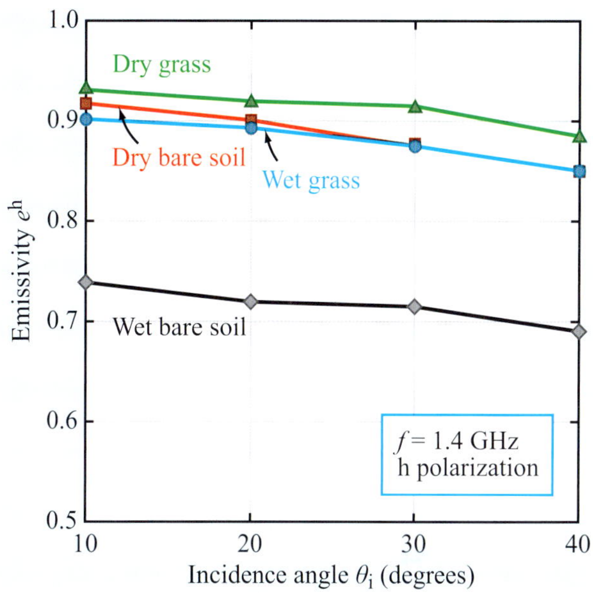

Figure 12-22: Measured emissivities at 1.4 GHz for bare and grass-covered soil, under wet and dry soil conditions [Jackson and Schmugge, 1991].

were selected such that the error between the model and the measured data is a minimum. For reference, we also show a computed plot of T_B versus m_v for a bare soil surface with the same value of h'.

If we approximate the variation of T_B with m_v over the range between 0.1 g/m^3 and 0.4 g/cm^3 by a straight-line equivalent, the slope of such a line defines the radiometric soil moisture sensitivity given by Eq. (12.54). That is,

$$S_{m_v} = \left| \frac{\partial e}{\partial m_v} \right| = \left| \frac{1}{T_0} \frac{\partial T_B}{\partial m_v} \right| .$$

For the bare-soil plot in Fig. 12-21, $S_{m_v} \approx 1$ (g/cm^3)$^{-1}$, compared with ≈ 0.35 (g/cm^3)$^{-1}$ for the vegetation-covered soil.

The masking effect of vegetation cover is illustrated by the data reported by Jackson and Schmugge (1991) at L-band; Fig. 12-22 shows angular plots for bare soil under wet and dry conditions, and grass-covered soil, also under wet and dry conditions. After measuring the emissivities of both fields using a truck-mounted radiometer (under dry-soil conditions), the two fields were sprinkler irrigated for two hours, and then their emissivities were measured again the next day. The plots in Fig. 12-22 indicate that it is difficult to distinguish the dry bare-soil emissivity from the grass-covered emissivity under either soil moisture condition.

To evaluate the applicability of the ZRT model, a defoliation experiment was conducted by Brunfeldt and Ulaby (1986) in which:

(1) brightness-temperature measurements were made for mature corn and soybeans canopies,

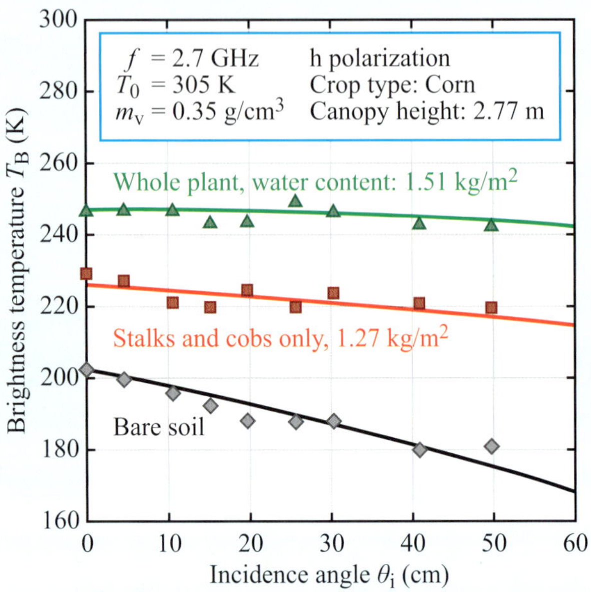

Figure 12-23: Emission by a corn canopy undergoing progressive stages of defoliation [Brunfeldt and Ulaby, 1984a, 1986].

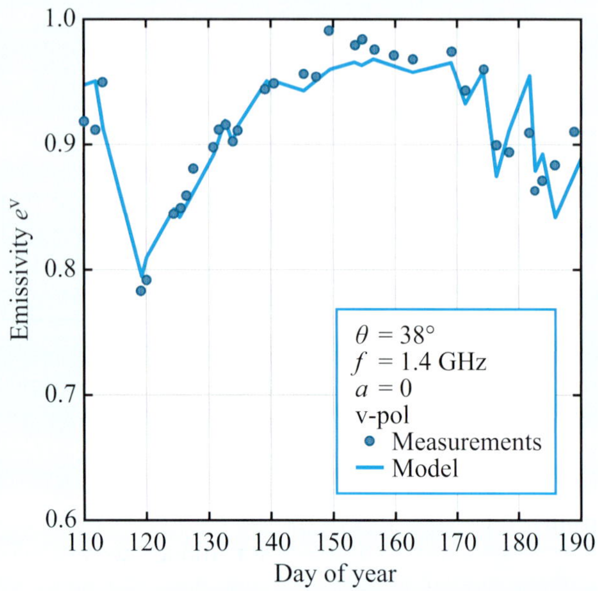

Figure 12-24: Comparison of simulated and measured wheat emissivities during the PORTOS-1993 at 1.4 GHz over the growing season [Wigneron et al., 1995].

(2) the leaves of the corn or soybeans plants were then removed and the brightness-temperature measurements were repeated, and then finally

(3) the stalks were removed and new brightness-temperature measurements were made of the remaining bare-soil surface.

An example of their results at 2.7 GHz is shown in Fig. 12-23. The bare-soil data provided good estimates of the soil surface reflectivity at every angle of incidence. Using these estimates, the ZRT model was then fitted to the full canopy and to the no-leaves canopy by treating the single scattering albedo a and the canopy attenuation τ_0 as free variables. The model appears to provide a good fit to the experimental data.

Using a multifrequency microwave radiometer mounted atop a 20 m high boom, the French group at INRA (Institut National de la Recherche Agronomique) made brightness-temperature measurements over a 2.5-month-long period for fields planted in wheat and soybeans (Wigneron et al., 1995). Figure 12-24 provides a comparison of the measured data and the values computed using the ZRT model. The value of the albedo used in the model is $a = 0$ and τ_0 was empirically related to the measured integrated canopy water content M_w (as discussed in Section 12-7.5). The model-calculated plot appears to track the experimental data very well.

12-7.4 Single-Scattering Albedo a

Assuming the ZRT model to be valid, several investigators generated estimates of the single-scattering albedo a by fitting the model to experimental observations of the emission from vegetation canopies (Mo et al., 1982; Brunfeldt and Ulaby, 1986; Pampaloni and Paloscia, 1986; Jackson, 1993; Wegmüller et al., 1995; Wigneron et al., 1995). The general consensus is that at 1.4 GHz, a varies over the range between 0.03 and 0.15, depending on vegetation type and structure. Even though for a given canopy structure the albedo a may have different values for h and v polarizations (Brunfeldt and Ulaby, 1986; Wigneron et al., 1995), the difference is small.

▶ Consequently, in most soil-retrieval studies, a is treated as polarization-independent. ◀

12-7.5 Vegetation Optical Thickness τ_0

The transmissivity is related to the optical thickness τ_0 by

$$\Upsilon = e^{-\tau_0 \sec\theta_i}.$$

In Section 11-8.6 we defined the ***integrated canopy water content*** M_w as the total mass of water contained in a vertical column of the canopy per unit ground surface area. Since a vegetation canopy is, in essence, a water cloud, it is reasonable to expect τ_0 to vary as a function of M_w. Values of τ_0 estimated from brightness-temperature measurements by matching the measurements to the ZRT model were used by Schmugge et al. (1986) and Brunfeldt and Ulaby (1986) to generate the results shown in Fig. 12-25 at 1.4 GHz and 2.7 GHz, respectively. Linear regression leads to

$$\tau_0 = 0.115 M_w, \qquad \text{at 1.4 GHz} \qquad (12.55a)$$

(for grass, corn, and soybeans)

and

$$\tau_0 = 0.36 M_w, \qquad \text{at 2.7 GHz for soybeans}. \qquad (12.55b)$$

In a similar study reported by Saleh et al. (2007) for a site dominated by clover grass, linear regression of the data shown in Fig. 12-26 led to

$$\tau_0 = 0.06 M_w - 0.003, \text{ at 1.4 GHz for clover grass}. \qquad (12.56)$$

The coefficient of M_w in Eq. (12.56) is approximately half the value of the coefficient determined by Schmugge et al. (1986) at the same microwave frequency [Eq. (12.55a)]. This large difference suggests that τ_0 is governed not only by the water content M_w, but also by the geometrical structure of the vegetation.

The frequency dependence of τ_0 was evaluated by Jackson and Schmugge (1991), who postulated that

$$\tau_0 = b M_w \qquad (12.57a)$$

and

$$b = b' f^x, \qquad (12.57b)$$

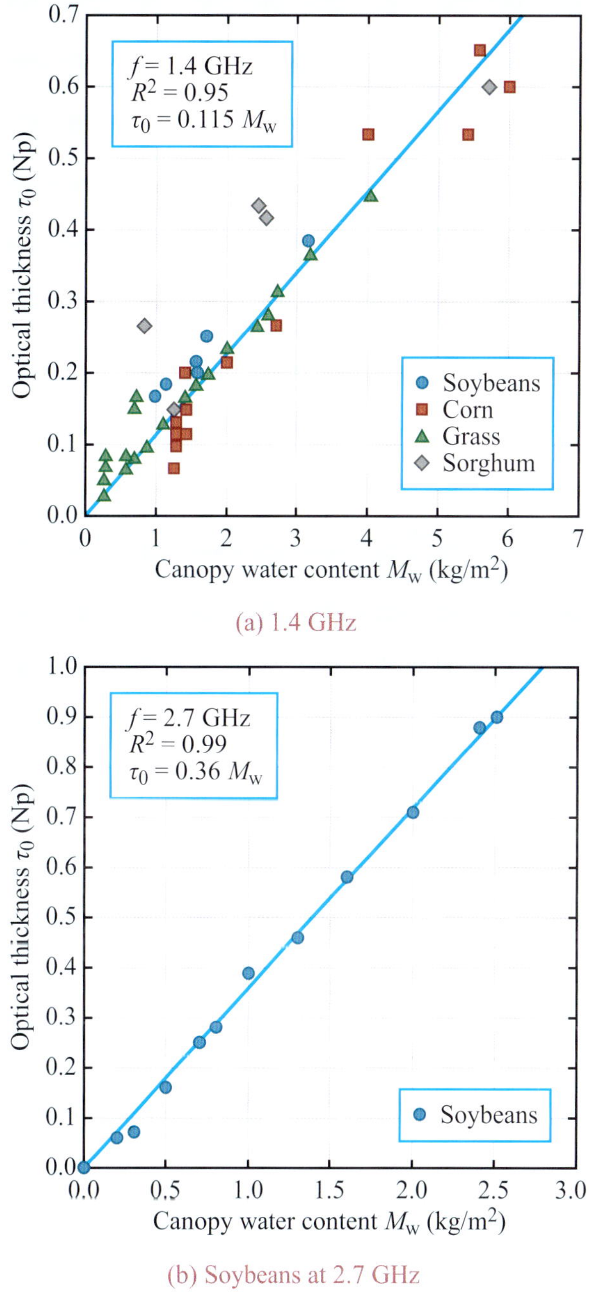

Figure 12-25: Variation of optical thickness τ_0 with vegetation water content for (a) corn-, grass-, and soybean-covered fields at 1.4 GHz and (b) soybeans at 2.7 GHz [Brunfeldt and Ulaby, 1986].

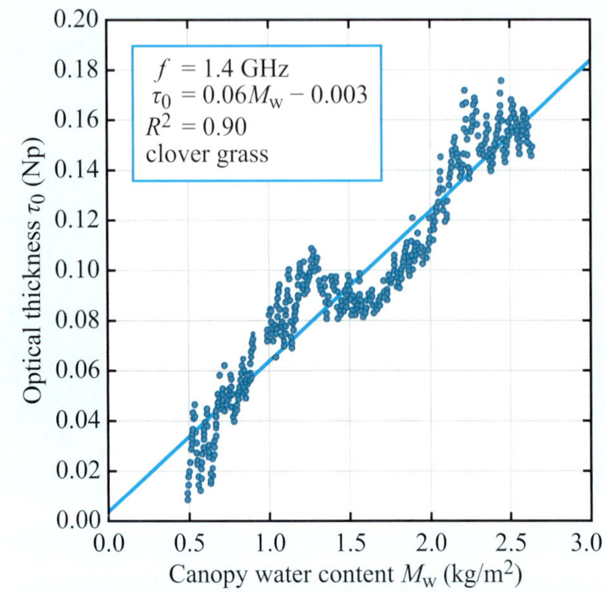

Figure 12-26: Attenuation (optical thickness) τ_0 versus vegetation water content for a site covered with clover grass [Saleh et al., 2007].

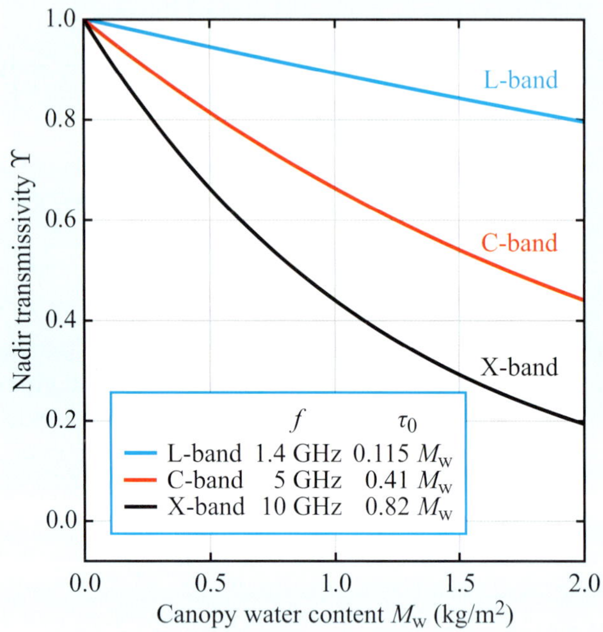

Figure 12-27: Representative canopy transmissivity for a grass canopy at three microwave frequencies.

where b is a frequency-dependent constant, b' is a vegetation-type specific constant and the exponent x describes the frequency dependence of b. To fit various data to the model, x varied between 1.08 for wheat to 1.38 for soybeans and 1.47 for oats. To illustrate the role of frequency, we show in Fig. 12-27 the variation of the nadir transmissivity Υ as a function of canopy water content at three microwave frequencies. The selected values of b are representative of vegetation as a class; the real transmissivity depends on the specific crop under consideration.

12-8 Estimation of Soil Moisture and Vegetation Water Content

In its simplest form, the ZRT model given by Eq. (12.53) consists of two soil-related parameters, namely the roughness parameter h' and the dielectric constant $\varepsilon_{\text{soil}}$, and two vegetation-related parameters, a and τ_0.

To estimate the volumetric soil moisture content m_v from the brightness temperature $T_{B_{\text{sat}}}$ measured by a satellite-borne radiometer, it is necessary to execute the following steps:

(1) Estimate vegetation brightness temperature $T_{B_{\text{veg}}}$

The brightness temperature measured by the satellite radiometer is given by

$$T_{B_{\text{sat}}} = \Upsilon_a [T_{B_{\text{veg}}} + \Upsilon^2_{\text{veg}} \Gamma_{\text{coh}} T_{\text{DN}}] + T_{\text{UP}}, \qquad (12.58)$$

where Υ_a is the atmospheric transmissivity and T_{DN} and T_{UP} are the downward-emitted and upward-emitted atmospheric brightness temperatures. All three atmosphere-related quantities can be estimated from atmospheric profile data provided by atmospheric satellite sensors (see Chapter 9). At L-band, $\Upsilon_a \approx 0.99$ and T_{DN} and T_{UP} are typically on the order of a few degrees. Since $\Upsilon_{\text{veg}} \leq 1$ and $\Gamma_{\text{coh}} \leq 1$, the term $\Upsilon^2 \Gamma_{\text{coh}} T_{\text{DN}}$ is very small in magnitude relative to $T_{B_{\text{veg}}}$, so it can be ignored altogether. The bottom line is: it should be possible to estimate $T_{B_{veg}}$ with excellent accuracy.

(2) Estimate vegetation emissivity e_{veg}

To convert $T_{B_{veg}}$ to e_{veg}, the physical temperature T_0 of the canopy (which is assumed to be the same as that of the soil) should be known or estimated. Several empirical algorithms are currently available for estimating T_0 on the basis of radiometric satellite observations made at 18.7 or 37 GHz (Owe and Van de Griend, 2001; Jones et al., 2007). The conversion leads to

$$e_{veg} = \frac{T_{B_{veg}}}{T_0}. \qquad (12.59)$$

(3) Determine ground cover

Using optical satellite sensors or other ancillary information, it is possible to determine whether the observed pixel is bare, vegetation-covered, or other (such as urban or water).

(a) If the scene is a bare-soil surface, the emissivity is governed by a minimum of two parameters, namely h' and ε_{soil}. Hence, two uncorrelated measurements are needed to invert the bare-soil emissivity model so as to estimate ε_{soil}, from which m_v can be estimated using dielectric models (Section 4-8). The two measurements may be made at two different frequencies, two different incidence angles, or two polarizations if θ_i is significantly larger than zero. Otherwise, the value of h' can be arbitrarily selected, in which case the estimated value of m_v might not be accurate.

(b) If the scene is vegetation-covered and the vegetation type is known (from optical imagery, for example), it is possible to assign a value to the scattering albedo a on the basis of past experiments, and to estimate τ_0 from

$$\tau_0 = bM_w = bcL, \qquad (12.60)$$

where b is a vegetation type-specific constant and c is a conversion factor between canopy water content M_w and leaf area index L (see Fig. 11-33). With such assignments, it may be possible to invert Eq. (12.53) to generate an estimate of m_v.

► Ideally, if 4 or more independent channels (multiple combinations of angles, frequencies, and polarizations) of radiometric data are available, it may be possible to invert the model without recourse to optical data or arbitrary assignment of parameter values. ◄

12-8.1 Single-Channel Soil Moisture Retrieval

The data presented in Fig. 12-28 are one of the earliest demonstrations comparing radiometer-retrieved values of m_v with *in-situ* measurements. They are based on airborne observations made at 1.4 GHz. Because only one channel of data was available, it was necessary to use the type of vegetation cover and other factors deduced from matching the ZRT model to the measured emissivities in the inversion process.

The ***PSR*** is a versatile, airborne microwave imaging radiometer that operates at multiple frequencies extend-

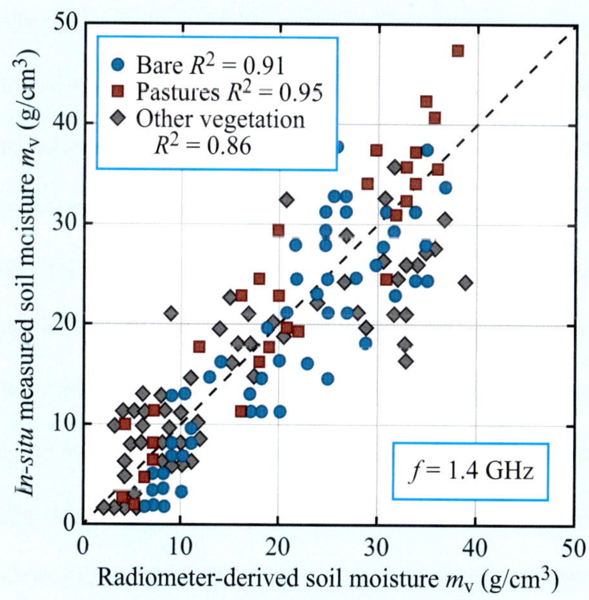

Figure 12-28: Measured values of soil moisture compared with those estimated from airborne observations of 1.4 GHz emissivities for a test site in central South Dakota [Schmugge, 1983].

ing between 6 GHz and 10.75 GHz (Piepmeier and Gasiewski, 2001b). One of its imaging modes involves conical scanning at a fixed incidence angle of 55°. The PSR was used in 2002 to image agricultural sites in Iowa, dominated by fields planted in corn and soybeans. Inversion of the ZRT model using specified values for the soil-surface roughness and vegetation coefficients led to the results shown in Fig. 12-29 (Bindlish et al., 2006) for the C-band (7.32 GHz) channel. Considering that these results are based on a single channel of radiometric data and that corn and soybean canopies are more lossy at 7.32 GHz than at lower frequencies, the significant amount of data scatter relative to the 1 : 1 line is not surprising.

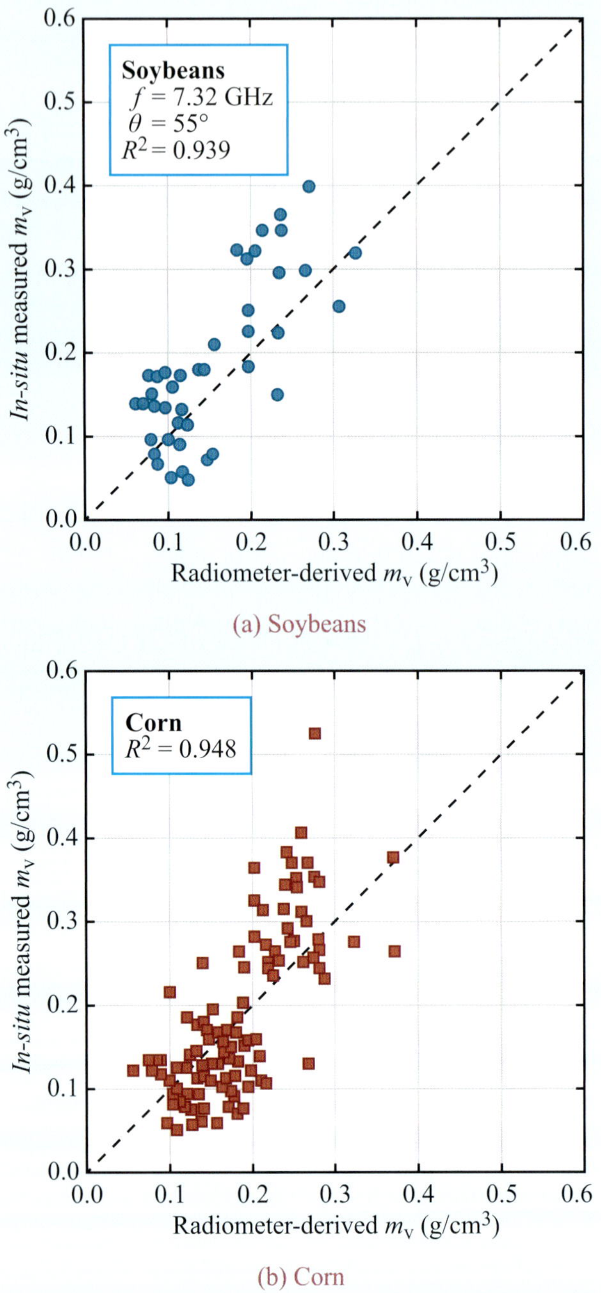

Figure 12-29: Comparison of *in-situ* measurements of m_v with radiometer-derived values based on C-band single-channel observations [Bindlish et al., 2006].

12-8.2 Multichannel Soil Moisture Retrieval

To overcome some of the limitations of single-channel radiometric retrieval of soil moisture, a multinational team of European and Australian researchers (Saleh et al., 2009) carried out an investigation in which two L-band radiometers were used to measure the emission from agricultural sites with low to moderate vegetation cover. The combination of the two radiometers provided observations with both h and v polarizations at six different angles of incidence. The multiangle data allowed the team to perform a two-step inversion process through which the roughness and vegetation parameters were estimated from the radiometric observations, rather than on the basis of ancillary ground observations. Their technique is illustrated by the results shown in Fig. 12-30. Retrieved values of m_v are highly correlated to *in-situ* values measured on the ground.

Instead of using multiple angles, Bolten et al. (2003) used multiple frequencies (L-band and S-band) and a combination of active and passive microwave sensors, aimed at evaluating how well m_v can be retrieved using various combinations of channels. Their study is based on data acquired by the ***Passive Active L- and S-band*** (PALS) instrument (Wilson et al., 2001), which they flew over a test site in Oklahoma. The site included fields planted in wheat, corn, alfalfa, as well as rangeland and bare-soil surfaces. The radiometers provided dual-polarization data at 1.4 and 2.69 GHz, and the radars provided multipolarization data at 1.26 and

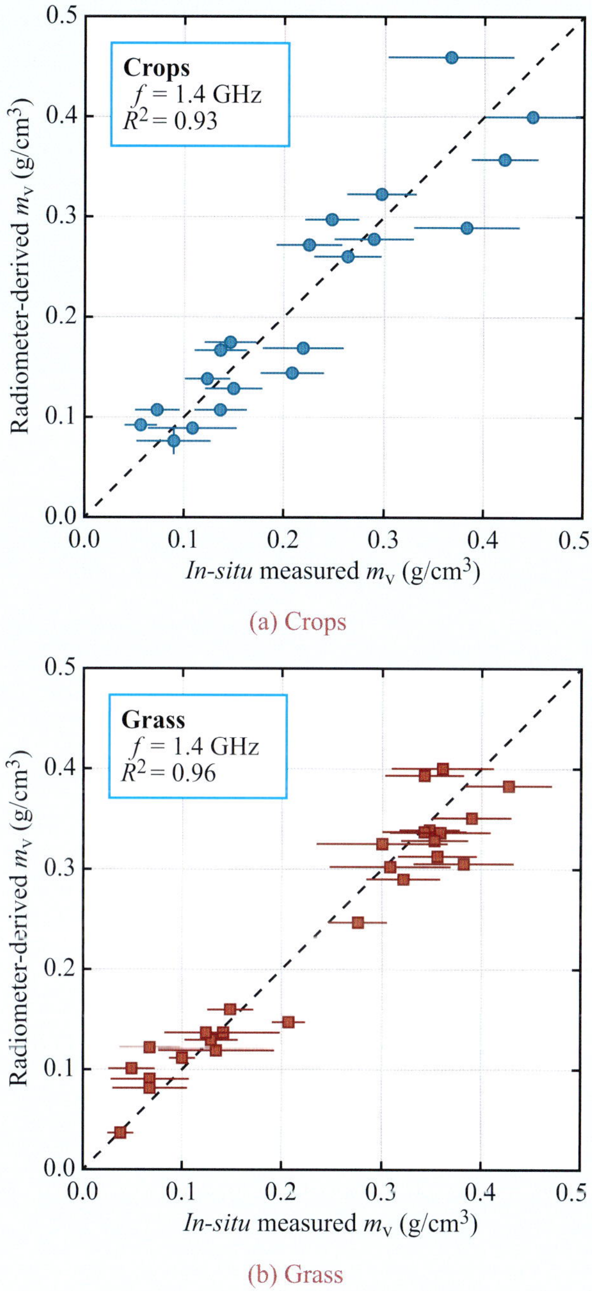

Figure 12-30: (a) Soil moisture retrievals for crops, and grass based on dual-polarization multiangle L-band observations [Saleh et al., 2009].

3.15 GHz. The retrieval results shown in Fig. 12-31 are based on multifrequency/multipolarization observations using the passive sensors alone and the active sensors alone. Unfortunately, insufficient data was available for evaluating the retrieval potential of combined active/passive sensing.

12-8.3 Change Detection

Using the PALS instrument (described in the preceding subsection) over the Iowa site, Narayan et al. (2006) examined the sensitivities of the L-band radar and radiometer to soil moisture changes between two dates, namely, July 5 and July 7 of 2002. The images in Fig. 12-32 represent the changes ΔT_{B} and $\Delta\sigma^0$ between image pairs recorded on July 5 and July 7. Given that the two dates are only two days apart, the presumption is that the vegetation cover was unchanged between the two dates, and therefore the observed changes in T_{B} and σ^0 are associated with corresponding changes in m_{v}. Overall, the pattern of ΔT_{B} in part (a) of Fig. 12-32 is similar to the pattern of $\Delta\sigma^0$ in part (b), except that the spatial resolution of the radiometer is 400 m $\times$ 400 m, compared with 30 m $\times$ 30 m for the radar. By ***aggregating*** the radar data (averaging multiple pixels) to the 400 m radiometer resolution, it was possible to compare ΔT_{B} with $\Delta\sigma^0$. The results, shown in Fig. 12-33, suggest that both instruments exhibit strong sensitivities to m_{v}.

12-9 Operational Satellite Missions

The European Space Agency (ESA) and NASA have each developed a satellite mission aimed at mapping the distribution of soil moisture content on a global scale. Brief descriptions of the two missions follow.

12-9.1 Soil Moisture and Ocean Salinity (SMOS) Mission

According to the performance review authored by Mecklenburg et al. (2012), ESA's ***Soil Moisture and Ocean Salinity*** (SMOS) mission was launched in 2009 to provide:

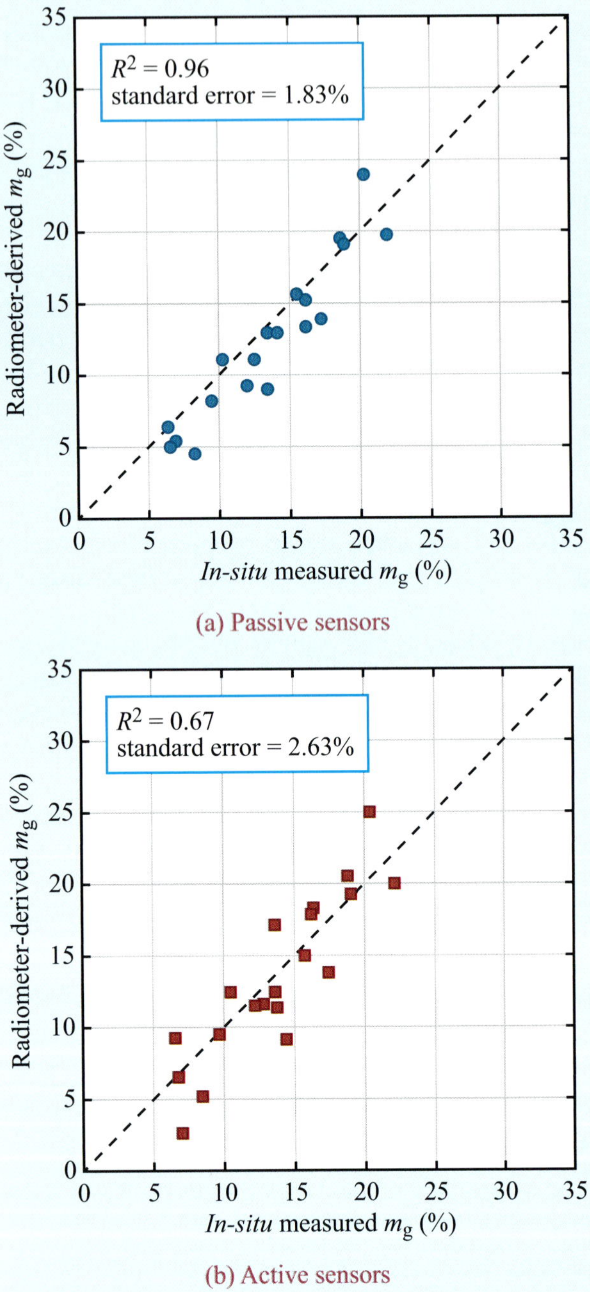

Figure 12-31: Predicted versus *in-situ* gravimetric soil moisture using (a) the passive channels and (b) the active channels on PALS [Bolten et al., 2003].

"**(a)** global volumetric soil moisture estimates with an accuracy of 0.04 g/cm^3 at a spatial resolution of 35–50 km and a temporal sampling of 1–3 days, and

(b) global ocean salinity estimates with an accuracy of 0.1 practical salinity scale units for a 10–30-day average for an open ocean area of 200 × 200 km^2."

The primary instrument on SMOS is ***MIRAS*** a Microwave Imaging Radiometer with Aperture Synthesis (McMullan et al., 2008) whose overall antenna consists of three deployable segments (Fig. 12-34) each containing 23 antenna elements. The antenna beam is created by applying aperture synthesis to the 69 elements (see Section 7-13). The radiometer operates at L-band (1.413 GHz) over an angular range from nadir to 55°, in dual and full polarization modes.

Soil moisture is retrieved by inverting the ZRT model described in earlier sections. To do so at a 40 km × 40 km cell size entails a great deal of guesswork regarding the distribution of scene classes (bare, vegetation, forest, water, urban, etc.), vegetation cover type, soil type, etc. Consequently, the retrieved values of m_v are of marginal accuracy. We show in Figs. 12-35 and 12-36 examples of the "best" results and the more "typical" results comparing SMOS-retrieved soil moisture with soil moisture estimates based on recordings made by a nationwide network managed by the U.S. Department of Agriculture. The plots and associated regression indicate general agreement between m_v (SMOS) and m_v (*in situ*), but the rms error is typically twice as large as the stated objective of 0.04 g/cm^3. Given the large size of the resolution cell, it is difficult to assess which of the various possible sources of error is the most responsible. The estimation process includes errors associated with the inversion algorithm and the instrument itself, as well as errors associated with the estimated average "*in-situ*" soil moisture of each 40 km × 40 km ground resolution cell.

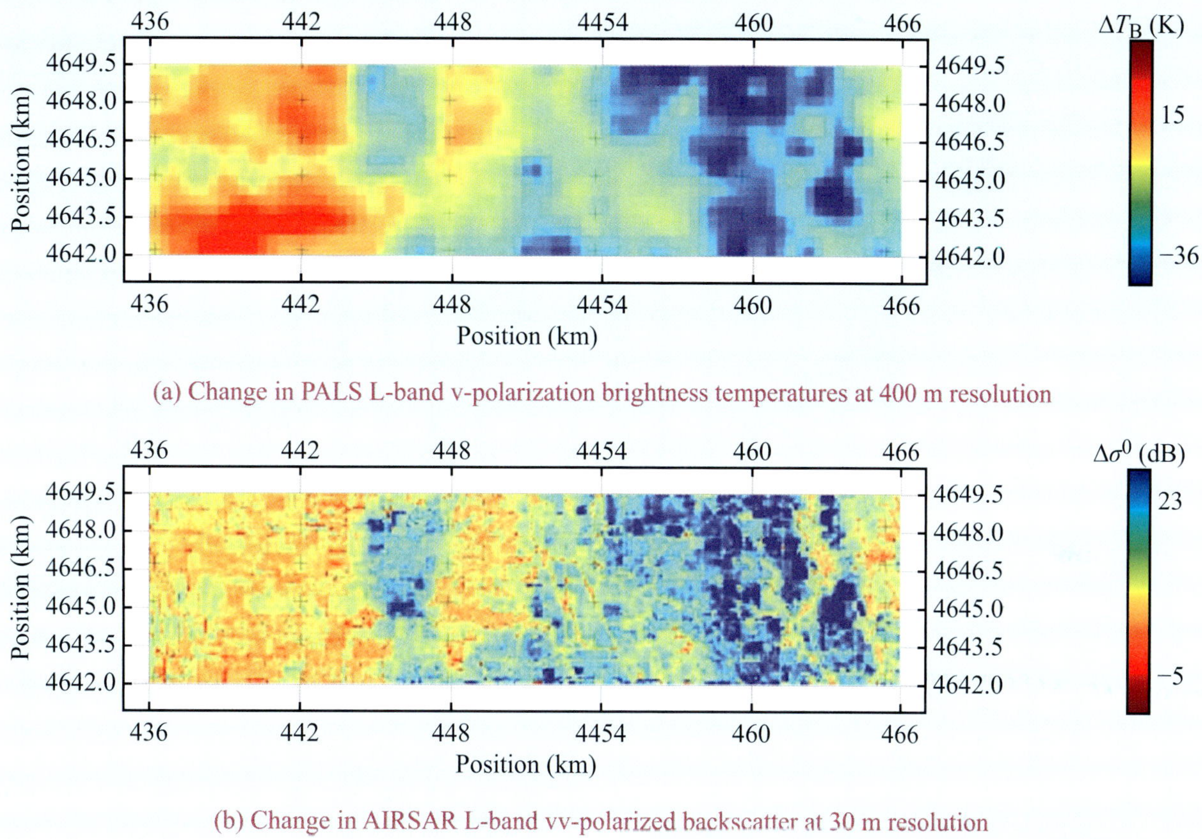

Figure 12-32: Difference images for change in PALS L-band, v-polarization brightness temperatures at 400 m resolution and change in AIRSAR L-band, vv-polarized backscatter at 30 m resolution for the period July 5 to July 7. The spatial patterns corresponding to wetting or drying are strikingly consistent in the two images [Narayan et al., 2006].

12-9.2 Soil Moisture Active Passive (SMAP) Mission

Like SMOS, NASA's Soil Moisture Active Passive (SMAP) mission, currently planned for launch in 2014, also includes an L-band radiometer with a 40 km spatial resolution, but it differs in two respects. Whereas SMOS uses aperture synthesis involving 69 antenna elements distributed along its three arms to achieve resolution, SMAP uses a 6 m diameter mesh reflector (Fig. 12-37) and imaging is realized by conically scanning the antenna at a constant incidence angle of 35.5° or 40° (Spencer et al., 2011). The second difference is that SMAP also includes an imaging L-band radar with a nominal single-look resolution of 250 m × 400 m. By degrading the radar image resolution to 1 km × 1 km, the number of independent samples is increased from 1 to 10. The basic concept behind SMAP is to generate soil-moisture maps at 10 km resolution, realized by combining an aggregated radar image at 10 km resolution with a disaggregated radiometer image at the same resolution. That is, the resolution of the radar image is degraded from 1 km to 10 km, and conversely, the radiometer resolution is *improved* from 40 km to 10 km. The algorithm proposed for realizing the intended soil-moisture product is described by Colliander et al. (2012). The feasibility of the SMAP mission and whether or not it can deliver the promised

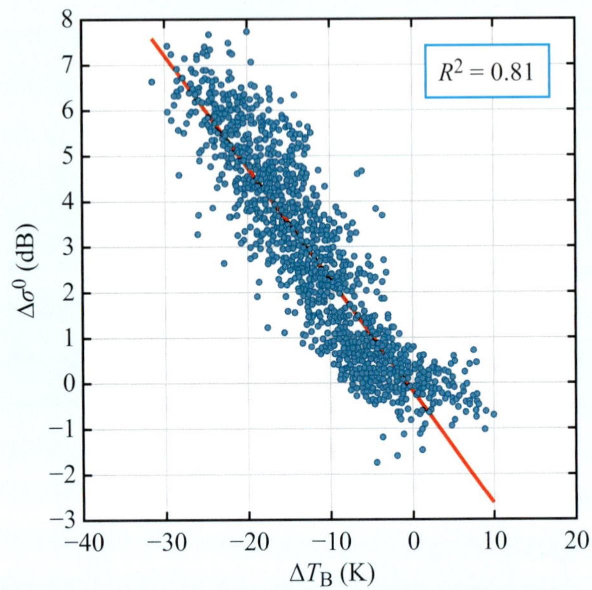

Figure 12-33: $\Delta\sigma^0$ versus ΔT_B corresponding to the images in Fig. 12-32, after aggregating the radar data to the 400 m resolution of the radiometer.

Figure 12-34: MIRAS 1.4 GHz radiometer on SMOS satellite.

0.04 g/cm^3 soil-moisture accuracy (Entekhabi et al., 2010) is uncertain at this time.

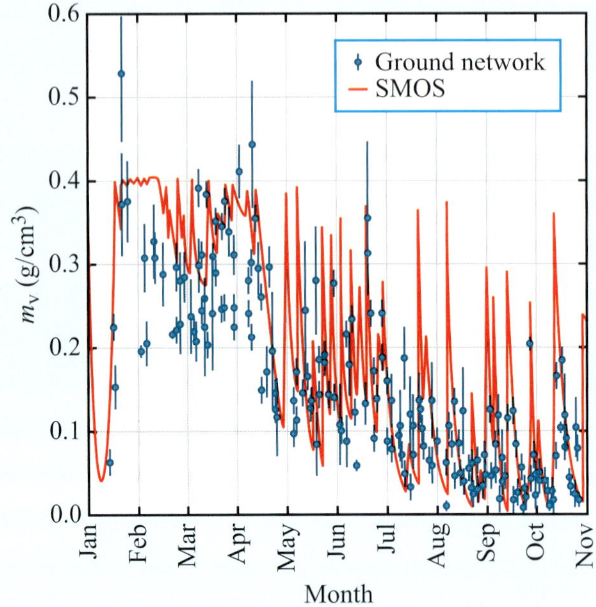

(a) 9-month history

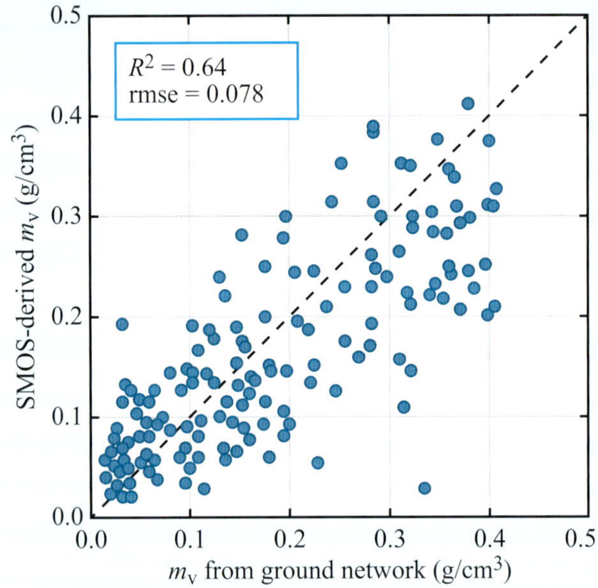

(b) Regression

Figure 12-35: Time series of SMOS-derived m_v and *in-situ* values estimated from a USDA network [Mecklenburg et al., 2012].

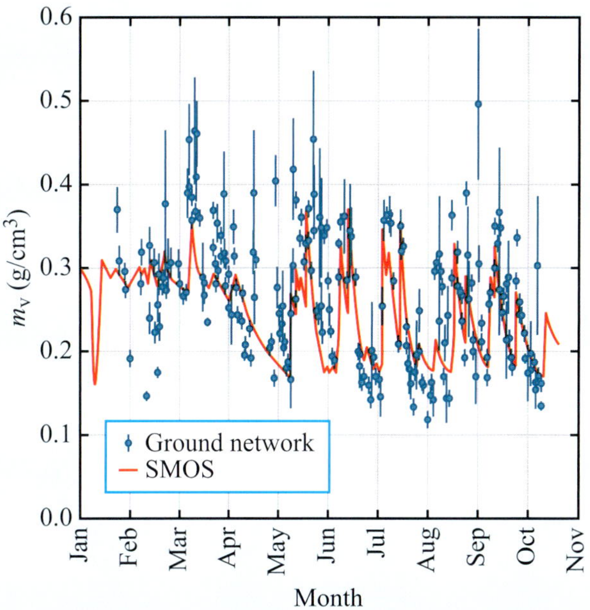

(a) 9-month history

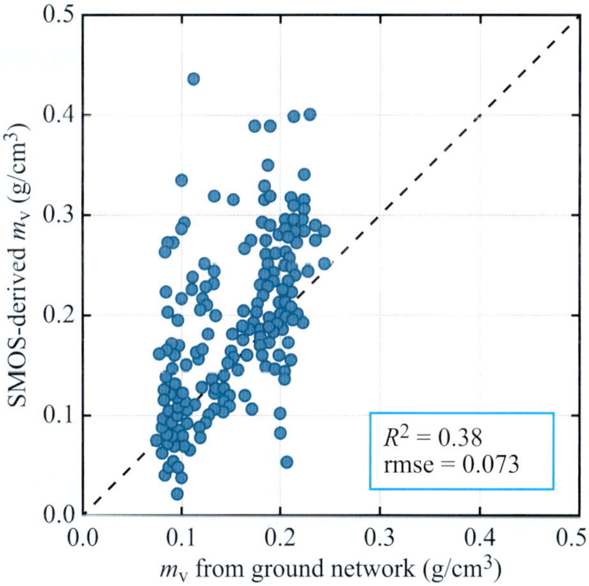

(b) Regression

Figure 12-36: Time series of SMOS-derived m_v and *in-situ* values estimated from USDA network [Bitar et al., 2012].

Figure 12-37: Artist's rendition of SMAP in orbit, showing the mesh antenna, antenna spot beam, and wide swath [Spencer et al., 2011].

12-10 Optical Depth and Emissivity of Forest Canopies

As noted earlier in connection with Fig. 12-25, the L-band optical thickness τ_0 of grasses and crops like corn and soybeans may vary between zero and about 0.8 Np, depending on the structure of the vegetation and its water content M_w. Owing to their greater height and biomass, forest canopies have larger optical thicknesses, lower transmissivities, and higher emissivities.

Using a five-channel ground-based radiometer system, Mätzler (1994) measured the optical thickness of a mature, 100-year-old beech tree under a variety of wind and temperature conditions. The radiometer system was pointed upward towards the middle of the tree. The measured variation of τ_0 with frequency is shown in Fig. 12-38 under both foliated and defoliated conditions. In the absence of leaves, $\tau_0 \approx 1$ Np across the entire measured spectrum. Under foliated conditions (with leaves), τ_0 increases markedly with increasing frequency f.

Another interesting result reported by Mätzler (1994) has to do with the variation of τ_0 with the physical temperature T_0 of the tree. A significant change

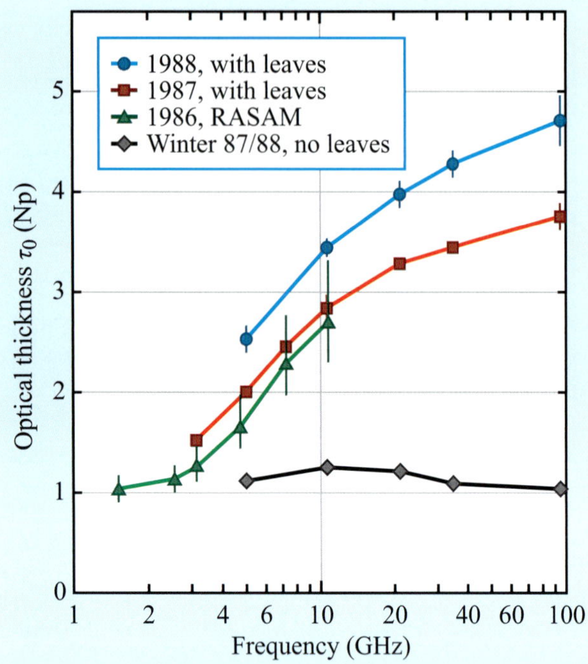

Figure 12-38: Spectra of optical depth measured along the vertical direction. The data for 1987 and 1988 are for the test beech tree in unfrozen, dry, defoliated winter conditions and for the foliated beech tree under calm wind conditions near 20 °C. The RASAM data of 1986 represent a collection of measurements with mean values and standard deviations obtained for a set of beech trees in another forest [Mätzler, 1994].

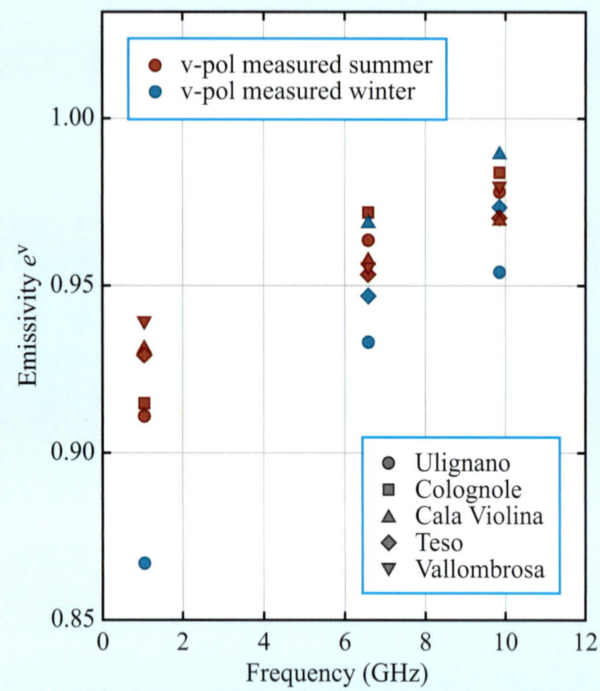

Figure 12-39: Measured emissivity as a function of frequency for five forests at $\theta_{\mathrm{i}} = 30°$ and vertical polarization [Vecchia et al., 2010].

in the magnitude of τ_0 occurs as the temperature changes between below-freezing temperatures and above freezing temperatures, consistent with the change expected to occur in the dielectric constant of any vegetation material as the temperature crosses the freezing temperature of the liquid contained in the vegetation (see Fig. 4-41).

As τ_0 increases, so does the emissivity, and in the limit, e approaches the constant value of $(1 - a)$. According to the data shown in Fig. 12-39 for the five different types of trees observed during both summer and winter conditions, the emissivity varies between about 0.9 at 1.4 GHz and 0.97 at 10 GHz (Vecchia et al., 2010).

12-11 Emission by Snow-Covered Terrain

Section 11-15 contains a thorough description of the propagation properties of dry and wet snow. Included are plots and formulations for the absorption, scattering and extinction coefficients as a function of the microwave frequency f, snow density ρ_{s}, ice crystal radius r, and liquid-water content m_{v}. Also displayed are plots of the single scattering albedo a and the penetration depth δ_{p} as a function of f and m_{v}. These properties and relationships are as relevant to the microwave emission from snow-covered terrain as it was then (Section 11-15) for radar scattering. The only additional relationship we wish to introduce is the following empirical expression developed by Hallikainen et al. (1987) for the extinction coefficient κ_e:

$$\kappa_e = 1.66 \times 10^{-3} f^{2.8} r^2 \qquad \text{(Np/m)}, \qquad (12.61)$$

(18–60 GHz)

where f is in GHz and r is the ice crystal radius in millimeters. The constant coefficient in Eq. (12.61) is different from that in the original publication because there the expression for κ_e was expressed in dB/m and the crystal size was represented by its diameter instead of its radius. The relationship described by Eq. (12.61) is based on an extensive set of transmission measurements made at 18, 35, 60, and 90 GHz for snow samples with mean crystal diameters extending between 0.2 mm and 1.6 mm. The samples included newly fallen snow and refrozen snow, and the range of snow density was from 0.17 to 0.39 g/cm^3.

12-11.1 Radiative Transfer Models

From Eq. (12.51), for a snow layer of depth d containing ice particles with a small albedo a, the zeroth-order radiative transfer (ZRT) solution of the radiative transfer equation assumes the form

$$T_{\mathrm{B_{snow}}} = \left(\frac{1-\Gamma_{\mathrm{as}}}{1-\Gamma_{\mathrm{as}}\Gamma_{\mathrm{sg}}\Upsilon_{\mathrm{s}}^2} \right) \cdot [(1+\Gamma_{\mathrm{sg}}\Upsilon_{\mathrm{s}})(1-\Upsilon_{\mathrm{s}})(1-a)T_{\mathrm{s}} + (1-\Gamma_{\mathrm{sg}})\Upsilon_{\mathrm{s}}T_{\mathrm{g}}] \qquad (12.62)$$

where $\Gamma_{\mathrm{as}} = \Gamma^p_{\mathrm{as}}(\theta_{\mathrm{i}})$ and $\Gamma_{\mathrm{sg}} = \Gamma^p_{\mathrm{sg}}(\theta'_{\mathrm{i}})$ are the (coherent) p-polarized reflectivities of the air-snow and snow-ground boundaries at incidence and refraction angles θ_{i} and θ'_{i}, respectively, Υ_{s} is the one-way transmissivity of the snow layer along direction θ'_{i}, and T_{s} and T_{g} are the physical temperatures of the snow layer and underlying ground surface, respectively. The ZRT approximation is valid for media with single scattering albedo $a \lesssim 0.2$. From the plots displayed in Fig. 11-76, we discern that for snow containing ice crystals with a mean radius of 0.5 mm, $a \approx 0.2$ at $f \approx 10$ GHz. Hence, the ZRT model is inapplicable at $f > 10$ GHz, and for snow with larger ice crystals, the model applicability is limited to even lower frequencies. Consequently, most computational models of snow emission rely on the use of numerical techniques and dense-media formulations (Chang et al., 1976; Zwally, 1977; Kong et al., 1979; Choudhury and Chang, 1981; Fung and Eom, 1981; Tsang et al., 1985; Ulaby et al., 1986; Fung, 1994). Nevertheless, the expression given by Eq. (12.62) remains useful even when $a > 0.2$, as its form is used to generate equivalent semiempirical models that describe the variation of the brightness temperature with physical parameters, such as the snow water equivalent W. Recall from Eq. (11.116) that W is the height in centimeters of equivalent liquid water contained in a vertical column of the snow layer.

Figure 12-40 provides comparisons of model calculations based on a numerical solution of the radiative transfer equation (Fung and Eom, 1981) with data measured at 5, 10.7, and 18 GHz for a 66 cm deep snowpack (Chang and Shiue, 1980). Overall, good agreement is observed between theory and experimental observations. To fit the model computation to the measured data, the radius of the equivalent ice spheres was selected to be 0.85 mm. The corresponding albedo is $a = 0.05$ at 5 GHz, 0.30 at 10.7 GHz, and 0.70 at 18 GHz. This means that the ZRT model is perfectly suitable at 5 GHz, but not at the higher frequencies.

12-11.2 Response of Dry Snow to Water Equivalent

The relative permittivity of dry snow, $\varepsilon'_{\mathrm{ds}}$, is typically between 1.2 and 2 depending on the snow density ρ_{s}, and $\varepsilon''_{\mathrm{ds}}/\varepsilon'_{\mathrm{ds}} \ll 1$ (Section 4-6.1). If the underlying soil medium is in a frozen state, its relative permittivity $\varepsilon'_{\mathrm{soil}}$ is around 3.0, and $\varepsilon''_{\mathrm{soil}}/\varepsilon'_{\mathrm{soil}} \ll 1$. These values are essentially independent of both temperature (below freezing) and frequency over the entire microwave region. The magnitude of Γ_{as} is a function of $\varepsilon_{\mathrm{ds}}$, the angle of incidence θ_{i}, and the polarization configuration $p = \mathrm{v}$ or h. Similarly, Γ_{sg} of the snow-ground boundary is governed by $\varepsilon_{\mathrm{ds}}$, $\varepsilon_{\mathrm{soil}}$, the refraction angle θ'_{i} and p. For the indicated values of $\varepsilon_{\mathrm{ds}}$ and $\varepsilon_{\mathrm{soil}}$, calculations of the Fresnel reflectivities lead to the following conclusion:

$$\Gamma_{\mathrm{as}}\Gamma_{\mathrm{sg}} < 0.01 \qquad \text{for } \theta \leq 70^\circ,$$

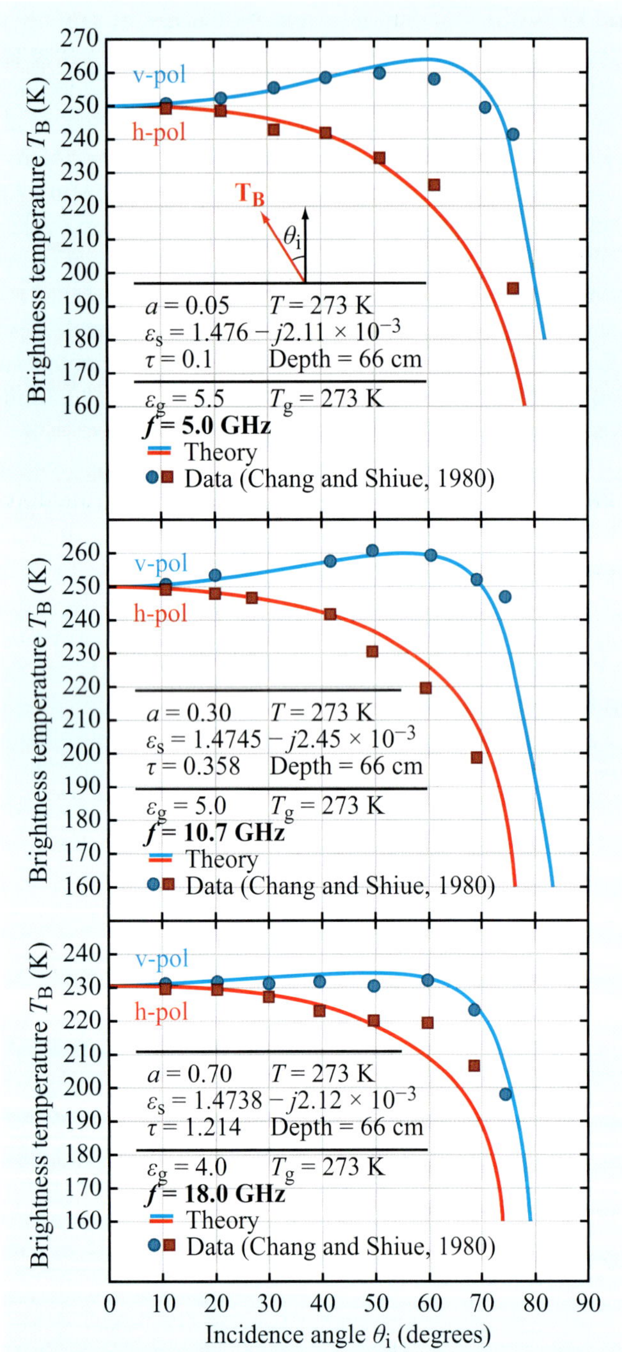

Figure 12-40: Comparison of theory with angular measurements for a snow layer with $\varepsilon_{\text{ice}} = 3.15 - j14 \times 10^{-3}$, snow density $\rho_s = 0.3$ g/cm^3, equivalent scatterer radius $r = 0.85$ mm, and wetness by weight $m_w = 0.06$ percent [Fung and Eom, 1981].

for both $p = \text{v}$ and $p = \text{h}$. Note that if either surface boundary is not smooth, its reflectivity is smaller than its specular reflectivity, in which case the product of the two reflectivities is even smaller than 0.01. Hence, the factor $(1 - \Gamma_{\text{as}}\Gamma_{\text{sg}}\Upsilon_s^2)$ in Eq. (12.62), which accounts for multiple reflections between the air-snow and snow-ground boundaries, may be set equal to unity with an error of less than 1% (remember that $\Upsilon_s \leq 1$). Further simplification may be made by neglecting $\Gamma_{\text{sg}}\Upsilon_s$ relative to 1, which is justified by the fact that $\Gamma_{\text{sg}} < 0.05$ for $\theta_i \leq 70°$ and $\Upsilon_s \leq 1$. Upon incorporating these simplifications in Eq. (12.62) and setting $T_s = T_g = T_0$, we obtain the following expression for the snow emissivity e_{snow}:

$$e_{\text{snow}} = \frac{T_{\text{B}_{\text{snow}}}}{T_0} = (1 - \Gamma_{\text{as}})[(1-a) + (a - \Gamma_{\text{sg}})\Upsilon_s]. \quad (12.63)$$

At a given incidence angle θ_i and polarization p, the three quantities Γ_{as}, Γ_{sg}, and a have constant values, and the snow layer transmissivity is given by

$$\Upsilon_s = e^{-\kappa_e d \sec\theta_i'} = e^{-\kappa_{\text{em}} d \rho_s \sec\theta_i'} = e^{-\kappa_{\text{em}} W \sec\theta_i'}, \quad (12.64)$$

where $\kappa_{\text{em}} = \kappa_e/\rho_s$ is called the ***mass extinction coefficient*** and $W = d\rho_s$ is the snow water equivalent. Using Eq. (12.64), we can express Eq. (12.63) in the form

$$e_{\text{snow}} = c_1 + c_2 e^{-c_3 W}, \quad (12.65)$$

with

$$c_1 = (1 - \Gamma_{\text{as}})(1 - a), \quad (12.66a)$$

$$c_2 = (1 - \Gamma_{\text{as}})(a - \Gamma_{\text{sg}}), \quad (12.66b)$$

$$c_3 = \kappa_{\text{em}} \sec\theta_i'. \quad (12.66c)$$

Of course, θ_i' is related to θ_i by Snell's law of refraction [Eq. (2.95)].

Even though Eq. (12.63) is based on the ZRT model, which is valid only if $a \lesssim 0.2$, the form of Eq. (12.65) has proven useful in relating e_{snow} to W

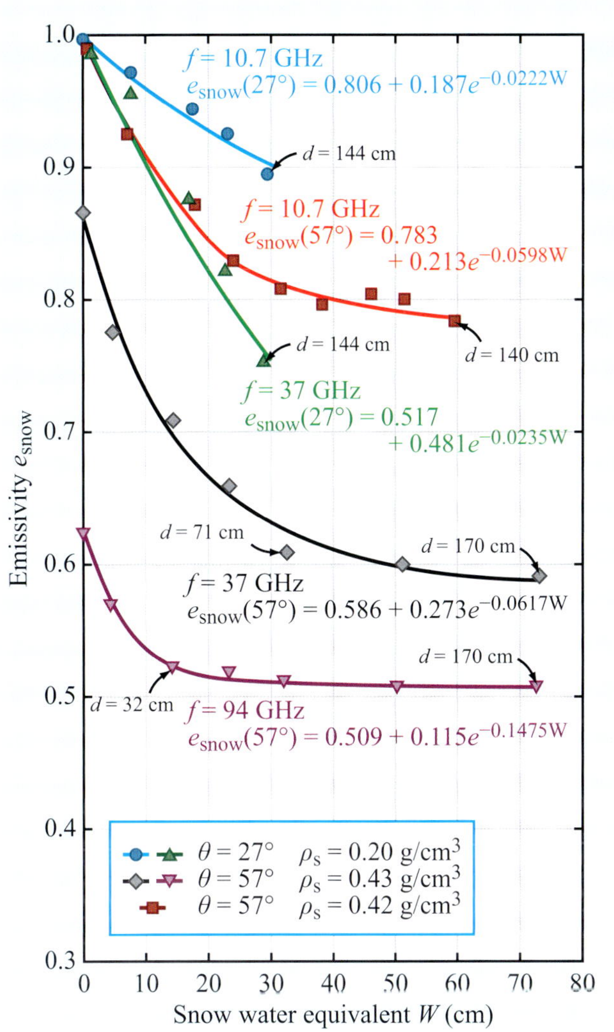

Figure 12-41: Variation of emissivity with snow water equivalent at 10.7, 37, and 94 GHz [from Stiles and Ulaby, 1980a].

even for $a > 0.2$. The data shown in Fig. 12-41 are part of three snow experiments conducted by Stiles and Ulaby (1980a) using truck-mounted radiometers with center frequencies at 10.7, 37, and 94 GHz. In one of the experiments, $\theta_i = 27°$ and in the other two $\theta_i = 57°$. In each experiment, the measurements were made by adding horizontal layers of snow in increments of 30–40 cm in depth and then making radiometric observations after each addition. All layers were composed of newly fallen snow, which resulted in an approximately uniform snowpack in terms of the density and size distribution of the ice crystals. In contrast to such an "artificial" snowpack, a natural snowpack usually consists of several layers having different densities and crystal-size distributions. The properties of the various layers are the result of the history of the snowpack, which usually is related to its location and elevation above sea level. The crystals contained in snow that has undergone one or more melting and refreezing cycles are very different in both shape and size from crystals of newly fallen snow.

Returning to our discussion of Fig. 12-41, at each incidence angle-frequency combination, the functional form of Eq. (12.65) was fitted to the measured data by selecting appropriate values for the constants c_1 to c_3. We observe that:

(a) The sensitivity of e_{snow} (its initial slope as a function of W) is steeper at 57° than at 27°.

(b) The slope is steeper at 94 GHz than at 37 GHz, and also steeper at 37 GHz than at 10.7 GHz.

(c) Since the magnitude of the h-polarized air-snow reflectivity is on the order of 0.02 for dry snow at $\theta_i = 27°$ and 57°, the factor $c_1 \approx (1 - a)$. Hence, at 57°, the effective value of the albedo is $a \approx 0.22$ at 10.7 GHz, 0.41 at 37 GHz and 0.49 at 94 GHz.

The range of W in Fig. 12-41 extends to slightly above 70 cm (corresponding to a layer depth of 1.7 m). Similar experimental measurements were conducted in Italy at 10 and 37 GHz (Macelloni et al., 2001), but over a narrower range of W, namely up to 20 cm. Their results, shown in Fig. 12-42, are in concert with those in Fig. 12-41.

The experimental observations presented in Figs. 12-41 and 12-42 corroborate predictions based on theoretical computations, an example of which is shown in Fig. 12-43. The plots were computed by Chang et al. (1987) to demonstrate the response of $T_{B_{snow}}$ to W at the five frequencies of the scanning multichannel microwave radiometer (SMMR) that was flown on the Nimbus-7 spacecraft. The plots are for $\theta_i = 50°$ and

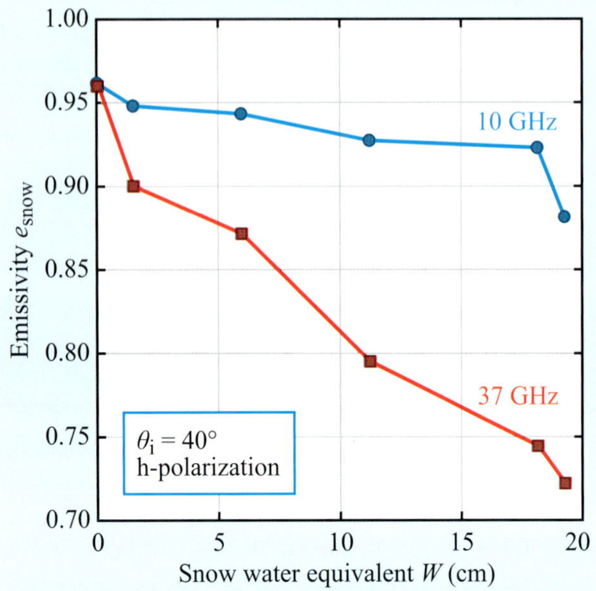

Figure 12-42: Measured emissivity at 10 and 37 GHz as a function of snow water equivalent [Macelloni et al., 2001].

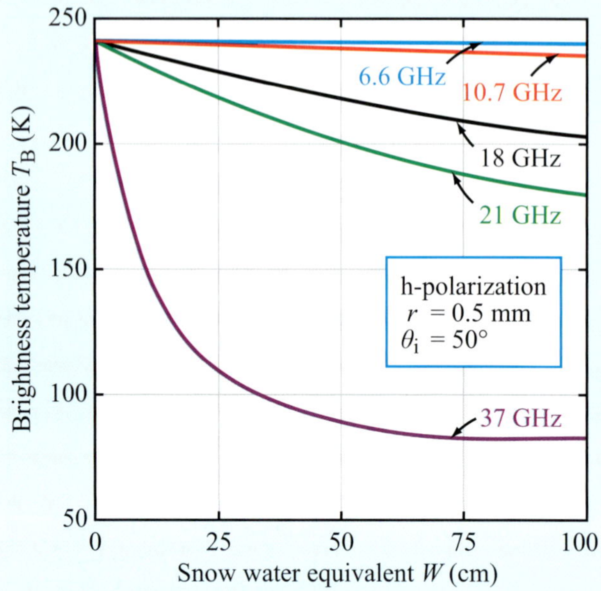

Figure 12-43: Calculated brightness temperatures versus snow water equivalent for five SMMR frequencies with horizontal polarization and 50° incidence angle [Chang et al., 1987].

h polarization, and a snowpack containing ice crystals with 0.5 mm radius. Increasing the crystal radius leads to steeper responses, lower plateau levels, and faster decay rates.

In contrast to the near-ideal, uniform snow conditions associated with the data and model computations shown in Figs. 12-41 to 12-43, the brightness-temperature measurements displayed in Fig. 12-44 are for a snowpack under natural conditions. The test field is located in Switzerland, 2450 m above sea level (Mätzler et al., 1982). The data covers a wide range of conditions for dry ***winter snow*** extending from the beginning of the snow season in October to the end of the season in March. ***Spring snow***, which usually occurs at that location between April and June, is discussed later.

According to the data in Fig. 12-44, for $W \leq 20$ cm (corresponding to a depth of 60–80 cm) $T_{B_{snow}}$ decreases with W much like the emissivity of the artificially generated snowpile responsible for the data in Figs. 12-42 and 12-43. However, at about $W = 20$ cm, the curves in Fig. 12-44 reverse direction, exhibiting upward trends with increasing W. This reversal raises two questions: (1) What causes the reversal in slope? and (2) will the fact that a single value of T_B corresponds to two very different values of W present an interpretation problem? We deal with the second question first: It is possible to resolve the ambiguity in the estimated value of W by using multichannel observations (Mätzler et al., 1982).

To answer the question regarding the reversal in the $T_{B_{snow}}$ versus W trend, we use the information contained in Fig. 12-45. Calculated values of the h- and v-polarized brightness temperatures are plotted as a function of the optical thickness $\tau_0 = \kappa_e d$ for each of several values of the scattering albedo a. These plots clearly show that a change in either τ_0 or a may cause a significant change in $T_{B_{snow}}$. For example, consider a snowpack characterized by large ice crystals such that the albedo $a \approx 0.87$. If $\tau_0 \geq 2.0$, we have $T_{B_{snow}} \approx 175$ K for horizontal polarization. Now, suppose that more snow is added to the surface. If the albedo of the new snow layer is smaller than 0.87, $T_{B_{snow}}$ decreases. For

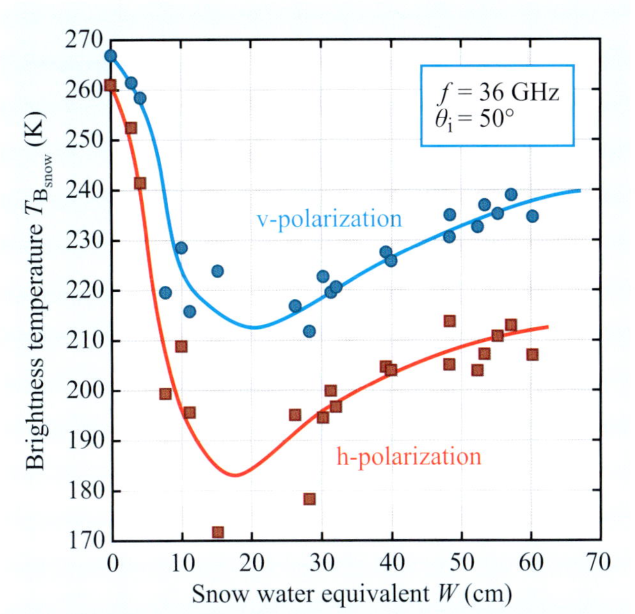

Figure 12-44: Brightness temperature at 36 GHz, with horizontal and vertical polarization, as a function of the water equivalent of dry snow from the beginning of the snow season through late winter conditions at the end of March. Data taken between March 1977 and December 1980 [from Schanda et al., 1983].

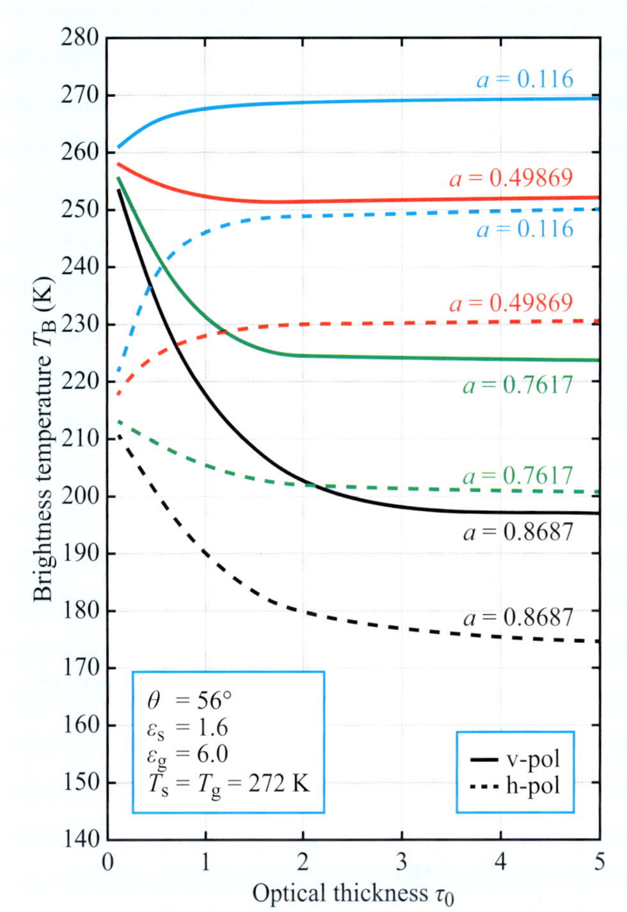

Figure 12-45: Calculated brightness temperature versus optical depth at an incidence angle of 56° for various values of the albedo a [from Lee, 1981].

dry snow, the albedo is governed by the wavelength and size distribution of the ice crystals. With regard to Fig. 12-44, one possible explanation for the reversal in the $T_{B_{snow}}$ versus W slope is that the ice crystals contained in the snowpack during the early part of the winter season (corresponding to $W \leq 20$ cm) are much larger in size (particularly in the depth-hoar part of the snowpack) than those contained in the layers that form after W has reached 20 cm. Clearly, there is nothing magical about the 20 cm figure; it simply happens to be characteristic of the cumulative weather history of the test site at which the data were obtained. According to Mätzler et al. (1982), the behavior of $T_{B_{snow}}$ depicted in Fig. 12-44, which includes data from four consecutive snow seasons, was characteristic of dry winter snow, even though very different snow histories were encountered during the four seasons.

Further discussion of the response of $T_{B_{snow}}$ to W is given in Section 12-11.6, in connection with satellite observations.

12-11.3 Snow Classes

Based on five years of extensive microwave observations of snow, the research gorup at the University of Bern (Schanda et al., 1983) devised the following classification scheme:

(1) ***Winter Snow:*** Snow that has not undergone any melting metamorphism. At high altitudes, this condition usually is satisfied during the period between November and March (in the Northern Hemisphere).

(2) ***Wet Spring Snow:*** The snowpack surface consists of thick (at least several centimeters deep) layers of wet, quasi-spherical ice crystals (1–3 mm in diameter) formed during the day at temperatures above the freezing point and usually associated with either the passage of warm fronts or sunny, clear-sky conditions.

(3) ***Dry (Refrozen) Spring Snow:*** The surface of the snowpack consists of a layer of refrozen, firn snow that forms during clear, cold nights and is several centimeters thick.

At high altitudes, spring-snow conditions extend over a period between early April and late June. The durations indicated for winter and spring snow pertain to the elevation (2450 m) of the Swiss test site referred to earlier. At lower or higher elevations, the duration of the snow season will be correspondingly shorter or longer; furthermore, the relative durations of winter and spring snows also depend on elevation. For example, at the lower elevations a snowpack might not contain any winter snow, in which case it would be classified as spring snow for the entire snow season.

Average spectra for $T_{B_{snow}}$ are shown in Fig. 12-46 for winter-snow conditions. All the measurements have been normalized to a snow water equivalent of 48 cm. For dry snow, an increase in the frequency f results in increases in both the extinction coefficient κ_e and the albedo a. This means that for a given snow-layer thickness d (or water equivalent W), $T_{B_{snow}}$ should decrease with increasing frequency. Indeed, that is the behavior exhibited by the spectra in Fig. 12-46 for the winter and spring dry-snow conditions. The two sets of spectra are significantly different in two respects: (1) the polarization difference $\Delta T_{B_{snow}} = T^{v}_{B_{snow}} - T^{h}_{B_{snow}}$ is much larger for winter snow than for spring snow, and (2) $T_{B_{snow}}$ of dry spring snow decreases more steeply with increasing frequency than $T_{B_{snow}}$ of winter snow. These differences are attributed to differences in the shapes and sizes of the ice crystals characteristic of the two snow conditions (Schanda et al., 1983). The stronger polarization dependence of winter snow is a result of

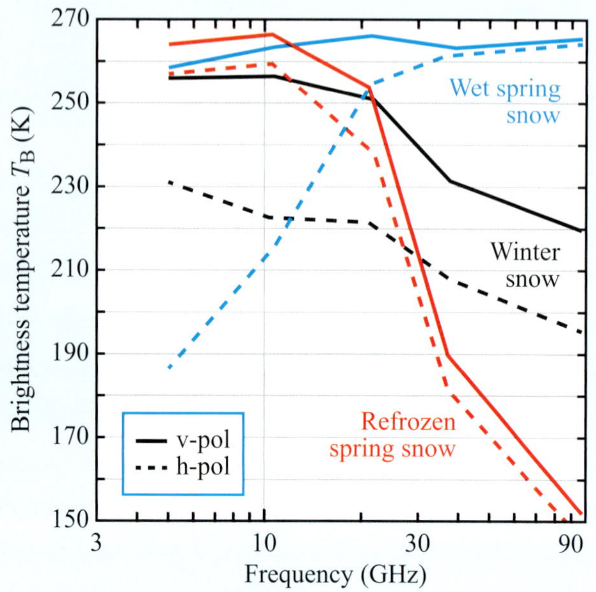

Figure 12-46: Average spectral brightness temperatures observed at horizontal and vertical polarizations with a 50-degree incidence angle during different stages of the snow season [from Schanda et al., 1983].

the fact that snow crystals are generally not spherical; however, they become so after several cycles of melting and freezing (which is the key attribute of spring snow). The melting and refreezing process also produces larger crystals, because the snow crystals tend to combine with each other. This in turn leads to more scattering, or a larger albedo, and is responsible for the steeper dependence of spring snow on f.

As was observed in connection with Fig. 11-78(a), the albedo a decreases with increasing snow wetness m_v. In the absence (or near absence) of scattering in the snow medium, the medium nearly becomes a blackbody radiator. Also, if $f \gg f_0$, where f_0 (≈ 9 GHz at $T = 0$ °C) is the relaxation frequency of water, the average permittivity of the snow medium is only slightly larger than that of air, which means that the transmissivity of the air-snow boundary is only slightly smaller than 1.0. Consequently, $T_{B_{snow}}$ approaches a value that falls between about $0.92T_0$ and $0.99T_0$ (where T_0 is the physical temperature of the snow's surface layer) for both polarization configurations. The spectral plots for wet spring snow in Fig. 12-46 are within the above range at frequencies above 21 GHz. At lower frequencies, the higher permittivity of water results in a higher value for the average permittivity of the snow's surface layer, which in turn means that the surface discontinuity is more significant, particularly for horizontal polarization.

12-11.4 Snow Wetness

We begin our examination of the effects of snow wetness on snow emission by comparing measured angular patterns of the brightness temperature $T_{B_{snow}}$ for dry and wet snow. Figure 12-47 shows such a comparison for horizontal polarization at 10.7 GHz and 37 GHz. By wet snow, we mean snow whose surface layer contains water in liquid form; however, the bulk of the snowpack may or may not be wet. The wetness values indicated in Fig. 12-47 and in the succeeding figure refer to the average volumetric water content of the surface 5 cm layer, as measured by a freezing calorimeter. The 10.7 GHz data indicate that $T_{B_{snow}}$ is almost insensitive to wetness variations at that frequency and polarization (h) combination. In contrast, a change of 1 percent in wetness results in a change of 70 K in $T_{B_{snow}}$ at 37 GHz. Also, in analogy with the case for the backscattering coefficient discussed in connection with Fig. 11-79, surface roughness exercises a stronger influence on $T_{B_{snow}}$ when the snow is wet than when it is dry, as the curves in Fig. 12-48 demonstrate.

An overall summary of the response to wetness is provided by the family of curves shown in Fig. 12-49. All the curves are based on measurements made at $\theta = 20°$ for a snow layer approximately 2 m deep; m_v represents the average volumetric liquid-water content of the 15 cm thick snow surface layer.

12-11.5 Diurnal Variations

An impressive example illustrating the importance of the frequency f with regard to the *sensitivity* of $T_{B_{snow}}$ to m_v is shown in Fig. 12-50. Observations of $T_{B_{snow}}$ made over a complete diurnal cycle indicate that when m_v changes from zero to between 1 and 2 percent, $T_{B_{snow}}$ increases by

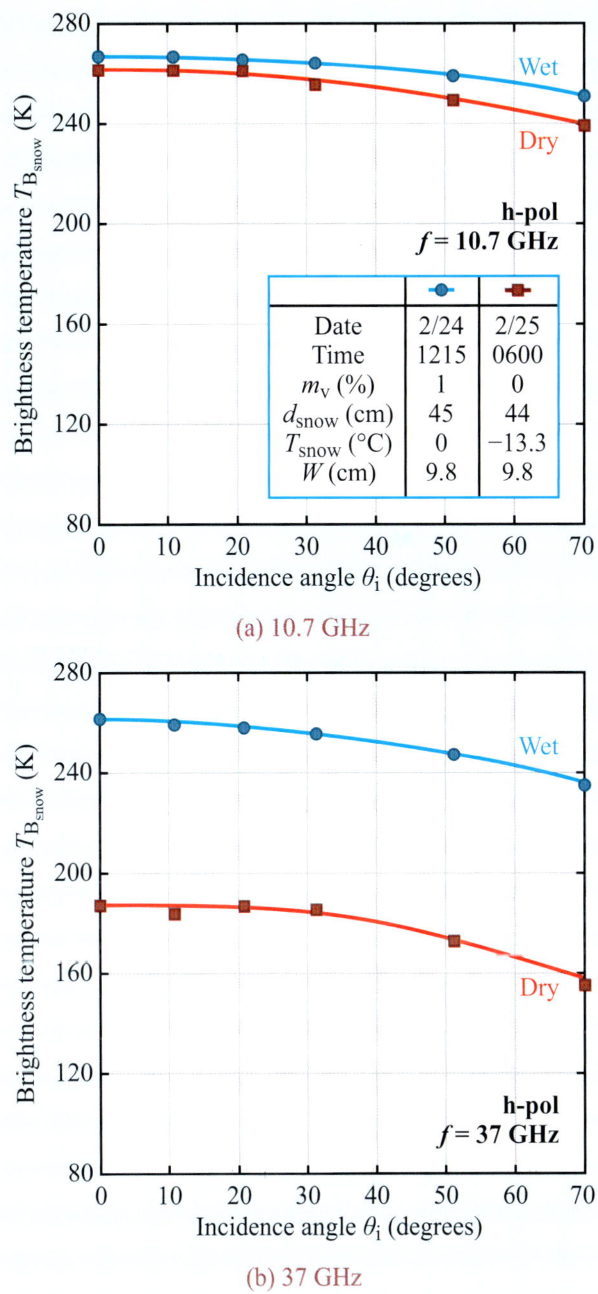

Figure 12-47: Angular response of $T_{B_{snow}}$ at (a) 10.7 GHz and (b) 37 GHz to wet and dry snow on 24 February 1977 and 25 February 1977 [from Stiles and Ulaby, 1980a].

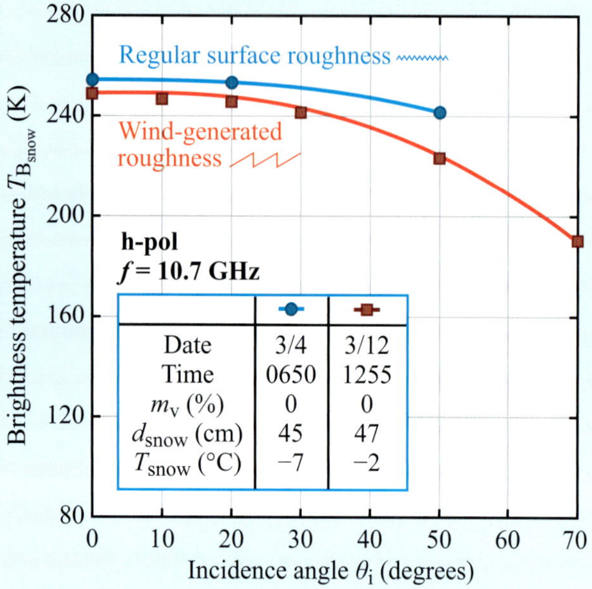

(a) Dry snow

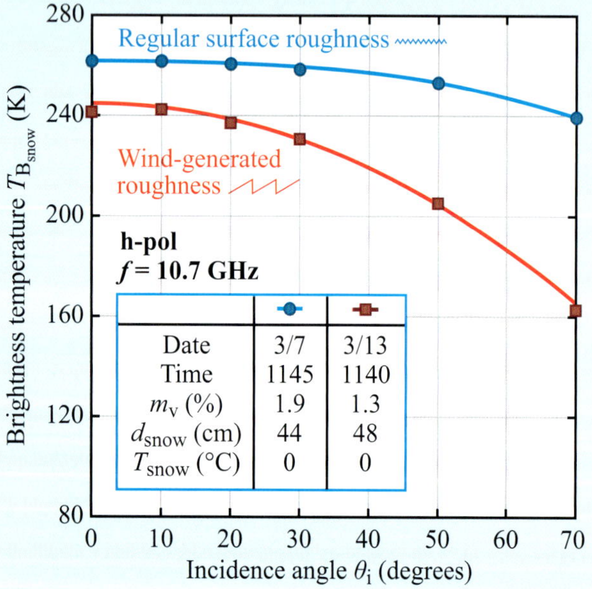

(b) Wet snow

Figure 12-48: The effect of surface roughness on $T_{B_{snow}}$ for (a) dry snow and (b) wet snow at 10.7 GHz [Stiles and Ulaby, 1980b].

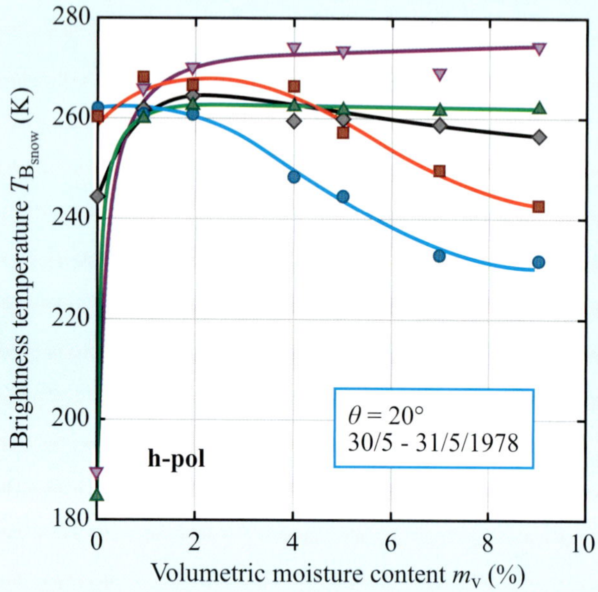

(a) h polarization

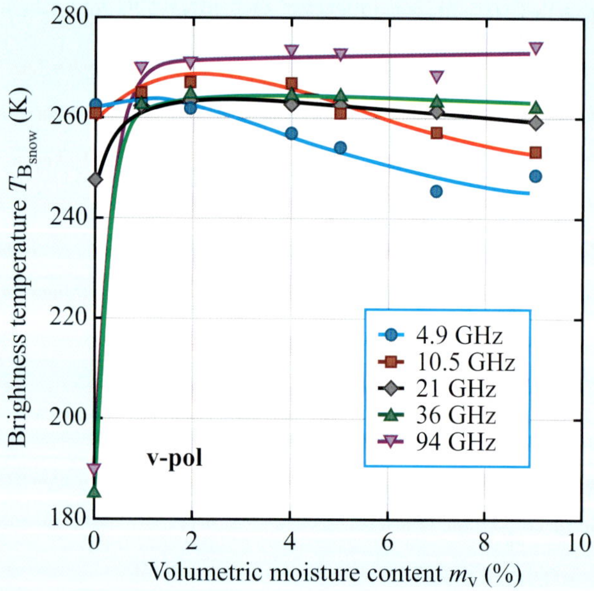

(b) v polarization

Figure 12-49: Measured variation of $T_{B_{snow}}$ with snow wetness during the melting cycle at five microwave frequencies. The wetness is the average of the top 15 cm surface layer [from Hofer and Mätzler, 1980].

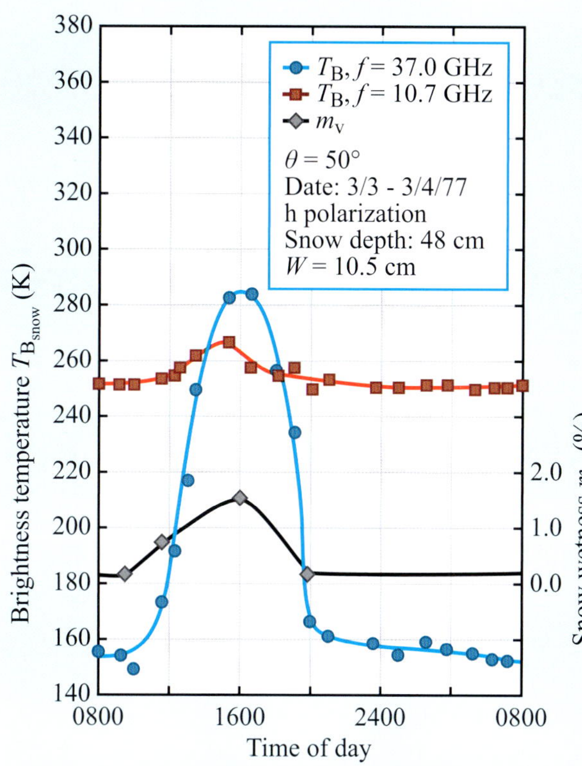

Figure 12-50: Measured diurnal variation of $T_{B_{snow}}$ at 10.7 and 37 GHz at 50° [from Stiles and Ulaby, 1980a].

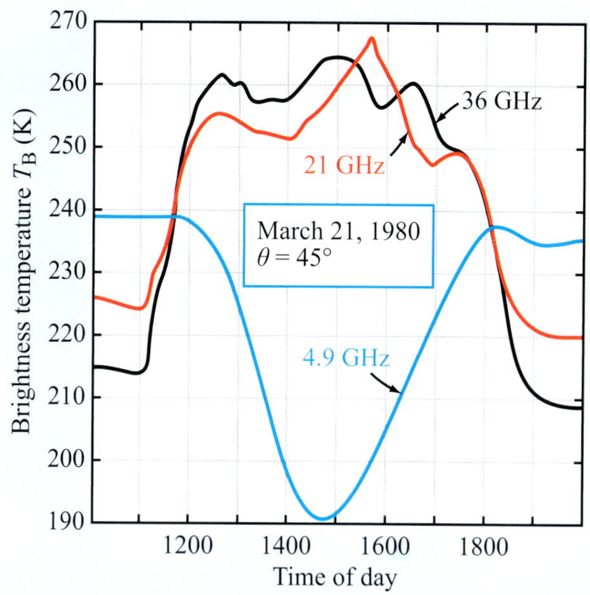

Figure 12-51: Temporal variation of brightness temperatures for horizontal polarization at 4.9, 21, and 36 GHz [Hofer and Mätzler, 1980].

10 K or less at 10.7 GHz and by over 100 K at 37 GHz. At 37 GHz, the peak value observed for $T_{B_{snow}}$ was within 1 K of the physical temperature of the snow surface. The absolute accuracy of the instrument was about ±2 K, which means that $T_{B_{snow}}$ could have been as high as 270 K. Assuming that the snow medium behaves as a perfect blackbody when wet, the above value for $T_{B_{snow}}$ implies that the snow-air transmissivity ≈ 0.99. Similar results have been reported by Schanda et al. (1982).

The sensitivity of the brightness temperature of snow to m_v exhibits a reversal in sign at a frequency f somewhere between 5 and 10 GHz. Above 10 GHz, $T_{B_{snow}}$ generally increases as m_v increases from zero to 2 percent. Observations at 4.9 GHz (Hofer and Mätzler, 1980; Schanda et al., 1982) show that $T_{B_{snow}}$ *decreases* with increasing m_v. An example comparing the behavior of $T_{B_{snow}}$ at 4.9 GHz with the behavior of $T_{B_{snow}}$ at 21 GHz and 36 GHz is given in Fig. 12-51.

12-11.6 Satellite Observations

The type of snow-cover information needed for hydrologic applications include: (1) *snow extent*, (2) the *water equivalent W* of snow-covered areas, and (3) the detection of the *onset of snow melt*. An evaluation of the operational usefulness of satellite-based snow-cover observations is given in Rango (1980), and a brief review of the methods used to generate snow-cover maps is available in the papers by Hall et al. (2001, 2002) and Armstrong and Brodzik (2002). The data used to generate the snow-cover maps are obtained from observations made by satellite-borne optical sensors and from ground-based observations. Optical observations are hampered by clouds and cannot provide information about snow depth, and ground observations must be extrapolated, often over long distances, in order to generate estimates for large areas. Thus, based on

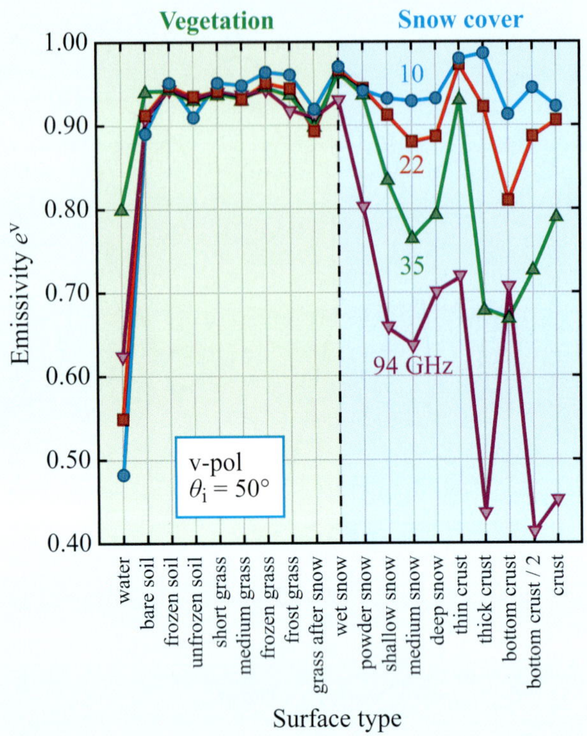

Figure 12-52: Ground-based v-polarized emissivity measurements at 10, 22, 35, and 94 GHz for different land surfaces. Surface types greater than 10 correspond to different classes of dry snow. The chart was compiled by Grody (2008) on the basis of data reported by Mätzler (1994).

an evaluation of the response of microwaves to snow parameters in general and on the results of studies using SMMR data in particular, we conclude that microwave sensors can offer a decided improvement, in terms of the accuracy of snow-covered information, over optical sensors.

The availability of multichannel data is the key ingredient necessary for the development of algorithms that can successfully distinguish between different classes of ground cover. The chart in Fig. 12-52 displays nominal values of the v-polarized emissivity for 20 classes of land surfaces under winter conditions, measured at four microwave frequencies: 10, 22, 35, and 94 GHz. The chart was compiled by Grody (2008) using data reported by Mätzler (1994) at $\theta_i = 50°$. When the temperature is below freezing, all snow-free classes (classes 2 through 9) exhibit emissivities of about 0.95 at all four frequency channels. In contrast, the emissivities of snow-covered classes vary widely between classes and among the four frequency channels. These large differences in emissivity are used to construct algorithms to distinguish between different cover classes and to estimate the associated amount of snow content. Examples comparing *in-situ* measured snow depth with T_B measured by the v-polarized 37 GHz channel of the SMMR instrument are shown in Fig. 12-53 for two different geographic regions (Canada and Russia). Using algorithms developed on the basis of such empirical observations (Chang et al., 1987; Hall et al., 2002) snow-depth and snow-water-equivalent maps with typical accuracies on the order of $\pm 10\%$ have been generated for over three decades. An example is shown in Fig. 12-54.

12-12 Coherent and Incoherent Emissivities

12-12.1 Coherent Emissivity

The configuration shown in Fig. 12-55(a) represents a dielectric layer of thickness d and dielectric constant ε_2, present above a homogeneous medium of dielectric constant ε_3. The top medium is air with $\varepsilon_1 = 1$. The upper and lower boundaries are electromagnetically flat and the spatial scale of variations within the volume of the middle layer is much smaller than λ, thereby qualifying it as an equivalent homogeneous medium. All media have the same physical temperature T_0.

Let us consider the differential volume labeled "emission source" in Fig. 12-55(a), and let us focus on the energy it contributes towards the total emission into the top medium along direction θ_1. Because of the parallel-plane configuration, the electric field of the energy contributed by the differential volume consists of an infinite number of contributions, all interrelated by specific phase relationships. The phase relationships are associated with the propagation delay between the two boundaries and the phase angles of the reflection coefficients at the two boundaries. Moreover, these phase relationships also apply to emission by every other differential volume within the middle layer, as well as by the emission originating in the bottom medium. Consequently, even though these emission sources are incoherent relative to one another, their sum acquires a coherent character. The expression for the emissivity of the two-layer structure can be obtained either by integrating over all energy sources contributing to the emission into the top medium along angle θ_1 or by invoking the law of conservation of energy. The latter approach leads to the following expression for the p-polarized ***coherent emissivity*** e^p_{coh}:

$$e^p_{\text{coh}} = 1 - \Gamma^p_{\text{coh}} = 1 - |\rho^p|^2, \tag{12.67}$$

(coherent reflectivity)[†]

where ρ^p is the p-polarized reflection coefficient given by Eq. (2.143), namely

$$\rho^p = \frac{\rho^p_{12} + \rho^p_{23} e^{-2\gamma_2 d \cos\theta_2}}{1 + \rho^p_{12}\rho^p_{23} e^{-2\gamma_2 d \cos\theta_2}}\,. \tag{12.68}$$

Hence, ρ^p_{12} is the p-polarized reflection coefficients of the upper boundary for incidence in medium 1 onto medium 2 at angle θ_1, and a similar definition applies to ρ^p_{23} for incidence in medium 2 onto the boundary with medium 3 at angle θ_2. Of course, angles θ_1, θ_2, and θ_3 are related by Snell's law and γ_2, the propagation constant of the middle layer, is defined in Section 2-4.

By way of illustration, we show in Fig. 12-55(b) a plot of e_{coh} versus d at normal incidence for a layer of pure ice of thickness d over a water surface. The frequency is 1 GHz. The oscillatory behavior of e_{coh} is a manifestation of the constructive and destructive interference associated with the coherent addition of multiply-reflected contributions.

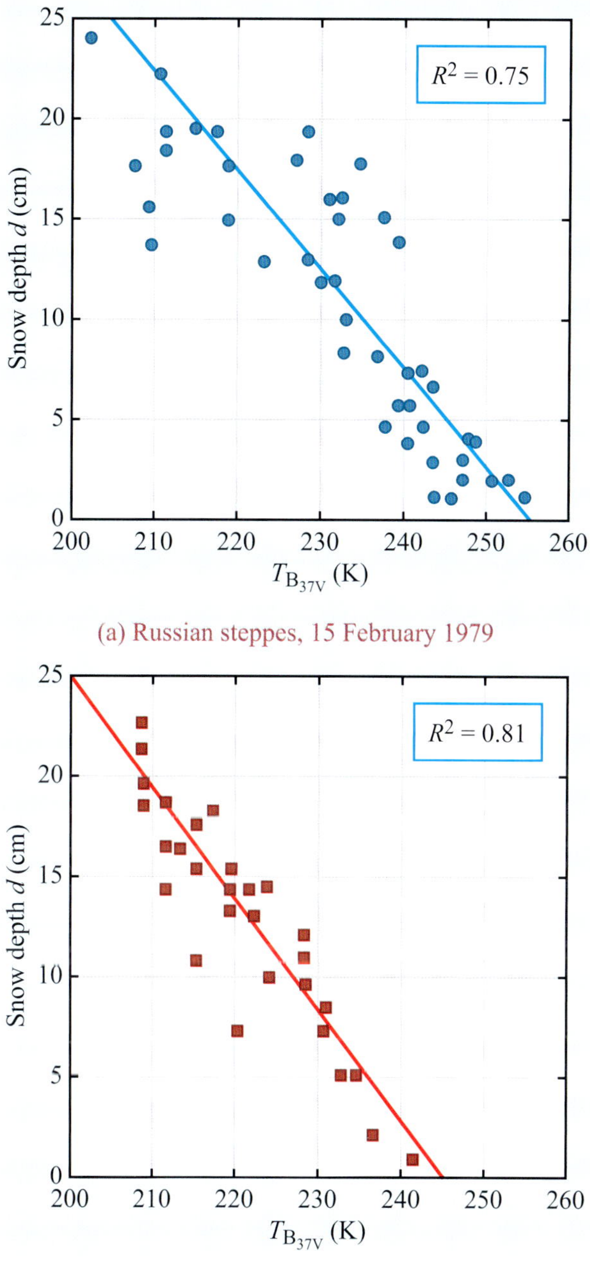

(a) Russian steppes, 15 February 1979

(b) Canadian high plains, 12 February 1983

Figure 12-53: SMMR brightness temperature versus snow water equivalent for (a) Russian steppes, 15 February 1979 and (b) Canadian High Plains, 12 February 1983 [Chang et al., 1987].

12-12.2 Incoherent Emissivity

If the middle layer in Fig. 12-56(a) has randomly distributed inhomogeneities whose dimensions are

[†]Computer Code 12.7.

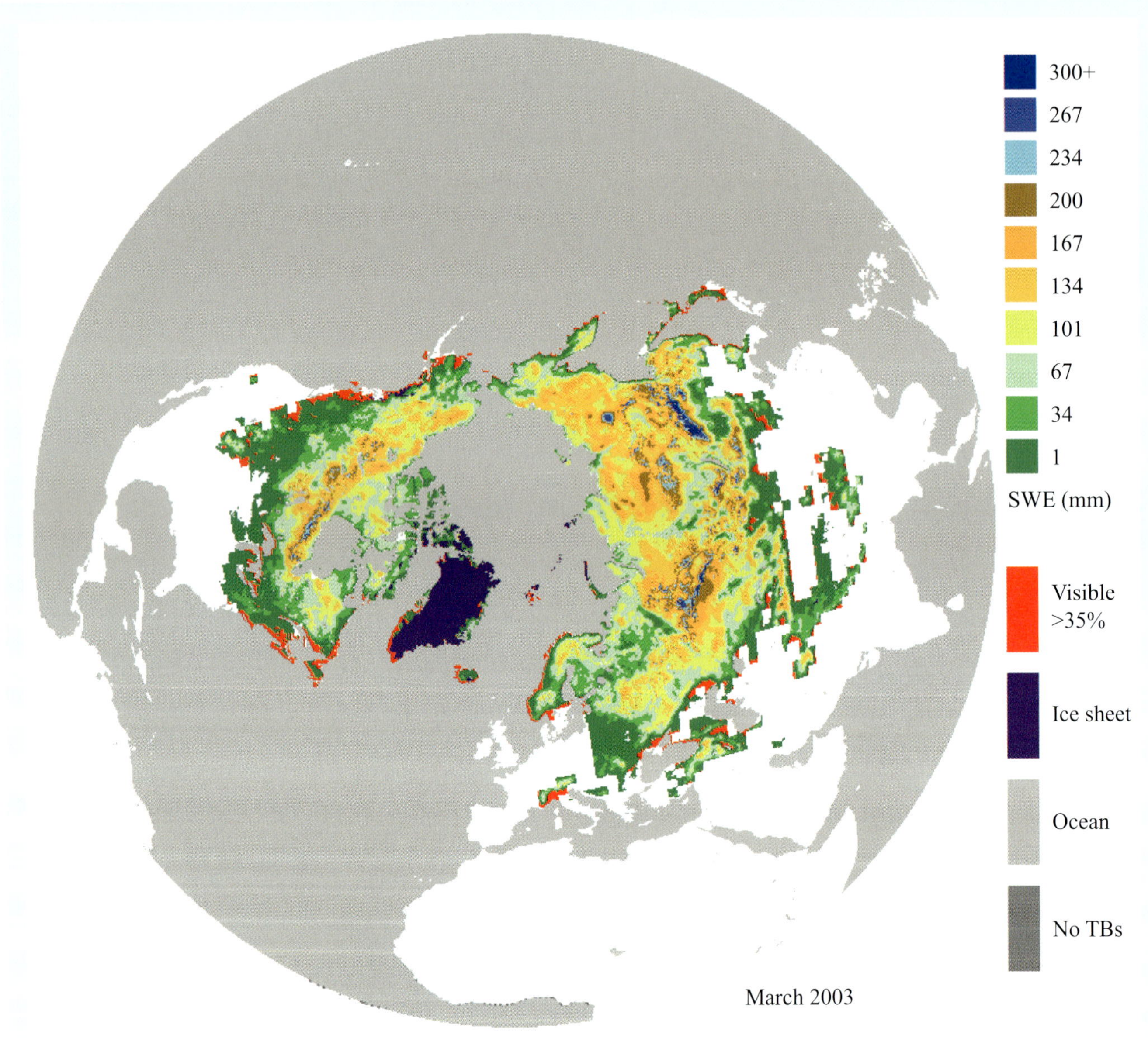

Figure 12-54: Color map of snow water equivalent [U.S. National Snow and Ice Data Center].

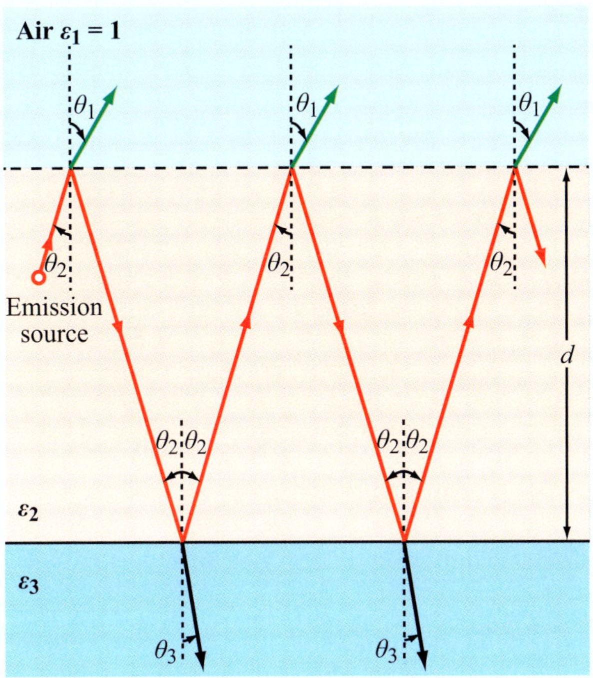

(a) Homogeneous middle layer

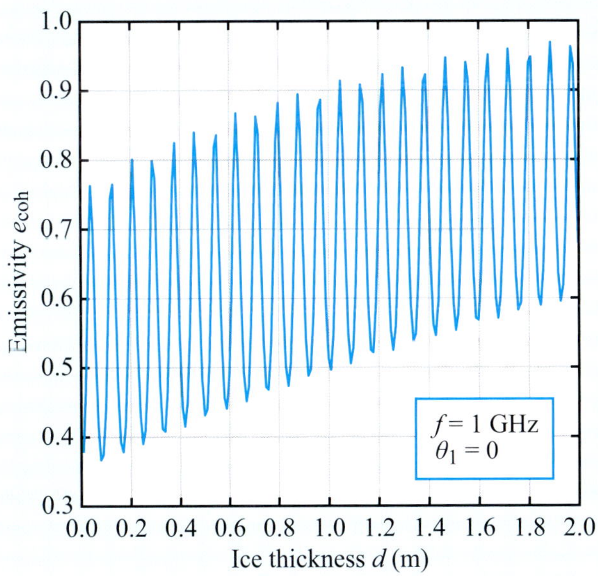

(b) e_{coh} versus d

Figure 12-55: Coherent emissivity of a two-layer structure consisting of a homogeneous layer of dielectric constant ε_2 and thickness d above a medium with ε_3: (a) emission contributions of a differential volume, and (b) variation of e_{coh} with d for a layer of ice over water at 1 GHz.

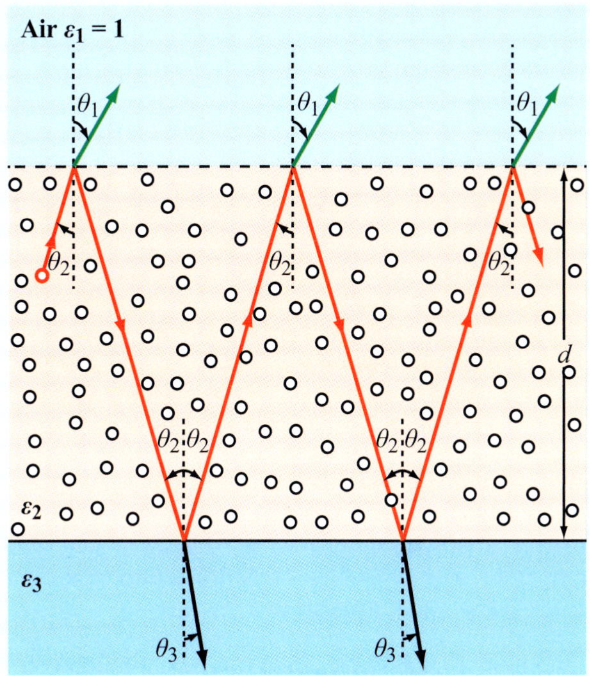

(a) Inhomogeneous middle layer

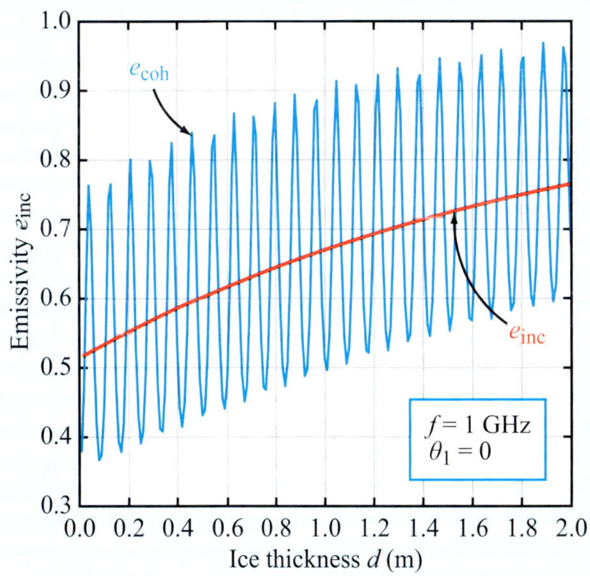

(b) e_{coh} and e_{inc} versus d

Figure 12-56: Incoherent emissivity (in red) of a two-layer structure consisting of an inhomogeneous layer of dielectric constant ε_2 and thickness d above a medium with ε_3: (a) emission contributions of a differential volume, and (b) variation of e_{inc} with d for a layer of ice over water at 1 GHz. Coherent emissivity (in blue) is shown for comparison.

larger than $\lambda/100$, the phase relationships among the multiply-reflected contributions are no longer preserved, in which case the emission by the middle and bottom layer can be added as powers (as opposed to electric fields), which leads to Eq. (12.51). Upon setting $T_2 = T_3 = T_0$ in Eq. (12.51) and then dividing the expression by T_0, we obtain the following expression for the p-polarized ***incoherent emissivity*** e^p_{inc}:

$$e^p_{\text{inc}} = \left(\frac{1-\Gamma^p_{12}}{1-\Gamma^p_{12}\Gamma^p_{23}\Upsilon^2}\right) \cdot [(1+\Gamma^p_{23}\Upsilon)(1-a)(1-\Upsilon)+(1-\Gamma^p_{23})\Upsilon], \tag{12.69}$$

(incoherent emissivity)†

where a is the single scattering albedo of the middle layer and

$$\Upsilon = e^{-\kappa_e d \sec\theta_2}. \tag{12.70}$$

If the random inhomogeneities in the middle layer are enough to destroy the phase coherence between multiple reflections, but simultaneously $a \ll 1$, Eq. (12.69) simplifies to

$$e^p_{\text{inc}} = \left(\frac{1-\Gamma^p_{12}}{1-\Gamma^p_{12}\Gamma^p_{23}\Upsilon^2}\right)(1-\Gamma^p_{23}\Upsilon^2). \qquad (a \ll 1) \tag{12.71}$$

In Fig. 12-56(b) we show plots of both the coherent and incoherent emissivities as a function of d. The two emissivities have the same general trend, but the coherent emissivity also exhibits the phase-interference oscillations discussed earlier.

12-13 Microwave Emission by Lake Ice

The presence of brine pockets in sea ice causes the mixture to behave like a lossy medium with significant absorption and scattering. In contrast, lake ice is a low-loss transmission medium. The dielectric properties of pure ice are summarized in Section 4-3. Because freshwater and lake water may contain impurities and small amounts of dissolved salts, the salinity is not exactly zero, but rarely exceeds a fraction of one part per thousand.

The multifrequency plots displayed in Fig. 12-57 were generated using the incoherent model given by Eq. (12.71). The 5 GHz curve is comparable to the experimental measurements shown in Fig. 12-58 over the range of d between approximately 20 cm and 65 cm. The sensitivity of T_{B} to d (i.e., the slope of T_{B} versus d) increases with increasing frequency, but the linear range becomes narrower. The model is based on near-ideal conditions, with uniform ice properties and no snow over the ice layer. If snow is present, we would need to develop and use a multilayer radiative transfer model.

To appreciate the complexity of real-world scenarios, consider the time series of the observations displayed in Fig. 12-59. The data pertains to Great Bear Lake in Canada's Northwest Territories (Kang et al., 2010). The time series at the bottom shows measurements of the mean air temperature from the middle of 2003 to the middle of 2004. Also shown are the temporal variations in ice thickness and snow depth above the ice. According to the authors, ***complete freeze-over*** (CFO) occurred during the ten-day interval centered around day 340 of 2003. Ice thickness increased monotonically until it reached a peak thickness of approximately 150 cm on approximately day 140 of 2004. ***Melt onset*** (MO) started shortly thereafter, and over the next 20 days the ice melted away and its thickness returned back to zero.

The paper by Kang et al. (2010) contains time series for T_{B} at 6.9, 10.72 and 18.7 GHz for both h and v polarizations. The measurements were recorded by the AMSRE system in 2003 and 2004. In Fig. 12-59, we show only the channels at the lowest and highest of the three frequencies. For all channels, T_{B} changes markedly upwards during the freezing stage centered at day 340, then T_{B} increases linearly over time until the ice (and its snow cover) starts to melt. Consistent with our observations in Section 12-11.4, wet snow causes T_{B} to increase further at melt onset. For the period between complete freeze-over and melt onset, T_{B} varies

†**Computer Code 12.8.**

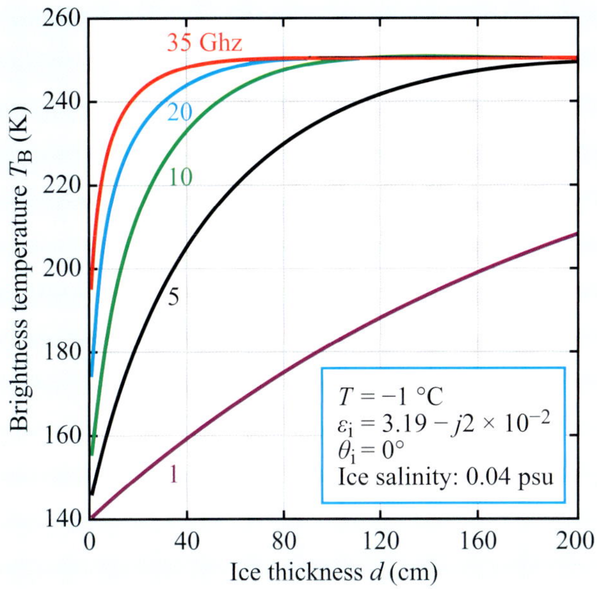

Figure 12-57: Calculated brightness temperature of lake ice as a function of ice thickness.

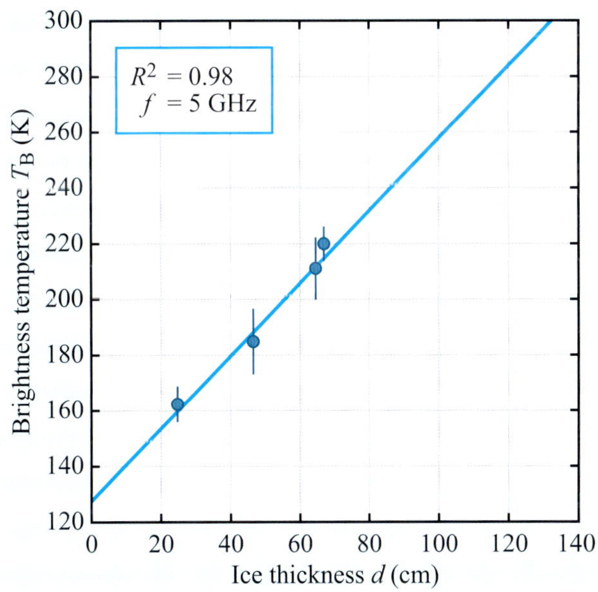

Figure 12-58: Observed relationship between T_B and ice thickness at 5 GHz [from Hall et al., 1981].

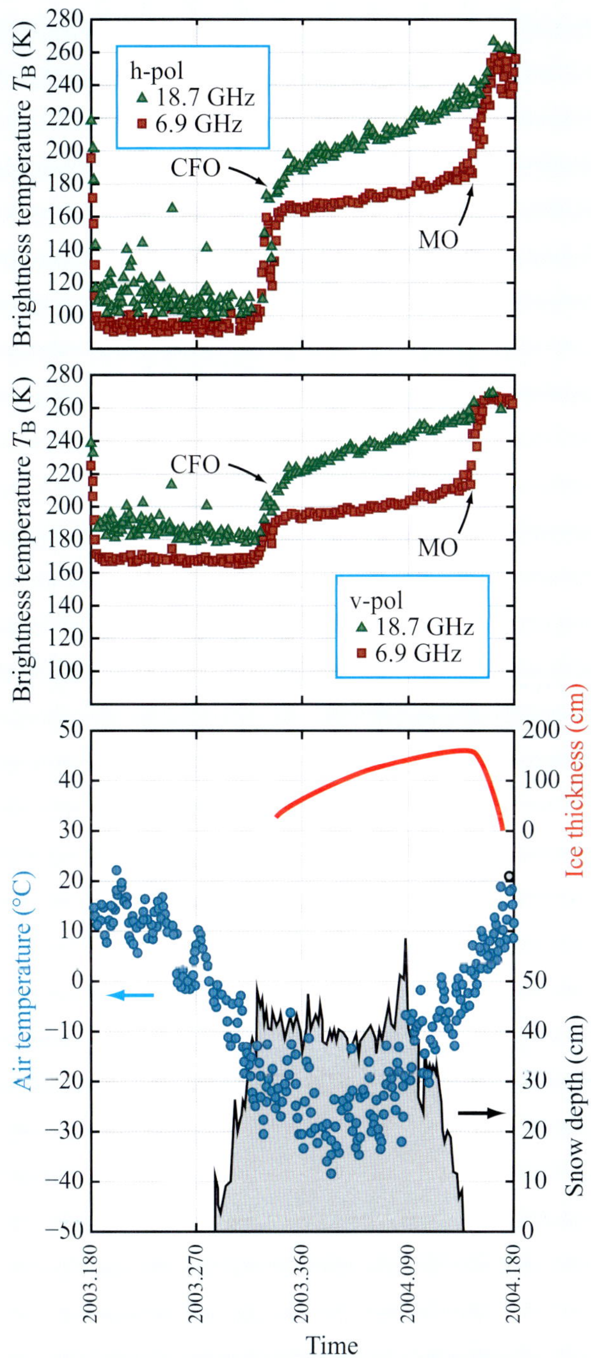

Figure 12-59: Time series of T_B at two frequencies for both h and v polarizations, mean air temperature, ice thickness, and snow depth, all for Great Bear Lake in Canada [Kang et al., 2010]. CFO = complete freeze-over and MO = melt onset.

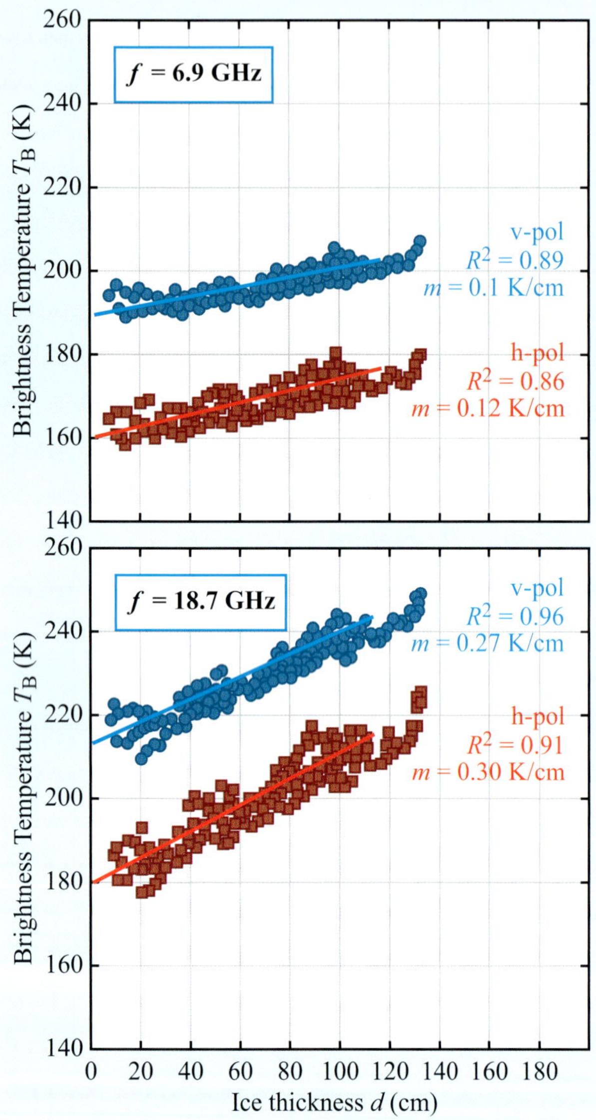

Figure 12-60: Linear regression plots between brightness temperature and lake-ice thickness [Kang et al., 2010].

approximately linearly with ice thickness d. Examples are shown in Fig. 12-60 for a total of 327 pairs of (T_B, d). The coefficient of determination (R^2) varies between 0.86 at 6.9 GHz to 0.96 at 18.7 GHz. More importantly, the slope $S_d = \partial T_B/\partial d$ increases from about 0.1 K/cm at 6.9 GHz to 0.28 K/cm at 18.7 GHz. These results are in concert with the general model predictions of Eq. (12.71) and those displayed in Fig. 12-57.

PROBLEMS

12.1 A relatively dry soil medium has a dielectric constant $\varepsilon_{soil} = 4 - j0.1$.

(a) Use Computer Code 12.1 to compute and plot the specular emissivities for h and v polarizations as a function of incidence angle from 0 to 70°.

(b) Use Computer Code 12.2 for I^{2}EM to generate similar plots, by assuming the surface to be very smooth. You can do so by setting $s = 0.1$ cm, $l = 20$ cm and $f = 1$ GHz. Compare the two sets of plots.

12.2 For a random surface with an exponential correlation function, rms height $s = 1$ cm and correlation length $l = 12$ cm, use Computer Code 12.2 to compute and plot the h-polarized emissivity at $\theta = 30°$ as a function of frequency from 1 to 10 GHz. Assume loamy soil with $m_v = 0.2$ g/cm^3.

12.3 Use Computer Code 12.3 to generate a family of plots for the 1 GHz h-polarized emissivity at $\theta_i = 30°$ as a function of m_v (0 to 0.3 g/cm^3). The plots correspond to the following values of the surface parameter $Z_s = s^2/l$: 0, 0.5, 1.0, 1.5.

12.4 A sinusoidal periodic surface with a peak-to-peak amplitude of 15 cm and a period of 75 cm has a soil moisture content of 0.25 g/cm^3. Assuming the surface to have no random roughness, generate and plot the emissivity for h and v polarizations as a function of incidence angle θ_0 for:

(a) $\phi_0 = 0$ (parallel look direction).

(b) $\phi_0 = 45°$.

(c) $\phi_0 = 90°$ (perpendicular look direction).

12.5 Repeat Problem 12.4 after adding to the surface of the sinusoidal variation a random roughness characterized by $s = 2$ cm and $l = 10$ cm. Use Code 12.4 at 1.4 GHz.

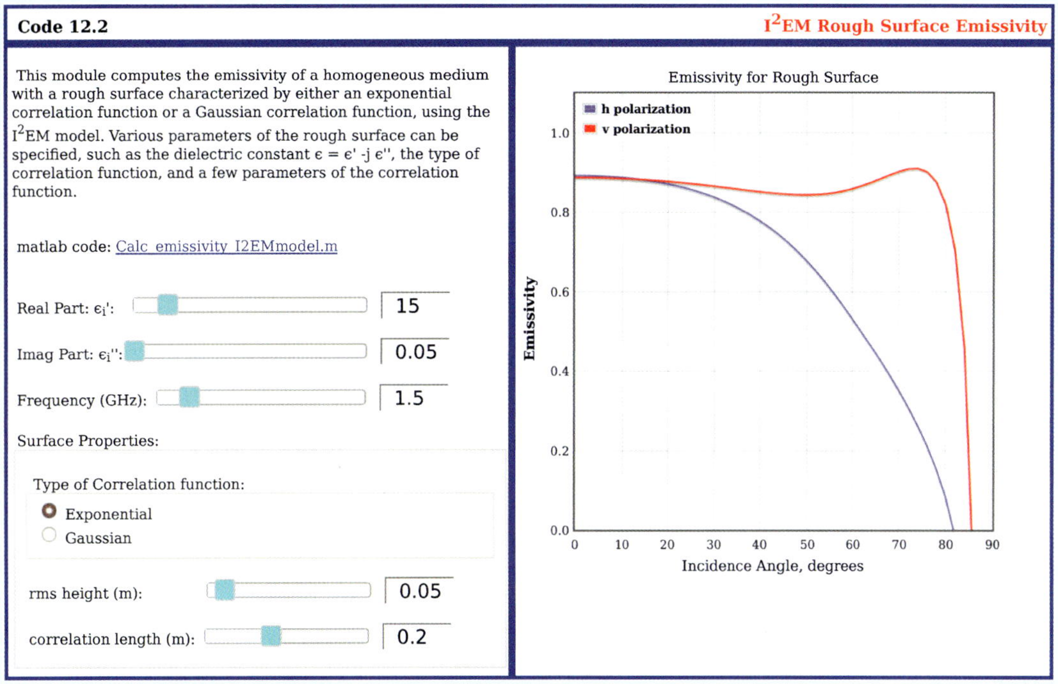

An example of one of the interactive modules available at the book website: mrs.eecs.umich.edu.

12.6 A vegetation canopy is characterized by the following parameters: albedo $a = 0.03f$ and extinction coefficient $\kappa_e = 0.3f$ (Np/m), with f in GHz ($1 \le f \le 10$ GHz). The underlying soil surface has an rms height $s = 2$ cm and correlation length $l = 10$ cm. Use Computer Code 12.5 to compute the ZRT emissivity at 1.4 GHz and $\theta_i = 30°$ for h polarization:

(a) as a function of canopy height d for $m_v = 0.05$ g/cm^3.

(b) as a function of canopy height d for $m_v = 0.3$ g/cm^3.

Vary d between 0 and 2 m.

12.7 Use the parameters of the canopy described in Problem 12.6 to generate and plot the h-pol ZRT emissivity at $\theta_i = 30°$ as a function of m_v (0 to 0.3 g/cm^3) at:

(a) L-band (1.4 GHz),

(b) X-band (10 GHz).

The canopy height is 1.2 m.

12.8 Use the parameters of the canopy described in Problem 12.6 to generate and plot the h-pol ZRT emissivity at $\theta_i = 30°$ and $f = 1.4$ GHz as a function of m_v (0 to 0.3 g/cm^3) for:

(a) $d = 0.2$ m,

(b) $d = 1.5$ m.

12.9 A layer of dry snow of depth d overlays a soil surface. The soil surface may be treated as specular with a dielectric constant $\varepsilon = 4 - j0.1$. At 10 GHz, snow has an albedo $a = 0.2$ and an extinction coefficient $\kappa_e = 1.1$ Np/m, and its density is 0.5 g/cm^3. Generate and plot the emissivity as observed by an h-pol downward-looking radiometer as a function of d.

12.10 Generate and plot the coherent emissivity versus thickness for an ice layer over water at normal incidence. Vary d between 0 and 1 m. The frequency is 3 GHz, the water is pure, the temperature is -5 °C, and the ice dielectric is $\varepsilon_{ice} = 3.2 - j2 \times 10^{-3}$.

CHAPTER
13

Radar Measurements and Scatterometers

SeaWinds Scatterometer on the QuikSCAT Satellite

CONTENTS

Overview

In microwave remote sensing, information about the earth's terrain, ocean, or atmosphere must be extracted from the received microwave signal emitted or reradiated by them. For remote sensing radar, the received signal is due to reflection of the transmitted signal by the subject area. Because the radar designer can control the transmitted signal, the transmitted signal can be tailored to extract more information about the subject than is possible from passive sensing. Fundamentally, a remote sensing radar can make measurements of the received signal (1) power or amplitude, (2) frequency or Doppler shift, and (3) time delay or range. Antenna pointing provides direction discrimination capability.

▶ Even though the received signal consists of the sum of echoes from different areas within the antenna footprint, range and Doppler information makes it possible to separate or discriminate the signal components arriving from different areas within the illuminated area. ◀

This means that a radar can achieve higher resolution than that dictated by the antenna pattern alone. This discrimination is possible by exploiting the properties of the transmitted signal and the coherent nature of the received signal. In order to develop the system design and signal processing concepts necessary to perform these tasks, we devote this chapter to a description of the basic radar system configurations used in remote sensing and delve into the theory of radar measurement and discrimination.

A ***scatterometer*** is a radar designed to measure the reflectivity of desired targets. It does this by transmitting a microwave signal and then observing the signal scattered from the target and received by the antenna. The simplest scatterometers measure only the power of the scattered signal return. However, more sophisticated systems can extract additional information from the received signal using time delay and Doppler frequency shift.

13-1 CW Radar

The simplest radar employs a ***continuous wave*** (CW) transmitter. As illustrated in Fig. 13-1, part of the signal transmitted by the radar reflects or ***scatters*** off the target toward the receive antenna. The radar may be ***monostatic***, wherein the same antenna is used for transmit and receive, or ***bistatic***, wherein different antennas (located at different positions but with overlapping footprints) are used for transmit and receive. In the radar signal flow diagrams illustrated in Fig. 13-1, the received signal is mixed with a copy of the transmitted signal. This is known as ***homodyne*** processing and is used to shift the frequency of the received signal down to baseband to simplify the signal processing. The receiver then amplifies the mixed-down signal and passes it on to an indicator or processor that extracts the characteristics of interest.

For a single stationary point target, the received signal is a time-delayed, attenuated copy of the transmitted signal. The time delay arises from the time-of-flight of the signal from the transmit antenna to the target and back to the receive antenna. The signal amplitude depends on the range, antenna gain, frequency, and the radar cross section (RCS) of the target, as given by the radar equation derived in Section 5-4. If the target is moving relative to the radar, the received signal has a slightly different frequency due to the Doppler shift caused by the motion of the target relative to the radar.

13-1.1 Target Stationary Relative to Radar

In a CW radar the transmit signal voltage $\upsilon_t(t)$ can be expressed as

$$\upsilon_t(t) = A\cos(2\pi f_0 t). \tag{13.1}$$

Limiting our attention to monostatic systems for simplicity,* the two-way signal ***time of flight*** T is related to the distance R between the radar and target by

$$T = \frac{2R}{c}. \tag{13.2}$$

*For a bistatic radar with separate transmit and receive antennas $2R$ is replaced by the sum of the distances between the target and transmit antenna R_t and target and receive antenna R_r.

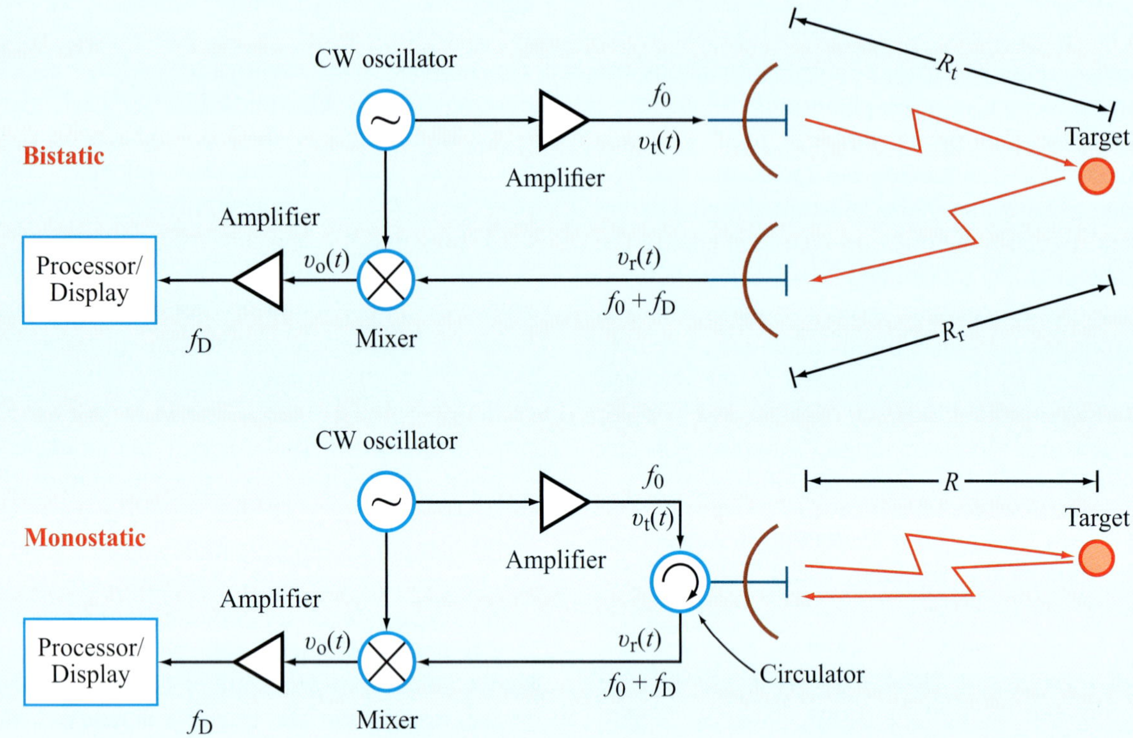

Figure 13-1: CW radar in bistatic and monostatic configurations.

The received signal $\upsilon_r(t)$, when the target is not moving relative to the radar, is

$$\upsilon_r(t) = K\cos[2\pi f_0(t-T)] = K\cos(2\pi f_0 t - \phi), \quad (13.3)$$

where K is a constant computed from the radar equation given by Eq. (5.29b):

$$K = A\,\frac{G\lambda}{(4\pi)^{3/2}R^2}\,\sqrt{\sigma_{pq}}\,, \quad (13.4)$$

and the signal phase shift ϕ due to the signal time-of-flight is

$$\phi = 2\pi f_0 T = 2\pi f_0 \cdot \frac{2R}{c} = \frac{4\pi R}{\lambda}\,. \quad (13.5)$$

In Eq. (13.4), G is the antenna gain in the direction of the target and σ_{pq} is the target RCS for a q-polarized transmit antenna and a p-polarized receive antenna. Note that Eq. (13.3) can be written as

$$\begin{aligned}\upsilon_r(t) &= K\cos[2\pi f_0(t-T)]\\ &= K\cos(2\pi f_0 t - 2kR),\end{aligned} \quad (13.6)$$

where $k = 2\pi/\lambda$ is the wave number. The phase ϕ, modulo 2π, can be determined by comparing the transmitted and received signals. This can be done with a mixer that, in effect, multiplies the two signals together. The mixer output $\upsilon_0(t)$ in Fig. 13-1 is

$$\begin{aligned}\upsilon_0(t) &= \upsilon_t(t)\;\upsilon_r(t)\\ &= [A\cos(2\pi f_0 t)][K\cos(2\pi f_0 - \phi)]\\ &= \frac{AK}{2}\,[\cos(4\pi f_0 t - \phi) + \cos\phi]\,.\end{aligned} \quad (13.7)$$

After lowpass filtering, only the second term, known as the ***baseband*** term, is retained and $\upsilon_0(t)$ becomes

$$\begin{aligned}\upsilon_0(t) = \frac{AK}{2}\cos\phi &= \frac{AK}{2}\cos 2kR \\ &= \frac{AK}{2}\cos\left(\frac{4\pi R}{\lambda}\right).\end{aligned} \tag{13.8}$$

(baseband signal)

Note that the output is a constant value that does not change with time if the relative range between the target and radar is fixed. It only changes if the range to the target changes. We note that for an ideal sine wave, only ϕ modulo 2π is significant, which corresponds to the fractional part of $2R/\lambda$. This means that the output voltage is very sensitive to small changes in range. The output amplitude can vary from $-AK/2$ to $+AK/2$, corresponding to a change of $\lambda/4$ in the range R. At X-band ($\lambda = 3$ cm), this is a change in R of only 0.75 cm!

13-1.2 Signal Scintillation

Consider what happens when there are two targets in the scene at slightly different ranges. The received signal voltage is the sum of the signal voltages from the two targets:

$$\begin{aligned}\upsilon_r(t) &= \upsilon_1(t) + \upsilon_2(t) \\ &= K_1\cos[2\pi f_0(t - T_1)] + K_2\cos[2\pi f_0(t - T_2)] \\ &= K_1\cos(2\pi f_0 t - 2kR_1) + K_2\cos(2\pi f_0 t - 2kR_2),\end{aligned} \tag{13.9}$$

where the subscripts indicate the target number. The corresponding baseband mixer output after lowpass filtering is

$$\upsilon_0(t) = \frac{AK_1}{2}\cos 2kR_1 + \frac{AK_2}{2}\cos 2kR_2. \tag{13.10}$$

(baseband signal for 2 targets)

For the purposes of analysis, we assume that the targets have similar radar cross sections and similar (but not identical) ranges, so $K_1 \approx K_2 = K$. Using this approximation and geometric relations, Eq. (13.10) can be written as

$$\upsilon_0(t) = AK\cos[k(R_1 + R_2)]\cos[k(R_1 - R_2)]. \tag{13.11}$$

For the next step, refer to Fig. 13-2(a), which shows the geometry of the two targets and the radar. The two targets are separated by a distance of d at a mean range of $R_0 = (R_1 + R_2)/2$. If the range is sufficiently large, the two rays to the targets are nearly parallel in the vicinity of the targets so that we can approximate R_1 and R_2 as

$$R_1 \approx R_0 + (d/2)\sin\theta, \qquad R_2 \approx R_0 - (d/2)\sin\theta. \tag{13.12}$$

Substituting these approximations in Eq. (13.11) leads to

$$\upsilon_0(t) = AK\cos(2kR_0)\cos(kd\sin\theta). \tag{13.13}$$

The magnitude of the baseband voltage has the form

$$|V| = V_0\,|\cos(kd\sin\theta)|, \tag{13.14}$$

where $V_0 = AK\cos 2kR_0$ is a constant.

A plot of $|V|$ as a function of $\sin\theta$ is shown in Fig. 13-2(b). The signals from the two targets interfere with each other to reinforce or cancel each other depending on the aspect angle θ between the baseline between the targets and the direction to the radar.

> ▶ This interference phenomenon arises due to the ***coherence*** of the transmit and receive signals. Signal coherence is a fundamental property of radars. ◀

It is instructive to consider the characteristics of this pattern. Note, for example, that the first null away from the origin occurs at $\theta = \theta_1$ where

$$\frac{2\pi d}{\lambda}\sin\theta_1 = \frac{\pi}{2}. \tag{13.15}$$

Here the subscript 1 is used to indicate the fact that it is the first null. This equation may be solved to obtain

$$\sin\theta_1 = \frac{\lambda}{4d},$$

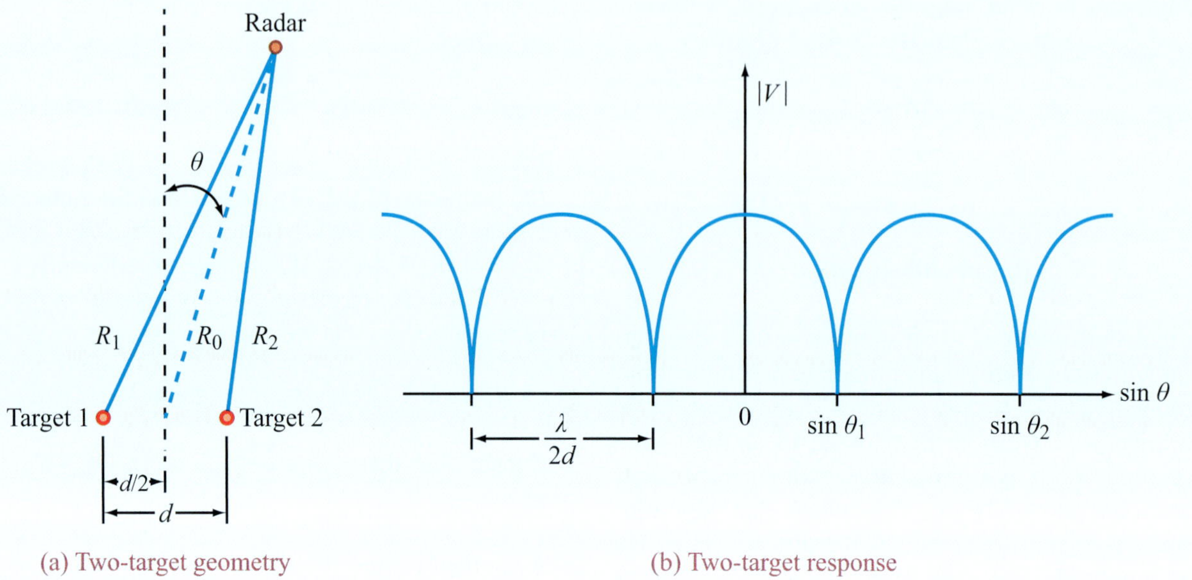

Figure 13-2: Scattering from two targets.

or

$$\theta_1 \approx \frac{\lambda}{4d}, \quad \text{if } d \gg \lambda. \tag{13.16}$$

The last result occurs when the scatterers are many wavelengths apart, so the null angle is close to zero. The second null is obtained from

$$\frac{2\pi d}{\lambda} \sin\theta_2 = \frac{3\pi}{2}, \tag{13.17}$$

which leads to

$$\sin\theta_2 = \frac{3\lambda}{4d},$$

or

$$\theta_2 \approx \frac{3\lambda}{4d}, \quad \text{if } d \gg \lambda. \tag{13.18}$$

The angular spacing between the two adjacent nulls can be derived using

$$\Delta \sin\theta = \sin\theta_2 - \sin\theta_1 = \frac{\lambda}{2d},$$

or

$$\Delta\theta \approx \theta_2 - \theta_1 = \frac{\lambda}{2d}, \quad \text{if } d \gg \lambda. \tag{13.19}$$

Thus the nulls in θ are spaced by approximately the inverse of twice the target spacing in wavelengths.

When θ is fixed, the baseband signal is fixed. However, if the targets move relative to the radar, θ may vary with time. This causes the baseband signal amplitude to vary so that the signal appears to "fade" in strength with time. This ***fading*** occurs due to the coherent nature of the scattered signals as discussed in detail in Section 5-6. Additional targets leads to more complicated fading patterns and signal fluctuations.

13-1.3 Target Moving Relative to Radar

When the target is moving relative to the radar, the received signal is also ***Doppler-shifted***. The ***Doppler frequency*** is the rate of change of the phase of the signal due to the changing distance between the radar and target. We note that the instantaneous frequency of a sine wave is the time derivative of the argument of the sine wave. Thus, the instantaneous frequency ω of the received signal in Eq. (13.6) is

$$\omega = \frac{d}{dt}(2\pi f_0 t - 2kR)$$

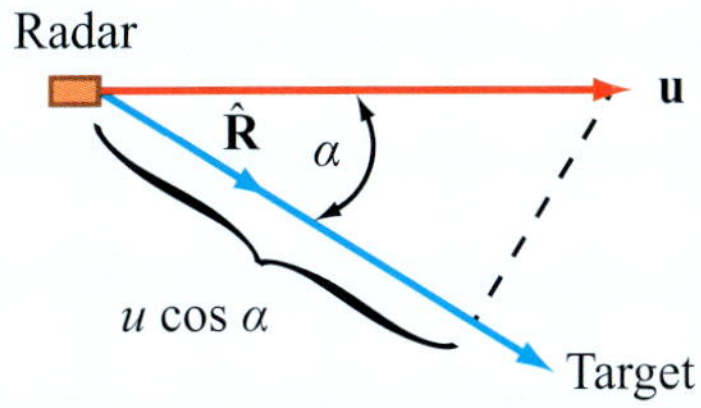

Figure 13-3: Doppler geometry.

$$= 2\pi f_0 - 2k\frac{d}{dt}R = 2\pi f_0 - \frac{4\pi}{\lambda}\frac{d}{dt}R. \qquad (13.20)$$

The frequency f of the received signal is then

$$f = \frac{\omega}{2\pi} = f_0 - \frac{2}{\lambda}\frac{d}{dt}R = f_0 + f_D, \qquad (13.21)$$

where f_D is the Doppler frequency shift of the carrier frequency f_0 caused by the change in R with time:

$$f_D = -\frac{2}{\lambda}\frac{d}{dt}R = -\frac{2u_R}{\lambda}, \qquad (13.22)$$

where u_R is the ***relative velocity*** between the radar and target in the radial direction. That is, the Doppler frequency shift is the negative of twice the radial speed in wavelengths per second.

▶ The negative sign comes about because the Doppler frequency is positive when the radar is approaching the target, so that R is decreasing, while the Doppler frequency is negative when the radar is receding from the radar when R is increasing. ◀

The radial component of the velocity u_R may be obtained from the total velocity vector **u** and the vector distance **R** by (Fig. 13-3)

$$u_R = -\mathbf{u} \cdot \frac{\mathbf{R}}{|\mathbf{R}|} = -u\cos\alpha, \qquad (13.23)$$

where α is the angle between the velocity vector and the radius vector from the radar to target. Hence, we may

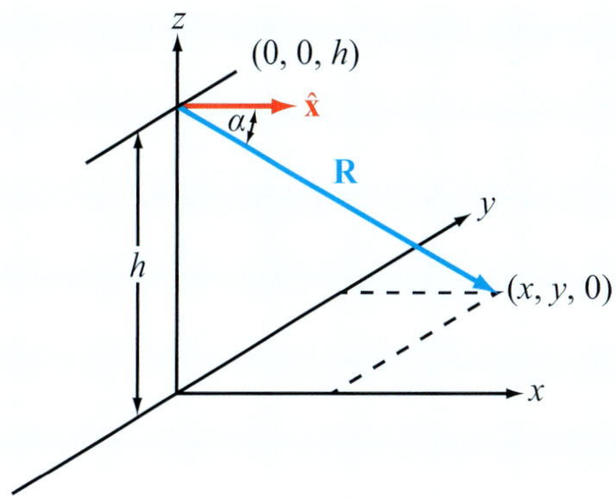

Figure 13-4: Geometry for Doppler calculation for horizontal motion above a plane surface.

write the Doppler frequency of Eq. (13.22) as

$$f_D = -\frac{2\mathbf{u} \cdot \mathbf{R}}{\lambda R}. \qquad (13.24)$$

If we assume that the radar is traveling in the positive x direction as shown in Fig. 13-4, then

$$\mathbf{u} = \hat{\mathbf{x}}u \qquad (13.25a)$$

and

$$\mathbf{R} = \hat{\mathbf{x}}x + \hat{\mathbf{y}}y - \hat{\mathbf{z}}h. \qquad (13.25b)$$

Using these definitions in Eq. (13.24) leads to

$$f_D = -\frac{2ux}{\lambda\sqrt{x^2 + y^2 + h^2}}. \qquad (13.26)$$

For a fixed value of f_D and h, this expression is a hyperbolic equation in x and y. The magnitude of the Doppler shift is largest when the radar is moving directly toward or away from the target; this value is defined as f_{D0},

$$f_{D0} = \pm\frac{2u}{\lambda}. \qquad (13.27)$$

Thus, the maximum range of Doppler frequencies extends from $-f_{D0}$ to $+f_{D0}$. Choosing various values within this range, we can plot lines of constant Doppler

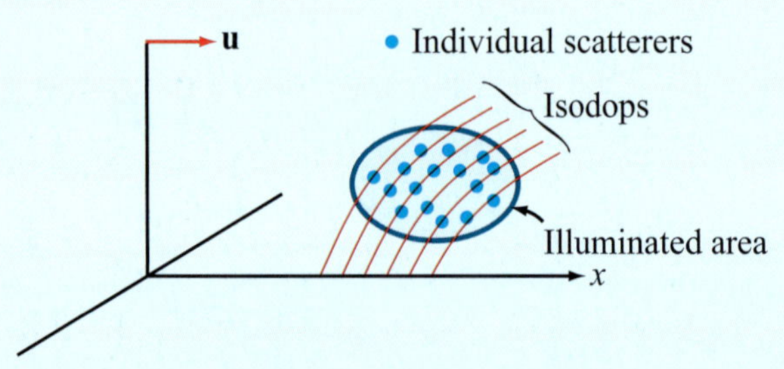

Figure 13-5: Multiple targets within an antenna footprint.

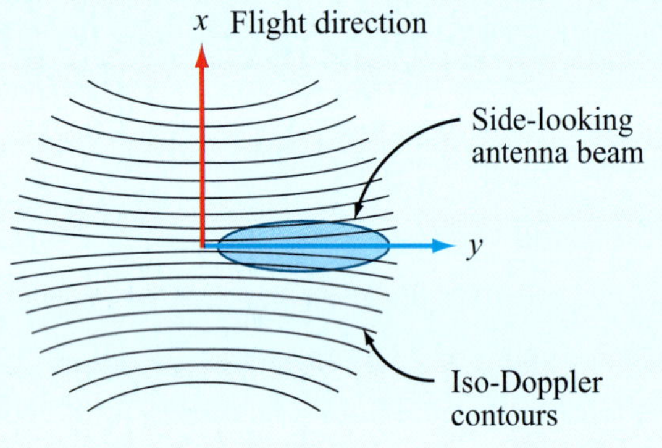

Figure 13-6: Iso-Doppler contours for horizontal motion above a plane surface. The oval represents the surface illumination area of the antenna mainlobe of a side-looking radar.

shift on the surface. These lines are known as ***iso-Doppler lines*** or ***isodops***. Figure 13-5 illustrates an evenly spaced set of isodops for a radar moving in the x direction over a stationary plane. Scatterers along the same isodop generate the same Doppler shift.

As an aside, an important special case of Eq. (13.26) occurs for a side-looking radar. For a side-looking radar traveling in the x direction, as shown in Fig. 13-6, the antenna is pointed to one side, along the y direction (or $-y$ direction). If the antenna beam is sufficiently narrow so that $x^2 \ll y^2$ over the mainlobe of the antenna on the surface, Eq. (13.26) can be approximated as

$$f_{\rm D} = -\frac{2ux}{\lambda\sqrt{x^2+y^2+h^2}} \approx -\frac{2ux}{\lambda\sqrt{y^2+h^2}} \,. \qquad (13.28)$$

This expression suggests that the Doppler frequency in the side-looking case is approximately linearly proportional to the along-track displacement x for a narrow-beam antenna.

When there are multiple targets in the illumination area shown in Fig. 13-5, the received signal is the linear sum of the returns from all targets. Since each target has a different x and y position, each target has a slightly different combination of Doppler shift and range or phase offset. This corresponds to the distributed target model introduced in Section 5-7. As the platform moves past the target, the relative phases and Doppler frequencies vary, resulting in time-varying signal fading and scintillation. This is discussed further later.

The isodop plot in Fig. 13-6 suggests a strategy whereby we can ***discriminate*** between different areas within the antenna illumination area because different areas of the antenna illumination pattern in Fig. 13-6 have different Doppler frequencies.

► By appropriately bandpass-filtering the received signal we can separate the return from one area (defined by the antenna pattern and the Doppler frequencies that fall within the bandpass filter) and another area similarly defined but for a different range of Doppler frequencies. ◄

This enables resolution of the antenna footprint into smaller areas by filtering the signal based on the Doppler frequency. Note that this is possible due to the Doppler shift of the received signal relative to the transmitted signal and requires that the radar be phase coherent.

13-2 Pulsed Radar

A CW radar transmits and receives continuously. However, signal leakage from the transmitter to the receiver limits its performance. As a result, most remote-sensing radars are pulsed. A ***pulsed*** radar transmits short signals, and then "listens" for the returns during the intervals between the transmit pulses. The transmit ***pulse length*** τ and ***pulse repetition frequency*** (PRF) $f_{\rm p}$ are chosen so that the desired return signal can

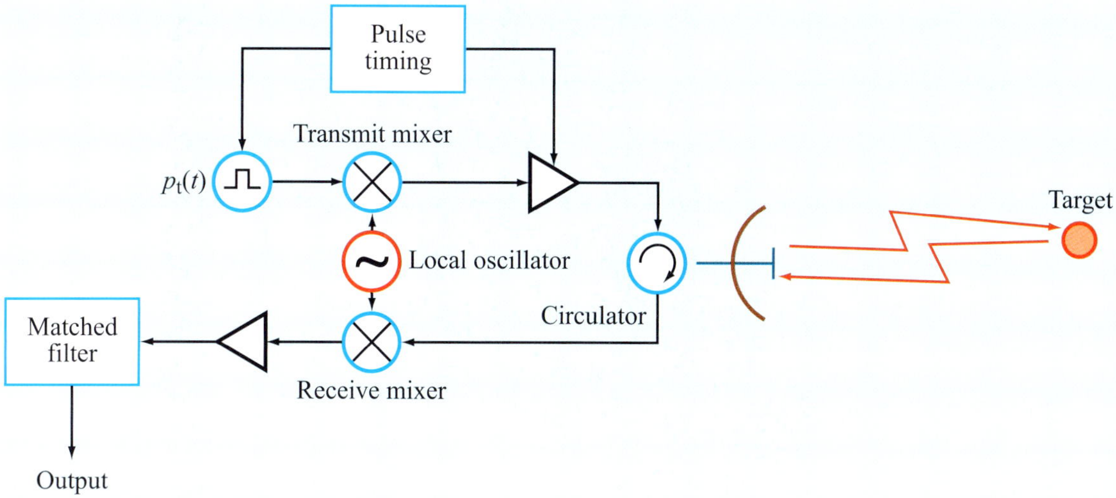

Figure 13-7: Pulsed-radar configuration.

be received without interference from the transmitted signal. On many radars, the timing is set so that there is only one pulse in flight at a time, though some radars employ multiple pulses in flight at a time.

Pulsed radars differ from CW radars by including a scheme to ***pulse-modulate*** the transmit signal as shown in Fig. 13-7. In this figure, the voltage of the transmitted pulse, represented by $p_t(t)$, is modulated or frequency-shifted from baseband to the transmit carrier frequency. Note that all radars generate baseband pulses; some radars generate transmit pulses by turning on and off a CW oscillator.

For a point target, the received pulse is a replica of the transmitted pulse delayed in time by an amount $T = 2R/c$:

$$\upsilon_r(t) = a(T)\, p_t(t-T) = a(R)\, p_t\left(t - \frac{2R}{c}\right), \quad (13.29)$$

where $a(T)$, and equivalently $a(R)$, is a factor that contains all system, target, and attenuation parameters, and $p_t(t)$ describes the transmitted pulse. For an area target containing n scatterers, the voltage received by the radar is the sum of all their contributions:

$$\begin{aligned}\upsilon_r(t) &= \sum_{i=1}^{n} a(T_i)\, p_t(t-T_i) \\ &= \sum_{i=1}^{n} a(R_i)\, p_t\left(t - \frac{2R_i}{c}\right),\end{aligned} \quad (13.30)$$

where $a(T_i)$ describes the amplitude of the backscattered voltage due to the ith scatterer at time delay T_i (or range R_i). For a ***rectangular pulse*** of duration τ,

$$p_t(t) = \begin{cases} V_0 \cos 2\pi f t & \text{for } 0 \le t \le \tau, \\ 0 & \text{otherwise.} \end{cases} \quad (13.31)$$

If the radar is moving relative to a point target i, the received voltage waveform is

$$p_r\left(t - \frac{2R_i}{c}\right) = \begin{cases} V_0 \cos\left[2\pi(f + f_{D_i})\left(t - \frac{2R_i}{c}\right)\right] & \text{for } T_i < t < T_i + \tau, \\ 0 & \text{otherwise,} \end{cases} \quad (13.32)$$

where T_i is the roundtrip time delay to target i, and f_{D_i} is the ***Doppler shift*** associated with that target.

Pulsing the transmit signal enables a capability not possible with passive sensors: ***range discrimination***—the ability to discriminate between targets at different ranges within the antenna illumination area. This capability is illustrated with the aid of Fig. 13-8. The figure shows $p_t(t)$ for a rectangular pulse approaching two targets at different ranges. In Fig. 13-8(a) targets 1 and 2 give radar echoes that cannot be separated or ***resolved*** because the return echoes overlap due to the fact that the two targets are too close together. At time t_1 the pulse has not returned from either target. At t_2 enough time has elapsed so that a return has started from target number 1. At t_3 a return has started from target 2, but target 1 is still providing a return. At t_4 no return is coming from target 1, but a return continues from target 2. And finally, at t_5 the pulse has passed both targets and no return is observed. The radar output shown separately exists for an extended period longer than the pulse, but there is no gap between the signals returned from targets 1 and 2 so they cannot be separated. The output as shown is constant, but in fact it may be different during the time when both targets are illuminated simultaneously. The return from the two targets together may be either larger or smaller than that from a single object, because the signals may either add coherently or cancel, depending upon their relative phases. The variation in amplitude, however, does not in general enable differentiation of the two targets, particularly in the presence of noise.

In Fig. 13-8(b) the two targets are far enough apart so that they can be separately resolved. That is, at t_2 the signal has started to return from target 1 and at t_3 it has ceased to return from target 1. Thus, a single unique return pulse is obtained from each target. The discrimination capability is therefore related to the duration of the pulse relative to the spacing between the two targets.

If target 1 is at a distance R_1 with corresponding round-trip time delay $T_1 = 2R_1/c$, the total time between the start of the transmitted pulse and the arrival of the tail

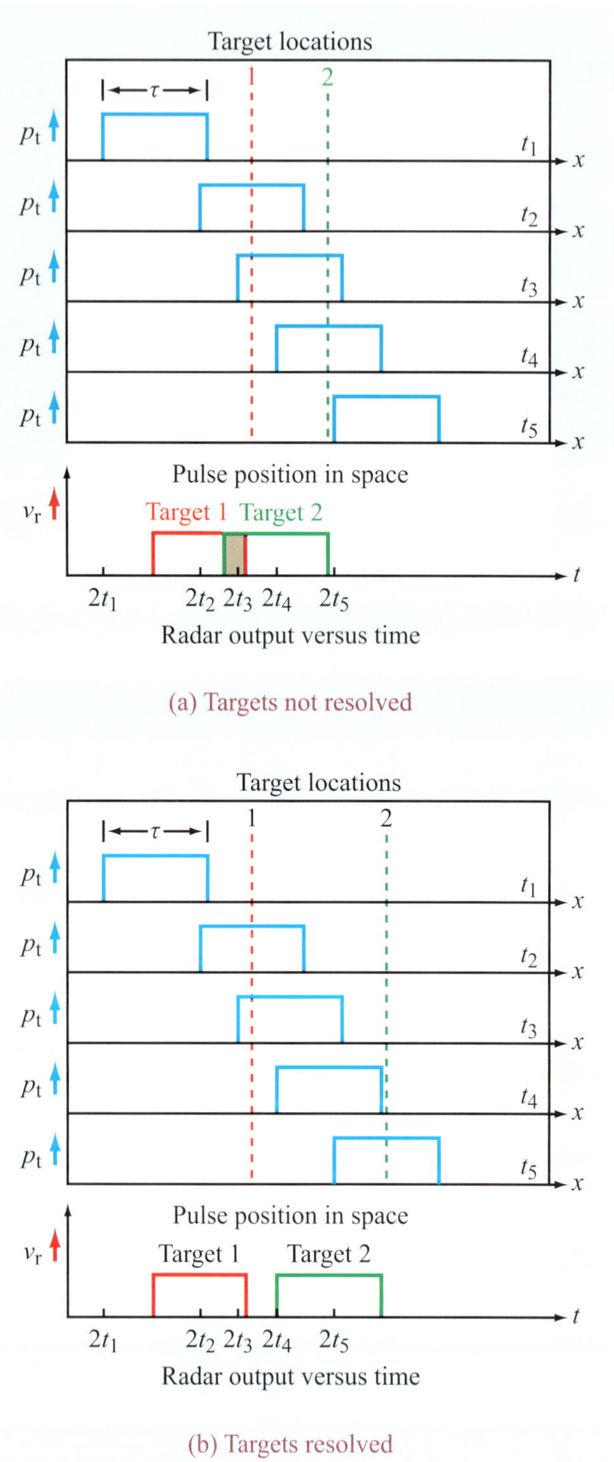

Figure 13-8: Examples of range resolution and target discrimination with a windowed pulse.

end of the returned pulse is

$$T_a = \frac{2R_1}{c} + \tau, \tag{13.33}$$

where τ is the pulse length. For target 2, the earliest response time, corresponding to the arrival of the front end of the returned pulse, is

$$T_b = \frac{2R_2}{c}\,. \tag{13.34}$$

To discriminate between the two targets, T_b must be greater than T_a. The condition $T_b = T_a$ represents the minimum spacing between R_2 and R_1 such that the two targets are distinguishable on the basis of their time delay. Setting $T_b = T_a$ leads to

$$\frac{2R_1}{c} + \tau = \frac{2R_2}{c}\,, \tag{13.35}$$

from which we obtain the ***range resolution***

$$\Delta R = R_2 - R_1 = \frac{c\tau}{2}\,. \tag{13.36}$$

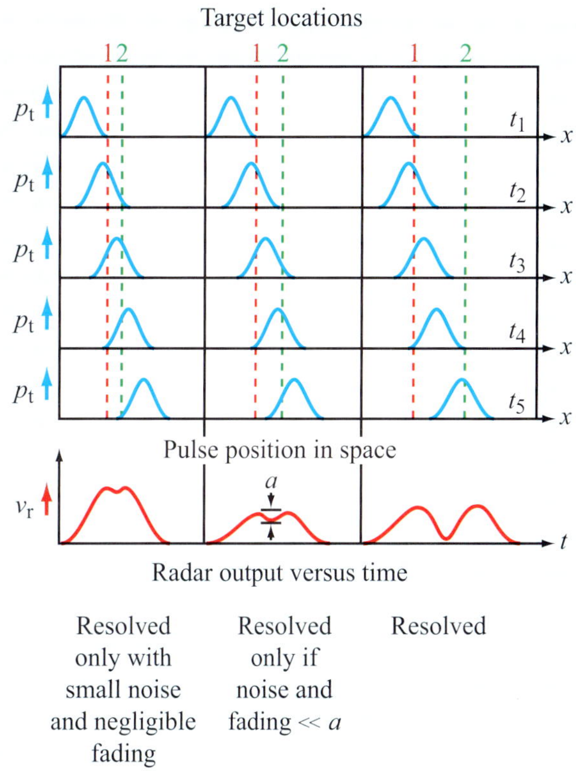

Figure 13-9: Examples of range resolution and target discrimination with a windowed pulse.

A more realistic pulse-shape example is shown in Fig. 13-9, where the pulse shape is no longer rectangular—either by design or due to unavoidable filtering in the hardware. In the left column, the two targets are so close together that the returns overlap. A small decrease in power at the midpoint is indicated, as would be found if the two returns were in phase with each other. However, the presence of two targets in this situation can only be established if the noise level is small enough that the dip between the two returns can be reliably interpreted. In the middle example the targets are somewhat farther apart, and have a correspondingly deeper dip between the two returns. In the right column, the targets are far enough apart so that the response goes essentially to zero between them, and one can clearly establish that two distinct targets are present.

We observe that the return signals from targets at different ranges or time delays can be separated based on the pulse shape and length. For a platform moving horizontally over a flat surface, Fig. 13-10 displays lines of constant range. These ***isorange lines*** define regions of the surface that generate return signals received by the radar at the same time.

Different areas of the antenna illumination pattern in Fig. 13-10 have different ranges, corresponding to different time delays. The return signals from these different regions within the footprint can be separated by ***time-filtering*** the data; i.e., only looking at the data within a short ***time window*** or ***gate***. Such filtering is known as ***range gating***. Range gating enables the radar to resolve (discriminate between) targets at different ranges.

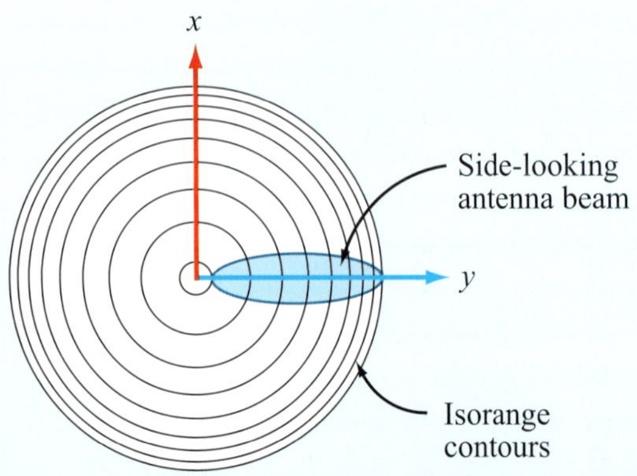

Figure 13-10: Isorange contours over a plane surface. The oval represents the surface illumination area of the mainlobe of an antenna for a side-looking radar traveling along the x direction. Isorange contours are equally spaced, but ground-range contours are not.

13-3 Range and Doppler Resolution

> ▶ In microwave sensing, ***resolution*** is the spacing (expressed in terms of angle, range, or speed as appropriate) between the points at which the response power is half the peak-power response. ◀

That is, the resolution is defined as the half-power width of the response. This is illustrated in Fig. 13-11, where the resolution w is shown as the width between the half-power points in the response from the target sensed.

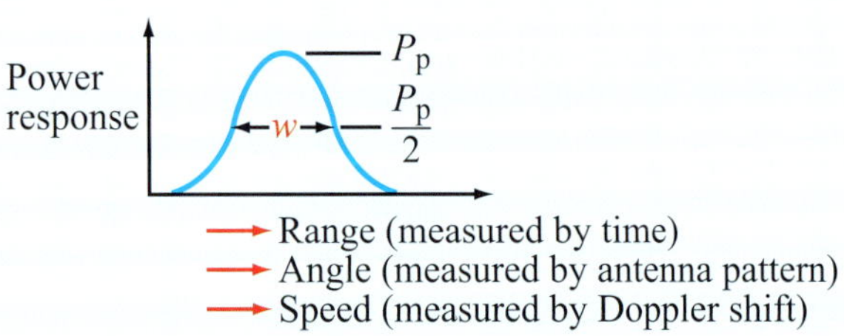

Figure 13-11: Radar definition of resolution—the half-power width.

The microwave system obtains ***discrimination*** and, hence, resolution by measurement of one or more of the three quantities: *angle, range,* and *speed.* Angular discrimination is defined by the beamwidth of the antenna. If the antenna has a narrower beam, the discrimination is better and the resolution distance on the ground is smaller: we say the resolution is ***finer***. Essentially, the ***angle resolution*** is set by the 3 dB (one-half power) beamwidth of the antenna illumination pattern on the surface.

Range resolution or distance resolution, is obtained through discrimination of time delay, which is equivalent to a range discrimination because of the known constant speed of the electromagnetic wave. Many different kinds of time delay techniques may be used in radar. The most common is pulsed operation as indicated in the examples shown in Figs. 13-8 and 13-9. ***Pulse compression*** or ***range compression***, discussed later, is an extension of this idea and can result in finer range discrimination.

Doppler frequency resolution and discrimination capability arise because the received frequency is proportional to the relative speed between the object sensed and the radar system. Thus, a filter discriminating the different frequencies received is capable of discriminating the different speeds. Different points on the earth have different relative speeds, so filters that distinguish relative speeds may be used to discriminate between signals from different parts of the surface of the earth.

Figure 13-12 illustrates the different techniques and their combinations for a microwave system carried on an aircraft traveling horizontally above the surface of the earth. In Fig. 13-12(a) angle measurement alone is used; that is, the beamwidth of the antenna establishes the resolution. The dark area on the ground lies within the half-power contours of the antenna pattern and therefore is defined as the resolvable area or ***resolution cell*** for an angle-only system. Passive microwave (radiometer) systems depend on angle resolution defined by the antenna pattern. Because both the range and speed measurement are based on comparison of the received signal with modulation imposed at a transmitter, and

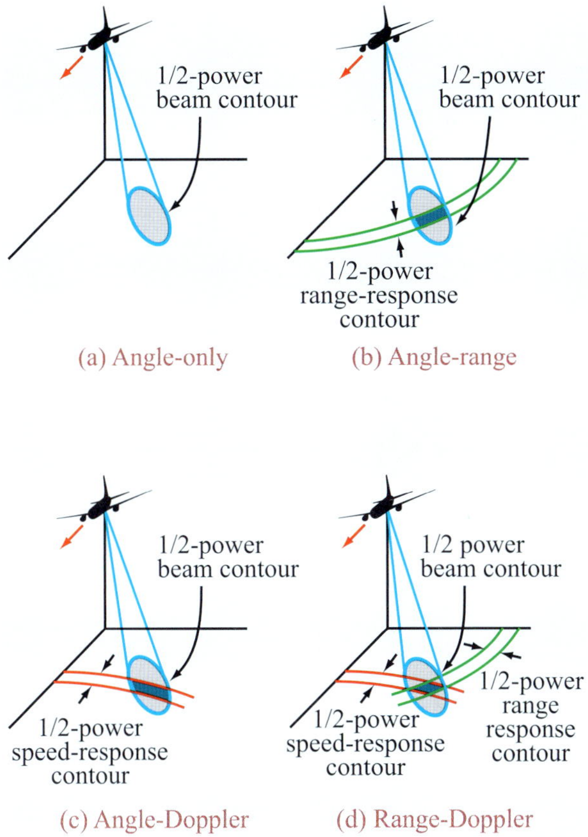

Figure 13-12: Methods for resolution in microwave sensing: (a) angle only, (b) angle and range, (c) angle and Doppler, (d) range and Doppler.

since the radiometer has no transmitter, these methods are not available to it.

In Fig. 13-12(b) the ***combination of range (distance) and angle*** resolution techniques is shown. If only range measurement is used (i.e., if the antenna beam illuminates the entire surface of the ground), the resolution cell is a ring lying between the two half-power range response contours. By combining the angle measurement and range measurement, the resolvable region or resolution cell is limited to the smaller dark area. In this case the antenna pattern is narrow in one direction and wide in the other direction. Range resolution allows breaking up the wide direction of the illuminated ground into smaller segments. The narrow direction of the antenna beam is used for resolution in angle.

Figure 13-12(c) shows a ***combination of angle and Doppler*** measurements used for resolution. Here, half-power Doppler response contours are illustrated on the surface. As previously noted, for horizontal motion above a plane surface, these contours are hyperbolas, so the resolution cell is a hyperbolic-shaped strip like that shown in the figure. By combining the angular and Doppler responses, a smaller resolvable area, as illustrated by the dark area, can be obtained. This method has been widely used in airborne and spaceborne scatterometer systems.

Figure 13-12(d) shows a ***combination of range and Doppler*** discrimination to establish the small resolution cell shown by the dark area as the intersection of the isorange and iso-Doppler contours.

▶ Since the half-power Doppler-response and half-power range-response contours may be made closer together than the antenna beamwidth, the resolution obtained by combining range and Doppler techniques is usually much finer than that obtained when an angle measurement is used. This idea is the basis for ***synthetic-aperture radar*** (discussed in Chapter 14) and provides the finest possible resolution. ◀

Since passive radiometers must generally depend on angular resolution alone, their ability to obtain a small resolution cell on the ground is limited. Hence, when fine resolution is required, radar is needed. For some applications where fine resolution is not required the selection between the sensor classes (active versus passive) may be made on the basis of the ability of one or the other (or a combination) to best determine the parameters of interest.

13-4 Frequency-Modulated Radar

A CW radar has no range- or timing-discrimination capability. This can be understood intuitively by noting that the time resolution of a CW radar is equivalent to an infinitely long pulse length. However, if ***frequency modulation*** (FM) of the transmit signal is employed, range resolution can become possible. In this section we examine the operational principles of FMCW and pulsed-FM radar.

In an ***FMCW radar*** the signal is continuously transmitted and received, but the signal frequency is varied as a function of time. While other modulation schemes are possible, ***linear FM*** (LFM) is most commonly used. Figure 13-13(a) illustrates the instantaneous transmit signal frequency for an LFMCW radar. The transmit signal is linearly swept across a band B of frequencies, from f_1 to f_2, in a time $T_R/2$, where T_R is the ***modulation period***. In this figure, the sweep or ***ramp rate*** α, which is the same for both rising and falling portions of the signal modulation, is

$$\alpha = \frac{2B}{T_R} \qquad \text{(Hz/s)}. \tag{13.37}$$

The upward or downward varying LFM signals are sometimes termed ***chirps***.

The echo from a point target is a time-delayed, attenuated copy of the transmit signal just as in the cases previously considered. Usually the swept bandwidth is much larger than the maximum Doppler; i.e., $B \gg f_{D0}$. Due to the time delay, the received signal lags the transmitted signal and as a result the two signals have different frequencies with the ***intermediate frequency difference*** f_I being directly proportional to the range to the target R, as indicated in Fig. 13-13(b). From the geometry of the similar triangles in Fig. 13-13(a), it is evident that the ratio of the time delay T to $T_R/2$ is the same as the ratio of f_I to B. Thus,

$$\frac{T}{T_R/2} = \frac{f_I}{B}\,. \tag{13.38}$$

Upon replacing T with $2R/c$ and solving for the frequency difference f_I, we have

$$f_I = \left(\frac{4B}{cT_R}\right) R = \frac{2\alpha}{c}\, R, \tag{13.39}$$

where α is the ramp rate defined by Eq. (13.37). Note the direct relationship between the intermediate frequency difference f_I and the range R; the proportionality constant consists of the velocity of light c and two system constants, the modulation bandwidth B and the modulation period T_R. Upon mixing the received signal with a copy of the transmitted signal and lowpass-filtering the mixer output signal, an LFMCW radar generates a signal whose frequency is f_I. The mixing process is known as ***dechirping***. Measuring f_I by a

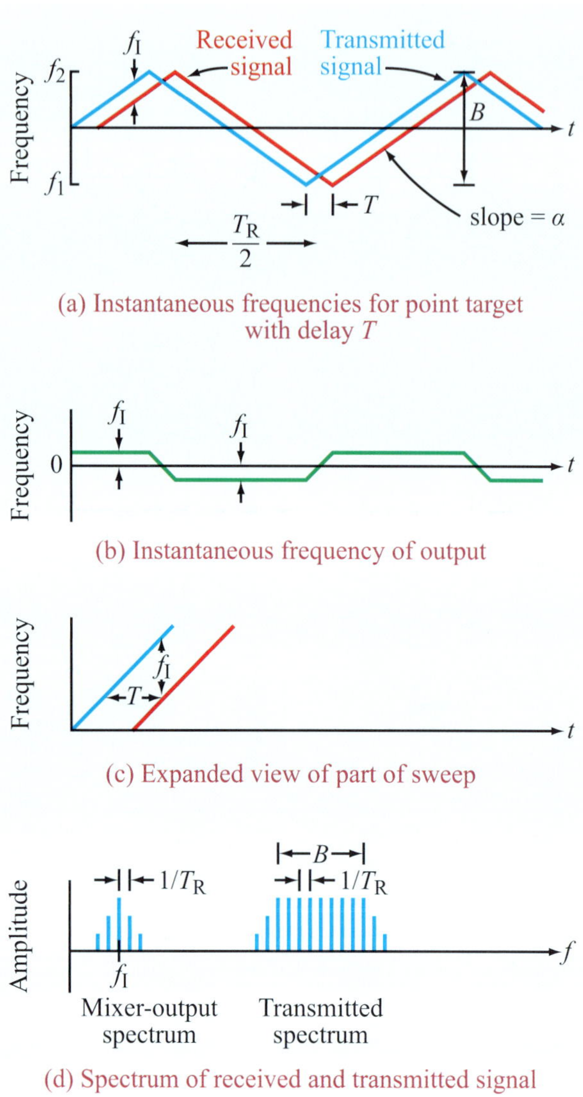

(a) Instantaneous frequencies for point target with delay T

(b) Instantaneous frequency of output

(c) Expanded view of part of sweep

(d) Spectrum of received and transmitted signal

Figure 13-13: Operation of an LFMCW radar.

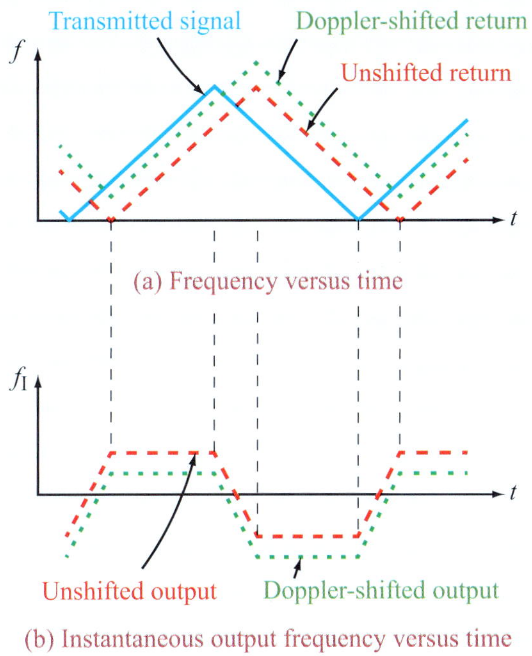

Figure 13-14: Effect of Doppler shift on an LFMCW radar: (a) frequency versus time; (b) instantaneous output frequency versus time. Note that the effect of Doppler is to shift the center frequency upward or downward depending on the sign of the Doppler shift.

frequency counter or a similar technique provides a measurement of R.

If the signal includes a Doppler shift, the time variation of f_I becomes unsymmetrical during the two halves of the modulation period T_R (Fig. 13-14). This asymmetry can provide an advantage through a processing circuit that can measure both the Doppler shift and the equivalent unshifted f_I, and then the latter is used to compute the range R.

We note that because the sweep is of finite length, the spectrum of the received signal is spread in frequency. The spectrum of a single, continuous sweep of finite duration is also continuous. However, the spectrum of the repetitive sweep is a line spectrum as illustrated in Fig. 13-13(d). So long as the sweep is of sufficient duration, the line spectrum closely approximates samples of the continuous spectrum. Mixing the transmit and receive signals and lowpass-filtering results in a signal spectrum centered at the difference frequency f_I, as shown in Fig. 13-13(d).

Over the time period of a single sweep and ignoring the transition periods at the beginning and end of the sweep, the ideal mixer output baseband signal $\upsilon_b(t)$ is

$$\upsilon_b(t) = K\cos(2\pi f_I t + \phi), \tag{13.40}$$

where ϕ is a target-dependent phase shift. When multiple targets at different ranges are present, the baseband mixer output is

$$\begin{aligned} \upsilon_b(t) &= \sum_n K_n \cos(2\pi f_{I_n} t + \phi_n) \\ &= \sum_n K_n \cos\left(\frac{8\pi B R_n t}{c T_R} + \phi_n\right), \end{aligned} \tag{13.41}$$

where we used Eq. (13.39) to relate f_{I_n} to range R_n. Note that targets at different ranges are at different frequencies. Thus, targets at different ranges can be discriminated by narrowband bandpass filters at different center frequencies corresponding to different ranges. The minimum bandwidth such filters can achieve determines the effective range resolution of the LFMCW radar.

At baseband, the signal return during a single frequency sweep is essentially a constant-amplitude cosine wave.

▶ From basic signal processing theory, the effective 3 dB bandwidth of a finite-duration CW signal of length $T_R/2$ is $2/T_R$. ◀

Hence, the effective frequency resolution Δf_I of the baseband signal is $2/T_R$; i.e., one frequency resolution element Δf_I extends from $f_a = f'_0 - 1/T_R$ to $f_b = f'_0 + 1/T_R$, where f'_0 is the center of the resolution element bandwidth. In effect, targets must be separated by at least Δf_I to be able to discriminate between them.

From Eq. (13.39), the conversion from baseband frequency to range is

$$R = \frac{c f_I T_R}{4B}. \tag{13.42}$$

The range resolution ΔR of the LFMCW system corresponds to $\Delta f_{\mathrm{I}} = 2/T_{\mathrm{R}}$:

$$\Delta R = \frac{cT_{\mathrm{R}}}{4B}\,\Delta f_{\mathrm{I}} = \frac{cT_{\mathrm{R}}}{4B}\cdot\frac{4}{T_{\mathrm{R}}} = \frac{c}{2B}\,. \tag{13.43}$$

(LFMCW range resolution)

▶ Thus, the effective range resolution depends only on the bandwidth of the LFMCW signal. It does not depend on the sweep length.‡ ◀

By implementing a bank of bandpass filters (or their digital equivalents), the LFMCW return signal can be resolved into multiple resolution elements. The result is that an LFMCW can have a range discrimination capability similar to pulsed radars.

We note that because the signal is continuously transmitted and received, the total signal energy is maximized compared with a pulsed radar with the same parameters. As discussed later, this maximizes the effective signal-to-noise ratio.

13-5 Matched Filtering

When processing the received signal, a radar filters the signal prior to detection. The filter that maximizes the ***signal-to-noise ratio*** (SNR) is derived from the waveform of the echo of the transmitted signal and is known as the ***matched filter***. The matched filter defines the radar's response, and hence its resolutions in time and Doppler frequency, or equivalently in range and speed. In this section we show that the matched filter indeed maximizes the measurement SNR.

A general filter can be described in terms of its ***impulse response***, which is the output of the filter for an impulse input, or its ***frequency response***, which is the Fourier transform of the impulse response. The frequency response describes the magnitude and phase of the filter output as a function of frequency. Figure

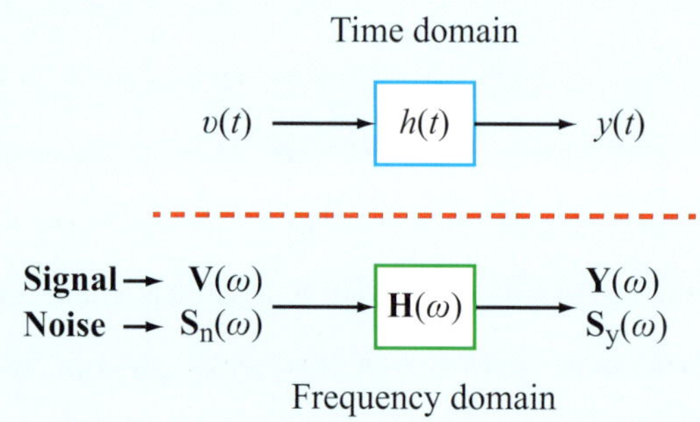

Figure 13-15: Filter inputs and outputs in the time and frequency domains.

13-15 illustrates the notation for analyzing the matched filter.

The ***Fourier transform*** of a general continuous signal $\upsilon(t)$ is defined as

$$\mathbf{V}(\omega) = \int_{-\infty}^{\infty} \upsilon(t)\, e^{-j\omega t}\, dt \equiv \mathcal{F}\{\upsilon(t)\}, \tag{13.44}$$

where $\omega = 2\pi f$, while the ***inverse Fourier transform*** of $\mathbf{V}(\omega)$ is

$$\upsilon(t) = \frac{1}{2\pi}\int_{-\infty}^{\infty} \mathbf{V}(\omega)\, e^{j\omega t}\, df \equiv \mathcal{F}^{-1}\{\mathbf{V}(\omega)\}. \tag{13.45}$$

The ***spectrum*** of a signal is defined as the magnitude squared of the Fourier transform. The total signal energy $\mathcal{E}_\upsilon$ in the time and frequency domains are related by ***Parseval's theorem***,

$$\mathcal{E}_\upsilon = \int_{-\infty}^{\infty} |\upsilon(t)|^2\, dt = \frac{1}{2\pi}\int_{-\infty}^{\infty} |\mathbf{V}(\omega)|^2\, d\omega. \tag{13.46}$$

When the signal $\upsilon(t)$ is passed through a linear time-invariant filter with an impulse response $h(t)$, the output $y(t)$ is the time-domain convolution of $h(t)$ and $\upsilon(t)$; i.e.,

$$y(t) = \upsilon(t) * h(t) = \int_{-\infty}^{\infty} \upsilon(t')\, h(t-t')\, dt'. \tag{13.47}$$

Using the properties of the Fourier transform, it can be shown that convolution in the time domain is equivalent

‡We note that this result holds for LFM signals for time-bandwidth products greater than about 10, i.e., $T_{\mathrm{R}}B > 10$. For lower time-bandwidth products, a more precise analysis may be required.

to multiplication in the frequency domain (Ulaby and Yagle, 2013). Thus, the output signal in the frequency domain $\mathbf{Y}(\omega) = \mathcal{F}\{y(t)\}$ can be written as

$$\mathbf{Y}(\omega) = \mathbf{H}(\omega)\,\mathbf{V}(\omega), \tag{13.48}$$

where the ***filter frequency response*** $\mathbf{H}(\omega) = \mathcal{F}\{h(t)\}$ is the Fourier transform of the filter's impulse response. The signal energy $\mathcal{E}_y$ at the output of the matched filter is

$$\mathcal{E}_y = \int_{-\infty}^{\infty} |y(t)|^2\,dt = \frac{1}{2\pi}\int_{-\infty}^{\infty} |\mathbf{Y}(\omega)|^2\,d\omega = \frac{1}{2\pi}\int |\mathbf{H}(\omega)\,\mathbf{V}(\omega)|^2\,d\omega. \tag{13.49}$$

Frequency analysis for random signals has to be treated somewhat differently by using the concept of ***power spectral density*** (PSD). The PSD of a random signal, such as noise, is the Fourier transform of the signal correlation function. For a wide-sense stationary random signal $n(t)$, the ***autocorrelation function*** $R_n(t')$ is defined as

$$R_n(t') = \langle n(t)\,n^*(t+t')\rangle = \int_{-\infty}^{\infty} n(t)\,n^*(t+t')\,dt, \tag{13.50}$$

where t' is the time shift and the brackets indicate the time average, which in this case is the same as the expected value. The signal PSD is then

$$\mathbf{S}_n(\omega) = \mathcal{F}\{R_n(t')\}. \tag{13.51}$$

Note that the average random signal power is

$$\langle |n(t)|^2\rangle = R_n(0) = \frac{1}{2\pi}\int_{-\infty}^{\infty} \mathbf{S}_n(\omega)\,d\omega. \tag{13.52}$$

When a random signal $n(t)$ with PSD $\mathbf{S}_n(\omega)$ is passed through the filter $\mathbf{H}(\omega)$, the output signal's PSD $\mathbf{S}_y(\omega)$ is given by

$$\mathbf{S}_y(\omega) = |\mathbf{H}(\omega)|^2\,\mathbf{S}_n(\omega). \tag{13.53}$$

The random signal power P_y is then

$$P_y = \frac{1}{2\pi}\int \mathbf{S}_y(\omega)\,d\omega = \frac{1}{2\pi}\int |\mathbf{H}(\omega)|^2\,\mathbf{S}_n(\omega)\,d\omega, \tag{13.54}$$

while the random signal energy over a time period τ is

$$\mathcal{E}_y = \tau P_y. \tag{13.55}$$

The received signal observed by the radar includes not only the echo of the transmitted signal, but also the noise-like signal emitted by the scene and generated in the radar hardware. For the radar, this is undesired noise. The total received signal $x(t)$ can be written as

$$x(t) = \upsilon(t) + n(t), \tag{13.56}$$

where $n(t)$ is the noise component and $\upsilon(t)$ is the radar signal. As described in Chapter 7, the power associated with noise is given by

$$P_\text{n} = kT_\text{SYS}B, \tag{13.57}$$

where T_SYS is the system noise temperature and B is the system noise bandwidth. Over practical microwave bandwidths, the radiometric signal can be treated as narrow-band Gaussian white noise. Over the bandwidth B (in Hz) centered at ω_0, the noise has a constant power spectral density (PSD) of $N_0 = kT_\text{SYS}$. Hence, the noise PSD $\mathbf{S}_\text{n}(\omega)$ is given by

$$\mathbf{S}_\text{n}(\omega) = \begin{cases} N_0, & \omega_0 - \pi B \leq \omega_0 \leq \omega_0 + \pi B, \\ 0 & \text{otherwise}. \end{cases} \tag{13.58}$$

For a radar, the noise and received signal can be treated as independent signals.

To derive the ***matched filter***, we want to determine the filter $h(t)$ (or, equivalently, $\mathbf{H}(\omega)$) that maximizes the SNR at the output of the filter. Since the filter is linear and signal and noise are independent, the matched-filtered signal and noise outputs are independent.

After filtering the noise with $\mathbf{H}(\omega)$, the noise output power P_n is given by

$$P_\text{n} = \frac{1}{2\pi}\int_{-\infty}^{\infty} |\mathbf{H}(\omega)|^2\,\mathbf{S}_\text{n}(\omega)\,d\omega. \tag{13.59}$$

Note that since the useful signal bandwidth *must* be contained within the receiver bandwidth, the integral is over the receiver bandwidth. Since the noise is assumed to be white; i.e., $\mathbf{S}_\mathrm{n}(\omega) = N_0$ over the receiver bandwidth,

$$P_\mathrm{n} = \frac{N_0}{2\pi} \int_{-\pi B}^{\pi B} |\mathbf{H}(\omega)|^2 \, d\omega. \tag{13.60}$$

At a time delay t_d, the matched filter output is

$$y(t_\mathrm{d}) = \frac{1}{2\pi} \int \mathbf{H}(\omega)\, \mathbf{V}(\omega)\, e^{j\omega t_\mathrm{d}} \, d\omega, \tag{13.61}$$

so the power of the matched filter is given by

$$|y(t_\mathrm{d})|^2 = \left| \frac{1}{2\pi} \int \mathbf{H}(\omega)\, e^{j\omega t_\mathrm{d}}\, \mathbf{V}(\omega) \, d\omega \right|^2. \tag{13.62}$$

The matched filter output noise power at a particular time is given by Eq. (13.60). Thus, at the time delay t_d, the matched filter output SNR is

$$\mathrm{SNR_{out}} = \frac{|y(t_\mathrm{d})|^2}{P_\mathrm{n}} = \frac{\left| \dfrac{1}{2\pi} \displaystyle\int \mathbf{H}^*(\omega)\, e^{j\omega t_\mathrm{d}}\, \mathbf{V}(\omega) \, d\omega \right|^2}{\dfrac{N_0}{2\pi} \displaystyle\int |\mathbf{H}(\omega)|^2 \, d\omega}. \tag{13.63}$$

To proceed, we use the ***Schwartz inequality*** given by

$$\left| \int f(x)\, g^*(x)\, dx \right|^2 \leq \left(\int |f(x)|^2 \, dx \right) \left(\int |g(x)|^2 \, dx \right). \tag{13.64}$$

Note that equality is achieved only when $g(x) = af(x)$ for some constant a.

The numerator of the SNR expression in Eq. (13.63) is then bounded by

$$\begin{aligned} &\left| \int \mathbf{H}(\omega)\, e^{j\omega t_\mathrm{d}}\, \mathbf{V}(\omega) \, d\omega \right|^2 \\ &\leq \left(\int \left|\mathbf{H}(\omega)\, e^{j\omega t_\mathrm{d}}\right|^2 d\omega \right) \left(\int |\mathbf{V}(\omega)|^2 \, d\omega \right) \\ &= 2\pi \mathcal{E}_\upsilon \int |\mathbf{H}(\omega)|^2 \, d\omega, \end{aligned} \tag{13.65}$$

where $\mathcal{E}_\upsilon$ is the signal energy prior to matched filtering. Note that $|\mathbf{H}(\omega)\, e^{j\omega t_\mathrm{d}}|^2 = |\mathbf{H}(\omega)|^2$. The SNR is thus bounded by

$$\mathrm{SNR_{out}} \leq \frac{\mathcal{E}_\upsilon \int |\mathbf{H}(\omega)|^2 d\omega}{N_0 \int |\mathbf{H}(\omega)|^2 d\omega} = \frac{\mathcal{E}_\upsilon}{N_0}. \tag{13.66}$$

Based on the Schwartz inequality, this bound, which gives the maximum signal-to-noise ratio, is achieved only when

$$\mathbf{H}(\omega) = a\, \mathbf{V}^*(\omega)\, e^{-j\omega t_\mathrm{d}}, \tag{13.67}$$

where a is an arbitrary complex constant. In the time domain this means that the optimum matched filter for time delay t_d is

$$h(t) = a\, \upsilon(t_\mathrm{d} - t). \tag{13.68}$$

▶ That is, the matched filter impulse response is a time-reversed copy of the signal waveform with the specified delay. ◀

We pause to note that this result applies to the signal after bandpass filtering by the receiver. If the receiver bandwidth (including any possible Doppler shift) is less than the transmit bandwidth, the matched filter corresponds to the receiver-filtered signal. Thus, the matched filter should correspond to the expected received signal rather than merely to the transmitted signal.

A matched filter can be implemented in a variety of ways. It can be implemented as an analog filter or a digital filter after sampling the data. Recalling the filter convolution relation in Eq. (13.47) and using $h(t) = a\upsilon(t_\mathrm{d} - t)$ as the matched filter, the output of the matched filter is

$$y(t) = a \int_{-\infty}^{\infty} \upsilon(t')\, \upsilon(t' + t_\mathrm{d} - t)\, dt'. \tag{13.69}$$

This is recognized as the autocorrelation of $\upsilon(t)$ evaluated at the lag time $(t_\mathrm{d} - t)$ [compare with the autocorrelation expression in Eq. (13.50)]. Thus, the matched filter can be computed by ***correlating*** the received signal with a time-delayed copy of the transmitted signal. This is typically done causally so that the matched filter output at time delay t_d can be written as

$$y(t_d) = \int_0^{t_d} \upsilon(t)\, p_t(t - t_d)\, dt, \tag{13.70}$$

where $p_t(t)$ is the voltage waveform of the transmit signal modulation.

For an ***interrupted continuous wave*** (ICW) radar, $p_t(t)$ is a rectangular pulse given by Eq. (13.31). After removing the carrier, the matched filter impulse response is given by $h(t) = p_t(-t)$. Since this leads to a noncausal filter, we modify the matched filter by implementing a fixed time delay T_p and setting $h(t) = p_t(T_p - t)$. The time shift has the effect of shifting time of maximum SNR from $t = 0$ to $t = T_p$.

Note that for this simple case of a rectangular pulse, the matched filter is also a rectangular pulse,

$$h(t) = \begin{cases} 1, & 0 \le t \le T_p, \\ 0 & \text{otherwise.} \end{cases} \tag{13.71}$$

Such a matched filter can be implemented as a gated integrator; i.e., the signal is integrated for a period T_p at which point the output of the integrator is the matched filter output corresponding to the signal. In this case, the integrator acts like a lowpass filter on the received signal.

After removing the carrier, the received signal is a delayed, Doppler-shifted, attenuated copy of the transmit signal plus noise. Ignoring the Doppler shift and amplitude, the received signal is

$$\upsilon(t) = p_t(t - T) + n(t), \tag{13.72}$$

where $T = 2R/c$ is the signal time of flight and $n(t)$ is the noise. The matched filter output is then

$$y(t) = h(t) * \upsilon(t) = h(t) * p_t(t - T) + h(t) * n(t). \tag{13.73}$$

In the noise-free case, $y(t)$ is the convolution of two boxcar functions, which results in a triangle function at the output of the matched filter; i.e., the matched filter output is

$$y(t) = \begin{cases} (t - T - \tau), & T - 2\tau \le t \le T + \tau, \\ (T + \tau - t), & T + \tau \le t \le T + 2\tau, \\ 0 & \text{otherwise.} \end{cases} \tag{13.74}$$

At $t = T + \tau$, the matched filter output achieves maximum SNR. This corresponds to the desired time delay and is where the output should be sampled for reporting the signal power. If the signal time-of-flight changes, the delay in the optimum point of the matched filter must also be changed. For this case, in the frequency domain, the signal, matched filter, and output spectra are

$$\mathbf{P}(\omega) = \mathcal{F}\{p(t)\} = e^{-j\omega\tau} \left(\frac{\sin \omega}{\omega} \right) \tag{13.75}$$

and

$$\mathbf{H}(\omega) = \mathcal{F}\{h(t)\} = e^{j\omega\tau} \left(\frac{\sin \omega}{\omega} \right), \tag{13.76}$$

so that

$$\mathbf{Y}(\omega) = \mathcal{F}\{y(t)\} = \left(\frac{\sin \omega}{\omega} \right)^2. \tag{13.77}$$

What if the matched filter does not match the received signal? The consequence is that at the output, the signal-to-noise ratio is less than optimum. Filter mismatch is considered in the discussion of the radar ambiguity function in Section 13-9.

The ***time resolution*** or ***range resolution*** corresponds to the width of the output of the matched filter between half-power points. From Eq. (13.74) for a rectangular pulse of length τ, the time resolution is τ. This yields a range resolution ΔR for an ICW pulse of

$$\Delta R = \frac{c\tau}{2}. \tag{13.78}$$

Since the 3 dB bandwidth of a rectangular pulse is approximately $1/\tau$, this result demonstrates that the radar time (and therefore range) resolution is related to the inverse of signal bandwidth. This is true for all pulse modulation schemes. Amplitude modulating the pulse shape acts like a window on the pulse spectrum. This reduces sidelobes, but widens the mainlobe of the pulse spectrum. Thus in general, the range resolution of a pulsed radar system can be written as

$$\Delta R = \alpha_0 \frac{c\tau}{2}, \tag{13.79}$$

where $\alpha_0 \ge 1$ is a parameter that depends on the pulse shape.

13-6 Pulsed-FM Radar

Some of the advantages of pulsed and LFMCW operation can be combined by generating frequency modulated pulses. Such radars are sometimes termed ***chirp radars***. In a chirp radar, the transmit signal is modulated in amplitude, frequency, and/or phase.

> ► Usually, the received signal is passed through a ***matched filter*** to maximize the signal to noise ratio at the receiver output. ◄

The most common pulse modulation is probably linear FM (LFM), but binary-encoded phase modulation is also used. Many chirp radars also amplitude modulate or shape the pulse envelope to control the frequency content of the transmitted signal. Since the received signal is a function of the transmit signal, modulating the transmit signal also controls the received signal properties.

The general principle of an LFM chirp radar is illustrated in Fig. 13-16(a). The transmit signal consists of a sequence of LFM signals, in this case upward sweeps, separated by a time gap dedicated to receiving signals. The pulse duration is τ, and its bandwidth after LFM modulation is B. Note that B is larger than $1/\tau$. There are various ways to do the modulation, but one commonly used approach is to generate a linear FM baseband chirp such as that shown in Fig. 13-17 and mix it with a fixed-frequency carrier to generate each transmitted pulse. In many modern systems, the transmit modulation signal is digitally generated.

The received signal from a point target consists of a time-delayed, attenuated, Doppler-shifted copy of the transmitted pulse. Using the concepts introduced in the previous section, the signal can be processed by ***dechirping*** the signal and using a bandpass filter (Section 13-4). These two operations can be combined into a single operation by designing a bandpass filter that has a constant amplitude, but a frequency-varying time delay as illustrated in Fig. 13-16(b) and (c). The filter time delay is chosen so that the frequency transmitted (and received) first is delayed long enough so that it arrives at the output of the filter at the same time as the frequency transmitted last. All the frequencies in between also arrive at this time, so they are superimposed at a single instant of time at the filter output. Of course, because of the finite pulse length and bandwidth, the actual signal has a sinc-like envelope as illustrated in Fig. 13-16(b). With a bandwidth B, the approximate width of the output pulse is $1/B$, and if the transmitted amplitude is constant during the pulse, it takes the form of a $(\sin x)/x$ pulse.

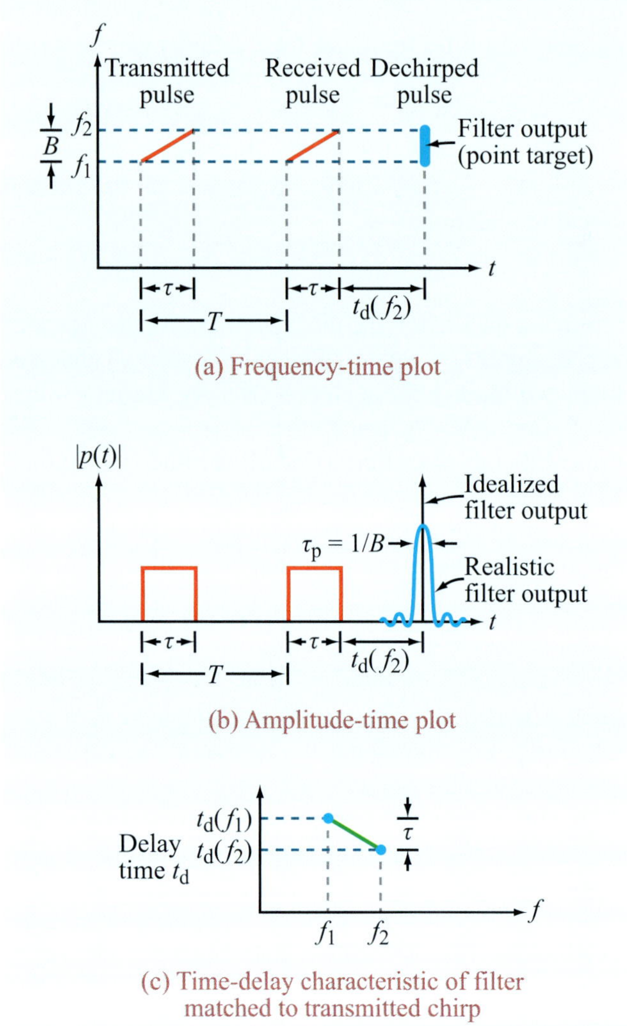

Figure 13-16: Principle of operation of an LFM chirp radar.

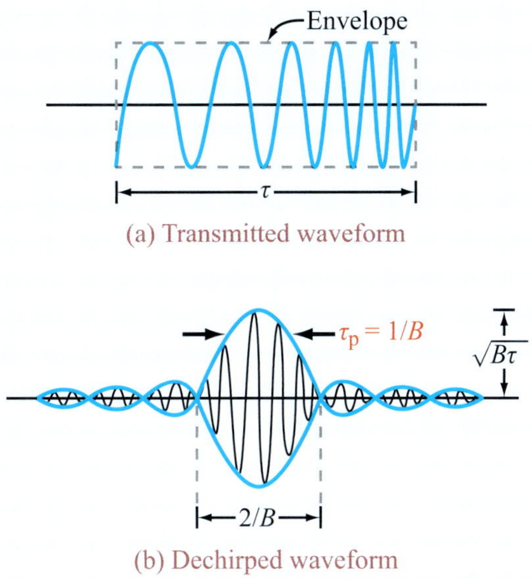

Figure 13-17: Waveforms for an LFM chirp radar with a point target. Note that the transmitted waveform has a boxcar envelope while the dechirped waveform has a sinc-function envelope.

Figure 13-16(c) shows the delay characteristics of the matched filter to the transmit pulse for the output shown in (b). Note that the filter time delay is the least for f_2, the last frequency transmitted, and most for f_1, the first frequency transmitted. The difference in time delay is just the duration of the pulse.

This time-delay filter is a particular implementation of the matched filter technique for this pulse. Other implementations are possible. For example, the signal can be digitized and the matched filter applied by multiplication (in the frequency domain) with the frequency response of the matched filter.

A key observation from Fig. 13-16(b) is that the matched filter "compresses" the pulse from length τ to a shorter effective length of $\tau_p = 1/B$. For this reason, matched-filter processing of the pulse signal is often termed ***pulse compression*** or ***range compression***. The ***compression ratio*** is

$$C = \frac{\tau}{\tau_p} = \tau B. \tag{13.80}$$

Note that the compressed pulse length τ_p only depends on the signal bandwidth B and not on the pulse length τ. The pulse length does, however, affect the signal-to-noise ratio of the matched filter output.

As previously shown, the signal-to-noise ratio of the matched filter output for white noise is given by Eq. (13.66); i.e., the output SNR is the ratio of the signal energy to the noise power spectral density:

$$\text{SNR}_{\text{out}} = \frac{\mathcal{E}_\upsilon}{N_0}. \tag{13.81}$$

On the other hand, the SNR at the input of the matched filter is given by the ratio of the signal power to the noise power prior to matched filtering:

$$\text{SNR}_{\text{in}} = \frac{P_\upsilon}{\displaystyle\int \mathbf{S}_{\text{n}}(\omega)\,d\omega} = \frac{\mathcal{E}_\upsilon}{N_0 \tau B}, \tag{13.82}$$

where $P_\upsilon = \mathcal{E}_\upsilon/\tau$ is the average signal power. The ***pulse compression gain*** G_{comp} is the ratio of the signal-to-noise ratio at the output of the matched filter to that at the input to the matched filter:

$$G_{\text{comp}} = \frac{\text{SNR}_{\text{out}}}{\text{SNR}_{\text{in}}} = \tau B. \tag{13.83}$$

Thus, the pulse compression gain for matched filtering is equal to the signal ***time-bandwidth product***. This means that the matched filter output signal SNR can be increased by lengthening the transmit pulse. Note that SNR_{out} depends only on the signal energy, not the details of the signal waveform, so the results can be applied to any pulse waveform.

Figure 13-17 shows the waveforms for the point-target response of the chirp radar. The transmitted waveform shown in (a) is modulated in frequency from a low frequency to a high frequency. In this example, its amplitude is constant throughout its duration. The ***dechirped*** waveform is shown in Fig. 13-17(b), where the familiar $(\sin x)/x$ shape appears. The null-to-null width for this case is $2/B$, but the effective 3 dB width is approximately $1/B$. The amplitude is increased from 1 for the input signal to $\sqrt{B\tau}$ for the dechirped waveform.

A convenient way to analyze a linear FM pulse signal is to use complex notation. Consider the upsweep chirp transmit pulse waveform shown in Fig. 13-17(a). This is the real part of the complex pulse waveform $p_{\text{LFM}}(t)$ that can be written as

$$p_{\text{LFM}}(t) = \begin{cases} e^{j2\pi\alpha t^2/2}, & |t| \leq \tau/2, \\ 0 & \text{otherwise}, \end{cases} \tag{13.84}$$

where α is the ***chirp rate*** $df/dt = B/\tau$. For a downsweep, the sign of α is negative. The exponential in Eq. (13.84) is a phase function of the form ωt, where ω is the angular frequency. The instantaneous frequency $f(t)$ of $p_{\text{LFM}}(t)$ can be computed by dividing the phase function by 2π and taking its time derivative. This gives the time-varying frequency $f(t)$ as

$$f(t) = \frac{1}{2\pi}\,\frac{d}{dt}\,(\pi\alpha t^2) = \alpha t, \tag{13.85}$$

which is a linear function of time. For the transmitted pulse plotted in Fig. 13-16(a), the minimum frequency $(-B/2)$ of the pulse modulation occurs at $t = -\tau/2$ while the maximum frequency $(B/2)$ occurs at $\tau/2$, so the bandwidth is B. A plot of the chirp spectrum for several different time-bandwidth products is shown in Fig. 13-18. In this plot the frequency axis for each plot has been normalized to simplify comparison. Note that as the time-bandwidth product increases, the spectrum becomes more and more like a boxcar function. The envelope of the Fourier transform $\mathbf{P}'_{\text{LFM}}(f)$ of $p_{\text{LFM}}(t)$ is approximately

$$\mathbf{P}'_{\text{LFM}}(f) = \begin{cases} 1, & 0 \leq f \leq \alpha\tau, \\ 0 & \text{otherwise}. \end{cases} \tag{13.86}$$

An approximation for the Fourier transform of this LFM chirp with a time-bandwidth product of $\tau B > 0$ is given by (Richards et al., 2010)

$$\mathbf{P}_{\text{LFM}}(f) \approx \mathbf{P}'_{\text{LFM}}(f)\; e^{-j[(\pi\tau/B)f^2 + \pi/4]}. \tag{13.87}$$

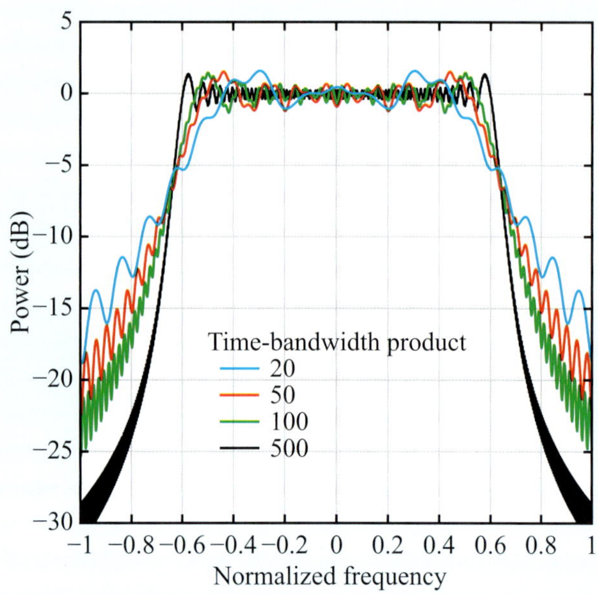

Figure 13-18: Comparisons of the spectra of linear FM chirps with different time-bandwidth products.

The normalized matched filter $h_{\text{LFM}}(t)$ corresponding to $p_{\text{LFM}}(t)$ has an impulse response that is the complex-conjugate time-reversed version of $p_{\text{LFM}}(t)$; i.e.,

$$h_{\text{LFM}}(t) = \begin{cases} e^{-j\alpha t^2/2}, & |t| \leq \tau, \\ 0 & \text{otherwise}. \end{cases} \tag{13.88}$$

The normalized matched filter output $y(t)$ is obtained by convolving $h_{\text{LFM}}(t)$ and $p_{\text{LFM}}(t)$, which leads to

$$y(t) = \left(1 - \frac{|t|}{\tau}\right) \frac{\sin\left[\left(1 - \frac{|t|}{\tau}\right)\pi Bt\right]}{\left(1 - \frac{|t|}{\tau}\right)\pi Bt}, \quad |t| \leq \tau. \tag{13.89}$$

This is the product of a triangle function (first term on the left) and a $(\sin x)/x$ function where x is

$$x = \left(1 - \frac{|t|}{\tau}\right)\pi Bt = \pi B\left(t + \frac{t^2}{\tau}\right). \tag{13.90}$$

Note that x is dominated by the quadratic function of time, which leads to a linear phase versus frequency, particularly near the peak of the $(\sin x)/x$ function.

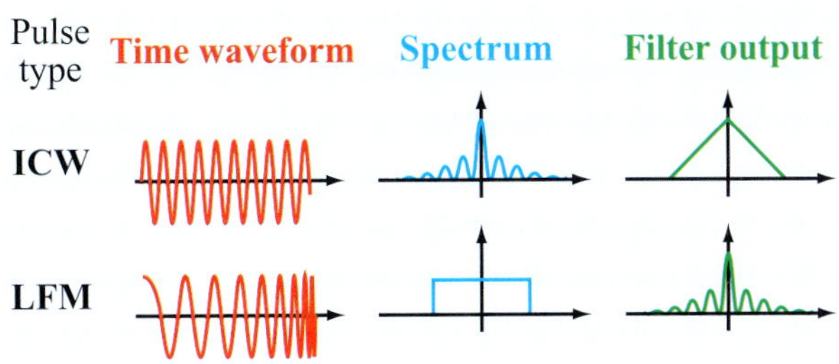

Figure 13-19: Comparison of ICW and pulsed linear FM waveforms, spectra and match filter outputs.

An intuitive explanation for the $(\sin x)/x$ function can be seen by examining the envelope of the Fourier transform of the LFM chirp given in Eq. (13.86). Assuming that the Doppler shift of the return signal is small compared with the chirp bandwidth, the envelope of the Fourier transform of the received signal is similar to the transmit signal. Thus, the matched filter output spectrum is approximately $|\mathbf{P}_{\text{LFM}}(f)|^2$, which is a boxcar function in frequency. The corresponding time-domain matched filter output is a $(\sin x)/x$ function. The width of the mainlobe is approximately $1/B$, while the peak height is τB. The peak occurs at the time delay corresponding to the pulse time-of-flight plus the time offset employed to ensure causality of the matched filter. When the received signal includes a Doppler shift the matched filter output includes a phase shift at the peak related to the Doppler shift.

For reference, Fig. 13-19 compares the properties of the ICW and pulsed LFM waveforms by displaying approximate spectra and matched filter outputs. Note that the pulsed LFM waveform has its power more uniformly spread in frequency than does the ICW pulse. For a given resolution, this enables more efficient spectrum utilization.

We note that the triangle function in Eq. (13.89) arises from the auto-convolution of the rectangular pulse envelope defined in Eq. (13.84) and acts as a window on the $(\sin x)/x$ function in the matched filter output. As previously noted, by amplitude modulating the pulse amplitude on transmit, the windowing function can be modified to control the matched filter output sidelobes. Shaping the pulse amplitude broadens the mainlobe, but suppresses the sidelobes. Because less power is transmitted during part of the pulse as a result of the pulse modulation, less total energy is transmitted and the effective SNR is reduced. Thus amplitude modulating the transmit pulse allows a trade-off between resolution (mainlobe width), SNR, and sidelobes. This trade-off applies for all pulse modulation schemes.

13-7 Pulsed Radar, General Modulation

So far, we have considered CW, ICW, pulsed, and LFM-chirp transmit modulation schemes. A variety of other modulation schemes can be used, including ***binary-coded phase modulation*** and ***frequency-hopping***. The former very rapidly varies the phase of the transmit signal according to a binary sequence, while the latter rapidly sequences through a set of frequencies in deterministic, usually cyclic order. Both approaches can be analyzed as special cases of more general pulsed signal modulation and both can be used with matched filtering to "compress" the pulse to improve the time or range resolution as well as improve the SNR.

For ***binary phase modulation***, the carrier phase is switched between 0 and π based on a binary sequence. Amplitude modulation is introduced by varying $p_t(t)$ to control sidelobes. A phase change of π is equivalent to changing the sign of $p_t(t)$. However, the phase shifts need not be limited to 0 and π.

For binary-sequence and frequency hopping modulation, $p_t(t)$ consists of a general amplitude term that multiplies a signal consisting of a sequence of much shorter signals called ***chips***. For binary phase modulation each chip is a complex constant, often $+1$ and -1, in a particular sequence. For frequency-hopping, each chip is a short complex sinusoid with the desired frequency offset for that chip. For example, the pulse waveform may have the functional form

$$p_t(t) = p_{\text{amp}}(t) \sum_{i=1}^{N_c} p_{\text{chip}}(i; t - iT_c), \tag{13.91}$$

Table 13-1: Barker codes.

Code length	Code	Sidelobe level (dB)
2	$-+, ++$	-6.0
3	$--+$	-9.5
4	$--+-, ---+$	-12.0
5	$+++-+$	-14.0
7	$---++-+$	-16.9
11	$---+++-++-+$	-20.8
13	$-----++--+-+-$	-22.3

where N_c is the ***number of chips per pulse***, $T_c = \tau/N_c$, $p_{amp}(t)$ is the amplitude modulation term defined over $t = 0$ to τ, and $p_{chip}(i;t)$ is the chip term for the ith chip.

Consider the binary phase modulation sequence with $N_c = 7$ and $p_{chip}(i;t) = a_i\, p_c(t)$ where

$$p_c(t) = \begin{cases} 1, & 0 < t < T_c \\ 0 & \text{otherwise,} \end{cases}$$

with $a_i = -1, -1, -1, 1, 1, -1, 1$. This is a particular example of a ***Barker code*** and the corresponding $p_c(t)$ chip sequence is illustrated in Fig. 13-20(b). Other Barker codes are shown in Table 13-1. ***Polyphase Barker codes*** that use more than two phase states also exist (Borwein and Ferguson, 2005).

For this binary modulation scheme the phase modulation is limited to 0° or π (180°) based on a binary sequence by multiplying by one or negative one. The matched filter for this waveform is, in essence, a cross-correlator between a stored sample of the transmitted code and the received signal. This can be implemented as suggested in Fig. 13-20 (alternate approaches can be used). If the transmitter uses a pulsed oscillator to produce pulses of duration τ, the binary code generator switches in and out a 180° phase shift, and the resulting signal is amplified and transmitted. This 180° phase shift can be inserted, as shown, in the RF chain ahead of an amplifier, or it can also be inserted between the amplifier and the antenna. The received signal comes in through the usual low-noise preamplifier and is mixed down to an IF signal. At the IF, a tapped delay line with total delay of τ is used to range compress the data. The outputs of the different taps are summed with or without phase shifting. The taps associated with the 180° phase shift on transmission have their output phase shifted by 180° on receive (in other words, the outputs are inverted). The taps with zero phase are sent directly to the adder and detector. When the time delay associated with reception is such that the first bit has just reached the end of the delay line, all of the taps in the delay line are (after suitable phase-shifting) producing signals that are in phase; hence the output of the adder is the coherent sum of the N_c chip signals. For any other position of the received pulse in the delay line, some of the pulse outputs add in phase while others add out of phase, so that the output of the summer is significantly less than N_c times the output of a single tap. This is illustrated in Fig. 13-20(d).

Table 13-2: Truth table for multiplier with $+1$ and -1 inputs and no carry.

Input 1	Input 2	Output
$-$	$-$	$+$
$-$	$+$	$-$
$+$	$-$	$-$
$+$	$+$	$+$

A simple algorithm for computing the output of Fig. 13-20(d) is shown here. Note that when the target is properly lined up so the chip signals add in phase, the output of the 7-bit summer is

$$\begin{aligned}(-1)(-1)+(-1)(-1)+(-1)(-1)\\ +1+1+(-1)(-1)+1=7.\end{aligned}$$

In effect, the outputs for the individual bits are the same as for a multiplier with $+1$ and -1 inputs, as indicated by the truth table of Table 13-2. Shown below are the results of applying this to different delays for the received pulse. On the top line is the adder phase. On the second and other even lines are the phases for the received signal, and on the third and other odd lines are the outputs for the individual bits, with the sums at the right (see Table 13-3).

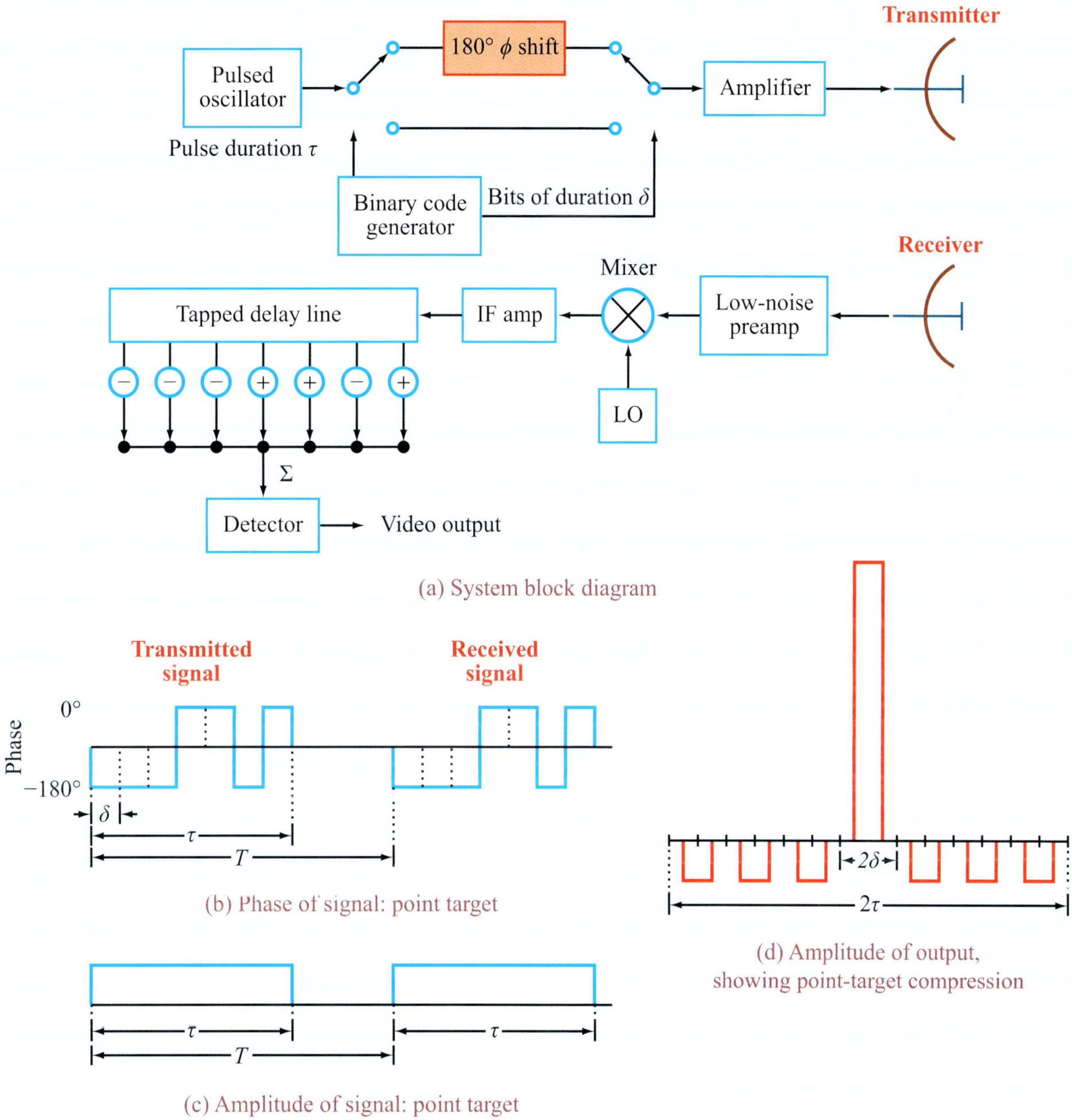

Figure 13-20: Example of 7-bit binary phase-code modulation and demodulation.

Table 13-3: Processing sequence.

Adder phase	−	−	−	+	+	−	+	
1. Properly aligned	−	−	−	+	+	−	+	
Output	1	1	1	1	1	1	1	$\Sigma = 7$
2. Misaligned by T_c		−	−	−	+	+	−	
Output		1	1	−1	1	−1	−1	$\Sigma = 0$
3. Misaligned by $2T_c$			−	−	−	+	+	
Output			1	−1	−1	−1	1	$\Sigma = -1$
4. Misaligned by $3T_c$				−	−	−	+	
Output				−1	−1	1	1	$\Sigma = 0$
5. Misaligned by $4T_c$					−	−	−	
Output					−1	1	−1	$\Sigma = -1$
6. Misaligned by $5T_c$						−	−	
Output						1	−1	$\Sigma = 0$
7. Misaligned by $6T_c$							−	
Output							−1	$\Sigma = -1$

For longer codes, one of the more common types used is the ***maximal-length sequence*** (Cook and Bernfield, 1967, pp. 247–251). These also are known as m-sequences and pseudorandom sequences. They may be readily generated using an n-stage shift register, as illustrated by the example of Fig. 13-21, which generates the same code used in the example, but with pluses instead of minuses. The shift register is connected with feedback to the input. The source of the feedback in this case is two inputs from different blocks in the shift register (here block 2 and block 3) to a modulo-two adder. The modulo-two adder has the truth table shown in the figure.

In the example illustrated in Fig. 13-21, we start with the shift register loaded with all pluses. The first output is a plus when the shift command is given, but the combination of the pluses in positions 2 and 3 gives a minus as the input (at that time) into position 1. The next shift also gives a negative input to position 1 and a positive output, but for the third shift, block 2 contains a minus and block 3 a plus, so that the feedback is positive, as shown. The first of the minus inputs has now been shifted to the output. This process continues as indicated, but in the seventh shift (eighth setting of the register) the sequence begins to repeat.

Maximal-length sequences have the following properties for an n-stage shift register:

(1) Length $= 2^n - 1$.

(2) Number of pluses and minuses equal within 1 unit (balance property).

(3) Last stage must be in feedback loop.

(4) Number of feedback taps must be even.

Maximal-length sequences have been extensively studied, and the results are available in the literature. The sequences are widely used not only in radar, but in secure communication systems and for generating pseudorandom numbers for computational purposes.

Table 13-4 gives examples of maximal-length sequences. Because the length of the sequence is $2^n - 1$, a

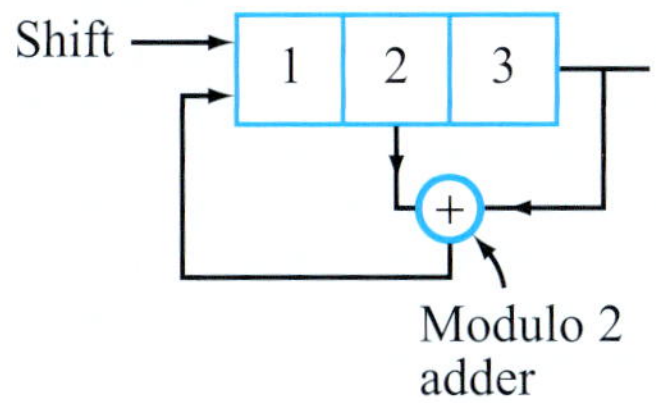

Truth table for modulo 2 adder

2	3	Output
−	−	−
−	+	+
+	−	+
+	+	−

Sequence

1	2	3	Output
+	+	+	+
−	+	+	+
−	−	+	+
+	−	−	−
−	+	−	−
+	−	+	+
+	+	−	−
+	+	+	+

(The first seven rows form 1 cycle; the last row: Repeat)

Figure 13-21: Example of generation of maximal-length sequence.

Table 13-4: Number of binary maximal-length sequences with an n-stage shift register.

Number of stages n	Length of sequence $2^n - 1$	Number of maximumal sequences	Example of feedback connections
3	7	2	3, 2
4	15	2	4, 3
5	31	6	5, 3
10	1023	60	10, 7

relatively short shift register may be used to obtain quite a long code. For instance, the ten-stage shift register achieves a 1023-bit code. The number of maximal sequences listed in the table depends upon the number of feasible feedback connections that will generate maximal-length sequences. Examples of feedback connections with just two connections are shown in the table. However, some of the codes are generated by use of feedback from 4, 6, or more of the blocks in the shift register.

Even though binary phase modulation is perhaps the most common discrete-step technique used for pulse compression, it is, of course, possible to use other techniques; for instance, amplitude modulation may be used, in which case the pluses represent a signal that is turned on, and the minuses a signal that is turned off. Frequency modulation may also be used, in which the plus represents one frequency and the minus another. The binary phase code is used so commonly because both the modulation and the pulse compression are very easy to achieve with this method.

Nonbinary phase modulation schemes also exist. These are widely used in communication but not in radar. For example, a quaternary rather than a binary system may be used in which four 90°-phase-shifted positions represent the numbers 0, 1, 2, and 3. The subject of such codes has also been extensively studied.

Discrimination in range can be achieved with short pulses, with frequency modulation, and with longer pulses that are modulated with either FM or binary phase codes. The binary-coded system may be used on a continuous basis like the FM system, and one may use nonlinear modulation techniques. In general, however, the time resolution of a well-designed system is always approximately $1/B$.

13-8 Measurement Precision

13-8.1 Effective Number of Samples

When a random signal is measured continuously, or when the samples of the random signal are very close together relative to the spacing between independent samples, Rice (1944–45) has shown that the variance s_T^2 of the signal integrated over a time T is given by

$$s_T^2 = \frac{2}{T} \int_0^T \left(1 - \frac{x}{T}\right) R_{\mathrm{vf}}(x)\, dx \tag{13.92}$$

where $R_{\upsilon f}(x)$ is the autocovariance function,

$$R_{\upsilon f}(\tau) = R_\upsilon(\tau) - \overline{P_\upsilon}^2, \tag{13.93}$$

derived from the autocorrelation function $R_\upsilon(\tau)$ of the voltage $\upsilon(t)$

$$R_\upsilon(\tau) = \lim_{T\to\infty} \frac{1}{T} \int_0^T \upsilon(t)\, \upsilon(t+\tau)\, dt, \tag{13.94}$$

and the average signal power $\overline{P_\upsilon}$ is defined by

$$\overline{P_\upsilon} = \lim_{T\to\infty} \frac{1}{T} \int_0^T \upsilon^2(t)\, dt. \tag{13.95}$$

The ***effective number of independent samples*** or ***number of looks*** may be obtained by relating the expression for the signal variance in Eq. (13.92) to the expression for the variance of N independent samples; i.e.,

$$s_T^2 = \frac{\overline{P_\upsilon}^2}{N}. \tag{13.96}$$

Then, the effective number of independent samples is given by

$$N = \frac{\overline{P_\upsilon}^2 T}{2\displaystyle\int_0^T \left(1 - \frac{x}{T}\right) R_{\upsilon f}(x)\, dx}. \tag{13.97}$$

The autocovariance function may be obtained either from the known signal or from the inverse Fourier transform of the signal PSD. For noise-like signals, the latter method is the more appropriate.

An important example is band-limited white noise, which has a power spectrum given by Eq. (13.58). The autocovariance function corresponding to this power spectrum is given by

$$R_{\upsilon f}(\tau) = \overline{P_\upsilon}^2 \left(\frac{\sin \pi B\tau}{\pi B\tau}\right)^2. \tag{13.98}$$

For this case, the integration in the variance expression in Eq. (13.92) can be approximated by

$$s_T^2 = \frac{2\overline{P_\upsilon}^2}{T} \int_0^T \left(\frac{\sin \pi Bx}{\pi Bx}\right)^2 dx, \tag{13.99}$$

since x/T varies slowly over the region where the integrand has a significant value. This expression can be integrated to obtain

$$s_T^2 = \frac{\overline{P_\upsilon}^2}{\pi BT}\left[\frac{\cos \pi BT - 1}{\pi BT} + 2\int_0^{2\pi BT} \frac{\sin t}{t}\, dt\right]. \tag{13.100}$$

Note that if the time-bandwidth product BT is much greater than one, the first term in the square brackets is negligible, while the second term is approximately π. Hence, if $BT \gg 1$

$$s_T^2 \approx \frac{\overline{P_\upsilon}^2}{BT}. \tag{13.101}$$

It follows from Eq. (13.96) that for band-limited noise with $BT \gg 1$

$$N \approx BT. \tag{13.102}$$

A more exact evaluation of Eq. (13.97) is plotted in Fig. 13-22 along with some other quantities of interest. In this figure the value of the normalized standard deviation is shown and compared with $1/\sqrt{BT}$. The ***normalized standard deviation*** is the standard deviation divided by the mean. Clearly, as BT approaches 10 the approximation $N \approx BT$ becomes fairly good.

Also shown in Fig. 13-22 is the approximation $N = BT$ compared with the result of evaluating Eq. (13.97) for this case. Note that the effective number of independent samples is always somewhat larger than BT. The effective number of independent samples must start with 1 because any measurement corresponds to at least one sample. It quickly approaches a line almost parallel with the BT line and displaced from it by about 0.6. Therefore, the effective number of independent samples for large BT is about 0.6 more than BT, a difference that is unimportant for large BT but may be significant for the smaller values, as indicated by the difference between the normalized-standard-deviation curve and the curve of $1/\sqrt{BT}$. Similar curves could be derived for other spectra and autocovariance functions, but this result is illustrative of what can be expected in most cases.

When discrete signal samples spaced in time by τ are collected, a summation rather than an integration is used.

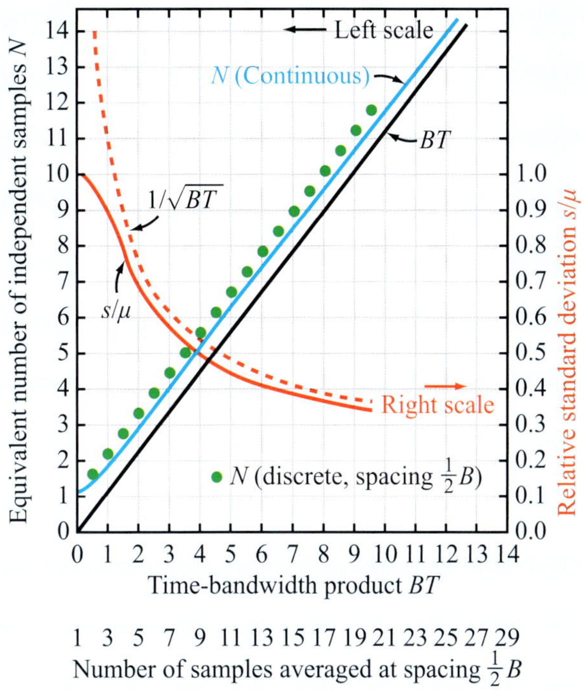

Figure 13-22: Equivalent number N of independent samples and normalized standard deviation as a function of time-bandwidth (BT) product.

In this case the variance of the sum can be shown to be (Davenport and Root, 1958, p. 80)

$$s_T^2 = \frac{1}{N_p^2}\left\{\sum_{i=1}^{N_p} s_i^2 + 2\sum_{k=1}^{N_p-1}\left(N_p - k\right) R_{\upsilon f}(kT_p)\right\}, \tag{13.103}$$

where s_i^2 is the standard deviation for a single sample, T_p is the time spacing between adjacent samples, and N_p is the total number of signal samples or pulses. Since for a stationary process all the s_i's are the same, we may replace s_i by just s. The signal correlation coefficients $\rho_{\upsilon f}(n)$ are then

$$\rho_{\upsilon f}(kT_p) = \frac{R_{\upsilon f}(kT_p)}{s^2}. \tag{13.104}$$

Substituting this value into Eq. (13.103) yields

$$s_T^2 = \frac{s^2}{N_p^2}\left\{N_p + 2\sum_{k=1}^{N_p-1}(N_p - k)\,\rho_{\upsilon f}(kT_p)\right\}. \tag{13.105}$$

Since the effective number of independent samples N is just the ratio of the variance for one sample to the variance of N_p samples for measurement over a time interval $T = N_p T_p$, it is given by

$$N = \frac{s^2}{s_T^2} = \frac{N_p}{1 + 2\sum_{k=1}^{N_p-1}(N_p - k)\,\rho_{\upsilon f}(kT_p)}. \tag{13.106}$$

Since the second term in the denominator is always positive, the maximum value of N is obtained when this term is zero. For the rectangular-spectrum signal, this happens when

$$T_p = 1/B,\ 2/B,\ 3/B, \ldots$$

because in this case

$$\rho_{\upsilon f}(kT_p) = 0$$

for all k, because

$$\sin k\pi BT_p = 0.$$

Then, for this case we have

$$N = N_p = BT + 1.$$

For any spacing except a multiple of $1/B$, the second term in the denominator of Eq. (13.106) is positive and N is less than the value given above. For example, consider a spacing of one-half the reciprocal of the bandwidth, that is,

$$T_p = \frac{1}{2B}.$$

The variance for different values of N_p is plotted in Fig. 13-22 as points. It is interesting to note that the number of independent samples obtained this way is larger than that obtained by continuous integration, also shown on the graph. For the larger values of N_p, the discrete sum variance is about 0.3 larger than the continuous-integration equivalent number of independent samples and about 0.9 larger than BT.

This is an interesting result, since it shows that the variance may be reduced better by taking samples at a spacing between $1/2B$ and $1/B$ than by taking samples closer together, which would approach the continuous case. Numerical calculation for samples at smaller spacing substantiates this; there is no particular advantage, and in fact some small disadvantage, if the PRF of the radar is made higher than $1/B$. For radiometers this never turns out to be a problem, because the time-bandwidth product is invariably very large.

13-8.2 Radiometric Precision

For radiometers, the radiometric precision is quantified by ΔT, the standard deviation of the brightness-temperature measurement. How do we best quantify the power measurement accuracy of a radar?

▶ The primary goal of a scatterometer is to make an accurate measurement of the signal amplitude or power. This need not require a high signal-to-noise ratio. ◀

Note that radiometers operate with a signal-to-noise ratio that may be much less than unity since the desired component of the radiometric signal is only a small part of the total received signal due to thermal noise added in the receiver and other parts of the signal path. Radiometers overcome the low signal-to-noise ratio by averaging large numbers of independent samples. This can be done for radar as well, which reduces the noise variance, but what about the signal variance?

We again note that when the target geometry remains fixed with respect to the radar (so there is no range variation and no Doppler shift), the radar obtains only a single look and no amount of averaging can reduce the signal variance because all of the samples are identical. Hence, averaging only reduces the noise variance. As previously noted, with multiple targets and changing range and/or Doppler, the signal varies with time (fading). In this case, averaging the signal in time can reduce the signal variance and thereby reduce the variance of estimates of the surface backscatter as well as reducing the noise variance.

A very useful approach to recording radar measurements of the signal power in the presence of noise is to make separate measurements of the radar signal power and the noise power. The latter can be collected by not transmitting pulses at some times, or by making power measurements over bandwidths outside of the radar echo bandwidth. Assuming the radar signal and the noise signal are independent, the mean value of the received power $\overline{P_r}$ is the sum of the mean values of the signal and the noise power because the signal and noise cross-terms average out. Thus we can write

$$\overline{P_r} = \overline{P_s} + \overline{P_n}, \tag{13.107}$$

where $\overline{P_s}$ is the mean signal power and $\overline{P_n}$ is the mean power of the noise. The received mean power can be expressed in terms of the mean signal power and the ***signal-to-noise ratio*** S_n as

$$\overline{P_r} = \overline{P_s}\left(1 + \frac{\overline{P_n}}{\overline{P_s}}\right) = \overline{P_s}\left(1 + \frac{1}{S_n}\right), \tag{13.108}$$

where

$$S_n = \frac{\overline{P_s}}{\overline{P_n}}. \tag{13.109}$$

As given in Section 5-7.3, if the receiver uses a square-law detector to measure the signal power, the standard deviation s_r of the signal without averaging (1 independent sample) is proportional to the mean and can be expressed as

$$s_r = \overline{P_r} = \overline{P_s}\left(1 + \frac{1}{S_n}\right). \qquad (N_r = 1) \tag{13.110}$$

When N_r independent samples of the received signal are averaged, the standard deviation s_{rN_r} for the average is given by

$$s_{rN_r} = \frac{\overline{P_s}}{\sqrt{N_r}}\left(1 + \frac{1}{S_n}\right). \qquad (N_r \geq 1) \tag{13.111}$$

The estimate of the received signal power is obtained by taking the difference between the estimate of the combination of signal-and-noise $\hat{\overline{P_r}}$ and the estimate of the noise alone $\hat{\overline{P_n}}$:

$$\hat{\overline{P_s}} = \hat{\overline{P_r}} - \hat{\overline{P_n}}, \qquad (13.112)$$

where the $\hat{}$ denotes an estimate. The estimate of the noise is obtained by averaging a set of N_n samples of the noise from separate measurements of the noise. The variance $s^2_{nN_n}$ of this noise estimate is

$$s^2_{nN_n} = \frac{\overline{P_n}}{\sqrt{N_n}}. \qquad (13.113)$$

The variance for either the sum or the difference of two independent random variables is the sum of the variances of the individual components; hence the variance for the signal estimate is the sum of the variances of the two measurements, namely the signal-plus-noise and noise-only powers, i.e.,

$$s^2_s = s^2_{rN_r} + s^2_{nN_n} = \frac{\overline{P_s}^2}{N_r}\left(1 + \frac{1}{S_n}\right)^2 + \frac{\overline{P_n}^2}{N_n}. \qquad (13.114)$$

Using

$$\overline{P_n} = \overline{P_s}\left(\frac{\overline{P_n}}{\overline{P_s}}\right) = \frac{\overline{P_s}}{S_n}, \qquad (13.115)$$

Eq. (13.114) becomes

$$s^2_s = \frac{\overline{P_s}^2}{N_r}\left(1 + \frac{1}{S_n}\right)^2 + \frac{\overline{P_s}^2}{N_n}\left(\frac{1}{S_n}\right)^2. \qquad (13.116)$$

The ***normalized standard deviation***, K_p, of the estimate of the signal power is then

$$K_p = \frac{s_s}{\overline{P_s}} = \sqrt{\frac{1}{N_r}\left(1 + \frac{1}{S_n}\right)^2 + \frac{1}{N_n}\left(\frac{1}{S_n}\right)^2}. \qquad (13.117)$$

When the signal and noise estimates use the same number of samples, Eq. (13.117) simplifies to

$$K_p = \frac{1}{\sqrt{N_r}}\sqrt{\left(1 + \frac{1}{S_n}\right)^2 + \left(\frac{1}{S_n}\right)^2}. \qquad (13.118)$$

Note that K_p is a function of both the signal-to-noise ratio of the measurements and the number of pulses used, which enables a trade-off between the number of pulses, the signal time-bandwidth product, and the signal-to-noise ratio. It may be easier to increase the number of independent looks by improving the signal time-bandwidth product than to increase the signal-to-noise ratio in a particular design. A general approach to computing K_p when filtering and detection are intermixed is given by Yoho and Long (2004).

Radiometers can obtain separate signal and noise measurements by alternately measuring (1) the output of a calibrated noise source of known temperature, combined with the noise in the receiver, and (2) the output of the antenna, also combined with the noise in the receiver. By suitable comparison of these, the effect of the receiver noise is calibrated out with a precision comparable to that indicated by Eq. (13.117). In some radar scatterometers used for oceanographic wind observation, noise measurements are obtained by skipping some transmit pulses. The resulting noise-measurement periods are interleaved with the measurements of the signal and noise.

As an example of the type of standard deviation that may be obtained with different signal-to-noise ratios for radar, consider a radar time-bandwidth product of 100, which means $N_r \approx 100$. If $S_n = 1$, the relative standard deviation is $K_p = \sqrt{5/100} = 0.22$. If $S_n = 10$, the normalized standard deviation is reduced by a factor of two to 0.11, and if S_n approaches infinity the normalized standard deviation (now due to fading alone) is 0.10. Therefore, almost no advantage is gained in having a signal-to-noise ratio exceeding $\sim$10, and relatively little advantage is obtained in having it much greater than unity. However, increasing the time-bandwidth product (or, equivalently, the number of independent looks) by 100 reduces the relative standard deviation at $S_n = 1$ from 2.2 to 0.22.

Consider, for example, a passive system such as a radio telescope with a bandwidth $B = 10^8$ Hz and an observation time $T = 100$ s. The equivalent number of independent samples is $N_r = BT = 10^{10}$. For unity signal-to-noise ratio, the relative standard deviation is $\sqrt{5/10^{10}} = 2.2 \times 10^{-5}$. For a much weaker signal-to-noise ratio, say $S_n = 0.001$, the normalized standard deviation is 1.4×10^{-2}, which still provides quite a precise measurement.

13-9 Ambiguities in Radar

Ambiguities can exist with pulse radars for both range and speed measurements. For pulse radars the ambiguity is often associated with the pulse repetition frequency (PRF) f_p.

13-9.1 Range Ambiguity

▶ Ambiguity in range measurement occurs at high PRF when the echo from a more-distant target appears to be the echo from a nearer target. ◀

Figure 13-23 illustrates the problem graphically by showing ambiguous range returns resulting from transmitting pulses too close together; i.e., the PRF f_p is too high. The transmitted pulses are shown at the top of the figure, and the next three lines illustrate returns associated with the first transmitted pulse for each of three targets whose spacings are such that the range delay is T_1 from the start of the first pulse for target 1, $T_1 + T_p$ for target 2 and $T_1 + 2T_p$ for target 3, where T_p is the pulse period. That is, $T_p = 1/f_p$. The first "return" line is for target 1 at a range corresponding to T_1; the second return line is for target 2 at a range corresponding to $T_p + T_1$, which places it just T_1 after the start of the second pulse. The third return line contains a comparable return that arrives a time T_1 after the start of the transmission of the third pulse. On the display all of these pulses appear together. That is, on the fifth line the return from target 1 appears at a time T_1 after the transmission of the first pulse. The returns from both target 1 and target 2 appear at a time T_1 after transmission of the second pulse, and the returns from all three targets are superimposed after the time of transmission of the third pulse and from then on. In the usual display, which shows the returns only for one interpulse period, as indicated in the sixth line, one cannot tell whether there are one, two, three, or even more targets present, and one cannot establish the range to the farther targets because one does not know which pulse caused the return. Thus, if the maximum range of the radar is the range to the third target, the pulse repetition period should be $4T_p$ or thereabouts, to assure that this ambiguity in interpretation of the returns does not exist. Increasing T_p results in a lower PRF, but this may be inconsistent with the requirement for a high

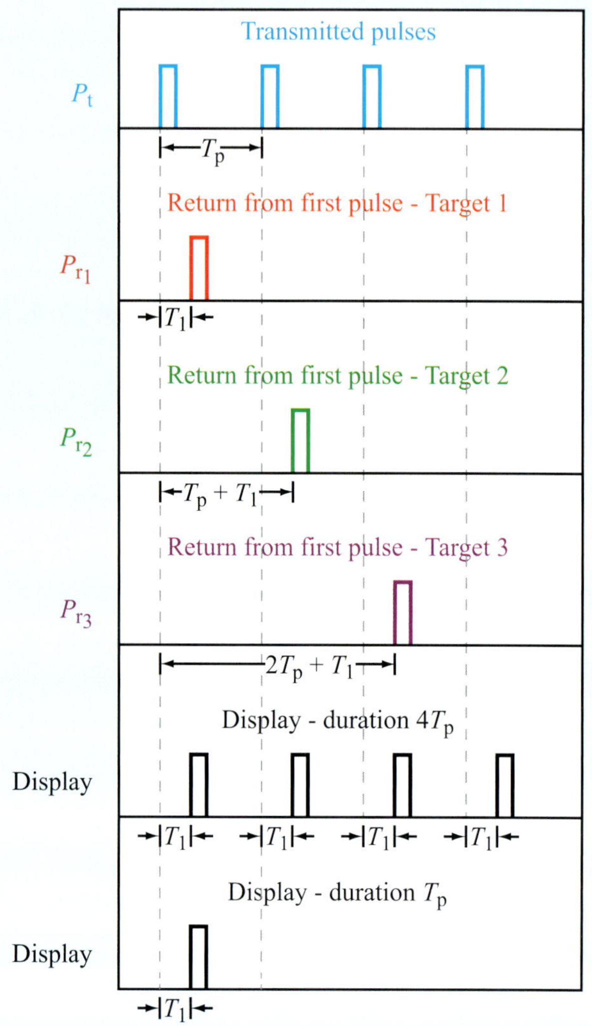

Figure 13-23: Ambiguous range returns.

enough PRF to meet Nyquist sampling requirements for the Doppler frequency.

13-9.2 Speed Ambiguity

▶ Ambiguity in speed measurement occurs when the sampling rate (PRF) is too low to satisfy the Nyquist sampling criterion, resulting in aliasing. ◀

When a single Doppler frequency is received, that is, when a single target is observed, one can see the ambiguity problem associated with the Nyquist sampling theorem by considering the waveforms in Fig. 13-24. In the second line, a single Doppler-frequency return is shown sampled at the Nyquist rate by the pulses transmitted in accordance with the first line. In the third line, however, a Doppler frequency three times that of the second line, that is, a Doppler frequency corresponding to a speed three times that associated with the first line, is shown, and the samples are identical with those for the second line; thus there is an ambiguity in the frequency of the waveform. One cannot tell from the samples whether the Doppler frequency is associated with the speed u_1 or with any odd multiple of u_1. If we are to avoid this ambiguity, and the maximum Doppler frequency is that shown in the third line, we must sample three times as often. That is, we must have a PRF three times that illustrated in the figure.

One way to resolve the ambiguities is illustrated in Fig. 13-25. Here the transmitted pulses are not emitted at a constant interpulse period. Rather, their spacing is ***jittered*** about a rate that allows adequate Doppler sampling, but would normally give range ambiguities. Note that a strong return occurs when all the pulses at appropriate delays are added together, but the other pulses appear to come from weaker targets at incorrect ranges. Unfortunately, with this type of system, the extra returns provide spurious artifacts and ***clutter***, which may or may not be tolerable, depending upon other system considerations.

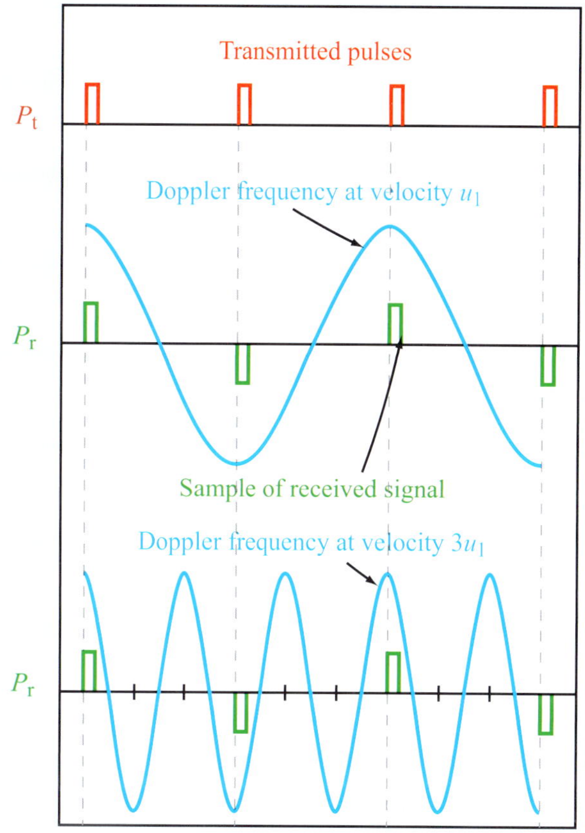

Figure 13-24: Ambiguous Doppler returns when sampling at the Nyquist rate.

13-9.3 Radar Ambiguity Function

▶ Ambiguity in timing or range measurement occurs when the PRF is too high so that the signal from a previous pulse appears during the expected return of the desired signal echo. ◀

Single-pulse ambiguity

The simple cases treated thus far illustrate the importance of Doppler and timing ambiguity in radar. Ambiguity also arises due to mismatch between the received signal and the reference matched filter used to process the signal. Choosing the appropriate transmit

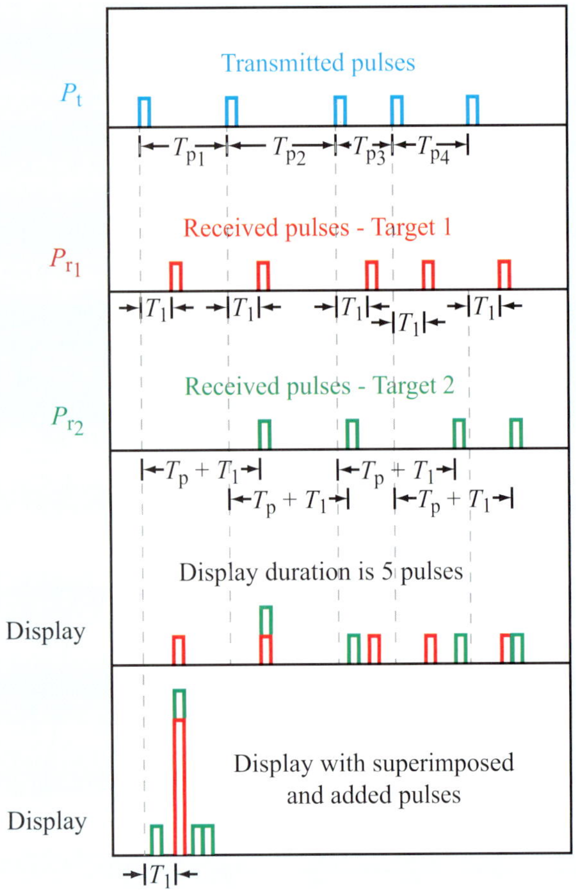

Figure 13-25: Reduction of ambiguous range returns by jittered pulse intervals.

waveform, PRF, and matched filter helps minimize ambiguity. A useful tool for this purpose is the ***radar ambiguity function***, which is derived from the matched filter for a single pulse, but which can also be used for analyzing coherent processing of multiple pulses. The ambiguity function is particularly useful for analyzing the properties of the system when different waveforms are used; however, a detailed study of waveform design is beyond the scope of this book, but is treated in other texts (e.g., Skolnik, 1980; Richards et al., 2010). In the following we illustrate the radar ambiguity function for several waveforms.

Consider the complex transmit function

$$v_t(t) = u(t)\, e^{j2\pi f_c t} \tag{13.119}$$

where $u(t)$ is the baseband complex modulated transmit waveform and f_c is the carrier frequency. Ignoring the amplitude and removing the carrier, the expected received signal is a time-delayed, Doppler-shifted replica of the of transmit signal and is given by

$$v_R(t) = u(t-\tau')\, e^{j2\pi f'_D (t-\tau')}, \tag{13.120}$$

where τ' is the time delay due to the time of flight and f'_D is the Doppler shift.

The matched filter output $y(t)$ is the convolution of the received signal with the filter impulse response given by the time-reversed complex conjugate of the reference signal $v_{REF}(t)$,

$$y(t) = \int_{-\infty}^{\infty} v_R(t')\, v^*_{REF}(t'-t)\, dt'. \tag{13.121}$$

In the ideal case, the reference function is $v_{REF}(t) = v_R(t)$ so that the matched filter output is

$$\begin{aligned} y(t) &= \int_{-\infty}^{\infty} u(t'-\tau')\, u^*(t'-t-\tau') \\ &\quad \cdot e^{j2\pi f'_D (t'-\tau')} e^{-j2\pi f'_D (t'-t-\tau')}\, dt' \\ &= e^{j2\pi f'_D t} \int_{-\infty}^{\infty} u(t'-\tau')\, u^*(t'-t-\tau')\, dt'. \end{aligned} \tag{13.122}$$

The output of the matched filter at time $t+\tau'$ is obtained by replacing t with $(t+\tau')$ and also t' with $(t'+\tau')$:

$$y(t+\tau') = e^{j2\pi f'_D t(t+\tau')} \int_{-\infty}^{\infty} u(t')\, u^*(t'-t-\tau')\, dt'. \tag{13.123}$$

We recognize the integral as the autocorrelation function of the signal modulation (Richards, 2005).

Let us examine what happens if the wrong reference function is used; i.e., what does the output of the matched filter change to when the filter and signal are mismatched? Suppose the reference function $v_{REF}(t)$ were to use a time delay τ_0 (instead of the time delay τ')

and a Doppler shift f_{D0} (instead of the true Doppler shift f'_D),

$$v_{REF}(t) = u(t-\tau_0)\, e^{j2\pi f_{D0}(t-\tau_0)}. \qquad (13.124)$$

The output of the matched filter is then

$$y(t) = \int_{-\infty}^{\infty} u(t'-\tau')\, u^*(t'-t-\tau_0) \cdot e^{j2\pi f'_D(t'-\tau')} e^{-j2\pi f_{D0}(t'-t-\tau_0)}\, dt'. \qquad (13.125)$$

Using the substitutions $t'' = t' - \tau'$, $\tau = \tau' - \tau_0$, and $f_D = f'_D - f_{D0}$, the matched filter output can be written as

$$y(t) = e^{j2\pi f_{D0}(t-\tau)} \int_{-\infty}^{\infty} u(t'')\, u^*(t''-t+\tau)\, e^{j2\pi f_D t''}\, dt''. \qquad (13.126)$$

The output of the matched filter at time $t+\tau$ is thus

$$y(t+\tau) = e^{j2\pi f_{D0} t} \int_{-\infty}^{\infty} u(t'')\, u^*(t''-t)\, e^{j2\pi f_D t''}\, dt'', \qquad (13.127)$$

which is similar in form to Eq. (13.123) but with Doppler frequency terms.

Note that now $\tau = \tau' - \tau_0$ is the ***error in the time delay*** between the actual delay and the time delay assumed in the reference function, and $f_D = f'_D - f_{D0}$ is the ***error in the Doppler shift*** between the received signal and the reference function. The magnitude of the matched filter output is maximized when the received and reference signals match, which occurs when $\tau = 0$ and $f_D = 0$, and falls off when they are mismatched. Thus, the magnitude of the matched filter output can be used to evaluate how well matched the received signal and the matched filter reference signal are.

To assist in this evaluation we define the generalized radar ambiguity function as

$$\chi(\tau, f_D) = \left| \int_{-\infty}^{\infty} u(t)\, u^*(t-\tau)\, e^{-j2\pi f_D t}\, dt \right|^2 \qquad (13.128)$$

or

$$\chi(\tau, f_D) = \left| \int_{-\infty}^{\infty} \mathbf{U}^*(f-f_c)\, \mathbf{U}(f-f_c-f_D)\, e^{-j2\pi f \tau}\, df \right|^2, \qquad (13.129)$$

where $\mathbf{U}(f)$ is the Fourier transform of $u(t)$. Some authors do not include the square in the definition of χ. Note that $\chi(0,0)$ corresponds to the total signal energy and that

$$\chi^2(\tau, f_D) \le \chi(0,0). \qquad (13.130)$$

The radar ambiguity function can be viewed as the convolution of the Doppler-shifted echo with a filter matched to an unshifted echo. *The variables f_D and τ describe the mismatch in Doppler and time delay, respectively,* between the actual signal and the signal that the matched filter is designed for. At $f_D = 0$ and $\tau = 0$ the signal and filter are matched, which produces the highest signal-to-noise ratio. Signal and filter mismatch reduces the signal-to-noise ratio primarily by reducing the signal output power.

Figure 13-26(a)–(c) illustrates the ambiguity function for several single-pulse modulation schemes, including a simple pulse, an LFM-pulse, and a binary phase-modulated pulse. Fig. 13-26(a) illustrates the ambiguity function for an ICW pulse. Note that for zero Doppler ($f_D = 0$) in Eq. (13.128), the ambiguity function has the form of the autocorrelation of the pulse. For an ICW pulse, this is triangular as derived in Eq. (13.74). If the time delay $\tau = 0$, the ambiguity function is a $|(\sin x)/x|^2$ function. A corresponding diagram of the region for which the ambiguity diagram for an ICW pulse exceeds the 1/2 power point is shown in Fig. 13-26(d). Everything within the red ellipse is essentially ambiguous; i.e., there is not enough information to "resolve" the signal any finer. Thus, this region defines the effective resolution in time delay and Doppler for a single pulse. The indicated size corresponds to the effective temporal and Doppler resolution of the pulse previously derived. Computation of the ambiguity function is thus an effective way of determining the temporal and frequency resolution of a particular transmit signal or pulse modulation scheme.

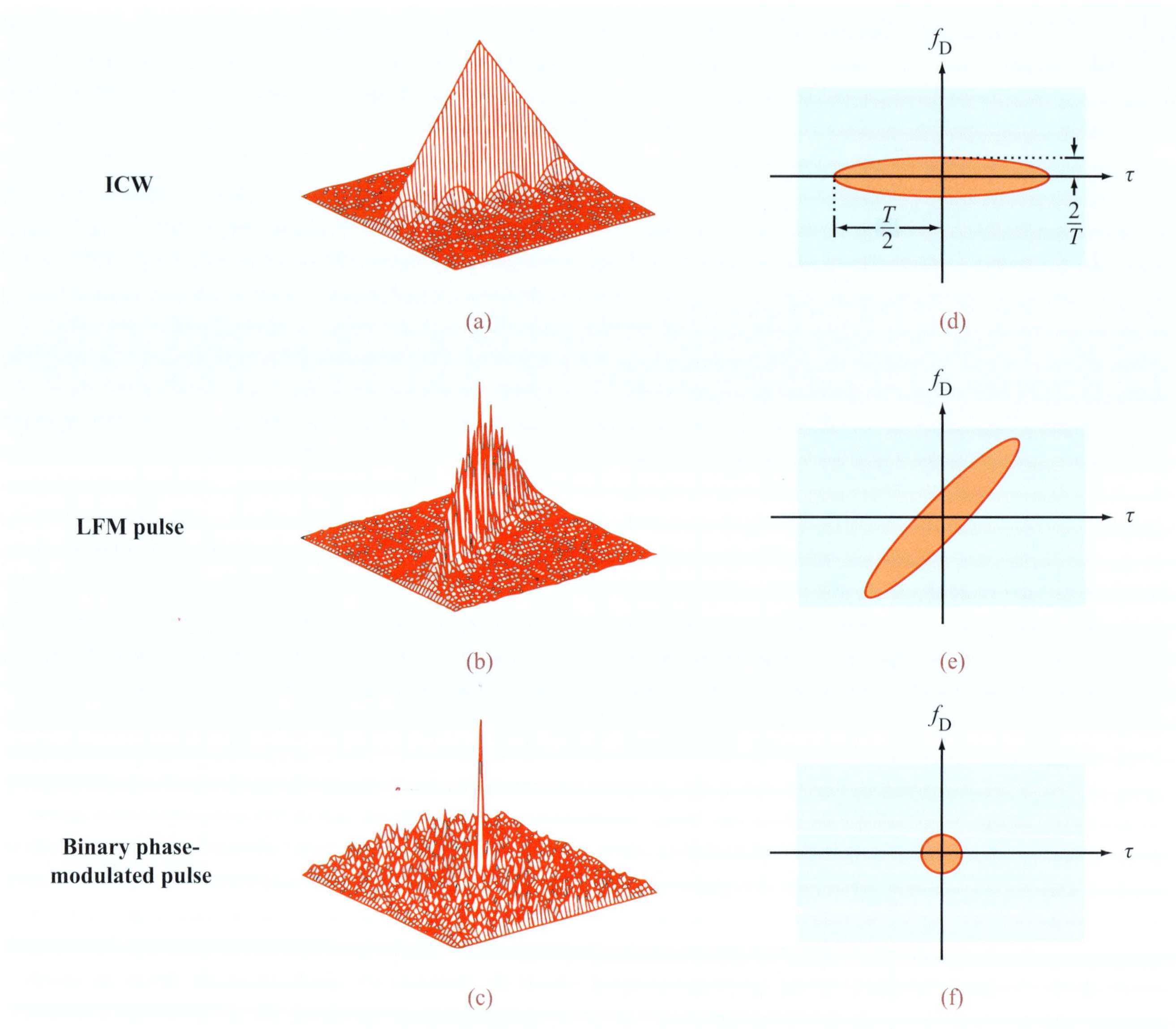

Figure 13-26: Examples of single-pulse ambiguity functions: (a) rectangular ICW pulse, (b) LFM pulse, (c) binary phase-modulated pulse and diagrams of their 1/2-power regions in time-Doppler space.

For the LFM pulse shown in Fig. 13-26(b) and (e), the "blade-like" shape of the ICW pulse is rotated at an angle. This couples the Doppler and range ambiguity. The LFM and ICW pulse shapes have very low "floors" off the side of the main peak. This means that the area resolved by the pulse is nicely isolated from the surrounding area and that strong signal returns from outside the resolved area have less impact on the measured power.

The ambiguity function shown in Fig. 13-26(c) and (f) for the binary phase-modulated pulse has the smallest central peak, and therefore potentially the finest resolution of the ambiguity functions shown. However, the ratio of the height of the central peak to the surrounding "floor" is much lower than for the other modulation schemes, which suggests that this modulation scheme is more susceptible to interference from strong signals near the desired resolution element. It can be shown that the total squared ambiguity function is fixed no matter what the shape of the modulation; i.e.,

$$\iint \chi^2(\tau, f_\mathrm{D})\, d\tau\, df_\mathrm{D} = \chi^2(0,0). \tag{13.131}$$

Thus, as the peak is narrowed by changing the pulse modulation the floor comes up to compensate. The ambiguity function is a useful tool for evaluating the sensitivity of the resolution element to the surrounding area.

Multiple-pulse ambiguity

The ambiguity function can be computed by coherently combining multiple pulses. Figure 13-27 illustrates the ambiguity function for a train of N pulses of length T and pulse repetition interval T_0. The total duration of the pulse train is NT_0. The plot illustrates numerous spikes of various heights (the height is suggested by the diameter of the circles). These correspond to possible time and frequency ambiguities. Combinations of delay and Doppler frequency that correspond to one of the spikes can be confused (i.e., they are ambiguous) with one another. Note that the peaks along the horizontal time axis are at the pulse spacing T_0, which give the ambiguities in range. Along the frequency axis the peaks are spaced $1/T_0$ apart. These are the ambiguities between the different Doppler frequencies associated with the sampling that occurs during pulsed operation. The width of the individual spikes in the Doppler domain is proportional to the reciprocal of the duration of the burst of pulses. By combining multiple pulses, the individual spikes are narrower, which yields higher resolution, but the existence of multiple spikes indicates multiple ambiguities. We note that a narrow antenna beam could resolve the ambiguities in Fig. 13-27 if the mainlobe illuminates only the area corresponding to one of the peaks.

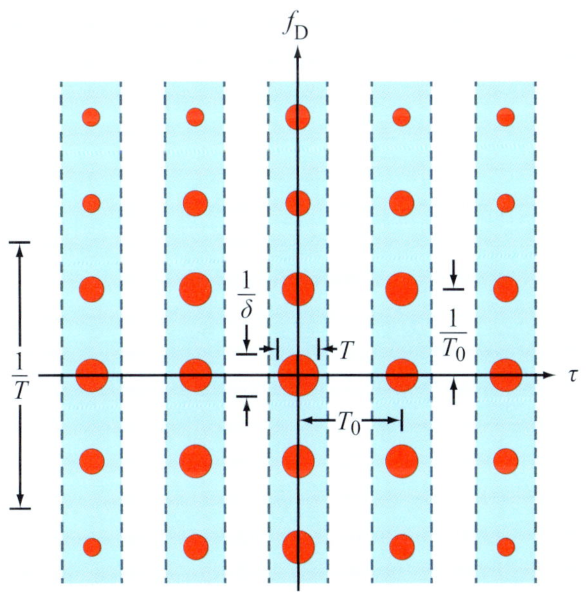

Figure 13-27: Ambiguity diagram for multiple pulses.

The subject of ambiguity functions is quite complex, and a detailed study of the different waveforms is beyond the scope of this book. Waveform design has been the subject of many papers and books or chapters in books on radar signals. For a more complete analysis, the reader is referred to Cook and Bernfield (1967), DiFranco and Rubin (1968), Berkowitz (1965), Skolnik (1980), and Richards et al. (2010).

13-10 Radar Calibration

▶ Internal calibration of a scatterometer system permits determination of *relative* scattering coefficients (or scattering amplitude S if the scatterometer is polarimetric), whereas external calibration permits absolute determination of the scattering coefficient σ^0 or the scattering amplitude S. ◀

13-10.1 Internal Calibration

Two different methods for internal calibration can be used—the method of independent or separate calibrations of different parts of the system, and the ***ratio method***, in which a sample of the transmitted signal is passed through the receiver and data-handling system. The latter is definitely a superior approach, because there is less chance for error and it may be conducted at very frequent intervals, whereas separate calibrations require interrupting the measurement series to a much greater degree. Furthermore, the calibration of each part separately results in many errors that can add up to decrease the precision of the overall calibration.

Recall from Eq. (5.37) that the scattering coefficient σ^0 of a distributed target is directly proportional to the received-to-transmitted power ratio, $P^{\mathrm{r}}/P^{\mathrm{t}}$. That is,

$$\sigma^0 = \left[\frac{P^{\mathrm{r}}}{P^{\mathrm{t}}}\right]\left[\frac{(4\pi)^3 R_0^4}{\lambda^2 G_0^2 A}\right]. \qquad (13.132)$$

Internal calibration allows us to measure $(P^{\mathrm{r}}/P^{\mathrm{t}})$, which includes all system gains and losses except for the antenna gain G_0.

The general concept of the ratio calibration method is illustrated by Fig. 13-28. A directional coupler is shown in the line from the transmitter to the antenna, and another in the line from the antenna to the receiver. Between the two is a path of known attenuation L_{c} so that a sample of the transmitter signal may be fed through to the receiver. Depending on the type of radar system, this may be done continuously or on an alternating basis, with the receiver sometimes connected to the incoming signal from the external target and sometimes connected to the calibration signal. In Fig. 13-28, the transmitter power at the antenna is P^{t}, and at the directional coupler, it is

$$P_0^{\mathrm{t}} = L_{\mathrm{t}} P^{\mathrm{t}}, \qquad (13.133)$$

where L_{t} is the loss factor of the transmitter directional coupler. The received power at the receiver directional coupler is either

$$P_0^{\mathrm{r}} = \frac{P^{\mathrm{r}}}{L_{\mathrm{r}}}, \quad \textbf{during target reception} \text{ (both switches in Position 1)} \qquad (13.134a)$$

or

$$P_0^{\mathrm{c}} = \frac{P_0^{\mathrm{t}}}{L_{\mathrm{c}} L_{\mathrm{DC_t}} L_{\mathrm{DC_r}}} = \frac{L_{\mathrm{t}} P^{\mathrm{t}}}{L_{\mathrm{c}} L_{\mathrm{DC_t}} L_{\mathrm{DC_r}}}, \quad \textbf{during calibration} \text{ (both switches in Position 2)} \qquad (13.134b)$$

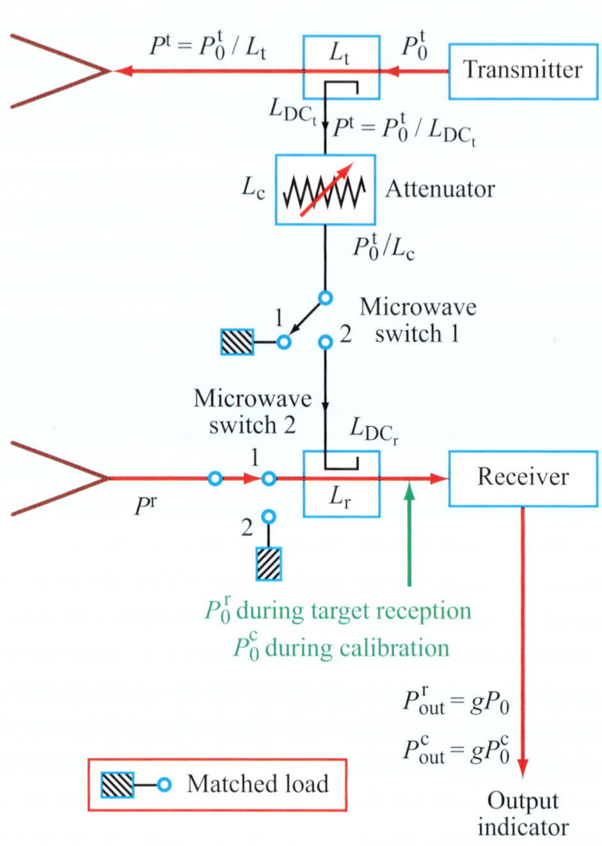

Figure 13-28: Ratio calibration technique. Switches 1 and 2 are synchronized.

where $L_{\mathrm{DC_t}}$ and $L_{\mathrm{DC_r}}$ are the loss factors of the two directional couplers for the internal path through the attenuator. With receiver gain g applied to both P_0^{r} and P_0^{c}, the ratio of the two final outputs is

$$\frac{P_{\mathrm{out}}^{\mathrm{r}}}{P_{\mathrm{out}}^{\mathrm{c}}} = \frac{gP_0^{\mathrm{r}}}{gP_0^{\mathrm{c}}} = \left(\frac{L_{\mathrm{c}} L_{\mathrm{DC_t}} L_{\mathrm{DC_r}}}{L_{\mathrm{r}} L_{\mathrm{t}}}\right) \frac{P^{\mathrm{r}}}{P^{\mathrm{t}}}, \qquad (13.135)$$

which can be inverted to give

$$\frac{P^{\mathrm{r}}}{P^{\mathrm{t}}} = K_{\mathrm{s}} \frac{P_{\mathrm{out}}^{\mathrm{r}}}{P_{\mathrm{out}}^{\mathrm{c}}}, \qquad (13.136)$$

where K_{s} is an ***overall calibration constant*** given by

$$K_{\mathrm{s}} = \frac{L_{\mathrm{r}} L_{\mathrm{t}}}{L_{\mathrm{c}} L_{\mathrm{DC_t}} L_{\mathrm{DC_r}}}. \qquad (13.137)$$

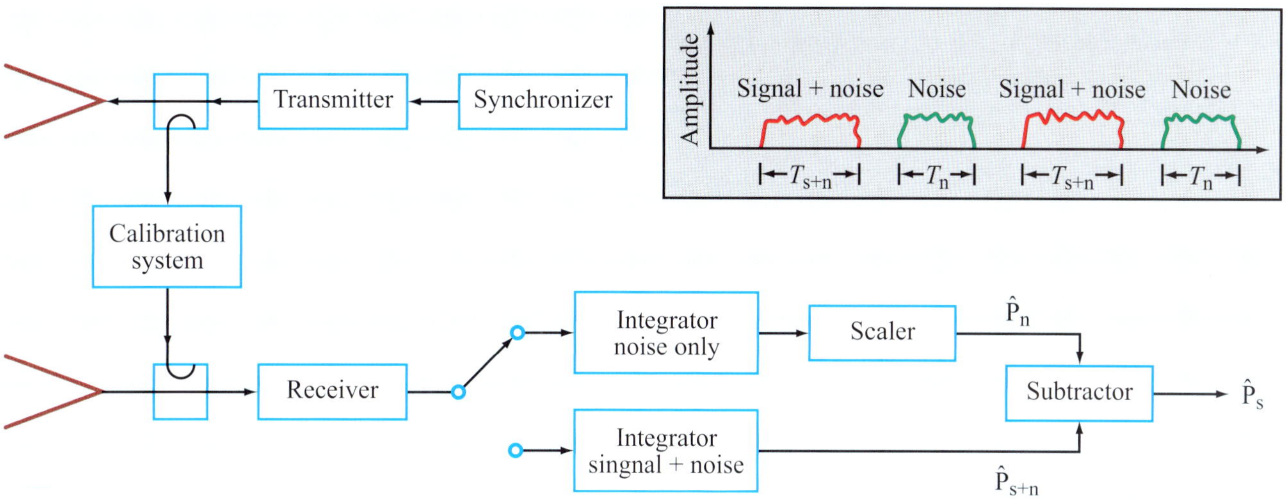

Figure 13-29: Noise measurements to allow radars to measure σ^0 at low SNRs.

Radars can be made very sensitive by using the same technique that is applied in radiometers. In most radiometer receivers, alternate measurements are made of the external scene viewed by the antenna beam and of a reference noise source. As discussed in Section 13-8, upon subtraction of the two measurements and then integrating the difference signal, much of the receiver noise is eliminated. When the concept is applied to a radar, the first measurement corresponds to the return from an external target plus the receiver noise, and the second signal is receiver noise alone. Subtraction followed by integration leads to significant improvement in signal-to-noise ratio (SNR) and K_p.

Figure 13-29 illustrates a switched system for realizing the SNR improvement technique. From time to time, the system is calibrated using the internal calibration method described earlier. During the interim period when it is assumed that the system is stable, the system is switched alternately between the sum of signal-and-noise and the noise alone. As indicated in the time sketch below the block diagram, this can be accomplished by transmitting a signal of finite duration, integrating the value received from this signal, and then integrating the noise during a waiting period before the receipt of the next signal. Since, as indicated in the sketch, the integration times for noise and the sum of signal-and-noise may be different, a scaler is required for one of the signals before they are subtracted to obtain the estimate of the signal alone. This type of system has been widely used in wind scatterometer systems. In the case of the Seasat scatterometer a precision of about 5 percent in the measurement was achieved with a signal-to-noise ratio of the order of -10 dB (Johnson et al., 1980).

13-10.2 External Calibration

The degree of success of the internal calibration techniques discussed in the preceding subsection depends in part upon the accuracy with which the antenna gain pattern and other system functions are known. It is often desirable to calibrate a radar imager or scatterometer by measuring the return power from a target of known radar cross section. Figure 13-30 illustrates calibration arrangements for ground-based and airborne scatterometers using a "hard" target (such as a corner reflector) to provide the necessary calibration level. As shown later, measurement precision is determined by the magnitude of the radar cross section of the calibration target relative to the radar cross section of the

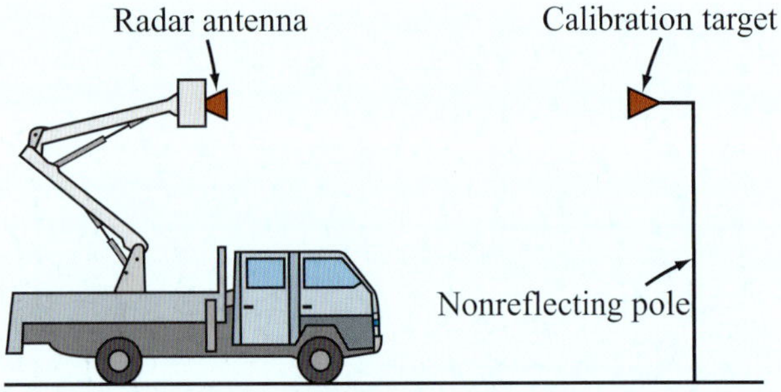

(a) Calibration of truck-mounted radar

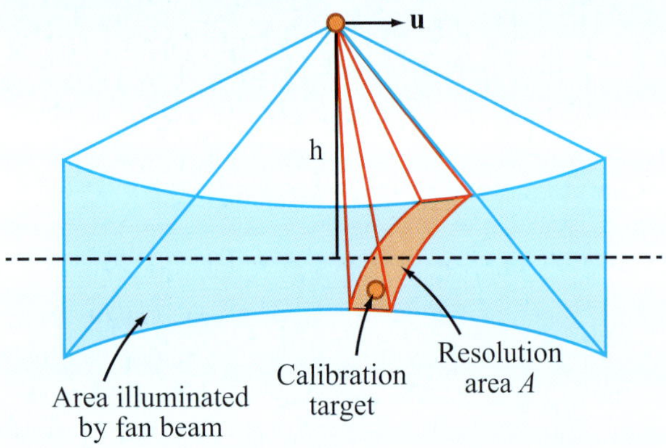

(b) Calibration of fan-beam Doppler scatterometer

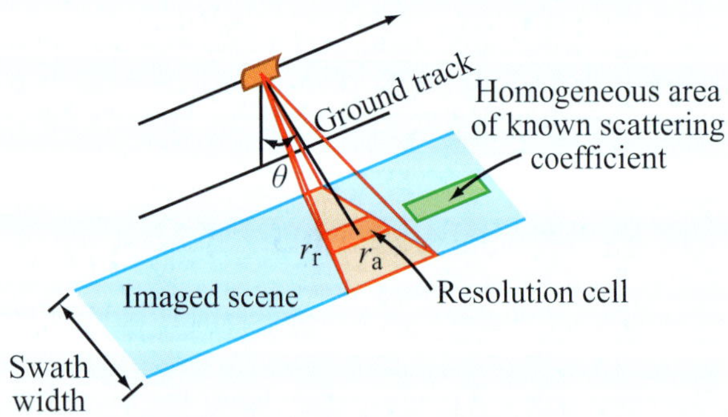

(c) Use of a homogeneous area of known scattering coefficient to calibrate an imaging radar

Figure 13-30: Configurations for calibrating (a) a ground-based (truck- or tower-mounted) scatterometer and (b) an airborne scatterometer or imager (here a fan-beam Doppler scatterometer) using a manmade calibration target, or (c) using a natural surface of known scattering properties.

background; in the case of Fig. 13-30(b), the background is the resolution area A. Alternatively, a "homogeneous" extended target of known scattering coefficient $\sigma^0(\theta)$ may be used [Fig. 13-30(c)]. This latter approach is particularly attractive for airborne and spaceborne systems because the background is itself the calibration target. Thus, the need for hard targets with large radar cross sections is avoided. The azimuth and range dimensions of the extended target should be much larger than the corresponding spatial resolutions of the radar, so that sufficient averaging can be performed to reduce the effects of signal fading. Ideally, the extended target should have a time-invariant scattering behavior (i.e., its dielectric properties and surface roughness remain constant over an extended period of time), and its $\sigma^0(\theta)$ should be a smooth, slowly varying function of θ over the angular range of the antenna elevation beamwidth (range direction). An example of an extended target is a concrete airport runway; its dimensions are suitable for a high-resolution imaging radar and its surface scattering characteristics remain essentially constant with time, as long as it remains dry. The major drawback of the (electromagnetically) smooth concrete surface, however, is that its scattering coefficient $\sigma^0(\theta)$ exhibits a strong dependence on θ, which means that the calibration accuracy becomes critically dependent on the accuracy with which θ can be determined and on the accuracy of $\sigma^0(\theta)$. It is assumed that $\sigma^0(\theta)$ is measured by a calibrated, truck-mounted or airborne scatterometer. A more suitable calibration surface is an artificially prepared or naturally existing, relatively fiat, rough surface located in an arid region. If the dimensions of the extended target are on the order of kilometers, an airborne imaging radar may be calibrated over several segments of its full angular range (in elevation) by flying the system several times so that the extended target is positioned at different ranges within the image swath width. For a spaceborne scatterometer or modest-resolution imager, the Amazon rain forest is the most homogeneous such target (Johnson et al., 1980; Kennett and Li, 1989; Long and Skouson, 1996; Tsai et al., 1999), and its σ^0 is more nearly independent of θ than that of a desert target.

13-10.3 Measurement Precision

When a calibration target of radar cross section σ_c is illuminated by a radar antenna, the backscattered electric field at the receiver is a combination of the desired field from the calibration target and the fields scattered by the background. Thus, in terms of the polarization components of those fields that are parallel to the polarization vector of the receiving antenna,

$$\mathbf{E}_m = \mathbf{E}_c + \mathbf{E}_b = \mathbf{E}_c \left[1 + \frac{\mathbf{E}_b}{\mathbf{E}_c}\right] = \mathbf{E}_c \left[1 + \left|\frac{\mathbf{E}_b}{\mathbf{E}_c}\right| e^{j\phi}\right], \tag{13.138}$$

where $\mathbf{E}_m$, $\mathbf{E}_c$, and $\mathbf{E}_b$ are respectively the receiver (measured) electric field, the electric field due to the calibration target, and the net electric field due to the background, and ϕ is the relative phase angle between $\mathbf{E}_b$ and $\mathbf{E}_c$. The measured radar cross section (RCS) σ_m is proportional to the received power:

$$\begin{aligned}\sigma_m = K|\mathbf{E}_m|^2 &= K|\mathbf{E}_c|^2 \left(1 + \frac{|\mathbf{E}_b|^2}{|\mathbf{E}_c|^2} + 2\left|\frac{\mathbf{E}_b}{\mathbf{E}_c}\right| \cos\phi\right) \\ &= \sigma_c \left(1 + \frac{\sigma_b}{\sigma_c} + 2\sqrt{\frac{\sigma_b}{\sigma_c}} \cos\phi\right),\end{aligned} \tag{13.139}$$

where K is a proportionality constant and σ_c and σ_b are the RCSs of the calibration target and background, respectively. If $\sigma_b \ll \sigma_c$, Eq. (13.139) reduces to $\sigma_m \approx \sigma_c$. In the general case, however, the measured radar cross section may differ from σ_c by a factor equal to the quantity inside the brackets in Eq. (13.139). In a measurement setup, ϕ is an unknown quantity that may take on any value between 0 and π. The corresponding ***calibration error*** falls between the maximum limits given by

$$\frac{\sigma_m - \sigma_c}{\sigma_c} = \frac{\sigma_b}{\sigma_c} \pm 2\sqrt{\frac{\sigma_b}{\sigma_c}} = S^2_{bc} \pm 2S_{bc}, \tag{13.140}$$

where

$$S^2_{bc} = \frac{\sigma_b}{\sigma_c} \tag{13.141}$$

is defined as the ***background-to-calibration-target radar-cross-section ratio***. Plots of the upper and lower limits of Eq. (13.140) are shown in Fig. 13-31 as a function of S_{bc}. Alternatively, σ_m and σ_c may be expressed in decibels, in which case the error limits are given by

$$\begin{aligned}\sigma_e(\text{dB}) &= \sigma_m(\text{dB}) - \sigma_c(\text{dB}) \\ &= 10\log(1 + S^2_{bc} \pm 2S_{bc}).\end{aligned} \tag{13.142}$$

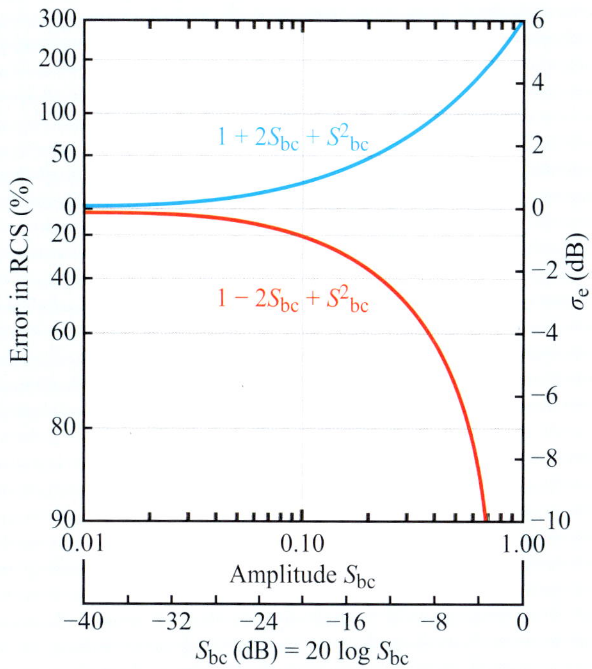

Figure 13-31: Maximum and minimum error bounds (in percent on left-hand scale and in dB on right-hand scale) associated with the measurement of a target with radar cross section σ_c against a background with radar cross section σ_b, plotted as a function of $S_{bc} = \sqrt{\sigma_b/\sigma_c}$ (or $S_b(\text{dB}) = 20\log S_b$) [adapted from Blacksmith et al., 1965].

As can be seen from Fig. 13-31, in order to insure that the error is bounded within the ± 20 percent interval (± 1 dB), we must have $\sigma_b/\sigma_c \leq 10^{-2}$ ($= -20$ dB). For an airborne radar viewing a ground-resolution cell of area A and average scattering coefficient σ^0_b, in which a calibration target of RCS σ_c is located, the preceding measurement-precision requirement of ± 1 dB translates into

$$A\sigma_b^0 \leq \frac{\sigma_c}{100}\,. \tag{13.143}$$

To illustrate this requirement with an example, consider a (fairly) high-resolution radar system with $A = 100$ m^2, and a bare soil background characterized by a $\sigma_b^0 = 0.1$. According to Eq. (13.143), the calibration target should have an RCS of $\sigma_c \geq 10^3$ m^2, or

$$\sigma_c(\text{dB}) \geq 30 \text{ dBsm},$$

where ***dBsm*** stands for the dB scale referred to a radar cross section of 1 m^2.

Obviously, calibration of ground-based radars [Fig. 13-30(a)] is a much easier task because the calibration target can be positioned at the top of a low-reflectivity structure (such as a pole whose surface is covered with absorbing material), thereby avoiding most of the backscattering from the ground. In that case, the background RCS is much smaller than for the ground-viewing case, and consequently, calibration targets with smaller RCS are required for a given level of measurement precision.

13-11 Passive Calibration Targets

A wide variety of calibration targets may be used in a laboratory environment, but for field operations where high directional accuracies are difficult to achieve, the calibration target should not only have a high RCS, but its RCS should also be insensitive to orientation (relative to the radar) over a wide range of angles. This characteristic may be described in terms of the half-power beamwidths of the RCS directional pattern, as discussed next for individual types of calibration targets. Calibration targets are of two types: (a) passive calibrators and (b) active calibrators. We summarize the properties of both types in this and the following sections.

13-11.1 Flat Rectangular Plate

When viewed from direction (θ, ϕ), the perfectly conducting rectangular plate shown in Fig. 13-32 exhibits a radar cross section $\sigma(\theta, \phi)$ given by Kerr and Goldstein (1951) as

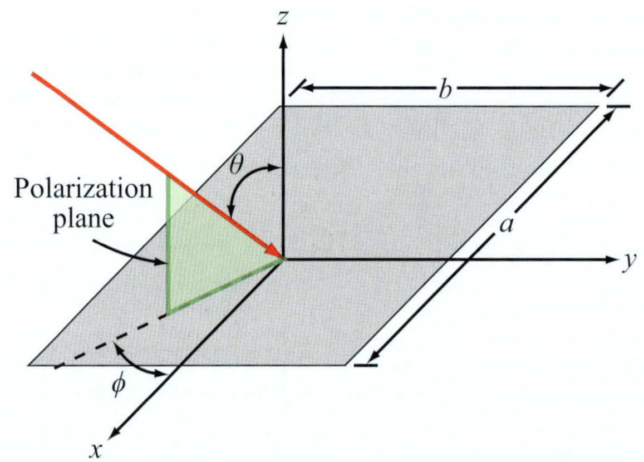

Figure 13-32: Plane of polarization of incident wave.

$$\sigma(\theta,\phi) = \frac{4\pi A^2}{\lambda^2}\left[\frac{\sin(ka\sin\theta\cos\phi)}{ka\sin\theta\cos\phi}\,\frac{\sin(kb\sin\theta\sin\phi)}{kb\sin\theta\sin\phi}\right]^2\cos^2\theta, \tag{13.144}$$

where a and b are the plate dimensions, $A = ab$ is its area, and $k = 2\pi/\lambda$. The above expression, which is polarization-independent, is in good agreement with experimental measurements (Fig. 13-33) for θ in the 0–30° range.

The maximum value of $\sigma(\theta, \phi)$ occurs at normal incidence ($\theta = \phi = 0°$) and is given by

$$\sigma_{\max} = \frac{4\pi A^2}{\lambda^2}\,, \tag{13.145}$$

and the ***half-power beamwidths*** may be computed using the procedure discussed in Section 3-5 for a rectangular antenna,

$$\beta \approx \begin{cases} 0.44\lambda/a & \text{in } xz \text{ plane } (\phi = 0), \\ 0.44\lambda/b & \text{in } yz \text{ plane } (\phi = \pi/2). \end{cases} \tag{13.146}$$

As may be observed from Fig. 13-33, the flat plate has very narrow beamwidths; in this case, $a = b = 16.5$ cm, $\lambda = 3.25$ cm, $\sigma_{\max} = 8.8$ m^2 (= 9.4 dBm^2), and $\beta \approx 0.086$ rad ($\approx 5°$).

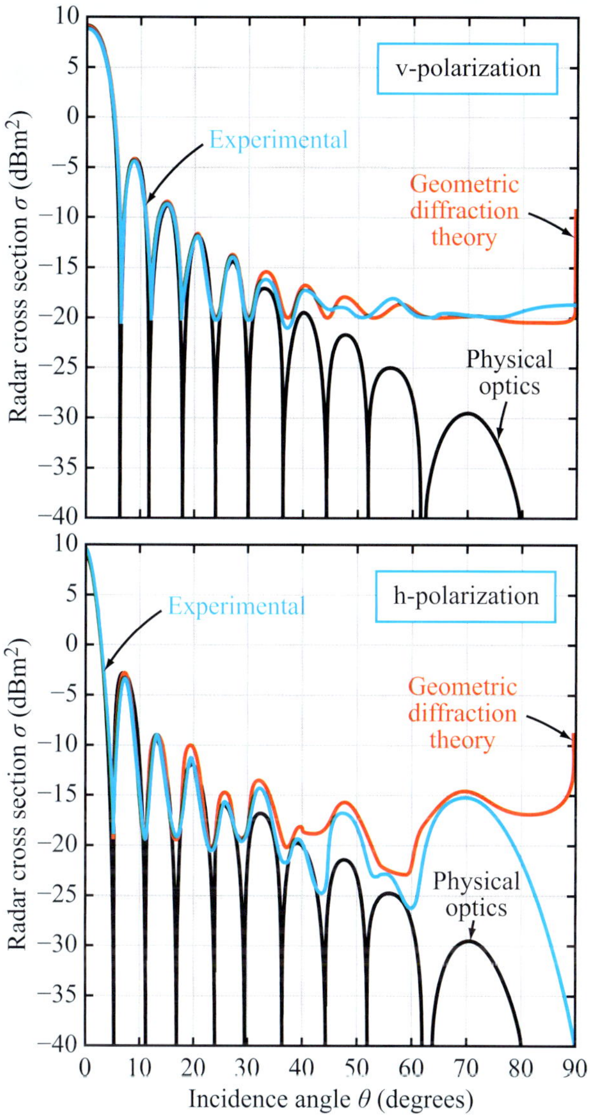

Figure 13-33: Radar cross section of a 16.5 cm by 16.5 cm flat plate at $\lambda = 3.25$ cm [from Ross, 1966].

13-11.2 Flat Circular Plate

The radar cross section of a circular disc of radius r is given by Kerr and Goldstein (1951) as

$$\sigma(\theta) = \frac{4\pi A^2}{\lambda^2}\left[2\,\frac{J_1(2kr\sin\theta)}{2kr\sin\theta}\right]^2\cos^2\theta, \qquad (13.147)$$

where A is the area of the plate and $J_1(\ )$ is the Bessel function of the first kind. Its maximum RCS $\sigma_{\max}$ is given by Eq. (13.145), and

$$\beta \approx \frac{\lambda}{4r} = \frac{\lambda}{2d}, \qquad (13.148)$$

where d is the disc diameter.

13-11.3 Sphere

Expressions for the backscattering cross section of a sphere are given in Section 8-5, and the frequency dependence of the backscattering efficiency of a metal sphere is plotted in Fig. 8-21. Above the Mie region ($r \geq \lambda$)

$$\sigma \approx A = \pi r^2. \qquad (13.149)$$

The most attractive feature of the metal sphere as a calibration target is the angular symmetry of its radar cross section [σ is independent of the observation direction (θ, ϕ)]. The magnitude of σ, however, is considerably smaller than the maximum radar cross section of a metal plate with the same physical cross section as the sphere. For example, if $r = 2\lambda$ is the radius of a metal sphere and also the radius of a circular metal plate, Eqs. (13.145) and (13.149) lead to

$$\frac{\sigma_{\max}(\text{plate})}{\sigma(\text{sphere})} \approx 160 \qquad (r = 2\lambda).$$

13-11.4 Corner Reflector

A ***dihedral corner reflector*** is an assembly consisting of two flat metal plates arranged at right angles to each other, as shown in Fig. 13-34(a). The process of reflection by a dihedral reflector may be considered in terms of ray theory; if a ray enters the corner reflector in a direction that is perpendicular to the line of intersection of the two plates, it experiences a specular reflection by one of the plates towards the other one, then reflects again off the second one, and finally returns in the direction of the transmitting source. This property results in a much wider beamwidth in the horizontal plane (perpendicular to the line of intersection of the two plates) than for a single plate, but has the same narrow-beamwidth limitation in the elevation plane.

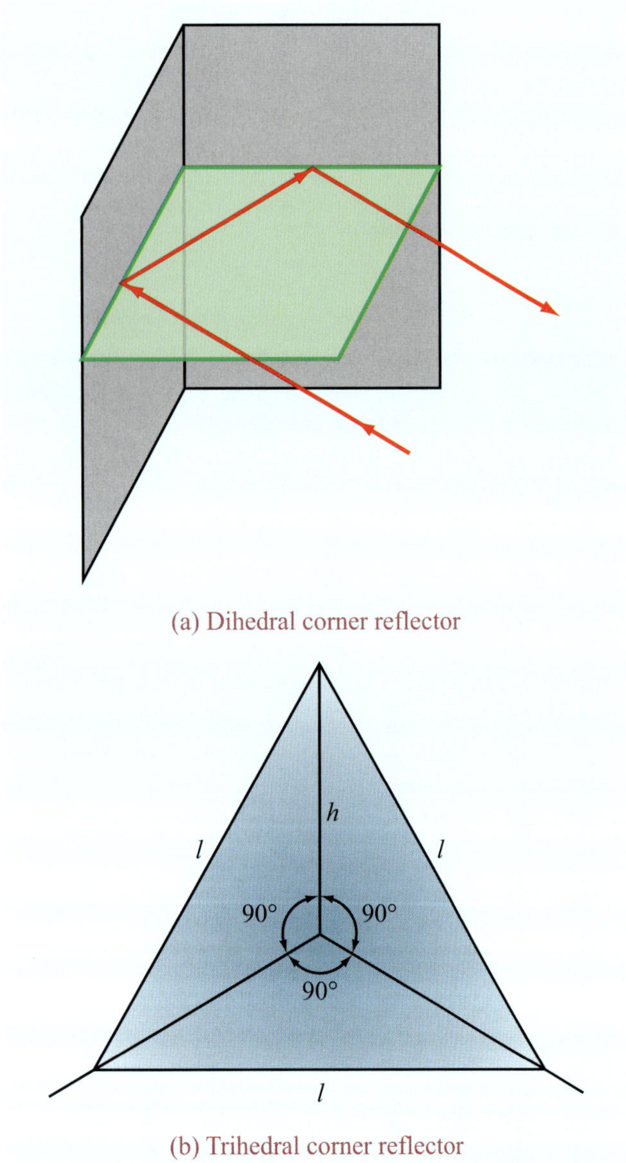

Figure 13-34: Dihedral and trihedral corner reflectors.

This restriction may be removed by adding a third plate that is orthogonal to the other two, as shown in Fig. 13-34(b). Such a reflector is called a ***trihedral corner reflector***. The directional variation of the radar cross section of the trihedral corner reflector is described in terms of the angles θ and ϕ, defined as the angles in the elevation and azimuth planes between the axis of symmetry of the reflector and the radar direction.

In Fig. 13-35(a), the plane containing the radar line of sight (line connecting radar to common corner of trihedral reflector) and the intersecting line connecting the reflector corner to point c is called the elevation plane. The reference plane is orthogonal to the elevation plane and contains the radar line of sight. The angle θ is measured between the radar line of sight and the projection of the reflector axis of symmetry onto the reference plane, and ϕ is the angle between the axis of symmetry and its projection onto the reference plane.

The size of the trihedral reflector is defined in terms of the length of its edges, l, or its height $h = l/\sqrt{2}$, as shown in Fig. 13-35(a). One of the attractive features of the trihedral corner reflector is that its RCS has broad patterns in both the azimuth and elevation planes. Examples of experimental measurements made at 9.5 GHz for a trihedral with 45 cm long edges (or equivalently, $h = 10\lambda$) are shown in Fig. 13-35(b) and (c). The half-power beamwidth is on the order of 30° in both planes.

The maximum RCS of the trihedral corner reflector occurs along its axis of symmetry ($\theta = 0$ and $\phi = 0$), and is given by

$$\sigma_{\text{max}} = \frac{4\pi}{\lambda^2} A_{\text{eff}}^2, \tag{13.150}$$

where A_{eff}, the effective area of the trihedral reflector, is related to l and h by

$$A_{\text{eff}} = \frac{l^2}{\sqrt{12}} = \frac{h^2}{\sqrt{3}} \,. \tag{13.151}$$

Because of the relatively broad beamwidth of its RCS in both the azimuth and elevation planes, the trihedral corner reflector has been the calibration target of choice for many radars, from marine radars used on ships to satellite imaging SARs flown in space. However, the trihedral corner reflector is vulnerable to the possible interaction between the reflector and its ground background. At incidence directions close to boresight, only a portion of each plate contributes to backscattering; for the incident ray to return to the radar,

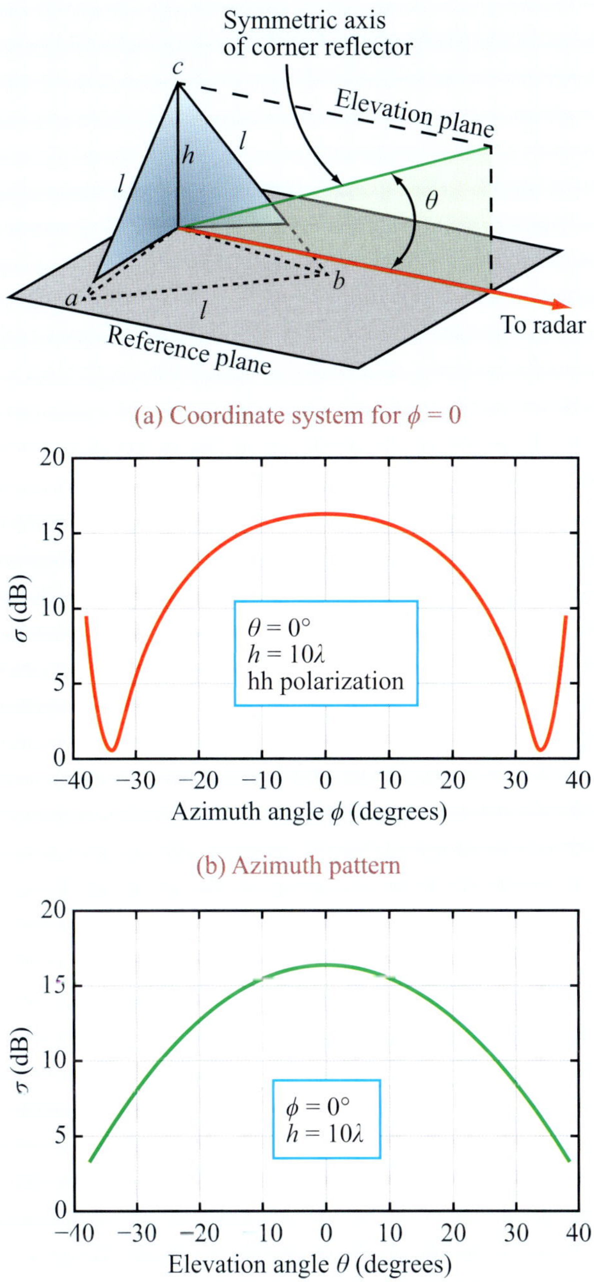

(a) Coordinate system for $\phi = 0$

(b) Azimuth pattern

(c) Elevation pattern

Figure 13-35: Trihedral corner reflector: (a) coordinate system showing the angle θ, and (b) and (c) are measured scattering patterns for $h = 10\lambda$ at 9.5 GHz [Sarabandi and Chiu, 1996].

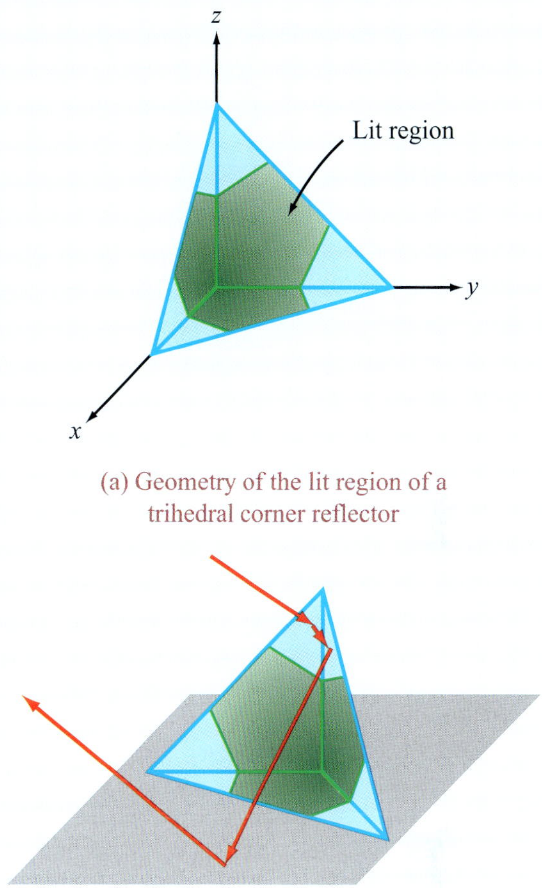

(a) Geometry of the lit region of a trihedral corner reflector

(b) Interaction between the reflector and the ground surface

Figure 13-36: Diagrams of the lit region of the trihedral reflector and of how undesirable components involving reflections by the unlit regions and the ground surface can occur.

it has to be reflected by all three plates. The part of each plate that satisfies this condition is called the ***lit region***. Ray analysis reveals that only 2/3 of the area of each plate satisfies the lit condition (Robertson, 1947; Knott, 1993; Sarabandi and Chiu, 1996), as illustrated in Fig. 13-36(a). This means that the unlit region can be eliminated without reducing the maximum RCS. By eliminating the unlit region, we avoid undesirable scattering contributions involving the combination of reflection by the unlit region and reflection by the ground [Fig. 13-36(b)]. These undesirable interactions between the reflector and the ground introduce uncertainties that end up compromising the quality and accuracy of the calibration process. In practice, instead of totally eliminating the unlit regions, they are replaced with light-weight absorbing material, thereby maintaining the physical shape of the trihedral, while simultaneously reducing its weight by about 30% and suppressing reflector-ground interactions. Sarabandi and Chiu (1996), who call such a reflector a ***pentagonal corner reflector***, have developed theoretical models for computing its RCS and have verified their results with experimental observations.

13-11.5 Luneburg-Lens Reflector

A spherical Luneburg lens is a dielectric sphere whose index of refraction $n(r)$ varies as a function of r, the distance from the center of the sphere (Fig. 13-37), according to the form defined by Luneburg (1964):

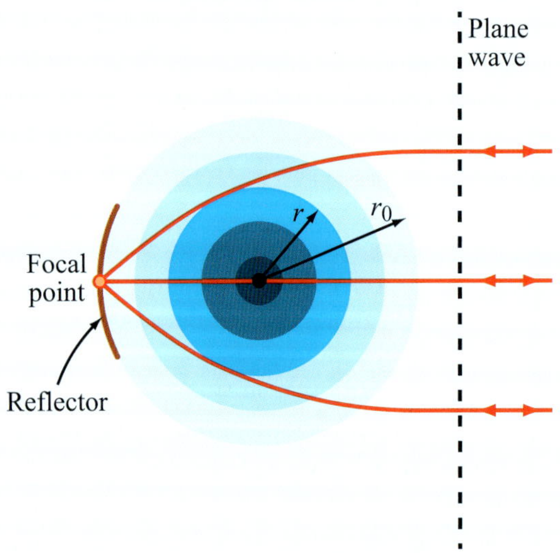

Figure 13-37: A Luneburg lens consists of many concentric dielectric shells, with the relative permittivity increasing from slightly greater than 1 for the outermost shell to 2 at the center. The permittivity variation causes an incident plane wave to refract towards the focal point.

$$n(r) = \sqrt{\varepsilon(r)} = \left[2 - \left(\frac{r}{r_0}\right)^2\right]^{1/2}. \qquad (13.152)$$

In other words, $n(r)$ varies between $n = \sqrt{2}$ at the center ($r = 0$) and $n = 1$ at the periphery ($r = r_0$). A plane wave incident upon such a lens is focused to a point on the opposite surface of the sphere. If a metallic reflecting surface is present on the back side of the sphere, the wave is reflected back in the direction from which it originated. The continuous variation of $n(r)$ is approximated in practice with discrete changes of $n(r)$; typically, 10 or more concentric spherical shells are used, with the outermost shell having a dielectric constant of about 1.1 (Buckley, 1960; Croney and Delany, 1963).

The angular coverage provided by a Luneburg-lens reflector is governed by the geometry of the reflecting surface, as illustrated by Fig. 13-38. Comparisons of measured angular patterns at 9.375 GHz are shown in Fig. 13-38(b) for three types of Luneburg-lens reflectors, in addition to a circular metal plate, and a metal sphere, all with the same diameter (12 inches $\approx$ 30 cm). Also shown is the pattern for a circular corner reflector whose dimensions are such that it can barely be contained within a 12-inch-diameter sphere. The Luneburg-lens reflector provides much wider beamwidths than the flat plate, yet it has approximately the same level of $\sigma_{\max}$. It is superior also to the corner reflector, in terms of both absolute level (for comparable dimensions) and angular coverage. Its disadvantages are its greater weight and much higher fabrication cost.

In the central angular region, $\sigma_{\max}$ of the Luneburg-lens reflector is equal (theoretically) to that of a circular plate of the same physical cross sectional area,

$$\sigma_{\max} = \frac{4\pi A^2}{\lambda^2} = \frac{4\pi^3}{\lambda^2}\, r_0^4, \qquad (13.153)$$

where r_0 is the radius of the spherical lens. For the reflectors shown in Fig. 13-38(b), $\sigma_{\max}$ of the Luneburg-lens reflector is about 2 dB lower than its theoretical value (compared with the flat plate) and about 28 dB higher than σ of the metal sphere; i.e., larger by a factor of 630. Measured values of $\sigma_{\max}$ usually are slightly smaller than the theoretical value given by Eq. (13.153).

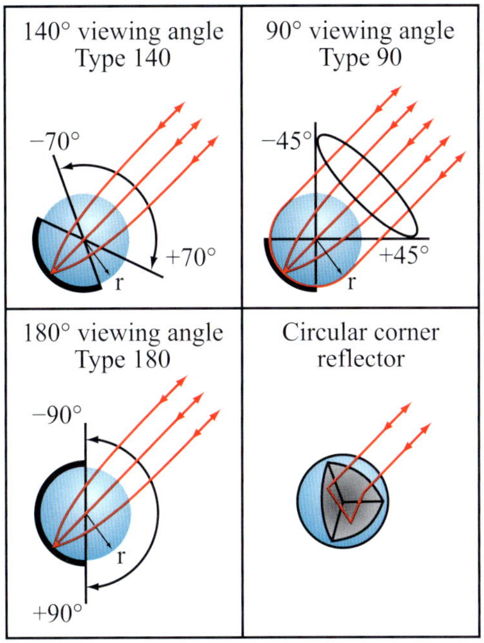

(a) Various types of Luneberg lens reflectors

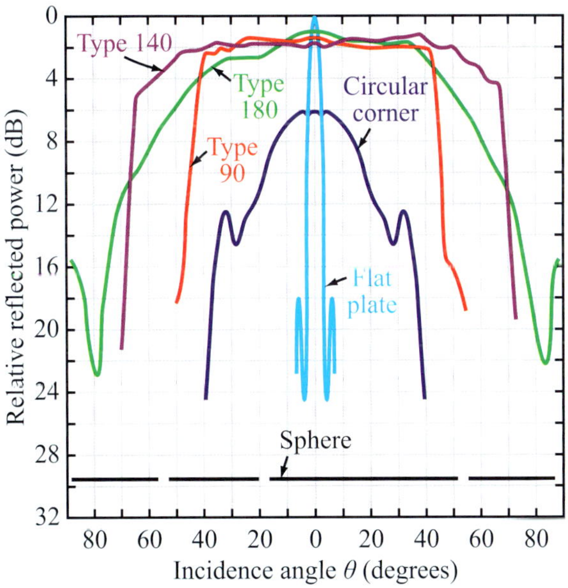

(b) Angular patterns

Figure 13-38: (a) Luneburg lens with three different metal reflector configurations, and (b) angular patterns of the radar cross sections of the three Luneburg reflectors, a circular metal plate and a metal sphere, all 12 inches in diameter, and the largest circular corner reflector that can be contained within a 12-inch-diameter sphere. All cross sections are relative to the peak value of the circular-plate radar cross section. Frequency is 9.375 GHz. [Courtesy of Emerson and Cumming, Inc.]

To account for this difference, $\sigma_{\max}$ of the Luneburg-lens reflector may be redefined as

$$\sigma'_{\max} = \eta \, \frac{4\pi A^2}{\lambda^2}, \qquad (13.154)$$

where η is an efficiency factor, $0 \le \eta \le 1$. The magnitude of η depends on the wavelength λ and on the number of discrete steps used in approximating the continuous dielectric variation given by Eq. (13.152). To illustrate with an example, measurements of $\sigma'_{\max}$ made over the 1 to 18 GHz region for a 23 cm diameter Luneburg-lens reflector showed that η decreases from about 0.99 at 1 GHz to 0.65 at 8 GHz and to 0.18 at 18 GHz (Stiles et al., 1979).

13-11.6 Comparison of Calibration Targets

Expressions for the maximum radar cross section, $\sigma_{\max}$, and for the angular coverage, or beamwidth, are summarized in Table 13-5 for the calibration targets discussed in the preceding subsections.

13-12 Active Radar Calibrators (ARCs)

The calibration targets discussed in the preceding subsections are collectively called ***passive*** calibrators, in contrast to a second group of calibrators that use ***active*** electronic amplification of the incident radar signal to boost its level and reradiate it back to the radar. The need for ***active radar calibrators*** (ARCs) was prompted by size and weight considerations, as well as the need to calibrate the cross-polarized channel of a radar scatterometer or imager. Since its invention in the early 1980s (Brunfeldt and Ulaby, 1984b), the ARC has evolved into a highly sophisticated calibrator and has become the standard tool for external calibration of scatterometers and imagers, including polarimetric SAR systems.

To appreciate the attractive features of the ARC as a calibration device, let us consider the size requirements for a trihedral corner reflector. The maximum RCS of a trihedral with edges of length l is $\sigma_{\max} = \pi l^4/3\lambda^2$. To build a trihedral with an RCS of 40 dBsm (i.e.,

Table 13-5: Characteristics of passive calibration targets.

Type of Target	σ_{max}[a]	Approximate Half-Power Beamwidth[b]	Comments
Rectangular plate	$4\pi A^2/\lambda^2$	$0.44\lambda/a, 0.44\lambda/b$	Large σ_{max}, very narrow beamwidths
Circular plate	$4\pi A^2/\lambda^2$	$0.25\lambda/r$	Large σ_{max}, very narrow beamwidths
Sphere	A	Omnidirectional	$r > 2\lambda$, small σ
Trihedral corner reflector	$\pi l^4/3\lambda^2 = 16\pi A_c^2/9\lambda^2$	$\approx 30°–40°$	σ_{max} about 3 dB lower than that of a flat plate with same aperture
Luneburg-lens reflector	$4\pi A^2/\lambda^2$	Up to 180° (140° in practice)	Large σ_{max}, wide beamwidth, efficiency decreases with frequency

[a]A = physical area for the flat plates and cross sectional area for the metal sphere and Luneburg lens. A_c = aperture area of corner reflector.
[b]a, b = sides of rectangular plate, r = radius of circular plate, and radius of sphere, l = edge of trihedral corner reflector.

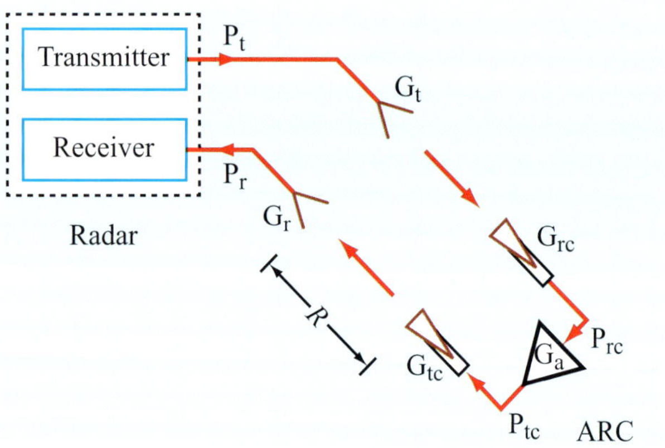

Figure 13-39: The active radar calibrator (ARC) consists of two antennas with an RF amplifier of gain G_a in between.

$\sigma_{max} = 10^4$ m^2) at 10 GHz, l should be 1.7 m, which makes such a trihedral fairly bulky and heavy. The problem is worse at lower frequencies; at 1 GHz, for example, l should be 5.4 m, and at UHF it should be larger still.

The basic concept of the active radar calibrator is very simple: it consists of two antennas separated by an RF amplifier (Fig. 13-39). To a radar, the ARC appears like a target with maximum RCS σ_{ARC} given by (Brunfeldt and Ulaby, 1984b)

$$\sigma_{ARC} = \frac{\lambda^2}{4\pi} G_{rc} G_{tc} G_a, \tag{13.155}$$

where G_{rc} and G_{tc} are the gains of the ARC's receive and transmit antennas and G_a is the ***loop gain***, which includes the gain of the RF amplifier as well as any mismatch or transmission line losses that may be present. The effective beamwidth of the ARC's RCS pattern is dictated by the size of its antennas relative to λ. To use the ARC to calibrate the hh-polarization channel of a scatterometer or imager, the ARC's antennas should have the same polarization orientation and they should be positioned such that their h-polarization direction matches that of the radar. To calibrate the hv-polarization channel, the ARC's receive antenna should be oriented such that its polarization vector is parallel to the v-polarized direction of the signal transmitted by the radar, and its transmit antenna should be oriented such that it radiates an h-polarized signal relative to the radar antenna.

Mutual coupling between the transmit and receive antennas of an ARC limits the maximum amplifier gain that can be used. If the product of the loop gain and coupling coefficient is comparable to or exceeds unity, the feedback system goes into oscillation, which is clearly undesirable. In practice, mutual coupling can be reduced to an acceptable level by insuring sufficient spatial separation between the two antennas.

Figure 13-40(a) is a diagram of an ARC designed to operate at 5 GHz. It consists of two standard-gain horns, each with a 16 cm × 22 cm aperture and gain $G_{tc} = G_{rc} = 75$ (or 18.75 dB), and a loop gain of 30.5 dB (or $G_a = 1122$). Use of Eq. (13.155) leads to

$$\sigma_{ARC} = \frac{\lambda^2}{4\pi} G_{rc} G_{tc} G_a = \frac{(6\times10^{-2})^2}{4\pi}(75)^2 \times 1122$$
$$= 1808, \text{ or } 32.57 \text{ dBsm}.$$

In part (b) of Fig. 13-40, we show azimuthal patterns of the RCS measured for a 30 cm × 30 cm flat plate, a 23 cm diameter Luneburg lens, and the ARC of Fig. 13-40(a). The ARC provides a broad pattern with a maximum RCS larger than that of the metal plate by a factor of 63 (18 dB).

The advantage of the ARC over a trihedral corner reflector is illustrated by the following example. Consider an L-band ($f = 1.275$ GHz) airborne or spaceborne imaging radar with a spatial resolution of 25 m × 25 m. The system is to be calibrated by using a calibration target whose RCS is 20 dB greater than the RCS of the terrain background. From Eq. (13.141) and Fig. 13-31, this condition corresponds to $S_{bc} = 0.1$ and a maximum measurement uncertainty of ±1 dB. The terrain background has a backscattering coefficient $\sigma^0 = -16$ dB. Hence, to satisfy the stated conditions, the calibration target should have an RCS of

$$\sigma \text{ (dBsm)} = -16 \text{ dB} + 10\log(25\times25) + 20 \text{ dB}$$
$$= 32 \text{ dBsm} \quad (\text{or } \sigma = 1584.9 \text{ m}^2).$$

Use of Eqs. (13.150) and (13.151) leads to the requirement that a trihedral corner reflector with such an RCS at 1.275 GHz should have sides $l = 3$ m, and a corresponding triangular aperture $A_{eff} = 2.6$ m^2.

The same RCS can be realized by an ARC with a loop gain of 48 dB and two microstrip antennas each $\lambda/2$ on the side (≈ 15 cm). The elevation beamwidth of the RCS of the ARC is 74°, compared with only 40° for the trihedral corner reflector (Brunfeldt and Ulaby, 1984b).

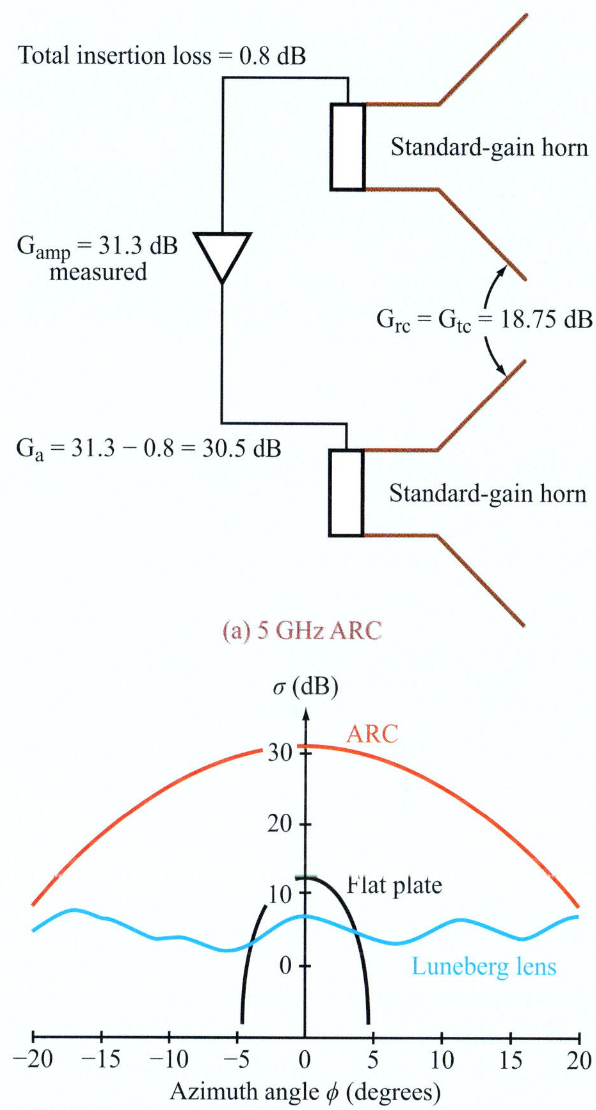

(a) 5 GHz ARC

(b) Measured RCS patterns

Figure 13-40: Measured RCS patterns for ARC, 30 cm × 30 cm flat plate and 23 cm Luneberg lens. Measurement frequency is 5 GHz [Brunfeldt and Ulaby, 1984b].

13-13 Polarimetric Active Radar Calibrator

A polarimetric radar transmits a v-polarized signal and then measures the v- and h-polarization components of the backscattered signal using a dual-polarized receiving antenna. The process leads to measuring S_{vv} and S_{hv}, the vv and hv components of the scattering matrix of the target. Upon switching the polarization of the transmitted signal from v to h polarization, repetition of the measurement process leads to S_{hh} and S_{vh} (Fig. 13-41). To calibrate a polarimetric radar system, the calibration target needs to accommodate all four polarization combinations. This can be realized by using an ARC whose antennas are orthogonally polarized relative to one another and whose orientations are at 45° (or 135°) relative to the polarization axes of the radar. Consider the configuration shown in Fig. 13-42. The radar transmit antenna is v polarized and the polarization direction of the ARC receive antenna is at 45° (halfway between the $\hat{\mathbf{v}}$ and $\hat{\mathbf{h}}$ directions). Consequently, the ARC receive antenna extracts a signal whose electric field is only $E^{\mathrm{t}}_{\mathrm{v}}/\sqrt{2}$ of the field $E^{\mathrm{t}}_{\mathrm{v}}$ of the incident wave. The corresponding power extracted by the ARC receive antenna is half of the power carried by the incident wave. After amplification by the RF amplifier, the signal is reradiated towards the radar by a transmit antenna whose polarization vector is oriented at 135° relative to the v-polarization direction. The radar uses two receive channels, one v polarized and the other h polarized. Each channel extracts one-half of the power carried by the reradiated signal. This allows calibration of the vv and hv channels. A similar process with a transmitted h-polarized wave $E^{\mathrm{t}}_{\mathrm{h}}$ leads to calibration of the hh and vh channels. Hence, the 45°-tilted ARC with orthogonally polarized antennas can provide known signals to calibrate all four channels of a polarimetric radar. Such an ARC is called a ***polarimetric ARC***, or PARC for short.

To reduce the overall size and weight of the PARC, Sarabandi et al. (1992a, b) developed a ***single-antenna PARC***, or ***SAPARC***. A block diagram is shown in Fig. 13-43(a), and the physical assembly of the antenna part is shown in Fig. 13-43(b). A square-horn antenna

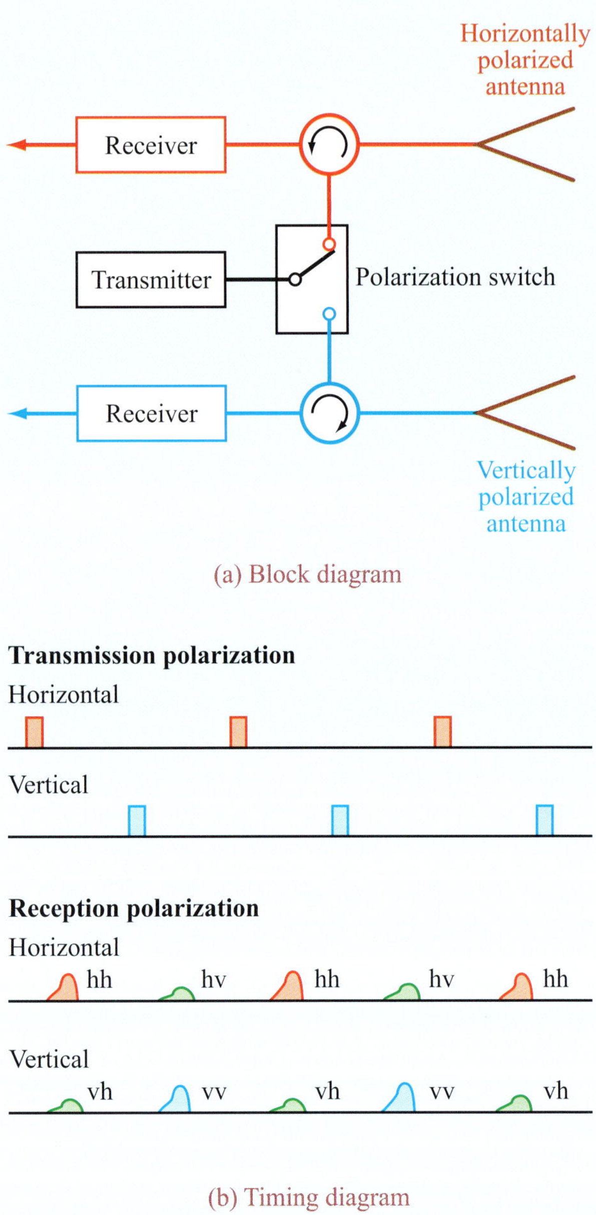

Figure 13-41: Calibration of polarimetric radar. A polarimetric radar is implemented by alternately transmitting signals out of horizontally and vertically polarized antennas and receiving at both polarizations simultaneously. Two pulses are needed to measure all the elements in the scattering matrix [van Zyl and Kim, 2011].

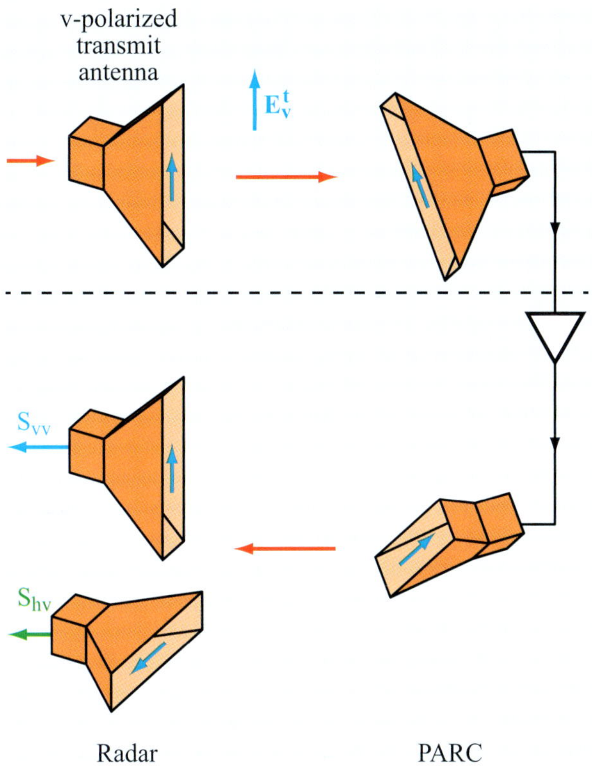

Figure 13-42: PARC with its antennas oriented at 45° relative to radar antennas.

that can accommodate both v and h polarizations is connected to a waveguide orthomode transducer (OMT) that uses wire grids placed inside the waveguide section to isolate the v and h ports. The isolation realized using this technique is on the order of 40 dB (Sarabandi et al., 1992b; 1995).

13-14 Polarimetric Scatterometers

Until the mid-1980s, the primary function of a ***magnitude-only scatterometer*** system had been to measure the backscattering coefficient σ^0 of distributed targets, usually as a function of the incidence angle θ. Often, the measurements are made for various configurations of receiving and transmitting antenna polarizations, such as hh, hv, vv, RR, RL and LL.

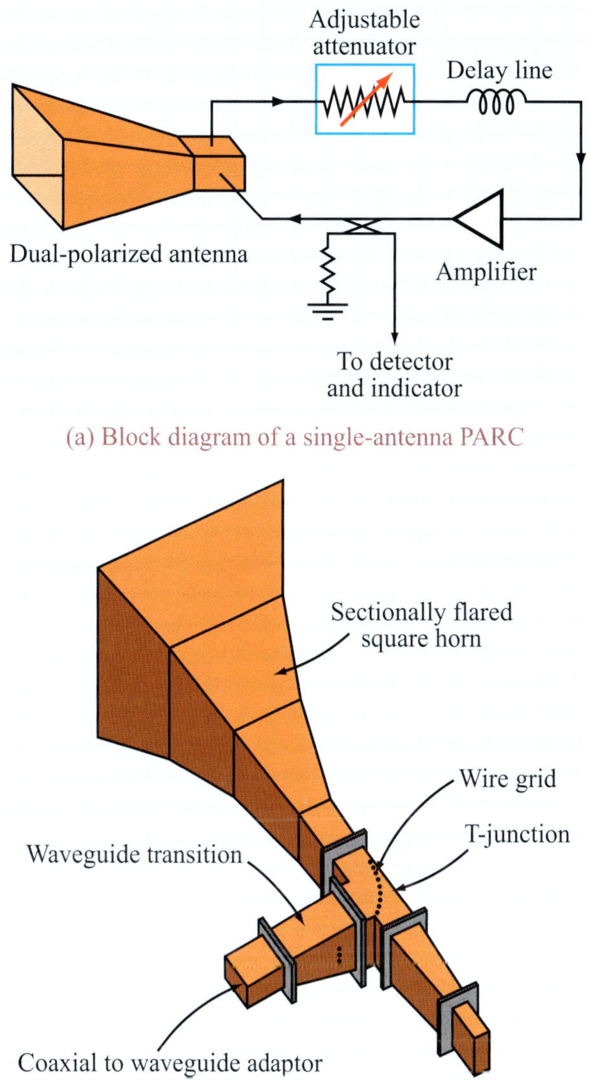

(a) Block diagram of a single-antenna PARC

(b) Antenna assembly of the L-band SAPARC

Figure 13-43: Block diagram and antenna assembly of a single-antenna PARC [Sarabandi et al., 1992b].

In all cases, however, the quantity measured by the scatterometer is the magnitude of the received signal, and no phase information is recorded. Recall that the first letter signifies the polarization of the receiving antenna and the second letter signifies the polarization of the transmitting antenna, with antenna polarization always defined for both transmitting and receiving antennas as if the antenna were in a transmitting mode; the letters h, v, R, and L denote horizontal, vertical, right circular, and left circular, respectively. In contrast to such a magnitude-only scatterometer, a ***polarimetric scatterometer*** is configured to measure the complete scattering matrix **S** of the target by measuring the magnitude and phase of the scattered electric field for all linear polarization combinations. The attractive feature of the polarimetric approach lies in the fact that given a matrix **S**, we can use the polarization synthesis technique (Section 5-10) to compute the scattering coefficient of the target for any desired receive-transmit polarization configuration.

To measure the scattering matrix **S** of a target, a radar must be able to transmit and receive two orthogonal polarizations (usually vertical (v) and horizontal (h)) and to measure both the magnitude and phase of the received signals. Although polarimetric scatterometers may be operated from aircraft or satellite platforms, we focus our consideration in this section on short-range scatterometers that typically measure targets at ranges of 20 m or less. This corresponds to a two-way time delay of roughly 133 ns. Because the target is at such a close range, leakage signals that are multiply reflected due to mismatch within the system can appear at the target range as noise with a level well above that of thermal noise. This problem, which is specific to short-range radars, must be seriously considered in the scatterometer design.

In principle, it should be possible to address short-range and noise-related considerations in the design of FM, pulse, and other types of radar systems, but the versatility of vector network analyzers makes them a natural basic building block for short-range polarimetric scatterometers (Ulaby et al., 1988b; Whitt and Ulaby, 1988; Blanchard and Rochier, 1987; Riegger et al., 1987). Hence, this section is concerned with the design and operation of basic network analyzer-based polarimetric scatterometers.

13-14.1 Network Analyzer Principles of Operation

Vector network analyzers measure the magnitude and phase characteristics of linear networks relative to some standard or reference. This is accomplished by making both transmission and reflection measurements to obtain specific characteristics of the network (Fig. 13-44). For transmission measurements, the signal transmitted through the test network is compared with the incident signal, which serves as the reference. For reflection measurements, the signal reflected from the input port is compared with the incident signal. The output port, in this case, is usually terminated with a matched load to remove reflections that travel back through the network and contribute to the reflected signal.

A basic vector network analyzer system consists of (1) an RF source, (2) an RF to IF converter, (3) an IF signal detector and analog-to-digital (A/D) converter, and (4) a digital microprocessor and display. In the context of Fig. 13-44, the RF source is used to supply the incident signal to the test network. The transmitted or reflected RF signal and a sample of the incident signal are then mixed down to the IF range, maintaining the magnitude and phase relationships between the two signals. The IF signals are then detected, converted to digital form, and processed to obtain information about the test network. If a frequency-swept incident signal is used, the result is a measurement of the frequency response of the network, and Fourier transform techniques can be used to obtain the time-domain response.

As an example, Fig. 13-45 illustrates the basic components of an HP 8510 vector network analyzer system used for making transmission-reflection measurements. In general, the principles are the same for other vector network analyzer systems (HP 8753, HP 8720, Wiltron 360, and similar systems by other manufacturers). The system of Fig. 13-45 consists of three pieces of equipment: (1) a synthesized sweeper, which serves as the RF source; (2) a frequency converter, which serves as the RF to IF converter; and (3) a network analyzer and display processor, which serves as the IF detector, A/D converter, and digital processor.

The frequency converter has four ports to which signals can be applied. Of these ports, the a_1 and a_2 ports

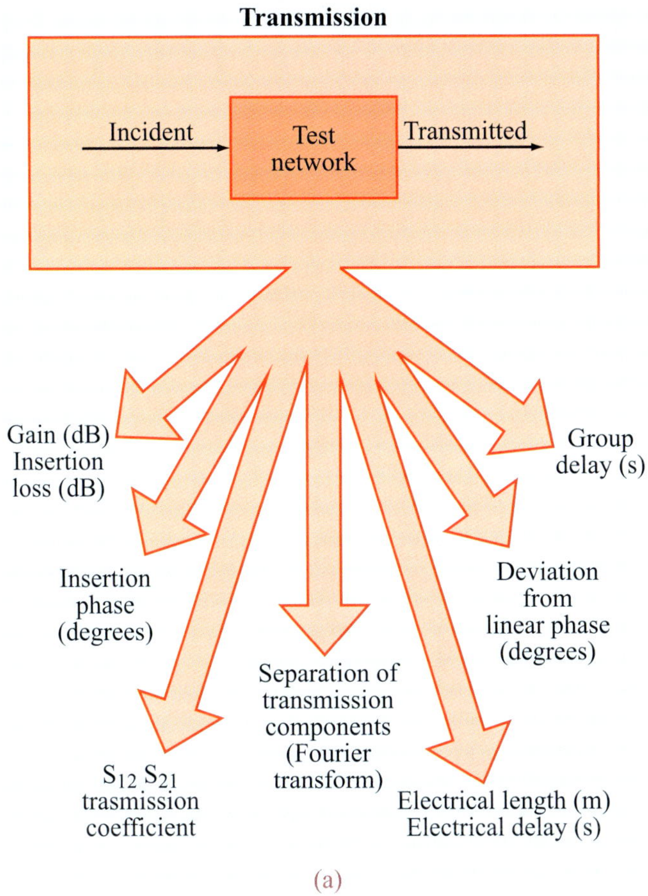

Figure 13-44: Transmission and reflection measurement capabilities of a vector network analyzer [Hewlett-Packard].

are used to input the reference signal, and the b_1 and b_2 ports are used to input the measurement signals. In the basic transmission-reflection configuration, a sample of the signal from the RF source is applied to the a_1 reference port, while the a_2 port is unused. The transmitted and reflected signals are then applied to the b_1 and b_2 measurement ports, respectively. The network analyzer can then provide the magnitude and phase of the reflected and transmitted signals relative to that of the incident signal by forming the ratios:

$$S_{11} = \frac{b_1}{a_1} \quad \text{and} \quad S_{21} = \frac{b_2}{a_1}, \tag{13.156}$$

respectively.

13-14.2 Network Analyzer Operation as a Scatterometer

The measurement system of Fig. 13-45 can be reconfigured to operate like a scatterometer. In the scatterometer configuration, the "test network" is simply some radar target separated from the measurement system by free space. In this case, the signal from the RF source is coupled to free space through an antenna and travels to the target (i.e., test network) through the air medium instead of a coaxial cable or waveguide. The signal reflected from the target is then received by another antenna and is compared with a sample of the transmitted signal from the RF source. The system

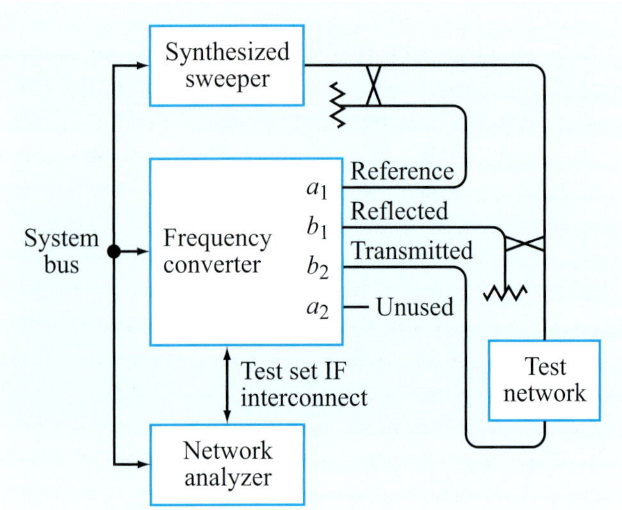

Figure 13-45: Configuration of a network analyzer for transmission and reflection measurements.

can be configured as either a dual-antenna system or a single-antenna system (using a circulator to separate the transmitting and receiving channels) as shown in Fig. 13-46.

There is a major difference between the transmission-reflection configuration and the scatterometer configuration. When operating as a scatterometer, there is a relatively long time delay between the returned and transmitted signals due to propagation to the target and back. During this delay, the reference signal has changed frequency by an amount Δf, which corresponds to a phase shift in degrees of $\Delta\phi = 360\,\Delta f\,\tau$, where $\tau = 2R/c$ is the two-way time delay. If a synthesized source is used, the transmitted signal can be stepped in frequency so as to keep the reference signal at the same frequency as the received signal.

The system configuration in Fig. 13-46 represents a traditional (magnitude-only) scatterometer, so it must be modified in order to operate it as a polarimetric scatterometer. Because it measures the complete scattering matrix of a given target, a polarimetric scatterometer must be able to transmit and receive a set of orthogonal polarizations. While other polarization schemes can be used, v- and h-linear polarizations are used here for illustrative purposes to remain consistent

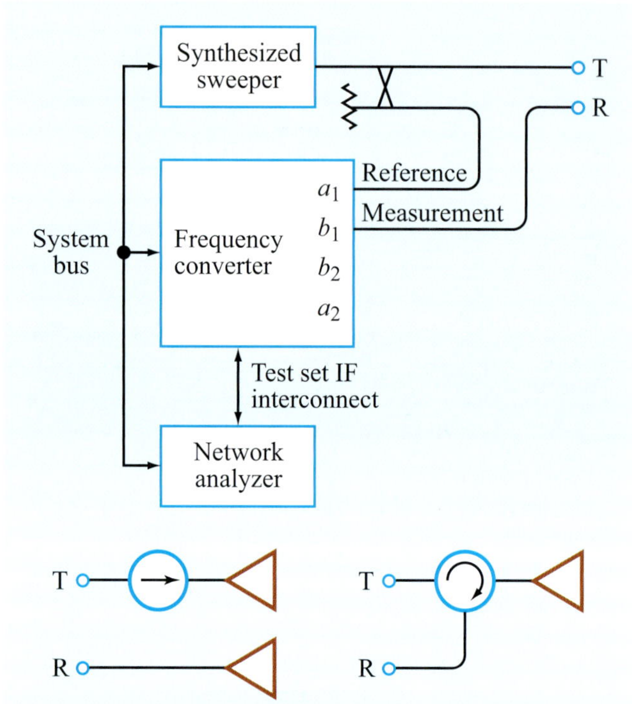

Figure 13-46: Configuration of a network analyzer for scatterometer measurements; (a) dual-antenna system, (b) single-antenna system.

with the rest of the text. For the single-antenna system, we add a second polarization channel and apply it to the b_2 port of the network analyzer. For the dual-antenna system, two options are available. An additional channel can be utilized, similar to the single-antenna system, or polarizers can be inserted before the antennas. These different configurations are shown in Fig. 13-47, where the network analyzer is assumed to be configured the same as in Fig. 13-46.

There are many advantages to using a vector network analyzer as the IF processor for a polarimetric scatterometer. The advantages all relate to the many features available with most network analyzers. One of these features is the ease with which the network analyzer can be programmed and controlled by computer over an interface bus. This feature is stressed because it reduces the hardware design to the

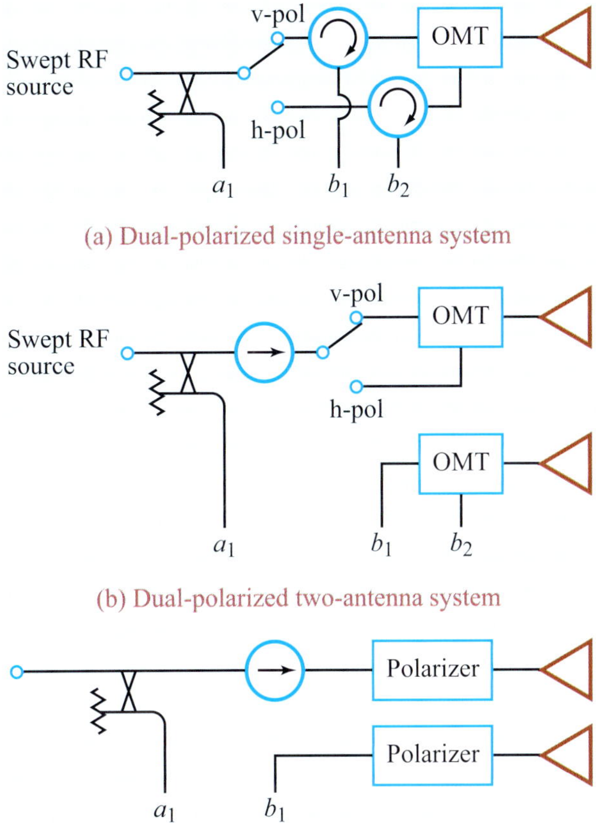

Figure 13-47: Polarimetric configurations of a network analyzer-based scatterometer: (a) dual-polarized single-antenna system, (b) single-antenna dual-polarized system, and (c) dual-antenna system using polarizers.

consideration of only RF components. The network analyzer supplies all the detection, processing, and data transfer options. To provide a complete description of all the options available with network analyzers is not possible, but we highlight some that are particularly useful in polarimetric scatterometer applications. One of the most useful processing options available with many network analyzers is time-domain operation. Because the returned signal is measured as a function of frequency, real-time Fourier transform techniques can be used to obtain the time-domain response. In scatterometer applications, the result is a measure of the radar return as a function of range. In addition, all other processing operations are available to both time and frequency domains. The signal, for example, may be displayed in the time domain where the data can be processed as a function of range. Time gating can then be used to isolate the return due to targets at a particular range. Suppose a measurement of the backscattered power is desired as a function of depth for a volume of scatterers. By using the time-domain option with time gating, the portion of the returned signal corresponding to a particular depth in the volume can be selected. The position and width of the time gate determines the range and range extent of the small volume within the entire volume of scatterers. The time-gated signal can then be transformed back to the frequency domain to obtain the frequency response of the small volume, as modified by the medium between it and the scatterometer. This process is illustrated in Fig. 13-48.

There are other features that can be used to remove nonvarying errors in the return signal. For example, scattering from objects other than the target can contribute to the noise of the measurement. Many network analyzers have data storage and complex math features that can remove these nonvarying errors. First, the response of the test setup without the target is stored and later subtracted from the response when the target is present, resulting in a measurement of the target response alone. The leakage signals caused by multiple reflections that appear at the target range can be removed in the same manner. An example illustrating this procedure is given in Fig. 13-49.

Another feature that is particularly useful in reducing random noise contributions is IF averaging. With this feature, a specified number of consecutive data traces are continuously averaged. The constant target signals add coherently, but the random noise adds incoherently, thereby improving the signal-to-noise ratio. This is similar to integration in traditional radar systems.

13-14.3 Microwave Polarimetric Scatterometers

Scatterometers must be able to measure targets over a wide range of radar cross sections. In practice, targets

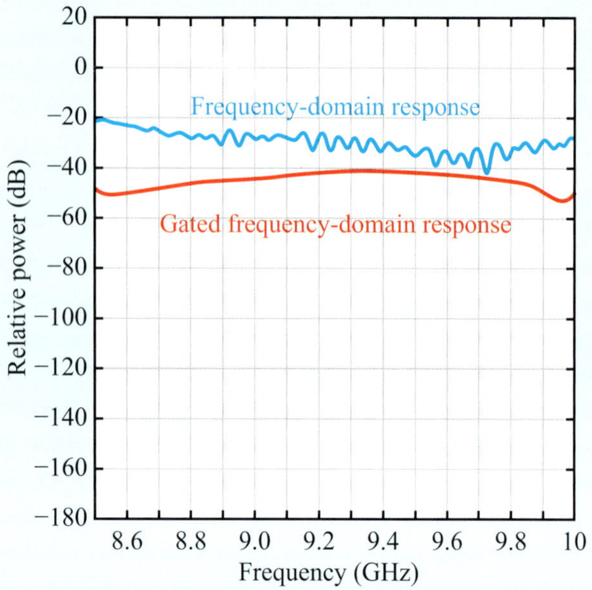

(a) Frequency- and gated frequency-domain response

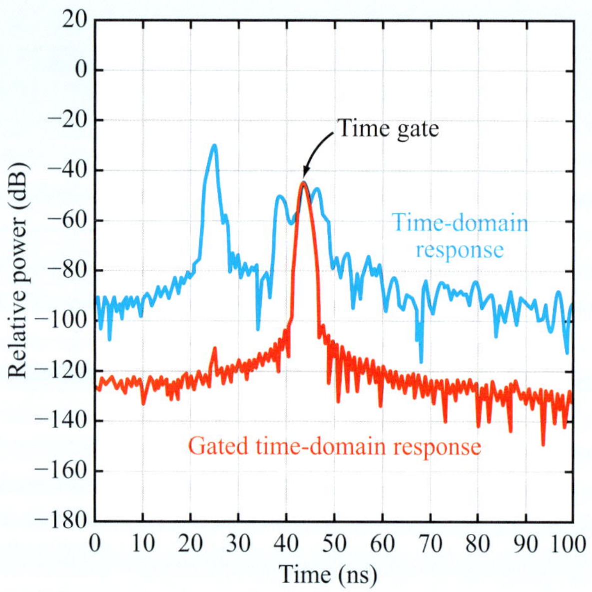

(a) Time- and gated time-domain response

Figure 13-48: Plots illustrating the use of gating with the time-domain option to measure the frequency response as a function of depth within a random medium: (a) frequency-domain response, (b) time-domain response.

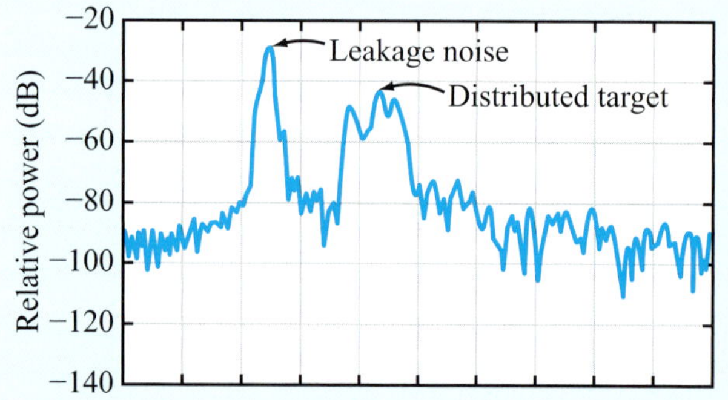

(a) Time-domain response with target

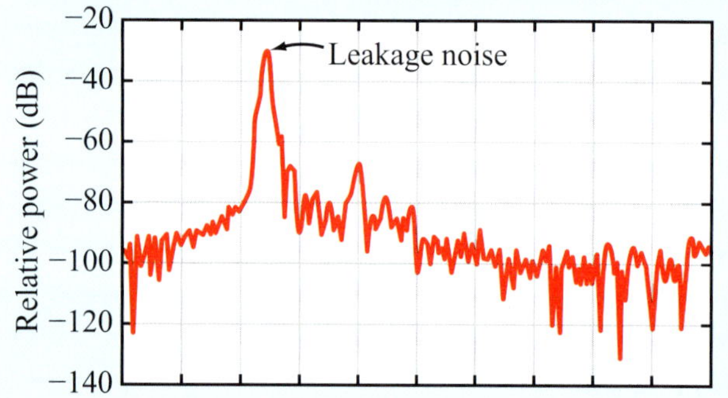

(b) Time-domain response without target

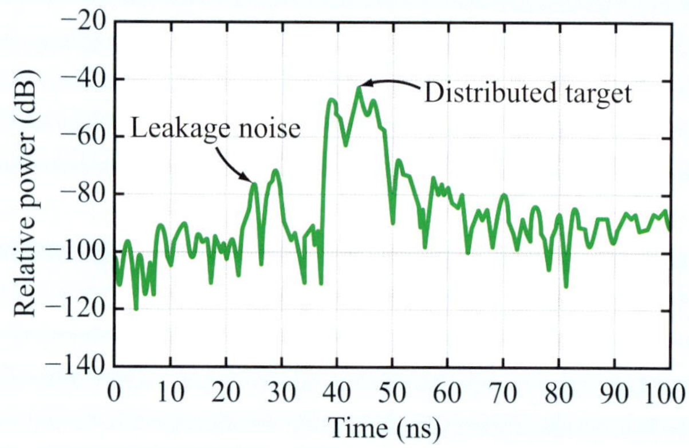

(c) Time-domain response of target alone

Figure 13-49: Plots illustrating the use of complex mathematical operations with trace storage to remove leakage signals: (a) time-domain response with target, (b) time-domain response without target, and (c) time-domain response after subtraction of (b) from (a).

with large radar cross sections are easy to measure, but there is a lower bound on the measurable RCS. This lower bound (***minimum detectable target***) for a given scatterometer at a fixed range from the target is limited by three major factors. The first factor is the thermal noise level, which is the absolute minimum detectable level and it is related to the system noise figure and the system bandwidth. The second limiting factor is the dynamic range of the receiver. This factor usually is dominant when a single antenna is used or if there is insufficient isolation between the transmitting and receiving antennas. In such cases, part of the transmitted signal returns to the receiver's sampler and sets the minimum detectable signal level. This minimum signal level can be determined by dividing the returned signal level by the dynamic range of the sampler. Finally, the third limiting factor is the effect of multiple internal reflections that appear at the target range. In a radar system, there usually exists impedance mismatches between different RF components and between the antenna and free space. A fraction of the transmitted signal corresponding to each mismatched interface returns to the receiver as noise. Multiple reflections between the mismatch interfaces also arrive at the receiver, but with different time delays corresponding to the total path length. The short-range reflections can be distinguished from the target signal in the time domain, but multiple reflections that appear at the target range cannot be separated from the target signal. Therefore, targets with a signal level that is less than the level of the multiple reflections are difficult to detect reliably.

13-15 Calibration of Polarimetric Radars

The advent of polarimetric scatterometers and imaging radars in the early 1980s led to the development of several techniques for calibrating them. The technique proposed by Barnes (1986) characterizes the errors introduced by the transmitter and receiver in terms of ***distortion matrices*** that alter the measured scattering matrix of the target. His calibration technique requires the use of three calibration targets with known radar cross sections, some of which must have zero elements in their scattering matrices. A technique similar to that of Barnes was used by Freeman et al. (1988) where the known target scattering matrices were realized by polarimetric active radar calibrators (PARCs). A technique introduced by Riegger et al. (1987) characterizes the system errors in terms of coupling coefficients between elements of the theoretical scattering matrix and elements of the measured scattering matrix. Riegger's is in essence the same model as that used by Barnes, except that Riegger has expanded the matrix product, which results in twice the number of unknowns. These techniques have been shown to be effective in practical applications, but are limited because the approaches require measurements of targets with a particular form of scattering matrix that is often difficult to realize.

Two additional polarimetric calibration techniques are of interest because they require only a single nondepolarizing target (such as a sphere or trihedral) to correct for co-polarized channel imbalance and absolute magnitude errors. The cross-polarization coupling (or crosstalk) errors are corrected by using targets with unknown scattering matrices. The first technique, proposed by Sarabandi et al. (1990) achieves calibration of the crosstalk errors by measuring any arbitrary depolarizing target. Knowledge of the scattering matrix of the arbitrary target is not required. A similar technique by van Zyl (1990) uses measurements of distributed natural targets to determine the crosstalk errors. The advantage of these techniques is their insensitivity to target positioning, which makes them particularly useful in field calibration.

In this section, we begin by introducing the concept of distortion matrices to model the errors introduced by polarimetric scatterometer systems. With this background, we then show how the technique by Sarabandi et al. (1990) can be used to calibrate polarimetric scatterometers if they have ***diagonal distortion matrices***. Two calibration targets, namely a metal sphere or trihedral reflector and any second target with strong cross-polarized RCS, are required to calibrate the scatterometer. The advantage of this method is that (1) it is relatively insensitive to orientation of the calibration targets with respect to the antenna coordinate system and (2) theoretical values for the scattering matrix of the second target are not required. Therefore, the calibration requires only the knowledge of the theoretical scattering matrices for a metal sphere or trihedral reflector, which are well known. The applicability of the technique, however, is limited to radars with diagonal distortion matrices.

For systems where no assumptions can be made about the distortion matrices, the technique proposed by Whitt et al. (1991) can be used. The technique requires the measurement of three known targets to determine the distortion matrices that characterize the effect of the measurement system on the transmitted and received waves. The scattering matrices for the known targets can be of any form, provided that a limited set of constraints is satisfied.

13-15.1 System Distortion Matrices

According to Eq. (5.19), the electric field $\mathbf{E}^{bs}$ backscattered by a target is related to the electric field incident upon it, $\mathbf{E}^{i}$, by

$$\mathbf{E}^{bs} = \left(\frac{e^{-jkR_r}}{R_r}\right)\mathbf{S}\mathbf{E}^{i}, \qquad (13.157)$$

where $k = 2\pi/\lambda$, R_r is the range between the radar and the target, and $\mathbf{S}$ is the ***scattering matrix*** of the target in the backscattering alignment convention,

$$\mathbf{S} = \begin{bmatrix} S_{vv} & S_{vh} \\ S_{hv} & S_{hh} \end{bmatrix}. \qquad (13.158)$$

For a polarimetric scatterometer observing a distributed area, $\mathbf{S}$ is the scattering matrix of the area illuminated by the antenna pattern, and for a polarimetric imaging radar, each pixel has a scattering matrix $\mathbf{S}$. The function of either radar is to measure $\mathbf{S}$. Because of errors introduced by the radar system, the measured scattering matrix is a distorted version of the true scattering matrix. The purpose of polarimetric calibration is to determine the distortion errors, so as to obtain an accurate estimate of $\mathbf{S}$.

v-Polarized transmit mode

The calibration problem can be modeled in terms of the generalized schematic representation shown in Fig. 13-50. Figure 13-50(a) depicts the condition when the polarization switch of the transmit antenna is set to transmit a v-polarized wave of amplitude E_0. Because the system is nonideal, the transmitted field $\mathbf{E}^{t}$ consists of a v-polarized component $T_{vv}E_0$ and an unintended

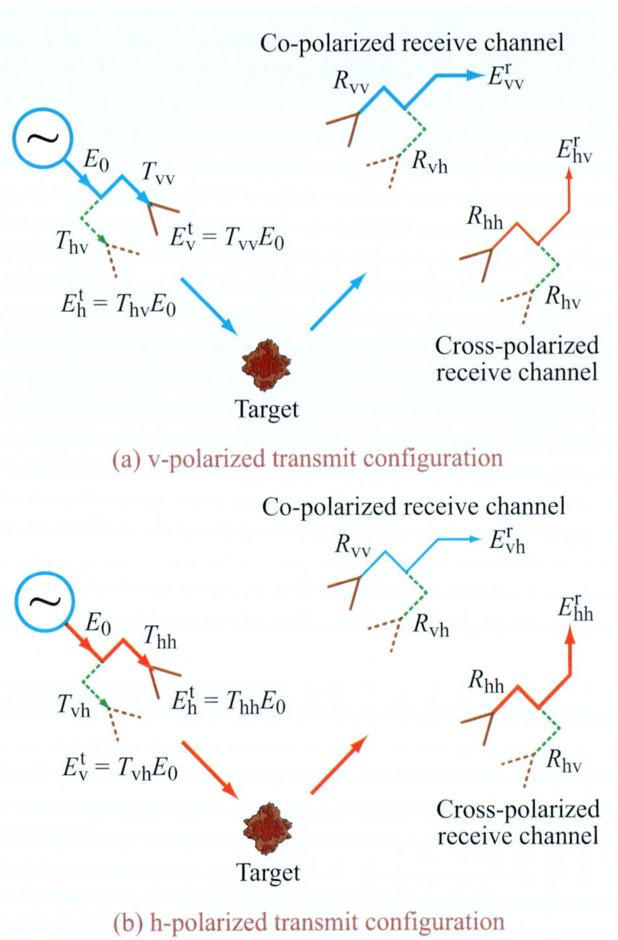

Figure 13-50: Schematic representation of transmitting and receiving antennas; (a) v-polarized transmitting configuration, (b) h-polarized transmitting configuration. The dashed lines represent polarization coupling. For an antenna with perfect polarization isolation, the coupling coefficients would be zero.

h-polarized component $T_{\rm hv}E_0$:

$$\mathbf{E}^{\rm t} = E_0 \begin{bmatrix} T_{\rm vv} \\ T_{\rm hv} \end{bmatrix}, \tag{13.159}$$

where the deviation of $T_{\rm vv}$ from 1 is due to amplitude and phase errors in the transmitted signal, and $T_{\rm hv}$ represents the coupling between the v and h ports of the antenna. For an ideal transmit antenna, $T_{\rm vv} = 1$ and $T_{\rm hv} = 0$. The incident field $\mathbf{E}^{\rm i}$ at the scatterer location is related to $\mathbf{E}^{\rm t}$ by

$$\mathbf{E}^{\rm i} = \left(\frac{G_{\rm t}}{4\pi R_{\rm r}^2}\right)^{1/2} e^{-jkR_{\rm r}}\mathbf{E}^{\rm t}, \tag{13.160}$$

where $G_{\rm t}$ is the nominal gain of the transmit antenna. Combining Eqs. (13.157), (13.159), and (13.160) leads to

$$\mathbf{E}^{\rm bs} = \frac{1}{R_{\rm r}^2}\left(\frac{G_{\rm t}}{4\pi}\right)^{1/2} e^{-j2kR_{\rm r}}\,\mathbf{S}\begin{bmatrix} T_{\rm vv} \\ T_{\rm hv} \end{bmatrix} E_0. \tag{13.161}$$

The receive antenna of a polarimetric radar has two channels, one intended to detect the v-polarized component of $\mathbf{E}^{\rm bs}$ and the other to detect its h-polarized component. The receive antenna also may have distortions, so the received electric field $\mathbf{E}^{\rm r}$ may be modeled as

$$\mathbf{E}^{\rm r} = K_1\left(\frac{G_{\rm r}\lambda^2}{4\pi}\right)^{1/2} \begin{bmatrix} R_{\rm vv} & R_{\rm vh} \\ R_{\rm hv} & R_{\rm hh} \end{bmatrix} \mathbf{E}^{\rm bs}, \tag{13.162}$$

where K_1 is a conversion constant related to the effective area of the receive antenna and transmission-line losses. The receive antenna parameters $R_{\rm vv}$ and $R_{\rm hh}$ account for amplitude or phase distortions introduced by the receive antenna, and $R_{\rm vh}$ and $R_{\rm hv}$ account for cross-polarization coupling. For an ideal, distortion-free receive antenna, $R_{\rm vv} = R_{\rm hh} = 1$ and $R_{\rm hv} = R_{\rm vh} = 0$.

Combining Eqs. (13.161) and (13.162) leads to

$$\mathbf{E}^{\rm r} = \begin{bmatrix} E^{\rm r}_{\rm vv} \\ E^{\rm r}_{\rm hv} \end{bmatrix} = e^{-j2kR_{\rm r}}\frac{K}{R_{\rm r}^2} \begin{bmatrix} R_{\rm vv} & R_{\rm vh} \\ R_{\rm hv} & R_{\rm hh} \end{bmatrix} \begin{bmatrix} S_{\rm vv} & S_{\rm vh} \\ S_{\rm hv} & S_{\rm hh} \end{bmatrix} \begin{bmatrix} T_{\rm vv} \\ T_{\rm hv} \end{bmatrix}$$

(v-pol transmit mode) (13.163)

with

$$K = K_1\left(\frac{G_{\rm t}G_{\rm r}\lambda^2}{(4\pi)^2}\right)^{1/2} E_0. \tag{13.164}$$

In Eq. (13.163), $E^{\rm r}_{\rm vv}$ and $E^{\rm r}_{\rm hv}$ are the electric fields measured by the v-polarized and h-polarized receive channels when the transmit antenna is set to transmit v polarization.

h-Polarized transmit mode

When the polarization switch of the transmit antenna is switched so that the antenna transmits a horizontally polarized wave, the applicable system configuration is as shown in Fig. 13-50(b). In this case, $\mathbf{E}^{\rm r}$ is given by

$$\mathbf{E}^{\rm r} = \begin{bmatrix} E^{\rm r}_{\rm vh} \\ E^{\rm r}_{\rm hh} \end{bmatrix} = e^{-j2kR_{\rm r}}\frac{K}{R_{\rm r}^2} \begin{bmatrix} R_{\rm vv} & R_{\rm vh} \\ R_{\rm hv} & R_{\rm hh} \end{bmatrix} \begin{bmatrix} S_{\rm vv} & S_{\rm vh} \\ S_{\rm hv} & S_{\rm hh} \end{bmatrix} \begin{bmatrix} T_{\rm vh} \\ T_{\rm hh} \end{bmatrix}.$$

(h-pol transmit mode) (13.165)

General form

Equations (13.163) and (13.165) can be combined into the compact form

$$\mathbf{E}^{\rm r} = e^{-j2kR_{\rm r}}\frac{K}{R_{\rm r}^2}\,\boldsymbol{\mathcal{R}}\,\mathbf{S}\,\boldsymbol{\mathcal{T}}\,\mathbf{p}^{\rm t}, \tag{13.166}$$

where

$$\mathbf{E}^{\rm r} = \begin{bmatrix} E^{\rm r}_{\rm vv} \\ F^{\rm r}_{\rm hv} \end{bmatrix} \quad \text{and} \quad \mathbf{p}^{\rm t} = \hat{\mathbf{v}} = \begin{bmatrix} 1 \\ 0 \end{bmatrix} \quad \textbf{(v-pol)} \tag{13.167a}$$

when the polarization switch of the transmitter is set to transmit vertical polarization, and

$$\mathbf{E}^{\rm r} = \begin{bmatrix} E^{\rm r}_{\rm vh} \\ E^{\rm r}_{\rm hh} \end{bmatrix} \quad \text{and} \quad \mathbf{p}^{\rm t} = \hat{\mathbf{h}} = \begin{bmatrix} 0 \\ 1 \end{bmatrix} \quad \textbf{(h-pol)} \tag{13.167b}$$

when it is set to transmit horizontal polarization. The ***receive*** and ***transmit distortion matrices*** $\boldsymbol{\mathcal{R}}$ and $\boldsymbol{\mathcal{T}}$ are given by

$$\boldsymbol{\mathcal{R}} = \begin{bmatrix} R_{\rm vv} & R_{\rm vh} \\ R_{\rm hv} & R_{\rm hh} \end{bmatrix}, \tag{13.168a}$$

and

$$\boldsymbol{\mathcal{T}} = \begin{bmatrix} T_{\rm vv} & T_{\rm vh} \\ T_{\rm hv} & T_{\rm hh} \end{bmatrix}. \tag{13.168b}$$

13-15.2 Distortionless Antennas

For an ideal radar with distortionless antennas, $\mathcal{R} = \mathcal{T} = \mathbf{I}$, where $\mathbf{I}$ is the identity matrix, in which case Eq. (13.166) reduces to

$$\mathbf{E}^{\mathrm{r}} = e^{-j2kR_{\mathrm{r}}} \frac{K}{R_{\mathrm{r}}^2} \mathbf{S}\,\mathbf{p}^{\mathrm{t}}. \tag{13.169}$$

The range R_{r} is proportional to one-half of the two-way-travel time delay measured by the radar, and K can be determined by calibrating the radar against a target of known RCS. By performing four measurements, one pair with $\mathbf{p}^{\mathrm{t}} = \mathrm{v}$ polarized and the second pair with $\mathbf{p}^{\mathrm{t}} = \mathrm{h}$ polarized, the four measurements of $\mathbf{E}^{\mathrm{r}}$ provide the four values of $\mathbf{S}$. This is the ideal case; the more realistic cases follow.

13-15.3 Reciprocal Distortion Matrices

With some single-antenna polarimetric radar systems, the transmit and receive distortion matrices are simply the transpose of one another. For such a ***reciprocal antenna*** system, $\mathcal{R} = \widetilde{\mathcal{T}}$, in which case Eq. (13.166) assumes the form

$$\mathbf{E}^{\mathrm{r}} = e^{-j2kR_{\mathrm{r}}} \frac{K}{R_{\mathrm{r}}^2} \widetilde{\mathcal{T}}\,\mathbf{S}\,\mathcal{T}\,\mathbf{p}^{\mathrm{t}}. \tag{13.170}$$

13-15.4 Matrix Inversion

Writing the components of $\mathbf{E}^{\mathrm{r}}$ and $\mathbf{p}$ in Eq. (13.166) explicitly for $\mathbf{p} = \hat{\mathbf{v}}$ and $\mathbf{p} = \hat{\mathbf{h}}$ gives

$$\begin{bmatrix} E^{\mathrm{r}}_{\mathrm{vv}} \\ E^{\mathrm{r}}_{\mathrm{hv}} \end{bmatrix} = e^{-j2kR_{\mathrm{r}}} \frac{K}{R_{\mathrm{r}}^2} \mathcal{R}\,\mathbf{S}\,\mathcal{T} \begin{bmatrix} 1 \\ 0 \end{bmatrix} \tag{13.171a}$$

and

$$\begin{bmatrix} E^{\mathrm{r}}_{\mathrm{vh}} \\ E^{\mathrm{r}}_{\mathrm{hh}} \end{bmatrix} = e^{-j2kR_{\mathrm{r}}} \frac{K}{R_{\mathrm{r}}^2} \mathcal{R}\,\mathbf{S}\,\mathcal{T} \begin{bmatrix} 0 \\ 1 \end{bmatrix}, \tag{13.171b}$$

which can be combined into

$$\begin{bmatrix} E^{\mathrm{r}}_{\mathrm{vv}} & E^{\mathrm{r}}_{\mathrm{vh}} \\ E^{\mathrm{r}}_{\mathrm{hv}} & E^{\mathrm{r}}_{\mathrm{hh}} \end{bmatrix} = e^{-j2kR_{\mathrm{r}}} \frac{K}{R_{\mathrm{r}}^2} \mathcal{R}\,\mathbf{S}\,\mathcal{T}. \tag{13.172}$$

Once the distortion matrices $\mathcal{R}$ and $\mathcal{T}$ have been determined through calibration (as discussed in the next section), the true scattering matrix $\mathbf{S}$ can be obtained from

$$\mathbf{S} = \left(\frac{R_{\mathrm{r}}^2}{K}\, e^{j2kR_{\mathrm{r}}} \right) \mathcal{R}^{-1} \begin{bmatrix} E^{\mathrm{r}}_{\mathrm{vv}} & E^{\mathrm{r}}_{\mathrm{vh}} \\ E^{\mathrm{r}}_{\mathrm{hv}} & E^{\mathrm{r}}_{\mathrm{hh}} \end{bmatrix} \mathcal{T}^{-1}. \tag{13.173}$$

For a reciprocal antenna with $\mathcal{R} = \widetilde{\mathcal{T}}$, characterization of only one matrix is required, simplifying the calibration task considerably. This representation assumes that the distortion matrices are invertible, which is generally the case for real radar systems. The phase factor $\phi = 2kR_{\mathrm{r}}$ is difficult to measure because the phase center of the target and the target position must be known exactly. Usually, only the phases of S_{hv}, S_{vh}, and S_{hh} relative to that of S_{vv} are needed for polarization synthesis or other applications.

13-15.5 Antennas with Diagonal Distortion Matrices

If the radar to be calibrated has excellent cross-polarization isolation, the distortion matrices are approximately diagonal. A ***diagonal distortion matrix*** corresponds to perfect isolation between the v and h ports of the radar antenna; thus,

$$T_{\mathrm{vh}} = T_{\mathrm{hv}} = R_{\mathrm{vh}} = R_{\mathrm{hv}} = 0. \tag{13.174}$$

Even though a polarimetric radar may use only a single antenna to provide both transmit and receive functions, for purposes of illustration we use the block diagram shown in Fig. 13-51 to represent a radar with perfect cross-polarization isolation.

Imposing the condition given by Eq. (13.174) onto Eq. (13.172) leads to

$$\begin{bmatrix} E^{\mathrm{r}}_{\mathrm{vv}} & E^{\mathrm{r}}_{\mathrm{vh}} \\ E^{\mathrm{r}}_{\mathrm{hv}} & E^{\mathrm{r}}_{\mathrm{hh}} \end{bmatrix} = e^{-j2kR_{\mathrm{r}}} \frac{K}{R_{\mathrm{r}}^2} \begin{bmatrix} R_{\mathrm{vv}}T_{\mathrm{vv}}S_{\mathrm{vv}} & R_{\mathrm{vv}}T_{\mathrm{hh}}S_{\mathrm{vh}} \\ R_{\mathrm{hh}}T_{\mathrm{vv}}S_{\mathrm{hv}} & R_{\mathrm{hh}}T_{\mathrm{hh}}S_{\mathrm{hh}} \end{bmatrix}. \tag{13.175}$$

This expression can be abbreviated as

$$E^{\mathrm{r}}_{pq} = e^{-j2kR_{\mathrm{r}}} \frac{K}{R_{\mathrm{r}}^2} R_{pp}T_{qq}S_{pq}, \qquad p,q = \text{h or v}. \tag{13.176}$$

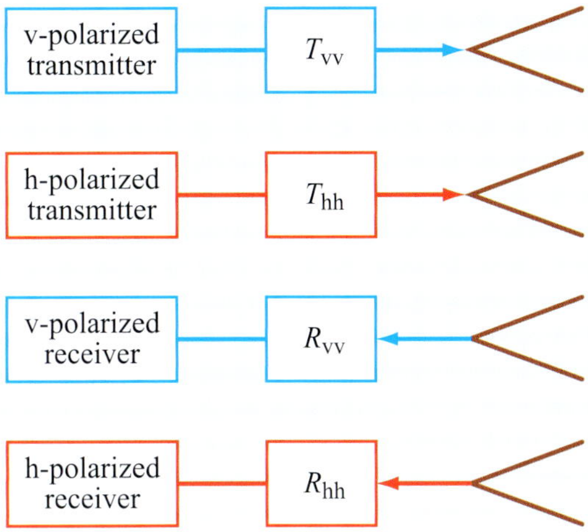

Figure 13-51: Simplified block diagram of a dual-polarized radar system with channel imbalances but no cross-channel leakage.

The standard calibration procedure involves the use of one reference target of known scattering matrix. Upon measuring E^{r}_{pq} with S_{pq} known, the quantity $(KR_{pp}T_{qq})$ can be determined in both amplitude and phase for all four polarization combinations, with the phase being relative to a selected reference.

The calibration procedure is simple and straightforward, but its accuracy depends on the accuracy with which the scattering matrix $\mathbf{S}$ of the calibration target is known. Of all possible calibration targets, the metal sphere is the easiest to align, and its scattering matrix can be computed exactly (Section 8-5.4). For downward-looking airborne or spaceborne radars, however, a metal sphere is not a practical calibration target because its RCS is not sufficiently large to dominate over the backscatter from the terrain background (see Section 13-10.3), but it is perfectly suitable for calibrating ground-based scatterometers if placed on a nonreflecting pole as depicted in Fig. 13-30(a).

A metal sphere can be used to calibrate only the vv and hh channels of a ground-based scatterometer, because the sphere is a nondepolarizing target: its $S_{\mathrm{hv}} = S_{\mathrm{vh}} = 0$. Hence, as we see later, we use the sphere to calibrate the co-pol channels and then we perform a second calibration against any depolarizing target (without the need to know its scattering matrix) to calibrate the cross-polarized channels. The calibration procedure is described in detail in the publication by Sarabandi et al. (1990). An abbreviated version follows.

The co-polarized scattering amplitudes of a metal sphere are $S_{\mathrm{vv}} = S_{\mathrm{hh}} = S_0$, with S_0 being a known quantity. If the radar is used to measure the received electric field $\mathbf{E}^{\mathrm{r}}$ for vv and hh polarizations with the sphere located at a range R_0 from the radar antenna, Eq. (13.176) yields

$$E^0_{\mathrm{vv}} = e^{-j2kR_0}\,\frac{K}{R_0^2}\,R_{\mathrm{vv}}T_{\mathrm{vv}}S_0, \tag{13.177a}$$

$$E^0_{\mathrm{hh}} = e^{-j2kR_0}\,\frac{K}{R_0^2}\,R_{\mathrm{hh}}T_{\mathrm{hh}}S_0, \tag{13.177b}$$

where the subscript and superscript "0" denote quantities associated with the metal sphere. Similarly, we use subscript and superscript "u" to denote quantities associated with measurements performed on an unknown target or surface of scattering matrix $\mathbf{S}^{\mathrm{u}}$:

$$E^{\mathrm{u}}_{\mathrm{vv}} = e^{-j2kR_{\mathrm{u}}}\,\frac{K}{R_{\mathrm{u}}^2}\,R_{\mathrm{vv}}T_{\mathrm{vv}}S^{\mathrm{u}}_{\mathrm{vv}}, \tag{13.178a}$$

$$E^0_{\mathrm{hh}} = e^{-j2kR_{\mathrm{u}}}\,\frac{K}{R_{\mathrm{u}}^2}\,R_{\mathrm{hh}}T_{\mathrm{hh}}S^{\mathrm{u}}_{\mathrm{hh}}. \tag{13.178b}$$

Combining the two pairs of equations yields

$$S^{\mathrm{u}}_{\mathrm{vv}} = \left(\frac{E^{\mathrm{u}}_{\mathrm{vv}}}{E^0_{\mathrm{vv}}}\right)\left(\frac{R_{\mathrm{u}}}{R_0}\right)^2 e^{-j2k(R_0-R_{\mathrm{u}})}\,S_0, \tag{13.179a}$$

$$S^{\mathrm{u}}_{\mathrm{hh}} = \left(\frac{E^{\mathrm{u}}_{\mathrm{hh}}}{E^0_{\mathrm{hh}}}\right)\left(\frac{R_{\mathrm{u}}}{R_0}\right)^2 e^{-j2k(R_0-R_{\mathrm{u}})}\,S_0. \tag{13.179b}$$

All quantities on the right-hand side of Eq. (13.179) are either known (such as k and S_0) or measured by the radar. Hence, Eq. (13.179) provides calibrated values of the co-polarized components of the scattering matrix of the ground cell illuminated by a scatterometer or the pixel area imaged by a SAR (when using a strong calibration target such as an ARC instead of the sphere).

Calibration of the cross-polarized channels calls for measuring the hv and vh channels of the received electric field with the radar pointed at a depolarizing calibration target with significant cross-polarized scattering amplitudes. A tilted cylinder or a PARC are perfect candidates. If we denote the measurements with the symbol "c," we have

$$E^{\mathrm{c}}_{\mathrm{hv}} = e^{-j2kR_{\mathrm{c}}} \frac{K}{R^2_{\mathrm{c}}} R_{\mathrm{hh}} T_{\mathrm{vv}} S^{\mathrm{c}}_{\mathrm{hv}}, \qquad (13.180\mathrm{a})$$

$$E^{\mathrm{c}}_{\mathrm{vh}} = e^{-j2kR_{\mathrm{c}}} \frac{K}{R^2_{\mathrm{c}}} R_{\mathrm{vv}} T_{\mathrm{hh}} S^{\mathrm{c}}_{\mathrm{vh}}. \qquad (13.180\mathrm{b})$$

The reciprocity theorem (Section 5-3.2) states that in the backscattering direction,

$$S^{\mathrm{c}}_{\mathrm{hv}} = S^{\mathrm{c}}_{\mathrm{vh}}$$

in the BSA convention. Next, we define the ratio of the two measurements as K_1:

$$K_1 = \frac{E^{\mathrm{c}}_{\mathrm{hv}}}{E^{\mathrm{c}}_{\mathrm{vh}}} = \frac{R_{\mathrm{hh}} T_{\mathrm{vv}}}{R_{\mathrm{vv}} T_{\mathrm{hh}}}. \qquad (13.181)$$

Also, from the co-polarized measurements given by Eq. (13.177) for the sphere or a PARC, we define the product K_2 as

$$K_2 = E^0_{\mathrm{vv}} E^0_{\mathrm{hh}} = \frac{K^2}{R^4_0}\, e^{-j4kR_0} R_{\mathrm{vv}} T_{\mathrm{vv}} R_{\mathrm{hh}} T_{\mathrm{hh}} S^2_0. \qquad (13.182)$$

Use of Eq. (13.180), in combination with Eqs. (13.181) and (13.182), leads to the solution

$$S^{\mathrm{u}}_{\mathrm{hv}} = \left(\frac{E^{\mathrm{u}}_{\mathrm{hv}}}{\sqrt{K_1 K_2}}\right) \left(\frac{R_{\mathrm{u}}}{R_0}\right)^2 e^{-j2k(R_0 - R_{\mathrm{u}})} S_0, \qquad (13.183\mathrm{a})$$

$$S^{\mathrm{u}}_{\mathrm{vh}} = \sqrt{\frac{K_1}{K_2}}\, E^{\mathrm{u}}_{\mathrm{vh}} \left(\frac{R_{\mathrm{u}}}{R_0}\right)^2 e^{-j2k(R_0 - R_{\mathrm{u}})} S_0. \qquad (13.183\mathrm{b})$$

According to the results reported by Sarabandi et al. (1990), this calibration method can be used to calibrate a ground-based scatterometer to an accuracy of ± 0.3 dB in magnitude and $\pm 5^\circ$ in phase.

13-15.6 Nonreciprocal Systems with Full Distortion Matrices

In the general case, the transmit and receive antennas may not be reciprocal of one another, and the isolation between the v and h ports may not be sufficient to justify setting the cross terms of the distortion matrices equal to zero [i.e., Eq. (13.174)]. From Eq. (13.172), the four channels of the received signal are represented by

$$\begin{bmatrix} E^{\mathrm{r}}_{\mathrm{vv}} & E^{\mathrm{r}}_{\mathrm{vh}} \\ E^{\mathrm{r}}_{\mathrm{hv}} & E^{\mathrm{r}}_{\mathrm{hh}} \end{bmatrix} = e^{-j2kR_{\mathrm{r}}} \frac{K}{R^2_{\mathrm{r}}} \begin{bmatrix} R_{\mathrm{vv}} & R_{\mathrm{vh}} \\ R_{\mathrm{hv}} & R_{\mathrm{hh}} \end{bmatrix} \begin{bmatrix} S_{\mathrm{vv}} & S_{\mathrm{vh}} \\ S_{\mathrm{hv}} & S_{\mathrm{hh}} \end{bmatrix} \begin{bmatrix} T_{\mathrm{vv}} & T_{\mathrm{vh}} \\ T_{\mathrm{hv}} & T_{\mathrm{hh}} \end{bmatrix}. \qquad (13.184)$$

To calibrate the radar, we need to determine K, the four elements of $\mathcal{R}$ and the four elements of $\mathcal{T}$. This can be accomplished by using three different calibration targets with known scattering matrices (Whitt et al., 1991). Alternatively, the necessary number of calibration targets can be reduced to fewer than three by taking advantage of reciprocity properties of both the calibration target and the terrain, and by invoking appropriate approximations (van Zyl and Kim, Chapter 4, 2011). Useful calibration-related references include Quegan (1994), Klein and Freeman (1991), van Zyl (1990), Cordey (1993), Zebker and Lou (1990), Freeman et al. (1995), and Sarabandi et al. (1995).

13-16 GNSS-R Bistatic Radar

The use of ***Global Navigation Satellite System*** (GNSS) signals of opportunity, such as those transmitted by the U.S. Global Positioning System (GPS), European ***Galileo***, and Russian ***Globalnaya Navigatsionnaya Sputnikovaya Sistema*** (GLONASS) satellites, for remote sensing purposes can be divided into two general classes. The ***radio occultation technique*** (GNSS-RO) discussed in Section 9-13 provides direct measurements of atmospheric refractivity profiles and, by inference, of the profiles of water-vapor density and air temperature. The ***reflectometry technique*** (GNSS-R) provides direct measurements of the bistatic radar cross section of the Earth surface and of the range to the surface, from which properties of the scattering target can be discerned. See Section 5-4 for a general discussion of bistatic radar. GNSS-R measurements are made using a forward scattering geometry, at which the radar cross section is generally highest and the scattered signal strength largest. Hall and Cordey (1988) first suggested the use of GNSS signals as a means of making bistatic scatterometer measurements of the ocean surface. A modified GPS navigation receiver could be used as the receive side of a bistatic radar link, and the magnitude of the power received from GPS signals scattered by the ocean could be related to the radar cross section of its surface. Martin-Neira (1993) later extended the possible applications to include ocean altimetry. GNSS signals are modulated by long pseudorandom noise (PRN) codes that enable highly accurate determination of the time-of-flight between the transmitter and receiver, in order to support precise position determination by triangulation. This feature can also be used for altimetry by measuring the difference in time-of-flight between a direct signal path from the transmitter to the receiver and a second signal path that is reflected by the ocean surface. If the locations of the transmitter and receiver are sufficiently well known, the height above sea level of the surface can be inferred from the difference in time-of-flight.

Early experimental demonstrations of the GNSS-R technique were made from stationary platforms and aircraft. Garrison et al. (1998) reported the first serious examination of how bistatic scattering from the ocean surface is related to the local sea state. The ocean surface roughness and, by inference, the near surface wind speed, are found to be highly correlated with the strength of the scattered signal at the specular point of reflection and with the extent of the diffuse scattering away from the specular point. As the winds and the roughness increase, the specular point scattering decreases in magnitude and the ***diffuse scattering region*** (also referred to as the ***glistening zone*** in optics) increases in extent. Both characteristics can be measured and used to estimate wind speed. Platform and airborne demonstrations of other applications of GNSS-R soon followed, including sea-ice mapping (Komjathy et al., 2000) and near-surface soil moisture content (Masters et al., 2004). In each case, the general approach taken is similar to that of conventional backscatter scatterometry, in which the received power is converted to radar cross section by inversion of the bistatic radar equation, which is then related to the dielectric and roughness properties of the target.

There are several important differences between a conventional backscatter scatterometer and a GNSS-R bistatic radar:

(1) The dependence of the radar cross section on target properties can be quite different between backscatter and forward scatter. For example, increasing ocean surface roughness will tend to increase the backscatter and decrease the forward scatter. For this reason, backscatter scatterometers are more amenable to the measurement of higher wind speeds, and GNSS-R bistatic radars have improved performance at lower wind speeds.

(2) The radar hardware is generally simpler and smaller and requires lower power for GNSS-R because there is no transmitter involved, and because GNSS receiver technology has become highly refined to support its mass commercial market.

(3) With GNSS-R systems, precise determination of the location of the scattering target does not require accurate antenna pointing control or knowledge. This is because if the locations of the transmitter and receiver are well known (as is typically the case with standard GNSS navigation capabilities), then the location of the specular point can be determined analytically from geometrical considerations. This can significantly simplify the requirements levied on the airborne or spaceborne platform on which the receiver is deployed.

The first spaceborne detection of an ocean-reflected GPS signal was made by Lowe et al. (2002), using calibration data from the SIR-C radar experiment that fortuitously happened to fall within the transmit passband of the GPS L2 signal at 1228 MHz. A more intentional effort to demonstrate the capabilities of spaceborne GNSS-R remote sensing was undertaken by the UK Space Agency's Disaster Monitoring Constellation (UK-DMC) program. UK-DMC-1 was launched in 2003 into a 680 km polar orbit with a

GNSS-R payload specifically designed to measure GPS signals scattered from the Earth surface. In addition to its GPS receiver, the payload included a nadir-pointing antenna with a peak gain of 11.8 dB and a 3 dB beamwidth of $20^\circ \times 70^\circ$. The mission also supported technology demonstrations for many other payloads, so the on-board GNSS-R processing capability was limited to measurements of up to three reflected signals simultaneously for 20 seconds at a time. The 20-second data records were telemetered to the ground, where they were processed into images of the scattered power received from the surface in the vicinity of the specular point. These images are referred to as ***Delay Doppler Maps***.

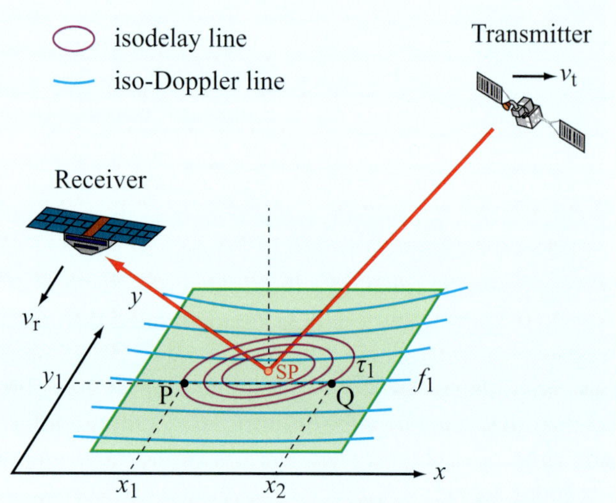

Figure 13-52: Coordinate transformation for the Delay Doppler Map measured by a GNSS-R scatterometer.The specular point (SP) represents the location on the surface with the shortest difference between direct and reflected signal propagation times from the transmitter to the receiver. Longer differences trace out elliptical lines on the surface. The hyperbolae represent lines on the surface of constant Doppler shift. A specific (delay, Doppler) coordinate corresponds to two points (e.g., P and Q) on the surface [Clarizia, 2012].

13-16.1 The Delay Doppler Map

The scattered power received by a GNSS-R sensor is measured by performing a lag correlation of the scattered signal with a time delayed version of the direct signal (or of a replica of the direct signal that is time synchronized to it). The form of the bistatic radar equation given by Eq. (5.29a) can be adapted for use in GNSS-R remote sensing by integrating over the distributed surface target and by explicitly including the effects of the lag correlator (Zavorotny and Voronovich, 2000). The received power can be written as

$$P^{\rm r}(\tau,f_{\rm D}) = \frac{P^{\rm t}\lambda^2}{(4\pi)^3} \iint_A \frac{G_{\rm t}(x,y)\, G_{\rm r}(x,y)}{R_{\rm t}^2(x,y)\, R_{\rm r}^2(x,y)}\, \Lambda^2(\tau,x,y) \cdot |S(f_{\rm D},x,y)|^2\, \sigma^0(x,y)\; dx\, dy, \qquad (13.185)$$

where τ is the difference in time-of-flight between the direct signal and the signal scattered by the surface, $f_{\rm D}$ is the Doppler shift of the scattered signal relative to the direct signal, $P^{\rm t}$ is the transmit power, λ is the signal wavelength, $R_{\rm t}$ and $R_{\rm r}$ are, respectively, the distances from the scattering target to the transmitter and receiver, $G_{\rm t}$ and $G_{\rm r}$ are the transmit and receive antenna gain patterns, Λ is the autocorrelation function of the PRN code for the GNSS signal, S is the response of the Doppler filter, and σ^0 is the bistatic radar cross section per unit area. The integration in Eq. (13.185) is performed over the entire region of diffuse scattering. In general, the received power is highest at the $(\tau, f_{\rm D})$ coordinates corresponding to the location of the specular point.

The minimum value of τ that produces a surface reflected signal is equal to the difference in time-of-flight between the direct signal and the specular reflection from the surface. If τ is increased above that, scattered power is measured from concentric regions of the glistening zone around the specular point. This is illustrated in Fig. 13-52, which shows a typical spaceborne measurement geometry. The ***specular point*** (SP) and concentric elliptical lines of constant isodelay are shown, together with hyperbolic lines of constant Doppler shift of the transmit carrier frequency. In addition to variations in the time lag of the correlator, the received signal can also be processed through a bank of Doppler filters to selectively measure scattering along the iso-Doppler lines. Two-dimensional variations of the delay and Doppler values produce a ***Delay Doppler Map*** (DDM) of the scattered power received from the surface. In practice, DDMs are typically sampled over a range of (τ, f) values that includes all significant regions of the glistening zone.

Examples of ocean DDMs measured by UK-DMC-1 are shown in Fig. 13-53 for three different surface conditions. The response of the DDMs to the wind-driven roughening, both in terms of the power scattered from the specular point and the extent of the glistening zone that surrounds it, is consistent with Eq. (13.135). The specular point with the highest scattered power is located at the top of the "horseshoe" shape in the figures. No scattered signal is present for measurements made at lower delay values (above the specular point in the figures). These measurements are typically used to monitor the noise floor of the receiver. Measurements at higher delay values (below the specular point in the figures) map out the glistening zone of the rough surface. Scatter from the specular point decreases, and the extent of the glistening zone increases, with increasing wind speed and roughness.

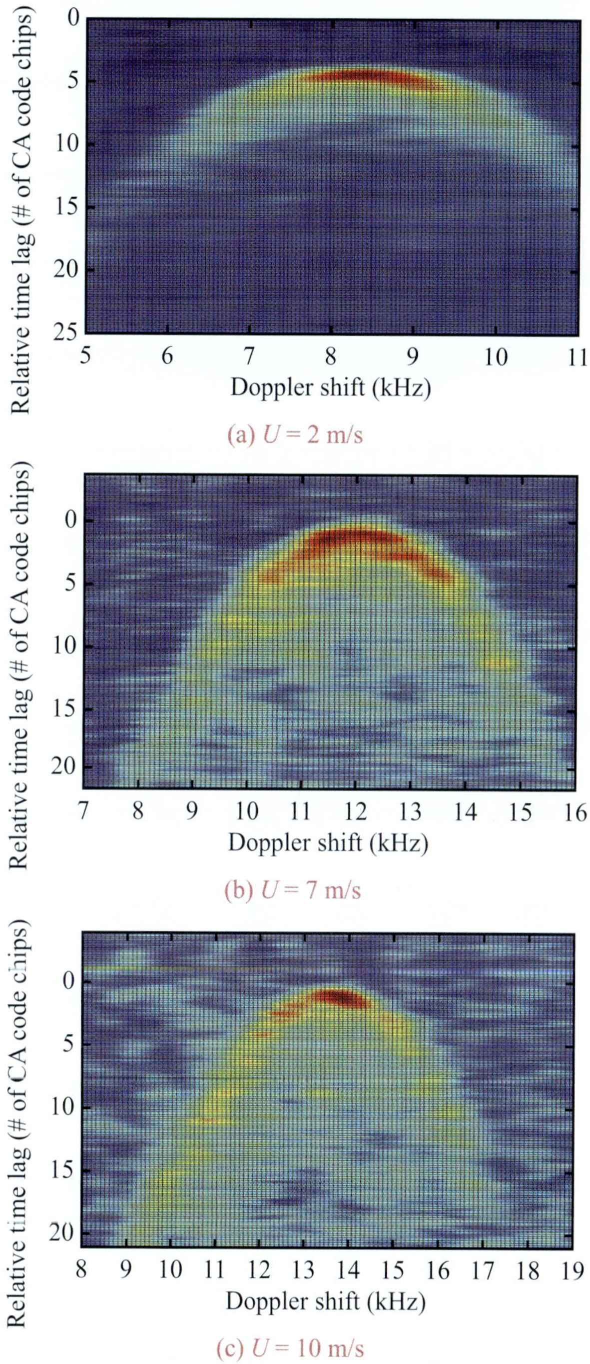

Figure 13-53: Delay Doppler Maps of the ocean surface measured by UK-DMC-1 at surface wind speeds of (a) 2 m/s, (b) 7 m/s, and (c) 10 m/s (referenced to 10 m height above the surface) [Gleason et al., 2005].

13-16.2 The Cyclone Global Navigation Satellite System (CYGNSS)

CYGNSS is a NASA spaceborne mission scheduled for launch in 2016. It consists of a constellation of eight microsatellites in low inclination orbit at 500 km altitude (Ruf et al., 2013). Each observatory carries an upgraded version of the UK-DMC-1 GNSS-R receiver. The upgrades permit continuous operation (i.e., 100% duty cycle) with the ability to sample up to 4 reflections simultaneously. The goal of CYGNSS is to better understand the coupling between surface winds and the moist atmosphere within tropical cyclones in order to improve the forecasting of storm genesis and rapid intensification. The ability to measure hurricane force winds with a GNSS-R bistatic radar has been demonstrated by an extensive history of airborne campaigns (Katzberg et al., 2001; Katzberg et al., 2006; Katzberg and Dunion, 2009). The use of a GNSS-R approach enables two key features of the

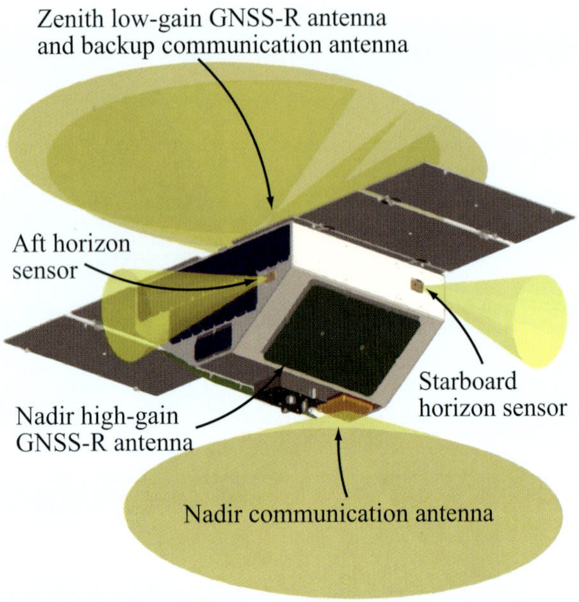

Figure 13-54: The CYGNSS observatory with fully deployed solar arrays measures $159 \times 51 \times 26$ cm, weighs 22 kg, and requires 52 W of power. The keep-out zones for communication antennas and horizon sensors are also shown. The large green panels are its nadir GNSS-R science antennas.

CYGNSS mission. Operation at the GPS L1 carrier of 1575 MHz allows the radar signal to penetrate through the extreme levels of precipitation typically found in the eyewall of a hurricane, which enables the measurement of surface wind speed in its inner core. GPS receivers, with their low power and easy antenna pointing accommodation requirements, allow for the use of small, simple satellites, which makes it affordable to deploy a constellation of eight. An illustration of one of the spacecraft is shown in Fig. 13-54. Use of a constellation produces very frequent (approximately every 3 hours) wind measurements throughout the tropics, which allows for time-resolved imaging of the rapid intensification phase of storm development. The launch of CYGNSS will mark the first time that a GNSS-R sensor is the primary payload on a spaceborne science mission.

PROBLEMS

13.1 A radar system has the following characteristics:

Peak transmitted power $P_t = 10^6$ W
Pulse duration $T_p = 1\,\mu$s
Bandwidth $B = 1/T_p$
Aperture (effective) area $= 3$ m^2
Wavelength $\lambda = 10$ cm
Noise figure $F = 10$ dB

Determine the received power and the signal-to-noise ratio for a target with a radar cross section of $\sigma = 5$ m^2 at distances of 10, 25, 50, and 100 km.

13.2 A SLAR has the following characteristics:

Peak transmitted power $P_t = 10$ kW
Antenna gain $G = 30$ dB
Wavelength $\lambda = 3$ cm
Horizontal beamwidth $= 0.01$ rad
Pulse duration $T_p = 100$ ns
Receiver bandwidth $B = 1/T_p$
Noise figure $F = 10$ dB

Determine the received power and single-pulse signal-to-noise ratio for $\sigma^0 = -20$ dB, at a height of 5 km and a slant range of 10 km. Assume the antenna boresight is pointed at the target area.

13.3 Three targets are spaced a distance of 5 m apart in a straight line parallel to the axis of motion. They have the same σ and are observed from very far away with a radar operating at a wavelength of $\lambda = 3$ cm.

(a) Sketch the received signal voltage as the radar passes the targets.

(b) Calculate the voltage versus time using the Doppler point of view for a velocity of $u = 50$ m/s.

13.4 Repeat Problem 13.3 with the line of targets perpendicular to the aircraft motion.

13.5 An experimental scatterometer flown on Skylab had the following characteristics:

Signal bandwidth $= 20$ kHz
Noise bandwidth $= 75$ kHz
$T_{S+N} = 1$ s
$T_N = 0.1$ s

Calculate the measurement K_p for the case when the signal-to-noise ratio is -10 dB, 0 dB, and 10 dB. Assume the signal bandwidth is centered in the noise bandwidth.

13.6 A LFMCW radar altimeter uses a chirp bandwidth of 10 MHz with a 100 ms period. Compute the instantaneous frequency of the return difference frequency. What range resolution is achieved?

13.7 Sketch the isorange lines for a radar with 10 m range resolution operation operating at a height of 3 km. The sketch should be to scale and include a minimium of 10 curves that are equally spaced in slant range.

13.8 Sketch the isodops for a radar traveling 200 m/s in level flight at a height of 3 km. Assume a radar operating frequency of 3 GHz. The sketch should be to scale and include a minimium of 10 curves that are equally spaced in Doppler frequency.

13.9 Determine the range-only ambiguity function for a Gaussian pulse envelope given by $p(t) = e^{-t^2/t_0^2}$, where t_0 is a constant. Based on the ambiguity function what is this pulse's equivalent duration and equivalent bandwidth?

13.10 Sketch the range-only ambiguity function for the binary 1101101 where the total pulse duration is T with equally spaced chips. Use continuous integration. What is the equivalent duration and bandwidth of this modulation function? In computing the ambiguity function a 0 should be treated as a -1.

CHAPTER 14

Real- and Synthetic-Aperture Side-Looking Airborne Radar

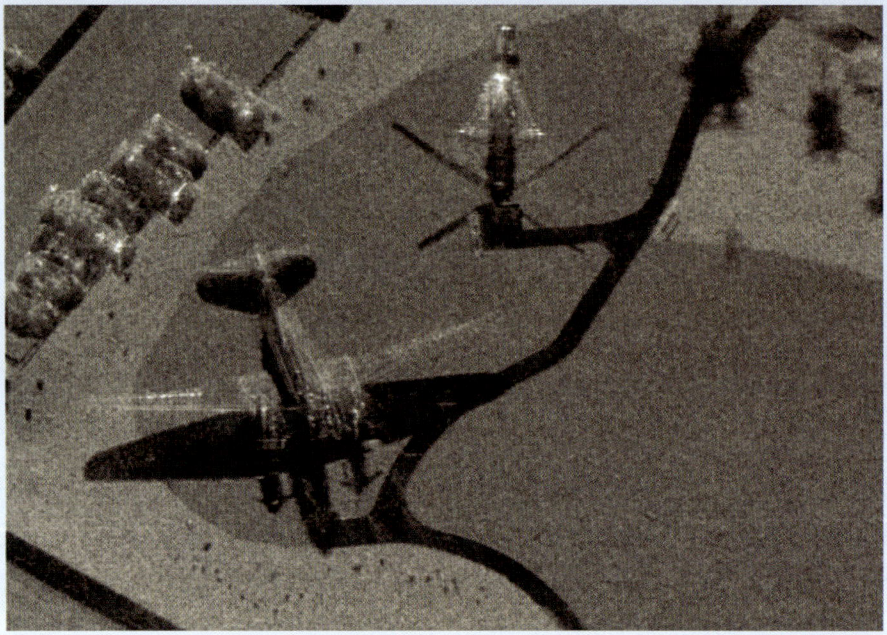

High-resolution SAR image of an airport runway with a plane and helicopter (Courtesy Sandia National Laboratories)

CONTENTS

Overview

Remote sensing radars can be either imaging or nonimaging. Nonimaging radars were discussed in Chapter 13 and now we consider the family of imaging radars generically known as ***side-looking airborne radars*** (SLARs). Two different types of SLAR are considered: ***real-aperture SLAR*** and ***synthetic-aperture radar*** (SAR). These two imaging radars differ primarily in how the signal's Doppler spectrum is processed. While technically SAR is a type of SLAR, by convention the term SLAR is generally only used to refer to real-aperture radars. We will follow this convention in this book. The goal of both SLAR and SAR systems is to generate maps or images of the radar backscatter from the illuminated area.

The basic theories of SLAR and SAR are discussed in this chapter, along with considerations relating to system design and construction, but the details of transmitters, receivers, and other hardware components are not treated in this book. Processing algorithms for SAR are only briefly considered. More detailed treatment of SAR processing algorithms can be found in separate texts (e.g., Cumming and Wong, 2005; Curlander and McDonough, 1991).

14-1 Introduction

Radar systems are much more complex than radiometer systems due to the addition of a transmitter. However, this complexity provides the designer with additional flexibility via modulation of the transmitted signal and demodulation and possible Doppler processing of the received signal. Radar receivers are similar to the radiometer receivers discussed in Chapter 7 and many radiometer techniques can be used for very sensitive radars that operate with signal-to-noise ratios close to or less than unity. However, radar receivers are primarily designed for receiving the reflected signal and often do not have to be as sensitive as radiometer receivers. In many cases the transmit power level may be large enough that rather noisy receivers can be used without degradation in system performance.

In a remote sensing context, the term ***target*** describes the surface or volume that scatters the radar signal illuminating it back to the receiver. Originally in radar, the word *target* referred to a specific object such as an aircraft or a ship, and the target background was referred to as ***clutter***. However, in modern usage *target* can refer to anything that reflects the radar signal, including areas on the surface. In most remote-sensing radar imaging applications, the target is the ground. For ***radar imaging***, we want to resolve an extended area target into smaller areas, sometimes termed ***pixels*** or ***resolution elements***. For SLAR, this is done using a combination of the motion of the radar, a narrow antenna beamwidth, and the radar's range resolution. SAR achieves finer resolution in the along-track direction by also using Doppler processing.

The received radar signal amplitude is proportional to the target area's scattering coefficient, which depends on the incidence angle, the frequency of operation, and the polarization states of the transmit and receive antennas. We note that the variation of the scattering coefficient with angle of incidence is different for different classes of targets. This variation can be used in target analysis and identification. With all imaging radars, a single pass that produces a single image provides scattering measurements at a single angle of incidence at each location. Multiple passes along different flight lines or at different altitudes can give information about the variation in scattering as a function of angle of incidence. Some radars are specifically designed to gather this information by employing one or more beams pointed ahead of or behind the platform track so that the scattering coefficient for a particular target may be measured at a variety of angles as the radar advances past the target.

Some targets can be identified or mapped by their characteristic shape, texture, and/or context in an image. While variations due to amplitude differences among different target classes is important, the use of shape, texture, and context does not necessarily require accurate amplitude calibration. Thus, some applications of imaging radar do not require accurate calibration, which can simplify their design. Photo analysts, for example, are accustomed to working with

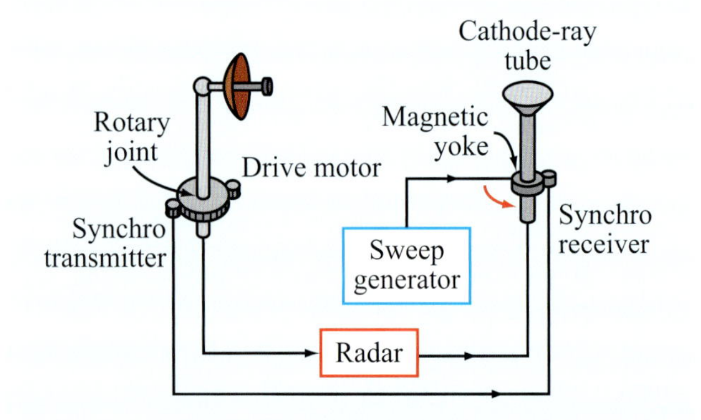

Figure 14-1: PPI radar system.

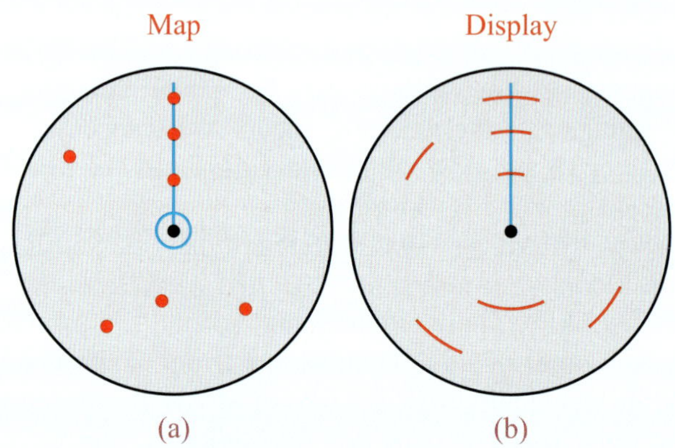

Figure 14-2: PPI display (360°) geometry: (a) surface view from above; the radar is at the center with bright targets indicated, and (b) corresponding PPI display.

aerial photographs in which calibration of amplitude is almost completely lacking. They are able to extract the information of interest from contrasting image brightness using shape, texture, and context, whether it is in photographs or radar images. In such cases the production of images with good geometric fidelity and good contrast may be more important than accurate calibration. On the other hand, radar imaging applications that relate the observed σ^0 to a geophysical property of interest may require accurate calibration on either an absolute or a relative basis.

The earliest imaging radar systems were based on real-aperture rotating antennas. The ***plan-position indicator*** (PPI) was invented early in World War II as a means of displaying rotating-antenna images. Figure 14-1 illustrates a PPI-type radar system. With PPI radar, the antenna is rotated through 360°, although in some systems the antenna rotates through a partial rotation sector. The PPI displays the radar echo power or voltage amplitude as a function of range and the angular position of the antenna. The rotation angle is sensed by a potentiometer, a two-phase "synchro" servo system, or other means. In an old cathode-ray tube system, the synchro receiver mechanically rotates the yoke used for magnetic deflection of the display tube in synchronism with the antenna. Thus, when the antenna is pointed straight ahead of the aircraft, the synchro positions the sweep so that the sweep line is vertical. When the antenna is pointed directly to the side, the synchro positions the sweep so that the sweep line is horizontal, and so on. By starting the sweep at the display center and finishing the sweep at the edge, the full 360° range may be displayed as indicated in Fig. 14-2. A simple target map is shown with a straight line directly ahead of the radar. With the PPI display, individual targets appear at approximately the correct geometric position, though they tend to appear on the display as segments of circles rather than as points. A PPI radar is capable of producing an image that represents reasonably faithfully the location of targets on the ground, and a PPI radar built with a large-enough antenna can achieve the same kind of results that can be achieved with a side-looking real-aperture radar having an antenna of the same length.

14-2 Real-Aperture SLAR

The general idea of side-looking airborne radar was introduced in Section 1-5.1. The aircraft (or spacecraft) carries a long, thin antenna that produces a fan beam pointed to the side of the flight track as illustrated in Fig. 14-3(a). In pre-1970 radar systems, the forward motion of the aircraft was synchronized with the motion of film in a recorder that passes the film in front of an intensity-modulated cathode-ray tube that produces a single line showing the radar echo amplitude versus

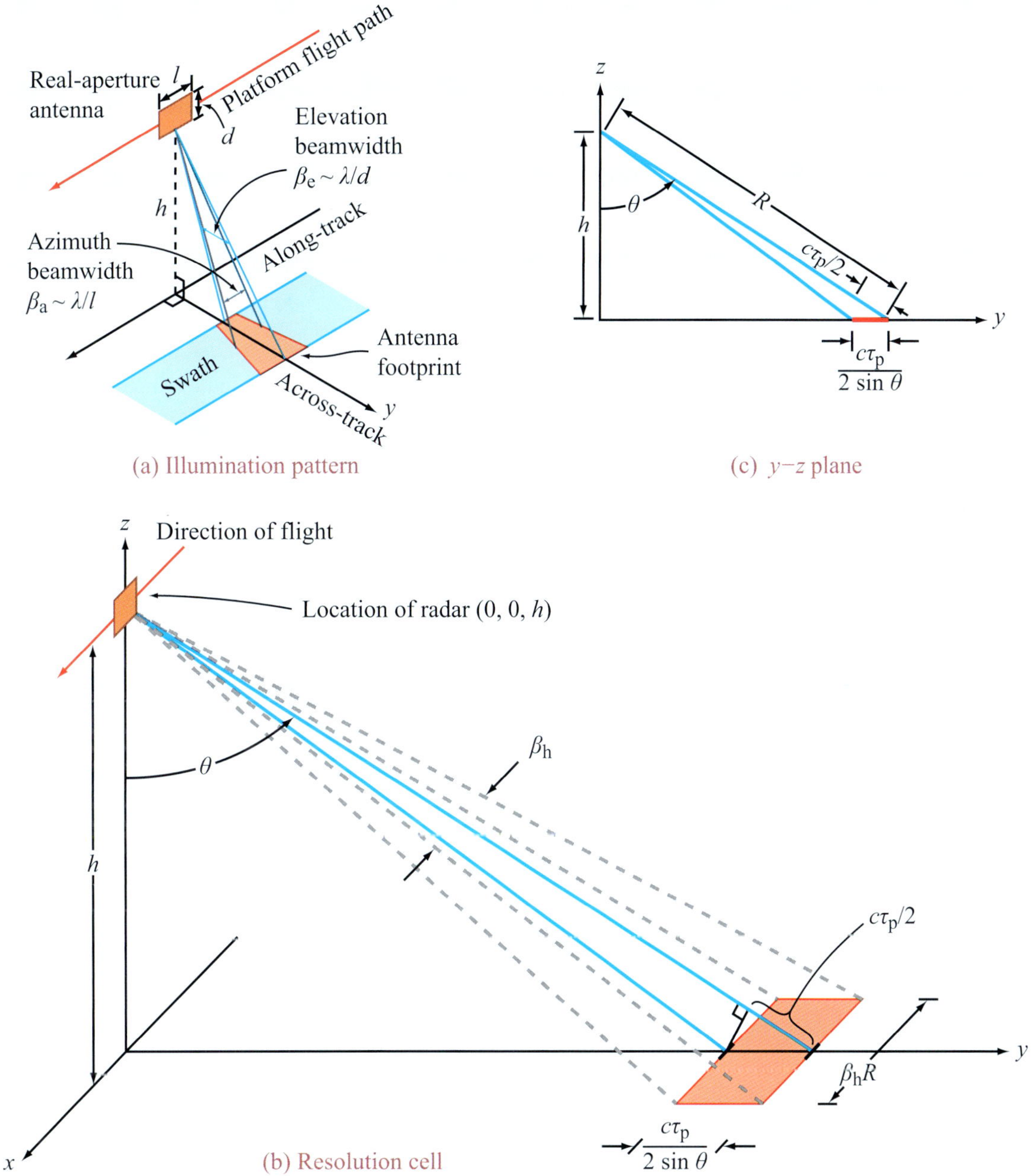

Figure 14-3: Geometry of real-aperture SLAR: (a) antenna illumination pattern on ground, (b) three-dimensional view of one resolution cell, and (c) view in the y–z (cross-track/vertical) plane.

time for each pulse transmitted. The motion of the film and aircraft produced an intensity image by this technique. In modern systems, cathode-ray tubes and film recordings are replaced by digital sampling and storage. This simple description applies to a real-aperture SLAR. For SAR, the recorded signal is further processed to obtain finer resolution in the along-track direction. In this section we focus on real-aperture SLAR; SAR systems are considered later.

14-2.1 SLAR Resolution

The spatial resolution of a real-aperture SLAR is dictated by the antenna beamwidth in the ***along-track*** or ***azimuth direction***, and by the effective pulse length (after range compression, if used—see Section 13-6) in the ***across-track direction***. Figure 14-3 gives the detailed geometry for resolution determination of the real-aperture SLAR.

The total round-trip time delay T experienced by the radar signal traveling from the radar to a point at ***slant range*** R is

$$T = \frac{2R}{c} . \tag{14.1}$$

For a pulse of length τ_p, the ***slant-range resolution*** r_r in the slant-range direction is

$$r_\text{r} = \frac{c\tau_\text{p}}{2} . \tag{14.2}$$

If range compression is used, τ_p is the effective pulse length after range compression. The ratio of the slant-range resolution to the total slant range is the same as the ratio of the pulse duration τ_p to the total time delay T. We are more interested in the resolution distance on the ground (i.e., the ground resolution) rather than in the slant-range resolution. The relation between the slant-range and ground-range resolutions can be determined with the aid of Fig. 14-3. The ***ground-range resolution*** r_y is given by

$$r_y = \frac{c\tau_\text{p}}{2\sin\theta} , \tag{14.3}$$

where θ is the angle of incidence.

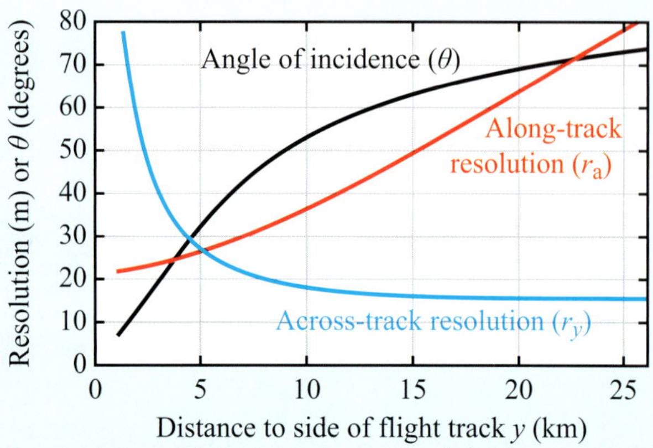

Figure 14-4: Example of SLAR resolution versus incidence angle for altitude $h = 7.5$ km, pulse length $\tau_\text{p} = 100$ ns, and along-track horizontal beamwidth $\beta_\text{h} = 3$ mrad (0.17°).

The ***along-track resolution*** r_a of a ***real-aperture*** SLAR is dictated by the antenna beamwidth at the ground; i.e., by the arc length corresponding to the ***horizontal beamwidth*** β_h:

$$r_\text{a} = \beta_\text{h} R = \frac{\beta_\text{h} h}{\cos\theta} , \tag{14.4}$$

where h is the height of the radar above the ground.

Because of the factor $\sin\theta$ in the denominator of the expression given by Eq. (14.3) for the ground-range resolution and the factor $\cos\theta$ in the denominator of the along-track resolution expression of Eq. (14.4), the along-track and ground-range resolutions are functions of the angle of incidence and, consequently, of the distance to the side of the flight track, as shown in Fig. 14-4.

► The across-track resolution r_y is degraded at short distances (corresponding to low incidence angles), and the along-track resolution r_a is degraded at long distances. ◄

Even with very long antennas, the along-track resolution degrades very quickly with incidence angle. For this reason, real-aperture SLARs are seldom used to produce wide-swath images and are rarely used from space. As an illustration, a real-aperture X-band SLAR used on a spacecraft at a 600 km altitude with a 10 m long antenna (beamwidth ~3 mrad or 0.17°) achieves a best-case along-track resolution of only 1.8 km.

Note that in general the SLAR image resolution cells (pixels) are not square. Figure 14-4 reveals that, for a given real-aperture SLAR, a pixel is square at only one range (in this example, at 5 km). At other ranges a pixel is rectangular, being elongated in the across-track direction at short ranges and in the along-track direction at long ranges. Thus, in specifying the resolution of a real-aperture SLAR, one should use both the beamwidth and slant-range resolution to describe the performance of the system, rather than a single value. From Eqs. (14.3) and (14.4), the pixel area $r_a r_y$ is a minimum at an incidence angle of 45°.

14-2.2 The SLAR Radar Equation

For a SLAR with the geometry of Fig. 14-3, the radar equation given by Eq. (5.40) becomes

$$P_r = \frac{P_t G^2 \lambda^2 \sigma^0}{(4\pi)^3 R^4} (\beta_h R) \left(\frac{c\tau_p}{2\sin\theta} \right) \tag{14.5}$$

or

$$P_r = \frac{P_t G^2 \lambda^2 \sigma^0 \beta_h c \tau_p}{2(4\pi)^3 R^3 \sin\theta} , \tag{14.6}$$

where P_r and P_t are the received and peak transmitted powers, G is the antenna gain, σ^0 is the scattering coefficient, and the illuminated area is written in terms of the product $r_a r_y$, using Eqs. (14.3) and (14.4). Proper correspondence between P_r and σ^0 requires P_r to be the ensemble average over many samples of the distributed target.

Often, the ***average transmitted power*** $P_{t_{av}}$, rather than the ***peak power*** P_t, is used. For a ***pulse-repetition rate*** f_p (in pulses/s), the average power $P_{t_{av}}$ is

$$P_{t_{av}} = P_t \tau_p f_p , \tag{14.7}$$

where τ_p is the pulse length. Accordingly, Eq. (14.6) may be written as

$$P_r = \frac{P_{t_{av}} G^2 \lambda^2 \sigma^0 \beta_h c}{2(4\pi)^3 R^3 f_p \sin\theta} . \tag{14.8}$$

From Chapter 7, the receiver noise power is

$$P_n = kT_0 BF, \tag{14.9}$$

where k is Boltzmann's constant, T_0 is the reference temperature (290 K), B is the bandwidth, and F is the receiver noise figure. Consequently, the ***signal-to-noise ratio*** S_n is

$$S_n = \frac{P_r}{P_n} = \frac{P_t G^2 \lambda^2 \sigma^0 \beta_h c \tau_p}{2(4\pi)^3 k T_0 B F R^3 \sin\theta} . \tag{14.10}$$

(SLAR SNR equation)

Note that, as discussed in Chapter 13, for a pulse with a realistic shape, one must use the actual form of the pulse $P_t(t)$ and the actual shape of the antenna pattern, so that Eq. (14.6) should be written as

$$P_r(t) = \iint_{\text{total area illuminated}} \frac{P_t(t-T)\, G^2(x,y) \lambda^2 \sigma^0(x,y)}{(4\pi)^3 R^4} \, dx\, dy. \tag{14.11}$$

For a fan-beam antenna, the gain pattern of the antenna can be separated into independent components in the θ (cross-track) and ϕ (along-track) directions, we can write

$$G(\theta,\phi) = G_\theta(\theta)\, G_\phi(\phi) = G_0\, g_\theta(\theta)\, g_\phi(\phi), \tag{14.12}$$

where $g_\theta(\theta)$ and $g_\phi(\phi)$ are pattern factors with maximum value of unity and G_0 is the peak gain. Note that for the narrow-beam case, the along-track differential distance may be expressed as

$$dx = d(R\phi) = R\, d\phi, \tag{14.13}$$

and $\sigma^0(\theta,\phi) \approx \sigma^0(\theta)$. Substituting Eqs. (14.12) and (14.13) in (14.11) leads to

$$P_r(t) = \int \frac{P_t(t-T)\, G_0\, g_\theta^2(\theta)\, \lambda^2\, \sigma^0(\theta)}{(4\pi)^3 R^4} \, dy \int g_\phi^2(\phi)\, d\phi. \tag{14.14}$$

An alternative form of this equation can be derived by noting that

$$R = cT/2 \tag{14.15}$$

and

$$P_t(t) = P_{t_{max}}\, p(t), \tag{14.16}$$

where $P_{t_{max}}$ is the maximum power value of the pulse and the pulse envelope $p(t)$ has a maximum value of one. Also note from the geometry of Fig. 14-3 that

$$\theta = \cos^{-1}\left(\frac{h}{R}\right) = \cos^{-1}\left(\frac{2h}{cT}\right). \tag{14.17}$$

Since

$$y = \sqrt{R^2 - h^2}\,,$$

dy in Eq. (14.11) can be replaced with

$$dy = dR\,\frac{R}{\sqrt{R^2 - h^2}}. \tag{14.18}$$

Making all of these substitutions in Eq. (14.14), we obtain

$$P_r(t) = \frac{P_{t_{max}} G_0^2 \lambda^2}{(4\pi)^3 (c/2)} \cdot \int \frac{p(t-T)\, g_\theta^2\left(\cos^{-1}\left(\frac{2h}{cT}\right)\right) \sigma^0\left(\cos^{-1}\left(\frac{2h}{cT}\right)\right) dT}{T^2 \sqrt{\left(\frac{cT}{2}\right)^2 - h^2}} \cdot \int g_\phi^2(\phi)\, d\phi, \tag{14.19}$$

which may be expressed in the form

$$P_r(t) = \frac{2P_{t_{max}} G_0^2 \lambda^2}{(4\pi)^3 c} \int p(t-T)\, b^2(T)\, dT, \tag{14.20}$$

where

$$b^2(T) = \frac{g_\theta^2\left(\cos^{-1}(2h)\,cT\right) \sigma^0\left(\cos^{-1}\left(\frac{2h}{cT}\right)\right) dT}{T^2 \sqrt{\left(\frac{cT}{2}\right)^2 - h^2}} \cdot \int g_\phi^2(\phi)\, d\phi. \tag{14.21}$$

In Eqs. (14.14) and (14.19) the along-track variation of the antenna beam (i.e., in its narrow direction) is in a separate integral, which is a great convenience because it means that the effect of the shape of the antenna pattern in the along-track direction is similar for all ranges. This simplifies processing and image pixel calibration.

The expressions given by Eqs. (14.19) and (14.20) describe a convolution of the pulse shape, the shape of the antenna pattern, the shape of the scattering-coefficient variation with angle, and the time delay or distance. For many applications, one can assume that the antenna pattern can be written in the form of Eq. (14.6) with half-power beamwidth β_h in the azimuth direction and equivalent half-power pulse duration τ_p in the range direction; however, for precise measurements, and in most cases near vertical, Eqs. (14.19) and (14.20) should be used instead.

14-2.3 SLAR Systems

▶ Unlike SAR systems, SLAR systems do not necessarily require pulse-to-pulse coherence. ◀

In this regard they are similar to pulse radars used for aircraft tracking. The primary differences between SLAR and aircraft tracking radars are associated with the method of scanning the antenna (by moving the aircraft rather than merely rotating the antenna) and the methods used to record and present the information. In the past, many SLAR systems were built using standard pulse radar receiver-transmitter units developed for other purposes such as marine navigation.

Figure 14-5 shows a conceptual block diagram of a 1960s SLAR that employs a film recorder for data storage. The entire system is controlled by a timing synchronizer that develops pulses used in controlling the transmitter pulse time, the gain control of the receiver, and the deflection circuits for the display. The synchronizer triggers a modulator that produces a pulse that modulates the power transmitter. In the past, magnetron oscillators were widely used for SLAR transmitters due to their ready availability, compact size, and reliability. While magnetrons produce short pulses with high peak power, their pulses do not always start with the same phase. Consequently, magnetrons are not suitable sources for fully coherent

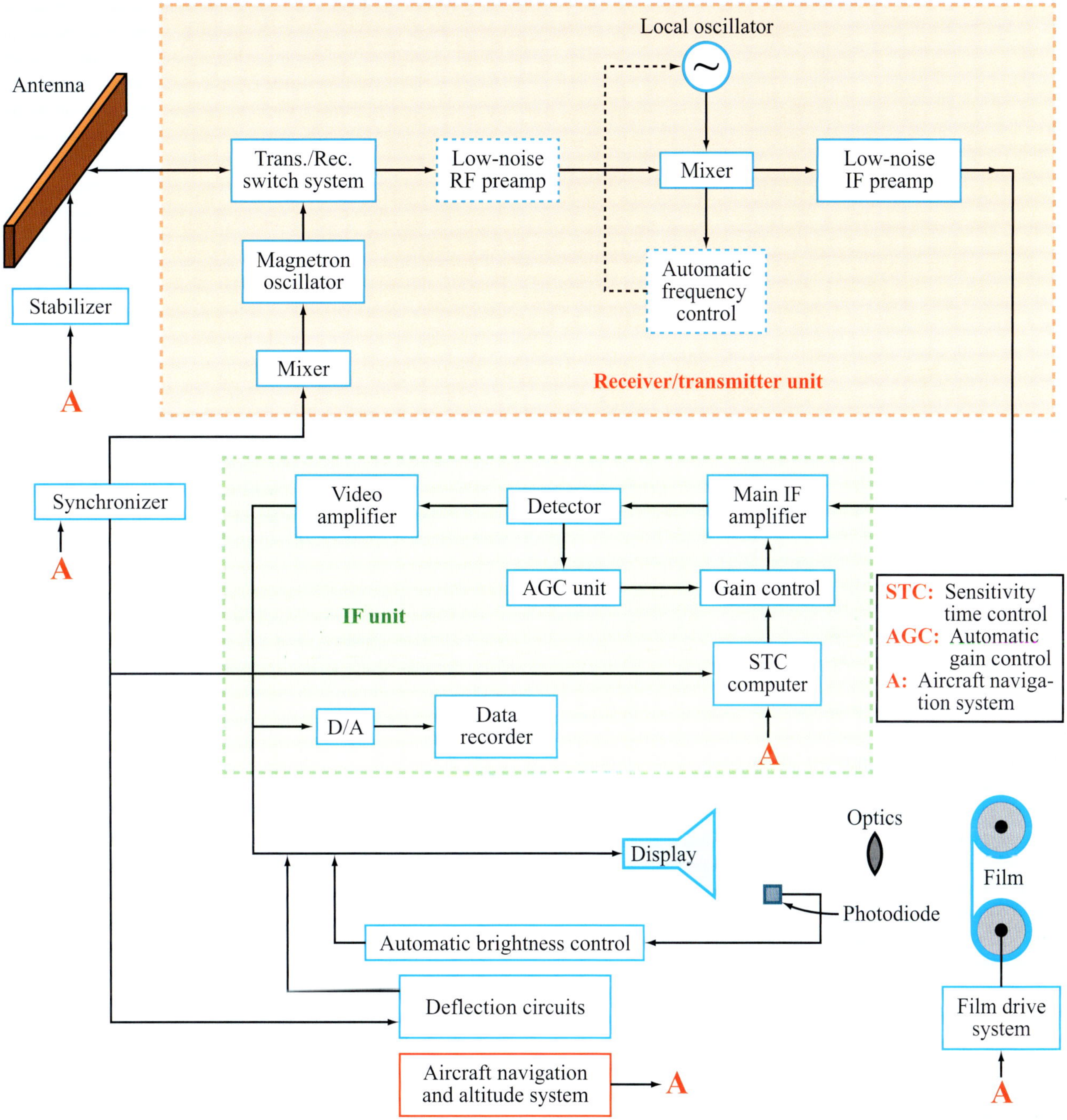

Figure 14-5: Example block diagram of a SLAR system.

radar systems. Modern systems often use ***traveling wave tube amplifiers*** (TWTAs) or ***solid-state transmitters***.

The output of the transmitter goes to a ***transmit-receive switching system***. This system involves ***circulators***, nonlinear receiver-protection devices, and so on, but its primary purpose is to provide a low-loss path between the transmitter and antenna during transmission while still protecting the receiver from the high peak power of the transmitter. During reception it must provide a low-loss path from the antenna to the receiver and reduce the amount of the quiescent noise output of the oscillator that reaches the receiver to prevent deterioration of the signal-to-noise ratio.

> ▶ A SLAR antenna is typically long and narrow so that it can achieve a narrow beam in the along-track direction and a wide beam in the across-track (or elevation) direction. ◀

The antenna may be rigidly mounted to the side of an aircraft when the application can permit avoidance of turbulent atmospheric conditions, but to compensate for attitude changes (of the aircraft) induced by ***atmospheric turbulence***, the antenna may be stabilized mechanically or electronically. When the antenna must be stabilized, the usual procedure is to allow for small motions about both the vertical axis and the horizontal axis. The small motions about the vertical axis compensate for minor variations in the roll of the aircraft, and those about the horizontal axis compensate for minor variations in yaw. Compensation for pitch variations is not usually attempted. The effect of yaw motions is to move the beam about on the ground and consequently cause the points imaged at a particular instant to be different from those assumed if a fixed antenna position relative to the flight line is assumed. Geometric variations in the elevation direction caused by roll of the aircraft are less significant, but aircraft roll affects the gain and image brightness across the swath. Since the antenna is typically long, mechanical compensation for yaw is more difficult than compensation for roll. A moving antenna must be enclosed in an aerodynamic ***radome*** and space must be left for the antenna's movement.

The transmitted signal travels to the target and returns to the antenna, and from there to the transmit-receive switching system, and on to the receiver. We note that radar design involves a trade-off between performance, on the one hand, and cost and complexity on the other. The receiver may or may not contain a low-noise RF preamplifier, depending upon the requirements for signal-to-noise ratio at the output, the expected value of the scattering coefficient, the gain of the antenna, and the power of the transmitter. Typically mixer-IF preamplifier systems may achieve noise figures of 8 to 10 dB, and sometimes better. Since the improvement associated with using a low-noise preamplifier in this situation is not likely to be more than about 6 dB, a preamplifier may not be required. The superheterodyne receiver is similar to the radiometer receivers discussed in Chapter 7. The primary differences between the radar receiver and the radiometer receiver relate to the greater need for a low noise-figure and greater gain-stability in the radiometer receiver, and the need for automatic frequency control and time-variable gain in the radar receiver.

When Doppler processing is not used, the transmitter need not be phase-coherent with the receiver. However, if transmit/receive IF frequency drift can occur, either ***excess receiver bandwidth*** or ***automatic frequency control*** (AFC) may be necessary to ensure capturing the full bandwidth of the received signals. To minimize RF losses, RF components are mounted in a receiver-transmitter unit close to the antenna. Other components can be located further away since losses in the transmission line connecting the low-noise IF preamplifier with the main IF amplifier have little impact on system performance. The IF unit contains the main IF amplifier (which usually has an adjustable gain in at least one stage), the detector, and the data recording system. Originally, data recording was done on film using cathode ray tubes but it is now done digitally by digitizing the signal versus time using an ***analog-to-digital*** (A/D) converter and storing the digital data.

Historically, to produce good real-time images with a SLAR, great care was taken to match the transfer characteristics of the different components of the system. SLARs have traditionally incorporated automatic-gain-control (***AGC***) and/or sensitivity-time-control (***STC***) circuitry to minimize dynamic range requirements in signal processing and recording. There is a trade-off in system dynamic range and complexity: dynamic range can be reduced by using antenna pattern and/or receiver gain correction at the expense of complexity.

14-3 Synthetic-Aperture Radar (SAR)

► More commonly referred to merely as a synthetic-aperture radar (SAR), a synthetic-aperture side-looking airborne radar system combines radar and signal processing to form high-resolution backscatter images. ◄

Multichannel SARs that make simultaneous (or nearly simultaneous) measurements at different polarizations are known as ***quad-pol*** or ***polarimetric*** SARs. SARs that simultaneously employ multiple antennas with specific geometries are termed ***interferometric*** SARs, or InSARs for short. InSARS are covered in the next chapter. In this section we consider the fundamental theory of SAR operation and associated trade-offs. We refer to other books for more details on SAR image formation processing (e.g., Cumming and Wong, 2005; Curlander and McDonough, 1991). We discuss the ambiguity problem, which is of extreme importance for spacecraft radars, along with motion compensation and image distortion. Signal-to-noise considerations for SARs are similar to those for real-aperture radars, but are not identical, and so the differences are highlighted.

14-3.1 Ways to Consider SAR

Fundamentally, SAR is based on movement of the radar antenna over a stationary target area. ***Inverse SAR*** fixes the radar in place and moves the target. An example is a target on a rotating platform. In the operation of a SAR, pulses are issued and received as the radar passes over the target area. The recorded received signals for each pulse are combined to generate a high-resolution radar image.

► The key to SAR is coherent recording and processing of the received signal over multiple pulses. ◄

SAR can be examined from several different points of view that can lead to different insights into its operation and implementation. Careful analysis of the mathematical descriptions of these views reveals that they are mathematically equivalent, as they must be. However, the fact that multiple methods can be used to describe the same process is an indication of its complexity.

The points of view listed below are explored in more detail in the subsections that follow:

1. Synthesized antenna aperture.
2. Doppler beam sharpening.
3. Correlation or matched filtering with a reference point-target response.
4. Dechirping of Doppler frequency shift.
5. Optical-focusing equivalent.

14-3.2 Synthesized Aperture

As suggested by its name, synthetic-aperture radar operates by "synthesizing" a long antenna aperture to produce finer resolution in the along-track direction. The synthesis occurs by recording the signal for each position of the antenna along a linear track, then combing the recorded signals as if they were simultaneously collected by an array antenna; in effect, SAR can be thought of as using time-storage techniques and the platform motion to create a long array antenna. Array-antenna analysis techniques similar to those developed in Chapter 3 can be applied to analyze the resolution of the SAR system.

Figure 14-6 illustrates the concept of a 1960s ***synthesized antenna aperture*** focused on a particular target. Note the difference between a focused antenna and the usual unfocused antenna. In unfocused antennas, the transmitted beam is confined as nearly as possible

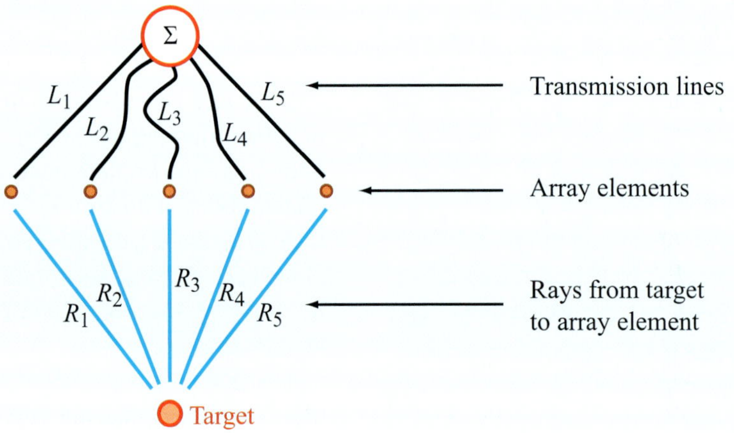

(a) Real-aperture equivalent array focused on target
$L_1 + R_1 = L_2 + R_2 = L_3 + R_3 = L_4 + R_4 = L_5 + R_5$

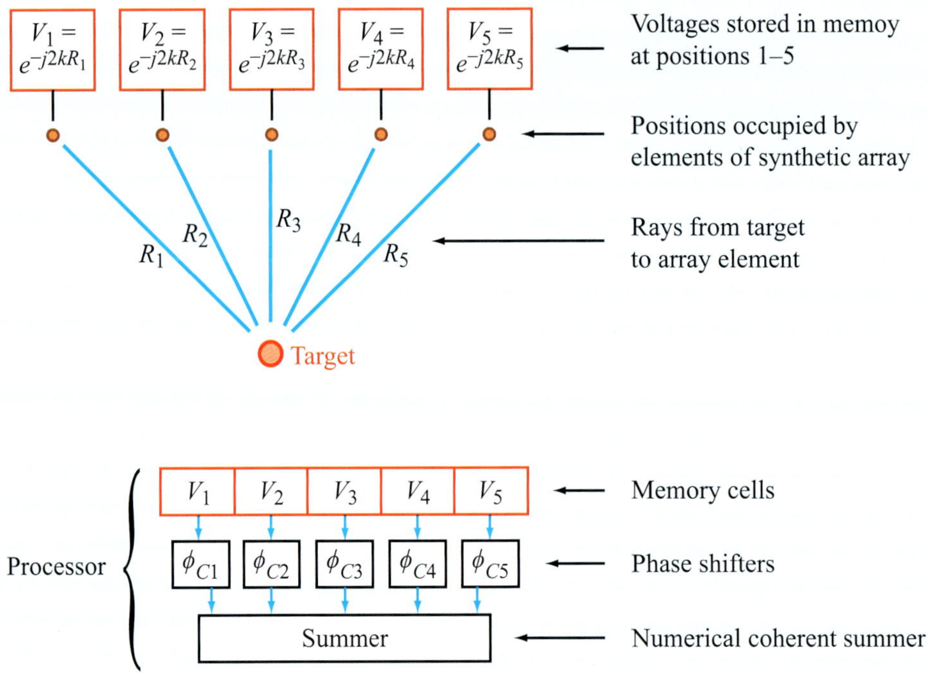

(b) Synthetic-aperture equivalent of (a)

Figure 14-6: Concept of synthetic aperture.

to parallel rays. For reception we regard the source of the signal to be so far away that the rays are essentially parallel. Antennas for communication, for real-aperture radar (RAR) systems, and for most other radars are designed with this in mind. However, a synthetic-aperture radar usually has an aperture that is so long that the beam must be focused at a point where the target is located, much as the lens in a camera is focused for nearby objects. In an analogy with the camera, we may say that in real-aperture systems, the "lens" (antenna) is focused at infinity. For a synthetic aperture the lens, or the "array" of antenna positions, is focused on a nearby object.

For clarity and intuition, Fig. 14-6(a) illustrates how one might build a real-aperture array of five antenna elements focused on the target T. Each element in the array is connected to a summing point, and the voltages from all the different elements are added at that point. The lengths of the transmission lines connecting each of the elements to the summing point are based on the distance to the target and the position of the particular element in the array.

▶ The total phase delay for a signal arriving at the summing point from the target must be the same for each element if the contributions of the different elements are to add in phase; when they do add in phase, the array is said to be focused at the target point. ◀

This means that the equivalent total distance from target to summing point must be the same for every element, so that

$$L_1 + R_1 = L_2 + R_2 = L_3 + R_3 = L_4 + R_4 = L_5 + R_5.$$

The lengths L_i are the "equivalent free-space lengths" of the transmission lines; that is, we choose the line length so that the time delay of the signal in the transmission lines corresponds to the travel time of the wave in space. On transmission, the signals all originate at the location that is the summing point for reception, so that for transmission the phase shift along each path is also the same as that along every other path. Consequently, for the location of target T, the total round-trip phase shift is the same for each element, and the electric fields from the different elements all add in phase. For other locations, the distances between target and summing point differ so that the contributions from the different array elements differ in phase. The summed output is smaller than it is for a target located at the focal point T. Note that the actual distances can be adjusted so long as the phase relationship is maintained.

Figure 14-6(b) shows the synthetic-aperture equivalent of the real-aperture focused array of Fig. 14-6(a). In SAR, each of the "elements" in the synthetic array is occupied at a different time as the SAR moves. When the element is occupied, both amplitude and phase of the received signal are recorded. When data from all element positions have been collected, the recorded signals can be summed to synthesize the aperture output.

For simplicity in our intuitive explanation here, we assume that the amplitudes are the same for each position and consider only the phase. The phases are as indicated in the boxes at the top of Fig. 14-6(b), with the round-trip phase shift for wave travel for element i being $2kR_i$, where $k = 2\pi/\lambda$. After all the target phases have been recorded—that is, after each position in the synthetic array has been occupied by the radar—the contents of memory cells 1 through 5 are phase-shifted by the amounts needed to compensate for the differing distances R_1 to R_5. In essence, V_1 is multiplied by e^{+j2kR_1}, and so on for the other elements. This phase correction $+2kR_1$ is indicated in the figure as ϕ_{C1}. Note that the phasor sum of all the corrected signals is 5 times the value for one signal.

It is significant to recognize that in this view of SAR, there is no requirement for continuous motion from array elements 1 to 2 to 3 to 4 to 5, that the elements be uniformly spaced, or that element positions be occupied in any particular order. The only requirement is that the position of each element relative to the focal point T be known and the appropriate phase correction applied in the processor (the summer in this case). It should be noted that this phase-correction scheme is applicable only when the system bandwidth is a small fraction of the operating frequency.

Mathematically we can designate the total phase shift

on path 1 of Figure 14-6(a) as ϕ_{T1}, the phase shift in space on path 1 as ϕ_{S1}, and the phase shift on the transmission line as ϕ_{L1}. Thus we may write

$$\phi_{\mathrm{T1}} = \phi_{\mathrm{L1}} + \phi_{\mathrm{S1}} = \phi_{\mathrm{T}i} = \phi_{\mathrm{L}i} + \phi_{\mathrm{S}i} = \phi_{\mathrm{T}} \qquad (14.22)$$

for $i = 1$ to 5, and the phase shift in space is

$$\phi_{\mathrm{S}i} = 2kR_i, \qquad (14.23)$$

and the phase shift on the transmission line is

$$\phi_{\mathrm{L}i} = 2k_{\mathrm{L}}R_i, \qquad (14.24)$$

where $k_{\mathrm{L}} = 2\pi/\lambda_{\mathrm{L}}$ is the transmit signal wave number for the transmission line. [Note: The wavelength of the signal in free space (λ) may be different from the wavelength (λ_{L}) of the same signal in the transmission line.]

The required phase shift in the transmission line is therefore

$$\phi_{\mathrm{L}i} = \phi_{\mathrm{T}} - \phi_{\mathrm{S}i}. \qquad (14.25)$$

From this and Eq. (14.24), the length of the transmission line can be determined as

$$L_i = \frac{\phi_{\mathrm{T}} - 2kR_i}{2k_{\mathrm{L}}}. \qquad (14.26)$$

The voltage received at element i is

$$V_i = e^{-j2kR_i} = e^{-j\phi_{\mathrm{S}i}}. \qquad (14.27)$$

Therefore, we need a phase-shift correction

$$\phi_{\mathrm{C}i} = \phi_{\mathrm{L}i} = \phi_{\mathrm{T}} - \phi_{\mathrm{S}i} \qquad (14.28)$$

to apply to the signal, as noted in Fig. 14-6(b). If we apply this correction, the resulting voltage at the summing point is

$$V = \sum_{i=1}^{5} e^{-j\phi_{\mathrm{T}}} = 5e^{-j\phi_{\mathrm{T}}} \qquad (14.29)$$

for a target at T. Since the absolute phase is essentially arbitrary, we set $\phi_{\mathrm{T}} = 0$ by choosing $\phi_{\mathrm{C}i} = -\phi_{\mathrm{S}i}$. More generally, to focus an N-element synthetic aperture on a target at position T, the expression

$$V = \sum_{i=1}^{N} a_i e^{j2kR_i} V_i \qquad (14.30)$$

is used, where V_i is the received signal for the ith aperture position, R_i is the distance between the ith aperture position and the target, and a_i is an optional weighting function that can be applied to control azimuth sidelobes.

14-3.3 Doppler Beam-Sharpening Approach

The original synthetic-aperture radar developed by Wiley in the early 1950s was called a "Doppler beam-sharpener." The use of the term "synthetic aperture" was adopted in the late 1950s. Figure 14-7 illustrates the view of a SAR as a focused Doppler beam-sharpening system, with Fig. 14-7(a) describing the geometry.

In Fig. 14-7(a) the radar is shown on an aircraft at a distance x_{r} back from the origin of coordinates at a height h; i.e, the aircraft coordinates are $(-x_{\mathrm{r}}, 0, h)$ and the coordinates of the target T are $(-x_{\mathrm{t}}, y_{\mathrm{t}}, 0)$. The ellipse shown in the figure represents the half-power contour of the antenna illumination on the ground of the real antenna carried on the aircraft. The width of the ellipse is exaggerated for clarity; typically, it is a narrow fan beam. The beamwidth in the along-track direction is β_{h}. The effective ground-range resolution r_y is indicated. The shaded area surrounding the target, and bounded in the x direction by the edges of the beam, shows the area contributing to the return at a particular instant of time. The dashed lines are constant Doppler frequency values (isodops) surrounding the target. Their spacing Δf_{D} is the bandwidth B_{Df} of the Doppler filter used for beam sharpening.

The Doppler frequency for the target in this case is given by

$$f_{\mathrm{dT}} = \frac{2u(x_{\mathrm{r}} - x_{\mathrm{t}})}{\lambda R}. \qquad (14.31)$$

Note that the target location is fixed, whereas the radar location x_{r} is varying. For the illustration in Fig. 14-7(a), the Doppler frequency is positive because both values

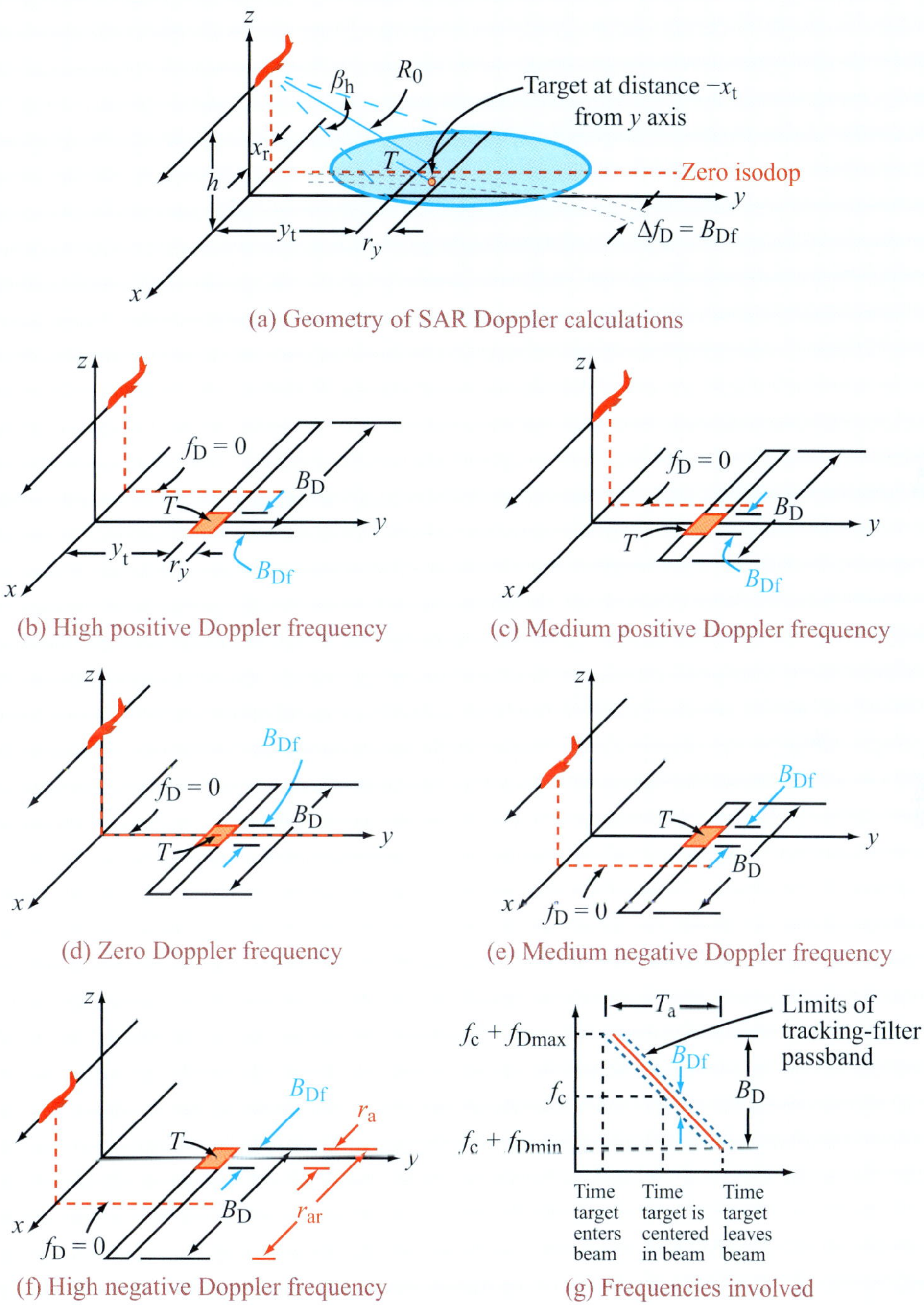

Figure 14-7: Doppler beam-sharpening SAR concept for a narrow-band Doppler filter.

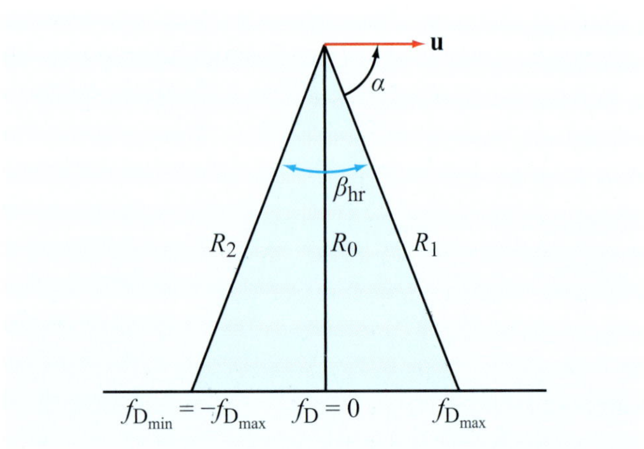

Figure 14-8: Real antenna with azimuth beamwidth β_{hr} traveling at speed **u** generates maximum and minimum Doppler shifts along R_1 and R_2.

of x are negative, with x_r having a larger magnitude. As x_r increases (it initially becomes less negative), the Doppler frequency decreases. If we neglect the variation in R (the slant distance from the radar to the target) during the passage of the beam over the target, the frequency decrease is linear as indicated in Fig. 14-7(g).

Figures 14-7(b)–(f) illustrate the effect of the "focused" ***Doppler beam-sharpening*** system as the target is successively illuminated first by the forward edge of the beam and later by other portions of the beam, until finally the beam passes it by. For illustration, the geometry of Fig. 14-7(a) is simplified by moving the coordinate system so that the target is located on the y axis at $(0, y_t, 0)$. The along-track length of the illuminated area on the ground is the resolution r_{ar} of a real-aperture radar of the same beamwidth; i.e.,

$$r_{ar} = \beta_{hr} R_0. \tag{14.32}$$

For a real-antenna aperture with azimuth beamwidth β_{hr} and velocity **u** (Fig. 14-8), the Doppler frequency shift is maximum for a target along the direction R_1 corresponding to the front edge of the beam,

$$f_{D_{max}} = \frac{2u}{\lambda} \cos\alpha, \tag{14.33}$$

where α is the angle between the velocity direction and the unit vector between the radar antenna and the ground. Upon replacing α with $(\pi/2 - \beta_{hr}/2)$, we have

$$\begin{aligned} f_{D_{max}} &= \frac{2u}{\lambda} \sin\left(\frac{\beta_{hr}}{2}\right) \\ &\approx \frac{2u}{\lambda}\left(\frac{\beta_{hr}}{2}\right) = \frac{u\beta_{hr}}{\lambda}, \end{aligned} \tag{14.34}$$

where we used the small angle approximation $\sin x \approx x$. The Doppler shift along R_2 is

$$f_{D_{min}} = -f_{D_{max}} = -\frac{u\beta_{hr}}{\lambda}. \tag{14.35}$$

Hence, the total Doppler bandwidth is

$$B_D = f_{D_{max}} - f_{D_{min}} = \frac{2u}{\lambda}\beta_{hr} = \frac{2ur_{ar}}{\lambda R_0}, \tag{14.36}$$

where r_{ar} is the azimuth resolution of the *real* antenna beam.

Doppler beam-sharpening is accomplished by using a narrow-bandwidth filter with a time-varying center frequency that "tracks" the Doppler-shifted signal frequency from the target. The limits on the ground defined by this filter are shown by the cross-hatched areas in Fig. 14-7(b)–(f) and have width r_a. The filter bandwidth is B_{Df}.

In Fig. 14-7(b), the beam begins to illuminate the target. The Doppler frequency is at its positive maximum, and the filter picks out a small section at the forward edge of the illuminated area. In Fig. 14-7(c), the aircraft has moved ahead about a quarter of the distance on the ground covered by the beam, and the Doppler filter has tracked the target signal down to a medium positive Doppler frequency. In Fig. 14-7(d), the aircraft has moved abreast of the target, whose Doppler frequency is now zero, so the filter is also centered on zero. In Fig. 14-7(e), the aircraft has moved forward to a point about another quarter beamwidth along the flight line, and the filter has tracked down to a medium negative Doppler frequency. Finally, in Fig. 14-7(f), the aircraft has moved forward to a point where the beam is about to leave the target. The filter is now centered on its maximum negative excursion.

Figure 14-7(g) shows this Doppler frequency shift as a function of time as the radar flies past the target. In this figure the dotted line shows the limits of the tracking filter passband as the tracking filter follows the Doppler shift associated with the point target. Radar returns from other parts of the beam are filtered out by the tracking filter. By analogy with Eq. (14.36), the bandwidth of the tracking filter is related to the along-track resolution r_a [Fig. 14-7(f)] by

$$B_{Df} = \frac{2ur_a}{\lambda R_0}. \tag{14.37}$$

This expression can be inverted to obtain the SAR azimuth resolution in terms of the Doppler filter bandwidth:

$$r_a = \frac{\lambda R_0}{2u} B_{Df}. \tag{14.38}$$

Note that a different center frequency-versus-time tracking filter is required for each cross-range image pixel. Although the use of a tracking filter, or a bank of them, may be a conceptually simple way to implement Doppler beam-sharpening, practical considerations often lead to alternative implementations. In addition, the width of the Doppler filter may be reduced, which impacts r_a, as discussed later.

14-3.4 Correlation or Matched Filtering with Reference Point-Target Response

SAR can be viewed as a two-dimensional correlation or matched-filtering operation. One dimension is range or cross-track. ***Range compression*** is done in ***fast-time***, the time scale of an individual pulse. The other dimension is cross-range or along-track, also known as the "azimuth dimension." By analogy with range compression, azimuth processing is called ***azimuth compression***. Azimuth compression uses multiple pulses. This time scale is termed ***slow-time***. Range compression was considered in Section 13-6. Here we consider azimuth compression by correlation or match filtering.

We note that the received signal for each pulse is a continuous-time signal. As a result, range compression is a continuous-time problem, even if the signal is digitally sampled and processed. On the other hand, the azimuth signal is a fundamentally discrete-time problem due to the fact that the radar uses discrete pulses as it travels in azimuth. This requires that the sample rate due to the PRF satisfies the Nyquist sampling criterion. This is addressed in greater detail later. While the azimuth signal is actually a discrete-time signal, for the purposes of comparison with range compression, we treat it as a continuous-time signal in this section.

The basic scheme for a continuous-time correlation device is indicated in Fig. 14-9(a). The phasor output of the receiver $V_s(t)$ is fed to a correlator along with a phasor reference function $V_r(t)$. When the two signals are coincident in time and have the same waveform, the correlation is maximum. Hence the correlator output is maximized for a target located at a point for which the reference function is designed, and targets located elsewhere are discriminated against by the filter.

To illustrate how this operates, suppose that the phasor received signal is

$$V_s(t) = e^{j\phi(t)}.$$

The phase of the received signal, which is the time integral of the angular frequencies associated with the carrier and the Doppler shift, is given by

$$\phi(t) = \int (\omega_c + \omega_D)\, dt - \omega_c t + \int \frac{4\pi ux}{\lambda R_0}\, dt, \tag{14.39}$$

where x is the along-track displacement between the radar and the target. Since this displacement varies linearly with time, that is,

$$x = ut,$$

where u is the aircraft velocity, the resulting expression for the phase is

$$\phi(t) = \omega_c t + \frac{2\pi u^2 t^2}{\lambda R_0}. \tag{14.40}$$

Ignoring the carrier-frequency term, the signal is

$$V_s(t) = e^{j2\pi u^2 t^2/\lambda R_0}. \tag{14.41}$$

This is the signal for a target whose x coordinate is zero; that is, the time origin is chosen as the instant that the radar comes abreast of the target. This radar position is known as the ***point of closest approach*** (PCA). The

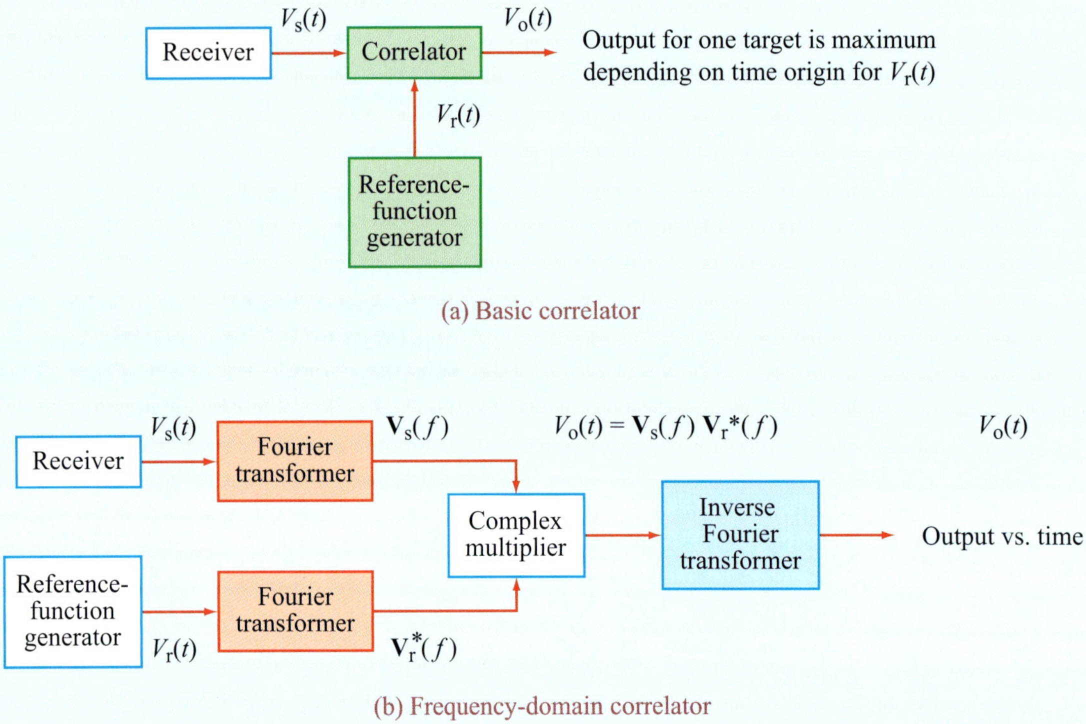

Figure 14-9: Correlation processor concept for SAR.

appropriate reference function for a point target at this location is given by

$$V_r(t) = e^{j2\pi u^2 t^2/\lambda R_0}, \tag{14.42}$$

which, of course, is the same as $V_s(t)$. The output of the correlator is

$$V_o(t_r) = \int_{-T_a/2}^{T_a/2} V_s(t)\, V_r^*(t+t_r)\, dt, \tag{14.43}$$

where t_r is the offset between the reference used and the reference for the target at $x = 0$. Substituting the values from Eqs. (14.41) and (14.42) into Eq. (14.43) yields

$$V_o(t_r) = \int_{-T_a/2}^{T_a/2} e^{-j(2\pi u^2/\lambda R_0)(2t_r t + t_r^2)}\, dt,$$

which may be simplified to

$$V_o(t_r) = e^{-j(2\pi u^2 t_r^2/\lambda R_0)} \int_{-T_a/2}^{T_a/2} e^{-j(4\pi u^2 t_r/\lambda R_0)t}\, dt. \tag{14.44}$$

Note that the limits that have been placed on the integral correspond to the ***time of illumination*** of the target by the antenna beam, with the total duration of this illumination being T_a. The duration of the reference function is assumed to be long enough that it does not affect the limits of the integral. When this integral is evaluated, the result is

$$V_o(t_r) = e^{-j(2\pi u^2 t_r^2/\lambda R_0)} \left[\frac{\sin\left(2\pi u^2 T_a t_r/\lambda R_0\right)}{2\pi u^2 T_a t_r/\lambda R_0} \right] T_a. \tag{14.45}$$

Thus the output of the correlator is a maximum when $t_r = 0$, which occurs when the reference function and the target are aligned with each other. Note that the

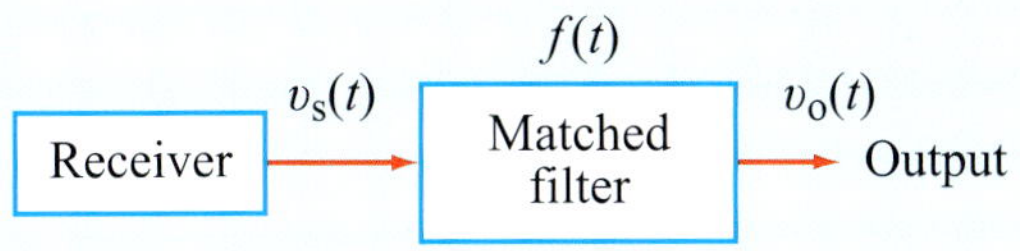

Figure 14-10: Matched-filter processor concept for SAR.

magnitude of the correlator output varies as $(\sin x)/x$. Hence. the output for a target located well away from the origin is small.

Clearly, if the SAR is to track multiple targets, many reference functions must be provided. A way to avoid this is to perform the equivalent of the correlation in the frequency domain, as indicated in Fig. 14-9(b). Here the reference function and receiver output are both Fourier-transformed, the Fourier transforms are multiplied, and the resulting frequency-domain function is converted by an inverse Fourier transform into the output waveform. For a single target, the time at which the output waveform has its maximum depends upon its location; for multiple targets, the output of the inverse transformer is a sequence, with the time for each target corresponding to its location. This frequency-domain approach is the basis of the ***range-Doppler algorithm*** (RDA) for processing SAR data, which we discuss in more detail in Section 14-9.4.

A similar but alternative view to correlation is ***matched filtering***. Figure 14-10 shows the elementary block diagram of the matched-filter processor. Here a filter, whose frequency response is the complex conjugate of the signal Doppler spectrum for a point target, is inserted in series with the input signal to obtain the output. The output signal is then the convolution of the input signal and the filter impulse response $f(t)$:

$$v_o(t) = v_s(t) * f(t), \tag{14.46}$$

where the filter frequency response is designed such that

$$\mathbf{F}(\omega) = \mathbf{V}_s^*(\omega). \tag{14.47}$$

Recall that the matched filtering and correlation approaches are identical for the target location when the matched-filter and correlation reference function are properly defined.

14-3.5 Dechirping Comparable with Range Pulse-Compression Dechirping

As discussed in Chapter 13, range compression (matched filtering) of a linear frequency-modulated transmit signal can be accomplished using a dechirping operation. The same concept can be applied to the Doppler signal. Recall the linear change in Doppler with time shown in Fig. 14-7. The dechirping view of SAR is illustrated with the aid of Fig. 14-11. Figure 14-11(a) shows the decreasing Doppler frequency from the target, which lasts for a time T_a during which the target is within the area illuminated by the antenna. Since the highest Doppler frequency is received first and the lowest Doppler frequency is last, the idea of the frequency selective delay filter is to delay the higher-frequency signal by an amount corresponding to the time required for the actual Doppler frequency to change to its lowest value. Thus, the desired frequency-dependent time delay characteristic of the filter is as shown in Fig. 14-11(b). The resulting idealized output superimposes the outputs corresponding to all the Doppler frequencies to a single frequency output at a single time as indicated in Fig. 14-11(c). In reality, the output has a sinc function-like envelope because the dechirped signal has a finite duration and bandwidth, similar to the output for the range pulse-compression signal. Note that for range dechirping, the signal typically has a bandwidth of many MHz and a pulse duration of microseconds; for the Doppler frequency the bandwidth is in kHz or hundreds of Hz, and the time delay is measured in tenths of seconds or seconds.

14-3.6 Optical-Focusing Equivalent of SAR

While SAR processing is now performed digitally, the original SAR processors used optics for processing the radar signal. An exception was Wiley's first Doppler sharpening system. ***Optical processing*** begins with recording the SAR data on photographic film, as shown in Fig. 1-16. To record the signal phase (i.e., to record both positive and negative signal amplitudes), the signal zero level is set to be midway in the film's grayscale characteristic, with positive signals being recorded towards saturation on the film and negative signals being

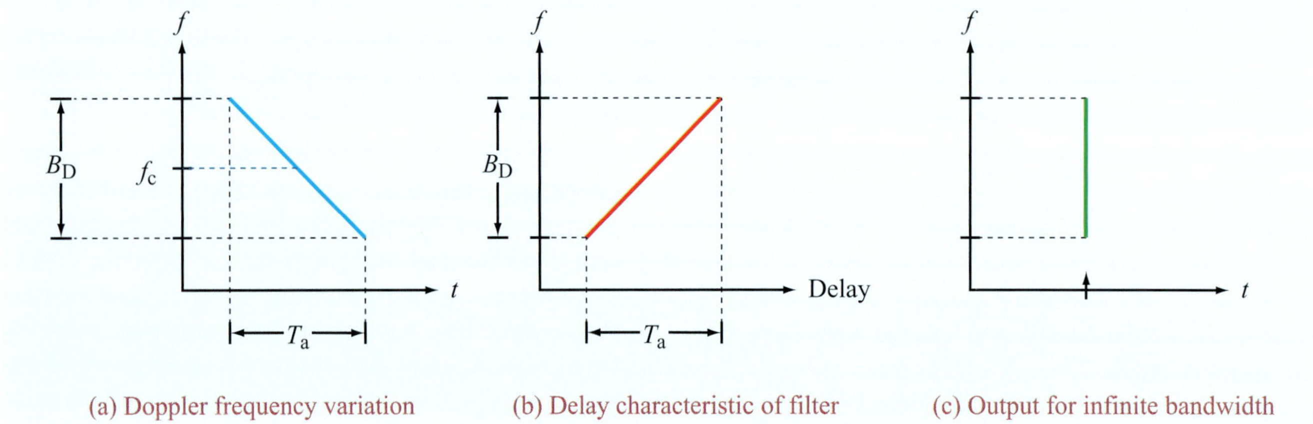

Figure 14-11: Delay-filter azimuth dechirp concept for SAR.

recorded toward the toe of the film response curve. The resulting recording on film is thus a grayscale rendition of the phase history of the received signal. Typically, a single pulse was recorded as a line on the film, as in the SLAR diagram illustrated in Fig. 1-16. Pulses were recorded sequentially as adjacent lines.

Figure 14-12 illustrates the principle of optical SAR processing. Recall the SAR concept shown in Fig. 14-6 and consider Fig. 14-6(a) for the real-antenna equivalent of the synthetic antenna as a model also for Fig. 14-12. In Fig. 14-12 the storage of the phase-coherent information is on film, rather than in a computer memory. The positions occupied by the elements of the actual antenna synthetic array extend over tens to hundreds of meters but are recorded over millimeters on the film and so the optics are closely spaced.

The optical processor is similar to building a real aperture as indicated in Fig. 14-6(a), except that the transmission lines are replaced by optical paths. This is indicated in Fig. 14-12(b) and (c). The bottom half of Fig. 14-12(b) is scaled with distances appropriate to the radar wavelength and radar geometry. The top half is scaled with distances appropriate to the optical wavelength and the geometry of the optical processor. Since the wavelength ratio is on the order of 10,000 or more in a typical case, the geometry is also scaled accordingly, although the actual optical wavelength is less important than the distances involved in recording on film. The radar signal goes from a point on the array out to the target and back, and then is recorded on film. The film is illuminated with collimated laser light, and the light passing through the film reproduces the strong and weak points of the equivalent phase front for the radar.

The signal is then focused at the optical image point with a lens as indicated in Fig. 14-12(c). Figure 14-12(b) illustrates the requirements for the lens geometry. The total phase shift between target and optical image point is the phase shift along path R_1 at the radar frequency and along path O_1 optically, and the same applies to all other path pairs R_i and O_i. The amount of excess radar distance at the ends of the aperture that must be corrected in the focusing is R_c. The corresponding amount of optical distance that must be corrected in the focusing is O_c. This focusing correction is achieved by the use of a convex lens, as indicated in Fig. 14-12(c). The shorter wavelength, or lower phase velocity, of the light in the glass of the lens causes additional phase shift where the lens is thick, relative to that where the lens is thin. The shape of the lens is thus selected to provide the desired correction,

The physical layout of the optical processor in its simplest form is shown in Fig. 14-12(d), where the

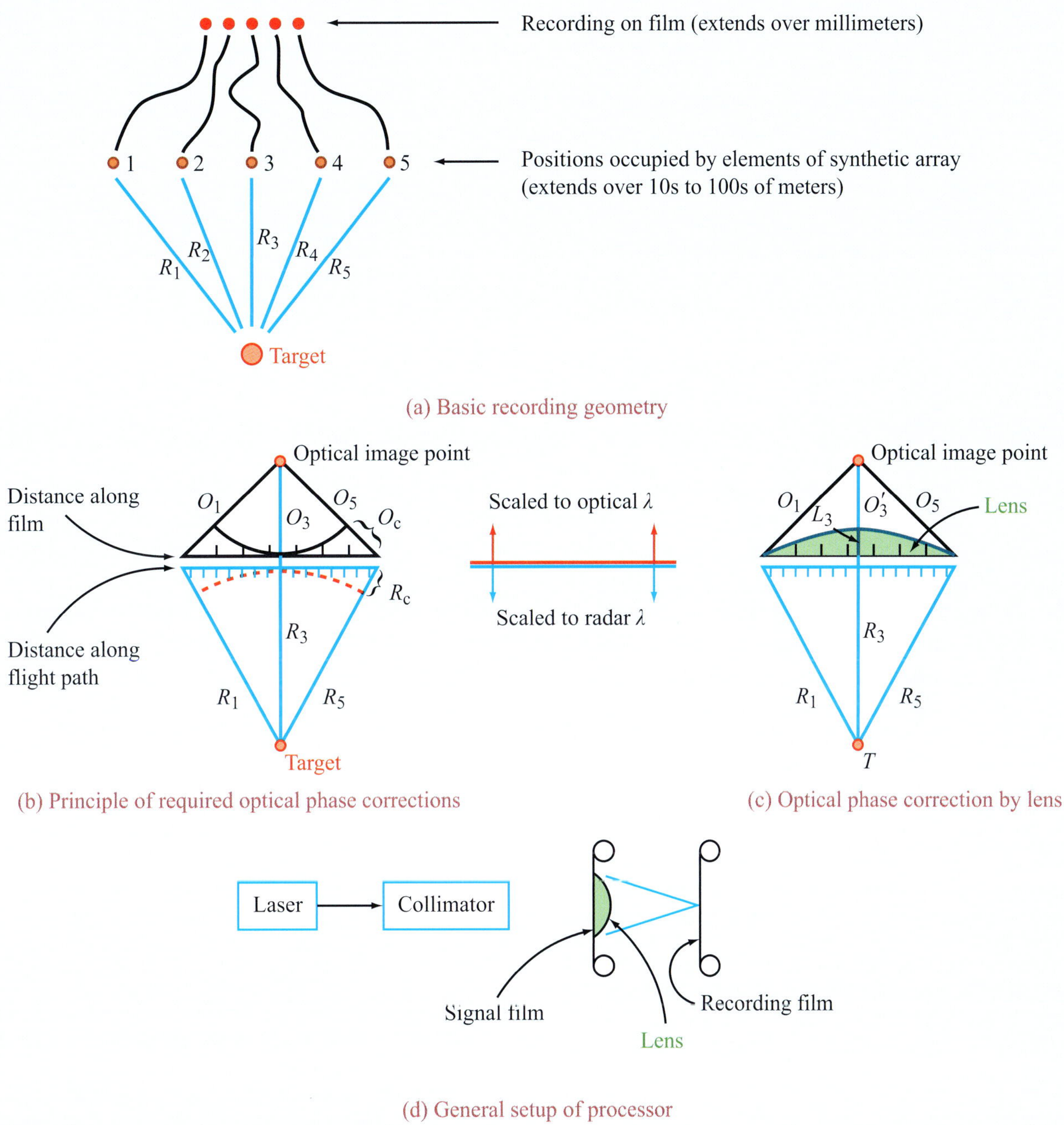

Figure 14-12: Optical processing for SAR.

laser light is collimated, fed through the film to the lens, and then focused on the film. Unfortunately, life is not this simple with optical processing, for the very process of focusing this signal on the film has the effect of defocusing in the range direction, so an additional lens must be added to compensate for that. Moreover, the correction computed for the center of the lens applies only for the particular radar range. Since the radar images a wide swath of the ground at varying slant ranges, the processor must accommodate these varying slant ranges by providing different corrections for different ranges for which the signal is recorded on film. As a result, optical processors were extremely complex to design and operate in practice. Optical processing was further complicated by the time delay required for processing the film, and the difficulties in accurately controlling the film processing.

14-4 SAR Resolution

The previous section introduced various points of view of synthetic-aperture radar. Here, we derive the SAR resolution equation by considering two of these points of view: namely, the synthesized aperture and the Doppler beam sharpener points of view. These were selected because of their importance in the understanding of the SAR image formation process; however, the equations for resolution are the same no matter what point of view is used.

14-4.1 Synthesized-Aperture Point of View

Figure 14-13(a) illustrates the along-track geometry of the SAR measurement of a target T. The radar antenna on the airborne vehicle is shown at three different locations where echo data is collected, and they are labeled A, B, and C. At location A, the forward edge of the beam is first to intercept the target; at location B, the antenna is abreast of the target; and at location C, the aft edge of the beam is just leaving the target. The ***total possible length of the synthetic aperture*** L_p defines the finest resolution that can be obtained with the real antenna in a real-aperture SLAR:

$$L_p = \beta_{hr} R_0, \qquad (14.48)$$

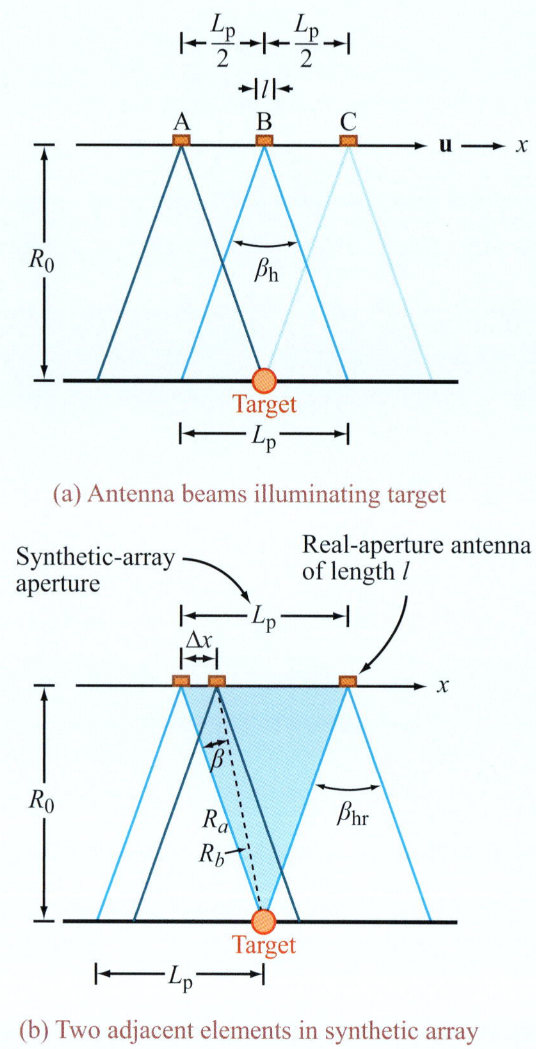

Figure 14-13: Synthesized-array concept for SAR.

where β_{hr} is the horizontal or along-track beamwidth of the real antenna and the subscript r is used to distinguish the real-aperture antenna beamwidth from the beamwidth of the synthetic array antenna.

In considering the beamwidth of a synthetic antenna, one must take into account the fact that the phase shift associated with a given target is due to two-way, rather than one-way, travel on the path between the antenna and the target. Figure 14-13(b) illustrates the ranges for two adjacent elements in an array. The beamwidth is determined by the relative phase shift along the paths between the target and the two elements. Two adjacent elements in the synthetic array, separated by a (small) distance Δx, are shown, with portions of the rays R_a and R_b going from these elements to the target. Since the two elements are close together, the rays may be considered essentially parallel. Hence, the difference in distance along the two rays is

$$R_a - R_b = \Delta x \sin\beta. \tag{14.49}$$

For the real-aperture antenna the phase shift associated with this distance is

$$\Delta\phi_\mathrm{r} = k(R_a - R_b) = k\,\Delta x \sin\beta. \tag{14.50a}$$

(real aperture)

For the synthetic-aperture array, the phase shift enters twice: once from antenna to target and again as it returns from target to antenna. Thus, in this case

$$\Delta\phi_\mathrm{s} = 2k(R_a - R_b) = 2k\,\Delta x \sin\beta. \tag{14.50b}$$

(synthetic aperture)

For a real (or synthetic) antenna of length $\Delta x = L$, where L is long enough so that the beamwidth is small, the first ***null*** in the azimuth pattern occurs at $\pm\beta_\mathrm{nr}$ (or β_ns) such that

$$kL\sin\beta_\mathrm{nr} \approx kL\beta_\mathrm{nr} = \pi \qquad \text{(real aperture)}, \tag{14.51a}$$

$$2kL\sin\beta_\mathrm{ns} \approx 2kL\beta_\mathrm{ns} = \pi \qquad \text{(synthetic aperture)}. \tag{14.51b}$$

The corresponding null-to-null beamwidths are

$$\beta_{\mathrm{null}_\mathrm{r}} = 2\beta_\mathrm{nr} = \frac{2\pi}{kL} \qquad \text{(real aperture)}, \tag{14.52a}$$

$$\beta_{\mathrm{null}_\mathrm{s}} = 2\beta_\mathrm{ns} = \frac{\pi}{kL} \qquad \text{(synthetic aperture)}. \tag{14.52b}$$

Hence, the effective length of the synthetic aperture is $2L$, not just L. Accordingly, *the beamwidth for the synthetic aperture is half that for a real aperture having the same length*.

The along-track beamwidth β_hs for the synthetic aperture is therefore

$$\beta_\mathrm{hs} = \frac{\lambda a_\mathrm{hs}}{2L}, \tag{14.53}$$

where a_hs is the aperture-illumination taper factor associated with the synthetic aperture. If the synthetic aperture has uniform weighting all along its length, then $a_\mathrm{hs} = 0.88$, as discussed in Chapter 3 for real antennas. However. most antennas, both real and synthetic, have nonuniform aperture illumination; so the value for a_hs is usually greater than 0.88. While the simplifying assumption that a_hs is 1 is common, we should keep in mind that actual antennas, both real and synthetic, have a factor closer to 1.5.

For the synthetic aperture, the along-track resolution is

$$r_\mathrm{a} = \beta_\mathrm{hs} R_0 = \frac{\lambda R_0}{2L} a_\mathrm{hs}. \tag{14.54}$$

This expression applies regardless of the actual length of the synthetic aperture used. If the entire ***potential length*** L_p is used, the ***finest possible resolution*** r_ap for the SAR is given by

$$r_\mathrm{ap} = \frac{\lambda R_0}{2L_\mathrm{p}} a_\mathrm{hs} = \frac{\lambda a_\mathrm{hs}}{2\beta_\mathrm{hr}}, \tag{14.55}$$

where we used the relationship $L_\mathrm{p} = \beta_\mathrm{hr} R_0$ relating the beamwidth of the real aperture β_hr to the along-track width L_p at the range R_0. Thus, *the resolution attainable is inversely proportional to the beamwidth of the real antenna used in the SAR*. This beamwidth of the real antenna is

$$\beta_\mathrm{hr} = a_\mathrm{hr}\frac{\lambda}{l}, \tag{14.56}$$

where a_hr is the aperture-illumination taper factor for the real antenna and l is its length. This value of β_hr may be substituted in Eq. (14.55) to obtain the potential synthetic-aperture resolution in terms of the length of the real antenna and the aperture taper factors as

$$r_\mathrm{ap} = \frac{\lambda a_\mathrm{hs} l}{2a_\mathrm{hr}\lambda} = \left(\frac{a_\mathrm{hs}}{a_\mathrm{hr}}\right)\frac{l}{2}. \tag{14.57}$$

Since the aperture-illumination taper factors are often about the same for the both real and the synthetic aperture, so that $a_{hs}/a_{hr} \approx 1$, the usual result is

$$r_{ap} \approx \frac{l}{2}. \tag{14.58}$$

(optimal focused resolution)

> ▶ Thus, the finest potential synthetic resolution is one half the length of the real aperture. Moreover, the SAR azimuth resolution is independent of range and also independent of wavelength. ◀

The range independence is a consequence of the fact that the length of the synthetic aperture L_p is directly proportional to the range, so a larger range results in a larger aperture, which results in a finer beamwidth for the synthetic aperture and maintains the effective resolution. The reason the resolution is independent of the wavelength is similar: if the wavelength is longer, the synthetic-aperture length in wavelengths is also longer, so that the net result is that the resolution remains the same because the equivalent synthetic beamwidth remains the same. The expression for the resolution of the synthetic aperture in Eq. (14.58) should be contrasted with the real-aperture along-track (azimuth) resolution

$$r_{ar} = \beta_{hr} R_0 = a_{hr} \frac{\lambda R_0}{l}, \tag{14.59}$$

where the resolution increases if either λ or R_0 increases.

The SAR azimuth resolution r_{ap} given by Eq. (14.57) is the best resolution attainable with a synthetic-aperture radar having an antenna fixed to an aircraft or spacecraft flying in a straight line. As described later, finer resolution can be obtained if a scanning antenna is used so that it may look ahead at the target, keeping the target ***spotlighted*** (i.e., the antenna is pointed at the target) as the radar flies by, and continues to look back at the target for some time. This lengthens the synthetic aperture. Of course, an antenna scanned in this fashion achieves finer resolution for a particular target at the expense of neglecting other targets. The scanning motion results in gaps in coverage along the flight path. As described later, flying in a circular path concentric with the target can also achieve fine resolution, though such a path eliminates the possibility of wide-area coverage.

Although the finest possible resolution is achieved by using the entire available synthetic aperture L_p, the designer may choose to reduce the effective resolution to reduce speckle (discussed in Section 14-8) or system complexity. If a shorter aperture is used, the resolution is that given by Eq. (14.54), rather than that given by Eq. (14.57).

14-4.2 Unfocused SAR

One can shorten the synthetic aperture to the point where the phase correction discussed in Section 14-3.2 can be neglected. When this is done, the radar is said to be ***unfocused***. For this situation, all the signals received at the different points in the synthetic aperture can be added in phase without correction. Figure 14-6 illustrates the basic idea for synthetic-aperture corrections. The corrections are needed to reach a focus at the target point T because, as illustrated in that figure, the different path lengths between the radar and the target result in in-phase shifts along the paths that are significantly different. The amount of correction needed at a particular point is

$$\phi_{Ci} = \phi_T - \phi_{Si}. \tag{14.60}$$

For unfocused SAR, we neglect corrections less than $\pi/4$ radians. That is, if

$$\phi_{Ci} \leq \pi/4,$$

ϕ_{Ci} is ignored and all contributions are added in phase. This simplifies processing at the expense of precise focus and resolution.

To derive the conditions under which the unfocused SAR operates, consider Fig. 14-14, where R_{mp} is the ***maximum possible slant range*** to the target, R_m is the ***maximum slant range to the target in the unfocused SAR***, and R_0 is the ***minimum range to the target***. The phase shift associated with the maximum possible range is

$$\phi_{Cmp} = 2k(R_{mp} - R_0). \tag{14.61}$$

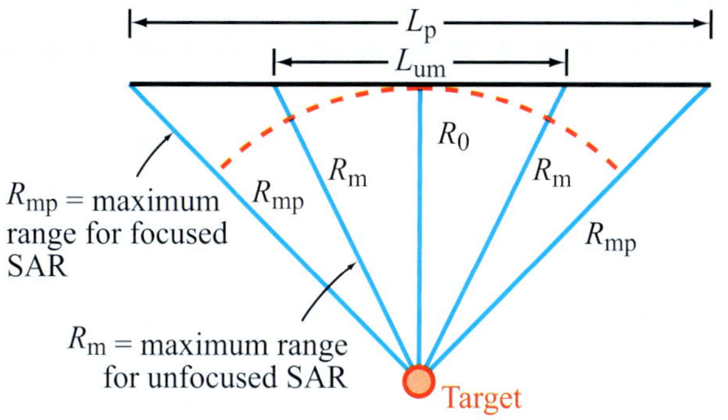

Figure 14-14: Illustration of needed phase compensation.

For the unfocused case,

$$\phi_{Cm} = 2k(R_m - R_0). \tag{14.62}$$

Setting $\phi_{Cm} = \pi/4$ and substituting $2\pi/\lambda$ for k yields

$$R_m - R_0 \leq \frac{\pi/4}{2(2\pi/\lambda)} = \frac{\lambda}{16}. \tag{14.63}$$

From Fig. 14-14, it can be seen that

$$R_m^2 = R_0^2 + (L_{um}/2)^2.$$

However, because R_m and R_0 are comparable in length, we can apply the Taylor series

$$R_m = R_0\sqrt{1+\left(\frac{L_{um}}{2R_0}\right)^2} \approx R_0\left[1+\frac{1}{8}\left(\frac{L_{um}}{R_0}\right)^2\right],$$

which leads to

$$R_m = R_0 \approx \frac{L_{um}^2}{8R_0}. \tag{14.64}$$

Using the limiting case of Eq. (14.63) in Eq. (14.64), we find that the ***maximum length of the unfocused synthetic aperture***, L_{um}, is given by

$$L_{um}^2 = \frac{8R_0\lambda}{16} = \frac{R_0\lambda}{2}.$$

Hence

$$L_{um} = \sqrt{R_0\lambda/2}. \tag{14.65}$$

(maximum unfocused length)

If we neglect the effect of taper in the unfocused synthetic aperture, the ***unfocused best possible resolution*** r_{apu} is given by

$$r_{apu} = \frac{\lambda R_0}{2L_{um}} = \sqrt{R_0\lambda/2} = L_{um}. \tag{14.66}$$

(optimal unfocused resolution)

Thus, the azimuth resolution for the unfocused SAR is equal to the length of the synthetic aperture involved. This can result in some simplification of processing for an unfocused system. On the other hand, the resolution for the unfocused case is dependent on wavelength and distance, while for the focused resolution case the resolution is independent of range and wavelength.

14-4.3 Doppler Point of View

From the Doppler beam-sharpening point of view, the along-track resolution of the SAR is given by Eq. (14.38), which can be written as

$$r_a = \frac{\lambda R_0}{2u} B_{Df}, \tag{14.67}$$

where B_{Df} is the Doppler sharpening filter bandwidth. To determine the limitations on the resolution, one must consider how narrow the filter bandwidth B_{Df} can be without loss of information.

The target is visible during the time T_a required for the antenna to travel the distance L_p. This is the time over which the filter operates. As an approximate rule of thumb, the filter bandwidth is greater than the inverse of its integration length, which suggests that

$$B_{Df} \geq 1/T_a. \tag{14.68}$$

The ratio between the aperture length L_p and the time T_a required to travel that length is related to the speed u of the platform:

$$L_p = uT_a. \tag{14.69}$$

Combining these results, the ***finest obtainable resolution*** is

$$r_{ap} = \left(\frac{\lambda R_0}{2u}\right)\left(\frac{u}{L_p}\right) = \frac{\lambda R_0}{2L_p}. \quad (14.70)$$

Using Eq. (14.48) for L_p and Eq. (14.56) for β_{hr} we can write

$$r_{ap} = \frac{l}{2a_{hr}}. \quad (14.71)$$

If the taper of the aperture distribution for the synthetic array is also considered, this becomes

$$r_{ap} = \left(\frac{a_{hs}}{a_{hr}}\right)\frac{l}{2}, \quad (14.72)$$

which, as expected, is the same as Eq. (14.57) derived for a synthesized aperture.

14-4.4 Comparison of Real-Aperture and Synthetic-Aperture Resolution

The along-track resolutions attained when the aperture taper factors are assumed to be 1 are given by Eq. (14.59) for SLAR, by Eq. (14.66) for unfocused SAR, and by Eq. (14.72) for focused SAR. That is,

$$r_{ar} = \frac{\lambda R_0}{l} \quad \textbf{(real-aperture radar)}, \quad (14.73a)$$

$$r_{apu} = \sqrt{\frac{R_0 \lambda}{2}} \quad \textbf{(optimal unfocused SAR)}, \quad (14.73b)$$

$$r_{ap} = \frac{l}{2} \quad \textbf{(optimal focused SAR)}. \quad (14.73c)$$

Note the differences in the variation of resolution with range, wavelength, and antenna length for each type of radar. For the real-aperture radar the finest resolution is attained with a small λ, a large l, and the smallest distance to the target R_0. For unfocused SAR, the antenna length l does not matter but the azimuth resolution variation with distance and wavelength is like that of real-aperture SLAR. In both cases, finer resolution comes with a longer antenna. For focused SAR, neither distance nor wavelength matters and the antenna should be made short to obtain the finest resolution. To illustrate, we provide a numerical comparison. Consider an X-band radar with $\lambda = 0.03$ m at two different antenna lengths, $l = 2$ m and $l = 10$ m. Three different ranges are considered: 5 km (a short-range airborne radar), 50 km (a long-range airborne radar), and 500 km (a spacecraft radar). The resulting values for the resolutions are given in Table 14-1.

Table 14-1: Comparison of azimuth resolutions for SLAR, fully focused SAR, and unfocused SAR.

	$l = 2$ m			$l = 10$ m		
Range (km)	5	50	500	5	50	500
Fully focused SAR	1	1	1	5	5	5
Unfocused SAR	8.7	27.4	86.8	8.7	27.4	86.6
Real-aperture radar	75	750	7500	15	150	1500

14-5 Ambiguity Considerations in SAR

► For aircraft SAR, timing ambiguity is seldom a limitation, but for spacecraft SAR systems it is a serious constraint. ◄

The problem is that the pulses must be far apart in time, and hence in along-track distance, to prevent range ambiguities; however, the large Doppler frequency spread for a high-speed spacecraft requires a high pulse repetition rate to be able to sample the Doppler bandwidth without aliasing. These conflicting requirements make ambiguity a problem.

Recall that the ***interpulse period*** T_p is defined as the reciprocal of the ***pulse repetition frequency*** (PRF) f_p:

$$T_p = 1/f_p. \quad (14.74)$$

Figure 14-15 illustrates the situation for a side-looking radar over a plane Earth. The maximum and minimum

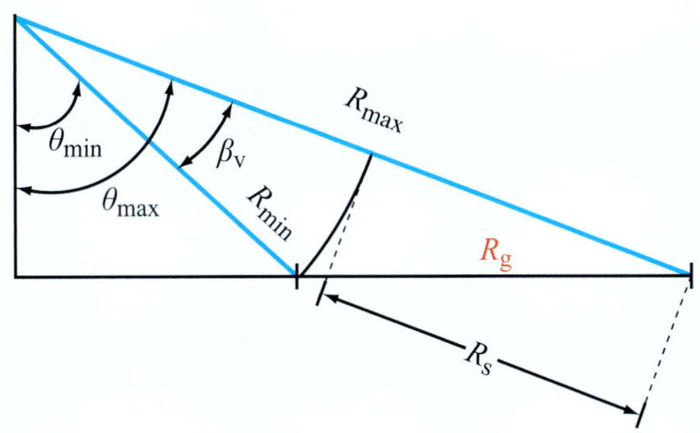

Figure 14-15: Range ambiguity diagram, side view. Radar is moving into or out of the page. Slant-range swath is R_s and ground-range swath is R_g.

ranges within the vertical beam (R_{max} and R_{min}) are shown, along with the vertical beamwidth β_v and the angles of incidence associated with the outer and inner edges of the beam (θ_{max} and θ_{min}). The slant swath width is the difference between maximum and minimum range and is given by R_s. The repetition period is determined by the condition that one cannot have two pulses contributing simultaneously to the received signal. The signals received from the ground from the two pulses would be from different ranges and consequently from different points on the ground. Mathematically this requirement is given by

$$cT_p/2 \geq R_s, \tag{14.75}$$

or in terms of the pulse repetition frequency,

$$f_p \leq c/2R_s. \tag{14.76}$$

Thus, the swath width places a constraint on the maximum possible pulse repetition frequency f_p.

A minimum constraint is placed on f_p by the requirement to sample the Doppler frequency properly. The magnitude of this constraint depends upon the specifics of processing used: ***zero-offset processing*** or ***azimuth-offset processing*** (Fig. 14-16). In zero-offset processing the carrier frequency is down-converted to zero frequency, so that the maximum video frequency is half the total bandwidth of the Doppler spectrum. For this type of processing, the constraint set by the Nyquist sampling theorem is that f_p must be twice this maximum bandwidth, or

$$f_{p0} \geq B_D. \tag{14.77}$$

The bandpass sampling theorem imposes this constraint if the sampling is actually done at a higher intermediate frequency.

Another type of processing, known as ***azimuth-offset processing***, down-converts the carrier frequency to approximately $B_D/2$, so that the maximum negative excursion of the Doppler frequency at baseband occurs at the zero-frequency (Fig. 14-16). For this situation the maximum video frequency is B_D, so the requirement on the pulse repetition frequency is

$$f_{pa} \geq 2B_D. \tag{14.78}$$

That is, the minimum sampling rate is $2B_D$.

We can combine Eqs. (14.77) and (14.78) by defining a ***Doppler processing constant*** a_D as 1 for zero-offset and 2 for azimuth-offset processing:

$$f_p \geq a_D B_D. \tag{14.79}$$

When zero-offset processing is used, both in-phase and quadrature channels are needed so that positive and negative Doppler frequencies may be distinguished from one another. Since two separate channels are used, one must collect two values [one ***in-phase*** (I) and one ***quadrature-phase*** (Q)] per sample at a sampling rate of B_D complex samples per second. Hence, the constraint on the PRF is different from that for the azimuth-offset processing, but in both cases $2B_D$ sample values per second are required.

Substituting Eq. (14.36) into Eq. (14.79) gives

$$f_p \geq 2a_D \frac{u}{\lambda} \beta_{hr}. \tag{14.80}$$

The expressions given by Eqs. (14.76) and (14.80) can be combined into

$$\frac{2a_D u}{\lambda} \beta_{hr} \leq f_p \leq \frac{c}{2R_s}. \tag{14.81}$$

Since the real-aperture azimuth beamwidth is

$$\beta_{hr} \approx (\lambda/l)a_{hr},$$

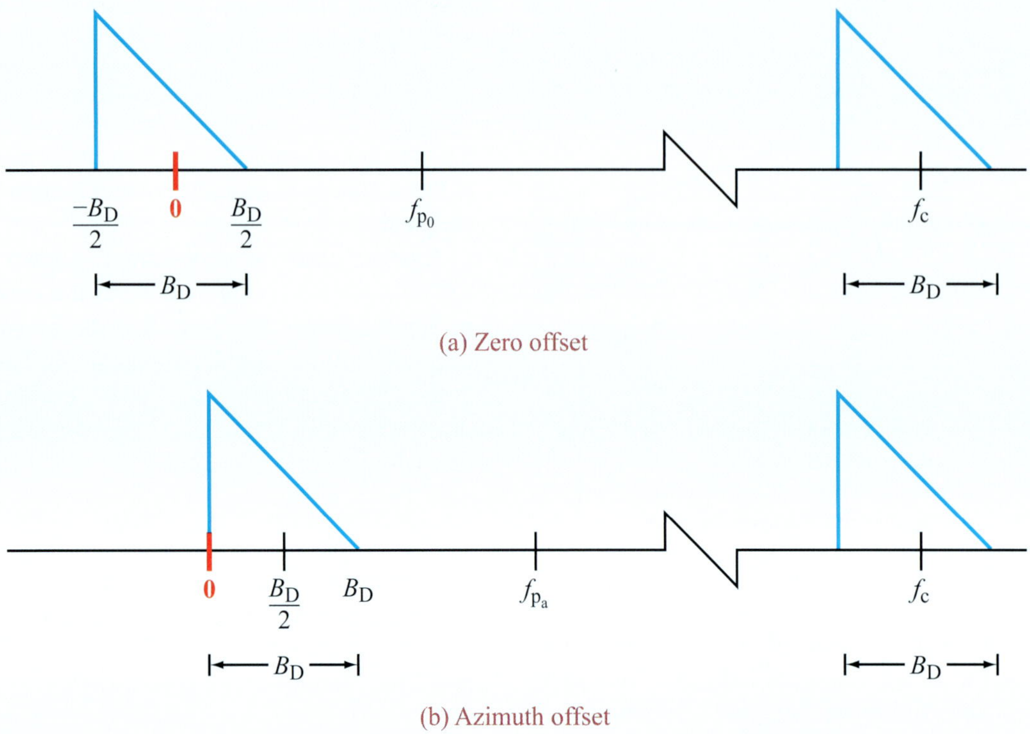

Figure 14-16: Doppler bandwidths for (a) zero-offset processing and (b) azimuth-offset processing.

Eq. (14.81) can be written as

$$a_{hr}\left(\frac{2a_D u}{l}\right) \le f_p \le \frac{c}{2R_s}. \tag{14.82}$$

▶ The ratio l/u is the amount of time it takes the radar to fly one antenna length. Hence, for $a_{hr} = 1$, the sampling constraint in Eq. (14.82) states that at least one sample (i.e., one pulse) must be taken for each half antenna length of travel for zero-offset processing ($a_D = 1$), and at least one sample for each quarter antenna length of travel for azimuth-offset processing ($a_D = 2$). ◀

Equation (14.82) may be considered a limitation on the slant swath width. That is,

$$R_s \le \frac{cl}{4a_D u a_{hr}}. \tag{14.83}$$

As a numerical example, consider an aircraft radar with $l = 2$ m and $u = 100$ m/s (about Mach 1). The slant swath is limited to

$$R_s \le 500 \text{ km } (a_D = 1) \qquad \text{or} \qquad 250 \text{ km } (a_D = 2)$$

for $a_{hr} = 1$. Since the aircraft cannot fly high enough to create a swath this wide, ambiguity is not a concern. For a spacecraft radar, however, the situation is quite different. A typical velocity for a low Earth-orbiting spacecraft is $u = 7.5$ km/s. For the antenna of length $l = 2$ m, we find a swath-width limitation of

$$R_s \le 20 \text{ km } (a_D = 1) \qquad \text{or} \qquad 10 \text{ km } (a_D = 2),$$

while for the antenna of length 10 m we find

$$R_s \le 100 \text{ km } (a_D = 1) \qquad \text{or} \qquad 50 \text{ km } (a_D = 2),$$

which represent significant limitations on the swath width that must be considered in the design. (Typically a safety factor is added.) Even without safety factors, the 2 m antenna gives a swath width that is too narrow to be usable from spacecraft altitudes, and the 10 m antenna gives one that is barely usable. Thus, the ambiguity constraint on spacecraft radars is severe.

Safety factors are necessary in the ambiguity constraints because antenna beams do not have perfect edges. In the range direction illustrated in Fig. 14-15, the slant swath shown is limited by the edges of the beam. If one uses the half-power points to represent the edges, this means that significant signals come back to the radar from pulses separated by a time somewhat greater than T_p calculated from Eq. (14.75). If one pulse strikes the ground near the outer edge of the illustrated swath, and another one strikes the ground at a value of θ slightly smaller than θ_{min}, the return from the ground associated with this second pulse may be quite strong because it is only a few dB down in the antenna pattern. Thus one must apply a safety factor so that the antenna pattern can be reduced far enough before another pulse strikes the ground. We call this ***slant-range safety factor*** k_r, and it modifies Eq. (14.76) to

$$f_p \le c/(2k_r R_s), \qquad k_r \ge 1. \tag{14.84}$$

Similarly, one may have problems in the along-track direction. In Fig. 14-16 we assume that the horizontal antenna pattern of the real aperture drops instantly to zero at the edge of the beam, so that the Doppler spectrum is sharply limited. In fact, the antenna pattern does not drop off so sharply, and consequently the Doppler spectrum is not so sharply limited. Therefore, a safety factor is also needed to assure adequate sampling of the Doppler spectrum. We call this ***along-track safety factor*** k_a. With this safety factor, Eq. (14.79) becomes

$$f_p \ge a_D k_a B_D, \qquad k_a \ge 1. \tag{14.85}$$

Equation (14.82) therefore becomes

$$\frac{2a_D u a_{hr} k_a}{l} \le f_p \le \frac{c}{2k_r R_s}. \tag{14.86}$$

Using this result, the constraint on the slant swath width becomes

$$R_s \le \frac{cl}{4a_D u a_{hr} k_r k_a}. \tag{14.87}$$

Since the factors k_r and k_a are always greater than 1 (typical values range from 1.3 to 1.6), the slant swath width is even less than the values given for the example above. The swath-width limitation problem may be alleviated by using a longer antenna, but a longer antenna is harder to use in space.

To see how the slant swath width relates to the ground swath width for the plane-Earth case (a spherical Earth should be used with spacecraft, but for simplicity here, we use a flat Earth), refer to Fig. 14-15 and note that

$$R_s = h(\sec\theta_{max} - \sec\theta_{min}), \tag{14.88}$$

which may be written as

$$R_s = h[\sec(\theta_{min} + \beta_v) - \sec\theta_{min}]. \tag{14.89}$$

For a small vertical beamwidth (required by the narrow swath in the spacecraft case), this result can be approximated by

$$R_s = \beta_v R_{av} \tan\theta_{av}, \tag{14.90}$$

where θ_{av} is the average value of θ_{max} and θ_{min} and R_{av} is the slant range at θ_{av}. Because of the flat-Earth approximation, the expression given by Eq. (14.90) should be regarded only as a first-order approximation for spacecraft radars.

We may use this expression along with Eq. (14.87) to find a ***minimum acceptable area*** for an antenna, from an ambiguity point of view. Substituting Eq. (14.90) in Eq. (14.87) gives

$$\beta_v R_{av} \tan\theta_{av} \le \frac{cl}{4a_D u a_{hr} k_r k_a}, \tag{14.91}$$

but the vertical beamwidth may be expressed as

$$\beta_v = \frac{\lambda}{H} a_{vr}, \tag{14.92}$$

where H is the height of the antenna aperture and a_{vr} is the ***real-aperture vertical taper factor***. Substituting

Eq. (14.92) into Eq. (14.91), using $f = c/\lambda$, and rearranging, the antenna area may be expressed by

$$\text{area} = lH \geq \frac{4a_D u R_{av} \tan\theta_{av}}{f} (a_{vr} a_{hr} k_r k_a). \qquad (14.93)$$

▶ Thus, there is a minimum area for the radar antenna. Any smaller antenna will not satisfy the ambiguity constraints. ◀

Note that the ambiguity constraints can be satisfied by antennas with differing ratios of H to l. That is, the area of the antenna is constrained, but not its shape, and consequently not the resolution and swath width—the resolution is specified by l and the swath width by H.

The minimum area constraint rarely impacts aircraft radar, but is important for spacecraft radar. As a numerical example, consider the following parameters:

$$u = 7500 \text{ m/s}, \qquad R_{av} = 1000 \text{ km},$$
$$\theta_{av} = 45^\circ, \qquad f = 10 \text{ GHz},$$
$$a_{vr} = a_{hr} = 1, \qquad k_r = k_a = 1.4.$$

Using Eq. (14.93), the minimum area of the antenna is

$$lH \geq 5.88 \text{ m}^2 \qquad (a_D = 1).$$

If the antenna length is 10 m, we find that the maximum slant swath width is $R_s \leq 51$ km. The ground swath width is then given by $R_g \approx 51/\sin 45^\circ = 72$ km. This is not a very wide swath; for a smaller angle of incidence, however, the swath is wider. For $\theta_{av} = 20^\circ$ with $R_{av} = 1000$ km, a smaller antenna can be used; i.e.,

$$lH \geq 2.14\text{m}^2 \qquad (a_D = 1).$$

At the smaller angle of incidence, R_s is still 51 km, but R_g is now $51/\sin 20^\circ = 149$ km, which is a more useful swath width.

14-5.1 Scanning Synthetic-Aperture Radar

Because of the ambiguity problem described in the previous section, the swath width for a spacecraft SAR is often less than desirable for some applications. One solution is to use a ***scanning synthetic-aperture radar*** (scanSAR), as illustrated in Fig. 14-17. Figure 14-17(a) illustrates a three-dimensional view of a scanSAR system in which the beam is scanned in elevation to three different positions. During the passage by the potential synthetic-aperture length L_p, one-third of this length is used for each of the three scan positions. In each scan position (subswath), Eq. (14.87) determines the swath width.

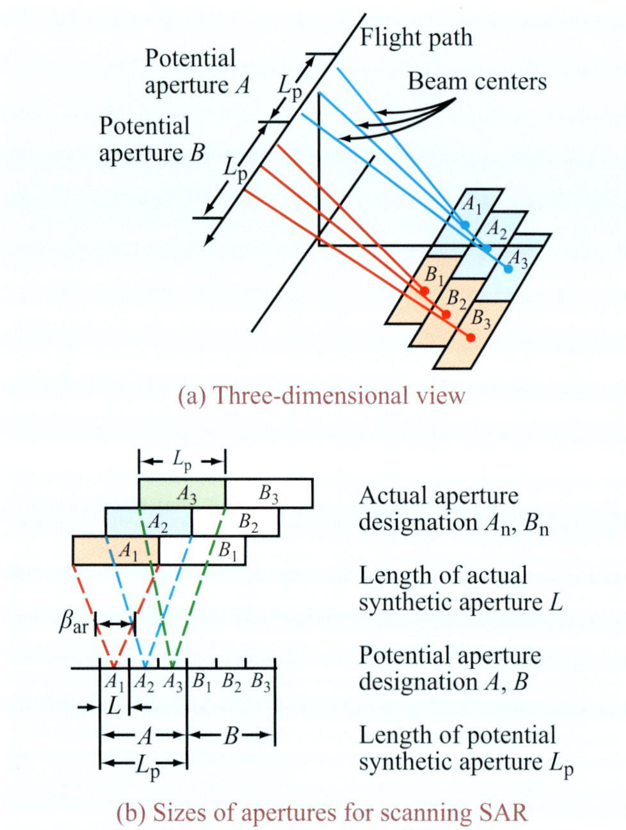

Figure 14-17: Scanning synthetic-aperture radar geometry: (a) three-dimensional view, (b) sizes of aperture for scanning SAR.

▶ For a given real-aperture antenna length, the swath width obtained by scanning in this example is three times as great as that without scanning. The penalty paid for this is that only one-third of the potential synthetic aperture may be used to obtain along-track resolution so the possible resolution r_a for this three-step scanning SAR is three times as large as that for a single-swath nonscanning SAR. That is, $r_a = 3r_{ap}$. ◀

Figure 14-17(b) illustrates the situation in a two-dimensional plot that allows better presentation of the dimensions. Two potential synthetic apertures A and B are shown, each one subdivided into three segments. The limits of the real-aperture beam are also shown. During the time for passing aperture A_1, the swath A_1 is imaged. At the end of this time the beam scans outward and the radar starts to image swath A_2, which it does during the aperture interval indicated as A_2. At the end of that time the beam scans out farther and swath A_3 is imaged. By the time the beam has completed imaging swath A_3, the radar is at the end of the first potential synthetic aperture A. This means that the beam must be swung back to the original position immediately so that the imaging of swath B_1 may start; otherwise a gap is created between swath A_1 and swath B_1. The procedure continues as the radar goes past the potential aperture B, and so on, so that a complete image is built up with three times the swath width that would be achieved without scanning.

One can see by analogy with the three-step example of Fig. 14-17 that the total distance available for the length of a synthetic aperture for a scanSAR with N_{sc} scan positions is

$$L \leq L_p / N_{sc}. \tag{14.94}$$

Consequently, the along-track resolution obtained with the scanSAR is

$$r_a \geq N_{sc} r_{ap}. \tag{14.95}$$

If one has a fixed antenna length and wishes to image a very wide swath, N_{sc} must be large and a significant sacrifice in resolution must be made. For some applications, more-frequently repeated coverage is more important than resolution, but the trade-off between swath width and resolution should be kept in mind when utilizing scanSAR.

14-5.2 Other SAR Observation Geometries

So far we have only considered SARs that look directly to the side of the spacecraft or the aircraft flight track, which is nominally straight as shown in Fig. 14-18(a). Such systems can create a continuous image strip and are known as ***stripmap-mode*** SAR systems. ScanSAR is a type of stripmap SAR that can be used for multiple swaths parallel to the flight path or diagonal paths as illustrated in Figs. 14-18(b) and (c). As suggested by the discussion in Section 14-4.1, other SAR observation geometries can be used. We consider a few of the most important of these in the following.

Squint-mode SAR

Many SAR systems, particularly aircraft radars for military purposes, look at some other azimuth angle, usually ahead of the aircraft as shown in Fig. 14-19. This is known as ***squint-mode SAR***. SARs operated in squint mode can be stripmap or scanSAR systems. In the example in Fig. 14-19, the antenna beam has been tilted ahead of the side-looking direction, or "squinted," so that its pattern on the ground is an ellipse with its major axis at about 45° to either of the coordinate axes. For comparison, a side-looking antenna is also shown. In both cases the iso-range contours associated with the range resolution are concentric circles that are locally orthogonal to the major axis of the ellipse. Note that in the squinted case the isodops and the iso-range lines are not perpendicular to each other. As a result the shape of the resolution cell is no longer approximately rectangular as it is for the side-looking case. Hence, the range and azimuth resolutions become coupled to one another in the squint-mode case. The ground-range resolution is denoted by the ***pulse range resolution*** r_p rather than r_y, since the ground range is no longer in the y direction. Similarly, the ***Doppler resolution*** r_d, which is in the direction perpendicular to the isodops and associated with the Doppler filter bandwidth B_{Df}, is now used rather than r_a.

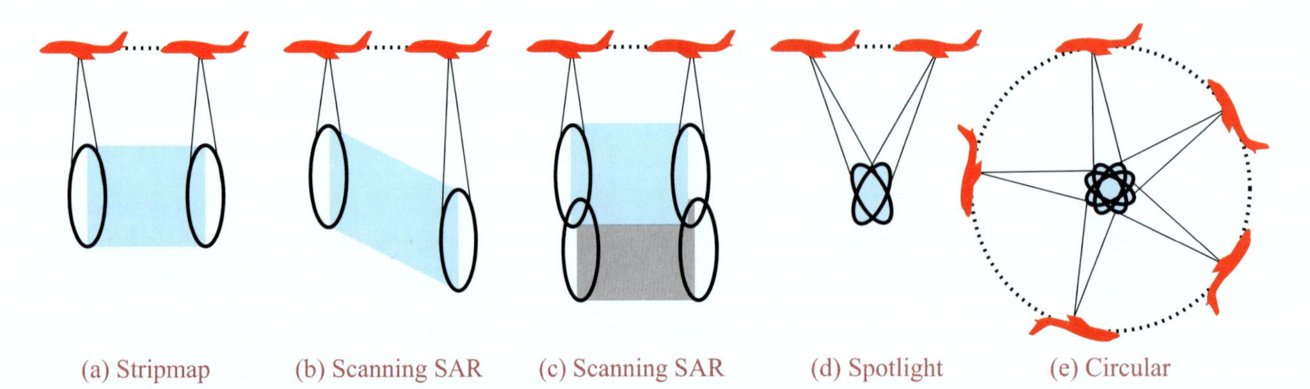

Figure 14-18: Comparison of SAR geometry modes: (a) stripmap, (b, c) scanning SAR, (d) spotlight, and (e) circular.

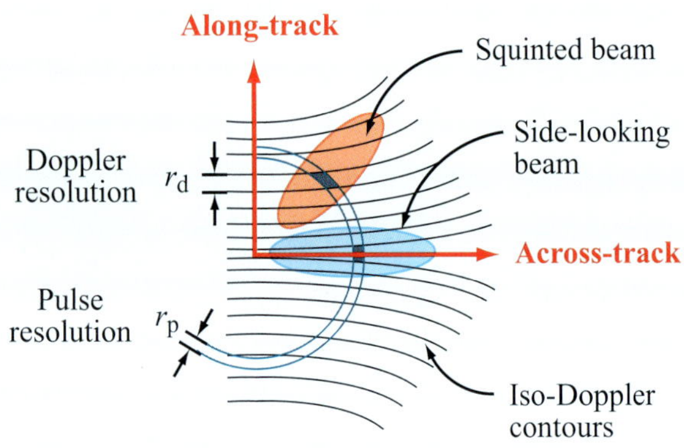

Figure 14-19: Squint-mode geometry.

Because the iso-Doppler and iso-range lines are not orthogonal, the commonly used range-Doppler processing algorithm cannot be used, and more sophisticated algorithms that explicitly account for the range and Doppler coupling must be used instead. We note that "***range walk***" is particularly important in squint-mode SARs. As discussed in Section 14-9.2, range walk occurs when the distance to the target migrates between range resolution bins or cells.

Spotlight-mode SAR

In ***spotlight-mode SAR***, a narrow-beam antenna remains steered toward the target as the aircraft passes by the target so that the target area is illuminated for an extended period, as shown in Fig. 14-18(d). This greatly lengthens the effective aperture, and thus improves the possible resolution at the expense of coverage. The narrow antenna beam has a higher gain than a fixed antenna that would produce the same synthetic aperture and thus provides higher signal-to-noise ratio. Otherwise, the analysis of spotlight SAR is similar to stripmap SAR. In both cases, the length of the synthetic array is limited by coherency over the azimuth angles the target is observed.

Circular SAR

The final SAR imaging geometry we consider is the ***circular SAR*** depicted in Fig. 14-18(e). In circular SAR, the aircraft flies a circular path about the target, which provides continuous observation of the target, which is a form of persistent surveillance. Figure 14-20 shows the geometry of circular SAR in more detail. For an aircraft moving in a circle about the origin $(0,0,0)$ at a speed u, radius r_0, and altitude h, the radar's angular position as a function of time is

$$\theta(t) = ut/2\pi r_0.$$

The time-varying x and y components of the distance from the target point (x_0, y_0) are

$$x(t) = r_0 \cos\theta(t),$$

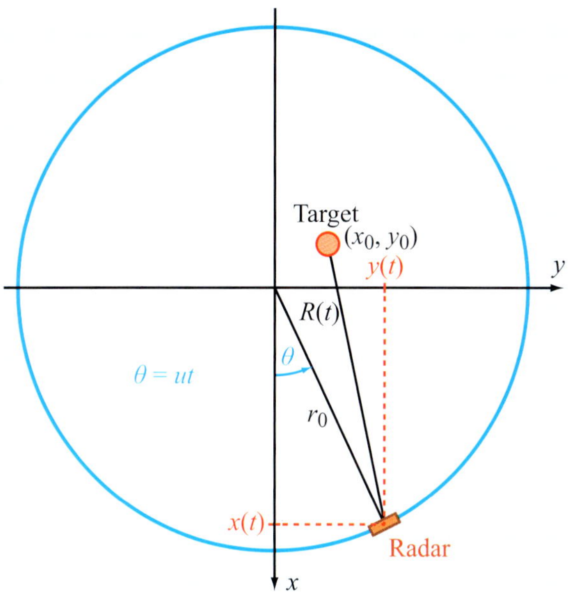

Figure 14-20: Spotlight-mode geometry.

and

$$y(t) = r_0 \sin\theta(t).$$

The distance R between the target and radar is

$$R(t) = \sqrt{(x(t)-x_0)^2 + (y(t)-y_0)^2 + h^2}\,. \qquad (14.96)$$

For simplicity, we consider only a small arc of the circle, which, without loss of generality, is centered at $\theta = 0$. For a small arc, $x(t) \approx r_0$ and $y(t) \approx r_0\,\theta(t) = ut/2\pi$ so that Eq. (14.96) becomes

$$R(t) = \sqrt{(r_0 - x_0)^2 + \left(\frac{ut}{2\pi} - y_0\right)^2 + h^2}. \qquad (14.97)$$

When the circle radius r_0 is significantly larger than the height h, we can use the Taylor's approximation $\sqrt{a^2+b} \approx a + b/2a$ valid for $a^2 \gg b$, which leads to

$$R(t) \approx (r_0 - x_0) + \frac{1}{2}\left(\frac{\left(\frac{ut}{2\pi} - y_0\right)^2 + h^2}{r_0 - x_0}\right)$$

$$\approx r_0 - x_0 + \frac{1}{2}\left(\frac{\left(\frac{u}{2\pi}\right)^2 t^2 - \left(\frac{uy_0}{\pi}\right)t + y_0^2 + h^2}{r_0}\right), \qquad (14.98)$$

where in the last step we used in the denominator the approximation $r_0 - x_0 \approx r_0$ for small angle θ. Grouping constants, this expression has the form

$$R(t) = h'_0 - \frac{uy_0}{2\pi r_0}\, t + \frac{u^2}{8\pi^2 r_0}\, t^2, \qquad (14.99)$$

where $h' = r_0 - x_0 + (y_0^2 + h^2)/2r_0$. The expression given by Eq. (14.99) is a quadratic function of time. The phase of the received signal is $4\pi R(t)/\lambda$, so the received signal will have a quadratic phase chirp whose DC value (h') is related to the x position while the linear phase term (frequency) is a function of the y position. This is similar to the azimuth chirp generated by stripmap-mode operation, and matched filtering can be employed to create an image of the targets within the illuminated area. Note that for a target at the center of the circle, the range is constant and so the matched filter has a constant impulse response. Specialized SAR processing algorithms that exploit the geometry of spotlight-mode and circular SAR have been developed (e.g., Jakowatz et al., 1996; Carrara et al., 1995), but they are not considered in this book.

14-6 SAR Power Considerations

Power considerations for synthetic-aperture radar are similar to those for real-aperture radar, although one can express them in many other ways. In terms of the along-track and across-track resolutions r_a and r_y, the power received by a real-aperture radar for a single pulse can be expressed as

$$P_r = \frac{P_t G^2 \lambda^2 \sigma^0 r_a r_y}{(4\pi)^3 R^4}\,. \qquad (14.100)$$

Since the SNR S_n is often of more interest than the received power *per se*, we equate P_r to the noise in the receiver as was done with SLAR in Eq. (14.10), obtaining

$$S_1 = \frac{P_t G^2 \lambda^2 \sigma^0 r_a r_y}{(4\pi)^3 R^4 k T_0 B F}. \tag{14.101}$$

Note the additional subscript here on S_1, which is used to refer to the SNR for a single pulse. In synthetic-aperture radar, coherent integration takes place over multiple pulses so one must also consider the signal-to-noise ratio for the summation of all of the pulses used in building the synthetic aperture.

The number of pulses in the synthetic aperture is N_s. Since the signals add coherently (and if we temporarily ignore the antenna gain variation with azimuth angle), the signal voltage after summing N_s pulses is related to the voltage received for one pulse V_1^s by

$$V_{N_s}^s = N_s V_1^s. \tag{14.102}$$

Because the noise voltages are random, they do not add coherently and we must use the ensemble average of the square of the noise voltage. Assuming the noise samples are independent, after adding N_s random squared noise voltages, the ensemble noise average is given by

$$\langle (V_{N_s}^n)^2 \rangle = N_s \langle (V_1^n)^2 \rangle. \tag{14.103}$$

▶ Hence, the signal-to-noise ratio after coherently adding N_s pulses is improved over the signal-to-noise ratio for one pulse by a factor N_s:

$$S_{N_s} = \frac{P^s}{P^n} = \frac{(V_{N_s}^s)^2}{\langle (V_{N_s}^n)^2 \rangle} = \frac{N_s^2 (V_1^s)^2}{N_s \langle (V_1^n)^2 \rangle} = N_s S_1. \tag{14.104}$$

◀

It is useful to express N_s in terms of radar-system parameters. First note that N_s is the ratio of the length of the synthetic aperture to the distance traveled between two pulses. That is,

$$N_s = \frac{L}{u T_p} + 1 \approx \frac{L}{u T_p}, \tag{14.105}$$

where L is the length of the aperture, u is the speed, and T_p is the pulse repetition period. Since L is not a fundamental parameter of the system, we use Eq. (14.54) to relate it to r_a to obtain

$$N_s = \frac{\lambda R a_{hs}}{2 r_a u T_p}. \tag{14.106}$$

Note that this expression includes the weighting term a_{hs}, which accounts for the antenna gain pattern.

Combining Eq. (14.106) with Eqs. (14.104) and (14.101) gives the signal-to-noise ratio after N_s pulses:

$$S_{N_s} = \frac{P_t G^2 \lambda^3 a_{hs} \sigma^0 r_y}{2(4\pi)^3 R^3 k T_0 B F u T_p}. \tag{14.107}$$

This equation suggests that the signal-to-noise ratio for a SAR pixel is independent of the along-track (azimuth) resolution r_a. We may process for the finest possible resolution r_{ap} or for some coarser resolution without changing S_{N_s}. This means that the power requirement for the synthetic-aperture radar is independent of r_a, and is determined only by the across-track (range) resolution r_y. On the other hand, the processor complexity is determined by the fineness of r_a. Thus, if one wishes to obtain a constant-area pixel as indicated in the discussion on speckle in Section 14-8, one can make a trade-off between r_y and r_a so long as their product remains the same.

Pulse compression is always used in SAR, so the average power may be more important than the peak power. For a rectangular pulse, the effective pulse duration is τ_p. For another pulse shape or type, we find an equivalent τ_p by noting that the energy in the pulse for a given shape of pulse may be equated to the energy in an equivalent rectangular pulse by

$$P_{t_{max}} \tau_p = \int P_t(t)\, dt. \tag{14.108a}$$

Using this definition for τ_p, one may replace the peak power P_t in Eqs. (14.100), (14.101), and, more particularly, (14.107) by

$$P_t = P_{av}/\tau_p f_p = \frac{T_p P_{av}}{\tau_p}, \tag{14.108b}$$

where f_p is the pulse repetition frequency, P_{av} is the average power, and $f_p T_p \equiv 1$. We may relate the bandwidth in Eq. (14.107) to τ_p through a pulse-taper factor a_B that relates B and T_p:

$$B = \frac{a_B}{\tau_p}. \quad (14.109)$$

Substituting these relations into Eq. (14.107), the SNR becomes

$$S_{N_s} = \frac{P_{av} G^2 \lambda^3 a_{hs} \sigma^0 r_y}{2(4\pi)^3 R^3 k T_0 F u a_B}. \quad (14.110)$$

14-6.1 SAR SNR Equation

According to Eq. (14.110), the SNR is strongly affected by the antenna gain, the wavelength, and the range to the target R.

Sometimes it is important to find the SNR for a system with a given set of specifications, in which case Eq. (14.110) is the most convenient form of the radar equation. However, in many cases it may be necessary to find the required average power for specified values of the SNR and other system parameters. For a synthetic-aperture radar, it is quite often more convenient to use the area of the antenna rather than the gain, because the individual components of the area relate to the resolution and the swath width. The effective area A_e for an antenna with gain G is

$$A_e = G\frac{\lambda^2}{4\pi}.$$

Using this expression in Eq. (14.110) and then solving for the average power P_{av} leads to

$$P_{av} = \frac{S_{N_s} 8\pi R^3 k T_0 F u \lambda a_B L_F}{A_e^2 a_{hs} \sigma^0 r_y}. \quad (14.111)$$

One additional important factor is included in Eq. (14.111), namely L_F, the ***loss factor*** that accounts for RF losses and the fact that the gain is not the same everywhere in the beam as it is at the beam center.

As a numerical example consider two radars, one on an aircraft and the other on a spacecraft, both with the same losses, noise figure, signal-to-noise ratio, and wavelength:

$$S_{N_s} = 4 \text{ (6 dB)}, \qquad F = 4 \text{ (6 dB)},$$
$$L_F = 4 \text{ (6 dB)}, \qquad \lambda = 0.03 \text{ m},$$
$$a_{hs} = a_B = 1, \qquad \sigma^0 = 10^{-2}.$$

The different speeds and altitudes of the aircraft and spacecraft radars lead to significantly different minimum area and power requirements:

	Aircraft	Spacecraft
R	20 km	1000 km
u	300 m /s	7.5 km /s
A_e	0.1 m^2	6 m^2
r_y	3 m	30 m
P_{av}	1.5 W	134 W

Note that the area of the spacecraft antenna, 6 m^2, is that required by ambiguity considerations in a middle angular range. Thus, the spacecraft's antenna cannot be as small as that of the aircraft radar. On the other hand, the aircraft radar can use an antenna as large as the spacecraft radar if, for some reason, one wanted to do so. If the aircraft antenna were to have an area of 6 m^2 and resolution r_y of 30 m, the resulting power requirement would be only 43 μW.

Aircraft radars, which operate at shorter ranges and lower speeds, usually use relatively small antennas to obtain fine resolutions, whereas spacecraft radars, operating at longer ranges and higher speed, are required to use larger antennas by both power and ambiguity considerations. Because of the high speed of the spacecraft and the consequent coverage rate for the radar mapping, a coarser resolution is also normally used.

14-6.2 Radiometric Resolution

The radiometric resolution of a SAR is often specified in terms of the ***noise equivalent*** σ^0 (NES). The NES is defined as the σ^0 value that yields an image pixel SNR of 1 (0 dB). Setting $S_{N_s} = 1$ in Eq. (14.110) and solving for σ^0, the NES is given by

$$\text{NES} = \frac{2(4\pi)^3 R^3 k T_0 F u a_B}{P_{av} G^2 \lambda^3 a_{hs} r_y}. \quad (14.112)$$

Note that the NES may vary in range since the SNR is range dependent.

While the requirements for the NES are application dependent, typical design values of the NES are −25 dB to −30 dB. Targets with σ^0 less than the NES cannot be reliably imaged since the pixel intensity value, which ideally is proportional to the σ^0 of the corresponding ground cell, is instead dominated by the system noise floor.

Two other parameters of interest in SAR system design are the spatial ***impulse response function***, sometimes abbreviated as IPR, and the ***integrated sidelobe ratio*** (ISLR). The IPR is the normalized system measurement response function for σ^0, and is a function of the SAR processing applied to the data, including any azimuth and range windows employed. The effective resolution of the SAR image is defined as the 3 dB dimensions of the IPR for one pixel. Note that signal processing windowing used in the SAR processing can be used to trade-off IPR sidelobes and the width of the IPR mainlobe: as sidelobes are reduced, the width of the mainlobe grows.

Typically, the IPR is computed in a small area centered over a particular pixel. It is typically normalized to a maximum value of one. The SAR processor processes and accumulates the pulses that illuminate the area. The pixel value is thus a weighted sum of the radar returns. The IPR is the result of the effective weighting in the SAR processor, the properties of the transmitted signal, and the measurement geometry.

> ▶ As the SAR processor impulse response, the IPR can be visualized as a two-dimensional plot of the output of the SAR processor for the pixel when the surface σ^0 is a delta function at the pixel center, as illustrated in Fig. 14-21. ◀

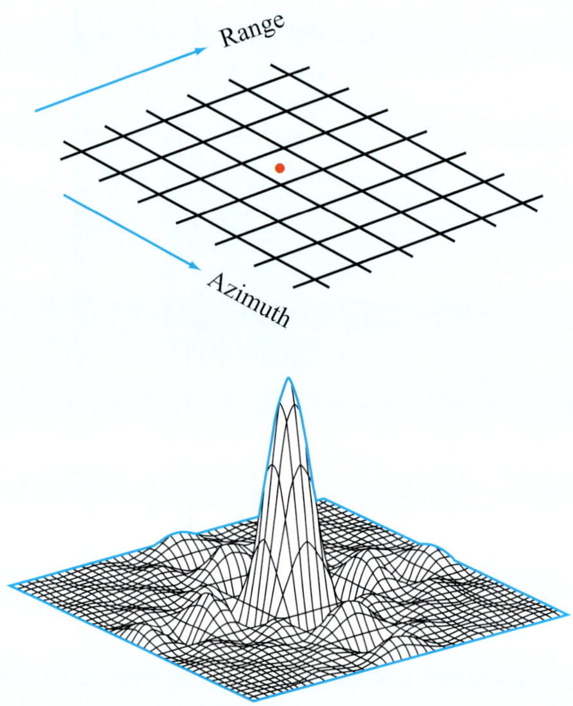

Figure 14-21: Illustration of an impulse response function for a SAR image.

The grid shown in the figure represents the SAR image pixel grid, with the red dot in the center being the location of a corner reflector. Below it is an example IPR function for this area.

In practice, the IPR is often measured by placing a corner reflector or an active radar calibrator in a low-backscatter region of the image and computing a SAR image centered about the reflector. A sufficiently large reflector provides a very bright point-like radar target. The image response is the IPR, which often is computed at a higher than normal resolution or interpolated to reveal details as shown in Fig. 14-22.

The ***integrated sidelobe ratio*** is defined as the ratio of the system gain over the sidelobes of the pixel response to the total integral of system response:

$$\text{ISLR} = \frac{\displaystyle\int_{\text{sidelobes}} \text{IPR}(x,y)\,dx\,dy}{\displaystyle\int_{\text{everywhere}} \text{IPR}(x,y)\,dx\,dy}. \qquad (14.113)$$

In a well-designed SAR, the ISLR is −13 dB or lower.

14-7 SAR System Configurations

SAR systems require coherent operation over the length of the synthetic aperture. The block diagram of a SAR

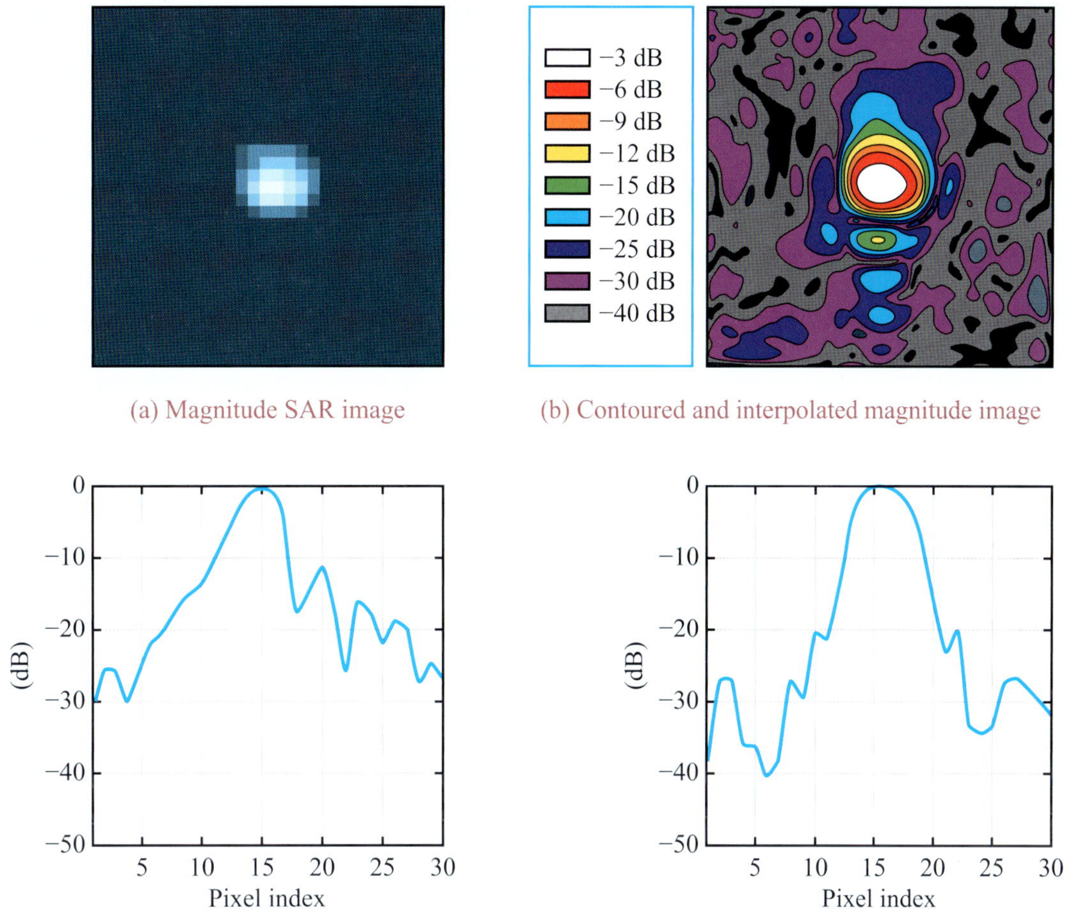

(a) Magnitude SAR image

(b) Contoured and interpolated magnitude image

(c) Vertical slice through peak response

(d) Horizontal slice through peak response

Figure 14-22: Example of a nonideal SAR IPR computed from a corner reflector in a dark scene: (a) magnitude image of area around reflector in a linear scale with the peak normalized to one, (b) contour plot of IPR, (c) plot of estimated IPR through a vertical (range) cut through the peak, which is the effective range compression response, (d) plot of estimated IPR through a horizontal cut through the peak, which is the effective azimuth response.

is similar to other pulse-Doppler radars such as ***moving target indicator*** (MTI) systems. Pulse compression is used for most applications of SAR so that the peak power may be kept down while obtaining a very fine range resolution. The key difference between SARs and other coherent radars lies in the processor rather than in the radar hardware.

Figure 14-23 shows a conceptual block diagram of a modern SAR. A key requirement of the hardware is maintaining phase and frequency coherence. All frequencies in the system are locked to a very stable oscillator at the intermediate frequency. For transmitting, a chirp generator produces a complex (I/Q) pulse with the desired timing and amplitude envelope. The pulse amplitude envelope may be shaped (windowed) or it may be a rectangular pulse. The chirp is fed to a ***single-sideband*** (SSB) mixer, where it is mixed with the local-oscillator signal to produce a pulse

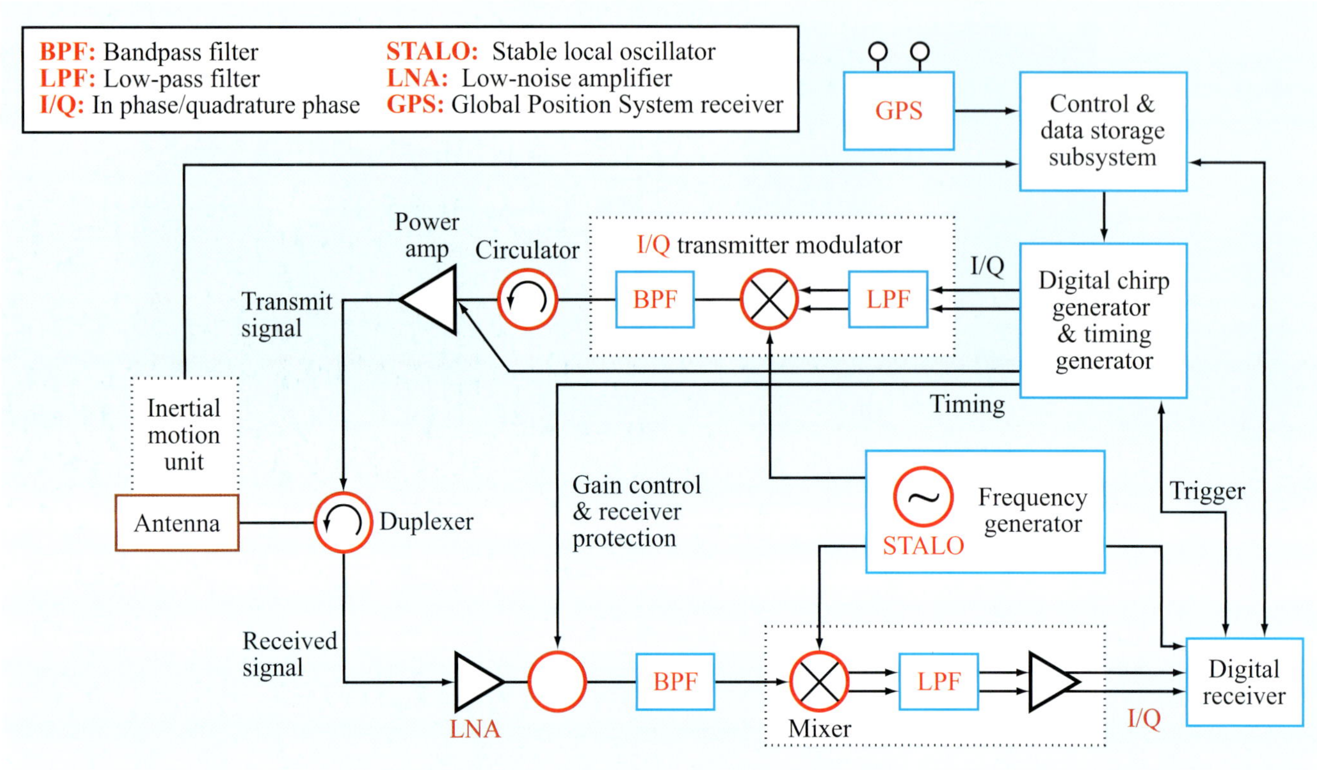

Figure 14-23: Basic block diagram of a SAR.

at the carrier frequency. The carrier-frequency output of the SSB mixer is then amplified to the required level in the transmitting RF power amplifier and sent to the duplexer. It passes through the duplexer to the antenna, and the signal is transmitted.

The received signal picked up by the antenna passes through the duplexer to the low-noise amplifier at the carrier frequency, and is mixed down to the intermediate frequency in the first mixer. The IF signal is amplified and digitally sampled. Information on antenna attitude and position is also collected.

The system shown is a zero-IF system, so it has SSB mixing with two mixers—an in-phase or I-mixer, and a quadrature or Q-mixer. These mixers combine the IF output of the receiver with an intermediate-frequency LO signal to beat the carrier signal down to zero frequency and the Doppler shifts down to frequencies near zero. In this regard the SAR system is just like any other coherent pulse radar. However, some SARs include circuitry for pre-correcting the IF due to motion errors, while others perform the correction as part of the processing. A more detailed SAR block diagram that includes such an error-correction system is shown in Fig. 14-24. The error-correction system operates on the second local oscillator (at IF) unit of the receiver. Such a correction system can also operate on the first receiver LO, and frequently the correction is accomplished in the processor.

For the system shown in Fig. 14-24, an error-frequency oscillator (typically at a very low frequency) has its output mixed with the IF signal from the stable IF oscillator. The result is an IF signal that has been shifted so that errors in velocity and attitude are converted to zero. The error-offset oscillator receives signals associated with both attitude and velocity errors. The velocity errors are obtained from integrating

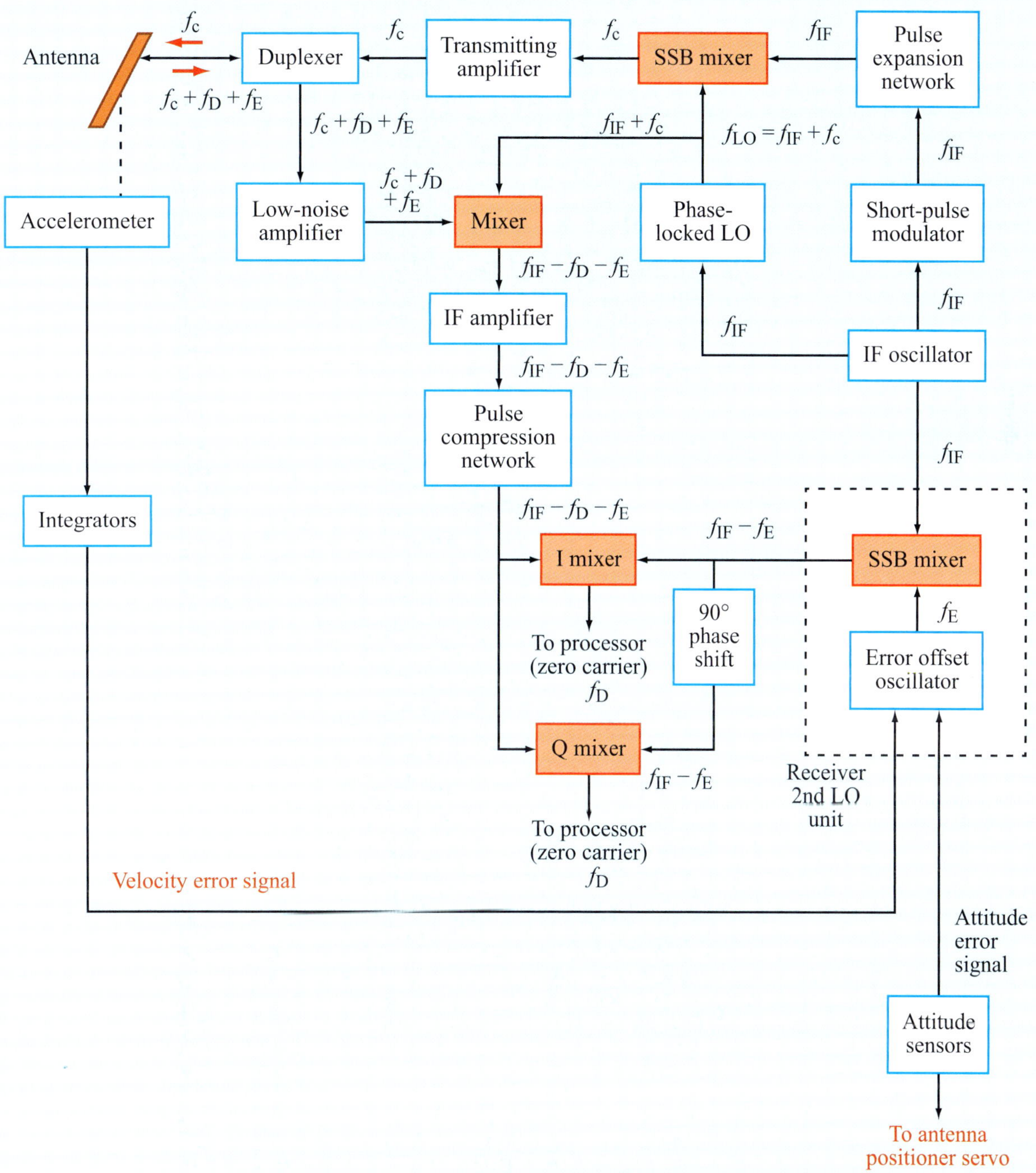

Figure 14-24: Block diagram of a SAR including IF error offset correction.

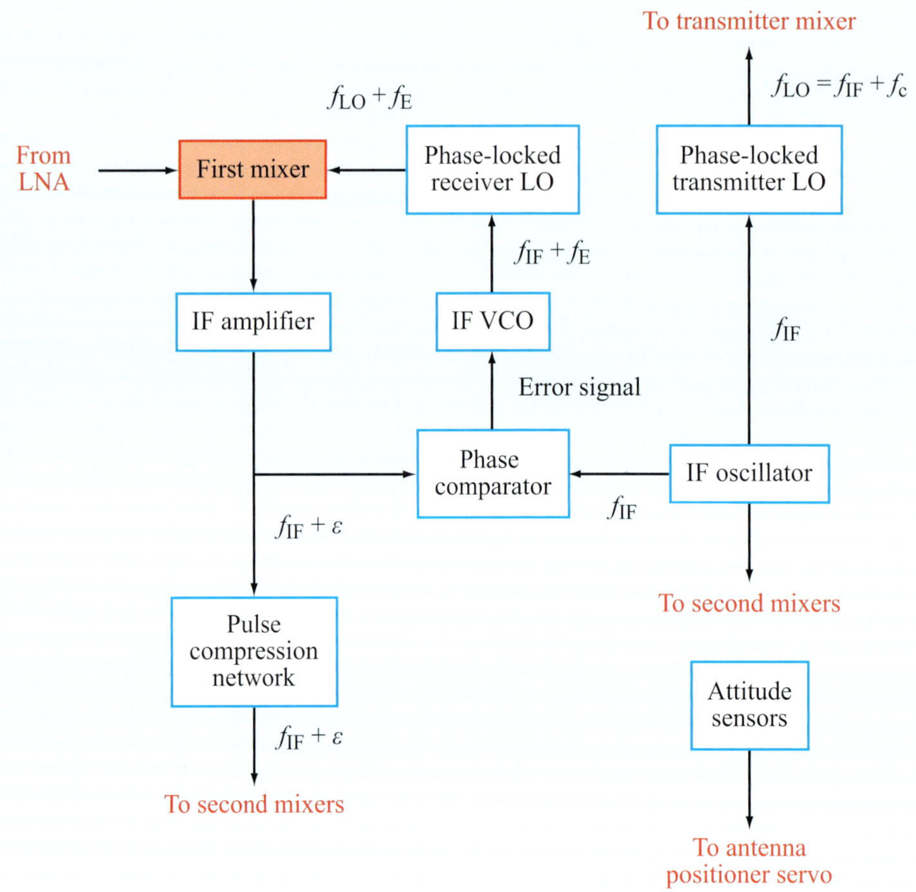

Figure 14-25: Clutter-lock system—IF implementation. f_E is the error offset frequency due to velocity and attitude errors, and ε is the frequency error in the phase-locked loop. For high loop gain, ε should be very small.

accelerometers mounted on the antenna, and the attitude errors are obtained from the aircraft navigation system. The attitude-error sensors produce signals that can make rapid corrections through the error-offset oscillator and slower corrections by repositioning the antenna using the antenna-positioner servo. Thus, the antenna-positioner servo serves the purpose of keeping the antenna attitude errors small, whereas the error-offset oscillator completes the correction. Note that the SSB mixer and error-offset oscillator are shown as the receiver second LO unit. When a clutter-lock system as described below is used, this second LO unit is not required.

The ***clutter-lock*** system uses the centroid of the received Doppler spectrum to adjust the LO frequency so that the IF carrier is located at the center of the IF Doppler spectrum rather than offset from it. Figure 14-25 illustrates such a system. The difference from the radar shown in Fig. 14-23 is in the form of the first local oscillator of the receiver. In Fig. 14-23 the receiver first LO is the same as that providing the transmitter signal. In Figure 14-25 a different receiver LO is used whose frequency is adjusted by a phase-locked loop so that the center of the IF Doppler spectrum is at the frequency of the IF oscillator, f_{IF}. This is accomplished by feeding the output of the IF amplifier to

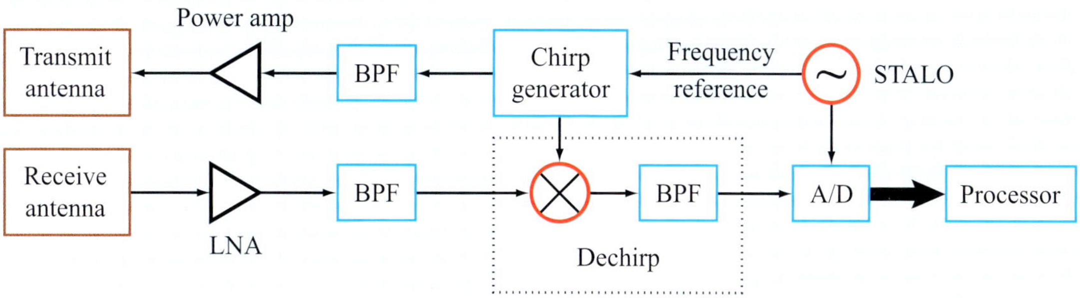

Figure 14-26: Functional block diagram of an LFMCW SAR.

a phase comparator along with the signal from the stable IF oscillator. An error signal is produced that adjusts the frequency of the IF ***voltage-controlled oscillator*** (VCO), which, in turn, adjusts the higher frequency of the phase-locked receiver LO. Adjusting this frequency in the proper manner reduces the difference between the centroid of the IF Doppler spectrum and the frequency f_{IF} to a minimum. Note that the clutter-lock phase comparator occurs before the pulse compression network, thereby allowing the relatively long pulse that contains information from many different ranges, to be used for adjusting the local oscillator. If the output of the pulse-compression network were used, the result would have to be integrated over many range gates to achieve the same stability and independence of individual-range target returns. If the pulse compression ratio is not large enough, such integration may be necessary even with the system shown in Fig 14-25. Of course, the time constant for operation of the phase-locked loop has to be adjusted to a reasonable level so it does not jitter with small variations in target-signal characteristics. A clutter-lock system can also be implemented completely in the processor. Details of the subsystems contained in the boxes of Figs. 14-23 through 14-26 are beyond the scope of this book.

For a low-cost, modest-performance SAR, a linear frequency-modulated continuous wave (LFMCW) system offers a simple design. A block diagram of an LFMCW SAR is shown in Fig. 14-26. Unlike other SAR designs, an LFMCW SAR continuously transmits and receives signals, though this complicates the receiver design due to leakage of the transmit signal into the receiver. Typically, separate transmit and receive antennas are used to maximize the transmit/receive isolation. By employing an LFMCW signal, the SNR is maximized, which can permit the use of a low-power transmitter. In Fig. 14-26 the transmitted LFM signal is mixed with the received signal to dechirp it. While intermediate stages can provide better performance, this simple homodyne design simplifies the hardware by lowering the baseband sampling frequency.

We note that most SAR systems operate in a pulsed mode in which the SAR transmits a short pulse, then waits for the echo. For an aircraft SAR, typically only a single pulse is in flight at a time. A spacecraft SAR, however, may have multiple pulses in flight. In both of these cases, due to the short pulse length, the movement of the SAR during signal transmission or during echo reception from the target is so small that it can be neglected. Thus, the system can be treated as stationary during the pulse (except for computation of the Doppler shift). This is known as the ***stop-and-go*** or ***stop-and-hop*** approximation. The stop-and-go approximation does not apply to LFMCW systems where transmission and reception are continuous and a more detailed analysis is required.

In a chirped LFMCW transmit signal, the frequency of the signal increases from a starting frequency f_0 and spans the bandwidth B at the ***chirp rate*** $\alpha = 2B/T_{\mathrm{R}}$, where T_{R} is the chirp repetition period. The frequency then ramps back down, as shown in Fig. 14-27. This up-down cycle is repeated at the pulse repetition interval

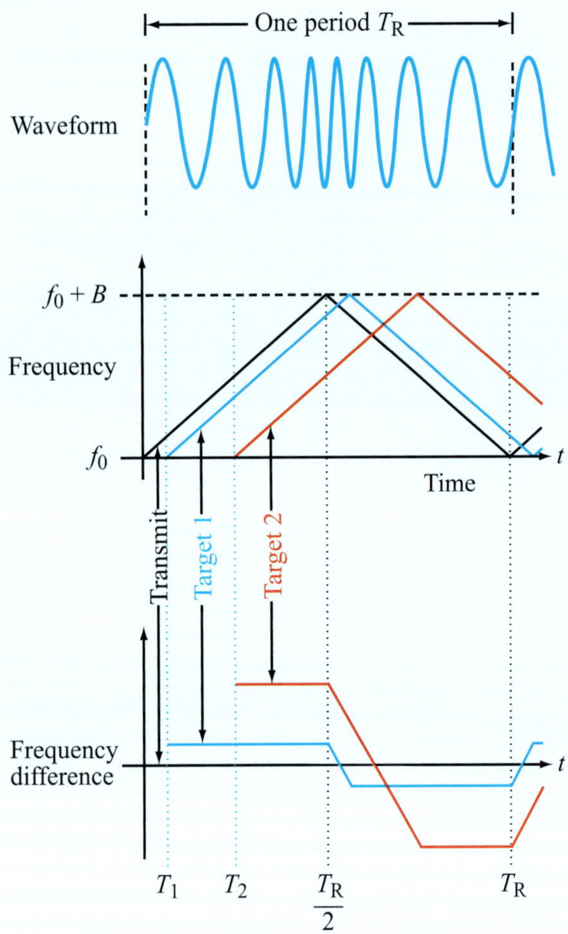

Figure 14-27: Frequency versus time for an LFMCW SAR.

of T_R. Using complex notation, the transmitted up-chirp signal can be expressed as

$$V_t(t) = e^{j(\phi + 2\pi f_0 t + \pi \alpha t^2)}, \tag{14.114}$$

where t is fast time and ϕ is the initial phase. This same signal is repeated for each up-chirp. The down-chirp signal can be expressed similarly to the up-chirp signal, but with $f_0 + B$ as the starting frequency and $-\alpha$ as the chirp rate.

The range R to the target is

$$R(t,\eta) = \sqrt{R_0^2 + u^2(t+\eta)^2}\,, \tag{14.115}$$

where R_0 is the range of the closest approach to the target, $\eta = NT_R$ with N being the slow-time chirp number, T_R the chirp interval, and u the along-track velocity. The received signal is delayed by the time-of-flight delay T:

$$T = \frac{2R(t,\eta)}{c} \tag{14.116}$$

so that the received signal for a single chirp can be written as

$$V_r(t,\eta) = e^{j(\phi + 2\pi f_0(t-T) + \pi\alpha(t-T)^2)}, \tag{14.117}$$

while remembering that T is a function of time.

The transmit signal is mixed with the received signal and low-pass filtered. This is equivalent to multiplying Eq. (14.117) by the complex conjugate of Eq. (14.114). This dechirping operation removes the carrier and results in the ***dechirped signal***

$$\begin{aligned} V_{dc}(t,\eta) &= e^{-j(2\pi f_0 T - \pi\alpha T^2)} \cdot e^{-j2\pi\alpha T t} \\ &= e^{-j\phi'} \cdot e^{-j2\pi f_{dc} t}, \end{aligned} \tag{14.118}$$

which is a sinusoid whose frequency $f_{dc} = \alpha T$ is proportional to range. Range compression is completed by taking the Fourier transform of the dechirped signal. This is typically done with a Fast Fourier Transform (FFT) after applying a signal processing window to suppress range sidelobes. Each output bin of the FFT corresponds to one range resolution. The range resolution is proportional to the inverse of the chirp bandwidth $1/\alpha T_p$. Motion of the platform during the time of flight delay has the effect of shifting the target in range, which can be corrected by applying a slow-time frequency chirp (de Wit et al., 2006).

14-8 Speckle in Radar Images

As discussed in Section 5-6 and illustrated in Section 5-8, fading and speckle are fundamental issues in radar. It is particularly important in radar imaging. The effect of the random fluctuations in the return signal observed from a distributed target is to produce noise-like speckle in the image. As the radar moves past a target pixel, it may only obtain a single independent "look" at the pixel, or multiple looks spaced far enough apart so that they represent independent samples. If only a single look is obtained at each pixel, the pixel brightness has a Rayleigh or exponential distribution, depending upon whether amplitude or square-law equivalent detection is used. Further, if the receiver is nonlinear in some way other than square-law, the distribution may take some other form. If only one independent sample is observed, the speckle makes it difficult to determine with any confidence the value of the scattering coefficient for that particular pixel. As demonstrated in Section 5-8, the adverse effects of speckle can be ameliorated to some extent by averaging nearby pixels, though this has the side effect of reducing the image resolution. If a target area is covered by multiple pixels, the pixel average has less variance, which results in a better scattering coefficient estimate. Pixel averaging or low-pass filtering can improve the image interpretability. In the following, we separately consider speckle in SLAR and SAR images.

14-8.1 Speckle in SLAR Images

Since SLAR images tend to have low resolution, in most cases an individual pixel in a SLAR image represents an average of several independent samples. As more independent samples are averaged for each pixel, the degree of speckle decreases. If enough independent samples are averaged into the reported pixel value, an individual pixel yields a good estimate of the scattering coefficient for the underlying terrain. Note that averaging employed to reduce the effects of speckle also averages the noise, which improves the SNR. Hence, precise estimates of the scattering coefficient can be made in the presence of noise if enough independent samples are averaged together, even if the noise level is high.

In this section, we treat two problems in SLAR: calculation of the number of independent samples per pixel and methods for increasing the number of independent samples averaged. To show the number of independent samples per pixel for a SLAR image, we note that the Doppler frequency shift for the backscatter from a target strip at a location x from the side-looking position (Fig. 14-28) is

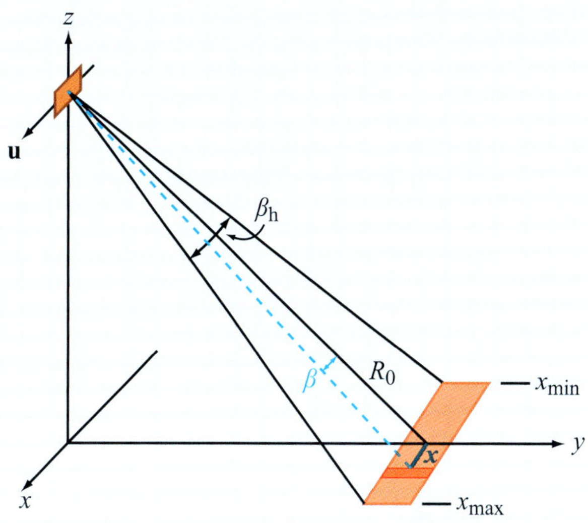

Figure 14-28: Geometry of SLAR fading.

$$f_D = \frac{2u}{\lambda}\sin\beta = \frac{2ux}{\lambda R_0}, \tag{14.119}$$

where u is the velocity of the aircraft, λ is the wavelength, and R_0 is the slant range. The maximum value taken on by x within the beam is

$$x_{max} = \beta_h R_0/2,$$

where β_h is the beamwidth of the antenna in the horizontal direction. Substituting this value into Eq. (14.119), the maximum Doppler frequency is

$$f_{Dmax} = \frac{u\beta_h}{\lambda} = -f_{Dmin}. \tag{14.120}$$

Note that this is also the magnitude of the minimum Doppler frequency f_{Dmin}, since the minimum value is equal in magnitude but opposite in sign to the maximum value. Hence, the ***Doppler bandwidth*** associated with the SLAR is

$$\Delta f_D = 2f_{Dmax} = \frac{2u\beta_h}{\lambda}. \tag{14.121}$$

The time for observing this fading signal, T, is the time that the target remains within the beam for a given point target, which is related to the along-track resolution r_a,

$$T = \frac{r_a}{u} = \frac{\beta_h R_0}{u}. \tag{14.122}$$

Since, as shown in Chapter 13, the number of independent samples is approximately equal to the product of the Doppler bandwidth and the time of observation, the number N_a of independent along-track samples is

$$N_a = T\,\Delta f_D = \frac{2\beta_h^2 R_0}{\lambda}. \tag{14.123}$$

This expression can be written as

$$\begin{aligned} N_a &= \left(\frac{2\beta_h}{\lambda}\right)(\beta_h R_0) \approx \left(\frac{2\lambda}{\lambda l}\right)(r_a) \\ &= \frac{2r_a}{l} = \frac{r_a}{l/2}, \end{aligned} \tag{14.124}$$

where for simplicity we have used $\beta_h = \lambda/l$, though we note that the effective beamwidth is a function of the antenna taper.

Equation (14.124) is very fundamental and applies to synthetic-aperture radar as well.

▶ This result states that the number of independent samples available to be averaged in the along-track direction is essentially equal to the ratio of the along-track resolution r_a to half the antenna length $(l/2)$ in the along-track dimension. ◀

Thus, if one can achieve a 30 m resolution with a 3 m antenna, the (approximate) number of independent samples N_a is 30/1.5, or 20. Figure 14-29(a) illustrates this averaging. For a pixel whose dimensions are $r_a \times r_r$, one can think of the along-track resolution r_a as being achieved by superimposing single independent-sample subpixels of width $l/2$, indicated by the vertical dashed lines in the figure.

As an example of the number of independent samples obtainable and the resulting relative standard deviation for the measurement for each pixel, consider the following SLAR numerical design example that operates at two frequencies with the following characteristics:

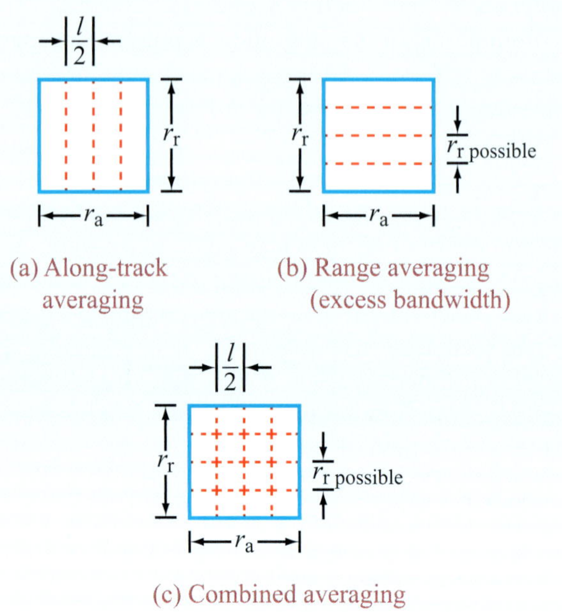

Figure 14-29: Ways to obtain averaging.

$$l = 5 \text{ m}$$

$$\begin{aligned} \lambda_1 &= 1 \text{ cm}, & \lambda_2 &= 3 \text{ cm}, \\ \beta_{h1} &= 0.002 \text{ rad}, & \beta_{h2} &= 0.006 \text{ rad}, \\ r_{a1} &= 2R_{km} \text{ m}, & r_{a2} &= 6R_{km} \text{ m}, \end{aligned}$$

where R_{km} is the slant range measured in kilometers. Using these values in Eq. (14.124), we find that

$$N_{a1} = \frac{2R_{km}}{2.5} = 0.8R_{km}$$

and

$$N_{a2} = \frac{6R_{km}}{2.5} = 2.4R_{km}.$$

From Eq. (5.85), the standard deviation-to-mean ratio of an N_a-look intensity image is given by

$$\frac{s}{\mu} = \frac{1}{\sqrt{N_a}}.$$

We may use these equations to calculate the resolution, the number of independent samples, and the relative standard deviation for the two radar wavelengths operating at two distances, 5 and 20 km:

	$\lambda_1 = 1$ cm		$\lambda_2 = 3$ cm	
R	5 km	20 km	5 km	20 km
r_a	10	40	30	120
N_a	4	16	12	48
s/μ	0.5	0.25	0.29	0.14

These values assume that all possible samples have been averaged. Note that the longer-wavelength system has a poorer along-track resolution than the shorter-wavelength system, but that the normalized standard deviation, and consequently the precision of measurement, is significantly better for the longer-wavelength system because of the larger number of independent samples.

Speckle can be further reduced by averaging independent samples in the range direction.

> ▶ ***Range averaging*** can be done after image formation, or the number of independent samples may be increased by transmitting more bandwidth than required for the final range resolution. This is known as ***excess range bandwidth***. ◀

Suppose a pulse of length $\tau_r = 1/B_r$ is transmitted. The possible slant-range resolution is then

$$r_{\text{r possible}} = \frac{c\tau_r}{2} = \frac{c}{2B_r} \,.$$

Now consider integrating N_r returns with this resolution, thereby obtaining a poorer resolution in range but averaging N_r independent samples. The equivalent pulse length associated with the integrated result is

$$\tau = N_r \tau_r .$$

Since the actual range resolution is $c\tau/2$, this resolution can be written as

$$r_r = \frac{cN_r}{2B_r} \,.$$

This situation is illustrated in Fig. 14-29(b), where the four possible range resolution cells are integrated to form one larger pixel with resolution r_r. It follows that one can express the number of independent samples in the range direction in terms of either the effective pulse duration or the bandwidth:

$$N_r = \frac{r_r}{r_{R\text{ possible}}} = \frac{\tau}{\tau_r} = \frac{B_r}{B} \,, \tag{14.125}$$

where B is the reciprocal of τ, the pulse length after integration. The bandwidth needed to achieve the resolution r_r is B, and the actual bandwidth used is B_r. Thus, N_r can be described as equal to the ***excess-bandwidth ratio***. Transmitting the excess bandwidth allows increasing the number of independent samples averaged.

The excess bandwidth ratio may be achieved by one of three methods:

1. Transmitting a short pulse of duration τ_r and then averaging multiple pixels in the range direction.
2. Transmitting a longer pulse of duration τ but modulating the pulse and detecting without correlation or dechirping.
3. Transmitting a long pulse with frequency modulation and dechirping only partially by using a filter whose delay-versus-frequency characteristic does not completely compensate for the chirp. This method is known as ***stretch processing*** (Skolnik, 1980).

In each of these methods, the averaging to achieve better standard deviation, and hence reduced speckle, is achieved by averaging in range (or, equivalently, range frequency) rather than in the along-track direction. If the averaging in frequency is performed continuously, as with methods 2 and 3 above, the resulting standard deviation depends upon the correlation on the frequency axis of the fading spectrum and upon the actual spectrum of the pulse.

Averaging in range and in azimuth may be combined to give a total number of independent samples

$$N = N_r N_a . \tag{14.126}$$

This situation is illustrated in Fig. 14-29(c), where a total of 16 samples is obtained, 4 by averaging along-track and 4 by averaging in range. In reality, one would not need to explicitly compute the pixels of the size of the subcells indicated in Fig. 14-29(c). The azimuth averaging is achieved naturally because of the wide beamwidth of the SLAR, and the range averaging may be attained by transmitting a wide bandwidth on a relatively long pulse and detecting it without dechirping. It should be noted that the transmission of excess bandwidth requires that the receiver have excess bandwidth; and if the receiver has excess bandwidth, the noise level is increased in proportion to the excess-bandwidth ratio. Hence, use of range averaging requires more power in the transmitter to overcome this noise, whereas use of along-track averaging requires no excess power.

14-8.2 Speckle in SAR Images

The problem of speckle is usually more severe with synthetic-aperture radars because SAR inherently produces pixels with one independent sample per pixel, which results in the worst-case speckle. To reduce speckle, the designer of the SAR processing system must specifically take into account the need for averaging multiple independent samples.

The speckle reduction schemes discussed previously for SLAR can be used for SAR. However, SAR processing provides additional options for speckle reduction, including high-resolution pixel averaging and ***subaperture processing***. Figure 14-30 illustrates some examples of the two approaches. In Fig. 14-30(a), a scale drawing compares the length of the effective antenna aperture and the along-track resolutions, ranging from 2 m for one independent look to 16 m for eight independent looks. In Fig. 14-30(b) the method of processing for the finest resolution $r_{\rm ap}$ and then averaging is illustrated. To do this, multiple along-track apertures must be processed simultaneously, each one resulting in a 2 m pixel for one of the four pixels shown. In Fig. 14-30(c) the synthetic aperture $L_{\rm p}$ is divided into four pieces or subapertures. Because the subapertures are shorter, the resolution of their pixels are coarser. Each of the subapertures is processed separately to obtain a pixel with an 8 m along-track dimension. The intensities (magnitude power) for the four pixels are then averaged. The processing involved in the method of (b) uses $4N_{\rm p}$ pulses, whereas each subaperture look in Fig. 14-30(c) uses $N_{\rm p}/4$ pulses for a total of $N_{\rm p}$ pulses. Thus, less processing is required for method of (c) than that of (b). Furthermore, the subaperture processing in (b) might be complicated by dealing with range-curvature and it is more susceptible to range-walk problems than (c). Finally, there is a difference in the ***depth of focus*** for the two cases (discussed in Section 14-9.3). Thus, one normally prefers to use shorter synthetic apertures and add the independent samples, rather than using long apertures and adding the smaller-size pixels. Some users, however, prefer to obtain the smaller pixels on the assumption that for certain applications the highest-resolution result may be superior. It allows for end users to make their own trade-off between resolution and speckle without returning to the raw SAR data.

Figure 14-31 illustrates subaperture processing from the Doppler point of view. A full resolution requires the full bandwidth $B_{\rm d}$. In subaperture images, separate images are created at lower resolution using bandpass filters that span the bandwidth of each subaperture. In this case, four nonoverlapping bands are used to cover the Doppler bandwidth. Incoherently summing them reduces speckle with $N_{\rm a} = 4$.

14-9 Introduction to SAR Processing

SAR processing is very complicated, partly because of the variety of options available and partly because of the complexity of the required calculations. In this section we describe a few of the processing algorithms, along with background information helpful in understanding SAR processing, notably the spectra that arise in the SAR system and the concept of depth of focus. As most SAR processing is now done digitally in computers, we consider only digital SAR processing algorithms. For the purpose of illustration, we present simplified examples of two SAR processing algorithms: the frequency-domain ***range-Doppler algorithm*** (RDA) and the ***time-domain backprojection*** (TDBP) algorithm.

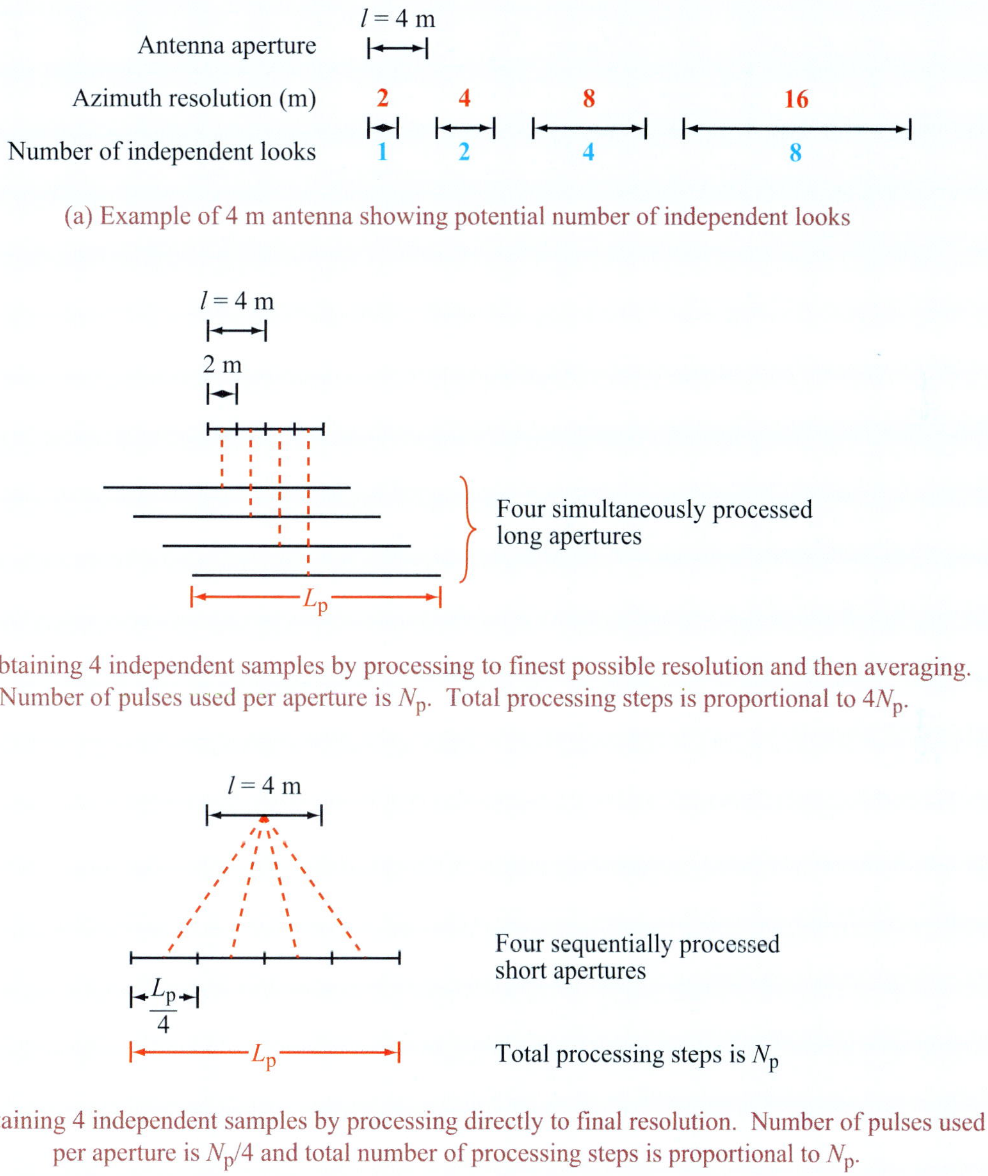

Figure 14-30: Obtaining independent looks with SAR.

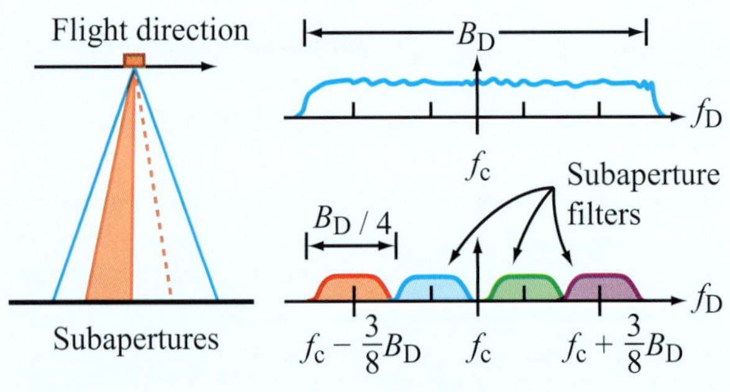

Figure 14-31: Subaperture processing from Doppler point of view.

Additional details on SAR processing are available in books dedicated to SAR processing algorithms (e.g., Cumming and Wong, 2005; Curlander and McDonough, 1991).

In line with common usage, we use the term azimuth instead of "along-track" to refer to the along-track direction and the processing associated with the formation of the synthetic beam. The term ***range processing*** refers to cross-track or perpendicular to along-track and ***azimuth processing*** refers to along-track or azimuth.

Figure 14-32 provides a block-diagram of the steps used in SAR processing. Here part (a) shows a general block diagram for the type of processing in which range compression occurs before azimuth compression, which is most common, but the alternative block diagram of (b) is also used with certain forms of processing. In (a), pulse compression is done first. Following range compression, and possibly using the Doppler-centroid information, the signal undergoes azimuth compression in which some method is used to process the phase-coherent signals to achieve fine azimuth resolution. After azimuth compression, the resulting complex images may be detected by computing the magnitude-square or amplitude of each pixel and then multilook averaging. In the alternative block diagram of Fig. 14-32(b), the only difference is that azimuth compression takes place before range compression.

14-9.1 SAR Signal Spectra

SAR has been described as the "marriage of radar and signal processing." Hence, it is instructive to understand the signal spectra in the synthetic-aperture system.

Different processors operate with different offsets of the carrier frequency from zero. Some processors beat the carrier frequency down to zero, and these must have means for distinguishing positive and negative frequencies. Others use a carrier frequency converted to near zero (***azimuth offset***), and still others use a carrier frequency somewhat higher (***range offset***).

The transmitted radar signal voltage is a sequence of repeating pulses that can be written in the form

$$\upsilon_t(t) = \mathfrak{Re}\left\{\sum_{i=-\infty}^{\infty} V_0\, p(t - iT_p)\, e^{j\omega_c t}\right\}, \qquad (14.127)$$

where V_0 is the peak magnitude and $p(t)$ is the modulation or shape factor (envelope) for the transmitted pulse. The carrier frequency f_c corresponds to an angular carrier frequency $\omega_c = 2\pi f_c$. In the frequency domain this time series may be represented by

$$\mathbf{V}_t(f) = V_0 \sum_{k=-\infty}^{\infty} \mathbf{P}(kf_p - f_c)\, e^{-j2\pi k/T_p}, \qquad (14.128)$$

where f_p is the repetition frequency $(1/T_p)$ and $\mathbf{V}_t(f)$ and $\mathbf{P}(k)$ are the Fourier transforms of $\upsilon_t(t)$ and $p(t)$, respectively. The line spectrum is illustrated in Figure 14-33(a), where both the baseband and RF spectra are shown. The amplitude of the individual lines is governed by the spectrum of the envelope of the transmitted signal $\mathbf{P}(f)$. The baseband signal is shown on the left (centered at zero), and the RF signal is shown on the right as a translation of $\mathbf{P}(f)$ from center frequency zero up to center frequency f_c. For an infinite number of pulses, each line in this spectrum is a continuous sine wave at a particular frequency. If a finite number of pulses are transmitted, each line is then representable by a sinc function $[(\sin x)/x]$ whose width is the reciprocal of the total duration of the pulse train.

Since the transmit signal has a relatively narrow bandwidth compared with the carrier frequency, the Doppler shift of the received signal is essentially the

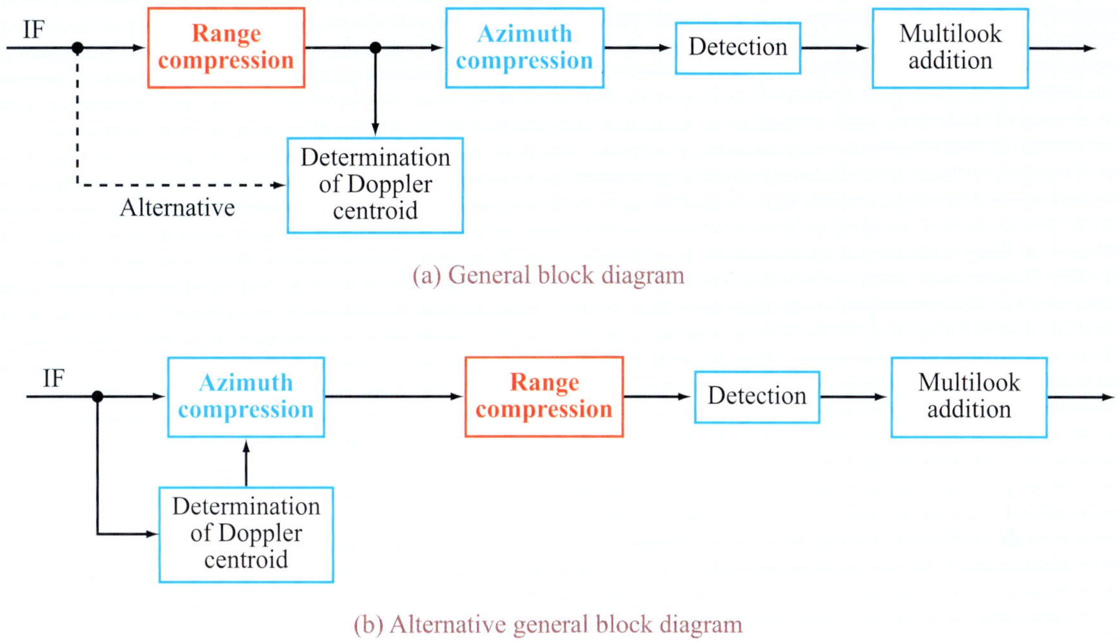

Figure 14-32: Steps in SAR processing.

same for each of the frequencies of interest. (In reality, the Doppler spectra for the different lines differ slightly, but this effect is generally negligible.) Figure 14-33(b) presents an expanded view of the received signals in the neighborhood of the carrier frequency. An asymmetric Doppler spectrum is indicated symbolically by triangles, since one normally expects that there will be different Doppler amplitudes ahead of the sideways direction (positive Doppler shift) and behind the sideways direction (negative Doppler shift).

The SAR data may be collected with different frequency offsets in hardware, but is frequently processed with either one of two different frequency offsets: zero-offset and azimuth-offset. Range-offset is yet another type of offset. We describe each briefly.

Zero-offset processing

Zero-offset processing is commonly used in digital SAR processors and requires complex (I and Q channel) signals and computation. For this case the signal is mixed with a local oscillator at the carrier frequency and the signal centered at zero frequency is retained as indicated in Figure 14-34(a). Zero-offset processing has the advantage of minimizing the sampling rate required for digitizing the data. In Figure 14-34(a) the Doppler-frequency spectra are now centered at zero compared with those in Figure 14-33(b). Each of these asymmetric spectra consists of summations of all of the different target Doppler chirps that are present at any particular instant. In zero-offset processing, normally only the spectral components centered about zero frequency are used, and low-pass filters are used to eliminate all of the components centered about harmonics of the PRF.

The PRF must be greater than or equal to the Doppler bandwidth, as indicated in the discussion on ambiguities in Section 13-9.2. This ensures that the spectral regions associated with the different harmonics of the PRF do not overlap and the Nyquist criterion is satisfied. Unfortunately, Doppler frequencies associated with sidelobes of the horizontal antenna pattern may extend to such high frequencies that they overlap the

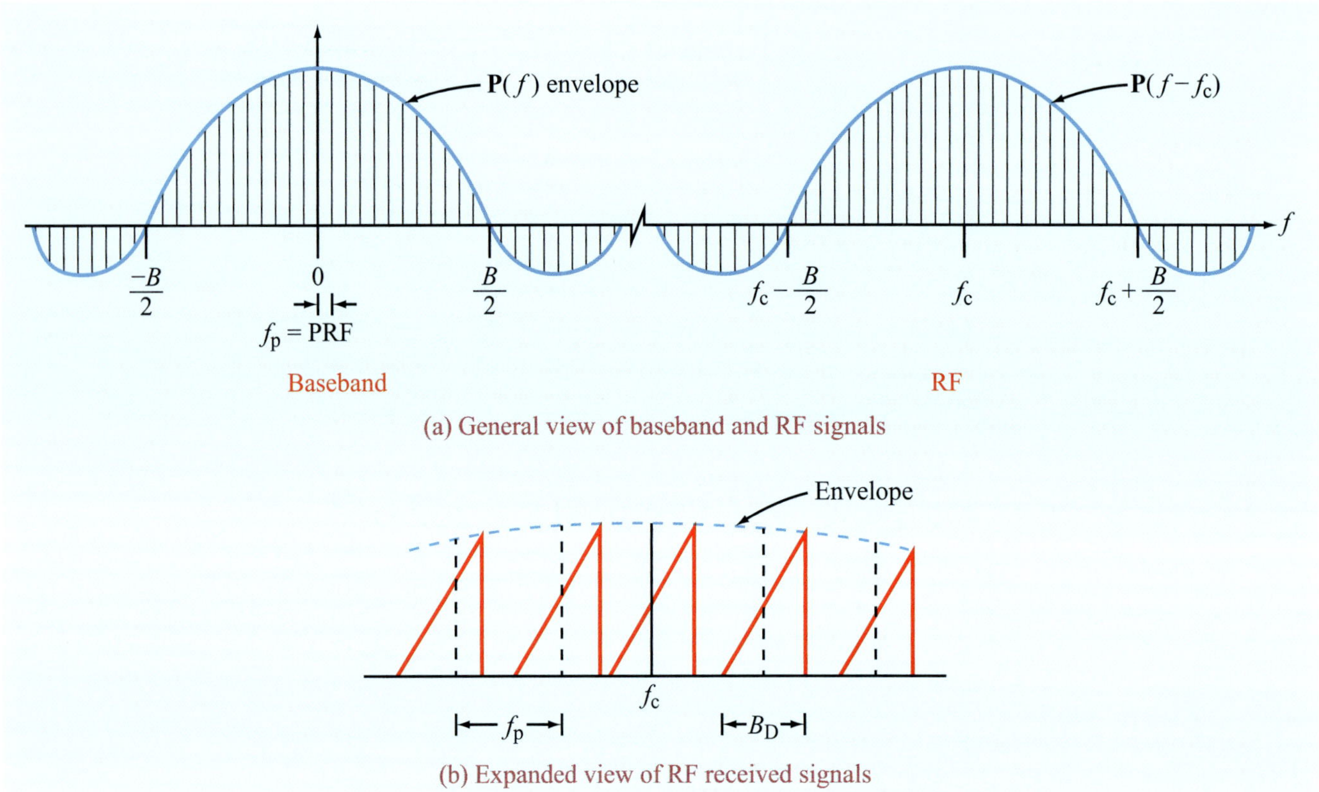

Figure 14-33: Spectra of transmitted and received signals.

frequency range of the mainlobe below and above multiples of f_p. This results in signal "aliasing"; aliased signals produce artifacts in product images.

Azimuth-offset processing

Figure 14-34(b) illustrates the case of azimuth-offset processing. In this situation the offset frequency is just high enough so that the entire Doppler spectrum associated with the carrier is above zero frequency. That is, the offset frequency is selected so that the maximum negative excursion of the Doppler spectrum just touches the zero-frequency point. Hence the offset frequency in this case is of the order of half the Doppler bandwidth. Since higher frequencies are present if the central Doppler spectral region is used, the sampling rate must be twice as high as with zero offset. That is, the Nyquist criterion requires a sampling rate of f_p to accommodate the maximum Doppler frequency, as indicated in Fig. 14-34(b). This type of processing has often been used for optical processing.

Range-offset processing

Figure 14-34(c) shows the third type of offset: range-offset processing. Here the offset carrier frequency is selected at approximately half the bandwidth of the transmitted pulse, so that the entire envelope of the transmitted pulse signal is above zero frequency. Range-offset processing is used when simultaneous processing of range and azimuth chirps is to be accomplished, as is often done in optical processing, and also when the azimuth processing is to be done in a range-sequential manner. The bandpass sampling theorem states that the PRF required to adequately sample this situation need only equal the Doppler bandwidth B_D, provided proper

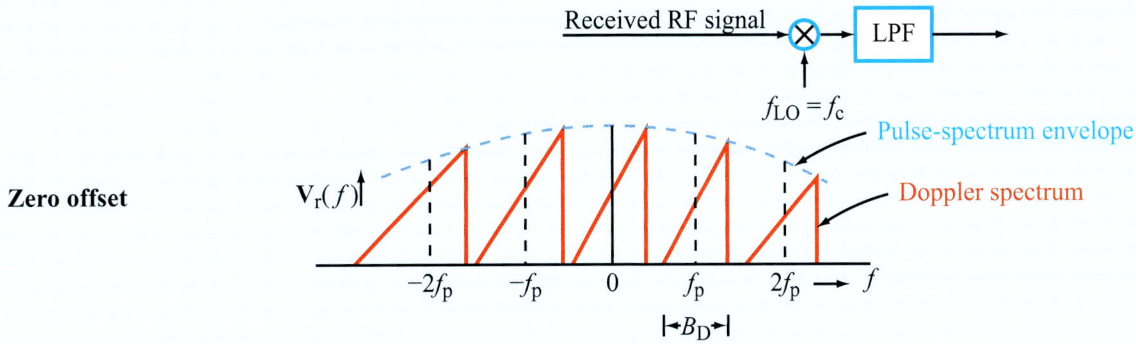

(a) Zero-offset spectrum after mixing. *I*- and *Q*-channels are needed to distinguish positive and negative Doppler frequencies. [$f_p \geq B_D$ to satisfy Nyquist criterion.]

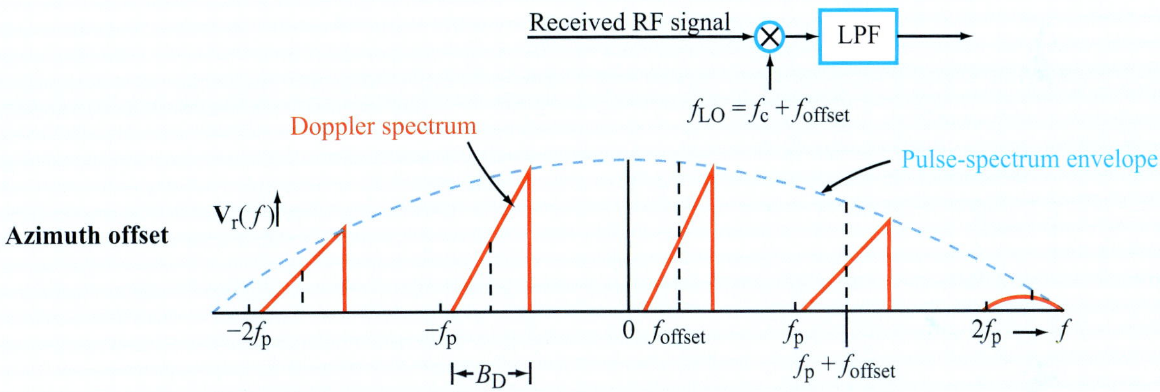

(b) Azimuth offset spectrum after mixing. Only one channel is needed. [$f_p \geq 2B_D$ to satisfy Nyquist criterion.]

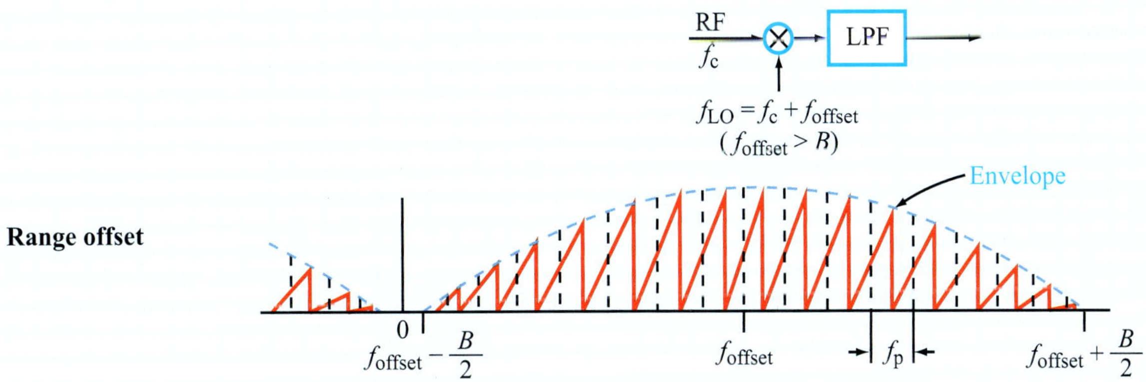

(c) Range-offset spectrum after mixing. The need for *I*- and *Q*-channels depends on processing method. [$f_p \geq B_D$ to satisfy Nyquist criterion.]

Figure 14-34: Spectra associated with frequency offsets.

relations are maintained between the sampling rate and the carrier and pulse-repetition frequencies. Thus the PRF for range offset may be the same as that for zero offset, whereas that for azimuth-offset processing must be twice as great.

14-9.2 Range Migration

Range migration is an effect important to both SLAR and SAR. In stripmap SAR, as the platform flies by the target, the slant range to the target varies with time. Referring to the geometry in Fig. 14-35(a), the range to the target as a function of time is

$$R(t) = \sqrt{y_t^2 + h^2 + x_t^2(t)} \,. \tag{14.129}$$

Thus, $R(t)$ is a hyperbolic function of time; i.e., $R(t)$ versus time traces one side of a hyperbola as illustrated in Fig. 14-35(b). This curve is sometimes called the "***range smile***." The amount of change in range to the target over the period of time of the synthetic aperture is known as the ***range curvature***.

In processing SAR data, we are primarily interested in the portion of time over which the mainlobe of the antenna observes the target. Two cases are shown in Fig. 14-35: side-looking and squint mode. The range varies from R_1, which is the range at the edge of the beam, to R_0, which is the point of closest approach, to R_2, which is the range at the other beam edge. The change in range is R_C for the side-looking mode and R_w for the squint-looking mode. These are the ***range curvatures*** for these cases. For narrow beamwidths or coarse range resolution, R_C may be less than r_y, which means that the target remains in the same range resolution bin. When R_C or R_w are greater than the effective range resolution r_y, the target migrates through multiple range resolution elements or bins. This is termed ***range migration*** or ***range walk***. While range migration affects all SLAR and SAR systems, it is most significant in SAR when the synthetic aperture is long compared with the distance to the target, and also in squint-looking antenna systems.

Some SAR processing algorithms compensate for range migration, while others do not. If the SAR processor assumes that the target remains in a single range bin but there is significant range migration, the image quality is degraded since the target becomes spread over multiple pixels in range.

14-9.3 Depth of Focus

The ***depth of focus*** describes the distance over which an image remains in adequate focus. In synthetic-aperture processing, it describes mismatch tolerance between the assumed target range and its actual range. Within the depth of focus, a fixed reference function corresponding to a point target at a particular range can be used in the processor.

> ▶ The minimum number of distinct reference functions required to focus the SAR image is given by the ratio of the swath width to the depth of focus. ◀

If the depth of focus is great enough, only one reference may be required for all the ranges used, which could greatly simplify the processor: if only a single reference function is required, the phase shifters used in Fig. 14-6(b) may be the same for all swath locations or only a single Doppler sharpening filter or correlation reference function is required. However, if the depth of focus is small, many different reference functions must be used in the processor to ensure the image is properly focused over the full swath.

The depth of focus is also useful for evaluating the sensitivity of the image to errors or uncertainty in the motion of the platform. If the change in range due to nonideal platform motion is less than the depth of focus, the image remains in focus. The idea behind the depth of focus is to define a level of error tolerance in the processing. The error tolerance is often defined in terms of the range curvature R_C, the change ΔR in range between the radar and target over the synthetic aperture. The depth of focus is a span of target ranges ΔR_D such that the difference between the actual target range curvature and the range curvature of the reference target is less than a phase tolerance threshold. The most common criterion is that the phase error over the synthetic aperture be less than $\lambda/4$ or $\lambda/8$.

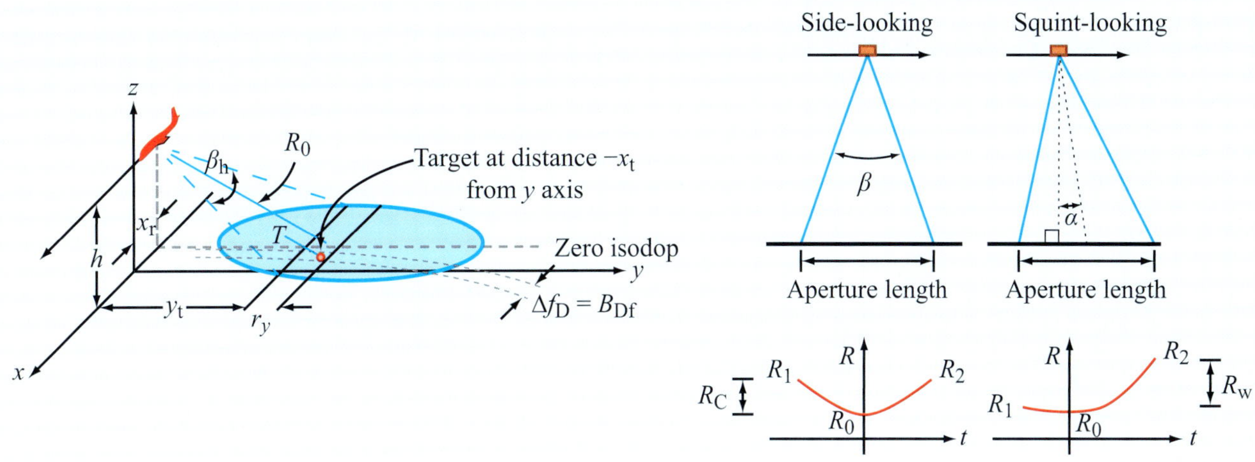

(a) Geometry of SAR Doppler calculations (b) Side-looking mode (c) Squint mode

Figure 14-35: Range curvature with side-looking and squint modes.

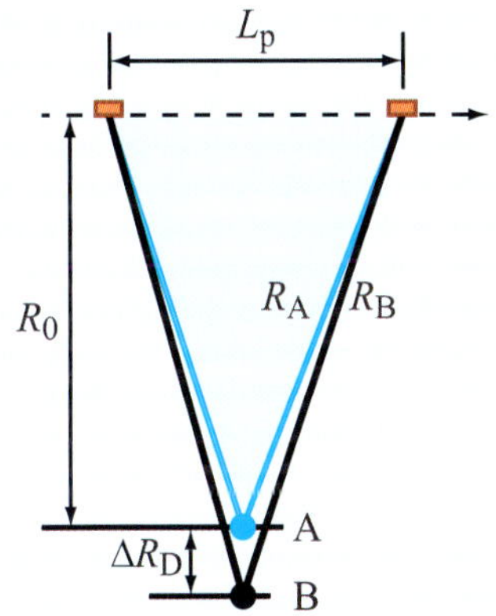

Figure 14-36: Depth of focus geometry. The actual target is at point A. The reference function is computed for the target at point B.

Figure 14-36 illustrates the depth of focus geometry, where one target (A) represents the actual target and the other (B) represents the reference target position. The worst-case range error occurs at the edges of the synthetic aperture. The range curvature for target A is R_{CA}:

$$R_{CA} = R_A - R_0, \tag{14.130}$$

where R_0 is the range of closest approach. In Fig. 14-36, L_p is the maximum possible length of the synthetic antenna. For a side-looking SAR, the point of closest approach occurs halfway through the synthetic-aperture length, so that

$$\begin{aligned} R_{CA} &= \left[\sqrt{\left(\frac{L_p}{2}\right)^2 + R_0^2}\right] - R_0 \\ &= \sqrt{R_0^2\left[\left(\frac{L_p}{2R_0}\right)^2 + 1\right]} - R_0 \\ &= R_0\left[\left(\sqrt{\left(\frac{L_p}{2R_0}\right)^2 + 1}\right) - 1\right]. \end{aligned} \tag{14.131}$$

Using a first-order Taylor series approximation for the square root,* it follows that

$$R_{CA} \approx \frac{L_p^2}{8R_0}. \tag{14.132}$$

*For $c \ll 1$, the first-order Taylor series expansion of $\sqrt{1+c}$ is $\approx 1 + (c/2)$.

The range to the point of closest approach for target B is $R_0 + \Delta R_D$, so R_{CB}, the range curvature for target B, is

$$R_{CB} = R_B - (R_0 + \Delta R_D). \tag{14.133}$$

Then, by the same approximation that led to Eq. (14.132),

$$R_{CB} \approx \frac{L_p^2}{8(R_0 + \Delta R_D)}. \tag{14.134}$$

To write this in a more convenient form, we first write

$$R_{CB} = \frac{L_p^2}{8R_0} \frac{1}{(1 + \Delta R_D/R_0)} \tag{14.135}$$

and use the first-order binomial-series approximation[†] for the second term, which then leads to

$$R_{CB} \approx \frac{L_p^2}{8R_0} - \frac{L_p^2\,\Delta R_D}{8R_0^2}. \tag{14.136}$$

The difference in range curvatures between the actual and assumed target ranges is then

$$R_{CA} - R_{CB} = \frac{L_p^2\,\Delta R_D}{8R_0^2}. \tag{14.137}$$

If the phase error threshold is set at $\lambda/4$, then $R_{CA} - R_{CB}$ should be $\leq \lambda/4$, or

$$\frac{L_p^2\,\Delta R_D}{8R_0^2} \leq \frac{\lambda}{4}. \tag{14.138}$$

The depth of focus is the maximum value of ΔR_D satisfying Eq. (14.138):

$$\Delta R_D = \frac{2\lambda R_0^2}{L_p^2}. \tag{14.139}$$

Using Eqs. (14.48), (14.53), and $r_a \approx l/2$, where l is the length of the real antenna, this can be written as

$$\Delta R_D = \frac{2\lambda}{\beta_{hr}^2} = \frac{2l^2}{\lambda a_{hs}^2} = \frac{8r_a^2}{\lambda a_{hs}^2}. \tag{14.140}$$

(depth of focus)

[†]The first-order binomial expansion of $(1+x)^{-1}$ (with $x^2 < 1$) is $\approx 1 - x$.

Thus the depth of focus is greater for longer real antenna lengths and shorter wavelengths. Degrading the azimuth resolution, which is equivalent to lengthening the antenna, increases the depth of focus.

As a numerical example, suppose that $r_a = 2$ m or 20 m and that $\lambda = 10$ cm, with $a_{hs} = 1$. The depth of focus is 320 m at 2 m resolution and 32 km at 20 m resolution.

14-9.4 SAR Image Processing: The Range-Doppler Algorithm

The range-Doppler algorithm (RDA) is perhaps the simplest and most commonly used algorithm for stripmap SAR processing. It is remarkably robust to motion and velocity errors, though these degrade and distort the image. Figure 14-37 shows a diagram of the signal processing for the RDA. The algorithm separately performs matched filtering in range and azimuth. By implementing the matched filtering using frequency-domain multiplication, the algorithm is computationally

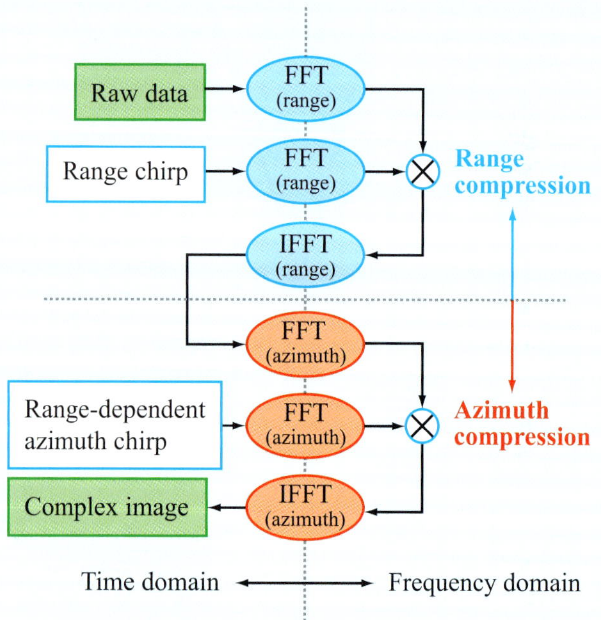

Figure 14-37: Signal-processing diagram for the range-Doppler algorithm.

more efficient than time-domain methods. RDA is an example of the generalized frequency domain processing algorithm (Zaugg and Long, 2009).

As shown in Fig. 14-37, RDA consists of two major steps: range compression and azimuth compression. Range compression is computed in fast time (i.e., within a single pulse) while azimuth compression is computed in slow time (i.e., across pulses). RDA is typically implemented on a block or sequence of pulses. Each pulse is first range-compressed by matched-filtering the digitized receive signal. This is done in the frequency domain by computing the Fast Fourier Transform (FFT) of the received signal and the range reference pulse (signal processing windows may be applied), which are then multiplied in the frequency domain and the inverse FFT (IFFT) is computed. This is equivalent to convolving the received signal and range reference pulse. The resulting signal contains the range-compressed signal magnitude and phase at each range bin or cell.

To illustrate RDA processing, Fig. 14-38(a) contains simulated raw SAR data collected over a scene containing two point targets at different cross-track ranges but the same azimuth position. For display purposes, the signal has been grossly over-sampled. Figure 14-38(b) illustrates the result of range compression. Note that in the range-compressed image, the signal is concentrated along two lines in azimuth at different ranges. Along these "smiles" the azimuth chirps are visible. The zero Doppler signal occurs about halfway along the azimuth axis. The range curvature or range migration is visible as the variation in the energy versus range along the azimuth axis.

RDA can be adapted for LFMCW SAR processing by modifying the range compression step. For dechirped LFMCW, range compression is accomplished by merely computing the FFT of the data.

For the azimuth compression step, RDA implicitly assumes that the pulses are evenly spaced in time and distance so that an FFT can be used. Figure 14-39 diagrams the isodop lines for a side-looking antenna at a given instant. These lines move with the radar as the platform flies in the along-track direction. The horizontal line reveals the track of a fixed target through

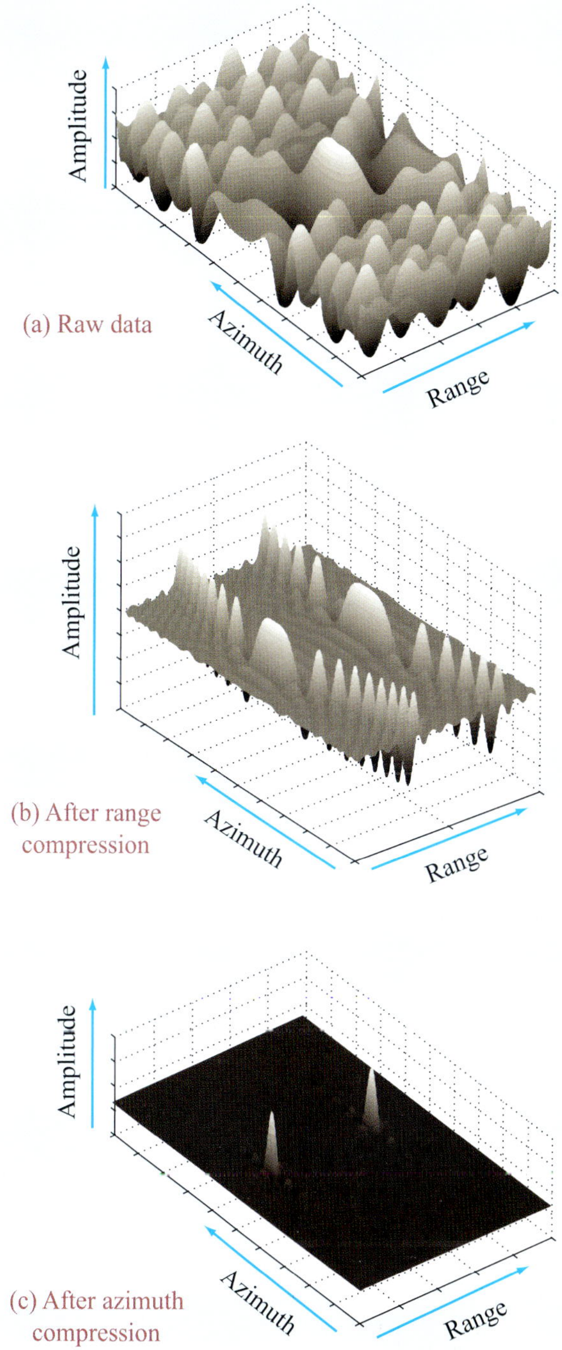

Figure 14-38: Illustration of range and azimuth compression showing the real part of the SAR signal from two point targets in (a) the raw data, (b) after range compression, and (c) after azimuth compression.

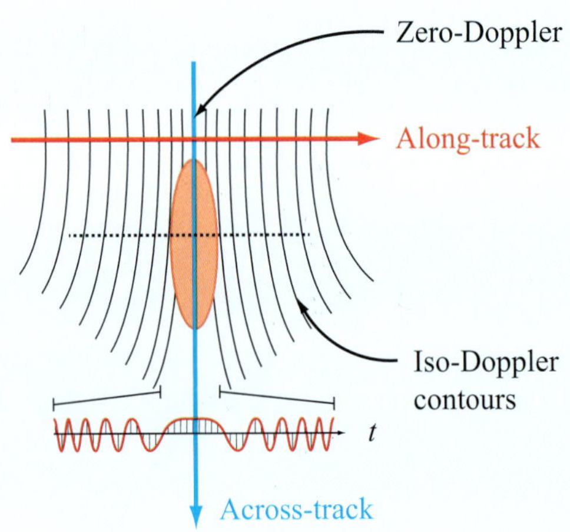

Figure 14-39: Azimuth sampling of the along-track Doppler chirp. The vertical lines in the bottom signal occur at the pulse sample times.

the isodop lines as the platform flies past. The diagram at the bottom shows a time-expanded plot of the real part of the Doppler-shifted signal as a function of time with the carrier removed. If there were only a single target and the signal were CW, this would be the received Doppler signal. Since the signal is pulsed, however, each pulse only provides a sample of this signal. The sample locations are illustrated with the vertical lines in the lower waveform, which shows the real part of the Doppler signal. The signal sample is the magnitude and phase (or I and Q values) of the range-compressed data corresponding to the range between the target and radar at each sample instant. Thus, the range-compressed data provides discrete-time samples of the Doppler signal. Assuming these samples are evenly spaced in time, frequency-domain techniques can be used to efficiently compute the azimuth-matched filter. This is the basis for RDA azimuth compression.

In RDA azimuth-compression, the reference azimuth chirp is computed for each range bin. The reference chirp is the expected Doppler signal for a point target in that range bin. (The number of reference chirps can be reduced by computing only one reference for each depth of focus.) The FFT of the reference chirp is then computed. For each range bin, the range-compressed data in the azimuth direction at the same bin is extracted and its FFT computed. The two frequency-domain signals are multiplied and the IFFT computed. This yields the azimuth-compressed SAR image. The distance between azimuth pixel spacing corresponds to the distance traveled between pulses, while the distance between range pixels is in the slant range and corresponds to the range bin spacing. Conversion to a ground-range image is considered in Section 14-10.2.

Note that the simple implementation of the RDA azimuth compression described above does not account for the range curvature, which limits the image quality when range walk is significant. Methods for accounting for the range curvature, known as ***range migration corrections***, are described in texts dedicated to SAR processing (e.g., Cumming and Wong, 2005).

Figure 14-38(c) shows the result of azimuth compression for the simulated signal example. Note that the energy spread in azimuth in Fig. 14-38(b) is now localized to the location of the point targets. We point out that the signals in the various panels have been normalized for display purposes. Due to the compression steps, the actual signal amplitude in Fig. 14-38(c) is much larger than (b), which is larger than (a).

The FFTs of reference range and azimuth chirps need to be computed only once, and can often be computed beforehand. The computational load for the remaining steps can be estimated from the number of range samples and azimuth pulses in a block. If there are N range samples and M pulses, the range-compression computation is proportional to $MN \log N$, while computation of azimuth compression is proportional to $NM \log M$ for a total of $MN(\log M + \log N)$. The log terms result from the computational efficiency of the FFT, which is $N \log N$ for a length N signal. Time domain methods for convolution would require computation proportional to $MN(M+N)$. The difference is significant for large M and N.

14-9.5 SAR Image Processing: the Backprojection Algorithm

The computational power of modern computer hardware makes time-domain SAR processing algorithms, such as ***backprojection***, practical. Though computationally more intense than RDA, the backprojection algorithm implements a two-dimensional (range and azimuth) matched filter and can produce higher-quality images.

The discrete backprojection algorithm is a deceptively simple algorithm for generating images from synthetic-aperture radar data. It can be applied to stripmap, circular, or spotlight-mode SAR data as well as 3-D SAR imaging. It requires accurate knowledge of the relative positions of the SAR antenna and the pixel locations on the ground for each pulse, and it intrinsically compensates for nonideal platform motion during the computational process. Though computationally taxing and requiring high-precision antenna position information, it can generate high-quality SAR images even under nonideal motion conditions so long as the antenna positions are known.

The algorithm computes the radar cross section over a grid of pixels on the surface. Each pixel is separately computed. The idea is to compute time-Doppler history that the radar should observe for a point target at that pixel. This is the reference function for the matched filter used to evaluate the matched-filter output for the particular pixel from the SAR data. Accurately computing the reference function requires knowing the distance between the antenna and the pixel over the length of the synthetic aperture. When the "stop-and-go" approximation applies, the antenna location at the start of the transmit pulse can be used.

Consider a point target centered in the pixel. After range-compression, the received signal at time $t_i = 2R_i/c$, corresponding to range R_i between the radar and the point target, is given by

$$V_i(t_i) = a_i e^{-j2kR_i}, \tag{14.141}$$

where i is the pulse index and a_i is the signal amplitude from the radar equation. While precision processing accounts for variations in a_i, here we ignore the amplitude for the reference function and write the ***phase-only reference function*** as

$$V_\mathrm{r}\left(\frac{2R_i}{c}\right) = V_i^*\left(\frac{2R_i}{c}\right) = e^{j2kR_i}. \tag{14.142}$$

The matched-filter output is the azimuth convolution of the this reference function with the observed range-compressed data. This is completed by correlating the range-compressed data evaluated at the pixel range with the phase reference function. The pixel value A is

$$A = \sum_i V_i\left(\frac{2R_i}{c}\right) V_\mathrm{r}\left(\frac{2R_i}{c}\right) = \sum_i a_i, \tag{14.143}$$

where the sum is computed over the pixels that define the synthetic aperture, i.e., those in which the pixel is in the mainlobe of the antenna beam. More generally, a pulse number-dependent signal processing window may be included in the summation to reduce sidelobes. The computation in Eq. (14.143) is computed for each pixel location in the SAR image. Note that this formulation of backprojection forms the image directly in the ground plane. Topographic variations can be explicitly accounted for by including the local pixel height when computing its range to the radar.

While the backprojection pixels may be arbitrarily spaced, a single-look image is formed with the pixels spaced $l/2$ in azimuth. In range, the pixels are spaced by the slant-range resolution of the SAR. A uniform ground-range grid has variable spacing in slant range.

The backprojection algorithm can also be used with LFMCW SARs. As noted previously, such systems differ from conventional pulse-mode SARs in that the pulse length is very long, so transmission and reception occur simultaneously. For an LFMCW SAR the stop-and-go approximation is not generally valid. The motion during the chirp can be compensated for by including an additional range-dependent chirp during range compression (Zaugg and Long, 2012).

The backprojection algorithm explicitly deals with range migration and can handle nonideal motion so long as it is accurately measured and reported. However, experience suggests it is much less robust than RDA when there are unknown position or motion errors. Because each pixel is computed independently, pixel computation can be done in parallel to speed computation. The required computation can be estimated from the number of range samples and azimuth pulses. For the simple version of the backprojection algorithm outlined above, range compression can be performed the same way as RDA. If there are N range samples and M pulses, computation in the range compression is then proportional to $MN \log N$. For an $M \times N$ image, the backprojection computation requires evaluating MN^2 terms that include computation of a square root (unless the R_i terms can be precomputed), a complex exponential, and a complex multiply and sum. There is also the expense of interpolating the range compressed data, which must be considered as well. Hence, the computational load for backprojection is substantially higher than it is for RDA.

14-10 Geometric Distortion in Radar Images

Some distortions occur naturally in SLAR and SAR images because of the radar imaging geometry, and others occur either due to nonideal motions of the vehicle or due to imperfect attempts to correct for geometric distortions or vehicle motion. We briefly consider some of these in the following subsections.

14-10.1 Elevation Distortion

Because a radar measures distance to the side of the aircraft by measuring slant range rather than ground range, a distortion occurs when the point being imaged is at a height other than the mean ground level. Consider the example shown in Fig. 14-40, where a tall flagpole target is used as an illustration. The radar is at height h above mean ground level, and consequently above the bottom of the flagpole, but the height of the radar from the top of the flagpole is, of course, different because of the flagpole height H. As seen in the figure, the slant range to the bottom of the flagpole is R_B, which is larger than the range to the top of the flagpole, R_T. Thus the top of the flagpole is closer to the radar than the bottom of the flagpole. As a result, in the radar image the top of the flagpole appears closer than the bottom. The angle of incidence associated with the top of the flagpole, θ_T, is larger than the angle associated with the bottom, θ_B. The top of the flagpole appears at a position to the side of the flight line by an amount y_T whereas its actual location to the side of the flight line is y.

Note that a ***radar shadow*** exists for the flagpole, or any other elevated object. This is the region behind the

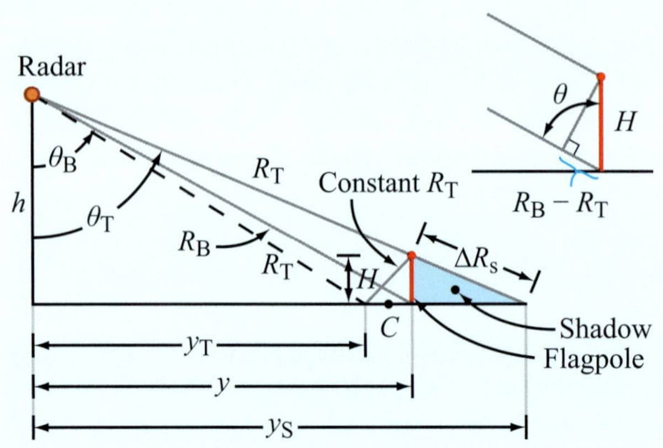

Figure 14-40: Radar imaging height distortion and shadow geometry.

object that is not illuminated by the radar (Fig. 14-40). The shadow extends from the bottom of the flagpole, y, to the end of the shadow at y_S. Thus the effect of the target height is both to distort the location of the top (and other elevated parts of the target) and to create a shadow region that cannot be imaged by the radar.

Optical imaging sensors determine the distance by the angle θ; that is, in an optical image, the top of the flagpole appears to be at a point y_s, whereas in a radar image it is much closer, at y_T. Thus, the image distortion due to height for radar is the opposite of that associated with an optical sensor. In both cases, the target shift is known as ***parallax*** displacement.

The magnitude of the displacement of the top of the flagpole is given by

$$R_B - R_T = \sqrt{y^2 + h^2} - \sqrt{y^2 + (h - H)^2}. \quad (14.144)$$

For a target with a relatively small elevation like the flagpole in this example, this may be approximated by

$$R_B - R_T \approx H \cos\theta, \quad (14.145)$$

where the approximation is obtained from the figure by using the chord instead of the arc of constant R_T.

This difference in slant range results in the apparent cross-range position of the top of the flagpole at

$$y_T = \sqrt{R_T^2 - h^2} = \sqrt{y^2 - 2hH + H^2}\,, \quad (14.146)$$

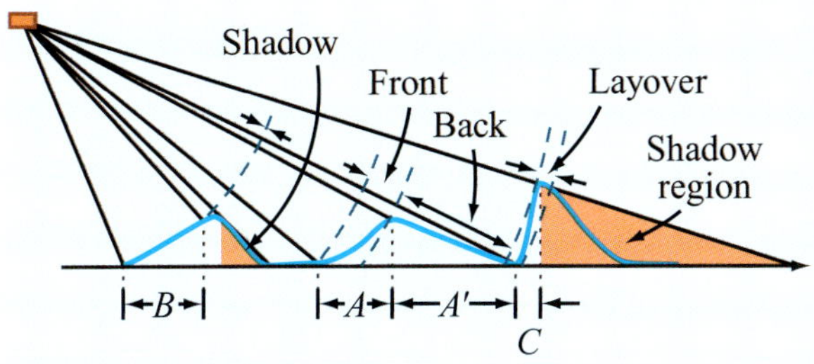

Figure 14-41: Examples of height distortion in SAR images.

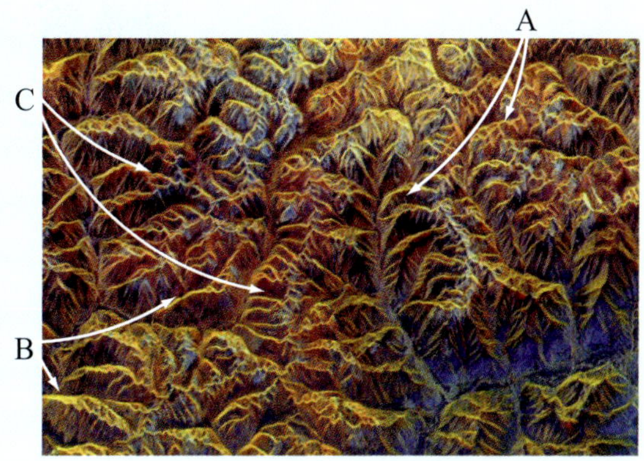

Figure 14-42: Foreshortening in mountainous terrain. The radar illumination direction is from the top of the image. [SIR-C/XSAR image courtesy of NASA/JPL.]

rather than at y. In other words, the top of the flagpole is apparent in the radar image at a cross-range distance that is closer to the radar than to the apparent location of the bottom of the image. This occurs because, in fact, the top of the flagpole *is* closer to the radar than the bottom is. This phenomenon is known as ***overlay***.

The effect of overlay can be significant in cities and steep mountainous areas. Overlay is an extreme example of the inherent ***foreshortening*** effect that occurs in radar imaging. Figure 14-41 provides several examples of different amounts of foreshortening. In this figure, the dashed arcs represent lines of constant slant range. For the center peak in Fig. 14-41, the horizontal distance A between the mountain base and its peak is much larger than the corresponding points in the slant-range image, while for the back-side positions, A' is stretched out in the slant plane compared with its horizontal displacement. Further, because the radar return from a large area ends up at the same distance in the radar image, the forward face of the mountain is often much brighter than the slant-range image of the back side. Peak A in Fig. 14-41 illustrates even more foreshortening where the entire forward slope of the mountain appears at a single slant range. The resulting SAR image has essentially no effective cross-range resolution at this point. We also note that the back side of the mountain is in the shadow of the peak and thus provides no reflection in the SAR image. Peak C in Fig. 14-41 illustrates when the foreshortening effect causes the top of the peak to appear closer to the radar than the bottom of the valley, creating overlay conditions. These effects are clearly evident in the SAR image of mountainous terrain shown in Fig. 14-42. Note that the mountains are "laid over" toward the radar and that their front sides appear brighter than the back sides. Elevation distortion is also visible in the distorted bridge under construction in Fig. 14-43.

As previously indicated, image shadows result from areas that do not provide any reflection for the radar. Shadows can lead to loss of information about the surface. However, the *shape* of the shadow can reveal information about the object creating the shadow. For example, the reflective metal of the aircraft in Fig. 14-44 reflects much of the radar signal away from the radar, providing limited information about the aircraft shape. However, the radar shadow on the flat ground clearly reveals the shape of the wings and the presence of the otherwise invisible helicopter blades. The shadows of trees in the upper right hint at their shape.

14-10.2 Range Distortion

The radar measures slant range, but to produce maps or ***plan view images***, the image must be converted to an image in ground range. The horizontal displacement y of the base of the flagpole in Fig. 14-40 is given by

$$y = \sqrt{R_{\rm B}^2 - h^2} = h \tan \theta_{\rm B}. \qquad (14.147)$$

Figure 14-43: Multilook SAR image with image distortions due to digital elevation model (DEM) errors. The bridge at the center is distorted since it is not included in the DEM. Streaks along the freeway are due to moving vehicles. Parallel lines at the bottom are railroad tracks.

Figure 14-44: Example of how a shadow can reveal the shape of the object creating the shadow. The radar illumination direction is from the top of the image. [Image courtesy of Sandia National Laboratories.]

Since y is not linearly related to R_{B}, the horizontal scale in the range direction is distorted in an image. If we denote the displacement on the image by Y and a scale factor for relating displacement on the image to displacement in space by a, we find

$$Y = a(R - R_0) = a\left(\sqrt{h^2 + y^2} - h\right), \qquad (14.148)$$

where we have deleted the subscript B from R_{B} because we are dealing with general relations rather than simply a single location. We also replaced R_0 with the value of R at $y = 0$, namely $R_0 = h$. The rate of change of displacement on the image relative to displacement on the ground is given by

$$\frac{dY}{dy} = a\,\frac{y}{R} = a\sin\theta. \qquad (14.149)$$

Since the rate of change of displacement on the image relative to that on the ground is proportional to y or to $\sin\theta$, the value of the derivative is small near vertical, where y is small, and grows larger across the swath. This means that the slant-range image compresses the horizontal variations near vertical, relative to those far from the flight line. This inherent distortion associated with the difference between slant range and ground range cannot be totally overcome in the radar processing.

14-10.3 SLAR Motion Distortion

SLAR images are generated by displaying the echo power versus time (or range) for one pulse in one dimension, and "stacking" the range lines in the other dimension to create an image. Nonideal motion and pointing can create distortions in the resulting image. These include the following:

1. **Varying speed:** The PRF is not adequately synchronized with the speed of the aircraft. This results in a nonlinear ***stretching*** or ***compression*** in the along-track direction.

2. **Lateral or vertical motions:** Deviations away from a straight-line flight path result in curvilinear distortions that make straight lines (such as roads) parallel to the flight track appear curved, often sinuous.

3. **Yaw of the aircraft:** The effect of yaw motion is to distort the direction of different points relative to others, depending upon their displacement from thc flight linc. Extrcmc yaw, such as occurs during a turn, can completely distort the image.

4. **Pitch of the aircraft:** The effect of a pitching motion is to move the intersection of the beam with the ground ahead of or back from the position directly to the side of the point beneath the aircraft.

5. **Roll of the aircraft:** The roll of the aircraft changes the antenna gain at different points in the image and therefore modulates the grayscale.

Items 1–4 are illustrated in Fig. 14-45. If these motions are adequately sensed by the aircraft navigation system, corrections can be made by including frequency shifting in the processing, by adjusting the timing or positioning of the pixels in the image, or by rectifying the image after formation.

14-10.4 SAR Motion Errors

Stripmap synthetic-aperture processing ordinarily depends upon the assumption that the radar is traveling horizontally in a straight line. Figure 14-46 illustrates the geometry for determining the effect of deviations from this straight-line path. The assumption is that the radar is traveling in the x direction at height h and is observing a ground cell at an angle θ from the vertical. The coordinates of the forward edge of the ground cell are $(r_{\mathrm{a}}/2, y, 0)$. The coordinates for the radar are $(0,0,h)$, and the range vector from the radar to the front edge of the cell is

$$\mathbf{R} = \hat{\mathbf{x}}\,\frac{r_{\mathrm{a}}}{2} + \hat{\mathbf{y}} R\sin\theta - \hat{\mathbf{z}} h.$$

If the radar velocity $\mathbf{u}$ has components in the y and z directions (in addition to the x direction),

$$\mathbf{u} = \hat{\mathbf{x}} u_x + \hat{\mathbf{y}} u_y + \hat{\mathbf{z}} u_z,$$

the total Doppler shift associated with the point $(r_{\mathrm{a}}/2, y, 0)$ is given by

$$\begin{aligned} f_{\mathrm{D}} &= \left(\frac{2\mathbf{u}}{\lambda}\right)\cdot\left(\frac{\mathbf{R}}{R}\right) \\ &= \frac{2}{\lambda R}(\hat{\mathbf{x}} u_x + \hat{\mathbf{y}} u_y + \hat{\mathbf{z}} u_z)\cdot\left(\hat{\mathbf{x}}\,\frac{r_{\mathrm{a}}}{2} + \hat{\mathbf{y}} R\sin\theta \;- \hat{\mathbf{z}} h\right). \end{aligned}$$

Hence,

$$f_{\mathrm{D}} = \frac{2}{\lambda}\left(\frac{u_x r_{\mathrm{a}}}{2R} + u_y\sin\theta - u_z\cos\theta\right). \qquad (14.150)$$

This Doppler frequency f_{D} can be expressed in terms of the desired Doppler frequency f_{D0} $(= B_{\mathrm{Df}}/2)$ and an error component f_{De} $(= f_{\mathrm{D}y} + f_{\mathrm{D}z})$:

$$f_{\mathrm{D}} = f_{\mathrm{D0}} + f_{\mathrm{De}} = \left(1 + \frac{f_{\mathrm{De}}}{f_{\mathrm{D0}}}\right) f_{\mathrm{D0}}.$$

If we define a limit bound ε_{r} to set the per-unit error allowed in the Doppler frequency, expressed as a fraction of B_{Df}, then Eq. (14.150) may be rewritten as

$$f_{\mathrm{D}} = (1 + 2\varepsilon_{\mathrm{r}}) f_{\mathrm{D0}}.$$

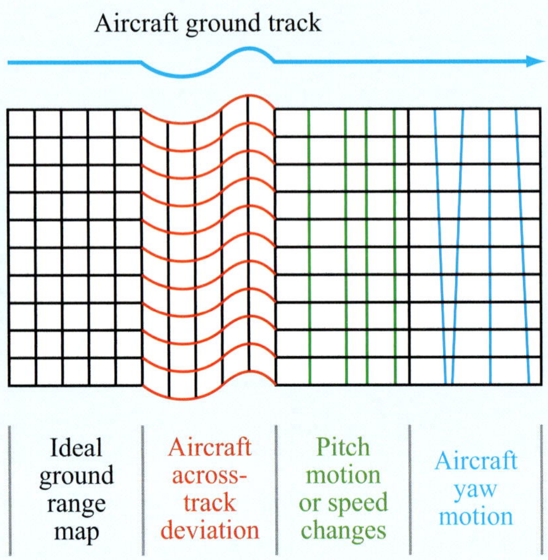

Figure 14-45: SLAR map of a square grid with uncompensated aircraft motion.

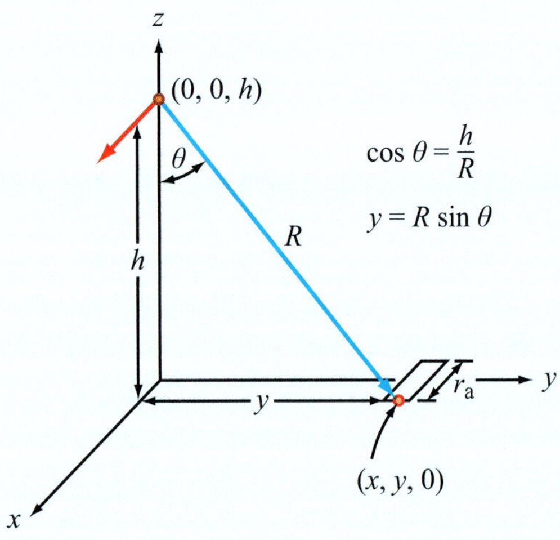

Figure 14-46: Geometry for SAR motion-error determination.

With this definition we have

$$\frac{f_{\rm De}}{f_{\rm D0}} \leq 2\varepsilon_{\rm r},$$

which sets the bounds on the size of the error velocities u_y and u_z as given in

$$u_y \sin\theta - u_z \cos\theta \leq \lambda f_{\rm D0}\varepsilon_{\rm r} = \frac{\lambda B_{\rm Df}\varepsilon_{\rm r}}{2}. \qquad (14.151)$$

If the velocity component errors are smaller than this, no compensation is required, but if they are larger, some sort of compensation is required to minimize adverse effects.

Consider an aircraft example with $\varepsilon_{\rm r} = 0.1$, $B_{\rm Df} = 2$ Hz, and $\lambda = 3$ cm. Substituting these values into Eq. (14.151) yields

$$u_y \sin\theta - u_z \cos\theta \leq 3 \times 10^{-3} \text{ m/s} = 3 \text{ mm/s}.$$

For a spacecraft, using the same $\varepsilon_{\rm r}$ and λ, $B_{\rm Df}$ is 25 Hz and the limits are

$$u_x \sin\theta - u_z \cos\theta \leq 3.75 \text{ cm/s}.$$

These are very small velocities when considered in terms of absolute values or values relative to the aircraft and spacecraft velocities.

With such tiny errors causing serious trouble for the SAR, some sort of compensation is absolutely necessary. It is unrealistic to imagine that either an aircraft or a spacecraft can be made to travel in a perfectly straight line. Although spacecraft are extremely stable, aircraft tend to bounce around in the turbulent air, and the corrections required for this nonideal movement can be very sizable.

To compensate for velocity and acceleration errors, one must first sense them. The sensing can be achieved by mounting integrating accelerometers directly on the antenna, or at least nearby. Typical aircraft navigation systems do not have the sensitivity required to sense and measure motion errors as small as required for SAR compensation. Further, motion measurement errors may be compounded by the long lever arm between the antenna mounting location and the aircraft motion sensors.

The effect of the uniform-velocity errors is to shift the center frequency for the expected Doppler spectrum away from the carrier frequency, as illustrated in Fig. 14-47(a). Here the expected spectrum is shown centered about the carrier frequency and the actual spectrum is shifted by the velocity errors. For a system in which the processing is performed at baseband, the signal is mixed down from the carrier frequency by beating the return signal with a local oscillator at the carrier frequency. One solution to the problem of velocity errors is to change the frequency of the local oscillator by just the right amount so that the LO frequency is centered in the passband shown in the solid blue lines in Fig. 14-47(a). This may be done either at the carrier frequency, at the intermediate frequency, or at baseband in the processor.

Correcting for the acceleration errors is, in effect, correcting for the resulting changes in velocity, so the correction method is similar. However, correcting for uniform-velocity shifts may be done with a system having a longer time constant than that required for the acceleration correction, so correction is easier to do.

The problem of compensating for motion errors is one of the primary limitations on the precision that can be achieved with a SAR.

14-10.5 SAR Attitude Errors

The aircraft or spacecraft may yaw, pitch, or roll, or do all three at the same time. Each of these causes an error in the SAR. The effect of the roll error is the same as for the SLAR; that is, it simply changes the gain for a particular point on the ground. If the antenna rolls too much this can be a serious problem, but it does not affect synthetic-aperture processing and its effect is not as severe as the yaw and pitch errors, which do affect the processing.

Figure 14-47(b) illustrates the geometry when the aircraft yaws to the right. This has the effect of moving the area illuminated by the real aperture away from the side-looking direction and rotating it slightly ahead of that direction. If the yaw were to the left it would result in the beam moving back of the side-looking position. This causes distortions in the image; but, more important for the SAR, it causes a shift of the Doppler frequencies

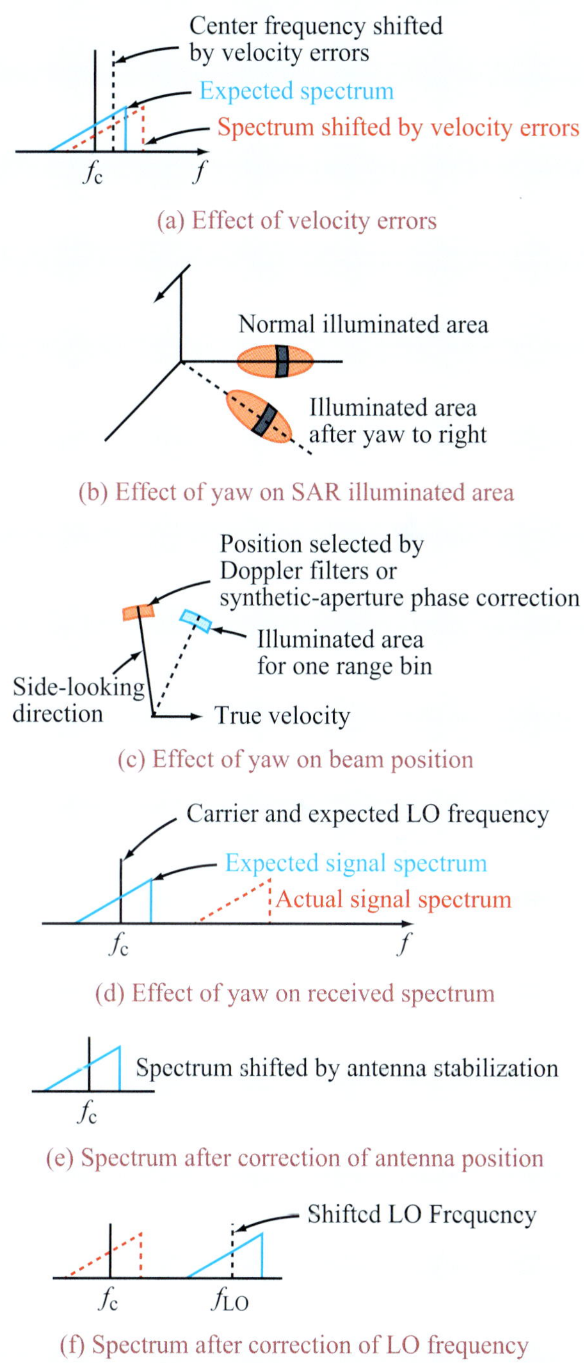

Figure 14-47: Illustration of velocity and attitude errors.

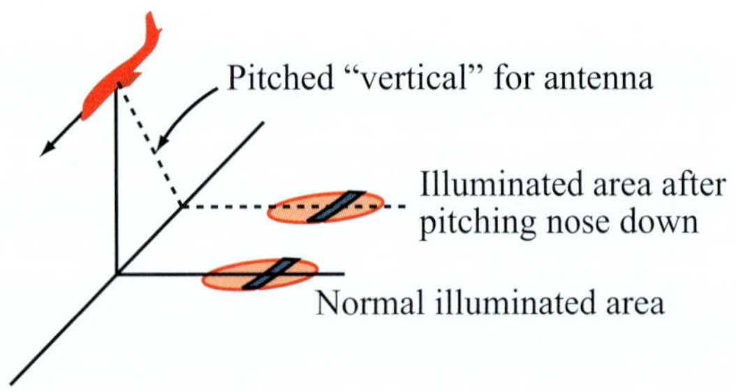

Figure 14-48: Effects of pitch on SAR illuminated area.

away from the beam. Figure 14-47(c) illustrates the situation looking down on the illuminated area for a given range position on the ground. If nothing is done about the problem, the Doppler filters select the dashed area shown, which is outside the area illuminated by the antenna beam, so the signal received cannot be correlated, as shown in Fig. 14-47(d). The signal spectrum is in the wrong place, just as it is for the velocity error.

Two types of correction are possible for this effect. Perhaps the best, if it can be done perfectly, is simply to stabilize the antenna so it always looks directly to the side. When this happens, there is no yaw of the antenna, even though the aircraft or spacecraft carrying it yaws. The result is that the spectrum is in the right place, as indicated in Fig. 14-47(e). The other solution is like that applied for velocity errors, as indicated in Figure 14-47(f); that is, one moves the local-oscillator frequency to the center of the passband for the Doppler frequencies. In practice, a combination of antenna stabilization and shifting of the LO is often required. This correction may be achieved with the same local-oscillator shifting that is done for the velocity errors. That is, the combination of velocity error and yaw error may be used to provide the error signal that shifts the local-oscillator frequency.

The effect of pitch is somewhat different, but the solution is the same. Figure 14-48 illustrates the effect of pitch on the illuminated area. Instead of rotating the beam about a vertical axis, the rotation is about a horizontal axis parallel to the side-looking direction. Thus, the two illuminated areas—the proper one and the one after pitching—are parallel to each other but displaced. The effect on the Doppler frequency, however, is the same as illustrated for yaw except that the shift may not be exactly the same at all ranges. Nevertheless, the correction can be most easily achieved by the same methods used for correcting for yaw, that is, stabilizing the antenna and shifting the local-oscillator frequency.

Note that even after the corrections have been made to the local-oscillator signal so that the SAR image can be produced, the image has errors in it. In the case of yaw, the errors are along-track displacements that are proportional to the distance from the point beneath the radar. In the case of pitch, they are approximately the same at all distances away from the radar track, but still represent along-track displacements of the pixels. When this occurs, the geometric fidelity of the image is degraded. It is possible to warp the final image to partially correct for this problem.

An inherent yaw effect is likely to occur with most spacecraft radars. This comes about because the spacecraft is aligned with the direction of travel along the orbit plane, but the zero-Doppler line is not quite perpendicular to this plane. Because the Earth is rotating beneath the spacecraft, the vector velocity relative to a point along the Earth contains a y component even if the spacecraft is traveling perfectly in the x direction in its orbit plane. The zero-Doppler line on the surface of the Earth is perpendicular to the vector sum of the velocity along the orbit direction and the velocity relative to the rotating Earth. Consequently, it is tilted from perpendicular to the orbit plane by an amount that varies up to 3.5° at the equator, where the Earth's linear velocity is the greatest. The inherent yaw built into most spacecraft systems may be corrected either by rotating the antenna about the spacecraft axis or by designing the spacecraft itself to be continuously reoriented perpendicular to the zero-Doppler line.

14-11 Elevations from SLAR and SAR

Elevations can be estimated using the interferometric techniques described in Chapter 15. Other approaches to height estimation that do not require the coherence of interferometric techniques include exploiting shadows in the image and image parallax with the aid of stereo techniques. These are briefly considered below.

14-11.1 Shadows

In the geometry shown in Fig. 14-40, a significant shadow exists behind a flagpole, and similar shadows may exist behind a mountain, a tree, or a building. In some cases, the shadow is long enough so that it may by used to estimate the height of the object. The length of the shadow, ΔR_S is given by

$$\Delta R_S = \frac{H}{\cos\theta} \,. \tag{14.152}$$

Since the denominator of the right-hand side of the equation is small for large θ, even relatively low objects may have their height determined by the shadow length. For example, consider a slant-range resolution r_r of 10 m. If the object has a height $H = 5$ m at $\theta = 80^\circ$ then $\Delta R_S = 28.8$ m, which is almost 3 pixels. This is enough to estimate the height of the object, although the estimate is not very good, because of the small number of pixels involved. Finer range resolution enables better height estimate accuracy.

14-11.2 Stereo with Radar

The stereoscopic effect is widely used with aerial photography for determining elevations. Most topographic maps are made by stereo interpretation of overlapping pairs of aerial photographs. Similar techniques can be used with radar. Radar stereo, however, is different from optical stereo, as suggested in the discussion of Fig. 14-40, since higher objects appear further away in a photograph but closer in a radar image. With photography, the stereo effect is obtained by taking successive overlapping photographs as the aircraft travels. With radar the normal procedure is to fly different flight lines with overlapping radar swaths and use the resulting images for stereo. Targets in the radar images have different incidence angles. The flight lines may be flown on the same side or the opposite side of the area being studied. Figure 14-49(a) illustrates opposite-side stereo. An object of height H is observed from the two indicated positions. It "leans" a distance of ΔR_1 toward the aircraft on flight line 1 and a distance of ΔR_2 toward the aircraft on flight line 2. When the images are superimposed and translated to the side until the two images of the same point on the ground coincide, the amount of translation required is the sum of ΔR_1 and ΔR_2. By knowing the location in the image and the spacing between the flight lines, one can compute the height above the mean ground level. When a pair of such images is observed through a stereo viewer, the effect of the parallax is to make the mountain appear in three dimensions, much as it would in an optical stereo view,

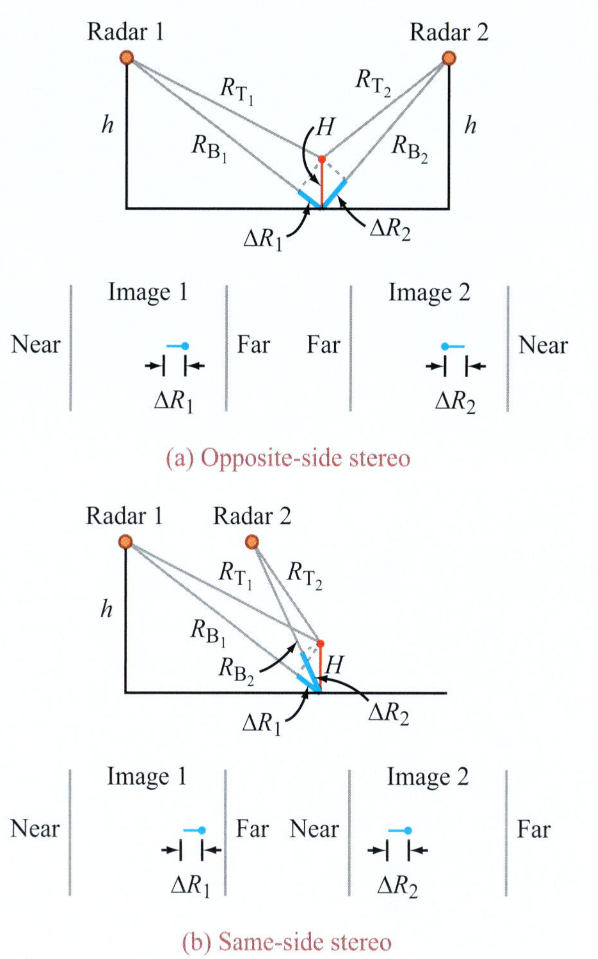

Figure 14-49: Two types of stereo imaging.

except that the displacements are the opposite of what they would be in the photographs.

An important difficulty with opposite-side stereo is shadows. Since the shadows from one image lie to one side of the object and from the other image to the other side of the object, objects are often difficult to recognize. In fact, a given object may be in the shadow of some taller object for one of the images and not for the other, so that it cannot even be seen on both images. The solution to this is to use same-side stereo, as indicated in Fig. 14-49(b). Two flight lines are also used for same-side stereo, but they are translated from each other, either horizontally, as illustrated, or vertically. Because in this case ΔR_2 also is different from ΔR_1, a stereo effect occurs. When the two images are translated so that a particular object is superimposed from one image to the other, the amount of translation can be used to establish the height above mean ground just as with opposite-side stereo or with photography. In this case, however, it is the difference of ΔR_2 and ΔR_1 that determines the height. Because the shadows are on the same side for both images, this method is easier to use than opposite-side stereo. It has been used rather widely in mapping remote areas where other elevation information is lacking.

14-11.3 Squint Stereo

A dual-beam SAR can produce stereo images in a single pass that may be used for a single flight line to determine altitudes. The situation is illustrated in Fig. 14-50. At a given point on the flight path, two beams are used—a forward beam and an aft beam. While a squint angle of 45° ahead of and behind the side-looking direction is used here, other squint angles can be used. Pulses are typically transmitted alternately on the two beams and two recording systems are used—one for each beam. In the figure, the direction of the scan lines is indicated first. Then the locations of the beams are shown for two positions at which a given point on the ground is imaged (position 1 on the forward beam, and position 2 on the aft beam). If the point on the ground represents a flagpole, its top is displaced from the bottom toward the radar on each image, but the direction toward the radar is different on the two images, as shown. The images may then be superimposed at the base of the flagpole, and the distance between the images of the tops indicates the height of the flagpole. Similar techniques can be used in which a given point at some elevation, such as the top, is superimposed on the two images, and the amount of translation required to align the relative positions of the images for a reference level determines the height. Squint-stereo systems require fan-beam antennas that are tilted relative to the axis of the aircraft.

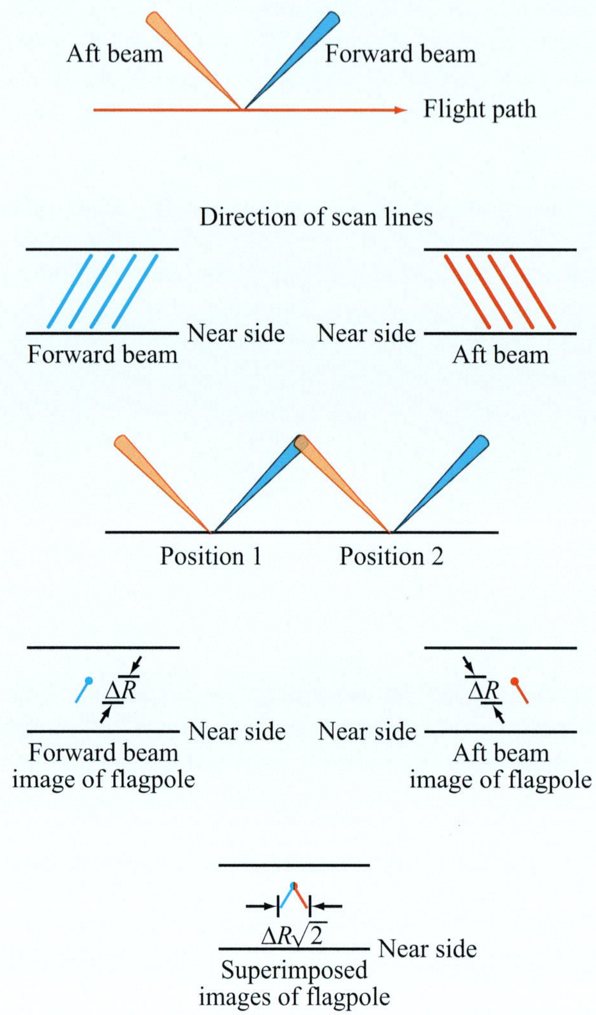

Figure 14-50: Dual squint-beam stereo radar with 45° squint angles.

14-11.4 Mountains and Buildings

Because of shadows behind mountains and buildings, images in mountainous terrain and cities with tall buildings necessarily fail to contain information about the areas that are in the shadow. In such regions, multiple flight lines are normally required to image all parts of the regions. Orthogonal flight lines, in which the region is observed from both sides along parallel flight lines and from both sides along parallel lines orthogonal to the first pair, are considered the best approach. When such images are superimposed, the shadows are generally filled in. Unfortunately, however, superimposing them in mountainous country can cause difficulties because the parallax displacement is different for the different images.

One way to accomplish part of this purpose with a single flight line is to use the dual squinted-beam imaging geometry illustrated in Fig. 14-50. The shadows are filled in to the degree that orthogonal looks can fill them in. Of course, under some circumstances, parts of a region may be shadowed from both positions on the flight line and a separate flight line on the other side may be required. Distortions due to deviation from a straight flight path and due to aircraft altitude changes still remain in this pair of images, and will differ between them, so that proper superposition of the images depends upon the maintenance of a straight flight line and a stable platform. This technique is viable for both real-aperture and synthetic-aperture systems.

14-12 Ionospheric Effects

The Earth's ionosphere, which extends from about 60 km above the Earth surface to as high as 1000 km depending on solar activity, consists of an ionized gas containing free electrons and positively charged ions (Kelley, 1989; Brekke, 1997; Hunsucker and Hargreaves, 2003). Under the influence of the Earth's magnetic field, a radio wave passing through this ionized medium may experience the ***Faraday effect*** that causes the wave's electric field vector to rotate, which, in turn, alters the polarization state of the radio wave. The amount of angular rotation, termed the ***Faraday rotation***, depends on the frequency of the wave, its direction of propagation relative to the direction of the Earth's magnetic field, and the state of the ionosphere (free electron density profile).

This Faraday rotation may have consequences to both radar and radiometric Earth observations if the observation platform is above 60 km, which includes all satellite orbits.

For a satellite-borne radar transmitting a v-polarized wave, the Faraday rotation alters the polarization state of the wave, so the wave illuminating the surface is no longer purely v polarized, but rather a combination of a v-polarized component and an h-polarized component. Consequently, the scattering measurements recorded by the radar no longer match the assumptions of the inversion algorithm used to convert the radar measurements into estimates of the biophysical or geophysical parameters of the observed scene. The mismatch usually translates into estimation errors. A similar scenario applies to the radiometer; the v-polarized energy captured by the radiometer antenna is a result of emission by the Earth's surface at a polarization direction other than vertical, such that after getting rotated by the ionosphere arrives as v-polarized at the radiometer antenna. Fortunately, the Faraday rotation can be ignored completely if $f \gtrsim 3$ GHz, and for frequencies in the range $0.5 \leq f \leq 3$ GHz, the rotation angle can be estimated and the measured radar observations can be corrected accordingly.

14-12.1 Rotation Angle

The Faraday rotation effect is illustrated by the simple diagram shown in Fig. 14-51, which depicts an EM wave traveling through an ionized layer (usually called a ***plasma***). Here, **B** represents the Earth's magnetic field and $B_{\|}$ is the component of **B** along the propagation direction. The rotation angle χ is given by Hunsucker and Hargreaves (2003) as

$$\chi = 2.365 \times 10^{-14} \frac{B_{\|}}{f^2} N_{\text{TECU}} \sec\theta \qquad \text{(rad)}, \quad (14.153)$$

where $B_{\|}$ is the magnetic flux density (in teslas) of the Earth magnetic field parallel to the direction

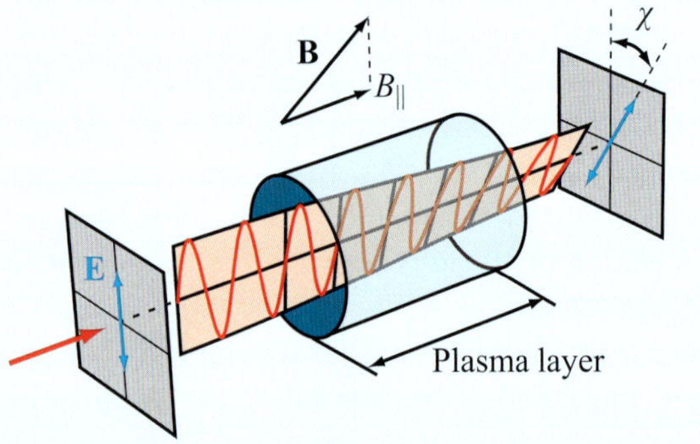

Figure 14-51: Illustration of the Faraday rotation effect resulting from EM propagation in a plasma under the influence of an external magnetic field **B**.

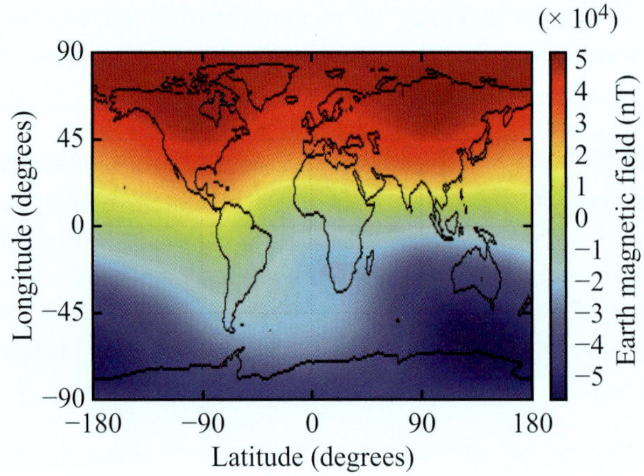

Figure 14-52: Vertical geomagnetic field as modeled by the IGRF10 for June 21, 2007 [Jehle et al., 2009].

of propagation (also called the ***line-of-sight*** (LOS) magnetic field), f is the frequency in GHz, θ is the incidence angle relative to nadir, and N_{TECU} is the integrated free-electron density for a vertical column through the entire ionosphere, measured in TECU, where TECU stands for ***total electron content unit*** and 1 TECU $= 10^{16}$ electrons/m^2. At night, N_{TECU} may be as low as 0.2 TECU, whereas in the daytime it may reach 100 TECU, and when sunspots are highly active N_{TECU} may reach 200. The magnitude of N_{TECU} also varies with season and geographic latitude. The magnetic field component $B_{\|} = B\cos\psi$, where ψ is the angle between the Earth's magnetic field and the propagation direction, and B can be computed from the International Geomagnetic Reference Field (IGRF) model (Macmillan and Maus, 2005), which varies with both latitude and longitude (Fig. 14-52). Using this model, we show in Fig. 14-53 plots for the one-way rotation angle χ as a function of f for a satellite sensor observing the Earth at an incidence angle $\theta = 40°$, at a midlatitude location. At 3 GHz, χ varies between $\approx 0.06°$ for $N_{\text{TECU}} = 1$ (typical nighttime condition) to only about 6° for $N_{\text{TECU}} = 100$. This means that the Faraday rotation effect may be ignored at $f \gtrsim 3$ GHz. At lower frequencies, the Faraday rotation may be more significant, as discussed next.

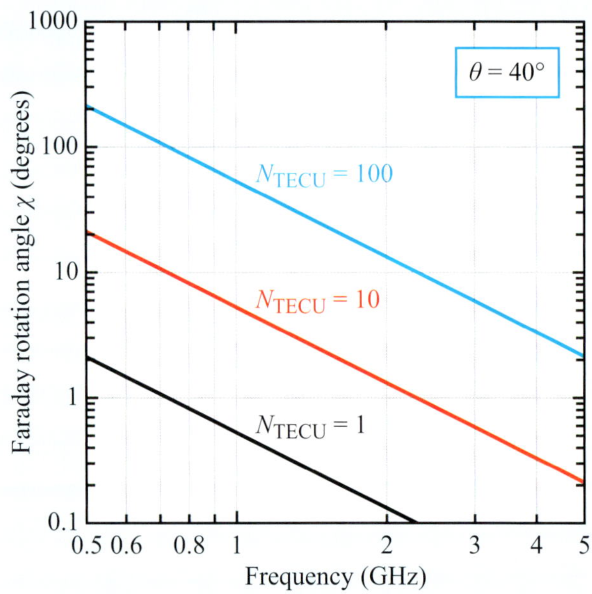

Figure 14-53: Computed variation of the one-way Faraday rotation angle with frequency for various values of N_{TECU}, assuming a midlatitude region and an incidence angle $\theta = 40°$.

14-12.2 Impact on SAR Data

To illustrate the impact of the Faraday rotation on SAR data, we show the results of two simulation studies. In the first study (Wright et al., 2003), the one-way Faraday rotation angle was computed at L-band (1.26 GHz) as a function of latitude and longitude for both low sunspot activity with N_{TECU} varying geographically between about 1 and 80, and high sunspot activity with N_{TECU} between 10 and 120. The incidence angle was set at 35°. The results of the study are displayed in Fig. 14-54. Under the low-sunspot-activity condition, $|\chi|$ seldom exceeds 5°, but when the sunspot activity is high, $|\chi|$ sometimes exceeds 15°. The impact of the Faraday rotation on the backscattering coefficient is illustrated by the computed values given in Table 14-2 for an agricultural area. Application of two-way Faraday rotation leads to insignificant errors (for $\chi \leq 10°$) for hh and vv polarizations, but the error for hv polarization is 1.61 dB for $\chi = 10°$ and as much as 4.23 dB at $\chi = 20°$. To provide a visual illustration of what such errors mean, the authors used the multipolarization image shown in Fig. 14-55(a) to simulate the same image after incorporating the Faraday rotation twice, once to simulate the transmission from the satellite to the ground and a second time for the return trip. Whereas the multipolarized image in Fig. 14-55(a) provided significant discrimination between different crops, the simulated images in parts (b) and (c) of the figure clearly demonstrate that much of the discriminating power has vanished because of the Faraday rotation effect.

The impact of Faraday rotation on SAR data was also evaluated by Jehle et al. (2009) for three SAR configurations with operating center frequencies of 9.65 GHz, 1.27 GHz, and 0.45 GHz. Their results, summarized in Table 14-3, suggest that whereas the impact of the Faraday rotation on the SAR parameters is negligibly small at X-band (9.65 GHz) and somewhat significant at L-band (1.27 GHz), it is very serious at P-band (450 MHz). For $N_{\text{TECU}} = 100$, the two-way Faraday rotation angle is $2\chi = 470°$! Even if the 2π-ambiguity problem can be resolved and χ can be estimated using the models described earlier, the errors associated with the estimate can pose serious problems

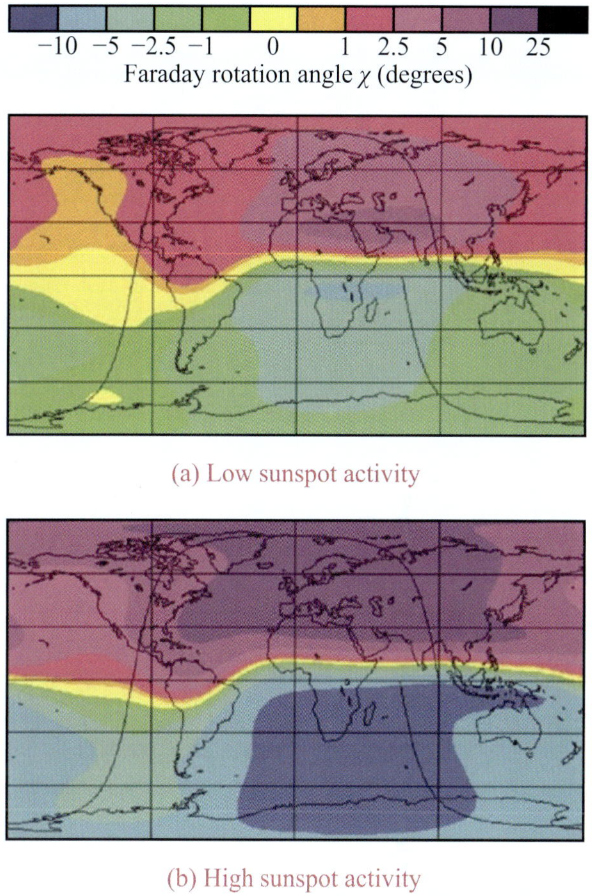

Figure 14-54: Predictions of one-way Faraday rotation angle under (a) low sunspot activity and (b) high sunspot activity, for 12:00 GMT, April and $\theta = 35°$ [Wright et al., 2003].

Table 14-2: Change in backscattering-coefficient magnitude for an agricultural scene observed at L-band due to the application of two-way Faraday rotation [Wright et al., 2003].

One-way χ	Change in mean hh (dB)	Change in mean vv (dB)	Change in mean hv (dB)
5°	−0.08	−0.09	0.51
10°	−0.35	−0.36	1.61
20°	−1.39	−1.49	4.23

(a) $\chi = 0°$ (b) $\chi = 10°$ (c) $\chi = 20°$

Figure 14-55: L-band SAR imagery of Feltwell, U.K. with one-way Faraday rotation of: (a) 0° (original image), (b) 10°, and (c) 20°. Blue = hh, green = vv, red = hv. The same color balance was used for the three images. [Wright et al., 2003.]

Table 14-3: Satellite sensor details and estimates of the two-way impact of 50 and 100 TECU on the path delay, chirp length, and Faraday rotation for TerraSAR-X, ALOS PALSAR, and a possible future spaceborne P-band sensor configuration. The Earth's magnetic field is modeled for June 21, 2007, 45° North and 0° East, at a height of 300 km and a nadir-looking sensor configuration [Jehle et al., 2009].

Sensor	TerraSAR-X		PALSAR		P-band	
Frequency (f_c) [GHz]	9.65		1.27		0.45	
Bandwidth (B) [MHz]	max. 300		max. 28		6	
Chirp duration [μs]	40		27		27	
Sampling rate [MHz]	max. 300		max. 32		8	
Chirp form	up		down		down	
Orbit (altitude) [km]	514		695		695	
TECU	50	100	50	100	50	100
Path delay at f_c [m]	0.5	1	27	54	218	436
Change in chirp length for B [ns]	−0.1	−0.2	4	8	19.4	38.7
Two-way Faraday rotation at f_c [°]	0.5	1	29.5	59	235	470
Δ Faraday rotation within chirp [°]	0.03	0.06	1.3	2.6	6.3	12.5

to both the interpretation of the measured data and the efficacy of correction algorithm applied to it.

Other noteworthy investigations of the impact of the Faraday rotation angle on SAR data include Rignot (2000), Gail (1998), Qi and Jin (2007), Freeman and Saatchi (2004), Lin et al. (2003), Kimura (2009), and Meyer and Nicoll (2008).

14-12.3 Impact on Radiometric Data

Using an ionospheric model similar to that discussed earlier in Section 14-12.1, Le Vine and Abraham (2002) modeled the effect of the Faraday rotation on radiometric emission from the ocean at L-band (1.4 GHz). The motivation was to determine the degree of error caused by the Faraday rotation relative to the requirement to measure sea surface salinity with an accuracy on the order of 0.1–0.2 psu (practical salinity unit), as discussed later in Section 18-2.4. An accuracy of 0.1 psu corresponds to a change of brightness temperature on the order of 0.05 K (Le Vine and Abraham).

Figure 14-56 displays simulated world maps of the Faraday rotation angle at L-band at $\theta = 30°$, for both 6 am and noon local time. The simulations, which were performed for high solar activity, show values for χ extending from 0 to greater than 15°. The change in brightness temperature, ΔT_B, due to the Faraday rotation is shown in Fig. 14-57 as function of the sunspot number R_s, which varies between 10 for a quiet sun to 150 for a highly active sun. We observe that at noon and $\theta = 50°$, ΔT_B increases from about 0.3 K at low sunspot activity ($R_s = 10$) to as much as 2.1 K at $R_s = 150$. These brightness-temperature errors are certainly much larger than the required precision of 0.05 K specified for measuring ocean salinity, which calls for the application of highly accurate correction algorithms (Le Vine and Abraham, 2002; Yueh, 2000).

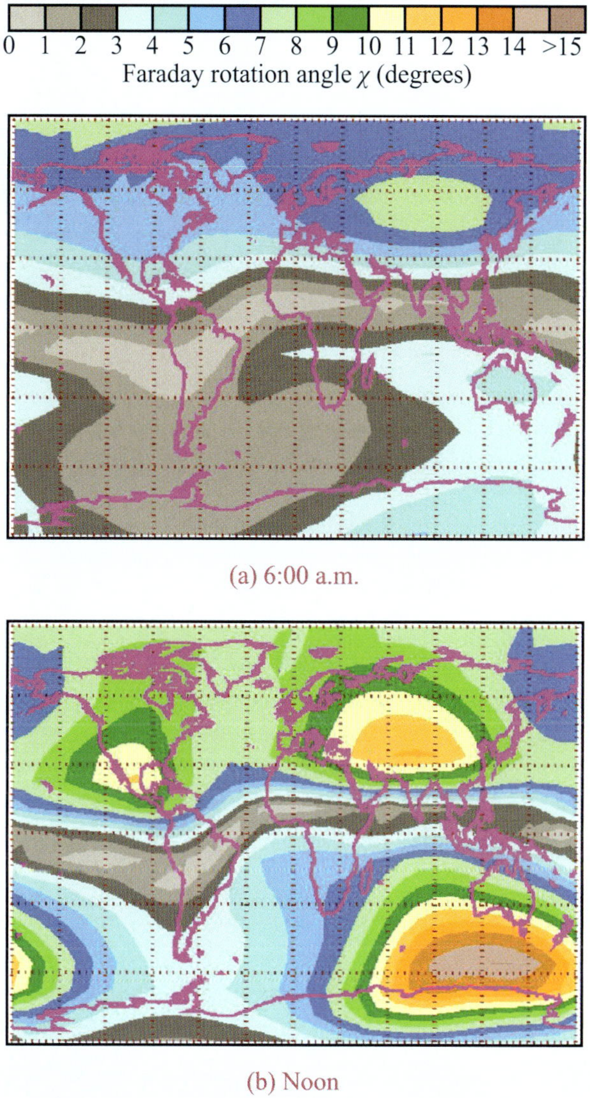

Figure 14-56: Global distribution of Faraday rotation for local time of 6:00 a.m. and noon. The data are for high solar activity (June 1989), an altitude of 675 km, and looking perpendicular to the satellite heading to the right at an incidence angle of 30°. [Le Vine and Abraham, 2002].

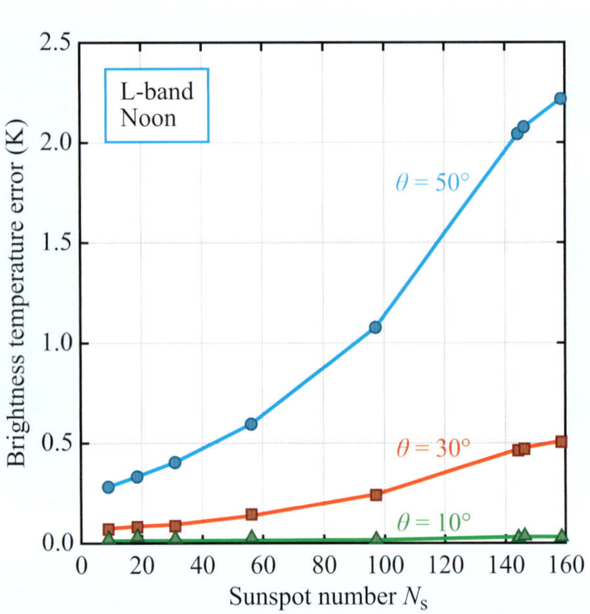

Figure 14-57: Error in brightness temperature due to neglecting Faraday rotation at noon at 30° N, 220° E for high solar activity (June 1989). The data are for a sensor at altitude of 675 km and looking perpendicular to the satellite heading to the right. The surface is ocean with $S = 35$ psu and $T_0 = 20$ °C. [Le Vine and Abraham, 2002.]

PROBLEMS

14.1 A SLAR uses an antenna with a $[(\sin x)/x]^2$ gain pattern with a 3 dB beamwidth of 30°. The antenna boresite is pointed at $\theta = 60°$ from a height of 100 m. For this problem assume that σ^0 over the ocean is given by

$$\sigma^0 = K \exp -\theta/6.5°,$$

and over land σ^0 is given by

$$\sigma^0 = K \exp -\theta/25°.$$

Calculate the relative signal power as a function of θ from 30° to 80° for land and ocean. Compare and discuss the effects of the antenna gain pattern and the incidence angle variation of σ^0 on the received signal power.

14.2 A SLAR operating at a frequency of 10 GHz uses 50 ns pulses and a 20 MHz receiver bandwidth. It has a 4 m long antenna with a uniform aperture distribution. Assuming a flight altitude of 5 km, determine the following as a function of across-track distance at 5 km intervals from 5 to 40 km:

(a) Azimuth resolution.

(b) Range resolution.

(c) Equivalent square pixel dimensions.

(d) Potential number of independent samples.

14.3 A SLAR has the following characteristics when flying at 100 m/s at a height of 2 km:

Peak power:	25 kW
Pulse duration:	100 ns
Peak antenna gain:	30 dB
Horizontal beamwidth:	0.01 mrad (two-way)
Noise figure:	10 dB
Bandwidth:	15 MHz
Frequency:	10 GHz
PRF:	2000 Hz

Calculate the following at 10 km ground range, assuming the antenna boresight is pointed at 10 km ground range:

(a) Ground-range resolution.

(b) Along-track resolution.

(c) Signal-to-noise ratio for $\sigma^0 = -20$ dB.

(d) Number of independent samples.

14.4 Separately sketch to scale the Doppler frequency versus time for a radar with 3 cm wavelength, 3 m antenna length, and speed of 200 m/s for targets at slant ranges of 5 km and 20 km. Indicate the frequency limits for an unfocused SAR in the sketch.

14.5 A SAR has a 3 m long antenna with $\lambda = 30$ cm and flies at 200 m/s. Plot the phase versus distance along a synthetic aperture for a point target at slant ranges of 5 and 20 km.

14.6 A radar operates at 10 GHz with a 3 m long antenna. Compare the azimuth resolution for SLAR, unfocused SAR, and fully focused SAR at slant ranges of 1, 5, 10, 50, 100, and 500 km.

14.7 A satellite SAR has the following characteristics:

$r_a = 25$ m	$u = 7.5$ km/s
$r_r = 25$ m	$\sigma^0 = -20$ dB
$l = 10$ m	$h = 500$ km
$G = 55$ dB	$\lambda = 3$ cm
$\theta = 30°$	range chirp $TB = 500$
$F = 5$ dB	$S_1 = 5$ dB

Calculate P_{av}, P_t and S_{N_s}.

14.8 A spaceborne radar with an aperture of $l = 10$ m at a 3 cm wavelength and elevation beamwidth of 7° is flown at 7.5 km/s at an altitude of 500 km and observes at $\theta = 20°$ directly to the side of the flight track. Plot isodops on the ground for this case. If the spacecraft is pitched forward 1°, locate the beam on the ground and plot isodops through it. Discuss frequency correction to compensate for the difference in mean Doppler shift. For simplicity, assume a flat Earth.

14.9 A dual-frequency SAR operates on an aircraft at frequencies of 1 and 10 GHz. Both frequencies use 2 m long antennas.

(a) For resolutions of 3 m in slant range and azimuth, compute and discuss the range curvature at ranges of 5 and 25 km.

(b) Repeat for 15 m resolution.

14.10 Calculate depth of focus for a SAR with $l = 6$ m, $r_a = 3$ m, $\lambda = 5$ cm, and $R = 600$ km. Repeat for $r_a = 30$ m and 60 m.

14.11 A spaceborne SAR has 3 cm wavelength and a 10 m long antenna. It operates at an orbital height of 500 km and a velocity of 7.5 km/s.

(a) Determine the swath width if the inner edge of the swath has an incidence angle of 20°. Use spherical-Earth geometry.

(b) If the resolution is to be 10 m × 10 m, determine the raw data rate in complex samples per second.

(c) Repeat part (b) for 50 m × 50 m resolution.

CHAPTER 15

Interferometric Synthetic-Aperture Radar

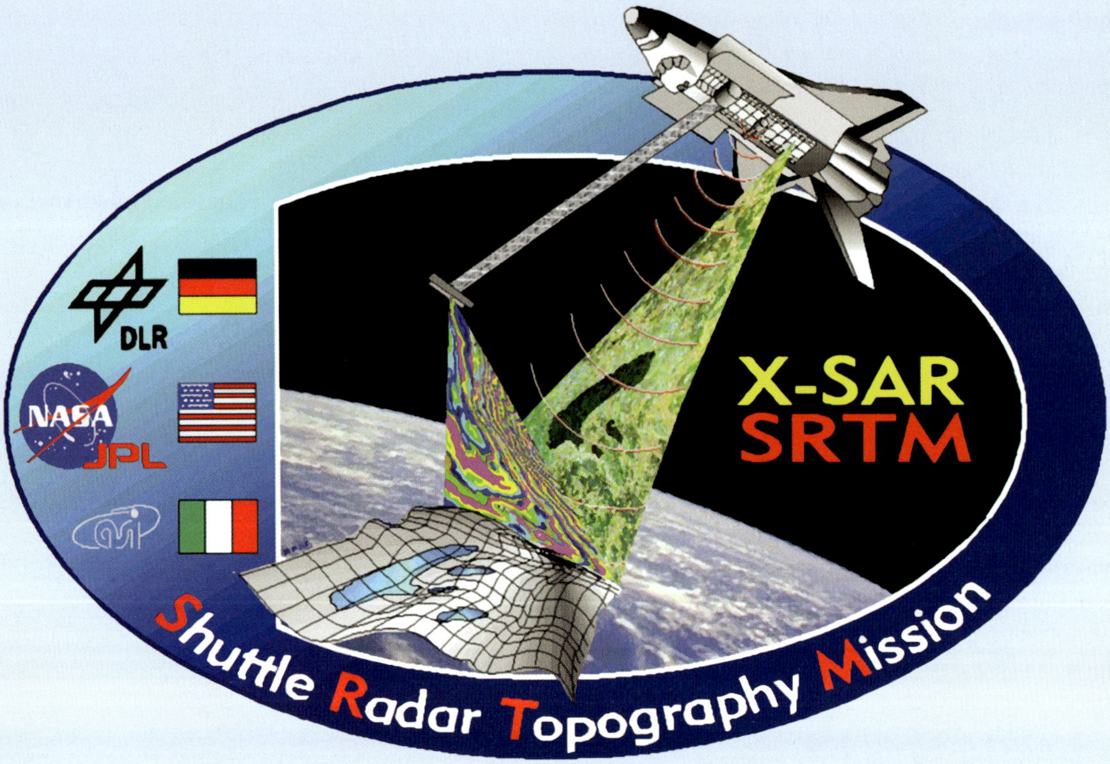

SRTM mission logo

CONTENTS

Overview

A radar echo contains both ***amplitude*** and ***phase*** information. Most imaging radar applications use algorithms that utilize only the amplitude to extract geophysical and biophysical information about the imaged scene. In some cases, the phase difference between the hh- and vv-polarized scattering amplitudes is used as well. In this chapter we consider how to exploit the ***phase difference*** between two radar echoes, associated with the same image pixel, but measured by two different systems along different directions. Measurement of such a phase difference is known as ***interferometry***, which is a well-developed technique in optics.

We can apply many of the same approaches used in optical interferometry to the modeling and analysis of radar signals. ***Radar interferometry*** allows us to measure the surface elevation of individual image pixels, which leads to the generation of ***topographic maps*** with height information. Radar interferometry also is used to measure surface motion. Because most of the applications of radar interferometry rely on the availability of high-resolution images of the terrain, the interferometric data are generally collected and processed using SAR techniques. Hence, such a radar is referred to as an ***interferometric SAR***, or InSAR or IFSAR for short.

In this chapter we introduce InSAR concepts and models, show how the interferometric radar data are used to extract useful geophysical information, and illustrate the InSAR technique with multiple examples.

15-1 Brief History of Radar Interferometry

Radar interferometry was used in Earth-based observations of Venus (Rogers and Ingalls, 1969), as a means to separate ambiguous range-Doppler echoes reflected simultaneously from Venus's northern and southern hemispheres. Subsequently, the technique was used to derive height information about the surfaces of the Moon (Zisk, 1972a, b) and Venus (Rumsey et al., 1974). Following these early planetary explorations, radar interferometric methods were adapted to measure the topography of the Earth's surface, from both aircraft and satellite platforms. Observations of interference fringes modulated by Earth surface topography were recorded by an aircraft interferometric radar and reported by Graham (1974). However, even though the recorded interference patterns clearly demonstrated that topography creates a predictable phase signal when observed interferometrically, they were difficult to analyze, and it proved impractical to extract topography information from the phase signatures.

In the early 1980s, Zebker and Goldstein (1986) implemented an airborne interferometric radar system on a NASA CV990 aircraft outfitted with a digital data-recording system. They recorded two radar channels, one from each of two separately-located radar antennas, and then processed the data using a JPL digital correlator system. They demonstrated that the phase difference between the two signals contained sufficient information so as to reconstruct the topography of the imaged radar scenes. Moreover, it was possible to correct the measured data for ever-present deviations of the aircraft flight path, which caused artifacts in the radar images. It was concluded that the digitally correlated radar interferometric technique can provide digital elevation information with an accuracy on the order of a few meters, which was far superior to the accuracy that was possible with the radar-stereo technique in use at the time.

The Goldstein-Zebker team extended the capabilities of radar interferometry by exploring the possibility of forming ***interferograms*** using radar antennas displaced *along* the flight path (rather than *across* the flight path), corresponding to temporal rather than spatial separation of the images. With this technique, they were able to map the velocities of ocean waves and currents with good accuracies (Goldstein and Zebker, 1987).

Through further work, JPL extended the use of radar interferometry to spaceborne platforms by interferometrically processing data acquired a decade earlier (1978) by the SEASAT SAR. The two major advances here were the ability to detect deformations at cm scale (Gabriel et al., 1988) and the ability to reconstruct continuous functions of radar phase, so-called ***phase***

unwrapping (Goldstein et al., 1988). The processed images detected surface deformations (between different dates) as small as a few centimeters and produced unambiguous maps of ground elevation, but despite these and other similar demonstrations, it was difficult to extend the capability of interferometric radar to mapping surface topography or deformation on a large scale, primarily due to the lack of a sufficiently large archive of interferometric radar data with supportable *in-situ* measurements.

The status of radar interferometry changed drastically after the launch of the ERS-1 satellite by the European Space Agency (ESA) in 1991. The satellite's C-band SAR acquired vast amounts of data on a global scale. At the French Center for Space Studies (CNES), Massonnet et al. (1993) developed an innovative method to isolate the interferometric deformation signature from radar interferograms of the surface, making it possible to readily view the surface displacements of geophysical processes. This development imparted a galvanizing effect on the radar community, and served to establish the credibility of radar interferometry. It also spurred the introduction of new methods and algorithms by numerous investigators (e.g., Zebker and Villasenor, 1992; Just and Bamler, 1994; Goldstein, 1995; Massonnet and Feigl, 1998; Zebker et al., 1997; Rosen et al., 2000) and led to two InSAR satellite missions: the ***Shuttle Radar Topography Mission*** (SRTM) in 2000 and the ***TanDEM-X*** mission in 2010.

The launch of the Envisat satellite by ESA in 2002 extended the continuous record of worldwide observations begun 10 years earlier with the ERS satellites. Other satellites, such as the Canadian Radarsat series, also added to the growing archive of available data. As data coverage became more and more routine, the InSAR technique matured into a standard analytical tool for the geophysics community (Peltzer et al., 2001; Bürgmann et al., 2002; Fielding et al., 2005; Pritchard and Simons, 2004). The technique further advanced by coupling InSAR analysis more closely with crustal deformation modeling (Amelung et al., 2000; Jonsson et al., 2002), by adding new capabilities such as vector deformation mapping (Bechor and Zebker, 2006), and also by moving into new application areas such as hydrological applications (Amelung et al., 1999; Hoffmann et al., 2003).

Since 2000, the focus of technological development in the radar interferometry arena has shifted towards the production of time series of interferograms so as to reconstruct surface-deformation histories, rather than snapshots at a single time of deformation. A number of new methods, including ***stacking*** (Lyons and Sandwell, 2003), ***persistent scattering*** (Ferretti et al., 2000), and ***small baseline subset analysis*** (Berardino et al., 2002), have become commonplace. The persistent-scattering approach was developed by Ferretti and colleagues (Ferretti et al., 2000, 2001, 2004), who applied it mainly to the analysis of urban structures with well-defined scattering centers. Its applicability to natural terrains was pioneered by Hooper et al. (2007), who showed that many surfaces or vegetation covers in fact exhibit scattering that does not scintillate in phase over many months or years. Improved persistent-scattering detection algorithms based on maximum likelihood methods (Shanker and Zebker, 2007) have led to more robust implementations exploiting these longer-lived pixels.

15-2 2-D versus 3-D Measurements

When we measure the returns of multiple pulses and then process them together to realize high-resolution capability in both the range and azimuth directions, we generate a radar image that displays the 2-D backscatter distribution of the imaged scene. Each pixel in the image has a time delay associated with its distance R from the radar, and a corresponding two-way ***phase delay*** given by

$$\phi = -\frac{4\pi}{\lambda} R, \tag{15.1}$$

where λ is the EM wavelength. The measured phase delay may include a noise contribution, but for simplicity we ignore it for the present time.

15-2.1 Interferometric Phase

> ▶ InSAR methods are based on combining radar returns from two different antennas, displaced from each other in space or time. ◀

If the antennas are displaced in space, we denote the spatial separation as the radar ***spatial baseline***. The two radar images of the scene differ due to the ***parallax*** of the two observations, generating a distortion of the one image compared with the other. The distortion is related to the surface topography.

The scenario shown in Fig. 15-1 depicts a SAR imaging a scene with topographic variations. The radar return signal from a particular range bin includes contributions from all equirange points on the surface corresponding to that range bin. Except for the few situations in which the radar shadow can be used to estimate the height of an object or feature, it is difficult to extract height information from single radar images, because the radar image is fundamentally two-dimensional in nature, whereas the scene is three-dimensional. Many surface points at different heights can return echoes simultaneously.

When two antennas are used to generate two separate images, a point P on the terrain surface, as depicted in Fig. 15-2, is at the intersection of two circles, one of radius R_1 with its center at antenna A_1—and another of radius R_2. If we know the positions of the two antennas, and if we have accurate measurements of R_1 and R_2, we can determine the 3-D coordinates of P unambiguously.

Prior to the advent of radar interferometry, the traditional method that was used to extract topographic information from pairs of radar images of the same scene relied on the radar-stereo technique described in Section 14-11. The stereo technique does not require or use phase information, but it has limited sensitivity to topography and it is inherently susceptible to the presence of speckle in the radar images. Furthermore, the two ranges R_1 and R_2 to a specific feature in the scene have to be significantly different in length in order for the stereo technique to provide a useful estimate of height.

The two receiver configurations commonly used in radar interferometry are displayed in Fig. 15-3:

(1) ***Cross-track with two independent radars***, each with its own transmit/receive antenna [Fig. 15-3(a)].

(2) ***Cross-track with common transmit antenna***, but separate receive antennas [Fig. 15-3(b)]. The transmit antenna also serves as the receive antenna for one of the radars and an auxiliary antenna is used as the receive antenna for the second radar.

(3) ***Along-track*** with two independent radars separated in time [Fig. 15-3(c)].

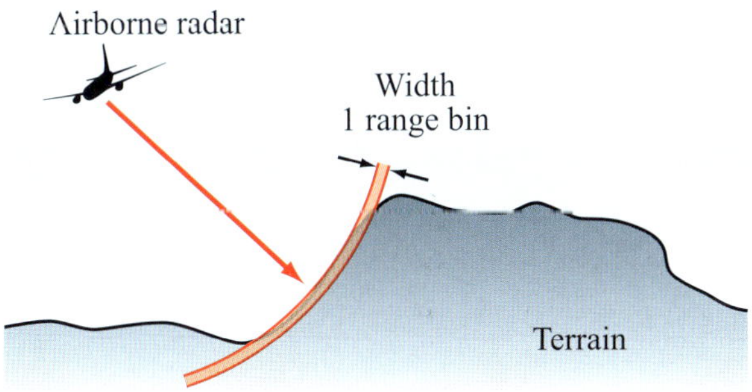

Figure 15-1: Radar returns from all points within any range bin are received at once. Thus conventional radar images cannot properly depict topographic information.

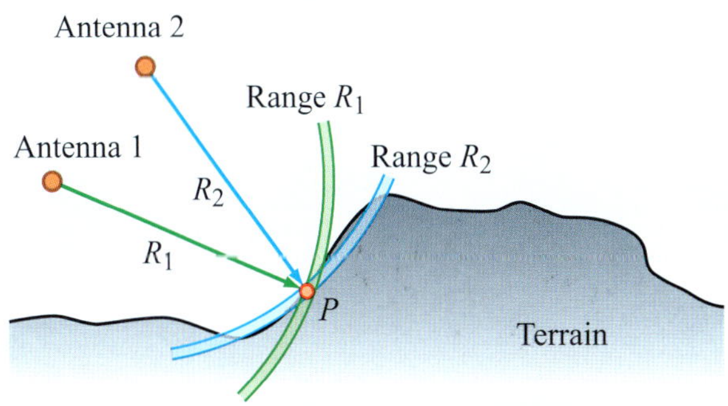

Figure 15-2: Distances to a point P from Antenna 1 and Antenna 2 are R_1 and R_2, respectively. The two arcs of these radii cross at the true position of the point in three-dimensional space.

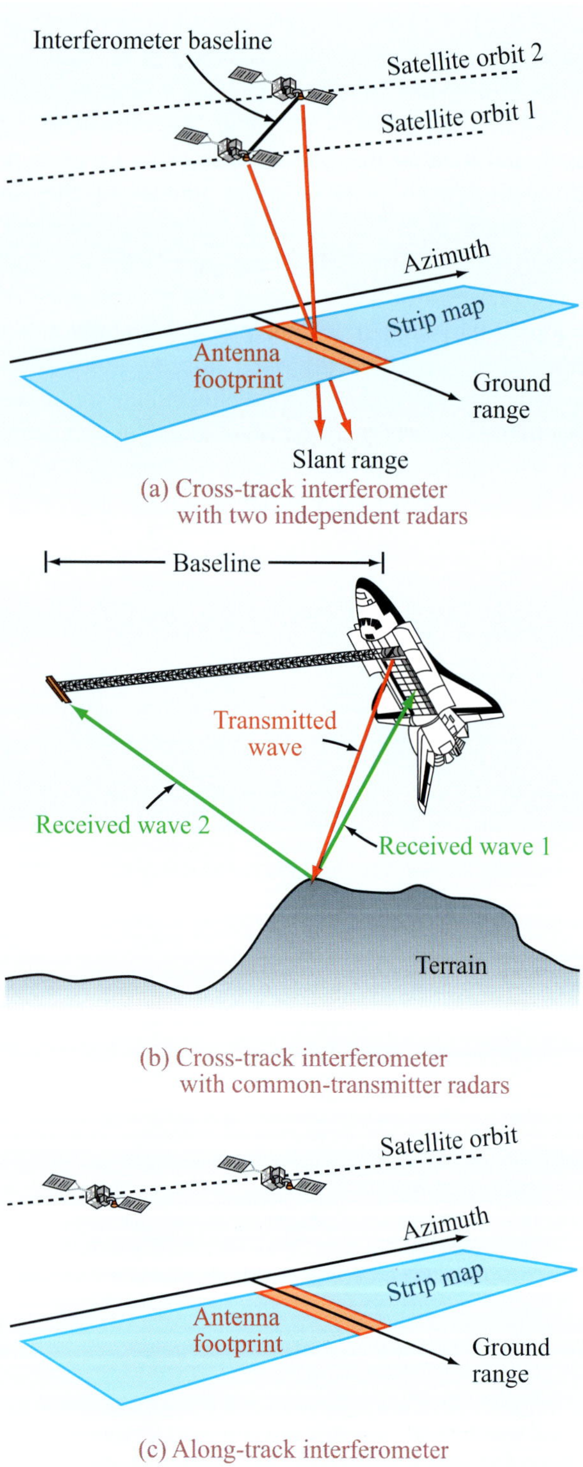

Figure 15-3: Cross-track interferometric configurations in (a) and (b) and along-track configuration in (c).

The two antennas may be be separated from one another in the cross-track direction, as shown in Fig. 15-3(a) and (b), which means that the two antennas comprising the interferometer are flown on separate, parallel tracks. Such an interferometer is called a ***cross-track interferometer***. Alternatively, the two antennas can be flown along the same track [Fig. 15-3(c)], with one displaced from the other by a certain distance or flown at a different time, and we refer to this geometry as an ***along-track interferometer***.

▶ As we see in later sections, the cross-track configuration provides topography (height) information, whereas the along-track configuration provides information about the movement of the surface between the observation times associated with the two images. ◀

The radar interferometric measurement technique relies on the use of phase information and it takes advantage of the ***parallel-ray approximation***, which requires ranges R_1 and R_2 in Fig. 15-4 to be different, but by only a very small amount. That is $|R_1 - R_2| \ll R_1$. The spacing between the two antennas—called the

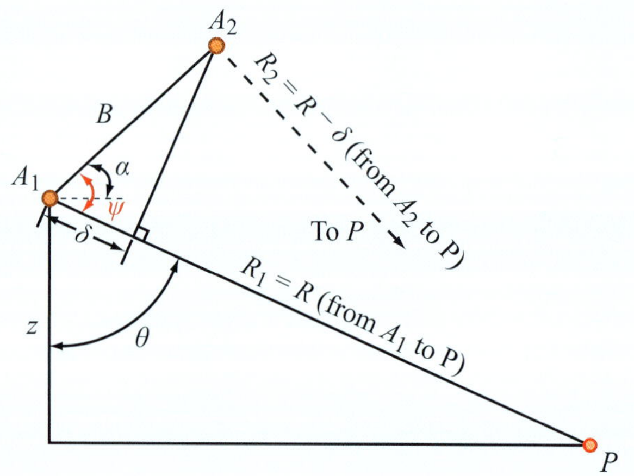

Figure 15-4: Parallel-ray approximation for InSAR geometry with $R_1 = R$ and $R_2 = R - \delta$. Note that R_1 is from A_1 to P.

baseline B—is greatly exaggerated in the figure relative to R_1 and R_2. For convenience, we label R_2 as $R - \delta$ and R_1 as R.

For the two-independent-radars configuration [Fig. 15-3(a)], the roundtrip path length between A_1 and P is $2R$ and for the second radar, it is $2R - 2\delta$. The corresponding ***interferometric phase difference*** (of radar A_1 relative to radar A_2) is

$$\phi_{\text{int}} = \phi_2 - \phi_1 = -2k\delta = -\frac{4\pi\delta}{\lambda}. \qquad (15.2a)$$

(independent radars)

When the same transmit antenna is used, as in Fig. 15-3(b), the two-way path length for receive antenna A_2 is only $2R - \delta$, in which case the phase difference is

$$\phi_{\text{int}} = -k\delta = -\frac{2\pi\delta}{\lambda}. \qquad (15.2b)$$

(common-transmit antenna)

The two expressions for ϕ_{int} can be combined into

$$\phi_{\text{int}} = -\frac{2\pi n\delta}{\lambda}, \qquad (15.2c)$$

where

$$n = \begin{cases} 1 & \text{for common-transmit antenna,} \\ 2 & \text{for two independent radars.} \end{cases} \qquad (15.2d)$$

In Fig. 15-4, we denote the orientation of the baseline B with respect to the horizontal axis by the angle α. For triangle A_1A_2P, angle ψ is given by

$$\psi = \frac{\pi}{2} - \theta + \alpha, \qquad (15.3)$$

and from the law of cosines

$$(R - \delta)^2 = R^2 + B^2 - 2RB\cos\left(\frac{\pi}{2} - \theta + \alpha\right). \qquad (15.4)$$

However, since $\delta \ll R$, we can apply the parallel-ray approximation which leads to

$$\delta = B\sin(\theta - \alpha). \qquad (15.5)$$

For a point P at height z relative to the height of A_1,

$$z = R\cos\theta. \qquad (15.6)$$

Knowledge of the range R is available from the radar data. Through Eqs. (15.2a) and (15.2b), δ is determined from the phase difference measurement, which in turn leads to a determination of θ from Eq. (15.5), and finally to a determination of the topographic height z from Eq. (15.6).

15-2.2 Height Measurement Precision

A critical parameter for the design of a topographic mapping system is the ***estimated height error*** (precision) associated with the measurement of z, to which we assign the standard deviation symbol s_z. The precision s_z is related to the precision with which δ can be measured. Application of the chain rule

$$\frac{\partial z}{\partial \delta} = \frac{\partial z}{\partial \theta}\,\frac{\partial \theta}{\partial \delta}, \qquad (15.7a)$$

and using Eqs. (15.5) and (15.6) to compute the derivatives

$$\frac{\partial z}{\partial \theta} = -R\sin\theta, \qquad (15.7b)$$

$$\frac{\partial \delta}{\partial \theta} = -B\cos(\theta - \alpha), \qquad (15.7c)$$

leads to

$$\frac{\partial z}{\partial \delta} = \frac{-R\sin\theta}{B\cos(\theta - \alpha)}. \qquad (15.8)$$

Thus, a change in δ of s_δ leads (approximately) to a change in z of s_z given by

$$s_z = \frac{R\sin\theta}{B\cos(\theta - \alpha)}\, s_\delta. \qquad (15.9)$$

The quantities s_z and s_δ represent the standard deviations associated with the inferred values of z and δ, respectively. The minus sign in Eq. (15.8) was not carried over to Eq. (15.9) because standard deviations are positive quantities representing deviations relative to mean values.

For typical values of θ and α, the factor $\sin\theta/\cos(\theta-\alpha)$ seldom deviates from 1 by more than a factor of 2 (larger or smaller), so we may approximate Eq. (15.9) as

$$s_z \approx \frac{R}{B}\, s_\delta. \tag{15.10}$$

The precision with which δ can be measured depends on the precision with which the interferometric phase difference ϕ_{int} can be measured. Using Eq. (15.2c),

$$s_\delta = \frac{\lambda}{2\pi n}\, s_{\phi_{\text{int}}}, \tag{15.11a}$$

which leads to

$$s_z = \frac{R}{B}\,\frac{\lambda}{2\pi n}\, s_{\phi_{\text{int}}}. \tag{15.11b}$$

Here, $n = 1$ for the common-transmitter interferometer and $n = 2$ for the two-independent-radar interferometer. In practice, it is possible to measure ϕ_{int} with an accuracy of a few degrees. If, for demonstration purposes, we choose $n = 2$ and we set $s_{\phi_{\text{int}}} = 7.2^\circ = 0.1257$ rad, we obtain

$$s_\delta = \frac{\lambda}{100} \tag{15.12a}$$

and

$$s_z = \frac{R}{B}\cdot\frac{\lambda}{100}. \tag{15.12b}$$

By way of illustration, the relevant parameters for the ERS SAR are: $R = 800$ km, $B = 1$ km (for two orbital passes separated by 1 km), and $\lambda = 6$ cm. Using Eq. (15.12b), we obtain

$$s_z \approx \frac{R}{B}\,\frac{\lambda}{100} \approx \frac{1}{2}\ \text{m}.$$

In contrast, the measurement height precision when only R is measured and no phase information is used—as in the radar stereo case—is 7.2 km! While this is a rather ideal noise-free scenario for the interferometric radar, it does illustrate the power of the interferometric approach to topography estimation.

To illustrate with a real example, we show in Fig. 15-5 SAR images at various stages of interferometric processing, leading to a 3-D topography map. The images and associated processing steps are discussed in the following sections.

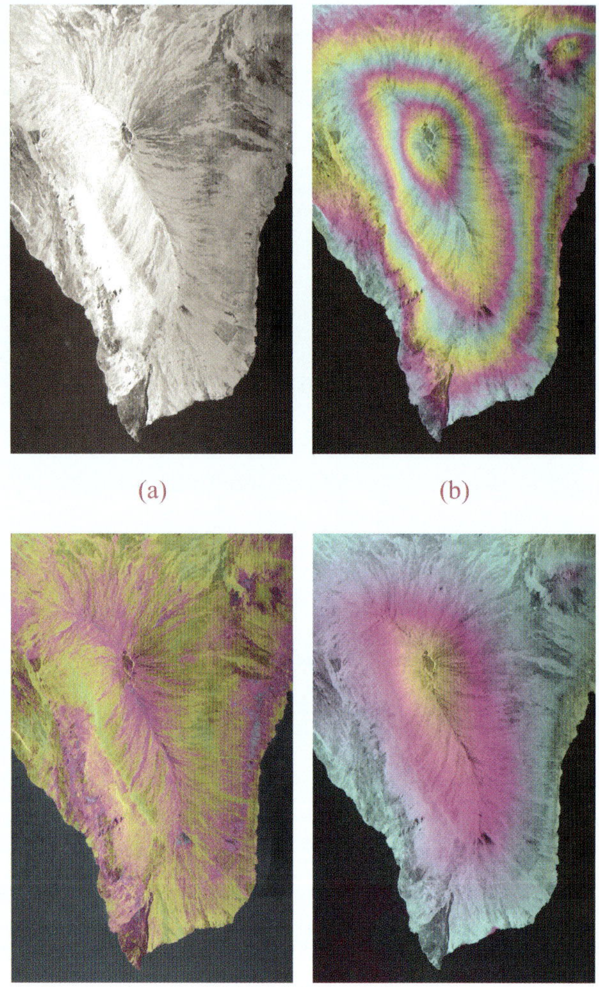

(a) (b) (c) (d)

Figure 15-5: Example of interferometric SAR processing: (a) SAR magnitude image, (b) phase difference (interferogram) between two images with flat-Earth phase removed, (c) correlation map between interferogram component scenes, and (d) inferred surface topography.

15-2.3 The Role of SNR

Clearly, interferometric performance depends critically on how well the interferometric phase difference ϕ_{int} can be measured, which in turn depends on the signal-to-noise ratio (SNR). We now consider how SNR impacts the phase-difference measurement uncertainty $s_{\phi_{\rm int}}$.

The variance of the measured phase difference, $s^2_{\phi_{\rm int}}$, is due to the random phase component introduced by the noise accompanying the signal, and it is approximately proportional to the inverse of the SNR,

$$s^2_{\phi_{\rm int}} \approx \frac{1}{\rm SNR} \,. \tag{15.13}$$

To secure the value $s_{\phi_{\rm int}} = 0.1257$ rad that we used in the example leading to Eq. (15.12), it is necessary that SNR = 63.29, or, equivalently, 18 dB.

Collecting Eqs. (15.9), (15.11b), and (15.13) leads to

$$s_z = \frac{\lambda}{2n\pi} \cdot \frac{R\sin\theta}{B\cos(\theta-\alpha)} \cdot \frac{1}{\sqrt{\rm SNR}} \,, \tag{15.14a}$$

$$n = \begin{cases} 1 & \text{for common transmitter,} \\ 2 & \text{for independent radar.} \end{cases} \tag{15.14b}$$

15-3 Cartographic Corrections

In order to further develop the interferometric technique for cartographic applications, we need to understand how topography affects the radar images. In the following, we briefly consider ***geocoding*** and ***georeferencing*** and their relationships to radar interferometry.

As described in Chapter 14, a typical ***slant range*** SAR image is presented in ***radar coordinates***, a coordinate system in which the range coordinate is the slant range and the azimuth coordinate corresponds to the along-track distance. In a ***ground range*** SAR format, the slant range is converted to ground range (Fig. 15-6). Surface topography often leads to foreshortening or even layover in the resulting image. This is caused by the nonlinear mapping of slant range into ground range, plus any effects from uncompensated topography.

Figure 15-7 illustrates various geometrical distortions that occur when the radar-coordinate data are mapped into a geographically corrected coordinate system. When the surface is flat, this transformation is straightforward, but it is more complicated when surface topography is present. As a consequence of the nonlinear mapping from slant to ground range, the shapes of objects are not preserved and ground range lines are not imaged as straight lines. This is illustrated in Fig. 15-8.

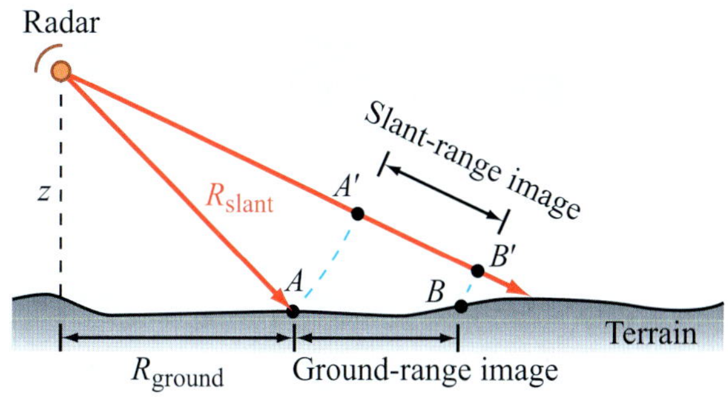

Figure 15-6: Relationship between slant-range and ground-range image presentations for a side-looking radar.

Multipath effects in radar reflection can lead to ghosting, artifacts, or replicated targets in the SAR image. Figure 15-9 illustrates how multipath can result in target reflections appearing at two different slant ranges. Consequently, the target appears at two different slant ranges in the image. Typically, the direct-path target image is well focused, but the longer multipath image is often less well focused due to the mismatch between the azimuth matched filter at the apparent range and the range migration of the target multipath.

These distortions result from the two-dimensional imaging of conventional SAR. Fortunately, by using interferometric techniques, InSAR generates three-dimensional information that allows us to estimate the terrain height, which allows us to correct for these distortion effects. Note that slant range $R_{\rm slant}$ and ground range $R_{\rm ground}$ are related to height by (Fig. 15-6)

$$R^2_{\rm ground} + z^2 = R^2_{\rm slant} \,, \tag{15.15a}$$

so if we resample a slant range image according to

$$R_{\rm ground} = \sqrt{R^2_{\rm slant} - z^2} \,, \tag{15.15b}$$

we can properly recover the undistorted ground-range image from the slant-range image. With the additional

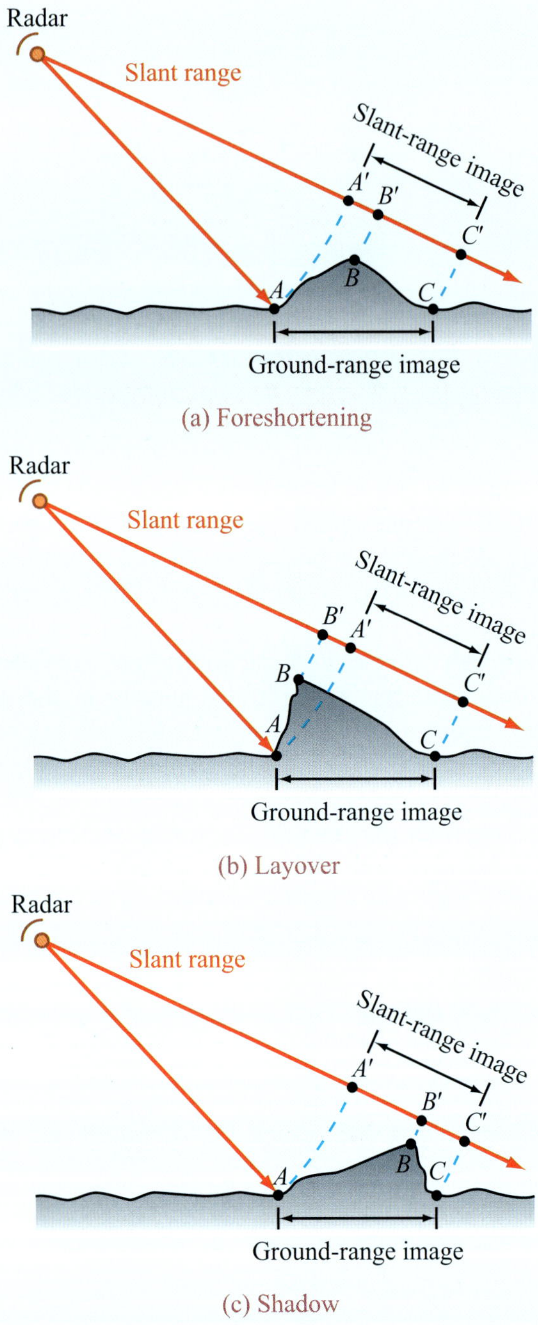

Figure 15-7: Geometric distortions: (a) foreshortening, which occurs when the distance between A' and B' in the slant range is less than the ground-range distance between A and B, (b) layover, where B' is imaged in front of A', and (c) shadow, where the unilluminated mountain back yields a dark region between B' and C'.

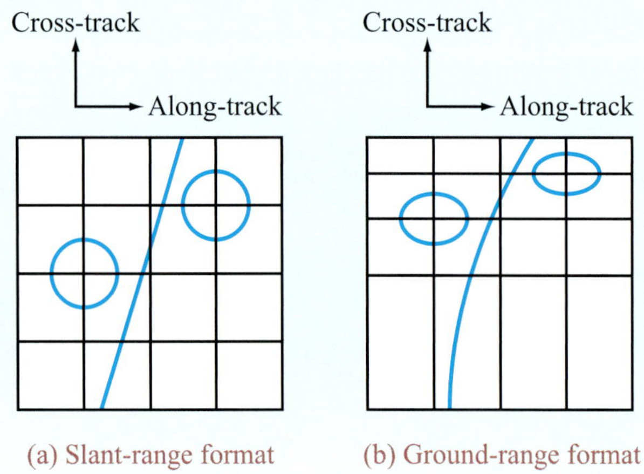

Figure 15-8: Distortions resulting from mapping (a) ground-range to (b) slant-range format. Note that slant-range images do not faithfully reproduce shapes and lines in ground-range images.

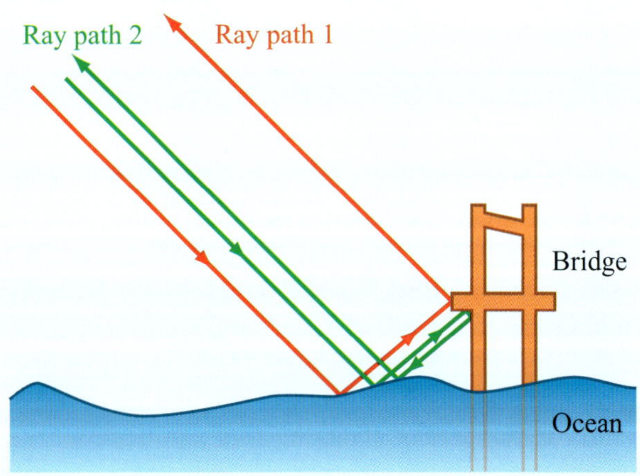

Figure 15-9: Multipath distortion. Two separate ray paths, with different total lengths, illuminate a bridge, causing it to appear twice in the radar image.

knowledge of height, a geometrically accurate ground range image can be formed from the slant range. This process is known as ***SAR georeferencing***, or SAR geocoding, although the latter is also concerned with properly locating each SAR image pixel in an absolute sense. Shadows and extreme foreshortening leading to layover still provide distortions in the georeferenced image. Backscatter variations due to topographic slope have to be corrected separately.

15-4 Forming the Radar Interferogram

A radar interferometer employs two antennas displaced from each other. In ***across-track interferometry***, the two antennas are displaced in the across-track direction, while in ***along-track interferometry*** they are displaced in the along-track direction. In ***single-pass interferometry***, the two antennas are mounted on a single platform and the data are collected simultaneously. In ***multiple-pass interferometry***, data collections are made at different times using either multiple platforms or multiple passes over the data collection area with the same platform.

In Section 15-2, we established that the phase difference between the signals received by the two antennas contains information about the surface topography. An image of the phase difference is known as an ***interferogram***. In the following, we examine in detail the formation of the interferogram.

15-4.1 Displacement versus Range

We start by restating Eqs. (15.5) and (15.6):

$$\delta = B\sin(\theta - \alpha), \tag{15.16}$$

and

$$z = R\cos\theta. \tag{15.17}$$

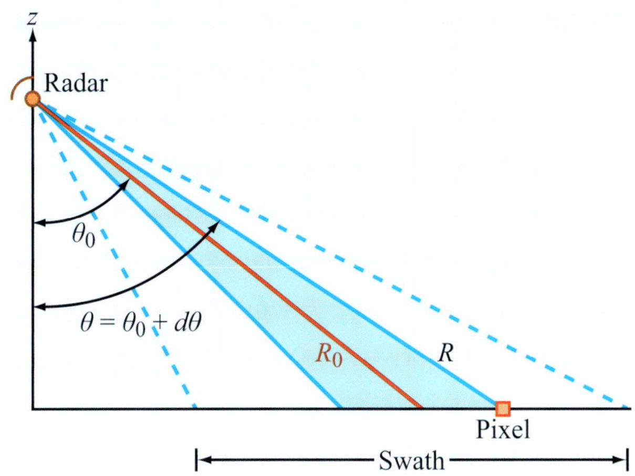

Figure 15-10: Illuminated swath is typically centered in angle about a mean look angle θ_0.

When creating and processing a radar interferogram, it is useful to reference the range location of a pixel to the center of the image swath. In Fig. 15-10, θ_0 represents the incidence angle along the center of the swath and the corresponding range is R_0. The elevation beamwidth of a satellite SAR is typically on the order of a few degrees (e.g., for ERS-1/2 SAR, θ extends from 20° to 26°, with θ_0 at 23°, and for JERS-1, θ extends from 32° to 38°, with θ_0 at 35°). This means that for the pixel P in Fig. 15-10, the angular deviation $d\theta$ from θ_0 is at most 3°. Replacing θ with $\theta_0 + d\theta$ in Eqs. (15.16) and (15.17) and using the approximations $\sin(d\theta) \approx d\theta$ and $\cos(d\theta) \approx 1$ gives

$$z = R\cos(\theta_0 + d\theta) \approx R[\cos\theta_0 - \sin\theta_0 \, d\theta] \tag{15.18}$$

and

$$\begin{aligned}\delta &= B\sin(\theta_0 - \alpha + d\theta) \\ &= B[\sin(\theta_0 - \alpha) + \cos(\theta_0 - \alpha)\, d\theta].\end{aligned} \tag{15.19}$$

Solving Eq. (15.18) for $d\theta$ yields

$$d\theta = \frac{R\cos\theta_0 - z}{R\sin\theta_0}, \tag{15.20}$$

and when this is inserted in Eq. (15.19), the latter becomes

$$\begin{aligned}\delta &= B\sin(\theta_0-\alpha)+B\cos(\theta_0-\alpha)\,\frac{\cos\theta_0 - z/R}{\sin\theta_0} \\ &= B\sin(\theta_0-\alpha)+\frac{B\cos(\theta_0-\alpha)}{\tan\theta_0} \\ &\quad -\frac{B\cos(\theta_0-\alpha)}{\sin\theta_0}\,\frac{z}{R}\,. \end{aligned} \tag{15.21}$$

Expanding z/R of the last term in Eq. (15.21) about R_0, where $R = R_0 + dR$, gives

$$\begin{aligned}\frac{z}{R} &= \frac{z}{R_0+dR} = \frac{z}{R_0}\left[\frac{1}{1+(dR/R_0)}\right] \\ &\approx \frac{z}{R_0}\left(1-\frac{dR}{R_0}\right) = \frac{z}{R_0} - \frac{z\,dR}{R_0^2}\,. \end{aligned} \tag{15.22}$$

Incorporating Eq. (15.22) into Eq. (15.21) leads to

$$\begin{aligned}\delta &= B\sin(\theta_0-\alpha)+\frac{B\cos(\theta_0-\alpha)}{\tan\theta_0} \\ &\quad -\frac{B\cos(\theta_0-\alpha)}{\sin\theta_0}\left(\frac{z}{R_0}-\frac{z}{R_0^2}\,dR\right). \end{aligned} \tag{15.23}$$

Upon using the approximation $z \approx R_0\cos\theta_0$, Eq. (15.23) simplifies to

$$\delta = B\sin(\theta_0-\alpha)+\frac{B\cos(\theta_0-\alpha)}{\tan\theta_0}\,\frac{dR}{R_0}\,. \tag{15.24}$$

We can define the baseline at the nominal incidence angle θ_0 in Fig. 15-11 as having a ***parallel*** component $B_{\|}$ and a ***perpendicular*** component $B_\perp$ given by

$$\begin{aligned} B_{\|} &= B\sin(\theta_0-\alpha), \\ B_\perp &= B\cos(\theta_0-\alpha). \end{aligned}$$

In other words, the total offset between the two images δ can be expressed as a function of the two ***InSAR baseline components*** $B_{\|}$ and $B_\perp$ as

$$\delta = B_{\|} + B_\perp\,\frac{dR}{R_0\tan\theta_0}\,. \tag{15.25}$$

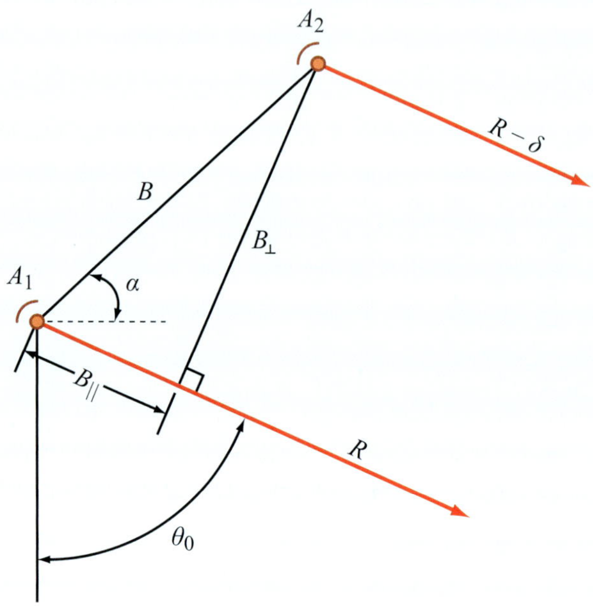

Figure 15-11: Parallel and perpendicular components of the interferometer baseline B as defined by incidence angle θ_0. Range R is from A_1 to target point.

This result suggests that δ has a constant shift equal to the parallel component of the baseline, and a linearly increasing shift (or ***stretch***) that varies inversely with the geometric factor $R_0\tan\theta_0$. Figure 15-12 schematically illustrates the spatial shift and scaling that occurs between the images created from the two antennas. Because of the scale and shift, a target observed at one pixel location in a particular image, is at a different pixel position in the second image

When forming the interferogram, the phase difference relates the phase in a pixel in image 1 to the corresponding phase of the same target in image 2, which may be at a different pixel location. Thus, it is necessary to establish a formatting relationship between each pixel in image 2 and its corresponding location in image 1.

In practice, this is accomplished by estimating the shift in image 2 relative to image 1 by identifying corresponding points in the two images and solving for

(a) Antenna 1 image (b) Antenna 2 image: shifted and scaled

Figure 15-12: (a) Image as seen by antenna 1 and (b) image as seen by antenna 2. The combination of angle-dependent mapping into slant range coordinates causes antenna 2 to see a shifted and scaled version of the image seen by antenna 1.

(a) Image 1 (b) Image 2

Figure 15-13: Schematic illustration of corresponding small regions in the two interferometer images: (a) image from antenna 1, with a particular small region identified, and (b) corresponding area in the image from antenna 2.

an equation for the offset. This is termed ***determining the interferometric offset field*** or ***image coregistration***.

15-4.2 Offset Determination

By comparing the positions of the same point scatterers in the two images, we can derive an empirical equivalent of the relation given Eq. (15.25). Because the radar data contain some noise, the offset determination process is imperfect, resulting in a somewhat noisy estimate. This is overcome by estimating the offsets for a large number of locations and solving for the parameters determining δ using a least-squares approach to find a well-fitting low-order approximate relation.

There are many ways to determine the offsets, but for the sake of brevity, only one recipe is considered here. Since in the general case there may be shifts both in range and azimuth, we adopt a 2-D approach to solving the shift estimation problem

▶ Offsets are found by choosing a small region in image 1, and cross-correlating it over a similar area in image 2 to find the best matching terrain. The position of the correlation peak then gives the offset at that location. ◀

The approach begins by identifying a small region within image 1, as shown in Fig. 15-13. From the imaging geometry, we know, approximately but not exactly, where the region should occur in image 2. To find the exact location, the cross-correlation of the region locations is computed. Let $\mathbf{C}_1(m,n)$ be the complex image in the selected region in image 1 at location (m,n), and let $\mathbf{C}_2(m_2,n_2)$ be the complex image of a region in image 2 centered at location (m_2,n_2). The ***cross-correlation function*** $\mathbf{K}_\mathrm{C}(i,k)$ is computed using

$$\mathbf{K}_\mathrm{C}(i,k) = \frac{\sum_i \sum_k \mathbf{C}_1(m,n)\, \mathbf{C}_2^*(m_2+i,\ n_2+k)}{\sqrt{\overline{\mathbf{C}_1}\ \overline{\mathbf{C}_2}}}\,, \tag{15.26}$$

with

$$\overline{\mathbf{C}_1} = \sum\sum \mathbf{C}_1(m,n)\, \mathbf{C}_1^*(m,n),$$

$$\overline{\mathbf{C}_2} = \sum\sum \mathbf{C}_2(m_2+i,\ n_2+k)\, \mathbf{C}_2^*(m_2+i,\ n_2+k),$$

and where the sums are computed over an area of the size of $\mathbf{C}_1$. In practice, a two-dimensional signal processing window is used in this computation. Because the initial offset is not known exactly, in practice $\mathbf{C}_2$ is typically selected to be larger than $\mathbf{C}_1$ as illustrated in Fig. 15-14.

The magnitude of the cross-correlation function $\mathbf{K}_\mathrm{C}(i,k)$ describes how similar the two image regions are when region 2 is spatially shifted by (i,k). The

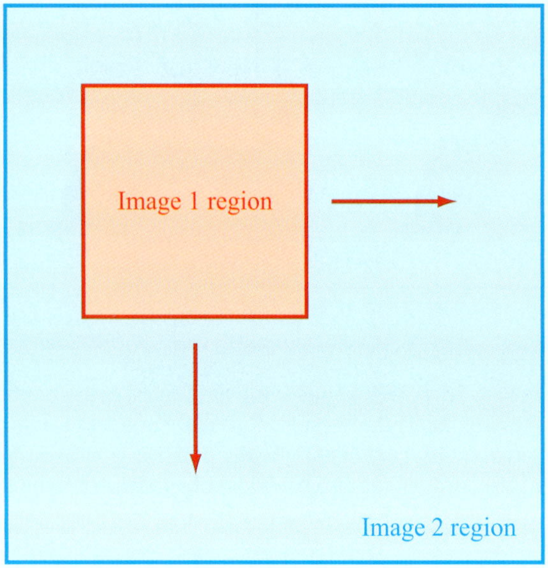

Figure 15-14: In computing the cross-correlation function, a small region from image 1 is cross-correlated against a larger piece of image 2 to enable a search range for the correlation peak.

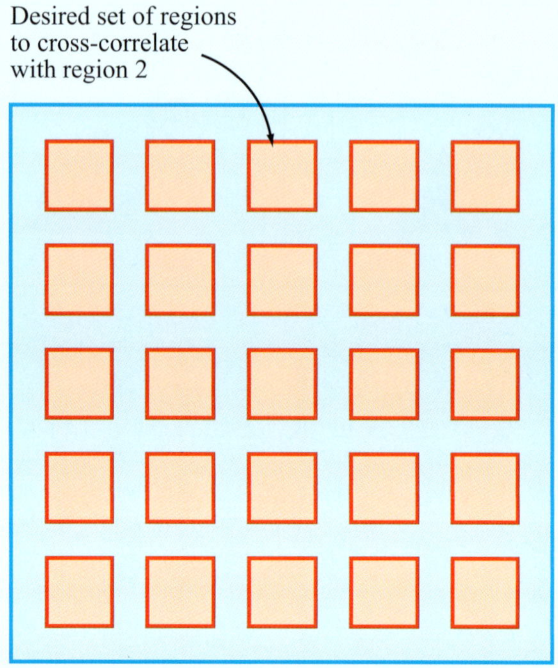

Figure 15-15: Grid of image 1 regions to cross-correlate with image 2.

cross-correlation is maximized at the shift that causes the complex regions to be the most similar. This peak, located at (i_0, k_0), is the image offset we seek.

Since the images are complex, the complex correlation described in Eq. (15.26) can be directly implemented to find the required shift. However, when the phase noise is large (due, for example, to low SNR in regions of the image with low σ^0), the complex correlation may lead to inaccurate estimates. It is useful to correlate the image powers (pixel magnitudes) rather than the complex pixel values. In this case the complex values in Eq. (15.26) are replaced with the pixel magnitude squared; i.e., by $|\mathbf{C}_1|^2$ and $|\mathbf{C}_2|^2$. This eliminates noise-related issues in regions with inadequate phase fidelity.

Note that because we are multiplying two detected images together when computing the cross-correlation, the bandwidth of the cross product (in the correlation process) effectively is twice the bandwidth of the original data. This can lead to aliasing if the original image data are not oversampled by at least a factor of 2. Hence, for best results we generally interpolate each region by at least a factor of 2 in each direction before performing the correlation.

The method described thus far is for a particular subregion of image 1. By analyzing multiple regions, the offset estimate can be improved. To obtain a good set of estimates for the next step, a good rule of thumb is to examine 10–20 locations in range and in azimuth as indicated in Fig. 15-15. For a given offset estimate, once the region correlations are computed, δ can be estimated empirically using a linear least-squares approach. Figure 15-16 is a typical example of a plot of the offset estimates for individual regions as a function of the range R. Also shown is the linear least-squares fit to the data. From the equation of the line, we can estimate both the parallel and perpendicular components of the baseline. The y intercept corresponds to $B_{\|}$ at the reference pixel defined by θ_0 and the slope of the line in Fig. 15-16 corresponds to $B_{\perp}$. The initial estimates of

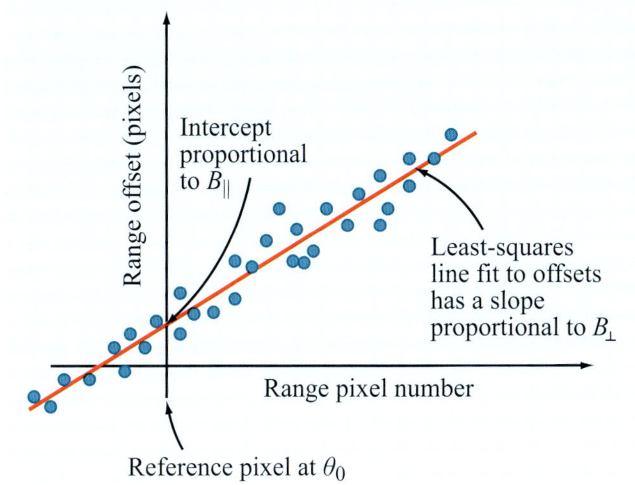

Figure 15-16: Plot of noisy measured offsets as a function of range location and the least-square fit of the data. The slope and intercept of the line yield the parallel and perpendicular baseline components.

$B_{\|}$ and $B_{\perp}$ can be refined later using known topographic corrections.

The next step is to resample image 2 to the coordinates of image 1 using the relation for δ given by Eq. (15.25) with the empirically determined values of $B_{\perp}$ and $B_{\|}$. In addition, if there is an azimuth shift due to timing considerations, this is removed at this time as well.

Finally, the interferogram $\mathbf{I}(x,y)$ is formed by computing

$$\mathbf{I}(x,y) = \text{image } 1(x,y) \\ (\text{image } 2(x+\delta,\ y+\alpha))^*, \qquad (15.27)$$

where the complex conjugate multiplication results in the desired phase differencing operation. In this equation x corresponds to the pixel index in the range direction, y is the pixel index in the azimuth direction, and α is the azimuth offset.

In summary, the steps involved in formatting the interferogram include:

(1) Processing the two SAR images measured by the two receive antennas.

(2) Measuring the offsets by:

(a) selecting a small region in image 1 and its approximate equivalent in image 2,

(b) interpolating by 2 or more,

(c) detecting complex values to obtain power images of several small regions,

(d) cross-correlating and recording of offset peak locations,

(e) repeating for many different locations,

(f) obtaining a least-square fit for the offset values as a function of range; also determining azimuth offset, if any, and

(g) solving for $B_{\perp}$ and $B_{\|}$ from the slope and intercept of the line.

(3) Resampling image 2 to image 1 coordinates.

(4) Cross-multiplying image 1 by the conjugate of image 2 to form the interferogram.

15-4.3 Multilooking

Having computed the interferogram, it is often desirable to ***multilook*** the data in order to improve the accuracy of the phase estimate, at the expense of spatial resolution. As described in Chapter 5, multilooking means that pixel intensities are averaged in one or both dimensions to reduce fluctuations. Multilooked data are much smoother than single-look imagery and can be downsampled without losing much information. This also reduces computation and the required data volume for storage.

For image data, recall from Chapter 5 that multilooking involves averaging power values rather than complex values. As a general rule, in most radar calculations, quantities representing second-order statistics are averaged in a multilook processor. For example, when considering the sum of two Gaussian variables, the variances, not the standard deviations, add. Thus when we average two pixels in a radar image, the mean power is preserved, not the mean standard deviation. It is this property that leads to the reduction of relative uncertainty for multilooked pixels. Because the interferogram pixels are derived from the product of two field quantities, these are essentially powers, even though they have a distinct phase associated with each of them.

> ▶ Hence, averaging the interferogram pixels directly approaches the mean power while preserving the average phase value. ◀

To help understand the difference from another perspective, note that in the multilooking case (when a single image is involved), the pixel phases are uniformly distributed and are essentially independent from pixel to pixel, so when they are added together to reduce noise, the powers add. For the interferogram (which is the phase difference between two images), the interferometric phase changes slowly from pixel to pixel, so coherent integration to reduce noise is very helpful.

15-4.4 Correlation

An important measure of interferogram quality is the ***degree of correlation*** between the two images, or how closely the phase in one image tracks that of the other. This quantity varies between 0 for images that are completely independent of each other to 1 for images with identical phases. The image correlation can be computed when generating the multilooked image by applying the following definition:

$$\rho = \frac{\sum \text{image } 1_{\text{i}}\text{image } 2_{\text{i}}^*}{\sqrt{\sum \text{image } 1_{\text{i}}\text{image } 1_{\text{i}}^*}\sqrt{\sum \text{image } 2_{\text{i}}\text{image } 2_{\text{i}}^*}}, \tag{15.28}$$

where image 1_{i} and image 2_{i} are the complex values of a corresponding point in image 1 and image 2 after image 2 has been resampled according to the estimated shift. Note that the numerator is the interferogram while the denominator is the product of the image *amplitudes*, not powers.

In general, the degree of correlation is a function of SNR and the local target characteristics (to be described later).

15-5 Decorrelation

> ▶ ***Correlation*** is quite fundamental to understanding and interpreting interferometric measurements and is closely coupled with our ability to estimate interferometric phase accurately. ◀

In general, accurate phase measurements require high correlation. Unfortunately, there are a number of mechanisms that can cause signal decorrelation. These are considered in the following subsections.

15-5.1 Speckle

Radar echoes are considered to be "correlated" with each other if the measured phases and amplitudes coincide and hence represent the "same" interactions of the radar signal with the scattering terrain. In an imaging radar, this means that the observed scattering characteristics or "speckle" patterns are nearly the same. As introduced in Chapter 5, ***speckle*** is an inherent consequence of the coherent nature of radar. It is observable as the "graininess" evident in single-look radar images. It arises due to EM interactions with multiple scattering centers within a single resolution element (or ***cell***) of the radar rather than actual variations in σ^0 of the surface.

Consider the backscattered voltage V from a given resolution cell when viewed by two radars from slightly different orientations. Since V is the coherent sum of the signals backscattered from the individual scatterers contained in the resolution cell, a change in orientation causes changes in the ranges to the individual scatterers, and hence a change in the magnitude and phase of V. This leads to ***geometric decorrelation***, and the degree of decorrelation depends on the amount of angular offset between the two antennas relative to the target.

15-5.2 Decorrelation Model

In this section we quantify the effects of decorrelation in InSAR radar images. Consider two complex radar signals, $\mathbf{V}_1$ and $\mathbf{V}_2$, acquired by two interferometric antennas, located very near each other, and passing through separate receivers. The signals can be modeled as consisting of a common signal component $\mathbf{C}$ and individual noise components $\mathbf{N}_1$ and $\mathbf{N}_2$:

$$\mathbf{V}_1 = \mathbf{C} + \mathbf{N}_1, \tag{15.29a}$$

$$\mathbf{V}_2 = \mathbf{C} + \mathbf{N}_2. \tag{15.29b}$$

Since the antennas are located very close to one another, the common signal component is essentially the same for both signals, but the receiver noise is different.

The received signal correlation is

$$\rho_{\text{thermal}} = \frac{\langle \mathbf{V}_1 \mathbf{V}_2^* \rangle}{\sqrt{\langle \mathbf{V}_1 \mathbf{V}_1^* \rangle \langle \mathbf{V}_2 \mathbf{V}_2^* \rangle}}, \tag{15.30}$$

where $\langle\ \rangle$ denotes ensemble averaging and the subscript "thermal" indicates that the noise is due to thermal properties.

Because $\mathbf{N}$ and $\mathbf{N}_2$ are uncorrelated random variables with zero means, it can be shown that

$$\rho_{\text{thermal}} = \frac{|\mathbf{C}|^2}{|\mathbf{C}|^2 + |\mathbf{N}|^2}. \tag{15.31}$$

Using SNR$= |\mathbf{C}|^2/|\mathbf{N}|^2$, the correlation is related to the measurement SNR by

$$\rho_{\text{thermal}} = \frac{1}{1 + \frac{1}{\text{SNR}}} = \frac{\text{SNR}}{1 + \text{SNR}}. \tag{15.32}$$

If SNR $\gg 1$, $\rho_{\text{thermal}} \approx 1$.

In InSAR, some geometric decorrelation occurs due to the difference in viewing angles of the two antennas. We assume for now that the viewing angle changes only slightly between the two images so that we can model the two signals as

$$\begin{aligned} \mathbf{V}_1 &= \mathbf{C} + \mathbf{D}_1 + \mathbf{N}_1, \\ \mathbf{V}_2 &= \mathbf{C} + \mathbf{D}_2 + \mathbf{N}_2, \end{aligned} \tag{15.33}$$

where $\mathbf{C}$ again is the correlated part of the signal, $\mathbf{N}_1$ and $\mathbf{N}_2$ are the thermal noise components, and $\mathbf{D}_1$ and $\mathbf{D}_2$ represent additive "noise" (which really is signal variability) due to the change in viewing direction. For reasons to be made apparent later, we call $\mathbf{D}_1$ and $\mathbf{D}_2$ the ***spatial decorrelation*** terms, since they involve spatial movement of the antennas.

Using the same arguments as previously, while ignoring thermal noise, leads to

$$\rho_{\text{spatial}} = \frac{|\mathbf{C}|^2}{|\mathbf{C}|^2 + |\mathbf{D}|^2}, \tag{15.34}$$

where the subscript "spatial" is used instead of "thermal."

If we include both the thermal and noise effects and treat them as uncorrelated random variables, the correlation is

$$\rho_{\text{spatial+thermal}} = \frac{|\mathbf{C}|^2}{|\mathbf{C}|^2 + |\mathbf{D}|^2 + |\mathbf{N}|^2}. \tag{15.35}$$

This result can be rewritten as

$$\rho_{\text{spatial+thermal}} = \frac{|\mathbf{C}|^2}{|\mathbf{C}|^2 + |\mathbf{D}|^2} \cdot \frac{|\mathbf{C}|^2 + |\mathbf{D}|^2}{|\mathbf{C}|^2 + |\mathbf{D}|^2 + |\mathbf{N}|^2}, \tag{15.36}$$

and, upon re-defining SNR as the ratio of all nonthermal to thermal powers, i.e.,

$$\text{SNR} = \frac{|\mathbf{C}|^2 + |\mathbf{D}|^2}{|\mathbf{N}|^2}, \tag{15.37}$$

we obtain the result

$$\begin{aligned} \rho_{\text{spatial+thermal}} &= \rho_{\text{spatial}} \cdot \rho_{\text{thermal}} \\ &= \frac{|\mathbf{C}|^2}{|\mathbf{C}|^2 + |\mathbf{D}|^2} \cdot \frac{1}{1 + \frac{1}{\text{SNR}}} \\ &= \frac{|\mathbf{C}|^2}{|\mathbf{C}|^2 + |\mathbf{D}|^2} \cdot \frac{\text{SNR}}{1 + \text{SNR}}. \end{aligned} \tag{15.38}$$

► Thus, the total correlation is the product of the two individual correlations, one due to signal correlation and the other due to thermal noise. ◄

Finally, we consider correlation when the two surfaces are imaged at different times. Even though the backscattered phase is a random quantity because it is the sum of echoes from n scatterers at distances R_1 to R_n, if the scene area is unchanged between observation times and the radar is at the same location, then R_1 to R_n are unchanged and the total echoes are completely correlated. However, if some of the scatterers move in location, even slightly, the echoes become slightly different. This is known as ***temporal decorrelation***.

There are a number of ways in which a surface might change between observations. The surface can undergo erosion, or if it is vegetated, the plants may grow, or some natural hazard such as an earthquake can disturb the scatterers. On short time scales, wind can alter the positions of leaves and stems. While the causes for temporal decorrelation are many, any disturbance of the locations of the scattering centers within a resolution element can produce temporal decorrelation in the radar signal.

The temporal decorrelation from slightly changed surfaces can be modeled by introducing a term in the backscatter equation corresponding to surface change. A little thought and repetition of the previous arguments leads to an expression of the form

$$\rho_{\text{total}} = \rho_{\text{thermal}} \cdot \rho_{\text{spatial}} \cdot \rho_{\text{temporal}}, \tag{15.39}$$

where ρ_{temporal} represents decorrelation due to a change in the surface condition between observations, ρ_{spatial} refers to a change in sensor location, and ρ_{thermal} is due to thermal noise inside the radar system.

Equation (15.39) is an approximation that only holds exactly for fairly high correlations.

15-5.3 Calculation of Spatial Baseline Decorrelation

This section quantifies ρ_{spatial} and determines just how close the two interferometer antennas need to be located to achieve accurate phase estimation. We begin by deriving a Fourier transform relation between the radar impulse response and the correlation coefficient as a function of the difference in viewing angles using the imaging geometry shown in Fig. 15-17.

Consider the phase of the reflection from the scatterer at P compared with the phase of the reflection from the center of the resolution cell. Using the parallel-ray approximation, the phase from P to A_1 is advanced by $y\sin\theta_1$ relative to the ray from O to A_1, and a similar phase advance of $y\sin\theta_2$ applies to antenna A_2. Assuming there are many scattering elements in the resolution cell, the total signal $\mathbf{V}_1$ at antenna 1 can be modeled as the integral of the signals reflected from all of the scatterers in the resolution element. That is,

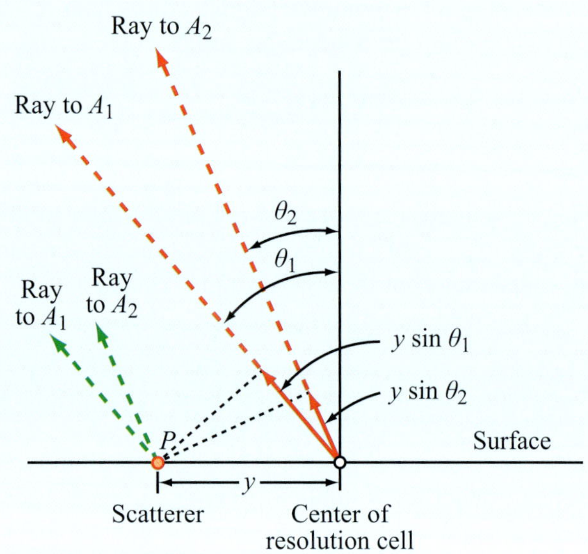

Figure 15-17: Imaging geometry for the decorrelation model. Scatterer P, at a distance y from the pixel center, is observed from two different incidence angles, so the phase of the wave reflected from P differs in each observation. The observed phase relative to the phase of the center of the pixel is that accumulated by a wave propagating a distance $y\sin\theta$ for each ray.

$$\mathbf{V}_1 = \iint \mathbf{S}(x-x_0,\, y-y_0)\exp\left[-j\frac{4\pi}{\lambda}\,(R+y\sin\theta_1)\right] \cdot \mathbf{p}(x,y)\,dx\,dy, \tag{15.40}$$

where $\mathbf{S}(x,y)$ represents the complex scattering amplitude from (x,y) on the surface and $\mathbf{p}(x,y)$ is the radar system impulse response function expressed in spatial coordinates. This is the sum of all of the contributions from the points in the resolution cell weighted by the impulse response.

Similarly, for antenna 2,

$$\mathbf{V}_2 = \iint \mathbf{S}(x-x_0,\ y-y_0) \exp\left[-j\frac{4\pi}{\lambda}\,(R+y\sin\theta_2)\right] \cdot \mathbf{p}(x,y)\ dx\,dy. \tag{15.41}$$

In both cases R is the range to the center of the resolution cell. In this analysis we assume R to be the same for both antennas. If this not the actual configuration, the mean phase of the return changes, but the conclusions derived below remain unchanged.

The interferogram is formed by computing $\mathbf{V}_1\mathbf{V}_2^*$:

$$\mathbf{V}_1\mathbf{V}_2^* = \iiiint \mathbf{S}(x-x_0,\ y-y_0)\, \mathbf{S}^*(x'-x_0,\ y'-y_0) \cdot \exp\left[-j\frac{4\pi}{\lambda}\,y(\sin\theta_1 - \sin\theta_2)\right] \cdot \mathbf{p}(x,y)\,\mathbf{p}^*(x',y')\ dx\,dy\,dx'\,dy'. \tag{15.42}$$

Making the assumptions that the surface scatterers are arranged randomly, distributed uniformly, and uncorrelated, it follows that

$$\left\langle \mathbf{S}(x,y)\,\mathbf{S}^*(x',y')\right\rangle = \sigma^0\,\delta(x-x',\ y-y'), \tag{15.43}$$

which enables reduction of the four-fold integral in Eq. (15.42) to

$$\langle \mathbf{V}_1\mathbf{V}_2^*\rangle = \sigma^0 \iint e^{-j(4\pi/\lambda)y\cos\theta\ d\theta}\,|\mathbf{p}(x,y)|^2\ dx\,dy\,, \tag{15.44}$$

where θ is the average of θ_1 and θ_2 and $d\theta$ is their difference. Since the exponential kernel is linear in y, it can be viewed as a scaled Fourier transform relating the correlation function $\langle \mathbf{V}_1\mathbf{V}_2^*\rangle$ to the power impulse response $|\mathbf{p}(x,y)|^2$.

Our objective is to compute the spatial correlation

$$\rho_{\text{spatial}} = \frac{\langle \mathbf{V}_1\mathbf{V}_2^*\rangle}{\sqrt{\langle \mathbf{V}_1\mathbf{V}_1^*\rangle\,\langle \mathbf{V}_2\mathbf{V}_2^*\rangle}}\,. \tag{15.45}$$

To that end, we approximate the impulse response as a sinc function in both dimensions, which is a good representation for high time-bandwidth chirped systems. That is,

$$\mathbf{p}(x,y) = \frac{1}{\sqrt{r_x r_y}}\left[\frac{\sin(\pi x/r_x)}{(\pi x/r_x)} \cdot \frac{\sin(\pi y/r_y)}{(\pi y/r_y)}\right], \tag{15.46}$$

where r_x and r_y are the ground resolutions of the radar system along the azimuth and range directions, respectively. Use of Eq. (15.46) leads to

$$\langle \mathbf{V}_1\mathbf{V}_1^*\rangle = \langle \mathbf{V}_2\mathbf{V}_2^*\rangle = \sigma^0\,\delta(x-x',\ y-y'). \tag{15.47}$$

After evaluating the Fourier transform of Eq. (15.44) and using the results in Eq. (15.45), we obtain the expression

$$\rho_{\text{spatial}} = 1 - \frac{2r_y\cos\theta}{\lambda}\,d\theta. \tag{15.48}$$

The differential angle $d\theta = \theta_1 - \theta_2$ in Fig. 15-17 can be related to $B_\perp$ and R as

$$d\theta = \frac{B_\perp}{R}\,, \tag{15.49}$$

in which case Eq. (15.48) can be cast in the form

$$\rho_{\text{spatial}} = 1 - \frac{2B_\perp r_y\cos\theta}{\lambda R}\,. \tag{15.50}$$

Note that this is a linearly decreasing function of $B_\perp$; ρ_{spatial} extends from 1 at $B_\perp = 0$ (perfect correlation) to zero at a ***critical value of $B_\perp$*** denoted B_c, with

$$B_c = \frac{\lambda R}{2r_y\cos\theta}\,. \tag{15.51}$$

If the perpendicular baseline $B_\perp$ approaches or exceeds B_c, no correlation is observed. This quantifies our earlier observation that the interferometer antennas must be located near each other in order for the two signals measured by them to be highly correlated.

Figure 15-18 shows the theoretical linear falloff to zero at the critical baseline. A sinc function spatial-response function is assumed. Measured data are often seen to fall off more quickly than predicted from the theoretical model. While the impulse response of the SEASAT radar used to collect the data of Fig. 15-18 was indeed a sinc function, its resolution may have been overestimated by the system model.

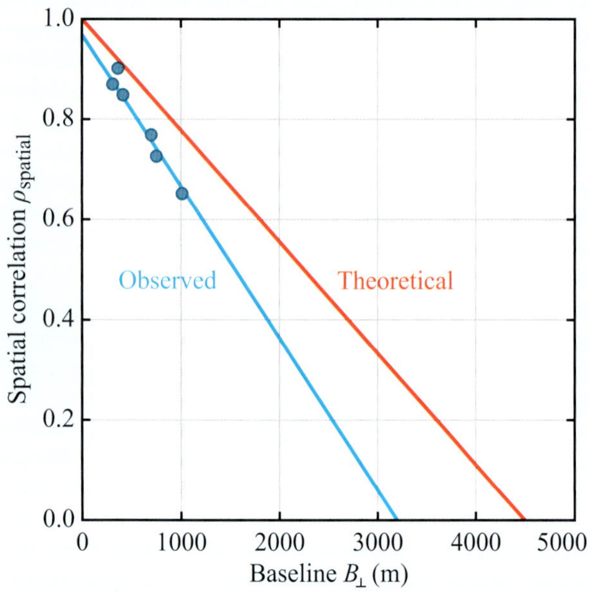

Figure 15-18: Baseline decorrelation for the sinc impulse function of the SEASAT SAR compared with actual data collected over Death Valley in 1978. The discrepancy in the rates of falloff between the curves is attributed to imprecise system models of the resolution. [Zebker and Villasenor, 1992.]

15-5.4 Rotational Decorrelation

In multipass interferometry, if the radar flight track of one pass is rotated relative to the track of the second one (i.e., they are not parallel), a second source of decorrelation due to the imaging geometry is introduced. The derivation of this correlation term is not provided here, but can be found in Zebker and Villasenor (1992). The rotational correlation can be expressed as

$$\rho_{\text{rotation}} = 1 - \frac{2r_x\, d\phi \sin\theta}{\lambda}, \qquad (15.52)$$

where r_x is the azimuth resolution and $d\phi$ is the rotation angular deviation between the two tracks. When applicable, this term should be included as a multiplicative factor in the correlation product expression in Eq. (15.39). Figure 15-19(a) compares Monte Carlo simulations and the predicted rotational correlation versus the path rotation angle for radars

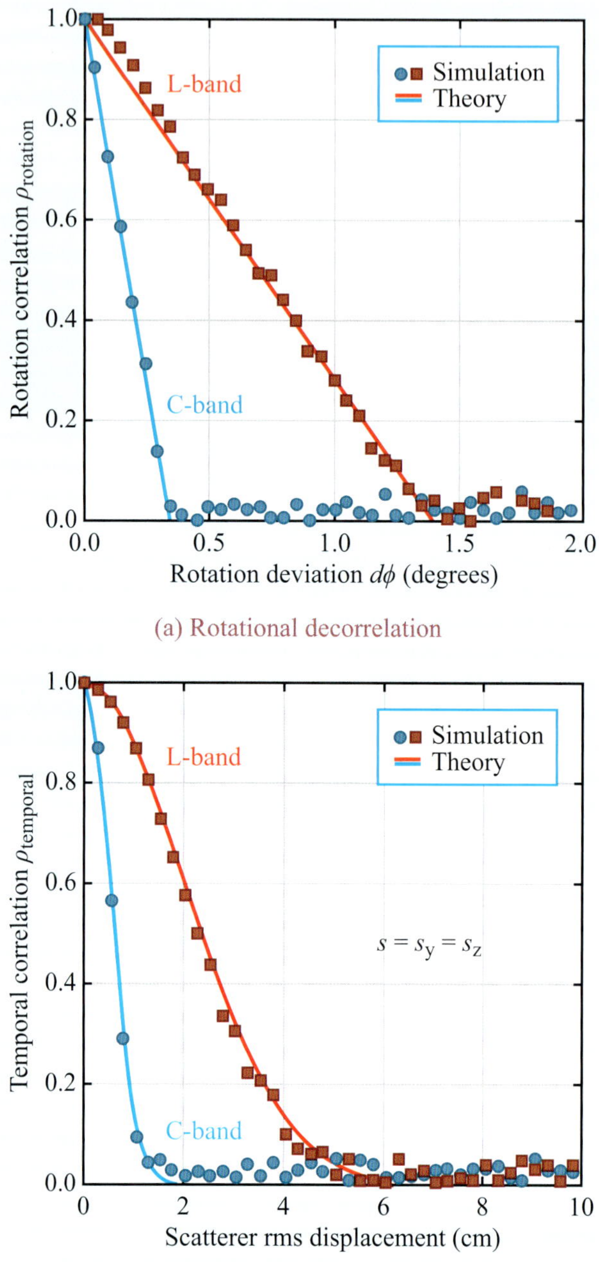

(a) Rotational decorrelation

(b) Motion decorrelation

Figure 15-19: Simulated and theoretical dependence of InSAR correlation on (a) track rotation as a function of rotation angle, and on (b) scatterer motion as a function of rms displacement of the scatterers. These curves are for a radar with 5 m resolution in both slant range and azimuth. [Zebker and Villasenor, 1992.]

operating at L- and C-band. Note that the correlation is high when the tracks are aligned (rotation angle zero) but falls off as the rotational angle grows. The fall-off with angle increases as the wavelength decreases.

Deviations from straight-line paths can also lead to decorrelation, but suitable models for evaluating the decorrelation are very specific to the details of the flight path difference, and must be computed in each instance following the methods described above.

15-5.5 Temporal Decorrelation

As previously noted, temporal decorrelation occurs if the scatterers move with respect to each other between observation times. The temporal decorrelation term can be modeled using 3-D motions of each of the individual scatterers, leading to the correlation product given by Zebker and Villasenor (1992) as:

$$\begin{aligned}\mathbf{V}_1\mathbf{V}_2^* = \iiiint\!\!\iint & \mathbf{S}(x-x_0,\ y-y_0,\ z-z_0) \\ & \cdot \mathbf{S}^*(x'-x_0,\ y'-y_0,\ z'-z_0) \\ & \cdot \exp\left[-j\frac{4\pi}{\lambda}\,(\eta_y \sin\theta + \eta_z \cos\theta)\right] \\ & \cdot \mathbf{p}(x,y)\,\mathbf{p}^*(x',y')\ dx\,dy\,dz\,dx'\,dy'\,dz',\end{aligned} \tag{15.53}$$

where η_y and η_z are incremental changes in the y- and z-locations of a scatterer. The computed average of this integral can be expressed as

$$\begin{aligned}\langle \mathbf{V}_1\mathbf{V}_1^* \rangle = \sigma^0 \iint & \exp\left[\ j\frac{4\pi}{\lambda}\,(\eta_y \sin\theta + \eta_z \cos\theta)\right] \\ & \cdot f_y(\eta_y)\ f_z(\eta_z)\ d\eta_y\ d\eta_z,\end{aligned} \tag{15.54}$$

where $f_y(\eta_y)$ and $f_z(\eta_z)$ are the probability distributions of motions in the y and z directions, respectively. By treating the pdfs as independent Gaussian processes, Zebker and Villasenor (1992) show that

$$\rho_{\text{temporal}} = \exp\left[-\frac{1}{2}\left(\frac{4\pi}{\lambda}\right)^2 (s_y^2 \sin^2\theta + s_z^2 \cos^2\theta)\right], \tag{15.55}$$

where s_y^2 and s_z^2 are variances of motions in y and z.

Figure 15-19(b) compares Monte-Carlo simulations with the predicted temporal correlation based on Eq. (15.55), with both plotted as a function of the rms displacement $s = s_y = s_y$ of the scatterers. The correlation function drops off at a much faster rate at the shorter wavelength, as expected.

The observed change in correlation as a function of the time difference between repeat passes of the SEASAT SAR is displayed in Fig. 15-20(a) for three geographic locations. Because the vegetation-free floor of Death Valley changes very little over the 18-day observation period, the temporal correlation remained at approximately 1 throughout. For the lightly vegetated lava flows in Eastern Oregon, the correlation decreased approximately linearly with time. Finally, for the heavily forested area, the rate of decrease is only slightly faster, but significant decorrelation is observed after only 2 days.

In Fig. 15-20(b), reported by Rosen et al. (2000), we note that the correlation is the lowest for the water surfaces and highest for urban areas.

15-6 Measurement of Topography

Before we examine the InSAR topographic measurement process in detail, let us highlight the primary reasons for measuring topography:

(1) Most of the world's surface is not well mapped, especially in digital form and at high resolution. InSAR is particularly suited for addressing this deficiency. In fact, the ***Shuttle Radar Topography Mission*** (SRTM), which flew for only 11 days in 2000, managed to map the topography of the entire segment of the globe between 56° S and 60° N at a spatial resolution of 90 m and a vertical precision of 16 m. SRTM used a baseline of 60 m between its two antennas. In many areas around the world, the SRTM-generated topography data is the best available today. However, there is need for improved mapping capability, and for mapping areas that were not mapped by SRTM, because they have very high relief or very low backscattering coefficients (such as some desert areas).

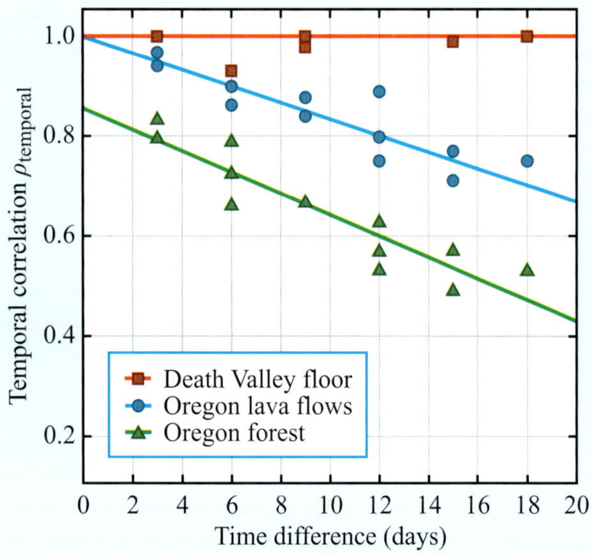

(a)

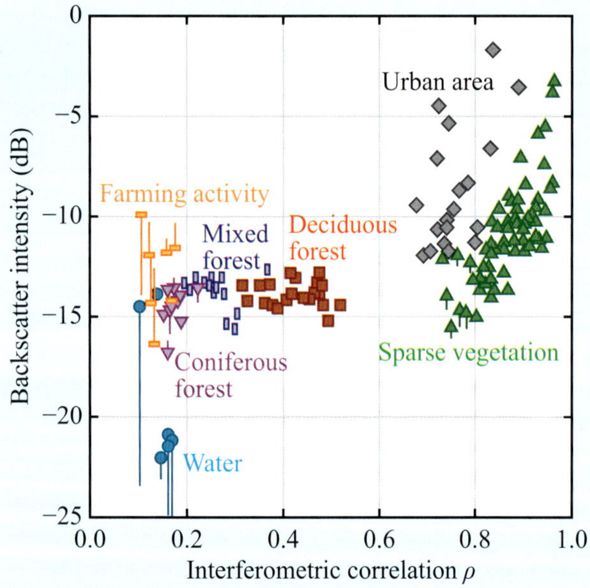

(b)

Figure 15-20: (a) Observed decorrelation with time for three different geographic regions, as measured by the SEASAT SAR in 1978; (b) scatter plot of relative backscatter intensity versus interferometric correlation for a water surface and various types of terrain [Rosen et al., 2000].

(2) Historically, different parts of the world were mapped on different ***datums***, or coordinate systems, making it difficult to combine them for use in global studies such as global change. InSAR processing is location dependent and can provide ***digital elevation maps*** (DEMs) on a single grid. SRTM demonstrated this by generating global DEMs with a consistent projection.

(3) Earthquakes, landslides, floods, and other events modify the local topography. InSAR-derived DEMs can detect topographic changes and help us understand the effects of such events and how they modify the terrain.

15-6.1 Inferring Topography from Interferometric Phase

In Section 15-2, we used the parallel-ray approximation to demonstrate the relationship between the measured phase difference ϕ_{int} and the height z of an image pixel. Precise implementation of the topographic mapping process, however, calls for the development of exact solutions. To that end, we use the observation geometry shown in Fig. 15-21 and reintroduce Eq. (15.4):

$$\begin{aligned}(R-\delta)^2 &= R^2 + B^2 - 2RB\cos\left(\frac{\pi}{2} - \theta + \alpha\right) \\ &= R^2 + B^2 - 2RB\sin(\theta - \alpha), \qquad (15.56)\end{aligned}$$

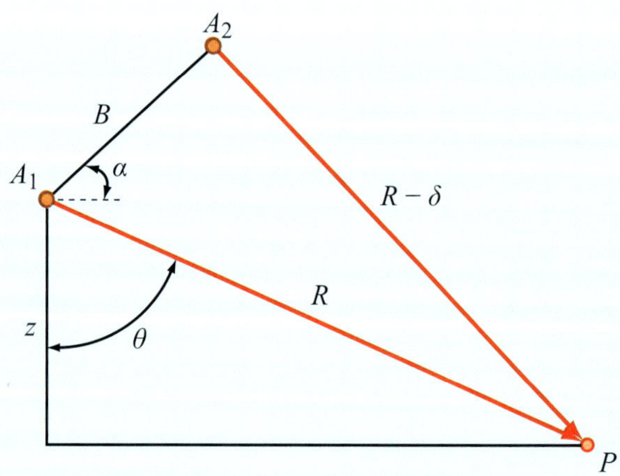

Figure 15-21: InSAR observation geometry.

from which it follows that

$$\sin(\theta-\alpha)=\frac{R^2+B^2-(R-\delta)^2}{2RB}\,. \qquad (15.57)$$

Phase error

From Eqs. (15.2a) and (15.6),

$$\delta=-\frac{\lambda}{4\pi}\,\phi_{\rm int} \qquad (15.58)$$

and

$$z=R\cos\theta, \qquad (15.59)$$

where $\phi_{\rm int}$ pertains to the two-independent-radars configuration, with the understanding that if the interferometer uses a common transmitter instead, the factor of 4 should be replaced with a factor of 2.

Hence, given B, α, and R we can solve for θ and use it to estimate z. To estimate the associated error, we use derivatives. In this case, a slightly different chain rule is used:

$$\frac{\partial z}{\partial\phi_{\rm int}}=\frac{\partial z}{\partial\theta}\frac{\partial\theta}{\partial\delta}\frac{\partial\delta}{\partial\phi_{\rm int}}\,, \qquad (15.60)$$

and we note that

$$\frac{\partial z}{\partial\theta}=\frac{\partial}{\partial\theta}(R\cos\theta)=-R\sin\theta\,. \qquad (15.61)$$

To obtain an expression for $\partial\theta/\partial\delta$, we differentiate Eq. (15.57) while treating only θ and δ as variables:

$$\frac{\partial}{\partial\delta}[\sin(\theta-\alpha)]=\frac{\partial}{\partial\delta}\left[\frac{R^2+B^2-(R-\delta)^2}{2RB}\right],$$

which yields

$$\cos(\theta-\alpha)\,\frac{\partial\theta}{\partial\delta}=\frac{2(R-\delta)}{2RB}\approx\frac{1}{B}\,. \qquad (15.62)$$

In the final step, we approximated $(R-\delta)/R$ as 1. Hence,

$$\frac{\partial\theta}{\partial\delta}=\frac{1}{B\cos(\theta-\alpha)}\,. \qquad (15.63)$$

Finally, from Eq. (15.58),

$$\frac{\partial\delta}{\partial\phi_{\rm int}}=\frac{-\lambda}{4\pi}\,,$$

so that we can write

$$\frac{\partial z}{\partial\phi_{\rm int}}=(-R\sin\theta)\left(\frac{1}{B\cos(\theta-\alpha)}\right)\left(\frac{-\lambda}{4\pi}\right)=\frac{\lambda}{4\pi}\,\frac{R\sin\theta}{B\cos(\theta-\alpha)}\,, \qquad (15.64)$$

which is consistent with the result given by Eq. (15.14a) (but without the SNR part) on the basis of the parallel-ray approximation.

Baseline-length error

If the two receive antennas of the SAR interferometer are physically connected to one another, such as if they are mounted on the two wings of an aircraft or at the two ends of a long mast—as was the case for the SRTM instrument, then the baseline distance B is a constant quantity with no significant variability. For repeat-pass systems, on the other hand, the knowledge of the value of B may have an element of uncertainty, either in terms of its length or, more likely, in terms of the angle α. Errors in the values of the baseline length or orientation contribute to height errors.

The contribution of the baseline-length error to the height error can be obtained by computing the appropriate partial derivatives:

$$\frac{\partial z}{\partial B}=\frac{\partial z}{\partial\theta}\cdot\frac{\partial\theta}{\partial B}\,. \qquad (15.65)$$

To derive an expression for $\partial\theta/\partial B$, we start with Eq. (15.57):

$$\frac{\partial}{\partial B}[\sin(\theta-\alpha)]=\frac{\partial}{\partial B}\left[\frac{R^2+B^2-(R-\delta)^2}{2RB}\right],$$

which leads to

$$\frac{\partial\theta}{\partial B}=\frac{B^2+\delta^2-2R\delta}{2RB^2\cos(\theta-\alpha)}\,. \qquad (15.66)$$

Since the range $R\gg\delta$ and also $R\gg B$, Eq. (15.66) can be simplified to

$$\frac{\partial\theta}{\partial B}=\frac{-\delta}{B^2\cos(\theta-\alpha)}\,. \qquad (15.67)$$

Combining this result with Eq. (15.61) leads to

$$\frac{\partial z}{\partial B} = \frac{\partial z}{\partial \theta} \cdot \frac{\partial \theta}{\partial B} = (-R\sin\theta)\left[\frac{-\delta}{B^2\cos(\theta-\alpha)}\right] = \frac{R\delta\sin\theta}{B^2\cos(\theta-\alpha)}. \quad (15.68)$$

Using the approximate relation $\delta = B\sin(\theta-\alpha)$, as given by Eq. (15.5), leads to

$$\frac{\partial z}{\partial B} \approx \frac{R}{B}\sin\theta\tan(\theta-\alpha). \quad (15.69)$$

Fortunately, the uncertainty in the length of B can be controlled to within a small value, so this error source usually is small for most practical systems.

Baseline-orientation error

We now consider the error due to the uncertainty associated with the baseline-orientation angle α, which depends on the knowledge uncertainty of the platform track or orbit and the platform altitude, plus any flexing of the platform itself. To compute $\partial z/\partial\alpha$, we use the chain rule

$$\frac{\delta z}{\partial \alpha} = \frac{\partial z}{\partial \theta} \cdot \frac{\partial \theta}{\partial \delta} \cdot \frac{\partial \delta}{\partial \alpha}. \quad (15.70)$$

The derivatives $\partial z/\partial\theta$ and $\partial\theta/\partial\delta$ are given by Eqs. (15.61) and (15.63), respectively. To compute $\partial\delta/\partial\alpha$, we differentiate Eq. (15.57) while treating only α and δ as variables:

$$\frac{\partial}{\partial\alpha}[\sin(\theta-\alpha)] = \frac{\partial}{\partial\alpha}\left[\frac{R^2+B^2-(R-\delta)^2}{2RB}\right],$$

which leads to

$$-\cos(\theta-\alpha) = \frac{2(R+\delta)}{2RB}\,\frac{\partial\delta}{\partial\alpha},$$

or

$$\frac{\partial\delta}{\partial\alpha} = \frac{-RB}{R+\delta}\cos(\theta-\alpha) \approx -B\cos(\theta-\alpha). \quad (15.71)$$

Using Eqs. (15.61), (15.63), and (15.71) in Eq. (15.70) gives

$$\frac{\partial z}{\partial \alpha} = (-R\sin\theta)\left[\frac{1}{B\cos(\theta-\alpha)}\right][-B\cos(\theta-\alpha)], \quad (15.72)$$

which simplifies to

$$\frac{\partial z}{\partial \alpha} = R\sin\theta. \quad (15.73)$$

▶ The fact that $|\partial z/\partial\alpha|$ and $|\partial z/\partial\theta|$ [as in Eq. (15.61)] are given by the same expression indicates that a tilt $d\alpha$ in the interferometer baseline is indistinguishable from a change $d\theta$ in the local slope of the surface. ◀

If we denote s_z, $s_{\phi_{\rm int}}$, s_B, and s_α as the standard deviations associated with their respective parameters, we can summarize the phase, baseline-length, and baseline-angle errors as follows:

Phase error:

$$s_z = \frac{\lambda}{4\pi}\,\frac{R\sin\theta}{B\cos(\theta-\alpha)}\,s_{\phi_{\rm int}}. \quad (15.74a)$$

Baseline-length error:

$$s_z = \frac{R\sin\theta\tan(\theta-\alpha)}{B}\,s_B. \quad (15.74b)$$

Baseline-angle error:

$$s_z = R\sin\theta\; s_\alpha. \quad (15.74c)$$

To gain insight into the magnitude of the error contributions, we consider the design parameters of the airborne NASA/JPL TOPSAR system:

$$R \approx 10 \text{ km}, \qquad \theta \approx 30^\circ,$$
$$B = 1.5 \text{ m}, \qquad \alpha = 63^\circ,$$
$$\lambda = 6 \text{ cm}, \qquad \text{SNR} \approx 100 \text{ (20 dB)}.$$

From Eq. (15.13), $s_{\phi_{\rm int}}$ for a 1-look image is

$$s_{\phi_1} = \frac{1}{\sqrt{\text{SNR}}} = 0.1 \text{ rad}.$$

This particular radar calculates about 10 looks in the interferogram before estimating the height. For N looks (Rodriguez and Martin, 1992),

$$s_{\phi_N} \approx \frac{s_{\phi_1}}{\sqrt{2N}}. \qquad (15.75)$$

Hence, using $N = 10$ for the TOPSAR,

$$s_{\phi_{10}} \approx 0.022 \text{ rad} \approx 1.25^\circ.$$

Using this result in Eq. (15.74a), along with the system parameters, leads to an rms height error (due to phase error) of

$$s_z = \frac{\lambda}{4\pi} \frac{R\sin\theta}{B\cos(\theta-\alpha)} s_{\phi_{10}} = 0.42 \text{ m} \qquad \textbf{(phase error)}.$$

For the baseline-length error, which for this system is primarily due to ***aircraft deformation***, s_B is estimated to be about 10^{-4} m. Using this value in Eq. (15.74b) yields

$$s_z = \frac{R\sin\theta\tan(\theta-\alpha)}{B} s_B = 0.216 \text{ m}, \qquad \textbf{(baseline-length error)}$$

which is about one-half of the phase-associated error.

Finally, for this system the baseline-orientation error is due to uncertainty in the ***aircraft attitude***, for which $s_\alpha \approx 0.01^\circ$. Using this value in Eq. (15.74c) gives

$$s_z = R\sin\theta\, s_\alpha = 0.88 \text{ m} \qquad \textbf{(baseline-angle error)},$$

which is about twice the phase-associated error. Thus we see that for this system the uncertainty in baseline orientation is the major source of error. This is true for most systems.

15-6.2 Phase Unwrapping

In the preceding sections, we developed a set of equations that relate the phase difference ϕ_{int} (measured by an interferometric SAR) for a particular pixel to the height z of that pixel. Hence, given an imaging geometry and absolute values of ϕ_{int} for each pixel, it should be possible to create a topographic map of the imaged scene. Phase, however, is only measured *modulo* 2π, which means that the phase must first be *unwrapped* before applying it to determine z. The measured phase modulo 2π is known as the ***wrapped phase***. The process of converting the wrapped phase into absolute phase is called ***phase unwrapping***. It entails determining how many multiples of 2π should be added to the wrapped phase to produce a continuous phase surface.

► Whereas the wrapped phase is limited to the range $[0, 2\pi]$, the unwrapped phase is not so limited. ◄

The problem of phase unwrapping for InSAR is fairly complicated, so in this chapter we detail only the separation of the total phase into the background, or flat-Earth, component, and the phase component due to topography or crustal deformation. The flat-Earth term is known from the imaging geometry, and its removal from the observed phase leaves only the more slowly varying signal of interest.

The phase unwrapping problem for InSAR was first addressed by Goldstein et al. (1988), who developed a theory for inconsistencies in the phase field, which they called residues. They showed that by properly joining residues with a series of curves that are not crossed during integration of measured pixel-to-pixel phase differences a consistent and physically plausible solution could be achieved. This was a conservative method that produced few incorrect unwrapped pixels, but often left large gaps in the resulting interferogram product. Ghiglia and Romero (1996) used variable-norm criteria to optimize the placement of the uncrossable boundaries and were able to retrieve more spatial coverage. Following this work, Costantini (1998) was able to show a duality between the placement of branch cut lines and the linear-programming methods used to optimize network accessibility problems, which yielded not only still greater coverage, but also incorporated network solver algorithms that permitted solution of complicated phase fields. The current state of the art in phase unwrapping follows Chen and Zebker (2001), who added weighting functions, based on statistics of radar signals from terrain slopes and deformations, to the network programming approach. Its combination of

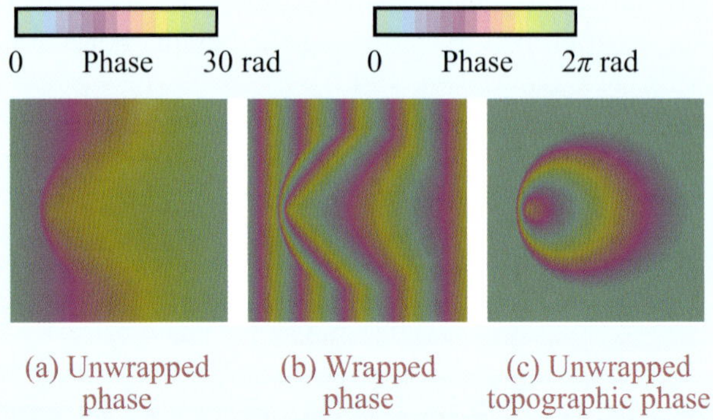

Figure 15-22: Simulation of InSAR phase difference: (a) true phase ϕ, (b) measured phase modulo 2π, and (c) tropospheric component of ϕ after removal of flat-Earth component.

accurately placed cut lines and robustness of solution leads to significant areal coverage and accuracy.

We start with Eq. (15.57), and then we simplify it as follows:

$$\begin{aligned}\sin(\theta-\alpha) &= \frac{R^2+B^2-(R-\delta)^2}{2RB} \\ &= \frac{R^2+B^2-R^2+2R\delta-\delta^2}{2RB} \\ &= \frac{B^2+2R\delta-\delta^2}{2RB} \\ &\approx \frac{B}{2R}+\frac{\delta}{B}\approx\frac{\delta}{B}, \qquad (15.76)\end{aligned}$$

where we used the knowledge that $\delta \ll R$ and $B \ll R$. This is the same expression given by Eq. (15.5) obtained using the parallel-ray approximation.

Next, we expand Eq. (15.76):

$$\begin{aligned}\delta &= B\sin(\theta-\alpha) \\ &= B[\sin\theta\cos\alpha-\cos\theta\sin\alpha]. \qquad (15.77)\end{aligned}$$

From Eq. (15.6), $\cos\theta = z/R$. It follows that

$$\sin\theta=\sqrt{1-\frac{z^2}{R^2}}$$

and

$$\delta = B\left[\sqrt{1-\frac{z^2}{R^2}}\cos\alpha-\frac{z}{R}\sin\alpha\right]. \qquad (15.78)$$

For an InSAR with two independent radars, the (absolute) phase difference is

$$\begin{aligned}\phi_{\text{int}} &= -\frac{4\pi}{\lambda}\,\delta \\ &= -\frac{4\pi B}{\lambda R}\left[\sqrt{R^2-z^2}\cos\alpha-z\sin\alpha\right]. \qquad (15.79)\end{aligned}$$

However, as we stated earlier, the *measured* phase is modulo 2π, whereas the true phase ϕ_{int} as given by Eq. (15.79) varies continuously with z/R. To illustrate, we simulate in Fig. 15-22(a) a continuous phase pattern that extends over a range of 30 radians. The measured phase difference, modulo 2π, is illustrated in part (b) of Fig. 15-22. The task of phase unwrapping is to convert the measured phase displayed in Fig. 15-22(b) into the true phase in part (a). An additional step of phase unwrapping involves segregating the two components of ϕ_{int}, namely the component ϕ_{topo} due to topographic (height) variations, from the component ϕ_{fe} due to the ***flat Earth*** reference. The latter is the phase pattern that the InSAR would measure if $z = z_0$ everywhere across the image, where z_0 is a constant. Figure 15-22(c) simulates the unwrapped phase after removal of the flat-Earth component. This is the true phase pattern ϕ_{topo} associated with the height variations that the interferometric process seeks to quantify.

To segregate the two phase components, we define z_0 as the mean height of the surface and dz as the deviation due to the topographic height relative to z_0. After replacing z with z_0+dz in Eq. (15.79), we have

$$\begin{aligned}\phi_{\text{int}} &= -\frac{4\pi B}{\lambda R}\left[\sqrt{R^2-(z_0+dz)^2}\,\cos\alpha-(z_0+dz)\sin\alpha\right] \\ &= -\frac{4\pi B}{\lambda R}\left[\sqrt{R^2-z_0^2-2z_0dz-(dz)^2}\cdot\cos\alpha\right. \\ &\qquad \left.-(z_0+dz)\sin\alpha\right]. \qquad (15.80)\end{aligned}$$

Neglecting $(dz)^2$ relative to the other terms, and then factoring out $\sqrt{R^2 - z_0^2}$ leads to

$$\phi_{\text{int}} = -\frac{4\pi B}{\lambda R}\left[\sqrt{R^2 - z_0^2}\cdot\sqrt{1 - \frac{2z_0\,dz}{R_0^2 - z_0^2}}\cos\alpha - z_0\sin\alpha - dz\sin\alpha\right]. \tag{15.81}$$

The second square root in Eq. (15.81) is of the form $\sqrt{1-a}$ with $a \ll 1$. Using the Taylor series expansion $\sqrt{1-a} \approx 1 - a/2$ leads to

$$\phi_{\text{int}} = -\frac{4\pi B}{\lambda R}\left[\sqrt{R^2 - z_0^2}\left(1 - \frac{z_0\,dz}{R_0^2 - z_0^2}\right)\cos\alpha - z_0\sin\alpha - dz\sin\alpha\right]. \tag{15.82}$$

The ***flat-Earth component*** is obtained by setting $dz = 0$:

$$\phi_{\text{fe}} = -\frac{4\pi B}{\lambda R}\left[\sqrt{R^2 - z_0^2}\cos\alpha - z_0\sin\alpha\right], \tag{15.83}$$

and the ***topographic component*** is given by

$$\begin{aligned}\phi_{\text{topo}} &= \phi_{\text{int}} - \phi_{\text{fe}} \\ &= \frac{4\pi B}{\lambda R}\left[\frac{z_0\cos\alpha}{\sqrt{R^2 - z_0^2}} + \sin\alpha\right] dz \\ &= \frac{4\pi B}{\lambda R}\left[\frac{\cos\alpha}{\tan\theta_0} + \sin\alpha\right] dz,\end{aligned} \tag{15.84}$$

where θ_0 is the incidence angle relative to the mean surface, and

$$\cot\theta_0 = \frac{z_0}{\sqrt{R^2 - z_0^2}}.$$

► The linear relationship between ϕ_{topo} and dz indicates that constant-phase contours appear at constant heights on the imaged surface. ◄

The wrapped phase repeats every 2π radians, with each ***fringe*** (a 0 to 2π transition) corresponding to a height difference Δz known as the ***ambiguity height***. We obtain Δz by setting $\phi_{\text{topo}} = 2\pi$ in Eq. (15.84), which leads to

$$\Delta z = \frac{\lambda R}{2B\left(\dfrac{\cos\alpha}{\tan\theta_0} + \sin\alpha\right)}. \tag{15.85}$$

By way of an example, for the TOPSAR with the system parameters listed in the preceding subsection,

$$\Delta z = \frac{6\times10^{-2}\times10^4}{2\times1.5\left(\dfrac{\cos 63^\circ}{\tan 30^\circ} + \sin 63^\circ\right)} = 119.24 \text{ m}.$$

This means that the phase contours repeat every approximately 120 m. We should note that this is based on single values for R and θ_0; in an exact topographic reduction process, the variations of R and θ_0 across the image should be properly incorporated. Examples of digital elevation model (DEM) maps acquired by airborne InSAR systems are shown in Fig. 15-23.

15-6.3 Curved-Earth Phase Pattern

For airborne InSAR observations at moderate incidence angles, the height of the radar above the surface is such that it is reasonable to treat the Earth surface as flat. Such an assumption, however, is not applicable for spaceborne SAR applications, nor for airborne SARs with high incidence angles. To achieve useful height accuracies from the InSAR phase measurements, we should replace the reference flat-Earth model with a curved-Earth model. The relevant geometry is shown in Fig. 15-24. Now, the incidence angle at point P relative to the mean curved-Earth surface is θ', not θ. Angles θ and θ' can be computed by applying the law of cosines, which yields

$$\theta = \cos^{-1}\left[\frac{(h+R_0)^2 + R^2 - R_0^2}{2R(h+R_0)}\right] \tag{15.86a}$$

and

$$\theta' = \pi - \cos^{-1}\left[\frac{R^2 + R_0^2 - (h+R_0)^2}{2RR_0}\right], \tag{15.86b}$$

where R_0 is the Earth radius.

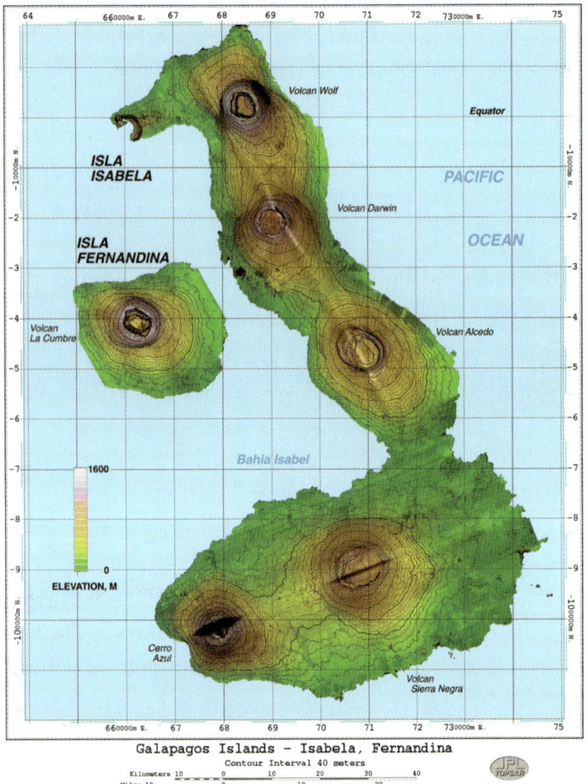

(a) Galápagos Islands: Fernandina and Isabela

(b) Askja, Iceland

Figure 15-23: (a) DEM of the Galápagos Islands Fernandina and Isabela, acquired by the NASA/JPL TOPSAR C-band interferometer. The swath width of the instrument is only 20 km, so many parallel flight lines were collected and combined. (b) DEM of Askja, Northern volcanic zone, Iceland, derived from the C-band EMISAR topographic mapping system of Denmark. The color variation in the image is derived from L-band EMISAR polarimetry [Christensen et al., 1998].

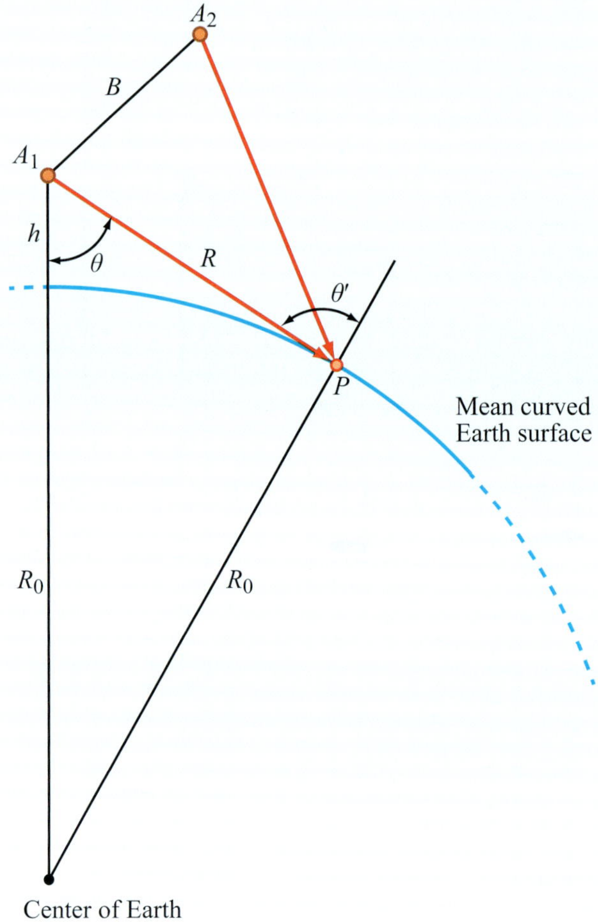

Figure 15-24: Geometry for computing the curved-Earth phase component.

15-7 Mapping Earth's Topography: The SRTM Mission

In February 2000, NASA, in partnership with the National Geospatial-Intelligence Agency and the German and Italian Space Agencies, launched and operated the Shuttle Radar Topography Mission (SRTM), producing the most complete, highest-resolution digital elevation model of the Earth to date. It used dual radar antennas attached to the space shuttle to acquire interferometric radar data, which were then processed to digital

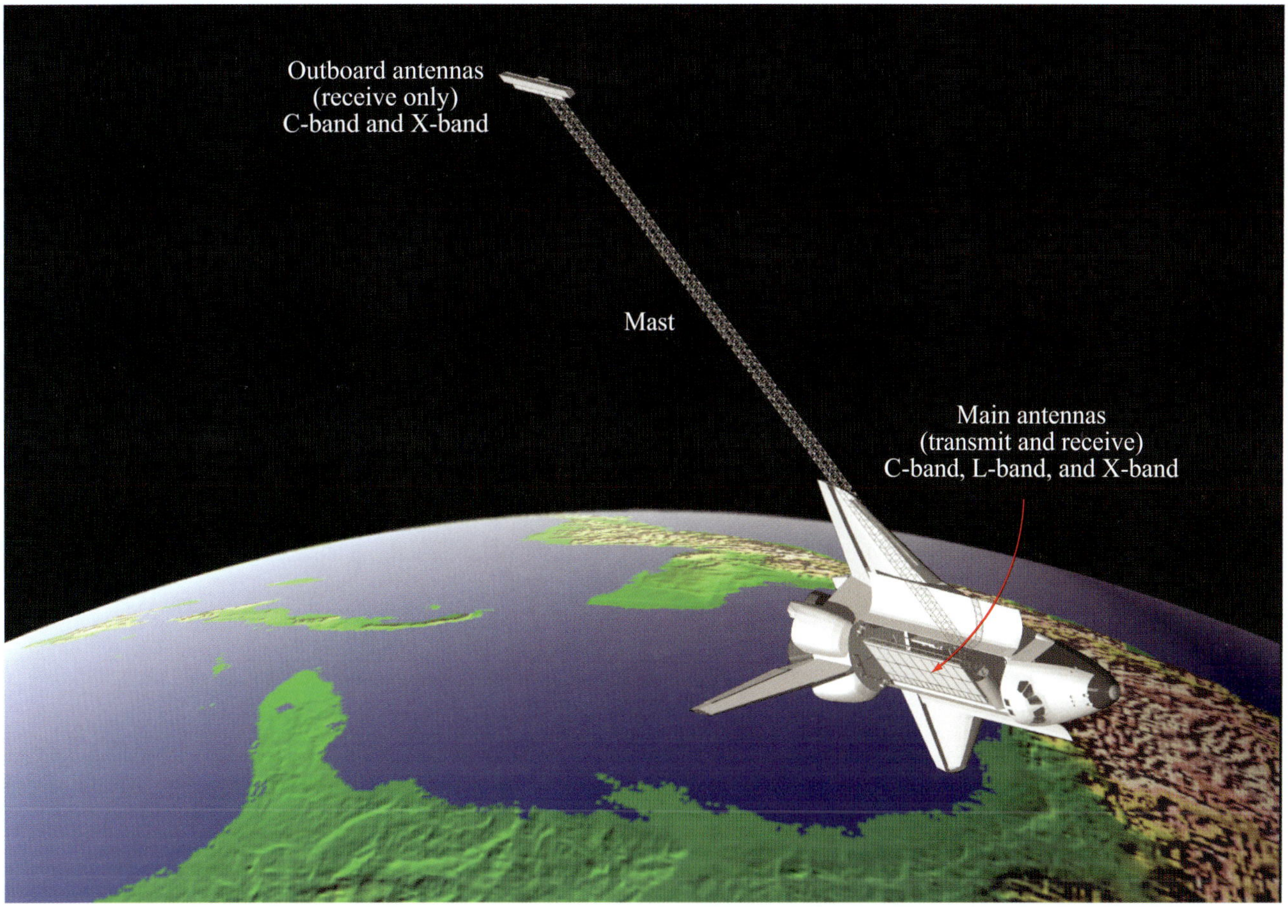

Figure 15-25: Cartoon illustrating the configuration of the SRTM instrument. Radar signals are transmitted from the main shuttle bay, and received simultaneously at that location and at an outboard antenna, forming an across-track baseline. [Courtesy NASA.] The mast is 60 m long.

topographic models at 1 arcsec resolution. SRTM was built on the hardware used by the previous NASA SIR-C mission, which included L-, C-, and X-band SAR systems. An illustration of the configuration of the instrument aboard the shuttle is shown in Fig. 15-25.

The mission objective was to acquire a digital elevation model of all land surfaces between about 60° north latitude and 56° south latitude, constituting about 80% of Earth's land surface. The data acquired by the system continue to be one of the most complete set of digital representations of the Earth available to date, and they are used by many different user communities.

The data-reduction process used a version of the steps outlined in the preceding sections. Because the system had a long boom holding the second set of antennas, and also because the shuttle itself exhibits significant attitude variations in flight, the main contributions to errors were from those components. Much of the processing was performed to address and compensate for attitude errors (Farr et al., 2007). Additional flight hardware systems were added to the shuttle avionics in order to better measure details of the craft's orbit and attitude.

The topographic products were derived from the C-band radar data, and were processed at JPL over a

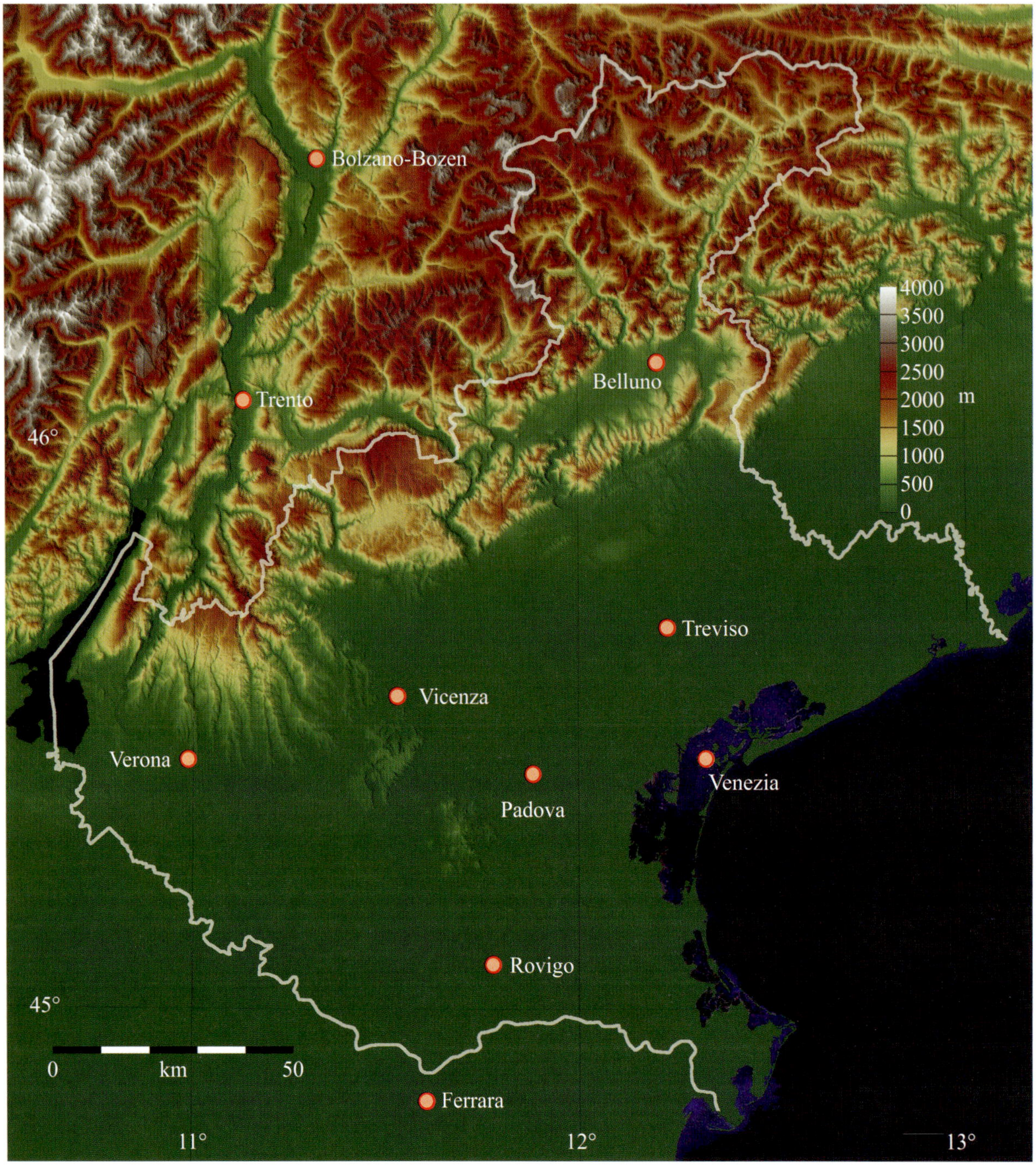

Figure 15-26: SRTM topographic map of part of Italy.

period of about 9 months. The terrain height data were produced in DTED-2 format at 1×1 arcsec spacing up to latitude 50° and 100 m (latitude) $\times$ 200 m (longitude) above 50°. All data are available free of charge from a server operated by the US Geological Survey. An example is shown in Fig. 15-26.

15-8 Along-Track Interferometry

The previous sections of this chapter were focused on the use of InSAR for measuring topography.

► By altering the imaging geometry to an ***along-track interferometer***, it is possible to use the InSAR to measure surface motion, rather than elevation. ◄

Many of the processing techniques described previously continue to be applicable, but in the new imaging geometry (Fig. 15-27) the baseline is described in terms of the temporal variable t, rather than in terms of spatial coordinates.

15-8.1 Temporal Baseline

The configuration depicted by Fig. 15-27 is an ***along-track radar interferometer*** in which antennas A_1 and A_2 are mounted on the front and back sections of the aircraft fuselage, rather than across-track on the two wings. The separation between the two antennas is the baseline B and the vehicle velocity is u. The two images generated by the two independent antennas are

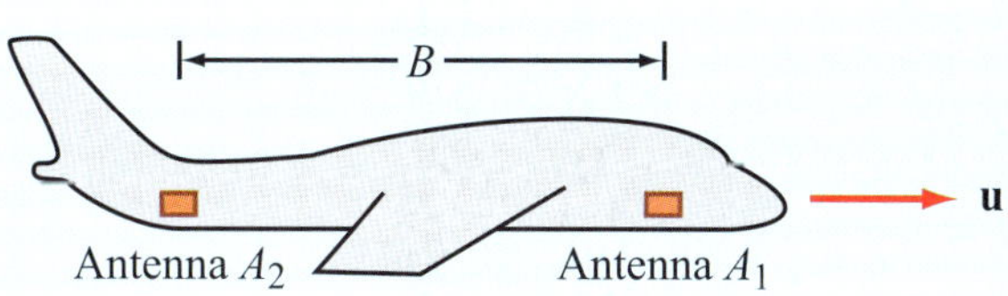

Figure 15-27: Interferometer geometry to measure along-track motions. Antennas mounted at the front and rear of an aircraft capture two radar images displaced in time by the interval it takes the plane to travel the baseline distance B.

essentially identical, except that the image generated by antenna A_2 is delayed in time by $\Delta t = B/u$, namely the amount of time it takes the aircraft to fly over the distance B. For a typical aircraft velocity, Δt is a fraction of a second. In Section 15-9 we consider the case where the two images are acquired by repeat pass satellite interferometry in which the same, almost identical, path is traversed by the same SAR at two times separated not by seconds, but by years! The complex voltage distributions $\mathbf{V}_1(x,y)$ and $\mathbf{V}_2(x,y)$ of the two images are related to each other by

$$\mathbf{V}_1(x,y) = \mathbf{V}_2(x+B,\, y), \tag{15.87}$$

where x and y are the along-track and cross-track coordinates, respectively. Alternatively, we can express the voltages in terms of (t,R), where t is time and R is the range to pixel (x,y). That is,

$$\mathbf{V}_1(t,R) = \mathbf{V}_2\left(t+\frac{B}{u},\, R\right). \tag{15.88}$$

The two images are identical, except for the shift in position in the along-track dimension. In Fig. 15-28, we show a top view of the imaging geometry, with particular interest in an object located at a distance y from the flight track. We first consider the interferogram for a stationary object. If image 1 is acquired at t_1, then when the second antenna acquires image 2 from the same position relative to the target, it acquires it at $t_2 = t_1 + B/u$. Since the ranges to the stationary object are identical, the interferometric phase difference is zero. But if the object is moving *away* from the flight path with a radial velocity component v_r, then

$$R_2 = R_1 + v_r\,\frac{B}{u}\sin\theta, \tag{15.89}$$

where θ is the incidence angle and $v_r \sin\theta$ is the component of v_r along the direction of the radar line of sight. The corresponding interferometric phase is

$$\begin{aligned}\phi_{\rm int} = \phi_1 - \phi_2 &= -\frac{4\pi}{\lambda}(R_1 - R_2) \\ &= \frac{4\pi}{\lambda u} B v_r \sin\theta. \end{aligned} \tag{15.90}$$

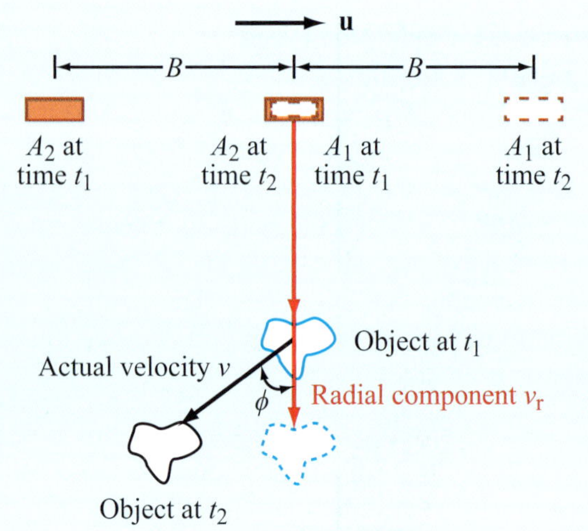

Figure 15-28: Observing a moving object receding from the flight path of the aircraft. In the time between observation, the object moves further away, and thus the phase of the radar signal increases.

The interferometric phase is directly proportional to the radial velocity of the object, v_r. Note that v_r is related to the true velocity v of the object by $v_r = v\cos\phi$ (Fig. 15-28).

15-8.2 Ocean Currents

In Section 15-5.5, we noted that in cross-track interferometry, the motion of scatterers in the resolution cell between the two observations leads to signal decorrelation, and therefore loss of interferometric phase information. In contrast, the along-track interferometry result encapsulated by Eq. (15.90) indicates that ϕ_{int} can be used to measure the radial velocity v_r, provided all scatterers in the resolution move together. This is the case for ocean currents (Fig. 15-29), which means that along-track interferometry can be used to measure v_r of the ocean surface.

To quantify how well the radial velocity v_r of the ocean surface can be measured, we replace ϕ_{int} in Eq. (15.90) with the phase measurement uncertainty $s_{\phi_{\text{int}}}$ and we also replace v_r with the radial velocity

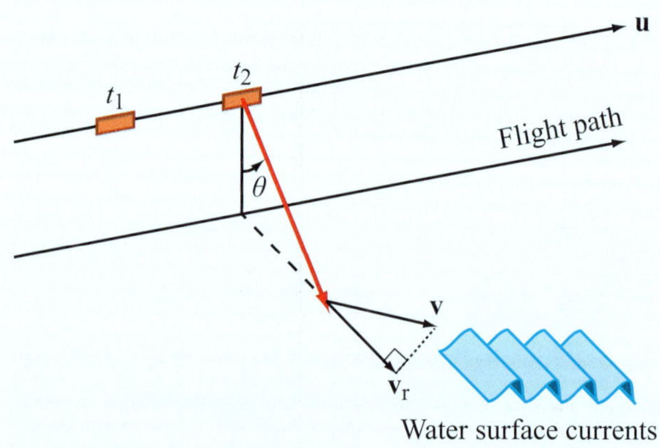

Figure 15-29: Along-track interferometer observing the velocity of the ocean surface.

measurement uncertainty s_{v_r}. The process gives

$$s_{v_r} = \frac{\lambda u}{4\pi B \sin\theta}\, s_{\phi_{\text{int}}}. \tag{15.91}$$

For an L-band SAR with $\lambda = 24$ cm, a baseline spacing $B = 20$ m along the belly of the aircraft, an aircraft velocity $u = 250$ m/s, and an incidence angle $\theta = 45^\circ$,

$$s_{v_r} = \frac{0.24 \times 250}{4\pi \times 20 \times 0.707}\, s_{\phi_{\text{int}}} = 0.34 s_{\phi_{\text{int}}}.$$

If ϕ_{int} can be measured with a precision $s_{\phi_{\text{int}}} = 3^\circ \approx 0.05$ rad, then $s_{v_r} \approx 1.7$ cm/s.

15-9 Measuring Surface Deformation

Satellites can be made to repeat their flight tracks with minimal cross-track deviation. The combination of two repeat-path SAR measurements constitutes an along-track interferometer. For a perfectly aligned flight track, such an interferometer loses all sensitivity to topography, but gains sensitivity to radial motions of the imaged surface.

Using a sequence of ERS-1 SAR images acquired with nearly identical paths over Antarctica, Goldstein et al. (1993) demonstrated that the flow velocity of the

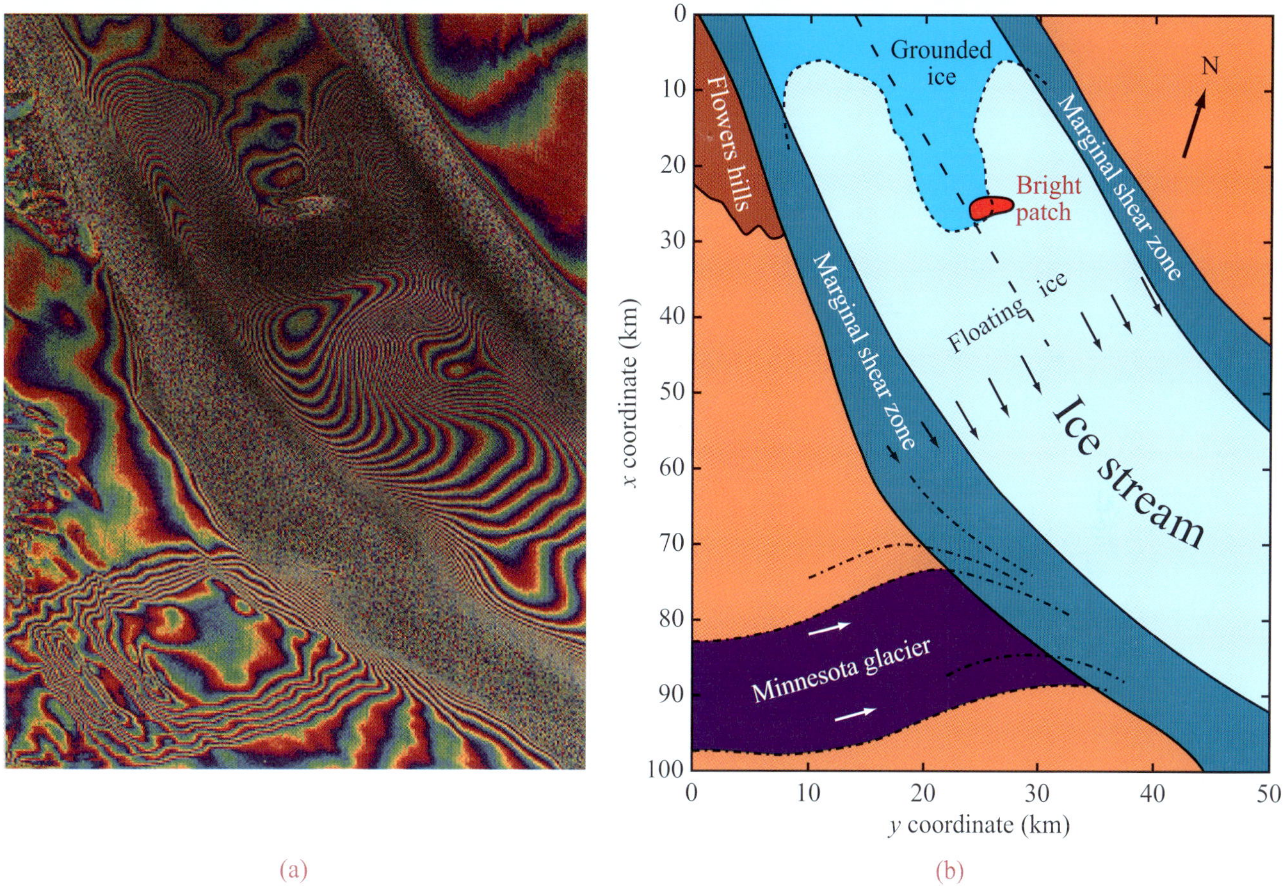

Figure 15-30: Radar interferogram of a portion of the Rutford ice stream in Antarctica, based on two ERS-1 images taken six days apart. The fringe pattern (color cycle) is essentially a map of ice-flow velocity, with one fringe representing 28 mm of range change along the radar line of site. [Image courtesy Jet Propulsion Laboratory, California Institute of Technology.]

ice stream can be measured with a precision of about 3×10^{-8} m/s. An interferometric image and associated ice-flow map are shown in Fig. 15-30.

In Eq. (15.91), the ratio B/u represents the time delay Δt between the two SAR observations. Making such a substitution gives

$$s_{v_r} = \frac{\lambda}{4\pi \sin\theta \, \Delta t} \, s_{\phi_{\text{int}}}. \tag{15.92}$$

If the two observations are separated by $\Delta t = 30$ days, repetition of the L-band example that follows Eq. (15.92) leads to

$$\begin{aligned} s_{v_r} &= \frac{0.24}{4\pi \times 30 \text{ days}} \times 0.05 \\ &= \frac{0.24 \times 0.05}{4\pi \times 0.707 \times 30 \text{ days}} \\ &= 4.5 \times 10^{-5} \text{ m/day} = 1.6 \text{ cm/year}. \end{aligned}$$

This measurement precision is comparable to the speed of tectonic motions of the major plates on the surface of the Earth, which means that InSAR has the capability to map even such very slow-motion processes.

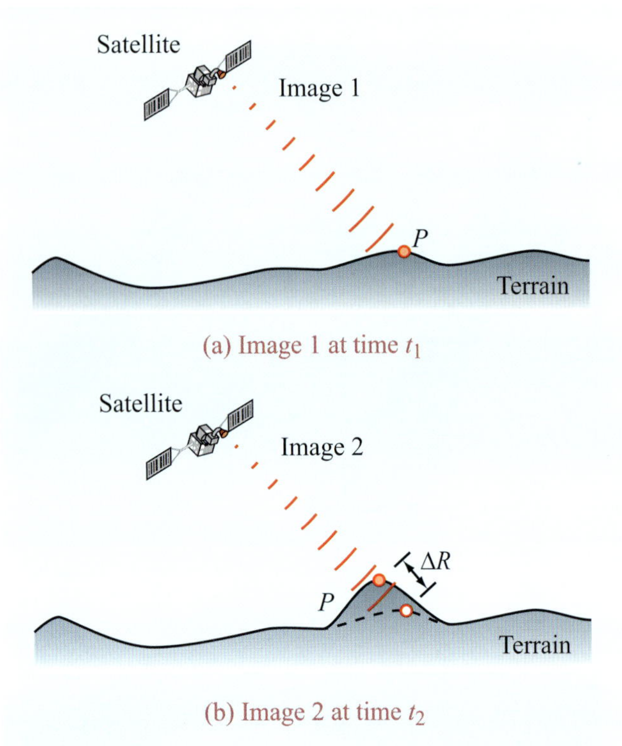

Figure 15-31: Surface deformation has displaced the location of point P by ΔR along the direction of the radar line of sight.

Actually, detecting displacements directly is a more practical approach than inferring velocities, in part because many tectonic motions are episodic and do not occur as smooth, continuous motions.

The scenario shown in Fig. 15-31 depicts a surface deformation that occurred between the times of acquisitions of two SAR images. The deformation causes the surface to change by ΔR along the direction of the radar line of sight. The phases of the pixel P in the two images are

$$\phi_1(t_1) = -\frac{4\pi}{\lambda}\, R_1 + \phi_{\text{scat}_1}, \tag{15.93a}$$

$$\phi_2(t_2) = -\frac{4\pi}{\lambda}\, (R_1 - \Delta R) + \phi_{\text{scat}_2}, \tag{15.93b}$$

where ϕ_{scat_1} is the phase of the scattering amplitude of the pixel at time t_1, and a similar definition applies to ϕ_{scat_2} at time t_2.

If the surface deformation moves the entire pixel, but does not change the relative locations of the individual scatterers contained in the pixel, then $\phi_{\text{scat}_1} = \phi_{\text{scat}_2}$. Consequently, the measured interferometric phase is

$$\phi_{\text{int}} = \phi_1 - \phi_2 = -\frac{4\pi}{\lambda}\, \Delta R.$$

Hence, in principle, by measuring the interferometric phase, we can quantify the line-of-sight deformation of the surface of every pixel in the image. This measurement capability, however, is compromised by two factors: (1) changes in the surface scatterers, causing $\phi_{\text{scat}_1} \neq \phi_{\text{scat}_2}$ and (2) deviation in the satellite orbit between the two observations. The first factor is related to the temporal decorrelation discussion of Section 15-5.5, and the second factor is associated with the role of topography. If the two tracks have a cross-track shift between them, ϕ_{int} becomes a sum of two components, one due to topography and another due to surface deformation. Through the application of more than two passes, together with the availability of precise track data, it is possible to separate the effects due to topography from those due to surface deformation.

The interferometric image in Fig. 15-32 shows surface displacements caused by the 6.3-magnitude L'Aquila earthquake in Italy, which occurred on April 6, 2009. The InSAR images were acquired by the ASAR system on ESA's ENVISAT. The indicated displacement values are the components of the true displacement along the direction of the radar line of sight.

15-10 Worldwide Dual Satellite InSAR Coverage: The TanDEM-X Mission

Most spaceborne InSAR experiments to date have been implemented using a single satellite in a repeat orbit to obtain the multiple images needed to form interferometric pairs. One major exception to this is SRTM, described in Section 15-7, in which two antennas on a single platform illuminate the ground simultaneously. Because the baseline length in this configuration is physically limited to structures that

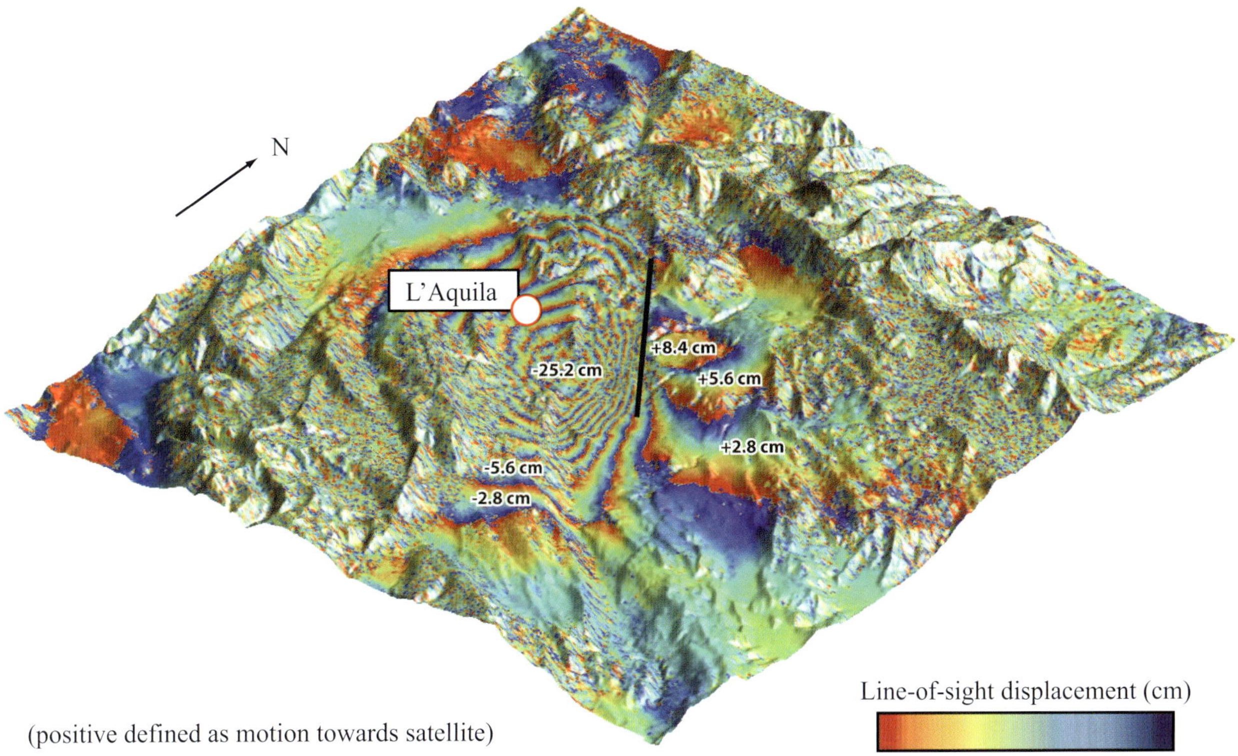

Figure 15-32: Interferogram due to surface deformation caused by the 6.3-magnitude L'Aquila earthquake in Italy [Walters et al., 2009].

can be reliably deployed in space, the performance of such a system is limited as described in Eq. (15.74). Mounting the two antennas on separate spacecraft with coordinated orbits so that simultaneity of observation is preserved, however, allows for a longer baseline separation in space (for topographic measurements) and in time (for velocity measurements, see Section 15-8). This approach was used by the German Aerospace Center, DLR, and Astrium consortium to build and fly the TanDEM-X (TerraSAR Digital Elevation Model - X) mission, which was launched in 2010. Over its expected three-year lifetime, the TanDEM-X mission will survey all 150 million square kilometers of Earth's land surface several times.

In this mission, each of two identical X-band radar satellites image the surface so that the same footprint is viewed by both satellites at the same time. Figure 15-33 is an artist's conception of the dual satellite configuration. This nearly eliminates the spatially and temporally varying atmospheric phase delay that is typically the largest statistical error source for InSAR. The ability to position the satellites independently allows for large spatial baselines, further reducing errors in the topographic product. The digital elevation models produced by the mission, with its 12 m spatial resolution and 2 m vertical accuracy, surpass even the best SRTM data sets acquired in 2000.

Furthermore, the two satellites may be positioned along track, rather than cross track, to form a temporal aperture and map surface velocities from space. This was first demonstrated using TanDEM-X data by Romeiser et al. (2010), who recorded images of currents

Figure 15-33: Artist's conception of twin TerraSAR satellites collecting simultaneous data to form the TanDEM-X system. The ability to achieve large cross-track baselines and greatly suppress atmospheric phase artifacts leads to digital elevation models with 2 m height accuracy at 12 m postings.

at the mouth of the Elbe river at varying tidal phases in order to visualize the flow and ebb of the water. They were able to achieve accuracies of several tenths of one m/s, as verified through ground measurements and predicted water velocities.

15-11 Time-Series InSAR Applications

Two-pass and three-pass interferometric methods for measuring surface deformation have evolved into ***multipass radar interferometry***. Additional applications are enabled if a long time series of observations from nearly repeating orbits are acquired. Multipass observations serve to improve the inferred deformation by averaging the predicted value, or to observe temporally changing deformation patterns, or both. Many existing SAR systems produce such data sets routinely. Through the use of time-series methods, such as ***stacking***, ***small baseline subset analysis*** (SBAS), and ***persistent scattering*** (PS), it is possible to improve the precision of the surface deformation estimate provided by a standard InSAR by an order of magnitude or more. Examples of InSAR analyses achieving millimeter-level precision with PS and SBAS techniques have been reported in numerous studies, including those by Ferretti et al. (2007), Johanson et al. (2009), Li et al. (2009), and Amelung et al. (2008).

15-11.1 Stacking

The stacking technique involves co-registering a ***stack of interferograms*** and adding them together (Lyons and Sandwell, 2003; Fialko, 2006). When implementing the stacking process, the interferograms may or may not be unwrapped in phase, and are often scaled by time separation in order to preserve deformation rates.

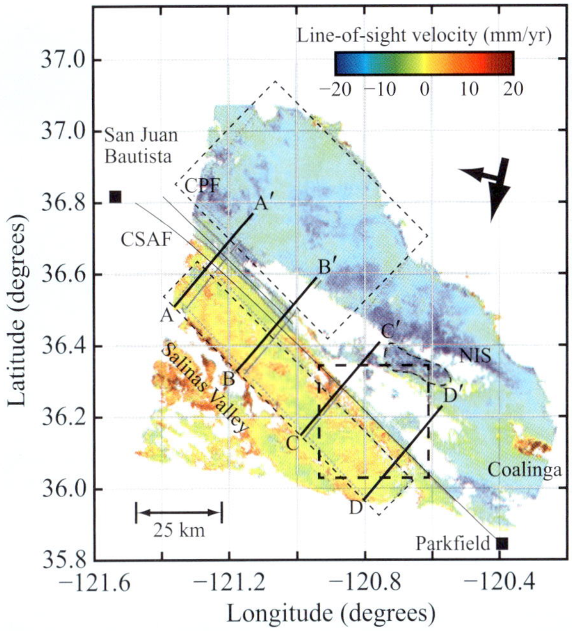

Figure 15-34: Stack of 12 descending ERS interferograms over the San Andreas Fault acquired between May 1992 and January 2001. The average creep rate in this section is 1.5–2.0 cm/yr, and it is easily seen in this image. Areas corrupted by spatial decorrelation, mainly because this is at a short, 6 cm wavelength, yield no useful displacement estimates and are white in this image. [Ryder and Bürgmann (2008).]

The main advantage of the stacking method is reduction of statistical errors, but it yields only a single average measurement and cannot easily retrieve time histories of deformation. An example of a stack is shown in Fig. 15-34, depicting average ***creep rate*** along the San Andreas Fault in California, from Parkfield northward to San Juan Batista. The fairly small displacement rate of about 1.5 cm/yr is easily discernible.

15-11.2 Small Baseline Subset Analysis (SBAS)

▶ The small baseline subset analysis (SBAS) technique can provide the *history* of the deformation of the surface crust, not just the mean deformation rate provided by the stacking technique. ◀

Successful implementation requires the selection of time-series interferograms with high correlation. The technique was first demonstrated by Berardino et al. (2002) who combined sets of high-correlation interferograms to obtain the time history for each correlated patch of ground over a site in southern Italy. The process, which includes spatial low-pass filtering and temporal high-pass filtering of the series of interferometric phase measurements for each imaged pixel, leads to the generation of an image displaying the measured rms deformation (Fig. 15-35) over the total observation time period (which in this case was close to 10 years), as well as a time series of the deformation at selected points in the image (Fig. 15-36). Other

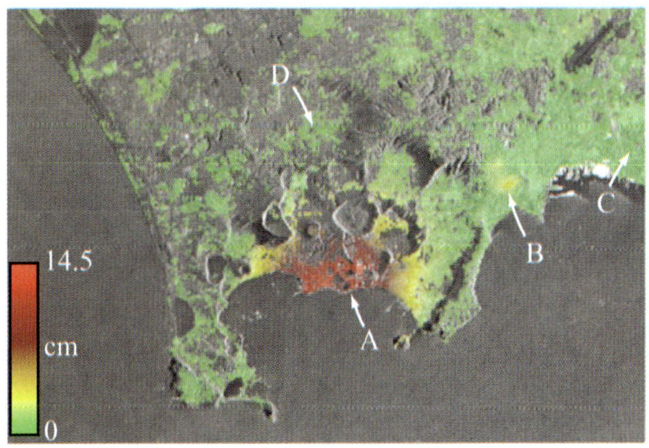

Figure 15-35: False-color map of the measured deformation rms superimposed on the SAR image amplitude of the investigated area. The temporal evolution of the deformations in the selected points identified by A, B, C are shown in Fig. 15-36(a)–(c), respectively [Berardino et al., 2002].

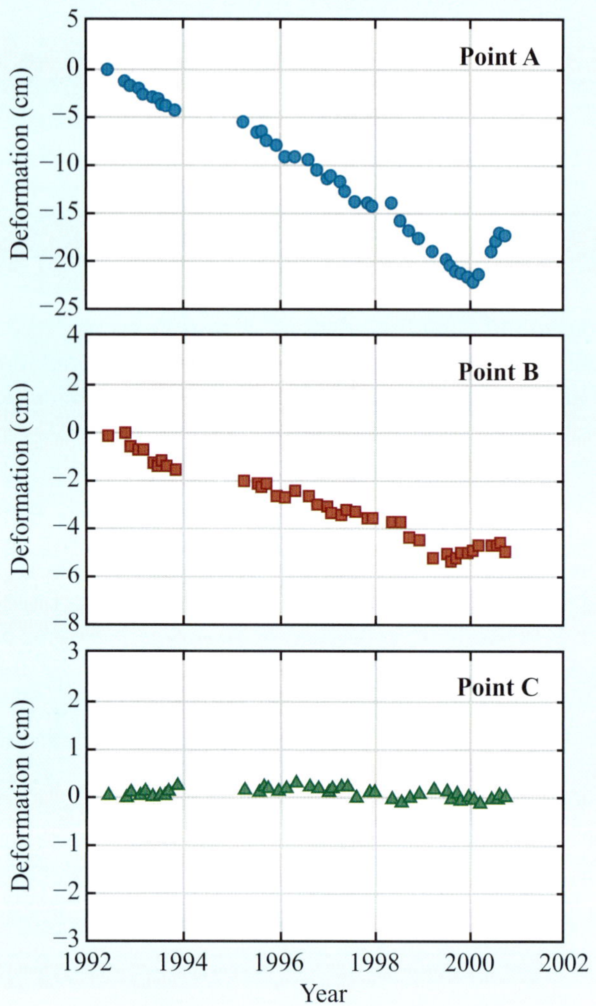

Figure 15-36: Time-series deformation measured at (a) point A, (b) point B, and (c) point C of Fig. 15-35 [Berardino et al., 2002].

examples of the SBAS technique are available in Pepe et al. (2005, 2011) and Reeves et al. (2011).

15-11.3 Persistent Scattering (PS)

The preceding techniques—stacking and SBAS—use a time series of conventional interferograms, and as such , they are subject to the decorrelation limitations of InSAR. Since decorrelation is dependent on the type of terrain and on changes that occur between one date and another, the SBAS technique can produce reliable deformation estimates over only those patches of the Earth's surface where the measured InSAR phase is reliable over some spatial extent. Consequently, decorrelation remains an obstacle to comprehensive coverage of surface deformation across the entire imaged scene.

One way to avoid the decorrelation problem is to identify individual pixels in the image that do not *scintillate*, thereby remaining highly correlated over the complete series of interferograms. Examples include buildings, bridges, and similar man-made structures. If a sufficient number of these ***persistent-scattering*** (PS) pixels are located in the imaged scene, a sufficiently dense geodetic network can be formed to allow deformation observations at small scales at fine (mm-level) precision. The PS technique was initially developed to track precise motions (subsidence) of man-made structures with mm-level resolution (Ferretti et al., 2000, 2001, 2004; Colesanti et al., 2003; Adam et al., 2003; Crosetto et al., 2003; Lyons and Sandwell, 2003; Werner et al., 2003).

The statistics of the field scattered from the many scatterers within a resolution cell are usually modeled by a Rayleigh fading model in which the power is exponentially distributed and the the phase is uniformly distributed (see Section 5-7). And whereas the Rayleigh fading model is appropriate for a cell containing a large number of randomly located scatterers, the model is not applicable when the total scattering is dominated by one or a few scatterers. Examples of dominant scatterers are corner reflectors formed by two walls or a wall and a ground surface, much like the corner reflectors used as radar calibration devices (see Section 13-11.4).

► Such persistent-scattering (PS) structures exhibit a *permanent* or nonscintillating behavior when viewed repeatedly by a SAR from the same perspective, which allows for very precise (mm-scale) measurements of their deformations. ◄

The PS technique was generalized and successfully applied to natural terrain to measure volcano deformation in vegetated areas (Hooper, 2006a; Hooper et al., 2007). This was accomplished by identifying persistent scatterers on the basis of their phase variations as a function of time, instead of their amplitude scintillation patterns.

The PS technique was further expanded in two major ways. First, by applying maximum-likelihood estimation methods, it is possible to improve the identification of PS points in the imaged scene, particularly in natural terrain (Shanker and Zebker, 2007). This leads to a denser geodetic network in space and time, which helps resolve more subtle and finer-scale deformations. The second expansion is related to the development of more accurate phase unwrapping methods (Shanker and Zebker, 2009, 2010). The improvement realized by these steps is illustrated in Fig. 15-37, which compares PS points identified using the original PS method (Ferretti et al., 2000), the more generalized approach introduced by Hooper (2006b), and the maximum-likelihood approach proposed by Shanker and Zebker (2007). Gray pixels in the images represent scintillating pixels and color pixels represent PS points. The maximum likelihood approach is able to identify not only all of the PS points identified by the other techniques, but also many others that are located in vegetated areas.

Application of the PS maximum-likelihood method to the San Andreas Fault area in California is demonstrated by the image in Fig. 15-38(a). The color of each identified PS point is associated with its displacement rate in mm/yr, measured along the line-of-sight of the radar. Hence, it is called the ***LOS displacement rate***. For the 20 km long segment noted in Fig. 15-38(a), the displacement rate as a function of distance from the fault is shown in Fig. 15-38(b). Pixels above the fault line exhibit an average displacement rate of about 6 mm/yr, compared with only about 1 mm/yr in the opposite direction for those pixels below the fault line.

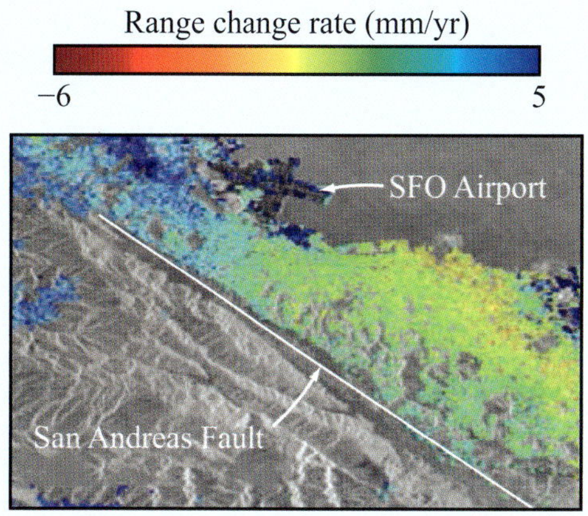

(a) Permanent scatterers amplitude dispersion (Ferretti et al., 2000)

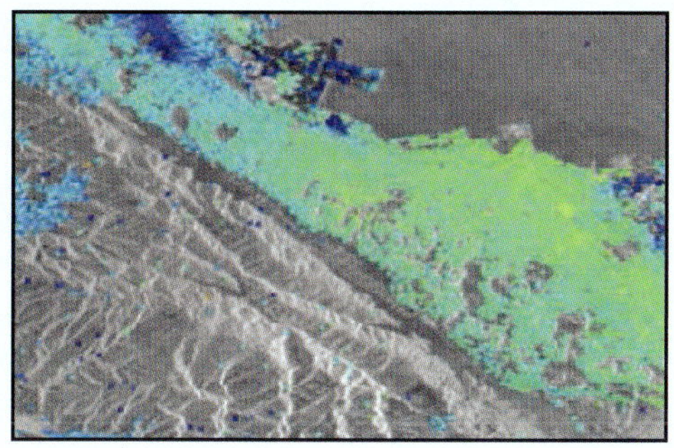

(b) Pixel phase and filtering (Hooper et al., 2004)

(c) Maximum likelihood (Shanker and Zebker, 2007)

Figure 15-37: Comparison of PS identification algorithms according to the methods by Ferretti et al. (2000), (b) Hooper et al. (2004), and (c) Shanker and Zebker (2007). Color pixels are persistent-scattering points.

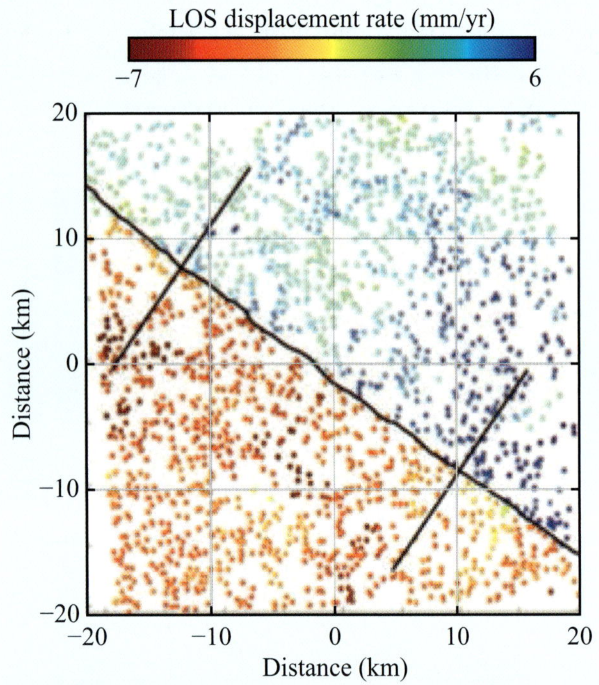

(a) Average LOS displacement rate image over San Andreas Fault estimated by using the edgelist phase unwrapping algorithm

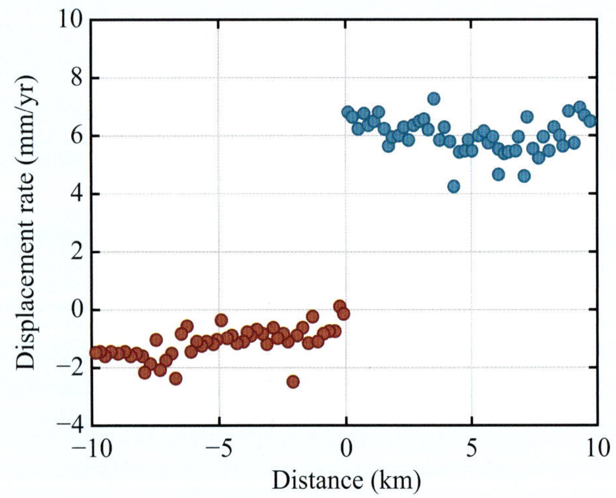

(b) Average LOS displacement rate as a function of distance from the fault

Figure 15-38: (a) Average LOS displacement rate image over the San Andreas Fault, (b) average LOS displacement rate as a function of distance from the fault. Assuming that all the displacement is purely due to ***strike slip motion*** across the fault, the estimated ***slip rate*** is 22 mm/yr. [Shanker, 2010.]

PROBLEMS

15.1 For an orbital topographic interferometer with an SNR of 15 dB, a perpendicular baseline length of 500 m, a range of 800 km, and operating at a wavelength of 6 cm, what is the error in the derived heights due to noise? The incidence angle is 23°.

15.2 Derive the slant to ground-range mapping function for a radar operating at a 35° incidence angle. Assume that the Earth is flat.

15.3 Repeat Problem 15.2 for a spherical Earth.

15.4 Given the following measurements of pixel offset in range versus pixel range deviation from the range to the pixel at the center of the beam of one of the images, what are the perpendicular and parallel baseline components for a radar with a range of 850 km, incidence angle of 25°, and slant-range pixel spacing of 7.5 m?

Range pixel deviation (m)	Measured offset (m)
100	−2.8
150	−1.8
200	−1.45
250	−1.1
300	−0.2
350	0.3
400	0.4
450	1.2
500	1.7
550	1.8
600	2.3
650	3.2
700	3.2
750	4.1
800	4.4

15.5 Calculate the critical baseline for a 12 cm wavelength radar with a sinc impulse response, a bandwidth of 80 MHz, a range of 700 km, and operating at 30° incidence angle?

15.6 A 1280 MHz radar operating from a range of 750 km has an SNR of 12 dB, an InSAR baseline of

2 km, and an incidence angle of 35°. Assuming that the baseline orientation angle is known to 0.01° and the baseline length to 2 m, what are the contributions to height error for topographic applications from phase, baseline length, and baseline orientation angle?

15.7 What is the uncertainty associated with the radial velocity of the surface when measured by a radar with 10 dB SNR, an along-track baseline of 10 m, and a platform velocity of 200 m/s, if the radar operates at 24 cm wavelength and an incidence angle of 45°?

For Problems 15.8–15.12, consider a radar system with the following description:

Orbit altitude:	696000 m
Wavelength:	0.236057 m
PRF:	2159.827 Hz
Sample rate f_s:	16.0 MHz
Chirp slope:	5.1851852×10^{11} Hz/s
Pulse length:	27.0 μs
Satellite velocity:	7179.4 m/s
R_0:	844768 m
f_D:	0 Hz

Copy two data files, image.1 and image.2, from *Images* in the book website mrs.eecs.umich.edu. Each image is a complex floating-point file consisting of 14336 lines of 3072 complex samples each. These files are used for the next five problems.

15.8 Form an interferogram by multiplying the two images together and display the phase, showing interferogram fringes. Calculate and submit a multilook version of the interferogram with 4 looks in range and 16 looks in azimuth.

15.9 Calculate the offsets over the original images, and plot the range offset versus range. Determine the average azimuth offset.

15.10 Calculate the baseline components $B_{\|}$ and $B_{\perp}$ from the offsets. Calculate the baseline length B and orientation angle α.

15.11 Resample image.2 to align with image.1 and form the new interferogram. If you need to resample in azimuth, be sure to account for the Doppler centroids of the images in some manner.

15.12 Calculate the correlation between the two images from step 1 and step 4, using 4×16 look averages for determining each correlation coefficient. Display the correlation images as byte files where zero correlation maps to 0, and unity correlation maps to 255.

CHAPTER 16

Radar Remote Sensing of the Ocean

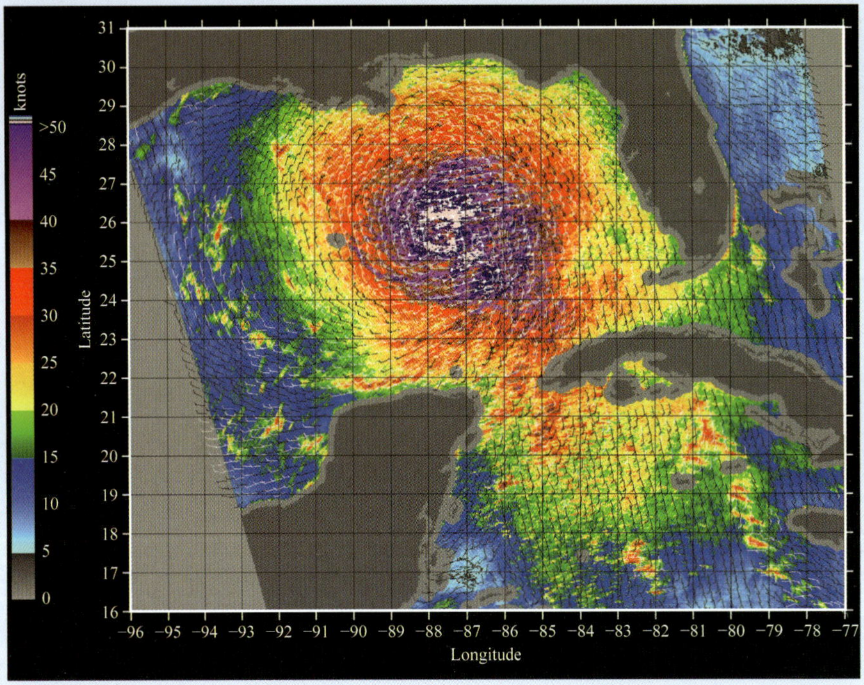

Near-surface winds of Hurricane Katrina based on data from SeaWinds scatterometer on board QuikSCAT on 28 August 2005

CONTENTS

Overview

The ocean covers more than three-fourths of the Earth's surface and dominates global weather and climate. Knowledge of environmental conditions over the oceans is critical for weather forecasting, navigation, and scientific studies. Satellite platforms can provide rapid global coverage of ocean parameters. Scatterometers and SARs can provide valuable information about the state of waves, wind over the ocean, rainfall rate, and sea-ice conditions. In this chapter we consider a number of wind scatterometer–based applications of microwave remote sensing over the ocean, with an emphasis on wind and rain measurements. Sea ice, which is of interest in the polar regions, is also considered. We also briefly discuss wind measurements by SAR. We close the chapter with an examination of how radar can be used to detect and map the distribution of oil slicks.

16-1 Wind-Vector Scatterometry

Knowledge of winds blowing over the oceans is crucial to accurate weather and wave prediction and to the understanding of air/sea interaction and gas exchange. Unfortunately, conventional *in-situ* techniques (ships of opportunity, buoys, island stations, aircraft, etc.) for measuring oceanic winds are hampered by their limited resolution, variable accuracy, and uneven sampling. Active microwave remote sensing offers a solution. Experience with a long series of satellite radar scatterometers has demonstrated that wind vectors over the ocean can be accurately measured from space. Radars can also measure rain over the ocean from space.

Under steady-wind conditions, the friction between the wind and the ocean surface causes the water surface to assume a periodic waveform composed of ***gravity waves*** with wavelengths on the order of 1 m or longer. Superimposed on the gravity waves are waves with much smaller wavelengths, on the order of 1 cm or shorter, called ***capillary waves***. The periodic waveform of the gravity waves alone is illustrated in 3-D in Fig. 16-1(a), and in part (b) of the same figure we show a 2-D version that includes both the gravity waves and the capillary waves on top of them. Also note that the periodic waveform is ***asymmetrical***; that is, it is skewed such that

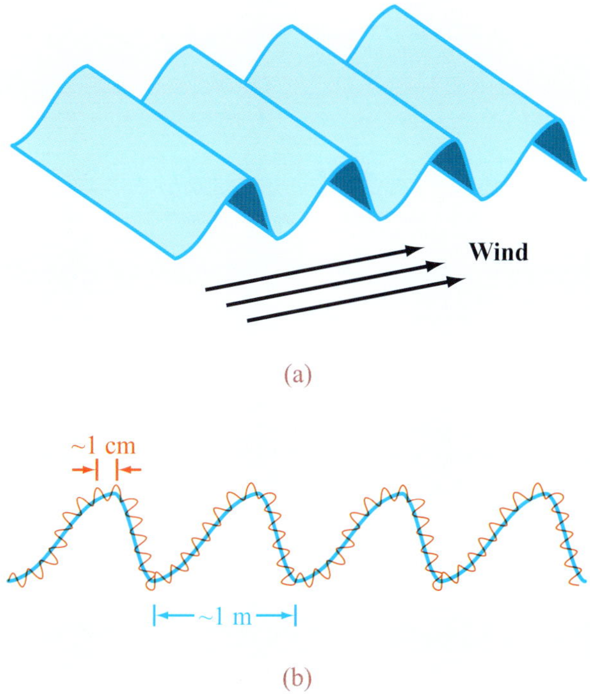

Figure 16-1: Ocean surface profile: (a) periodic gravity waves and (b) small capillary waves riding on top of the gravity waves.

the slope of the upswing part of the wave is smaller than that of the downswing part. Greater discussion of wave spectrum and skewness is given in Section 16-2.

Steady wind action leads to a sea state known as a ***fully developed sea***. Under such conditions, the backscattering coefficient σ^0 is a function of not only the wind speed U, but also the wind direction relative to the radar look direction, denoted by

$$\chi = \phi_{\text{wind}} - \phi_{\text{radar}}, \qquad (16.1)$$

where ϕ_{wind} and ϕ_{radar} are the azimuth angles of the wind vector and radar-to-surface vector, both defined relative to the north direction as shown in Fig. 16-2. These directions are defined in the ***oceanographic convention***, in which $\phi_{\text{wind}} = 0^\circ$ corresponds to a wind flowing from the south toward the north with positive angles measured CW as seen from above. Angle χ is the ***relative azimuth*** angle between the wind vector and the radar look direction.

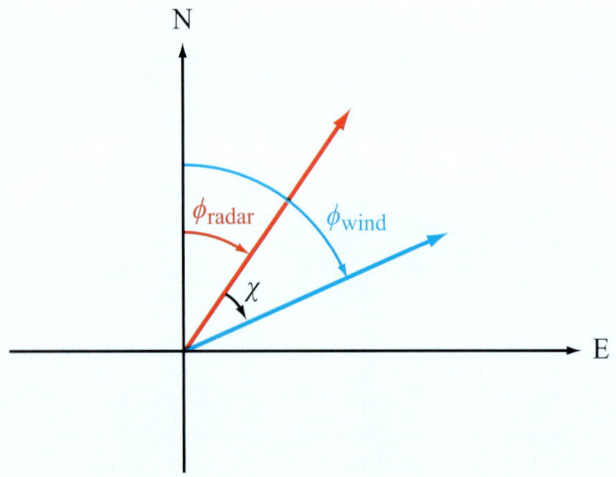

Figure 16-2: In the oceanographic convention, the azimuth angle of the radar look direction, ϕ_{radar}, and the wind azimuth ϕ_{wind} angle are defined relative to north. The angle between them is the relative azimuth angle $\chi = \phi_{wind} - \phi_{radar}$.

As discussed in later sections, the periodic sea-surface pattern leads to a backscattering coefficient with an azimuthal dependence of the form

$$\sigma^0(\chi) = A + B\cos\chi + C\cos 2\chi, \tag{16.2}$$

where coefficients A, B, and C are functions of four parameters, namely the wind speed U and the three radar wave parameters: incidence angle θ, radar frequency f, and receive-transmit polarization configuration pq. For a specified set of radar wave parameters, σ^0 is a function of U and χ. Hence, to measure the wind velocity vector (speed and direction), it is necessary to perform two or more different observations of the area on the ocean surface. The most common approach is to measure the sea surface from multiple azimuth angles. Figure 16-3 shows the illumination patterns of four spaceborne wind scatterometers. All four patterns include observations at 45° and 135° relative to the satellite track, and three of them also include a third observation, at either 90° or 115°. This means that as the satellite flies along its path past a resolution cell of sea surface, the cell gets observed along two or three azimuth directions. The multiple measurements are used in an algorithm that infers both U and χ. Table 1-2 provides a summary of the radar wave parameters for the systems whose illumination patterns are displayed in Fig. 16-3. As discussed later, scanning pencil-beam scatterometers can provide observations at multiple azimuth angles.

16-2 Wind and Wave Modeling

16-2.1 Wind

The atmosphere is in constant motion in response to temperature and pressure differentials and is highly variable at all scales. Kinetic energy in the atmosphere is dissipated by frictional forces. Large scale motions force lesser-scale motions in the so-called ***energy cascade***. The kinetic energy of the smaller scales is ultimately dissipated through viscosity effects. Instability and irregularities in the Earth's surface also produce small scale eddies and disturbances that dissipate energy from larger-scale wind fields. This energy cascade is important in defining the spectrum of the wind fields.

Long-term empirical observations of the statistics of atmospheric turbulence have shown that motion statistics remain relatively stable, which is a distinctive characteristic of ***turbulent flows***. Although the details of the turbulent flow appear random, the statistics of the flow parameters are stable in time and can be used to characterize the flow. Because turbulent flow is random, deterministic models for the flow cannot model it adequately. This has led to the development of ***turbulence models*** of atmospheric motion (Charney, 1971).

> ► ***Wind***, the flow of air, is a vector quantity with a magnitude (speed) and direction. ◄

Wind has both horizontal and vertical components. Temperature and moisture content play a particularly important role in vertical air motion. When a volume of air is colder or contains more moisture, it is denser than the surrounding air mass and sinks. Conversely, when a volume of air is less dense than the surrounding air due to a lower moisture content or warmer temperature, it is more ***buoyant*** than the surrounding air and rises. When this occurs, the atmosphere is termed ***unstable***. As an air mass rises it cools due to expansion and becomes more dense than the surrounding air. This tends to

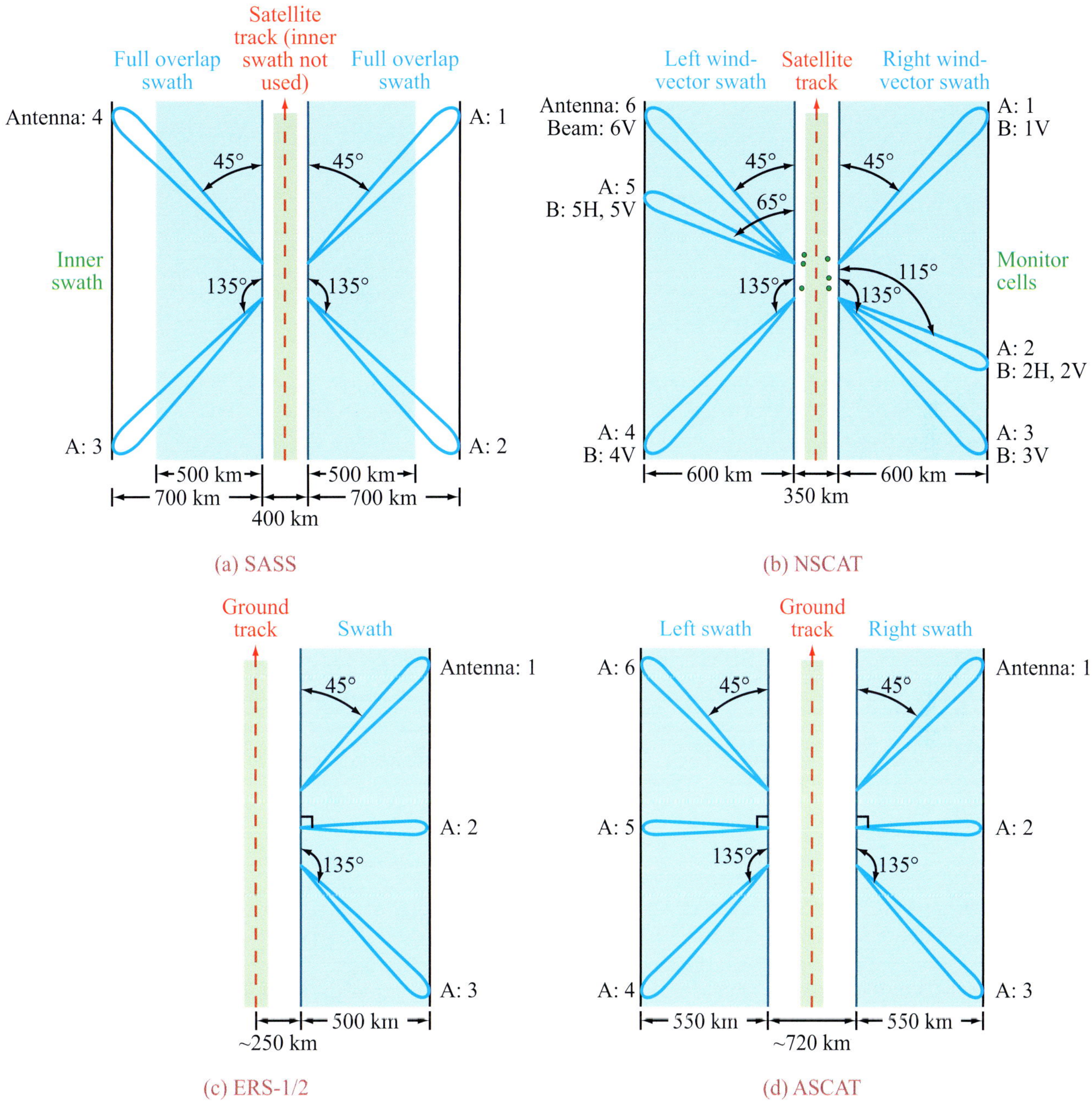

Figure 16-3: Configurations for the antenna footprints for fan-beam wind scatterometers flown to date. (a) Seasat Scatterometer (SASS), 1978. (b) NASA Scatterometer (NSCAT), 1996–1997. (c) ERS-1/2 AMI, 1992–2001. (d) Advanced Scatterometer (ASCAT), 2007–.

cancel the effects of buoyancy. When an air mass is less buoyant than the surrounding atmosphere and sinks, the conditions are termed ***stable***. When there is no net buoyancy, the atmosphere is said to be ***neutrally stable***.

For wind blowing horizontally, friction with the Earth's surface tends to slow the wind at the surface level. The lowest region of the troposphere, the portion most affected by the surface, is known as the ***surface boundary layer***, the top of which may be from tens to many hundreds of meters above the surface depending on local stability conditions.

> ▶ Friction with the wind blowing over the ocean's surface transfers momentum from the atmosphere to the ocean. This momentum transfer is known as ***wind stress***, which is a vector quantity with magnitude and direction. ◀

Wind stress is often the parameter of greatest interest to atmospheric and oceanic modelers. Wind stress depends on the ***wind shear***, the difference between the wind vector and the ocean surface current vector, as well as the ocean surface roughness, and the air/sea temperature difference.

Within the surface boundary layer the profile of the wind speed $U(h)$ as a function of height h above the surface can be expressed approximately as (Pedlosky, 1979)

$$U(h) = \frac{u_*}{0.4}\left(\ln\frac{h}{Z_0} + \psi\right) \qquad \text{(m/s)}, \tag{16.3}$$

where u_* is the ***friction velocity*** of the wind at the ocean surface, ψ is a function of the atmospheric stability of the surface boundary layer, and Z_0 is the ***roughness length*** of the surface. Friction velocity and the ***wind stress*** τ (in N/m^2) are related by (Phillips, 1985)

$$u_* = \sqrt{\frac{\tau}{\rho}} \qquad \text{(m/s)}, \tag{16.4}$$

where ρ is the air density in Kg/m^3.

> ▶ Note that ψ is positive in a stable atmosphere and negative in an unstable atmosphere. When $\psi = 0$ the conditions are neutrally stable and the air and ocean temperatures are the same. ◀

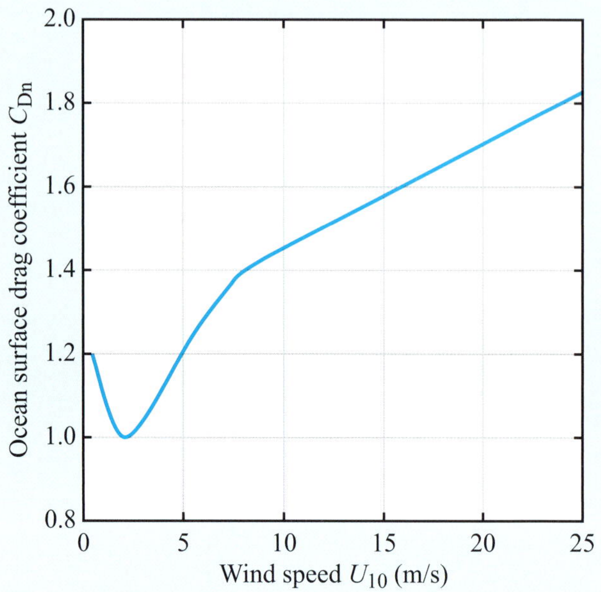

Figure 16-4: Empirical ocean surface drag coefficient C_{Dn} versus wind speed at 10 m above the ocean surface.

When wind blows over long distances of the ocean, heat transfer between the ocean and atmosphere tends to produce neutral-stability conditions.

Alternatively, the stress can be written in terms of the wind speed at a fixed reference height, U, as (Kinsman, 1965)

$$\tau = \rho C_D (U - U_s)^2, \tag{16.5}$$

where U_s is the surface current and C_D is a unitless stability-dependent and reference height-dependent ***drag coefficient***. The friction velocity u_* also can be written as

$$u_* = \frac{\sqrt{C_{Dn}}\,(U - U_s)}{1 - \sqrt{C_{Dn}}\,\psi/0.4}, \tag{16.6}$$

where C_{Dn} is the ***drag coefficient for a neutral atmosphere***. Under the conditions of neutral stability ($\psi = 0$), Eq. (16.6) reduces to

$$u_* = \sqrt{C_{Dn}}(U - U_s). \tag{16.7}$$

The most accurate estimates of the drag coefficient C_{Dn} come from field observations and are averaged over wave conditions with no surfactants. Fig. 16-4

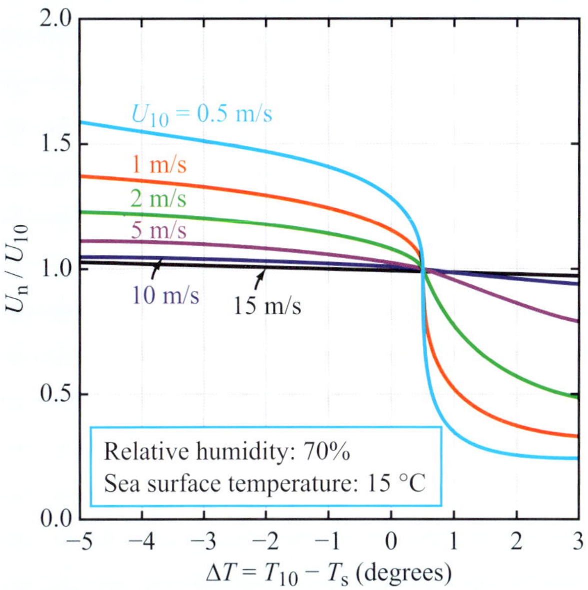

Figure 16-5: The ratio of the equivalent neutral stability wind speed U_n to the actual wind speed U_{10} at a height of 10 m as a function of the air-sea temperature difference ΔT using the model of Liu et al. (1979) with a sea surface temperature of 15 °C and relative humidity of 70%.

illustrates measured values of C_{Dn} versus wind speed at a standard reference height of 10 m, which is labeled U_{10}. Generally, the drag coefficient increases with wind speed as the surface becomes rougher.

For moderate to strong winds over the ocean, the effects of atmospheric stability are small. However, stability can be important in low-wind conditions. As illustrated in Fig. 16-5, for moderate to high winds, the ratio of the neutral-stability wind speed U_{n} to U_{10} is nearly unity. However, at lower wind speeds the ratio is temperature dependent and thus stability dependent. The wind stress, which dictates the momentum transfer between the ocean and atmosphere, is thus temperature dependent.

Winds over the ocean are turbulent over all spatial scales. Experimentally verified turbulence models predict that at spatial scales ranging from 10 to 1000 km, the wind energy spectrum approximately follows a power-law dependence on the spatial wavenumber. The power-law model takes the form αK^{-b}, where $K = 2\pi/L$ is the spatial spectrum wavenumber, L is the spatial scale, and $b \approx 5/3$ is a constant (Charney, 1971; Leith, 1971; Freilich and Chelton, 1986). The parameter α is proportional to the squared mean wind speed. The actual wind spectrum varies from this ideal because of unmodeled effects. This turbulent energy drives ocean motion through ***air-sea interaction***.

16-2.2 Waves

As previously noted, the friction of the wind blowing over the ocean's surface generates atmospheric and oceanic turbulence, which leads to the development of small ***capillary waves***, waves of a centimeters or less in length (Kinsman, 1965). These waves interact nonlinearly to generate longer waves. Waves 1 cm to 1 m or so in length are known as ***capillary-gravity waves***, while longer waves (greater than ~1 m) are known as ***swell*** or ***gravity waves***. Once produced, gravity waves can propagate long distances.

▶ Unlike electromagnetic waves, which interact linearly, interactions between water waves are nonlinear. When two water waves interact, they generate new waves at both higher and lower wavelengths. ◀

This leads to an ***energy cascade*** in the water wave spectrum that transfers wave energy to both smaller and larger wavelengths. ***Viscosity*** causes dissipation of energy at the smallest waves. Energy at the largest wavelengths is lost through ***wave breaking*** and through ***wave-wave interactions*** that propagate energy to shorter waves. The energy input from the wind friction with the surface thus generates a spectrum of water wavelengths. For steady winds, these processes result in an equilibrium wave spectrum known as a ***fully developed sea***. An idealized plot of the wind-speed-dependent wave spectrum is shown in Fig. 16-6.

The precise characteristics of the shape of the wave spectrum depend on the ***fetch***, or distance over which the wind has been blowing, the surface viscosity of the water (and thus its temperature and, to a lesser degree, its salinity), and the air/sea temperature difference. The surface friction is affected by surfactants, such as oil or biological material, if any is present.

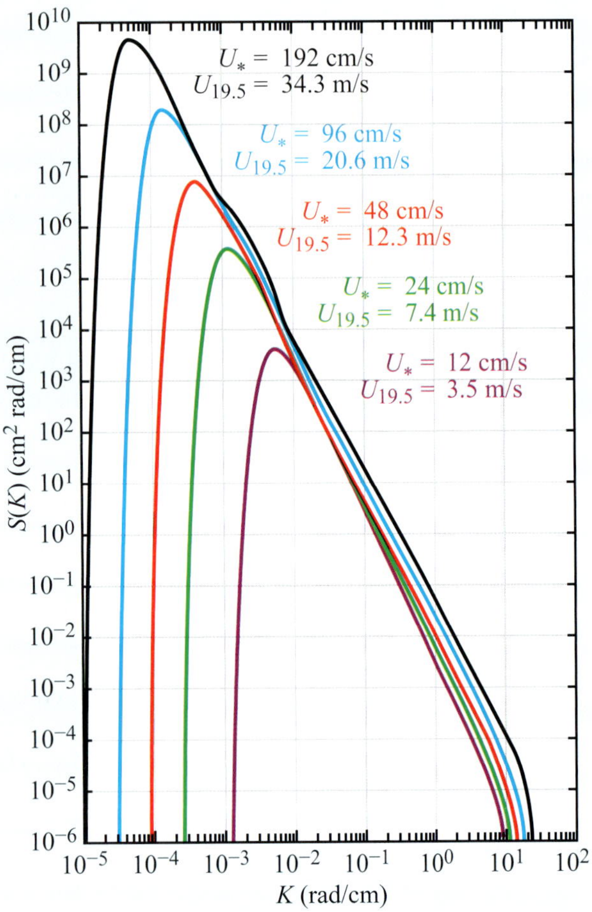

Figure 16-6: Wavenumber spectra for fully developed seas versus wind stress [Bjerkaas and Riedel, 1979].

16-3 Radar Scattering

Radar backscatter from the ocean's surface depends on the surface roughness and the water's dielectric properties.

> ▶ Since the wind is responsible for most of the surface roughness, the backscatter is related to the wind vector, though other factors play a role as well. ◀

In the following we consider how radar backscatter at moderate incidence angles (20°–60°) is related to the near-surface wind vector. This relationship can be exploited to infer wind speed and direction from backscatter measurements.

Before we begin, we note that as described in Chapter 4, the dielectric constant ε of the ocean's surface is dependent on the water temperature, the presence of ***surface films***, ***salinity***, etc. However, ε is treated as a constant whose value is computed using the expressions given in Section 4-2 at the specific radar wavelength under consideration and for the expected values of the ocean salinity and water temperature.

In considering scattering from the sea surface, we can model it as a time-varying, anisotropic, and skewed rough surface whose roughness is due to local wind conditions and waves propagating from other areas into the region of interest. Clearly, it is very different from and much more complex than the rough surfaces discussed in Chapter 10.

An examination of a sea surface under wind-driven conditions reveals that there are many scales of waves present on the ocean's surface. In particular, there are capillary waves of many sizes riding on larger-scale waves of varying sizes. As previously noted, large waves can propagate long distances and thus may not be related to the local wind conditions. On the other hand, capillary waves tend to be in equilibrium with the local wind. Hence for wind sensing, the radar wavelengths that are particularly sensitive to capillary wavelengths are the most appropriate for measuring local wind conditions. Further, since at near-nadir and near-grazing incidence angles the backscatter has a significant component due to large waves, moderate incidence angles (20°–60°) are desirable for inferring surface roughness due to capillary waves.

We note that the physical mechanisms that contribute to backscattering are diffraction and incoherent scattering. These mechanisms are activated by the waves that are comparable to or smaller than the incident wavelength.

> ▶ Thus, in microwave backscattering from rough surfaces at moderate incidence angles, only roughness scales comparable to or smaller than the incident radar wavelength are important. ◀

Within a short time frame and a localized region, the roughness of the sea surface can be thought of as a stationary random process.

16-3.1 Ocean Surface Statistics

To calculate the backscattering coefficient σ^0 we need to know either the ***surface correlation function*** $\rho(r,\phi)$ or its Fourier transform, the ***surface spectrum*** $S(r,\phi)$, for the roughness scales in the centimeter range. Recall from Section 10-2.2 that the surface correlation function is a measure of how the surface heights of two points on the surface are statistically correlated when the two points are separated by a distance r and the distance vector between them makes an angle ϕ relative to a reference direction. In the present context, ϕ is defined relative to the upwind direction. An investigation of the correlation function of a multiscale rough surface by Fung and Chen (2009) reveals that regardless of the correlation behavior of the individual roughness scales, the correlation function of the combined multiscale surface has an exponential appearance. To demonstrate this idea, we assume that each of five roughness scales are Gaussian-correlated and that they are statistically independent of each other. Under these conditions the ***surface correlation function*** along a given azimuth direction ϕ may be written as

$$\rho_s(r) = \sum_{i=1}^{5} s_i^2 \exp\left(-\frac{r^2}{L_i^2}\right), \tag{16.8}$$

where L_i is the scale length and s_i is the standard deviation of the surface roughness at scale length L_i for the ith scale. To continue with the illustration we somewhat arbitrarily let $L_5 = 0.1L_1$, $L_4 = 0.14L_1$, $L_3 = 0.2L_1$, $L_2 = 0.45L_1$, $s_5^2 = 0.06s_1^2$, $s_4^2 = 0.15s_1^2$, $s_3^2 = 0.3s_1^2$, and $s_2^2 = 0.8s_1^2$. A plot of the resulting normalized $\rho_s(r)$ with $s_1 = 0.8$ cm and $L_1 = 10$ cm is shown in Fig. 16-7.

Even though all five roughness scales are Gaussian-correlated, the total surface correlation function appears

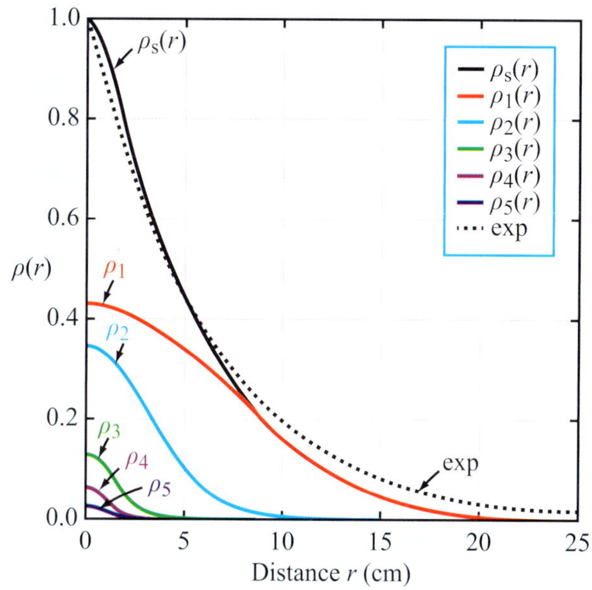

Figure 16-7: Plot of the normalized correlation function of a multiscale rough surface. $\rho_1(r)$ to $\rho_5(r)$ are the correlation functions of the individual terms in Eq. (16.8), while $\rho_s(r)$ is the total correlation. Note that the total correlation function resembles the exponential function, even though the individual terms are Gaussian correlated. This is the result of sums of multiple, smaller resolution elements with one or more larger-scale terms.

exponential, except near the origin. It is the addition of several much smaller components that cause the overall correlation to have a sharp peak near the origin, thus creating the exponential look. An exponential function $\exp[-r/(0.6L_1)]$ is plotted for comparison. Except for a small region near the origin and a slight deviation in the tail, there is good agreement between the exponential and the multiscale correlation function. Since the sea surface has many small roughness scales, we can expect an exponential function to be a good fit to its correlation function in the capillary region.

Since the correlation function for any surface must be centrosymmetric, an approximate representation of the surface correlation function for the short wave portion of the sea surface is given by

$$\rho_s(r,\phi) = s^2 \exp\left(-\frac{r}{L_t}\right), \tag{16.9}$$

where $L_t = L_u \cos^2\phi + L_c \sin^2\phi$ is the correlation length along the ϕ direction; L_u is the correlation length in the ***upwind*** direction (along $\phi = 0$); L_c is the correlation length in the ***crosswind*** direction ($\phi = 90°$), and s^2 is the surface height variance. The three parameters in the surface correlation are functions of windspeed. It is expected that $L_c > L_u$ and that both of them are decreasing functions of windspeed, while s^2 is an increasing function of windspeed. As discussed in Section 16-3.4 these parameters can be related to the wind speed U through empirically determined functions.

We now consider the surface ***skewness***. In a stationary random process the surface moments are related to its second-order and third-order ***cumulant functions***, also known as correlation and ***bicoherence*** functions, respectively. The Fourier transform of the correlation and bicoherence functions are the spectrum and ***bispectrum*** (Mendel, 1991). The second-order C_2 and third-order C_3 cumulants of a zero-mean, stationary random process $x(t)$ are given by

$$C_2(\tau) = E\{x(t)x(t+\tau)\}, \tag{16.10a}$$

$$C_3(\tau_1,\tau_2) = E\{x(t)x(t+\tau_1)x(t+\tau_2)\}, \tag{16.10b}$$

where $E\{\cdot\}$ is the statistical expectation and τ, τ_1, and τ_2 are time delays. $C_2(\tau)$ is recognized as the conventional ***temporal correlation function***, which describes the process covariance. The third-order cumulant function $C_3(\tau_1,\tau_2)$ is useful for describing the skewness of the process.

Early studies of cumulant functions were carried out by many investigators (Longuet-Higgins et al., 1963; Hasselmann et al., 1963; Srokosz and Longuet-Higgins, 1986; Masuda and Kuo, 1981). Application of these functions to radar scattering were reported by Chen and Fung (1990), Fung and Chen (1991), and Chen et al. (1992). An investigation of the form and properties of the bicoherence function was carried out by Amar (1989). The bicoherence function has vanishing first and second derivatives at the origin and can be written as the sum of symmetric and antisymmetric functions. The symmetric function is responsible for vertical skewness and the antisymmetric function accounts for horizontal skewness. Since the bicoherence function is a third-order product [see Eq. (16.10b)], the impact of the symmetric component on scattering is small compared with the impact of the temporal surface correlation function, which is a second-order product (Fung, 1994). Hence, the symmetric component of the bicoherence function can be ignored in scattering calculations. Our major interest is in the antisymmetric function.

As we see shortly, the antisymmetric component of the bicoherence function causes the upwind backscatter to be higher than the downwind backscatter. It has no influence on σ^0 in the crosswind direction. We refer to the antisymmetric component of the bicoherence function as the ***skewness function***. Its Fourier transform, the bispectrum, is purely imaginary. A summary of Amar's work on skewness is available in Appendix 7A of Fung (1994), where a suggested form for the skewness function is

$$S_s(r,\phi) = s^3 \left(\frac{r\cos\phi}{s_0}\right)^3 \exp\left(-\frac{r^2}{s_0^2}\right), \tag{16.11}$$

where s_0 is the skewness parameter (which depends on windspeed), r is the separation distance, and ϕ is the azimuth angle relative to the wind direction.

16-3.2 IEM Scattering Model

Before the 1990s, the available radar surface scattering models were restricted mainly to the small perturbation model for small roughness scales (Fung and Lee, 1982) and the Kirchhoff model for large roughness scales (Fung and Chen, 1991). The Kirchhoff model can be applied when there is no wind and consequently no small-scale roughness. Unfortunately, wind induces surface waves that are smaller than can be accommodated by the small perturbation model, which is limited to roughness scales longer than the EM wavelength. Two-scale scattering models have been developed to overcome this limitation (Wentz, 1975; Plant, 1986; 2002).

A more general approach that can accommodate roughness scales comparable to the incident wavelength was developed by Fung et al. (1992). They use the integral equation governing the electrical surface current to better estimate the surface current induced by the incident wave. Their model bridges the gap between the small perturbation and the Kirchhoff models, and includes these models as the low and high frequency limits of their model, respectively. This ***integral equation model*** (IEM) employs an algebraic form involving the surface spectrum and its higher orders (cumulants) for both vertical and horizontal polarizations. For backscattering from the sea surface, the IEM backscattering coefficient *without the skewness component* is given by (Fung, 1994)

$$\sigma_{pp}^{(1)}(\theta,\chi) = \frac{k^2}{2}\exp(-2k^2s^2\cos^2\theta)\sum_{n=1}^{\infty}\frac{|I_{pp}^n|^2}{n!}W_n(K,\chi), \tag{16.12a}$$

where $p = \mathrm{v}$ or h stands for vertical or horizontal polarization, θ is the incidence angle, χ is the azimuth angle of the wind vector relative to the radar look direction (Fig. 16-2), and

$$I_{pp}^n = (2k\cos\theta)^n f_{pp}\exp(-k^2s^2\cos^2\theta) + \frac{(k\cos\theta)^n}{2}F_{pp},$$

$$f_{\mathrm{vv}} = \frac{2\rho_{\mathrm{v}}}{\cos\theta},$$

$$f_{\mathrm{hh}} = -\frac{2\rho_{\mathrm{h}}}{\cos\theta},$$

$$F_{\mathrm{vv}} = \frac{2\sin^2\theta}{\cos\theta}\left[\left(1-\frac{\varepsilon\cos^2\theta}{\varepsilon-\sin^2\theta}\right)(1-\rho_{\mathrm{v}})^2 + \left(1-\frac{1}{\varepsilon}\right)(1+\rho_{\mathrm{v}})^2\right],$$

$$F_{\mathrm{hh}} = \frac{2\sin^2\theta}{\cos\theta}\left[4\rho_{\mathrm{h}} - \left(1-\frac{1}{\varepsilon}\right)(1+\rho_{\mathrm{h}})^2\right].$$

Here, k is the EM wavenumber, s is the surface rms height, ρ_{v} and ρ_{h} are the Fresnel reflection coefficients, ε is the surface dielectric constant, and

$$W_n(K,\chi) = \frac{1}{2\pi}\int_0^{2\pi}\int_0^{\infty}\rho_{\mathrm{s}}^n(r,\phi)\,e^{-jKr\cos(\chi-\phi)}\,r\,dr\,d\phi, \tag{16.12b}$$

where $\rho_{\mathrm{s}}(r,\phi)$ is the surface correlation function given by Eq. (16.9), and K is the sea-surface wavenumber.

For the computation in the next section we choose a simple approximate form for $W_n(K,\chi)$ instead of going through the details of the general Fourier transform computations. This approximate form for $W_n(K,\chi)$ is related to the exponential correlation function and has been successfully used in surface scattering computations (Fung, 1994):

$$W_{\mathrm{n}}(K,\chi) = \left(\frac{sL_{\mathrm{t}}}{\mathrm{n}}\right)^2\left[1+\left(\frac{KL_{\mathrm{t}}}{\mathrm{n}}\right)^2\right]^{-3/2}, \tag{16.13}$$

where $L_{\mathrm{t}} = L_{\mathrm{u}}\cos^2\phi + L_{\mathrm{c}}\sin^2\phi$. It is important to note that the spectrum corresponding to the function given by Eq. (16.13) only approximates the portion of the sea spectrum representing wave sizes in the range of centimeters, which is the spectrum relevant to radar scattering in the microwave range. It does not include either the lower or the higher end of the wavenumber spectrum of the sea surface. The rms height s, correlation length L_{u} in the upwind direction, and correlation length L_{c} in the crosswind direction are determined empirically from radar measurements.

The portion of the backscattering coefficient due to surface skewness is given by (Fung, 1994)

$$\begin{aligned}\sigma_{pp}^{(2)}(\theta,\chi) = {}& \frac{k^2}{2}|f_{pp}|^2\exp(-4k^2s^2\cos^2\theta)\cdot\sum_{n=1}^{\infty}\frac{(-8k^3\cos^3\theta)^n}{n!}B_n(K,\chi)\\ &+\frac{k^2}{2}\mathfrak{Re}(f_{pp}^*F_{pp})\exp(-3k^2s^2\cos^2\theta)\cdot\sum_{n=1}^{\infty}\frac{(-3k^3\cos^3\theta)^n}{n!}B_n(K,\chi)\\ &+\frac{k^2}{8}|F_{pp}|^2\exp(-2k^2s^2\cos^2\theta)\cdot\sum_{n=1}^{\infty}\frac{(-k^3\cos^3\theta)^n}{n!}B_n(K,\chi),\end{aligned} \tag{16.14}$$

where

$$B_n(K,\chi) = \frac{-j}{2\pi}\int_0^{2\pi}\int_0^{\infty}S_{\mathrm{s}}(r,\phi)^n\,e^{-jKr\cos(\chi-\phi)}r\,dr\,d\phi \tag{16.15}$$

with $S_{\mathrm{s}}(r,\phi)$ as given by Eq. (16.11). The total σ^0 is the sum of Eqs. (16.12a) and (16.14):

$$\sigma_{pp}^0(\theta,\chi) = \sigma_{pp}^{(1)}(\theta,\chi) + \sigma_{pp}^{(2)}(\theta,\chi). \tag{16.16}$$

16-3.3 Comparison with Measurements

Unlike land surfaces, the sea surface varies continuously with time. For a short time over a small area, it is reasonable to assume that its roughness follows a stationary process, which enables us to use the models described previously. However, the assumption of stationarity is only an approximation because in reality the wind vector fluctuates spatially and may vary during measurements. Nevertheless, the simple scattering model still holds well.

Extensive measurements of ocean σ^0 have been collected by radars mounted on towers, aircraft, and satellites. Analysis of the data reveals consistent trends for the backscattering coefficient as a function of both the azimuth windspeed and the angle χ at a wide variety of frequencies and incidence angles.

To illustrate use of the IEM approach, consider the data sets collected by Schroeder et al. (1985) at Ku-band and Masuko et al. (1986) at X-band. In the following discussion, the surface scattering model given by Eq. (16.16) is first compared with the data of Masuko et al. (1986), which were collected at 10 GHz, 52° incidence angle, and hh polarization. The radar measurements were made at three windspeeds: 3.2, 9.3, and 14.5 m/s.

The dielectric constant of sea water used in the model is determined by frequency and temperature using the model of Chapter 4. Assuming a water temperature of 20° C, the dielectric constant of the sea surface at X-band is estimated to be $\varepsilon = 56.2 - j37.4$. The scales of roughness responsible for the scattering are determined by frequency, incidence angle, and windspeed. In the scattering model the signal level at upwind is controlled mainly by the correlation length L_u. Beyond $\theta = 30°$, an increase in L_u causes σ^0 to decrease. At crosswind, σ^0 is controlled by L_c in a similar way. The quantity that determines the upwind-downwind difference is s_0. A larger value causes a larger increase in upwind and a larger decrease in downwind, simultaneously. Usually, the value of s_0 is around 0.1 or smaller. Finally, the scattering level at all azimuth angles rises or falls with the rms height s.

A comparison between the model and the radar measurements is shown in Fig. 16-8. The IEM model parameters are estimated from the data for each condition (i.e., wind speed and direction, incidence and azimuth angles, and polarization configuration). For this case, the surface rms height s increases with windspeed from 0.13 cm at $U = 3.2$ m/s to 0.29 cm at $U = 9.3$ m/s and to 0.35 cm at $U = 14.5$ m/s. Simultaneously, L_u decreases from 30 cm through 13 cm to 9 cm; while L_c changes from 58 cm through 27 cm to 25 cm. All three parameters have a clear upward or downward trend with windspeed. The skewness parameter, however, does not show a clear trend. For both the high and the low windspeeds, $s_0 = 0.06$ cm provides the best fit while $s_0 = 0.08$ cm is best at the mid-windspeed. Note that a higher s_0 value at a higher windspeed does not necessarily mean larger upwind-downwind difference because the backscattering coefficient is larger at a higher wind speed. The data show an upwind-downwind difference of 3.5 dB at 3.2 m/s, 3 dB at 9.3 m/s and 0.5 dB at 14.5 m/s. The minimum radar return does not occur at crosswind locations; instead, it is shifted slightly towards the downwind location. This is most apparent at low-wind conditions. Overall, the IEM model is able to match the data behavior very well. Hence, Fung's function for the surface spectrum given by Eq. (16.13) can adequately represent the intended portion of the sea spectrum.

16-3.4 Empirical Fits for IEM Parameters

The discussion regarding Fig. 16-8 suggests that if a more complete data set, in terms of windspeed, is available at a given frequency and incidence angle, it is possible to establish empirical expressions for the parameters s, L_u, L_c, and s_0 as a function of windspeed. Schroeder et al. (1985) acquired data sets for many windspeeds at 13.9 GHz and 40° incidence angle, which enabled computation of error bars for each model fit. The measured data indicate that for a given condition, there is more than 1 dB of fluctuation in the observed σ^0, which is not unexpected for scattering from ocean surfaces.

Using data from Schroeder et al. for vv polarization, the IEM model is matched to each data set at windspeeds of 4.2, 5.5, 7.5, 12, 15, and 19.4 m/s to determine the values of the model parameters s, L_u, L_c, and s_0. At

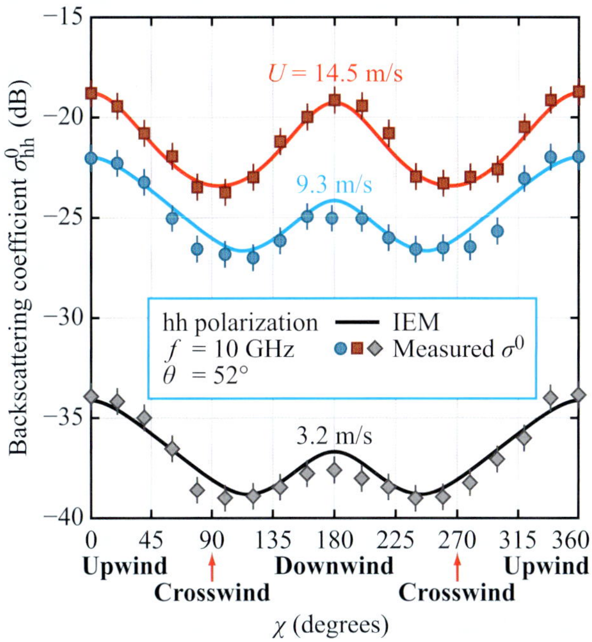

Figure 16-8: Comparison between the IEM surface scattering model for the sea surface and a data set collected by Masuko et al. (1986). The results indicate clear trends for σ^0, L_u, L_c and s with windspeed. Data show an upwind-downwind difference of 3.5 dB at 3.2 m/s, 3 dB at 9.3 m/s, and 0.5 dB at 14.5 m/s. Note that there is a clear shift of the minimum towards the downwind direction in the data especially at low-wind conditions.

13.9 GHz and 20° C, the sea water dielectric constant is $\varepsilon = 46.59 - j38.98$. The matching parameters found by fitting the model to the data using a curve fitting program are given in Table 16-1.

A curve is then fit to each model parameter versus wind speed to determine its wind-speed dependence. Through a conventional curve-fitting process, the following relations are obtained:

$$s = -0.00768 + 0.0237U - 0.000587U^2, \quad (16.17a)$$

$$L_u = 24.25 - 1.56U + 0.0302U^2, \quad (16.17b)$$

$$L_c = 33.89 - 0.225U + 0.0453U^2, \quad (16.17c)$$

$$s_0 = 0.0901 - 0.0061U + 0.000346U^2. \quad (16.17d)$$

Table 16-1: Scattering model coefficients for the IEM model at 13.9 GHz, $\theta = 40°$, and vv polarization.

Windspeed (m/s)	s (cm)	L_u (cm)	L_c (cm)	s_0 (cm)
4.2	0.09	19	32	0.085
5.5	0.098	17	31	0.075
7.5	0.165	14	30	0.055
12	0.19	10	26	0.07
15	0.215	9	20	0.08
19.4	0.23	6	13	0.1

All parameters are measured in cm and U is in m/s.

Applying these relations to the IEM model leads to the azimuthal pattern displayed in Fig. 16-9. Of the six windspeeds, five of them fit the IEM model quite well; at 7.5 m/s, the measured data is consistently higher than the IEM model. Upon detailed examination, it appears likely that the true wind speed is greater than 7.5 m/s, or else the radar measurements might have a calibration error.

Figure 16-10 provides plots of the backscattering coefficient versus windspeed at upwind, downwind and crosswind locations. The difference between upwind and crosswind σ^0 increases slightly with increasing wind speed, and so does the upwind-downwind difference.

16-3.5 The Wind Geophysical Model Function

As we have seen, the ocean surface σ^0 is a complicated function of the near-surface wind vector. The dominant scattering mechanism at low incidence angles is specular scattering while the dominant mechanism for microwave backscattering for moderate (20°–60°) incident angles is Bragg scattering, i.e., the return signal is proportional to the roughness of the ocean surface on the scale of the radar wavelength. In general, as wind speed increases, surface roughness increases and so does σ^0, although both may saturate under some wind-speed conditions and radar frequencies.

In general we can write the surface σ^0 as a function of the radar parameters, observational geometry, and

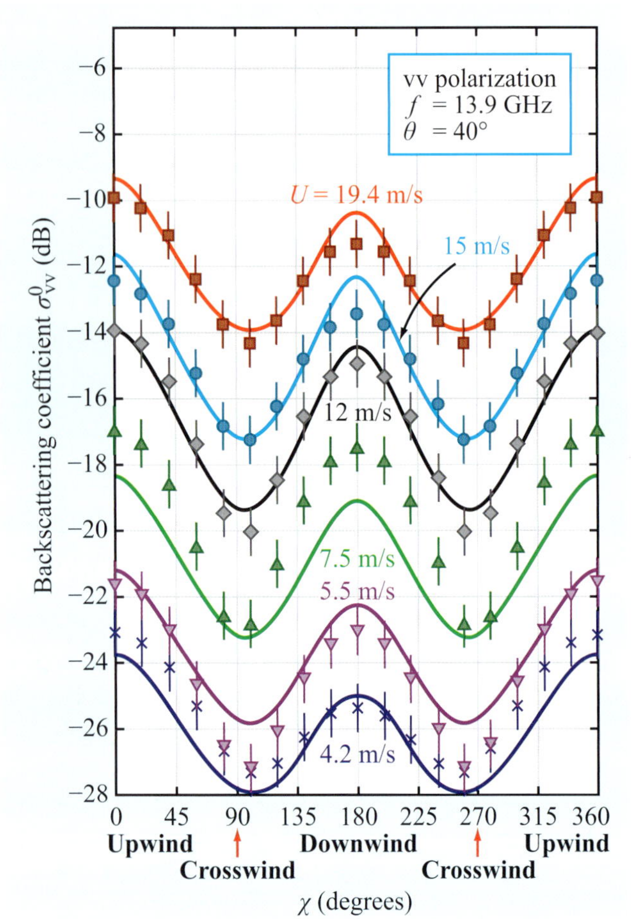

Figure 16-9: Backscattering coefficient versus windspeed and azimuth angle at 13.9 GHz and 40° incidence. Note that upwind backscatter is always larger than downwind and crosswind and that the backscattering coefficient always rises with windspeed. Measured data is from Schroeder et al. (1985).

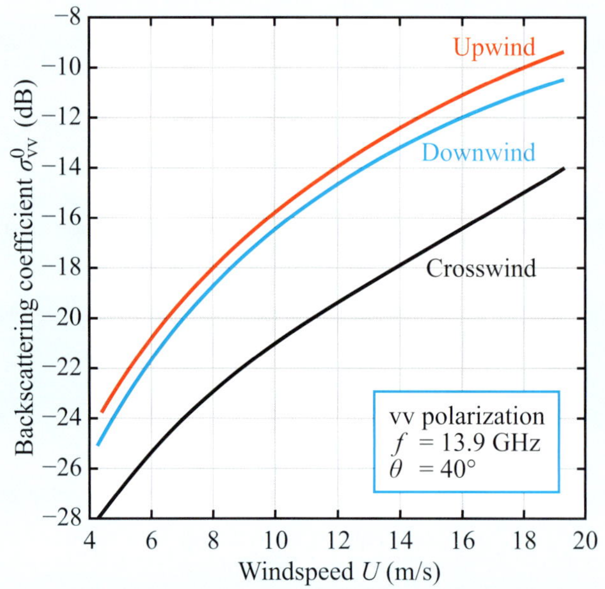

Figure 16-10: Backscattering versus windspeed at 13.9 GHz. Note that the largest increase with windspeed occurs along the upwind direction.

geophysical properties as

$$\sigma^0 = \mathcal{M}(U, \chi, pq, \theta, f, \ldots) \qquad (16.18)$$

where $\mathcal{M}$ is termed the forward ***geophysical model function*** (GMF), U is the wind speed, $\chi = \phi_{\text{wind}} - \phi_{\text{radar}}$ is the relative azimuth angle between the wind direction ϕ_{wind} and the incident radar signal look direction ϕ_{radar}, pq is the polarization, θ is the incidence angle, f is the radar frequency, and $\ldots$ denotes other geophysical parameters such as water temperature, air temperature, etc. In common practice, U is chosen to be the ***neutral-stability equivalent wind speed*** at either 19.5 m or 10 m, above the sea surface.

An example of some of the properties of the GMF at Ku-band is shown in Fig. 16-11, while an illustrative plot of the GMF at C-band at a 40° incidence angle is shown in Fig. 16-12. Even though the details of the GMF vary with radar frequency, its general properties are similar over most of the microwave band. We note that the σ^0 of the ocean's surface depends *primarily* on the observation azimuth angle, incidence angle, wind direction, and wind speed.

A key property of the GMF is that σ^0 is not a unique function of wind direction: as shown in Fig. 16-11(b), there is a $\cos 2\chi$ dependence of σ^0 on the wind direction χ such that the upwind σ^0 has a slightly higher value than the downwind σ^0. The variation in σ^0 as a function of χ can be expressed as

$$\sigma^0(\chi) = A + B\cos\chi + C\cos 2\chi \qquad (\text{m}^2/\text{m}^2). \quad (16.19)$$

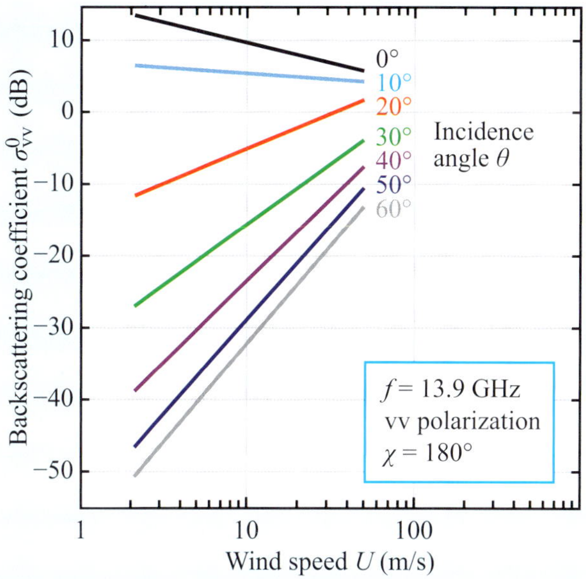

(a) σ^0 versus U at $\chi = 180°$

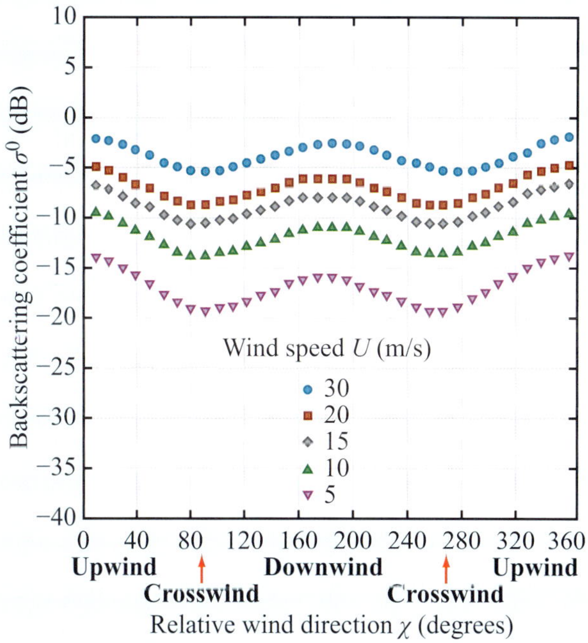

(b) σ^0 versus χ

Figure 16-11: The Ku-band SASS-1 geophysical model function relating ocean surface σ^0 to the near-surface wind speed: (a) σ^0 versus incidence angle for when the wind is blowing toward the radar (downwind). (b) σ^0 versus relative azimuth angle χ between the radar look direction and the wind direction at a fixed incidence angle.

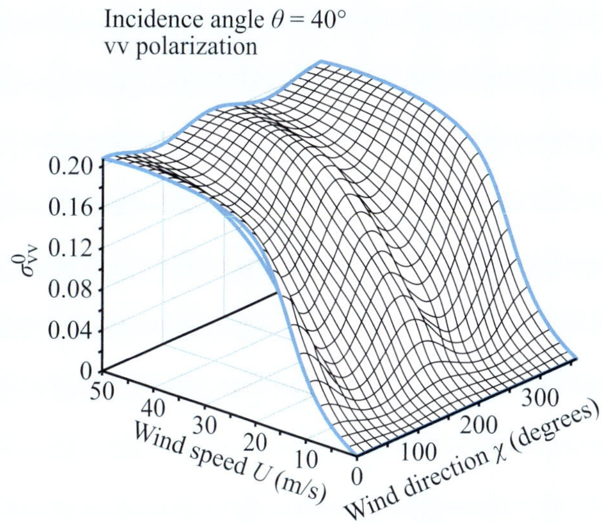

Figure 16-12: The C-band CMOD5 geophysical model function relating ocean surface σ^0 to the near-surface wind at an incidence angle of 40° and vv polarization.

The coefficients A, B, and C depend on the wind speed and the three radar wave parameters: θ, f, and polarization pq, and they exhibit a power-law dependence on wind speed. That is, at a particular incidence angle, radar frequency and polarization configuration, the three coefficients vary as

$$A \approx c_a U^{\gamma_a},$$
$$B \approx c_b U^{\gamma_b},$$
$$C \approx c_c U^{\gamma_c},$$

where γ_a, γ_b, γ_c vary from ~1 to ~2.4, depending on frequency, polarization, and incidence angle. Coefficients c_a, c_b, and c_c also depend on the radar parameters. The dependence of γ_a, γ_b, and γ_c on θ is responsible for the different slopes of the lines plotted in Fig. 16-11(a).

The scattering model introduced in the previous section describes the general behavior of the observed GMF, but a complete analytic formulation of the GMF remains elusive. However, empirical methods for deriving $\mathcal{M}$ have enjoyed significant success in practice. An approximate GMF can be derived from temporally and spatially collocated radar backscatter measurements and wind vector measurements recorded *in situ* by buoys, ships, and towers. By fitting the measured data to appropriate functional forms, empirical expressions are obtained that relate σ^0 to U and χ. The goodness of fit of the empirical approach depends on the availability of accurate *in-situ* wind-vector measurements that adequately span the desired range of wind conditions. Unfortunately, this is seldom the case because wind conditions cannot be controlled, and so individual experiments are limited to the range of wind conditions prevailing at the time. Nevertheless, a number of model functions have been developed based on such an empirical approach (Bracalente et al., 1980; Schroeder et al., 1982).

The first widely used Ku-band GMF, known as ***SASS-1***, was derived by fitting the two-scale model described above to aircraft scatterometer measurements. Unique values for the model parameters were found by fitting the model to the observations in a least-squares sense (Wentz, 1977; 1978). The derived parameters were found to have a small residual rms error of 0.75 dB between the model and aircraft backscatter observations. The final SASS-1 model (Schroeder et al., 1982) is a refinement of these results in which the tabular model-function coefficients were adjusted so as to produce smooth variations with incidence angle. For a given polarization, the SASS-1 model function expresses σ^0 (in dB) as a function of the wind speed U as follows:

$$\sigma^0\,(\text{dB}) = G(\theta,\chi) + H(\theta,\chi)\log_{10} U. \qquad (16.20)$$

The SASS-1 model coefficients $G(\theta,\chi)$ and $H(\theta,\chi)$ are tabulated every 2° in θ and 10° in χ, and exhibit bi-harmonic variation with χ as in Eq. (16.19).

Wentz et al. (1984; 1986) developed a Ku-band model function derived primarily from the statistics of the Seasat scatterometer (SASS) data, based on the assumption that over a sufficiently long time (3 months in the case of SASS) the wind vector exhibits a high degree of variability relative to the antenna look direction so that the probability distribution of two orthogonal components of the wind vector can be modeled by a bi-variate normal distribution. This yields a Rayleigh distribution for the wind speed. The single parameter of the Rayleigh distribution (the mean global wind speed) is obtained from a climatological atlas.

The SASS-2 GMF model (Wentz et al., 1984) uses the same functional form of Eq. (16.19), but the coefficients are modeled as

$$A = a_0 U \alpha_0,$$
$$B = A(a_1 + \alpha_1 \log_{10} U),$$
$$C = A(a_2 + \alpha_2 \log_{10} U),$$

where the coefficients a_0, a_1, a_2, α_0, α_1, and α_2 are tabulated functions of incidence angle. Coefficients with subscripts 0 and 2 were derived directly from the statistics of SASS measurements while coefficients with a subscript 1 were derived from aircraft scatterometer data (Wentz, 1991).

More recently, GMFs have been developed based on collocated radar backscatter measurements with winds predicted from ***numerical weather prediction*** (NWP) models (Freilich and Dunbar, 1993). GMFs constrained to particular forms, purely tabular forms, and mixed-form GMFs have been developed this way (Chen et al., 1992; Wentz et al., 1984; Wentz and Smith, 1999; Hersbach et al., 2007). This approach requires extremely large collocated data sets for model generation but can easily incorporate other geophysical parameters of interest. Since scatterometer-derived winds are often used as input to NWP models, the GMF tends to be consistent with the NWP model from which it is derived. The use of synoptic scale NWP wind-field analyses can produce a GMF without the need for an analytical form, although using a functional form can smooth the model function.

On the other hand, NWP models have their own errors and limitations, including limited spatial resolution, spatial biases, and mislocated extreme weather events. As the NWP surface-wind velocities are used to assign each backscatter measurement to a wind speed and azimuth angle bin, NWP errors that are a systematic function of wind speed result in misassignment of backscatter measurements, with corresponding errors in the calculation of bin-averaged values that form the initial model function estimate. Random NWP velocity errors result in misassignment of backscatter measurements; however, as the random errors are not systematic function of wind speed, to lowest order their effects are smaller. Random NWP errors, however,

Table 16-2: Coefficients of the C-MOD5 model function (Hersbach et al., 2007) for C-band (5.3 GHz) and vv polarization.

Coefficient	Value	Coefficient	Value
C_1	−0.688	C_{15}	0.007
C_2	−0.793	C_{16}	0.33
C_3	0.338	C_{17}	0.012
C_4	−0.173	C_{18}	22.0
C_5	0.0	C_{19}	1.95
C_6	0.004	C_{20}	3.0
C_7	0.111	C_{21}	8.39
C_8	0.0162	C_{22}	−3.44
C_9	6.34	C_{23}	1.36
C_{10}	2.57	C_{24}	5.35
C_{11}	−2.18	C_{25}	1.99
C_{12}	0.4	C_{26}	0.29
C_{13}	−0.6	C_{27}	3.80
C_{14}	0.045	C_{28}	1.53

can result in systematic under-predictions of the GMF directional dependence. Importantly, the accuracy of individual NWP analyses is of lesser importance than the overall systematic accuracy of the analyses; in the absence of systematic wind-speed-dependent or geography-dependent errors, analysis errors (such as missing fronts and misplacing the centers of lows, etc.) add only random errors. An NWP-based GMF was used in developing a number of tabular Ku-band GMFs (Wentz and Smith, 1999; Lungu, 2006) and for refining C-band GMFs based on a functional form (Hersbach et al., 2007). The NWP-GMF methodology is not limited to just wind vectors, but also can be applied to model functions for synoptic pressure as well (Patoux et al., 2003).

Like Ku-band, C-band model functions have an extensive history. An example of a widely-used mixed-form GMF is the C-band C-MOD5 model function (Hersbach et al., 2007). With the numerical coefficients defined in Table 16-2, the functional form for C-MOD5 is given by

$$\mathcal{M}(U,\chi,\theta) = B_0(1+B_1\cos(\chi)+B_2(2\cos^2\chi-1))^{1.6}, \tag{16.21}$$

where the B_i coefficients are empirically determined using the following complicated sequence of equations:

$$X = \frac{\theta-40}{25},$$

$$A_0 = C_1 + C_2X + C_3X^2 + C_4X^3,$$

$$A_1 = C_5 + C_6X,$$

$$A_2 = C_7 + C_8X,$$

$$\gamma = C_9 + C_{10}X + C_{11}X^2,$$

$$S_0 = C_{12} + C_{13}X,$$

$$S = A_2U,$$

$$S_2 = \begin{cases} S_0 & S_0 > S, \\ S & S_0 \le S, \end{cases}$$

$$A_4 = \frac{1}{1+e^{-S_2}},$$

$$A_3 = A_4\begin{cases} 1 & S > S_0, \\ \left(\dfrac{S}{S_0}\right)^{S_0(1-A_4)} & S \le S_0, \end{cases}$$

$$U_0 = C_{21} + C_{22}X + C_{23}X^2,$$

$$D_1 = C_{24} + C_{25}X + C_{26}X^2,$$

$$D_2 = C_{27} + C_{28}X,$$

$$A = C_{19} - \frac{C_{19}-1}{C_{20}},$$

$$B = \frac{1}{C_{20}(C_{19}-1)^{C_{20}-1}},$$

$$B_0 = A_3^{\gamma}10^{A_0+A_1U},$$

$$B_3 = C_{15}U\left(\frac{1}{2}+X-\tanh 4\left[X+C_{16}+C_{17}U\right]\right),$$

$$B_1 = \frac{C_{14}(1+X)-B_3}{e^{0.34(U-C_{18})}+1},$$

$$U_2 = \begin{cases} \dfrac{U}{U_0}+1 & \dfrac{U}{U_0}+1 \ge C_{19}, \\ A+B\left(\dfrac{U}{U_0}\right)^{C_{20}} & \dfrac{U}{U_0}+1 < C_{19}, \end{cases}$$

$$B_2 = (D_2U_2 - D_1)e^{-U_2}.$$

As before, U is the wind speed in m/s, $\chi = \phi_{\text{wind}} - \phi_{\text{radar}}$

is the relative azimuth angle between the wind direction ϕ_{wind} and radar look angle ϕ_{radar}, and all angles are in degrees. C-MOD5 computes vertically polarized σ^0 at a particular radar incidence angle and azimuth angle at C-band for a neutral-stability wind speed and direction at a height of 10 m above the sea surface. Note that the wind direction for C-MOD5 is defined in ***meteorological convention***[a] where 0° corresponds to a wind flowing from the north with positive angles CW seen from above. Most Ku-band GMFs have used oceanographic convention for wind directions. Associated with the GMF is a modeling uncertainty factor, typically expressed as a normalized standard deviation, K_{pm}, which may be a function of the true wind. K_{pm} describes the expected variability of σ^0 for a given set of wind and radar conditions and can be thought of as an error bar. Typical values of K_{pm} vary from 0.14 to 0.17 for C- and Ku-band, respectively. Wind retrieval precision for the ERS-1/2 and ASCAT scatterometers is about 1.8 m/s.

Figure 16-13 is an image of near-surface winds for Hurricane Katrina. The radar data were acquired by the SeaWinds scatterometer on board QuikSCAT on 28 August, 2005.

16-4 Rain

As previously noted, in a rain-free environment the scatterometer-observed σ^0 of the ocean's surface is due to scattering from wind-generated gravity-capillary waves. Rain affects the observed σ^0 in a number of ways. Figure 16-14 illustrates wind waves and rain effects over the ocean.

> ▶ First, falling rain attenuates the radar signal by scattering it in directions away from the radar. ◀

This two-way attenuation affects both the signal headed toward the surface as well as the signal scattered from the surface toward the radar.

> ▶ Second, some of the transmitted signal from the radar is scattered toward the radar by raindrops. ◀

As described below, backscatter due to rain can be confused with surface backscatter. Rain scattering and attenuation are treated in detail Chapter 8. Such effects are frequency dependent and are less important at lower microwave frequencies.

> ▶ Third, rain striking the surface alters the surface roughness. The rain impacts the surface by generating ripples and splash products such as stalks, crowns, and other roughening artifacts that tend to enhance the surface backscatter. ◀

The contribution of these splash products to the backscattering varies with radar wavelength, incidence angle and polarization (Bliven and Giovanangeli, 1993; Bliven et al., 1997). For C- and Ku-band at vv polarization, backscatter from rain-generated ring-waves is a dominant component of rain-induced radar backscattering at all incidence angles. At hh polarization, the radar backscatter from ring-waves decreases with increasing incidence angle while the radar backscatter from nonpropagating splash products increases (Bliven et al., 1989).

The impacting raindrops generate turbulence in the upper water layer that attenuates the gravity-wave spectrum. This modification depends on the wavelength of the water waves, the wind speed, and the rain rate (LeMehaute and Khangaonkar, 1990). At high rain rates, the net effect of the impinging raindrops on the sea surface is a decrease in the amplitudes of those water waves whose wavelengths are longer than a transition wavelength of about 10 cm, and an increase in the amplitudes of water waves whose wavelengths are shorter than 5 cm (Melsheimer et al., 2001).

The transition wavelength between the two spectral regions is not well defined and depends on rain rate, drop-size distribution, wind speed, and the temporal evolution of the rain event. In the transition wavelength region, raindrops impinging on the sea surface may increase or decrease the amplitude of the Bragg waves,

[a]In ***oceanographic convention*** 0° corresponds to a wind flowing toward the north. Both widely used, oceanographic and meteorological conventions differ by 180°.

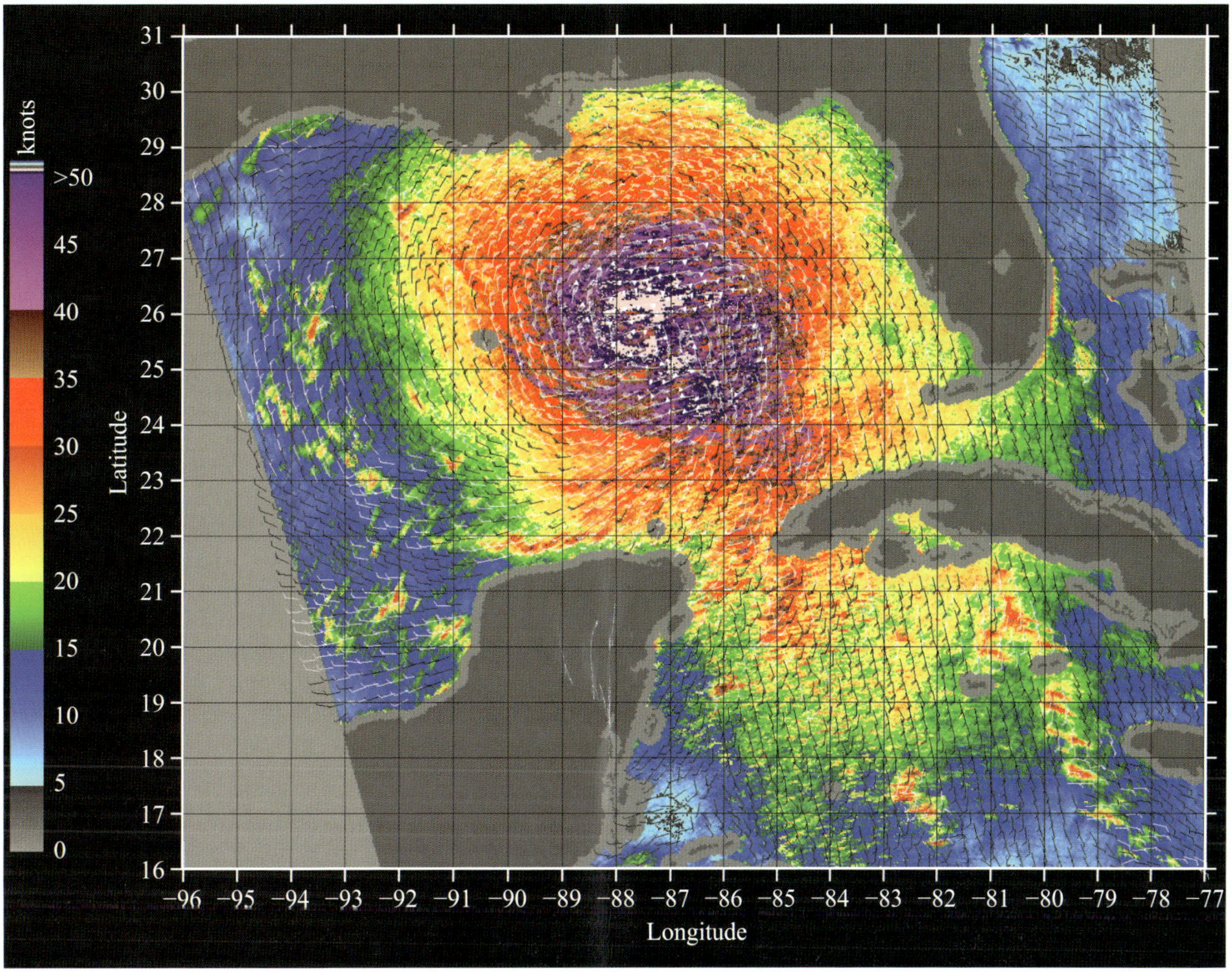

Figure 16-13: Near-surface winds of Hurricane Katrina, based on backscatter data recorded by the SeaWinds scatterometer on board QuikSCAT on 28 August 2005. Pixel size is 2.5 km×2.5 km.

with a corresponding increase or decrease in σ^0. Thus the presence of rain increases the uncertainty or variability in σ^0 of the ocean's surface.

Finally, as illustrated in Fig. 16-15, ***downdrafts*** associated with falling rain alter the local near-surface wind-vector field, which affects σ^0. This is particularly important for ***convective rain events*** because the rain is concentrated in relatively small (typically of the order of 1 km) rain cells associated with strong downdrafts and ***microbursts***. Such effects are highly dependent on local storm conditions. An example of a convective rain event observed with a C-band SAR is shown in Fig. 16-16. Notice the complex effects of rain exhibited in the image. Relative to the surrounding gray level due to the mean wind flow, the bright areas are due to enhanced backscatter due to scattering from the rain and the rain-roughened surface. Low backscatter is due

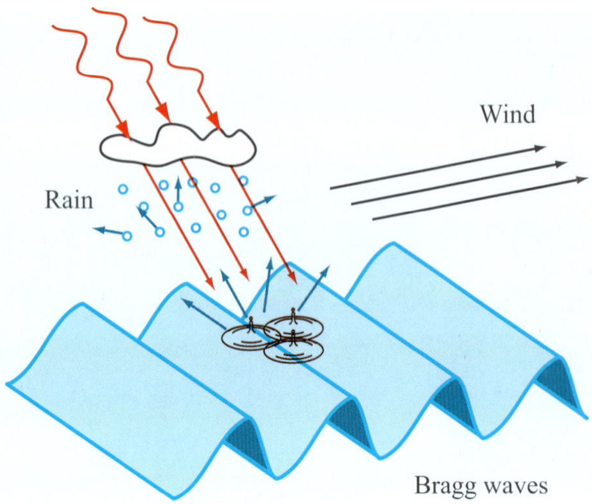

Figure 16-14: Diagram illustrating the radar observation of wind and rain over the ocean's surface. Rain droplets striking the surface produce ring waves, stalks, and splash products. These attenuate the wind-generated Bragg wave spectrum. Scattering from falling rain causes attenuation of the surface signal and adds a backscatter component of its own.

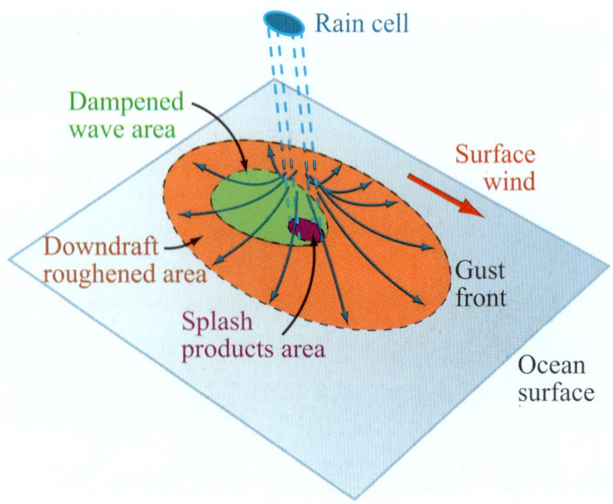

Figure 16-15: Schematic diagram of airflow associated with a rain cell. The red arrow indicates the mean wind direction outside of the downdraft-affected area. The blue arrows show the downdraft air flow. The green area indicates the rain-impacted area at the surface. The purple area within the green area shows the region whose surface wave spectrum is dampened by the effects of rain and the moving rain cell.

to dampening of the surface spectrum and from signal attenuation by the falling rain.

Interpretation of SAR backscatter images of rain is complicated by the vertical structure of the rain and the radar slant viewing geometry. Figure 16-17 schematically illustrates the vertical geometry of a convective rain event. Even in the tropics, precipitation is often first generated as snow at high altitudes. As it falls, it melts within a narrow altitude range. Because this melt phase has an enhanced radar backscatter compared with lower and higher altitudes, this part of the atmosphere is often known as the ***radar bright band***. From this layer downward, rain falls toward the surface. In Fig. 16-17, the radar observes the surface at an incidence angle $\theta = 45^\circ$. The oblique blue lines represent lines of constant range whose width is the radar's range resolution. The projection of this slant range resolution yields a ground resolution of x_0. All backscattering within a single range bin appears as if it comes from the surface. However, this includes not only the surface backscatter but also scattering from rain above the surface within the range bin. Rain scattering can increase the apparent backscatter in the corresponding SAR pixel, which can be confused with scattering from the surface. Because radar is a range-measuring instrument, scattering by rain layers above the sea surface gets displaced horizontally on the radar image. The total rain-induced scattering can be significant, particularly at high rain rates.

16-4.1 Modeling the Surface Effects of Rain

As the previous discussion suggests, rain can have a very complicated effect on the wind-induced ocean σ^0, either increasing or decreasing the σ^0 measured by the radar depending on the rain rate, local wind and wave conditions, and radar characteristics including wavelength, incidence angle, and azimuth angle relative to the wind direction. In wind scatterometry, the variability of rain effects degrades the accuracy of

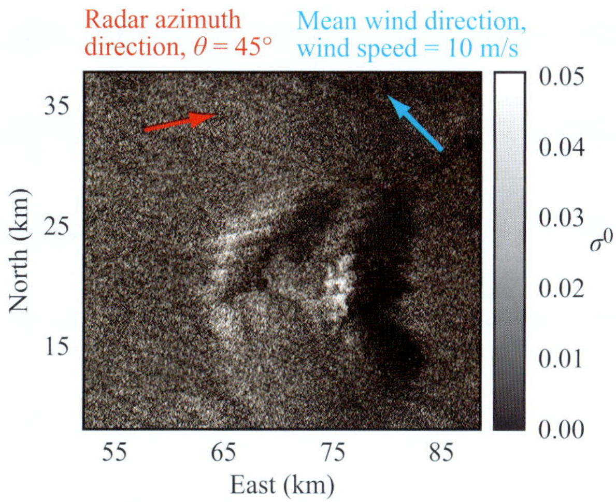

Figure 16-16: C-band SAR image of a rain cell. The red arrow shows the radar azimuth direction. The blue arrow shows the mean wind direction in the surrounding area. The corresponding wind speed is 10 m/s. The incidence angle of the radar signal at the surface is approximately 45°.

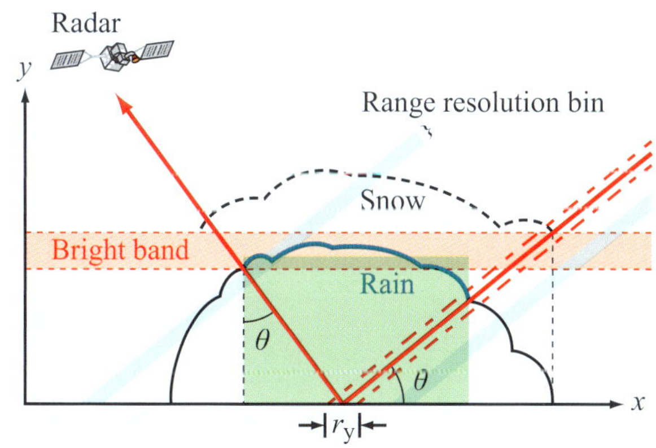

Figure 16-17: Schematic diagram of the radar scattering geometry for a rain cell. The blue bars approximate the effective length of the radar pulse assuming plane wave propagation.

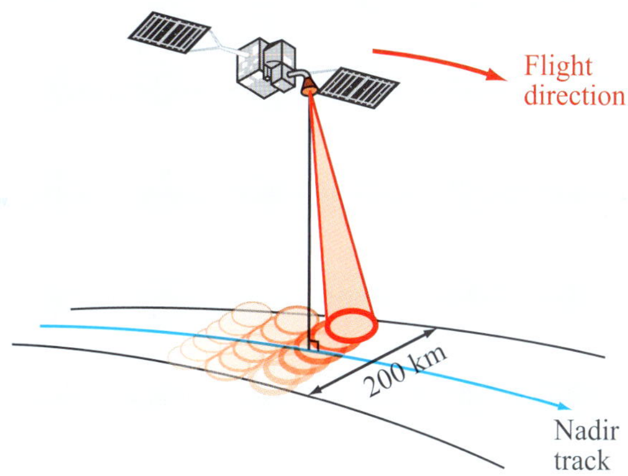

Figure 16-18: An illustration of the TRMM-PR measurement geometry. Switched spot beams span the swath at different incidence angles. These are sequentially scanned to image the swath as the spacecraft moves along-track.

wind estimates made from backscatter measurements and has typically led to discarding measurements that are excessively rain contaminated. This is often termed ***rain flagging***. Rain flagging may be accomplished using separate radiometer measurements, weather radar measurements, or just scatterometer backscatter measurements. Backscatter measurements that are rain flagged may be discarded, corrected to equivalent rain-free σ^0 values, or used to estimate the rain rate.

Spaceborne radar sensors designed for rain measurements such as the ***Tropical Rain Measuring Mission Precipitation Radar*** (TRMM-PR) operate at low incidence angles (less than 17°) to help minimize ambiguity between backscatter from rain above the surface and the surface itself (Meneghini and Kozu, 1990; Stephens and Kummerow, 2007). In particular, the Ku-band TRMM-PR uses electronic switching of 128 antenna elements in a 2 m × 2 m antenna array to generate 49 fixed ∼ 0.5° pencil beams at different incidence angles across a 215 km wide swath in a push-broom configuration, with a horizontal resolution of approximately 4 km (Fig. 16-18). The TRMM-PR employs pulse compression to provide 250 m vertical resolution measurements of the scattering volume viewed by each of its antenna beams. For range bins not containing the surface reflection, the vertical rain profile can be directly estimated from the range bin backscatter measurement (Iguchi et al., 2000). For range bins with surface reflection, the rain profile helps provide corrections for the atmospheric scattering and attenuation to improve the accuracy of the surface backscatter measurements (Meneghini and Kozu, 1990; Stephens and Kummerow, 2007).

In the following we restrict our attention to large-footprint (10 km or more) scatterometers. For footprints of that size, a simple but effective approach to model the effects of rain on σ^0 is to treat the rain as a perturbation of the backscatter value that would be obtained from the wind-generated wave field if the rain cell were not present. This approach is appropriate when wind scatterometer–derived wind estimates are to be incorporated into global ***numerical weather prediction*** (NWP) models. Such models often do not explicitly include rain and thus prefer nonraining wind estimates.

We consider a simple rain perturbation model that includes attenuation and backscatter from the falling rain and accounts for the rain effects on the surface. It does not explicitly include wave-spectrum dampening. A simple phenomenological model for rain effects is expressed as

$$\sigma^0_{\mathrm{m}} = (\sigma^0_{\mathrm{w}} + \sigma^0_{\mathrm{sr}})\Upsilon^2_{\mathrm{r}} + \sigma^0_{\mathrm{r}}, \tag{16.22}$$

where σ^0_{m} is the total backscatter, σ^0_{w} is the wind-only backscatter, σ^0_{sr} is the contribution to the backscatter due to rain impacting the surface, Υ^2_{r} is the two-way path-integrated transmission factor for rain, and σ^0_{r} is the effective backscatter augmentation due to falling rain. This model is often simplified to

$$\sigma^0_{\mathrm{m}} = \sigma^0_{\mathrm{w}}\Upsilon^2_{\mathrm{r}} + \sigma^0_{\mathrm{eff}} \tag{16.23}$$

with

$$\sigma^0_{\mathrm{eff}} = \Upsilon^2_{\mathrm{r}}\sigma^0_{\mathrm{sr}} + \sigma^0_{\mathrm{r}}, \tag{16.24}$$

and σ^0_{eff} is called the ***effective rain backscatter***. With this rain perturbation model, a general ***wind and rain GMF*** can be written as

$$\sigma^0_{\mathrm{m}} = \mathcal{M}(U,\chi,p,\theta,f)\,\Upsilon^2_{\mathrm{r}}(R'_{\mathrm{r}},p,\theta,f) + \sigma^0_{\mathrm{eff}}(R'_{\mathrm{r}},p,\theta,f), \tag{16.25}$$

where $\mathcal{M}$ is the wind-only GMF and R'_{r} is the ***integrated rainfall***. In scatterometer models, two rain-related quantities are used:

R_{r}: surface rainfall rate in mm/hr,
R'_{r}: ***integrated rainfall*** in km-mm/hr.

The latter is the rainfall rate integrated over the vertical extent of the rain layer. Plots of Υ^2_{r} and σ^0_{eff} for vertical polarization for this model are shown in Fig. 16-19 for Ku-band. At C-band Υ_{r} is ≈ 1. A plot of σ^0_{eff} as a function of incidence angle is shown in Fig. 16-20 for C-band.

The various parameters in Eqs. (16.22) and (16.23) can be determined empirically from temporally and spatially collocated measurements of the vertical rain profile, NWP wind, and σ^0. While *in-situ* (i.e., from buoys or ships) wind measurements may seem preferable, there are several issues with them. First, such measurements are very rare when coupled with the requirement of collocated vertical rain profile and backscatter measurements. Second, such measurements are affected by downdrafts and thus violate the underlying assumptions behind the rain perturbation model.

Alternatively, NWP winds can be generated that predict local wind conditions without the rain, and thus describe the expected wind for the perturbation model. A complicating factor for this modeling approach, however, is commonly known as irregular ***beam filling***. Figure 16-21 illustrates the beam filling problem. Rain, particularly in convective conditions such as thunderstorms, occurs in localized “cells” a few kilometers in extent. Such rain cells may be much smaller than the radar footprint and thus do not completely cover or “fill” the scatterometer footprint. In such cases, the rain effects vary over the radar footprint. This adds to the variability and uncertainty in the rain model. Area and gain weighting of the observed rain profile, along with accounting for the range bin effect previously considered, help mitigate beam filling effects (Owen and Long, 2011).

Since rain effects depend on frequency, it is instructive to contrast the characteristics of the rain perturbation model at C-band and Ku-band. According to the rain perturbation model, the total backscatter is the sum of

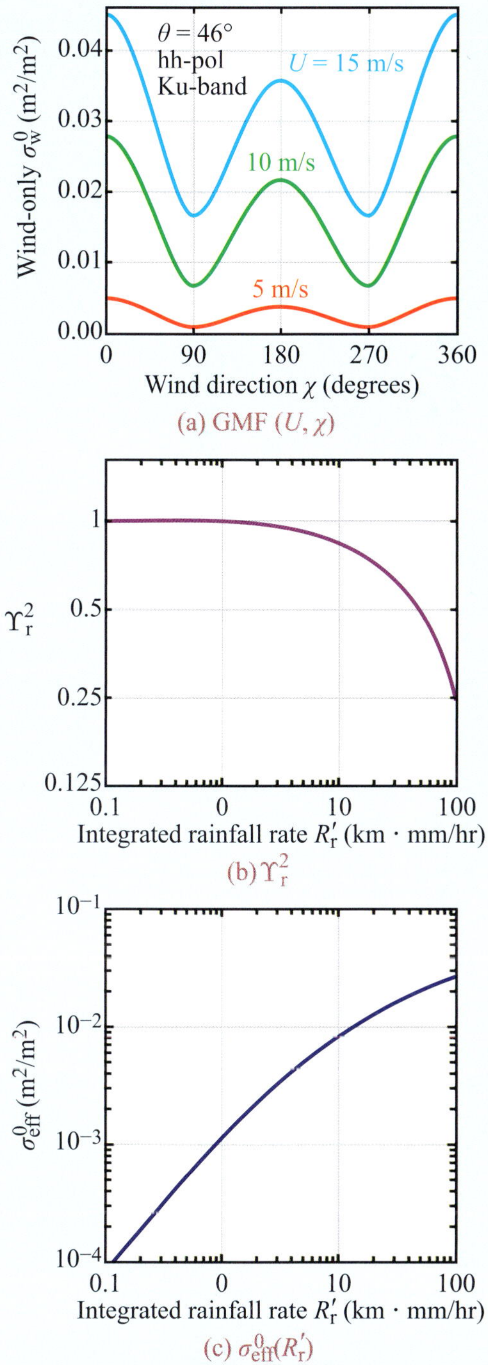

Figure 16-19: Illustration of a Ku-band wind/rain GMF (a) wind (conventional wind-only GMF), (b) rain attenuation and (c) rain effective backscatter model function [Draper and Long, 2004a, b].

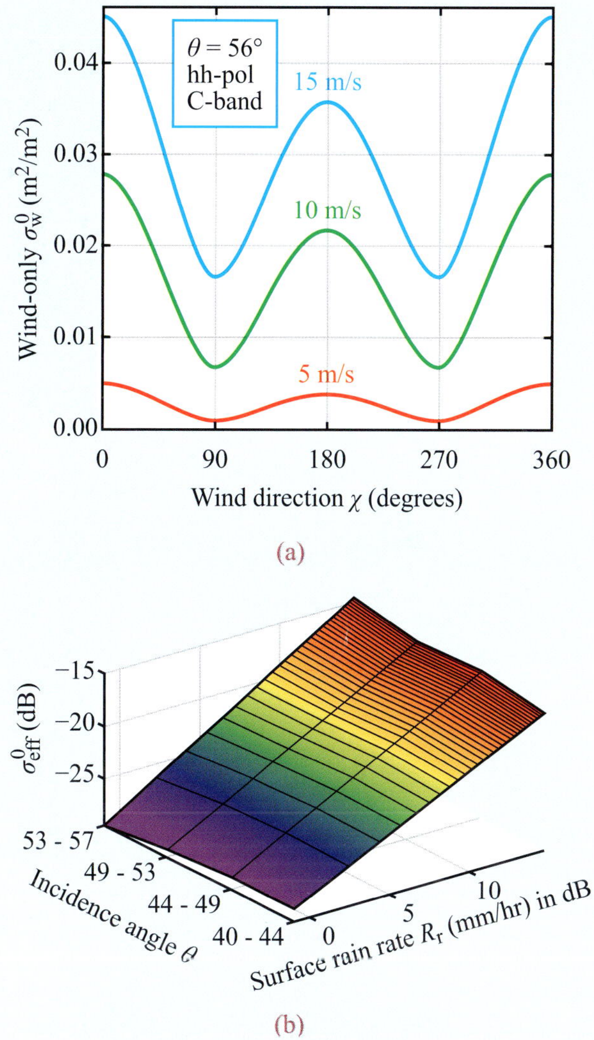

Figure 16-20: Illustration of a C-band wind/rain GMF (a) wind (conventional wind-only GMF) and (b) rain effective backscatter model function [Nie and Long, 2007]. R_r (dB) $= 10\log R_r$.

the surface backscatter due to the wind, attenuated by the rain in the atmosphere, plus the rain perturbation effects. As suggested by Fig. 16-19, at Ku-band, the signal attenuation can be significant even at lower rain rates. However, at C-band, signal attenuation is very small, even at high rain rates. At C-band, only the surface effects due to rain are important, while at Ku-band rain

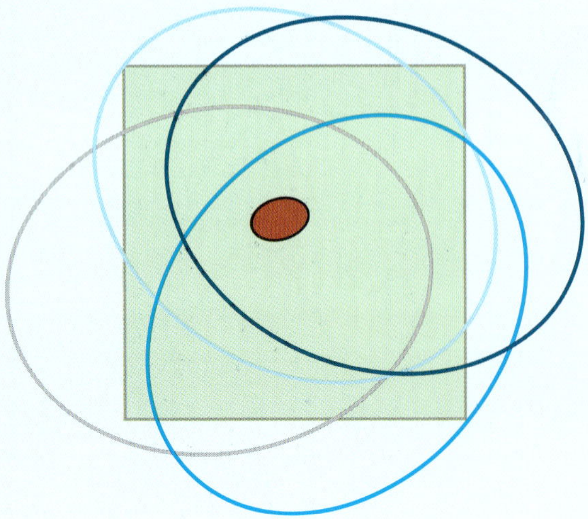

Figure 16-21: Illustration of the beam filling problem. The red area represents a rain cell. The ellipses represent the illuminated area for different antenna beams looking at the same square area representing a wind scatterometer wind vector cell.

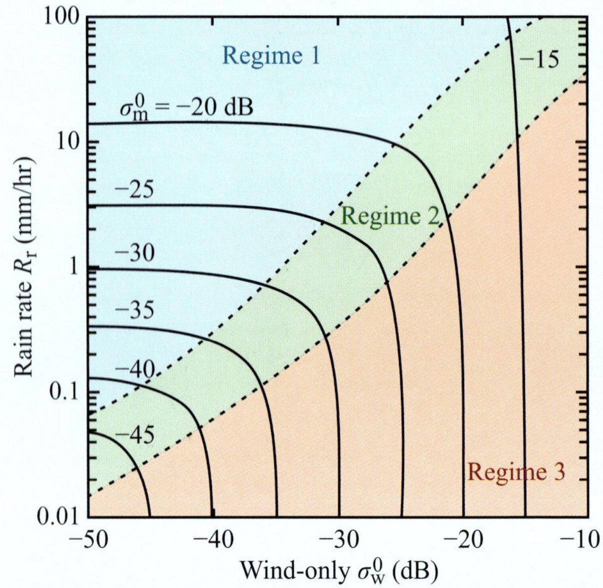

Figure 16-22: Contour plots of vv-pol Ku-band rain perturbation model displaying total σ^0_m with respect to wind-induced backscatter σ^0_w and rain rate. Three backscatter regimes are identified as colored areas. Incidence angle is 54°. Model derived from SeaWinds and TRMM-PR data [Draper and Long, 2004b]. Regime 1: $\sigma^0_{eff}/\sigma^0_m > 0.75$; regime 2: $0.25 \leq \sigma^0_{eff}/\sigma^0_m < 0.75$; regime 3: $\sigma^0_{eff}/\sigma^0_m < 0.25$.

is significant and factors such as the height of the bright band, which dictates the thickness of the rain layer, also are important (Nielsen and Long, 2009; Stephens and Kummerow, 2007).

16-4.2 Perturbation Model Regimes

For further comparison between C- and Ku-band, we examine the relative contribution of rain scattering to the wind-induced surface scattering. Figure 16-22 illustrates a contour plot of σ^0_m, the total backscattering coefficient, at Ku-band and $\theta = 54°$ as a function of the wind-only contribution and the rain rate R_r. As illustrated in the plot, there are three distinct backscatter regimes. These regimes can be identified by thresholding the ratio $\sigma^0_{eff}/\sigma^0_m$. For regime 1: $\sigma^0_{eff}/\sigma^0_m > 0.75$; for regime 2: $0.25 \leq \sigma^0_{eff}/\sigma^0_m < 0.75$; while for regime 3: $\sigma^0_{eff}/\sigma^0_m < 0.25$. In regime 1, rain-induced backscatter dominates the total backscatter while in regime 3 rain effects are negligible. In regime 2, the rain-induced and wind-induced backscatter have comparable magnitudes. For regime 3, rain may be ignored when estimating wind, but since rain dominates the backscattering in regime 1, it it not possible to estimate the wind from the backscatter measurements. In regime 2 it is possible to simultaneously estimate wind and rain from the backscatter measurements (Draper and Long, 2004b). According to the study by Draper and Long (2004a), at Ku-band and for rain rates greater than 0.1 km-mm/hr, approximately 14% of SeaWinds backscatter measurements fell in regime 1, 41% in regime 2, and 44% in regime 3, of the total, less than 8% had rain rates greater than 0.1 km-mm/hr.

Similar regimes can be defined at C-band, but due to the longer radar wavelength, rain has a smaller effect on the total backscatter and is important only at high rain rates. Figure 16-23 plots vertically polarized C-band σ^0_m as a function of σ^0_w and rain rate for various narrow

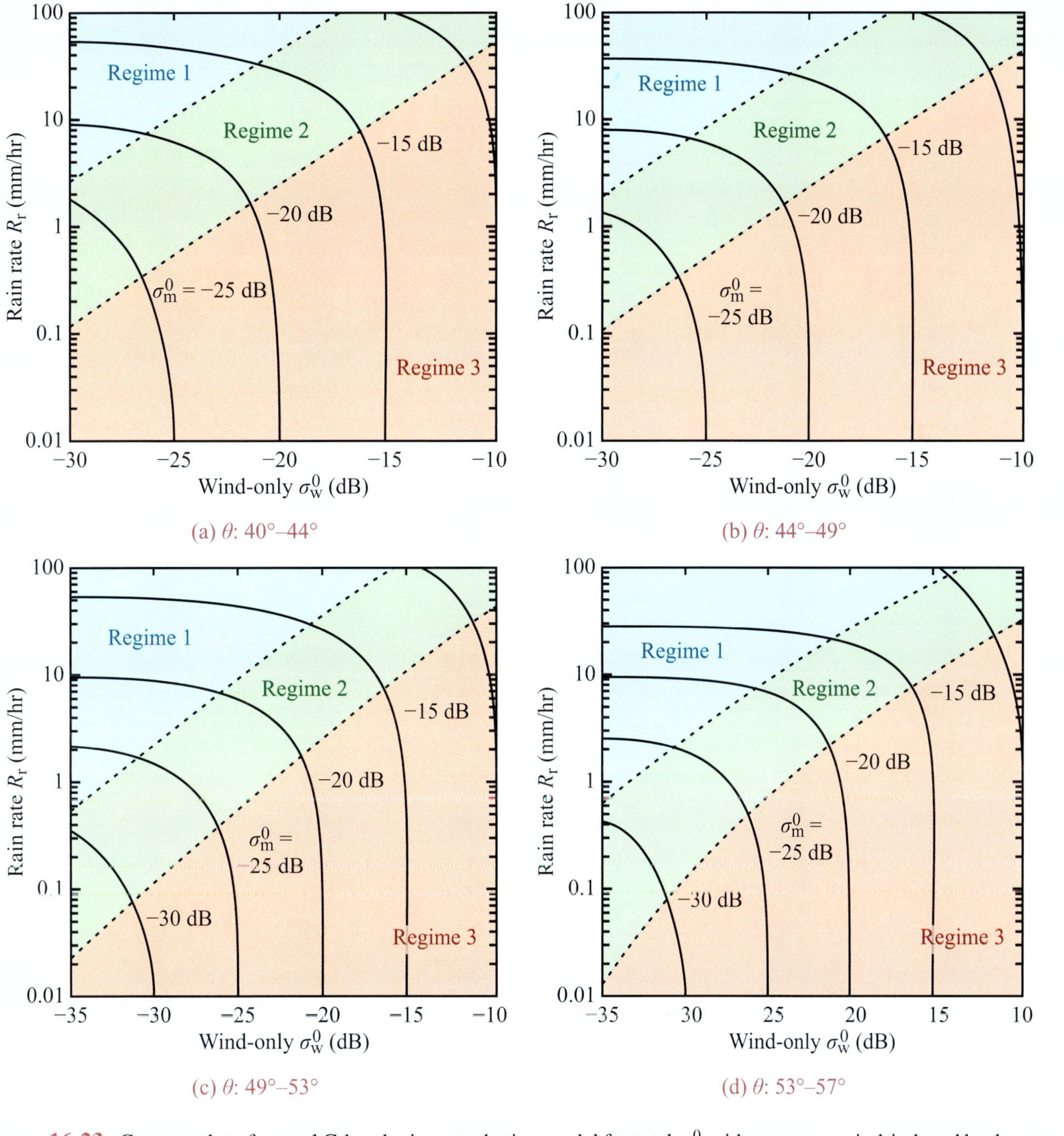

Figure 16-23: Contour plot of vv-pol C-band rain perturbation model for total σ^0_m with respect to wind-induced backscatter σ^0_w and rain rate R_r for several incidence angle ranges. Three backscatter regimes are identified as colored areas. Model derived from ERS-1/2 and TRMM-PR data [Nie and Long, 2006]. Regime 1: $\sigma^0_\mathrm{eff}/\sigma^0_\mathrm{m} > 0.75$; regime 2: $0.25 < \sigma^0_\mathrm{eff}/\sigma^0_\mathrm{m} < 0.75$; regime 3: $\sigma^0_\mathrm{eff}/\sigma^0_\mathrm{m} < 0.25$.

ranges of incidence angles. As the incidence angle is increased, the area of regime 1 reduces while the area of regime 3 increases, suggesting that rain has more significant impact at higher incidence angles (Nie and Long, 2006). For significant rain at C-band ($>$ 0.8 km-mm/hr), which occurs less than 3% of the time, Nie and Long (2006) found that approximately 1.67% of ERS scatterometer backscatter measurements that were within the angular range 53° to 57° fall within regimes 1 or 2.

16-5 Wind Scatterometers

A satellite radar scatterometer is designed to accurately measure the normalized radar cross section σ^0 of the Earth's surface. A ***wind scatterometer*** is a specialized scatterometer *system* designed for measuring near-surface ocean wind speed and direction. Its design and observation geometry is driven by the general behavior of the GMF relating ocean backscattering to wind speed and direction, which dictates the need to obtain multiple measurements of σ^0 at each point of the swath. A wind scatterometer system includes the processing of the raw measurements into calibrated σ^0 measurements and ultimately to wind vector.

Estimation of the wind velocity (speed and direction) from scatterometer backscatter measurements involves several steps:

1. The development of an accurate σ^0 model (GMF) over the range of viewing geometries and expected wind conditions. This was considered in previous sections.

2. The acquisition of accurate measurements of σ^0 for each desired sample point of the swath from several different viewing directions.

3. The implementation of a method to identify and/or correct rain-contaminated σ^0 measurements.

4. The application of an algorithm for estimating ("retrieving") wind velocity for each sample point using the GMF.

5. The implementation of a method to resolve potential directional ambiguities in the estimated winds. Typically, this requires selecting which of several directional ambiguities are closest to the true wind.

Errors in the retrieved wind velocities are dependent not only on the model function, but also on errors and noise in the backscatter measurements and on details of the wind-retrieval and ambiguity-selection algorithms. While designed primarily for estimating ocean winds, wind scatterometers have utility in land and ice applications, including severe weather monitoring, vegetation studies, freeze/thaw mapping, sea-ice monitoring, iceberg tracking, and long-term climate studies in the polar regions. In this section we focus primarily on ocean wind retrieval.

16-5.1 Scatterometer Viewing Geometry

As noted earlier, σ^0 exhibits a $\cos 2\chi$ dependence on the wind direction χ such that the upwind σ^0 is slightly higher than the downwind σ^0, and the minimum σ^0 occurs at (approximately) crosswind. As a result of this behavior, a single measurement of σ^0 is not sufficient to estimate both wind speed and wind direction. In order to estimate the wind vector, σ^0 must be measured from several azimuth and/or incidence angles. As discussed later in Section 16-8, the $\cos 2\chi$ dependence on direction also leads to multiple possible directions for the wind. These possible directional "ambiguities" (also sometimes known as "aliases") must be resolved with the aid of an ***ambiguity selection*** step to produce a unique estimate of the wind vector.

Two different approaches have been used in satellite wind scatterometers to measure σ^0 at multiple azimuth and/or incidence angles. The difference is related to the antenna type: fan-beam versus pencil-beam. In addition, different signal modulation schemes can be used to resolve the antenna footprint into finer resolution elements. As might be expected, there are advantages and disadvantages to each approach, requiring trade-offs in the design process.

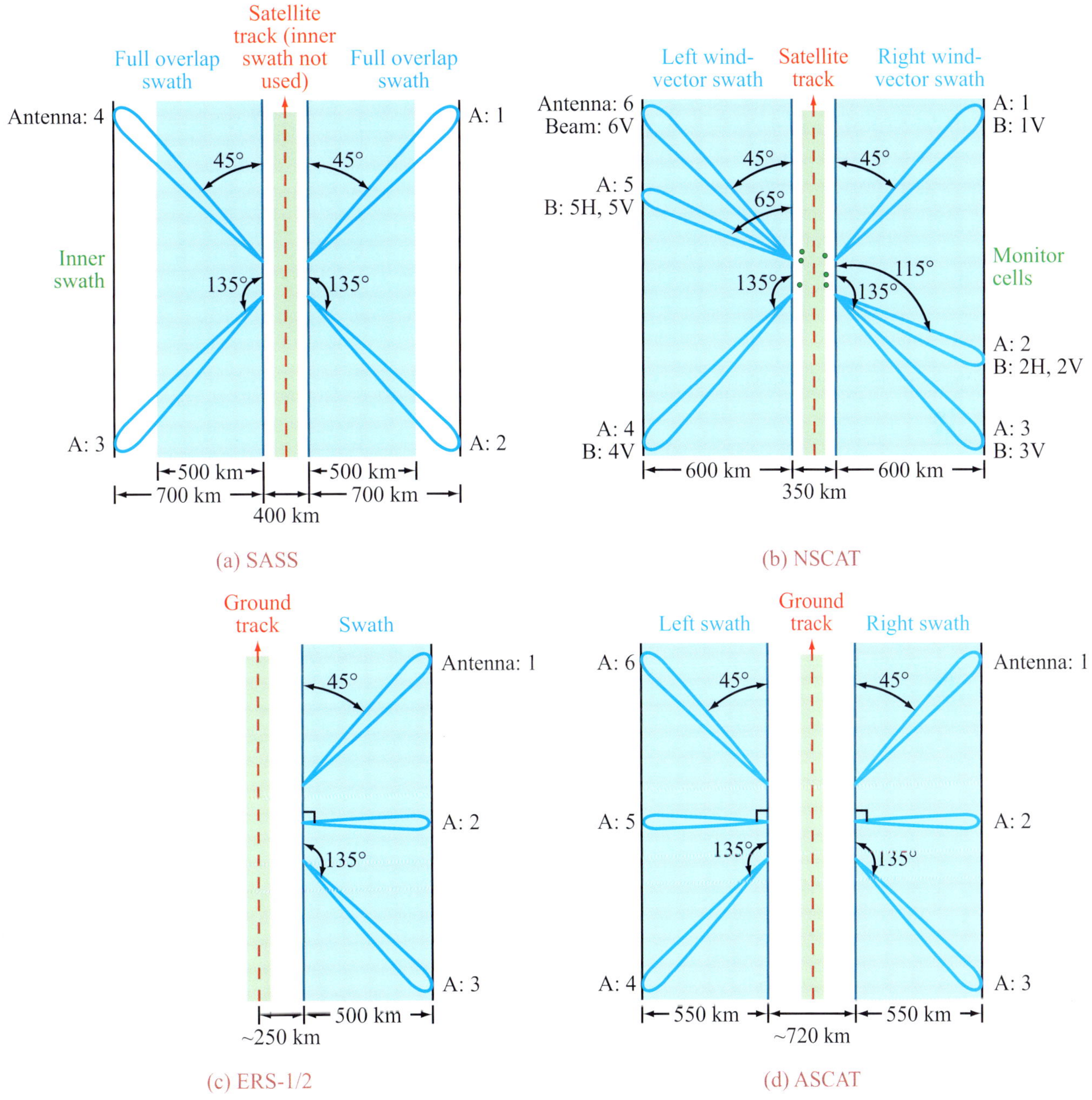

Figure 16-24: Configurations for the antenna footprints for fan-beam wind scatterometers flown to date. (a) Seasat Scatterometer (SASS), 1978. (b) NASA Scatterometer (NSCAT), 1996–1997. (c) ERS-1/2 AMI, 1992–2001. (d) Advanced Scatterometer (ASCAT), 2007–.

16-5.2 Fan-Beam Wind Scatterometers

In a ***fan-beam scatterometer***, the surface is illuminated by one or more fan-beam antennas producing long, thin footprints that are resolved into smaller areas or "resolution cells" using range compression or Doppler filtering. While rotating fan-beam concepts have been proposed (Dong et al., 2010; Lin et al., 2012), most fan-beam wind scatterometers flown to date have used fixed antennas providing illumination patterns of various configurations as shown in Fig. 16-24. The history of satellite wind scatterometers is summarized in Table 1-2.

> ▶ Fan-beam scatterometer systems illuminate the surface across a wide swath (hundreds of kilometers), usually on both sides of the subsatellite track. ◀

As described later, there is a coverage gap near nadir.

The narrow antenna beam, coupled with pulse timing, yields along-track resolution. To obtain cross-track resolution, the antenna footprint is divided into along-beam segments as illustrated in Fig. 16-25. Each of the two major approaches for obtaining along-beam resolution are indicated, namely (a) Doppler filtering and (b) range resolution.

Figure 16-25(a) superimposes an antenna pattern for a forward-looking antenna over the iso-Doppler contours resulting from the motion of the scatterometer over the surface. As discussed in Chapter 13, Doppler filtering of the return echo enables separating the signal from different parts of the antenna beam. By using multiple Doppler filters, multiple cross-track resolution elements or cells can be obtained simultaneously. Adjusting the ***antenna azimuth angle*** (ϕ_a) relative to the cross-track direction affects the Doppler bandwidth and swath width. Due to the configuration of the iso-Doppler lines, the Doppler filtering approach is effective only for azimuth angles in the range $15^\circ \leq |\phi_a| \leq 75^\circ$ on either side of the sub-satellite track.

Figure 16-25(b) superimposes the antenna pattern over isorange contours. As discussed in Chapter 13, range gating or range-compression of the return echo can be used to resolve the antenna illumination pattern into sub-elements or cells in the along-beam direction.

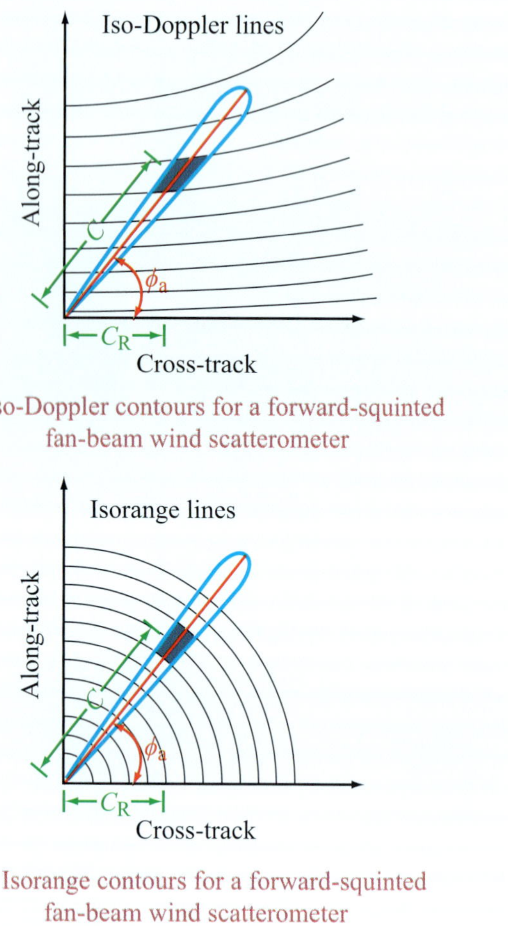

(a) Iso-Doppler contours for a forward-squinted fan-beam wind scatterometer

(b) Isorange contours for a forward-squinted fan-beam wind scatterometer

Figure 16-25: (a) Iso-Doppler contours for a forward-squinted fan-beam wind scatterometer. (b) Isorange contours for a forward-squinted fan-beam wind scatterometer.

Unlike Doppler filtering, range processing can be used for any antenna azimuth angle. As discussed later, the antenna azimuth angle affects the wind estimation accuracy and couples with the illumination pattern length in setting the useable swath width.

A minimum of two azimuth angles are required for wind estimation, but as discussed later the addition of more azimuth angles greatly reduces directional ambiguity (Schroeder et al., 1985; Shaffer et al., 1991). Referring to Fig. 16-24, a given point within the

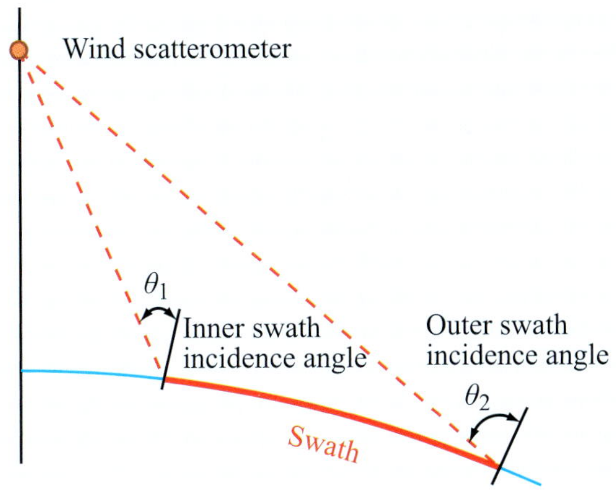

Figure 16-26: Incidence angle variation over the swath of a wind scatterometer.

observation swath is first observed by a forward-looking antenna. As the spacecraft moves along-track, this same point is then observed by the central antenna (if present), and finally by the aft-looking antenna. This provides the capability to measure σ^0 for each point within the swath from 2 or 3 different azimuth angles. While the first Doppler scatterometer (SASS) used only two antenna azimuth angles[b] on each swath side, later scatterometers employed three beams, which improve estimation of the ocean wind direction, as discussed later. Wind estimation performance is optimized when this third azimuth angle is nearly at broadside (Lin et al., 2012), which is not possible with a Doppler system because the Doppler shift is zero at broadside. The attraction to using a Doppler system over a ranging system is that the Doppler systems usually requires less transmit power for the same backscatter measurement performance level.

We observe from the sketch in Fig. 16-26 that the measurement incidence angle varies over the swath width for a fan-beam system and that for wide swath systems the curvature of the Earth must be considered. Since the GMF used to estimate the wind vector is a function of the incidence angle θ, it is necessary to adjust the estimation algorithm across the swath. Also, the sensitivity of σ^0 to wind direction is zero at nadir, and increases monotonically with θ. Hence, in order to estimate the wind direction, the σ^0 measurements must be acquired at incidence angles sufficiently greater than zero.

The minimum useful incidence angle for measurement of the wind vector varies with frequency, but is approximately 15°. In a typical fan-beam configuration (Fig. 16-24), this results in a gap in the satellite track over which winds can not be estimated. However, early Ku-band Doppler scatterometers included observations at lower incidence angles as calibration aids (Jones et al., 1982; Naderi et al., 1991).

► On the high end of the useable range of θ, the maximum incidence angle is limited by SNR considerations due to the long slant range and fall-off of σ^0 at high incidence angles. ◄

Typically a ranging scatterometer has a fixed *slant range* resolution. Due to the illumination geometry (Fig. 16-27), this translates into a variable resolution at the surface. Assuming a locally spherical Earth with radius $R_{\rm E}$ and a scatterometer operating at an altitude $H_{\rm s}$, the along-beam distance C between the subsatellite nadir point and the point of the swath at a slant range R can be computed using the ***law of cosines***. The distance C is related to the angle α by

$$C = \alpha R_{\rm E}, \tag{16.26}$$

and α is the solution to the quadratic equation governing triangle SEB in Fig. 16-27:

$$R^2 = (R_{\rm E} + H_{\rm s})^2 + R_{\rm E}^2 - 2(R_{\rm E} + H_{\rm s})R_{\rm E}\cos\alpha. \tag{16.27}$$

Solution of Eq. (16.27) for α gives

$$\alpha = \cos^{-1}\left[\frac{(R_{\rm E} + H_{\rm s})^2 + R_{\rm E}^2 - R^2}{2(R_{\rm E} + H_{\rm s})R_{\rm E}}\right]. \tag{16.28}$$

Similarly, angle ψ in Fig. 16-27 is governed by the quadratic equation

$$(R_{\rm E} + H_{\rm s})^2 = R_{\rm E}^2 + R^2 - 2RR_{\rm E}\cos\psi, \tag{16.29a}$$

[b]Owing to the fixed Doppler filters used on SASS, some parts of its swath were covered by only a single antenna.

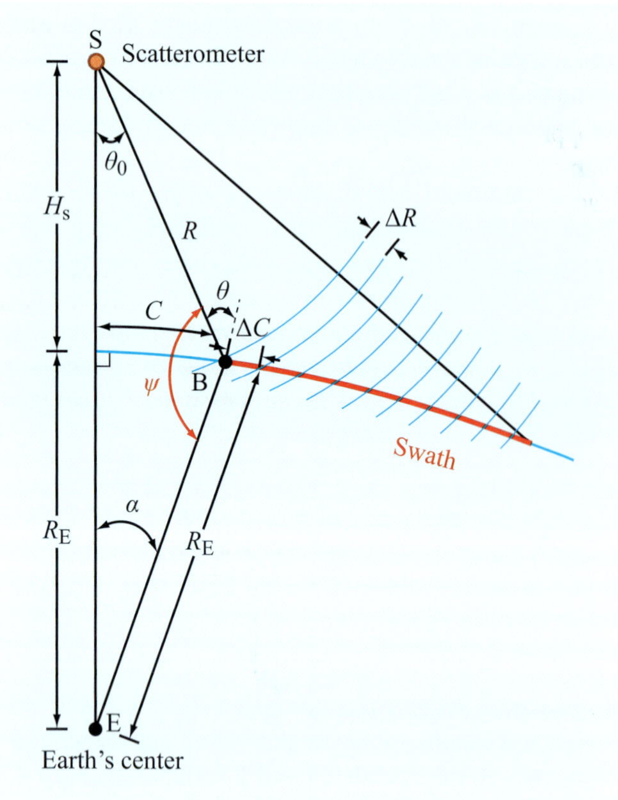

Figure 16-27: Slant-range geometry relative to ground-range geometry for a spherical Earth.

and its solution is

$$\psi = \cos^{-1}\left[\frac{R_E^2 + R^2 - (R_E + H_s)^2}{2RR_E}\right]. \quad (16.29b)$$

The local incidence angle θ is then given by

$$\theta = \pi - \psi. \quad (16.30)$$

Note that slant ranges with $R > \sqrt{(R_E + H_s)^2 + R_E^2}$ extend beyond the observable surface.

From Fig. 16-25 the cross-track distance C_R is computed from the along-beam distance C and the antenna azimuth angle ϕ_a by

$$C_R = C\cos\phi_a. \quad (16.31)$$

It follows that for a given C_R, the slant range R is

$$R = \sqrt{(R_E + H_s)^2 + R_E^2 - 2(R_E + H_s)R_E\cos\left(\frac{C_R}{R_E\cos\phi_a}\right)}. \quad (16.32)$$

To support wind estimation on a uniform grid, it is desirable to have measurements of σ^0 at a set of fixed cross-track distances. Using Eq. (16.32), we can compute the corresponding slant ranges at which the measurements should be made to satisfy the uniform-grid criterion. A more precise formulation of this result factors in the variation of the Earth's oblateness along the antenna beam, which causes R_E to vary with latitude, i.e., the local Earth radius. The more precise value of R_E is

$$R_E = R_{EQ}(1 - f\sin^2 l), \quad (16.33)$$

where l is the latitude angle, f is the ***flatness*** of the Earth, and R_{EQ} is the equatorial radius of the Earth. For the WGS84 geoid model $f = 1/298.257$ and $R_{EQ} = 6378.1778$ km. Note that R_E varies by up to 10 km from equator to pole.

The along-beam resolution ΔC depicted in Fig. 16-27 is related to the slant range resolution ΔR by

$$\Delta C = \Delta R\frac{dC}{dR}. \quad (16.34)$$

Computing the derivative,

$$\frac{dC}{dR} = \frac{R_E R}{(R_E + H_s)R_E\sin\alpha}, \quad (16.35)$$

it follows that

$$\Delta C = \Delta R\,\frac{R}{(R_E + H_s)\sin\alpha}. \quad (16.36)$$

The cross-track resolution ΔC_R is related to the along-beam resolution by

$$\delta C_R = \Delta C\cos\phi_\alpha = \Delta R\frac{R\cos\phi_\alpha}{(R_E + H_s)\sin\alpha}. \quad (16.37)$$

Note that at low inclination angles, which occur at small cross-track distances, the cross-track range resolution is very coarse, but it becomes finer at larger cross-track distances. This implies that if it is desirable to maintain a fixed cross-track size and spacing of resolution elements on the surface, the span of slant ranges integrated into each resolution element changes across the swath width.

A similar analysis can be performed for a Doppler scatterometer using Eq. (10.24), which gives the Doppler shift as a function of the relative velocity of the scatterometer relative to the pixel location on the Earth's surface. Calculating the Doppler shift is rather complicated because of the orbit geometry and rotation of the Earth under the moving spacecraft, and so it is not presented here. We note that the alignment and spacing of the iso-Doppler lines illustrated in Fig. 16-25(a) varies along the orbit track.

16-6 σ^0 Measurement Precision

▶ A general design goal of a wind scatterometer is to minimize the ***normalized standard deviation*** K_p of the measured signal power or energy. ◀

To first order, K_p of a fan-beam scatterometer is given by the approximate expression (Long et al., 1988)

$$K_p = \frac{1}{\sqrt{N_{\text{looks}}}} \left[1 + \frac{2}{\text{SNR}} + \frac{1}{\text{SNR}^2} \right], \tag{16.38}$$

where N_{looks} is the number of independent looks. For a Doppler scatterometer employing a ***digital processor***,

$$N_{\text{looks}} = N_p \tau B_c, \tag{16.39}$$

where N_p is the number of pulses averaged together, τ is the transmit pulse length, and B_c is the bandwidth corresponding to the resolution element. As discussed later, the value of B_c is set by the desired spatial resolution. To obtain good radiometric resolution (small K_p), N_p is made large and τ long. The longest pulse that can be transmitted without interference corresponds to the length of the pulse (T_p) plus the time required for the leading edge of the pulse to propagate from the transmitter to the closest point of the swath and back to the radar. The next pulse cannot be transmitted until the full echo from the farthest part of the swath is received. Also, delays associated with switching the transmitter must be included.

Figure 16-28 illustrates the timing diagram of the Doppler NSCAT scatterometer. The scatterometer transmits 25 pulses and measures the received energy for each pulse. The received signal includes both the return echo from the surface and system noise. A ***signal+noise*** energy measurement is accumulated over the 25 pulses. Then, the system "listens" without transmitting for a duration equivalent to 4 pulses to collect a separate ***noise-only*** energy measurement. This simplifies the hardware design and calibration by enabling use of the same hardware for computing the signal+noise and noise-only measurements. In the ground processor, the scaled noise-only measurement is subtracted from the signal+noise measurements to generate estimates of the return signal energy. The process is outlined in Fig. 16-29.

The number of pulses that can be accumulated (i.e., the dwell time) is set by the PRF and the orbital speed of the satellite. Typically, a σ^0 measurement is made from each of the individual beams in sequence. If the pulse repetition period is T and the ground-track velocity is u, the maximum number of pulses allowed for collecting an integrated energy measurement is

$$N_p \leq \left\lfloor \frac{D}{uN_a T} \right\rfloor, \tag{16.40}$$

where N_a is the number of antenna beams and D is the along-track measurement spacing. Noting from Eqs. (16.38) and (16.39) that K_p is inversely proportional to the number of pulses, which is constrained in Eq. (16.40) by the along-track spacing of the measurements, we see that there is a trade-off between K_p and the along-track spacing: reducing the along-track spacing has the side effect of increasing K_p, while integrating more pulses reduces the resolution and K_p, but increases the along-track spacing.

For NSCAT, an along-track spacing of 25 km was desired. The timing diagram shown in Fig. 16-28 results in 25 signal pulses and 4 noise pulses for each of NSCAT's eight antenna beams and yields a 25 km along-track spacing.

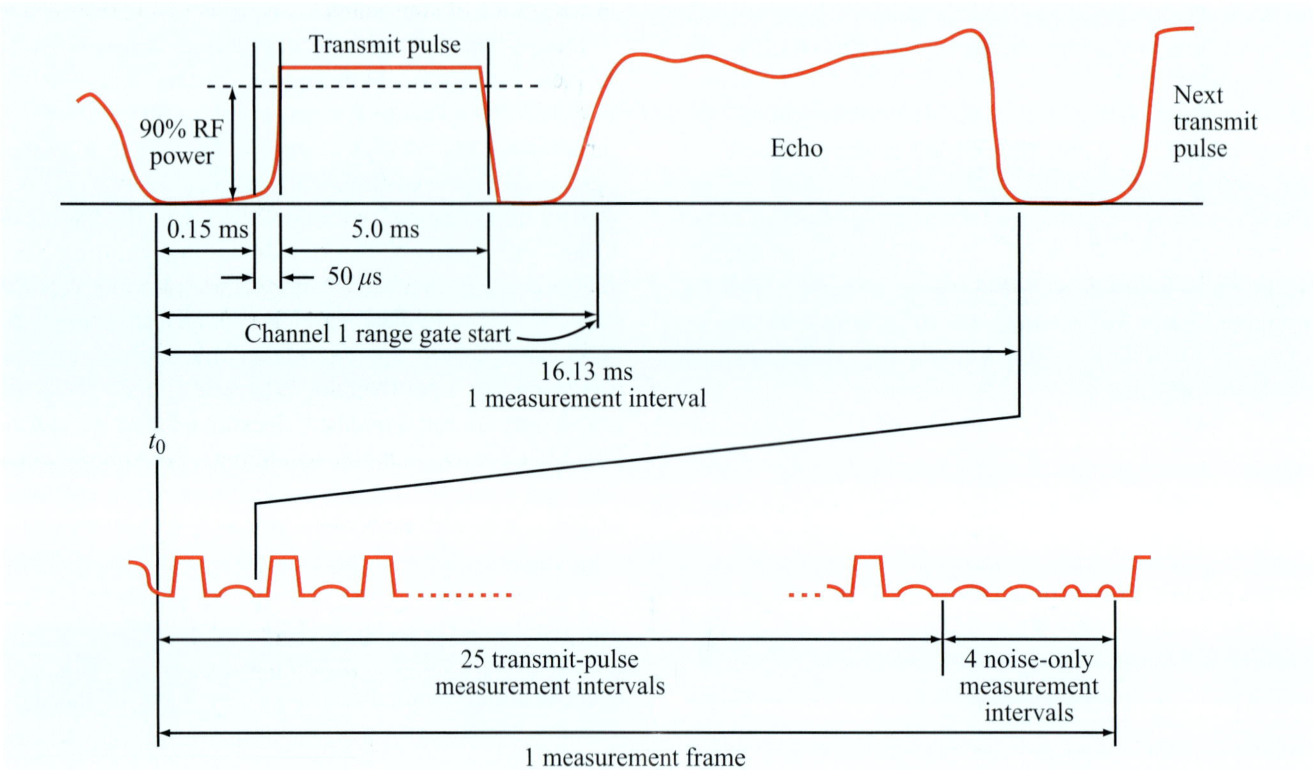

Figure 16-28: The transmit and receive timing used by the NASA Scatterometer (NSCAT). For a given beam, the signal power measured for each of 25 transmit pulses is accumulated. A separate noise-only measurement is made by processing the noise energy during 4 transmit pulses when the transmitter is disabled. This timing is repeated for each of the antenna beams (4 single polarization antennas and 2 dual-polarization antennas).

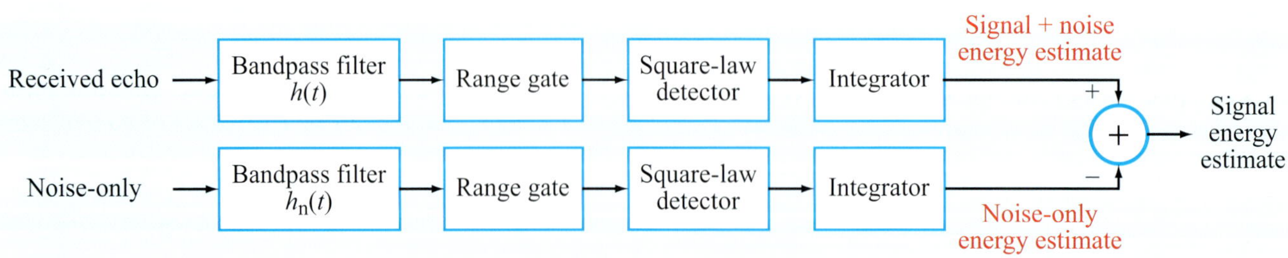

Figure 16-29: Signal-processing flow diagram for estimating the scatterometer signal energy. Not shown is the scaling factor applied to the noise-energy measurement prior to its subtraction from the signal+noise energy estimate.

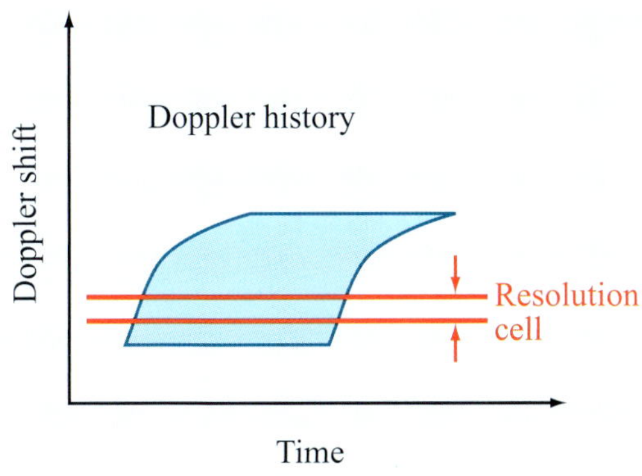

(a) Doppler shift versus time of the return echo for a long-pulse fan-beam scatterometer

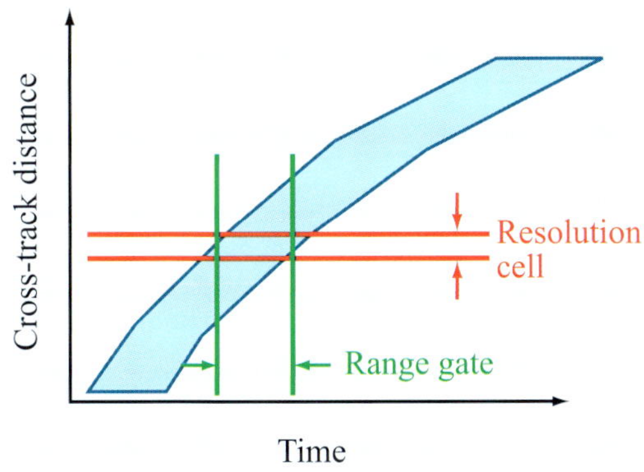

(b) Cross-track distance versus time for a short-pulse fan-beam scatterometer

Figure 16-30: Wind scatterometer signal history for (a) a long-pulse Doppler scatterometer and (b) a short-pulse Doppler scatterometer.

16-6.1 Doppler-Filtering Scatterometers

For a Doppler system, the incidence angle, slant range, and Doppler frequency all are interrelated. Figure 16-30 schematically illustrates the time-Doppler history for short- and long-pulse systems. The shaded areas display the Doppler shift of the echo from the illuminated swath. Because of the very long slant ranges from the satellite to the sea surface, the variation in time-of-flight from the near-swath to the far-swath can be very long, resulting in a very pronounced curvature in the time-Doppler history. For the fan-beam systems under consideration, a given Doppler shift typically corresponds to a particular narrow range of slant ranges, and thus to a narrow range of delay times. In Fig. 16-30(a), at a particular Doppler frequency the horizontal extent of the ***time-Doppler history*** corresponds to the pulse length plus this narrow variation in time-of-flight. For a long-pulse system, the variation in time-of-flight at a given Doppler shift is very small compared with the pulse length, while for a short pulse system, the variation may be much larger than the pulse length. While ranging systems can use short ($\leq 100\ \mu$s) pulses, Doppler scatterometer systems typically use long (> 1 ms) pulses, which provide capability for fine Doppler frequency discrimination. For example, NSCAT used 5 ms long pulses, which yielded a 200 Hz Doppler resolution. The fine Doppler resolution helps increase the number of looks, thereby reducing the measurement variance represented by K_{p}.

As previously noted, Doppler filtering of the return signal can provide cross-track resolution. To select the center frequencies and bandwidths of the Doppler filters, we specify a nominal cross-track resolution. Typically this is the same as the along-track measurement spacing. For each cross-track location, the Doppler frequency corresponding to the center of the antenna beam illumination pattern on the surface is determined using the methods discussed in Chapter 13. The Doppler bandwidth for each resolution element then corresponds to the spacing between the Doppler center frequencies.

As shown in Fig. 16-31, the relative motion of the spacecraft and Earth's surface has the effect of "tilting" the iso-Doppler contours slightly off axis compared with the along-track/cross-track coordinate system aligned with the spacecraft's ground track. The result is that each antenna requires a different Doppler center frequency for each resolution element to ensure that the same area is measured by the forward-facing and aft-facing antennas. Further, to maintain the resolution elements at fixed cross-track distances throughout the orbit, their Doppler center frequencies must be adjusted continuously as a function of time.

The plots in Fig. 16-32(a) display the Doppler center frequencies corresponding to resolution elements at fixed cross-track distances for the forward-looking antenna beam of NSCAT, which flew in an 830 km

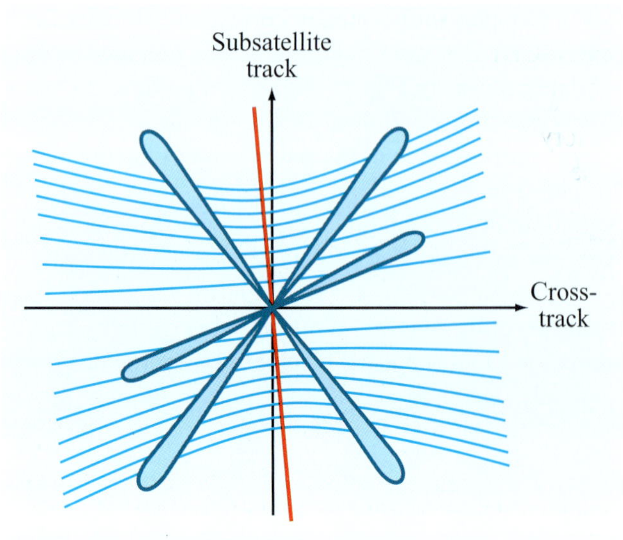

(a) Ascending

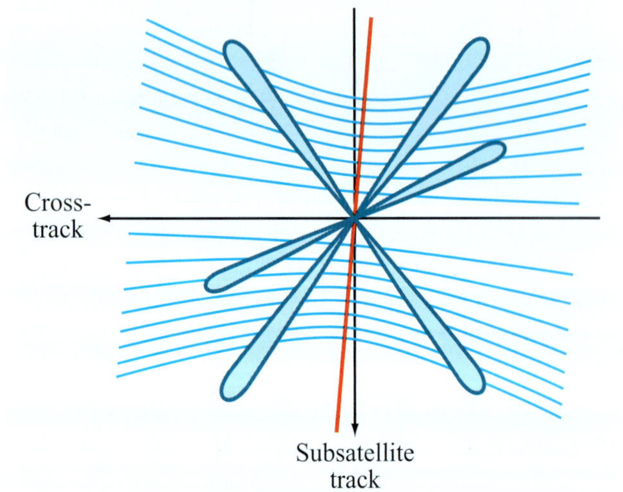

(b) Descending

Figure 16-31: Iso-Doppler "tilt" for a spaceborne Doppler scatterometer for (a) ascending (northbound) and (b) descending (southbound) equator crossings.

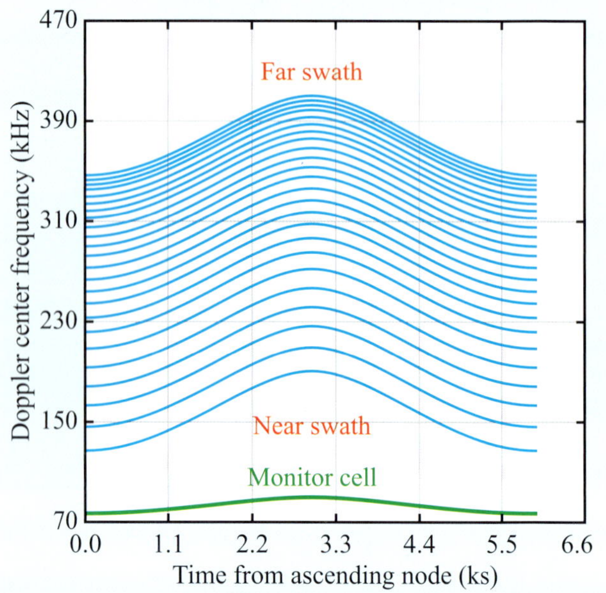

(a)

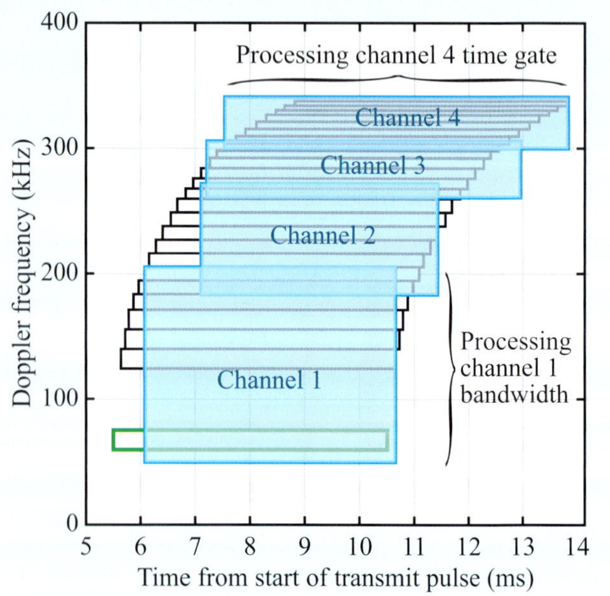

(b)

Figure 16-32: (a) Center frequency for each 25 km resolution cell for a forward-looking antenna versus orbit position in time from equator crossing, (b) Doppler history for each resolution cell at a particular orbit time with the time/bandwidth of the digital processing channels superimposed.

nearly circular orbit. The resolution elements are spaced 25 km apart in cross-track, starting (after the first element, which is closer and used for calibration) at 185 km. The variation with time from the northern equator crossing, known as the orbit ***ascending node***, is due to orbit geometry and the oblateness of the Earth. At the far swath, the variation in center frequency can be nearly 100 kHz while the variation at the near swath is less than 20 kHz. Note that the maximum Doppler shift is over 420 kHz while the minimum Doppler shift is approximately 75 kHz. Aft-facing antennas have the same general characteristics, but with a negative Doppler shift. The Doppler shift for the center antennas have a smaller range. Because it is difficult to build well-calibrated analog filters that can support this much variation in center frequencies, a digital processor is used instead. Digital processors also offer the flexibility of changing the Doppler filters on-orbit to support changes in orbital parameters or adjusting the swath (Long et al., 1988). Digital processors can easily change bandwidth and center frequencies of the resolution elements to maintain them at a constant cross-track position throughout the orbit.

For each channel and for each pulse, after range-gating and digitizing the received signal, the digital processor computes the FFT of the sampled signal, applies a signal processing window to minimize spectral leakage by convolution, then performs power detection for each FFT bin by squaring its value to compute the signal periodogram. The result is an estimate of the signal energy in a narrow bandwidth. Based on the desired center frequency and bandwidth, the periodogram bins are summed into a single energy value for each resolution element. The energy measurements are accumulated for multiple pulses and transmitted to the ground for further processing. The onboard processing and accumulation minimizes the downlink data volume.

The approximate number of looks N_{looks} for a particular σ^0 measurement made by a Doppler scatterometer is given by

$$N_{\text{looks}} = N_{\text{p}} \tau B_{\text{c}}. \qquad (16.41)$$

Using the NSCAT parameters ($N_{\text{p}} = 25$, $\tau = 5$ ms, and $5\,\text{kHz} \leq B_{\text{c}} \leq 20$ kHz), we find that NSCAT measurements provide a large number of looks, namely $625 < N_{\text{looks}} < 2500$, which permitted operation at even very low (-15 dB) SNR while maintaining an acceptable K_{p} at low wind speeds. By operating at low SNR, the transmit power can be kept very small (120 W for NSCAT).

In a Doppler system the curvature in the time-Doppler history can be exploited to reduce the system noise when estimating the return energy by using range-gating and Doppler filtering. Since the signal for a particular resolution element is present for only a small fraction of the receive interval, the received signal can be range-gated so that only the return echo over the time period for which the desired return is present is used for measuring the signal power integrated over the pulse length to estimate the signal energy. Optimally, this is done separately for each resolution element. However, in some cases it may be advantageous to combine the processing of multiple resolution elements into one of several processing channels as illustrated in Fig. 16-32(b). In this figure the time-Doppler history of each of the scatterometer resolution elements is indicated with a box. Note that resolution elements closer to the sub-satellite swath have smaller Doppler center frequencies but larger Doppler bandwidths than the same-sized resolution elements on the ground at the far swath. The processing bandwidth is split into four frequency channels with overlapping bandwidths. A digital processor implements Doppler filters for each resolution element. The center frequencies and bandwidths of the filters are based on an uploaded table (Long et al., 1988). We note that the channel range gates indicated in Fig. 16-30(b) are shorter than the transmit pulse. Chosen based on hardware constraints, the short range gates have the effect of slightly degrading K_{p} over having full-length range gates. Range gates longer than the signal length include more noise and thus also degrade K_{p}. Ideally, to optimize K_{p} the range gate should be the same length as the signal. Compared with processing all of the resolution elements within a single channel, multichannel processing has the advantage of minimizing K_{p} by more closely matching the range gate lengths to the signal.

Figure 16-33 illustrates in block-diagram form the RF hardware and the digital signal processing used to process the signal for NSCAT. The block diagram also shows the RF switching hardware to select between each of the multiple antennas used in the design. The beam

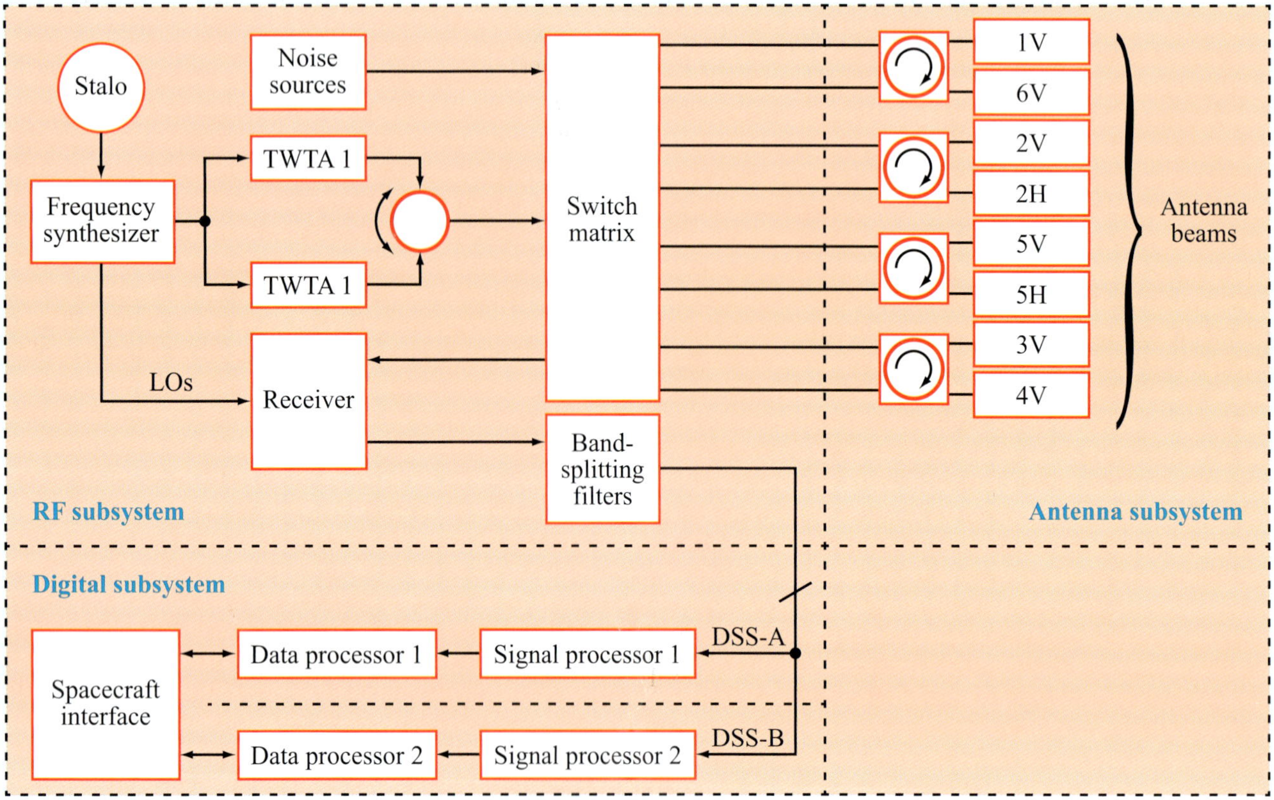

(a) NSCAT RF subsystem

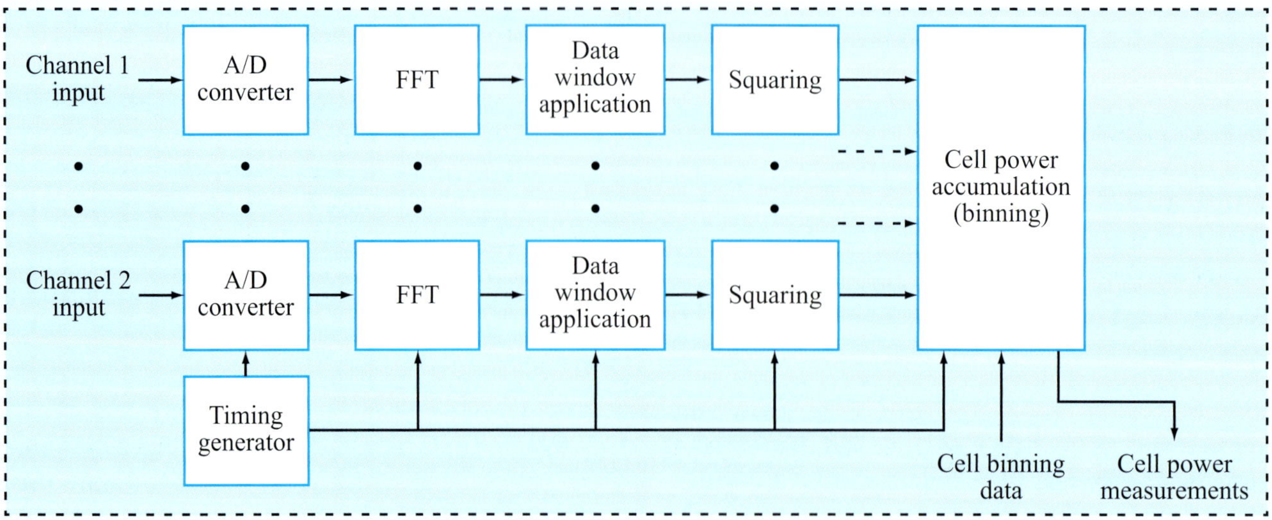

(b) NSCAT digital signal processing

Figure 16-33: (a) Block diagram of the NSCAT RF subsystem, (b) signal processing flow diagram for the NSCAT digital processor.

switching is controlled by the central timing processor. A set of signal energy measurements are acquired from each beam, which are cycled in sequence. The cycling time is synchronized with ground track velocity so that measurements of a given beam are collected with a fixed along-track spacing, typically 25 km.

16-6.2 Range-Resolution Wind Scatterometers

Range-resolution fan-beam scatterometers share some of the same features with Doppler systems, but are less-restrictive when it comes to selecting the center-beam azimuth angle, which can be chosen to point directly to the side if desired. The first range-resolution wind scatterometer on ERS-1/2 used interrupted continuous-wave (ICW) pulses, each 100 μs long with power levels of several kilowatts. Later systems (e.g., ASCAT) used longer pulses (10 ms) and linear frequency modulation (LFM) with dechirp range compression (discussed in Chapter 13) to enable reduction of the peak transmit power (to 120 W) while maintaining a low K_p. In both cases, the signal bandwidth is large compared with the range of Doppler shifts over the swath, which reduces the need to compensate for the Doppler shift in processing the data into resolution elements.

As in a Doppler scatterometer, a number of pulse measurements are summed into each energy measurement for a given beam, and pulsing is sequenced between the beams so that measurements from a given beam are uniformly spaced along the spacecraft ground track. The range bins accumulated into each measurement are adjusted as a function of orbit position to maintain fixed cross-track locations of the measurements since the spacecraft height above the surface varies over the orbit.

After mixing the received signal to baseband, the received signal is digitized. In the long-pulse case, range compression is applied. Power measurements are formed by summing the range bins corresponding to the desired resolution, and the accumulated power measurements are downlinked to the ground for further processing (Wilson et al., 2005).

The antenna pattern and signal processing filters combine to create a spatial response pattern for measuring the surface σ^0. To better understand the spatial response pattern for both Doppler and range scatterometers, Fig. 16-34 shows a zoomed-in view of the resolution elements in Fig. 16-25. Note that in both cases the cross-beam response is dictated by the antenna gain pattern in the narrowbeam (cross-beam) direction. For the Doppler filtering system, the Doppler filter response defines the along-beam response with the slowly varying widebeam antenna gain pattern superimposed. The iso-Doppler contours define the along-beam 3 dB response, while the antenna beamwidth in the narrow direction defines the cross-beam 3 dB response. The total spatial response function is the product of the antenna and Doppler filter gain responses.

Similarly, for the range scatterometer the antenna gain pattern defines the cross-beam gain pattern while the range resolution filter or range compression gain response defines the along-beam response function. This response includes the effects of summing multiple range bins into a single measurement. The variation in the Doppler shift over the footprint can play a role by slightly tilting the range filter response lines.

When multiple pulses are integrated into a single measurement, the 3 dB gain response area of the integrated response is thought of as the "resolution" of the measurement, even though the integrated footprint areas may not completely cover the desired region. Using the parameters of the radar equation, estimates of the surface σ^0 are computed from the signal energy measurements. While the σ^0 measurements can be used directly in wind estimation, the measurements can be spatially interpolated with the aid of a smoothing window to enable reporting of σ^0 on a uniform spatial grid at the surface (Wilson et al., 2005). This greatly simplifies collocating σ^0 measurements from the various azimuth beams when performing wind estimation, but imposes limits on reconstruction processing (Long, 2003). For land and ice applications, where σ^0 evolves slowly compared with the ocean wind field, multiple passes over an area can be combined to fill gaps between the spatial response functions. Using knowledge of the spatial response function with reconstruction techniques, it is possible to generate fine-resolution

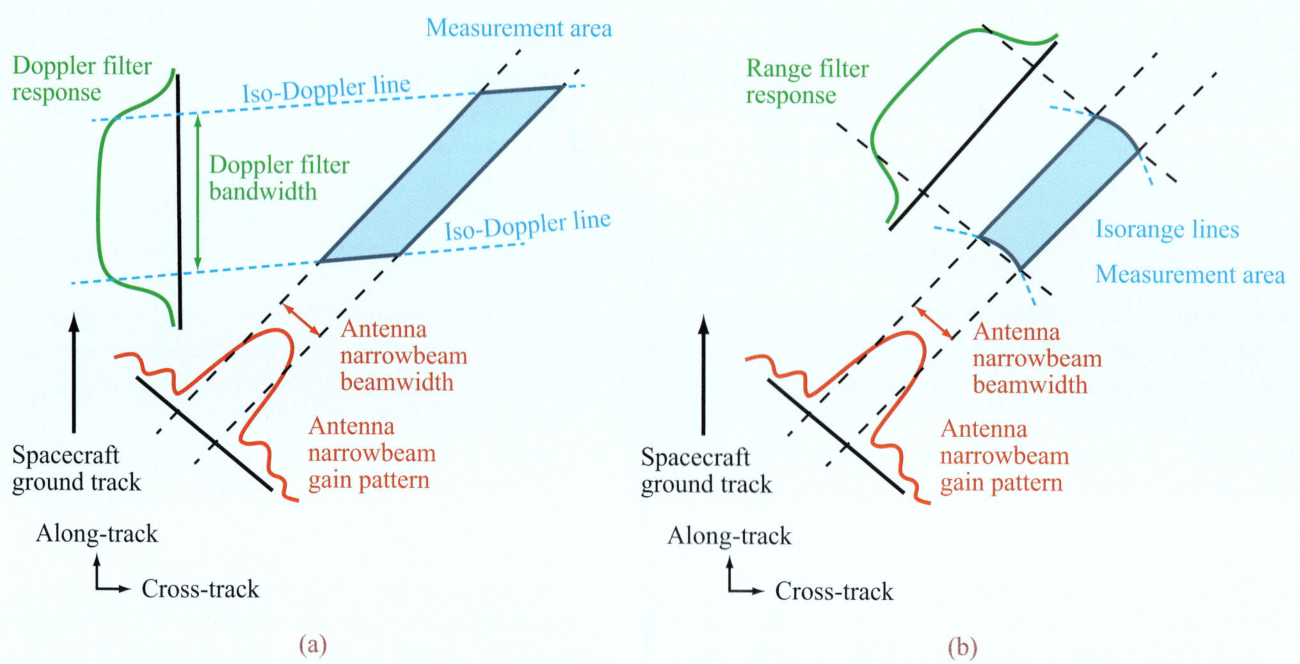

Figure 16-34: Illustration of how the spatial response function for a fan-beam scatterometer measurement is determined for (a) a Doppler system and (b) a ranging system for a single pulse. The shaded areas correspond to the 3 dB response region. The 3 dB response area of an integrated measurement is computed by averaging the spatial responses of the pulses integrated into the measurement.

estimates of the surface σ^0 (Early and Long, 2001; Williams and Long, 2011b).

16-7 Scanning Pencil-Beam Wind Scatterometers

In a ***scanning pencil-beam scatterometer***, a narrow pencil-beam antenna is rapidly scanned over the observation swath. This approach has the advantage of higher SNR for a given transmit power, but has reduced dwell time due to the scanning. The reduced dwell reduces the number of available looks compared with a fan-beam design. As described shortly, a scanning scatterometer has a much different distribution of azimuth angles for its σ^0 measurements compared with the 2 or 3 azimuth angles typical of a fan-beam system, all of which remain fixed over the swath.

To date, scanning pencil-beam wind scatterometers (Table 1-2) have employed two antenna beams at different incidence angles rotating about the nadir axis, as illustrated by Fig. 16-35. The rotation rate is set so that the spacecraft propagates along the subsatellite track a distance equal to the desired along-track resolution distance during each rotation. The two beams trace out a helix on the surface as the spacecraft moves along its orbit. For simplicity, this is illustrated with a single beam in Fig. 16-36. As shown in this figure, each point of the swath width is first observed at a forward-looking azimuth angle ϕ_f, and then shortly thereafter (seconds to a few minutes later) by at an aft-looking azimuth angle ϕ_a. With two beams at different incidence angles, over most of the swath, each point within the swath is observed four times at different azimuth and incidence angles. This diversity of azimuth and incidence angles improves the wind estimate performance. Since the

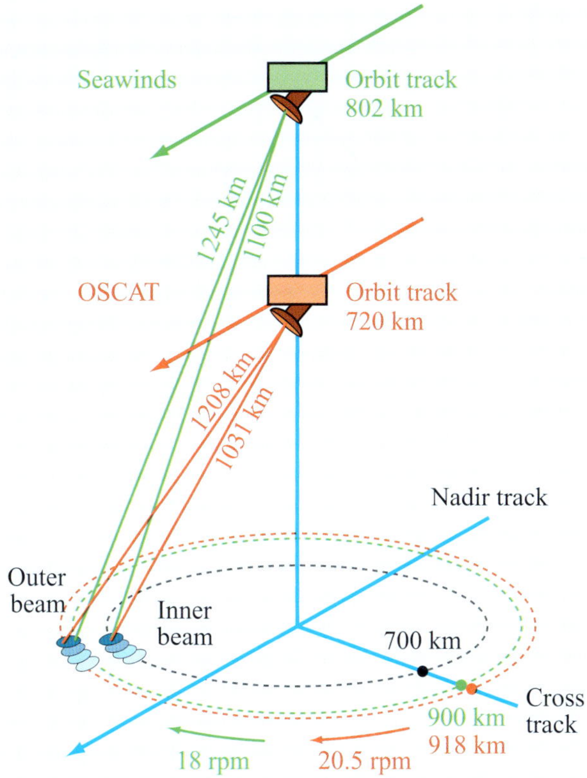

Figure 16-35: Superimposed geometry for two scanning pencil beam scatterometer systems, the NASA SeaWinds and ISRO Oceansat-2 scatterometer (OSCAT).

GMF varies with polarization, diversity is enhanced by mixing antenna polarizations.

Pencil-beam wind scatterometer systems rely on a narrow-beam antenna that must be scanned over the swath. While this could be done electronically, it has traditionally been done using a mechanically rotated antenna. This requires a separate momentum compensation system since, like all nadir-facing sensors, the spacecraft must be rotated once per orbit to keep the antennas facing earthward, as shown in Fig. 16-37.

While a narrow-beam antenna has high gain, which produces high system SNR, the need to scan the beam over the swath reduces the dwell time per measurement. Pulse timing and integration must be varied as a function of the antenna azimuth angle if measurements are to

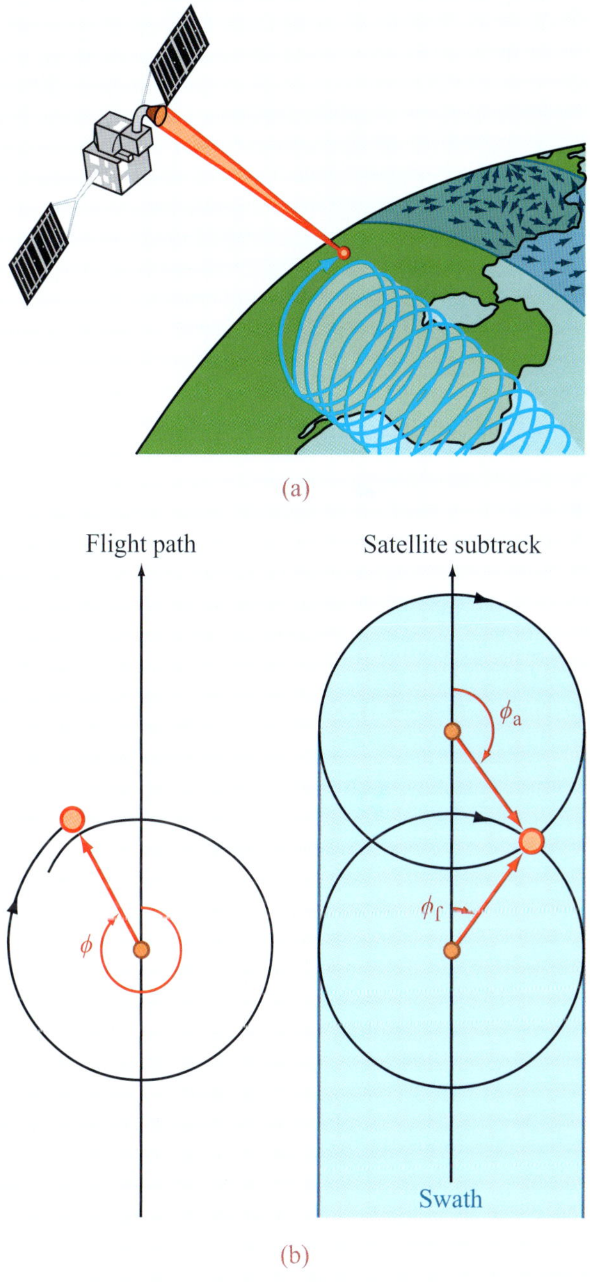

Figure 16-36: Scanning concept for a single beam scanning pencil-beam scatterometer, and geometry of multiple azimuth angles from a single beam.

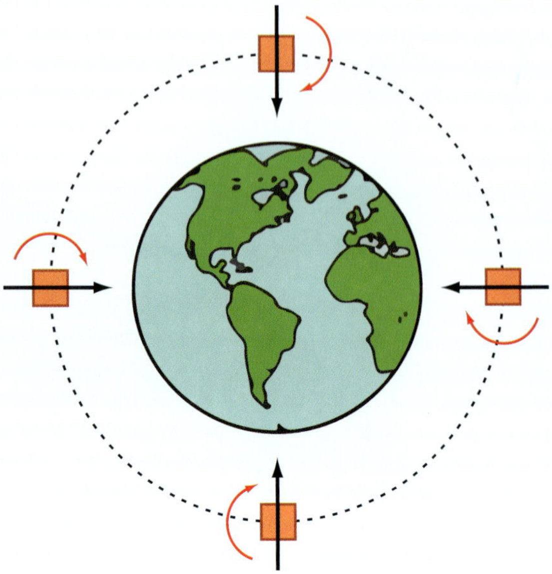

Figure 16-37: Illustration of how a spacecraft must be rotated once per orbit to maintain its downward pointing.

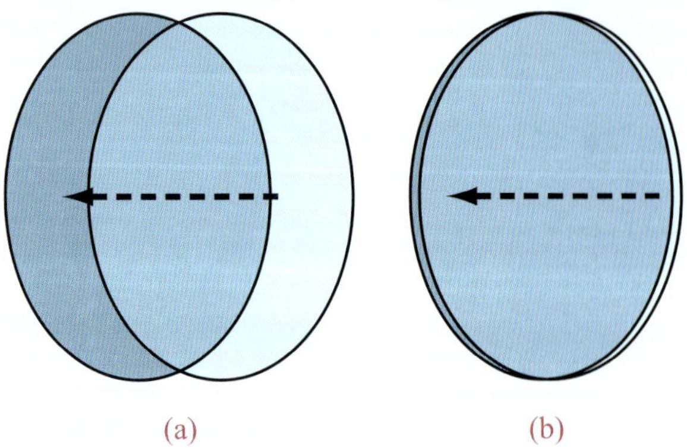

Figure 16-38: Illustration of scanning loss using 3 dB antenna illumination patterns on the surface. The dashed arrow indicates the scanning direction. (a) Misaligned antenna footprints, (b) nearly perfectly aligned footprints.

be collected at fixed cross-track distances or else the measurements have to be gridded after collection.

16-7.1 Scanning Loss

In spinning antenna systems, due to the long (> 10 ms) two-way travel time, the antenna can rotate a significant fraction of the beam width between the issuance of the transmit pulse and reception of the received signal. This causes misalignment of the transmit and receive antenna patterns, resulting in loss of signal gain known as ***scanning loss*** (Fig. 16-38). The ***two-way gain*** pattern is the product of the transmit and receive patterns including the rotation offset. This can be illustrated by writing the integral form of the radar equation as given by Eq. (5.34) with separate antenna gain patterns for transmit and receive,

$$P_p^{\rm t}(\theta) = \iint\limits_A \frac{P_q^{\rm t}\, G_{\rm t}(\theta_{\rm a}^{\rm t}, \phi_{\rm a}^{\rm t})\, G_{\rm r}(\theta_{\rm a}^{\rm r}, \phi_{\rm a}^{\rm r})\, \lambda^2}{(4\pi)^3 R_{\rm a}^4} \cdot \sigma^0\, dA, \tag{16.42}$$

where $G_{\rm t}(\theta_{\rm a}^{\rm t}, \phi_{\rm a}^{\rm t})$ is the transmit antenna pattern with its geometry $(\theta_{\rm a}^{\rm t}, \phi_{\rm a}^{\rm t})$ applicable during the transmit pulse and $G_{\rm r}(\theta_{\rm a}^{\rm r}, \phi_{\rm a}^{\rm r})$ is the receive antenna gain pattern with its geometry $(\theta_{\rm a}^{\rm r}, \phi_{\rm a}^{\rm r})$ applicable during the receive interval. In writing this expression, we have assumed that the transmit pulse and the receive integral are sufficiently short so that the geometry can be assumed to be fixed during these time periods. The two-way effective antenna gain pattern G is

$$G(\theta_{\rm a}, \phi_{\rm a}) = G_{\rm t}(\theta_{\rm a}^{\rm t}, \phi_{\rm a}^{\rm t})\, G_{\rm r}(\theta_{\rm a}^{\rm r}, \phi_{\rm a}^{\rm r}). \tag{16.43}$$

The integrated antenna gain $G_{\rm Int}$ is

$$G_{\rm Int} = \iint\limits_A G_{\rm t}(\theta_{\rm a}^{\rm t}, \phi_{\rm a}^{\rm t})\, G_{\rm r}(\theta_{\rm a}^{\rm r}, \phi_{\rm a}^{\rm r})\, dA. \tag{16.44}$$

When transmit and receive antenna gain patterns are properly aligned, $G_{\rm Int}$ is maximized. Let $G_{\rm Int}^{\rm max}$ be this maximum gain. The scanning loss $G_{\text{scan loss}}$ is then the ratio (difference in dB) between the maximum gain and the gain achieved; i.e.,

$$G_{\text{scan loss}} = \frac{G_{\rm Int}^{\rm max}}{G_{\rm Int}}. \tag{16.45}$$

Note that when the transmit and receive antenna gain patterns are similar and the footprints are properly aligned, $G(\theta_{\rm a}, \phi_{\rm a}) \approx G_{\rm t}^2(\theta_{\rm a}^{\rm t}, \phi_{\rm a}^{\rm t})$ and the scan loss is 0 dB.

To minimize scanning loss, the receive gain pattern at the time of the signal return needs to be aligned with the direction that the return signal is returning from. The required azimuth angle spacing $\Delta\phi_{\text{ant}}$ in radians to accomplish this can be computed from the antenna rotation rate ρ_{rot} in radians per second and the nominal slant range R to the center of the footprint using

$$\Delta\phi_{\text{ant}} = \frac{2R}{c}\,\rho_{\text{rot}}. \qquad (16.46)$$

Note that the receive beam "lags" the transmit beam by $\Delta\phi_{\text{ant}}$ because the antenna will have rotated by the time the signal echo returns from the surface. Variations in the nominal slant range due to oblateness of the Earth and orbital eccentricity cause variations in the scanning loss that must be accounted for when computing σ^0 from the signal energy measurements.

The SeaWinds scatterometer included separate feeds for transmission and reception, which were spaced in azimuth angle but shared the same reflector. The azimuth angle spacing was chosen to minimize scanning loss.

Using separate transmit and receive feeds can eliminate the transmit/receive switch common to most radars. A simplified schematic diagram of a single-beam scanning scatterometer system using separate feeds is illustrated in Fig. 16-39. Note, however, that separate feeds sharing the same reflector are possible only if there is sufficient room between them. Otherwise, an array feed requiring a transmit/receive switch may be needed.

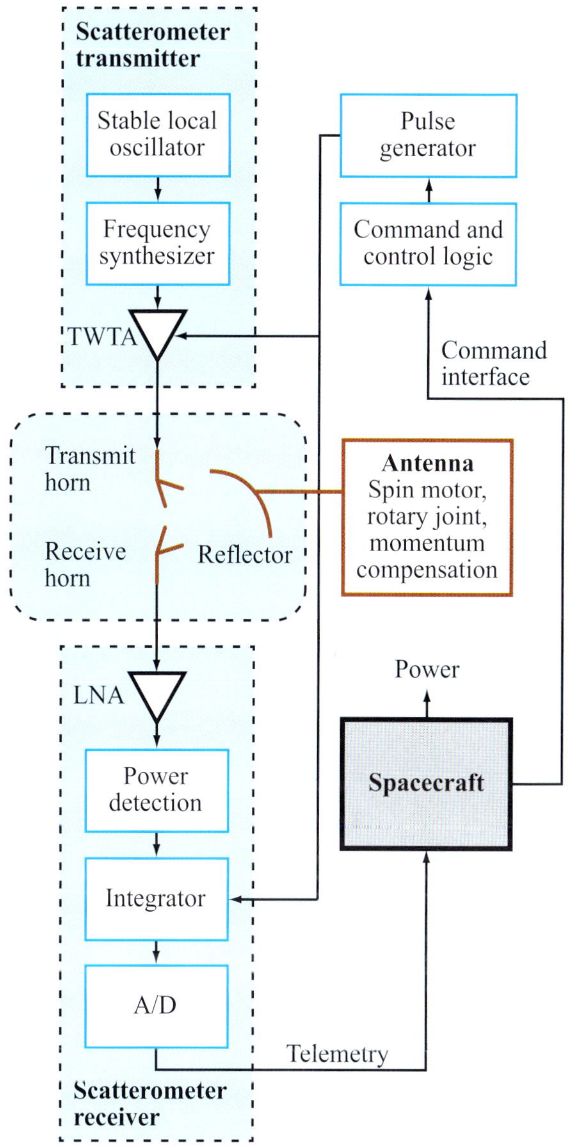

Figure 16-39: Schematic diagram of a simple single-beam pencil-beam scatterometer system with separate transmit and receive feeds.

16-7.2 Measurement Precision

The spatial resolution of a pencil-beam system depends on the footprint size. However, higher resolution can be obtained by using range and Doppler processing. To gain insight into what is possible, consider Fig. 16-40, which illustrates the arrangement of isorange and iso-Doppler lines observed by the pencil-beam antenna footprint during part of its rotation. When looking straight-ahead, the isorange and iso-Doppler lines essentially coincide. In this forward-looking case range resolution and Doppler filtering techniques yield essentially the same results. This geometry yields the fewest independent looks. As the antenna rotates, the isorange and iso-Doppler lines separate and are nearly orthogonal at a 90° azimuth angle. This case has the highest number of looks. As discussed in Chapter 13, the pulse bandwidth sets the range resolution dr while the Doppler resolution $d\omega$ is inversely related to the

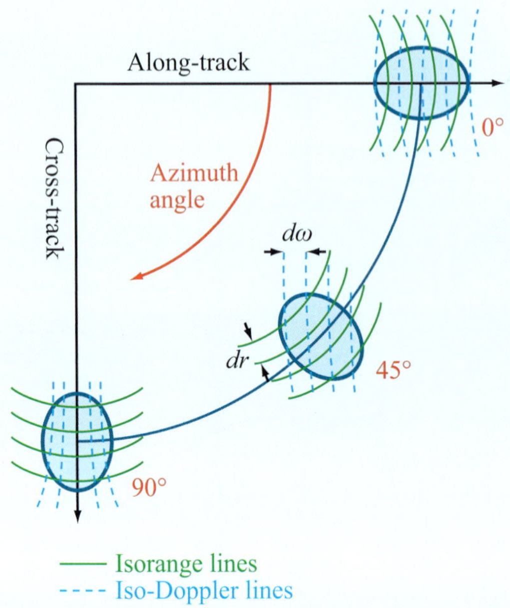

Figure 16-40: Relationship of the iso-Doppler and isorange lines for a scanning pencil-beam scatterometer as a function of azimuth angle. Symmetry can be applied for the other quadrants.

pulse length. The range/Doppler geometry results in a variation in the measurement K_p with azimuth angle. The details of the processing do have an impact on the effective K_p (Yoho and Long, 2004). In the case of SeaWinds, the transmit signal was LFM to a bandwidth of 377 kHz with a dechirp range-compression followed by periodogram processing. Multiple range bins are accumulated into separate range measurements known as ***slices***. Like a Doppler scatterometer, the number of slices and their center frequency is adjusted as a function of orbit time, as well as rotation angle, in order to maintain a fixed spatial resolution.

For a general pencil-beam system, K_p can be expressed as (Long and Spencer, 1997)

$$K_p = \frac{1}{\sqrt{N_p \tau B_c}} \left[\alpha + \frac{\beta}{\text{SNR}} + \frac{\gamma}{\text{SNR}^2} \right], \qquad (16.47)$$

where N_p is the number of pulses averaged together, τ is the effective pulse length, and B_c is the effective signal bandwidth. The precise coefficients $\alpha \approx O(1)$, $\beta \approx O(2)$, and $\gamma \approx O(1)$ [see Eq. (16.38)] have complicated dependencies on the modulation characteristics, filtering, and signal bandwidth. The effective pulse length may be shorter than the transmit pulse length if part of the signal is clipped by the range gate. The signal bandwidth depends on both the Doppler signal bandwidth over the footprint, the transmit modulation bandwidth, and the receiver bandwidth. Since multiple range slices can be collected simultaneously, the footprint is "sliced" into sub-elements. These yield higher resolution at the expense of higher K_p.

Figure 16-41 illustrates the spatial response function for a pencil-beam system (Ashcraft and Long, 2003). Two cases are shown: the spatial response for the energy accumulated over the full footprint is shown in Fig. 16-41(a), while the case when the signal energy is accumulated over only a single range slice is shown in Fig. 16-41(b). Figure 16-41(c) illustrates the 3 dB footprint and slice contours, where all the slices are shown. While this figure is specific to SeaWinds, OSCAT employs a similar approach.

As previously discussed in Chapter 13, to determine the signal energy, separate measurements of the signal+noise and the noise-only energy measurements are collected and subtracted. The noise-only signal corresponds to the radar operating as a radiometer; the radiometric signal plays the role of noise in the scatterometer. As discussed in Chapter 7, the radiometric signal is due to the receiver thermal noise plus thermal emission by the scene. Thus, the radar noise power is scene dependent. For a fan-beam system, the scene extends over a large area and changes more slowly than the scene brightness temperature for scanning pencil-beam system. Due to the rapid motion of the scanning, the background scene temperature can change significantly from pulse to pulse. While not transmitting a signal and measuring the noise-only energy during the receive interval can be used with only minimal adverse impact (due to not transmitting) in the fan-beam case, it cannot be used for a pencil-beam system. Instead, the noise measurement must be made simultaneously with the signal+noise measurement. This can be accomplished using a two-channel baseband

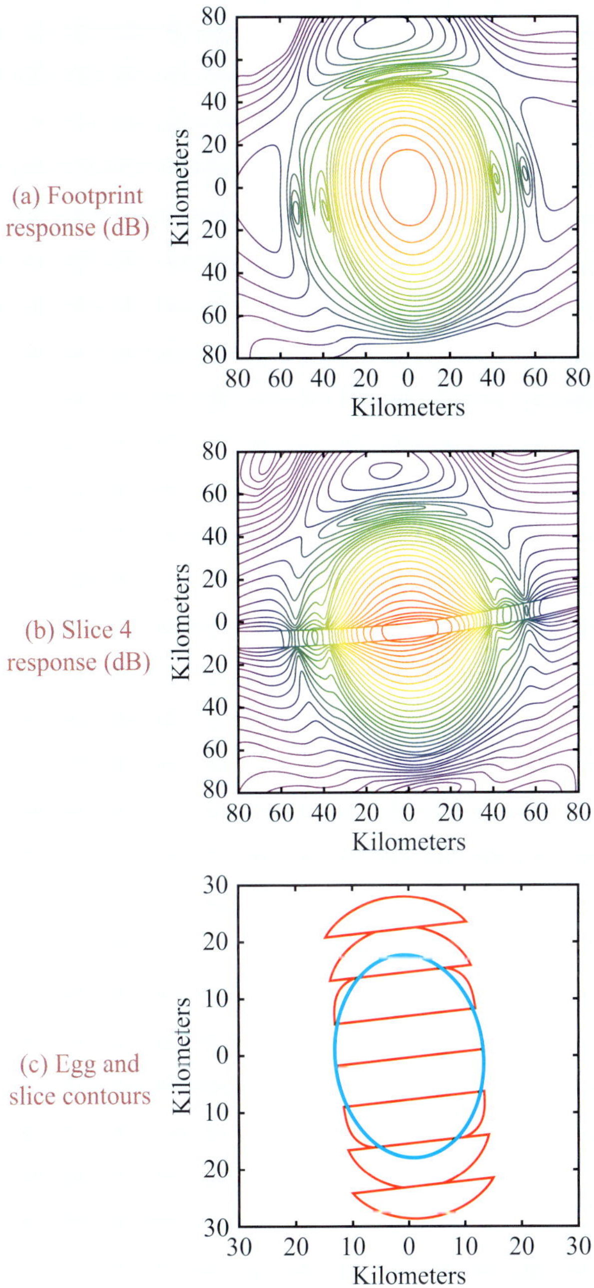

Figure 16-41: Spatial response functions for the SeaWinds pencil-beam scatterometer: (a) antenna beam-limited, (b) Doppler-filter slice with contour lines spaced at 3 dB intervals, and (c) diagram of the 3 dB footprint (blue ellipse) and Doppler slices (red) contours.

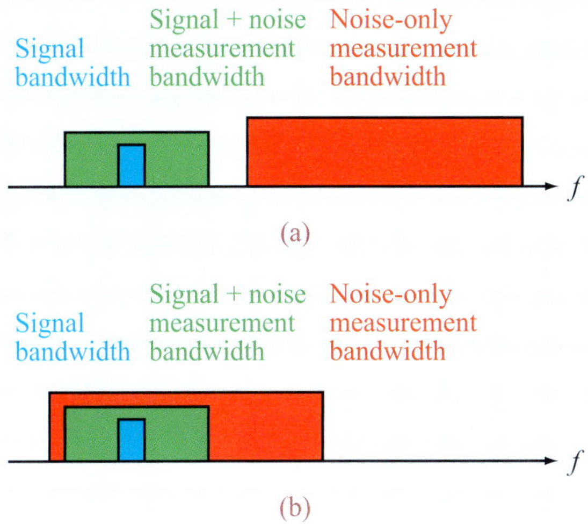

Figure 16-42: Diagram for frequency plans for measuring signal power: (a) disjoint bandwidths, and (b) overlapping bandwidths.

receiver with the noise-only channel operating in a disjoint bandwidth, as illustrated in Fig. 16-42(a). Note that the sidelobes of the transmit signal have to be suppressed so that undesired signals in the noise-only measurement bandwidth are minimized. The noise-only measurement bandwidth should be set to be as large as possible in order to minimize the uncertainty of the noise-only energy.

Let $C_{\rm sn}$ be the signal+noise energy measurement collected over an effective signal bandwidth $B_{\rm r}$ and $C_{\rm no}$ be the noise-only energy measurement collected over an effective noise bandwidth $B_{\rm n}$, where $B_{\rm n} \gg B_{\rm r}$. Since separate channels for the signal+noise and noise-only may have separate gains, denoted by $g_{\rm sn}$ and $g_{\rm no}$, respectively, the channel gains must be carefully measured and accounted for to minimize measurement biases. For disjoint bandwidths, the signal energy measurement $C_{\rm s}$ is

$$C_{\rm s} = \frac{1}{g_{\rm sn}}\, C_{\rm sn} - \frac{B_{\rm r}}{g_{\rm no} B_{\rm n}}\, C_{\rm no}. \qquad (16.48)$$

Alternatively, is it possible to allow the noise-only bandwidth to overlap the signal bandwidth, as illustrated in Fig. 16-42(b). In that case, Eq. (16.48) is modified to

$$C_s = \frac{1}{g_{sn}(B_n - B_r)} C_{sn} - \frac{B_r}{g_{no}(B_n - B_r)} C_{no}. \quad (16.49)$$

Critical to the performance of scatterometer wind estimation is minimization of bias in the noise subtraction. Bias errors in σ^0 lead to biases in the estimated wind. Long-term calibration and stability are essential in a wind scatterometer.

The radar equation for a narrow-beam scatterometer is first introduced in Section 5-5.1. This equation relates the transmitted signal power to the received signal power and includes the effects of the antenna pattern through the illumination integral. The signal energy is the integral of the signal power, which can be related to the surface σ^0, assumed to be constant over the resolution element, as

$$C_s = \frac{1}{X_{cal}} \sigma^0, \quad (16.50)$$

where X_{cal} is the illumination integral, including range gating and filtering. The precise formulation depends on the details of the signal processing employed (Long and Spencer, 1997; Spencer et al., 1997; 2003). The general idea can be seen from the radar power equation given in Eq. (16.42). Using τ for the ideal signal length, the ideal integrated signal energy $C_s(n)$ for a single pulse n can be computed as

$$C_s(n) = \int_{t=0}^{\tau} \iint_A \frac{P_q^t\, G_t(\theta_a^t, \phi_a^t)\, G_r(\theta_a^r, \phi_a^r)\, \lambda^2}{(4\pi)^3 R_a^4} \cdot \sigma^0 \, dA\, dt. \quad (16.51)$$

The total ideal integrated signal energy C_s is the sum over the accumulated pulses,

$$C_s = \sum_{n=1}^{N_p} C_s(n). \quad (16.52)$$

Assuming that σ^0 is constant over the footprint, it can be factored out. This enables us to express the signal energy in terms of the ***system spatial response function*** $S(x,y)$ as

$$C_s = \sigma^0 \iint_A S(x,y)\, dA, \quad (16.53)$$

where

$$S(x,y) = \sum_{n=1}^{N_p} \int_{t=0}^{\tau} \frac{P_q^t\, G_t(\theta_a^t, \phi_a^t)\, G_r(\theta_a^r, \phi_a^r)\, \lambda^2}{(4\pi)^3 R_a^4}\, dt$$
$$\approx \tau \sum_{n=1}^{N_p} \frac{P_q^t\, G_t(\theta_a^t, \phi_a^t)\, G_r(\theta_a^r, \phi_a^r) \lambda^2}{(4\pi)^3 R_a^4}, \quad (16.54)$$

where the dependence of the various terms on the pulse number due to motion is implied. Examples of $S(x,y)$ for particular SeaWinds measurements are shown in Fig. 16-41.

Using these results, X_{cal} in Eq. (16.50) is given by

$$X_{cal} = \iint_A S(x,y)\, dA$$
$$\approx \tau \sum_{n=1}^{N_p} \iint_A \frac{P_q^t\, G_t(\theta_a^t, \phi_a^t)\, G_r(\theta_a^r, \phi_a^r)\, \lambda^2}{(4\pi)^3 R_a^4}\, dA. \quad (16.55)$$

The precise computation for X_{cal} must include the effects of signal clipping due to range gates, signal loss due to filtering, and channel gain variation. Note that motion between pulses must be considered in computing X_{cal}.

Finally, given a noisy measurement $\widehat{C_s}$ of C_s, the estimated σ^0 is then

$$\widehat{\sigma^0} = \frac{1}{X_{cal}} \widehat{C_s}. \quad (16.56)$$

The measurements of $\widehat{\sigma^0}$ from different azimuth angles and/or polarizations and incidence angles form the input to the wind-scatterometer wind-estimation process. Associated with these σ^0 measurements are the geometry and K_p parameters required for wind estimation.

In deriving Eq. (16.53), it was assumed that σ^0 was constant. However, if σ^0 varies over the surface, the observed σ^0 can be written in terms of the system spatial response function as

$$\widehat{\sigma^0} = \frac{1}{X_{cal}} \iint_A S(x,y)\, \sigma^0(x,y)\, dA, \quad (16.57)$$

where $\sigma^0(x,y)$ is the spatially varying backscatter. Thus, the observed or measured σ^0 is a weighted spatial average of $\sigma^0(x,y)$ over the illuminated area where the weighting function is the spatial response function. The spatial variation in σ^0 may arise due to gradients in the wind field, wind fronts, and the combination of the variation in incidence angle over the footprint and the GMF dependence on incidence angle.

While concentrated in a local area (the measurement footprint), the spatial response function can be nonzero over a larger area. The observed σ^0 value can include attenuated contributions from areas some distance away from the measurement center. This can be important for σ^0 measurements collected near icebergs or sea ice due to the large σ^0 values of ice. Measurements near coastal regions can also be affected since the σ^0 of land surfaces may be much larger than the σ^0 of the ocean, particularly under low wind speed conditions. In such cases the observed value is a weighted average of the ocean and land contributions,

$$\widehat{\sigma^0} = \frac{\iint_{\text{ocean}} S(x,y)\,\sigma^0_{\text{ocean}}(x,y)\,dA}{\iint_{\text{ocean}} S(x,y)\,dA} + \frac{\iint_{\text{contam}} S(x,y)\,\sigma^0_{\text{contam}}(x,y)\,dA}{\iint_{\text{contam}} S(x,y)\,dA}, \tag{16.58}$$

where the contamination may be due to ice or land. Contaminated σ^0 measurements are those whose ice or land contribution is sufficiently large as to cause a significant error in the wind estimate when the σ^0 measurements are used in wind estimation. Such measurements must either be discarded or corrected prior to wind estimation. A general mask can be defined to discard measurements within a fixed distance from the coast or sea ice, or the expected contamination can be evaluated and the measurements discarded if the estimated contamination exceeds a wind-dependent threshold. An appropriate threshold can be computed by evaluating the wind error versus the contamination (Owen and Long, 2009).

16-7.3 Dealing with Rain

Globally, approximately 7% of wind scatterometer measurements over the ocean may have some influence from rain. Rain affects the estimated surface σ^0. Since most wind scatterometers to date have operated at Ku- or C-band, we focus on the effects of rain at those frequencies. While Ku-band is more sensitive to rain, measurements at C-band also are affected. If the effects of rain on σ^0 are sufficiently large, the wind estimation accuracy is degraded, and it may be impossible to estimate the wind vector with useable accuracy when the rain effects overwhelm the wind signal in the σ^0 measurement. At both C- and Ku-band, the sensitivity of the wind error due to rain is dependent on incidence angle, with reduced sensitivity at lower incidence angles (Figa and Stoffelen, 2000; Nie and Long, 2008).

There are four major options for dealing with rain effects in wind estimation:

(1) Detect the presence of rain and provide a "flag" to indicate it. The wind estimate can then be discarded as rain contaminated, or rain-contaminated σ^0 measurements can be discarded prior to estimating the wind.

(2) Estimate the rain parameters and "correct" the σ^0 measurements to the values that would be observed without the presence of rain so that conventional wind estimation can be used.

(3) Simultaneously estimate wind and rain using a wind/rain GMF.

(4) Use an empirical approach that estimates the winds without explicitly estimating rain parameters.

All of these approaches can benefit from collocated radiometer measurements. Such radiometers have been available on some wind scatterometer platforms, but not on all. Note that scanning pencil-beam wind scatterometer systems can use their noise-only measurements as a radiometer, which can help with rain detection (Ahmad et al., 2005, 2006). Rain can be detected in collocated radiometer measurements because it has a higher brightness temperature than that of the ocean. Rain can also be detected in wind retrieval as poor fits to the GMF, which can be determined by examining likelihood values (Stoffelen and Anderson, 1997; Figa and Stoffelen, 2000; Portabella and Stoffelen, 2001, 2002; Mears et al., 2000; Anderson et al., 2012).

Approach (1) avoids rain by discarding data. Assuming accurate identification of rain, discarding measurements avoids the issue of rain contamination altogether. Unfortunately, rain is often associated with meteorological conditions of interest such as hurricanes and other storms and so discarding rain contaminated data limits the study of these cases (Huddleston and Stiles, 2000).

Approach (2), σ^0 correction, requires auxiliary data, such as collocated radiometer measurements, in order to infer the rain parameters to compute correction factors due to signal attenuation, rain backscatter, and surface scattering (Huddleston and Stiles, 2000; Tournadre and Quilfen, 2003; Hilburn et al., 2006; Hristova-Veleva et al., 2006). The integral radiometer measurements of scanning pencil-beam wind scatterometer systems can provide independent rain estimates (Ahmad et al., 2005; 2006)

Approach (3), ***simultaneous wind/rain estimation***, can be done with scatterometer-only data. It has the advantage of estimating both rain and wind. The rain estimates can be used as a rain flag. The method requires sufficient azimuth diversity for the σ^0 measurements and so it cannot be applied over the full swath of scanning pencil-beam systems (Draper and Long, 2004a, b), and for C-band, it only can be applied at incidence angles greater than about 40° (Nie and Long, 2008). Due to noise in the scatterometer measurements, low rain rates cannot be accurately estimated and are discarded (Nielsen and Long, 2009).

Approach (4) attempts to estimate the wind vector under all conditions. Neural nets have been employed to implement this approach, which is still in an early stage of development (Stiles et al., 2006; Stiles and Dunbar, 2010).

16-8 Wind-Vector Retrieval

Once the wind scatterometer has collected multiple σ^0 measurements from a diverse geometry at each sample point within the swath, the measurements are used to estimate the near-surface wind vector. This is sometimes termed ***wind retrieval***. In effect, the GMF is inverted to find the wind vector that "explains" the observed σ^0 values.

Typically, the σ^0 measurements are collected on, or gridded into, a uniform rectilinear grid aligned with along-track and cross-track. A grid spacing of 25 km is commonly used. Gridding may be the "drop in the bucket" type where a given σ^0 measurement is assigned to the grid element in which its center falls. Alternatively, signal reconstruction techniques may be employed (Williams and Long, 2011a), or the measurements may be spatially interpolated to the grid element centers (Wilson et al., 2005).

Ideally, within each grid element, known as a ***wind-vector cell*** (WVC), multiple σ^0 measurements are collected, each measured from a different azimuth angle and/or incidence angle and/or polarization. This diversity, the details of which may vary over the swath, is critical to the performance of the wind estimation algorithm. In practice, the σ^0 measurements are not perfectly aligned with the grid and provide an irregular sampling of the surface σ^0.

An ideal measurement $\widehat{\sigma^0}_{\text{ideal}}$ of σ^0 exactly covers the WVC area so that

$$\widehat{\sigma^0}_{\text{ideal}} = \frac{\displaystyle\iint_{\text{WVC area}} \sigma^0(x,y)\, dA}{\text{WVC area}} \,. \tag{16.59}$$

That is, the ideal observed σ^0 is the average σ^0 over the WVC area. However, the measured σ^0 given by Eq. (16.57) is a weighted area average. Further, the spatial response footprint area does not generally match the WVC area. For example, Fig. 16-41 illustrates the 3 dB footprints of the measurements for a particular WVC for the SeaWinds scatterometer. When σ^0 varies spatially over the WVC area, the ideal and measured σ^0 values may not match. This can lead to small errors in the estimated wind in the presence of spatial wind gradients.

16-8.1 Noise-Free Retrieval Algorithm

In order to provide insight into the wind estimation process, we first consider the problem of determining the wind speed and direction from noise-free σ^0 measurements. At a given WVC, several σ^0 measurements are collected. These are related to the wind vector by the GMF. Given these measurements and their geometry and polarization, we want to infer the wind vector that "explains" the observed σ^0 values, i.e., that yields the same σ^0 values as the observations when the GMF is evaluated at the same geometry and polarization as the observations for the wind speed and direction. This is illustrated for several common geometries in Figs. 16-43 through 16-45. For example, Fig. 16-43 considers the two-beam case used by SASS (Fig. 16-25) where the azimuth angles are 45° apart. In Fig. 16-43 each of the four panels considers a different wind direction. To generate a particular panel, the σ^0 observations for each antenna beam are determined by evaluating the GMF at a 10 m/s wind speed and the particular wind direction considered. These are the noise-free σ^0 measurements. Using these measurements, the curves of wind speeds and directions that yield the same σ^0 are computed using the GMF and geometry. This sinusoidal-like curve representing speed versus direction is plotted for each σ^0 measurement. The wind speed and direction where the curves corresponding to the two azimuth measurement directions intersect, yields a consistent estimate of the wind vector. Note, however, that with only two azimuth angles there are 2–4 intersections. Given only the σ^0 measurements, we cannot determine which of the multiple intersections is the correct one: there is an intrinsic ***ambiguity*** in the wind estimate.

To help identify the location of the ambiguities, the dashed line illustrates the mean-squared difference between the curves. This is the same as the ***least-squares objective function*** L_{LS} described later. Ambiguities occur at the local minima of the objective function and the height of the objective function at these minima can be used to evaluate the "goodness" or ***likelihood*** that the particular ambiguity corresponds to the true wind.

It is important to note that the number and distribution of the ambiguities varies with the number of measurements and their geometry. While Fig. 16-43 illustrates the two-azimuth angle case, Figs. 16-44 and 16-45 consider three-azimuth angle cases. Figure 16-44 uses an NSCAT-like geometry (see Fig. 16-24) where the center beam is at an azimuth angle of 60°, while Fig. 16-45 considers a center beam at a 90° azimuth angle, which is the geometry used by the ERS-1/2 and ASCAT scatterometers (see Fig. 16-24). The additional antenna beam reduces the number of exact intersections, sometimes to one. However, as suggested by the objective functions, there are still near-intersections corresponding to local minima of the objective function. We observe that there are fewer local minima when the center beam is at 90° rather than at 60°, which suggests that the 90° center beam azimuth angle is superior for wind retrieval (Stoffelen and Anderson, 1997). As previously noted, this cannot be done with a fan-beam Doppler scatterometer; it requires a ranging scatterometer

Although not illustrated, a scanning pencil-beam system results in a variable azimuth geometry over the swath. This azimuth geometry also yields multiple intersections of the solution curves with similar characteristics as observed in the fan-beam case, but with different distribution of intersections.

16-8.2 Retrieval in the Presence of Noise

The analysis presented so far does not consider the effects of noise in the σ^0 measurements. As a result, the simple approach described in the earlier subsection cannot be used as is. Intuition can be gained by noting that noise added to the true σ^0 values has the primary effect of shifting the curves shown in Figs. 16-43 through 16-45 up or down, with a lessor impact on the curve amplitude. Depending on the number of beams, shifting the curves up or down has the effect of shifting the intersections to different directions and speeds. Shifting vertically can also eliminate and create intersections. Further, the values of the local minima of the objective function are altered, which leads us to retain all of the local minima of the objective function

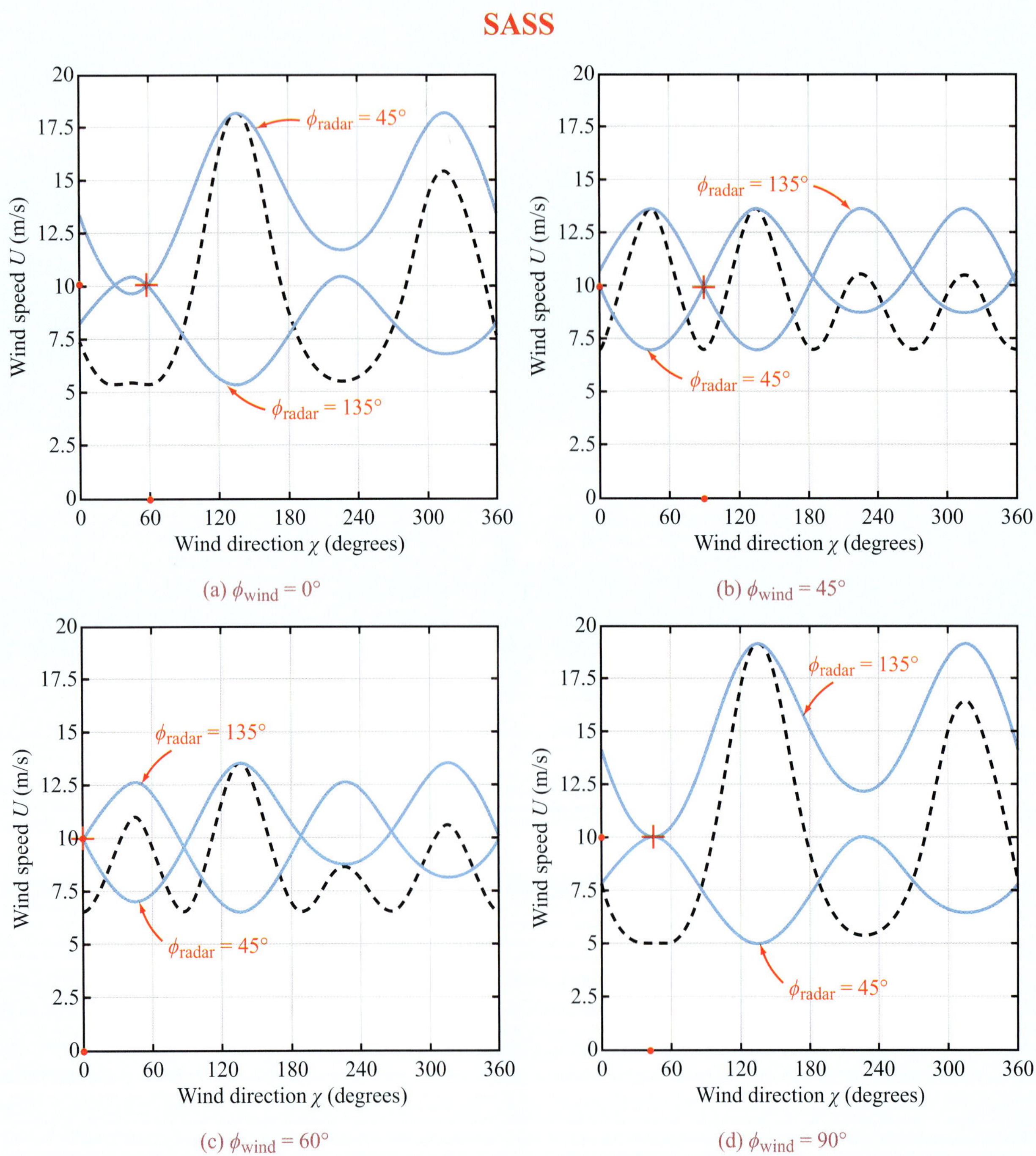

Figure 16-43: Simplified wind estimation examples using the SASS GMF. Solid lines are wind speed versus relative wind direction for noise-free σ^0 values corresponding to a 10 m/s wind speed at various wind directions as observed from each antenna beam at a 20° incidence angle. The dashed line is the arbitrarily scaled L_{LS} objective function versus wind direction evaluated at the minimum speed. This case is for a two-measurement geometry similar to SASS with two antennas at azimuth angles of 45° and 135°.

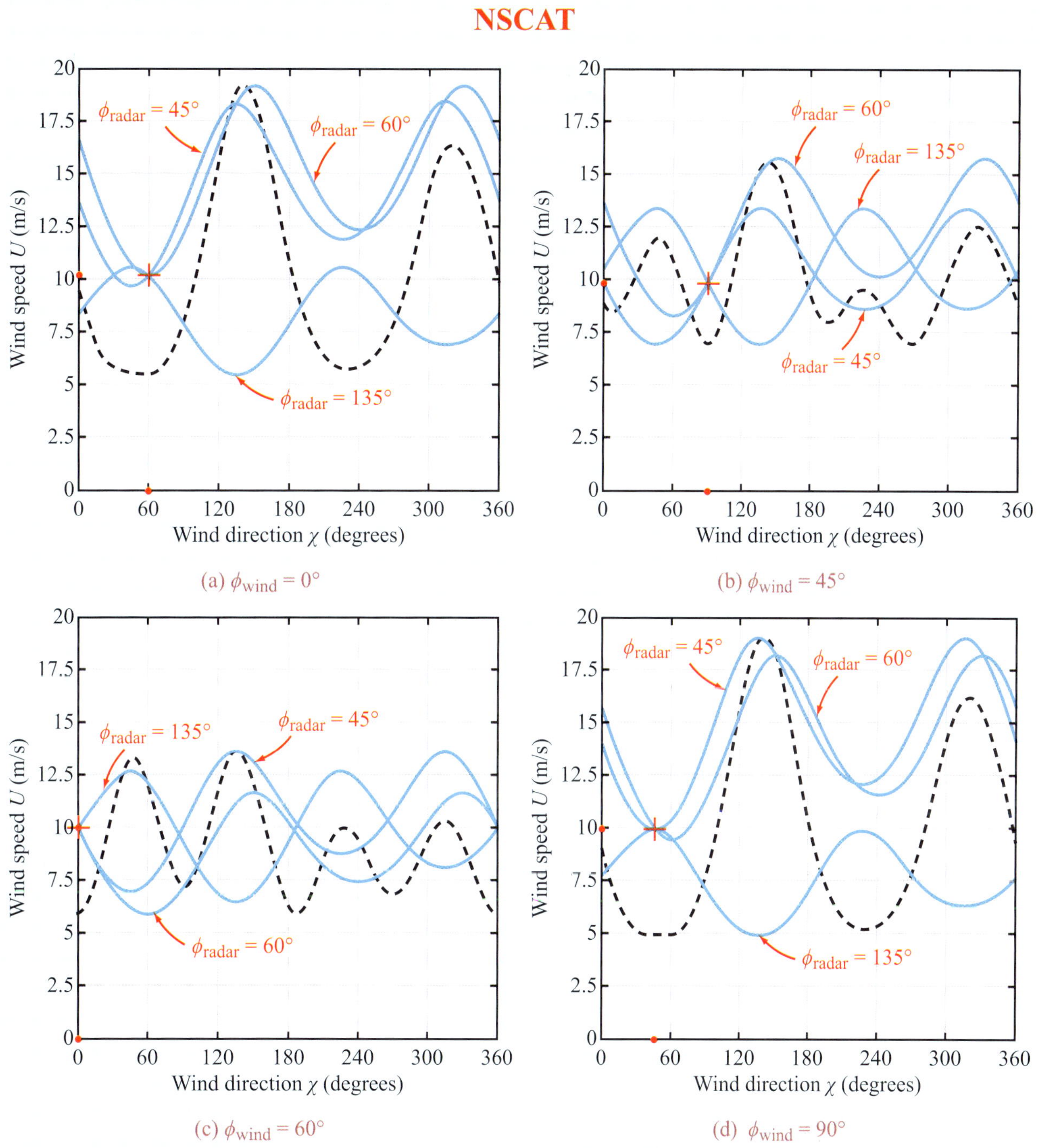

Figure 16-44: Same as Fig. 16-43 but for an NSCAT-like geometry: three antennas at azimuth angles of 45°, 60°, and 135°.

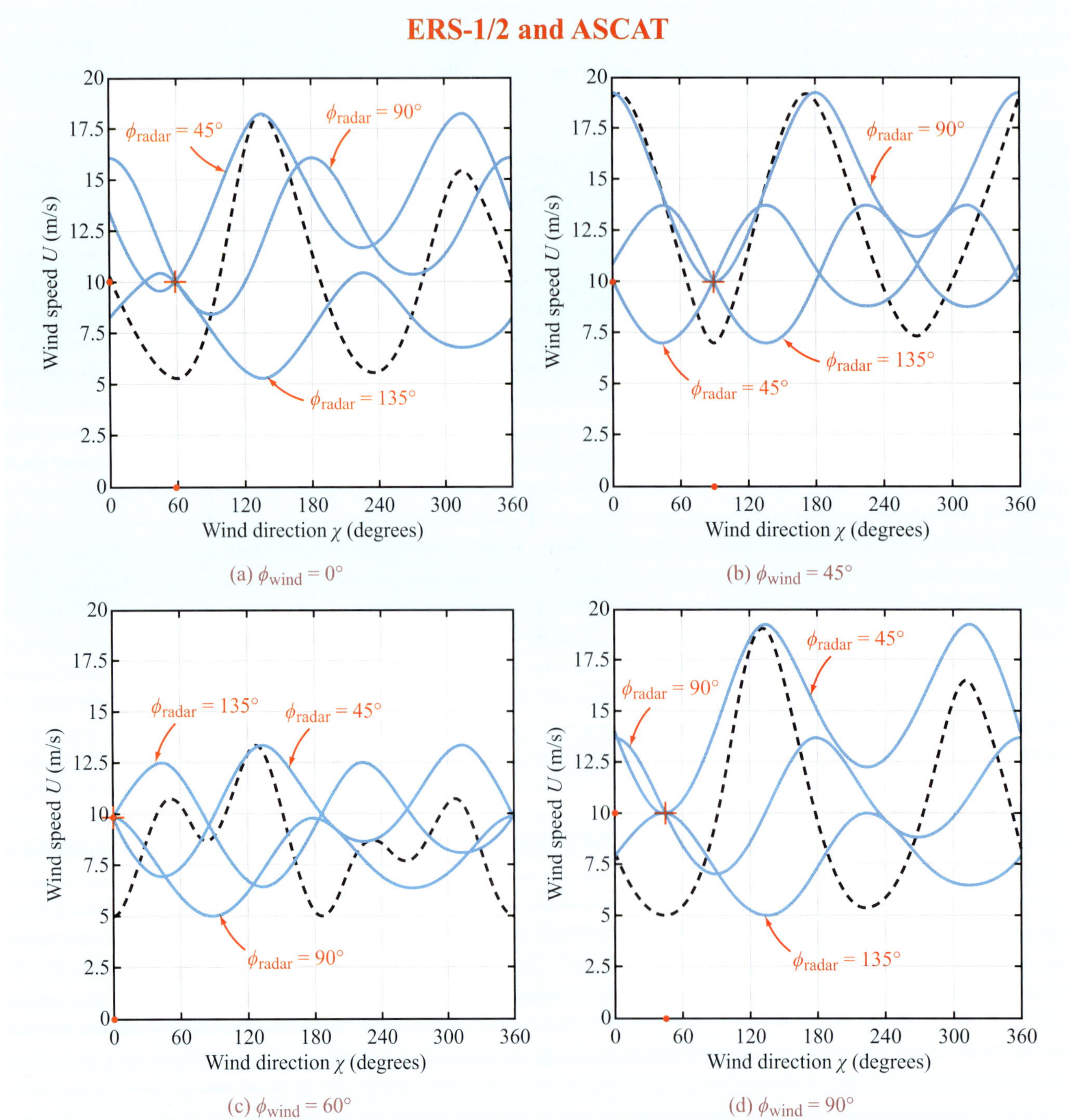

Figure 16-45: Same as Fig. 16-43 but for an ERS-1/2 and ASCAT geometry: three antennas at 45°, 90°, and 135°.

as possible wind solutions (ambiguities)—even those far from the global minimum.

Over the years, a number of approaches to estimating the wind from the σ^0 measurements have been developed based on minimizing an objective function comparing the measurements with the GMF (Chi and Li, 1988). Most of these are based on maximum-likelihood or least-squares estimation techniques. In the least-squares approach, an objective function $L_{\rm LS}(U,\chi)$ given by the sum-squared error between the measurements and the GMF evaluated at a particular wind speed U and direction χ is computed,

$$L_{\rm LS}(U,\chi)=\sum_{i=1}^{N_{\rm wvc}}(\sigma_i^0-\mathcal{M}(U,\chi,\phi_i,\theta_i))^2,$$

where $N_{\rm wvc}$ is the number of σ_i^0 measurements available at a particular location. Examples of $L_{\rm LS}(U,\chi)$ are shown in Figs. 16-43 to 16-45.

Because the objective function has multiple minima, termed ***wind ambiguities***, we cannot uniquely report the wind, but must retain all the solutions. Scatterometer wind estimation is thus a two-step process: first, optimization of the objective function to determine the wind ambiguities and second, selection of a unique ambiguity known as the ***unique wind estimate***. This step is termed ***ambiguity selection***.

Maximum-likelihood techniques rely on statistical models for the σ^0 measurement variability (Fischer, 1972; Long et al., 1988; Pierson, 1989; Long and Mendel, 1991; Yoho and Long, 2003; Portabella and Stoffelen, 2004) from which an objective function is formed (Chi and Li, 1988; Portabella and Stoffelen, 2006). The wind estimates are the wind vectors that minimize the objective function. Like the least-squares approach, maximum-likelihood estimation produces multiple wind solutions that must be resolved using an ambiguity selection algorithm (Schroeder et al., 1985; Shaffer et al. 1991; Stiles et al., 2002; Portabella and Stoffelen, 2004; Hoffman, 1982, 1984; Buehner, 2002).

The details of the signal models, wind estimation techniques, and ambiguity selection algorithms are outside of the scope of this book; the reader is referred to the cited literature.

16-9 SAR Imaging of Ocean Winds

Imaging the ocean using SAR is a complex process and has led to many theoretical attempts to explain the observed images. The fundamental electromagnetic scattering problem has been largely solved; the difficulties that remain are related to the modeling of the motion of the ocean surface and its two-dimensional description.

SAR images of the ocean clearly exhibit waves and wind-generated features, and because of their high resolution there is interest in using SAR data to study meso-scale wind features. Unlike wind scatterometers, which are designed to measure the backscatter of the surface from multiple azimuth angles, SAR backscatter is observed along essentially a single direction. Since the wind GMF exhibits a nonunique dependence of backscatter on azimuth angle, determining the wind speed and direction using SAR data entails a different approach from that used with wind scatterometers.

Fortunately, the high resolution of SAR images can often permit inference of the wind direction from wind-generated features in the image such as wind streaks, wind shadows, boundary layer rolls, or wave patterns. Wind direction can also be inferred from the estimated directional spectrum, though this generally results in a 180° directional ambiguity. Scatterometer-derived wind directions can be used and wind direction can be derived from the storm phenomenology such as in a hurricane when the local wind direction is known relative to the location of the eye. Once the wind direction is determined, the wind speed is estimated from the SAR backscatter image using a wind GMF. Accurate wind measurements require careful calibration of the SAR backscatter measurements since the observed σ^0 is very sensitive to wind speed and direction.

Figures 16-46 through 16-48 illustrate wind features in SAR images. In Fig. 16-46 the wind flow is from upper-left to lower right. Islands block the flow, resulting in low wind speed regions in their lee, and accelerated wind speeds between the islands. The wave-like patterns are the results of boundary layer turbulence modifying the surface σ^0. Figure 16-47 illustrates different σ^0 values across an atmospheric front due to different wind stresses on either side of the front. An example of a SAR image of a hurricane is shown in

Figure 16-46: C-band SAR image over the Aleutian Islands. Wind flows from upper left to lower right. Wind shadows result in dark areas in the lee of the islands with accelerated winds between mountains and islands. Boundary layer turbulence produces a wave-like signature in the surface backscatter downwind of the islands. [Data courtesy of Canadian Space Agency.]

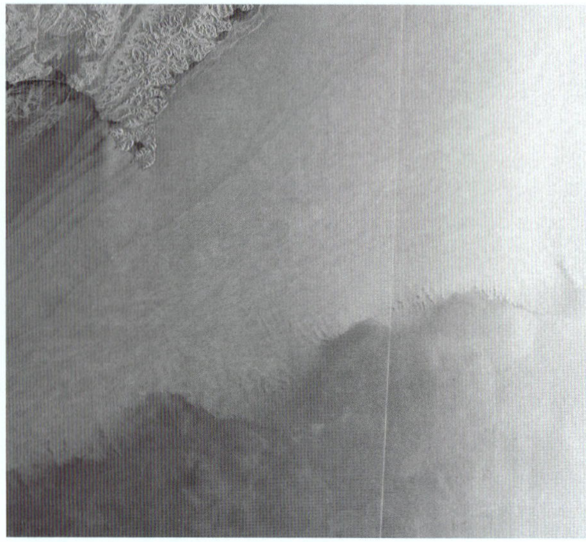

Figure 16-47: C-band SAR image of the ocean near the coast. Wind streaks extend from upper right to lower left. Note the wind shadows from coastal features. An atmospheric front extends horizontally in the lower half of the image, with different wind conditions on each side, resulting in a backscatter brightness difference in the image. Turbulence near the boundary generates atmospheric waves that produce a wave-like signature in the surface backscatter. [Data courtesy of Canadian Space Agency.]

Fig. 16-48. The hurricane eye is visible as a low σ^0 region, with higher σ^0 values surrounding it. The effects of rain on the SAR signal due to attenuation and surface modification are visible as an irregular arc at the image center. To the right, the surface is darker due to lower speeds. Land is contained at the upper right.

An example of an operational wind product derived from SAR data is shown in Fig. 16-49, which shows wind speeds derived from C-band hh-pol RADARSAT data. The wind direction used to retrieve the speed is derived from numerical weather prediction products, shown as arrows in the image. Alternatively, wind directions are determined from scatterometer data. At each image pixel, the SAR-observed σ^0 is used with the assumed wind direction to "lookup" (invert) the wind speed from the model function. In this image, the general wind flow is from the northwest toward

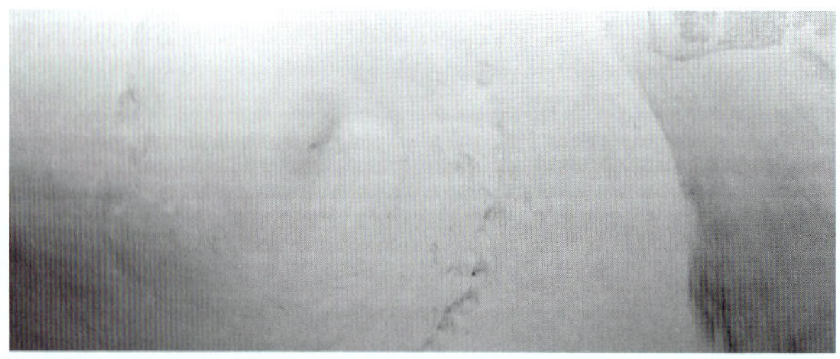

Figure 16-48: C-band SAR image of Hurricane Isabel off the East Coast of the U.S. The hurricane eye is visible at the center-left. The dark cloud-like features are convective rain events. Wind streaks are visible in the ocean to the right. The wind direction is counterclockwise around the eye. [Data courtesy of Canadian Space Agency.]

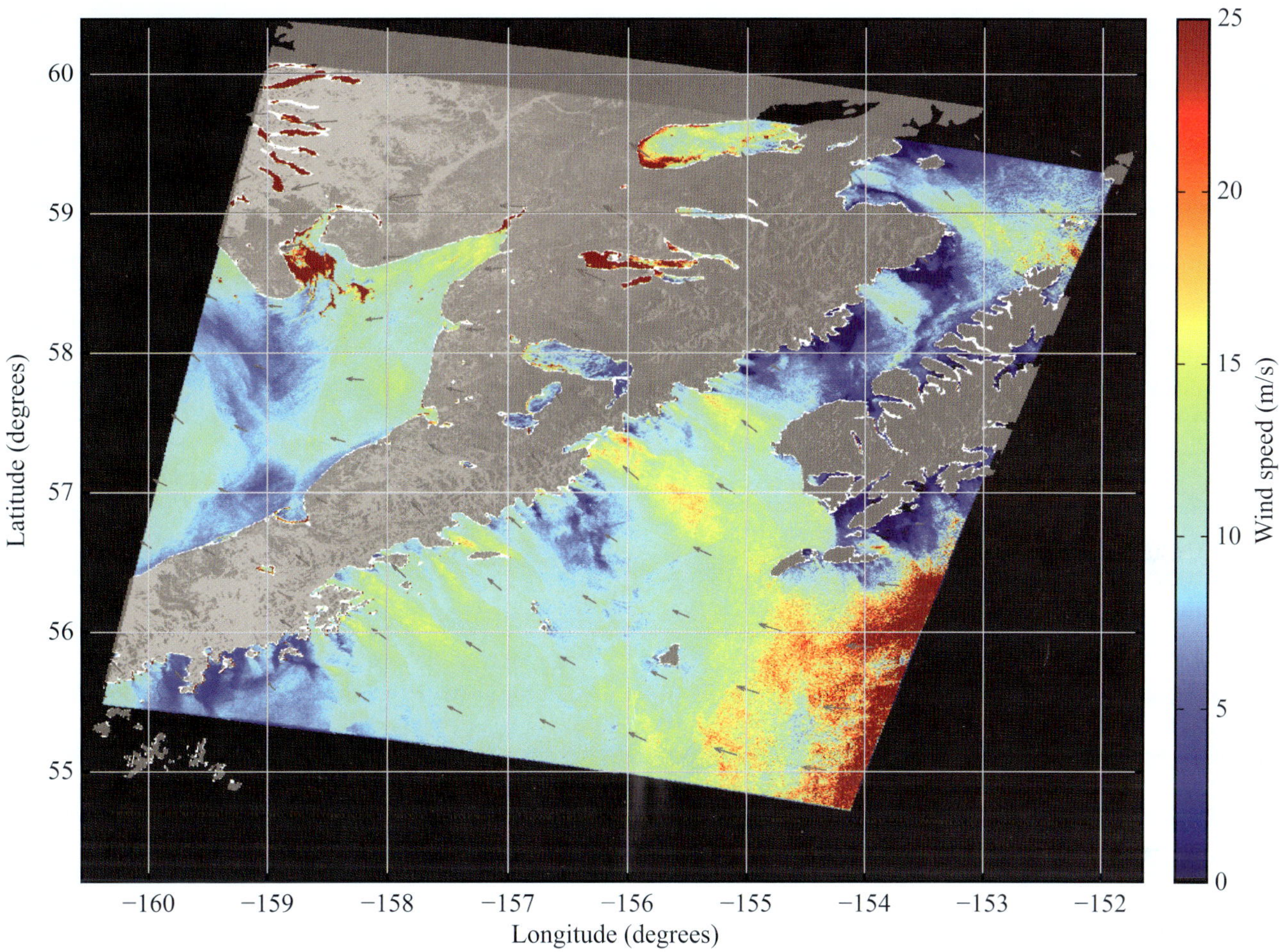

Figure 16-49: Winds derived from C-band SAR data south of Alaska. Black arrows (whose directions are according to the meteorological wind convention for which the arrow points in the direction from which the wind is blowing) indicate numerical prediction winds that have been interpolated to infer the wind direction. Speeds are derived from SAR σ^0 values and the assumed wind direction.

the southeast. Mountains along the Alaskan Peninsula block the wind in some areas, leading to low-wind speed shadows, and funnel the wind through passes, resulting in high-speed wind jets. Such images are regularly produced from several operational SARs such as RADARSAT-2 and TerraSAR-X.

16-10 Properties of Sea Ice

The ***cryosphere***, which includes sea ice, ice sheets, snow, ***glaciers***, and ***permafrost***, is the least known and least understood part of the Earth's surface (Campbell and Gudmansen, 1981). Approximately one-ninth of the world's oceans are covered by sea ice. Monitoring sea ice is important for both science applications and industries involved in exploration for and transportation

Table 16-3: Ice parameters of importance in various operations and research areas [Weeks, 1981].

Area of Interest	Pertinent Ice Parameters
Offshore operations	Extent, type, thickness, drift velocity, internal stress, properties (air temperature, atmospheric pressure, wind velocity, current velocity)
Climate	Extent, thickness
Albedo	Extent, type, snow cover
Insulation	Type, thickness, snow cover (air temperature, wind velocity)
Latent heat export	Thickness, drift velocity
Surface stress	Drift velocity, top and bottom ice roughness (wind velocity, current velocity)
Ocean mixed layer	Ice growth and ablation rates, drift velocity (current velocity, water-column stability)

Parameters in parentheses are also important, but are not directly related to ice.

of mineral and petroleum resources in the polar regions. Vessels navigating in polar seas are likely to be both safer and more efficient if they have access to information about the types and extent of sea ice they will encounter. The scientific community is particularly interested in mapping the distribution of sea ice because it is a sensitive indicator of climatic change and controls the transport of heat between the ocean and atmosphere (Weeks, 1981; Carsey et al., 1982).

A brief summary of the information requirements for sea-ice sensing is given in Table 16-3.

16-10.1 The Nature of Sea Ice

Sea ice and its formation are complicated, making the remote-sensing problem complicated. The main factors that influence the formation of sea ice are the ***brine content*** of the surface water (i.e., the water ***salinity*** and ***density***), the vertical distribution of salinity, the surface temperature, and the depth of the water. Because ice forms earlier in shallow water than in deep water, ice formation often starts near coastlines. If the brine density exceeds 24.7 percent, the water column must be cooled from top to bottom to temperatures below the normal freezing point of water before freezing begins at the top. This is a result of ***convection*** that takes place because the maximum density of water containing less than 24.7 percent brine occurs at a temperature above freezing. Once freezing starts, the initial ice cover forms without regard to salinity. However, ice formation and its rate are influenced by factors such as wind speed, ocean currents and waves, and the rate of cooling.

When freezing takes place depends on environmental conditions. If the surface is calm and cooling is rapid, the ice forms in small crystals known as ***frazil*** ice. According to Weeks and Assur (1969), the first crystals of pure ice to form are minute spheres, which rapidly change into circular disks. The maximum diameter to which these disks grow varies, but it is of the order of a few mm. The disks then change to dendritic hexagonal stars, and the star crystals grow rapidly and join together to form a relatively uniform sheet of ***young ice***.

When the surface is disturbed by waves or currents, the process is somewhat different. Initially, the cooling water surface takes on an oily, opaque appearance, at which point ***grease ice*** is said to exist. Upon further freezing, grease ice develops into ***nilas*** or ***ice rinds***, the former appearing as an elastic crust and the latter as a brittle, shiny crust. When the wind is blowing, the ice develops into masses as a result of water motion. Freezing occurs in the center of the masses, and then spreads outward to form circular, flat, ice disks up to 10 cm in diameter. These disks typically have raised edges and rounded shapes resulting from collisions. As the disks grow together, they form ***pancake ice***. In due time, they merge to form continuous sheets less than 30 cm thick. During the growth phase the thickness may exceed 10 cm in less than 48 hours, after which the growth rate slows. Weeks and Assur (1969) state that the main body of an ice sheet then grows in columnar form known as ***congelation ice***, which is composed of vertical ice crystals that gradually increase in size with depth. Wind, ocean currents, waves, and pressure usually cause some deformation and drifting of the ice, resulting in a crowding together of the ice in some places and dispersal in others. Horizontal pressure may cause the sheets of ice to pile up in ***ice ridges*** and ***hummocks***. Moreover, one piece of ice may override another, a process known as ***rafting***. ***Ridging*** occurs when the sea-ice sheet bends or fractures and piles up on itself forming a ***ridge line*** up to 6 m above the surface, which is supported by a ***keel*** up to 25 m below the surface consisting of blocks of ice pushed downward in the ridging process.

During the growth of the ice crystals, brine is expelled. The brine is colder and denser than the surrounding seawater and much of it is expelled below and sinks. Some of the brine is trapped in the forming ice and concentrates into tiny vertically oriented pockets within the ice. The brine in the pockets is much more concentrated than the brine in sea water, and the freezing point of the pockets is far lower than that of sea water. During the summer, the ice melts near the surface, and the brine dissipates in the upper ice layers, so second-year and multiyear ice generally have much lower salinities near the surface than does newer ice. ***Multiyear*** (MY) ice is generally much thicker than ***first-year*** (FY) ice.

Selected examples of photographs of various sea ice types are shown in Fig. 16-50. A more comprehensive atlas of ice types organized by Bushuyev et al. (1974) shows more photographs of the various types of ice. Schematic diagrams of some of the principal types of sea ice are shown in Fig. 16-51. Note the summer deformation that occurs in MY ice.

While the terminology for describing sea ice is complex, fortunately for remote sensing most of the semantic complication occurs with regard to very thin types that are of little importance to navigation and offshore operations. The most important impact on the energy transfer between ocean and atmosphere is the mere occurrence of a sea-ice cover. In many

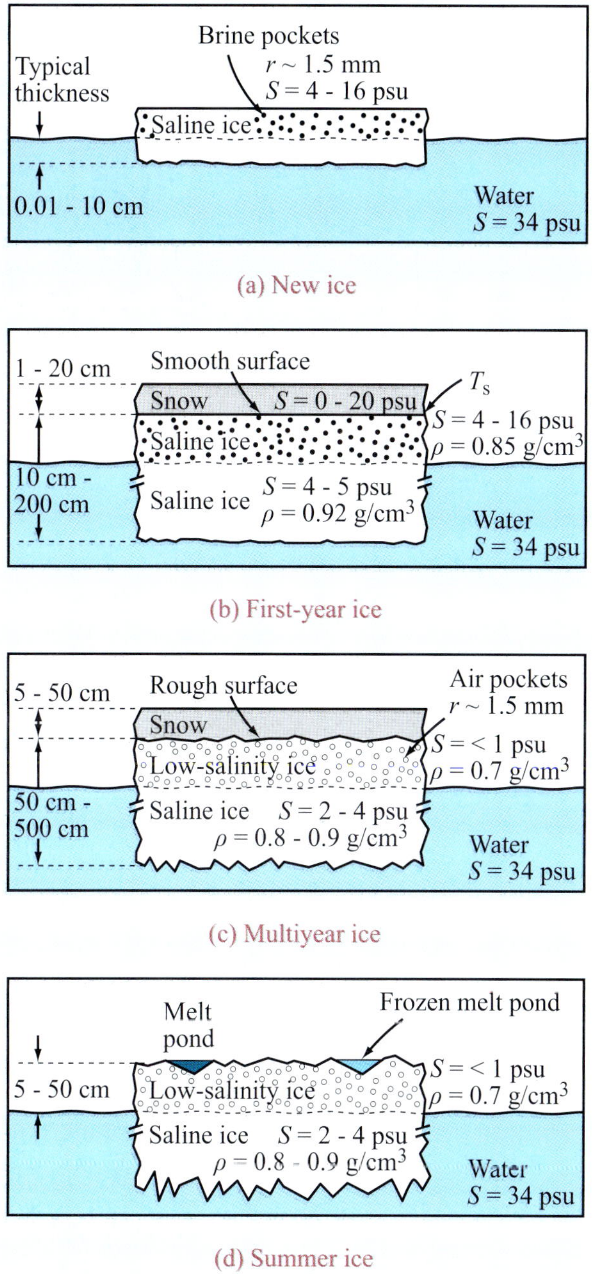

Figure 16-51: Schematic representation of some principal ice types [from Zwally et al., 1983].

(a) Pancake ice (b) Nilas and smooth ice (c) Overlapping nilas

(d) Iceberg surrounded by pancake ice (e) Snow-covered sea ice with melt ponds and sastrugi (f) Open water and various ice types

Figure 16-50: Photographs of various types of sea ice. [Courtesy of the National Snow and Ice Data Center: Ted Scambos, Terry Haran, Alice O'Connor, Rob Bauer.]

applications, it is often not essential to be able to distinguish between the many types of thin ice so long as the remote sensor can recognize the presence of ice. Ice thickness is an important parameter for navigation. Discrimination between the two major classes of sea ice, namely FY ice and MY ice, is adequate for many purposes.

Characterizing sea ice is complicated by its large area, inaccessibility, harsh weather, clouds and fog, and winter darkness. Furthermore, driven by wind and currents, most sea ice is in constant motion. Typical sea ice movements vary from remaining virtually stationary to traveling as much as 50 km per day (Allen and Long, 2006). As ice moves, it deforms, and creates ridges and cracks. ***Leads*** (open water between patches of floating ice) open and close. As ice sheets distort, ice ***floes*** form, rotate, break up, and reform.

The summer sea-ice situation is outlined in Figs. 16-52 and 16-53 for FY and MY ice. In Fig. 16-52, the first diagram depicts FY sea ice prior to melting, with the ice covered by a thin layer of snow. Initially, early summer heating by the sun moistens the snow near the surface. Liquid water that forms percolates into the snow and begins to form ***melt pools*** in and under the snow. The downward percolating water spreads out and freezes on contact with the cold lower ice, forming a rough layer of superimposed ice. As the air temperature rises, the snow melts completely, leaving a surface of superimposed ice that is much rougher than the winter surface of FY ice. In time, the surface of the ice melts, leaving extensive pools of water on the surface known as ***melt ponds***, along with small mounds of ice above the surface, where the superimposed ice is thickest. In late summer, ***thaw holes*** form that allow the water to drain away.

The MY-ice summer situation is illustrated in Fig. 16-53. MY ice usually is rough and hummocky during the winter as a result of the effects of the previous summer. Its surface contains relatively large bumps interspersed with flat surfaces where the previous summer's melt pools and ponds have refrozen. The entire surface is covered with snow of varying depths, which masks depressions in the ice containing the refrozen melt pools.

In early summer, the snow melts off the top of the hummocks, and water accumulates in the melt pools that had formed the year before. Later, the snow disappears and the melt pools deepen. Because of the premelt roughness of the MY ice, a smaller fraction of the area is covered in midsummer by melt pools than for FY ice; indeed, FY ice may be almost completely covered. This difference can have a significant effect on the relative changes in the radar response of the two classes of summer ice.

Pressure ridges are of great significance to Arctic operations, because of their large masses. A pressure ridge is defined by the World Meteorological Organization as a line or wall of broken ice forced up by pressure. The upper part is called the ***sail*** and the submerged part is called the ***keel***; keels are generally much larger than sails. A new ridge is usually made up of broken fragments of ice, and the slope of its sides usually equals or exceeds 40°. Weathered (particularly multiyear) ridges are more rounded, having side slopes between 20° and 40°. Voids exist between blocks in the newer ridges but are filled in by the melting process during the summer, so multiyear ridges are more solid. Keel depths to 25 m and sail heights to just under 6 m have been measured using submarine sonar and airborne lidar (Wadhams, 1979).

Total precipitation over sea ice is small, and most of the Arctic and Antarctic is a virtual desert. Nevertheless, snow does fall, and by spring, sea ice is covered with a layer of snow that varies from essentially no snow at all to 10 cm or so, with some tens of cm in wetter areas. Snow can have an important effect on radar backscattering, particularly during early summer, when

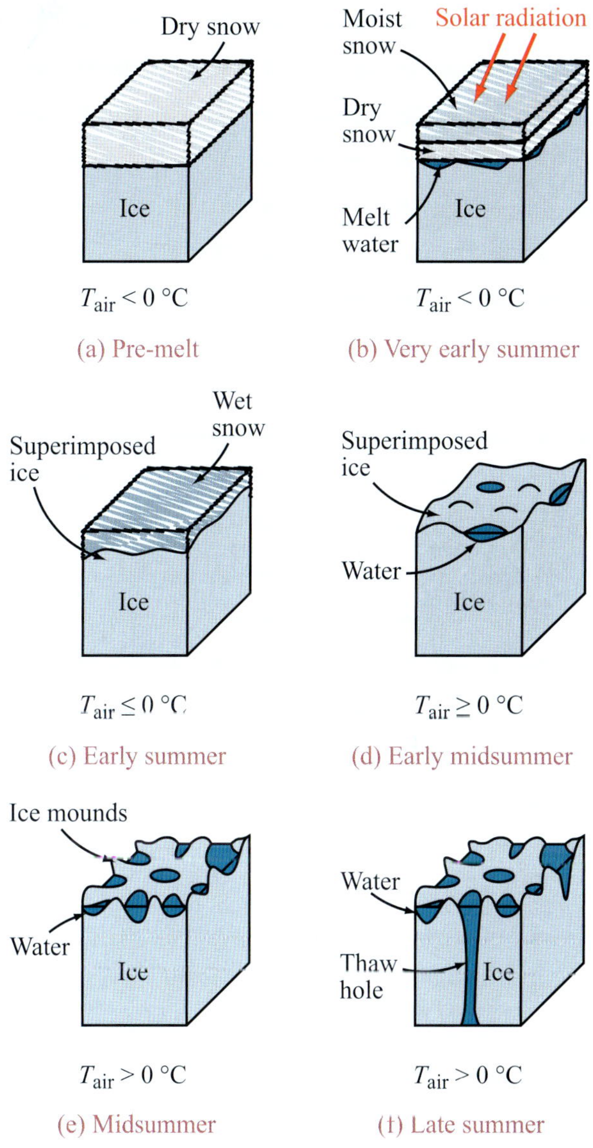

Figure 16-52: Schematic diagram of the changes in first-year ice during the summer melting season [from Gogineni, 1984].

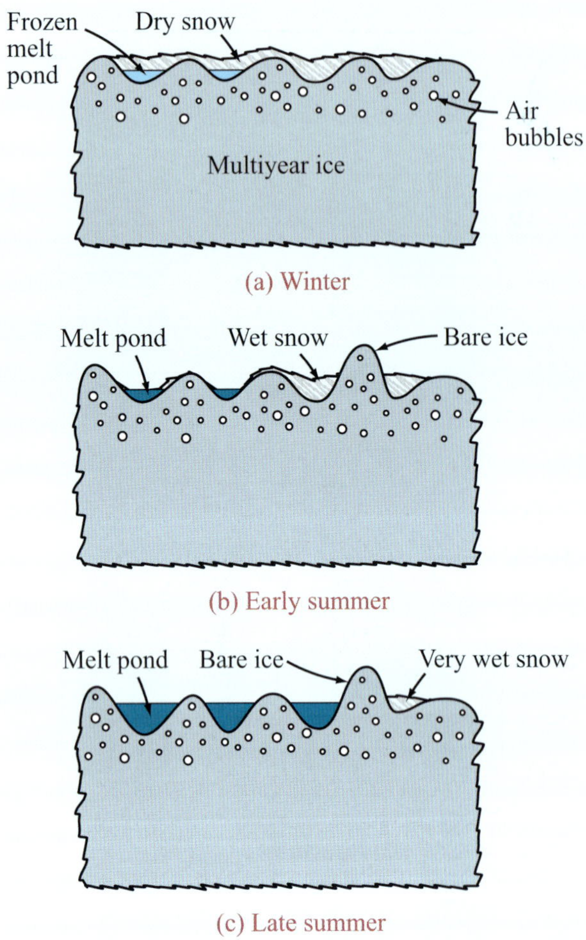

Figure 16-53: Schematic diagram of the changes in multiyear ice during the summer melt season [from Gogineni, 1984].

the snow begins to melt. The result can be confusion in the radar and passive-microwave signatures of ice types because of the obscuration of volume scattering by snow.

16-10.2 Physical Properties

The physical properties of sea ice are quite variable, as expected on the basis of the complex formation and modification processes involved. Nevertheless, some representative profiles provide an indication of the nature of the different kinds of ice. Figure 16-54 shows density profiles for first-year and multiyear ice based on *in-situ* observations north of Alaska. MY ice is much less dense near the surface than FY ice; this is in part because during the transformation into MY ice, the brine pockets are replaced with air-filled voids. The effect of the summer melting period (between the MY and the MY+1 measurement) is also clear.

The difference in salinity between FY and MY ice is apparent in the salinity profiles shown in Fig. 16-55. The salinity in the surface layer of FY ice is orders of magnitude higher than that of MY ice—or of FY ice at the end of its first melting season, when it has just become second-year ice.

16-10.3 Penetration Depth in Sea Ice

Because it is a mixture of ice, salt, brine pockets, and air bubbles, sea ice exhibits a complicated dielectric behavior (see Section 4-5). The relative dielectric constant of sea ice, ε_{si}, varies with several sea-ice

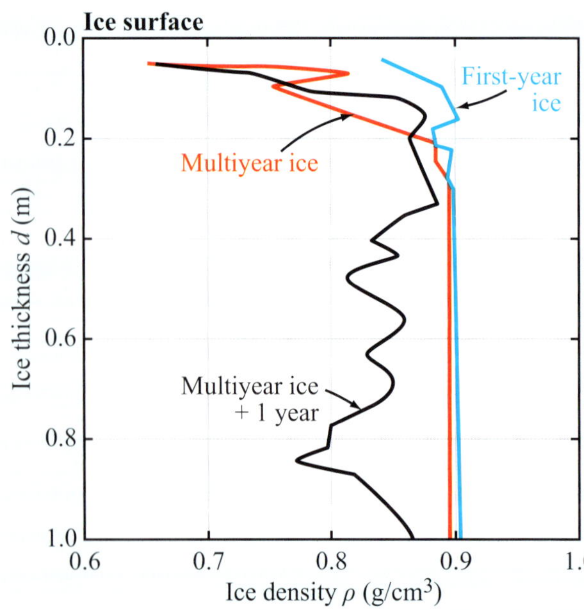

Figure 16-54: Density versus depth below the surface for FY and MY ice. The MY+1 curve was measured in the fall [from Campbell et al., 1978].

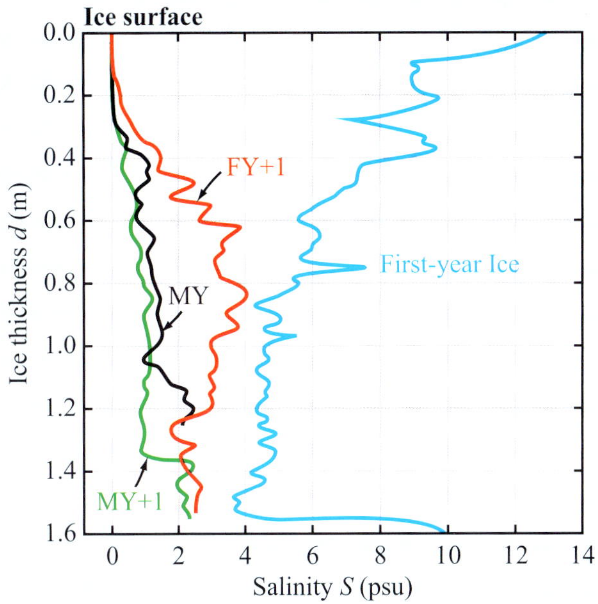

Figure 16-55: Salinity versus depth below the surface for FY and MY ice [Campbell et al., 1978].

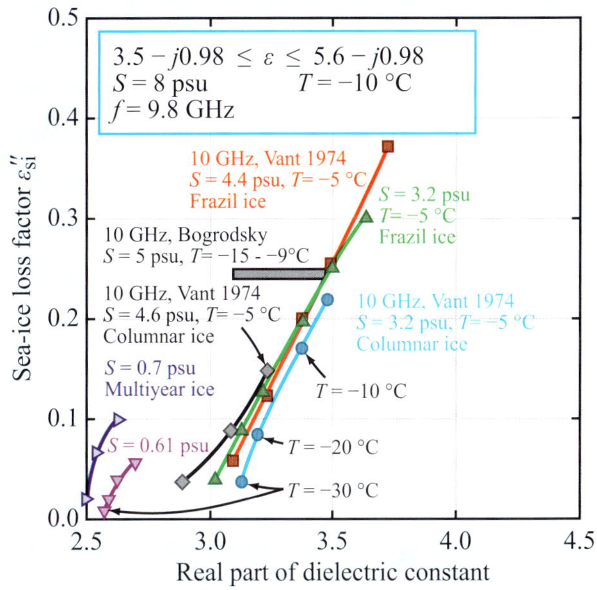

Figure 16-56: Measured values of the complex dielectric constant of sea ice [Kim, 1984].

parameters, most notably salinity and temperature. Furthermore, it is frequency dependent and influenced by the size, shape, and orientation distributions of the brine inclusions. It may also depend on the polarization configuration and direction of propagation of the incident wave (i.e., it may be anisotropic). Over the frequency range between 1 and 40 GHz, most measured values of the relative permittivity of sea ice fall in the relatively narrow range $2.5 \leq \varepsilon_{si} \leq 8$. In contrast, the dielectric loss factor ε_{si} covers a wide range of values, from less than 0.01 to more than 1.0. In general, ε_{si} (1) decreases rapidly with increasing negative temperature, (2) increases with increasing ice salinity, and (3) is typically 3 to 10 times larger for FY ice than for MY ice, because the salinity of FY ice is typically 5 times that of MY ice. Dielectric data measured at or near 10 GHz is summarized in Fig. 16-56.

The penetration depth of radar signals in sea ice is shown as a function of frequency in Fig. 16-57. This plot provides a general idea of the significant differences to be expected for FY and MY ice, and of the decrease in the penetration of the signal with increasing radar frequency. The dielectric properties of sea ice are shown in more detail in Figs. 4-12 through 4-16.

Penetration into the snow cover is illustrated versus the wetness of the snow in Fig. 16-58 for C-band and X-band frequencies. Clearly, penetration into dry snow is so great that the snow has little effect on sea-ice radar measurements made in winter. In the summer, however, the high attenuation of wet snow can cause the surface of the ice to be masked completely.

16-11 Sea-Ice Radar Scattering

From the description of sea ice given previously, it is apparent that sea ice has a complicated physical structure that varies considerably from winter to summer. As a result, its scattering properties exhibit considerable seasonable variation. In particular, the presence of liquid water in the ice or in its overlaying snow cover has a dramatic effect on the scattering characteristics.

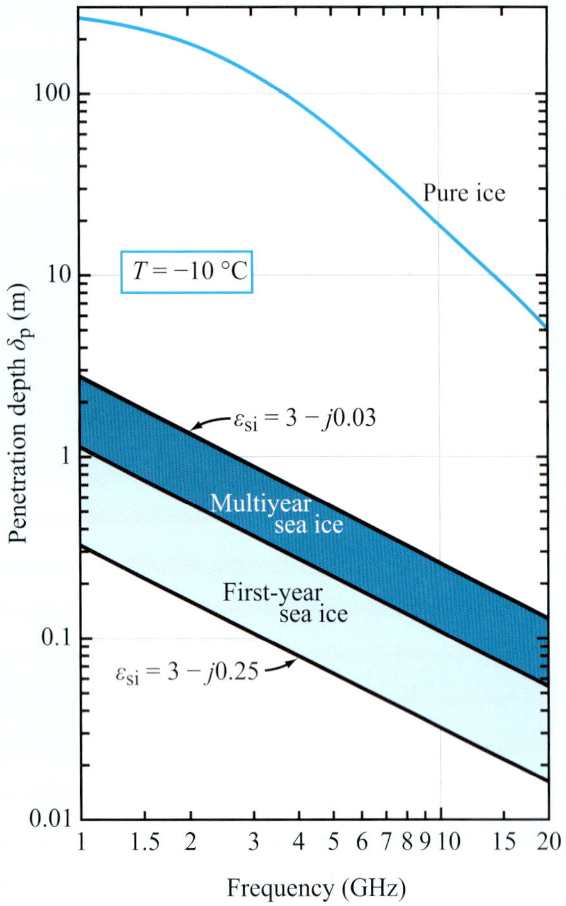

Figure 16-57: Penetration depth in pure ice, first-year sea ice, and multiyear sea ice computed using the model described in Chapter 4.

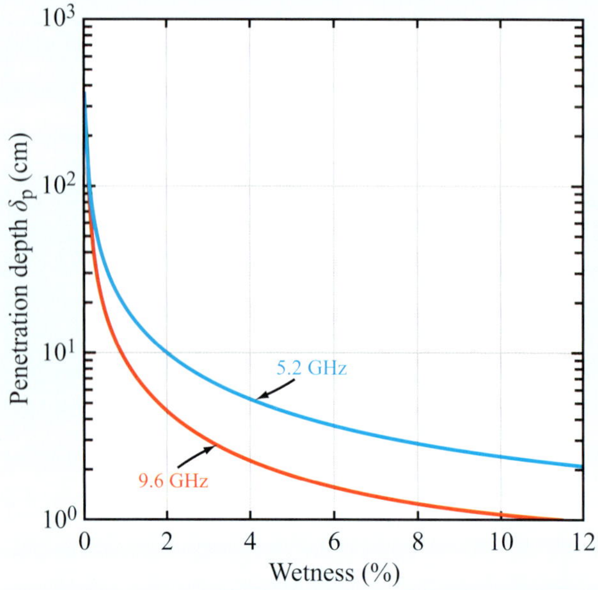

Figure 16-58: Depth of penetration in snow at C- and X-band frequencies as a function of snow wetness [Gogineni, 1984].

A few general comments about sea-ice scattering in different seasons are in order. Schematic diagrams of winter ice conditions were shown earlier in Fig. 16-51. New ice is very salty, with many brine pockets. It may or may not be smooth when it first forms, but usually it is quite smooth after it grows to a few cm thick. The result is a relatively large σ^0 for the very thinnest ice, where sharp and plentiful edges contribute strongly to volume scattering. As the ice grows beyond a few centimeters thick, off-nadir σ^0 decreases because the high conductivity and smooth surface of the ice cause most of the energy to be reflected away from the radar. Older ice tends to become roughened somewhat, so σ^0 increases with time. The high salinity in first-year ice assures that penetration is small, so essentially all of the scattering is from the surface. While the snow cover that accumulates during the winter contributes some volume scattering, and because the snow is dry, its low attenuation has little effect on the signal received from the ice surface.

Multiyear ice contains many small air bubbles or pockets in its upper layer in the spaces that were occupied by brine during its first year. The low salinity allows greater penetration, and volume scattering from the brine inclusions. At the higher frequencies, where the bubbles are not too much smaller than a wavelength, volume scattering is much larger than surface scattering. Moreover, the undulating surface of MY ice contributes significant scattering. Thus, the σ^0 of MY ice at the higher frequencies is usually much larger than that of FY and thinner types of ice. Because smooth surfaces do not depolarize much, the cross-polarized surface return from first-year types is quite low. However, the contrast between FY and MY ice at cross-polarization is enhanced because the depolarizing effect in the volume scattering present for MY ice is not matched by the surface scattering of FY ice.

In the summer, the presence of liquid water prevents the signal from penetrating either FY or MY ice, and a layer of wet snow can mask the ice return and contribute a return of its own. Initially, as shown in Fig. 16-51, the melting of the upper part of the snow on FY ice causes some increase in attenuation by, and backscattering from, the snow, but primarily it causes the surface of the ice to be roughened by the introduction of superimposed ice. During this period, the attenuation of the snow reduces the σ^0 of MY ice slightly.

When the snow on FY ice disappears, the rough surface of the superimposed ice is exposed. At this point, the σ^0 of the FY ice can temporarily exceed the σ^0 of MY ice. As melting progresses, the superimposed ice largely melts into sheets of water (melt ponds), resulting in a large decrease in σ^0 for the FY ice. Even though melting also occurs on the MY ice, as shown in Fig. 16-52, the water runs into the flat areas of previous years' melting, and most of the ice surface remains exposed; this maintains a large σ^0 for the MY ice until late summer when most of the surface becomes covered with melt ponds with low σ^0.

As melting progresses during the summer, the FY/MY scattering contrast returns to its normal order, with MY ice having a larger σ^0 than FY, though the contrast is less than during winter. At the start of winter, ice that survives the summer freezes again. The ice that had been FY ice has lost its salinity and behaves like MY ice, and a new FY ice begins to form.

First-year pressure ridges have steep slopes, high salinity, sharp corners, and voids in which multiple scattering can take place. The ridges show up reasonably well on radar images because, presumably, of the corner-reflector and multiple-scattering effects and because of their steep slopes. At low grazing angles, shadows are also apparent. Multiyear pressure ridges tend to be more rounded, have fewer large voids, and are not as steep-sided. Their scattering cross sections are more like those of the surrounding MY ice than of FY ridges. Accordingly, they provide a lower contrast, and they are more difficult to identify on radar images.

Open water (polynyas and leads) in areas of sea-ice cover vary in size and extent. As discussed earlier in this chapter, backscatter from open water depends on local wind conditions, but in general, ocean backscatter is lower than that for sea ice. When the radar sensor is unable to resolve individual open-water features, resulting in resolution cells containing a mixture of open water and sea ice, the observed backscatter may be increased or decreased compared with when the area is covered completely by sea ice.

Backscatter from sea ice has been studied extensively, using scatterometers and spectrometers, as well as SAR. Observations have confirmed radar's ability to detect sea ice, discriminate between major classes of sea ice (e.g., FY and MY), measure ice thickness, and map snow cover. This section provides illustrative examples of these applications.

16-11.1 Ocean versus Sea-Ice Discrimination

We first consider the use of radar for mapping sea ice against the open ocean. In Figs. 16-59 and 16-60 we display typical plots of the radar backscatter for FY ice, MY ice, and open water under calm conditions, for winter and summer conditions, respectively. Generally, backscattering from water is lower than from sea ice during the winter. In summer, when sea ice is often covered with melt ponds and wet snow, the distinction between sea ice and open water is lessened. We also note that at Ku-band frequencies, the angular roll-off of σ^0 with incidence angle is much steeper for water than for sea ice. The difference between sea ice and ocean backscatter is greatest at large incidence angles, which suggests that radars operating at large incidence angles can best differentiate sea ice and open water. Comparing the summer and winter plots, it is apparent that FY ice and MY ice have different signatures in winter than during the summer while the calm ocean σ^0 is similar year round.

A complication in differentiating ice from open water results from the fact that high winds roughen the water's surface and raise the water σ^0 for $\theta > 10°$. Windy conditions reduce the backscatter contrast between the

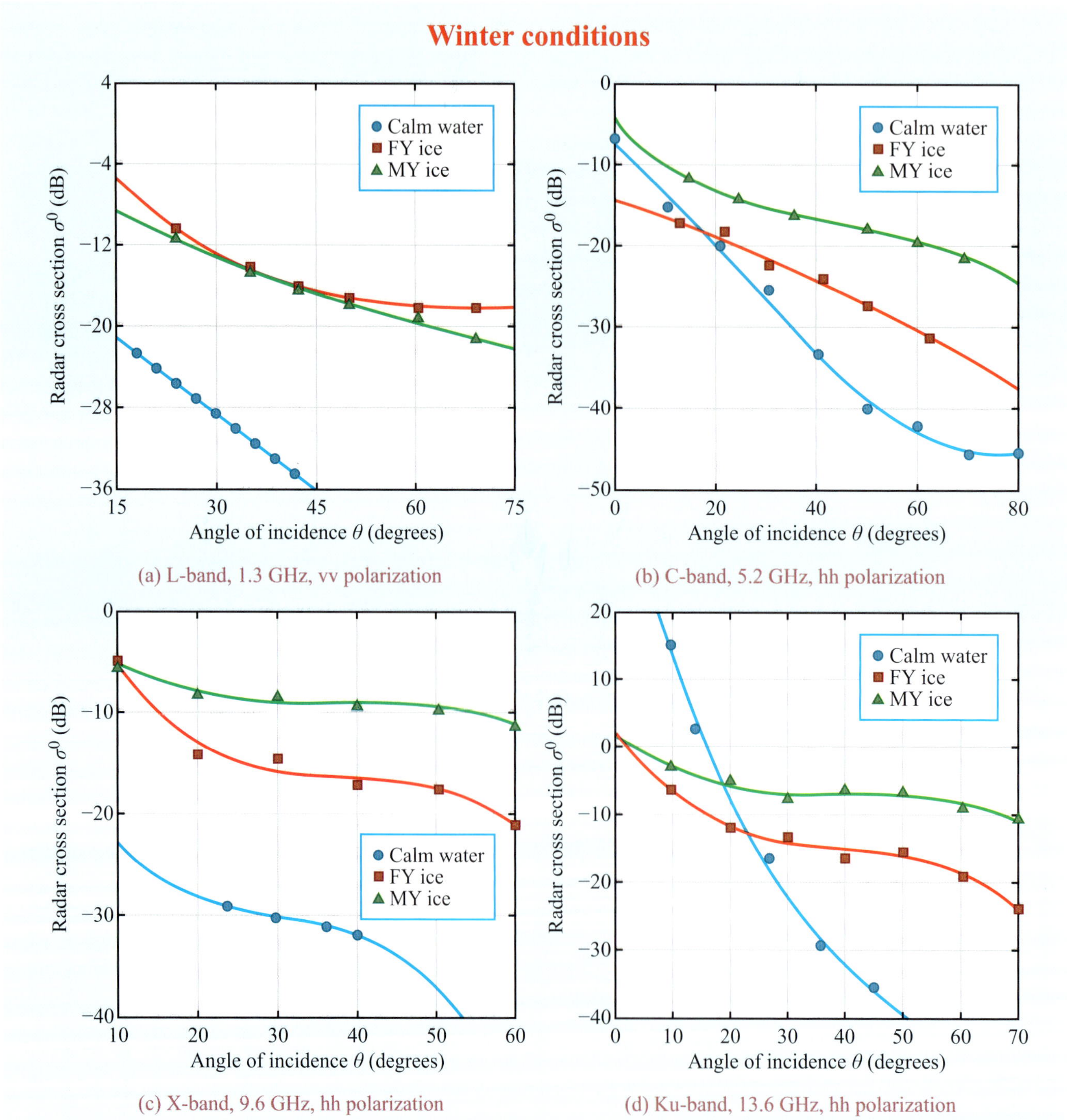

Figure 16-59: Radar backscatter cross sections of sea ice and open water under calm conditions versus incidence angle at various frequencies during winter (Onstott, 1992).

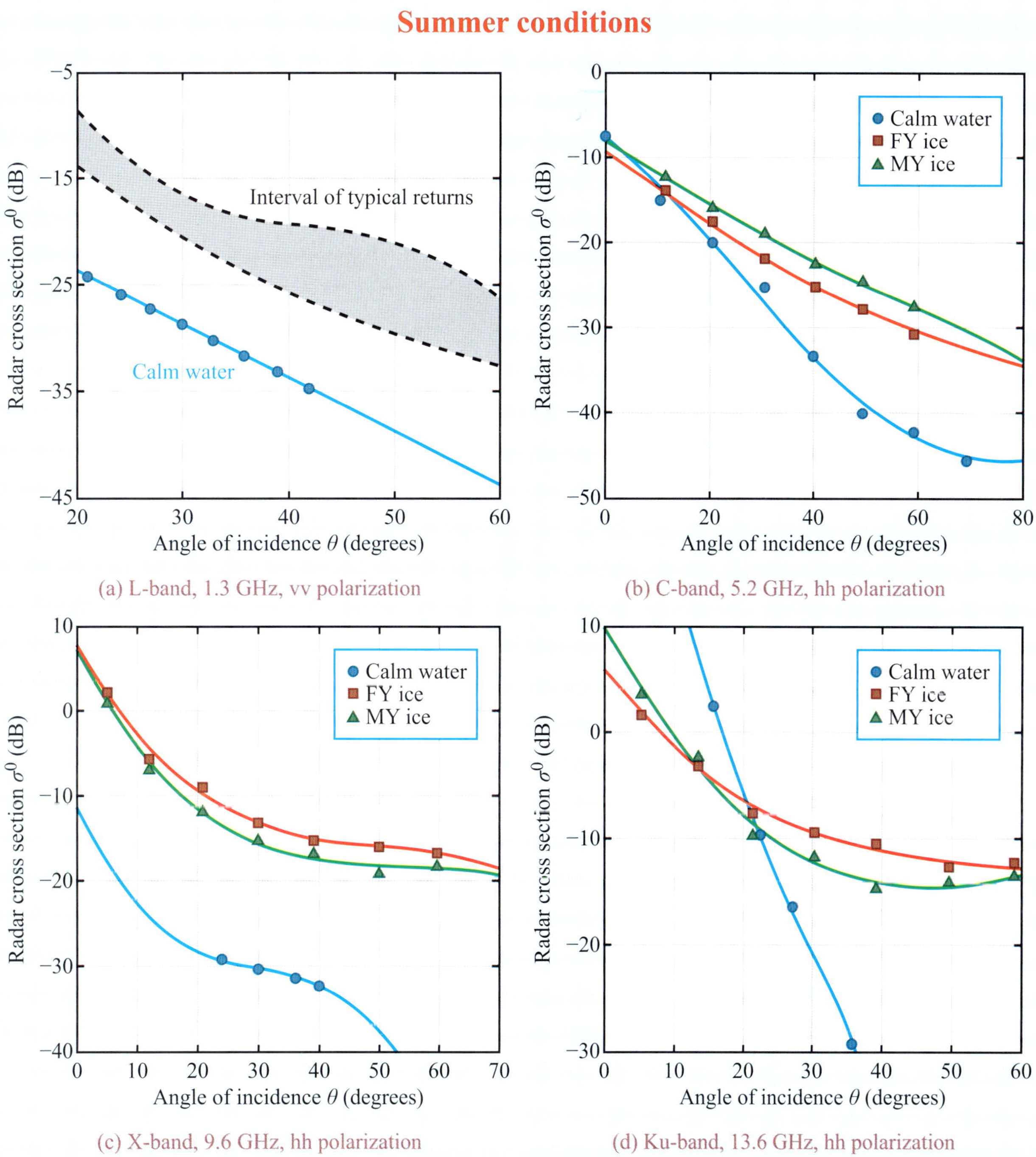

Figure 16-60: Radar backscatter cross sections of sea ice and open water under calm conditions versus incidence angle at various frequencies during summer [Onstott, 1992].

ocean and sea ice, and complicate the use of a simple σ^0-threshold to discriminate between sea ice and open ocean.

To improve the discrimination between sea ice and open water under a wide range of sea conditions, Wakabayashi et al. (2004, 2013) explored the use of polarimetric data acquired by airborne and spaceborne SAR systems. They used the data to compare histograms of σ^0 for water and sea ice at each of several polarization combinations (hh, vv, and hv), as well as in terms of the radar ***target entropy*** H_T defined by Eq. (5.174). An example of their results using PALSAR data is shown in Fig. 16-61. Of the three scattering coefficients, vv is the best discriminator and hv is the poorest, but H_T is superior to all; the H_T histograms for water and sea ice exhibit no overlap between them. An important observation made by Wakabayashi et al. (2013) is that whereas it is difficult to separate thin sea ice from open water on the basis of the radar backscatter alone, the entropy contrast between thin sea ice and open water is even larger than between thick sea ice and water. This improves the performance of ocean and sea ice classification considerably.

16-11.2 Discriminating between Different Ice Types

As previously noted, discriminating between FY and MY ice is of great interest for navigation and climate research. Under appropriate conditions radar measurements can be used to discriminate between FY and MY sea ice with a high degree of success.

We first consider low-frequency scattering during winter conditions. Figure 16-62 shows σ^0 versus θ for thick FY, MY, and lake ice, along with a measurement of a pressure ridge in the spring season (Onstott et al., 1982). Except at 60° and 70°, no significant difference appears between σ^0 for thick FY and MY ice. Note that very smooth lake ice produces a significantly weaker signal than sea ice, while the small pressure ridge gives a much stronger signal at most θ. The lower backscatter from lake ice may be attributed to (1) its smoother surface and (2) the lack of brine pockets. At low radar frequencies, such as L-band, the signal penetrates deeper into the ice than higher frequencies, which results in

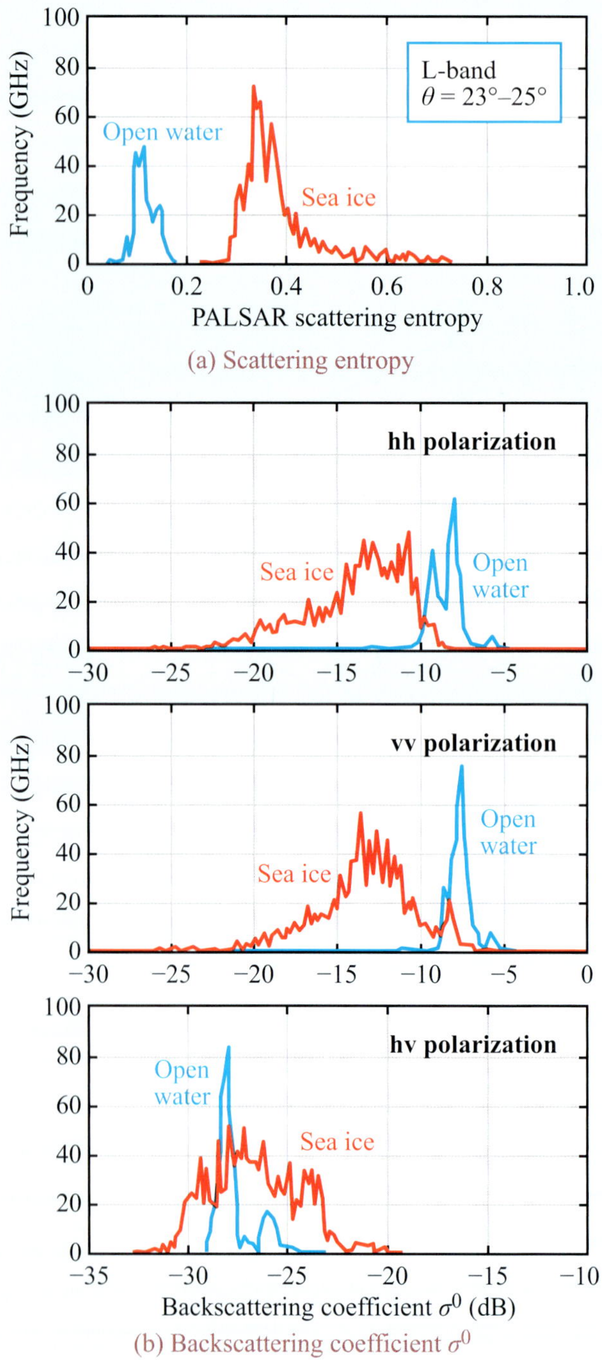

Figure 16-61: Histograms of scattering entropy and backscattering at L-band. (a) target entropy, (b) hh-polarization σ^0, (c) vv-polarization σ^0, (d) hv-polarization σ^0 [Wakabayashi et al., 2013].

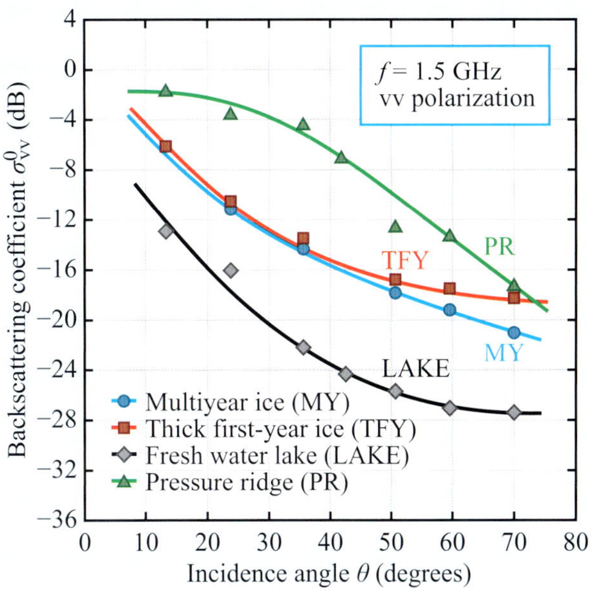

Figure 16-62: Scattering from thick FY and MY sea ice, lake ice, and a pressure ridge, at 1.5 GHz with vertical polarization (from Onstott et al., 1982).

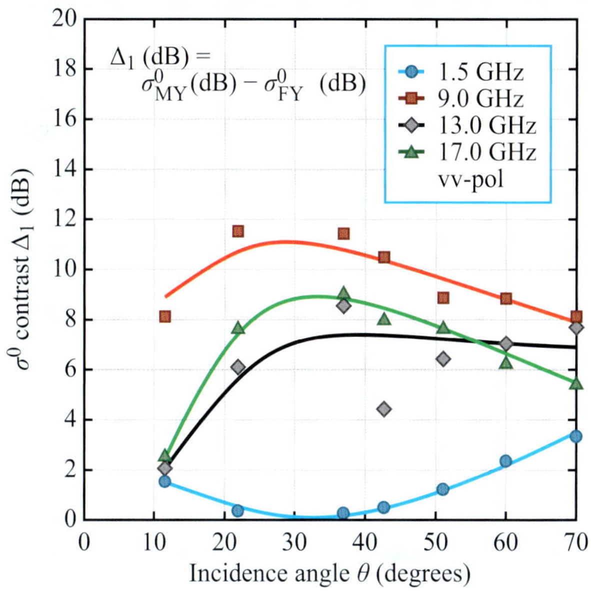

Figure 16-63: Contrast between MY and thick FY ice during spring at different frequencies as a function of angle of incidence for vertical polarization (from Onstott et al., 1982).

more surface-type scattering from ice layers, possibly including the subsurface ice/ocean interface. Further, the scattering contrast between FY and MY is smaller at low frequencies because the cross section of the internal voids and bubbles that contribute to volume scattering are proportionally smaller compared with the wavelength.

At higher frequencies, sea-ice scattering is often dominated by volume scattering (Gray et al., 1982). This contributes to the relatively flat roll-off of σ^0 with incidence angle of sea ice, as seen in Figs. 16-59 and Figs. 16-60 at X-band and Ku-band. In the winter at Ku-band, the scattering from MY ice greatly exceeds that from FY ice at all angles (10°–45°). For the summer date of the measurements shown in Fig. 16-60, the FY ice return has not changed much from the winter condition, but the volume scattering in the MY ice is suppressed by attenuation in the wet upper layer, so the MY return is comparable to that from the FY ice. We note that scattering measurements made in summer are influenced strongly by the particular point in the melting cycle at which the measurements were made.

The ability of different-frequency radars to distinguish among major ice types during spring is illustrated in Figs. 16-63 through 16-65. In these figures, the contrast is computed by taking the difference between the measured values of σ^0 in dB for the two indicated types of ice. The figures compare different thicknesses of MY and FY ice. The contrast between MY and thick FY ice in Fig. 16-63 is small at 1.5 GHz, but increases with frequency. The maximum contrast at higher frequencies is at approximately $\theta = 30°$, which corresponds to the minimum contrast at L-band. For cross-polarized signals at higher angles of incidence, L-band provides minimal contrast between thick-FY ice and MY ice, as shown in Fig. 16-65.

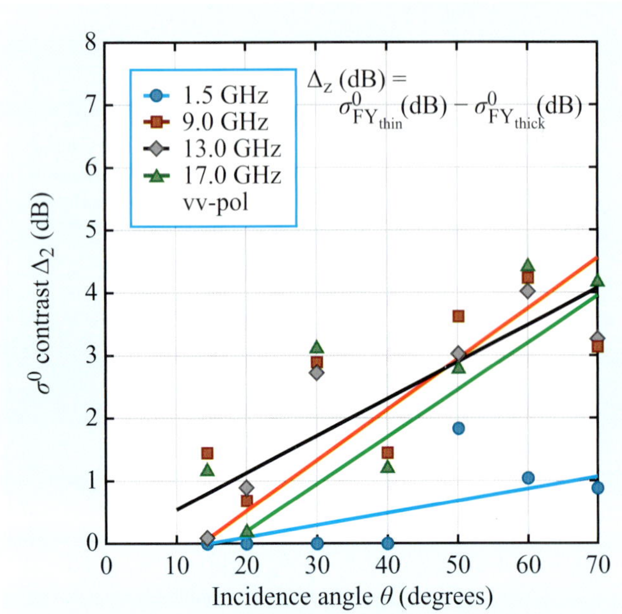

Figure 16-64: Contrast between thin FY and thick FY ice during spring at different frequencies as a function of angle of incidence for vertical polarization (from Onstott et al., 1982).

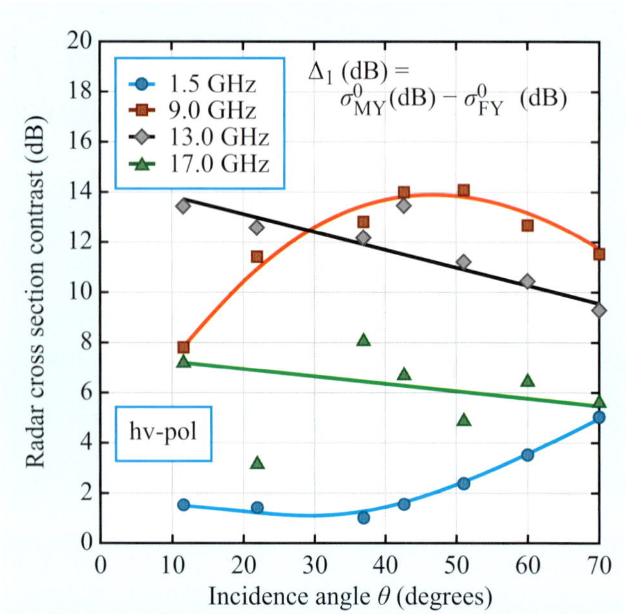

Figure 16-65: Contrast between MY and thick FY ice during spring at different frequencies as a function of angle of incidence for cross polarization (Onstott et al., 1982).

16-11.3 Measuring Ice Thickness

For sea ice grown under laboratory conditions at CRREL (the U.S. Army Cold Regions Research and Engineering Laboratory in Hanover, New Hampshire), its C-band backscattering coefficient exhibited a dynamic range of about 8 dB as the ice thickness increased between 2 cm and 12 cm (Kwok et al., 1998). The plots in Fig. 16-66 show similar trends for all three polarization combinations and all four incidence angles. These trends suggest that the backscatter is dominated by volume scattering, and that it may be possible to estimate ice thickness from the measured σ^0 values. The results of a more recent investigation conducted under natural sea conditions, also at C-band, are displayed in Fig. 16-67. In this case, the ice thickness is plotted along the vertical axis and the cross-polarization ratios ($\sigma^0_{hv}/\sigma^0_{hh}$ in part (a) and $\sigma^0_{hv}/\sigma^0_{vv}$ in part (b)) are along the horizontal axis. Also, the polarization ratios are in natural units, not dB. The measured data is primarily for MY ice and the range of sea-ice thickness is from about 20 cm to 50 cm.

16-11.4 Snow on Sea Ice

During winter, snow accumulates on sea ice, forming a layer that insulates the sea ice from the atmosphere. During summer, this layer melts and often completely disappears. Dry snow is essentially transparent at low microwave frequencies, but contributes to volume scattering at higher frequencies. Barber et al. (1998) measured the effects of dry snow over sea ice at multiple frequencies. At very high (38 GHz) frequencies, volume scattering from the snow layer dominates the radar scattering return as shown in Fig. 16-68. In this plot, measurements are compared against modeled scattering cross sections using a radiative transfer model that includes both surface and volume scattering. We note that volume-scattering effects are larger at larger values of θ.

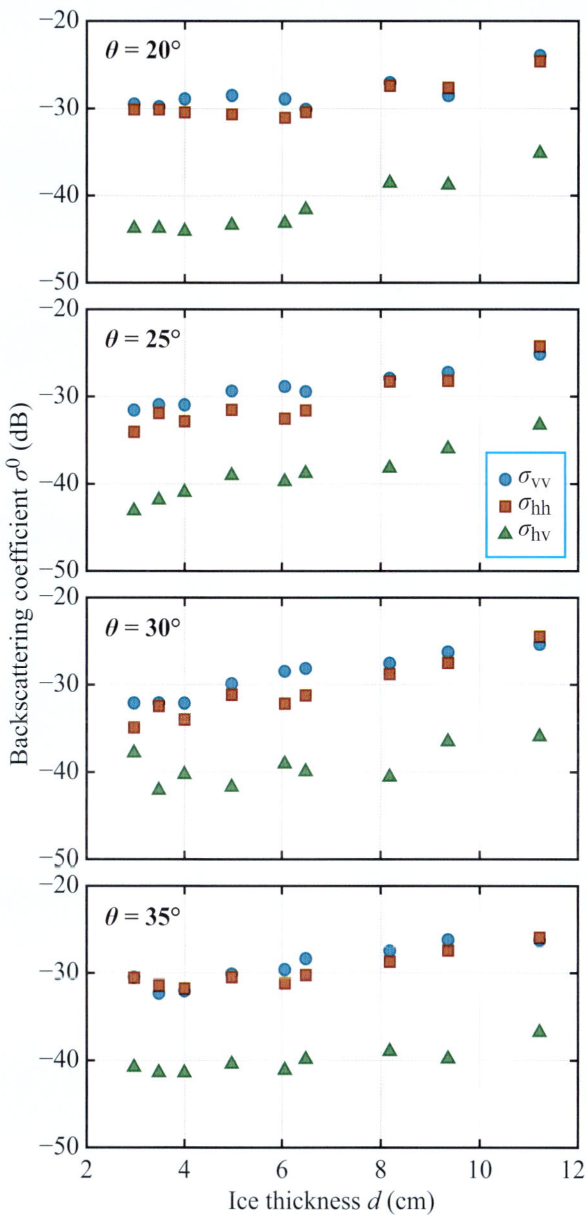

Figure 16-66: C-band σ^0 measurements of thin ice grown under laboratory conditions [Kwok et al., 1998].

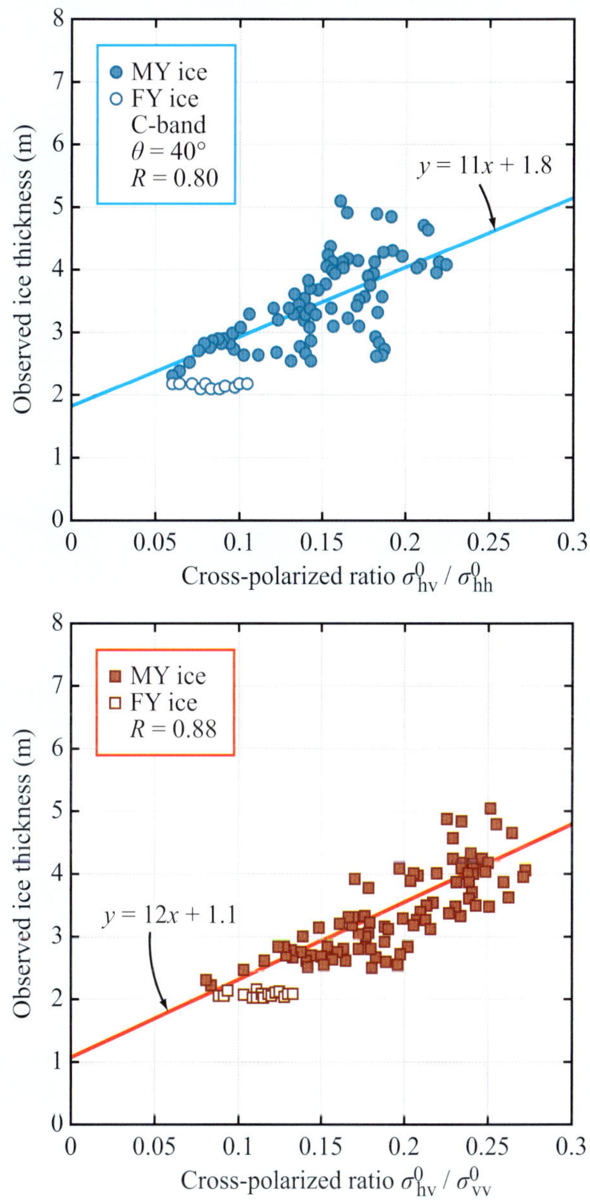

Figure 16-67: Sea-ice thickness versus two cross-polarization ratios at C-band [Kim et al., 2012].

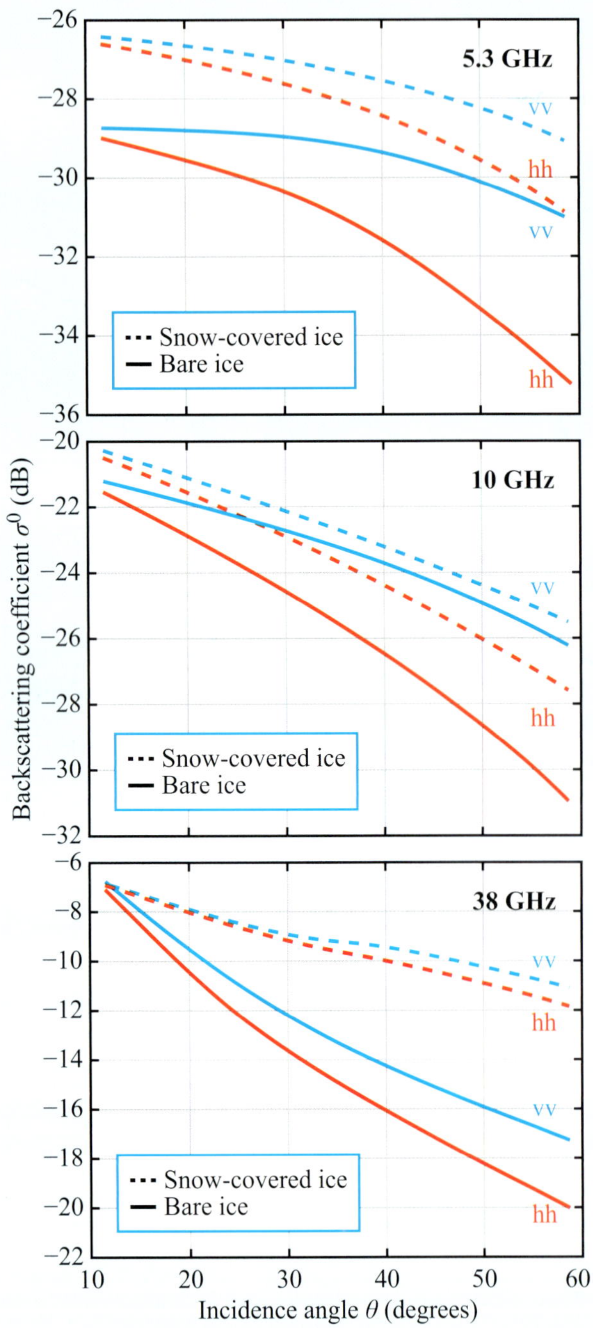

Figure 16-68: Comparison of modeled and measured scattering versus θ for snow-covered saline ice and bare ice from a controlled experiment [Barber et al., 1998].

Because liquid water has a substantially larger dielectric constant than does ice, the presence of liquid water in snow significantly alters the effective dielectric constant of the snow and hence the emission and radar scattering. Radar backscatter is thus very sensitive to the freeze/thaw transition of snow and ice. To illustrate, Fig. 16-69 shows multiyear time-series plots of Ku-band radar backscatter near a free-floating buoy that also measured the daily average air temperature. As the air temperature climbs above freezing, the surface snow and ice layer begin to melt, initiating a strong drop in the backscatter for MY ice, but an increase in backscatter for FY ice, which melts by the end of the summer to reform again in the fall. The strong microwave scattering freeze/thaw signal also occurs in the great ice sheets of Greenland and Antarctica. This microwave scattering freeze/thaw signature has been exploited by a number of investigators to map freeze/thaw transitions over both land and ice (Mortin et al., 2012; Kunz and Long, 2006; Nghiem et al., 2001).

Backscatter images from satellite scatterometer images are useful for visualizing the spatial and temporal variability of the backscatter of sea ice over large areas and time periods. Figures 16-70 and 16-71 present midwinter and summer backscatter images over the Arctic and Antarctic regions at, respectively, Ku- and C-band. During winter, MY year ice appears lighter than FY. In these images, ocean backscatter is dependent on local wind speed conditions, and generally appears brighter at C-band than at Ku-band. The differences in sea ice coverage from winter to summer are apparent. The next section considers how wind scatterometer data can be used for mapping sea ice extent and sea ice type.

16-11.5 Scatterometer Mapping of Sea-Ice Extent

As previously noted, sea ice is very dynamic, both intra-annually and interannually. Maps of sea-ice extent are of value for fishing, shipping, and scientific studies of climate. Wind scatterometers have been operationally used for mapping sea-ice extent since the 1990s. A number of approaches to mapping sea-ice extent have been developed (Ezraty and Cavanié, 1999; Remund and Long, 1999; Haarpainter et al., 2004; Anderson

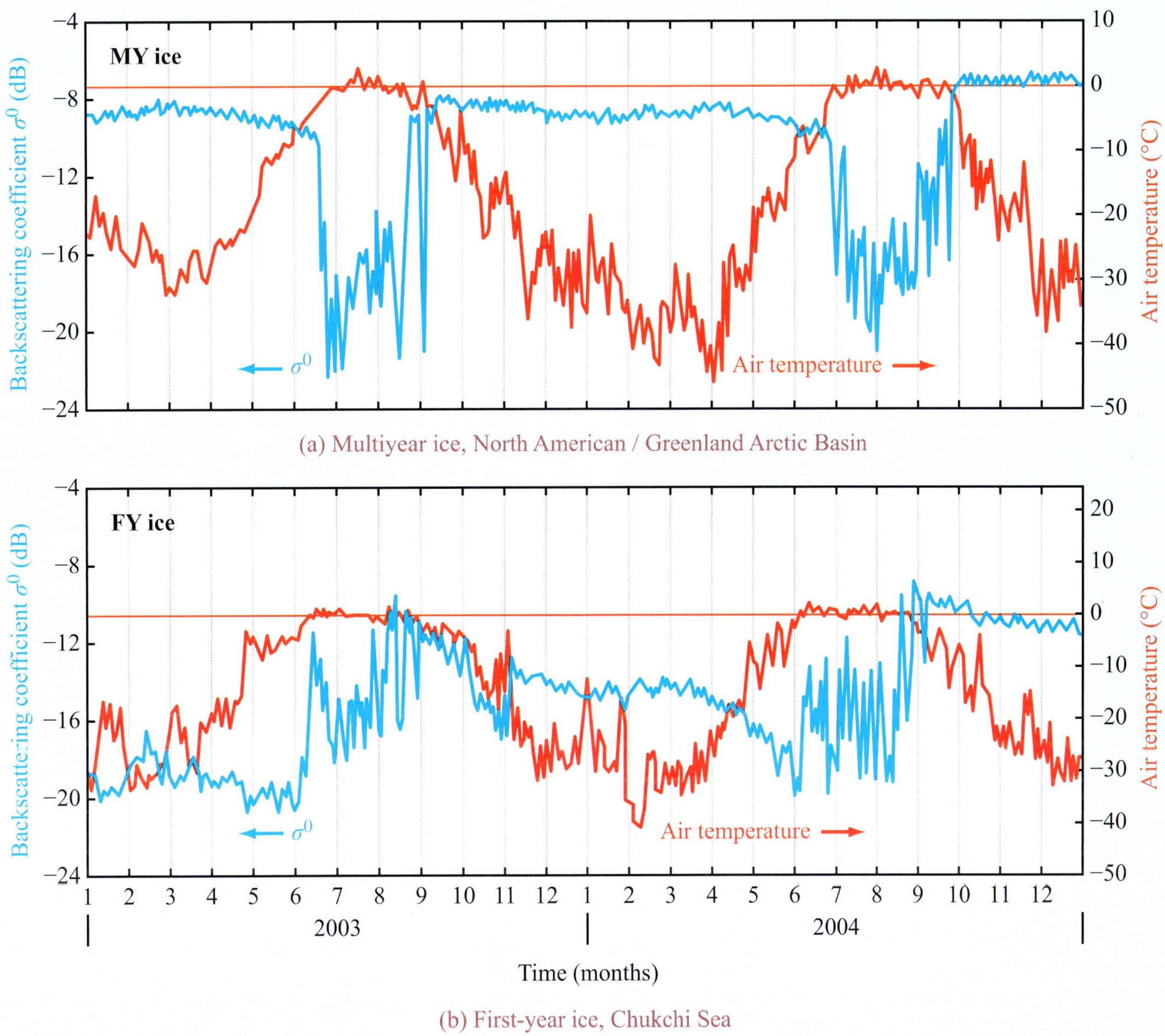

Figure 16-69: Time series of air temperature and Ku-band backscatter for (a) MY ic and (b) FY ice. Temperature data were obtained from buoys while backscatter data were measured by the QuikSCAT scatterometer [Mortin et al., 2012].

and Long, 2005; Haarpainter and Spreen, 2007; Nghiem et al., 2006). These approaches have many features in common, some of which we consider below. Even though illustrated with Ku-band wind scatterometer data, similar results are obtained for C-band. The methods are based on the different empirical microwave scattering signatures of sea ice and open ocean.

Because satellite-borne wind scatterometers operate in polar orbit, and because they have broad swaths, they can provide backscatter measurements over essentially the entire polar regions, with many locations observed

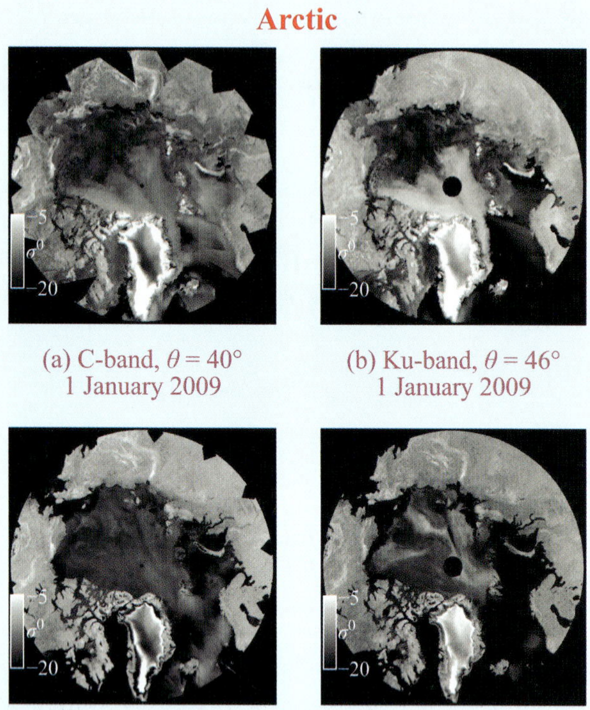

(a) C-band, $\theta = 40°$ 1 January 2009

(b) Ku-band, $\theta = 46°$ 1 January 2009

(c) C-band, $\theta = 40°$ 5 July 2009

(d) Ku-band, $\theta = 46°$ 5 July 2009

Figure 16-70: σ^0 images over the Arctic observed at C- and Ku-band based on a single day of wind scatterometer measurements recorded in 2009. Notches and circular region at the pole are areas without coverage. Bands are artifacts due to temporal change during the day.

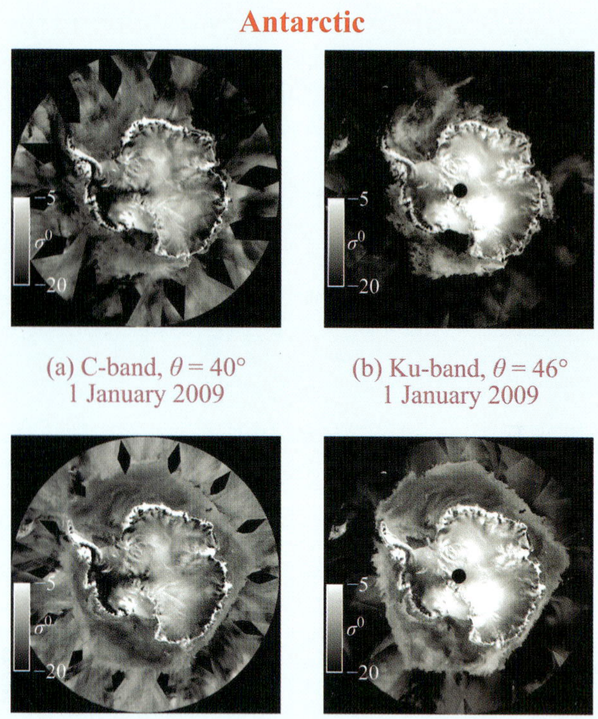

(a) C-band, $\theta = 40°$ 1 January 2009

(b) Ku-band, $\theta = 46°$ 1 January 2009

(c) C-band, $\theta = 40°$ 5 July 2009

(d) Ku-band, $\theta = 46°$ 5 July 2009

Figure 16-71: σ^0 images over the Antarctic observed at C- and Ku-band based on a single day of wind scatterometer measurements in 2009. Black diamonds and circles at the poles are areas without coverage. Bands are artifacts due to temporal change during the day.

more than once per day. Fan-beam scatterometers measure σ^0 over a range of incidence angles, while scanning scatterometers typically measure σ^0 at only two incidence angles.

In the fan-beam case, multiple passes over a given pixel over several days can be used to estimate the slope of σ^0 versus incidence angle, normalized to σ^0 at a particular incidence angle, say 40°. For this purpose over the incidence angle range 20° to 60°, the dependence of σ^0 on incidence angle is often modeled as

$$\sigma_p^0(\mathrm{dB}) = A_p + B_p(\theta - 40°), \tag{16.60}$$

where p is the polarization, A_p is σ^0 for pp polarization normalized to σ^0 at 40°. For multipolarized scatterometers A_p and B_p are computed separately for each polarization. The ***copol ratio*** γ is defined as

$$\gamma = \frac{A_\mathrm{v}}{A_\mathrm{h}}\,. \tag{16.61}$$

Backscatter maps can be generated at the intrinsic sensor resolution, or at a higher resolution using reconstruction techniques (Early and Long, 2001). Figure 16-72 illustrates an example of A_p and B_p for both polarizations derived from 6 days of NSCAT data. Note that in these images, A_p and B_p for sea ice are

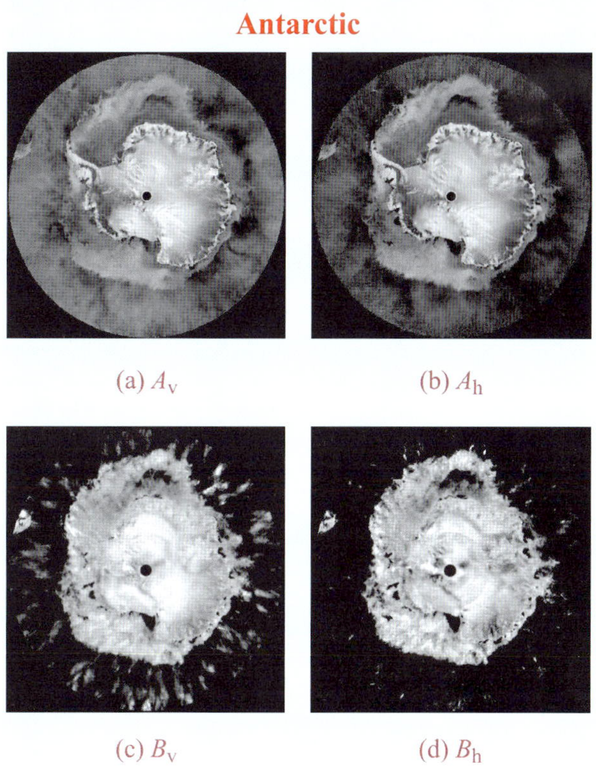

Figure 16-72: Backscatter images derived from NSCAT data. (a) A_v, (b) A_h, (c) B_v, and (d) B_v of Eq. (16.60) [from Remund and Long, 1999].

generally higher than for the ocean. We can exploit this to map the extent of sea ice.

To classify each pixel as either sea ice or ocean, a ***feature vector*** V_{fea} is formed for each image pixel as the combination of γ, A_h, B_h:

$$V_{fea}(n) = [\gamma(n)\ A_h(n)\ B_h(n)], \tag{16.62}$$

where n is the pixel index. The feature vector may be augmented by the variances v_h and v_v of the observed measurements over the imaging period.

For a scanning scatterometer, B_p cannot be estimated since σ^0 is measured at only two different incidence angles, and at different polarizations. In this case the ***pseudo copol ratio*** γ' is defined as

$$\gamma' = \frac{\sigma^0_{vv}}{\sigma^0_{hh}}. \tag{16.63}$$

The feature vector is then defined as

$$V_{fea}(n) = [\gamma'(n)\ \sigma^0_{hh}]. \tag{16.64}$$

For both fan-beam and scanning-beam scatterometers, a statistical approach can be used to classify each pixel as either sea ice or open ocean. Here, we focus only on mapping sea-ice extent, although Haarpainter and Spreen (2007) were also able to extract sea-ice concentration from the image data.

A simple ***classification scheme*** can be based on using histograms of the feature vectors and discrimination techniques. This is illustrated with the aid of Fig. 16-73, which shows two-dimensional histograms of γ versus B_v, two of the elements of the feature vector. For the purposes of discussion, only a two-element feature vector is used in order to simplify the presentation. Using a land mask, measurements over land are excluded so that the histogram includes only σ^0 measurements of ocean or sea ice. The two peaks in the histogram correspond to the signatures of open ocean and sea ice. The ocean peak at the lower right exhibits more spatial variability and so is more spread out, while sea ice tends to have a narrower peak.

The normalized histogram is an estimate of the ***probability density function*** (pdf) of the feature vector $p_{tot}(V_{fea})$. When there are two classes, for a given pixel the total pdf can be written as

$$p_{tot}(V_{fea}) = p_{ice}(V_{fea}|\text{ice})\ p(\text{ice}) + p_{ocean}(V_{fea}|\text{ocean})\ p(\text{ocean}), \tag{16.65}$$

where $p_{ice}(V_{fea}|\text{ice})$ is the ***conditional probability*** of V_{fea} given that the pixel is ice, and $p_{ocean}(V_{fea}|\text{ocean})$ is the ***conditional probability*** of V_{fea} given that the pixel is ocean. Of course, $p(\text{ocean})$ and $p(\text{ice})$ are the ***prior probabilities*** of the pixel being ocean and ice, respectively. Note that $p(\text{ocean}) = 1 - p(\text{ice})$. In some models, the pdfs are assumed to be spatially constant, while in others they have been treated as spatially

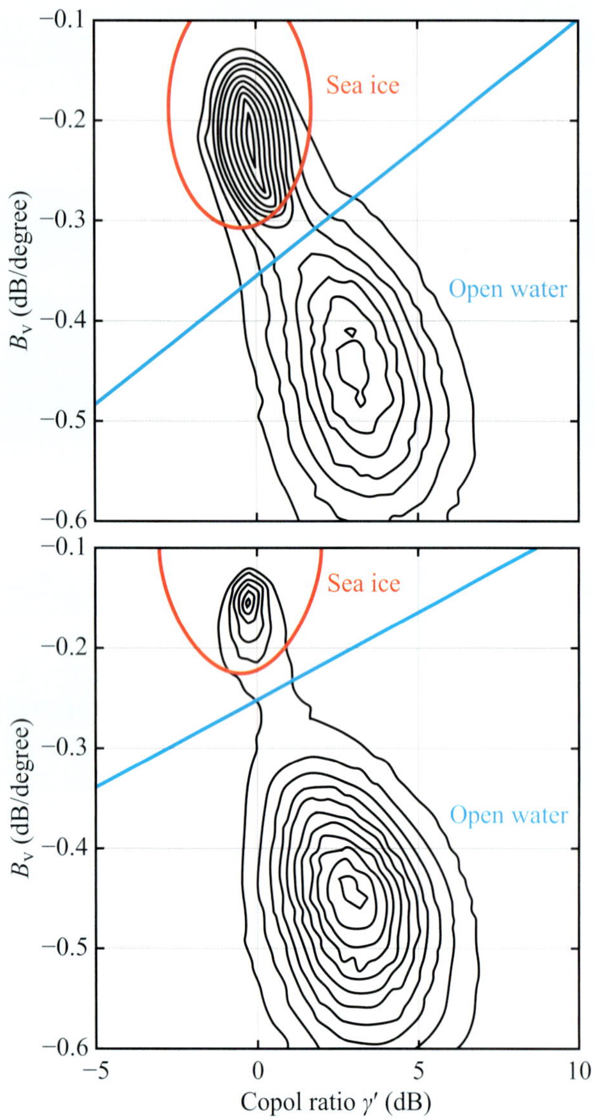

Figure 16-73: Contour plots of two-dimensional histograms of B_v versus γ' for two different days in spring and midsummer in Antarctica. The curved and linear lines represent the decision boundaries for linear and Mahalanobis discrimination.

varying. The priors have been treated differently as well. When no prior information is available, the ***uninformative prior*** is selected as $p(\text{ocean}) = 1/2$.

To ***classify*** a pixel as ocean or ice, we compute the distance between its feature vector and the peak centers (Duda and Hart, 1973). In a ***linear discriminator***, this results in a straight line, plane, or hyperplane to divide the space into two classes, each containing a peak. The pixel is then classified according to the region it falls in. The location of the line is selected to minimize the probability of classification error. In Fig. 16-73, the line is approximately orthogonal to the line connecting the two peaks. If the conditional pdfs are assumed to be Gaussian, the peaks correspond to the mean of the conditional pdf. When the priors have similar value, the minimum probability of error occurs when the ***Mahalanobis distance*** between the observed feature vector and the Gaussian means is minimized (Duda and Hart, 1973).

Pixel classification results in a binary image that indicates the location of sea ice and ocean. Noise in the measurements and overlapping conditional pdfs result in misclassified pixels. Many errors can be corrected using binary image processing techniques (Remund and Long, 1999). For example, region-growing can eliminate the isolated misclassification patches in the ocean and ice. Image erosion and dilation techniques (Rush, 1995) can be used to filter the sea-ice edge. Figure 16-74 illustrates a noisy initial classification and the result after binary processing.

The scatterometer-derived sea ice extent can be validated through comparison with high-resolution SAR imagery. Two examples are provided in Fig. 16-75. In both images, sea ice fills the left side of the image, with open ocean to the right. Note the wind-related storm feature in Fig. 16-75(a).

In Fig. 16-75(b), the growing ice edge is very diffuse and thus poorly defined. The higher resolution of the scatterometer data provides a finer resolution edge estimate than the coarse (25 km) resolution SSM/I radiometer sea-ice edge. In Fig. 16-75(a) the scatterometer edge is contained with the 30% ice concentration, while it lies between the 5% and 30% ice concentration lines in Fig. 16-75(b). This suggests that the active and passive signals are sensitive to slightly different sea ice conditions.

This becomes more apparent when comparing the sensors over the annual seasonal cycle. In Fig. 16-76,

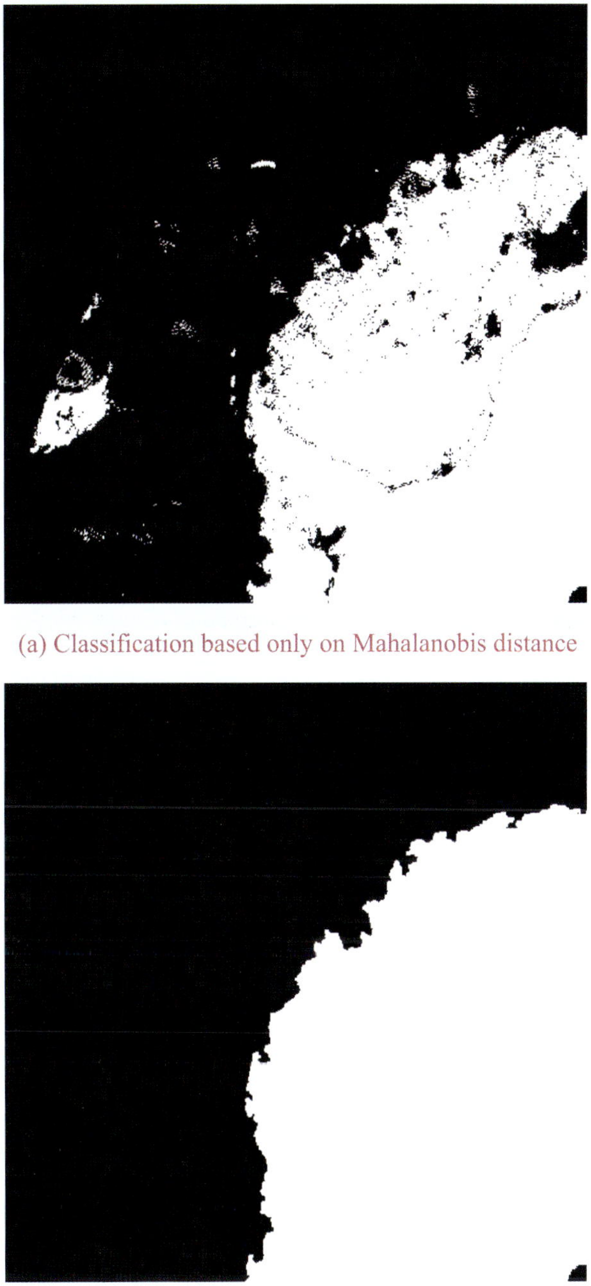

(a) Classification based only on Mahalanobis distance

(b) After post-classification processing to reduce spurious classifications

Figure 16-74: Example of ice/ocean discrimination where white pixels indicate locations classified as ice and black pixels are pixels classified as ocean. (a) Classification based only on Mahalanobis distance. (b) After post-classification processing to reduce spurious classifications [Remund and Long, 2003].

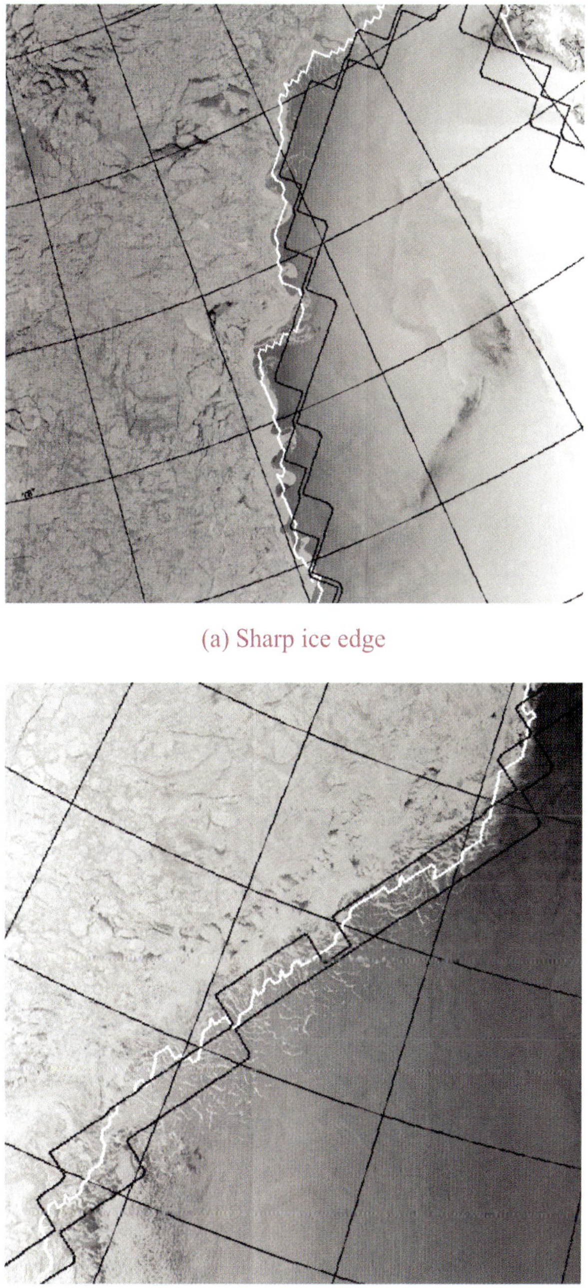

(a) Sharp ice edge

(b) Diffuse ice edge

Figure 16-75: Radarsat SAR images with overlays of (white) SeaWinds on QuikSCAT-derived sea ice extent and (black) SSMI-derived NASA Team 5% and 30% ice concentrations.

the mean sea-ice extent measured over a 10-year period as a function of day of the year are compared for the SeaWinds scatterometer and the SSM/I radiometer. Note that during the summer, when sea-ice extent is at a minimum, the scatterometer-derived sea-ice extent corresponds to a small radiometer-derived sea-ice concentration, whereas in the summer at peak extent the scatterometer extent corresponds to a higher radiometer sea-ice extent (Meier and Stroeve, 2008). The precise reasons for the difference is uncertain but may be due to the seasonally varying contrast between sea ice and ocean for the two sensors.

16-11.6 Scatterometer Mapping of Sea-Ice Type

We noted earlier that FY and MY have different σ^0 signatures (at least during the winter). This difference can be exploited to map FY and MY sea ice on a daily basis using wind scatterometer data. The statistical approach to mapping MY ice using Ku-band wind scatterometer data has been considered by several investigators (Swan and Long, 2012; Kwok et al., 2009; Nghiem et al., 2006). Like the sea-ice extent mapping described in the previous section, this statistical approach exploits the differences in the scattering of MY and FY sea ice but does not need to explicitly model the scattering signatures.

Earlier, we illustrated through Figure 16-70 the winter backscatter of the Arctic sea ice at C- and Ku-band. In the images MY ice is brighter than FY ice. One approach to mapping MY ice is to start with daily backscatter images and create histograms of σ^0 for each day. Figure 16-77 illustrates the evolution of the daily backscatter histograms of σ^0_{vv} values at $\theta = 54^\circ$ within the area covered by sea ice over the course of a winter season. The regions near the ice edge, known as the ***marginal ice zone*** (MIZ), have been excluded. The vertical height of the histogram is converted to color and shown as an image in Fig. 16-78(a). Note that during the winter, there are two distinct populations of ice with different σ^0 values. In Fig. 16-78(b) these are identified as MY ice, which has a higher σ^0 value, and FY, which has a lower σ^0 value. Only MY ice survives the summer. FY ice begins forming in late September, grows through

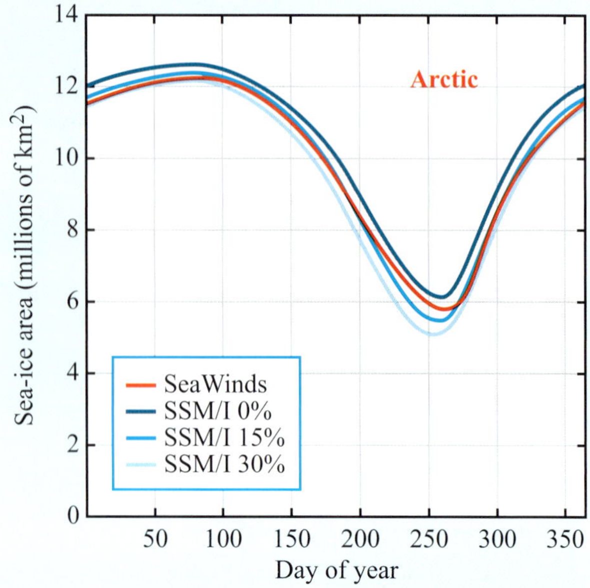

(a) Arctic mean ice area above 60° N latitude

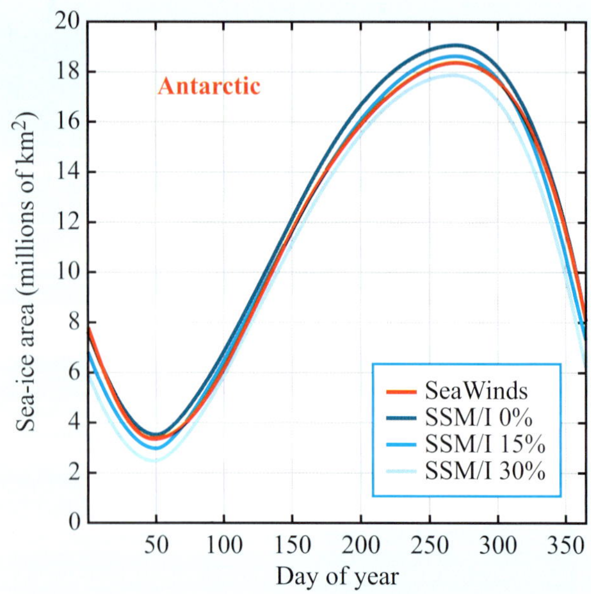

(b) Antarctic mean ice area

Figure 16-76: Mean daily total sea-ice area during the prime QuikSCAT operational mission 1999–2009. SeaWinds on QuikSCAT-derived sea-ice extent and SSM/I NASA Team 0%, 15%, and 30% ice concentration extents are shown for the (a) Arctic and (b) Antarctic.

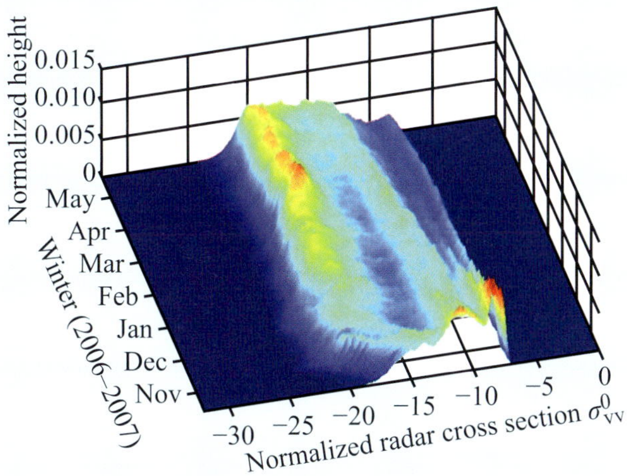

Figure 16-77: Time series of histograms of Ku-band σ^0_{vv} over the 2006–2007 winter. Color indicates normalized histogram height [Swan and Long, 2012].

winter, and disappears due to melting by the end of summer.

Thus, from late fall through spring (October through May), the ice can be classified as MY or FY on the basis of its σ^0 value. During the summer melt, σ^0 of all ice types drops, and MY and FY ice can no longer be distinguished. A fixed threshold (at approximately −13 dB) has been used (Kwok et al., 2009; Nghiem et al., 2006), but using an adaptive threshold can provide discrimination (Swan and Long, 2012).

Figure 16-79 summarizes MY/FY areal coverage derived over a 7-year-long period based on QuikSCAT measurements. According to the time series of Ku-band σ^0_{vv} values shown in Fig. 16-79(b), the plot in Fig. 16-79(a) shows the total area of sea-ice coverage compared with the MY areal coverage. It should be noted that the minimum ice area during summer maybe inaccurate due to summer melt conditions. However, note that there is a gradual shift in the ice cover from MY to FY ice during winter. The MY ice mode becomes progressively weaker and dispersed during this period, while the FY ice mode becomes progressively stronger. Since the total area of sea ice under consideration does not change significantly from winter to winter, it is

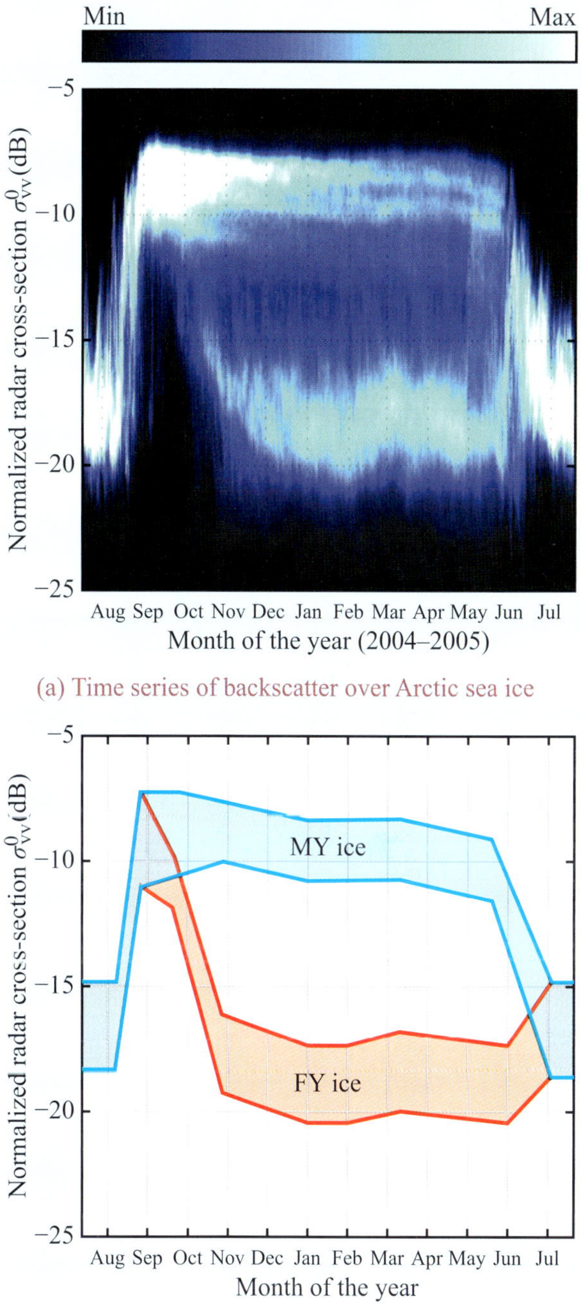

(a) Time series of backscatter over Arctic sea ice

(b) Diagram of approximate backscatter ranges

Figure 16-78: Seasonal Ku-band backscatter evolution. (a) Time series of histograms of backscatter over Arctic sea ice for the winter of 2004–2005. (b) Cartoon showing approximate ranges of backscatter for FY and MY ice for each year between 2003 and 2009. The range of backscatter for each ice type is seasonally dependent. [Swan and Long, 2012.]

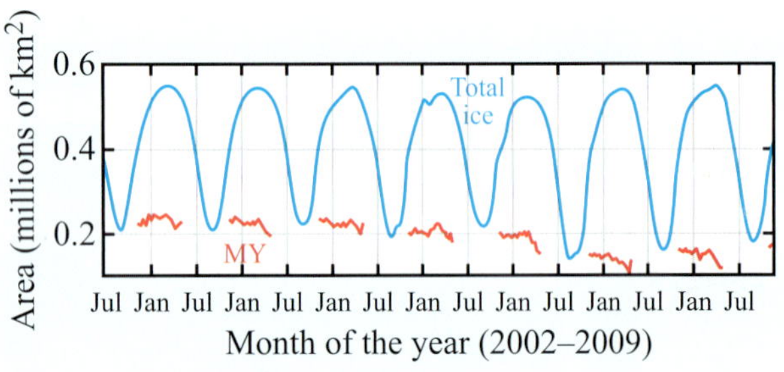

(a) Total area of ice and area of MY ice in winter

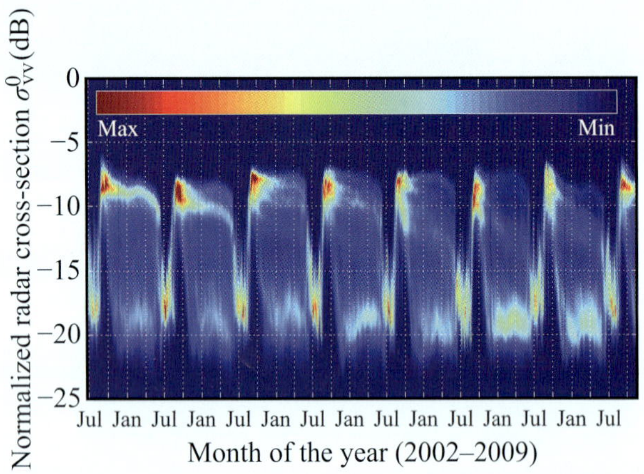

(b) Time series of QuikSCAT σ^0_{vv}

Figure 16-79: (a) Total area of ice for each day and area of MY ice during the winter, (b) time series of histograms of QuikSCAT σ^0_{vv} [Swan and Long, 2012].

apparent that MY ice was gradually replaced with FY ice over this period.

A limitation of this classification method is that MY ice can sometimes "look like" FY ice, particularly in the MIZ, which is why the MIZ was excluded in the analysis. In some instances, FY ice may resemble MY ice if it has undergone significant rafting and ridging and/or brine exclusion.

16-12 Radar Imaging of Sea Ice

SAR imaging from satellites has been used for monitoring sea ice since the early 1990s. Radars can image the ice features and by creating a time series of images, we can extract motion information about the ice. This became apparent from the first satellite SAR data. Using Seasat SAR imagery, Leberl et al. (1983) stated: "(1) There is no difficulty in identifying homologue features on sequential radar images; (2) The accuracy is entirely controlled by accuracy of the orbit data and the geometric calibration of the sensor; (3) Conventional radargrammetric methods serve well for satellite radar ice mapping; (4) Ground control points can be used to calibrate the ice location and motion measurements where orbit data and sensor calibration are lacking." In their study, ice motion was found to be about 6.4 km ± 0.5 km per day. Follow-up investigations have developed automated tracking of ice features over extended periods, which enables observation of motion and deformation of the ice. The reader is referred to the extensive literature for more information on the theory and application of SAR-derived sea-ice motion, including Kwok et al. (1995, 2003, 2009), Liu et al. (1997), Kwok (2004), and Kwok and Cunningham (2012).

Many examples of SAR images of sea ice could be selected, but we provide only one example, shown in Fig. 16-80. This image offers a sense of the capability of radar to image details in sea ice. Higher-frequency images allow better ice-type discrimination based on grayscale contrasts, but much variability shows within each type. The use of shape and edge information allows ice identification even at those frequencies where the σ^0 contrast is low. Lower frequencies give somewhat better location of ridges, but ridges can be seen at all frequencies. Summer conditions produce images that may be much lower in contrast.

16-13 Iceberg Tracking

Icebergs are formed by the ***calving*** of large chunks of ice from glaciers. Glacial ice is formed from layers of accumulated snow, and is thus composed of freshwater ice. Volume scattering from the layers of snow on the surface of the iceberg causes them to appear very bright in radar images unless the surface is undergoing melt, in which case the contrast in σ^0 between an iceberg and

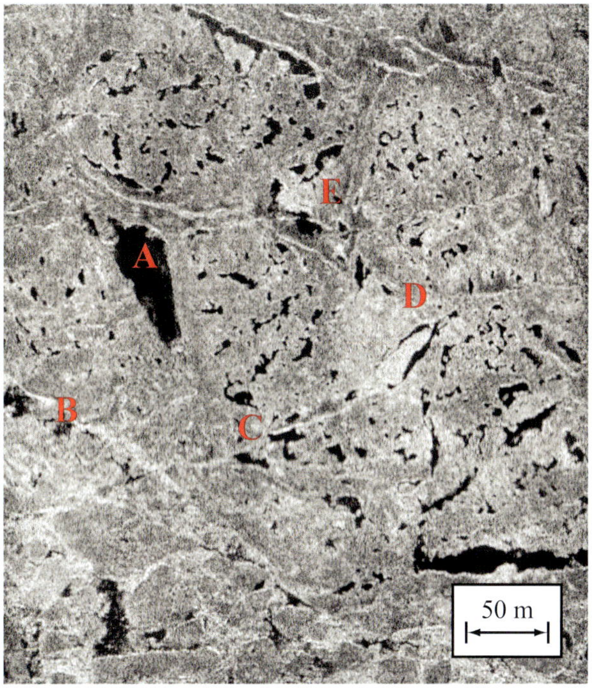

(a) C-band hh-polarization SAR image

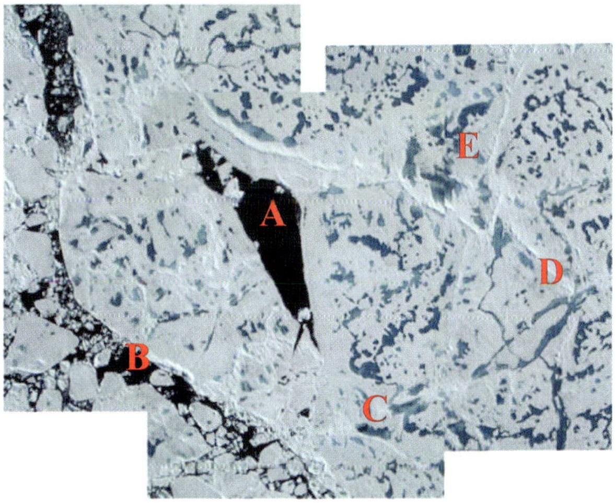

(b) Optical image

Figure 16-80: Annotated comparison of a C-band hh-pol SAR image and an optical image. Note the light blue melt ponds on the surface of ice floes appear black in the SAR image. Labels are A: open-water polynya. B: open-water lead. C: ridge with ponds. D, E: melt ponds in relatively high-backscatter areas.

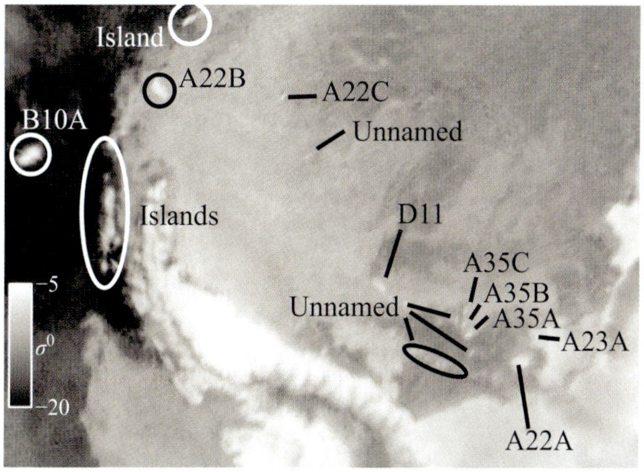

Figure 16-81: Ku-band σ^0_{hh} image ($\theta = 46°$) generated from one day of QuikSCAT data from day 101 of 1999. Annotation indicates the locations and names of large tabular icebergs in the image. Glacial ice covering the Antarctic land mass shows up as very bright, while sea ice is somewhat darker. Open ocean at the upper left is dark.

the surrounding sea ice or wind-roughened sea may be small.

Icebergs come in a wide variety of sizes from extremely large (tens of km) ***tabular icebergs*** found primarily in Antarctica, to small (few m) fragments known as "***growlers***." Tabular icebergs greater than 5 km in size are large enough to be visible and tracked in reconstructed σ^0 images generated from satellite scatterometer data, as shown in Fig. 16-81. Icebergs generally have better contrast with respect to sea ice at Ku-band than at C-band. In 1999, the 65-mile-long B10A iceberg was relocated in a QuikSCAT σ^0 image after having been lost to optical trackers for many months due to cloud cover. Once found again, optical and SAR images were collected, and these are compared in Fig. 16-82. Figure 16-83 illustrates daily positions of icebergs tracked with satellite scatterometers for over three decades.

With their higher resolution and their ability to see through clouds and operate at night, SARs are useful for locating icebergs. Large icebergs are easily visible in SAR images, but small icebergs and fragments of

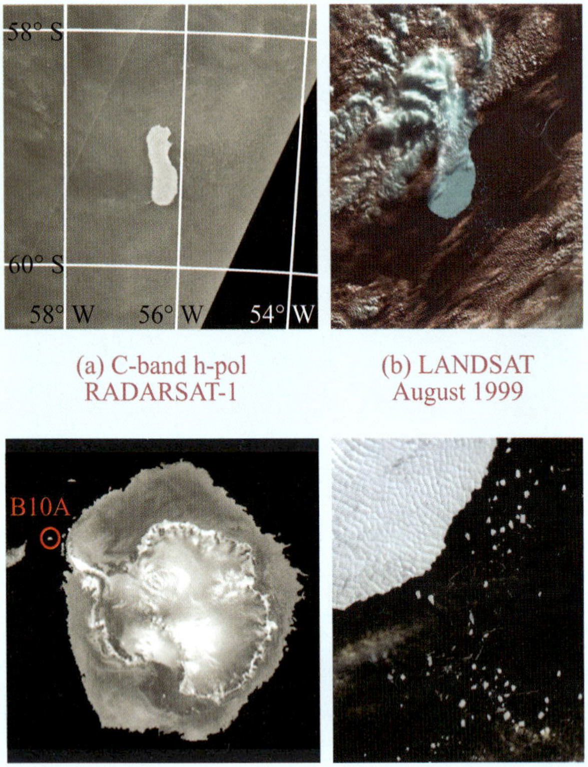

(a) C-band h-pol RADARSAT-1

(b) LANDSAT August 1999

(c) Ku-band v-pol QuikSCAT July 1999

(d) Zoom of (b) showing 1 km fragments

Figure 16-82: Images of Antarctic iceberg B10A. (a) C-band hh-pol RADARSAT-1, (b) LANDSAT August 1999, (c) Ku-band vv-pol QuikSCAT with ocean masked off, July 1999, (d) Zoomed-in view of part (b) showing 1 km fragments. [Images (a), (b), and (d) courtesy of the U.S. National Ice Center.]

icebergs may be difficult to discriminate in images. A spectacular RADARSAT image of a large tabular iceberg calved off the end of an Antarctic glacier is shown in Fig. 16-84. The large-scale surface texture of the iceberg is the result of its parent glacier flowing over subsurface topography. Note the surrounding sea ice. In this image the prevailing current and wind is in the upward direction. The blocking effect allows for the formation of new sea ice and open water in the lee of the iceberg.

Since icebergs have a different polarimetric signature

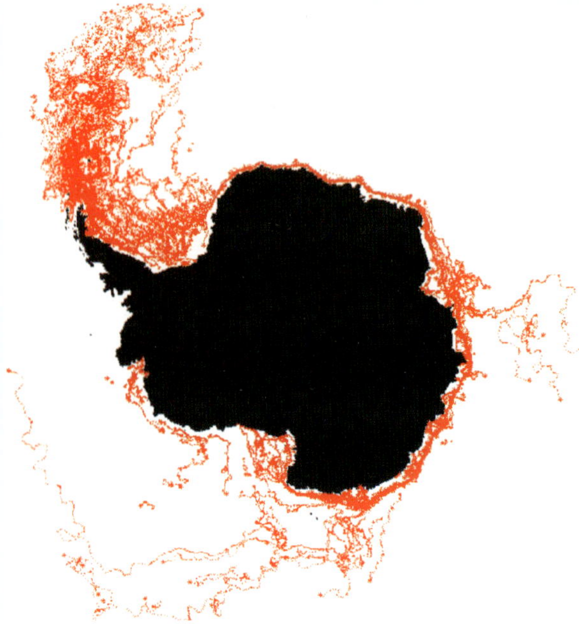

Figure 16-83: Daily positions of Antarctica icebergs detected in scatterometer images collected from 1978 to 2010 [Stuart and Long, 2011].

Figure 16-84: C-band hh-pol RADARSAT image of an Antarctic tabular iceberg. The iceberg is grounded, and has not moved far from the glacier tongue from which it calved. [Image courtesy of Canadian Science Agency.]

than sea ice, polarimetric SAR images are effective in discriminating icebergs from sea ice, at least if the iceberg is sufficiently large. For reliable detection of small icebergs, a combination of optical and SAR images may be required.

16-14 Radar Detection of Oil Spills

Radar sensors are often used for remote detection of oil slicks by virtue of their all-weather capability and independence of day and night conditions. Because of their high resolution, SAR has been widely used for oil slick detection and mapping (Alpers, 2002; Brekke and Solberg, 2005; Girard et al., 2005; Migliaccio et al., 2007, 2011; Solberg et al., 2007; Zhang et al., 2011), although for extremely large oil spills, they also can be monitored with satellite scatterometers (Lindsley and Long, 2012).

16-14.1 SAR Observation of Oil Slicks

The presence of oil alters the ocean's radar scattering properties, resulting in a difference between the measured backscatter and the backscatter that would be measured if oil were not present. Generally, this is evident as a reduction in the surface σ^0, which appears as a dark area in a SAR image, surrounded by a brighter ocean surface. Figures 16-85 and 16-86 illustrate examples of SAR images containing oil slicks. The slick in Fig. 16-85 is the result of a naturally occurring oil seep from a geologic source, while the oil slick visible in the upper right section of the image in Fig. 16-86 is from spills introduced by oil rigs. Oil slicks can also originate from deliberate dumping from ships.

Radar-based detection of an oil slick requires that a moderate wind be blowing to roughen the surface by generating capillary waves. Wind that is too low does not generate waves, while high winds mix the oil into the water. Based on the scattering theory discussed earlier, the backscattered radar power is approximately proportional to the spectrum of the short ocean waves present on the surface. Oil dampens or suppresses the generation of these small waves. This dampening is more effective at shorter wavelengths (Alpers and Hühnerfuss, 1989), so shorter-wavelength radars (e.g., X- and C-band radars) are more efficient at detecting oil slicks than longer-wavelength radars (L- or P-band). This efficiency has been confirmed

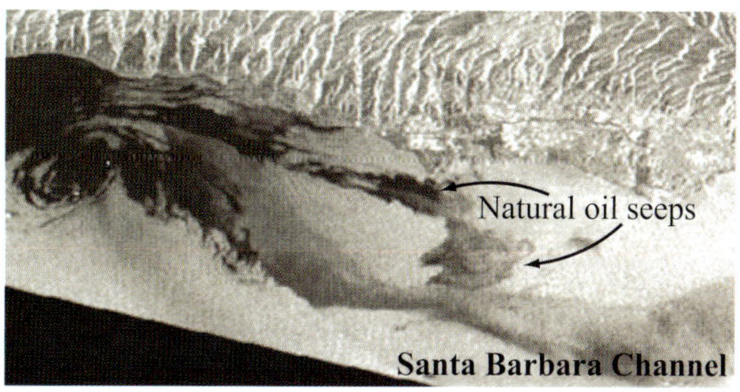

Figure 16-85: C-band SAR image of natural oil slicks in Jan. 1996. [Courtesy of ESA.]

Figure 16-86: SAR image of an oil spill in the Caspian Sea. [Courtesy of ESA.]

experimentally (Johannessen et al., 1995; Gade et al., 1998).

We note that while oil slicks are associated with low backscatter of the water surface, there are other causes of low backscatter that can create dark patches. These "look-alike" conditions can be confused with oil slicks. Such sources include rain cells, very low wind conditions, and the surface expression of internal waves that alter the surface wave spectrum. A trained human operator is often able to discriminate between oil slicks and look-alikes based on experience, visual spill properties, and the presence of man-made sources such as ships and oil platform. Automated algorithms have also been developed and this remains an active area of research. SAR polarimetry is particularly effective in enabling discrimination between man-made oil slicks and look-alikes (Zhang et al., 2011; Migliaccio et al., 2007, 2011).

16-14.2 Scatterometer Observation of Oil Slicks

The 2010 Deepwater Horizon oil spill in the Gulf of Mexico covered a sufficiently large area that it could be observed by the European Space Agency Advanced Scatterometer (ASCAT) on MetOp-A when processed with reconstruction algorithms. In oil-free conditions the σ^0 is used to determine the near-surface wind vector. However, when oil is present, the observed σ^0 is reduced for given wind conditions relative to the oil-free conditions. By comparing the observed σ^0 with that predicted by numerical weather prediction, an incidence-angle–dependent threshold can be selected to detect and map the spatial extent of the oil (Lindsley and Long, 2012). While the scatterometer has much lower resolution than a SAR, its wide swath provides frequent coverage, enabling time-series analysis, as illustrated by Fig. 16-87.

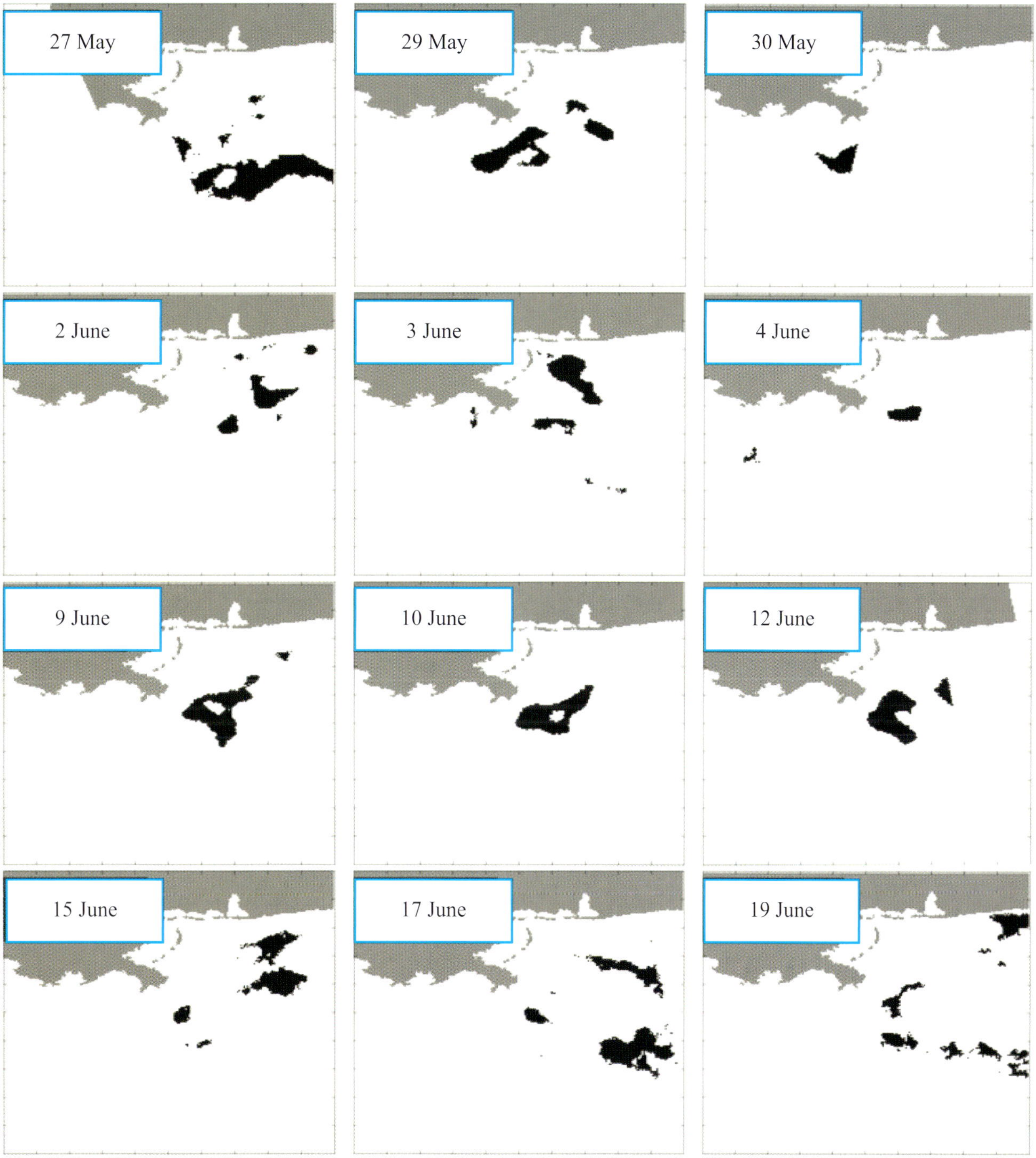

Figure 16-87: Time series of the 2010 Deepwater Horizon oil spill in the Gulf of Mexico derived from the ASCAT scatterometer. Each image is the thresholded backscatter error for the date indicated in the upper left corner. Since the an ASCAT pass may not completely cover the region on a given day, a diagonal crop occurs in some images [Lindsley and Long, 2012].

PROBLEMS

16.1 Reproduce Fig. 16-8 using the C-band CMOD5 model function.

16.2 Using the tabulated Ku-band NSCAT-1 model function available on the book website and the C-band CMOD5 model function, plot σ^0 versus wind speed at 45° incidence angle and 0° azimuth angle. Calculate the GMF wind-speed ***sensitivity*** U_S, defined to be

$$U_S(U,\sigma^0) = \left.\frac{dU}{d\sigma^0}\right|_{U,\sigma^0}.$$

Note that $\sigma^0(U)$ is given by the GMF. For a fixed uncertainty in σ^0, the value of $\Delta\sigma^0$ due to noise and calibration error is related to the corresponding wind-speed error by approximately $\Delta U = U_S(U,\sigma^0)\,\Delta\sigma^0$. For $K_p = 0.02$, calculate ΔU versus wind speed U and comment on the speed sensitivity.

16.3 Rain alters the measured surface σ^0 due to attenuation and added backscatter. Using the sensitivity function developed in Problem 16.2, determine the approximate wind-speed error versus rain rate at 5, 10, and 15 m/s at Ku-band. Comment on the sensitivity of the estimated wind speed to rain.

16.4 For a radar at Ku-band with a fixed rain rate of 10 km-mm/hr at a wind speed of 10 m/s, compute the beam-filling error in the observed σ^0 when the size of the rain cell is one-half the size of the antenna footprint.

16.5 Calculate the 3 dB antenna footprint size for a Ku-band (14 GHz) pencil-beam scatterometer with a 0.5° circular beamwidth operating at an 800 km altitude and antenna incidence angle of 55°. Assume a spherical Earth and an ideal antenna pattern.

16.6 Repeat Problem 16.5 for C-band (5.3 GHz), assuming the antenna size is the same.

16.7 For Problem 16.5, sketch the geometry and compute the slant range to the surface.

16.8 For Problem 16.5, what is the differential slant range between the portions of the antenna footprint closest to and furthest from the radar? For a 100 μs transmit pulse, what is the shortest possible transmit-pulse/receive cycle? Comment on the longest transmit-pulse/receive cycle.

16.9 For Problem 16.5, what is the highest PRF possible with only a single pulse in flight?

16.10 For the scatterometer parameters described in Problem 16.5, and assuming a ground-track velocity of 7 km/s, how long does it take for the nadir track to move 25 km?

16.11 For Problem 16.10, what is the required rotation rate so that the antenna rotates once during the time required for the spacecraft to move 25 km along-track? Compute the change in azimuth angle of the antenna beam during the time it takes for the radar to move 25 km along-track.

16.12 Plot the Doppler shift as a function of azimuth angle for a pencil-beam scatterometer operating at an incidence angle of 45°, an altitude of 800 km, and a ground-track velocity of 7.5 km/s. Assume a nonrotating spherical Earth. Consider both C- and Ku-band.

16.13 Repeat Problem 16.12 but take into account the effects of the rotation of the Earth. Assume the orbit inclination angle is 96.5°. Plot the result for both ascending and descending orbit overpasses at the equator.

16.14 Using the tables and computer code on the book website, write a simple computer program to compute the C-band σ^0 as a function of wind speed and direction for each fan-beam antenna at a fixed across-track distance where the antenna azimuth angles are 45°, 90°, and 135° relative to the scatterometer ground track. Assume a northward-moving track. Then plot, for each measurement, the wind speed versus wind direction that gives the same observed σ^0 for the corresponding antenna beam.

16.15 Using the code developed for Problem 16.14, at a particular wind speed of 10 m/s and for two directions 0° and 45° (meteorological convention), add simulated ***Monte Carlo*** noise with $K_p = 0.05$ to σ^0. Compute and plot the $L_{LS}(U,\psi)$ objective for both cases. How many local optima (wind aliases) are found for each case?

Compute the wind-vector error between the true wind vector and the wind alias closest to the true wind vector.

16.16 Extend the results for Problem 16.15 to compute the root mean square (rms) error over 100 realizations of the Monte Carlo noise for each true wind vector. Repeat for wind directions 0° through 360° at 10° increments. This technique is known as ***compass simulation*** and is useful for predicting the scatterometer wind performance at particular wind speed and swath location. Plot the rms wind speed and wind direction error as a function of the true wind direction. Comment on the rms error versus direction behavior.

16.17 Repeat Problem 16.15 using only two antennas at 45° and 135° azimuth angles. Comment on the results.

16.18 In a scanning scatterometer, the relative azimuth angles of the σ^0 measurements are not fixed as they are in a fan-beam antenna. Compute and plot the azimuth angles for a single-beam scanning scatterometer as a function of cross-track distance. Assume a constant incidence angle of 45° and a radius of scan on the surface of 1200 km.

16.19 Repeat the compass simulation analysis described in Problem 16.16 using the antenna azimuth geometry derived in Problem 16.18. Comment on the results.

CHAPTER 17

Spaceborne Altimetry

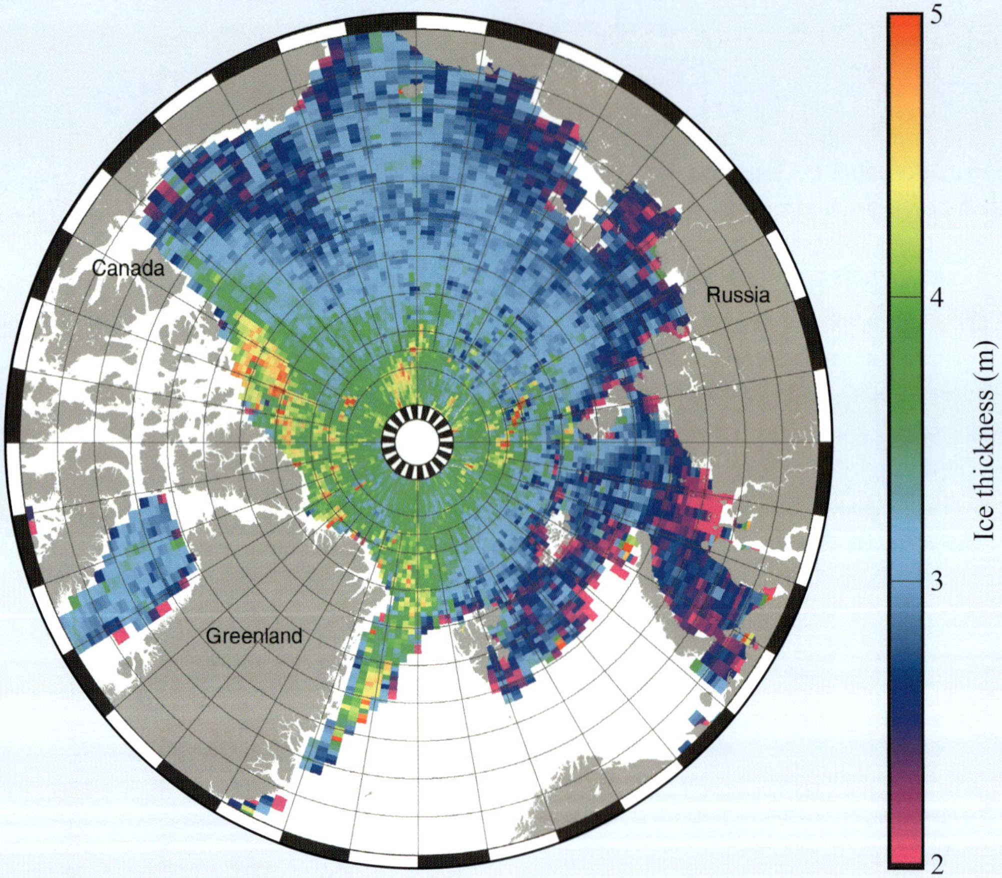

Arctic sea-ice thickness from CryoSat altimeter data

CONTENTS

Overview

A ***spaceborne altimeter*** is a specialized ranging radar designed to measure the distance from a spacecraft to the surface. Coupled with precision orbit information the surface topography can be determined with a high degree of precision. Also used for measuring the topography of cloud-covered planets and moons (Pettengill et al., 1980; West et al., 2009), spaceborne altimeter systems excel in Earth applications where ocean topographical height accuracies to a few centimeters or less have been demonstrated (Fu and Cazenave, 2001). Moreover, the reflected altimeter waveform can be used to extract information about ocean wave height and near-surface wind speed that can in turn be used to track the evolution of ice sheets (Martin et al., 1983) and measure the distribution of icebergs in the southern hemisphere (Tournadre et al., 2008).

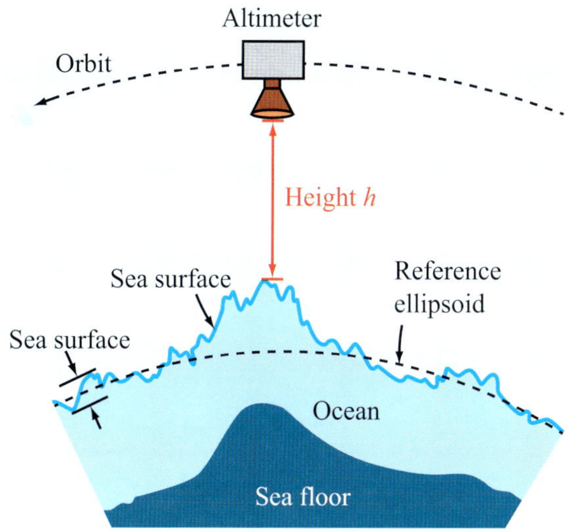

Figure 17-1: Spaceborne altimeter.

17-1 Introduction

A spaceborne altimeter is a ***system*** consisting of a nadir-facing radar, a satellite, an orbit determination system, and a data processing and analysis subsystem. Figure 17-1 depicts a satellite altimeter system in operation. In this chapter we consider some of the subsystems, but our focus is on the radar subsystem. To date, most spaceborne altimeters have been real-aperture systems, though synthetic-aperture altimeters also have been proposed (Elachi et al. 1990; Stephens et al., 2002).

The essential idea in altimetry is to use the time of flight of the radar signal to measure height. The altimeter transmits a modulated pulse toward the surface and measures the ***time delay*** t_d from pulse transmission to receipt of its surface echo. The height h of the altimeter above the surface is then

$$h = \frac{ct_\mathrm{d}}{2} . \tag{17.1}$$

Typically, the pulse is modulated using linear frequency modulation (LFM) and the received signal is range-compressed using a matched filter (see Chapter 13). Based on peak detection, the time delay accuracy Δt_d in measuring t_d is inversely related to the transmit bandwidth B; i.e.,

$$\Delta t_\mathrm{d} = \frac{1}{B} , \tag{17.2}$$

which leads to a fundamental ***height uncertainty*** in measuring the height, namely

$$\Delta h = \frac{c\, \Delta t_\mathrm{d}}{2} = \frac{c}{2B} . \tag{17.3}$$

Additional range errors arise from uncertainties in the system clock used to compute t_d, noise that affects the pulse detection, variations in the propagation speed of the signal through the atmosphere from the radar to the surface, and uncertainty in detecting the leading edge of the echo.

The expression given by Eq. (17.3) suggests that high-precision time measurement requires wide bandwidth. For scientific applications the desired height precision is a few centimeters or less, which would require a bandwidth of many GHz to achieve. But international frequency allocation regulations limit the bandwidth of altimeter signals to 320 MHz or less, depending on the operating band. *Given this restriction, how do we achieve the desired precision?* As described below, to

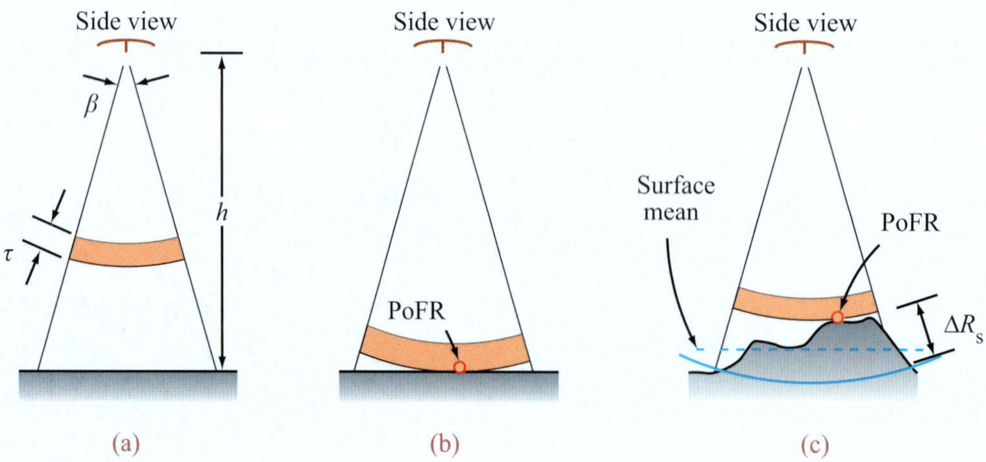

Figure 17-2: Illustration of the effective pulse length τ and the point of first reflection (PoFR) for (a) a pulse in midflight, (b) a flat surface just as the pulse first impacts it, and (c) a rough surface with significant height variation over the footprint. In (c), ΔR_s is the difference between the maximum and minimum slant range within the footprint.

achieve the desired level of precision, a combination of radar design, signal modeling, and signal processing of the received waveform is used to exploit the scattering characteristics of the surface to more accurately estimate the propagation time t_d. The received echo depends on the antenna beamwidth, the antenna (platform) pointing attitude, and the surface roughness and associated radar reflectivity.

In order to provide intuition about how to exploit the return signal characteristics, we delve into the details of the return. In the following, we consider the ***effective*** pulse length τ *after range compression* (see Chapter 13 for details on range compression), which corresponds to the range resolution of the pulse; i.e., $\Delta h = c\tau/2$. Figure 17-2(a) illustrates a transmit pulse in flight. The red arc represents the effective pulse length. The signal transmitted from the altimeter propagates from the transmit antenna to the surface, where it is reflected back toward the satellite. For a flat surface this occurs at the nadir point below the spacecraft (Fig. 17-2(b)). This is the distance we desire to measure. However, if the surface is tilted, or has topography, the ***point of first reflection*** (PoFR) may not be at nadir, as illustrated in Figure 17-2(c), so in general, the desired location point cannot be determined. Our primary interest is in the distance to the ***mean surface height*** rather than in the peak height, so merely identifying the leading edge of the return signal does not meet our needs.

Figure 17-3 illustrates the evolution of the illumination of a circular antenna footprint as a function of time after the signal first strikes the surface. When the surface is flat, the illuminated area maybe either a filled circle or an annulus. Initially, the pulse illuminates a circular area (Fig. 17-3(a)). Note that the range from the altimeter varies from h at nadir to $h_m = h/\cos(\theta_B/2)$ at the edge of the beam, where θ_B is the 3 dB two-way antenna beamwidth. If $\tau > h_m - h$, the pulse completely illuminates the antenna footprint and the return is from the full antenna footprint (Fig. 17-3(b)). This is known as ***beam-limited altimetry***. If $h_m - h > \tau$, the pulse never completely illuminates the footprint, resulting in a return from an annular region (Fig. 17-3(c)). This is ***pulse-limited altimetry***. Figures 17-4(a) and 17-4(b) illustrate idealized profiles of power-versus-time waveform for the return echo for these two cases.

The situation is more complicated for rough surfaces. Figures 17-3(d)–(f) show the evolution of the illuminated area for the pulse-limited case. The topography variations within the footprint lead to slower ramp-up and ramp-down of the power-versus-time profile, as well

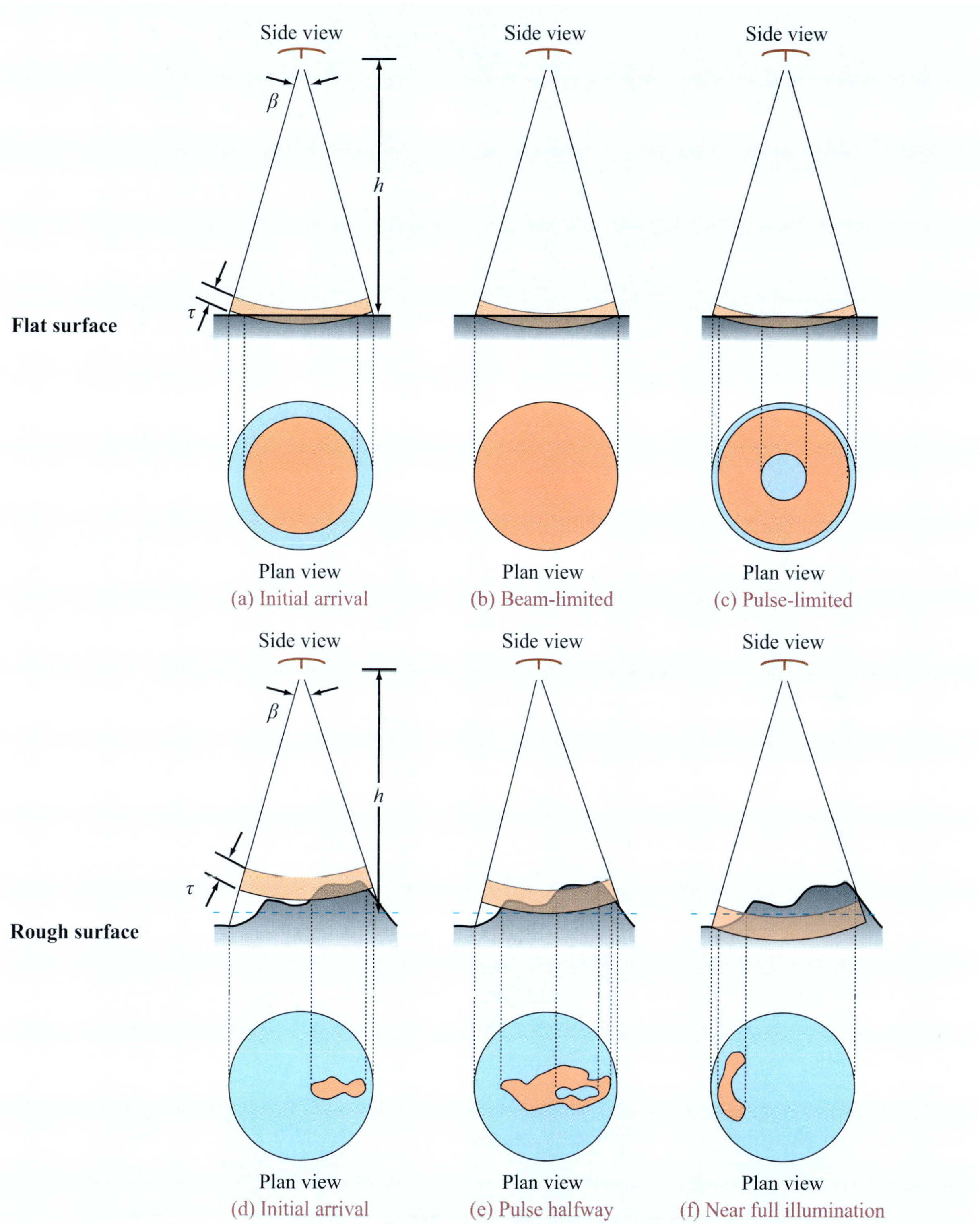

Figure 17-3: Time evolution for a pulse of effective length τ transmitted from height h above the surface by an antenna of beamwidth β: (a) flat surface shortly after pulse first begins reflecting from surface, (b) ***beam-limited altimeter*** case when the pulse fully illuminates the surface, and (c) ***pulse-limited altimeter*** case when the pulse partially illuminates the surface. Parts (d)–(f) pertain to a pulse-limited altimeter over a rough surface.

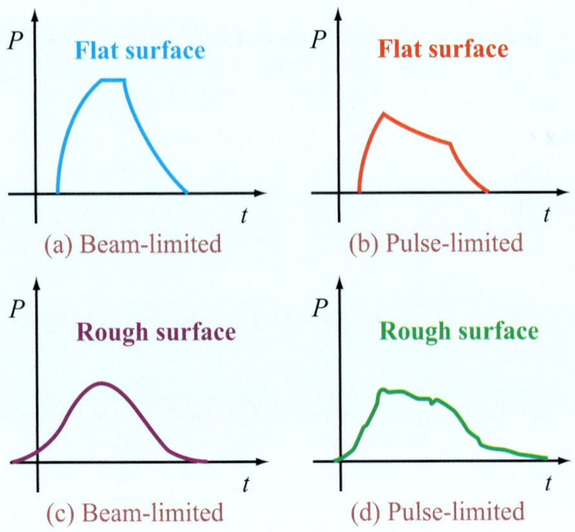

Figure 17-4: Idealized power versus time histories for (a) a beam-limited altimeter over a flat surface, (b) a pulse-limited altimeter over a flat surface, (c) a pulse-limited altimeter over a rough surface, and (d) a beam-limited altimeter over a rough surface.

as noise-like variability. A notional example of the power-versus-time profile for the beam-limited case is illustrated in Fig. 17-4(c). In both the beam-limited and pulse-limited cases, the topography variations within the footprint create noise-like variations in the return echoes that must be dealt with when estimating the pulse delay time, and hence when computing the mean surface height.

17-2 Signal Modeling

Modeling the altimeter echo signal enables high-precision estimation of timing information. The models depend on the surface. We start with the ocean surface model, noting that other surfaces also have been successfully modeled for use in spaceborne altimetry, including land and glacial ice.

17-2.1 Ocean Surface Signal Model

To a first approximation, the ocean's surface is flat and is roughened by waves that dominate its surface roughness. Wind blowing over the surface generates short (wavelength < 1 cm) ***capillary waves*** with small amplitudes that interact to produce longer, higher-amplitude waves known as ***gravity waves***. (A longer discussion on ocean waves is contained in Chapter 16.) Figure 17-5 illustrates an idealized ocean wave spectrum. Gravity waves have more energy than capillary waves, yielding a "red" spectrum for the height displacement. For a well-developed sea, the location of the long-wavelength peak depends on the mean wind speed. Gravity waves can be many meters high and are an important factor in measuring the mean surface height.

For small wave slopes, the ***surface displacement*** $H(x,t)$ of a water wave is given by the approximate

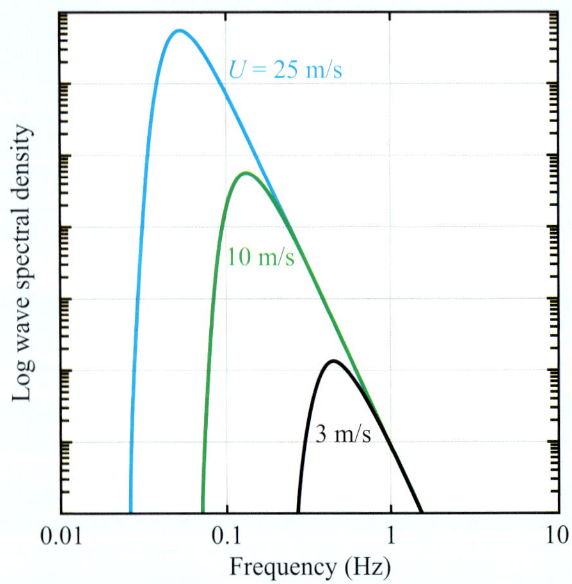

Figure 17-5: The Pierson and Moskowitz (1964) idealized ocean wave spectrum for several wind speeds. Since there is more energy at lower wavenumbers than higher wavenumbers, the spectrum shape is sometimes referred to as a "red" spectrum.

sinusoidal expression

$$H(x,t) = H_0 \cos(Kx - \Omega t), \tag{17.4}$$

where $K = 2\pi/\lambda_w$ is the water wave number, λ_w is the water wavelength, and Ω is the angular frequency of the water wave. The quantities Ω and K are related by the dispersion relations

$$\Omega = \sqrt{gK \tanh(KD)} + \hat{\mathbf{K}} \cdot \mathbf{U} \tag{17.5}$$

for gravity waves and

$$\Omega = \sqrt{BK^3} + \hat{\mathbf{K}} \cdot \mathbf{U} \tag{17.6}$$

for capillary waves, where g is the acceleration of gravity, D is the water depth, $B = 73$ cm^3/s^2 is the ***surface tension to water density ratio***, $\mathbf{U}$ is the ***mean surface current***, and $\hat{\mathbf{K}}$ is a unit vector denoting the propagation direction. The wave spectrum is the Fourier transform of $H(x,t)$. The Fourier transform with respect to x is known as the ***wavenumber spectrum*** and is denoted by $F(K)$ while the Fourier transform with respect to t is the ***frequency spectrum*** denoted $S(\Omega)$. By Parseval's theorem, $F(K)\,dK = S(\Omega)\,d\Omega$. The mean ocean height $\bar{h}$ is the spectrum at $\Omega = 0$ while the height variance $\overline{h^2}$ is the integral of the spectrum,

$$\overline{h^2} = \iint F(K)\,dK = \int S(\Omega)\,d\Omega. \tag{17.7}$$

Since the radar pulse is very short compared with the motion of the surface, we can assume that the surface is stationary during the pulse echo period. An intuitive initial analysis of the received signal from a fixed surface of height $H(\theta,\phi)$ at time t can be obtained with the aid of a binary-valued indicator function $I(\theta,\phi,t)$. This function has value one if the height at a given location in θ and ϕ falls within an equivalent "pulse length" distance $c\tau/2$ at time t and is zero otherwise. Formally,

$$I(\theta,\phi,t) = \begin{cases} 1, & R_\theta - c(t-\tau)/2 \le H(\theta,\phi) \le (R_\theta - ct), \\ 0 & \text{otherwise}, \end{cases} \tag{17.8}$$

where θ and ϕ are the elevation and azimuth angles within the antenna beam and R_θ is the slant range from the antenna to the mean surface height. For a smooth surface, the indicator function outlines a circular or annular region at a given moment as shown in Figs. 17-3(a)–(c). For a rough surface, the indicator function defines a more complicated area, as illustrated in Figs. 17-3(d)–(f). Using the indicator function, the instantaneously illuminated area $A_I(t)$ at a given time is given by

$$A_I(t) = \int_0^{2\pi} \int_0^{\pi} I(\theta,\phi,t)\, R_e^2 \sin\theta\, d\theta\, d\phi, \tag{17.9}$$

where R_e is the local radius of the Earth.

Note that because the surface is random, $I(\theta,\phi,t)$ should be averaged over many possible realizations of the random surface. This average is given by

$$\langle I(\theta,\phi,t)\rangle = \int_{-\infty}^{\infty} H(x,t)\, I(\theta,\phi,t)\, dx \tag{17.10}$$

$$= \int_{R_\theta - c(t-\tau)/2}^{R_\theta - ct/2} H(x,t)\, dx. \tag{17.11}$$

The corresponding average value of the illuminated area is

$$\langle A_I(t)\rangle = 2\pi R_e^2 \int_0^{\pi} \int_{R_\theta - c(t-\tau)/2}^{R_\theta - ct/2} H(x,t) \sin\theta\, d\theta\, d\phi. \tag{17.12}$$

For historical reasons the wave-height distribution is characterized by the average crest-to-trough height of the upper one-third of the highest waves, which is known as the ***significant wave height*** and is denoted by $H_{1/3}$. In satellite altimetry, *$H_{1/3}$ is traditionally defined to be four times the standard deviation of the surface height* (see Chelton et al., 2001).

Figure 17-6 illustrates $\langle A_I(t)\rangle$ for various values of $H_{1/3}$ for a particular set of altimeter parameters. Note that as the significant wave height increases, the rate of change of area with time flattens out. The power received is approximately proportional to the illuminated area and so the rise time of the leading edge of the pulse changes with significant wave height. This sensitivity of the pulse rise time to $H_{1/3}$ makes it possible to estimate the significant wave height from the pulse rise time.

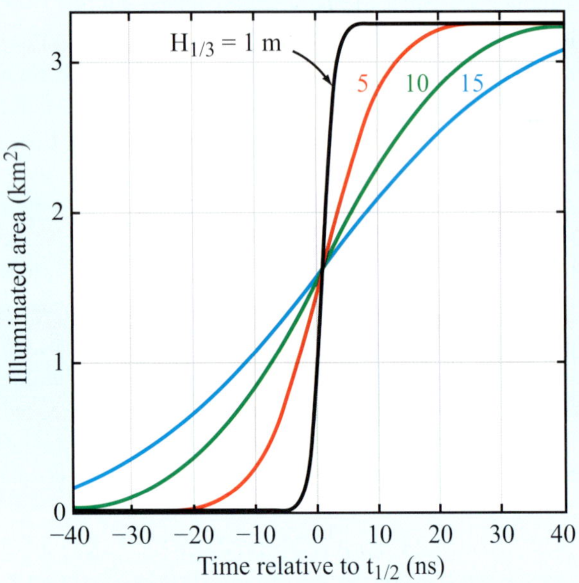

Figure 17-6: An example of the average illuminated area versus time. Time is displayed in nanoseconds relative to $t_{1/2}$, the midpoint of the pulse reflected from the mean sea surface.

Equation (17.12) calculates the average illuminated area for a fixed random surface, but does not include the height probability density function $p_{\rm H}(x)$. For a more precise computation we need to incorporate the height pdf. Over a sufficiently large footprint the central limit theorem suggests that the pdf of sea surface height can be treated as a Gaussian function:

$$p_{\rm H}(x) = \frac{1}{(2\pi)^{1/2} s_{\rm H}} \exp\left\{-\frac{1}{2}\left(\frac{x}{s_{\rm H}}\right)^2\right\}, \qquad (17.13)$$

where $s_{\rm H}$ is the ***standard deviation of the sea surface height***. Since $s_{\rm H}$ is one-fourth of the significant wave height $H_{1/3}$, the surface height distribution of Eq. (17.13) can be written as

$$p_{\rm H}(x) = \frac{4}{(2\pi)^{1/2} H_{1/3}} \exp\left\{-\frac{1}{2}\left(\frac{4x}{H_{1/3}}\right)^2\right\}, \qquad (17.14)$$

where $s_{\rm H}$ is replaced with $(H_{1/3})/4$.

The expected value of the average illumination area is equal to the convolution of the height probability density function with the range pulse response, which can be expressed as

$$\langle A_I(t)\rangle = p_{\rm H}(t) * p_\tau(t) * u(t - t_{1/2}), \qquad (17.15)$$

where $*$ denotes convolution, $p_{\rm H}(t)$ is the probability density function for the sea surface height expressed in terms of two-way travel time $t = 2x/c$, $p_\tau(t)$ is the ***effective two-way system range response function***, $u(t)$ is the unit step function, and $t_{1/2} = 2R/c + t_{\rm d}$ is the two-way time of flight corresponding to the half-power point. The effective system range response function after range compression is typically sinc-like (see Chapter 13), i.e., $p_\tau(t) \approx {\rm sinc}^2(t)$, where ${\rm sinc}(x) = \sin(\pi x)/(\pi x)$, rather than the "box-car" or "rect" response used in computing Eq. (17.12). For a particular range response function, the expression in Eq. (17.15) can be evaluated numerically.

Using the radar equation, the average received power $P(t)$ versus time is approximately

$$P(t) = \frac{P_t \lambda^2\, G^2(t)\, \sigma^0}{(4\pi)^3\, R^4(t)} \langle A_I(t)\rangle, \qquad (17.16)$$

where $G(t)$ is the antenna gain expressed as a function of the two-way travel time and σ^0 is the average normalized radar cross section over the footprint. Over the ocean, σ^0 is related to the near-surface wind speed, as discussed later in Section 17-3.4.

Since the nominal platform height $\bar{H}$ is much greater than the variation in range due to the topography within the footprint, we can use the nominal range to nadir $\bar{R}$ in the denominator of Eq. (17.16) with little error. We can then write $P(t)$ as (Chelton et al., 2001)

$$P(t) = P_0\, p_{\rm H}(t) * p_\tau(t) * P_{\rm FS}(t), \qquad (17.17)$$

where $P_{\rm FS}(t) = G(t)\, u(t - t_{1/2})$ and the constant P_0 is given by

$$P_0 = P_t \lambda^2 \sigma^0/(4\pi)^3 \bar{R}^4. \qquad (17.18)$$

Thus, the received power-versus-time signal is the combined convolution of the surface height (expressed in two-way travel time), the effective system range response, and the antenna pattern. Figure 17-7 illustrates

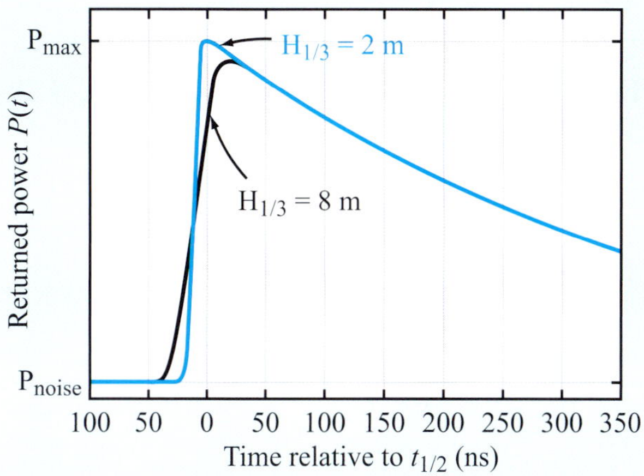

Figure 17-7: An example of $P(t)$ versus time. Time is displayed in nanoseconds relative to midpoint of the pulse reflected from the mean sea surface. [Adapted from Chelton et al. (2001).]

$P(t)$ for two values of $H_{1/3}$ for the system configuration corresponding to the TOPEX/Poseidon altimeter. From the waveform shape, we can extract the slope of the leading edge to estimate $H_{1/3}$ and the half-power point to estimate the sea surface height, as discussed in more detail in Section 17-4.

Equation (17.17) describes the average power versus time for a pulse. However, realizations of individual pulses are noisy due to thermal noise and the particular realization of the random surface. To mitigate the effects of noise, multiple pulses are transmitted and averaged to extract estimates of the height and leading edge slope. As more pulses are averaged, the noise level of the average decreases, but movement of the platform during the averaging period degrades the along-track resolution due to averaging of the individual pulse footprints, leading to a trade-off between noise level and along-track resolution. (We note that cross-track resolution is determined by the antenna beamwidth.) The average thermal SNR versus time is $P(t)/kT_{\text{REC}}B$ where T_{REC} is the receiver noise temperature and B is the receiver bandwidth. The SNR for a particular pulse depends on the surface height realization. Averaging multiple pulses improves the SNR. For example, the TOPEX/Poseidon altimeter averages 128 pulses.

17-2.2 Land-Surface Signal Modeling

Signal modeling for land is more complicated than the ocean since very large changes in surface height can occur over short distances. Over land, the statistical model developed for ocean waves is not generally applicable and comparatively less has been done on land signal modeling. Notable exceptions are the relatively smooth regions of glaciated Greenland and Antarctica. Most of the effort has focused on the development of "re-tracking" algorithms of altimeters optimized for ocean observation in order to improve performance over land (see Section 17-4.2).

In addition to more height variability, land surfaces also exhibit considerably more variation in backscatter.

▶ Vegetated areas can have very low backscatter, resulting in unreliable height estimates due to low SNR. On the other hand, glacial ice exhibits very high radar backscatter due to volume scattering. ◀

Altimeter measurements over the great ice sheets are used to study surface elevation changes and mass balance to better understand climate change. Also, while ocean scattering occurs only at the surface, microwave radar signals can penetrate to significant depths in vegetation, snow, and ice. This causes a spread in the observed pulse profile. Sea-ice backscatter is highly variable depending on thickness, roughness, snow cover, and melt conditions.

A key difference between ocean and glacial ice-sheet scattering is related to surface slope: while the mean slope of the ocean's surface is small, surface slope is significant over the great ice sheets of Greenland and Antarctica. Accordingly, we revisit our discussion of pulse-limited altimetry (introduced in Section 17-2) to consider the effects of mean surface slopes. Compare Figs. 17-8(a)–(c), which illustrate pulse-limited operation from a smooth, flat surface, with Figs. 17-8(d)–(f), which illustrate pulse-limited operation on a surface with nonzero slope. For a surface

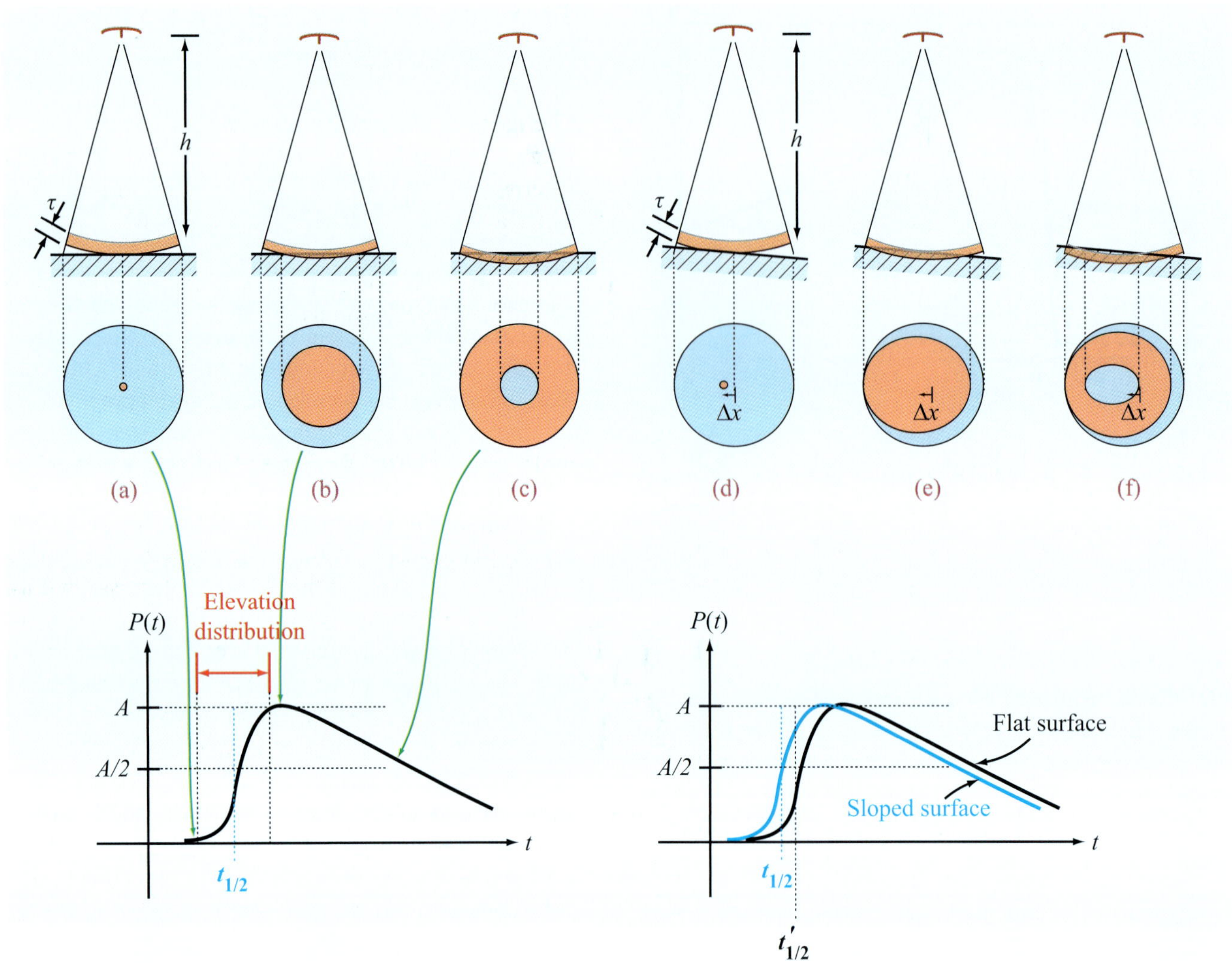

Figure 17-8: Comparison of pulse-limited altimetry for a flat surface (left) and a sloped surface (right). A schematic of the timing of the illumination and mean pulse waveform is indicated for each case. At the bottom, the black line corresponds to the waveform for the flat surface, while the blue line shows the waveform for the sloped-surface case.

with a nonzero mean slope, the initial return and mean power point no longer correspond to the point directly below the altimeter, but are shifted horizontally. As illustrated in Fig. 17-9, the vertical displacement Δh of the *observed* height is offset from the nadir point by

$$\begin{aligned} \Delta h &= \Delta \ell \sin\alpha \\ &= (h\sin\alpha)\sin\alpha = h\sin^2\alpha, \end{aligned} \tag{17.19}$$

where α is the mean surface slope. For small slopes,

$$\Delta h \approx h\alpha^2. \tag{17.20}$$

The horizontal displacement Δx of the reflection point is

$$\Delta x = h\cos\alpha\sin\alpha. \tag{17.21}$$

The direction of the displacement cannot be determined from a single pulse. An initial digital elevation map

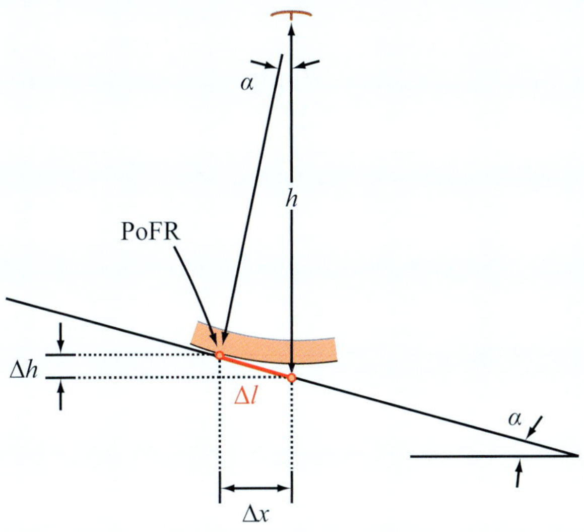

Figure 17-9: Geometry of sloped-surface height error.

(DEM) is required to estimate the displacement and make corrections. Because the slope and displacement vary along the altimeter track and must be compensated for when estimating the height, a glacial ice-height estimation algorithm is significantly more complicated than ocean-height estimation. As in ocean altimetry, the distribution of surface heights within the beam footprint dictates the average slope of the leading edge of the pulse, as seen in Fig. 17-8. However, the observed scattering in not just from the air/ice interface, but also includes volume scattering from ice grains buried beneath the surface. Volume scattering, and thus the depth of the effective scattering surface, is a function of ice grain size and density, which varies spatially and temporally as a function of season. The formation of liquid water in the top layer of the snow or ice dramatically reduces the penetration depth. Thus care must be taken when estimating altimeter-derived sea-ice surface heights.

17-3 Height Estimation Errors and Corrections

The goal of a satellite altimeter is to measure the height of the surface by subtracting the measured radar-determined range to the surface from the platform height (refer to Fig. 17-1). The latter requires using precision orbit determination methods. As discussed below, there are a number of considerations affecting the accuracy of the radar range measurement.

17-3.1 Precision Orbit Determination

An essential element of the system for spaceborne altimetry is orbit determination since any errors in orbit determination directly affect the estimated surface height. The details of the orbit determination process are outside the scope of this book, but the essential idea is to precisely model the orbit trajectory in the Earth's gravitational field, including the effects of atmospheric drag and radiation pressure, and to use measurements of the spacecraft velocity and/or position to constrain the orbit estimate. Since the gravitational field varies with tides and surface topography (which is what we are trying to measure), iterative methods may be needed. Part of the original motivation for flying altimeters was to better determine the gravity field to facilitate better spacecraft orbit prediction for other missions. This effort continues today.

Dynamic modeling of spacecraft orbits and orbit determination are well-developed fields (Seeber, 2004; Tapley et al., 2004). High-precision orbit models employ high-order spherical harmonic models for the Earth's gravity field and include many hundreds of terms. Factors such as atmospheric drag and radiation pressure from solar illumination are included in the models. These factors depend on spacecraft attitude, solar activity, and atmospheric density. The latter, while tenuous at orbital altitudes, varies over the year and from day to night and is an important factor at lower orbits. Higher orbits minimize the effects of the high-order gravity field terms and atmospheric drag. Partially for these reasons, the TOPEX/Poseidon altimeter employed a 1336 km orbit altitude rather than the lower polar

Table 17-1: Typical midafternoon values of the corrections due to atmospheric and sea-state effects for the 13.6 GHz TOPEX/Poseidon satellite altimeter (adapted from Chelton et al., 2001).

	Mean adjustment (cm)		Time-variation (cm)	
Source	Equatorial	Midlatitude	Equatorial	Midlatitude
Tropospheric refraction				
Dry	226	226	0.5	2
Wet	24	10	6	5
Ionospheric refraction	12	6	5	2
Sea-state bias	4	6	1	3

Equatorial latitudes: 30S–30N deg Midlatitudes 60S–30S deg & 30N–60N

orbits (700–800 km) more commonly used for remote sensing satellites.

Spacecraft motion measurements are made with the aid of multiple global networks of ground stations, which measure spacecraft velocity using lasers and/or the Doppler shift of beacon signals transmitted from the spacecraft. The Global Positioning System (GPS) is yet another tool. GPS systems employ conventional position measurements along with carrier phase measurement for precision Doppler determination.

17-3.2 Atmospheric Effects

The radar measures the time of flight of the transmitted signal. This is converted to distance using the speed of light. However, the precise wave velocity in the atmosphere varies based on atmospheric conditions. In practice, a constant speed of light is generally assumed, and a correction factor for atmospheric effects is applied to the resulting estimated time-of-flight range. Other factors affect the accuracy of the range estimate and are compensated for by estimating a bias correction.

Complicating the altimeter signal model previously presented in Section 17-2 are the effects of signal propagation through the atmosphere. As discussed in Chapter 9, variations in atmospheric density and thickness, as well as the presence of water vapor in the lower part of the atmosphere and free electrons in the ionosphere, alter the wave velocity compared with that in vacuum, resulting in signal refraction and delay. At the level of precision (~ 1 cm) required for satellite altimetry, these effects, which vary with position and time, must be accounted for. A detailed analysis of all of these effects is outside of the scope of this chapter, but Table 17-1 summarizes the magnitudes of these effects on height measurement for a spaceborne altimeter. Note that tropospheric effects vary with atmospheric moisture and temperature profiles, while ionospheric conditions vary with location, sunspot activity, time of day, and season.

Since ionospheric delay varies with frequency, the difference in height measured at different frequencies can be exploited when estimating a range correction term for ionospheric delay. For example, a three-step procedure was used for computing range estimates for the TOPEX/Poseidon altimeter, which collected range estimates at both Ku-band and C-band. First, range estimates were collected without taking into account atmospheric variations. These range estimates were then separately corrected for sea-state bias (see next section). Finally, ionospheric and tropospheric range corrections were computed and applied.

The range correction ΔR_{atm} due to atmospheric delay can be written as

$$\Delta R_{\text{atm}} = \frac{ct_{1/2}}{2} - \frac{c}{2}\int_0^{t_{1/2}} \frac{1}{n}\,dt = \frac{c}{2}\int_0^{t_{1/2}} \frac{n-1}{n}\,dt, \tag{17.22}$$

where c is the speed of light in free space and n is the real part of the index of refraction (which may vary with height, and thus with t, as discussed in Section 9-8). Two separate ***range-delay corrections*** are applied, one for propagation through the ionosphere and another for propagation through the troposphere:

$$\Delta R_{\text{atm}} = \Delta R_{\text{ion}} + \Delta R_{\text{trop}}. \tag{17.23}$$

Propagation in the troposphere is nondispersive, so the index of refraction n can be treated as a constant across the signal bandwidth. In contrast, the ionosphere is a dispersive medium, which means that n does vary with frequency. This difference between these two layers necessitates the use of different models for n.

Ionospheric path length

At microwave frequencies, the real part of the ionospheric index of refraction, $n'_{\text{ion}}(z)$ at altitude z, is related to the number density of electrons $n_{\text{e}}(z)$ (in electrons/m^3) by the approximate expression (Chelton et al., 2001)

$$n'_{\text{ion}}(z) \approx 1 + \frac{40.3}{f^2}\, n_{\text{e}}(z), \tag{17.24}$$

where f is the frequency in Hz. We note that in the referenced source, the expression includes a factor of 10^6 because in there n_{e} is expressed in electrons/cm^3. The plasma frequency f_{p}, which is the natural frequency of oscillation of the electrons in the ionospheric layer, is related to n_{e} by

$$f_{\text{p}} = (80.6 n_{\text{e}})^{1/2}. \tag{17.25}$$

The electron density is highest during daylight hours ($n_{\text{e}} \approx 10^{12}$ electrons/m^3) and the corresponding plasma frequency is $f_{\text{p}} \approx 9$ MHz.

For vertical propagation through an ionospheric layer of height R, the ***ionospheric excess path length*** ΔR_{ion} is the incremental path length associated with the deviation of n'_{ion} from unity,

$$\begin{aligned}\Delta R_{\text{ion}} &= \int_0^R (n'_{\text{ion}} - 1)\, dz \\ &= \frac{40.3}{f^2} \int_0^R n_{\text{e}}(z)\, dz = \frac{40.3}{f^2}\, N_{\text{ed}}(z),\end{aligned} \tag{17.26}$$

where N_{ed} is the ***integrated electron density*** in electrons/m^2. A standard unit for N_{ed} is the TECU,

Table 17-2: Typical low and high ionospheric path lengths, corresponding to $N_{\text{ed}} = 1$ TECU and 100 TECU, respectively.

Frequency	Low ΔR_{ion}	High ΔR_{ion}
C-band (5.3 GHz)	1.43 cm	143.1 cm
Ku-band (13.6 GHz)	0.2 cm	21.8 cm
Ka-band (35 GHz)	0.03 cm	3.3 cm

which stands for ***total electron content unit*** and it is defined such that 1 TECU $= 10^{16}$ electrons/m^2. Hence,

$$\Delta R_{\text{ion}} = \frac{40.3}{f^2} \times 10^{16} N_{\text{TECU}}, \tag{17.27}$$

where N_{TECU} is N_{ed} measured in TECUs. Over a diurnal cycle, N_{ed} may vary between a nighttime low of 0.2 TECU and a daytime high of about 100 TECU. However, N_{ed} may reach as high as 200 when sunspots are highly active. As noted in Table 1-3, most altimeters operate at C-band (5.3 GHz), Ku-band (13.6 GHz), or Ka-band (35 GHz). The corresponding typical ranges of the ionospheric path length ΔR_{ion} are summarized in Table 17-2. Because ΔR_{ion} is inversely proportional to f^2, it is much smaller at 35 GHz than at the lower frequencies.

Single-frequency altimeters must rely on independent measurements of the integrated electron density N_{TECU} in order to determine ΔR_{ion}. Global maps of N_{TECU} can be derived from GPS measurements. Alternatively, ΔR_{ion} can be determined without knowledge of N_{TECU} by operating a dual-frequency altimeter.

If R_1 and R_2 are the ranges measured by a two-frequency altimeter operating at frequencies f_1 and f_2, respectively, then

$$R_1 = R_{\text{t}} + \Delta R_{\text{ion}_1} \tag{17.28a}$$

and

$$R_2 = R_{\text{t}} + \Delta R_{\text{ion}_2}, \tag{17.28b}$$

where R_{t} is the true range and ΔR_{ion_1} and ΔR_{ion_2} are the excess ionospheric path lengths. The difference between

R_2 and R_1 is

$$R_2 - R_1 = \Delta R_{\text{ion}_2} - \Delta R_{\text{ion}_1} = a\left(\frac{1}{f_2^2} - \frac{1}{f_1^2}\right), \tag{17.29}$$

where $a = 40.3 \times 10^{16} N_{\text{TECU}}$. Also,

$$\Delta R_{\text{ion}_1} = \frac{a}{f_1^2}\,. \tag{17.30}$$

Using Eq. (17.29) in Eq. (17.30) to replace a leads to

$$\Delta R_{\text{ion}_1} = \frac{1}{f_1^2}(R_2 - R_1)\left(\frac{1}{f_2^2} - \frac{1}{f_1^2}\right)^{-1} = \left(\frac{\delta_{\text{f}}^2}{1-\delta_{\text{f}}^2}\right)(R_2 - R_1), \tag{17.31}$$

where $\delta_{\text{f}} = f_2/f_1$. Using Eq. (17.31) in Eq. (17.28a) leads to an estimate of the true range R_{t} as

$$R_{\text{t}} = R_1 - \Delta R_{\text{ion}_1} = R_1 - \left(\frac{\delta_{\text{f}}^2}{1-\delta_{\text{f}}^2}\right)(R_2 - R_1). \tag{17.32}$$

Tropospheric path length

According to the discussion in Section 9-8, the ***tropospheric excess path length*** ΔR_{tropo} consists of a dry component ΔR_{d} and a wet component associated with water vapor, ΔR_{v}:

$$\Delta R_{\text{tropo}} = \Delta R_{\text{d}} + \Delta R_{\text{v}}. \tag{17.33}$$

Given an accurate measurement of atmospheric pressure at the surface, ΔR_{d} can be estimated with an uncertainty of less than 1 cm, and ΔR_{v} also can be estimated with an uncertainty of 1 cm through measurements by a downward-looking satellite radiometer with a channel in the 20 GHz range (Section 9-8).

17-3.3 Sea-State and EM Bias

The altimeter echo is the result of specular backscattering from the surface. The shape of the echo is

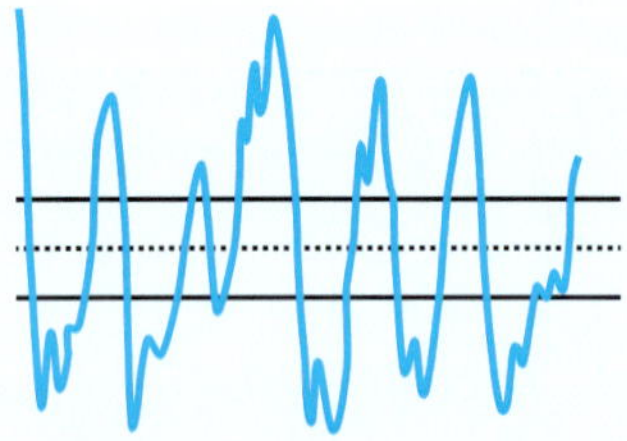

Figure 17-10: Illustration of mean surface heights.

determined by the height distribution of the specular echoes from the wave-roughened surface. For a given specular facet on the surface, the backscattered power is proportional to the local radius of curvature of the surface. As a result, the backscattered signal from the wave troughs is larger than that from the wave crests. This has the effect of biasing the measured surface height towards a lower estimate; i.e., the mean scattering surface appears lower than the mean surface height, as depicted in Fig. 17-10. This height bias, which depends on wind conditions, is frequently termed the ***EM bias***. In addition, the shapes of ocean waves are not quite symmetric but are skewed; i.e., the mean and median surface heights differ. This results in a ***skewness bias***. The sum of the EM and skewness biases is termed the total ***sea-state bias***. To accurately estimate the actual mean surface height, the sea-state bias factor must be compensated for.

The EM bias depends on many properties of the surface. However, since a correction for the EM bias must be estimated from the altimeter return, the EM bias correction has traditionally been expressed in terms of the significant wave height $H_{1/3}$; e.g.,

$$\Delta R_{\text{EM}} = -bH_{1/3}, \tag{17.34}$$

where b is a unitless parameter. It has been experimentally determined that the values of the parameter b depends on wind speed and $H_{1/3}$. Figure 17-11 illustrates the variation of b with these parameters at Ku-band (Chelton et al., 2001).

A number of theoretical approaches to determining b have been developed (e.g., Elfouhaily et al., 2000), but empirical estimates of b have proven superior in

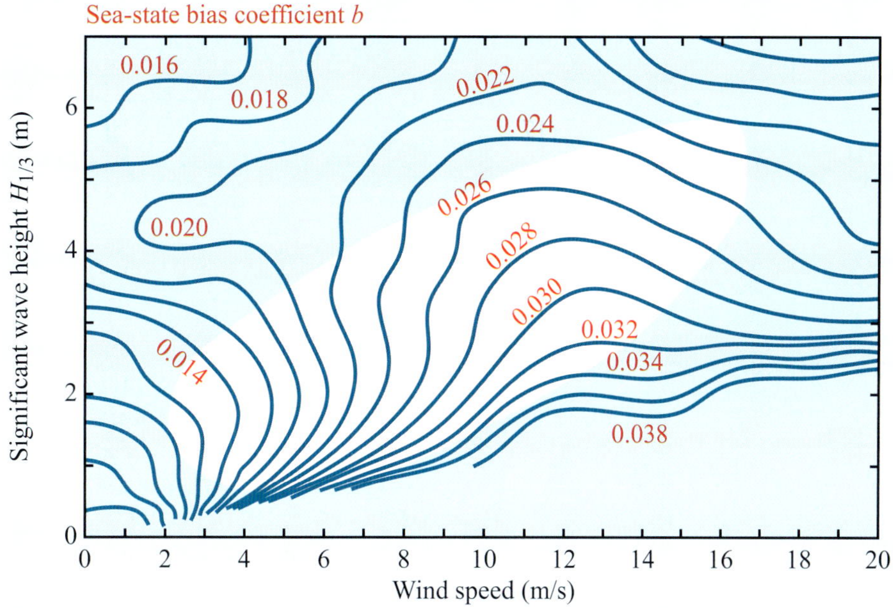

Figure 17-11: Empirically estimated coefficient b (unitless) for sea-state bias using Eq. (17.34). The white area contains greater than 99% of wind speed/$H_{1/3}$. [From Chelton et al. (2001).]

practice—in part because they include all effects linearly related to $H_{1/3}$ and so they can incorporate skewness bias corrections. Empirical estimates are determined by comparing the scattering surface height measured with an altimeter with mean surface heights determined by other sensors such as buoys.

17-3.4 Significant Wave Height

As we have seen, the shape of the altimeter echo is related to the significant wave height $H_{1/3}$, which can be inferred from the slope of the leading edge (see Fig. 17-7). Significant wave height is a measure of large waves on the surface and is used to parameterize some of the height bias terms. The value of $H_{1/3}$ inferred from the leading edge slope is dependent on the assumptions of the signal model. In practice $H_{1/3}$ is estimated on-board the spacecraft as part of the range tracking system (see Section 17-4.2); however, improvements to the $H_{1/3}$ estimate can be made with ground-based processing where the dependence of the waveform shape on additional model parameters can be considered.

Altimeter-derived $H_{1/3}$ values are validated against open-ocean buoy measurements. Validation is complicated by the fact that the altimeter measurements are essentially instantaneous spatial averages over the footprint while measurements recorded by a buoy are temporal averages at a single point. Various validation studies have suggested a 25–45 cm rms error for the altimeter-estimated value of $H_{1/3}$.

17-3.5 Wind Speed

The backscattered power observed in the altimeter signal depends on the surface σ^0 in accordance with using the radar equation [Eq. (17.16)]. Computing the surface σ^0 from the measured signal requires accurate calibration of the transmit power and antenna gain, as well as compensating for the receiver gain. As we have seen, the average area can be determined from the shape of the waveform. Signal attenuation due to atmospheric effects

Table 17-3: Scales of ocean topography variations.

Source	Height Magnitude	Spatial Scale (km)	Time Scale
Seamounts/bathymetry	±1–20 m	100–200	—
Tides	1–16 m depending on location	100–1000	hours
Large scale currents	1 m	100	months
Mesoscale currents	1 m	100	days
Coastal winds	few m	10–100	days
Large scale winds (El Niño/La Niña)	10 cm	10000	months
Pressure field	few cm	100	hours
Tsunami (deep ocean)	few cm	1–2	seconds

must be compensated for as well. With careful design and calibration, a σ^0 measurement-precision capability of a few tenths of a dB can be achieved.

As discussed in Chapter 16, over the ocean the surface σ^0 is related to the near-surface wind speed. For a nadir-looking altimeter, σ^0 is largest at low wind speeds when the surface is smooth and is reduced as the wind roughens the surface, causing scattering of the signal away from the altimeter direction. The relationship between the near-surface wind speed and σ^0 is termed the ***geophysical model function*** (GMF). Theoretical and empirical approaches for deriving altimeter GMFs have been developed. The Freilich and Challenor (1994) Ku-band GMF can be written as

$$\sigma^0 = 12.40 - 0.2459U + 8.956\exp(-0.9593U), \tag{17.35}$$

where U is the ***neutral-stability wind speed*** at a 10 m height above the surface. This model is in close agreement with other GMFs. The wind speed U is estimated from measurements of σ^0 by inverting this equation using least-squares or maximum-likelihood techniques. Wind speed accuracies of 1.67–1.75 m/s (rms error compared with anemometers on ocean buoys) have been reported (Chelton et al., 2001).

17-3.6 Topography

While widely used, the term "sea level" is a misnomer. The sea surface has significant height variations over a variety of scales. At an appropriate spatial scale, the sea-surface topography can written as

$$h = h_{\text{geod}} + h_{\text{dyn}} + h_{\text{tide}} + h_{\text{atm}}, \tag{17.36}$$

where h_{geod} is the ***surface undulation of the geoid*** relative to the reference ellipsoid, h_{dyn} is the ***dynamic surface height*** due to ocean currents, h_{tide} is the ***tidal height***, and h_{atm} is the ***surface height due to atmospheric loading***. Table 17-3 summarizes the magnitude of ocean height perturbations over spatial and time scales. Figure 17-12 reproduces a well-known plot of sea-surface height over a transect of the Gulf Stream in the Atlantic Ocean measured from a spaceborne altimeter. Key features in the altimeter track are identified.

The ocean's surface follows an equipotential surface based on the local gravity field, modified by tides, winds and waves. This equipotential surface defines the geoid and corresponds to the mean sea level. To first order, the geoid is an ellipsoid, but includes higher order terms resulting from inhomogeneities in the Earth's gravitation field and local mass distribution. These higher order terms, called ***undulations***, cause the geopotential surface to vary by up to 100 m from an ideal ellipsoid.

Temporal variations result from tidal pull from the moon and sun, as well as wind and currents. Ocean tides vary significantly from place to place and with time and can be many meters in some areas.

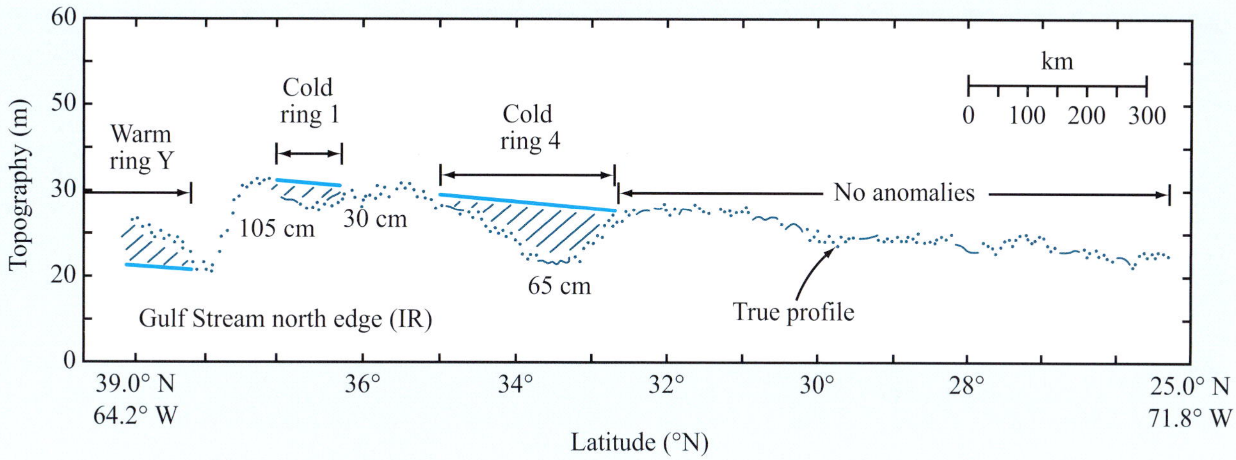

Figure 17-12: Ocean topography across the Gulf Stream. Small-scale eddies known as cold rings produce small-scale variations. The Gulf Stream creates the slope. [Adapted from Cheney and Marsh, 1981.]

Winds and atmospheric-pressure differential also shift the surface, and internal and ocean currents lift or depress the surface. For example, winds blowing over long fetch distances push the water, piling against coastlines. This can cause height changes of a few meters in shallow water. Wind-induced ocean topography changes of a few tens of cm are a signature of periodic El Niño/La Niña events. Ocean currents can cause variations of up to a meter. Differential atmospheric pressure variations over large scales can depress or raise the sea surface.

A key goal of spaceborne altimetry is to accurately measure the dynamic sea-surface topography, from which currents can be extracted. Essential to this is proper spatial and temporal sampling, which is dictated by the orbit (Section 17-4.3). An example of an image of derived ocean topography from TOPEX/Poseidon is shown in Fig. 17-13. The dynamic topography is estimated from the measured height after subtracting the geoid, tidal, and atmospheric effects. Uncertainties in these terms currently limit the spatial scales of dynamic sea-surface topography to fairly large scales.

17-3.7 Bathymetry

Bathymetry is measuring seafloor topography. Satellite altimeters have been remarkably successful at observing seafloor topography, and have located previously unknown seamounts and other undersea features. The basic theory for predicting seafloor topography from satellite altimeter measurements is summarized by Dixon et al. (1983). A detailed analysis is outside the scope of this chapter; however, the essential idea is to relate the observed gravity field to variations in seafloor topography. Since the ocean has essentially a constant density, the primary concern is the variation in crustal density, thickness and sediment density and load. Ship-based depth soundings are used to constrain the long-wavelength depth with shorter-wavelength topography derived from the satellite gravity measurements (Smith and Sandwell, 1994; Sandwell and Smith, 2001).

17-4 Practical Sensor Considerations

Cost and spacecraft considerations limit the available power, weight, size, and data downlink capacity of spaceborne altimeter instruments, requiring complicated

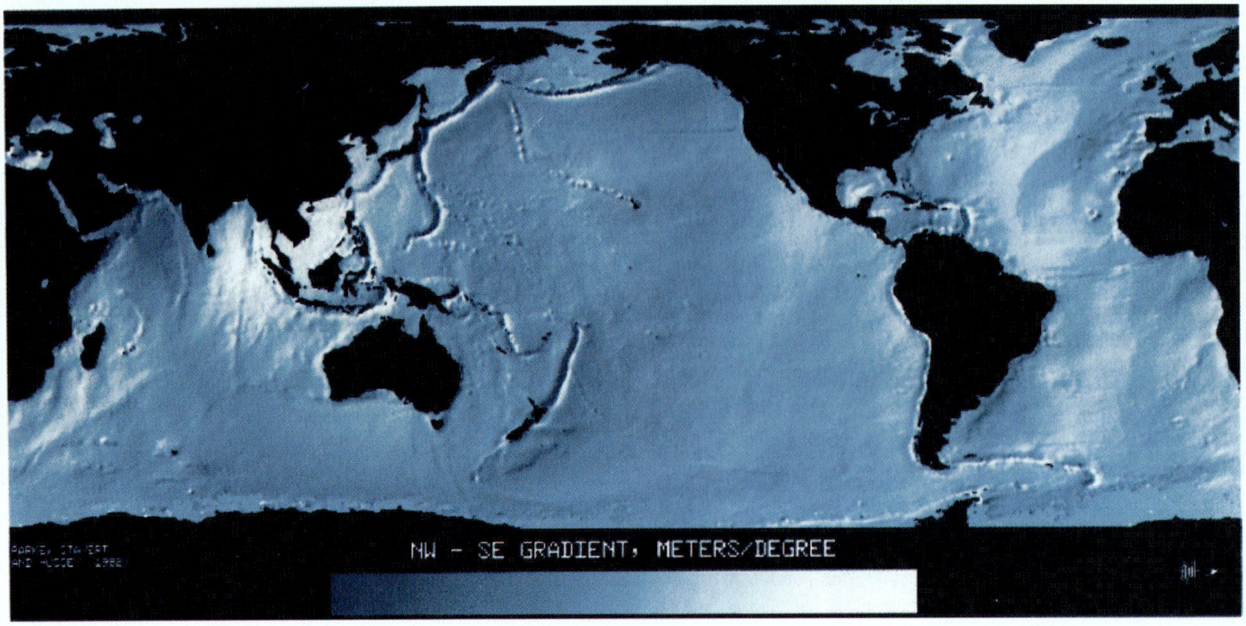

Figure 17-13: Ocean topography derived from the Seasat altimeter. [Courtesy of NASA.]

trade-offs in altimeter implementation, capability and performance.

17-4.1 Range Compression and Deramping

The altimeter signal model presented in previous sections was described in terms of the range-compressed signal. In practice, altimeters typically use linear frequency modulation (LFM) for the transmit pulse. Such transmit signals are known as ***chirps***. The received signal is multiplied by a delayed version of the transmit LFM signal and filtered. This process, known as ***deramping*** or ***dechirping***, serves to reduce the bandwidth of the signal. Range compression of the dechirped signal is then accomplished by sampling the signal and computing the discrete Fourier transform (DFT) (see Chapter 13). The output of the DFT is often described as "in the frequency domain"; however, this is true only of the dechirped signal. The DFT is merely completing the range compression of the signal prior to deramping and thus the output is actually in the range-compressed time domain. As shown in Chapter 13, the effective range resolution is inversely related to the bandwidth of the LFM signal. For example, a transmit frequency bandwidth B of 320 MHz yields an effective two-way travel time resolution of $\Delta t = 1/B = 3.125$ ns.

In LFM signal modulation the ramp rate α for a transmit signal of bandwidth B and a transmit pulse length τ is

$$\alpha = \frac{B}{\tau} \,. \tag{17.37}$$

When deramping is employed, the timing delay t_d between the start of the transmit pulse chirp and the start of the deramping chirp determines the center frequency of the deramped signal, which corresponds to a two-way travel distance of $R_\mathrm{d} = ct_\mathrm{d}$. A range $R_0 = R_\mathrm{d} + \Delta R$ (where $\Delta R > 0$) corresponds to the frequency f_0 in the deramped signal where

$$f_0 = \alpha(R_0 - R_\mathrm{d}). \tag{17.38}$$

Thus R_0 can be computed from f_0 as

$$R_0 = R_\mathrm{d} + \frac{f_0}{\alpha} \,. \tag{17.39}$$

The discrete frequencies of the range-compression DFT yield a set of f_0 values. The output is the return signal versus range. The discrete DFT bins, often termed range bins, are the waveform samples. The DFT output is the signal waveform modeled in Section 17-2. Note that by adjusting t_d, the base range R_0 and the corresponding frequency of the deramped signal can be altered. The DFT of the deramped signal yields the range-compressed signal versus range.

17-4.2 Range Tracking

Practical considerations for the design of on-board digital processors limit the time the deramped signal can be sampled and the number of samples that can be collected and analyzed for each pulse. By dynamically adjusting t_d, the range R_0 to the surface can be maintained within a narrow range, thereby limiting the length of the deramped signal that needs to be sampled and the computational requirements for the DFT and other processing steps. The timing is set with the aid of an ***automatic tracking unit*** (ATU), which implements an algorithm that analyzes the return echoes.

In some cases the timing adjustment is done using a coarse adjustment of t_d in discrete steps, with a fine adjustment applied by frequency shifting the deramped signal prior to computing the DFT. This approach was used for the TOPEX/Poseidon altimeter (Chelton et al., 2001).

As individual pulses are quite noisy, multiple pulses are averaged together prior to analysis by the ATU. To maximize the SNR, altimeters typically use very high PRFs, as high as 20 kHz. Averaging pulses improves the SNR compared with a single pulse and reduces the data volume that must be processed.

Downlink data volume limitations preclude sending the averaged pulse to the ground. Instead, the timing determined by the ATU is downlinked, and after further ground-based processing it is used as the basis of the height measurement reported by the altimeter. However, selected portions of averaged pulses are periodically downlinked to the ground for further processing. To reduce storage, the waveform sample resolution is reduced by averaging adjacent range bins. Ground re-analysis of the downlinked averaged pulses is termed ***re-tracking***. Because more sophisticated analysis can be performed on the ground, re-tracking can yield improved topography estimates. Specialized re-tracking algorithms are used for glacial ice (Martin et al., 1983) and identifying the signature of icebergs (Tournadre et al., 2008).

The on-board ATU is designed to run in real time; hence, the analysis it provides is not very sophisticated, which means that it may not perfectly track the surface height, particularly during land/ocean transitions. This can result in loss of surface track. The ATU must then re-acquire and re-center the surface echo in its analysis window. During this period, the reported timing is inaccurate and is not used for topography measurement. Ground re-tracking of the downlink may not be able to recover estimates during this period either.

A detailed description of ATU and ground re-tracking algorithms is outside the scope of this book. A detailed description of the TOPEX/Poseidon on-board ATU algorithm is given by Zieger et al. (1991) and the issues in ground re-tracking for TOPEX/Poseidon are covered by Rodriguez and Martin (1994). Here, we discuss only some basic principles of satellite ATU.

Initially, the ATU sweeps over a wide set of possible ranges to crudely select the range. A set of *early*, *middle*, and *late* gates are defined, as shown in Fig. 17-14. These span different widths of waveform samples to help localize the half-power point $t_{1/2}$. The value for each gate is the average of the waveform samples it covers.

The slope of the leading edge of the waveform, which as previously noted is related to the significant wave height, is computed from the difference of the late and early gates. The slope is first computed from the widest gate, then narrower gates depending on the estimated value of $H_{1/3}$. When properly centered, the middle gate value should average to one-half of the peak signal. If the value is too high or two low, the timing is adjusted to center the waveform within the fixed pattern of gates. The index of the middle gate is selected from the index of the early/late gates and used to determine $H_{1/3}$. To minimize noise, the gate-derived timing updates are smoothed by a filter whose coefficients depend on the design and desired sensitivity. Actual updates are applied up to 20 times per second.

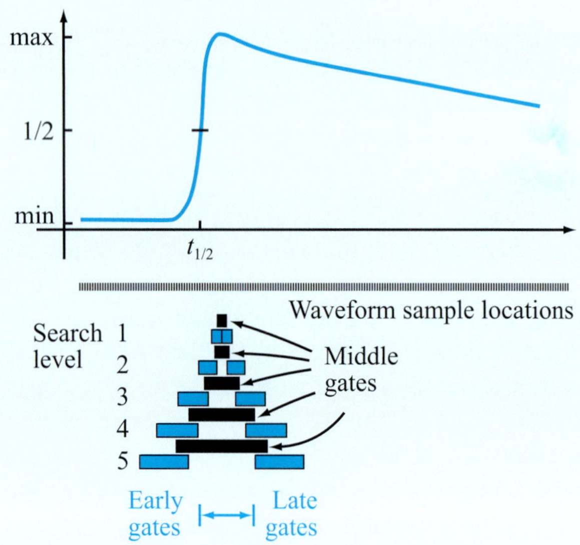

Figure 17-14: Notional illustration of averaged "gates" used by an altimeter ATU. The top shows an expected altimeter waveform shape with the locations of the waveform samples indicated by the vertical lines. The span of several search levels of early, middle, and late gates are shown. The waveform samples covered by the span are included in the gate average.

As discussed earlier, clouds and rain can introduce delay and attenuation in the signal. This alters the signal time of flight. Small or periodic attenuation figures can create extra "noise" in the tracker and degrade its performance. Antenna pointing errors alter the shape of the expected waveform and can bias the estimated t_d by shifting the middle gate balance point.

17-4.3 Orbit Considerations

Altimeters measure the distance from the spacecraft to the sub-satellite point. In order to collect global measurements, it must rely on the movement of the satellite in its orbit. The spacecraft orbit dictates the track of the altimeter over the Earth's surface and thus its coverage. Spacecraft orbits are constrained by the equations describing orbital motion and by the need of the spacecraft to have sufficient solar illumination to provide power for spacecraft and altimeter operations. Other orbit considerations include the radiation environment and atmospheric drag. Higher orbits result in less atmospheric load but the increase in the measurement distance requires higher-power transmitters to maintain sufficient SNR. The orbit altitude also affects the ground speed, with lower altitudes having higher ground speed.

Altimeter orbit design is treated in detail by Parke et al. (1987). For our purpose, the primary factors that affect the orbit geometry are the ***orbit altitude***, ***orbit eccentricity***, and ***orbit inclination angle***. In general, circular orbits with low eccentricity (< 0.001) are desired. The orbit inclination angle controls the range of latitudes that can be observed. For a nadir-looking sensor such as an altimeter, the maximum latitude that can be sampled is related to the orbit inclination angle i by $|\text{lat}_{\text{max}}| = 90^\circ - |90^\circ - i|$. Note that unless the altimeter is an imaging type, it provides height measurements only over a very narrow swath directly below the satellite orbit. The surface is not observed by the sensor elsewhere.

Certain combinations of the orbit parameters enable ***repeat orbits*** that precisely repeat their track periodically, which is highly desirable in most altimeter applications as it can provide long-term monitoring of fixed points. The repeat orbit creates a repeating track that samples the surface in a fixed pattern, which includes locations where the orbit tracks cross each other, as shown in Fig. 17-15. These crossover points are useful for altimeter calibration and in estimating tides. For an altimeter in an exact repeat orbit, a given location is only sampled every few days. This means that local tidal height variations, which have dominant cycles at once and twice per day, are undersampled. The tidal height is thus aliased in the altimeter height measurement at each point. The aliasing can be resolved using the fact that the tidal components are harmonically related to the Earth's rotation rate and lunar period. The tidal height variation can be estimated if a sufficiently long time series is sampled so that the tidal aliases are separable and do not occur near zero, corresponding to the mean sea-level height. Insuring this is the case provides another set of constraints for selecting the orbit parameters, the details of which are not considered here.

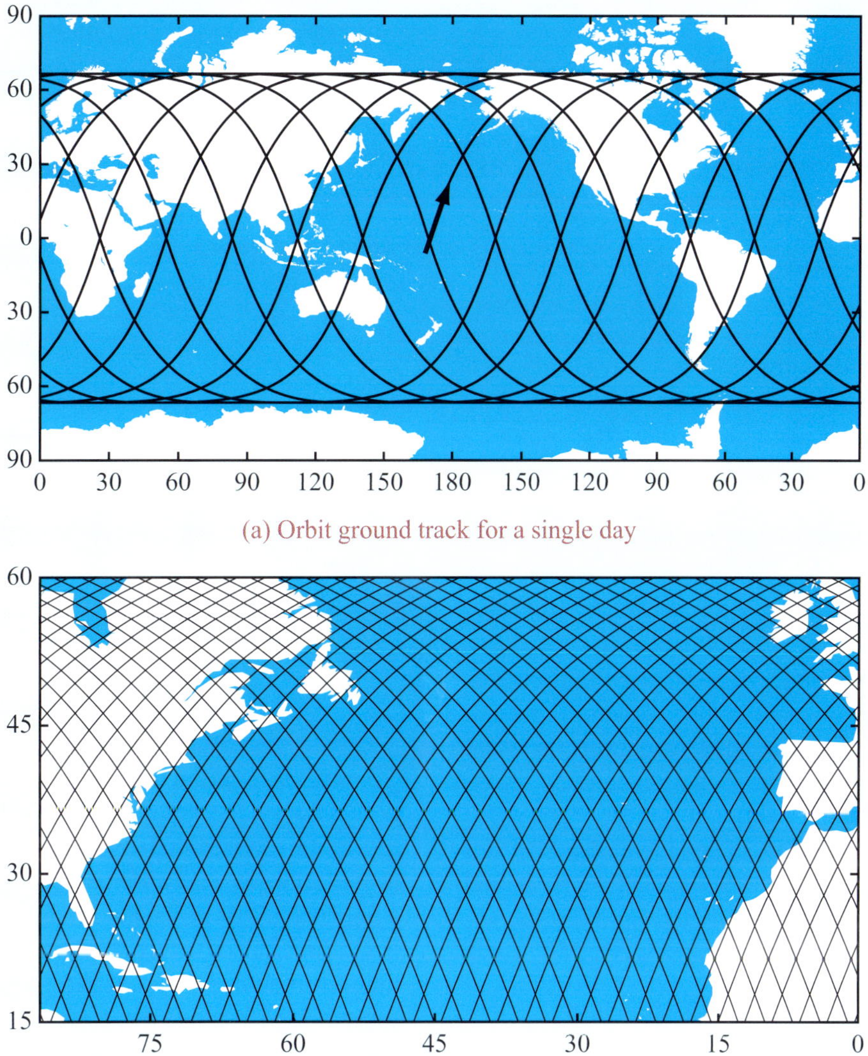

(b) Zoomed-in view of the ground tracks for one 10-day exact repeat cycle

Figure 17-15: Plots of the spatial ground coverage of the TOPEX/Poseidon orbit: (a) orbit ground track for a single day, (b) zoomed-in view of the ground tracks for one 10-day exact repeat cycle [from Chelton et al. (2001)].

Altimeters have flown in a variety of orbits, with the most common being the 1336 km altitude orbit used by TOPEX/Poseidon and the Jason follow-on missions. This orbit was selected to provide the time/space sampling required for high-precision measurement of tides while meeting the desired height and coverage considerations. The estimated tides can then be subtracted from the height measurements to improve the quality of the dynamic sea-level topography.

17-5 Synthetic-Aperture Altimetry

The along-track resolution of an altimeter is dictated by the antenna beamwidth or, equivalently, its along-track dimension, plus the distance traveled during the integration of pulses to estimate the height. Finer along-track resolution can be obtained using synthetic-aperture techniques. This enables a reduction in the length of the antenna, in exchange for signal processing complexity. The general theory developed in Chapter 14 for SAR can be applied to ***synthetic-aperture radar altimetry*** (SARA). SARA improves the along-track resolution but does not affect the cross-track resolution.

For SARA we must reconsider the PRF limits to include the surface topography variation. To ensure that the signals from two different pulses are not received at the same time, the upper limit on the PRF f_p is

$$f_p < \frac{c}{2\Delta R_s}, \tag{17.40}$$

where ΔR_s is the maximum slant-range variation over the footprint, including topographic variation and and slope [Fig. 17-2(c)]. Note that variations in slant range over the full footprint extent, including cross-track, must be considered in computing ΔR_s. The slope induced by off-nadir viewing angles increases ΔR_s. For convenience we neglect the cross-track variations and the Earth's curvature to enable us to approximate ΔR_s by the expression

$$\Delta R_s = H_s + h\alpha\beta + \frac{h\beta^2}{8}, \tag{17.41}$$

where H_s is the topographic variation around the mean slope, β is the antenna beamwidth, h is the satellite height, and α is the mean surface slope angle. Further, ignoring H_s because it is much smaller than the other terms and using $\beta \approx \lambda/l$, where l is the antenna along-track length, we have

$$R_s \approx \frac{h\lambda}{l}\left(\alpha + \frac{\lambda}{8l}\right). \tag{17.42}$$

This leads to the PRF constraint

$$f_p < \frac{cl}{2h\lambda(\alpha + \lambda/8l)}. \tag{17.43}$$

As shown in Chapter 13, the PRF must be at least one pulse every time the antenna moves one-half its length; i.e., PRF $\geq 2u/l$, where u is the platform velocity. Using the smallest value of PRF and solving for l, we find that for small α,

$$l^3 > \frac{4h\lambda^2 u}{c} \tag{17.44}$$

while for large α,

$$l^2 > \frac{4h\lambda u\alpha}{c}. \tag{17.45}$$

These expressions dictate the minimum antenna size. As a numerical example for a low altitude, low frequency altimeter with $\lambda = 0.2$ m, $h = 300$ km, and $u = 7$ km/s with $\alpha = 0.4$ yields a minimum antenna length of $l > 0.1$ m from the two expressions. Note that using $l = 0.1$ m requires a PRF of 70 kHz.

17-6 Wide-Swath or Imaging Altimetry

Conventional altimeters provide height measurements only along the subsatellite track at a cross-track resolution determined by the antenna beamwidth. To provide a wider measurement swath, multiple beams or a single beam that is scanned in the cross-track dimension is required. Alternately, topography can be provided by interferometry. In this section we briefly consider the former. Interferometry is discussed in Chapter 15.

Multiple simultaneous beams can be provided using separate antennas, multiple feeds with a single reflector, or active antennas. Scanning can be done mechanically or with an electrically scanned antenna. Mechanical scanning schemes such as those used for radiometers can be used (see Chapter 7). A variety of electronic beam steering schemes are possible. For all these schemes the spatial resolution is determined by the antenna illumination area on the ground. Figure 17-16 illustrates the wide-swath altimeter geometry.

Scanning a single beam introduces an additional consideration in the design because the beam must be scanned to all-cross-track positions during the time that it takes to the spacecraft to move along-track by

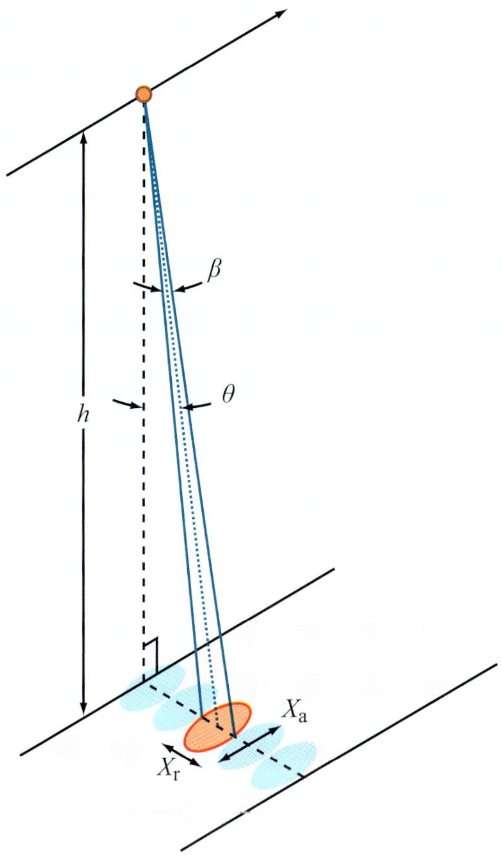

Figure 17-16: Geometry for a wide-swath scanning altimeter. Only one side of the satellite track is shown. Practical sensors observe to both sides.

the distance corresponding to the along-track resolution of the antenna beam. The approximate along-track resolution X_a for a real-aperture beam is given by

$$X_a = 2h \tan \beta/2 \approx h\beta \approx \frac{h}{\lambda/l} . \qquad (17.46)$$

If there are N_B beam positions across the swath, each of these must be visited during the time required for the spacecraft to move a distance along-track equal to X_a. The maximum dwell time per beam t_{dwell} is thus

$$t_{dwell} < \frac{X_a}{uN_B} . \qquad (17.47)$$

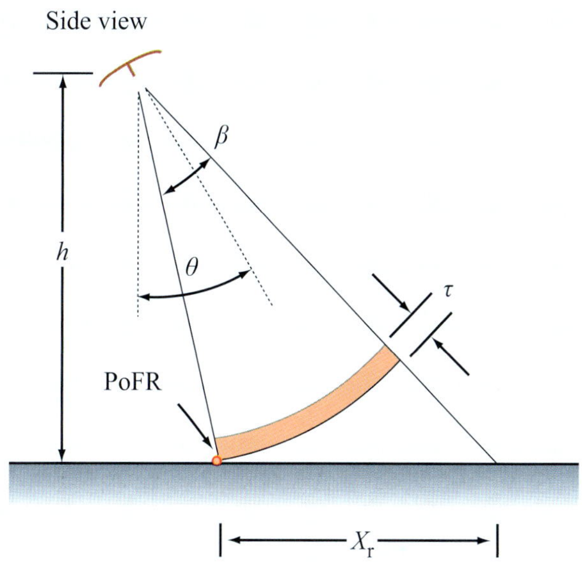

Figure 17-17: Off-nadir altimeter geometry.

Note that this constraint does not include beam switching time. For fine-resolution systems with small X_a and large N_B, t_{dwell} could be comparable to t_d so that a carefully designed scanning and timing control scheme is needed due to the long delay between transmit and receive times. Note that the constraint in Eq. (17.47) can be relaxed by using synthetic-aperture techniques to improve the along-track resolution as discussed in the next section.

Note that off-nadir pointing of the altimeter antenna beam has the effect of introducing a slope in the topography seen by the beam as illustrated in Figure 17-17. The slope analysis considered in Section 17-2.2 applies and the signal model must be adapted for each beam location. Off-nadir pointing requires better spacecraft attitude knowledge to avoid generating a slope artifact in the estimated heights.

17-7 CryoSat-2 Mission

The frozen-water part of Earth is called the ***cryosphere***. CryoSat-2 is a satellite mission whose objective is to map and measure the thickness of the ice sheets that overlie Greenland and Antarctica and to measure the thickness of ice floating in the polar oceans (eoportal.org). After a failed launch of its predecessor in October of 2005, CryoSat-2 was successfully launched in April 2010, and has made a number of discoveries about Antarctica including the identification of a vast crater that probably was left behind when a lake lying under 3 km of ice suddenly drained (ESA CryoSat Bulletin, 2 July 2013).

The primary instrument on CryoSat-2 is a combined SAR/two-antenna interferometer/altimeter. It is called ***SIRAL***: the SAR Interferometric Radar Altimeter. Figure 17-18 is an artist's conception showing its 15 km footprint and 250 m along-track resolution. Its operating frequency is 13.575 GHz and its range resolution is 45 cm. Table 17-4 provides a summary of the instrument specifications, and Fig. 17-19 displays a map of Arctic sea-ice thickness for March, 2011. The thickness of the ice layer above the water surface can be as much as 5 meters, whereas the penetration depth at SIRAL's operating frequency is only a few centimeters, so it is not possible for the sensor to measure the depth of the sea ice directly. The radar altimeter measures the ice freeboard, which is the height of the ice above the water surface. From knowledge of the ratio of the freeboard to the thickness of the ice submerged below the water surface, it is possible to estimate the total sea-ice thickness. Through repeat coverage, CryoSat-2 will provide monthly distributions of thickness, allowing climate scientists to track the evolution of ice content in both the Arctic and Antarctic regions.

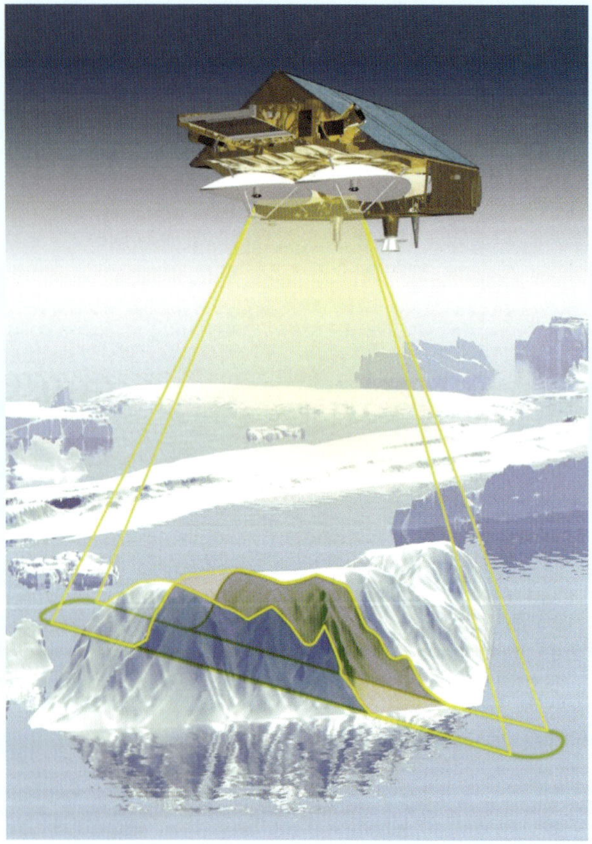

Figure 17-18: Artist's conception of SIRAL on CryoSat-2 [European Space Agency].

Table 17-4: Specifications of the SIRAL instrument on CryoSat-2 [European Space Agency].

Parameter	Value
RF frequency	13.575 GHz (single-frequency Ku-band radar)
Pulse bandwidth	320 MHz (40 MHz for tracking only in SARIn)
PRF (Pulse Repetition Frequency)	1.97 kHz in LRM, 17.8 kHz in SAR and in SARIn; coherent pulse transmission for Doppler processing
Burst mode PRF	1970 Hz in LRM, 85.7 Hz in SAR, and 21.4 Hz in SARIn
Pulses/burst	N/A in LRM, 64 in SAR and in SARIn
Pulse duration	50 μs
Timing	Regular PRF in LRM, burst mode in SAR/SARIn
Samples/pulse	128 in LRM and SAR, 512 in SARIn
RF peak power	25 W
Antenna size	2 reflectors 1.2 m $\times$1.1 m, side by side
Antenna beamwidth (3 dB)	1.08° (along-track) $\times$1.2° (cross-track)
Antenna footprint	15 km
Range resolution	About 45 cm
Along-track resolution	250 m (SAR/SARIn)
Data rate	50 kbit/s for LRM, 12 Mbit/s in SAR, 2 $\times$ 12 Mbit/s in SARIn
Instrument mass (with antennas)	70 kg nonredundant
Instrument power	149 W

Figure 17-19: Arctic sea-ice distribution generated from SIRAL/CryoSat-2 data [European Space Agency].

PROBLEMS

17.1 Sketch the beam-limited footprint for a altimeter operating in a 820 km orbit using a 1 m diameter antenna for both Ku-band (14.6 GHz) and C-band (5.4 GHz). What is the effective range resolution of each case?

17.2 In Problem 17.1, if a 320 MHz bandwidth LFM pulse is used, what is the pulse-limited range resolution? For the Ku-band case, sketch to scale the spatial coverage of each of a few range bins on the surface, assuming a smooth, flat surface.

17.3 For Problem 17.2, if a maximum of 64 range bins are retained, what is the maximum range of surface heights that can be estimated? Assume the 64 bins are fixed.

17.4 For the Ku-band radar used in Problems 17.1 and 17.2, plot the height error Δh as a function of spacecraft tilt angle.

17.5 For Problem 17.1, what is the maximum spacecraft tilt permissible if we require that the nadir point for a smooth flat surface be contained within the footprint?

CHAPTER

18

Radiometric Remote Sensing of the Ocean

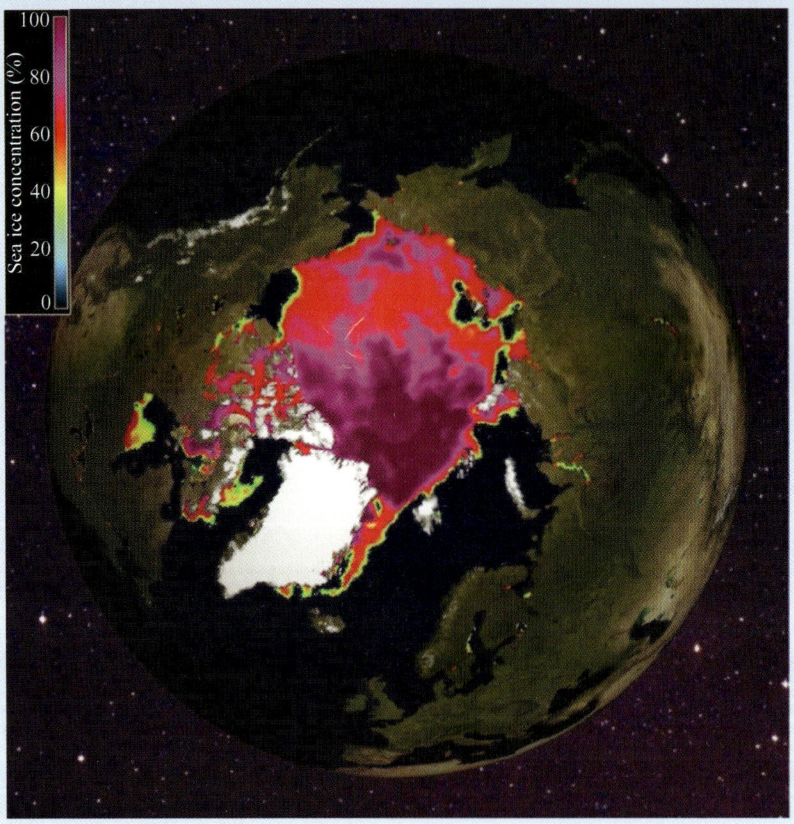

Arctic sea-ice concentration from radiometer data

CONTENTS

Overview

The Nimbus-7 satellite, launched in 1978, afforded investigators their first opportunity to conduct comprehensive evaluations of the use of microwave radiometric techniques to monitor the dynamics of the ocean surface from space. The satellite carried an instrument called the ***Scanning Multichannel Microwave Radiometer*** (SMMR) whose purpose, as articulated by the Nimbus-7 SMMR Experiment Team (Gloersen et al., 1984), was to provide "nearly all-weather observations of the sea-surface temperatures, near-surface winds over the oceans, water vapor over oceans, sea-ice concentration, sea-ice age, sea-ice radiating temperature, and snow cover over terrain." The SMMR instrument included radiometers at five microwave frequencies (6.6, 10.7, 18, 21, and 37 GHz) and used an oscillating offset parabolic reflector to generate images of the Earth's surface. In later years, instruments with SMMR-like radiometers were launched on DMSP Block 5D in 1987, 1990, 1991, 1995, 1997, 1999, and 2003, and also on TRMM in 1997, Aqua in 2002, ADEOS in 1996, and Coriolis in 2003 (see Table 1-4). The purpose of this chapter is to examine the relationship between the emitted microwave radiation and the various ocean parameters listed earlier, and how to retrieve them from multifrequency/multipolarization observations. Absent among the listed ocean parameters is ocean salinity. To measure ocean salinity, it is necessary to use a radiometer at an operating frequency that is much lower than the lowest of the SMMR frequencies. To that end, the European Space Agency launched in 2009 the ***Soil Moisture and Ocean Salinity*** (SMOS) satellite, with a 1.4 GHz imaging radiometer as its primary payload.

18-1 Brightness Temperature of the Sea Surface

The p-polarized brightness temperature observed by a satellite-borne radiometer viewing the Earth's surface at an incidence angle θ through a horizontally stratified atmosphere is given by Eq. (6.113), which we rewrite in the form:

$$T_{\rm B}^{p}(\text{satellite}) = \Upsilon_{\rm a}(T_{\rm B}^{p} + T_{\rm SS}^{p}) + T_{\rm UP}, \tag{18.1}$$

where $T_{\rm B}^{p}$ is the p-polarized brightness temperature of the sea surface in the absence of the atmosphere, $T_{\rm SS}^{p}$ is the p-polarized, upward-reflected, *surface-scattered* fraction of the downward-emitted sky brightness temperature $T_{\rm SKY}$, $T_{\rm UP}$ is the *upward-emitted* atmospheric temperature, and $\Upsilon_{\rm a}$ is the one-way transmissivity of the atmosphere. The most important quantity in the context of the sea surface is $T_{\rm B}^{p}$ because it is directly related to the sea-surface parameters. All quantities on the right-hand side of Eq. (18.1) are functions of θ.

The atmosphere-related quantities—$\Upsilon_{\rm a}$, $T_{\rm SKY}$, and $T_{\rm UP}$—can be estimated from knowledge of the state of the atmosphere above the sea surface, with said knowledge provided by atmospheric satellite sensors. Also, $T_{\rm SS}^{p}$ can be estimated with reasonable accuracy from the estimated value of $T_{\rm SKY}$ and an estimate of the surface reflectivity. By way of illustration, let us consider a clear-sky scenario in which an 18 GHz satellite-borne radiometer is viewing the sea surface at $\theta = 50^\circ$ and the sea-surface temperature is $T_{\rm s} = 290$ K. As shown later, the h-polarized brightness temperature of the sea might vary between about 80 K and 100 K due to the combination of wind speed and wind direction relative to the antenna beam direction. At $T_{\rm B}^{\rm h} = 80$ K, the emissivity is

$$e^{\rm h} = \frac{T_{\rm B}^{\rm h}}{T_{\rm s}} = \frac{80}{290} = 0.276.$$

Under clear-sky conditions, $T_{\rm SKY} \approx 15$ K. The surface scattered contribution is

$$\begin{aligned} T_{\rm SS}^{\rm h} &\approx (1 - e^{\rm h})T_{\rm SKY} \\ &\approx (1 - 0.276)15 = 10.9 \text{ K}. \end{aligned}$$

If we were to repeat the calculation for the maximum possible value of $T_{\rm B}^{\rm h} = 100$ K, we would end up with $T_{\rm SS}^{\rm h} = 9.83$ K. Hence, the maximum possible error is on the order of 1.1 K.

▶ The point of the example is to illustrate that it is indeed possible to convert T_B^p(satellite), the brightness temperature measured by the satellite-borne radiometer, into an estimate of T_B^p, the brightness temperature of the sea surface, within an error of about 1 K. ◀

Next, we focus our attention on how T_B^p is related to the geophysical parameters of the ocean surface, including its physical temperature, salinity, roughness spectrum (which is related to the wind vector), and the presence or absence of ice or oil.

18-1.1 Spectral Sensitivity

The brightness temperature is a function of several geophysical parameters of the atmosphere and the ocean surface. In terms of spatial and temporal variability, the most important among these parameters are the sea surface temperature T_s, the surface wind speed u, the salinity S, the integrated precipitable water vapor h_v, and the cloud liquid-water path h_L (Section 9-6). We may write T_B^p in the following functional form:

$$T_B^p = T_B^p(f, \theta;\; T_s, u, S, h_v, h_L), \qquad (18.2)$$

which consists of two sets of variables: (1) a sensor set comprising the microwave frequency f, the angle of incidence θ, and the antenna polarization p, and (2) a geophysical set comprising the five geophysical parameters listed earlier. The relationships between T_B^p and the atmospheric variables were examined in Chapters 8 and 9 as a function of f and θ, and the relationships between T_B^p and the three ocean parameters are examined in forthcoming sections as a function of all three sensor variables. Through such examinations, it is possible to make judicious selections of sensor parameters in order to optimize the retrieval capability of the radiometric observation technique. Wilheit et al. (1980) illustrate this process in terms of the frequencies selected for the SMMR instrument. Figure 18-1 shows a spectral plot of the ***sensitivity*** $\partial T_B^p/\partial g_i$ for each of the five geophysical parameters, denoted g_i. These spectral sensitivity curves, which

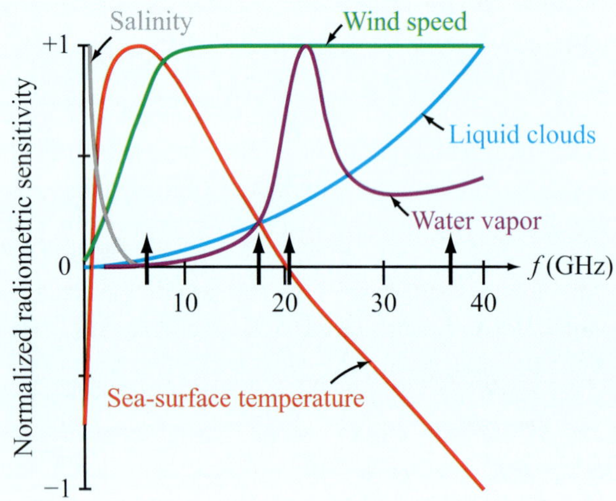

Figure 18-1: Spectra of the normalized radiometric sensitivity $Q_{g_i}/(Q_{g_i})_{max}$, where $Q_{g_i} = \partial T_B^p/\partial g_i$, for various geophysical parameters g_i. The arrows indicate the SMMR frequencies. The signs have been chosen to be positive in the frequency range of primary importance to the given parameter [from Wilheit et al., 1980].

are normalized for the purpose of presentation on the same scale, depict the incremental change in T_B^p at the frequency f caused by an incremental change in the given parameter, g_i. The arrows shown on the frequency axis indicate the frequencies of the SMMR channels. The curve depicting sensitivity to salinity indicates that the effect of salinity on T_B^p becomes negligible at frequencies above 4 GHz, which is one of the reasons why ocean salinity cannot be measured by SMMR. The surface temperature sensitivity, $\partial T_B^p/\partial T_s$, exhibits a peak at approximately 5 GHz, which is in close proximity to SMMR's 6.6 GHz channel. Similarly, the 10.7 GHz, 21 GHz, and 37 GHz channels may be regarded as the primary channels for measuring wind speed, integrated precipitable water vapor (IPWV), and cloud liquid-water path (LWP), respectively. Actually, each of the channels (including the 18 GHz channel) contains some information about each of the geophysical parameters under consideration. Hence, in practice the algorithm designed to retrieve a given parameter utilizes measurements provided by several of the available channels rather than by the primary channel alone. Actually, the SMMR has 10 channels—a horizontally polarized channel and a vertically polarized channel for each of the five frequencies—and both are used in the retrieval algorithms.

18-1.2 Brightness Temperature of a Smooth Surface

Before we examine the effects of wind-induced surface roughness, we consider the simple case of a smooth, calm water surface. For a specular surface, the brightness temperature T_B^p and the scattered temperature T_{SS}^p are defined in terms of the ***specular Fresnel reflectivity*** Γ^p:

$$T_B^p = e^p T_s = (1 - \Gamma^p) T_s, \tag{18.3a}$$

and

$$T_{SS} = \Gamma^p T_{SKY}. \tag{18.3b}$$

The expressions for Γ^v and Γ^h are available in Section 2-8.

The angular variation of the brightness temperature of a specular sea surface is shown in Fig. 18-2 for four microwave frequencies. The plots were computed using the dielectric model for saline water given in Section 4-2. The general pattern is characteristic of the emissivity of any specular surface; i.e., for horizontal polarization, e^h decreases with increasing angle, while for vertical polarization e^v increases with increasing θ until θ reaches the Brewster angle θ_B, and then e^v decreases rapidly between θ_B and 90°.

The spectral variation of T_B^p is shown in Fig. 18-3 for three observation configurations: $\theta = 0$, $\theta = 50°$ with $p = $ h, and $\theta = 50°$ with $p = $ v. For purposes of comparison, two sets of plots are included, one set for pure water and the other for water with a salinity of 36 psu,* which is a typical value for sea water. The difference between T_B^p for pure water and that for sea water is quite significant at 1 GHz, but decreases rapidly with increasing frequency and becomes negligibly small above approximately 4 GHz. This behavior is in

*psu (practical salinity unit) is equivalent to 1‰ (mass of solid salt dissolved in 1 kg of water).

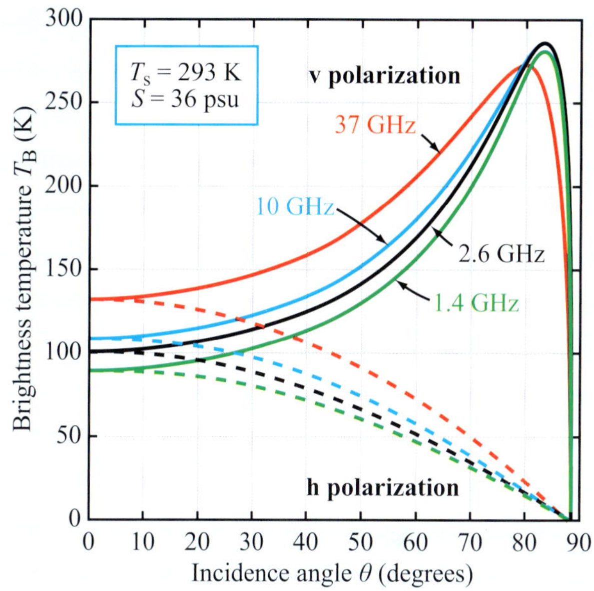

Figure 18-2: Brightness temperature of a specular sea surface at four microwave frequencies.

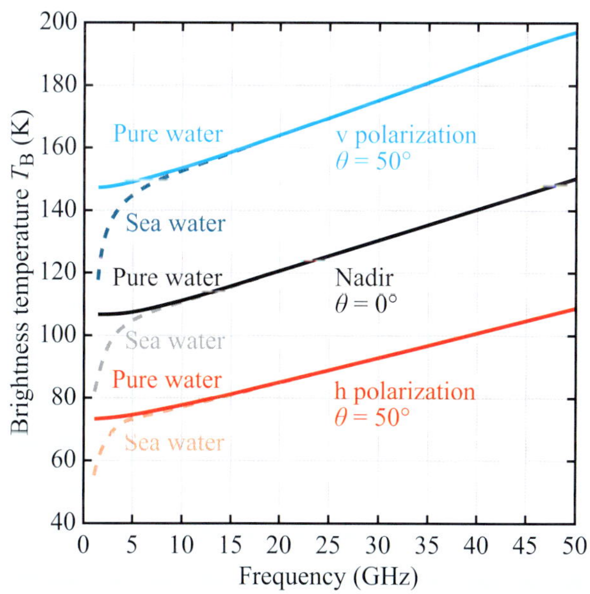

Figure 18-3: Spectral variation of the brightness temperature of pure water and sea water at 20 °C.

response to the frequency dependence of the dielectric constant ε_w of sea water. Further discussion of the relation between T_B^p and water salinity is given in Section 18-2.

18-1.3 Penetration Depth in Sea Water

The expression given by Eq. (18.3) is for a semi-infinite water medium with a uniform temperature profile. In the strictest sense, sea water does not have a uniform temperature profile, but the surface layer responsible for most of the emission by the water medium does approximate this condition quite well.

▶ The reason is simple: microwaves penetrate sea water to a depth of only about 1 cm. ◀

Part (a) of Fig. 18-4 displays a plot of the penetration depth δ_p at 1.4 GHz as a function of water salinity, and part (b) shows δ_p as a function of frequency for sea water at 20 °C. These graphs are based on the dielectric models given in Sections 4-1 and 4-2. For sea water, δ_p decreases from 0.5 cm at 1 GHz to 1 mm at 10 GHz. For fresh water, δ_p is about 5 cm at 1 GHz, but decreases rapidly with increasing frequency and becomes close to that for sea water by 10 GHz.

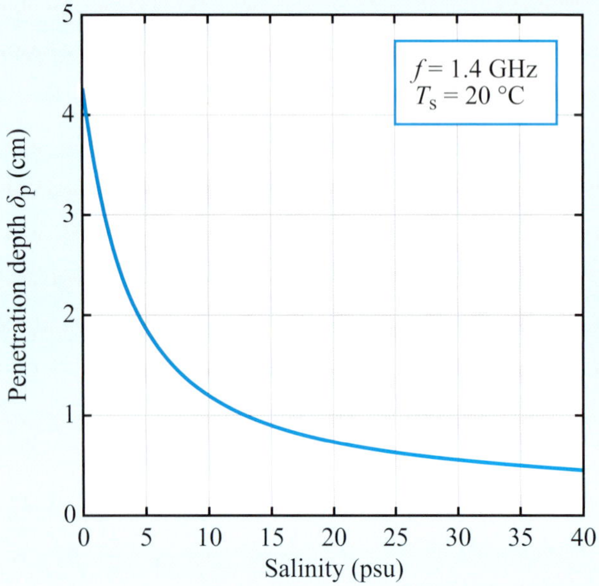

(a) δ_p versus salinity

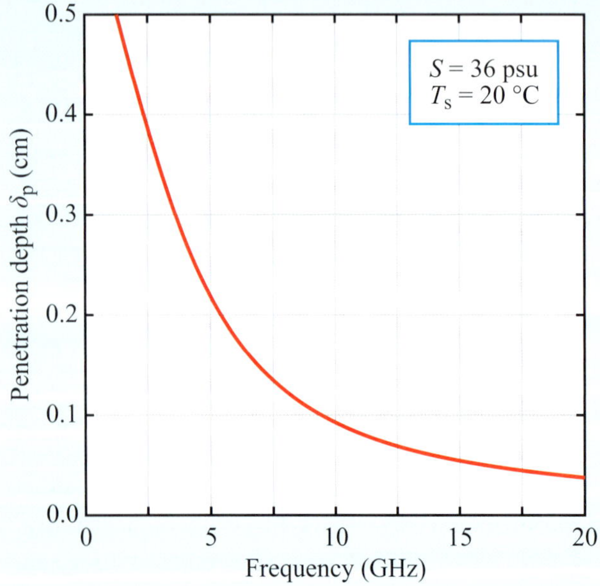

(b) δ_p versus frequency

Figure 18-4: Penetration depth of water as a function of (a) salinity at 1.4 GHz and (b) frequency at a salinity of 36 psu.

18-2 Measurement of Sea-Surface Temperature and Salinity

The ability to remotely monitor the temporal and spatial variations of sea-surface temperature and salinity is useful in the study and modeling of ocean circulation and energy exchange between the ocean and atmosphere. As pointed out in Blume et al. (1977), satellite-borne thermal infrared radiometers (TIR) have been providing maps of sea-surface temperature since the late 1960s. The accuracy of the inferred temperatures, however, is dependent upon prevailing cloud-cover conditions. In contrast, cloud attenuation effects are much weaker in the microwave region, particularly at frequencies below 10 GHz.

The brightness temperature of the sea surface, T_B^p, is a function of the ***thermometric sea surface temperature*** T_s (often referred to as SST), the salinity S, the surface roughness, and the presence of foam on the surface. The sensitivity of T_B to variations in surface roughness and foam, both of which are wind-generated, increases with increasing frequency up to 10 GHz and is essentially flat above that (see Fig. 18-1).

► For measurements of sea-surface temperature and salinity, it is desirable to operate at low microwave frequencies so as to depress the impact of wind-related factors on T_B, thereby decreasing the uncertainty (caused by these factors) associated with the values of T_s and S inferred from the measured values of T_B. ◄

18-2.1 Sensitivity Analysis for Salinity S

If we ignore the effects of surface roughness for the time being, the radiometric brightness temperature of the ocean is given by Eq. (18.3a) as

$$T_B^p = (1 - \Gamma^p) T_s, \tag{18.4}$$

and the expressions for the specular reflectivity Γ^p are available in Table 2-5 for $p = \text{v}$ and $p = \text{h}$. The reflectivity is specified by the incidence angle θ and the dielectric constant of the water ε_w. The latter is, in turn, dependent on the sea-surface temperature T_s, the water salinity S (often called the ***sea-surface salinity*** (SSS)), and the microwave frequency f. The relevant relationships are available in Section 4-2. By holding three of the four variables (θ, f, T_s, S) constant, we can compute the variation of T_B^p with respect to the fourth one. An example is shown in Fig. 18-5, which displays T_B as a function of S for various values of T_s, all at normal incidence and $f = 1.4$ GHz.

The p-polarized ***salinity sensitivity*** Q_S^p is defined as

$$Q_S^p = \frac{\partial T_B^p}{\partial S}, \tag{18.5}$$

which corresponds to the local slope of the curve relating T_B^p to S. In general, T_B^p exhibits a nonlinear

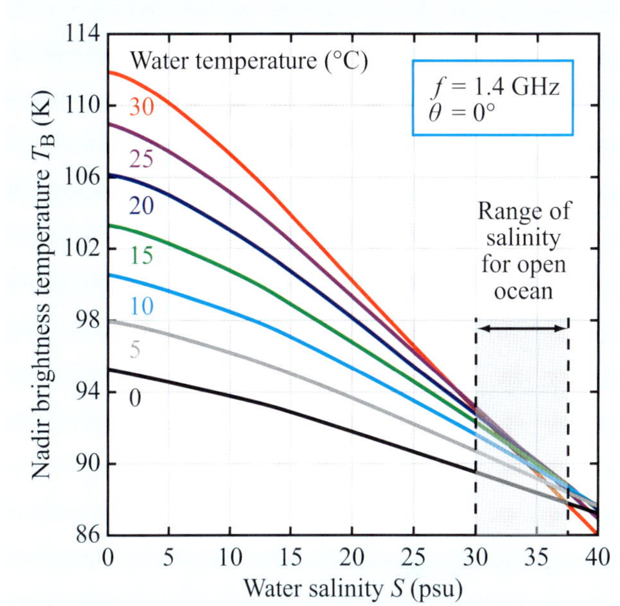

Figure 18-5: Variation of nadir brightness temperature with water salinity at 1.4 GHz under calm sea conditions.

variation with S, and its salinity sensitivity Q_S^p is strongly dependent on T_s. For the curves shown in Fig. 18-5, the slope Q_S is smallest at $T_s = 0$ °C and largest at $T_s = 30$ °C.

For the open ocean, the average salinity is around 32.5 psu, and its range is seldom below 30 psu or above 38 psu. Let us consider the case where $T_s = 20$ °C and let us define Q_S^p as the linear slope of T_B^p versus S over the narrow range between $S = 30$ psu and $S = 38$ psu. Figure 18-6 displays plots of Q_S^p as a function of f for $\theta = 0$, $\theta = 40°$ with $p = \text{h}$, and $\theta = 40°$ with $p = \text{v}$. We observe that for all three curves, $|Q_S^p|$ is largest at 1 GHz and decreases rapidly with increasing frequency. In fact, above 5 GHz, $Q_S^p \approx 0$. Also, among the three sensor configurations, the v-polarized 50° configuration provides the strongest sensitivity to salinity variations.

The preceding analysis assumed calm sea conditions with no surface roughness, which is not realistic. As we see later, to infer S from T_B^p data, it is necessary to use correction factors that account for variations in sea-surface temperature T_s and windspeed u. These

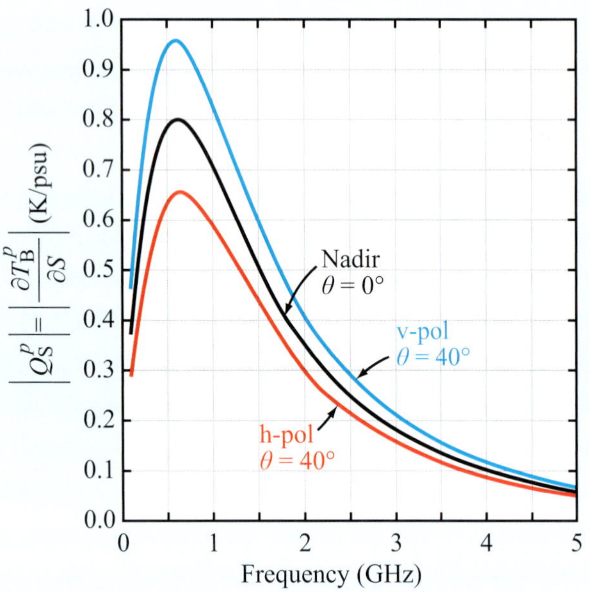

Figure 18-6: Spectral variation of the absolute value of the salinity sensitivity, evaluated for a surface with $T_s = 20$ °C and $S = 35$ psu [Le Vine et al., 2010].

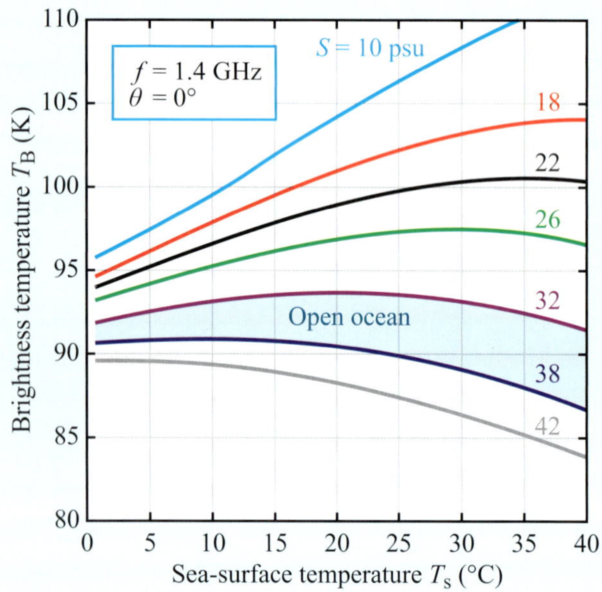

Figure 18-7: T_B versus T_s at L-band [Le Vine et al., 2010].

correction factors use estimated values of T_s and u provided by other microwave radiometer channels.

18-2.2 Sensitivity Analysis for Sea-Surface Temperature T_s

The brightness temperature T_B^p is dependent on the physical temperature T_s, not only explicitly as indicated by Eq. (18.4), but also indirectly through the dependence of Γ^p on T_s. The variation of T_B with T_s is shown in Fig. 18-7 at each of several values of S. The plots were generated by Le Vine et al. (2010) using a dielectric model by Klein and Swift (1977), which is in excellent agreement with the models presented in Section 4-2 (at $f = 1.4$ GHz). If we set S at 32.5 psu and approximate T_B^p as a linear function of T_s over the range between $T_s = 0$ °C and $T_s = 30$ °C, we obtain the spectral plots shown in Fig. 18-8. The slope is the ***sea-surface temperature sensitivity***

$$Q_{T_s}^p = \frac{\partial T_B^p}{\partial T_s} \,. \tag{18.6}$$

With both T_B^p and T_s measured in kelvins, $Q_{T_s}^p$ is dimensionless. According to the results displayed in Fig. 18-8, $Q_{T_s}^p$ may take on positive or negative values or may be equal to zero, depending on the sensor configuration (f, θ, p). Based on our earlier discussion of the salinity sensitivity Q_S^p, we can nearly eliminate the dependence on S by operating at frequencies above 5 GHz. This is the primary reason for selecting 6.6 GHz as one of the channels of the SMMR instrument and its subsequent replacements.

► The primary purpose of the 6.6 GHz channel is to measure T_s. In Fig. 18-8, $Q_{T_s}^p$ has a broad maximum centered in the neighborhood of 6.6 GHz, and of the three sensor configurations, v polarization at $\theta = 50°$ has the highest sensitivity. ◄

We also note that $Q_{T_s}^p \approx 0$ in the neighborhood of 1.4 GHz.

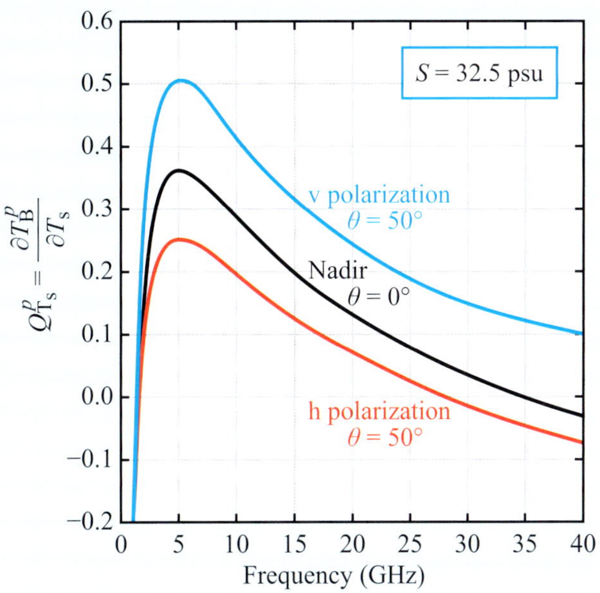

Figure 18-8: Variation of $Q^p_{T_s}$ with frequency for $S = 32.5$ psu.

18-2.3 Satellite Measurement of SST

The launch of the SMMR instrument aboard Nimbus-7 in 1978 provided the first opportunity to develop and test algorithms for retrieving sea-surface temperature from microwave observations by satellite-borne radiometers. As noted earlier, in connection with Fig. 18-8, the sensitivity of the brightness temperature to sea-surface temperature has a broad maximum around 6.6 GHz. Prior to 1978, the only satellite instruments that made radiometric observations of the Earth's oceans at frequencies below 10 GHz were the 3.5 GHz radiometer aboard Cosmos 243 in 1968 and Cosmos 384 in 1970, and Skylab's 1.4 GHz radiometer in 1973. Because the Cosmos radiometer acquired data for only a few days, it was not possible to conduct a thorough quantitative evaluation of its sensitivity to sea-surface temperature. The measurements made by the Skylab radiometer in 1973–74 showed that the brightness temperature T_B is essentially independent of the sea surface temperature T_s (Lerner and Hollinger, 1977). This result confirms the theoretical calculations shown in Fig. 18-8, which indicate that the sensitivity $Q_{T_s} \approx 0$ at 1.4 GHz for sea water with a salinity of about 35 psu.

The SMMR's five frequencies extend from 6.6 to 37 GHz. The 6.6 GHz channels (horizontal and vertical polarization) are perfectly suitable for providing information on sea-surface temperature. Algorithms to retrieve sea-surface temperature (SST) from SMMR observations were the subject of numerous papers and reports in the early 1980s (Wilheit and Chang, 1980; Hofer and Njoku, 1981; Lipes, 1982; Pandey and Kakar, 1983; Milman and Wilheit, 1983; Prabhakara and Wang, 1983). The first among these algorithms was developed by Wilheit and Chang (1980) by applying ***statistical inversion techniques*** (Section 9-3.5) to a *simulated* data set. The data set was generated using an ensemble of radiative transfer models for a variety of oceanic and atmospheric conditions. The algorithm was developed using pre-launch calibration constants for the SMMR instrument. Comparison of the SST values estimated by the SMMR algorithm with data collected by ships showed "that there was a large (about 20 °C) bias in the SMMR SSTs, and there was a large variance with respect to the ship measurements" (Milman and Wilheit, 1983). The algorithm was later modified by using improved calibration constants for SMMR and adapting the physical models used to generate the simulated data set. This new version, which Milman and Wilheit (1985) refer to as ***Version II***, is given by

$$T_s = T'_s - 1.34 - 0.2[7.5 - T'_s(1 - 0.025T'_s)], \quad (18.7a)$$

$$\begin{aligned} T'_s = {} & 1.7T^v_B(6.6) - 0.37T^h_B(6.6) + \frac{56(285 - T^h_B(10.7))}{285 - T^v_B(10.7)} \\ & - \frac{245(285 - T^h_B(18))}{285 - T^v_B(18)} - 326\ln(280 - T^v_B(18)) \\ & + 370\ln(280 - T^h_B(18)) - 11\ln(280 - T^h_B(21)) \\ & - 3\theta_i - 73.15, \end{aligned} \quad (18.7b)$$

where T_s (°C) is the inferred sea-surface temperature, T'_s (°C) is the sea-surface temperature prior to correcting for the dependence of emissivity on temperature, T^v_B and T^h_B refer to the vertically or horizontally polarized components of the measured brightness temperature, and θ_i is the incidence angle, which is nominally equal to 50° and changes only a fraction of a degree across the conical scan pattern of the SMMR antennas. Among the SMMR's ten channels, the 6.6 GHz vertically polarized channel is the most important for sensing sea-surface temperature.

► This frequency (6.6 GHz) was chosen for its proximity to the peak in the sensitivity of the brightness temperature to T_s (see Fig. 18-8), and at SMMR's 50° incidence angle the vertically polarized emissivity is approximately wind-speed independent. ◄

The approach used by Wilheit and Chang (1980) in the development of the first SST algorithm later became the fundamental structure around which other algorithms were built. The main differences among the various SST algorithms pertain to (1) the specific methods used to sequentially select the most significant channels from among SMMR's 10 channels, and (2) the type of correction used to eliminate systematic errors, which may be dependent on the time of day and may be different for different time periods (Milman and Wilheit, 1985).

Figure 18-9 compares SMMR-derived values of T_s with ship measurements for 18,702 coincidences (surface reports that are within about 75 km and 24 hours of a given SMMR observation). The algorithm used is a modified version of that given by Eq. (18.7), which Milman and Wilheit (1985) refer to as ***Version IV***. The rms difference was found to be 1.79 °C. Accuracies between 1 °C and 1.5 °C were reported in other studies (Hofer and Njoku, 1981; Lipes, 1982; Pandey and Kakar, 1983) involving fewer observations, covering narrower time periods and/or more restricted geographic regions. Today, global SST maps are generated using either satellite infrared observations provided by AVHRR (***Advanced Very High Resolution Radiometer***)—which includes 6 infrared channels extending in wavelength between 0.6 μm and 12 μm, and whose nadir spatial resolution is ≈ 1 km (NOAASIS.NOAA.gov)—or a combination of observations made by infrared and microwave radiometers. An example of a global SST map is shown in Fig. 18-10.

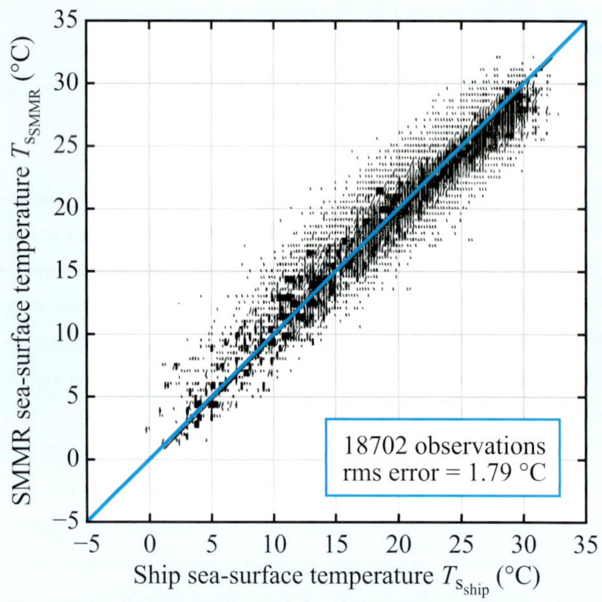

Figure 18-9: Scatter plot of SMMR-retrieved SST versus ship SST [from Milman and Wilheit, 1985].

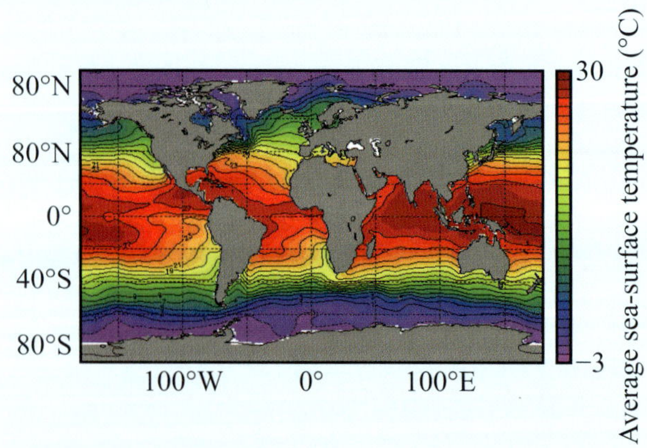

Figure 18-10: Global SST map. [Courtesy of NOAA.]

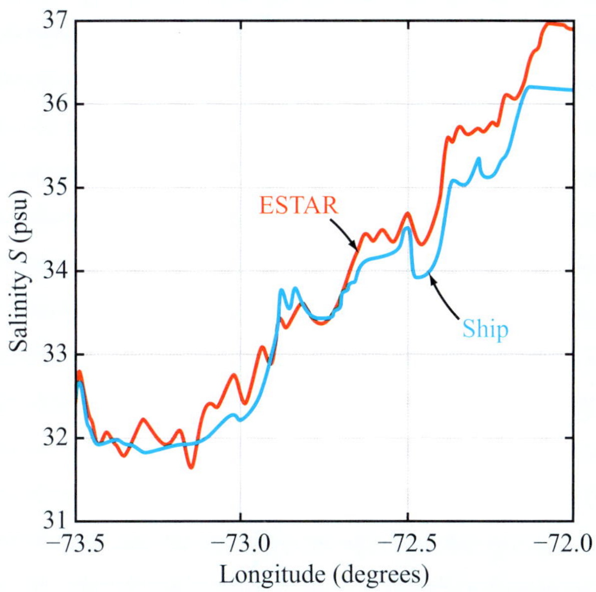

Figure 18-11: Comparison of salinity measured *in situ* by a ship with salinity derived from ESTAR measurements along a track where the aircraft flew over the ship [Le Vine et al., 2010].

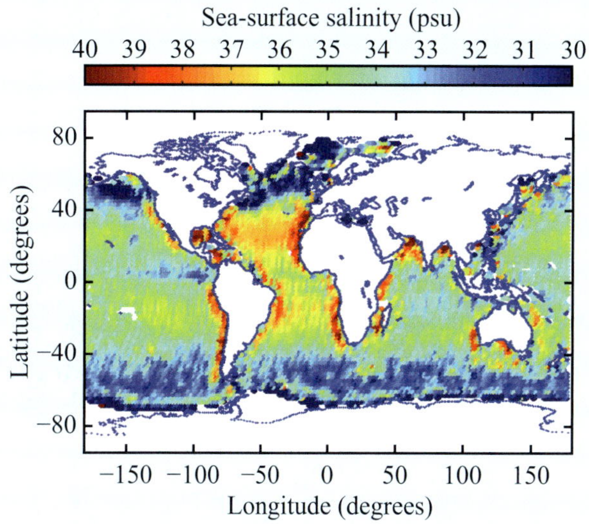

Figure 18-12: Example of a weekly global SSS map (27 April – 4 May 2010) generated by SMOS [Font et al., 2010].

18-2.4 Satellite Measurement of SSS

As noted earlier in connection with Figs. 18-6 and 18-8, the brightness temperature T_B^p at L-band (1.4 GHz) has a very strong sensitivity to sea-surface salinity (SSS) and, simultaneously, a very weak sensitivity to sea-surface temperature. Figure 18-11 compares temporal profiles of salinity measured by a ship's sensor and inferred from brightness temperatures measured at 1.4 GHz by ***ESTAR*** [the ***Electronically Scanning Thinned Array Radiometer*** (Le Vine et al., 2001)] as it followed the path of the ship while flying on a NASA aircraft (Le Vine et al., 2010). The radiometer-inferred profile tracks the *in-situ* profile with high fidelity.

Two satellite L-band radiometers have been flown to date to map sea-surface salinity. The first is part of the ***SMOS*** (Soil Moisture and Ocean Salinity) mission (Font et al., 2004; 2010), which was launched in 2009, and the second is part of the ***Aquarius*** mission (Lagerloef et al., 2008), which was launched in 2011. A third mission—the ***SMAP*** (Soil Moisture Active Passive), which is planned for launch in 2014, also includes an L-band radiometer. The stated goal of salinity missions is to map the spatial distribution of sea-surface salinity of the open ocean with an accuracy of 0.2 psu. The SMOS algorithm uses an iterative convergence technique that strives to minimize the error between the measured brightness temperature and the brightness temperature computed using a forward model based on an assumed initial salinity. The assumed value of the salinity is adjusted iteratively until convergence is realized (Zine et al., 2008). An example of a salinity map is shown in Fig. 18-12.

18-3 Measurement of Near-Surface Wind Vector

18-3.1 Azimuth Variation

As noted earlier in Section 16-1.3, wind action causes the water surface to assume an antisymmetric wave pattern, as a consequence of which the backscatter of the

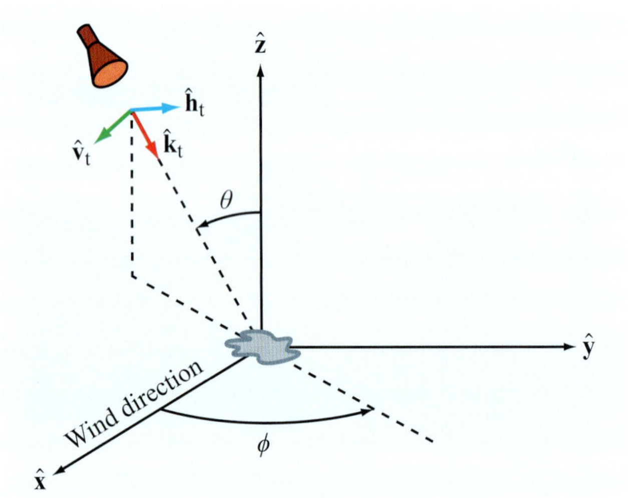

Figure 18-13: Azimuth angle ϕ is defined such that $\phi = 0$ corresponds to when the antenna beam direction is along the wind direction (upwind), and $\phi = 180°$ corresponds to downwind direction.

ocean surface when viewed along the upwind direction is greater than when viewed along the downwind direction. The azimuth angle ϕ in Fig. 18-13 is defined relative to the wind direction; $\phi = 0$ corresponds to when the antenna beam points along the direction of the wind vector **u** (i.e., upwind), and $\phi = 180°$ corresponds to when it points along the opposite direction (downwind).

▶ The surface asymmetry of wind-driven ocean waves that leads to a directional dependence for the radar backscatter also leads to a directional dependence for the brightness temperature. ◀

Experimental observations reported in the early 1990s by Dzura et al. (1992) and by Wentz (1992) demonstrated that both the h-polarized and v-polarized brightness temperatures exhibit periodic variations with ϕ, and that the angular patterns are different for the two polarizations, even at normal incidence. Since those early reports, numerous theoretical and experimental investigations have been conducted to examine the directional dependence for all four Stokes parameters of the brightness temperature vector $\mathbf{T}_\mathrm{B}$ (Yueh et al., 1994; 1995; 1997; 1999; 2010; Gasiewski and Kunkee, 1994; Germain et al., 2002; Piepmeier and Gasiewski, 2001a, b; Yueh, 1997; Yueh and Chaubell, 2012). Recall from Section 6-9 that $\mathbf{T}_\mathrm{B}$ is defined as

$$\mathbf{T}_\mathrm{B} = \begin{bmatrix} T_\mathrm{B}^\mathrm{v} \\ T_\mathrm{B}^\mathrm{h} \\ T_\mathrm{B}^3 \\ T_\mathrm{B}^4 \end{bmatrix} = K \begin{bmatrix} \langle |E_\mathrm{v}|^2 \rangle \\ \langle |E_\mathrm{h}|^2 \rangle \\ 2\mathfrak{Re}\,\langle E_\mathrm{v} E_\mathrm{h}^* \rangle \\ 2\mathfrak{Im}\,\langle E_\mathrm{v} E_\mathrm{h}^* \rangle \end{bmatrix}, \qquad (18.8)$$

where $K = \lambda^2/(k\eta_0 B)$, k is Boltzmann's constant, η_0 is the intrinsic impedance of free space and B is the radiometer bandwidth. The plots in Fig. 18-14 represent the deviation $\Delta\mathbf{T}_\mathrm{B}$ from the value of $\mathbf{T}_\mathrm{B}$ at a reference direction. The variations of the four components of $\mathbf{T}_\mathrm{B}$ with ϕ are illustrated in Fig. 18-14 for a wind speed of 11 m/s at 5 m elevation above the surface. The plots compare measured observations made by a 19 GHz aircraft-borne radiometer along an incidence angle of 30° with model computations using a two-scale model developed by Yueh (1997). The measured data were generated by flying the aircraft in a circular pattern while maintaining the incidence angle constant at 30°. Note that the plots in Fig. 18-14 correspond to

$$\Delta T_\mathrm{B}^\mathrm{v} = T_\mathrm{B}^\mathrm{v}(\phi) - T_\mathrm{B}^\mathrm{v}(45°), \qquad (18.9a)$$

$$\Delta T_\mathrm{B}^\mathrm{h} = T_\mathrm{B}^\mathrm{h}(\phi) - T_\mathrm{B}^\mathrm{h}(45°), \qquad (18.9b)$$

$$T_\mathrm{B}^3 = T_\mathrm{B}^3(\phi), \qquad (18.9c)$$

$$T_\mathrm{B}^4 = T_\mathrm{B}^4(\phi). \qquad (18.9d)$$

Under calm sea conditions, $T_\mathrm{B}^3 = T_\mathrm{B}^4 = 0$, whereas T_B^v and T_B^h may be anywhere between 50 K and 250 K, depending on the values of T_s and θ (over the range between 0° and 65°). Hence, to illustrate the azimuth variation under windy conditions, the plots of T_B^v and T_B^h in Fig. 18-14 are referenced to their values at $\phi = 45°$.

In addition to indicating very good agreement between the measured and modeled values of the four Stokes parameters of $\mathbf{T}_\mathrm{B}$, the angular plots in Fig. 18-14 assert that:

(a) The v-polarized brightness temperature T_B^v is at its maximum value at $\phi = 0$ (upwind) and at its minimum value at $\phi = 90°$ (crosswind), and it has a second peak at

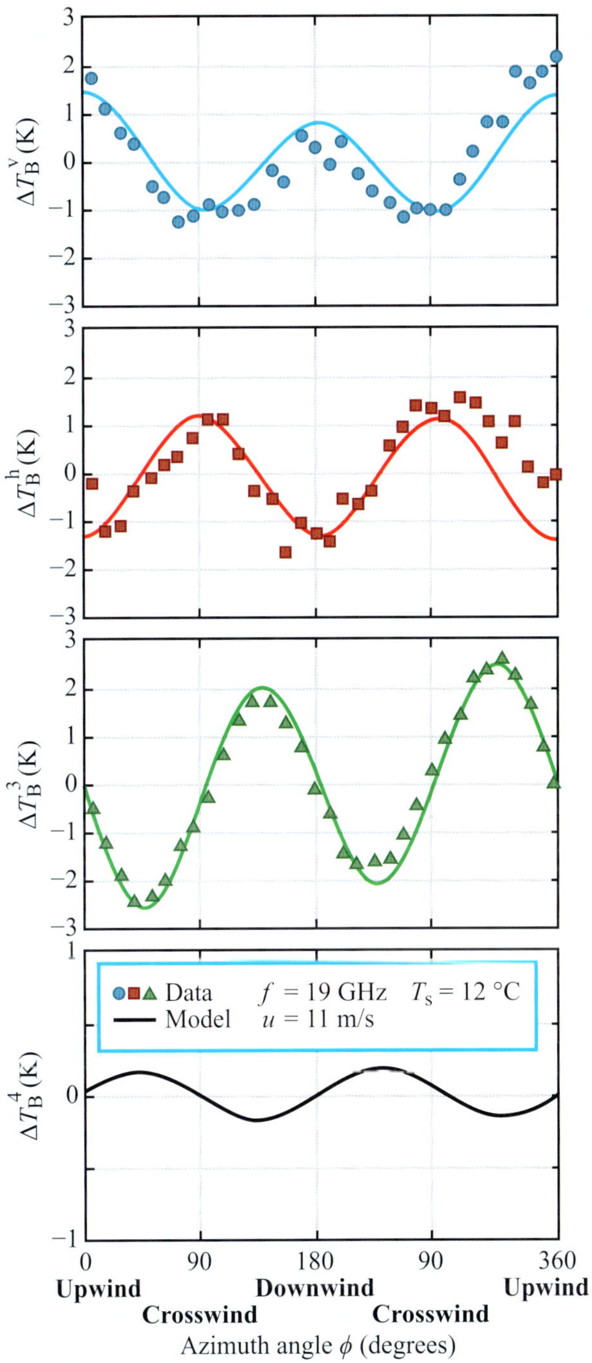

Figure 18-14: Comparison of theoretical Stokes parameters with the aircraft 19 GHz radiometer data at 30° incidence as a function of the azimuth angles ϕ [Yueh, 1997].

$\phi = 180°$ (downwind) but it is smaller than the upwind peak.

(b) The pattern of T_B^h is approximately a mirror-image of that of T_B^v.

(c) The pattern of T_B^3 exhibits unequal maxima at $\phi = 135°$ and $315°$ and unequal minima at 45° and 225°.

(d) The calculated pattern of T_B^4 has a very small amplitude of about 0.2 K (no measurements were made in the reported experiment).

In an independent study reported by Germain et al. (2002), aircraft observations made by a 19.35 GHz radiometer were compared with model computations based on Yueh's two-scale model (1997), but with the effects of shadowing incorporated in the model, which are not important at angles of incidence close to nadir, but may influence the emission at high angles of incidence. Their results are shown in Fig. 18-15 for $\theta = 55°$ and $u = 9$ m/s.

18-3.2 Wind-Speed Dependence

Using 19 GHz and 37 GHz SSMI satellite data observed at $\theta = 53°$, in conjunction with buoy reports supplied by the U.S. National Data Buoy Center, Wentz (1992) developed a semiempirical model for T_B^v and T_B^h as a function of the magnitude of the wind speed u and its direction ϕ. He also incorporated corrections to account for atmospheric attenuation and emission contributions. The model provides very good fits to the experimental data at both frequencies and over a wide range of wind speeds (Fig. 18-16). By inverting his model to predict wind speed from SSMI measurements of T_B^v and T_B^h at 37 GHz, he was able to demonstrate that the wind speed u can be inferred from the radiometric data with an rms error of 1.6 m/s, if no correction due to wind direction is applied, and that the error can be reduced to 1.3 m/s if the wind direction inferred from the data is incorporated into the inversion algorithm. Wentz's results are based on 3321 comparisons. An updated version of Wentz's 1992 model was reported by Meissner and Wentz (2002).

In 1998, Chang and Li (1998) developed a neural-network retrieval model to infer u and ϕ. An example of their results based on 37 GHz SSMI data is shown in Fig. 18-17. The reported retrieval rms error is 0.965 m/s for u and 19.55° for ϕ.

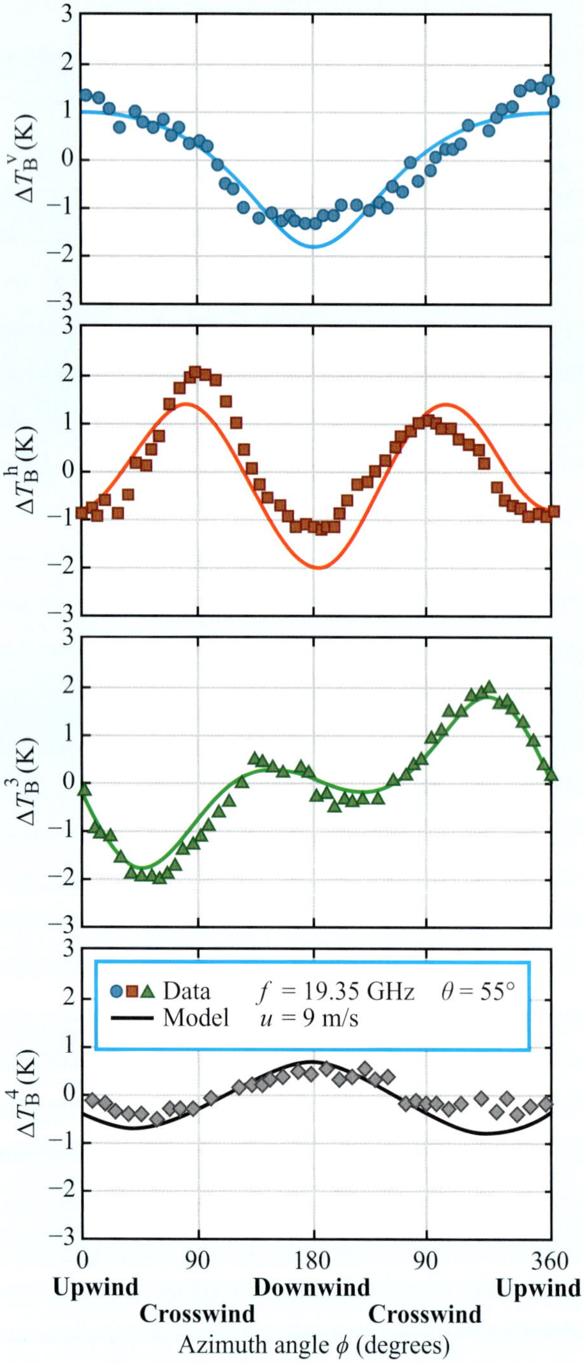

Figure 18-15: Comparisons of 19.35 GHz aircraft measurements with theoretical model predictions at $\theta = 55°$ for $u = 9$ m/s at 5 m above the sea surface [Germain et al., 2002].

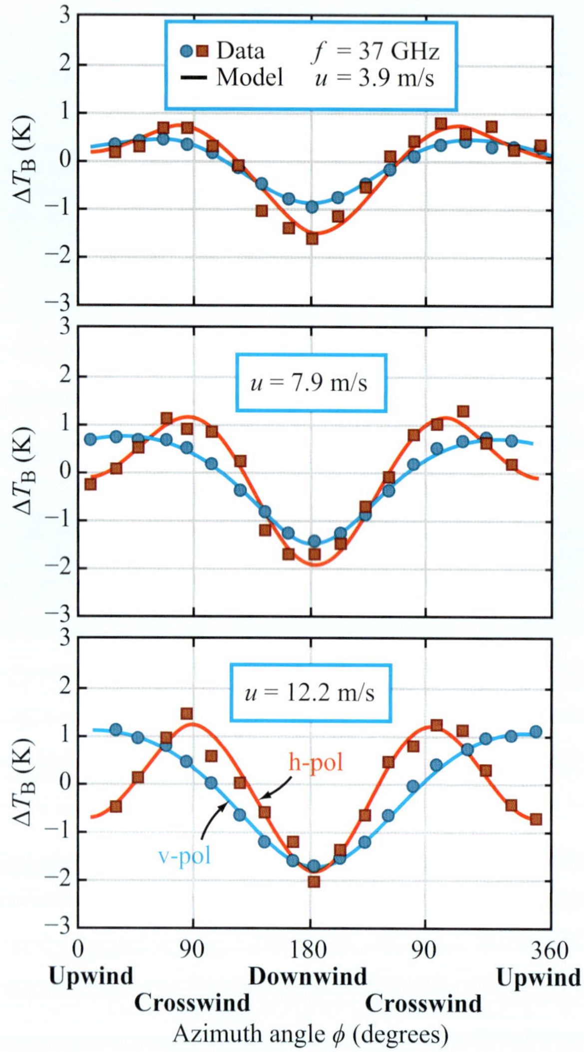

Figure 18-16: Comparison of semiempirical model for ΔT_B^v and ΔT_B^h with measured SSMI data at three different wind speeds [Wentz, 1992].

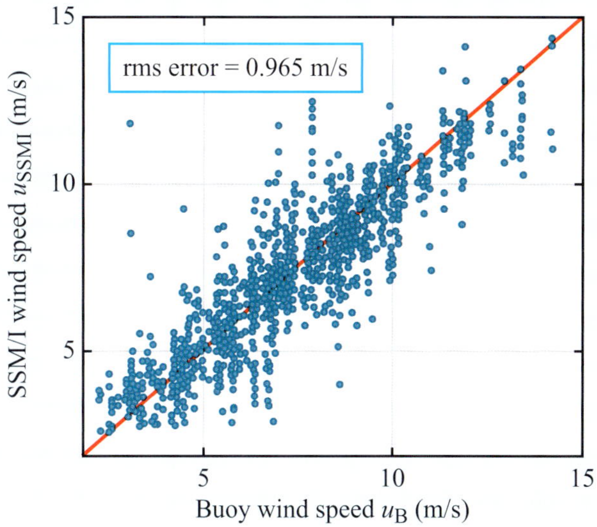

(a) Wind speed u

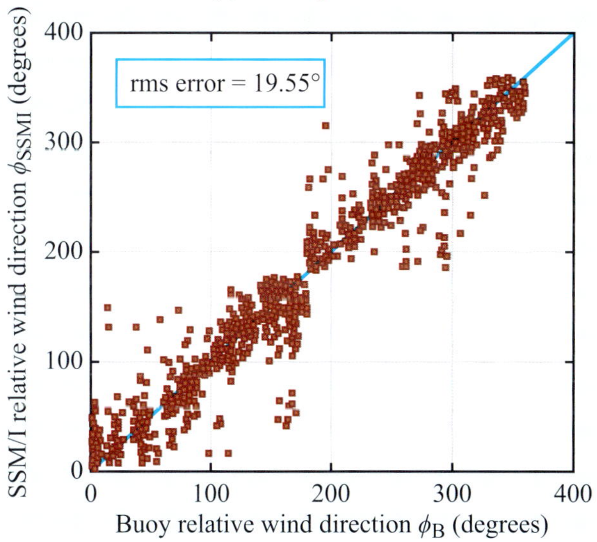

(b) Wind direction ϕ

Figure 18-17: Scatter plot of buoy wind vectors versus retrieved wind vectors from SSMI 37 GHz T_B^v and T_B^h observations using a neural-network algorithm [Chang and Li, 1998].

18-3.3 WindSat Retrieval Algorithm

The Coriolis satellite, launched in January of 2003, carries a new generation version of the SMMI and AMSR sensors called *WindSat*. Its radiometers operate at five frequencies between 6.8 GHz and 37 GHz, it uses conical scanning at an approximately constant incidence angle of 53°, and three of its frequency channels (namely, 10.7, 18.7, and 37 GHz) can measure all four Stokes parameters of $\mathbf{T}_B$. Spatial resolution is on the order of 30 km. More details are available in the paper by Gaiser et al. (2004).

The dependence of the four Stokes parameters on ϕ is given by the model (Yueh et al., 2006):

$$T_B^v = T_{B0}^v + T_{B1}^v \cos\phi + T_{B2}^v \cos 2\phi, \qquad (18.10a)$$

$$T_B^h = T_{B0}^h + T_{B1}^h \cos\phi + T_{B2}^h \cos 2\phi, \qquad (18.10b)$$

$$T_B^3 = T_{B1}^3 \sin\phi + T_{B2}^3 \sin 2\phi, \qquad (18.10c)$$

$$T_B^4 = T_{B1}^4 \sin\phi + T_{B2}^4 \sin 2\phi, \qquad (18.10d)$$

where ϕ is the angle between the radiometer beam look direction and the wind direction. Coefficients T_{B0}^v and T_{B0}^h represent the v-polarized and h-polarized brightness temperatures when $u = 0$ (no wind). The remaining coefficients in Eq. (18.10) are functions of the wind speed u. Several empirical models have been proposed for relating these coefficients to u, and for inverting the overall model to infer u and ϕ (Yueh et al., 2006; Brown et al., 2006; Meissner and Wentz, 2006; Smith et al., 2006). The algorithms are based on WindSat radiometer data, together with validation data sets reported by buoys and other sources. The estimated retrieval error using WindSat data is on the order of 1 m/s for u. The error for ϕ depends on u; it is large at low wind speeds, but less than 20° for $u > 7$ m/s.

The official WindSat retrieval algorithm uses various combinations of frequency and polarization channels to retrieve the wind vector, and includes correction for atmospheric effects due to water-vapor and cloud-cover variations. The only channel not used for this application is the 6.6 GHz channel, as it is the primary channel for measuring the sea-surface temperature T_s. The global image shown in Fig. 18-18 displays the magnitude and

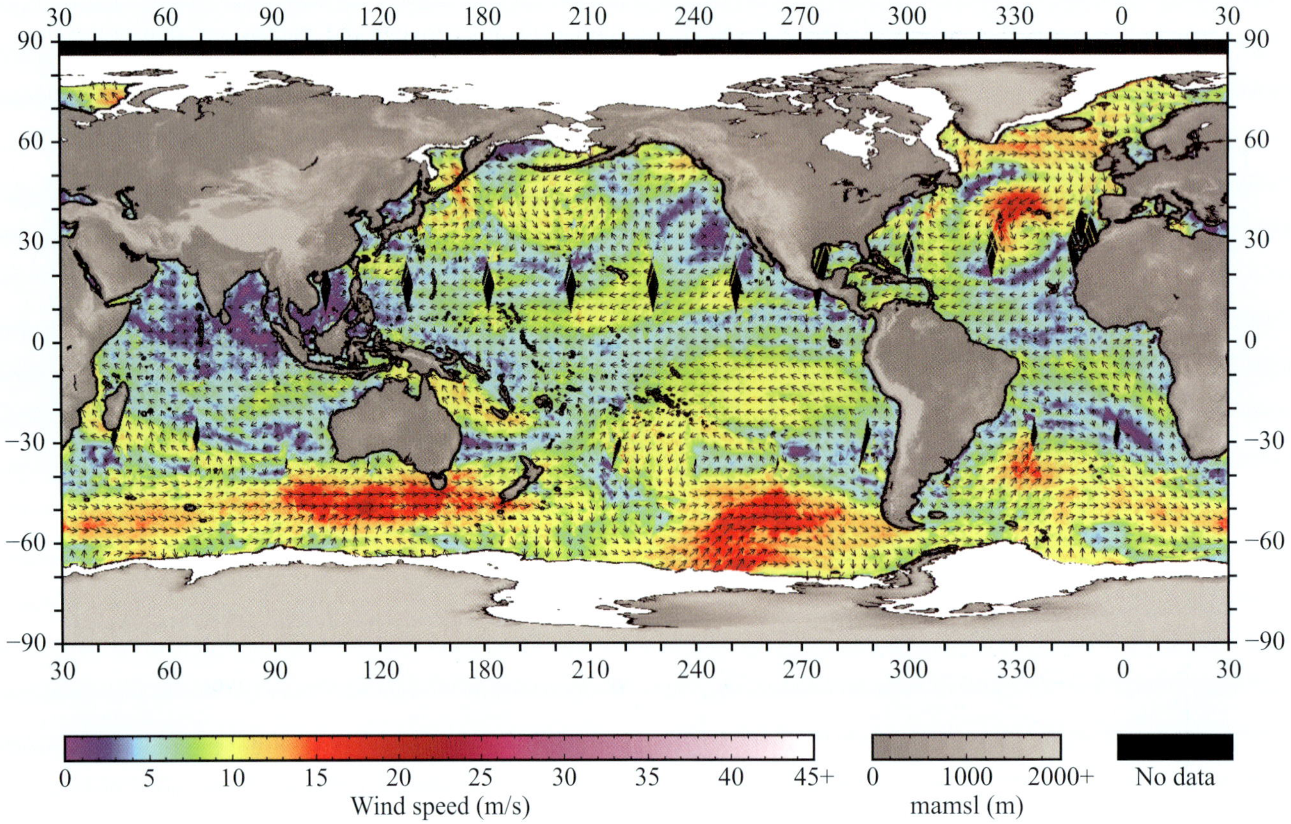

Figure 18-18: Global image of wind magnitude and direction over the oceans.

direction of the wind vector, inferred from three days of WindSat radiometric data.

Both the Aquarius and SMAP missions use a combination of radar and a radiometer, both operating at L-band (1.26 GHz for the radar and 1.41 GHz for the radiometer), which has spurred interest in examining the dependence of σ° and T_B on wind speed and direction. Using aircraft data, Yueh and Chaubell (2012) developed models for both σ° and T_B based on the angular function given in Eq. (18.10a). Figure 18-19 depicts the azimuth variation for T_B^v, T_B^h, σ_vv°, and σ_hh° at each of five different wind speeds.

18-4 Mapping Sea-Ice Type and Concentration

The SMMI radiometer instruments that flew onboard the DMSP satellites beginning in 1987 and their successors in the form of the AMSR radiometer suite aboard the Aqua and ADEOS-II satellites (see Table 1-4) have provided credible information about the spatial distribution of sea ice in both the Arctic and Antarctic regions of the world. The two primary pieces of information extracted from the multifrequency radiometric data are sea-ice type [***first-year*** (FY) versus ***multiyear*** (MY)] and sea-ice concentration. The physical properties of sea ice are described in Section 16-10 in connection with the radar backscatter response

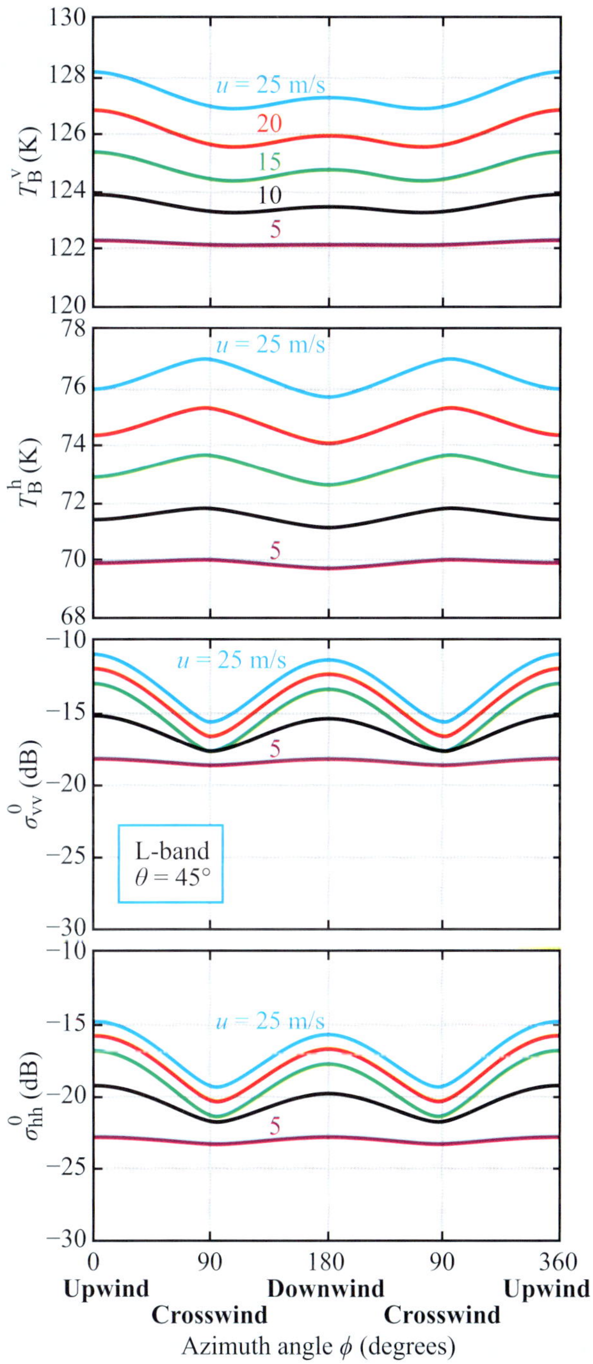

Figure 18-19: L-band model azimuth patterns for T_B and σ° at $\theta = 45^\circ$ [Yueh and Chaubell, 2012].

to sea ice. In the present section, we examine the relationship between the microwave emissivity and the relevant physical and dielectric parameters of sea ice, and we do so as a function of the microwave frequency f because, as we see later, the emissivity frequency gradient (de/df) has proven to be a most important parameter for discriminating between ***open water*** (OW) and ice, and between FY ice and MY ice. We then conclude the section with a brief overview of inversion algorithms and an image example of sea ice distribution.

18-4.1 Coherent versus Incoherent Emissivity

Consider the air/ice configuration shown in Fig. 18-20. The emissivity measured by a nadir-looking radiometer depends on the following parameters:

(a) $\varepsilon_i = \varepsilon_i' - j\varepsilon_i''$ of the sea-ice layer,

(b) $\varepsilon_w = \varepsilon_w' - j\varepsilon_w''$ of the water,

(c) d, the thickness of the ice layer,

(d) f, the microwave frequency,

(e) a, the albedo of the ice layer,

(f) the roughnesses of the air-ice and ice-water boundaries.

If we treat the boundaries as planar interfaces, we can proceed with computing the emissivity along one of three directions:

(1) Coherent Emissivity: If the spatial inhomogeneity within the ice volume (due to air or liquid brine pockets) is at a scale much smaller than λ in the medium, then the medium may be assumed to be homogeneous with an albedo $a = 0$, in which case we can use the ***coherent emissivity*** model of Section 12-12.1 to compute e_{coh}.

(2) Incoherent Emissivity with $a < 0.2$: If the wavelength is such that the albedo is not zero, but smaller than about 0.2, we can use Eq. (12.69) of Section 12-12.2 to compute the ***incoherent emissivity*** e_{inc}.

(3) Incoherent Emissivity with $a > 0.2$: If $a > 0.2$, we have to resort to a numerical solution of the radiative transfer equation.

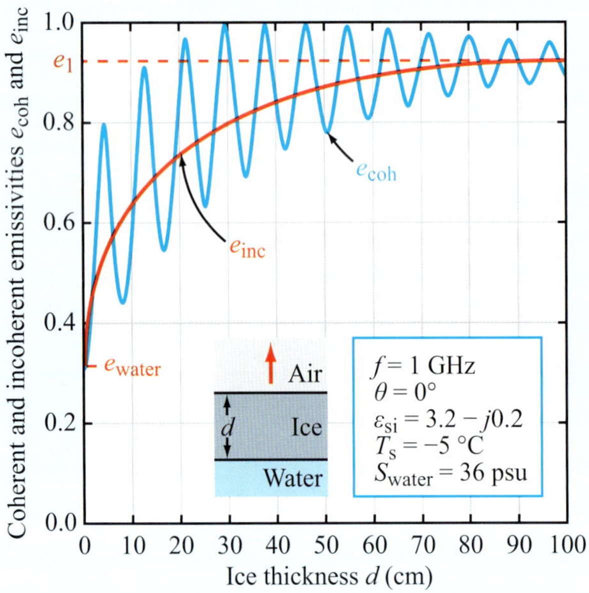

Figure 18-20: Calculated variation of the coherent and incoherent emissivities of sea ice over water as a function of ice thickness at 1 GHz.

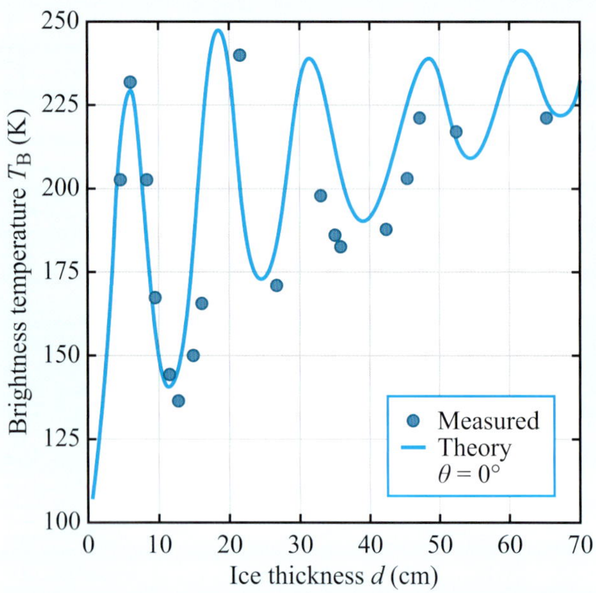

Figure 18-21: Comparison of experimental low-salinity results and the layered model at 610 MHz [Hallikainen, 1983].

In Fig. 18-20 we display plots for e_{coh} and e_{inc} (with $a \approx 0$) as a function of the ice thickness d. The frequency is 1 GHz. As expected, e_{coh} exhibits a damped oscillatory variation with d, whereas e_{inc} does not. In both models, the energy emitted by the ice and water media undergo multiple reflection between the two boundaries. However, in the coherent model the contributions of the multiple reflection are added coherently by adding their electric fields first, before computing the powers, whereas in the incoherent model the contributions are added incoherently by adding their power components directly.

A somewhat more complicated (but also more realistic) model was used by Hallikainen (1983) to analyze radiometric measurements made at frequencies in the 0.5 to 1 GHz band. The model divides the ice layer into a large number of discrete layers, allows each layer to have its own dielectric constant and temperature, and includes a snow layer on top of the ice. Comparison of Hallikainen's model calculations with the measured brightness temperature at 0.61 GHz is shown in Fig. 18-21. The good overall agreement between theory and measurement supports the applicability of the coherent emissivity model at this low microwave frequency.

To examine the dependence of emissivity on ice thickness at higher microwave frequencies, we use the incoherent emissivity model because the inhomogeneities in the ice volume are likely to be much more significant at the shorter wavelengths. The results—shown in Fig. 18-22 for 10.7, 18.7, 36.5, and 89 GHz (all of which are AMSR channels)—are for ice with a salinity of 1 psu. For all practical purposes, at frequencies greater than 10.7 GHz, the ice behaves like an opaque medium if $d > 15$ cm. The indicated ice salinity is characteristic of MY ice, which is far less lossy than younger types of ice. Consequently, unless we operate a radiometer at a frequency much lower than 1 GHz, it is not possible to determine the ice thickness from the observed brightness temperature. And since frequencies below 1 GHz are not practical for radiometry, it is not realistic to consider such an

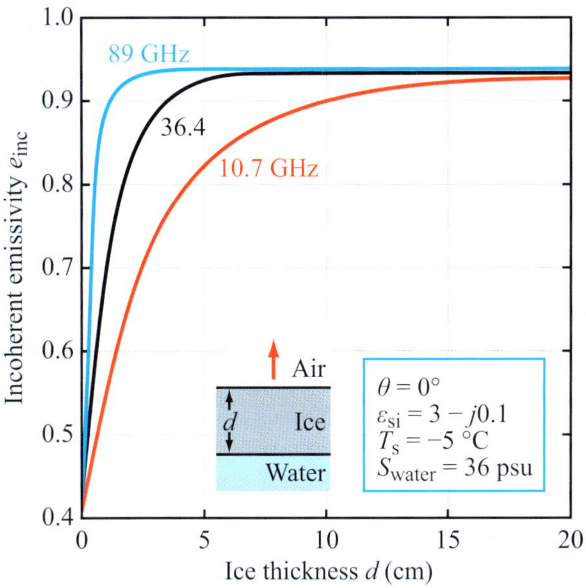

Figure 18-22: Incoherent emissivity of sea ice as a function of ice thickness at multiple frequencies.

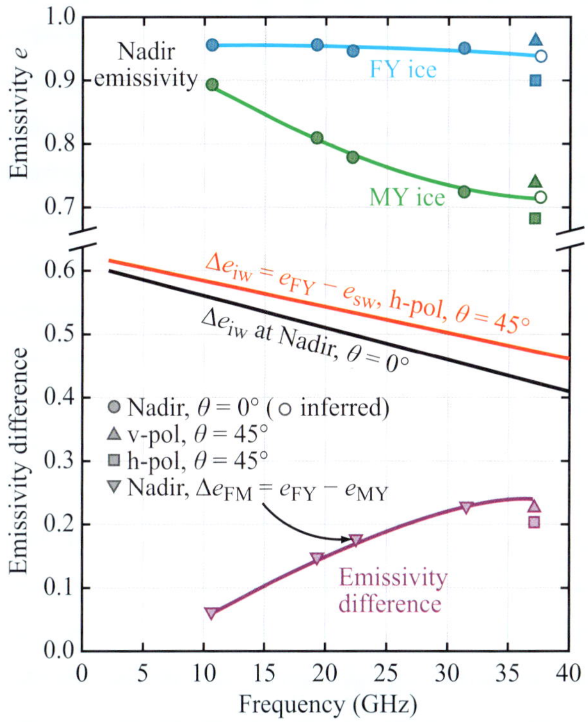

Figure 18-23: Measured nadir emissivities of first-year ice and multiyear ice [Wilheit et al., 1971].

application from a satellite platform. Nevertheless, it is possible to distinguish different types of ice on the basis of their polarization ratio at 18.7 GHz and the spectral gradient ratio of the 18.7- and 36.5 GHz channels (both parameters are defined later in Section 18-4.5). This makes it possible to map the location of the generally thicker MY ice from orbital altitudes using the emissivity measurements.

18-4.2 Aircraft Observations

Among the first investigations of sea-ice microwave emission were those conducted in 1967 and 1970 using the NASA Convair 990 airborne observatory (Wilheit et al., 1971). The two missions were conducted over the Arctic Sea north of Point Barrow, Alaska, and the microwave observations were made from an altitude of 10 km. In the 1967 mission, only one radiometer, operating at 19.35 GHz, was used. However, nine radiometers were added for the 1970 mission. The major conclusions derived from the microwave observations were (1) that there is a large contrast between the emissivities of sea ice and water, and (2) that the contrast between the emissivities of "new" ice (FY ice) and "old" ice (MY ice) increases as frequency increases. The top part of Fig. 18-23 shows the ***nadir emissivity*** measured for FY ice and MY ice at five microwave frequencies. The values shown at 37 GHz are estimates based on dual-polarization measurements made at a nadir angle of 45°. The bottom part of the figure shows a plot of $\Delta e_{iw} = e_{FY} - e_{sw}$, representing the difference in emissivity between FY ice and sea water, and a plot of $\Delta e_{FM} = e_{FY} - e_{MY}$, representing the difference in emissivity between FY ice and MY ice.

According to Fig. 18-23, e_{FY} is approximately frequency-independent. In contrast, e_{MY} decreases rapidly with frequency, from 0.89 at 10.69 GHz to 0.72 at 31.4 GHz. These trends were confirmed by additional measurements made by the NASA group in 1971 (Gloersen et al., 1973), and by a Naval Research

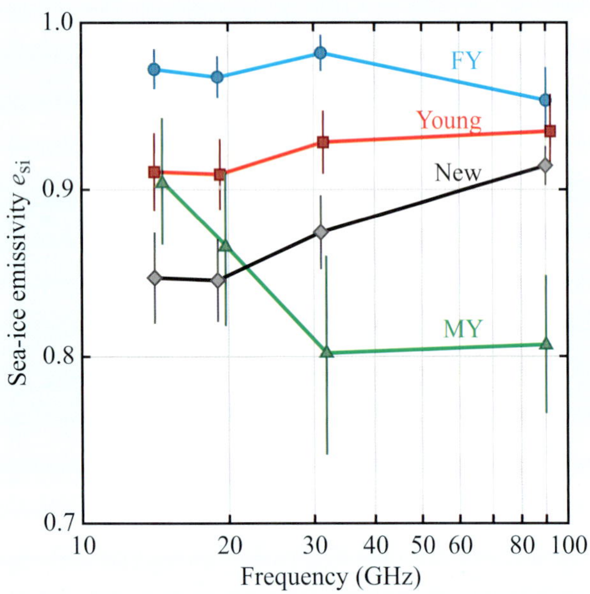

Figure 18-24: Nadir emissivity in the 14 to 90 GHz range for four ice types. Each vertical bar shows ±1 standard deviation in the data. Symbols have been displaced slightly to avoid overlapping [from Troy et al., 1981].

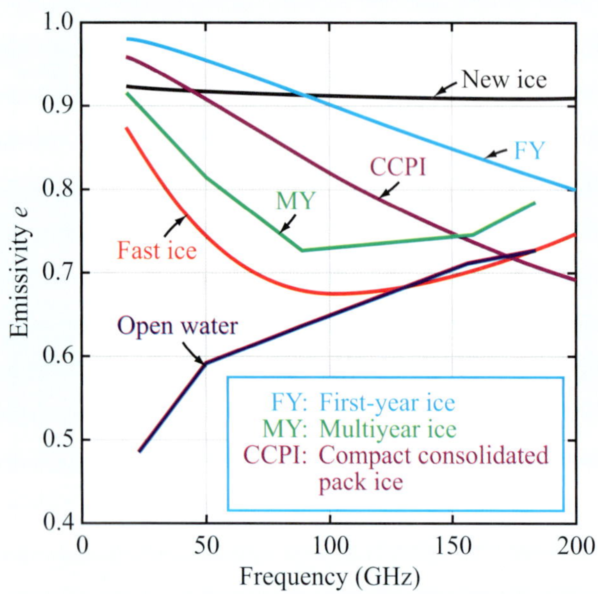

Figure 18-25: Emissivity spectrum for open water and five types of sea ice [Hewison et al., 2002].

Laboratory (NRL) group in 1973 (Tooma et al., 1975) and 1977 (Troy et al., 1981). The results of the 1977 mission, which involved multifrequency measurements made from an altitude of 200 to 600 m, are shown in Fig. 18-24. The two sets of measurements, namely those shown in Figs. 18-23 and 18-24, are in very close agreement for the FY ice category. For MY ice, the values given in Fig. 18-23 do not agree as well with the mean values of the measurements given in Fig. 18-24, but they do fall within the indicated vertical bars, each representing one standard deviation relative to the measured mean. The only exception is the 31 GHz emissivity, for which the measurements by Troy et al. (1981) give $e_{\text{MY}} = 0.81 \pm 0.05$, and the average given by Wilheit et al. (1971) is $e_{\text{MY}} = 0.72$. This discrepancy may be explained as being partly the result of the presence of second-year ice patches in the MY ice areas observed by Troy et al.; the emissivity of second-year ice is intermediate between the emissivities of FY and MY ice.

An aircraft-mounted four-frequency radiometer with center frequencies at 24, 50, 89, and 157 GHz was flown in 1995 and 1997 over Baltic sea ice to measure the nadir emissivity over this wide range of the millimeter-wave spectrum. The system belonged to the Remote Sensing Branch of the UK Meteorological Office. The results, which were reported by Hewison and English (1999) and Hewison et al. (2002), are summarized in Fig. 18-25. These measurements confirm the overall character of the spectral responses measured in earlier campaigns.

18-4.3 Models for the Emissivity of Sea Ice

The major factors governing the emission behavior of sea ice are (1) sea-ice thickness, (2) the dielectric constant and temperature profiles of the ice layer, (3) the thickness of any snow layer that may be present on top of the ice, and whether it is dry or wet, and (4) the degree of inhomogeneity and anisotropy of the ice volume. We previously examined the dependence of the emissivity

Table 18-1: Representative physical and electrical parameters for four sea-ice types.

Parameter	New Ice	Young Ice	First-Year Ice	Multiyear Ice
Depth range (cm)	<10	10–30	30–100	>100
Ice salinity (psu)	15	10	5	1
Ice surface temperature (°C)	−5	−8	−13	−18
Ice density ρ_i (g/cm^3)	0.87	0.87	0.87	0.77
Brine salinity (psu)	86	126	164	197
Brine volume fraction	0.152	0.067	0.023	0.004
Ice dielectric constant at 10 GHz	$5.65 - j2.25$	$4.03 - j0.81$	$3.32 - j0.23$	$2.77 - j0.03$
Ice dielectric constant at 30 GHz	$4.09 - j1.52$	$3.46 - j0.58$	$3.15 - j0.17$	$2.74 - j0.02$
Ice dielectric constant at 100 GHz	$3.39 - j0.67$	$3.17 - j0.26$	$3.06 - j0.08$	$2.73 - j0.01$
Absorption coefficient κ_a at 10 GHz (Np/m)	194	84	27	3.8
Absorption coefficient κ_a at 30 GHz (Np/m)	466	195	60	8.1
Absorption coefficient κ_a at 100 GHz (Np/m)	755	309	92	12.0

on ice thickness in Section 18-4.1 and concluded that at frequencies above 10 GHz, the ice layer appears (electromagnetically) semi-infinite in depth. Now we examine the other factors.

Ice type

In general, different types of ice are characterized by different dielectric properties. This is in part because of differences in salinity and density, and in part because the physical temperature of the ice surface is often different for different types of ice. The surface of thick, multiyear ice is better insulated from the warm water beneath it than the surface of new ice, and hence the surface temperature of multiyear ice is likely to be colder than that of new ice. Furthermore, temperature differences affect the brine volume fraction of ice, which in turn affects the dielectric constant of the sea-ice mixture.

Four types of ice are considered: new ice, young ice, first-year (FY) ice, and multiyear (MY) ice. While the first two types are technically subdivisions of FY ice, in the following each is treated as a distinct type of ice because of their significantly different emissivities. Table 18-1 gives typical values for the physical temperature of the ice surface T_s, the ice salinity S_i, and the ice density ρ_i for each of the four types of ice. These parameters are used in the model results presented shortly. Because of spatial resolution considerations, we are interested in frequencies higher than 10 GHz. Therefore, we are interested only in the surface and near-surface regions of the ice layer because, as was demonstrated previously, the penetration depth in sea ice is on the order of centimeters or less, depending on the type of ice and the microwave frequency. Consequently, emission contributions from the sea-water medium beneath the ice can be neglected. For the present, we assume that the ice surface is smooth and bare (i.e., without snow cover) and that the ice volume is homogeneous. These are reasonable assumptions for new ice and young ice, but not for FY or MY ice. Well-established FY ice is generally covered by a 10 to 20 cm snow layer, which modifies the emission behavior of the ice (the effects of snow on emissivity are examined later). Multiyear ice also may be snow-covered. Furthermore, MY is not homogeneous. The density of MY ice is lower than that of the other types of ice (Table 18-1) because MY ice contains many air bubbles left behind when the brine drains from the ice during the summer season. The presence of the air bubbles causes volume scattering.

To derive a simple model for sea-ice emission, we assume a smooth, bare surface and a homogeneous, semi-infinite ice volume, in which case the emissivity is given by

$$e^p(\theta) = 1 - \Gamma^p(\theta), \tag{18.11}$$

where $\Gamma^p(\theta)$ is the specular reflectivity. To compute $\Gamma^p(\theta)$ for each of the four ice types listed in Table 18-1, we use the refractive dielectric mixing model given by Eq. (4.45) with $\alpha = 1/2$. That is,

$$\sqrt{\varepsilon_{\mathrm{si}}} = v_{\mathrm{b}}\sqrt{\varepsilon_{\mathrm{b}}} + v_{\mathrm{i}}\sqrt{\varepsilon_{\mathrm{i}}} + (1 + v_{\mathrm{i}} + v_{\mathrm{b}})\sqrt{\varepsilon_{\mathrm{air}}}\,, \tag{18.12}$$

where v_{b} is the volume fraction of brine and v_{i} is the volume fraction of ice. The dielectric constants ε_{b}, ε_{i}, $\varepsilon_{\mathrm{si}}$ and $\varepsilon_{\mathrm{air}}$ refer to brine, pure ice, sea ice, and air, respectively. This simple mixing model provides good agreement with the measured values of the dielectric constant of FY ice (Gloersen and Larabee, 1981). Note that the ice and brine volume fractions v_{i} and v_{b} are related to each other through the sea-ice density ρ_{si},

$$\rho_{\mathrm{si}} = \rho_{\mathrm{i}} v_{\mathrm{i}} + \rho_{\mathrm{b}} v_{\mathrm{b}} \tag{18.13}$$

where $\rho_{\mathrm{i}} = 0.916$ g/cm^3 is the density of pure ice and $\rho_{\mathrm{b}} \approx 1$ g/cm^3 is the density of brine. Given the temperature and salinity of the ice, the brine volume fraction v_{b} can be determined using the models in Section 4-5.2, and the expression for ε_{b} is given by Eq. (4.47). Naturally, $\varepsilon_{\mathrm{air}} = 1 + j0$. For ice we take $\varepsilon_{\mathrm{i}} = 3.15 + j0$ (the imaginary part of ε_{i} is actually on the order of 10^{-3}, but neglecting it does not have any significant effect on the calculated value of $\varepsilon_{\mathrm{si}}$). The above procedure leads to the specifications given in Table 18-1 for the brine salinity S_{b}, the brine and ice volume fractions, and the absorption coefficient κ_{a} at 10, 30, and 100 GHz. The dielectric constants calculated from Eq. (18.12) are shown as a function of frequency in Fig. 18-26. The relative levels of the dielectric-constant curves among the four ice types is a direct result of their relative salinities.

Figure 18-27 displays spectral curves for the nadir emissivity of new ice, young ice, and FY ice, calculated according to Eq. (18.11) and the dielectric curves given in Fig. 18-26. The points and associated vertical bars shown in Fig. 18-27 are the mean values and the standard deviations (± 1) of the emissivities measured by Troy et al. (1981) at 14, 19, 31, and 90 GHz. The calculated curves are in good agreement with the measured data for new ice and young ice. The FY-ice curve delineates a

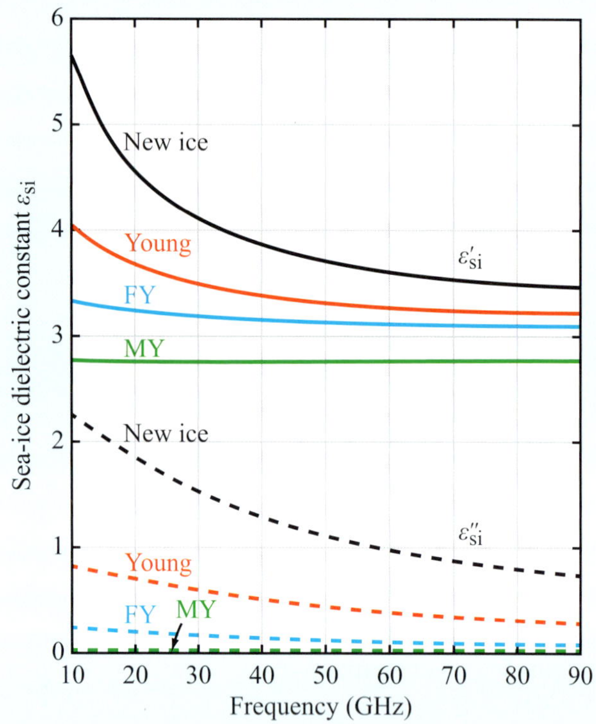

Figure 18-26: Dielectric constants of four types of sea ice as a function of frequency, computed according to Eq. (18.12) and the specifications given in Table 18-1.

behavior that is approximately independent of frequency and is in agreement with the trend indicated by the data. Its level, however, is around 0.92, as compared with 0.97 for the general level of the measured data. As we shall see in the case considered next, this difference in level can be reconciled by adding a layer of snow to the ice when calculating the emissivity of FY ice.

Snow layer over FY ice

The presence of a snow layer affects the observed brightness of sea ice. To understand this, consider a snow layer of thickness d over a layer of FY ice. The combination of two factors, namely an uneven snow surface (i.e., fluctuations in d as a function of spatial position) and a small degree of volume scattering by the ice crystals in the snow layer, justifies the use

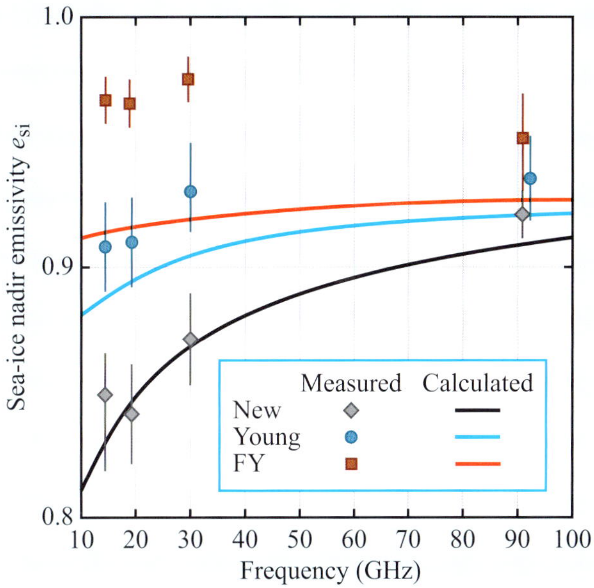

Figure 18-27: Calculated spectra of nadir emissivities of new ice, young ice, and first-year ice (without snow cover). For comparison, the emissivities measured by Troy et al. (1981) are also shown.

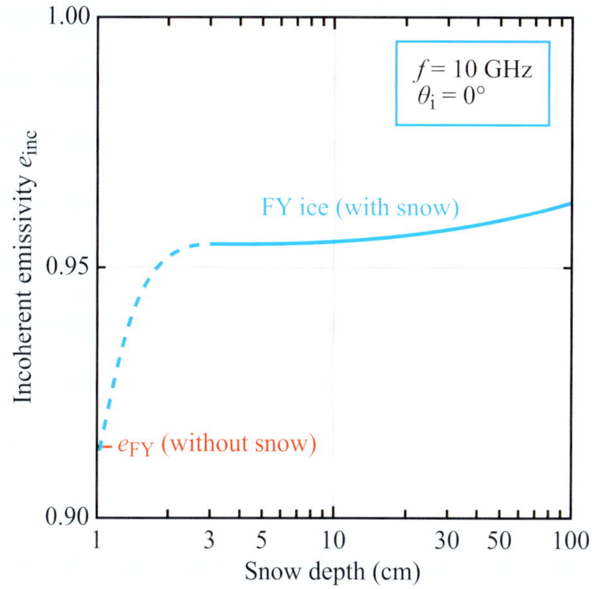

Figure 18-28: Computed variation of emissivity of snow-covered first-year ice with snow depth at 10 GHz.

of the incoherent emissivity model for computing the emissivity of the snow-over-ice composite.

From Eq. (12.71), the incoherent emissivity at nadir ($\theta = 0$) is given by

$$e = \frac{(1-\Gamma_{12})(1-\Gamma_{23}\Upsilon^2)}{1-\Gamma_{12}\Gamma_{23}\Upsilon^2}, \tag{18.14}$$

where Γ_{12} and Γ_{23} are the reflectivities of the air-snow and snow-ice interfaces and Υ is the one-way transmissivity of the snow layer. For normal incidence,

$$\Gamma_{12} = \left|\frac{\sqrt{\varepsilon_s}-1}{\sqrt{\varepsilon_s}+1}\right|^2, \tag{18.15a}$$

$$\Gamma_2 = \left|\frac{\sqrt{\varepsilon_{si}}-\sqrt{\varepsilon_s}}{\sqrt{\varepsilon_{si}}+\sqrt{\varepsilon_s}}\right|^2, \tag{18.15b}$$

and

$$\Upsilon = e^{-\kappa_a d}, \tag{18.16}$$

where κ_a is the absorption coefficient of snow. The dielectric constant ε_{si} of FY sea ice is given in Fig. 18-26. The snow is assumed to be dry and to have a density of 0.3 g/cm^3, which leads to $\varepsilon_s' \approx 1.6$ (see Section 4-6.1). From Fig. 4-18, the corresponding dielectric loss factor at 9.375 GHz is $\varepsilon_s'' = 4.8 \times 10^{-4}$. For simplicity, we assume that both parts of ε_s are independent of frequency. Since $\varepsilon_s''/\varepsilon_s' \ll 1$, the approximate relationship

$$\kappa_a \approx \frac{2\pi}{\lambda_0}\frac{\varepsilon_s''}{\sqrt{\varepsilon_s'}} \tag{18.17}$$

may be used, where λ_0 is the wavelength in free space.

Figure 18-28 shows the computed nadir emissivity of FY ice when covered by a layer of snow, plotted as a function of the snow depth d. The variation is almost binary in nature: at $d = 0$, $e = e_{FY} = 0.915$, the emissivity of snow-free sea ice, and for $d \geq 2$ cm, $e \approx 0.955$ and increases very slowly with increasing depth d. The dashed curve is a free-hand extrapolation because Eq. (18.14) does not apply for d less than about

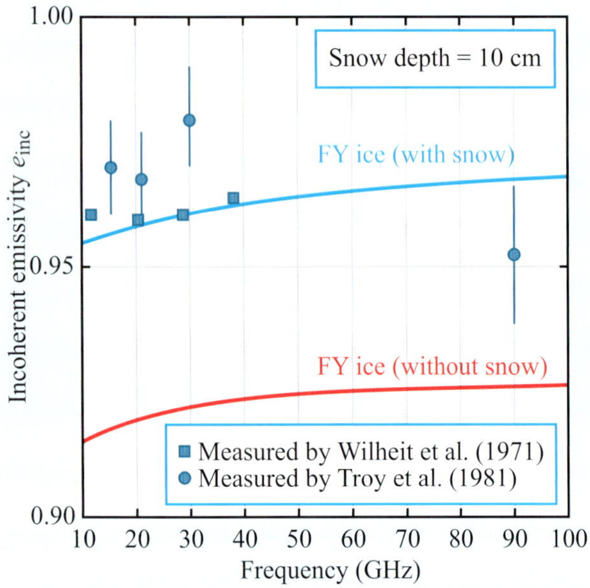

Figure 18-29: Spectral variation of emissivity of first-year ice with and without snow cover.

$\lambda_0/2$. Because e is essentially independent of d, we can choose a reasonable thickness of, let us say, 10 cm when calculating the emissivity as a function of frequency. The result is shown in Fig. 18-29, where it is compared with the measured values reported by Troy et al. (1981) and Wilheit et al. (1971). Overall, the calculated curve is in good agreement with the measured data. It is especially significant that (1) the model used is quite simple, (2) the results are approximately independent of snow depth for 2 cm $\leq d \leq$ 30 cm, and (3) the results are approximately independent of snow density for a wide range of densities extending from 0.2 to 0.7 g/cm^3. More sophisticated models for snow cover over sea ice have been developed. The reader is referred to Tjuatja et al. (1992), Nghiem et al. (1995), and Golden et al. (1998) for more details.

Multiyear ice

Modeling the emission from MY ice can be more difficult than for FY ice. Multiyear ice is characterized by a greater penetration depth than FY ice, and it contains air bubbles of a size comparable to the wavelength. These factors call for modeling the ice both as a scattering and an absorbing medium. For simplicity, we model MY sea ice as an inhomogeneous medium with a plane upper boundary. Assuming a Rayleigh phase function for the scattering produced by the air bubbles, the curves shown in Fig. 18-30 are obtained using a numerical solution of the radiative transfer equation. The nadir emissivity curves in Fig. 18-30(a) appear to be in fairly good agreement with the data reported by Wilheit et al. (1971) and Troy et al. (1981), and the same can be said for the horizontal and vertical polarization curves shown in Fig. 18-30(b) for $\theta = 50°$.

The comparisons between measurements and theoretical calculations given in this section are intended to illustrate some of the approaches to understanding and interpreting the radiometric observation of sea ice.

► Because the physical and dielectric parameters of a given type of ice vary widely, it is not possible to provide a unique set of curves to describe the emission properties of sea ice. ◄

Except for the clear distinction between FY ice and MY ice, well-defined criteria useful in separating one subtype from another are difficult to formulate. This is partly because the physical characteristics of sea ice vary as a function of location and time, which leads to variations in the physical parameters of the ice (e.g., temperature, salinity, density) and in variations in the dielectric and propagation parameters as well. Furthermore, the ice may be covered by a snow layer whose thickness, density, and average crystal size may vary, and which may be dry or may contain water in liquid form (wet snow). The net result of these variations is that it may not be possible to define, by measurement or by theory, the unique radiative properties that can consistently distinguish different types of ice under all conditions.

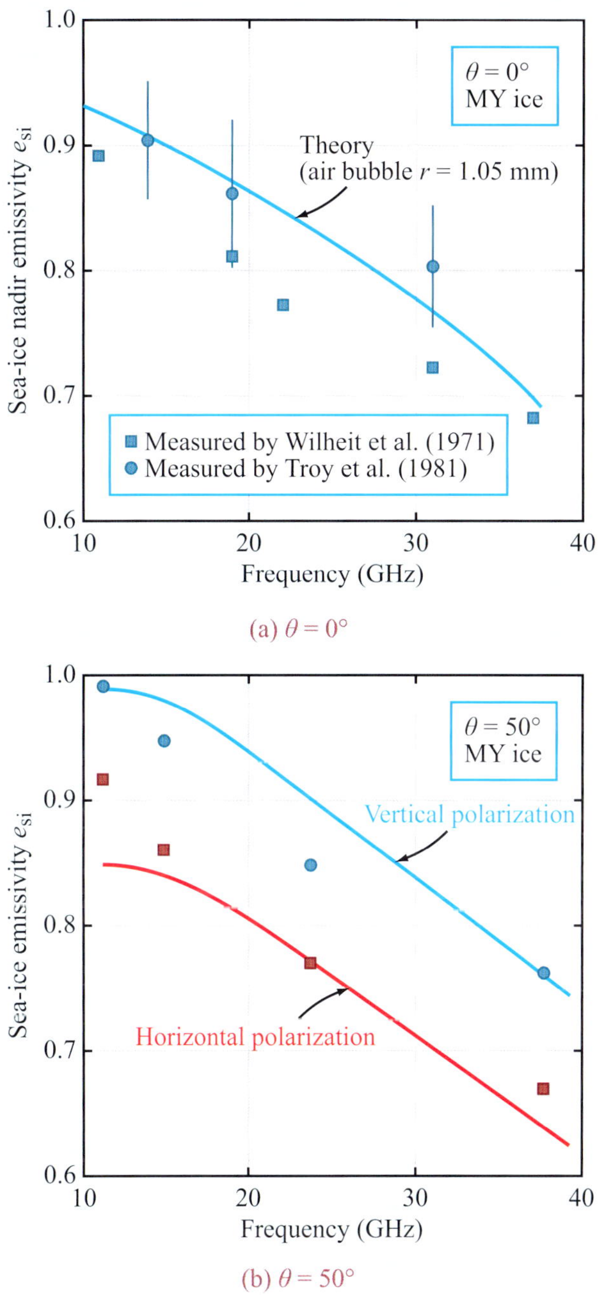

Figure 18-30: Comparison of measured nadir emissivity of multiyear ice with theoretical calculations.

▶ If the ice is covered with a layer of wet snow, it is not possible to distinguish among the different ice types because the penetration depth of wet snow is on the order of only a few centimeters at frequencies above 10 GHz. Surface ponds have even less penetration. ◀

18-4.4 Satellite Observations

The ***cryosphere***, which includes sea ice, ice sheets, snow, ***glaciers***, and ***permafrost***, is the least known and least understood part of the Earth's surface (Campbell and Gudmansen, 1981). Furthermore, because it is both difficult and expensive to conduct surface investigations in ice-bound areas such as the Arctic and Antarctic, these regions are in effect surrounded by what Campbell and Gudmansen (1981) refer to as an "observational barrier." In order to penetrate this barrier, it is necessary to acquire synoptic-scale imagery from satellite platforms on a repetitive basis. However, visible and infrared sensors are of limited use in polar regions because these areas are dark for half the year and are often cloud-covered. Fortunately, active and passive microwave systems have proven to be reliable sensors of sea ice.

The brightness temperature of the sea surface when observed from a point outside the Earth's atmosphere is given by

$$T_B(\text{satellite}) = T_B e^{-\tau\sec\theta} + \Gamma T_{SKY} e^{-\tau\sec\theta} + T_{UP}, \tag{18.18}$$

where τ is the total atmospheric opacity, T_{SKY} and T_{UP} are the downward and upward atmospheric brightness temperatures, Γ is the surface reflectivity, and T_B is the brightness temperature of the sea surface, averaged over the antenna's ***field of view*** (FOV). In the polar regions, where the atmosphere is relatively dry, τ is small enough to permit the linearization $e^{-\tau} \approx 1 - \tau$.

The expression for T_B(satellite) reduces to

$$T_B(\text{satellite}) = T_B + [\Gamma T_{SKY} + T_{UP} - \tau\sec\theta(T_B + \Gamma T_{SKY})]. \tag{18.19}$$

At 19.35 GHz, a commonly used satellite radiometer frequency, the magnitude of the second term varies from about 4 K over ice to about 14 K over open ocean, and the range of T_B extends between 120 K over the ocean and about 240 K over FY ice (Carsey, 1982).

In the general case, the brightness temperature corresponding to a certain field of view may consist of contributions from open water, FY ice, and MY ice,

$$T_B = a_{sw} e_{sw} T_{sw} + a_{fy} e_{fy} T_{fy} + a_{my} e_{my} T_{my}, \tag{18.20}$$

where the subscripts sw, fy, and my refer to sea water, FY ice, and MY ice, respectively; T_{sw}, T_{fy}, and T_{my} are the physical temperatures;[†] and a denotes the areal concentrations (fractions of the FOV) of the three constituents with

$$a_{sw} + a_{fy} + a_{my} = 1. \tag{18.21}$$

In some cases, it may be feasible to consider more than three categories if sufficient multifrequency and multipolarization observations are available to allow separation of the various categories.

An extensive analysis of the spectral, temporal, and spatial variations of sea-ice emissivity was conducted by Comiso (1983) using SMMR observations of the Arctic region for a period of approximately one year. For each SMMR channel, the emissivity was calculated by ratioing the brightness temperature T_B measured by that channel to the ice temperature T_i; i.e.,

$$e = \frac{T_B}{T_i}.$$

In Comiso's (1983) paper, the ice temperature T_i was estimated from the brightness temperature measured by the ***temperature-humidity infrared radiometer*** (THIR) that accompanied the SMMR onboard the Nimbus-7 satellite. To establish effective values for the emissivities of first-year (FY) and multiyear (MY) ice, Comiso generated histograms for a midwinter day using data from a large area containing specific areas known to be consistently first-year or multiyear-ice areas. These histograms are shown in Fig. 18-31. The labels FY and MY identify the effective emissivity values for FY and MY ice, based on the emissivity measurements for the specific areas of known ice type. For all except the 10.7 GHz channels, an emissivity data point was excluded from the histogram when the corresponding h-polarized emissivity of the 10.7 GHz channel of the same pixel was lower than 0.7. This condition was used to exclude or minimize any water-contaminated data points, thereby limiting the distribution to FY, MY, and ***MIX*** (mixture of FY and MY) ice. For the 10.7 GHz histograms, water-contaminated pixels were excluded by requiring the corresponding h-polarized emissivity of the 18 GHz channel to be higher than 0.7. The label ***I/W***, which refers to ice-water mixtures, identifies the emissivity range of those pixels not excluded by this emissivity "filter."

▶ The important role of the microwave frequency f is clearly evident in the histograms in Fig. 18-31. We observe that (1) the total dynamic range of the emissivity e increases with increasing frequency, (2) the difference $\Delta e_{FM} = (e_{FY} - e_{MY})$ increases from about 0.05 at 6 GHz to 0.21 at 37 GHz, and (3) the dynamic ranges and emissivity differences do not exhibit statistically significant differences with regard to polarization. ◀

To examine the spatial variability of emissivity, Comiso generated cluster diagrams, an example of which is shown in Fig. 18-32. Other combinations of frequencies and polarizations are available in Comiso (1983). The cluster diagram illustrates how well the various ice types can be distinguished by using a particular set of channels. Some of the data from the cluster, labeled MIX, should actually correspond to second-year ice because previous determinations of second-year ice emissivities (e.g., Tooma et al., 1975) fall within the enclosed ellipse. The sensitivity of the snow cover, especially during the onset of spring, has been examined by Comiso et al. (1984) using this cluster technique. The study was made in a region where surface ice temperatures were available. Figure 18-33 shows the temporal variability of the FY-ice emissivity cluster caused by changes in the characteristics of the snow cover. Comiso et al. (1984) have shown that a

[†]Lowercase subscripts are used to represent physical temperatures here, since uppercase subscripts on T have previously signified radiometric temperature.

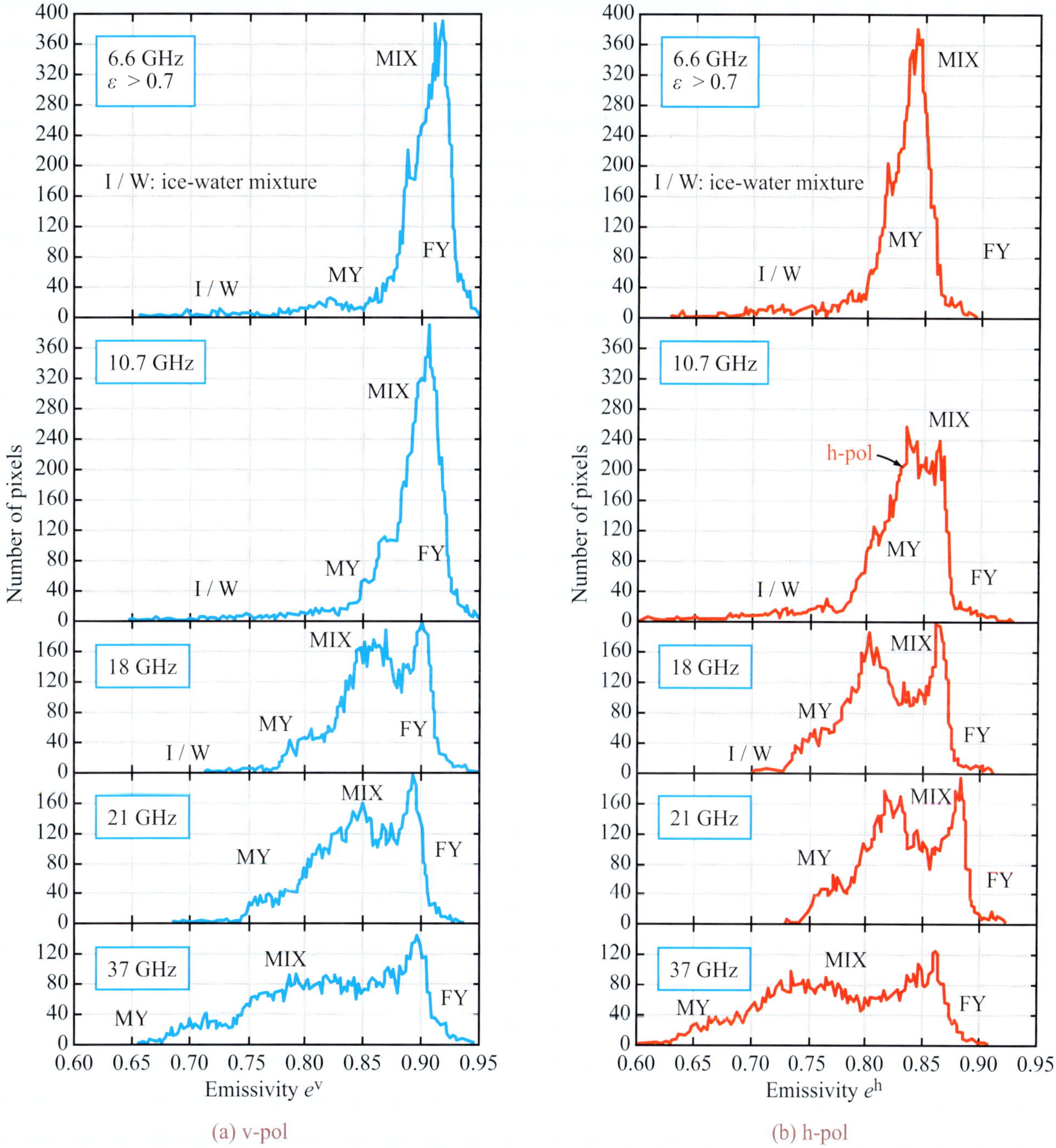

Figure 18-31: Areal distribution (in pixels) of effective microwave sea-ice emissivities at all SMMR channels over approximately a year (from Comiso, 1983).

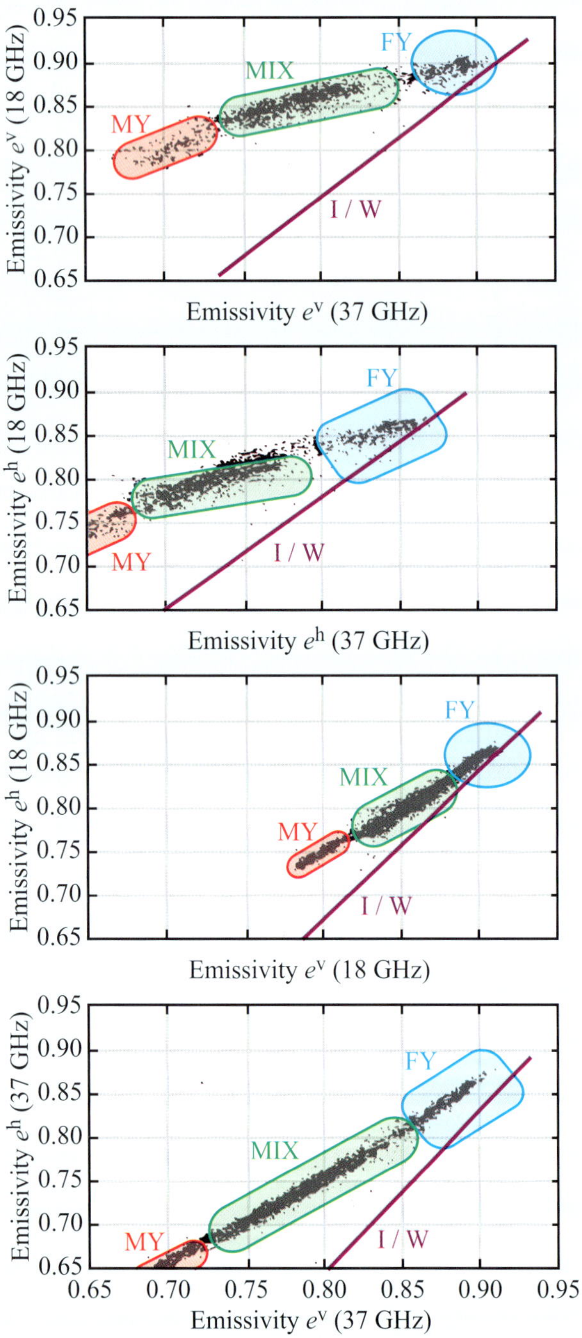

Figure 18-32: Multispectral analysis of emissivities, using the 18 GHz and 37 GHz channels for both polarizations (from Comiso, 1983).

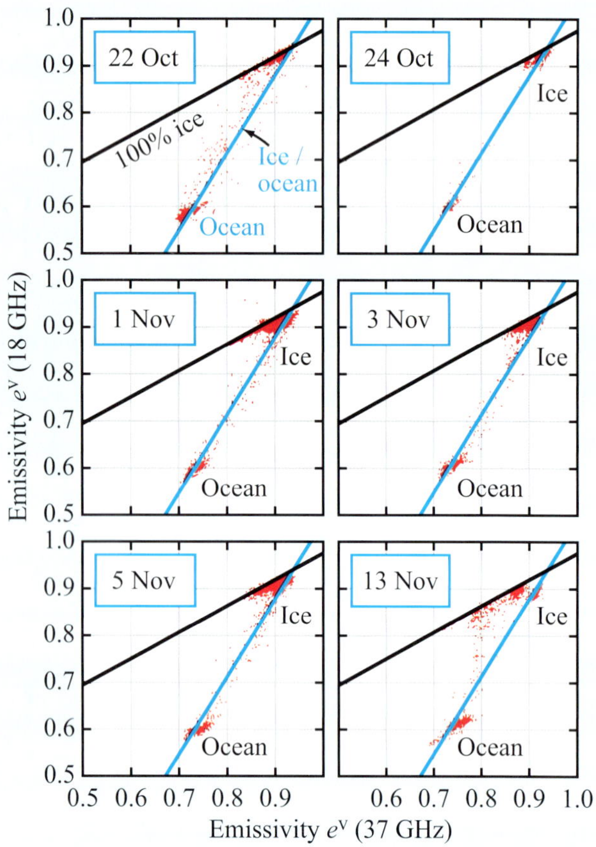

Figure 18-33: Multispectral cluster analysis using the emissivities at 18 GHz and 37 GHz, vertical polarization, during October 22, 24 and November 1, 3, 5, and 13, 1981 (from Comiso et al., 1984).

more accurate retrieval of ice concentration can be made by taking into account the variations in the emissivity of first-year ice than by using either the 18 GHz or the 37 GHz channel alone.

18-4.5 Sea-Ice Algorithms

Although the brightness temperature of FY ice is generally much higher than that of MY ice, especially at frequencies higher than 30 GHz, and the brightness temperature of any type of ice is much higher than that of open water, the interpretation of radiometer measurements is complicated by the fact that the field of view (FOV) of the radiometer often comprises more than one type of surface. With single-frequency observations, the poor spatial resolution can lead to ambiguous estimates of sea-ice concentration (the fraction of an area covered with ice). For example, a certain concentration of high-brightness-temperature FY ice and open water can exhibit the same total brightness temperature as a markedly different concentration of MY ice and open water. If conditions are such that both types of ice exist at a specific time and geographical location in an observed FOV, the ambiguity problem becomes even greater.

Tooma et al. (1975) argued that it should be possible to reduce the degree of ambiguity of sea-ice type and concentration by capitalizing on the differences in the spectral behavior of the brightness temperatures of the different types of ice. Using nadir-looking airborne radiometers with frequencies of 19.34 GHz and 31 GHz, they collected radiometer data over a wide variety of ice types and developed a simple, but effective discriminator of FY ice, MY ice, and open-water/new ice using a two-dimensional vector whose components are the average of and difference between the two brightness temperatures measured at 19.34 and 31 GHz (Fig. 18-34). They also observed that there are two regions of FY ice, namely, a region with an average brightness temperature of about 245 K that usually is correlated with ridged FY ice, and a region with an average brightness temperature of about 235 K, which usually is correlated with smooth FY ice. Similarly, two forms of MY ice were differentiated by the microwave observations. These regions were associated with second-year (SY) ice and older MY ice.

The rectangles representing the five categories shown in Fig. 18-34 are based on typical values for homogeneous or "near-homogeneous" sections of ice (Tooma et al., 1975). For more complex ice conditions, where the FOV may contain more than one category, the average and difference brightness temperatures will depend on the relative concentrations of the various categories. If the FOV contains only two of the four major categories (open-water, new ice, FY ice, and MY ice), the two frequency brightness temperature vector produces points along the perimeter of the dashed triangle. The location of a given point depends on the identities and relative concentrations of the ice categories. Although many points stray from the perimeter of the triangle, the overall shape of the distribution and the density of the data points support the conclusions reached by Tooma et al. (1975). As discussed next, this general approach forms the basis of many sea-ice mapping algorithms.

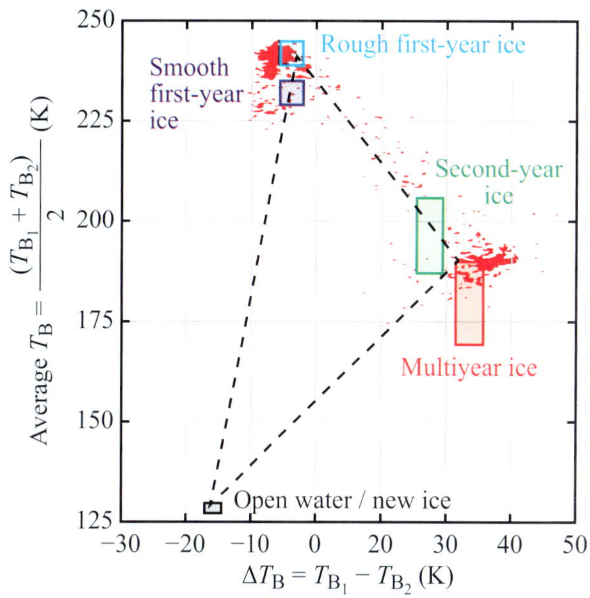

Figure 18-34: Idealized diagram depicting the general areas of homogeneous ice forms delineated by dual-frequency passive microwave radiometry. T_{B1} and T_{B2} are nadir brightness temperatures at 19.3 GHz and 31.0 GHz, respectively (from Tooma et al., 1975).

In 1994, Cavalieri (1994) proposed an algorithm based on two emissivity-derived parameters:

$$\text{PR} = \frac{T_B^v(f_1) - T_B^h(f_1)}{T_B^v(f_1) + T_B^h(f_1)}, \quad \text{(18.22a)}$$

(18.7 GHz polarization ratio)

and

$$\text{SG} = \frac{T_B^v(f_2) - T_B^v(f_1)}{T_B^v(f_2) + T_B^v(f_1)} \quad \textbf{(spectral gradient)}, \quad \text{(18.22b)}$$

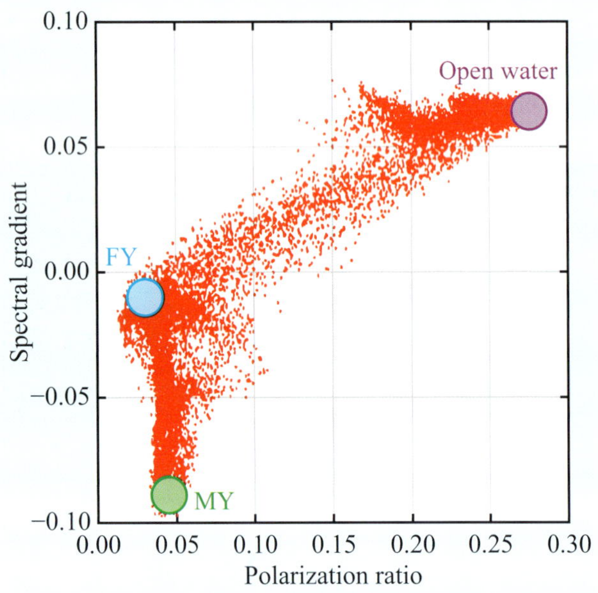

Figure 18-35: Sea-ice scatter plot.

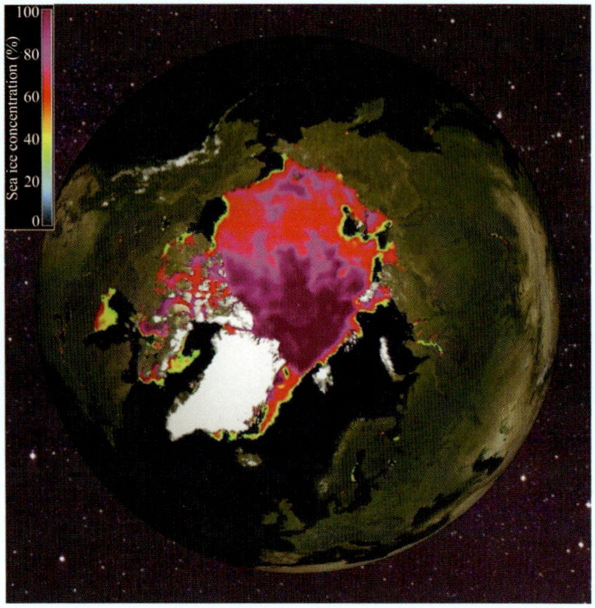

Figure 18-36: Arctic sea-ice concentration map derived from microwave radiometric observations. [Courtesy of University of Illinois.]

where $f_1 = 18.7$ GHz and $f_2 = 36.5$ GHz. Using the ***polarization ratio*** (PR) and ***spectral gradient*** (SG) as axes, Fig. 18-35 displays a scatter plot for the Arctic measured on 15 February 1996 (Markus and Dokken, 2002). The algorithm defined by Eq. (18.22) and modified versions thereof (Markus and Dokken, 2002; Cavalieri et al., 2010; Kongoli et al., 2011) are used to generate sea-ice concentration maps of the type shown in Fig. 18-36.

18-5 Oil-Slick Detection

In dictionary terms, an ***oil slick*** is "a film of oil floating on water." Oil slicks are sometimes generated naturally by organic substances in the ocean environment or they may be the result of man-made oil pollution, usually from accidental spills or deliberate dumping. The U.S. Coast Guard is the agency responsible for enforcing antipollution laws and regulations in the navigable waters of the United States. To meet this responsibility, the Coast Guard has developed an ***Airborne Oil Surveillance System*** (Maurer et al., 1977) designed to be "...a real-time, all weather, day-night sensor system with the capability to (1) detect oil slicks, (2) indicate the magnitude of spills—areal extent and approximate thickness, (3) identify and document the source(s) of discharge, (4) assess cleanup operations, and (5) gather data regarding the frequency and magnitude of significant spills." This, and most commercial oil-surveillance systems, use a combination of multiple sensors including optical, radar, and microwave radiometers. The article by Leifer et al. (2012) provides an excellent review of airborne and spaceborne sensing tools for the detection and surveying of oil spills.

18-5.1 Emissivity of an Oil-Covered Water Surface

The presence of an oil layer may affect the emissivity of the sea surface in two ways: (1) it may alter the transmission of the energy emitted by the water medium;

and (2) if the sea surface is rough, the oil film may dampen the smaller wave structures, thereby reducing the roughness of the surface, which leads to a drop in the magnitude of the emissivity (Edgerton et al., 1971; Hollinger, 1974). For now, let us consider the case of a smooth water surface covered by a uniform layer of oil. Because the oil material is homogeneous in composition, the coherent emissivity model given by Eq. (12.67) is suitable for calculating the emissivity of the oil-covered water surface.

The increase in the p-polarized brightness temperature attributed to the oil slick is given by

$$\Delta T_{\mathrm{B}}^{p} = (e_{\mathrm{oil}}^{p} - e_{\mathrm{w}}^{p})T_{\mathrm{s}}, \qquad (18.23)$$

where e_{oil}^{p} is the p-polarized coherent emissivity of the oil-water composite and e_{w}^{p} is the p-polarized emissivity of the water alone. The expression for e_{oil}^{p} is given by the combination of Eqs. (12.67) and (12.68) with the middle layer being the oil film with dielectric constant $\varepsilon_2 = \varepsilon_{\mathrm{oil}}$ and the bottom layer is the water with $\varepsilon_3 = \varepsilon_{\mathrm{w}}$. For the water alone, $e_{\mathrm{w}}^{p} = 1 - \Gamma^{p}$, where Γ^{p} is the reflectivity of water. According to the review reported by Hollinger (1974), "the real part of the dielectric constant ($\varepsilon'_{\mathrm{oil}}$) of the majority of oils is near 2.1 but ranges from about 1.8 to 2.6. The imaginary part of the dielectric constant ($\varepsilon''_{\mathrm{oil}}$) of the oils measured is generally very small, usually less than 0.02, except for aged 40 gravity crude and Bunker C fuel oil where it may be as large as 0.3." Hollinger's assessment was based on data published by Edgerton and Trexler (1970), Howard et al. (1967), and von Hippel (1954).

Figure 18-37 displays the increase in brightness temperature ΔT_{B} at 20 GHz as a function of oil thickness for several combinations of the angle of incidence θ and the antenna polarization p. The pattern is ***oscillatory***, and its spatial period is a function of the wavelength in the oil layer $\lambda \approx \lambda_0/\sqrt{\varepsilon'_{\mathrm{oil}}}$ and the angle θ. The magnitude of ΔT_{B} increases with increasing incidence angle for horizontal polarization, and exhibits the opposite behavior for vertical polarization. The stronger sensitivity of the horizontally polarized brightness temperature to the presence of an oil slick was observed in several experimental investigations (Edgerton and Trexler, 1970; Edgerton et al., 1971; Brown et al., 1998).

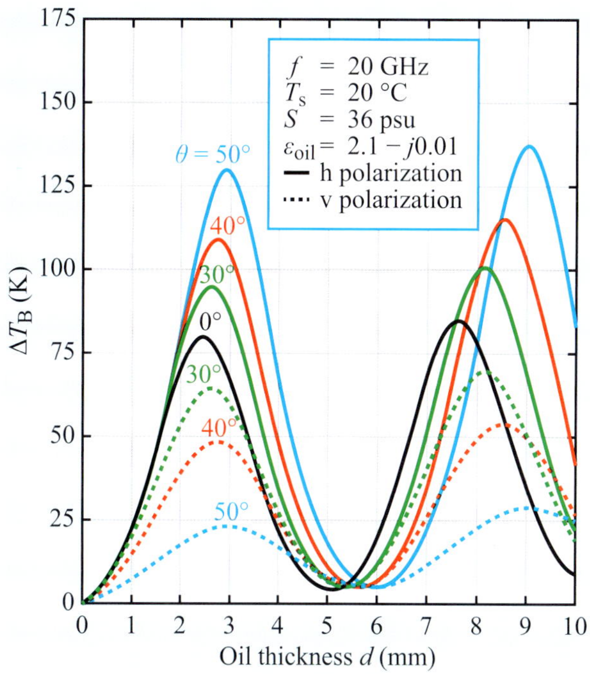

Figure 18-37: Computed increase in brightness temperature as a function of oil thickness at 20 GHz for various combinations of nadir angle and polarization configuration.

Laboratory confirmation of the coherent emissivity model described above was obtained by Hollinger (1974), who measured the brightness temperature of a water tank after adding oil to the surface in increments of known volume. Examples of his results are shown in Fig. 18-38 for 19.34 and 69.75 GHz. The measured values of the increase in ΔT_{B} are in excellent agreement with the calculated curves at both frequencies. Similar results were reported by Lääperi and Nyfors (1982) at 4.95, 11.6, and 37 GHz. This type of measurement suggests that microwave radiometric observations can be used not only to detect the presence of an oil slick, but to estimate its thickness as well. The ambiguity associated with the periodic dependence of ΔT_{B} on the thickness d can be removed by using two or more frequencies. This is illustrated in Fig. 18-39, which presents a graph of T_{B} at 4.95 GHz versus T_{B} at 16.5

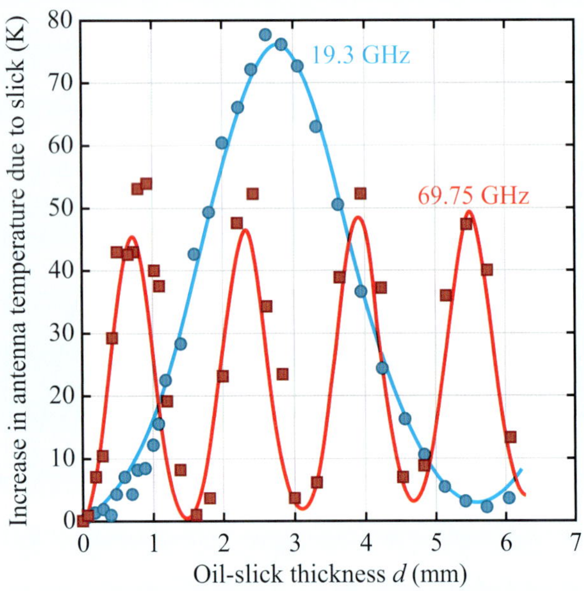

Figure 18-38: Measurements made at 19.3 and 69.75 GHz of the increase in brightness temperature due to No. 2 fuel oil spread over a smooth water surface in a test tank as a function of film thickness. The curves represent the calculations that best fit the measurements [from Hollinger, 1974].

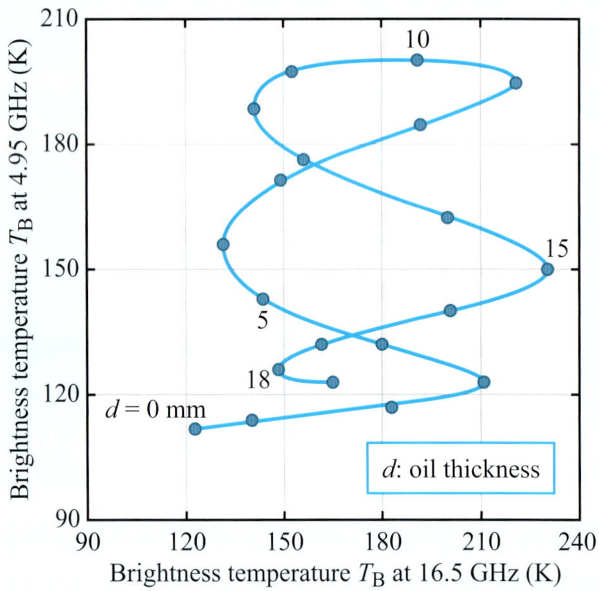

Figure 18-39: Ambiguity chart for the frequency combination 4.95 GHz and 16.5 GHz [Lääperi and Nyfors, 1982].

GHz, drawn as a continuous plot with d as parameter, starting at $d = 0$ mm and ending at $d = 19$ mm. For $d \leq 15$ mm, only one ambiguous pair of points exists: the value $d = 7.2$ mm and $d = 13$ mm give the same combination of brightness temperatures (Lääperi and Nyfors, 1982).

18-5.2 Airborne Observations

To demonstrate the feasibility of mapping oil thickness with microwave radiometers, a series of airborne observations of controlled oil spills were made by NRL in cooperation with the U.S. Coast Guard using multifrequency imaging radiometers (Hollinger and Mennella, 1973; Hollinger, 1974; Troy and Hollinger, 1977). Figure 18-40 illustrates one form of data presentation along with corresponding estimates of the oil-thickness contours.

The German company Optimare (www.optimare.de) uses a three-frequency (18.7, 36.5, and 89 GHz) scanning radiometer consisting of six channels. The stated measurement range of oil thickness is from 0.05 mm to 3 mm. When flown at an attitude of 300 m, its 4° instantaneous field of view has a radius of about 20 m at nadir and increases in size as the beam scans out in angle.

Instead of a multifrequency system, Brown et al. (1998) considered the use of a wide-band radiometer that operates over the continuous range from 26 GHz to 40 GHz. Figure 18-41 compares measurement spectra of T_B for oil-covered water with theoretical computations based on the coherent emissivity model. Their results confirm the earlier conclusion that h polarization affords a greater contrast between water and oil-covered water than v polarization.

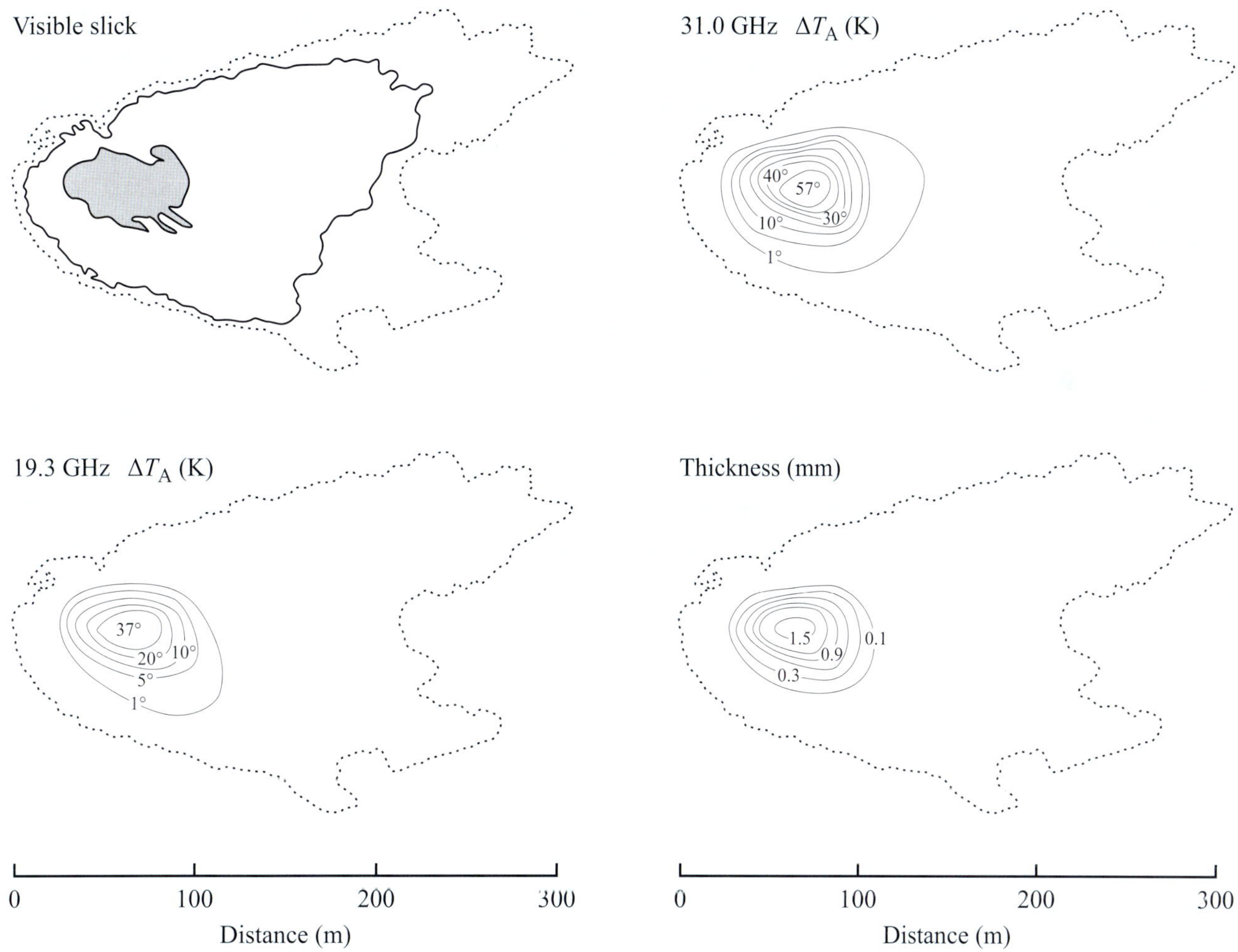

Figure 18-40: The drawing at upper left is a tracing of a color photograph of the oil slick resulting from a controlled spill of 630 gallons of No. 2 fuel oil. The oil had been dyed red to allow the thick regions of oil to be identified visibly. The outer line is the extreme edge of the visible slick, the next inner line is the region of color fringing, and the crosshatched area is the region of thick oil. The antenna temperatures measured at 19.3 and 31.0 GHz are shown, superimposed on the outline of the visible slick. The thickness contours derived from the microwave data are also shown [from Hollinger and Mennella, 1973].

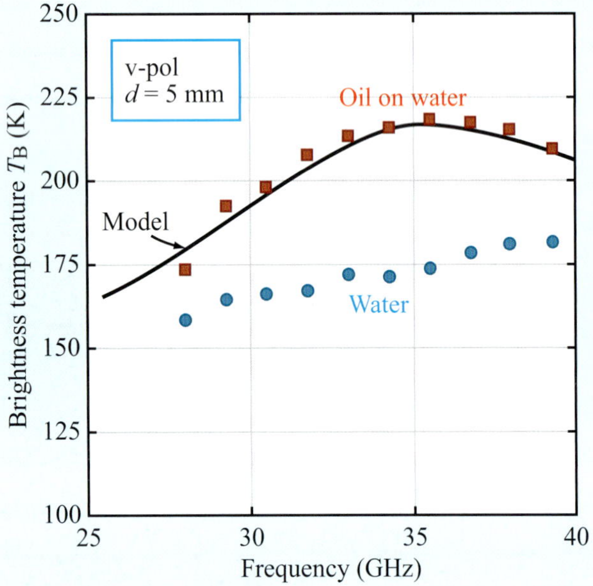

(a) v-pol

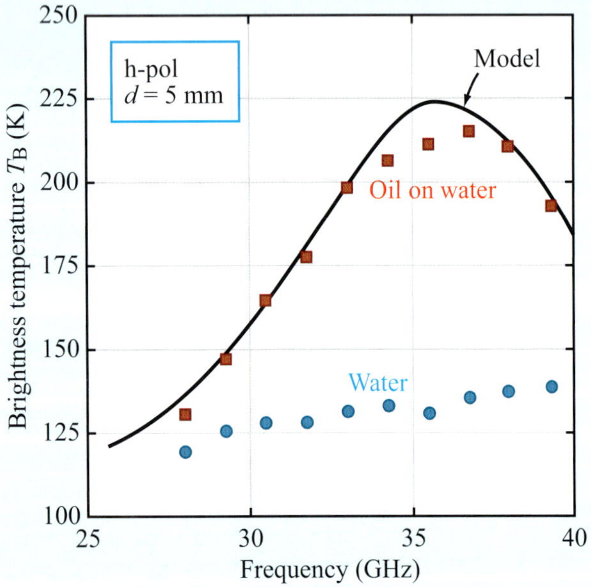

(b) h-pol

Figure 18-41: Comparison of experimental and theoretical T_B spectra for (a) v polarization, and (b) h polarization [Brown et al., 1998].

Appendix A Symbols, Quantities, and Units

Symbol	Quantity	SI Unit
A	Area	m^2
A	Attenuation rate	—
a	Absorptivity	—
a	Temperature lapse rate	K/km
a	Single scattering albedo	—
A_i	Complex feeding coefficient	V
A_u	Depolarization factor	—
B	Bandwidth	Hz
B	Baseline length	m
B	Biomass	kg/m^2
c	Velocity of light in vacuum	m/s
$\mathbf{C}$	Scattering covariance matrix	m^2
$C(\,)$	Fresnel cosine integral	—
D	Directivity (antenna)	—
d	Distance or length	m
$\mathcal{E}$	Energy (work)	J
$\mathbf{E}$	Electric-field intensity	V/m
e	Emissivity	—
F	Radiation intensity (normalized)	—
f	Frequency	Hz
$\mathcal{F}$	Fourier transform	
f_D	Doppler frequency	Hz
g	Conductance	S
G	Gain (power)	—
h, H	Height	m
h	Planck's constant	J·s
$h(t)$	Impulse response	—
h	Horizontal polarization	—
h_v	Integrated precipitable water vapor	cm
h_L	Cloud liquid water path	cm

Symbol	Quantity	SI Unit
$\mathbf{H}$	Magnetic field intensity	A/m
$\mathbf{H}(\omega)$	System frequency response	—
H_T	Target entropy	—
I, i	Current	A
I	Intensity	$Wm^{-2}sr^{-1}$
$J(\,)$	Bessel function	—
$\mathbf{J}$	Current density (volume)	A/m^2
J_a	Absorption source function	$Wm^{-2}sr^{-1}$
J_s	Scattering source function	$Wm^{-2}sr^{-1}$
k	Wavenumber	rad/m
k	Boltzmann's constant	J/K
$\hat{\mathbf{k}}$	Wave direction	—
l, L	Length	m
l	Correlation length	m
L	Loss factor	—
$\mathcal{L}$	Propagation factor	—
m	rms slope	—
$\mathbf{M}$	Mueller matrix	m^2
M_v	Total water vapor mass	kg/m^2
m_v	Cloud water content	g/m^3
m_v	Snow or soil wetness	—
M_w	Vegetation water content	kg/m^2
n	Index of refraction	—
N	Number	—
N	Refractivity	—
p, P	Power	W
p	Co-polarization ratio	—
P_H	Polarimetric pedestal	—
p	Pressure	N/m^2
$\mathbf{p}$	Antenna polarization vector	V/m
$P(x < x')$	Cumulative probability	—

Symbol	Quantity	SI Unit
$p(t)$	Pulse waveform	V
$p(x)$	Probability density function	(units of x)$^{-1}$
q	Charge	C
q	Cross-polarization ratio	—
Q	Cross section	m^2
R	Resistance	Ω
R	Range	m
$\mathbf{R}$	Distortion matrix	—
r	Radial distance, radius	m
r	Cross correlation	—
r	Spatial resolution	m
$\mathbb{R}$	Boundary reflectivity matrix	—
S	Scattering amplitude	m
S	Salinity	psu
$S(\ )$	Fresnel sine integral	—
S_n	Signal-to-noise ratio	—
$\mathbf{S}$	Scattering matrix	m
$\mathbb{S}$	Power density	W/m^2
s	rms height	m
s_x	Standard deviation of x	units of x
T	Period	s
T	Tone (digital number)	—
T	Temperature	K or °C
$\boldsymbol{T}$	Distortion matrix	—
T_A	Antenna temperature	K
T_B	Brightness temperature	K
T_N	Noise temperature	K
Υ	Layer transmissivity	—
$\mathbb{T}$	Transmissivity (boundary)	—
$\boldsymbol{\mathbb{T}}$	Boundary transmissivity matrix	—
$\mathbf{u}$	Velocity	m/s
u_p	Phase velocity	m/s
v	Vertical polarization	—
V	Visibility function	Wm^{-2} sr^{-1}
υ, V	Voltage	V
W	Snow water equivalent	cm
W_x	Weighting function of x	m^{-1}
Z	Reflectivity factor	m^3
$\mathbf{Z}$	Impedance	Ω
Z_x	Slope of $z(x)$	—
α	Attenuation constant	Np/m
α	Inclination angle	rad or °

Symbol	Quantity	SI Unit
α	Auxiliary angle	rad or °
β	Beamwidth	rad or °
β	Phase constant (wavenumber)	rad/m
Γ	Reflectivity	—
γ	Propagation constant	m^{-1}
δ_p	Penetration depth	m
δ	Phase angle	rad or °
$\delta(\ )$	Impulse function	—
δ_s	Skin depth	m
ε	Dielectric constant	—
ε_0	Free-space permittivity	F/m
η	Intrinsic impedance	Ω
η_a	Aperture efficiency	—
η_b	Antenna beam efficiency	rad or °
θ	Incidence or elevation angle	rad or °
κ_e	Extinction coefficient	m^{-1}
κ_s	Scattering coefficient	m^{-1}
κ_a	Absorption coefficient	m^{-1}
λ	Wavelength	m
μ, μ_0	Permeability	H/m
μ_x	Mean value of x	units of x
ξ	Antenna radiation efficiency	—
ξ	Efficiency	—
ρ	Reflection coefficient	—
ρ	Density	g/cm^3
ρ_v	Charge density (volume)	C/m^3
ρ_v	Water vapor density	g/m^3
σ	Conductivity	S/m
σ	Radar cross section	m^2
σ^0	Radar scattering coefficient	—
τ	Transmission coefficient	—
τ	Optical depth	—
τ	Relaxation time	s
τ	Pulse length	s
ψ	Rotation angle	rad or °
ψ	Phase angle	rad or °
$\psi(\mathbf{R}, \mathbf{R}_i)$	Scattering phase function	m sr^{-1}
ϕ	Azimuth angle	rad or °
χ	Ellipticity angle	rad or °
Ω	Solid angle	sr
ω	Angular frequency	rad/s
ω	Angular velocity	rad/s

Appendix B Glossary of Names and Acronyms

A/D Analog to digital

ac Alternating current

ACM Association for Computing Machinery

ADEOS Advanced Earth Observing Satellite

AGARD Advisory Group for Aerospace Research and Development

AGC Automatic gain control

AIRS Atmospheric Infrared Sounder

AIRSAR Airborne synthetic-aperture radar

ALOS Advanced Land Observation Satellite

AMAP Architectural plant model

AMSR Advanced Microwave Scanning Radiometer

AMSU Advanced Microwave Sounding Unit

AN/APQ-97 Model number: K-band side-looking radar by Westinghouse Electric

ANFIS Adaptive-Network-Based Fuzzy Inference Systems

ARC Active radar calibrator

ARTEMISS An Algorithm to Retrieve Temperature and Emissivity

ASAR Advanced Synthetic Aperture Radar

ASCAT Advanced Scatterometer aboard Metop satellites

ASI Agenzia Spaziale Italiana (Italian Space Agency)

ATMP Precipitation retrieval algorithm developed for ATMS

ATMS Advanced Technology Microwave Sounder

ATU Automatic tracking unit

AVHRR Advanced Very High Resolution Radiometer

AWG Arbitrary waveform generator

BPF Bandpass filter

BSA Backscatter alignment convention

BW Bandwidth

C-MOD5 A widely used C-band GMF

CARABAS Coherent All-Radio Band Sensing

CCIR Comité Consultatif International de Radio Communications (French: International Radio Consultative Committee)

CFO Complete freeze-over

CHAMP Challenging Minisatellite Payload, German small satellite mission

CNES Centre National d'Etudes spatiales (French National Center for Space Studies)

COBE Cosmic Background Explorer

COHO Coherent oscillator

ColdFET Cold field-effect transistor

Comstar Series of U.S. domestic telephone satellites

CONAE Comisión Nacional de Actividades Espaciales (Argentine National Space Activities Commission

COSMIC Constellation Observation System for Meteorology, Ionosphere, and Climate

COSMO-SkyMed Constellation of small satellites for the Mediterranean basin observation

CRC Chemical Rubber Company (scientific reference publisher)

CRESDA China Center for Resources Satellite Data and Application

CrIMSS Crosstrack Infrared and Microwave Sounding Suite

CrIS Crosstrack Infrared Sounder

CryoSat-2 ESA satellite dedicated to ice-cap research

CSA Canadian Space Agency

CYGNSS Cyclone Global Navigation Satellite System

CW Continuous-wave radar

CZCS Coastal Zone Color Scanner

D/A Digital-to-analog converter

D3M Double-Debye dielectric model

dc Direct current

DDM Delay Doppler map

DEM Digital Elevation Model *or* Digital Elevation Measurement

demux Demultiplexer

DFT Discrete Fourier transform

DLR German Aerospace Center (Deutsches Zentrum für Luft- und Raumfahrt e.V.)

DMSP Defense Meteorological Satellite Program

DOF Degrees of freedom

DQD Digital quadrature demodulator

DSP Digital signal processing

DSS Digital subsystem

E-SAR P- and L-band airborne SAR system operated by DLR

EADS European Aeronautic Defence and Space Company

EM Electromagnetic

EMSL European Microwave Signature Laboratory

ENR Excess noise ratio

Envisat Environmental Satellite (ESA)

EOS Earth Observing System

ERB Earth radiation budget

ERIM Environmental Research Institute of Michigan

ERS European Remote Sensing satellites *or* Earth Resources Satellite

ESA European Space Agency

ESCAT ESA ERS scatterometer aboard ERS-1 and -2

ESMR Electrically Scanning Microwave Radiometer

ESTAR Electronically Steered Thinned Aperture Radiometer

ETM Enhanced Thematic Mapper

FEM Finite element method

FET Field-effect transistor

FFMLP Feedforward Multilayer Perceptron

FFT Fast Fourier transform

FIA Forest inventory and analysis

FM Frequency modulation

FMCW Frequency-modulated continuous-wave radar

FOV Field of view

FSA Forward Scattering Alignment convention

FY / FYI First-year ice

GACM Global Atmospheric Composition Mission

GEOS-3 Geodynamics Experimental Ocean Satellite altimeter

Geosat Geodetic Satellite

GeoSTAR Geostationary Synthetic Thinned Array Radiometer demonstrator instrument

GIFTS Geosynchronous Imaging Fourier Transform Spectrometer

GLONASS Russian Globalnaya Navigatsionnaya Sputnikovaya Sistema satellites

GMF Geophysical model function

GNSS Global Navigation Satellite System

GNSS-R GNSS Reflectometry

GNSS-RO GNSS Radio Occultation

GOM Geometric-optics model

GPS Global positioning system

GRACE Gravity Recovery and Climate Experiment

GSR Ground-based scanning radiometer

GUI Graphical user interface

HAMSTRAD $H_2$2O Antarctica Microwave Stratospheric and Tropospheric Radiometer

HES Hyperspectral Environmental Suite

HIRS High-Resolution Infrared Radiation Sounder

HJ-1A, -1B, -1C Huanjing series of Chinese satellites for disaster and environmental monitoring

HPA High-power transmit amplifier

HRRR High Resolution Rapid Refresh model

I / Q In-phase and quadrature-phase

I / W Ice-water mixtures

I^2EM Improved IEM scattering model

ICW Interrupted continuous wave

IEM Integral equation method

IF Intermediate frequency

IFFT Inverse fast Fourier transform

IFOV Instantaneous field of view

IfSAR Interferometric synthetic-aperture radar

IGRF International Geomagnetic Reference Field

INS Inertial navigation system

InSAR Interferometric synthetic-aperture radar

IP Internet protocol suite

IPR Impulse response function

IPWV Integrated precipitable water vapor

IR Infrared

ISA International Standard Atmosphere

ISAR Inverse synthetic-aperture radar

ISLR Integrated sidelobe ratio

ISO International Standards Organization

ISRO Indian Space Research Organisation

ITU International Telecommunication Union

JAXA Japanese Aerospace Exploration Agency

JERS Japanese Earth Resources Satellite

JPL Jet Propulsion Laboratory

JPSS Joint Polar Satellite System

KARI Korea Aerospace Research Institute

KOMPSAT Korean multipurpose satellite (a.k.a. Ariring)

LAI Leaf area index

LANDFIRE Landscape Fire and Resource Management Planning Tools Project

Landsat Series of satellites dedicated to acquisition of satellite imagery of Earth

LCX L-band, C-band, X-band

LEOMAC Low-Earth Orbit Multi-spectral Atmospheric Composition mission

LFM Linear FM

LFMCW Linear FMCW

LHC Left-hand circular

LIMS Limb Infrared Monitor of the Stratosphere

LNA Low-noise receive amplifier

LO Local oscillator

LOS Line of sight

LPF Low-pass filter

LUT Lookup tables *or* load under test

LWP Liquid-water path

MAP Maximum a posteriori despeckling

MetOp Meteorological Operational Satellite Program of Europe

MIMICS Michigan Microwave Canopy Scattering Model

MIMO Multiple-input and multiple-output

MIR Midinfrared *or* microwave interferometric radiometer

MIRAS Microwave Imaging Radiometer with Aperture Synthesis

MIT Massachusetts Institute of Technology

MIX Mixture of FY and MY ice

MIZ Marginal ice zone

MLS Microwave limb sounder

MMSE Minimum mean-square error

MO Melt onset

MODIS Moderate Resolution Imaging Spectroradiometer

MPM Millimeter-wave propagation model

MSS Multispectral scanners

MSU Microwave sounding unit

MTI Moving Target Indicator

mux Multiplexer

MWR Microwave radiometer

MY Multiyear (ice; has survived one melt season)

NASA National Aeronautics and Space Administration

NAST-M NPOESS Airborne Sounder Testbed–Microwave

NBCS National Biomass and Carbon Dataset

NEMS Nimbus E Microwave Spectrometer

NES Noise equivalent σ^0

NESDIS NOAA National Environmental Satellite, Data, and Information Services

NEXRAD Next-Generation Radar (WSR-88D: Weather Surveillance Radar, 1988, Doppler)

NGAS Northrop Grumman Aerospace Systems

Nimbus U.S. meteorological research satellite series

NIR Near-infrared

NLCD USGS National Land Cover Dataset

NMM3D Numerical method-of-moment 3-D

NOAA National Oceanic and Atmospheric Administration

NODS NASA Ocean Data System, at JPL

NPOESS National Polar-Orbiting Operational Environmental Satellite System

NRL U.S. Naval Research Laboratory

NRSCC National Remote Sensing Center of China

NSCAT NASA scatterometer aboard Japanese satellite, Midori (a.k.a. ADEOS)

NSIDC U.S. National Snow and Ice Data Center

NSOF NOAA Satellite Operations Facility

NWP Numerical weather prediction

Oceansat Indian satellite series

OCM Ocean Colour Monitor instrument aboard Oceansat

OMT Orthomode Transducer

OSCAT Oceansat-2 Scatterometer

OW Open water

PA Power amplifier

PALS Passive active L-band system

PALSAR Phased Array type L-band Synthetic Aperture Radar

PARC Polarimetric ARC

PATH Precipitation and All-Weather Temperature and Humidity mission

PAZ Spanish radar satellite

PCA Point of closest approach *or* principal components analysis

pdf Probability density function

PiSAR Japanese airborne, polarimetric, L-band SAR

PoFR Point of first reflection

POLARSCAT LCX Polarimetric Scatterometer

PolSCAT Polarimetric Ku-band Scatterometer

PORTOS Five-frequency radiometer on French ARAT aircraft and associated datasets

Poseidon CNES solid-state nadir-pointing radar altimeter

PPI Plan-position indicator

PR Polarization ratio

PRF Pulse repetition frequency

PRISM Polarimetric Radar Inversion for Soil Moisture

PRN Pseudorandom noise

PS Persistent scattering

PSD Power spectral density

PSR Polarimetric scanning radiometer

PTOST Pacific THORpex Observing System Test

QuikSCAT NASA Quick Scatterometer satellite

RADARSAT Pair of Canadian remote sensing satellites

RADSCAT Radiometer / Scatterometer flown on Skylab (a.k.a. S-193)

RAM Random-access memory

RAR Real-aperture radar

RASAM radio-scatterometer

RCS Radar cross section

RDA Range-Doppler algorithm

RF Radio frequency

RFI Radio frequency interference

RHC Right-hand circular

RISAT Radar Imaging Satellite: Series of Indian radar imaging reconnaissance satellites

rms Root-mean-square error/deviation

rmsd Root-mean-square deviation

rmse Root-mean-square error

RT Radiative transfer models

RVI Radar Vegetation Index

S193 Skylab Microwave Radiometer/Scatterometer and Altimeter (a.k.a. RADSCAT)

S194 Skylab L-band Radiometer

S^{2}RT Single-scattering radiative transfer model

S^{2}RT/R S^{2}RT model with Rayleigh particles

SAM Stratospheric aerosol measurement

SAMS Stratospheric and Mesospheric Sounder

SAOCOM Satélite Argentino de Observación con Microondas (Argentine Microwaves Observation Satellite)

SAPARC Single-antenna PARC

SAR Synthetic-aperture radar

SARA SAR altimetry

SASS Seasat-A Scatterometer System

SBAS Small baseline subset analysis

SBUV/TOMS Solar Backscatter Ultraviolet / Total Ozone Mapping Spectrometer

SCAMS Scanning Microwave Spectrometer

scanSAR Scanning SAR

SD^{2}M Single-Debye dielectric model

Seasat NASA / JPL first satellite designed for remote sensing of the Earth's oceans

SeaWinds NASA scatterometer aboard QuikSCAT satellite

SEE Standard error of estimate

Sentinel ESA satellite series

SFO Code for San Francisco International Airport

SG Spectral gradient

SIR Shuttle Imaging Radar

SIR-C/X-SAR Spaceborne Imaging Radar-C/X-band Synthetic-Aperture Radar

SIRAL SAR Interferometric Radar Altimeter

SLAR Side-looking airborne radar

SLC Scan line corrector

SMAP Soil Moisture Active Passive mission and instrument

SMART Soil Moisture Assessment Radar Technique

SMEI Solar Mass Ejection Imager

SMILES Superconducting Submillimeter-Wave Limb Emission Sounder

SMMR Scanning Multichannel Microwave Radiometer

SMOS Soil Moisture and Ocean Salinity mission

SNR Signal-to-noise ratio

SPIE International Society for Optics and Photonics

SPM Small-perturbation model

SRTM Shuttle Radar Topography Mission

SSB Single-sideband

SSM/I or SSMI Special Sensor Microwave / Imager

SSM/T Special Sensor Microwave Atmospheric Temperature Profiler

SSMIS Special Sensor Microwave Imager / Sounder

SSS Sea-surface salinity

SST Sea-surface temperature

SSU Stratospheric Sounding Unit

STALO Stable local oscillator

STC Sensitivity time control

Suomi NPP U.S. Suomi National Polar-Orbiting Partnership

SY Second-year ice

TanDEM-X TerraSAR-X add-on for Digital Elevation Measurement

TBSCAT Two-stream radiative transfer model

TCP Transmission Control Protocol

TDBP Time-domain backprojection

TECU Total electron content unit

TEM Transverse electromagnetic

TerraSAR-X Earth observation satellite (DLR and EADS)

THIR Temperature / Humidity Infrared Radiometer

THORpex The Observing-System Research and Predictability Experiment

TIR Thermal infrared

TIROS Television Infrared Observation Satellite

TMR TOPEX Microwave Radiometer

TOPEX NASA nadir-pointing radar altimeter

TOVS TIROS Operational Vertical Sounder

TR or T/R Transmit-receive switch

TRM Transmit / receive module

TRMM Tropical Rain Mapping Mission

TRMM-PR TRMM Precipitation Radar

TropSat Earth observation satellite mission dedicated to global tropics

TVB Tinga-Voss-Blossey formulas

TWTA Traveling wave tube amplifier

UARS Upper Atmosphere Research Satellite

UAV Unmanned aerial vehicle

UAVSAR Uninhabited aerial vehicle synthetic-aperture radar

UK-DMC U.K. Space Agency's Disaster Monitoring Constellation

UPS Uninterruptible power supply

URSI International Union of Radio Science

USDA United States Department of Agriculture

USGS United States Geological Survey

VCO Voltage-controlled oscillator

VHF Very High Frequency

VIWV Vertically integrated water vapor

WindSat Wind microwave radiometer

WMAP Wilkinson Microwave Anisotropy Probe

WVC Wind vector cell

XFMR Transformer

ZRT Zeroth-order radiative transfer

Appendix C Mathematical Formulas

Trigonometric Relations

$$\sin(x \pm y) = \sin x \cos y \pm \cos x \sin y$$
$$\cos(x \pm y) = \cos x \cos y \mp \sin x \sin y$$
$$2 \sin x \sin y = \cos(x - y) - \cos(x + y)$$
$$2 \sin x \cos y = \sin(x + y) + \sin(x - y)$$
$$2 \cos x \cos y = \cos(x + y) + \cos(x - y)$$
$$\sin 2x = 2 \sin x \cos x$$
$$\cos 2x = 1 - 2 \sin^2 x$$
$$\sin x + \sin y = 2 \sin\left(\frac{x+y}{2}\right) \cos\left(\frac{x-y}{2}\right)$$
$$\sin x - \sin y = 2 \cos\left(\frac{x+y}{2}\right) \sin\left(\frac{x-y}{2}\right)$$
$$\cos x + \cos y = 2 \cos\left(\frac{x+y}{2}\right) \cos\left(\frac{x-y}{2}\right)$$
$$\cos x - \cos y = -2 \sin\left(\frac{x+y}{2}\right) \sin\left(\frac{x-y}{2}\right)$$
$$\cos(x \pm 90^\circ) = \mp \sin x$$
$$\cos(-x) = \cos x$$
$$\sin(x \pm 90^\circ) = \pm \cos x$$
$$\sin(-x) = -\sin x$$
$$e^{jx} = \cos x + j \sin x \qquad \text{(Euler's identity)}$$
$$\sin x = \frac{e^{jx} - e^{-jx}}{2j}$$
$$\cos x = \frac{e^{jx} + e^{-jx}}{2}$$
$$\sqrt{j} = \frac{1+j}{\sqrt{2}}$$
$$d\Omega = \sin\theta \, d\theta \, d\phi$$

$$\left|1 + e^{jx} + e^{j2x} + \cdots + e^{j(N-1)x}\right|^2 = \left[\frac{\sin(Nx/2)}{\sin(x/2)}\right]^2$$

Approximations for Small Quantities

For $|x| \ll 1$,

$$(1 \pm x)^n \approx 1 \pm nx$$
$$(1 \pm x)^2 \approx 1 \pm 2x$$
$$\sqrt{1 \pm x} \approx 1 \pm \frac{x}{2}$$
$$\frac{1}{\sqrt{1 \pm x}} \approx 1 \mp \frac{x}{2}$$
$$e^x = 1 + x + \frac{x^2}{2!} + \cdots \approx 1 + x$$
$$\ln(1 + x) \approx x$$
$$\sin x = x - \frac{x^3}{3!} + \frac{x^5}{5!} + \cdots \approx x$$
$$\cos x = 1 - \frac{x^2}{2!} + \frac{x^4}{4!} + \cdots \approx 1 - \frac{x^2}{2}$$
$$\lim_{x \to 0} \frac{\sin x}{x} = 1$$

Appendix D Useful Reference Books

Balanis, C. A. (2008). *Modern Antenna Handbook*. Wiley & Sons.

Barton, D. K. (1975). *Radars, Vol. 5: Radar Clutter*. Artech House.

Blackwell, W. J. and F. W. Chen (2009). *Neural Networks in Atmospheric Remote Sensing*. Artech House.

Born, M. and E. Wolf (1965). *Principles of Optics: Electromagnetic Theory of Propagation, Interference and Diffraction of Light*. Oxford: Pergamon Press.

Burke, B. F. and F. Graham-Smith (2002). *An Introduction to Radio Astronomy*. Cambridge University Press.

Carrara, W. G., R. S. Goodman, and R. M. Majewski (1995). *Spotlight Synthetic Aperture Radar: Signal Processing Algorithms*. Artech House.

Cloude, S. (2009). *Polarisation: Applications in Remote Sensing*. Oxford University Press.

Cumming, I. G. and F. H. Wong (2005). *Digital Processing of Synthetic Aperture Radar Data: Algorithms and Implementation*. Artech House.

Curlander, J. C. and R. N. McDonough (1991). *Synthetic Aperture Radar: Systems and Signal Processing*. Wiley-Interscience.

Elachi, C. (1988). *Spaceborne Radar Remote Sensing: Applications and Techniques*. IEEE Press.

Fu, L.-L. and A. Cazenave (2001). *Satellite Altimetry and Earth Sciences: A Handbook of Techniques and Applications*. Academic Press.

Fung, A. K. (1994). *Microwave Scattering and Emission Models and Their Applications*. Artech House.

Fung, A. K. and K. S. Chen (2010). *Microwave Scattering and Emission Models for Users*. Artech House.

Hanssen, R. F. (2001). *Radar Interferometry: Data Interpretation and Error Analysis. Remote Sensing and Digital Image Processing*. Kluwer Academic.

Ishimaru, A. (1978). *Wave Propagation and Scattering in Random Media*. Academic Press.

Ishimaru, A. (1991). *Electromagnetic Wave Propagation, Radiation, and Scattering*. Prentice Hall.

Janssen, M. A. (ed.) (1993). *Atmospheric Remote Sensing by Microwave Radiometry*. Wiley & Sons.

Lee, J.-S. and E. Pottier (2009). *Polarimetric Radar Imaging: From Basics to Applications*. CRC Press.

Massonnet, D. and J.-C. Souyris (2008). *Imaging with Synthetic Aperture Radar*. CRC Press.

Mätzler, C. (2006). *Thermal Microwave Radiation: Applications for Remote Sensing*. IET.

Meischner, P. (2004). *Weather Radar: Principles and Advanced Applications*. Springer.

Oliver, C. and S. Quegan (2004). *Understanding Synthetic Aperture Radar Images*. SciTech.

Raemer, H. R. (1997). *Radar Systems Principles*. CRC Press.

Richards, M. A. (2005). *Fundamentals of Radar Signal Processing*. McGraw-Hill.

Richards, M. A., J. A. Scheer, and W. A. Holm (2010). *Principles of Modern Radar*. SciTech.

Sihvola, A. H. (1999). *Electromagnetic Mixing Formulas and Applications*. IEE.

Skolnik, M. I. (1980). *Introduction to Radar Systems*. McGraw-Hill.

Thompson, A. R., J. M. Moran, and G. W. Swenson (1994). *Interferometry and Synthesis in Radio Astronomy*. Krieger Publishing Company.

Tinbergen, J. (1996). *Astronomical Polarimetry*. Cambridge University Press.

Tsang, L., J. A. Kong, and R. T. Shin (1985). *Theory of Microwave Remote Sensing*. Wiley & Sons.

Ulaby, F. T. and M. C. Dobson (1989). *Handbook of Radar Scattering Statistics for Terrain*. Artech House.

Ulaby, F. T. and C. Elachi (1990). *Radar Polarimetry for Geoscience Applications*. Artech House.

Ulaby, F. T., E. Michielssen, and U. Ravaioli (2010). *Fundamentals of Applied Electromagnetics*. Prentice Hall.

Ulaby, F. T., R. K. Moore, and A. K. Fung (1981). *Microwave Remote Sensing: Active and Passive, Vol. I—Fundamentals and Radiometry.* Artech House.

Ulaby, F. T., R. K. Moore, and A. K. Fung (1982). *Microwave Remote Sensing: Active and Passive, Vol. II—Radar Remote Sensing and Surface Scattering and Emission Theory.* Artech House.

Ulaby, F. T., R. K. Moore, and A. K. Fung (1986). *Microwave Remote Sensing: Active and Passive, Vol. III—Volume Scattering and Emission Theory, Advanced Systems and Applications.* Artech House.

Ulaby, F. T. and A. E. Yagle (2013). *Engineering Signals and Systems*. NTS Press.

Van Zyl, J. J. and Y. Kim (2011). *Synthetic Aperture Radar Polarimetry*. Wiley & Sons.

Bibliography

Adam, N., B. Kampes, M. Eineder, J. Worawattanamateekul & M. Kircher (2003). The development of a scientific permanent scatterer system. In *International Society for Photogrammetry and Remote Sensing Hannover Workshop*. Hannover, Germany.

Ahmad, K. A., W. L. Jones, T. Kasparis, S. Wiechecki Vegara, I. S. Adams & J. D. Park (2005). Oceanic rain rate estimates from the QuikSCAT Radiometer: A Global Precipitation Mission pathfinder. *Journal of Geophysical Research*, *110*(D11), D11101.

Ahmad, K. A., W. L. Jones & T. Kasparis (2006). QuikSCAT Radiometer (QRad) Rain Rates Level 2B Data Product. In *IEEE International Geoscience and Remote Sensing Symposium* (pp. 4126–4129). 31 July–4 August, Denver, CO.

Allen, C. T. & F. T. Ulaby (1984). Modeling the Polarization Dependence of the Attenuation in Vegetation Canopies. In *IEEE International Geoscience and Remote Sensing Symposium*. 27–30 August, Strasbourg, France.

Allen, C. T., B. Brisco & F. T. Ulaby (1984). *Modeling the Temporal Behavior of the Microwave Backscattering Coefficient of Agricultural Crops*. Lawrence, KS: Technical Report 360-21, Remote Sensing Laboratory, University of Kansas Center for Research, Inc.

Allen, J. R. & D. G. Long (2006). Microwave observations of daily Antarctic sea-ice edge expansion and contraction rates. *IEEE Geoscience and Remote Sensing Letters*, *3*(1), 54–58.

Alpers, W. (2002). Remote sensing of oil spills. In *Maritime Disaster Management Symposium* (pp. 19–23).

Alpers, W. & H. Hühnerfuss (1989). The damping of ocean waves by surface films: A new look at an old problem. *Journal of Geophysical Research*, *94*(C5), 6251–6265.

Amar, F. (1989). *Directional Random Sea Surface Generation*. M.S. Thesis, University of Texas at Arlington, Arlington, TX.

Ambach, W. & A. Denoth (1980). *Dielectric behaviour of snow: a study versus liquid water content*. Ft. Collins, CO: NASA Conference Publication 2153.

Amelung, F., D. L. Galloway, J. W. Bell, H. A. Zebker & R. J. Laczniak (1999). Sensing the ups and downs of Las Vegas: InSAR reveals structural control of land subsidence and aquifer-system deformation. *Geology*, *27*(6), 483–486.

Amelung, F., S. Jónsson, H. A. Zebker & P. Segall (2000). Widespread uplift and "trapdoor" faulting on Galápagos volcanoes observed with radar interferometry. *Nature*, *407*(6807), 993–996.

Amelung, F., N. Gourmelen & S. Baker (2008). InSAR time-series: Results from Kilauea volcano, Hawaii, and the Eastern California Shear Zone. In *Fall Meeting of the American Geophysical Union*.

Anderson, C., J. Figa, H. Bonekamp, J. J. W. Wilson, J. Verspeek, A. Stoffelen & M. Portabella (2012). Validation of Backscatter Measurements from the Advanced Scatterometer on MetOp-A. *Journal of Atmospheric and Oceanic Technology*, *29*(1), 77–88.

Anderson, H. S. & D. G. Long (2005). Sea ice mapping method for SeaWinds. *IEEE Transactions on Geoscience and Remote Sensing*, *43*(3), 647–657.

Arcone, S. A., A. J. Gow & S. McGrew (1986). Structure and dielectric properties at 4.8 and 9.5 GHz of saline ice. *Journal of Geophysical Research*, *91*(C12), 14281–14303.

Armstrong, R. L. & M. J. Brodzik (2002). Hemispheric-scale comparison and evaluation of passive-microwave snow algorithms. *Annals of Glaciology*, *34*(1), 38–44.

Ashcraft, I. S. & D. G. Long (2003). The spatial response function of SeaWinds backscatter measurements. *Proceedings of SPIE*, *5151*, 609–618.

Askne, J. I.. H. & B. G. Skoog (1983). Atmospheric Water-Vapor Profiling by Ground-Based Radiometry at 22 and 183 GHz. *IEEE Transactions on Geoscience and Remote Sensing*, *21*(3), 320–323.

Assur, A. (1960). *Composition of sea ice and its tensile strength*. Wilmette, IL: U.S. Army Corps of Engineers Snow, Ice and Permafrost Research Establishment.

Atkinson, P. M. & A. R. L. Tatnall (1997). Introduction to neural networks in remote sensing. *International Journal of Remote Sensing*, *18*(4), 699–709.

Atlas, D. (1964). Advances in Radar Meteorology. In H. E. Landsberg & J. van Mieghem (Eds.), *Advances in Geophysics* (Vol. 10). New York: Elsevier.

Attema, E. P. W. (1991). The Active Microwave Instrument on-board the ERS-1 satellite. *Proceedings of the IEEE*, *79*(6), 791–799.

Attema, E. P. W. & F. T. Ulaby (1978). Vegetation modeled as a water cloud. *Radio Science*, *13*(2), 357–364.

Aumann, H. H., M. T. Chahine, C. Gauter, M. D. Goldberg, E. Kalnay, L. M. McMillin, ... J. Susskind (2003a). AIRS/AMSU/HSB on the Aqua Mission: Design, Science Objectives, Data Products, and Processing Systems. *IEEE Transactions on Geoscience and Remote Sensing*, *41*(2), 253–264.

Aumann, H. H., F. W. Chen, L. Chen, S. Gaiser, D. Hagan, T. Hearty, ... D. C. Tobin (2003b). *Validation of AIRS/AMSU/HSB Core Products for Data Release Version 3.0* (E. Fetzer, Ed.). Technical Report D-26538, NASA JPL, Caltech, Pasadena, CA.

Backus, G. E. & F. Gilbert (1970). Uniqueness in the inversion of inaccurate gross earth data. *Philosophical Transactions of the Royal Society of London. Series A, Mathematical and Physical Sciences*, *266*(1173), 123–192.

Baghdadi, N., F. Saba, M. Aubert, M. Zribi & F. Baup (2011). Evaluation of Radar Backscattering Models IEM, Oh, and Dubois for SAR Data in X-Band Over Bare Soils. *IEEE Geoscience and Remote Sensing Letters*, *8*(6), 1160–1164.

Balanis, C. A. (1989). *Advanced Engineering Electromagnetics*. New York: Wiley & Sons.

Balanis, C. A. (2008). *Modern Antenna Handbook* (Constantine A. Balanis, Ed.). Hoboken, NJ: Wiley & Sons.

Barber, D. G., A. K. Fung, T. C. Grenfell, S. V. Nghiem, R. G. Onstott, V. I. Lytle, ... A. J. Gow (1998). The Role of Snow on Microwave Emission and Scattering over First-Year Sea Ice. *IEEE Transactions on Geoscience and Remote Sensing*, *36*(5), 1750–1763.

Barnes, R. M. (1986). *Antenna Polarization Calibration Using In-Scene Reflectors*. Report TT-65, MIT Lincoln Laboratory, Lexington, MA.

Barrett, A. H. & E. Lilley (1963). Mariner-2 Microwave Observations of Venus. *Sky & Telescope*, *25*(4), 192–195.

Barrick, D. E. & C. T. Swift (1980). The Seasat microwave instruments in historical perspective. *IEEE Journal of Oceanic Engineering*, *5*(2), 74–79.

Barrow, W. L. & L. J. Chu (1939). Theory of the Electromagnetic Horn. *Proceedings of the IRE*, *27*(1), 51–64.

Barthel, J., K. Bachhuber, R. Buchner & H. Hetzenauer (1990). Dielectric spectra of some common solvents in the microwave region. Water and lower alcohols. *Chemical Physics Letters*, *165*(4), 369–373.

Barton, D. K. (1975). *Radars Vol. 5 - Radar Clutter*. (D. K. Barton, Ed.). Norwood, MA: Artech House.

Basharinov, A. E., A. S. Gurvich, S. T. Yegorov, A. A. Kurskaya, D. T. Matvyev & A. M. Shutko (1971). The results of microwave sounding of the Earth's surface according to experimental data from the satellite Cosmos 243. *Space Research*, *11*.

Batelaan, P. E., R. M. Goldstein & C. T. Stelzried (1973). *Improved Noise-Adding Radiometer for Microwave Receivers*.

Batlivala, P. P. & F. T. Ulaby (1976). Radar Look Direction and Row Crops. *Photogrammetric Engineering & Remote Sensing*, *42*(2), 233–238.

Beal, R. C. (1980). Spaceborne Imaging Radar: Monitoring of Ocean Waves. *Science*, *208*(4450), 1373–1375.

Bechor, N. B. D. & H. A. Zebker (2006). Measuring two-dimensional movements using a single InSAR pair. *Geophysical Research Letters*, *33*(16), L16311.

Beckmann, P. & A. Spizzichino (1963). *The scattering of electromagnetic waves from rough surfaces*. Oxford: Pergamon Press.

Berardino, P., G. Fornaro, R. Lanari & E. Sansosti (2002). A New Algorithm for Surface Deformation Monitoring Based on Small Baseline Differential SAR Interferograms. *IEEE Transactions on Geoscience and Remote Sensing*, *40*(11), 2375–2383.

Bergen, K. M., M. C. Dobson, L. E. Pierce & F. T. Ulaby (1998). Characterizing Carbon Dynamics in a Northern Forest Using SIR-C/X-SAR Imagery. *Remote Sensing of Environment*, *63*(1), 24–39.

Berkowitz, R. S. (Ed.) (1965). *Modern Radar: Analysis, Evaluation, and System Design*. New York: Wiley & Sons.

Bespalova, Y. A., V. M. Veselov & V. Y. Gershenzon (1982). Determining surface wind velocity by measurements of polarization anisotropy of natural and scattered microwave radiation. *Issledovanie Zemli iz Kosmosa*, *1*, 87–94.

Best, A. C. (1950). The size distribution of raindrops. *Quarterly Journal of the Royal Meteorological Society*, *76*(327), 16–36.

Bindlish, R., T. J. Jackson, A. J. Gasiewski, M. Klein & E. G. Njoku (2006). Soil moisture mapping and AMSR-E validation using the PSR in SMEX02. *Remote Sensing of Environment*, *103*(2), 127–139.

Birkeland, P. W. (1974). *Pedology, weathering, and geomorphological research*. New York: Oxford University Press.

Bishop, C. M. (2007). *Pattern Recognition and Machine Learning*. Springer.

Bitar, A. A., D. Leroux, Y. H. Kerr, O. Merlin, P. Richaume, A. Sahoo & E. F. Wood (2012). Evaluation of SMOS Soil Moisture Products Over Continental U.S. Using the SCAN/SNOTEL Network. *IEEE Transactions on Geoscience and Remote Sensing*, *50*(5), 1572–1586.

Bjerkaas, A. W. & F. W. Riedel (1979). *Proposed model for the elevation spectrum of a wind-roughened sea surface*. Technical Report APL-TG-1328-I-31, Applied Physics Laboratory, Johns Hopkins University, Laurel, MD.

Black, H. D. (1978). An Easily Implemented Algorithm for the Tropospheric Range Correction. *Journal of Geophysical Research*, *83*(B4), 1825–1828.

Blacksmith, P., R. E. Hiatt & R. B. Mack (1965). Introduction to radar cross-section measurements. *Proceedings of the IEEE*, *53*(8), 901–920.

Blackwell, W. J. (2011). Hyperspectral microwave atmospheric sounding using neural networks. In *Signal and Image Processing for Remote Sensing*. CRC Press.

Blackwell, W. J. & F. W. Chen (2009). *Neural Networks in Atmospheric Remote Sensing*. Norwood, MA: Artech House.

Blanchard, A. J. & J. Rochier (1987). Bistatic Radar Cross Section Measurement Facility. In *IEEE International Geoscience and Remote Sensing Symposium* (Vol. 1, pp. 545–548). 18–21 May, Ann Arbor, MI.

Blanchard, B. J. & A. T. C. Chang (1983). Estimation of Soil Moisture from Seasat SAR Data. *Water Resources Bulletin*, *19*(5), 803–810.

Bliven, L. F. & J. P. Giovanangeli (1993). An experimental study of microwave scattering from rain- and wind-roughened seas. *International Journal of Remote Sensing*, *14*(5), 855–869.

Bliven, L., J. P. Giovanangeli & G. Norcross (1989). Scatterometer Directional Response During Rain. In *IEEE International Geoscience and Remote Sensing Symposium* (Vol. 3, pp. 1887–1890). 10–14 July, Vancouver, Canada.

Bliven, L. F., P. W. Sobieski & C. Craeye (1997). Rain generated ring-waves: Measurements and modelling for remote sensing. *International Journal of Remote Sensing*, *18*(1), 221–228.

Blume, H. J. C. (1977). Noise Calibration Repeatability of an Airborne Third-Generation Radiometer (Letters). *IEEE Transactions on Microwave Theory and Techniques*, *25*(10), 852–855.

Blume, H. J. C., M. Kendall & J. C. Fedors (1977). *Sea-surface temperature and salinity mapping from remote microwave radiometric measurements of brightness temperature*. NASA Technical Paper 1077, Langley Research Center, Hampton, VA.

Boerner, W. M. (1981). Use of Polarization in Electromagnetic Inverse Scattering. *Radio Science*, *16*(6), 1037–1045.

Bolten, J. D., V. Lakshmi & E. G. Njoku (2003). Soil Moisture Retrieval Using the Passive/Active L- and S-Band Radar/Radiometer. *IEEE Transactions on Geoscience and Remote Sensing*, *41*(12), 2792–2801.

Born, M. & E. Wolf (1965). *Principles of Optics: Electromagnetic Theory of Propagation, Interference and Diffraction of Light*. Oxford: Pergamon Press.

Borwein, P. & R. Ferguson (2005). Polyphase sequences with low autocorrelation. *IEEE Transactions on Information Theory*, *51*(4), 1564–1567.

Bouvet, A., T. Le Toan & L.-D. Nguyen (2009). Monitoring of the Rice Cropping System in the Mekong Delta Using ENVISAT/ASAR Dual Polarization Data. *IEEE Transactions on Geoscience and Remote Sensing*, *47*(2), 517–526.

Bracalente, E., D. Boggs, W. Grantham & J. Sweet (1980). The SASS^1 scattering coefficient $\sigma°$ algorithm. *IEEE Journal of Oceanic Engineering*, *5*(2), 145–154.

Bradley, G. A. & F. T. Ulaby (1981). Aircraft radar response to soil moisture. *Remote Sensing of Environment*, *11*, 419–438.

Breit, G. & M. A. Tuve (1926). A Test of the Existence of the Conducting Layer. *Physical Review*, *28*(3), 554–575.

Brekke, A. (1997). *Physics of the Upper Polar Atmosphere*. Chichester, UK: Wiley & Sons.

Brekke, C. & A. H. S. Solberg (2005). Oil spill detection by satellite remote sensing. *Remote Sensing of Environment*, *95*(1), 1–13.

Brown, E. R., O. B. McMahon, T. J. Murphy, G. G. Hogan, G. D. Daniels & G. Hover (1998). Wide-Band Radiometry for Remote Sensing of Oil Films on Water. *IEEE Transactions on Microwave Theory and Techniques*, *46*(12), 1989–1996.

Brown, S. T., C. S. Ruf & D. R. Lyzenga (2006). An emissivity-based wind vector retrieval algorithm for the WindSat polarimetric radiometer. *IEEE Transactions on Geoscience and Remote Sensing*, *44*(3), 611–621.

Brunfeldt, D. R. & F. T. Ulaby (1984a). Measured microwave emission and scattering in vegetation canopies. *IEEE Transactions on Geoscience and Remote Sensing*, *22*, 520–524.

Brunfeldt, D. R. & F. T. Ulaby (1984b). Active Reflector for Radar Calibration. *IEEE Transactions on Geoscience and Remote Sensing*, *22*(2), 165–169.

Brunfeldt, D. R. & F. T. Ulaby (1986). Microwave Emission from Row Crops. *IEEE Transactions on Geoscience and Remote Sensing*, *24*(3), 353–359.

Buckley, E. F. (1960). Stepped-Index Luneburg Lenses. *Electronic Design*, *8*, 86–89.

Buckreuss, S., R. Werninghaus & W. Pitz (2008). The German satellite mission TerraSAR-X. In *IEEE Radar Conference* (pp. 1–5).

Buehner, M. (2002). Assimilation of ERS-2 scatterometer winds using the Canadian 3D-var. *Atmosphere-Ocean*, *40*(3), 361–376.

Bürgmann, R., M. E. Ayhan, E. J. Fielding, T. J. Wright, S. McClusky, B. Aktug, … A. Turkezer (2002). Deformation during the 12 November 1999 Düzce, Turkey, Earthquake, from GPS and InSAR Data. *Bulletin of the Seismological Society of America*, *92*(1), 161–171.

Burke, B. F. & F. Graham-Smith (2002). *An Introduction to Radio Astronomy*. Cambridge, UK; New York: Cambridge University Press.

Burke, W. J., T. J. Schmugge & J. F. Paris (1979). Comparison of 2.8- and 21-cm microwave radiometer observations over soils with emission model calculations. *Journal of Geophysical Research*, *84*(C1), 287.

Bush, T. F. & F. T. Ulaby (1975). Fading Characteristics of Panchromatic Radar Backscatter from Selected Agricultural Targets. *IEEE Transactions on Geoscience Electronics*, *13*(4), 149–157.

Bush, T. F., F. T. Ulaby, T. Metzler & H. Stiles (1976). *Seasonal Variations of the Microwave Scattering Properties of Deciduous Trees as Measured in the 1-18 GHz Spectral Range*. Technical Report 177-60, Remote Sensing Laboratory, University of Kansas, Lawrence, KS.

Bushuyev, A. V., N. A. Volkov & V. S. Loshchilov (1974). *Atlas of Ice Formations*. Leningrad, USSR: Gidrometeoizdat.

Bussey, H. E. (1967). Measurement of RF Properties of Materials: A Survey. *Proceedings of the IEEE*, *55*(6), 1046–1053.

Campbell, C. K. (1978). Free-Space Permittivity Measurements on Dielectric Materials at Millimeter Wavelengths. *IEEE Transactions on Instrumentation and Measurements*, *27*(1), 54–58.

Campbell, M. J. & J. Ulrichs (1969). Electrical Properties of Rocks and Their Significance for Lunar Radar Observations. *Journal of Geophysical Research*, *74*(25), 5867–5881.

Campbell, W. J. & P. Gudmansen (1981). The application of microwave remote sensing for snow and ice research. In K. R. Carver (Ed.), *1st IEEE International Geoscience and Remote Sensing Symposium* (pp. 951–957). 8–10 June, Washington, DC.

Campbell, W. J., J. Wayenberg, J. B. Ramseyer, R. O. Ramseier, M. R. Vant, R. Weaver, … T. G. Farr (1978). Microwave remote sensing of sea ice in the AIDJEX Main Experiment. *Boundary-Layer Meteorology*, *13*(1-4), 309–337.

Carrara, W. G., R. S. Goodman & R. M. Majewski (1995). *Spotlight Synthetic Aperture Radar: Signal Processing Algorithms*. Norwood, MA: Artech House.

Carsey, F. D. (1982). Arctic Sea Ice Distribution at End of Summer 1973–1976 From Satellite Microwave Data. *Journal of Geophysical Research*, *87*(C8), 5809–5835.

Carsey, F. D., R. O. Ramseier & W. F. Weeks (1982). *Sea Ice Mission Requirements for the U.S. FIREX and Canada RadarSat Programs*. Report based on Bilateral Ice Study Team Workshop, 11-13 February, 1981, Comwall, Ontario, Canada.

Carver, K. R. (1975). *Antenna and Radome Loss Measurements for MFMR and PMIS*. Technical Report PA00817, Physical Science Laboratory, New Mexico State University, Las Cruces, NM.

Castel, T., A. Beaudoin, N. Floury, T. Le Toan, Y. Caraglio & J.-F. Barczi (2001). Deriving Forest Canopy Parameters for Backscatter Models Using the AMAP Architectural Plant Model. *IEEE Transactions on Geoscience and Remote Sensing*, *39*(3), 571–583.

Cavalieri, D. J. (1994). A microwave technique for mapping thin sea ice. *Journal of Geophysical Research*, *99*(C6), 12561–12572.

Cavalieri, D. J., T. Markus, D. K. Hall, A. Ivanoff & E. Glick (2010). Assessment of AMSR-E Antarctic Winter Sea-Ice Concentrations Using Aqua MODIS. *IEEE Transactions on Geoscience and Remote Sensing*, *48*(9), 3331–3339.

CCIR (1981). *Attenuation by Precipitation and Other Atmospheric Particles*. ITU; Draft Report 721 (MOD F), Document 5/5046-E, Committée Consultative Internationale de Radio, Geneva, Switzerland.

Chandrasekhar, S. (1960). *Radiative Transfer*. New York: Dover.

Chang, P. S. & L. Li (1998). Ocean surface wind speed and direction retrievals from the SSM/I. *IEEE Transactions on Geoscience and Remote Sensing*, *36*(6), 1866–1871.

Chang, A. T. C. & J. C. Shiue (1980). A comparative study of microwave radiometer observations over snowfields with radiative transfer model calculations. *Remote Sensing of Environment*, *10*(3), 215–229.

Chang, A. T. C., P. Gloersen, T. J. Schmugge, T. T. Wilheit & H. J. Zwally (1976). Microwave Emission from Snow and Glacier Ice. *Journal of Glaciology*, *16*(74), 23–39.

Chang, A. T. C., J. L. Foster, D. K. Hall, A. Rango & B. K. Hartline (1982). Snow water equivalent estimation by microwave radiometry. *Cold Regions Science and Technology*, *5*(3), 259–267.

Chang, A. T. C., J. L. Foster & D. K. Hall (1987). Numbus-7 SMMR Derived Global Snow Cover Parameters. *Annals of Glaciology*, *9*, 39–44.

Charney, J. G. (1971). Geostrophic Turbulence. *Journal of the Atmospheric Sciences*, *28*(6), 1087–1095.

Chelton, D. B., J. C. Ries, B. J. Haines, L.-L. Fu & P. S. Callahan (2001). Satellite Altimetry. In L. Fu & A. Cazenave (Eds.), *Satellite Altimetry and Earth Sciences*. Academic Press.

Chen, C. W. & H. A. Zebker (2001). Two-dimensional phase unwrapping with use of statistical models for cost functions in nonlinear optimization. *Journal of the Optical Society of America A*, *18*(2), 338.

Chen, F. W. & D. H. Staelin (2003). AIRS/AMSU/HSB Precipitation Estimates. *IEEE Transactions on Geoscience and Remote Sensing*, *41*(2), 410–417.

Chen, K. S. & A. K. Fung (1990). A Bragg Scattering Model for the Sea Surface. In *Ocean 90 Conference* (pp. 249–252).

Chen, K. S., A. K. Fung & D. A. Weissman (1992). A backscattering model for ocean surface. *IEEE Transactions on Geoscience and Remote Sensing*, *30*(4), 811–817.

Chen, K. S., T.-D. Wu, L. Tsang, Q. Li, J. Shi & A. K. Fung (2003). Emission of rough surfaces calculated by the integral equation method with comparison to three-dimensional moment method simulations. *IEEE Transactions on Geoscience and Remote Sensing*, *41*(1), 90–101.

Chen, L., J. Shi, J.-P. Wigneron & K.-S. Chen (2010). A Parameterized Surface Emission Model at L-Band for Soil Moisture Retrieval. *IEEE Geoscience and Remote Sensing Letters*, *7*(1), 127–130.

Cheney, R. E. & J. G. Marsh (1981). SEASAT altimeter observations of dynamic topography in the Gulf Stream region. *Journal of Geophysical Research*, *86*(C1), 473–483.

Chi, C.-Y. & F. K. Li (1988). A comparative study of several wind estimation algorithms for spaceborne scatterometers. *IEEE Transactions on Geoscience and Remote Sensing*, *26*(2), 115–121.

Choudhury, B. J. & A. T. C. Chang (1981). On the angular variation of solar reflectance of snow. *Journal of Geophysical Research*, *86*(C1), 465.

Choudhury, B. J., T. J. Schmugge, A. Chang & R. W. Newton (1979). Effect of Surface Roughness on the Microwave Emission From Soils. *Journal of Geophysical Research*, *84*(C9), 5699–5706.

Christensen, E. L., N. Skou, J. Dall, K. W. Woelders, J. H. Jørgensen, J. Granholm & S. N. Madsen (1998). EMISAR : An Absolutely Calibrated Polarimetric L- and C-band SAR. *IEEE Transactions on Geoscience and Remote Sensing*, *36*(6), 1852–1865.

Chu, L. J. (1940). Calculation of the Radiation Properties of Hollow Pipes and Horns. *Journal of Applied Physics*, *11*(9), 603–610.

Chu, T. S. & D. C. Hogg (1968). Effects of precipitation on propagation at 0.63, 3.5, and 10.6 microns. *Bell System Technical Journal*, *47*(5), 723–759.

Chuah, H. T., K. Y. Lee & T. W. Lau (1995). Dielectric Constants of Rubber and Oil Palm Leaf Samples at X-Band. *IEEE Transactions on Geoscience and Remote Sensing*, *33*(1), 221–223.

Cimini, D., E. R. Westwater, A. J. Gasiewski, M. Klein, V. Y. Leuski & J. C. Liljegren (2007a). Ground-Based Millimeter- and Submillimeter-Wave Liquid Water Contents. *IEEE Transactions on Geoscience and Remote Sensing*, *45*(7), 2169–2180.

Cimini, D., E. R. Westwater, A. J. Gasiewski, M. Klein, V. Y. Leuski & S. G. Dowlatshahi (2007b). The Ground-Based Scanning Radiometer: A Powerful Tool for Study of the Arctic Atmosphere. *IEEE Transactions on Geoscience and Remote Sensing*, *45*(9), 2759–2777.

Cimini, D., G. Visconti & F. S. Marzano (2010). *Integrated Ground-Based Observing Systems*. Berlin, Germany: Springer-Verlag.

Clarizia, M. P. (2012). *Investigating the Effect of Ocean Waves on GNSS-R Microwave Remote Sensing Measurements*. Ph.D. Thesis, 258 pp, University of Southhampton, UK.

Cloude, S. R. (1992a). Recent Developments in Radar Polarimetry: A Review. In *AMPC Asia-Pacific Microwave Conference* (Vol. 2, pp. 955–958).

Cloude, S. R. (1992b). Uniqueness of target decomposition theorems in radar polarimetry. In W. M. Boerner, L. A. Cram, W. A. Holm, D. E. Stein, W. Wiebeck, W. Keydel, … H. Brand (Eds.), *Direct and Inverse Methods in Radar Polarimetry, Part 1, NATO-ARW* (pp. 267–296). Norwell, MA: Kluwer Academic.

Cloude, S. R. & E. Pottier (1995). The concept of polarization entropy in optical scattering. *Optical Engineering*, *34*(6), 1599–1610.

Cloude, S. R. & E. Pottier (1996). A review of target decomposition theorems in radar polarimetry. *IEEE Transactions on Geoscience and Remote Sensing*, *34*(2), 498–518.

Cobb, S. (2004). Weather Radar Development Highlight of National Severe Storms Laboratory's First 40 Years. *NOAA Magazine Online*. NOAA. Retrieved from http://www.magazine.noaa.gov/stories/mag151.htm

Cofield, R. E. & P. C. Stek (2006). Design and Field-of-View Calibration of 114-660-GHz Optics of the Earth Observing System Microwave Limb Sounder. *IEEE Transactions on Geoscience and Remote Sensing*, *44*(5), 1081–1166.

Colesanti, C., A. Ferretti, C. Prati & F. Rocca (2003). Monitoring landslides and tectonic motions with the Permanent Scatterers Technique. *Engineering Geology*, *68*(1-2), 3–14.

Colliander, A., S. Chan, S.-B. Kim, N. Das, S. H. Yueh, M. Cosh, … E. G. Njoku (2012). Long term analysis of PALS soil moisture campaign measurements for global soil moisture algorithm development. *Remote Sensing of Environment*, *121*, 309–322.

Collin, R. E. (1959). Properties of Slotted Dielectric Interfaces. *IRE Transactions on Antennas and Propagation*, *7*(1), 62–73.

Collins, M. J. & J. M. Allan (2009). Modeling and Simulation of SAR Image Texture. *IEEE Transactions on Geoscience and Remote Sensing*, *47*(10), 3530–3546.

Collins, M. J. & J. Huang (1998). Uncertainties in the estimation of ACF-based texture in synthetic aperture radar image data. *IEEE Transactions on Geoscience and Remote Sensing*, *36*(3), 940–949.

Colvin, R. S. (1961). *A Study of Radio Astronomy Receivers*. Report 18, Radioscience Laboratory, Department of Electrical Engineering, Stanford University, Stanford, CA.

Comiso, J. C. (1983). Sea ice effective microwave emissivities from satellite passive microwave and infrared observations. *Journal of Geophysical Research*, *88*(C12), 7686–7704.

Comiso, J. C., S. F. Ackley & A. L. Gordon (1984). Antarctic sea ice microwave signatures and their correlation with in situ ice observations. *Journal of Geophysical Research*, *89*(C1), 662–672.

Compton, R. T. & R. E. Collin (1969a). Open Waveguides and Small Horns. In R. E. Collin & F. J. Zucker (Eds.), *Antenna Theory*. New York: McGraw-Hill.

Compton, R. T. & R. E. Collin (1969b). Slot Antennas. In R. E. Collin & F. J. Zucker (Eds.), *Antenna Theory*. New York: McGraw-Hill.

Conrath, B. J. (1972). Vertical Resolution of Temperature Profiles Obtained from Remote Radiation Measurements. *Journal of the Atmospheric Sciences*, *29*(7), 1262–1271.

Cook, C. E. & M. Bernfield (1967). *Radar Signals: An Introduction to Theory and Application*. New York: Academic Press.

Cooper, B. F. C. (1970). Correlators with two-bit quantization. *Australian Journal of Physics*, *23*, 521–527.

Corbella, I., N. Duffo, M. Vall-llossera, A. Camps & F. Torres (2004). The visibility function in interferometric aperture synthesis radiometry. *IEEE Transactions on Geoscience and Remote Sensing*, *42*(8), 1677–1682.

Cordey, R. A. (1993). On the accuracy of crosstalk calibration of polarimetric SAR using natural clutter statistics. *IEEE Transactions on Geoscience and Remote Sensing*, *31*(2), 447–454.

Costantini, M. (1998). A novel phase unwrapping method based on network programming. *IEEE Transactions on Geoscience and Remote Sensing*, *36*(3), 813–821.

Crane, R. K. (1971). Propagation phenomena affecting satellite communication systems operating in the centimeter and millimeter wavelength bands. *Proceedings of the IEEE*, *59*(2), 173–188.

Crane, R. K. (1981). Fundamental Limitations Caused by RF Propagation. *Proceedings of the IEEE*, *69*(2), 196–209.

Croney, J. & W. D. Delany (1963). A New Type of Omniazimuthal Radar-Echo Enhancer. *Microwave Journal*, *6*, 105–109.

Crosetto, M., A. Arnaud, J. Duro, E. Biescas & M. Agudo (2003). Deformation monitoring using remotely sensed radar interferometric data. In *11th FIG Symposium on Deformation Measurements*. Santorini, Greece.

Croswell, W. F. (2007). Slot Antennas. In J. L. Volakis (Ed.), *Antenna Engineering Handbook* (4th ed.). New York: McGraw-Hill.

Cumming, I. G. & F. H. Wong (2005). *Digital Processing of Synthetic Aperture Radar Data: Algorithms and Implementation*. Norwood, MA: Artech House.

Cumming, W. (1952). The Dielectric Properties of Ice and Snow at 3.2 cm. *Journal of Applied Physics*, *23*, 768–773.

Curlander, J. C. & R. N. McDonough (1991). *Synthetic Aperture Radar: Systems and Signal Processing*. New York: Wiley-Interscience.

Cutrona, L. J. (1970). Synthetic Aperture Radar. In M. I. Skolnik (Ed.), *Radar Handbook*. McGraw-Hill.

Cutrona, L. J. & G. O. Hall (1962). A Comparison of Techniques for Achieving Fine Azimuth Resolution. *IRE Transactions on Military Electronics*, *6*(2), 119–133.

Cutrona, L. J., W. E. Vivian, E. N. Leith & G. O. Hall (1961). A High-Resolution Radar Combat-Surveillance System. *IRE Transactions on Military Electronics*, *5*(2), 127–131.

Davenport, W. B. & W. L. Root (1958). *Introduction to Random Signals and Noise*. New York: McGraw-Hill.

Davies, H. & G. G. Macfarlane (1946). Radar echoes from the sea surface at centimetre wave-lengths. *Proceedings of the Physical Society*, *58*(6), 717–729.

De Grandi, G. D., J.-S. Lee & D. L. Schuler (2007). Target Detection and Texture Segmentation in Polarimetric SAR Images Using a Wavelet Frame: Theoretical Aspects. *IEEE Transactions on Geoscience and Remote Sensing*, *45*(11), 3437–3453.

de Loor, G. P. (1956). Dielectric properties of heterogeneous mixtures with a polar constituent (Rijksuniversiteit te Leiden Thesis). *Applied Scientific Research*, *11*(1), 310–320.

de Loor, G. P. (1968). Dielectric Properties of Heterogeneous Mixtures Containing Water. *Journal of Microwave Power and Electromagnetic Energy*, *3*(2), 67–73.

De Roo, R. D. & F. T. Ulaby (1994). Bistatic specular scattering from rough dielectric surfaces. *IEEE Transactions on Antennas and Propagation*, *42*(2), 220–231.

De Roo, R. D., Y. Du, F. T. Ulaby & M. C. Dobson (2001). A semi-empirical backscattering model at L-band and C-band for a soybean canopy with soil moisture inversion. *IEEE Transactions on Geoscience and Remote Sensing*, *39*(4), 864–872.

De Wit, J. J. M., A. Meta & P. Hoogeboom (2006). Modified range-Doppler processing for FM-CW synthetic aperture radar. *IEEE Geoscience and Remote Sensing Letters*, *3*(1), 83–87.

Decker, M. T., F. O. Guiraud & E. R. Westwater (1973). Correction of Electrical Path Length by Passive Microwave Radiometry. In *Conference on Propagation of Radio Waves at Frequencies above 10 GHz*. 10-13 April, London, UK.

Decker, M. T., E. R. Westwater & F. O. Guiraud (1978). Experimental evaluation of ground-based microwave radiometric sensing of atmospheric temperature and water vapor profiles. *Journal of Applied Meteorology*, *17*(12), 1788–1795.

Decker, M. T., R. G. Strauch & E. R. Westwater (1982). Ground-based remote sensing of atmospheric temperature, water vapor, and wind. In *IEEE International Geoscience and Remote Sensing Symposium* (Vol. 1). 1–4 June, Munich, West Germany.

Deepak, A. (Ed.) (1977). *Inversion Methods in Atmospheric Remote Sounding* (p. 638). New York: Academic Press.

Deepak, A. (Ed.) (1980). *Remote Sensing of the Atmospheres and Oceans*. New York: Academic Press.

Deirmendjian, D. (1969). *Electromagnetic Scattering on Spherical Polydispersions*. New York: Elsevier.

Deitz, P. & J. G. Constantine (1979). Passive MMW Detection in Snow. In *IEEE National Aerospace and Electronics Conference (NAECON)* (pp. 371–381). Dayton Convention Center, Dayton, OH.

Deschamps, G. & H. Cabayan (1972). Antenna synthesis and solution of inverse problems by regularization methods. *IEEE Transactions on Antennas and Propagation*, *29*(3), 268–274.

Desnos, Y. L., C. Buck, J. Guijarro, J. L. Suchail, R. Torres & E. Attema (2000). ASAR - Envisat's Advanced Synthetic Aperture Radar - Building on ERS achievements towards future earth watch missions. *ESA Bulletin*, (102), 91–100.

Dicke, R. H. (1946). The Measurement of Thermal Radiation at Microwave Frequencies. *Review of Scientific Instruments*, *17*(7), 268–275.

DiFranco, J. V. & W. L. Rubin (1968). *Radar detection. Information theory series*. Englewood Cliffs, NJ: Prentice-Hall.

Dobson, M. C., F. T. Ulaby, M. T. Hallikainen & M. A. El-Rayes (1985). Microwave Dielectric Behavior of Wet Soil-Part II: Dielectric Mixing Models. *IEEE Transactions on Geoscience and Remote Sensing*, *23*(1), 35–46.

Dobson, M. C., F. T. Ulaby, L. E. Pierce, T. L. Sharik, K. M. Bergen, J. Kellndorfer, … P. Siqueira (1995a). Estimation of Forest Biophysical Characteristics in Northern Michigan with SIR-C / X-SAR. *IEEE Transactions on Geoscience and Remote Sensing*, *33*(4), 877–895.

Dobson, M. C., F. T. Ulaby & L. E. Pierce (1995b). Land-cover classification and estimation of terrain attributes using synthetic aperture radar. *Remote Sensing of Environment*, *51*(1), 199–214.

Dobson, M. C., L. E. Pierce & F. T. Ulaby (1996). Knowledge-Based Land-Cover Classification Using ERS-1/JERS-1 SAR Composites. *IEEE Transactions on Geoscience and Remote Sensing*, *34*(1), 83–99.

Doerry, A. W., D. F. Dubbert, M. E. Thompson & V. D. Gutierrez (2005). A portfolio of fine resolution Ka-band SAR images: part I, SPIE Defense and Security Symposium, 28 March–1 April. Sandia National Laboratories. Retrieved from http://www.sandia.gov/RADAR/imageryka.html

Dong, X., D. Zhu & W. Lin (2010). A Ku-Band Rotating Fan-Beam Scatterometer: Design and Performance Simulations. In *IEEE International Geoscience and Remote Sensing Symposium* (pp. 1081–1084). 25–30 July, Honolulu, HI.

Draper, D. W. & D. G. Long (2004a). Evaluating the effect of rain on SeaWinds scatterometer measurements. *Journal of Geophysical Research*, *109*(C12), C02005.

Draper, D. W. & D. G. Long (2004b). Simultaneous wind and rain retrieval using SeaWinds data. *IEEE Transactions on Geoscience and Remote Sensing*, *42*(7), 1411–1423.

Dubois, P. C., J. J. van Zyl & T. Engman (1995a). Measuring Soil Moisture with Imaging Radars. *IEEE Transactions on Geoscience and Remote Sensing*, *33*(4), 915–926.

Dubois, P. C., J. J. van Zyl & T. Engman (1995b). Corrections to "Measuring Soil Moisture with Imaging Radars." *IEEE Transactions on Geoscience and Remote Sensing*, *33*(6), 1340.

Duda, R. O. & P. E. Hart (1973). *Pattern Classification and Scene Analysis*. New York: McGraw-Hill.

Durden, S. L., J. J. van Zyl & H. A. Zebker (1990). The Unpolarized Component in Polarimetric Radar Observations of Forested Areas. *IEEE Transactions on Geoscience and Remote Sensing*, *28*(2), 268–271.

Dzura, M. S., V. S. Etkin, A. S. Khrupin, M. N. Pospelov & M. D. Raev (1992). Radiometers-Polarimeters: Principles of Design and Applications for Sea Surface Microwave Emission Polarimetry. In *IEEE International Geoscience and Remote Sensing Symposium* (Vol. 2, pp. 1432–1434). 26–29 May, Houston, TX.

Early, D. S. & D. G. Long (2001). Image Reconstruction and Enhanced Resolution Imaging from Irregular Samples. *IEEE Transactions on Geoscience and Remote Sensing*, *39*(2), 291–302.

Ebert, E. E., J. Janoviak & C. Kidd (2007). Comparison of near real time precipitation estimates from satellite observations and numerical models. *Bulletin of the American Meteorological Society*, *88*(1), 47–59.

Edgerton, A. T. & D. T. Trexler (1970). Oceanographic Applications of Remote Sensing with Passive Microwave Techniques. In *6th International Symposium on Remote Sensing of Environment* (Vol. 2, pp. 767–788). 13-16 October, Ann Arbor, MI.

Edgerton, A. T., D. Meeks & D. Williams (1971). Microwave emission characteristics of oil slicks. In *Joint Conference on Sensing of Environmental Pollutants*. Palo Alto, CA, November 8-10, 1971: American Institute of Aeronautics, Paper No. 71-1071.

Ehrlich, M. & J. Short (1954). Mutual Coupling Considerations in Linear-Slot Array Design. *Proceedings of the IRE*, *42*(6), 956–961.

El-Rayes, M. & F. T. Ulaby (1987). Microwave Dielectric Spectrum of Vegetation—Part I: Experimental Observations. *IEEE Transactions on Geoscience and Remote Sensing*, *25*(5), 541–549.

Elachi, C. (1988). *Spaceborne Radar Remote Sensing: Applications and Techniques*. New York: IEEE Press.

Elachi, C., K. E. Im & E. Rodríguez (1990). Global digital topography mapping with a synthetic aperture scanning radar altimeter. *International Journal of Remote Sensing*, *11*(4), 585–601.

Elfouhaily, T., D. R. Thompson, B. Chapron & D. Vandemark (2000). Improved electromagnetic bias theory. *Journal of Geophysical Research*, *105*, 1299–1310.

Emerson, W. (1973). Electromagnetic wave absorbers and anechoic chambers through the years. *IEEE Transactions on Antennas and Propagation*, *21*(4), 484–490.

Entekhabi, D., E. G. Njoku, P. E. O'Neill, K. H. Kellogg, W. T. Crow, W. N. Edelstein, … J. Van Zyl (2010). The Soil Moisture Active Passive (SMAP) Mission. *Proceedings of the IEEE*, *98*(5), 704–716.

Eom, H. J. & A. K. Fung (1984). Backscattering coefficient of a perturbed sinusoidal surface. *IEEE Transactions on Antennas and Propagation*, *32*(3), 291–294.

Escorihuela, M. J., Y. H. Kerr, P. de Rosnay, J.-P. Wigneron, J.-C. Calvet & F. Lemaître (2007). A Simple Model of the Bare Soil Microwave Emission at L-Band. *IEEE Transactions on Geoscience and Remote Sensing*, *45*(7), 1978–1987.

Etkin, V. S., M. D. Raev, M. G. Bulatov, A. V. Smirnov & Y. A. Militsky (1991). *Radiohydrophysical Aerospace Research of Ocean*. Technical Report IIp-1749, Akademiya Nauk SSSR, Institut kosmicheskikh issledovaniĭ (Space Research Institute, Russian Academy of Science), Moscow, Russia.

Evans, S. (1965). Dielectric Properties of Ice and Snow—A Review. *Journal of Glaciology*, *5*(42), 773–792.

Ezraty, R. & A. Cavanié (1999). Construction and evaluation of 12.5-km grid NSCAT backscatter maps over Arctic sea ice. *IEEE Transactions on Geoscience and Remote Sensing*, *37*(3), 1685–1697.

Fante, R. L. (1981). Relationship between radiative-transport theory and Maxwell's equations in dielectric media. *Journal of the Optical Society of America*, *71*(4), 460–468.

Farr, T. G., P. A. Rosen, E. Caro, R. Crippen, R. Duren, S. Hensley, … D. Alsdorf (2007). The Shuttle Radar Topography Mission. *Reviews of Geophysics*, *45*(2), RG2004.

Fenner, R. G., S. C. Reid & C. H. Solie (1981). A Multifrequency Evaluation of Active and Passive Microwave Sensors for Oilspill Detection and Assessment. In *1st IEEE International Geoscience and Remote Sensing Symposium* (pp. 1252–1267). 8–10 June, Washington, DC.

Ferraro, R. R., F. Weng, N. C. Grody, L. Zhao, H. Meng, C. Kongoli, … C. Dean (2005). NOAA operational hydrological products derived from the Advanced Microwave Sounding Unit. *IEEE Transactions on Geoscience and Remote Sensing*, *43*(5), 1035–1048.

Ferretti, A., C. Prati & F. Rocca (2000). Nonlinear subsidence rate estimation using permanent scatterers in differential SAR interferometry. *IEEE Transactions on Geoscience and Remote Sensing*, *38*(5), 2202–2212.

Ferretti, A., C. Prati & F. Rocca (2001). Permanent scatterers in SAR interferometry. *IEEE Transactions on Geoscience and Remote Sensing*, *39*(1), 8–20.

Ferretti, A., M. Bianchi, C. Prati & F. Rocca (2004). Higher-order permanent scatterers analysis. *EURASIP Journal on Applied Signal Processing*, *2005*(20), 3231–3242.

Ferretti, A., A. Monti-Guarnieri, C. Prati, F. Rocca & D. Massonnet (2007). *InSAR Principles: Guidelines for SAR Interferometry Processing and Interpretation* (K. Fletcher, Ed.). Noordwijk, The Netherlands: European Space Agency.

Fialko, Y. (2006). Interseismic strain accumulation and the earthquake potential on the southern San Andreas fault system. *Nature*, *441*(7096), 968–971.

Fielding, E. J., M. Talebian, P. A. Rosen, H. Nazari, J. A. Jackson, M. Ghorashi & R. Walker (2005). Surface ruptures and building damage of the 2003 Bam, Iran, earthquake mapped by satellite synthetic aperture radar interferometric correlation. *Journal of Geophysical Research*, *110*(B3), B03302.

Figa, J. & A. Stoffelen (2000). On the assimilation of Ku-band scatterometer winds for weather analysis and forecasting. *IEEE Transactions on Geoscience and Remote Sensing*, *38*(4), 1893–1902.

Figa-Saldaña, J., J. J. W. Wilson, E. Attema, R. Gelsthorpe, M. R. Drinkwater & A. Stoffelen (2002). The advanced scatterometer (ASCAT) on the meteorological operational (MetOp) platform: A follow on for European wind scatterometers. *Canadian Journal of Remote Sensing*, *28*(3), 404–412.

Finkelstein, M. I., V. G. Glushnev, A. N. Petrov & V. Y. Ivashchenkov (1970). Anisotropic attenuation of radio waves in sea ice. *Izvestiya, Atmospheric and Oceanic Physics*, *6*(3), 1975–1976.

Fischer, R. E. (1972). Standard Deviation of Scatterometer Measurements from Space. *IEEE Transactions on Geoscience Electronics*, *10*(2), 106–113.

Font, J., Y. H. Kerr & M. Berger (2000). Measuring Ocean Salinity from Space: the European Space Agency's SMOS Mission. *Backscatter*, *11*(3), 17–19.

Font, J., G. S. E. Lagerloef, D. M. Le Vine, A. Camps & O.-Z. Zanife (2004). The determination of surface salinity with the European SMOS space mission. *IEEE Transactions on Geoscience and Remote Sensing*, *42*(10), 2196–2205.

Font, J., J. Boutin, N. Reul, P. Spurgeon, J. Ballabrera, A. Chuprin, … S. Zine (2010). Overview of SMOS Level 2 Ocean Salinity Processing and First Results. In *IEEE International Geoscience and Remote Sensing Symposium* (pp. 3146–3149). 25–30 July, Honolulu, HI.

Frankenstein, G. & R. Garner (1967). Equations for Determining the Brine Volume of Sea Ice from −0.5 °C to −22.9 °C. *Journal of Glaciology*, *6*(48).

Franklin, J. N. (1968). *Matrix Theory*. Englewood Cliffs, NJ: Prentice-Hall.

Franklin, J. N. (1970). Well-posed stochastic extensions of ill-posed linear problems. *Journal of Mathematical Analysis and Applications*, *31*(3), 682–716.

Fraser, K. S., N. E. Gaut, E. C. Reifenstein & H. Sievering (1975). Interactions Mechanisms—Within the Atmosphere. In R. G. Reeves (Ed.), *Manual of Remote Sensing: Interpretation and Applications*. American Society of Photogrammetry.

Frater, R. H. & D. R. Williams (1981). An Active "Cold" Noise Source. *IEEE Transactions on Microwave Theory and Techniques*, *29*(4), 344–347.

Freeman, A. & S. L. Durden (1998). A Three-Component Scattering Model for Polarimetric SAR Data. *IEEE Transactions on Geoscience and Remote Sensing*, *36*(3), 963–973.

Freeman, A. & S. S. Saatchi (2004). On the detection of Faraday rotation in linearly polarized L-band SAR backscatter signatures. *IEEE Transactions on Geoscience and Remote Sensing*, *42*(8), 1607–1616.

Freeman, A., C. Werner & Y. Shen (1988). Calibration of Multipolarisation Imaging Radar. In *IEEE International Geoscience and Remote Sensing Symposium* (Vol. 1, pp. 335–339). 13–16 September, Edinburgh, Scotland.

Freeman, A., M. Alves, B. Chapman, J. Cruz, Y. Kim, S. Shaffer, … K. Sarabandi (1995). SIR-C Data Quality and Calibration Results. *IEEE Transactions on Geoscience and Remote Sensing*, *33*(4), 848–857.

Freilich, M. H. & P. G. Challenor (1994). A new approach for determining fully empirical altimeter wind-speed model functions. *Journal of Geophysical Research*, *99*(C12), 25051–25062.

Freilich, M. H. & D. B. Chelton (1986). Wavenumber spectra of Pacific winds measured by the SEASAT scatterometer. *Journal of Physical Oceanography*, *16*(4), 741–757.

Freilich, M. H. & R. S. Dunbar (1993). Derivation of satellite wind model functions using operational surface wind analyses—An altimeter example. *Journal of Geophysical Research*, *98*(C8), 14633–14649.

Fu, L.-L. & A. Cazenave (2001). *Satellite Altimetry and Earth Sciences: A Handbook of Techniques and Applications*. *International Geophysics Series*. San Diego, CA: Academic Press.

Fu, L. L., E. J. Christensen, C. A. Yamarone, M. Lefebvre, Y. Menard, M. Dorrer & P. Escudier (1994). TOPEX/POSEIDON Mission Overview. *Journal of Geophysical Research*, *99*(C12), 24369–24381.

Fukunaga, K. & W. L. G. Koontz (1970). Application of the Karhunen-Loève Expansion to Feature Selection and Ordering. *IEEE Transactions on Computing*, *19*(4), 311–318.

Fung, A. K. (1994). *Microwave Scattering and Emission Models and Their Applications*. Norwood, MA: Artech House.

Fung, A. K. & K. S. Chen (1991). Kirchhoff Model for a Skewed Random Surface. *Journal of Electromagnetic Waves and Applications*, *5*(2), 205–216.

Fung, A. K. & K. S. Chen (2009). *Microwave Scattering and Emission Models for Users*. Norwood, MA: Artech House.

Fung, A. K. & K. S. Chen (2010). *Microwave Scattering and Emission Models for Users*. Norwood, MA: Artech House.

Fung, A. K. & H. J. Eom (1981). A Theory of Wave Scattering from an Inhomogeneous Layer with an Irregular Interface. *IEEE Transactions on Antennas and Propagation*, *29*(6), 899–910.

Fung, A. K. & K. Lee (1982). A semi-empirical sea-spectrum model for scattering coefficient estimation. *IEEE Journal of Oceanic Engineering*, *7*(4), 166–176.

Fung, A. K., Z. Q. Li & K. S. Chen (1992). Backscattering from a randomly rough dielectric surface. *IEEE Transactions on Geoscience and Remote Sensing*, *30*(2), 356–369.

Fung, A. K., W. Y. Liu, K. S. Chen & M. K. Tsay (2002). An Improved IEM Model for Bistatic Scattering from Rough Surface. *Journal of Electromagnetic Waves and Applications*, *16*(5), 689–702.

Gabriel, P., S. Lovejoy, D. Schertzer & G. L. Austin (1988). Multifractal analysis of resolution dependence in satellite imagery. *Geophysical Research Letters*, *15*(12), 1373–1376.

Gade, M., W. Alpers, H. Hühnerfuss, H. Masuko & T. Kobayashi (1998). Imaging of biogenic and anthropogenic ocean surface films by the multifrequency/multipolarization SIR-C/X-SAR. *Journal of Geophysical Research*, *103*(C9), 18851–18866.

Gail, W. B. (1998). Effect of Faraday rotation on polarimetric SAR. *IEEE Transactions on Aerospace and Electronic Systems*, *34*(1), 301–307.

Gaiser, P. W., K. M. S. Germain, E. M. Twarog, G. A. Poe, W. Purdy, D. Richardson, ... P. S. Chang (2004). The WindSat spaceborne polarimetric microwave radiometer: sensor description and early orbit performance. *IEEE Transactions on Geoscience and Remote Sensing*, *42*(11), 2347–2361.

Galin, N., A. Worby, T. Markus, C. Leuschen & P. Gogineni (2012). Validation of Airborne FMCW Radar Measurements of Snow Thickness Over Sea Ice in Antarctica. *IEEE Transactions on Geoscience and Remote Sensing*, *50*(1), 3–12.

Garrison, J. L., S. J. Katzberg, & M. I. Hill (1998). Effect of sea roughness on bistatically scattered range coded signals from the Global Positioning System. *Geophysical Research Letters*, *25*(13), 2257–2260.

Gasiewski, A. J. (1992). Numerical Sensitivity Analysis of Passive EHF and SMMW Channels to Tropospheric Water Vapor, Clouds, and Precipitation. *IEEE Transactions on Geoscience and Remote Sensing*, *30*(5), 859–869.

Gasiewski, A. J. & D. B. Kunkee (1993). Calibration and applications of polarization-correlating radiometers. *IEEE Transactions on Microwave Theory and Techniques*, *41*(5), 767–773.

Gasiewski, A. J. & D. B. Kunkee (1994). Polarized microwave emission from water-waves. *Radio Science*, *29*(6), 1449–1466.

Gasiewski, A. J. & J. R. Piepmeier (1996). Polarimetric scanning radiometer for airborne microwave imaging of ocean thermal emission. In *Fifth Specialist Meeting on Microwave Radiometry and Remote Sensing of the Environment*. 4-6 November, URSI, Boston, MA.

GE (1973). *S193 Microwave Radiometer/Scatterometer and Altimeter Calibration Data Report, Flight Hardware*. General Electric Space Systems Division, Valley Forge, PA.

Germain, K. M. S., G. A. Poe & P. W. Gaiser (2002). Polarimetric Emission Model of the Sea at Microwave Frequencies and Comparison with Measurements. *Progress in Electromagnetics Research*, *37*, 1–30.

Ghiglia, D. C. & L. A. Romero (1996). Minimum L^p-norm two-dimensional phase unwrapping. *Journal of the Optical Society of America A*, *13*(10), 1999–2013.

Girard-Ardhuin, F., G. Mercier, F. Collard & R. Garello (2005). Operational Oil-Slick Characterization by SAR Imagery and Synergistic Data. *IEEE Journal of Oceanic Engineering*, *30*(3), 487–495.

Glaister, J., A. Wong & D. A. Clausi (2013). Despeckling of Synthetic Aperture Radar Images Using Monte Carlo Texture Likelihood Sampling. *IEEE Transactions on Geoscience and Remote Sensing*, 1–11.

Gleason, S., S. Hodgart, C. Gommenginger, S. Mackin, M. Adjrad & M. Unwin (2005). Detection and Processing of bistatically reflected GPS signals from low Earth orbit for the purpose of ocean remote sensing. *IEEE Transactions on Geoscience and Remote Sensing*, *43*(6), 1229–1241.

Gloersen, P. & J. K. Larabee (1981). *An Optical Model for the Microwave Properties of Sea Ice*. NASA Technical Memorandum 83865, Goddard Space Flight Center, Greenbelt, MD.

Gloersen, P., W. Nordberg, T. J. Schmugge, T. T. Wilheit & W. J. Campbell (1973). Microwave Signatures of First-Year and Multiyear Sea Ice. *Journal of Geophysical Research*, *78*(18), 3564–3572.

Gloersen, P., D. J. Cavalieri, A. T. C. Chang, T. T. Wilheit, W. J. Campbell, O. M. Johannessen, ... R. O. Ramseier (1984). A summary of results from the first Numbus-7 SMMR observations. *Journal of Geophysical Research*, *89*(ND4), 5335–5344.

Glushnev, V. G., B. D. Slutsker & M. I. Finkelstein (1976). Measurement of 8-mm radio-wave attenuation in sea and freshwater ice and in snow. *Radiophysics and Quantum Electronics*, *19*(9), 916–918.

Goggins, W. B. (1967). A Microwave Feedback Radiometer. *IEEE Transactions on Aerospace and Electronic Systems*, *3*(1), 83–90.

Gogineni, S. P. (1984). *Radar Backscatter from Summer and Ridged Sea Ice, and the Design of Short-Range Radars*. Ph.D. Thesis, University of Kansas, Lawrence, KS.

Golden, K. M., M. Cheney, K. Ding, A. K. Fung, T. C. Grenfell, D. Isaacson, ... D. P. Winebrenner (1998). Scattering Models for Sea Ice. *IEEE Transactions on Geoscience and Remote Sensing*, *36*(5), 1655–1674.

Goldfinger, A. D. (1980). Refraction of microwave signals by water vapor. *Journal of Geophysical Research*, *85*(C9), 4904–4912.

Goldstein, R. M. (1995). Atmospheric limitations to repeat-track radar interferometry. *Geophysical Research Letters*, *22*(18), 2517–2520.

Goldstein, R. M. & H. A. Zebker (1987). Interferometric Radar Measurement of Ocean Surface Currents. *Nature*, *328*(6132), 707–709.

Goldstein, R. M., H. A. Zebker & C. L. Werner (1988). Satellite radar interferometry: Two-dimensional phase unwrapping. *Radio Science*, *23*(4), 713–720.

Goldstein, R. M., H. Engelhardt, B. Kamb & R. M. Frolich (1993). Satellite Radar Interferometry for Monitoring Ice Sheet Motion: Application to an Antarctic Ice Stream. *Science*, *262*(5139), 1525–1530.

Goodberlet, M. A. (2000). Improved Image Reconstruction Techniques for Synthetic Aperture Radiometers. *IEEE Transactions on Geoscience and Remote Sensing*, *38*(3), 1362–1366.

Goodberlet, M. A., C. T. Swift & J. C. Wilkerson (1989). Remote sensing of ocean surface winds with the special sensor microwave/imager. *Journal of Geophysical Research*, *94*(C10), 14547–14555.

Gorelik, A. G., V. V. Kalashnikov, L. S. Raykova & Y. A. Frolov (1973). Radiothermal Measurements of Atmospheric Humidity and the Integral Water Content of Clouds. *Izvestiya, Atmospheric and Oceanic Physics*, *9*, 527–530.

Gower, J. F. R. (1979). The computation of ocean wave heights from GEOS-3 satellite radar altimeter data. *Remote Sensing of Environment*, *8*(2), 97–114.

Graham, L. C. (1974). Synthetic interferometer radar for topographic mapping. *Proceedings of the IEEE*, *62*(6), 763–768.

Grantham, W. L., E. M. Bracalente, W. L. Jones & J. W. Johnson (1977). The SeaSat-A satellite scatterometer. *IEEE Journal of Oceanic Engineering*, *2*(2), 200–206.

Gray, A. L., R. K. Hawkins, C. E. Livingstone, L. D. Arsenault & W. M. Johnstone (1982). Simultaneous scatterometer and radiometer measurements of sea-ice microwave signatures. *IEEE Journal of Oceanic Engineering*, *7*(1), 20–32.

Greffet, J.-J. (1992). Theoretical model of the shift of the Brewster angle on a rough surface. *Optics Letters*, *17*(4), 238–240.

Grody, N. C. (1978). Microwave Radiometry Applied to Synoptic Meteorology and Climatology. In D. H. Staelin & P. W. Rosenkranz (Eds.), *High Resolution Passive Microwave Satellites*. Cambridge, MA: Applications Review Panel Report, Research Laboratory of Electronics, MIT.

Grody, N. C. (1993). Remote sensing of the atmosphere from satellites using microwave radiometry. In M. A. Janssen (Ed.), *Atmospheric Remote Sensing by Microwave Radiometry*. New York: Wiley & Sons.

Grody, N. C. (2008). Relationship between snow parameters and microwave satellite measurements: Theory compared with Advanced Microwave Sounding Unit observations from 23 to 150 GHz. *Journal of Geophysical Research*, *113*(D22), D22108.

Guiraud, F., J. Howard & D. Hogg (1979). A Dual-Channel Microwave Radiometer for Measurement of Precipitable Water Vapor and Liquid. *IEEE Transactions on Geoscience Electronics*, *17*(4), 129–136.

Gustavsson, A., L. M. H. Ulander, L. E. Andersson, P. O. Frolind, H. Hellsten, T. Jonsson, ... G. Stenstrom (1996). The experimental airborne VHF SAR sensor CARABAS. A status report. In *IEEE International Geoscience and Remote Sensing Symposium* (Vol. 3, pp. 1877–1880). 27–30 May, Lincoln, NE.

Haarpaintner, J. & G. Spreen (2007). Use of Enhanced-Resolution QuikSCAT/SeaWinds Data for Operational Ice Services and Climate Research: Sea Ice Edge, Type, Concentration, and Drift. *IEEE Transactions on Geoscience and Remote Sensing*, *45*(10), 3131–3137.

Haarpaintner, J., R. T. Tonboe, D. G. Long & M. L. van Woert (2004). Automatic detection and validity of the sea-ice edge: an application of enhanced-resolution QuikScat/SeaWinds data. *IEEE Transactions on Geoscience and Remote Sensing*, *42*(7), 1433–1443.

Hach, J.-P. (1966). Proposal for a continuously calibrated radiometer. *Proceedings of the IEEE*, *54*(12), 1016–2015.

Hach, J.-P. (1968). A Very Sensitive Airborne Microwave Radiometer Using Two Reference Temperatures. *IEEE Transactions on Microwave Theory and Techniques*, *16*(9), 629–636.

Hagen, J. B. & D. T. Farley (1973). Digital-correlation techniques in radio science. *Radio Science*, *8*(8-9), 775–784.

Hajj, G. A., C. O. Ao, B. A. Iijima, D. Kuang, E. R. Kursinski, A. J. Mannucci, … T. P. Yunck (2004). CHAMP and SAC-C atmospheric occultation results and intercomparisons. *Journal of Geophysical Research*, *109*(D6), D06109.

Hall, C. D. & R. A. Cordey (1988). Multistatic Scatterometry. In *IEEE International Geoscience and Remote Sensing Symposium* (Vol. 1, pp. 561–562). 13–16 September, Edinburgh, Scotland.

Hall, D. K., J. L. Foster, A. T. C. Chang & A. Rango (1981). Freshwater Ice Thickness Observations Using Passive Microwave Sensors. *IEEE Transactions on Geoscience and Remote Sensing*, *19*(4), 189–193.

Hall, D. K., J. L. Foster, V. V. Salomonson, A. G. Klein & J. Y. L. Chien (2001). Development of a Technique to Assess Snow-Cover Mapping Errors from Space. *IEEE Transactions on Geoscience and Remote Sensing*, *39*(2), 432–438.

Hall, D. K., R. E. J. Kelly, G. A. Riggs, A. T. C. Chang & J. L. Foster (2002). Assessment of the relative accuracy of hemispheric-scale snow-cover maps. *Annals of Glaciology*, *34*(1), 24–30.

Hallikainen, M. T. (1977). *Dielectric properties of NaCl ice at 16 GHz*. Report S 107, Helsinki University of Technology, Espoo, Finland.

Hallikainen, M. T. (1978). Measured Permittivities of Snow and Low-Salinity Sea Ice for UHF Radiometer Applications. In *URSI National Conference on Radio Science*. Stockholm, Sweden.

Hallikainen, M. T. (1980). Dielectric Properties and Passive Remote Sensing of Low-Salinity Sea Ice at UHF Frequencies. *Acta Polytechnica Scandinavica*, *45*.

Hallikainen, M. T. (1983). New low-salinity sea-ice model for UHF radiometry. *International Journal of Remote Sensing*, *4*(3), 655–681.

Hallikainen, M. T. (1992). Review of the Microwave Dielectric and Extinction Properties of Sea Ice and Snow. In *IEEE International Geoscience and Remote Sensing Symposium* (Vol. 2, pp. 961–965). 26–29 May, Houston, TX.

Hallikainen, M. T., F. T. Ulaby & M. Abdelrazik (1983). Modeling the Dielectric Behavior of Wet Snow in the 4-18 GHz Range. In *IEEE International Geoscience and Remote Sensing Symposium*. 31 August–2 September, San Francisco, CA.

Hallikainen, M. T., F. T. Ulaby & M. Abdelrazik (1984a). The Dielectric Behavior of Snow in the 3 to 37 GHz Range. In *IEEE International Geoscience and Remote Sensing Symposium* (pp. 169–176). 27–30 August, Strasbourg, France.

Hallikainen, M. T., F. T. Ulaby, M. C. Dobson & M. A. El-Rayes (1984b). Dielectric Measurements of Soils in the 3 to 37 GHz Band Between −50 °C and 23 °C. In *IEEE International Geoscience and Remote Sensing Symposium* (pp. 163–168). 27–30 August, Strasbourg, France.

Hallikainen, M. T., F. T. Ulaby, M. C. Dobson, M. A. El-Rayes & L. K. Wu (1985). Microwave dielectric behavior of wet soil - Part I: Empirical models and experimental observations. *IEEE Transactions on Geoscience and Remote Sensing*, *23*(1), 25–34.

Hallikainen, M. T., F. T. Ulaby & M. Abdelrazik (1986). The Dielectric Properties of Snow in the 3 to 37 GHz Range. *IEEE Transactions on Antennas and Propagation*, *34*(11), 1329–1340.

Hallikainen, M. T., F. T. Ulaby & T. E. van Deventer (1987). Extinction Behavior of Dry Snow in the 18- to 90-GHz range. *IEEE Transactions on Geoscience and Remote Sensing*, *25*(6), 737–745.

Hallikainen, M. T., M. V. O. Toikka & J. M. Hyyppa (1988). Microwave Dielectric Properties Of Low-salinity Sea Ice. In *IEEE International Geoscience and Remote Sensing Symposium* (Vol. 1, pp. 419–420). 13-16 September, Edinburgh, Scotland.

Hansen, R. C. (1964). Aperture Theory. In R. C. Hansen (Ed.), *Microwave Scanning Antennas* (Vol. 1). New York: Academic Press.

Hanssen, R. F. (2001). *Radar Interferometry: Data Interpretation and Error Analysis*. *Remote Sensing and Digital Image Processing*. Dordrecht; Boston; London: Kluwer Academic.

Hanway, J. J. (1971). *How a Corn Plant Develops*. Special Report No. 48, Cooperative Extension Service, Iowa State University, Ames, IA.

Hardy, W. N., K. W. Gray & A. W. Love (1974). An S-Band Radiometer Design with High Absolute Precision. *IEEE Transactions on Microwave Theory and Techniques*, *22*(4), 382–390.

Hasselmann, K., W. Munk & G. MacDonald (1963). Bispectra of Ocean Waves. In M. Rosenblatt (Ed.), *Time Series Analysis* (pp. 126–139). New York: Wiley & Sons.

Hasted, J. B. (1973). *Aqueous Dielectrics*. London: Chapman and Hall.

Hauck, B., F. T. Ulaby & R. D. De Roo (1998). Polarimetric bistatic measurement facility for point and distributed targets. *IEEE Antennas and Propagation Magazine*, *40*(1), 31–41.

Hayes, D., U. H. W. Lammers, R. Marr & J. McNally (1979). Millimeter Wave Backscatter from Snow. In *Workshop on Radar Backscatter from Terrain*. U.S. Army Engineer Topographic Laboratories, Ft. Belvoir, VA, 9-11 January: Remote Sensing Laboratory Technical Report 374-2, University of Kansas Center for Research, Inc.

Haykin, S. (2008). *Neural networks and learning machines* (3rd ed.). Prentice Hall.

Healy, S. B., J. Wickert, G. Michalak, T. Schmidt & G. Beyerle (2007). Combined forecast impact of GRACE-A and CHAMP GPS radio occultation bending angle profiles. *Atmospheric Science Letters*, *8*(2), 43–50.

Heise, S., J. Wickert, G. Beyerle, T. Schmidt & C. Reigber (2006). Global monitoring of tropospheric water vapor with GPS radio occultation aboard CHAMP. *Advances in Space Research*, *37*(12), 2222–2227.

Hering, W. S. (1965). Atmospheric Composition, Section 6.2. In S. L. Valley (Ed.), *Handbook of Geophysics*. Office of Aerospace Research, Cambridge Research Laboratories, United States Air Force.

Hersbach, H., A. Stoffelen & S. de Haan (2007). An improved C-band scatterometer ocean geophysical model function: CMOD5. *Journal of Geophysical Research*, *112*(C3), C03006.

Hewison, T. J. & S. J. English (1999). Airborne retrievals of snow and ice surface emissivity at millimeter wavelengths. *IEEE Transactions on Geoscience and Remote Sensing*, *37*(4), 1871–1879.

Hewison, T. J., N. Selbach, G. Heygster, J. P. Taylor & A. J. McGrath (2002). Airborne Measurements of Arctic Sea Ice, Glacier and Snow Emissivity at 24-183 GHz. In *IEEE International Geoscience and Remote Sensing Symposium* (Vol. 5, pp. 2851–2855). 24–28 June, Toronto, Canada.

Hilburn, K. A., F. J. Wentz, D. K. Smith & P. D. Ashcroft (2006). Correcting Active Scatterometer Data for the Effects of Rain Using Passive Radiometer Data. *Journal of Applied Meteorology and Climatology*, *45*(3), 382–398.

Hoekstra, P., & P. Cappillino. (1971). Dielectric properties of sea and sodium chloride ice at UHF and microwave frequencies. *Journal of Geophysical Research*, *76*(20), 4922–4931.

Hoekstra, P. & A. Delaney (1974). Dielectric Properties of Soils at UHF and Microwave Frequencies. *Journal of Geophysical Research*, *79*(11), 1699–1708.

Hofer, R. & C. Mätzler (1980). Investigations on Snow Parameters by Radiometry in the 3- to 60-mm Wavelength Region. *Journal of Geophysical Research*, *85*(C1), 453–460.

Hofer, R. & E. G. Njoku (1981). Regression Techniques for Oceanographic Parameter Retrieval Using Space-Borne Microwave Radiometry. *IEEE Transactions on Geoscience and Remote Sensing*, *19*(4), 178–189.

Hoffman, R. N. (1982). SASS wind ambiguity removal by direct minimization. *Monthly Weather Review*, *110*(5), 434–445.

Hoffman, R. N. (1984). SASS Wind Ambiguity Removal by Direct Minimization. Part II: Use of Smoothness and Dynamical Constraints. *Monthly Weather Review*, *112*(9), 1829–1852.

Hoffmann, J., D. L. Galloway & H. A. Zebker (2003). Inverse modeling of interbed storage parameters using land subsidence observations, Antelope Valley, California. *Water Resources Research*, *39*(2), SBH 5-1 – SBH 5-13.

Hogg, D. C. (1980). Ground-based measurements of microwave absorption by tropospheric water vapor. *Proceedings of the International Workshop on Atmospheric Water Vapor, Vail, CO, 11-13 September, 1979*, 219–228.

Hollinger, J. P. (1974). *The Determination of Oil Slick Thickness by Means of Multifrequency Passive Microwave Techniques*. Department of Transportation Report No. CG-D-31-75, U.S. Coast Guard, Office of Research and Development, Washington, DC.

Hollinger, J. P. & R. A. Mennella (1973). Oil Spills: Measurements of Their Distributions and Volumes by Multifrequency Microwave Radiometry. *Science*, *181*(4094), 54–56.

Hooper, A. J. (2006a). *Persistent Scatterer Radar Interferometry for Crustal Deformation Studies and Modeling of Volcanic Deformation*. Ph.D. Thesis, Department of Geophysics, Stanford University, Stanford, CA.

Hooper, A. J. (2006b). Multiple Acquisition InSAR Analysis: Persistent Scatterer and Small Baseline Approaches. In *Eos, Transactions, American Geophysical Union* (Vol. 87).

Hooper, A. J., H. A. Zebker, P. Segall & B. Kampes (2004). A new method for measuring deformation on volcanoes and other natural terrains using InSAR persistent scatterers. *Geophysical Research Letters*, *31*(23), L23611.

Hooper, A. J., P. Segall & H. Zebker (2007). Persistent scatterer interferometric synthetic aperture radar for crustal deformation analysis, with application to Volcán Alcedo, Galápagos. *Journal of Geophysical Research*, *112*(B7), B07407.

Hopfield, H. S. (1971). Tropospheric Effect on Electromagnetically Measured Range: Prediction from Surface Weather Data. *Radio Science*, *6*(3), 357–367.

Hornik, K. M., M. Stinchcombe & H. White (1989). Multilayer Feedforward Networks are Universal Approximators. *Neural Networks*, *4*(5), 359–366.

Hottel, H. C., A. F. Sarofim, W. H. Dalzell & I. A. Vasalos (1971). Optical Properties of Coatings. Effect of Pigment Concentration. *AIAA Journal*, *9*(10), 1895–1898.

Howard, D. D., N. A. Thomas & M. C. Licitra (1967). *Microwave Monitoring of Sea Water Contamination of Navy Fuel Oils*. Report No. 6552, Naval Research Laboratory, Washington, DC.

Hristova-Veleva, S. M., P. S. Callahan, R. S. Dunbar, S. H. Yueh, B. W. Stiles, J. N. Huddleston, ... R. W. Gaston (2006). Revealing the SeaWinds ocean vector winds under the rain using AMSR. Part I: The physical approach. In *14th Conference on Satellite Meteorology and Oceanography*. 29 January–2 February, Atlanta, Georgia.

Huang, S., L. Tsang, E. G. Njoku & K. Chen (2010). Backscattering Coefficients , Coherent Reflectivities , and Emissivities of Randomly Rough Soil Surfaces at L-Band for SMAP Applications Based on Numerical Solutions of Maxwell Equations in Three-Dimensional Simulations. *IEEE Transactions on Geoscience and Remote Sensing*, *48*(6), 2557–2568.

Huddleston, J. N. & B. W. Stiles (2000). A multidimensional histogram rain-flagging technique for SeaWinds on QuikSCAT. In *IEEE International Geoscience and Remote Sensing Symposium* (Vol. 3, pp. 1232–1234). 24–28 July, Honolulu, HI.

Hülsmeyer, C. (1904). Hertzian-wave Projecting and Receiving Apparatus Adapted to Indicate or Give Warning of the Presence of a Metallic Body, such as a Ship or a Train, in the Line of Projection of such Waves. British.

Hunsucker, R. D. & J. K. Hargreaves (2003). *The High-Latitude Ionosphere and its Effects on Radio Propagation*. New York: Cambridge University Press.

Huynen, J. R. (1970). *Phenomenological theory of radar targets*. Ph.D. Thesis, Drukkerij Bronder-Offset, N.V., Rotterdam, Netherlands.

IEEE (1979). *IEEE Standard Test Procedures for Antennas*. New York: IEEE Press.

Iguchi, T., T. Kozu, R. Meneghini, J. Awaka & K. Okamoto (2000). Rain-Profiling Algorithm for the TRMM Precipitation Radar. *Journal of Applied Meteorology*, *39*(12), 2038–2052.

Inoue, Y., T. Kurosu, H. Maeno, S. Uratsuka, T. Kozu, K. Dabrowska-Zielinska & J. Qi (2002). Season-long daily measurements of multifrequency (Ka, Ku, X, C, and L) and full-polarization backscatter signatures over paddy rice field and their relationship with biological variables. *Remote Sensing of Environment*, *81*(2), 194–204.

Ioannidis, G. A. & D. E. Hammers (1979). Optimum antenna polarizations for target discrimination in clutter. *IEEE Transactions on Antennas and Propagation*, *27*(3), 357–363.

Ishimaru, A. (1978). *Wave Propagation and Scattering in Random Media*. New York: Academic Press.

Ishimaru, A. (1991). *Electromagnetic Wave Propagation, Radiation, and Scattering*. Englewood Cliffs, NJ: Prentice Hall.

Ishimaru, A. & R. L. T. Cheung (1980). Multiple scattering effects on wave propagation due to rain. *Annales des Telecommunications*, *35*(11/12), 373–379.

Ishimaru, A. & Y. Kuga (1982). Attenuation constant of a coherent field in a dense distribution of particles. *Journal of the Optical Society of America*, *72*(10), 1317–1320.

Ishimaru, A., R. Woo, J. W. Armstrong & D. C. Blackman (1982). Multiple scattering calculations of rain effects. *Radio Science*, *17*(6), 1425–1433.

Jackson, D. R. & A. A. Oliner (2007). Leaky-Wave Antennas. In C. A. Balanis (Ed.), *Modern Antenna Handbook* (pp. 325–367). Hoboken, NJ: Wiley & Sons.

Jackson, T. J. (1993). Measuring surface soil moisture using passive microwave remote sensing. *Hydrological Processes*, *7*(2), 139–152.

Jackson, T. J. & T. J. Schmugge (1991). Vegetation effects on the microwave emission of soils. *Remote Sensing of Environment*, *36*(3), 203–212.

Jakes, W. C. (1951). Gain of Electromagnetic Horns. *Proceedings of the IRE*, *39*(2), 160–162.

Jakes, W. C. (1961). Horn Antennas. In H. Jasik (Ed.), *Antenna Engineering Handbook*. New York: McGraw-Hill.

Jakowatz, C. V., D. E. Wahl & P. A. Thompson (1996). Ambiguity resolution in SAR interferometry by use of three phase centers. *Proceedings of SPIE*, *2757*(1).

Janssen, M. A. (Ed.) (1993). *Atmospheric Remote Sensing by Microwave Radiometry*. New York: Wiley & Sons.

Jao, J. K. (1984). Amplitude distribution of composite terrain radar clutter and the K-distribution. *IEEE Transactions on Antennas and Propagation*, *32*(10), 1049–1062.

Jarnot, R. F., V. S. Perun & M. J. Schwartz (2006). Radiometric and Spectral Performance and Calibration of the GHz Bands of EOS MLS. *IEEE Transactions on Geoscience and Remote Sensing*, *44*(5), 1131–1143.

Jasik, H. (Ed.) (1961). *Antenna Engineering Handbook*. New York: McGraw-Hill.

Jehle, M., M. Rüegg, L. Zuberbühler, D. Small & E. Meier (2009). Measurement of Ionospheric Faraday Rotation in Simulated and Real Spaceborne SAR Data. *IEEE Transactions on Geoscience and Remote Sensing*, *47*(5), 1512–1523.

Jezek, K. C. (2003). Observing the Antarctic Ice Sheet Using the RADARSAT-1 Synthetic Aperture Radar. *Polar Geography*, *27*(3), 197–209.

Jin, Y.-Q. (1993). *Electromagnetic scattering modelling for quantitative remote sensing*. Singapore; River Edge, NJ: World Scientific.

Jin, Y.-Q. & C. Liu (1997). Biomass retrieval from high-dimensional active/passive remote sensing data by using artificial neural networks. *International Journal of Remote Sensing*, *18*(4), 971–979.

Johannessen, J. A., D. R. Lyzenga, R. Shuchman, H. Espedal & B. Holt (1995). Multifrequency SAR observations of ocean surface features off the coast of Norway. In *IEEE International Geoscience and Remote Sensing Symposium* (Vol. 2, pp. 1328–1330). 10–14 July, Florence, Italy.

Johanson, I. A., R. Bürgmann, A. Ferretti & F. Novali (2009). Variable Creep on the Concord fault from PS-InSAR and SBAS. In *Fall Meeting of the American Geophysical Union*. Abstract #G23B-0693.

Johnson, J. T. & Y. Y. Cai (2002). A theoretical study of sea surface up/down wind brightness temperature differences. *IEEE Transactions on Geoscience and Remote Sensing*, *40*(1), 66–78.

Johnson, J., L. A. Williams Jr., E. M. Bracalente, F. B. Beck & W. L. Grantham (1980). Seasat-A satellite scatterometer instrument evaluation. *IEEE Journal of Oceanic Engineering*, *5*(2), 138.

Joint Research Centre (n.d.) European Microwave Signature Laboratory (EMSL). Retrieved from http://sta.jrc.ec.europa.eu/index.php/emsl

Jones, L. A., J. S. Kimball, K. C. McDonald, S. T. K. Chan, E. G. Njoku & W. C. Oechel (2007). Satellite Microwave Remote Sensing of Boreal and Arctic Soil Temperatures From AMSR-E. *IEEE Transactions on Geoscience and Remote Sensing*, *45*(7), 2004–2018.

Jones, R. G. (1976). Precise Dielectric Measurements at 35 GHz Using an Open Microwave Resonator. *Proceedings of the Institution of Electrical Engineers (IEE)*, *123*(4), 285–290.

Jones, W. L., L. C. Schroeder, D. H. Boggs, E. M. Bracalente, R. A. Brown, G. J. Dome, ... F. J. Wentz (1982). The SEASAT-A satellite scatterometer: The geophysical evaluation of remotely sensed wind vectors over the ocean. *Journal of Geophysical Research*, *87*(C5), 3297.

Jónsson, S., H. A. Zebker, P. Segall & F. Amelung (2002). Fault Slip Distribution of the 1999 Mw 7.1 Hector Mine, California, Earthquake, Estimated from Satellite Radar and GPS Measurements. *Bulletin of the Seismological Society of America*, *92*(4), 1377–1389.

Jordan, R. L. (1980). The Seasat-A synthetic aperture radar system. *IEEE Journal of Oceanic Engineering*, *5*(2), 154–164.

Just, D. & R. Bamler (1994). Phase statistics of interferograms with applications to synthetic-aperture radar. *Applied Optics*, *33*(20), 4361–4368.

Kanagaratnam, P., T. Markus, V. Lytle, B. Heavey, P. Jansen, G. Prescott & S. P. Gogineni (2007). Ultrawideband radar measurements of thickness of snow over sea ice. *IEEE Transactions on Geoscience and Remote Sensing*, *45*(9), 2715–2724.

Kang, K., C. R. Duguay, S. E. L. Howell, C. P. Derksen & R. E. J. Kelly (2010). Sensitivity of AMSR-E Brightness Temperatures to the Seasonal Evolution of Lake Ice Thickness. *IEEE Geoscience and Remote Sensing Letters*, *7*(4), 751–755.

Kashihara, H., K. Nakada, M. Murata, M. Hiroguchi & H. Aiba (1984). A study of amplitude distribution of space-borne synthetic aperture radar (SAR) data. In *International Symposium on Noise and Clutter Reduction (ISNCR-84)*. Tokyo.

Katzberg, S. J. & J. Dunion (2009). Comparison of reflected GPS wind speed retrievals with dropsondes in tropical cyclones. *Geophysical Research Letters*, *36*(17), L17602.

Katzberg, S. J., R. A. Walker, J. H. Roles, T. Lynch & P. G. Black (2001). First GPS signals reflected from the interior of a tropical storm: Preliminary results from Hurricane Michael. *Geophysical Research Letters*, *28*(10), 1981–1984.

Katzberg, S. J., O. Torres & G. Ganoe (2006). Calibration of reflected GPS for tropical storm wind speed retrievals. *Geophysical Research Letters*, *33*(18), L18602.

Kelley, M. C. (1989). *The Earth's Ionosphere: Plasma Physics and Electromagnetics*. San Diego, CA: Academic Press.

Kellndorfer, J., W. Walker, K. Kirsch, G. Fiske, J. Bishop, L. LaPoint, … J. Westfall (2013). NACP Aboveground Biomass and Carbon Baseline Data, V. 2 (NBCD 2000). Data set. Oak Ridge, TN: ORNL DAAC. http://dx.doi.org/10.3334/ORNLDAAC/1161

Kennaugh, E. M. (1951). *Effects of the type of polarization on echo characteristics*. Antenna Laboratory Report 389-9, Ohio State University, Columbus, OH.

Kennett, R. G. & F. K. Li (1989). Seasat over-land scatterometer data, Part II: Selection of extended area and land-target sites for the calibration of spaceborne scatterometers. *IEEE Transactions on Geoscience and Remote Sensing*, *27*(6), 779–788.

Kerker, M. (1969). *The Scattering of Light and Other Electromagnetic Radiation. Physical Chemistry*. New York: Academic Press.

Kerr, D. E. (1951). *Propagation of Short Radio Waves. MIT Radiation Laboratory Series*. New York: McGraw-Hill.

Kerr, D. E. & H. Goldstein (1951). Radar Targets and Echoes. In D. E. Kerr (Ed.), *Propagation of Short Radio Waves*. New York: volume 13 of MIT Radiation Laboratory Series. McGraw-Hill.

Kerr, Y. H., P. Waldteufel, J. P. Wigneron, S. Delwart, F. Cabot, J. Boutin, … S. Mecklenburg (2010). The SMOS Mission: New Tool for Monitoring Key Elements of the Global Water Cycle. *Proceedings of the IEEE*, *98*(5), 666–687.

Kidder, S. Q. & A. S. Jones (2007). A Blended Satellite Total Precipitable Water Product for Operational Forecasting. *Journal of Atmospheric and Oceanic Technology*, *24*, 74–81.

Kikuchi, K., T. Nishibori, S. Ochiai, H. Ozeki, Y. Irimajiri, Y. Kasai, … M. Shiotani (2010). Overview and early results of the Superconducting Submillimeter- Wave Limb-Emission Sounder (SMILES). *Journal of Geophysical Research*, *115*(D23).

Kim, S., B. Kim, Y. Kong & Y. Kim (2000). Radar backscattering measurements of rice crop using X-band scatterometer. *IEEE Transactions on Geoscience and Remote Sensing*, *38*(3), 1467–1471.

Kim, Y. S. (1984). *Theoretical and experimental study of radar backscatter from sea ice*. Ph.D. Thesis, University of Kansas, Lawrence, KS.

Kim, Y. & J. J. van Zyl (2001). Comparison of forest parameter estimation techniques using SAR data. In *IEEE International Geoscience and Remote Sensing Symposium* (Vol. 3, pp. 1395–1397). 9–13 July, Sydney, Australia.

Kim, Y. & J. J. van Zyl (2009). A Time-Series Approach to Estimate Soil Moisture Using Polarimetric Radar Data. *IEEE Transactions on Geoscience and Remote Sensing*, *47*(8), 2519–2527.

Kim, Y., T. J. Jackson, R. Bindlish, H. Lee & S. Hong (2012). Radar Vegetation Index for Estimating the Vegetation Water Content of Rice and Soybean. *IEEE Geoscience and Remote Sensing Letters*, *9*(4), 564–568.

Kimura, H. (2009). Calibration of Polarimetric PALSAR Imagery Affected by Faraday Rotation Using Polarization Orientation. *IEEE Transactions on Geoscience and Remote Sensing*, *47*(12), 3943–3950.

Kindt, J. T. & C. A. Schmuttenmaer (1996). Far-infrared dielectric properties of polar liquids probed by femtosecond terahertz pulse spectroscopy. *Journal of Physical Chemistry*, *100*(24), 10373–10379.

King, A. P. (1950). The Radiation Characteristics of Conical Horn Antennas. *Proceedings of the IRE*, *38*(3), 249–251.

Kinsman, B. (1965). *Wind Waves*. Englewood Cliffs, NJ: Prentice-Hall.

Klein, J. D. & A. Freeman (1991). Quadpolarization SAR calibration using target reciprocity. *Journal of Electromagnetic Waves and Applications*, *5*(7), 735–751.

Klein, L. A. & C. T. Swift (1977). An Improved Model for the Dielectric Constant of Sea Water at Microwave Frequencies. *IEEE Journal of Oceanic Engineering*, *2*(1), 104–111.

Knott, E. F. (1993). *Radar Cross Section Measurements*. New York: Van Nostrand Reinhold.

Komjathy, A., J. Maslanik, V. U. Zavorotny, P. Axelrad & S. J. Katzberg (2000). Sea ice remote sensing using surface reflected GPS signals. In *IEEE International Geoscience and Remote Sensing Symposium* (Vol. 7, pp. 2855–2857). 24–28 July, Honolulu, HI.

Kong, J. A., R. Shin, J. C. Shiue & L. Tsang (1979). Theory and Experiment for Passive Microwave Remote Sensing of Snowpacks. *Journal of Geophysical Research*, *84*(B10), 5669–5673.

Kongoli, C., S.-A. Boukabara, B. Yan, F. Weng & R. Ferraro (2011). A New Sea-Ice Concentration Algorithm Based on Microwave Surface Emissivities—Application to AMSU Measurements. *IEEE Transactions on Geoscience and Remote Sensing*, *49*(1), 175–189.

Kostinski, A. B. & W. M. Boerner (1986). On foundations of radar polarimetry. *IEEE Transactions on Antennas and Propagation*, *34*(12), 1395–1403.

Kozu, T., T. Kawanishi, H. Kuroiwa, M. Kojima, K. Oikawa, H. Kumagai, … K. Nishikawa (2001). Development of precipitation radar onboard the Tropical Rainfall Measuring Mission (TRMM) satellite. *IEEE Transactions on Geoscience and Remote Sensing*, *39*(1), 102–116.

Kramer, M. A. (1991). Nonlinear principal component analysis using autoassociative neural networks. *AIChE Journal*, *37*(2), 233–243.

Kraszewski, A. (1978). A Model of the Dielectric Properties of Wheat at 9.4 GHz. *Journal of Microwave Power and Electromagnetic Energy*, *13*(4), 293–296.

Kraus, J. D. (1966). *Radio Astronomy* (2nd ed.). New York: McGraw-Hill.

Kraus, S. P., J. E. Estes, S. G. Atwater, J. R. Jensen & R. R. Vollmers (1977). Radar Detection of Surface Oil Slicks. *Photogrammetric Engineering & Remote Sensing*, *43*(12), 1523–1531.

Kseneman, M. & D. Gleich (2009). Soil Moisture Estimation with TerraSAR-X. In *16th International Conference on Systems, Signals and Image Processing* (pp. 1–4).

Kuga, Y., F. T. Ulaby, T. F. Haddock & R. D. De Roo (1991). Millimeter-wave radar scattering from snow: Part 1—Radiative transfer model. *Radio Science*, *26*(2), 329–341.

Kummerow, C. D., S. Ringerud, J. Crook, D. Randel & W. Berg (2011). An observationally generated a priori database for microwave rainfall retrievals. *Journal of Atmospheric and Oceanic Technology*, *28*(2), 113–130.

Kunkee, D. B, G. A. Poe, D. J. Boucher, S. D. Swadley, Y. Hong, J. E. Wessel & E. A. Uliana (2008). Design and Evaluation of the First Special Sensor Microwave Imager/Sounder (SSMI/S). *IEEE Transactions on Geoscience and Remote Sensing*, *46*(4), 863–883.

Kunz, L. B. & D. G. Long (2006). Melt Detection in Antarctic Ice Shelves Using Scatterometers and Microwave Radiometers. *IEEE Transactions on Geoscience and Remote Sensing*, *44*(9), 2461–2469.

Kuo, C. C., D. H. Staelin & P. W. Rosenkranz (1994). Statistical iterative scheme for estimating atmospheric relative humidity profiles. *IEEE Transactions on Geoscience and Remote Sensing*, *32*(2), 254–260.

Kuria, D., H. Lu, T. Koike, H. Tsutsui & T. Graf (2006). Multi-Frequency Microwave Response to Periodic Roughness. In *IEEE International Geoscience and Remote Sensing Symposium* (pp. 1744–1747). 31 July–4 August, Denver, CO.

Kursinski, E. R., G. A. Hajj, J. T. Schofield, R. P. Linfield & K. R. Hardy (1997). Observing the Earth's atmosphere with radio occultation measurements using the Global Positioning System. *Journal of Geophysical Research*, *102*(D19), 23429–23465.

Kurum, M., R. H. Lang, P. E. O'Neill, A. T. Joseph, T. J. Jackson & M. H. Cosh (2009). L-Band Radar Estimation of Forest Attenuation for Active / Passive Soil Moisture Inversion. *IEEE Transactions on Geoscience and Remote Sensing*, *47*(9), 3026–3040.

Kwok, R. (2004). Annual cycles of multiyear sea ice coverage of the Arctic Ocean: 1999–2003. *Journal of Geophysical Research*, *109*(C11), C11004.

Kwok, R. & G. F. Cunningham (2012). Deformation of the Arctic Ocean ice cover after the 2007 record minimum in summer ice extent. *Cold Regions Science and Technology*, *76-77*, 17–23.

Kwok, R., D. A. Rothrock, H. L. Stern & G. F. Cunningham (1995). Determination of the age distribution of sea ice from Lagrangian observations of ice motion. *IEEE Transactions on Geoscience and Remote Sensing*, *33*(2), 392–400.

Kwok, R., S. V. Nghiem, S. Martin, D. P. Winebrenner, A. J. Gow, D. K. Perovich, … E. J. Knapp (1998). Laboratory Measurements of Sea Ice: Connections to Microwave Remote Sensing. *IEEE Transactions on Geoscience and Remote Sensing*, *36*(5), 1716–1730.

Kwok, R., G. F. Cunningham & W. D. Hibler (2003). Sub-daily sea ice motion and deformation from RADARSAT observations. *Geophysical Research Letters*, *30*(23), 2218.

Kwok, R., G. F. Cunningham, M. Wensnahan, I. Rigor, H. J. Zwally & D. Yi (2009). Thinning and volume loss of the Arctic Ocean sea ice cover: 2003–2008. *Journal of Geophysical Research*, *114*(C7), C07005.

Lääperi, A. & E. Nyfors (1982). Microprocessor Controlled Microwave Radiometer System for Measuring the Thickness of an Oil Slick. In *IEEE International Geoscience and Remote Sensing Symposium* (Vol. 1). 1–4 June, Munich, West Germany.

Lagerloef, G. S. E., F. R. Colomn, D. M. Le Vine, F. J. Wentz, S. H. Yueh, C. S. Ruf, … C. T. Swift (2008). The Aquarius / SAC-D mission: Designed to meet the salinity remote-sensing challenge. *Oceanography*, *21*(1), 68–81.

Lahtinen, J., A. J. Gasiewski, M. Klein & I. S. Corbella (2003). A calibration method for fully polarimetric microwave radiometers. *IEEE Transactions on Geoscience and Remote Sensing*, *41*(3), 588–602.

Lambrigtsen, B. H., S. T. Brown, T. C. Gaier, L. Herrell, P. P. Kangaslahti & A. B. Tanner (2010). Monitoring the Hydrologic Cycle With the PATH Mission. *Proceedings of the IEEE*, *98*(5), 862–877.

Landau, L. D., E. M. Lifshitz, J. B. Sykes, J. S. Bell & M. J. Kearsley (1975). *Electrodynamics of Continuous Media*. New York: Pergamon Press.

Lane, J. A. & J. A. Saxton (1952). Dielectric Dispersion in Pure Polar Liquids at Very High Radio Frequencies. III. The Effect of Electrolytes in Solution. *Proceedings of the Royal Society of London. Series A, Mathematical and Physical Sciences*, *214*(1119), 531–545.

Lawrence, H., J. Wigneron, F. Demontoux, A. Mialon & Y. H. Kerr (2013). Evaluating the Semiempirical H – Q Model Used to Calculate the L-Band Emissivity of a Rough Bare Soil. *IEEE Transactions on Geoscience and Remote Sensing*, *51*(7), 4075–4084.

Laws, J. O. & D. A. Parsons (1943). The relation of raindrop size to intensity. *Transactions of the American Geophysical Union*, *24*, 452–460.

Le Vine, D. M. (1999). Synthetic aperture radiometer systems. *IEEE Transactions on Microwave Theory and Techniques*, *47*(12), 2228–2236.

Le Vine, D. M. & S. Abraham (2002). The effect of the ionosphere on remote sensing of sea surface salinity from space: absorption and emission at L band. *IEEE Transactions on Geoscience and Remote Sensing*, *40*(4), 771–782.

Le Vine, D. M., C. T. Swift & M. Haken (2001). Development of the synthetic aperture microwave radiometer, ESTAR. *IEEE Transactions on Geoscience and Remote Sensing*, *39*(1), 199–202.

Le Vine, D. M., G. S. E. Lagerloef & S. E. Torrusio (2010). Aquarius and Remote Sensing of Sea Surface Salinity from Space. *Proceedings of the IEEE*, *98*(5), 688–703.

Leberl, F., J. Raggam, C. Elachi & W. J. Campbell (1983). Sea ice motion measurements from SEASAT SAR images. *Journal of Geophysical Research*, *88*(C3), 1915.

Lee, J.-S. & E. Pottier (2009). *Polarimetric Radar Imaging: From Basics to Applications*. Boca Raton, FL: CRC Press.

Lee, K. K. (1981). *Emission from a Layer of Mie Scatterers. Electrical Engineering*. Thesis, Electrical Engineering, University of Kansas, Lawrence, KS, Lawrence, KS.

Lehner, S., J. Schulz-Stellenfleth, B. Schattler, H. Breit & J. Horstmann (2000). Wind and wave measurements using complex ERS-2 SAR wave mode data. *IEEE Transactions on Geoscience and Remote Sensing*, *38*(5), 2246.

Leifer, I., W. J. Lehr, D. Simecek-Beatty, E. Bradley, R. Clark, P. Dennison, … J. Wozencraft (2012). State of the art satellite and airborne marine oil spill remote sensing: Application to the BP Deepwater Horizon oil spill. *Remote Sensing of Environment*, *124*, 185–209.

Leitao, C. D., N. E. Huang & C. G. Perra (1978). *Final Report of GEOS-3 Ocean Current Investigation Using Radar Altimeter Profiling*. Technical Momorandum 73280, NASA Wallops Flight Center, Wallops Island, VA.

Leith, C. E. (1971). Atmospheric Predictability and Two-Dimensional Turbulence. *Journal of the Atmospheric Sciences*, *28*(2), 145–161.

LeMehaute, B. L. & T. Khangaonkar (1990). Dynamic interaction of intense rain with water waves. *Journal of Physical Oceanography*, *20*(12), 1805–1812.

Lerner, R. M. & J. P. Hollinger (1977). Analysis of 1.4 GHz Radiometric measurements from Skylab. *Remote Sensing of Environment*, *6*(4), 251–269.

Leslie, R. V. & D. H. Staelin (2004). NPOESS Aircraft Sounder Testbed-Microwave: Observations of Clouds and Precipitation at 54, 118, 183, and 425 GHz. *IEEE Transactions on Geoscience and Remote Sensing*, *42*(10), 2240–2247.

Li, J. & P. Stoica (2009). *MIMO Radar Signal Processing*. Hoboken, NJ: Wiley & Sons.

Li, L., P. Gaiser, M. R. Albert, D. G. Long & E. M. Twarog (2008). WindSat Passive Microwave Polarimetric Signatures of the Greenland Ice Sheet. *IEEE Transactions on Geoscience and Remote Sensing*, *46*(9), 2622–2631.

Li, Z., E. J. Fielding & P. Cross (2009). Integration of InSAR Time-Series Analysis and Water-Vapor Correction for Mapping Postseismic Motion After the 2003 Bam (Iran) Earthquake. *IEEE Transactions on Geoscience and Remote Sensing*, *47*(9), 3220–3230.

Liebe, H. J. (1981). Modeling attenuation and phase of radio waves in air at frequencies below 1000 GHz. *Radio Science*, *16*(6), 1183–1199.

Liebe, H. J. (1985). An updated model for millimeter wave propagation in moist air. *Radio Science*, *20*(5), 1069–1089.

Liebe, H. J., G. Hufford & M. Cotton (1993). Propagation modeling of moist air and suspended water/ice particles at frequencies below 1000 GHz (Presented at the Electromagnetic Wave Propagation Panel Symposium, Palma de Mallorca, Spain, 17–20 Mat 1993). *AGARD Conference Proceedings 542*.

Lin, C.-C., M. Betto, M. Belmonte Rivas, A. Stoffelen & J. de Kloe (2012). EPS-SG Windscatterometer Concept Tradeoffs and Wind Retrieval Performance Assessment. *IEEE Transactions on Geoscience and Remote Sensing*, *50*(7), 2458–2472.

Lin, Y. C. & K. Sarabandi (1999a). Retrieval of forest parameters using a fractal-based coherent scattering model and a genetic algorithm. *IEEE Transactions on Geoscience and Remote Sensing*, *37*(3), 1415–1424.

Lin, Y. C. & K. Sarabandi (1999b). A Monte Carlo coherent scattering model for forest canopies using fractal-generated trees. *IEEE Transactions on Geoscience and Remote Sensing*, *37*(1), 440–451.

Lindsley, R. D. & D. G. Long (2012). Mapping Surface Oil Extent From the Deepwater Horizon Oil Spill Using ASCAT Backscatter. *IEEE Transactions on Geoscience and Remote Sensing*, *50*(7), 2534–2541.

Lipes, R. G. (1982). Description of SEASAT radiometer status and results. *Journal of Geophysical Research*, *87*(C5), 3385–3395.

Liu, A. K., S. Martin & R. Kwok (1997). Tracking of Ice Edges and Ice Floes by Wavelet Analysis of SAR Images. *Journal of Atmospheric and Oceanic Technology*, *14*(5), 1187–1198.

Liu, Q. & F. Weng (2005). One-dimensional variational retrieval algorithm of temperature, water vapor, and cloud water profiles from advanced microwave sounding unit (AMSU). *IEEE Transactions on Geoscience and Remote Sensing*, *43*(5), 1087–1095.

Liu, W. T. (2002). Progress in Scatterometer Application. *Journal of Oceanography*, *58*(1), 121–136.

Liu, W. T., K. B. Katsaros & J. A. Businger (1979). Bulk Parameterization of Air-Sea Exchanges of Heat and Water Vapor Including the Molecular Constraints at the Interface. *Journal of the Atmospheric Sciences*, *36*(9), 1722–1735.

Livesey, N. J., M. L. Santee, P. C. Stek, J. W. Waters, P. Levelt, P. Veefkind, … A. Roche (2008). LEOMAC: A future Global Atmospheric Composition Mission (CACM) concept. In *IEEE Aerospace Conference* (pp. 1–13).

Long, D. G. (2003). Reconstruction and Resolution Enhancement Techniques for Microwave Sensors. In C. H. Chen (Ed.), *Frontiers of Remote Sensing Information Processing*. World Scientific.

Long, D. G. & J. M. Mendel (1991). Identifiability in wind estimation from scatterometer measurements. *IEEE Transactions on Geoscience and Remote Sensing*, *29*(2), 268–276.

Long, D. G. & G. B. Skouson (1996). Calibration of spaceborne scatterometers using tropical rain forests. *IEEE Transactions on Geoscience and Remote Sensing*, *34*(2), 413–424.

Long, D. G. & M. W. Spencer (1997). Radar backscatter measurement accuracy for a spaceborne pencil-beam wind scatterometer with transmit modulation. *IEEE Transactions on Geoscience and Remote Sensing*, *35*(1), 102–114.

Long, D. G., C.-Y. Chi & F. K. Li (1988). The design of an onboard digital Doppler processor for a spaceborne scatterometer. *IEEE Transactions on Geoscience and Remote Sensing*, *26*(6), 869–878.

Long, D. G., R. Milliff & E. Rodríguez (2009). The TropSat mission. An observatory for mesoscale convective system processes in the global tropics. In *IEEE Radar Conference*. Pasadena, CA.

Long, M. W. (1975). *Radar Reflectivity of Land and Sea*. Lexington, MA: Lexington Books.

Longuet-Higgins, M. S., D. E. Cartwright & N. D. Smith (1963). Observations of the directional spectrum of sea waves using the motions of a floating buoy. In *Ocean Wave Spectra* (pp. 111–136). Easton, MD.

Lopes, A. (1983). *Etude expérimental et théorique de l'atténuation et de la rétrodiffusion des micro-ondes par un couvert de blé. Application a la télédétection (An Experimental and Theoretical Study of Microwave*

Attenuation and Backscattering from a Wheat Canopy). Thesis, Sabatier University, Toulouse, France.

Lowe, S. T., J. L. LaBrecque, C. Zuffada, L. J. Romans, L. E. Young & G. A. Hajj (2002). First spaceborne observation of an Earth-reflected GPS signal. *Radio Science*, *37*(1), 7-1–7-28.

Luneburg, R. K. (1964). *Mathematical Theory of Optics*. Berkeley & Los Angeles, CA: University of California Press.

Lungu, T. (2006). *QuikSCAT Science Data Product Users Manual Overview and Geophysical Data Products*. Pasadena, CA: NASA JPL, Caltech.

Lutz, R., T. T. Wilheit, J. R. Wang & R. K. Kakar (1991). Retrieval of atmospheric water vapor profiles using radiometric measurements at 183 and 90 GHz. *IEEE Transactions on Geoscience and Remote Sensing*, *29*(4), 602–609.

Lyons, S. & D. T. Sandwell (2003). Fault creep along the southern San Andreas from interferometric synthetic aperture radar, permanent scatterers, and stacking. *Journal of Geophysical Research*, *108*(B1), 2047.

MacDonald, F. C. (1956). The Correlation of Radar Sea Clutter on Vertical and Horizontal Polarizations with Wave Height and Slope. *IRE Convention Record, Part 1*, 29–32.

MacDonald, H. C. & W. Waite (1978). *Significance of Dual-Polarized Long Wavelength Radar for Terrain Analysis*. Contract No. 954940, JPL, University of Arkansas, Fayetteville, AR.

Macelloni, G., S. Paloscia, P. Pampaloni, F. Marliani & M. Gai (2001). The relationship between the backscattering coefficient and the biomass of narrow and broad leaf crops. *IEEE Transactions on Geoscience and Remote Sensing*, *39*(4), 873–884.

Machin, K. E., M. Ryle & D. D. Vonberg (1952). The Design of an Equipment for Measuring Small Radio-Frequency Noise Powers. *Proceedings of the Institution of Electrical Engineers (IEE)*, *99*, 127–134.

Macmillan, S. & S. Maus (2005). International Geomagnetic Reference Field — the tenth generation. *Earth, Planets and Space*, *57*(12), 1135–1140.

Maity, S., C. Patnaik, M. Chakraborty & S. Panigrahy (2004). Analysis of Temporal Backscattering of Cotton Crops Using a Semiempirical Model. *IEEE Transactions on Geoscience and Remote Sensing*, *42*(3), 577–587.

Manninen, A. T. & L. M. H. Ulander (2001). Forestry parameter retrieval from texture in CARABAS VHF-band SAR images. *IEEE Transactions on Geoscience and Remote Sensing*, *39*(12), 2622–2633.

Marconi, S. G. (1922). Radio Telegraphy. *Proceedings of the IRE*, *10*, 215–238.

Markus, T. & S. T. Dokken (2002). Evaluation of late summer passive microwave Arctic sea ice retrievals. *IEEE Transactions on Geoscience and Remote Sensing*, *40*(2), 348–356.

Marliani, F., S. Paloscia, P. Pampaloni & J. A. Kong (2002). Simulating coherent backscattering from crops during the growing cycle. *IEEE Transactions on Geoscience and Remote Sensing*, *40*(1), 162–177.

Marshall, J. S. & W. M. Palmer (1948). The Distribution of Raindrops with Size. *Journal of Meteorology*, *5*(4), 165–166.

Martin, T. V., H. J. Zwally, A. C. Brenner & R. A. Bindschadler (1983). Analysis and Retracking of Continental Ice Sheet Radar Altimeter Waveforms. *Journal of Geophysical Research*, *88*(C3), 1608–1616.

Martin-Neira, M. (1993). A Passive Reflectometry and Interferometry System (PARIS): Application to ocean altimetry. *ESA Journal*, *17*, 331–355.

Massonnet, D. & K. L. Feigl (1998). Radar interferometry and its application to changes in the Earth's surface. *Reviews of Geophysics*, *36*(4), 441.

Massonnet, D, M. Rossi, C. Carmona, F. Adragna, G. Peltzer, K. Feigl & T. Rabaute (1993). The displacement field of the Landers earthquake mapped by radar interferometry. *Nature*, *364*(6433), 138–142.

Masters, D., P. Axelrad & S. J. Katzberg (2004). Initial results of land-reflected GPS bistatic radar measurements in SMEX02. *Remote Sensing of Environment*, *92*(4), 507–520.

Masuda, A. & Y.-Y. Kuo (1981). A note on the imaginary part of bispectra. *Deep Sea Research Part A. Oceanographic Research Papers*, *28*(3), 213–222.

Masuko, H., K. Okamoto, M. Shimada & S. Niwa (1986). Measurement of Microwave Backscattering Signatures of the Ocean Surface Using X-Band and Ka-Band Airborne Scatterometers. *Journal of Geophysical Research*, *91*(C11), 13065–13083.

Mathur, N. C., M. D. Grossi & M. R. Pearlman (1970). Atmospheric Effects in Very Long Baseline Interferometry. *Radio Science*, *5*(10), 1253–1261.

Matsuo, S. (1938). A Direct-Reading Radio-Wave-Reflection-Type Absolute Altimeter for Aeronautics. *Proceedings of the IRE*, *26*(7), 848–858.

Matsuoka, T., S. Fujita & S. Mae (1996). Effect of temperature on dielectric properties of ice in the range 5–39 GHz. *Journal of Applied Physics*, *80*(10), 5884.

Mattioli, V., S. Bonafoni, P. Basili, G. Carlesimo, P. Ciotti, L. Pulvirenti & N. Pierdicca (2010). Neural Network for the Satellite Retrieval of Precipitable Water Vapor over Land. In *IEEE International Geoscience and Remote Sensing Symposium* (pp. 2969–2963). 25–30 July, Honolulu, HI.

Mätzler, C. (1994). Passive microwave signatures of landscapes in winter. *Meteorology and Atmospheric Physics*, *54*(1-4), 241–260.

Mätzler, C. (1998). Microwave properties of ice and snow. In *Solar System Ices* (pp. 241–257). Dordrecht, Netherlands: Kluwer Academic.

Mätzler, C. (2006). *Thermal Microwave Radiation: Applications for Remote Sensing*. Stevenage, UK: Institution of Engineering and Technology.

Mätzler, C. & E. Schanda (1984). Snow mapping with active microwave sensors. *International Journal of Remote Sensing*, *5*(2), 409–422.

Mätzler, C. & U. Wegmüller (1987). Dielectric properties of fresh-water ice at microwave frequencies. *Journal of Physics D Applied Physics*, *20*, 1623–1630.

Mätzler, C., E. Schanda & W. Good (1982). Towards the Definition of Optimum Sensor Specifications for Microwave Remote Sensing of Snow. *IEEE Transactions on Geoscience and Remote Sensing*, *20*(1), 57–66.

Maurer, A. T., A. T. Edgerton & D. C. Meeks (1977). U.S. Coast Guard Airborne Oil Surveillance Systems Status Report. In *11th International Symposium on Remote Sensing of Environment* (pp. 1639–1640). 22-29 April, Ann Arbor, MI.

McDonald, K. C. & F. T. Ulaby (1993). Radiative Transfer Modeling of Discontinuous Tree Canopies at Microwave Frequencies. *International Journal of Remote Sensing*, *14*(11), 1097–2128.

McDonald, K. C., M. C. Dobson & F. T. Ulaby (1991). Modeling multi-frequency Diurnal backscatter from a walnut orchard. *IEEE Transactions on Geoscience and Remote Sensing*, *29*(6), 852–863.

McGillem, C. D. & T. V. Seling (1963). Influence of System Parameters on Airborne Microwave Radiometer Design. *IEEE Transactions on Military Electronics*, *7*(4), 296–302.

McMullan, K. D., M. A. Brown, M. Martín-Neira, W. Rits, S. Ekholm, J. Marti & J. Lemanczyk (2008). SMOS: The Payload. *IEEE Transactions on Geoscience and Remote Sensing*, *46*(3), 594–605.

Mears, C. A., F. J. Wentz & D. K. Smith (2000). *SeaWinds on QuikSCAT normalized objective function rain flag*. Product Description Verson 1.2, Remote Sensing Systems, Santa Rosa, CA.

Mecklenburg, S., M. Drusch, Y. H. Kerr, J. Font, M. Martín-Neira, S. Delwart, ... R. Crapolicchio (2012). ESA's Soil Moisture and Ocean Salinity Mission: Mission Performance and Operations. *IEEE Transactions on Geoscience and Remote Sensing*, *50*(5), 1354–1366.

Meeks, M. L. & A. E. Lilley (1963). The microwave spectrum of oxygen in the Earth's atmosphere. *Journal of Geophysical Research*, *68*(6), 1683–1703.

Meier, W. N. & J. Stroeve (2008). Comparison of sea-ice extent and ice-edge location estimates from passive microwave and enhanced-resolution scatterometer data. *Annals of Glaciology*, *48*(1), 65–70.

Meischner, P. (2004). *Weather Radar: Principles and Advanced Applications*. Berlin; New York: Springer.

Meissner, T. & F. J. Wentz (2002). An updated analysis of the ocean surface wind direction signal in passive microwave brightness temperatures. *IEEE Transactions on Geoscience and Remote Sensing*, *40*(6), 1230–1240.

Meissner, T. & F. J. Wentz (2004). The complex dielectric constant of pure water and sea water from microwave satellite observations. *IEEE Transactions on Geoscience and Remote Sensing*, *49*(9), 1836–1849.

Meissner, T. & F. J. Wentz (2006). Ocean Retrievals for WindSat. In *IEEE MicroRad* (pp. 119–124).

Melbourne, W. G. (2004). *Radio Occultations Using Earth Satellites: A Wave Theory Treatment* (J. H. Yuen, Ed.). Pasadena, CA: Monograph 6, Deep Space Communications and Navigation Series, Deep Space Communications and Navigation Systems, Center of Excellence, NASA JPL, Caltech.

Melon, P., J. M. Martinez, T. Le Toan, L. M. H. Ulander & A. Beaudoin (2001). On the retrieving of forest stem volume from VHF SAR data: observation and modeling. *IEEE Transactions on Geoscience and Remote Sensing*, *39*(11), 2364–2372.

Melsheimer, C., W. Alpers & M. Gade (2001). Simultaneous observations of rain cells over the ocean by the synthetic aperture radar aboard the ERS satellites and by surface-based weather radars. *Journal of Geophysical Research*, *106*(C3), 4665–4677.

Mendel, J. M. (1991). Tutorial on higher-order statistics (spectra) in signal-processing and system-theory: Theoretical results and some applications. *Proceedings of the IEEE*, *79*(3), 278–305.

Meneghini, R. & T. Kozu (1990). *Spaceborne weather radar*. Boston, MA: Artech House.

Merzouki, A., H. McNairn & A. Pacheco (2011). Mapping Soil Moisture Using RADARSAT-2 Data and Local Autocorrelation Statistics. *IEEE Journal of Selected Topics in Applied Earth Observations and Remote Sensing*, *4*(1), 128–137.

Meyer, F. J., & J. Nicoll. (2008). The Impact of the Ionosphere on Interferometric SAR Processing. In *IEEE International Geoscience and Remote Sensing Symposium* (Vol. 2, pp. 391–394). 8-11 July, Boston, MA.

Mie, G. (1908). Beiträge zur Optik trüber Medien, speziell kolloidaler Metallösungen. *Annalen der Physik*, *330*(3), 377–445.

Migliaccio, M., A. Gambardella & M. Tranfaglia (2007). SAR Polarimetry to Observe Oil Spills. *IEEE Transactions on Geoscience and Remote Sensing*, *45*(2), 506–511.

Migliaccio, M., F. Nunziata, A. Montuori & W. G. Pichel (2011). A Multifrequency Polarimetric SAR Processing Chain to Observe Oil Fields in the Gulf of Mexico. *IEEE Transactions on Geoscience and Remote Sensing*, *49*(12), 4729–4737.

Miller, C. K. S., W. C. Daywitt & M. G. Arthur (1967). Noise standards, measurements, and receiver noise definitions. *Proceedings of the IEEE*, *55*(6), 865–877.

Milman, A. S. & T. T. Wilheit (1983). Sea surface temperatures from the Nimbus-7 scanning multichannel microwave radiometer. In *1st Workshop on Satellite-Derived Sea Surface Temperature*. Pasadena, CA, January 27-28, 1983: NASA JPL, Caltech, May 1, 1983, A1-A20 (JPL Publication 83-34).

Milman, A. S. & T. T. Wilheit (1985). Sea surface temperatures from the scanning multichannel microwave radiometer on Nimbus 7. *Journal of Geophysical Research*, *90*(C6), 11631.

Mo, T., B. J. Choudhury, T. J. Schmugge, J. R. Wang & T. J. Jackson (1982). A Model for Microwave Emission From Vegetation-Covered Fields. *Journal of Geophysical Research*, *87*(C13), 11229–11237.

Moncrief, F. J. (1980). Side-Looking Radar Will Spot Oil Slicks. *Microwaves*, *19*, 25–26.

Moore, R. K. & J. D. Young (1977). Active microwave measurement from space of sea-surface winds. *IEEE Journal of Oceanic Engineering*, *2*(4), 309–317.

Moran, J. M. & B. R. Rosen (1981). Estimation of the propagation delay through the troposphere from microwave radiometer data. *Radio Science*, *16*(2), 235–244.

Moreira, A., P. Prats-Iraola, M. Younis, G. Krieger, I. Hajnsek & K. P. Papathanassiou (2013). A Tutorial on Synthetic Aperture Radar. *IEEE Geoscience and Remote Sensing Magazine*, 6–43.

Moreira, Alberto & R. Bamler (2010). Foreword to the Special Issue on TerraSAR-X: Mission, Calibration, and First Results. *IEEE Transactions on Geoscience and Remote Sensing*, *48*(2), 603–604.

Mortin, J., T. M. Schrøder, A. W. Hansen, B. Holt & K. C. McDonald (2012). Mapping of seasonal freeze-thaw transitions across the pan-Arctic land and sea ice domains with satellite radar. *Journal of Geophysical Research*, *117*(C8), C08004.

Mott, H. (2007). *Remote Sensing with Polarimetric Radar*. Los Alamitos, CA; Hoboken, NJ: IEEE Press; Wiley-Interscience.

Naderi, F. M., M. H. Freilich & D. G. Long (1991). Spaceborne radar measurement of wind velocity over the ocean-an overview of the NSCAT scatterometer system. *Proceedings of the IEEE*, *79*(6), 850–866.

Napier, P. J., A. R. Thompson & R. D. Ekers (1983). The very large array: Design and performance of a modern synthesis radio telescope. *Proceedings of the IEEE*, *71*(11), 1295–1320.

Narayan, U., V. Lakshmi & T. J. Jackson (2006). High-Resolution Change Estimation of Soil Moisture Using L-Band Radiometer and Radar Observations Made During the SMEX02 Experiments. *IEEE Transactions on Geoscience and Remote Sensing*, *44*(6), 1545–1554.

Narvekar, P. S., T. J. Jackson, R. Bindlish, L. Li, G. Heygster & P. Gaiser (2007). Observations of land surface passive polarimetry with the WindSat instrument. *IEEE Transactions on Geoscience and Remote Sensing*, *45*(7), 2019–2028.

NASA (1972). *Nimbus 5 User's Guide*. Greenbelt, MD: The Electrically Scanning Microwave Radiometer (ESMR) Experiment, NASA Goddard Space Flight Center.

Nash, R. T. (1964). Beam Efficiency Limitations of Large Antennas. *IEEE Transactions on Military Electronics*, *8*(3), 252–257.

Nashashibi, A. Y. & F. T. Ulaby (2007). MMW Polarimetric Radar Bistatic Scattering From a Random Surface. *IEEE Transactions on Geoscience and Remote Sensing*, *45*(6), 1743–1755.

Nelson, S. O. (1973). Microwave Dielectric Properties of Grain and Seed. *Transactions of the American Society of Agricultural Engineers*, *16*(5), 902–905.

Nelson, S. O. (1976). Microwave Dielectric Properties of Insects and Grain Kernels. *Journal of Microwave Power and Electromagnetic Energy*, *11*(4), 299–304.

Nelson, S. O. (1978). Electrical Properties of Grain and Other Food Materials. *Journal of Food Processing and Preservation*, *2*(2), 137–154.

Nelson, S. O. (1979). RF and Microwave Dielectric Properties of Shelled, Yellow-Dent Field Corn. *Transactions of the American Society of Agricultural Engineers*, *22*(6), 1451–1457.

Nelson, S. O. & L. E. Stetson (1976). Frequency and moisture dependence of the dielectric properties of hard red winter wheat. *Journal of Agricultural Engineering Research*, *21*(2), 181–192.

Newton, R. W. & J. W. Rouse (1980). Microwave Radiometer Measurements of Soil Moisture Content. *IEEE Transactions on Antennas and Propagation*, *28*(5), 680–686.

Nghiem, S. V., R. Kwok, S. H. Yueh, J. A. Kong, C. C. Hsu, M. A. Tassoudji & R. T. Shin (1995). Polarimetric scattering from layered media with multiple species of scatterers. *Radio Science*, *30*(4), 835–852.

Nghiem, S. V., K. Steffen, R. Kwok & W.-Y. Tsai (2001). Detection of snowmelt regions on the Greenland ice sheet using diurnal backscatter change. *Journal of Glaciology*, *47*(159), 539–547.

Nghiem, S. V., Y. Chao, G. Neumann, P. Li, D. K. Perovich, T. Street & P. Clemente-Colón (2006). Depletion of perennial sea ice in the East Arctic Ocean. *Geophysical Research Letters*, *33*(17), L17501.

Nicolson, A. M. & G. F. Ross (1970). Measurement of the Intrinsic Properties of Materials by Time-Domain Techniques. *IEEE Transactions on Instrumentation and Measurements*, *19*(4), 377–382.

Nie, C. & D. G. Long (2006). The Effect of Rain on ERS Scatterometer Measurements. In *IEEE International Geoscience and Remote Sensing Symposium* (pp. 4119–4121). 31 July–4 August, Denver, CO.

Nie, C. & D. G. Long (2007). A C-Band Wind/Rain Backscatter Model. *IEEE Transactions on Geoscience and Remote Sensing*, *45*(3), 621–631.

Nie, C. & D. G. Long (2008). A C-band scatterometer simultaneous wind/rain retrieval method. *IEEE Transactions on Geoscience and Remote Sensing*, *46*(11), 3618–3631.

Nielsen, S. N. & D. G. Long (2009). A Wind and Rain Backscatter Model Derived From AMSR and SeaWinds Data. *IEEE Transactions on Geoscience and Remote Sensing*, *47*(6), 1595–1606.

Njoku, E. G. & J. A. Kong (1977). Theory for passive microwave remote sensing of near-surface soil moisture. *Journal of Geophysical Research*, *82*(20), 3108–3118.

Njoku, E. G., W. J. Wilson, S. H. Yueh, S. J. Dinardo, F. K. Li, T. J. Jackson, … J. D. Bolten (2002). Observations of soil moisture using a passive and active low-frequency microwave airborne sensor during SGP99. *IEEE Transactions on Geoscience and Remote Sensing*, *40*(12), 2659–2673.

Nyfors, E. (1982). *On the Dielectric Properties of Dry Snow in the 800 MHz to 13 GHz Range*. Report S-135, Radio Laboratory, Helsinki University of Technology, Espoo, Finland.

Nyquist, H. (1928). Thermal Agitation of Electric Charge in Conductors. *Physical Review*, *32*(1), 110–113.

Oguchi, T. (1973). Attenuation and phase rotation of radio waves due to rain: Calculations at 19.3 and 34.8 GHz. *Radio Science*, *8*(1), 31–38.

Oh, Y. (2004). Quantitative Retrieval of Soil Moisture Content and Surface Roughness From Multipolarized Radar Observations of Bare Soil Surfaces. *IEEE Transactions on Geoscience and Remote Sensing*, *42*(3), 596–601.

Oh, Y., K. Sarabandi & F. T. Ulaby (1992). An empirical model and an inversion technique for radar scattering from bare soil surfaces. *IEEE Transactions on Geoscience and Remote Sensing*, *30*(2), 370–381.

Oh, Y., K. Sarabandi & F. T. Ulaby (2002). Semi-Empirical Model of the Ensemble-Averaged Differential Mueller Matrix for Microwave Backscattering From Bare Soil Surfaces. *IEEE Transactions on Geoscience and Remote Sensing*, *40*(6), 1348–1355.

Oh, Y., S. Hong, Y. Kim, J. Hong & Y. Kim (2009). Polarimetric Backscattering Coefficients of Flooded Rice Fields at L- and C-Bands: Measurements, Modeling, and Data Analysis. *IEEE Transactions on Geoscience and Remote Sensing*, *47*(8), 2714–2721.

Ohm, E. A. & W. W. Snell (1963). A Radiometer for a Space Communications Receiver. *Bell System Technical Journal*, *42*(5), 2047–2080.

Oliver, C. & S. Quegan (2004). *Understanding Synthetic Aperture Radar Images*. Raleigh, NC: SciTech.

Olsen, R., D. V. Rogers & D. B. Hodge (1978). The aR^b relation in the calculation of rain attenuation. *Antennas and Propagation, IEEE Transactions on*, *26*(2), 318–329.

Onstott, R. G. (1990). *Polarimetric Radar Measurements of Artificial Sea Ice During CRRELEX'88*. Technical Report 196100-23-T, Environmental Research Institute of Michigan.

Onstott, R. G. (1992). SAR and scatterometer signatures of sea ice. In F. D. Carsey (Ed.), *Microwave Remote Sensing of Sea Ice (Geophysical Monograph 68)* (pp. 73–104). Washington, DC: American Geophysical Union.

Onstott, R. G., R. K. Moore, S. P. Gogineni & C. Delker (1982). Four years of low-altitude sea ice broad-band backscatter measurements. *IEEE Journal of Oceanic Engineering*, *7*(1), 44–50.

Orhaug, T. & W. Waltman (1962). A Switched Load Radiometer. *Publications of the National Radio Astronomy Observatory*, *1*, 179–204.

Otoshi, T. Y. (1968). The Effect of Mismatched Components on Microwave Noise-Temperature Calibrations. *IEEE Transactions on Microwave Theory and Techniques*, *16*(9), 675–686.

Owe, M. & A. A. van de Griend (2001). On the relationship between thermodynamic surface temperature and high-frequency (37 GHz) vertically polarized brightness temperature under semi-arid conditions. *International Journal of Remote Sensing*, *22*(17), 3522–3532.

Owen, M. P. & D. G. Long (2009). Land-Contamination Compensation for QuikSCAT Near-Coastal Wind Retrieval. *IEEE Transactions on Geoscience and Remote Sensing*, *47*(3), 839–850.

Owen, M. P. & D. G. Long (2011). Simultaneous Wind and Rain Estimation for QuikSCAT at Ultra-High Resolution. *IEEE Transactions on Geoscience and Remote Sensing*, *49*(6), 1865–1878.

Page, D. F. & R. O. Ramseier (1975). Application of radar techniques to ice and snow studies. *Journal of Glaciology*, *15*(73), 171–191.

Pampaloni, P. & S. Paloscia (1986). Microwave Emission and Plant Water Content: A Comparison Between Field Measurements and Theory. *IEEE Transactions on Geoscience and Remote Sensing*, *24*(6), 900–905.

Pampaloni, P. & S. Paloscia (2000). *Microwave Radiometry and Remote Sensing of the Earth's Surface and Atmosphere*. Zeist, Netherlands: VSP BV.

Pandey, P. C. & R. K. Kakar (1983). Selection of optimum frequencies for atmospheric electric path length measurement by satellite-borne microwave radiometers. *IEEE Transactions on Antennas and Propagation*, *31*(1), 136–140.

Papoulis, A. (1965). *Probability, Random Variables, and Stochastic Processes*. New York: McGraw-Hill.

Parke, M. E., R. H. Stewart, D. L. Farless & D. E. Cartwright (1987). On the choice of orbits for an altimetric satellite to study ocean circulation and tides. *Journal of Geophysical Research*, *92*(C11), 11693.

Parmar, R. M., R. K. Arora, M. Venkata Rao & K. Thyagarajan (2006). OCEANSAT 2: mission and its applications. *Proceedings of SPIE*, *6407*(1).

Patoux, J., R. C. Foster & R. A. Brown (2003). Global pressure fields from scatterometer winds. *Journal of Applied Meteorology*, *42*(6), 813–826.

Payne, V. H., J. S. Delamere, K. E. Cady-Pereira, R. R. Gamache, J.-L. Moncet, E. J. Mlawer & S. A. Clough (2008). Air-Broadened Half-Widths of the 22- and 183-GHz Water-Vapor Lines. *IEEE Transactions on Geoscience and Remote Sensing*, *46*(11), 3601–3617.

Pazmany, A. L. (2007). A Compact 183-GHz Radiometer for Water Vapor and Liquid Water Sensing. *IEEE Transactions on Geoscience and Remote Sensing*, *45*(7), 2202–2206.

Pedlosky, J. (1979). Finite amplitude baroclinic waves in a continuous model of the atmosphere. *Journal of the Atmospheric Sciences*, *36*(10), 1908–1924.

Peltzer, G., F. Crampé & P. A. Rosen (2001). The Mw 7.1, Hector Mine, California earthquake: surface rupture, surface displacement field, and fault slip solution from ERS SAR data. *Comptes Rendus de l'Academie des Sciences Series IIA Earth and Planetary Science*, *333*(9), 545–555.

Peng, J. & C. S. Ruf (2008). Calibration Method for Fully Polarimetric Microwave Radiometers Using the Correlated Noise Calibration Standard. *IEEE Transactions on Geoscience and Remote Sensing*, *46*(10), 3087–3097.

Pepe, A., E. Sansosti, P. Berardino & R. Lanari (2005). On the Generation of ERS / ENVISAT DInSAR Time-Series Via the SBAS Technique. *IEEE Geoscience and Remote Sensing Letters*, *2*(3), 265–269.

Pepe, A., P. Berardino, M. Bonano, L. D. Euillades, R. Lanari & E. Sansosti (2011). SBAS-Based Satellite Orbit Correction for the Generation of DInSAR Time-Series: Application to RADARSAT-1 Data. *IEEE Transactions on Geoscience and Remote Sensing*, *49*(12), 5150–5165.

Peplinski, N. R., F. T. Ulaby & M. C. Dobson (1995a). Dielectric Properties of Soils in the 0.3–1.3 GHz Range. *IEEE Transactions on Geoscience and Remote Sensing*, *33*(3), 803–807.

Peplinski, N. R., F. T. Ulaby & M. C. Dobson (1995b). Corrections to "Dielectric Properties of Soils in the 0.3–1.3-GHz Range." *IEEE Transactions on Geoscience and Remote Sensing*, *33*(6), 1340.

Perry, J. W. & A. W. Straiton (1972). Dielectric Constant of Ice at 35.3 and 94.5 GHz. *Journal of Applied Physics*, *43*(2), 731–733.

Pettengill, G. H., D. F. Horwood & C. H. Keller (1980). Pioneer Venus Orbiter Radar Mapper: Design And Operation. *IEEE Transactions on Geoscience and Remote Sensing*, *18*(1), 28–32.

Petty, G. W. (1994). Physical retrievals of over-ocean rain rate from multichannel microwave imagery. Part I: Theoretical characteristics of normalized polarization and scattering indices. *Meteorology and Atmospheric Physics*, *54*(1-4), 101–121.

Phillips, D. L. (1962). A Technique for the Numerical Solution of Certain Integral Equations of the First Kind. *Journal of the ACM*, *9*(1), 84–97.

Phillips, O. M. (1985). Spectral and statistical properties of the equilibrium range in wind-generated gravity waves. *Journal of Fluid Mechanics*, *156*, 505–531.

Piepmeier, J. R. & A. J. Gasiewski (2001a). Digital Correlation Microwave Polarimetry: Analysis and Demonstration. *IEEE Transactions on Geoscience and Remote Sensing*, *39*(11), 2392–2410.

Piepmeier, J. R. & A. J. Gasiewski (2001b). High-Resolution Passive Polarimetric Microwave Mapping of Ocean Surface Wind Vector Fields. *IEEE Transactions on Geoscience and Remote Sensing*, *39*(3), 606–622.

Pierson, W. J. (1989). Probabilities and statistics for backscatter estimates obtained by a scatterometer. *Journal of Geophysical Research*, *94*(C7), 9743–9759.

Pierson, W. J. & L. Moskowitz (1964). A proposed spectral form for fully developed wind seas based on the similarity theory of S. A. Kitaigorodskii. *Journal of Geophysical Research*, *69*(24), 5181–5190.

Plant, W. J. (1986). Two-scale model of short, wind-generated waves and scatterometry. *Journal of Geophysical Research*, *91*(C9), 10735–10749.

Plant, W. J. (2002). A stochastic, multiscale model of microwave backscatter from the ocean. *Journal of Geophysical Research*, *107*(C9), 3120.

Poe, G. A., A. P. Stogryn & A. T. Edgerton (1972). *A Study of the Microwave Emission Characteristics of Sea Ice*. Final Technical Report 1749R-2, Contract No. 2-35340, Aerojet Electrosystems, Azusa, CA.

Polder, D. & J. H. van Santen (1946). The effective permeability of mixtures of solids. *Physica*, *12*(5), 257–271.

Pollard, B. D., E. Rodríguez, L. Veilleux, T. Akins, P. Brown, A. Kitiyakara, ... A. Prata (2002). The wide swath ocean altimeter: radar interferometry for global ocean mapping with centimetric accuracy. In *IEEE Aerospace Conference* (Vol. 2, pp. 1007–1020). Big Sky, MT.

Portabella, M. & A. Stoffelen (2001). Rain Detection and Quality Control of SeaWinds. *Journal of Atmospheric and Oceanic Technology*, *18*(7), 1171–1183.

Portabella, M. & A. Stoffelen (2002). Characterization of residual information for SeaWinds quality control. *IEEE Transactions on Geoscience and Remote Sensing*, *40*(12), 2747–2759.

Portabella, M. & A. Stoffelen (2004). A probabilistic approach for SeaWinds data assimilation. *Quarterly Journal of the Royal Meteorological Society*, *130*(596), 127–152.

Portabella, M. & A. Stoffelen (2006). Scatterometer Backscatter Uncertainty Due to Wind Variability. *IEEE Transactions on Geoscience and Remote Sensing*, *44*(11), 3356–3352.

Pozar, D. M. (1997). *Microwave Engineering*. New York: Wiley & Sons.

Prabhakara, C. & I. Wang (1983). Statistical method to sense sea surface temperature from the Nimbus-7 scanning multichannel microwave radiometer. In *1st Workshop on Satellite-Derived Sea Surface Temperature*. Pasadena, CA, January 27-28, 1983: NASA JPL, Caltech, May 1, 1983, B1-B7 (JPL Publication 83-34).

Prakash, R., D. Singh & N. P. Pathak (2012). A Fusion Approach to Retrieve Soil Moisture With SAR and Optical Data. *IEEE Journal of Selected Topics in Applied Earth Observations and Remote Sensing*, *5*(1), 196–206.

Pritchard, M. E. & M. Simons (2004). An InSAR-based survey of volcanic deformation in the southern Andes. *Geophysical Research Letters*, *31*(15), L15610.

Qi, R. & Y. Jin (2007). Analysis of the Effects of Faraday Rotation on Spaceborne Polarimetric SAR Observations at P-Band. *IEEE Transactions on Geoscience and Remote Sensing*, *45*(5), 1115–1122.

Quegan, S. (1994). A unified algorithm for phase and cross-talk calibration of polarimetric data—theory and observations. *IEEE Transactions on Geoscience and Remote Sensing*, *32*(1), 89–99.

Racette, P., R. F. Adler, A. J. Gasiewski, D. M. Jakson & D. S. Zacharias (1996). An airborne millimeter-wave imaging radiometer for cloud, precipitation, and atmospheric water vapor studies. *Journal of Atmospheric and Oceanic Technology*, *13*(3), 610–619.

Racette, P. E., E. R. Westwater, Y. Han, A. J. Gasiewski, M. Klein, D. Cimini, … P. Kiedron (2005). Measurement of low amounts of precipitable water vapor using ground-based millimeterwave radiometry. *Journal of Atmospheric and Oceanic Technology*, *22*(4), 317–337.

Raemer, H. R. (1997). *Radar Systems Principles*. Boca Raton, FL: CRC Press.

Rango, A. (1980). Operational Applications of Satellite Snow Cover Observations. *Water Resources Bulletin*, *16*(6), 1066–1073.

Rawson, R. F. & F. L. Smith (1974). 4-Channel Simultaneous X-L Imaging SAR Radar. In *9th International Symposium on Remote Sensing of the Environment* (Vol. 1, pp. 251–271). University of Michigan, Ann Arbor, MI.

Reeves, J. A., R. Knight, H. A. Zebker, W. A. Schreüder, P. Shanker & T. R. Lauknes (2011). High quality InSAR data linked to seasonal change in hydraulic head for an agricultural area in the San Luis Valley, Colorado. *Water Resources Research*, *47*(12), W12510.

Remund, Q. P. & D. G. Long (1999). Sea ice extent mapping using Ku band scatterometer data. *Journal of Geophysical Research*, *104*(C5), 11515–11527.

Remund, Q. P. & D. G. Long (2003). Large-scale inverse Ku-band backscatter modeling of sea ice. *IEEE Transactions on Geoscience and Remote Sensing*, *41*(8), 1821–1833.

Reutov, A. P. & B. A. Mikhaylov (1970). Radiolokatsionnyye Slantsii Bokovogo Obzora [Side-Looking Radar, English translation available from National Technical Information Service, U.S. Department of Commerce, Catalog Number AD-787070]. *Soviet Radio*.

Ricaud, P., B. Gabard, S. Derrien, J.-P. Chaboureau, T. Rose, A. Mombauer & H. Czekala (2010a). HAMSTRAD-Tropo, A 183-GHz Radiometer Dedicated to Sound Tropospheric Water Vapor Over Concordia Station, Antarctica. *IEEE Transactions on Geoscience and Remote Sensing*, *48*(3), 1365–1380.

Ricaud, P., B. Gabard, S. Derrien, J.-L. Attie, T. Rose & H. Czekala (2010b). Validation of Tropospheric Water Vapor as Measured by the 183-GHz HAMSTRAD Radiometer Over the Pyrenees Mountains, France. *IEEE Transactions on Geoscience and Remote Sensing*, *48*(5), 2189–2203.

Rice, S. O. (1944). Mathematical Analysis of Random Noises. *Bell System Technical Journal*, *23*, 282–352.

Rice, S. O. (1945). Mathematical Analysis of Random Noises. *Bell System Technical Journal*, *24*(1), 46–156.

Rice, S. O. (1951). Reflection of electromagnetic waves from slightly rough surfaces. *Communications on Pure and Applied Mathematics*, *4*(2-3), 351–378.

Richards, M. A. (2005). *Fundamentals of Radar Signal Processing*. New York: McGraw-Hill.

Richards, M. A., J. A. Scheer & W. A. Holm (2010). *Principles of Modern Radar*. Raleigh, NC: SciTech.

MIT Radiation Lab (*1947–1953*). *MIT Radiation Laboratory Series*. New York: McGraw-Hill.

Riegger, S., W. Wiesbeck & A. J. Sieber (1987). On the Origin of Cross-polarization in Remote Sensing. In *IEEE International Geoscience and Remote Sensing Symposium* (Vol. 1, pp. 577–580). 18–21 May, Ann Arbor, MI.

Riendeau, S. & C. Grenier (2007). RADARSAT-2 Antenna. In *IEEE Aerospace Conference* (pp. 1–9).

Rignot, E. J. M. (2000). Effect of Faraday rotation on L-band interferometric and polarimetric synthetic-aperture radar data. *IEEE Transactions on Geoscience and Remote Sensing*, *38*(1), 383–390.

Robertson, S. D. (1947). Targets for Microwave Radar Navigation. *Bell System Technical Journal*, *26*(4), 852–869.

Rocken, C., R. Anthes, M. Exner, R. Ware, D. Feng, M. Gorbunov, … X. Zou (1997). Analysis and validation of GPS/MET data in the neutral atmosphere. *Journal of Geophysical Research*, *102*(D25), 29849–29866.

Rodgers, C. D. (1976a). The Vertical Resolution of Remotely Sounded Temperature Profiles with a priori statistics. *Journal of the Atmospheric Sciences*, *33*(4), 707–709.

Rodgers, C. D. (1976b). Retrieval of Atmospheric Temperature and Composition From Remote Measurements of Thermal Radiation. *Journal of Geophysical Research*, *41*(7), 609–624.

Rodgers, C. D. (1977). Statistical principles of inversion theory. *United States National Aeronautics and Space Administration, Wash., D.C., Conference Publication NASA-CP-114*, 117–134.

Rodgers, C. D. (1996). Information content and optimisation of high spectral resolution measurements. *Proceedings of SPIE*, *2830*(1), 136–147.

Rodgers, C. D. (2000). *Inverse Methods for Atmospheric Sounding: Theory and Practice*. Singapore; River Edge, NJ: World Scientific.

Rodríguez, E. & J. M. Martin (1992). Theory and design of interferometric synthetic aperture radars. *IEE Proceedings-F Radar and Signal Processing*, *139*(2), 147–159.

Rodríguez, E. & J. M. Martin (1994). Assessment of the TOPEX altimeter performance using waveform retracking. *Journal of Geophysical Research*, *99*(C12), 24957.

Rodríguez, E., B. D. Pollard & J. M. Martin (1999). Wide-Swath Ocean Altimetry Using Radar Interferometry. *IEEE Transactions on Geoscience and Remote Sensing*.

Rogers, A. E. E. & R. P. Ingalls (1969). Venus: mapping the surface reflectivity by radar interferometry. *Science*, *165*(3895), 797–799.

Romeiser, R., S. Suchandt, H. Runge, U. Steinbrecher & S. Grünler (2010). First analysis of TerraSAR-X along-track InSAR-derived current fields. *IEEE Transactions on Geoscience and Remote Sensing*, *48*(2), 820–829.

Rønne, C., L. Thrane, P.-O. Åstrand, A. Wallqvist, K. V. Mikkelsen & S. R. Keiding (1997). Investigation of the temperature dependence of dielectric relaxation in liquid water by THz reflection spectroscopy and molecular dynamics simulation. *The Journal of Chemical Physics*, *107*(14), 5319.

Rosen, P. A., S. Hensley, I. R. Joughin, F. K. Li, S. N. Madsen, E. Rodríguez & R. M. Goldstein (2000). Synthetic Aperture Radar Interferometry. *Proceedings of the IEEE*, *88*(3), 333–382.

Rosenkranz, P. W. (1995). A rapid atmospheric transmittance algorithm for microwave sounding channels. *IEEE Transactions on Geoscience and Remote Sensing*, *33*(5), 1135–1140.

Rosenkranz, P. W. & D. H. Staelin (1988). Polarized thermal microwave emission from oxygen in the mesosphere. *Radio Science*, *23*(5), 721–729.

Rosenkranz, P. W., K. D. Hutchison, K. R. Hardy & M. S. Davis (1997). An Assessment of the Impact of Satellite Microwave Sounder Incidence Angle and Scan Geometry on the Accuracy of Atmospheric Temperature Profile Retrievals. *Journal of Atmospheric and Oceanic Technology*, *14*(3), 488–494.

Ross, R. A. (1966). Radar cross section of rectangular flat plates as a function of aspect angle. *IEEE Transactions on Antennas and Propagation*, *14*(3), 329–335.

Roth, K., R. Schulin, H. Flühler & W. Attinger (1990). Calibration of time domain reflectometry for water content measurement using a composite dielectric approach. *Water Resources Research*, *26*(10), 2267–2273.

Rouse, J. W. (1969). Arctic ice type identification by radar. *Proceedings of the IEEE*, *57*(4), 605–611.

Rowe, A. P. (1948). *One Story of Radar*. New York: Cambridge University Press.

Ruck, G. T., D. E. Barrick, W. D. Stuart & C. K. Krichbaum (1970). *Radar Cross Section Handbook* (G. T. Ruck, Ed.). New York: Plenum Press.

Ruf, C. S. (1993). Numerical annealing of low redundancy linear arrays. *IEEE Transactions on Antennas and Propagation*, *41*(1), 85–90.

Ruf, C. S. (1995). Digital correlators for synthetic aperture interferometric radiometry. *IEEE Transactions on Geoscience and Remote Sensing*, *33*(5), 1222–1229.

Ruf, C. S. & J. Li (2003). A Correlated Noise Calibration Standard for Interferometric, Polarimetric and Autocorrelation Microwave Radiometers. *IEEE Transactions on Geoscience and Remote Sensing*, *41*(10), 2187–2196.

Ruf, C. S. & C. Principe (2003). X-Band Lightweight Rainfall Radiometer First Light. In *IEEE International Geoscience and Remote Sensing Symposium* (Vol. 3, pp. 1701–1703). 21–25 July, Toulouse, France.

Ruf, C. S., C. T. Swift, A. B. Tanner & D. M. Le Vine (1988). Interferometric synthetic aperture microwave radiometry for the remote sensing of the earth. *IEEE Transactions on Geoscience and Remote Sensing*, *26*(5), 597–611.

Ruf, C. S., Y. Hu & S. T. Brown (2006a). Calibration of WindSat Polarimetric Channels With a Vicarious Cold Reference. *IEEE Transactions on Geoscience and Remote Sensing*, *44*(3), 470–475.

Ruf, C. S., S. M. Gross & S. Misra (2006b). RFI detection and mitigation for microwave radiometry with an agile digital detector. *IEEE Transactions on Geoscience and Remote Sensing*, *44*(3), 694–706.

Ruf, C., A. Lyons, M. Unwin, J. Dickinson, R. Rose, D. Rose & M. Vincent (2013). CYGNSS: Enabling the Future of Hurricane Prediction. *IEEE Geoscience and Remote Sensing Magazine*, *1*(2), 52–67.

Rumsey, H. C., G. A. Morris, R. R. Green & R. M. Goldstein (1974). A radar brightness and altitude image of a portion of Venus. *Icarus*, *23*(1), 1–7.

Rush, J. C. (1995). *The Image Processing Handbook*. Boca Raton, FL: CRC Press.

Ryder, I. & R. Bürgmann (2008). Spatial variations in slip deficit on the central San Andreas Fault from InSAR. *Geophysical Journal International*, *175*(3), 837–852.

Ryle, M. (1952). A New Radio Interferometer and Its Application to the Observation of Weak Radio Stars. *Proceedings of the Royal Society of London. Series A, Mathematical and Physical Sciences*, *211*(1106), 351–375.

Ryle, M. & A. Hewish (1960). The synthesis of large radio telescopes. *Monthly Notices of the Royal Astronomical Society*, *120*, 220–230.

Saatchi, S., K. Halligan, D. G. Despain & R. L. Crabtree (2007). Estimation of Forest Fuel Load From Radar Remote Sensing. *IEEE Transactions on Geoscience and Remote Sensing*, *45*(6), 1726–1740.

Saillard, M. & D. Maystre (1990). Scattering from metallic and dielectric rough surfaces. *Journal of the Optical Society of America*, *7*(6), 982–990.

Saleh, K., J.-P. Wigneron, P. Waldteufel, P. de Rosnay, M. Schwank, J.-C. Calvet & Y. H. Kerr (2007). Estimates of surface soil moisture under grass covers using L-band radiometry. *Remote Sensing of Environment*, *109*(1), 42–53.

Saleh, K., Y. H. Kerr, P. Richaume, M. J. Escorihuela, R. Panciera, S. Delwart, … J. P. Wigneron (2009). Soil moisture retrievals at L-band using a two-step inversion approach (COSMOS / NAFE'05 Experiment). *Remote Sensing of Environment*, *113*(6), 1304–1312.

Sandberg, G., L. M. H. Ulander, J. E. S. Fransson, J. Holmgren & T. Le Toan (2011). L- and P-band backscatter intensity for biomass retrieval in hemiboreal forest. *Remote Sensing of Environment*, *115*(11), 2874–2886.

Sandwell, D. T. (1991). Geophysical applications of satellite altimetry. *Reviews of Geophysics Supplement*, *29*, 132–137.

Sandwell, D. T. & W. H. F. Smith (2001). Bathymetric Estimation. In L.-L. Fu & A. Cazenave (Eds.), *Satellite Altimetry and Earth Sciences* (pp. 441–457). Academic Press.

Sarabandi, K. (1992). Derivation of phase statistics from the Mueller matrix. *Radio Science*, *27*(5), 553–560.

Sarabandi, K. & T.-C. Chiu (1996). Optimum Corner ReflectOrs for Calibration of Imaging Radars. *IEEE Transactions on Antennas and Propagation*, *44*(10), 1348–1361.

Sarabandi, K., T. B. A. Senior & F. T. Ulaby (1988). Effect of Curvature on the Backscattering from a Leaf. *Journal of Electromagnetic Waves and Applications*, *2*(7), 653–670.

Sarabandi, K., F. T. Ulaby & M. A. Tassoudji (1990). Calibration of polarimetric radar systems with good polarization isolation. *IEEE Transactions on Geoscience and Remote Sensing*, *28*(1), 70–75.

Sarabandi, K., L. E. Pierce & F. T. Ulaby (1992a). Calibration of a polarimetric imaging SAR. *IEEE Transactions on Geoscience and Remote Sensing*, *30*(3), 540–549.

Sarabandi, K., Y. Oh & F. T. Ulaby (1992b). Performance characterization of polarimetric active radar calibrators and a new single antenna design. *IEEE Transactions on Antennas and Propagation*, *40*(10), 1147–1154.

Sarabandi, K., L. E. Pierce, M. C. Dobson, F. T. Ulaby, J. M. Stiles, T. C. Chiu, … A. Freeman (1995). Polarimetric Calibration of SIR-C Using Point and Distributed Targets. *IEEE Transactions on Geoscience and Remote Sensing*, *33*(4), 858–866.

Schaber, G. G., C. Elachi & T. G. Farr (1980). Remote sensing data of SP Mountain and SP Lava flow in North-Central Arizona. *Remote Sensing of Environment*, *9*(2), 149–170.

Schaber, G. G., J. F. McCauley, C. S. Breed & G. R. Olhoeft (1986). Shuttle Imaging Radar: Physical Controls on Signal Penetration and Subsurface Scattering in the Eastern Sahara. *IEEE Transactions on Geoscience and Remote Sensing*, *24*(4), 603–623.

Schanda, E., C. Mätzler, K. Künzi, S. Patil & H. Rott (1982). Microwave Signatures of Mapping of Snow. In *International Society for Photogrammetry and Remote Sensing, Commission VII Exhibition*. Toulouse, France.

Schanda, E., C. Mätzler & K. Künzi (1983). Microwave Remote Sensing of Snow Cover. *International Journal of Remote Sensing*, *4*(1), 149–158.

Schaper, L. W., D. H. Staelin & J. W. Waters (1970). The estimation of tropospheric electrical path length by microwave radiometry. *Proceedings of the IEEE*, *58*(2), 272–273.

Schelkunoff, S. A. (1943). *Electromagnetic Waves*. New York: D. Van Nostrand Co.

Schelkunoff, S. A. & H. T. Friis (1952). *Antennas: Theory and practice*. New York: Wiley & Sons.

Schleher, D. C. (1976). Radar Detection in Weibull Clutter. *IEEE Transactions on Aerospace and Electronic Systems*, *12*(6), 736–743.

Schmugge, T. J. (1978). Remote Sensing of Surface Soil Moisture. *Journal of Applied Meteorology*, *17*(10), 1549–1557.

Schmugge, T. J. (1983). Remote Sensing of Soil Moisture: Recent Advances. *IEEE Transactions on Geoscience and Remote Sensing*, *21*(3), 336–344.

Schmugge, T. J. & B. J. Choudhury (1981). A comparison of radiative transfer models for predicting the microwave emission from soils. *Radio Science*, *16*(5), 927–938.

Schmugge, T. J., P. E. O'Neill & J. R. Wang (1986). Passive Microwave. *IEEE Transactions on Geoscience and Remote Sensing*, *24*(1), 12–22.

Schroeder, L. C., W. L. Grantham, J. Mitchell & J. Sweet (1982). SASS measurements of the Ku-band radar signature of the ocean. *IEEE Journal of Oceanic Engineering*, *7*(1), 3–14.

Schroeder, L. C., W. L. Grantham, E. M. Bracalente, C. L. Britt, K. S. Shanmugam, F. J. Wentz, ... B. B. Hinton (1985). Removal of Ambiguous Wind Directions for a Ku-Band Wind Scatterometer Using Three Different Azimuth Angles. *IEEE Transactions on Geoscience and Remote Sensing*, *23*(2), 91–100.

Seeber, G. (2004). *Satellite Geodesy: Foundations, Methods and Applications*. Berlin, Germany: Walter de Gruyter & Co.

Sekhon, R. S., & R. C. Srivastava (1970). Snow Size Spectra and Radar Reflectivity. *Journal of the Atmospheric Sciences*, *27*(2), 299–307.

Seling, T. V. (1964). The application of automatic gain control to microwave radiometers. *IEEE Transactions on Antennas and Propagation*, *12*(5), 636–639.

Senior, T. B. A., K. Sarabandi & F. T. Ulaby (1987). Measuring and modeling the backscattering cross section of a leaf. *Radio Science*, *22*(6), 1109–1116.

Shaffer, S. J., R. S. Dunbar, S. V. Hsiao & D. G. Long (1991). A median-filter-based ambiguity removal algorithm for NSCAT. *IEEE Transactions on Geoscience and Remote Sensing*, *29*(1), 167–174.

Shanker, P. (2010). *Persistent scatterer interferometry in neutral terrain*. Ph.D. Thesis, Electrical Engineering, Stanford University, Stanford, CA.

Shanker, P. & H. A. Zebker (2007). Persistent scatterer selection using maximum likelihood estimation. *Geophysical Research Letters*, *34*(22), L22301.

Shanker, P. & H. A. Zebker (2009). Sparse Two-Dimensional Phase Unwrapping Using Regular Grid Methods. *IEEE Geoscience and Remote Sensing Letters*, *6*(3), 519–522.

Shanker, P. & H. A. Zebker (2010). The Edgelist Algorithm for Constraining Phase Unwrapped Solutions with Additional Geodetic Information. In *IEEE International Geoscience and Remote Sensing Symposium*. 25–30 July, Honolulu, HI.

Shannon, C. E. (1948). A Mathematical Theory of Communication. *Bell System Technical Journal*, *27*(3), 379–423.

Sherwin, C. W., J. P. Ruina & R. D. Rawcliffe (1962). Some Early Developments in Synthetic Aperture Radar Systems. *IRE Transactions on Military Electronics*, *6*(2), 111–115.

Shi, J., L. Jiang, L. Zhang, K.-S. Chen, J.-P. Wigneron & A. Chanzy (2005). A Parameterized Multifrequency-Polarization. *IEEE Transactions on Geoscience and Remote Sensing*, *43*(12), 2831–2841.

Sihvola, A. H. (1999). *Electromagnetic Mixing Formulas and Applications*. London: Institution of Electrical Engineers.

Silver, S. (1949). *Microwave Antenna Theory and Design*. New York: McGraw-Hill.

Skolnik, M. I. (1980). *Introduction to Radar Systems* (2nd ed.). New York: McGraw-Hill.

Skoog, B. G., J. I. H. Askne & G. Elgered (1982). Experimental determination of water vapor profiles from ground-based radiometer measurements at 21.0 and 31.4 GHz. *Journal of Applied Meteorology*, *21*(3), 394–400.

Skou, N. & B. Laursen (1998). Measurement of ocean wind vector by an airborne, imaging polarimetric radiometer. *Radio Science*, *33*(3), 669–675.

Slater, P. N. (1980). *Remote Sensing: Optics and Optical Systems*. Reading, MA: Addison-Wesley.

Slone, A. J. (1995). Improved remote sensing data analysis using neural networks. Electrical Engineering and Computer Science, MIT, Cambridge, MA.

Smith, C. K., M. Bettenhausen & P. W. Gaiser (2006). A Statistical Approach to WindSat Ocean Surface Wind Vector Retrieval. *IEEE Geoscience and Remote Sensing Letters*, *3*(1), 164–168.

Smith, E. K. (1982). Centimeter and millimeter wave attenuation and brightness temperature due to atmospheric oxygen and water vapor. *Radio Science*, *17*(6), 1455–1464.

Smith, E. K. & S. Weintraub (1953). The Constants in the Equation for Atmospheric Refractive Index at Radio Frequencies. *Proceedings of the IRE*, *41*(8), 1035–1037.

Smith, W. H. F. & D. T. Sandwell (1994). Bathymetric prediction from dense satellite altimetry and sparse shipboard bathymetry. *Journal of Geophysical Research*, *99*(B11), 21803–21824.

Snider, J. B. (1972). Ground-Based Sensing of Temperature Profiles from Angular and Multi-Spectral Microwave Emission Measurements. *Journal of Applied Meteorology*, *11*(6), 958–967.

Snider, J. B., F. O. Guiraud & D. C. Hogg (1980). Comparison of cloud liquid content measured by two independent ground-based systems. *Journal of Applied Meteorology*, *19*(5), 577–579.

Solberg, A. H. S., C. Brekke & P. O. Husøy (2007). Oil Spill Detection in Radarsat and Envisat SAR Images. *IEEE Transactions on Geoscience and Remote Sensing*, *45*(3), 746–755.

Sollner, M. & H. Suss (1996). Design and first experimental results of a full-polarimetric quasioptical radiometer system at 90 GHz with high spatial, radiometric and polarimetric resolution. In *Fifth Specialist Meeting on Microwave Radiometry and Remote Sensing of the Environment*. URSI, Boston, MA.

Spencer, M. W., C. L. Wu & D. G. Long (1997). Tradeoffs in the design of a spaceborne scanning pencil beam scatterometer: Application to SeaWinds. *IEEE Transactions on Geoscience and Remote Sensing*, *35*(1), 115–126.

Spencer, M. W., C. L. Wu & D. G. Long (2000). Improved resolution backscatter measurements with the SeaWinds pencil-beam scatterometer. *IEEE Transactions on Geoscience and Remote Sensing*, *38*(1), 89–104.

Spencer, M. W., W.-Y. Tsai & D. G. Long (2003). High-resolution measurements with a spaceborne pencil-beam scatterometer using combined range/doppler discrimination techniques. *IEEE Transactions on Geoscience and Remote Sensing*, *41*(3), 567–581.

Spencer, M. W., K. Wheeler, S. Chan, J. R. Piepmeier, D. Hudson & J. Medeiros (2011). The planned Soil Moisture Active Passive (SMAP) mission L-band radar/radiometer instrument. In *IEEE International Geoscience and Remote Sensing Symposium* (pp. 2310–2313). 24–29 July, Vancouver, BC, Canada.

Srokosz, M. A. & M. S. Longuet-Higgins (1986). On the Skewness of Sea-Surface Elevation. *Journal of Fluid Mechanics*, *164*, 487–498.

Staelin, D. H. (1977). Inversion of Passive Microwave Remote Sensing Data from Satellites. In A. Deepak (Ed.), *Inversion Methods in Atmospheric Remote Sounding* (pp. 361–394). New York: Academic Press.

Staelin, D. H. & F. W. Chen (2000). Precipitation Observations Near 54 and 183 GHz Using the NOAA-15 Satellite. *IEEE Transactions on Geoscience and Remote Sensing*, *38*(5), 2322–2332.

Staelin, D. H., A. H. Barrett, J. W. Waters, F. T. Barath, E. J. Johnston, P. W. Rosenkranz, ... W. B. Lenoir (1973). Microwave Spectrometer on the Nimbus 5 Satellite: Meteorological and Geophysical Data. *Science*, *182*(4119), 1339–1341.

Stephens, G. L. & C. D. Kummerow (2007). The remote sensing of clouds and precipitation from space: A review. *Journal of the Atmospheric Sciences*, *64*(11), 3742–3765.

Stephens, G. L., D. G. Vane, R. J. Boain, G. G. Mace, K. Sassen, ... Z. Wang (2002). The CloudSat Mission and the A-Train. *Bulletin of the American Meteorological Society*, *83*(12), 1771–1790.

Stiles, B. W. & R. S. Dunbar (2010). A Neural Network Technique for Improving the Accuracy of Scatterometer Winds in Rainy Conditions. *IEEE Transactions on Geoscience and Remote Sensing*, *48*(8), 3114–3122.

Stiles, B. W., B. D. Pollard & R. S. Dunbar (2002). Direction interval retrieval with thresholded nudging: a method for improving the accuracy of QuikSCAT winds. *IEEE Transactions on Geoscience and Remote Sensing*, *40*(1), 79–89.

Stiles, B. W., J. N. Huddleston, S. M. Hristova-Veleva, R. S. Dunbar, M. H. Freilich, B. A. Vanhoff, ... W.-Y. Tsai (2006). Revealing the SeaWinds ocean vector winds under the rain using AMSR: Part II, the empirical approach. In *14th Conference on Satellite Meteorology and Oceanography*.

Stiles, W. H. & F. T. Ulaby (1980a). The Active and Passive Microwave Response to Snow Parameters: Part 1—Wetness; Part 2—Water Equivalent of Dry Snow. *Journal of Geophysical Research*, *85*(2), 1037–1049.

Stiles, W. H. & F. T. Ulaby (1980b). *Microwave Remote Sensing of Snowpacks*. Technical Report 340-3, Remote Sensing Laboratory, University of Kansas, Lawrence, KS.

Stiles, W. H., D. R. Brunfeldt & F. T. Ulaby (1979). *Performance Analysis of the MAS (Microwave Active Spectrometer) Systems: Calibration, Precision and Accuracy*. Technical Report 360-4, Remote Sensing Laboratory, University of Kansas, Lawrence, KS.

Stiles, W. H., F. T. Ulaby, A. K. Fung & A. Aslam (1981). Radar Spectral Observations of Snow. In *1st IEEE International Geoscience and Remote Sensing Symposium* (pp. 654–668). 8–10 June, Washington, DC.

Stoffelen, A. & D. Anderson (1997). Scatterometer Data Interpretation: Measurement Space and Inversion. *Journal of Atmospheric and Oceanic Technology*, *14*(6), 1298–1313.

Stogryn, A. (1971). Equations for Calculating the Dielectric Constant of Saline Water. *IEEE Transactions on Microwave Theory and Techniques*, *19*(8), 733–736.

Stogryn, A. (1989a). The magnetic field dependency of brightness temperatures at frequencies near the O2 microwave absorption lines. *IEEE Transactions on Geoscience and Remote Sensing*, *27*(3), 279–289.

Stogryn, A. (1989b). Mesospheric Temperature Sounding with Microwave Radiometers. *IEEE Transactions on Geoscience and Remote Sensing*, *27*(3), 332–338.

Stogryn, A. P., H. T. Bull, K. Ruayi & S. Iravanchy (1996). *The microwave permittivity of sea and freshwater*. Aerojet internal report, Sacramento, CA.

Straiton, A. W., C. W. Tolbert & C. O. Britt (1958). Apparent Temperatures of Some Terrestrial Materials and the Sun at 4.3 Millimeter Wavelengths. *Journal of Applied Physics*, *29*(5), 776–782.

Strand, O. N. & E. R. Westwater (1968). Statistical Estimation of the Numerical Solution of a Fredholm Integral Equation of the First Kind. *Journal of the ACM*, *15*(1), 100–114.

Strang, G. (1980). *Linear Algebra and Its Applications* (2nd ed.). New York: Academic Press.

Stratton, J. A. (1941). *Electromagnetic Theory* (pp. 464–469). New York: McGraw-Hill.

Stuart, K. M. & D. G. Long (2011). Tracking large tabular icebergs using the SeaWinds Ku-band microwave scatterometer. *Deep Sea Research Part II: Topical Studies in Oceanography*, *58*(11-12), 1285–1300.

Stuchly, M. A. & S. S. Stuchly (1980). Coaxial Line Reflection Methods for Measuring Dielectric Properties of Biological Substances at Radio and Microwave Frequencies—A Review. *IEEE Transactions on Instrumentation and Measurements*, *29*(3), 176–183.

Sun, G. & K. J. Ranson (1995). A three-dimensional radar backscatter model of forest canopies. *IEEE Transactions on Geoscience and Remote Sensing*, *33*(2), 372–382.

Surussavadee, C. & D. H. Staelin (2006). Comparison of AMSU Millimeter-Wave Satellite Observations, MM5/TBSCAT Predicted Radiances, and Electromagnetic Models for Hydrometeors. *IEEE Transactions on Geoscience and Remote Sensing*, *44*(10), 2667–2678.

Surussavadee, C. & D. H. Staelin (2008). Global Millimeter-Wave Precipitation Retrievals Trained With a Cloud-Resolving Numerical Weather Prediction Model, Part I: Retrieval Design, Part II: Performance Evaluation. *IEEE Transactions on Geoscience and Remote Sensing*, *46*(1), 109–118.

Surussavadee, C. & D. H. Staelin (2009). Satellite Retrievals of Arctic and Equatorial Rain and Snowfall Rates using Millimeter Wavelengths. *IEEE Transactions on Geoscience and Remote Sensing*, *47*(11), 3697–3707.

Surussavadee, C. & D. H. Staelin (2010a). NPOESS Precipitation Retrievals using the ATMS Passive Microwave Spectrometer. *IEEE Geoscience and Remote Sensing Letters*, *7*(3), 440–444.

Surussavadee, C. & D. H. Staelin (2010b). Global precipitation retrieval algorithm trained for SSMIS using a Numerical Weather Prediction Model: Design and evaluation. In *IEEE International Geoscience and Remote Sensing Symposium* (pp. 2341–2344). 25–30 July, Honolulu, HI.

Surussavadee, C. & D. H. Staelin (2010c). Global Precipitation Retrievals Using the NOAA/AMSU Millimeter-Wave Channels: Comparison with Rain Gauges. *Journal of Applied Meteorology and Climatology*, *49*(1), 124–135.

Surussavadee, C. & D. H. Staelin (2011). Evaporation Correction Methods for Microwave Retrievals of Surface Precipitation Rate. *IEEE Transactions on Geoscience and Remote Sensing*, *49*(12), 4763–4770.

Swan, A. M. & D. G. Long (2012). Multiyear Arctic Sea Ice Classification Using QuikSCAT. *IEEE Transactions on Geoscience and Remote Sensing*, *50*(9), 3317–3326.

Swartz, A. A., H. A. Yueh, J. A. Kong, L. M. Novak & R. T. Shin (1988). Optimal polarizations for achieving maximum contrast in radar images. *Journal of Geophysical Research*, *93*(B12), 15252–15260.

Tan, S. & M. Mavrovouniotis (1995). Reducing data dimensionality through optimizing neural network inputs. *AIChE Journal*, *41*(6), 1471–1480.

Tanelli, S., S. L. Durden, E. Im, K. S. Pak, D. G. Reinke, P. Partain, … R. T. Marchand (2008). CloudSat's Cloud Profiling Radar After Two Years in Orbit: Performance, Calibration, and Processing. *IEEE Transactions on Geoscience and Remote Sensing*, *46*(11), 3560–3573.

Tanner, A. B. & C. T. Swift (1993). Calibration of a Synthetic Aperture Radiometer. *IEEE Transactions on Geoscience and Remote Sensing*, *31*(1), 257–267.

Tapley, B. D., B. E. Schutz & G. H. Born (2004). *Statistical Orbit Determination*. San Diego, CA: Academic Press.

Tassoudji, M. A., K. Sarabandi & F. T. Ulaby (1989). *Design consideration and implementation of the LCX polarimetric scatterometer (POLARSCAT)*. Technical Report 0022486-T-1, College of Engineering, University of Michigan, Ann Arbor, MI.

Taylor, A. H., L. C. Young & L. A. Hyland (1934). Patent 1981884, System for Detecting Objects by Radio. United States.

Thompson, A. R., J. M. Moran & G. W. Swenson (1994). *Interferometry and Synthesis in Radio Astronomy*. Malabar, FL: Krieger Publishing Company.

Tikhonov, A. (1963). On the solution of incorrectly stated problems and a method of regularization. *Doklady Akademii Nauk SSSR*, *151*, 501–504.

Tinbergen, J. (1996). *Astronomical Polarimetry*. New York: Cambridge University Press.

Tinga, W. R., W. A. G. Voss & D. F. Blossey (1973). Generalized approach to multiphase dielectric mixture theory. *Journal of Applied Physics*, *44*(9), 3897–3902.

Tiuri, M. E. (1964). Radio Astronomy Receivers. *IEEE Transactions on Military Electronics*, *8*(3), 264–272.

Tjuatja, S., A. K. Fung & J. Bredow (1992). A scattering model for snow-covered sea ice. *IEEE Transactions on Geoscience and Remote Sensing*, *30*(4), 804–810.

Tooma, S. G., R. A. Mennella, J. P. Hollinger & J. R. Ketchum (1975). Comparison of sea-ice type identification between airborne dual-frequency passive microwave radiometry and standard laser/infrared technique. *Journal of Glaciology*, *15*(73), 225–239.

Tournadre, J. & Y. Quilfen (2003). Impact of rain cell on scatterometer data: 1. Theory and modeling. *Journal of Geophysical Research*, *108*(C7), 3225.

Tournadre, J., K. Whitmer & F. Girard-Ardhuin (2008). Iceberg detection in open water by altimeter waveform analysis. *Journal of Geophysical Research*, *113*(C8), C08040.

Touzi, R., W. M. Boerner, J. S. Lee & E. Lueneburg (2004). A review of polarimetry in the context of synthetic aperture radar: concepts and information extraction. *Canadian Journal of Remote Sensing*, *30*(3), 380–407.

Trebits, R. N., R. D. Hayes & L. C. Bomar (1978). mm-wave reflectivity of land and sea. *Microwave Journal*, *21*(8), 49–53.

Troy, B. E. & J. P. Hollinger (1977). *The Measurement of Oil Spill Volume By a Passive Microwave Imager*. Final Report No. CG-D-55-77, National Technical Information Service, Springfield, VA.

Troy, B. E., J. P. Hollinger, R. M. Lerner & M. M. Wisler (1981). Measurement of the microwave properties of sea ice at 90 GHz and lower frequencies. *Journal of Geophysical Research*, *86*(C5), 4283.

Truong-Loï, M.-L., A. Freeman, P. C. Dubois-Fernandez & E. Pottier (2009). Estimation of Soil Moisture and Faraday Rotation From Bare Surfaces Using Compact Polarimetry. *IEEE Transactions on Geoscience and Remote Sensing*, *47*(11), 3608–3615.

Tsai, W.-Y., J. E. Graf, C. Winn, J. N. Huddleston, S. Dunbar, M. H. Freilich, … W. L. Jones (1999). Postlaunch sensor verification and calibration of the NASA Scatterometer. *IEEE Transactions on Geoscience and Remote Sensing*, *37*(3), 1517–1542.

Tsang, L., J. A. Kong, E. G. Njoku, D. H. Staelin & J. W. Waters (1977). Theory for microwave thermal emission from a layer of cloud or rain. *IEEE Transactions on Antennas and Propagation*, *25*(5), 650–657.

Tsang, L., J. A. Kong & R. T. Shin (1985). *Theory of Microwave Remote Sensing*. New York: Wiley & Sons.

Turchin, V. F., V. P. Kozlov & M. S. Malkevich (1971). The use of mathematical statistics methods in the solution of incorrectly posed problems. *Soviet Physics Uspekhi*, *13*(6), 681–703.

Twomey, S. (1965). The application of numerical filtering to the solution of integral equations encountered in indirect sensing measurements. *Journal of the Franklin Institute*, *279*(2), 95–109.

Ulaby, F. T. (1974). Radar measurement of soil moisture content. *IEEE Transactions on Antennas and Propagation*, *22*(2), 257–265.

Ulaby, F. T. & J. E. Bare (1979). Look-Direction Modulation Function of the Radar Backscattering Coefficient of Agricultural Fields. *Photogrammetric Engineering & Remote Sensing*, *45*, 1495–1506.

Ulaby, F. T. & M. C. Dobson (1989). *Handbook of Radar Scattering Statistics for Terrain*. Norwood, MA: Artech House.

Ulaby, F. T. & M. El-Rayes (1987). Microwave Dielectric Spectrum of Vegetation - Part II: Dual-Dispersion Model. *IEEE Transactions on Geoscience and Remote Sensing*, *25*(5), 550–557.

Ulaby, F. T. & C. Elachi (1990). *Radar Polarimetry for Geoscience Applications*. Norwood, MA: Artech House.

Ulaby, F. T. & R. P. Jedlicka (1984). Microwave Dielectric Properties of Plant Materials. *IEEE Transactions on Geoscience and Remote Sensing*, *22*(4), 406–415.

Ulaby, F. T. & W. H. Stiles (1981). Microwave response of snow. *Advances in Space Research*, *1*(10), 131–149.

Ulaby, F. T. & A. E. Yagle (2013). *Engineering Signals and Systems*. Allendale, NJ: National Technology and Science Press.

Ulaby, F. T., J. Cihlar & R. K. Moore (1974). Active microwave measurement of soil water content. *Remote Sensing of Environment*, *3*(3), 185–203.

Ulaby, F. T., P. P. Batlivala & M. C. Dobson (1978). Microwave Backscatter Dependence on Surface Roughness, Soil Moisture, and Soil Texture: Part 1—Bare Soil. *IEEE Transactions on Geoscience Electronics*, *16*(4), 286–295.

Ulaby, F. T., G. A. Bradley & M. C. Dobson (1979). Microwave Backscatter Dependence on Surface Roughness, Soil Moisture, And Soil Texture: Part 2—Vegetation-Covered Soil. *IEEE Transactions on Geoscience Electronics*, *17*(2), 33–40.

Ulaby, F. T., R. K. Moore & A. K. Fung (1981a). *Microwave Remote Sensing: Active and Passive, Vol. I—Fundamentals and Radiometry*. Norwood, MA: Artech House.

Ulaby, F. T., M. Razani, A. K. Fung & M. C. Dobson (1981b). *The Effects of Vegetation Cover on the Microwave Radiometric Sensitivity to Soil Moisture*. Technical Report 460-6, Remote Sensing Laboratory, University of Kansas, Lawrence, KS.

Ulaby, F. T., R. K. Moore & A. K. Fung (1982a). *Microwave Remote Sensing: Active and Passive, Vol. II—Radar Remote Sensing and Surface Scattering and Emission Theory*. Norwood, MA: Artech House.

Ulaby, F. T., A. Aslam & M. C. Dobson (1982b). Effects of Vegetation Cover on the Radar Sensitivity to Soil Moisture. *IEEE Transactions on Geoscience and Remote Sensing*, *20*(4), 476–481.

Ulaby, F. T., F. Kouyate, A. K. Fung & A. J. Sieber (1982c). A Backscatter Model for a Randomly Perturbed Periodic Surface. *IEEE Transactions on Geoscience and Remote Sensing*, *20*(4), 518–528.

Ulaby, F. T., W. H. Stiles, A. K. Fung, H. J. Eom & M. Abdelrazik (1982d). *Observations and Modeling of the Radar Backscatter from Snowpacks*. Technical Report 527-4, Remote Sensing Laboratory, University of Kansas, Lawrence, KS.

Ulaby, F. T., C. T. Allen & A. K. Fung (1983). Method for Retrieving the True Backscattering Coefficient from Measurements with a Real Antenna. *IEEE Transactions on Geoscience and Remote Sensing*, *21*(3), 308–313.

Ulaby, F. T., C. T. Allen, G. Eger & E. Kanemasu (1984). Relating the microwave backscattering coefficient to leaf area index. *Remote Sensing of Environment*, *14*(1), 113–133.

Ulaby, F. T., R. K. Moore & A. K. Fung (1986a). *Microwave Remote Sensing: Active and Passive, Vol. III—Volume Scattering and Emission Theory, Advanced Systems and Applications*. Norwood, MA: Artech House.

Ulaby, F. T., F. Kouyate, B. Brisco & T. H. L. Williams (1986b). Textural Information in SAR Images. *IEEE Transactions on Geoscience and Remote Sensing*, *24*(2), 235–245.

Ulaby, F. T., A. Tavakoli & T. B. A. Senior (1987). Microwave Propagation Constant for a Vegetation Canopy With Vertical Stalks. *IEEE Transactions on Geoscience and Remote Sensing*, *25*(6), 714–725.

Ulaby, F. T., T. F. Haddock & R. T. Austin (1988a). Fluctuation statistics of millimeter-wave scattering from distributed targets. *IEEE Transactions on Geoscience and Remote Sensing*, *26*(3), 268–281.

Ulaby, F. T., T. F. Haddock, J. R. East & M. W. Whitt (1988b). A millimeterwave network analyzer based scatterometer. *IEEE Transactions on Geoscience and Remote Sensing*, *26*(1), 75–81.

Ulaby, F. T., K. McDonald, K. Sarabandi & M. C. Dobson (1988c). Michigan Microwave Canopy Scattering Models (MIMICS). In *IEEE International Geoscience and Remote Sensing Symposium* (p. 1009). 13–16 Septmber, Edinburgh, Scotland.

Ulaby, F. T., T. E. van Deventer, J. R. East, T. F. Haddock & M. E. Colluzi (1988d). Millimeter-wave bistatic scattering from ground and vegetation targets. *IEEE Transactions on Geoscience and Remote Sensing*, *26*(3), 229–243.

Ulaby, F. T., T. H. Bengal, M. C. Dobson, J. R. East, J. B. Garvin & D. L. Evans (1990a). Microwave Dielectric Properties of Dry Rocks. *IEEE Transactions on Geoscience and Remote Sensing*, *28*(3), 325–336.

Ulaby, F. T., M. W. Whitt & M. C. Dobson (1990b). Measuring the Propagation Properties of a Forest Canopy Using a Polarimetric Scatterometer. *IEEE Transactions on Antennas and Propagation*, *38*(2), 251–258.

Ulaby, F. T., T. F. Haddock, R. T. Austin & Y. Kuga (1991). Millimeter-wave radar scattering from snow: Part 2—Comparison of theory with experimental observations. *Radio Science*, *26*(2), 343–351.

Ulaby, F. T., K. Sarabandi & A. Nashashibi (1992). Statistical properties off the Mueller matrix off distributed targets. *IEE Proceedings-F Radar and Signal Processing*, *139*(2), 136–146.

Ulaby, F. T., E. Michielssen & U. Ravaioli (2010). *Fundamentals of Applied Electromagnetics* (6th ed.). Boston, MA: Prentice Hall.

Ulich, B. L. (1977). A radiometric antenna gain calibration method. *IEEE Transactions on Antennas and Propagation*, *25*(2), 218–223.

Vaccaneo, D., L. Sambuelli, P. Marini, R. Tascone & R. Orta (2004). Measurement System of Complex Permittivity of Ornamental Rocks in L Frequency Band. *IEEE Transactions on Geoscience and Remote Sensing*, *42*(11), 2490–2498.

Valenzuela, G. R. & M. B. Laing (1972). Nonlinear energy transfer in gravity-capillary wave spectra, with applications. *Journal of Fluid Mechanics*, *54*(3), 507–520.

van Beek, L. K. H. (1967). Dielectric Behavior of Heterogeneous Systems. In J. B. Birks (Ed.), *Progress in Dielectrics*. London: Heywood.

van Cittert, P. H. (1934). Die Wahrscheinliche Schwingungsverteilung in Einer von Einer Lichtquelle Direkt Oder Mittels Einer Linse Beleuchteten Ebene. *Physica*, *1*, 201–210.

van de Griend, A. A. (2001). The effective thermodynamic temperature of the emitting surface at 6.6 GHz and consequences for soil moisture monitoring from space. *IEEE Transactions on Geoscience and Remote Sensing*, *39*(8), 1673–1679.

van de Hulst, H. C. (1957). *Light Scattering by Small Particles*. New York: Wiley & Sons.

van Zyl, J. J. (1985). *On the importance of polarization in radar scattering problems*. Ph.D. Thesis, Antenna Laboratory Report No. 120, 152 pp., Caltech, Pasadena, CA.

van Zyl, J. J. (1990). Calibration of polarimetric radar images using only image parameter and trihedral corner reflector responses. *IEEE Transactions on Geoscience and Remote Sensing*, *28*(3), 337–348.

van Zyl, J. J. & Y. Kim (2011). *Synthetic Aperture Radar Polarimetry*. Hoboken, NJ: Wiley & Sons.

van Zyl, J. J., C. H. Papas & C. Elachi (1987a). On the optimum polarizations of incoherently reflected waves. *IEEE Transactions on Antennas and Propagation*, *35*(7), 818–825.

van Zyl, J. J., H. A. Zebker & C. Elachi (1987b). Imaging radar polarization signatures: Theory and observation. *Radio Science*, *22*(4), 529–543.

Vant, M. R. (1976). *A combined empirical and theoretical study of the dielectric properties of sea ice over the frequency range 100 MHz to 40 GHz*. Ph.D. Thesis, Carleton University, Ottawa, Canada.

Vant, M. R., R. B. Gray, R. O. Ramseier & V. Makios (1974). Dielectric properties of fresh and sea ice at 10 and 35 GHz. *Journal of Applied Physics*, *45*(11), 4712–4717.

Vant, M. R., R. O. Ramseier & V. Makios (1978). The complex dielectric constant of sea ice at frequencies in the range 0.1–40 GHz. *Journal of Applied Physics*, *49*(3), 1264–1280.

Vasalos, I. A. (1969). *Effects of Separation Distance on the Optical Properties of Dense Dielectric Particle Suspensions. Chemical Engineering*. Ph.D. Thesis, Chemical Engineering, MIT, Cambridge, MA.

Vecchia, A. D., P. Ferrazzoli, L. Guerriero, R. Rahmoune, S. Paloscia, S. Pettinato & E. Santi (2010). Modeling the Multifrequency Emission of Broadleaf Forests and Their Components. *IEEE Transactions on Geoscience and Remote Sensing*, *48*(1), 260–272.

Viksne, A., T. C. Liston & C. D. Sapp (1969). SLR Reconnaissance of Panama. *Geophysics*, *34*(1), 54–64.

Vila, D., R. R. Ferraro & R. Joyce (2007). Evaluation and improvement of AMSU precipitation retrievals. *Journal of Geophysical Research*, *112*(D20), D20119.

Vincente, G. A., R. A. Scofield & W. B. Menzel (1998). The operational GOES infrared rainfall estimation technique. *Bulletin of the American Meteorological Society*, *99*(9), 1883–1898.

von Hippel, A. R. (1954). *Dielectric Materials and Applications*. Cambridge, MA: MIT Press.

Wadhams, P. (1979). The Estimation of Sea Ice Thickness from the Distribution of Pressure Ridge Heights and Depths. In *International Workshop on Remote Estimation of Sea Ice Thickness* (pp. 53–76). 25-26 September, St. Johns, Newfoundland, Canada.

Wakabayashi, H., T. Matsuoka, K. Nakamura & F. Nishio (2004). Polarimetric Characteristics of sea ice in the sea of Okhotsk observed by airborne L-band SAR. *IEEE Transactions on Geoscience and Remote Sensing*, *42*(11), 2412–2425.

Wakabayashi, H., Y. Mori & K. Nakamura (2013). Sea Ice Detection in the Sea of Okhotsk Using PALSAR and MODIS Data. *IEEE Journal of Selected Topics in Applied Earth Observations and Remote Sensing*, *6*(3), 1516–1523.

Walters, R. J., J. R. Elliott, N. D'Agostino, P. C. England, I. Hunstad, J. A. Jackson, … G. Roberts (2009). The 2009 L'Aquila earthquake (central Italy): A source mechanism and implications for seismic hazard. *Geophysical Research Letters*, *36*(17), L17312.

Wang, J. R. & B. J. Choudhury (1981). Remote Sensing of Soil Moisture Content Over Bare Field at 1.4 GHz Frequency. *Journal of Geophysical Research*, *86*(6), 5277–5282.

Wang, J. R., J. C. Shiue & J. E. McMurtrey (1980). Microwave remote sensing of soil moisture content over bare and vegetated fields. *Geophysical Research Letters*, *7*(10), 801–804.

Wang, J. R., P. E. O'Neill, T. J. Jackson & E. T. Engman (1983). Multifrequency Measurements of the Effects of Soil Moisture, Soil Texture, And Surface Roughness. *IEEE Transactions on Geoscience and Remote Sensing*, *21*(1), 44–51.

Wang, J. R., W. C. Boncyk & A. K. Sharma (1993). Water vapor profiling over ocean surface from airborne 90 and 183 GHz radiometric measurements under clear and cloudy conditions. *IEEE Transactions on Geoscience and Remote Sensing*, *31*(4), 853–859.

Wang, J. R., J. D. Spinhim, P. Racette, L. A. Chang & W. Hart (1997). The effect of clouds on water vapor profiling from the millimeter-wave radiometric measurements. *Journal of Applied Meteorology*, *36*(9), 1232–1244.

Watanabe, M., M. Shimada, A. Rosenqvist, T. Tadono, M. Matsuoka, S. A. Romshoo, … T. Moriyama (2006). Forest Structure Dependency of the Relation between L-band σ^0 and Biophysical Parameters. *IEEE Transactions on Geoscience and Remote Sensing*, *44*(11), 3154–3165.

Waters, J. W. (1971). *Ground-based microwave spectroscopic sensing of the stratosphere and mesosphere*. Ph.D. Thesis, Electrical Engineering and Computer Science, MIT, Cambridge, MA.

Waters, J. W. (1993). Microwave limb sounding. In M. A. Janssen (Ed.), *Atmospheric Remote Sensing by Microwave Radiometry*. New York: Wiley & Sons.

Waters, J. W., L. Froidevaux, R. S. Harwood, R. F. Jarnot, H. M. Pickett, W. G. Read, … M. J. Walch (2006). The Earth observing system microwave limb sounder (EOS MLS) on the Aura Satellite. *IEEE Transactions on Geoscience and Remote Sensing*, *44*(5), 1075–1092.

Watson-Watt, R. (1957). *Three Steps to Victory: A Personal Account by Radar's Greatest Pioneer*. London: Odhams Press.

Weeks, W. F. (1981). Sea ice: the potential of remote sensing. *Oceanus*, *24*(3), 39–48.

Weeks, W. F. & A. Assur (1969). *Fracture of lake and sea ice*. Research Report 269, Cold Regions Research and Engineering Laboratory, Office of Naval Research, Department of the Navy, Washington, DC.

Weger, E. (1960). Apparent Sky Temperatures in the Microwave Region. *Journal of Meteorology*, *17*(2), 159–165.

Wegmüller, U. (1986). *Signaturen zur Mikrowellen-Fernerkundung: Bodenrauhigkeit und Permittivität von Eis*. Thesis, Institut für angewandte Physik der Universität Bern, Germany.

Wegmüller, U. (1993). Signature research for crop classification by active and passive microwaves. *International Journal of Remote Sensing*, *14*(5), 871–883.

Wegmüller, U., C. Mätzler & E. G. Njoku (1995). Canopy Opacity Models. In B. J. Choudhury, Y. H. Kerr, E. G. Njoku & P. Pampaloni (Eds.), *Passive Microwave Remote Sensing of Land-Atmosphere Interactions* (pp. 375–388). Utrecht, The Netherlands: VSP.

Weinreb, S. (1961). Digital radiometer. *Proceedings of the IRE*, *49*(6), 1099.

Weinstock, W. W. (1965). Radar Cross-Section Target Models. In R. S. Berkowitz (Ed.), *Modern Radar Analysis, Evaluation and System Design*. New York: Wiley & Sons.

Weir, W. B. (1974). Automatic measurement of complex dielectric constant and permeability at microwave frequencies. *Proceedings of the IEEE*, *62*(1), 33–36.

Wells, J. S., W. C. Daywitt & C. K. S. Miller (1964). Measurement of Effective Temperatures of Microwave Noise Sources. *IEEE Transactions on Instrumentation and Measurements*, *13*(1), 17–28.

Wentz, F. J. (1975). Two-scale scattering model for foam-free sea microwave brightness temperatures. *Journal of Geophysical Research*, *80*(24), 3441–3446.

Wentz, F. J. (1977). *A two-scale scattering model with application to the JONSWAP '75 Aircraft Microwave Scatterometer Experiment (NASA contractor report)*. NASA Langley Research Center.

Wentz, F. J. (1978). Forward scattering of microwave solar radiation from a water surface. *Radio Science*, *13*(1), 131–138.

Wentz, F. J. (1991). A simplified wind vector algorithm for satellite scatterometers. *Journal of Atmospheric and Oceanic Technology*, *8*(5), 697–704.

Wentz, F. J. (1992). Measurement of oceanic wind vector using satellite microwave radiometers. *IEEE Transactions on Geoscience and Remote Sensing*, *30*(5), 960–972.

Wentz, F. J. & D. K. Smith (1999). A model function for the ocean-normalized radar cross section at 14 GHz derived from NSCAT observations. *Journal of Geophysical Research*, *104*(C5), 11499–11514.

Wentz, F. J., S. Peteherych & L. A. Thomas (1984). A model function for ocean radar cross sections at 14.6 GHz. *Journal of Geophysical Research*, *89*(C3), 3689.

Wentz, F. J., L. A. Mattox & S. Peteherych (1986). New algorithms for microwave measurements of ocean winds: applications to SEASAT and special sensor microwave imager. *Journal of Geophysical Research*, *91*(C2), 2289–2307.

Werner, C., U. Wegmüller, T. Strozzi & A. Wiesmann (2003). Interferometric point target analysis for deformation mapping. In *IEEE International Geoscience and Remote Sensing Symposium* (Vol. 7, pp. 4362–4364). 21–25 July, Toulouse, France.

West, R. D., Y. Anderson, R. Boehmer, L. Borgarelli, P. Callahan, C. Elachi, ... H. A. Zebker (2009). Cassini RADAR Sequence Planning and Instrument Performance. *IEEE Transactions on Geoscience and Remote Sensing*, *47*(6), 1777–1795.

Westwater, E. R. (1965). Ground-based passive probing using microwave spectrum of oxygen. *Journal of Research of the National Bureau of Standards Section D - Radio Science*, *D69*(9), 1201.

Westwater, E. R. (1978). Accuracy of water vapor and cloud liquid determination by dual-frequency, ground-based microwave radiometry. *Radio Science*, *13*(4), 677–685.

Westwater, E. R. (1979). Ill-posed problems in remote sensing of the Earth's Atmosphere by microwave radiometry. In *International Symposium on Ill-Posed Problems: Theory and Practice*. 2-6 October, University of Delaware, Newark, DE.

Westwater, E. R. & A. Cohen (1973). Application of Backus-Gilbert inversion technique to determination of aerosol size distributions from optical scattering measurements. *Applied Optics*, *12*(6), 1340–1348.

Westwater, E. R. & M. T. Decker (1977). Application of statistical inversion to ground-based microwave remote sensing of temperature and water vapor profiles. *United States National Aeronautics and Space Administration, Wash., D.C., Conference Publication NASA-CP-114*, 395–425.

Westwater, E. R., & N. C. Grody (1980). Combined Surface- and Satellite-Based Microwave Temperature Profile Retrieval. *Journal of Applied Meteorology*, *19*(12), 1438–1444.

Westwater, E. R. & O. N. Strand (1968). Statistical Information Content of Radiation Measurements used in Indirect Sensing. *Journal of the Atmospheric Sciences*, *25*(5), 750–758.

Westwater, E. R., J. B. Snider & A. V. Carlson (1975). Experimental Determination of Temperature Profiles by Ground-Based Microwave Radiometry. *Journal of Applied Meteorology*, *14*(4), 524–539.

Wexler, R. (1948). Rain Intensities by Radar. *Journal of Meteorology*, *5*(4), 171–173.

Wexler, R. & D. Atlas (1963). Radar Reflectivity and Attenuation of Rain. *Journal of Applied Meteorology*, *2*(2), 276–280.

Whitt, M. W. & F. T. Ulaby (1988). Millimeter-wave polarimetric measurements of artificial and natural targets. *IEEE Transactions on Geoscience and Remote Sensing*, *26*(5), 562–573.

Whitt, M. W. & F. T. Ulaby (1994). Radar Response of Periodic Vegetation Canopies. *International Journal of Remote Sensing*, *15*(9), 1813–1848.

Whitt, M. W., F. T. Ulaby, P. Polatin & V. V. Liepa (1991). A general polarimetric radar calibration technique. *IEEE Transactions on Antennas and Propagation*, *39*(1), 62–67.

Wickert, J., G. Beyerle, T. Schmidt, C. Marquardt, R. König, L. Grunwald & C. Reigber (2002). First CHAMP mission results for gravity, magnetic and atmospheric studies. In C. Reigber, H. Lühr & P. Schwintzer (Eds.), *First CHAMPS Science Meeting*. Potsdam, Germany: Springer-Verlag.

Wickert, J., T. Schmidt, G. Beyerle, R. König, S. Heise & C. Reigber (2006). GPS radio occultation with CHAMP and GRACE: Recent Results. In U. Foelsche, G. Kirchengast & A. Steiner (Eds.), *Atmosphere and Climate: Studies by Occultation Methods*. Berlin, Germany.

Wigneron, J.-P., A. Chanzy, J.-C. Calvet & N. Bruguier (1995). A Simple Algorithm to Retrieve Soil Moisture and Vegetation Biomass Using Passive Microwave Measurements over Crop Fields. *Remote Sensing of Environment*, *51*(3), 331–341.

Wigneron, J.-P., L. Laguerre & Y. H. Kerr (2001). A simple parameterization of the L-band microwave emission from rough agricultural soils. *IEEE Transactions on Geoscience and Remote Sensing*, *39*(8), 1697–1707.

Wigneron, J.-P., A. Chanzy, Y. H. Kerr, H. Lawrence, J. Shi, M. J. Escorihuela, ... K. Saleh-Contell (2011). Evaluating an Improved Parameterization of the Soil Emission in L-MEB. *IEEE Transactions on Geoscience and Remote Sensing*, *49*(4), 1177–1189.

Wilheit, T. T. (1978). Radiative Transfer in a Plane Stratified Dielectric. *IEEE Transactions on Geoscience and Remote Sensing*, *16*(2), 138–143.

Wilheit, T. T. (1979). Effect of wind on the microwave emission from the ocean's surface at 37 GHz. *Journal of Geophysical Research*, *84*(8), 4921–4926.

Wilheit, T. T. (1986). Some comments on passive microwave measurement of rain. *Bulletin of the American Meteorological Society*, *67*(10), 1226–1232.

Wilheit, T. T. (1990). An algorithm for retrieving water vapor profiles in clear and cloudy atmospheres from 183 GHz radiometric measurements: Simulation studies. *Journal of Applied Meteorology*, *29*(6), 508–515.

Wilheit, T. T. & A. T. C. Chang (1980). An algorithm for retrieval of ocean surface and atmospheric parameters from the observations of the scanning multichannel microwave radiometer. *Radio Science*, *15*(3), 525–544.

Wilheit, T. T., W. Nordberg, J. Blinn, W. J. Campbell & A. T. Edgerton (1971). Aircraft measurements of microwave emission from Arctic Sea ice. *Remote Sensing of Environment*, *2*, 129–139.

Wilheit, T. T., A. T. C. Chang, M. S. V. Rao, E. B. Rodgers & J. S. Theon (1977). A Satellite Technique for Quantitatively Mapping Rainfall Rates Over the Oceans. *Journal of Applied Meteorology*, *16*(5), 551–560.

Wilheit, T. T., A. T. C. Chang & A. S. Milman (1980). Atmospheric corrections to passive microwave observations of the ocean. *Boundary-Layer Meteorology*, *18*(1), 65–77.

Wilheit, T. T., A. T. C. Chang & L. S. Chiu (1991). Retrieval of monthly rainfall indices from microwave radiometric measurements using probability distribution functions. *Journal of Atmospheric and Oceanic Technology*, *8*(1), 118–136.

Wilheit, T. T., C. D. Kummerow & R. Ferraro (2003). Rainfall Algorithms for AMSR-E. *IEEE Transactions on Geoscience and Remote Sensing*, *41*(2), 204–213.

Williams, B. A. & D. G. Long (2011a). Reconstruction From Aperture-Filtered Samples With Application to Scatterometer Image Reconstruction. *IEEE Transactions on Geoscience and Remote Sensing*, *49*(5), 1663–1676.

Williams, B. A. & D. G. Long (2011b). A Reconstruction Approach to Scatterometer Wind Vector Field Retrieval. *IEEE Transactions on Geoscience and Remote Sensing*, *49*(6), 1850–1864.

Wilson, J. J. W., J. Figa-Saldaña & E. O'Clerigh (2005). *ASCAT Level 1 Production Generation*. EUMETSAT document number EUM.EPS.SYS.SPE.990009.

Wilson, W. J., S. H. Yueh, S. J. Dinardo, S. L. Chazanoff, A. Kitiyakara, F. K. Li & Y. Rahmat-Samii (2001). Passive Active L- and S-Band (PALS) Microwave Sensor for Ocean Salinity and Soil Moisture Measurements. *IEEE Transactions on Geoscience and Remote Sensing*, *39*(5), 1039–1048.

Wohlleben, R., D. Fiebig, A. Prata & W. V. T. Rusch (1991). Beam squint in axially symmetric reflector antennas with laterally displaced feeds. *IEEE Transactions on Antennas and Propagation*, *39*(6), 774–779.

Wright, P. A., S. Quegan, N. S. Wheadon & C. D. Hall (2003). Faraday rotation effects on L-band spaceborne SAR data. *IEEE Transactions on Geoscience and Remote Sensing*, *41*(12), 2735–2744.

Wu, S.-C. (1979). Optimum frequencies of a passive microwave radiometer for tropospheric path-length correction. *IEEE Transactions on Antennas and Propagation*, *27*(2), 233–239.

Yagi, H. (1928). Beam transmission of ultra-shortwaves. *Proceedings of the IRE*, *16*, 715–740.

Yang, L., H. Du, J. Zhao & Q. Liu (2011). Global Vegetation Dynamic Monitoring Using Multiple Satellite Observations. In *IEEE International Geoscience and Remote Sensing Symposium* (pp. 771–774). 24–29 July, Vancouver, BC, Canada.

Yoho, P. K. & D. G. Long (2003). An improved simulation model for spaceborne scatterometer measurements. *IEEE Transactions on Geoscience and Remote Sensing*, *41*(11), 2692–2695.

Yoho, P. K. & D. G. Long (2004). Correlation and covariance of satellite scatterometer measurements. *IEEE Transactions on Geoscience and Remote Sensing*, *42*(6), 1179–1187.

Yueh, S. H. (1997). Modeling of wind direction signals in polarimetric sea surface brightness temperatures. *IEEE Transactions on Geoscience and Remote Sensing*, *35*(6), 1400–1418.

Yueh, S. H. (2000). Estimates of Faraday rotation with passive microwave polarimetry for microwave remote sensing of Earth surfaces. *IEEE Transactions on Geoscience and Remote Sensing*, *38*(5), 2434–2438.

Yueh, S. H. & J. Chaubell (2012). Sea Surface Salinity and Wind Retrieval Using Combined Passive and Active L-Band Microwave Observations. *IEEE Transactions on Geoscience and Remote Sensing*, *50*(4), 1022–1032.

Yueh, S. H., R. Kwok, F. K. Li, S. V. Nghiem, W. J. Wilson & J. A. Kong (1994). Polarimetric passive remote sensing of ocean wind vectors. *Radio Science*, *29*(4), 799–814.

Yueh, S. H., W. J. Wilson, F. K. Li, S. V. Nghiem & W. B. Ricketts (1995). Polarimetric measurements of sea surface brightness temperatures using an aircraft K-band radiometer. *IEEE Transactions on Geoscience and Remote Sensing*, *33*(1), 85–92.

Yueh, S. H., W. J. Wilson, F. K. Li, S. V. Nghiem & W. B. Ricketts (1997). Polarimetric Brightness Temperatures of Sea Surfaces Measured with Aircraft K- and Ka-Band Radiometers. *IEEE Transactions on Geoscience and Remote Sensing*, *35*(5), 1177–1187.

Yueh, S. H., W. J. Wilson, S. J. Dinardo & F. K. Li (1999). Polarimetric microwave brightness signatures of ocean wind directions. *IEEE Transactions on Geoscience and Remote Sensing*, *37*(2), 949–959.

Yueh, S. H., W. J. Wilson, S. J. Dinardo & S. V. Hsiao (2006). Polarimetric Microwave Wind Radiometer Model Function and Retrieval Testing for WindSat. *IEEE Transactions on Geoscience and Remote Sensing*, *44*(3), 584–596.

Yueh, S. H., S. J. Dinardo, A. Akgiray, R. West, D. W. Cline & K. Elder (2009). Airborne Ku-Band Polarimetric Radar Remote Sensing of Terrestrial Snow Cover. *IEEE Transactions on Geoscience and Remote Sensing*, *47*(10), 3347–3364.

Yueh, S. H., S. J. Dinardo, A. G. Fore & F. K. Li (2010). Passive and Active L-Band Microwave Observations and Modeling of Ocean Surface Winds. *IEEE Transactions on Geoscience and Remote Sensing*, *48*(8), 3087–3100.

Yunck, T. P. (2002). An overview of atmospheric radio occultation. *Journal of Global Positioning Systems*, *1*(1), 58–60.

Yunck, T. P., C. H. Liu & R. Ware (2000). A history of GPS sounding. *Terrestrial, Atmospheric and Oceanic Sciences*, *11*(1), 1–20.

Zaugg, E. C. & D. G. Long (2009). Generalized Frequency-Domain SAR Processing. *IEEE Transactions on Geoscience and Remote Sensing*, *47*(11), 3761–3773.

Zaugg, E. C. & D. G. Long (2012). Generalized LFM-CW SAR Processing. *IEEE Transactions on Geoscience and Remote Sensing*, in press.

Zavorotny, V. U. & A. G. Voronovich (2000). Scattering of GPS signals from the ocean with wind remote sensing application. *IEEE Transactions on Geoscience and Remote Sensing*, *38*(2), 951–964.

Zebker, H. A. & R. M. Goldstein (1986). Topographic Mapping From Interferometric Synthetic Aperture Radar Observations. *Journal of Geophysical Research*, *91*(5), 4993–4999.

Zebker, H. A. & Y. L. Lou (1990). Phase calibration of imaging radar polarimetric stokes matrices. *IEEE Transactions on Geoscience and Remote Sensing*, *28*(2), 246–252.

Zebker, H. A. & J. Villasenor (1992). Decorrelation in interferometric radar echoes. *IEEE Transactions on Geoscience and Remote Sensing*, *30*(5), 950–959.

Zebker, H. A., J. J. van Zyl & D. N. Held (1987). Imaging Radar Polarimetry From Wave Synthesis. *Journal of Geophysical Research*, *92*(1), 683–701.

Zebker, H. A., P. A. Rosen & S. Hensley (1997). Atmospheric effects in interferometric synthetic aperture radar surface deformation and topographic maps. *Journal of Geophysical Research*, *102*(B4), 7547–7563.

Zernike, F. (1938). The concept of degree of coherence and its application to optical problems. *Physica*, *5*(8), 785–795.

Zhan, X., P. R. Houser, J. P. Walker & W. T. Crow (2006). A Method for Retrieving High-Resolution Surface Soil Moisture From Hydros L-Band Radiometer and Radar Observations. *IEEE Transactions on Geoscience and Remote Sensing*, *44*(6), 1534–1544.

Zhang, B., W. Perrie, X. Li & W. G. Pichel (2011). Mapping sea surface oil slicks using RADARSAT-2 quad-polarization SAR image. *Geophysical Research Letters*, *38*(10), L10602.

Zhang, M. & J. T. Johnson (2001). Comparison of modeled and measured second azimuthal harmonics of ocean surface brightness temperatures. *IEEE Transactions on Geoscience and Remote Sensing*, *39*(2), 448–452.

Zieger, A. R., D. W. Hancock, G. S. Hayne & C. L. Purdy (1991). NASA radar altimeter for the TOPEX/POSEIDON Project. *Proceedings of the IEEE*, *79*(6), 810–826.

Zine, S., J. Boutin, J. Font, N. Reul, P. Waldteufel, C. Gabarró, ... S. Delwart (2008). Overview of the SMOS Sea Surface Salinity Prototype Processor. *IEEE Transactions on Geoscience and Remote Sensing*, *46*(3), 621–645.

Zisk, S. H. (1972a). A new, earth based radar technique for the measurement of lunar topography. *The Moon*, *4*(3-4), 296–306.

Zisk, S. H. (1972b). Lunar topography: first radar-interferometer measurements of the alphonsus-ptolemaeus-arzachel region. *Science*, *178*(4064), 977–980.

Zwally, H. J. (1977). Microwave Emissivity and Accumulation Rate of Polar Firn. *Journal of Glaciology*, *18*(79), 195–215.

Zwally, H. J. & P. Gloersen (1977). Passive Microwave Images of the Polar Regions and Research Applications. *Polar Record*, *18*(116), 431–450.

Zwally, H. J., J. C. Comiso, C. L. Parkinson, W. J. Campbell, F. D. Carsey & P. Gloersen (1983). *Antarctic Sea Ice, 1973-1976: Satellite Passive-Microwave Observations* (pp. 21–25). NASA Report SP-459, Washington, DC.

Index